Lecture Notes in Civil Engineering

Volume 264

Series Editors

Marco di Prisco, Politecnico di Milano, Milano, Italy

Sheng-Hong Chen, School of Water Resources and Hydropower Engineering, Wuhan University, Wuhan, China

Ioannis Vayas, Institute of Steel Structures, National Technical University of Athens, Athens, Greece

Sanjay Kumar Shukla, School of Engineering, Edith Cowan University, Joondalup, WA, Australia

Anuj Sharma, Iowa State University, Ames, IA, USA

Nagesh Kumar, Department of Civil Engineering, Indian Institute of Science Bangalore, Bengaluru, Karnataka, India

Chien Ming Wang, School of Civil Engineering, The University of Queensland, Brisbane, QLD, Australia

Lecture Notes in Civil Engineering (LNCE) publishes the latest developments in Civil Engineering—quickly, informally and in top quality. Though original research reported in proceedings and post-proceedings represents the core of LNCE, edited volumes of exceptionally high quality and interest may also be considered for publication. Volumes published in LNCE embrace all aspects and subfields of, as well as new challenges in, Civil Engineering. Topics in the series include:

- Construction and Structural Mechanics
- Building Materials
- Concrete, Steel and Timber Structures
- Geotechnical Engineering
- Earthquake Engineering
- Coastal Engineering
- Ocean and Offshore Engineering; Ships and Floating Structures
- Hydraulics, Hydrology and Water Resources Engineering
- Environmental Engineering and Sustainability
- Structural Health and Monitoring
- Surveying and Geographical Information Systems
- Indoor Environments
- Transportation and Traffic
- Risk Analysis
- Safety and Security

To submit a proposal or request further information, please contact the appropriate Springer Editor:

- Pierpaolo Riva at pierpaolo.riva@springer.com (Europe and Americas);
- Swati Meherishi at swati.meherishi@springer.com (Asia—except China, Australia, and New Zealand);
- Wayne Hu at wayne.hu@springer.com (China).

All books in the series now indexed by Scopus and EI Compendex database!

More information about this series at http://www.springer.com/bookseries/15087

Yun Li · Yaan Hu · Philippe Rigo ·
Francisco Esteban Lefler ·
Gensheng Zhao
Editors

Proceedings of PIANC Smart Rivers 2022

Green Waterways and Sustainable Navigations

Set 1

 Springer

Editors

Yun Li
Nanjing Hydraulic Research Institute,
Ministry of Water Resources, Ministry of
Transport and National Administration of
Energy
Nanjing, China

Yaan Hu
Nanjing Hydraulic Research Institute,
Ministry of Water Resources, Ministry of
Transport, National Administration of
Energy
Nanjing, China

Philippe Rigo
Liege University
Liège, Belgium

Francisco Esteban Lefler
PIANC
Brussels, Belgium

Gensheng Zhao
Nanjing, China

Nanjing Hydraulic Research Institute This work was supported by Nanjing Hydraulic Research
Institute

ISSN 2366-2557 ISSN 2366-2565 (electronic)
Lecture Notes in Civil Engineering
ISBN 978-981-19-6137-3 ISBN 978-981-19-6138-0 (eBook)
https://doi.org/10.1007/978-981-19-6138-0

This Springer imprint is published by the registered company Springer Nature Singapore Pte Ltd.
The registered company address is: 152 Beach Road, #21-01/04 Gateway East, Singapore 189721,
Singapore

Contents

Inland Navigation Structure

Smart Shipping

River System Management

Logistics

Future Challenges for Waterway Hydraulic Structures

Claus Kunz[✉]

Bundesanstalt Fuer Wasserbau, Kussmaulstrasse 17, 76187 Karlsruhe, Germany
claus.kunz@baw.de

Keywords: Climate change · Sustainability · Existing hydraulic structures · Gray energy

1 Introduction

Looking into the future, (massive) hydraulic structures face a not inconsiderable number of challenges that need to be overcome. Almost like a bracket, climate change encompasses several issues, of which, in addition to adaptation to climate change and climate change mitigation, examples of newly constructed hydraulic structures, existing hydraulic structures and construction methods for hydraulic structures are mentioned in more detail. Due to the short time available, the topics can only be briefly touched upon. It can be assumed that several contributions to PIANC SmartRivers 2022 will already address individual of these topics. Otherwise, future exchanges on the most pressing issues will be sought.

2 Climate Change

The consequences of climate change are increasingly being felt. In particular, the world faces more extreme weather events, like heat waves, forest fires and droughts, heavier precipitation and an increased risk of flooding and erosion, Fig. 1. The impacts of climate change will further grow in the coming decades and will threat human lives and nature. Impacts affect water management and hydraulic engineering.

Y. Li et al. (Eds.): PIANC 2022, LNCE 264, pp. 1–9, 2023.
https://doi.org/10.1007/978-981-19-6138-0_1

Fig. 1. Flood affecting urban areas

The challenge of climate change makes adaptation to climate change necessary on the one hand and efforts to protect the climate (climate change mitigation) on the other.

2.1   Climate Change Adaption

In order to prepare water bodies and structures planned or located in them for climate change, climate change surcharges can be helpful, which mean a surcharge on the runoff. In Germany, either flat-rate surcharges (e.g. 20%) or surcharges differentiated

T [years]	Climatic factors				
	1	2	3	4	5
2	1,25	1,50	1,75	1,50	1,75
5	1,24	1,45	1,65	1,45	1,67
10	1,23	1,40	1,55	1,43	1,60
20	1,21	1,33	1,42	1,40	1,50
50	1,18	1,23	1,25	1,31	1,35
100	1,15	1,15	1,15	1,25	1,25
200	1,12	1,08	1,07	1,18	1,15
500	1,06	1,03	1,00	1,08	1,05
1000	1,00	1,00	1,00	1,00	1,00

Fig. 2. Climate factors for 5 regions of Baden-Württemberg to be considered for discharges of different annuality (source: LUBW 2015)

by river basin and annuality are known (LAWA 2017), Fig. 2. For the sea level rise on the German North Sea coast, a water level rise of about 0.7 m is to be foreseen for an observation period from now on over the next approx. 100 years. This rise has to be taken into account for constructional facilities.

In the future, building structures must be able to withstand higher temperatures in summer. These temperatures mean higher thermal gradients and consequently greater constraints in the structure. Standardization activities and implementation of the standardization request on adaptation to climate change has been intiated by CEN under the mandate M/526 (CEN 2021).

The management of water, whether low water or high water, can be done by dams or weirs. Depending on the river regime, this requirement may or may not be reconciled with inland navigation. Navigation itself can adapt to climate change, e.g. by developing and maintaining shallow draft vessels for low water periods.

2.2 Climate Change Mitigation

Climate change is attributed to the increase in greenhouse gases. Of the total emissions of the construction industry, about a quarter and thus about 10% of the global CO_2 emissions are attributable to the construction of buildings (Global Allience 2020). It is therefore necessary to record the CO_2 emissions caused by projects in the form of CO_2 equivalents (corresponding to GWP = Global Warming Potential) and to reduce them, cf. in particular Sect. 3. Hydraulic engineering should contribute to the reduction of harmful greenhouse gases through low-emission building materials and construction methods and continue to build durable structures. However, the preservation and extension of the service life of existing (hydraulic) structures also makes a significant contribution to climate change mitigation, cf. Sect. 4.

As an energy-efficient - and moreover safe - mode of transport, inland shipping can contribute to reducing greenhouse gases and thus to climate protection simply by taking over transports. This also includes further optimization and improvement of propulsion technology, alternative drives, ship controls and automation in ship operation. In Germany at least, the potential for transporting goods by waterway could be exploited to a much greater extent.

3 New Structures

3.1 Building Materials

The building materials concrete and steel are predominant in structural hydraulic engineering. Table 1 shows the CO_2 equivalents of the construction materials predominant in hydraulic engineering.

Table 1. Values from (BMI 2022)

Material	Unit	CO_2-equivalents in [kg]
Concrete C30/37	m^3	299
Reinforcement steel	t	684
Construction steel	t	1.127
Wood	m^3	89

In the production of concrete, the main component, clinker, is emission-intensive. Approx. One third of the CO_2 emissions result from the operation of the kiln and approx. Two thirds from the chemical reaction to deacidify the limestone to quicklime (Feldmann et al. 2022). Reducing the proportion of clinker and increasing the proportion of natural additives, such as ground limestone or calcined clays, would be possible solutions. Steel production is also very CO_2-intensive in the production of pig iron and further in the processing into crude steel. Recycling steel scrap into secondary steel can reduce CO_2 emissions by more than half (Arcelor Mittal 2021). However, due to limited supply and strong demand, only around 20% of global steel production can be covered by steel scrap.

In this context, it is important to use the materials optimally within the framework of structural design and to consider the life cycle in order to be able to make comparative considerations and decisions. Thus, the durability of the building materials or even the construction methods also plays a role (Westendarp and Kunz 2020). Durability would also have to be evaluated in the development of new materials. While experience often proves the long-term durability of known materials, suitable performance tests are often lacking for new materials.

3.2   Construction Methods

Alternative construction methods may be considered for some hydraulic structures. For example, a concrete lock could be evaluated over a sheet pile sluice, where the sheet pile steel chamber would generally be constructed with anchorages, if the foundation soil is suitable. A combined construction method for locks would be possible if the excavation pit enclosure were solidly constructed (diaphragm wall construction) and integrated into the final lock construction. Another promising approach is the construction of lock chambers with precast elements (Lühr et al. 2020), Fig. 3, which the BAW is dealing with in a research project (Hasselder 2021).

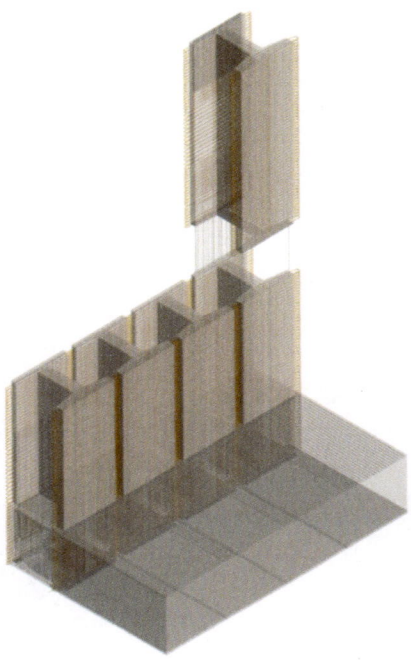

Fig. 3. Cut-out of a lock chamber wall planned with prefabricated cellular concrete elements (Source BAW)

In addition to an optimization of the concrete cross-section, recycled materials could provide for an appropriate dead weight in the cavities in a cellular construction, whereby other construction material properties could be in the lower limit range.

Due to the high corrosion susceptibility of steel weir gates, the BAW is currently investigating the use of inflatable elastomer structures. Inflatable structures have already been in use internationally for decades, but not on shipping lanes and not for dam heights up to approx. 5 m (PIANC 2018). A Eurocode-compliant verification procedure for weir gates of this type planned for German inland waterways has already been developed (BAW 2019), guidelines for hydraulic dimensioning, material requirements and material testing are in preparation. Although the inflatable weir membrane has a shorter service life of approx. 30 years compared to a steel closure, it does not require corrosion protection and replacement is easier.

3.3 Sustainability Assessment

In the future, for new structures CO_2 emissions must be limited and resource consumption reduced. A comprehensive concept is offered by a sustainability assessment, which, however, includes even more far-reaching aspects than the reduction of greenhouse gases, Fig. 4. In principle, it encompasses the 5 criteria of ecological quality, economic quality, sociocultural quality, technical quality and process quality (BMI 2019). Sustainability assessments have already been developed and applied in building construction, but not yet in hydraulic engineering. A partial section considers ecological balancing by means of life cycle assessment (LCA) to quantify the environmental impact (Haller et al. 2022). BAW has recently started a research project "Sustainability assessment for new construction and rehabilitation of massive transport water structures under consideration of established and alternative building materials and construction methods". The project aims to develop a sustainability assessment system for the construction and rehabilitation of massive hydraulic structures. The use of new alternative building materials and construction methods is to be taken into account in the sustainability assessment.

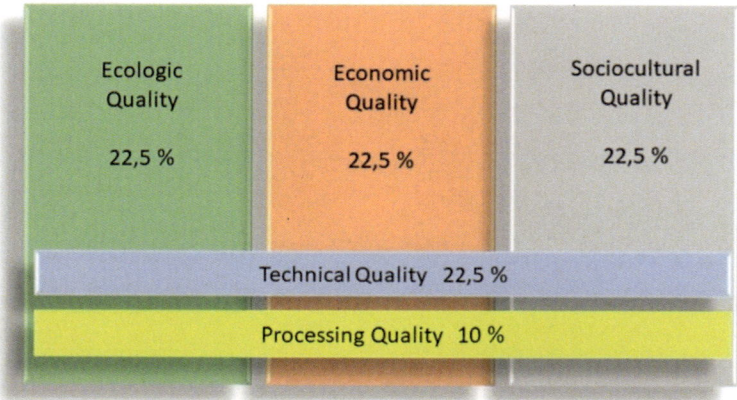

Fig. 4. Five criteria for assessing sustainability and their weighting in [%] (Source: based on (BMI 2019))

However, the most effective measure for reducing CO_2 emissions and resource consumption is to avoid new constructions. This makes the preservation of existing building fabric enormously more important, cf. Sect. 4.

4 Existing Structures

4.1 Static Verification

Hydraulic structures are long usable infrastructure assets that have been designed for a long service life since time immemorial. Age statistics in Germany with approx. 30% of hydraulic structures that have exceeded a service life of 100 years to be applied today should rather fill with pride than indicate the obsolescence of the structures, Fig. 5. Nevertheless, structural verifications are necessary for existing hydraulic structures, because some structures show damage, others have changed their boundary conditions in the course of time, and others were built according to earlier standards, in whose verifications changes and further developments were made (conceptual ageing).

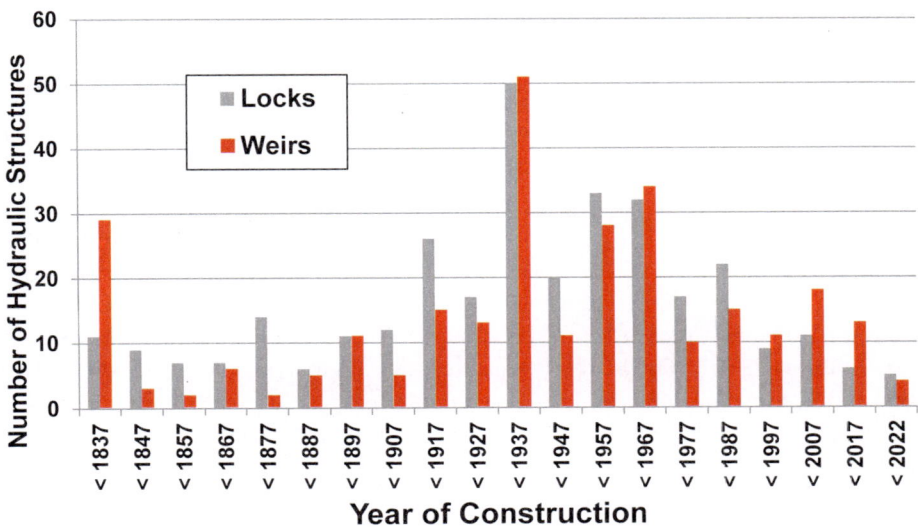

Fig. 5. Age distribution of German locks and weirs within the waterway system (Source: BAW)

In many cases, normative safety can no longer be verified with the relevant standards. From the discrepancy that structures can no longer be verified numerically, i.e. they are "unsafe", and the fact that they have been operated inconspicuously for decades in some cases, a frame-work was developed to get to the bottom of this discrepancy and to clarify the facts leading to the uncertainty. Basis is the verification guideline (BAW 2016). The outcome may lead to an adjustment or change of procedures and standards.

The framework for a more realistic assessment provides for the following three categories:

1. observations, experimental tests and monitoring,
2. revision of actions and verification formats in current standards,
3. considerations of the safety concept and reliability.

For category 1, the application of test loads that exceed normal actions and a subsequent post-calibration can be mentioned. Further a minimum safety level could be calculated, from which a remaining service life can be determined. Within category 2, the application of tensile strength in massive concrete cross-sections is investigated, which is not yet permitted in the current codes. Tensile strength helps to reduce the effect of crack water pressure and to make the verifications more reliably. Within category 3 a reliability level BETA for existing hydraulic structures may be defined different from that for buildings. A risk-based methodology may be used for verifications and for decision making for necessary upgrading measures.

The safe and economical preservation of structures, especially those with a high cement content, by means of adapted verification procedures preserves the "gray energy" contained in the structures and thus prominently serves climate change mitigation. Gray energy is the energy that must be expended for the manufacture, transportation, construction, operation, and demolition of the structure and its components. For a new building, gray energy accounts for over 50% of the energy consumption in the life cycle of the structure, with gray energy representing around 80% of CO_2 emissions.

4.2 Maintenance

In addition to the verification, however, the preservation of existing structures on site, and especially maintenance, is also important. Timely maintenance can effectively prevent exponentially increasing damage, and its cost-effectiveness has been proven. Corrosion is known to be the dominant damage mechanism in hydraulic steel structures. In the field of hydraulic steel structures, the BAW was concerned with repair products for corrosion protection that can be implemented at short notice, so-called SmartRepair (BAW 2020). Products whose rapid on-site application is possible repair damage for a short time so that the progress of corrosion can be halted. BAW is also currently researching maintenance options for the concrete of massive hydraulic structures (BAW 2021).

5 Outlook

In this article, current and near future challenges for hydraulic structures were presented, partly from a German point of view. These are essentially the adaptation to climate change for hydraulic structures and inland navigation, climate change mitigation in the construction of new hydraulic structures and here the assessment of the sustainability of building materials and construction methods. Furthermore the preservation of hydraulic structures through adapted verification formats and above all as an effective contribution to climate change mitigation has been adressed. Some aspects may already be covered by contributions in SmartRivers 2022, others still require discussion and further development. The author would be pleased about a feedback or contact from the circle of the participants by e-mail to one or the other aspect. A further bilateral or international discussion is not excluded.

References

ArcelorMittal (2021) Climate Action Report 2, July 2021
BAW (2016) Merkblatt Bewertung der Tragfähigkeit bestehender, massiver Wasserbauwerke (TbW). Bundesanstalt für Wasserbau, Karlsruhe
BAW (2019) Merkblatt Schlauchwehre (MSW) – Teil B: Nachweis der Tragfähigkeit von Membranen wassergefüllter Schlauchwehre an Binnenwasserstraßen. Bundesanstalt für Wasserbau, Karlsruhe
BAW (2020) FuE-Abschlussbericht "Smart Repair": Reparatur bzw. Ersatz von Korrosionsschutzmaßnahmen zum Erhalt des Korrosionsschutzes und der Stahlkonstruktion, BAW-Nr. B3951.02.04.70009. Bundesanstalt für Wasserbau, Karlsruhe

BAW (2021) Wartung massiver Wasserbauwerke. https://izw.baw.de/publikationen/forschung-xpress/0/BAWFoX_2021_69.pdf. Accessed 30 Aug 2022

BMI (2019) Leitfaden Nachhaltiges Bauen: Zukunftsfähiges Planen, Bauen und Betreiben von Gebäuden (3. Auflage). Bundesministerium des Innern und für Heimat. Berlin

BMI (2022) Ed. Bundesministerium des Innern und für Heimat. https://oekobaudat.de/Oekobau. DAT. Accessed 26 Aug 2022

CEN (2021) Climate Change. https://www.cencenelec.eu/areas-of-work/cen-cenelec-topics/environment-and-sustainability/climate-change/. Accessed 30 Aug 2022

Feldmann A, Dombrowski M, Nearchou N, Grün S (2022) Die Klimakrise – transformation der gebauten Umwelt: Entwurfsgrundsätze bei der Tragwerksplanung. In: Ingenieurbau, June 2022, pp 40–44

Global Alliance for Buildings and Construction (2020) 2020 Global status report for buildings and construction

Haller JI, Appelaniz D, Nowak J, Wrede C (2022) Die Klimakrise – transformation der gebauten Umwelt: Präzisere Einordnung bei der Ökobilanzierung. In: Ingenieurbau, May 2022, pp 46–49

Hasselder M (2021) Einsatz von Fertigteilen im massiven Verkehrswasserbau. In: BAW-Kolloquium Angewandte Forschung, von der Forschung in die Praxis. Karlsruhe/virtual, 03 November 2021

LAWA (2017) Auswirkungen des Klimawandels auf die Wasserwirtschaft – Bestandsaufnahme, Handlungsoptionen und strategische Handlungsfelder. Bund-/Länderarbeitsgemeinschaft Wasser (Hrsg.)

LUBW (2015) Abfluss – BW: Regionalisierte Abfluss-Kennwerte Baden-Württemberg. Landesanstalt für Umwelt und Messen Baden-Württemberg, Karlsruhe

Lühr S, Westendarp A, Stephan C, Kunz C (2020) Einsatz von Fertigteilen im massiven Verkehrswasserbau. Bautechnik 97(6):404–414

PIANC (2018) Report n° 166 – 2018: Inflatable Structures in Hydraulic Engineering. PIANC InCom, Brussels

Westendarp A, Kunz C (2020) Massige Betonbauteile von Wasserbauwerken im (Klima-) Wandel. In: Beton- und Stahlbetonbau, vol 115. Verlag W. Ernst & Sohn, Berlin

Development of Inland Shipping and Construction of New Three-Gorges Ship Channel

Xinqiang Niu[✉], Jundong Wu, Yachao Wang, Xiaowei Wang, Hao Zhou, and Yu Zheng

Chang Jiang Survey, Planning, Design and Research CO., LTD., 1863 Jiefang Avenue, Jiang'an District, Wuhan, China
niuxinqiang@cjwsjy.com.cn

Keywords: Inland shipping of China · The Yangtze River shipping · New three gorges channel · Freight volume

1 Development and Goals of China Inland Shipping

1.1 Development Status

Inland waterways in China flows from west to east, affects the economy of South and North China, communicates with the ocean. It has the comparative advantages of large capacity, less pollution, less land occupation, low cost and low energy consumption. By the end of 2020, the navigable mileage of inland waterways in China was about 127,700 km (Fig. 1), and the inland water freight volume was 3.815 billion tons, ranking first in the world's inland rivers.

1.2 Development Goals

During the "Thirteenth Five-Year Plan (2016–2020) period", China's inland water transportation (IWT) has played an important supporting role in building a comprehensive transportation system and serving the implementation of the national strategy (National Development and Reform Commission 2021). During the "14th Five-Year Plan period", China has entered a stage of high-quality development. The "14th Five-Year Plan" and the strategy of building a transportation powerhouse require IWT to strengthen its role as an important channel and hub in the comprehensive transportation system. The overall goal of IWT at this stage is to basically build a national high-grade waterway of 25,000 km with "four vertical, four horizontal and two networks" by 2035; at the same time, promote the integrated planning and construction of port hubs, improve the collection and distribution system, and vigorously develop rail-water combined transport, water and water transfer, and promote the high-quality development of intermodal transportation (Table 1).

Y. Li et al. (Eds.): PIANC 2022, LNCE 264, pp. 10–15, 2023.
https://doi.org/10.1007/978-981-19-6138-0_2

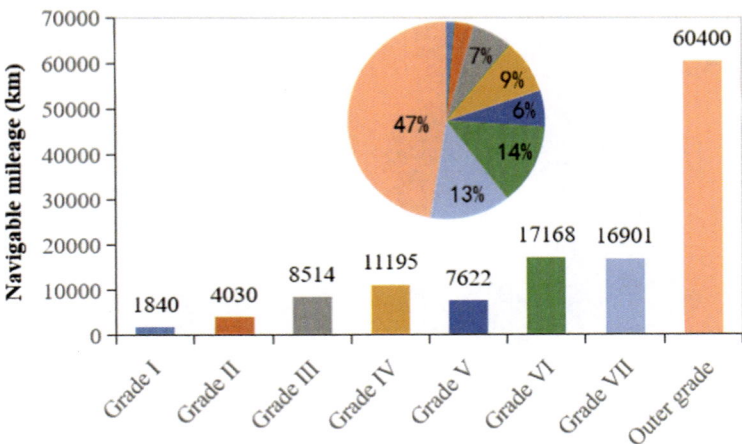

Fig. 1. In 2020, the mileage (km) and proportion of various grades of inland waterways in China

Table 1. Main indicators of China water transport development during "14th Five-Year Plan period" (MOT 2022)

Indicator	Year 2020	Year 2025	Increase
New and improved inland waterway mileage (km)	/	/	About 5000
Newly added national high-grade waterway (km)	/	/	About 2500
Adaptability of passing capacity of large specialized coastal wharfs	>1.0	>1.1	/
Railway entry rate of major coastal ports (%)	>90		/
Average annual growth rate of containerized rail-water combined transport (%)	15		/

2 Main Problems of the Yangtze River Shipping Development

2.1 Development Status

The Yangtze River system is an important part of China's inland shipping network, with a navigable mileage of about 64,700 km, accounting for 50.7% of the national inland shipping mileage. At the same time, it is the golden waterway with the busiest IWT and the largest water transportation volume in the world. The freight volume has ranked first in the world for 17 consecutive years since 2005. In 2021, the freight volume of the Yangtze River mainstream will reach 3.53 billion tons (Fig. 2).

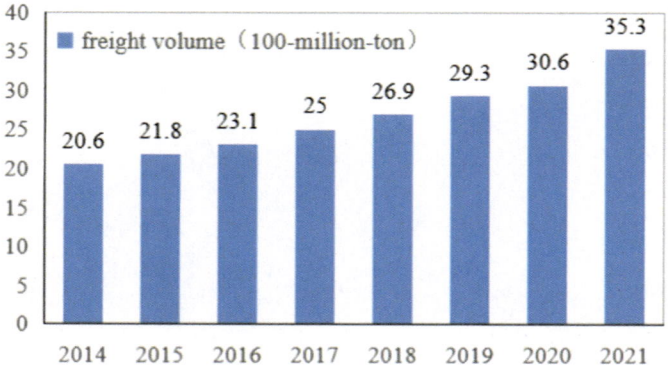

Fig. 2. Development of freight volume in the main stream of the Yangtze River

2.2 Main Problems

In order to implement the concept of the Yangtze River Protection and green development, and pursue the goals of carbon peaking and carbon neutrality, the shipping of Yangtze River needs to comprehensively promote the systematization governance of mainstream waterways in accordance with the general idea of "deep downstream, smooth middle reaches, and extended upstream of the Yangtze River", improve the scale of the waterway and the passing capacity to create a smooth, efficient, safe and green golden waterway. Three main problems need to be solved to achieve the construction goals and promote the high-quality development of Yangtze River Shipping. The first is that the shipping technology system such as channel standards, ship size, and navigable building scale cannot meet the strategic requirements of the golden waterway. Second, there are bottlenecks in the Yangtze River waterway, including the four upper cascade high dams that block the further extension of the Yangtze River waterway to the upstream; the water depth of the middle reaches is relatively deeper compared with the upper and lower reaches, making it difficult to rectify and restricting the development of 10,000-ton river-sea direct ships; the channel conditions of the downstream river are not stable enough, and ship density of local regions is relatively high. The third is that the demand for passing through the Three Gorges Ship Lock has exceeded the design capacity, and the backlog of ships is serious. The efficient and smooth shipping of the Yangtze River needs to focus on solving the problem of insufficient passing capacity of the Three Gorges Hydro-Junction because it is a throat hub of the Yangtze River mainstream.

3 Construction of New Three Gorges Channel

3.1 Development Status and Freight Volume Forecast

Freight volume passing through Three Gorges ship locks has grown explosively after the completion of the Three Gorges Project. In 2011, the traffic volume exceeded 100 million tons, reaching the planned traffic volume 19 years ahead of schedule; in 2021, the total traffic volume reached 151 million tons (Fig. 3). The Three Gorges Hydro-Junction has become the core issue that restricts the further development of the Yangtze River shipping. According to shipping development of the Mississippi River and the Rhine River, it can be seen that the demand for IWT is closely related to the stage of economic development and urbanization development, and will continue to grow until the end of post-industrialization and urbanization (Fig. 4). At present, China is in the transition stage from the middle to the late stage of industrialization and the middle stage to the late stage of urbanization. According to forecast of several domestic authoritative institutions, freight demand of the Three Gorges Hydro-Junction will continue to increase in the medium and long term (Fig. 5). It is predicted that the freight demand will reach 300 million tons in 2050, exceeding the passing capacity of the existing ship locks (CIECC 2021). Therefore, it is proposed to build a new Three Gorges waterway to solve the problem of insufficient passing capacity of the Three Gorges Hydro-Junction.

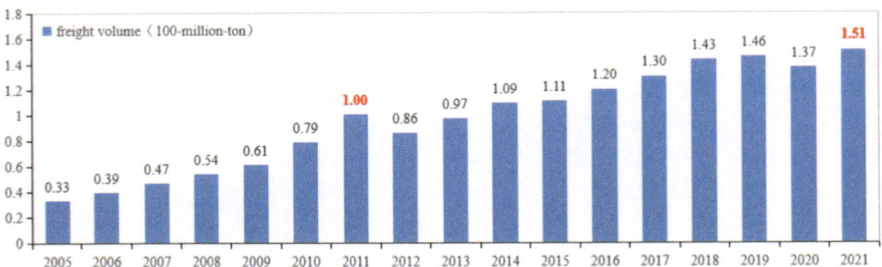

Fig. 3. Freight volume of the three gorges ship lock over the years

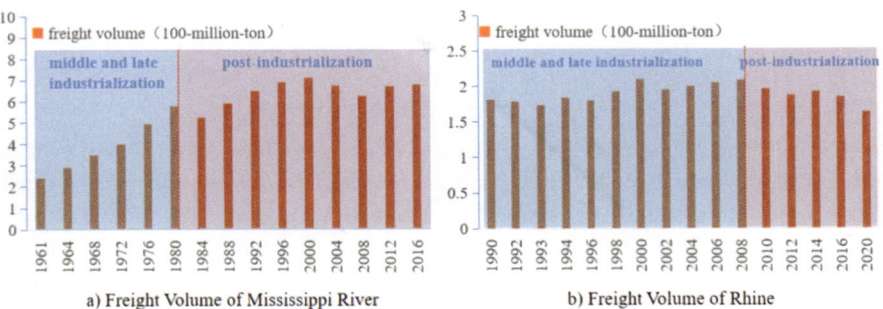

a) Freight Volume of Mississippi River   b) Freight Volume of Rhine

Fig. 4. Freight volume of mississippi river and rhine over the years

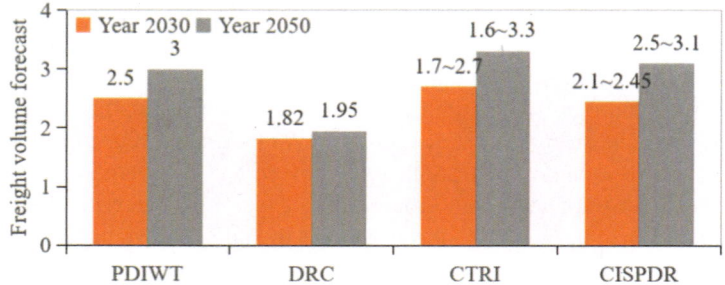

Fig. 5. Freight volume forecast of the three gorges hydro-junction by authoritative institutions

3.2 Construction Standards

The New Three Gorges Waterway Project plans to build two-line ship locks at the Three Gorges Project, and dismantle the original No. 3 ship lock of the Gezhouba Project and build a new two-line large ship lock. The effective size of the new ship lock chamber is 280 m × 40 m × 8.0 m (length × width × minimum threshold water depth), and the maximum designed ship shape is 130 m × 22 m × 5.5 m (length width × depth of immersion). The one-way throughput capacity of the Three Gorges Hydro-Junction will reach about 180 million tons, which meets the demand of the predicted transportation volume in 2050 after the completion of the project, and leave a certain space for development.

3.3 Key Research Questions

The construction of the New Three Gorges Waterway has large-scale, high technical indicators, and many constraints, and there are four issues that need to be studied, which include navigation hydraulics, large-scale long-distance navigation tunnels, super-wide and high water-head herringbone door and hoist equipment, and navigation during construction. **Key issue 1:** the New Three Gorges Channel has a high navigable water head. It has broken through the technical indicators of ship locks built at home and abroad in terms of ship lock water delivery efficiency, energy dissipation of ship lock chamber, and valve air defense. Higher requirements have been put forward for these indicators. **Key issue 2:** a excavated navigation tunnel of the new shipping channel line IV lies on the left bank of the Three Gorges Hydro-Junction. The navigation tunnel has long distance, large cross-section scale, large fluctuation of the upstream water level, and high navigation standards. Since it lacks practical experience of large-scale navigation tunnels, researches on key issues such as navigation safety, fire safety and ship passage way are essential. **Key issue 3:** technical indicators such as minimum water depth on shiplock sill, submerged water depth of the herringbone door operation, and the width of shiplock chamber exceed the built shiplocks, which puts forward higher requirements for normal operation of the herringbone door and hoist equipment. It is necessary to conduct a comprehensive study on the key technical problems such as the structural force and operation mode of the herringbone door, and propose a technical solution for the herringbone door and hoist that meets the operation requirements. **Key Issue 4:** During the construction period, the No. 3 ship lock of the Gezhouba Hydro-Junction needs to be dismantled. In addition, the deep excavation of the Sanjiang Channel will occupy part of the navigation width, which will affect passing capacity. At the same time, considering further increase in the transportation volume during the construction period, it is necessary to combine the transportation system of overturning the Gezhouba Hydro-Junction, focusing on the navigation problems during the construction period.

References

National Development and Reform Commission (2021) Making national development plans for high-quality economic and social development during the 14th five-year plan period. QIUSHI 2021(2020-6):25–35

MOT (2022) The 14th five-year development plan for water transport. Ministry of Transport of the People's Republic of China, Beijing

CIECC (2021) General report on preliminary demonstration and research of deepening the new three gorges waterway. China International Engineering Consulting Corporation, Beijing

Inland Navigation, a Priority for PIANC/InCom

Philippe Rigo$^{(\boxtimes)}$

University of Liege, Allée de La Découverte, B52, 4000 Liege, Belgium
Ph.rigo@Uliege.be

Keywords: Inland navigation · Waterway · Infrastructure · INCOM

1 Introduction

This keynote paper presents the PIANC Inland Navigation Commission (INCOM; http://incomnews.org/). INCOM is one of the 4 technical commissions of PIANC, which is targeting to Inland Navigation, and particularly to the waterway infrastructure and management.

In this paper, we review few recent INCOM reports (already published) and the currently running INCOM Working Groups (WG), which will release soon their reports (2023–2025).

In August 2022, the active InCom members are:

Chairman
Philippe RIGO (Belgium)

Secretary
Jasna MUSKATIROVIC (Serbia)

Members
Leonel TEMER (Argentina)
Jürgen TRÖGL (Austria)
Katrin HASELBAUER (Austria)
Stefan DEVOCHT (Belgium)
Catherine SWARTENBROEKX
(YP, Belgium)
Jose Renato Ribas FIALHO
(Brasil)
Eduardo Pessoa de QUERIOZ
(Brazil)
Mitko TOSHEV (Bulgaria)
Denitsa MATEVA (Bulgaria)

Raphaël WISSELMANN
(Central Rhine Commission)
Luc BOISCLAIR (Canada)
Dianguang MA (China),
Yun LI (China),
Gensheng ZHAO (YP, China)
Fabio ZAPATA (Colombia)
Tero SIKIÖ (Finland)
Benoit DELEU (France)
Fabrice DALY (France)
Gabriele PESCHKEN (Germany)
Katja RETTEMEIER (YP,
Germany)
Takashi KADONO (Japan)
Choonghyo LEE (South Korea)
Otto KOEDIJK (Netherlands)
Hugo LOPES (Portugal)
Ignacio SANCHIDRIÁN VIDAL
(Spain)
Rodrigo Garcia ORERA (Spain)

Houston HETHER (United
Kingdom)
José SANCHEZ (USA)
Brian TETREAULT (USA)
Jeremy GIOVANDO (PIANC
Climate Change expert).

Former active members:
Gernot PAULI (Germany)
Jean-Michel HIVER (Belgium)
Jim Stirling (UK)
Wu PENG (China)

Experts:
Jorge Enrique SÁENZ-SAMPER
(Col.)
Eduardo Nery MACHADO
(Brazil)

Y. Li et al. (Eds.): PIANC 2022, LNCE 264, pp. 16–34, 2023.
https://doi.org/10.1007/978-981-19-6138-0_3

2 Recent Published INCOM WG Report

Here are some recent INCOM/PIANC reports, published and released by PIANC (available on https://www.pianc.org/publications/inland-navigation-commission).

WG 125: Permanent WG on River Information Services (RIS)

- WG 125/I – Guidelines and Recommendations for River Information Services;
- WG 125/II – Technical Report on the Status of River Information Services;
- WG 125/III – RIS Related Definitions 2019.

Chair: Cas Willems (NL)
https://www.pianc.org/publications/inland-navigation-commission/wg125-1

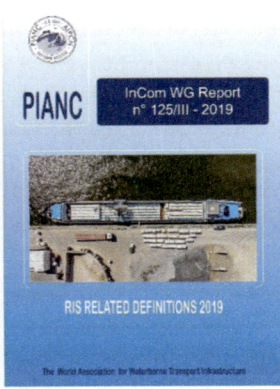

WG 179 – Standardization of Inland Waterways – Proposal for the Revision of the ECMT 1992 Classification;

Chair: Ivo Ten Broeke (RWS, NL)
https://www.pianc.org/publications/inland-navigation-commission/wg179

In 2016, the WG 179sent a questionnaire to all PIANC National Sections to collect information about the applied classification for inland waterways in their countries and their suggestions for modernizing the current ECMT (1992) and UN/ECE classifications.

Based on the results of the questionnaire, the WG concluded that in Europe the connected waterways share a common classification. However, inland navigation and the inland waterways fleet in other countries is rather different from that in Europe. The WG started an investigation into developments in the inland waterways fleet in

Europe. The overall waterway network was studied, and the specific dimensions of existing locks and bridges have been considered in drafting a new classification proposal.

The report includes:

– a proposal for new EU classification and the differences between the ECMT and UNECE classifications are highlighted and explained.
– a description of fleet characteristics and of waterway characteristics.
– enhances the fleet and waterway characteristics findings through a market analysis.
– the synthesis of the information presented in the report to develop the proposal for a revised classification.

WG 189 – Fatigue of Hydraulic Steel Structures;
Chair: Dirk Jan Peters (RHDHV, NL),
https://www.pianc.org/publications/inland-navigation-commission/wg189

This report contains a detailed analysis of the current engineering practice and offers guidelines for a more uniform, systematic approach to fatigue related issues. It provides a summary of the appropriate design tools, analysis methods, technical codes, other guidelines and best practices. It gives examples of both correct and incorrect solutions, provides the discussion of crucial issues and presents the lessons learned from fatigue failures of hydraulic structures. Apart from the design, the report also provides proper recommendations and best practices for the repair of different fatigue damages and for the management (particularly monitoring and assessment) of structures exposed to fatigue.

The existing guidelines and norms that handle fatigue of structures in other fields have thoroughly been reviewed and recommended if and where appropriate.

The matters that have been investigated include: Nature of fatigue in hydraulic structures, significance and specific character of fatigue damage; Identification of fatigue loads, their sources, characters and correlations.

WG 191 – Composites for Hydraulic Structures

Chair: Hota Gangarao (WVU, USA),
https://www.pianc.org/publications/inland-navigation-commission/wg191
Webinar:
Part 1: https://www.youtube.com/watch?v=siPPp9ub88g
Part 2: https://www.youtube.com/watch?v=bR5jcortYH8

Composites have been evolving over the years and are making major in-roads into the marine, aviation and other industries where corrosions and self-weight are the major impediments to advancing the state-of-the-art. Civil Works engineers have been reluctant to take advantage of these composite materials and systems, partially because of the absence of well documented success stories, accepted design and construction practices or specifications, limited understanding of composite system behavior, absence of training in design, construction, evaluation and repair, higher initial costs in some applications and others including unfavorable reputation for recycling. A few navigational structures using fiber reinforced polymer (FRP) composites have recently been designed, manufactured and installed in the United States of America, France, United Kingdom, the Netherlands, and other countries. US Army Corps of Engineers is embarking on higher volume applications of composites for navigational structures.

This report summarizes the state of the art of FRP composites for hydraulic structures including design, construction, evaluation and repair. For clarity and brevity, only essential concepts related to composites, major manufacturing methods, key structural characteristics and engineering science issues of composites are briefly included in the report, while more in-depth general discussions related to composites are directed for deeper exploration by readers through an extensive set of references provided in this report. Emphasis is placed on applications of composites in waterfront, marine, navigational structures including lock gates, gates and protection systems. Design of composite hydraulic structures is presented or referenced for the cases available, such as design of FRP Recess Panel, Wicket Gates, Miter Gates, FRP gates and repair of corroded Steel Piles. This is followed by discussions on operation and maintenance guidance including nondestructive inspection ad evaluation techniques. Cost considerations are discussed in Chapter 7. The report concludes with summary remarks and recommendations.

WG 192 – Report on the Developments in the Automation and Remote Operation on Locks and Bridges

Chair: Lieven Dejonckheere, Vlamse Waterweg, BE

https://www.pianc.org/publications/inland-navigation-commission/wg192

Since 2008, there are a lot more waterways that have implemented or in the process of implementing remote operation technology. At the same time, events around the world have led to a much tighter security posture for marine transportation. These have a significant impact on remote operation of locks and bridges.

The objective of this WG concerns the automation and remote operation of locks and bridges to reflect technological advancement and new considerations related to remote operation.

The WG have collected recent development and case studies from different countries on remote operation of structures. The standards, guidelines and best practices in this field have been reviewed critically. The matters that have been investigated include:

- New development in remote operation of structures
- Physical security including perimeter protection, intrusion detection technology, video analytic and access control
- Network security including protection of data, intrusion prevention/detection (hackers), etc.
- Integration of SCADA and Process control with other systems such as traffic management, RIS, ERP

- Scanning & video technology including High Definition cameras, thermal cameras and advanced image processing such as facial recognition
- Human Factor Engineering
- Simulation technology for training & certification of operators
- Big Data Analysis
- Self-learning technology.

WG 197 – Small Hydro Power Plant in Waterways

Chair: Nicholas Crosby, KCAL-Global, UK
https://www.pianc.org/publications/inland-navigation-commission/wg197
Webinar: https://www.youtube.com/watch?v=V0jJJ5QRgiM

Hydropower structures are rarely built for a single purpose. Hydropower is usually incorporated in a multipurpose system used for water storage (irrigation and drinking water), flood attenuation and water management, navigation, and amenity. In most fully developed economies, all the large commercially viable hydropower potentials have been developed. Even in developing economies, hydropower is often well developed with most of the larger schemes having been developed or under development. However, there is considerable potential in all countries to increase hydro capacity using small, mini- or micro-sized turbines on smaller water courses, rivers, and even man-made canals.

PIANC   InCom WG Report n° 197 - 2021

SMALL HYDRO POWER PLANT IN WATERWAYS
The World Association for Waterborne Transport Infrastructure

Any organization that controls or manages a water course can utilize the potential of moving water to generate renewable energy and inland navigations are an obvious possibility with existing infrastructure creating differences in level and water movement.

In the past, the developers of navigations paid little attention to the effect on the environment of the creation of the navigation. Rivers and water courses were blocked with weirs and dams to facilitate the passage of vessels, preventing the long-distance migration of diadromous fish to/from the sea and even the localized potamodromous movement of fish within the freshwater river system. The transportation of silt downstream during floods, often a source of land fertilization for deltas in the lower reaches, can be blocked by the dams, weirs and other control structures causing land degradation a long way downstream.

WG 198 – Saltwater Intrusion Mitigation in Inland Waterways
Chair: Tom O'MAhoney (NL) and Ruifeng Ray Liang (Co-chair)
https://www.pianc.org/publications/inland-navigation-commission/wg198
Webinar: https://www.youtube.com/watch?v=
siPPp9ub88g

This report provides recommendations for the study of saltwater intrusion in inland waterways and, where necessary or required, its mitigation. Mitigation methods are summarised as well as measurement and modelling techniques that can be used to predict or determine the effectiveness of various measures. Attention is given to both inland waterways (i.e. waterways that are enclosed via dams with shipping locks) and to open river estuaries.

The objectives of the WG were to provide guidance and insight into the measuring, modelling and mitigating saltwater intrusion:

- Quantify salt propagation (speed, distance and concentration);
- Monitor waterway salinity intrusion,
- Measurement: how saltwater intrusion can be measured and monitored;
- Modelling: Physics based multi-dimensional modelling, and physical modelling for insight and solution discovery for salinity intrusion issues,
- Mitigation: Methods for arresting salinity intrusion.

WG 201 – Framework for an Inland Waterway Classification in South America

Chairs: Philippe Rigo (ULiege, BE), Ricardo Sánchez and Azhar Jaimurzina (ECLAC-CEPAL, UN)

https://www.pianc.org/publications/inland-navigation-commission/wg201

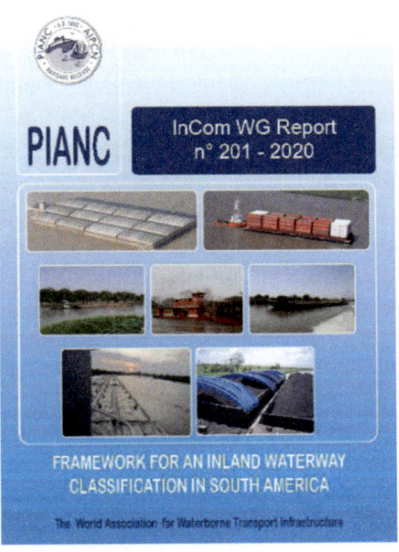

The river network in South America (S.A.) is extensive and consists of some of the largest river basins in the world. However, despite these natural features, inland navigation is underutilized and still plays a marginal role in the transport of commercial goods in the region.

In South America, there are several independent inland waterway systems, which currently have varying levels of development and investment. From a macro perspective, the uses of the inland waterways in the S.A. regions/countries are challenged by various factors. These factors include: a low level of investment in the construction and maintenance of waterway infrastructures and inland ports; incomplete, outdated or absent national and regional norms and regulatory frameworks; poor administrative structures and institutional capacities; limited use of navigational services and technologies.

These challenges have limited the potential of inland navigation, affecting not only the direct use of this mode of transport, but also its integration with other modes.

A common inland waterway classification for South America could be a tool to support the development of inland navigation in South America. Given the experiences in other regions of the world (EU, USA, China, etc.), such classifications can be a powerful and dynamic way to support and implement inland waterway policies and projects as it enables the identification of the limitations and the economic potential of navigable waterways in the region and can encourage and monitor the development of their capacity for transport of goods and people.

In Conclusion, the WG Proposes an Inland Waterways Classification for South America

WG 204: Awareness Paper on Cybersecurity in Inland Navigation

https://www.pianc.org/publications/inland-navigation-commission/tg204.

Chair: Gernot Pauli

Cyberspace is understood as a complex environment where people, software and services interact, supported by information and communications technology (ICT) devices and interconnected networks, especially the internet. Cybersecurity is the protection of this environment and its elements from theft or damage (hardware, software, information), as well as from disruption or misdirection of the services they provide.

Since the end of the last century the number and the complexity of navigational and information equipment installed onboard inland navigation vessels have increased dramatically. ICT is transforming shipping, bringing enhanced monitoring, communication and connection capabilities. One key example are the RIS, which make inland waterway transport safer and more efficient. At the same time, RIS have also reinforced the dependency on ICT and networks. This trend looks set to continue, as there are many initiatives to merge information (related to infrastructure, traffic, vessel, cargo, people, etc.) from various stakeholders (shipping companies, waterways manager, classification societies, ports and governments …), and thereby facilitating the development of new generations of intelligent transport systems, including automatic sailing inland navigation vessels.

In this context, inland navigation should be seen extensively, as a complex system in its own right, including the vessels, waterways, ports, shipping companies and cargo, linked by ICT services (such as RIS) and subject to cybersecurity risks.

WG 210: Smart Shipping on Inland Waterways

Chair: Ann-Sofie Pauwelyn (Vlaamse Wateeweg, BE); Lea Kuiters (NL)

https://www.pianc.org/publications/inland-navigation-commission/wg210

PIANC does not ignore the innovations in the field of autonomous transportation. Autonomous driving and truck platooning are expected to reduce the costs of road transport and to increase its flexibility, while new rail corridors and the next generation of freight trains are expected to lower the technical and organizational barriers for rail freight. It is therefore of paramount importance that technologically innovative initiatives like smart shipping are in the focus of the IWT sector to improve efficiency, safety and sustainability. These improvements will also counter the potentially

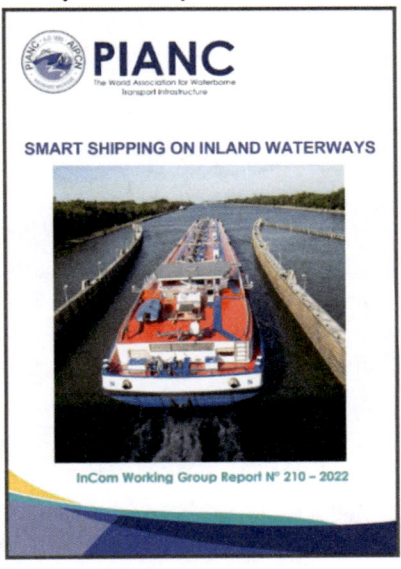

competitive advantage provided by technological evolutions in the other transport modes. When smart shipping is actually used, this will have an impact. Therefore, we need to establish a framework that allows the deployment of smart shipping in a safe and reliable way.

PIANC is aware that Smart developments in other transport modes have been reaching a mature status (as train and automobile). So, this WG refers the methodologies implemented in these modes and focuses on smart shipping with a specific interest on the waterborne infrastructure.

3   The Running INCOM WGs

Here is the list of the running INCOM WGs.

WG 125 – Guidelines and Recommendations for River Information Services (RIS)

Chair: Piet Creemers, VlamseWaterweg, Belgium.

http://incomnews.org/wg?reposname=WG+125+-+River+Information+Services+%28I+-+II%29

Since the last technical report of PIANC on River Information Services the development in the implementation of River Information Services has been considerably.

Since 2010 studies have been conducted on RIS enabled Corridor Management. The concept of Corridor management is recognised as the next step in the deployment of RIS.

"Corridor Management is defined as information services among waterway authorities mutually and with waterway users and related logistic partners in order to optimise use of inland navigation corridors within a network of waterways"

Enhancing inland navigation with the concept of IWT Corridor Management will lead to the benefits for inland waterborne transport in the logistic chain e.g.:

- Reliable voyage planning to improve the operation of skippers, terminal and port operators;
- Improved added value of Vessel Traffic Management Services in the logistic chain;
- Simplification of the administration procedures by the usage of an intelligent information management.

The PIANC RIS guidelines are essential for the further development of River Information Services and as such the development and implementation of RIS enabled Corridor Management.

Based on these and other recent developments, the PIANC RIS guidelines are going to be updated by the Permanent Working Group 125.

WG 128 - Alternative Technical-Biological Bank Protection Methods for Inland Waterways

Chair: Bernhard   Soehngen,   BAW, Germany.
http://incomnews.org/wg?reposname=WG +128+-+Alternative+Technical-Biological

+Bank+Protection+Methods+for+Inland+Waterways

The objective of the InCom WG128 is to understand, evaluate and report on the effectiveness of best practice examples **of innovative (alternative) bank protection measures**, as related to different impact influences and boundary conditions, to fulfil the technical purposes and additionally to improve the ecological conditions.

WG 190 – Corrosion Protection of Lock Equipment

Chair: Rebekah Wilson, USACE, USA
http://incomnews.org/wg?reposname=WG+190+-+Corrosion+Protection+of+Lock +Equipment.

In today's competitive environment, maintenance costs are a crucial and very significant part of a structure's life cycle cost and its ability to deliver value to its owner. Corrosion, whether stress induced and/or caused by the environmental conditions, can be a major degradation factor responsible for significant maintenance costs. One of the most effective strategies to prolong the life of steel structures and equipment is thus a high performance corrosion protection system. There is currently very little research and documentation available for corrosion protection systems as they pertain to lock equipment such as gates and valves which operate in an aggressive aquatic environment and are subjected to hydro-mechanical forces. If such knowledge was easily available, it would allow facility owners and operators to make durable and sustainable decisions, from original construction of lock equipment to long term maintenance strategies that prolong the life of the assets.

WG 199 – Health Monitoring for Port and Waterway Structures
Chair: Brian Eick, Mathew Smith, USACE, USA
http://incomnews.org/wg?reposname=WG+199+-+Health+Monitoring+for+Port+and +Waterway+Structures.

Structural health monitoring (SHM) principles, a damage prognosis strategy, and technology adoption can provide continuous measurements of aging infrastructure to support real-time operations, provide alerts concerning imminent failures, and provide longer-term monitoring to accurately quantify asset and component condition, including remaining service life, risk assessment, and maintenance requirements.

These strategies are built upon a foundation of sensor and inspection measurement data and utilize physical models, numerical simulations, and statistical models to provide a probabilistic measure of condition and probability of failure along with confidence estimates of this quantity. Use of such a probabilistic measure of likelihood of failure will substantially improve confidence in the risk measures used to decide upon infrastructure maintenance and capital expenditures, while also providing defensible evidence as a basis for those decisions.

The main goal of structural health monitoring of Waterway and Port structures is to provide quantified probabilistic measures of risk and reliability necessary to make operational and financial decisions concerning the functionality and safety of those structures.

WG 203 – Sustainable Inland Waterways – A Guide for Waterways Managers on Social and Environmental Impacts
Chair: Andreas Dohms, WSV, Germany.
http://incomnews.org/wg?reposname=WG+203+-+Sustainable+Inland+Waterways+ +A+Guide+for+Waterways+Managers+on+Social+and+Environmental+Impacts
WG203 focuses on:
- Multifunction of Inland Waterways – Chances and Challenges for IW Managers
- Social and Environmental Awareness of Waterborne Infrastructure Managers

This will have a first part focusing of a general concept "Social and Environmental Awareness of Waterborne Infrastructure Managers", also called CSR (**Corporate Social Responsibility**). The purpose is to raise the global awareness of the PIANC community, pushing to change education and mentality towards a more sustainable world.

Then, in a second part subtitled "Multifunction of Inland Waterways – Chances and Challenges for IW Managers" will give case studies showing how these concepts of "Social and Environmental Awareness" and "multifunction of IW (also called Co-

Creation)" have been applied by some managers. The public authorities responsible for common welfare associated with IW infrastructure should consider the application of CSR concepts and approaches in the execution of their responsibilities.

WG 206 – Update of the Final Report of the International Commission for the Study of Locks

Chair: John Clarkson, USACE, USA

http://incomnews.org/wg?reposname=WG+206+-+Update+the+Final+Report+of+the+International+Commission+for+the+Study+of+Locks

The main objective of the WG 206 is to update the PIANC 1986 Report of the International Commission for the Study of Locks. It has been over 30 years since this benchmark document was produced and much has evolved and an updated report, second version, is needed for the navigation community. The new lock design textbook (not for academic education but young professional in lock design) will be a valuable instrument to promote PIANC and the Inland Navigation industry. This publication will serve the navigation community for years and will solidly place PIANC as the preeminent inland water transport organization.

The original document was an outstanding document, 445 pages, in its time how-ever much of it is simply outdated and now is of limited value. Many of the designs presented simply are not used as more efficient, reliable, cost effective, and environ-mentally friendly solutions are favored. There are multiple areas to update.

As a second volume, it is envisioned the basic outline of the book will be retained, updated with new chapters or headings for subjects that were not common at the time for such items such as sustainability. Many countries now have mature water transport infrastructure and it is becoming clear the driving force for design are new efficient rehabilitation strategies when expanding or building a lock and maintain existing traffic in an overcrowded waterway. Other strategies such as in-the-wet construction can allow for much smaller footprint since a full scale cofferdam is not needed.

WG 207 – Innovations in Shiplift Navigation Concepts

Chair: Hu Yaan, Gensheng Zhao, NHRI, China.

http://incomnews.org/wg?reposname=WG+207+-+Innovations+in+ShipLift+Navigation+Concepts

Since the 1990's, shiplift technol-ogy has been developing rapidly in the world and particularly in China, UK, Germany and Belgium. Many different types of shiplifts have been built or are being designed, e.g., Strépy-Thieu shiplift in Belgium, Three Gorges shiplift and Jinghong shiplift in China, new Niederfinow shiplift in Germany, Falkirk Wheel in UK. Many advanced and innovational construction techniques and design concepts have been used in these projects.

It was therefore required to establish a new PIANC report. It discusses the future development direction of shiplifts and give guidance on construction, management and maintenance of new and old shiplifts for the coming 20–30 years.

WG 216 – Best Practices in Planning Inland Waterways Multimodal Platforms
Chair: Philippe Rigo (Uliege, BE); Roberto Zanetti (NL).
http://incomnews.org/wg?reposname=WG+216+-+Best+Practices+in+Planning +Inland+Waterways+Multimodal+Platforms

Planning the development of ports in general, but also of inland ports, is not only a matter of infrastructures but also of intermodality, logistics and service given to costumers:

– Logistics and intermodality are nowadays strongly linked with an efficient transport (goods and passengers). To develop the economy and a sustainable mobility of a region (city), good accessibility (transport) is a key issue. Having a river is not enough, it should be a waterway, allowing cheap, safe, reliable and just on time transportation (import and export) of goods of different types (containers, bulk, construction material, fluid, fuel,..) to the inland ports.
– Service given to costumers: Identifying the customers' needs is also required to identify the needed infrastructures and the relevant facilities as a multimodal platform.

To integrate the infrastructure with the various transportation modes, making a river an efficient mean of transport (an Inland Waterway) the "requested" facility is a multimodal platform. This will contribute to move from RIVERS to WATERWAYS.

This WG focuses on multi modal platform along inland waterways which have other specificities than sea ports (smaller dimensions, different equipment's, other constraints and geographical implementation, traffic density, …and other governances).

WG 219 – Guidelines for IW Infrastructure to Facilitate Tourism
Chair: Rudy Van Der Ween, Port of Gent, BE (in Collaboration with IWI)
http://incomnews.org/wg?reposname=WG+219+-+Guidelines+for+IW+Infrastructure +to+Facilitate+Tourism.

Recreational navigation is a growing activity, also in the managed inland waterway systems. The increase in demand for IW recreational of activities has led to development of infrastructure which should be sustainable and well-integrated with transportation systems.

A sustainable model for navigation recreation infrastructure aiming to encourage sustainable initiatives and measures in the natural spaces, where fluvial tourism activities take place, should be technically and economically feasible but also environment friendly and have to a positive social impact.

Tourism and recreation navigation have the potential to develop synergies with ecosystem restoration, natural protection, and urban waterfront redevelopment providing also social benefits and promoting cultural and historical heritage.

Management measures should not only include immediate actions such as efficient waste management and responsible use of resources (energy), but generate awareness among all actors involved in fluvial tourism activities (SMEs, managers and owners of river ports, professionals, local community, etc.).

WG 228 – Extended Values of "LOW-USE" Inland Waterways

Chair: Arjan de Heer, WitteWeenBos, NL (in collaboration with IWI)
http://incomnews.org/wg?reposname=WG+228+-+Extended+Values+of+Low-Use+Inland+Waterways

Historically, navigable rivers and canals have been an important feature of human society through transportation, water supply, agricultural management, economic and societal benefits. With increasing population and the concomitant development of market economy many rivers have been transformed to inland waterways (IW) operated as navigation corridors by use of structures (locks, weirs,…) and the design of specific vessels to support rapid and efficient transport of bulk materials (e.g. ore, petroleum and coal). Common river training (including dredging, groins, …) was also used to maintain the navigability of the IW across a range of hydrologic conditions. The result has been a complex infrastructure and training management integrated to support navigation during all the year (as much as possible).

This navigation infrastructure requires also governing bodies (as CCNR, USACE, …) to establish parameters for its effective use such as safety and cost-effective use while protecting populations from floods and droughts. In addition, ecosystem services and recreation were often neglected in the past along waterways whereas nowadays the importance of ecological (and social) functions of waterways are emphasized and ask for special care.

Nowadays, while large inland waterways remain economically viable and continue to support substantial commercial navigation, competition from rail and truck has resulted in less commercial use of some smaller inland waterways. Concurrently, large waterways are emphasized with new effective and modern infrastructure and management improvements, and, on the contrary, we observe less incentive for investment in less economically viable inland waterways. These are the so-called "low-use inland waterways" (see definition in Sect. 2).

In many cases, decommissioning the low-use IW and their infrastructure is not feasible and not desirable, thus we need to identify new beneficial values of these inland waterways in terms other than only the commercial transport navigation. Nevertheless maintaining the use of IW by commercial vessels still remains a decisive focus as the developments in Smart Shipping and logistics might provide a strong and sustainable economically viable base.

Economic sectors such as recreational navigation (INCOM WG 219) are significant opportunities to reevaluate society's investment in navigation infrastructure. Indeed, the economic benefits of waterborne tourism need to be considered in the economic analyses.

WG 229 – Guidelines for Sustainable Performance Indicators for Inland Waterways

Chair: Klaas Visser, TUD, NL.

http://incomnews.org/wg?reposname=WG+229+-+Guidelines+for+Sustainable
+Performance+Indicators+for+Inland+Waterways

This WG has as main objective to tackle "sustainable performance indicators", with a specific target on the **hazardous emissions and greenhouse gas production aspects** induced by the IW navigation. PIANC is aware that performance indicators should also relate to technical performance, economic performance, maintenance performance, etc. But these are not in the scope of this WG.

In addition, IW performance depends on:

- The characteristics the waterway (blockage coefficient of confined water, channel design, etc.)
- The shape of the ship hull for propulsion efficiency (hydrodynamics optimization performed during early design stage of a vessel.

These are not in the scope of this WG. The WG targets alternate propulsion technology (e.g. LNG, electric, and others instead of typical fossil fuels).

WG 234 – Infrastructure for the Decarbonisation of IWT

Chair: Mark Van Koningsveld, TUD, NL
http://incomnews.org/wg?reposname=WG+234+TG+Infrastructure+for
+Decarbonisation+of+Inland+Navigation

In 2020, InCom concluded that decarbonisation is of existential importance for inland navigation. Without decarbonisation, IWT will lose all its political support and will become as transport mode "non grata" for freight forwarders. Global and European societal pressure is growing to keep climate change and air pollution within acceptable limits; as illustrated by:

- "*The European Green Deal*" (Dec. 2019), to ensure that Europe will be the first climate-neutral continent, and making Europe a prosperous, modern, competitive and climate neutral economy, as envisaged in the Commission Communication "A Clean Planet for All: A European strategic long-term vision for a prosperous, modern, competitive and climate neutral economy" (November 2018);
- The Paris Agreement Objectives (COP21);
- Accordingly, political and regulatory attention has been increasingly directed towards IWT, as in many cases this transport mode's environmental and climate impact is not negligible. See for example
- The Central Commission for Navigation of the Rhine (CCNR)'s Ministerial Mannheim declaration (October 2018) and the calls from the European Council and European Parliament to enhance the environmental track record of inland waterway transport;
- The Sustainable Development Goals (SDG) of the United Nations' Development Program (UNDP), particular SDG 9 (Industry, Innovation and Infrastructure); SDG 13 (Climate Action) and SDG 14 (Life Below Water).

The availability of zero-emission fuels infrastructure, including onshore electric power supply, will be key to enable zero-emission vessels and increase the competitiveness of IWT as a whole, at a time when other modes of transport are reducing their ecological footprint.

WG 236 – Sustainable Management of the Navigability of Free Flow Rivers
Chair: Calvin Creech, USACE, NL

This WG concerns only free-flowing currents or rivers in their natural state, entirety or partially, in which the flow is not constrained by any hydraulic infrastructure. We shall refer to these as "natural rivers".

Most navigable natural rivers are "large rivers". They are alluvial rivers, with hard points (rock outcrops) in the low-flow and middle riverbed, with specific bottom morphologies. As opposed to regulated rivers, the evolutions of morphological variables (bottom slope, width, cross sections and plan view) of natural rivers makes it difficult to fix a navigation channel in a permanent alignment.

Therefore, there needs to be a change in paradigm and methodologies for navigation in natural rivers.

This WG is dedicated to the improvement of navigability natural rivers that are not intended to be regulated. The goal is finding ways to guaranty/improve navigability without conventional training structures, concrete, rock etc. inside the natural rivers.

WG237 - Bottlenecks and Best Practices of Transport of Containers on Inland Waterway
Chair: Krämer Iven,SWH Bremen, Germany
http://incomnews.org/wg?reposname=WG+237+-+Bottlenecks+and+Best+Practices+of+Transport+of+Containers+on+Inland+Waterways

Making our economies climate neutral will lead to a decline in the volume of fossil fuels, often the most important cargo for inland navigation. In Europe, the market for coal is already shrinking and with the further electrification of road transport, the market for liquid fossil fuels will follow. Thus, inland navigation needs urgently to find other cargoes and to expand its market share in the transport of cargoes, which will stay relevant in the future.

In the Rhine region, the transport of containers is seen for inland navigation, perhaps, the most promising market segment of the future. However, road and rail are strong competitors, having often the advantage of being faster and requiring fewer transshipments of the containers. Thus, inland navigation must become more efficient, more reliable, and more customer oriented to increase its market share or just being able to defend the status quo of container transport.

In other parts of the world, in particular in China, container transport on inland waterways is growing rapidly. Thus, the challenge is less the competition with other modes of transport, rather than managing the growth.

A common feature of large scale container transport on inland waterways is its interdependence with seaports. This is not a surprise, as seaports are generally the largest hubs for container transport. In Europe, container transport on inland waterways is by far most developed in the wider Rhine region, where Europe's busiest seaports are situated. In China the container transport on inland waterways concentrate mainly on the Pearl river and the Changjiang river. The estuaries of these rivers are home to the largest container ports in the world. In the port of Shanghai, which is the largest container port in the world, about half of the container volume is handled at terminals at the river mouth which can be accessed by inland vessels. Another half is handled at the terminals out of the river mouth (Yangshan port area), which are served by river/sea vessels. This type of transport is growing rapidly along with increased waterway dimensions of the Changjiang river.

WG241 - Handling Accidents and Calamities in Hydraulic Structures
Chair: Yves Masson, CNR, France, with Richard Daniel (NL), and Tim Paulus (USACE, USA)
http://incomnews.org/wg?reposname=WG+241+-+Handling+Accidents+and +Calamities+in+Navigation+Hydraulic+Structures

While the prior objective of hydraulic structures (such as lock gates, navigation river weirs and storm surge barriers) is to remain in service, engineers must also be capable to adequately handle their failures. Despite the ongoing development of expertise, design tools, norms, and construction methods, there are still a considerable number of accidents and calamities that happen to such structures. In addition, the costs of losses as result of these so-called "upset events" are growing due to the growing complexity of waterborne infrastructure.

Accidents to hydraulic structures, like locks, navigation weirs and flood barriers, happen not only as result of strength excess. Other possible causes are, e.g., unforeseen conditions, lack of maintenance, improper operation, and navigation errors. There are often combinations and complex sequences of events that lead to disastrous results.

Various PIANC WGs have, so far, provided guidance for preventing accidents from happening, particularly the accidents resulting from strength excess. While this should remain the engineer's main concern, there is also a demand for guidance how to effectively handle the accidents and calamities that have actually happen. This is a matter of combined effort of not only engineers. Nevertheless, engineers can and should contribute to the solutions in such cases.

WG242 -Permanent Floating Houses Along IW Banks and Infrastructure
Chair: Heiner Haass, D-Marin Consult, Germany
In many large cities (Paris, London, Amsterdam, ..) a significant ratio (%) of the inland waterway banks (and sometime also the infrastructure) are used by moored ships and floating houses as fixed and permanent residences or accommodations (for private and commercial uses). These (*often old*) former cargo ships often are not able to navigate anymore and have been fully been refurbished for a comfortable live (with a garden/terrace on the upper deck, large living room in the cargo holds, ..).

The floating houses are for a large part the result of a societal problem, namely the shortage of (affordable) houses, and/or a trend to live close to the nature.

The floating houses for residential or hospitality uses are connected to land with permanent gangways and water and electricity supply and sewage treatment systems.

The WG will make recommendations.

4   Conclusions

PIANC/INCOM is a key worldwide scientific and technical Association in charge of the dissemination of the knowledge in the field of inland waterways, inland navigation and inland infrastructure.

http://incomnews.org

Welcome to the PIANC INCOM Commission

InCom (Inland Navigation Commission) is one of the 4 international technical commissions of PIANC (Permanent International Association of navigation Congresses)

The InCom Commission focusses :

- Inland Navigation.
- Inland Waterways.
- River and Port Infrastructures.
- Inland Waterway transport and Logistics, ..

Please, feel free to surf this web-page in order to get more information about our Inland Navigation Commission

Contact: Prof Ph RIGO. , Univ of Liege, InCom Chairman; Tel: +32 4 366 93 66 (office), or use the CONTACT FORM

If you have expertise in this field, contact PIANC to join an INCOM WG to share your expertise and get benefit of the PIANC network. **YOU ARE WELCOME!**

References

INCOM. http://incomnews.org
PIANC. https://www.pianc.org/
INCOM WGs. https://www.pianc.org/publications/inland-navigation-commission

Innovations in Shiplift Navigation Concepts

Yaan Hu[1](✉), Gensheng Zhao[1](✉), Claus Kunz[2], Zhonghua Li[1],
Jan Akkermann[3], Marc Michaux[4], Fabrice Daly[5], Jim Stirling[6],
Weili Zheng[7], Jean-Michel Hiver[8], Michael Thorogood[9],
Jianfeng An[1], Xin Wang[1], Shu Xue[1], and Chao Guo[1]

[1] Nanjing Hydraulic Research Institute, 223, Guangzhou Road, Nanjing 210029,
China
{yahu, gszhao}@nhri.cn

[2] Bundesanstalt fuer Wasserbau, Kussmaulstrasse 17, D-76187 Karlsruhe,
Germany

[3] University of Applied Sciences Karlsruhe, Moltkestraße 30,
D-76133 Karlsruhe, Germany

[4] Service public de Wallonie mobilité infrastructures, Rue Verte, 11,
7000 MONS, Beglium

[5] Cerema, 12 Rue Léon Teisserenc de Bort, 78190 Trappes, France

[6] Project Solutions, Solutions 4U Ltd Hamilton, Rue Verte, 11,
South Lanarkshire ML3 6JT, UK

[7] General Navigation Administration of the Three Gorges Dam of Yangtze
River, NO. 12, Shangdaodi Road, Xiling District, Yichuang City, China

[8] Université Libre de Bruxelles Ecole Polytechnique - BATir - CP 194/250,
avenue Franklin Roosevelt-1050, Bruxelles, Belgium

[9] Eadon Consulting Limited Advance Manufacturing Park, Technology Centre,
Catcliffe, Rotherham, UK

Keywords: Shiplift · Innovation · Navigation · PIANC WG 207

1 Introduction

Shiplifts are one of the main types of navigation structures in canals and high dams in natural rivers. In the 21st PIANC International Navigation Congress, the development of shiplifts was emphasized, and it is recognized that shiplifts have several advantages over navigation ship locks when the lift height is over 40 m. In 1984, PIANC established a "Study Commission of ship lift" and published a technical report in 1989 named "Ship lifts", which introduced and summarized the experiences in design and management of shiplifts from 1950 to 1986.

Since the 1990's, shiplift technology has been developing rapidly in the world and particularly in China, UK, Germany and Belgium. Many different types of shiplifts have been built or are being designed, e.g., Strépy-Thieu Shiplift in Belgium, Three Gorges Shiplift and Jinghong Shiplift in China, new Niederfinow Shiplift in Germany, Falkirk Wheel in UK. Many advanced and innovational construction techniques and design concepts have been used in these projects.

The tasks of the PIANC WG 207 report provides guidelines and recommendations to persons involved in shiplift research, design, construction, management and maintenance, with a specific focus on the historical ship lifts. Database of shiplifts, including

© The Author(s) 2023
Y. Li et al. (Eds.): PIANC 2022, LNCE 264, pp. 35–51, 2023.
https://doi.org/10.1007/978-981-19-6138-0_4

brief lists of the types, dimensions and technical parameters of representative shiplifts in the world. Evaluation of research, design, construction, management and maintenance of Shiplift and approaches used for operational, engineering, financial and policy decision-making. WG 207 report includes conceptual design, design research, analytical models, numerical models, desktop and physical models, prototype survey and test, which are used to address the new developments of shiplifts in the world.

2   Layout of Shiplifts

General the primary layout and components of vertical ship lift contain the upper approach, tower, chamber, drive system, safety mechanism, balance system, lower approach and etc., which given below and shown schematically in Fig. 1.

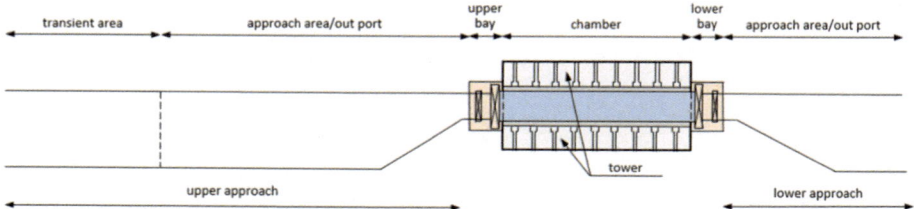

Fig. 1. Scketch of shiplift layout

The determination of the ship lift type should take into account the impact of the flood discharge, sediment, power station operation and etc. When there are no engineering cases to reference, it should be determined through special studies.

Normal the shiplift should be designed single stage. When restricted by terrain, geological conditions or the lifting height is too large, a multi-step scheme can be adopted. Open channels, water bridges, water tunnels and other types of intermediate channels can be used between the stages of the multi-step shiplift.

When the variability of navigable water level is small, the all balanced vertical ship lift should be used; when the hydrological water level fluctuations or short-term operational water level fluctuations is large is large, the partial balanced vertical ship lift or special upper/ lower docking station should be adopted.

3   Innovation of Typical Shiplifts

3.1   Strépy-Thieu Shiplift

The Strépy-Thieu Shiplift lies on a branch of the Canal du Centre in the municipality of Le Rœulx, Hainaut, Belgium. With a height difference of 73.15 m between the upstream and downstream reaches, it was the tallest shiplift in the world upon its completion, and remained so until the Three Gorges Dam ship lift in China was completed in January 2016.

Side elevation

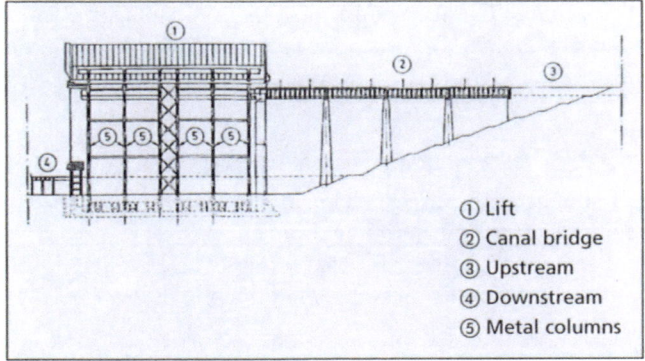

① Lift
② Canal bridge
③ Upstream
④ Downstream
⑤ Metal columns

Fig. 2. Side view of Strépy-Thieu Shiplift

The structure at Strépy-Thieu Shiplift consists of two independent counterweighted caissons which travel vertically between the upstream and downstream sections. Because of Archimedes' Principle, the caissons weigh the same whether they are laden with a boat or simply contain water (Fig. 2). In practice, variations in the water level mean that the mass of each caisson varies between 7200 and 8400 tonnes. The caissons have useful dimensions of 112 m × 12 m and a water depth of between 3.35 and 4.15 m.

Each caisson is supported by 112 suspension cables (for counterbalance) and 32 control cables (for lifting/lowering), each of 85 mm diameter (Fig. 3). The mass of the counterbalance was calculated to keep the tension in each of the control cables below 100 kN at all times. The suspension cables pass over idler pulleys with a diameter of 4.8 m. Four electric motors power eight winches per caisson via speed-reduction

Fig. 3. Machine hall of Strépy-Thieu Shiplift

gearboxes and the 73.15-m lift is completed in seven minutes. The structure is massively reinforced to provide rigidity against torsional forces during operation and has a mass of approximately 200,000 tonnes. The vertically moving watertight gates are designed to withstand a 5 km/h (3.1 mph) impact from a 2000-tonne vessel.

3.2 Three Gorges Ship Lift

The Three Gorges ship lift is the largest rack and pinion climbing ship lift in the world, with a lifting height of up to 113 m, internal dimensions of 120 × 18 × 3.5 m (useable space) and moving mass of approx. 15500 tons. The technical level of complete equipment manufacturing, installation and commissioning of the Three Gorges ship lift have reached the world advanced level.

Four sets of rack and pinion climbing drive mechanisms of the Three Gorges ship lift are installed on the ship compartment, and four sets of accident safety mechanisms are arranged adjacent to the drive mechanism. The safety mechanism adopts the long nut column short rotating screw type, and uses the friction self-locking condition of the screw and nut trapezoidal thread to realize the safe locking of the ship's cabin. The driving mechanism and safety mechanism of the Three Gorges ship lift have complex forms, and the upstream and downstream water levels change greatly and rapidly during operation. There are many technical problems as follows:

- Shifting control of cabin equipment and tower structure.
- Clearance design of thread pair of safety mechanism.
- Rack, BSO support and force transmission design.
- Super large cabin structure design.
- Influence of water level variation and rate of change on gate arrangement and gate.
- Concrete tower column construction and major equipment installation.

The ship chamber is the core equipment of the ship lift, and the design of other structures, equipment and facilities is centered on the ship carriage. Self-supporting ship carriage is generally composed of two main girders, two safety beams, two driving beams, a chamber panel structure and a head of ship carriage structure and an accessory structure. Both ends of the safety crossbeam and the driving crossbeam are suspended to the outside of the main longitudinal beam to form a flank platform structure (Fig. 4).

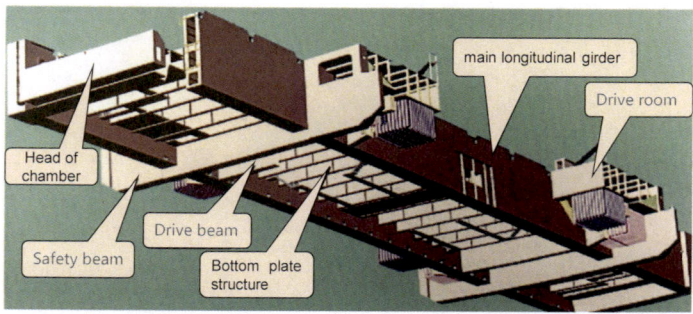

Fig. 4. Layout of self-supporting ship chamber

The Three Gorges Ship Lift adopts the scheme of stacking beams with the same gate slot and lifting flat gate. The scheme has outstanding advantages in gate sealing reliability and equipment operation safety, and has good applicability for large navigable water level variation. The upper gate of the Three Gorges Ship Lift has 7 working overlapping beams and 1 working flat gate. The upper gate of the Three Gorges Ship Lift has 7 working overlapping beams and 1 working flat gate. The auxiliary water retaining gate is arranged in the form of the same slot overlapping beam and the lifting flat gate, which is located upstream of the working gate. The auxiliary gate is used as the accident maintenance gate. The auxiliary overlapping beam has 8 sections (Fig. 5).

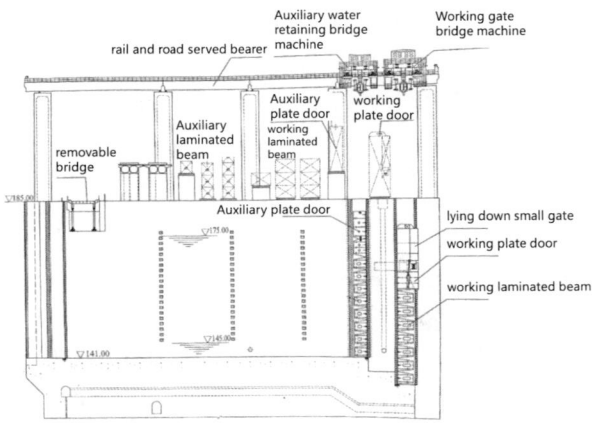

Fig. 5. Layout of upper ship lift gate and equipment

The water level of the downstream approach channel of the ship lift is generally affected by the negative adjustment of the unit, the flood discharge of the junction and the downstream flood, and the change rate is generally fast. The change amplitude of the downstream water level of the Three Gorges ship lift is about ±0.5 m/h. In order to adapt to the change of water level, the working gate of the lower lock head of the Three Gorges ship lift is a super large double leaf flat sinking gate with pressure adjustment, inflatable water stop, step-by-step locking, and a small lying down gate (Fig. 6).

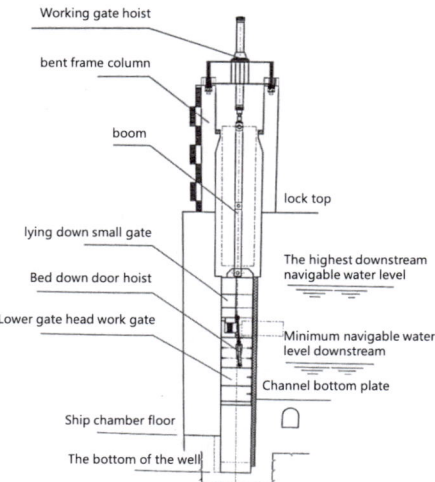

Fig. 6. Layout of lower ship lock gate and equipment

3.3 Jinghong Hydro-Floating Shiplift

Jinghong Hydro-Floating shiplift is located in the Jinghong hydropower project on Lancanjiang River in Yunnan Province (China), which is the first hydro-floating Shiplift in the world. It is also the biggest navigation structures in Yunnan province. The lift's preliminary study is at 2004, and has gone into operation in Nov, 2016.

Hydro-floating ship lift (HFSL) is a new type of ship lift invented by Chinese Engineers in 2000's, which main components are consist of ship chamber, counter-weight, shafts, F/E system, mechanical synchronization system and so on. The lift uses water energy to drive ship chamber moving up and down.

The principle of Hydro-floating lift is similar with lock, water filling or emptying from 16 different shafts which longitudinal arranged in tower both side of the chamber, the counterweights in the shafts are ascend or descend (filling process or emptying process) with the variations of shaft water level. The ship chamber connected to the counterweights with steel cables will lifting with the movement of the counterweights (Fig. 7).

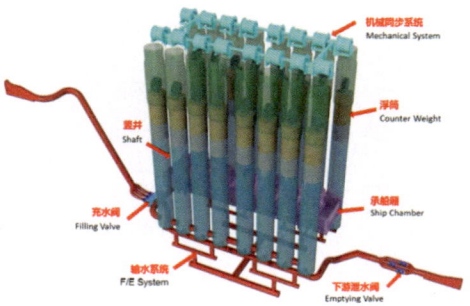

Fig. 7. Components of HFSL

Fig. 8. Filling valves of F/E system

HFSL is different with the motor-driven ship lift, which using the water power to drive the chamber, the lift weight of hydro-floating ship lift is no longer limited by the motor capacity. Therefore there is possibility to design larger-sized ship chamber.

Filling/emptying valves are main control devices of HFSL, which is comprised of 3 piston valves with diameter of 1.6 m (Fig. 8). The piston valves can precisely control the operation of ship lift by regulating flux in F/E system. F/E system adopted equal inertia layout. The water flow in the main pipe with diameter of 2.5 m is distributed both horizontally and vertically twice to 16 branch pipes with diameter of 1.6 m which connected to 16 shafts.

Hydro-floating ship lift applies the advantages of simple maintenance and high safety standard of ship lock, and the advantages of high efficiency of motor-driven ship lift. The hydro-floating ship lift has high-level safety standard. When the load of chamber changes, the draught of counterweight is auto adjusted by the buoyancy, the chamber and counterweight are keep balanced again, which resolved the shiplift's safety in accidents such as chamber leaking.

3.4 Goupitan Multi-step Shiplift

Goupitan multi-step ship lift is located in the main stream of Wujiang River in Guizhou Province, China, which is part of Goupitan Hydropower Station with a dam height of 225 m. The maximum water difference needs to be overcome for ships is 199 m. The variance of the upstream navigable water level is up to 40 m. It is the current navigable building with the highest head and maximum water level variation in the world. Three-step vertical ship lifts are built for the first time at home and abroad. In order to fit the large variation of the water level both the upstream and the downstream, the first and third step lifts take the rope hoist type with chamber entering water. So, they are also the largest ship lifts with chamber entering water. The lifting height of the second step is up to 127 m, which is the highest lift. The three ship lifts are connected with navigable tunnels and aqueducts, and the operating conditions are complex. A number of technical indicators have exceeded the domestic and foreign built ship lift (Fig. 9).

Fig. 9. Goupitan three-step ship lift in China

The main buildings include upstream approach, first step vertical lift, first step inter-mediate channel (including navigable tunnel, aqueduct), second stage vertical lift, second step intermediate channel (including aqueduct), third step vertical lift and downstream approach (Fig. 10). Among them, the first and third steps adopt the wire rope hoisting vertical lift with chamber entering the water. The second step adopts the wire rope hoisting full balance vertical lift. The total length of the three-step vertical lift line is 2181.7 m.

The first step lift is arranged in the upstream reservoir to adapt to the change of navigable water level from 590.0 m to 630.0 m in the upstream. The lower gate head is connected with the first step intermediate channel through the aqueduct. The water level of the first step intermediate channel is 637.0 m, the maximum lifting height of the first step lift is 47 m. The second step lift is arranged between the two intermediate

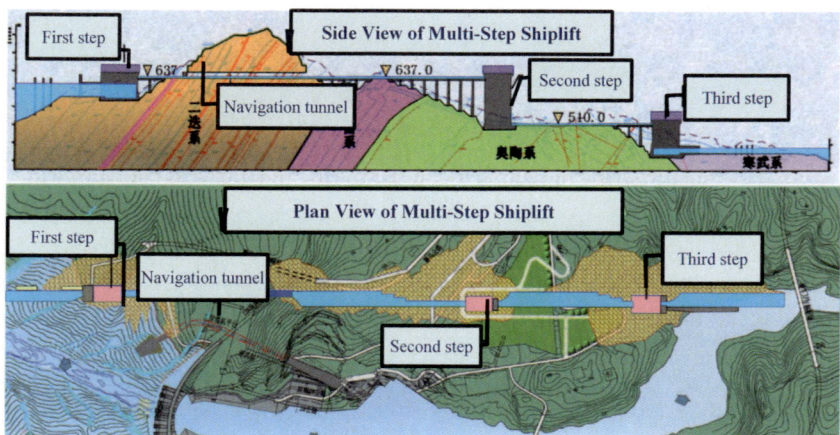

Fig. 10. Layout of Goupitan shiplift

channels to overcome the head drop between the two channels, and the upper and lower gate heads are respectively connected with the two intermediate channels through the aqueduct, the water level of the second intermediate channel is 510.0 m. The lifting height of the second step ship lift is 127 m. The third step lift is arranged in the downstream, which is used to adapt to the change of downstream navigable water level from 431.0 m to 445.82 m of the hub. The upper gate head is connected with the downstream intermediate channel through the aqueduct, and the lower gate head is connected with the downstream approach, and the maximum lifting height is 79 m.

The chambers of the first step and third step ship lifts are directly entering the water to connect the upstream and downstream without docking with gate heads. This ship type has several advantages. For instance, it could adapt the large variation of the water level to solve the problem of the distribution of the gate head. Also, many devices in the normal gate head are saved and the operation procedure is simplified. The transporting ability is promoted as well.

3.5   The Falkirk Wheel

The chambers of the first step and third step ship lifts are directly entering the water to connect the The Falkirk Wheel, opened in 2002, is the only rotating ship lift in the world. It is located in central Scotland at Falkirk and was conceived as the centerpiece of a large economic and social regeneration project known as The Millennium Link that aimed to restore the Forth and Clyde Canal and the Union Canal, both of which had been closed to navigation in the 1960's (Fig. 11).

Whilst most shiplifts are built to support commercial goods transport, the Falkirk Wheel was envisaged as a visitor attraction to move leisure traffic and to boost tourism.

Fig. 11. Layout of Falkirk wheel and union canal

Aesthetics thus played a huge role in the design and layout of the Wheel and the approach aqueduct with many of the design decisions being guided by aesthetics rather than purely technical reasoning.

The wheel has an overall diameter of 35 m and consists of two opposing arms extending 15 m beyond the central axle and taking the shape of a Celtic-inspired, double-headed axe. Two sets of these axe-shaped arms are connected to a 3.8 m diameter central axle of length 28 m. Two diametrically opposed water-filled caissons, each with a capacity of 250,000 L, are fitted between the ends of the arms.

The caissons or gondolas always carry a combined weight of 500 tonnes (490 long tons; 550 short tons) of water and boats, with the gondolas themselves each weighing 50 tonnes (49 long tons; 55 short tons). Care is taken to maintain the water levels on each side, thus balancing the weight on each arm. According to Archimedes' principle, floating objects displace their own weight in water, so when the boat enters, the amount of water leaving the caisson weighs exactly the same as the boat (Fig. 12). This is achieved by maintaining the water levels on each side to within a difference of 37 mm (1.5 in) using a site-wide computer control system comprising water level sensors, automated sluices and pumps. It takes 22.5 kW (30.2 hp) to power ten hydraulic motors, which consume 1.5 kilowatt-hours (5,100 BTU) per half-turn, roughly the same as boiling eight kettles of water.

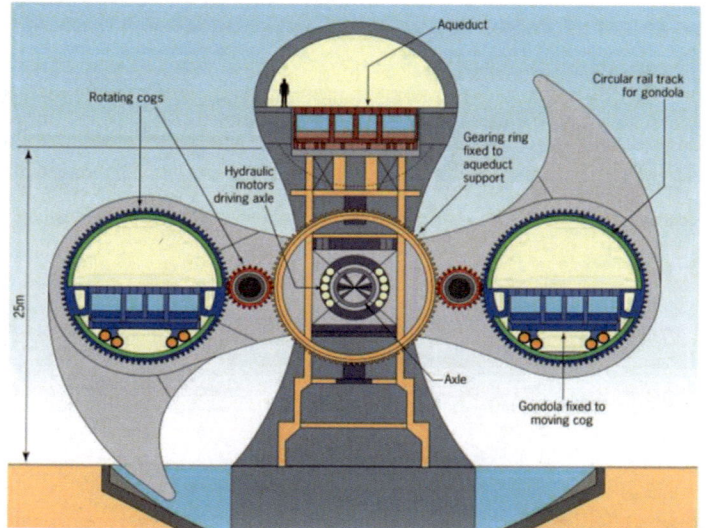

Fig. 12. Cross section of the wheel's arms

The caissons are required to turn with the wheel in order to remain level. Whilst the weight of the caissons on the bearings is generally sufficient to rotate them, a gearing mechanism using three large identically sized gears connected by two smaller ones ensures that they turn at precisely the correct speed and remain correctly balanced.

Each end of each caisson is supported on small wheels, which run on rails on the inside face of the 8 m diameter holes at the ends of the arms (Fig. 13).

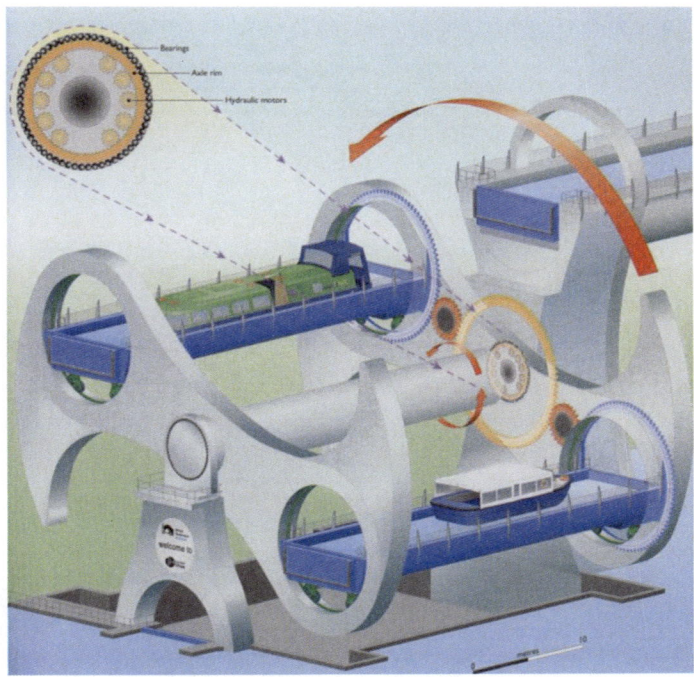

Fig. 13. Operation mechanism of Falkirk wheel

3.6   New Niederfinow Shiplift

The capacity of the existing ship lift is not big enough to manage the prospected traffic volume. Also after seventy years of operation, the costs of maintenance are increasing. Therefore the WNA Berlin was assigned the task to plan and erect a new construction to surmount the 36 m height incline in Niederfinow (Fig. 14). It takes navigability, environment, and costs into consideration (Fig. 15).

The ship lift consists of a ship chamber with counterweights and a supporting concrete construction. The main dimensions of the construction amount to length/width/height 133 m/46.4 m/54 m (Fig. 16). The foundation is constructed as an 11 m deep concrete trough. The construction is symmetrical and consists of four u-shaped towers, one in each quarter, and 12 columns. The lift operates with 14 sets of concrete counterweights balancing the deadweight of the ship chamber and the water volume up to the operation level. 224 wire cables run from the chamber to the top of the lift, over pulleys, and down to concrete weights. These vertical loads of 2 × 8,500 t are transferred into the concrete structure by two cable pulley beams, each of them placed along the side of the construction on top of two towers and six of the columns.

Four engines with total 1280 kW move the chamber by a rack-and-pinion drive; each engine comprises of two electric motors and corresponding gears (Fig. 17). They are situated in engine rooms at the sides of the chamber. The engines are synchronized by shafts and also electronically. Because of the counterbalanced system, only a small amount of power is required to overcome the friction in the bearing and guidance

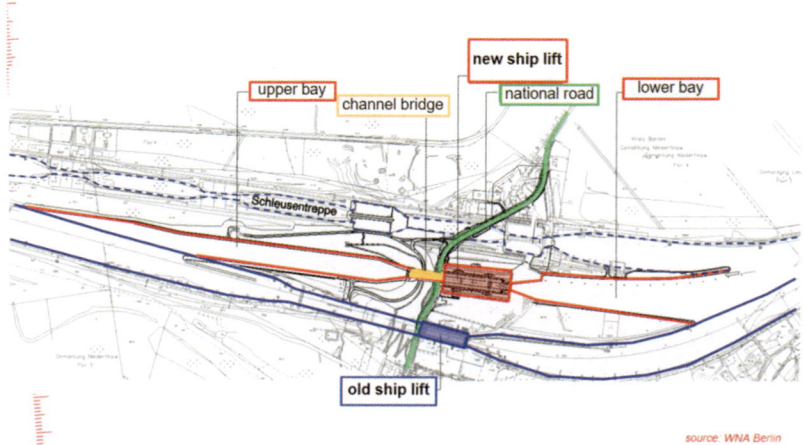

Fig. 14. Layout of New Niederfinow Ship Lift and Old Niederfinow Shiplift

Fig. 15. New Niederfinow Ship Lift and Old Niederfinow Shiplift

systems and to accelerate the chamber. In the case of imbalance, for example caused by an empty chamber, the rack-and-pinion drive cannot stand the load. For this purpose the ship chamber safety system was developed. The system consists of a rotary-lock-bar, shaped like a screw with a diameter of 1085 mm, embedded in a 36 m long splitting side-thread. The rotary-lock-bar is connected with the chamber across the split (Fig. 18).

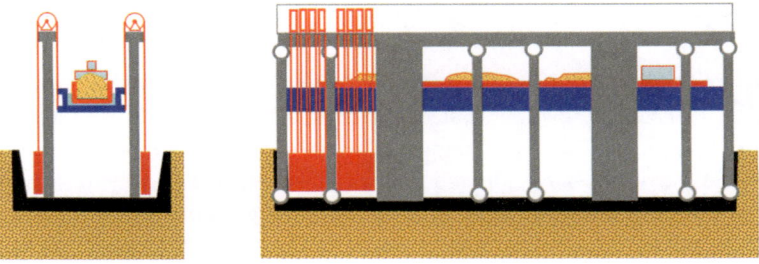

Fig. 16. Layout of counterweights and a supporting concrete

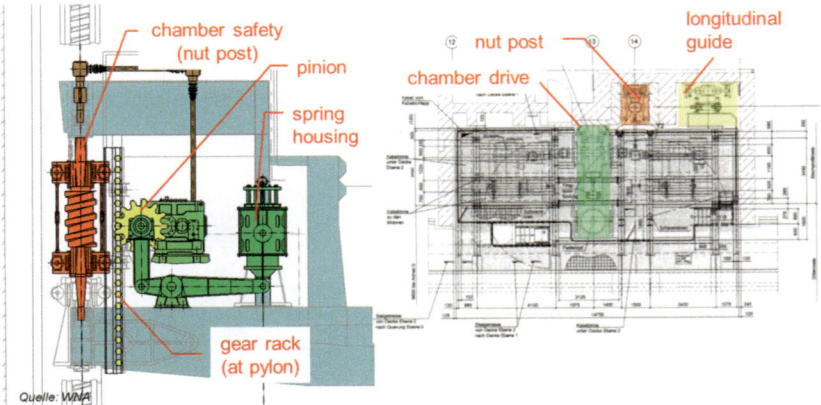

Fig. 17. Layout of chamber drives

Fig. 18. Photo of gear rack and nut post

4   Analysis Methodologies of Shiplifts

The construction of ship lift is a systematic project integrating hydrodynamics, civil construction, mechanical, hydraulic, steel structure, electric engineering, etc. Its overall operating characteristics are very complex. The overall physical model test of the ship lift can reflect the actual operation and force condition of the prototype ship lift more objectively, so the physical model test is its main research method. Numerical simulation has the advantages of low cost, short cycle time, and easy model modification, etc. It can also simulate accidents and extreme conditions that cannot be performed in physical model tests, guarantee the safety of test personnel and instruments, and provide technical guarantee for the safe operation of ship lifts under various conditions. With the rapid development of computer technology and numerical calculation methods, numerical simulation technology is playing an increasingly important role in the field of ship lift research. Besides, the prototype observation can find the problems existing in the operation of a new ship lift, deepen the understanding of the operation characteristics of ship lifts, and obtain a large number of measured data to guide the operation and management of ship lifts, it has also become an essential research tool after the construction of large ship lifts.

4.1   Physical Model

Shiplift research covers hydraulics, structural mechanics, electromechanical control and other subjects. The technical problems are very complex. It is difficult to reflect the complex interaction of "ship-water-ship chamber-driving system" of ship lift by partial physical model (Fig. 19).

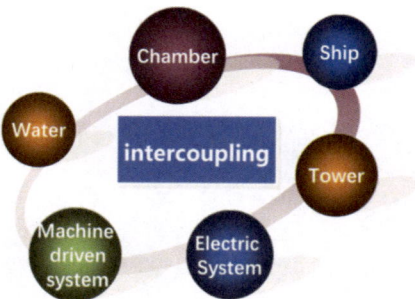

Fig. 19. Complex interaction of ship lift

Due to different research objects and scientific problems in different parts of the ship lift, the similarity rate and design method of each part are also different (Fig. 20).

- The scale of physical model is generally 1:10–1:20;
- Research on mechanical synchronous system focuses on synchronous shaft torque and torsional deformation during operation, so the model design should meet similitude of motion, stiffness and geometric deformation;

- Ship chamber hydrodynamics consist of the load changes caused by slamming force and adsorption force when the bottom of the ship chamber falling into or rising out of water, longitudinal tilting moment of chamber and mooring force of ships caused by water level fluctuation when the ship chamber's gate being in operation, the stern squat while ship driving into and out of chamber. Model design should meet hydrodynamic factors similarity.
- The running characteristics of ship chamber in normal operation and accident condition should meet gravity and motion similarity.

a) Rack and pinion ship lift（3000t）   *b) Hydro-floating ship lift（2×500t）*

c) Hydro-floating ship lift（2×1000t）   *d) Wire Rope hoist ship lift（2×1000t）*

Fig. 20. Ship lift physical model of different types

4.2 Numerical Model

Numerical simulation has the advantages of low cost, short research period and easy modification of the model. At the same time, it can also simulate accidents and extreme working conditions that cannot be carried out in the physical model test, so as to ensure the safety of test personnel and instruments and provide technical support for the safe operation of ship lift under various conditions. With the rapid development of computer technology and numerical calculation methods, numerical simulation technology is playing an increasingly important role in the research field of shiplift.

4.3 Integrated Simulation Technology

In view of the shiplift research involving multidisciplinary characteristics, the dynamic 3D virtual prototype of ship lift's overall characteristics is studied by means of data collection, theoretical analysis, model development and computer experiment. The subsystem of ship lift 3d virtual prototype (Fig. 21) mainly includes the flow-solid coupling simulation model of ship chamber entering/leaving water, ship chamber gate opening/closing, ship passing through the chamber, the dynamics simulation of mechanical lifting system, and tower structure.

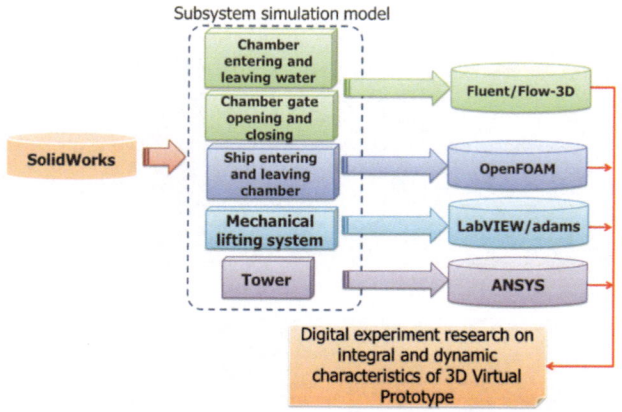

Fig. 21. Concept of ship lift 3D virtual prototype

4.4 Prototype Monitoring Technolgoy

Ship chamber hydrodynamics, ship berthing conditions, the forces of steel wire rope and other structures are important bases to determine the normal operation parameters of the ship lift, evaluate the safety of accident conditions, and guide the electrical control design. Therefore, it is very necessary to carry out the systematic prototype debugging and observation of the ship lift work characteristics and safety evaluation after the completion of the ship lift. The prototype observation can find the problems existing in the operation of a new ship lift, deepen the understanding of the operation characteristics of ship lifts, and obtain a large number of measured data to guide the operation and management of ship lifts.

5 Outlook

There are many types of ship lifts, and vertical ship lifts are the main development direction in the future. Among them, rope and winch type, rack and pinion climbing type and hydro-floating ship lift will be the mainstream type. Under some special topographical condition, inclined ship lifts also have certain application space, and water slope ship lift will gradually withdraw from history. In some tourist waterways and urban landscape water systems, there will be some new ship lift types, and the functions of hydraulic and floating ship lift will gradually transform from navigation to tourism, cultural heritage display and other functions.

In the future, ship lift will have a spread application prospect in artificial canals, water shortage areas and mountainous areas, especially in ship navigation of crossing the watershed in interconnected river system, and navigation of 200–300 m large hydropower hub, ship lift will be the main navigation type.

From the technical indicators and scale of ship lifts, the lifting height of large ship lift in the future will be increased from 100 m to 200 m, the lifting weight of ship lifts will be more than 20000 T, and 4 ships of 2000–3000 T can pass at one time. At present, some countries and institutions have carried out preliminary research work in this regard. The rapid development of civil construction level and mechanical manufacturing capacity has provided the conditions and ability to build large ship lift.

The construction of safe, efficient, green and intelligent ship lift is the overall development trend in the future, and the health monitoring, safety warning and operation guarantee technology of ship lift is one of the key concerns. Moreover, with the development of 5G, big data, artificial intelligence and other emerging technologies, more attention will be paid to the green and intelligent aspects of ship lift construction in future. Through a series research and development of innovative technology and intelligent equipment, the operation and maintenance of ship lift would form an organic integration with emerging technologies. The intelligence operation of ship lift is an important developing trend in the future.

References

PIANC (1989) Supplement to Bulletin No 65. Technical report, Ship Lifts

Hu Y, Xuan G, Li Z (2008) Key technology research of hydraulic floating ship lift for Jinghong hydropower station on Lancangjiang River in Yunnan. Nanjing Hydraulic Research Institute, Nanjing

Hu Y (2011) Basic research and application of hydro-floating ship lift. Doctor thesis, Nanjing Hydraulic Research Institute

Waterway Infrastructure

Construction of Cai Mep International Container Terminal

Koji Suzuki[✉] and Yoshinori Sumita

International General Head Quarter, Toa Corporation, Tokyo, Japan
{ko_suzuki, y_sumita}@toa-const.co.jp

Abstract. Cai Mep International Container Terminal was constructed 80 km southeast of Ho Chi Minh city, Vietnam, where a thick and soft clay deposit was widely found from the top of the ground. To minimize consolidation settlement after construction, soil improvement by PVD was extensively applied. Sand filling including preload up to +11.8 mCD was required on the existing ground of +3.5 mCD. Stability of this sand filling on river side was the first concern in the construction work. According to soil investigation to study shear strength profile of the soft clay, original design of sand filling was revised to maintain stability of the filling on river side. This report presents the original and the revised design, as well as the results of soil investigation.

Keywords: Container terminal · Soft soil · Slope stability

1 Introduction

Cai Mep international container terminal was constructed as a part of Cai Mep-Thi Vai International Port Construction Project funded by Japanese ODA loan. The location is 80 km south-east of Ho Chi Min city, Vietnam, and very close to the river mouth of Thi Vai river. Quay structure for berthing, supported by pile foundation, is 600 m wide and equipped with 6 gantry cranes. Reclamation for terminal area behind the quay is approximately 38 ha. Construction of the terminal was started in Oct/2008 and completed in Oct/2012.

Due to wide distribution of a very soft clay deposit (more than 30 m thick) on the top of the ground in the reclamation area, application of prefabricated vertical drain (PVD) with preloading sand fill was planned to accelerate consolidation settlement. The height of the scheduled preloading fill exceeded 5 m because the expected settlement under preloading fill was more than a couple of meters. This height of preloading fill causes some concern about stability of the slope of the preloading fill on the river side.

Soil investigation was carried out in the early stage of construction to determine engineering properties of the soft clay deposit in not only the terminal area, but also under the river bed. Then, stability of the river side slope was evaluated. According to the stability analysis, a slope inclination milder than the original design was adopted together with a 30 m set-back of terminal area.

This report presents revision of the original design of reclamation work, including the results of soil investigation carried out in the early stage of the construction.

© The Author(s) 2023
Y. Li et al. (Eds.): PIANC 2022, LNCE 264, pp. 55–61, 2023.
https://doi.org/10.1007/978-981-19-6138-0_5

2   Location of the Site

Figure 1 presents the location of the site by a circle. It is 80 km from HO Chi Minh city and 20 km to Vung Tau city. The container terminal subjected to this report is shown by an arrow in the figure. As can be seen in the figure, the terminal is facing Thi Vai River, and it has the same kind of facilities on both sides. Since the location is very close to the river mouth, water level of the river is affected by the tide of the sea. HWL and LWL are +3.97 mCD and +0.58 mCD, respectively.

3   Original Design

Original design of berthing facilities is presented in Fig. 2. The width of the quay structure is 55 m and the top elevation is +5.0 mCD. Distance from face line of the quay to the retaining wall is 140 m, including 85 m long trestle.

The site is widely covered with soft clay deposit (SPT N values are mostly 0 to 2). Figure 3 shows the ground condition of the river bank. A low cliff is formed by a soft clay along river bank. This soft clay deposit continues to the depth of −38 mCD in the river and to the depth of −32.5 mCD in terminal area (24 m thick in river side and 36 m thick in land side), as indicated in Fig. 2.

Figure 4 demonstrates the shape of preload filling and counter weight mound in river side. Top of sand filling is +10.8 mCD at the river bank, and +11.5 m in terminal area. Soil improvement by prefabricated vertical drain (PVD) was extensively applied to accelerate consolidation settlement of the soft clay in the terminal area behind the retaining wall.

Due to wide distribution of soft and thick clay deposit, stability of sand mound presented in Fig. 4 was the first main concern to be considered. Since filling of counter weight mound was scheduled very early stage of construction, immediate study of stability was required.

Fig. 1.   Location of the site

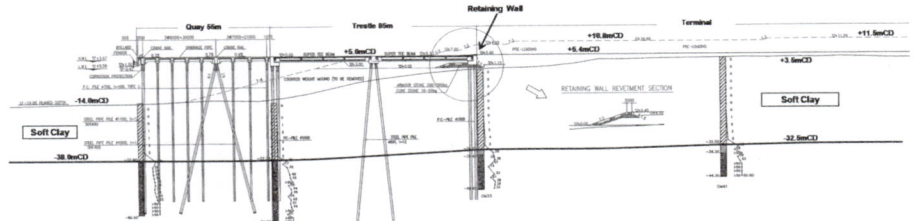

Fig. 2. Cross section of quay structure, trestle and reclamation

Fig. 3. Condition of river bank

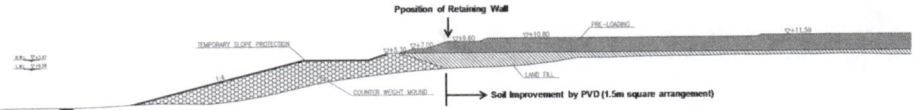

Fig. 4. Cross section of original design for preload fill and counter weight mound on the river bed

4   Engineering Properties of the Soft Clay

Soil investigation was carried out to study engineering properties of the soft clay deposit and to evaluate stability of sand filling presented in Fig. 4. Boreholes were made on the river bed as presented in Fig. 5, as well as land side boreholes. Undisturbed sampling was performed with hydraulic piston sampler to take soil samples in high quality.

Figure 6 demonstrates water content, Atterberg limits and shear strength profile given by an on-land borehole. Natural water content is very close to or slightly higher

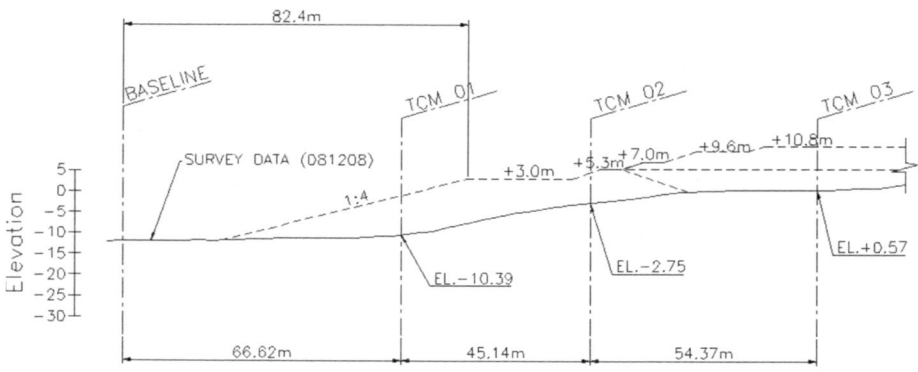

Fig. 5. Boreholes (TCM01 to 03) on the river bed

than liquid limit, suggesting that the clay deposit has high compressibility. Direct shear test (DST) was performed with recompression method, in which test specimens (6 cm diameter and 2 cm height) were consolidated at in-situ overburden stress before shear. The factor of 0.85 (Hanzawa, 1992) was applied to consider the effect of deformation speed during shear on strength. The picture on right side of Fig. 5 shows advanced type DST apparatus used for the test. This apparatus can achieve constant volume condition (equivalent to undrained condition) during shear precisely.

The results of field vane test (FVT) in Fig. 6 are almost the same as DST results through the depth. The results of unconfined compression test ($q_u/2$) are smaller than other two tests, except shallow part of the deposit.

Figure 7(a) presents undrained shear strength profile obtained by hand vane for riverbed boreholes (TCM01 and 02). Picture on the right side of Fig. 7 shows hand vane devise. This test was carried out for the soil sample in the sampling tube immediately after soil samples were retrieved from the ground. The factor of 0.85 was also applied as in the case of DST. As can be found in the figure, these strength values given by hand vane are slightly smaller than strength on land shown in Fig. 7(b).

Asada et al. (2005) studied the effect of release of overburden on shear strength of naturally deposited clay. They used DST apparatus shown in Fig. 6 and found that the effect can be expressed by Eq. (1), where R_{su} is the ratio of s_u after swelling due to stress release to that before swelling, and OCR is the ratio of effective vertical stress before swelling to that after swelling.

$$R_{su} = OCR^{-0.141} \qquad (1)$$

The solid lines in Fig. 7(a) are derived by Eq. (1) from strength profile on land shown in Fig. 7(b). There is good agreement between the lines given by Eq. (1) and measured values. This strongly suggests that the river bank presented in Fig. 3 was formed by erosion, and riverbed soil was subjected to stress release.

For boreholes in the river, drilling machine was place on a platform supported by two small boats. Although the use of floating platform is efficient for drilling on the

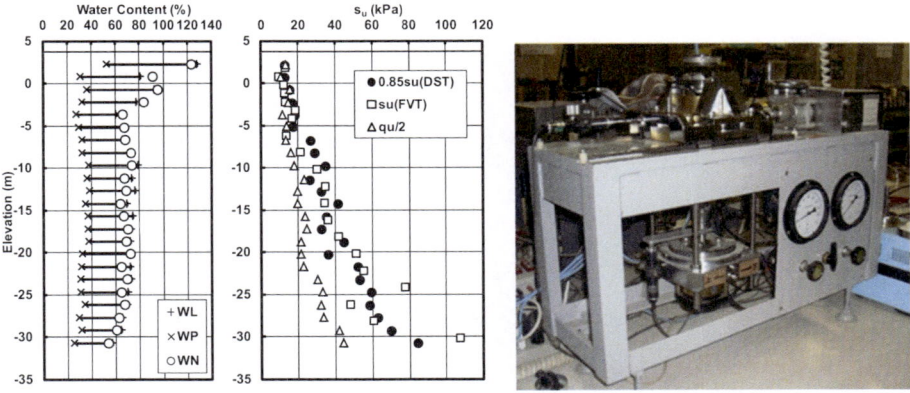

Fig. 6. Test results obtained from a on land borehole and DST apparatus used for the study

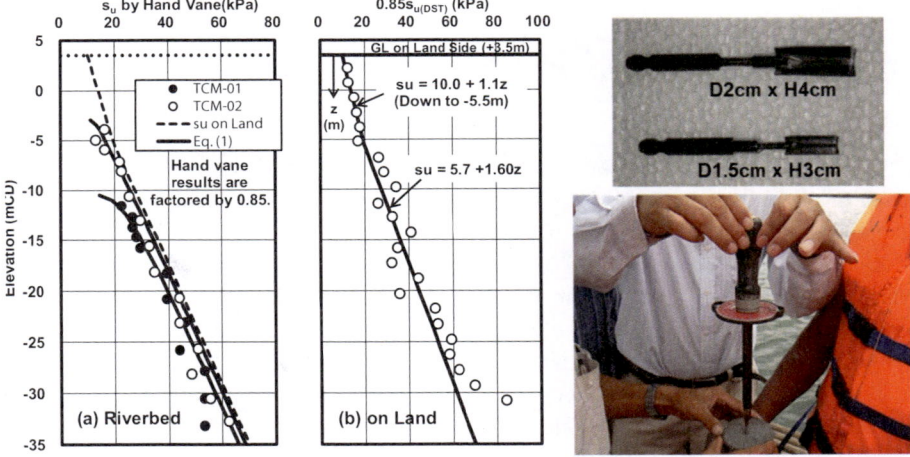

Fig. 7. Shear strength determined by hand vane for river boreholes and picture of hand vane

water, application of field vane test is difficult because of movement of platform due to tide and waves. This was the reason to use hand vane instead of field vane.

Figure 8 shows reliability of hand vane test. In the figure, hand vane results are compared with triaxial compression and extension test. Specimens for triaxial tests were first consolidated anisotropically, then sheared under undrained condition. As can be seen in the figure, hand vane results (factored by 0.85) are agree with the average strength of compression and extension triaxial tests.

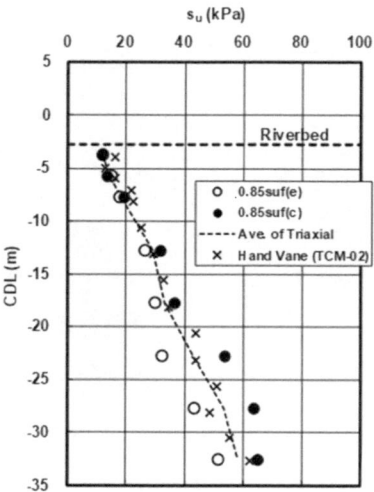

Fig. 8. Comparison of test results from hand vane test and triaxial test

5 Revised Design

According to the results of soil investigation, many cases of stability analysis were carried out, and cross section of preload fill and counter weight mound was revised as presented in Fig. 9. The slope inclination was reduced from 1:4 to 1:5, and the position of retaining wall was set-back 30 m. Figure 10 demonstrated quay and trestle of the revised design. There is no change in quay structure. The length of trestle was increased from 85 m to 115 m due to a 30 m set-back of retaining wall.

Concrete beam for trestle was also revised at the same time from prestressed concrete to reinforced concrete. Due to this revision, weight of concrete beams became much lighter than the original ones. This change made it possible to use smaller construction machine, of which preparation is much easier in the region than larger one.

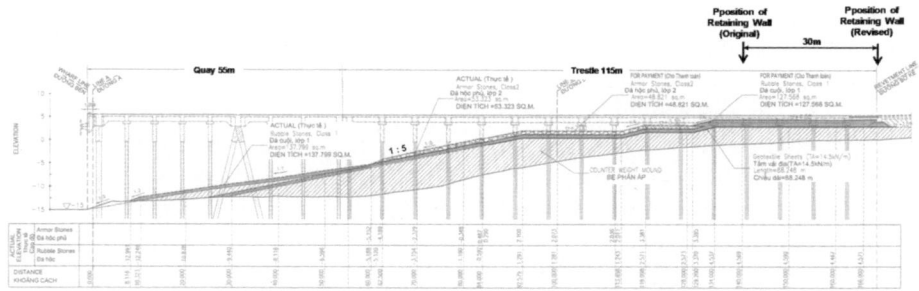

Fig. 9. Revise design of trestle

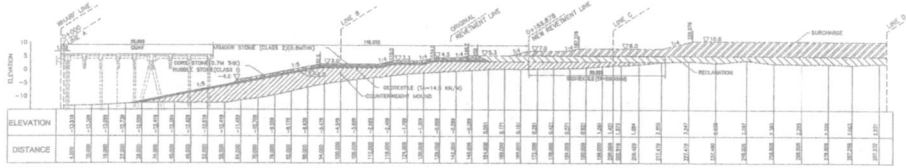

Fig. 10. Revised design of preload fill and counter weight mound on the river bed

6 Conclusions

Filling work in the project including counter weight mound and preload was successfully completed without any indication of sliding failure. Since soil data given by direct shear test and triaxial test presented in Figs. 6, 7 and 8 were not available at the time of design revision, shear strength profile of the soft clay applied to stability analysis was slightly smaller than that presented in the figures. This may give some contribution to successful completion of the filling work.

References

Hanzawa H (1992) A new approach to determine soil parameters free from regional variations in soil behavior and technical quality. Soils Found 32(1):71–84

Asada H., Tanaka Y., Suzuki K., Watabe Y.: Shear strength characteristics of marine clay subjected to swelling by direct shear test. In: 40th Annual Conference of Japanese Geotechnical Society, pp. 323–324 (2005). (in Japnese)

Continuous Management of the Channel Design to Optimize Navigability in the Middle Paraná River Waterway: "Santa Fe – Confluence". Argentina

M. L. Marpegan$^{(\boxtimes)}$ and L. A. Temer

EMEPA S.A, Buenos Aires, Argentina
mlmarpegan@gba-emepa.com.ar

Abstract. This paper exposes how HIDROVIA S.A. (Inland Waterway Concession company) worked in the Middle Paraná River inland waterway "Santa Fe – Confluence" to maintain the navigation channel without dredging works by means of four steps: frequent bathymetric surveys, study of AIS vessels tracks, a special coordination of logistic works in the river with the technical team and buoy tenders crew and effective communication plans with involved stakeholders. This is a stretch of the well-known Paraguay – Paraná Inland Waterway in South America that reaches deep waters in the Río de la Plata. Physical characteristics and depths along this section of Paraná River determine what kind of vessels can sail according to its draught up to 10 ft. With 2 ft. For under keel clearance. From October 2010 to September 2021, dredging works and installation and maintenance of modern aids to navigation system were granted by the concession company. Given the intense morphological dynamics and thalweg migration of the river in this section, the contract stated that it was no suitable to keep a fixed channel design by means of permanent dredging works. It established that dredging works should only be done when and where it is not possible to maintain the channel design by readjustment of its profile with partial changes in the trace and direction of the waterway axis. The Aids to Navigation Department of HIDROVIA S:A. succeeded in doing that by implementing the continuous management of the channel design to optimize sustainable navigation in the Santa Fe – Confluence waterway.

Keywords: Middle Paraná River · Thalweg migration · Trace changes · Sustainable waterway

1 Introduction

The Middle Paraná River inland waterway "Santa Fe – Confluence" flows entirely through h Argentina country in South America. This is a stretch of the well-known regional Paraguay – Paraná Inland Waterway (from now on HPP, for its acronym in Spanish) which flows through another four countries: Brazil, Bolivia, Paraguay, and Uruguay and finally reaches deep waters in the Río de la Plata next to the open sea (Fig. 1). Average discharge of this stretch of the Parana River is about 15000 m^3/s (mean waters condition).

© The Author(s) 2023
Y. Li et al. (Eds.): PIANC 2022, LNCE 264, pp. 62–73, 2023.
https://doi.org/10.1007/978-981-19-6138-0_6

Fig. 1. The Santa Fe – Confluence waterway. (Modify from Hidrovía SA 2021).

Physical characteristics and depths along this section of Paraná River determine what kind of vessels can navigate and the allowed draught is determined consequently. From Confluence downstream to Santa Fe Port navigation is allowed for vessels loaded up to 10 ft. And fluvial traffic is developed mainly by tug pushing barges convoy and, in some few cases, by self-propelled. Design depth of the waterway also includes additional 2 ft. For under keel clearance.

From October 2010 to September 2021, both dredging works and installation and maintenance of modern aids to navigation system were granted by HIDROVIA S.A. Company. The Aids to Navigation Department was in charge of the installation, maintenance and management of the aids to navigation system that included buoys and beacons according IALA-B guidelines, a network of automatic water level measuring stations and the installation of antennae for the reception of AIS: Automatic Identification System signals. Given the intense morphological dynamics of the river in the Santa Fe – Confluence section, the contract stated that it was no suitable to keep a fixed channel design by means of permanent dredging works. It established that dredging works should only be done when and where it is not possible to maintain the channel design by readjustment of its profile with partial changes in the trace and direction of the waterway axis. That was how HIDROVIA S.A. succeeded in implementing the continuous management of the channel design to optimize navigability in the Santa Fe – Confluence waterway. The purpose of this solution was not only to reduce costs but also to minimize environmental impacts in the river system reaching a sustainable management of the waterway. This requirement could be reached following the river thalweg migration with the channel axis, but the fluvial activity was so important that it demanded a continuous labor-intensive management to succeed. It required about 50 trace changes of the channel design a year, as an average, that were materialized successfully by HIDROVIA S.A.

This paper exposes the works carried out in the Paraná waterway Santa Fe – Confluence to maintain the designed navigational channel without dredging works by means of four basic steps: frequent bathymetric surveys, study of AIS vessels tracks, an accurate coordination of logistics works in the river with the technical team and buoy tenders crew and effective communication plans with involved stakeholders. Since 2019, the channel design in this waterway has been sustainably guaranteed with no dredging works.

2   Santa Fe – Confluence Waterway

The Paraná River section of HPP is in one of the most populated and industrialized area of South America. It is a strategic link to facilitate trade between the south of Brazil, Bolivia and Paraguay and Rosario city (Argentina) from where it can connect to deep draught vessels with the Atlantic Ocean. It is considered the most important integration way of MERCOSUR since it is one of the most important transport routes needed to facilitate physical integration for the five countries. Its commercial area of direct influence (hinterland) is estimated at about 720.000 km^2 and about 3.500.000 km^2 of indirect influence with a population of more than 40 million inhabitants (Escalante 2015). Santa Fe – Confluence waterway is 650 km long.

2.1   Commerce

Main cargo for this route is iron ore and grains with some share of containers and fuel. Important to highlight is that two countries, both Paraguay and Bolivia have this waterway as the only international trade connection due to their condition of land-locked countries. The hinterland of the waterway has a very big agriculture potential as well as reserves of iron ore and manganese that have of worldwide importance. Soybean and by-products, iron ore and fuels gather at least 88% of total freight (*Op. Cit.* 2015). Table 1 shows tonnage passed through the waterway in: 2015–2019 (Hidrovía SA. 2021).

Table 1.   Cargo tons in the Santa Fe - Confluence waterway. 2015–2019.

Products	Total cargoes [tons]				
	2015	2016	2017	2018	2019
Soybeans	7.721.289	8.322.094	9.034.433	9.837.442	8.106.063
Other grains	4.214.646	633.464	1.318.628	887.298	1.766.175
Iron ore	4.126.000	3.564.751	3.807.790	3.574.748	3.792.662
Liquid cargo	4.064.111	3.679.247	3.599.026	4.402.984	3.939.994
General cargo	1.460.559	1.075.810	994.310	1.253.193	1.379.994
Total	21.586.605	17.275.366	18.754.188	19.955.665	18.984.772

The volume of transported cargo along the stretch Santa Fe – Confluence has been growing up since 3,7 million tons in 1996 up to 22 million tons in 2019. In 2019 there were more than 11.300 travels, and in 2021 about 12.300 trips (Marpegan and Pérez, 2020).

2.2   Navigation Along the Waterway

The Santa Fe–Confluence waterway allows both domestic and international commercial river traffic to and from Argentina, Uruguay, Paraguay, Bolivia and south of Brazil. Cargo is transported by a fleet composed mainly by barges and pushers and some self-propelled vessels. These barges are arranged in convoys of different sizes pushed by tugs of adequate power (Escalante 2015). Usual arrays applied to carry bulk cargo, are convoys from 16 to 30 barges: 4 × 4; 4 × 5; 5 × 5 and 5 × 6 all of them with a tug pusher and the biggest convoy with a pusher of more than 7,000 HP (*Op. Cit.* 2015; Hidrovía SA. 2021). Physical dimensions are shown in Table 2 considering Mississippi type barges 60 m long and a tug pusher length of 50 m (Hidrovía S.A. 2021). The longer ones have maneuvering restrictions along straight areas and or where the river turns in curves of the channel.

Table 2. Dimensions of convoys usually used in the Santa Fe - Confluence waterway.

Vessel type	Barges [units]	Length [m]	Breath [m]
General Cargo	1	139/200	18
Convoy 4 × 4	16	290	44
Convoy 4 × 5	20	290	55
Convoy 5 × 5	25	350	55
Convoy 5 × 6	30	350	66

3   Technical and Sustainable Channel Management

The required waterway management must keep safe navigation conditions, for all kind of vessels along the Santa Fe – Confluence channel for 24 h all day of the year in any fluvial and hydrometric condition: high, mean or low waters. The contract demanded a level service higher than 97% that meant a permanent coordinated work, in agreement with authorities (Subsecretary of Ports, Waterways and Merchant Marine, Naval Hydrographic Service, Argentine Naval Prefecture) and users (Captains, boat skippers and shipowners). This also requires planned regular services campaigns in the river to attend each buoy and beacon, a permanent availability of technical and human resources, analysis of recent bathymetric surveys and the study of convoys or ships tracks up and downriver.

To succeed in offering proper response for each necessary change or adjustment in any buoy or beacon, two operational bases were installed in both ends of this waterway, downstream in Santa Fe port and at the upper end in Barranqueras port (Fig. 2). Between these two stations, other strategic locations were Goya, Reconquista, La Paz and Paraná ports.

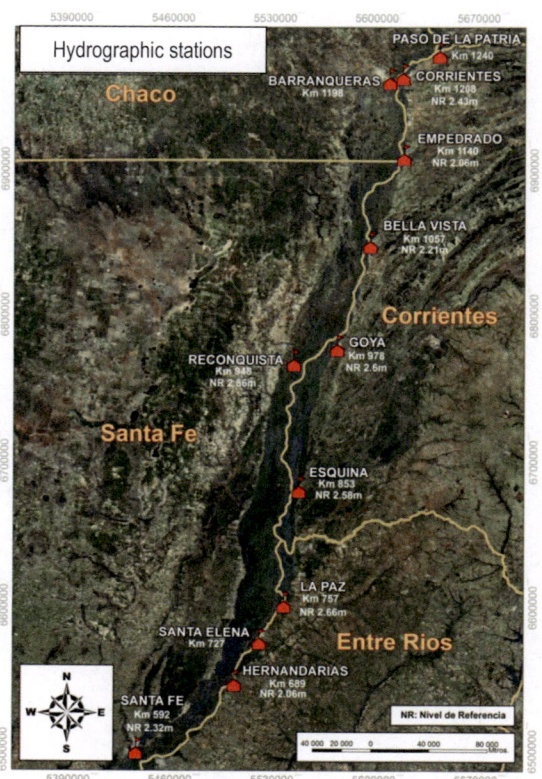

Fig. 2. Hydrographic stations and Hidrovía SA operational bases in The Santa Fe – Confluence waterway. (Hidrovía SA 2021b).

3.1 Bathymetric Surveys

Bathymetric surveys were regularly done for different purposes with hydrographic boats. Surveys were carried out along the channel axis and its edges with high density of profiles (each 200 m) where it is difficult to sail the river (shallow and stretch waters) in the "Pasos"; along the channel axis and less density of cross sections (each 1000 m) between subsequent Pasos (in what is called "Entrepasos") and along secondary branches of the river. The annual survey reached a total profiles length of 9250 km in the Santa Fe – Confluence waterway.

Pasos and Entrepasos are stretches of the Middle Paraná River where the contract requires to keep 3,66- and 4,00-meters depth respectively. Pasos are the places where a strong sedimentation takes place close to the channel and there is adjacent intense activity of sandbanks and dunes. Entrepasos are the more stable hydro sedimento-logical parts of the waterway. The Pasos are about 43 % of the Santa Fe – Confluence waterway length and have hold 65 % of channel axis changes up to September 2020 (Marpegan and Pérez, 2020).

3.2   Channel Axis Changes

With the bathymetric surveys, the trace of the channel can be monitored to detect when a change in direction is necessary. The fluvial movements of sandbanks and dunes of the Middle Paraná River change the river morphology almost in a continuous way so that secondary branches of the river and some areas of its principal course that were initially unfit to commercial navigation turn to be deep o wide enough to allow development of this activity safely in agreement with technical contract conditions.

Table 3 shows how much changes in channel axis were done annually from 2011 to July 2020 in Pasos, Entrepasos and in areas combined by Pasos and Entrepasos. 413 changes were carried out in that period.

Table 3.   Annual changes in direction of the channel axis in the Santa Fe - Confluence waterway (Marpegan and Pérez, 2020).

Year	Pasos	Entrepasos	Combination sites: Pasos + Entrepasos	Total
2011	12	0	7	19
2012	33	9	9	51
2013	23	16	1	40
2014	33	16	3	52
2015	34	16	14	64
2016	28	17	4	49
2017	47	9	2	58
2018	29	16	1	46
2019	2	11	1	14
To June 2020	6	4	10	20
Total:	247	114	52	413

Figure 3 shows the changes done in the Santa Fe – Confluence waterway along the river, from upriver in 1230 km at Confluence to 584 km in Santa Fe port where 3 sections with different morphological activity can be identify. From the lower end: Santa Fe – Santa Elena stretch with low activity, Santa Elena – Goya section is the critical zone with the most frequent talweg migration and Goya – Confluence section with moderate activity. These characteristics of each stretch determine the number of changes in the channel axis that are necessarily made to comply with the contract of keeping safe navigation conditions with less o no dredging works.

3.3   Tracs from Automatic Identification System: AIS

Information about fluvial traffic in the Santa Fe – Confluence waterway get from AIS records was studied and correlated with the vessel type, the direction of each trip, stretch of the river, local hydrometric levels and type and amount of transported cargo. This analysis was particularly meaningful in sections where the Paraná River offers alternative branches as in Paso *Raigones – San Juan* Branch shown in Fig. 4 and consequently allowed to:

- Make a real time monitoring of navigation in the waterway
- Supervise and oversee the boats in charge of bathymetric surveys and technical assistance to the buoying system.
- Generates warm maps of different traffic densities
- Know how the convoys and vessels use the channel and identify collisions between vessels and or with buoys.
- Validate actual channel design and installed aids to navigation quality of service.

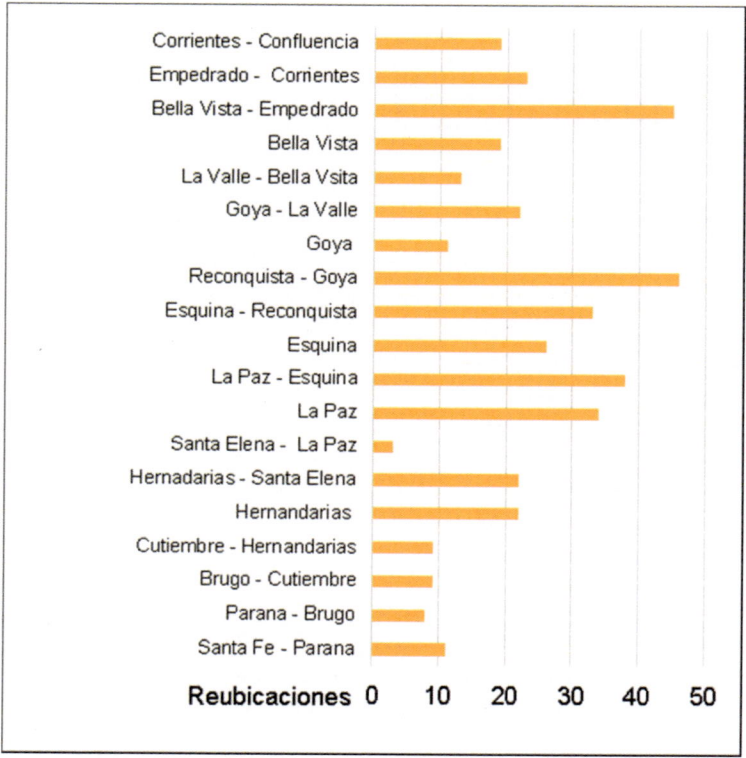

Fig. 3. Spatial distribution of channel axis changes in Santa Fe – Confluence waterway. Period: 2012 – July 2020

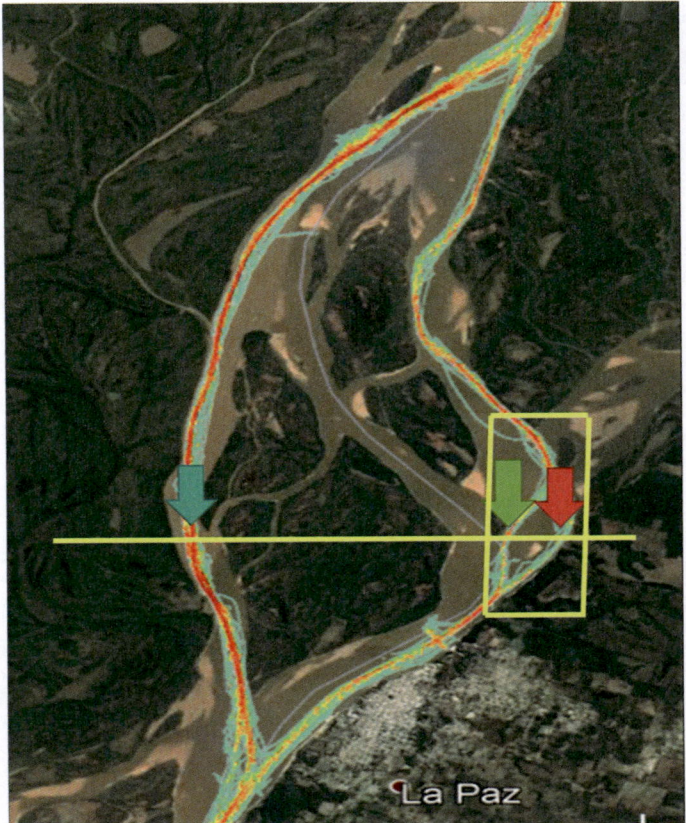

Fig. 4. Traffic nautical density. Paso *Raigones/San Juan* Branch case. Santa Fe – Confluence waterway.

3.4   Interaction with Users and Stakeholders

Since the beginning of the concession in 2011, Hidrovía S.A established and kept a frequent interaction with stakeholders, users and interested public by means not only of available official tools but also implementing new ones in order that user of the channel always access and know technical information about the in the Santa Fe – Confluence waterway status. For example:

- List of signals installed
- List of signals with any fault or misfunction and its type
- Minimum depth in each Paso
- Planed area for the survey campaigns
- Preliminary design of channel axis changes to be proposed to the authorities
- Other interested and related topics.

4 Case Study: Paso Raigones and San Juan Branch

18% of the 413 axis changes carried out between 2011–2020 were in the area where the fluvial morphology and thalweg migration are more active. It is close to La Paz port in Santa Fe – Confluence waterway between 740 km and 785 km including 4 Pasos: *Cortada Arroyo Seco, Riacho Raigones, Curuzú Chalí* and *Garibaldi* and additionally the Entrepasos between them (Fig. 5). This area is about 50 km long and 765 km was the most critical place where more thalweg migration was registered.

In 783 km the Middle Paraná River opens its principal course into 3 other ones, like a "trifurcation" that gives place to *San Juan* and *San Juancito* branches and additionally subsequent Pasos *Raigones* and *Curuzú Chalí* (Fig. 5). The morphologic intense activity in this area demands almost permanent bathymetric surveys that were done by Hidrovía S. A, to verify and or rectify navigational channel axis trace and the visual aids to navigation installed to maintain safe nautical conditions. It could be statistically reported that the interval between successive changes was 70 days as an average.

Fig. 5. Middle Paraná River close to La Paz Port.

4.1   AIS Data

AIS data allow an offline analysis to understand and follow the trace evolution. As a result of that, the company was able to (Fig. 6):

- Study temporal deviation from vessels path from the designed channel
- Identify areas for bathymetric surveys
- Evaluate use and frequency of change for the existing alternatives
- Report about navigation habits along shallower areas (kind of vessels)

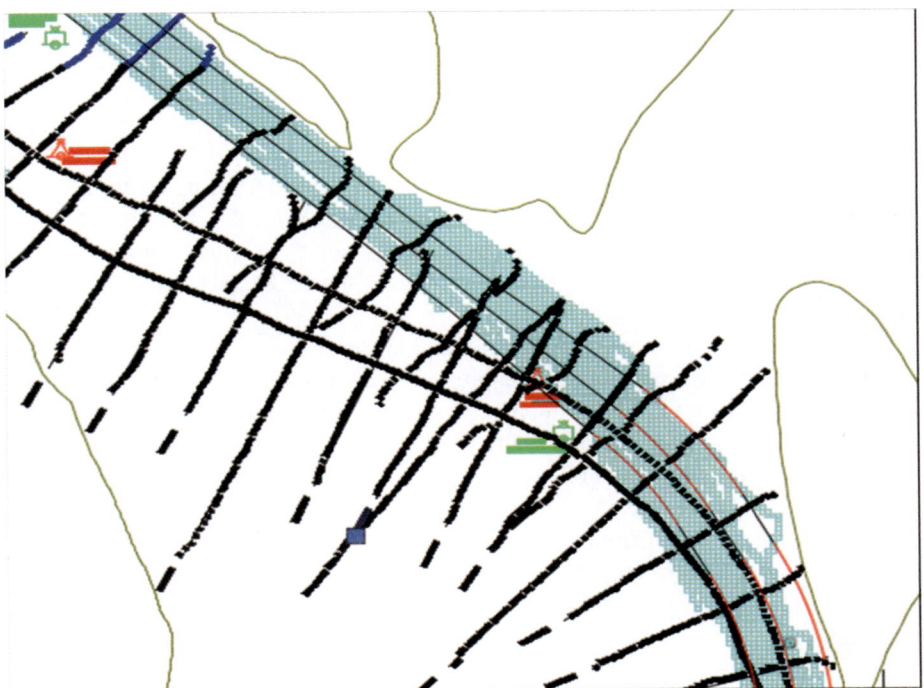

Fig. 6. Channel shift to left margin.

4.2   Bathymetric Surveys and Alternatives Analysis

For this example, bathymetric surveys along Paso *Riacho Raigones* were done to look for bigger depths that could justify a new axis change. Some other surveys were also carried out along secondary branches *San Juan* and *San Juancito*. In all of them the navigability conditions were observed and analyzed, and the critical depth were identified. It allowed to keep Paso *Riacho Raigones* as the deepest option for the channel.

The final design was then validated with traffic density analysis got from all the vessels tracks in the area and additionally where also new aids to navigation designs were determined.

5 Conclusions

The intense morphological activity of the Middle Paraná River with continuous bed forms and dunes migration is a strong characteristic of the fluvial system that determines how the navigational channel can be defined along its course. This was considered in the contract that required a channel design maintenance with reduced or no dredging works, tracing continuous thalweg migration with the axis changes.

The work done by Hidrovía S.A. in the Santa Fe – Confluence waterway between 2011 and June 2020, shows clearly that no unique channel design is possible, it cannot be fixed in time and demands a permanent adequation.

That fluvial complex scenery is complemented by maneuvering restrictions to barges convoys that sail this waterway, specially the longer ones. Sometimes the best nautical option (or a shorter straight path) for the channel axis does not match the proper morphological option (the deepest). Different ships and barges convoys have its own preferred alternative.

It is important to promote stakeholders, authorities and users participation in the channel and the aids to navigation design process that is necessary not only to optimize the waterway facilities but also to improve navigation safety avoiding groundings and collisions between vessels or against the buoys. This collective work also leads to continuous improvement methodology for the intervention in Santa Fe – Confluence waterway.

The work carried out to follow natural migration of the river thalweg moving the channel axis was a heavy task for Hidrovía S.A., and the achieved success shows that sustainable navigation is possible with soft intervention in the river and minimum or no dredging works.

References

Escalante, R.S.: Parana – paraguay rivers inland waterway. Smart Rivers Proceedings. Buenos Aires Argentina (2015)

Hidrovía, S.A.: Estudio de Tráfico y su Proyección. Año 2019. Segundo Semestre. Informe HDRV-SFN/104/2021a. Buenos Aires. 11 de Mayo (2021a)

Hidrovía, S.A.: Informe Ambiental Anual. Año 2021b. Plan de Gestión Ambiental. Vía Navegable Troncal Sección Santa Fe - Confluencia. Informe HDRV-SFN/103/2021b. Buenos Aires. 5 de Agosto (2021b)

Marpegan, M.L., Pérez, E.: Gestión Continua de la traza de la Vía Navegable Troncal para garantizar la navegabilidad en el tramo Santa Fe – Confluencia. XI Congreso Argentino de Ingeniería Portuaria. Buenos Aires. Argentina. 09 al 11 de septiembre (2020)

Experimental Study on Flow Characteristics Around a Submerged Half-Buried Pipeline

Zishun Yao[1], Lidi Shi[2], Shoupeng Xie[3], Peng Li[4],
and Dawei Guan[5(✉)]

[1] Department of Civil and Environmental Engineering,
The University of Auckland, Auckland, New Zealand
zyao132@aucklanduni.ac.nz
[2] State Key Laboratory of Hydroscience and Engineering, Tsinghua University,
Beijing, China
shild18@mails.tsinghua.edu.cn
[3] West Africa Corporation of China Harbor Engineering Co.,
Abidjan, Côte d'Ivoire
spxie@chec.bj.cn
[4] China Satellite Maritime Tracking and Control Department, Jiangyin, China
1823633508@qq.com
[5] Key Laboratory of Ministry of Education for Coastal Disaster and Protection,
Hohai University, Nanjing, China
david.guan@hhu.edu.cn

Abstract. This paper describes the flow characteristics around a half-buried pipeline exposed to different current conditions by flume experiment. Particle Imaging Velocimetry (PIV) technique was used in the experiment to reveal the flow structure. The experiment results indicate that the hydrodynamic parameters, including average kinetic energy, vorticities, Reynolds shear stress, and kinetic turbulence energy, increased with the Renolds numbers. Furthermore, it is found that the vortex at the upstream side of the half-buried pipeline vanishes gradually with increasing Renolds numbers. However, the two vortices at the locations downstream of the half-buried pipeline exist all the time, in which the vortex closed to the downstream of the pipeline becomes unstable, the vortex at the pipeline wake remains unchanged. In the meantime, the turbulence intensity at the upstream side of the half-buried pipe near the bed surface strengthens significantly under Reynolds number conditions, which may accelerate the potential scour process in front of the pipeline.

Keywords: Half-buried pipeline · PIV · Flow patterns · Turbulence structure

1 Introduction

Pipelines are usually buried or half-buried in rivers for oil and gas transmission or cable conduits. The flow past the pipelines underwater has long been a subject of study and research due to the complicated interaction between the structure and the hydrodynamics. Numerous papers have been written on this subject (Ahmed and Rajaratnam 1998; Bearman and Trueman 2016; Jensen et al. 1990; Lin et al. 2009; Okajima 1990;

© The Author(s) 2023
Y. Li et al. (Eds.): PIANC 2022, LNCE 264, pp. 74–81, 2023.
https://doi.org/10.1007/978-981-19-6138-0_7

Roshko 1960; Tritton 1959; Yang et al. 2021). The flow characteristics, including patterns and turbulent structures around the submerged pipelines, a typical cylindrical structure underwater, have also been extensively investigated. Jensen et al. (1990) studied the flow patterns around a submerged pipeline placed on a flat, erodible bed under current-only conditions. The flow features around the pipeline and the vortex shedding pattern were presented using the two-color flow visualization technique. Zhu et al. (2013) investigated the flow fields around the pipeline on a scoured bed by numerical simulation. The numerical results indicated that the flow field around the pipeline is subjected to the gap between the pipeline and the scoured bed. Guan et al. (2019) investigated the flow field around the vibrating pipeline within a scour hole and presented a visualization method to obtain the detailed flow fields. These studies have revealed a general principle of the flow structure around the pipeline. However, the studies concerning the flow structure around a submerged half-buried pipeline are not available yet.

The purpose of this study is to investigate the flow structure and its features around a submerged half-buried pipeline by physical experiment. Because this study focuses on the flow structure, the scour process is not considered, and the bed is set as fixed and smooth in the experiments. To visualize the vortex development and to manifest the features of the flow structure, Particle Imaging Velocimetry (PIV) technique as a contactless measurement has been adopted in hydrodynamic experiments in recent years (Apsilidis et al. 2015; Jenssen and Manhart 2020; Yang et al. 2021), was used in this study. The experiment results are expected to enhance the knowledge of the flow structure around the submerged half-buried pipelines and facilitate channel improvement and river-crossing pipeline construction design.

2 Methodology

2.1 Experiment Set-up

The experiment set-up comprises three main parts: a semi-cylinder model, a water circulating flume, and a PIV system. As shown in Fig. 1, the flume dimensions were length = 11 m, width = 0.6 m, height = 6 m. An aluminum semi-cylinder model with a length of 0.6 m and a radius of 0.025 m, used as the model of the half-buried pipeline. The model was fixed in the flume at a distance of 5 m away from the outfall, as shown in Fig. 2(a). The PIV system consists of a highspeed camera (Photron FASTCAM Mini WX50) and a laser device. The laser device was mounted on the flume at the position above the pipeline model so that the laser light plane (as shown in Fig. 2(b)) could be cast into the flume. The laser light plane was perpendicular to the lengthwise direction of the pipeline model and parallel to the flow direction in the flume. The highspeed camera was placed at the side of the flume aiming at the laser light plane with the axis view perpendicular to the laser light plane.

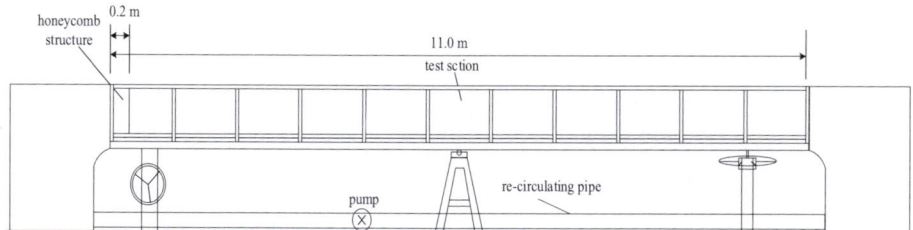

Fig. 1. The dimensions of the flume

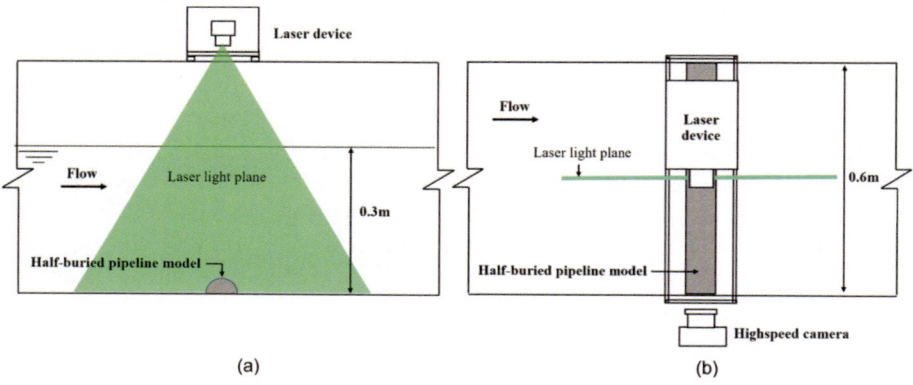

Fig. 2. Experiment set-up: (a) Front view; (b) Top view.

2.2 Experimental Method

Three tests were taken in this study, and the test conditions are shown in Table 1. The water depth was kept constant in the tests at 0.3 m. The mean velocity of the flow was calculated based on the flow discharge in each test.

Table 1. Test conditions.

No.	Flow discharge Q $(10^{-3}$ m^3/s)	Mean velocity U_m (m/s)	Reynolds number Re	Water depth h (m)
L1	5.8	0.032	8891	0.3
L2	28.8	0.16	47952	0.3
L3	50.4	0.28	83916	0.3

The procedure of each test is described below:

(1) Running the water pump, and adjusting the flow discharge to the designated value in each test.

(2) Turning on the laser device, injecting tracking particles slowly into the water at the inlet of the flume, then taking photos by highspeed camera when the tracking particles pass through the laser light plane.
(3) Saving the images taken during the test for flow field calculation, then starting the next run.

A total of 3000 images of each test were used to obtain the time-averaged flow field around the pipeline model by the PIV technique. The details of the PIV technique and the calculation principle can be found in Guan et al. (2019). It is worth noting that, as shown in Fig. 3., the origin of coordinates adopted in this study is located at the center of the semi-cylinder bottom; the X-axis and the Y-axis of the coordinates are parallel to the longitudinal and vertical axis of the flume, respectively. u and v are the horizontal velocity and vertical velocity of the flow, respectively.

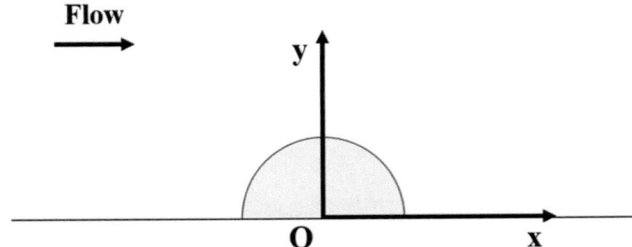

Fig. 3. Two-dimensional coordinates

3 Experimental Results and Discussion

3.1 Characteristics of Time-Averaged Flow Field

The time-averaged flow field around the half-buried pipeline model in three tests was calculated by the PIV technique. The time-averaged flow features of each test, including the distribution of streamlines, the average kinetic energy (AKE), and the vorticities, were presented in Fig. 4.

As shown in Fig. 4(a), three vortices could be observed around the pipeline model. To better distinguish the vortices, these vortices were defined as:

(1) V_{up}: the vortex closed to the upstream side of the pipeline model;
(2) V_{dp}: the vortex closed to the downstream side of the pipeline model;
(3) V_{wp}: the relatively big vortex situated at the wake of the pipeline model, the size of this vortex is much larger than that of the two vortices stated above and is nearly the same as the pipeline model.

By comparing the streamline distributions of the three tests, it can be found that as the Reynolds number increases, V_{up} gradually disappears, V_{dp} becomes unstable, but V_{wp} exists all the time. It can be deduced that with the increase of the approaching flow velocity, the backflow zone in front of the pipeline is compressed, causing the V_{up} becomes smaller and disappear eventually. In the meantime, the area of the V_{dp} is contracted to cause a unstable status. Moreover, with the increase of the Reynolds number of the flow, the V_{wp} exists all the time during the experiment, and its position remains unchanged.

The different distributions of AKE, as shown in Fig. 4(b), indicate that with the increase of Reynolds number of the flow, the wake zone is compressed and becomes smaller, especially in the vertical direction. In the meantime, as the Reynolds number increases, the difference between the AKE of pipeline wake and the AKE of undisturbed flow grows significantly.

The time-averaged vorticities at different flow Reynolds number are shown in Fig. 4(c). The two-dimensional distribution of the vorticities shows that the area with relatively large vorticities mainly appears downstream of the pipeline model near the V_{dp} and V_{wp}. The maximum vorticity appears on the top of the pipeline model. With the increase of flow Renolds number, the maximum vorticity increases, and the area with relatively large vorticities extends downstream. It may because that the pipeline's existence constrains the vertical development of the area of relatively large vorticities above the pipeline, but the horizontal extension of this area receives fewer restrictions. Thus, the area of relatively large vorticities tends to extend significantly downstream in the horizontal direction rather than in the vertical direction.

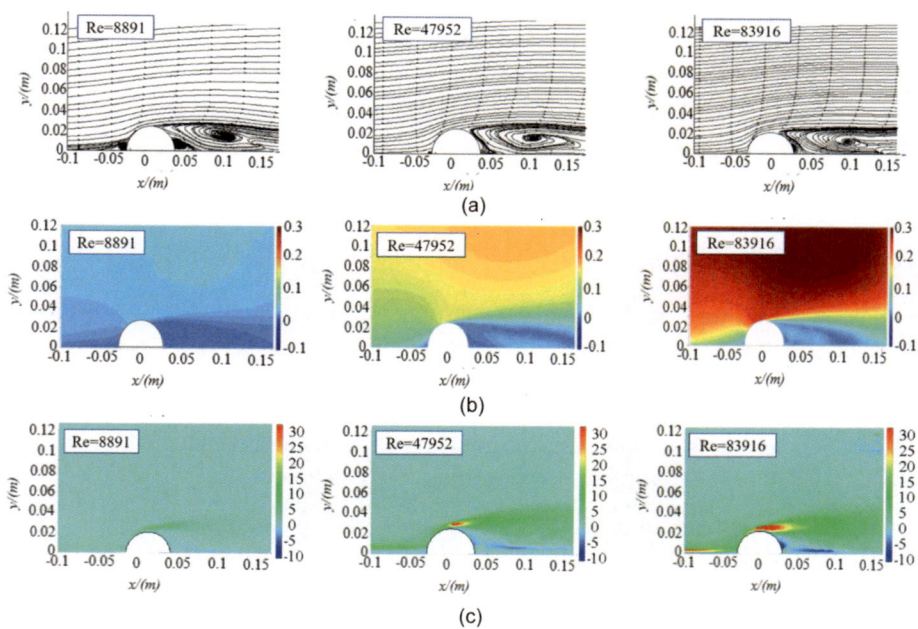

Fig. 4. Time-averaged flow field around the half-buried pipeline at different Reynolds number. (a) Streamlines distributions; (b) AKE distributions; (c) Vorticities distributions.

3.2   Characteristics of Wake Turbulence Structure

Figure 5 shows the characteristics of the wake turbulence structure in three tests, including the Reynolds shear stresses, turbulence intensities, and turbulence kinetic energy (TKE).

As shown in Fig. 5(a), the areas with considerable Reynolds shear stress, situated downstream of the half-buried pipeline model, can be classified as the upper region behind the top of the pipeline and the lower region near the bed surface. It is found that the Reynolds shear stress in the upper region is much greater than that in the lower region, which is different from the Reynolds shear stresses of flow around the single-cylinder, where the Reynolds shear stresses of the upper and lower shear layers are almost equal in magnitude. In addition, the comparison of the Reynolds shear stresses at different Reynolds number indicates that as the flow Reynolds number increases, the maximum value of Reynolds shear stress as well as the kinetic energy of the flow increase.

Figure 5(a) and Fig. 5(c) show the turbulence intensity at different Reynolds number in the axis-X direction (TI_x) and in the axis-Y direction (TI_y), respectively. It can be observed that with the increase of the flow Reynolds number, TI_x and TI_y in the half-buried pipeline wake increase, and the triangular-shaped region with relatively high TI_x and TI_y value at the pipeline wake is expanded. TI_x and TI_y of the area upstream of the pipeline above the bed surface developed as the flow Reynolds number increased. Moreover, similar to the turbulence intensity distribution around submerged weir in the downstream direction and vertical direction (Guan et al. 2014), TI_x are more than twice TI_y by comparing turbulence intensities of the same point at the same Reynolds number. The triangular-shaped region of relatively high TI_x is broader than that of TI_y. The main reason is that the flow velocity in the horizontal direction (u) is much higher than that in the vertical direction (v), resulting in a more significant turbulent intensity TI_u. Furthermore, the flow above the half-buried pipeline is constrained vertically. u is more affected than v in the wake flow of the pipeline.

The distribution pattern of TKE in the flow field of each test, as shown in Fig. 5(d), is similar to that of TI_u. The outline of the high TKE zone and the high TI_u zone at the wake of the pipeline are almost the same. Similar to the development rules of TI_x and TI_y, with the increase of the flow Reynolds number, the maximum TKE increases, and the area of the high TKE region develops. Especially the turbulence intensity at the upstream side of the half-buried pipe near the bed surface increases significantly when the Reynolds number increases. It indicates that the flow structure of the upstream side of the half-buried pipeline becomes turbulent at high flow velocity conditions, which may lead to the potential scour in front of the pipeline, and may eventually result in the suspension of the submerged pipeline above the scoured bed threatening the safety of the pipeline. The scour problem of the submerged half-buried pipeline needs to be further studied in the future.

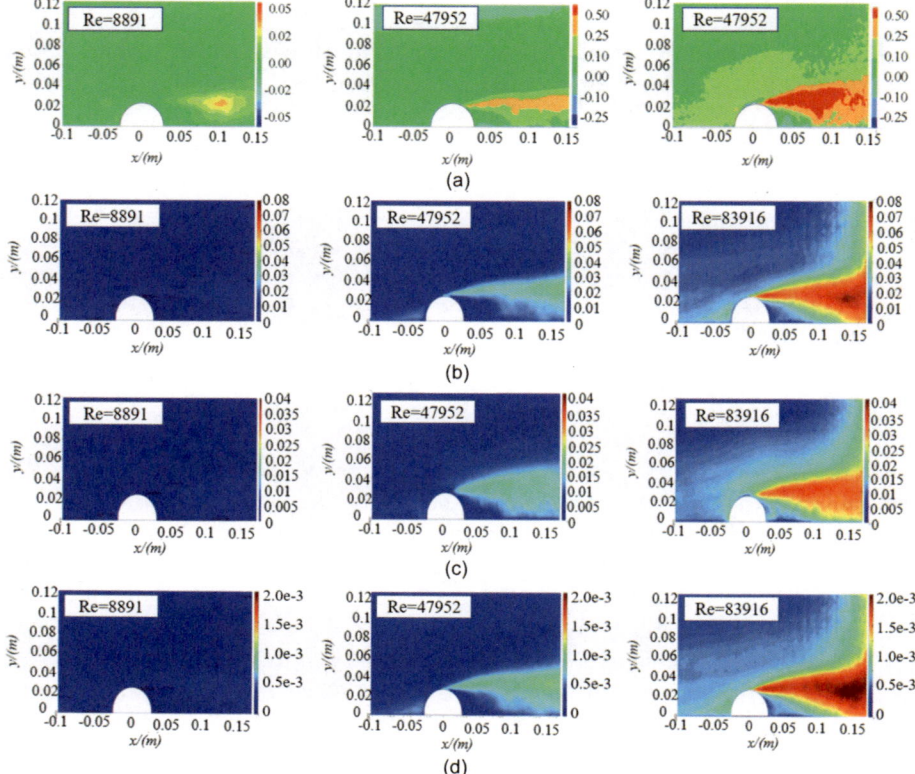

Fig. 5. Turbulence structure around the half-buried pipeline at different Reynolds number. (a) Reynolds shear stresses distributions; (b) TI_x distributions; (c) TI_y distributions; (d) TKE distributions.

4   Conclusion

This experimental study investigated the flow field patterns and the turbulent structures around the half-buried pipelines at different Reynolds number using the PIV technique. The flow structures studied include the streamlines, the average kinetic energy, the vorticities, the Reynolds shear stress, the turbulence intensities, and the turbulence kinetic energy. The main conclusions are summarized as follows:

(1)   As the approaching flow Reynolds number increases, the vortex at the upstream side of the half-buried pipeline vanishes gradually. In the meantime, the vortex closed to the downstream of the pipeline is compressed and becomes unstable. However, another vortex positioned at the pipeline wake remains unchanged.

(2)   The increase of the approaching flow Reynolds number results in the increase of AKE, the vorticities, the Reynolds shear stress, and TKE of the flow structure at the half-buried pipeline wake.

(3)   The turbulence intensity at the wake of the half-buried pipeline develops dramatically with the increase of the approaching flow Reynolds number. Furthermore,

the turbulence intensity at the upstream side of the half-buried pipe near the bed surface strengthens significantly under high Reynolds number conditions.

Acknowledgements. This work was supported by the National Natural Science Foundation of China (Project No. 52122109), and the first author would like to thank China Scholarship Council (CSC) for the financial support of this research.

References

Ahmed F, Rajaratnam N (1998) Flow around bridge piers. J. Hydraul. Eng. 124(3):288–300

Apsilidis N, Diplas P, Dancey CL, Bouratsis P (2015) Time-resolved flow dynamics and Reynolds number effects at a wall–cylinder junction. J. Fluid Mech. 776:475–511

Bearman PW, Trueman DM (2016) An investigation of the flow around rectangular cylinders. Aeronaut Q 23(3):229–237

Guan, D., Chiew, Y.M., Wei, M., Hsieh, S.C.: Visualization of flow field around a vibrating pipeline within an equilibrium scour hole. J. Vis. Exp. (150) (2019)

Guan D, Melville BW, Friedrich H (2014) Flow patterns and turbulence structures in a scour hole downstream of a submerged weir. J. Hydraul. Eng. 140(1):68–76

Jensen BL, Sumer BM, Jensen HR, Fredsoe J (1990) Flow around and forces on a pipeline near a scoured bed in steady current. J. Offshore Mech. Arct. Eng. 112(3):206–213

Jenssen U, Manhart M (2020) Flow around a scoured bridge pier: a stereoscopic PIV analysis. Exp. Fluids 61(10):1–18. https://doi.org/10.1007/s00348-020-03044-z

Lin W-J, Lin C, Hsieh S-C, Dey S (2009) Flow characteristics around a circular cylinder placed horizontally above a plane boundary. J. Eng. Mech. 135(7):697–716

Okajima A (1990) Numerical simulation of flow around rectangular cylinders. J. Wind. Eng. Ind. Aerodyn. 33(1–2):171–180

Roshko A (1960) Experiments on the flow past a circular cylinder at very high Reynolds number. J. Fluid Mech. 10(3):345–356

Tritton DJ (1959) Experiments on the flow past a circular cylinder at low Reynolds numbers. J. Fluid Mech. 6(4):547–567

Yang Y, Qi M, Li J, Ma X (2021) Experimental study of flow field around pile groups using PIV. Exp. Therm. Fluid Sci. 120:110223. https://doi.org/10.1016/j.expthermflusci.2020.110223

Zhu H, Qi X, Lin P, Yang Y (2013) Numerical simulation of flow around a submarine pipe with a spoiler and current-induced scour beneath the pipe. Appl. Ocean Res. 41:87–100

Field Measurements of Flow Velocities
in Propeller Jets

Irene Cantoni[1], Arne Van Der Hout[2(✉)], Erik Jan Houwing[3],
Alfred Roubos[4], and Michel Ruijter[3]

[1] Civil Engineering and Geosciences, TUDelft, Delft, The Netherlands
[2] Deltares, Harbour, Coastal and Offshore Engineering, Delft, The Netherlands
Arne.vanderHout@deltares.nl
[3] Rijkswaterstaat, GPO, Utrecht, The Netherlands
[4] Port or Rotterdam Authority, Port Development, Rotterdam, The Netherlands

Abstract. Propellers of ships generate high velocities adjacent to quay walls, jetties and locks. Generally, a bottom protection is installed in order to prevent instability due to scour. Although design guidance exist, propeller-induced loads are far from fully understood and have predominantly been derived on the basis of model tests. The validation of the existing design methods is lacking, especially for specific types of bow thrusters. In this research, field measurements of flow velocities induced by a 4-channel bow thruster system against a vertical quay wall have been performed. Test results showed a flow characterized by low mean velocities and large fluctuations, with the extent of reflected flow limited to few meters from the quay wall and inflow beneath the suction points playing a role.

Keywords: Propeller jet · Bow thruster · Quay wall · Field measurement · Bed protection

1 Introduction

Use of transverse thrusters, such as bow thrusters, is common during berthing procedures. Bow thruster-induced wash can cause scour (Roubos, et al., 2014), creating therefore the need for a bottom protection. Due to the interaction between propeller, ship hull and lateral restrictions such as quay walls and sea bed, the flow pattern of the jet is quite complex and is characterized by turbulence and high flow velocities near the bed. An accurate understanding of the propeller-induced flow field is fundamental for engineers, in order to quantify scour and to design an adequate bed protection (Hamill and Kee, 2016).

While an unconfined jet has received attention from several studies, fewer research has been focused on the flow field induced by propellers when confinement elements, such as vertical quay walls or proximity to the sea bed, are present (Wei and Chiew, 2019). Most of the research on propeller induced flow has been based on Albertson *et al.*, who used axial momentum theory to investigate the flow field of a plain-water jet (Albertson et al., 1950).

© The Author(s) 2023
Y. Li et al. (Eds.): PIANC 2022, LNCE 264, pp. 82–100, 2023.
https://doi.org/10.1007/978-981-19-6138-0_8

General features of jets are: diffusion, mixing layers and a great amount of turbulence derived by decelerating flow (Verhagen, 2001). Using the results of Albertson et al., several others developed semi-empirical equations to describe the velocity field within a propeller jet, slightly modifying Albertson's equations. Fuehrer and Römisch (1987), Blaauw and van de Kaa (1978), Hoffmans and Verheij (2011) and Schmidt (2000) based their research on scale models, while Blokland (1996) conducted field measurements. Similar to the present research, Schmidt (2000) and Blokland (1996) investigated reflection of a propeller jet against a vertical quay wall, although they did consider a different type of (bow) thruster than investigated here. Based on Blaauw and van de Kaa, and Fuehrer and Römisch research, BAW (2010) and PIANC (2015) design guidelines have been developed.

In the most recent years, a series of studies identified discrepancies between expected theoretical values as computed using general design guidelines and scale or numerical models. Location of the maximum flow velocities and flow patterns were different than expected (Deltares, 2015). Another study, conducted in Hamburg by Danish Hydraulics (Köppen and Best, 2014), pointed the attention on current guidelines to be overly conservative. Furthermore, guidelines don't provide a clear indication with regard to the width of a bottom protection, which is often designed based on vessel characteristics instead on the extent of the flow velocities (PIANC, 2015).

This has created the need to increase knowledge about propeller jets, especially when reflecting from a vertical quay wall. Both the research community and industry show interest in the topic, with the final goal of achieving optimized bottom protection designs. This study is the result of a collaboration between Port of Rotterdam Authority, Rijkswaterstaat, Deltares, TU Delft, Rotterdam Municipality, Boskalis, Deme and CROW.

In November 2018, pilot measurements on this topic were conducted using a tug at the Antarcticakade, a quay wall for inland barges with a nautical water depth of approximately 7–9 m in the Port of Rotterdam (Deltares, 2018). During that measurement campaign, flow velocities in the propeller jet have been measured using a single upward looking Acoustic Doppler Current Profiler (ADCP). Measurement results indicated the applied set-up was adequate to measure propeller jets, being practical and applicable. Measurements carried out in the present research represent a follow-up study, using a similar test setup and taking place at the same location, with objective of obtaining a better understanding of the flow velocities that occur near the bed and to verify the applicability of design methods presently in use.

2 Field Measurements Methodology

For conducting the measurements, an inland motor vessel, MTS Vorstenbosch, was utilized (see Fig. 1, left panel). It was considered representative for ultimate design situations, since this inland vessel is the largest in the Netherlands. The quay wall at the Antarcticakade consists a sheet pile vertical quay wall, and it has a bed protection of a 20 m wide and 0.85 m thick layer of loose rocks 10–60 kg. For the first 10 m from the quay wall, the protection is penetrated with colloidal concrete (see Fig. 2).

Fig. 1. Measurement location with the Vorstenbosch (left); Schematic representation of a 4-channel bow thruster, illustrating the working principle of the system (Right, source: vethpropulsion.com).

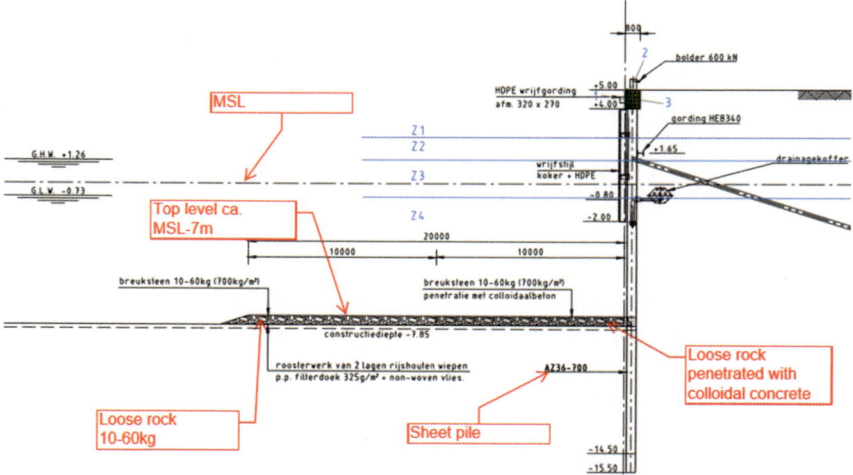

Fig. 2. Cross section of the Antarcticakade, with bottom protection characteristics highlighted.

The MTS Vorstenbosch is equipped with two 4-channel bow thruster systems (see Fig. 1, right panel). This vessel was selected, since for this type of thruster no clear design guidelines are yet available. Due to the shape of the bow of the vessel, the length of the thruster channel and the bow thruster inlet and outlet locations are different for each bow thruster. By applying different combinations of power for both bow-thrusters, it was therefore possible to test a wide variety of conditions. The characteristics of the vessel are listed in Table 1. For the most backward placed propeller (bow thruster 1), the wall clearance was equal to 4.23 times the propeller diameter Dt, and for the most forward propeller (bow thruster 2) the wall clearance was 2.32 times Dt.

Table 1. Characteristics of the Vorstenbosch

	Symbol	Parameter	Value
Ship	L	Length (m)	147.5
	B	Beam (m)	22.8
	D	Draught (m)	5.4
Veth Jet	Pt	Maximum power (kW)	618
4K-1400A	Dt	Propeller diameter (mm)	1420

To measure flow velocities, two Acoustic Doppler Current Profilers (ADCP) and two Acoustic Doppler Velocimeters (ADV) were installed (Nortek, 2017). While ADCPs measures an array of flow velocities, ADV sensors have a fairly small measurement volume (cylinder of 14 mm diameter and 14 mm height) with high sampling rate of 64 Hz. The ADCP measures the flow velocity in an array of cells by combining information of 4 acoustic beams, assuming a uniform flow within each measurement cell. The size of these measurement cells increases with increasing distance from the instrument. Due to the highly turbulent flow within a propeller jet, it was assessed that the assumption of uniform flow would not be met when the distance from the instrument would be too large. Therefore, here only the measurement results of the second cell of the ADCP are presented (located 0.5 m from sensor head), with a measurement volume that can be visualized as a pyramid trunk with a lateral area of 0.23 m base1 × 0. 41 m base2 × 0.20 m height (total approximate volume: 0.016 m^3) and a sampling rate of 8 Hz.

The instruments were mounted on a frame that was positioned on the sea bed, perpendicularly to the quay wall. One ADCP and one ADV were situated close to each other at the end of the frame near the quay, where maximum velocities are expected. One ADV was located in the middle of the frame, with the objective of being close to the bow thruster inlets and monitor the inflow. Finally, the last ADCP has been mounted at the external end of the frame, at ∼ 14 m from the quay wall. ADCPs were mounted horizontally, therefore looking underneath the ship's hull.

Three different ship positions have been considered. Firstly, the ship has been positioned with the axis of the bow thrusters symmetrical with respect to the measurement frame. Then, tests have been performed with the ship positioned 5 m ahead in respect to the first position, and 5 m behind it. Figure 3 (left panel) shows the three different adopted ship positions and their relative position to the measurement frame. A general reference system is adopted, with the positive x axis pointing perpendicularly away from the quay wall, y axis along the quay wall (positive towards the bow of the vessel), and z representing the vertical dimension in the water column (positive upwards). To acquire insight in the spatial distribution of the flow, it is assumed that mean flow patterns are stationary and that the results of individual tests at different ship positions can be combined. Since moving the vessel was more efficient than installing additional instruments, this approach was adopted as a practical compromise. An illustration of the measurement points with respect to the vessel is depicted in Fig. 3 (right panel).

As a general measurement protocol, each bow-thruster was activated for 2 minutes for 4 applied power steps (25%, 50%, 75% and 100%), then both bow-thrusters were activated simultaneously also applying the same 4 power steps. The three tests were repeated for three different ship positions. Tests were then repeated both a low and high tide, to investigate influence of different keel clearances. To assess the influence of the changing water level, and identify eventual recirculation patterns, one test with a longer duration of 30 min was performed during the rising tide. In addition, one test where the ship is de-berthing was performed using both bow-thrusters at 100% of power, to allow a comparison of stationary tests with a realistic maneuver. Table 2 lists an overview of all performed tests.

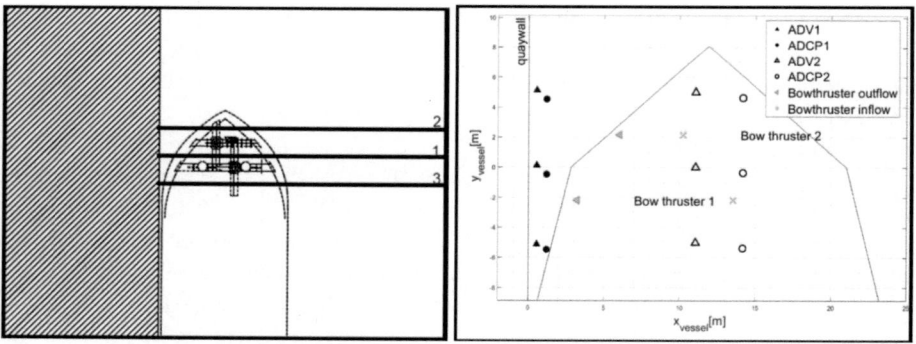

Fig. 3. The horizontal lines indicate the position of the frame on which the instrument where mounted with respect to vessel for the different considered ship positions (*left*); Illustration of the position of the instrument. ADVs are shown as triangles and cell 2 of ADCPs as circles. Inflow and outflow coordinates of both bow thrusters are illustrated, as well as a schematic outline of the ship (*right*).

Table 2. Measurement protocol for tests

Test	Subtest			Bow thruster	Ship position	Average keel clearance (m)
	Steps of applied power	Duration of each subtest (min)	Total duration of test (min)			
1	25%, 50%, 75%, 100%	2	8	1	1	1.03
2	25%, 50%, 75%, 100%	2	8	2		1.03
3	25%, 50%	2	4	Both		1.04
4	25%, 50%, 75%, 100%	2	8	2	2	1.06
5	25%, 50%, 75%, 100%	2	8	1		1.10
6	25%, 50%, 75%, 100%	2	8	Both		1.12
7	25%, 50%, 75%, 100%	2	8	1	3	1.30
8	25%, 50%, 75%, 100%	2	8	2		1.38
9	25%, 50%, 75%, 100%	2	8	Both		1.50
10	50%	30	30	Both	1	2.07
11	25%, 50%, 75%, 100%	2	8	1		2.57
12	25%, 50%, 75%, 100%	2	8	2		2.79
13	25%, 50%, 75%, 100%	2	8	Both		2.88
14	25%, 50%, 75%, 100%	2	8	1	2	2.82
15	25%, 50%, 75%, 100%	2	8	2		2.76
16	25%, 50%, 75%, 100%	2	8	Both		2.69
17	25%, 50%, 75%, 100%	2	8	1	3	2.58
18	25%, 50%, 75%, 100%	2	8	2		2.54
19	25%, 50%, 75%, 100%	2	8	Both		2.51
20	25%, 50%, 75%, 100%	2	8	1	1	1.33
21	25%, 50%, 75%, 100%	2	8	2		1.30
22	25%, 50%, 75%, 100%	2	8	Both		1.26
23	100%	7	7	Both	Moving	1.21

3   Test Results

3.1   General Flow Patterns

To investigate general flow patterns, the mean flow velocity and standard deviation for each step of applied power was calculated. Test 1, where bow thruster 1 was used at low tide for ship position 1, is taken as a base case; results are shown in Fig. 4 (left panel).

Observing the measured flow velocities for each instrument, it is possible to notice some general flow patterns. Instruments located near to the quay wall (ADCP1 and ADV1) recorded an increase in mean flow velocity with increase in applied power. Mean horizontal flow velocity appeared to be in the same order of magnitude for these two instruments, however the ADV recorded larger fluctuations. Also, the instrument located beneath the suction points (ADV2) showed an increase in mean flow velocity with an increase in applied power, but velocity magnitude was lower than recorded by the two instruments near the quay wall, and the trend seems different. The last instrument (ADCP2), located 14 m from the quay wall, recorded low mean flow velocities and the increase of applied power did not correspond to an increase in mean flow velocities.

These general observations are valid for most of the tests: mean flow velocities were usually in the order of magnitude of 1 m/s. The instruments near the quay wall presented a clear relation between an increase in applied power and an increase in velocities, whereas the instrument located beneath the suction points, despite recording an increase of velocity with increase of applied power, showed lower mean flow velocities. ADCP2 did not show a relation between use of bow thrusters and recorded velocities.

However, the data from the two instruments near the quay wall was not always consistent: the ADV often recorded higher mean flow velocities than the ADCP, and a larger variability, expressed by higher standard deviations. This behavior is especially evident in tests where the flow had more space to develop (larger keel and/or wall clearance). It is hypothesized that the relative low flow velocities measured by the ADCP is caused by the relatively large size of measurement cell compared to the turbulent structures present in the flow. However, this could not be confirmed during data processing and should be investigated further.

Concerning the spatial distribution of the flow: by assuming that placing the ship at a different position and combining individual test results is equivalent to have extra measurement points in the corresponding location, it is possible to gain an insight on the spatial distribution of the flow. This assumption was based on the uniformity of the measuring location. Test in different ship positions were performed with comparable water levels. A check conducted during post-processing gave confidence that the flow during the test was stationary enough to capture the small variations, as well as it did the absence of recirculation patterns during test 10. The only uncertainty on the validity of this assumption is represented by the variability of the sheet pile wall. Nonetheless, the results from the tests conducted with the ship located in position 2 and 3, consistently shown flow velocities significantly lower than the ones recorded for tests with the ship in position 1, giving a clear indication of the extent of the flow along the quay wall.

As observable in Fig. 4 (right panel), measurements do not show a clear return flow underneath the ship as there is a large variability in flow direction. Measurement points further away from the outlet do not seem to measure the reflected jet but show that the flow near the inflow point was directed towards the suction points of the bow thrusters. A comparison of flow direction recorded by the instrument located between the two suction points (ADV2), confirms that at this location (~ 10 m from the quay wall) the flow was dominated by the inflow and not by the reflected jet from the quay wall. This became particularly evident when comparing tests where different bow thrusters were used. Depending on which one is activated, the flow was directed towards the respective suction point (Fig. 5).

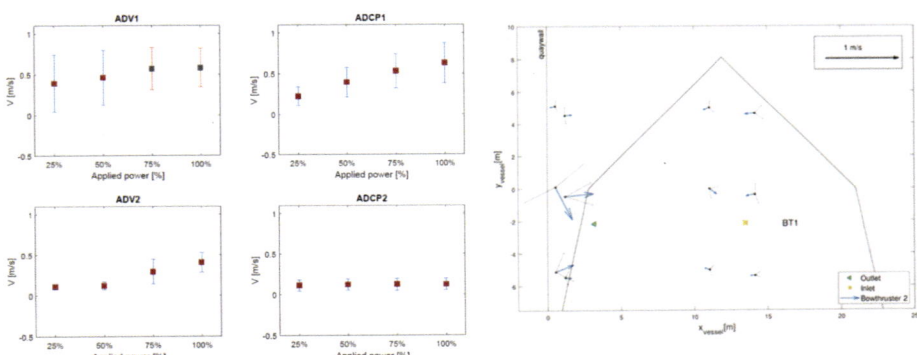

Fig. 4. Results from test 1 (use of bow thruster 1, ship position 1, low water). Horizontal mean velocity and standard deviation for each subtest are presented for each instrument. Non reliable data is colored in grey (*left*); Mean horizontal velocity magnitude and dominant direction for bow thruster 1 induced flow field at 50% of applied power. Tests 1, 5 and 7 (low water) have been combined. Dotted lines: range in direction, corresponding to ±st. Dev of direction (*right*).

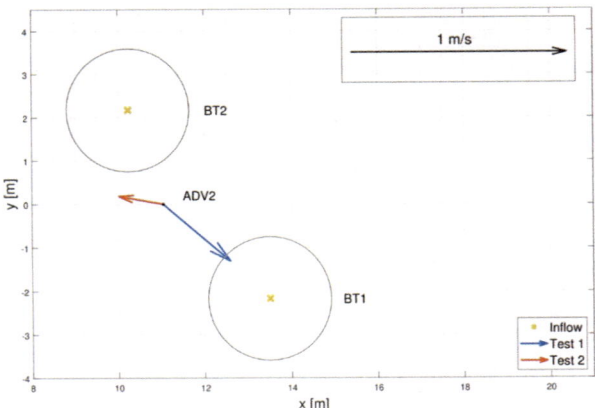

Fig. 5. Comparison of mean velocity and dominant direction measured by ADV2 for 100% of applied power in test 1 (bow thruster 1) and test 2 (bow thruster 2). Suction points of both bow thruster projections on the bed are shown.

3.2   Influence of Wall Clearance

To investigate impact of different wall clearances, the tests using one or the other bow thrusters were compared. The two bow thrusters differ in channel length, and hence in the resulting wall clearance. Wall clearances of the bow thrusters are, relative to the propeller diameter, respectively 2.32 D_t for bow thruster 1 and 4.23 D_t for bow thruster 2. For both bow thrusters however, also the relative position of the instruments with respect to the outlet was different. It was assumed that for this set-up, this difference was not relevant. Figure 6 shows that when low water tests are compared, no significant differences can be observed between the use of one or the other bow thruster, despite the asymmetric shape of the bow. For this condition, the influence of wall clearance seems small. When assessing the influence of under keel clearance however, the results are different for bow thruster 1 and bow thruster 2. This suggests that the impact of the under keel clearance is depending on the wall clearance and vice versa (Fig. 6).

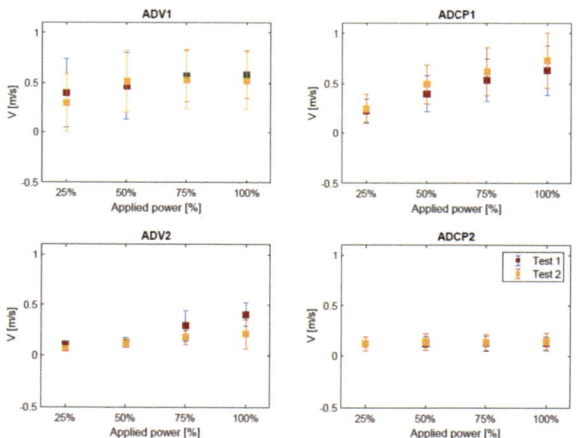

Fig. 6. Comparison of horizontal flow velocity mean magnitude and standard deviation for each step of applied power for test 1 (bow thruster 1) and test 2 (bow thruster 2), at low water.

3.3   Influence of Under Keel Clearance

Comparing the tests conducted at low or high tide, it is possible to observe the influence of under keel clearance on bed velocities. The results, as it can be observed in Fig. 7, vary depending on which bow thruster is used. Observing the data collected by ADV1, the use of bow thruster 1 seems to result in lower values of mean horizontal flow velocities with a larger under keel clearance. On the contrary, when bow thruster 2 is used, ADV1 records larger mean horizontal flow velocities with a larger under keel clearance, as well as an increase in turbulence.

This is though not reflected by measurements from ADCP1, which recorded similar velocities for both cases. It is also worth noting how, for bow thruster 2, the difference in mean horizontal flow velocity between low and large under keel clearance recorded by ADV1 increased with the increase in applied power (Fig. 7, right). In contrast for bow thruster 1, the difference remained fairly constant (Fig. 7, left). From this result, it therefore appears that the influence of under keel clearance on the bed velocities also depends on wall clearance.

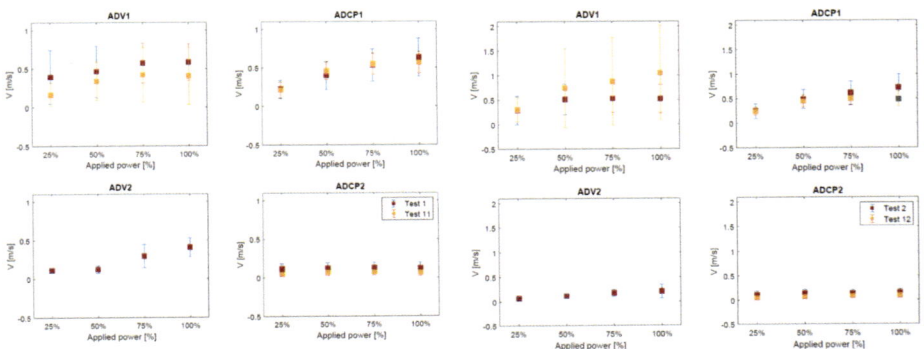

Fig. 7. Comparison for bow thruster 1 of horizontal mean flow velocity and standard deviation for each step of applied power for test 1 (low water) and test 11 (high water) (*left*); Same for bow thruster 2 for test 2 (low water) and test 12 (high water) (*right*).

3.4   Use of Multiple Bow Thrusters

Simultaneous use of the bow thrusters was also investigated during this research. Figure 8 (left panel) compares the results of tests 1, 2 and 3. In these tests, bow thruster 1, bow thruster 2 and both bow thrusters have been used during low tide. It seems that the use of both bow thruster simultaneously resulted in significantly higher flow velocities as could be expected. It should be noted that test 3 has been conducted only for the first two steps of applied power. This comparison is also possible for tests 11, 12 and 13, where respectively bow thruster 1, 2 and both bow thrusters simultaneously were used, but during high tide (Fig. 8, right panel). While tests 1, 2 and 3, presented above, were conducted at low water and with a fairly constant water depth, tests 11, 12 and 13 were carried out at high water and with larger variation in water level. There was, in fact, a difference of approximately 30 cm between water depth in test 11 and in test 13.

Figure 8 (right panel) shows that differences between use of bow thruster 1 and bow thruster 2 were more significant than the difference between the use bow thruster 2 or both bow thrusters at the same time. It is also worth noting the discrepancies in data recorded by the two instruments located near the quay wall (ADV1 and ADCP1), and the fact that velocities recorded by ADV1 during test 13 did not increase constantly with the increase in applied power, differently than in the other tests. The reason for this was unclear. Consequently, simultaneous use of both bow thrusters do not seem to lead to differences in mean flow velocities larger than the ones between the use of each bow thruster singularly for large keel clearances.

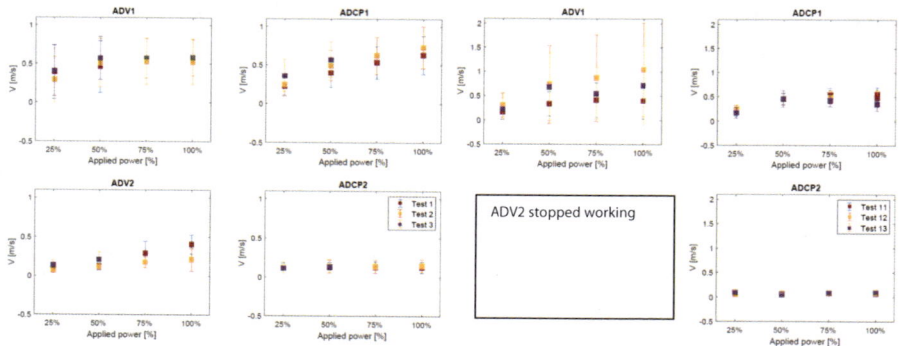

Fig. 8. Comparison of horizontal flow velocity mean magnitude and standard deviation for each step of applied power for test 1 (bow thruster 1) and test 2 (bow thruster 2), and test 3 (both bow thrusters) at low water (*left*); Same for test 11 (bow thruster 1) and test 12 (bow thruster 2), and test 13 (both bow thrusters) at high water (*right*).

4   Comparison with Guidelines

Results from field measurements were compared to the theoretical predictions using the formulae suggested by commonly used guidelines.

4.1   Maximum Velocities at the Bottom

The maximum velocities at the bottom have been estimated on the basis of both the Dutch and the German method suggested by PIANC. Equations suggested by the guidelines are based on the efflux velocity V_0, which depends on the power of the bow thruster P_t, water density ρ_w, and propeller diameter D_t as:

$$V_0 = 1.15 \left(\frac{P_t}{\rho_w D_t^2} \right)^{0.33} \tag{1}$$

Furthermore, the maximum velocities V_{max}, which can be expected at the corner between the quay and the bed, can be calculated according to German method from:

$$V_{max} = a_L 1.9 \, V_0 \left(\frac{L}{D_p} \right) \tag{2}$$

where a_L is an empirical coefficient chosen based on wall and under keel clearance and L is the wall clearance. Following Dutch method, V_{max} is calculated using the following equations, with h_t the height of the thruster above the bed:

$$V_{max} = V_0 \frac{D_t}{h_t} \qquad for \frac{L}{h_t} < 1.8 \qquad (3)$$

$$V_{max} = 2.8\, V_0 \frac{D_t}{L + h_t} \qquad for \frac{L}{h_t} > 1.8 \qquad (4)$$

The data obtained from measurements are compared to the theoretical predictions in Fig. 9 (left and right panel) for both bow thrusters and respectively for a keel clearance of ~ 1 m, and ~ 2.7 m. This comparison shows that both formulae are generally conservative compared to the measured velocities, with the exception of data measured during use of bow thruster 2 at high water (test 12, see Fig. 9, right panel). Furthermore, it is possible to observe how sensitivity the methods are to variations in wall and under keel clearance. This could also be because some of the considered wall and under keel clearance ratio falls outside the range of values for which the formulae have been developed and validated.

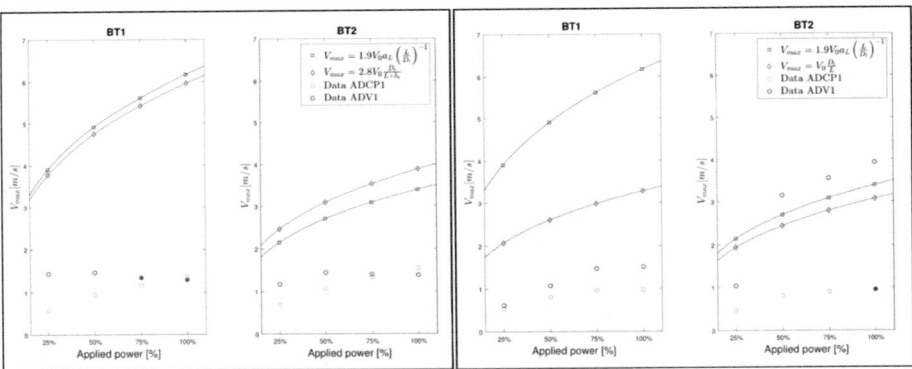

Fig. 9. Comparison between maximum velocities on the bed, as calculated with Dutch and German method, and measured maximum velocities from ADV1 and ADCP2 with a small UKC in Test 1 and Test 2 (*left*); Same for a larger UKC in Test 11 and Test 12 (*right*). Unreliable data (low correlation) depicted in grey.

4.2   Relative Turbulence Intensity

Relative turbulence intensity is a parameter that is often used to give an indication of turbulence level and can be calculated as $r = \frac{\sqrt{u'^2}}{\bar{u}}$, ratio between the root-mean-square of turbulent fluctuations u', and mean velocity $\bar{u}$. . Previous literature focused mainly on unrestricted flow; moreover, several methods (full scale measurements, scale models, analytical relations) were used. This resulted in a quite large typical range for relative turbulence intensity. Current guidelines indicate 0.25–0.4 as a typical range of values for r in propellers jet (PIANC, 2015). Figure 10 shows the data collected during the field measurements and shows that values are generally larger. This is mainly since the measured flow was characterized by low mean flow velocities and large fluctuations. The instruments near the quay wall record relative turbulence intensity levels even in the order of magnitude of 1.

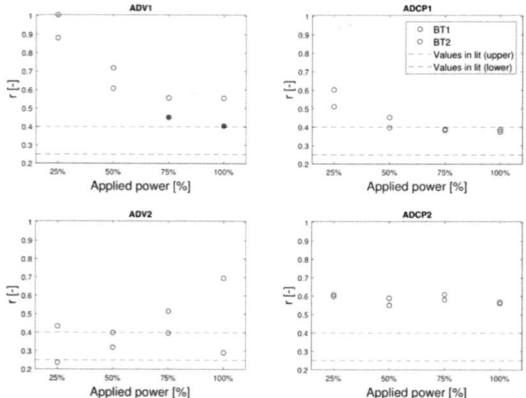

Fig. 10. Relative turbulence intensity measured by each instrument for each step of applied power. Comparison between bow thruster 1 (test 1) and bow thruster 2 (test 2). Unreliable data are depicted in grey. Reference lines indicate lower- and upper-bound values expected from literature.

4.3 Use of Multiple Propellers

According to literature, two approaches were adopted to take into account the use of multiple propellers: either linear superposition (PIANC, 2015) or proportionality to square root of number of propellers used (Schiereck, 2016). As observable from Fig. 11 both theories result to be generally conservative. Data collected during test 22 by the ADV is an exception (Fig. 11, right panel) and measurement results seem to be close to theoretical values obtained by multiplying theoretical velocities of one propeller by $\sqrt{2}$.

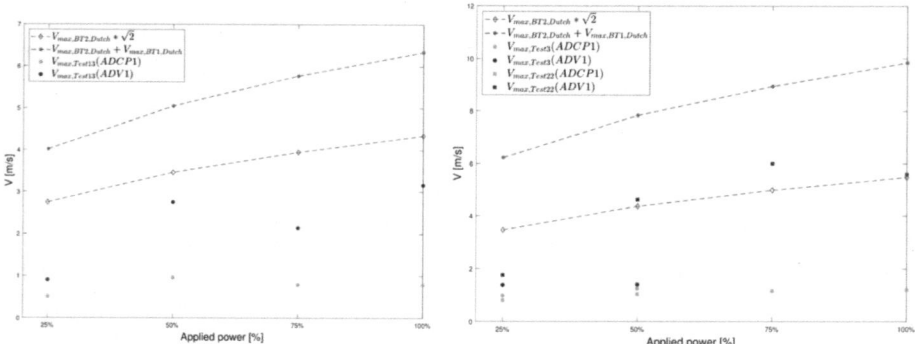

Fig. 11. Comparison of linear superposition theory and proportionality to sq. Root of n° of used propellers, with data from ADV1 and ADCP1 during test 13 (both bow thrusters, high water) (*left*). Same during test 3 and test 22 (both bow thrusters, low water) (*right*). Unreliable data in grey.

5 Discussion

5.1 Assessment of the Measurement Set-Up and Protocol

The measurement set-up used includes some limitations. Firstly, the efflux velocity of the thruster V_0 could not be measured directly with the used measurement setup. In the comparison of the measurement results with the design guidelines, the theoretical value of V_0 has been estimated based on Eq. (1) and the applied power as read from the instrument panel in the ship cabin. It is possible however that the actual efflux velocity in the field tests was lower than this theoretical value and that part of the conservatism shown in Section 4 is originating from the uncertainty in actual efflux velocity. It is therefore advised to measure V_0 directly in future research to verify the validity of Eq. (1) for these types of 4-channel bow thrusters.

Secondly, the limited amount of measuring points and the low spatial resolution result in a fragmented knowledge of the spatial patterns of the flow. ADCPs should have provided more insight on the extent of the reflected flow, but due to the high level of turbulence caused by the reflected jet, the cells more distant from the instrument were considered not reliable due to their relatively large measurement volume. Furthermore, in some tests there is a clear discrepancy between the two instruments placed near the quay wall. Despite their vicinity, ADV1 and ADCP1 are often showing different results. Namely, the ADV records higher mean flow velocities and larger fluctuations than the ADCP. Moreover, velocities corresponding to pressure fluctuations recorded by the ADCP appear to be larger than the recorded flow velocities. This could raise the suspicion that the ADCP, that averages the flow over measuring volume, it is not able to capture the turbulent flow caused by the bow thrusters. Therefore, we consider the flow measured by the ADVs to be more reliable, although the ACDP data can still valuable in a qualitative way, e.g. for relative comparison between different tests and showing trends in the data.

Despite the discrepancies and uncertainties above mentioned, the used measurement set-up was able to clearly determine three different zones in the bow thrusters-induced flow on the bed. These zones fit in the framework identified by Schmidt and we adapted it for 4-channel type bow thrusters. Figure 12 illustrates these zones of the flow induced by the reflection of bow thrusters against a vertical quay wall.

- Outflow zone (zone 1): differently than Schmidt, who distinguished an establishment flow zone and an established flow zone, for these tests only an outflow zone was identified. This is chosen on basis of the comparison between tests where bow thruster 1 was used (flow should have not been developed yet when it impinges the quay wall) and bow thruster 2 (flow should have been developed at the impingement). Since no major differences were observed, only one zone is assumed for the considered wall clearances;
- Impingement zone (zone 2) and spread along the quay wall (zone 3) were the zones identified by Schmidt in his experiments. During the present study, the focus was on the near bed velocities. Therefore, these zones haven't been investigated;
- A reflection zone (zone 4), where the flow is characterized by the highest recorded velocities and presents a clear relation with the use of bow thrusters;

- An inflow zone (zone 5), where the flow presents lower velocities than in the reflection zone, but still correlates to the use of bow thrusters. The flow is directed towards the 4-channel system suction point(s);
- A zone (zone 6) where the use of the bow thrusters is not noticed by the instruments anymore. At 14 m from the quay wall, flow velocities are low and there is not an increase in velocity following the increase of applied power from the bow thrusters.

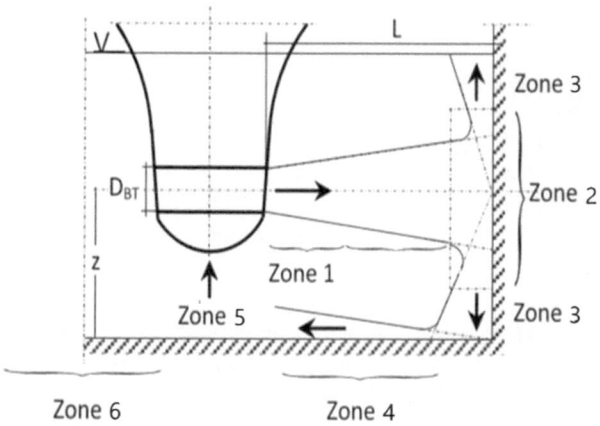

Fig. 12. Zones of reflected jet identified by the tests analyzed in this paper. Adapted by the authors from Schmidt (2000).

Furthermore, the used set-up confirmed the two assumptions that were made during testing. The first assumption was that activating the bow thrusters with the ship moored at the quay would be representative of realistic mooring operations. To test this hypothesis, a test was performed with the ship de-berthing, both bow thrusters activated at 100% of power. Results from this test show that mean velocities of the dynamic situation are comparable with the ones from the static one (see Fig. 13). Since the guidelines provide calculations for the mean velocity, the focus on the dynamic test was on the moving mean. These values were comparable with the ones in the static tests. The higher instantaneous velocities recorded by ADV1 shown in Fig. 13 may be partly influenced by measurement errors and should be investigated further. Stationarity checks conducted on the static tests ensured that the mean over each subtest is representative. Therefore, the results from tests 1–9 and 11–22 are considered representative for a realistic berthing or de-berthing situation.

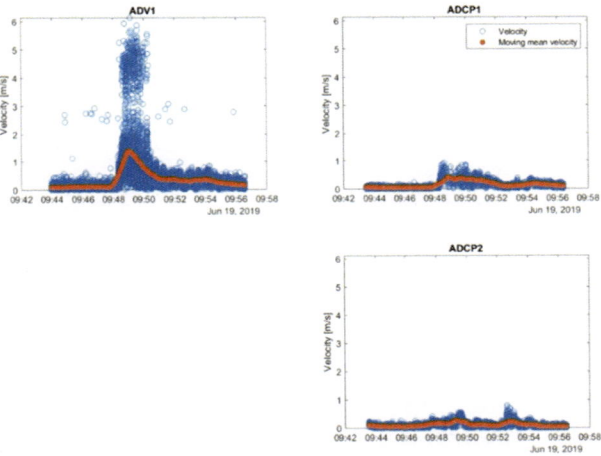

Fig. 13. Time series of recorded horizontal flow velocities during test 23 (de-berthing maneuver, both bow thrusters activated at maximum power). Both instantaneous and moving mean over 1 min are shown.

5.2 Contextualization in the Theoretical Framework

From comparison with Dutch and German calculation methods of near-bed flow velocities recommended by PIANC guidelines, data collected in this study identified the guidelines as conservative in most cases, especially the ones were the maximum absolute velocities were recorded and are, therefore, determinant for design. On the other side, data recorded by ADV1 in some tests, namely test 12 and 22, resulted in maximum velocities at the bottom slightly larger than the expected ones. Both in test 12 and test 22, bow thruster 2 was used: in test 12 on its own, in test 22 simultaneously with bow thruster 1. Bow thruster 2 is characterized by a larger wall clearance, equal to 4.23 D_t, while bow thruster 1 is located at 2.32 D_t from the quay wall. Most of the research upon which the guidelines have been based considered situations where the flow was less restricted than in the case presented in the present study: (Blaauw and Van de Kaa, 1978) conducted physical scale modelling for ducted and non-ducted propellers, measuring the velocities as the ship was moving in relatively unrestricted waters. (Schmidt, 2000) studied propeller jets against a quay wall in a stationary situation, evaluating different wall and keel clearances. Wall clearance studied by Schmidt were between 7.3- and 4-times D_t, while height of the thruster in the water column was either 2.2- or 2.4-times D_t. In the present study, height of the bow thruster changed more gradually within the tests, covering values between 1.21 D_t and 2.45 D_t. The results of the present study for the situation with a height of the bow thruster of 2.45 D_t and a wall clearance of 4.23 D_t fall within the range explored in Schmidt experiments. As Fig. 9 illustrates, maximum velocities recorded by the ADV present a similar trend to the one forecasted following the German method. However, the maximum velocities recorded by the ADV were slightly larger than expected. For the test with an under keel and wall clearance outside the range taken into account by the studies upon which the guidelines were drafted, the results showed the guidelines to be over conservative.

In combination with the aforementioned smaller wall clearances associated with the bow thruster system, the results of the present study show how the flow develops in a more restricted environment compared to previous studies and shows that for these conditions the design guidelines are conservative. Already a physical scale model research performed by Deltares (2015) found that formulae in literature being generally conservative for small wall clearances, with the exception of larger distances between quay wall and ship (9.5 D_t). Blokland (1996) as well, using field measurements, found higher flow velocities than expected from calculations for wall clearances between 3.2 D_t and 16 D_t.

These results support the hypothesis that currently generally used guidelines do not reflect influence of wall and keel clearance accurately enough for restricted situations, which are outside the range of tested conditions on which the theoretical formulations used in design guidelines are based. These conclusions are relevant, since the situation of small wall clearance considered in the present field measurements is representative for most inland vessels, which are usually characterized by a blunt hull shape that results in a small distance between the wall and the bow thruster's outlet.

5.3   Impact on Bottom Protection Design

When using measured flow velocities in the commonly used stability formulae to determine a suitable rock size required for bed protection, results vary greatly according to which tests are chosen as representative for design conditions. Concerning stability formulas such as Izbash and Pilarczyk, one should consider that these formulations have not been specifically developed for propeller jets. For propeller-jet induced damage, and stability in general, turbulence has been previously identified as a mechanism which might be even more important than mean flow velocity, for instance by Verhagen (2001) and Hofland (2005). As observed in Paragraph 4, low values of mean flow velocities and high relative turbulence intensities were present near the bed. The resulting calculations of rock stability for these conditions suggested that some damage would occur at the Antarcticakade for the most severe conditions. However, surveys conducted after the measurements did show an absence of damage (no displaced rocks) and also the instruments did not show any signs of impact due to moving loose material. This leads to think that a more precise representation of the physical phenomena should be sought.

6   Conclusions

From the measurements conducted at the Antarcticakade quay wall, bow thruster-induced flow on the bed seems to be characterized by fairly low mean flow velocities in combination with a high relative turbulence level. In contrast to research that was previously conducted, no clear return flow underneath the vessel was observed. The extent of the reflected flow from the quay wall seemed to be limited to a few meters. In addition, the flow pattern below the vessels seems to be largely influenced by the suction points of the 4-channel bow thruster system. Furthermore, the measured flow velocities for small wall and under keel clearances showed to be much lower than predicted by methods that are presently used in the design of bottom protections. For larger keel and wall clearances, a better match with general design guidelines was found, but the maximum velocities were slightly larger than expected. Consequently, the design guidelines for bottom protection seem to be conservative for berths facilitating inland vessels in situations with small under keel clearances (approximately < 2 D_t) and small wall clearances (approximately < 4 D_t). It is however important to note how more restricted situations are more relevant for design, since they cause the largest absolute flow velocities. These results are in agreement with previous scale model tests conducted by Deltares (2015). It is highly recommended that additional tests be performed to derive more insight into the relation between the flow velocities near quay walls and small under keel and wall clearances.

Acknowledgements. On behalf of the Delft University of Technology, Deltares, Port of Rotterdam, and Rijkswaterstaat the authors of this paper would like to thank all companies involved for their support, funding and hospitality. In particular, VT and Shell are gratefully acknowledged for their support and for providing a fully laden Vorstenbosch for this test.

References

Albertson ML, Dai YB, Jensen RA, Rouse H (1950) Diffusion of submerged jets. Trans. Am. Soc. Civ. Eng. 115(1):639–664. https://doi.org/10.1061/TACEAT.0006302

BAW: Principles for the design of bank and bottom protection for inland waterways. Code of Practice, Karlsruhe, March 2011, ISSN 2192-9807 (2010)

Blaauw HG, Van de Kaa EJ (1978) Erosion of bottom and sloping banks caused by the screw race of manoeuvring ships. In: 7th International Harbour Congress, May 22–26, 1978, Antwerp

Blokland T (1996) Schroefstraal Tegen Kademuur - Stroomsnelheid en Erosie Meeting (in Dutch), Gemeentewerken Rotterdam, Project Delta 2000-8

Deltares (2015) Propeller jets. Knowledge Gap 7 - Reflection of Transverse Jets by Vertical Quay Walls. Ref: 1210140-000-HYE-0008. 20 Nov 2015

Deltares (2018) Veldmetingen schroefstraalbelastingen (memo in Dutch), project number 11202175. 21 Dec 2018

Fuehrer M, Römisch K (1987) Propeller jet erosion and stability criteria for bottom protection of various constructions. PIANC Bullettin No. 58

Hamill G, Kee C (2016) Predicting axial velocity profiles within a diffusing marine propeller jet. Ocean Eng. 124:104–112

Hoffmans G, Verheij H (2011) Jet scour. Inst. Civ. Eng. Proc. Marit. Eng. 164(4):185–193. https://doi.org/10.1680/maen.2011.164.4.185

Hofland B (2005) Rock and Roll - Turbulence-induced damage to granular bed protection. PhD dissertation, TU Delft

Köppen J, Best J (2014) Messung von Schiffsinduzierten Belastungen auf das Bubendey Ufer Naturmessungen von Uferbelastungen. DHI

Nortek (2017) The Comprehensive Manual. Nortek

PIANC (2015) Report No. 180. Guidelines for protecting berthing structures from scour caused by ships

Roubos A, Blokland T, Van der Plas T (2014) Field Tests Propeller Scour Along Quay Wall. PIANC World Congress San Francisco, USA

Schiereck GJ (2016) Introduction to Bed, Bank and Shore Protection. Delft Academic Press/VSSD, Delft

Schmidt E (2000) Belastungen Durch Bugstrahlruder. Dresdner Wasserbauliche. Mitteilungen 18, Technische Universität Dresden, Institut für Wasserbau und Technische Hydromechanik, pp. 145–157. https://hdl.handle.net/20.500.11970/104017

Verhagen HJ (2001) Bowthrusters and the stability of a riprap revetment. https://www.researchgate.net/publication/238606205

Wei M, Chiew YM (2019) Impingement of propeller jet on a vertical quay wall. Ocean Eng. 183:73–86

Field Observation and Experimental Study on the Interaction Between Ship Waves And Vertical Wave Dissipation Revetment

Liehong Ju, Jingxin Huang, and Junning Pan[✉]

Laboratory of Hydrology-Water Resource and Hydraulic Engineering,
Nanjing Hydraulic Research Institute, Nanjing, China
jnpan@nhri.cn

Abstract. With China's economic development, many inland waterways adopt vertical revetment structure to save land resources, but this structure is not conducive to wave attenuation. According to the measurement results of the ship wave data of typical ship types in Sunan Canal sailing at different speeds collected in field observation, the maximum wave height in front of the vertical revetment wall was less than 1.0 m under normal navigation conditions, while the navigation administration boats can form a wave height of nearly 1.8 m in front of the vertical revetment wall at the maximum speed. In channels with heavy freight and fast ship speed, the ship waves formed by ships, superposed with the reflected waves by vertical impervious revetment, would result in severe water surface fluctuation, affecting the safety of ship navigation and the stability of revetment structure. Based on the regulation project of Sunan Canal, this paper put forward two types of vertical wave dissipation revetment, namely round-hole caisson structure and grid-type structure. And their wave dissipation effects are compared and optimized through 2-D wave test. Finally, the 3-D physical model tests of ship wave and revetment were carried out, and the effects of reducing ship wave height of round-hole caisson structure and grid structure are compared. The results show that these two vertical wave dissipation structures can reduce the maximum ship wave height in front of the revetment wall by about 20–25%. The relevant research results of this paper can serve as a reference for similar projects.

Keywords: Wave dissipation revetment · Ship wave · Field observation · Physical model test

1 Introduction

Jiangsu Province boosts abundant canals and developed water transportation since ancient times (Sun 2007), occupying about one fourth of China's shipping mileage, tonnage of transportation vessels, annual cargo transportation volume and turnover. In response to the economic development and shipping needs in recent years, many canals have carried out waterway regulation by dredging channels and building revetments to improve the efficiency of canals.

© The Author(s) 2023
Y. Li et al. (Eds.): PIANC 2022, LNCE 264, pp. 101–112, 2023.
https://doi.org/10.1007/978-981-19-6138-0_9

The interaction between ship waves and revetment is a traditional hydrodynamic issue. Revetment structure is one of the important facilities of inland waterway engineering, which can alleviate the scouring of bank slope by water flow and ship waves, thus playing an important role in ensuring the stable and unobstructed waterways. Choosing an appropriate revetment structure is the primary task of revetment engineering design. The revetment structures are generally divided into slope type and vertical type. Slope revetment structure is more commonly used and has a long history, which tends to be more ecological and diversified in recent years (Li et al. 2019). Compared with the slope structure, the vertical revetment requires less land resources but has weaker ability of dissipating ship waves. In order to reduce wave reflection and dissipate waves, people set holes in vertical structures on the wave-ward side, often with perforated caissons. The caisson type breakwaters was first proposed by Jarlan in the 1960s. Subsequently, many domestic and foreign scholars have carried out a lot of experimental research and theoretical analysis on the wave dissipation effect and mechanical characteristics of perforated caissons, and achieved fruitful results (Chen et al. 2001; Shi et al. 2011) while the researchers are focusing more numerical simulation in recent years (Cai et al. 2022). The main mechanism of wave dissipation by using perforated caissons is to introduce part of the wave energy into the caissons through the holes opened on the wave-ward side of the vertical structure, and weaken the reflected waves through the dissipation of waves in the caisson and the influence of phase difference, so as to realize the wave dissipation effect.

The main hydrodynamic force in the canal is ship waves. Throughout the long history of studies on ship waves, many achievements have been made in theoretical research on Kelvin's ship-wave pattern (Ursell 1959) and empirical formulas have been developed through field observation and experiments (Blaauw et al. 1984; Pilarczyk 1984; Robert and Sorensen 1986; Zhou and Chen 1995). In recent years, the research on ship waves mostly focuses on numerical simulation (Zhou et al. 2013).

Relying on the regulation project of Sunan Canal, the field observation was carried out to collect the ship wave under different conditions. Then, 2-D wave tests were carried out to explore different vertical revetment wave dissipation structures and propose the reasonable types of structure accordingly. Finally, 3-D tests of ship waves and different wave dissipation revetments were carried out to verify the effect of vertical revetment on dissipating ship waves.

2   Field Observation of Ship Waves

2.1   Basic Information About Field Observation of Ship Waves

The field observation on the ship waves in Sunan Canal was mainly conducted at two sites: ① Zhenjiang section of Sunan Canal. The revetment spacing on both sides of the channel is 90 m, and the water level during observation was 2.3 m; ② Yixing beltway section of Wushen Line. The revetment spacing on both sides of the channel is 70 m, and the water level during observation was 2.1 m. The upstream and downstream shorelines of the two observation locations are straight. The canal section of the ship wave field observation sites is shown in Fig. 1.

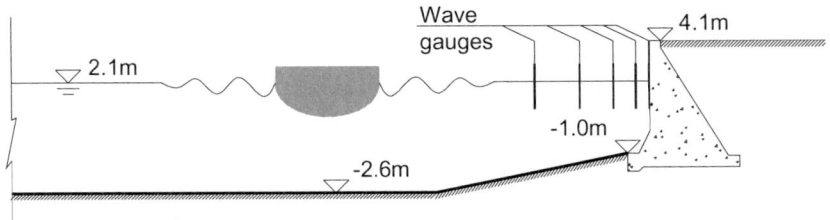

Fig. 1. Schematic diagram of ship wave field observation

The wave gauges which adopted for field ship wave surface measuring are 2 m high. During wave surface measuring, the computer-controlled acquisition system was used, the data collection frequency is 20 20 Hz. The layout of wave height gauges is shown in Fig. 2. The distance between the ship and the shoreline was measured by infrared rangefinder.

Ship types at two sites: 500 ton–1000 ton cargo ships, navigation administration boats.

Ship waves were measured mainly in two scenarios: ① when the canal is under temporary control, and individual vessels pass the canal separately at a constant speed, record the water surface fluctuation process when the vessel passes through the observation points; ② when the canal is open for normal navigation without control measures, record the water surface fluctuation process for a period of time (0.5 hours).

2.2   Main Results of Ship Wave Field Observation

Figure 2 shows the typical water surface fluctuation process in field under normal navigation conditions, while Fig. 3 shows the same process when the navigation administration boat (NA boat) passes through at high speed. Among them, wave gauge 1# was deployed adjacent to the revetment wall, while the distance between wave gauges 2#, 3#, 4# and 5# and the revetment wall were 0.6 m, 1.6 m, 3.1 m and 5.1 m, respectively. The results of the maximum wave height and the maximum runup on the revetment wall measured by wave gauges under normal navigation conditions are shown in Table 1, while those obtained when the NA boat passes through are shown in Table 2.

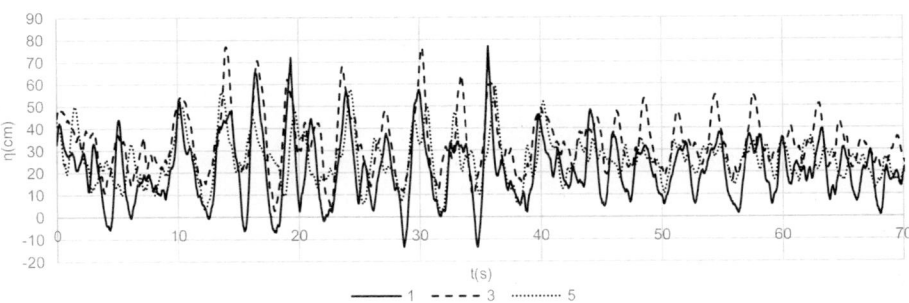

Fig. 2. Typical water surface fluctuation process under normal navigation condition

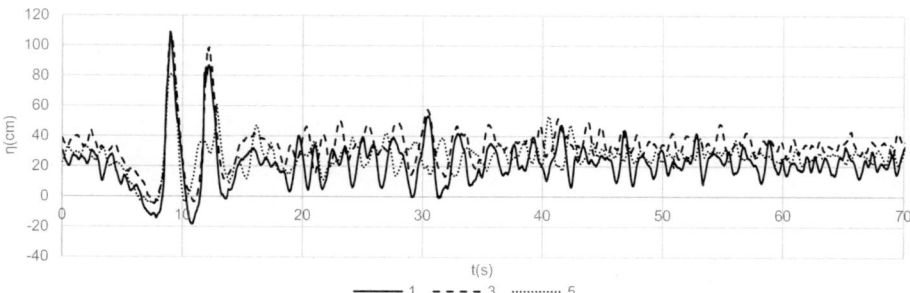

Fig. 3. Typical water surface fluctuation process when the NA boat passes through at a high speed

Table 1. Maximum wave height and runup on the revetment wall measured at two sites (normal navigation)

Measurement no.	Maximum wave height, Hmax (m)					Max runup on the revetment wall (m)	Period (s)
	Gauge 1#	Gauge 2#	Gauge 3#	Gauge 4#	Gauge 5#		
1	0.34	0.29	0.25	0.28	0.33	0.21	1.36
2	0.72	0.70	0.68	0.54	0.54	0.46	1.52
3	0.43	0.31	0.25	0.24	0.32	0.27	1.54
4	0.44	0.40	0.33	0.26	0.32	0.22	1.88
5	0.36	0.29	0.22	0.27	0.27	0.24	2.00
6	0.38	0.35	0.27	0.31	0.32	0.24	1.72
7	0.47	0.33	0.36	0.27	0.31	0.25	1.60
8	0.84	0.73	0.68	0.55	0.50	0.53	2.64

Table 2. Maximum wave height and runup on the revetment wall measured at two sites (NA boat)

Measurement no.	Offshore distance s (m)	Average speed v (m/s)	Hmax (m)					Max runup on the revetment wall (m) (m)	Period (s)
			Gauge 1#	Gauge 2#	Gauge 3#	Gauge 4#	Gauge 5#		
1	15	4.0	0.71	0.64	0.67	0.60	0.65	0.45	2.85
2	17	3.8	0.57	0.50	0.46	0.41	0.35	0.37	2.00
3	15	5.3	1.73	1.61	1.11	0.53	0.84	1.17	3.60
4	15	4.6	1.27	1.14	1.11	1.01	0.89	0.87	3.02

Under normal navigation conditions, the average value of the maximum wave height of different measurement times at the measuring points on the revetment wall was 0.49 m and the maximum wave height was 0.84 m; the average runup of wave on the revetment wall was 0.30 m, and the maximum runup was 0.53 m. The wave height formed by a single cargo ship passing through the observation point was small due to its small speed; when the NA boat passed through the observation point at a high speed, the wave was much higher. The maximum wave height of the observation points on the revetment wall was 1.73 m and the maximum runup height of the wave surface was 1.17 m.

3 2-D Test of Vertical Wave Dissipation Revetment

3.1 Section of Vertical Revetment Wave Dissipation Structure

According to the hydrological, geological and topographic data of the regulated channel, two types of vertical wave dissipation structures were adopted: round-hole caisson structure and grid-type structure. See Fig. 4 for the section and front of the round-hole caisson wave dissipation structure (the elevation unit in the drawing is m, and other marked units are mm). The top elevation of round-hole caisson structure was +4.1 m and the bottom elevation was −1.9 m. The circular holes were 150 mm in diameter, arranged in three rows, with a transverse spacing of 800 mm and a longitudinal spacing of 1000 mm. The elevation of the center point of the top row was +2.6 m and that of the bottom row was +0.6 m. See Fig. 5 for the section and front of the grid-type wave dissipation structure. The upper and lower elevations of grid openings were 3.7 m and 1.3 m respectively, the grid width was 400 mm, and the column width between grids was 600 mm.

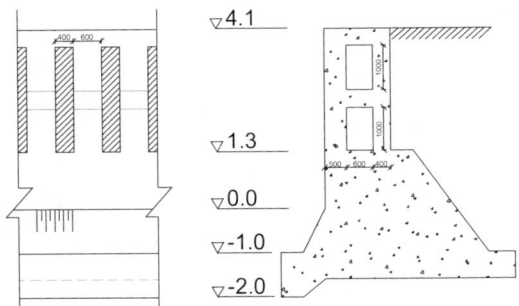

Fig. 4. Round-hole caisson wave dissipation structure

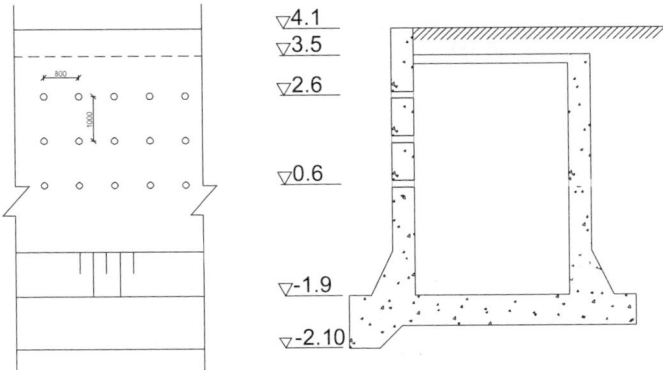

Fig. 5. Grid-type wave dissipation structure

3.2   Test Conditions

The physical model test of wave section was carried out in the wave tank, which was 170 m long, 1.2 m wide and 1.6 m deep. One end of the flume was equipped with a wave maker and the other end a wave dissipation slope. The test was designed according to Froude's law of similarity, and the geometric scale of the model was 1:5. Regular wave is adopted for the section test wave, with the test wave period including 2.0 s and 3.0 s, and the test wave height including 0.3 m and 0.5 m. During the test, capacitive sensors were adopted to measure the wave surface, which was automatically collected by computer control at the sampling frequency of 40 Hz; the reflection coefficient was analyzed by multipoint method.

3.3   Test Results and Analysis

(1) Test results of round-hole caisson wave dissipation structure
   The opening ratio of round-hole caisson wave dissipation structure has a great influence on its wave dissipation effect. In the section test, the reflection coefficients of perforated caisson under different opening ratio were measured to analyze the influence of opening ratio on the reflection coefficient. The caisson opening ratio (n) is equal to the ratio of the hole area in the opening area to the total area. When n = 0.15, the transverse spacing of holes was 40 cm, the longitudinal spacing was 30 cm and the hole diameter was 15 cm. When n = 0.22, the transverse spacing of holes was 40 cm, the longitudinal spacing was 20 cm and the hole diameter was 15 cm. When n = 0.26, the transverse spacing of holes was 40 cm, the longitudinal spacing was 30 cm and the hole diameter was 20 cm. When n = 0.39, the transverse spacing of holes was 40 cm, the longitudinal spacing was 20 cm and the hole diameter was 20 cm. See Fig. 6 for comparison of reflection coefficient test results of wave dissipation structure with different opening ratio.
   The results show that when the water level is 2.5 m, the wave period is 2 s and the wave height is 0.3 m, the reflection coefficients of the round-hole caisson revetment wave dissipation structure are 0.94, 0.71, 0.61, 0.59 and 0.57 respectively under opening ratio of 0.02, 0.15, 0.22, 0.26 and 0.39, indicating that the decreasing trend of reflection coefficient slows down when the opening ratio is greater than 0.26.

(2) Test results of grid-type wave dissipation structure
   Considering the structural design of the grid-type structure, the test was made with two kinds of opening structures. One structure was with an opening ratio of about 0.40, grid width of 90 cm (solid part), grid spacing of 60 cm (cavity part), while the other was with an opening ratio of about 0.33, grid width of 100 cm (solid part) and grid spacing of 50 cm (cavity part). The top and bottom elevations of the grid structure are shown in Fig. 5. The test results of reflection coefficients of the two grid-type structures under the action of different waves with the water level of 2.5 m are shown in Table 3.

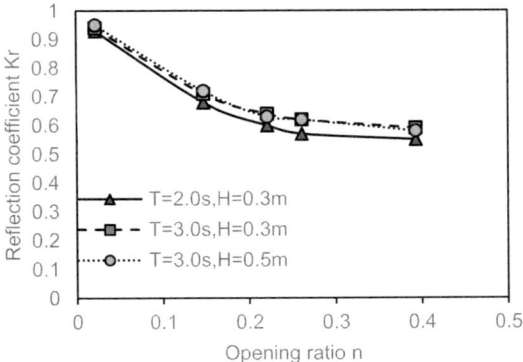

Fig. 6. Variation of wave reflection coefficient of round-hole caisson under different opening ratio

Table 3. Test results of wave reflection coefficient of different grid-type structures

Period (s)	Incident wave height (m)	n = 0.40	n = 0.33
2.0	0.3	0.75	0.78
3.0	0.3	0.74	0.77
	0.5	0.66	0.67

The test results show that the two grid structures have similar wave dissipation effects. Under the action of forward waves, the wave reflection coefficient of the grid-type wave dissipation structure was larger than that of the round-hole caisson wave dissipation structure.

4   3-D Test of Ship Waves and Revetment

4.1   Test Conditions

The bottom width of the test channel is 46 m, and the spacing of the fronts of the revetment wall is 90 m. See Fig. 7 for the section. The test water level is the average navigable water level, i.e. +2.5 m.

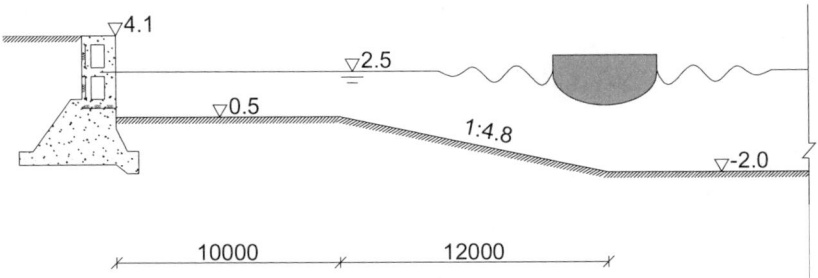

Fig. 7. Channel section of the 3-D test of ship waves and revetment

The test ship includes 1000 t cargo ship and 500 t cargo ship. See Table 4 for the type and size of ship for the 3-D test.

Table 4. Ship type and size for the 3-D test of ship waves and revetment

Ship type	Length (m)	Width (m)	Draft depth (m)
500 t cargo ship	41.6	8.3	3.0
1000 t cargo ship	44.6	8.8	3.4

Five types of different revetment structures were tested, including:

REVETMENT 1: vertical impervious structure;
REVETMENT 2: round-hole caisson structure (opening ratio n = 0.02);
REVETMENT 3: round- hole caisson structure (opening ratio n = 0.26);
REVETMENT 4: grid-type structure (opening ratio n = 0.4);
REVETMENT 5: riprap slope structure (slope gradient 1:2).

4.2   Model Test Design

The test was carried out in a water tank with a length of 50 m, a width of 6.0 m and a height of 0.8 m. The revetment structure is simulated within a 20 m section in the middle. The ends of the water tank are the acceleration and deceleration areas of the test ship, as shown in Fig. 8. The geometric scale of the 3-D ship waves and revetment wave dissipation test model is 1:20.

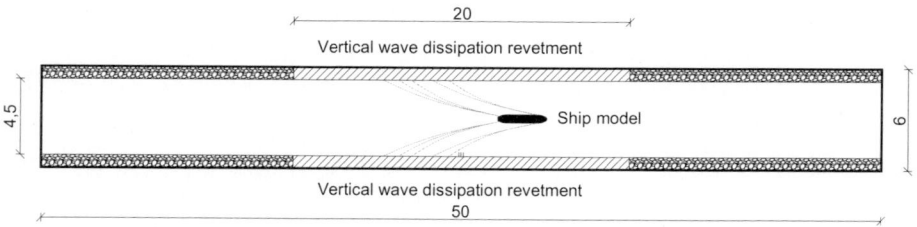

Fig. 8. Layout of the 3-D ship wave and revetment test model

The model ship is self-propelled with remote control, with the designed maximum speed of 2.5 m/s.

The wave height is measured by using capacitive wave surface sensors, and sample collection is controlled by computer at the sampling frequency of 40 Hz. The layout position of wave gauge is the same as that in the field observation.

Particle imaging analysis technology is adopted for the measurement of ship speed trajectory: install position identification devices on the ship model, erect high-definition camera devices above the water tank, and calculate the motion trajectory and speed of the ship through video analysis software.

During the test, accelerate the model ship to a certain speed before the test section and stabilize it, and then drive it into the model test section at a uniform speed. Measure the ship speed by using the particle imaging system and collect the wave surface process of each measuring point by the wave surface acquisition system. See Fig. 9 for test photos.

Fig. 9. Photos of 3-D ship wave and revetment test

4.3   Model Test Results

(1) Comparison between the test results and field observation results

The fourth condition of field observation of NA boats in Table 5 was re-demonstrated in the 3-D test. During the ship wave re-demonstration test, the revetment wall is a vertical impervious structure. See Fig. 10 for the comparison between the waveform change at the revetment wall measured in the laboratory and the field observation results. In general, the waveform changes of the ship waves were basically identical, from the larger "bow waves" caused by the ship to the "stern waves" with relatively small wave height. The peak shape and interval of the "bow waves" simulated in the laboratory were basically consistent with those measured during field observation. However, the water reduction phenomenon of the bow wave front was greater than that in the field observation, which was mainly due to the difference between the near shore terrain of the channel at the field observation site and the simulated terrain in that the water at the field was deeper while that in the near shore of the laboratory was shallow. The comparison between the maximum wave height at the revetment wall generated by 1000 t cargo and 500 t cargo ship sailing at different speeds in the laboratory and the field observation results is shown in Fig. 11 (22 m from the center line along the edge of the channel). It can be seen that the trend of maximum wave height is close to the speed variation.

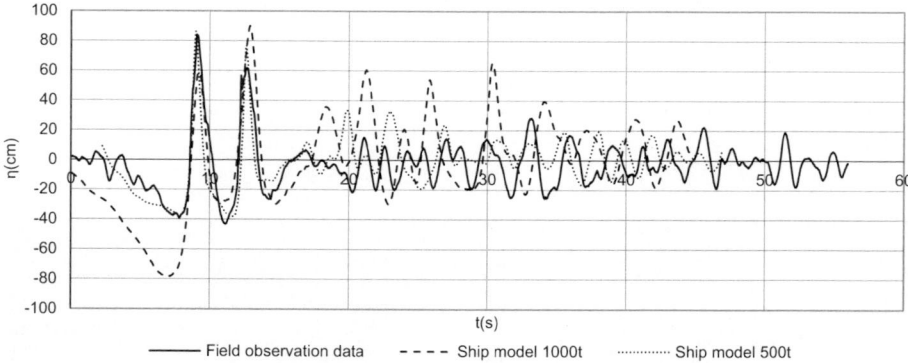

Fig. 10. Comparison of wave surface fluctuation process between field observation and test

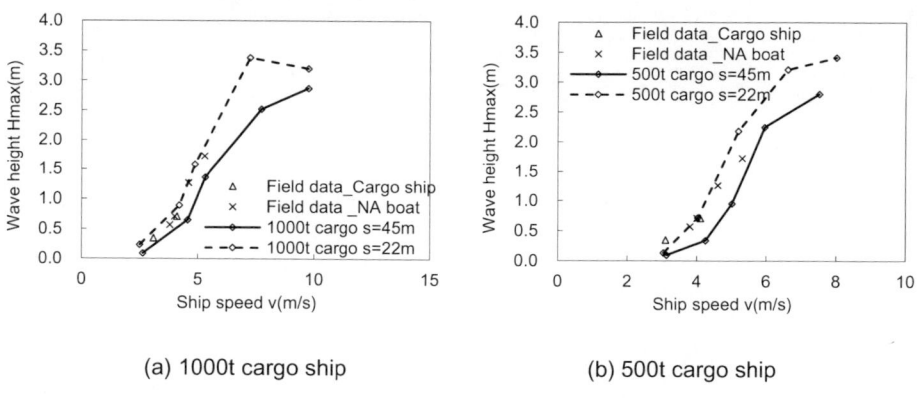

(a) 1000t cargo ship   (b) 500t cargo ship

Fig. 11. Comparison of ship wave height between test and field observation

(2) Comparison of wave dissipating effect of different revetment structures
In the test, the wave height in front of the revetment wall was measured under different ship speeds and offshore distance. The comparison of the maximum wave height in front of the revetment wall in different routes is shown in Fig. 12

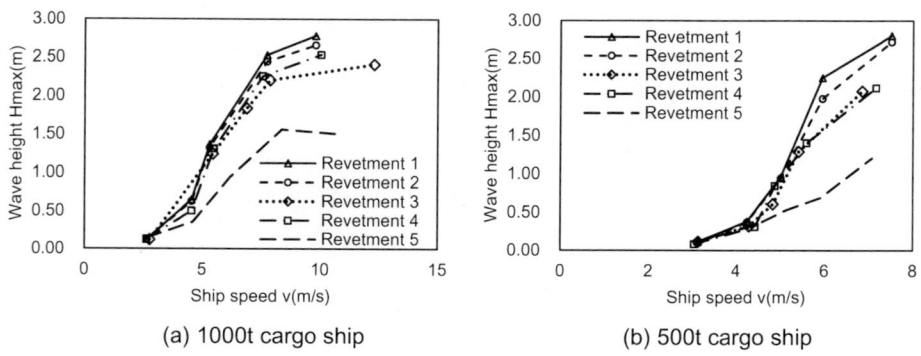

(a) 1000t cargo ship   (b) 500t cargo ship

Fig. 12. Comparison of the maximum ship wave height in front of the revetment wall of different revetment structures when the vessel is sailing in the middle of the channel

The test results show that the maximum wave height was similar in the case of vertical impervious revetment structure and round-hole caisson structure (opening ratio n = 0.02), followed by grid-type structure (opening ratio n = 0.4) and round-hole caisson structure (opening ratio n = 0.26), and the smallest height was generated with riprap slope structure. The maximum wave height in front of the revetment wall of the round-hole caisson structure (opening ratio n = 0.26) and the grid-type structure (opening ratio n = 0.4) was reduced by about 20–25%, while that of the riprap slope structure was reduced by about 40%.

5   Main Conclusions

Vertical revetment is a structure that saves land resources, but it can hardly reduce the wave reflection. In the waterway with busy shipping and fast vessel speed, the interaction between vertical impervious revetment structure and ship waves often leads to severe water surface fluctuation, which will affect navigation safety. In addition, the larger wave height is also unfavorable to the stability and safety of revetment. Through field observation, 2-D wave test and 3-D ship wave and revetment test, this study came to the following conclusions:

(1) The field observation on ship waves in Sunan Canal shows that under normal navigation conditions, the maximum ship wave height formed by cargo ships on the canal was less than 1.0 m, but the maximum height of the waves formed in front of the vertical revetment wall can reach 1.73 m when the NA boats sail at the maximum speed.
(2) The 2-D wave test results show that the wave dissipation effect of the round-hole caisson structure was continuously improved with the increased opening ratio of the wave-ward side. When the water level was 2.5 m, the wave period was 2 s, the wave height was 0.3 m, and the opening ratio of the round-hole caisson wave dissipation structure were 0.02, 0.15, 0.22, 0.26 and 0.39, the wave reflection coefficients were 0.94, 0.71, 0.61, 0.59 and 0.57, respectively.
(3) The 3-D test of the interaction between ship waves and revetment shows that the round-hole caisson structure (opening ratio n = 0.26) and grid-type structure (opening ratio n = 0.4) could effectively reduce the front wave height of the revetment wall by about 20–25%.

References

Chen X, Chen R (2001) Experimental study on the interaction between waves and perforated caissons. China Offshore Platform (5):6

Cai Y et al (2022) Numerical study on the influence of opening ratio on the regular wave reflection coefficient of perforated caisson on open foundation bed (1)

General Secretariat of PIANC (1987) Guidelines for the design and construction of flexible revetments incorporating geotexitiles for inland waterways. Supplement to Bulletin No. 57

Blaauw HG, van der Knaap FCM, de Groot MT, Pilarczyk KW (1984) Design of Bank Protection of Inland Navigation Fairways. Delft Hydraulic Laboratory. Publication No. 320

Pilarczyk KW (1984) Prototype tests of slope protection systems. Hartel Canel, The Netherlands

Li J, Shi B (2019) Research progress of ecological revetment of inland waterway based on ship wave dissipation. China Water Transp: Later Half Month 19(3):4

Liu J et al (2016) Stability analysis of interlocking concrete block slope protection surface under ship waves. J Water Resour Water Eng 27(6):8

Robert M, Sorensen F (1986) Bank protection for vessel generated waves. IMBT Hydraulic Laboratoty Report No. THL-117-86

Shixiaodi et al (2011) Experimental study on wave dissipation performance of perforated caisson. Water Transport Eng (03):16–20

Sun J (2007) Study on the countermeasures of accelerating the development of inland shipping in Jiangsu. Modern Econ Res (4):4

Ursell F (1959) On Kelvin's ship-wave pattern

Zhou J, Chen W (1995) Overview of ship waves and river bank slope protection project. Jiangsu Traffic Eng (1)

Zhou M, Roelvin D, Verheij H, Ligteringen H (2013) Study of passing ship effects along a bank by Delft3D-FLOW and XBeach. In: International Workshop on Next Generation Nautical Traffic Models. Delft, The Netherlands

Meshed Remote Operation as the Default Mode: From Technical Challenge to Society Opportunnities

Anges Peil[✉] and Michiel Coopman

De Vlaamse Waterweg, Hasselt, Belgium
{agnes.peil,michiel.coopman}@vlaamsewaterweg.be

Abstract. This paper describes the choice of the Flemish waterway authority – De Vlaamse Waterweg – for a new vision on remote operation of its inland waterway infrastructure. From now on, remote control is the default operation mode. As of 2019, a new vision was approved to remotely control all our movable structures (ca.330) by 2032 from 3 remote control centres. The innovative aspect is that they will function as a fully interconnected meshed network and can be used as a back-up for each other. At first glance this may seem a purely technical challenge. It is however shown that by pursuing this goal, it compels us to rethink large parts of our organisation in regards with change management, business processes, roles and responsibilities, training and technical management. At the same time, we see society opportunities revealing themselves. This paper is aimed at sharing our experiences and as a call-to action for other waterway authorities, academic institutions and private companies to help envision and built the future waterway-systems.

Keywords: Future of inland waterway governance · Remote operation · Change management · Innovation · Traffic management

1 Introduction

De Vlaamse Waterweg NV is an Inland Waterway Authority in Belgium that manages almost all inland waterways and infrastructure, excluding the ports, in Flanders, the northern region of Belgium. Flanders has approximately 1,000 kms of navigable waterways within an area of 13,625 km². The organization of shipping- and water management requires approx. 330 bridges, locks, weirs, pumping stations, etc.

De Vlaamse Waterweg has a long history developing remote control systems for bridges, locks and weirs. As with most inland waterway authorities, these systems were implemented using a project-based approach, clustering the operation of infrastructure per waterway or section of a waterway.

We have come to a point that we have outgrown this stage. From now on, remote control is the default operation mode. As of 2019, a new vision was approved to remotely control all our movable structures (ca.330) by 2032 from 3 remote control centres. The innovative aspect is that they will function as a fully interconnected meshed network and can be used as a back-up for each other.

Y. Li et al. (Eds.): PIANC 2022, LNCE 264, pp. 113–121, 2023.
https://doi.org/10.1007/978-981-19-6138-0_10

What started as a purely technical story, has evolved into a future full of opportunities. We found that by formulating such a progressive vision and actually getting to work with it, it opened up a lot of new sometimes unexpected challenges but most of all, a lot of unexpected new opportunities. Opportunities for our organisation, for our employees, but above all for society, the logistics sector and shipping in general.

This paper is aimed at sharing our experiences and as a call-to action for other waterway authorities, academic institutions and private companies to help envision and built the future waterway-systems.

2   Towards a Meshed Network

Up until 2019, De Vlaamse Waterweg had a vision of clustering the operation of movable infrastructures per waterway or specific sections of waterways. A total of 13 stand-alone Remote Operation Centers (ROC's) were envisioned. Some of them have already been built and in operation for decades, some of them are in the finale stage of construction and some of them will forever remain on the drawing table.

During 2019, a number of evolutions made that De Vlaamse Waterweg adjusted this vision: the logistics sector is evolving towards supply chain approach and there is a desire to sail more efficiently (fuel consumption, loading and unloading,....); the 13 corridors have different opening hours and traffic intensity which makes the workload for the employees very different and which increases the risk of less readability of the waterway for the skipper; the Flemish Government imposed staff reduction, ... Based on these evolutions, the Board of Directors chose at the end of 2019 for the new vision where De Vlaamse Waterweg will operate its infrastructure from 3 major ROC's, aiming at 2032 to accomplish this.

From a technical viewpoint, the vision of evolving towards 3 major ROC's starting from several pre-existing stand-alone minor ROC's sparked the idea to approach this in a different way than before. First off, there is the question of availability, redundancy and disaster recovery. By centralising the operations, the impact of an incident in one of the ROC's scales along with the size of the ROC. This will become a major concern in the final state. Secondly, the path towards this finale state of 3 major ROC's, will have to accommodate for the current situation of several minor ROC's and allow for a flexible and smooth migration, preferably with some sort of fall-back plan and stable intermediate situation, all while ensuring continuous service.

These considerations lead to the idea to make both the existing ROC's and the future major ROC's, interoperable and redundant such that they form a meshed network. The idea is that all the operating desks and involved technical systems are interoperable, no matter where they are located.

In doing so it will become possible during the transition stage, to remain in operation until the day of migration and then to turn the switch as a matter of speech. This allows to detach the operational and organizational decision from the technical timeline. And in case of incident or emergency, it will be just as easy to fall-back or to move operations towards one of the other ROC's.

Lessons Learned and Recommendations:

i. The society and the world we live in are constantly changing and technological developments are having an impact on how we do things and vice-versa.
ii. As waterway authority, we have a duty towards society and to adjust to these changes.
iii. It is the challenge to find the balance between continuity of service, building for the future and allowing for change and innovation.

3   Change Management

The new vision from 2019 (to serve 330 infrastructure works from 3 control centres) introduced technical, organisational, communicative, … challenges within De Vlaamse Waterweg. Between 2019 and 2022, this new vision was translated into a new approach in various areas: ensuring redundancy across the 3 control centres, building long-distance stable network connections, extensive standardization, improving cybersecurity, drawing new organizational charts and business processes, developing a central approach to water management, providing professional training to operators, drafting an implementation plan with a view up to 2032, …

3.1   Working with Ambassadors

We think that supportive communication and a precise insight in the stakeholders of your change is essential for success. Therefore we make extensive use of various forms of communication and interaction with our stakeholders.

At the end of 2021, an community of ambassadors was set up to extend the diversity in our guiding coalition to Remote control. The community of ambassadors are a group of volunteers within our organization committed to the transformation effort. They promote remote control and share feedback of the business engagement they experience.

This community officially started at a digital (because of COVID-19 measures) seminar attended by approximately 1/10th of the employees of De Vlaamse Waterweg. The Managing Director and the Operations Director shared the change story with the participants to create a sense of urgency. Besides the importance of remote control for society and shipping, the importance for their own employees, the professionalization of the operation, the new opportunities created for collaboration within the organization, the opportunities to standardize processes,… Were discussed in an interactive way.

3.2   Communication

This was followed with an intensive communication through various media. A one-pager was created to promote the change story on various ways throughout the organization explaining the 'what-why-how'. A seperate page Remote control was created on the intranet of De Vlaamse Waterweg. Here colleagues can easily find all the latest information about remote control an get access to the digital workplaces.

3.3   Focus Groups

We use co-creation and interaction to activate our organisation for Remote control. Dedicated 'focus groups are regularly organized to capture expertise of our operators and their managers. This on numerous strategic subjects such as the organizational chart within a control centre, on the manuals with the operational guidelines, on solving personnel related bottlenecks, etc. The unique concept of collaboration across different functions ensures support and acceptance of the entire transition among staff members and valuable insight we otherwise would miss. The game rules are: a focus group delivers input and advice from the field, makes proposals for solutions, … But does not have decision-power. The existing decision-making bodies, both at management level and for consultations with the trade unions, will be retained. In the focus groups, the proposals made by the employees are treated with respect, they are explored in depth, and the "why" question is the most frequently asked question in such a focus group: "why are you saying this", "why is this important to you", etc.

Lessons Learned and Recommendations:

i. The implementation of remote control can be worked out as a pure technical story. However, it offers added value to an organization if it is worked out as a program that operates as a catalyst for other developments within the organization.
ii. It is important that management recognizes and supports the change story and that employees perceive that management makes decisions based on the input that operators provide from their real-world experience.
iii. It is important to "plant the flag" far enough into the future. That way, space and time can be made to work on the picture of the future and to determine the steps towards it.

4   Tools

4.1   Training Simulator

De Vlaamse Waterweg NV has set up the AWATAR project. AWATAR stands for Automation of Waterways: Training and Reference. The 3 main goals of AWATAR are, to establish and maintain the technical and operational reference, to allow for standardized and professional training, facilitate, professionalize and mature the way remote control is implemented.

We made virtual 3D models of several lock and bridges, together with models of ships, cars, pedestrians, … inside a gaming engine. We connected these virtual models with real PLC and SCADA software and simulated different camera-viewpoints, traffic situations and weather conditions.

This allows us to discuss and try very different HMI designs, camera-view points, functional behaviour, operational procedures, … all without disturbing any real-life operation and in great detail. For the development of the simulator, we organized multiple participation sessions, workshops and feedback-loops with both engineers and operators. We used Virtual Reality Glasses to design the operator desk.

The simulator offers the possibility to employees to train various types of infrastructure (e.g. lift bridge, swing bridge, rolling bridge, ...) and various scenarios in a short period of time. In addition to the normal traffic flow, various weather conditions can be rolled out (rain, fog, etc.), incidents can occur, etc. This allows the operator to practice his skills in a safe environment on generic structures. On the work floor, training is then given on the specific structures in the remote control center.

4.2   Learning Environment

A digital learning platform was set up, tailored to operators. this learning platform uses 12 clickable tiles. The information offered depends on the theme. sometimes a short explanation is given and a link is made to existing tools or internet pages, e.g. for water management, a link is made to the website www.waterinfo.be of the Flemish Government.

In the case of more complex subjects, such as 'navigation regulations and signaling', a video film provides a verbal guide to help colleagues navigate through the complex regulations. Basic information can be found, for example, behind the tile 'naming of waterways' or 'languages' where the most common nautical terms are translated into the languages mainly used by skippers in Flanders. The digital learning platform also offers a 'do the test' tile where colleagues can see on the basis of a test on which subjects they should best learn more.

The website https://www.visuris.be/ developed by De Vlaamse Waterweg, in accordance with a number of agreements at the scale of the European Union, is also a source of information.

Lessons Learned and Recommendations:

i. Developing a digital learning platform takes the necessary time, also keeping it up to date by supplementing the content and adding new questions to the self-test requires discipline and attention.
ii. The ownership of the learning platform lies within the business, because the colleagues were allowed to develop this themselves, they are also the most important ambassadors of this learning platform.

5   Technical Implementation

The formulation of the new vision towards large, meshed ROC's requires a mature technology management.

5.1   Research and Development Center

The changing functional needs of the organization require an specific approach from both the technology and the technical organization. The technical realization evolves towards a continuous-life-cycle-approach with a more centrally coordinated organizational unit. To achieve all this, the need for a proper and in-company research and development centre (R&DC) was identified. Beforehand, the technologies implemented in a ROC were mostly contractor-developments and as such, contractor-specific. These technical systems used to be integrated on the spot, in a live environment.

As of the creation the R&DC, contractors are obliged to develop their systems in cooperation with the waterway authority. A systematic development cycle is used, going from functional requirements to technical requirements to proof-of-concept, prototyping and testing, with multiple feedback loops. Only when this cycle is successfully complement, real-life integration is allowed. Furthermore, a development-version is kept, to test future alternations and updates. The idea is to gradually implement and standardize all systems for the whole infrastructure and keep them evolving at the same pace of technological development, in contrast with just using the new technologies in new projects.

5.2   Innovation

New functionalities are requested to support the operation of and between the remote control centres. A different balance is sought between decentralization and centralization of the infrastructure, between standardization and customization. Demands on availability are increasing, but can be partly mitigated by the network synergies when compared to classic redundancy solutions. The scalability and multi-deployability of the systems are a critical point of attention. Technical systems evolve towards an almost completely network-driven operation.

In 2022, in our R&DC, we successfully accomplished a fully functional prototype of 2 ROC's who function redundant for each other. We found that almost all technical systems to we used this date, like the Supervisory Control And Data Acquisition-system (SCADA), Video-management-system (VMS), Automatic Identification System (AIS) now have the capability to function within a network, as they are becoming more and more IP-based. As a general design approach, we make sure to make as little adaptations as possible on the local infrastructure. It is the overarching systems that offer these functionalities. The only system that we have found that poses problems in realizing the needed new functions, is the VHF-system. Different sections of waterway have different settings of the VHF-equipment and the number of sections this equipment can handle is limited. However this is not problematic for now, since the goal of the network of ROC's is to operate mainly predefined sections of waterways and only switch-over in case of major incidents. In such a case, whole sections of waterways will be redistributed across the remaining ROC's and not isolated specific structures. The work-around for now, is to work with redundant VHF equipment within the ROC to allow for some flexibility on this part and to go for manual configuration in case of larger incidents and thus switch-overs.

5.3   Procurement Management

In order to realize a performant and uniform technological standard across all our infrastructure and across the whole life-cycle of our systems, we had to rethink our procurement management. Contracts are no longer aligned with geographical borders (a specific waterway, a specific ROC, ..) or made to measure for a specific project. Contracts are now framework-contracts focussing on technological-system-borders and include often not only new implementations but also updates, replacements, maintenance etc. during several years. This allows to attract specialised contractors and built a long-term cooperation. The challenge of course, is to coordinate between these different framework contracts while realizing large projects and making sure these different technical systems can fully cooperate. Another reason why the R&DC is a crucial piece of the puzzle.

5.4   Project Management

In order to achieve the goal of remote control as a default mode for our ca. 330 works, a complex task is set out for our organisation as a whole, in regards of project management. As of now, all projects on the movable assets become part of the larger puzzle. The full potential is only reached when all pieces of the puzzle come together and the picture becomes clear. Following the timeline of one project can already be a challenging task but now the timeline of all these projects influence one another. A dedicated team is assigned this daunting task.

Lessons Learned and Recommendations:

  i. Most of the needed technology to accomplish meshed remote operation is available today
 ii. Because of the links on both a technical level as a governance level, meshed remote operation has a tendency to make things more complex and in need of a mature technical management

6   Opportunities

6.1   Service

Remote control offers the opportunity to let the service model of De Vlaamse Waterweg evolve in a future-oriented way. In this matter, too, the involvement of the sector is of great importance. In its service developments, the Flemish Waterway wants to take into account the various players and their possibilities. Companies, logistics players, independent skippers, ... Should be given time and the opportunity to adapt and organize their way of working. By organizing specific board groups, the input of the sector is captured.

Today, the operating hours of navigable waterways in Flanders are very diverse. It must be noted that there is also a large variety of vessels on these waterways. On some waterways there is mainly commercial shipping, other waterways have mainly recreational boating.

It seems logical to state that operating hours anywhere can expand to 24/7 once the remote control centers are up and running. However, for various reasons, it seems that it is not appropriate to decide this as such and requires further investigation. On the side of the waterway manager, for example, an adapted intervention strategy will have to be worked out if waterways are navigated 24/7, which means that more technicians need to be available to solve technical failures. Against this desired higher availability, there is also a higher cost of operation. On the side of the skipper, attention should be paid to the possible shift in the competitive position of independent skippers who have to take into account rest periods versus companies with many skippers who can sail in shifts.

6.2   Integrating with RIS Smart Shipping and Smart Logistics

By moving from a fragmented approach to an overarching integrated strategy and state-of-the-art overall remote control system, full integration with RIS, Smart Shipping and Smart Logistics, and the physical internet becomes possible. Data from the infrastructure will be gathered and turned into useful information.

The opportunities for the logistics sector and shipping are created by embracing innovative technologies and integrating them within the day-to-day operations. We provide the necessary tools to shipping to plan their journey as effective as possible and the systems in the remote control centres support the operators in offering an optimal locking process.

6.3   Water Management

Remote operation of water-controlling structures is somewhat different form operating locks and bridges. The former focusses on a large area and tries to optimise the water management across this whole area. The operating time of each structure is rather short. The latter focusses on one specific structure and takes a larger amount of time (not taking into account traffic management). The main advantage of remote operation on water-controlling structure is exactly that you can oversee the whole area by gathering and using all available data from all the involved structures and basins. It allows to build a decision-support system based on real-time information and predictions. Climate-change is putting more stress on our infrastructure resulting from more extreme weather and water conditions. Making sure we use the full potential of our infrastructure and basins is a way to mitigate the effects with fairly reasonable effort.

Lessons Learned and Recommendations:

i. Similarly, so-called "logical decisions" in a service delivery model require both internal research and discussions with stakeholders.
ii. It is important not to view the technical activities within a remote control program as stand-alone activities but to frame them within larger cross-border projects and programs. Waterways do not stop at a country's borders.

7 Conclusions

The strategy towards all-out meshed remote-control makes a digital evolution possible. As such, this originally technical story will contribute to prosperity and well-being in Flanders. Organizations must dare to dream and dare to seize technical evolutions as a basis for the many opportunities to improve the future for the shipping and logistics sector, while at the same time responding efficiently and effectively to the impact of climate change.

Acknowledgements. The authors would like to thank and honor all operators, colleagues, management, contractors and stakeholders of De Vlaamse Waterweg for their work and contributions in the development and realization of the new remote operation vision.

Presenting the Work of PIANC TG234 "INfrastructure for the Decarbonisation of IWT"

Mark Van Koningsveld[1(✉)] and Gernot Pauli[2]

[1] Delft University of Technology, Delft, The Netherlands
m.vankoningsveld@tudelft.nl
[2] Self employed, Berlin, Germany
gernot.pauli@outlook.de

Abstract. PIANC provides guidance and technical advice for sustainable waterborne transport infrastructure. To address the challenge of making inland navigation infrastructure sustainable, a Task Group "Infrastructure for the decarbonisation of Inland Water Transport" (TG234) was set up in January 2021. The objective of the TG was to identify knowledge gaps and major challenges that need to be urgently addressed and advise PIANC on further actions, such as setting up a working group. A report was to be delivered in early 2022. The TG involved a range of international experts who met every three months on-line. They jointly participated in the following activities: sharing experiences in their area of expertise; gathering, organising, and discussing literature; discussing the perspective to be taken while reporting; contributing an overview of the developments either in their country or organisations and of course in the various discussions. A final report was produced that took the perspective of the waterway manager, highlighting a number of key questions that need to be answered in the transition to reduced/zero emissions. With the state-of-art knowledge gathered by the group, it became possible to identify the existing knowledge gaps and the major challenges that need to be addressed. The report can guide PIANC in evaluating the best way forward to address the decarbonisation of IWT infrastructure.

Keywords: Inland waterway transport · Decarbonisation · IWT infrastructure · Alternative fuels

1 Introduction

The availability of zero-emission fuels infrastructure, including onshore electric power supply, will be key to enable zero-emission vessels and increase the competitiveness of IWT as a whole, at a time when other modes of transport are reducing their ecological footprint. To address the challenge of making inland navigation infrastructure sustainable, PIANC Task Group TG234 "Infrastructure for the decarbonisation of Inland Water Transport" was set up in January 2021. The objective of the TG is in line with the declaration of PIANC, namely developing approaches to decarbonise the operation of port and navigation infrastructure (i.e. move to net-zero emissions), whilst at the

Y. Li et al. (Eds.): PIANC 2022, LNCE 264, pp. 122–134, 2023.
https://doi.org/10.1007/978-981-19-6138-0_11

same time enabling the reduction of greenhouse gas emissions from vessels by providing the necessary facilities, infrastructure and, where appropriate, incentives. For this purpose TG234 was tasked to identify knowledge gaps and major challenges that need to be urgently addressed and advise PIANC on further actions. The findings of TG234 have been written down in a final report that has been submitted to INCOM (PIANC 2022). The report serves as a coarse knowledge base to guide further steps towards a rational approach to developing infrastructure for the decarbonisation of IWT.

2 Approach

The steps taken by the TG to identify knowledge gaps and challenges are listed here briefly. As suggested in the Terms of Reference (TOR dated October 1st 2020) and to provide first insights into the decarbonisation of IWT, TG234 compiled key developments per country/organisation that participated in the TG. Similarly, main developments per energy carrier were also compiled. These compilations or briefing notes form an integral part of the report and are summarized in Chapter 3. Next, attention was given to the questions that an actor striving for decarbonisation would have. A comprehensive list of questions was drawn up in Chapter 4 and an approach was suggested over how to answer them in Chapter 5. Some conclusions are drawn in Chapter 6.

3 Key Developments Related to Decarbonisation of Infrastructure

A feasible starting point for the members of TG234 was to create briefing notes on key developments on zero emission IWT as they observed them in their own countries/organisations. In the following, short summaries per briefing note are included. Next to the key developments per organisation or country, briefing notes were also made of a number of promising new energy carriers. It is clear that the work done so far is not yet comprehensive at global scale, but it provides an inspiring first step.

3.1 Key Developments per Country or Organization

Short Report Decarbonisation IWT Europe. In 2019, the European Union presented the European Green Deal with the aim of ensuring that the continent is greenhouse gases (GHG) emission-free by 2050. In July 2021, the European Commission adopted a set of proposals (Fit for 55 package) to make the EU's climate, energy, transport and taxation policies fit for reducing net GHG emissions fit for reducing net GHG emissions by at least 55% by 2030. The initiatives include Alternative Fuels Infrastructure Regulation (AFIR), the Renewable Energy Directive (RED) and the Energy Taxation Directive (ETD). AFIR supports the deployment of alternative fuels infrastructure, including refueling points for natural gas and hydrogen. Member States are required to set up national policy frameworks to establish markets for alternative fuels and report their progress. RED deals with the promotion of energy from renewable sources and has set a binding target to produce 40% of energy from renewable sources by 2030. The ETD aims to ensure the proper functioning of the EU internal market by ensuring that energy taxation is aligned with climate objectives.

Short Report Decarbonisation IWT CCNR. The Central Commission for the Navigation of the Rhine (CCNR) has drawn up a roadmap (Adopted on 9 December 2021) to lay the foundation for a common approach to the energy transition and emissions reduction by all stakeholders. This roadmap should be understood as the primary CCNR instrument for climate change mitigation and setting transition pathways for the fleet (new and existing vessels), suggesting, planning, and implementing measures directly adopted or not by the CCNR, and monitoring intermediate and final goals set by the Mannheim Declaration. In 2021, CCNR published the results of in-depth studies over financial instruments to be seen as part of a broad discussion process at Rhine, European and international level. It mandated its committees to feed the study results into the PLATINA3 project, desiring an action plan for the further development of a European funding and financing instrument to be drawn up and detailed. CCNR regularly organizes workshops on innovative technologies.

Short Report Decarbonisation IWT Austria. The current political ambitions to decarbonise IWT in Austria are higher than those on European level. The Mobility Masterplan 2030 and Government Programme 2021–2027 make concrete recommendations while also committing to endeavours such as installing shore power units. Implementation projects prepared by the Austrian waterway company, Viadonau, include the installation of shore power supply for cargo vessels at selected existing and future berths, and implementation plans for cruise vessels are underway. CCNR and Viadonau have set up an international conference aiming at international harmonisation of technical standards and the operational and billing systems of a future shore power system along the European waterways.

Short Report Decarbonisation IWT France. In order to meet the GHG emissions reduction targets in the transport sector, policies have been set up for the inland navigation sector. The "Mobility orientation Act" of 2019 eased the establishment of low emission zones (ZFEs) while the "National hydrogen plan" of 2018 aims at achieving mass-production of green hydrogen as a fuel for mobility. A bill entitled "Delivery of a vessel certificate for a restricted navigation", that was introduced in 2019 allowing green vessels to derogate from the EU technical regulations if they operate on a limited journey in an area of local (national) interest, has proven efficient. As a result, the French inland fleet will welcome hydrogen and GNC powered vessels in the coming years.

Short Report Decarbonisation IWT Germany. The Federal Climate Protection Law (Bundes-Klimaschutzgesetz – KSG), amended by the German Federal Parliament in 2021, aims to achieve GreenHouse Gas (GHG) neutrality in Germany by 2045. The Federal Ministry for Digital and Transport (BMDV) will support climate friendly inland navigation with subsidies for decarbonisation and development of inland waterway infrastructure as well as with research and development. It has commissioned work, to be published in 2023, to develop energy efficiency indices for inland navigation together with a proposal for their practical implementation. German IWT companies are already investing in climate neutral vessels. The report concludes that IWT in Germany has a chance to survive, when it will be innovative and when there will be an adequate regulation for GHG emissions from transport, including carbon prizing, that honours the inherent energy efficiency of IWT.

Short Report Decarbonisation IWT Netherlands. The targets for emissions reduction in the Netherlands have been drafted in the Dutch Green Deal on Maritime and Inland Shipping and Ports (2019), signed by various governmental authorities, trade associations, ports, sector representatives and research institutes, each with a list of planned actions. Numerous initiatives have been launched e.g., investing in shore power facilities for around 500 state berths, a national ban on degassing while sailing (to be introduced step by step), examining blending biofuel obligation in inland shipping vessels. The Dutch national government has set up a supporting system for innovations such as fully emission-free powered ships for the inland shipping sector. Funding schemes have been put in place for greening of the Dutch fleet. Another initiative to stimulate the decarbonization of the fleet is a new labelling system for inland vessels' emissions performance. Attention will be given to the necessary bunkering infrastructure and the safety requirements and legal framework to facilitate the introduction of new energy carriers in the inland waterway sector in the coming year.

3.2   Key Developments Per Carrier

Short Report Hydrogen for Propulsion. Pressurized hydrogen storage is currently furthest developed for mobile applications (inland shipping) and is the most applied method in current hydrogen vessel projects. Liquid hydrogen could be an option as a mid-term solution when liquefaction plants are built and the fuel price comes down. Bunkering can take place via four different configurations: truck-to-ship, ship-to-ship, bunker stations and swapping of tank-containers and depends on the physical state in which hydrogen (pressurized, liquid or hydrogen carrier) is stored on board inland navigational vessels. The most feasible scenario for the short-term is swapping pressurized hydrogen in swappable containerized containment systems (tube-containers) at container terminals. Regulations for the use of hydrogen on board of-, and bunkering of hydrogen to inland navigational vessels are still under development. The availability of hydrogen as a fuel for vessels relates to hydrogen fuel production as well as to provision of bunkering infrastructure in a sufficient number of ports in the operating area. Strategic engagement of a large industrial player (gas producer, utility company, oil or energy major), who is not only aiming at supplying (moderate amounts of) green hydrogen to inland waterway vessels but also to large consumers along the Inland Waterways, is required for a breakthrough.

Short Report Biofuel for Propulsion. Rapeseed methyl ester, also known as OLEO100, is a biofuel produced exclusively from rapeseed oil. It can be used in its pure form and does not need to be mixed with a fossil fuel. It has an energy density comparable to that of diesel (slightly lower). Similar to diesel/gasoil, OLEO100 is used in internal combustion engines and can be mixed with diesel, it is therefore compatible with existing conventional propulsion systems. It is mainly used by heavy road vehicles, but it is being tested for application on inland vessels. Currently, refuelling is done either by refuelling trucks or directly by drums delivered to the refuelling station. It can be regarded as a conventional fuel when it comes to existing rules. OLEO100 is not considered harmful for human nor the environment, no specific policies are needed. Infrastructure changes required are minor and costs are therefore negligible compared to other alternative fuels. If available on location, OLEO100 refuelling specifics are comparable to those of conventional fuels; since comparable energy density and viscosity means comparable volumes and refuelling times. The main challenge is the long-term availability of this fuel if it is widely adopted.

Short Report Methanol for Propulsion. Methanol is a climate neutral fuel, when it is produced from renewable energy and can be used as fuel for combustion engines or for fuel cells. Methanol has a low energy density compared to gasoil/diesel fuel. Otherwise, it is rather similar to diesel/gasoil and be used for all applications. Bunkering is possible from bunkering vessels, tank trucks and fixed tank stations. The required safety distances are also similar to diesel/gasoil. Safety risks during methanol transport are well understood and safety measures in place. Technical requirements and standards for methanol as fuel on inland navigation vessels are under development in Europe. Infrastructure costs are on the same level as for diesel/gasoil and low in comparison to other alternative/climate neutral fuels. More refuelling (bunkering) stops are needed because of low energy density. The main challenge is the high cost for methanol itself, when it is produced from renewable energy. Otherwise, methanol could become a standard fuel for inland navigation.

Short Report Battery Electric Propulsion. A battery electric propulsion system consists in general of rechargeable batteries, electric switch board and an electric propulsion system. Because of low energy density, battery electric propulsion is most suitable for ships that travel short distances (between stops). Fixed batteries require electric charging points at mooring places and exchangeable batteries require cranes, e.g., on container terminals with nearby charging point. Infrastructure costs are high as many charging points are needed and as rechargeable batteries for inland navigation vessels require a high-capacity power supply. Battery fires are rare, but hard to control. Technical requirements and standards for rechargeable batteries exist or are under development. In Europe, national and EU policies support the implementation of charging points at suitable locations of the inland waterway network. Battery costs are expected to further decrease, and energy density will increase, allowing battery electric propulsion becoming a technical and economically feasible alternative for certain inland navigation tasks. According to the latest RWS studies on the safety aspects of new energy carriers, the surrounding safety zones between the location of the battery containers and the surrounded buildings can be as low as 5 m. The most important recommendation is the location awareness for energy services. This is particularly important in case of shipping accidents.

4 Relevant Questions for Waterway Managers

4.1 General

From Chapter 3 it can be concluded that the path to decarbonisation of IWT is different for different corridors and in different countries. While a relatively short list of potential energy carriers appears to emerge when discussing decarbonisation, which carrier (or a mix of carriers) is likely to emerge as preferred depends on a whole range of local situations. In practice, it is seen that vessels owners and bunker station operators have a strong influence on the alternative energy carriers they would like to use (bottom-up). But at the same time, the question whether a selected alternative energy carrier is going to be successful at the corridor scale can depend on a range of policy measures and subsidy schemes (top town). While momentum for change appears to be stronger bottom-up, TG234 considered that a top-down approach could provide a stronger rational framework. So while fully aware of the bottom-up as well as the top-down perspective, TG234 decided to take on the perspective of a waterway manager that faces the need for decarbonisation of his/her waterway. By discussing step-by-step the kind of questions that arise, a structured approach to decarbonisation emerges.

4.2 List of Questions

TG234 foresees that a waterway manager that seeks to decarbonise his/her network encounters the questions listed below:

    i. What are the most promising technologies (or energy carriers) for the decarbonisation of IWT?

    ii. What is the overall transport challenge in my network (amount of cargo, number of passengers, from where to where now and in the future)?

    iii. What is the state of the water transport network and of the fleet that operates on it (proportion of vessels of given type/classification, now and in the future, alternative transport modes)?

    iv. What is the energy consumption that is associated with the transport challenge, given the current and future state of the network as well as of the fleet and the waterway conditions in the future considering impacts of climate change? (Emission hotspots?)

    v. What type of energy carriers can replace the current ones, what quantities of fuel are needed where, and how will these fuels affect range, payload, velocity, etc.?

    vi. Where should we position bunkering points or refuelling stations? What are the charging/fuelling times and the waiting times at refuelling stations?

    vii. How can the estimated demand for alternative fuels (electricity, hydrogen, methanol, etc.) be supplied over the network?

    viii. What are the standards or existing regulations that must be followed?

What are the most promising technologies (or energy carriers) for the decarbonisation of IWT?

The impact on the infrastructure for the decarbonisation varies substantially with the different energy carriers. As was shown in the short reports of the previous chapter, biofuel for decarbonisation allows for continuous use of the existing refuelling infrastructure, whereas a switch to electric propulsion would require building a totally new infrastructure. Furthermore, none of the future energy carriers is suitable for all transport tasks. Therefore, the waterway manager is well advised to get a good understanding of the different technologies or energy carriers for the decarbonisation of IWT. This will also help him or her to efficiently consider the following questions.\

What is the overall transport challenge in my network (amount of cargo, number of passengers, from where to where, now and in the future)?

It is important to consider what the transport challenge is in the network. The type and amount of cargo that needs to be transported, in combination with the origin and destination of this cargo, determines the demand for transport and also the location of the bunkering infrastructure of the new energy carriers. It is also important to assess whether or not there are alternative transport modes that are likely to compete with inland shipping.

What is the state of the water transport network and of the fleet that operates on it (proportion of vessels of given type/classification, now and in the future, alternative transport modes)?

When it is clear what the transport demand is depending on future traffic flows and the vessels required to transport it, it becomes important to assess the potential for transport over water.

Looking at the state of the water transport network will reveal the vessel classes (PIANC 2020) that will be able to fulfil the transport demand. The maximum vessel class that can operate on a waterway is typically restricted by a maximum available air draught (e.g., due to the presence of fixed bridges), a maximum allowable width, length and draught (e.g., due to the presence of locks) and the presence of other width and depth bottlenecks (Van Dorsser et al. 2020; CCNR 2021c; Vinke et al. 2022). Other aspects that can come into play are traffic intensity and environmental aspects like wind and current. The state of the waterway, the available water depth and the ambient current conditions influence the amount of energy that is associated with the transport function.

The waterway classification determines the maximum size of the vessels that can use it. Beyond that it is important to know the composition of the fleet that is available to perform the transport function on the waterway network. Not all vessels that are part of the fleet will be of the maximum size. Smaller vessels will need more trips to transport the same amount of cargo compared to larger ships. Older vessels might still have older engines that may perform less when it comes to emissions. Also, it is important to assess the availability of alternative transport modes, e.g., road, rail, pipeline. When alternatives are available *and* capable to accommodate a significant modal shift, this will put more pressure on the inland shipping sector to adopt/convert to other energy carriers.

What is the energy consumption that is associated with the transport challenge, given the current and future state of the network as well as of the fleet and the waterway conditions in the future considering impacts of climate change? (Emission hotspots?)

When the transport demand (volumes, origins, destinations), the state of the waterway network (e.g., water depths, currents), and the state of the fleet (composition, engine ages, etc.) are known, the associated energy demand for transport can be estimated using vessel resistance algorithms (Bolt 2003; Vehmeijer 2021; Segers 2021; Van Koningsveld et al. 2021; Rijkswaterstaat, 2022a, b) (Fig. 1).

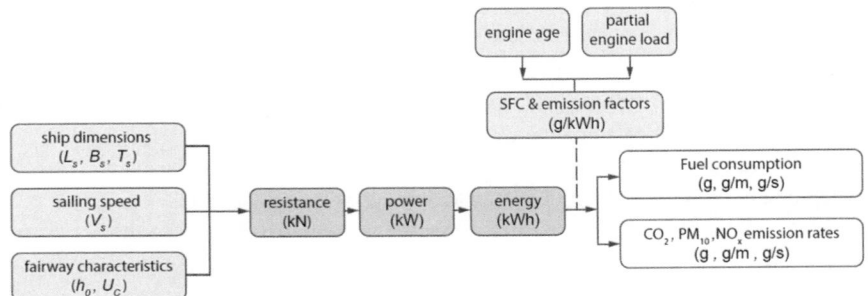

Fig. 1. Methodology for estimating emissions for IWT vessels (image modified from Segers, 2021, by TU Delft Ports and Waterways is licenced under CC BY-NC-SA 4.0)

For modern waterway networks that are already actively used, the availability of Automatic Identification System (AIS) data can be an important source of input. Depending on the country of origin, digital information on the state of the waterway network may also be openly available.

AIS data, combined with waterway network data, can be utilised to provide a promising first estimate of the energy demand that is associated with the transport function. To estimate future demands, growth/shrinkage scenarios can be of use.

Table 1. General SFC and emission factors of CO_2, PM10 and NO_x for diesel fuel and different engine construction year classes (source: Ligetink et al. 2019 and modified based on the emission standards described at DieselNet (2021)).

construction year classes	weight class	Fuel consumption [g/kWh]	CO_2 [g/kWh]	PM10 [g/kWh]	NO_x [g/kWh]
1900–1974	L1–L3	235	746	0.6	10.8
1975–1979	L1–L3	230	730	0.6	10.6
1980–1984	L1–L3	225	714	0.6	10.4
1985–1989	L1–L3	220	698	0.5	10.1
1990–1994	L1–L3	220	698	0.4	10.1
1995–2002	L1–L3	205	650	0.3	9.4
2003–2007 CCR-1	L1–L3	200	635	0.3	9.2
2008–2018 CCR-2	L1–L3	200	635	0.2	7
2019–2019 CCR-2	L1–L3	200	635	0.2	7
2019–20XX STAGE V	L1	205	650	0.1	2.1
2020–20XX STAGE V	L2 and L3	190	603	0.015	1.8

It is useful to take the energy consumption (kWh) as a basis for analysis, since empirical information is typically available to estimate the associated fuel consumption, via so-called Specific Fuel Consumption factors (g/kWh) (Table 1). When assumptions are made on partial engine loads and engine ages also CO_2 and environmental pollutant emissions can be estimated (Hulskotte 2013; Smart Freight Centre 2019; Wijaya et al. 2020).

It may be worthwhile to validate these coarse estimates with actual energy consumption, fuel use and emissions. Also, it will be useful to document the current locations and capacities of bunker facilities.

With the above method so-called energy consumption, fuel use and emission footprints can be generated for individual vessels. Overall patterns can be generated by aggregating the footprints of individual vessels that together represent the traffic on a corridor (Jiang et al. 2022).

Such heatmaps, created from individual contributions, can be used to identify hotspots and identify root causes. This information can be used to design policies to reduce emissions (Segers 2021). On the long term such policies will probably involve zero emission energy carriers, but in the years before those other measures may be necessary in an effort to reduce emissions.

More coarse methods to estimate energy consumption, fuel use and emissions may also be used. Various methods are available that estimate fuel use per tonne kilometre (tkm) based on empirical data. While these methods are easier to use, especially in situations of limited data availability, they are less useful to test new situations. The most practical way forward as such is a trade-off.

What type of energy carriers can replace the current ones, what quantities of fuel are needed where, and how will these fuels affect range, payload, velocity, etc.?

Once the total energy demand, fuel use and emissions, as well as the locations and capacities of current bunker facilities are known, it becomes possible to estimate the required volumes in case alternative energy carriers would be considered.

Alternative energy carriers will have a different energy content than more traditionally available options. Also, other energy carriers may involve alternative energy conversion systems. For each energy carrier/energy conversion combination it should be investigated what the potential influence on sailing range, payload amount and sailing velocity is. If only the sailing range is affected, an increased number of bunker stops is the main transport efficiency impact. If the range remains the same but the amount of payload is affected, the main transport efficiency impact is an increased number of trips required to transport the same amount of cargo.

Where for the previous question AIS data could be used as a basis for quantification of the current state, testing the effectiveness of alternative policies requires simulation. A common approach for this these days is the use of agent-based meso-scale simulation models (Van Koningsveld and Den Uijl 2020; Jiang et al. 2022). With such models the effect of changes to the vessels (the agents) can be assessed beforehand.

A typical question is of course what performance indicators are most suitable. Given that a known amount of cargo needs to be transported a typical measure of transport performance is the unit of tonne kilometre (tkm), or the tons of cargo times the km of distance over which it needs to be transported. Obviously, an important indicator is the cost of transport. When we are interested in energy, fuel and emission efficiency respective units of kWh/tkm, g fuel/tkm and g emission/tkm become relevant.

It is good to realise that in the cost and emission units the efficiency of the transport chain becomes visible. Let's imagine that 3.000 tons of cargo need to be transported over 100 km. Then the transport performance can be expressed as $3.000 \times 100 = 300.000$ tkm. This performance is irrespective of vessel size. But now let's assume this cargo is transported with a vessel that has a capacity of 3.000 tons or a vessel that has a capacity of 1.500 tons. In the first case the cargo can be transported by 1 full trip to the destination and 1 empty trip back to the origin. In the second case 2 full trips and 2 empty trips are needed to transport the same amount of cargo.

Depending on local circumstances and vessel properties this will result in different emission patterns: there is good chance that the second option will have a poorer total efficiency in terms of e.g., g CO_2/tkm, at the same time the emission source in terms of g CO_2/s or g CO_2/km can be lower since the emission will be spread out over time. While for CO_2 the totals are likely to be of interest, the actual peak values may be of interest for other environmental pollutants like fine particle emissions such as PM10. It will also be interesting to see what the cost effects are the situation will be complicated further when aspects like ambient current and available water depth are included. It is clear that the total performance of the IWT mode is complex. An increasingly popular approach these days is that the effects of policies are tested in simulation models or digital twins. It is necessary to do this since relying on intuition or coarse empirical data may yield unreliable results. Especially since the use of alternative energy carriers can affect things like sailing range (refuelling more often, and possibly taking longer/shorter), amount of payload that can be carried (more trips required) and perhaps the velocity profile that can be achieved. Information on refuelling/charging times and waiting times will influence the cost competitiveness of a suggested solution.

Where should we position bunkering points or refuelling stations? What are the charging/fuelling times and the waiting times at refuelling stations?
Insight in the total energy demand over the network tells the waterway manager something about the capacity requirements of the bunker stations/charging stations on the network. Insight in the range of vessels for different energy carrier/energy converter combinations will tell the waterway managers something about the maximum inter-distance of the bunker stations/bunker vessel.

How this all works out in detail will depend on the current vessel mix, and scenarios for possible future vessel mixes as well as on scenarios for the energy carrier mix that is assumed to be used on the network.

It is good to realise that also developments in other transport modes will be of interest, as well as developments in other corridors. In the end the selected solution (or mix of solutions) must be price competitive compared to available alternatives. Unless the other transport modes lack the capacity to accommodate a modal shift, poor price competitiveness will lead to the decline of the IWT mode.

How can the estimated demand for alternative fuels (electricity, hydrogen, methanol, etc.) be supplied over the network?
Insights in potential locations and capacities of alternative fuel bunker points are already an important step forward. But the availability (and cost) of alternative energy

carriers may also be a deciding factor. When the supply of sufficient amounts of a given energy carrier is problematic, a preferred energy carrier, while potentially suitable, might not become the implemented solution. Availability of certain energy carriers can vary significantly from one location to the next. This is at least one of the reasons why it is not possible to point to any one energy carrier as a preferred solution that fits all.

What are the standards or existing regulations that must be followed?
Next to demand for and the potential supply of alternative energy carriers, another important factor for likely success or failure of an energy carrier is the presence/absence of regulations. Mandatory safety margins for example, may pose inhibiting restrictions on the implementation of a given energy carrier.

5   Suggested Approach to Answer the Questions

The questions posed in Chapter 4 are key for any waterway manager to contemplate while decarbonising the IWT mode. TG234 suggests that these questions should be addressed first before detailed guidance can be provided on the actual dimensions of the required energy-related infrastructure components. It makes a big difference if you are dealing with a very busy shipping corridor that supports a wide range of vessels and substantial cargo flows, or a much smaller waterway that caters to a limited number of vessels and only one cargo type. Or if you are dealing with a waterway system that has substantial current vs one that has more calm conditions. In any case, waterway managers need a good understanding of the most promising technologies and energy carriers for the decarbonisation of IWT. Therefore, PIANC could compile the information that allows waterway managers to gain this understanding without having to do their own research.

6   Conclusions

The report concludes that PIANC could consider setting up a WG that focuses on the methodological approaches in the field of decarbonisation of inland waterway infrastructure that are at different development stages at an international level. A subsequent PIANC WG could work on coupling guidelines with the currently lacking standards in the field. A complicating factor is that insights on decarbonization are still very much under development. As a result it may not be easy to collate best practices since these are continuously being developed and updated. An alternative could be to approach the decarbonisation challenge with a Permanent Task Group/Strategic Initiative, such as the Permanent Task Group on Climate Change.

Acknowledgements. The report on which the paper is based has been produced by an international Task Group convened by InCom, comprising the following members (in alphabetical order): Cees Boon, Peng Chuandsheng, Nathaly Dasburg-Tromp, Turi Fiorito, Man Jiang, Mark van Koningsveld (chair), Hugo Lopes, Ulf Meinel, Hyumin Oh, Baptiste Panhaleux, Gernot Pauli (co-chair) and Poonam Taneja.

References

Bolt, E.: Schatting energiegebruik binnenvaartschepen. Technical Report. Rijkswaterstaat Adviesdienst Verkeer en Vervoer, Rotterdam (2003)

CCNR: CCNR roadmap for reducing inland navigation emissions. https://www.ccr-zkr.org/12090000-en.html (2022a)

CCNR: CCNR study on energy transition towards a zero-emission inland navigation sector. https://www.ccr-zkr.org/12080000-en.html#01 (2021a)

CCNR: Workshop "Shore power at berths". https://www.ccr-zkr.org/13020155-en.html (2022b)

CCNR: Workshop Alternative energy sources for electrical propulsion systems in inland navigation. https://www.ccr-zkr.org/13020154-en.html (2021b)

CCNR: Reflection paper "Act now!" on low water and effects on Rhine navigation. https://www.ccr-zkr.org/files/documents/workshops/wrshp261119/ien20_06en.pdf (2021c)

DieselNet: Summary of worldwide engine and vehicle emission standards. https://dieselnet.com/standards/ (2021)

European Commission: Energy efficiency first: a guiding principle of the european energy and climate governance. https://energy.ec.europa.eu/topics/energy-efficiency/energy-efficiency-targets-directive-and-rules/energy-efficiency-first_en (2021a)

European Commission: Commission proposes new EU framework to decarbonise gas markets, promote hydrogen and reduce methane emissions. https://ec.europa.eu/commission/presscorner/detail/en/ip_21_6682 (2021b)

Huang HX et al (2020) Effect of seasonal flow field on inland ship emission assessment: a case study of ferry. Sustainability 12:7484. https://doi.org/10.3390/su12187484

Hulskotte, J.H.J.: Toelichting Rekenapplicatie PRELUDE versie 1.1. Technical Report. TNO (2013).

Jiang, M., Segers, L.M., Van der Werff, S.E., Baart, F., Van Koningsveld, M.: OpenTNSim (v1.1.2). https://github.com/TUDelft-CITG/OpenTNSim (2022). https://doi.org/10.5281/zenodo.6447088

Ligterink, N.E., et al.: Emissiefactoren wegverkeer - Actualisatie 2019. Technical Report. TNO (2019)

PIANC: Standardisation Of Inland Waterways: Proposal For The Revision Of The ECTM 1992 Classification. PIANC WG 179 Report (2020)

PIANC: Infrastructure for the decarbonisation of IWT. InCom TG Report 234 (2022)

Rijkswaterstaat: BIVAS: applicatie. https://bivas.chartasoftware.com/Home (2022a)

Rijkswaterstaat: BIVAS: emissies. https://bivas.chartasoftware.com/Home/BIVASApplicatie/Documentatie/Emissies/ (2022b)

RVW: Richtlijnen Vaarwegen 2020. Rijkswaterstaat, Ministerie van Infrastructuur en Waterstaat. https://puc.overheid.nl/rijkswaterstaat/doc/PUC_632307_31/1/ (2020)

Segers, L.: Mapping inland shipping emissions in time and space for the benefit of emission policy development: a case study on the Rotterdam-Antwerp corridor. Master's thesis. Delft University of Technology, Delft, The Netherlands. http://resolver.tudelft.nl/uuid:a260bc48-c6ce-4f7c-b14a-e681d2e528e3 (2021)

Smart Freight Centre: Global Logistics Emissions Council Framework for Logistics Emissions Accounting and Reporting. URL: https://www.smartfreightcentre.org/en/how-to-implement-items/what-is-glec-framework/58/ (2019)

Van Dorsser C, Vinke F, Hekkenberg R, van Koningsveld M (2020) Effect of low water on loading capacity of inland ships. Eur. J. Transp. Infrast. 20:47–70. https://doi.org/10.18757/EJTIR.2020.20.3.3981

Van Koningsveld, M., Den Uijl, J.: OpenTNSim (Version 1.0.0). https://github.com/TUDelft-CITG/OpenTNSim (2020). https://doi.org/10.5281/zenodo.3813871

Van Koningsveld, M., Verheij, H., Taneja, P., de Vriend, H.: Ports and Waterways: Navigating the changing world. Delft University of Technology, Hydraulic engineering, Ports and Waterways, Delft, The Netherlands (2021). https://doi.org/10.5074/T.2021.004

Vehmeijer, L.: Measures for the reduction of CO_2 emissions, by the inland shipping fleet, on the Rotterdam-Antwerp corridor. Master's thesis. Delft University of Technology, Delft, The Netherlands. http://resolver.tudelft.nl/uuid:1abd88e0-9ab6-47fd-a503-2f19ba13bbff (2021)

Vinke F, Van Koningsveld M, Van Dorsser C, Baart F, Van Gelder P, Vellinga T (2022) Cascading effects of sustained low water on inland shipping. Climate Risk Manag. 35:100400. https://doi.org/10.1016/j.crm.2022.100400

Wijaya ATA, Ariana IM, Handani DW, Abdillah HN (2020) Fuel oil consumption monitoring and predicting gas emission based on ship performance using Automatic Identification System (AISITS) data. IOP Conf. Series: Earth Environ. Sci. 557:012017. https://doi.org/10.1088/1755-1315/557/1/012017

The Application of Beidou High-Precision Positioning Technology in the Deformation Monitoring of Ship Locks

Haizun Huang[1](✉), Shihua Qin[2], Chihua Wang[1], Weixun Fang[2], Yulong Wang[1], Hao Gong[3], Dongjie Ning[1], and Jian Shi[3]

[1] Guangxi Beibu Gulf Port Big Data Technology Co., Ltd., Nanning, China
huanghaizun@139.com, {wangch,ningdj}@bbwport.com
[2] Guangxi Xijiang Development and Investment Group Co., Ltd., Nanning, China
[3] Shanghai Huace Navigaition Technology Ltd., Shanghai, China
{hao_gong,jian_shi}@huace.cn

Abstract. Accompanying the construction of inland navigation projects is the generation of high slopes, whose stability has always been an important safety issue of great concern during the process. The safety incidents involving high slope instability often cause huge economic losses and even casualties. Therefore, it is of particular importance to monitor high slopes, to identify safety hazards and to predict the occurrence of safety accidents in advance through the tendency of changing. Once the project is completed, effective monitoring is also essential during the operation of the locks in order to avoid safety accidents such as the collapse of ship locks and other main structures.

Based on the BeiDou high-precision positioning technology, this paper investigates the application of the automated displacement monitoring system in the construction and operation of navigation projects. The system mainly consists of sensor subsystem, data transmission subsystem, data processing and control subsystem as well as other auxiliary support subsystems. The system collects static satellite data from fiducial points and monitoring points, carries out baseline vector solution to realize millimeter-level displacement monitoring, and effectively monitors high slope displacement during construction according to the storage, management, query, statistics and analysis of the monitoring data, as well as timely detects abnormal situations about the displacement. When there is slope instability, early warning will be issued by grading so that measures could be implemented in advance to avoid safety accidents. When applied during the operation of ship locks, it can monitor the displacement of the lock chamber's main work persistently to ensure its normal operation.

This paper takes the construction site of certain ship lock project as the object and sets up 1 fiducial point and 10 monitoring points. The real-time displacement monitoring of the site's high slopes throughout the construction period is carried out, the main functions of the monitoring system are tested and the monitoring data and results are briefly analyzed. After the completion of the project, the monitoring system is applied during the operation of the ship lock and the deformation of the lock chamber is constantly monitored and analyzed.

© The Author(s) 2023
Y. Li et al. (Eds.): PIANC 2022, LNCE 264, pp. 135–143, 2023.
https://doi.org/10.1007/978-981-19-6138-0_12

Keywords: BeiDou high-precision positioning technology · Ship lock · Automated monitoring system

1　Background

Deformation monitoring is the periodic and repeated measurement of the observation points set on deformed bodies in order to determine the changes of their spatial location and the characteristics of their internal structure over time [1]. Throughout the construction and during the operation, the engineering structures should be continuously monitored in order to keep track of the deformation, identify problems in time and ensure their security.

The deformation monitoring of the lock includes the main work such as the lock head, the lock chamber [2], and the foundation pit [3]. The fill behind the lock chamber walls gradually increases in height during the construction, covering the original observation points. Therefore, they must be increased in height as well to enable continuous monitoring. Due to the phased construction of the lock chamber wall, its surrounding area is excavated and back filled every now and then, making it necessary to frequently change the monitoring route for settlement measurements and thus it would compromise the continuity and accuracy of the monitoring. What's more, the release of water within the lock chamber during the operation phase will affect the observation points during the construction phase, also creating difficulties for deformation monitoring.

Since its establishment, the BeiDou system has been widely used in various fields like navigation and positioning. Its advantages of high accuracy, high efficiency, all-weather service and no need for line of sight in static relative positioning have led many researchers to choose it over the conventional methods such as triangulation, trilateration and triangulateration. And it has brought fruitful results both in theory and practice. It is also increasingly widely used in the deformation monitoring of precision engineering.

Therefore, this paper carries out research on the automated displacement monitoring in the main work of ship locks based on the BeiDou high-precision positioning technology.

2　The Principle of BeiDou High-Precision Measurement

Four satellites are required in the BeiDou positioning technology. The distance from each of the two satellites to the user machine is the radius of two spheres intersecting at two points. Suppose the distances from the 4 satellites to the user machine are R1, R2, R3 and R4 respectively [4].

$$R_1 = c(t_r + dt_r - t_{s1}) = \sqrt{(x_1 - x)^2 + (y_1 - y)^2 + (z_1 - z)^2} \tag{1}$$

$$R_2 = c(t_r + dt_r - t_{s2}) = \sqrt{(x_2 - x)^2 + (y_2 - y)^2 + (z_2 - z)^2} \tag{2}$$

$$R_3 = c(t_r + dt_r - t_{s3}) = \sqrt{(x_3 - x)^2 + (y_3 - y)^2 + (z_3 - z)^2} \qquad (3)$$

$$R_2 = c(t_r + dt_r - t_{s4}) = \sqrt{(x_4 - x)^2 + (y_4 - y)^2 + (z_4 - z)^2} \qquad (4)$$

where xi, yi and zi are the coordinates of the i-th satellite; c is the speed of light; tr is the signal arrival time measured by the receiver; dtr is the clock bias between the receiver and the satellite; tsi is the signal emission time measured by the satellite.

Four unknowns can be solved with the above four equations to obtain the coordinates of the station. To improve the accuracy of the BeiDou measurement, RTK technology is used where one receiver is placed on the reference station while other receiver(s) on a carrier (called mobile station). Both the reference station and the mobile station receive signals from the same BeiDou satellite synchronously. The reference station sends the carrier phase measurement, pseudo-range measurement and the coordinates of the reference station to the mobile station in real time by radio transmission. The monitoring station starts the differential processing based on the simultaneously received data from the reference station and the BeiDou satellite to obtain the baseline vectors (Δx, Δy, Δz) of the reference station and the mobile station. The baseline vectors are combined with the reference station coordinates to obtain the WGS84 coordinates for each point of the mobile station. And finally, the plane coordinates x, y and the normal height h of each point of the mobile station are converted and obtained by coordinate transformation. The BeiDou displacement measuring system [5], based on the RTK principle, satellite positioning and post-processing of measured data, has certain lags in the post-processing of measured data due to the large amount of data. The BeiDou monitoring equipment requires its reference station to be located in a stable zone, generally on an unobstructed and steady bedrock. The layout of BeiDou monitoring equipment is shown in Fig. 1. The accuracy of static displacement measurement of the BeiDou monitoring equipment reaches the millimeter level [6].

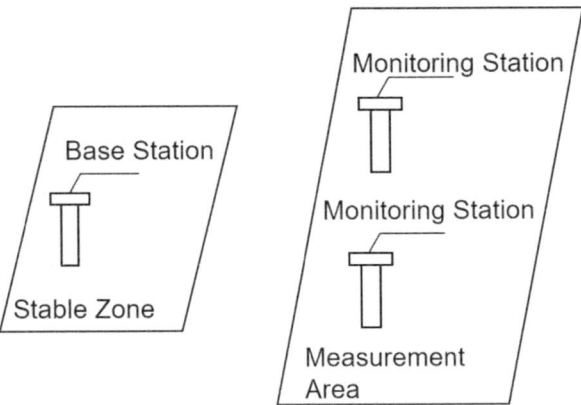

Fig. 1. Layout of the BeiDou monitoring equipment

3 Deformation Monitoring System for Ship Lock Projects Based on BeiDou High-Precision Positioning Technology

Through the equidistant deployment of professional BeiDou-based displacement monitoring system between the upper and lower lock heads on the lock chamber wall, coupled with the existent building deformation monitoring and early warning system platform for real-time displacement monitoring of stress points, the lock chamber monitoring and early warning demonstration system is built, which could allow dynamic monitoring on the displacement of the lock chamber, as well as store, manage, inquire, count and analyze the monitoring data. When the monitoring data reaches the threshold, the early warning system will be activated. Through the development of this project, we'll gradually improve and enrich the application scenarios of the BeiDou system on shipping hubs and water conservancy projects, providing technical support for lock chamber safety monitoring. The technical route is shown in Fig. 2.

The monitoring system consists of four main components, namely, the sensor subsystem, the data transmission subsystem, the data processing and control subsystem and the auxiliary support system.

Sensor subsystem: it is responsible for data acquisition. This is mainly a BeiDou receiver that monitors the horizontal displacement, vertical displacement and the rate of change of the lock chamber. What's more, the sensor subsystem can also be extended with strain gauges, rain gauges, video dome machines and other equipment.

Data transmission subsystem: it is responsible for transmitting the data collected by the sensor system to the monitoring centre server. Most of the sensors have built-in transmission modules, which can transmit data via 3G/4G/Lora/NB and other communication methods.

Data processing and control subsystem: it is the data processing and analysis centre of the whole monitoring system and the monitoring centre of the system. The monitoring centre server collects, processes, stores, analyzes, displays and alarms all kinds of sensor data in real time.

Auxiliary support system: it consists of equipment that assists in the normal operation of the entire monitoring system, including power supply, lightning protection, comprehensive wiring and off-site cabinets and other subsystems.

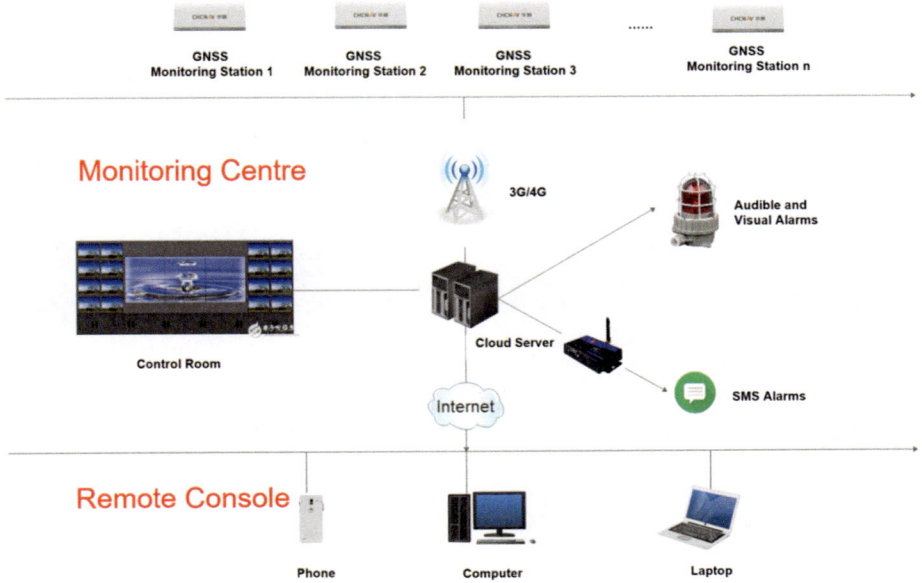

Fig. 2. Architectural diagram of the monitoring system

4 Application of Deformation Monitoring System for Ship Lock Projects Based on High Precision Positioning Technology

4.1 Introduction to the Project

The monitoring project is the second-line ship lock of Guigang Shipping Hub of Xijiang Main River, which is located in Xintang Town, Gangnan District, Guigang City, Guangxi. It aims to monitor the horizontal and vertical displacements and the rate of change of displacements of the lock chamber walls between the upper and lower lock heads of the second-line ship lock of Guigang Shipping Hub of Xijiang Main River. A BeiDou receiver was used to receive signals from BeiDou satellite system. And the differential between the reference station and the monitoring station will be used to calculate horizontal and vertical displacements. In the early phase of the project, system testing was carried out for slope monitoring during the construction. And it will eventually be deployed to the main body of the lock chamber.

4.2 Monitoring Points Layout

According to field situations and technical requirements, a total of 10 BeiDou monitoring stations were laid out on both sides of the lock chamber at equal intervals to monitor the overall deformation activities and changes on the surface of the force bearer on both sides of the lock chamber. Another BeiDou reference station was laid on the roof of the central building of the engine room, which is not affected by the potential energy of the water level, to improve the accuracy of deformation monitoring. The points are shown in Fig. 3 and Fig. 4.

Fig. 3. Map of monitoring points on the top chamber wall

Fig. 4. Map of monitoring points on the lower chamber wall

4.3   Analysis of Monitoring Accuracy

During the test period, total station prisms were installed at the columns of 2 BeiDou monitoring points simultaneously. Comparing the BeiDou monitoring data with the total station monitoring data, it can be seen that the basic deformation trend shown by the 2 groups of data is the same. The comparison between the monitoring data from BD1 and BD2 equipment and that from total station equipment is shown in Fig. 5.

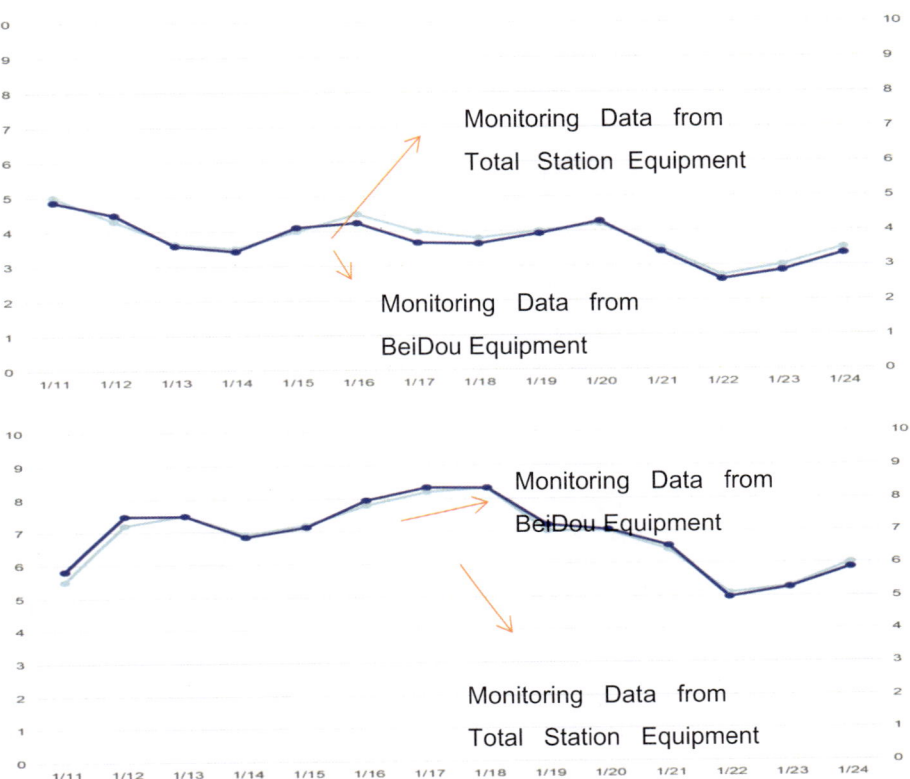

Fig. 5. Comparison of the monitoring data from BD1 and BD2 equipment and the total station equipment

As can be seen from Fig. 5, the deformation trends of the locks are basically the same, excluding the manual observation errors. In order to verify the usability of the BeiDou displacement monitoring, a hypothesis test was conducted on the two monitoring results of BD1 and BD2 [11]. The original hypothesis is that the two kinds of observation means, BeiDou monitoring and total station monitoring, have no significant effect on the displacement data. The results of the significance calculation are shown in Table 1.

As can be seen from Table 1, the ANOVA results for BD1 and BD2 were 0.8516 and 0.7487, both of which were greater than 0.05. Therefore, the original hypothesis was accepted as there was no significant difference between the observations of the two monitoring methods.

Table 1. Calculation results of significance

Significance analysis indicators	BD1		BD2	
	Inter-group	Intra-group	Inter-group	Intra-group
S 11. 53		14 329 0.10	92. 47	37 160. 40
f	1	44	1	42
F-value	11.0	325. 7	92. 47	884.70
P-value	0.851 6		0. 748 7	
F threshold value 4. 06			4. 07	

Note: S is the sum of squares of the sample data; f is the degree of freedom; F is the ratio of the sum of squares between groups to the mean square of the sum of squares of the error; P is the probability of the F value at the corresponding significant level.

5 Conclusion

This paper briefly outlines the technical method of applying BeiDou technology to monitor the deformation of ship locks, and successfully applies it to the displacement monitoring of the second-line of ship lock at Guigang Shipping Hub of Xijiang Main River. The conclusions are as follows:

1) It is feasible to apply the automated monitoring method of BeiDou monitoring equipment to monitoring ship lock projects, which has the advantages of being all-weather, unaffected by bad weather, continuous observation and high observation efficiency, compared with manual observation by means of total stations and so on.
2) BeiDou monitoring and total station monitoring data are in good agreement, and there is no significant difference between the data. The comparison with the data from total station monitoring verifies the reliability of BeiDou monitoring data, and basically removes the error of manual observation;
3) The extreme weather conditions of rainstorm and typhoon during the field test had little impact on the fully automated monitoring by BeiDou. Compared with the manual monitoring of total station, automated monitoring made by BeiDou featured stronger usability.
4) The monitoring system can be applied to the construction and operation of lock project, so it is meaningful to promote the application of such projects, and it can be extended to the monitoring of breasting dolphins at the same time after the completion of the lock project.

References

Yin X, Li B (2007) Deformation Monitoring Technology and Its Application. Yellow River Water Conservancy Press, Zhengzhou

Wu X, Zheng M (2010) Monitoring of horizontal displacements and deformations of three gorges permanent locks and result analysis. Yangtze River 41(20):16–18

Zhum X, Chen Y (2011) Foundation pit monitoring of Zhaojiagou ship lock. Technol. Manag. Port Harbor Eng. 34(3):7–10

Liu Z (2014) GNSS Slope Monitoring and Deformation Analysis. Survey and Mapping Press, Beijing

Zhou W (2013) Research and Realization on Theories and Methods of Precise Positioning Based on BeiDou Navigation Satellite System. PLA Information Engineering University, Zhengzhou

Yao Y, Ruan Y, Liu B et al (2015) Control technology of construction quality for high filled airport based on Beidou satellite navigation system. Chin. J. Geotech. Eng. 37(Suppl. 2):6–10

Research on Key Parameters Selection of Lock Capacity Simulation

Yingfei Liu[1]([⊠]) and Fengshuai Cao[2]

[1] China Institute of Water Resources and Hydropower Research,
Beijing 100038, China
yfliu@mwr.gov.cn

[2] CCCC Water Transportation Consultants Co., Ltd., Beijing 100007, China
caofengshuai@pdiwt.com.cn

Abstract. There are many factors affecting the lock capacity, including ship size and ship type combination, navigation speed of ships entering and leaving the lock, ship spacing, lock filling and discharging time, gate opening and closing time, dimension of lock chamber, navigation days, ship loading rate, traffic volume imbalance, daily working time, etc. In the simulation model, the ship's random arrival at the lock and the distribution proportion of ship type can be simulated, so as to realize the random process of ship passing through the lock and simulate the ship passing scheduling rules. Compared with the methods of manual calculating the lock capacity according to the code, the simulation method is more objective and scientific. In the simulation analysis of lock capacity, the two ship operation parameters of ship entering and exit speed and start interval have strong randomness. This paper studies the rationality of the parameters by constructing the simulation model of lock capacity system. The simulation test shows that with the increase of randomness of ship entering and exit speed, the passing capacity of ship lock decreases gradually, but this decrease can be ignored. The randomness of the start interval has no obvious effect on the lock capacity. Therefore, it is reasonable to take the average value of the model parameters when using the system simulation method to determine the navigation capacity of the lock, and considering all factors comprehensively, it is recommended to take the average value of the parameters in the modeling.

Keywords: Lock capacity · System simulation model · Random parameters

1 Introduction

The lock capacity is an important technical and economic parameter to determine the scale of lock. It is an important research content in the planning and design of lock. The scientificity of its calculation results will directly affect the rationality of lock design scale. China's code for Master Design of shiplocks (2001) defines the lock capacity as the total deadweight tonnage of ships passing through the lock and the cargo volume passing through the lock in the design level year. Among them, the total deadweight tons of ships passing the lock refer to the sum of deadweight tons of all ships passing the lock in a year under the conditions of predicted ship type combination, ship passing efficiency and navigation days, which is the capacity under ideal conditions. The

© The Author(s) 2023
Y. Li et al. (Eds.): PIANC 2022, LNCE 264, pp. 144–154, 2023.
https://doi.org/10.1007/978-981-19-6138-0_13

calculation of cargo volume passing through the gate should also consider non cargo factors, ship loading rate and operation imbalance, so as to obtain the sum of the actual cargo volume passing through the lock in one year. There are many factors affecting the lock capacity, including ship type scale and ship type combination, navigation speed of ships entering and leaving the lock, ship spacing, lock filling and emptying time, gate opening and closing time, plane scale of lock chamber and approach channel, navigation days, ship loading rate, traffic volume imbalance, daily working time, etc. so the calculation of lock capacity is a very complex work.

The simulation research of lock capacity refers to the method of analyzing the lock capacity by building a simulation model based on the system simulation theory and using the scheduling algorithm and optimization algorithm. In the simulation model, the random arrival of ships and the distribution proportion of ship types could be considered, so as to realize the random arrangement of ships passing through the lock and simulating the dispatching rules of ships passing through the lock. Compared with the methods of manual arrangement shifting and calculating the lock capacity according to the code for overall design of lock, it is more objective and scientific.

Many scholars have applied the research method of system simulation to the determination of the lock capacity, and the optimization problems such as the joint scheduling of multi-level and multi-line locks and upstream and downstream navigation buildings (Shang et al. 2018). Shang et al. (2011) constructed the simulation model of the joint scheduling of the four line lock of Changzhou hydro junction project, and analyzed the lock capacity and operation effect under different joint scheduling schemes. Zhang et al. (2015) constructed the shipping scheduling simulation model of multi lock and bottleneck reach of Yangzhou section of Yanshao line to simulate and optimize the shipping scheduling. Adhikari et al. (2014) used the research method of sensitivity analysis on demand change and waiting time to evaluate the Ohio Hanoi river shipping system composed of 9 ship locks.

When the system simulation method is applied to study the lock capacity and other related problems (Shang et al. 2018; Wu et al. 2013), and when the lock capacity is scientifically and reasonably determined in engineering practice, the parameters of the simulation model are generally taken as the average value. However, in reality, some data in the process of ship passing the lock change within a certain range and have a certain randomness. Therefore, whether the average value of the corresponding parameters in the simulation model, rather than the random value within a certain range, will have a significant impact on the results of lock capacity. So this needs further research. This paper analyzes and studies the rationality of taking the average value of parameters when using the system simulation method to study lock capacity.

2 Calculation and Simulation Model of Lock Capacity

2.1 Scope of Model Research

The whole process of ship passing through the lock is simulated by the lock capacity simulation model, which includes upstream and downstream anchorages, upstream and downstream approach channels, berthing structures and lock chambers.

2.2　Ship Passing Simulation Process

The simulation flow of ship passing through the lock (taking the downstream direction as an example) is shown in Fig. 1. The ship's passing arrangement algorithm is based on the two-dimensional packing algorithm (Shang et al. 2011).

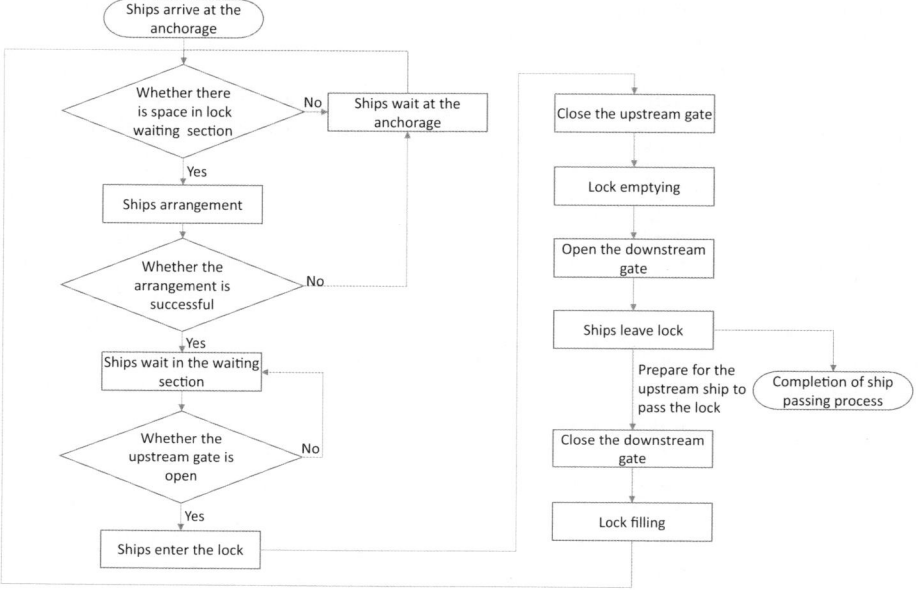

Fig. 1. Simulation flow for ship passing the lock.

2.3　Model Input Parameters

- Lock dimensions. It mainly includes the dimensions of the lock chamber, approach channel and other dimensions related to the ship arrangement and navigation in and out of the lock.
- Passing ship. It mainly includes the scale of the ship passing the lock, the composition of the ship and the relevant characteristics of the ship's entering and exiting operation.
- Lock operation. It mainly includes the operation time of relevant facilities and equipment of the lock and the time when the lock can work normally.

Among the above parameters, the length and width of ship arrangement in the lock chamber, the water depth on sill and the operation distance entering and exiting the lock have been determined when designing the lock. The ship type combination and the distribution of ship arrival law are determined after economic analysis. The annual operation days and daily working hours are determined by considering the factors that may affect the operation in the design stage.

Generally, these parameters will not change in the simulation process. In the real scene, the speed of entering and leaving the lock and the start interval between the front and back ships are not only affected by the objective conditions such as the ship's operating performance, the weight and type of loaded goods, the layout of the lock and the water flow conditions, but also affected by the subjective factors such as the captain's experience and driving style. Therefore, there are great differences between ships. The opening and closing time of the gate is mainly determined by the mechanical equipment and water flow conditions. The filling and emptying time is mainly determined by the layout of the lock culvert and water flow conditions. Generally, there will be little change during normal operation.

Therefore, this paper mainly analyzes the rationality of average value for the ship entering and exiting speed and start interval in the simulation model.

2.4   Output Results

Through the simulation model, the utilization rate of lock chamber, the number of ships passing through the lock and the lock capacity can be obtained. Among them, the utilization rate of lock chamber is the average value of the ratio of the sum of the lock chamber area occupied by ships in lock chamber area during the operation period of the simulation model. The number of ships passing through the lock is the sum of the number of ships passing through the lock in the running time period of the simulation model; the lock capacity of the ship lock is the sum of the cargo tonnage of each ship passing through the lock in the running time period of the simulation model.

3   Selection of Probability Distribution

3.1   Typical Probability Distribution

Typical probability distributions include normal distribution, Irish distribution, exponential distribution and triangular distribution. Normal distribution is the most important distribution in probability theory and mathematical statistics. The central limit theorem has proved that many random variables, such as measurement error, product weight, human height and annual rainfall, can be described or approximately described by normal distribution. The distribution is also widely used in engineering field. Irish distribution and exponential distribution are widely used in queuing theory, including port service system. The analysis of a large number of port statistical data shows that the arrival law of ships and the time of ships occupying berths generally accord with the exponential distribution and Irish distribution. Triangular distribution is usually used to calculate reasonable common values according to the known minimum and maximum values in the absence of sampling data.

3.2   Random Values and Assumptions of Parameters

In conclusion, in the absence of statistical data, a simple probability distribution, namely triangular distribution, should be adopted as the probability distribution of random value of parameters. Normal distribution or Irish distribution can also be selected when there are certain statistical data.

In the simulation model of this study, the random values of ship entering and exiting speed and start interval are assumed as follows:

i. The start intervals of each ship and its previous ship in a lock are different, and the starting intervals of the same ship when entering and exiting the lock are the same, which meets the triangular distribution;

ii. The speed of each ship entering and exiting the lock is different, and the speed meets the triangular distribution, while the ratio of the speed of the same ship entering and leaving the lock to the corresponding average value is the same;

iii. If the calculated speed of ship is greater than the front ship, in order to avoid the possibility of ship overtaking or collision in the model, set the speed of the ship behind to be the same as the front ship, so that the distance between the two ships remains unchanged.

4   Case Analysis

4.1   Test Parameters and Simulation Conditions

4.1.1   Dimension of Lock Chamber

In the simulation model, the length of the lock chamber is 280 m and the width is 34 m and 40 m respectively, without considering the influence of the water depth on sill when the ship passes the lock.

4.1.2   Ship Type Distribution Proportion

The statistical data of ship types passing through the lock in a certain year are selected as the distribution proportion of current ship types (see Table 1). Considering the improvement of the scale of the lock and the navigation conditions of the channel in the future, it is expected that the ship type passing the lock will continue to develop towards large-scale, and the proportion of large ships (ship width greater than 16.3 m) will gradually increase and become the main ship type. Therefore, it is predicted that the proportion of large ships is 50%, 60%, 70% and 80% respectively as combination 1–4 (see Table 2). Large ships of current ship types account for about 20.2%.

Table 1. Distribution proportion of current ship type

Loading capacity of ship (t)	Ship length (m)	Ship width (m)	Proportion
392–392	44.6–44.6	8.5	0.1%
829–829	54.6–54.6	9.53	1.2%
500–2708	60–79.5	11	10.0%
1298–2838	67–81.5	13	8.8%
100–3647	68–89.6	13.8	24.7%
150–4938	87.6–91.3	15	14.8%
2431–6931	88.7–104.4	16.3	20.1%
1142–7767	100–112	17.2	13.7%
2000–8890	110–118	19.2	6.5%

Table 2. Predicted ship type distribution proportion

Ship cargo capacity (t)	Ship length (m)	Ship width (m)	Combination 1	Combination 2	Combination 3	Combination 4
2000–2500	72–88	13.8	5%	5%	2%	2%
2000–3000	82–88	15.0	2%	2%	3%	2%
2500–3500	82–88	16.3	3%	3%	5%	3%
3500–4500	82–88	16.3	10%	10%	5%	3%
4000–6500	90–105	16.3	15%	10%	7%	5%
5500–7500	125–130	16.3	15%	10%	8%	5%
7500–9000	110–135	19.2	20%	30%	25%	30%
8500–10,500	110–135	22.0	20%	20%	25%	30%
9000–12,000	110–135	25.0	10%	10%	20%	20%

4.1.3 Parameter Random Value

Take the average value of the ship's entering and exiting speed and the start interval between the front and back ships as the basic working condition. When the above parameters are taken at random, considering the range of probability distribution from small to large, the following three groups of triangular distribution forms are adopted: Form 1 - the minimum and maximum values of distribution are 90% and 110% of the average value respectively, form 2 - the minimum and maximum values of distribution are 80% and 120% of the average value respectively, and form 3 - the minimum and maximum values of distribution are 70% and 130% of the average value respectively. From taking the average value to taking the random value in the form of triangular distribution 1–3, the randomness of parameter value gradually increases.

4.1.4 Lock Operation

Considering the weather, maintenance, overhaul and other reasons, the annual operation days are taken as 335d. The daily operation hours are taken as 24 h.

4.1.5 Test Conditions

A total of 40 groups are selected according to different lock chamber dimensions and random parameters. See Table 3 for details.

Table 3. Test conditions

Dimension of lock chamber	Current ship type combination	Predicted ship type combination 1	Predicted ship type combination 2	Predicted ship type combination 3	Predicted ship type combination 4
280 m × 34 m 280 m × 40 m	The random parameters take the average value or three triangular distribution forms (4 test conditions)				

4.2   Analysis of Simulation Results

4.2.1   Test Condition Analysis

The lock capacity is affected by many factors, such as the total number of ships passing through the lock, the distribution proportion of ship types and so on. For the 34 m wide lock, although with the increase of the proportion of large ships, the total number of ships passing through the lock decreases and the utilization rate of the lock chamber decreases, the lock capacity decreases first and then increases; For the 40 m wide ship lock, with the increase of the proportion of large ships, the total number of ships passing through the lock decreases, the utilization of the lock chamber first increases and then decreases, while the lock capacity first increases and then decreases; See Figs. 2, 3 and 4 for details. With the increase of the proportion of large ships, the complex change trend of lock capacity shows that different factors have successively become the main factors affecting the lock capacity. Therefore, it can be considered that the test condition can comprehensively reflect the combination of different influencing factors.

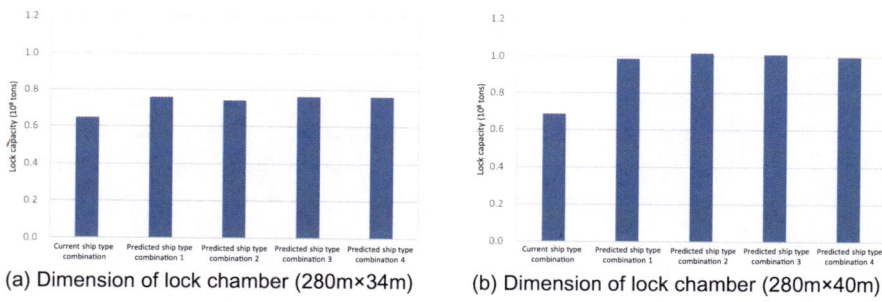

Fig. 2. Lock capacity.

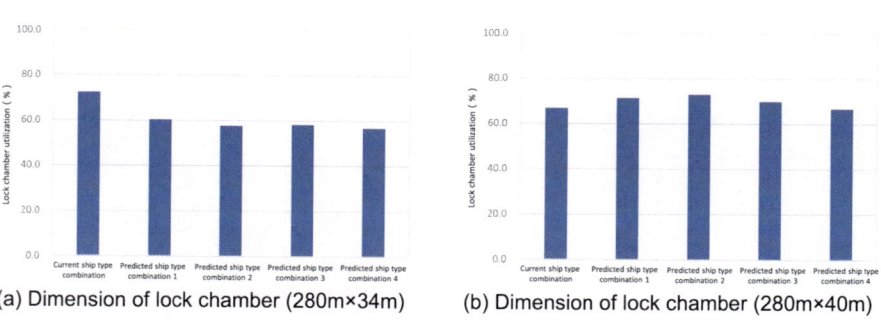

Fig. 3. Lock chamber utilization.

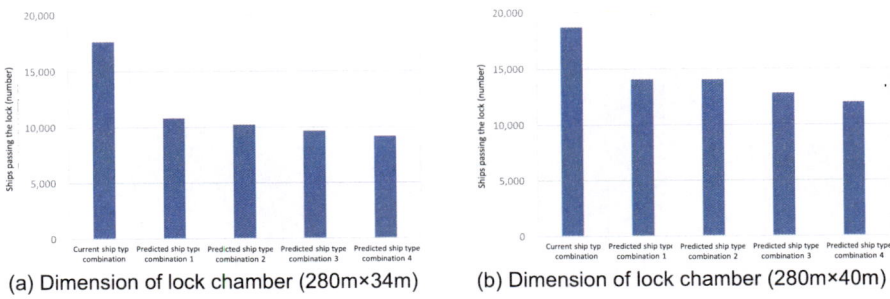

(a) Dimension of lock chamber (280m×34m)   (b) Dimension of lock chamber (280m×40m)

Fig. 4. Numbers of ships passing the lock.

4.2.2   Influence of Random Value of Entering and Exiting Speed on the Lock Capacity

Keeping the average start interval unchanged, the lock capacity under the condition of random entering and exiting speed are shown in Fig. 5 and Fig. 6. It can be seen that under the condition of the same lock chamber size and ship type distribution combination, the lock capacity decreases with the increase of the randomness of the entering and exiting speed. Under the same ship type distribution combination, compared with the average value method, the maximum reduction of the lock capacity is only 1.5%–1.7%, which can be considered as no effect on the lock capacity.

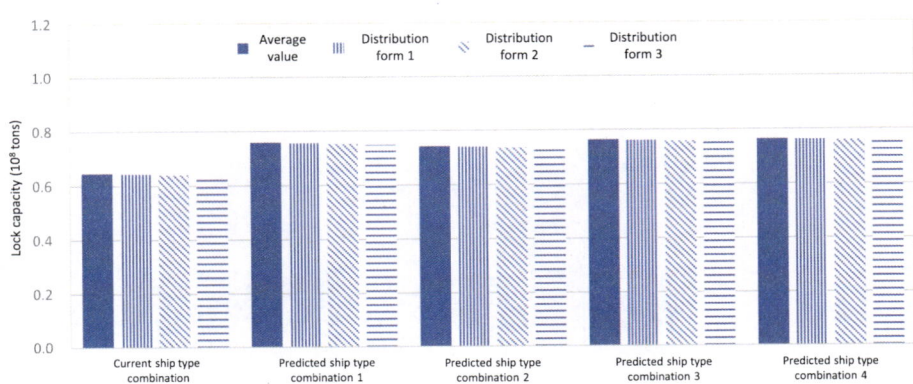

Fig. 5. The lock capacity under the different ship type combinations, when the entering and exiting speed is taken as different distribution (dimension of lock chamber, 280 m × 34 m).

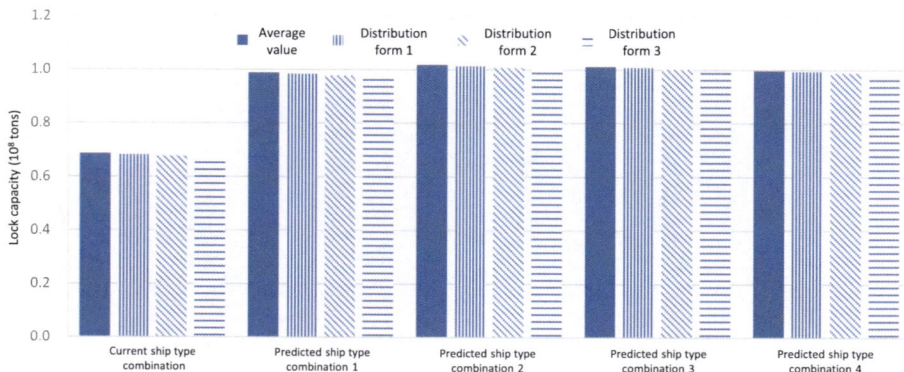

Fig. 6. The lock capacity under the different ship type combinations, when the entering and exiting speed is taken as different distribution (dimension of lock chamber, 280 m × 40 m).

4.2.3   Influence of Random Value of Start Interval Between Front and Back Ships on the Lock Capacity

Keeping the average value of the entering and exiting speed unchanged, the lock capacity under the condition of random value of the start interval between the front and back ships are shown Fig. 7 and Fig. 8. It can be seen that under the same lock chamber scale and ship type distribution combination, with the increase of the randomness of the value of start interval, there is no obvious change law of ship lock capacity. Under the same ship type distribution combination, compared with the average value method, the maximum reduction of the lock capacity is only 0.1%, which has no effect.

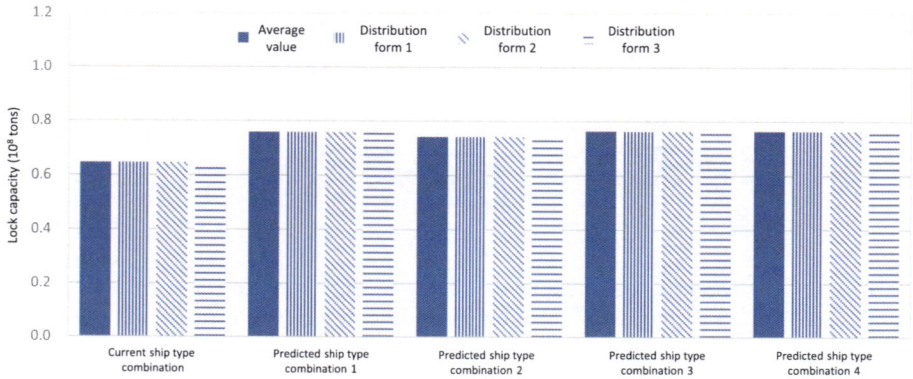

Fig. 7. The lock capacity under the different ship type combinations, when the start interval of the ship is taken as different distribution (dimension of lock chamber, 280 m × 34 m).

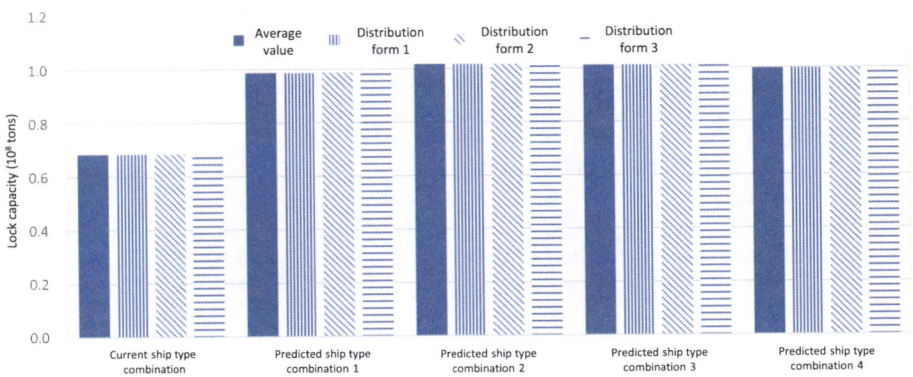

Fig. 8. The lock capacity under the different ship type combinations, when the start interval of the ship is taken as different distribution (dimension of lock chamber, 280 m × 40 m).

5   Conclusions

i. The parameters with certain randomness in the simulation study of lock capacity mainly include ship entering and exiting speed and start interval.

ii. Under the condition of the same lock chamber scale, ship type distribution combination and start interval, with the increase of the randomness of the value of the entering and exiting speed, the lock capacity shows a decreasing trend, but it can be ignored.

iii. Under the condition of the same lock chamber scale, ship type distribution combination and average entering and exiting speed, the lock capacity does not change significantly with the increase of the randomness of the starting interval.

iv. It is reasonable and credible to take the average value of each parameter in the simulation model when carrying out the simulation research of lock capacity and applying the system simulation method to scientifically and reasonably determine the ship lock capacity in engineering practice.

v. The random value of parameters in the simulation model will increase the difficulty of parameter acquisition and simulation modeling in the research process, and may reduce the operation efficiency of the model. Therefore, considering the above factors, it is suggested to take the average value of the parameters in the model when using the system simulation method to determine the lock capacity.

References

Adhikari G., et al.: Simulation analysis of the ohio river waterway transportation system. In: Proceedings of the 2014 Industrial and Systems Engineering Research Conference (2014)

CCCC water transportation planning and Design Institute: Code for Master Design of shiplocks (JTJ 305–2001). Ministry of Communications of the People's Republic of China. People's Communications Press, Beijing (2001)

Shang JP et al (2018) Overview of researches on lock throughput capacity. Port & Waterway Engineering 2018(7):109–113

Shang, J.P., Wu, P., Tang, Y.: On multiple-lane lock's joint scheduling scheme plan based on computer simulation. Port & Waterway Engineering **2011**(9), 199–204 (2011)

Wu, P., Shang, J.P.: Research on computer simulation analysis method of lock capacity. In: Proceedings of standardization and technological innovation of port & waterway engineering, People's Communications Press, Beijing, pp. 354–363 (2013)

Zhang W, Gu DP, Wang QM (2015) Analysis and Simulation of navigation in Yangzhou segment of Yanshao waterway. Port & Waterway Engineering 2015(5):122–127

Research on Theoretical Framework and Implementation Path of Green Maintenance of Inland Waterway

Zhaoxing Han[1,2,3], Chaohui Zheng[1,3], Wenxi Jiang[1,3],
Jinxiang Cheng[1,3(✉)], and Liguo Zhang[1,3]

[1] Transport Planning and Research Institute, Ministry of Transport,
Beijing 100028, China
chengjx@tpri.org.cn
[2] Tsinghua University, Beijing 100084, China
[3] Ministry of Transport, Laboratory of Transport Pollution Control
and Monitoring Technology, Transport Planning and Research Institute,
Beijing 100028, China

Abstract. Inland waterway maintenance could guarantee channel dimension and maintain navigation conditions. With the continuous increase of channel mileage, inland waterway maintenance gradually became more frequent and scale-up, as well. The maintenance activities would inevitably cause adverse environment impacts on waters, and disturb the ecosystem. Therefore, it was necessary to implement the green maintenance to reduce the negative impacts in the maintenance process. However, the research on green maintenance of inland waterway was not systematic and impeccable. This paper proposed a solution for the problem above. Firstly identified the environmental impacts of maintenance based on the analysis of maintenance process. Secondly the notion and principles of green maintenance are put forward, according to the concept of green transportation and green channel. Thirdly, the study constructed the technical framework of green maintenance and summarized the major measures for green maintenance. An last, implementation path of green maintenance was addressed including standard-setting, establishing system management, technology research and development, etc.

Keywords: Inland waterway · Green maintenance · Environment impact · Theoretical framework · Implementation path

1 Introduction

Inland navigation shipping has the comparative advantages of large traffic volume, low freight rate, low energy consumption, less land occupation and light pollution (Yuan Yuan et al. 2017). It is an important part of the comprehensive national transport network (the comprehensive national transport network planning, 2021) and the main path for the transportation industry to achieve carbon peak (carbon peak action plan before 2030, 2021). Inland waterway is the main support for the development of inland navigation. Since the 13th five year plan, the construction of inland water transport

© The Author(s) 2023
Y. Li et al. (Eds.): PIANC 2022, LNCE 264, pp. 155–168, 2023.
https://doi.org/10.1007/978-981-19-6138-0_14

infrastructure has been strengthened. And an inland waterway network including the Yangtze River main line, Xijiang River shipping trunk line, Beijing Hangzhou canal, Yangtze River Delta and Pearl River Delta high-grade waterway network has been basically established, with the connection between trunk and branch lines, the river and the sea. (Ministry of Transport 2021). By the end of 2020, the navigation mileage of inland waterways in China reached 127700 kilometers, including 14400 kilometers of high-grade waterways of class III and above, accounting for 11.3% of the total mileage (Fig. 1) (2020 statistical bulletin on the development of transportation industry, 2021). According to the outline of comprehensive national transport network planning, it is expected that by 2035, the national high-grade waterway will reach about 25000 km.

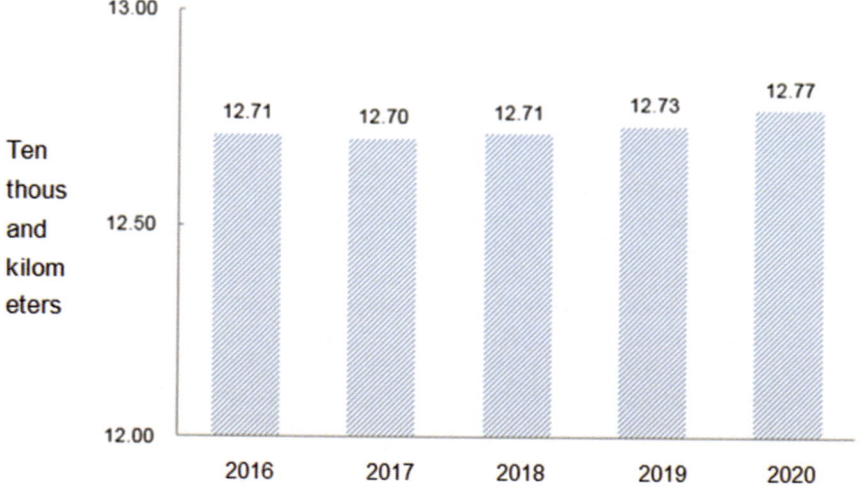

Fig. 1. Mileages of inland waterway in China from 2016 to 2020

Inland waterway maintenance is the basic work to keep the navigation dimensions and ensure the safety of ship navigation. With the continuous increase of navigation mileage of inland waterway, inland waterway shipping has gradually shifted from large-scale construction to large-scale maintenance, and the work of waterway maintenance is becoming more and more arduous. Generally speaking, inland waterway maintenance includes regular measurement of water depth, adjustment and maintenance of navigation aids, temporary dredging of local navigable river sections, protect and repair of inland waterway regulation buildings, etc. (technical code for channel maintenance, 2021). The above work involves many engineering and technical measures, which resulting in construction operations and inevitably producing energy consumption, sewage and waste gas emission and ecological disturbance. Therefore, in June 2021, the Ministry of transport issued the technical guide for green maintenance of inland waterway to guide the ecological and environmental protection in waterway maintenance. However, the current theoretical framework of green maintenance is not clear, the energy-saving and environmental protection technology system in inland channel maintenance has not been established, and the specific engineering and technical measures are still in the initial research and development stage. To solve this problem, based on the specific content of channel maintenance, this paper identifies its impact on the ecological environment, puts forward the main measures of inland waterway maintenance, and clarifies the implementation path of green maintenance.

2 Impact of Inland Waterway Maintenance on River Ecosystem

2.1 Inland Waterway Maintenance

Channel maintenance usually includes maintenance dredging, regulating structures maintenance and navigation aids maintenance. Table 1 combs the typical construction processes of different waterway maintenance works.

Table 1. Construction process of typical channel maintenance method

Construction content	Main process
Maintenance dredging	construction preparation → pre dredging survey → dredging operation → dredged material transportation → dredged material treatment
Waterway regulating structures maintenance (spur dike maintenance)	construction preparation → fixed-point riprap → demolition of damaged spur dike part
Waterway regulating structures maintenance (revetment and bottom protection)	construction preparation → sinking mattress → water riprap
Navigation aids maintenance	site inspection → navigation acids position adjustment → navigation aids cleaning, derusting and painting → parts replacement

2.2 Environmental Impact

1) aquatic environment impact

In terms of water regime impact, dredging and excavation will cause hydrodynamic variation and thus brings aquatic habitats changes. In terms of water quality impact, it will be deteriorated by the underwater construction work such as sinking mattress, riprap and dredging which increase the concentration of suspended solid. Feng (2017) simulated the diffusion of suspended solids during the dredging works of Taipingkou waterway in the middle reaches of the Yangtze River. It was shown that the diffusion range of which increment of suspended solids was greater than 10mg/L was no longer than 3km. The field monitoring of typical waterway regulation projects shows that the influence of sinking mattress and riprap can be control within the 200m of the construction site. Suspended solids will also be increased by the sediment overflow during the dredged material transportation. Field observation and experiments showed that the sediment concentration in the turbidity zone formed around the vessel is 3 ∼

10 times higher than before. However, due to the rapid settling velocity, the sediment concentration has decreased to about twice as much as before 20 minutes after construction completion. (Dai et al. 1997). In the maintenance of navigation aids, the process of cleaning, derusting and painting is carried out in water. Toxic chemicals produced during the rust cleaning and painting of navigation aids may cause water pollution.

2) Ecological impact

Sinking mattress and riprap directly occupy the habitat of benthic organisms, resulting in their direct burial and coverage. Moreover, surrounding fish will be dispersed due to the disturbance. However, studies have shown that the complex flow conditions formed by sinking mattress and enrockment have a contribution to the formation of artificial reef effect, which can ensure the living space of fish and providing them with spawning grounds, wintering grounds and feeding grounds (Chang et al. 2019). The increasing concentration of suspended solid caused by wading construction brings adverse impact on plankton, benthos and fishery resources. However, the water quality will recover to the original level very soon as sediment can quickly settle down. So this kind of impact could be regarded as short-termed and temporary (Xie and Ding 2020). In addition, excavation directly destroys the habitat of benthos and causes them to be buried (Ma Yi 2014). Organic matters and heavy metals in bottom sludge will be released to water bodies, and endanger human health through food chain transfer and biological amplification effect (Zuo et al. 2014). Currently the utilization rate of dredged soil is low and most of it is directly dumped to the designated dumping area (Fu et al. 2011), which will cause resuspension of sediment and bury benthos in the short term. Land filling of dredged soil occupies land resources, and pollutants lead to potential ecological risks through rainwater runoff and infiltration (Zuo et al. 2014).

3) Other effects

The construction wastes, batteries, solar panels, rust and other wastes generated from the replacement of navigation aids will poses a threat to the environment if they are improperly disposed (Yan 2017). In addition, regular sailing inspection triggers huge air pollutants and greenhouse gas emission as a result of fuel consumption, which put great pressure on the goal of carbon emission reduction (Yan 2017). In addition, the mechanical vibration noise generated by vessels will give rise to the hearing damage of fish and affect the foraging behavior (Liu Dan et al. 2021), especially pose a great threat to the communication and hearing of the Yangtze finless porpoise (Zhang Tianci et al. 2018). Fuel oil leakage、improper disposal of vessel domestic sewage, oily sewage and garbage will bring about water pollution and damage the aquatic ecology as well.

The environmental impact of inland waterway maintenance was showed in Fig. 2.

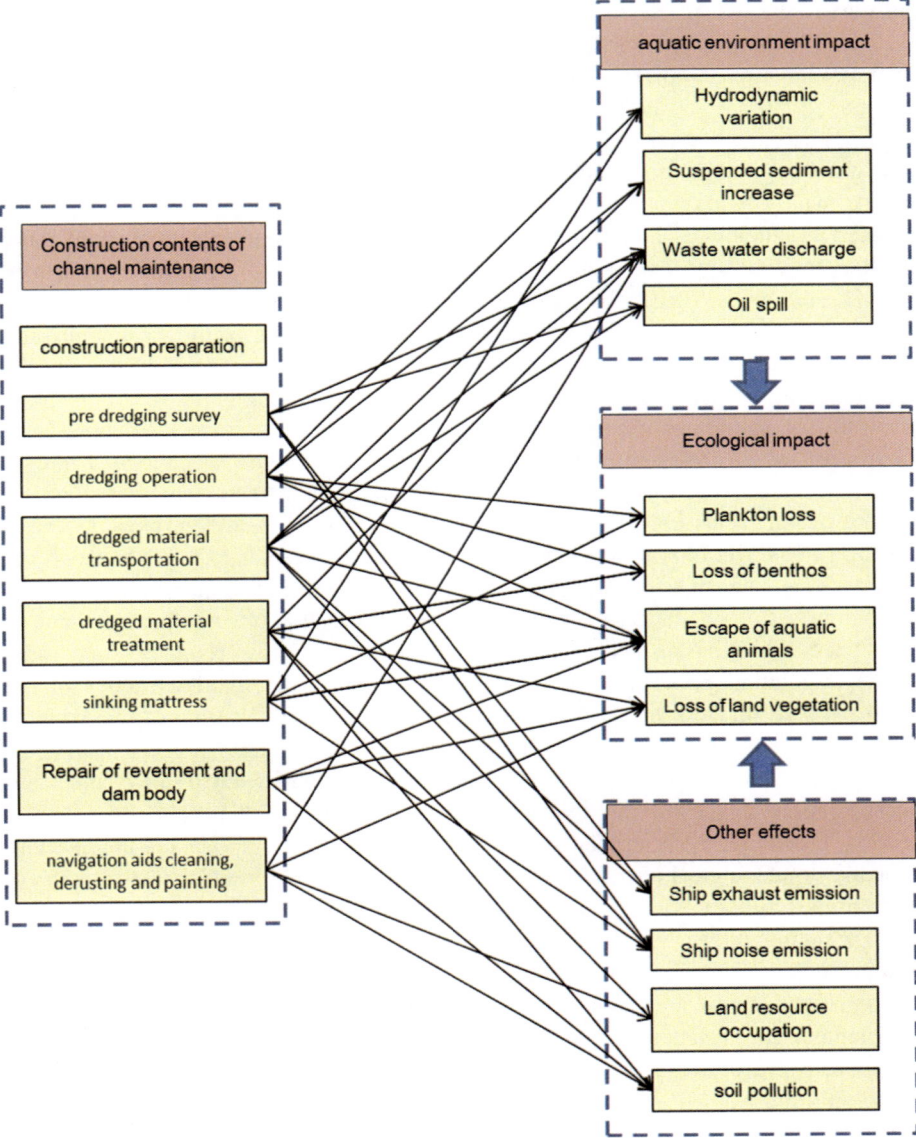

Fig. 2. Environmental impact of inland waterway maintenance

3 Theoretical Framework of Green Conservation of Inland Waterway

3.1 Notion of Green Conservation of Inland Waterway

The demand for waterway transport has always been a very important part of river comprehensive development all over the world. For example, Germany puts waterway transport in the first place in river development, and France implements the comprehensive development plan of Rhone River, which is led by waterway transport, power generation and irrigation. The same situation is also true in the southern water system of China. In order to further improve the water transport capacity and economic benefits, many large rivers in the world are carrying out or have been carried out large-scale waterway regulation projects. The traditional waterway construction and maintenance mainly focus on effectiveness and stability of the projects, and do not much consider the ecological and environmental impacts. As people pay more and more attention to the environment, the industry began to explore the notion of `green channel'. Yan et al. (2018) believe that ecological channel refers to integrating the navigation function of channel with the function of river ecosystem, and maintaining the structure and function of navigation river ecosystem on the basis of playing the basic function of channel. Liu Xinyi (2017) believes that combining the navigation function of channels with the original water ecosystem can not only hold the basic function of safe, convenient, smooth and efficient water transport, but also have a reasonably balanced ecosystem structure, stable self-regulation and recovery ability and pleasant overall landscape effect, so as to finally realize green and sustainable development. Liu Junwei (2015), Lei Guoping (2016) and Liu huaihan (2020) focus on the Yangtze River, which is the longest and highest navigable river in China, with discussion of the theory and key technology for the construction of ecological channel. In recent years, China has also carried out positive practices in green channels, such as the channel regulation project of Jingjiang river section in the middle reaches of the Yangtze River, the 12.5 m deep-water channel project below Nanjing of the Yangtze River, the Beijing Hangzhou canal, etc.

Channel maintenance is an important link in the whole life cycle of the channel and a necessary guarantee for the channel to continue to give full play to its transport function. Clarifying the scientific connotation of ecological channel is the premise and foundation of ecological channel research. 'The technical requirements for assessment of green transportation facilities Part 3: green channel' (JT/T 1199.3-2018), formulated in 2018, defines green channel as: within the whole life cycle of the channel, take sustainable development as the concept, carry out technical and economic analysis and environmental impact assessment; through reasonable planning, design, construction and maintenance management, reduce resource and energy consumption, reduce pollution emissions, protect the ecological environment to the greatest extent, pay attention to the improvement of quality construction and operation efficiency, and build a channel for harmonious development with resources, environment, ecology and society. Among them, specific evaluation indicators is set up for channel maintenance, covering whether the construction and materials used for channel maintenance are environmental-friendly, whether the maintenance frequency is appropriate, whether the waste disposal is reasonable, etc. Since this standard focuses on design and construction stages, the requirements for maintenance are not integrated. It defines whether the specific channel maintenance operation "reaches green", but it does not answer "how to achieve green". In order to make up for these deficiencies, the Ministry of transport issued the technical guide for green maintenance of inland waterways in

2021, which gives the definition of green maintenance of waterways, that is, on the basis of ensuring the functions of waterways, implement the concept of environmental protection, energy conservation and carbon reduction, and apply new materials, equipment, processes and technologies to reduce the impact on the ecological environment.

In general, the discussion and practices of green channel have developed rapidly in recent ten years. In general, the discussion and practices of green channel have promoted rapidly in recent ten years. The understanding of green channel has also experienced many changes. One is altering from simply doing good environmental management of channel construction to paying attention to the coordination of construction and environment, and afterwards to integrating the concept of ecological civilization and sustainable development into construction. Another is altering from only optimizing the construction scheme to concerning the harmony between shipping and river ecosystem in the whole life cycle of channels, which involves maintaining the biodiversity of aquatic species, maximizing the ecological services of river ecosystem, ensuring the health and sustainability of rivers, and improving the safety carrying capacity of the waterway.

3.2 Principles of Green Maintenance of Inland Waterway

Green channel maintenance is an important embodiment of green channel, and the sustainable development concept of harmonious coexistence with natural ecology should be fully implemented. Therefore, the overall principle of green maintenance is to meet the needs of channel use to the greatest extent with the least resource consumption and the least impact on environment. Specific principles include (Fig. 3):

1) Greening maintenance operation. For example, construction activities such as maintenance dredging and obstacle removal should be arranged without long-term and severe impact on river water quality and ecological function.

2) Ecologicalization of structures and materials. On the premise of meeting the use function, new structures and materials can be used to make up for the interference of the channel to the ecosystem.

3) Minimizing material and energy consumption. It includes reducing the direct consumption of consumables, materials & equipment, and selecting energy-saving appliances.

4) Reducing pollution discharge from maintenance activities. For example, activities such as patrol, mapping and inspection, and the use of ships and vehicles also produce some pollution, which should be minimized.

5) Under the goal of "double carbon", the ecological function of the channel should be better through conservation as far as possible, so as to enhance the positive role of the river in climate change.

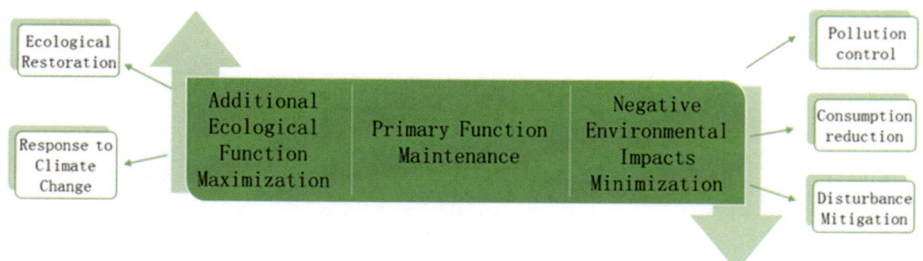

Fig. 3. Principles of green maintenance of inland waterway

3.3   Research Framework for Green Conservation of Inland Waterway

The research on the green maintenance of inland waterways can be divided into four aspects: basic theoretical research, ecological impact mechanism research, green conservation technology research & development, and implementation path research (Fig. 4.).

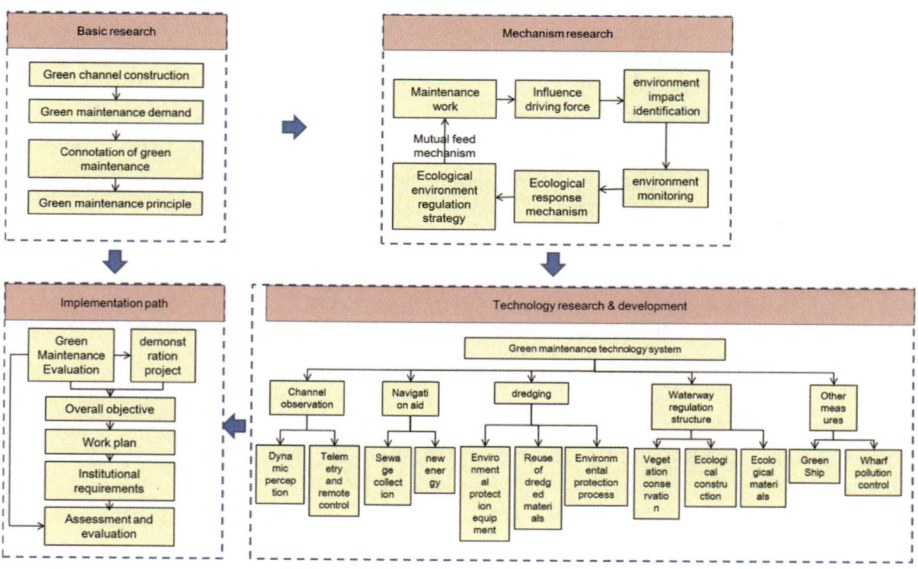

Fig. 4. Theoretical framework of green conservation of inland waterway

The basic research focuses on analyzing the specific needs of channel green maintenance according to the relevant definitions and contents of green channel, so as to put forward the concept and connotation of green maintenance and the basic principles of green maintenance. Mechanism research starts with the content of maintenance work, analyzes the main driving forces of various channel maintenance behaviors on the impact on the ecological environment, so as to identify the ways, scope and degree of the impact on the ecological environment, and studies the response mechanism of the ecological environment to channel maintenance. Based on the environmental monitoring and ecological investigations, mitigation measures and regulation strategies for maintenance work was put forward.

Technology research & development will focus on forming a green maintenance technology system according to the ecological environment impact and response mechanism, including channel observation, navigation aid maintenance, maintenance dredging, channel regulation, and other measures. In terms of implementation path, according to the results of basic research, mechanism research and technological research & development, we will build green maintenance evaluation standards and demonstration projects, so as to put forward the overall goal of green maintenance and incorporate it into the maintenance work plan. With the increase of technical maturity, the relevant maintenance technical requirements are gradually incorporated into the maintenance management system and technical specifications. According the result of green maintenance assessment, the green evaluation standard of green maintenance is dynamically improved.

4   Major Measures for Green Maintenance of Inland Waterway

Three maintenance activities that have the most significant impacts on the environment is discussed, including dredging, regulating structure rectification and navigation aid maintenance. The major measures for these activities are listed in Table 2.

4.1   Dredging

In general, the main adverse environmental impact of dredging is the shapely increased concentration of suspended solids (SS). The diffused SS may consequently pollute the water sources of residential areas and interfere the growth of aquatic organisms. Green measures so aim on SS control. Compared with traditional dredgers, grab dredgers, drag suction dredgers, suction cups and cutter suction dredgers are more environmental friendly. For example, the cutter suction dredger cuts the underwater medium by the cutter and then pumps the mud to the designated place; the drag suction dredger generates mud by joint force of dredger movement and underwater rake head, and the mud is also pumped away (Liao Yongtao 2016). The positive pressure of the mud pump would greatly reduce the diffusion of mud. The use of environmental rake head would further enhance the effectiveness of SS control (Ren et al. 2017). For example, drag suction dredgers adopt the rake head with eddy current protective cover or the rake head installed with a closed plate; cutter suction dredgers adopt disc type, shovel suction type and screw type reamers; grab dredgers adopt fully enclosed leak proof grabs. Dredging ships, vehicles and equipment powered by fuel oil will have air pollution. Using the ships, vehicles and equipment that meet emission standards or electric power is an effective measure to reduce these air pollution.

The additional environmental problem of dredging is the disposal of dredged soil. Randomly discarded dredged soil is a kind of pollution, but it is possible to recycle

Table 2. Main measures for green maintenance

	Air pollution control measures	Water pollution control measures	Ecological protection and restoration measures	Resource conservation and intensive utilization	Other measures
Maintenance dredging	-Use ships, vehicles and equipment that meet the exhaust emission limit standards	-Select dredging ships or dredging equipment with environmental protection function; -Dredging vessels use machines and tools to manufacture relatively few suspended solids and control the diffusion range of suspended solids; -As far as possible, the loading shall not overflow. The operation with overflow can adopt the form of underwater overflow and other relatively environmental protection -Anti pollution enclosure for water source		-Classified utilization of dredged soil -Dredged soil shall be transported out without being thrown and sealed;	-Control plane scale, silt preparation depth and operation time, and strictly control over excavation; -Optimize the dredging scheme to avoid environmentally sensitive areas; -Positioning technology and visualization technology are used to assist precise dredging -Equipped with pollution emergency equipment
Renovation and maintenance of buildings			-Increase environmental protection or ecological functions for general renovation buildings, if possible -The function or local improvement of repairing defects for ecological regulation buildings		-Select plants with barren resistance, flooding resistance, drought resistance, easy reproduction, high survival rate, developed roots and easy sprouting; -The selection of exotic plants shall be confirmed to have no potential ecological hazards
Aids to navigation maintenance	-Use ships, vehicles and equipment that meet the exhaust emission limit standards -Navigation aids of remote control and telemetry system are adopted to reduce on-site cruise inspection;	-If possible, the buoy shall be ashore cleaned, painted and repaired -Control the use of paint -Collect and clean the sewage, waste and oil pollution of navigation aids by categories;	-Only trim the trees around the shore mark that block the intervisibility range;	-choose environmentally friendly products and materials with high strength, corrosion resistance and aging resistance when replacing batteries and other consumables -Energy saving lamps and solar energy are used for power supply, when the lighting duration and brightness of navigation aids are enough;	-Local materials, or natural degradable raw materials; -Dispose the collected solid wastes (such as old batteries, attachments, etc.) according to the requirements of national regulations

dredged soil. The actual use of dredged soil is closely related to its physical and chemical elements (Xie and Ding 2020). According to the technical status, the clean dredged soil can be used for land reclamation, ecological beach consolidation, sand pillow filling, artificial island construction or wetland construction, etc. For example, in the regulation of the north channel of the Yangtze River Estuary, the throwing and blowing process of secondary transportation is adopted to blow and fill the dredged soil to the water area of Hengsha east beach for siltation promotion. Most of developed countries have formulated a complete set of policies and technical standards related to the utilization of dredged soil (Zhao et al. 2013). The contaminated dredged soil shall be transferred and treated on shore. The pre-treatment shall be carried out according to the degree of pollution. The dredged soil that is identified as hazardous waste shall be delivered to a qualified treatment enterprise. When the dredged soil has to be discarded partly, the setting of the dumping area should avoid the ecologically sensitive areas such as nature reserves, fish spawning grounds, feeding grounds, wintering grounds and scenic spots, and set up anti-pollution screens and anti-pollution curtains to reduce the scope of influence.

Optimizing the dredging scheme prior to operation is the basic work. A scientific dredging scheme should avoid environmentally sensitive areas and plan a reasonable dredging scale, silt preparation depth and operation time. Auxiliary technologies such as positioning technology and visualization technology can ensure the accurate implementation of dredging scheme and control overbreak to the greatest extent (Xie and Ding 2020).

4.2 Regulating Structure Rectification

Regulating structures are permanent, which will not continuously generate water pollution or air pollution. Rectification of regulating structure focus on enhancing the effectiveness of ecological protection. For general dams, revetments, slope protectors, beach protectors, bottom protectors, the structural form can be transformed to improve the ecological function, with the consideration of original structure, flows, scouring and silting, damage degree, environmental requirements, habitat needs, etc.. For ecological regulating structures, it is mainly to maintain the original structure, materials and vegetation coverage. When it is found that the function has obvious defects, partly rectification is possible. Vegetation coverage is the most commonly used way to reflect the ecological function of regulating structures (Cao et al. 2018; Chen 2017). In order to coordinate with the surrounding ecosystem, native plants or exotic plants without potential ecological hazards should be selected. In addition, the selected plants have the advantages of barren resistance, flood resistance, drought resistance, easy reproduction, high survival rate, developed roots and easy germination, which can reduce the loss of plants, but they should still be inspected and replanted after the flood season.

4.3 Maintenance of Navigation Aids

In terms of environmental pollution, buoy maintenance is the most likely to cause adverse environmental impact on the river. Therefore, it is primary to transfer the buoy to the shore maintenance base for centralized cleaning, painting and repair. The sewage, oil stain, garbage, etc. produced during operation can be conveniently collected and sent for appropriate treatment. Local materials or natural degradable raw materials are also advocated to reduce the difficulty of subsequent treatment and long-term adverse impact on the environment. When on-shore maintenance is not feasible, it is necessary to take some measures to prevent pollutants from entering the water body. Similar to dredging, it is better to use cleaner ships, vehicles and equipment for navigation aids maintenance, in order to reduce air pollution. Another measure is to promote the new technology of remote control and telemetry instead of on-site inspection, resulting in the less use of vehicles and ships. The increasingly mature remote control and telemetry technology of navigation aids enables the management department to monitor the location and energy status of navigation aids in real time. Compared with the timing incandescent lamps and batteries are mainly used, it is feasible to change the "monthly inspection" rule (Yan 2017).

Navigation aids continuously consume energy and modules (batteries, lamps, standards, calibration, reflective films, etc.). Correspondingly, environmental-friendly products and materials with high strength, corrosion resistance and aging resistance can decrease consumption. For example, the standard body made of non-metallic materials such as FRP and high-strength polyethylene can be free from maintenance within $5 \sim 10$ (Yan 2017). In addition, the use of energy-saving lamps and solar power supply can reduce daily energy consumption.

5   Implementation Paths on Green Maintenance of Inland Waterway

5.1   Develop Green Maintenance Evaluation Standards

With the gradual deepening of the research on the mutual feeding mechanism between the construction & operation of inland waterway and the river ecological environment, the evaluation standard of green maintenance of inland waterway can be established according to the ways, degrees and mitigation measures of ecological environment impact caused by waterway maintenance. The green maintenance work of the channel can be scored to determine the green level, according to the characteristics of the maintenance practice, the application of green maintenance technology, the degree of technological innovation, etc. The introduction of green maintenance evaluation standards can make horizontal and vertical comparison of green maintenance practices in different work contents and regions, so as to effectively guide and promote the green maintenance of inland waterways.

5.2   Carry Out Green Maintenance Demonstration Project

The channel maintenance work has obvious regional characteristics. The maintenance priorities and main contents of mountainous channels, plain river network channels and tidal channels are quite different. The channel maintenance of natural channels and artificial canals are also significantly diverse. Therefore, it is suggested to build green conservation demonstration projects, establish industry typical benchmarks, and build a typical case base of channel green conservation for different regions and different types of typical segments, so as to provide reference for other channels to carry out green conservation.

5.3 Form Green Maintenance System Specification

With the increase of technical maturity and the standardization of operation, it is necessary to gradually integrate the concept, technology and measures of green maintenance into the daily maintenance of inland waterways, clarify the technical requirements and specific measures of green maintenance in the annual maintenance plan, and ensure the capital investment. At the same time, it is necessary to revise and improve the existing industry standard technical specification for waterway maintenance JTS/T 320-2021), incorporate the technical measures related to green maintenance into the technical specification, and institutionalize and daily green maintenance of inland waterways.

5.4 Carry Out Research and Development of Green Maintenance Technology

In view of the maintenance needs of inland waterway, it is necessary to continuously improve and optimize the existing green maintenance technical measures, further develop and apply the maintenance technology with low pollution emission, low energy consumption and low ecological interference, so as to continuously enrich and expand the green maintenance technology system of inland waterway. For example, we can develop and build intelligent channel sensing system, and dynamically sense the change of channel water depth in combination with mobile facilities (such as UAVs) and fixed facilities (such as navigation aids and channel hydrological stations), so as to reduce energy consumption and pollution emission caused by channel survey. It is necessary to develop low impact dredging equipment and processes according to the characteristics of maintenance dredging sediment, so as to meet the water quality requirements of ecologically sensitive river sections such as water sources and protected areas.

References

Cao MX, Shen X, Ying HH (2018) Study on ecological structures of waterway regulation in the Yangtze River below Nanjing. Port & Waterway Engineering 1:1–11

Chang LH, Xu B, Zhang P, Tang W (2019) The influence of spur dikes for deep-water channel regulation on fish habitat. Journal of Water Conservancy 515(09):1086–1094

Chen DQ (2017) Study on green channel regulation buildings in the middle and lower reaches of the Yangtze River. The Fortune Times 10:144

Chen YR (2021) Practice and reflection on green waterway construction on jingjiang river section in the middle reaches of yangtze river. Yangtze River Technology and Economy 15 (02):51–53

Dai CL, Zhang JQ, Zheng YG, Zeng FT (1997) Experimental analysis on the impact of loading overflow construction of self-propelled trailing suction dredger on the environment of nearby waters. Water Transportation Engineering 10:59–62

Feng TH (2017) Numerical simulation of suspended solids diffusion in Yangtze River Channel Dredging. Green Technology 08:47–51

Fu G, Zhao DZ, Cheng HF (2011) Comparison and analysis of comprehensive utilization of dredged materialsat home and abroad. Port & Waterway Engineering 451(03):90–96

Ju T, Zhang TC, Wang ZT, Xie Y, Zheng CH, Wang KX, Wang D (2017) Characteristics of riprapping underwater noise and its possible impacts on the Yangtze finless porpoise. Acoustic Technology 36(06):580–588

Liao YT (2015) Exploration on Energy Conservation and Emission Reduction of Dredging Ships Based on the Improvement of Dredging Machines and Tools. Dalian Maritime University

Liu HH, Lei GP, Yin SR, Kuang HW (2016) Ecological measures and technical prospect of Yangtze River trunk channel regulation. Water Transportation Engineering 511(01):114–118

Ma Y, Cheng TJ (2014) Influence of waterway regulation engineering on water ecological environment and countermeasures. Port & Waterway Engineering 497(11):115–119

Ministry of transport. In 2035, China will basically build a "four vertical, four horizontal and two networks" national high-grade waterway of 25000 kilometers. [EB/OL].

Ministry of transport: JTS / T # 320-2021 technical specification for channel maintenance. People's Communications Press, Beijing (2021)

Ren XL, Li X, Xiang GS (2017) Thoughts on the development of new environmental protection technology of dredger. Ship Materials and Market 147(05):59–64

Wang KX, Wang D (2015) Analysis of Impact of Waterway Adjustment Activities on Yangtze Finless Porpoise and Mitigation Measures. Environmental Impact Assessment 216(03):13–17

Xie PJ, Ding ZY (2020) Analysis of ecological impact and Countermeasures in the construction of inland waterway dredging. China Water Transport 20(01):128–129

Yan CZ (2017) On strengthening environmental protection in navigation mark management. Navigation 228(03):59–60

Yan DH, Dou P, Cui BS, Xie T, Wang H (2018) Theoretical framework kand key issues in cological inland waterway construction. Journal of Beijing Normal University (Natural Science) 54(6):755

Yuan Y, Zhao L (2017) Comparative analysis of inland waterway maintenance development at home and abroad. China Water Transport 17(01):42–45

Zhao DZ, Liu J, Cheng HF, Wang ZZ (2013) Current situation and future prospect of dredged material disposal in the Yangtze estuary deepwater navigation channel. Hydro-Science and Engineering 2:26–32

Zuo JP, Chen YM, Zhou JX (2014) Discussion on ecological dredging of inland waterway based on ecological protection. China Water Transport 14(03):176–178

Shipper Response Surveys and their Importance in the Evaluation of U.S. Inland Waterways

Kevin P. Knight[✉]

Institute for Water Resources, US Corps of Engineers, Alexandria, VA, USA
Kevin.P.Knight@usace.army.mil

Abstract. Shipper response surveys have been growing in favor in recent years and if administered correctly, can help quantify shipper's behaviors and opinions which present themselves as transportation demand functions by mode of delivery (water, rail or truck). Demand functions, defined by economists as "numeric demand elasticities" or changes in quantity demanded as a result of changes in rates/duration/reliability, have enabled to the Corps to analyze (1) lock reliability and component failures; (2) demand projections vs. system capacity to determine equilibrium waterway traffic and (3) selection and sequence of replacement, repair or modernization efforts over the planning horizon.

This paper highlights this innovative approach in predicting shipper behavior using shipper response surveys that were recently applied to two major waterways in the U.S.– the Upper Mississippi River and Ohio Rivers) and how the resulting outputs could be used in to analyze lock investment priorities as well as impacts due to closures and other events.

Keywords: Inland navigation · Economic impact analysis · Shipper behavior · Lock and dam investments

1 Introduction

The U.S. Corps of Engineers (USACE) oversees the operations and maintenance of the U.S. Inland River System, which includes 27 rivers, 12,000 miles of inland river channels, 207 Lock chambers at 171 lock sites, all of which carry at least 850 billion tons of a wide variety of commodities each year while benefitting shippers, receivers, and communities (Fig. 1).

Fig. 1. The U.S. navigation system

Y. Li et al. (Eds.): PIANC 2022, LNCE 264, pp. 169–178, 2023.
https://doi.org/10.1007/978-981-19-6138-0_15

In the late 1990s and early 2000s, following several controversies in USACE's analyses of the Upper Mississippi River locks, the National Academy of Sciences was sought out to review USACE's draft analysis and recommended that they develop an approach to estimate the structure of transportation demands for use in USACE planning models, particularly by analyzing shipper behavior to changes in conditions to the waterways.

Following that time, USACE completed several studies that examined actual and hypothetical behavior on the Upper Mississippi and Illinois River Basins (Train and Wilson, 2004, 2005, 2007a, 2019), the Columbia-Snake Waterway (Train and Wilson, 2006a), the Ohio River (Train and Wilson, 2008c), and the Calcasieu River (Wilson et al., 2011) and more recently the Upper Mississippi River and Ohio River Restudy (Wilson, 2021).[1] In every case, survey methods focused on shippers of commodities that had a historical presence on the waterway and on shippers of varying distance from the waterway to capture the effects of space that are central to the decision to use the waterway. Using these survey data, demand models have been estimated that yield significant evidence that shippers do respond to changes in rates, time and reliability. The responsiveness is two-fold: shippers' discrete decisions (where and how to ship the product) and continuous decisions (the volume of shipments) are both embedded in most of the studies. In all cases, the analyses proved that shippers respond to changes in attributes that are affected by USACE infrastructure decisions.

The results of the survey were used to calculate transportation demand functions by mode of delivery which allow for the generation of demand elasticities used in waterway investments. Moreover, these elasticities were then incorporated into various models, such as the USACE Navigation Investment Model, which provided the following outputs:

- Event trees and hazard functions to model lock failures, closures, and repairs
- Demand curves by origin, destination, and commodity to define shipping forecasts
- Optimal configuration of tows for each movement, including selection of towboat, number of barges, and reconfiguration by river segment to minimize cost
- Congestion-based costs on a lock-by-lock basis and annual systemwide traffic equilibriums
- Automated calibration of the model to historical traffic using optimization routines
- Optimal selection and time sequencing of alternatives to improve capacity and reliability
- Determination of optimal lockage fees
- All these features help analyze (1) lock reliability and component failures; (2) demand projections versus system capacity to determine equilibrium waterway traffic; and (3) selection and sequence of replacement, repair, or modernization efforts over the planning horizon.

---

[1] These reports have also led to several papers published in academic outlets. There have also been numerous surveys conducted using these data, including Train and Wilson, 2006b, 2007b, 2008a, 2008b, and2009, and Sitchinava et al. 2005. The citations enumerated in the text contain the primary reports for each of the surveys conducted.

2 Role of Elasticities in Inland Navigation Benefits Analysis

The heart of transportation demand models involves the computation of demand elasticity, which is an economic principle that measures the extent of response to changes in quantity demanded because of a change in price. A variable is said to be elastic (having an absolute elasticity value greater than 1) when it responds more than proportionally to changes in price. In contrast, an inelastic variable (with an absolute elasticity value less than 1) is one which changes less than proportionally in response to changes in price. When there are substitutes available, the elasticity is likely to be higher, as one could switch from one good to another even if the price change is minor. By the same token, the more necessary a good is, the lower its demand elasticity, as people will attempt to buy it no matter the price. Common examples include insulin or a heart transplant. The same economic principle has been applied to carrier when faced with making decisions on various modes of transport (truck, rail or barge) to move its commodities.

Numerous studies have applied the principle of elasticity for various modes of public transportation including automobile usage where the estimated elasticity is between 0.01 and 1.26, urban transit with an estimated elasticity between 0.01 and 1.32, airline travel with an estimated elasticity between 0.36 and 4.60, and rail travel with an estimated elasticity between 0.12 and 1.54 (1). Equally important has been the estimation of freight transportation, i.e. the movement of commodities. The elasticity estimates of these various modes are found to depend heavily upon the commodity being transported but with general elasticity estimates of: rail transportation with an estimated elasticity between 0.02 and 3.50 and motor carrier transportation with an estimated elasticity between 0.14 and 2.96 (1). By examining the response to changes in cost, the cost of alternative movements (in other words) benefits to the waterways can be estimated (Wilson, 2004).

According to the Corps' Planning Guidance Notebook ER1105-2-100, National Economic Development benefits are *"the difference in costs of mode transport between the without-project condition (when rails, trucks or different waterways or ports are used) and the with-project condition (improved locks, waterways or channels). The economic benefit to the national economy is the savings in resources from not having to use more costly mode or point of transport. (Additional benefits can be realized by reductions in costs incurred from trip delays (e.g. reduction in lock congestions), reduction in costs associated with the use of larger or longer tows, and reduction in costs due to more efficient use of barges."*

3 Survey Description

All data used in the recent Ohio and Upper Mississippi River analyses were generated from a survey of agricultural shippers using a survey instrument approved by the U.S. Office of Management and Budget and supplemented with information from the Surface Transportation Board's Carload Waybill Statistics, USACE river distances, and Google map distances. The goal of the research was to gather data that pertain to all agricultural shippers that could conceivably ship on Ohio River and Upper Mississippi River Basin waterways. Figures 2 and 3 show the shipper locations and destinations.

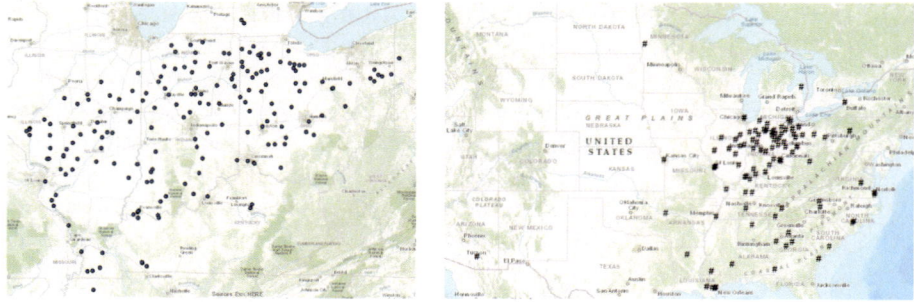

Fig. 2. Shipper locations for the ohio study shipper destinations for the ohio study

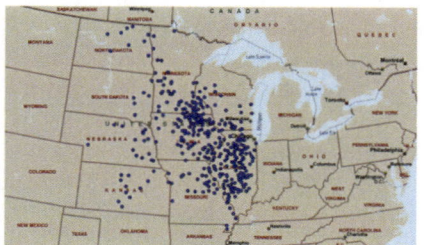

Fig. 3. Shipper locations - upper miss study shipper destinations – upper miss study

More specifically, several thousand shippers were sampled. The sample was stratified by distance from the waterway, with a higher percentage for those located closest to the waterway. For both river systems, corn shipments dominated the sample, representing 65% and 47% for the Upper Mississippi River and Ohio River System, respectively. The remaining commodities included but were not limited to wheat, soybeans, beans, and grain.

Survey questions included grain elevator age, loading capacity, shipment information, last shipment, and shipping alternatives. The meat of the survey contained questions on alternate modes: For example, "If rates went up by X percent, how likely would you switch from water to rail or truck, etc.? (Fig. 4)" Questions were repeated for changes in time and changes in reliability.

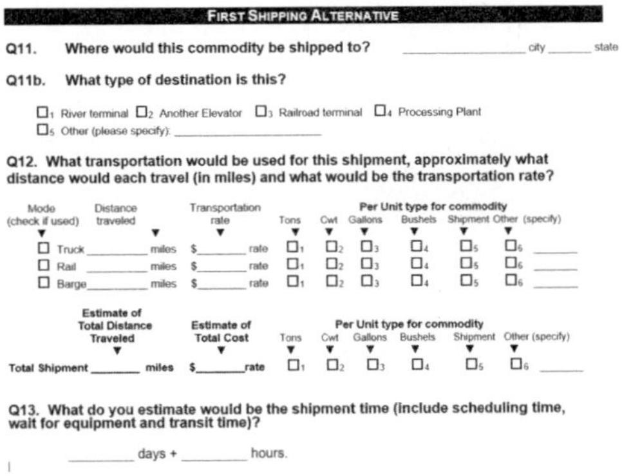

FIRST SHIPPING ALTERNATIVE

Q11. Where would this commodity be shipped to? ______ city ______ state

Q11b. What type of destination is this?

☐1 River terminal ☐2 Another Elevator ☐3 Railroad terminal ☐4 Processing Plant
☐5 Other (please specify): ______

Q12. What transportation would be used for this shipment, approximately what distance would each travel (in miles) and what would be the transportation rate?

Mode (check if used)	Distance traveled		Transportation rate		Per Unit type for commodity: Tons	Cwt	Gallons	Bushels	Shipment	Other (specify)	
☐ Truck	______	miles	$______	rate	☐1	☐2	☐3	☐4	☐5	☐6	______
☐ Rail	______	miles	$______	rate	☐1	☐2	☐3	☐4	☐5	☐6	______
☐ Barge	______	miles	$______	rate	☐1	☐2	☐3	☐4	☐5	☐6	______

	Estimate of Total Distance Traveled		Estimate of Total Cost		Per Unit type for commodity: Tons	Cwt	Gallons	Bushels	Shipment	Other (specify)	
Total Shipment	______	miles	$______	rate	☐1	☐2	☐3	☐4	☐5	☐6	______

Q13. What do you estimate would be the shipment time (include scheduling time, wait for equipment and transit time)?

______ days + ______ hours.

Fig. 4. Portion of shipper response survey

4 Revealed Vs. Stated Preference

Revealed data reflect actual decisions made by shippers is very useful in predicting shipper behavior. However, many believe that revealed data's attributes do not provide a large enough range of data to identify the parameters of interest. Rates per ton-mile, times-in-transit, and reliability each have similar values between the *chosen* and *next-best* alternative. Because of limited variation in such statistics, there has been a growing literature on *stated preference* modeling. A stated preference survey presents survey respondents with a set of hypothetical states, and then solicits a preference. This approach considerably simplifies analysis and the difficulty of collecting survey responses to confidential information. However, it is criticized as being based on hypothetical situations instead of real-world decision-making. This *stated preference* approach differed from the standard approach in that the stated preference questions were grounded in the revealed decisions made. In particular, survey recipients were asked what they did and what they would do if the chosen alternative were not available. This was taken as their next best alternative. The stated preference questions ask if the shipper would stay with their choice, switch to a different option, or shut-down if an attribute change. For example, for the last shipment, if the attribute changed x percent, would you continue with the original mode and destination or switch to your best alternative choice?). This framing of the question grounds the decision- making not to hypothetical alternatives, but rather to alternatives commonly confronted by the individual making the decision.

In the survey, three such questions related to **rate, time, and reliability.** The percentage change was randomly offered to each and ranged from 10 to 60%, generating a range of values over which to identify the parameters of the profit-function on which decisions are made. In addition, if the shipper did not switch, they were asked what level of the attribute would induce a switch with outcomes presented in Tables 1, 2, and 3.

For the Ohio River Shipper Survey Report (and similarly applied to the Upper Mississippi River Report), 6 rate changes, from a 10 to 60% increase in rates, were used in the survey. A total of 319 responses were observed. At low values of rate changes, 66% of responses indicated they would not switch to the alternative. As the rate change increased, this proportion fell. However, even for large rate increases, 45% (30 of 66) of respondents report they would still not switch. If they would switch, there were two alternatives selected. First, they could switch to their next best mode/destination. At various rate changes, there were a total of 128 such switches. Second, they could switch to *shut-down*. Shut-down is and has been a major factor in all of the surveys conducted discussed in the literature. In this sample, 22 of 319 (7%) reported that they would shut-down at the rate increase prompt. As expected, switching to an alternative tends to increase with the magnitude of the rate change.

Table 1. Shipment stated preference – rate responses

% Rate Change	No Switch	Switch	Shut-down	Total	% No	% Switch	% Shut-down
10	54	26	2	82	66	32	2
20	42	23	6	71	59	32	8
30	22	28	4	54	41	52	7
40	14	14	2	30	47	47	7
50	8	7	1	16	50	44	6
60	29	30	7	66	44	45	11
Total	169	128	22	319	53	40	7

The same information with respect to increases in transit time was examined, with transit times defined, again, as including the setup and waiting times and the time once loaded to reach the final destination. There were 316 responses. If time changes, shippers report that a total of 177 (56%) shipments would not change regardless of the time change. As with rates, switch rates generally increase with progressively higher changes in transit times.

The same information as in Tables 1 and 2 with respect to reliability is presented in Table 3, with a total of 412 responses. The same general pattern as with rate and time is indicated (as expected). For decreases in reliability, the switch rate increases with the percentage change in reliability. Generally, Tables 10, 11, and 12 each follow expectations. Further, shippers appear to be more responsive to rates than to time and reliability, particularly for large rate changes.

Table 2. Shipment stated preference – time responses

% Time Change	No Switch	Switch	Shut-down	Total	% No Switch	% Switch	% Shut-down
10	45	16	5	66	68	24	8
20	49	20	2	71	69	28	3
30	30	22	5	57	53	39	9
40	21	22	4	47	45	47	9
50	19	23	3	45	42	51	7
60	13	13	4	30	43	43	13
Total	177	116	23	316	56	37	7

Table 3. Shipment stated preference – reliability responses

% Reliability Change	No Switch	Switch	Shut-down	Total	% No	% Switch	% Shut-down
10	26	4	1	31	84	13	3
20	43	23	8	74	58	31	11
30	34	17	3	54	63	31	6
40	13	9	4	26	50	35	15
50	23	20	2	45	51	44	4
60	33	42	9	84	39	50	11
Total	172	115	27	314	55	37	9

5 Major Report Findings

1. The choice models indicated statistically important responses of shippers to changes in the rates, reliability, and distance. These responses also differed by shipper attributes that included rail car loading capacity and storage capacity.
2. There were important differences in the responses for truck, rail, and barge shipments.
3. Many firms reported limited alternatives in their choice of mode and destination, and many reported that they would shut down in the presence of rate increases or if the chosen alternative was taken away. Unlike previous studies conducted under NETS, the effect of a shut-down alternative was reflected in the choices and explicitly captured in the models of switching behavior.
4. Arc elasticities were calculated for each mode and shipment attribute. Demand was found to be inelastic; that is, the arc-elasticities were all less than one in magnitude.
5. The rate demand elasticities were all inelastic. Barge elasticities ranged from −0.41 to −0.48; rail elasticities ranged from −0.22 to −0.12, and truck elasticities ranged from −0.29 to −0.26.
6. The time demand elasticities were all inelastic, and smaller than rate elasticities. Barge time elasticities ranged from −0.287 to −1.067; and truck elasticities ranged from −0.19 to −0.76.
7. The reliability elasticities were all inelastic and rest between those of rate and time elasticities (in magnitude). Barge reliability elasticities ranged from 0.29 to 1.067; rail elasticities ranged from 0.16 to 0.36 and truck elasticities ranged from 0.19 to 0.76.
8. Annual volume demand elasticities were also estimated for rate, time and reliability. The responses of shippers often pointed to no change in annual volumes from a change in an attribute. A Heckman (1979) model was, therefore, used to estimate the model. The results suggest that shippers with large storage capacities were more likely to adjust volumes in response to rate changes than for smaller capacities. Given a change does occur, the change is driven largely by the level of the change in the attribute. That is, the elasticities conditioned on a change occurred did not vary with shipper attributes or commodity, but, whether or not a change occurred depended on shipper attributes.

9. The Heckman model allows the calculation of two different elasticities. These are a conditional elasticity (given a shipper's volume changes) and an unconditional elasticity (where a shipper's volume may or may not change). By definition, the former is larger in magnitude than the latter for each attribute. In some cases, annual volumes, given a change in volume, were responsive to changes in attributes. However, generally the unconditional elasticities were less than one in magnitude, pointing to relatively inelastic demands.

10. Two different rate elasticities are presented – one in which the shipper and its competitors face the same rate change, and one in which the shipper but not its competitors face a rate change. The elasticities calculated from the former are much smaller in magnitude than those calculated from the latter, and in the latter case, there are some unconditional elasticities greater than one in magnitude for the median shipper. For some rate change levels, the conditional elasticities are greater than one in magnitude. This suggests that if there is a rate change that induces a volume change, the change is relatively responsive.

11. Both time in transit and reliability elasticities are nonzero, a finding that suggests shippers do adjust annual volumes to these shipment attributes. As with rates, the unconditional elasticities are less than one in magnitude.

6   Conclusion

This report continues a series of demand studies aimed at providing shipper-level information that can be used by USACE to evaluate the benefits of waterway improvements. The shipper-based surveys that have been developed and modified over the last several years are designed to collect information on shipper and shipments. These data, in turn, are used to estimate the responsiveness of mode, destination choices, and annual volumes to changes in rates, time in transit, and reliability.

The choice models were estimated with a logit methodology applied to both revealed and stated preference data. The results suggest that while demands are responsive to changes in rates, time in transit, and reliability, the response is somewhat small and points to relatively inelastic demands (i.e., demand elasticities less than one in magnitude). The annual volume models were estimated with a Heckman selection model using stated preference data. Generally, the results suggest that shippers respond to rates and time in transit, but as with the choice models, the response is somewhat small, with most elasticities less than one in magnitude.

The demand functions appear to be reasonably steep and point to a large degree of captive shippers (i.e., shippers that do not switch to alternatives even for large changes in the attributes). While this result points to relatively large benefits to infrastructure investments, there are limits. A novelty of this research is the incorporation of the option of no longer shipping (i.e., shutting down). This finding has been a consistent theme throughout this line of research. In the present case, the option to shut down is explicitly represented in the choice model. Hence, attributes, particularly rates, cannot increase without bound because eventually shippers will opt out of the market. This reaction places limits on the benefit calculations necessary for USACE planning models. The resulting elasticities are used widely in the analysis of lock closures, delayed investments and other challenges to the inland waterway system.

References

Heckman JJ (1979) Sample selection bias as a specification error. Econometrica 47:153–161

National Research Council: Inland navigation system planning: the upper Mississippi River-Illinois Waterway. National Academies Press (2001)

National Research Council: Review of the US Army Corps of Engineers Restructured Upper Mississippi River-Illinois Waterway Feasibility Study: Second Report. National Academies Press (2005a)

National Research Council: Water Resources Planning for the Upper Mississippi River and Illinois Waterway. National Academies Press (2005b)

Sitchinava, N., Wilson, W., Burton, M.: Stated Preference Modeling of the Demand for Ohio River Shipments. IWR draft report (2005). Available at http://www.nets.iwr.usace.army.mil/docs/other/OhioStatedPreferenceDemand.pdf

Train K (2003) Discrete Choice Methods with Simulation. Cambridge University Press

Train K, Wilson WW (2004) Shippers' Responses to Changes in Transportation Rates and Time: The Mid-American Grain Study. IWR Report 04-NETS-R-02. Available at https://planning.erdc.dren.mil/toolbox/library/IWRServer/04-NETS-R-021.pdf

Train K, Wilson WW (2005) Econometric Analysis of Stated-preference Experiments Constructed from Revealed preference Choices. IWR Report 05-NETS-P-08. Available at https://planning.erdc.dren.mil/toolbox/library/IWRServer/05-NETS-P-08.pdf

Train K, Wilson WW (2006a) Transportation Demands in the Columbia-Snake River Basin. IWR Report 06-NETS-R-03. Available at https://planning.erdc.dren.mil/toolbox/library/IWRServer/06-NETS-R-03.pdf

Train K, Wilson WW (2006) Spatial demand decisions in the pacific northwest: mode choices and market areas. Transportation Research Record: Journal of the Transportation Research Board 1963:9–14

Train K, Wilson WW (2007a) Transportation Demands for the Movement of Non-Agricultural Commodities Pertinent to the Upper Mississippi and Illinois River Basin. IWR Report 07-NETS-R-1

Train K, Wilson WW (2007) Spatially generated transportation demands. Research in Transport Economics: Railroad Economics 20:97–118

Train K, Wilson WW (2008) Estimation on stated-preference experiments constructed from revealed preference choices. Transportation Research Part B: Methodological 42(3):191–203

Train K, Wilson WW (2008) Transportation demand and volume sensitivity: a study of grain shippers in the upper mississippi river valley. Transportation Research Record 2062:66–73

Train K, Wilson WW (2008c) The Demand for Transportation in the Ohio River Basin. Report submitted to the USACE NETS program

Train K, Wilson WW (2009) Monte carlo analysis of SP-off-RP data. Journal of Choice Modeling 2(1):101–117

Train K, Wilson WW (2019) Demand for the Transportation of Agricultural Products: An Application to Shippers in the Upper Mississippi and Illinois River Basins. Report to the USACE

USACE (2000) Planning Guidance Notebook. Engineer Regulation (ER) 1105-2-100. Available at https://www.publications.usace.army.mil/Portals/76/Publications/EngineerRegulations/ER_1105-2-100.pdf

Wilson WW, Campbell M, Gleasman W (2011) 2010 Shipper Response Models for the Calcasieu Lock and GIWW-WEST. Report submitted to USACE

Winston C (1983) The Demand for Freight Transportation: Models and Application. Transportation Research – Part A **17A**, 419–427

Winston C (1985) Conceptual developments in the economics of transportation: an interpretive survey. Journal of Economic Literature 23(1):57–94

Shoreside Power at Berths for Inland Navigation Vessels – How to Make Available a Harmonised System of Shoreside Power Access on the Rhine to Reduce Air and Noise Pollution

Raphaël Wisselmann[(✉)] and Kai Kempmann

Central Commission for the Navigation of the Rhine, Strasbourg, France
{R.Wisselmann, k.kempmann}@ccr-zkr.org

Abstract. The Central Commission for the Navigation of the Rhine (CCNR) is an international organisation that exercises an essential regulatory role in the navigation of the Rhine. It is active in the technical, legal, economic and environmental fields. In all its areas of action, its work is guided by the efficiency of transport on the Rhine, safety, social considerations, and respect for the environment. Many of the CCNR's activities now reach beyond the Rhine and are directly concerned with European navigable waterways more generally.

So as to take into particular account the challenges of climate change, in 2018 the transport ministers of the Member States of the Central Commission for the Navigation of the Rhine (Germany, Belgium, France, the Netherlands, Switzerland) signed a declaration, the so-called Mannheim Declaration, to reassert the objective of largely eliminating greenhouse gases and other pollutants by 2050 and to task the CCNR to develop a roadmap to achieve these goals. These goals that are intended to protect the environment and the climate concern not only inland navigation vessels' propulsion systems but also the on-board power supply for operating machinery, for example when at berth.

Last but not least, the conflicts over berths on the Rhine in city centres demonstrate that joint efforts are required to reduce or largely eliminate both pollutant and noise emissions. Supplying inland navigation vessels with shore power can play an important role in reducing emissions and noise and helps achieve the objectives of the Mannheim Declaration, while also securing attractive city centre berths for future generations of boatmen. Together with its stakeholders, the CCNR works on a regular basis to identify technical and regulatory gaps in standards and provisions, and proposes activities that aim at a harmonised implementation of shoreside power infrastructure on the Rhine.

Keywords: Inland navigation · Shore power · Berth · Infrastructure · Emission reduction

Y. Li et al. (Eds.): PIANC 2022, LNCE 264, pp. 179–185, 2023.
https://doi.org/10.1007/978-981-19-6138-0_16

1   Shoreside Power and the Significance for Emission Reduction

As with all other transport modes, inland navigation is confronted with the changing climate and the required adaptation measures to support climate change mitigation and adaptation. Addressing the issue of climate change is a political top priority.

In their Declaration signed in Mannheim on 17 October 2018, the so-called Mannheim Declaration [1], the transport ministers of the Member States of the Central Commission for the Navigation of the Rhine (CCNR - Germany, Belgium, France, the Netherlands, Switzerland) also reasserted the objective of largely eliminating greenhouse gases and other pollutants by 2050.

To further improve the environmental sustainability of inland navigation on the Rhine and inland waterways, the Mannheim Declaration tasked the CCNR to develop a roadmap to:

- reduce greenhouse gas emissions by 35% compared with 2015 by 2035,
- reduce pollutant emissions by at least 35% compared with 2015 by 2035, and
- largely eliminate greenhouse gases and other pollutants by 2050.

However, these targets for protecting the environment and climate concern not only inland navigation vessels' propulsion systems, but also the on-board power supply for operating machinery, for example when at berth. Furthermore, conflicts over berths in city centres demonstrate that joint efforts are required to reduce or largely eliminate both pollutant and noise emissions.

2   The Role of Berths for Inland Navigation

A berth is a designated location in a port, terminal or along the waterway used for mooring vessels. Berths provide safe mooring and facilitate embarking and disembarking for the crews and family members, as well as the loading and unloading of their cars. They are thus important elements of the inland navigation system. In addition to the economic aspects, they also have an important role for the social sustainability of inland navigation. Contrary to maritime navigation, inland vessels are often operated by entrepreneurs who live with their family on board the vessel in private accommodation. To ease social life participation for the crews and their families, berths should not only be equipped with gangways and car dropping facilities, but also be connected to public transport and offer services such as the internet, communication and shore power as well fresh water supplies and waste disposal at strategic locations.

3   The Role of Electricity in Rhine and Inland Navigation

3.1   Electricity for Propulsion

The CCNR roadmap [3] outlines transition pathways for the fleet (existing and newbuilds), addressing the roles that will be played by different technological solutions in the energy transition challenge. For the propulsion of the vessel, it is expected that battery electric propulsion will become important, for example for ferries, but also for smaller motor cargo and tanker vessels. Providing shore power at berths is thus one element among others in the CCNR roadmap towards eliminating greenhouse gas emissions and air pollutants from inland navigation by 2050. Indeed, measures must be taken on shore to deliver these new energy carriers, notably electricity!

3.2   Electricity for Vessel Operation at Berth

The CCNR noted that, at more and more berths along the Rhine, energy generation with on-board diesel generators are no longer welcome for pollutant and noise emission reasons. The CCNR has responded by amending the Rhine Police Regulations (RPR) with a new mandatory sign, thus enabling competent local authorities to target and avoid possible emissions (Fig. 1).

Fig. 1. New mandatory sign in article 7.06 of the RPR

Vessels berthed where the sign is displayed are required to use either emission-free onboard systems such as battery packs or connect to shore power. While at berth, the boat master is required to cover all the electrical power requirements from this shore-based power system. Whether this sign is displayed or not is for the competent local authorities to decide. These authorities can provide for variations whereby, for example, this obligation only applies at night.

The requirements for shore power facilities differ depending on which types of vessels are connected. Passenger vessels, in particular cabin vessels, have the highest energy demand and therefore require other installations than, for example, dry cargo vessels. The difference in energy demand is high as are the requirements on the shore power connection infrastructure.

4   Standards and Gaps – Important Aspects to Address

The CCNR has assessed the status of the development of shore-side and ship-side European standards. Stakeholders have highlighted that not only an international standardisation of the shore power connection is necessary, but equally a standardisation of the operating and payment system, and a closing of the standardisation gaps that still exist in their view, for example for electrical connections with currents between 125 and 250 amperes.

Currently the following standards and gaps exist in Europe:

- Connection systems (EN 15869-1:2019 up to 125 amps and EN 16840:2017 from 250 amps) - but there is no standard for currents between 125 and 250 amps. Other proprietary standards also exist, for example, the Powerlock system used for supply of vessels with higher energy demand, such as cabin vessels at berth.
- Operation of shoreside power connections - no uniform concept exists. The standard EN 15869-1:2019 only stipulates that an instruction manual must be attached externally. However, the shoreside power connections differ in their form and handling. There is a need to agree on a harmonized concept at least for handling of shoreside power infrastructure.
- Payment system - standards exist but no uniform concept for paying or respectively billing. Various payment systems exist in parallel, such as prepaid cards, debit and credit cards, fleet cards with RFID technology as well as apps and websites. There is a need to agree on a commonly accepted payment method (Figs. 2 and 3).

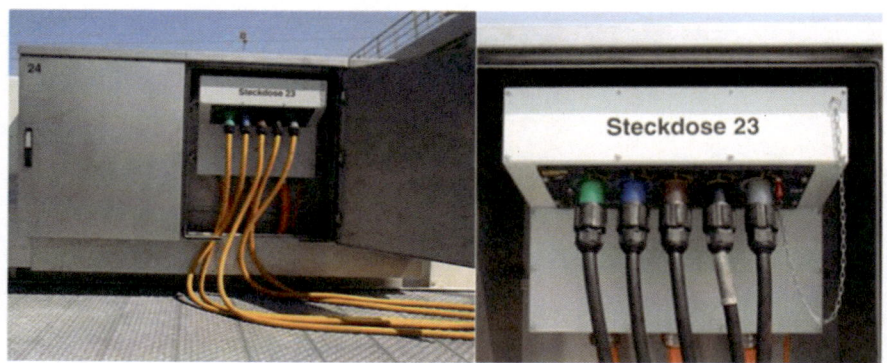

Fig. 2. Powerlock system (© Port of Switzerland)

Fig. 3. EN 15,869 type system (© North Sea Port)

5 Implementation of Shoreside Power – Important Aspects to Consider

There is a need for public policy to put infrastructure into place. This policy needs to address funding and financing issues to support the implementation, but also the location of such facilities to achieve an appropriate distribution along a corridor. In addition, the formation of partnerships for the development of a harmonised network of shoreside electricity connections has been proposed. Hence, activities should be organised to collect and share experiences from implemented pilot projects. The dialogue between providers and users should be continued in order to achieve a coordinated implementation, especially at river basin or, respectively, at corridor level.

There is a need for clarification on operational, technical and practical aspects, not only among users, but also among providers of shore power:

- Exchange of information at the ship/shore interface. The ship-side requirements and boundary conditions must be mutually considered and jointly developed with those of the shore-side infrastructure.
- It might often be unclear whether sufficient electricity can be offered at the berth. This depends crucially on the shoreside electricity grid and the energy supplier.
- Availability of shoreside power infrastructure is not only the mere presence of a shoreside power connection, but also its reliable functioning and provision of services, for example help in fixing system malfunctions.
- Systems for shoreside power connections should be designed in such a way that they might also be used as a platform for other services, e.g. to manage drinking water, waste disposal and access to the internet.

Focus shall also be also placed on the users of the shoreside electricity infrastructure. Special requirements for the qualification of crew members (e.g. minimum qualification regarding electrical equipment, training in first aid) should be defined if not already stipulated as, for example, in the European standard ES-QIN. Connecting and operating the shoreside power system is an activity that the personnel must be introduced to by means of instruction or training. Occupational health aspects must be taken into account in advance when drawing up a job description. Cables are not only laid and connected during the day, but also in bad weather conditions, e.g. in darkness, rain and cold.

6 Conclusions – Important Elements for the Way Ahead

In 2022, the CCNR organised a workshop [4], in cooperation with viadonau, to jointly advance the interests of Rhine navigation and European inland navigation, on the topic of regional coordinated implementation. As a platform for dialogue with stakeholders, the workshop provided important impulses on the various issues and initiated discussions among all those involved in waterway transport.

There is consensus among the representatives of the inland navigation sector that berths must be thought of as a part of the overall inland navigation system. Berths have an important role for navigation, especially for the crew. In this context, people's needs should be placed at the centre of the inland navigation system.

As an important milestone, the CCNR drafted an action plan. It focusses on the above identified elements of standardisation, availability, implementation, provider as well as user aspects and proposed short-, medium- and long-term supporting measures, such as:

- Clarification of the identified gap in the standards in coordination with competent standardisation institutions.
- Collection of good practice examples of operating concepts to be added to CCNR's collection of examples of berths [2].
- Continuation of the dialogue in the form of workshops or round tables to identify further standardisation needs and exchange with standardisation institutions with regard to the ship/shore interface.

Further developments in inland navigation will require the implementation of a monitoring of the required electric currents, in particular related to the emerging battery-electric propulsion systems since the charging of batteries has a higher energy demand than for the operation of vessels at berth. There are two possible solutions to charge batteries, either at shore (exchange of battery containers at berth) or quick charging at locks or berths. The latter creates a high load for the energy grid and is thus to be developed in close coordination with energy producers and grid providers.

Battery-electric propulsion systems and accumulators for self-sufficient power supply bear the risk that shoreside power connections might become a bridging technology. The vessels energy demand for the operation at berth could be met by the onboard batteries. Thus, a shoreside power infrastructure may no longer be required.

To avoid dead-end investments, shoreside power infrastructure should be planned in the most flexible way and as service-oriented as possible to allow adaptation to future needs, such as a service platform. These service platforms could then not only be used for shoreside power but also for giving access to water, internet, communication and other services.

A great deal of complex work lies ahead for the decision-makers and planners of the Rhineriver. The challenges should be tackled as a whole and the solutions coordinated at international and interdisciplinary levels. Inland navigation plays an important role in globally mitigating climate change and must therefore be fully supported.

References

1. CCNR: Mannheim Declaration. 150 years of the Mannheim Act – The driving force behind dynamic inland navigation (2018). https://www.zkr-kongress2018.org/files/Mannheimer_Erklaerung_en.pdf
2. CCNR: Collection of examples on the need and equipment of berths. Collection of communications from CCNR delegations (2020). https://www.ccr-zkr.org/13020600-en.html

3. CCNR: CCNR roadmap for reducing inland navigation emissions(2021). https://www.ccr-zkr.org/files/documents/resolutions/ccr2021-II-36en.pdf
4. CCNR: Workshop on "Shore power at berths" (2022). https://www.ccr-zkr.org/13020155-en.html

Simulation Study to Assess the Maximum Dimensions of Inland Ships on the River Seine in Paris

Marc Mansuy[1(✉)], Maxim Candries[1], Katrien Eloot[2], and Sebastien Page[3]

[1] Maritime Technology Division, Ghent University, Ghent, Belgium
{marc.mansuy, maxim.candries}@ugent.be
[2] Flanders Hydraulics (FH), Antwerp, Belgium
katrien.eloot@mow.vlaanderen.be
[3] IMDC, Antwerp, Belgium
sebastien.page@imdc.be

Abstract. Traversing the river Seine in Paris is challenging for inland navigation vessels due to the density and diversity of local traffic and the variety of manoeuvres to be encountered in a confined environment. The waterway authority, Voies Navigables de France (VNF), commissioned a study to assess the relevance of the current regulations when present and future vessels of varying types and dimensions cross Paris. This paper describes the use of fast time and real time simulations to assess the maximum dimensions of ships crossing the Seine in Paris. In a first phase, fast time simulations were executed with a track controller, which allowed to identify bottlenecks on the full length (12 km) of the river Seine in Paris. Based on those results, critical scenarios were selected to be tested on a full mission bridge simulator by skippers familiar with the crossing of Paris. Inspired by PIANC INCOM WG 141 Detailed Design and Safety and Ease Approaches, the main challenges related to the simulation setup and the assessment methodology are presented and discussed in this paper. The simulations have shown that the main bottlenecks are related to the succession of passages under narrow bridges with non-aligned openings. The maximum water levels for which safe passage is possible, were determined for each ship type and compared with the existing regulations. Finally, recommendations were formulated, which were then discussed with VNF, end users and stakeholders.

Keywords: Transport of containers · River · Bridges · Manoeuvring simulator

1 Introduction

The river Seine is a major axis of the French inland waterway transport network with a high traffic density. Traversing the city of Paris is challenging for inland navigation vessels due to the variety of manoeuvres involved. In the area of the two isles, the main artery of the waterway is passing in between the two isles (south of the Ile Saint-Louis and north of the Ile de la Cité), so that larger ships have to deal with sharp bends in

© The Author(s) 2023
Y. Li et al. (Eds.): PIANC 2022, LNCE 264, pp. 186–200, 2023.
https://doi.org/10.1007/978-981-19-6138-0_17

between the two islands. On top of that, there is a high number of historically important bridges, where traffic has to pass underneath narrow arches, while taking into account delicate current conditions on a bending trajectory (see Fig. 1).

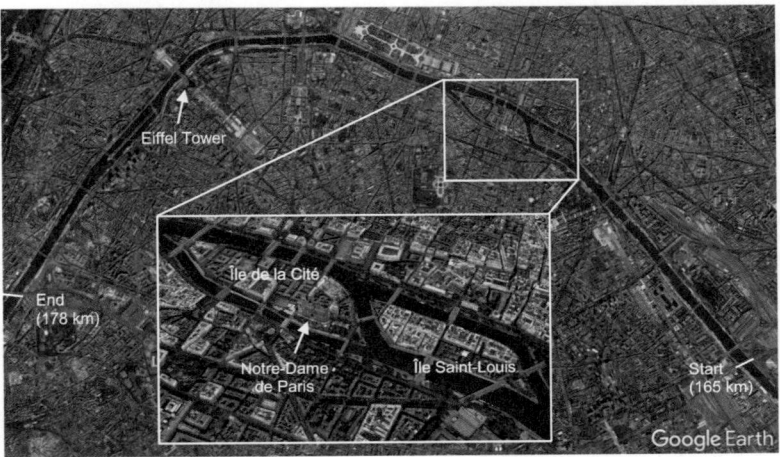

Fig. 1. Study area (12 km): river seine crossing the city of Paris, France.

Some decades ago, regulations concerning the maximum ship dimensions depending on the water level of the river were put in place to ensure the safety of navigation. However, with increasing capacity demand and with new types of ships, the question arose whether the regulations are still up to date and whether the safety is sufficient to increase the traffic and ensure the competitiveness of inland waterways transportation.

Specific nautical studies executed in the past on short stretches of the Seine river have been consulted. Some recommendations about the navigation of 135 m × 11.4 m ships between the two islands are given by CETMEF (2010) based on full-scale tests. Recommendations for 105 to 125 m long and 11.4 m wide ships in the same area are given by DNT (2016) based on navigation simulations. Navigation simulations executed by CETMEF (2013) provided information about the navigation conditions around Alexandre III bridge. Moreover, DST (2013) gave some recommendations about turning manoeuvres of 135 m long passenger ships downstream of Grenelle island, in the southwest of Paris. However, a comprehensive study assessing the maximum length of ships able to cross Paris under different hydraulic conditions had never been conducted. To this end, the waterway authority (VNF) commissioned a study to assess the navigation in Paris under different hydraulic conditions for the largest ships that are expected to cross Paris. VNF therefore requested to concentrate on the influence of length and ship type for 11.4 m wide ships.

Design guidelines based on three different approaches were recently published by PIANC (2019) with the vision of optimization of inland waterways dimensions based on local constraints, present and future fleet. A first step in the design or upgrade of an existing waterway is to use national guidelines. If no national guidelines are applicable, the PIANC guidelines provide recommendations for the dimensions of fairways (Concept Design) depending on the so-called Safety & Ease level which is aimed by the waterway authority.

This first approach has some limitations, e.g., it is not applicable in rivers with high flow velocities. Existing examples can then be used as a reference if the situation is comparable to the one studied (Practice Approach). When the situation is too different and large uncertainties remain or if environmental, local constraints limit the dimensions of the waterway, a third approach (Detailed Design) is recommended.

Under bridges, PIANC recommends guaranteeing a minimum height on the total width of the fairway with additional safety distance for collision risk. However, most of the waterways in France were designed before any regulations about air draft had been put in place (Deplaix and Daly 2019) and the fleet has significantly changed over the last decades. Therefore, the minimum height under arched bridges would only be guaranteed over a very narrow width if applied in Paris. Moreover, no guidelines are given for rivers with significant flow velocities.

A ship manoeuvring simulator is a useful tool to reproduce the passage under bridges in specific hydro-meteorological conditions. Söhngen and Butterer (2015), for example, simulated the passage of a 135 m long motor vessel within the width delimited by two piles of a flat bridge. Real time simulations are also used extensively to evaluate the accessibility in bends, as for example is the case for the Nord-Pas-de-Calais network in France (Adams et al. 2019) and the Upper-Seascheldt in Belgium (Eloot 2015).

This paper describes the use of simulations to assess the operational limits based on ship length on the river Seine inside the city of Paris and is illustrated by a selection of the main scenarios of navigation with motor ships in bends and under non-aligned narrow and low arched bridges. Section 2 describes the simulation setup. Section 3 presents the methodology applied to assess the safety of the manoeuvres and the main findings. The challenges and limitations of the methodology are discussed in Section 4. In Section 5, conclusions are given.

2 Simulations Setup

2.1 Manoeuvring Simulators

Several hundreds of scenarios with different ship types (fret and passengers) were simulated in Paris on three full mission manoeuvring simulators at Flanders Hydraulics Research. The simulators are dedicated for research studies and training. The main simulator is composed of a bridge with 360° aerial view of the surroundings projected on a cylindrical screen as shown on Fig. 2. A second simulator was used to simulate the navigation of large passenger ships and coupled to the first one to simulate ship meetings in bends. A third simulator was used to simulate the navigation of a small hydrofoil craft. The bridge of all simulators is equipped with:

- ECDIS and radar;
- Controllable camera views;

- Controllable bridge height;
- Propulsion and steering controls adapted to each ship type.

Fig. 2. Main full mission bridge simulator with 360° view at FH.

2.2 Waterway Environment

A total length of 12 km of the river Seine crossing Paris was modelled in 3D for this study. The 3D environment is divided into two independently designed parts:

- 3D external view: this is the visible part of the environment above the waterline (see Fig. 3). This part is projected on screens and allows the skipper to orient himself. The visual aspect of the external environment created is of medium resolution except for bridges which are accurately reproduced (± 10 cm) from original plans.

Fig. 3. 3D external view of the non-aligned bridges.

- Bathymetry: this is the part under the waterline. It was reproduced from the bathymetric data and influences the hydrodynamic behaviour of the ship.

2.3   Hydraulics Conditions

The current was implemented with TELEMAC, which is a software package developed by IMDC that resolves the 2D shallow water equations to model the water flow (Breugem 2020). The mesh has a resolution of 10 m with a refinement of 5 m close to the banks and 2 m around the bridge piles. To obtain approaching current velocities at the depth corresponding to the draft of the ships, a correction factor based on a logarithmic distribution of the velocities was applied. Hydraulic conditions from low water (0.82 m measured at the reference station Austerlitz) to the maximum water level for which navigation is currently allowed (4.30 m measured at Austerlitz) were modelled with an increment of 0.10 m. The significant variation of the water level along the study area (i.e. longitudinal profile) was also implemented. The modelled flow fields were later tested and validated on the manoeuvring simulator by experienced skippers.

2.4   Ship Models

The ship models listed in Table 1 were all implemented in the simulators used in the study. The manoeuvring behaviour of each ship model is determined by a mathematical model which computes:

- hydrodynamic forces, propulsion and steering forces, shallow water effects, restricted water effects;
- aerodynamic forces;
- interaction with encountering and overtaking target vessels.

Table 1. Ship models

Ship type	Transport	ECMT class	Length	Beam	Draft	Air draft
Push convoy	bulk	Vb	180	11.4	1.70 / 2.80	2.90 / 4.00
Push convoy	bulk		150	11.4	1.70 / 2.80	2.90 / 4.00
Motorship	2 layers cont. / bulk		135	11.4	1.70 / 2.80	2.90 / 4.00 / 3.65 / 4.75
Motorship	2 layers cont. / bulk		125	11.4	1.70 / 2.80	2.90 / 4.00 / 3.65 / 4.75
Motorship	2 layers cont. / bulk	Va	110	11.4	1.70 / 2.80	2.90 / 4.00 / 3.65 / 4.75
Motorship	passengers		125	11.4	2.10	6.00
Motorship	passengers		60	10	1.48	4.00
Motorship	passengers		5	2.5	1.62 / 0.9	1.62 / 2.28

The mathematical model of the container vessels and bulk carriers shown in Table 1, is based on tests performed with a 1:25 scale physical model of a ECMT class Va vessel (110 m x 11.4 m), which was validated by full scale measurements (Verwilligen et al. 2015). Mathematical models for longer vessels (135 m x 11.4 m) were also derived and used for various nautical studies on the simulators at Flanders Hydraulics Research. For this study, models of lengths 105 m and 125 m were derived from the ECMT class Va model.

The models at a draft of 2.8 m were derived by interpolating the available drafts in the database. Models with a draft of 1.7 m (corresponding to a moderately loaded vessel) were deduced by extrapolation.

The models of the 60 m x 10 m and 125 m x 11.4 m passenger ships were derived from a 110 m x 17 m ship model equipped with rotating pods (Z-Drive thrusters) in the simulator database (Delefortrie et al. 2014).

Various ECMT class Vb/VIb/VIa pushed convoy combinations have been tested in the towing tank for shallow water for several drafts and water depths (see Fig. 4). The Vb model, available for two loading conditions, was used to develop two new models for 150 m and 180 m pushed convoys (Delefortrie et al. 2020). The models at a draft of 2.8 m were interpolated from the two available loading conditions and the models at a draft of 1.7 m were developed by CFD calculations (Van Hoydonck et al. 2020).

Fig. 4. Towing tank tests: ECMT class Vb.

The 5 m long passenger hydrofoil craft was derived from a simplified model of Riva boat type scaled to the desired dimensions and tuned based on full scale trials (only flying mode was modelled).

2.5   Track Controller for Fast Time Simulations

Flanders Hydraulics (FH) and Ghent University use five different fast time simulation techniques to systematically study ship manoeuvres in shallow and confined waters (Lataire et al. 2018). For this study, "Fast-time Track Controller" simulations (referred to as "fast time simulations" in this paper) were used taking full use of the mathematical model. The controls of the ship (rudder and propeller) are changed in time by a Prescience Model based Track Controller called PMTC by Chen et al. (2021) to follow a predefined path at a predefined speed. The input of this type of simulation is the desired trajectory and the margin for deviations that are allowed over the trajectory (Eloot et al. 2009).

2.6 Skippers

The real time simulations in this study were executed by professional skippers who have ample experience with navigation in Paris. One skipper is particularly familiar with 110 to 180 m long bulk convoys with a beam of 11.4 m, another skipper is particularly familiar with container ships of 86 m × 9 m and a third skipper is particularly familiar with passenger ships sailing in Paris. Prior to the actual simulations, the skippers spent a day on the simulators, during which they could test the modelled environment and provide feedback on the realism of both the mathematical manoeuvring model and the hydraulic model. The skippers shared their experience before, during and after each simulation. The human factor is taken into account by repeating the scenarios with two different skippers at the water level identified as possible limit. The scenarios are also assigned to skippers based on their particular experience (push convoy, container ships, passenger ship…) so that the different nuances in sailing different ship types are taken into account.

3 Detailed Study

3.1 Methodology

3.1.1 Simulation Protocol

Due to a large number of different parameters involved in the study, the simulations were carried out in two phases. A first phase consisted of fast time simulations to investigate the influence of each parameter and to identify the most critical conditions and bottlenecks. A second phase of the study consisted of real time simulations which allowed a more detailed analysis of the critical scenarios identified in the first phase.

3.1.2 Debriefing and Skippers Feedback

After each real time simulation, the skippers are invited to the control room to give their opinion about the manoeuvres that were performed. The difficulty as well as the safety of the manoeuvre is rated on a scale from 1 to 6. At this point, the nautical expert can already judge the accessibility level based on what was observed and discussed in the control room. However, the final results depend on the comparative and objective analysis of all the parameters conducted after a detailed post-processing of the simulation runs (based on the safety criteria described in 3.1.3). Preliminary results are nevertheless useful to drive the protocol and select the testing conditions in an optimized way.

3.1.3 Safety Criteria

Different criteria are used to evaluate the difficulty and safety of the manoeuvres. At low water levels, the most critical parameter is the distance between the ship and the depth line corresponding to the draft of the vessel. At high water levels, the most critical parameters are the horizontal distance between the ship and the line corresponding to the air draft of the ship and the vertical distance between the ship and the bridge. Three other parameters are monitored as well: the reserve of the propeller, the reserve of the bow thruster and the reserve of the rudder. In general, the reserve of a control parameter n, written as R_n, is the reserve that is available in case a problem occurs and is defined by Eq. (1) in function of the mean value $\hat{n}$ and the maximum value n_{max} of the parameter n over the duration of a simulation.

$$R_n = 1 - \frac{\hat{n}}{n_{max}} \tag{1}$$

For the three criteria mentioned above, the control parameter n is equal to the number of revolutions of the main propeller, the number of revolutions of the bow thruster and the rudder angle respectively.

Another parameter that is used as a criterion to assess the safety margin of a manoeuvre is the number of rudder variations (in °/s) derived from the mean rate of turn. This parameter is a good indication of the level of stress that the pilot experiences during the manoeuvre. Three other parameters are also considered in the analysis: under keel clearance (UKC), the vertical distance between the ship and a bridge and the distance to other ships.

The manoeuvrability of the ship can then easily be evaluated based on the criteria using a colour code. Figure 5 shows the colour code used in this study and indicates for which values for each of the parameters a colour changes.

	UKC	Reserve main propeller	Reserve bow thruster	Reserve rudder	Rudder variations	Horizontal distance ship – ground	Horizontal distance ship – bridge	Vertical distance ship - bridge	Distance to other ships
No constraints	> 0,84 m	≥ 40%	≥ 70%	≥70%	≤ 2 °/s	≥ 1 m	≥ 50 cm	≥ 50 cm	≥ 5 m
Acceptable	0,47 m - 0,84 m	25% - 40%	50% - 70%	50% - 70%	2 °/s – 4 °/s	0 m – 1 m			1 m – 5 m
Inacceptable	< 0,47 m	≤ 25%	≤ 50%	≤50%	> 4 °/s	0 m	< 50 cm	< 50 cm	< 1 m

Fig. 5. Safety criteria and colour codes used for the evaluation.

3.1.4   Nautical Expert Evaluation

Finally, an accessibility level is attributed and commented by the nautical expert based on the safety criteria defined in Sect. 3.1.3, the analysis of the trajectories as well as feedback from the skippers. A manoeuvre is considered as impossible when at least one of the safety criteria turns red.

3.2   Analysis

3.2.1   Fast Time Simulations

A first phase of fast time simulations was executed for every ship model at different water levels varying every 10 cm. The reference trajectory is based on the navigation axis found on the ECDIS chart and adapted based on skippers' experience provided during the validation day as well as from trajectories observed in previous studies which were executed for other purposes in this study area. The results were compared with previous studies executed in the city centre of Paris in order to validate the threshold of the different criteria. The results were also compared to actual regulations and skippers' feedback on reference cases for which the safety and ease level is well known. Tuning of the Track Controller parameters was then necessary to increase the realism of the manoeuvres as well as tuning of the threshold levels of the criteria.

The main bottlenecks were highlighted from fast time simulations and the waterway was divided into different critical sections. The passage under the Pont-Neuf was identified as the most critical section for the ships with a beam of 11.4 m at every water level due to a reduced beam. At the lowest water level, the base of the pile located underwater is a bottleneck for fully loaded vessels while the reduced width due to the arch of the bridge becomes the bottleneck at high water levels, especially for empty ships.

3.2.2   Real Time Simulations

Based on the results of the fast time simulations, a protocol was established with logical links to prioritize the scenarios to be tested with the skippers on the full mission bridge simulators. The number of scenarios could thus be estimated in advance but the final protocol had to be decided upon during the simulation day based on a rapid analysis by the skipper and the nautical expert immediately after each run.

Detailed post-processing was carried out after each simulation day and the results were grouped together in different data sheets, providing an overview per ship and per sailing direction Fig. 6 gives an example of what such a sheet looks like for a 135 m × 11.4 m container ship. It can be seen that the analysis indicates that navigation is impossible in the second section of the simulated trajectory (red color).

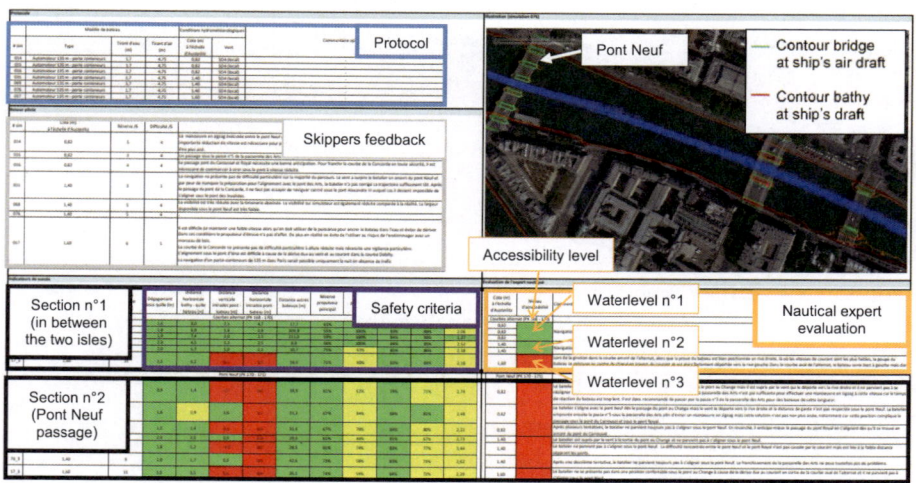

Fig. 6. Extract from a simulations sheet for a 135 m × 11.4 m container ship sailing downstream with a draft of 1.7 m. Analysis of the section n°1 and n°2 of the waterway.

3.2.3 Synthesis

The results of the fast time simulations and real time simulations were combined to recommend a level of accessibility for each of the sections of the 12 km long trajectory. The main bottleneck for 11.4 m wide ships was sailing under the Pont Neuf, where the ship must arrive perfectly aligned due to the restricted width. This is especially difficult to achieve when sailing downstream after passing the bends in between the two islands and the non-aligned bridges.

During the simulations, the skippers preferred to sail fast in order to improve manoeuvrability and attempt to pass safely under the Pont Neuf. As this involved repeating the same passage on the simulator and a certain advance knowledge of the problems involved, the manoeuvre cannot be considered as acceptable unless some measures are taken.

Some scenarios were only feasible after repeating the simulations or by using a specific technique, such as speeding up. Those scenarios show that only a well-trained skipper familiar with the navigation in high water levels could manage to pass safely. The risk is that a change of regulations would open the navigation to new skippers with no experience of crossing Paris, who might not anticipate the bottlenecks. Therefore a certification system (where the waterway authority makes an exception for a ship exceeding the maximum dimensions allowed), training strategy (e.g. by using ship handling simulators) and other recommendations were formulated when the navigation could not immediately be validated based on simulation results.

When all conditions have been tested, a final accessibility level can be recommended for the full length of the trajectory for each design ship in order to visualize easily the operational limits (i.e. water levels) of the different ships, as shown in Fig. 7. This helps the waterway authority in the decision making for an adaptation of the current regulations, which are based on ship length. The results were then discussed with VNF and stakeholders.

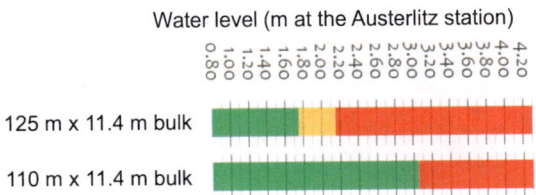

Fig. 7. Extract from the synthesis of simulation results showing. the operational limits (i.e. water levels) for two different bulk carriers (green = possible, orange = very difficult, red = impossible).

4   Challenges and Limitations of the Methodology

4.1   Safety Margins

At very high water levels, the headroom under bridges is so reduced that the skippers take advantage of the shape of the ship to pass underneath a bridge, for example by using the reduced air draft of the bulk carriers in front of the ship bridge. Figure 8 shows that the skipper brings the bow very close to the left side of the bridge (orange shape) and the aft of the ship close to the centre of the bridge (dark blue shape). However, in the analysis, a rectangular cuboid (bounding box) is considered around the ship (see Fig. 9) and some scenario can be rejected even though the skipper thought he had safely passed under the bridge based on what he saw during the simulations. Due to uncertainties on the actual shape of the superstructure of the ship crossing Paris, the analysis uses a conservative approach to make the conclusions applicable to any shape of superstructure.

Videos of the simulations as well as the 3D visualization software developed by Siradel, in which simulations can be replayed, can be used as a support for a more detailed discussion with the stakeholders about safety margins in order to take a final decision on the accessibility level.

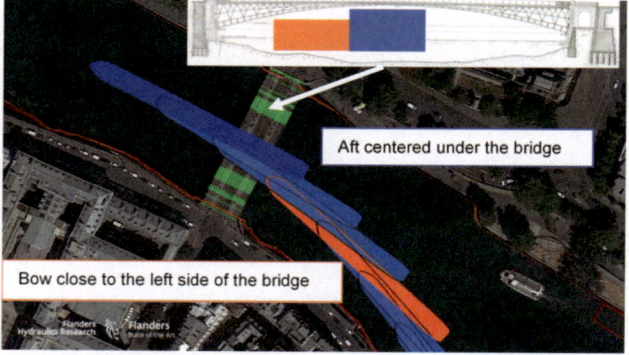

Fig. 8. 125 m ship passing the Pont d'Arcole in Paris.

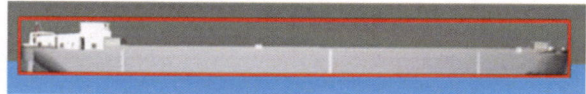

Fig. 9. Bounding box used to compute safety distances.

4.2   Human Factor

Several scenarios were executed by different skippers who shared their experience and techniques to tackle the bottlenecks with the operator. For example, one skipper would slow down, quickly using reverse engine at the exit of the second bend in between the islands, while another one would accelerate and reach high speeds (as shown in Fig. 10). Other skippers, equipped with a 360° bow thruster, would use it to slow down the ship while maintaining high main engine power to get a high rudder efficiency. However, such feedback needs to be balanced with the fact that the scenarios tested on the simulators differ from what is experienced in real life since, in reality, most skippers are used to sail on a ship with a beam that is smaller than 11.4 m. The feedback of the skippers was taken into account to provide recommendations at the end of the study such as investigating scenarios which were not considered initially, for example with a smaller ship beam, or to look for solutions so that navigation could be allowed, for example by using a certificate system or by requiring training on simulators.

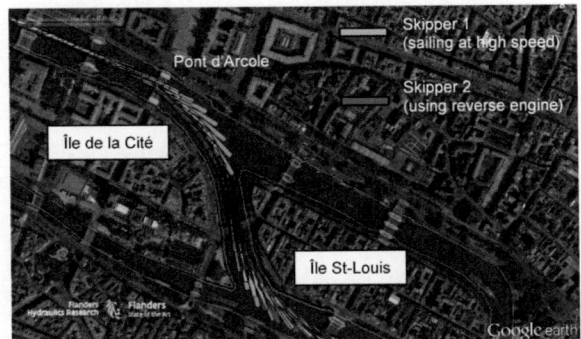

Fig. 10. Simulations with a 125 m long ship sailing in current.

4.3   Realism of the Simulation Tool

In this study, high level of detail for the modelling of the waterway was requested, especially around the bridges for which an accuracy of 10 cm was respected. This level of detail is important for the analysis of the simulation runs, but also for the immersive experience of the skipper and has an impact on the results of the simulations.

Moreover, distances can be more difficult to estimate on a simulator than in reality due to blind spots and the fact that the skipper only sees a virtual image centred on a fixed eye point. During the validation days, the simulators had to be adapted to avoid blind spots, therefore a 360° simulator was selected, and an extra feature allowed the skipper to move the view to the side of the ship to have a better visibility on the aft part of the ship (as shown in Fig. 11). In reality, the skipper is assisted by crew members who can move around the ship to evaluate distances.

Under low bridges, the skippers would usually lower the ship bridge by the maximum amount and steer the ship by passing their head through a hatch. This is not possible on the simulator; therefore the eye point has been placed 20 cm above the roof of the ship (see Fig. 2). This indicates that the level of realism clearly needs to be commented on in the simulation report, as discussed in Mansuy et al. (2021) and PIANC (2019).

Fig. 11. View moved to the side on Sim 360°.

The level of realism of the mathematical model also has an impact on the results. To tackle this, the feedback of the skippers is directly included in the datasheets where the real time simulations are synthesized (cf. Sect. 3.2). The datasheets were presented to VNF and stakeholders and after discussion balanced conclusions were drawn. One of the main conclusions was that it would most likely be possible to sail at higher water levels with ships having a beam smaller than 11.4 m. To validate this, additional simulations with ships with a beam smaller than 11.4 m were ordered by VNF. Moreover, it is recommended to have a real-life tests phase on the river Seine if the operational limits obtained from simulations need to be refined even further as the diversity of the fleet and the experience of the skippers is large.

5   Conclusions

A study was carried out to optimize the operational limits for which present and future vessels of varying types and dimensions cross Paris. Fast time and real time simulations were executed for different water levels of the river with experienced skippers at Flanders Hydraulics Research. The simulations showed that the main bottleneck is located at the narrow width under the Pont Neuf where 11.4 m wide ships have difficulties aligning in order to pass safely.

A critical water level up to which ships can safely sail could be successfully identified for the different ship models that were tested. Moreover, recommendations on the optimization of the operational limits by means of further measures were formulated (certification system, training on simulators, progressive tests in real conditions…).

The methodology applied, inspired by PIANC WG 141 recommendations, shows that in a detailed study, not only objective evaluation criteria should be considered but also the skippers' feedback and the recommendations from the waterway authority, from the end users and from the stakeholders who have experience with the conditions in reality. The nautical expert can then discuss both objective and subjective results to help the waterway authorities with the decision making.

Finally, in conditions which have never been experienced by the skippers or which are at the limits of feasibility, the manoeuvrability of ships is difficult to calibrate. However, the simulator should be considered as a tool which allows testing the response of a ship in unknown situations from which the different stakeholders can learn. Such use should be distinguished from the use of a simulator for training purposes during which a skipper is taught to be become familiar with a well-known situation.

Based on the results of the first study, it has been shown that regulations based on length only may be too restrictive for ships with smaller beams. Therefore, additional simulations have been ordered to assess the influence of ship's beam on the operational limits. New simulations have been executed with ships of reduced beam closer to the actual fleet characteristics.

Acknowledgements. The study was commissioned by Voies Navigables de France (VNF) and the Direction Régionale et Interdépartementale de l'Environnement, de l'Aménagement et des Transports d' Île-de-France (DRIEAT Île-de-France). The study was coordinated by International Marine and Dredging Consultants (IMDC) in collaboration with Ghent University, Flanders Hydraulics and Siradel. The results of the analysis were discussed on several occasions with representatives of the inland navigation sector and with skippers familiar with navigation in Paris, whose valuable input is acknowledged.

References

Adams R, et al. (2019) A comprehensive characterization of the nautical accessibility and traffic flow of the ECMT class Va inland waterway network of Nord-Pas-de Calais (France). In: Smart Rivers 2019 Conference, September 30 - October 3, 2019 Cité Internationale / Centre de Congrès Lyon, France

Breugem AW (2020) Ongoing developments in TELEMAC and TOMAWAC at IMDC. In: Online proceedings of the papers submitted to the 2020 TELEMAC-MASCARET User Conference October 2020, pp. 115–121. Antwerp: International Marine & Dredging Consultants (IMDC)

CETMEF (2010) Étude de la navigabilité des bateaux Va+ dans le bief de Paris Va+. Rapport d'étude

CETMEF (2013) Rapport provisoire de l'étude de manoeuvrabilité au niveau du pont Alexandre III. Rapport d'étude

Chen C, Verwilligen J, Mansuy M, Eloot K, Lataire E, Delefortrie G (2021) Tracking controller for ship manoeuvring in a shallow or confined fairway: Design, comparison and application. Applied Ocean Research 115:102823

Delefortrie G, Eloot K, Lemmens M (2020) Étude des conditions de navigation pour la traversée de Paris (VNF): sub report 2. Set-up of push model convoys. version 4.0. FHR reports, 19_012_2. Flanders Hydraulics Research: Antwerp. VII, p. 16

Deplaix JM, Daly F (2019) Evolution du dimensionnement des voies navigables. In: Smart Rivers 2019 Conference, September 30 - October 3, 2019 Cité Internationale / Centre de Congrès Lyon. France

DN&T (2016) Mission d'études de trajectographie pour la mise en place de bateaux entre le pont d'Arcole et le débouché du tunnel des Tuileries sur la Rive Droite de la Seine, Paris - Rapport final

DST (2013) Simulation of turning-manoeuvres of a 135 m inland passenger vessel

Eloot K, Verwilligen J, Vantorre M (2009) Safety assessment on head on encounters and overtaking manoeuvres with container carriers in confined channels through simulation tools. In: MARSIM 2009. Panama City

Eloot K (2015) Workshop Design Guidelines for Inland Waterways – Application of WG 141 approach including full bridge ship handling simulators for Class Va-vessels to the Upper-Seascheldt. In: PIANC-SMART Rivers 2015. Buenos Aires

Lataire E, et al. (2018) Systematic techniques for fairway evaluation based on ship manoeuvring simulations. In: Proceedings of 34th PIANC World Congress, pp. 1–13

Mansuy M, Candries M, Eloot K (2021) Nautical access study based on real time Bird's Eye ViewSimulations. TRANSNAV Journal 15:53–61

PIANC (2019) Design Guidelines for Inland Waterway Dimensions. PIANC, InCom WG, p. 141

Söhngen B, Butterer R (2015) Workshop design guidelines for inland waterways. In: Application of WG 141 approach including elaboration of field data and fast time simulation for Class Va-vessel passing narrow Jagstfeld bridge in the German Neckar River, PIANC-SMART Rivers 2015. Buenos Aires

Van Hoydonck W, Delefortrie G, Lemmens M, Mostaert F (2020) Étude des conditions de navigation pour la traversée de Paris (VNF): CFD computations of a barge at 90° drift for 18 combinations of draft and under-keel clearance. Version 2.0. FHR reports, 19_012_1. Flanders Hydraulics Research: Antwerp. VII, 27 pp. + 10 p. app

Verwilligen J, Delefortrie G, Vos S, Vantorre M, Eloot K (2015) Validation of mathematical manoeuvring models by full scale measurements. In: Proceedings of the International Conference on Marine Simulation and Ship Manoeuvrability (MARSIM), 8–11 September 2015, Newcastle upon Tyne, pp. 1–16. International Marine Simulator Forum (IMSF), UK

Solar Parks and Wind Farms Along Inland Waterways – Mitigating Measures Concerning Hindrance for Vessel Traffic

Otto C. Koedijk[1,2]

[1] Rijkswaterstaat WVL, Rijswijk, Netherlands
otto.koedijk@rws.nl
[2] Technical University, Delft, Netherlands
o.c.koedijk@tudelft.nl

Abstract. In the search for space for producing renewable energy, possible negative effects of solar parks and wind farms along inland waterways can easily be overseen. This paper provides an exploratory description of effects for navigation like blinding of helmsmen, disturbance of radio communication and exaggeration of vessel's radar images and concludes with a chapter on mitigating measures.

Keywords: Solar parks · Wind farms · Effects for navigation · Inland waterways

1 Introduction

To stop or at least slow down climate change, a world wide pursuit to produce renewable energy instead of using fossiles is high on the agenda. Two ways to produce green energy are booming: solar parks and wind farms. Being quite space consuming, they are often planned along inland waterways as the banks are often free of population, buildings and growth.

The possible negative effects of solar parks and wind farms along inland waterways for vessel traffic can easily be overseen. In the case of solar parks, the installation can cause electromagnetic hindrance (disturbance of radio communication). Also, reflection of the sun in the solar panels can lead to blinding of the helmsman (visual hindrance). Both hindrances can have a negative effect on nautical safety, as recent research from the Dutch Organization for applied scientific research (TNO) shows (Emmerik et al. 2022). This research was ordered by the Dutch main waterway authority Rijkswaterstaat.

In the case of wind farms, each wind turbine reflects the echo of radars, that are installed aboard vessels or ashore, as a part of a VTS system. The wind turbines can produce false echos in the fairway at the radar-screen and thus lead to misunderstanding by the skipper or VTS-operator. Again, this hindrance can have a negative effect on nautical safety.

Y. Li et al. (Eds.): PIANC 2022, LNCE 264, pp. 201–209, 2023.
https://doi.org/10.1007/978-981-19-6138-0_18

2   Solar Parks

2.1   Visual Hindrance

Possible visual hindrance of solar parks for helmsmen starts with the reflection of sunlight by the panels. Reflection is highly dependent on the angle of incidence. In a perpendicular situation, resulting from diffractive indexes of air and glass reflection of sunlight is only 4%, but at larger angles reflection increases up to 100%.

2.1.1   Human Aspect of Reflection of Sunlight

The first definition of nuisance is based on the point when it becomes uncomfortable to perceive reflected sunlight. This can be expressed in many ways, but common reactions are a tendency to turn the head away from the light source or hold the hand above the eyes or in front of the light source.

A second way to define this hindrance is to rely on the process that takes place in the eye. The Solar Glare Hazard Analysis Tool (SGHAT) from Sandia National Laboratories in the US is based on this (Ho et al. 2011). This tool calculates the amount of light that falls on the retina. The angular size of the light source in the visual field of the observer is also calculated. Based on these two measures, it is then calculated how much light falls on the retina and how large the light source is depicted on the retina.

2.1.2   Application of the SGHAT Model on the Dutch Waterway Network

The effects of solar parks along Dutch inland waterways were calculated by making use of an adapted SGAHT model. Parameters of this model are azimuth, angle of inclination, (viewing) direction and orientation of the solar panels.

Being a relatively small country (300 x 200 km), the outcome of the model is the same in the whole of the Netherlands. The outcome of the model is the hindrance of the reflected sun, expressed in the number of hours a year. Emmerik et al. (2022) recommended a somewhat arbitrary upper limit of 100–150 hours annualy.

As an example of this research, Figure 1 below gives an expression of the hindrance for northerly directions of view.

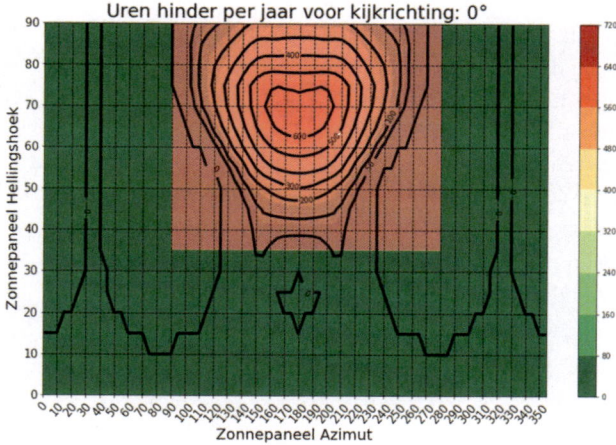

Fig. 1. Riskfull solar panels for northerly directions of view.

Special fairway situations as junctions and trajectories served by operators of locks and bridges can ask for a stricter number of hours of hindrance (up to zero), to be determined by the waterway authority concerned.

The research carried out by Emmerik et al. (2022) is not suitable for movable solar panels, following the sun; it can be assumed that those cause more hindrance than predicted by the SGHAT model.

2.2   Radio Disturbance

Solar parks along inland waterways can possibly cause disturbance of radio communication between vessels or between vessel and operator of lock, bridge or Vessel Traffic Service. Most important communication systems involved in the Netherlands are maritime radiotelephone (VHF), automatic identification system (AIS) and the system used by first responders (C2000).

2.2.1   Cause of Disturbance

The radio disturbance can be caused by the combination of the inverter(s) and the electric wires, connecting the panels. Together they can cause significant high-frequent emissions. In the past, severe degradation of the C2000 network has shown in Dutch residential areas. A recurrence in the VHF, AIS and C2000 networks should be prevented.

Figure 2 below shows emissions of a solar installation -measured by the Dutch governemental Radio Agency- which can reach a level of 14 dB. Emmerik et al. (2022) limit this increase of radio noise of solar systems however to 3 dB.

2.2.2   Rules for Electromagnetic Compatibility (EMC)

In Europe, Guideline 2014/30/EU was published to prevent all kind of electric appliances from radio noise. This EMC-Guideline refers to European harmonized standards NEN-EN 55011:2016 and EN61000-6-4/A1, agreed on by manufacturers and radio users. Appliances that comply to these European standards, are supposed not to disturb and get a CE-label.

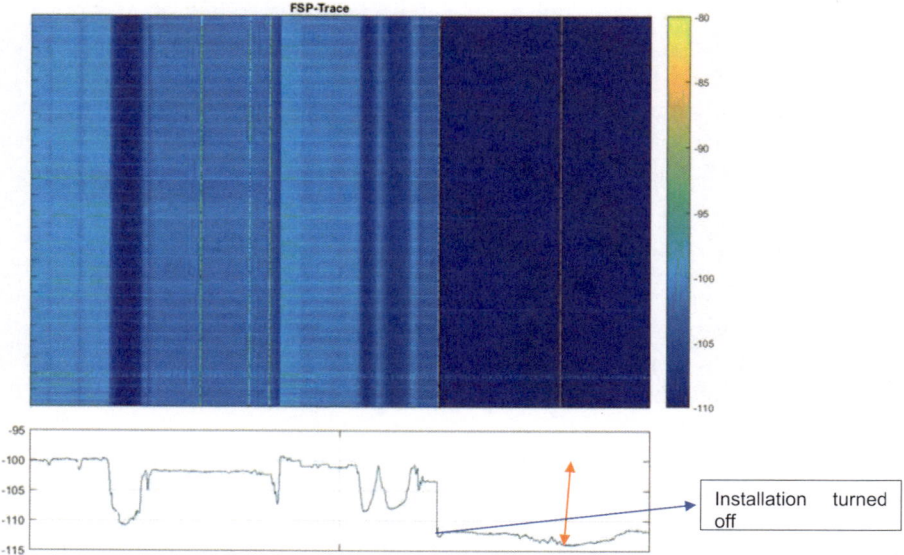

Fig. 2. Noise level of 14 dB (orange arrow), produced by a solar installation. Source: Radio Agency Netherlands.

Emmerik et al. (2022) however, showed that a complete solar installation whose appliances apply to the European standards, can still produce serious radio disturbance and violate the so called essential requirements of the European Guideline 2014/30/EU. The same message comes from Keyer et al. (2014), who state that nearly all European national agencies, responsible for enforcing the law in this matter, are not able to stop this fast growing number of violations.

3 Wind Farms

3.1 Projection of a Windturbine at a Radarscreen

Windturbines are projected at the radar screen of an inland waterway commercial vessel deviant from the real situation in a number of ways. Most important deviation is the exaggerated projection of their width, as a result of the width of the radar bundle. Such radars are certified for an observing distance of 1200 m. At this distance, this wider projection of the windturbine is illustrated in Fig. 3 below.

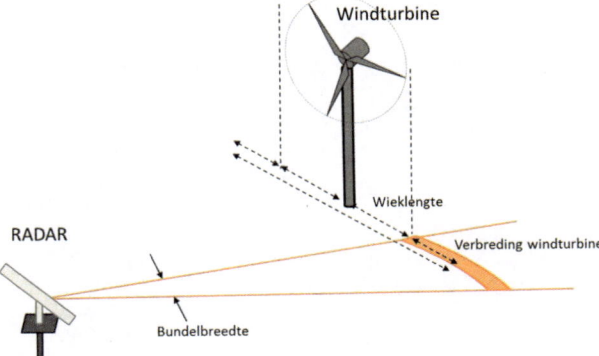

Fig. 3. Principle of the exaggerated width of the windturbine on the radar screen. Source: Rob van Heijster et al. 2016.

3.2 Problem of Exaggerated Width

As a result of the exaggerated with, a windturbine (to be) placed close to the fairway can be projected on the radarscreen as an object laying partly in the fairway. Note that this projected object is at maximum size when -due to the momentary wind direction- the rotor of the windturbine is perpendicular positioned to the fairway. This projection can lead to all kind of misinterpretations of the radar observer (helmsman) and subsequently to riskfull behavior, like navigating around or stopping in front of the would be obstacle in the waterway. An illustration of this deviant image of a windturbine on the vessel's radar screen can be seen in Fig. 4 below, giving the impression of a largely blocked fairway, indicated by the white circle.

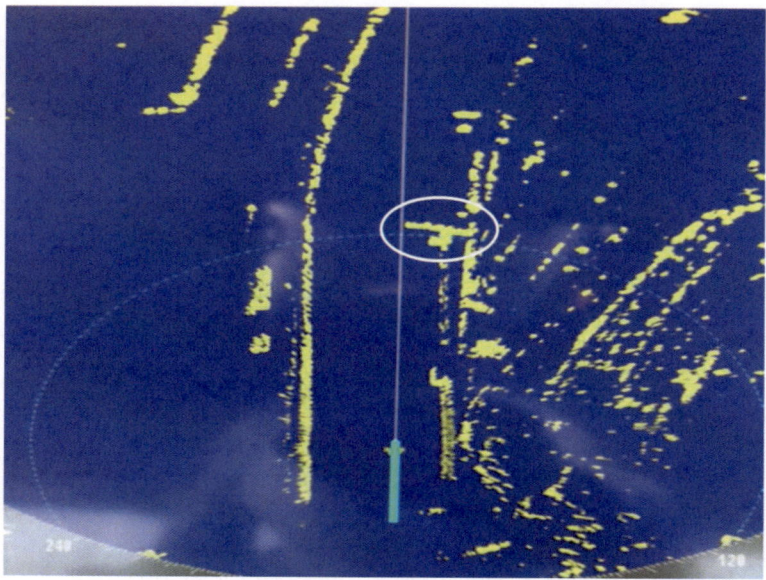

Fig. 4. Image on the radar of a commercial vessel, navigating the Dutch Hartel canal. Source: Rob van Heijster et al. 2016

3.3   Current Dutch Policy on Minimal Distance

In order to avoid images of windturbines at vessel's radar screens in the fairway like in Fig. 4 above, the Dutch Minister of Infrastructure (2015) published his policy concerning applications for permission to place windturbines. The minimal distance from the edge of the fairway should be at least 50 m and in all cases at least half the size of the rotor of the windturbine, to avoid possible collision with vessels.

3.4   Windfarms at Sea

At sea, the number of windfarms is growing rapidly, also in areas of marine traffic. Various effects of the windturbines on the radar at seagoing vessels are addressed recently by the National Academies of Sciences, Engineering and Medicine (2022), but they are out of the scope of this paper, which focusses on inland waterways.

4   Mitigation

This chapter provides points of interest and possible measures to reduce the hindrances previously discussed. The first part concerns diminishing hindrance of solar parks, the second part of wind farms.

4.1   Mitigating Hindrance of Solar Parks

4.1.1   Mitigating Visual Hindrance

The reflection of solar panels can be reduced by the use of coatings or textured glass. For an idea of the reflection by different types of solar panels, see Fig. 5 below. The lowest curve in Fig. 5 concerns deeply textured panels.

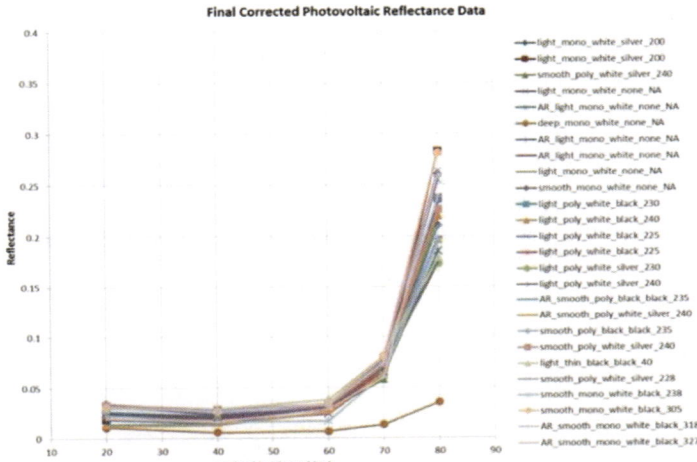

Fig. 5.   Source: Ho et al. 2011.

4.1.2   Mitigating Electromagnetic Hindrance

Various measures can contribute to diminish electromagnetic effects and avoid radio disturbance. First of all, electric wires should not form a string and can be put side by side in iron bars or cable trays. Secondly, the right type of EMC/EMI-filters and ferrite beads can be applied (see Fig. 6 below); the latter is often seen in the case of computer devices.

Nonetheless these measures, specialized measurements before and after realization of a solar park are necessary to be certain that no radio hindrance is produced.

Fig. 6. EMC/EMI filter (left) and ferrite beads (right). Source: Dutch Agency for Telecommunication.

4.1.3   Mitigating Hindrance of Windfarms

Based on the fact, that a windturbine's reflection point is relatively small and concentrated, Heijster et al. (2016) starting from the then current largest windturbines determined the exaggerated width at 30 m at both sides of the windturbine at a distance of 1200 m. This led to the recommendation for the Dutch Minister of Infrastructure to change his policy (see §3.3) into a minimum distance to the edge of the fairway of half the rotor size of the windturbine plus 30 m.

Another direction for a possible contribution to nautical safety is providing a windturbine with an AIS transponder, thus informing the helmsman on the presence of the windturbine, thereby avoiding misunderstanding of it's image at the vessel's radar screen.

Acknowledgements. The author wishes to express his gratitude to TU Delft colleague ir.dr. Ernst Schrama, board member of the Rotterdam division of the Dutch Association for experimental radio research (Veron), for providing practical information and scientific literature for free.

References

van Emmerik ML, Hoefsloot PC, van Sanden KPHM (2022) Radio en visuele hinder door zonneparken langs vaarwegen (Hindrance of solar parks along fairways for vessel traffic). Dutch Organisation for applied scientific research (TNO). published by Rijkswaterstaat WVL, 2022. Download: https://puc.overheid.nl/doc/PUC_707355_31

Ho CK, Ghanbarri CM, Diver RB (2011) Methodology to assess potential glint and glare hazards from concentrating solar power plants: analytical models and experimental validation. Am. Soc. Mech. Eng.

2014/30/EU Guideline from the European Parliament and Council concerning the harmonization of the rules of member states about electromagnetic compatibility (rearrangement)

NEN-EN 55011:2016 - Industrial, scientific and medical equipment - Radio-frequency disturbance characteristics - Limits and methods of measurement

EN61000-6-4/A1: Electromagnetic compatibility (EMC) - Part 6-4: Generic standards - Emission standard for industrial environments

Keyer C, Timens R, Buesink F, Leferink F (2014) In-situ measurement of high frequency emission caused by photo voltaic inverters. https://research.utwente.nl/en/publications/in-situ-measurement-of-high-frequency-emission-caused-by-photo-vo

Van Heijster R, van Gent O, Tan R, Theil A (2016) Effecten van windturbines op binnenvaartscheepsradars. Een voorstel tot een nieuwe nationale regelgeving. https://puc.overheid.nl/rijkswaterstaat/doc/PUC_712434_31/1/

Minister of Infrastructure Netherlands (2015) Beleidsregel voor het plaatsen van windturbines op, in of over rijkswaterstaatswerken. https://wetten.overheid.nl/BWBR0013685/2015-11-21

National Academies of Sciences, Engineering, and Medicine (2022) Wind Turbine Generator Impacts to Marine Vessel Radar. The National Academies Press, Washington, DC. https://doi.org/10.17226/26430

Study on Flow Distribution on Diversion Surface at the Head of Sandbar in Bifurcated Reach

Wanli Liu[✉] and Weiyan Xin

Tianjin Research Institute for Water Transport Engineering,
M.O.T, 1, Tianjin, China
tjtgliuwanli@163.com

Abstract. The rule of flow distribution in the diversion region at the head of sandbar of bifurcated reach is studied. The results show that there are diversion points and stagnation points in the diversion area at the head of sandbar in straight-bifurcated reach, and there exists diversion surface due to the difference between flow directions of surface and bottom in the diversion area at the head of sandbar in meandering-bifurcated reach. The concept of diversion surface is put forward through the study. It refers to the curved surface along vertical direction calculated according to the diversion ratio of two branches in the local reach of the head of sandbar in diversion area. Natural flow characteristics at the head of typical bifurcated reach in the middle reaches of the Yangtze River are studied based on the concept of diversion surface, and the layout principle of regulation project at the head of sandbar in bifurcated reach is put forward, that is, the project at the head of sandbar should be set near the relatively stable diversion surface if the diversion ratio of two branches is not planned to be changed, while the purpose can be achieved by deflecting the starting point of the project to one side if it is made to play the role of changing the diversion ratio of two branches. The proposal of the concept and calculation method of diversion surface provides a theoretical basis for the design and construction of location of regulation works in bifurcated reach.

Keywords: Bifurcated reach · Flow characteristics · Diversion surface

1 Introduction

Daijiazhou reach in the middle reaches of the Yangtze River is selected as a typical bifurcated reach for our study. The total length of Daijiazhou reach is about 34 km (see Fig. 1), and it is a slightly curved bifurcated reach. The bar of Daijiazhou in the river divides the channel into two branches, and the left branch is a circular channel while the right branch is a straight one (Liu 2015). The outline of the head of bar of Dai Jiazhou's is a long upward extending "wedge" shape, which bends on the plane and gradually becomes higher along the longitudinal direction.

The analysis shows that the head of Daijiazhou is constantly changing, moving left or right, or lifting up or falling down (Cai and Li 2011). Its instability is a direct reflection of the instability of channel conditions of Daijiazhou reach in dry season. The

© The Author(s) 2023
Y. Li et al. (Eds.): PIANC 2022, LNCE 264, pp. 210–218, 2023.
https://doi.org/10.1007/978-981-19-6138-0_19

head of Daijiazhou is constantly changing under the action of distinct flow and sediment conditions (Fig. 2), resulting in the alternation of main channels in dry season of two branches.

The straight channel of right branch is the main navigable channel of the reach. The study shows that most straight channels will develop into single channels whose channel conditions are good when the head of Daijiazhou extends upward while the multiple transitions in deep pool in the straight reach is easy to form whose channel conditions are poor when the position of the head of Daijiazhou is lower. Obviously, taking engineering measures to stabilize the head of Daijiazhou and promoting its siltation to extend upward is very important to improve the channel conditions of straight channel. However, the project layout of the head of sandbar has a direct impact on the effect of the regulation project. Therefore, it is necessary to study the rule of flow distribution of diversion area at the head of sandbar in bifurcated reach to direct the project layout of the head of bar to be reasonable (Liu 2020).

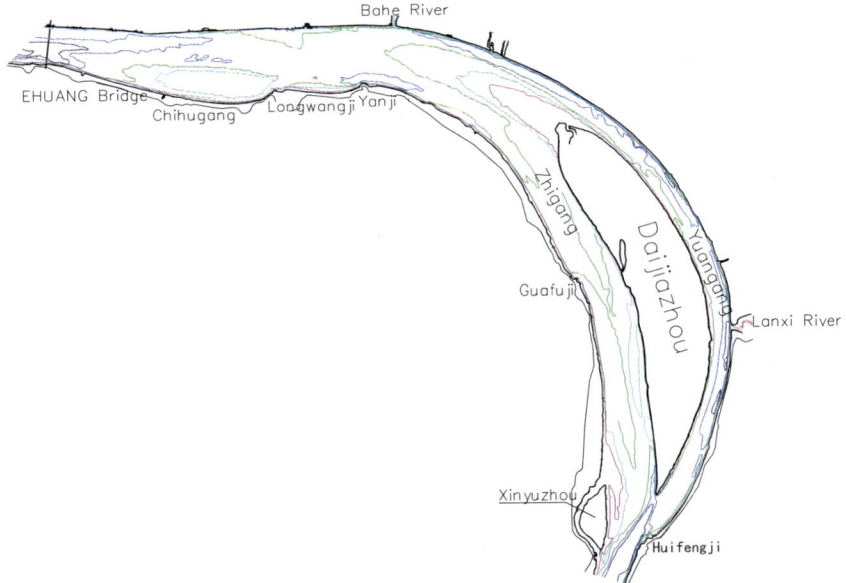

Fig. 1. Sketch of the Daijiazhou reach

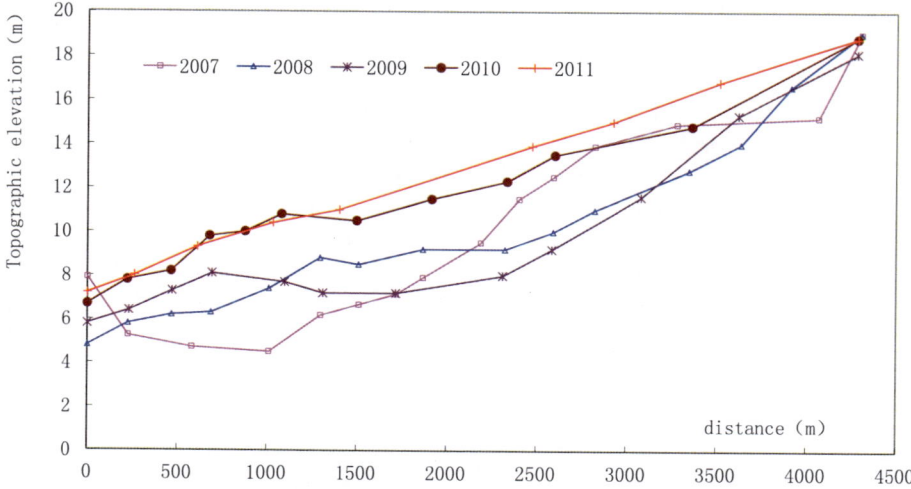

Fig. 2. The variation of of topographic longitudinal profile of the head of the Daijiazhou sandbar

2   The Diversion Point and Confluence Point of Distributed Reach

For the bifurcated reach, there are generally diversion points in diversion area at the head of bar and confluence points in confluence area at the tail of bar. The study shows that the position of the diversion points of Daijiazhou reach move upstream in dry season and moved downstream in flood season (Fig. 3), and there is little difference between the locations of confluence points at various discharge. The reason is that the position of water edge varies greatly under distinct discharge because of flat terrain at the head of Daijiazhou. The position of diversion point moves up and down due to that the larger the discharge is the greater the flow momentum is and it has a large variation range. The tail of Daijiazhou is steep, and the position of water edge under different discharge has little difference, while the position of the flow dynamic axis under all levels of discharge has little displacement here. Therefore, the change of the position of confluence point is small.

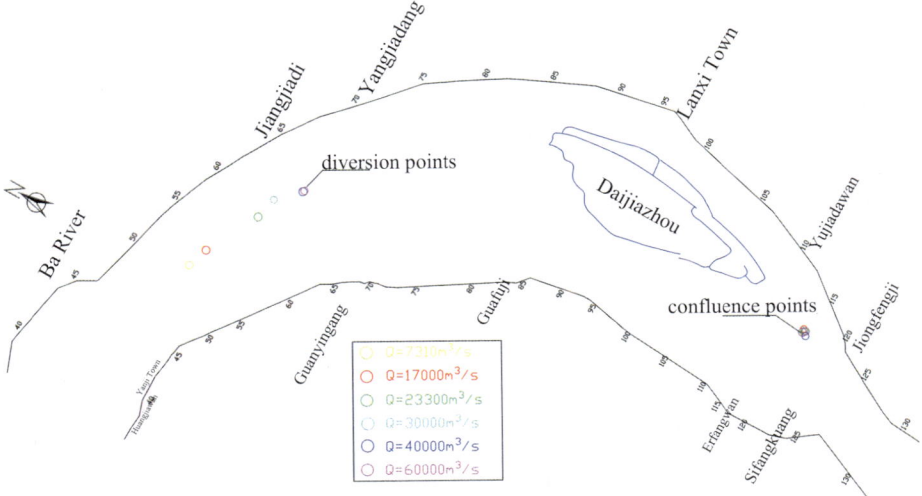

Fig. 3. The location of diversion and confluence point under various flows

3   Flow Distribution Characteristics in Diversion Area at the Head of Sandbar

The flow distribution characteristics in diversion area at the head of sandbar are analyzed through numerical simulation calculation and physical model experiment, and the rules of flow movement near the head is revealed comprehensively. The change of flow direction at the head of Daijiazhou is obviously subject to the amount and direction of incoming flow, the terrain at the head of sandbar, the topographic change of two branches, and the comparison relationship of resistance between the two branches (Fig. 4).

(1) The flow at the head of Daijiazhou is obviously left biased given the discharge less than 20000 m³/s, and a less discharge leads to a greater angle to the left. This is the phenomenon that the incoming flow under the basic straight river regime is affected by the downstream curved river regime and the flow carrying capacity of the straight channel is limited. This kind of deflection may be closely related to the landform created by the straight channel during the flood period and the difficulty of forming a well connected deep groove in the falling water period. And the deflected flow will have a certain oblique scouring effect on the submerged bar at the head of Daijiazhou. In addition, surface flow deflects more frequently due to that the deflected flow is affected by the topography of the head of Daijiazhou. Therefore, more sediment from the upstream may enter the straight channel, which has negative effect on low-flow scouring of the shoals in straight channel (Fig. 5).

(2) The upstream flow of Daijiazhou head has no obvious deflection angle given the discharge between 20000 m³/s and 30000 m³/s. This is because the stagnation

point at the head of Daijiazhou enters the bend due to the rise of water level when the discharge is large, and the diversion area is already located in the upper reach of the bend. The horizontal distribution of flow matches the flow capacity of the two branches to achieve the balance. Therefore, the direction of flow at the head of sandbar shall not biase to one side obviously.

(3) The flow at the head of Daijiazhou tends to the right given the discharge between 30000 m³/s and 50000 m³/s. With the further increase of the mount of flow and the further downward movement of the stagnation point at the head of Daijiazhou, the wetted area of the straight channel increases with the increase of water level, and it is obviously better than circular channel. Therefore, the increase of the discharge capacity of straight channel is greater than that of circular channel, and the cross-sectional transverse flow velocity distribution in the diversion area is more affected by the curved river regime, which is also limited to the discharge capacity of circular channel to the right.

(4) The water overflows Xinzhou and the flood flow is straight given the discharge greater than or equal to 50000 m³/s, and there is no obvious biase to one side near the head of bar.

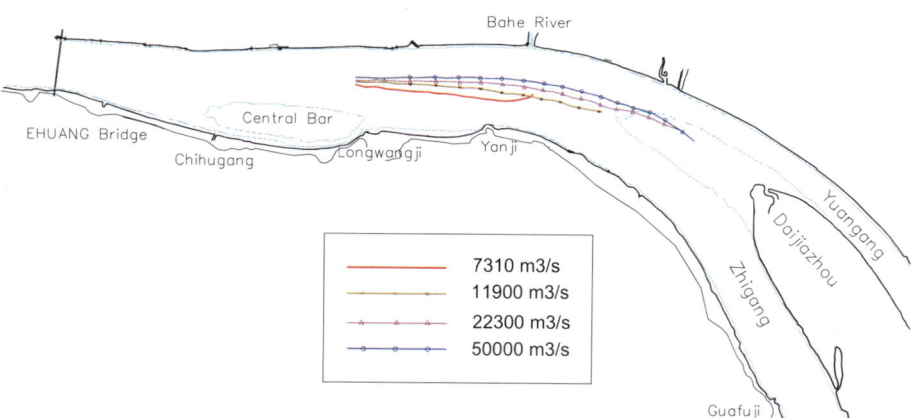

Fig. 4. The diversion at the head of Daijiazhou under distinct levels of flow under the current terrain

Fig. 5. Local flow pattern at the head of Daijiazhou sandbar under low flow (Q = 7310 m³/s)

4 Diversion Surface and the Principle of Project Layout of Head of Sandbar

Taking Daijiazhou reach of the middle reaches of the Yangtze River as an example, the concept of diversion surface is put forward creatively based on the study of the flow distribution characteristics of diversion surface at the head of sandbar. The diversion surface of the head in bifurcated reach is studied and its layout of regulation project is guided through numerical simulation calculation (Liu et al. 2009).

The diversion surface refers to the curved surface along the vertical line calculated according to the diversion ratio of two branches in the local reach of the head of diversion area (Fig. 4). That is some sections in the reach upper the head of sandbar should be taken and the cumulative discharge between sections taken and the bank shall be calculated respectively. The points on each section where the cumulative discharge is equal to the diversion amount of this branch are connected and the longitudinal surface formed by the connecting lines at distinct vertical positions of the point is the diversion surface. The location of the diversion surface is different under different levels of discharge. For example, there are obvious difference between low-flow and flood locations of diversion surface of Daijiazhou (Fig. 6).

Based on the concept of diversion surface, the layout principle of the regulation works at the head of sandbar in bifurcated reach is put forward combined with the natural flow characteristics at the head of sandbar in bifurcated reach. The regulation works should deviate from the side of low-flow diversion surface as far as possible, such as the left or right side if the regulation works are arranged at the head to increase the low-flow diversion ratio of the straight channel (main branch). The sandbar-head project should be arranged near the flood diversion surface in order to ensure that the regulation project does not affect the flood characteristics of two branches. The sandbar-head project shall be set on the low-flow diversion surface if the regulation project does not significantly change the diversion ratio of the two branches. The above principles can play an important role in the layout of bar-head regulation works in bifurcated reach.

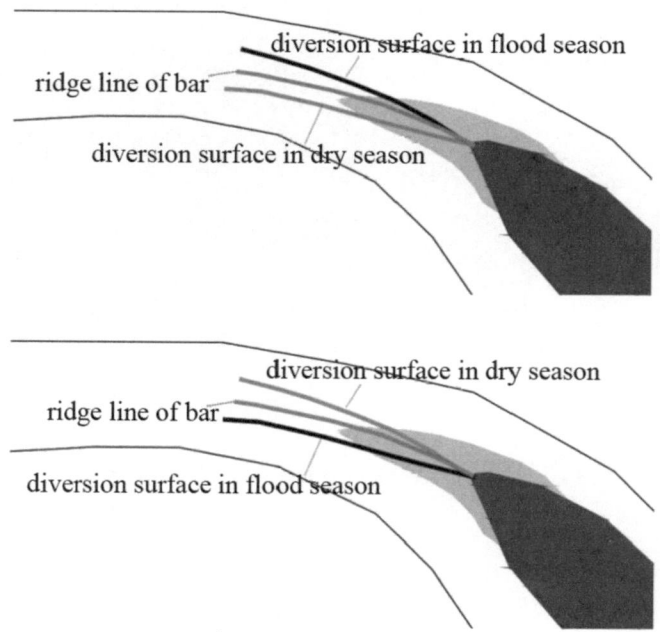

Fig. 6. The diagram of the relation between sandbar-head diversion surface and layout of regulation works

5   The Project Layout in Sandbar-Head Diversion Area

As the diversion area at the head of sandbar is an inundated area, which shows characteristics of single section not bifurcated section. Therefore, the change of single section to bifurcated section will bring special changes in flow conditions. First, the diversion ratio of two branches will change. The sandbar-head project is more manifested in changing the inflow conditions of the branch to make the diversion ratio of the two branches adjusted to a certain extent due to that the diversion ratio is affected by many factors, including the length of the branch (on-way resistance), section profile, local resistance, sand wave resistance and inlet inflow conditions, etc. Second, the sandbar-head project changes the flow velocity on both sides of the project area significantly, and the flow velocity on one side (the side to which the flow at the head is biased before the project) increases because more water replenished along the way through the diversion surface bypasses the project head and enters the artificial branching reach, which shows that the discharge and flow velocity of this reach increases from the lower to upper gradually, while the other side is opposite, that is, the discharge and velocity in this reach decreases from the lower to upper gradually. This kind of phenomenon is more obvious in the early stage of project implementation and will weaken to a certain extent with the continuous development of the influence of sandbar-head project on riverbed deformation.

The fish mouth type dividing dike project is often set in order to stabilize or change the diversion condition of flow and sediment of two branches (Li 2007). Stabilizing the diversion condition of flow and sediment of the two branches or increasing the diversion ratio of the main channel, and narrowing the width of river in dry season (the river width in dry season in diversion area of bifurcated reach is often larger, which will lead to small depth inevitably, and extending upstream with "fish mouth" project can create a sandbar of a certain scale so as to improve channel depth condition in sandbar-head area) are two main functions of "fish mouth" project. The project of sandbar-head diversion area of Daijiazhou reach is arranged on the left side of diversion surface in dry season to increase diversion ratio of main channel in dry season. And the project acts on shaping sandbar pattern, narrowing river width and increasing depth (Fig. 7).

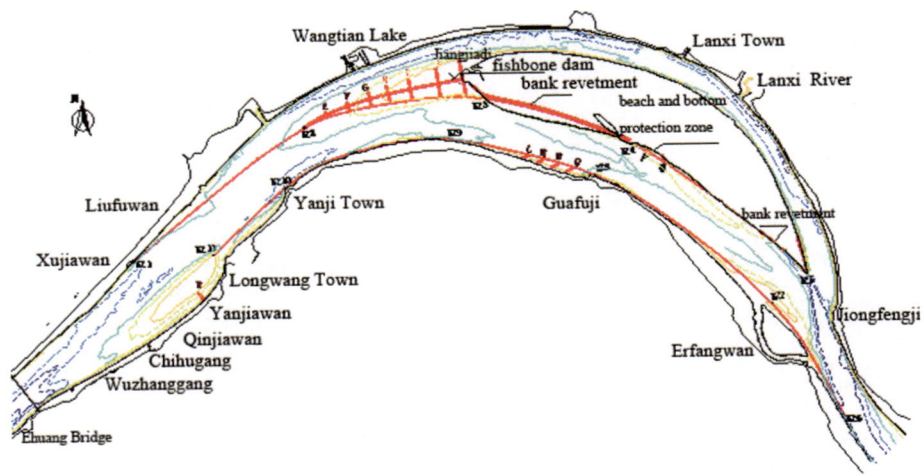

Fig. 7. The diagram of fish mouth type dividing dike project at the head of sandbar

6 Conclusions

(1) The study shows that there are diversion points in the diversion area at the head of sandbar in bifurcated reach generally. The terrain of the head of Daijiazhou is flat. The position of flow edge under distinct discharges varies greatly, and the larger the discharge is the greater the flow momentum is. Therefore, the position of diversion point moves up and down with a large change amplitude.

(2) The change of flow direction at the head of Daijiazhou is obviously restricted by factors such as the amount and direction of incoming flow, the topography of the head of Daijiazhou of the reach, the topographic change of the two branches and the resistance comparison relationship between the two branches. There are many phenomena such as obvious left deviation, no obvious deviation angle, right deviation of the flow at the head of Daijiazhou.

(3) There is a diversion surface in the diversion area at the head of bar in meandering bifurcated reach due to the difference between directions of surface flow and the bottom flow. The concept of diversion surface is put forward through study. The location of the diversion surface varies under distinct discharges. For example, there are obvious differences in the locations of diversion surface of dry season and flood season in Daijiazhou reach.

(4) The layout principle of the regulation project at the head of bar of bifurcated reach is put forward based on the concept of diversion surface and combined with the natural flow characteristics at the head of bar of bifurcated reach. The sandbar-head project shall be set near the relatively stable diversion surface if the diversion ratio of the two branches shall not be changed significantly while the purpose can be achieved by shifting the starting point of the project to one side to make the project play its role of changing the diversion ratio of the two branches.

Acknowledgements. This work was financially supported by National key R & D Plan (2018YFB1600400), and Basic research funding of national commonweal research institutions (TKS20210403).

References

Cai D, Li Q (2011) River channel evolution and characteristics of navigation-obstruction in Daijiazhou section in the middle Yangtze River. J. Sediment. Res. 2:47–54

Liu WL (2015) Systematic Treatment Technology of Bifurcated Reach in the Middle and Lower Reaches of the Yangtze River. People's Communications Publishing House Co. Ltd, pp 110–112

Liu WL (2020) Guiding Restoration Technology and Practice of Main Navigable Branches in the Bifurcated Reach of the Middle and Lower Reaches of the Yangtze River. People's communications Publishing House Co. Ltd, pp 83–85

Liu WL, Li WSH, Li YB, Zhu YD (2009) Discussion on channel regulation of Daijiazhou reach in the middle reaches of the Yangtze River. Water Transportation Engineering 30(1):31–36

Li WSH (2007) Some thoughts on the technical problems of waterway regulation in the middle and lower reaches of the Yangtze River. Waterway Port 12:418–424

Study on Prediction Method for Compression Scour Depth of River-Crossing Bridge

Qianqian Shang[1(✉)], Hui Xu[1], and Jian Zhang[1,2]

[1] Nanjing Hydraulic Research Institution, Nanjing, China
{qqshang, huixu}@nhri.cn
[2] Hohai University, Nanjing, China

Abstract. Riverbed deformation caused by river-crossing bridge construction can be divided into compression scour and local scour. Compared with local scour, fewer studies have been made on the compression scour caused by bridge piers. It is noteworthy that, the compression scour can lead to riverbed scour of the whole cross section along a bridge site, which is obviously detrimental to the bridge foundation safety. Based on a summary of existing research findings, a prediction model for the compression scour of bridge piers is constructed, and the model is applied in predicting the compression scour depth of Shiyezhou River Bridge in the lower reaches of the Yangtze River. Firstly, the pier boundary treatment methods at different spatial scales are discussed. Subsequently, the selection method of flow and sediment processes is proposed from the engineering safety point of view, according to the flow and sediment characteristics on the lower reaches of the Yangtze River. Finally, the depth of compression scour around the upstream and downstream of Shiyezhou Bridge piers are predicted, and comparisons were made between the prediction depth of Shiyezhou Bridge and other existing bridges in the lower reaches of the Yangtze River. Comparisons show that the compression scour depth of Shiyezhou Bridge was basically equivalent to that of other bridges downstream the Yangtze River. The results indicate that the method for predicting the compression scour depth of bridge piers is reasonable and feasible, and the prediction of compression scour depth can provide technical basis for determining the embedment depth of the bridge pier foundation.

Keywords: Bridge construction · Pier treatment · Flow and sediment process · Compression scour · Prediction model

1 Introduction

In the lower reaches of the Yangtze River, many river-crossing bridges are built to strengthen the communication between cities on both river sides, and to mitigate the river-crossing traffic pressure. Due to the narrowing action of the bridge substructure on the water flow, the riverbed deforms accordingly (WUHES 1981). The riverbed changes caused by bridge can be divided into compression scour and local scour, where the former refers to the scour occurring after the sectional contraction of piers, including natural riverbed scour and the scour arising from the contraction of water flow. As for local scour, it is the erosion, i.e., scour hole, just around piers formed by

Y. Li et al. (Eds.): PIANC 2022, LNCE 264, pp. 219–231, 2023.
https://doi.org/10.1007/978-981-19-6138-0_20

flow vortex (Zhang et al. 1993). At present, studies mainly focus on local pier scour (Olsen and Kjellesvig 1998, Karim and Ali 2000, Roulund et al. 2005, Liu and Garcia 2006, Chen 2008 and Sumer et al. 2001), while much less attention has been paid on compression scour (Yong and Blair 2010; Fenocchi and Natale 2016; Guo and Qi 2011). In most case, if a pier experiences compression scour apparently, the cross-section of the bridge site will be subjected to overall scour along the bridge, which is disadvantageous for the safety of the building foundation (Guo 2013). Especially on the middle and lower reaches of the Yangtze River, the sediment concentration declines sharply after the impoundment of the Three Gorges Reservoir, which accelerates the downstream riverbed scour. Therefore, the foundation depth of bridge piers should be chosen with an overall consideration. In the design of river-crossing bridges, the prediction of compression scour of piers is one of the key technical problems that must be solved.

Previously, empirical prediction formulas for the compression scour of piers have been acquired through model tests or field measurements. There are some commonly used formulas, including formulas published in railway and highway specifications (MRPRC 1999; MTPRC 2002), American HEC-18 formula (Shirole and Holt 1991), Soviet Baldakov, Lestervan, and Andreev formulas (Lu and Gao 1996). However, those formulas can only reflect the equilibrium value of compression scour under the prolonged action of specific flow and sediment conditions, along with the disadvantage of coefficient uncertainty. In recent years, domestic and foreign scholars have started predicting the compression scour of piers through numerical models. For instance, Liu et al. (1993) simulated the scour of railway bridge piers in the midstream of the Yellow River during a once-in-a-century catastrophic flood, using a method of adjusting the local head loss coefficient of grids. Results showed that the scouring-silting trend on the riverbed surface basically accorded with the test, but the measured values and calculated values were different. Lai et al. (2010), Guo and Qi (2011), Guo (2013), and Due and Rodi (2008) verified rectangular flume experiments on long contraction segments by using numerical simulation, and the model could simulate the variation of water level along the river and the scour distribution in the contraction segment, but sediment process and suspended load were not considered. For the lower reaches of the Yangtze River, due to the impoundment of the Three Gorges Reservoir, flow and sediment processes are obviously different from the natural situation, leading to much complex scouring-silting conditions. Compared with river-crossing tunnels, more factors should be considered for river-crossing bridge (Zhang et al. 2011; Wei et al. 2016; Yue et al. 2010). Therefore, predicting the compression scour of piers, a model should meet the following conditions: (a) flow and sediment processes should be reasonably selected; (b) parameters should be calibrated and verified so that annual and interannual scouring-silting changes of riverways can be rechecked; (c) models should reflect the flow and sediment movements before and after bridge construction well.

Taking a newly built bridge as an example, a mathematical model for compression scour of bridge piers was constructed in this study. The model parameters were verified, the pier boundary was proved, the flow and sediment processes were selected, and then the scour depth was predicted, expecting to provide a reference for solving similar problems.

2   Project Overview

The newly built Lianyungang-to-Zhenjiang Railway, which is 249 km in overall length, continues Xinyi-Changxing Railway in the north, connects Nanjing-Qidong Railway in the middle, and meets Beijing-Shanghai High-Speed Railway and Shanghai-Nanjing Intercity Railway in the south, being the main north-south longitudinal railway passage linking northern, central, and southern Jiangsu regions. This railway is planned to span the Yangtze River at the river reach from Shiyezhou to Wufengshan. Two bridge positions were initially designed, i.e., Shiyezhou bridge site scheme and Wufengshan bridge site scheme. And at last, the bridge scheme crossing over the central bar named Shiyezhou was chosen. The suspension bridge scheme was adopted, where the left-branch main span was 406 m and the right-branch main span was 910 m. The plane layout and elevation layout are as shown in Fig. 1 and Fig. 2.

Fig. 1. Plan layout of the scheme of Shiyezhou bridge location

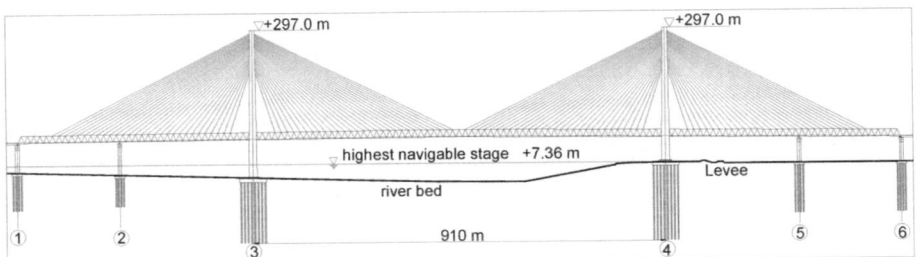

Fig. 2. Elevation layout of the right branch

3   Recent Evolution of the River Reach

Shiyezhou river reach is gently bent and branched, which is divided into left and right branches by Shiyezhou in the middle of the river (Fig. 1). The left branch is the secondary branch, with the measured water diversion ratio of 38.8% in December 2012, and the right branch is the main branch. The overall river regime remains relatively stable due to the levees and the nodes alongside the riverway. But the local evolution is still obvious, which is mainly manifested by the change in the diversion pattern of branches and the adjustment of riverbed.

(1) The changes in the diversion pattern of branches are mainly embodied in the increasing diversion ratio of the left branch and the declining diversion ratio of the right branch. From the middle 1970s to the early 1990s, the water diversion ratio of the left branch grew slowly, with an annual average increase amplitude of 0.1%. After 1995, the growth rate was evidently elevated because of continuous floods, and the annual growth rate reached 2.8% during 1997–1999. But the growth rate was slowed down again after 2000. The water diversion rate of the left branch in Shiyezhou rose from 20% in the 1970s to about 40% in recent years.

(2) As for the adjustment of riverbed, the low elevated beach of Shiyezhou experienced sustained scouring and regression; the left branch was subjected to overall scouring, with the local scour depth reaching 4 m; the entrance section of the right branch in Shiyezhou was characterized by side beach scouring at the left side and deep channel silting at the right side, while the middle-lower section of the right branch was manifested as side beach silting at the right edge of Shiyezhou and deep channel scouring. Variation of the transverse profiles at the bridge site are showed in Fig. 3.

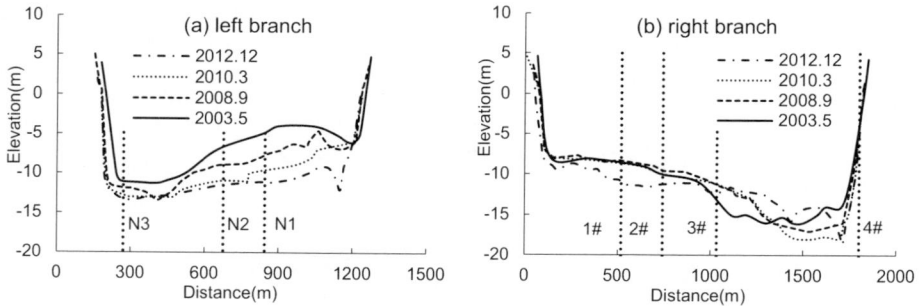

Fig. 3. Transverse profiles at the Shiyezhou bridge site

4 Modeling and Verification

4.1 Governing Equations Set

The governing equations set for the 2D numerical planar flow and sediment model under the orthogonal curvilinear coordinate system is as follows:

$$\frac{\partial Z}{\partial t} + \frac{1}{C_\xi C_\eta}\left[\frac{\partial(C_\eta Hu)}{\partial \xi} + \frac{\partial(C_\xi Hv)}{\partial \eta}\right] = 0 \tag{1}$$

A. Continuity equation of water flow:

$$\frac{\partial(Hu)}{\partial t} + \frac{1}{C_\xi C_\eta}\left[\frac{\partial}{\partial \xi}(C_\eta Huu) + \frac{\partial}{\partial \eta}(C_\xi Hvu) + Hvu\frac{\partial C_\xi}{\partial \eta} - Hv^2\frac{\partial C_\eta}{\partial \xi}\right]$$
$$= -\frac{gu\sqrt{u^2+v^2}}{C^2} - \frac{gH}{C_\xi}\frac{\partial Z}{\partial \xi} + \frac{1}{C_\xi C_\eta}\left[\frac{\partial}{\partial \xi}(C_\eta H\sigma_{\xi\xi}) + \frac{\partial}{\partial \eta}(C_\xi H\sigma_{\eta\xi}) + H\sigma_{\xi\eta}\frac{\partial C_\xi}{\partial \eta} - H\sigma_{\eta\eta}\frac{\partial C_\eta}{\partial \xi}\right]$$

$$(2)$$

B. Motion equation of water flow:

$$\frac{\partial(Hv)}{\partial t} + \frac{1}{C_\xi C_\eta}\left[\frac{\partial}{\partial \xi}(C_\eta Huv) + \frac{\partial}{\partial \eta}(C_\xi Hvv) + Huv\frac{\partial C_\eta}{\partial \xi} - Hu^2\frac{\partial C_\xi}{\partial \eta}\right]$$
$$= -\frac{gv\sqrt{u^2+v^2}}{C^2} - \frac{gH}{C_\eta}\frac{\partial Z}{\partial \eta} + \frac{1}{C_\xi C_\eta}\left[\frac{\partial}{\partial \xi}(C_\eta H\sigma_{\xi\eta}) + \frac{\partial}{\partial \eta}(C_\xi H\sigma_{\eta\eta}) + H\sigma_{\eta\xi}\frac{\partial C_\eta}{\partial \xi} - H\sigma_{\xi\xi}\frac{\partial C_\xi}{\partial \eta}\right]$$

$$(3)$$

where (ξ, η) represents the coordinates of the curvilinear coordinate system; (x, y) stands for physical coordinates; u and v are the flow velocity components in the directions ξ and η, respectively; Z means the water level; t is time; H refers to the water depth; C signifies the Chezy coefficient ($C = \frac{1}{n}H^{1/6}$); n denotes the roughness coefficient; $\sigma_{\xi\xi}$, $\sigma_{\eta\eta}$, $\sigma_{\xi\eta}$ and $\sigma_{\eta\xi}$ indicate turbulent shear stresses; v_t is the turbulent viscosity coefficient.

C. Unbalanced sediment transport equation of suspended loads

$$\frac{\partial(HS_i)}{\partial t} + \frac{1}{C_\xi C_\eta}\left[\frac{\partial}{\partial \xi}(C_\eta HuS_i) + \frac{\partial}{\partial \eta}(C_\xi HvS_i)\right]$$
$$= \frac{1}{C_\xi C_\eta}\left[\frac{\partial}{\partial \xi}\left(\frac{\varepsilon_\xi}{\sigma_s}\frac{C_\eta}{C_\xi}H\frac{\partial S_i}{\partial \xi}\right) + \frac{\partial}{\partial \eta}\left(\frac{\varepsilon_\eta}{\sigma_s}\frac{C_\xi}{C_\eta}H\frac{\partial S_i}{\partial \eta}\right)\right] + \alpha_i\omega_i(S_i^* - S_i) \qquad (4)$$

where S_i is the sediment content of suspended loads in group i; S_i^* represents the sediment transport capacity of suspended loads in group i; ε_ξ and ε_η stand for the sediment diffusion coefficients in the directions ξ and η, respectively; σ_s is a constant, taken as 1.0; α_i denotes the saturation recovery coefficient of suspended loads in group i; ω_i means the sediment deposition velocity in group i.

D. Unbalanced transport equation of bed loads:

$$\frac{\partial HS_{bL}}{\partial t} + \frac{1}{C_\xi C_\eta}\left[\frac{\partial}{\partial \xi}(C_\eta HuS_{bL}) + \frac{\partial}{\partial \eta}(C_\xi HvS_{bL})\right] = \alpha_{bL}\omega_{bL}(S_{bL}^* - S_{bL}) \qquad (5)$$

where S_{bL} denotes the sediment concentration of bed loads in group L; S_{bL}^* is the sediment transport capacity of bed loads in group L, $S_{bL}^* = g_{bL}^*/(\sqrt{u^2+v^2}h)$, and g_{bL}^* refers to the sediment transport rate of different bed load groups; α_{bL} is the saturation recovery coefficient of bed loads in group L; ω_{bL} stands for the deposition velocity of bed loads in group L.

E. Riverbed deformation equation:

Riverbed deformation caused by suspended load scouring-silting: $\gamma_{0i}\frac{\partial Z_i}{\partial t} = \alpha_i\omega_i(S_i - S_i^*)$.

Riverbed deformation arising from bed load scouring-silting: $\gamma_{0bL}\frac{\partial Z_{bL}}{\partial t} = \alpha_{bL}\omega_{bL}(S_{bL} - S_{bL}^*)$.

Total riverbed scouring-silting thickness: $Z = \sum\limits_{i=1}^{m} Z_i + \sum\limits_{L=1}^{n} Z_L$.

Where m and n represent the numbers of suspended load groups and bed load groups, respectively; γ_{0i} denotes the dry bulk density of sediments in suspended load group i; Z_i is the riverbed scouring-silting thickness induced by the suspended loads in group i; γ_{0bL} stands for the dry bulk density of sediments in bed load group L; Z_L is the total riverbed scouring-silting thickness.

4.2   Model Discretization and Solving

To numerically solve the water flow motion Eqs. (1–3), the equations were discretized using the control volume method with integral conservation, the coupled equations were solved via SIMPLER formula, and such a process was repeatedly iterated until the convergence of the flow field. After discretization, the suspended load transport Eq. (4) and bed load transport Eq. (5) were implicitly solved through the underrelaxation technology and progressive scanning time division multiple address (TDMA) technology. Next, the riverbed deformation Eq. (6) was explicitly solved through finite difference discretization.

4.3   Relevant Parameter Settings

The main parameters of water flow motion equations include roughness coefficient n and turbulent viscosity coefficient v_t, where the former reflects the resistance of natural river, which can be determined by measuring the water level and verifying the flow velocity. The turbulent viscosity coefficient is calculated through $v_t = k_t u_* H$, wherein u_* is the friction velocity and k_t is a constant.

The main parameters of sediment motion equations include sediment transport capacity S_i^* of different suspended load groups and that (g_{bL}^*) of different bed load groups. $S_i^* = P_* S_*$, where P_* is the gradation of sediment transport capacity in different groups and S_* is the total sediment transport capacity of water flow, which are generally determined through the following Zhang R.J. formula:

$$S_* = k\left(\frac{\sqrt{u^2 + v^2}}{gh\omega}\right)^m \tag{6}$$

where ω is the average sediment deposition velocity of suspended loads; k and m are constants; P_* can be determined through the method rendered in Appendix B of Literature (Zhang et al. 2011). $g_{bL}^* = P_{bL}\eta_L g_b^*$, where P_{bL} is the percentage of sediments (group L) in riverbed sediments, η_L is the camouflage coefficient, and g_b^* is the total sediment transport rate of bed loads, which can be calculated through the formulas listed in Appendix A of Literature (Zhang et al. 2011).

4.4   Considerations in Pier Calculation

In the past calculations, piers are generalized mainly using local terrain correction, additional roughness coefficient, *etc.*, which can only simulate the water-resisting effect of piers while failing to give the detailed flow field changes near piers or calculate the compression scour of piers. As computer performance is enhanced and numerical methods are improved, piers can be processed by densifying the grids so that the grids are consistent with the pier scale. How to process the pier boundary is the key to accurately simulating the flow and sediment motions near piers, and the models can be divided into small-scale and large-scale models according to the spatial scale simulated.

Small-scale models are usually applied to laboratory pier simulation, with a small grid size (mm level). In this process, the pier boundary can be processed by combining the large coefficient method and wall shear rate, *i.e.*, the grids within the pier (internal grids encircled by grid nodes 1–16 in Fig. 4) are processed through the large coefficient method (if the source item is a large coefficient, its calculation result is taken as 0), while those adjacent to the pier wall (grid nodes 1–13) are processed using wall shear rate.

Large-scale models are generally used for pier simulation in natural riverways with complicated riverway boundaries and meter-level grid scale. When wall shear ratio is used, the distance from the first computing node on the wall surface to the wall surface is $z^+ = \frac{z_l u_*}{v}$, $30 < z^+ < 100$; for the lower reaches of the Yangtze River, z_l is 0.001 m under the flow velocity of 2 m/s, water depth of 10 m, and roughness coefficient of 0.025, so the scale needed in the calculation is much smaller than the pier scale. Therefore, the wall shear ratio is rarely used in the actual calculation, but instead, the inaccessible conditions and non-slip conditions are directly stipulated on the wall surface of the pier. The practice has proved that this method can simulate the detouring flow phenomenon nearby the pier very well.

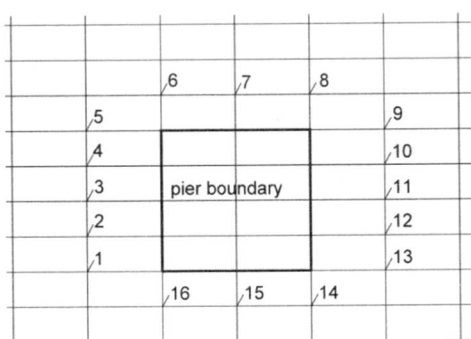

Fig. 4. Schematic diagram of bridge pier boundary treatment

4.5   Model Verification

In this study, a 2D planar flow and sediment model of Shiyazhou river reach was constructed, which reached the three rivers estuary upward and extended Liuxu estuary downward. The overall length of the river section simulated was about 40 km. To improve the model prediction accuracy, the model parameters were repeatedly calibrated and verified as seen in Table 1. The verification results of water level along the river and typical cross-sectional flow velocity are presented in Fig. 4. Restricted by the length of this paper, only the medium and long-term verification results of riverbed scouring-silting were given. In this model, the terrain in October 2011 was taken as the initial terrain and that in November 2015 as the verification terrain. The inlet flow rate and sediment content were the measured data from Datong Station, the outlet water level was acquired by the water-level-flow rate relation curves from Zhenjiang Station, and then this model verified the riverbed scouring-silting of this river reach after 4 complete hydrological years. The calculation results accorded well with the measured scouring-silting position and distribution.

Table 1. Calibration results of 2D flow and sediment model parameters for Shiyezhou

Verification type	Verification content	Verification data	Calibration parameter
Water flow	Flow velocity, water level and water diversion ratio	(1) Low water verification: December, 2012 (flow rate during measurement: 19,950 m^3/s) (2) Flood verification: July, 2005 (flow rate during measurement: 39,500 m^3/s)	Riverway roughness coefficient n
Sediment	Riverbed scouring-silting position, amount and distribution	(1) Annual scouring-silting: flood season (2010.3–2010.8), dry season (2010.8–2012.12) (2) Short-term (1 or 2 years) scouring-silting: 2010.3–2012.12, 2013.7–2014.7 (3) Medium and long term (3 or 5 years) scouring-silting: 2011.10–2015.11	Sediment transport capacity coefficient, saturation recovery coefficient, sediment transport rate of bed loads, etc

5 Prediction of the Compression Scour of Piers

5.1 Selection of Flow and Sediment Processes

The rationality of prediction results is closely related to the selection of flow and sediment processes. In this study, the determination method of flow and sediment processes was proposed from the partially safe angle according to the inflow flow and sediment characteristics on the lower reaches of the Yangtze River and considering the flow and sediment changes of a downstream hydrological control station (Datong) after the impoundment of the Three Gorges Reservoir.

(1) Flow and sediment processes in the years of the catastrophic flood: The year of once-in-three hundred years flood was selected as the year of catastrophic flood according to the design criteria (once-in-a-century flood design and once-in-three hundred years flood check) of the Shiyezhou bridge site scheme. The flow process was based on the year 1998 (once-in-a-century), and the peak discharge was amplified according to the frequency. The sediment process corresponding to the year of the catastrophic flood was derived (Fig. 5) using lower envelopes from the angle of engineering safety according to the flow rate-sediment transport rate relation curves of Datong Station after the impoundment of the Three Gorges Reservoir.

(2) Flow and sediment processes in serial hydrological years: Typical serial years were adopted, and the combination of the years of catastrophic floods was considered. According to the flow and sediment data of Datong Station during 1950–2014, the flow rate of Datong Station changed little since the impoundment of the Three Gorges Reservoir, the years of medium and small floods played a dominant role, but the sediment discharge was evidently smaller than that before the impoundment, which reflected the impounding and sediment trapping effects of the Three Gorges Reservoir. Before the impoundment of the Three Gorges Reservoir, the multi-year sediment discharge was 427 million tons in Datong Station. After the impoundment, the annual average sediment discharge was 143 million tons during 2004–2014, which was 66% lower than that before the impoundment, as shown in Fig. 6. The period of 2007–2010 represented the flow and sediment characteristics during 2004–2014 very well after the construction of the Three Gorges Reservoir. According to statistics, the runoff was 837,200 million m^3 during 2007–2010, which was basically equivalent to the average runoff (832,100 million m^3) after the construction of the Three Gorges Reservoir, with an annual maximum flow rate of 64,600 m^3/s and minimum flow rate of 10,000 m^3/s (Fig. 6). Considering the influence of the 1998 flood, the serial hydrological years were finally determined as 2007–2010+1998 (the sediment process was expressed by the lower envelope in Fig. 5(a)).

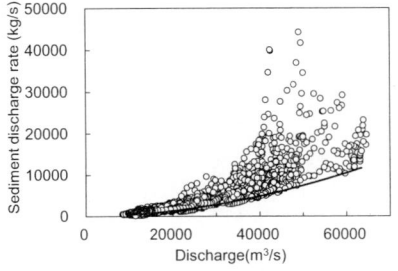

(a) Discharge and sediment discharge rate

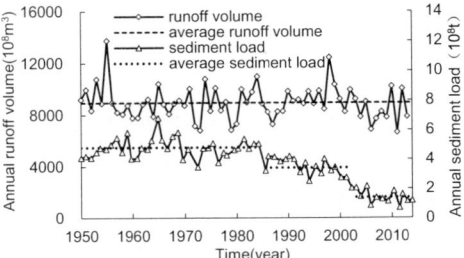

(b) Annual runoff and sediment load

Fig. 5. Runoff and sediment of Datong Station

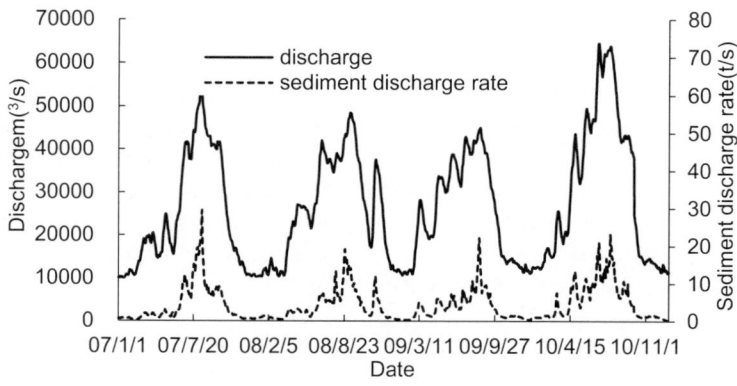

Fig. 6. Discharge and sediment discharge rate

5.2 Cross-Sectional Hydrodynamic Changes of the Bridge Site After Contraction of Piers

Due to the bridge construction, the water level and flow velocity distribution within a certain range of the bridge area were changed, which would have a direct bearing on the cross-sectional flow rate per unit width of the bridge site. Generally, the discharge per unit width between piers was elevated somehow after the bridge construction (Fig. 7). Impacted by water damming and covering of pier studs, the upstream and downstream flow rate per unit width was reduced. At the left side of the right branch in Shiyazhou, the flow rate per unit width between piers was reduced due to the small pier spacing and large resistance, and meanwhile, the increased amplitude of the flow rate per unit width of the right riverway was large.

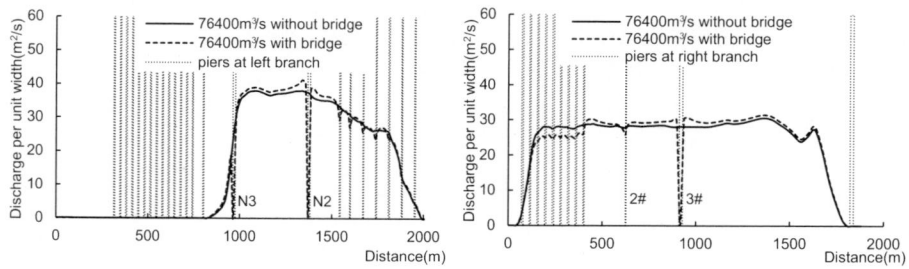

Fig. 7. Transverse distribution of the unit discharge of bridge site sections

5.3 Prediction of the Compression Scour of Piers

After encountering the year of a catastrophic flood, the upstream flow velocity was decelerated due to the water-resisting effect of piers, and riverbed silting dominated; the downstream riverbed experienced both scouring and silting due to the water flow extruding and covering effects of piers; given the presence of piers on the cross-section of this bridge site, the discharge area was narrowed, the flow rate per unit width was elevated, compression scour took place, and the scour amplitude was greater at the left-branch, e.g., N3# pier (9.46 m) and N2# pier (5.92 m), while the cross-sectional scour amplitude was relatively smaller at the right branch, e.g., 2# auxiliary bridge pier (2.81 m) and 3# main pier (3.70 m) (see Fig. 8 and Fig. 9).

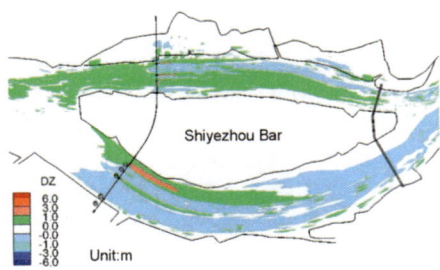
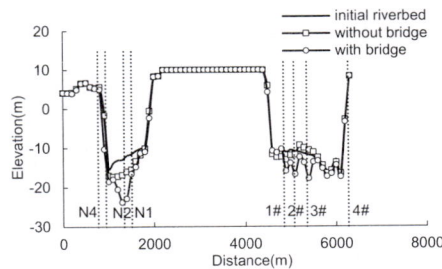

Fig. 8. Erosion and siltation caused by bridge piers after a catastrophic flood year

Fig. 9. Changes of transverse section after a catastrophic flood year

DaTong-WuSong Estuary sailing charts of the Yangtze River in 1998 and 2013 (Lu 2016) were used to calculate the cross-sectional compression scour (Table 2) at typical bridge sites on the lower reaches of the Yangtze River. It could be seen that due to the different bridge locations and cross-sectional morphologies, the depth of compression scour varied from the bridge to bridge, and the scour amplitude ranged from 2.6 m to 8.0 m, which was equivalent to the prediction result of the Shiyezhou bridge site model.

Table 2. Statistical measured maximum depth of cross-sectional compression scour at typical bridge sites on the lower reaches of the Yangtze river (1998–2013)

	Tongling Yangtze River Highway Bridge	Wuhu Yangtze River Bridge	Dashengguan Yangtze River Bridge	Nanjing Yangtze River Bridge	Nanjing Second Yangtze River Bridge	Nanjing Fourth Yangtze River Bridge
Bridge width (km)	1.1	2.1	1.4	1.2	1.6	1.9
Cross section	V shaped	U shaped	V shaped	V shaped	U shaped	V shaped
Maximum depth of compression scour (m)	4.0	4.2	8.0	2.8	2.6	6.0

6   Conclusions

(1) A prediction model for the compression scour of piers was constructed. Small-scale models can be processed by combining the large coefficient method and wall shear rate. Wall shear rate is inapplicable to natural large-scale models, and thus inaccessible conditions and non-slip conditions can be directly stipulated on the wall surface of piers. Repeated calibration and validation allow the model parameters to apply to recheck the annual and interannual scouring-silting deformation of riverways.

(2) The determination method of flow and sediment conditions was proposed and the adverse flow and sediment processes were determined from the angle of engineering safety based on the flow and sediment characteristics on the lower reaches of the Yangtze River and the changes in the flow and sediment conditions of Datong Station since the impoundment of the Three Gorges Reservoir.

(3) Impacted by bridge piers, the cross-section of the bridge site was subjected to compression scour, where the scour amplitude was great at left-branch N3# pier (9.46 m) and N2# pier (5.92 m), while the cross-sectional scour amplitude was relatively smaller at the right branch. Through the statistical cross-sectional compression scour at the already built bridge sites on the lower reaches of the Yangtze River, the scour depth ranged from 2.6 m to 8.0 m, which was basically equivalent to the model prediction result.

Acknowledgements. This research is supported by the Subsequent Work of the Three Gorges Project (No. SXHXGZ-2020-3) and Central Public-Interest Scientific Institution Basal Research Fund (No. Y222013).

References

The river sediment Engineering Department of Wuhan University of Hydraulic and Electric Science 1981 River sediment engineering. Conservancy Press, Beijing

Zhang HW, Ma JY, Zhang JH, et al (1993) The design river and transfer. China Architecture Building Press, Beijing (in Chinese)

Olsen NRB, Kjellesvig HM (1998) Three-dimensional numerical flow modeling for estimation of maximum local scour depth. J Hydraulic Res 36(4):579–590

Karim OA, Ali KHM (2000) Prediction of flow patterns in local scour holes caused by turbulent water jets. J Hydraulic Res 38(4):279–287

Roulund A, Sumer BM, Fredsoe J, Michelsen J (2005) Numerical and experimental investigation of flow and scour around a circular pile. J Fluid Mech 534:351–401

Liu X, Garcia MH (2006) Numerical simulation of local scour with free surface and automatic mesh deformation. World Environmental & Water Resource Congress. Omaha, NE, (CD-ROMs)

Chen XL (2008) Study on mechanics and numerical simulation of flow and local scour around hydraulic structures. Tsinghua University, Beijing (in Chinese)

Sumer BM, Withehouse RJS, Torum A (2001) Scour around coastal structures: a Summary of recent research. Coast Eng 44(2):153–190

Yong GL, Blair PG (2010) Predicting contraction scour with a two-dimensional depth-averaged model. J Hydraul Res 48(3):383–387

Fenocchi A, Natale L (2016) Using numerical and physical modeling to evaluate total scour at bridge piers. J Hydraul Eng 142(3):06015021

Guo H, Qi ML (2011) Numerical simulation study on the contraction scour of bridge cross the river. China Railway Sci 32(5):43–49 (in Chinese)

Guo H (2013) Numerical study of contraction scour at bridge crossings. Beijing Jiaotong University (in Chinese)

Ministry of Railways of the People's Republic of China (1999) TB10017-99 Code for survey and design on hydrology of railway engineering. China Railway Publishing House, Beijing (in Chinese)

Ministry of Transport of the People's Republic of China (2002) JTG C30-2002 hydrological specifications for survey and design of highway engineering. China Communications Press, Beijing (in Chinese)

Shirole AM, Holt RC (1991) Planning for comprehensive bridge safety assurance program (Transport Research Report No. 1290). Transportation Research Board, Washington, D.C., pp 137–142

Lu H, Gao DG (1996) Bridge Hydraulics. China Communications Press, Beijing (in Chinese)

Due BM, Rodi W (2008) Numerical simulation of contraction scour in an open channel. J Hydraul Eng 134(4):367–377

Liu YL, Chu YD (1993) Numerical modelling of scouring and silting of full sediment in the bridge reach. J China Railway Soc 4:96–102 (in Chinese)

Zhang W, Li YT, Yuan J (2011) Prediction of maximum bed erosion depth near a crossing tunnel under the lower reach Yangtze River. J Hydroelectric Eng 30(4):90–97 (in Chinese)

Wei S, Li GL, Chen S (2016) Mathematical model studies on maximum bed erosion depth near Shiyezhou river-crossing tunnel. Hydro-Sci Eng 1:1–8 (in Chinese)

Yue HY, Gu LH, Zhang J (2010) Fluvial process and prediction of maximum erosion depth of riverbed in tunnel location across Hanjiang River. Yangtze River 41(6):35–39 (in Chinese)

Ministry of Transport of the People's Republic of China (1999) JTJ/T232-98 technical regulation of medelling for flow and sediment in inland waterway and harbour. China Communications Press, Beijing (in Chinese)

Lu XJ (2016) Research on the local scour at bridge piers in the tidal reach of the Changjiang river. East China Normal University (in Chinese)

Sustainable Management of the Navigability of Natural Rivers (PIANC WG 236)

Calvin Creech[1]($\boxtimes$), Erik Mosselman[2]($\boxtimes$), Jean-Michel Hiver[3]($\boxtimes$), and Nils Huber[4]($\boxtimes$)

[1] US Army Corps of Engineers, Mobile, USA
Calvin.T.Creech@usace.army.mil
[2] Deltares, TU Delft, Delft, The Netherlands
Erik.Mosselman@deltares.nl
[3] Université Libre de Bruxelles, Brussels, Belgium
jean-michel.hiver@ulb.be
[4] BAW, Karlsruhe, Germany
nils.huber@baw.de

Abstract. The PIANC InCom/EnviCom Working Group 236 was established in early 2021 to develop PIANC guidelines for improving navigability conditions on natural or quasi-natural rivers, while maintaining morphological processes and natural river form and function. Its key objectives include: 1) development of guidelines to improve and maintain the navigability in natural rivers; 2) assess the sustainability of river training works designed to improve the navigability; 3) assess the sustainability of dynamic river management (monitoring and shifting of navigation aids to adapt the navigation channel to the river dynamics); 4) highlight the technical, operational, economic and environmental considerations for navigation in natural rivers compared to that in regulated rivers and canals; and 5) improve the understanding of the physical processes in natural rivers, developed with or without river training works. The developed guidance includes a planning framework for developing a navigability improvement masterplan for a natural or quasi-natural river system, and the integrated and adaptive management strategies that can be applied at a system scale. Specific interventions and measures have been identified to meet the dual goals of maintaining morphological river function and improving navigability conditions. These measures include dynamic charting; morphological dredging and disposal management; Temporary, Adaptable, and Flexible Training Structures (TAFTS); riverbed armoring and sediment nourishment; rock excavation; meander cutoffs and oxbow development; localized traditional river training structures; and channel closure structures. The impacts and strategies for mitigation associated with some of the measures are analyzed and discussed. Finally, the continual monitoring, management, and operational tools available for improving navigability in a morphologically active river system is presented. It is recognized that natural and quasi-natural rivers will typically be more fluvially active and dynamic than systems that have used traditional methods for navigability improvements including heavily trained rivers or systems with locks and dams. These unrestricted and unconfined river systems, therefore, will require new and innovative strategies to monitor the fluvial and geomorphic changes of the system in order to inform managers and navigators

© The Author(s) 2023
Y. Li et al. (Eds.): PIANC 2022, LNCE 264, pp. 232–242, 2023.
https://doi.org/10.1007/978-981-19-6138-0_21

of the river. Case studies are presented that include the Madeira River (Brazil); Magdalena River (Colombia); Niger Delta (Nigeria); Yangtze River (China); the Brahmaputra-Jamuna River (India); and the Red River (Vietnam).

Keywords: Nature-based · Inland navigation · Management

1 Introduction

Inland Waterway Transport (IWT) include three types of waterway features: 1) rivers and estuaries, 2) lakes and reservoirs, and 3) canals. In the riverine systems, there are two possible situations: natural rivers and rivers that are regulated or trained by hydraulic infrastructure.

The primary application of the PIANC Working Group 236 lies within natural or quasi-natural river systems – systems that are not constrained or significantly impacted by dams or river training structures. The navigable natural rivers are often some of the "large" rivers of the world. Due to significant depths and widths of these large rivers, navigation opportunities are naturally present in these systems.

A natural river system exhibits two important characteristics – unregulated hydrology and unconstrained morphology. In other words, both the hydrology and the morphology of the system is "natural", and the river can freely respond to the environmental boundary conditions provided by the watershed. These are the systems that have not been subjected to significant engineering interventions. This unconstrainted and unregulated condition results in dynamic and natural river evolutionary processes within the riverine corridor that are increasingly valued from viewpoints of ecology and natural heritage. Therefore, understanding these system-specific natural river processes are important in developing recommendations for navigability improvement on these systems.

In many natural river systems, it is not technically feasible nor environmentally desirable to improve navigability through river training works. The only solution is then to assist the river in maintaining a navigable channel through specific actions – for example morphological dredging or adaptive management of the navigation channel itself.

It is within this context that the PIANC Working Group 236 was formed. The focus of the Working Group is on best practices that can be implemented within the natural system, which will maintain natural riverine processes following implementation. This approach results in innovative navigability improvement and management strategies while maintaining the natural river system behavior (see the lower right quadrant in Table 1).

Table 1. Resulting guidance based on existing system conditions and future end state

	River State	Existing System Conditions	
		Regulated or constrained by Hydraulic Infrastructure	Natural or Quasi-Natural
Proposed System Conditions	Regulated or constrained by Hydraulic Infrastructure	*Resulting Guidance: Management and maintenance of existing systems that are controlled by hydraulic infrastructure.*	*Resulting Guidance: Traditional methods of river engineering.*
	Natural or Quasi-Natural	*Resulting Guidance: River restoration of systems that have been trained or significantly regulated.* **Recommended for a future PIANC Working Group**	**Resulting Guidance: Innovative navigability improvements that maintain natural river system behavior. Focus of WG 236.**

2   Planning and Design Strategies to Improve Navigability in Natural Rivers

A planning process is first developed as a framework for the navigability improvement masterplan associated with a natural river system. This begins with developing an understanding of the natural processes followed by an integrated and adaptive management strategy at the system scale within the context of the dynamic morphology of the system. This requires specific steps and guidance for implementing a planning study on a natural river, which include data acquisition (hydrographic surveying, remote sensing, sediment data collection), geology and fluvial geomorphology studies, hydrology and hydraulic analysis, sediment transport analysis, economic analysis, and socio-environmental studies.

On a strategic level, navigability in a natural or quasi-natural river system can be improved either by means of classical and dedicated river engineering projects or by means of continuous improvement processes of the daily management of the river. Whereas river engineering projects have a defined project framework (set of river engineering measures, concrete timeline, defined project budget, planning and construction phase, acquisition of necessary permissions) the improvement of the daily river management is a continuous and more fluent task, which can be described by means of following the river management cycle (see Fig. 1).

This river management cycle displays the basic elements (planning and execution of works, monitoring, and evaluation of continuous information on fairway status) in the daily work of a river manager aiming at maintaining or improving navigability in a natural free river stretch.

The continuous improvement of these management elements is a dynamic process, which has to take into consideration the specific financial and human resources of the respective waterway authority, the specific annual discharge pattern of the considered natural free river stretch, the morphological and ecological opportunities and limitations of the respective river system as well as the specific needs of the users of the fairway (navigation sector).

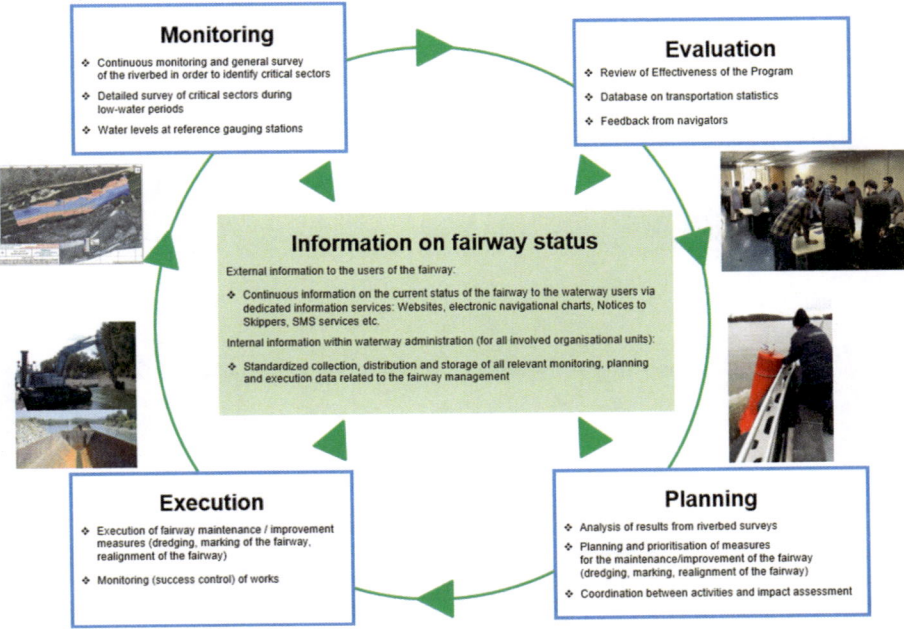

Fig. 1. The river management cycle

3 Interventions and Measures to Improve Navigability in Natural Rivers

Following the development of a masterplan, it is necessary to identify the measures that can be implemented to improve navigability in natural rivers given the system processes and constraints. Several interventions have been identified by the Working Group and are presented in the following sections.

3.1 Dynamic Fairway Management

Rivers that are highly dynamic, with high sediment transport loads and a very strong seasonal variation in water level and discharge, require an innovative approach to enable navigation. For these types of pristine rivers, it is not desirable, nor environmentally or economically justified to enhance navigation by providing structural interventions along the full stretch of the river. Moreover, on some free-flowing and braiding rivers, it is also not viable economically to establish and maintain one static channel. Therefore, because of their nature, this type of rivers can only be feasibly controlled by applying flexible, adaptable or dynamic management strategies. Dynamic

(Fairway) Management should be considered as an alternative solution to enable navigation in otherwise difficult to manage environments.

For highly-dynamic, fluvial rivers, like the Ayeyarwady and the Jamuna, where Inland Water Transport (IWT) is highly restricted by the available water depth and as a result IWT is yet to be developed, the only cost-effective method to develop the navigation channel is to use Dynamic Fairway Management techniques including (see Fig. 2):

- Hydrographic surveying and (electronic) chart updating,
- Aids to Navigation and their repositioning.
- Dredging to eliminate remaining navigation constraints.

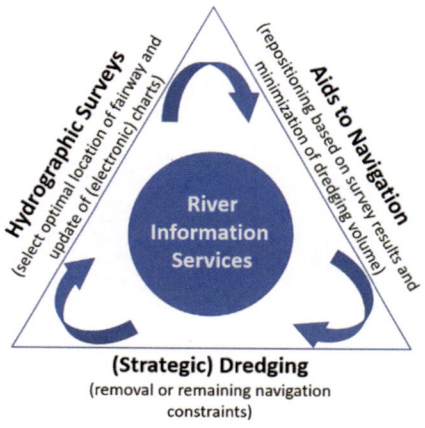

Fig. 2. Dynamic (fairway) management

3.2   Navigability Forecasts and Alerts

Navigation charts need to be updated regularly when the river bed changes. This regular updating can be enhanced by making use of the echosounder data of the vessels that are plying the river. This is the background of the CoVadem initiative (www.covadem.com) which gathers and combines echosounder data from a large set of inland vessels. Currently, over 250 Dutch ships participate by measuring, logging and exporting data from every river and canal where they sail in the Netherlands, Belgium, Germany and Switzerland. During their trips, they continuously measure underkeel clearance using conventional echosounder equipment as well as location using a GPS meter. Underkeel clearances are translated into water depths by correcting for draught, squat and trim. The draught is taken from the logged loading gauge just before a trip starts. The squat and trim are calculated using an empirical model. The results are made available in the form of an operational water depth chart for 600 km of the Rhine in Germany and the Netherlands. This chart is constructed in two steps. First, measured water depth data are used to derive an up-to-date river bed topography. Second, this topography is combined with hydrological predictions of present and near-future water levels. The resulting operational chart helps to maximize loading capacity, to reduce fuel consumption by sailing more efficiently, to guarantee a reliable estimated time of arrival (ETA), and to plan survey and dredging operations (Van der Mark and Lemans 2020). The challenge is to develop similar forecast systems for natural and quasi-natural rivers (Fig. 3).

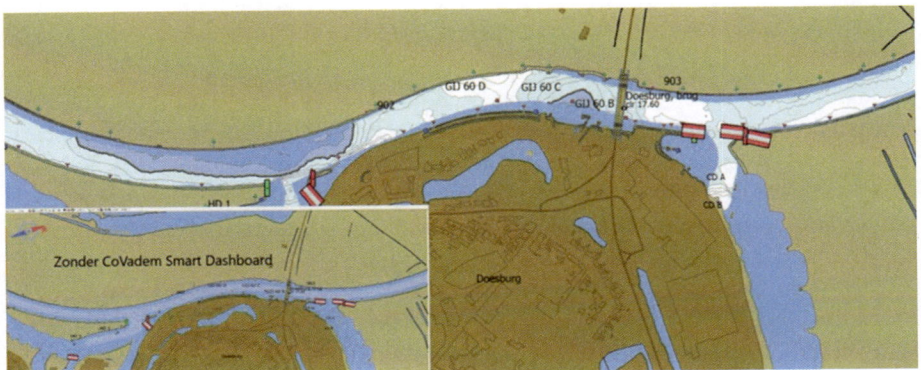

Fig. 3. Up-to-date water depth chart from CoVadem (www.covadem.com)

3.3 Morphological Dredging and Disposal Management

Morphological dredging also called smart dredging makes it possible to perpetuate the measurer's management of the river to maintain or improve the navigability based mainly on continuously monitoring and prediction of the morphological evolution of the river. The morphological dredging effectiveness relies on good understanding of the morphological evolution of the hydrofluvial system and on following the principles of working with nature.

3.4 Removal of Obstacles in the Navigation Fairway (Rock Excavation, Clearing of Snags and Obstructions)

For some natural channels, the presence of rock will serve as the limit for available channel depth in the navigation fairway under low-water conditions. This rock may present an unavoidable obstacle if there are constraints to the fairway location (i.e. the channel cannot be re-aligned such that the rock does not pose an obstacle). In this case, the governing entity is posed with a choice: pursue rock removal, or exhaust channel limitation options (first light-loading, and eventually channel closure). This is driven by the economics of the rock removal (the volume needing removal and the cost per volume unit) vs. the cost to shipping of reduced channel efficiency or routing to another means of transportation. To determine the economics of rock removal, the volume needing removal and the competency of the material must be determined. The removal volumes is determined with extensive multibeam surveying, as sediment transport often covers rock features. The volume needing removal can vary tremendously, as at some locations, only rock pinnacles or minor patches infringe on the desired low water navigation channel geometry, whereas other locations have rock infringe on that geometry for considerable height for entire reaches. The competency of the rock drives decisions on the removal methodology, as more competent rock requires more extreme removal measures, elongating the schedule and increasing environmental permitting requirements. Soft or loose rock can be removed via excavator. Harder rock will require more intensive means, such as a hydraulic hammer, rock grinder, or explosives. Geologists should be consulted to discuss the appropriate means of removal. Some of these alternatives (e.g. hydraulic hammers) come in different sizes with different rates and capabilities to break up rock. Specialized equipment such as a hydraulic hammer or grinder head for excavator will be expensive. However, this expense may be competitive with blasting when factoring in production rates, the cost of channel closures due to safety, and the increased environmental monitoring and mitigation measures that may be required.

Once the feasibility and the economics of rock removal have been determined, a detailed coordination plan is required to manage the removal activities, surveying for quality control or quality assurance on successful removal and removal quantities, and environmental monitoring. This plan needs to be highly flexible, as removal activities often have a high dependence on river stage at the removal location. The quality control or quality assurance process establishes needs to be capable of quick confirmation, as sediment transport into the site will obscure cleared rock sites with mobile material. Thus, breaking a removal site down into smaller units that can be surveyed and analyzed quickly is recommended.

3.5   Temporary, Adaptable, and Flexible Training Structures (TAFTS)

The traditional usage of training structures on untrained rivers has largely focused on forcing a new, rigid organization on a system through robust, permanent structures for the singular goal of navigability. These rigid structures are optimized for a certain discharge, not taking account the dynamics of natural or quasi-natural river systems throughout the year. This philosophy has changed, as understanding of what was lost through the traditional methodology has grown, and re-imagining and re-framing of historic practices have presented ways to better align with the natural flexibility of untrained systems. This section seeks to highlight temporary, adaptable, and flexible structures such as:

- Temporary placement of vessels or jacks for channel constriction
- Low-cost woody debris structures allowed to fail
- Base extensions and notching to increase or reduce constriction

These structures can be altered easily and at low cost to adjust for the lateral or longitudinal migration of the channel. Such temporary and flexible structures can also contribute to improve the ecological functions of the river and flood protection.

3.6   Armouring and Sediment Nourishment

Improving the navigability of a river possibly comes along with managing the sediment transport capacity. In erosive environments, in combination with a bedload deficit, e.g., downstream of dams or induced by extensive mining, this can result in progressing riverbed incision. An erosive system cannot only worsen river ecosystems, but also deteriorate navigability conditions, e.g., by destabilizing buildings along the river, exacerbating the access to harbors and non-fluvial parts of the IWT network, or revealing obstacles like rocks. Therefore, methods are necessary to prevent these effects and to control a degrading system by stabilizing the riverbed.

Riverbed stabilization can be accomplished by many different interventions and measures whereby three different approaches can be mentioned: Riverbed armouring aims for increasing the particle erosion resistance of the riverbed. Sediment nourishment tends to raise the sediment supply. The third approach focuses on interventions decreasing the capacity of the flow to transport sediment (expressed by flow velocity, bed shear stress or stream power).

The objective of this section is to provide a flexible and dynamic tool as part of a river basin wide sediment management strategy. If properly designed, it can also promote natural river morphology and have positive effects on environmental conditions such as habitat diversity. For this purpose, riverbed armouring and sediment nourishment are suitable methods, while the former acts locally and the latter also has impact on downstream river reaches.

Riverbed armouring as a method describes the adding of coarse material to the riverbed to control ongoing erosion by increasing the erosion resistance of the upper bed layer. In contrast to sediment nourishment, the added sediments are intended to remain in place and form an armour layer rather than be transported. This procedure can be considered as a direct method to stabilize the river bed. It is suitable for locations with locally high bed shear stresses and high transport capacity in combination with low particle erosion resistances.

The stability of riverbed armouring depends on the type of riverbed armouring, in particular its grain size. If the riverbed armouring is much coarser than the existing riverbed, this may result in a completely immobile armour layer. Although this may be efficient, it is also a risk in the long term due to resulting sediment starvation downstream. Likewise, in dynamic natural or quasi-natural rivers, the relative position of the armour layer within the river may change over time. For example, an armour layer placed in an outer bend may end in the inner bend after meander migration. Also, in case of overall riverbed degradation, the armour layer may form an obstacle for navigation on the long term (examples in the Rhine). Riverbed armouring should therefore be applied with care and with knowledge about the long-term development of the river.

3.7   River Training Works

River works for regulation of the low-water bed may include the construction of bank revetments, longitudinal dikes, and groynes or spur-dikes. There are a great variety of these structures, in type and dimensions as well as in the materials of construction. These types of structures include:

- Spur dikes
- Bottom vanes
- Longitudinal dikes with or without transverse dikes
- Revetments (Fig. 4).

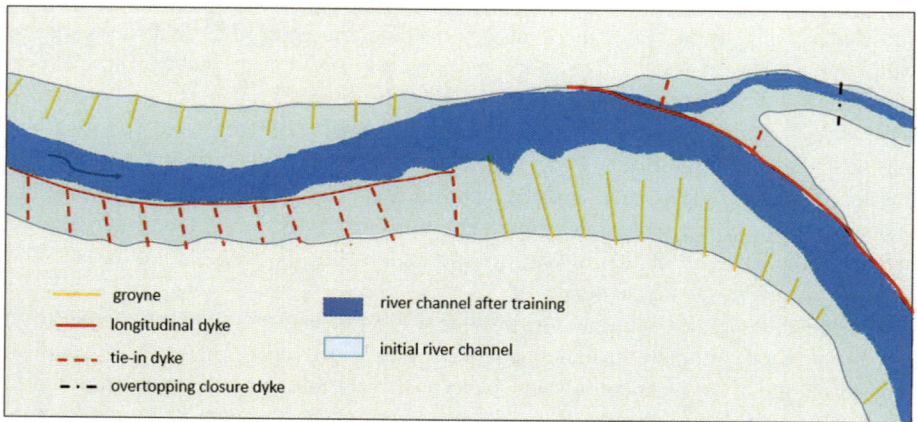

Fig. 4. Various types of river training structures

4   Future Work

4.1   Impacts and Mitigation of Navigability Improvement Interventions

The impacts and mitigation of navigability improvements will then described in the Working Group Report. Here the focus will be on river training works as they have the largest impacts. Nonetheless, impacts associated with other measures will be included too, as applicable. These impacts include mitigation of bank erosion processes as well as sediment nourishment. The impacts on reaches of the river that have been adapted by river training structures are necessary to analyze within the context of navigation safety. The Working Group also will include descriptions of potential impacts from river training structures (or other physical navigability improvement measures) to other areas including flood risk and ecosystem resources. This section will also include potential measures to mitigate against these impacts and discuss the long-term navigability of the system following the implementation of these measures.

4.2   Monitoring, Management, and Operations Tools

Reasoned work planning methods may be described in a cycle leading to continuous navigability quality improvement, the service offered to users and any economic, environmental or other objective that has been clearly identified and defined.

These tools are dependent on structural measures, management, institutional, technical, human and economic resources, such as:

- structural measures (river training works, bank protection, …)
- morphological dredging
- programmed inspections and frequency
- marking
- technical and technological resources (hydrographic boats, measuring equipment, …)
- efficiency of the hydrographic service (management and informatics resources, …)
- human resources and their qualification
- financial resources devoted to the management of the river
- institutional measures

Consequently, it is important to have a set of indicators - efficiency, pertinence, expectation, satisfaction – (i.e. respect of the navigability requirements, evolution of the river during the hydrological cycle, annual duration of navigation, proactive intervention management) to judge the quality of each of these tools and to consider possible actions to improve them.

For example, the following considerations are included:

- critical analysis of information collected from pilots regarding problems, alerts or accidents due to navigability failures
- improvement of corrective interventions on local structural measures, morphological dredging (i.e. optimisation of trends analyses based on data base) and marking management
- improvement of protocol field measurements, data analysis and integration of new technology
- improvement of the activities of a hydrographic service (i.e. informatics resources and continuous training of the personnel
- where asset performance relates to the ability of the assets to meet target levels of service, inspections allow the monitoring of that performance with time.

4.3 Summary and Conclusions

The current version of the Working Group 236 report is the first compiled document containing all the contributions of the WG's members. It is a working document allowing the Working Group to continue its work according to the ToR.

The Working Group will conduct the following recommendations:

1. Gather information under the different themes listed in "Monitoring, Evaluation, Planning and Execution" in order to concatenate and issue specific recommendations for them.
2. Gather relevant information specific to the sustainability of the effects of measures to improve navigability and their technical and socio-environmental impacts.
3. There are three specific recommendations that are being advanced by the working Group for inclusion in the PIANC WG 236 Report. These include:

a. Establish or create a hydrography service (institutional aspects, human resources, innovations technologies, financial aspects, etc.). It is an important point to guarantee continuously monitoring.
b. Operations and Maintenance of the Aids to Navigation
c. Continuous training of the staff (local and/or outside training)

Reference

Van der Mark C, Lemans M (2020) Operational 2D water depth prediction using echo sounder data of inland ships. River Flow 2020. CRC Press

The Design of Safety Monitoring System for Navigation Hubs Built in Plain Area

Yutang Ding[✉], Yangyang Lu, and Guowen Xiong

Nanjing Hydraulic Research Institution, Nanjing, China
ytding@nhri.cn

Abstract. Navigation hubs built in plain area usually have large space span and compact structure layout. Safety monitoring performance, especially the surface deformation monitoring performance, is sensitive to the monitoring distance and structure features. Thus the monitoring system should be elaborately designed to meet the requirement of accuracy and robustness. Conventional approaches for surface deformation monitoring are time-consuming and suffering from low accuracy. In this paper, the designing strategies and monitoring techniques for surface deformation monitoring in Jiepai Navigation Hub, a typical navigation-power hub built in plain area, are demonstrated. The monitoring result shows that the designed monitoring system has good performance both in accuracy and robustness. The designing strategy can serve as a reference for other similar projects built in plain area.

Keywords: Safety monitoring · Navigation hub · Plain area · Designing strategy

1 Introduction

Surface deformation monitoring is an essential part of safety monitoring system for waterway infrastructures. The deformation data serves as one of the most important bases for assessing structure running state. The accuracy and robustness of surface deformation monitoring system, however, depend strongly on monitoring approaches and structure features. Manual observation, GNSS and sensor monitoring are the three main methods for surface deformation monitoring in hydraulic engineering (Ding et al. 2022; Zhou et.al. 2022).

Manual observation using the total station and the electronic level is the most widely used method due to the low cost. However, navigation hubs built in plain area usually have large space span and complex structure layout. Monitoring accuracy can be easily reduced due to error accumulation along the long monitoring distance. Intervisibility between monitoring points and datum points may be destroyed by the complex structure layout. Moreover, manual observation is time-consuming and labor intensive, thus the real-time monitoring cannot be achieved.

GNSS, i.e. Global Navigation Satellite System, is a rapidly developing method in deformation monitoring area (Huang et al. 2018). This method is based on satellite

© The Author(s) 2023
Y. Li et al. (Eds.): PIANC 2022, LNCE 264, pp. 243–250, 2023.
https://doi.org/10.1007/978-981-19-6138-0_22

positioning technology, deformation data is automatically computed from the distance between satellites and monitoring stations. The accuracy can hardly be influenced by space span or structure features. The cost of GNSS, however, is relatively expensive. Thus it is uneconomical to design a large number of GNSS monitoring stations for navigation hubs with large space span.

Sensor monitoring is another widely adopted monitoring approach (Yan and Li 2002; Li and Li 2002). The surface deformation is automatically measured by a set of sensors and then collected by the monitoring platform software, thus the real-time monitoring can be achieved. The cost for sensor monitoring is lower than GNSS method while the accuracy is relatively higher. Installation space, however, must be ensured for facility installation and maintenance.

Jiepai Navigation Hub, which serves as a key waterway project on Xinjiang River in east China, is a typical inland navigation hub built in plain area. Diverse structures including the navigation lock, the overflow dam, the flat slab dam, the sluice gates, the concrete dam, the earth dam and the hydro-power stations are assigned compactly along the long engineering axis. Difficulties are experienced when designing the surface deformation monitoring system. This paper introduces the designing strategies of surface deformation monitoring system for Jiepai Navigation Hub. Solutions to the difficulties including the low deformation monitoring accuracy for long-axis structures and the lack of construction space are stressed. Some representative monitoring results are also exhibited and analyzed. The designing strategy can be used as a reference for other similar navigation projects.

2 The Designing Strategy of Surface Monitoring System

2.1 Layout of the Project

The layout of the project can be seen from Fig. 1. Diverse structures are assigned across Xinjiang River between the left bank and the right bank. A road bridge is also built on the top of water retaining structures. The overall length of engineering axis is about 1.0 km. The project layout is complex and the monitoring distance is much longer than ordinary hydraulic infrastructures.

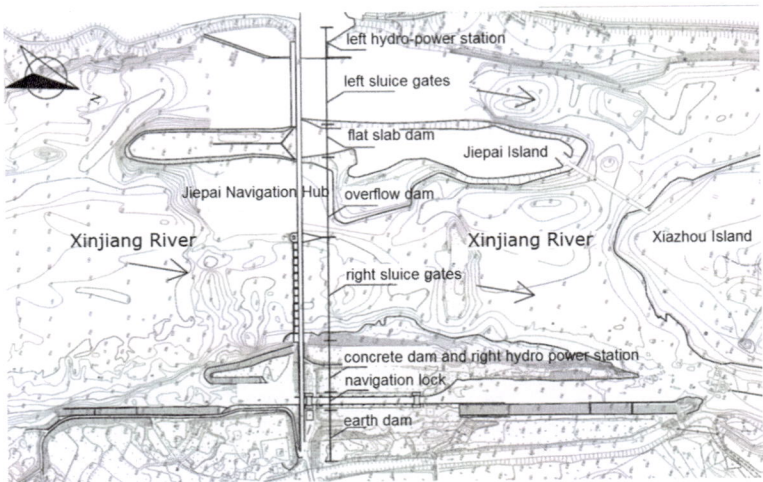

Fig. 1. The layout of Jiepai Navigation Hub.

2.2 The Designing Strategy of Surface Deformation Monitoring System

According to the discussion in Sect. 1, GNSS method and manual observation method have significant disadvantages in economical cost or monitoring accuracy. Then sensor monitoring method turns to be an appropriate choice as the cost and the accuracy are satisfying. Two kinds of sensor systems, hydrostatic level system and tension wire alignment system are thus applied for vertical deformation monitoring and horizontal deformation monitoring respectively.

The navigation lock and the main water retaining structures, including the sluice gates, the flat slab dam and the overflow dam, are the primary monitoring objects. As the measuring points in tension wire alignment system should be built along a straight line, the monitoring system is divided into 5 independent sub-systems according to the structure feature. The sketch of the system can be seen in Fig. 2.

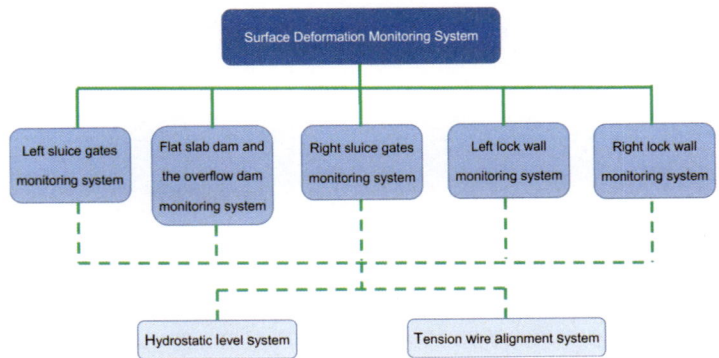

Fig. 2. Sketch of surface monitoring system.

To give a detailed illustration, the designing detail for right lock wall is shown in Fig. 3. The right lock wall consists of 11 independent blocks. The water level in lock chamber varies significantly and frequently as ships keep driving through the navigation lock. The deformation of each block is of great concern as the blocks are in complex loading conditions. Thus sensors, including hydrostatic level sensors and tension wire alignment sensors, are placed near the center of block surface. The inverse plummet device and double-metal pipe device are employed as the horizontal displacement datum point and vertical displacement datum point respectively. All the sensors are connected into the safety monitoring platform, thus the deformation can be measured and collected automatically.

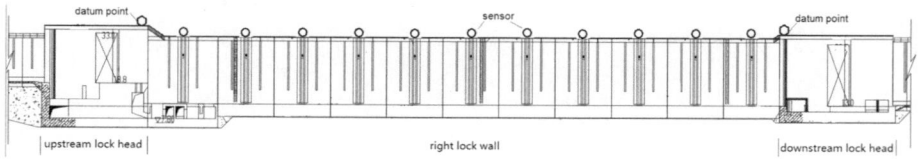

Fig. 3. Layout of measuring points for right lock wall.

2.3   Solution to the Lack of Construction Space

The above designing strategy successfully guarantees the feasibility and accuracy of the monitoring system as the sub-system are independent and the measuring distances are reduced to no more than 200 m. Difficulties, however, such as lack of construction space still exist. Auxiliary facilities are thus needed. The implementation of overflow dam monitoring system is a typical example. As can be seen from Fig. 4(a), sensors are placed on the top of the pier with narrow space. No passageway is available for sensor installation. To solve the above problem, a steel auxiliary bridge is designed. The sketch can be seen from Fig. 4(b). U-bars are fixed on the both sides of the piers and the auxiliary bridge is placed on the U-bars. Bearing capacity is examined by calculation and in-situ testing. Figure 4(c) shows the completed bridge. The passageway for installation and maintenance is thus ensured.

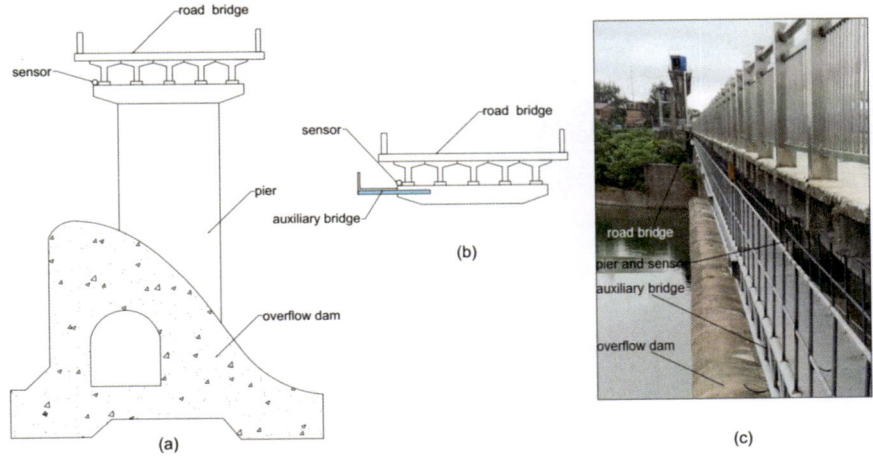

Fig. 4. Layout of sensors and auxiliary bridge for overflow dam.

3 Working Principles of the Monitoring System

3.1 The Hydrostatic Level System

The hydrostatic level system is built according to the principle of connected vessels. As can be seen from Fig. 5, the sensors are connected by tubes and filled with fluid. At the initial moment, the surface of the fluid in each sensor are at the same elevation. The fluid height is measured and recorded as $h_0, ..., h_i, ..., h_n$ respectively. Thus we have:

$$H_0 + h_0 = H_i + h_i = H_n + h_n \tag{1}$$

where H stands for the structure elevation. At the current moment, the structure experienced some deformation, and we have:

$$H_0^* + h_0^* = H_i^* + h_i^* = H_i^* + h_i^* \tag{2}$$

where the superscript "*" stand for the current moment. Eq. (2) can be rewritten as:

$$(H_0 + \Delta H_0) + (h_0 + \Delta h_0) = (H_i + \Delta H_i) + (h_i + \Delta h_i)$$
$$= (H_n + \Delta H_n) + (h_n + \Delta h_n) \tag{3}$$

With:

$$\Delta H = H_0^* - H_0; \ \Delta h = h_0^* - h_0 \tag{4}$$

Thus the vertical deformation of the ith measuring point can be obtained:

$$\Delta H_i = \Delta H_0 + \Delta h_0 - \Delta h_i \tag{5}$$

where ΔH_0 is measured by double-metal pipe device (datum point), Δh_0 & Δh_i are measured by "sensor 0" and "sensor i" respectively.

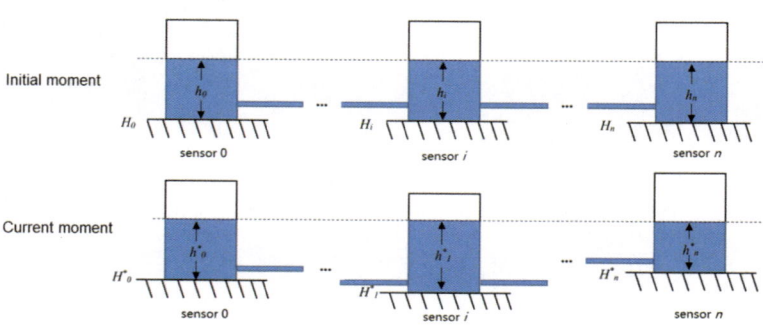

Fig. 5. Sketch of hydrostatic level system.

3.2   The Tension Wire Alignment System

The tension wire alignment system is made of deformation sensors, datum points (inverse plummet device), and tension wire device. The system is assigned into a straight line and the tensioned wire goes through each sensor. As can be seen from Fig. 6, the relative horizontal displacement (ΔS_i) between the wire and the structure is measured by the ith sensor. The true deformation then can be obtained with the correction of datum points (ΔS_{j1}, ΔS_{j2}).

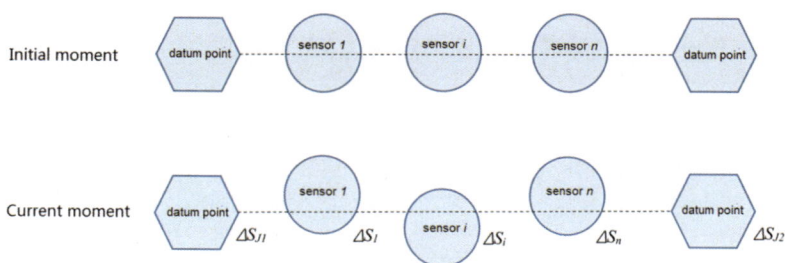

Fig. 6. Sketch of tension wire alignment system.

4   Monitoring Performance

To evaluate the performance of the designed system, some typical monitoring results are demonstrated and analyzed. The monitoring data of sensor 5–7, which locate on the middle part of the right lock wall blocks, are chosen. Deformation curves are plotted and shown in Fig. 7. As can be seen from the figure, the deformation curves are smooth with slight perturbations. The vertical displacements of the 3 sensors are very close, indicating that no inhomogeneous vertical deformation occurs between blocks. The horizontal displacements, however, are different. The maximum difference between neighboring block can reach 4 mm. The result implies that the horizontal stiffness of the lock wall blocks have slight difference though they have the same structure formation.

The deformation increments between adjacent measuring times, as can be seen, are no more than 1 mm. The accuracy of GNSS method, however, is about 2.5 mm and 5 mm for horizontal deformation measuring and vertical deformation measuring respectively. Thus the designed monitoring system exhibits its advantage in accuracy.

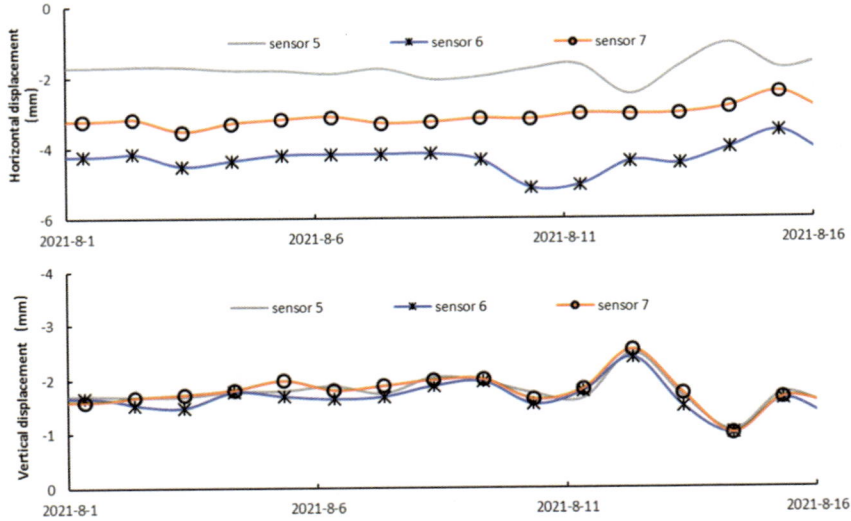

Fig. 7. Deformation curves of typical blocks in right lock wall.

5   Conclusions

Surface deformation monitoring system designing is challenging for navigation hubs built in plain area, due to the long engineering axis and the complex structure layout. In this paper the monitoring system designing strategy of Jiepai Navigation Hub is demonstrated. The whole system is divided into 5 independent sub-systems according to the structure feature. The monitoring distances are thus reduced to appropriate values. By comparing different monitoring methods, the sensor method is chosen as the main monitoring approach. The problem of lacking construction space is solved introducing auxiliary facilities. The monitoring results indicate satisfying performance of the designed system. The designing strategy can serve as a reference for other similar projects built in plain area.

Acknowledgements. The financial support of NHRI research foundation (No. Y321011) is gratefully acknowledged by the authors.

References

Ding Y, Zhou W, Chen Z, Wang C, Lu Y (2022) Monitoring and analysis on micro-amplitude deformations of a concrete face rock-fill dam with long-term service conditions. Hydro-Sci. Eng. 1–11 (in Chinese)

Zhou J, Mao H, Xu W, Xie G, Shi L (2022) Application research on deformation monitoring technology of substation on soft foundation in mountainous area under strong electromagnetic condition. Rock Soil Mech 1–10 (in Chinese)

Huang K, Chen Q, Ju X (2018) Study on prediction method of dam deformation for GNSS automatic monitoring system. Bull Surv Mapp (01):147–150 (in Chinese)

Yan J, Li S (2002) Optimization of deformation monitoring system designing for Three Gorges Dam.Yangtze River (06):36–38

Li Z, Li G (2002) Review of dam exterior deformation monitoring methods in China. Bull Surv Mapp (10):19–21 (in Chinese)

The Navigable Frontier Canal
of the African Rift

J. M. Deplaix[✉]

AFTM, Paris, France
aftmjmd@hotmail.com

Abstract. This Multipurpose Canal is one of the biggest infrastructure projects for Africa in the 21st Century. It aims at linking 3 river basins, Nile, Congo, and Zambezi. It will eventually connect all the Great Lakes of the African Rift: Albert, Edward, Kivu, Tanganyika and Malawi, as well as Victoria.

It is Multipurpose, as it will involve Hydropower, Navigation, as well as Security and Peace aspects.

Here, are described only Navigation aspects.

This project presents many challenges to the designer, mostly because of the huge difference of elevation between the lakes, and of the divides to climb, reaching up to 2 000 m between the Nile and Congo basins.

The Report presents the design criteria, as well as tentative alignments, possible canalisation steps and techniques to be used over this route, in particular Waterslope, Blue Wave and Pulsating canalisation.

This project is promoted by CSP-REGLA, a Congolese NGO. The first step, Integrated Development of Semliki River, is being studied by the Ministry of Regional Integration of DRC. More details are given about this particular project.

Keywords: Congo · Semliki · Lake Albert · Lake Edward · Lake Victoria

1 Introduction

This Multipurpose Canal is one of the biggest infrastructure projects for Africa in the 21st Century. It aims at linking 3 river basins, Nile, Congo, and Zambezi. It will eventually connect all the Great Lakes of the African Rift: Albert, Edward, Kivu, Tanganyika and Malawi, as well as Victoria.

It is Multipurpose, as it will involve Hydropower, Navigation, as well as Security and Peace aspects.

Here, only Navigation aspects are discussed.

This project presents many challenges to the designer, mostly because of the huge difference of elevation between the lakes, and of the divides to climb, reaching up to 2 000-m above sea level (masl) between the Nile and Congo basins.

Here are presented tentative alignments, as well as the design criteria, possible canalisation steps and techniques to be used over this route, in particular Waterslope, Blue Wave and Pulsating canalisation. These 3 techniques have been invented by members of PIANC and presented in the Bulletin of the organisation of their time.

© The Author(s) 2023
Y. Li et al. (Eds.): PIANC 2022, LNCE 264, pp. 251–267, 2023.
https://doi.org/10.1007/978-981-19-6138-0_23

This project is promoted by CSP-REGLA, a Congolese NGO, and is known under the local name "Sula ya Amani" (the Face of Peace). The first step, Integrated Development of Semliki River, is being studied by the Ministry of Regional Integration of DRC. More details are given about this particular project.

2 The Proposed Route of the Project

2.1 A 3 000 km Link Towards the Ocean

The route shall start in the Nile Basin, on Lake Albert, elevation 616 (masl), reach Lake Edward at elevation 913 through the Semliki River, and join with Lake Victoria, elevation 1 137, through Lake George, elevation 915, and a divide canal at elevation 1 197.

To cross into the Congo Basin, the route will leave Lake Edward and pass along 2 active volcanoes, with a divide reach around 2 068 m, then descend onto Lake Kivu, elevation 1 463. From there, it will reach Lake Tanganyika, elevation 768, through Ruzizi River and connect with the Congo-Lualaba, elevation 548, through Lukuga River.

Crossing into Zambezi Basin can be through Tanzania (divide at 1 430 m) or Zambia (divide at 1 630 m), to reach Lake Malawi, elevation 477. Then, through Lake Malombe and Shire River, the route will reach Zambezi River, elevation 28, and finally the Indian Ocean.

2.2 Nile Basin

2.2.1 Lake Albert-Lake Edward Link

Navigation on Lake Albert is a classical Lake navigation, with short waves and possible fetch effect when the wind blows in the direction of the main axis of this body of water, measuring some 150 km by 40 km.

The project shall start, from North to South, at the border between RDC and Uganda, in the middle of the lake, at the tri-point between RDC, Northern and Northwestern Regions of Uganda, approximately 2°09′14.96N 31°18′21.97. From there, one branch of Victoria Nile is some 8 km away, in Uganda, the start of the Albert Nile.

Elevation of Lake Albert is given by Google Earth at 616 masl.

Along the direct route, navigation on Lake Albert is 144 km to reach the opposite shore, at the point where it is proposed to locate the outlet of the Navigation Canal of the Semliki.

This Navigation Canal is designed to avoid the numerous and tight bends of the Semliki found in the part which constitutes the border with Uganda. From the Lake, it would go up to opposite Bundibundu, the last RDC village on the right bank of the Semliki, a stretch of some 68 km. At this place, elevation is given around 674 masl, so slope/gradient is practically 1 m/km.

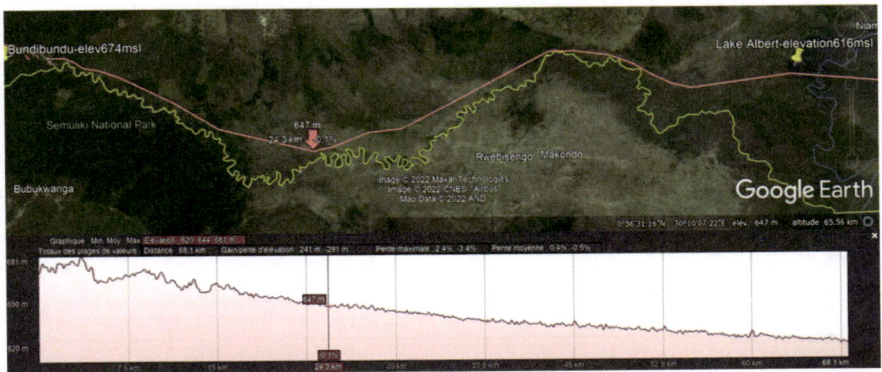

To compensate for this fall of 58 m, a succession of 6 locks could be devised, while it is also feasible to use a Blue Wave system, with one gate every 500 m. Such system would avoid an elevated waterway in the middle of a flat plain, or 10 m deep cuts.

Upstream from the border, the river itself will be developed for navigation, with a succession of dams, sited so as to damage to the least extent the virgin nature of the flood plain, which is included into the Virunga National Park up to Lake Edward, elevation 913 masl.

The proposed solution to overcome the 230m+ difference of elevation between Bundibundu and Lake Edward is studied later in this paper. It has to be noted that up to Lake Edward the route is fully in RDC territory.

2.2.2   Lake Edward-Lake Victoria Link

From Lake Edward, another link could be developed towards Lake Victoria. This could be a great interest for Uganda tourism, as it is fully in Ugandan territory and would pass through an area already developed for tourism, the Kazinga channel, connecting Lake Edward to Lake George.

From Lake George, 915 masl, the route would climb the hill up to 1197 masl, thanks to 2 or 3 Waterslopes.

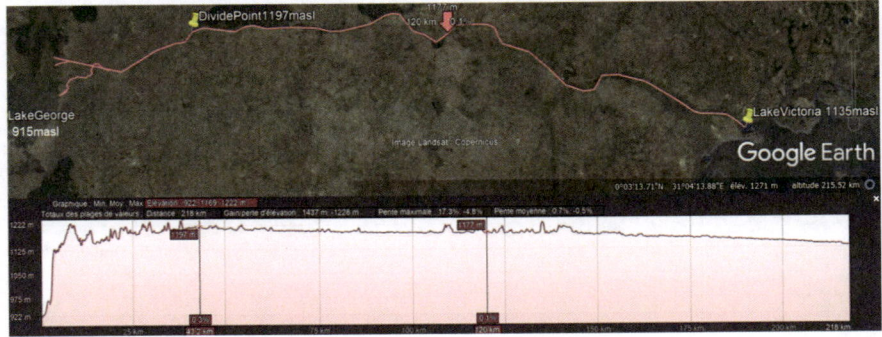

From this divide point, the descent towards Lake Victoria will be easy, only 62 m fall in 175 km, possibly using Blue Waves, 3 gates every 3 000 m.

The working of Waterslopes and Blue waves is explained later in this paper.

2.3   Congo Basin

From Lake Edward to Lake Kivu, and hence to Lake Tanganyika, the proposed route will fall into the Congo Basin. It is the part which poses the highest challenge to the designer, with a divide around 2068 masl, a 1 150 m difference to climb!

2.3.1   Lake Edward-Lake Kivu Link

This link can be designed either on the East side of the volcanoes, or on the West side.

Here is one possible alignment of the East version, North being to the right of this picture:

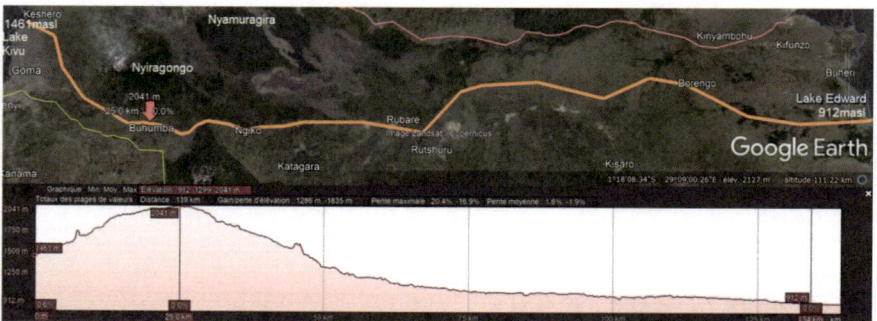

This route would be exposed to lava from Nyiragongo, but at the same time it may help to divert the flow of lava from inhabited areas.

The West version can be better described in two parts:

The rising part from Lake Edward till the divide, and the divide reach plus the descending part towards Lake Kivu.

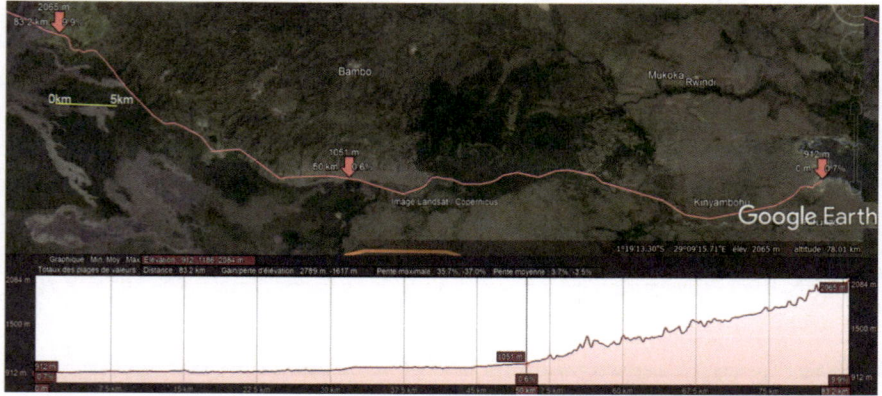

Rising from 912 masl on Lake Edward, it takes 50 km to reach elevation 1050 m. From there, it takes 33 km to reach the divide at 2068 masl, with an average slope less than 4%, which leaves the possibility to site there a succession of 13 Waterslopes, 2 km long each, climbing 80 m each. The last 160 m will be tricky to arrange, because the slope of the natural terrain there is more like a cliff, and the structure will have to be sited askew.

On the first 50 km it is possible to use either 8 locks, or a long Blue Wave with some 140 gates, or a combination of both.

From the divide, there will be a 11 km long divide reach at elevation 2068 masl, then a more or less regular descent towards Lake Kivu, down to elevation 1461 masl, according to Google Earth measurements.

This 600 m fall will be compensated by Waterslopes, 2 with a 2.3% slope, between 2068 and 1908 masl over 7 km, and 7 with a 4% slope over 14 km.

This route would be exposed to lava from the Nyamuragira. The Waterslopes would be sited out of the historical domain of this very fluid lava, but the divide reach is likely to be destroyed now and then. Since it is only a canal, it will be easy to reconstruct, on average something like every 15 years.

Another point of interest is that this route follows in many places the limits of the Virunga National Park, except for the divide reach. This would facilitate the protection from encroachment and bandits.

This combined route is 116 km long, 18 km shorter than the East version.

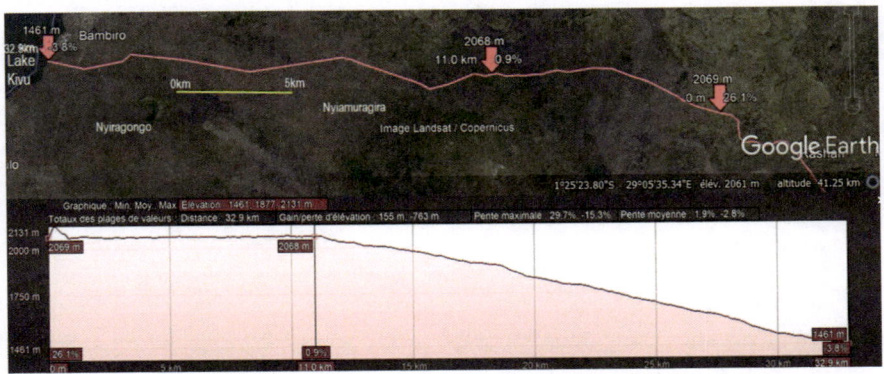

Thanks to the use of Waterslopes, the use of water will be minimal, and transfers from one basin to the other are likely to be negligible.

Both versions fall entirely in RDC territory.

2.3.2   Lake Kivu-Lake Tanganyika Link

To Reach Lake Tanganyika, waters of Lake Kivu follow a river, the Ruzizi, which marks the border between RDC, Rwanda and Burundi. This Ruzizi River is 159 km long and has a head of 645 m.

This river is already being developed for hydropower, thanks to treaties, and navigation could use this diplomatic way in order to be established. Thus we shall not study this link in detail, since it will be a matter of diplomacy rather than technique.

Nevertheless, a number of short-circuit can be devised, the river being replaced by a lateral canal in RDC territory. In particular, a large part of the 110 km where the river makes the border with Burundi is easy to canalise, the terrain being rather flat. A good location for a 28 m dam has been found at 83.7 km, and could provide a 17 km navigable reach. Further, the final 10 km could be short-circuited, since it falls entirely in Burundi.

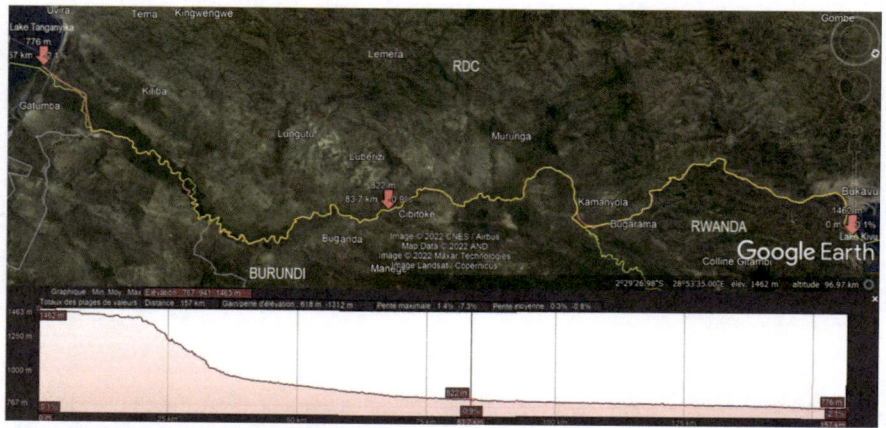

As far as the upper part is concerned, it could be developed with a combination of locks and Waterslopes, the final design being subject to the negotiations between RDC and Rwanda.

2.4   Zambezi Basin

Passing from Congo basin to Zambezi basin will also be challenging, with again a 1 150 m difference to overcome. Two routes can be envisioned, one through Zambian territory, the other through Tanzania.

Through the Zambian territory, there will be a first 900 m to climb up to the divide, then a slow descent down to the Malawi border, and a 1 150 m descent towards Lake Malawi, as can be seen in the profile.

This route would have the advantage of not impacting the virgin character of Lake Rukwa, which is recommended to be protected from invasion of exogenous species.

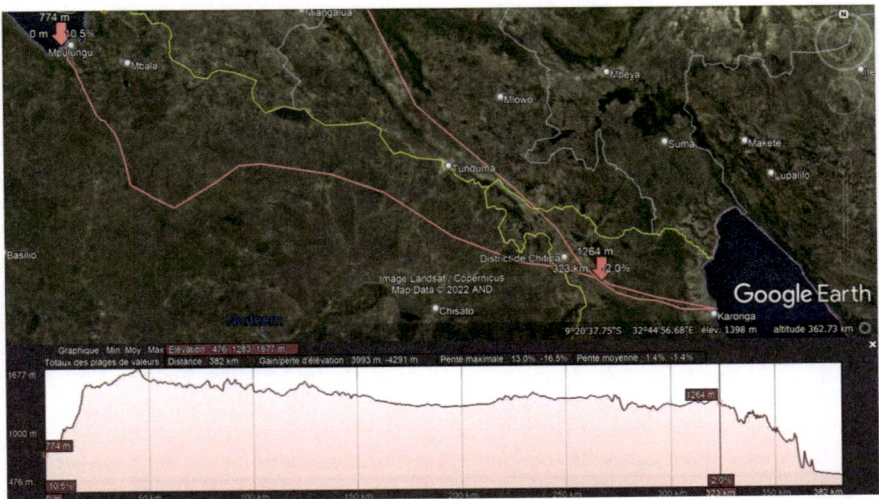

The other route would be through Tanzanian territory, with 2 mountain ranges to cross and using Lake Rukwa over nearly 150 km. Although it would seem more obvious, ecological considerations and the 4 slopes to negotiate, rather than 2 in the other option, put this route as a worse solution.

Lake Malawi, 465 masl, is fully in the Zambezi basin, yet reaching the Zambezi River proper is not an easy task, because the river leading to it does not always flow year-round, and is so rocky and narrow that finding a navigable alignment will be difficult. Further, Malawi government has deferred its plans to develop an inland port on its territory, on the banks of the Shire River, some 100 km from the Zambezi River.

Similarly, the Mozambique government does not have plans to create a major harbour at the mouth of the Zambezi. Thus, inland vessels would be unable to unload their cargo in sea ships there. However, to navigate these big lakes, the vessels must be of a stronger scantling than a pure inland ship, and most of the vessels reaching the mouth of the Zambezi would also be able to navigate along the coast up to the existing ports of Beira and Quelimane, or even up to Maputo and Nacala, and the existing small port of Chinde would cater for the other, smaller boats. However, such big developments shall be done with the full support of the governments, and that will be obtained only when a convincing development of traffic is seen over the stretch Lake Albert-Lake Tanganyika.

2.5   Lukuga/Lualaba Development

Another route was studied, the link between Lake Tanganyika and the Congo River. The Lualaba is the name given to that part of the Congo River upstream Kisangani, coming from Katanga. Taking into consideration the mineral riches of Katanga, it would seem sensible to develop a link between Lake Tanganyika and Katanga, to complement the rail link towards the Indian Ocean. At the same time, it would enable to draw a navigable connection with the Congo River proper across the various rapids, and put Kinshasa in contact with the Navigable Frontier Canal of the African Rift. At a later stage, if navigation through the Inga cataracts is made possible, it would be the seed of a cross-continent navigable connection from the Atlantic Ocean to the Indian Ocean.

As a first step, it means to render the Lukuga River navigable. This river is the outlet of Lake Tanganyika, and its flow is very variable, depending on the rains over the Lake catchment. A variation of 11.6 m of the lake level is on record, since the times of Stanley and Livingstone, when the Lake was endorheic. Thus, the first reach of this project will be a "divide reach" starting in the Lake, with a way to overcome a huge variation.

Due to the topography and the position of human settlements, this divide reach may be broken in 2 parts. The higher may be used only when Lake levels are high, to protect human settlements and the existing railway line.

Assuming existing Lake level of 768 masl as on Google Earth[1] and the entrance of Lukuga at 333 km from the Lualaba, a lock & dam would be sited at 303 km and be level with the lowest elevation of the Lake, assumed at 766 masl. The second lock & dam would be sited at 323 km, and compensate for every variation, between 766 and 777 masl. The bottom of both parts would be sited at 762 masl, so as to provide 4 m depth even at low water.

The beam of the largest vessel on the Lake being 17 m, locks would be 18 m wide, and length could be 100 m.

Below this divide, the river can be divided in 3 parts, as far as slope is concerned:

- From 230 km to 303 km, a little less than 1 m per kilometer (67 m in 73 km)
- From 165 km to 230 km, about 1.6 m per kilometer (101 m in 65.5 km)
- From the Origin in the Lualaba till 165, a very slight slope, about 0.33 m per kilometer (51 m in 163.5 km). The last 86 km are even flatter, since the Lukuga only descends 19 m in 80 km (0.24 m/km).

---

[1] This elevation is some 7m lower than that given by water gauges in Burundi.

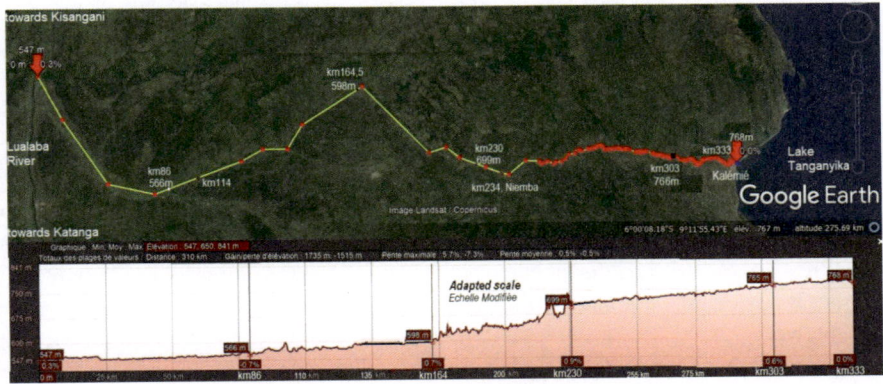

The discharge is extremely variable, from 23 to 2 300 m³/s, and a long-term average has not been computed. It is assumed to be in the range of 300 m³/s below Niemba, where a large tributary meets the Lukuga, and 200 m³/s out of the Lake.

The first part is strewn with rapids, and the proximity of the railway line precludes the use of long reservoirs. We find the same problem as on the divide reach. For all tentative dams up to the divide reach, the great variability of the flow from year to year makes it difficult to make profitable a hydroelectric development upstream of Niemba, with floods one year, minute flow the next, while falls can be only no more than 10 or 13 m because of the railway. To pass the rapids, Blue Waves may be used. An example is shown later in this paper. Small lock & dams may also be used. Insufficient knowledge of the terrain forbids to be more precise.

The second part would be a very good case for hydroelectric development, with 116 m fall in 64 km. 6 locations have been identified, with heads varying between 10 and 25 m. These rather low heads have been selected so that dams be not too wide, inundated area be restricted and that they can be passed by single locks.

As for the dam at 716 masl (width = 815 m), it is set at the highest point so as not to flood the town of Niemba. Its reservoir extends into the first part, and it will even be necessary to plan for a Blue Wave on the last kilometers upstream, or else a deep rock cut so that the boats can reach a possible next dam. The presence of the railway line would make it necessary to lower the level of this next dam, or to rebuild the line over several tens of kilometres.

Below 170 km, the plain is too flat to enable siting large dams. It will be studied at a later stage whether a free-flowing navigation would be possible, or whether some navigation barrages are necessary.

3   Proposed Techniques to Be Used in This Project

3.1   Design of Low-Head Developments

To render navigable an area of rapids, 3 solutions can be proposed depending on the height of the fall of the rapids:

- For a fall of approximately 1 m, a secure pass solution, with a length of less than 500 m.
- For a fall of 2 to 3 m, a Blue Wave, with a length that can exceed 1 km.
- For a higher fall, 5 m and above, a lock.

3.1.1 Secure Pass

The secure pass is well known in Europe as the "Passe Marinière" and was used to cross water mills on rivers. It then required a means of level control, to guarantee the fall used by the mills. It has disappeared, replaced little by little by locks. Two remain in Prague. It has been applied in particular to the Zinga cataract, on the Ubangui, at the border between the CAR and the DRC, and allows the continuity of navigation between Kinshasa or Brazzaville, and Bangui.

A possible application would be at 290 km of Lukuga: a safe route may be devised, where the weak current flows without preventing the ascent. Most of the current follows the cataract, sucked up by the steepest fall.

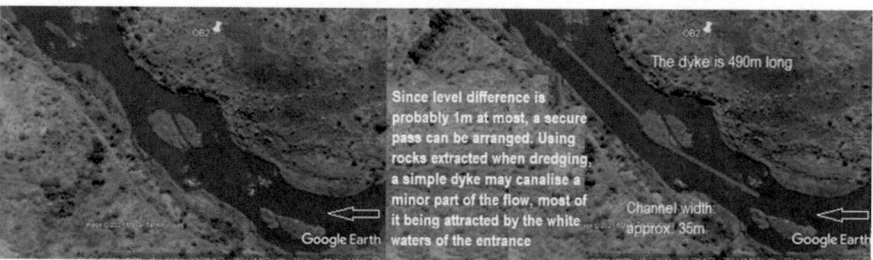

3.1.2 Blue Wave

The Blue Wave is a patent revealed to the general public by an article in the PIANC Bulletin in 2001. It was there that it was found by the team of Sula Ya Amani. Depending on the project boat and the project speed, the length of each intermediate reach will be calculated for each location, here probably around 200 m. A notable advantage of the Blue Wave, its walls can even be curved and do not have to be straight, when it reuses an old channel of the rapids for example. A possible application would be right at the end of the dividing reach, but the precision of Google Earth's Digital Terrain Model does not allow exact values to be given: the water at the top of certain rapids is lower than downstream. These will only be simple examples.

At the downstream end of the divide reach, a lock is necessary, because the fall can exceed 6 m. 766 masl is the lowest known altitude of Lake Tanganyika, while the other reach of the "divide reach" serves to compensate for the fall during high waters, up to 777 masl. If we take the "Amani", the largest boat sailing on Tanganyika (90 × 17 m), as a reference boat, locks should be 18 m wide and 100 m long.

Downstream of the divide reach, falls are much lower, and we enter the Blue Wave area of expertise. The Blue Wave itself is made up of a number of cells of about 200/500 m each. As a lot of water is available, we can use the upstream and downstream cells as buffers, and save a cell, both downhill and uphill. Here, to cross the first rapids, the canal is 800 m, we have therefore only drawn 4 cells, each 200 m long, whereas in a canal we would need 5, or even 6. Another paper in this Congress gives more details on Blue Wave, with an example.

Below are the successive levels that a boat would encounter when descending this Blue Wave. In blue (figures and lines), these are the starting levels, the 2 m drop being divided into 3 stages of 66 cm. In green, we find the figures and levels of equilibrium, when we open the valve between 2 cells. It is 33 cm lower than the upper reach, and 33 cm higher than the next cell. Here, 765.00 masl for the first cell.

The boat having penetrated from one cell to the other, the intermediate door is closed, and the valve is opened towards the 3rd cell. There is a 1 m drop here, with equilibrium at 764.50. The boat enters the 3rd cell. The u/s door of the 3rd cell closes, the valves feeding the 4th cell, 1.17 m lower, open and the levels equalize at 763.915; the d/s door open and the boat moves forward in the 4th cell. The door closes behind her, and a bottom valve open to lower the level of this cell to the downstream level, that is to say 763.33. The downstream door then opens, and the boat can exit the Blue Wave.

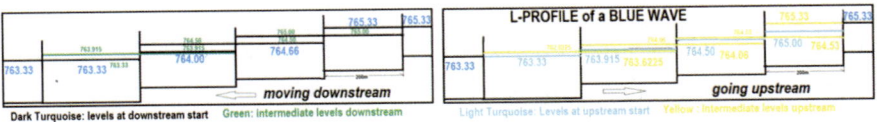

On the ascent, we start from altitude 763.33, in yellow this time. By equalizing with the level of the 3rd cell (green), we reach 763.6225 (yellow), and the boat can pass in the 3rd cell. By equalizing with the level of the 2nd cell (green), we reach 764.0613 (yellow), and the boat can pass in the 2nd cell. By equalizing with the level of the 1st cell (green), we reach 764.53 (yellow), and the boat can pass in the 1st cell. Once the door is closed behind her, an upstream valve is opened to bring the level of the cell back to that of the u/s forebay, in order to be able to open the door and allow the boat to enter this u/s forebay.

The bottom of each cell should be 4 m below the lowest level found in the cell, either blue figure or yellow figure. Each cell is as wide as a normal canal, thus there is the same amount of earthworks[2] than a bypass canal with a lock.

By integrating the average of a possible wait, the passage with a lock would last 2 h, and only 1h32 with the Blue Wave.

3.1.3   Lock

This technique is well proven and well known. It has the advantage of experience and reliability. Thus, we shall not give details about it.

Its domain ranges between 5 m and 30 m of head and is best applicable between 10 m to 25 m. Lower falls are costly for a difference of height manageable by other techniques, and higher falls are in the domain of high-head devices.

---

[2] In actuality, 2 out of 4 cells require less digging than a by-pass canal and its lock. Some 20 000m³ are saved, here.

3.2 Design of High-Head Developments

For high-head developments, there are 4 techniques available, which will be studied here, keeping in view the extreme difference of levels to be covered, up to 1 150 m:

- Locks, single or in succession
- Inclined planes
- Elevators
- Waterslopes

3.2.1 Lock

The highest lock is located in the former USSR. Today, it is in Kazakhstan, the lock of Ust-Kamenogorsk, renamed Oskemen, on the Irtych. It has a drop of about 40 m. The 3-Gorges dam locking system in China is higher, 118 m, but it is made of 2 lines of 5 cascading locks. Itaipu, in South America, has the same drop, but there is no lock yet. In Brazil, the Tucurui dam (73 m drop) is crossed by 2 locks of 35 m drop, at each end of an intermediate reach.

It is usually considered that the cheapest lock per metre of head is that of 25 m head. Combination of locks enable to cover any difference of height, but strong earthworks are necessary when the terrain is gently sloping.

3.2.2 Incline Planes

The domain of Incline Planes ranges from 35 to 150 m+. Examples are found starting from 13 m, like in Elblag (Poland, 5 units totaling 99 m), with the highest being 110 m at Krasnoiarsk (Russia), which has 2 railway tracks of opposite slopes connected by a huge turntable. The caisson is a self-propelled system.

There are 2 types, longitudinal (Ronquières, 68 m, and Krasnoiarsk) or transversal (St Louis-Arzviller, 44.5 m). Some longitudinal carry the boats in the dry (Biwako, Japan, and Elblag), which is difficult for boats carrying goods. Most have counterweights, but Krasnoiarsk is self-propelled, and could have any length. All examples are in straight line, which makes it at times difficult to place in the existing terrain.

Their disadvantage is the weight of the steel caisson containing the water on which the boat floats, and the difficult foundations of the steel tracks, which carry the water, the caisson and the possible counterweight. The pressure on the underlying ground is extremely high and concentrated. It is usually a very costly proposition.

3.2.3 Elevators

The same problem, cost, is found with elevators, because here the weight is even more concentrated, due to the weight of the structure which guides the caisson full of water. It took 20 years to finish Strepy-Thieu, due to problems of cost and probably of foundations.

The highest on record is at the 3 Gorges dam, 118 m. The difficulty there was to cope with the variation of the upstream reservoir. A sensible solution has been found.

Another difficulty is to find a location where the head is sufficiently concentrated, and this usually obliges to make a deep cut downstream and provide a canal-bridge upstream, both adding to the cost.

3.2.4   Waterslopes

The problem of difficult foundations does not exist with Waterslopes[3].

Their principle is to push a wedge of water, on which a boat can float, using a Shield, sort of movable dam attached to propelled wheels, powered by electric motors. This is called Shield ("Bouclier") or Pusher.

As can be seen on the figure, the weight of the wedge of water is spread on a long surface, and the most weight is found under the shield and its pusher, which must be heavy for adherence (grip) purposes.

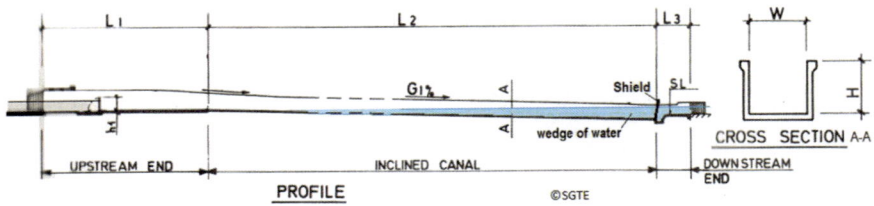

As with self-propelled inclined plane, the maximum elevation to cover is practically infinite, 100 m, 500 m, etc. For traffic considerations, a head of 1 150 m will be compensated by 14 Waterslopes, some 80 m each, rather than by a single one of 1 150 m of elevation. The big one would be totally feasible, and much cheaper, but would take some 8 h to climb, so only 3 boats per day could pass, rather than 36 with 14 Waterslopes. Besides, it is unlikely to find a 28 km slope regular and straight. 14 Waterslopes would better adjust to the terrain.

The necessary power is not linked to the difference of elevation, only to the mass of water to move upwards, which itself is linked to the width, the length and draught of the boat, and the slope.

It would seem that, with a slope of 4% and a speed of 1 m/s, Waterslopes should better be 2 km long, with a head of 80 m each. For a width of 18 m and a draught of 2.8 m, the necessary power for the pusher is below 5 000 kW, less than a big electric locomotive. This was found feasible on most of the routes analyzed. At times, a slope of 5% is necessary, at times only 2.3%, as in the descent towards Lake Kivu, and the length may be tailored to what the terrain requires.

Although it is best with a straight alignment, a Waterslope has some possibility to provide curves and counter-curves down to a radius of 1 300 m approximately, for 95 m long boats, when, as proposed, boats have a beam of 17 m and the Waterslope is 18 m wide. This flexibility may be very useful to find alignments with the least excavations. Thus, locating a set of Waterslopes will be rather easy on most of the routes experienced by the Frontier Canal.

---

[3] The Waterslope is a technique invented by Professor Aubert and SGTE, using an idea by Engineer Bouchet.

Here are pictures of Montech Waterslope, which worked for 35 years until traffic became too small to run it. This 6 m wide structure used 2 locomotives on tyres, with a 1 500 kW diesel-electric propulsion for 3% slope.

4 Application to Semliki River Development

4.1 General Concept

The main issue is the steep fall of about 160 m, concentrated over a few kilometres. A preliminary design for a hydroelectric dam has been found at 105 km, with a reservoir extending over 34 km. But how can navigation pass it? The second issue is that this part of the river falls entirely within the limits of the Virunga National Park, thus it is necessary to tamper to the least extent with the virgin forest. This led to multiply dams, to less inundate the Park. The third issue was the large variation of Lake Edward over the ages, with an extreme of 15 m, which may oblige to deepen the upper segment of the river, in order navigation be possible even at low stages.

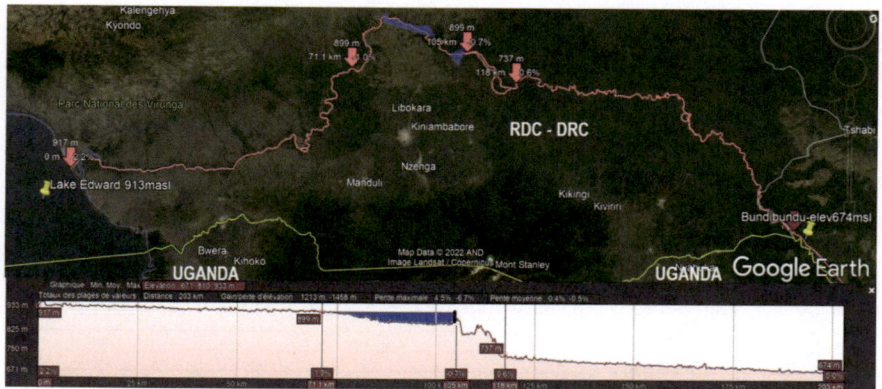

4.2 Navigating a 162 m Fall

The 162 m drop on the Semliki between the headrace and the downstream compensation reservoir is currently the highest navigable height difference studied in the world. Overcoming it is possible in a number of ways, with 7 or 8 high head locks for instance. Here is only shown the solution using 3 Waterslopes.

The trick is to use a contour canal, following elevation 899 masl, branching out from the reservoir at the 105 km dam. This 324 m^2 canal, 36 m wide, 9 m deep, will be navigable for about 7 km, and this headrace doubles as a feeder canal to bring the flow of the river to the 3.3 km underground penstocks feeding the turbines.

At the end of this contour canal, there would be a first Waterslope, from 899 to 850 masl, then a 1.4 km embankment at 850 masl to cross one tributary of the Semliki, the Butawu, then a second, bended Waterslope from 850 to 794 masl, which follows more or less the border of the Virunga National Park, then 2 options, depending on whether the Park insists on protecting the Luusilubi River, in which case the third Waterslope will be again bended, or this river can be developed for navigation, with a straight Waterslope leading to it.

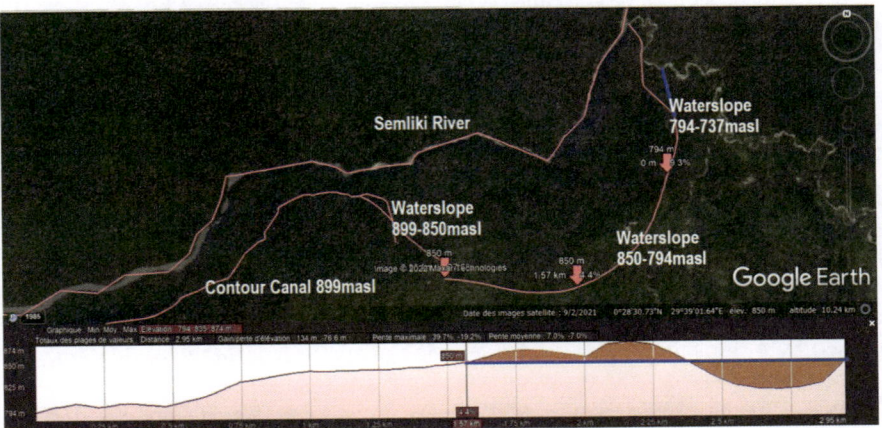

By the way, this Luusilubi is the highest source of River Nile, vying with the Butawu for this title, these brooks starting above elevation 4 650 m.

4.3   Developing the River Below the Dam

Attempts were made to design barrages between elevation 737 and the border. It seems that eight are needed, avoiding a ninth, too close to the border and which would risk, by reducing the solid flow, modifying the course of the border river, which could be the source of many territorial conflicts.

The highest barrage in this provisional design is only 12 m high, most are 7 to 10 m high, and 2 are only 4 m high. This was necessary, because the land is very flat, and 1 m more of head increases the flooded area by several tens of hectares. Despite the low drop, it is difficult to imagine the use of Blue Waves; the river is, in fact, very wide and there are no rapids on which a Blue Wave could be grafted. Thus, dams would be necessary in any case.

Due to an imprecise knowledge of the topography and the small discharge of the Semliki (200 m^3/s), the possibility of a Pulsating Canalisation has not been thought possible either.

For navigation, there are a large number of steep curves, radius below 200 m, and if not covered by reservoirs, they will have to be short-circuited.

4.4 Developing the River Above the Dam

Between Lake Edward and the proposed dam, only 2 barrages are necessary: one to cope with Lake Edward variations, the other to serve as the end of the "divide reach" when Lake Edward is extremely low. There is evidence of historical variations of Lake Edward down to 15 m lower than present. However, it was most probably endorheic in those dry years, so the lowest navigable level will be said to be 908 masl only. Presently, Google Earth gives an elevation of 913 for Lake Edward, and the highest high level will be said to be 917 masl. Gauges around the lake give figures some 6 m higher, but since our sole reference is Google Earth, we shall use its values.

Both parts of the divide reach may only flood land likely to be covered by an exceptional flood. An interesting location is proposed for the first barrage, with a controlling valve culminating at 917 masl, in case Lake Edward rises to the extreme and floods all the reservoir, but used mostly around 913 masl. A lock would compensate for these variations and lower vessels at elevation 908 masl. This reach, as well as the second, would be dug at 904 masl so that navigation is possible even in the driest year. In such case, the lock would not be used.

The second barrage would have a head of only 9 m and be dredged at 904 masl, but to avoid inundating one extra square kilometre the 908 masl reach will have to be shortened by 4 km and the 899 masl reservoir extended accordingly by dredging over 2 kms.

In the reservoir, a 4 m channel will have to be provided, marked and maintained.

5 Conclusion

This paper provides the proof of a technical feasibility for the Navigable Frontier Canal of the African Rift, as well as its junction both with Lake Victoria and with the Congo (Lualaba) River.

Its hydroelectrical potential may be sufficient to fund the cost of the first element of this Canal, the Semliki River Development, an International waterway entirely in RDC territory.

It may also be the occasion to test in full scale the Waterslope and Blue Wave techniques, which offer an interesting potential and may render workable new navigations in mountainous terrain. They are further very well adapted to autonomous navigation, prepared for the future.

References

Deplaix JM (2001) The Blue Wave, a new concept of waterways enabling to increase velocity or to reduce impact on environment. PIANC Bull 108, Autumn 2001. French version. http://aftm.free.fr/PromoL.htm

Aubert J (1965) La Canalisation Pulsatoire, Brevet 1437528, 22 mars 1965, INPI
Aubert J (1971) La pente d'eau remplacera-t-elle l'écluse? Navigation, Ports et Industries, 16–17, 541–545, 591–596
Aubert J (1979) La pente d'eau va-t-elle remplacer l'écluse? Navigation, Ports et industries

Inland Navigation Structure

A Modernized Safety Concept for Ship Force Evaluations During Lock Filling Processes

Fabian Belzner[1(✉)], Carsten Thorenz[1(✉)], and Mario Oertel[2(✉)]

[1] Federal Waterways Engineering and Research Institute, Karlsruhe, Germany
`{fabian.belzner, carsten.thorenz}@baw.de`
[2] Helmut-Schmidt-University Hamburg, University of the Federal Armed Forces, Hamburg, Germany
`mario.oertel@hsu-hh.de`

Abstract. A ship in a lock chamber is exposed to forces acting on the hull during the filling and emptying processes. These forces accelerate the ship and lead to a displacement. To avoid a collision of the ship with the lock structure, it is moored with mooring lines, which can be strained up to a certain breaking load. The force acting in the mooring lines is called the mooring line force and must be distinguished from the ship force. If the mooring line force exceeds the breaking load, the mooring line will fail and the tension energy will abruptly be transformed into kinetic energy. A snap-back of the mooring line ends can produce great forces and the mooring staff is at risk for major injuries. Furthermore, the ship will start to move and could damage the structure and itself. Thus, the mooring line force must be limited during the locking process. The mooring line force depends on the ship force and, furthermore, on the properties of ship and mooring lines. Due to the number of possible parameter combinations the given ship force alone might not be sufficient to judge on the mooring line safety. In this paper a statistical approach to determine the relations between mooring line configuration, mooring line forces and ship force based on Monte Carlo simulations is shown.

Keywords: Navigation lock · Ship · Safety · Inland navigation

1 Introduction

Inland navigation locks are mostly located between sections of impounded rivers or canals, at connection points within the waterway network or at tidal rivers. They enable ships to overcome water level differences. The safe and efficient operation of these locks is essential for the transport of goods on the waterways.

Inside the navigation locks ships are connected to the structure by mooring lines to avoid a collision of the ship with the lock structure. During the leveling process the ship in a lock chamber is exposed to forces acting on the hull. The mooring lines must absorb these forces to avoid any unwanted displacement of the ship. The lines can be strained up to a certain breaking load. If a mooring line fails (breaks) a snap-back of the mooring line ends can produce great forces and the mooring staff is at risk for major

Y. Li et al. (Eds.): PIANC 2022, LNCE 264, pp. 271–280, 2023.
https://doi.org/10.1007/978-981-19-6138-0_24

injuries. Furthermore, the ship will start to move and can damage the lock or the ship itself. Thus, the mooring line force must be limited during the locking process.

The forces acting on the ship depend on the hydraulic effects during the filling process. A major factor is the gradient of the filling discharge. A faster opening of the filling facilities (e.g. valves) will cause higher forces but reduce the filling time and vice versa. Another relevant factor for the forces is the effect of jets on the ship. These jets can be reduced by adding energy dissipation objects (i.g. breakers) at the culvert outlets, but these hinder the flow. The hydrodynamics of the lock filling process are described in Belzner *et al.* (2018).

Thus, to reach a short filling time and to reduce the forces acting on a ship are contrary objectives. During the planning process of a lock a compromise between the filling time and acceptable forces must be found. The aim is to fill the lock as fast as possible and as slowly as needed.

2   Force Criteria

The force acting on the ship is the sum of external forces acting the hull. It depends on the hydrodynamics in the space between lock chamber floor and walls and the ship hull. During the hydraulic design process analytical, numerical or physical models are used to reproducibly determine this force. The mooring line force, on the other hand, depends on the temporal variation of the amplitude of the ship force and the configuration of the mooring line system: number, length, diameter, material and pretention of the mooring lines. Thus, it can only be reproduced for exactly the same setup, which is unlikely to occur in reality.

Figure 1 shows a principle sketch of a ship in a lock chamber. The force acting on the hull can be decomposed in longitudinal and transversal components. For through the head filling systems the longitudinal force component usually dominates. The mooring line force is the force that acts in the mooring lines, which are illustrated with dashed lines in Fig. 1.

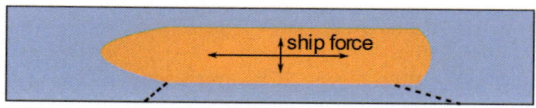

Fig. 1. Principle sketch of a ship (yellow) in a lock chamber with mooring lines (dashed lines)

Today the ship force is a commonly used criteria to assess the safety of the locking process: in practice it is assumed, that the mooring lines will not break if the ship force does not exceed a certain level. In the literature this certain level is often called the "ship force criteria" or the "hawser force criteria" and is mostly based on a mixture of semi-deterministic considerations and experience.

The ship force criterion used in Germany is based on estimations of Partenscky (1986) combined with practical considerations of the Federal Waterways Research and Engineering Institute (BAW). Partenscky (1986) analytically analyzed one specific (worst) case and generalized his results to the fleet of the 1980s. For a ship with a length of 110 m (CEMT Va) a force of 23.5 kN is considered acceptable (PIANC 2015), which is approximately 0.80 ‰ of the weight force of the displaced volume. In the Netherlands this criterion is defined as 0.85 ‰ of the weight force of the displaced volume and 1.10 ‰ of the weight force of the displaced volume for locks with floating bollards (Rijkswaterstaat 2000). These criteria are based on considerations of Vrijburcht (1994), who analyzed a range of configurations which fitted to the Dutch fleet in those times. Experience shows that almost no accidents with breaking mooring lines occur in locks due to the locking process itself. Thus, the approaches of Partenscky (1986) and Vrijburcht (1994) seem to be on the safe side.

The requirements for the mooring line equipment on board of the ships are specified for the European Union in the transnational ES-TRIN regulation (European committee for drawing up standards in the field of inland navigation 2021) and depend on the size of the ship (Eq. (1)):

$$R_S = \begin{cases} 60 + \frac{L_V \cdot W_V \cdot y_V}{10} & \text{for } L_V \cdot W_V \cdot y_V \leq 1000\,\text{m}^3 \\ 150 + \frac{L_V \cdot W_V \cdot y_V}{100} & \text{for } L_V \cdot W_V \cdot y_V > 1000\,\text{m}^3 \end{cases} \tag{1}$$

where,
R_S minimum breaking load [kN]
L_V length [m]
W_V width [m]
y_V draught [m].

3 Physics of the Mooring Line System

In a simplified way, the ship can be considered as a mass and the mooring lines as springs. Thus, the interaction between ship force, the reaction of the ship and the mooring lines can be described by the mass-spring equation (Eq. (2)):

$$m_V \cdot a + k \cdot \Delta x = F_V \tag{2}$$

where,
m_v mass of the ship [kg]
a acceleration of the ship [m/s^2]
k spring constant [N/m], $k = 0$ for $\Delta x \leq 0$
Δx elongation of the spring [m].

Angular effects, the number and mass of mooring lines and damping effects also influence the behavior of the system. For convenience these parameters are neglected in Eq. (2). Later the model is enhanced with all these parameters to get a more realistic approximation.

The mooring line force $F_{m.l.}$ can be derived from the elongation by Hook's law:

$$F_{m.l.} = k \cdot \Delta x \tag{3}$$

where the spring constant has to be calculated from the mooring line's material properties:

$$k = \frac{E \cdot A}{l_0} \tag{4}$$

where,
E Young's modulus [N/m^2]
A cross sectional area [m^2]
l_0 original length of the mooring line [m].

The mooring line force is calculated by Hook's law (Eq. (3)) from the elongation, which depends on the dynamic behavior of the mooring system and the spring constant. The ES-TRIN regulation requires a minimum breaking load for the mooring lines but there are no requirements concerning the stress-strain response of the lines, expressed by Young's modulus. The spring constant is calculated from the mooring lines' material and geometric properties, which are not regulated as long as the mooring fulfills the requirements for the minimum breaking load. A study from the Netherlands shows that the ES-TRIN requirements are over-fulfilled by most of the currently circulating mooring lines (PIANC 2015), indicating that these lines will also be more stiff. The considerations of Partenscky (1986) are based on mooring lines made of steel. Today it can be observed that the predominant part of the mooring lines is made of fiber-synthetic materials like polyester or polypropylene, which shows significantly different material properties compared to steel. Under certain circumstances, however, tankers must moor with steel hawsers. Furthermore, the spring constant scales inversely proportional with the length of the mooring line, which is also hard to predict.

In summary, the properties of the mooring line system cover a wide spectrum between rather "soft" and noticeably "stiffer" systems. Figure 2 illustrates the reactions of two mooring systems with significantly different characteristic. The forces are illustrated on the vertical axis in a normalized form where the forces are divided by the maximum amplitude of the ship force. The blue line shows a ship force acting on a ship with a mass of about $m_v = 4200$ t in longitudinal direction. Following the thoughts of Partenscky (1986) the acting external force is applied in shape of a ramp. Its peak amplitude is at about 1 ‰ of the ship's weight force. The mooring line force is calculated by Eq. (2) considering that even pretensioned lines will have a certain slack allowing the ship to accelerate before the ship force and the ship's kinetic energy is absorbed by the mooring line. Two cases are illustrated: a rather stiff system (short steel rope) and a rather soft system (long synthetic rope).

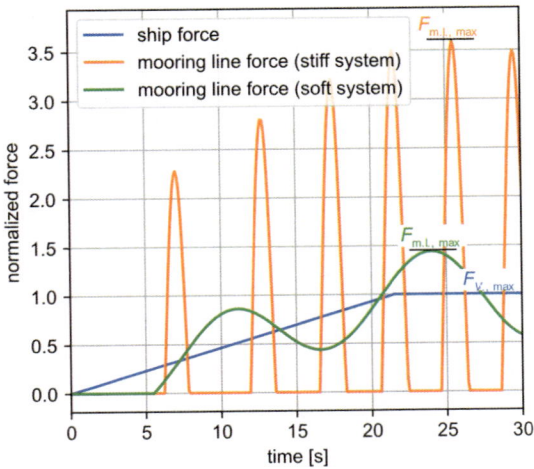

Fig. 2. Reaction force in the mooring lines in comparison with the ship force for a soft system and a stiff system

At the beginning the ship force acts on the ship, which accelerates until the line slack is reduced and the horizontal line force due to dead load and the ship force are in equilibrium, respectively. When the line is strained, the ship's kinetic energy and the ship force are converted into line tension. The mooring line force is the result of the elongation of the line and the spring constant and acts as a reset force accelerating the ship in the opposite direction. Depending on the configuration of the mooring line system, the line will slack again or remain strained. The example illustrated in Fig. 2 shows the difference in mooring line forces depending on the so called "mooring system configuration": the peak forces in the stiff and the soft system differ significantly. Softer or stiffer systems result from a combination of the material of the lines, the length and the diameter. One may assume that the stiffer system will be able to absorb higher forces without failing, but the only reliable comparative value is still given in the ES-TRIN standard (Eq. (1)).

This example shows in a very simplified way, that the mooring line force cannot be deterministically predicted on the base of the ship force alone. It depends on the magnitude and the temporal development of the ship force and, furthermore, on the configuration of the mooring line system. Even if the timeline of the ship force can be a known parameter, the configuration of the mooring system must be considered as a random variable.

Figure 2 shows an oversimplified example where only one mooring line is considered. A more realistic case is the application of two mooring lines: a bow line and a stern line. This amplifies the shown effect due to the interaction of two slacking and straining lines.

4 Probabilistic Approach

4.1 Amplification Factor

For a given ship, we can assume that the ship force depends on the characteristics of the lock filling system in a deterministic way. Then for this ship, the mooring line force depends on the ship force and the characteristic of the mooring line system, which is an unknown random variable. With considerations like shown in Fig. 2 an amplification factor can be defined, which is the ratio between the maximum of the mooring line force and the maximum of the ship force:

$$V = \frac{F_{\text{V, max}}}{F_{\text{m.l., max}}} \tag{5}$$

where,

V	amplification factor [-]
$F_{\text{V, max}}$	maximum of ship force [N]
$F_{\text{m.l., max}}$	maximum of mooring line force [N]

The amplification factor can be interpreted as an extremal value of every experiment.

4.2 Semi-probabilistic Approach

The current generation of the standards for structural engineering (e.g. ISO 22111:2019 2019) uses semi-probabilistic design approaches to determine impacts on structures. Structural loads and resistance are treated as probabilistic distributions from which partial safety factors can be calculated. This allows the planning engineers to guarantee a certain failure probability by considering a distributed quantity without having detailed knowledge of its statistical distribution. For example: the snow load on the rooftop of a building depends on its geographical position. It follows a statistical distribution based on long-standing observations. To avoid overloading and as consequence failure of the roof, a sufficient snow load must be used for the structural calculation. Using the maximum value of the observation will be uneconomic, using the mean value will underestimate the load with a probability of 50%. Absolute safety does not exist but a certain level of safety with a known failure probability (e.g. 10^{-6}) can be guaranteed. Therefore, the structure must be designed to sustain a certain quantile of the snow load with a chosen reliability.

4.3 Monte Carlo Simulations

For wind or snow loads long-standing observations exist, allowing statistical analysis with the aim to choose a "safe" load for the structural calculation. For mooring line forces observations also exist, but the quantity is not sufficient for statistical analysis. That might be the reason why Partenscky (1986) did a kind of "worst case" calculation and Vrijburcht (1994) considered a limited range of situations, which covers the situation in the Netherlands and probably also on the waterway network in Belgium and Germany.

Another approach to overcome this lack of information is to perform Monte Carlo simulations with a high variety of randomly chosen parameters for the configuration of the mooring line systems. With this approach a large sample set of mooring line loads is produced where the amplification factor V can be calculated from.

The main influence factors like length of the mooring lines, diameter, Young's modulus and pretention were randomly chosen from a range of uniform distributed values between certain limits that seem reasonably from an engineering point of view. Equation 2 was enhanced to consider more than one spring, angular effects, damping, rope characteristics and the fact that mooring lines cannot absorb pressure. As a consequence, Eq. 2 becomes a discontinuous differential equation, which cannot be solved analytically. Thus, the solution was performed numerically with a leap-frog algorithm (Skeel 1993).

After a sufficient number of simulations to obtain a statistically significant result the amplification factor will follow a Gumbel distribution, also called LogWeibull or Fisher-Tippetts distribution. An example is given in Fig. 3(left) where the results are shown compared to a fitted Gumbel probability density function (pdf). With this distribution function a chosen probability of exceedance P can be assigned to a corresponding amplification factor V_p. An example is given in Fig. 3(right) where a chosen probability on the vertical axis is assigned to an amplification by the cumulative distribution function (cdf). Note: this example is showing an exceedance probability of 1:10, which is very unsecure. This example was chosen to improve the readability. In reality a probability in the order of $1:10^6$ would be chosen.

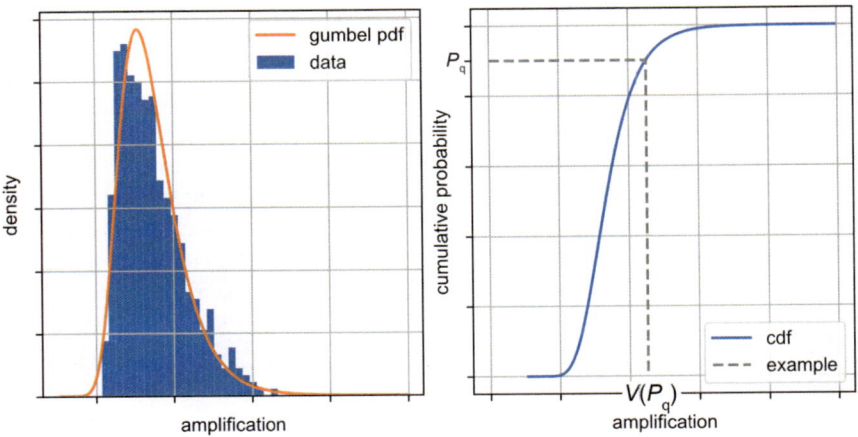

Fig. 3. Qualitative example for the probability density function (pdf, left) and cumulative distribution function (cdf, right) of the amplification factor

Now the correlation between the maximum ship force and the mooring line force in one experiment or locking process with a defined probability is known. Additionally, the minimum breaking load of the mooring lines R_s is known from the ES-TRIN standard or other regulation. From this relationship, the maximum allowable ship force can be calculated:

$$F_{V, max} = \frac{R_S}{V_p} \tag{6}$$

where,

V_P	amplification factor with the probability P [-]
$F_{V, max}$	maximum of ship force [N]
R_S	minimum breaking load of the mooring line [N]

The amplification factor V_p is the result from a probabilistic approach. For further usage, the design of the filling system and the valve schedules depend only this factor or furthermore the resulting maximum acceptable ship force are used without doing any statistics. This approach is based on probabilistic considerations but due to its reduction to one factor representing the probability, it is called a semi-probabilistic approach.

5 Outlook

Here, the ship in the lock chamber and the mooring line system are treated as a spring-mass system (Eq. (2)). Even if this approach is enhanced by several influence factors to make it more realistic, it does not consider the interaction between the ship and the water in the lock chamber. The authors perform numerical simulations with the CFD toolbox OpenFOAM® like presented in Thorenz and Schulze (2021) and additionally physical experiments with a scale model to investigate the coupling between the ship and the water in the lock chamber. As a result, the coupling will either be neglected or must be considered in the model.

Most semi-probabilistic design concepts consider both, the impact and the resistance as distributed values. The safety is in-between, where the impact does not exceed a certain value and the resistance does not fall below a certain value. Currently, the resistance (minimum breaking load) is typically treated as a deterministic value. In the future, the breaking load of mooring lines has to be described statistically to be able to assess probabilities on the resistance side.

An acceptable probability of exceedance of the mooring line force must be chosen. This will be a measure of how many locking processes can (statistically) be conducted without failing of the mooring lines. This may be a political decision regarding also other disciplines. It must be taken into account, that not every mooring line failure will have lethal consequences but it cannot be ruled out.

6 Conclusions

The authors showed a probabilistic approach that aims at an updated concept for planning engineers to determine the maximum acceptable ship forces on the base of a chosen exceedance probability.

The main findings of this paper in a short summary:

- During the design process of a lock or the lock filling process a criterion for a maximum acceptable ship force is needed.
- The ship force can reproducibly be determined with physical or numerical models. It must be distinguished from the mooring line force. Actually, the mooring line force is relevant for the safety but depends furthermore on the parameters of the ship and the mooring line configuration, which are random variables and cannot be forecasted. Thus, an approach to determine a maximum acceptable ship force considering various mooring line configurations is needed.
- Existing approaches are ether based on outdated assumptions or do not enable to consider a chosen reliability. Nevertheless, these approaches seem to be rather conservative.
- Existing observations of ship forces and mooring line forces are not sufficient for statistical analyses and forecast.
- Monte Carlo simulations with a high variety of randomly chosen parameters for the configuration of the mooring line systems were made to produce a sample set which allows statistical analysis of the expectable mooring line forces depending on a given ship force.
- Bases on statistical analysis, an amplification factor can be defined, which allows to determine the maximum acceptable ship force depending on the required mooring line equipment.

In this paper the procedure was summarized without giving hard values for target probabilities or acceptable forces. At the present time the procedure is in a proof-of-concept state. The German Federal Waterways Engineering and Research Institute and the Helmut Schmidt University are currently further developing this approach to update the criteria for the hydraulic design of navigation locks based on a scientific base.

References

Belzner F, Simons F, Thorenz C (2018) An application-oriented model for lock filling processes. In: Proceedings of the PIANC-world congress Panama City, Panama, 7–11 May 2018. https://henry.baw.de/handle/20.500.11970/107576

European committee for drawing up standards in the field of inland navigation (ed.) (2021) Technical Requirements for Inland Navigation ships: ES-TRIN

International Organization for Standardization (ISO) (ed.) (2019) ISO 22111:2019: Bases for design of structures—General requirements

Partenscky H-W (1986) Schleusenanlagen. Springer, Berlin

PIANC (ed.) (2015) PIANC InCom WG 155: Ship Behaviour in Locks and Lock Approaches

Rijkswaterstaat (ed) (2000) Design of locks 1

Skeel RD (1993) Variable step size destabilizes the Störmer/leapfrog/verlet method. BIT 33 (1):172–175.https://doi.org/10.1007/bf01990352

Thorenz C, Schulze, L (2021) Numerical investigations of ship forces during lockage. J Coastal Hydraulic Struct (1). https://doi.org/10.48438/jchs.2021.0005. Accessed 4 Dec 2021

Vrijburcht A (1994) Troskrachtcriteria van schutsluizen. Onderzoek. https://puc.overheid.nl/rijkswaterstaat/doc/PUC_96549_31/. Accessed 2 Nov 2021

Analysis of Piano Key Weir Drainage Characteristics

Zixiang Li[1,2], Jiayi Xu[1,2], Yanfu Li[1,2], and Changhai Han[1,2(✉)]

[1] Nanjing Hydraulic Research Institute, Nanjing 210029, China
chhan@nhri.cn
[2] State Key Laboratory of Hydrology Water Resources
and Hydraulic Engineering, Nanjing 210098, China

Abstract. The piano key weir is a new type of labyrinth weir structure form, its advantages such as stable structure, upstream and downstream inverted overhang and small base area greatly improve the scope of application, and it has high discharge capacity and overflow efficiency, which is considered as a very effective method to solve the shortage of reservoir and dam discharge flow, and has more applications abroad. In this paper, the experimental observation of the hydraulic characteristics of the discharge from the hydraulic physics model of the piano key weir confirms its good overflow capacity, but also limits the study of its complex flow. Combined with the numerical simulation software, the three-dimensional flow field of the piano key weir is numerically simulated based on the RNG k-ε turbulence model and the free-surface VOF technique to analyze the changes of hydraulic characteristics such as flow patterns, flow lines and flow velocities, and to clarify the hydraulic characteristics of the spillway and the mechanism of the improved spillway capacity by combining the experimental results of the physical model. In addition, a comparative analysis of the overflow capacity and flood discharge efficiency between the piano key weir and the traditional thin-walled weir is carried out to better reflect the advantages of the piano key weir in flood discharge by using the over-discharge ratio, and to discuss the application of the piano key weir in actual diseased reservoir and dam projects.

Keywords: Piano key weir · Discharge capacity · Discharge flow · Over-discharge ratio

1 Introduction

In recent years, global climate change has intensified and extreme weather has emerged frequently, resulting in severe flooding, frequent occurrence of mega-floods and super-standard floods, and the emergence of extreme hydrological events has raised higher requirements for flood safety of reservoirs and dams. Piano key weir is a new type of labyrinth weir, which is structurally wide at the top and narrow at the bottom, with upside down weir walls hanging upside down, invented by French Hydroccop Association in 2003, the first piano key weir was constructed in the summer of 2006 in Goulours, Southwest France (Guo et al. 2014). As the weir axis is elongated along the longitudinal direction, several connected side weirs are formed and the overflow front

© The Author(s) 2023
Y. Li et al. (Eds.): PIANC 2022, LNCE 264, pp. 281–293, 2023.
https://doi.org/10.1007/978-981-19-6138-0_25

is increased accordingly. When the head on the weir is the same, the overflow capacity of the piano key weir is increased several times compared with the ordinary linear weir, which can significantly improve the discharge capacity and flood discharge efficiency.

The piano key weir is wide at the top and narrow at the bottom, resembling the piano keys in appearance. The piano key weir is a high super-stationary structure from the structural mechanics point of view, with good overall balance and it has a wide range of applications (Lempérière and Ouaman 2003) due to its small footprint and upside down design structure. The symbols and meanings of the geometric parameters of the piano key weir are as follows: L is the total length of the overflow leading edge of the piano key weir; W is the weir width of the piano key weir, W_i is the weir inlet width, W_o is the weir outlet width; B is the total upstream and downstream length of the piano key weir, the side weir length, B_i is the downstream inverted overhang length of the inlet; B_o is the upstream inverted overhang length of the outlet; B_b is the length of the base; P is the weir height, P_d is the base height; S_i is the inlet inverted overhang angle, S_o is the inverted overhang angle of the outlet (Neelakantan et al. 2019). The geometric schematic diagram of the piano key weir is shown in Fig. 1.

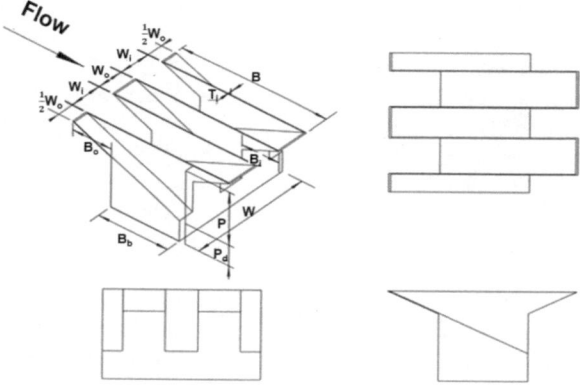

Fig. 1. Geometric diagram of the piano key weir

In this paper, we observe the over-weir flow pattern and drainage capacity of the piano key weir through physical model tests, and analyze the advantages of the piano key weir in terms of drainage compared to traditional thin-walled weirs (Jiang et al. 2019). Due to the complex flow pattern of the weir, the model test alone cannot fully analyze its superior drainage efficiency mechanism. numerical simulation is used as a supplement to study the drainage characteristics in terms of flow pattern and flow velocity, and research the mechanism for improving the drainage of the weir, so as to provide a reference for the optimization of the weir body shape and parameter sensitivity analysis.

2   Physics Model Experiment

2.1   Experimental Setup

The experimental device was built in the hall of Tiexinqiao test base of Nanjing Hydraulic Research Institute, and the experimental device consisted of water measuring weir and water tank, as shown in Fig. 2 below. During the experiment, the water from the groundwater reservoir is pumped to the upstream water measuring weir, and the water flow is stabilized by the water measuring weir free overflow sink, and finally the water flow is stabilized again by the elimination wall to the downstream sink, and the water flows through the sink to reach the model of the piano key weir, and then the overflow flows through the model and finally flows into the groundwater reservoir through the sink, forming a circulatory system throughout.

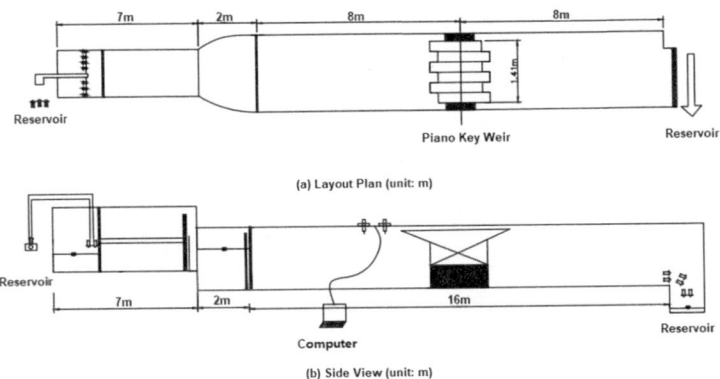

(a) Layout Plan (unit: m)

(b) Side View (unit: m)

Fig. 2. Experiment layout

The model of the piano key weir is made of 3 mm organic plastic board, and the specific parameters are shown in Table 1. The model is 6 inlet chambers and 5 outlet chambers with inverted A-type inverted structure upstream and downstream, and the model is installed in the middle of the 16 m-long downstream flume, and the model centerline coincides with the flume centerline in the incoming flow direction, and the model is spaced 0.81 m from both ends of the flume to the left and right. The experiments were conducted to study the hydraulic characteristics of the weir and to compare the discharge capacity with that of the traditional thin-walled weir, and reasonable body parameters were selected according to the conclusions of previous studies by experts and scholars (Lemperiere and Guo 2005): the spreading ratio L/W = 5.34, the tangent of the upstream and downstream inversion angle Tan Si = 0.53, and the tangent of the upstream and downstream inversion angle Tan S_o = 0.53. Tan S_o = 0.53, upstream/downstream inversion ratio B_i/B_o = 1.06, inlet/outlet width ratio W_i/W_o = 1.23, and weir height to weir width ratio P/W = 0.16.

Table 1. Dimensions of the piano key weir model (Unit: m)

Parameters	L	W	W_i	W_o	B	B_i	B_o	B_b	P	P_d	Tan S_i	Tan S_o
Size	7.53	1.41	0.135	0.11	0.63	0.17	0.16	0.30	0.23	0.10	0.53	0.53

2.2 Experimental Results

In order to observe the flow pattern and discharge capacity of the piano weir through physical model tests and to compare it with the traditional thin-walled weir, 12 sets of water crossing experiments were conducted throughout the experiment, and data such as flow rate, syringe reading, and travel velocity were measured and recorded, and the head on the weir was obtained from the water level syringe reading, and the total head on the weir was calculated from the measured travel velocity. The ratio of the total head on the weir to the weir height can be obtained by combining the total head on the weir with the weir height, which is known from the previous measurements of piano key weir. The specific data are shown in Table 2 below.

Table 2. Experimental data table

Groups	Q (L/s)	Reading #1 (cm)	h_0(cm)	Reading #2 (cm)	h_0(cm)	V (cm/s)	H (cm)	H/P	C_d
1	15.4	15.87	1.09	32.41	0.95	1.43	1.09	0.05	2.16
2	25.0	16.33	1.55	32.87	1.41	2.90	1.55	0.07	2.07
3	40.0	16.92	2.14	33.46	2.00	5.52	2.16	0.09	2.02
4	60.0	17.71	2.93	34.20	2.74	8.80	2.97	0.13	1.88
5	75.0	18.25	3.47	34.77	3.31	9.72	3.52	0.15	1.82
6	90.0	18.92	4.14	35.41	4.05	11.23	4.20	0.18	1.67
7	100	19.32	4.54	35.91	4.67	13.26	4.63	0.20	1.61
8	115	19.92	5.14	36.47	5.28	14.46	5.25	0.23	1.53
9	125	20.33	5.55	36.94	5.89	16.15	5.68	0.25	1.48
10	150	21.33	6.55	37.94	6.48	19.65	6.75	0.30	1.37
11	170	22.23	7.45	38.82	7.36	22.89	7.72	0.34	1.27
12	200	23.49	8.71	39.11	7.65	23.91	9.00	0.39	1.19

In order to compare the overflow capacity of the piano key weir and the traditional linear weir, it is assumed that the position of the piano key weir in the flume is replaced by a thin-walled weir of equal width and height (Qi et al. 2018), and the overflow flow of the thin wall is projected with the same total head on the weir, and the ratio of the flow of the piano key weir to the flow of the thin-walled weir is expressed as the over-discharge ratio: r, $r = \frac{Q_{PKW}}{Q_S}$ and the over-discharge ratio of the two is shown in Table 3 below.

The discharge expression of the piano key weir is similar to the thin-walled weir flow equation through the dimensional analysis, and the flow equation of both is as follows.

$$Q_{PKW} = C_d W \sqrt{2g} H^{3/2} \tag{1}$$

$$Q_S = m_0 W \sqrt{2g} H^{3/2} \tag{2}$$

where: C_d is the piano key weir flow coefficient, W is the piano key weir width, H is the total head on the piano key weir, $H = h_0 + \frac{V^2}{2g}$, h_0 is the head on the piano key weir, V is the traveling flow rate, m_0 is the thin walled weir flow coefficient, $m_0 = 0.4034 + 0.0534 \frac{H}{P_1} + \frac{0.0007}{H}$, P_1 is the sum of the weir height and the base height, $P_1 = P + P_d$.

Table 3. Over-discharge ratio of piano key weir and thin-walled weir

Groups	H(cm)	Q_{PKW}(L/s)	Q_S(L/s)	r
1	1.09	15.4	3.33	4.62
2	1.55	25.0	5.43	4.60
3	2.16	40.0	8.69	4.60
4	2.97	60.0	13.77	4.36
5	3.52	75.0	17.64	4.25
6	4.20	90.0	22.86	3.94
7	4.63	100	26.40	3.79
8	5.25	115	31.80	3.62
9	5.68	125	35.74	3.50
10	6.75	150	46.20	3.25
11	7.72	170	56.51	3.01
12	9.00	200	71.24	2.81

2.3 Analysis of Drainage Capacity

In order to compare the discharge capacity of the piano key weir and the traditional thin-walled weir, the flow rate of the piano key weir and the thin-walled weir at the same head on the weir, and the flow coefficient of the piano key weir and the thin-walled weir at the same head on the weir are made as follows, and the relationship between the over-discharge ratio and the head on the weir and the weir height ratio is made as follows (Barcouda et al. 2006) which can visually show the comparative relationship between the discharge of the piano weir and the thin-walled weir.

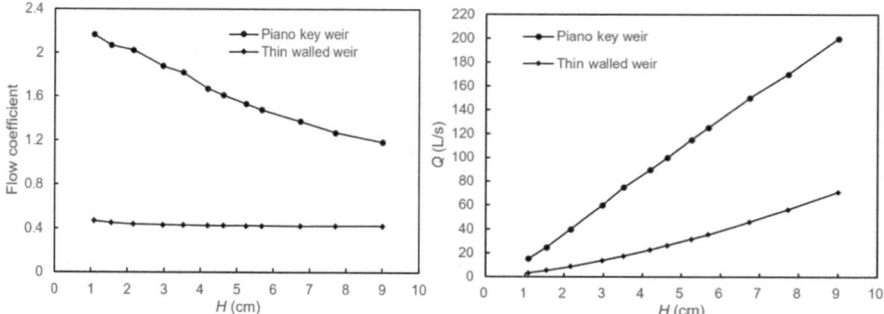

Fig. 3. The relationship between the piano key weir and thin-walled weir flow coefficient, flow rate and total water head on the weir

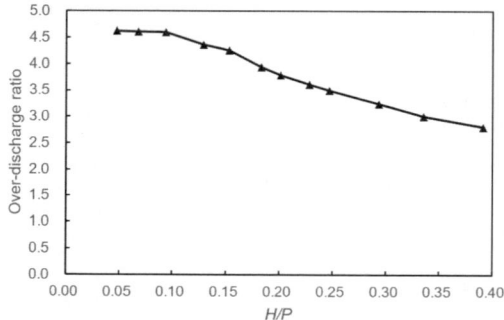

Fig. 4. The relationship between over-discharge ratio and H/P

As can be seen from Fig. 3, the flow coefficient and discharge volume of the piano key weir are significantly higher than those of the thin-walled weir at the same total head on the weir, and the piano key weir has a better discharge capacity than the thin-walled weir. As can be seen from Fig. 4, the piano key weir has a high discharge efficiency than the thin-walled weir, and the overflow ratio is basically maintained at 3–4 times, with the increase of upstream flow, the overflow ratio of the piano key weir shows a decreasing trend, the decrease of the piano key weir discharge capacity is because the overflow effect of the outlet chamber and the side weir of the inlet chamber receives a weakening when the upstream flow increases, but overall the piano key weir has a better overflow capacity compared with the thin-walled weir.

3   Mathematical Model Setup and Result Analysis

3.1   Model Building and Numerical Processing Methods

3.1.1   Model Setup

In order to study the hydraulic characteristics of the discharge of the piano key weir in more detail, the above physical model was used as a prototype and modeled 1:1 using software, and finally imported into numerical simulation software for simulation calculations. The simulated calculation model of the piano key weir is as follows, which consists of upstream channel, weir body and downstream channel (Li et al. 2015). The length of the piano key weir body in the direction of water flow is 630 mm, the inlet court is 135 mm, the outlet court is 110 mm, and the width of the weir body is 785 mm. In order to ensure the smooth flow over the weir and better simulate the water tongue as well as the actual situation, the length of the upstream channel is 800 mm and the length of the downstream channel is 800 mm in the calculation model 2000 mm.

The grid quality, quantity and division method have an important influence on the accuracy of the simulation results. In the numerical simulation of pian key weir, the total area grid number is 387.7946 million, the calculation file size asks 12 GB, the finish time is 50 s, and the calculation time is about 60 h, among which the grid number of PKW area is 107.2537 million and the grid size is 0.005 m.

The simulated area is divided into the upstream channel area, the pian key weir area and the downstream channel area according to the overall area and the piano key parts. The upstream channel area and the downstream channel area have relatively regular flow patterns, so the grid is relatively sparse. The weir area and the area behind the weir are the piano key parts of the study, so a denser grid is generated to ensure the accuracy of the simulation (Li et al. 2016).

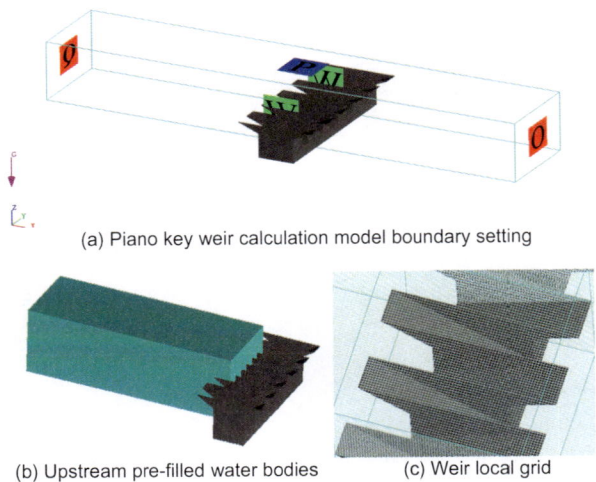

(a) Piano key weir calculation model boundary setting

(b) Upstream pre-filled water bodies   (c) Weir local grid

Fig. 5. The mathematical model setting of the piano key weir

3.1.2 Turbulent Flow Equation and VOF

Combined with the flow characteristics of the experimental piano key weir, the turbulent flow mathematical model is selected as the RNG k-ε equation model (Jiang 2020). The RNG k-ε equation is a mathematical model derived using the statistical method of reformed group theory for the instantaneous N–S equation, compared with the standard k-ε model, the RNG k-ε model adds a term to the ε equation that reflects the mainstream time-averaged strain rate and improves the accuracy of high-speed motion (Zhao et al. 2020). In addition, the RNG k-ε model takes into account the effect of vortices on turbulent flow and eliminates the singularity of the standard k-ε model, which combined makes the RNG k-ε equation more accurate and reliable in simulating jet diffusion, etc. The k and ε equations are as follows.

$$\frac{\partial(\rho k)}{\partial t} + \frac{\partial(\rho k u_i)}{\partial x_i} = \frac{\partial}{\partial x_i}\left[\frac{\mu_t}{\sigma_k}\frac{\partial k}{\partial x_i} + \mu\right] + G_k - \rho\varepsilon \tag{3}$$

$$\frac{\partial(\rho\varepsilon)}{\partial t} + \frac{\partial(\rho\varepsilon u_i)}{\partial x_i} = \frac{\partial}{\partial x_j}\left[\frac{\mu_t}{\sigma_\varepsilon}\frac{\partial\varepsilon}{\partial x_j} + \mu\right] + \rho C_1 S\varepsilon - \rho C_2 \frac{\varepsilon^2}{k + \sqrt{v\varepsilon}} \tag{4}$$

where: ρ is the fluid density, t is time, k is the turbulent kinetic energy, ε is the turbulent kinetic energy dissipation rate, u_i is the velocity component, x_i, x_j are the coordinate components, v is the viscosity, $v = \frac{\mu}{\rho}$, μ is the coefficient of viscosity, $\sigma_k = 1.0$, $\sigma_\varepsilon = 1.2$, G_k is the turbulent kinetic energy generation term due to the average velocity gradient, $G_k = \mu_t S^2$, $S = \sqrt{(2E_{ij}E_{ij})}$, $E_{ij} = \frac{1}{2}\left(\frac{\partial u_i}{\partial x_j} + \frac{\partial u_j}{\partial x_i}\right)$, the coefficient $C_1 = max\left[0.43, \frac{\eta}{\eta+5}\right]$, $\eta = \frac{Sk}{\varepsilon}$, $C_2 = 1.9$, μ_t is the turbulent viscosity, $\mu_t = \rho C_\mu \frac{k^2}{\varepsilon}$, $C_\mu = \frac{1}{A_0 + \frac{A_s U^* k}{\varepsilon}}$, $A_0 = 4.0$, $A_s = \sqrt{6}\cos\phi$, $\phi = \cos^{-1}(\sqrt{6}\omega)$, $\phi = \frac{E_{ij}E_{jk}E_{ik}}{\sqrt{(E_{ij}E_{ij})}}$, $U^* = \sqrt{E_{ij}E_{ij} + \widetilde{\Omega}_{ij}\widetilde{\Omega}_{ij}}$, $\widetilde{\Omega}_{ij} = \Omega_{ij} - 2\varepsilon_{ijk}\omega_k$, $\Omega_{ij} = \widetilde{\Omega}_{ij} - \varepsilon_{ijk}\omega_k$.

The flow pattern is complex during the release of the piano key weir, and the method VOF (Fan et al. 2022) is used in the capture of the free liquid surface, which is constructed using the volume ratio function $F(x_i, t)$, and the $F(x_i, t)$ function is as follows.

$$\frac{\partial F}{\partial t} + u_i \frac{\partial F}{\partial x_i} = 0 \tag{5}$$

When: $F(x_i, t) = 0$, it means the place is water phase; when $0 < F(x_i, t) < 1$, it means the place is water-gas phase intersection; when $F(x_i, t) = 1$, it means the place is gas phase.

3.1.3 Model Boundary Settings

Set the boundary conditions of the model before simulation, set the specified pressure at the upstream inlet, and set the fluid elevation; set the outflow at the outlet: the two walls and the bottom are set as non-sliding wall; the top is set as atmospheric pressure inlet boundary condition, see Fig. 5 above for the specific settings (Li 2020).

3.2   Numerical Simulation Results and Analysis

3.2.1   Drainage Capacity Model Validation

In order to verify the numerical simulation of the discharge capacity of the piano key weir, by giving different working conditions of the head on the weir, the respective discharge volume was calculated by simulation, and the error between the simulated and experimental values was calculated, the experimental data and the data derived from the simulation are shown in Table 4 below, and Fig. 6 can more visually reflect the error between the experimental and simulated values.

Table 4. Data table of experimental and simulated values

Groups	H(cm)	$Q_{\text{experiment}}$(L/s)	$Q_{\text{simulation}}$(L/s)	Error value %
1	1.09	15.4	15.92	3.38
2	1.55	25.0	26.30	1.56
3	2.16	40.0	39.46	−1.35
4	2.97	60.0	58.61	−2.32
5	3.52	75.0	73.18	−2.43
6	4.20	90.0	87.92	−2.31
7	4.63	100	99.74	−0.26
8	5.25	115	113.77	−1.07
9	5.68	125	124.13	−0.69
10	6.75	150	151.64	1.09
11	7.72	170	165.77	−2.49
12	9.00	200	197.28	−1.36

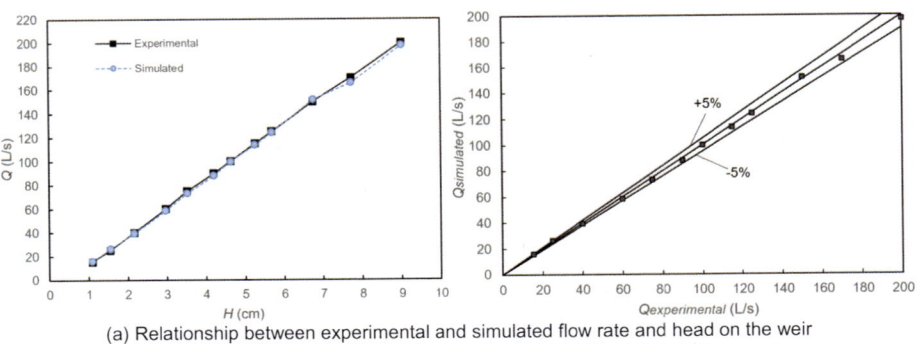

(a) Relationship between experimental and simulated flow rate and head on the weir

(b)Flow rate experimental value versus simulated value

Fig. 6. Relationship between the experimental and simulated values of the piano key weir

By comparing the flow value and simulation value of this model experiment, it can be found that the flow value calculated by numerical simulation of head on the same weir is basically linearly distributed with the experimental value and the values of the two curves are very close, and the error values of both are distributed around the straight line with slope k = 1, and the error values are between the two straight lines of −5% and +5%, and the results of numerical simulation are more reliable and meet the requirements of numerical simulation of hydraulics.

3.2.2   Flow Analysis

The following figure shows the profiles of the inlet chamber and outlet chamber of the weir and four profiles in the direction of the incoming flow of the weir. The flow pattern has the characteristics of ternary flow. When the water flows through the piano key weir, the water flows partly from the inlet, and the water flows downstream along the bottom plate in the inlet chamber and freely overflows to the downstream sink at the outlet, and the water flows in the inlet chamber also overflows from the side weir to the outlet chamber. Some of the water flows in from the exit overflow, collide and mix with the water flowing down the side weir of the incoming palace chamber, flow downstream along the bottom plate of the outgoing palace chamber, and are ejected at the exit. The overall flow over the weir is a mixture of forward and lateral flow, the flow pattern is complex, and the water flows collide and mix with each other, the flow pattern of the whole process can be seen in Fig. 7 below.

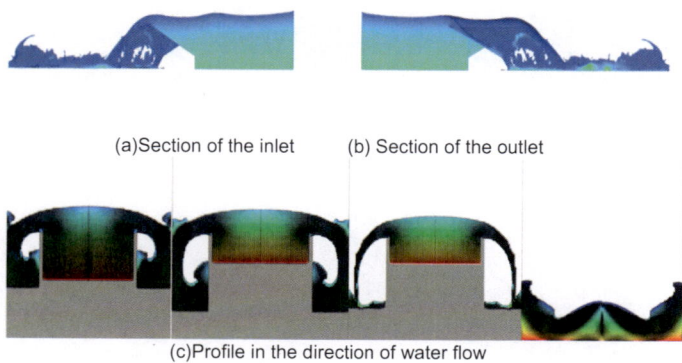

(a)Section of the inlet   (b) Section of the outlet

(c)Profile in the direction of water flow

Fig. 7. Schematic diagram of the flow pattern of water over the weir of the piano key weir

During the simulation, it was found that when the flow rate of water over the weir was small, the water surface line upstream of the piano key weir was basically the same as the water surface line of the flume, and the water flowed smoothly through the piano key weir, producing only a small drop in water level at the entrance and exit of the palace, and it could be found from the profile feature map in the direction of incoming flow that the water tongue overflowing from the side weir to the exit palace chamber was detached from the side wall, and the lateral water flow was free to flow out, and the lateral water tongue was thin, and the adjacent side walls were free to flow out Tongue of water does not interfere with each other into the exit palace room, in the middle of the front of the exit palace room to stimulate the final convergence of water outflow. The flow pattern at the exit of the exit chamber is more complex, the water jet at the exit while the water will spread down and left and right, and again collide and mix with the water flowing down from the overflow of the incoming palace, the water diffusion at the exit of the exit chamber is intense, and the flow pattern is chaotic.

3.2.3 Flow Rate Distribution

The flow pattern of the water over the weir is complex, and the velocity of the water will change when passing the weir. The velocity contour clouds of the inlet and outlet chambers are derived from the numerical simulation, which can reflect the magnitude of the fluid velocity visually. From the simulation results, it can be seen that the upstream of the weir is smooth and the velocity is small. The water flows into the inlet chamber first and the velocity starts to increase, and then picks up at the end of the inlet chamber in the form of a water tongue (Fig. 8).

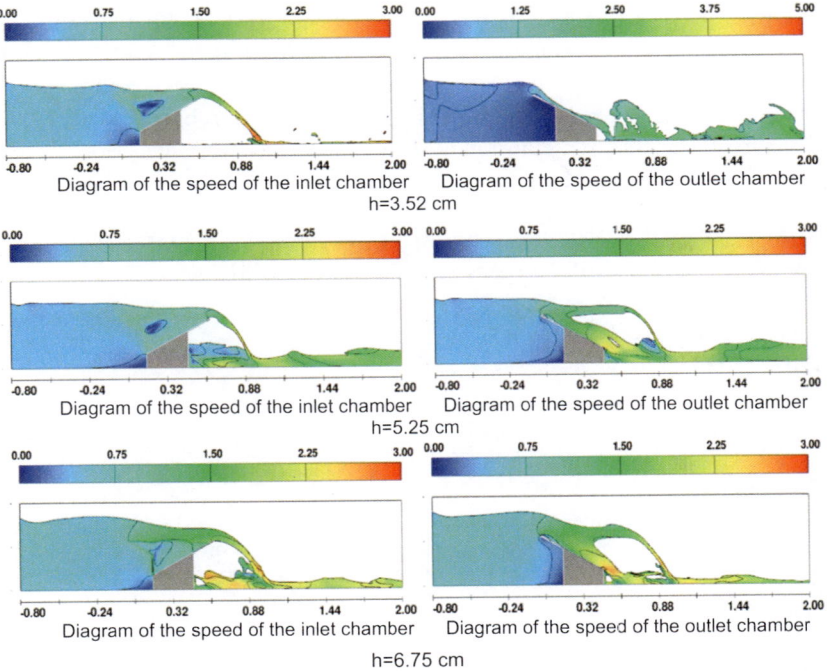

Fig. 8. Schematic diagram of the inlet and outlet speed of the water head on different weirs

From the above figure, it can be seen that the overall flow velocity of the piano key weir is small at low head, and the water can flow smoothly and evenly along the palace chamber and side walls, and finally overflows out from the weir at the outlet, with good overall flow. With the increase of the head on the weir, the flow velocity of the water over the weir also increases, indicating that the increase of the head on the weir can increase the flow velocity of the water over the weir. However, while the head on the weir increases, some areas with lower flow velocity appear at the entrance and exit of the palace, and these areas obstruct the water flow under the action of the inertia of the water flow, which reduces the drainage capacity of the piano key weir, which is the reason why the head on the piano key weir increases and the flow velocity increases but the drainage capacity decreases.

4   Conclusion

As a new type of labyrinth weir, the piano key weir has a good application prospect in flood control and drainage. In this paper, the flow pattern and discharge capacity of the piano key weir are studied through physical model tests, and the discharge capacity of the piano key weir is compared with that of the thin-walled weir and the over-discharge ratio of the piano key weir. The numerical simulations of the discharge capacity of the piano key weir were compared with the physical simulations, and the flow patterns matched with the experimental flow patterns, which verified the reliability of the simulations. The simulation analysis of the flow velocity of the piano key weir shows that the increase of head on the weir can increase the flow velocity of the water over the weir, but the increase of head on the weir also decreases the discharge efficiency, which provides a reference for the optimization of the piano key weir shape to improve the discharge efficiency in the future.

Acknowledgements. This study is supported by Central-level Public Welfare Research Institutes Basic Research Business Expenses Special Funds (Y121005).

References

Barcouda M et al (2006) Cost-effective increase in storage and safety of most dams using fuse gates or P.K. weirs. Proceeding of the 22nd

Guo XL, Yang KL, Xia QF, Fu H (2014) Discharge capacity characteristics of piano key weir. J Hydraul Eng 45(7):867–882 (in Chinese)

Fan HH, Li Z, Li BB, Li L, Shang S, Wang JY (2022) Numerical investigation of gas-assisted sludge atomization and breakup based on VOF-DPM coupled model. Chin J Process Eng 1–10 (in Chinese)

Jiang D, Li GD, Li SS (2019) Experimental study on discharge characteristics of different upstream-downstream overhang ratios of piano key weir. Water Res Hydropower Eng 50 (7):124–130 (in Chinese)

Jiang D (2020) Study on parameter optimization of piano key weir upside down ratio and numerical simulation of downstream scour. Xi'an University of technology, Thesis of Master (in Chinese)

Lempérière F, Ouaman A (2003) The piano keys weir: a new cost-effective solution for spillways. Hydropower Dams 7(5):144–149

Lemperiere F, Guo J (2005) Low cost increase of dams storage and flood mitigation: the piano keys weirs. J Hydraul Res 35(3):76–79

Li GD, Miao Z, Gao B, Chen G (2015) Numerical study on discharge capacity of piano-key weir at its different overflow edges. J Hydroelectric Eng 34(8):77–84 (in Chinese)

Li SS, Li GD, Miao Z, Chen G (2016) Numerical simulation study on discharge characteristics of piano-key weir with various heights. Water Resour Hydropower Eng 47(5):60–64 (in Chinese)

Li SS (2020) Study on hydraulic characteristics and geometric. Xi'an University of technology, Thesis of Master (in Chinese)

Qi YY, Li GD, Li SS, Mi T (2018) Numerical simulation of hydraulic characteristics of piano-key weir. South-to-North Water Transfers Water Sci Technol 16(1):164–169 (in Chinese)

Neelakantan TR, Rajeshwaran T, Renganathan GN (2019) Hydraulic advantage of piano-key weir over ogee weir. Int J Innov Technol Exploring Eng (IJITEE) 9(2s2)

Zhao QY, Liu SH, Liao WJ (2020) Study of total flow control equations and energy loss characteristics of steady turbulent flow in open channel. Adv Water Sci 31(02):270–277 (in Chinese)

Analysis on Throughput Capacity of Water-Saving Ship Lock in Simulation Method

Ying Tang$^{(\boxtimes)}$, Chunze Liu, Fengshuai Cao, and Jianping Shang

CCCC Water Transportation Consultants Co., Ltd., Beijing, China
{tangying, liuchunze}@pdiwt.com.cn

Abstract. When the double-line water-saving ship locks operate in the mode of filling and emptying mutually, the filling water process of one lock chamber and emptying water process of the other one are carried out at the same time, thus the situation of chambers waiting for each other is inevitable. Besides, the variation of ship navigating speed and safety interval time between two adjacent ships may have a more significant impact on the operation efficiency of the ship lock, which cannot be ignored. Combined with the actual condition of a planned lock on a canal in China, the throughput capacity and operation efficiency of the double-line water-saving ship locks under the mode of filling and emptying mutually are studied. A coupled operation simulation model of double-line locks is established, in which the mode of filling and emptying water mutually or independently is adopted. In the model, the whole operation processes of the locks are simulated. The simulation results show that the throughput capacity of the planned ship lock is reduced by 5.6%–7.8%, and the average lockage time is increased by about 10.8%, when comparing the mode of filling and emptying water mutually with independently. Besides, using average value of speed and safety interval time of ships is acceptable when doing research on the throughput capacity and operation efficiency of ship locks. The achievements can provide technical support for the analysis of the throughput capacity and operation efficiency of the water-saving ship lock.

Keywords: Water-saving ship lock · Throughput capacity · Operation process of ship lock · System simulation

1 Introduction

The operation of ship locks on the inland river and canal in China is often restricted by the shortage of water resources, so it is necessary to build water-saving ship locks in some places. There are four main types of water-saving ship locks: the ship lock with a water-saving pool, the multi-level ship lock without waiting for navigation, the double-line ship locks filling and emptying water mutually, and the ship lock with an intermediate channel (Chen et al. 2021). At present, the water-saving ship locks that have been put into operation in China mainly include the third and fourth line ship locks in Changzhou and the second and third lines in Feilaixia. A number of new water-saving ship locks are under planning and construction (Dongle et al. 2020).

© The Author(s) 2023
Y. Li et al. (Eds.): PIANC 2022, LNCE 264, pp. 294–304, 2023.
https://doi.org/10.1007/978-981-19-6138-0_26

The operation experience of the third and fourth line ship locks in Changzhou shows that the throughput capacity and operation efficiency of water-saving ship locks are affected by filling and emptying water mutually.

When the double-line water-saving ship locks operate in the mode of filling and emptying mutually, the filling water process of one lock chamber and emptying water process of the other one are carried out at the same time. On the premise that other operating parameters of locks are the same, because the number of ships passing through the two locks may be different in one lockage, the situation of chambers waiting for each other is inevitable, which affects the operation efficiency of the two locks. In addition, in the mode of filling and emptying mutually, the operation of the double-line water-saving ship locks are coupled with each other. Compared with the mode of filling and emptying independently, the fluctuation of ship navigating speed and interval time may have a more significant impact on the operation efficiency of the ship lock, which cannot be ignored.

The current research of water-saving ship locks focuses on the hydraulics of water conveyance system (Chen et al. 2021), the calculation of water-saving rate (Dongle et al. 2020) and benefit analysis about water-saving (Yang and Chen 2013). But there is almost no research on the impact of water-saving mode on the throughput capacity and operation efficiency of ship lock.

Combined with the condition of a planned lock on a canal in China, the throughput capacity and operation efficiency of the double-line water-saving ship locks under the mode of filling and emptying mutually are studied. Based on the simulation model of double-line locks which filling and emptying independently, a coupled operation simulation model of double-line locks is established, in which the mode of filling and emptying mutually is adopted. The whole operation processes of the locks are simulated, under two set of ship types and proportions. Indexes such as throughput capacity, chamber occupy rate and one lockage time are collected in the model, statistic data are compared and analyzed.

2 Building, Verifying and Validating Simulation Model

2.1 Model Scope and Boundary

The model includes the anchorage, connecting section, approach channel, berthing structure, lock head at the upstream and downstream and lock chamber of the double-line locks. The upper and lower boundaries of the model are the upstream and downstream anchorages of the ship lock.

2.2 Lockage Process

Lockage process in the mode of filling and emptying water independently and mutually are shown separately in Fig. 1 and Fig. 2.

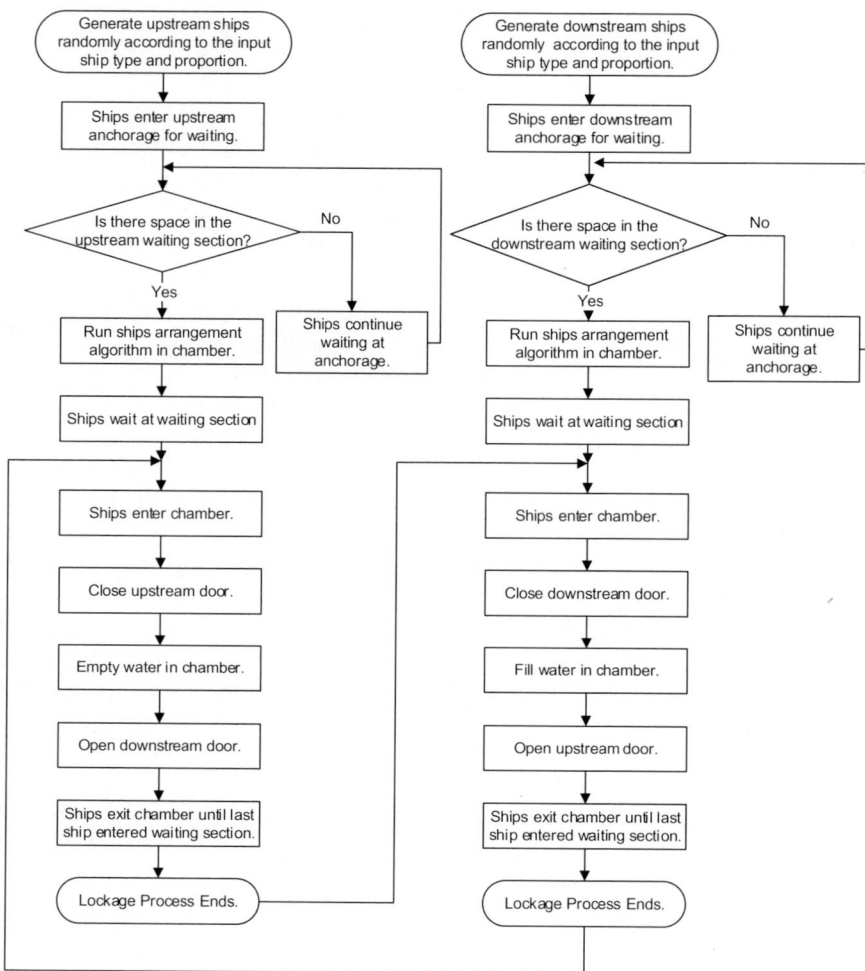

Fig. 1. Lockage process in the mode of filling and emptying water independently

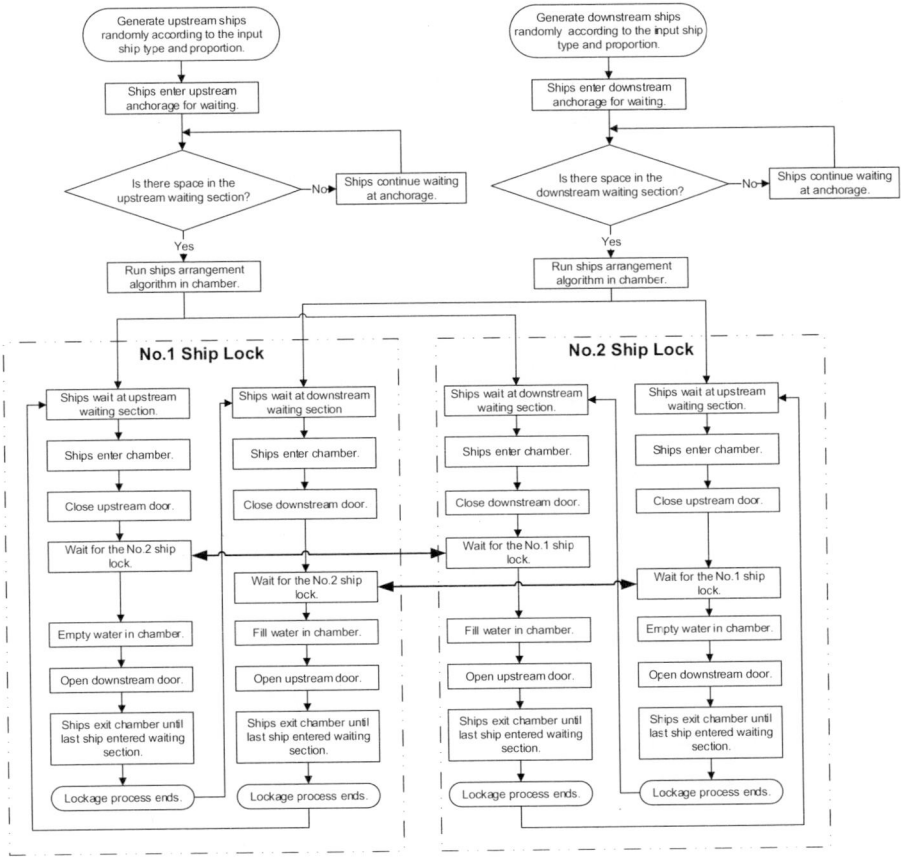

Fig. 2. Lockage process in the mode of filling and emptying water mutually

2.3   Main Algorithms and Basic Assumptions

The model specifies the ship lock for a ship, according to the state of the double-line ship locks (upgoing or downgoing). When the ship lock allows ship to wait at waiting section, the position of the ship in the lock chamber and order in the queue are determined according to the ship arrangement algorithm, and then ships of one same lockage are sent to the waiting section. After that, the whole process of ships passing through lock are simulated. The ship arrangement algorithm is shown in Fig. 3, which is based on the two-dimension packing algorithm. For details, please refer to the ship arrangement algorithm for single lock adopted by Shang et al. (2011) and Liu et al. (2020).

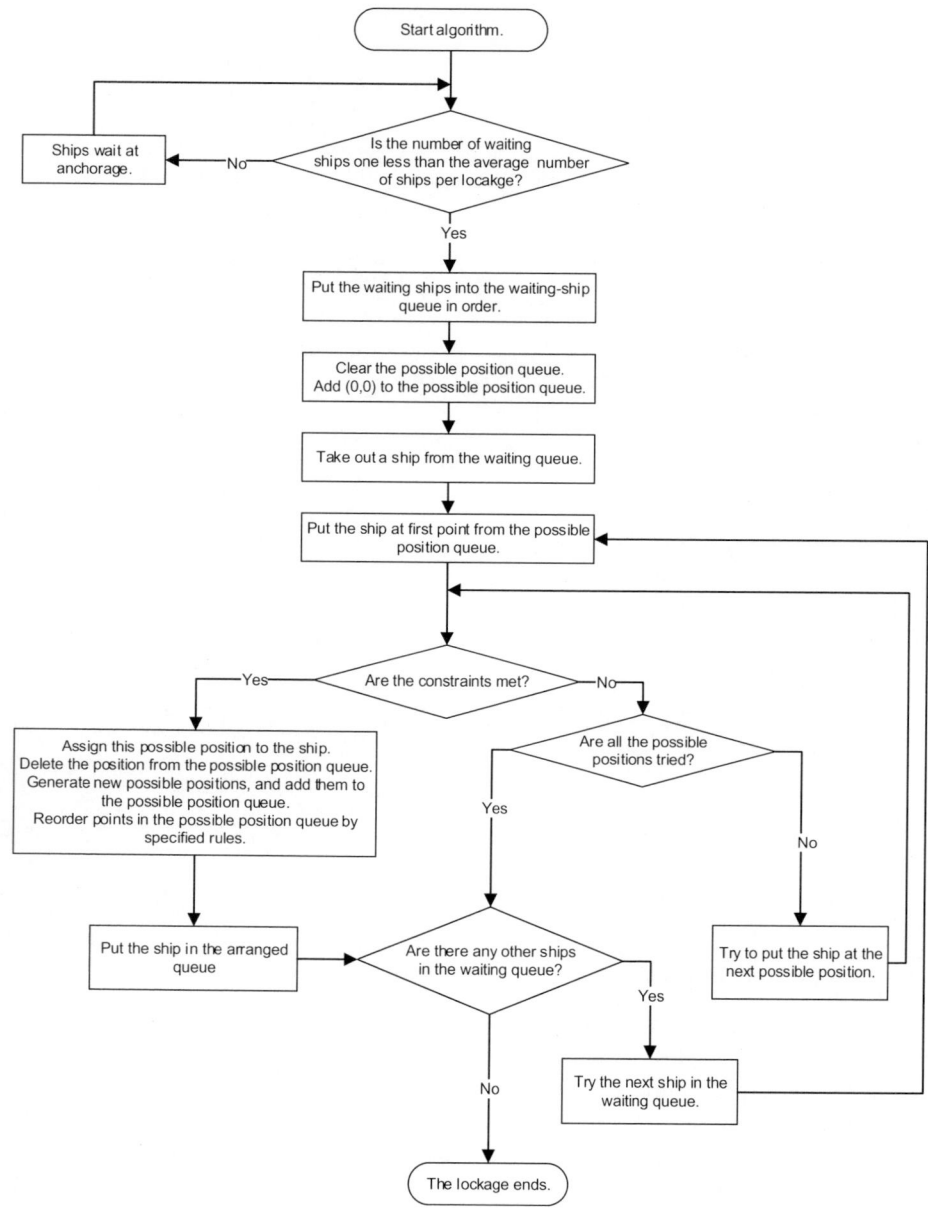

Fig. 3. Ship arrangement algorithm

The basic assumptions of model are listed below:

(1) The target of the model is to calculate the capacity of the ship lock. In order to ensure the full load operation of the ship lock, there should be enough ships waiting for lock in the anchorage.

(2) Ships are generated according to the input ship type and proportion.
(3) The acceleration and deceleration process of the ship navigation are not simulated in the model. Ships move at a fixed speed, which will change dynamically and immediately. When this ship has speed less than the speed of the front ship, or is more than 1.5 m away from the front ship, it will move at the preset speed. When this ship has higher speed than the front ship, and is within 1.5 m away from the stern of the front ship, it will move at the same speed as the front ship, to keep a safety distance.
(4) When arrange ships in chamber, the lock chamber shall be filled as full as possible.

2.4 Statistic Data and Key Performance Indicators

These data are collected: total number of ships passing through single lock upstream or downstream, total number of lockages, total number of bulk ships, total number of container ships, total DWT of bulk ships, total container capacity of container ships, average DWT of one lockage, the occupy rate of chamber, average number of ships per lockage, average lockage time, and average number of lockages per day.

These key performance indicators are calculated: total DWT of ships passing through one-line lock upstream or downstream per year (A TEU is treated as 15 ton), one-way annual capacity of one-line ship lock, which equals to the total DWT of ships passing through one-lock upstream or downstream per year × ship loading coefficient/unbalanced coefficient of traffic volume, and one-way capacity of double-line ship locks.

2.5 Model Verification and Validation

The model is verified and validated using the upgoing data of the Three Gorges double-line five-level locks in 2013. The plane dimension of lock chamber of each level in Three Gorges is 280 m × 34 m (length × width). The length of lock head is 50 m. The locks operate in one-way and ships wait at waiting section just before the lock head. Assume that ships enter and exit chamber in turn one by one, and the entering speed is set to 0.4 m/s, the exiting speed is set to 1.0 m/s. The safety interval time between two ships is set to 2 min. Ships are arranged in chamber according to the first-come-first-serve rule, and ships are generated randomly following the same rule within a year. The unbalanced coefficient of traffic volume is 1.1, which is used to amend the total one-direction cargo ton or the total DWT of ships passing through one-line lock collected in the model.

The results of model verification and validation show that the simulation model has correct logic and can get reasonable results, which can be used for the next research.

3 Simulation Experiment

3.1 Experiment Scheme

The model does not simulate the process that the ship lock cannot operate (accident, maintenance, flood and dry season, etc.), so the simulation model runs 340 days per year and 22 h per day. The warmup period of the model is 10 days. The experiment scheme is shown in Table 1 below.

Table 1. Experiment scheme

Type of water conveyance system	The variation amplitude of ship speed and safety interval time when entering and leaving lock, relatively to base value (input)
Filling and emptying water mutually	No variation, variation ±30%, variation ±50%
Filling and emptying water independently	No variation, variation ±30%, variation ±50%

In Table 1, "no variation" means that the ship enters and exits the lock at the base value of speed and safety interval time. And "variation ±30%" "variation ±50%" means that ship speed and safety interval time are randomly sampled from 70%–130% of the input basic value and 50%–150%, which are decided when ships are generated. Those random values are input in the model, but can only be treated as expected, because the actual speed and safety interval time of a ship will be affected by the speed of the front ship. When this ship has speed less than the speed of the front ship, or is more than 1.5 m away from the front ship, it will move at the preset speed. When this ship has higher speed than the front ship, and is within 1.5 m away from the stern of the front ship, it moves at the same speed as the front ship.

As a result, if input the same entering and leaving speeds in "variation" experiments as in "no variation" experiments, the average speeds of those two type of experiments are not the same for sure. To study the influence of variation, the average value of entering and leaving speeds are adjusted to be the same among all experiments, thus the input value is a little higher in "variation" experiment schemes.

3.2 Input Parameters

3.2.1 Plane Layout

The lock chamber is 300 m in length and 34 m in width. Considering the safe distance between the ship in chamber and the chamber wall and between ships, the length and width of the lock chamber for arranging ships are 280 m and 32.8 m (Fig. 4).

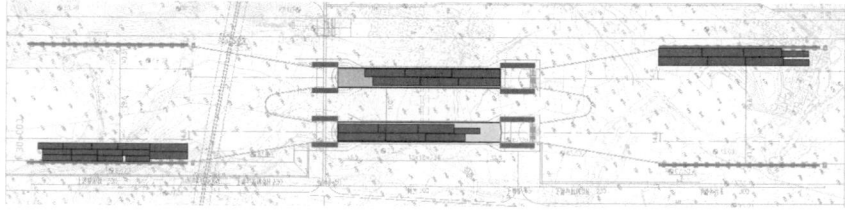

Fig. 4. Ship lock layout in the model

3.2.2 Ship Type and Proportion

According to the history statistic data, length and width of ships of each type are ranges. As input in the model, each level of tonnage and container capacity are subdivided into three grades: ship length and width taking the lower limit of its range, taking the average value and taking the upper limit, and the corresponding proportion of each grade are 33%, 34% and 33% in its tonnage level or container capacity level. Besides, the DWT value of bulk ships is equal to its tonnage, the DWT value of container ships is equal to the container capacity (in TEU) × 15t.

Some parameters of ships in the model are not considered, such as the height above the waterline and the loading rate (Table 2).

Table 2. Ship type and proportion input in the model

Ship type	Tonnage/Container capacity	DWT (ton)	Length (m)	Width (m)	Ship number proportion	
					Near future	Future
Bulk ship	1000	1000	45.0–50.0	10.8–11.0	39.7%	24.0%
	2000	2000	68.0–74.0	13.8–14.0	10.4%	18.1%
	3000	3000	74.0–80.0	15.5–15.8	15.4%	13.3%
	5000	5000	88.0–90.0	15.5–15.8	6.4%	7.7%
Container ship	70TEU	1050	54.0–60.0	10.8–11.0	15.5%	14.1%
	160TEU	2400	68.0–74.0	13.8–14.0	4.1%	10.6%
	200TEU	3000	68.0–74.0	15.5–15.8	6.0%	7.8%
	250TEU	3750	88.0–90.0	15.5–15.8	2.5%	4.5%
Total	-	-	-	-	100%	100%

3.2.3 Time Related Parameters and Other Inputs

The time of opening and closing gate are all 4 min. And the time of filling and emptying water are all 14 min.

The basic values of ship speed entering and leaving locks are 0.8 m/s and 1.2 m/s in "variation" experiments, 1.0 m/s and 1.4 m/s in "no variation" experiments. The actual speed of the ship entering and leaving the lock is randomly determined by the basic value and variation range. The basic value of the safety interval time between two adjacent ships entering and leaving the lock is 2 min. The actual value of a ship entering and leaving the lock are the same, and are also randomly determined by the basic value and variation range.

According to the ship lock design code in China, the ship loading coefficient is taken as 0.8 when there is no relevant statistic. The unbalanced coefficient of traffic volume is taken as 1.1.

4 Simulation Experiment Results and Analysis

In this study, the speed of the ship entering and leaving the lock and the safety interval time between two adjacent ships are the two main factors that may influence the throughput capacity and operation efficiency of ship locks, under the condition of filling and emptying water mutually or independently.

After preliminary analysis, the above factors mainly have impact on total number of ships passing through one-line lock upstream or downstream, total number of lockages, one-way annual capacity of one-line ship lock, one-way capacity of double-line ship locks and average lockage time, but have no impact on average DWT of one lockage, the occupy rate of chamber, average number of ships per lockage. The one-way capacity of double-line ship locks and the average lockage time can fully reflect the influence, so the two indexes are mainly analyzed. See Table 3 and Table 4 for simulation results.

Table 3. Simulation results of the one-way capacity of double-line ship locks

The variation amplitude of ship speed and safety interval time when entering and leaving lock, relative to base value (input)		No variation		Variation ±30%		Variation ±50%	
Ship type proportion		Near future	Future	Near future	Future	Near future	Future
One-way capacity of double-line ship locks of different kinds of water conveyance system (million ton)	Filling and emptying water mutually	72.77	79.64	71.35	77.78	69.70	75.91
	Filling and emptying Water independently	78.47	84.32	76.95	82.50	75.61	80.97
The reduction ratio of Filling and emptying water Mutually relative to independently		7.3%	5.6%	7.3%	5.7%	7.8%	6.2%

Table 4. Simulation results of the average lockage time in "no variation" experiments

Type of water conveyance system	Filling and emptying water independently			Filling and emptying water mutually		
Ship type proportion	Near future	Future	Index	Near future	Future	Index
Average number of ships per lockage	5.6	4.7		6.1	4.9	
Ships average waiting time in chamber (minute)	0.00	0.00		1.44	1.27	
Ships maximum waiting time in chamber (minute)	0.00	0.00		16.00	14.25	
the average lockage time (minute)	53.89	49.21	a	59.76	54.47	A
A-a (minute)	-	-		5.87	5.26	B
B/a	-	-		10.9%	10.7%	

(1) The throughput capacity of the double-line ship locks will be reduced comparing filling and emptying water mutually with independently. The range of capacity reduction is related to the composition of ships, and the variation amplitude of ship speeds entering and leaving locks. Different composition of ships may lead to different value of average ship number per lockage, and the greater the average ship number per lockage, the greater the reduction of capacity. The greater the variation of speeds and safety interval time, the greater the reduction of capacity.

(2) According to the boundary conditions and input parameters in the simulation experiments, the throughput capacity is reduced by 5.6%–7.8%, and the average lockage time is increased by about 10.8%, when comparing the mode of filling and emptying water mutually with independently. It should be noted that the filling and emptying water time of the ship lock in this study are the same in the two filling and emptying modes. But generally, it is slightly longer in the mode of filling and emptying water independently. If then, it is hard to tell whose average lockage time is longer, and further research is needed.

(3) In terms of the influence of the variation of the ship speed entering and leaving locks and safety interval time, if the variation is 30%, the throughput capacity will be reduced by 1.9%–2.2%, comparing the mode of filling and emptying water mutually with independently. And if the variation is 50%, the throughput capacity will be reduced by 3.6%–4.0%. Those reductions are not significant. The results of this case shows that it is acceptable to use average value of navigation speed and safety interval time when doing research on the throughput capacity and operation efficiency of ship locks.

5 Conclusions

(1) On the premise of the same time of filling and emptying water, the throughput capacity will be reduced if the double-line ship locks operate in the mode of filling and emptying water mutually compared with independently. The range of

capacity reduction is related to the composition of ships, and the variation amplitude of ship speeds entering and leaving locks and safety interval time between two adjacent ships.

(2) Under the boundary conditions and input parameters of this study, the throughput capacity is reduced by 5.6%–7.8%, the average lockage time is increased by about 10.8%, comparing the mode of filling and emptying water mutually with independently.

(3) The speed of ships entering and leaving locks and safety interval time with no variation or different degrees of variation, have little impact on the throughput capacity of ship lock. Therefore, it is acceptable to use average value of speed and safety interval time of ships when doing research on the throughput capacity and operation efficiency of ship locks.

Acknowledgements. We appreciate the support from China Communications Construction Company (CCCC), which is our superior company. The CCCC pays great attention to scientific research and application, and try his best to support young research staffs in the company.

References

Chen M, Lv SP, Liu Y, Huang HJ (2021) Review of hydraulics investigations of filling and emptying system in ship locks with water-saving basins. Port Waterway Eng 586(09):106–112

Dongle SY, Wang Q, Zhang N, He LD, Li H (2020) Calculation method of water saving rate of multi-level ship lock. China Water Transp (12):87–89

Liu CZ, Cao FS, Tang Y, Shang JP (2020) Comparison of plane dimension schemes for capacity expansion project of Gezhouba junction. J Waterway Harbor 41(2):172–178

Shang JP, Wu P, Tang Y (2011) On multiple-lane lock's joint scheduling plan based on computer simulation. Port Waterway Eng 457(9):199–204

Yang ZC, Chen MD (2013) Efficiency and application of water-saving ship lock in inland river navigation. Port Waterway Eng 486(12):131–135

Applicability of the Blue Wave to the Canal from Villiers to Beaulieu

J. M. Deplaix$^{(\boxtimes)}$

AFTM, Paris, France
aftmjmd@hotmail.com

Abstract. The Blue Wave is a locking structure that allows simultaneous navigation in both directions of navigation, without stopping in most cases, and at a speed depending on the local topography.

Its principle has been described in an issue of PIANC Bulletin, N°108, september 2001.

It could be used, in lieu of a set of locks, on a Canal which is presently being improved to CEMT Class Va, far upstream from Paris, the Canal from Villiers to Beaulieu (Petite Seine, France).

It is possible to form 10 reaches of 1,000 m each, separated by doors (mitre or valve). The depth available in the existing borrow pits is sufficient for the navigation of the project boats, which have 31.9 m^2 of wetted section.

By using the existing pits, which were dug in advance for improvement of this waterway, the structure very faithfully respects the current levels of these bodies of water, which avoids costly measures to restore the water table. In addition, the 2 upstream gates being non-overtopping, the non-overtopping dykes only total 2 km, instead of 16 km in the lock option.

These two points confirm that the proposed solution is more respectful of the environment and landscapes than the lock option. Thus, since the set speed of the concept is practically the optimum speed permitted by the cross-section of the canal, it can be said that the site of the Canal from Villiers to Beaulieu is optimal for the application of the Blue Wave principle.

Keywords: Blue-Wave · Canal · Lock · Seine River · Speed

1 Introduction

The Blue Wave is a locking structure that allows simultaneous navigation in both directions of navigation, without stopping in most cases, and at a speed depending on the local topography.

Its principle has been described in an issue of PIANC Bulletin, N°108, september 2001.

It could be used, in lieu of a set of locks, on a Canal which is presently being improved to CEMT Class Va, far upstream from Paris, the Canal from Villiers to Beaulieu (Petite Seine, France).

This paper will show that the proposed solution is more respectful of the environment and landscapes than the lock option. Thus, since the set speed of the concept is

Y. Li et al. (Eds.): PIANC 2022, LNCE 264, pp. 305–310, 2023.
https://doi.org/10.1007/978-981-19-6138-0_27

practically the optimum speed permitted by the cross-section of the canal, it can be said that the site of the Canal from Villiers to Beaulieu is optimal for the application of the Blue Wave principle.

2   The Blue Wave Concept

In cities, traffic lights are synchronised, so as to allow vehicles, running at a set speed, to travel without finding a red light. This is called Green Wave.

Similarly, on a canal, the Blue Wave concept enables craft to never stop, while encountering differences of elevation. If the Lock is akin to an elevator, the Blue Wave is an escalator. The difference of elevation is broken in a number of steps, each usually some 1 m high, and the reach varies up or down while the craft is moving, albeit at slow speed. In contrast, the craft does not move in a lock, and needs to have, between 2 locks, a large canal to ply sufficiently fast to compensate for this stop.

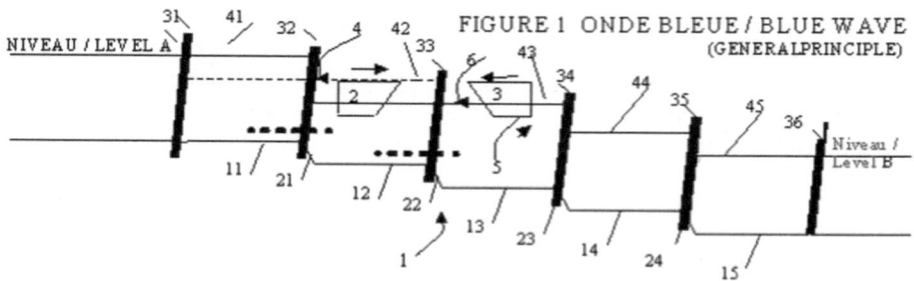

FIGURE 1 ONDE BLEUE / BLUE WAVE (GENERAL PRINCIPLE)

As can be seen on the figure, boats can navigate up and down at the same time, provided sufficient passing space is provided. Locks cannot do that. Another advantage, on shallow terrain, is that the Blue Wave can stick to the slope of the natural terrain, while, for a lock, there must be deep trenches or high embankments which, apart from being costly, are very conspicuous in the landscape. This is less and less accepted.

In natural rivers with rapids (less than 5 m fall), a Blue Wave can enable to navigate on a parallel canal across the rapids. This will be of use on African, South American, and Asian rivers, where there are many shallow rapids.

Similarly, where a bend is short-circuited, for hydraulics reasons the main flow has to continue to use the bend, and there is a slight difference of level between the two ends of the by-pass. With a set of Blue Wave gates at each end that alternatively open, it is possible to compensate for this slight fall, while the natural hydraulics of the river remain unchanged.

In canals, the cross section of the canal can be smaller with a Blue Wave than with a set of locks, and yet retain the same average speed on the whole voyage. A canal designed for 6.7 km/h is much cheaper to excavate than the one designed for 10 km/h.

Let us assume a 10 km canal designed for 10 km/h, with a 10 m high lock at the end. It takes roughly 1.5 h to cross both. 6.7 km/h on average, or 1.85 m/s.

A Blue Wave needs 10 km to overcome 10 m. At 1.85 m/s, this is 5400 s or 1.5 h, with a canal designed for 6.7 km/h only, thus a much smaller cross-section.

If the 12 gates needed for a Blue Wave are not more expensive than a 10 m lock, with all its concrete, then a Blue Wave is cheaper than a classical waterway.

The design speed depends on the local topography and the length of the craft.

3 Application to Bray-Nogent Project

3.1 The Bray-Nogent Project History

In the eighties, the improvement of the Bray-Nogent waterway was already planned. It was thought possible to seize an opportunity and excavate a large gauge canal parallel to the existing Villiers-Beaulieu canal, the excavated material being needed to create the platform of a Nuclear Power Plant nearby. There was actually 8 borrow pits, so that the water table would not be impaired, but the Bray-Nogent waterway was deferred, and the borrow pits were left unused.

40 years later, the project is on the verge to be implemented, and a classical canal with one lock has been designed parallel to the Villiers-Beaulieu canal. This project is using the borrow pits, but the 4.44 m fall at the lock obliges to raise the banks of the new canal over 12 km (2 × 6 km), to deepen some borrow pits (4 km) and to take the actions necessary to stabilise the water table, a very touchy problem in this area.

3.2 The Blue Wave Solution

It was thought possible to fruitfully apply the Blue Wave technique there:

By using the borrow pits, the structure very faithfully respects the current levels of these bodies of water, which avoids costly measures to restore the water table. In addition, only the banks of the 2 upstream cells need to be raised, and by 1 m only, against 16 km in the lock option, by up to 4 m!

These two points confirm that the proposed solution is more respectful of the environment and landscapes than the lock option.

Here, the set speed may be 2 m/s, i.e. 7.6 km/h, very close to the commercial speed permitted by the wet section of the canal.

It is possible to form 10 reaches of 1,000 m each, separated by collapsible doors (mitre or valve) with a feeding system from one reach to the other. The depth available in the borrow pits is sufficient for the navigation of the project boats (31.9 m^2 of wetted section). For the economy of the system, it would be preferable, but not essential, for the width of the different reaches to be identical, and not to exceed 50 m at the water surface. In this case, the volume exchanged from reach to reach will be 25,000 m^3, since it is on average a section of 0.5 m that thus passes from one reach to the other.

The first two reaches are used to dampen level variations of the Seine in the reach of Beaulieu, between 60.60 (Normal Pond Level-RN/NPL) and 61.57 (Reference Water Level-LER/RWL). The following figures, extrapolated from VNF documents, show the kinematics for a downhill boat.

3.2.1   Downstream Movement at Normal Pond Level

At NPL, at the green light, the boat (or the group of boats) enters reach 1 towards reach 2, both being at the upstream pond level, and while it travels in it at 2 m/s, the door pk0 closes. It then enters reach 2, since gate pk1 has remained open, and, as soon as it enters, this gate pk1 closes, while in synchronization (as soon as it is completely closed) reach 2 begins to empty into reach 3 with a flow rate between 50 and 100 m³/s.

When the equality of levels is reached (duration approximately 360 s) the gate pk2 opens, and the boat passes into reach 3 which is then at elevation 60.10.

If a rising boat were in this reach, it can enter reach 2 before gate pk2 closes. This crossing operation can be carried out simultaneously, if the doors are wide enough (30 m), or in succession, under the control of lights, if they are narrower (18 m).

For the downstream boat, the gate pk2 closes as soon as any rising boat has crossed it, reach 3 empties into reach 4 until the levels are equal (59.60), at which time the gate pk3 opens to allow the passage of the boat in reach 4.

The operations are repeated step by step, with 5 crossing possibilities over the entire route.

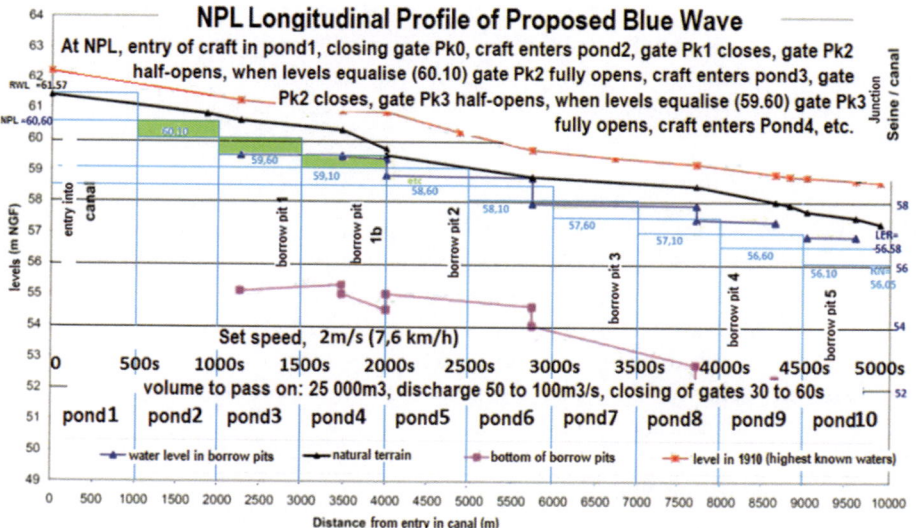

3.2.2   Downstream Movement in Flood Condition

If the Beaulieu reach is higher than the NPL, up to the RWL (or the Highest High Water), the first reach is used to dampen this variation:

At RWL, at the green light, the boat (or the group of boats) enters reach 1, and while it travels through it at 2 m/s, the gate pk0 closes, while in synchronization (as soon as it is completely closed) reach 1 begins to empty into reach 2 with a flow between 50 and 100 m³/s. The excess volume is rejected downstream by dumping in the reach 3 above or around the gate pk2.

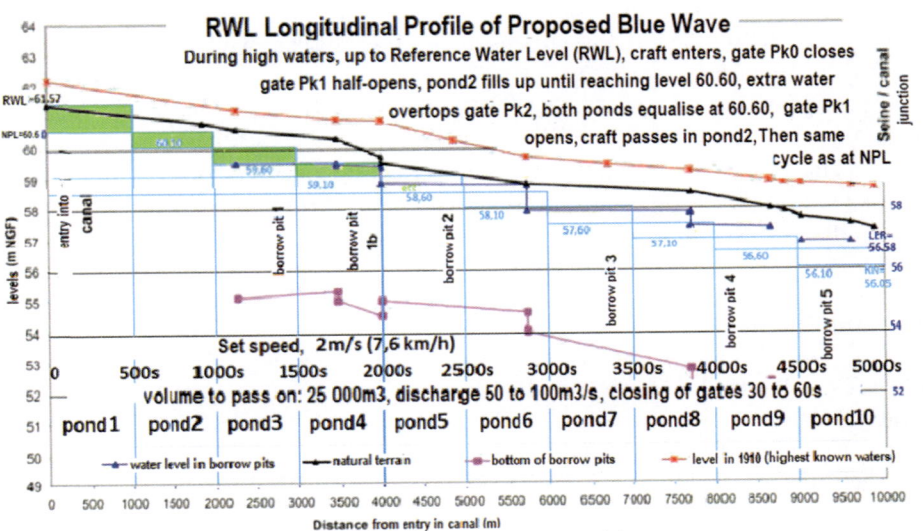

When the equality of levels is reached (duration approximately 360 s) the door pk1 opens, and the boat passes in the reach 2 which is then at the elevation 60.60. We are then in the same situation as before, see above the NPL conditions.

3.2.3   Upstream Movement at Normal Pond Level

For rising boats, the principle is similar, and is explained within the figure for NPL conditions, see below:

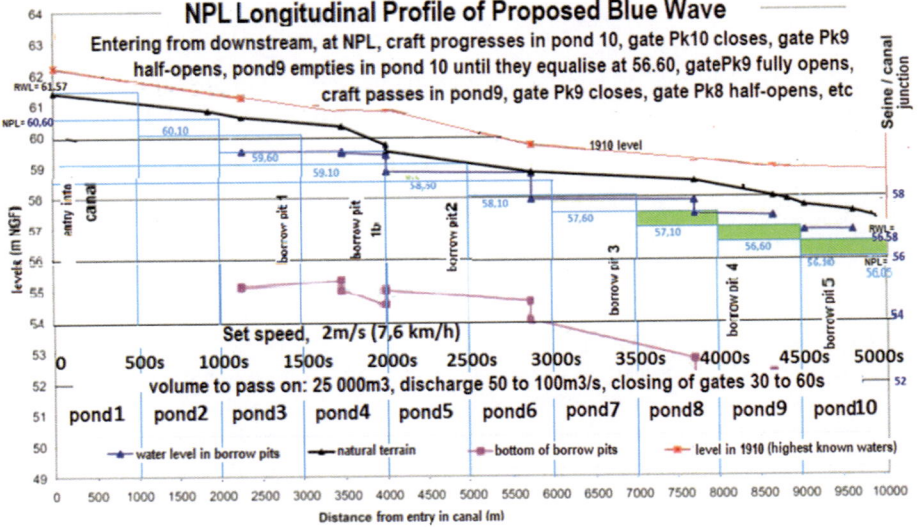

If the Seine is at RWL or in full flood, gates pk10 and pk9 are open or anyway already at the same flood level, since the level in reach 9 is practically identical to that of the Seine RWL (56.58). In extreme flood, both reaches will be at flood level, possibly even reach 8.

At the green light, reach 10 and reach 9 are travelled at set speed by the upbound boat. As soon as gate pk9 is crossed, it closes and while the boat travels along reach 9, reach 8 empties into reach 9 until the levels are equal (57, 10, or higher), which then allows the gate pk8 to be opened in front of the rising boat, which can enter reach 8, and so step by step until pk0.

4 Conclusion

We can see that the site of the Canal de Villiers in Beaulieu is optimal for the application of the Blue Wave principle, since the set speed is practically the commercial speed allowed by the wet section of the canal.

The structure would be crossed in 5,000 s (1 h 23), while crossing the same canal with passing through a lock would last at best 5,400 s (1 h 30).

Regarding the waiting time in front of the structure, the maximum duration of a possible red light before entering the structure would only be 1,000 s (16 min 40 s), while the mathematical expectation of waiting in front of the lock would be 30 min (duration of ½ cycle), according to the Operation Programme.

By integrating the average of a possible wait, the passage with lock would last 2 h, and only 1 h 39 with the Blue Wave.

Thus, using a Blue Wave in Canal de Beaulieu would seem to be a better proposition than a Lock.

Reference

Deplaix JM (2001) The Blue Wave, a new concept of waterways enabling to increase velocity or to reduce impact on environment. PIANC Bulletin, 108, Autumn 2001. French version. http://aftm.free.fr/PromoL.htm

Benchmark of Turning Basin Options for the ECMT Class V Network of Nord-Pas-de-Calais (France)

Sebastien Page[1(✉)], Marc Mansuy[2], Katrien Eloot[3], Maxim Candries[2], and Roeland Adams[1]

[1] International Marine and Dredging Consultants (IMDC) NV, Antwerp, Belgium
{sebastien.page,roeland.adams}@imdc.be
[2] Maritime Technology Division, Ghent University, Ghent, Belgium
{marc.mansuy,maxim.candries}@ugent.be
[3] Flanders Hydraulics, Antwerp, Belgium
katrien.eloot@mow.vlaanderen.be

Abstract. The Nord-Pas-de-Calais Division of Voies navigables de France (VNF), the French Waterways authority, is preparing a master plan for the development of turning basins for its ECMT class V network. In this frame, IMDC has been asked to carry out a comparative analysis (benchmark) of the different designs of turning basins in France and elsewhere in the world and to propose design criteria adapted for future developments (for ships of 110 m and 135 m) on the Nord-Pas-de-Calais network. The first part of the study consisted in a desktop review of available national and international guidelines or recommendations on the subject of turning basins design. This helped building a better understanding of the needs in terms of geometry for safe and/or comfortable turning maneuvers. Technical studies of built or projected turning basins on the Nord-Pas-de-Calais network were also analyzed to refine the knowledge on local conditions. Based on the findings of the desktop analysis, interviews of European waterways authorities, international experts and skippers were carried out to further improve the understanding of the problem, both from the infrastructure side as well as from the users' side. On top of geometry concerns, the interviews tackled hydrometeorological constraints, operational and environmental aspects, auxiliary equipment, etc. From this, a draft technical reference system was proposed for future turning basins on the Nord-Pas-de-Calais network. It proposes two options, depending on whether the basin is located on two banks or on one bank only, and aims at offering each option in two safety and ease levels: a comfort level version, on which the turn must always be relatively easy, and a safety level version, on which the turn must always be possible, but with less ease. The designs were then tested using real time navigation simulations. Optimizations of some of the designs were also proposed to adapt to severe hydro-meteorological conditions. Finally, a decision tree has been developed to facilitate the selection of a design solution according to the local characteristics of the projected basin location on the network. This tool allows the selection of the geometry allowing the easiest turn possible according to the available space and the hydrometeorological conditions present on the envisaged site.

© The Author(s) 2023
Y. Li et al. (Eds.): PIANC 2022, LNCE 264, pp. 311–323, 2023.
https://doi.org/10.1007/978-981-19-6138-0_28

Keywords: Waterway infrastructure · Turning basins · Navigation · Hydraulics · Ship simulator

1 Introduction

The Nord-Pas-de-Calais ECMT Class V network will play a major role in linking the Canal Seine-Nord Europe to the ports of Dunkirk and the inland navigation network of Flanders and Wallonia (Belgium), and connected ports in the Scheldt-Rhine delta such as Ghent, Antwerp or Rotterdam.

The future commissioning of the Seine Nord-Europe Canal and the prospect of 24-h navigation should result in a significant increase in traffic, particularly of vessels of 110 m and 135 m.

The Nord-Pas-de-Calais Division of Voies navigables de France (VNF), the French waterways manager, is therefore preparing a master plan for the development of turning basins for ships of up to 135 m on the ECMT class V network. Within this masterplan, VNF has hired IMDC and its subcontractors Flanders Hydraulics (FH) and Ghent University to propose well suited geometries that can be applied in all locations and conditions on the network (Fig. 1).

Fig. 1. The ECMT class V Nord-Pas-de-Calais network (circled in red) in the Seine-Scheldt environment (source: VNF)

The network has a total length of 240 km and, on most of its length, has a width of 34 m at 3.5 m depth. It is situated in a rather highly urbanized environment. From a hydrometeorological point of view, the canalized rivers of the network (Deûle, Lys and

Scheldt Rivers) can be host to significant currents while strong wind conditions (5 Bf and above) are more frequent at the coast (Dunkerque area) than more inland.

The proposed turning basin geometries shall be suited to all envisaged locations, with varying hydrometeorological conditions (current and wind), availability of (both or only one) banks for construction, environmental or land use constraints while guaranteeing satisfactory operational conditions and for limited costs. For each of the proposed designs, the study aims to define a minimum safety geometry allowing safe turning considered and a geometry with larger dimensions, allowing turning under better comfort conditions, always whatever the location and conditions considered on the ECMT class V network of Nord-Pas-de-Calais.

In a first phase, a desktop review of available national and international guidelines on the subject of turning basins design as well as of technical studies of built or projected turning basins on the Nord-Pas-de-Calais network is performed. Based on the findings of the desktop analysis, interviews of European waterways managers, international experts and skippers are carried out to refine the understanding of the constraints at stake. From this, propositions of well-suited geometries of turning basins are made. Finally, these geometries are tested on real-time navigation simulators and, in some cases, optimized.

2 Design Ships

The study focuses on ECMT Va class vessels of 110 m and 135 m (the latter is named "Va+" below). The vessels studied are container ships since this type of vessel is the most constraining when turning in windy conditions. The draft in loaded condition is taken equal to that of a loaded bulk carrier (2.5 m) in order to guarantee the validity of the simulations for loaded bulk carriers as well (since container ships generally have smaller drafts) (Table 1 and Fig. 2).

Table 1. Main dimensions of the design ships

Class	Length [m]	Beam [m]	Draft loaded/empty [m]	Air draft loaded/empty [m]
Va	110	11.4	2.5/1.7	3.4/4.2
Va+	135	11.4	2.5/1.7	3.4/4.2

Fig. 2. A 110 m-long ship on the ECMT class V Nord-Pas-de-Calais network

It should be noted here that on canals or canalized rivers with relatively weak current (which is the case of the study area), a ship can always turn empty except in exceptional cases. Exceptions are either rare (diversion of deliveries, unexpected and prolonged closures of the channel or of a lock) or limited to bulk carriers and avoidable (bulk loading generating dust settling in the deckhouse).

In terms of steering aids, both design ships are equipped with bow thrusters. On European waterways, this type of equipment is mandatory on any vessel longer than 110 m (Va+ class in this case) and it is present on the majority of current Va units.

3 Literature Review

The first step of the literature review consisted in studying recommendations related to turning basin design in national and international guidelines and recommendations.

The sections related to turning basins in the Dutch (RWS 2017), German (BMVBS 2011), French (Ministère de l'Équipement 1995), Chinese (PIANC 2019), Russian (PIANC 2019) and American (USACE 2006) guidelines were studied as well as in the PIANC InCom WG 141 recommendations for the design of inland waterways (PIANC 2019). It is worth noting that the main features of national guidelines in relation to turning basins design are recapitulated (and translated) in the annexes to the WG 141 report.

The guidelines which are more applicable to the studied case are the Dutch and the German guidelines as they both target the design ships of the study. The French directive, which is of application in the present case, does not make recommendations pertaining to turning basins. Other consulted guidelines are less applicable in the present case as, for instance, some consider sea going ships only (US guidelines) or ships with much lower maneuverability or sailing on rivers with much stronger current (Chinese guidelines).

For waterways of medium traffic density, which is the case of the studied network, the Dutch guidelines recommends circular geometries with a diameter of 1.2 L (L = length of the ship). It must be noted that this is applicable to canals or rivers with a current lower than 0.5 m/s which is the case in more than 95% of the time in the study area.

On the other hand, the German guidelines recommend using basins of trapezoidal shapes. The case of 110 m and 135 m inland ships is specifically addressed in the guidelines. For both ships, a distance from the waterway axis to the small base of the trapeze equal to the length of the ship (1 L) is proposed (see Fig. 3). No indication is given for the dimension of the larger base or for the angles of the trapeze.

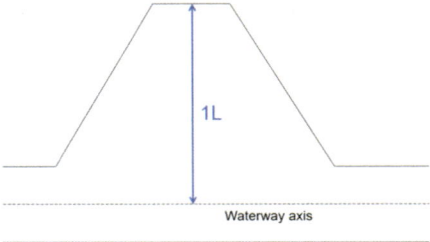

Fig. 3. Turning basins dimension for 110 m and 135 m ships in the German guidelines (L: length of ship)

The PIANC recommendations build on the above-mentioned guidelines and propose a dimension of 1.2 L (for a quality of driving C[1]) with no indication on the shape of the basin (circular or trapezoidal) (see Table 2). The size of the basins should be chosen according to the targeted level of accessibility (quality of driving). The report however acknowledges the presence of turning basins of 1.15 L (and less) in Europe.

Table 2. Minimum width of turning basins (including safety allowances) as a factor of ship dimension from existing guidelines (PIANC 2019). Minimum widths for sheltered water and canals are circled in red.

	Open water	Sheltered water and canals	Quality of driving
	Free-flowing river	Canals	
China	2.5 L (4.0 L)*	2.0 L	A-B
Dutch	case by case	1.2 L	C
Germany	case by case	1.2 L	C
US	1.5 L**	1.2-1.5 L**	A-B
Recommendation of WG 141	1.5 L free turn and 1.2 L fixed turn	1.2 L free and fixed turn	C

(*flow velocity >1.5 m/s), (** US figure is for sea-going ships, no figure is given for IWT, flow velocity < 0.8 m/s)
Remarks concerning WG 141 recommendations: 1.5 L was chosen according to the US guidelines, 1.2 L according to geometric aspects. Because turning is generally a challenging manoeuvre, the S&E quality will always be C

The PIANC recommendations (PIANC 2019) also indicate that hydrometeorological conditions (wind and current) must be taken into account. A formula is suggested for lengthening the basins of rivers with significant current. This formula is based on the hypothesis of a drift not compensated by the pilot, which is generally not the case during a turning maneuver in a confined environment such as a canal or river.

---

[1] Quality of driving C corresponds to a strongly constricted drive, as opposed to quality of driving A (nearly unrestricted drive) or B (moderate to strongly restricted drive) (PIANC, 2019).

The formula is therefore not applicable in the context of the study as it would lead to excessive basins lengths.

In a second phase, existing or studied turning basins on the ECMT class V network of Nord-Pas-de-Calais were studied. This helped better understand the constraints to be taken into account for the design of turning basins in the study area (wind, current and lack of available space, mainly). In the reviewed studies, the proposed geometries were either circular or trapezoidal with dimensions ranging from 1.1 L to 1.3 L.

It is worth noting that, although most studies acknowledge the interest of keeping the turning basins outside of the navigation channel as much as possible, all proposed designs integrate the entire channel (in order to limit the excavation costs, and to avoid additional expropriation costs). This is also true for most of the existing turning basins on the Nord-Pas-de-Calais network.

Amongst the reviewed turning basins studies, the report of a turning basin project in Arques (FH & Ghent University 2017) for ships of 135 m and 143 m on either circular or trapezoidal geometries supported by navigation simulations helped better evaluate safety and ease level of the turns on such geometries and under the hydrometeorological conditions of Nord-Pas-de-Calais.

Finally, a short analysis of the existing turning basins (geometries, dimensions, etc.) was also carried out, to serve as input when considering upgrades (enlargement) of existing turning basins.

4   Interviews of Skippers, Waterway Managers and Experts

With the knowledge acquired in the previous phase, a questionnaire was drafted to refine the understanding of the problem by means of interviews of waterways authorities, international experts, and skippers. In total, seven employees of waterways authorities (VNF in France, DVW and SPW in Belgium and RWS in the Netherlands), six international experts and six skippers, from various countries have been interviewed. This enabled a better understanding of the problem from both the side of the user as well as the side of the infrastructure provider. Information related to bank protection for turning basins, equipment (signaling, lighting, mooring, etc.), waterway maintenance and management (dredging, impact from and on traffic, etc.) was also collected.

Most experts or waterway managers referred to Dutch, German or PIANC guidelines and to circular or trapezoidal geometries with dimensions of 1.2 L to 1.3 L being the most commonly cited options. Many also pointed to the fact that ship maneuvering simulations are recommended by most guidelines in case of strong current or wind. There was a wide consensus on having a depth equal to the depths of the waterway, with many pointing to the fact that large under keel clearances (UKC) have a positive impact on the ease of the turn.

Skippers tend to prefer circular geometries for ease of maneuvering but confirm that trapezoidal geometries are also interesting options. Moreover, they are often obliged to turn in much narrower basins and mention that basins of 1.1 L width or less are not uncommon.

5 Proposition of Geometries Suited for the ECMT Class V Network of Nord-Pas-de-Calais

From the information collected above, two main designs have been proposed: a circular shape for situations where both banks of the canal or canalized river can be used and a trapezoidal shape when the turning basin can only be built on one bank of the waterway. For both options, two sizes are proposed based on information collected in the previous phases of the study: a larger one which is expected to allow comfortable turns in all conditions and a smaller one, which is designed to be sufficient to allow turns, though with less ease.

5.1 Option N°1: Circular Geometry, Using Both Banks of the Waterway

A circular geometry has the advantages of allowing simpler (circular) maneuvers while keeping excavation volumes limited. It also is the most commonly encountered geometry on the Nord-Pas-de-Calais network and its users are therefore very well accustomed to it. The main disadvantage of this solution is the total interruption of traffic when a turn is performed, regardless of the size of the turning ship. If such interruption is deemed unacceptable, a single bank solution (trapezoidal) should be considered.

Both small and large sizes of the proposed circular geometry are shown in Fig. 4.

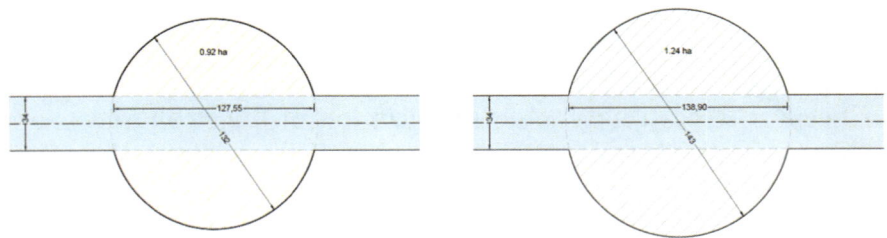

Fig. 4. Proposed circular geometry for 110 m ships using both banks of the waterway, small size (left) and large size (right). The channel is shown in blue and the area to be excavated in hatching

5.2 Option N°2: Trapezoidal Geometry, Using Only One Bank of the Waterway

The advantages of this option are mainly land use (and budget) related. Indeed, it is the solution that requires the least amount of space when the basin must be located on a single bank and therefore potentially minimizes excavations and land acquisitions. Also, by being positioned off-center, this geometry does not imply a systematic interruption of the traffic during turns, at least not during the turns of ships of smaller dimensions than those of the design ship.

On the downside, the two-stage maneuver, which must be carried out on this type of basin, is more complex and more time-consuming than the one-stage circular maneuver, but it is nevertheless commonly practiced by all skippers. Finally, this trapezoidal solution is not yet widely used on the Nord-Pas-de-Calais network and the services in charge of waterway maintenance might have to get accustomed to this new design. The skippers should not encounter major difficulties in using such basin, since they are already used to turning in various types of geometries (wide portions of the waterway, junctions, etc.).

Both small and large sizes of this geometry are shown in Fig. 5.

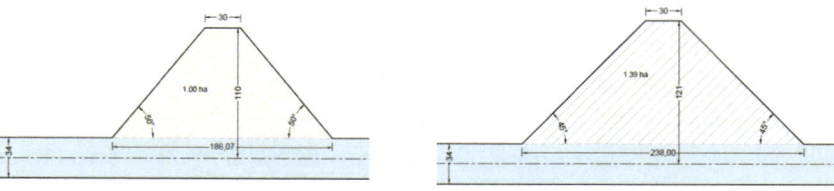

Fig. 5. Proposed trapezoidal geometry for 110 m ships using only one bank of the waterway, small size (left) and large size (right). The channel is shown in blue and the area to be excavated in hatching

The base angles of the trapezoid are chosen based on results of navigation simulations on ships of 135 m on similar geometries in the study of the turning basin in Arques cited above (FH & Ghent University 2017).

6 Check of Accessibility and Optimization of the Proposed Geometries by Means of Real-Time Navigation Simulations

All proposed geometries have then been tested on the real-time navigation simulators of Flanders Hydraulics (FH) in Antwerp, Belgium.

The parameters of the study (ship classes, geometries, hydrometeorological conditions, basin positioning, etc.) are numerous and the number of possible scenarios (several thousands) greatly exceeded the material possibilities of the study. A two-phased approach was therefore used for the real-time simulations: a first phase of exploratory simulations, during which the impacts of the different criteria were analyzed, and a second phase, during which the most relevant scenarios (key scenarios) were studied.

6.1   Results of the Exploratory Phase

6.1.1   Analysis of the Impact of the Simulation Parameters (Current, Wind, Ship Draft, etc.)

6.1.1.1 Impact of the Hydrometeorological Conditions

The selected current velocities and wind speeds for the simulations correspond to strong but not extreme conditions (values not exceeded 95% of the time). Currents up to 0.4 m/s and wind speed up to 5 Bf depending on locations were computed and introduced in the simulators.

In the case of trapezoidal basins, turning in the upstream direction is more difficult than turning in the downstream direction. This is because the stern of the ship is in a current that opposes the rotation of the ship, causing it to drift toward the edge of the basin. On the other hand, turning after approaching in the downstream direction is easier because the current drags the stern of the ship and makes it turn in the desired direction. The result is a smooth and centered maneuver in the basin (Fig. 6).

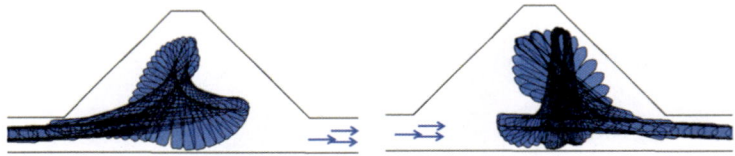

Fig. 6. Large trapezoid for a 110 m ship, sailing downstream (left) and upstream (right), with a current of 0.4 m/s. In the case of the upstream turn (right), the shorter distance to the banks and the lower fluidity are clearly visible.

In the case of circular basins, the direction of the current has no influence because of the symmetry of the layout.

Transverse wind was identified as the most unfavorable regardless of the type of basin (circular or trapezoidal).

The unfavorable effect of the wind is further illustrated in Fig. 7 where the maneuver is performed very close to the side opposite to the wind direction for instance.

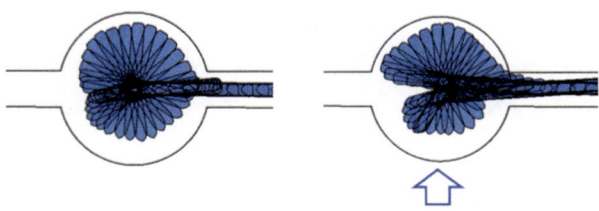

Fig. 7. Influence of the wind in a circular basin. Condition without wind (left) and with crosswind (blue arrow) 5 Bf (right)

6.1.1.2 Impact of Ship Draft

In the case of the circular basins, empty ships (i.e. ships with empty containers) are the more critical ships in windy situations, as shown in Fig. 8. Indeed, empty ships lead to an increased effect of the wind which is not counterbalanced by the better underwater maneuverability.

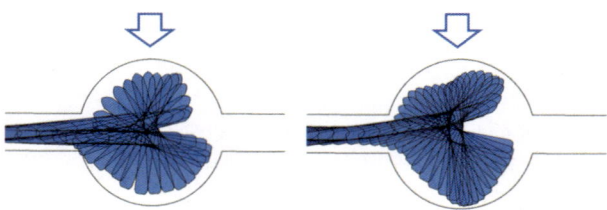

Fig. 8. Impact of the draft on turns on circular basins in unfavorable wind condition (5 Bf transverse), empty ship (left) and loaded ship (right). Note: the contact in the empty ship maneuver (left) is due to reduced visibility on the simulator and could have been avoided.

In the absence of wind, the loaded condition is the most critical, as shown in Fig. 9, since maneuverability is reduced by the high draft.

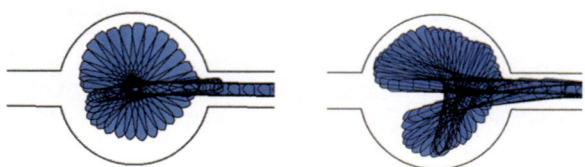

Fig. 9. Influence of loading on turns on circular basins in the absence of wind. Empty ship (left) and loaded ship (right).

In the case of trapezoidal basins, the loaded condition is more critical even in the presence of wind, as shown in Fig. 10. Indeed, in a trapezoidal basin, the ship makes a turn to enter the basin. This maneuver is more difficult in the loaded condition since the reduction in maneuverability of a loaded ship compared to a light ship outweighs the difficulty associated with the increase in windage area of an empty ship (empty containers) during the turn.

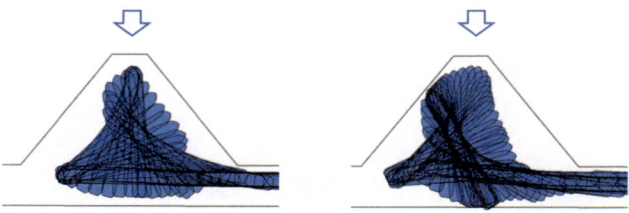

Fig. 10. Influence of draft on trapezoidal basins, empty ship (left) and loaded ship (right) in wind condition (5 Bf transverse)

The impact of other parameters, such as basin positioning (left or right bank, in a straight line or a curve, etc.) were studied but are not described here.

6.1.2   Optimization of Certain Geometries

The tests performed in the exploratory phase revealed a too constrained space in the small trapezoids and optimizations of this geometry were proposed, the objective remaining to guarantee the turns in unfavorable hydrometeorological conditions in a geometry of reduced size.

Two optimized geometries have been proposed depending on whether the solution is limited to a single bank or whether it can occupy both banks (e.g. the case of the enlargement of a circular basin occupying both banks) The optimizations were performed using an overlay of the simulation trajectories and are shown at Fig. 11.

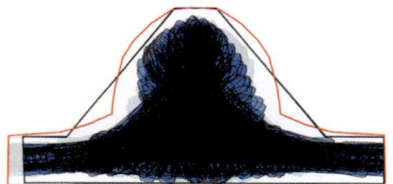 

Fig. 11. Principle of the optimization of the small trapezoid (in red), optimized geometry on one bank only (left): rounded top and opening from the channel following the approach paths and optimized geometry using both banks (right): widening of the channel on the opposite bank

6.2   Results of the Second Phase of Simulations

Over both phases of simulations, 97 simulations were performed, and 85 results were deduced directly from simulations results.

Results showed that the proposed large circle allows for an unconstrained turn (of empty ships) while the small circle allows for a constrained turn (of the same ships). These geometries therefore meet the desired high and low quality of driving objectives as defined in the introduction. It is also worth noting here that turns of loaded ships are possible in all hydrometeorological conditions, though with less ease, on these basins.

Results for the trapezoidal geometries (large trapezoid and small optimized trapezoid) showed similar results for empty ships at the exception of the case of a construction on a river (with current) and when the smaller base would be oriented towards the direction of the prevailing winds. In such case, turns proved to be impossible for all trapezoidal basins. Tests on optimized geometries showed the improved ease of driving compared to that of the original small trapezoid. It must be noted here that on the numerous stretches with no current on the Nord-Pas-de-Calais network (canals), the large and small optimized trapezoids can be used regardless of the orientation and meet the desired high and low quality of driving objectives.

To facilitate the decision making of VNF, a decision tree was proposed. It allows to determine the most appropriate geometry (the one that allows the easiest turn) according to the availability banks (both or only one), the available space, the presence or absence of current and the possibility of arranging the basin so that the top of the trapezoid is not oriented in the direction of the prevailing winds.

7   Conclusion

A study was performed to identify turning basin geometries which allow turns of ships up to 135 m in any location of the Nord-Pas-de-Calais network, with varying hydrometeorological conditions (current and wind), availability of (both or only one) banks for construction, environmental or land ownership constraints while guaranteeing satisfactory operational conditions and for limited costs.

The first phase of the study consisted in the review of national and international guidelines in relation to turning basins design as well as the analysis of past turning basins studies and existing turning basins on the Nord-Pas-de-Calais network. In a second phase, interviews of waterways authorities' staff, international experts, and skippers were performed to better understand how the information collected in the first phase should be applied to the case of the studied network.

From this, geometries were proposed for the future turning basins of the Nord-Pas-de-Calais. Two options were proposed, depending on whether the basin can be located on two banks (circular geometry) or on one bank only (trapezoidal geometry), and are given in a comfort level version, for which the maneuver must always be relatively easy, and a safety level version, for which the turn must always be possible, potentially with less ease.

These draft geometries were then tested on real-time navigation simulators in order to demonstrate their accessibility and to potentially propose improvements. Optimizations of the trapezoidal geometries were proposed to adapt to severe hydrometeorological conditions, for a single or double bank implementation.

The results show that circular geometries allow the turn in all circumstances and can be retained according to the available space between a comfort level (diameter 1.3 L) and a safety level (diameter 1.2 L). Optimized trapezoids allow for easier turns but are not suitable when the small base is oriented in prevailing wind direction when applied on rivers (with current). On the canals sections of the network however, the proposed trapezoidal geometries did give the desired results, in other words: comfortable turns on the large trapezoid and safe turns on the small (optimized) geometries.

Finally, the results have been compiled in a decision tree which will facilitate the selection of a design solution according to the local characteristics of the projected basin location, keeping in mind that a certain flexibility will be required in the application of the proposed designs since local constraints often complicate the application of generic solutions.

Acknowledgements. The study was commissioned by Voies navigables de France (VNF) The study was coordinated by International Marine and Dredging Consultants (IMDC) in collaboration with its subcontractors: Ghent University and Flanders Hydraulics. The results of the analysis were discussed on several occasions with representatives of the inland navigation sector, whose valuable input is acknowledged.

References

BMVBS (2011) Richtlinien für Regelquerschnitte von Binnenschifffahrtskanälen

FH & Ghent University (2017) Etude de trajectographie du bassin de virement d'Arques – Simulations complémentaires

Ministère de l'Équipement (1995) Circulaire 76.38 modifiée par la circulaire 95.86 relative aux caractéristiques des voies navigables

PIANC (2019) Design Guidelines for Inland Waterway Dimensions. InCom WG 141

Rijkswaterstaat (2017) Richtlijnen Vaarwegen RVW 2017

USACE (2006) EM 1110-2-2602. Hydraulic design of deep-draft navigation projects

Comparative Study of the Hydraulic Characteristics of Stratified Energy Dissipators in In-Chamber Longitudinal Culvert Systems

Xin Ma[1,2(✉)], Yaan Hu[1], and Zhonghua Li[1]

[1] State Key Laboratory of Hydrology-Water Resources and Hydraulic Engineering, Nanjing Hydraulic Research Institute, Nanjing 210029, China
{xma, yahu, zhli}@nhri.cn
[2] State Key Laboratory of Hydraulics and Mountain River Engineering, Sichuan University, Chengdu 610065, China

Abstract. The study of energy dissipation characteristics of stratified energy dissipators in the lock chamber is carried out by numerical simulation. Discussed the hydraulic characteristics of different energy dissipators in the lock chamber and open channel area (e.g., flow field in the lock chamber, flow distribution in the branch holes and energy dissipation mechanism). The results show that the stratified energy dissipation makes the water flow in the open ditch area more fully mixed; the flow distribution in the branch holes is more balanced; the maximum flow velocity into the bottom of the lock chamber is reduced. The in-chamber longitudinal culvert system with open channel + cover has the best flow conditions, followed by open channel + grating, both of the above are better than the traditional open ditch energy dissipator.

Keywords: Numerical simulation · Locks · Filling and emptying system · Energy dissipator

1 Introduction

With the rapid development of China's economy, a large number of world-class high-head water conservancy projects have been completed and put into use one after another. The establishment of high headlocks to enable ships to travel unimpeded on the rivers has become another new national strategy. However, high headlocks are required to meet two conditions at the same time - cargo capacity and protection of locks and ship safety. The instantaneous large flow rate will endanger the safety of ships moored in the lock. To meet the safety of ships mooring in the lock, it is necessary to set up the energy dissipator in the filling and emptying system, so as to reduce the energy of the water around the ship and enhance the uniformity of water flow to achieve the purpose of reducing the mooring ship bollard force (Zong and Yang 1989).

In the 1930s, the United States in some locks used the open ditch energy dissipator type: Lower Granite, Bay Spring, New Bankhead and other locks had been applied. In addition, in the 1960s, the U.S. Army Corps of Engineers had also placed open ditch energy dissipators at ship locks experiment stations on the Columbia River and Snake River, and many experiments and studies had been done for this purpose. (Perkins and

© The Author(s) 2023
Y. Li et al. (Eds.): PIANC 2022, LNCE 264, pp. 324–333, 2023.
https://doi.org/10.1007/978-981-19-6138-0_29

Chanda 1975; Richardson 1969; Richardson and Webster 1960). Similarly, Marmet Lock (Hite 1999) with the open ditch energy dissipator on both sides of the increased in force dissipation threshold to increase the smoothness of the flow to optimize mooring conditions. In China, the open ditch energy dissipation type was first used in Gezhouba 2 locks, after which the energy dissipation type has been widely used, such as Sichuan Fuzhou River Lotus Temple, Santai locks, Xijiang River Guiping a line of locks. Recently, China's researchers on the basis of single ditch research, for this reason, the domestic energy dissipation on the basis of single ditch, proposed a double ditch, three ditch arrangement type (Li et al. 2021).

For higher heads, such as a 60 m single-stage ship lock, the open ditch energy dissipator will be difficult to reach the lock design requirements. By adding grilles or covers, the single layer open trench energy dissipator, is changed to a "double layer energy dissipator". Whether the innovative types of ditch + grille and ditch + cover are better than single layer is to be demonstrated. This study investigates the hydraulic characteristics of single-layer open ditch energy dissipator and layered open ditch energy dissipation (ditch + grille and ditch + cover form) by numerical simulation, aiming to understand the effect of different energy dissipation arrangements on the flow regime of the lock chamber, the flow distribution of branch holes and energy dissipation effect.

2 Numerical Simulation

2.1 Calculation Area and Grid

Without affecting the calculation accuracy, reduce the calculation volume and improve the calculation efficiency. According to the characteristics of the arrangement of the outlet water transmission system in the second section of the open channel, half of the entire lock chamber area was selected for the simulation study in the calculation area. At the same time, to ensure that the flow velocity and flow direction on the inlet boundary is relatively smooth, the corridor inlet was extended horizontally, and the calculation model is shown in Fig. 1.

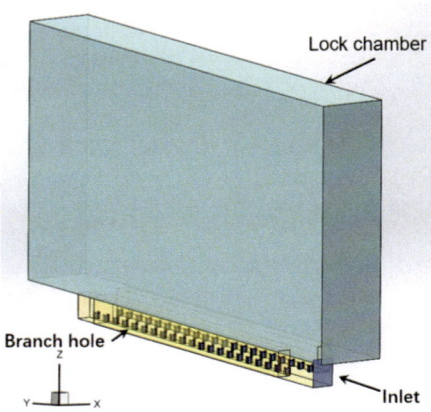

Fig. 1. Calculation area schematic

The calculation area has meshed with a hexahedral structure. Mesh encryption is applied to the side branch holes, open trenches and their nearby areas. The total number of grid cells is about 2.54 million and the total number of nodes is about 1.65 million. The grid division diagram is shown in Fig. 2.

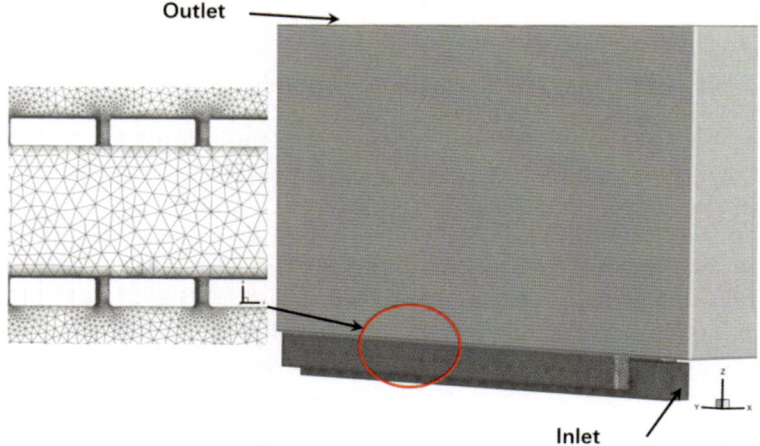

Fig. 2. Grid profile diagram

2.2 Control Equations and Boundary Conditions

In terms of model selection, the volume of fluid method (VOF) was chosen for the free surface tracking (Hirt and Nichols 1981); the renormalization group (RNG) k - ε turbulence model was used for the turbulence model (Yakhot and Orszag 1986). According to the actual lock gate opening-closing flow variation under the lock chamber inlet, the inlet flow rate is adopted as variable flow rate that is, the incoming flow is non-constant. The highest head, the lowest initial head and the shortest opening time in the physical model test were selected (Design water head $H = 60$ m, Initial lock chamber water level $h_0 = 6$ m, Gate opening time $t_v = 8$ min, and the inlet flow velocity variation is shown in the following figure (Fig. 3).

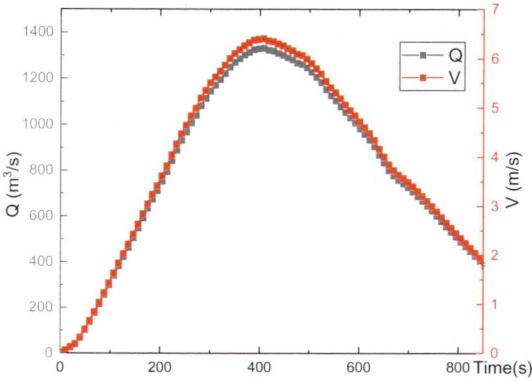

Fig. 3. Variation of inlet flow rate and inlet flow rate

The boundary settings are shown below:

(1) Inlet boundary: the velocity inlet boundary condition is used, and the UDF is applied for the input of variable flow velocity.
(2) Outlet boundary: the pressure outlet boundary condition is adopted, and the normal gradient of all variables is 0.
(3) Wall boundary: using no-slip boundary conditions and standard wall functions for the viscous bottom layer.
(4) Free water surface: the free water surface is treated as pressure boundary, and the relative pressure value on the pressure boundary is P = 0.

2.3 Model Validation

In the physical model experiments, the pressure measurement points of the side branch holes and the monitoring points of the water level change in the lock chamber are compared respectively (Fig. 4).

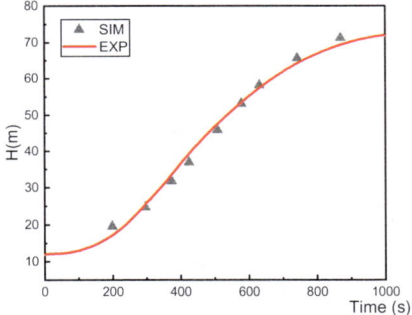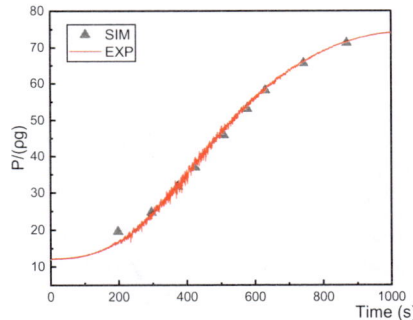

Fig. 4. Comparison of numerical simulation results with physical experimental data: a. Water level change in the lock chamber; b. Pressure change at the top of the main corridor.

According to the change of water level in the lock chamber and the change of pressure in the branch hole, we could get that the simulation was validated against the experimental data which were within a reasonable range.

3 Results and Analysis

3.1 Lock Chamber Flow Field

It can be seen from Fig. 5, at the maximum flow rate of $Q = 330$ m^3/s ($T = 400$ s), the jet velocity of the traditional open channel branch hole reaches 12.21 m/s; the flow velocity of the open channel + grille branch hole is 9.42 m/s; the flow velocity of the open channel + cover plate branch hole is 11.32 m/s. From the Fig. 5, it can be seen that after the water flows through the open ditch, there is still a high-speed water flow of more than 3 m/s into the lock chamber. And no larger flow velocity is found in the open ditch + grille and open ditch + cover. It means that the open ditch cannot fully kill the energy of water flow in the branch hole.

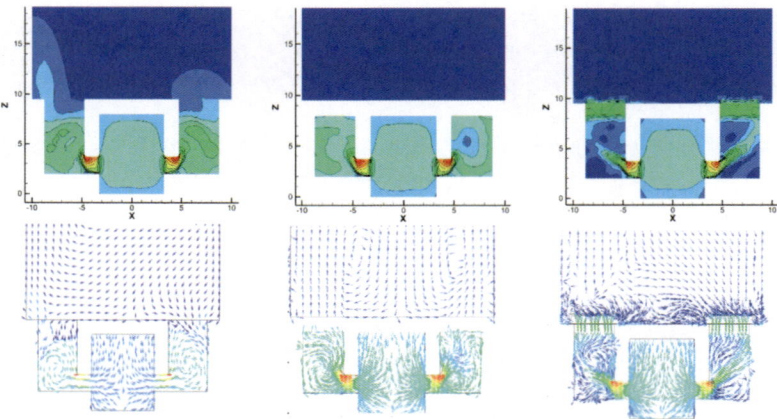

Fig. 5. Three different energy dissipator transverse flow velocity diagram and vector diagram

From the vector distribution diagram (Fig. 6), it can be seen that, compared with the conventional open channel energy dissipators, the addition of the grille makes the water in the open ditch dissipator area mixed more strongly, the spin roll is enhanced, energy dissipation is more adequate. In addition, the addition of the grille also adjusts the direction of the main jet of the lock chamber flow, so that the speed into the lock chamber is further reduced, and the lateral diffusion is more rapid. The upper branch hole and the addition of the cover plate (ditch + cover energy dissipator), so that the bright ditch area whirlpool intensity more and more intense and in the cover plate out of the flow formed after the second spin roll, into the lock chamber inside the maximum flow velocity of about 2.24 m/s, only 47% of the traditional bright ditch energy dissipators, bright ditch + cover plate layered energy dissipation water transfer system.

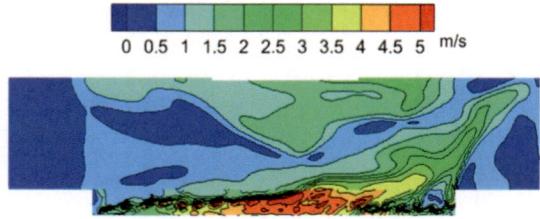

Fig. 6. Longitudinal chamber flow velocity distribution at the moment of maximum water transfer (with ditch energy dissipator)

From the longitudinal flow field diagram at the time of the maximum flow rate (x = 6.75 m cross section), it can be seen that the maximum inclination of the water flow in the vertical direction at the time of the maximum flow rate is about 45°, of which the inclination of the upstream side is about 52°, that is, part of the jet is shot obliquely to the water body of the lock chamber and the maximum flow velocity into the lock chamber is 4.97 m/s. This makes the whole lock chamber form a huge transverse vortex, and the main jet decays along the course to maintain a certain velocity near the water surface. The flow velocity is maintained near the water surface. This indicates that when the design head reaches 60 m, the dissipation effect in the open channel area slows down significantly, and more energy needs to be dissipated by mixing and diffusion in the water body of the lock chamber (Fig. 7).

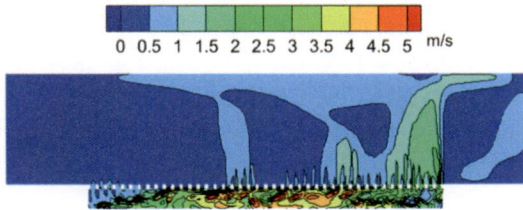

Fig. 7. Longitudinal chamber flow velocity distribution at the moment of maximum water transfer (with ditch + grille energy dissipator)

The velocity of water entering the lock chamber on the upstream side is obviously reduced, but at the end, there is still a large flow velocity (maximum velocity out of the lock chamber is 2.64 m/s) due to the blockage at the end of the branch corridor and the inertia of the water. However, after the water flows through the grille, the water flows make a local impact due to the sudden narrowing of the outlet aperture, which restrains the flow direction, and the flow direction after the lock chamber is almost vertical into the lock chamber. This leads to the lock chamber without a wide range of vortex, and because of the grille barrier, the water flows out of the lock chamber and the water body inside the lock chamber through the role of volume suction, rapid mixing and elimination of energy, combined with a faster reduction in velocity (Fig. 8).

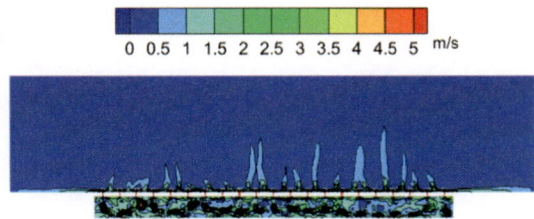

Fig. 8. Longitudinal chamber flow velocity distribution at the moment of maximum water transfer (with ditch + cover energy dissipator)

At the time of the maximum water transfer of the open channel + cover, there is still a lower velocity water body around the branch hole outflow in the middle section for absorption and mixing. It means that the setting of the open channel + cover plate has caused some influence on the outflow of the branch hole. It can be found that, due to the value-added cover plate, the dissipation caused by the impact of water flow on the cover plate is very large, plus the skirt of the cover plate to increase the outflow length, adjust the uniformity, the outflow are more uniform and flow velocity is not large (maximum velocity out of the lock chamber is 1.67 m/s). The middle main shot in the low water depth for the emergence of convergence, increasing the respective lateral diffusion space, while the upstream and downstream ends on both sides also bear a certain dissipation. Resulting in the entire flow is very uniform, near the surface combined velocity does not exceed 0.28 m/s.

3.2   Branch Hole Flow Distribution

The energy dissipator are different, but the outflow law is roughly similar: at the beginning of the lock transmission, the outflow velocity of each branch hole is decreasingly distributed along the flow direction of the branch corridor, that is, the flow velocity of the branch hole near the inlet of the branch corridor is large (4#, 10#), and the flow velocity of the branch hole at the end of the branch corridor is small (16#, 19#); over time, the outflow velocity of each branch hole is increasingly distributed along the branch corridor, that is, the flow velocity of the branch hole near the inlet of the branch corridor is small, and the flow velocity of the branch hole at the end of the branch corridor is large. The flow velocity of each branch is increasing along the branch corridor (Fig. 9).

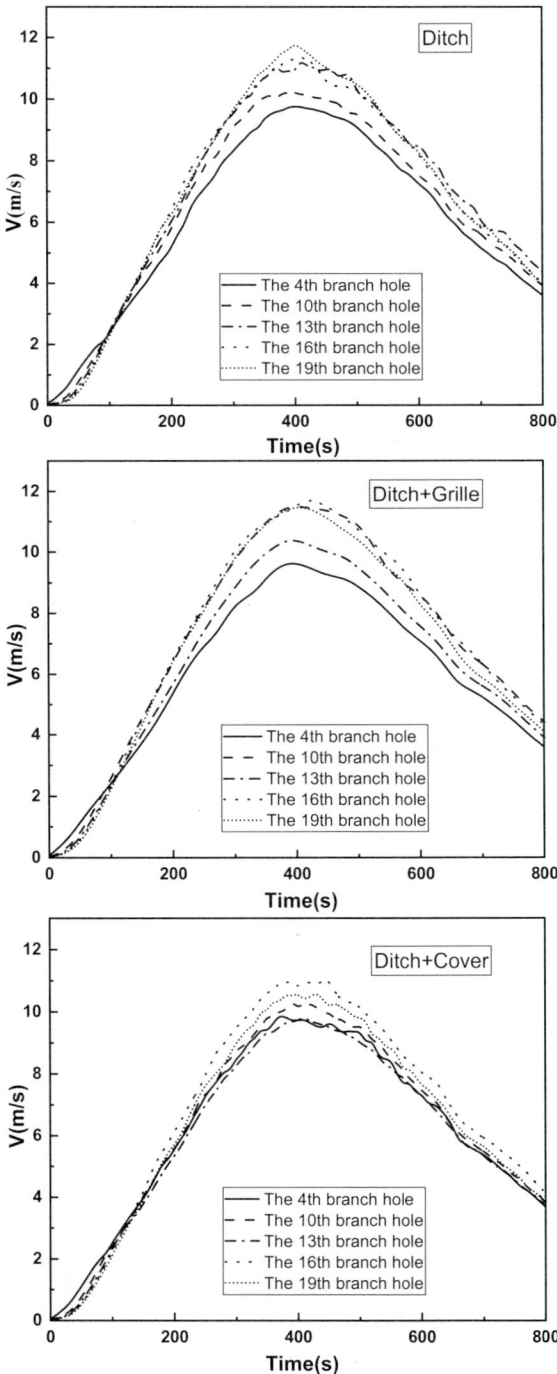

Fig. 9. Three different energy dissipators side branch hole flow velocity distribution: a) ditch dissipator; b) ditch + grille dissipator; c) ditch + cover dissipator.

The difference is that, when the flow rate is larger, the uniformity of the traditional open channel energy dissipator branch hole outflow is poor ($Q = 330$ m^3/s moment each branch hole flow rate statistics standard deviation σ is 0.75). Adding the equal spacing grille, the maximum flow rate moment ($Q = 330$ m^3/s) standard deviation σ reduced to 0.47, the flow peak is obviously reduced. The addition of the cover, the water flow can be more fully in the open ditch energy dissipation area mixing, so that the water pressure near each branch hole to get re-equilibrium, resulting in the flow of water outflow flow rate does not appear large differences, regardless of the initial filling or maximum water transfer time, branch hole outflow uniformity is the best among the three (the maximum flow time standard deviation σ only 0.44).

3.3   Energy Dissipation Analysis

The magnitude of turbulence energy can reflect the dissipation state of the open ditch area:

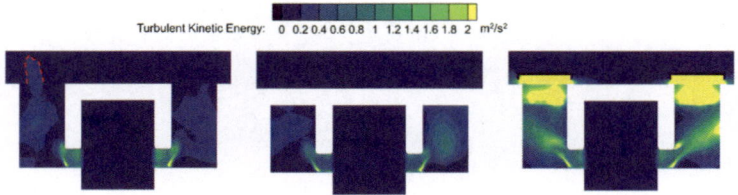

Fig. 10. Turbulent energy values corresponding to the internal flow structures of the three body types ($T = 400$ s, cross-section at the longitudinal symmetry center of the lock chamber)

From the Fig. 10, the turbulence is most intense in the open trench area of the open ditch + cover, with large turbulence energy values near the orifice, in the middle of the open trench and near the cover. The addition of the grille allows the more turbulent flow to be concentrated in the middle of the orifice and open trench. The turbulent energy values in the ditch energy dissipator are relatively discrete except at the orifice, and there is still some mixing at the bottom of the lock chamber (marked in red). It shows that the effect of energy dissipation by stratified-energy dissipation is obvious, and the energy dissipation is concentrated in the open ditch area.

4   Conclusions

In this study, the hydraulic characteristics of the stratified energy dissipator open channel + grate and open channel + cover are studied by numerical simulation. The flow characteristics, branch hole flow distribution and energy dissipation mechanism of the two new energy dissipators are discussed in detail. Based on these studies, the following conclusions were obtained.

1 Single-layer open channel energy dissipators due to the uneven flow of branch holes, as well as into the open channel, the lock chamber after the diffusion and mixing efficiency are low, resulting in the lock chamber flow pattern is poor.
2 Ditch + grille energy dissipator although the side branch hole flow and out of the hole flow pattern and open ditch energy dissipator similar, but due to the addition of

grille, the water in the open ditch mixes more fully and adjust the flow of water into the lock chamber flow pattern, so that with the surrounding water diffusion more fully, the main jet to the bottom of the lock chamber near the energy is lower.

3 Ditch + cover because the upper branch hole area is smaller so that the entire branch corridor side branch hole inlet pressure distribution is more uniform. The cover greatly enhances the strength of the water turbulence in the open ditch area. The flow pattern of water entering the lock chamber is the best among the three.

In summary, for 60 m head, stratified energy dissipators is more superior than traditional energy dissipators.

Acknowledgements. The research reported herein is funded by the China National Key R&D Plan (Grant No.: 2016YFC0402001).

References

Hirt CW, Nichols BD (1981) Volume of fluid (VOF) method for the dynamics of free boundaries. J Comput Phys 39:201–225

Hite JE (1999) Model Study of Marmet Lock Filling and Emptying System, Kanawha River, West Virginia. Army Engineer Waterways Experiment Station Vicksburg MS Coastal and Hydraulics Lab

Li Z, An J, Ma X, Huang W (2021) The key technology investigation of the high-efficient filling and emptying system of lock based on hierarchical energy dissipation concept. Nanjing Hydraulic Research Institute

Perkins LZ, Chanda AJ (1975) Filling and emptying systems, Little Goose Lock, Snake River, Washington: hydraulic model investigation. U.S. Department of Defense, Department of the Army, Corps of Engineers, North Pacific Division, Hydraulic Laboratory

Richardson GC (1969) Filling system for lower granite lock. J Waterways Harbors Div 95:275–289

Richardson GC, Webster MJ (1960) Hydraulic design of Columbia river navigation locks. Trans Am Soc Civ Eng 125:345–364

Yakhot V, Orszag SA (1986) Renormalization group analysis of turbulence. I. Basic theory. J Sci Comput 1:3–51

Zong M, Yang M (1989) Hydraulic design of a filling emptying system. Nanjing Institute of Water Resources Science

Corridor Scale Planning of Bunker Infrastructure for Zero-Emission Energy Sources in Inland Waterway Transport

Man Jiang[1]($\boxtimes$), Fedor Baart[1,2], Klaas Visser[1], Robert Hekkenberg[1], and Mark Van Koningsveld[1,3]

[1] Delft University of Technology, Delft, The Netherlands
{m.jiang-3, K.Visser, m.vankoningsveld}@tudelft.nl
[2] Deltares, Delft, The Netherlands
Fedor.Baart@deltares.nl
[3] Van Oord Dredging and Marine Contractors B.V., Rotterdam, The Netherlands

Abstract. The availability of supporting bunker infrastructure for zero-emission energy sources will be key to accommodate zero-emission inland waterway transport (IWT). However, it remains unclear which (mix of) zero-emission energy sources to prepare for, and how to plan the bunker infrastructure in relative positions and required capacity at corridor scale. To provide insight into the positioning and dimensions of bunkering infrastructure we propose a bottom-up energy consumption method combined with agent based network simulation. In the method, we first produce a two-way traffic energy consumption map, aggregated from the energy footprint of individual vessels on the transport network. Next we investigate the potential sailing range of the vessels on the network if they would sail the same routes, but with alternative energy carriers. Based on the sailing range of the vessels for different energy carriers, the maximum inter-distance between refuelling points can be estimated. By aggregating the energy consumptions of all the vessels on the network, we can estimate the required capacity of a given refuelling point. To demonstrate the basic functionality we implement the method to four representative corridor scale inland shipping examples using zero-emission energy sources including hydrogen, batteries, e-NH3, e-methanol and e-LNG. The application in this paper is limited to four abstract cases. A recommended next step is to apply this approach to a more realistic network.

Keywords: Inland waterway transport · Zero-emission · Bunkering infrastructure · Sustainable energy sources · Energy consumption

1 Introduction

The world's economy relies heavily on waterborne supply chains. Approx. 80% of all global trade is shipped by marine transport; according to UNCTAD (2021) subdivided into tanker trade (2020: 2,918 106 tons loaded), main bulk (2020: 3,181 106 tons loaded) and other dry cargo (2020: 4,549 106 tons loaded, of which a little over 40% is attributed to container transport). Overall efficiency of global supply chains is to a great

Y. Li et al. (Eds.): PIANC 2022, LNCE 264, pp. 334–345, 2023.
https://doi.org/10.1007/978-981-19-6138-0_30

extent determined by the inland transport networks they are connected with. Approximately one third of the operating cost of vessels is related to energy use.

The Paris Climate Agreement, and its subsequent implementation, a.o. by the International Maritime Organization (IMO), requires significant changes in power systems on board (engine room, energy storage). These changes will not only affect the performance of individual vessels (e.g. loading capacity, range, velocity), but also the performance of ports and waterway networks (e.g. allocation of bunker stations, development of the new fuel supply network, potential modal shift, network inter-competitiveness).

A major challenge that is currently hindering the energy transition is the lack of insight in how alternative power source strategies on board of individual vessels cascade through a ports and waterways system, ultimately impacting its overall competitive performance. A recent first step was the development of a method enabling corridor scale estimation of inland shipping related energy consumption, fuel use and emission patterns (Jiang et al. 2022b). A logical next step is to develop a method that builds on these energy consumption and fuel use patterns to support the rational design of bunkering networks (per energy carrier estimate a logical maximum inter-distance between bunkering stations and their respective required total capacity).

2 Method

When the transport demand (volumes, origins, destinations), the state of the waterway network (e.g., water depths, currents), and the state of the fleet (composition, engine ages, etc.) are known, the associated energy demand for transport can be estimated using vessel resistance algorithms (Bolt 2003; Hekkenberg 2013; Vehmeijer 2019; Segers 2021; Van Koningsveld et al. 2021, Rijkswaterstaat 2022a; Rijkswaterstaat 2022b).

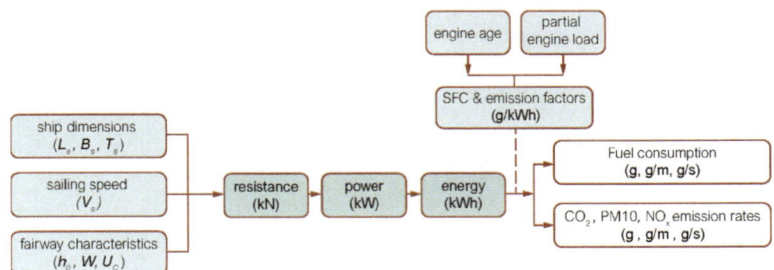

Fig. 1. Methodology for estimating emissions for IWT vessels (image modified from Segers 2021, by TU Delft Ports and Waterways is licenced under CC BY-NC-SA 4.0)

Figure 1 describes the methodology for estimating emissions for IWT vessels. Starting point of the analysis are the ship dimensions (length at the waterline (L_s), beam (B_s) and actual draught (T_s)), the vessel sailing speed (V_s) relative to the water, and the waterway characteristics (water depth (h_0), waterway width (W), current (U_c)). With this information, based on Holtrop and Mennen's method (1982) with Zeng et al.'s (2018) shallow water effect correction, we can estimate the total resistance a vessel experiences while sailing at a given velocity with respect to the water. Once the total resistance (kN) is calculated we estimate the total power (kW) that is required to overcome this resistance, which includes the power for propulsion and hotel system. Next we calculate the energy (kWh) that is consumed by multiplying this total power with the duration of its application. The energy consumption estimate can then be translated to fuel use and emissions.

Segers (2021), showed how this approach can be used to estimate corridor scale energy consumption, fuel use and emission patterns; both to estimate current patterns, using position information from the Automatic Identification System (AIS) that vessels need to have on board, and future patterns under various scenarios and policies. Jiang et al. (2022b) further generalized the approach by Segers (2021) to come to an approach that in principle is world-wide applicable (provided the required input data can be provided of course). A logical next step is to use this method to estimate energy consumption and fuel use to inform decision making on bunker infrastructure, both placement and capacity. Figure 2 displays the method we propose in this paper to plan a corridor scale bunkering infrastructure in terms of relative positions and required capacity for IWT.

The relative position of a refueling point is related to the sailing range (m) of the fleet, which depends on the total amount of energy storage (kWh) on board and the energy consumption per meter (kWh/m). The required capacity of the refueling points can be estimated based on the total energy consumption (kWh) of the ships between the refueling points.

As a fundamental step, the energy consumption calculation algorithm for a single ship and the whole corridor network has been written into the Python package Open source Transport Network Simulation (OpenTNSim) version v1.1.2 (Jiang et al. 2022a), which enables the further determination of the relative position and required capacity of refueling points in corridor scale.

The following sections describe (1) OpenTNSim simulation with an energy module and how the energy consumption algorithm can be applied using OpenTNSim, (2) how energy consumption maps can be used to rationalize the relative positions of bunker stations, and (3) how energy distribution over the shipping network can be used to define the required capacity of bunker stations.

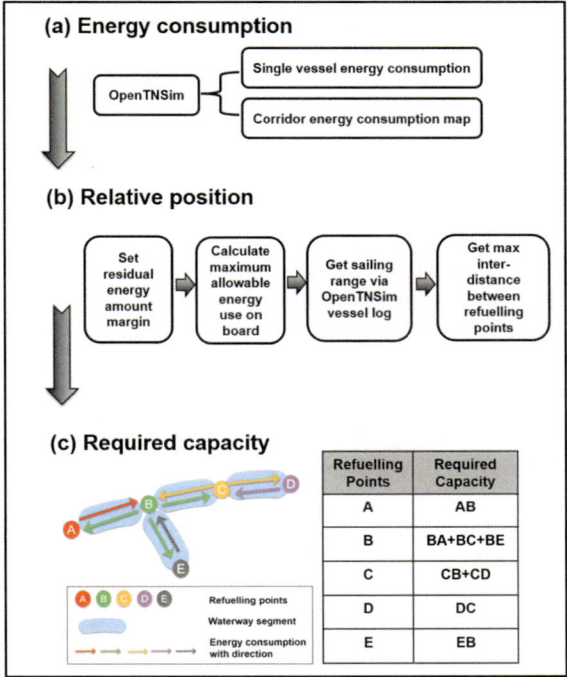

Fig. 2. Schematic diagram of the method

2.1 Energy Consumption

2.1.1 OpenTNSim

OpenTNSim is a python package for the investigation of traffic behaviour on networks. It can be used to investigate how water transport chains interact with the waterway network and its infrastructure. Simulations can be used to compare the consequences of traffic scenarios and network configurations (Van Koningsveld and Den Uijl 2020). In this paper we use OpenTNSim version v1.1.2 (Jiang et al. 2022a) (which includes an energy module) to perform energy consumption estimation and determine the relative positions and required capacity of refueling points. The energy module contains 'resistance, power and energy consumption estimation' algorithm, that can be applied to consecutive dx/dt events from either actual position data (AIS data, trip logs, etc.) or simulated position data (e.g. discrete event simulation output). This enables us to resolve footprints as a function of space and time. The model simulation mainly includes three components:

- Vessel objects with properties (including sailing log information).
- A graph that contains nodes and edges to represent the waterway network. The nodes linked by edges, contain geo-locations (longitude and latitude) of the waterways. The edges are made bi-directional to allow for two-way traffic, and contain waterway characteristics in the edge information.
- A simulation environment for sailing event simulation.

2.1.2   Energy Consumption Estimation in OpenTNSim

The general steps for the energy consumption estimation for a single vessel in OpenTNSim are as follows. First we create a vessel with its vessel properties, mainly include vessel dimensions, vessel sailing speed relative to the water, vessel installed engine power, etc. Then we create a graph with nodes linked by edges composing a network. We use nodes position to represent the geo-locations (longitude and latitude) of waterways. The waterway characteristics which cover the water depth, waterway width, and current speed are contained on each linking edge as edge information. Then we define the route (which is 'path' in the graph) that the vessel will sail. The path is defined by providing the origin and destination nodes in Dijkstra's path algorithm to find the shortest paths. Next we make an environment and add the created graph to the environment. Then we add the created vessel, to which we will append the environment and the route. Lastly, we give the vessel the process of moving from the origin to the destination of the defined path. We incorporate the energy module to this moving vessel, and via this energy module, the resistance, power and energy consumption of the moving vessel are successively calculated per edge along the path. Summing up the energy consumption of all the edges, the total energy consumption along the route can be obtained. Energy consumption in time and space of the vessel can also be mapped via displaying the energy consumption of all the edges along the route together.

The energy consumption of the corridor network can be mapped by aggregating the results of multiple vessels that together represent the corridor's traffic. The corridor energy consumption map is composed of the geographic waterway network, the energy consumption of ships per sailing edge aggregated in space along the waterways, the energy consumption directions according to the ship sailing directions, and the energy consumption in different time scales such as daily, weekly, monthly, seasonally and yearly (for the determination of required capacity for refueling points in different time scales).

2.2   Relative Position

The refueling points are positioned according to the minimum sailing range of the representative fleet (PROMINENT 2016; MOT 2022) in the corridor. The minimum sailing range provides an indicator for the maximum inter-distance between two refueling points, as this minimum value allows all types of ships in the corridor to arrive at the next refueling point. For the minimum sailing range determination, in OpenTNSim, we define vessels with a 'fuel volume', each discrete event we calculate how much energy is consumed and how much fuel. This volume can then be subtracted from the fuel volume. The fuel tank slowly depletes along with the sailing. When a certain 'buffer' is reached this can be defined as the sailing range by that vessel for the given fuel and given the volume of the fuel tank. Then by doing this for various vessels we can find the minimum range. The inter-distance between bunker points should not be larger than this minimum range. It should be noted that the maximum allowable energy use amount is smaller than the total amount of energy storage on board, since we set a residual energy margin which is 5–10% of the total energy storage on board, to prevent the ships running out of all the fuels on board before reaching the next refueling point. The total amount of energy storage can be determined by knowing the mass or volume of zero-emission energy sources on board, energy density and energy conversion system efficiency.

2.3 Required Capacity

Once refueling points are positioned on the corridor energy consumption map, the required capacity of each refueling point can be estimated based on the total energy consumption at various time scales. As shown in Fig. 2(c), the waterway network is divided into waterway segments by the refueling points. As a first scenario we put one refueling station is 'in charge' of all the waterway segments it directly connects. However, to prevent overlapping energy supply for the same segment by the two refueling points at both ends, we allocate the required energy supply amount according to the energy consumption direction, as shown in the refueling points capacity table in Fig. 2(c).

3 Results

Zero-emission energy sources in this paper refers to (green) hydrogen, e-fuels, and batteries. Among which, e-fuels are produced from electricity, water and carbon dioxide or nitrogen. When using electricity from renewable sources and circular carbon dioxide (e.g. from biomass or direct capture from the air), net emissions are near zero (Van Kranenburg et al. 2020). The e-fuels analyzed here include e-NH3, e-methanol and e-LNG.

Although there are increasing number of vessels powered by zero-emission energy sources in recent years (Arief and Fathalah 2022), the energy storage and conversion systems for zero-emission energy sources on board for wide use are still in designing stages (de Vos et al. 2018; de Vos 2020; van Kranenburg et al. 2020; Huang et al. 2021). Therefore, there is no certain value of the total storage amount for zero-emission energy sources on board to refer to. This forms an obstacle to get the sailing range of a zero-emission energy source powered vessel, which in turn hinders the determination of the relative positions and required capacities of the refuelling points. However, with the method implemented in OpenTNSim, we could calculate the required amount of energy (in kWh) and zero- emission energy sources (in kg and m^3) of a vessel sailing on a route to give some insight on designing the energy storage volume on board and preparing the energy amount for the whole route and corridor network.

Via the investigation of inland waterway characteristics (MOT 2022; RVW 2020; CCNR 2021), we have simulated the energy consumption of a motor vessel (M8 type (RVM 2020)) sailing in the corridor network along four distinct routes using Open-TNSim. The vessel properties and waterway network with characteristics are shown in Fig. 3. With the same total sailing distance of 450 km, Route 1 starts from Node 0, through Nodes 1, 2, 3 to Node 4, representing an 'unrestricted waterway' from sea port to the hinterland. Route 2 starts from Node 0 through Nodes 1, 2, 5 to Node 6,

representing a route from sea port to the hinterland that includes a 'shallower section (150 km)'. Route 3 starts from Node 0 through Nodes 1, 2, 7 to Node 8, representing a route from sea port to the hinterland that includes a 'very shallow section
(150 km)'. Route 4 starts from Node 0, through Nodes 1, 2, 9 to Node 10, representing
a route from sea port to the hinterland that includes a 'very shallow section (25 km)'.
At routes 1 and 2, the M8 vessel is able to sail with its maximum draught. However, at
routes 3 and 4, the M8 vessel has to reduce cargo to gain a smaller draught to pass the
limited water depth waterway sections.

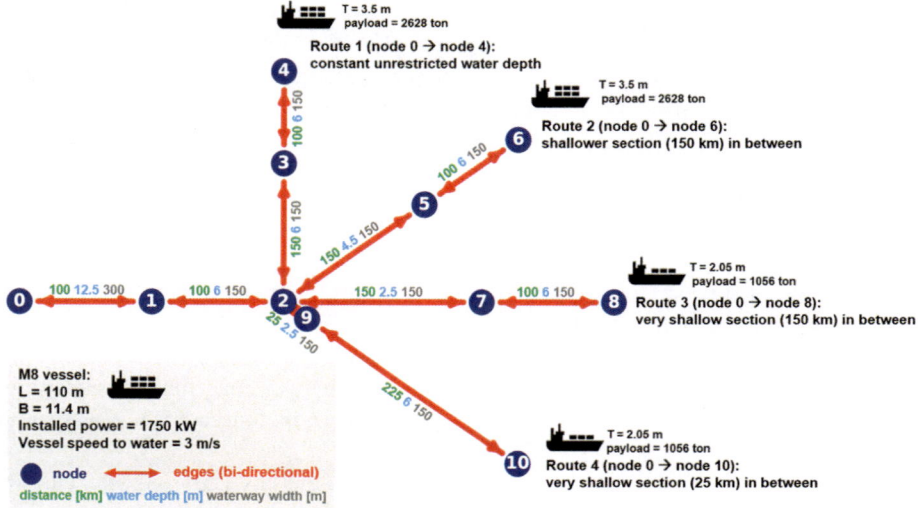

Fig. 3. OpenTNSim input: vessel properties and graph information.

Knowing the net energy gravimetric density (kWh/kg) and net energy volumetric
(kWh/m^3) density for zero emission energy sources and the corresponding energy
conversion system efficiency, the amount of energy consumption on board can be
translated to the amount of zero emission energy sources consumption in mass and
volume on board.

The required amount of energy and zero-emission energy sources (in mass and
volume) along the route of the M8 vessel estimated via OpenTNSim, are shown in
Table 1.

Table 1. The required amount of energy and zero emission energy sources (in mass and volume) along the route

	Required energy amount (MWh)	Required amount of zero emission energy sources (for fuel only, excludes storage system)					
		Hydrogen (liquid, -253°C)	E–NH3 (liquid, -34°C)	E–methanol (liquid)	E–LNG (liquid)	Battery 2 MWh (20ft Containers)	Diesel
Route 1: constant unrestricted water depth	13.12	Mass (ton) 1.02	6.55	6.13	2.35		3.48
Payload: 2628 ton		Volume (m³) 13.26	8.65	7.72	5.84	10.4 containers	4.77
Route 2: with	13.35	Mass (ton) 1.03	6.66	6.22	2.39		3.53
shallower section (150km) Payload: 2628 ton		Volume (m³) 13.47	8.79	7.84	5.93	10.6 containers	4.85
Route 3: with very shallow section (150km)	13.18	Mass (ton) 1.02	6.57	6.14	2.36		3.49
Payload: 1056 ton		Volume (m³) 13.28	8.67	7.74	5.85	10.5 containers	4.78
Route 4: with very shallow section (25 km)	11.36	Mass (ton) 0.89	5.74	5.37	2.06		3.05
Payload: 1056 ton		Volume (m³) 11.61	7.58	6.76	5.12	9.2 containers	4.18

As shown in Table 1, with the same sailing distance (450 km (Fig. 3)) and same payload (2628 ton), the required energy amount and the required amount of each zero-emission energy source for Route 1 and Route 2 are different. The required energy amount for Route 2 with a shallower section is 0.23 MWh higher than it for Route 1, as more energy is needed to overcome higher sailing resistance in the shallower section. Similarly, the required energy amount for Route 3 is 1.82 MWh higher than Route 4 though with the same sailing distance (450 km) and reduced payload (1056 ton), which is due to the longer 'very shallow section' in Route 3.

In general, for each route, the required mass of zero-emission energy sources in order from smallest to largest are: hydrogen (liquid, −253 °C), E–LNG (liquid), E-methanol (liquid), E–NH3 (liquid, −34 °C), battery with 2 MWh capacity (20ft Containers). The required volume of zero-emission energy sources in order from smallest to largest are: E–LNG (liquid), E- methanol (liquid), E–NH3 (liquid, −34 °C), hydrogen (liquid, −253 °C), battery with 2 MWh capacity (20ft Containers).

For Route 1, it requires 1.02 ton of hydrogen (liquid, −253 °C), which is the lowest mass among other zero emission energy sources. However, the practical constraints on board of a ship are not so much mass, but volume. The required volume of hydrogen (liquid, −253 °C) is the highest among others, 13.26 m^3, which is 2.78 times than required diesel volume. Considering its packing factor, 2, on a ship, the actual required space would be 26.52 m^3, 7.7 times higher than diesel (Van Kranenburg et al. 2020). If its storage space is designed to be of the same size as a diesel tank, with the consideration of 10% residual energy margin on board, then at least 9 bunker points along Route 1 are needed for the vessel to refuel. If its storage space is designed the double size as diesel tank (meanwhile less cargo on board can be taken), with the consideration of 10% residual energy margin on board, then at least 5 bunker points along Route 1 are needed for the vessel to refuel.

The required amount of E–NH3 (liquid, −34 °C) and E- methanol (liquid) are similar both in mass and volume, with E–NH3 is slightly higher. Considering the packing factors, which are 1.1 and 1 respectively, for Route 1, the actual required storage space on board of E–NH3 (liquid, −34 °C) and E- methanol (liquid) are 9.51 m^3 and 7.72 m^3, respectively.

There is only a small difference between the required fuel amount of E–LNG (liquid) and diesel. The required fuel amount of E–LNG (liquid) for Route 1 in mass is 1.13 ton less than diesel, in volume is 1.07 m^3 more than diesel. The packing factor of E– LNG (liquid) is 2 times higher than diesel, which leads to 2.14 m^3 of extra required storage space than diesel on board.

For Route 1, the required number of battery containers with 2 MWh capacity (in 20ft container), considering 10% residual energy margin, is 12. If the M8 vessel takes 2 battery containers on board, at least 6 docking stations along the route are needed; if the M8 vessel takes 4 battery containers on board, then at least 3 docking stations along the route are needed. It should also be noted that the payload capacity onboard is reduced due to more battery containers.

4 Discussion

To estimate the energy amount for the whole route and corridor network, the reliable quantification of energy demand (kWh) is needed. It should be observed that even the same vessel type with the same payload (ton) and sailing distance (km) oftentimes results in different energy consumption patterns along the route due to different sailing situations such as water depth variation as shown in the Results. Therefore, it is essential to consider waterway characteristics along the route to get reliable quantification of energy demand of an individual vessel and the whole corridor.

The data availability may hinder the wider use of the method, as it requires quite some input data: on the state of the IWT network (water depth, ambient currents, waterway classes, available routes), on the state of the vessel fleet (vessel speeds to the water, vessel dimensions, payload levels, actual draughts, installed engine power), origin - destination information, etc. However, recently much of this information is gathered and disseminated by authorities, which greatly enhances the data availability. Changjiang Waterway Bureau (China), for example, provides Changjiang Waterway electronic map including a comprehensive description of waterway characteristics for public free use. Pearl River Administration of Navigational Affairs (China) publishes daily number of ships passing through locks in Pearl River with sailing directions and payload information. Rijkswaterstaat (The Netherlands) hosts Vaarweginformatie.nl, where a lot of crucial information on the water transport network is disseminated and kept up to date. Other countries undertake similar efforts, a.o. in European projects. Realistic vessel speeds (relative to the ground), for the sailing events, can be obtained from AIS data (ship geographic position time series). Combined with data on ambient currents, vessels speeds relative to the water can be estimated. When detailed information is hard to come by, averages or probability distributions may be useful to provide insight.

As this method implemented in OpenTNSim is able to quantify the required energy amount of both an individual vessel and the whole corridor network with various water depths, current, payload, and sailing speed, vessel types and amount, etc., it can be used for designing bunker infrastructure with various scenarios including extreme discharge scenarios, concerning environment and economy changes. However, as the bunker locations in practice is selected not merely based on energy demand but also other societal considerations, this method should not be used to really find the exact locations of bunker points, but rather to make sure that the bunker points are located at a reasonable spacing given the vessels, their fuel tanks and energy carriers.

5   Conclusions and Recommendations

This paper proposes a bottom-up method implemented in the agent-based transport simulation, OpenTNSim, for corridor scale planning of bunker infrastructure for zero-emission energy sources (hydrogen, batteries, e-NH3, e-methanol and e-LNG) in IWT. It focuses on the positioning and dimensions of bunkering points based on vessel energy consumption in the corridor.

Taking vessel properties and waterway characteristics into account for energy consumption estimation, the method is applied to four distinct IWT routing examples, in which the variation of the energy demand (MWh) for an individual M8 vessel sailing at routes with same sailing distance (km) and payload (ton) conditions is revealed. The total required amount for each zero-emission energy source in mass and volume (number of containers for batteries) is also calculated and analysed. The application examples provide insight into designing the energy storage volume on board, preparing the number of bunkering points and total energy amount for the whole route and corridor network.

The method is recommended to be applied to map corridor two-way traffic energy consumption in time and space with actual data such as AIS data, trip log and the depth, width, and current information of waterway network. Then with the value of suitable energy storage space on board for different vessel types as input, derive the sailing range via OpenTNSim for the determination of relative positions of bunkering points, and finally estimate the required capacity in various time scales of bunkering points in the corridor network.

Acknowledgements. This paper is part of Man Jiang's PhD project "Greenpower – Assessing the Consequences of the Energy Transition on Waterborne Supply Chains", that is funded by SmartPort. This project is supervised a.o. by Klaas Visser, Robert Hekkenberg and Mark van Koningsveld. The work done in this paper is assisted by Fedor Baart. Finally we thank Nanjing Hydraulic Research Institute for organizing the PIANC Smart Rivers 2022 Conference.

References

Arief IS, Fathalah AZM (2022) Review of alternative energy resource for the future ship power. In: IOP conference series: earth environmental science, vol 972, p 012073

Bolt E (2003) Schatting energiegebruik binnenvaartschepen. Technical report. Rijkswaterstaat Adviesdienst Verkeer en Vervoer, Rotterdam

CCNR (2021) "Act now!" on low water and effects on Rhine navigation. https://www.ccr-zkr.org/files/documents/workshops/wrshp261119/ien20_06en.pdf

De Vos P (2020) AmmoniaDrive: a solution for zero-emission shipping?! SWZ Maritime 141 (3):36–37

De Vos P, Stapersma D, Duchateau E, van Oers B (2018) Design space exploration for on-board energy distribution systems: a new case study. In: Bertram V (ed) Proceedings of the 17th international conference on computer and IT applications in the maritime industries (COMPIT 2018), pp 463–481. Technische Universität Hamburg-Harburg

Hekkenberg R (2013) Inland ships for efficient transport chains. Ph.D. thesis. Delft University of Technology

Holtrop J, Mennen GGJ (1982) An approximate power prediction method. Int Shipbuilding Prog 29:166–170

Huang MY, He W, Incecik A, Cichon A, Królczyk G, Li ZX (2021) Renewable energy storage and sustainable design of hybrid energy powered ships: a case study. J Energy Storage 43:103266. https://doi.org/10.1016/j.est.2021.103266

Jiang M, Segers LM, Van der Werff SE, Baart F, Van Koningsveld M (2022a) OpenTNSim (v1.1.2). Zenodo. https://doi.org/10.5281/zenodo.6447088

Jiang M, Segers LM, Van der Werff SE, Hekkenberg R, Van Koningsveld M (2022b) Quantification of energy consumption, fuel use, and emissions for sustainable inland shipping (submitted)

MOT (Ministry of Transport of the People's Republic of China) (2022) Navigation Standard of Yangtze River Trunk Line. https://wtis.mot.gov.cn/syportalapply/sysnoticezl/indexzl (in Chinese)

PROMINENT (2016) Identification of the fleet, typical fleet families & operational profiles on European inland waterways. https://www.prominent-iwt.eu/wp-content/uploads/2015/06/2015_09_23_PROMINENT_D1.1-List-of-operational-profiles-and-fleet-families.pdf

Rijkswaterstaat (2022a) BIVAS. https://bivas.chartasoftware.com/Home

Rijkswaterstaat (2022b) BIVAS: emissies. https://bivas.chartasoftware.com/Home/BIVASAppl icatie/Documentatie/Emissies/

RVW (2020) Richtlijnen Vaarwegen 2020. Rijkswaterstaat, Ministerie van Infrastructuur en Waterstaat. https://puc.overheid.nl/rijkswaterstaat/doc/PUC_632307_31/1/

Segers L (2021) Mapping inland shipping emissions in time and space for the benefit of emission policy development: a case study on the Rotterdam-Antwerp corridor. Master's thesis

UNCTAD (2021) Review of Maritime Transport 2021. https://unctad.org/webflyer/review-maritime-transport-2021

Van Koningsveld M, Den Uijl J (2020) OpenTNSim (Version 1.0.0). https://github.com/TUDelft-CITG/OpenTNSim/. https://doi.org/10.5281/zenodo.3813871

Van Koningsveld M, Verheij H, Taneja P, de Vriend H (2021) Ports and waterways: navigating the changing world. Delft University of Technology. https://doi.org/10.5074/T.2021.004

Van Kranenburg K et al (2020) E-fuels: towards a more sustainable future for truck transport, shipping and aviation. TNO report

Vehmeijer L (2019) Measures for the reduction of CO_2 emissions, by the inland shipping fleet, on the Rotterdam-Antwerp corridor. Master's thesis. Delft University of Technology

Zeng Q, Thill C, Hekkenberg R, Rotteveel E (2018) A modification of the ITTC57 correlation line for shallow water. J Mar Sci Technol 24(2):642–657. https://doi.org/10.1007/s00773-018-0578-7

Design of Longitudinal Air Bubbler System Inside Ship Lock

Iqbal Singh Biln[1]([✉]) and Chandler Engel[2]([✉])

[1] St. Lawrence Seaway Management Corporation, St. Catharines, ON, Canada
ibiln@seaway.ca
[2] US Army Cold Regions Research and Engineering Laboratory,
Hanover, NH, USA
chandler.s.engel@usace.army.mil

Abstract. This study investigated the expected behavior of installing a longitudinal air bubbler system in the St Lambert Lock, Quebec, Canada to bring "warm" water from the depths of the lock chamber and reduce ice buildup on the lock walls during the winter navigation. The bubbler manifold would be installed near the bottom of the lock chamber and extend along the entire wall of the lock chamber. Compressed air would be provided to the manifold and the air would be released through a series of holes (orifices) installed along the length of the manifold. A vertical water current would be induced in the lock chamber by the rising air bubbles. The continuous supply of water to the surface should reduce ice formation along the wall of the chamber. This use of a longitudinal air bubbler system to reduce ice buildup along lock walls during the winter season is a novel use of an air bubbler system in North America in a navigation lock.

Keywords: Ice · Air screen · Bubbler · Lock

1 Introduction

The St Lawrence Seaway Management Corporation (SLSMC) contemplates installing a pilot-level longitudinal air bubbler system in the St. Lambert Lock to bring "warm" water from the depths of the lock chamber and reduce ice buildup during the winter season. Figure 1 below shows the ice buildup on the lock walls which can become significant enough to impede the clear passage of vessels without intervention. Figures 2 and 3 show the current methods of ice removal from the lock walls with a customized ice scraper tug and a long arm excavator.

Y. Li et al. (Eds.): PIANC 2022, LNCE 264, pp. 346–362, 2023.
https://doi.org/10.1007/978-981-19-6138-0_31

Fig. 1. Ice buildup on the lock wall surface

Fig. 2. Ice removal with a customized ice scraper tug

Fig. 3. Ice removal with long arm excavator

The bubbler manifold would be installed near the bottom of the lock chamber and extend along the entire south wall of the chamber where ice buildup is most pronounced due to reduced solar exposure. Compressed air would be provided to the manifold and the air would be released through a series of holes (orifices) installed along the length of the manifold. A vertical water current would be induced in the lock chamber by the rising air bubbles. The continuous supply of water to the surface should reduce ice formation along the walls of the chamber.

This use of a longitudinal air bubbler system to reduce ice buildup along lock walls during the winter season is a novel use of an air bubbler system in North America at a navigation lock. In all other cases known to the authors, air bubbler systems have been installed at locks primarily to create horizontal flow velocities at the water surface to move or deflect floating surface ice. Those systems are relatively high flow bubblers that require large air flow rates (see for example Rand 1988; Haehnel 2016). The SLSMC became aware of an application of a longitudinal air bubbler system to create vertical currents in Finland at locks located along the Saimaa Canal.

The proposed longitudinal air bubbler system would not be required to create substantial horizontal flow velocities at the water surface. Rather the goal of the proposed system would be to provide a constant supply of relatively "warm" water to the surface in order to suppress or a least reduce ice formation on the lock wall surface. Here, "warm" refers to the temperature of water that is greater than 0 °C with no specification of an actual temperature.

Unfortunately, nothing is known about the vertical temperature structure of the water contained in the St. Lambert Lock during the winter, and little is known about the environmental conditions in the lock at the water surface during the winter season. Therefore it is not possible to determine the required supply of water that must be brought to the surface. This conceptual design will estimate the longitudinal air bubbler performance over a range of practical orifice diameters, orifice spacings, manifold sizes, and air compressor sizes. A best design will then be selected based on the estimated performance, practicality of installation, likelihood of success, and potential for future modification, if required.

2 Background

There are many publications describing the mechanics of air bubble plumes that are relevant to the use of air bubbler systems at navigation structures (See for example, Kobus (1968), Ashton (1974), Tuthill and Stockstill (2005), and Haehnel (2016)). Ashton (1974) specifically addresses the use of air bubblers systems to suppress surface ice. These authors address the use of "high flow" bubblers. The proposed longitudinal air bubbler system that is the focus of this study would be used to suppress ice formation using a "low flow" bubbler. The basis of this approach rests on the change of density of water with temperature. Fresh water is an unusual substance in that its maximum density occurs at 4 °C (approximately 39 °F). Because of this, during the winter season, the water in the lock is likely to be stratified with the coldest water at the surface. The air bubbler is used to induce an upward vertical current to move the "warm" water from the depths of the lock chamber to the surface. This continuous supply of warmer water should suppress the formation of ice.

As mentioned above, a longitudinal air bubbler system in Finland has been applied in locks located along the Saimaa Canal. (Private communication from Tero Sikiö, Head of Unit, Finnish Transport Infrastructure Agency, Inland Waterways) The manifold is located about 3 ft above the bottom of the lock and attached directly to the wall of the lock chamber. The manifold diameter is 20 mm (3/4″ in.), the orifice diameter is quite small, only 0.7 mm (0.02756 in.) and the orifice spacing is 2.0 m.

3 Air Bubbler System Design

The air bubbler system is required to provide sufficient vertical upward water flow to effectively suppress ice formation along the length of the lock wall. Given that the water temperature structure of the lock is not known it is not possible to calculate the ice suppression based on heat transfer calculations. Rather a range of possible designs was determined based on the estimated performance, practicality of installation, likelihood of success, and potential for future modification, if required. The of the longitudinal bubbler system for the St Lambert Lock was investigated using the computer program BUB300 (USACE 2006). The input data for BUB300 include: manifold length and diameter, supply line length and diameter, orifice diameter and spacing, nominal compressor pressure, and submergence depth.

3.1 Bubbler System Dimensions

The relevant system dimensions used in the analysis are described here.

Manifold Length: The required length of the manifold is 768 ft to provide coverage along the entire length of the south wall of the lock.

Supply Line Length and Diameter: The air supply line length was estimated to be 100 ft and the supply line diameter was assumed to match the manifold diameters, discussed below.

Manifold Submergence Depth: The submergence depth varied between 25 ft and 50 ft depending on the operation of the lock and the upstream and downstream water levels (Fig. 4). This compares to the maximum possible depth of the lock, from the bottom of the lock to the top of the wall, which is 60 ft.

Sizing Manifold Diameter to Provide Uniform Flow Out of All Orifices. The first step in the manifold design is to compare the orifice sizes and manifold size to ensure that the flow rates out of the orifices are as nearly uniform as possible along the length of the manifold. A commonly used rule of thumb is that the flow rates out of the orifices will be nearly uniform if the sum of the area of all the orifices is less than 25% of the cross-sectional area of the manifold. Flow rates through each orifice will be calculated directly with BUB300 and the actual uniformity of a proposed system geometry can be assessed for various input flow rates.

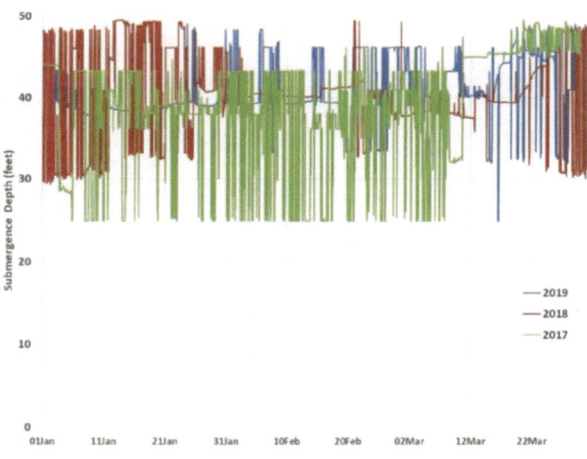

Fig. 4. Estimated Submergence depth of manifold in feet for three winter seasons

A manifold diameter of 3″ was selected as the best choice to provide practicality of installation, likelihood of success, and potential for future modification. The number of orifices required is equal to the total length of the manifold, 768 ft., divided by the orifice spacing. Table 1 shows the required manifold diameter, calculated using the 25% area criteria discussed above, for various orifice spacings (4, 8, 12, 16 foot) and various orifice diameters (0.7 mm, 1.0 mm, and 1/16″, 1/8″, 3/16″, 1/4″, 5/16″, 3/8″, 7/16″, 1/2″). The orifice diameters of 0.7 mm and 1.0 mm are included to match the orifice diameters used in the Finnish locks. A 3″ diameter manifold will provide sufficient flow capacity to maintain uniform flow across the manifold only for the cases shaded in yellow in Table 1. The detailed analysis of the 3″ manifold was limited to these cases.

3.2   Bubbler System Performance

The manifold and supply line dimensions used in the analysis are shown in Table 2. The range of manifold depths, orifice spacings, and orifice diameters analyzed are listed in Table 3. The largest orifice analyzed was 3/16″ (4.76 mm) because it was found that uniform flow could not be maintained out of all orifices at a larger orifice diameter and 8′ orifice spacing. BUB300 estimated the required air discharge through each of the orifices based on the air pressure supplied by the compressor. The sum of the individual orifice flow rates represents the total airflow to operate the bubbler at the specified pressure. The characteristics of the vertical plume created by the air bubbles were analyzed using the description provided by Ashton (1974) using the calculated air flow rate from each orifice and the depth of submergence of the manifold.

The vertical plume carries a volume of water to the surface. The vertical water discharge carried by the plume increases with depth of submergence of the manifold due to flow entrainment and the plume expands as it rises. An important point is the length of the lock wall protected by the plume reaching the surface.

Protection requires that the plume expands enough to cover the spacing between the orifices. The ratio of the plume width at the surface to orifice spacing was calculated to provide insight into portion of the wall that would be reached by the plume. It must be noted that the equations of Ashton assume that the bubbler orifices are located far from any vertical wall. In the case of the St Lambert Lock application, the bubbler will be located immediately next to the lock wall. It is not thought that this will make a significant difference in the results.

Table 1. Required manifold diameters for 768 ft. length, with various orifice spacings (4, 8, 12, 16 foot) and various orifice diameters. A 3″ diameter manifold will provide sufficient flow capacity for cases shaded in yellow.

Orifice Diameter			Orifice Spacing			
			4'	8'	12'	16'
			Number of orifices			
mm	Inches	Inches	192	96	64	48
			Required Manifold Diameter (inches)			
0.7		0.02756	0.76	0.54	0.44	0.38
1		0.03937	1.09	0.77	0.63	0.55
1.59	1/16	0.0625	1.73	1.22	1.00	0.87
3.18	1/8	0.1250	3.46	2.45	2.00	1.73
4.76	3/16	0.1875	5.20	3.67	3.00	2.60
6.35	1/4	0.2500	6.93	4.90	4.00	3.46
7.94	5/16	0.3125	8.66	6.12	5.00	4.33
9.52	3/8	0.3750	10.39	7.35	6.00	5.20
11.11	7/16	0.4375	12.12	8.57	7.00	6.06
12.70	1/2	0.5000	13.86	9.80	8.00	6.93

Table 2. Overall dimensions of Manifold and Supply Line

Component	Value
Manifold length	768 ft
Manifold diameter	3 in.
Manifold roughness	Hydraulicly smooth
Supply line length	100 ft
Supply line diameter	3 in.

Several trial geometries were evaluated to determine the optimal combination of orifice spacing and orifice diameters for the 3-in. manifold. Three scenarios were evaluated:

Scenario 1. This scenario investigated the impact of changing the orifice diameter. The orifice spacing was set at 8 ft and the submergence depth at 60 ft. The manifold diameter was 4 in. in this case which limited the effects of pressure drop in the manifold due to friction. Orifice diameters from the first five rows of Table 1 were evaluated.

Scenario 2. This scenario investigated the impact of changing the orifice spacing. The orifice diameter was set at 1/16 in. (1.59 mm) and the submergence depth at 60 ft. The manifold diameter was 3 in. Spacing of orifices included 4, 8, 12 and 16 ft.

Scenario 3. This scenario investigated the impact of changing manifold submergence depth. The orifice diameter was set at 1/16 in. (1.59 mm) and the orifice spacing at 8 ft. The manifold diameter was 3 in. Performance at depths of 20, 30, 40 and 50 ft were evaluated.

4 Results

Scenario 1: Varying orifice diameter. This scenario investigated the impact of changing the orifice diameter.

The air discharge per orifice is shown in Fig. 5. It should be noted that the air discharge across all orifices along the manifold was very uniform for all cases. The difference from completely uniform was only a few percentages for the largest orifice case.

The air discharge out of an orifice is proportional to the square of the orifice diameter. It can be seen in Fig. 5 that the air discharge per orifice increases rapidly for larger sized orifices. The total air supply required for all the 96 orifices in the manifold is shown in Fig. 6. Again, it can be seen that the total air supply required increases rapidly for larger sized orifices.

The ratio of surface spread of plume to orifice spacing is shown in Fig. 7. An orifice diameter of 1/16 in. or larger should provide good coverage of the lock wall for an orifice spacing of 8 ft (ratio is ~1 or larger). An orifice diameter of 1/16-in was selected to proceed with because it is relatively small which will help limit the airflow rate per orifice and therefore the vertical velocity. The size is a bit larger than the

0.7 mm holes used in the Finnish canals. The slightly larger size will also help to reduce the potential for biofouling.

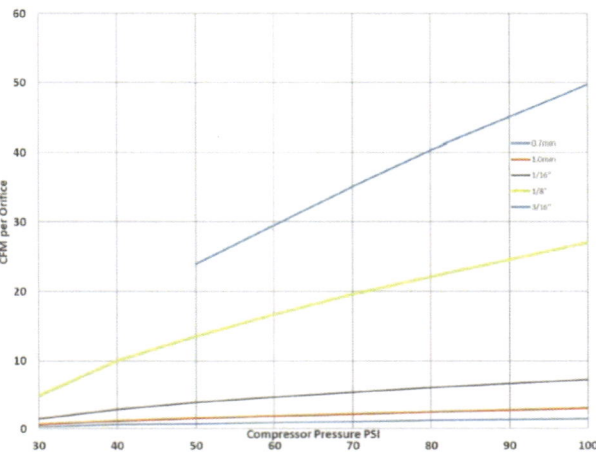

Fig. 5. Air discharge per orifice for a range of orifice diameters (Depth 60′, orifice spacing 8′, manifold diameter 4″))

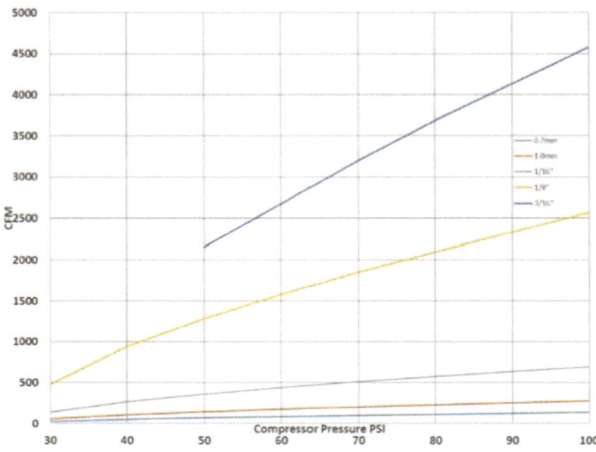

Fig. 6. Total air supply required for a range of orifice diameters (Depth 60′, orifice spacing 8′, manifold diameter 4″))

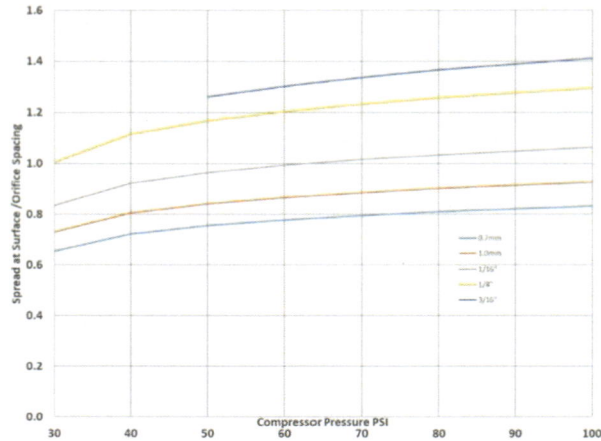

Fig. 7. Ratio of surface spread of plume to orifice spacing for a range of orifice diameters (Depth 60′, orifice spacing 8′, and manifold diameter 4″))

Scenario 2: Varying orifice spacing.

This scenario investigated the impact of changing the orifice spacing. The orifice diameter was set at 1/16 in., the manifold diameter at 3 in. and the submergence depth at 60 ft.

The air discharge per orifice is shown in Fig. 8. It can be seen that the orifice spacing does not have a significant impact on the discharge per orifice over the range of compressor pressures. This is due to the relatively minor pressure drops through the 3 in. manifold.

The total air supply required is shown in Fig. 9. At a given pressure, the total air supply required is linearly proportional to the number of orifices. The ratio of surface spread of plume to orifice spacing is shown in Fig. 10. The 4 foot spacing provides extra coverage with a ratio greater than 2.0. The 12 foot and 16 foot spacing provide poor coverage with ratios of 0.6 or less. The 8 foot spacing has a ratio just over 0.8 for most pressures evaluated, meaning that at least 80% of the wall would be covered by the bubble plume at the surface.

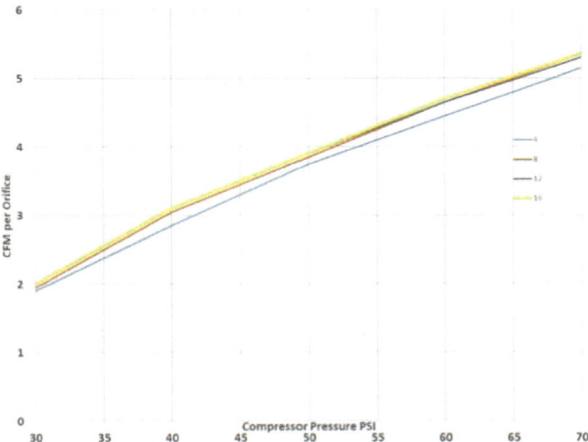

Fig. 8. Air discharge per orifice for a range of orifice spacings (Depth 50′, orifice diameter 1/16″, manifold diameter 3″))

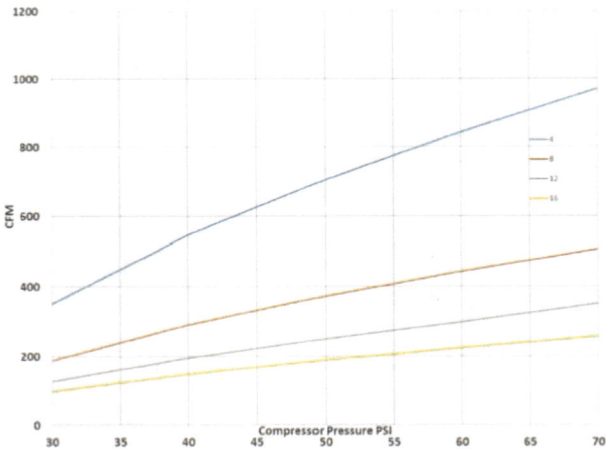

Fig. 9. Total air supply required for a range of orifice spacings (Depth 50′, orifice diameter 1/16″, manifold diameter 3″)

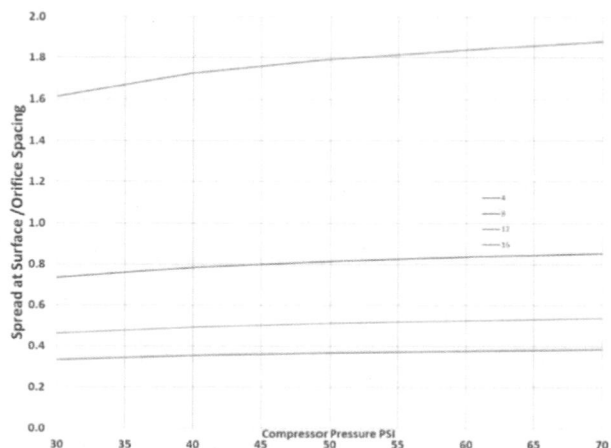

Fig. 10. Ratio of surface spread of plume to orifice spacing for a range of orifice spacings (Depth 50′, orifice diameter 1/16″, manifold diameter 3″)

Scenario 3: Varying the manifold submergence depth this scenario investigated the impact of changing manifold submergence depth.

The air discharge per orifice is shown in Fig. 11. At lower compressor pressures the discharge per orifice increases as the submergence depth decreases. However, at higher pressures, the increase in orifice discharge increases less as the submergence decreases. When the ratio of the manifold pressure to the outside pressure is 0.528 or less, the orifice discharge does not change at all with decreasing submergence. At this point the velocity from the manifold is choked, that is, the orifice discharge velocity is no longer sensitive to the pressure outside the manifold and depends only on the pressure inside the manifold. Choked flow is a common occurrence with compressible flows.

The total air supply required is shown in Fig. 12. Similar to the previous figure, the impact of choked flow is evident on the total air supply required.

The ratio of surface spread of plume to orifice spacing is shown in Fig. 13. The spread of the plume is linearly proportional to the depth of submergence. Therefore less submergence means that the plume has less opportunity to spread. It can be seen that manifolds with less submergence do not protect the entire lock wall.

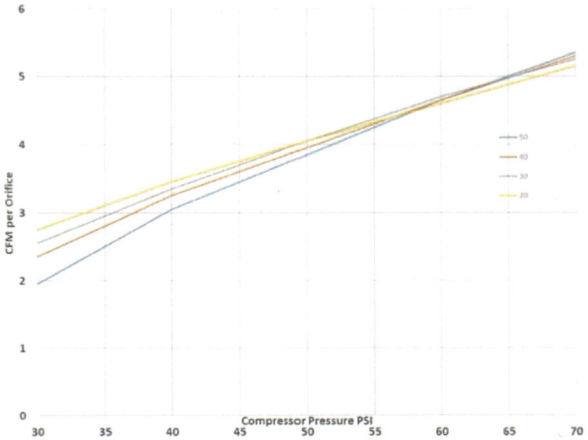

Fig. 11. Air discharge per orifice for a range of submergence depths (Orifice spacing 8′, orifice diameter 1/16″, manifold diameter 3″)

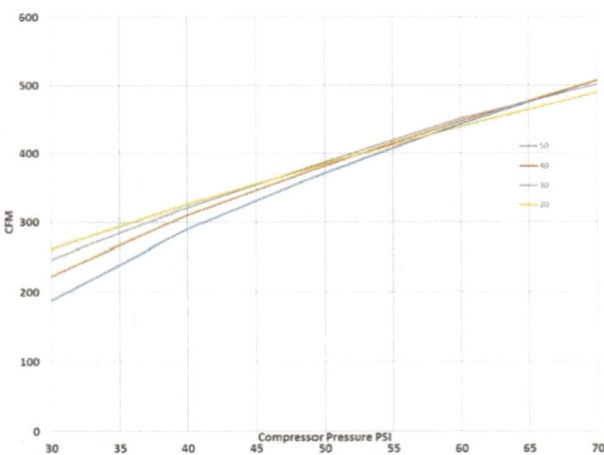

Fig. 12. Total air supply required for a range of submergence depths (Orifice spacing 8′, orifice diameter 1/16″, manifold diameter 3″)

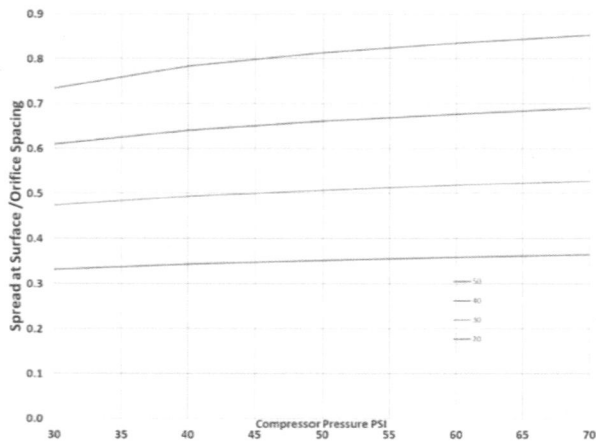

Fig. 13. Ratio of surface spread of plume to orifice spacing for a range of submergence depths (Orifice spacing 8′, orifice diameter 1/16″, manifold diameter 3″)

5 Discussion

The performance of a conceptual longitudinal air bubbler system for the St Lambert Lock was estimated based on BUB300 (USACE 2006) and the approach of Ashton (1978). The air flow through each orifice, the total air supply required for all the orifice, and the fraction of the lock wall protected by the bubbler plume was estimated for a range of orifice diameters, orifice spacing, and manifold submergence depths.

The air flow out of each orifice is strongly influenced by the orifice diameter. In general, for a given pressure difference across the orifice, the air flow out of the orifice is proportional to the square of the velocity. It is not thought likely that a large air flow is required to suppress ice formation along the lock wall. Therefore a relatively small orifice will likely provide sufficient air flow for this task. The use of a relatively small orifice is supported by orifice sizes used for longitudinal bubblers at locks located in Finland along the Saimaa Canal. (Private communication from Tero Sikiö), where orifice diameters of 0.07 mm are used. This size is much smaller than any orifices used at navigation locks in the US (however, bubblers at locks in the US are generally used to create large horizontal currents at the water surface to move surface ice.) This study focused on an orifice diameter of 1.59 mm (1/16″). This is larger than the Finish orifices but smaller than any currently in use at navigation locks. There are potential practical problems with very small orifice sizes – especially blockage of the orifice caused by marine debris, rust, and other small objects.

A 3 in. manifold provides very uniform flow through all the orifices – within about 3% variation for an 8 foot orifice spacing at 40 PSI pressure. A 3 in. manifold can also provide uniform distribution even if the orifice size is doubled to 1/8″. This provides flexibility if it is found that a larger orifice is required from field testing.

A range of orifice spacings were investigated. It was found that a spacing of 8 foot or less is best for providing protection along the entire length of the lock wall. Spacings greater than 8 foot do not protect the entire wall length because the vertical plumes do not spread out enough to provide flow to the entire distance between the orifices. This is especially true at small submergences (less than 40 ft of water depth). As discussed above, the equations used to estimate the bubbler spreading assume that the bubbler orifices are located far from any vertical wall. In the case of the St Lambert Lock application, the bubbler will be located immediately next to the lock wall. It is not clear if this would make a significant difference in the result. An orifice spacing of 8 ft is reasonable first try. If it is found that a closer spacing is required to provide complete surface coverage, the 3 in. manifold could provide uniform distribution to orifices with 1/8 in. diameter and 4 foot spacing.

Compressor pressures ranging from 30 psi to 70 psi were investigated. In general, the results at lower pressures were quite satisfactory and there was not any significant increase in air flow rate per orifice or in providing protection along the entire length of the lock wall. Note that pressure losses due to couplings, bends, flow valves, and other fixtures in the air supply line and manifold were not included in the calculations. There should be an allowance for extra capacity in the compressor to account for these losses.

6 Field Testing

The SLSMC constructed the pilot low flow air system at the St. Lambert Lock in 2020 using the findings of this analysis. The final dimensions of the system include a 3-in diameter manifold with 1/16-in diameter orifices spaced at 7-ft. The manifold was fastened to the lock wall in a position where it is less prone to effects from sedimentation, impacts from vessels and does not obstruct filling/emptying ports. An example of the manifold fastening method is shown in Fig. 14. System tests performed showed that the system provided continuous wall coverage and moderate flow up the wall but without a visible fountain, or excess disturbance of the surface which could cause splashing and incidental ice formation (Fig. 15).

The system was shown to be effective, and also found to have an incidental benefit from the gentle horizontal current which pushed any floating ice towards the center of the lock where it was easier to flush with vessel traffic. The 1/16 in. orifice diameter provided sufficient airflow such that the need to potentially increase the orifice diameter was determined to be unnecessary. This allowed a modification of the design to use a smaller diameter manifold of 2-in. while still providing uniform flow along the system length.

This design, with 2-in. diameter manifold, was used for the installation two additional bubblers (one on each wall) in the Upper Beauharnois Lock 4 (Fig. 16).

Fig. 14. Example of manifold attachment to lock wall at Upper Beauharnois

Fig. 15. Pilot longitudinal bubbler system in operation at St. Lambert Lock. Note bubble plumes visible along the lock wall.

Fig. 16. Installation of one of a pair of longitudinal bubblers at the Upper Beauharnois Lock 4 in Quebec, CA. Note the installation location above the filling ports along the bottom of the lock.

7   Conclusions

This study investigated the expected behavior of installing a longitudinal air bubbler system in the St Lambert Lock, Quebec, Canada to bring "warm" water from the depths of the lock chamber and reduce ice buildup on the lock walls during the winter navigation. The specialized software BUB300 written by engineers at the US Army Corps of Engineers Cold Regions Research and Engineering Laboratory was used to analyze potential system designs and find an optimal geometry for the subject location. Based on this analysis the SLSMC installed a pilot system was found to perform very well. A minor refinement to the design, a reduction of the manifold diameter from 3 to 2 in. provided some cost savings for future installs while still maintaining the performance capability of the pilot system. These systems are an important tool which provide timely and effective mitigation of ice buildup on lock walls, reducing the need for labor-intensive and potentially hazardous ice removal techniques.

References

Ashton GD (1974) Air Bubbler Systems to Suppress Ice. Cold Regions Research and Engineering Laboratory (CRREL) Special Report 210. September 1974

Ashton GD (1978) Numerical simulation of air bubbler systems. Can J Civ Eng 5(1978):231–238

Haehnel RB (2016) Review of Ice-Control Methods at Lock 8, Welland Canal, Port Colborne, Ontario. U.S. Army Engineer Research and Development Center (ERDC) Cold Regions Research and Engineering Laboratory (CRREL) Special Report 16-1. May 2016

Kobus HE (1968) Analysis of the flow induced by air-bubble-systems. Chapter 65 of Part 3. Coastal Structures. In: Proceedings eleventh conference on coastal engineering, London, England, vol II, pp 1016–1031. ASCE New York

Rand J (1988) High-flow air screens reduce or prevent ice-related problems at navigation locks. In: Proceedings of the international association of hydraulic research, ice symposium 1988, Sapporo, Japan, pp 34–43

Tuthill AM, Stockstill RL (2005) Field and laboratory validation of high-flow air bubbler mechanics. J Cold Reg Eng 19(3):85–99

USACE (U.S. Army Corps of Engineers) (2006) Ice Engineering. EM 1110-2-1612. U.S. Army Corps of Engineers, Washington, DC

Development of a Port Facility Diagnostic System that Utilizes Data Measured by Strong Motion Seismographs

Hiroaki Matsunaga[1]([✉]), Junichi Kyouda[1], Sou Itakura[1],
Tatsuru Yamamoto[2], Akito Sone[3], and Takashi Kadono[4]

[1] Ministry of Land, Infrastructure, Transport and Tourism, Chubu Regional
Development Bureau, Nagoya Port and Airport Research Office, Aichi, Japan
{matsunaga-h852a,kyouda-j852a,
itakura-s852a}@mlit.go.jp
[2] Port, Harbors and Coastal Engineering Group, NEWJEC Inc., Osaka, Japan
yamamotott@newjec.co.jp
[3] Port, Harbors and Coastal Engineering Group, NEWJEC Inc., Tokyo, Japan
sonekt@newjec.co.jp
[4] NEWJEC Inc., Osaka, Japan
kadonotk@newjec.co.jp

Abstract. The Authors developed a damage assessment system for mooring facilities. This system uses information on seismic motion observed by strong motion seismographs when an earthquake occurs in order to determine whether or not mooring facilities can be used based on the degree of damage suffered. The system creates maps estimating damage based on the results. Two methods are available for determining usability—immediate diagnoses and detailed diagnoses—and all calculations are performed automatically, from the acquisition of the seismic motion information to the determination of usability. The resulting maps estimating damage can be used to select surveyed facilities for on-site damage assessments, making this system an effective tool for supporting initial response systems after a disaster strikes.

Keywords: Development of a port facility diagnostic system · Large-scale earthquakes · Utilization of seismographs' data · Efficiency improvement of field surveys · Transporting emergency supplies

1 Introduction

In Japan, there is concern that large-scale earthquakes will occur in the near future. When large-scale earthquakes occur, road and railway networks can become disrupted in many sections. Waterborne transport, including IWT, is therefore expected to play a crucial role in transporting emergency supplies. Prompt resumption of trunk line waterborne transport is also important for the economic recovery in disaster-affected areas. In order for waterborne transport to function properly in the event of a disaster, it is necessary to know which facilities at what ports can be utilized. Conventionally, this information has been obtained solely through field surveys after an earthquake. The

© The Author(s) 2023
Y. Li et al. (Eds.): PIANC 2022, LNCE 264, pp. 363–371, 2023.
https://doi.org/10.1007/978-981-19-6138-0_32

Authors developed a diagnostic system for assessing the degree of damage to port facilities with the aim of streamlining on-site work at quays and facilitating the early resumption of waterborne transport.

This paper presents an example of a study on ports in Japan. The examples presented are expected to prove beneficial in the field of waterborne transport, which mainly involves large-scale canals. It is the Authors' hope that this report will be helpful in the development of future disaster response measures in the field of waterborne transport.

2   Three Challenges with Field Surveys Following Large-Scale Earthquakes

Generally, the decision on whether or not mooring facilities can be used following a large-scale earthquake are based on the results of field surveys. However, there are major challenges to conducting these field surveys.

(1) For piers and sheet pile quays, the degree of damage to underground and underwater areas cannot be ascertained with field surveys (simple visual surveys) immediately after a disaster occurs. For example, steel pipe piles at a facility that appeared sound had buckled, creating extremely dangerous conditions; members had no residual bearing capacity (Fig. 1).

(2) The survey of damage to mooring facilities at larger ports requires considerable time and person-hours, demanding greater efficiency.

(3) When a tsunami warning or advisory is issued, field surveys cannot be conducted immediately because people cannot approach coastal areas until the warning or advisory is lifted.

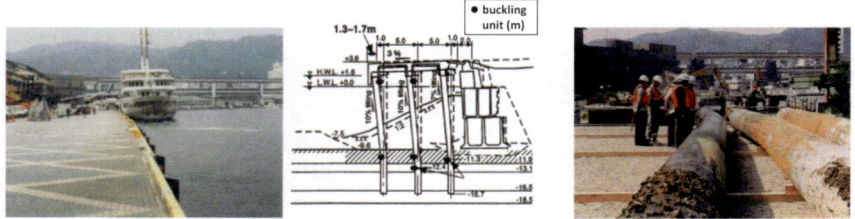

Fig. 1. Example of damage to a pier following the 1995 Hyogo-ken Nanbu earthquake

3   Development of a Port Facility Diagnostic System

3.1   Outline of Port Facility Diagnostic System

Strong motion seismographs are installed at critical ports across Japan (Fig. 2), allowing seismic tremors to be observed. The overall process behind a Port Facility

Diagnastic System (PFDS) is shown in Fig. 3. When an earthquake occurs, the data observed by the strong motion seismographs is stored in a data server at the National Institute of Maritime, Port and Aviation Technology, Port and Airport Research Institute (PARI). An e-mail notification is sent out indicating that an earthquake has occurred ((1), (2) in Fig. 3). After this email is distributed, PFDS accesses the PARI data server to obtain the observed seismic data ((3), (4) in Fig. 3). The obtained seismic waveforms are those observed on the surface or underground. The seismic motion used in the field of seismic design of port facilities in Japan is generally assumed to be the waveform in the engineering bedrock. In seismic engineering, the engineering bedrock refers to the boundary layer between the surface (soft ground), which has a significant influence on vibration characteristics, and the deeper area (hard ground), which has less influence on vibration characteristics. It is established for convenience. For the evaluation of the mooring facilities, a one-dimensional seismic response analysis is conducted using the equivalent linearization method (Yoshida 1996), which can consider the frequency dependence using the ground model for the observation point. The observed waveform data is used after being converted into an earthquake waveform in the engineering bedrock. The facility diagnosis registered in PFDS is performed using the obtained seismic data on the engineering bedrock, ((5), (6) in Fig. 3). There are two types of facility diagnostic methods. The first is an immediate diagnosis, where the diagnosis is performed immediately after an earthquake occurs. The second is a detailed diagnosis, where a detailed diagnosis is performed by conducting a two-dimensional seismic evaluation of the facility in question. Details are described below. After facility diagnoses are performed, a map estimating the damage is created for each registered facility at an individual port, and the map is distributed to the emergency operations center ((7), (8) in Fig. 3). PFDS performs these steps automatically.

Note that in the Japanese port sector, the ground vibrations are studied and divided into zones based on similar vibration characteristics. In the example of Shimizu Port in Japan shown in Fig. 4, the applicable area of strong motion seismograph observation data ranges from about 3 to 10 km.

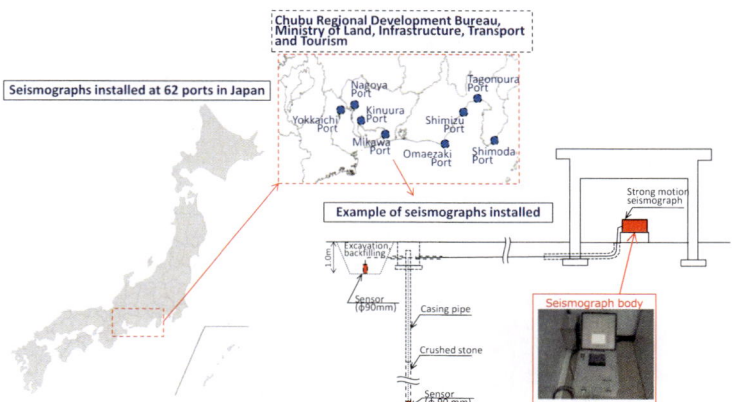

Fig. 2. Strong motion earthquake observation points in port areas and an example of strong-motion seismograph installations

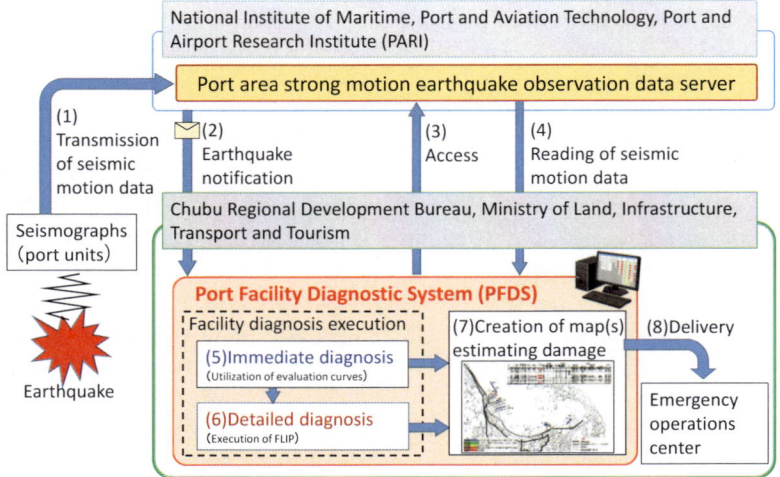

Fig. 3. The overall process behind the Port Facility Diagnostic System

Fig. 4. Zones with similar seismic characteristics (Example: Shimizu Port)

3.2   Development of Diagnostic Methods

In the Japanese port sector, FLIP, a two-dimensional seismic evaluation code (Iai et al. 1990), is used as a seismic inspection and design tool for mooring facilities assuming large-scale earthquakes. FLIP is a seismic evaluation program developed to predict structural damage due to liquefaction. This analysis code is an analysis program based on the finite element method. It enables deformation analysis considering the dynamic interactions between the ground and the structure by modeling the port structures and the ground in an integrated manner. In addition, the analysis can reproduce the damage to port facilities caused by large-scale earthquakes in the past with a high degree of accuracy. The analysis code is extremely effective in performing detailed seismic inspections of facilities. A diagnostic method for facilities that utilizes FLIP was therefore developed in PFDS.

3.2.1   Immediate Diagnostic Method

Performing a FLIP analysis—used in detailed diagnoses—can provide detailed estimates of damage to each member component comprising a facility. However, doing so requires several to more than ten hours of analysis time, whereas immediacy is demanded following an earthquake, Therefore, FLIP analyses were conducted for all facilities registered in PFDS beforehand, looking at seismic motions of various sizes with earthquakes of different magnitudes. The results of these analyses were compiled into evaluation curves that show the relationship between the magnitude of the earthquake and the degree of damage to the main member components of the facility structure (Fig. 5 and Fig. 6). The immediate diagnostic method uses these evaluation curves to perform diagnoses. The PSI of the velocity is used as an indicator of earthquake magnitude. The PSI of the velocity is defined by Eq. (1) (Nozu and Iai 2001) and is a seismic motion index that is highly correlated with the deformation of port structures and can be easily calculated from observed seismic data. A typical example of the extent of damage to main member components of pier or sheet pile quay structures is the maximum curvature ratio of steel pipe members, as defined by Eq. (2). For example, according to the technical standards and explanations of Japanese port facilities (Japan Port Association 2018), if the maximum curvature ratio of a sheet pile quay is less than 1.0, the mooring is structurally stable and can be judged to have residual bearing capacity. In other words, if the maximum curvature ratio exceeds 1.0, this indicates the steel pipe members are in an extremely dangerous condition, as they have lost their residual bearing capacity and have plasticized. The steel members are judged to be unusable. In addition to this structural judgment, as a utilization judgment, it is assumed that vessels can berth if the residual horizontal displacement at the top of the facility is 2.0 m or less. Judging using the evaluation curve shown in Fig. 6, a PSI of the velocity of 61 cm/s$^{0.5}$ or less is considered provisionally usable, while 61 cm/s$^{0.5}$ or greater is considered unusable.

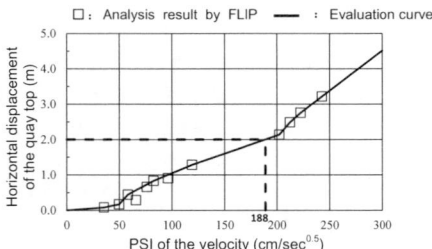

Fig. 5. Example of an evaluation curve used in the immediate diagnostic method (1) Relationship between PSI of the velocity and residual horizontal displacement

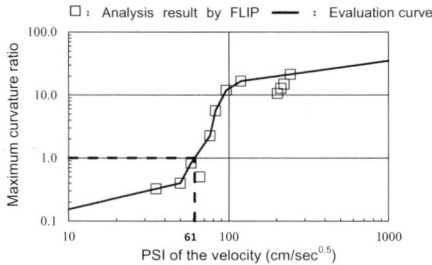

Fig. 6. Example of an evaluation curve used in the immediate diagnostic method (2) Relationship between PSI of the velocity and maximum curvature ratio

$$\text{PSI of the velocity} = \sqrt{\int_0^\infty v^2(t)dt} \tag{1}$$

where, v(t): velocity of earthquake motion at each time (cm/sec)
    dt: integration time (sec)

$$\text{Maximum curvature ratio} = \frac{\varphi_{max}}{\varphi_u} \tag{2}$$

where, φ_{max}: maximum curvature generated in steel pipe members based on FLIP analysis results (1/m), φ_u: limit curvature of steel pipe members (1/m).

3.2.2   Detailed Diagnostic Method

Detailed diagnostics is a method of analysis using FLIP based on pre-created analysis models. The analysis uses seismic waveforms converted from observed seismic waveform data to waveforms in the engineering bedrock. A residual deformation diagram is shown in Fig. 7 as an example of the analysis results. In addition to residual deformation, the degree of damage to each of the other main components can be estimated. For mooring facilities with container cranes, the container cranes are also modeled and analyzed using FLIP. This gives the maximum response acceleration at the center of gravity mass of a crane. This maximum response acceleration can be used to verify the crane's lifting limit acceleration and design seismic intensity.

3.3   Creation of Maps Estimating Damage

An example of output results of immediate and detailed diagnoses as a damage estimation map is shown in Fig. 8. It is color-coded according to the usability classification. The map is intended to be used for reporting to the emergency operations center
in the event of a disaster.

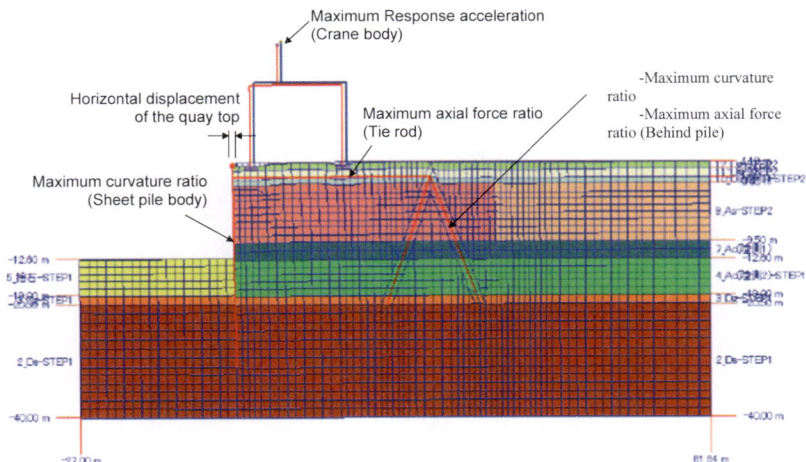

Fig. 7. Example of FLIP analysis results (added to the residual deformation diagram)

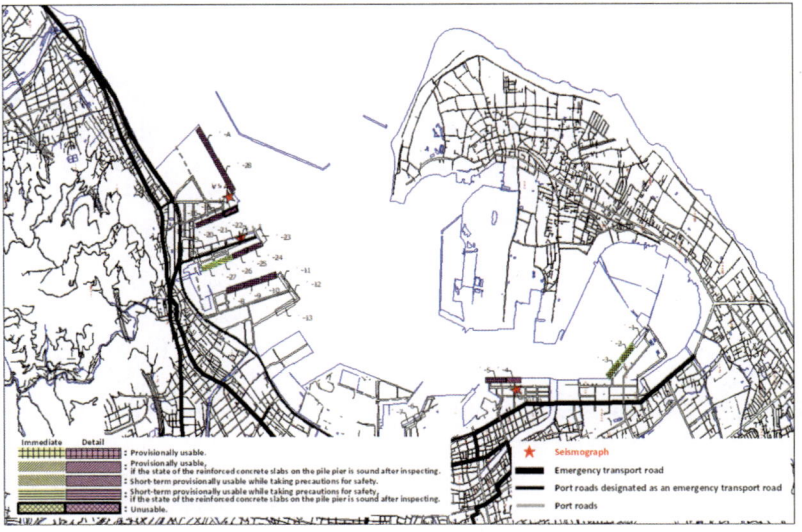

Fig. 8. Creation of Maps Estimating Damage (Example: Shimizu Port)

3.4   Addressing Challenges (Effectiveness of PFDS)

The challenges facing field surveys described in 2. were solved through the development of PFDS as follows.

(1) Ascertaining damage underground, where judgements are not possible with visual surveys: PFDS, using FLIP analyses, has made it possible to ascertain the extent of damage to underground components.
(2) Field surveys require considerable time and person-hours: the maps estimating damage created by PFDS makes it possible to improve the efficiency of field surveys, shorten the time required, and save labor.
(3) When a tsunami warning or advisory is issued, field surveys cannot be conducted for a certain span of time: immediate and detailed diagnoses can be performed by PFDS during these spans of time. Therefore, the time lost due to tsunami warnings, etc. can be reduced.

4   Conclusions

PFDS brings together, and was developed based on, current technologies, including strong motion seismographs installed in port areas, records of damage to port facilities to date, a two-dimensional effective stress seismic evaluation program (FLIP) that can accurately reproduce the extent of damage to mooring facilities caused by earthquake motion, and the latest research results on the damage mechanisms seen in port facilities. Although a comprehensive decision on whether or not a facility can be used must be made after confirming the damage on site, PFDS can be used to narrow down the list of facilities that have a high possibility of being usable from among many facilities, enabling more efficient and effective field surveys.

Even in countries where there are insufficient seismographs installed, it may be possible to develop a system with the contents presented in this paper by collaborating with local weather stations and existing observatories. It is the Authors hope that our study will help the development of waterborne transportation in times of disaster.

References

Nozu A, Iai S (2001) A consideration on the seismic motion index used for immediate damage estimation of the quay. 28th Kanto branch technology research presentation lecture summary pp 18–19

Yoshida N (1996) DYNEQ, A computer program for Dynamic response analysis of level ground by Equivalent linear method

Iai S, Matsunaga Y, Kameoka T (1990) Strain space plasticity model for cyclic mobility. Rep Port Harbour Res Inst 29(4):27–56

Japan Port Association (2018) Technical Standards and Commentaries for Port and Harbour Facilities in Japan

Development of a Verification Procedure of Partial Loading on Existing Solid Hydraulic Structures - Probabilistic Assessment for 3D Material Variations

Sophie Rüd$^{(\boxtimes)}$, Hilmar Müller, Helmut Fleischer, and Christoph Stephan

Federal Waterways Engineering and Research Institute, Karlsruhe, Germany
sophie.rued@baw.de

Abstract. One of the challenges in assessing the load-bearing capacity of existing solid hydraulic structures is the formal verification of concentrated loads for plain concrete. Due to the age bandwidth of such structures in Germany, this applies to hundreds of cases and especially to older structures of rammed concrete. Typical examples of components subjected to partial loads are found at weir pillars: e.g. support niches of inspection closures. Although they cannot be formally verified using the current regulations, the BAW Code of Practice "Evaluation of the load bearing capacity of existing solid hydraulic structures" (TbW) allows more detailed investigation methods to be applied, e.g. the use of non-linear probabilistic calculations. The principle research motivations are a higher loading capacity by numerical simulations with a more realistic material model compared to the usual linear calculations and a higher loading capacity by reproducing a "natural" bandwidth of material characteristics in these simulations. The aim of the current research project is the development and standardisation of the numerical simulations for such a verification procedure and its underlying safety concept by a classification of structural markers. As a result, the necessity of complex reinforcements for such structures could be assessed. The paper introduces the research concept and addresses the investigation steps regarding measured and generic 3D material distributions and FEM representation specifics as the material model. Furthermore, the preparation of the stochastic analysis is introduced by a demonstration model: The resulting hundreds to thousands of simulations of individual cases enable the stochastic analysis of metamodels to deduce general probabilistic results. Prospectively, the demonstration model will be transferred to further component measures and compressive strength classes.

Keywords: Verification · FEM · 3D material distribution · Probabilistic · Existing solid hydraulic structures

© The Author(s) 2023
Y. Li et al. (Eds.): PIANC 2022, LNCE 264, pp. 372–383, 2023.
https://doi.org/10.1007/978-981-19-6138-0_33

1   Introduction

Assessing the load-bearing capacity of existing solid hydraulic structures is generally based upon stylized assumptions in the calculation methods, including a severely simplified model of the material behaviour. Additionally, the assessment assumes an at least per section homogeneous construction material and therefore constant material parameters. Such simplifications are balanced in the assessment process, e.g. by safety factors and characteristic values as material parameters. For example, the broad scattering of the composition, processing and mixing of the building materials is addressed in the hypothesis of a normed material distribution and a 5% quantile approach of the characteristic material parameters. Furthermore, the standard-compliant zeroing of the tensile strength under bending stress prevents formal verification of the load-bearing capacity of plain concrete structures components under partial loading. This may result in complex reinforcements such as anchoring or additional constructions as sheathing of the loading areas. Figure 1 illustrates this kind of loading case for a weir pillar and the corresponding anchor reinforcements constructed at a weir pillar at the Donau-barrage Kachlet. In Germany, there are currently circa 700 weirs and locks in the portfolio of the Federal Waterways and Shipping Administration (WSV), of which almost 200 structures are over 100 years old and about 300 between 50 and 100 years, therefore this applies to hundreds of cases and especially to older structures build of rammed concrete. To address this challenge not on an individual case basis but to develop a verification procedure, the numerical simulations require a scientific foundation and standardisation in their simulation scenarios as well as their analysis in accordance to an underlying safety concept, e. g. by a classification of structural markers. As a result, the appraisal of subsequent and complex reinforcements for such structures could be supported in terms of necessity and efficiency. The classification of revision recesses and their loads in a decision matrix, according to their geometric and material-specific/construction time period, corresponds with a reduction in the investigation effort per individual case. The BAW Code of Practice "Evaluation of the load bearing capacity of existing solid hydraulic structures" (TbW) offers an assessment process set in three stages A, B and C. The stages mirror a rising accuracy in the assessment and the corresponding exploration of possible reserves in loading capacity: In the calculations of the first two stages, characteristic material values are either proposed according to the relevant material or construction period (stage A) or result from measurement campaigns (stage B) as 5% quantile values. In the third stage in accordance with TbW, FEM simulations are used to represent the non-linear material behaviour as realistically as possible, while probabilistic methods account for the anticipated range of the building material.

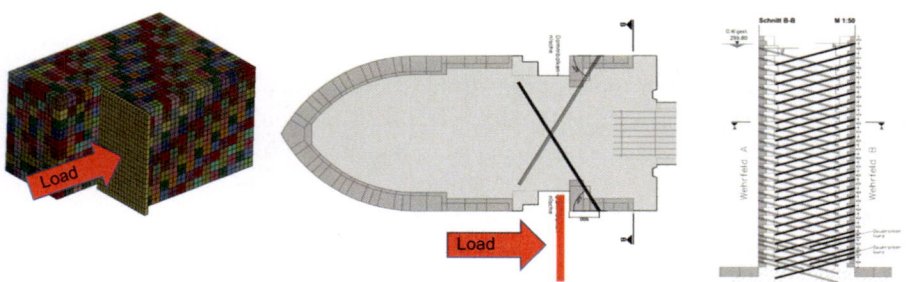

Fig. 1. Left: loading scenario partial loading, middle: loading scenario revision recess, right: reinforcements realized at Kachlet. Drawings modified from KW011A-SG01-4 (2016)

Regarding the FEM simulations, the main motivations are twofold. First, a higher loading capacity compared to the linear calculations by numerical simulations with a more realistic material model. The appeal of this aspect includes not only an extension of the linear material behaviour stipulated in typical verification calculation, but notably the load case independent inclusion of tensile strength. Second, reproducing a heterogeneous realistic 3D bandwidth of material characteristics averages to a higher overall loading capacity than the homogenous applied 5% quantile. However, the 5% quantile is part of the safety concept in EC0 (2002) and based upon model assumptions as well as material uncertainties, e.g. the broad spread of compositions and 3D localisation of the materials used in concrete construction. Furthermore, the material distribution of the loading areas is not known in such a detail as the simulations require. A representative appraisal therefore requires an appropriate probabilistic analysis of FEM simulations with 3D distribution fields of the material characteristics to account for the influence of material distribution and spatial permutations. The resulting hundreds to thousands of individual case simulations can be merged in parameterised result analysis and thereby enable a stochastic analysis. By building and sampling metamodels, general probabilistic results can be deduced at a defined safety level. Figure 2 introduces the FEM simulation preparation by illustrated process steps. The 3D distribution fields are spatial layout variations of generically produced distributions based on literature research and evaluation of measurements. Mapped on a section of a structural component and infused with a nonlinear material model, they constitute variations in otherwise constant FEM simulation scenarios. Each scenario yields one set of result values, e.g. a maximum force or number and failed elements prior to the maximum force. Approximating a sufficient amount of result sets via metamodels enables a probabilistic analysis of further scenarios. In the next sections these process steps are discussed and the preparation of the stochastic analysis is showcased by a demonstration model.

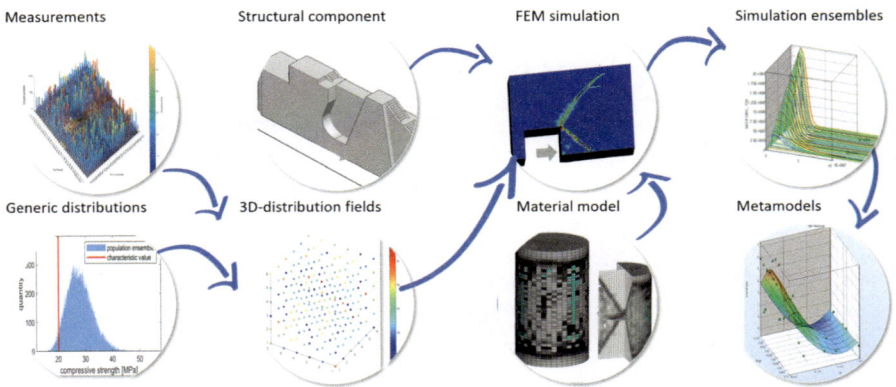

Fig. 2. Creation process of metamodels for reliability analysis based on FEM simulations of recesses with three-dimensionally distributed material parameters.

2   3D Distribution Fields of Material Parameters

2.1   Measurements

The influence of the spatial variation of material properties on the load-bearing behaviour of a structure depends not only on the distribution ensemble of the examined material properties, but also on the distribution pattern and its characteristics. Depending on its properties, the 3D variation can furnish decisive insights into the load-bearing capacity of the component in question and thus constitute a design parameter itself. Generating 3D distribution fields requires choices about the type and parameters of the 3D arrangement, e.g. random, structured or partially structured. A key attribute is the (self-) correlation length, which describes the distance of influence for a specific property. Regarding the correlation lengths of concrete, current studies point to a range of meters. The Joint Committee on Structural Safety (2002) for example invokes a correlation length of 5 m, based on Kersken-Bradley (1985) with 5 m for slim and 1 m for massive structures. Bouhjiti (2018) concurringly stipulates a correlation length of 1 m based on findings of De Larrard (2010), whose experiments and analysis however cautioned regarding scattering and the dependence per investigated parameter. Correlation lengths less than 1 m were obtained in a non-destructive measurement campaign by Borosnyói (2015) conducting spatial analysis with different rebound hammers. Rebound measurements offer a higher resolution than typical destructive testing methods and yield a velocity based rebound quotient (Q value), correlating to the compressive strength. Turning to older plain concrete, and especially to rammed concrete, the construction process per layer and successive condensing by manual or mechanical rammed implies a vertically dominated layout of the concrete density and composition within each layer and across layers. This layout is visually detectable in cut-outs and the height per layer usually ranges from 0.15 to 0.5 m (Rehm 2019; DIN 1045:1925-09). However, in view of the recess area size, the numerical simulations aim for a spatial and material field discretising in the range of centimetres. In order to obtain realistic distribution fields of correspondingly high resolution of rammed concrete, three series of rebound measurements were carried out on recent cut-out sections of two existing hydraulic structures and a test wall in 2020/21. The first structure is a double lock at the barrage Kachlet, situated at Donau-km 2.230 and in operation since 1928. Measurements were taken inside the massive lower head of the lock in the shaft partition during a repair measure. As the shaft excavations exposed relatively smooth vertical sections of quasi-unreinforced concrete, two measurement areas of about 2 and 3 m^2 at different height levels inside a massive hydraulic structure were available. The measurements at the Koblenz weir (Mosel-km 1.944, 1951) formed an edge area counterpart, offered by the construction of an inspection recess in vicinity to the first weir pillar. The rammed concrete test wall, courtesy of another research project from 2015 at the BAW headquarters in Karlsruhe, represents a relatively slim construction under laboratory conditions and encompasses three measurement areas. Although their horizontal extent is limited to less than half a meter, as for the recess, the cut-out section of the test wall enabled a horizontal comparison over a meter (cut-out length) as well as for the decimetre range (homogenisation depth per respective cut-out face). Due to a significant amount of non-valid measurements at the first measurements in Kachlet, the rebound hammer was switched and, contrary to DIN EN 12504-2:2012-12, the presented tests have been carried out with a rebound hammer of low impact energy to include low compressive strength values of less than 10 MPa. In order to record several individual layers and layer transition areas, measuring grid sizes of 0.3 m (Koblenz, Karlsruhe) to 0.5 m (Kachlet) match the spatial discretisation requirements of the numerical simulations in accordance with the minimal distance requirement (EN 12504-2:2012) of 2.5 cm between the individual measuring points. Due to the grid measurement, there was no pre-selection according to DIN EN 12504-2:2012-12, so areas were tested that would normally have been avoided due to rock grain, roughness, porosity, etc. Equally, test results were included even if the test caused subsiding of the test point. All in all, the measurements amounted to ~4000 single point values, each visually categorised (unannotated, stone, subsided on testing etc.) to form subsets per measurement area. For each area, the analysis differentiated ensembles by these subsets and their combinations. An example of the rebound results mapped on the measurement area is given in Fig. 3 on the left side with the face colour corresponding to the measured Q-values of the rebound hammer. Their edge colouring and line style indicate the visual category.

The main investigation objectives are the distribution characteristics, correlation length in vertical as well as horizontal direction respectively patterns or trends tailored to layers induced by the construction process. Furthermore, differences between the measurement objects are discussed. Analysing eventual layer patterns included the line wise comparison of median, mean and standard deviation and horizontal and vertical direction per (sub-)sets as well as semi-variograms, yielding a vertical correlation length in the decimetre range per area. Apart from outliers at rock grains and subsided points, the measurement areas support a horizontal layering as they yield a rather more homogeneous curve of mean values, standard deviations and medians in the vertical direction than per horizontal grouping. The repetition of a vertically sloped layer pattern is most pronounced in the measurements of the test wall, for which it is not only possible to match the measurement positions to the current component pictures, but also to validate the stipulated layers via matchup to the actual ones of the construction process as documented in the previous research project. The least pronounced layering was found at Kachlet, where the visual overlay in Fig. 3 suggests a division into an upper and lower subarea. Each subarea encompasses local clusters as well as distinctive counter layers of higher or lower means, highlighted in the slopes or stretched C-shapes in the respective horizontal means on the right side of Fig. 3.

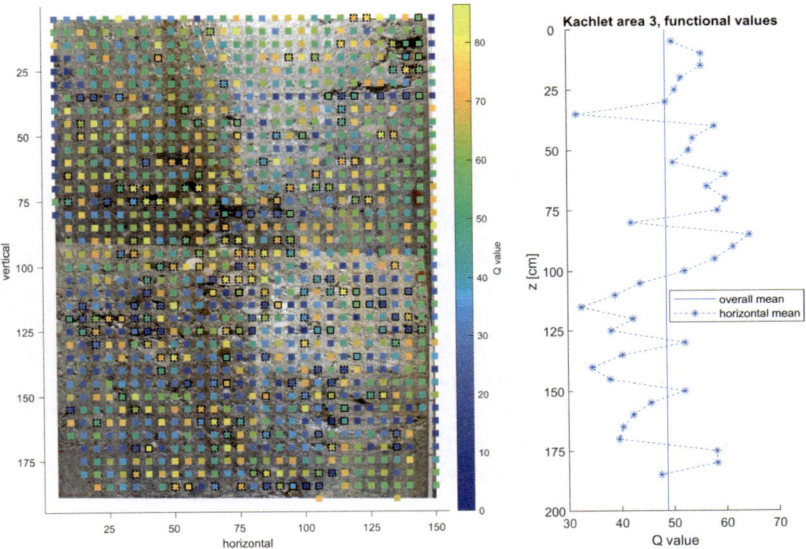

Fig. 3. Q-values of the rebound tests at Kachlet, left: colour-coded results with subsets indicated by line style (no line: unannotated, dotted line: stone, continuous line: subsided), right: horizontal mean values

Histograms and distribution fits of the (sub-)sets formed a starting point regarding the distribution shape, whereas Anderson-Darling-Tests and probability plots assessed whether the (sub-)sets follow rather a normal or lognormal distribution (types as supposed by Eurocode). Additionally, they visualised the ranges and the extent of deviations. Mostly, a lognormal distribution type similarity could only be attributed to unannotated ensembles and in case of the Kachlet measurements would classify only in rather lognormal than normal, especially due to higher values. The regularity within the measurement areas was approached by varying the sample ensembles for the median based classification in compressive strength classes detailed in DIN EN 13791/A20:2020-02.

Figure 4 compares histograms per subsets for the recess area at Koblenz with an inside area at Kachlet and a measurement area of the test wall, suggesting a logarithmic-like distribution for the functional values in the former case and a less obvious one at the latter. In all cases, the majority of the tests (Kachlet: $\sim 80\%$, Koblenz: $\sim 64\%$, test wall $\sim 92\%$) displayed no visual peculiarities. The contra intuitive differences between the visual markers and the test values can be ascribed to the remaining homogenisation in depth, e.g. if a rock grain was covered, it was not recorded as unannotated, even if the level of the rebound number suggests a rock grain directly underneath.

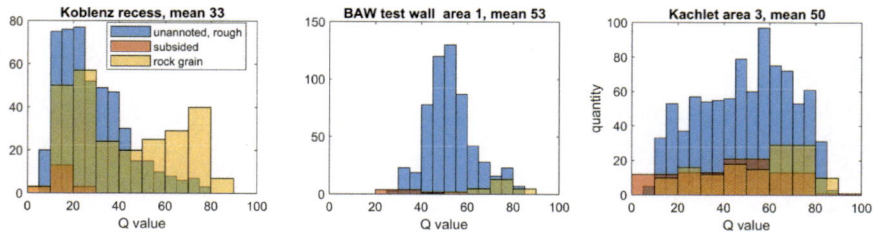

Fig. 4. Q-value histograms of the rebound tests at Koblenz (left), test wall (middle) and Kachlet (right) by subsets of functional, subsided and stone

2.2   Generic Distributions and FEM Implementation

The modelling of the spatial distribution of the material properties can be decisive for the calculated load-bearing capacity by prearranging probable load paths, and consequently resistance extent to specific loading conditions, or by inducing material weaknesses. For example, nests of gravel in the concrete can be represented by a cluster of higher values with reduced connectivity of the elements within and construction joints by a localisation in levels. The selection, implementation and evaluation of the distribution functions and parameters are therefore a focus of the research project. By generating random realisations of a distribution type with specific design parameters (log-normal distribution, characteristic value, standard deviation and population size), so-called populations are obtained as ensemble sets of material characteristics with different value compositions. In accordance with normative assumptions and the procedure assumptions to obtaining material characteristics in TbW, the base population of the compressive strength per structure is set as a logarithmic distribution. Furthermore, the strength classes of DIN EN 1992-1-1 imply a mean value corresponding to its 5% characteristic value plus 8 MPa. Each base population ensemble yields a large number of 3D distribution fields for the FEM model by random or structured mapping of the material property to the modelled component. These combinations of spatial allocation and material properties represent sub-ensembles per material distribution. The creation of these (sub-)ensembles as input files for the FEM calculation was automated in Matlab for random mapping and, based on literature and the rebound measurement analysis, a yet more realistically structured set-up is proposed as visualised in Fig. 5: The compressive strength population of the analysed component (e.g. loading area) follows the lognormal distribution of the left side, with the red bar indicating the characteristic value. The mapping divides into two population subsets. Firstly, the main pattern, constituted of the layers and layer transition areas, and secondly the strength outliers as deviations. As a simplified example, one possible division of the population is shown on the right side by color-coding. The main pattern without the deviations consists of two aspects: a vertical gradient in groups per cluster level and a horizontally uncorrelated distribution of these clusters per level.

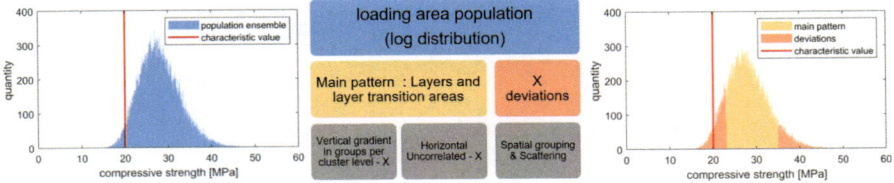

Fig. 5. Spatial mapping structogram for rammed concrete simulation, visualised by color-coded partitions of a histogram of one distribution

3   Numerical Simulation Set-Up

3.1   Non-linear Material Model

Various material models for geotechnical or concrete applications are already implemented in LS-Dyna (LSTC 2020). A first plausibility check regarding their suitability for the simulation of plain concrete behaviour included their capacity to reproduce specific damage for tensile and compressive loading. In further benchmarks, the material model "Continuous Surface Cap Model" (CSCM) was selected according to criteria relating to the tensile-compression behaviour as well as fracture patterns. The former behaviour was studied for single element tests and multi-element tests of cylinder and cube specimens, and the latter by comparing the damage and failure of the multi-element compound tests to general fracture patterns required for testing (DIN EN 12390-3:2009). CSCM offers a material parameterisation to be generated from compressive strength and maximum aggregate size. However, this primarily addresses the compressive strength range from 20 to 58 MPa, whereas hydraulic engineering and in particular its historical solid hydraulic structures also addresses significantly lower values. Parameter studies with the optimisation program LS-Opt facilitated the derivation of a correlation of the CSCM-generated compressive strengths for values down to and below 1 MPa. In numerical aspects, specifics as hourglassing parameters have been determined by sensitivity studies. Furthermore, series of simulations were carried out on cylinder models, beam models and a simplified model of a generic recess for the investigation of the 3D distribution influence. In these benchmarks, the selected material model, variations of the spatial distribution and distribution characteristics are evaluated regarding their damage behaviour and compared with standard calculations. Figure 6 depicts results of such a benchmark for a $2 \times 2 \times 10$ m beam under bending load by symmetrical single forces, represented in the simulation counterparts as two slowly advancing rigid cylinders. The dotted lines represent the analytical maximal force obtained by formula with the material values as in DIN EN 1992-1-1, whereas the histograms and their respective logarithmic fits visualise the simulation results of the matching 3D distributions. Although the number of simulations is rather low (50 per compressive strength), their bandwidth matches well in between of their encompassing quantile-based values.

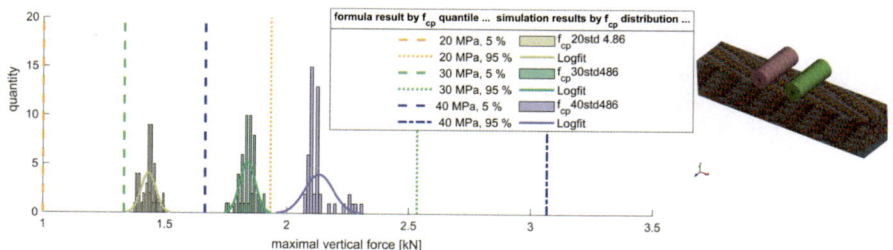

Fig. 6. Left: Calculated maximal vertical force of beams under bending for f_{cp} as in DIN EN 1992-1-1 compared to simulation result histograms and logarithmic fits of 3D distributions. Right: FEM model example.

3.2   FEM Simulations and Stochastic Analysis

For the example of a specific characteristic compressive strength, the FEM steps start with a distribution population generated with specific design parameters as shown in Fig. 5. This population contains the total amount of compressive strength values to be mapped as part properties with the respective generic material model parameters on the component's geometry in hundreds of spatially different variations, as visualised for a recess section in the left side of Fig. 7 for two variations. Each of these variations then experiences an identical loading scenario in LS-Dyna and fails somewhat differently due to their spatial strength configuration, with the differences encompassing damage patterns and scope of the failing elements as well as the failure mode itself in case of changing load and damage paths as depicted in Fig. 7 (middle). The scenario includes the embedding of the analysed part into the homogenous component with boundary conditions and gravity loading on the modelled structure as well as due to overlying structure parts. The loading itself models a continuously rising, corresponding to a rise in fictional water level, normal force application as depicted in Fig. 1. Key histories (time or value-based value curves, e.g. force-time, stress-strain) and virtual sensors (e.g. number of failed elements in a specific area) are defined as part of the FEM input for numerical dependencies and in preparation of the postprocessing. So far, the process described above would require the configuration of hundreds to thousands of scenarios and individual result analysis. As for the case of the material benchmarks simulations, the optimisation software LS-Opt facilitates the configuration and result analyses in order to evaluate all these variations in a parameterised manner. Such a set-up in LS-Opt includes on the pre-processing side the definition of the design space (parameters, their type and range), the sampling procedure, the parameter based spatial configuration and the LS-Dyna simulation control. On the post-processing side, histories are defined and extracted from the simulation runs for subsequent processing, as visualised for force trajectories per failed element number in Fig. 7 on the right. As a further step in this processing, single value responses are obtained, e.g. maximal value at a specified time or value. The approximation of such resulting value sets by meta-modelling methods enable a propagation between the individual simulation cases, e.g. between two recess depths or loading positions, and thereby the evaluation of a higher number of scenarios than actual simulations could deliver due to cpu and time limits.

These approximation methods span from cpu-friendly sequential approaches with linear basis functions up to Radial Basis Functions and Feedforward Neural Networks and yield quantitative and qualitative descriptions of the scenarios. In combination with failure classification, e.g. by response values, the probability of failure and the level of accuracy and reliability of the load cases can be determined in the overall context of the safety concept.

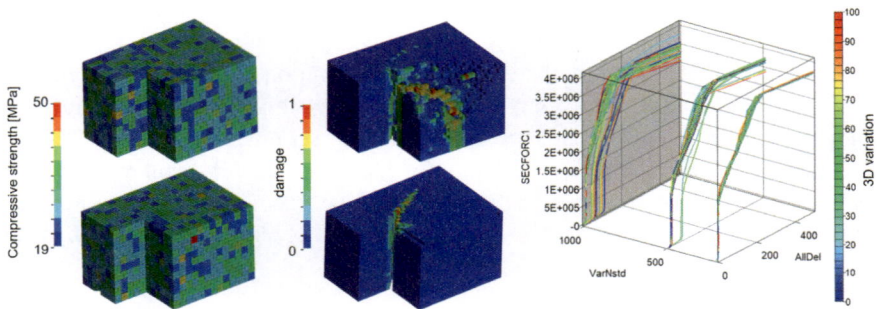

Fig. 7. Left to right: Color-coded compressive strength visualisations of two random 3D configurations of the same population. Failure variations differing only due to their 3D configurations. Section force per failed (deleted) elements with color-coding per 3D configuration.

3.3   Failure Classification

Contrary to the usually determined utilisation ratio or to individual case assessment, the proposed simulations require specific and automatable failure classification. For example, a partial damage per element could be acceptable, as it implies a change in bearing capacity without equalling a loss of bearing capacity. Due to robustness aspects as possible load path redundancies in the massive structures, even failure and thereby loss of bearing capacity of a few elements leads not unconditionally to the failure of the structural system. However, tipping points may be breached and transfer the structural system into a non-acceptable susceptibility. The requirements of failure classification are therefore divided in two parts: first the definition of qualitative indicators and secondly their combined quantitative description into an automated classification scheme, e.g. as the classifier approach implemented in LS-Opt. Indicators encompass testing related types as cracks (cohesion, element failure), movement at defined points etc. as well as numerically typical aspects as changes in energy types or element damage accumulation. The current simulations serve to validate the modelling approaches and to estimate the applicability of indicators for evaluation and cate-gorisation for the recess simulations. Furthermore, the demonstrator runs constitute a base of the overall set-up. For example, their results enable to identify combinations of the failure indicators, which are functional for constructing classifier-based sampling to reduce simulations scenarios in less susceptible design areas. This addresses the amount of simulation cases necessary to deduct the structural response over the design space depending on sensitivity analysis per parameter of the design space.

4 Conclusion and Outlook

Failure type and resistance capacity are both influenced by the material values ensembles and their 3D distributions. Reserves in bearing capacity are affirmative, however, simulations mirror the limitations of the 3D fields assumptions and the approximations due to material model and simulation set-up. This emphasises the relevance of the literature and measurement-based generation of 3D material parameter fields and of tailoring these fields for the rammed concrete typical for the existing solid hydraulic structures. Based upon showcase simulation set-ups, direct response values were successfully used as failure indicators for failure probability analysis. The next step in the project addresses the classification based upon combined failure indicators for the analysis of extended recess models reproducing component measures and compressive strength classes.

References

BAW (2016) BAW-Merkblatt: Bewertung der Tragfähigkeit bestehender, massiver Wasserbauwerke [BAW guideline: Assessment of the bearing capacity of existing, massive hydraulic structures]. Bundesanstalt für Wasserbau (BAW), Karlsruhe

Borosnyói A (2015) NDT assessment of existing concrete structures: spatial analysis of rebound hammer results recorded in-situ. Eng Struct Technol 7(1):1–12

Bouhjiti DE-M (2018) Analyse probabiliste de la fissuration et du confinement des grands ouvrages en béton armé et précontraint. Université Grenoble Alpes

DIN 1045:1925-09. Bestimmungen für Ausführung von Bauwerken aus Eisenbeton. Deutsche Fassung

DIN EN 12390- 3:2009. Prüfung von Festbeton – Teil 3: Druckfestigkeit von Probekörpern. Deutsche Fassung EN 12390-3:2009

DIN EN 12504-2:2012-12. Prüfung von Beton in Bauwerken – Teil 2: Zerstörungsfreie Prüfung – Bestimmung der Rückprallzahl. Deutsche Fassung EN 12504-2:2012

DIN EN 137912020-02. Bewertung der Druckfestigkeit von Beton in Bauwerken und in Bauwerksteilen. Deutsche Fassung EN 13791:2019

DIN EN 1992-1-1: Eurocode 2. Bemessung und Konstruktion von Stahlbeton- und Spannbetontragwerken

Teil 1-1: Allgemeine Bemessungsregeln und Regeln für den Hochbau. Deutsche Fassung EN 1992-1-1:2004 + AC:2010

EC0 (2002) EN 1990:2002–04: Eurocode 0: Basis of structural design. European Committee for Standardization, Brussels

Fib (2012) Model Code 2010 - Final draft, vol 1. International Federation for Structural Concrete (fib), Lausanne, p 350

JCSS (2002) JCSS probabilistic model code. Part 3: Resistance models. Joint Committee on Strutural Safety. https://www.jcss.byg.dtu.dk/Publications/Probabilistic_Model_Code

Kersken-Bradley M (1985) Sicherheit von Baukonstruktionen. Handbuch der Sicherheitstechnik, Band 1, O. Peters and A. Meyna, Hanser, München, pp 253–334

KW011A-SG01-4 (2016) Ausführungsplan Anker Wehrfeld B, Zeichnungsnr. KW011A-SG01-4, Wasser und Schifffahrtsamt Regensburg

De Larrard T (2010) Variabilité des propriétés du béton: caractérisation expérimentale et modélisation probabiliste de la lixiviation. École normale supérieure de Cachan

LSTC (2020) LS-DYNA® Keyword User's Manual, Volume II: Material Models. LS-DYNA R12, Livermore Software Technology (LST), California

Rehm J (2019) Eisenbeton im Hochbau bis 1918: Dokumentation und Analyse realisierter Bauwerke im Raum München. TUM University Press

Effect of River Regulation on Navigable Flow Conditions for River Bend

Honghao Fan[1], Jianjun Zhao[2], Jianguo Ye[3], Xianzhong Chen[4], Jintao Fang[2], Yu Wang[2(✉)], and Xingwei Zheng[1]

[1] Quzhou Jujiang River Shipping Construction and Development Co., Ltd., Quzhou, China

[2] State Key Laboratory of Hydrology-Water Resources and Hydraulic Engineering, Nanjing Hydraulic Research Institute, Nanjing, China
wangy@nhri.cn

[3] Quzhou Communications Investment Group, Quzhou, China

[4] Changshan River Waterway Development and Construction Headquarters, Quzhou, China

Abstract. The layout of the navigable junction on the curved river was relatively difficult. The combination of the curved flow and the oblique flow would increase lateral flow velocity in the entrance area, which affected the navigable safety of the ship. There was a bend in the entrance area of the Zhaoxian Junction downstream and a large area of flood plain on the right bank of the downstream channel, which contributed to hardly meet the navigation requirements due to flow coming straight towards waterway affected by flood plain squeezed and river bend. In this paper, normality hydraulic model with 1/80 scale was employed to experimentally optimize the influence of different river regulation measures on navigable flow conditions considering the characteristics of river bend and another situation. The effective river regulation measures were proposed by moving sluice to the left, dredging flood plain on the right bank of sluice downstream and adjusting the operation mode of sluice gate. It was verified that dredging the downstream flood plain to 65 m elevation combined with the operation of left sluice can meet navigable flow conditions requirements. The research results could be used as a reference for relevant junction projects.

Keywords: River regulation · Navigable flow conditions · River bend

1 Introduction

The river bend section is a common plane form of mountainous rivers. The navigable safety problems of the junctions located on the narrow and the curved river sections are relatively serious. The main difficulty in junction layout lies in the complex navigable flow conditions of entrance area, especially river regime with large areas of flood plain. River regulation is an important technical means to ensure safe navigation (Guo et al. 2015).

The navigable junction was generally arranged on the straight transition section or curve in the middle of the two curves for curved river. The entrance area of upstream and downstream approach channel was located on the convex or concave bank of the river thanks to the limitation of terrain, which caused field concern to the layout of the

© The Author(s) 2023
Y. Li et al. (Eds.): PIANC 2022, LNCE 264, pp. 384–393, 2023.
https://doi.org/10.1007/978-981-19-6138-0_34

navigable buildings. The comparison of flow conditions between concave bank and convex bank demonstrated that it was advisable to adopt a dispersed arrangement on the side of the convex bank when the junction was located in the downstream bend segment for the narrow continuous bend river segment with sharp bends in the upstream and reverse bends in the downstream (Pu et al. 2012). For mountainous rivers, some studies noted that navigable buildings should be arranged on concave banks (Cao and Zhou 2012). On basis of the comparison of the ship locks arrangement on the left or right banks of Naji Junction, it was more appropriate to be located on the mainstream side of the concave bank (Zheng and Chen 2005). Furthermore, improvement measures of navigable flow conditions should be taken according to the type of river bending (Yu et al. 2014). Han et al. (2014) proposed the junction position should be adjusted if the entrance area of approach channel was set on the convex bank on basis of Yongning Project. Li et al. (2016) carried out the optimization experiment of flow conditions in the upstream and downstream approach channels for ship lock arranged on the concave bank, and proposed the optimization measures of the boundary excavation and the setting of the permeable guidance wall.

In fact, the method of sluice operation had a significant impact on the navigable flow conditions (Wu et al. 2016). For the junctions which the ship lock was built on the convex bank, opening of the remote sluice hole could effectively reduce the backflow and oblique flow in the entrance area of the approach channel (Zhang et al. 2021). The simulation of navigable flow conditions adopted physical model test and numerical simulation (Ahmed et al. 2010; Zhang et al. 2018). On basis of the overall hydraulic model experiment of Zhaoxian Project, this paper analyzed flow regime of entrance area and connecting section on the concave bank of curved river section downstream, and proposes the engineering improvement measures.

2 Project Summary

Zhaoxian Junction, the fourth cascade project of the Changshan River cascade development in the upper reaches of the Qiantang River, was located in Changshan Country, Quzhou. The engineering layout from left to right was: ship lock, sluice and power station, as shown in Fig. 1. The left section of sluice set 13 holes with the baseboard elevation of 66.00 m, and left section of sluice set 8 holes with the baseboard elevation of 68.00 m. The net width of sluice was 252 m.

The lock chamber of Zhaoxian Junction was located on the left bank. The length of ship lock was 256 m, of which the head of the upper lock was 36 m, the lock chamber was 190 m, and the head of the lower lock was 30 m. The upstream approach channel was open without separation levee, and the layout was "straight in and curved out". The length of straight section of upstream approach channel was 330 m, of which the navigation adjustment section was 140 m, the breasting section was 190 m, and the bottom width of the upstream approach channel was 60 m. The downstream approach channel adopted the layout of "straight in and curved out". The total length of the approach channel was 330 m containing 140 m navigation adjustment section and 190 m breasting section, and the bottom width of the downstream approach channel was 60 m.

The characteristics of Zhaoxian Junction embodied: 1) the sluice and power station protruded to the right bank, and flood channel was relatively narrow. 2) There was a large area of flood plain on the right bank with the 66.30–73.80 m elevation, which was not conducive to discharge flood and should be partially dredged (see Fig. 2). 3) There are bends in vicinity of downstream entrance area and the Wuli High-speed Bridge, therefore, flow conditions were more complicated.

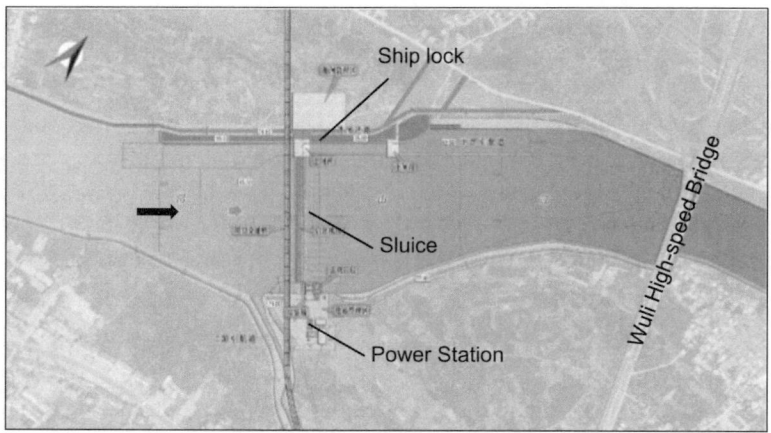

Fig. 1. Layout of Zhaoxian Junction

Fig. 2. Flood plain of Zhaoxian Junction

3   Experimental Setup and Methodology

3.1   Experimental Setup

The test model was designed at the geometric scale of 1/80 on basis of Froude similarity criterion according to the similarity law of model, the terrain conditions of junction upstream and downstream and the laboratory site. The Zhaoxian Junction was located the gentle S-shaped river section in the middle of the slightly curved river channel, therefore, the upper boundary of physical model was arranged at 1600 m upstream of sluice taking into account the smooth inflow and navigable flow conditions.

Wuli High-speed Bridge, situated in Zhaoxian Junction downstream with 40 m navigable net width, adjoined neighbor the bend of downstream approach channel, which contributed to complex flow conditions. Consequently, the lower boundary of physical model should be taken to 2160 m downstream of sluice considering the length of sufficient flow adjustment.

The section method is used for terrain production, and the triangulation wire system is employed for plane stakeout where the triangle closure error hardly exceeded ±1'. The model elevation was measured by a level and checked during the production process, and the installation elevation error of the section was controlled within ±1 mm. For complex terrain, the intensive section was added and processed separately in order to improve the production accuracy. The junction, including sluice, ship lock and power station, was made of PMMA. The surface of the river was made of cement mortar. The overall model was photographed in Fig. 3.

The flow velocity measurement adopted the acoustic Doppler flow meter produced by Nortek Company with a range of 0.1–400 cm/s. The flow discharge was measured by a standard rectangular thin-walled weir with an error of less than 1%; the velocity was measured by a water level stylus with an accuracy of 0.02 mm.

Fig. 3. Model photo of Zhaoxian Junction

3.2   Experimental Cases

Table 1 listed the experimental cases for navigable flow conditions of the Zhaoxian Junction. The water level of downstream of sluice was controlled on basis of the numerical simulation of stage-discharge relation curve. Preliminary experiment ensure the method of sluice operation.

Table 1. Experimental cases of Zhaoxian Junction

Cases no.	Flow discharge (m³/s)	Water level (m)	Sluice operation
1	1000	70.30	2.32 m opening for six holes
2	800	69.83	1.92 m opening for five holes
3	600	69.27	1.52 m opening for four holes
4	400	68.65	2.08 m opening for two holes
5	200	67.96	1.68 m opening for one hole

4   Results and Discussions

4.1   Downstream Flow Conditions of Initial Layout

For $Q = 1000$ m³/s with the opening of #1–#6 holes in the left section, there would be backflow within the range of 200 m downstream from the entrance area of the downstream approach channel. The longitudinal velocity of the approach channel generally exceeded 2.0 m/s within the range of 240 m to 320 m downstream from the entrance area of approach channel, and the lateral velocity in the left section was relatively large, which seldom met the navigable requirements.

The reason for the above phenomenon was that the downstream level of river channel was lower than 70.3 m, and the flood plain (higher than 71 m elevation) on the right bank was not completely submerged, which resulted in concentrated flow discharged between ship lock and flood plain. The concentrated flow rushed straight towards entrance area and connecting section of downstream approach channel, where developed a large-scale backflow in vicinity of the entrance area of the downstream approach channel, as shown in Fig. 4. Additionally, the mainstream in concentrated on the channel in vicinity of the Wuli High-speed Bridge area due to the high terrain on the right bank of the river channel, which led to large longitudinal velocity and large flow resistance of upward ships. Therefore, the rudder efficiency was poor, and the control difficulty increased.

Fig. 4. Downstream flow regimes of initial layout ($Q = 1000$ m³/s)

4.2   Downstream Navigable Flow Conditions for Sluice Moving to Left

The sluice was moved 32 m to left due to the flood plain on downstream right bank. Table 2 listed downstream navigable flow conditions. The data demonstrated lateral velocity exceeded navigation requirements in 200 m–360 m downstream from entrance for $Q = 600$ m³/s–1000 m³/s. In general, navigable flow conditions had improved with sluice moving to left, whereas navigation requirements remained unsatisfied. It is mainly manifested in the following aspects: 1) flow out of sluice was not smooth, and the overflow on the floodplain was small. The blocking effect of the floodplain on the right bank precipitated flow to the left, which easily caused scouring of the riverbed in vicinity of lock wall and separation levee. 2) Concentrated flow rushed straight towards entrance area and connecting section of downstream approach channel as similar to Fig. 4. Therefore, the dredging on the right bank of the river channel was mainly considered in the optimization of downstream navigable flow conditions.

Table 2. Navigable flow conditions with sluice moving 32 m to left

Discharge (m³/s)	Range (Distance downstream from entrance)	Lateral velocity (m/s)	Longitudinal velocity (m/s)	Backflow velocity (m/s)
1000	160 m	<0.30	<1.54	<0.37
	200 m–360 m	0.38–0.80	>2.00	<0.32
600	160 m	<0.32	<1.45	<0.27
	200 m–360 m	0.31–0.77	<1.92	<0.29

4.3   Dredging Downstream Flood Plain to 67 m Elevation

Fig. 5. Flow regimes of downstream entrance area with dredging flood plain to 67 m elevation ($Q = 800$ m^3/s)

Figure 5 photographed flow regimes of downstream entrance area with dredging downstream flood plain to 67 m elevation, and Table 3 listed the data of navigable flow conditions. It could be seen that the dredging of flood plain partly decreased flow velocity and improved flow regimes, however, the main trough was still close to ship lock. The flow diffusion into entrance area brought about relatively large flow velocity. Moreover, blocking effect on the right bank upstream from Wuli High-speed Bridge still existed due to the insufficient dredging range of flood plain, which narrowed the flow cross-section in entrance area downstream and mainstream of the river in vicinity of entrance area was biased towards the left bank. From the data point of view, the lateral velocity still exceeded navigable requirements in a certain range, and we could continue to further deepen the dredging range.

Table 3. Navigable flow conditions with dredging flood plain to 67 m elevation

Discharge (m^3/s)	Range (Distance downstream from entrance)	Lateral velocity (m/s)	Longitudinal velocity (m/s)	Backflow velocity (m/s)
1000	160 m	0.30–0.53	>2.00	<0.40
	200 m–280 m	<0.29	<1.71	No backflow
800	160 m	0.31–0.49	>2.00	<0.24
	200 m–280 m	<0.19	<2.00	No backflow
600	160 m	<0.41	<2.09	<0.29
	200 m–360 m	<0.21	<1.82	No backflow
400	160 m	<0.29	<1.93	<0.24
	200 m–360 m	<0.17	<1.55	No backflow

4.4 Dredging Downstream Flood Plain to 67 m Elevation

Figure 6 photographed flow regimes of downstream entrance area with dredging downstream flood plain to 67 m elevation, and Table 4 listed the data of navigable flow conditions. It could be seen that for $Q = 800$ m^3/s–1000 m^3/s with dredging downstream flood plain to 65 m elevation, the longitudinal velocity and backflow velocity in entrance area and connecting section of the approach channel met navigable requirements, and lateral velocity slightly exceeded navigable requirements within the range of 80 m–200 m downstream from the entrance. Nevertheless, the maximum lateral velocity was less than 0.39 m/s, and the scope of influence was slight. Therefore, the navigable flow conditions of the entrance area and connecting section basically met the requirements under this condition.

Table 4. Navigable flow conditions with dredging flood plain to 65 m elevation

Position	Discharge (m^3/s)	Maximum lateral velocity (m/s)	Maximum longitudinal velocity (m/s)	Maximum backflow velocity (m/s)
Mooring area	200	0.03	0.09	0.02
	400	0.13	0.30	0.13
	600	0.10	0.25	0.25
	800	Overall 0.17, local 0.23	0.28	0.19
	1000	Overall 0.18, local 0.25	0.19	0.24
Entrance area and connecting section	200	0.22	0.58	0.28
	400	Overall 0.22, local 0.34	1.17	0.32
	600	Overall 0.25, local 0.33	1.56	0.33
	800	Overall 0.30, local 0.36	1.81	0.35
	1000	Overall 0.30, local 0.39	1.83	0.36
Wuli High-speed Bridge	200	0.13	0.96	No backflow
	400	0.11	1.41	
	600	0.15	1.21	
	800	0.18	1.89	
	1000	0.18	1.85	

Fig. 6. Flow regimes of downstream entrance area with dredging flood plain to 65 m elevation ($Q = 1000$ m³/s)

5   Conclusions

1) Zhaoxian Junction was located on the gentle S-shaped river section in the middle of the slightly curved river channel. The downstream approach channel was located on the concave bank, and there was flood plain on downstream right bank. Relatively large intersection existed between the centerline of the approach channel and the mainstream direction, which resulted in large longitudinal velocity and lateral velocity and affected the safety of ship navigation. It was necessary to take engineering measures to improve the navigable flow conditions in the entrance area to ensure the safety of ships entering and leaving the lock.

2) By moving the sluice to the left and dredging the downstream flood plain, longitudinal velocity, lateral velocity and backflow velocity in the entrance area could be significantly reduced, and the navigable flow conditions could be effectively improved. The experimental results showed that the maximum navigable flow discharge of the initial and left-moving layout could only reached 400 m³/s. By means of effective measures, For $Q \leq 1000$ m³/s, hydraulic indexes of the entrance area and connecting section of the upstream and downstream approach channels met the navigation requirements, and navigable flow conditions had been significantly improved.

Acknowledgements. This research has been financially supported by the National Natural Science Foundation of China (Grant no. 52009083), Natural Science Foundation of Jiangsu Province (Grant no. BK20200159).

References

Ahmed HS, Hasan MM, Tanaka NT (2010) Analysis of flow around impermeable groynes on one side of symmetrical compound channel: an experimental study. Water Sci Eng 3(1):56–66

Cao YF, Zhou HX (2012) Analysis of navigation condition of the lack located along the concave and convex banks of bend channel. Hydro-science Eng (4):77–81 (in Chinese)

Guo HM, Xia XB, Cao GC, Cai LM (2015) River remediation study about improving navigable flow condition of Shipai sharp turn. Water Resour Power 33(3):86–88 (in Chinese)

Han CH, Yang Y, Li YF, Li QX (2014) Layout of navigable channel of navigation-power junction on the convex bank with slightly bended river section. Port Waterway Eng (10):121–125 (in Chinese)

Li J, Zhao JJ, Hong J, Xuan GX, Wang XD (2016) Flow condition optimization for upstream and downstream approach channels when ship lock lies on concave bank. Port Waterway Eng (12):101–105 (in Chinese)

Pu XG, Li M, Li JT, Hao YY (2012) Study on layout plan of navigation-power junction in narrow continuous meandering river. J Waterway Harbor 33(1):39–44 (in Chinese)

Wu ZY, Jiang CB, Chen J, Deng B, Yang W (2016) Influence of sluice gate opening mode on navigation flow condition. Adv Sci Technol Water Resour 36(3):73–77 (in Chinese)

Yu ZG, Han CH, Peng CB, He TY, Fan LE (2014) Overview of bent waterway navigation hub layout research. Pearl River 35(6):102–105 (in Chinese)

Zhang Y, Du QJ, Li ZP, Zhao JY (2021) Experimental study on influence of discharge lock dispatching mode on navigable flow conditions of convex-bank ship lock. Port Waterway Eng (4):97–101 (in Chinese)

Zhang S-H, Wu Y, Jing Z, Yi Y-J (2018) Navigable flow condition simulation based on two-dimensional hydrodynamic parallel model. J Hydrodyn 30(4):632–641. https://doi.org/10.1007/s42241-018-0062-1

Zheng BY, Chen B (2005) Test on navigation flow condition of curving connect section of upstream lock entrance. J Waterway Harbor 26(2):99–102 (in Chinese)

Evolution of Undular Surges in a Navigation Channel

Feidong Zheng[1(✉)], Xueyi Li[1], Fan Zhang[1], and Ping Mu[2]

[1] College of River and Ocean Engineering, Chongqing Jiaotong University, Chongqing, China
feidongzheng@126.com, xy_lee@cqjtu.edu.cn
[2] Key Laboratory of Hydraulic and Waterway Engineering of the Ministry of Education, Chongqing Jiaotong University, Chongqing, China
muping@cqjtu.edu.cn

Abstract. An undular surge is a secondary wave characterized by free-surface undulations over the body of a positive wave. The formation of undular surges is strongly linked to the departure of pressure distribution from hydrostatic due to the combined effects of wave nonlinearity and dispersion. In navigation channels, undular surges may be generated by the filling or emptying operation of lock chambers. The propagation of undular surges is associated with the periodic flow variations, and therewith they can cause impulse motion responses of navigation ships and even navigation accidents in some instances. In the present study, the free-surface and hydrodynamic properties of undular surges induced by the emptying operation of lock chambers were experimentally investigated. Detailed free-surface and velocity measurements were performed with a series of capacitance wave gauges and an acoustic Doppler velocimeter in a horizontal rectangular water wave channel. Both nonbreaking and breaking surges were recognized and analyzed. The results demonstrated that the wave face of a nonbreaking undular surge could be accurately described by solitary wave theory. However, the velocity distributions beneath the surge deviated significantly from the solitary wave solution. Based on the experimental results, a formula relating the longitudinal velocity component to free-surface elevation was established. Overall, this study provided new insights into the analogy of an undular surge and a solitary wave.

Keywords: Undular surge · Secondary wave · Navigation channel · Lock chamber · Solitary wave

1 Introduction

An undular surge is a secondary wave characterized by a train of undulations over the body of a positive wave (Peregrine 1966). The occurrence of undular surges is strongly linked to the departure of pressure distribution from hydrostatic due to the combined effects of wave nonlinearity and dispersion (Castro-Orgaz and Chanson 2020; Soares Frazao and Zech 2002). In navigation channels, undular surges may be generated by the filling or emptying operation of lock chambers (Treske 1994). The propagation of undular surges is associated with the periodic flow variations, and therewith they can

© The Author(s) 2023
Y. Li et al. (Eds.): PIANC 2022, LNCE 264, pp. 394–405, 2023.
https://doi.org/10.1007/978-981-19-6138-0_35

cause impulse motion responses of navigation ships and even navigation accidents (Zheng et al. 2021; Zheng et al. 2018). Moreover, these surges may lead to unexpected interaction events with downstream structures (e.g., lock gates, sluices), which may cause significant runup and impose additional impact loads on these structures (Zheng and Li 2021).

In light of the potential impact posed by undular surges, their evolution mechanism has been widely investigated in recent decades. The generation and development of undular surges advancing in still water were studied by several researchers, for example, Favre (1935), Benet and Cunge (1971), Treske (1994), Zheng et al. (2018), Zheng et al. (2021). Recently, most of the research effort has been devoted to the evolution of undular surges propagating against a steady flow. Herein, we can cite Koch and Chanson (2009, 2008), Chanson (2010a, 2010b), Gualtieri and Chanson (2012, 2011), Leng and Chanson (2017). However, most of previous studies were limited to qualitative analysis of the free-surface and velocity characteristics, quantitative information is quite sparse. Furthermore, the inherent relationship between the wave field and velocity field is still not fully understood. Nevertheless, the identification and quantification of these aspects are essential for engineering practice. The objective of this study is to quantitatively analyze the evolution of undular surges advancing in still water.

2 Experimental Methodology

Figure 1 shows the experiment configuration in a right-handed coordinate system $oxyz$. The experiments were performed in a rectangular horizontal water wave tank ($L \times W \times H = 38$ m $\times$ 0.3 m $\times$ 0.35 m), which was made of smooth polyvinyl chloride bed and plexiglass walls. The tank was connected to a reservoir through a pressurized pipe, where a plate gate and its shaft were equipped. When opening the gate at a given time t_v, the water in the reservoir flowed into still water in the tank and subsequently generated a solitary-like long wave. During the propagation of the wave, undulations grew at the wave front. In this study, the experiment conditions were: still water depth in the tank $d_0 = 0.08$ m; drop height $H_u = 0.35$ m; gate lifting time $t_v = 40$ s. These parameters were selected to generated both nonbreaking and breaking undular surges.

Two measurement systems were used to record the wave properties along the tank centerline. The tank was instrumented with capacitance wave gauges at thirteen locations to capture the entire evolution of an undular surge. The wave probes were located at $x = 3, 6, 9, 12, 15, 18, 21, 24, 27, 30, 33, 36$ and 37 m. An acoustic Doppler velocimeter NortekTM Vectrino + (serial number VCN9235) was used to determine the instantaneous velocity measurements at $x = 12, 21$ and 30 m. For each longitudinal location, velocity data were recorded at five vertical elevations, that is, $z/d_0 = 0.08$, 0.13, 0.41, 0.71 and 0.91. The unsteady measurements were performed with a sampling frequency of 200 Hz.

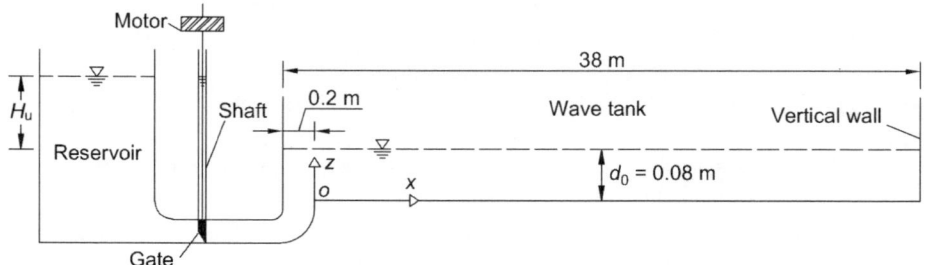

Fig. 1. Sketch of the experimental facility.

3   Results and Discussion

3.1   Basic Flow Patterns

The measured wave profiles during wave propagation are presented in Fig. 2, where the normalized water depth, d/d_0, was plotted against the normalized wave propagation time from gate lifting $t(g/d_0)^{0.5}$. g denotes the gravitational acceleration. Note that the wave profiles measured at $x \geq 12$ m are vertically offset by 0.8 for clarity. The emptying operation of the reservoir was associated with temporal changes in the discharge. The discharge increased up to a maximum value and subsequently decreased proportionally to the difference in water elevation between the tank and the reservoir. The increase in discharge generated a positive surge, whereas the subsequent decrease in discharge resulted in a negative surge after the former surge. Hence, the emptying of the reservoir led to the formation of a solitary-like wave with a single wave crest (e.g., $x = 3$ m). During the wave downstream propagation, the wave face of the positive surge tended to steepen and subsequently disintegrated into a series of well-formed undulations at $x = 12$ m (i.e., a nonbreaking undular surge). The undular surge underwent a significant amplification over propagation distance (e.g., $x = 18$ and 24 m) and finally evolved into a breaking undular surge at $x = 27$ m. Note that the wave amplitude of a breaking undular surge was much smaller that of the nonbreaking undular surge immediately before the apparition of wave breaking at the leading wave crest (e.g., $x = 30$ m).

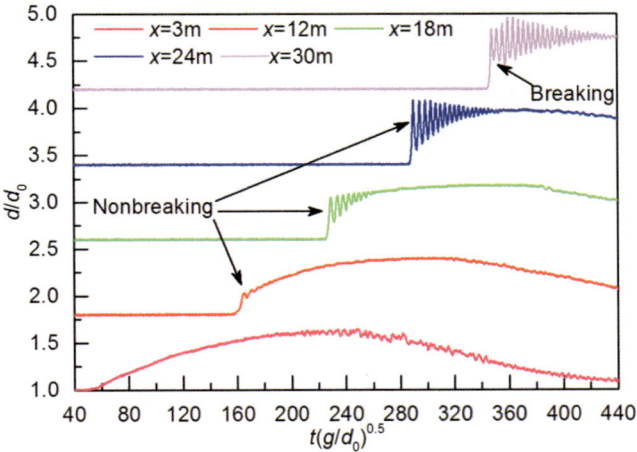

Fig. 2. Time evolution of wave profiles.

3.2   Free-Surface Characteristics

As the leading wave of an undular surge exhibited a solitary-like profile, an analogy technique with a solitary wave was used to characterize the spatial evolution of the leading wave over propagation distance. If the leading wave front was seen as a part of a solitary wave of the same amplitude, its wave shape could be depicted using the following mathematical expression (Montes 1998):

$$d = (d_{1c} - d_0)\mathrm{sech}^2\left(\sqrt{\frac{3(d_{1c} - d_0)}{4d_0^3}}(x - c_s t)\right) + d_0 \tag{1}$$

where,

$$c_s = \sqrt{gd_0}\left(1 + \frac{1}{2}\frac{(d_{1c} - d_0)}{d_0} - \frac{3}{20}\left(\frac{(d_{1c} - d_0)}{d_0}\right)^2\right) \tag{2}$$

d_{1c} denotes the water depth at the leading wave crest. The measured undular surges were compared to theoretical solutions in Fig. 3, where the abscissa was the normalized time relative to the leading wave crest. Notably, the leading wave front of a non-breaking undular surge (e.g., $x \leq 27$ m) could be satisfactorily described by the solitary wave theory. It can be also seen from Fig. 3 that the wavefront of a breaking undular surge was close to a solitary wave of the same amplitude (e.g., $x \leq 30$ m).

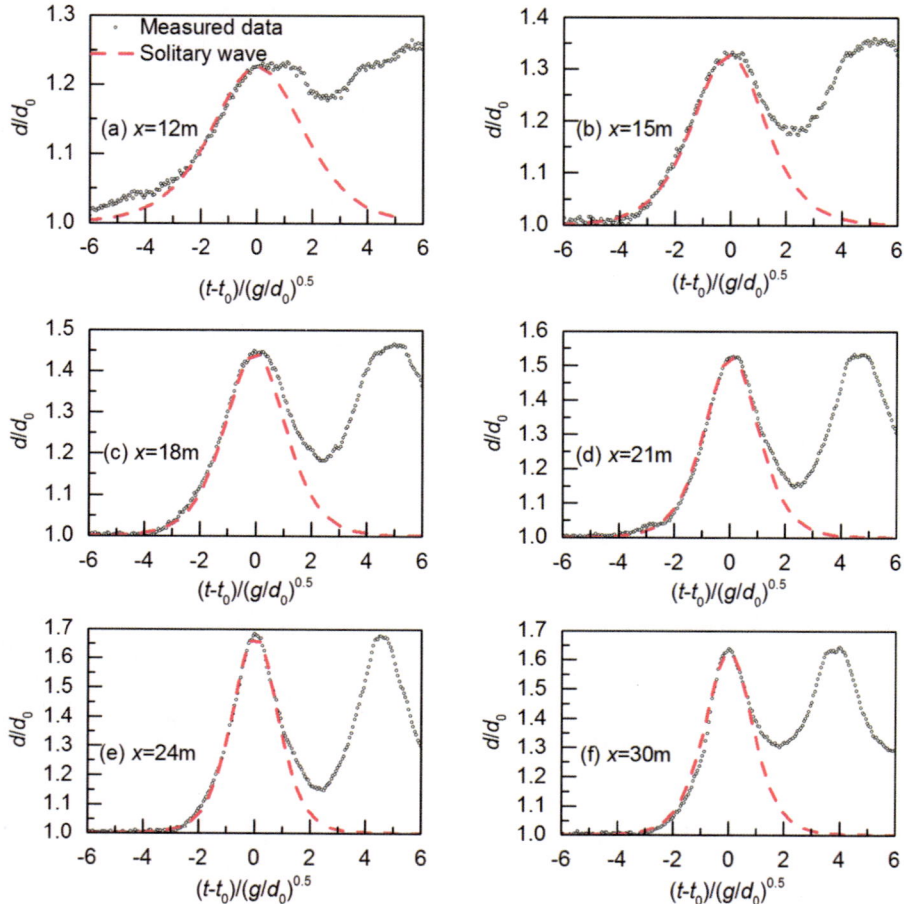

Fig. 3. Comparison of measured wave profiles with solitary wave solutions.

In undular surge flows, the mean water depth behind the leading wave d_m was usually defined as the average depths at the leading wave crest d_{1c} and its following wave trough d_{1t}. In Fig. 4, the normalized leading wave amplitude $(d_{1c} - d_0)/d_0$ was plotted against the normalized mean depth d_m/d_0 for two types of undular surges. For nonbreaking undular surges, the leading wave amplitude exhibited a linear increase with an increasing mean water depth. However, the leading wave height underwent a sharp decrease immediately after the occurrence of wave breaking and subsequently exhibited less variations during breaking surge propagation. It is believed that the maximum leading wave amplitude took place shortly before wave breaking at the wave crest. Noteworthy, the value of $(d_{1c} - d_0)/d_0$ observed at $x = 24$ m was close to the breaking limit of a solitary wave defined by McCowan (1894) (i.e., $(d_{1c} - d_0)/d_0 = 0.78$).

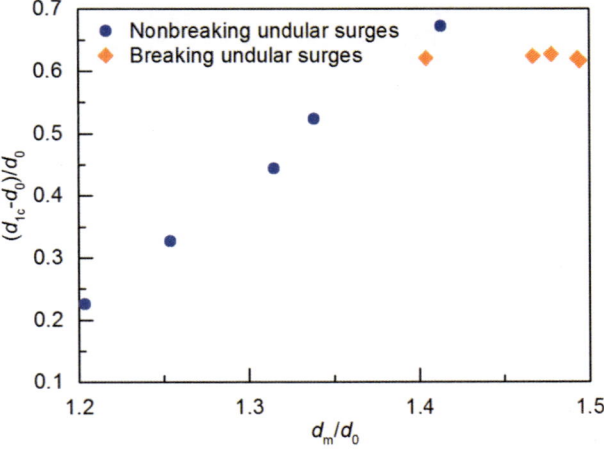

Fig. 4. Leading wave amplitude evolution during surge propagation.

The wave amplitude of an undular surge a_w was defined as half of the difference between the water depths at the leading wave crest and its following trough. The dependence of the normalized wave amplitude a_w/d_0 on the normalized mean depth is presented in Fig. 5. In this figure, the analytical solution of Lemoine (1948) based on the linear wave theory was also exhibited for comparison. As d_m/d_0 increased, a_w/d_0 increased for nonbreaking undular surges, whereas it decayed for breaking undular surges. Within the range of present investigations, the values of a_w/d_0 were significantly lower than the theoretical values predicted by the linear wave theory. Notably, while the experimental trend for nonbreaking undular surges was close to the linear wave theory, the data tended to suggest a significantly faster increase in a_w/d_0 with d_m/d_0.

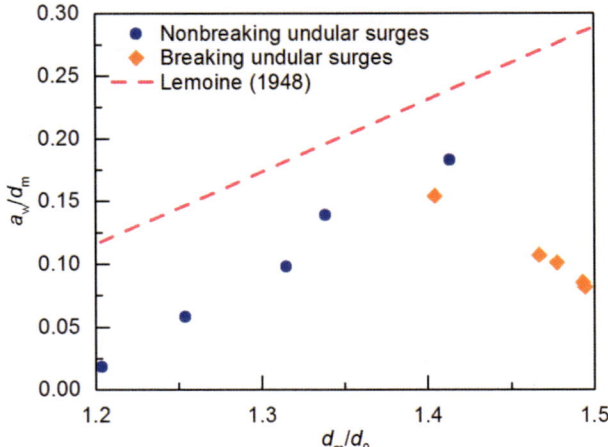

Fig. 5. a_w/d_m as a function of d_m/d_0.

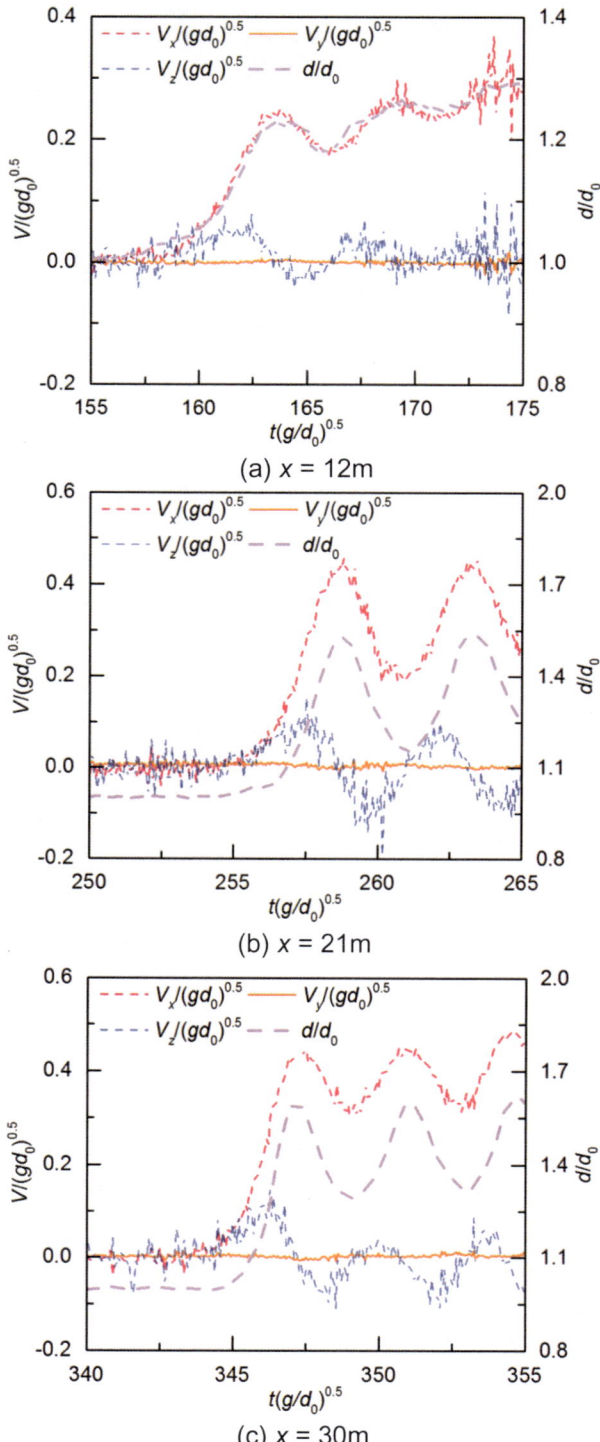

(a) $x = 12$m

(b) $x = 21$m

(c) $x = 30$m

Fig. 6. Variations in the instantaneous velocity components induced by undular surges.

3.3   Instantaneous Velocity

Figure 6 illustrates the typical time-variations of the instantaneous velocity components beneath undular surges at three longitudinal locations. Each graph includes the normalized velocities $V_x/(gd_0)^{0.5}$, $V_y/(gd_0)^{0.5}$, and $V_z/(gd_0)^{0.5}$, and the normalized water depth, d/d_0, as functions of the normalized time, $t(g/d_0)^{0.5}$. Note that the undular surges observed at $x = 12$ and 21 m were in the nonbreaking surge region, whereas the undular surge at $x = 30$ m was in the breaking surge region. Typical flow properties were clearly observed. The instantaneous longitudinal velocity component showed an oscillating pattern with free-surface undulations with minimum and maximum velocities obtained below the wave troughs and crests, respectively. The instantaneous transverse velocity component tended to fluctuate with the free-surface undulations. The instantaneous vertical velocity component tended to oscillate with the free surface and was very close to its null value below the wave crests and troughs.

The present experimental results demonstrated that the front face of a nonbreaking undular surge matched the solitary wave theory. Therefore, a detailed investigation on the longitudinal velocity component beneath the leading wave front was performed. At each normalized vertical elevation z/d_0, velocity data were recorded for eight series of characteristic moments, denoted $(t - t_0)/(g/d_0)^{0.5} = -3.5, -3.0, -2.5, -2.0, -1.5, -1.0, -0.5, 0$. Note that $(t - t_0)/(g/d_0)^{0.5} = 0$ indicated the arrival of the leading wave crest. The normalized longitudinal velocity component beneath the leading wave front was plotted against the free-surface elevation in Fig. 7. In this figure, the present data was also compared to the second-order solitary wave solution derived by Laitone (1960). It is evident that the experimental results were in qualitative but not in quantitative agreement with the solitary wave theory. Within the range of present investigations, the longitudinal velocity component beneath the leading wave front could be described by (Fig. 8):

$$\frac{V_x}{\sqrt{gd_0}} = 0.65 \left(\frac{d - d_0}{d_0} \right)^{0.73} \tag{3}$$

This finding indicated that the wavefront of an undular surge could not be treated as a portion of a solitary wave of the same wave amplitude.

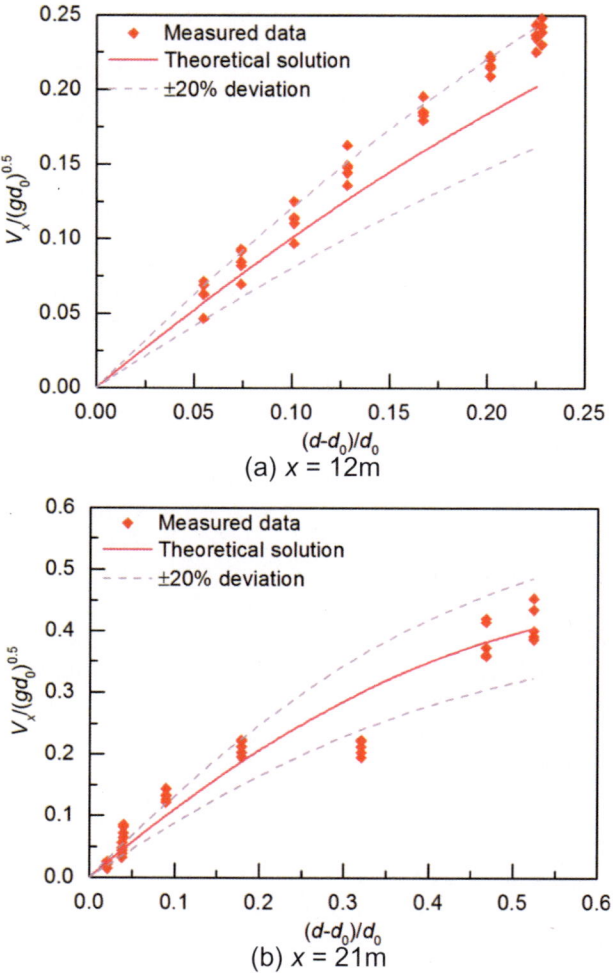

Fig. 7. Variation of $V_x/(gd_0)^{0.5}$ with $(d-d_0)/d_0$ for nonbreaking undular surges.

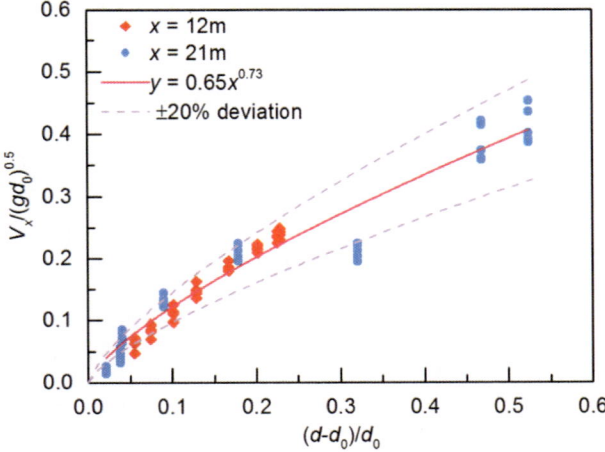

Fig. 8. $V_x/(gd_0)^{0.5}$ as a function of $(d - d_0)/d_0$ for nonbreaking undular surges.

4   Conclusions

In navigation channels, undular surges generated by the operation of lock chambers may significantly increase the degree of difficulty encountered maneuvering a ship and even cause navigation accidents in some instances. In this study, both nonbreaking and breaking undular surges were generated by the emptying of a reservoir. Detailed free-surface and velocity measurements were performed with a series of capacitance wave gauges and an acoustic Doppler velocimeter in a horizontal rectangular water wave channel. The experimental results indicated that the leading wave shapes of non-breaking undular surges during wave evolution closely matched the theoretical solitary wave shapes. Moreover, the maximum wave amplitude of the leading wave prior to wave breaking was close to breaking limit of a solitary wave. The maximum wave amplitude of an undular surge during its entire evolution occurred immediately before wave breaking at the leading wave crest. For nonbreaking undular surges, although the experimental trend was close to the linear wave theory, the data tended to suggest a significantly faster increase in a_w/d_0 with d_m/d_0. Furthermore, the present data demonstrated that the longitudinal velocity beneath the wave front of a nonbreaking undular surge was only in qualitative agreement with the solitary wave solution. Based on the experimental results, a formula relating the longitudinal velocity component to free-surface elevation was proposed. Overall, the present finding highlighted the difference between a nonbreaking undular surge and a solitary wave of the same wave amplitude.

Acknowledgements. This study was supported by the Natural Science Foundation of Chongqing, China (Grant No. cstc2020jcyj-bshX0043); Key Laboratory of Hydraulic and Waterway Engineering of the Ministry of Education, Chongqing Jiaotong University (Grant No. SLK2021B07).

References

Benet F, Cunge JA (1971) Analysis of experiments on secondary undulations caused by surge waves in trapezoidal channels. J Hydraul Res 9(1):11–33

Castro-Orgaz O, Chanson H (2020) Undular and broken surges in dam-break flows: a review of wave breaking strategies in a Boussinesq-type framework. Environ Fluid Mech 20(6):1383–1416

Chanson H (2010a) Unsteady turbulence in tidal bores: effects of bed roughness. J Waterway Port Coast Ocean Eng 136(5):247–256

Chanson H (2010b) Undular tidal bores: basic theory and free-surface characteristics. J Hydraul Eng 136(11):940–944

Favre H (1935) Etude théoretique et expérimentale des ondes de translation dans les canaux découverts (Theoretical and experimental study of travelling surges in open channels). Dunod, Paris

Gualtieri C, Chanson H (2011) Experimental study of a positive surge. Part 2: comparison with literature theories and unsteady flow field analysis. Environ Fluid Mech 11(6):641–651

Gualtieri C, Chanson H (2012) Experimental study of a positive surge. Part 1: basic flow patterns and wave attenuation. Environ Fluid Mech 12(2):145–159

Koch C, Chanson H (2008) Turbulent mixing beneath an undular bore front. J Coastal Res 24 (4):999–1007

Koch C, Chanson H (2009) Turbulence measurements in positive surges and bores. J Hydraul Res 47(1):29–40

Laitone EV (1960) The second approximation to cnoidal and solitary waves. J Fluid Mech 9:430–444

Lemoine R (1948) Sur les ondes positives de translation dans les canaux et sur le ressaut ondulé de faible amplitude (on the positive surges in channels and on the undular jumps of low wave height). J La Houille Blanche 2:183–185

Leng X, Chanson H (2017). Upstream propagation of surges and bores: Free-surface observations. Coast Eng J 59(1):1750003-1–1750003-32

McCowan J (1894). XXXIX. On the highest wave of permanent type. London Edinburgh Dublin Philos Mag J Sci 38(233):351–358

Montes JS (1998) Hydraulics of open channel flow. ASCE Press, New York

Peregrine DH (1966) Calculations of the development of an undular bore. J Fluid Mech 25 (2):321–330

Soares Frazao S, Zech Y (2002) Undular bores and secondary waves - experiments and hybrid finite-volume modelling. J Hydraul Res 40(1):33–43

Treske A (1994) Undular bores (Favre-waves) in open channels—experimental studies. J Hydraul Res 32(3):355–370

Zheng F, Li Y, Xuan G, Li Z, Zhu L (2018) Characteristics of positive surges in a rectangular channel. Water 10(10):1473

Zheng F, Wang P, An J, Li Y (2021) Characteristics of undular surges propagating in still water. KSCE J Civ Eng 25(9):3359–3368

Zheng F, Li X (2021) Undular surges interaction with a vertical wall. Marine Georesour Geotechnol

Experimental Investigation of Hydrodynamics on Abrupt-Expansion Pipe Behind Control Valve of Hydro-Driven Shiplift

Jiao Wang[1(✉)] and Yaan Hu[2]

[1] Southwest Hydroengineering Research Institute for Waterway,
Chongqing Jiaotong University, Chongqing 400016, China
1004612079@qq.com
[2] Nanjing Hydraulic Research Institute, Nanjing 210029, Jiangsu, China

Abstract. Control valve is the special core equipment of hydro-driven ship lift, it is similar to the electric motor of the electric driven ship lift, which is the key to the safe and efficient operation of ship lift. Hydro-driven ship lift requires very high performance of control valve. It not only needs to control the flow accurately, but also to meet the needs of large flow. It is difficult to avoid cavitation for control valve to work under high pressure difference. How to control the erosion and destruction of valves and pipelines caused by cavitation is related to the operation safety and efficiency of hydro-driven ship lift. Sudden expansion of pipe behind the valve is a simple and efficient valve cavitation suppression technology. Aiming at typical control valve (plunger valve), this paper uses physical model tests and theoretical analysis to study the effect of sudden expansion of pipe behind the valve on the flow resistance characteristics of valves, and comprehensively consider the effects of valve type, flow pattern, and flow pulsation. According to the analysis, the relationship between sudden expansion ratio of the pipe and valve flow coefficient is obtained, and the test results are in good agreement with the theoretical analysis. Based on the theoretical analysis and considered the effect of cavitation defense, the expansion ratio for this type of abrupt expansion pipe is suggested to be 3.00. This study has guiding significance for the anti-cavitation technology of industrial valves, and can be used as a reference for the design of pipelines for water delivery and pressure regulation projects.

Keywords: Control valve · Cavitation · Flow coefficient · Sudden expansion ratio

1 Introduction

Hydraulic driven shiplift (HDSL) is a new attempt on high dam navigation technology in the world. It mainly uses water power as the lifting power and security measure. The ship reception chamber moves with the rise and fall of the buoys in the vertical shafts, which are driven by the filling and emptying system. When the ship chamber load is changed, the submerged depth of the buoys will have a corresponding change to make the ship chamber reach a new equilibrium between chamber and buoys, and solve the

© The Author(s) 2023
Y. Li et al. (Eds.): PIANC 2022, LNCE 264, pp. 406–417, 2023.
https://doi.org/10.1007/978-981-19-6138-0_36

serious leakage problem in ship chamber, which is an incomparable technical advantage than traditional electric driven shiplift (Hu 2011). It is especially suitable for high dam navigation, and has a broad application prospect. Jinghong shiplift is the first HDSL in the world (Fig. 1).

The filling and emptying system of HDSL is similar to the electric drive system of traditional shiplift, which is the source of power. Its running speed and acceleration are controlled through the precise regulation of water flow. Hence, the performance of the valve plays an important role in the safety and stability of HDSL.

Cavitation is a transition phase between liquid and cavitation bubbles which occur within low pressure zone and crush with pressure recovery (Pan et al. 2013). Cavitation is a very common hydraulic phenomenon in hydraulic machinery such as propellers (Pennings et al. 2016), venturi tubes (Bertoldi et al. 2015), vane pumps (Li 2016), and control valves (Hubballi et al. 2013). Studies of valve cavitation always focus on noise, vibration, and performance reduction (Shirazi et al. 2012). Most research still relies on experiments (Long et al. 2018) with a variety of measurement methods such as high-speed photography (Wang et al. 2015), PIV (Kravtsova et al. 2014), probes (Pham et al. 1999), and observation window (Osterman et al. 2009). With the fast development of CFD (computational fluid dynamics), numerical simulations have also been used to study cavitation inside the valves and other situations beyond the capability of experiments (Liu et al. 2006).

Cavitation should be controlled for the safety and efficiency of HDSL. Based on the mechanism of cavitation, the cavitation defense measures for control valves are mainly divided into two categories: aeration (Tomov et al. 2015) and structure optimization (Gholami et al. 2015). Chern studied the globe valve with or without a cage by cavitation model (Chern et al. 2004). The results indicated that the cage could limit the cavitation to the vicinity of cage and prevent cavitation to erode valve body and downstream pipe. Hubballi performed analytic and experimental research by using cavitation index to predict the cavitation effects for hydraulic control valves. Jo made a numerical study for reducing cavitation in butterfly valve (Jo et al. 2013), and the perforated plate was thought to be effective to suppress the cavitation inside of the pipe. Gholami studied the needle valve by 3-D numerical simulation. A circular row of vanes was used at the end section of the needle valve. From their study, they concluded that cavitation was suppressed, but the flow coefficient also was decreased. Lee made a shape design of bottom plug used in a 3-way reversing valve to minimize the cavitation effect (Lee et al. 2016). Han selected three kinds of typical structures of poppet valves to research (Han et al. 2017). The results revealed that two-stage throttle valve (TS valve) can effectively suppress the occurrence of cavitation while the flow force of TS valve was much bigger than other valves.

The present researches about industrial valve cavitation, mainly concentrate on the cavitation mechanism and the structure optimization of valve. In the conditions of high pressure and large flow rate, there have few researches about useful measures which can be applied to practical engineering for efficiently restraining cavitation. In the current research, experimental study on abrupt expansion pipe behind plunger valve was carried out, the flow regime and cavitation phenomena were observed through glass tube, and the influence rules of abrupt-expansion on the hydraulic characteristics of valve were analyzed.

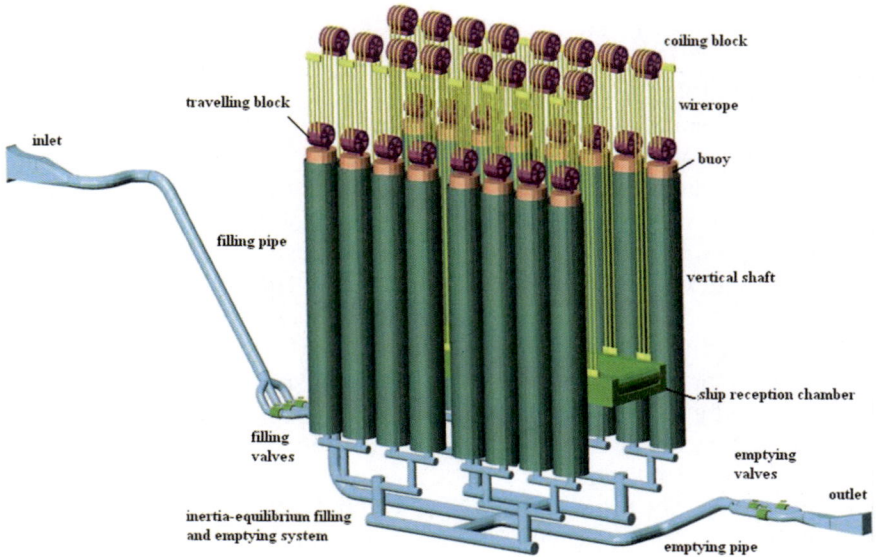

Fig. 1. Jinghong HDSL perspective view.

2   Materials and Methods

In this paper, as seen in Fig. 2, the test valve is a plunger valve, DN = 150 mm, the type of sleeve is SZ20–30% (made by VAG Germany). The experiments were carried out in the multi-functional cavitation experiment hall of Nanjing Hydraulic Research Institute (Nanjing, China). This laboratory is automated with the automatic control of pumps, valves, pressure and flow monitoring. The maximum capacity of water pressure supply system is 1.5MPa and the maximum flow rate is 0.15m³/s. The experimental model is mainly composed by test valve and glass pipe (Fig. 3).

The influence rules of hydraulic characteristics of valve with abrupt-expansion were observed in this study. It mainly includes the properties of flow capacity, wall pressure characteristics and cavitation characteristics. The representative parameters are as follows: flow coefficient (μ), pressure characteristics of pipe after valve including time-average pressure (P_{ta}) and fluctuating pressure (P_{rms}) and cavitation index (K). The main physical quantities required to be observed are as follows: flow rate (Q), cavitation noise, wall pressure, upstream steady pressure (P_u) and downstream steady pressure (P_d). The locations of the sensors are shown in Fig. 4.

In this research, the experimental studies of three kinds of abrupt-expansions were investigated. The specific dimensions of parameters were shown in Table 1, and the definition of expansion ratio (R) was shown in Eq. 1. *ID* and *OD* denote the inside diameter and outside diameter of glass pipe.

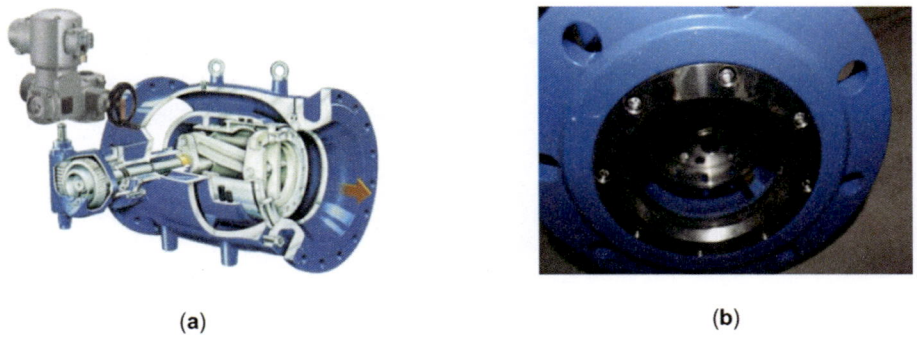

(a)   **(b)**

Fig. 2. (a) Perspective view of plunger valve; (b) Picture of plunger valve.

1 pump; 2 control valve; 3 bypass valve; 4 electromagnetic
flowmeter; 5 pressure stabilizing box; 6 electronic pressure gauge;
7 aeration ring; 8 test valve; 9 glass tube; 10 hydrophone; 11
pressure sensor; 12 data acquisition system; 13 computer; 14
camera; 15 pool

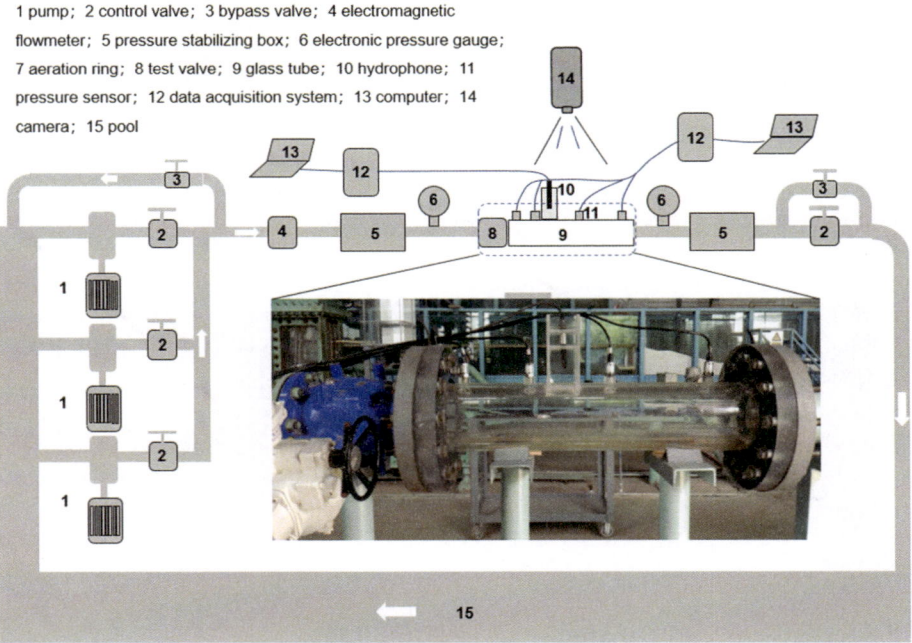

Fig. 3. Experimental setup.

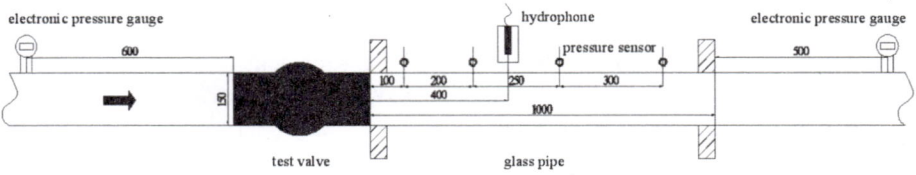

Fig. 4. Sensor layout (mm).

Table 1. Experiment content.

R	ID (mm)	OD (mm)	L (mm)
1.00	75	105	1000
2.15	110	145	1000
4.00	150	195	1000

$$R = \frac{A_1}{A_2} \tag{1}$$

A_1 denotes the reference section area of abrupt-expansion (m^2); A_2 denotes the reference section area of pipe (m^2).

3   Results and Discussion

3.1   Flow Phenomena

The typical cavitation phenomena of plunger valve are shown in Fig. 5. The cavitation appearances in three types of abrupt-expansions are same. It is mainly misty cavitation. Due to the particular structure of plunger valve, the misty cavitation can be combined into vortex rope cavitation at opening degree of 0.3. Enlarging of the abrupt-expansion size, a thickness of water cushion is added between the cavitation and the pipe wall. For R = 1.00, the pipe is filled with misty cavitation, the collapse of bubbles will directly erode the pipe wall. For R = 2.15, the pipe is basically filled with misty cavitation, but occasionally there is a layer of water cushion between bubbles and pipe wall. For R = 4.00, there is a stable water cushion between cavitation and pipe wall. The process of bubble collapse is confined within the water body, so the impact of collapse on pipe can be buffered, and the pipe vibration and cavitation damage could be reduced significantly.

3.2   Flow Coefficient

In this research, μ is defined as follow:

$$\mu = \frac{Q}{A\sqrt{2\frac{(P_u - P_d)}{\rho}}} = \frac{1}{\sqrt{1 + \sum \zeta}} \tag{2}$$

A denotes the reference section area (m^2); ρ denotes the density of water; g denotes the acceleration of gravity; $\sum \zeta$ denotes the resistance of the test system. Figure 6 and Table 2 show the difference of μ for various R.

From the definition of μ, it can be said that flow loss is the root cause of the change of μ. The resistance of the test system is composed with valve resistance (ζ_v) and abrupt-expansion resistance (ζ_e). The ζ_e includes resistance of the expansion section

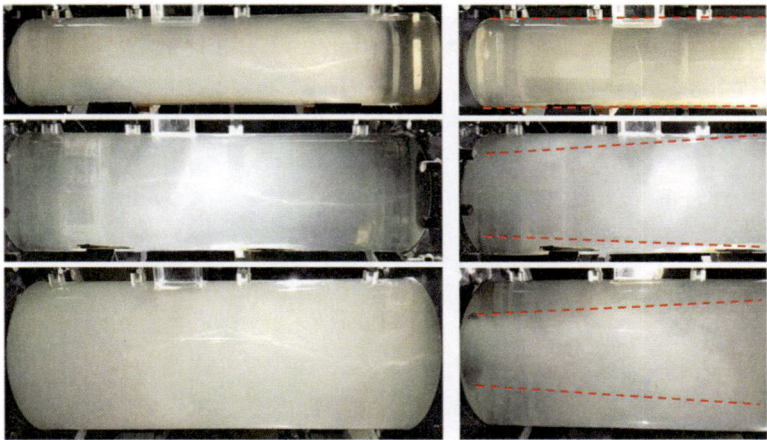

Fig. 5. Cavitation phenomena.

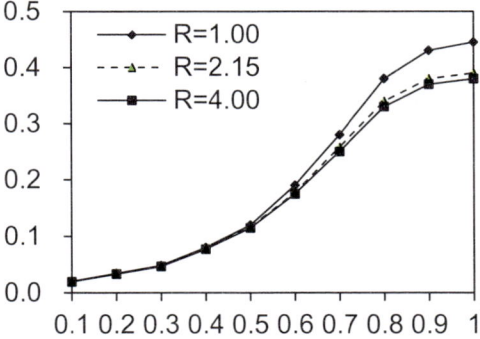

Fig. 6. Flow coefficient for various R.

(ζ_{ex}), resistance along the expansion body (ζ_λ) and resistance of the contraction section (ζ_c), and shown as follows:

$$\sum \zeta = \zeta_v + \zeta_e = \zeta_v + (\zeta_{ex} + \zeta_\lambda + \zeta_c) \tag{3}$$

Because the l/d of abrupt-expansion is large enough (Du et al. 2015), ζ_λ is small and it can be negligible. Therefore, ζ_e can be simplified as:

$$\zeta_e \approx \zeta_{ex} + \zeta_c \tag{4}$$

Section 1 and 2 were defined as shown in Fig. 7. Because the two sections were in the same elevation, so there was no need to consider the potential energy. The relationship of R and ζ_c could be analyzed by theoretical method as follows:

Table 2. Flow coefficient for various R.

n	μ			Decline (%)	
	$R = 1.00$	$R = 2.15$	$R = 4.00$	$R = 2.15$	$R = 4.00$
0.1	0.020	0.0198	0.0195	1.0	2.5
0.2	0.034	0.0335	0.0330	1.5	2.9
0.3	0.048	0.0470	0.0465	2.1	3.1
0.4	0.080	0.0780	0.0770	2.5	3.8
0.5	0.120	0.1165	0.1150	2.9	4.2
0.6	0.190	0.1770	0.1750	6.8	7.9
0.7	0.280	0.2580	0.2500	7.9	10.7
0.8	0.380	0.3400	0.3300	10.5	13.2
0.9	0.430	0.3800	0.3700	11.6	14.0
1	0.445	0.3900	0.3800	12.4	14.6

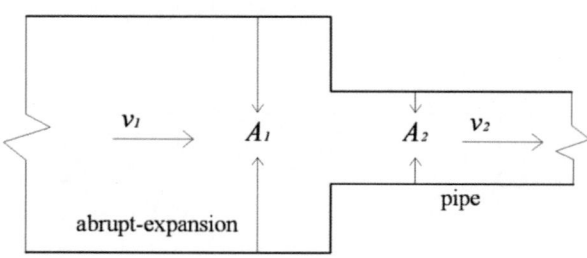

Fig. 7. Abrupt-contraction section.

The Bernoulli's equation considered flow loss:

$$\frac{v_1^2}{2g} + \frac{p_1}{\rho g} = \frac{v_2^2}{2g} + \frac{p_2}{\rho g} + \zeta_c \frac{v_2^2}{2g} \tag{5}$$

The continuity equation:

$$A_1 v_1 = A_2 v_2 \tag{6}$$

The simultaneous solution of Eq. (1)–(6):

$$\mu = \frac{1}{\sqrt{1 + \frac{R^2}{R^2-1}\zeta_c}} \tag{7}$$

In the same way, the influence of R on ζ_{ex} can be analyzed, and combined with Eq. 7:

$$\mu = \frac{1}{\sqrt{1 + \frac{R^2}{R^2-1}\zeta_e}} \tag{8}$$

The empirical formulas of ζ_{ex} and ζ_c are as (Li et al. 2006):

$$\zeta_{ex} = (1 - \frac{1}{R})^2 \tag{9}$$

$$\zeta_c = 0.5(1 - \frac{1}{R}) \tag{10}$$

μ parameter can be obtained as:

$$\mu = \frac{1}{\sqrt{1 + \frac{R^2}{R^2-1}\zeta_e}} = \frac{1}{\sqrt{1 + (1.5 - \frac{2.5}{R+1})}} \tag{11}$$

The impact of R on ζ_e shows in Fig. 8: with the increase of R, ζ_e is increasing, and μ is reducing, but the reducing extent is dropped. For $R = 3$, $\zeta_e \approx 0.9$; for $R = 8$, ζ_e is only 1.2. It can be seen that, when R increases to a certain extent, the further enlargement of R will not have a significant effect on ζ_e. Because the abrupt expansion pipe wall is too far from the mainstream to make a significant impact.

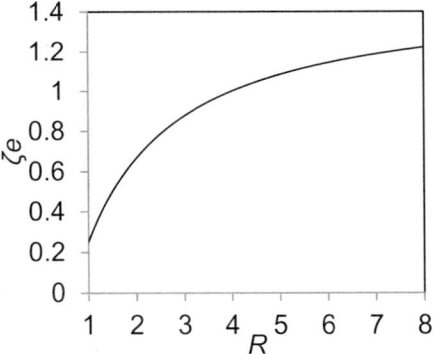

Fig. 8. Relationship of R and ζ_e.

Based on Eq. 11, the formula of μ on the whole test section with R could be obtained:

$$\mu = \frac{1}{\sqrt{1 + \zeta_v + 1.5 - \frac{2.5}{R+1}}} \tag{12}$$

The comparison between theoretical value and measured data of μ is shown in Fig. 9.

It can be seen that the measured data are in good agreement with the theoretical values. Based on the results of above analysis, the variation law of μ can be fully explained.

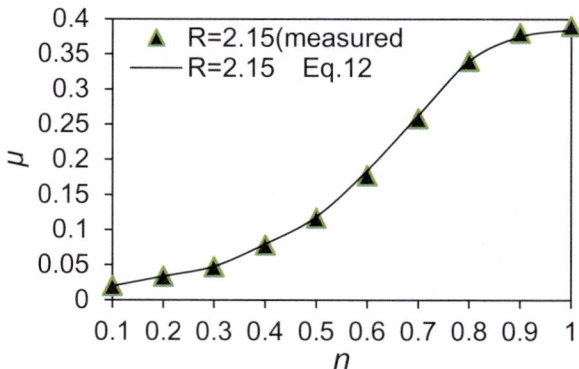

Fig. 9. Comparison of theoretical value and measured data.

For $n = 0.1$–0.5, the variations of μ are very small. Because of ζ_v is much bigger than ζ_e. The influence of abrupt-expansion could be neglected. For $n = 0.6$–1.0, with the increase of n, the ζ_v declined dramatically, and the influence of abrupt-expansion increased remarkably. For R has increased from 2.15 to 4.00, the decrease of μ was 5%, which shown in both of theoretical and experimental results.

3.3   Pressure Characteristics

Figure 10 shows the pressure characteristics for various R values of typical condition. For R increased from 1.00 to 2.15, P_{ta} increased and its maximum promotion was 20 kPa; P_{rms} near valve port decreased significantly and the drop was 2 kPa; For R increased from 2.15 to 4.00, the change of P_{ta} and P_{rms} were very small.

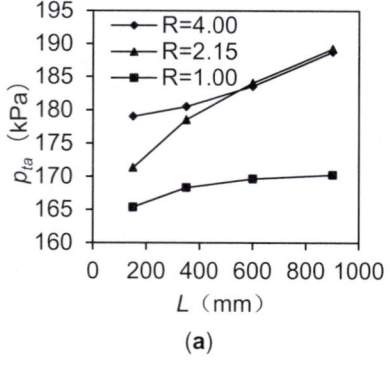

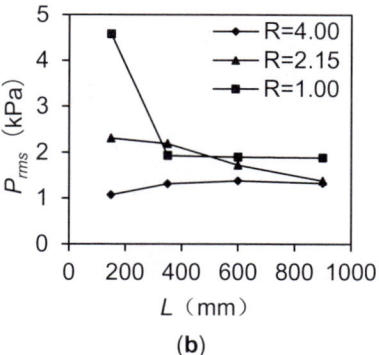

(a)   (b)

Fig. 10. Pressure characteristics for various R of typical condition. (a) Time-average pressure for various R; (b) Fluctuating pressure for various R.

Combined with pictures of cavitation phenomena, there is a layer of water subfill between pipe wall and mainstream, which is beneficial to increase the P_{ta} after valve, the collapsing of cavitation can be limited within the water body and the P_{rms} can be reduced significantly. After the stable water cushion is formed, the further increase of R, which means the further increase of the thickness of water cushion, has marginal effect on the pressure characteristics of pipe wall.

3.4   Noise Characteristics

As shown in Fig. 11, For R increased from 1.00 to 2.15, time-average sound pressure (SP) dropped rapidly by about 30 Pa, the sound pressure level (SPL) decreased about 40 db in the dominant frequency (6 kHz), and cavitation had been significantly inhibited. For R increased from 2.15 to 4.00, SP and SPL changed slightly. The comparisons of cavitation noise characteristics for various R values in different working conditions are shown in Table 3, it is indicated that when the stable water cushion is formed, the further increase of R wouldn't increase the cavitation suppression effect obviously.

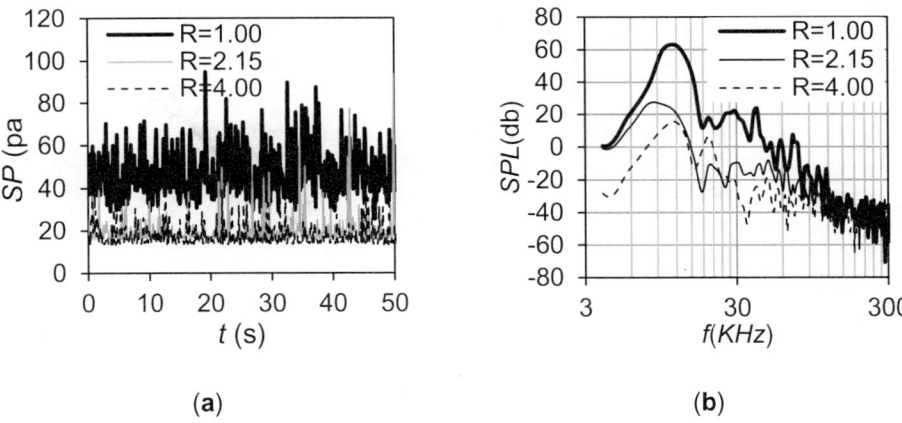

(a)   (b)

Fig. 11. Noise characteristics for various R of typical condition. (a) Sound pressure for various R; (b) Sound pressure level for various R.

Table 3. Cavitation noise characteristics (*CNC*) under different *R*.

Conditions (P_u-P_d, kPa)	CNC	R			Decline (%)	
		1.00	2.15	4.00	1.00 → 2.15	2.15 → 4.00
229-180	SP(Pa)	44.0	15.5	14.0	64.8	9.7
	SP_{rms}(Pa)	10.0	1.0	0.5	90.0	50.0
	SPL(db)	60.0	−15.0	−20.0	125.0	33.3
275-180	SP(Pa)	47.2	20.0	19.0	57.6	5.0
	SP_{rms}(Pa)	11.1	7.6	5.6	31.5	26.3
	SPL(db)	60.0	25.0	20.0	58.3	20.0
325-180	SP(Pa)	205.2	140.5	137.0	31.5	2.5
	SP_{rms}(Pa)	52.5	35.0	24.2	33.3	30.9
	SPL(db)	100.0	70.0	60.0	30.0	14.3

4   Conclusions

(1) Abrupt-expansion reduced the discharge capability of valve. For $n = 0.1$–0.5, the influence was very small. Because of ζ_v was much bigger than ζ_e. The influence of abrupt-expansion could be neglected. For $n = 0.6$–1.0, the influence increased remarkably, because ζ_v decreased with n obviously. The variation of μ for various R values was smaller for R increased from 2.15 to 4.00, because ζ_e decreased slightly with the further increase of R.

(2) Abrupt-expansion could improve the pressure characteristics on pipe wall behind valve and inhibit cavitation significantly. There was a layer of water subfill between the wall of abrupt-expansion and mainstream, which could increase the P_{ta}, limit the collapse of cavitation and reduce the P_{rms} significantly.

(3) In order to minimize the influence on the flow coefficient and ensure the formation of stable water cushion, it is suggested that the expansion ratio for this type of abrupt-expansion is suitable for 3.00.

Acknowledgements. The authors sincerely acknowledge the financial support received from "Science and Technology Research Project of Chongqing Education Commission of China (Grants No. KJQN202000722)".

References

Bertoldi D, Dallalba CCS, Barbosa JR (2015) Experimental investigation of two-phase flashing flows of a binary mixture of infinite relative volatility in a Venturi tube. Exp Therm Fluid Sci 64:152–163

Chern MJ, Wang CC (2004) Control of volumetric flow-rate of ball valve using V-port. J Fluids Eng 126:471–481

Du HY (2015) Influence of changed fuel injection pulse width and pressure on discharge coefficient in diesel engine. Trans Chin Soc Agric Eng 31:71–76

Gholami H, Yaghoubi H, Alizadeh M (2015) Numerical analysis of cavitation phenomenon in a vaned ring-type needle valve. J Energy Eng 141:04014053

Han MX, Liu YS, Wu DF et al (2017) A numerical investigation in characteristics of flow force under cavitation state inside the water hydraulic poppet valves. Int J Heat Mass Transf 111:1–16

Hu YA (2011) Basic theory research on hydraulic driven shiplift application. Ph.D. thesis of Nanjing Hydraulic Research Institute, Nanjing, China

Hubballi B, Sondur V (2013) A review on the prediction of cavitation erosion inception in hydraulic control valves. Int J Emerg Technol Adv Eng 3:110–119

Jo SH, Kim HJ, Song KW (2013) A numerical study for reducing cavitation in a butterfly valve with a perforated plate. In: Proceedings of Korean society for fluid machinery, pp 308–308

Kravtsova AY, Markovich DM, Pervunin KS et al (2014) High-speed visualization and PIV measurements of cavitating flows around a semicircular leading-edge flat plate and NACA0015 hydrofoil. Int J Multiph Flow 60:119–134

Lee MG, Lim CS, Han SH (2016) Shape design of the bottom plug used in a 3-way reversing valve to minimize the cavitation effect. Int J Precis Eng Manuf 17(3):401–406. https://doi.org/10.1007/s12541-016-0050-8

Li W (2006) Handbook of hydraulic computation, 2nd edn. China Water Conservancy and Hydropower Press, Beijing

Li WG (2016) Modeling viscous oil cavitating flow in a centrifugal pump. J Fluids Eng 138:011303

Liu YS, Yang YS, Li ZY (2006) Research on the flow and cavitation characteristics of multi-stage throttle in water-hydraulics. Proc IMech Part E: J Process Mech Eng 220:99–108

Long XP, Wang J, Zhang JQ et al (2018) Experimental investigation of the cavitation characteristics of jet pump cavitation reactors with special emphasis on negative flow ratios. Exp Thermal Fluid Sci 96:33–42

Osterman A (2009) Characterization of incipient cavitation in axial valve by hydrophone and visualization. Exp Therm Fluid Sci 33:620–629

Pan SS, Peng XX (2013) Physical mechanism of cavitation. National Defense Industry Press, China

Pennings P, Westerweel J, Terwisga TV (2016) Cavitation tunnel analysis of radiated sound from the resonance of a propeller tip vortex cavity. Int J Multiph Flow 83:1–11

Pham TM, Larrarte F, Fruman DH (1999) Investigation of unsteady sheet cavitation and cloud cavitation mechanisms. J Fluids Eng 121:289–296

Shirazi NT, Azizyan GR, Akbari GH (2012) CFD analysis of the ball valve performance in presence of cavitation. Life Sci J Acta Zhengzhou Univ Overseas Edn 9:1460–1467

Tomov P, Khelladi S, Ravelet F, et al (2015) Experimental study of aerated cavitation in a horizontal venturi nozzle. Exp Therm Fluid Sci 8–18

Wang ZY, Huang BG, Wang Y et al (2015) Experimental and numerical investigation of ventilated cavitating flow with special emphasis on gas leakage behavior and re-entrant jet dynamics. Ocean Eng 108:191–201

Experimental Study on the Relationship Between the Height of Submerged Dam at the Entrance of Pit-Type Pool and the Treatment Effect

Qiang Ying[✉], Xinnong Zhang, Dongdong Jia, and Changying Chen

Nanjing Hydraulic Research Institute, Key Laboratory of Port,
Waterway and Sedimentation Engineering of Ministry of Transport,
Nanjing, China
qying@nhri.cn

Abstract. The submerged dam is one of the commonly used schemes for the prevention and treatment of the pit-type collapse, which is often used in the pit-type collapse emergency treatment in Anhui and Jiangsu sections of the Yangtze River and has achieved good silt-promoting effects. However, the height of submerged dam lacks theoretical and experimental basis in the design and has been determined by experience in the past. Through the physical model of the pit-type collapse of Guizhou Village in the Yangzhong Reach, five different heights of the submerged dam were tested, and the flow velocity, flow direction and surface flow pattern of 12 points in the pond were measured. The test results show that the surface velocity of the inner surface of the submerged dam in the lower part of the pond increases due to the influence of the submerged dam, while the velocity of the bottom layer decreases, and the velocity of the inner side of the upper part mainly decreases; the average velocity in the pond decreases with the increase of the height of the submerged dam and there is a linear relationship between the two; considering that the engineering volume of the submerged dam has a square relationship with the height of the submerged dam, it is believed that the submerged dam should not be too high as long as it meet the sedimentation in the pond.

Keywords: Pit-type pool · Control · Submerged dam · Mean velocity

1 Introduction

The pit-type collapse is the most dangerous form of riverbank collapse in the lower reaches of the Yangtze River. In a period of 1–2 days, a pit-type pool of several hundred meters in length, width and depth; volume of one million cubic meters can be formed. Ports, docks, flood control dike and houses in the dike on the collapsed body will also be destroyed (Luo et al. 2020; Jin et al. 1998; Guo 1996; Qi et al. 2002), which will bring huge losses to people's lives and properties. The pit-type collapse involves the interaction between water and soil, which is affected by many factors and has the characteristics of sudden and rapid development. There is no unified understanding of the cause of pit collapse (Chen et al. 1985; Yu 2007; Ding et al. 1985;

© The Author(s) 2023
Y. Li et al. (Eds.): PIANC 2022, LNCE 264, pp. 418–426, 2023.
https://doi.org/10.1007/978-981-19-6138-0_37

Zhang et al. 2008;). There are only a few studies on early warning and prevention before pit-type collapse (Cao et al. 2019; Liu et al. 2017). In the aspects of emergency protection and post-treatment after the pit-type collapse, based on the practice of the water conservancy scientific and technical personnel, the principle of "Guarding the shoulders, consolidating the periphery, first promoting the siltation, and then filling the entrance" is put forward (Zhong et al. 2011; Yao et al. 2009; Zhang et al. 2011; Luo et al. 2019). But this principle lacks theoretical support. In practice, there are still problems such as how to determine the size of each project in order to optimize the overall effect and save the project investment.

The two-shoulder guarding works of Pit-type pool include the riprap of upper and lower shoulders and the submerged dam at the entrance. The submerged dam has better effect of promoting siltation. Materials for constructing submerged dams include riprap (Jin et al. 2001), caisson (Sun et al. 1996), sand filling woven bag, tree head stone, etc. The Crest Height of the submerged dam is determined by experience or material characteristics, so it is necessary to study the relationship between the crest height of the submerged dam and the engineering effect.

2 Model Design and Test

2.1 Model Design

Model design: The 1.1 km upstream and the 0.9 km downstream of the pit-type collapse at the Zhinan village in Yangzhong reach is chosen as the research objects. The width of the riverbed including the pit-type pool and the upper and lower reaches of the deep trough (−20 m contour) is used as the scale of the model, which is simulated by a 1:100 normal model. The scope and layout of the model are shown in Fig. 1 (Yuan et al. 2021).

2.2 Model Validation

The verification flow is 28500 m^3/s in the dry season period. The water level verification point is located at the No. 1 flow measurement point on the inlet section. Four flow test verification points are shown in Fig. 1. The results are shown in Table 1.

Table 1. Verification of velocity of measuring point in physical model (m/s)

Model	Discharge/(m^3/s)	Point number			
		1	2	3	4
Verification value	28500	0.678	1.158	0.705	0.750
Required value		0.657	1.098	0.671	0.792

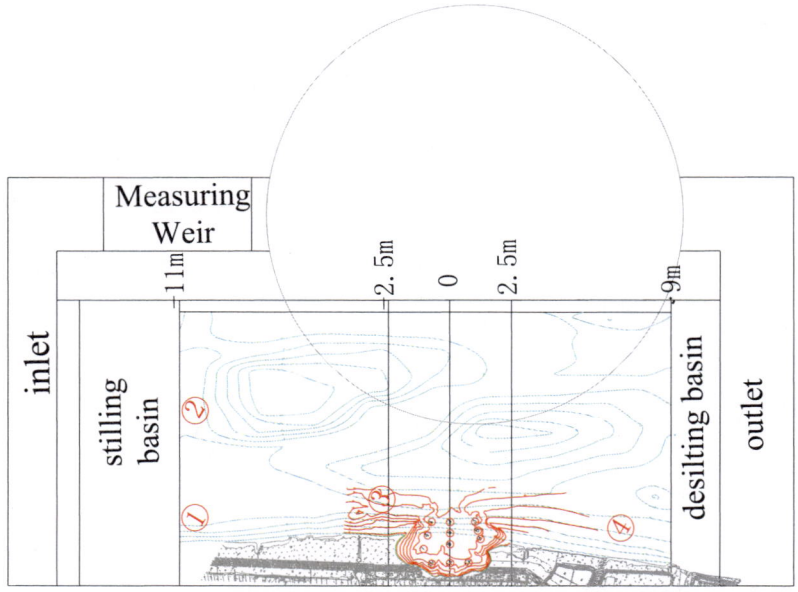

Fig. 1. Scope of physical model

2.3   Layout of Measuring Points

Considering that the pit-type collapse mainly occurs in the dry season, the flow rate $Q = 28500$ m^3/s is selected as the test discharge. The propeller current meter is used to measure the flow velocity. 12 Flow measuring points were arranged in the pit-type pool. The position distribution of each measuring point is shown in Fig. 2.

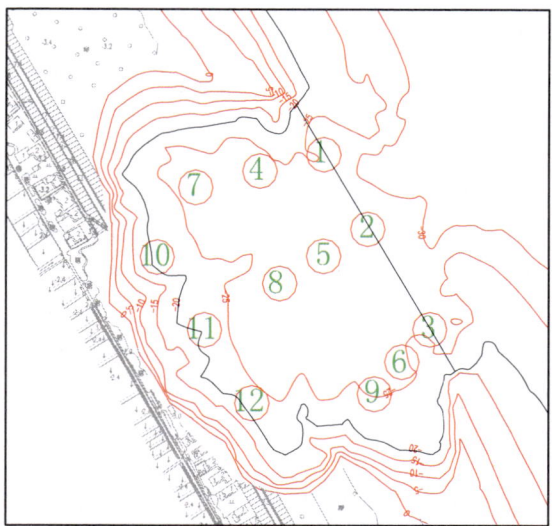

Fig. 2. Arrangement of velocity measuring points

2.4   Velocity Measurement

Each measuring point is arranged with a propeller measuring Rod, and each vertical line is measured by three-point method (0.2h, 0.6h, 0.8h, h is the depth of the measuring point). Each point is measured four times, each sampling time is 10 min, four times average value is taken as the velocity of this point. The surface flow field is measured by particle image velocimetry (PIV).

2.5   Simulation of Submerged Dam

The submerged dam is made of concrete. In the model, the submerged dam is a vertical dam with a top width of 3 cm; According to the maximum water depth of the section where the submerged dam is located, the height of the submerged dam is set to be 0.1, 0.2, 0.3, 0.4 and 0.5 times of the maximum water depth respectively. The height of each submerged dam is made in layers, which is convenient for assembly and disassembly during the test. See Fig. 3 for details.

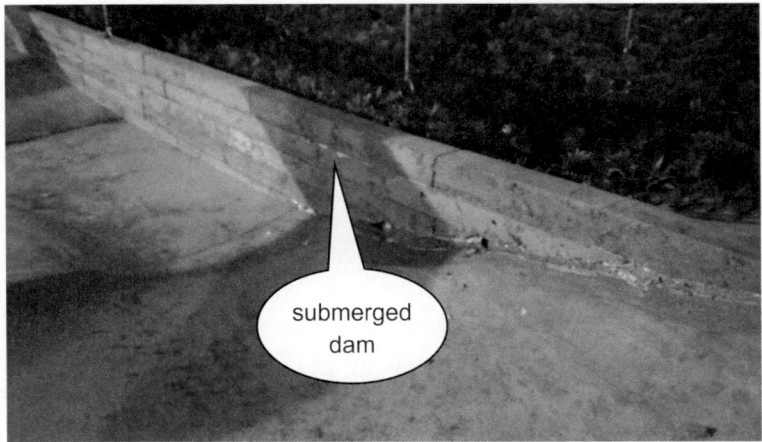

Fig. 3. Schematic diagram of submerged dam for model

3   Test Results and Analysis

3.1   Surface Flow Pattern

The surface flow pattern of the model without engineering is shown in Fig. 4. It can be seen that the main flow velocity is the highest in the deep trough. The surface velocity at the entrance of the pit-type pool is about 1/2 of the main flow velocity, and the velocity in the pit-type pool is even smaller. The flow in pit-type pool moves clockwise and the reflux center changes in the plane position at different times, but all of them are centered at 8 #. The surface flow pattern does not change much when the submerged dam is arranged at the entrance of the pit-type pool.

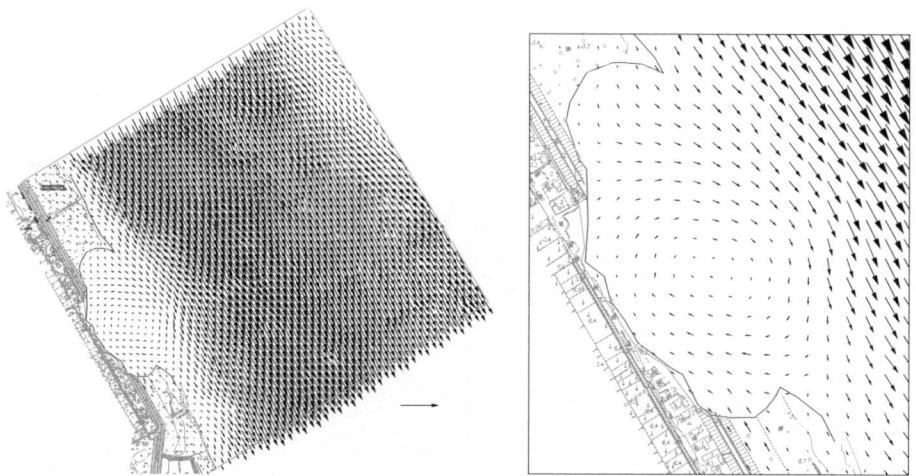

Fig. 4. Surface flow field near the pit pool in model without engineering (PIV)

3.2   Vertical Velocity Distribution

When there is no project, the distribution of the velocity of the upper, middle and lower layers varies greatly in different positions in pit-type pool. The flow direction of the upper, middle and lower layers of 1#, 2# points point all point to the downstream; the velocity of the upper layer of 3# measuring points still points to the downstream; but the flow direction of the middle and lower layers has pointed to the pit-type pool, as shown in Fig. 5. The velocity distribution in the pit-type pool is mainly affected by the flow inertia when entering the pit-type pool and the underwater topographic conditions in the pit-type pool. The upper velocity (flow direction) is mainly affected by the inertia, while the lower ones (flow direction) is greatly affected by the topography. The flow velocity of the surface, middle and bottom layers tends to converge, and some even have the maximum phenomenon of the bottom layer.

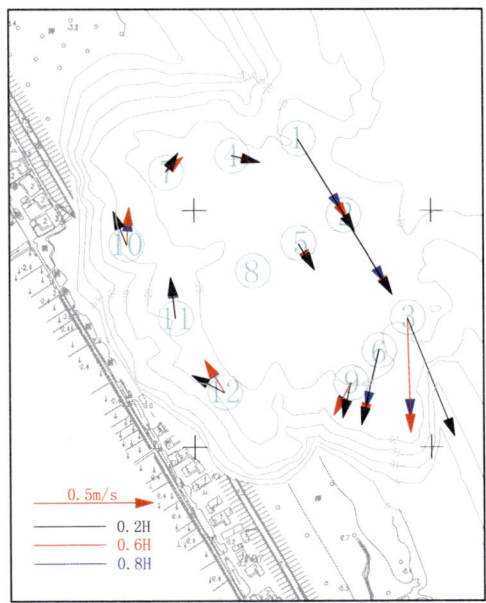

Fig. 5. Velocity distribution in the pit-type pool without Engineering

After the submerged dam is set, the movement form of the surrounding water body is changed. The velocity of the water body affected by the water resistance of the submerged dam decreases, while the flow velocity of the water body above the dam crest increases. For example, when the height of submerged dam is 0.5 times the water depth, the velocity at 0.2 h of 6#, 9# measuring points before and after the project is 0.326–0.411 m/s, 0.151–0.176 m/s respectively; The velocity at 0.6h before and after the project is 0.309–0.286 m/s and 0.155–0.180 m/s respectively. It can be seen that the velocity at the 6# measuring point close to the dam body decreases due to the obstruction of the submerged dam, and the velocity at the 9# measuring point far from the dam body increases. The flow velocity at 0.8 h decreases at both measuring points; At the 4# measuring points at the upstream of pit-type pool, the velocity after the project at 0.2 h, 0.6 h and 0.8 h is lower than that before the project. See Fig. 6 for details.

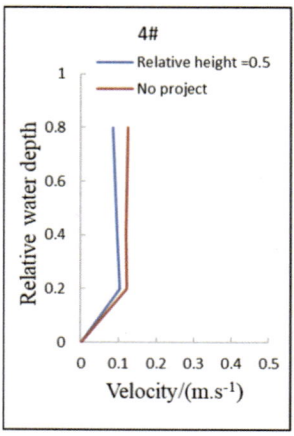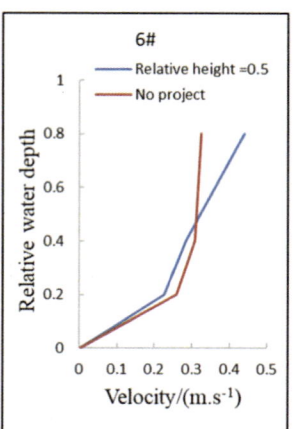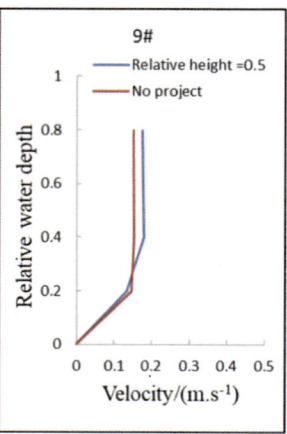

Fig. 6. Velocity distribution at 4 #, 6 # and 9 # before the project and 0.5H relative dam height

3.3 Average Velocity Distribution

The three-point method is used to calculate the average velocity of the measuring points. The average velocity of each measuring point under different schemes is shown in Fig. 7. Because the 1#, 2# and 3# measuring points are on the top or outside of the submerged dam, they are not considered in the analysis. It can be seen from the figure that the change trend of velocity values at different measuring points in the same scheme is not very obvious. In addition to the fact that the velocity in the pool is low and not easy to measure, there is also the random oscillation of the flow in the pool. Therefore, when analyzing the effect of submerged dam, average the velocity of 9 measuring points in the pool and compare the change of average velocity in the pool. The results are shown in Fig. 8. It can be approximated that the average velocity $\frac{V}{V_0}$ of measuring points in the pool has a linear relationship with the relative height $\frac{H_d}{H}$ of submerged dam, which can be expressed as:

$$\frac{V}{V_0} = 1 - 0.2195 \frac{H_d}{H} \tag{1}$$

where: v - average velocity of measuring points in the pool when there is a submerged dam; v_0 - average velocity of measuring points in the pit when there is no submerged dam; H_d - Maximum Height of submerged dam section; H - Maximum depth of submerged dam section.

This shows that with the increase of submerged dam height, the average velocity of water in the Pit-type pool decreases linearly. The main reason is that the dam body prevents part of the water body of the river from flowing into the Pit-type pool. A higher submerged dam is beneficial to reduce the flow movement in the Pit-type pool and stabilize the Pit-type pool. However, with the increase of dam height, the cross-sectional area of dam body increases with the square of dam height (as shown in Eq. 2) that is, the volume of submerged dam increases with the quadratic power of dam height and dam length. At the same time, considering that the dam length will also increase with the increase of dam height, the engineering quantity and engineering cost of submerged dam will increase rapidly.

For the dam body with trapezoidal section, the section area is:

$$s = \frac{(m_1 + m_2)}{2} H_d^2 + B H_d \tag{2}$$

where: H_d - dam height; B - dam crest; m_1, m_2 -side slopes.

Because the decrease of flow velocity in the nest pond is only linear with the dam height, and the engineering cost is 2–3 power with the dam height. Therefore, under the condition of ensuring the effect of reducing flow and promoting sedimentation in the Pit-type pool, it is not appropriate to select a higher dam height from an economic point of view.

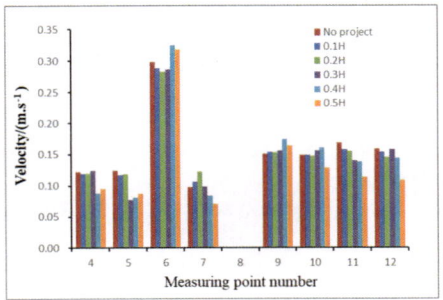

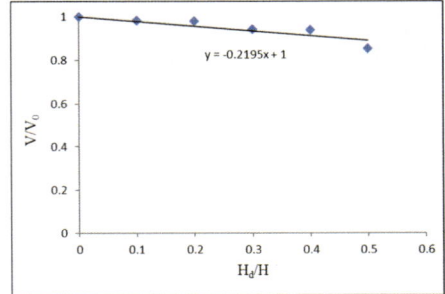

Fig. 7. Velocity values in the pit pool under different submerged dams

Fig. 8. Relationship between relative dam height and relative velocity

4   Conclusions

Submerged dam project is a common method used in the treatment of pit-type collapse. In this paper, a 1:100 normal model of pit-type collapse in Zhinan village of the Yangtze River is made, and five relative submerged dam heights are tested. The results show that the decrease of relative velocity in the pit-type pool is in a linear relationship with the increase of the relative height of the submerged dam. Considering the multi-power relationship between the dam volume and the dam height. Therefore, under the condition of ensuring the safety of the pit-type pool, it is recommended to take a lower dam height.

Acknowledgements. This work was supported by National Natural Science Foundation of China (U2040215, 52079080, 51779148) and the National Key R&D Program of China (2021YFC3200403).

References

Cao S, Cai L, Liu P (2019) Research and application of comprehensive evaluation method for bank collapse early warning. EWRHI 40(8):21–28

Chen YC, Peng HY (1985) Occurrence and protection of pit collapse in lower Yangtze River. In: Symposium of bank protection in middle and lower Yangtze River, III edn. Yangtze River Institute of Water Resources and Hydropower Research, Wuhan, pp 112–117

Ding PY, Zhang JY (1985) Discussion on the relationship between liquefaction and collapse of bank soil. In: Symposium of bank protection in middle and lower Yangtze River, III edn. Yangtze River Institute of Water Resources and Hydropower Research, Wuhan, pp 104–109

Guo Y (1996) Occurrence and prevention of ω—caving bank in Yangzhou port. J Yangzhou Teach Coll (Nat Sci Edn) 16(4):53–58

Jin LH, Wang NH, Fu QH (1998) Analysis of topography of bank slides and its affecting factors in Mahu reach of the Yangtze River. J Sediment Res (2):67–71

Jin LH, Shi XQ, Wang NH (2001) Research on the mechanism and control measures of dike pit-type slide in the Yangtze river. J Sediment Res (2):39–43

Liu DF, Lu P (2017) Research and application of bank collapse early warning technology of the Yangtze river in Anhui Province. EWRHI. 38(11):91–95

Luo LH, Su CC, Ying Q, et al (2020) Emergency treatment and effect analysis of arc collapsing in Zhinan Village, Yangzhong Reach of the Yangtze river. Jiangsu Water Resour (2):25–28

Luo LH, Su CC, Ying Q, et al (2019) Analysis on the causes of arc collapsing in Zhinan Village Yangzhong Reach of the Yangtze river. Jiangsu Water Resour Suppl.2:65-69, 80

Qi JP, Deng XD (2002) On harnessing the collapse and dangerous section of Yongan embankment in Yangtze river. J Nanchang Coll Water Conserv Hydroelectr Power 21(3):49–55

Sun XD, Cai YY, Yao WQ (1996) Practice of using sunk tree head-stone to control bank collapse of Yangtze river. Water Conserv Constr Manag (4):45–47

Yao F (2009) Formation and treatment of bank collapse in Nanjing section of the Yangtze River. China Water Transp 9(9):244–245

Yuan WX, Ying Q, Luo LH, et al (2021) Experimental study on the water blocking effect of tree head–stone in the treatment of pit collapse at the Yangzhong reach of Yangtze river. Hydro-Sci Eng (3):119–125

Yu WC (2007) Preliminary study on mechanism of formation of bag type pit collapse in middle and lower Yangtze river. Yangtze River 38(6):40–42

Zhang XN, Jiang CF, Chen CY et al (2008) Types and features of riverbank collapse. Adv Sci Technol Water Resour 28(5):66–70

Zhang ZQ, Zang YP, Zhong L (2011) Pit collapse and emergency protection at the Sanjiangkou riverbank. Hydro-Sci Eng (2):71–75

Zhong L, Zang YP, Qian HF, et al (2011) Methods and typical case analysis of bank collapse control. China Water Resour (16):31–34

Grounding Risk Estimation in Inland Navigation with Monte Carlo Simulations and Squat Estimation

Juan Carlos Carmona[(✉)], Raúl Atienza, Raúl Redondo,
and José R. Iribarren

Siport21, Madrid, Spain
{jcarlos.carmona, raul.atienza, raul.redondo,
jose.r.iribarren}@siport21.es

Abstract. In inland ports, where access is done navigating along an estuary, river or artificial canal, the operation may be strongly conditioned by the tide (in case it has enough wide run) or the water level in the river. The variations in water level imply restrictions on the draft of the vessels that can access such ports.

Siport21 has been working for several years in ports of these characteristics, where there is no possibility to dredge the inland waterway. The alternative is to develop synchronization analysis tools, which allow identifying the "operational windows" and maximizing the draft of the vessels in transit operations. The result takes advantage of the tidal run by means of adequate planning, so that there is always enough underkeel clearance safety margin.

Grounding risk estimation is elaborated applying Monte Carlo method. A failure (grounding) function is defined, considering the propagation of the tidal wave (water level and current), ship speed along the waterway, wind conditions, squat, and other variables. Probability distributions of all variables involved are considered, so that thousands of random navigation conditions can be simulated. This allows to estimate the failure probability.

This methodology is applied to a practical case of a port that is carrying out actions to improve and optimize its operations. To do this, AIS data and tide data along the entire waterway, obtained from measurement sensors and a calibrated numerical prediction model, have been used.

Keywords: Squat · Measurements · Monte-Carlo · Risk · Inland

1 Introduction

The objective of the paper is to present the methodology used to determine the risk of contact with the bottom in the access and exit navigations to the inland port. With this aim, a model for calculating the probability of contact with the bottom during an individual transit has been created. The maximum permissible contact probability for each type of operation can be established from the acceptable global risk in this failure mode for the entire design lifetime of the port.

© The Author(s) 2023
Y. Li et al. (Eds.): PIANC 2022, LNCE 264, pp. 427–439, 2023.
https://doi.org/10.1007/978-981-19-6138-0_38

The model obtains the grounding probabilities of individual operations, so that they can be compared with previously established limit values. Several fundamental factors are involved in the calculations to consider:

- ship type
- speed
- tidal wave propagation
- synchronization with tidal wave
- bathymetry
- squat
- climate conditions

The risk model serves as a tool for optimizing the maximum draft allowed in port access and exit maneuvers.

2   Input Variables

2.1   Water Level and Current

The port has developed a very precise mathematical model to obtain predictions of water levels and currents throughout the waterway. During the navigation simulation of the ships, the tide and current curves allow to know precisely both the water level available and the current (intensity and direction) at any time and position of the ship. In this work, five tidal ranges are studied, corresponding to exceedance probabilities of 20%, 40%, 60%, 80% and 100% (Fig. 1).

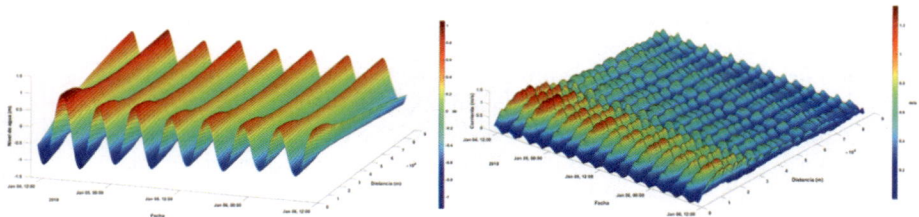

Fig. 1. Water level and current tidal waves (time-distance graphs)

2.2   Bathymetry

To obtain the total water column, it is necessary to know the depth of the waterway. The Port Authority has provided detailed bathymetric data obtained during measurement campaigns carried out on a scheduled basis for the maintenance of the waterway. From these bathymetric data, the profiles of the river sections along the different kilometer points are obtained (Fig. 2).

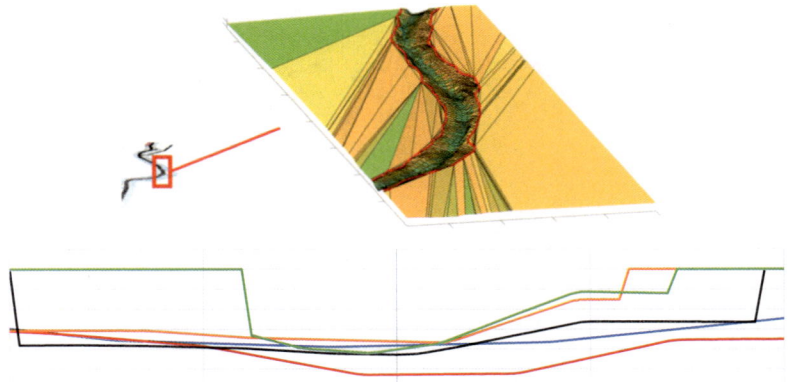

Fig. 2. 3D bathymetry and example sections

2.3 Ship Type

The type of the ship is a very important factor since, depending on U or V hull forms, the squat will vary significantly. In the project, ships are classified into three types, according to the block coefficient assigned to ship classes: cruise ship, container ship and bulk carrier. Types and main particulars of the ships studied have been obtained from AIS traffic data provided by the port, so that the usual types and dimensions were described (Table 1).

Table 1. Considered ships main particulars

Ship type	Container	Bulk carrier	Cruise
L_{OA} (m)	145.0	169.4	198.2
L_{BP} (m)	134.0	160.0	179.5
B (m)	22.0	27.2	26.0
C_B	0.7	0.8	0.6
Draft (m)	7.1 -7.5, each 0.1	7.0 -7.2, each 0.1	6.4

2.4 Speed

Navigation speed is critical in the simulation, as it determines the position of the ship on the waterway at any time. With position and time, both water level and current acting on the ship are obtained from the prediction model. The speed, intensity and direction of the current are fundamental parameters to compute ship squat. Depending on the water level, navigation is carried out with a different speed profile. These speed profiles are obtained from the provided AIS traffic data (Fig. 3).

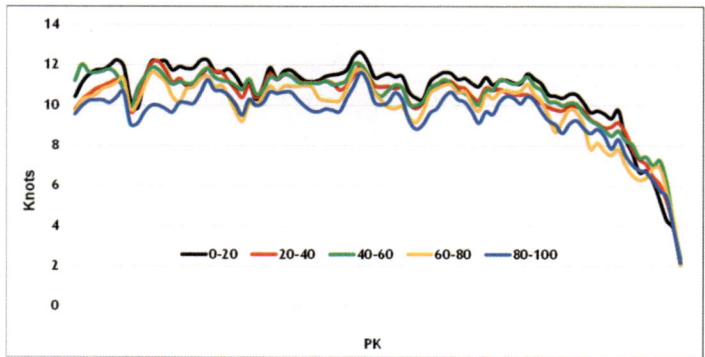

Fig. 3. Average speed profiles for different tide level steps

2.5   Environmental Conditions

The risk model includes the probability distributions for different climate agents, including joint distribution of wind and waves outside the estuary, wind distribution at the estuary and wind distribution in the interior zone. These distributions are used to generate random environmental conditions that are used in the Monte Carlo simulations (Fig. 4).

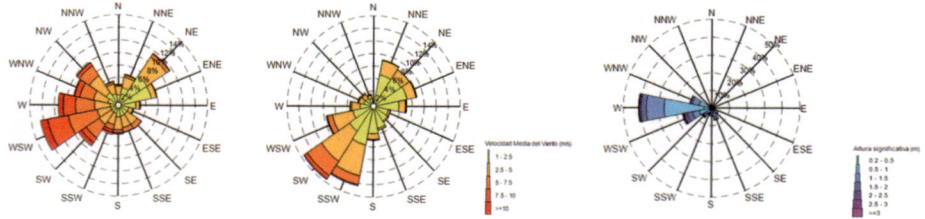

Fig. 4. Wind and wave distribution roses

3   Squat Estimation

To obtain a precise enough estimate of squat, calculations are made using empirical formulations and these calculations are compared with the results of simulations carried out using a numerical panel model. This comparison of results provides higher reliability and quality to the empirical estimates. The results of the numerical model are taken as a reference to correct the results of the empirical formulations because it is a more reliable and less conservative calculation method. All calculations are made for the three ship types listed above.

3.1 Empirical Calculations

Squat calculation through empirical formulations is carried out following the methods described in PIANC report No. 121-2014 "Harbour Approach Channels. Design Guidelines". This document recommends empirical formulas for squat prediction, suitable for several types of ships and channels. The different formulations for estimating the maximum squat presented in PIANC report analyzed are: Tuck (1966), Huuska/Guliev (1976), Barrass3 (2004), Yoshimura (1986) and Römisch (1989) (Table 2).

Table 2. Considered squat formulations

Formulation	Type of channel			Restrictions						
	U	R	C	F_{nh}	C_B	B/T	h/T	h_T/h	L_{pp}/B	L_{pp}/T
Tuck	x	x	x	-	-	-	-	-	-	-
Huuska/Guliev	x	x	x	≤0.7	0.6-0.8	2.19-3.5	1.1-2.0	0.22-0.81	5.5-8.5	16.1-20.2
Barrass3	x	x	x	-	0.5-0.85	-	1.1-1.4	-	-	-
Yoshimura	x	x	x	-	0.55-0.8	2.5-5.5	≥1.2	-	3.7-6.0	-
Römisch	x	x	x	-	-	2.6	1.19-2.25	-	3.5-9	-

3.2 Numerical Calculations

The numerical model used, based on a potential panel method, allows obtaining the forces and moments generated on a ship sailing along a restricted channel. In this way, from the hydrodynamic interaction forces and moments between sailing ships and the environment (bathymetry, slopes, confinements, …) it is possible to assess the suction and repulsion forces and, therefore, the sinkage and trim of ships. The model considers the specific hull shape of each ship, in a given load condition, as well as representative channel cross sections and allows the calculation of the hydrodynamic interaction (Fig. 5).

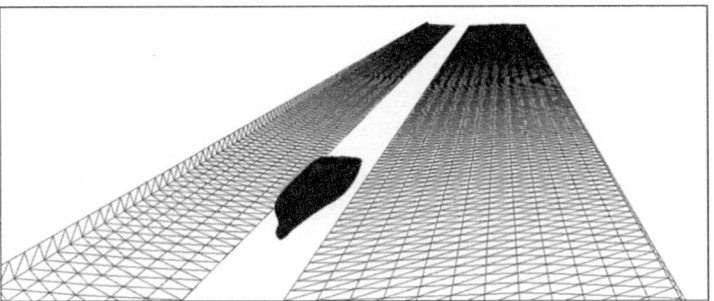

Fig. 5. Numerical model panel example

3.3   Calculation Results

Five sections of the waterway are chosen to make the comparison between empirical and numerical calculations. The sections are chosen to cover the three generic navigation channel typologies (U, R and C) and both symmetric and asymmetric sections are analyzed. The following figure shows the comparison of results obtained with empirical formulations (color bars) and numerical model (black dot) for one of the sections. The x-axis shows first water relative navigation speed in knots, critical speed

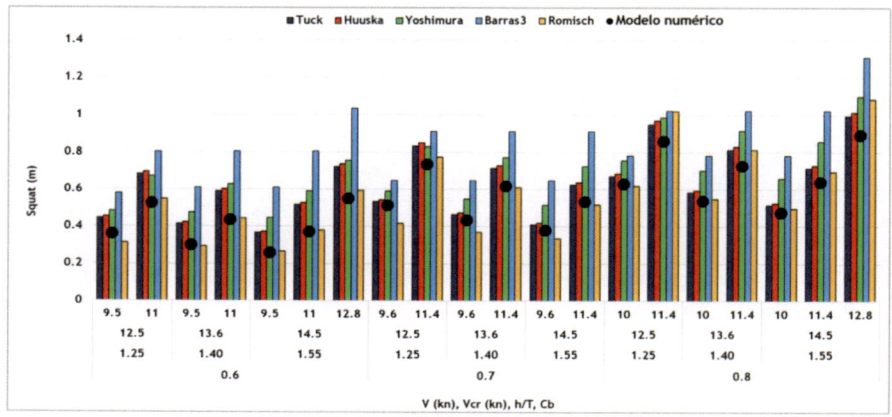

Fig. 6. Example of squat estimation results

in knots, h/T ratio, and finally Cb (Fig. 6).

The formulations with the highest stability in terms of deviation from numerical calculations are Tuck and Huuska/Guliev. It is proposed to use the Huuska/Guliev formulation, since it is somewhat simpler than Tuck's. The previous results allow to obtain correction coefficients for Huuska formulation depending on the section and ship type.

3.4   Full Scale Measurements

In order to validate the squat estimation method described, full scale squat measurements were conducted during a complete navigation onboard a container ship. Next table shows the main particulars of the vessel (Table 3).

Table 3. Measured ship main particulars

L_{OA} (m)	137.5
L_{BP} (m)	126.0
B (m)	22.0
Draft (m)	6.7 aft, 6.4 fore

The instrumentation consisted of a GPS unit with RTK precision (2 cm) for linear position (in plan and vertical) and an IMU for angle measurement. Next figure shows the instrumentation used (Fig. 7).

Fig. 7. Measurement instrumentation

The following plots show speed recorded by GPS, tide level during navigation as well as roll and pitch angles (Fig. 8):

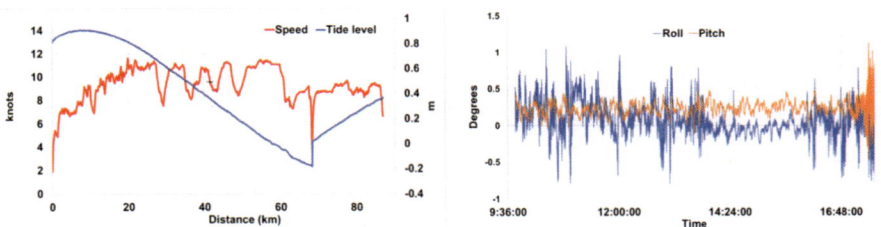

Fig. 8. Speed, Tide level and angles during navigation

The maneuver included an intermediate stop at a deeper section to let the low tide pass safely, therefore, a step in tide level can be found in the left plot. The following plot shows the comparison of the vertical movement recorded by GPS (zero reference was set at zero speed) and the squat estimate for the navigation (Fig. 9).

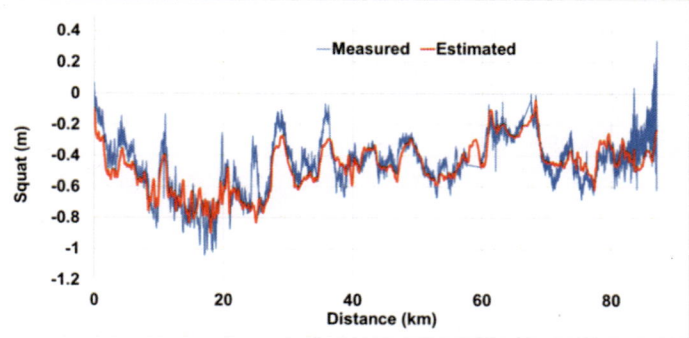

Fig. 9. Comparison of measured and estimated Squat

The previous plot shows that squat profile in general follows the speed profile as expected. There are some sections in which the difference between measurements and estimation reaches about 0.2 m, but in general the estimation is very close to the measurements. Risk estimations were carried out with a squat calibration based on previous data.

4   Risk Estimation

4.1   Acceptable Risk and Probability

Admissible risk is expressed by the probability of a failure occurring during the design lifetime of the facility. The following table shows admissible risk levels depending on the consequences of the failure (Table 4):

Table 4. Admissible risk levels based on the consequences (ROM 3.1-99, Spanish Port Authority).

RISK OF FAILURE INITIATION		
Economic impact	Possibility of human losses	
	REDUCED	EXPECTED
LOW	0.5	0.3
MEDIUM	0.3	0.2
HIGH	0.25	0.15
RISK OF TOTAL FAILURE		
Economic impact	Possibility of human losses	
	REDUCED	EXPECTED
LOW	0.2	0.15
MEDIUM	0.15	0.1
HIGH	0.1	0.05

Considering the type of operations carried out in this case, the allowable risk level is set at 0.5 (low economic impact in case of grounding in soft bottom together with limited risk of victims). To obtain the admissible probability of failure in a single

Table 5. Recommended return periods in risk assessment (PIANC 121 2014)

Type of Channel	Channel Bed Condition		
	Hard	Medium	Soft
General Navigation Channel	E1: 50	E1: 35	E1: 25
	E2: 250	E2: 150	E2: 100
	E3: 800	E3: 520	E3: 400
Specific Industrial Channel	E1: 35	E1: 25	E1: 15
	E2: 150	E2: 100	E2: 50
	E3: 500	E3: 350	E3: 250
E1= Low risk of loss of human life or environmental damages			
E2= Medium risk of loss of human life or environmental damages			
E3= High risk of loss of human life or environmental damages			

transit, it is necessary to establish the return period, or design lifetime that must be considered in the application of the admissible risk level. The following table shows the recommended return periods depending on the type of channel, the type of bottom and the risk of human life loss (Table 5):

In the case of the analyzed port, the navigation channel is general type, with soft bottom and low risk of human losses and environmental damage (E1). Therefore, the recommended return period is 25 years. As indicated above, the risk level is associated with the probability of an accident or failure occurring during the complete lifetime, in this case 25 years. The relationship between this overall probability and the probability of a bottom contact occurring during an individual operation is obtained using the long-term Poisson distribution:

$$P_{UKC} = 1 - \exp(-P_P \cdot N_Y \cdot Y_L) \tag{1}$$

where,

P_{UKC}: Failure probability in the return period considered

Y_L: Return period

N_Y: Operations per year

P_P: Failure probability per transit

Solving P_P from the previous equation, we obtain:

$$P_P = \frac{-\log(1 - P_{UKC})}{N_Y \cdot Y_L} \tag{2}$$

Expression by which the maximum admissible probability for an operation can be obtained.

AIS data statistical analysis allows to obtain the acceptable probability value of 0.0001 ($1*10^{-4}$).

4.2   Failure Probability Estimation

To obtain the probability of contact with the bottom due to vertical ship movements, the probabilistic method is used through Monte Carlo simulations. In this method, a failure function is defined, which depends on a series of variables with random variability. Once the behavior of this randomness of the variables is known, a series of simulations is carried out for the failure function, introducing random values in the variables based on their known probability distribution functions. In this case, two failure modes are considered, therefore two failure functions are defined: UKC and maneuver (sailing outside the channel) failures.

The UKC failure function (bottom contact) used is defined as follows:

$$\overline{UKC(PK,1\ldots N)} = \overline{H(PK,1\ldots N)} + \overline{TL(PK,1\ldots N)} - \overline{T(PK,1\ldots N)} - \overline{Sq(PK,1\ldots N)} - \overline{HEE(PK,1\ldots N)}$$

$$(3)$$

where,

$\overline{UKC(PK,1\ldots N)}$ = Under Keel Clearance, in m

$\overline{H(PK,1\ldots N)}$ = Channel depth, in m

$\overline{TL(PK,1\ldots N)}$ = Tide level, in m

$\overline{T(PK,1\ldots N)}$ = Ship draft, in m

$\overline{Sq(PK,1\ldots N)}$ = Ship squat, in m

$\overline{HEE(PK,1\ldots N)}$ = Environmental agents and heading effects

PK = Distance (km)

N = Number of random simulations.

This failure function (UKC) depends on five main variables (H, TL, T, Sq and HEE), which at the same time are a function of other variables. These variables will have a statistical distribution which must be defined to perform the random simulations of the Monte Carlo method.

The manoeuvre failure function (grounding because of exceeding channel width) is defined as follows:

$$\overline{WC(PK,1\ldots N)} = \overline{AW(PK,1\ldots N)} - \overline{OW(PK,1\ldots N)} \tag{4}$$

where,

$\overline{WC(PK,1\ldots N)}$ = Width clearance, in m

$\overline{AW(PK,1\ldots N)}$ = Available width, in m

$\overline{OW(PK,1\ldots N)}$ = Occupied width, in m

PK = Distance (km)

N = Number of random simulations.

The available width AW(PK, 1…N) includes the effect of the randomness of all the variables of UKC failure mode, therefore this variable depends on the variables of UKC failure mode.

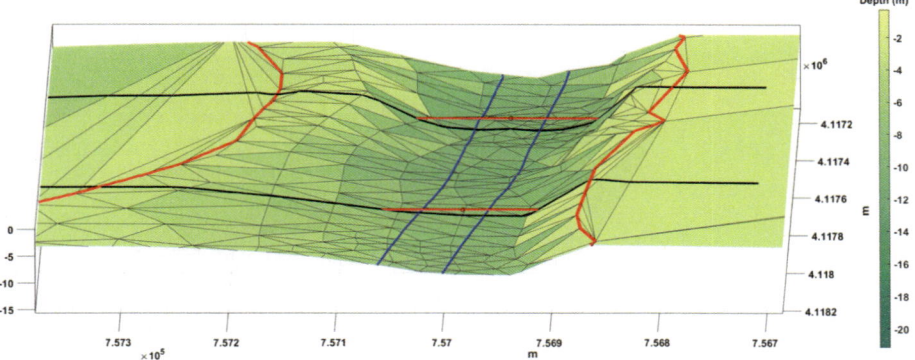

Fig. 10. Example of available UKC and width during navigation

On the other hand, the occupied width OW(PK,1...N) depends on environmental variables, navigation speed and ship maneuverability variables. The influence of the maneuverability variables is obtained through previous calculations applying SHIPMA autopilot maneuvering software. These calculations were calibrated with maneuvers carried out in SIPORT21 Real Time Bridge Simulator. The following figure illustrates the definitions of both failure functions. The available width and UKC in two sections of the river for the lowest point of the ship during navigation can be appreciated (Fig. 10).

5   Results

Navigation simulations were carried out varying the synchronization with the tide, from 2 h before high tide (HT−2) at the beginning of the manoeuvre to 0.5 h after high tide (HT+0.5). This allows to assess the operational windows. The following figures show different ship trajectories over the tide wave, depending on synchronization with high tide (red indicates high water level, blue indicates low water level) (Fig. 11).

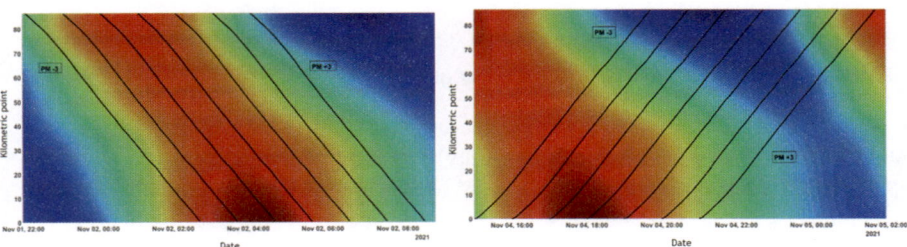

Fig. 11. Example of trajectories over tide wave for access (left) and exit (right) operations

Table 6. Probability results in access operations

Ship type	Draft (m)	Exceedance (%)	HT-2	HT-1.5	HT-1	HT-0.5	HT 0	HT +0.5
Container	7.1	100	1.E+00	6.E-01	2.E-02	<1E-4	<1E-4	<1E-4
	7.2	80	1.E+00	4.E-02	<1E-4	<1E-4	<1E-4	<1E-4
	7.3	60	1.E+00	1.E+00	7.E-02	<1E-4	<1E-4	<1E-4
	7.4	40	1.E+00	1.E-01	<1E-4	<1E-4	<1E-4	<1E-4
	7.5	20	1.E+00	1.E+00	9.E-03	<1E-4	<1E-4	<1E-4
Bulk carrier	7.0	100	2.E-02	2.E-03	<1E-4	<1E-4	<1E-4	<1E-4
	7.1	80	5.E-03	<1E-4	<1E-4	<1E-4	<1E-4	<1E-4
	7.2	60	1.E+00	4.E-02	<1E-4	<1E-4	<1E-4	<1E-4
	7.2	40	8.E-04	<1E-4	<1E-4	<1E-4	<1E-4	<1E-4
	7.2	20	4.E-01	<1E-4	<1E-4	<1E-4	<1E-4	<1E-4
Cruise	6.4	100	<1E-4	<1E-4	<1E-4	<1E-4	<1E-4	<1E-4
		80	<1E-4	<1E-4	<1E-4	<1E-4	<1E-4	<1E-4
		60	<1E-4	<1E-4	<1E-4	<1E-4	<1E-4	<1E-4
		40	<1E-4	<1E-4	<1E-4	<1E-4	<1E-4	<1E-4
		20	<1E-4	<1E-4	<1E-4	<1E-4	<1E-4	<1E-4

The model obtains the failure probability for each case considered (ship, tide range and tide synchronization). As an example, next table shows results for access operations. Cases in which probability is lower than the limit are highlighted in green and in red if the probability is higher (excessive risk) (Table 6).

This allows to detect the safe combinations of ship type-draft-tidal range and the most adequate planning of the access transit (when to start the maneuver in order to keep safe UKC margins).

6 Conclusions

Applying the methodology developed it is possible to obtain the following information about operations in the river:

- Maximum acceptable failure probabilities for each operation
- Failure probability for access and exit operations for a specific vessel type and draft
- Operational windows (relative to high water)
- Water level along the waterway
- Squat along the waterway
- UKC available along the waterway
- Critical points (km) in which failure occurs.

All this information makes possible to establish the maximum access and exit drafts based on the tide level (forecasted every day). To achieve this, it is necessary to have available a tool for predicting the tidal wave propagation along the waterway (time-distance curves). The port studied has developed such tool, validated with water level and current measurements. It is also necessary to elaborate a squat estimate with adequate precision, validated by field measurements on board ships during operations.

On the other hand, in access channels with soft and changing bottoms, as in this case, a precise and updated knowledge of navigation channel bathymetry allows the optimization of maximum access and exit drafts. These values can be kept up-to-date in case of getting new bathymetric data periodically.

Finally, the methodology also provides information about critical sections, which is a very relevant indication for the maintenance of the waterway.

References

Albers T, Reiter B, Treuel F (2018) Measurement and analysis of ship's squat on the river Elbe, Germany

Briggs M, Debaillon P (2014) Comparisons of PIANC and numerical ship squat predictions for rivers Elbe and Weser

Harkin A, Harkin J, Suhr J, Tree M, Hibberd W, Mortensen S (2018) Validation of a 3D underkeel clearance model with full scale measurements

PIANC (1997) Approach channels. A guide for design

PIANC (2014) Harbour approach Channels design guidelines

Puertos del Estado (2000) ROM 3.1-99 proyecto de la configuración marítima de los puertos; Canales de Acceso y Áreas de Flotación

Verwilligen J, Marc Mansuy M, Vantorre M, Eloot K (2018). Full-scale measurements to assess squat and vertical motions in exposed shallow water

Handling Accidents and Calamities in Hydraulic Structures – Objectives of Pianc Working Group WG-241

Ryszard A. Daniel[1(✉)], Timothy M. Paulus[2], Linda Petrick[3], and Yves Masson[4]

[1] RADAR Structural, Gouda, The Netherlands
r.a.daniel@xs4all.nl

[2] U.S. Army Corps of Engineers, Washington, DC, USA
timothy.m.paulus@usace.army.mil

[3] WEHNER Beratende Ingenieure, Bremen, Germany
l.petrick@wehner-ingenieure.de

[4] Compagnie Nationale du Rhône, Lyon, France
y.masson@cnr.tm.fr

Abstract. While the prior objective of hydraulic structures (such as lock gates, navigation river weirs and storm surge barriers) is to remain in service, engineers must also be capable to adequately handle their failures. Despite the ongoing development of expertise, design tools, norms, and construction methods, there are still a considerable number of accidents and calamities that happen to such structures. In addition, the losses and costs of damages as result of these so-called "upset events" are growing due to the growing complexity of waterborne infrastructure, intensity of navigation or other utilization of inland waters.

Accidents to hydraulic structures happen not only when their loads exceed the design strength. Other possible causes are, for example, unforeseen conditions, lack of inspection and maintenance, improper operation, and navigation errors. These other causes of accidents are often less controlled by technical norms than the relations between loads and resistances of structures. In addition, there are often combinations and complex sequences of events that may lead to disastrous results.

So far, various PIANC Working Groups have provided guidance for preventing accidents from happening, e.g. PIANC (2019) and PIANC (2020), including the accidents resulting from ship collision, e.g. PIANC (2014) and PIANC (2018). While this should remain the engineer's main concern, there is also a demand for more guidance how to effectively handle the accidents and calamities that actually happen. This is a matter of combined effort of not only engineers. Nevertheless, engineers can and should contribute to the solutions in such cases. Therefore, a new PIANC InCom Working Group has been established to investigate the existing practices in handling accidents and calamities; and to provide guidance in this field for professionals involved. This paper presents the objectives of the Working Group, selected investigation approach, some preliminary investigation results, and the envisioned contents of the final report.

© The Author(s) 2023
Y. Li et al. (Eds.): PIANC 2022, LNCE 264, pp. 440–453, 2023.
https://doi.org/10.1007/978-981-19-6138-0_39

Keywords: Navigation · Waterway · Accidents · Calamities · Upset events · Handling accidents

1 Main Objective

The main objective of the WG is to investigate the accidents happening and to develop recommendations for mitigating the effects of accidents. This task has been undertaken by an international group of structural and mechanical professionals, specialized in hydraulic structures such as navigation locks, weirs in navigable rivers and canals, shipyard docks, flood and tide barriers. The group started its proceedings in November 2021. It aims to collect, assess and systemize the existing know-how on handling accidents and calamities in hydraulic structures, including failure mechanisms, their assessments, and the rules of effective handling. The causes of accidents, such as unforeseen conditions, lack of maintenance, improper operation and navigation errors will be evaluated. The investigation will be focused on the engineering aspects but in correlation with the multi-task and multi-disciplinary actions to be taken after the accident has actually happened. The WG is aware that handling accidents and calamities is an interdisciplinary matter involving the actions like:

- Stabilizing the situation, limiting the damage;
- Rescuing endangered people and property;
- First reassessment of preventive measures;
- Preventing failures of related structures;
- Informing the stakeholders (like services involved, local residents, navigation);
- Restoring control of water flow;
- Documenting relevant events, collecting data and evidence;
- Investigation, choice of repair strategy;
- Temporary measures prior to actual repair;
- Planning and contracting of the repair;
- Performing the repair, commissioning, resuming normal operation;
- Implementing lessons learned.

Most of these actions can substantially be optimized when the root causes of accidents are properly investigated, identified and communicated. These root causes are often related to engineering and operation, which explains the focus of the WG. An example is the accident at the Rhône River Sablons Lock in France on February 18, 2020 (Fig. 1), which was caused by a number of technical issues with the downstream gate. As a result, 2 central gate elements collapsed under the pressure of 8 m of differential water head. A barge carrying 2200 tons of toxic, highly inflammable Vinyl Chloride Monomer (VCM) was flushed downstream and severely damaged. This accident, further discussed in the WG Case Study by Masson (2022), resulted in a number of local evacuations and a navigation interruption of one and a half months.

Fig. 1. Barge *the Pampero* after accident at the Sablons Lock, France, photo by Vivre Ici Vallée du Rhône Environnment

The Working Group will, occasionally, also address other than technical backgrounds of accidents and calamities. However, the WG does not intend to deliver a comprehensive guideline for the management of accidents and calamities, and particularly not for the issues like the division of responsibilities, mitigating the potential losses, evacuation plans, communication procedures etc. The focus remains on the technology and on the interaction between the crews and the users of navigation sites on the one hand, and the affected structures and systems on the other hand.

2   Classification of Accidents

One of the first steps after the accident happens is the classification of its severity. This classification usually determines the organizational level and the procedures of further handling. The Working Group will investigate the regulations in this matter followed various countries, including those of the WG members, compare the actual field experiences with these regulations, and draw general conclusions, possibly with recommendations based on lessons learned.

The most global differentiation is that between "routine" *accidents* and *calamities*. Merriam Webster defines a calamity as "*a disastrous event marked by great loss and lasting distress and suffering*". In this view, any accident or failure that endangers life or health should be categorized as a calamity. Any classification – including the differentiation between accident and calamity – may, however, also depend on the perspective. For example, the damage to a lock gate caused by the ship collision shown in photo (a) of Fig. 2 will in the first instance be a calamity for the waterway administration, while the sunken vessel in photo (b) of Fig. 2 will harm more the shipping company, see Daniel and Paulus (2019/2).

 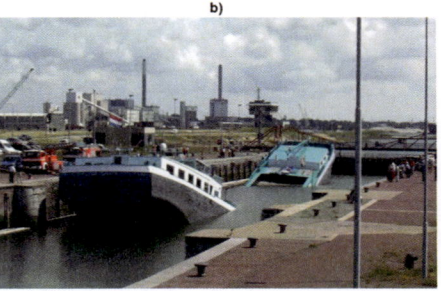

Fig. 2. a) Ship collision with Kiel Canal lock gate, Germany, photo by F. Behling, Kieler Nachrichten; b) sunken barge in IJmuiden Locks, Netherlands, photo by Rijkswaterstaat

The Working Group intends to determine objective, well-balanced criteria of accident classification, taking into account the interests as well as damages and losses suffered by all parties involved.

3   The WG-241 Plan of Action

A general plan of the WG investigations was laid out in the "Terms of Reference" (TOR), in accordance with the PIANC standard procedures. This document was discussed and approved by the PIANC Inland Navigation Commission (InCom) in March 2021. It describes the intended final product (report), emphasizing that it should draw clear distinctions between emergencies and regular procedures in face of a calamity. The report should by no means be received by the engineering community as imposing bureaucratic or other limits to the methods of accident handling. It should rather provide awareness of the consequences of different approaches; and recommend good practices.

On its kick-off meeting in November 2021, the Working Group decided that an effective approach to meet these PIANC InCom Terms of Reference will be to first perform a number of case studies of real-life accidents and calamities, and the manners in which they had been handled. The factual material collected in this way will allow for comparisons, assessments of effectiveness of various actions, and – finally – conclusions and recommendations of good practices. At the time of writing this paper, a number of case studies have already been performed. In addition, various accident investigation reports by other parties have been collected.

Simultaneously, the WG identified and reviewed the existing publications in the field of accident handling, including those of PIANC. Although accidents and calamities in hydraulic structures have not, so far, been covered by PIANC in a single comprehensive report, some specific aspects of this subject were discussed as part of other concerns. The following existing PIANC reports appeared to contain helpful information:

- WG 112: Mitigation of Tsunami Disasters in Ports;
- WG 119: Inventory of inspection and underwater repair techniques of navigation structures (concrete, masonry, and timber) both underwater and in the dry;
- WG 137: Navigation Structures – their Role within Flood Defense Systems;
- WG 151: Design of Lock Gates for Ship Collision;
- WG 155: Ship Behavior in Locks and Approaches;
- WG 175: A Practical Guide to Environmental Risk Management for Navigation Infrastructure Projects;
- WG 192: Developments in the Automation and Remote Operation of Locks and Bridges;
- TG 193: Resilience of the Maritime and Inland Waterborne Transport Systems.
- In addition, the following still active PIANC Working Groups might potentially be of interest:
- WG 182: Underwater acoustic imaging of waterborne transport infrastructure;
- WG 199: Health monitoring for port and waterway structures;
- WG 215: Accidental impact of ships on fixed structures – Update of PIANC WG 19;
- WG 233: Inspection, maintenance, and repair of waterfront facilities.

All these studies resulted in the intended layout of the Working Group report as globally indicated below. This layout is now a framework for further activities of the Working Group.

1. **Summary and Terms of Reference:**

   - Introduction, objectives, Terms of Reference;
   - Summary, references to Appendices.

2. **Context and classification of accidents:**

   - As far as helpful to activate appropriate procedures, no bureaucratization;
   - Simple classification, there is normally no time to loose.

3. **Identifying and reducing the risks:**

   - Although the subject is handling accidents that actually happen (i.e. not preventing them from happening), the general links to risk assessment are desired;
   - Focus on construction, operation, maintenance, not only the strength of structure.

4. **Investigation of accidents:**

   - When, by whom, and how deep? Relation with classification of accidents;
   - Knowledgeable, interdisciplinary, impartial, specific, detailed, …
   - Why the design did not prevent the accident from happening?

5. **Handling life safety risks:**

   - Life safety as an absolute priority;
   - Life safety in design, operation, maintenance, …

6. **Recovering from the damage:**

   - Differences from 'routine' projects;
   - Discuss (further develop?) the "fish diagram" (see drawing in Fig. 5), global command structure: who is managing what and when.

7. **Evaluation, recommendations and lessons learned:**

   - Focus on the handling of accident, do not repeat the investigation;
   - When, by whom, how deep? Distance from the "issues of the day".

4   Investigations of Accidents

In principle, all accidents to hydraulic structures should be investigated and evaluated, although the level and depth of investigations may differ from case to case. The prior objective of investigations is to determine the direct causes and the factors that contributed to the accident. Waterway administrations may have specific regulations that determine the level and scope of investigations depending on the severity of the accident or calamity. Table 1 below has been extracted from such regulations by the U. S. Army Corps of Engineers that administrates and develops the U.S. waterways, see USACE (2010).

Table 1. Classification of accidents and their investigations according to USACE regulations

	Class A	Class B	Class C	Class D
Cost of USACE property damage	$\geq$ \$ 2,000,000	$\geq$ \$ 500,000 but < \$ 2,000,000	$\geq$ \$ 50,000 but < \$ 500,000	$\geq$ \$ 2,000 but < \$ 50,000
Injuries to USACE civilian personnel or contractor crews	An injury that results in a fatality or total permanent disability	Permanent partial disability, or when 3 or more personnel are hospitalized as inpatients	Nonfatal injury that causes at least one day away from work beyond the day/shift on which it occurred	Nonfatal injury that causes absence from work shorter than in Class C
Required documents and investigations	▪ a preliminary accident notification ▪ a report of serious accident ▪ an accident investigation report ▪ a Board of Investigation (BOI)	▪ a preliminary accident notification ▪ a report of serious accident ▪ an accident investigation report ▪ a Board of Investigation (BOI)	▪ a preliminary accident notification ▪ an accident investigation report	▪ a preliminary accident notification ▪ an accident investigation report

It can be observed that the scope of required investigations is wider for the accidents causing high losses in terms of costs and injuries to personnel (Class A and B) than for the accidents causing low losses (Class C and D). In particular, for Class A or B the Board of Investigation (BOI) will be launched, which includes experts from universities, research institutes and industry; while for Class C or D a routine, internal investigation will normally be sufficient. In both cases, however, the key instructions to be followed by personnel involved are as listed below, Daniel and Paulus (2019):

- Start the investigation timely, desirably within 1–2 days after the accident. Delays may result in forgotten details or removed evidence.
- Investigation needs to be impartial.
- Only knowledgeable, experienced personnel should be members of the investigation team.
- Ensure that all reports are timely completed, preferably within 30 days after accident.
- Determine who or what caused the accident, why it happened, where it happened, and how it could have been prevented.
- Identify all circumstances regarding the event. Was there high water? What time of day was it? What time of year was it? What was the temperature? Was it raining, storming etc.?
- Describe details, lessons learned, steps to prevent similar accidents or failures from happening again.

The regulations in other countries represented in the WG are similar, although often integrated in a wider context of handling accidents and calamities. For example in the Netherlands, the so-called 3-level approach has been recommended. It comprises: 1) Prevention, 2) Mitigation, and 3) Actual handling. The details can be found in ref. Rijkswaterstaat (2011).

5   Handling Life Safety Risks

Modern management of waterways and hydraulic structures is largely based on risk analysis. Life safety risks are normally given the most prominent place in this analysis. Yet, the analysis of life safety risks is often limited to those hydraulic structures, which – in the case of failure – result in the inundation of the downstream areas. These are mainly the closures in river dams, flood barriers and storm surge barriers. Like other risks, the life safety risks in such structures are plotted, analyzed and evaluated on fault trees and risk matrices. In the Netherlands, the risk analysis – loss of life risk in particular – was a base for dividing the country into the so-called "embankment circles", with legally limited probabilities of inundation, ref. Overheid (2008). The Dutch methodology to estimate the loss of life in flood risk management has, for example, been discussed by Jonkman (2007).

However, the existing legislation and risk analysis methods do not usually capture the life safety risks of accidents such as a push-tow getting stuck under a dam gate, crew losing control of the vessel nearby the dam, maintenance crews losing control of their equipment, and the like. In all such cases, life safety risks concern the users or crews of hydraulic sites rather than the population of downstream areas. The challenge for the designers and managers of hydraulic structures is to identify such risks, to minimize them, and to provide appropriate actions when the accidents happen.

An example is the tragic accident at the Ohio River Montgomery Dam in January 2005 (Fig. 3) that cost the lives of four crewmembers of towboat *Elisabeth M*, as discussed by Sullivan (2016) and Daniel and Paulus (2019/1). Although the main cause was the extraordinary flow conditions, the incorrect assessments in field and the resulting human errors contributed to this accident as well. This can be seen from the description in image (a) of Fig. 3, while image (b) shows the sunken towboat. The tragedy resulted in technical measures and restrictions for the navigation through the lock.

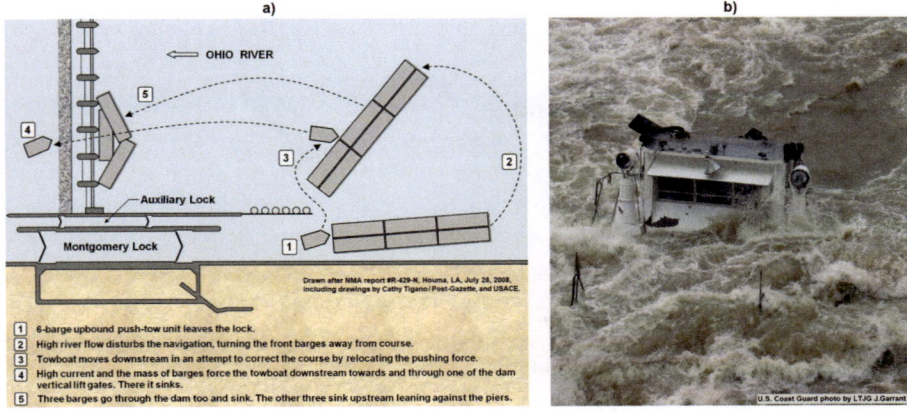

Fig. 3. Accident at Ohio River Montgomery Dam: a) subsequent stages, Daniel & Paulus (20219/1); b) sunken towboat, photo by US Coast Guard

Life safety risks do not only appear in large navigation structures. They may also occur at relatively small, recreational or other hydraulic sites. This should not be underestimated, considering that such sites are usually unmanned or remotely controlled, so there is nobody to undertake a rescue action. The least that should be done is then to discourage improper use of such structures by the public by placing barriers, clear warning signs and the like. Figure 4 presents two examples of such measures in small hydraulic structures in the Netherlands. The Working Group will also address such issues.

Fig. 4. Prohibition signs for swimmers and boaters on small river weirs, photos by authors

6   Recovering from the Damage

It is expected that the large number of case studies referring to actual accidents will enable the Working Group to distinguish typical stages in handling accidents and calamities. The identification of these stages will, in turn, help the professionals involved to effectively manage all the necessary activities and particularly the recovery from the damage. It is important, however, that such stages are not seen as a management's tool to limit the scope and depth of the required activities, but rather as a guideline to control the process. Accidents and calamities are, after all, "upset events" that happen and proceed irregularly, which requires a great deal of individual focus rather than standard approach.

The preliminary work has already been done in this matter. An example is the USACE Regulation 385-1-99 quoted earlier in this paper, ref. USACE (2010). The WG will also review relevant regulations and practices by other waterway administrations. An idea for discussion and possible further development is the so-called "fish diagram" (Fig. 5), proposed by Daniel and Paulus (2019/1, 2, 3). It distinguishes 5 typical stages of handling accidents and calamities. It also indicates the different urgencies of these stages, as well as the demand for personnel, material and other resources. The depicted relations are briefly commented below.

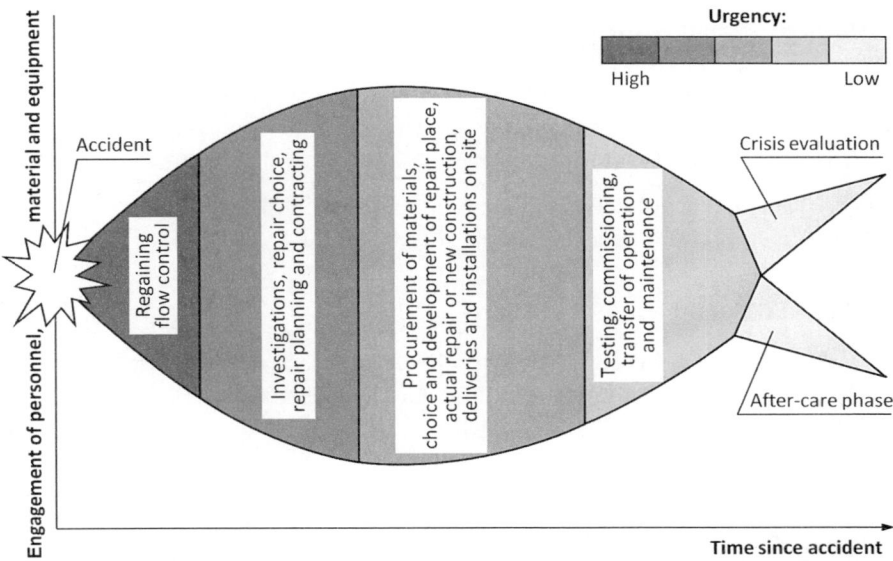

Source: R.A. Daniel, T.M. Paulus, *Lock Gates and Other Closures in Hydraulic Projects*, Butterworth-Heinemann, Oxford UK, Cambridge MA, 2019.

Fig. 5. Stages and engagement of resources when handling accidents

- The urgency to act is usually the highest immediately after (if possible even during) the accident. The emerging situation is then out of control. The priority is to regain that control and to prevent or at least limit any further damage.
- Despite the high urgency, the available means (personnel, material, equipment) to meet the emerging needs are then the lowest. Organizations may exercise freeing these resources at short notice, but it always takes time to effectively mobilize them.
- Once the control has been restored, one can inventory the damage, investigate its causes and plan the repair. The urgency decreases at this stage but is still high. The engaged resources are already substantial but the main work still needs to be done. It should begin quickly and efficiently.
- The repair can be done either by the in-house forces or by contractors or by a joined effort of both. USACE largely relies on the in-house forces. The administrations that have outsourced such forces (like Rijkswaterstaat) will need to contract all works. In both cases, the result should be a repaired or replaced hydraulic structure that can safely operate for many years to come.
- Upon delivery and installation of the repaired or replaced structure, the system will undergo a series of tests. Then it will be commissioned and handed over to the operation and maintenance crews. At this point, the work begins to resemble a regular project, although the urgencies are usually higher.
- The last stage is the so-called after-care phase when various "teething problems" may occur. They are solved in accordance with concluded agreements. This is also the time to evaluate the accident and to draw lessons learned. The evaluating team should include all main actors, but keep distance from the daily issues. This is marked by a split in the fish's tail in Fig. 5.

7 Example Case Studies

As already mentioned, the Working Group largely relies on real-life case studies while investigating the accidents in hydraulic structures and drawing appropriate conclusions. By means of examples, two of these case studies are presented below in more detail:

1. Ohio River Markland Lock gate failure in Sept., 2009, Paulus (2022);
2. Meuse River Grave Weir gate damage in Dec. 29, 2016, Daniel (2022).

7.1 The Markland Lock Gate Failure

The Markland Lock and Dam on the Ohio River is located near Warsaw, Kentucky, United States. The accident in question took place on September 27, 2009, and concerned the lower miter gate of the main lock chamber with the clear dimensions of 33.5 m by 365.7 m. The lock has also a parallel auxiliary chamber of the same width but a twice shorter length. Under normal conditions, both chambers and the dam carry the water level difference of 10.6 m.

The main reason of the lower gate failure was the uncontrolled filling of the lock chamber while the gate was improperly mitered. The flow entered the chamber too soon through the filling valves, overpowered the gate hydraulic drives and slammed the gate two leaves together in an incorrectly mitered position. In that condition, the gate was not capable of carrying the growing differential water head; and was forced by it over the sill. The strut arms broke away and the gate hinge anchors failed. One leaf broke entirely off and fell into water (Fig. 6a), while the other leaf remained upright far away from its operating position (Fig. 6b).

The accident caused a severe damage to both gate leaves, their drive struts, anchors and supporting structures. A cruise ship that was supposed to be locked upbound, broke off her moorings but did not suffer further damage. The navigation shut-down as result of necessary repairs lasted over 5 months.

Fig. 6. Markland Lock gate failure: a) anchors of sunken leaf; b) upright leaf forced off its position, photos by USACE

As the property loss due to the Markland Lock gate failure was over $9 million, the Board of Investigation (BOI) was called upon to investigate the accident, see Table 1 above. The BOI investigation report exposed additional circumstances and shortcomings that contributed to the lock gate failure. All these factors have been collected in the case study and will be analyzed by the Working Group. It is expected that this case will generate a number of lessons learned to be provided in the WG final report.

7.2   The Grave Weir Gate Damage

The Meuse River Grave weir damage of December 2016 was caused by a ship collision. This weir is of a so-called "bridge-weir" type, of which only two exist in the world. Hydraulic loads are carried by liftable water retaining panels, supported by vertical beams (posts) that, in turn, are hinged to bridge spans above the weir bays. During floods, the beams are hoisted out of water to the horizontal position under the bridge decks, in order to facilitate high water discharge. The Grave Weir has two such bridge spans. Under normal conditions, it controls a water level difference of 3 m.

On December 29, 2016, a downbound tanker *Maria Valentine*, carrying 2000 tons of benzene, rammed into the northern span at full speed. This happened early in the morning in a dense fog. The ship damaged several vertical beams and panels. Then she dove and passed the weir under the beams. It was considered a wonder that this did not cause any injuries or fatalities on board. Despite the damage on deck, the crew managed to anchor the vessel downstream of the weir. No benzene leaked into the river. Nevertheless, the weir structures were severely damaged (Fig. 7). This in combination with the late closing of the locks in the lateral canal caused the loss of navigation and substantial other damage in a wide area. The tanker itself suffered large damage to the appliances and equipment on deck, but no significant structural damage.

Fig. 7. Damage to Grave Weir after ship collision, photo by Rijkswaterstaat

Due to the large scale of damage, and the risks involved (e.g. in case that the benzene on board caught fire), the final stage of investigations was carried out by the Dutch Safety Board, which is an independent, highest level institution entitled to carry such investigations. The Dutch Safety Board decides by itself which incidents it will investigate and may do so when asked or on its own initiative. The Board decided to carry an investigation in January 2017, and the final report was released in May 2018. This report, like the BOI report on the Markland Lock accident, revealed a large number of organizational and other issues that contributed to the calamity and might have potentially caused still more damage. The Working Group will certainly consider the findings of this report in their further studies.

8 Conclusions

Since the Working Group has only recently been established, it is still too soon to derive conclusions from the investigations performed. The first substantial conclusions and recommendations as regards the handling of accidents and calamities can be expected mid 2023. Nevertheless, the discussions during WG meetings so far allow for a few general notes.

One general note is that while accidents and calamities in hydraulic structures are infrequent events, the consequences of these events are usually harmful for many parties. It is, therefore, highly desirable to develop not only the policy of handling these events but also the effective tools, like the structure of command, means of communication, legal framework for emergency actions, ways to mobilize the required expertise, supplies of materials and equipment etc.

Some administrations of, for example, waterways or harbors do already have such tools, but the real-life examples indicate that the handling of accidents or calamities is often far from optimal. It seems important that the organizations in charge of hydraulic structures:

a) maintain the readiness to properly act when accidents happen, for example by frequent exercising;
b) keep improving this readiness, for example by applying lessons learned from actual accidents.

These first observations will still be subject to further investigations, assessments and specifications by the Working Group. Should anybody from the PIANC community be willing to share their own experience, expertise, good practices or only ideas in this matter with the Working Group, then please feel encouraged to contact any of the co-authors of this paper.

References

Daniel RA, Paulus TM (2019/1) Lock gates and other closures in hydraulic projects. Elsevier Butterworth-Heinemann, Oxford – Cambridge, pp 924–926

Daniel RA, Paulus TM (2019/2). Handling accidents and calamities in hydraulic structures. In: 2019 IABSE congress "the evolving metropolis", New York City, 6 p

Daniel RA, Paulus TM (2019/3) Managing accidents and failures of hydraulic structures. In: International conference on structural failures, Międzyzdroje, Poland, 13 p

Daniel RA (2022) Case study: ship collision to the grave weir gate structures on the Meuse river. PIANC working group 241 case studies (in progress), 10 p

Jonkman SN (2007) Loss of life estimation in flood risk assessment. Delft University of Technology, 354 p. http://library.wur.nl/ebooks/hydrotheek/1875249.pdf

Masson Y (2022) Case study: sablons downstream gate failure. PIANC Working group 241 case studies (in progress), 8 p

Overheid (2008) Wet op de waterkering (in Dutch). De Minister van Verkeer en Waterstaat, The Hague. First published in 1995. https://wetten.overheid.nl/BWBR0007801/2008-07-01

Paulus TM (2022) Case study: markland lock gate failure. PIANC working group 241 case studies (in progress), 12 p

PIANC (2019) A practical guide to environmental risk management (ERM) in infrastructure projects. WG-175 report, PIANC EnviCom, Brussels, 70 p

PIANC (2020) Resilience of the maritime and inland waterborne transport system. TG-193 report, PIANC EnviCom, Brussels, 80 p

PIANC (2014) Design of lock gates for ship collision. WG-151 report. PIANC InCom, Brussels, 52 p

PIANC (2018) Accidental impact from ships on fixed structures. WG-215 report (in draft). PIANC InCom & MarCom, Brussels, in progress

Rijkswaterstaat (2011) Ruimtelijke kwaliteit in relatie tot Sociale veiligheid en security (in Dutch). Kort Tielens Architecten, Rijkswaterstaat DI, Utrecht, 52 p

Sullivan TM (2016). Historical perspective: navigation project incidents, periodic assessment workshop. U.S. Army Corps of Engineers, Institute for Water Resources, Risk Management Center, Louisville KY

USACE (2010) Accident Investigation and Reporting, engineering regulation ER 385-1-99, U.S. Army Corps of Engineers, Washington DC

Improvement Measures of Navigable Flow Conditions in the Baihutan Hydro-Junction in Changshanjiang River, China

Rongrong Wang[1], Aiping Sun[2], Shouyuan Zhang[1], Hongyu Cheng[3], and Gensheng Zhao[1(✉)]

[1] State Key Laboratory of Hydrology-Water Resources and Hydraulic Engineering, Ministry of Water Resources, Nanjing, China
gszhao@nhri.cn
[2] Quzhou Highway, Navigation and Transportation Management Center, Quzhou, China
[3] College of Harbour Coastal and Offshore Engineering, Hohai University, Nanjing, China

Abstract. The Baihutan hydro-junction of Changshanjiang River is located in the straight section of S-shaped channel, and the upstream approach channel is located in the concave bank of the bendy section of river. Navigable flow conditions in the entrance area of upstream and the bend connection area are complicated and disordered under the influence of circumfluence. Ships are prone to drift due to the high-speed transverse flow, which makes it difficult to meet the requirements of safe navigation. In the present study, the navigable flow conditions were evaluated by building a 1:80 hydraulic model and the remote-control ship model test were carried out. Besides, flow conditions in the entrance area, the approach channel and the bend connection area were investigated, and optimization and improvement measures were also proposed. Researches indicated that, affected by the water flow at the top of the bend and the narrowed river channel, the transverse and longitudinal flow velocity in the entrance area and the bend connection area were faster. Adjusting the layout of channel to reduce the angle between the mainstream and the axis of the channel, and expanding the cross-sectional area of the approach channel could effectively improve the navigable flow conditions and ensure the navigation safety. The results of this study can provide references for similar rivers.

Keywords: Baihutan hydro-junction · Navigable flow conditions · Entrance area · Physical hydraulic model · Ship model

1 Introduction

It commonly happens when building a new hydro-junction construction on a curved river of mountain area, while there are too many bends but few straight channels. The main characteristics of this type of river included shallow water, numerous beaches, more sharp bands and large surface slope (Jiang et al. 2022). This makes the flow conditions of approach channel and the entrance area more complicated, which is

Y. Li et al. (Eds.): PIANC 2022, LNCE 264, pp. 454–463, 2023.
https://doi.org/10.1007/978-981-19-6138-0_40

unfavorable for navigations (Yan 2021; Zhang 2020; Li 2019). Therefore, before constructing a new hydro-junction with a ship lock on curved rivers of mountain area, the navigation flow conditions should be studied deeply (He 2021; Sun 2016; Zhou 2008). Wang et al. (2021) compared how several engineering measures, including adjusting the height of the partition dam, adding partition walls and channel dredging, improve the navigable flow of a double-lane ship lock through the flood discharge variation test. Results showed that the comprehensive measure of adding partition walls and dredging the channel was effective in improving the flow conditions of the approach channel. There have been a large number of researches proposing several improvement measures for the navigable flow conditions. Shen et al. (2022) established a moving-bed partial model to predict the effects of sedimentation on the navigation and optimize the navigation condition of channels by adopting non-engineering measures such as dispatching the scouring sluice. Han et al. (2021) analyzed the flow conditions of the upstream and downstream approach channels and the connecting sections in the newly-built ship lock, and offered partial dredging as the improvement engineering measure. Besides, Wang et al. (2019) improved the flow conditions in the entrance area by adding spur dike complex and submerged dike complex. This paper aims to simulate the upstream flow of the montanic hydro-junction by building a physical hydraulic model, and improve the navigable conditions by proposing specific engineering measures.

Baihutan hydro-junction is the first hydro-junction built in the upper reaches of the "Changshanjiang River Development Key Projects" approved by the State Council. The navigation capacity of Baihutan is quite significant to the traffic volume of Changshanjiang River and Qujiang River. However, the bend in the upstream may limit and affect the navigation (Zhang 2020). This present study took the upstream of Baihutan hydro-junction as the research object, and built a normal physical hydraulic model to evaluate the flow conditions of upstream approach channel, entrance area and the bend connection area. Moreover, the remote-control ship model test is also applied for supplementary analysis. To sum up, engineering measures, including expanding the approach channel to concave bank to increase the cross-sectional area of the riverway, and adjusting the channel to avoid the top of the bend and accordingly reduce the angle between the mainstream and the axis of the channel, were proposed in this paper, which provide references for improving the navigation conditions of other hydro-junctions.

2 Hydro-Junction Information

The Baihutan hydro-junction is located in the Changshanjiang drainage basin in Zhejiang Province, China. The entire hydro-junction is lied in the straight section of the S-shaped riverway. Figure 1 shows the location and overall layout of the Baihutan hydro-junction. The layout of buildings from left to right successively includes the power station, sluice gates and ship lock. There are 17 sluice gates and the total net width of the flood discharge is 204 m with an 83-m bottom. The gates are divided into two parts by the longitudinal navigation walls. There are 8 gates on the ship lock side and 9 gates on the power station side, which are numbered 1#–17# from the power station to the ship lock by turns. The upper lock bay of the Baihutan Ship Lock is aligned with the axis of the sluice gates. Besides the lock chamber and the lower lock bay are located in the downstream of the dam axis. The available size of the lock chamber is 180 m × 23 m × 4 m (effective length × effective width × submergence), which is designed for navigating 500-ton ships. The levels of the ship lock and the navigation channel are Class IV. The approach channel of upstream is located in the upstream reservoir and connected by a curved section between the anchorage section and the navigation section. The anchorage section of upstream is 220 m long and 60 m width. The navigation section of upstream is 100 m long. In the initial design of upstream, the central angle of the bending section is 30°, the radius of centerline is 360 m, and the length of the centerline in the bending section is 172.8 m.

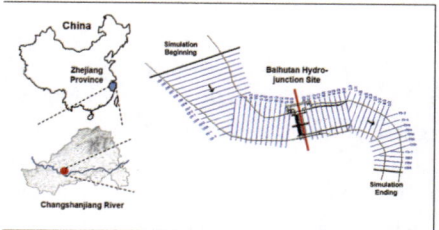

Fig. 1. Location of the Baihutan hydro-junction

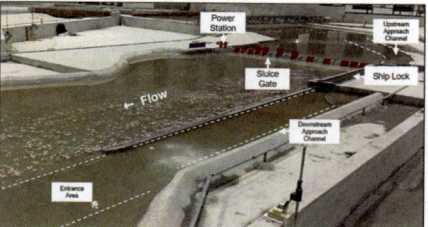

Fig. 2. Hydraulic model

The entrance area of the approach channel in upstream of the Baihutan ship lock is located in the concave bank with a typical large-angle sharp bend. When the water flow enters the bend, the water level in the concave bank is higher than that in the convex bank and forms a banked-up water level on the concave bank under the effect of centrifugal force. This will make the surface of water body flows to the concave bank, because its centrifugal force is greater than the pressure in water. While the bottom of water body flows to the convex bank since its pressure is greater than the centrifugal force. As a result, the main circulation is formed on the profile side. The transverse flow velocity of the water surface in the entrance area near the top of the bend may be large and the navigable flow conditions may be more complicated under the influence of the circulation flow. Excessive flow velocity may cause drifting or twisting to the ship. In order to avoid adverse flow conditions such as oblique flow, backflow and whirlpools affecting the navigation safety, it is necessary to further demonstrate whether the ship can navigate safely and propose specific engineering measures to improve the navigable flow conditions.

3   Experiment Design

3.1   Hydraulic Model

The normal fixed-bed hydraulic model is designed on the basis of Froude similarity criterion, and its geometric scale is 1:80. The model simulates the whole hydro-junction building, the upstream channel, the upstream approach channel, the bend connection area and the downstream river channel. The layout of the hydro-junction, the distance of the model flow adjusting from turbulent flow to subcritical flow and the length of the start-up distance of the remote-control ship model are fully considered when delineating the simulation boundaries. The upper boundary of the model is set at the straight section from the upstream of the bend, 1400 m away from the hydro-junction building. The boundary of downstream is set 1700 m away from the building. The model is built on the basis of the measured topographic data using the terrain profile analysis method. The model elevation is controlled less than ±1 mm, the closure difference of triangle or polygon is within ±1′, and the channel roughness is 0.030–0.035. The surface of the model is made of cement mortar. The discharge is measured and controlled by using the YC101E type electromagnetic flowmeter produced by SYCIF company, and its deviation is less than 1%. The water level is measured by water level point gauges with 0.01 mm precision. The flow velocity is measured by the Vector type acoustic doppler velocimetry produced by Nortek company, whose range is 0.1 cm/s–400 cm/s. The overall hydraulic model is shown in Fig. 2.

3.2   Remote-Control Ship Model

The design of ship model is consistent with the hydraulic model, which can be controlled wirelessly. According to the actual operation status of the Changshanjiang River waterway, ship with a size of 67.5 m × 10.8 m × 2.2 m (length × width × designed draft) is chosen for simulation. The outline of the ship, the blades and the rudder blades are scaled down to build the ship model. The draft of ship model and the headway speed in still water are calibrated before the experiment to ensure the consistency of maneuverability between the model and the real ship.

3.3   Experiment Cases

The upstream approach channel of the Baihutan ship lock is designed in an open type, which has no navigation walls in the upstream. This will make it easier and faster for ships going in and out of the lock, and more economical for the construction cost. However, if the sluice gates near the ship lock side are opened, the navigation of ships will be affected. Under the condition of meeting the normal discharge demand for the power station, the rest flow rate should be discharged through the left sluice gates, which is equidistant opened, so as to guarantee positive navigable conditions. Based the layout of the sluice gates, the opening distance regulation of the sluice gates and the opening sequence of the sluice gates, balancing the navigable conditions of upstream and downstream as well as the scouring of the downstream river channel, a variety of opening operation schemes of sluice gate were compared before the experiment (Zhang 2020; Zhang et al. 2019). Finally, five different flood discharge cases and sluice gate schedule were determined (see Table 1).

Table 1. Experiment cases of hydraulic model.

Case	Flood discharge (m^3/s)	Discharge through turbine (m^3/s)	Sluice gates schedule	Power station state	Description
1	102	102	All closed	Switch on	Power station operated alone
2	600	102	6#–8# opened	Switch on	Sluice gates and power station operated together
3	800	102	6#–8# opened	Switch on	
4	1000	102	6#–10# opened	Switch on	
5	1200	102	6#–11# opened	Shut off	Sluice gates opened alone

4   Navigable Flow Conditions in Initial Design

The riverbed topography of the upstream approach channel and the entrance area in Baihutan hydro-junction is complicated. The river narrows 110 m inward in the concave bank, which is 560 m away from the hydro-junction. When the ship goes down and enters the upstream approach channel, the maneuverability of the ship may be affected by the bed topography and the water flow in the bend. According to national standards of navigation conditions, the longitudinal flow velocity in the entrance area of the approach channel in Class IV should be less than 2.0 m/s, the transversal velocity should be less than 0.3 m/s, and the backflow velocity should be no more than 0.4 m/s. Moreover, the longitudinal flow velocity of the braking section and the anchorage section should be not more than 0.5 m/s, the transversal flow velocity should be not more than 0.15 m/s, the navigation section and the adjustment section of the approach channel should be still water.

Table 2. Measured flow velocity in initial design.

Flood discharge (m^3/s)	Maximum measured surface flow rate (m/s)								
	Bend connection area			Entrance area			Approach channel		
	Longitudinal flow	Transversal flow	Backflow	Longitudinal flow	Transversal flow	Backflow	Longitudinal flow	Transversal flow	Backflow
102	0.10	0.07	0.02	0.10	0.05	0	0.09	0.02	0
600	1.11	0.19	0	0.91	0.16	0.02	0.41	0.07	0.04
800	1.27	0.22	0	1.08	0.17	0.02	0.59	0.11	0.04
1000	2.00	0.37	0	1.80	0.33	0	0.77	0.17	0.07
1200	2.21	0.48	0	2.10	0.45	0	0.80	0.18	0.08

Results showed that when the flood discharge was 102 m³/s–800 m³/s, the maximum velocity in the navigation section and the adjustment section of approach channel is 0.04 m/s, and the water surface is the still water without fluctuations (see Table 2). The maximum velocity in the anchorage section is 0.60 m/s, and most of the flow velocity is within the range of 0.45 to 0.53 m/s. The maximum longitudinal velocity is 0.59 m/s, and the maximum transversal velocity is 0.11 m/s. Compared with the above-mentioned areas, the water flow velocity in the entrance and the connection area of the bend, is larger. The maximum longitudinal velocity in the bend is 1.27 m/s, the maximum transversal velocity is 0.22 m/s, and the transversal velocity in all measuring points is less than 0.30 m/s. Therefore, when the flood discharge is less than 800 m³/s, the flow velocity in the entrance area of the upstream approach channel and the bend connection area can both meet the requirements for navigation.

When the flood discharge is 1000 m³/s, the navigation section and the course adjustment section of the approach channel are still water areas. The maximum velocity of the anchorage section is 0.83 m/s, the maximum longitudinal velocity is 0.77 m/s, and the maximum transversal velocity is 0.27 m/s. In the bend connection area, the maximum longitudinal flow velocity is 2.00 m/s and the maximum transverse flow velocity is 0.37 m/s. When the discharge reaches 1200 m³/s, the flow velocity in the navigation section, the course adjustment section and the anchorage section of the approach channel can still guarantee the navigation safety. But the transverse flow velocity in the bend connection area is larger, and the transverse flow velocity in some individual measuring points can reach 0.48 m/s. From the perspective of the characteristics of riverway layout, the cross-sectional area of water body decreases, and the flow velocity increases under the effect of the shrinkage in the bend connection area. In addition, the location of the shrinkage is too closed to the top of the bend. It can be observed that the angle between the mainstream and the axis of the designed channel is about 10° when the flood discharge is 1000 m³/s, and the angle is about 14° when the flow discharge reaches 1200 m³/s. Larger angle makes the projection of the mainstream flow vector velocity in the normal direction of the channel centerline bigger, which causes larger transverse flow velocity in this area.

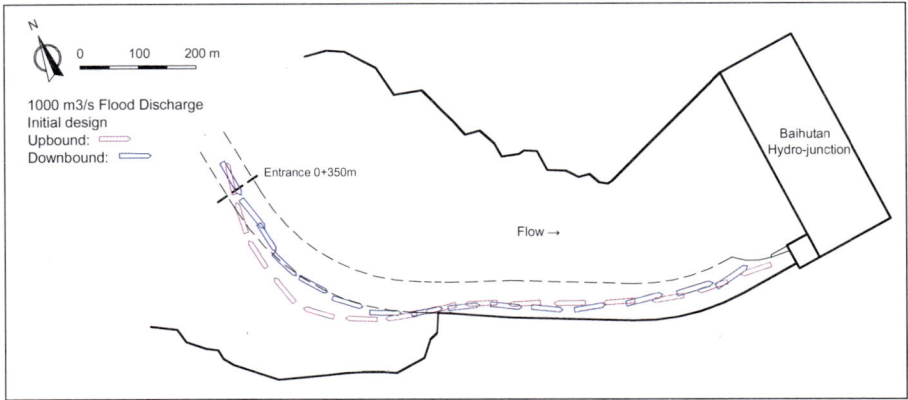

Fig. 3. The upbound and downbound sailing tracks in initial design.

In order to explore the influence of the large transverse flow velocity on the actual maneuvering of the ship in this area, a remote-control ship model was built to conduct the simulated tests under the 1000 m³/s flood discharge. Figure 3 shows the sailing tracks of the upbound and downbound of the upstream ships in the Baihutan ship locks when the flood discharge meets 1000 m³/s. Figure 3 indicated that when the ship goes up, the ship can be well controlled in all the navigation section, the course adjustment section and the anchorage section. However, the ship was found to drift toward the concave bank in the entrance area and the connecting section of the bend. Thus, the ship needs to be steered the full rudder in advance to avoid collision with the concave bank and deviation from the predetermined route. Ships need to be frequently steered in 0+100 m to 0+250 m away from the entrance. The transverse flow velocity in this area is about 0.25–0.42 m/s and exceeds the standard value. The ship enters the predetermined route in 350 m away from the entrance. The sailing track has large radian in the in the bend connection. The speed of the ship relative to the bank is about 2.6 m/s, and the maximum drift angle in the entrance area is 23°, which is larger than the limited value of 15°. Besides, the transversal swing angle of the ship is less than ±1.5° and has no obvious longitudinal tilt. When the ship goes down, the left rudder should be turned ahead to enter the approach channel. The speed of the ship relative to the bank is about 3.2 m/s, the maximum drift angle in the entrance area is 18°, and the drift angle is about 10°. The sailing track deviates to the concave bank, which is dangerous for sailing. Therefore, the transverse flow velocity exceeded the standard value in the entrance area and the bend connection area in the initial design, which make it infeasible for ships to navigate safely.

5   Optimized Navigable Flow Conditions

Adjusting the layout of channels, excavating local riverbed and adjusting shorelines are effective measures to promote the navigable flow conditions. There are two main reasons for the excessive flow velocity in upstream of the initial design. Firstly, the entrance area of the upstream approach channel is located in the sharp bend section of river, and the connection between the entrance area and the bend is too close to the top of the bend. Thus, the smooth navigation of ships was restricted by the bend. Secondly, the narrowed channel at the top of the bend leads to faster flow velocity, and the elevation of the bed in the concave bank is higher, which causes deflexion of water flow when passing through the top of the bend, and increase the angle between the mainstream and the axis of the channel. To reduce this angle and slow down the flow velocity, the following measures were taken: 1) Change the angle of the bending section between the anchorage section and the navigation section into 15°, and extend the approach channel toward the concave bank to expand the cross-sectional area of the riverway. 2) Extend the channel 150 m to the south, and connect the channel and the approach channel with a 72° curve to avoid the top of the bend. 3) Widen the inner side of the channel from the original 35 m on both sides to 35 m on the outer side and 70 m on the inner side. The differences between these two channel designs were demonstrated in Fig. 4.

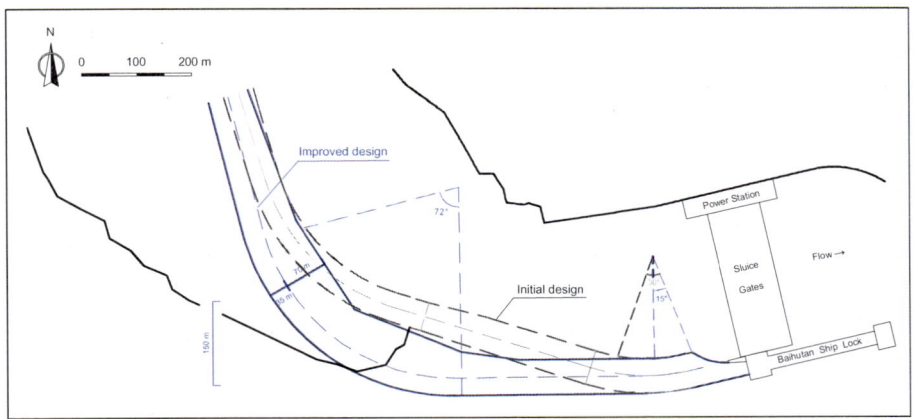

Fig. 4. Comparison of channel layout between improved design and initial design.

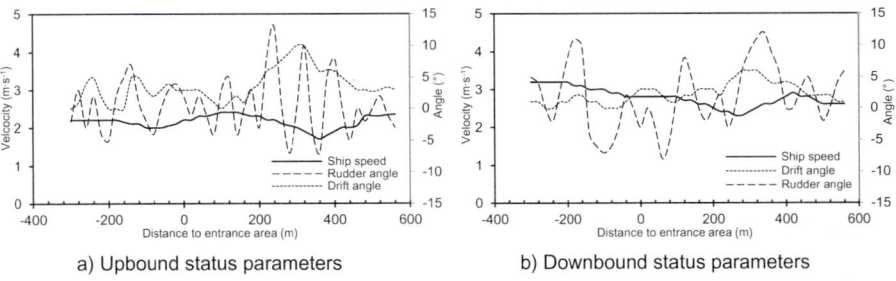

a) Upbound status parameters   b) Downbound status parameters

Fig. 5. Rudder angle, drift angle and ship speed process curves of upbound and downbound.

Table 3. Measured flow velocity in improved design.

Flood discharge (m³/s)	Maximum measured surface flow rate (m/s)								
	Bend connection area			Entrance area			Approach channel		
	Longitudinal flow	Transversal flow	Backflow	Longitudinal flow	Transversal flow	Backflow	Longitudinal flow	Transversal flow	Backflow
102	0.09	0.03	0.01	0.07	0.05	0.01	0.05	0.02	0
600	0.83	0.07	0	0.69	0.07	0.02	0.27	0.06	0.02
800	1.00	0.10	0	0.78	0.09	0	0.59	0.08	0.02
1000	1.54	0.15	0	0.97	0.15	0	0.77	0.10	0.03
1200	1.75	0.22	0	1.24	0.20	0	0.39	0.11	0.05

The flow velocity in improved design is shown in Table 3. Results indicated that when the flood discharge is 1200 m³/s, the navigation section and the course adjustment section is still water. The maximum flow velocity in the anchorage section is 0.42 m/s, the maximum longitudinal flow velocity is 0.39 m/s, and the maximum transverse flow velocity is 0.11 m/s. Besides, in the connecting area between the entrance area and the bend, the maximum longitudinal flow velocity is 1.75 m/s, and the maximum transverse flow velocity is 0.22 m/s. The angle between the mainstream and the axis of the channel reduced significantly, and the transverse velocity dropped about 56% in the same position and the same flow rate. The flow velocity in the entrance area, the approach channel and the bend connection area meet the requirements for navigation. The upbound and downbound sailing status parameters of the ship model are shown in Fig. 5. Results of the ship model test revealed that when the flood discharge is 1200 m³/s, the sailing track of upbound was smooth. Moreover, the ship was able to enter the upstream channel successfully under simple manipulations, among which the maximum drift angle throughout is 13°, the maximum drift angle of the entrance area is 10°, the average frequency of steering is 2 times/min, and the drift angle between the entrance area and the bend connection area is within 10°. What's more, the sailing track of downbound was also smooth, among which the maximum drift angle throughout is 12°, the maximum drift angle throughout is 6°, and the average frequency of steering is 3 times/min. Furthermore, no obvious tilt was observed and the ship could navigate easily.

To sum up, by adopting engineering measures including adjusting the layout of channels appropriately, changing the angle of bending section, expanding the approach channel and et al., the angle between the mainstream and the axis of the channel was reduced, and the navigable flow conditions in upstream were improved, which made it feasible to meet the navigation requirements under the flood discharge of 102–1200 m³/s.

6 Conclusions

This study evaluated the navigable flow conditions in the upstream of Baihutan hydro-junction by building a normal physical hydraulic model and a remote-control ship model. The shrinkage of the riverway in the entrance area leads to a large angle between the mainstream and the axis of the channel, which is the main reason for the high flow velocity in the entrance area and the bend. In the initial design with 1000 m³/s flood discharge, the maximum longitudinal flow velocity in the entrance area and the bend is 2.00 m/s, the maximum transverse flow velocity is 0.37 m/s, and the angle is about 10°. This affects the normal navigation, and engineering measures need to be taken to improve the flow conditions.

Reducing the angle between the mainstream and the axis of the channel is the main method to decrease the transverse flow velocity. What's more, other measures including adjusting the channel to avoid the top of the bend, reducing the central angle of the bending section from 30° to 15° and extending the concave bank 150 m away from the approach channel to increase the cross-sectional area, were all found effective in improving the flow conditions.

The maximum navigation discharge of the upstream in the initial design is 800 m³/s. After conducting the engineering measures mentioned above, the maximum navigation discharge increases to 1200 m³/s. In summary, the status parameters of the ship model indicated the effect of water flow on the ship. The sailing tracks and process curves of rudder angle, drift angle and ship speed were also proved to play an auxiliary role in the analysis of navigation conditions.

References

He YH (2021) Research on navigable flow conditions and improvement measures of shipai acute bend reach between the two dams of the three gorges. D. Chongqing Jiaotong University, Chongqing

Han K, Yu KW, Zhao JJ, Han CH (2021) Experimental study on navigation flow condition and scheme optimization of Gedi junction in Changshan River. J Port Waterway Eng 08:139–144

Jiang PF, Tong SC, Xu GX, Huang GX (2022) Review and application of determination methods on navigable hydraulic index of rapids in mountainous river. J. Ocean Eng 110830:1–12

Li JT (2019) Research on technological improvements with regard to navigable flow conditions at the entrance area of ship lock. In: Calvo L (ed) E-proceedings of the 38th IAHR world congress, IAHR, Panama, pp 4635–4640

Sun GD (2016) Study on improvement measures of navigation flow condition in the entrance area of second line ship lock of Mengli project. D. Chongqing Jiaotong University, Chongqing

Shen LQ, Chen L (2022) Experimental study on optimization model of navigation condition of Nianpanshan hydropower station. J/OL China Rural Water Hydropower 03:1–11

Wang JP, Ying FL, Chen YF (2019) Optimization of navigable flow condition of ship lock entrance area in curved river. J Port Waterway Eng 11:86–91

Wang B, He FF, Wang XY, Xu JC (2021) Study on navigable flow conditions of multi-lane lock in curved narrow channel section. J Port Waterway Eng 04:108–115

Yan ZQ, Zhao JJ, Gu JD, Wang Y (2021) Study on navigable flow conditions of S-shaped abrupt bend. J Port Waterway Eng 06:128–134

Zhou SQ (2008) Research on navigation conditions of approach channel. D. Chongqing Jiaotong University, Chongqing

Zhang M (2019) Numerical simulation of one-dimensional flow in the whole section of the Changshanjiang River navigation project in the upper reaches of the Qujiang River. R. Nanjing Hydraulic Research Institute, Nanjing

Zhang SY (2020) Navigable flow conditions of under approach channel—a case study of Baihutan hydraulic project. D. Hohai University, Nanjing

Innovative Mooring in Locks Using Shoretension: Density and Mooring Force Measurements in the North Lock Ijmuiden

Arne Van Der Hout[1(✉)], Allert Schotman[2], Thijs Hoff[2],
Jan Wim Van Der Veen[2], and Arnout Quax[3]

[1] Deltares, Delft, The Netherlands
Arne.vanderHout@deltares.nl
[2] Dutch Pilot Association, Loodswezen Amsterdam-IJmond,
IJmuiden, The Netherlands
[3] ShoreTension, Rotterdam, The Netherlands

Abstract. A field measurement campaign was conducted in the North Lock of IJmuiden, in which the density distribution in the lock chamber and the resulting mooring forces were measured. The innovative mooring system ShoreTension was applied in a navigation lock for the first time to investigate possible ways to increase safety of the mooring process. Density measurements provided valuable insight in the governing hydraulic processes during levelling and lock-exchange and most influential external drivers for density differences over the lock complex could be identified. Potential benefits and limitations of the ShoreTension system were identified, and valuable practical experience was gained during the test period.

Keywords: Navigation lock · ShoreTension · Innovative mooring · Density current · Field measurements

1 Introduction

1.1 Background

In the city of IJmuiden in the Netherlands, recently a new navigation lock "Sea Lock IJmuiden" is built, providing larger vessels access to the Port of Amsterdam. This new lock is currently the largest navigation lock in the world. In preparation to the operation this new large lock, more insight in nautical procedures and mooring line forces due to levelling and lock-exchange process was required. Therefore, prior to its opening on 26th of January 2022, a field measurement campaign was conducted in the North Lock of IJmuiden in 2020, which at that time was the largest operational lock of the IJmuiden lock complex.

The IJmuiden lock complex forms a boundary between fresh (North Sea Canal side) and salt water (North Sea side), see Fig. 1. For the IJmuiden Locks, density currents generate the most dominant loads on moored vessels during the lockage process. The physical process of density currents in navigation locks has been already extensively studied and described by e.g. Vrijburcht (1991). These effects were

© The Author(s) 2023
Y. Li et al. (Eds.): PIANC 2022, LNCE 264, pp. 464–478, 2023.
https://doi.org/10.1007/978-981-19-6138-0_41

carefully investigated in physical scale model research to assess the design of the Sea Lock IJmuiden (Van der Hout 2017; Kortlever 2018; Van der Hout 2018; Nogueira 2018). However, in order to translate theoretical force criteria used in the design of a lock to practical nautical procedures, more information on the present-day operation was needed. During the field test campaign described in the present paper, mooring line forces and the density distribution in the lock chamber were measured simultaneously to provide better understanding of the governing hydraulic processes.

Fig. 1. IJmuiden lock complex, with the names of the most important elements indicated (Google Earth).

In anticipation of mooring forces that were expected to be larger in the new Sea Lock IJmuiden, due to its larger dimensions and expected vessel sizes, than presently in the North Lock (van der Hout 2018), it was investigated whether innovative mooring solutions could be applied in a lock and whether it would help to increase safety of the lock operation. For this, the ShoreTension system (see Sect. 1.3) was tested in a field test campaign. Although it is a proven system in port settings, it was not known yet how such a dynamic mooring system would perform in a navigation lock.

1.2   North Lock of IJmuiden

All tests described here have been performed in the North Lock of IJmuiden. The lock chamber of the North lock has a length of 400 m between the outer gates and a width of 50 m. The bed level is located at 15 m below NAP (Amsterdam Ordnance Datum, which is almost equal to mean sea level). The lock is equipped with short culverts for levelling (see Fig. 3). At the canal side the water level remains relatively constant and varies around NAP-0.4 m. At the sea side of the lock the water level varies with the tide, with typical tidal ranges of NAP $\pm$ 1 m.

1.3   Shore Tension System

ShoreTension is a dynamic mooring system that has been developed for port operations and is already successfully applied worldwide at several terminals. The working principle of the ShoreTension system is that a cylindrical shaped unit, attached to the bollards on a quay wall, exerts a constant tension to the ship's mooring lines. This requires no energy except for an external hydraulic system which only needs to be used once to initialize ShoreTension at the correct pretension. After that, the cylinder of ShoreTension hydraulically moves along with the force variations in the mooring line. When the force in the mooring line becomes larger than a certain selected threshold value, the cylinder extends, thereby reducing the force in the line. Once the force in mooring line drops below a preset minimum value, the cylinder contracts again, thereby increasing the line force. Because of this dynamic operation, much control over pretension is achieved. The ShoreTension units are equipped with own, relatively stiff (Dyneema) mooring lines that need to be secured on board of the ship. This implies that a different mooring protocol needs to be followed than usual, as normally the ship's mooring lines are given out to be connected to the bollards on the lock wall. The used ShoreTension system has a Safe Working Load (SWL) of 150 Ton in fully extended position and a SWL of 100 Ton during normal operation. ShoreTension is equipped with sensors measuring the forces in the mooring rope and measurement data can be saved autonomously to a data file.

1.4 Objectives of the Test Campaign

The test campaign aimed to provide insight in the possibilities to increase safety of the mooring operation in a sea lock using an innovative mooring system like ShoreTension. It was meant as a pilot to gain real-life experience with the performance of the system. Next to insight in the applicability of the working principles of such a dynamic mooring system, also practical aspects were investigated during the tests. So, practical questions like "How long does it take to connect a ShoreTension unit?", "What protocol should the lines men follow to have a safe operation?" or "What mooring configuration works best?" have been addressed during the test campaign.

This paper provides an overview of the performed tests and will present insight in the most important hydraulic processes leading to mooring loads in sea locks. It also shows possible benefits and limitations of applying the ShoreTension system in lock operations. The knowledge obtained within this project and presented in this paper aims to make future operation of navigation locks safer. First, some general information on the density of fresh and salt water that is present on both sides the IJmuiden lock complex is provided, and it is discussed how external factors may cause variation in density differences. Next, the measurement setup using the ShoreTension system is presented. Lastly, the results of the measurements are provided and the suitability of the ShoreTension system in lock operation is discussed.

The research project was performed as part of TKI-Delta Technology program and was supported by Deltares, Dutch Pilot Association, ShoreTension, Arcadis and Rijkswaterstaat.

2 Density Differences Over the Lock

In preparation of the mooring force measurement campaign it was important to have a better understanding of actual density differences that are present over the lock in daily conditions. Although much historic monitoring data is already available in the North

Sea Canal -mostly measured somewhat further away from the IJmuiden lock complex-
and some data was available for the outer approach harbor on the sea side, these
datasets spanned different time periods and simultaneous measurements on both sides
of the lock were never performed before. Therefore, it was still unknow how variable
the density close to the lock would be and how the density difference over the lock
itself would be influenced by several external processes. As the strength of the density
current during a lock-exchange depends on the density difference and since density
currents generate the dominant hydraulic forces on moored ships in the North Lock, it
is important to know how large the density difference over the lock complex can be.
The density difference over the lock complex provides an upper limit for the actual
density difference that can be present over a lock gate when opening.

To get more understanding of the local density variations over the lock, a monitoring campaign of three weeks was conducted prior to the measurements with the
ShoreTension system (Van der Veen 2021). At two locations, one in the outer and one
in the inner approach harbor of the North Lock, the density of the water was measured
close to lock gates. At each location CTD sensors (sea Sect. 3.2) were deployed at
several water depths, varying from 3 to 5 sensors in the vertical depending on measurement period. The vertical distribution of sensors was chosen such that a reasonable
estimate of vertical density profiles was obtained and variations in average densities
could be captured adequately. Such stationary density measurements give valuable
information in temporal variations of the density, but due to the limited number of
sensors available, the spatial resolution that could be obtained was quite coarse.
Therefore, the stationary density measurements were complemented with manual
profile measurements, by lowering and raising a CTD sensor on a rope in the water at
specific time instances. These density profiles were mainly used to verify the parameters derived from the stationary measurements, but of course, instantaneous profile
measurements provide less information on larger scale temporal changes in density.

The main conclusions of the monitoring campaign can be summarized as follows:

- The mean density difference that was present over the lock was around 14 kg/m^3,
  which was in line with expectations (Kortlever 2018). The density difference over
  the lock varied between a minimum value of 11–12 kg/m^3 and maximum value of
  17–18 kg/m^3, with somewhat larger density variations on the sea side than on the
  canal side.
- The wind had the largest effect on the density distribution in the approach harbors,
  with a stronger influence for the larger wind speeds. For wind blowing in offshore
  direction (from east to west), the mean density in the outer approach harbor was
  much higher than for onshore (westerly) wind directions. The main reason for this
  was thought to be that the upper fresh water layer is pushed towards the sea by the
  easterly wind, while westerly winds push the upper fresh water layer towards the
  lock complex and thereby inducing a larger mixing layer in the upper part of the
  water column. Large scale upwelling and downwelling processes could also play a
  role, although that could not be conclusively confirmed by the obtained dataset.
- Smaller influences on local density differences were identified for tidal variations
  and water discharged through the nearby discharge sluices. The levelling process

itself mainly causes short scale variations in the density distributions in the approach harbors, due to the levelling and opening of the lock gates (in the order of 30 min – 1 h).

Within the monitoring campaign, that had a relatively short duration, already quite some variation in density difference over the lock was observed. Although the obvious influence of the wind was well captured by the measurements, longer scale variations due to e.g. varying discharges of the North Sea Canal related to wet/dry periods, water availability and water control decisions could not be captured. As the density balance in this water system is known to be quite complex, it is expected that variations in density on the canal side will to a large extent be influenced by these processes with longer time scales. It was therefore surprising that the largest density differences that were measured in this monitoring campaign (18 kg/m^3) were already close to the upper value of the density difference (20 kg/m^3) that was chosen as design value for the new Sea Lock IJmuiden (Kortlever 2018).

The density monitoring campaign provided valuable background information on the magnitudes of forces to be expected during the measurement campaign and it provided insight in the main parameters influence forces during a lock-exchange most.

3   Measurement Setup

3.1   Mooring Force Measurements

The mooring force measurement setup has been described by Hoff (2021) in more detail; here the most important elements are reproduced. All vessels that were measured were moored with 4 mooring lines (forward and aft hawser and forward and aft spring). The forces in the mooring lines were measured using two traditional load cells with a capacity of 100 Ton and two ShoreTension units. The load cells were not available during the full test campaign, so in some tests not all mooring lines loads were measured. Figure 2 shows the four mooring layouts that were (planned to be) used during the test campaign.

In configuration A, the ShoreTension units are connected to both hawsers. In configuration B, ShoreTension is used on both springs. In configuration C, ShoreTension is used on the forward spring and the aft hawser, and in configuration D, the ShoreTension units are connected to the forward hawser and aft spring.

The units can be operated in different operation modes, in which specific settings can be chosen to ensure optimal performance. In the performed tests, a constant pretension of 10 Ton was selected. If the force in the mooring line would drop below this value, the cylinder would heave-in the line until the required pre-tension was obtained again. If the force in the mooring line increases however, an upper limit was specified at which the ShoreTension cylinder starts to pay out, thereby reducing the force in the line again. In the performed tests, 40 Ton was selected as upper working limit, ensuring that the SWL of bollards and hawse-holes would not be exceeded during the tests.

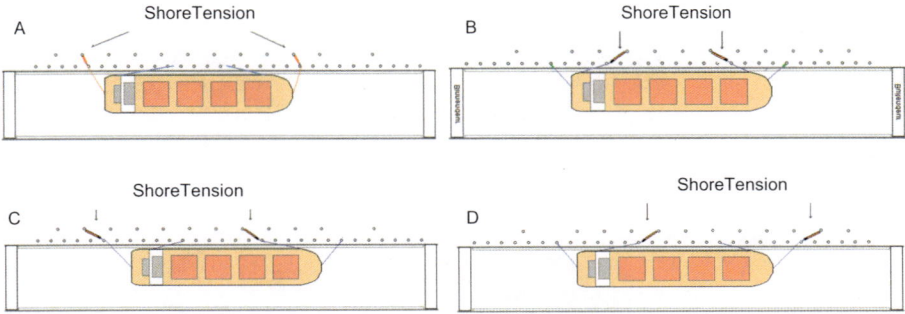

Fig. 2. Schematic measurement setup mooring line loads for the four mooring configurations (A-D), reproduced from (Hoff 2021).

To secure the ShoreTension unit on the quay wall, it was connected to a storm bollard present at the North Lock, located approximated 15 m away from the lock chamber wall. On the opposite site of the unit, a mooring line is connected to the movable part of the cylinder and this line was secured to the deck of the ship. However, the mooring line is not directly tied from the ShoreTension unit to the ship but guided via a "standard" bollard located close to the lock wall. A sleeve was used to reduce friction and wear and tear on the lines. A picture of the used setup is shown in Fig. 3.

Because the ShoreTension mooring line is not directly connected from the cylinder to the ship but guided via a bollard as shown in Fig. 3, the force measured by the ShoreTension cylinder will be lower than the actual mooring force exerted on the bollard due to friction at that bollard. To assess these friction losses, dedicated calibration tests have been performed. In those tests, the ShoreTension line was connected directly to a load cell and different line angles around a guiding bollard were tested. By comparing the loads measured by the ShoreTension unit and by the load cell, a relation between line angle and friction was derived. During all mooring tests the line angles were estimated and documented, and in post-processing the measured line forces are corrected for the friction loss. In the presented results, the forces that are exerted on the bollards are shown, which are a factor 1.1–1.7 higher than the forces measured by the ShoreTension unit depending on the applied line angle.

Fig. 3. Setup of the ShoreTension system (orange cylinder). Note that the number of people in the picture is not representative for the personnel needed to operate the system, as this picture was taken during one of the first "instruction" measurements.

3.2   Density Measurements

To measure density currents in the lock chamber during the operation of the lock, six CTD-divers were mounted in the lock chamber. These CTD-divers measure Conductivity, Pressure and Temperature simultaneously (Van Essen 2016). The measured conductivity and temperature are converted to density, using the UNESCO-1981 sea water formulation (UNESCO 1981). The divers logged autonomously, storing the data in their internal memory. The divers were taken out of the water after approximately 4 days to download the data, to verify operation and accuracy using a calibration liquid, to time-synchronize, to clean the sensors and to restart the and remount the sensors in the lock chamber again. A sample interval of 10 s or 30 s was used, depending on the type of sensor (two types were available: a somewhat older type and newer type with larger internal memory capacity, hence a shorter sample interval could be used). A sample interval of 30 s was thought to be sufficient to capture the large-scale density currents during a lock-exchange process, which has a typical time scale of 30 min - 1 h.

On each side of the lock chamber three sensors were attached to a vertical rope that was kept in place by a weight on the bottom. On the west (sea) side of the lock chamber the sensors were positioned 17 m from the gate and on the east (canal) side of the lock the sensors were positioned 5.5 m from the gate (Fig. 4). The vertical positions of the divers are given in Table 1.

By measuring at different heights in the water column, insight is obtained in density gradients in the lock chamber. Although the spatial resolution is limited, the difference in (mean) density between the east and the west side of the lock chamber can be used as indication for the density difference over the length of the ship. This density difference is expected to have a direct correlation with the longitudinal force acting on the ship.

The measured pressure was used to determine the water level in the lock chamber. The pressure measurements were corrected for atmospheric pressure variations, using a separate barometric pressure sensor mounted on the lock chamber wall above the water.

Table 1. Position of the CTD divers

	Horziontal position	Vertical position
D1	West	NAP-2.2 m
D2	West	NAP-8.2 m
D10	West	NAP-14.3 m
D11	East	NAP-2.0 m
D12	East	NAP-8.0 m
D13	East	NAP-13.3 m

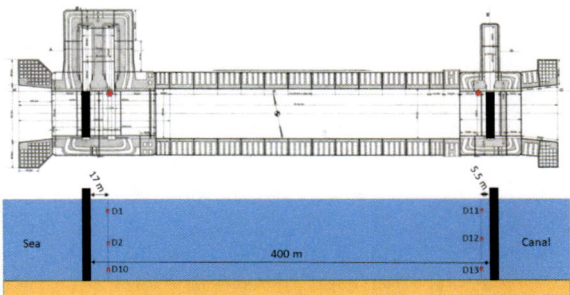

Fig. 4. Schematic measurement setup density sensors.

4 Measurement Results

4.1 Mooring Forces

The measurement campaign had a duration of 15 days. Within that period, mooring line forces were measured of 21 large vessels that passed through the North Lock. Ten vessels were inbound, i.e. sailing west to east from sea to canal, and eleven vessels were outbound, i.e. sailing east to west from canal to sea. All vessels were moored on the north side of the lock, as the ShoreTension units were only deployed at that side of the lock.

Table 2 provides an overview of all vessels and maximum measured line forces. The highest forces were generally measured as a results of density currents during the lock-exchange after opening the lock gate. In some tests however, largest peak loads in the lines occurred during the mooring procedure when fastening the mooring lines. The mooring configuration that is used follows from the definition provided in Fig. 2. The load cells and ShoreTension units are labelled "$_{east}$" and "$_{west}$", referring to their deployment location. Depending on the sailing direction of the vessel, this refers to aft or bow of the vessel, i.e. "$_{east}$" refers to forward lines for inbound vessels and to aft lines for outbound vessels (and vice versa for "$_{west}$").

4.2 Discussion on the Measured Forces Using Mooring Configuration A

The forces in the ShoreTension system are generally higher than in the conventional mooring lines measured with load cells. Since the ShoreTension mooring lines are much stiffer than the conventional mooring lines, most forces on the vessel are absorbed by the former lines. Next to that, also the pretension in the ShoreTension lines was typically larger. In several tests, the maximum line force in the conventional mooring lines even stayed below the set pretension in the ShoreTension lines (>10 Ton). During previous measurements by Meijering (2018) in the North Lock of IJmuiden, it was already observed that pretension in conventional mooring lines can vary strongly and that pretension is generally lower than theoretically desired. He also observed that largest vessel motions during levelling typically occurred for vessels with low pretension.

Table 2. Performed tests, with *in/out* indicating inbound or outbound sailing direction. In the grey columns the maximum line forces are shown, with the forces measured by the ShoreTension units marked *ST*, the line forces measured by the load cells marked *LC*.

Test No	Ship name	Sailing dir.	Mooring config.	Length [m]	Beam [m]	Draft [m]	ST_{east} [Ton]	ST_{west} [Ton]	LC_{east} [Ton]	LC_{west} [Ton]
1	Baker River	in	A	190	32.3	5.22	20	-	-	-
2	Yeoman Bontrup	out	A	250	38.0	7.68	63	-	-	-
3	Team cavatina	in	A	183	32.2	7.60	19	-	-	-
4	Baltic favour	out	A	183	27.3	11.00	25	18	-	-
5	Andrea victory	out	A	183	32.2	12.00	28	26	-	-
6	Aiolos	out	A	184	27.4	6.3	-	16	-	-
7	Katelina	in	ST_{east} only	118	18.5	5.28	-	-	-	-
8	Haruka	in	A	169	27.2	9.97	35	38	-	-
9	Niagara highway	in	B	200	37.2	8.35	21	15	-	-
10	Maersk Cirrus	in	A	183	32.2	7.00	23	14	-	-
11	Velos leo	out	A	183	32.2	11.1	37	31	-	-
12	Hellas avatar	in	A	183	32.2	11.4	36	33	-	-
13	Leon Apollon	out	A	229	32.2	13.00	38	22	8	16
14	Bow Pioneer	in	A	228	37.0	12.6	18	38	11	16
15	Cielo di New York	in	A	183	32.2	7.80	43	13	12	4
16	Med Atlantic	out	A	170	25.4	11.5	32	23	5	6
17	Minerva Zen	out	A	183	32.2	11.8	30	16	11	16
18	AC Splendor	out	B	177	28.4	8.50	36	48	13	3
19	Bontrup Amsterdam	out	B	224	32.2	6.13	18	21	12	9
20	Friedrich Russ	out	A	154	20.6	6.10	31	20	3	11
21	Sea Helios	in	A	180	32.2	6.75	29	19	-	14

With some exceptions, highest forces are measured on the east side. This was expected (Van der Hout 2018), since for inbound vessels the dominant density forces during a lock-exchange is directed backwards (towards sea) with a strong bow-out moment, pushing the bow of the vessel away from the lock wall. This creates largest line loads in the bow hawser (ST_{east}). For outbound vessels, the dominant density force is in longitudinal direction, pushing the vessel forward (also towards sea). This creates highest forces in the front spring and aft hawser. Because the ShoreTension line was typically stiffer than the conventional mooring line and because of the higher pretension, also for outbound vessels largest forces were measured by the east ShoreTension system.

Most tests are performed using mooring configuration A, in which ShoreTension is connected to the bow and aft hawser of the vessel. This configuration worked quite well and provided most flexibility in the placement of the ShoreTension units. Due to the used line angles of the hawsers, both transversal and longitudinal forces exerted on the vessel could be absorbed by the ShoreTension lines.

In almost all tests with mooring configuration A, resulting vessel motions during levelling and lock-exchange were absent or small, i.e. less than 0.5 m. The only exception is test 2, in which the setting of the east ShoreTension unit was accidently incorrect, and consequently the cylinder already started to pay out the line at a force lower than 40 Ton. As a result, driven by the density current of the lock-exchange, the vessel started to move towards the outer approach harbor, leading to a strong increase of the force in the aft hawser (up to 63 Ton). Although test 2 is not representative for the correct working of the ShoreTension system, it shows how high line forces can become once a ship starts moving, thereby easily exceeding safe working load of conventional mooring lines.

4.3   Discussion on Other Mooring Configurations

In three tests, mooring configuration B was used, in which ShoreTension was connected to the spring lines. This configuration gave also good control over vessel motions and line forces but was less flexible than configuration A. Due to the line angles in configuration B, the ShoreTension lines could absorb longitudinal forces well, but transversal forces less effectively.

In test 7, only one ShoreTension unit was connected to the bow hawser, since for inbound vessels the highest force due to the lock-exchange is expected on this bow line. The setup with one ShoreTension unit showed not to be practical, because the pretension exerted by the ShoreTension unit was so high that it pulled the aft of the ship off the fenders. Apparently, the crew on board of the vessel was not expecting such (large) pretension in the mooring lines and slacked the aft lines during tensioning of the ShoreTension system, or the hold capacity of the winches was not high enough on this relatively small vessel to keep the aft lines under proper tension. As a result, the vessel started to move forward during the test and the test was aborted.

In theory, mooring configuration C and D could work very well for outbound and inbound vessels respectively. In those situations, both ShoreTension units would work in the same direction, opposing the direction of the dominant density force, i.e. which is always directed towards the (saltier) sea side. However, after the experience in test 7 (and test 2) it was decided not to apply these mooring configurations, as the risk existed that the deck crew would slack the conventional mooring lines too soon, potentially leading to unwanted, uncontrolled vessel motions.

4.4   General Observations

In all tests, the forces measured by the ShoreTension units stayed below the limit of 40 Tons, so the system did not have to automatically pay out the line to reduce line forces. Due to friction at the guiding bollard, the actual forces in the mooring line itself could higher than measured at the cylinder and did exceed 40 Ton (as shown in Table 2). It means that, would a vessel have been moored with conventional mooring lines only, with similar pretension values, the SWL of the conventional mooring line would be exceeded. (In lock design, typically a value of 50%–55% of the minimum breaking load of a mooring line, MBL, is used as SWL). In 10 of the 21 tests, the SWL of conventional mooring lines would have been exceed with line forces reaching up to 181% of the SWL of the conventional mooring lines on board of the vessel. It should be mentioned however that for test purposes the mooring lines were intentionally released later than normal to be able to measure the development of forces during the lock-exchange. In normal operation, mooring lines might have been released earlier for the vessel to sail out of the lock. Also, when forces in mooring lines increase, the crew may decide to pay out the lines or which breaks may start slipping, thereby reducing line forces. This may however result in unwanted ship motions.

4.5   Relation to Density Differences

In 7 tests (Test 13 – Test 19), density measurements were available simultaneous with force measurements. Based on the three sensors on each side of the lock chamber, the density distribution in the vertical could be estimated at those positions. The depth averaged density was also determined for both sides, and hence the mean density difference between the west and the east side of the lock could be quantified.

Figure 5 shows a typical measurement result. It shows the results of test 15, the handysize tanker *Cielo di New York*, an inbound sailing vessel, using mooring configuration A. Around 11:50 the vessel sails into the initially salty lock chamber from the west side, visible by the small disturbance in upper density on the west side of the lock. After the west gate closes (red rectangle in the upper two panels), the water level in the lock chamber is adjusted to the lower canal level. During this levelling process, mooring lines forces of the aft hawser (ST_{west}) and forward spring (LC_{east}) reduce slightly, the force in the bow hawser remains constant (ST_{east}). The force in the aft spring (LC_{west}) was too small to be accurately measured by the load cell and remained below 2 Ton during levelling and lock-exchange.

When the east gate opens (grey rectangle in the upper two panels), fresh water flows into the lock chamber; first only visible at the east side of the lock, reaching the west side of the lock approximately 10 min later at 12:20. The incoming fresh water initially concentrates in the upper part of the water column propagating over the outflowing saltier bottom layer. Only after reflection at the west gate, the density distribution becomes more homogeneous spread over the vertical. The increase in density difference over the length of the lock chamber (purple dotted line in the bottom panel of Fig. 5), is followed by a strong increase of the force in the bow hawser (ST_{east}). The forces in the other mooring lines remain similar or decrease.

Around 12:24 the mooring lines are released, and the vessel can sail out of the lock chamber. Around 12:35 a disturbance of the density measured by the lowest sensor on the east side is observed. This is thought to be the moment that the vessel passes that sensor location, thereby inducing mixing over the vertical.

The strong increase of the bow hawser force as a result of the lock exchange in test 15 was in line with expectations from previous research (Van der Hout 2018). The mean density difference between both approach harbors in this test was around 12 kg/m^3, with a mean density of 1020 kg/m^3 in the outer approach harbor and 1008 kg/m^3 in the inner approach harbor. During the lock-exchange, the mean density difference over the lock chamber reached up to 8 kg/m^3, which is a good indication for the density difference that developed over the length of the vessel. As hydrodynamic forces during a lock-exchange scale with the density difference over the vessel (Van der Hout 2018), mooring line forces may increase in conditions with larger density differences. As described in Sect. 2, density difference larger than 12 kg/m^3 are not unusual and may occur frequently in IJmuiden.

In all tests, a clear relation between the density difference over the lock chamber and mooring lines forces was found, confirming that the density current is the dominant forcing mechanism in the North Lock. For conciseness, not all measurement results could be reproduced here, and only test 15 is provided as illustration for the quality of data that has been obtained during the measurement campaign.

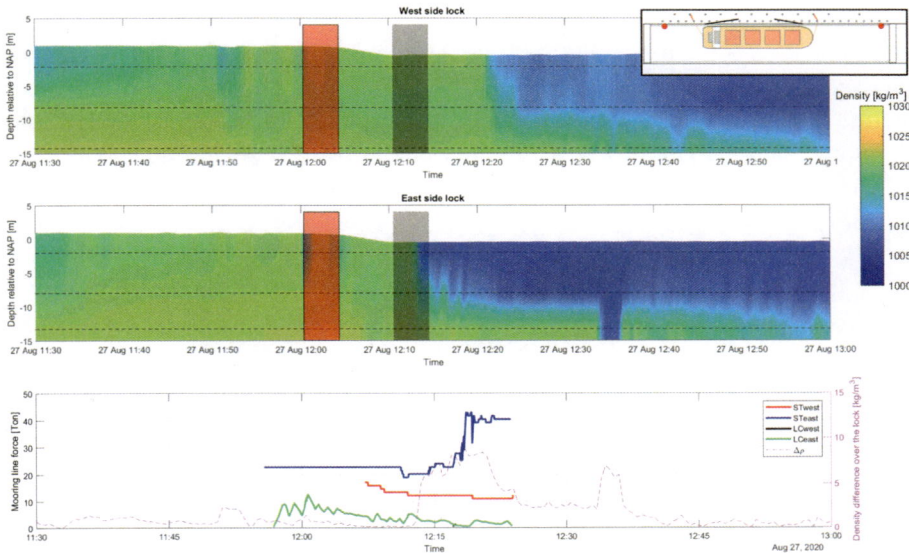

Fig. 5. Combined density and mooring force results for test 15, Cielo di New York, inbound, mooring configuration A. *Top panel*: contour plot of the density distribution on the west side of the lock. Density values are linearly interpolated between the sensor positions, indicated with the horizontal black dotted lines. Constant extrapolation to water surface and lock bottom is applied. The red rectangle indicates the period that the west gate closes, the grey rectangle the period that the east gate opens. *Middle panel*: idem, for the east side of the lock chamber. *Bottom panel*: measured forces (solid lines, left axis) and difference in mean density between east and west side (purple dotted line, right axis).

4.6   Use of the ShoreTension System in a Navigation Lock

An important objective of the test campaign was to gain practical experience with the ShoreTension system in navigation lock operation. As described above, during the test campaign it could be clearly identified which mooring configuration worked best: configuration A, with ShoreTension units connected to both hawsers. After connection of the lines, vessel motions were minimized due to proper pretension, and mooring line forces stayed comfortably within safe working load limits of the system. The people operating the system and performing the tests indicated that they experienced a positive effect on mooring safety, mainly because it provided more control over desired pre-tension in the mooring lines. Experience during the test period learned that line handling by vessel crews during lock passage is not always optimal and that it is often the cause for potentially dangerous situations, e.g. high line loads or unwanted vessel motions, as also reported by Meijering (2018).

The use of a system like ShoreTension however comes at a cost, because it takes more time to connect mooring lines than using conventional mooring lines. This is partly because the lines men and the vessel crew do not have experience with the system yet and need to adjust to new working procedures, and partly because the system was not optimized for lock operation yet. During the test period it was monitored how much (extra) time was needed to connect the ShoreTension lines. The results are shown in Fig. 6. For the first 4 tests no information was recorded, but for the following tests a learning curve was observed. This is illustrated by an indicative trendline, which shows that after a few tests to build experience with the system, the process went quicker. For the last 10 vessel passages, the time needed did not decrease much further, and remained on average around 10–11 min. If further time reduction is desired, the system should be optimized for lock operations.

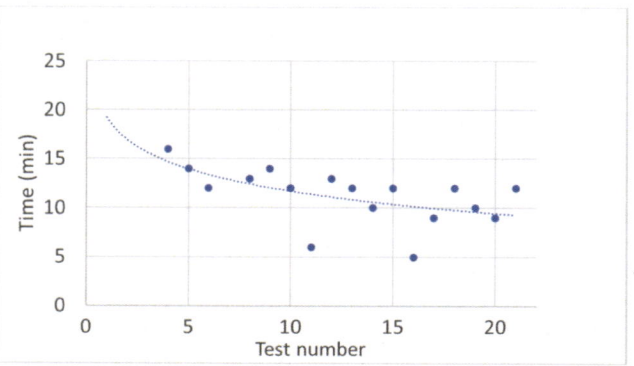

Fig. 6. Time needed to connect the ShoreTension lines

5   Conclusions

The results of the measurement campaign show that hydrodynamic forces during a lock-exchange in North Lock IJmuiden are large and that under daily conditions safe working load limits of conventional mooring lines may be exceeded, potentially leading to line breakage or unwanted vessel motions. In normal operation, this will usually not lead to large problems since mooring lines are released quickly after opening of the lock gate and the vessel is able to sail out of the lock. However, if something goes wrong, e.g. a vessel must wait, or a mooring line is stuck in a winch, this can lead to dangerous situations.

The ShoreTension system was tested for the first time in lock operation. Using this system showed to have several benefits: proper pretension could be applied to the mooring lines, thereby preventing large vessel motions during levelling and lock-exchange. Since it is a dynamic system, the risk of exceeding safe working loads is much smaller than for conventional mooring lines as lines loads are reduced automatically by extending the hydraulic cylinder. Furthermore, the hold capacity of the ShoreTension unit is larger than the break hold capacity of the winches on board of vessels passing the North Lock IJmuiden, thereby providing better control of the moored vessel behavior. Especially under critical conditions, e.g. for large vessels or large density differences, the use of the ShoreTension system can therefore improve safety of lock operations. During the test weeks, the number of ships was limited, and mostly relatively small vessels with a beam up to 32 m were tested. It is expected that for larger vessels, the ShoreTension system could even be more beneficial, since hydrodynamic forces will increase for larger vessels.

The downside of using the ShoreTension system is that it cost more time to connect the mooring lines than using conventional mooring lines. In the test campaign the additional time needed for connecting the system was approximately 10 min per vessel. This might however be reduced in the future if the system configuration is optimized for the use in locks. Also, when using the ShoreTension system a different mooring procedure needs to be used, since the mooring lines now need to be fastened on board of the vessel instead of on the lock wall. During the test period, the system could be deployed flexibly, but not all locks will have a similar bollard configuration as the North Lock, so mooring layout may need to be optimized for other navigation locks.

All in all, a successful test period in the North Lock of IJmuiden was performed. The dynamic mooring system ShoreTension was successfully tested in a practical lock operation for the first time, showing its potential to increase safety of lock operations. The density measurements provided valuable insight in the governing hydraulic processes during levelling and lock-exchange, and the most influential external drivers for density differences over the lock complex were identified.

Acknowledgements. The authors would like to thank TKI-Delta Technology and contributing partners for providing the necessary funds to conduct the measurements. Special appreciation goes out to ShoreTension, for making their system available and providing support during the test period. Furthermore, the lines men of Corps van Vletterlieden and the Pilots of Amsterdam-IJmond are gratefully acknowledged for their support during this research project.

References

Hoff T (2021) ShoreTension in de noordersluis (in Dutch). Thesis Master in Maritime Piloting, STODEL, 10 January 2021

Kortlever WCD, van der Hout AJ, O'Mahoney T, de Loor A, Wijdenes T (2018) Leveling the new sea lock in The Netherlands; including the density difference. In: 34th PIANC world congress, Panama

Meijering B (2018) Krachtopbouw trossen Noordersluis (In Dutch). Thesis Master in Maritime Piloting, STODEL, 17 December 2018

Nogueira HIS, van der Ven P, O'Mahoney T, de Loor A, van der Hout AJ, Kortlever W (2018) Effect of density differences on the forces acting on a moored vessel while operating navigation locks. J Hydraul Eng 144(6):04018021. https://doi.org/10.1061/(ASCE)HY.1943-7900.0001445

UNESCO (1981) The practical salinity scale 1978 and the international equation of state of seawater 1980. Tenth report of the Joint Panel on Oceanographic Tables and Standards (1981), (JPOTS), Sidney, B.C., Canada. UNESCO technical papers in marine science, No. 36

Van der Hout AJ (2017) New lock of IJmuiden – physical scale model of the world's largest lock. In: PIANC smart rivers conference, Pittsburgh, PA, 18–21 September 2017

Van der Hout AJ, Nogueira HIS, Kortlever WCD, Schotman AD (2018) Scale model research and field measurements for two new large sea locks in The Netherlands, In: 34th PIANC world congress, Panama

Van der Veen MJW (2021) Dichtheidsverschil Noordersluis (in Dutch). Thesis Master in Maritime Piloting, STODEL, 10 January 2021

Van Essen (2016) Product manual. https://www.vanessen.com/products/data-loggers/ctd-diver/
Vrijburcht A (1991) Forces on ships in a navigation lock induced by stratified flows. Doctoral
    thesis, TU Delft, Delft, Netherlands

Innovative Salt-Freshwater Separation System at the Krammer Locks, The Netherlands. Hydraulic Modelling to Balance Functional Requirements

Tom O'Mahoney[1]([✉]), Martin De Jonge[2], René Boeters[2], and Tjerk Vreeken[1]

[1] Deltares, Delft, The Netherlands
{tom.omahoney, tjerk.vreeken}@deltares.nl
[2] Rijkswaterstaat Zee and Delta, Middelburg, The Netherlands
{martin.de.jonge, rene.boeters}@rws.nl

Abstract. This paper presents research on the impact of a plan to change the salt-fresh water separation system at the Krammer locks in the Netherlands. This lock complex forms the connection for inland navigation between the Eastern Scheldt estuary and the hinterland. The new salt-fresh system will include bubble screens at both lockheads and an additional flushing discharge through the lock chambers and a neighbouring flushing sluice. The interaction between water management, salt management, ecology requirements and navigational requirements has been accounted for with the aid of WANDA-Locks and Delft 3D simulations. The effect of expected sea level rise on the performance of the new proposed new system is also modelled. The modelling efforts show that the new system should be able to keep the salt concentration of the fresh water lake Volkerak-Zoom within the required limits for the duration of the system's technical lifespan. The modelling approach is able to tune the operation of the lock to the hydraulic conditions of each lockage and within the seasonal restrictions on the water management of the region, giving for each situation the required air discharge for the bubble screens and the flushing discharge. This operation has also been designed to remain safe for the vessels passing through the lock complex.

Keywords: Salt intrusion · Shipping locks · WANDA-Locks · Bubble screens

1 Introduction

The Krammer locks (see Fig. 1) are set of shipping locks in the Netherlands between the fresh water lake Volkerak (Volkerak Zoommeer) and the saltwater Eastern Scheldt estuary. They form part of the Dutch Deltaworks in the South-West Delta of the Netherlands and have been operational since the 1980s. They currently use a salt-freshwater separation system incorporating salt water substitution (van der Kuur 1985).

The complex consists of two commercial shipping locks capable of accommodating class CEMT VIb vessels and two smaller recreational shipping locks. The target water

© The Author(s) 2023
Y. Li et al. (Eds.): PIANC 2022, LNCE 264, pp. 479–493, 2023.
https://doi.org/10.1007/978-981-19-6138-0_42

level in the lake Volkerak-Zoom is around NAP+0.0 m, with a normal variation of NAP ± 0.25 m, where NAP is the Normaal Amsterdam Peil or approximately Mean Sea Level. On the Eastern Scheldt side, the water levels are tide dependent, for which a maximum operational water level was set to NAP +2.75 m and a minimum operational water level of NAP −2.50 m.

The commercial locks are 280 m long between the gates and 26 m wide. The available width between the fender is 24.1 m. The lock floor is at a depth of NAP −6.25 m. This floor is perforated to allow the substitution of salt water for freshwater during lockage. The salt water is then transferred to a chamber below the lock which has a floor at depth NAP −11.0 m. The density difference across the lock complex can reach as high as 20 kg/m^3.

The concept of salt water substitution was developed in the 1960s for the lock at Dunkirk and has come to be known as the 'Dunkirk system'. This system has been optimized and applied at several locations, such as the Krammer and Kreekrak in the Netherlands. The improved system of the large locks at Krammer in the Netherlands, is considered the state of the art for this type of locks (PIANC 2021).

In the Dunkirk system, the saltwater is introduced and withdrawn vertically through the perforated lock floor while the fresh water is supplied and evacuated in horizontal transversal direction through wall openings, located high in the lock chamber walls. The openings in the walls and on the lock floor are distributed along the entire length of the lock to reduce mooring forces.

Fig. 1. Aerial view of the Krammer locks complex. The approach harbor on the Eastern Scheldt side is shown to the left and lake Volkerak-Zoom to the right. (Source Rijkswaterstaat).

Owing to tidal variations in the waterlevel on the sea side relative to the lake side, different locking cycles are possible, as illustrated in Fig. 2. The first illustrated situation is when a ship has entered the lock from the sea side. Once the salt water gate is closed (phase 1), the lock is levelled via the lock floor to or from a salt water basin (phase 2 - levelling), depending on the tidal phase. Afterwards, the boundary plane between salt and fresh water in the lock chamber is lowered (phase 3 - exchanging) by simultaneously withdrawing salt water through the lock floor and introducing fresh water through the wall openings. This operation is ended once the boundary plane is below the perforated lock floor (phase 4). The lock gates can then be opened without a density current being initiated as the density is equal across the gate.

The overall time required for a lockage is larger (more than twice as long) than in conventional locks with levelling valves in the lock gates. While levelling through the lock floor is more efficient (lower hawser forces of vessels), a lot of time is lost by exchanging salt and fresh water. This additional time is governed by the allowable mixing of salt and fresh water on the one hand and by allowable hawser forces on the other hand.

At the large Krammer locks, the lateral wall openings are in direct contact with the fresh water basin through short ducts. The Krammer lock complex is designed in such a way that the fresh water basin enfolds the lock chamber entirely, see Fig. 1. This is an important improvement of the original Dunkirk system, because this ensures an even distribution of the fresh water entering through the lock chamber walls.

The actual lock floor is located below the perforated floor. The space in between the perforated and the actual lock floor is connected to several salt water culverts in the middle of the lock chamber. The required salt water is supplied or withdrawn through these culverts. The supply through the culverts can be either gravitational or mechanical (by means of pumps). In the case of the large Krammer locks, several water saving basins have been constructed next to the lock complex in which salt and/or brackish water are stored. The basins and the outer salt water basin are interconnected with culverts that are equipped with valves and pumps to allow different filling and emptying scenarios.

The salt – fresh separation system is very effective with reductions up to 90% to 98% of the salt load on the fresh water basin. The system leads to an additional fresh water loss of up to 30% to 80%, depending on the tidal cycle.

The main disadvantages of the Krammer system are the high initial capital expenditure and ongoing operating expenditure, the additional time required for locking, the transversal forces on the ships, the necessity of pumping and the complicated control system.

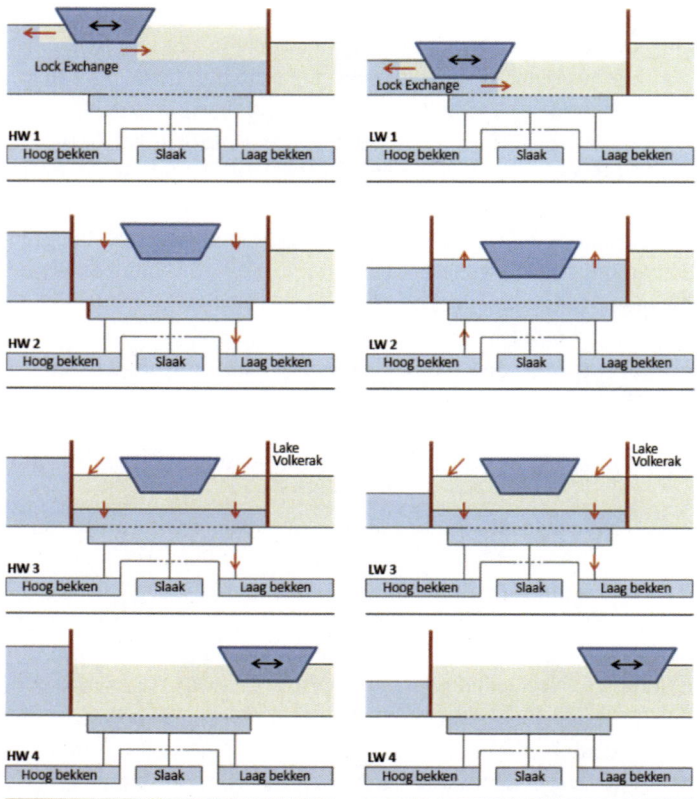

Fig. 2. Schematic representation of the current (Dunkirk) system of salt-freshwater separation.

2   The Innovative Fresh-Salt Separation System (IZZS)

Rijkswaterstaat plan to replace the salt-freshwater separation system at the Krammer locks with a new innovative system, the Innovatieve Zout-zoet Scheiding in Dutch (IZZS). The motivation for considering a change in the system are long lockage times and high maintenance costs of the current system. The basis of the new system is the installation of a bubble screen (Van der Burgh) to allow the lock gates to be opened directly after levelling. The lock-exchange flow, as a result of density currents, is reduced by the bubble screen. In this way, the exchange of salt and freshwater through the perforated floor of the lock is skipped and the total lockage time is shorter. In addition, the valves along the length of both lock chambers can be decommissioned. The bubble screen does not have the same effectiveness as the Dunkirk system so in addition a flushing discharge though the locks is needed. An option of an additional water screen has also been investigated but will not be used in the final system (see Fig. 3 for a schematization of all components). The adapted lockage cycle of the IZZS is given in Fig. 4.

Fig. 3. Schematic representation of the options which have been investigated for the Krammer locks, including bubble screen, water screen, flushing discharge and raised lock sill. (From Keetels et al. 2011)

The Innovative Salt-Fresh Separation system has been designed by a series of research projects. In 2009 the first investigations were made for the system. Laboratory tests and Computational Fluid Dynamics (CFD) simulations were made of bubble screens followed by a field measurements at the Stevin locks in the Netherlands (Keetels et al. 2010). These tested a new design of a bubble screen to improve the effectiveness and robustness when compared to older designs (Abraham van der Burgh, van der Kuur). These first investigations were made for a possible system at the Volkerak locks, at the inland end of lake Volkerak-Zoom owing to plans (since put on hold by the Dutch government) to make the lake Volkerak-Zoom a saltwater body. As the two lock complexes are rather similar, the first designs of the system could be easily transferred to the Krammer locks. Subsequently, in 2014 a pilot field campaign was made at the smaller Krammer recreational locks. The hydraulic model WANDA-Locks was then developed for the Krammer system, having been validated against the Stevin lock measurements (Deltares 2010). This model was used from 2015 until the present to study effect of the IZZS on the salt load into lake Volkerak-Zoom and to weigh up all the options of operation of the new system for many of the different functions of the lock complex.

2.1 Functions of the Krammer Locks

The Krammer locks perform many functions in the water system of the Dutch South West Delta;

- Safe and fast shipping; the lock complex provides a shipping route between the Scheldt and the Rhine.

- Flood safety; excess water from upstream is mostly discharged via the Haringvliet but the Krammer locks also have a flushing function in extreme cases.
- Water management; management of water levels in lake Volkerak-Zoom is provided together with Bathse Flushing Sluice
- Fish migration; an additional fish migration route through the new flushing sluice at the Krammer locks is part of the future renovation.
- Salt management; the Krammer locks are the principle point of salt intrusion into lake Volkerak-Zoom and mitigation is therefore most effective at this location.

2.1.1   Restrictions on Water Usage

In many parts of the Netherlands (and elsewhere (Mausshardt and Singleton 1995) saltwater intrusion at locks is effectively mitigated through flushing of the locks. However, the Krammer locks have several restrictions on its water usage which limit this option.

- Water availability owing to drought or agriculture; lake Volkerak-Zoom acts as an important freshwater provision for agriculture in Brabant and the South West Delta. It is maintained by freshwater discharges from the Hollands Diep and the rivers of Brabant. In times of reduced discharges restrictions are in place in on the discharges available for flushing of the locks.
- Saltwater ecology in the Eastern Scheldt; near the Krammer locks on the Eastern Scheldt side there are ecologically valuable saltwater habitats and a number of aquaculture areas for mussel and oyster farming. Too much freshwater discharge through the locks would adversely affect these areas. It is therefore limited to average daily discharges of 9 m^3/s in the growing season. Outside the growing season the discharges can be up to 3 times more.
- Nautical safety; flushing through the locks can create currents in the approach harbours which have an effect on the maneuverability of vessels entering (and exiting) the locks. Rijkswaterstaat maintains guidelines on limiting longitudinal and transverse velocities in the fairway.
- Energy usage for pumping; in order to minimize energy use at the locks flushing of water is performed as much as possible under the gravitational head available at low water on the Eastern Scheldt.

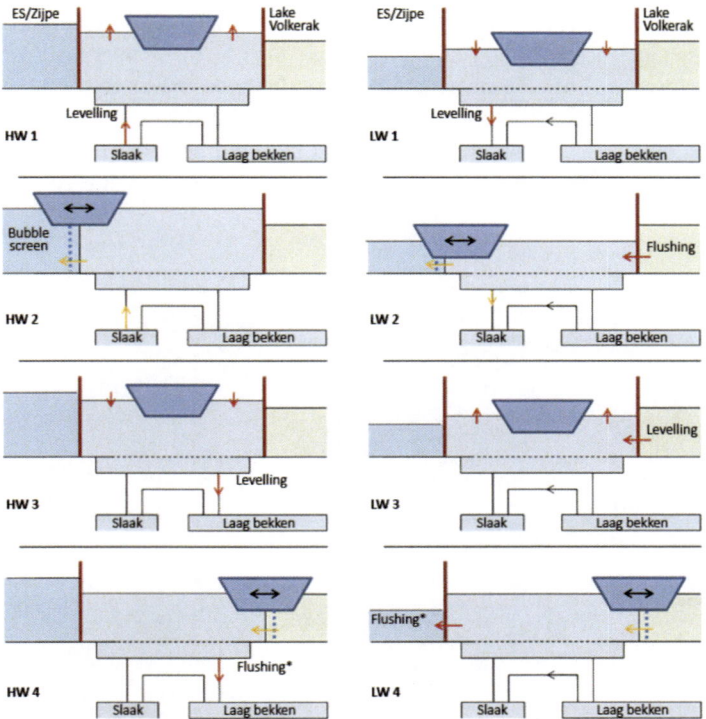

Fig. 4. Schematic representation of the proposed Innovative Salt-Freshwater separation system.

These multiple functions and numerous restrictions on water use make the design of a salt-fresh separation system and its planned operation complicated. The model instrument developed in this project has been used to balance all the conflicting requirements of the operation. Some of the results of this modelling are highlighted in the following section.

3   Hydraulic Modelling of the Krammer System

3.1   WANDA-Locks Model

WANDA (see reference) is a one-dimensional computational program developed by Deltares. The origin of the program is found in calculations on flow through pipeline systems. Various hydraulic components, such as pumps and valves, are included in its library. The WANDA-Locks library (see reference) allows the determination of the salt mass in the reservoirs. The formulae used are based on Deltares (2010). The lock chamber is a component included in the WANDA-Locks component library. Approach harbours can be modelled using the boundary condition for water level and salinity. The lock chamber and approach harbour are connected via valves that can comprise both the levelling system and the doors.

In case of advective flow the computation can be thought of as bookkeeping: the incoming discharge has a known salinity and can thus be taken as a load of salt being added to the reservoir. In case of baroclinic flow the hydrodynamic formulae simulate the density current using a fourth component in the WANDA-Locks library. These formulae are extended to include the impact of mitigating measures such as water and/or bubble screens by defining a so-called salt transmission factor, see Weiler et al. (2015).

Fig. 5. The schematization of lake Volkerak-Zoom in the WANDA-Locks model The Krammer locks are modelled in detail and provide fluxes and concentrations into the box at the west (left) side.

The model also schematises lake Volkerak-Zoom as different basins of water (see Fig. 5) which exchange salt through diffusion. The diffusion coefficients are an input of the model and have been calibrated and validated against measurements in the Eastern Scheldt and lake Volkerak-Zoom, see Fig. 6. The calibration was made against measurements at a number of locations (seen in Fig. 5) for a relatively dry year (2003). Only the January to June months were used for the calibration and subsequently the model was validated against the remaining months of the year (July to December).

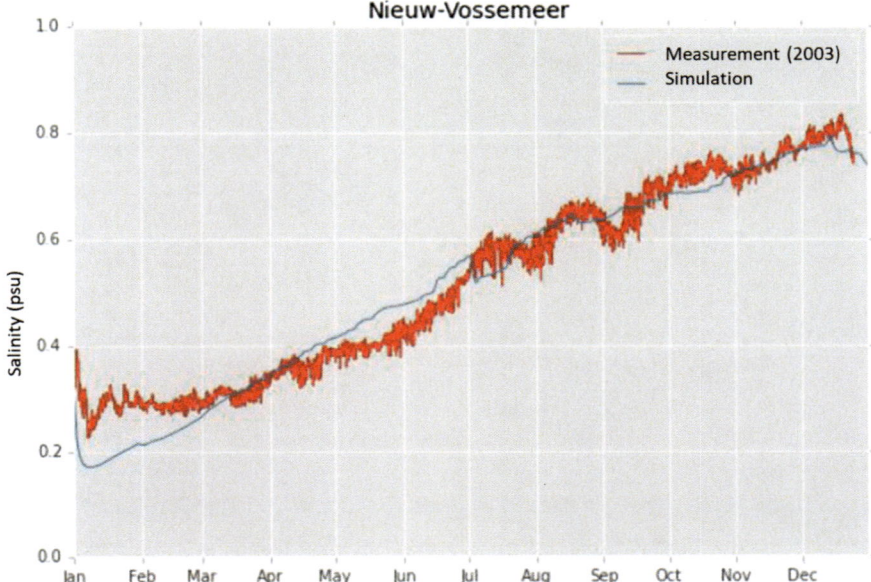

Fig. 6. A comparison of the predicted concentrations from the WANDA-Locks model and measured values in 2003. The model is calibrated on the data January to June and validated on the data July to December.

3.2   Effect of Lockage Operations

Previous work (Weiler et al. 2015) has shown the effect of gate open times (the amount of time that the gate is open for vessel sailing in and out of the lock) on salt intrusion through lock-exchange. The lock-exchange component of WANDA-Locks models the lock-exchange as a time dependent transfer of salt from lock chamber to approach harbour (or vice versa). The traffic handling is managed via a control module (van der Ven 2015).

For the Krammer locks the gate open times are not of importance for the current Dunkirk system as the gates are only opened after the saltwater in the lock chamber has been exchanged with fresh water, so there is no density difference over the lock gate. For the IZZS, the gates are opened directly after levelling and the bubble screens are operated. Their effectiveness is dependent on the gate open times (see Fig. 7).

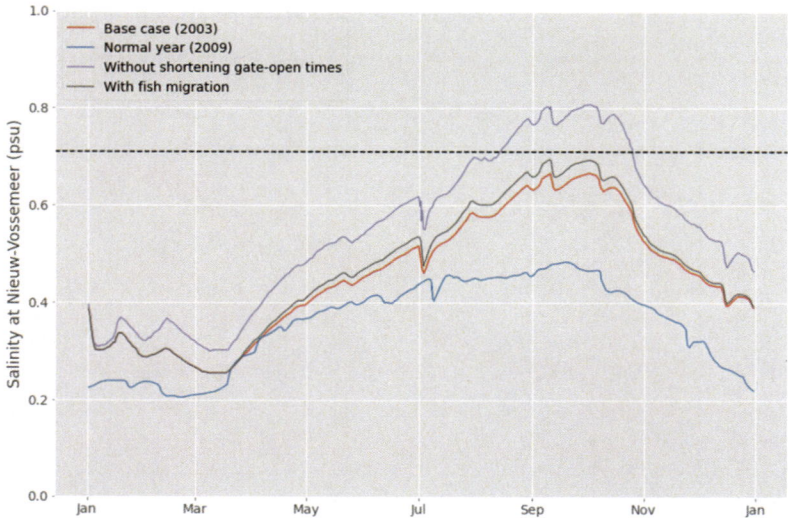

Fig. 7. WANDA-Locks simulations of the IZZS for a representative dry year (2003) and a representative normal year (2009)[1]. The effect of gate open times and fish migration measures at the locks is also included. The dashed black line is the threshold value for salinity maintained by Rijkswaterstaat.

3.3   Effect of Sea Level Rise (SLR)

Simulations have been run of the base case scenario with additional Sea Level Rise (SLR). A maximum 80 cm SLR has been simulated (see Fig. 8). In the base case a representative dry year (based on 2003) was simulated. In this case the lower basin, which stores the salt lockage prisms during high tide, is emptied to the Eastern Scheldt during low tide under gravity. As the sea level rises, the tidal window for emptying becomes shorter and smaller and the lower basin cannot be fully emptied. Some lockage prisms must then be discharged onto lake Volkerak-Zoom, causing salt intrusion. The threshold value for allowable salt concentration in lake Volkerak-Zoom is reached after only 20 cm of sea level rise.

To mitigate this, it is possible to install a pump in the culvert leading from the lower basin to the Eastern Scheldt to assist (when necessary) with emptying the lower basin. When this is done (see Fig. 9) the salt intrusion can again be controlled within the threshold value until a higher level of sea level rise. At that point the capacity of the pump is reached, and other factors in the operation make control of salt concentration within the limits difficult (see flushing discharge below). This required only 1 pump in the 4 available culverts.

---

[1] The simulation for this representative normal year was performed at an early stage in the project. Some of the aspects of the operation used for this simulation are not fully compatible with those used for the definitive simulations of the dry year. The normal year is included therefore only as a rough reference level.

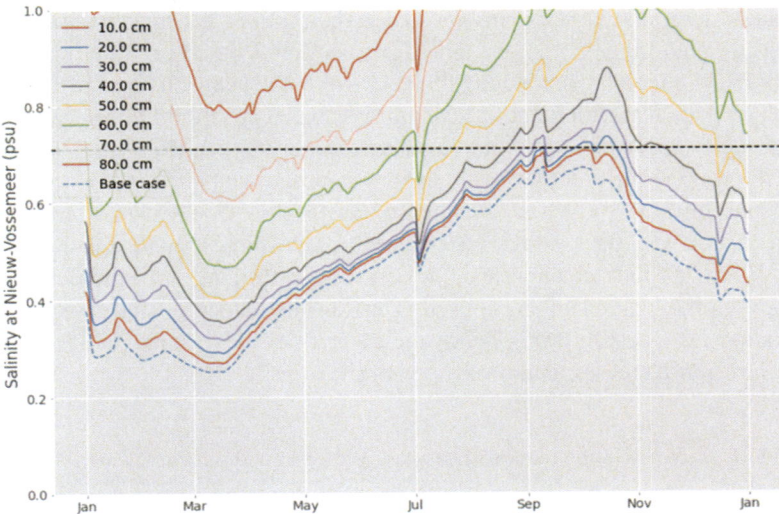

Fig. 8. WANDA-Locks simulations of the IZZS for a representative dry year including the effect of SLR. The dashed black line is the threshold value for salinity maintained by Rijkswaterstaat.

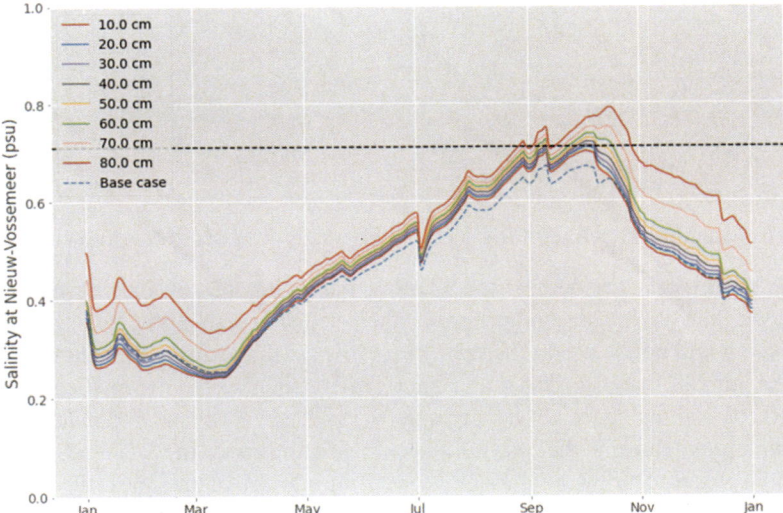

Fig. 9. WANDA-Locks simulations of the IZZS for a representative dry year including the effect of SLR and an additional pump from the lower storage basin (Laagbekken) to the Eastern Scheldt. The dashed black line is the threshold value for salinity maintained by Rijkswaterstaat.

An additional factor of the IZZS is the flushing discharge through the lock chambers. This is added when one of the lock gates is open for vessels sailing in and out. This flushing is done under gravity and without pumps. For reason of nautical safety, the flushing discharge is only included when it does not make maneuvering of the vessels unsafe (see next section). There are also limits based on water management (see Sect. 2.1.1). Finally, there are the limits based on the available head of the inactive (closed) lock gate at the moment of flushing. For flushing to the Eastern Scheldt, this head is reduced (on average) by sea level rise. Also, the available time window for flushing is reduced. In order to maintain the same amount of freshwater usage the instantaneous flushing discharge is increased as the low tide window shortens (see Table 1). In this way the average water usage for flushing is maintained as long as possible. When the lock chamber cannot be used for flushing, an additional flushing sluice between the lock chambers is used. This flushing sluice is will also be used for fish migration. At a certain level of SLR the capacity of the flushing sluice is reached. The salt intrusion increases as a result (see Fig. 8).

Table 1. Increase in the required instantaneous flushing discharge owing to SLR.

Sea level rise [m]	Flushing discharge [m³/s]	
	Through lock chamber	Through flushing sluice
0	9.00	2.72
0.1	9.54	3.82
0.2	10.07	5.15
0.3	10.62	6.78
0.4	11.22	9.58
0.5	11.82	max
0.6	12.48	Max
0.7	13.19	Max
0.8	14.06	Max

3.4   Safe Flushing Discharge for Vessels Entering and Exiting the Lock

Although an important aspect of the IZZS is the simultaneous flushing of the lock chamber with the bubble screen operations, there remain some moments in the lockage cycle flushing will not be used. The reasons for this are threefold: there is not always an available hydraulic head for flushing under gravity; it may not be nautically safe to flush; and, the sailing speed of vessels may be reduced too much by flushing.

The first point above is the case during sailing in and out at the Eastern Scheldt side during high water. During low water vessels sailing out of the lock should not be adversely affected by a flushing discharge in the direction of movement, however vessels sailing into the lock will also be sailing into the flushing discharge. This will reduce their speed and when the blockage of the vessel is large (such as when entering the lock and when the under keel clearance is lower at lower water levels) it has been decided to impose a maximum flushing discharge which is dependent on the water level (and available under keel clearance), see Table 1. This is the third point from above. Also, for the largest vessels it is recommended to limit the flushing discharge in all cases to 10 m³/s. This is equal to the maximum flushing discharge during the dry season, which is also limited for reasons of water management in lake Volkerak-Zoom.

When the gate is open in the lake Volkerak-Zoom side flushing will be used even at high water. The flushing discharge is stored in the Laagbekken (lower basin), which is kept at the lower water level to allow for a hydraulic head between the lake and this basin. At low water the chamber can be flushed directly into the Eastern Scheldt. For outgoing vessels, sailing into the flushing discharge there should always be sufficient under keel clearance available at this side (lake water levels at NAP 0.0 m) for the flushing discharge to not cause excessive delays, except for the largest vessels for which the maximum discharge of 10 m^3/s is used. For ingoing vessels the flushing discharge could cause an unsafe stopping procedure for ships in the lock. Therefore, the flushing will be stopped before the ships enter the lock. The moment at which the flushing discharge will be stopped has been chosen to allow for the longest flushing time. Vessels will be allowed to exit the lock and the incoming vessels will be allowed to approach the lock during flushing. The discharge will be stopped as the incoming vessels reach the approach harbour (Table 2).

Table 2. Maximum flushing discharge while vessels are entering the lock from the Eastern Scheldt.

Waterlevel Zijpe	Maximum flushing discharge
>NAP − 0.93 m	20,0 m^3/s
NAP – 1,00 m	19,1 m^3/s
NAP – 1,25 m	15,8 m^3/s
NAP – 1,50 m	12,5 m^3/s
NAP – 1.75 m	12,4 m^3/s
NAP – 2,00 m	12,2 m^3/s
NAP – 2,25 m	12,0 m^3/s
NAP – 2,50 m	11,9 m^3/s

3.5   Effect of Fish Migration Sluice on the Salt Intrusion

The IZZS envisions an addition of a flushing sluice between the lock chambers. This will have multiple functions: ; for salt management it will be used as an extra flushing discharge next to that through the lock chambers and it will also be used for fish migration; and, to increase the discharge capacity of the entire complex for extreme river discharge situations for which lake Volkerak-Zoom is used as storage basin for the Meuse and Rhine rivers.

The sluice gate will be kept open at for a short period after the end of low water to allow additional fish migration as the current reverses. This may allow some salt intrusion if salt water is present near the opening. The WANDA-Locks model does not account for possible density currents in the opening during this operation but the volumes of water exchange from the fish friendly operation are sufficiently small not to warrant additional mitigation measures (see Fig. 7).

4 Conclusions

For the Krammer locks the multiple functions and restrictions on fresh water usage complicate the design of an alternative salt-fresh separation system. The Innovative Salt-Fresh Separation System (IZZS) has been designed, combining bubble screens at both lock heads with an intermittent flushing discharge through the locks, using a combination of laboratory experiments, field measurements and CFD simulations. In addition, a hydraulic model of the lake Volkerak-Zoom system which accurately predicts the salt intrusion at the locks and the effect of operation and mitigation measures there, has been essential in weighing all the different and myriad functions and requirements. The model developed here, a box model with WANDA-Locks, can capture the complexity of the operation and give confidence that the planned operation will keep the fresh water availability in lake Volkerak-Zoom within the required limits of salinity for the foreseeable future.

Acknowledgements. The authors would like to acknowledge the many colleagues who contributed to the work summarized in this paper. In particular, the work of colleagues Otto Weiler and Arne van der Hout has contributed greatly to this paper.

References

Deltares (2010) Voorstudie: ontwerpstudie en praktijkproef zoutlekbeperking Volkeraksluizen – Model voor zoutvracht-berekeningen. Technical report 1201226-011-ZKS-0002 (in Dutch)

Deltares (2016) Validatierapport WANDA-Locks, het nieuwe zoutlekmodel. Technical report 1209463-000-HYE-0002 (in Dutch)

Keetels G, Uittenbogaard R, Cornelisse J, Villars N, van Pagee H (2011) Field study and supporting analysis of air curtains and other measures to reduce salinity transport through shipping locks. Irrig Drain 60(Suppl. 1):42–50

Mausshardt S, Singleton G (1995) Mitigating salt-water intrusion through Hiram M. Chittenden locks. J Waterways Port Coast Ocean Eng 121(4):224–227

Uittenbogaard RE, Cornelisse JM, O'Hara K (2015) Water – air bubble screens reducing salt intrusion through shipping locks. In: Proceedings of the 36th IAHR world congress, The Hague

PIANC (2021) Saltwater Intrusion and Mitigation in Inland Waterways. PIANC Report 198

der Kuur V (1985) Lock with devices to reduce salt intrusion. J Waterway Ports Coast Ocean Eng 111(6):1009–1021

Van der Ven PPD, De Groot I, Vreeken DJ, Weiler OM (2015) Simulating lock operation in the generic salt intrusion model WANDA-locks. In: PIANC SMART rivers 2015, Buenos Aires, Argentina, Paper 137

WANDA. https://www.deltares.nl/en/software/wanda/

WANDA-Locks. https://www.deltares.nl/en/software/module/wanda-locks/

Weiler OM, van de Kerk AJ, Meeuse KJ (2015) Preventing salt intrusion through shipping locks: recent innovations and results from a pilot setup. In: Proceedings of the 36th IAHR world congress

Live Digital Twin for Hydraulic Structures Fatigue Estimation

Cyril Condemine[1(✉)], Loic Grau[1], Yves Masson[2],
and Sebastien Aubry[2]

[1] MORPHOSENSE, Grenoble, France
{cyril.condemine, loic.grau}@morphosense.com
[2] Compagnie Nationale du Rhone (CNR), Lyon, France
{y.masson, s.aubry}@cnr.tm.fr

Abstract. Maintaining hydraulic structures such as dams, penstocks, or water lock gates in operating conditions and optimizing their maintenance costs are key issues for energy production or river navigation. The ultimate objective is to know the real state of fatigue and damage of the structure and identify any related anomalies. In this paper, we introduce a digital twin, for fatigue evaluation merging measured data obtained with an embedded sensor network and a 3D numerical model that converts in real time measured data into fatigue. After 3 years of R&D collaboration between CNR and Morphosense in the maintenance of navigation lock gates or dam gates, this presentation exposes how the proposed Live Digital Twin solution contributes to fatigue evaluation and more generally to global structural monitoring in dealing with fundamental issues of hydraulic structures: risk assessment, maintenance in operating conditions and maintenance costs optimization. After a context and state of the art introduction, the second part will detail the system overview. In the third part, the monitoring system will be addressed.

Keywords: Digital Twin · Fatigue · Monitoring · Sensor

1 Background

The lock gate is a key point in river transport ecosystem for which an unanticipated damage can have a high financial impact. The failure of only one of it can lead to several million euros losses (Eick et al. 2017; Treece 2015). Old gates reach their limit lifetime and needs for predictive maintenance. But there is no standards or regulations to monitor these structures and follow damages on a lock gate.

In addition, lock gate access is highly constrained by navigation traffic continuity of service. Only few navigation stops dedicated to lock gate maintenance and inspection are planned over the year. The traditional way to manage these actions is to organize periodic onsite inspections of the structure. However, this method provides visual inspections of a limited set of areas of the structure, sometimes using instrumentation but often purely visual checks. Onsite inspection cannot access all parts of the structure; including the underwater areas of the structure where there are also potential areas of structural vulnerability. Even more, this implies a constrained time to detect defects and

Y. Li et al. (Eds.): PIANC 2022, LNCE 264, pp. 494–505, 2023.
https://doi.org/10.1007/978-981-19-6138-0_43

to provide an efficient action plan. To establish a predictive maintenance strategy, the use of monitoring solutions contributes to continuous and real time information, between navigation stops. The proposed approach aims at reducing the cost of the inspection and maintenance with a Digital Twin process to assist remote control of the structure when it is needed instead of a prescriptive uninformed calendar-based inspections.

CNR operates and maintains 18 development schemes on the Rhône River in France and takes a very close interest in the hydromechanics equipment that equip its main structures: hydroelectric power plants, mobile dams, and navigation lock-gates. The hydromechanics assets include more than 300 large-sized vantellerie structures: width between 10 m and 45 m, height between 6 m and 21 m and weight between 50 t and 200 t. The large number of these equipment, as well as their strategic importance in terms of safety and security of the installations, make them a priority asset. Among these vantellerie bodies, the lock-gates are, together with the downstream safety valves of the production units, the equipment that are both the most stressed and that which has undergone the greatest number of operating cycles. Half of operational lock-gates is over 50 years old, keeping in mind that this equipment are generally designed for a 70 years of service lifetime as detailed in DIN 19704-1 2014-11 (2014).

Infrastructure performance expressed either in terms of risk assessment or in terms of cost-effectiveness is established using specific indicators. The ultimate objective is to know either the real remaining lifetime, the damage on the structure, and identification of any related anomalies and the comparison between design, simulation, and real behavior for design optimization of the structure.

Digital twins are an emerging technology. Structural Digital Twins which simulate and monitor structures in near real time are now being considered by significant players in energy producing structures. Some companies provide Digital Twin products, a broad term covering a range of disparate technologies. But there is no commercial product today which can provide a complete structural integrity solution. An operator that wants a structural integrity monitoring tool today will need to build a costly set of customized hardware and software that will only partially answer the requirements and in addition will have the following limitations:

- Current sensor technologies provide limited data by the sensors number, position, or static and dynamic precision, and are unsuitable for installation on large structure.
- Current numerical modelling methods cannot process the structural complexity in real-time without gross simplifications or major economic trade-off.

In addition, there is no solution combining the two elements into a holistic model. Some solutions based on machine learning propose behavior tracking and alert generation, but they are based on patterns of known events and cannot anticipate unknown events or structurally based fatigue failure. The weakness of big data with Machine Learning to identify outliers is that there is no physical understanding of the cause of the problem when there is a deviation of indicators, nor 3D representation of the areas with corresponding stresses. It is thus difficult to plan an efficient maintenance response.

A Digital Twin for structural monitoring is a twin numerical model of an instrumented structure providing a physical understanding of phenomenon which allows better decision making. Based on merging data captured by the sensors with a numerical model of the instrumented structure, the proposed Digital Twin allows to get additional structure indicators (like stress) not directly accessible to the measurement. Then it became possible to process the real fatigue and so the remaining lifetime. In case of behavioral modification, the numerical model can immediately identify this change and send a notification to the duty holder/operator specifies. In addition, our methodology includes of a fully numerical model from the design stage so that the digital twin can support inspection planning and predictive maintenance support.

In the second part of this paper, we will present the global system overview. In the third part, the methodology applied to monitor the lock-gate using the NEURON system and water level measurements is introduced. A relevant indicator combining the maximum deformation of the lock-gate and the water level will be presented in detail. The fourth part will focus on the Live Digital Twin building, combining the data from the previous indicators, the 3D numerical model of the lock-gate and the continuous data flow from the monitoring system. Finally operational implementation and results of the Digital Twin for Avignon water lock gate located on the Rhône River. The estimation of the fatigue and the service lifetime of the lock-gate with the Digital Twin is emphasized and the use for the maintenance plan optimization of the owner is explained.

2 Solution Overview

Digital twins are of most use when object characteristics are changing over time or may be difficult to evaluate directly, thus making the initial model of the object invalid, and when measurement data that can be correlated with this change can be captured. To ensure real fatigue, damage and remaining lifetime estimation allowing decisions based on physical understanding of the observed phenomena, the proposed system is based onto 3 parts:

- A monitoring system which allows to get all the needed static or dynamic indicators of the structure as well as the environmental or loads indicators
- A 3D numerical model of the monitoring structure based on the knowledge of the structure which considers the environmental loads, the structural behavior, and the boundary conditions and fast enough to integrate sensor data, and operate in real-time physics-based simulations
- A data hosting and processing platform architecture, between the measurement system and the physics-based numerical model of the structure, enabling the data-model convergence by dynamically updating the digital twin model and structural key indicators estimation to support operations and maintenance.

Based on the monitoring system, we proposed a Digital Twin updating in real time according to the endured loads by the structure and its integrity. The Digital Twin model is based on a 3D numerical model of the structure and is continuously feed with structural indicators (static, dynamic, modal, …) and environmental loads measurement (water height, wind states, sea state, temperature, …). The digital Twin delivers, from stress calculations, an estimate of fatigue, damage, and residual life at all points of the structure.

To provide an efficient Digital Twin, the developed solution simultaneously controls the measurement technologies (sensors, acquisition, transmission, processing), the 3D modelling of the structures (accurate, quick, and continuously updated, interfaces, boundary conditions), the structure load sources and their measurements (wind, swell, pedestrians, etc.), and a specific IT architecture to aggregate all the collection of data and simulation (Fig. 1).

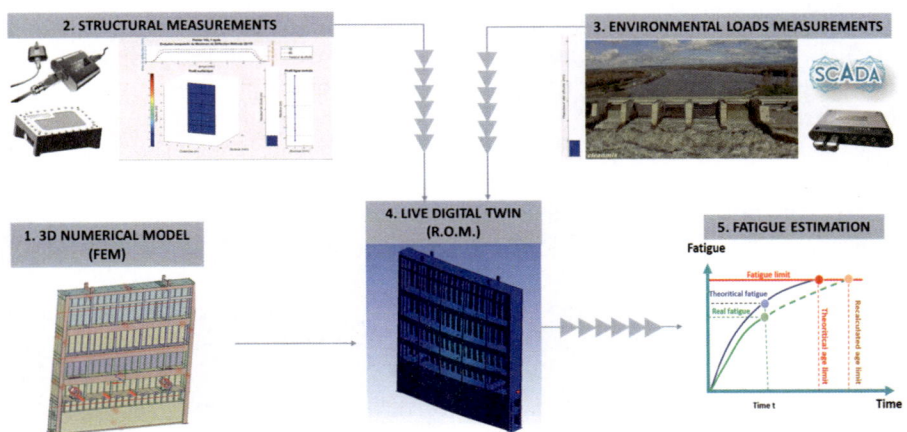

Fig. 1. Digital Twin overview

2.1   Industrial IoT Sensor Networks

Structural monitoring is based on two main types of indicators: structural and environmental. To allow qualitative correlation analyses of these measurements, the I-IoT system (Neuron) acquires and synchronizes data on the instrumented structure. Those data can then be correlated and merged into the calculations and digital models. The proposed monitoring I-IoT sensor network is comprised of a "plug-and-play" single cable hard-wired network of measurement nodes (synchronized high resolution tri-axis accelerometers, gyros, magnetometers, and a temp sensor) collecting static and dynamic deformation at a spatial density adapted to the structure and a software solution processing raw data into to specific indicators meaningful for the application (Fig. 2).

This network is suitable for large size structure (up to 30 nodes and 1km long), is available in different housings (IP65, 100m Subsea and ATEX Z1), and provide external sensors interoperability (strain gauges, analog and digital sensors) for load measurement. The mechanical indicators, static (tilts, roll, norm, torsion, deformation, convergence…) and dynamic (modal analysis, frequencies analysis and tracking, dynamic deformation), express the structural response to external loads and would be compared with Digital Twin simulated outputs. The environmental indicators measure external loads (stress, environment) applied on the structure: wind on wind turbines, hydrostatic load for lock gate, vehicles on a road bridge, trains on a railway bridge, the temperature for metal structures, humidity, corrosion. For each type of structure, only the environmental indicators which are significant for structural fatigue estimation are relevant for observation.

Data are pre-processed on the embedded gateway and send by 4G or Ethernet connection to an AWS cloud platform for data processing, long-term storage, and visualization.

2.2   Numerical Model

The Digital Twin provides an accurate description of objects that change over time: a precise, up-to-date copy of certain properties and states of the physical structure, such as its shape, position, state, and movements.

The model used in a Digital Twin is not a data-driven model, but it should produce results that are directly equivalent to measured quantities (so that the model updating process is data-driven), and it is likely that the model will take in other measured quantities as boundary conditions, loads, or material properties (Wright and Davidson 2020).

Digital Twins can use any type of model that is a sufficiently accurate representation of the physical object that is being twinned. In an ideal world, where computation would be instantaneous, and accuracy would be perfect, Digital Twins would use models derived directly from physics that took all phenomena likely to affect the quantities being measured and updated into account.

Digital models are mainly based on the finite element, boundary element or finite difference method, considering an almost infinite variety of types of structural geometry integrating all types of loads, material and boundary conditions. However, the drawback of a very complex model is that calculations may then take considerable time and means that real-time updating is not possible. Simplifying models to reduce these calculation times is therefore a real challenge. A model for a Digital Twin should be sufficiently physics-based in to manage meaningful updated parameters, sufficiently accurate to manage application useful updated parameters, sufficiently quick to run in comparison with the observed physical phenomenon. But the barrier of computational cost at high accuracy does not mean that physics-based approaches should be discarded altogether. Different solutions can be used to adapt or replace the computationally expensive model:

- Some applications of Digital Twins do not require high-speed computation, because the time frame over which the twin is to be updated is hours rather than seconds.
- Some applications of a Digital Twin can use local models of key parts of a structure or an object rather than considering the complete system.
- Generate a surrogate model or metamodel based on a set of reliable results within the known operating parameter envelope of the physical object (pure data-driven model with Kriging).
- Generate a reduced order model (ROM) by seeking to characterize the system being modelled in terms of a small number of functions or "modes".

In the proposed solution, we choose to use FEM model based on the structural geometry and on ANSYS simulation tools with linear elastic assumption. To process unsymmetric events or dynamic approach, a full structure model is usually used. Load acting on the structure (like wind, ocean waves and marine current, water level, temperature, ...) and physical parameters (cracks, erosion, ...) will also be modelled as described in the Digital Twin section, to generate a specific reduced order model. This reduced order model will be used to simulate tilts for each node (for calibration and data-model convergence algorithm) and to provide stress estimation (for fatigue estimation).

3 Monitoring: From Tilts Measurement to Deformation Indicator

The maximum of deflection is a fundamental indicator of the static behavior of the structure, especially a water lock gate. This indicator is also correlated with other environmental phenomenon such as temperature, top and bottom level of water. Standard solutions, as detailed in Eick et al. (2017), propose indicators that focus on the structural behavior, especially the correlation between the constraints and the water level. These indicators can be generalized by using the maximum deformation instead of the local stress. But measuring stress on a lock gate is limited by the robustness of a stress gauges in water environment and measuring the deflection of the gate is tricky with optical external devices due to the presence of obstacles, water, and 'dirty' environment. The proposed solution is based on a high accurate accelerometer sensors networks distributed on the structure allowing to capture the deformation in real time and on the significant area of the lock gate as detailed in Carmona et al. (2019). For the chosen use case (Avignon located lock gate from CNR on Rhône River) the monitoring network is comprised of 9 nodes, fixed with magnets, and a software interoperability with CNR SCADA providing upstream water level.

Fig. 2. I-IoT Sensor network and water lock gate instrumentation

Then, using tri-axis acceleration measurement, we estimate the tilt and then process, thanks to patented algorithms, the static deformation of the lock gate for each node.

The static information (tilts) leads to the estimation of spatial deformation, and dynamic data (accelerations) allow the modal behavior estimation. Using those data from all the nodes distributed along a vertical central line on the lock gate, our monitoring system can reconstruct the global deformation of all the lock gate, regardless sensor position, during different phases of the lock gate operation.

Using the global deformation measurement of the lock gate for each cycle of water charge/discharge we generate the maximum deformation indicator, in correlation with the water level. So based on this indicator, the structural behavior is monitoring in continuous, the measure data are easily accessible and depending on the measure phenomenon. In addition, loads and environmental indicators are available thanks to monitoring system interoperability. Data-driven model (Machine learning) can be derived from available measurement to detect prior to the Digital Twin deviant behavior and thus generate alarms. To compute the ROM (Reduced Order Model) model for the Digital Twin, we need tilts and water level during a long period, typically several days or weeks. All those data were then merged to process sensitivity studies leading to a reduction order of the FEM model (Fig. 3).

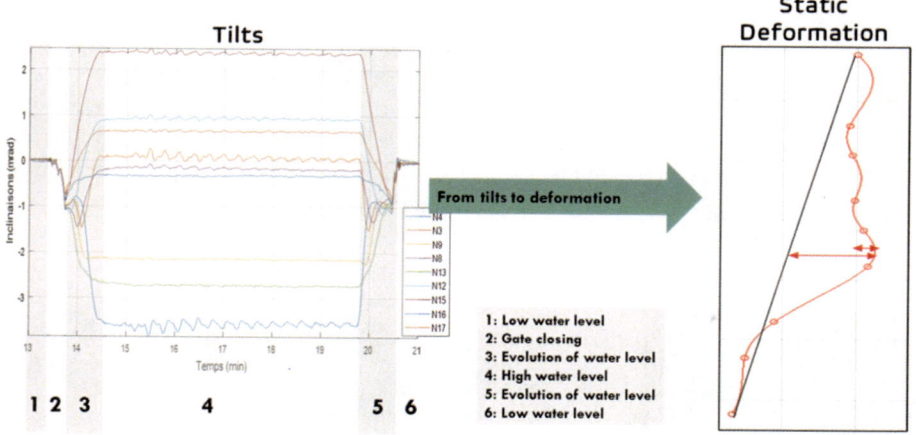

Fig. 3. Lock gate operation phases and maximum lock gate global deformation

4   Digital Twin

4.1   Numerical Model

Digital Twin building starts with the CAD model and design study of the structure. The 3D numerical model will fulfill the requirements described in the previous model section. To define sensor position, a first simulation is done to determine maximum deformation areas. As we use acceleration to estimate tilts and global deformation, our sensors will be installed on these specific areas where the lock gate deformation is at a local maximum value.

As described in the next figure, the digital twin design process is divided into to 5 main steps (Fig. 4).

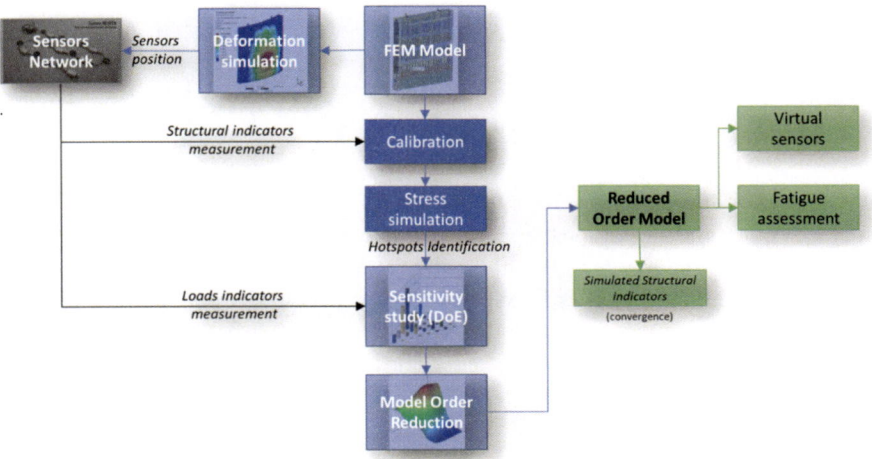

Fig. 4. Digital Twin design process

4.1.1   Model Calibration

The calibration of the numerical model based on measurement is an essential step to provide estimation of the boundary condition and load correction. This step will allow the structural behavior to be validated with an acceptable error independently from the load applied on it. AS we are in static applications, calibration will be made on tilt angles indicators. After the installation of the monitoring system on the structure, comparison of simulated tilts with measurement is performed, to tune the simulation model to best fit of the measurement data. Needed inputs are sensor node position on the structure and measured structural indicators. The next figure shows model calibration based on measured and simulated tilt angle comparison for each node and for different boundaries conditions: number of contacts between the gate and the support wall, type of upper and lower blockage of the gate, and the stiffness coefficient of supports (Fig. 5).

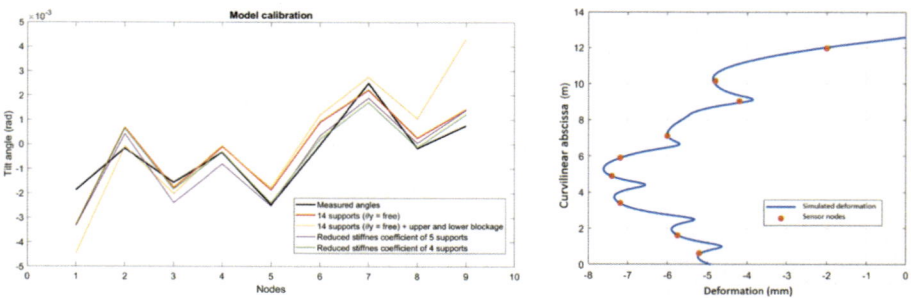

Fig. 5. Model calibration based on tilt angles and estimated deformation

4.1.2   Hotspots Identification

Hotspot identification aims at localizing the position of the most important stress on the structure where a fatigue assessment of the structure should be performed. The localization of the most important stress may be a welded or a non-welded area, regardless of the sensor position. The identification of this hotspot is often validated in collaboration with experts of the studied structure to provide the most possible feedback and then have a realistic hotspot. With a full picture of the stress distribution of the asset it is then possible to establish the actual level of the fatigue of the structure. In the case of Avignon located CNR lock Gate, we have identified 36 critical hotspots.

4.1.3   Loads Sensitivity Study and Reduced Order Model (ROM) Generation

A sensitivity study will also allow to define the uncertainties linked to the model/measurement correlation. This variational analysis is performed using: (i) external indicators as parametric inputs, (ii) tilts at sensor positions and stress at hotspots as parametric outputs. The main objective is to condense the relation between inputs and outputs to reduce the order of the simulated FEM model. As a result, simulations taking hours to days with FEM models decreased to seconds to minutes with a ROM model (Zhao et al. 1992; Chinesta et al. 2018).

The ROM is a collection of response surfaces with water levels (up and downstream) as inputs and with outputs such as stress on chosen hotspots and tilts angles for each node. Stresses on hotspot will be inputs for fatigue estimation and tilts angles will be inputs for data-model convergence monitoring.

The next figure introduces the ROM model for sensor node 1 between upstream and downstream water levels and tilt angles and for one specific hotspot, the weld number 28 near the edge of the lock gate between upstream and downstream water levels and stress amplitude (Fig. 6).

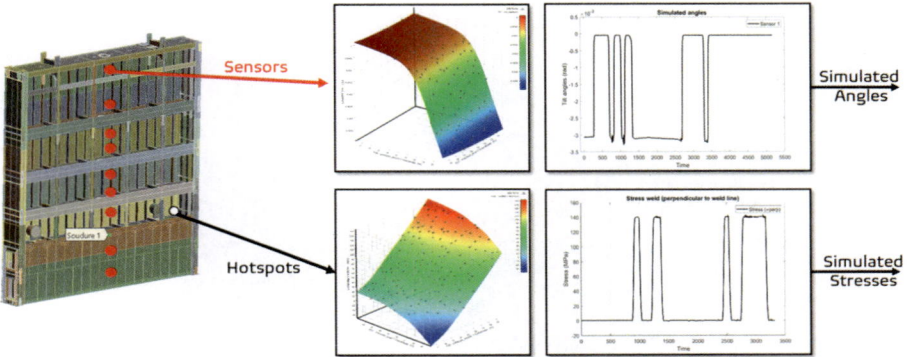

Fig. 6. Response surfaces examples: Sensor n°1 and Weld n°28

4.1.4 Fatigue Estimation

After calibration of the load and structure, the fatigue assessment requires to extract the stress from the model which can be performed on the calibrated Reduced Order Model. The fatigue assessment, based on linear damage accumulation model, provides the remaining lifetime of the structure based on stress amplitude and cycling within the structure. The stress, provided by the ROM model is then counted using a standard Rainflow counting method (amplitude and cycling) and applied to S-N curves for damage accumulation estimation. The fatigue assessment implementation is based on standards and guidelines such as Eurocode 3, DNV or FKM (Forschungskuratorium Maschinenbau 2020). The delivered output of this module is the degree of structural performance and the remaining useful lifetime based on the historical usage (Fig. 7).

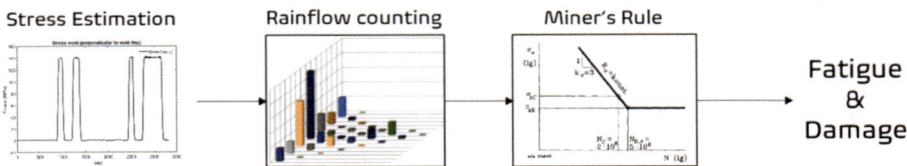

Fig. 7. From stress to fatigue estimation

4.2 Digital Twin Processing

This phase corresponds to the operational deployment of the Digital Twin with an automatic and real time estimation of the fatigue depending on the real stress of the Lock gate. The structural estimation of the fatigue and the remaining lifetime are the main attending indicators. The ROM takes as inputs loads measurements (environment indicators) and deflections/tilts of the structure at sensor node positions and constraints/stress at hotspots as outputs. The model is updated when the difference between the simulated and measured tilts or deflections is higher than a selected threshold. In the same way, variations in physical parameters such as cracks size or erosion factor can lead to an update of the model (Fig. 8).

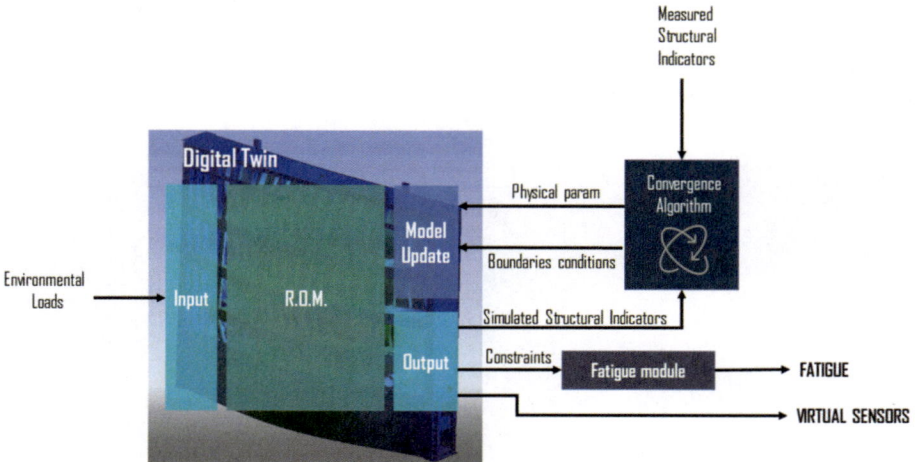

Fig. 8. Digital Twin process

Based on this approach, the proposed digital Twin provides residual lifetime for a list of 36 critical hotspots on the Avignon water lock gate. The residual lifetime estimation based on loads history shows:

- 20 uncritical welds
- 6 welds are identified to be repaired (plastic deformation, zero residual lifetime)
- 10 critical welds with low residual lifetime (<10 years).

Those results are in high correlation with the maintenance operations on the gate realized by CNR.

The next figure shows the remaining lifetime for these 10 welds and the model behavior. After a learning phase, the remaining lifetime evolves with the real and measured load cycles. A new value is calculated each day based on measures indicators (Fig. 9).

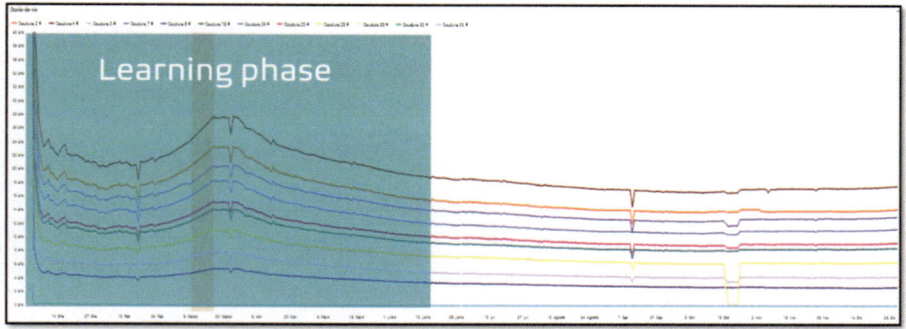

Fig. 9. Remaining lifetime estimation for critical hotspots

5   Conclusion and Perspectives

To provide information on the remaining lifetime of every part of the structure, we have designed, tested, and validated a Digital Twin with a real-time, physics based numerical model of a full-scale asset to give greater insight into performance and operation. We deliver a Digital Twin of a whole complex structure which can track the loading, via sensors, and accurately predict potential failures and areas of fatigue.

Thanks to a collaboration with CNR on the Avignon Lock gate on Rhône River, our approach has been validated in real environment, and since this first demonstration, 3 others Digital Twins have been deployed on other hydraulic structures.

Residual lifetime estimation was a first application of our solutions, but with the same tools, we can also provide design validations on new structures or virtual sensors to monitor inaccessible areas of the structure or physical phenomena difficult to measure.

This approach is currently validating on other structures such as floating offshore wind turbine, large size bridge or nuclear plant.

References

Carmona M, et al (2019) Innovative solution for navigation locks monitoring. Smart Rivers 2019

Chinesta F, Cueto E, Abisset-Chavanne E, Duval JL, El Khaldi F (2018) Virtual, digital and hybrid twins. Arch Comput Methods Eng. https://doi.org/10.1007/s11831-018-9301-4

DIN 19704-1 2014-11 $ 7-6. Fatigue (2014)

Eick BA, et al (2017) Automated damage detection in miter gates of navigation locks. Wiley

Forschungskuratorium Maschinenbau (2020). Analytical strength assessment of components: FKM guideline. Book 7th edn. EAN 9783816307457

Treece ZR (2015) USACE smart gate SHM to preserve America's infrastructure. In: IWSHM 2015 Stanford

Wright L, Davidson S (2020) How to tell the difference between a model and a digital twin. Adv Model Simul Eng Sci 7(1):1–13. https://doi.org/10.1186/s40323-020-00147-4

Zhao W, Baker MJ (1992) On the probability density function of rainflow stress range. Int J Fatigue 14(2):121–135

Mechanism and Variation Characteristics of Longitudinal Tilt of Ship Chamber of Hydro-Floating Ship Lift (HFSL)

Shu Xue$^{(\boxtimes)}$, Yaan Hu, Zhonghua Li, and Ying Jin

Key Laboratory of Navigation Structure Construction Technology,
Ministry of Transport, PRC, No.225 Guangzhou Road, Gulou District,
Nanjing, Jiangsu, China
sxue@nhri.cn

Abstract. The mounting clearance and deformation of the mechanical synchronization system of the HFSL directly affects the attitude of ship chamber, especially the longitudinal tilt of the ship chamber. In this paper, a generalized model is established to derive the analytical solution of the longitudinal tilt of the ship chamber according to the kinetic equations and differential equation theory of the rigid body with fixed axis rotation, and to reveal the mechanism of the longitudinal tilt of the ship chamber. According to the moment balance equation, the design conditions of the synchronous shaft rigidity of the HFSL are obtained. In addition, a mathematical model is established to simulate the change process of the longitudinal tilt of the ship chamber and analyze its variation characteristics.

Keywords: HFSL · Mechanical synchronization system · Longitudinal tilt of ship chamber · Mechanism · Change characteristics

1 Introduction

HFSL and rope winch vertical ship lift (RWSL) are the same in the following aspects: the ship chamber lifted through wire rope suspension, in the process of operation, the ship chamber will be inevitably subject to many unbalanced load. Local instability caused by unbalanced load resulting in the ship chamber deviated from the balance position, and then lead to the flow of the water in the ship chamber. The shift of the water gravity center extra imposed on the ship chamber a capsize moment, so that the amount of tilt of the ship chamber increased, and boost the water further flow, resulting in the tilt of the ship chamber continue to enlarge. If there is no sufficient anti-tilt moment, it will eventually lead to safety accident.

For the RWSL, the closed-loop mechanical synchronization system forces each winding drum to run synchronously which ensures each lifting point of the ship chamber lift almost synchronously. The system of "motor- synchronous shaft- winding drum" is equipped with reducer to realize the transmission, see Fig. 1. In the process of operation, the synchronous shaft mainly bears the additional torque generated by the

Y. Li et al. (Eds.): PIANC 2022, LNCE 264, pp. 506–515, 2023.
https://doi.org/10.1007/978-981-19-6138-0_44

unbalanced output of the motor, or when a driving motor or the main dragging equipment fails, the strength of the synchronous shaft is sufficient to transmit the rated output torque of a motor and maintain the synchronous lifting of each lifting point.

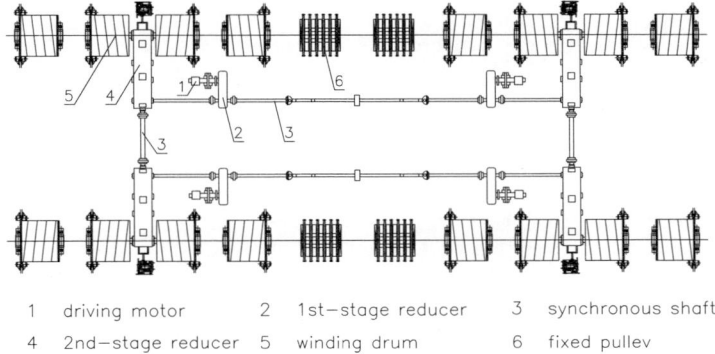

| 1 | driving motor | 2 | 1st–stage reducer | 3 | synchronous shaft |
| 4 | 2nd–stage reducer | 5 | winding drum | 6 | fixed pulley |

Fig. 1. Schematic diagram of mechanical synchronization system of RWSL

HFSL is driven by water energy and the structure of its mechanical synchronous system without motor and reducer, which is a closed-loop structure directly connected by synchronous shaft to each winding drum in series, using synchronous shaft rigidity to guarantee the synchronous rotation of winding drums, seen Fig. 2. Once the ship chamber is subjected to unbalanced load and starts to tilt, the winding drums rotates in sequence from upstream to downstream, producing a cumulative effect. Due to the installation gap between synchronous shafts and winding drums, there is no force on the synchronous shaft at the beginning. The relative rotation between the winding drums only eliminates the installation gap, then the drums become truly rigidly connected through the synchronous shafts, and the rigidity of the synchronous system comes into play to produce torsional deformation to resist the tilting moment of the ship chamber.

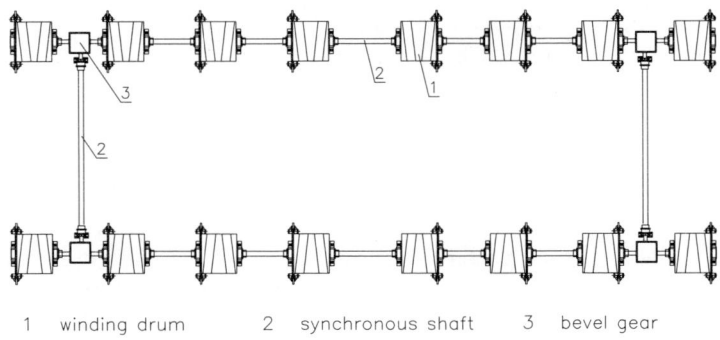

| 1 | winding drum | 2 | synchronous shaft | 3 | bevel gear |

Fig. 2. Schematic diagram of mechanical synchronization system of HFSL

In summary, for RWSL, the mechanical synchronization system can eliminate the installation gap through the motor preload, and the deformation produced by the synchronous shaft under the action of torque is reduced substantially through the reducer and then transferred to the ship chamber by winding drums. Thus, the installation gap of the mechanical synchronization system and the synchronous shaft torsional deformation have less influence on the ship chamber tilt. However, the installation gap of mechanical synchronous system and synchronous shaft torsional deformation of HFSL are directly reflected to the longitudinal tilt of the ship chamber. The longitudinal tilt is more sensitive to the installation gap and synchronous shaft rigidity, which directly affects the safe operation of the ship lift. If the synchronous shaft rigidity is large enough, the tilt of the vessel converges, otherwise the ship

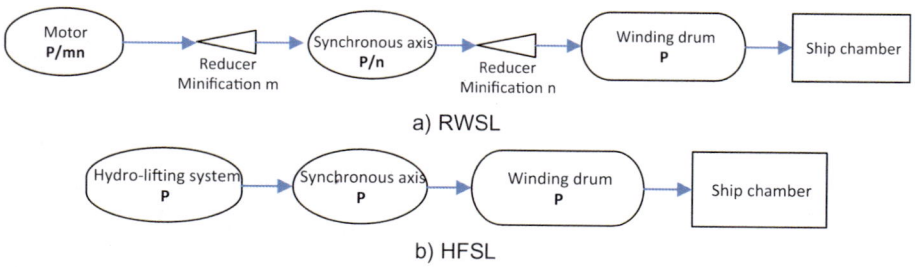

a) RWSL

b) HFSL

Fig. 3. Mechanical synchronization system transmission mechanism

chamber will be overturned (Fig. 3).

2   Generalized Model of Mechanical Synchronization System of HFSL

The mechanical synchronization system of HFSL is a hyper-static structure and it is difficult to determine the torque of each synchronous shaft by static balance condition. Therefore, a generalized model of mechanical synchronization system is established for analysis, and the most unfavorable situation that the load is concentrated on either end of synchronous shaft is considered.

Assume that the ship chamber is a rigid body, the left and right sides of synchronous axis is completely symmetrical and the influence of the modulus of elasticity of the wire rope is ignored. The force on synchronous axis is shown in Fig. 4. The tension of the ship chamber are F_1 and F_2, side tension of balance weight are F_1' and F_2'. When the ship chamber is in the ideal horizontal state, the ship chamber and water load gravity center is in the center. The load acting on each lifting point is equal ($F_1 = F_2$). But when the ship chamber is tilted, gravity center of the water load on the ship chamber will shift which leading to a overturn moment, and the synchronous shafts would produce a torque to resist chamber tilting.

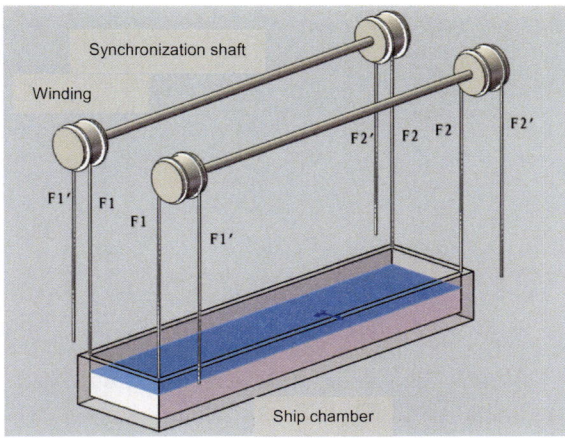

Fig. 4. Generalized model of mechanical synchronization system and ship chamber

3   Design Condition of Synchronous Shaft Torsional Rigidity

The relative rotation angle of the winding drums caused by the installation gap of the mechanical synchronization system is defined as θ_a, the longitudinal tilt of the ship chamber is defined as Δh_a, the relative rotation angle of the reel caused by the torsional deformation of the synchronous shaft is defined as θ_b, the longitudinal tilt of the ship chamber is defined as Δh_b.

The relative rotation angle θ of the winding drums at both ends of the longitudinal synchronous shaft is as follows:

$$\theta = \theta_a + \theta_b \tag{1}$$

The total tilt of ship chamber Δh is as follows:

$$\Delta h = \Delta h_a + \Delta h_b \tag{2}$$

The moment analysis is carried out with the ship chamber as the object of study, and the moment of force on the ship chamber consists of two parts:

1) Overturning moment of the water on the ship chamber M_G

$$M_G = 2 \int_0^{L/2} \rho g B \frac{\Delta h}{L} x^2 dx (1+k) = X \cdot \Delta h (1+k) \tag{3}$$

In which, $X = \frac{1}{12}\rho g B L^2$, B is the width of ship chamber, L is the length of ship chamber, k is the amplification factor of dynamic water moment in the chamber which reflects additional effect on the force of moment by the acceleration of the water and water surface fluctuation in the chamber.

2) Ship chamber anti-tilt moment M_K

$$M_K = M_I + M_B = 2\frac{\Delta h_b}{R^2\frac{L}{GI_p}}L + Y\Delta h(1+q) \tag{4}$$

$$M_I = 2\frac{\Delta h_b}{R^2\frac{L}{GI_p}}L \tag{5}$$

$$M_B = Y\Delta h(1+q) \tag{6}$$

In above formula, M_I is the anti-tilt moment generated by synchronous shaft, I_P is the cross-sectional polar inertial moment of synchronous shaft, G is the shear modulus of synchronous shaft, R is the radius of reel, M_B is the anti-tilt moment, Y is balance weight static anti-tilt moment $Y = \tau X$, τ is the balance weight anti-tilt coefficient which is related to bottom area of balance weight and lifting point position. q is amplification factor of vertical shaft water surface fluctuation on the static anti-tilt moment of balance weight, reflecting the additional effect of vertical shaft water surface fluctuation.

According to the ship chamber longitudinal moment convergence condition, the anti-tilt moment should be greater than the tilt torque generated by the water in the ship chamber before the ship chamber reaches the maximum permissible tilt.

$$2\frac{[\Delta h] - \Delta h_a}{R^2\frac{L}{GI_p}}L + Y[\Delta h](1+q) > X[\Delta h](1+k) \tag{7}$$

According to above formula, the design condition of synchronous shaft rigidity is as follows:

$$K = \frac{GI_p}{L} > \frac{[X(1+k) - Y(1+q)]R^2}{2L\left(1 - \frac{\Delta h_a}{[\Delta h]}\right)} \tag{8}$$

4 Analytical Solution of Longitudinal Tilt of Ship Chamber

The analytical solution of the tilt amount of ship chamber can be derived from the kinetic equations for the fixed-axis rotation of a rigid body. Define I as the rotational inertia of the chamber (the center of the line connecting the two longitudinal lifting points of the ship chamber is the axis of rotation), φ is the longitudinal tilt angle of ship chamber, $\ddot{\varphi}$ is the angular acceleration of the ship chamber longitudinal ends of the rotation.

The relation between θ and φ is:

$$\varphi = \frac{\theta R}{L} \tag{9}$$

1) When $\theta < \theta_a$, $X\Delta h(1+k) - Y\Delta h(1+q) = I\ddot{\varphi} = I(\Delta h/L)''$, the formula can be simplified to

$$\Delta h'' - [X(1+k) - Y(1+q)]\frac{L}{I}\Delta h = 0 \tag{10}$$

General solution of above homogeneous equation is:

$$\Delta h = \frac{\omega_0 L}{2\alpha}e^{\alpha t} - \frac{\omega_0 L}{2\alpha}e^{-\alpha t} \tag{11}$$

In which, $\alpha = \left\{[X(1+k) - Y(1+q)]\frac{L}{I}\right\}^{1/2}$, ω_0 is the initial disturbance angular velocity.

2) When $\theta > \theta_a$, $X\Delta h(1+k) - Y\Delta h(1+q) - 2\frac{\Delta h_p}{R^2\frac{L}{GI_p}}L = I\ddot{\varphi} = I(\Delta h/L)''$, the formula can be simplified to

$$\Delta h'' + \frac{L}{I}\left[\frac{2GI_p}{R^2} + Y(1+q) - X(1+k)\right]\Delta h = \frac{2GI_pL}{R^2I}\Delta h_a \tag{12}$$

General solution of above equation is

$$\Delta h = \frac{C}{\Omega^2} + \left(m - \frac{C}{\Omega^2}\right)\cos[\Omega(t - t_a)] + \frac{n}{\Omega}\sin[\Omega(t - t_a)] \tag{13}$$

In which, $\Omega = \left\{\frac{L}{I}\left[\frac{2GI_p}{R^2} + Y(1+q) - X(1+k)\right]\right\}^{\frac{1}{2}}$, $C = \frac{2GI_pL}{R^2I}\Delta h_a$, t_a is the time to eliminate the installation gap, $m = \theta_a R$, $n = \frac{\omega_0 L}{2}e^{\alpha t_a} + \frac{\omega_0 L}{2}e^{-\alpha t_a}$.

According to the above derivation, in the process of elimination of synchronous shaft rotation gap ($\theta < \theta_a$), the tilt amount of the ship chamber (Δh) is a hyperbolic sine function of t increasing monotonically. After the synchronous shaft rotation gap eliminated ($\theta > \theta_a$), the basic form of the equation is the forced motion of spring type, which reflects the forced torsional vibration of the "synchronous shaft system with suspended ship chamber and balance weight". Ω is the inherent frequency of the torsional vibration of the system, which is related to the synchronous shaft rigidity, the arrangement type of the ship chamber and balance weight, etc.

5 Characteristics of the Longitudinal Tilt of the Ship Chamber

Relying on a HFSL engineering machinery synchronization system and the characteristic dimensions of the ship chamber, 3D numerical simulation is carried out. Calculation parameters are shown in the Table 1.

Table 1. Calculation parameters of typical working condition

No	Items	Parameter
1	The length of water area in the ship chamber $L(m)$	66
2	The width of water area in the ship chamber $B(m)$	12
3	The radius of reel $R(m)$	2.125
4	Rotational inertia of ship chamber $I(\text{kg} \cdot \text{m}^2)$	3.46e8
5	Initial disturbance angular velocity of ship chamber longitudinal tilt $\omega_0(\text{rad/s})$	0.0001
6	Rotation gap of mechanical synchronous shaft $\theta_a(\text{rad})$	0.05
7	Shear modulus of synchronous shaft $G(\text{GPa})$	80
8	Torsional rigidity of synchronous shaft $K(\times 10^3 \text{kN} \cdot \text{m/rad})$	14.55

The process lines of the longitudinal tilt amount of the ship chamber with time are shown in Fig. 5. The change law of the longitudinal tilt of the ship chamber is as follows:

1) The torsional rigidity of synchronous shaft meets the design requirements, then the longitudinal tilt of the ship chamber converges.
2) During the elimination process of the synchronous shaft rotation gap, the longitudinal tilt of the ship chamber increases monotonically.
3) After the synchronous shaft gap is eliminated, the longitudinal tilt of the ship chamber oscillates and decays slowly at the balance position of the longitudinal moment of the ship chamber.
4) After the mounting gap is completely eliminated, the longitudinal tilt of the ship chamber can be expressed as the superposition of two different frequencies of the trigonometric functions. The longitudinal tilt of the ship chamber contains two dominant frequencies: one is in the mounting gap elimination moment, there is a high frequency vibration with a circular frequency of about 5.5rad/s which reflects the inherent frequency of the system torsional vibration. Due to the large damping created by inertia of the water in the ship chamber, it decays quickly. The other is the circular frequency of about 0.2rad/s of low frequency vibration.

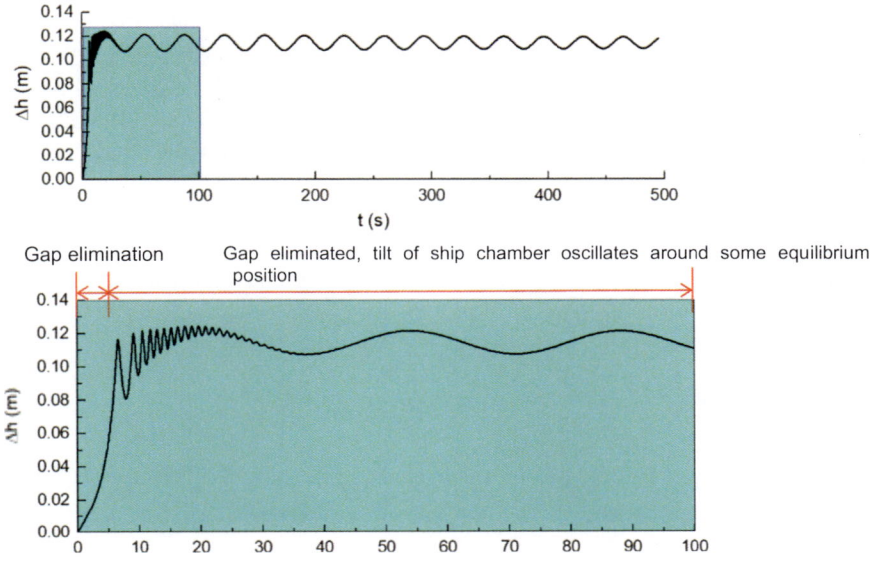

Fig. 5. Convergence of ship chamber longitudinal tilt

By analyzing the low frequency, the ship chamber is in the equilibrium position after stabilization, the fluctuation period is about 30s, converted into a circular frequency of 0.20 rad/s. Ship chamber can be regarded as a rectangular container. Self-oscillation circular frequency of water in a rectangular container has the analytical solution of:

$$\omega_{mn}^2 = gk \cdot \tanh(kh)$$

$$k^2 = \pi^2 \left(\frac{i^2}{a^2} + \frac{j^2}{b^2} \right) \tag{14}$$

In which, a, b, h are the length, width and depth of the water in the vessel, g is the acceleration of gravity. i and j are integers greater than or equal to zero, by taking different values, the longitudinal m-order self-oscillation circular frequency and the transverse n-order self-oscillation circular frequency of the liquid in the vessel can be calculated. The longitudinal first-order self-oscillation circular frequency of the water body in the vessel is 0.21 rad/s by substituting the dimension of the water body in the vessel, which is basically the same as the longitudinal tilting circular frequency of the vessel calculated by the numerical model in this paper. The low frequency reflects the longitudinal first-order self-oscillation frequency of the water body in the vessel. Due to the inability to accurately simulate the effects of system damping and water viscosity in the mathematical model, the decay is very slow. The sway of the water body decays much faster in the actual project.

Changing the calculation parameters by only reducing the synchronous shaft torsional rigidity, the longitudinal tilt change is investigated under the conditions of insufficient synchronous shaft rigidity by setting the synchronous shaft torsional rigidity $K = 0.5 \times 10^3 \text{kN} \cdot \text{m/rad}$. The calculation results show that if the torsional rigidity of the synchronous shaft is not enough, the tilt of the vessel diverges rapidly. The change process of synchronous shaft torque and ship chamber tilt under two working conditions is shown in Fig. 6. It shows that the synchronous shaft torque is zero before the gap is eliminated, and after the gap is eliminated, if the synchronous shaft rigidity is insufficient, the torque will increase sharply, and then will cause the synchronous shaft fracture damage; if the synchronous shaft rigidity is sufficient, the synchronous shaft torque and ship chamber tilt gradually converge and stabilize.

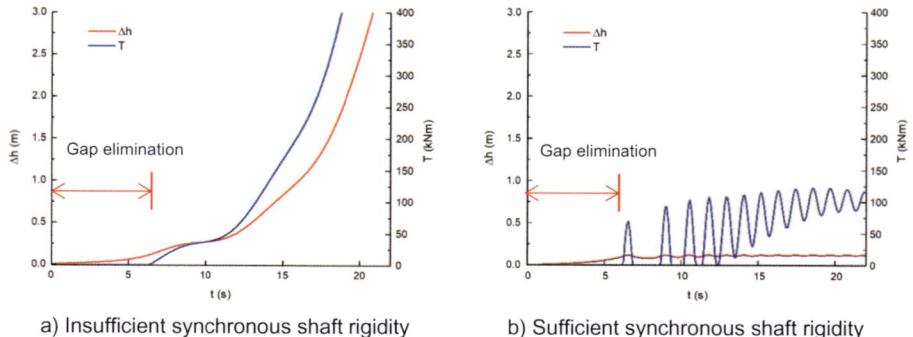

a) Insufficient synchronous shaft rigidity   b) Sufficient synchronous shaft rigidity

Fig. 6. Characteristics of ship chamber tilt and synchronous shaft torque variation

6 Conclusions

The conclusion of this paper is as follows:

1) The installation gap and synchronous shaft rigidity of the mechanical synchronization system of the HFSL directly affect the longitudinal tilt of the ship chamber. The design condition of synchronous shaft rigidity is $K = \frac{GI_p}{L} > \frac{[X(1+k)-Y(1+q)]R^2}{2L\left(1-\frac{\Delta h_a}{[\Delta h]}\right)}$.

2) The change of the longitudinal tilt of the HFSL ship chamber is divided into two stages. During the mechanical synchronous system installation gap elimination process, the longitudinal tilt of the ship chamber is a hyperbolic sine function of time, monotonically increasing; after the synchronous shaft installation gap eliminated, the change of the longitudinal tilt of the ship chamber is expressed as the superposition of two different frequency triangular functions, the high frequency vibration reflects the inherent frequency of synchronous shaft torsional vibration, the low frequency vibration reflects the longitudinal first order self-oscillation frequency of the water body in the ship chamber. Due to the influence of system damping and water body viscosity, the oscillation amplitude gradually decaying, and finally stabilizing in the ship chamber moment balance position.

References

Shu X (2017) Hydraulics of hundred-meter scale hydro-floating ship lift. Doctor thesis, Hohai University

Lamb H (1992) Theoretical fluid dynamics. The Science Press, Beijing

Su TC, Kang SY (2015) Numerical simulation of liquid sloshing. In: Engineering mechanics in civil engineering. ASCE, pp 1069–1072

Bogaers AEJ, Kok S, Reddy BD, Franz T (2016) An evaluation of quasi-Newton methods for application to FSI problems involving free surface flow and solid body contact. Comput Struct 173(C):71–83

Su Y, Liu ZY (2016) Numerical model of sloshing in rectangular tank based on boussinesq-type equations. Ocean Eng 121:166–173

Numerical Investigation of an Inland 64 TEU Container Vessel in Restricted Waters

Jinyu Kan[1], Lizheng Wang[1], Jialun Liu[2,3,4(✉)], Xuming Wang[2,3,4,5], and Bing Han[4,6]

[1] School of Naval Architecture, Ocean and Energy Power Engineering, Wuhan University of Technology, Wuhan, China
[2] Intelligent Transportation Systems Research Center, Wuhan University of Technology, Wuhan, China
jialunliu@whut.edu.cn
[3] National Engineering Research Center for Water Transport Safety, Wuhan University of Technology, Wuhan, China
[4] Academician Workstation of COSCO SHIPPING Group, Shanghai, China
[5] Yangtze River Delta Shipping Development Research Institute (Jiangsu) CO., LTD, Nanjing, China
[6] Shanghai Ship and Shipping Research Institute, Shanghai, China

Abstract. Compared with sea-going ships, inland vessels mostly sail in restricted waters, which may cause resistance and ship motions to change greatly. To ensure the safety of navigation, it is of great importance to study the hydrodynamic performance of inland vessels navigating in restricted waters. A 64 Twenty-feet Equivalent Unit (TEU) container vessel is numerically simulated at different speeds and water depth draft ratios (water depth/draft = 2, 2.5, 3, 4, 5, 16). The numerical methods are firstly verified and then applied to systematic simulations. The resistance components, ship motions, and details of flow fields are calculated and analyzed. Generally, the total, frictional and residual resistance coefficients increase with a decrease of water depth as expected. However, at relatively low speeds (Fr = 0.1129 and 0.1135) of h/T = 2, 2.5, 3, the resistance components change conversely that they decrease as the water depth gets shallower. This special phenomenon may be caused by the design of the ship hull or the use of the turbulence model that may not be appropriate. The residual resistance has the same trend as the total resistance and the lines are nearly parallel, which shows that the residual resistance is dominant in the component of total resistance. The ship squat phenomenon happens but is not severe in the shallowest condition (h/T = 2). With the water depth decreasing, the wave amplitude becomes larger and the wave crests near the ship bow and stern also increase, while the troughs change slightly at different water depths.

Keywords: Computational fluid dynamics · Numerical simulations · Ship resistance · Ship squat

© The Author(s) 2023
Y. Li et al. (Eds.): PIANC 2022, LNCE 264, pp. 516–528, 2023.
https://doi.org/10.1007/978-981-19-6138-0_45

1   Introduction

Water transport is the most essential transport in many countries of the world. Its ability to undertake large quantities and long-distance carriage makes it a valuable portion of transportation. Inland vessels play an important role and the demand is increasing rapidly. Though the bulk carriers still account for a large proportion of the quantity of inland water transport, container ships are more suitable for intelligent and autonomous shipping since containers are standardized to be preferable for modular transportation. Container ships mostly sail at a relatively high speed in the open sea compared with inland container vessels. Due to the depth and width of the inland waterways, the container vessels have to lower their speeds and the flow fields are becoming fairly complex. At present, there are lots of small container vessels like 36 TEU and 64 TEU, sailing in the Zhejiang inland waterways. A typical 64 TEU inland container vessel (here and after use 64 TEU instead) is shown in Fig. 1 and a novel 64 TEU is designed based on the common 64 TEU ones. The hydrodynamic performance is of great importance at the preliminary stage of ship design. Its prediction can guide the subsequent power selection of the main engine or ship type optimization etc. Hence, how to accurately determine the hydrodynamic forces acted on the ship is meaningful.

Nowadays, Computational Fluid Dynamics (CFD) methods have been developed rapidly. Although model testing is considered to be a very important tool for determining the resistance and power requirements of ship hull forms, CFD can be used efficiently for the same purpose (Ahmed et al. 2009). Compared with Experimental Fluid Dynamics (EFD), CFD can get details of the flow field faster and cheaper. Nonetheless, as for EFD, the credibility of CFD simulations requires the assessment of the modeling and numerical uncertainties to avoid the risk of taking erroneous conclusions (Pereira et al. 2017). For the vessels sailing in inland waterways, the researchers pay more attention to the ship resistance, ship-generated waves, and squat varying with the speed and water depth using the CFD method. Senthil Prakash et al. (2013), Pacuraru et al. (2017), and Ammar et al. (2019) performed simulations to predict the shallow-water resistance of different large block coefficient ships respectively. With the application of commercial code Fluent, Jachowski (2008) validated the CFD method with the results obtained by the existing methods of squat identification and searched the influence of the water depth and ship speed on the squat and wave profile. Raven (2012) focused on the shallow water effects on viscous resistance by simulating double body viscous flow for four different ships. Raven (2019) presented the shallow-water ship model testing and the determination of water depth effect on ship resistance. Mousaviraad et al. (2015) applied Unsteady Reynolds Averaged Navier-Stokes (URANS) method to study the shallow water effect of high-speed planning crafts while Tafuni et al. (2016) put attention on the bottom pressure and wave elevation using the Smoothed Particle Hydrodynamics method. Zeng et al. (2017) used both numerical and experimental methods to investigate the resistance extrapolation of an inland ship model in shallow water. As Zeng et al. (2017) mentioned, there is an unavoidable discrepancy with the ITTC 57 line. Zeng et al. (2018) proposed a numerical friction line for correcting shallow water effects on a ship's bottom using CFD calculations.

Based on the previous research, this paper aims to predict the ship resistance performance of a newly designed 64 TEU in restricted water. Six water depths and five Froude numbers have been considered in this study and simulations are performed using a Reynolds-Averaged Navier-Stokes (RANS) based CFD model with commercial code STAR-CCM+. The ship resistance and squat performances are analyzed combining with the details of flow fields. Section 1 presents the research background and status. Section 2 shows the details of the ships and tank tests. The applied CFD method is described and verified in Sect. 2.5. Section 3 shows the simulated conditions, results, and discussions from different aspects of resistance, ship motions, wave profiles, pressure, and velocity distributions. The conclusions are drawn in Sect. 4.

Side

Stern

Fig. 1. Common 64 TEU container ships sailing in the Zhejiang inland waterways.

2   Numerical Method and Verification

2.1   Model of KCS and 64 TEU

The 64 TEU is a newly designed electrically-driven vessel to operate in the waterway from Huzhou to Shanghai in China. It is proposed as green and intelligent inland vessel with multiple advanced technologies, such as a rim-driven thruster propulsion system, containerized batteries, and ship-shore coordination assisted navigation. From the appearance, the 64 TEU has a blunt bow and flat stern with a large block coefficient, which is different from the traditional 64 TEU in Fig. 1.

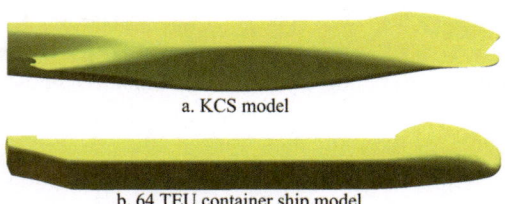

a. KCS model

b. 64 TEU container ship model

Fig. 2. Side views of the KCS and 64 TEU model.

To ensure the accuracy of the CFD method used in this paper, the KCS model is selected for verification. The KCS is an international standard model which has plentiful test data to be compared and verified with. It is a modern container ship with a

Table 1. Main particulars of the KCS and 64 TEU at model scale.

	KCS	64 TEU
SCALE	1/31.599	1/34.5
L_{PP} (m)	7.279	2.000
B_{WL} (m)	1.019	2.066
T (m)	0.342	0.075
S (m^2)	9.659	0.970
Δ (m^3)	1.649	0.050
C_B (-)	0.651	0.874
C_M (-)	0.985	0.984
Fr (-)	0.260	0.107

bulbous bow and stern which has a more complex flow field than the 64 TEU. The three-dimensional ship geometric models of KCS and 64 TEU are built using Solidworks. The side views of KCS and 64 TEU are illustrated in Fig. 2. Table 1 shows the main particulars of KCS and 64 TEU at the model scale. The KCS is also chosen for its large block coefficient and midship section coefficient which are close to the 64 TEU as a container ship.

2.2 Governing Equations of the Numerical Model

The RANS equations are solved to simulate the flow field around the ship. The continuity and momentum equations in incompressible flow are shown as follows:

$$\frac{\partial \rho}{\partial t} + \frac{\partial(\rho u_i)}{\partial x_i} = 0, \tag{1}$$

$$\frac{\partial}{\partial t}(\rho u_i) + \frac{\partial}{\partial x_i}(\rho u_j u_i) = -\frac{\partial p}{\partial x_i} + \frac{\partial \sigma_{ij}}{\partial x_j} + \frac{\partial}{\partial x_j}\left(-\rho u_i' u_j'\right), \tag{2}$$

where the u_i and u_j are the Reynolds average velocity components in x_i and x_j directions, omitting the average sign. The u_i' and u_j' represent the fluctuating velocities. The ρ, p and σ_{ij} are fluid density, static pressure, and stress tensor component respectively. Due to the increase of the Reynolds stress term (the last term in Eq. 2, the RANS equation is not closed. So the turbulence model needs to be established. Common turbulence models include $k - \varepsilon$ and $k - \varpi$. According to previous research (Querard et al. 2008), the $k - \varepsilon$ turbulence model can save nearly 25% CPU time compared with the $k - \varpi$ SST turbulence model for instance. It is also the most widely used turbulence model at present (Tezdogan et al. 2016; Mousaviraad et al. 2015; Castiglione et al. 2014; Jachowski 2008). To better simulate the characteristics of turbulence, the Realizable $k - \varepsilon$ turbulence model is applied in this paper.

The High-Resolution Interface Capturing (HRIC) scheme is used for Volume Of Fluid (VOF) simulations to maintain sharp interfaces between the participating liquid phases. In the HRIC settings, the lower limit and upper limit of Courant number are set to 500, so that the result can converge quickly and the Kelvin waves can be generated clearly without a smaller time step in the STAR-CCM+ code. The Dynamic Fluid Body Interaction (DFBI) model can solve the equation of a 6-Degree of Freedom (DOF) rigid body motion. In this paper, it is activated to get a stable posture of the ship hull with two degrees of freedom (heave and pitch). Accordingly, ship resistance is calculated.

2.3   Computational Domain and Boundary Condition

Because the studied ships in this paper are geometrically symmetric, only half a ship is researched. The coordinate system of the computational domain is right-handed: the x-axis is oriented against the direction of ship navigation; the y-axis points to the starboard side of the ship; the z-axis is vertical to the sea level upward; the origin is at the point of intersection of aft perpendicular plane, symmetry plane and still water plane. To make the flow field fully developed but save computing resources, the mid-ship section is 1.5 L_{PP} to the inlet plane and 2.5 L_{PP} on the contrary. The side plane is 2 L_{PP} away to ensure the wake is fully developed. The ship is 1 L_{PP} to the upper plane and 2 L_{PP} to the bottom plane for verification. In the following cases for different water depths, except for the distance to the bottom, the rest remains the same.

The initial conditions must be defined depending on the physics of the problem to be solved (Tezdogan et al. 2015). Boundary conditions are settled as Fig. 3. The computational domain is divided into the background mesh region and the overset mesh region. For the background mesh region, the upper plane and the negative x-direction plane are set as velocity inlet. The positive x-direction plane is settled as pressure outlet to prevent backflow. Since only half of the ship is taken for study, the *xoz* plane is set as symmetry. The ship wall is set as a non-slip stationary wall. So is the bottom plane considering the shallow water effect. The boundary conditions of the overset mesh region are set as overset except the *xoz* plane which is set as same as the background mesh region.

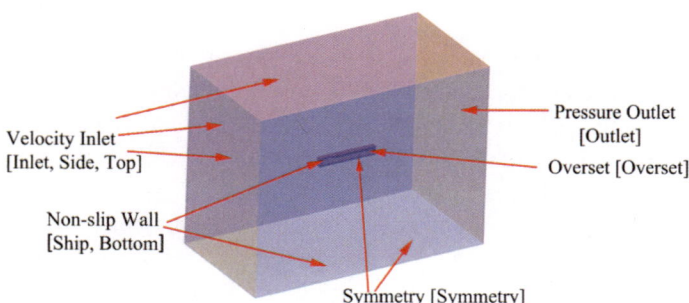

Fig. 3. Boundary conditions of the computational domain.

2.4   Discrete Grid

Commercial code STAR-CCM+ is used to perform the mesh generation, which can automatically generate unstructured hexahedral trimmed meshes with high quality. The KCS model is divided into multiple parts defined with different mesh sizes according to their surface curvature. Three boxes are set surrounding the ship to refine the mesh. The size of the outer box mesh is doubled to the previous box. The mesh near the waterplane needs to be finer to catch the Kelvin waves, which also uses three control boxes of different sizes. The control boxes refine the mesh in different degrees at x, y, z directions. The match between overset and background mesh is also considered.

The boundary layer, which was firstly proposed by Prandtl (1904), has a great influence on the accuracy of the resistance and flow field prediction. In the STAR-CCM + code, the prism layer mesh is applied to individually define the mesh near the ship. The most important parameter is the thickness near the ship wall, which is described by a non-dimensional number y^+. When applying the wall function method, the y^+ is generally required to be between 30 to 300. In this paper, the full y^+ wall treatment is used, and the y^+ is chosen to 80 with 8 prism layers after trials. Also, the bottom plane is set with 4 prism layers for catching the complex flow near the bottom wall at shallow water conditions and a good transition to the upper mesh. Mesh

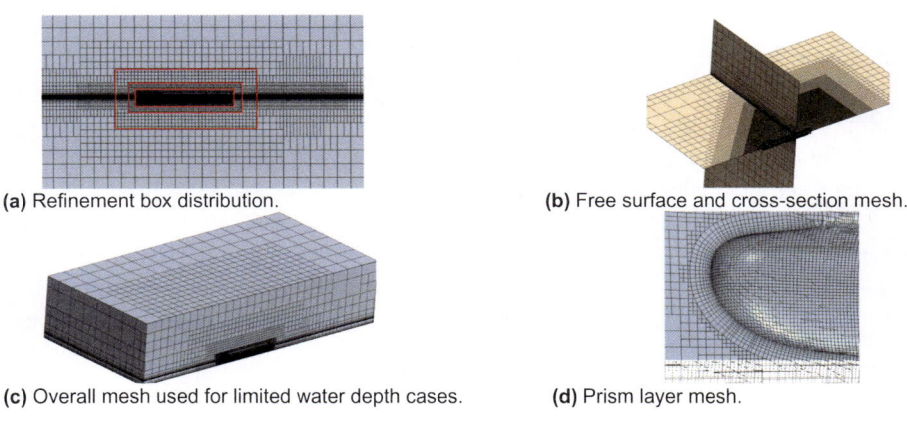

(a) Refinement box distribution.   (b) Free surface and cross-section mesh.

(c) Overall mesh used for limited water depth cases.   (d) Prism layer mesh.

Fig. 4.   Mesh distributions from different views.

distributions from different views are shown in Fig. 4.

2.5   ITTC Verifications of KCS and 64 TEU

It is desirable in CFD computations to perform an Uncertainty Analysis (UA) to evaluate the accuracy of the results (Castro et al. 2011). The recommended UA procedures (The Specialist Committee on UA 2008) include verification and validation, only the former is researched in this paper. In the uncertainty analysis, the mesh size is the most influential parameter compared to others like time step, iteration number, etc. Series of mesh with three different densities are used to simulate the KCS model at the $Fr = 0.1516$ condition and 64 TEU at $Fr = 0.2032$ condition. The base size of the three sets of mesh is diminishing by the scale $\sqrt{2}$.

Firstly, mesh convergence is studied. Then the procedures described in The Specialist Committee on UA (2008) are used for verification. The mesh parameters and results of the KCS and 64 TEU model are listed in Table 2, where the R_G, P_G, C_G, U_G,

Table 2. Mesh convergence study and verification of KCS ($Fr = 0.1516$) and 64 TEU model ($Fr = 0.2032$).

	KCS	64 TEU
COARSE MESH C_t (*10^{-3})	3.901	7.270
MIDEUM MESH C_t (*10^{-3})	3.654	6.386
FINE MESH C_t (*10^{-3})	3.573	6.244
EFD C_t (*10^{-3})	3.64	–
R_G (-)	0.3279	0.1606
P_G (-)	3.2132	5.2709
C_G (-)	2.0453	5.2137
U_G (%C_t_FINE MESH)	3.43%	4.15%
δ_G^* (%C_t_FINE MESH)	2.27%	2.29%
S_C (*10^{-3})	3.492	6.101

δ_G^* and S_C are convergence factor, order-of-accuracy, correction factor, the uncertainty of mesh, error of mesh, and corrected solution.

For the KCS model, as the mesh becomes finer, the calculated resistance is decreasing and gradually going to converge with a small deviation to the EFD data. The U_G and δ_G^* are small and the corrected solution has a deviation within 5% to the EFD data. For the 64 TEU model, the result shows a convergent trend and the U_G and δ_G^* are also small. Hence, it can be considered that the CFD method is verified. The verified method is applied to the other Froude number conditions of the KCS model.

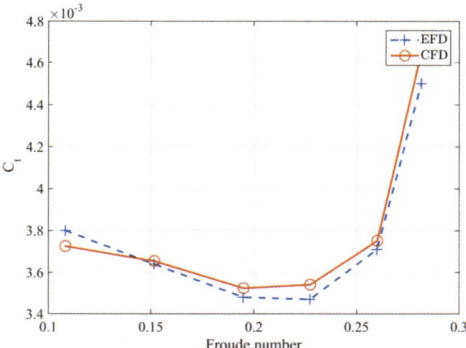

Fig. 5. Comparison of numerical and experimental C_t results of KCS model at different Froude numbers.

Compared with the EFD data, the calculated total resistance coefficient is shown in Fig. 5.

It is shown that the numerical results are in good agreement with the experimental results in the figure. The maximum deviation is around 4%. And the mesh convergence index shows uncertainties around 4%. Thus, all the results fall within the uncertainty area and all results can be considered validated.

3 Numerical Results

For inland vessels, which commonly navigate in shallow waters, the shallow water effect is a significant part of the research. Based on the previous research (Vantorre 2003), shallow water is usually defined as the water depth draft ratio (h/T) is smaller than 1.5. The waterway from Shanghai to Huzhou is a three-class channel and the minimum depth is 3.2 m. Hence, h/T=1.2 is the shallowest condition for the 64 TEU. However, for the simulation at h/T = 1.5 condition, the ship squats severely, and the overset mesh region exceeds the background mesh region which causes the unexpected stop in the first few seconds of the iteration.

After trials, the simulations are performed at h/T = 2, 2.5, 3, 4, 5, 16 and five Froude numbers within 0.1129 to 0.2032. Two degrees of freedom (heave and pitch) are considered. Details of the studied cases are listed in Table 3. The computed results are analyzed from the aspects of

Table 3. Details of simulated cases.

Fr	V (m/s)	h/T	Number of cases
0.1129	0.5	2, 2.5, 3, 4, 5, 16	6
0.1355	0.6	2, 2.5, 3, 4, 5, 16	6
0.1580	0.7	2, 2.5, 3, 4, 5, 16	6
0.1806	0.8	2, 2.5, 3, 4, 5, 16	6
0.2032	0.9	2, 2.5, 3, 4, 5, 16	6

3.1 Ship Resistance

The ship resistance components against the speed and water depth are shown in Fig. 6. For the cases from V = 0.7m/s to V = 0.9m/s, that the Froude numbers vary from 0.1580 to 0.2032, the total and residual resistance increase with the speed at different water depths while the frictional resistance decreases. In some water depths, like h/T = 2, 2.5,4, the total and residual resistance slightly decreases then increases with the speed. Overall, the highest speed condition has the highest total and residual resistance at different water depths.

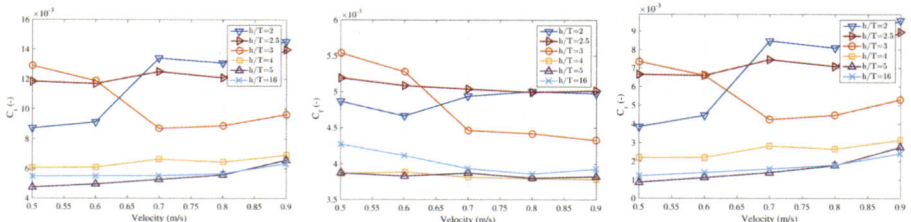

Fig. 6. Ship resistance coefficients (C_t, C_r, and C_f) change with the speed and water depth.

With the water getting shallower, the total, residual, and frictional resistance are gradually rising. These three resistance components are close between $h/T = 4$ and $h/T = 16$, which means the water depth has little effect on the resistance in this range. However, while $h/T=4$ shows higher residual resistance, $h/T = 16$ shows higher frictional resistance. There are significant resistance increases at $h/T = 2.5$ and $h/T = 3$. Then the rise becomes small from $h/T = 2.5$ to $h/T = 2$. The resistance components at V=0.5m/s and V=0.6m/s have different or even opposite laws of change in restricted water conditions like $h/T = 2, 2.5, 3$. It seems that at these two speeds, the total, residual, and frictional resistance firstly increase with a decrease of the water depth. There is a sudden change at the $h/T = 3$ condition. Then, the resistance components decrease when the water gets shallower.

The trends of the residual and total resistance are nearly parallel, which can be observed in Fig. 7. This figure compares the ship resistance components at $h/T = 2$ condition. From this figure, the frictional resistance keeps at the same level at different speeds. It is shown that the change of velocity has little effect on the frictional resistance at the shallowest water depth. The lines of residual and total resistance are almost parallel which means that the residual resistance dominates the change of total resistance.

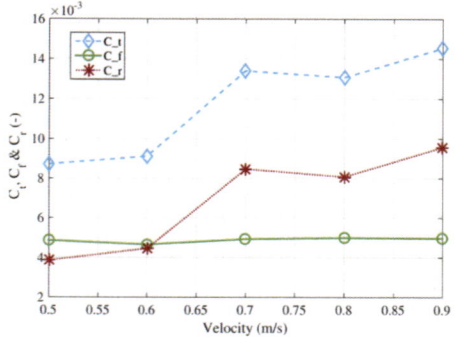

Fig. 7. Comparison of ship resistance components at $h/T = 2$ condition.

This special phenomenon may be caused by the design of the ship hull. Different from the common 64 TEU in Fig. 1, this ship has a blunt bow and the transition between bow and bottom is not smooth enough. The particularity of the ship hull may lead to this phenomenon of resistance performance at low speeds. According to the bottom view of Reynolds number distributions in Fig. 8, most regions of the ship bottom at V = 0.5m/s and V = 0.6m/s have Reynolds numbers that are smaller than 10^6. The turbulent flow near the bottom of the ship is not yet fully developed. Hence, the turbulence model may not be suitable to perform the simulations under these conditions.

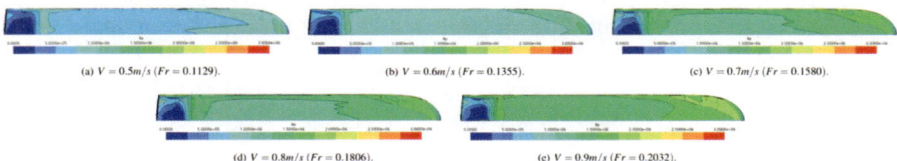

(a) $V = 0.5m/s$ $(Fr = 0.1129)$.   (b) $V = 0.6m/s$ $(Fr = 0.1355)$.   (c) $V = 0.7m/s$ $(Fr = 0.1580)$.

(d) $V = 0.8m/s$ $(Fr = 0.1806)$.   (e) $V = 0.9m/s$ $(Fr = 0.2032)$.

Fig. 8. The bottom view of Reynolds number distributions near the ship hull at $h/T = 2$ condition.

3.2 Ship Squat

When a ship proceeds the water, it generally drops vertically and trims forward or aft. The overall decrease in the static under keel clearance, forward or aft, is called ship squat Derrett 2006). The ship squat is a common phenomenon that can be severe when navigating in restricted waters and cause safety issues. So the study of squat has great significance. Figure 9 shows the ship pitching and heaving against the speed and water depth. The positive angle represents the ship trimming by stern and the negative value of sinkage represents the ship moving downwards. The sinkage is non-dimensionalized by dividing the water depth

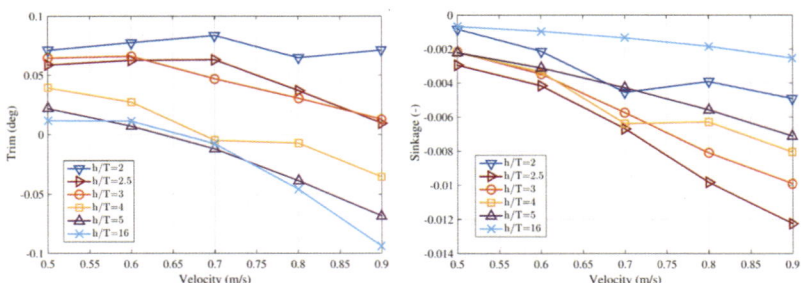

Fig. 9. Ship trim and sinkage change with the speed and water depth.

In the relatively deep water conditions, the ship trims by the stern and gradually becomes trimmed by the ship bow. At the medium-deep condition like $h/T = 3$, the ship trims by the stern at all considered speeds but the angle is getting smaller. When the water gets shallower, the ship keeps trimming by the stern, and the angle changes in a small range. The mechanism is that, at the most restricted condition, the water is accelerated at the ship's bottom. The flat bottom under the ship's bow makes it have a larger high-speed region than that at the stern. The uneven velocity distribution leads to a bigger sinkage at the bow than at the stern. Hence, the ship keeps trimming by the ship bow at the shallowest condition. The non-dimension sinkage refers to the movement of the center of gravity relative to water depth. According to the previous study (Linde et al. 2016), the sinkage is more significant when the water depth is restricted. As expected, its absolute value increases with an increase in speed and a decrease in water depth.

3.3   Wave Profiles

The free-surface cuts at $y/L_{PP} = 0.1$ are shown in Fig. 10. The x-coordinate is 0 which represents the middle of the ship, with the positive x-axis pointing towards the ship bow. At relatively low speeds, the section wave profiles are close to each other at different water depths in Fig. 10(a) and Fig. 10(b). The effect of limited water depth becomes more obvious and the wave amplitude becomes larger with the speed. It can be observed that the $h/T = 2$ condition has the highest bow and stern wave crest among all the water depth conditions in Fig. 10(c). Also, the wave near the ship side is higher than other water depth conditions. But with the speed increasing, the bow wave crests of the $h/T = 2.5$ and $h/T = 3$ are also increasing and gradually become close to the $h/T = 2$ condition in Fig. 10(e).

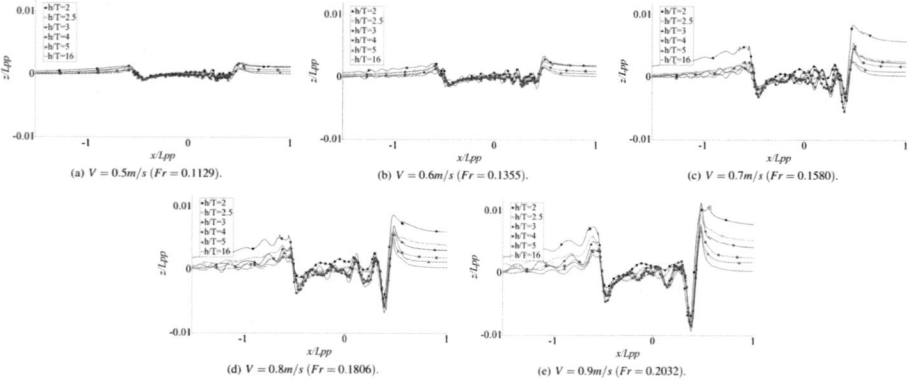

Fig. 10. Free-surface cuts computed at $y/L_{PP} = 0.1$.

4   Conclusions

This paper focuses on the resistance and flow field of a newly designed 64 TEU sailing in restricted waters at the model scale. The CFD method is verified according to ITTC recommended procedures with the KCS and 64 TEU model. After that, it is applied to different speeds and water depths of the 64 TEU model. Conclusions are drawn as follows.

1. The ship's total resistance at higher speeds increases with an increase in speed and a decrease in water depth as expected.
2. At lower speeds, the resistance performance is unusual at restricted water conditions. There is a sudden change at $h/T = 3$ condition. With the water getting shallower, the total resistance decreases, which is different from that at high speeds. This phenomenon may be caused by the blunt ship bow and the low Reynolds number near the hull at the model scale. The mechanism of this phenomenon needs further study.
3. The ship squat is not severe at $h/T = 2$ condition at all considered speeds, which guarantees safety at shallow water conditions.
4. As the water depth decreases, the wave amplitude becomes larger and the wave crests near the ship bow and stern also increase, while the troughs change slightly at different water depths.

Acknowledgements. Supported by Southern Marine Science and Engineering Guangdong Laboratory (Zhuhai) (SML2021SP101).

References

Ahmed Y, Guedes Soares C (2009) Simulation of free surface flow around a VLCC hull using viscous and potential flow methods. Ocean Eng 36(9):691–696

Ammar NR, Elgohary MM, Zeid A, Elkafas AG (2019) Prediction of shallow water resistance for a new ship model using CFD simulation: case study container barge. J Ship Prod Design 35(2):198–206

Castiglione T, He W, Stern F, Bova S (2014) URANS simulations of catamaran interference in shallow water. J Mar Sci Technol 19(1):33–51. https://doi.org/10.1007/s00773-013-0230-5

Castro AM, Carrica PM, Stern F (2011) Full scale self-propulsion computations using discretized propeller for the KRISO container ship KCS. Comput Fluids 51(1):35–47

Derrett DR (2006) Ship stability for masters and mates. Elsevier Butterworth-Heinemann, pp 286–292

Jachowski J (2008) Assessment of ship squat in shallowwater using CFD. Arch Civ Mech Eng 8 (1):27–36

Linde F, Ouahsine A, Huybrechts N, Sergent P (2016) Three-dimensional numerical simulation of ship resistance in restricted waterways: effect of ship sinkage and channel restriction. J Waterway Port Coast Ocean Eng

Mousaviraad SM, Wang Z, Stern F (2015) URANS studies of hydrodynamic performance and slamming loads on high-speed planing hulls in calm water and waves for deep and shallow conditions. Appl Ocean Res 51:222–240

Pacuraru F, Domnisoru L (2017) Numerical investigation of shallow water effect on a barge ship resistance. IOP Conference 227(1):012088

Pereira F, Vaz G, Eca L (2017) Verification and validation exercises for the flow around the KVLCC2 tanker at model and full-scale Reynolds numbers. Ocean Eng 129(1):133–148

Prandtl L (1904) Motion of fluids with very little viscosity. Technical report Archive and Image Library

Querard A, Temarel P, Turnock SR (2008) Influence of viscous effects on the hydrodynamics of ship-like sections undergoing symmetric and anti-symmetric motions, using RANS. In: Proceedings of the ASME 27th international conference on offshore mechanics and arctic engineering (OMAE)

Raven HC (2012) A computational study of shallow water effects on ship viscous resistance. In: Proceedings of the 29th symposium on naval hydrodynamics

Raven HC (2019) Shallow-water effects in ship model testing and at full scale. Ocean Eng 189:106343

Senthil Prakash MN, Chandra B (2013) Numerical estimation of shallow water resistance of a river-sea ship using CFD. Int J Comput Appl 71(5):33–40

Tafuni A, Sahin I, Hyman MC (2016) Numerical investigation of wave elevation and bottom pressure generated by a planing hull in finite-depth water. Appl Ocean Res 58:281–291

Tezdogan T, Incecik A, Turan O (2015) A numerical investigation of the squat and resistance of ships advancing through a canal using CFD. J Mar Sci Technol 21(1):86–101. https://doi.org/10.1007/s00773-015-0334-1

Tezdogan T, Incecik A, Turan O (2016) Full-scale unsteady RANS simulations of vertical ship motions in shallow water. Ocean Eng 123(1):131–145

The Specialist Committee on UA (2008) ITTC recommended procedures and guidelines. International towing tank conference

Vantorre M (2003) Review of practical methods for assessing shallow and restricted water effects. In: International conference on marine simulation and ship maneuverability, pp 1–11

Zeng Q, Hekkenberg R, Thill C, Rotteveel E (2017) A numerical and experimental study of resistance, trim and sinkage of an inland ship model in extremely shallow water. In: ICCAS 2017

Zeng Q, Thill C, Hekkenberg R, Rotteveel E (2018) A modification of the ITTC57 correlation line for shallow water. J Mar Sci Technol 24(2):642–657. https://doi.org/10.1007/s00773-018-0578-7

On the Numerical Modelling of Ship Forces During Lockage

Carsten Thorenz[✉] and Fabian Belzner

Federal Waterways Engineering and Research Institute, Karlsruhe, Germany
{carsten.thorenz, fabian.belzner}@baw.de

Abstract. When designing navigation locks, several parameters must be considered: Apart from the construction and operation costs also the operation speed and lockage safety must be considered. The safety of the lockage is often expressed in terms of the acceptable ship forces. Traditionally, apart from physical models different kinds of numerical models are used to determine the speed and safety of operation: During the predesign phase often simple analytical approaches are used, while in later design phases more sophisticated one- or three-dimensional numerical models or physical models are used independently or in a hybrid modelling approach. With these models the flow rates, the development of the water levels and the resulting forces on the ship can be computed or measured. While the application of three-dimensional numerical models for the lock filling process was in the past limited to very few and extraordinary projects, nowadays their application becomes more and more engineering practice. Here we are giving an overview of the numerical modelling methodologies and are presenting the state-of-the-art for three-dimensional numerical models.

Keywords: Lock filling · Ship forces · Numerical model · Three-dimensional · CFD

1 Introduction

For a valid hydraulic design of a lock, several parameters must be considered and be balanced with the costs for construction and operation. This includes the physical properties of the flow (e.g. pressures or velocities) as well as the speed and safety of the lockage. The safety is mostly expressed in terms of the allowable ship forces (PIANC 2015; Belzner 2018). Numerical models are often used to determine the speed and safety of operation during the hydraulic design. The level of sophistication is typically coupled to the planning stages. In early phases simple models are used, because they are fast to apply, but possibly rough in terms of accuracy. In later phases, more accurate models are required to validate and optimize the design. With these models the flow rates, the development of the water levels and the resulting forces on the ship can be computed or measured.

While the application of three-dimensional numerical models for the lock filling process was in the past limited to very few and extraordinary projects, nowadays their application becomes more and more engineering practice. For the design of locks it is

© The Author(s) 2023
Y. Li et al. (Eds.): PIANC 2022, LNCE 264, pp. 529–541, 2023.
https://doi.org/10.1007/978-981-19-6138-0_46

very helpful to evaluate specific features (e.g. pressure or velocity distributions) of the flow with these models. One must be aware that considering all aspects of the lock filling process offers several pitfalls which might be unexpected, particularly for modellers unexperienced with this specific topic. This is especially problematic if the modeller is already familiar with numerical modelling concepts in general (e.g. from finite element analysis of structures) and tries to transfer this knowledge to the process of lock filling: Unfortunately, with the available numerical modelling packages it is possible to produce valid looking results for the lock filling simulations even if the physical meaning is limited or if the results are plainly wrong. In this situation it is necessary that the modeller has an in-depth knowledge of the expected behaviour of the flow and of the ship in the chamber, in order to judge the quality of the numerical results.

2   Classes of Commonly Used Numerical Models

2.1   Introduction

Over the last decades, numerical models have developed into an irreplaceable tool for the hydraulic design of locks. Depending on the planning stage, the complexity and importance of the project, they are applied either alone, in combination with each other or with physical models. In Thorenz (2009) and De Mulder (2011) an overview of the state-of-the-art in that time was given. Looking back at these considerations, the methods presented then are still valid and in use. But with the further advance of computer technology, the application of three-dimensional numerical models has become more standard for the engineering practice. This is currently typically limited to singular aspects of the flow field, but promising research results are available now for the complete locking process. The simulation of the whole lock filling system, with all of its aspects (i.e. with opening of the valves and moving of the ship) is still challenging and thus must be based on a proper validation with physical model data.

2.2   Zero-Dimensional Numerical Models and Analytical Approaches

Based on the governing analytical equations for the lock filling process, it is possible to compute the temporal development of the water levels in the lock chamber for different valve schedules, configurations etc. by hand or with a spreadsheet program. This requires that the hydraulic parameters of the filling system are known with sufficient accuracy. These parameters can be determined either on the basis of physical model measurements, high-quality three-dimensional numerical models or by prototype measurements. If additional physical effects like wave propagation or jet effects on the vessel are regarded with analytical approaches, even the order of resulting longitudinal ship forces can be computed.

It must be pointed out that this kind of model is typically used only for through-the-head filling systems. The results depend heavily on the chosen parameters and thus on the availability of calibration data. The big advantage of this kind of models is that the results are available more or less instantaneously and thus make parameter studies easy.

All relevant effects must be described by analytical approaches, which requires a deep understanding of the hydromechanics of the lock filling process.

2.3 One-Dimensional Numerical Models

In the last decades, one-dimensional numerical models have become a de-facto standard in the pre-dimensioning phase of the design process. Three types of models can be distinguished:

- Models, that represent the flow in the lock chamber by a set Saint-Venant equations, i.e. by discretizing the chamber in the longitudinal direction.
- Models, that represent a complex filling system by a network of pipes, connectors and loss coefficients, but represent the chamber only as basin with no spatial discretization.
- Models which combine the two aforementioned approaches.

With the one-dimensional chamber models, only effects on the vessel in the longitudinal direction can be computed. The models are typically used to compute filling times and longitudinal forces on the vessel etc. They are useful to pre-dimension valve sizes and valve schedules for locks with through the head filling systems. The popular "LockFill" program (De Loor et al. 2013; available at https://oss.deltares.nl/web/lockfill/) is a hybrid program, which contains zero-dimensional and one-dimensional model parts. It is specifically tailored for through the head filling systems. Sometimes, a coupling of the one-dimensional chamber model with a network model for the filling system is incorporated ("LockSim", Schohl 1999). This facilitates the usage for more complex filling systems like longitudinal culvert systems, but is also still restricted to the longitudinal processes in the lock chamber. In other programs ("LoMo", Belzner et al. 2018; available at https://github.com/baw-de/lomo), the results of externally computed flow distributions can be fed in any spatial distribution and with variable momentum impact into the one-directional computational core which computes the flow in the lock chamber, thus enabling more complex flow situations.

For all of these programs it must be pointed out, that a proper usage requires understanding of the hydraulic behaviour of the lock filling system. The results depend heavily on the chosen coefficients, which can roughly be estimated for simple systems by expert knowledge. For more complex or new systems, parameter determination by physical or three-dimensional numerical modelling or by prototype measurements is necessary. Thus, the used parameters must be chosen on a reliable basis, as they have a fundamental impact on the results.

As an example, for the lock chamber modelling class of programs, Fig. 1 presents a typical result of LoMo, showing the user interface in the left panel and the computed flow rates (green), water levels (blue) and ship forces (yellow) in the right panel. The user has to set the correct values in the left panel. While the geometry parameters are easily chosen, the parameters under the "Filling" tab require deeper knowledge in order to describe the hydraulic behaviour correctly.

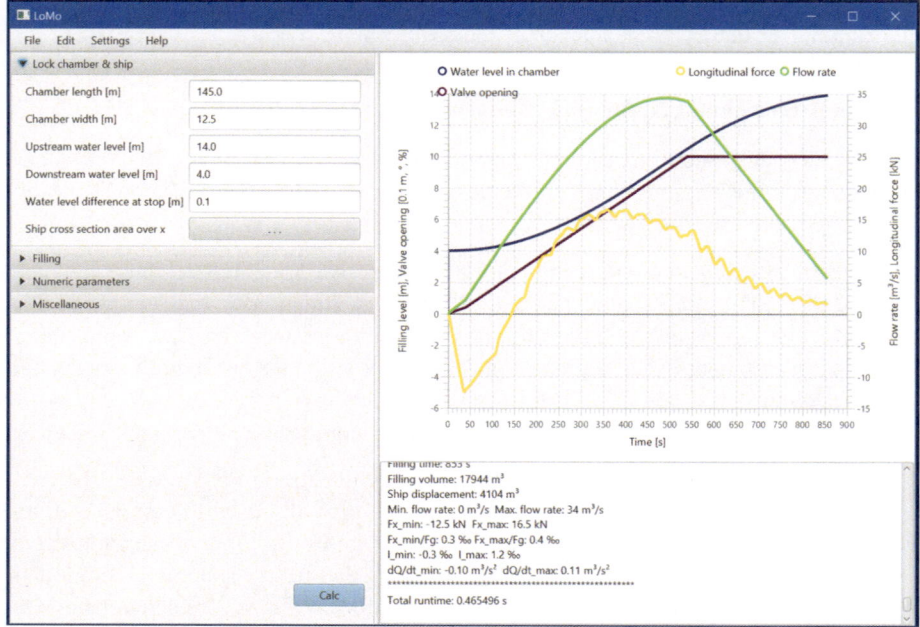

Fig. 1. Example computation with the one-dimensional lock model "LoMo".

For the computation of the flow in filling system networks, different tools are available. Figure 2 shows a network for a complex filling system with a lock chamber in the middle of the network, longitudinal culverts along the chamber and a set of water saving basins connected to the filling system (Thorenz 2010). This model was used in a hybrid fashion in conjunction with three-dimensional models for the calculation of the necessary coefficients for the parameterization of the one-dimensional model.

Summarizing one can say, that one-dimensional models are a very helpful tool for fast parameter studies in the design process. But it must be pointed out that the chosen parameter sets are a key factor for the results and thus must be evaluated with care.

2.4   Two-Dimensional Numerical Models

Two-dimensional numerical models are rarely used to model the lock filling process. Their main use case are computations of the far field in the inlet and outlet areas. Thus, further elaborations are omitted here.

2.5   Three-Dimensional Numerical Models

2.5.1   Overview

In the last two decades, three-dimensional numerical models have evolved from specialized tools which were only applicable for research purposes to a set of tools for engineering practice. Still the required knowledge level to achieve reliable results is high due to the high sensitivity of these models. In-depth knowledge of the numerical

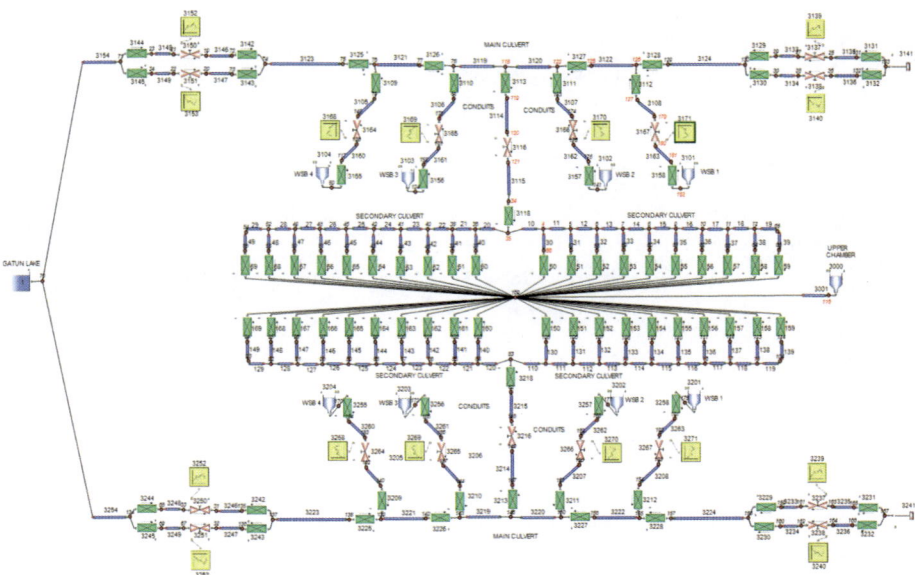

Fig. 2. Network for the one-dimensional modelling of a complex lock filling system (Thorenz 2010)

model is required in order to judge on the required grid and time step resolution, the used numerical schemes, the chosen turbulence and wall friction laws etc. The required computer power is acceptable today for local studies of the flow field, while still substantial computer power is required for more difficult questions. But for some aspects of the hydraulic design process, three-dimensional numerical modelling is no longer out of scope.

2.5.2   Local Models for Hybrid 1D-3D Approaches

Starting around 2010, the combination of three-dimensional models for the estimation of parameters with one-dimensional models for the filling system and/or the chamber became common (i.e. Roux et al. 2010, Thorenz 2010, Thorenz and Strybny 2012, O'Mahoney et al. 2018). This approach requires to set up detailed numerical models for all critical points in the filling system, where more complex geometric shapes make it difficult to predict the loss coefficients. Figure 3 presents a complex connection point, where flow from the lower left branch is distributed into the upper two branches. The time and space averaged results of this models can then be used to compute loss coefficients for a one-dimensional numerical model. For this type of models, the required computational resources are nowadays acceptable for engineering practice. Care must be taken for the choice of the turbulence model, an adequate grid resolution and sufficiently accurate numerical schemes.

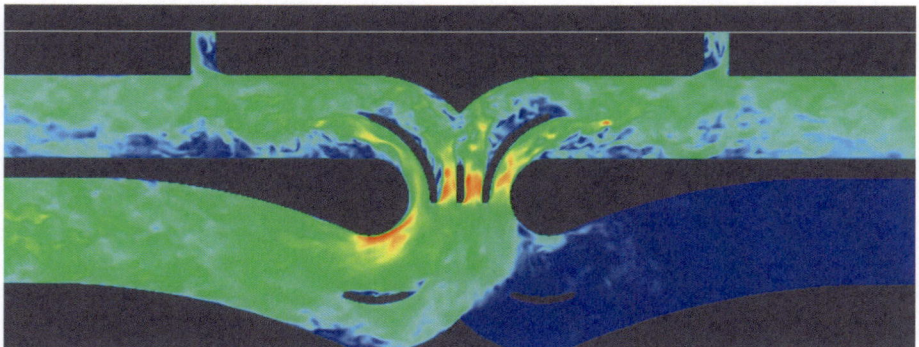

Fig. 3. Horizontal cross-section through the culvert system of a lock. The colormap indicates velocities from low (blue) to high (red). The flow direction is from the lower left branch into the two top branches.

2.5.3 Simplified Three-Dimensional Numerical Models for the Lock Chamber

Due to the complexity of modelling the complete locking process in a three-dimensional numerical model, it can be useful to limit the scope of the model. For example, it can be useful to compute the rate of flow into the chamber with a one-dimensional model and to use these results as a flowrate boundary condition in a three-dimensional model of the chamber (i.e. Roux et al. 2010, Verelst et al. 2018). Thus, the modelling of the moving valve and the feeding culverts would be omitted in the three-dimensional model. But it must be pointed out that this simplification comes with a price: The interaction between water motions in the lock chamber and the flowrate is broken by this approach. Thus, the dampening effect of the interaction between sloshing water and flowrate is not considered and the sloshing in the chamber can be overestimated for through-the-head systems.

Another feasible solution can be the simplification of the ship movement. If the vertical movement of the ship is small in relation to the under keel clearance, it can be favourable to use a stationary ship in the simulations. This reduces the computational complexity and costs significantly, but again comes with the price of a reduced accuracy. This approach can be a useful simplification for maritime locks with large water depth and small lift heights. A possible further simplification would be to omit the ship completely and just study the flow field and water level variations in the chamber (Stockstill 2009).

2.5.4 Full Three-Dimensional Models

The modelling of the complete locking process, including moving valves, the movement of the ship and of the water body from the upstream reach to the chamber is still a challenging task. It requires a significant understanding of the relevant hydraulic processes and of the required numerical modelling tools and allowable simplifications. Though some numerical models are seemingly able to produce results, these can be horribly wrong, even if they look plausible on the first glance. Thus, expert hydraulic knowledge is required to separate reasonable results from implausible ones.

When setting up full three-dimensional models for the lock filling process, the modeler should validate his toolchain and methodology first. This should be done by simulating a case for which the results are known from a physical model or from prototype data. While the simulation of the hydraulics (i.e. flow rates and water level development) is feasible today with reliable results, the reaction of the ship in terms of ship forces can be very hard to capture correctly. In Thorenz and Schulze (2021) a benchmark example is given, were a lock with a complex filling system was analysed both with a physical model and with a full three-dimensional model. The used validation datasets are freely available for further use. The example is presented in short form in Sect. 3.

2.5.5   Consideration of Density Currents and Air Entrainment

The modelling can be even more complicated, if the regarded fluid is not pure water, but is mixed with salt or air. The problem of mixing fresh and salt water occurs for locks in coastal regions. The salt can have a significant impact on the resulting ship forces, both during the filling process of the lock chamber and when opening the gates (Nogueira et al. 2018, Kortlever et al. 2018). This is due to the fact, that salt and fresh water do not mix immediately, but instead density currents are establishing in the lock chamber. These transport a higher momentum compared to currents under conditions with constant density. When filling a lock chamber, which was initially filled with saltwater, with freshwater, another not so obvious effect occurs: The freshwater can pool up in front of the bow of the vessel, while the saltwater remains behind the ship's aft. In this situation the forces on the vessel can be quite high, even if the flow velocities are very low. This is solely due to the hydrostatic pressure difference between salt- and freshwater columns.

Within the three-dimensional numerical modelling, the salt is regarded as a tracer in the water, and as such is transported by advection and dispersion. But in contrast to a traditional tracer, it has an impact on the density of the water. A critical point can be the turbulence modelling. A stable stratification of salt and fresh water will reduce turbulent mixing, thus both the diffusion of salt and the spreading of momentum will be reduced. For an instable stratification, mixing and spreading of momentum will be increased instead. This must be considered in the model. It can either be incorporated in the turbulence model and in the transport equation for the salt or be accounted for by using a Large-Eddy-Simulation approach, to directly capture the behaviour of turbulence and salt transport in the large turbulent structures. An example for this is given in Thorenz (2019).

Generally, the entrainment of air is avoided when designing lock filling systems. But for locks with through-the-head systems it can be attractive from an economic point of view to use filling systems with valves which are located above the downstream water level. In that case, air entrainment is unavoidable. In numerical modelling, air entrainment is even more difficult to handle than salt transport. This has several reasons. First of all, the impact on the fluid density can be much bigger. Secondly, the movement of air in the water has a dynamic of its own, i.e. bubbles are rising to the surface. From a numerical point of view, several approaches exist. Models used in the chemical industry have very detailed physical representation of the bubble dynamics, but are limited to their scope. For the purpose described here, these models are not easily suitable. Recent research in hydraulic structures modelling are focused on simplifications of these models, in order to make them applicable for the flow around hydraulic structures. Schulze et al. (2018) showed simulations for a solver which incorporates air bubble movements and evaluated the impact on a ship in the chamber. Thorenz and Grefenstein (2022) recently compared a further model simplification with physical model results and evaluated the possible impact of entrained on the flow behind a weir.

3   Example for Full Three-Dimensional Modelling of the Locking Process

3.1   Introduction

The three-dimensional modelling of the complete lock filling process, including moving valves and floating ship, is still a daunting task due to its numerical requirements. It requires both substantial resources and specialist knowledge in hydraulics and the numerical tools. What makes things even worse is the fact, that numerical modelling toolboxes can produce accurate looking results, because the flow computed field looks plausible, even if the results are far from reality. On the other hand, the modelling process in terms of accurate ship forces is fragile, so the validation of the modelling methodology is obligatory. Basic requirements are the same as for the modelling of parts of the filling systems. Additionally, the dynamics of the ship must accurately be captured. A minimum requirement is the correct modelling of an initial calm period before opening the valves. In this period, which must last at least as long as the wave sloshing time in the chamber (e.g. 30 s–60 s), the modelled ship forces should be equal to zero in both longitudinal and transversal direction and the movements of the ship must be negligible. Practically the forces will be larger than zero due to inevitable numerical errors. In that case, the resulting non-zero forces must be compared to the acceptable errors for the simulations and, if too high, the model must be enhanced. Furthermore, it is necessary to analyse the recorded forces during the locking process. They should be evaluated in terms of the temporal development of the forces, which follows a typical pattern for each filling-system. In order to judge on the quantity of the forces, the modelling process should be validated with physical model data or prototype measurements. In the following a very short extract of the modelling and validation procedure as performed by Thorenz and Schulze (2021) is presented.

3.2   Model Setup

The regarded system is a navigation lock with a through-the-bottom filing system consisting of a large pressure chamber, fed by side culverts from lateral water saving basins and by culverts at the heads for the residual filling. Figure 4 presents a 3D-sketch of a part of the hydraulic system, showing only one of the three water saving basins. The culverts leading to the other water saving basins are cut at the valves.

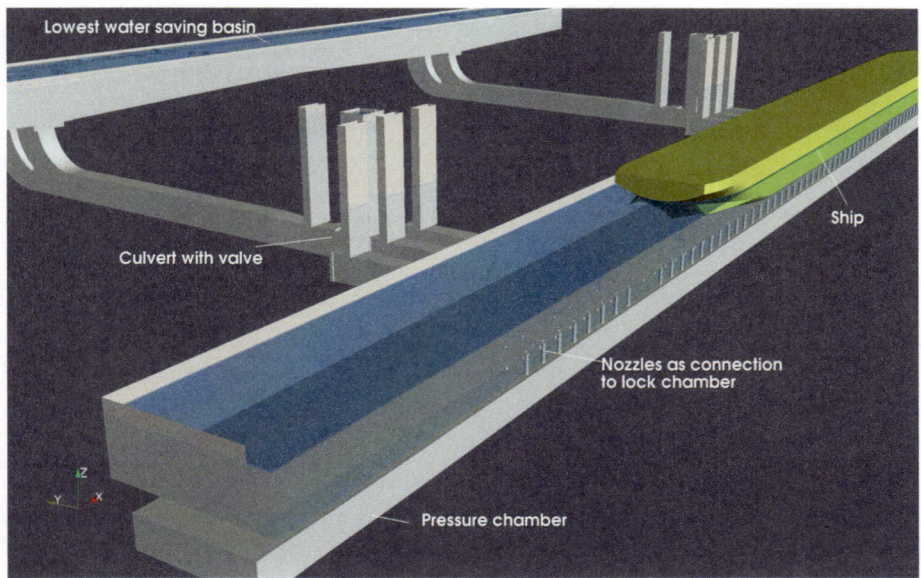

Fig. 4. Regarded parts of the filling system with the ship in the chamber (yellow) as they were used in the comparison study.

For this setting a three-dimensional numerical model has been built based with the OpenFOAM® software package. The generated computational grid uses several areas of local refinements (e.g. at the valves, for the jet behind the valves, at the nozzles of the bottom filling system and for the jets in the chamber) and has 28 million computational cells. The numerical model was not calibrated in any way to fit to model data. Instead, the numerical approximation was performed as good as regarded as necessary by the modellers.

3.3 Assessment of Results

For the quality assessment of the numerical model, a physical scale model in 1:25 was available. A thorough comparison of numerical and physical model data was performed in order to judge on the quality of the numerical model (Thorenz and Schulze 2021). The comparison of flow rates and water levels showed a very good agreement between numerical and physical model. As expected, the impact of friction (i.e. roughness) was low in the numerical model (Fig. 5, roughness height is given in prototype scale).

It must be pointed out that the flowrates cannot actually be measured in the physical model, but are computed from the chamber water level development. Thus, small deviations can be expected. The further evaluation of the longitudinal forces showed an excellent agreement with the physical model data (Fig. 6, again the roughness height is given in prototype scale). Please note that the height and shape of the first peak is most relevant and that the further development is of lower importance.

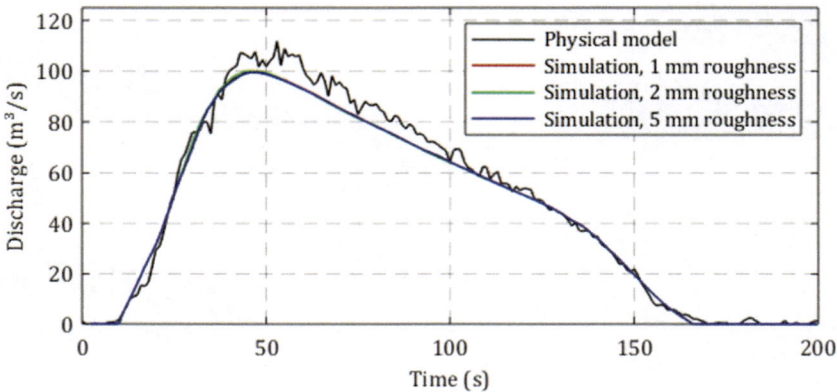

Fig. 5. Comparison of flowrates computed from physical model results with numerical results (flowrate and roughness given in prototype scale).

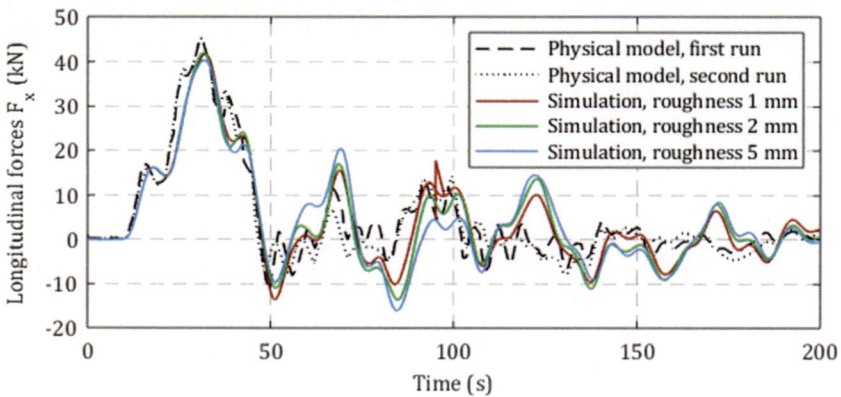

Fig. 6. Comparison of longitudinal ship forces from physical model results with numerical results (forces and roughness given in prototype scale).

The comparison of the transversal forces showed the correct behaviour in terms of temporal development, but the absolute value was overestimated by the numerical model. Thorenz and Schulze (2021) present an analysis, where several possible causes are investigated. This includes also an analysis of possible errors in the physical model, because the analysis has also shown discrepancies between planned and realized physical model.

4   Conclusions

Over the last decades, numerical models have become a non-substitutable tool in the hydraulic design process for locks. Ranging from one-dimensional models, which can produce results almost immediately, to full blown three-dimensional models of the complete locking process, which will compute days on a cluster computer system, a large toolkit has been developed which can be used for a broad scope of analyses.

Analytical calculations, zero- and one-dimensional numerical models are seemingly simple, but they require to have calibration data available for the not resolved hydrodynamic processes. This data can be generated either by physical model runs or by three-dimensional numerical models for the regarded parts. To transfer it from one setup to a seemingly similar setup, expert knowledge is required. Based on the achieved parameterization, the lock filling process e.g. for different valve schedules can be quickly evaluated. This is a fast and appropriate method to evaluate significant parameters like filling time, ship force and flow velocities. But one must consider, that for these methods the results are strongly dependent on parameters which have to be evaluated with experiments or with models of higher complexity. Furthermore, more complex flow phenomena cannot be considered with these approaches.

Three dimensional models became more feasible over the last years. Unfortunately, even poor three-dimensional models will produce "realistic" looking results. Thus, professional experience is necessary to assess the results. The modeller should understand the expected hydraulic behaviour in order to judge on the qualitive results of the flow field. In correctly built three-dimensional numerical models, calibration of the model should not be necessary in most cases, as the only unknown quantity which is suitable for calibration is the wall friction, which is typically of lower importance. The quality of the modelling strategy is the most relevant factor to achieve reliable results. This includes the choice of a capable numerical model, an accurate representation of the geometry, choice and validation of turbulence models, suitable grid and timestep resolutions and the setup of the whole model in terms of physical correctness. Expert knowledge in numerical models is required to set up the models and expert knowledge in lock hydraulics is required to evaluate the quality of the results.

For the validation of any modelling methodology, it is necessary to test the modelling strategy first on a known case, before using it for an unknown setup. For the known testcase, the relevant data (i.e. flowrates and ship forces) should have been measured in a physical model or in a prototype. This offers the possibility to check the modelling methodology before applying it to other cases.

Summarizing one can say that numerical and physical modelling cannot replace expert knowledge. Models are only a tool, supporting professional work. A good compromise between accurate and economic engineering work can be reached with hybrid modelling, where different types of modelling approaches are combined.

References

Belzner F, Simons F, Thorenz C (2018) An application-oriented model for lock filling processes. In: Proceedings of the 34th PIANC-world congress, 7–11 May 2018, Panama City, Panama. https://henry.baw.de/handle/20.500.11970/107576

De Loor A, Weiler O, Kortlever W (2013) LOCKFILL: a mathematical model for calculating forces on a ship while levelling through the lock head. In: Proceedings of the PIANC SMART rivers conference. PIANC Association, Brussels

De Mulder T (2011) Computational fluid dynamics (CFD) in lock design: progress and challenges. In: "What's new in the design of navigation locks" 2nd international workshop, PIANC - New-Orleans, 13–14 September 2011

Kortlever W, Van der Hout A, O'Mahoney T, de Loor A (2018) Levelling the new sea locks in the Netherlands: including the density difference. In: 34th PIANC world congress, Panama

Nogueira HIS, van der Ven P, O'Mahoney T, de Loor A, van der Hout A, Kortlever W (2018) Effect of density differences on the forces acting on a moored vessel while operating navigation locks. J Hydraul Eng 144(6)

O'Mahoney T, Heinsbroek A, De Loor A, Kortlever W Verelst K (2018) Numerical simulations of a longitudinal filling system for the new lock at Terneuzen. In: 34th PIANC world congress, Panama

PIANC (2015) InCom WG 155. Ship Behaviour in locks and lock approaches. Edited by PIANC, Brussels, Belgium

Roux S, Roumieu P, De Mulder T, Vantorre M, De Regge J, Wong J (2010) Determination of hawser forces using numerical and physical models for the third set of Panama locks studies. Paper 151 In: Proceeding of the PIANC MMX congress, Liverpool, UK, 10–14 May 2010

Schohl GA (1999) User's manual for LOCKSIM: hydraulic simulation of navigation lock filling and emptying systems. Report CHL-99-1, U.S. Army Engineer WaterwaysExperiment Station, Vicksburg, USA

Schulze L, Stamm J, Thorenz C (2018) A new two-phase flow model for the investigation of the effect of entrained air in navigation locks. In: E-proceedings of the 38th IAHR world congress, 1–6 September 2019, Panama City, Panama

Stockstill R (2009) Computational model of a lock filling system. Report ERDC/CHL CHETN-IX-18, US Army Corps of Engineers

Thorenz C (2009) Computational fluid dynamics in lock design - state of the art. Paper 10, international workshop on "innovations in navigation lock design", PIANC Brussels, 15–17 October 2009

Thorenz C (2010) Numerical evaluation of filling and emptying systems for the New Panama canal locks. In: Proceedings of the 32nd PIANC Congress 125th anniversary PIANC – setting the course, Liverpool, UK, 10–14 May 2010, Brussels, Belgium, pp 568–589

Thorenz C (2019) Can better turbulent mixing reduce density induced ship forces during lockage?. In: 34th PIANC world congress, Panama

Thorenz C, Grefenstein A (2022) Numerical modeling of air bubble transport behind a model scale weir. In: 39th IAHR world congress, 19–24 June 2022, Granada, Spain

Thorenz C, Strybny J (2012) On the numerical modelling of filling-emptying systems for lock. In: 10th international conference on hydroinformatics, HIC 2012, Hamburg, Germany

Thorenz C, Schulze L (2021) Numerical investigations of ship forces during lockage. J Coast Hydraul Struct. https://doi.org/10.48438/jchs.2021.0005

Verelst K, Vercruysse J, de Mulder T (2018) Comparison of software for computation of longitudinal forces on a ship in a lock chamber during levelling with openings in the lock gate. In: Bung, D, Tullis B (eds) 7th IAHR international symposium on hydraulic structures, Aachen, Germany, 15–18 May 2018. https://doi.org/10.15142/T3KD2H

Optimizing Upstream Approach Wall to Navigation Lock in Narrow Rivers

Didier Bousmar[1(✉)], Catherine Swartenbroekx[1], Geoffrey Pierard[2],
and Emmanuel Van Hees[3]

[1] Hydraulic Research Lab., Service Public de Wallonie Mobilité et
Infrastructures, Châtelet, Belgium
didier.bousmar@spw.wallonie.be
[2] Université Catholique de Louvain, Louvain-la-Neuve, Belgium
[3] ECAM, Brussels Engineering School, Brussels, Belgium

Abstract. In canalized rivers, navigation locks are often located aside a weir. River currents are concentrated towards the weir. As a result, vessels face transverse currents and significant outdraft forces when approaching the lock. An approach wall can be designed to isolate the final lock approach from the weir flow. This wall can be a plain wall, or a perforated wall. Depending on the design of the wall openings, the flow velocity gradient and the outdraft forces are translated upstream at the entrance of the port. At this location, the vessel keeps a larger velocity and can manoeuvre more easily. An optimal design of the opening should minimize the transverse forces acting on the vessel. Guidelines for the design of such approach walls may be found in the literature but cover only large and very large rivers. The present contribution focuses on narrower rivers, with a low ratio between port entrance width and river width.

Keywords: Navigation lock · Approach wall · Numerical modelling · Physical modelling · Cross current

1 Introduction

The current developments of fluvial transport in Europe often imply the navigation of larger vessels on existing waterways. Larger vessels contribute to reduce transport costs and to a larger modal share of the waterborne transport. These developments may rely on local improvement of the waterways, like local enlargement, but also on improved manoeuvrability of modern vessels. New larger locks are often needed to accommodate those larger vessels. These new locks must be built on confined space as existing amenities have to be maintained both on the river (weir, hydropower plant, etc.) and on land (roads, railways, urbanization, etc.). Accordingly, designing the approaches for these new locks results from a compromise between the extension on the approach area and the remaining space for river flow.

A lock approach can be defined as the necessary navigational area to support a safe entry and exit of the lock chamber (PIANC 2019). The vessels in the lock approaches are protected from the flow in the river and can manoeuvre safely to the lock. Entering the upstream lock approach is usually the most critical phase. River currents are

Y. Li et al. (Eds.): PIANC 2022, LNCE 264, pp. 542–552, 2023.
https://doi.org/10.1007/978-981-19-6138-0_47

concentrated towards the weir and downbound vessels face significant cross currents while they should slow down. Sufficient width and length of the approach area is required to enable safe navigation. Various recommendations and guidelines on the require size can be found in the literature (see PIANC 2019 for a synthesis).

Designing openings (ports) in the guard wall may also reduce cross currents at the port entrance (PIANC 2015). Different parameters influencing the flow pattern and the intensity of cross currents at the approach entrance can be identified: (1) the ratio between approach zone entrance width and the whole river width; (2) the ratio of the wall openings area and the approach entrance cross-sectional area; (3) the shape of the openings (rounded, sharp, on the whole depth, etc.); and (4) the plan form of the river upstream of the lock and weir.

Apart from case studies specific to a given site, only limited general guidelines for the design of guard wall can be found in the literature. Stockstill et al. (2004, 2005) conducted a systematic investigation of different guard wall geometries: solid wall, multi-cell wall and floating wall. Two test series covered approach width equal to 0.18 and 0.33 times the total river width. For the narrower approach and the multi-cell wall, the optimal ratio between the total wall opening area and the approach entrance area was found to be equal to 0.9. For the wider approach, the optimal value of this ratio dropped to 0.5 and larger transverse forces were observed.

Bousmar et al. (2010, 2014) and Swartenbroekx and Bousmar (2018) conducted specific case studies on narrower rivers where larger locks were planned. The ratio of approach width to total river width grew up to 0.4 in those cases. The optimal values found by Stockstill et al. for wall opening design were no more appropriate for these extreme cases. Specific studies concluded that lower opening ratio were required to minimize transverse forces.

These latest studies also highlighted the need for systematic studies on the design of guard wall for wider approach in narrower rivers. This paper presents preliminary results of such study, covering width ratio from 0.50 to 0.20. A guard wall with circular shaped openings similar to the multi-cell design by Stockstill et al. was tested. The area of the openings was systematically investigated, using physical and numerical modelling, to determine the optimal ratio.

2 Definition of the Investigated Layout

An idealized upstream lock approach is defined, in a straight channel configuration. Figure 1 defines the main dimensions of this layout, in both model and prototype values. The prototype dimensions are fixed considering a ECMT Class Va vessel of 110 m × 11.40 m. The model scale is 1:33.3. The approach width Wa is around 5 times the vessel width. It may correspond to either an approach for a single lock (navigation axis B, located in the middle of the approach) or an approach for a pair of locks (navigation axis A and C, located at first and last quarter). Flow depth in both approach and river is fixed to H = 4.5 m (150 mm in the model).

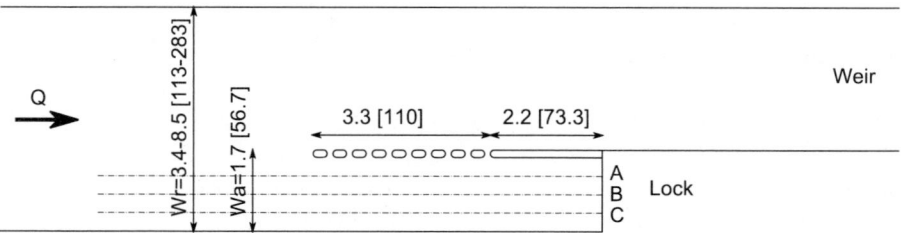

Fig. 1. Layout of the investigated upstream approach. Dimensions in meters. Physical model values [prototype values in brackets].

Different ratio of the approach width to the river width are investigated. This ratio is defined as

$$AP = \frac{Wa.H}{Wr.H} \tag{1}$$

where Wa = 1.7 m is the approach width, Wr is the river width, and H = 150 mm is the flow depth. Different values of the river width are tested, as listed in Table 1, to cover AP values in the range 0.50 to 0.20. The upstream discharge Q is adjusted the obtain a constant upstream specific discharge and velocity.

Table 1. Value of the width ratio AP investigated.

AP	River width Wr (model)	River width Wr (prototype)	Discharge Q (model)
0.50	3.40 m	113.3 m	120 l/s
0.40	4.25 m	141.7 m	150 l/s
0.33	5.10 m	170.0 m	180 l/s
0.25	6.80 m	226.7 m	240 l/s
0.20	8.50 m	283.3 m	300 l/s

For each AP value, different guard walls are tested. The wall thickness in model dimensions equals 50 mm. In all tests, the 2.2 m part closest to the lock is kept as a solid wall. The upstream 3.3 m part is fitted with rounded openings. In all cases, the size of the openings is kept constant to 50 mm, flanked by two 50 mm diameter half circles, on the whole flow depth. The adjustment parameter is the number of openings, to obtain different values of the ratio M between the opening area and the approach entrance area:

$$M = \frac{\sum L_i.H}{(Wa - Wg).H} \tag{2}$$

where Li = 50 mm is the opening width, and Wg = 50 mm is the wall thickness. The investigated values of M are summarized in Table 2. The different tested configurations will be referred by APxxxMyyy where xxx is the AP value in percent and yyy the M value in percent.

Table 2. Value of the opening ratio M investigated.

M	1.00	0.60	0.50	0.33	0.24	0.15	0.00
# openings	33	20	16	11	8	5	0
Σ Li (m)	1.65	1.00	0.80	0.55	0.40	0.25	0.00

3   Physical Model

The physical model used for this investigation is illustrated on Fig. 2. The whole configuration is set in a flat bottom basin. The left bank can be translated to obtain different AP configurations. Upstream discharge is controlled through an electromagnetic flowmeter and downstream water level is adjusted with a flap gate. Water level is controlled with two ultrasonic distance probes located up- and downstream.

Velocities are recorded with an electromagnetic velocity probe. An automated displacement carriage moves the probe on a grid with 200 mm longitudinal intervals and 100 mm transverse intervals. Preliminary investigations confirmed that a measurement duration of 30 s per station was sufficient and that measurement at a depth of 80 mm was representative of the depth-averaged velocity.

In the present study, only configurations AP050M000, M033, M050, M100, AP040M033 and M100 are tested. Measurements are used for calibration and validation of the numerical model.

Fig. 2. General view of the physical model, configuration AP040. Upstream and downstream views.

4   Numerical Model

Depth-averaged numerical modelling of the flow is performed using Telemac2D software. The layout was modelled with the same dimensions as the physical model to facilitate comparisons. The unstructured mesh covers an area extending 8.5 times the approach width Wa in the upstream direction from the approach entrance and 6 times this width in the downstream direction, so that flow in the approach is not influenced by the boundary conditions. The mesh was refined in the wall opening area (with cell size around 10 mm), and in the approach entrance (cell size around 30 mm). As the openings extend on the whole flow depth, no other specific treatment is required. A fixed specific discharge is imposed upstream and a constant water level downstream. Bed roughness is modelled using Manning' equation, with n = 0.013. Turbulence is modelled using either a depth-averaged k-epsilon model, or a constant eddy viscosity $v_t = 10^{-5}$ m²/s. Figure 3 shows a typical velocity field. The numerical model was eventually validated by comparison with physical model results (see Fig. 4).

Fig. 3. Numerical model results, configuration AP050M033 (blue: low velocities, orange: large velocities).

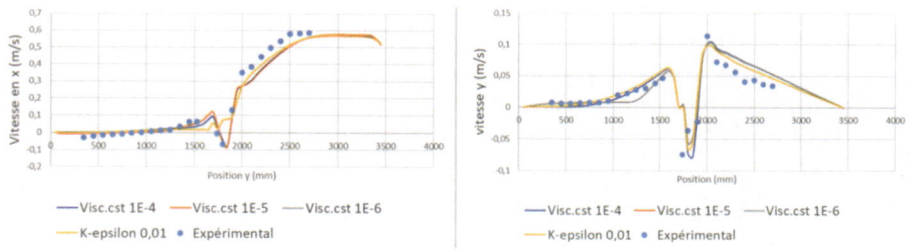

Fig. 4. Configuration AP050M033. Numerical and experimental results at cross section X = 18 m, located at the downstream end of the opened part of the guard wall. Longitudinal (left) and transverse (right) velocities.

5 Results

5.1 Velocities

The global flow pattern is depicted on Fig. 3. At the entrance of the lock approach, the flow converges towards the contracted river and the weir. Transverse velocities are observed on a distance extending from upstream the approach entrance to inside the approach. A velocity gradient area develops in diagonal from the right bank to the guard wall edge. Inside the approach area, only low velocities are observed, with a smooth recirculation zone extending on the whole area. In the river channel, due to the converging flow, maximum velocities are observed along the left bank and a recirculation area develops along the solid part of the guard wall.

The velocities in and upstream the approach area are analyzed in more details by extracting the longitudinal and transverse velocity components along the navigation axis identified on Fig. 1. Figure 5 shows the longitudinal velocities along navigation axis C (close to the bank) and A (close to the wall) for configuration AP050 and various M values. Vertical black lines indicate the extension of the openings in the wall. The velocity gradient is clearly observed from full velocity in the upstream river to quasi-zero in the approach zone. When comparing axis C and A, it is observed that the flow penetrates farther in the approach zone close to the wall, as highlighted on the global flow pattern on Fig. 3. It is also clear that the flow penetrates more the approach zone when the guard wall permeability increases with larger M values. With M100, the flow gradient is almost similar to the pattern observed with a plain wall (M000) just shifted 3.3 m downstream. Lastly, the recirculation zone in the quiet area of the approach zone (X > 18 m) can be identified with positive longitudinal velocities along the wall and negative values along the bank.

The benefit of the wall openings is more clearly highlighted by the analysis of transverse velocities plotted on Fig. 6. The transverse velocities corresponding to the flow contraction develop from upstream the approach zone to inside the zone. Transverse velocities along the wall (axis A) are almost three time larger than along the bank (axis C). Further analysis will therefore focus on axis A. The largest velocities are observed for the solid wall M000 and for the highest permeability M100. The velocity profiles are similar and just shifted 3.3 m as highlighted for the longitudinal velocities. This shows that M100 wall is quasi transparent to the transverse flow and can be expected to have no impact in terms of transverse forces reduction. For intermediate permeabilities, transverse velocity maximum decreases and the flow extends on a longer distance. Minimum values are obtained for M033 and M024 cases, with a (double) peak shifting progressively from inside the approach to upstream the approach.

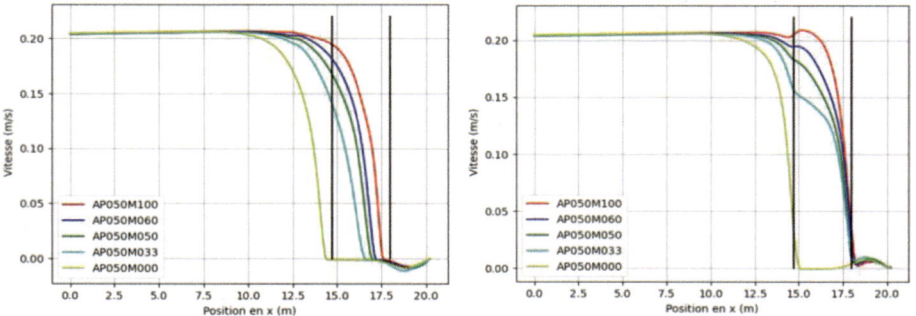

Fig. 5. Configuration AP050. Longitudinal velocity along navigation axis C (left) and A (right).

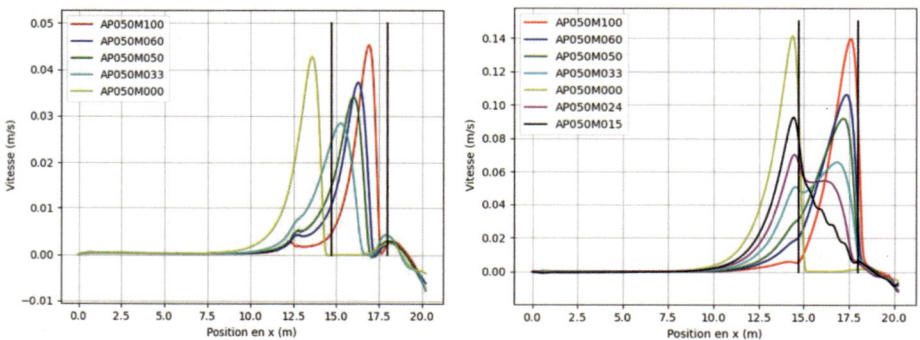

Fig. 6. Configuration AP050. Transverse velocity along navigation axis C (left) and A (right).

5.2   Forces and Momentum

For an improved analysis of the impact of the guard wall configuration on the vessel navigation, forces and momentum acting on the vessel are estimated from the flow pattern. As a first approximation, the transverse force is estimated from the unperturbed flow distribution. Transverse force is estimated as proportional to the square of local transverse velocity, integrated along the vessel hull length (drafted at 3.00 m). Figure 7 shows the evolution of the force for successive positions of the vessel (given as the center of the vessel) along navigation axis A. Yaw momentum is estimated similarly, considering the center of the vessel as rotation center.

Figure 7 shows the transverse force profiles along axis A for configurations AP025 and AP050. Conclusions drawn from the analysis of velocity profiles are confirmed here. The vessel is submitted to large transverse forces during its transit along navigation axis, from upstream to inside the approach. Largest values are observed for solid wall configuration M000 and large permeability configuration M100, just shifted by the length of the opened part of the wall. Configuration M100 has accordingly too many openings to be efficient. For intermediate configurations (M060 to M015), smaller maximum forces are observed. The vessel is also submitted to the transverse currents on a slightly longer distance. When reducing the permeability, the force peak progressively moves from inside to upstream the approach. Lastly, when comparing AP050 to AP025 configurations, it appears that transverse forces for the narrower channel AP050 are around 30% larger than for AP025, due to the stronger flow contraction.

Quite similar conclusions can be drawn from the analysis of yaw momentum profile as shown on Fig. 8. In the case of momentum, two peak are observed. A negative peak occurs when the vessel enters the transverse current zone, and the bow is drifted towards the river. A positive peak appears when the vessel leaves the transverse current zone, and the stern is now drifted toward the river. Again, largest momentum values are observed for case M000 and M100, shifted by 3.3 m. Intermediate values are observed for intermediate permeabilities (M060 to M015), with the peak location moving progressively from inside to upstream the approach. And again, peak values for configuration AP050 are around 30% larger than for AP025.

Peak values of transverse forces and yaw momentum are summarized on Fig. 9 for all the configurations investigated. This graph shows that the optimal wall configuration is globally M033. For the narrower channel AP050, a smaller permeability M024 could be an alternative considering maximum force and maximum momentum. For the larger channel AP025 and AP020, a larger permeability M050 could also be considered regarding minimum momentum.

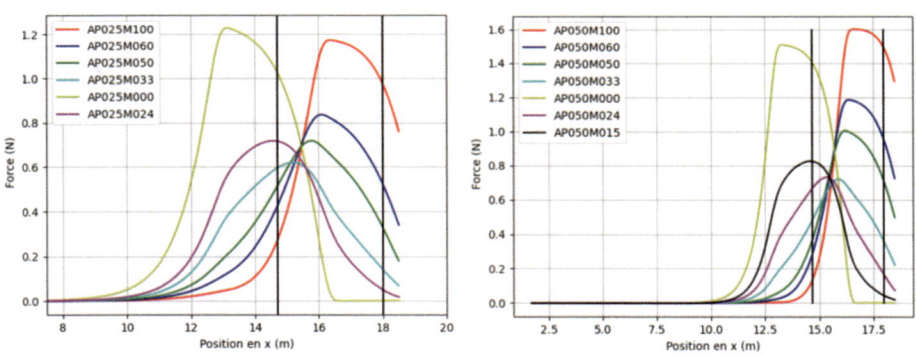

Fig. 7. Transverse force along navigation axis A. Configuration AP025 (left) and AP050 (right).

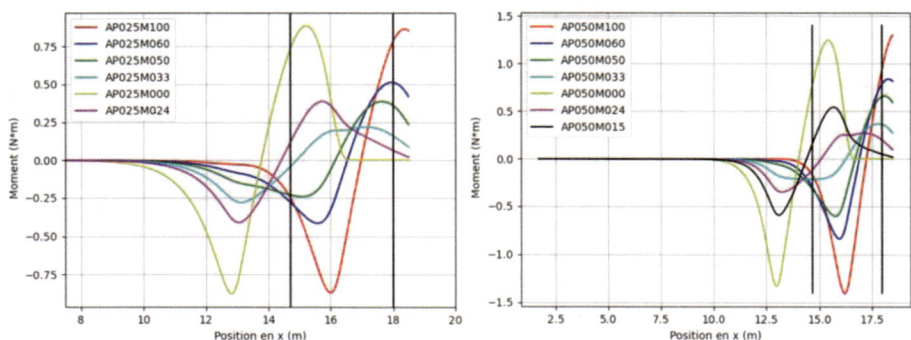

Fig. 8. Yaw momentum along navigation axis A. Configuration AP025 (left) and AP050 (right).

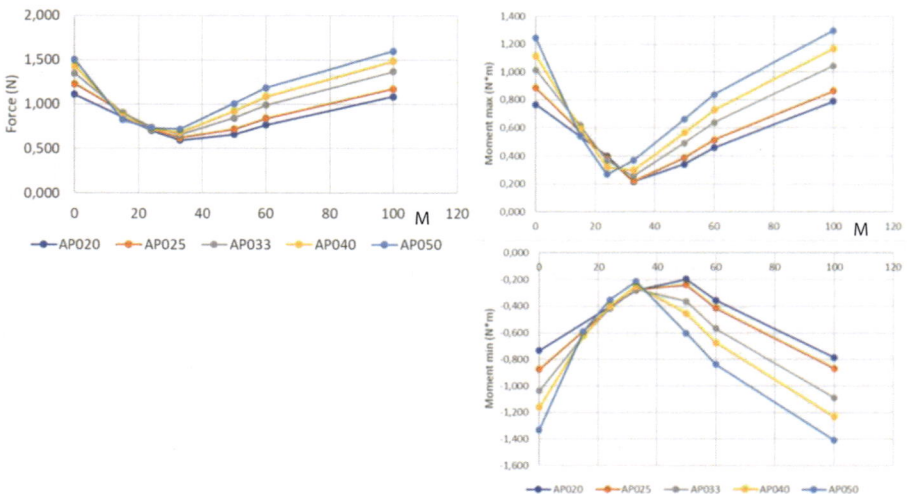

Fig. 9. Maximum transverse force (left) and maximum and minimum yaw momentum (right) as a function of M, along navigation axis A.

As the analysis of minimum and maximum yaw momentum values can lead to undefined results, a further analysis is performed. The integral of the yaw momentum supported by the vessel is calculated along its trajectory on the navigation axis, considering in a first approximation a constant sailing velocity. This integral is supposed to represent the overall steering effort along the entering course of the vessel, to compensate alternatively negative and positive yaw momentum. Results plotted on Fig. 10 confirm that configuration M033 is optimal, with M024 as an alternative for the narrower channel AP050 and M050 as an alternative for the largest channel AP020.

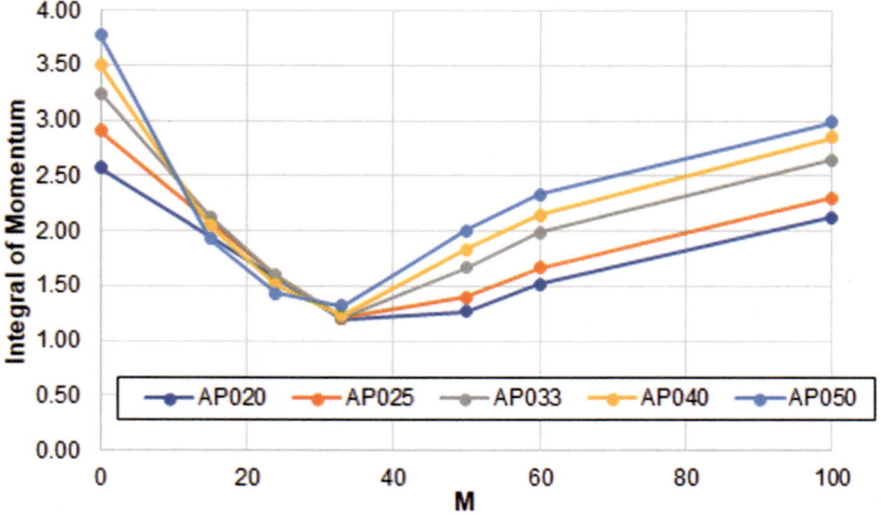

Fig. 10. Integral of yaw momentum along navigation axis A.

5.3   Discussion

Comparison with previous results by Stockstill et al. (2004, 2005) confirms the effectiveness of permeable guard wall. The need for lower permeability for narrower channels is also confirmed but smaller permeability values are obtained in this study compared to Stockstill et al. For AP020, the present study suggests a permeability of 33 to 50%, while Stockstill suggested 90%. For lower AP033, the present study recommends also a permeability of 33%, instead of 50%. These differences could be due to the different wall configurations. In the present study, openings are present on the whole wall depth, while in Stockstill et al. a beam forced the water to contract on the river bottom. Also, the exact AP value for Stockstill et al. study differs as their river channel was deeper than their approach channel.

Another point of interest to be further analyzed is the flow contraction and the associated recirculation zone in the river channel downstream the approach. This contraction can be severe for the narrowest channel and generate a significant head loss. This head loss may impact the water surface profile upstream and increase flood risk for surrounding areas. An alternative to round shaped openings in the wall is inclined straight openings. The latter will force the orientation of the flow in the river channel and may significantly reduce or even eliminate the recirculation zone (see e.g. Bousmar et al. 2010). Preliminary results depicted on Fig. 11 for AP050M033 configuration with inclined openings at an angle of 45° show than velocity pattern and forces on the vessel in the approach are expected to be quite similar, while the contraction in the river is significantly reduced.

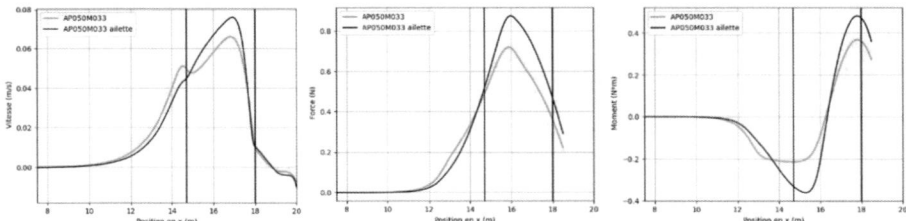

Fig. 11. Configuration AP050M033, navigation axis A. Comparison between round shaped and 45° inclined openings. Transverse velocity, transverse force and yaw momentum.

6   Conclusions

The proper design of an upstream lock approach and guard wall guarantee a safe and easy entrance of the vessels to the lock. The present study focused on narrow rivers on which larger vessels are nowadays allowed as a results of waterborne transport development. Larger vessels and larger lock imply larger flow contraction upstream the lock, with approach width to river width ratio sometimes up to 50%. Existing guidelines for the design of guard walls did not cover such ratio.

Systematic tests were performed on physical and numerical models, covering width ratio from 20 to 50% and opening ratio from 0 to 100%, with round shaped openings. Results show the effectiveness of a proper design of the wall openings. When openings are too large, the flow behaves as if there was no wall. When reducing the opening, the peak velocity and associated transverse force and yaw momentum decreases and is shifted upstream. If the openings are too much reduced, the peak grow again and moves upstream the approach zone, as if the wall was a solid wall. For all the configurations tested, a permeability of 33% seemed to be optimal. A smaller permeability may be convenient for the narrowest channel (AP050) and a slightly larger permeability could be used for the largest channel (AP020).

References

Bousmar D, Bayart P, Zimmerman N, Gronarz A (2014) Real-time navigation simulations for improving navigation on existing waterways: river Meuse in Belgium. In: PIANC world congress, San Francisco, USA

Bousmar D, Bertrand G, Hiver JM, Barlet S, Boogaard A, Elzein R, Veldman JJ (2010) The approaches for the new class VIb lock at Ivoz-Ramet, River Meuse, Belgium. In: PIANC MMX congress, Liverpool, UK

PIANC (2015) Ship behavior in locks and lock approaches. Report n°155. Brussels, Belgium

PIANC (2019) Design guidelines for inland waterway dimensions. Report n°141. Brussels, Belgium

Stockstill RL, Park HE, Hite Jr JE, Shelton TW (2004) Design considerations for upper approaches to navigation locks. Report ERDC/CHL TR-04-4, USACE, USA

Stockstill RL, Hite JE Jr, Park HE (2005) Hydraulic design of upper approach walls to navigation locks. PIANC Mag 118:17–25

Swartenbroekx C, Bousmar D (2018) Physical design of upper harbor at Auvelais lock, Belgium. In: 7th international symposium on hydraulic structures, IAHR, Aachen, Germany

Prototype Monitoring of Cavitation in Valve Culvert of Qianwei Shiplock

Xiujun Yan[✉], Zhonghua Li, and Lin Chen

Key Laboratory of Navigation Structures, Nanjing Hydraulic Research Institute, Nanjing, China
229073561@qq.com, xiaolinzi@tju.edu.cn

Abstract. The working condition of water valve is an important sign of the success for navigation lock design, and the valve cavitation characteristics and suppression technology is the key technical challenge in the design of high head navigation lock, in the view of the "flat bottom & top spreading" with the condition of "less project quantities, convenient construction" simple type corridor with the head of 19 m in Qianwei ship lock, the method of combining the physical model test and prototype observation are both adopt to this research. Through the physical model test, the cavitation position, cavitation characteristics and the effect of natural ventilation on cavitation suppression are comprehensively determined by analyzing the change law of cavitation noise intensity from several cavitation noise sensors under different opening conditions during the opening process, also the visual observation of flow state and auditory observation. Prototype observation focuses on the effect of self-aerated technology of valve lintel. When the air pipe outside the value lintel position is closed, there are evident in the cavitation pulse signal, and the maximum strength can reach to 182Pa, the "crackling" of cavitation collapse could be clearly felt at the top of the value, occasionally with 2–3 times slight "muffled thunder", The results show that the valve section has stronger cavitation during the opening process of value. On the contrary, when self-aerated of valve lintel was realized, the maximum ventilatory capacity was 0.169 m³/s, which was close to the 0.170 m³/s calculated by the physical model in the design stage, and no cavitation pulse signal was detected among the valve wells on both sides, and the process line was "stable" with the maximum noise intensity only about 10 Pa, no sound of cavitation collapse was heard at the top of the valve section during the whole valve opening process. The comparison of value steeve vibration also verifies the effect of self-aerated technology of valve lintel at the same times. The results show that self-aerated technology of valve lintel can significantly inhibit the cavitation of valve segment of this form.

Keywords: Cavitation · Self-aerated technology · "Flat bottom & top spreading" shape · Prototype observation · Physical model test

Y. Li et al. (Eds.): PIANC 2022, LNCE 264, pp. 553–564, 2023.
https://doi.org/10.1007/978-981-19-6138-0_48

1 Introduction

The valve plays an important role as the throat of the ship lock's water delivery system. Moreover, the operating condition of the valve is directly related with the safety and the efficiency of the ship lock. At present, The development of the ship lock projects in China leads to the construction of several high head locks, such as the Shuikou three-step ship lock, the Wuqiangxi three-step ship lock, the three-gorge five-step ship lock, Wan'an second-line lock, Changzhou lock and Banghai lock (Hu 2009; Bian 2000; Xue 2021).

The valve lintel gap is inevitable between the valve panel and the corridor in the navigation lock filling and the discharging valve's valve lintel. This gap is narrow, and the hydraulic head acting upon it is the actual working head of the lock valve (Luo et al. 2016). When a high-speed water flow pass through the lintel gap, the pressure in the gap reduces considerably according to the Bernoulli's law, and a strong cavitation will then occur subsequently not only to erode the flow passage but also to increase the pulsation of the valve opening and closing forces. The valve cavitation was widely studied (Ceccio 2010), which has several types, including the top gap cavitation, the bottom edge cavitation, the step-down floor cavitation, and the valve slot cavitation.

A new cavitation resistance technology for high head valves is developed to address the urgent need problem of the high head valve cavitation (Wu et al. 2013; Simpson et al. 2019). This technology involves a combined measure, including a new valve culvert shape model, a reasonable initial submergence depth, and the natural aeration (Li et al. 2019). The cavitation problem of the high head valves is properly resolved. The natural aeration measure can suppress the top gap cavitation with a good inhibitory effect on the bottom edge cavitation (Lindau et al. 2005). Moreover, this measure enables the aerated flow to reach the bottom edge of the valve (Paik et al. 2008; Rhee et al 2010). Consequently, the natural aeration has become a necessary measure for the protection against cavitation of high head valves. (Aydin and Ozturk 2009; Bhosekar et al. 2012; Zhang et al. 2011). Compared with forced aeration, self-aeration does not require an additional air compressor or control system and is more suitable for engineering applications. As a kind of self-aerated equipment, the self-aerated technology of valve lintel (SATVL) has the advantage of no moving parts, self-aeration, reliable operation, high efficiency of aeration, and corrosion reduction. Yan and Thorpe (Yan et al. 1990) pioneered the investigation of the cavitation flows, particularly, those of the choked cavitation flows and the choked cavitation conditions, and the choked cavitation number was proposed to describe the choked cavitation flows. He pointed out that the CVs could provide a constant mass flow rate of liquid while operating under the choked cavitation conditions, and the choked flow regime could be used to control the mass flow rate.

The key point of SATVL should focus on whether the valve lintel can be aerated or not. 1:1 full-scale slice physical model and CFD numerical simulations were established to investigate the sensitivity of structural parameters of the valve lintel to the critical self-aerated conditions, to optimize the valve lintel structural parameters to achieve the optimal self-aerated performance (Wang et al. 2017; Wang et al. 2020; Wu et al. 2020; Wu et al. 2021).

The dimension of Qianwei lock is 200 m × 34 m × 4.5 m (effective length & effective width & threshold depth), and the valve is classified as medium and high head lock with maximum working head 19.0 m. In order to reduce the amount of work and facilitate the construction, the hydraulic model of the valve in the design period recommended that the simple type of "flat bottom & expansion at the top gradually" would be adopted in the corridor of the valve. Natural aeration of the lintel is a necessary measure to suppress the cavitation of the gap and bottom edge of the lintel. The model test showed that lintel aeration can restrained cavitation of valve segment, but its effect needs to be verified by prototype results.

2 Physical Model

2.1 Model Arrangement

Figure 1 shows the shape of the corridor behind the door and the arrangement of aeration pipes used in the hydraulic physical model test of Qianwei ship lock. Two aeration pipes(φ = 100 mm) are set in the distance of 10.0 m and 16.0 m behind the door, and the control valve is set at the top of the reversed radical valve, the purpose is to eliminate the air in the corridor when water is filled for the first time and remove the air bag formed on the top of the corridor during the process of filling and discharging in the lock easily. Forced aeration is necessary depending on the prototype debugging as a backup measure to suppress cavitation of the corridor in the end.

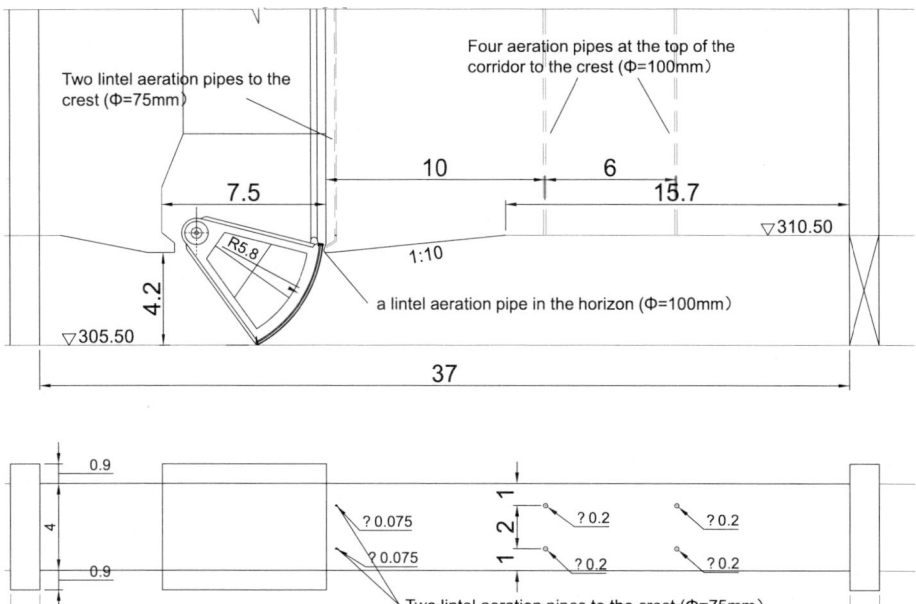

Fig. 1. Shape of corridor and layout of aeration pipe

2.2   Measuring Points Arrangement

The hydrophone layout in the physical model is shown in Fig. 2. The 1# hydrophone mainly receives the signal of lintel cavitation and strong cavitation noise at the bottom edge, The bottom edge cavitation noise signal is mainly received by the 2# and 3# hydrophone, The 4 # hydrophone receives cavitation noise signals transmitted to the lower access door slot and downstream access door slot itself when strong cavitation occurs at the bottom edge. The cavitation characteristics of the corridor of the valve section were studied by closing the natural aeration pipe of the lintel in a pressure reducing compartment.

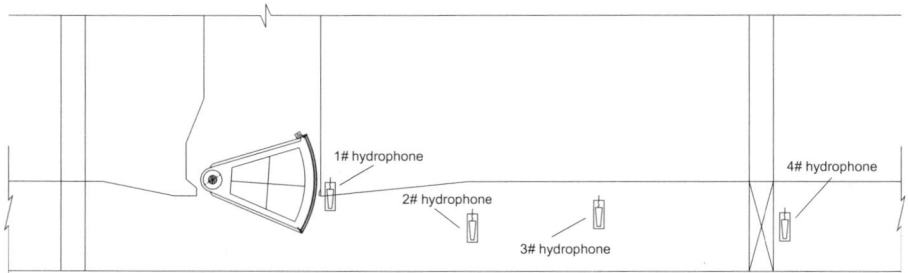

Fig. 2. Hydrophone arrangement among the valve

2.3   Model Test Results

Bottom edge cavitation and lintel cavitation is existed in the corridor shape of valve section in the absence of natural aeration measures for lintel, Cavitation morphology in typical opening of the value is shown in Fig. 3. The bottom edge cavitation occurs in the vortex at the tip of the bottom edge. Due to the strong turbulent shear action on the interface between the main flow behind the door and the rolling zone, the bottom edge cavitation is strengthened and developed in the shear layer, and its type is vortex cavitation.

The cavitation collapse zone is generally limited to 20.0 m behind the valve well., and the bottom edge cavitation collapse zone does not reach the lower access well even if the cavitation is particularly intense. The opening range of bottom edge cavitation is n = 0.2–0.8. The opening range of n = 0.4–0.7 is relatively strong, and the noise intensity is also large. The opening of n = 0.9 has no cavitation at the bottom edge.

Figure 4 shows the lintel shape and aeration pipe layout of Qianwei lock. The test shows that the noise intensity pulse is large and intensive when the door lintel is aerated, the noise intensity decreased significantly after aeration, and the measured noise intensity process line has a small amount of pulse. The pulse signal of 2# hydrophone is basically disappeared when the lintel single width aeration volume is only 0.018 m^3/s/m and the lintel total aeration volume is 0.072 m^3/s. The pulse signal of the 2# hydrophone is completely disappeared when the single width air flow rate of the lintel is only 0.044 m^3/s/m and the total air flow rate of the lintel was 0.176 m^3/s, The results showed that strong bottom margin cavitation (including lintel cavitation) is suppressed to a large extent after lintel ventilation, and lintel ventilation effect is remarkable.

Fig. 3. Bottom edge cavitation morphology when opening of valve is 0.6 (n = 0.6)

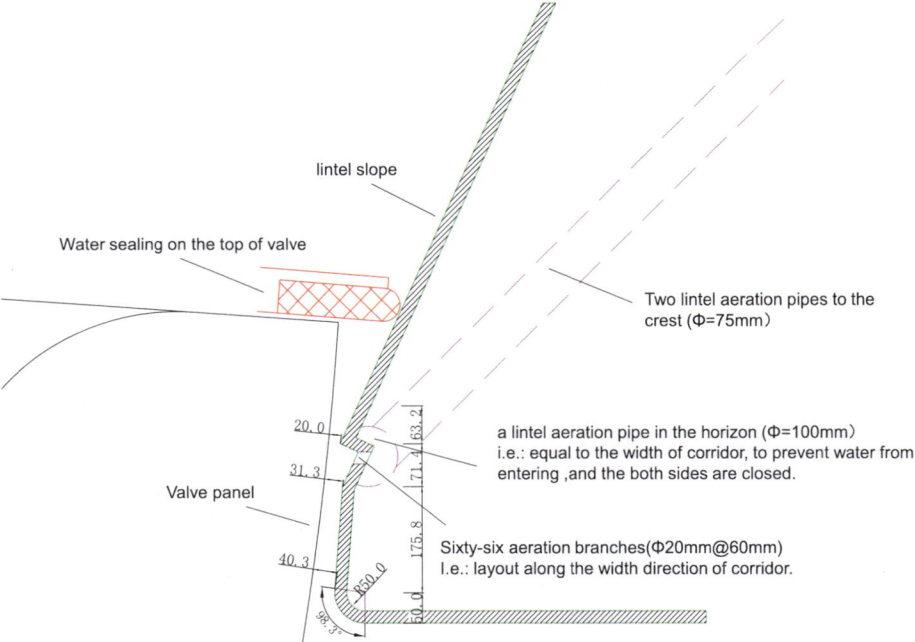

Fig. 4. Lintel shape of Qianwei Lock (unit: mm)

3   Prototype Observation

3.1   Effect of Natural Aeration of Lintel Inhibits Cavitation in the Section of Bilateral Filler and Drainage Valve

3.1.1   Section of Filler Valve

Water level during debugging is when the upstream water level is 334.4 m, and the initial water level of lock chamber is 318.2 m, working head of lock reach 16.2 m.

Operation mode: the filler valve is fully opened at the rate of $T_V = 4$ min, the valve is dynamic closed to $0.3°$ when the remaining water head is 0.6 m, When the water level inside and outside of the gate chamber is equal, Miter gate of the head bay is opened immediately.

If the aeration pipes of the valve lintel on both sides are blocked and the lintel is not aerated, the cavitation noise intensity in the upper left valve well and the upper right valve well is shown in Fig. 5, Early observation results show that hydrophone arranged in the position of valve well detected a relatively large pulse signal at the time when valve opening is in the early stages.

There are two reasons: seal head cavitation and the valve opening and closing system overcome the impact of the Static load changes to dynamic load at the beginning of the top valve seal separated from lintel, this kind of phenomenon is common in other high head lock, also is inevitable. The background elevation of cavitation noise signal of the left valve ($n = 0.05$–0.2) increases gradually, but the pulse is less, indicating that the bottom edge cavitation is not obvious before the opening is 0.2, when $n = 0.2$–0.8, the cavitation noise pulse is obvious and the intensity is high. After $n = 0.8$, the intensity decreases, only individual pulses when $n = 0.9$. The range of opening and the characteristics of the bottom edge cavitation in the prototype are similar to those in the model. The cavitation noise trend of the right valve is basically the same as that of the left valve, but the numerical value is slightly different. There is a significant cavitation pulse signal throughout the valve operation, the maximum is about 200 Pa. The "crackling sound" of cavitation collapse with two or three imperceptible low "muffled thunder sound" can be obviously felt at the top of the valve, which indicates strong cavitation is occurred in the water-filled valve section.

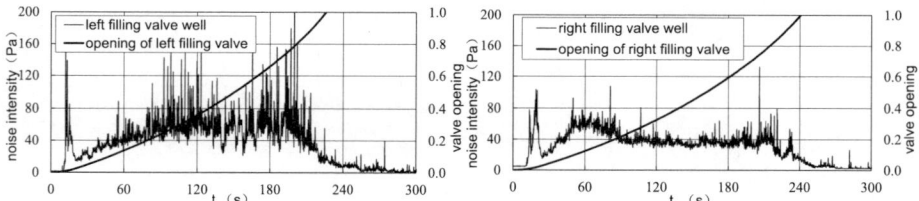

Fig. 5. Cavitation noise intensity in valve wells on both sides of bilateral water filler lock

The natural aeration of the lintel is realized by removing the plug of the pipe, The aeration flow is shown in Fig. 6. The lintel is well aerated with the maximum wind speed of the aeration pipe about 27.9 m/s and the maximum aeration flow of 0.169 m^3/s approximately, which is slightly larger than the aeration flow calculated by the physical model of the valve. The intensity of cavitation noise in the upper left valve well and the upper right valve well is shown in Fig. 7. There is a relatively obvious cavitation pulse signal when the valve is just opened, the maximum is about 93.1 Pa., It is caused by the seal head cavitation and the valve opening and closing system overcome the impact of the Static load changes to dynamic load at the beginning of the top valve seal separated from lintel as mentioned above. Subsequently, no cavitation pulse signal was detected in the valve wells on both sides during the valve operation process, and the process line was "stable". The average noise signal during valve opening was 7 Pa, and the maximum noise signal was no more than 15 Pa, Cavitation of the water-filled valve section was fully inhibited after the lintel is aerated, and no sound of cavitation collapse was appeared on the top of the valve section during the whole valve opening process.

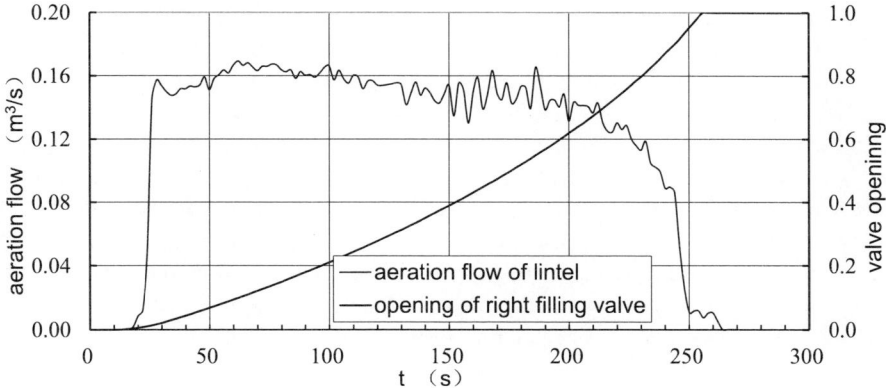

Fig. 6. Aeration flow of lintel of bilateral water filling valve of ship lock (natural aeration)

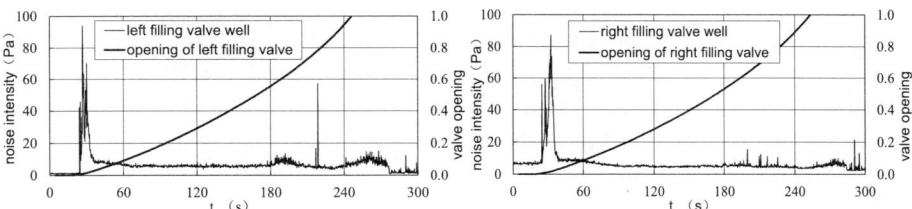

Fig. 7. Cavitation noise intensity with bilateral water filling process (natural aeration)

3.1.2   Drain Valve

Cavitation noise intensity contrast in the location of drain valve well is shown in Fig. 8 when lintel aeration or not, hydrophone arranged in the position of valve well detected a relatively large pulse signal at the time when valve opening is in the early stages (The reason is the same as above).

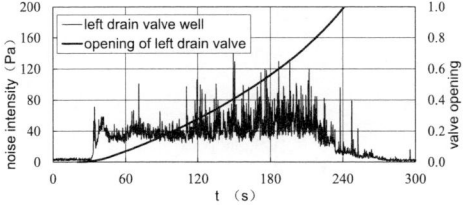

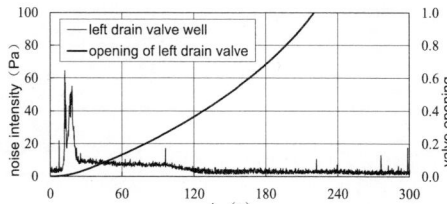

Fig. 8. Cavitation noise intensity on the left side

Fig. 9. Cavitation noise intensity (natural aeration)

During the entire valve operation process, the cavitation pulse signal is evident and obvious, cavitation noise signal background raised gradually increases on the left side of the valve when n = 0.05–0.2, n = 0.2–0.3 remain stable, n = 0.3–0.75 have an increasing trend and signal density, n = 0.7–0.8 signal gradually decreased, n = 0.9 only received individual signal pulse, the maximum noise intensity is about 142.0 Pa. The "crackling sound" of cavitation collapse can be obviously felt at the top of the valve, and 2–3 low "muffled sound" can be detected occasionally, and there is strong cavitation in the drainage valve section. The right side is basically similar to the left.

By removing the plug of the natural aeration pipe of the door lintel, the valve door lintel is aerated naturally with a maximum flow of 0.150 m³/s. The cavitation noise intensity of the lower left valve well is shown in Fig. 9. When the valve is just opened, there is an obvious cavitation pulse signal, about 64.1 Pa at most (The reason is the same as above). During the operation of the valve, no cavitation pulse signal was detected in the valve wells on both sides, and the process line was "stable". No abnormal cavitation sound was detected at the valve top section, and the cavitation of the drainage valve section was fully restrained after the lintel aeration.

3.2   Cavitation Characteristics of One Side Valve

3.2.1   Water Filling Valve Section

When water is filled on one side, the aeration flow of the valve lintel is shown in Fig. 10, and the maximum is 0.199 m³/s. The intensity of cavitation noise in the upper left valve well and the upper right valve well is shown in Fig. 11. The observation results show that when the valve is just opened, there is a relatively obvious cavitation pulse signal, about 125.2 Pa at most. In the process of valve operation, before n = 0.5 opening, there is basically no pulse signal detected in the valve well. Between n = 0.5 and 0.7, there is weak cavitation noise signal, and the whole process line is relatively "stable". During the valve opening process, no abnormal cavitation sound was detected on the top of the valve section. After natural aeration of the lintel, the original strong cavitation in the unilateral water filled valve section was fully restrained.

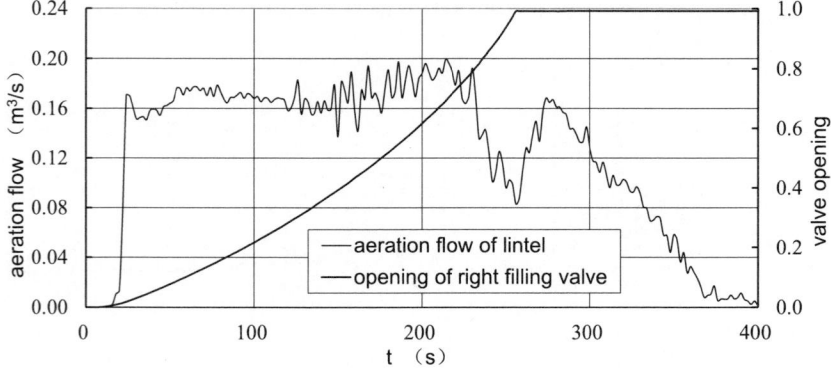

Fig. 10. Lintel aeration flow with unilateral water filling valve of ship lock

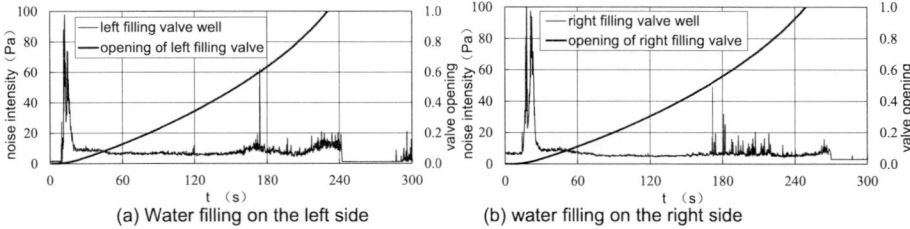

(a) Water filling on the left side   (b) water filling on the right side

Fig. 11. Cavitation noise intensity in valve wells on both sides of unilateral water-filled lock

3.2.2 Drain Valve Section

In the case of unilateral drainage, the aeration flow of the lintel is shown in Fig. 12, and the maximum is 0.160 m³/s. The intensity of cavitation noise in the lower left valve well is shown in Fig. 13. The observation results show that, except for the head cavitation of seal at the beginning of opening, no cavitation pulse signal is detected in the valve well during the operation of the valve, and its process line is "stable". During the whole valve opening process, no abnormal cavitation sound was detected at the top of the valve section. After the natural aeration of the lintel, the original strong cavitation in the unilateral drainage valve section was fully restrained.

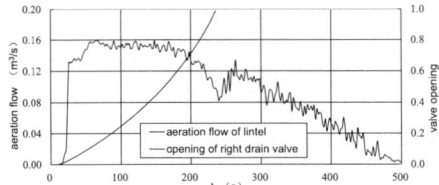

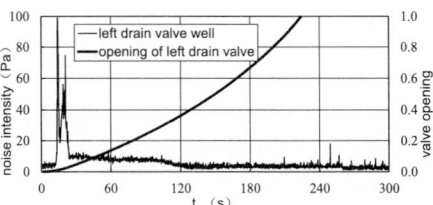

Fig. 12. Lintel aeration flow with unilateral discharge

Fig. 13. Cavitation noise intensity on the left side (unilateral discharge)

3.3 Effect of Natural Aeration of Lintel on Inhibiting Valve Hanger Rod Vibration

Figure 14 shows the longitudinal vibration acceleration of the suspender of the right water-filled valve when the lintel is aerated or not. The longitudinal vibration acceleration of the suspender decreases after the lintel is aerated. After natural aeration of the lintel, the original strong cavitation in the water-filled valve section was fully suppressed and the vibration of the valve hanger rod was improved.

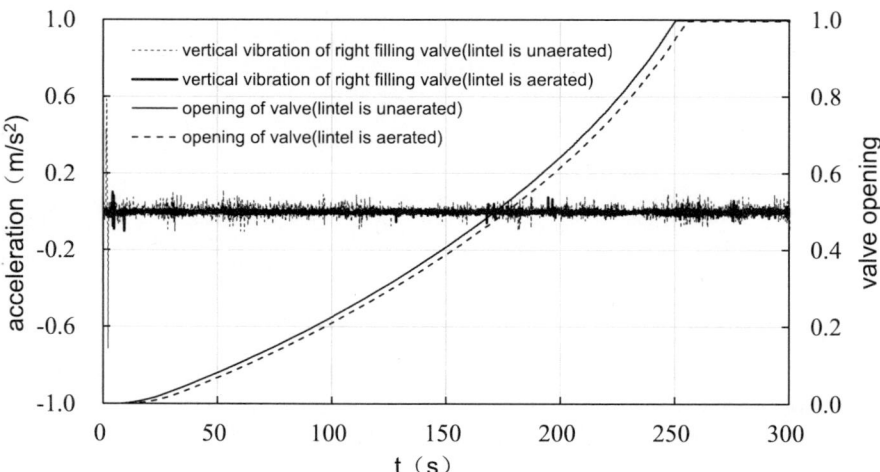

Fig. 14. Vertical vibration of the hanger rod of the right water-filled valve on lintel aeration or not

4   Conclusions

(1) Based on the hydro-dynamic model test of the valve, it is recommended that Qianwei lock with the water head of 19.0 m adopt the simple corridor shape of "flat bottom + gradual expansion at the top" with "small amount of work and simple construction", and give the natural aeration facilities and aeration pipe arrangement of lintel adapted to its cavitation characteristics.

(2) The model test showed that strong bottom margin cavitation (including lintel cavitation) was suppressed to a large extent after lintel aeration. Lintel aeration effect is remarkable. In ventilation, the noise intensity pulse is large and dense, After aeration, the noise intensity decreased significantly, and the measured noise intensity process line had a small amount of pulse. When the lintel single width aeration flow was only 0.018 m³/s/m and the lintel total aeration flow was 0.072 m³/s, the pulse signal of 2# hydrophone basically disappeared. When the single width air flow rate of the lintel was only 0.044 m³/s/m and the total aeration flow rate of the lintel was 0.176 m³/s, the pulse signal of the 2# hydrophone completely disappeared.

(3) Prototype observation shows that the effect of natural ventilation of lintel is significant, and cavitation of valve segment is fully inhibited after natural aeration of lintel. The natural aeration pipe of the door lintels was intentionally blocked when the two sides were filled with water. During the whole operation of the valve, there was a relatively obvious cavitation pulse signal, the maximum was about 182.3Pa. The "crackling sound" of cavitation collapse can be obviously felt at the top of the valve, and there is 2–3 imperceptible low "muffled thunder sound" occasionally, and the valve section has strong cavitation. After the natural aeration of the valve lintel, no cavitation pulse signal was detected in the valve wells on both sides during the operation of the valve, and its process line was "stable". During the

whole valve opening process, no sound of cavitation collapse was detected on the top of the valve section. The effect of natural aeration of lintel on inhibiting cavitation of valve segment is also significant.

(4) The comparison of suspender vibration of lintel ventilation or not shows that the vibration of valve suspender is improved after the natural aeration of lintel, because the original strong cavitation in the water-filled valve section is fully suppressed.

Acknowledgements. This research is supported by the National Key R& D Program of China (Grant No. 2016YFC0402006/04), Fundamental Research Funds for The Central Public-Interest Scientific Research Institutes (Major Project, Grant No. Y121009; Youth Foundation, Grant No. Y122009), China Postdoctoral Science Foundation (General Programs, Grant No.2021M701752), the Science and Technology Project of China Huaneng Group Co., Ltd. "Key Equipment Development and Engineering Demonstration (the first stage) of Megawatt Gravitational Compressed-Air Energy Storage System" (Grant number: HNKJ21-H33). Support from the funding agency is sincerely acknowledged.

References

Aydin MC, Ozturk M (2009) Verification and validation of a computational fluid dynamics (CFD) model for air entrainment at spillway aerators. Can J Civ Eng 36(5):826–836

Bhosekar VV, Jothiprakash V, Deolalikar PB (2012) Orifice spillway aerator: hydraulic design. J Hydraul Eng 138(6):563–572

Bian ZS (2000) Study on cavitation and acoustic vibration of water delivery valve section of Gezhouba Lock. Port Waterway Eng (07):34–37+58. (in Chinese)

Ceccio SL (2010) Friction drag reduction of external flows with bubble and gas injection. Annu Rev Fluid Mech 42:183–203

Hu YA (2009) Innovative technology and practice of cavitation-defense for high head lock valve. Nanjing Hydraulic Research Institute. Jiangsu Province. (in Chinese)

Li M, Bussonnière A, Bronson M et al (2019) Study of Venturi tube geometry on the hydrodynamic cavitation for the generation of microbubbles. Miner Eng 132:268–274

Lindau JW, Boger DA, Medvitz RB et al (2005) Propeller cavitation breakdown analysis. J Fluids Eng 127(5):995–1002

Luo X, Ji B, Tsujimoto Y (2016) A review of cavitation in hydraulic machinery. J Hydrodyn 28 (3):335–358. https://doi.org/10.1016/S1001-6058(16)60638-8

Paik BG, Kim KY, Ahn JW (2008) Influence on the rudder gap cavitation by the scaling of its clearance. Ocean Eng 35(17–18):1707–1715

Rhee SH, Lee C, Lee HB et al (2010) Rudder gap cavitation: Fundamental understanding and its suppression devices. Int J Heat Fluid Flow 31(4):640–650

Simpson A, Ranade VV (2019) Modeling hydrodynamic cavitation in venturi: influence of venturi configuration on inception and extent of cavitation. AIChE J 65(1):421–433

Wang X, Hu YA, Yan XJ, Wu B, Qian WX (2017) Experimental study on cavitation slice of top Slot of high head lock Valve. J Hydro-Sci Eng 04:14–19 (in Chinese)

Wang X, Hu YA, Zhang JM (2020) Experimental study of anti-cavitation mechanism of valve lintel natural aeration of high head lock. J Hydrodyn 32(2):337–344

Wu B, Hu Y, Wang X, Yan X-J (2020) Experimental and CFD investigations of choked cavitation characteristics of the gap flow in the valve lintel of navigation locks. J Hydrodyn 32 (5):997–1008. https://doi.org/10.1007/s42241-020-0065-6

Wu B, Hu YA, Wang X, Yan XJ (2021) Experimental and CFD study on the optimization of valve lintel's structural parameters under critical self-aerated conditions. Eng Appl Comput Fluid Mech 15(1):1629–1644

Wu J-H, Gou W-J (2013) Critical size effect of sand particles on cavitation damage. J Hydrodyn 25(1):165–166. https://doi.org/10.1016/S1001-6058(13)60350-9

Yan Y, Thorpe RB (1990) Flow regime transitions due to cavitation in the flow through an orifice. Int J Multiph Flow 16(6):1023–1045

Xue S, Yan XJ, Xiang MM (2021) Cavitation characteristics and defense measures of delivery valve section for 60-meter water-saving ship lock. Port Waterway Eng (02):12–16+43. (in Chinese)

Zhang J, Chen J, Xu W, Wang Y, Li G (2011) Three-dimensional numerical simulation of aerated flows downstream sudden fall aerator expansion-In a tunnel. J Hydrodyn 23(1):71–80

Propagation and Development of Nonlinear Long Waves in a Water Saving Basin

Xueyi Li$^{(\boxtimes)}$, Feidong Zheng, Duoyin Wang, and Ming Chen

College of River and Ocean Engineering, Chongqing Jiaotong University, Chongqing, China
{xy_lee,wdy}@cqjtu.edu.cn

Abstract. The water saving lock layout plays a key role in addressing the navigation hydraulic problems with high dams. However, the study of the propagation and development of nonlinear long waves induced by ship-lock operation in a water saving basin has received less attention so far. Specially, the mechanisms governing the formation of secondary waves and the impact of these waves on the impounds of the basin are still not fully understood. In the present study the entire evolution of a nonlinear long wave in a water saving basin was numerically simulated. The wave shape, wave celerity and wave force were analyzed. It was found that the leading edge of a long wave propagated along the water saving basin with a celerity which varied with time and space. Two distinct stages could be recognized and defined: a rapid acceleration phase characterized by a sharp increase in the celerity with propagation distance and a gentle acceleration phase where the long wave propagated in a more gradual manner. Moreover, the water surface slop of a long wave front equal to 0.045 could be used as an estimate of the occurrence of secondary waves. Furthermore, the present results highlighted the wave force on the impounds of a water saving basin was controlled by wave nonlinearity. These results may provide theoretical guidance and technical support for the hydraulic design and operation of water saving locks.

Keywords: Water saving lock · Nonlinear long waves · Secondary waves · Wave reflection · Wave dispersion

1 Introduction

Separated and integrated water saving basins are two typical layouts for water saving locks. A flexible water level division scheme is the main advantage of the former layout although it always needs a large usage of land. In contrast, a relatively small land area is needed in the integrated layout. However, the determination of water level division scheme is a rather complicated and must consider many factors, such as the fluctuations of upstream and downstream water level, the area ratio of water saving basins and lock chamber, the thickness of water saving basin floor, the minimum depth of water saving basin, and the minimum height of water saving basin.

In previous studies, the water levels at different locations in water saving basins were assumed to rise or drop in the same speed, thus the free surface was treated as a

© The Author(s) 2023
Y. Li et al. (Eds.): PIANC 2022, LNCE 264, pp. 565–577, 2023.
https://doi.org/10.1007/978-981-19-6138-0_49

plane. In fact, the operation of lock chambers is associated with temporal changes in the discharge, which increases up to a maximum value and subsequently decreases proportionally to the difference in water elevation between the basin and the lock chamber. Therefore, the operation of lock chambers leads to the formation of a long wave with a single wave crest or trough in the basin (Maeck and Lorke 2014). During the wave propagation, the positive part of the wave can evolve into secondary surges due to the combined effects of wave nonlinearity and dispersion (Castro-Orgaz and Chanson 2020; Soares Frazao and Zech 2002; Treske 1994; Zheng et al. 2018). As the secondary waves are superimposed on the original change in water level, they thus give rise to far larger maximum water level elevations (Benet and Cunge 1971). This phenomenon must be taken into account for the determination of the thickness of water saving basin floor, the minimum depth of water saving basin, and the minimum height of water saving basin (Treske 1994). Moreover, these waves will impose large impact loads on structural components (Zheng and Li 2021). In this study, the propagation of long waves and secondary waves in saving basins induced by the emptying operation of lock chambers were numerically investigated. The present results have the potential to provide some guidance for the design of water saving basins in integrated lock layout.

2   Numerical Method

2.1   Modeling

Based on a certain project, the length and width of the water saving basin were 337.56 m and 30.4 m, respectively. A rectangular water inlet was arranged in the longitudinal center of the basin with a length of 13.5 m and width of 4.5 m. As the longitudinal scale of the basin was much larger than the transverse scale, the basin was simplified as a two-dimension model in the vertical plane. Herein, only a half part of the basin was modeled by using a symmetric boundary. In the numerical model, the length and height of the basin were respectively 168.78 m and 6 m. The width and height of water inlet boundary were 1.78 m and 25 m, respectively.

After several mesh size distributions in the preliminary tests, the following grid scheme was used to achieve balance between the computing time and results quality. The maximum size of longitudinal x grid of water saving basin corridor is 0.02 m, and the maximum vertical grid is 0.02 m. The grid adopts quadrilateral grid, and the total number of grids is about 1.268 million, which is shown in the Fig. 1.

A specified mass inflow process is imposed at the bottom of the corridor, and the specific value is shown in the Fig. 2. The pressure inlet boundary condition P = Pa is adopted at the top of the basin. The symmetrical boundary is set at the symmetrical surface, and the remaining boundary is set as the wall. No-slip boundary conditions were applied to the walls in the simulation domain. The initial water depth in the water saving basin was selected on the basis of the actual engineering operation conditions, that is, $d_0 = 0.6$ m.

In conjunction with the renormalization group k–ε turbulence model, the numerical model solved the Reynolds-averaged Navier-Stokes (RANS) equations to predict the evolution of nonbreaking undular surges in a water–air system. The interface between water and air was captured using the volume-of-fluid method. An explicit scheme and second-order upwind scheme were applied for time and spatial discretization, respectively. The SIMPLE algorithm was adopted for pressure and velocity coupling.

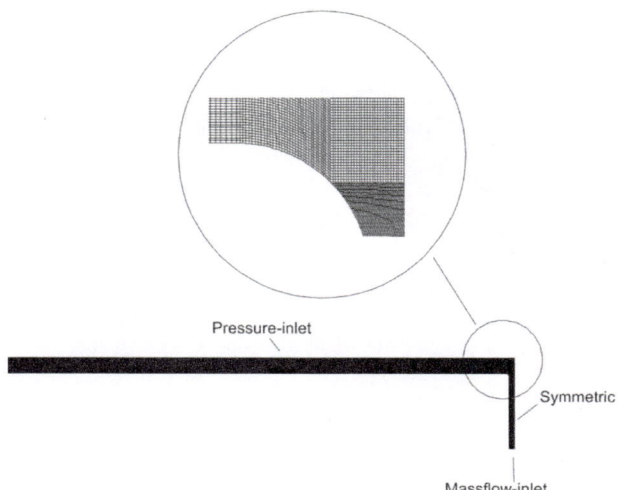

Fig. 1. Numerical model.

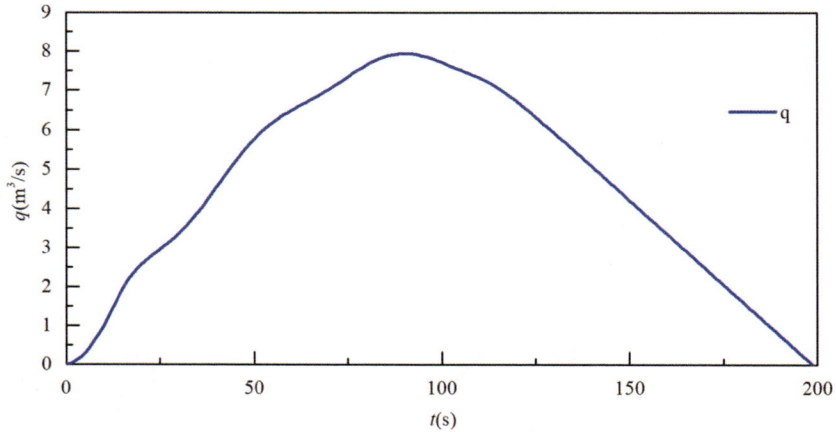

Fig. 2. Average discharge per unit width.

2.2   Validation

The formation of secondary waves is due to the continuous effect of wave nonlinearity in the long wave propagation, which leads to an enhanced water wave dispersion. Hence, the key to modelling secondary waves is therefore the ability of the numerical method used to effectively account for both wave nonlinearity and dispersion. Zheng and Li (2021) devised an experimental setup to validate a similar numerical model. Figure 3 provided a comparison between measured wave profiles and calculated results. Both the propagation of long waves and the evolution of secondary waves are well produced.

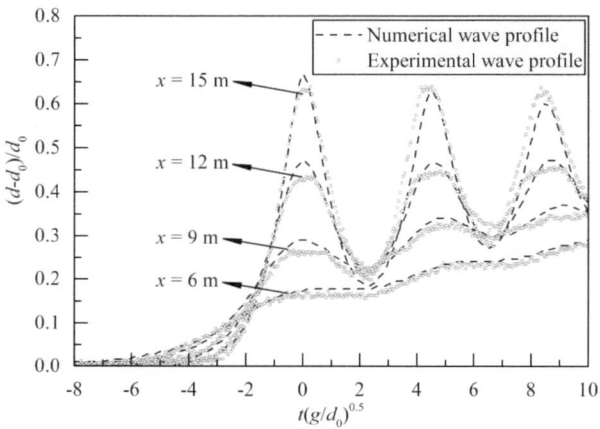

Fig. 3.   Comparison the results of experimental and numerical

3   Results and Discussion

3.1   Wave Profile

The emptying of the lock chamber generated a nonlinear long wave comprising a positive wave and a subsequent negative wave in the water saving basin (Maeck and Lorke 2014). When the long wave advancing in still water, its positive part tended to steepen due to wave nonlinearity and subsequently disintegrated into a series of well formed free-surface undulations (Soares Frazao and Zech 2002). The propagation of the secondary wave was accompanied by a significant wave amplification (Zheng et al. 2021a). After that, the secondary wave transferred into a breaking regime (Pelinovsky et al. 2015). During the further propagation of the secondary wave, its free-surface undulations disappeared and the wave evolved into a breaking bore (Chanson

2010a, Koch and Chanson 2009). The long wave and secondary wave were subject to positive reflection at the upstream and downstream impoundments of the water saving basin, which resulted into increased wave heights (Zheng et al. 2021b). After the first wave reflection, these waves propagated either in the flow direct or against an unsteady flow in the impounded basin. It should be noted that in this process, the change of the secondary wave type occurred due to the combined effect of wave nonlinearity and dispersion. Therefore, in the subsequent analyses, the investigation on the free-surface characteristics was restricted to the wave propagation prior to the secondary wave reflection at the impoundment.

The wave profile evolution before wave reflection is presented in Fig. 4. At $t = 5.17$ s, the surface sarcomport appeared at the top of the inlet of the water saving basin. At $t = 10.12$ s, the surface sarcomport region extended. At $t = 15.1$ s, a nonlinear long wave propagated to the left impoundment with the maximum water surface slope of the wave front of 0.022. At $t = 20.11$ s, the maximum slope of water surface slope reached 0.045. At $t = 25.1$ s, a series of well-formed undulations appeared at the wave front of the long wave. At $t = 30.11$–35.1, the propagation of the secondary wave was characterized by a considerable wave amplifications. At $t = 45.2$–55.2 s, the secondary wave showed some wave breaking and eventually evolved into a breaking bore. At this moment, the wave front of the long wave was characterized by a steep water wall. At $t = 60.2$s, the wave front arrived at the upstream impoundment.

Figure 5 exhibits the development of wave profiles after the first wave reflection. The secondary wave located at the wave front of the long wave showed an undular appearance immediately after wave reflection. At $t = 95.2$ s, the secondary wave exhibited some wave breaking immediately after its wave amplitude reached the maximum value. At $t = 115.2$ s, the free-surface undulations disappeared. The wave profiles after the long wave arrived at the downstream symmetrical boundary are presented in Fig. 6. Notably, the collisions of long waves facilitated the occurrence of undular secondary waves. In this context, the wave length of the secondary wave increased and the slope of the wave front decreased. The evolution of these waves after the second wall reflection is presented in Fig. 7. It can be seen that the wave amplitude of the secondary wave during wave propagation showed obvious wave attenuation due to the combined effect of wall friction and dispersion (Chanson 2010b).

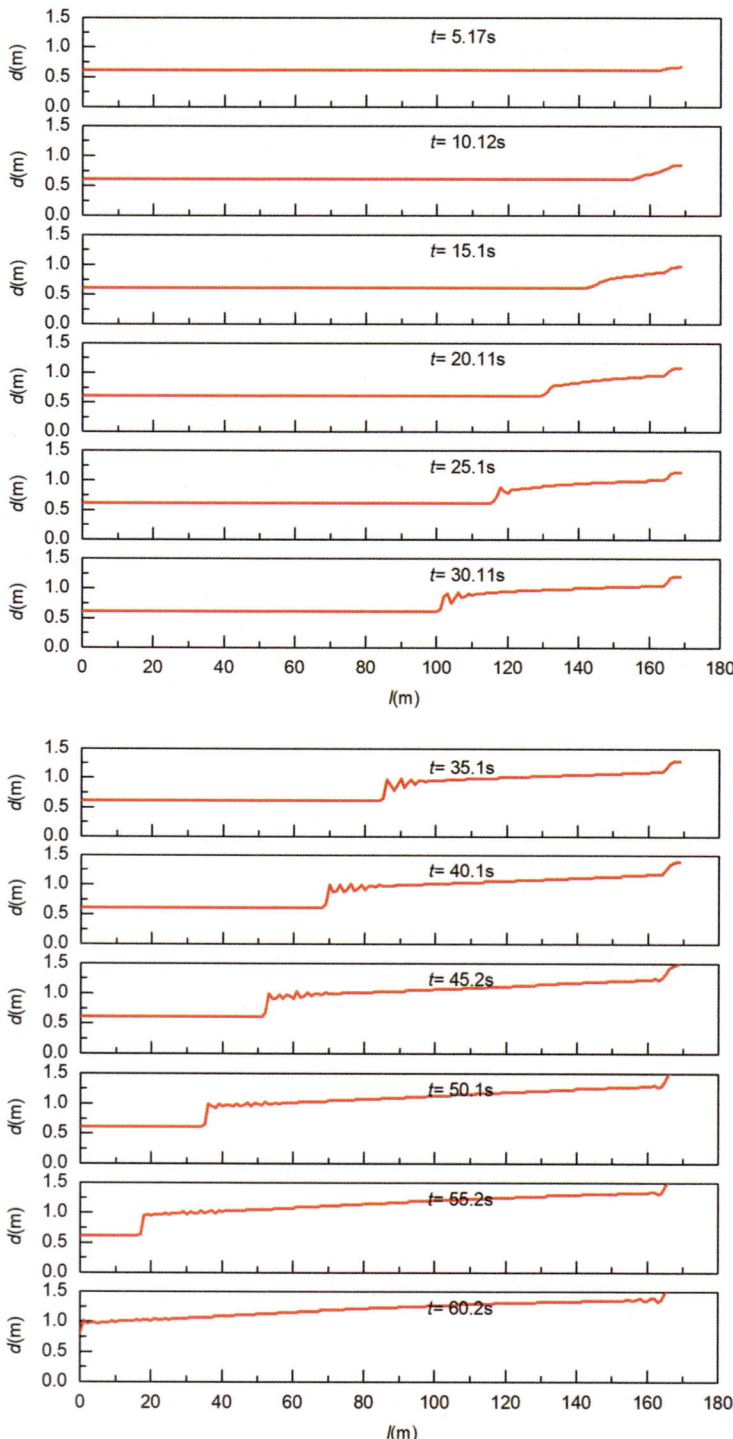

Fig. 4. Wave profile evolution before wall reflection.

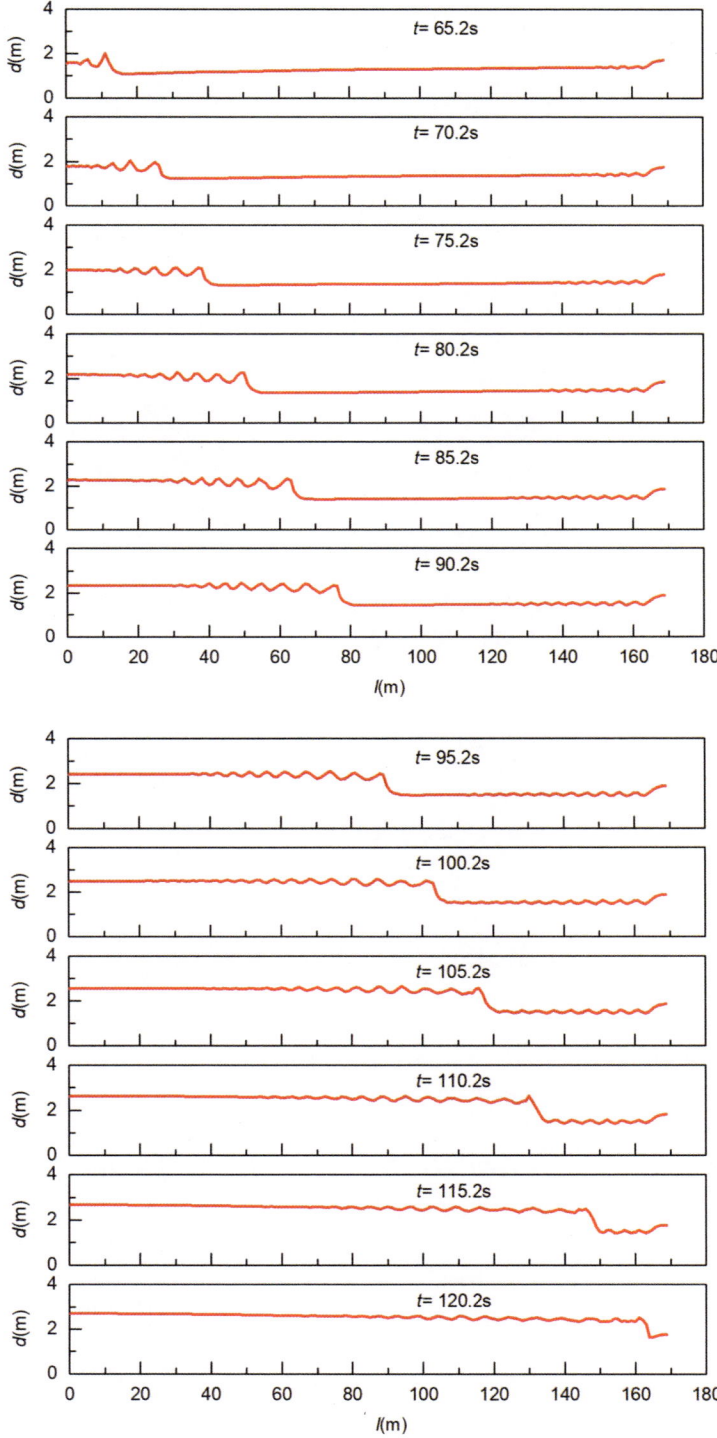

Fig. 5. Wave profile evolution after the first wall reflection.

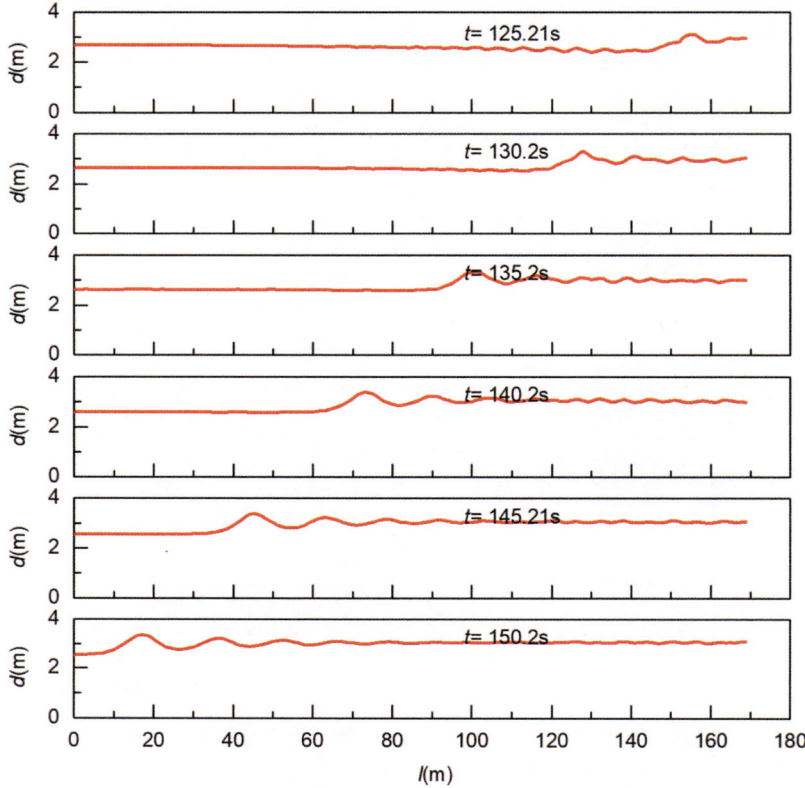

Fig. 6. Wave profile evolution after wave collision.

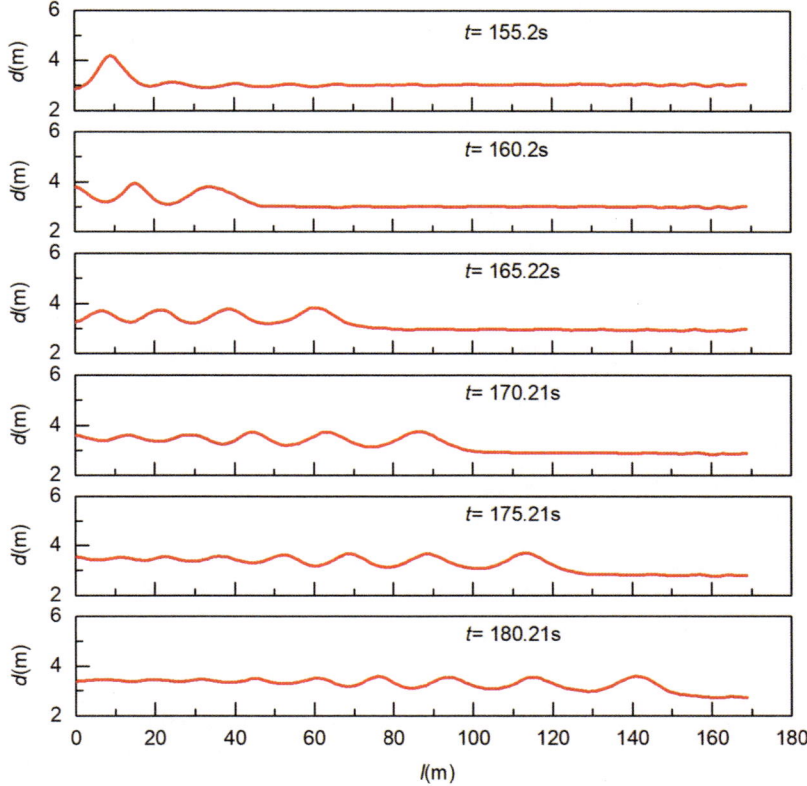

Fig. 7. Wave profile evolution after the second wall reflection.

3.2   Wave Celerity

The leading edge of a long wave propagated along the water saving basin with a celerity which varied with time and space. The evolution of long wave celerity in the water saving basin is shown in Fig. 8. C1 referred to the wave celerity of the long wave propagating in still water. C2 denoted the wave celerity of the long wave propagating from the upstream impoundment to the symmetric boundary immediately after the first wall reflection. C3 was the wave celerity of the long wave advancing from the symmetric boundary to the upstream impoundment. C4 was defined as the wave celerity after the second wall reflection. It can be observed that two distinct phases existed in the propagation process of the long wave in still water: a rapid acceleration phase characterized by a sharp increase in the celerity with propagation distance up to $x = 16.78$ m, and a gentle acceleration phase where the long wave propagated in a more gradual manner with a maximum wave celerity equal to 3.5 m/s. It is worth noting that the wave celerity at the phase transition was 2.5 m/s, which was close to the shallow wave celerity (i.e., 2.4 m/s). A sharp drop in the wave celerity was observed shortly immediately after the first wall reflection, i.e., C2 = 2.6. After that, the wave celerity showed some fluctuations during the subsequent wave propagation. C3 and C4 exhibited some irregular fluctuations over propagation distance with average values of 5.6 m/s and 5.2 m/s, respectively.

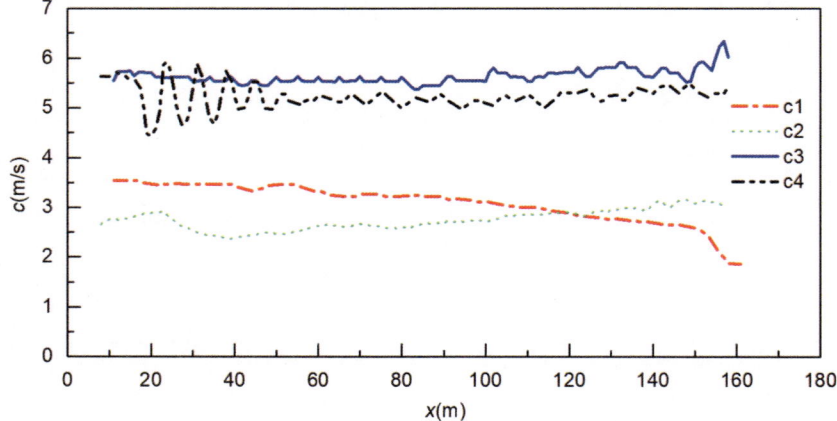

Fig. 8. Wave celerity evolution during wave propagation.

3.3 Wave Force

The temporary evolution of wall force at the upstream impoundment is present in Fig. 9. Considering the first wall reflection, the wave force profile exhibited a double maximum structure due to the nonhydrostatic effects resulted from wave nonlinearity (Cooker et al. 1997). The wall force increased significantly with the water level and reached the first maximum value before the maximum wave run up at the impoundment. Subsequently, the force profile presented a local minimum value near the maximum wave run up and the second maximum value. For the second wall reflection, the wall force tended to oscillated in phase with the free-surface undulations. In this situation, the wave nonlinearity was rather weak and only single peak in force distribution was exhibited (Chen et al. 2015).

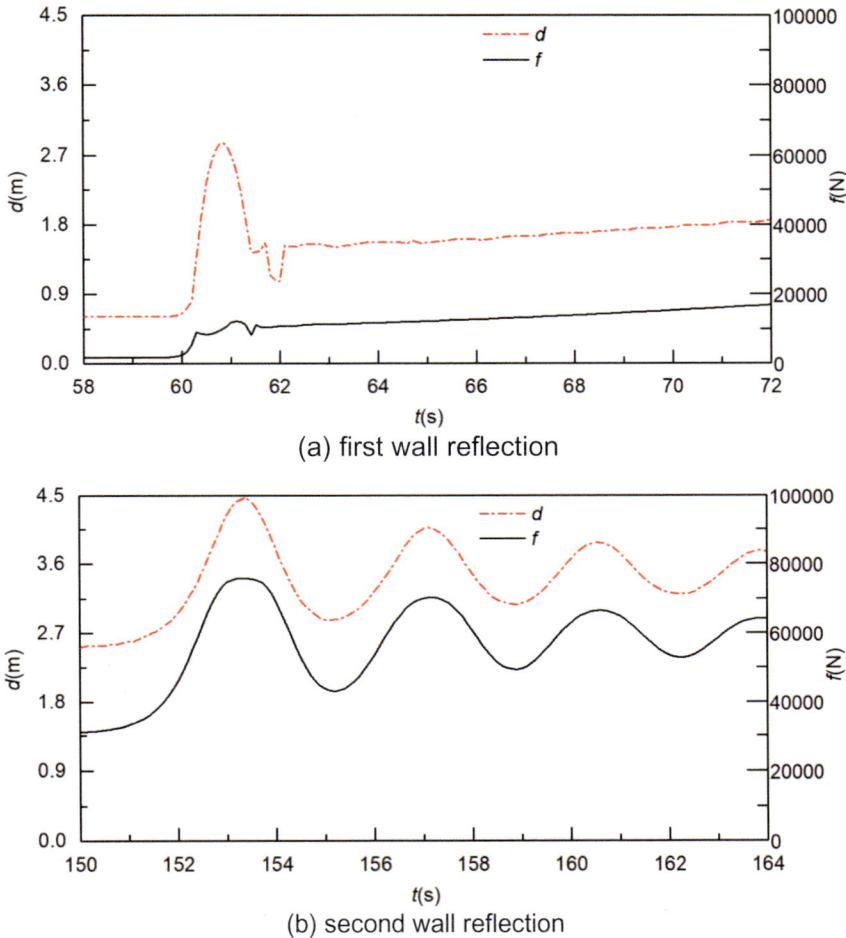

Fig. 9. Time evolution of free surface and wall force.

4   Conclusions

The basin in a water saving lock is an impounded channel where long waves generated by the filling and emptying operation of lock chambers exist. During the propagation of the waves along the basin, several complicated hydraulic phenomena can be observed, such as the appearance of secondary waves over the body of the long waves, the wave type transition of secondary waves, and the wave reflection at the upstream and downstream impoundments. In this study, the propagation and evolution of long waves induced by ship-lock operation in a water saving basin was numerically investigated. The results showed the leading edge of a long wave propagated along the water saving basin with a celerity which varied with time and space. Two distinct stages were observed in the propagation process of the long wave in still water: a rapid acceleration phase characterized by a sharp increase in the celerity with propagation distance up to 16.78 m, and a gentle acceleration phase where the long wave propagated in a more gradual manner. It was found secondary waves occurred on the wave front of the long wave when its slop exceeded 0.045. In addition, the present results demonstrated that the impact of water waves in a water saving basin on impoundments was strongly dependent on wave nonlinearity. The present results have the potential to provide some guidance for the optimum design of water saving basins in integrated lock layout.

Acknowledgements. This study was supported by the China Postdoctoral Fund (Grant No. 2021M700620); Natural Science Foundation of Chongqing, China (Grant No. cstc2021jcyj-bshX0049; cstc2020jcyj-bshX0043); Natural Science Funds of Chongqing (No. cstc2019jcyj-msxmX0759).

References

Benet F, Cunge JA (1971) Analysis of experiments on secondary undulations caused by surge waves in trapezoidal channels. J Hydraul Res 9(1):11–33

Castro-Orgaz O, Chanson H (2020) Undular and broken surges in dam-break fows: a review of wave breaking strategies in a Boussinesq-type framework. Environ Fluid Mech 20(6):1383–1416

Chanson H (2010) Undular tidal bores: basic theory and free-surface characteristics. J Hydraul Eng 136(11):940–944

Chanson H (2010) Unsteady turbulence in tidal bores: Effects of bed roughness. J Waterw Port Coast Ocean Eng 136(5):247–256

Chen YY, Kharif C, Yang JH, Hsu HC, Touboul J, Chambarel J (2015) An experimental study of steep solitary wave reflection at a vertical wall. Eur J Mech - B/Fluids 49:20–28. https://doi.org/10.1016/j.euromechflu.2014.07.003

Cooker MJ, Weidman PD, Bale DS (1997) Reflection of a high-amplitude solitary wave at a vertical wall. J Fluid Mech 342:141–158

Koch C, Chanson H (2009) Turbulence measurements in positive surges and bores. J Hydraul Res 47(1):29–40

Maeck A, Lorke A (2014) Ship-lock induced surges in an impounded river and their impact on subdaily flow velocity variation. River Res Appl 30(4):494–507

Pelinovsky EN, Shurgalina EG, Rodin AA (2015) Criteria for the transition from a breaking bore to an undular bore. Izv Atmos Ocean Phys 51(5):530–533. https://doi.org/10.1134/S0001433815050096

Soares Frazao S, Zech Y (2002) Undular bores and secondary waves - experiments and hybrid finite-volume modelling. J Hydraul Res 40(1):33–43

Treske A (1994) Undular bores (Favre-waves) in open channels—Experimental studies. J Hydraul Res 32(3):355–370

Zheng F, Li Y, Xuan G, Li Z, Zhu L (2018) Characteristics of positive surges in a rectangular channel. Water 10(10):1473

Zheng F, Wang P, An J, Li Y (2021) Characteristics of undular surges propagating in still water. KSCE J Civ Eng 25(9):3359–3368. https://doi.org/10.1007/s12205-021-0858-3

Zheng F, Wang P, Wang M, Zhang J (2021) The evolution and runup of nonbreaking undular surges. Mar Georesour Geotechnol 40(7):774–781

Zheng F, Li X (2021) Undular surges interaction with a vertical wall. Mar Georesour Geotechnol

Reliability Based Rehabilitation of Existing Hydraulic Structures

Arslan Tahir$^{(\boxtimes)}$ and Claus Kunz

Department of Structural Engineering, Bundesanstalt für Wasserbau, Karlsruhe, Germany
arslan.tahir@baw.de

Abstract. Existing hydraulic structures may show damage with increasing age and operation, so structural verification is crucial. In case of structural deficits, repair measures must be planned, and their effectiveness demonstrated. The advent of improved structural analysis methods and subsequent standardization processes facilitate the verification of existing structures to ensure sufficient reliability of infrastructure. Among the existing inland navigation hydraulic structures, older ship locks had been constructed with primitive construction materials such as damped plain concrete. At times, the structure exhibited neither any severe damages nor an indication of failure but failed to satisfy the limit states prescribed by the latest standards. This contribution considers a similar ship lock built in 1922 as a case study. The ship lock has a half-frame structural system with plain concrete gravity walls and a lightly transverse reinforced base slab. Cross-section based static verification revealed that the structure does not provide sufficient resistance in case of sliding and overturning limit states which could be attributed to crack and pore-water pressures in the cross-section. Consequently, rehabilitation of the lock walls with a vertical anchoring system was proposed to conform to required standards. Similar problems are expected for other existing locks in the German waterway system. Therefore, a methodology was developed to verify and to optimize the structural reliability of similar structures using full probabilistic methods while considering standard-based limit state functions. This involved uncertainty quantification of parameters for relevant loads (self-weight, water pressure, earth/ groundwater pressure, temperature, etc.) and materials (concrete, steel). To calculate the probability of failure and reliability indexes First Order Reliability Methods (FORM) was applied, considering its computational efficiency and more suitable for the presented Reliability-Based Design Optimization (RBDO) scheme. The contribution provides a probabilistic framework to study the influence of three aspects on the reliability of existing hydraulic structures, crack and pore-water pressures, operational conditions and lastly, the effect and optimization of rehabilitation in the form of anchoring.

Keywords: Existing hydraulic structures · Structural reliability · Rehabilitation · Optimization · Crack/pore-water pressures

© The Author(s) 2023
Y. Li et al. (Eds.): PIANC 2022, LNCE 264, pp. 578–590, 2023.
https://doi.org/10.1007/978-981-19-6138-0_50

1 Introduction

Hydraulic structures are critical elements of the inland navigation network, and the safe functioning of these elements directly influences the system's reliability. A considerable number of these structures are passing their planned service lifetime, i.e. more than 100 years, and based on safety reassessment, require minor repairs, rehabilitation or replacement. Since constructing these structures, several advancements have been made in codes, standards and limited state functions for verification. Three issues need to be reviewed during the safety reanalysis of these structures.

Firstly, construction-based limit state functions, i.e. gravity-based unreinforced concrete hydraulic structures, differ significantly from slender reinforced concrete structures DIN 19702 (2013) and BAW (2016), a guideline for existing structures, indicate several of these differences and provide recommendations to deal with them.

Secondly, compared to ancient approaches, Crack and Pore-Water Pressures (CPWP) have to be considered at a section level and hydraulic structure-specific resistance models developed. (Eugene and Victor 1995a; 1995b) concluded through a series of experimentations that CPWP exists and water penetration into a concrete crack is time-dependent, and water under pressure penetrates the fracture process zone of a concrete crack. This creates additional uplifting pressures against the gravity principle and destabilizes the structure. During reassessment of existing hydraulic structures, it was found that the stresses and forces due to CPWP were neither considered in the original design BAW (2016) and Kiesel (2018). Therefore, future safety assessments require an appropriate CPWP analysis method for solid hydraulic structures. Thirdly, in case of insufficient safety, the selection, modelling and application of efficient rehabilitation methods for the extension of service life (Brown 2015; USACE 2016).These methods include pre-stressed anchoring, local bolting anchors or passive anchoring.

We adapted the probabilistic methods rather than the code-based partial safety factor method to evaluate the safety. Employing a full probabilistic approach provide two significant advantages firstly, parametric uncertainty can be considered through probabilistic modelling of field data and secondly, its flexibility to adapt the design to a designer's desired target reliability. We adapted the First Order Reliability Method for our framework, primarily due to its computational efficiency.

Not much research or recommendations exist that consider these three aspects discussed earlier, and almost none exist that consider the reliability-based design under parametric uncertainty associated with hydraulic structures. The current study provides a practical solution with a primary focus on unreinforced concrete gravity-based systems experiencing CPWP and requiring rehabilitation. We developed a probabilistic framework that provides reliability-based optimized rehabilitation considering hydraulic structure-specific failure modes. Additionally, the framework allows the designer to design rehabilitation based on his desired extension in service life and risk acceptance rather than code dictated fixed service design life, i.e.100 years.

We provide a framework for the analysis model at the section level, which incorporates the CPWP and models the steel reinforcement based on the anchoring principle.

Section 2 of this paper presents the methodology for evaluating crack and pore-water pressures, followed by hydraulic structure-specific limit state functions in Sect. 3. The formulation of a reliability-based rehabilitation optimization problem is discussed in Sect. 4 in combination to target reliability for existing hydraulic structures. All presented methods are sewed together, and their application is presented through a case study of a typical 100-year-old structure in Sect. 5. Finally, the conclusion of the current study and discussion of the possible future outlook of work is presented.

2   Crack and Pore-Water Pressures (CPWP) in Rehabilitated Sections

DIN 19702 (2013) advises the analyst to consider crack and porewater pressures (CPWP) for hydraulic structures but provides a basic crack length and stress calculation method. The standard was further extended by BAW (2016) for existing hydraulic structures. The algorithm extends to cases with water on both sides and incorporates its effect on the effective normal forces and bending moment. Additionally, it provides calculations for the compressive stresses. Neither DIN 19702 (2013) nor BAW (2016) recommend the consideration of the tensile strength of concrete. Therefore, a section under tensile stresses is considered cracked and effective CPWP is applicable. Eventually, in cases where the resultant force acts outside the section base, the structure is considered unsafe. This is indicated in Fig. 1, part A, where the evolution of cracking is shown for an unreinforced section with increased tensile stresses (bending moment) and concrete tensile strength is ignored. Figure 1, part B, indicates the cases where the destabilizing tensile stresses are transferred through steel anchor and provide safety even in cases where the section is completely cracked. Figure 1, part C, shows all the sectional stress distribution and balance of forces for all possible scenarios from parts A and B. The essential contribution of this study is the extension of the BAW (2016) based analytical method for cases where anchoring is provided and evaluating the influence of CPWP in cases where the system is unstable.

This study considers the anchor as a point load contributing with its full strength to counter the destabilizing bending moments and sliding forces. Other designs and analyses to model steel reinforcement and CPWP are possible. The anchoring principle's major advantage is that no significant difference is seen in limit state functions and its probabilistic parameters in cases with/without CPWP and with/without rehabilitation, hence making the comparison considerably transparent.

3   Limit State Functions for Concrete Section Under CPWP

The stress distribution on a section level considering CPWP is further used in limit state functions to verify the safety of the structure. We consider three limits state function for the analysis of section considering different sections and operational load conditions. These include compression failure limit state function, eccentricity verification and sliding in construction joint. The last two are Ultimate Limit States.

3.1 Compression Failure

Most of the existing structures were constructed using non-standardized plain concrete, which has considerably variable and low compressive strengths compared to modern concrete. Additionally, the CPWP compounds the compressive forces increasing the probability of failure due to compressive failure. For all load combinations in the ultimate limit state (ULS) and in according to Code of Practice BAW (2016), the concrete compressive stresses shall be limited as follows:

$$\sigma_{cd}^* = \frac{3}{4} \cdot \sigma_{cd} < f_{cd} \tag{1}$$

with: σ_{cd} – Maximum stress value of the triangular compressive stress distribution.
σ_{cd}^* - Stress value of the simplified rectangular compressive stress distribution.
In case the entire section in under compression and no crack is present the stress in the system is calculated using the following equations

$$\sigma_{cd} = \frac{N_d}{h} + \frac{6 \cdot N_d \cdot e_d}{h^2} - H_2 \cdot \gamma_w \tag{2}$$

With water pressure on both sides, the effective concrete compressive stress σ_{cd} can be determined according to BAW (2016), Eq. (8), as follows:

$$\sigma_{cd} = \sigma_{abs} - H_2 \cdot \gamma_w \tag{3}$$

$$\sigma_{abs} = H_2 \cdot \gamma_w + \frac{4}{3} \cdot \frac{(N_{Ed} - h \cdot H_1 \cdot \gamma_w)^2}{[h \cdot (N_{Ed} - h \cdot H_1 \cdot \gamma_w) - 2 \cdot N_{Ed} \cdot e_d]} \tag{4}$$

with: H_1 - Water level on the side with compressive stresses in [m].
H_2 - opposite water level height from H_1 in [m].
γ_w - Weight of water in [kN/m^3].
e_d - load eccentricity under the design internal forces M_{Ed} und N_{Ed} in [m].
The final limit state function for reliability evaluation becomes

$$G(x) = \theta_R \cdot (\alpha_{cc} \cdot f_{cm}) - \theta_E \cdot \sigma_{cd}^* \tag{5}$$

where α_{cc} the coefficient to account for long-term effects on the concrete compressive strength.

θ_R is the coefficient of uncertainty in the resistance model and θ_E is the coefficient of uncertainty in the action model.

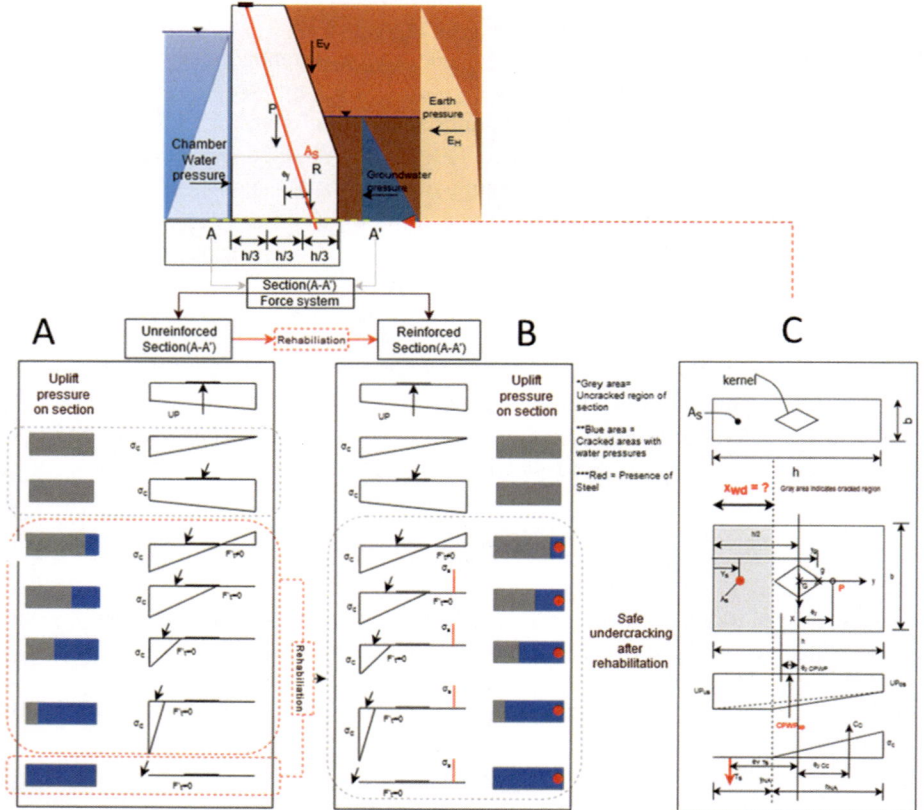

Fig. 1. Different possible states with CPWP in a rehabilitated concrete section

3.2   Eccentricity Based Design Check

Although not a standard-based verification, eccentricity determines the section's cracked state, which is vital in the calculation of CPWP and is also a geometrical indicator of the system stability. This limit state only includes loads with no consideration for material resistance. However, in case concrete tensile strength is not ignored, further modifications of the Limit State Function (LSF) can be transformed into a material based LSF. Furthermore, CPWP is a direct function of eccentricity, and in case the limit of h/2 as the half of the cross-section is breached, the section can no longer be verified by the methods discussed above.

The resultant force appears in three limit states corresponding to different tensile states and cracked based. Limit-1 is when the applied load eccentricity(e_d) to the section length (h) is $e_d/h < 1/6$, considering the case where a full section is under compression. Limit-2 represents cracked sections with $1/6 < e_d/h < 1/2$ and Limit-3 is when the section is fully cracked with $e_d/h > ½$. Furthermore, this aspect directly influences the effective area, an important aspect for calculation of CPWP and consequently N_{wd}, M_{wd} (revised normal/moments with CPWP).

Therefore, the equation for σ_{abs} for calculation of compression states is valid only if the following application limit is met:

$$\frac{e_{wd}}{h} = \frac{e_d/h}{1 - \overline{\sigma}_{wd}} < 0.5 \tag{6}$$

At an eccentricity $e_{wd}/h > 0.5$ there is a stability problem (no compression zone any more, load resultant N_{Ed} outside the cross section). Thus the limit state function for reliability evaluation becomes

$$G(x) = 0.5 \cdot \theta_R - \theta_E \cdot \frac{e_d/h}{1 - \overline{\sigma}_{wd}} \tag{7}$$

3.3   Sliding Failure in Construction Joints

As it is expected that the solid concrete hssydraulic structures were constructed in concreting steps. It can be assumed that construction joints were provided while construction. For these joints, the following verification must be performed with regard to ensuring the sliding force transmission:

$$\Sigma(V_E) \leq A_{cc} \cdot v_{Rd} \tag{8}$$

In this case, the design value of the sliding capacity in the joint v_{Rd} has to be determined following DIN EN 1992–1-1, by

$$v_{Rd} = c \cdot f_{ctd} + \mu \cdot \sigma_n \leq 0,5 \cdot v \cdot f_{cd} \tag{9}$$

where, c, μ and v are coefficients as a function of the joint roughness.

For usual cross sections the verification is performed assuming a "rough" joint $(c = 0,4 \, / \, \mu = 0,7/v = 0,5)$ according to the standard. But for the investigated lock lab tests for the evaluation of the joint roughness has been performed which lead to higher than the expected values. In the case of unreinforced concrete cross-sections, the load-bearing component of cohesion bond strength is completely omitted. The shear resistance is thus exclusively composed of the static friction (joint roughness) and normal stress components. Tensile strength of concrete is ignored in this contribution. Addition of the anchor leads to a modification of the above presented set of equations, since anchor is considered as a point load hence sliding resistance it provides is

$$V_{rd,anchor} = 0.5 \cdot A_{anchor} \cdot f_y \tag{10}$$

where f_y is the strength of the anchor and A_{anchor} is the cross sectional area of anchor. The final limit state function for reliability evaluation becomes

$$G(x) = \theta_R \cdot \left(A_{cc} \cdot \mu \cdot \sigma_n + 0.5 \cdot A_{anchor} \cdot f_y\right) - \theta_E \cdot \Sigma(V_E) \tag{11}$$

where θ_R is the coefficient of uncertainty in the resistance model and θ_E is the coefficient of uncertainty in action model. It must be noted that the anchor is expected to provide simultaneously sliding/shearing resistance and normal resistance to bending stresses.

4   Reliability Based Design Optimization

This phase has three components: reliability analysis of a structural system; secondly, computationally efficient optimization scheme for design variables; and thirdly, integration of the two aspects. Several methods exist for the first two components for reliability analysis (first/second-order reliability methods, subset simulations, importance sampling). Since an analytical set of equations for limits state functions has been derived, FORM due to its computation efficiency can be used. The probability of a failure and reliability analysis is essentially solving the following integral.

$$P_f = \Pr[g(\underline{\mathbf{X}}) \le 0] = \int_{g(\underline{\mathbf{X}}) \le 0} f_{\underline{\mathbf{X}}}(\underline{\mathbf{x}}) d\underline{\mathbf{x}} \tag{12}$$

where X is a random vector of input parameters with joint probability density function $f_X(x)$ and $g(X)$ is the limit state function (LSF).

Furthermore, Reliability Based Design Optimization (RBDO) model can be articulated as:

$$\min_{d,X} F(d, X) \tag{13}$$

$$\text{s.t.} \begin{cases} \Pr[L_i(d, X) \le 0] \le P_f^{\text{Target}}(i = 1, 2, \cdots, n) \\ C_j(d) \le 0 (j = 1, 2, \cdots, m) \end{cases} \tag{14}$$

where d is the deterministic design vector, X random vector is uncertain design vector, $F(d, X)$ is the optimization function, $L_i(d, X) \le 0$ is the limit state function for failure, P_f^{Target} is the target failure (allowable), $\Pr[L_i(d, X) \le 0]$ is the probability of failure under the given operational conditions and state, n is the number of uncertain constraints, $C_j(d) \le 0$ are deterministic constraint with m number of deterministic constraint.

Dissociated double loop approach was used. Where reliability analysis is conducted within the inner loop, we employed First Order Reliability Method FORM (Rackwitz and Fiessler 1978). Whereas the exterior loop, we used the golden search optimization algorithm to find the optimal point for design parameters. The approach was adopted for two reasons; firstly, the optimization problems can be reduced to one essential parameter, i.e. the area of steel, since we fixed the steel cover based on some dimensional requirements. Secondly, implementing the golden search algorithm with the FORM algorithm is computationally efficient in combination to multi limit state multi-operational conditions-based system reliability.

5 Target Reliability

The entire process depended on the P_f^{Target}, i.e. the allowable failure probability, which can be also translated into the reliability index β through the relation $P_{f_{\text{Target}}} = \Phi(-\beta_{\text{Target}})$ where Φ is the cumulative density function (CDF) of the standard normal distribution. Several aspects and approaches influence the evaluation of the target reliability. (DIN EN 1990, 2010) provides a target reliability index of 3.8 for 50 years service lifetime for new constructions. It has been adopted for solid hydraulic structures and 100 years service lifetime by DIN 19702 (2013). The discussion on modified reliability target becomes even more important as several of the existing structures do not fulfill partial safety factor design in codes for new structures. In BAW (2016), for example, modified partial safety factors for existing hydraulic structures are already considered.

6 Case Study

The considered structure is representative of a broad portfolio of similar structures in the German inland water navigation network. Lock Oldenburg, built in 1922, has a clear width of 12.0 m and a usable length of 105 m to accommodate ships with 1350 tons bearing capacity. The lock walls are constructed as a gravity wall from unreinforced rammed concrete. On the ground side, the walls are stepped upwards so that the chamber walls are 4.50 m thick at the bottom and reduced like stairs to a thickness of 1.50 m as shown in the Fig. 2. The geotechnical assessment carried out in 2007 indicated two layers of soil. The first layer from NN + 5.90 m (ground level) to NN −2.00 m has fine/medium sands of very low strength. The second layer from NN −2.00 to NN −4.00 m has fine/medium sands of low to medium strength.

The bearing capacity deficits in the northern chamber wall and the bottom, even using the favorable design approach EQU according to DIN EN 1990 /NA (2010). The lack of safety is primarily due to the crack and pore water pressure, usually not considered during construction. Therefore, a rehabilitation of the structure was necessary in this case through steel reinforcement. Figure 2 shows the old structure without rehabilitation and the designed rehabilitation configuration.

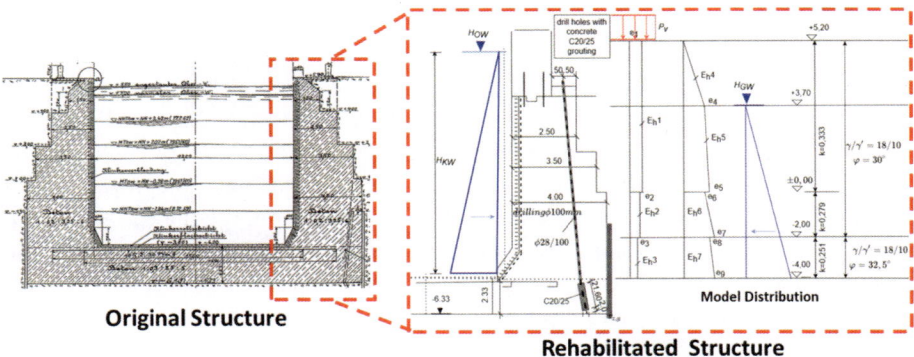

Original Structure

Rehabilitated Structure

Fig. 2. Structural configuration of a component pre and post rehabilitation

6.1   Probabilistic Design Parameters

Information regarding parameters of the models of CPWP and limit state function was gathered through different sources. For collected data, probabilistic modelling was conducted and appropriate distributions and their parameters were calculated. The following table indicates the selected distributions and parameters for the limit state functions and CPWP evaluation system.

Table 1. Probabilistic models for parameters of different limit state functions

	Parameters	Distribution	Mean	Std dev	Source
Concrete	Density of concrete wall	Lognormal	21.5	1.95	Test/Data
	Friction coefficient (tested)	Lognormal	2.59	1.09	Test/Data
	Concrete strength fcm	Lognormal	17.4	7.60	Test/Data
	Modulus of elasticity concrete	Lognormal	28300	2795	JCSS
	Surcharge loads	Gumbel	2.5	0.24	Literature
Soil layer1	Dry soil density	Lognormal	18	0.90	Test/Data
	Saturated soil density	Lognormal	10	0.50	Test/Data
	Angle of internal friction	Lognormal	30	2.65	Test/Data
Soil layer 2	Dry soil density	Lognormal	18	0.90	Test/Data
	Saturated soil density	Lognormal	10	0.50	Test/Data
	Angle of internal friction	Lognormal	32.5	3.19	Test/Data
Water levels	Ground water level	Gumbel	7.88	0.59	Drawing
	Case 1 no water in chamber	Constant	0		
	Case 2 water in chamber (tailwater)	Uniform	Min = 3.6 Max = 3.9		Drawing
	Case 3 water in chamber (headwater)	Uniform	Min = 8.5 Max = 9.5		Drawing
	Model uncertainty resistance	Lognormal	1	0.15	JCSS
	Model uncertainty actions	Lognormal	1	0.20	JCSS
Steel	Steel tensile mean strength	Lognormal	560	28.63	JCSS

6.2   Results

Reliability assessment was conducted for lock Oldenburg northern wall considering the CPWP and stress distributions shown in Fig. 1 and load models shown in Fig. 2, although with the probabilistic parameters indicated in Table 1. Table 2 indicates the reliability assessment conducted with three classifications. Class 1 is based on the significant operational conditions in a typical ship such as Oldenburg, i.e. revision with no water in the chamber, tail water level and head water level under operational conditions. Class 2 considers limit state functions, i.e. eccentricity, compression failure and sliding failure. Class 3 indicates the three different states of section analysis, i.e. State-1 represents the original design condition of the structure, which neither considered CPWP nor anchoring in the structure. This represents the safety level at the time of construction in the 1920ies. State-2 is when the structure is reanalyzed as an unreinforced section with CPWP; this state validates if the rehabilitation is required or not. State-3 represents cases where the structure is rehabilitated with anchor under CPWP. It must be noted that only state 3 is further used to optimize rehabilitation, whereas other states and their corresponding reliabilities are only used for comparative analysis. The anchor steel area for rehabilitation was designed according BAW (2016) with $a_s = 6$ cm^2/m and cover of 0.9 m from the earth side.

Table 2. Reliability levels achieved for different levels and anchor steel area of 6 cm^2/m

Level 1	Level 2	Level 3	Results	Crack base
Operation condition	**Limit State function**	**State of section** ***NA = no anchor *NP = no CPWP** ***YA = yes anchor *YP = yes CPWP**	**Reliability Index β**	
Case 1 −4 NNm (Revision situation)	Eccentricity	G (eccentricity-Case1-YA-YP)	3.89	
		G (eccentricity-Case1-NA-YP)	Not possible	
		G (eccentricity-Case1-NA-NP)	2.46	
	Compression	G (compression-Case1-YA-YP)	9.13	29%
		G (compression-Case1-NA-YP)	Not possible	100%
		G (compression-Case1-NA-NP)	8.09	Not required
	Sliding	G (sliding-Case1-YA-YP)	4.58	
		G (sliding-Case1-NA-YP)	Not possible	
		G (sliding-Case1-NA-NP)	5.83	
Case 2 −0.36 m + NN (3.64 m water) (Minimum Operational Water level, tail water)	Eccentricity	G (eccentricity-Case2-YA-YP)	3.76	
		G (eccentricity-Case2-NA-YP)	Not possible	
		G (eccentricity-Case2-NA-NP)	2.36	
	Compression	G (compression-Case2-YA-YP)	5.01	83%
		G (compression-Case2-NA-YP)	Not possible	100%
		G (compression-Case2-NA-NP)	8.43	Not required
	Sliding	G (sliding-Case2-YA-YP)	8.33	
		G (sliding-Case2-NA-YP)	2.66	
		G (sliding-Case2-NA-NP)	5.99	
Case 3 + 5,0 m NN (9 m water) (Maximum Operational Water level, head water)	Eccentricity	G (eccentricity-Case3-YA-YP)	8.91	
		G (eccentricity-Case3-NA-YP)	3.77	
		G (eccentricity-Case3-NA-NP)	1.87	
	Compression	G (compression-Case3-YA-YP)	14.41	0%
		G (compression-Case3-NA-YP)	14.69	0%
		G (compression-Case3-NA-NP)	14.83	Not required
	Sliding	G (sliding-Case3-YA-YP)	11.66	
		G (sliding-Case3-NA-YP)	6.84	
		G (sliding-Case3-NA-NP)	8.53	

From the above analysis, it can be concluded that water level operations significantly influence the reliability of the structure. The most critical load case is case 2, which represents the chamber's minimum operational water level. In cases where eccentricity and compressive failure without anchors and considering CPWP was considered, no results were possible. The primary reason being the entire section is under tensile stresses, and the entire section is cracked. Consequently, the method presented in Fig. 1 and their corresponding equations provide numerical instabilities. This is indicated by the state of cracking in the last column in Table 2, where the percentage of base crack is shown for each load case, using the deterministic mean values. This further supports why eccentricity or overturning should be considered as a limit state function for structures under CPWP.

The results presented in Table 2 indicate results for only one steel design dimensions. We use the optimization routine discussed in the section earlier and conclude that for a supposed target reliability index $\beta = 3.8$ the optimized anchor steel required for eccentricity LSF is 6.25 cm^2/m, compression LSF is 4.75 cm^2/m, and sliding LSF is 4.92 cm^2/m, and consequently, therefore for the entire system 6.25 cm^2/m.

The following graph in Fig. 3 indicates the minimum reliability and probability of failure achieved for different operational conditions and limits state functions and systems for a variable steel content in the cross section. At the lower steel content, the difference between the different limit state functions is low. A lower steel content makes compression failure a dominant failure condition, and an anchor area of 3 cm^2/m and lesser has no significant influence on the reduction of reliability.

Because of a code-based ignorance of tensile strength eccentricity check remains the most critical limit state function hence the dominant failure mode for system reliability. A lower steel content structure is safer in sliding than compression failure. This could be attributed to the resistance doubling effect of the anchoring to the system. Firstly, due to the reduction of crack, CPWP and increased effective normal forces. Secondly, the anchor provides shear resistance to the horizontal applied forces. As the steel content increases in the cross section, the crack in the section closes, and applied compression is reduced exponentially, making compression the least probable failure mode.

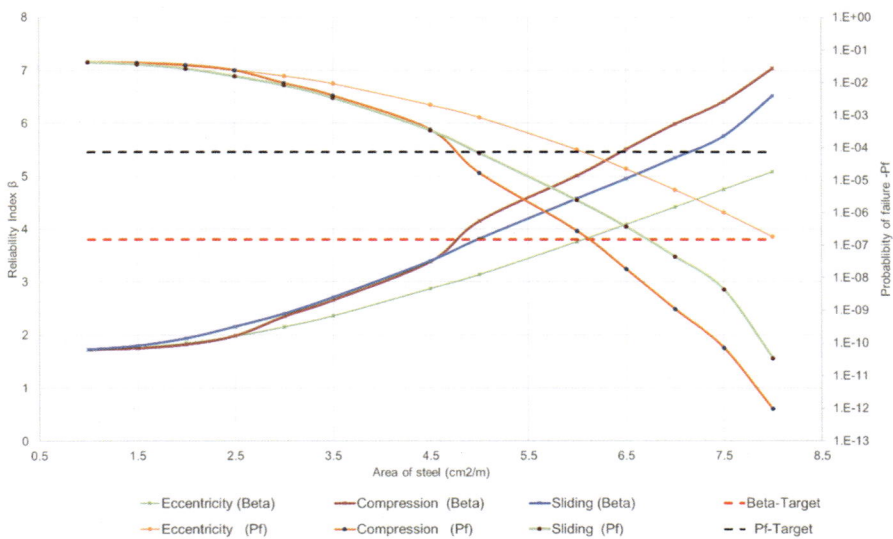

Fig. 3. Evolution of reliability of different limit state functions with area of steel in anchor

7   Conclusions

This contribution presents a practical solution through a framework with its field application for the impending problem of deficits in the reliability of existing concrete hydraulic infrastructure. The framework provides an agile reliability-based approach which provides the flexibility on an extension of service life and simultaneously provides a stepping stone for risk-based planning of assets at the network level. An example of lock Oldenburg represents structures that are a significant part of the existing inland navigation infrastructure throughout the world. These structures are nearing their planned service lifetime, and with good performance, reanalysis and reliability based minor rehabilitation they can extend their lifetime hence saving financial and environmental costs. The novelty of the contribution is the solution it provides to address the emerging issue of crack and porewater pressures in solid hydraulic structures, considering structure-specific uncertainties and integrating the anchoring system into the framework. Furthermore, the optimization scheme momentarily only for a single parameter since the primary target of the study was validation of the developed concepts and application of the framework was presented. However, the number of optimization parameters can be increased for complex problems and different structural components and systems. The authors believe that three aspects are of high importance and require further considerations. However, concrete split tensile tests indicate that the concrete has some tensile strength, but following the standards and codes, tensile strength is ignored, leading to a conservative verification. The presented method can be further modified to the inclusion of tensile strength of concrete. Sliding friction coefficients achieved during the testing of this exemplary lock structure was considerably higher than the literature-based values. Therefore, further investigating of the appropriate evaluation of the coefficient, which might significantly influence the results of the presented work is needed. Lastly, the anchoring principle presented in this contribution is one of the methods to model and assess the reliability of rehabilitated sections under crack and porewater pressures. The authors dealt with two further methods for the reliability assessment which will be presented in the future.

References

Kiesel A (2018) Ansatz von Riss- und Porenwasserdruck bei massiven, unbewehrten Wasserbauwerken gemäß BAWMerkblatt TbW. BAWBrief 1

BAW (2016) BAWMerkblatt: Bewertung der Tragfähigkeit bestehender, massiver Wasserbauwerke (TbW). Bundesanstalt für Wasserbau (BAW), Abteilung Bautechnik, Karlsruhe

Brown ET (2015) Rock engineering design of post-tensioned anchors for dams – a review. J Rock Mech Geotechnical Eng 7:1–13. https://doi.org/10.1016/j.jrmge.2014.08.001

DIN 19702 (2013) Massivbauwerke im Wasserbau - Tragfähigkeit, Gebrauchstauglichkeit und Dauerhaftigkeit. DIN 19702, Beuth Verlag, Berlin

DIN EN 1990 (2010) Eurocode: Grundlagen der Tragwerksplanung. DIN EN 1990, Beuth Verlag, Berlin

Eugene B, Victor S (1995a) Water fracture interaction in concrete part 1 fracture properties. ACI Mater J 92

Eugene B, Victor S (1995b) Water fracture interaction in concrete part ii hydrostatic pressure in cracks. ACI Mater J 92

Rackwitz R, Fiessler B (1978) Structural reliability under combined random load sequences. Comput Struct 9:489–494. https://doi.org/10.1016/0045-7949(78)90046-9

USACE (2016) Technologies to Extend the Life of Existing Infrastructure, vol 1. USACE Engineer Research and Development Center, Georgia, USA

Research Developments in Hydrodynamics of Ships Entering and Leaving the Tank of a Ship Lift

Luzhidan Fu[✉], Yaan Hu, and Zhonghua Li

Key Laboratory of Navigation Structure Construction Technology,
Nanjing Hydraulic Research Institute, Nanjing, China
503867214@qq.com

Abstract. In the design process of a ship lift, the size of the tank and the depth of water inside the tank are always minimized under the premise of satisfying the standard of ship size and freight volume, in order to reduce the power of electrical driving system, the construction cost of the project and the difficulty of equipment manufacturing, as a result, the cross-section coefficient of the tank is small. Therefore, the up-and downstream docking between the tank and the reaches of canal involves many related hydraulic problems, for instance, opening and closing of the vertical lift gates, entering and leaving of the tank, are very complicated. The docking process is an important link in the whole operations of the ship lift. At the same time, the change of unsteady flow of the hub will also affect the safety and efficiency of the docking process. This paper first briefly introduces the docking process of different types of ship lift, and systematically summarizes the scientific problems involved in the docking process when the ship enters and leaves the tank. Secondly, combined with the research results obtained from a variety of methods such as theoretical analysis, physical model test, mathematical simulation calculation and real ship test of previous researchers, the hydrodynamic research progress of ships entering and leaving the tank is introduced. In summary, the design criteria for the water depth of the ship lift's tank obtained on the basis of the progress of the hydrodynamic research of the ship entering and leaving the tank are introduced above.

Keywords: Ship lift · Research development · Hydrodynamics

1 Introduction

The ship lifts play an important role in river channelization and communication of different water systems in inland water transportation. Compared with ship locks, ship lifts have obvious advantages, especially suitable for high dam navigation, canal navigation between different water systems. The process of ship entering and exiting the ship lift chamber is one of the core links of the whole process of the ship lift operation. Due to the high blockage ratios of the ship lift chamber, the hydraulics problem becomes a complex three-dimensional water flow problem coupling with ship operation. To prevent the ship from touching bottom it is necessary to formulate reasonable ship draught control standards and ship navigation methods based on the

© The Author(s) 2023
Y. Li et al. (Eds.): PIANC 2022, LNCE 264, pp. 591–598, 2023.
https://doi.org/10.1007/978-981-19-6138-0_51

ship squat and the water fluctuation of the ship lift chamber, so as to ensure the safety of ship navigation and ship lift docking.

Some scholars proposed the practical methods for calculating the squat with the general ship characteristics (Tuck 1973; Husska 1976; Eryzulu 1978; Icorels 1980; Barrass 1981; Norrbin 1985; Millward 1990; Millward 1992; Eryzulu, 1994; Gourlay 2008;). Bao (Bao 1991) presented the formula of maximum squat in ship exiting the chamber according to the previous studies in China. The German Federal Waterway Design and Research Institute carried out the field investigations and the numerical calculation on the extra-large ships entering and exiting the Lüneburg ship lift. NHRI (Hu et al. 2011) advanced the formula of Bao based on the experiments of the Three Gorges ship lift.

The current studies at home and abroad mainly combine with the concrete engineering lack of the universality. In addition, most of them uses the formulas of ordinary restricted channels to make calculations, without considering the plane dimension influence of narrow navigation, enclosure and so forth.

In conclusion, this paper introduces the docking process of different types of ship lift, and systematically summarizes the scientific problems involved in the docking process when the ship enters and leaves the tank. Explain in detail the variation law of the ship's sailing resistance characteristics during the process of entering and leaving the tank. Describe the variation law of water surface fluctuation in the tank during the process of ships entering and leaving, analyze the relationship between the fluctuation characteristics and the ship speed, cross-section coefficient, entering or leaving, the water level in up-and downstream docking process. Summarizes the variation characteristics of water surface in the tank. Introduce the main influencing factors of ship squat, including ship size, ship draught, ship speed, tank size, tank water depth, etc. Research results and the relevant empirical formulas are summarized in chronological order according to different research methods. Outline the impact of the ship's entering and leaving of the tank on the characteristics of tank's load change and the force characteristics of the ship lock mechanism.

2 The Docking Process of Ship Lifts

Ship lifts come in a variety of classifications. According to whether the ship tank going into the navigation pool or not, it can be divided into two types: tank-launching ship lift and non-tank-launching ship lift. For the non-tank-launching ship lift, the upstream and downstream docking processes are consistent. For the tank-launching ship lift, the upstream docking process is the same as that of the non-tank-launching ship lift, but the downstream docking process is special.

The tank of the tank-launching ship lift enters the downstream approach channel directly, eliminating the need of the lower lock heads and its corresponding equipment, which not only saves the quantity of work, but also eliminates some auxiliary equipment such as retaining and sealing mechanism; meanwhile, the procedure of the process is greatly reduced, the time ships take to pass the dam is shortened, the operational reliability, passing capacity and operational efficiency are all increased a lot, compare to the non-tank-launching ship lift. The main factors affecting the hydrodynamic characteristics of tank-launching ship lift's process which tank going in and out the navigation pool is: the shape of the ship lift's tank, the size of the navigation pool, the speed of the tank going in and out, and the size of the approach channel.

The docking process between the ship and the upstream and downstream is an important part of the whole process of the ship lift operation. It involves problems like the hydrodynamic characteristics of the ship lift's tank when the tank's door is opened and closed; the ship navigation conditions and hydrodynamic characteristics in the tank during the process of ships entering and leaving; the change affections of downstream unsteady flow on the operation of the ship lift and other related hydrodynamic problems.

3   The Process of Ships Entering Leaving the Tank of a Ship Lift and its Hydraulic Problems

During the ship entering the tank, the bow kept pushing the water flow into the tank, which leads to the water level of tank gradually rises, and the water level of the stern lowers, forming a fore and aft water level difference, causing reverse flow of water on the ship's side and the bottom of the ship. The ship trim and sink then appears. When the ship completes the entrance and parks in the tank, the water surface decreases due to the inertia of the flow in the tank, yet the stern water is connected to the channel, with the supplement by the channel water, the water level drops less, and the stern sinks. During the ship leaving the tank, the bow water is connected to the channel, the bow kept pushing the water flow into the channel and the water level in the tank gradually decreases. Although the approach channel replenishes the water in the tank, but because of the ships blocking while the water in the stern is small, water supply insufficient, the water level drops more obviously, and the stern sinks is also larger.

The safety of the ship in the tank is one of the most concerned issues in the design and research of the ship lift as the cross-section coefficient of the tank is small. It is necessary to determine the reasonable water depth of the tank and the way of ships' navigation to prevent the bottoming and the drastically changing of the tank' load when the ships entering and leaving the tank, and to ensure the safety of the ship and the tank. Using the combination method of mathematical model and physical model and real ship test, the hydrodynamic characteristics of the ship entering and leaving the ship lift at different speeds are studied to determine the reasonable water depth of the tank and provide a basis for reasonable navigation.

Using the combination method of mathematical model and physical model and real ship test, the hydrodynamic characteristics of the ship entering and leaving the ship lift at different speeds are studied to determine the reasonable water depth of the tank and provide a basis for reasonable navigation (Figs. 1 and 2).

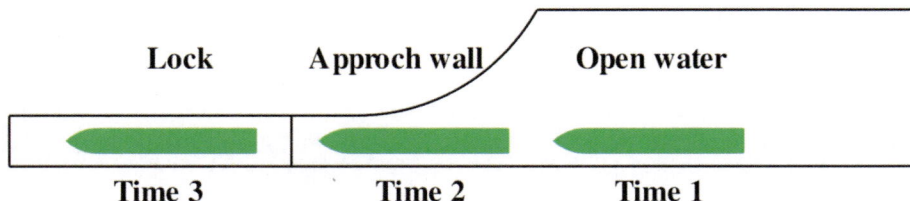

Fig. 1. Schematic diagram of three typical moments of ship entering the tank

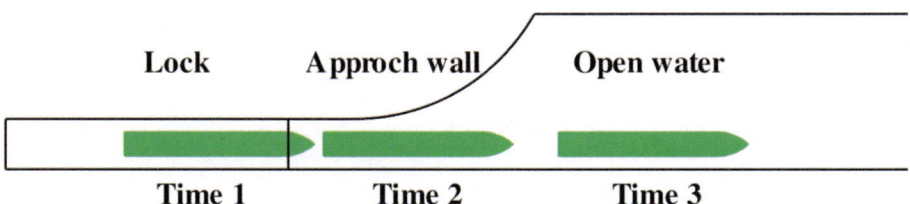

Fig. 2. Schematic diagram of three typical moments of ship leaving the tank

4   Hydrodynamics Study of Ships Entering and Leaving the Tank

4.1   Characteristics of Ship's Sailing Resistance

During the process of ship entering and leaving the tank, the ship's navigation environment is narrow and shallow, which lead to the flow field around the hull presents obvious bank effect and shallow water effect. The total resistance, lateral force, pitching moment and turning moment of the ship increase as the speed of the ship increases, and the speed of the ship has a significant impact on the force of the ship. Therefore, from a safety perspective, it is recommended that the ships enter and leave the tank at a lower speed. Meanwhile, when the ship enters the approach channel from the open water, the asymmetry of the flow field on both sides of the ship increases due to the narrowing of the water on both sides of the ship, which result in the rapid increase of lateral force and the turning moment. Therefore, it is more likely hit the wall or turn when ship sailing at this moment. Anti-collision measures can be considered in practical work.

4.2   Characteristics of Water Surface Fluctuation in the Tank During the Process of Ships Entering and Leaving

When the ship leaves the tank, the water pushed out by the ship in the rear of the stern can only be replenished by the narrow space around the ship and the tank, resulting in a drop in the depth of the tank and the channel behind the ship. Obviously, the closer to the approach channel, the greater the maximum drop in the water surface, and the maximum drop in the water surface in the tank basically occurs before the bow completely leaves the tank. When the ship enters the tank, the distribution of the fluctuation of the water surface in the tank is opposite to that of the ship's leaving. The closer to the closed end of the tank, the greater the fluctuation of the water level, and the maximum drop in the water surface of the tank appears after the ship enters the tank. The ship speed in and out of the tank has a greater impact on the maximum reduction of the tank's water surface. The maximum decrease in the water surface increases as the speed of ship increases.

4.3 Characteristics of Ship Squat During the Process of Ships Entering and Leaving

The ship's speed has a significant impact on the ship squat. At the same speed, the squat of the ship when leaving the tank is larger than that when entering the tank. Therefore, the ship squat when sailing out of the tank is the safe control condition for the ship to enter and leave the tank without the bottoming. As the tank's water depth decreases, the ship squat also decreases, the influence of the tank's water depth is less than the impact of the ship's speed on the squat. Under the same depth of the tank and the speed of the ship in and out, the larger the ship's tonnage, the smaller the section coefficient of the tank, and the greater the squat of the ship (Fig. 3).

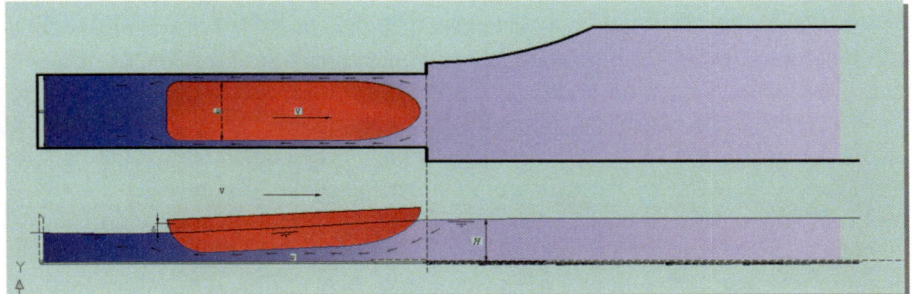

Fig. 3. Schematic diagram of ships leaving chamber

The relationship between the maximum squat of the ship (δ), entering and leaving the tank, and the water depth (H), section coefficient (n), and speed of the ship (V) can be expressed by two dimensionless quantities:

$$P = \frac{\delta}{H}, \; K = \frac{V^2}{2gH} \times \left[\left(\frac{n}{n-1}\right)^2 - 1\right] \tag{1}$$

in the range of 0.3–1.0 m/s, the maximum squat of the ship is basically linear with the square of the ship speed out of the tank (Fig. 4).

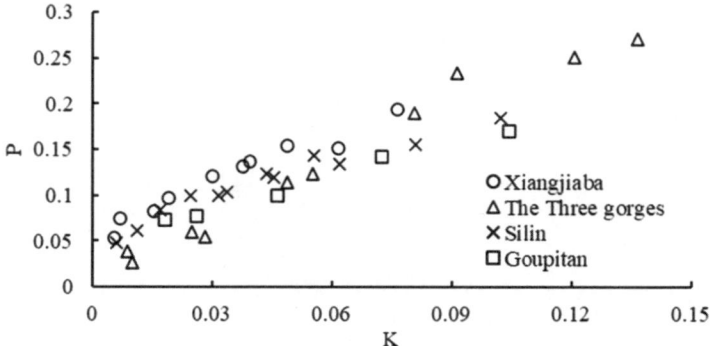

Fig. 4. Relationship of ship squat of different ship lifts

4.4   Force Characteristics of the Ship Lock Mechanism

During the process of upstream and downstream docking, ship's entering and leaving the tank, the two groups of locks at the same end of the tank are subject to fluctuations in force and amplitude, and the amplitudes of the lock mechanisms at the upper and lower ends are greatly different, the fluctuation near the closed end of the tank is large, and on the other end is small. It is necessary to pay attention to the impact of downstream water level fluctuation on the load of the ship. Because the downstream water level changes too fast, when the ship is docked with the downstream, the water level between the tank and the downstream waterway may form a difference, resulting in the water loss or full water of the tank. At this time, the four locking mechanisms are significantly more stressed than the docking state of the upstream side, and may even exceed its acceptable range. Therefore, before opening the door of tank, it is necessary to confirm whether there is a water level difference between the downstream and the tank.

4.5   The Design Criteria for the Water Depth of the Ship Lift's Tank

China's "Inland Navigation Standards" (GB50139-2004) stipulates the standard water depth (H) of the tank as calculated by the following formula:

$$H = T + \Delta H \tag{2}$$

In the formula: T- the standard draught of the ship; ΔH- the additional depth, which is the sum of the ship squat and safety margin without bottoming.

In the "Inland Navigation Standards", the value of the safety margin without bottoming is not clearly stated. Factors that should be considered for the additional depth of the ship: the ship will increase the draught when sailing in shallow water; the unevenness of the tank bottom; the sinking of the stern when the ship leaves the tank. Therefore, the safety margin of the tank should generally be not less than 0.5–0.6 m.

For the 500t class ship lift (designed ship draught 1.6 m) and the 1000 t class ship lift (designed ship draught 2.0 m) to ensure that the minimum normal design water depth of the corresponding tank is 2.5 m or 3.0 m, the safety margin of the tank is 0.5–0.6 m, the design speed of the ship out of the tank should be less than 0.6–0.8 m/s. Considering the misloaded water depth of the tank, which the limit should not exceed −0.2 m, the speed of the ship out of the tank should be controlled at 0.6–0.7 m/s.

After the ship's speed exceeds 0.7 m/s, the ship's tilting moment caused by the fluctuation of the water surface in the tank increases rapidly. When the ship's entering speed is around 0.8–0.9 m/s, the ship's tilting moment peaks. Therefore, considering the safety margin, the longitudinal tilting moment of the tank, the maximum speed of the ship entering the tank should not exceed 0.7 m/s.

5 Conclusion

After years of research, the general formula of calculating the maximum ship squat when ships entering and leaving the tank is established by NHRI. Meanwhile, the mooring force in ship lift tank when the reclining door is opened and closed is established which reveals the unsteady flow influence on safe operation of ship lift mechanism. Putting forward ship lift docking control method to assure the safety of ships and the ship lift which include the reclining door opening and closing speed and water depth control standards, solved the hydrodynamic safety issues when ship entering and leaving the tank of a ship lift.

References

Tuck EO (1973) Sinkage and trim in shallow water of finite width. Schiffstechnik 1973:4

Huuska O (1976) On the evaluation of under keel clearances in finite waterways report no. 9. Helsinki University of Technology, Ship Hydrodynamics Laboratory, Otaniemi

Eryuzlu N, Hausser R (1978) Experimental investigation into some aspects of large vessel navigation in restricted waterways. In Symposium proceedings on aspects of navigability of constraint waterways

Icorels (1980) Report of Working Group IV, No. 35. Bulletin

Barrass CB (1981) Squat-a reply to a comparison of 2 prediction methods. Nav Archit 6:268–271

Norrbin N (1985) Bank clearance and optimal section shape for ship canals. In: 26th PIANC international navigation congress, Belgium

Millward A (1990) A preliminary design method for the prediction of squat in shallow water. Mar Technol SNAME News 27(1):10–19

Millward A (1992) Comparison of the theoretical and empirical prediction of squat in shallow water. Int Shipbuild Prog 39(417):69–78

Eryuzlu NE (1994) Under keel requirements for large vessels in shallow waterways. In: 28th international navigation congress, PIANC, Paper, vol 2, pp 17–25

Gourlay T (2008) Slender-body methods for predicting ship squat. Ocean Eng 35(2):191–200

Bao GJ (1991) Formula of maximum sinkage for ships sailing in the chamber. J Nanjing Hydraulic Res Inst 03:279–282

Hu YA, Luo SZ, Li ZH (2011) Research on key technology of draft control standard for three gorges ship lock. Nanjing Hydraulic Research Institute

Research on Influence from Ship Navigating in the Intermediate Channel Between Ship Lifts on Hydraulic Characteristics

Yingying Chen[1,2]($\boxtimes$), Yaan Hu[1,2], and Zhonghua Li[1,2]

[1] Nanjing Hydraulic Research Institute, Nanjing, People's Republic of China
yychen@nhri.cn
[2] Key Laboratory of Navigation Structure Construction Technology,
Ministry of Transport, Nanjing, People's Republic of China

Abstract. Through the method of physical model test, tests of ship navigating in the intermediate channel of the navigation structure of Longtan Hydropower Station were carried out to study the hydraulic characteristics, such as the water surface fluctuation in the ship chamber and the intermediate channel, the heave of ship bow and stern, and the mooring force of the berthing ship under different navigational conditions. The results indicate that from the perspective of ship lift operational efficiency, docking safety, ship navigating safety and berthing safety, the design navigable water depth of 3.6 m in the intermediate channel is reasonable and the navigational speed of the ship in the intermediate channel should not exceed 1.5 m/s. The research results can provide scientific basis and technical reference for the design and operation of the navigation structure of Longtan Hydropower Station.

Keywords: Longtan hydropower station · Ship lift · Intermediate channel · Hydraulic characteristics · Physical model

1 Introduction

Navigation structures of high dams mainly include ship lift and multistage ship lock (Li et al. 2014). In the construction of navigation structures for high dams and ultra-high dams, ship lift has been more and more applied in China because of its characteristics of fast operation speed, low investment and water saving (Hu et al. 2016). So far, several different types of vertical ship lifts have been built and operated in Geheyan, Yantan, Shuikou, Silin, Shatuo, Three Gorges, Jinghong, Xiangjiaba and other projects. When the working head of the hydropower station is higher, the multistage ship lift with intermediate channels can be adopted, for instance, the navigation structure of Goupitan Hydropower Station adopted three vertical ship lifts linked by two intermediate channels which has been put into trial operation (Chen et al. 2018).

The intermediate channel between ship lifts is an artificial channel that links multistage ship lifts and is used for the navigation of upstream and downstream ships. Its two ends are both closed, and the channel flow velocity is almost zero. Due to the limitation of engineering terrain conditions, the cross section scale of the intermediate

© The Author(s) 2023
Y. Li et al. (Eds.): PIANC 2022, LNCE 264, pp. 599–610, 2023.
https://doi.org/10.1007/978-981-19-6138-0_52

channel is usually smaller than the requirements of restricted channel scale in navigation standard of inland waterway. It is a special restrictive channel. At present, there is no specification or standard for its scale design.

When ship navigates in the narrow closed channel in shallow water like the intermediate channel, its stern may sink obviously and even rub the bottom due to the blocking effect of the ship. And meanwhile, it produces a large surge wave. The surge wave will be reflected after being transmitted to the closed end, superimposed with the advancing surge wave, and oscillate back and forth in the intermediate channel and the ship chambers to form a complex long wave motion, which takes a long time to make the water surface stable. This kind of long wave motion of water flow has certain harm to ships in the intermediate channel, which will cause the ships to sway and shake, and has an adverse effect on the docking of the ship lift.

The navigation structure of Longtan Hydropower Station is a two-stage vertical ship lift with intermediate channel. The maximum lifting height of the two-stage ship lift is 62.4 m and 93.6 m respectively. The two-stage ship lift adopts fully balanced winch lifting system (Chen et al. 2020). The structure of the ship chamber is totally the same, the total length of the ship chamber is 88 m, and the design effective size is 73.0 × 12.2 × 3.5 m (length × width × water depth). The design navigable water depth of the intermediate channel is 3.6 m, and the size of the design ship is 68.0 m × 11.0 m × 2.4 m (total length × shaped width × design draft). In this paper, through the method of physical model test, the influence from the ship's navigation in the intermediate channel between ship lifts on the hydraulic characteristics is studied, and the control standards such as the limit speed of the ship's navigation in the intermediate channel is proposed.

2   Test Methods and Research Conditions

2.1   Physical Model Design

This paper mainly studies the characteristics of the water surface fluctuation in the intermediate channel and the ship chambers, the heave of the navigating ship, and the mooring force of the berthing ship, etc. Therefore, the model design does not consider the lifting equipment and the itinerary of the ship chamber, and only considers the actual water area between the two-stage ship lift of Longtan Hydropower Station. The actual water area between the two-stage ship lift includes the two ship chambers and the upper head of the first stage ship lift, aqueduct 3 one-way channel and other sections of the intermediate channel between the two-stage ship lift. The plane layout of the two-stage ship lift with intermediate channel of the navigation structure of Longtan Hydropower Station is shown in Fig. 1.

The prototype of the test ship is a 1000t-class single ship. One test ship is the navigating ship and the other one is the berthing ship. The ship model is mainly designed according to the gravity similarity criterion, and the geometric scale is consistent with the hydraulic physical model. In addition to meeting the size and linear geometry similarity of the ship model, the speed of ship model should be similar to that of the real ship, and the displacement of the ship model and the real ship should also be similar.

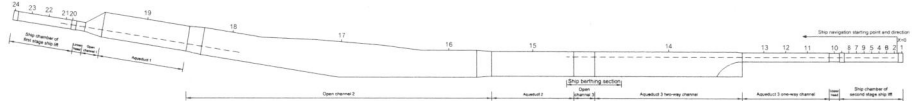

Fig. 1. Plane layout of the two-stage ship lift with intermediate channel of the navigation structure of longtan hydropower station and the ship navigation starting point and direction

2.2 Measuring Point Arrangement and Measuring Method

In the test, the capacitance liquid level meter was used to measure the water surface fluctuation of the two ship chambers and the intermediate channel between the two-stage ship lift, and 24 measuring points were arranged, as shown in Fig. 1. The ultrasonic range finders ware fixed on the front and rear traction device to measure the heave of the navigating ship. The total station instrument measured the speed and distance of the ship navigation. The pressure sensors measured the mooring force of the berthing ship. The test data were automatically collected and preliminarily processed by a multi-channel high-speed data acquisition system.

2.3 Research Conditions

The navigation method of ships in the intermediate channel should ensure the safety of the ships and navigation buildings. On this premise, the navigating time should be shortened as much as possible to improve the passing efficiency. Therefore, the more reasonable navigation method is to navigate along the route with variable speed.

According to previous research, when the ship navigates out of the ship chamber, a large amount of water in the ship chamber is pushed out, which may cause the ship to rub the bottom, affecting the safety of the ship and the ship chamber, and also cause water surface fluctuations in the ship chamber and the intermediate channel, affecting the docking safety of the ship lift and the berthing safety of ship in the intermediate channel (Li et al. 2016). In addition, considering the factors of preventing the ship from collision with the side wall and stabilizing the course, the navigational speed of the ship entering and leaving the upper and lower heads and the aqueduct 3 one-way channel (the narrow section of the intermediate channel) should not be too high. It is therefore decided that the ship should navigate out of the ship chamber of the second stage ship lift (hereinafter referred to as "ship chamber 2"), the upper head and the aqueduct 3 one-way channel at the maximum speed of 0.6 m/s in the test to ensure safety. After the ship leaves the aqueduct 3 one-way channel, the ship navigates in the aqueduct 3 two-way channel, open channel 3 and other intermediate channel sections at the maximum navigational speed accelerating to 1.0, 1.5, 2.0, 2.5 and 3.0 m/s (high speed in the intermediate channel, hereinafter referred to as "high speed").

There were two ship models in the test. One was towed by the traction system and navigated with variable speed in the ship chamber and the intermediate channel, and the other was berthed in the right ship berthing section on the upstream side of the second-stage ship lift (as shown in Fig. 1), and its mooring force was measured to analyze the influence from ship navigating in the ship chamber and the intermediate channel on the berthing conditions of the ship berthing in the intermediate channel.

At the same time, in the test, it is necessary to study the influence from ship navigation on the flow in the intermediate channel and the berthing conditions of the ship when the lower chamber gate of ship chamber of the first-stage ship lift (hereinafter referred to as "ship chamber 1") is open (docking with the intermediate channel) or closed. In particular, it is necessary to measure the water level change outside the downstream gate of ship chamber 1, and analyze whether the water surface fluctuation affects the smooth docking of the first stage ship lift and its downstream intermediate channel. The test groups are shown in Table 1.

Table 1. Test groups

Test groups	Speed out of the chamber (m/s)	High speed (m/s)	State of lower chamber gate of ship chamber 1
LZ1	0.6	1.0; 1.5; 2.0; 2.5; 3.0	Open
LZ2	0.6	1.0; 1.5; 2.0; 2.5	Closed

3 Water Surface Fluctuation in Intermediate Channel and Ship Chamber

The water level change at typical measuring points and the ship navigation distance with time under the condition that ship navigates at a high speed of 1.5 m/s in the intermediate channel in test group LZ1 are shown in Fig. 2. Table 2 shows the maximum water surface fluctuation at typical measuring points in different test groups.

Combining with the Fig. 1 and Table 2, due to the blocking effect of the ship, when the ship starts, a large amount of water in ship chamber 2 is pushed to the intermediate channel and ship chamber 1 during the ship navigation out of ship chamber 2, upper head of ship chamber 2 and aqueduct 3 one-way channel. The water level of ship chamber 2, ship chamber one-way channel, upper head of ship chamber 2 and aqueduct 3 behind the ship stern decreases, the ship stern sinks obviously and the ship may rub the bottom. The water level of aqueduct 3 two-way channel and its upstream intermediate channel and ship chamber 1 rises, but the amplitude of water surface rise is basically less than that of the water surface decline in upper head of ship chamber 2 and ship chamber 2. When the stern of the navigating ship navigates out of aqueduct 3 one-way channel and enters aqueduct 3 two-way channel, a large amount of water suddenly enters ship chamber 2, upper head of ship chamber 2 and aqueduct 3 one-way channel from the upstream side intermediate channel of aqueduct 3 one-way channel, and forms a relatively large surge wave. This kind of surge wave is reflected after being transmitted from the upstream end of aqueduct 3 one-way channel to the downstream end of ship chamber 2 and superimposed with the surge wave advancing from the upstream. Then the complex long wave motion is formed in the intermediate channel and the ship chambers, which takes a long time for the water surface to restore stability. In the process of ship navigation, aqueduct 3, open channel 2 and aqueduct1 have relatively small water surface fluctuation due to wide section size and large section coefficient.

The navigation structure of Longtan Hydropower Station is a two-stage vertical ship lift with intermediate channel. It is necessary to consider the influence of ship navigating from chamber 2 to the intermediate channel on the downstream docking of the first stage ship lift and the upstream docking of the second stage ship lift. If the water surface at lower head of the first stage ship lift fluctuates greatly and the water level difference between the inside and outside of the ship chamber gate is too large, it may be difficult to open the ship chamber gate for docking. When the water level outside the ship chamber gate is higher than that inside the ship chamber, a large amount of water in the intermediate channel will flow into the ship chamber after the ship chamber gate is opened. When the water depth outside the ship chamber gate is lower than that inside the ship chamber, a large amount of water in the ship chamber will flow out. The water flowing into or out of the ship chamber causes large changes in the weight of water and the longitudinal tilt torque in the ship chamber in a short period of time, which not only affects the operation efficiency of the ship lift, but also threatens the locking mechanism of the ship lift and the safety of navigation or berthing of ships in the ship chamber and the intermediate channel.

Referring to the design requirements of the Three Gorges ship lift, when the ship chamber gate is opened, the water level difference between the inside and outside of the ship chamber is required to be no more than 0.1m. During the docking process of the ship lift, the water surface fluctuation in the approach channel should be no more than 0.1 m too. Therefore, during the docking process of the Longtan ship lift, the water surface fluctuation at lower head of ship chamber 1 and upper head of ship chamber 2 can also be controlled by not more than 0.1 m.

From the process line of water level change and the maximum statistical value of the measuring point at the lower head of ship chamber 1, it can be seen that when the high speed of the navigating ship reaches 2.0 m/s or above, the maximum value of the water surface fluctuation at lower head of ship chamber1 exceeds 0.1 m which will adversely affect the downstream docking safety of the first stage ship lift. Therefore, in order to ensure the downstream safe docking of the first stage ship lift, the navigational speed of the ship in the intermediate channel should not exceed 1.5 m/s.

In the process of ship navigating from ship chamber 2 to the intermediate channel, the water surface decline value is greater than the rise value. The maximum water surface fluctuation at upper head of ship chamber 2 is greater than 0.25m. When the high speed is 1.0–1.5 m/s, the water surface fluctuation at upper head of ship chamber 2 is relatively small which is basically less than 0.1 m after about 15 min. When the high speed is greater than 2.0 m/s, the water surface fluctuation at upper head of ship chamber 2 needs to wait more than 40 min until it is less than 0.1 m. The greater the high speed is, the longer the waiting time for water surface to stabilize is, and the lower the operation efficiency of the ship lift is. Therefore, from the perspective of improving the safety and operation efficiency of the upstream docking of the second stage ship lift, the navigational speed of the ship in the intermediate channel should be lower than or equal to 1.5 m/s.

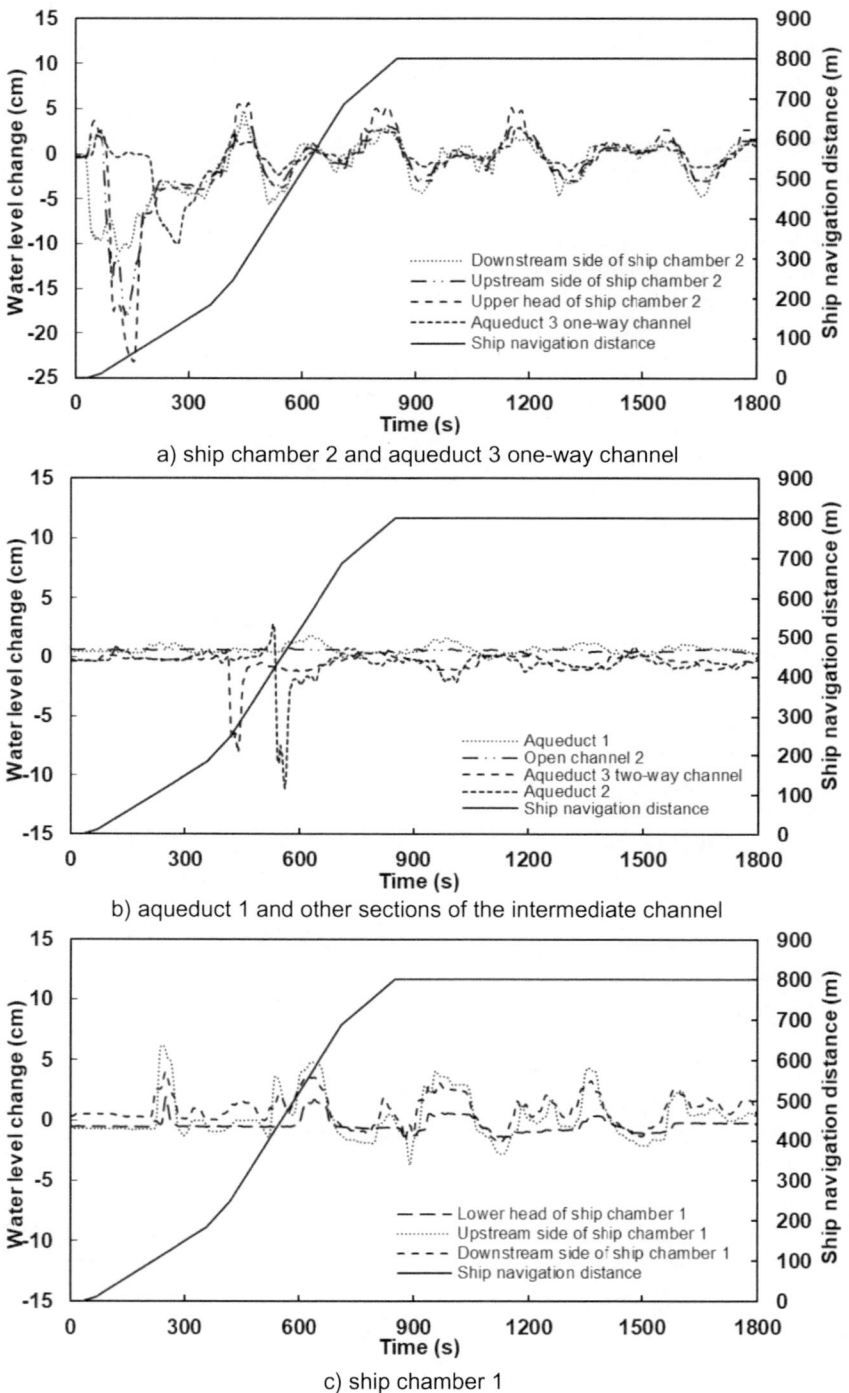

a) ship chamber 2 and aqueduct 3 one-way channel

b) aqueduct 1 and other sections of the intermediate channel

c) ship chamber 1

Fig. 2. Process line of water level change and ship navigation distance with time

Table 2. The maximum water surface fluctuation at typical measuring points (Unit: m)

Test groups	High speed (m/s)	Downstream side of ship chamber 2	Upper head of ship chamber 2	Aqueduct 3 one-way channel	Aqueduct 3 two-way channel	Aqueduct 1	Lower head of ship chamber 1	Upstream side of ship chamber 1
LZ1	1.0	0.17	0.31	0.17	0.03	0.03	0.07	0.14
	1.5	0.16	0.29	0.18	0.08	0.02	0.05	0.10
	2.0	0.23	0.35	0.22	0.17	0.07	0.20	0.26
	2.5	0.58	0.51	0.36	0.42	0.25	0.39	0.54
	3.0	1.03	0.97	0.80	0.82	0.50	0.96	1.33
LZ2	1.0	0.16	0.30	0.17	0.03	0.03	0.05	/
	1.5	0.14	0.29	0.18	0.10	0.03	0.10	/
	2.0	0.22	0.33	0.24	0.19	0.10	0.15	/
	2.5	0.65	0.43	0.38	0.23	0.26	0.47	/

4   Heave of Ship Bow and Stern

When the ship navigates out of the ship chamber and navigates in the intermediate channel, the water flow pushed away by the forward process of the ship produces a blocking effect under the influence of boundary conditions. The water in front of the ship bow moves to the stern, forming a movement of water around the ship. This backflow movement of water is accompanied by a loss of velocity, resulting in a water level difference. The drawdown of the water level will cause the ship to sink. When the ship sinks a large amount and the water depth of the channel is insufficient, it may cause the ship to rub the bottom, which will have an adverse impact on the safety of the ship and the ship lift. Accordingly, it is necessary study the heave of the ship bow and stern during the ship navigation, so as to determine the reasonable channel water depth and ship speed limits.

The heave of ship bow and stern with ship navigation distance under the condition that ship navigates at a high speed of 1.5 m/s and 2.5 m/s in the intermediate channel in test group LZ1 are shown in Fig. 3. The maximum sinkage and minimum safety margin of the navigating ship and the corresponding ship navigation distance and position in different test groups are listed in Table 3.

The figure and table show that in the process of ship navigation, the ship sinkage value is greater than the rising value. When the high speed of the navigating ship is 1.0–2.0 m/s, the maximum sinkage of the ship occurs when the stern is still in the ship chamber 2, and the maximum sinkage is mainly affected by the speed of the navigating ship out of the ship chamber 2. In the test, the maximum navigational speed of the ship out of the ship chamber is 0.6 m/s, so the maximum sinkage of the ship is 0.27–0.29 m, and the minimum safety margin is 0.93–0.91 m, which is relatively close. When the speed of the sailing ship is 2.5–3.0 m/s, the maximum sinkage occurs when the ship stern is located in aqueduct 2, and the maximum sinkage is mainly affected by the high speed of the navigating ship. When the high speed is 2.5 m/s, the maximum sinkage is 0.43–0.53 m, the minimum safety margin is 0.78–0.67 m. And when the high speed is 3.0 m/s, the maximum sinkage is 1.08 m, the minimum safety margin is only 0.12 m.

In order to ensure the safety of ship navigation, the safety margin at the bottom of the 1000 t-class design ship with a draft of 2.4 m for the Longtan ship lift should not be less than 0.7 m. Therefore, considering that the minimum safety margin of the ship is 0.7 m, the navigational speed of the ship in the intermediate channel should not be greater than 2.0 m/s.

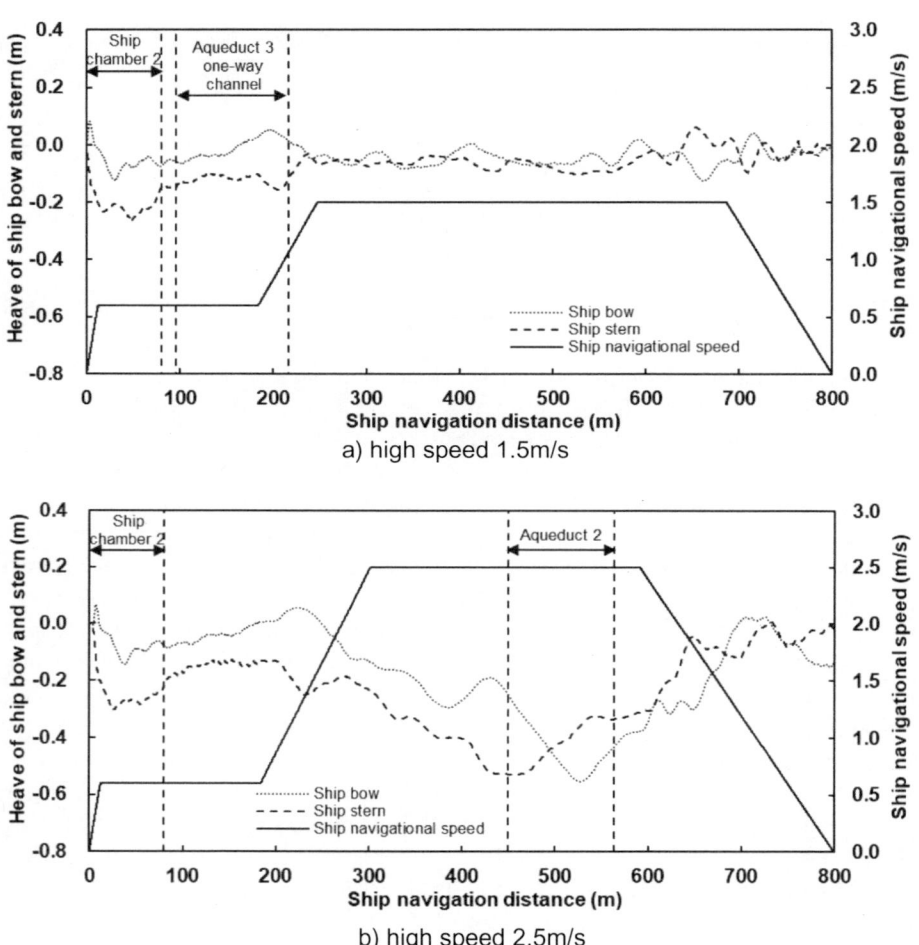

a) high speed 1.5m/s

b) high speed 2.5m/s

Fig. 3. Process line of heave of ship bow and stern with ship navigation distance

Table 3. The maximum sinkage and the minimum safety margin of the navigatiing ship

Test groups	High speed (m/s)	Maximum sinkage (m)	Minimum safety margin (m)	Ship navigation distance and position
LZ1	1.0	−0.29	0.91	X = 49.42 m, Ship chamber 2
	1.5	−0.27	0.93	X = 48.46m, Ship chamber 2
	2.0	−0.27	0.93	X = 47.98m, Ship chamber 2
	2.5	−0.53	0.67	X = 526.80m, Aqueduct 2
	3.0	−1.08	0.12	X = 466.53m, Aqueduct 2
LZ2	1.0	−0.28	0.92	X = 49.18, Ship chamber 2
	1.5	−0.27	0.93	X = 51.58, Ship chamber 2
	2.0	−0.27	0.93	X = 47.98, Ship chamber 2
	2.5	−0.43	0.78	X = 504.02, Aqueduct 2

Note: The maximum sinkage of the navigating ship is represented by negative value, and the minimum safety margin is represented by positive value, the ship navigation starting point is represented by X = 0 and the X value gradually increases along the ship navigation distance.

5   Ship Berthing Condition

The complex long wave motion oscillates back and forth in the intermediate channel and ship chambers will cause the ships to sway and shake and may easily cause the ship's mooring force to exceed the standard, which will do certain harm to the ships berthing in the intermediate channel and produce potential safety hazards.

The mooring force with time and ship navigation distance under the condition that ship navigates at a high speed of 1.5 m/s in the intermediate channel in test group LZ1 are shown in Fig. 4 and Fig. 5. The maximum mooring force of the berthing ship in different test groups are listed in Table 4.

From the analysis of the figures and table, it can be seen that the maximum longitudinal mooring force of the berthing ship is greater than the maximum lateral mooring force of the corresponding test condition. The maximum mooring force of the

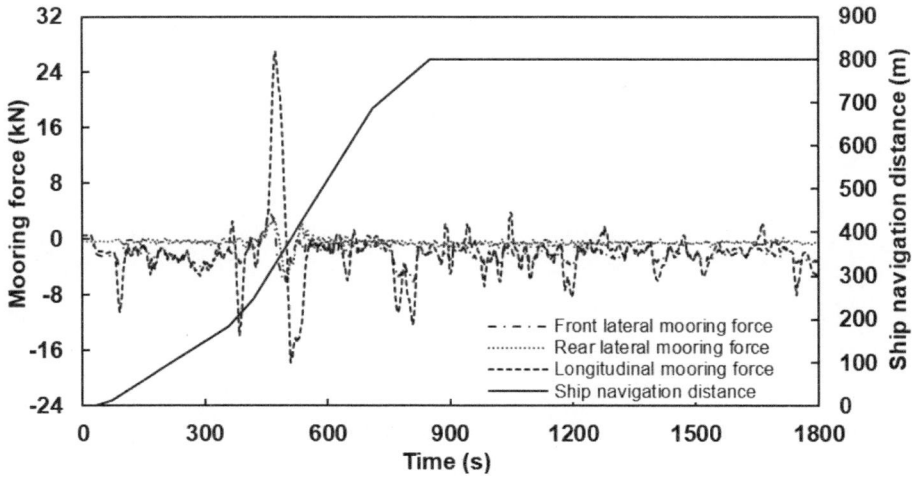

Fig. 4. Process line of mooring force and ship navigational distance with time

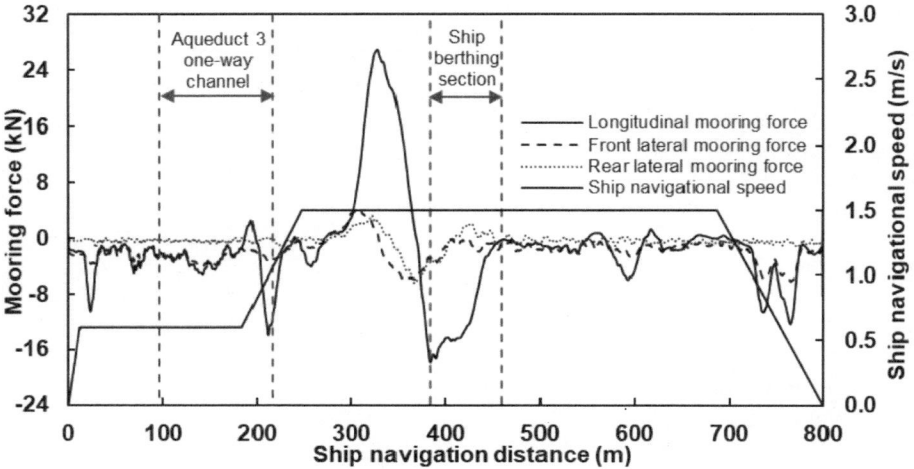

Fig. 5. Process line of mooring force with ship navigation distance

berthing ship basically increases with the increase of the high speed of the navigating ship, and the increase is more obvious when the high speed is greater than or equal to 2.0 m/s.

The design ship of the navigation structure of Longtan Hydropower Station is 1000t-class. The maximum limit value of longitudinal mooring force allowed by the code is 32 kN (GB 51177-2016, China), and the maximum limit value of lateral mooring force is 16 kN. Therefore, starting from controlling the mooring force of the berthing ship to meet the requirements of the code, the navigational speed of the ship in the intermediate channel should not exceed 1.5 m/s to ensure the safety of berthing ship in the intermediate channel.

Table 4. The maximum mooring force of the berthing ship (Unit: kN)

Test groups	High speed (m/s)	Longitudinal mooring force		Front lateral mooring force		Rear lateral mooring force	
		+	−	+	−	+	−
LZ1	1.0	13.9	−13.5	1.6	−5.2	1.8	−3.3
	1.5	28.4	−16.3	5.8	−4.8	3.2	−6.4
	2.0	70.0	−44.6	8.8	−7.2	4.7	−12.1
	2.5	137.2	−78.4	9.8	−5.8	8.2	−18.4
	3.0	229.6	−101.3	13.1	−20.7	19.0	−41.2
LZ2	1.0	14.9	−13.2	3.5	−4.7	1.6	−2.7
	1.5	26.5	−12.7	5.8	−4.7	4.2	−6.3
	2.0	65.5	−38.3	7.6	−8.3	4.7	−11.8
	2.5	92.7	−76.5	9.1	−12.4	6.3	−18.5

Note: The longitudinal mooring force of the berthing ship points to the upstream side as " +" and points to the downstream end as "−", the front lateral and the rear lateral mooring force point to the left as " +" and point to the right as "−".

6 Discussion on Ship Navigational Speed

According to the analysis of the above test results, the navigating process of the ship causes the water surface fluctuation in ship chambers and the intermediate channel, the heave of the navigating ship, and the mooring force change of the berthing ship in the intermediate channel. Considering these factors, the navigational speed control standard of the ship in the intermediate channel is discussed.

The process of ship navigation causes the water surface fluctuation in the ship chambers and the intermediate channel, which affects the safety and efficiency of the ship lift docking. From the perspective of docking safety, the water surface fluctuation of lower head of ship chamber 1 shall be less than 0.1m, and the high speed of the navigating ship in the intermediate channel should not be greater than 1.5 m/s. From the perspective of improving the efficiency of the upstream docking operation of the second stage ship lift, the high speed of the navigating ship in the intermediate channel should not exceed 1.5 m/s.

During the navigation of the ship, the stern of the navigating ship sinks obviously. In order to prevent the ship from rubbing the bottom, it is necessary to ensure that there is a certain safety margin at the bottom of the ship. From the perspective of ship navigation safety, the minimum safety margin at the bottom of the ship is 0.7 m, and the high speed of the navigating ship in the intermediate channel should not exceed 2.0 m/s.

The process of ship navigation causes the long wave motion in the intermediate channel and ship chambers which will do certain harm to the ship berthing in the intermediate. Considering the requirement that the mooring force of the berthing ship does not exceed the allowable limit of the code (the maximum allowable limit of longitudinal mooring force is 32 kN and the maximum allowable limit of lateral mooring force is 16 kN), the navigating speed of the ship in the intermediate channel should not be greater than 1.5 m/s.

In addition, after the navigating ship entering the intermediate channel, its acceleration also needs to be controlled. The speed increase should not be too fast, especially when there are berthed ships in the channel of the speed increase section.

7 Conclusions

Through the ship navigation test in the intermediate channel of the Longtan ship lift, this paper analyzes the variation characteristics of the water surface fluctuation, the heave of the bow and stern of the navigating ship, and the mooring force of the berthing ship in the intermediate channel. The influence of ship navigation on the docking of the ship lift, the navigating ship and the berthing ship is studied, and the navigational speed control standard of the ship in the intermediate channel is discussed.

Considering the operation safety and efficiency of the ship lift, the design navigable water depth of 3.6m in the intermediate channel is reasonable and the navigational speed of the ship in the intermediate channel should not exceed 1.5 m/s.

Under the conditions of the navigable water depth of 3.6 m in the intermediate channel, the maximum navigational speed of 0.6 m/s of the ship out of the ship chamber, and the high speed of 1.5 m/s of the ship in the intermediate channel, the maximum water surface fluctuation of the lower head of the first stage ship lift is 0.1m, and the water surface fluctuation of the upper head of the second stage ship lift is also less than 0.1 m after about 15 min. The maximum sinkage of the navigating ship is 0.27 m and the minimum safety margin is 0.94 m. The maximum longitudinal mooring force of the berthing ship in the intermediate channel is 28.4 kN and the maximum lateral mooring force is 6.4 kN. The research results can provide scientific basis and technical reference for the design and operation of Longtan ship lift.

References

Chen Y, Hu Y, Li Z, Fu L (2020). Comprehensive research on hydraulic characteristics of the second stage ship lift of Longtan. Port Waterway Eng 11:18–25, 70
Chen M, Zhang X, Xu G, Chen M (2018) Navigation structures type and run mode of Pengshui dam in Wujiang River. Port Waterway Eng 11, 85–90
Hu Y, Li Z, Li Y, Xuan G (2016) Research developments in the field of major ship lift in China. Port Waterw Eng 12:10–19
Li Z, Hu Y, Liu K (2016) Water depth standard for Xiangjiaba ship lift's chamber. Port Waterw Eng 12:153–157
Li Y, Liu J (2014). Analysis of navigational speed and channel scale and test of navigation conditions in the intermediate channel between ship lifts of Baise Power Station. J Waterw Habor 4:393–398
Ministry of Water Resources of the People's Republic of China. (2016). Design code for shiplift GB 51177-2016, p 27

Resistance of Plane Lock Gates Subjected to Ship Impact

Sara Echeverry Jaramillo[1](✉), Marine Geers[2], Loïc Buldgen[3],
Jean-Philippe Pecquet[4], and Philippe Rigo[5]

[1] Centre de Recherches Des Instituts Groupés de La Haute École Libre Mosane
(CRIG-HELMo), Liège, Belgium
s.echeverry@helmo.be

[2] University of Liège, Liège, Belgium
marine.geers@student.uliege.be

[3] Haute École Libre Mosane (HELMo), Liège, Belgium
l.buldgen@helmo.be

[4] SBE Engineering, Namur, Belgium

[5] University of Liège, Department ARGENCO-ANAST, Liège, Belgium
ph.rigo@uliege.be

Abstract. This paper presents the analysis of lock gates submitted to ship impacts. This problem is generally approached in two different ways: firstly, by the use of an equivalent static method, in which the impact is modeled by a quasi-static force, or by the use of dynamic numerical simulations, during which the progress of the ship and the temporal evolution of the impact force is taken into account. The second approach requires extensive calculation and modeling efforts, which are generally prohibitive in the early design stage.

A simplified analytical method is presented to evaluate the resistance of such structures when impacted by a ship. The principle is based in the super-element method, firstly evaluating the resistance of the lock gate in local deformation mode, assuming crushing only of certain structural elements in a limited zone, located in the close vicinity of the impact. Secondly, the entire resistance of the lock gate is calculated, considering the global deformation mode, assuming a bending of the entire structure.

The scientific objective is to extend this approach to lock gates which cannot currently be treated with this method, in particular those where the layout of the stiffening elements can be highly irregular. This implies to study thoroughly the global deformation modes of the structure and the plastic mechanisms involved in the energy dissipation, in order to predict a correct displacement field and resistance force.

Keywords: Lock gate · Super elements · Elastic-plastic structural analysis · Ship collision

Y. Li et al. (Eds.): PIANC 2022, LNCE 264, pp. 611–622, 2023.
https://doi.org/10.1007/978-981-19-6138-0_53

1   Introduction

The design of lock gates requires to account for accidental limit states, in which ship impacts shall be considered. Such analyses are not always straight forward. Dynamic numerical simulations in which the temporal evolution of the impact force is taken into account, require extensive calculation and modeling efforts, which are generally prohibitive in the early design stage. The objective of this research is to develop a calculation tool based on the super-element method to circumvent these difficulties and allow the consideration of ship collisions in the early design phase of lock gates.

The analytical approach of the super-element method assumes that the failure modes of the gate subjected to the impact are known. But it is obviously difficult to imagine in a generic and exhaustive way all the failure modes which can cover all conceivable lock gates.

In order to identify the failure modes of the plane lock gates, it appeared necessary to quickly carry out finite element analyzes using the LS-DYNA software, and compare the results of the analytical method based on the use of super-elements. Such preliminary simulations allowed to model the lock gates in a simplified manner, with only vertical and horizontal frames and girders (minimizing the structural details in some parts, such as the openings in the horizontal girders, who don't contribute to the collision resistance). These simplifications can reduce the number of elements, while reducing significantly the computational time.

In addition, the stresses caused by a ship impact on a lock gate depend strongly on the type of ship (more precisely the shape of the bow). It was observed that the most dangerous situation (in terms of deformation of the gate) appears during an impact with a raked bow (in contrast with a collision with a barge). Indeed, in this case, it was observed that the plastic deformations of the gate remain mainly concentrated around the point of impact, the rest of the gate being little mobilized. This reflects a high stress on only a few structural elements, so that the impact resistance will be relatively low. This is why in this article only the cases of collision with a raked bow are highlighted.

This paper presents an analytical method to evaluate the resistance of plane lock gates impacted by a ship. Such approach is an approximation, as the goal is not to replace finite element analyses, but to have a complementary tool that could be used at the pre-design stage. At this point several collision cases can be considered in order to find an optimal configuration of the gate or to identify the most damaging scenario. This latter to be further analyzed in detail with a finite element simulation for example.

Some studies have already been presented in this matter, as for instance Le Sourne (2007), who integrated developments to evaluate the crushing resistance of ship components in a program to evaluate the resistance of two colliding ships. On the other hand, Buldgen et al. (2015, 2013 and 2012) achieved the developments for analysis of plane and miter gates impacted by a raked bow and also by a barge. Other authors have advanced in the developments of super-elements to analyze crashworthiness of other type of structures, such as offshore wind turbines, namely Pire (2018) for jacket structures, Echeverry (2021) for a spar buoy floating wind turbine and Marquez et al. (2021) for a concrete floating wind turbine.

2 Description of the Problem

The collision scenario considered in this research is depicted in Fig. 1, where the ship advances with an initial velocity V_0 in the X direction of the global coordinate system. For simplicity of the model, only the bow is modelled but the entire mass of the ship is taken into consideration in the energy balance. The transversal girders are simply supported at each side of the gate, therefore it is free to rotate once the collision happens. It is assumed that the ship is colliding the gate from the upstream, impacting directly the plate of the downstream gate of the lock.

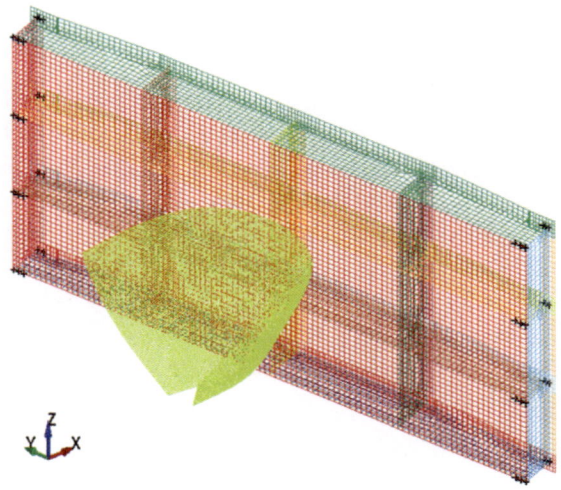

Fig. 1. Collision scenario, FEM model

The general principles of the problem and methodologies used in this research are presented more in detail in the subsequent sections. The dimensions of the gate and each of the stiffening members are not presented for confidential reasons.

2.1 General Principles

As the ship is characterized by its velocity V_0 and mass (including its added mass M_∞ in the direction of collision) $M_0 = M_{ship} + M_\infty$. The initial kinetic energy to be dissipated by the collision is assumed to be $M_0 V_0^2/2$.

The gate is constituted of steel, therefore all the simulations and analytical calculations in this paper deal with this material, represented by a bilinear elastoplastic material law. The elastic phase is characterized by a Young modulus **E** of 210 GPa and a yield stress σ_0 of 255 MPa. A linear strain hardening is considered by a tangent modulus obtained as $E_T = 0{,}5\%\,E$.

For the type of lock gate analyzed, it is assumed that a ship of mass 4500 tons is sufficiently large to induce important damage to the structure. The ship speed is then chosen according to the energy limits to study. For instance, a collision in the order of

2 m/s for such a vessel can, a priori, induce high deformations because the energy to be dissipated is 9 MJ. Having a relation between the collision resistance P_t and the ship penetration δ, the energy dissipated by the gate can then be calculated by simple integration.

The theoretical basis for the methodology used in this research is the so-called upper-bound theorem, which states that a system of applied loads on a structure will cause collapse of it, if the work rate of the system of applied loads equal to the corresponding internal energy dissipation rate. In the case of lock gate crashworthiness, it is then obvious that the external energy rate $\dot{E}_{ext}$ is produced by the force P_t applied by the ship on the gate. This way, we can obtain:

$$\dot{E}_{ext} = P_t V_0 = P_t \dot{\delta} \tag{1}$$

If we neglect the dynamic effects in the structure and assuming an infinitely rigid ship, all this external energy is to be dissipated by deformation of the gate, this is $\dot{E}_{ext} = \dot{E}_{int}$. By integration over the total volume of the structure, it is possible to obtain a relation between the stress and strain rate tensors. Finally, using the Green-Lagrange tensor within the upper-bound theorem, it is possible to find a link between the deformation of the gate and the ship penetration δ in a kinematically admissible displacement field. More details in this theoretical approach are described by Buldgen et al. (2012).

2.2   Super-Element Method

The integration of the force P_t over the entire volume of the gate is quite a laborious task to be performed analytically, this is why it is assumed that the structure can be split into super-elements. Pre-existing analytical models, i.e. Buldgen (2014) and Lützen et al. (2000), are used as a basis for analyzing plane lock gates impacted by a ship. In order to obtain an analytical expression of the local resistance, the gate structure is modeled with a limited number of nodes and some large structural components (girders and frames).

Three types of super-elements (namely *SE1*, *SE2* and *SE3*) are defined to form the structure of the gate and allow to evaluate its local resistance. These three components (*SE1*, *SE2*, and *SE3*) are shown in Fig. 2. Each super-element achieves a different failure mode, and the combination of all failure modes defines the total strength of the structure. These failure modes may include bending or buckling mechanisms after local deformation by crushing of the bow.

When the structure is struck by a perfectly rigid ship, the ship continues to move forward until the total initial kinetic energy has been fully dissipated (either as internal energy or kinetic energy). At the start of the collision, for low values of penetration δ, the damage caused to the structure remains mainly localized in a rather restricted zone confined around the initial point of contact. This region is subject to significant plastic deformations which can sometimes lead to the rupture of certain structural elements. Simultaneously, an elastoplastic global bending movement of the whole structure is superimposed on this localized indentation and is responsible for quite small out-of-plane displacements affecting the whole structure. In such a case, the structure is said to be resistant by a local mode of deformation.

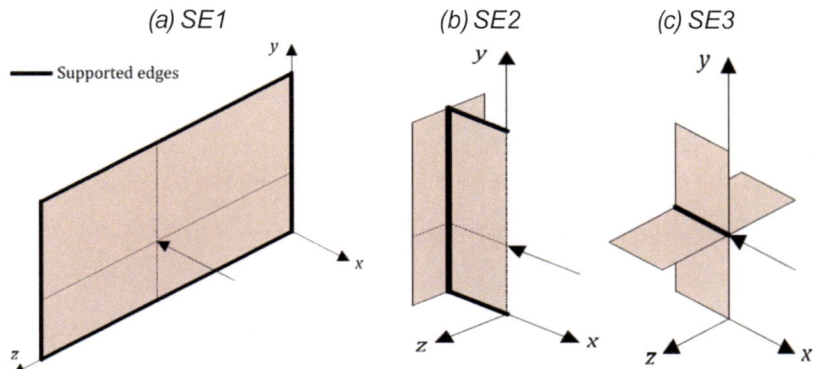

Fig. 2. Super-elements developed for studying lock gates (Buldgen 2014)

On the other hand, the impact can lead to large out-of-plane displacements affecting the entire structure. In this situation, a generalized rigid-plastic mechanism develops over the whole grid and this second process is called global deformation mode.

In order to analytically derive the internal energy E_{int}, the structure is first assumed to resist through a local deformation mode. Nevertheless, with increasing values of the ship displacement δ, it becomes more and more difficult for the ship to continue moving. Therefore, for a given penetration δ, the collision force reaches a sufficient level to activate a global plastic mechanism and a switchover is assumed to occur from the local mode to the global mode. An example of transition from local to global deformation modes of a lock gate is shown in Fig. 3.

As it is hard to take the coupling between local and global modes into account, it is assumed that a sudden switch between the two modes happens at certain penetration δ_t. At the beginning, when the ship first impacts the gate, only the local mode is activated and provides a resistance force P_t. This is true as long as the penetration does not exceed the value δ_t, which activates then the global mode. This means that at this penetration, the force P_t applied by the ship is sufficient to cause an overall bending of the whole structure.

In the local mode, each super-element is evaluated separately as they might present different dissipation energy mechanisms. For instance, the plate element (SE1) presents a deformation shape which is influenced in high manner by the shape of the bow (or the bulbous bow) striking directly into it. The beam elements (SE2) present a localized folding mechanism and a global bending that contribute to the resistance of the element. And the cross elements (SE3) present a folding mechanism that can be represented as the combination of folding mechanisms for each of the wings connected.

A representation of such deformation patterns is depicted in Fig. 4. More details in the calculation of P_t for the local modes (for each activated super-elements) and the global mode, are presented by Buldgen et al. (2012, 2014).

(a) Local and global deformation modes   (b) Collision resistance and internal energy

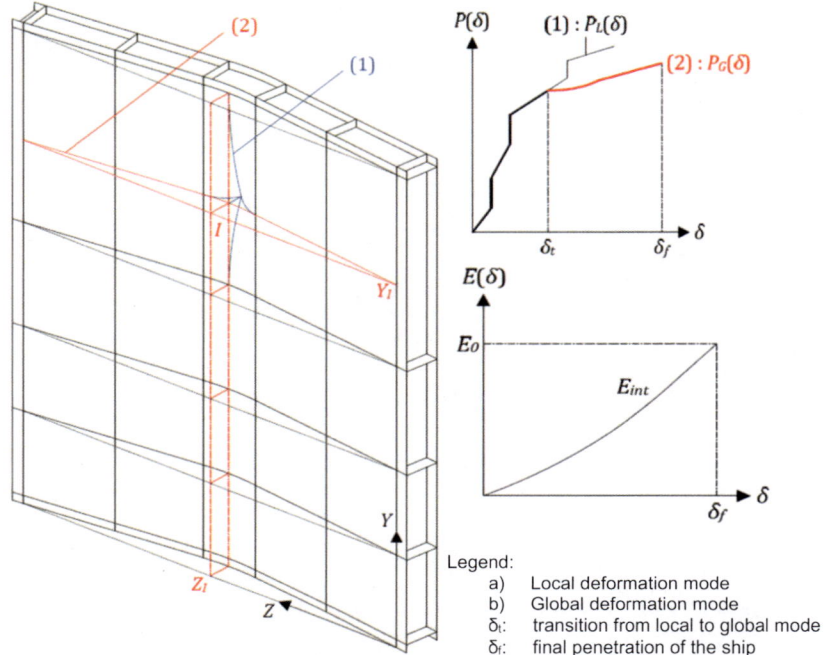

Legend:
a)   Local deformation mode
b)   Global deformation mode
δ_t: transition from local to global mode
δ_f: final penetration of the ship

Fig. 3. Example of deformation modes (local and global) of a lock gate (Buldgen 2014)

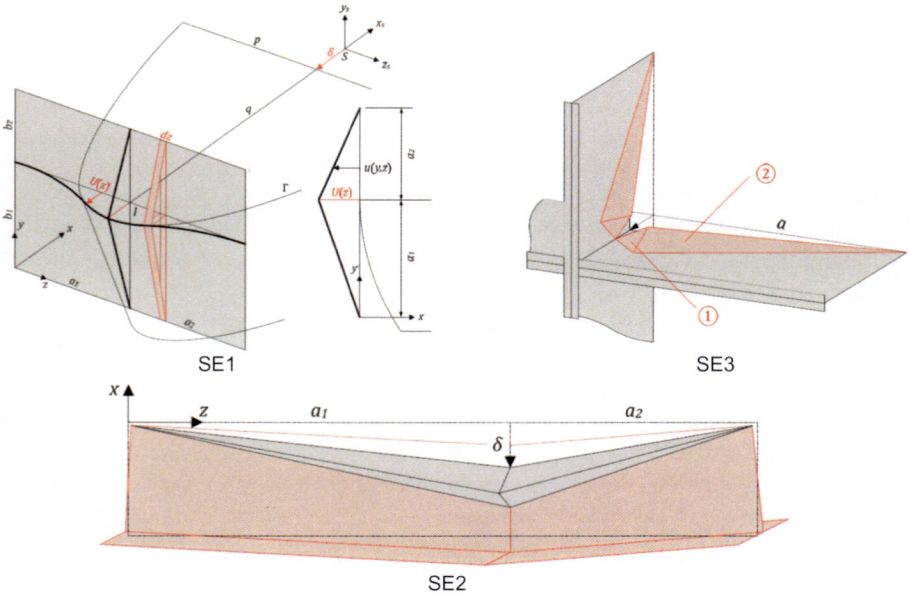

Fig. 4. Deformation patterns of the super-elements of a lock gate (Buldgen 2014)

2.3   Finite Element Simulations

Several numerical simulations have been performed using the finite element software LS-Dyna. A preliminary analysis has shown that a raked bow can cause important damage to the gate, presenting a more prominent localized deformation of the impacted area in comparison with a barge.

For the FEM simulations, the gate is entirely modeled with Belytschko- Tsay shell elements of size 12 cm, which is sufficient to see the folding mechanisms of some of the impacted elements at the collision speeds analyzed. The material law used is a stainless steel already described in Sect. 2.1. The bow of the ship is modelled with same elements but with an infinitely rigid material, in order to save some computational time and focus on the deformation of the gate at this stage of the research. The mass of the ship is considered in the simulations for the sake of energy conservation. The models were presented previously in Fig. 1.

In preliminary simulations, the importance of some parameters of the simulation were observed. Such analyses concerned the material law, collision speed and selection of boundary conditions of the gate. Also the rupture and size of elements can play an important role if the collision happens near a vertical or horizontal reinforcement element, causing folding of the web. These results are not presented in this paper, but it is important to have in mind that such parametric analyses would lead to a more or less reliable result of the FEM simulations, which are crucial for the validation and verification of the analytical calculations.

Several collision cases were analyzed, but four of them are highlighted as specific stiffening elements are activated in each case: a collision on an intersection, a horizontal girder, a vertical frame or a plate. These collision scenarios are presented in Fig. 5. In all of the cases, the most stressed element is the plate, as well as the horizontal girder close to the point of impact. These stresses obviously appear depending on the phase of the collision, since the plating is activated in the local deformation phase in each case.

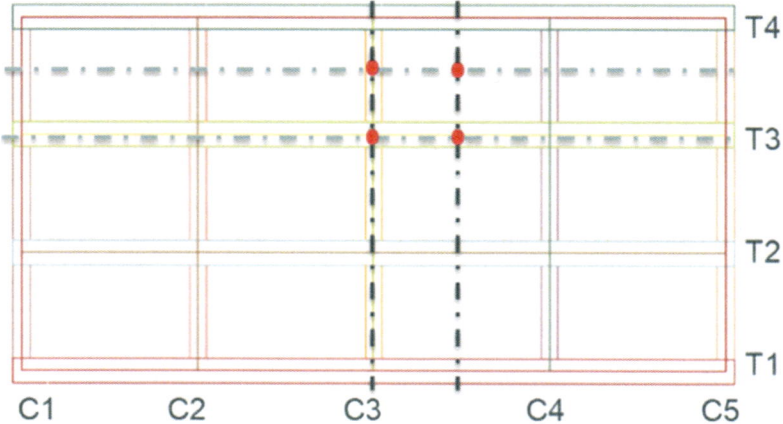

Fig. 5. Collision scenarios for specific super-element activation

It was observed, that it is important to analyze the gate crashworthiness in a plastic regime in order to correctly identify the failure modes. Such failure modes are directly related to the deformation patterns described before and depicted in Fig. 3, where a transition from local to global mode can be assumed in the analytical calculation.

Choosing adequate mass and speed parameters is not always easy. The mass must be chosen according to the vessels currently navigating on the waterway where the lock in question is located. Regarding the speed, there is no particular recommendation on how it should be chosen if it is not contractually specified. It is obvious that at a lower speed and for a lower mass, the energy to be dissipated also decreases, which can certainly reduce the effects due to plasticity, which is an important limit to consider in an analysis of impact.

Finally, it is noted in Fig. 6 a) that a collision of a ship of 4500 tons (added mass included) at a speed of 0.5 m/s is not sufficient to observe an overall deformation of the gate, because the structural response is clearly in the local deformation regime (only the horizontal girder is activated locally), therefore the energy dissipated is much lower compared to a higher velocity collision (for instance 2 m/s in Fig. 6 b).

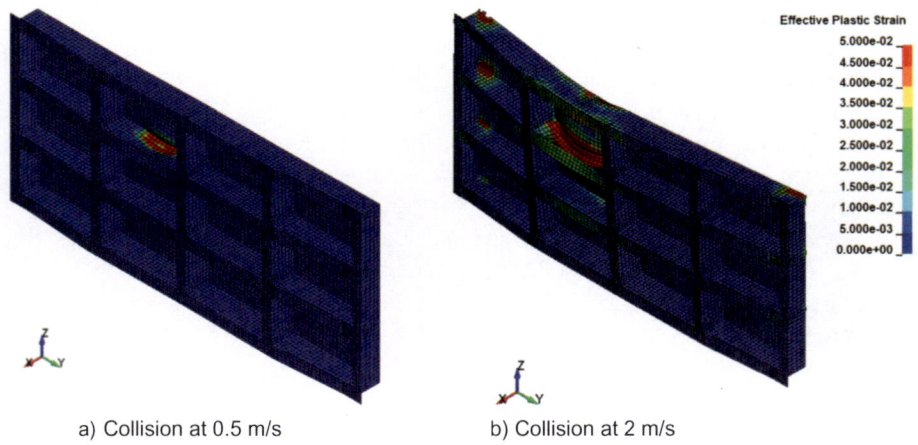

a) Collision at 0.5 m/s   b) Collision at 2 m/s

Fig. 6. Effective plastic Strain for a collision on a horizontal girder

The vertical deformation modes observed for the four collision cases at the impact point are presented in Fig. 7, where it is observed in all cases a localized deformation due to the shape of the bow and a global vertical deformation of the gate, the latter tending to be almost linear. This first approach of linear displacement field of the gate is of vital importance for the analytical developments, where the energy dissipation depends on such known displacement field. This is a different assumption as the one from Buldgen et al. (2012, 2014), who considered a bilinear displacement field.

It is important to note that for each case of collision (on an intersection, horizontal girder, vertical frame or plate), the most stressed element is the plate, as well as the horizontal girder close to the point of impact. These stresses obviously appear depending on the phase of the collision, as mentioned.

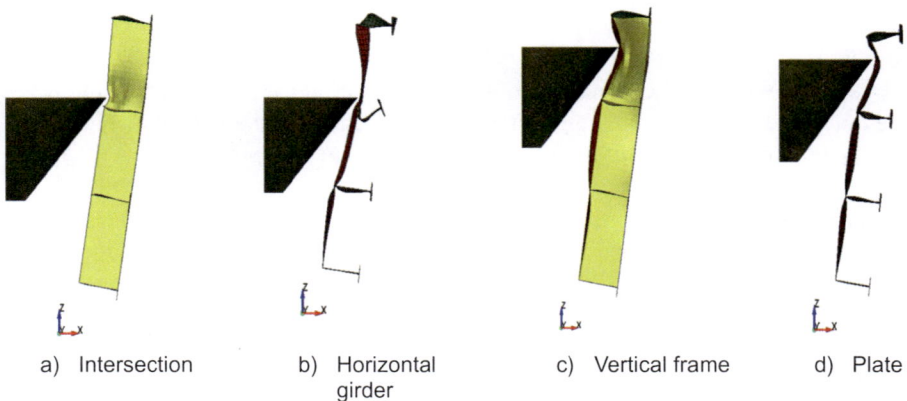

a) Intersection   b) Horizontal   c) Vertical frame   d) Plate
                               girder

Fig. 7. Vertical displacement field of the impacted area for four collision cases

Another method for designing lock gates against collisions mentioned by Buldgen (2014), could be to consider an equivalent static force to be applied to the structure. For example, in the European standards (EN 1991-1-7), the design values of the force are given according to the class of waterways. They were derived for a collision speed of about 2 m/s and assuming that the energy dissipated by the structure was negligible. This approach is not very realistic for lock gates, therefore extreme care should be taken before applying an equivalent static force method.

Using other methods and/or analytical theories (such as a stiffness matrix or an orthotropic plate analysis) would also permit to predict the displacement field of the gate in a quasi-static-elastic regime, including simple static simulations where a constant (known) force is applied. These studies are not in the scope of this paper but were also considered during this research for further analyses in this topic.

3   Results and Discussion

From the preliminary analyses and using the linear displacement field observed for the global deformation of the gate, an analytical calculation was performed, including the local and global deformation of the super-elements activated in each case. Main results for the collision force with respect to the ship penetration are presented in Fig. 8, where the blue line depicts the gate resistance calculated with FE simulation (LS-Dyna) and the orange line presents the analytical calculation. The collision cases concern a ship of 4500 tons (added mass included) at a speed of 2 m/s.

Some differences are observed due to assumptions regarding to the membrane effects in the case of a collision on a horizontal girder, which makes the analytical model overestimate the gate resistance after 0.2 m penetration. Similarly, for the case of collision on a vertical frame or an intersection, at the beginning of the collision, the force calculated with the LS-Dyna simulation presents an interesting hump, that might be related to the element size or collision speed. Considering this, the analytical method underestimates the force in global bending of the gate, which needs to be further analyzed as this might be directly related to the displacement field chosen.

Despite these differences observed, the model is promising to calculate the resistance of a plane lock gate in more general cases such as collision on the plate. In further analyses for the current case study, other methodologies to predict the vertical displacement field can be evaluated, looking forward to predict the displacement field for each collision scenario in a prior analytical calculation (in the elastic regime). Moreover, the effects of membrane strains should be analyzed in detail for some of the scenarios, where the gate resistance force might be overestimated due to a miss conception of the membrane effects.

Similar analyses are to be performed in lock gates with a more irregular stiffness system, in order to see the applicability of the method in such cases.

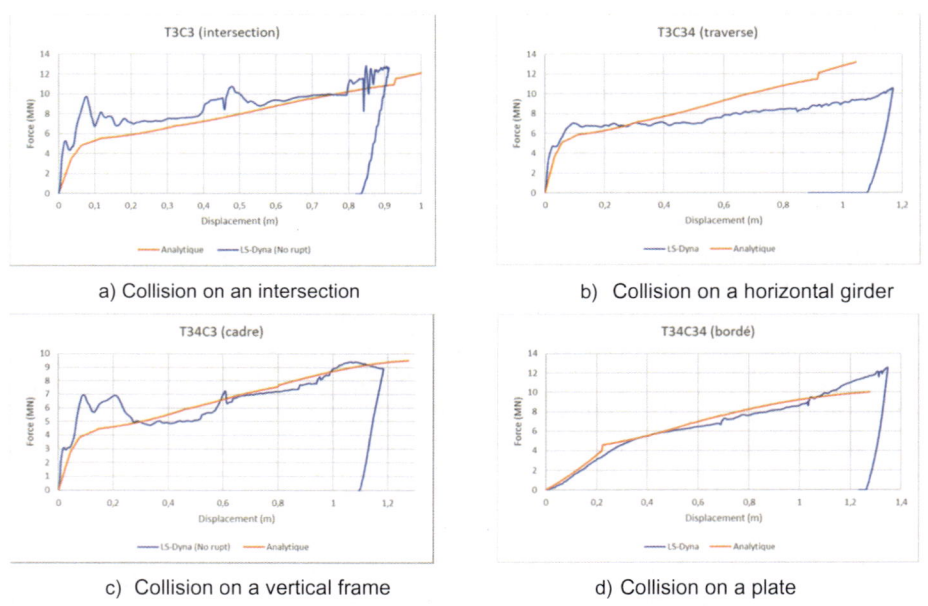

a) Collision on an intersection

b) Collision on a horizontal girder

c) Collision on a vertical frame

d) Collision on a plate

Fig. 8. Force vs displacement, results for four collision cases

4   Conclusion

This paper exposes a simplified method for assessing the crashworthiness of plane lock gates, based on the super-element method proposed by Buldgen (2014). At this stage of the project a plane gate with simple plating and classical orthogonal stiffness system has been considered. The deformation of the impacted structure is assumed to happen in two phases: a local indentation of the impacted elements and a global bending of the whole gate after a certain penetration of the ship inside the gate.

It is important to remember that the upper bound energy theorem systematically provides an estimate by excess of the resistance to impact. Then, the accuracy of the analytical results depends on the adequacy of the displacement field of the gate assumed prior any calculation. It is therefore crucial that this displacement field is chosen as carefully as possible. For this, the use of prior numerical simulations is currently the only option allowing reasonable assumptions to be made.

It was found, compared to the situations studied in the context of Buldgen (2014), that the lock gate configuration considered in this project presents significant differences. An important modification of the displacement field, in particular the introduction of a linear displacement field $g(Y,\delta)$ in the analytical model (and not bilinear as the hypothesis presented in Fig. 3) made it possible to obtain more accurate results. This confirms that the assumed displacement field remains a rather important condition to know before making the crashworthiness analyses using the super element method.

For the prediction of displacement field of different lock gates, a proposal of using other theories (such as the stiffness matrix method or the orthotropic plate) in the elastic regime, is a relevant option for future implementation and discussion of the current method. Furthermore, for a better understanding of the plastic global deformation of the gate, the membrane effects (related to the lateral supports of the gate) should be analyzed in detail, in order to predict a correct gate resistance for some specific collision cases. Finally, gates with an irregular stiffness system could be analyzed to validate and improve the current method.

Acknowledgements. This research is included within the framework of the Belgian COLLFOWT project from the Walloon Region, plan Marshall- GreenWin-Belgium, 2021–2023, in partnership with the University of Liège and SBE.

References

Buldgen L, Le Sourne H, Rigo P (2012) Simplified analytical method for estimating the resistance of lock gates to ship impacts. J Appl Math 2012:1–39

Buldgen L, Le Sourne H, Rigo P (2013) Fast strength assessment of miter gates to ship impact. Int J Crashworthiness 18:423–443

Buldgen L (2014) Simplified analytical methods for the crashworthiness and the seismic design of lock gates. PhD thesis, Faculty of applied sciences, University of Liege, Belgium

Buldgen L, Le Sourne H, Rigo P (2015) A simplified analytical method to estimate the resistance of plane lock gates impacted by river barges. Mar Struct 43:61–86

Echeverry S (2021) Numerical and analytical study of a spar-like floating offshore wind turbine impacted by a ship. PhD. thesis, Faculty of applied sciences, University of Liege, Belgium

Le Sourne H (2007) A ship collision analysis program based on super-element method coupled with large rotational ship movement analysis tool. ICCGS-2007. In: International conference on collision and grounding of ships. pp 131–8

Lützen M, Simonsen BC, Pedersen PT (2000) Rapid prediction of damage to struck and striking vessels in collision event. In: Proceedings of ship structure symposium for the new millennium: supporting quality in shipbuilding, Arlington, United States

Marquez L, Le Sourne H, Rigo P, (2021) Ship collision events against reinforced concrete offshore structures. In: Proceedings of the 8th international conference on marine structures, MARSTRUCT 2021, 7–9 June 2021, Trondheim, Norway

Pire T (2018) Development of a code based on the continuous element method to assess the crashworthiness of an offshore wind turbine jacket. PhD thesis, Faculty of applied sciences, University of Liege, Belgium

Risk-Based Maintenance of Lock Gates Based on Multiple Critical Welded Joints

Thuong Van Dang[1]([☒]) and Philippe Rigo[2]

[1] Thuyloi University, 175 Tay Son, Dong Da, Hanoi, Vietnam
thuongdv@tlu.edu.vn
[2] Department ArGEnCo, ANAST, University of Liege, Liege, Belgium
ph.rigo@uliege.be

Abstract. In many countries, inland waterway transport plays a significant role in the overall transport system. Navigation locks and dams are critical for inland waterways which regulate water and allow vessels to navigate. These infrastructures normally utilize large hydraulic steel structures, which are primarily of welded steel structures. Many of the critical welded joints of navigation lock gates have been in service for decades and are experiencing varying degrees of deterioration, mainly from fatigue. Inspection and maintenance of lock gates are expensive, generally requiring the complete closure of locks. Therefore, innovative strategies for the inspection and maintenance of lock gates are required. Fatigue cracks reflect the inherently poor fatigue performance of welded joints. Due to the cyclic loading nature of lock gates the fatigue of critical details requires assessment methods based on the reliability methods. Risk-based inspection planning are used for marine structures but seldom applied to inland navigation lock gates. This paper presents methods to update the failure probability of welded joints considering crack inspection data by using Dynamic Bayesian Network. Optimal inspection and repair plans can be evaluated by risk analysis, combining failure probabilities and associated expected costs for different events. In this study, a numerical example of the procedure of risk-based maintenance of a lock gate based on multiple critical welded joints is described. This reference case is a lock gate fabricated with five critical welded joints corresponding to different equivalent stress ranges. The conclusion is that risk-based maintenance of lock gates based on the optimization of the total expected cost is recommended.

Keywords: Fatigue · Lock gates · Risk-based inspection planning · Dynamic Bayesian networks · Welded joints

1 Introduction

Hydraulic steel structures are very diverse. Depending on the type of structures and environmental conditions, their components and structures may be suffered from different degradations (fatigue damage). Fatigue damage usually occurs in stress-concentrated areas where localized stress is high. Lock gates are large hydraulic steel structures fabricated by welded vertically and/or horizontally members and steel plates. Due to complex geometry, there are imperfections and residual stresses during the manufacturing process. Welded components are particularly prone to fatigue rather

© The Author(s) 2023
Y. Li et al. (Eds.): PIANC 2022, LNCE 264, pp. 623–631, 2023.
https://doi.org/10.1007/978-981-19-6138-0_54

than the base metal. Therefore, the fatigue strength analysis of welded structures at the design stage and operation stage as well as is of high practical interest. In this paper, a numerical example of the procedure of risk-based maintenance of a lock gate with multiple critical welded joints is described. This reference case is a lock gate fabricated with five critical welded joints corresponding to different equivalent stress ranges. These components are vulnerable to fatigue and are designed according to the "category E" of AASHTO.

2   Deterioration Modeling

Navigation lock gates, are subjected to cyclic loads. The primary fatigue load is the differential water head between the sides of a lock gate (called water head). It stands for the difference between water levels on the upstream and downstream sides of the lock gate. Since the water head is not exactly the same for each lockage due to seasonal flows of the river, the different stress ranges can be represented by an unique "equivalent stress range" (EUROCODE 2005). In the United States, the standard method for fatigue design and assessment of hydraulic steel structure is the nominal stress approach. The nominal stress approach is based on S–N curves when the capacity of welded steel joints related to the fatigue strength is represented by these S-N curves. There is no specific S-N curve available for hydraulic steel structures. The United States Army Corps of Engineers has recommended the code ETL 1100-2-584 (USACE-ETL-1110-2-584 2014) to perform fatigue assessment with the American Association of State Highway and Transportation Officials (AASHTO) code. Table 1 summarizes the parameters of the S-N.

Table 1. Input data of welded joints for S-N model

Parameters	Distribution	Mean	Cov
Δ	Lognormal	1.0	0.3
m_{SN}	Deterministic	3	
C_{SN}	Lognormal	$3.61 \cdot 10^{11}$	
$\Delta\sigma_A$[Mpa]	Deterministic	90	
$\Delta\sigma_B$ [Mpa]	Deterministic	85	
$\Delta\sigma_C$[Mpa]	Deterministic	80	
$\Delta\sigma_D$[Mpa]	Deterministic	75	
$\Delta\sigma_E$ [Mpa]	Deterministic	70	
a_c[mm]	Deterministic	22	
$\Delta\sigma$ [Mpa]	Deterministic	57	
B_s	Lognormal	1	0.25
n	Deterministic	3600	
T	Deterministic	100	

The fatigue crack growth model is utilized for incorporating crack inspection results in assessing the failure probability. The most widely used model is the Paris law (Paris & Erdogan 1963). Table 2 presents the parameters of the fatigue crack growth model. Assuming that the geometry function (Y) is constant, crack size a_t time t can be calculated as shown in Eq. (1).

$$a_t = [a_{t-1}^{1-\frac{m}{2}} + \left(1 - \frac{m}{2}\right) CB_s^m \Delta\sigma^m Y^m n\pi^{m/2}]^{2/(2-m)} \tag{1}$$

where a_0 (t = 1) is the initial crack size, C and m are material parameters. $\Delta\sigma$ is the equivalent nominal stress range used for fatigue analysis, n is the number of cycles per year and B_s is the load uncertainty.

Table 2. Input data of critical welded joints for the FM model

Parameters	Distribution	Mean	Cov
a_{0A}[mm]	Exponential	0.1778	
a_{0B} [mm]	Exponential	0.1716	
a_{0C} [mm]	Exponential	0.1590	
a_{0D} [mm]	Exponential	0.1450	
a_{0E}[mm]	Exponential	0.1390	
a_c [mm]	Deterministic	22	
C	Normal	$2.3247 \cdot 10^{-12}$	0.1063
m	Deterministic	3.0	
B_s	Lognormal	1.0	0.25
B_y	Lognormal	1.0	0.1
Y	Deterministic	1.0	
n	Deterministic	3600	
T	Deterministic	100	

Figure 1 shows the fatigue reliability curves for the five critical welded joints.

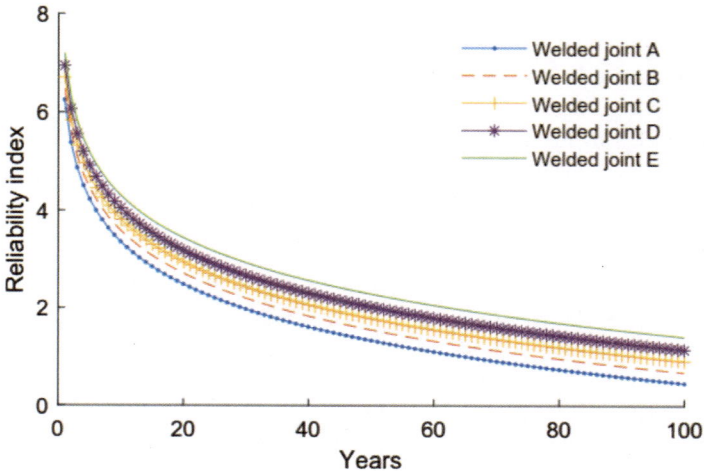

Fig. 1. Fatigue reliability of critical welded joints

3   Dynamic Bayesian Network (DBN)

A Dynamic Bayesian Network (DBN) is a special class of Bayesian network, that represents the temporal evolution of variables over time. DBN was developed in the early 1990s by extending static belief-network models to more general dynamic forecasting models (Dagum et al. 1992). A DBN framework for stochastic modeling of deterioration process and updating the failure probability is proposed by Straub (Straub 2010). DBN is used to update the failure probability for critical welded joints of the lock gate. For a conservative way, it is assumed the lock gate fails when a critical welded joint fails.

The variable $q = \left(1 - \frac{m}{2}\right) CB_s^m \Delta\sigma^m Y^m n\pi^{m/2}$ (see Eq. (1)) is defined in order to reduce the dimension of the joint distribution and consequently the computational time. The crack depth at the end of each year can be expressed recursively as a function of the crack depth in the previous year as shown in Eq. (2).

$$a_t = [a_{t-1}^{1-\frac{m}{2}} + q\,]^{2/(2-m)} \tag{2}$$

By instantiating the inspection variables I_t in the DBN with the observed events at the times of inspection, the failure probability is updated considering the inspection outcomes.

The DBN representation including inspection results I_t for critical welded joints are shown in Fig. 2.

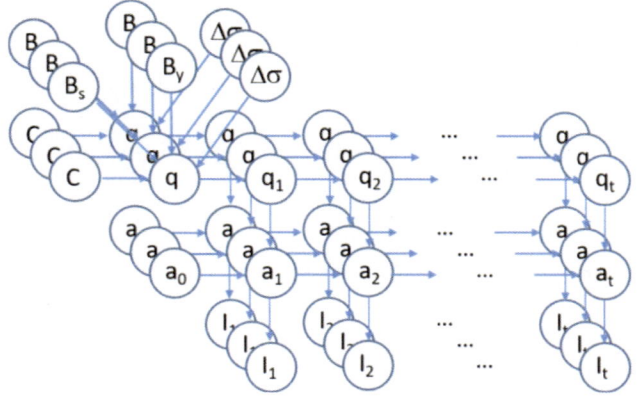

Fig. 2. DBN of multiple components

4 Risk-Based Decision Analysis

Risk-Based Inspection (RBI) is an attractive issue for hydraulic steel structures due to the increase of aging structures and that many failures are detected. The aim of RBI is to find a balance between the benefit of inspection, repair schedule versus failure cost. The failure probabilities using DBN are combined with the cost model.

Risk-based maintenance of the lock gate is performed according to three scenarios:

- Optimal inspection and repair schedule of the lock gate based on the optimization of the expected cost of the independently critical welded joints.
- Optimal inspection and repair schedule of the lock gate based on the optimization of the expected cost of the combined critical welded joints.
- Optimal inspection and repair schedule of the lock gate based on the optimization of the expected cost of groups of combined critical welded joints.

The different unitary costs utilized in this numerical example are shown in Table 3. Because there is a difference of maintenance cost between a single welded joint and multiple welded joints, it is assumed that inspection cost and repair cost for an individual welded joint is equal to 50% of the inspection cost and repair cost that are used for multiple welded joints.

Table 3. Relative cost characteristics of critical welded joints

Cost	Value (money unit)	
	multiple welded joints	welded joint separately
Failure cost, C_f	10^6	10^6
Inspection cost, C_{insp}	$0.003\ C_f$	$50\% \cdot 0.003\ C_f$
Repair cost, C_r	$0.045\ C_f$	$50\% \cdot 0.045\ C_f$
Discounting rate, α_r	0.03	0.03

In the first scenario, the optimal schedule is performed for each welded joint independently.

As result, the expected cost for the lock gate is calculated by Eq. (3).

$$E_{totgate} = E_{totA} + E_{totB} + E_{totC} + E_{totD} + E_{totE} \qquad (3)$$

Figure 3 shows the result of the optimization of the total expected cost for the critical welded joints, A to E, independently.

The result of the optimization cost for each critical welded joint (A to E) is given in Table 4 and Fig. 3.

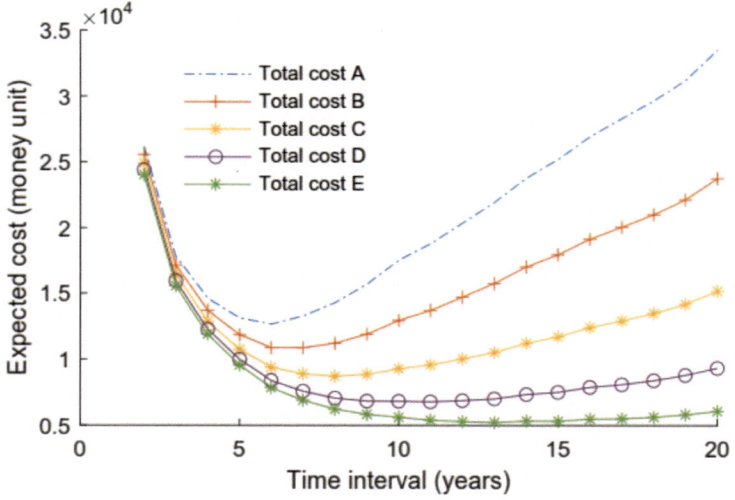

Fig. 3. Results of optimizing the total expected cost of single joint cost considered separately

Table 4. Result of optimizing for each critical welded joint

Welded joint	E_{tot}	ΔY (year)
A	$1.27.10^4$	6
B	$1.09.10^4$	7
C	$0.87.10^4$	8
D	$0.68 \cdot 10^4$	11
E	$0.52 \cdot 10^4$	13

In the second scenario (combined critical welded joints), the inspection and repair of the five critical welded joints are performed with a similar time period to figure out the optimal schedule. Repair cost for each component is equal to the repair cost of the entire gate divided by the number of investigated critical welded joints. The failure probability of the lock gate P_{fgate} is calculated by Eq. (4).

$$P_{fgate} = P\left(\sum_{i=1}^{v} A_i\right) = \sum_{i=1}^{v} P(A_i) - \sum_{i<j} P(A_i A_j) +$$

$$\sum_{i<j<k} P(A_i A_j A_k) - \sum_{i<j<k<l} P(A_i A_j A_k A_l) + \ldots + (-1)^{v-1} P(A_1 A_2 \ldots A_v)$$

$$(4)$$

The total expected cost for the lock gate of the second approach is then calculated by Eq. (5)

$$E_{totgate} = \sum_{t=1}^{T} C_f P_{fgate}(t) \frac{1}{(1+\alpha_r)^t} + \sum_{i=1}^{T_{insp}} \frac{\sum_{j=1}^{v} (C_{rj} P_{rj})(T_i) + C_{insp}}{(1+\alpha_r)^{T_i}} \quad (5)$$

where:
  C_{insp} inspection cost,
  C_{rj} repair cost for a critical welded joint,
  C_f failure cost,
  α_r annual discounting rate,
  P_f probability of failure,
  P_{rj} probability of repair for a critical welded joint,
  T_i year of inspection.
  v number of critical welded joint.

Figure 4 shows the outcome of the optimal schedule the lock gate with multiple critical welded joints. The optimal expected cost of the gate is equal $2.38 \cdot 10^4$. The inspection and repair interval is of 6 years.

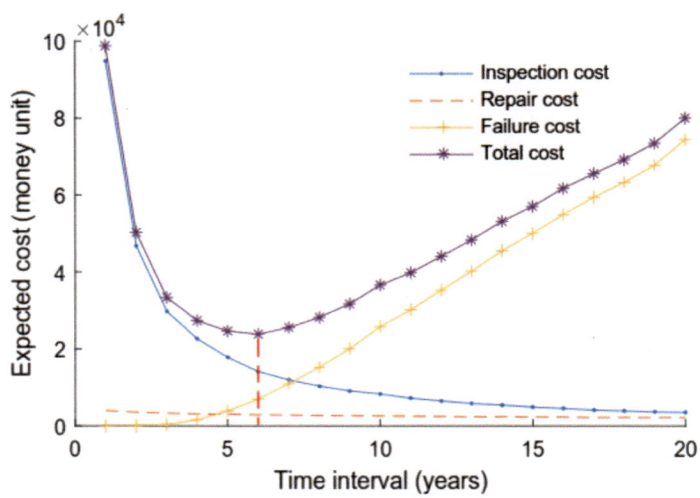

Fig. 4. Results of risk-based maintenance the lock gate with multiple critical welded joints

Thirdly, we consider two groups of combined critical welded joints. Critical welded joints having simultaneously inspection and repair time are combined together.

In this example, five critical welded joints are divided into two groups:

- group 1 includes welded joint A and B with inspection and repair time interval is 7 years
- group 2 includes welded joint C, D and E with time interval is 13 years.

Depending on the groups, the cost of inspection and repair are different. It is assumed that inspection and repair cost for group 1 and group 2 are equal to 80% and 90% of the inspection and repair cost that are used for the multiple welded joints (Table 3).

The procedure of the calculation in the third scenario is then similar to the second scenario (Eq. (4) and Eq. (5)) and the result is provided in Table 5.

Table 5. Total expected cost of the lock gate (100 years)

	Multiple joints	Single joint	Two group of multiple joints
$E_{totgate}$	2.38 ·104	4.18 104	3.92 104
No. of insps	16	47	20
ΔY (year)	6	(6, 7, 8, 11, 13)	(7, 13)

In the comparison of the three scenarios, we see that, the optimized total expected cost for the lock gate based on multiple critical welded joints gets better result (Table 5), with smaller cost ($\approx$50% and 60%) than the optimized total expected cost based on the failure probability of independently critical welded joints and the groups of multiple joints.

5   Conclusions

This paper presents an innovative methodology to establish inspection and maintenance of lock gates considering multiple critical welded joints. The risk-based maintenance of a lock gate performed based on the optimization of the total expected cost of the combined critical welded joints is recommended.

Acknowledgements. The authors acknowledge the financial support provided by the Wallonie-Bruxelles International (WBI), Belgium for this research.

References

Dagum P, Horvitz E, S U (1992) Dynamic network models for forecasting section on medical informatics stanford university school of medicine abstract we have developed a probabilistic forecasting methodology through a synthesis of belief network models and classical time-series analys. pp 41–48

EUROCODE (2005) European standard EN 1993.1.9. vol 7, no 2006

Paris P, Erdogan F (1963) A critical analysis of crack propagation laws. J Basic Eng 85(4):528. https://doi.org/10.1115/1.3656900

Straub D (2010) Stochastic modeling of deterioration processes through dynamic Bayesian networks. J Eng Mech 135(10):1089–1099

USACE-ETL-1110-2-584 (2014) ETL 1110-2-584 Design of Hydraulic Steel Structures. June. https://doi.org/10.1016/S0389-4304(97)00006-4

Ship Impact for Suederelbe Bridge Crossing in Hamburg

Claus Kunz[1]([⊠]) and Jan Schülke[2]

[1] Bundesanstalt fuer Wasserbau (BAW), Karlsruhe, Germany
claus.kunz@baw.de
[2] Deutsche Einheit Fernstrassenplanung- und bau GmbH (DEGES), Berlin,
Germany

Abstract. In the harbour of Hamburg a new bridge crossing the Suederelbe at
Elbe km 620 has been planned. The bridge is part of the BAB26 motorway.
Contracted by DEGES, a German company for highway planning and design,
BAW determined the ship impact loads on the basis and methodology of EN
1991-1-7 (2006) on a site - specific basis, BAW (2016). The Suederelbe Bridge
crossing will have a length of 695,60 m with main span bridging the fairway
over 350 m. The pylon height of the main piers is planned to be 150 m. The
clearance height of the Suederelbe crossing should be 53 m above sea level. The
two pylons and main piers of the Suederelbe bridge crossing, which are to be
positioned close to the bank, are at risk of ship impact. About 6,000 seagoing
vessels up to 33,000 dwt are passing the future bridge. Fleet structure, ship
passages, speeds, accident rates and nautical conditions were analysed and ship
impact loads for the piers and protective structures were determined using a load
model on one hand and a collision model on the other hand. Load and collision
model are probabilistic and based on corresponding distributions of the decisive
influencing parameters, whereby this approach represents state-of-the-art technology. Based on the collision model, the average time that the eastern pylon is
being impacted is expected to be 470 years, while the average time for the
western pylon is expected to be 1,240 years.

The impact loads for the east pylon are dynamic loads with FF_{dyn} = 17.5
MN for frontal impact and FL_{dyn} = 4.0 MN for lateral impact. Out of the
models the impact load can be determined as dependent on the distance of the
piers (pylons) to the fairway center line. The variation of the span distance of the
pylons show the plausibility of the developed and used modelling.

Keywords: Ship collision · Seagoing vessels · Load model · Collision model ·
Probabilistic ship impact

1 Introduction

DEGES GmbH is planning the new Suederelbe Bridge crossing Hamburg-Moorburg as
part of the BAB26 motorway at Elbe km 620 for the Hanseatic City of Hamburg. For
the planned bridge, bridge piers close to the bank are planned on both sides of the
fairway. Ship impact loads for the service lifetime condition cannot be derived directly
from the introduced relevant code EN 1991-1-7 (2006) and DIN EN 1991-1-7/NA

© The Author(s) 2023
Y. Li et al. (Eds.): PIANC 2022, LNCE 264, pp. 632–643, 2023.
https://doi.org/10.1007/978-981-19-6138-0_55

(2010) from a technical-economic point of view. This is due, on the one hand, to the insufficiently classified and blanket ship impact loads for seagoing vessels specified in EN 1991-1-7 and, on the other hand, to the pier positions provided in the bank area, so that a case-by-case investigation has been carried out with regard to ship impact loads.

On behalf of DEGES, Bundesanstalt fuer Wasserbau (BAW) determined the ship impact loads within the scope of a case-by-case consideration on the basis (methodology) of EN 1991-1-7 and adjacent German National Annex for the final condition and specified object-specific requirements regarding the proof for ship impact (e.g. geometries, e.g. deformation proofs,…). Traffic pattern, ship passages, speeds, accident occurrence, nautical conditions have been analysed and the impact loads for piers or protective bank structures are determined by a link of a load model with a collision model. In the absence of data, comparative observations have been made.

2 Bridge and Waterway Specifications for Suederelbe

2.1 Bridge Crossing Suederelbe

The new Suederelbe Bridge, Figs. 1 and 2, will be part of the BAB26, which in the eastern section will serve as a cross-connection between the motorways BAB7 and the BAB1 to accommodate supra-regional east-west traffic, improve the accessibility to Hamburg port, bundle long-range port traffic and thus relieve inner-city residential quarters in the south of Hamburg. Currently, work on the construction of the BAB26 from the west to the BAB7 is in full swing. In preparation for the work up to the BAB1, the approval procedures have been initiated in three sections and the designs are being planned. The total length of the new bridge infrastructure amounts to about 5 kms.

Fig. 1. Animation Suederelbe Bridge crossing, view from west to east; Source: DEGES

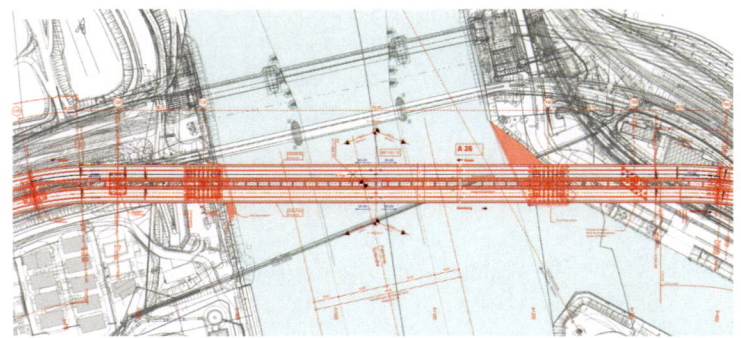

Fig. 2. Top view of the planned Suederelbe Bridge crossing; Source: DEGES

The new Süderelbe crossing in Hamburg will be designed as a cable-stayed bridge with two high piers (pylons), Figs. 1 and 3. The winning design of the competition comes from a German-Danish planning consortium. The construction is scheduled to begin in 2026.

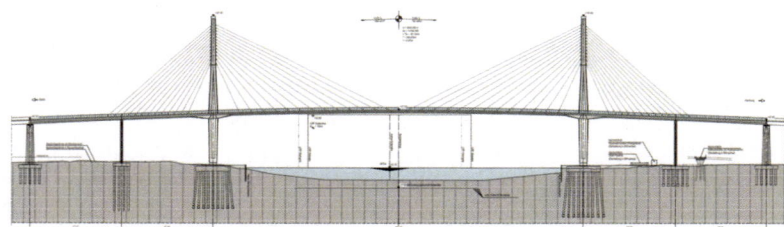

Fig. 3. Longitudinal section/view of the planned Suederelbe crossing; Source: DEGES

The current bridge of the new Suederelbe crossing is a five-span symmetrical cable-stayed bridge, with centrally arranged double cable levels in fan shape. The structure is designed as a semi-integral construction with spans of 86.05–86.75 - 350.00 - 86.75–86.05 m, total length is 695,6 m. The clearance under the bridge has a height of 53 m and thus approximately 51 m above the mean high tide, with a width of 150 m. The planned construction depth is - 17 m + NN corresponding to a water depth of 15 m plus 2 m for tolerances and scouring. The cable-stayed bridge is supported by pylon stems arranged centrally in the route axis with a height of about 150 m next to the Suederelbe and piers in the foreland areas. The superstructure consists of a one-piece cross-section consisting of two partial cross-sections coupled by cross girders, separated by a centrally arranged light gap, Fig. 4. The anchoring of the cables takes place within the inner cells of the superstructure cross-section adjacent to the light gap.

The western pier (pylon) stands on the edge of the Suederelbe and is protected by a bank wall parallel to the course of the river. On the east side, the pier is located directly at the entrance to the Hohe Schaar harbour a harbor for inland vessels, roughly in line with the eastern course of the river bank. To protect the piers additionally to the impact

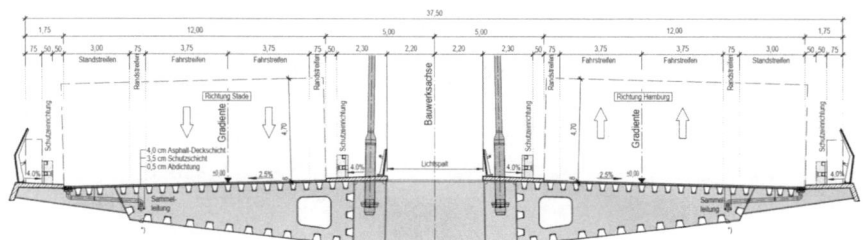

Fig. 4. Cross section of the planned Suederelbe crossing; Source: DEGES

design, bank walls and impact protection measures will be provided in front of the deep foundations.

The Suederelbe is tidally influenced, so that water levels of up to 6.9 m + NN can be expected at the bridge site. The planned heights of the ground surfaces in the area of the pier cap are 6.0 m + NN on the west and east sides. A bank wall in front is also being considered for the west side.

The locations of the planned piers (pylons) are in the bank or embankment area, so that direct ship impacts are not very likely to occur. However, an impact can be indirectly transmitted to the bridge pier foundations via the bank walls, which must be taken into account in the planning. Due to the nearby Kattwyk Bridge with a main opening of 108.5 m, which is directly adjacent to the planned Suederelbe Bridge crossing, the vehicle flows from the north are bundled to a lane width of B = 94 m.

Endangered by ship impact are therefore:

- the pylon pier on the east side from the south by seagoing and inland vessels and from the north only by inland vessels due to the shielding for seagoing vessels by a bridge pier of the Kattwyk Bridge,
- the pylon pier on the west side from the south by seagoing and inland vessels and from the north only by inland vessels because of the shielding for seagoing vessels by a bridge pier of the Kattwyk Bridge,
- the superstructure of the Suederelbe Bridge crossing.

2.2   Suederelbe Waterway

The Suederelbe is an approximately 16 km long section of the Lower Elbe in the area of the city of Hamburg, Fig. 5. The Suederelbe bypasses the historic port areas of Hamburg and instead passes Harburg and some newer port areas. Like the entire Lower Elbe, the Suederelbe is also subjected to the tides. The mean tidal range is about 3.60 m and leads to strong currents, Faltboot (2016).

On average, about 725 seagoing vessels sail per year and direction. The ships come repeatedly or regularly. The displacement ranges from approx. 200 m^3 to 107,000 m^3, the average value is 10,550 m^3; 98% of the transits for seagoing vessels remain below 50,000 m^3, Table 1. The most significant shares in the transit volume are accounted for by the general cargo ship with approx. 42% (length < 120 m), the chemical tanker with approx. 36% (length < 180 m), the tanker with approx. 13% (length < 201 m) and the

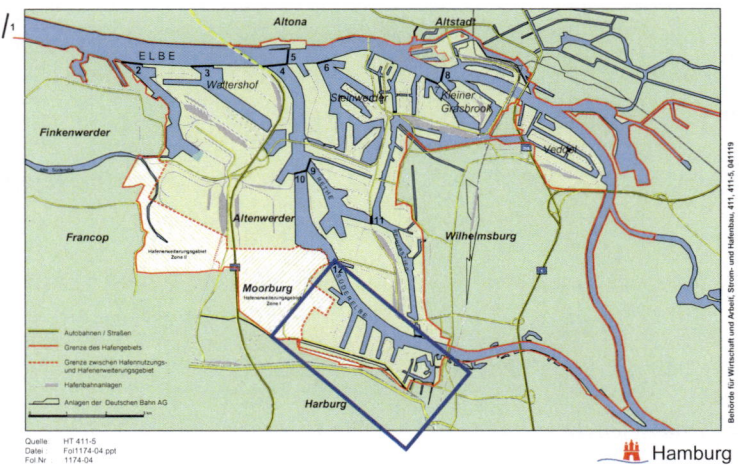

Fig. 5. Overview of the port of Hamburg and the Suederelbe; Source: Hamburg Port Authority

bulk carrier with approx. 8% (length < 204 m). Tug assistance is provided to vessels with a length of more than 200 m, in addition to wind forces of more than 8 Beaufort and visibility of less than 3 ship lengths. Every year, about 70 coal ships with up to 100,000 tdw for the Moorburg power plant station with a maximum discharge T = 12.8 m have been towed aft ahead.

There are about three times as many inland vessels as seagoing vessels, so that these are additionally taken into account with a vessel number in the size of 17,400 inland vessels in the smallest displacement class <5,500 m^3. For the planning horizon, which is the design service life of the new bridge from the start of construction, the updated vessel passages and distribution were retained after discussion with the client. The updated size distribution, which was divided into m^3-classes with a class width of 5,500 m^3, is shown in Table 1.

Table 1. Class frequencies of ship displacements for the Suederelbe crossing; Source: BAW (2016)

Displacement Class [m^3]	[%]	Sum [%]	Displacement Class [m^3]	[%]	Sum [%]
<5,500	89.10	89.10	<60,500	0.03	97.28
<11,000	3.09	92.19	<66,000	0.01	97.29
<16,500	0.84	93.03	<71,500	0.01	97.30
<22,000	0.73	93.76	<77,000	0.01	97.31
<27,500	0.61	94.36	<82,500	0.00	97.31
<33,000	0.83	95.19	<88,000	0.00	97.31
<38,500	0.80	95.99	<93,500	0.00	97.32
<44,000	0.57	96.57	<99,000	0.00	97.32
<49,500	0.52	97.09	<104,500	2.36	99.68
<55,000	0.16	97.25	<110,000	0.32	100.00

Table 1 shows an annual passage rate of about 3,000 ships per direction and year, a total of about 6,000 ships per year. Approx. 90% of all ships, including the inland waterway vessels, have a displacement of less than 11,000 m^3, approx. 95% less than 33,000 m^3.

3 Determination of Ship Impact

3.1 General

In general, the impact action is characterised by two load directions which do not occur simultaneously and which are assumed to act perpendicularly to the surface of the structural member:

– The frontal impact generally acts in the direction of travel (often in the longitudinal axis of a pier or abutment). In the case of frontal impact, the ship in distress is usually completely stopped.
– The lateral impact generally acts perpendicular to the direction of travel (often perpendicular to the longitudinal axis of a pier or abutment) and at the same time a frictional impact acts in parallel, i.e. in the longitudinal direction.

The point of application of the lateral impact should be set in such a way that the most unfavourable effect is achieved (moveable action).

According to EN 1991-1-7 (2006), C.4.4(2), probabilistic models of the basic variables determining the deformation energy or the impact behaviour of the ship can be used for the determination of the impact loads. A collision model according to EN 1991-1-7 (2006), B.9.3.3, has been used to determine the collision rate. For the treatment of the ship impact the impact load model was linked with the collision model in order to get a distribution of probable impact loads, with which the design value is determined using acceptable risk criteria.

3.2 Impact Mechanics

The physical determination of the impact load is based on the deformation energy in the impacting vessel. For the frontal impact the complete dissipation of the kinetic energy is assumed, for the lateral impact the deformation energy is calculated on the basis of the initial kinetic energy from an impact impulse calculation at collision. The determination of the impact loads is based on an empirically-analytically determined load-deformation relationship for seagoing vessels, Pedersen and Zhang (1998), EN 1991-1-7 (2006). The load-energy relationship for seagoing vessels is calculated according to equation (C.11) in EN 1991-1-7 (2006), here Eq. (1a/b). In contrast, the impacted structure is assumed to be a rigid structure, which leads to maximum loads in terms of structural dynamics. For deformable structures, this assumption of a rigid structure is no longer applicable and an adjustment of the contact force must be made according to the respective stiffnesses of the ship and the structure. For this project impact load was determined by:

$$F_{bow} = \begin{cases} F_o \cdot \overline{L} \left[\overline{E}_{imp} + (5.0 - \overline{L}) \, \overline{L}^{1.6} \right]^{0.5} & for \overline{E}_{imp} \geq \overline{L}^{2.6} \\ 2.24 \cdot F_o \left[\overline{E}_{imp} \overline{L} \right]^{0.5} & for \overline{E}_{imp} < \overline{L}^{2.6} \end{cases} \qquad (1a/b)$$

where as

$\overline{L} = L_{pp}/275\,m$, $\overline{E}_{imp} = E_{imp}/1425\,MNm$ and $E_{imp} = \frac{1}{2} m_x v_o^2$
and

F_{bow}	maximum bow impact force [MN];
F_o	reference value for the impact force = 210 MN;
E_{imp}	energy to be dissipated by plastic deformation;
L_{pp}	length oft he vessel [m];
m_x	mass including (hydrodynamic) added masses for longitudinal motion [10^6 kg];
v_o	impact velocity [m/s].

Equation (1b) applies to the traffic pattern under consideration. This also subsumes the inland navigation vessel quantity according to Table 1, which is on the safe side. The largest ship deformation smax.
is determined via

$$s_{max} = \frac{\pi \, E_{imp}}{2 \, F_{bow}} \qquad (2)$$

and the associated impact time T_0 is calculated with

$$T_o \approx 1.67 \frac{s_{max}}{v_o}, \qquad (3)$$

which can be used for dynamic analysis.

3.3   Impact Load Model

For the probabilistic determination of the effect of ship impact, an impact load distribution function is calculated taking into account the traffic and other data. The methodology is described in Kunz (2011). For this calculation, data are evaluated in terms of the frequentist probability concept.

Masses and frequencies of ships passing the bridge site are determined according to Table 1. Ships have been assumed as "fully loaded" which therefore is a conservative assumption. A ship frequency of 3,000 ships per year and direction can be expected. A change of these data for the future over the service lifetime of the planned bridge crossing is not seen. The speed distribution was set for southbound and northbound traffic with an average speed v = 10 km/h, the standard deviation was chosen from experience to be σ_v = 2.0 km/h. The speed distribution is not expected to change over the time periods considered. From comparative calculations a distribution for the impact angle is assumed, in which normal forces are acting on the impacted structure dependent on the impact angle and considering a sliding friction impact. Statistical evaluations of accidents at sea and on inland waterways allow the interpretation that not every ship collision is associated with the mechanically possible maximum impact. Therefore a log-normal impact severity function based on the above mentioned statistics has been applied. Combining the afore mentioned parameters leads to an impact distribution function, which indicates the probability of an impact load given a collision.

3.4   Collision Model

The probability of a collision is determined by a collision model, which considers the geometry of the waterway and the possibly impacted structure, the sailing line as well as the stopping ability of the ships. A specific collision model for inland waterways is shown in Fig. 6 and is mathematically described by Kunz (2011):

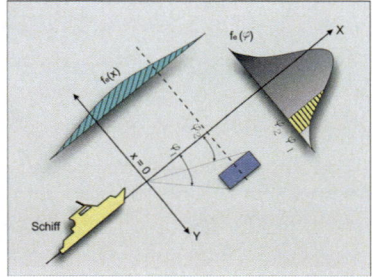

Fig. 6. Collision model for waterways; $f_s(x)$: avoidance probability; $f_s(\varphi)$: collision probability, Kunz (2011)

$$\lambda = \sum N_i * \int (d\lambda x/ds) * W_1(s) * W_2(s)\ ds \qquad (4)$$

where as:

λ the annual collision rate.

$\sum N_i$ the annual number of passing ships, if necessary according classes.

$(d\lambda x/ds)$ the stretch-referred accident rate.

$W_1(s)$ $= F\varphi\ (\varphi_1) - F\varphi(\varphi_2)$, the conditional probability of a collision way,

$W_2(s)$ $= 1 - F_x\ (s)$, the conditional probability that the collision cannot be avoided.

The number N of passing ships or vessels per year and direction at the bridge crossing was determined to 3,000 passages per year and direction. The stretch-referred accident rate $(d\lambda x/ds)$ is statistically evaluated from observations, whereby e.g. only collisions which are relevant for bridge accidents are considered. Based on accident statistics for the years 2010 to 2015 in Hamburg harbour, adjacent maritime accidents were determined to be 32 accidents per year on average. During this period, an average of 18,699 seagoing vessels (passages) were in service. The average distance travelled by seagoing vessels entering and leaving the port of Hamburg is calculated to be 15 km. Thus, a mean accident distance rate URSKM $= 32/(18,699 * 15) = 1.14 * 10^{-4}$ [accidents/(passage * km)] can be calculated. Since the evaluation is statistically an estimate of events that vary over the years, a 95% confidence interval for URSKM is determined. This leads to a URSKM of $URSKM_{KONF,95\%} = 1.2 * 10^{-4}$[accidents/(passage * km)]. However, an URSKM $= 1.5 * 10^{-4}$ [accidents/(passage * km)] was used by expert judgement because of certain imponderables (data basis, possible trend).

Concerning the collision probability $W_1(s) = F_{\varphi}(\varphi_1) - F_{\varphi}(\varphi_2)$ and $W_2(s) = 1 - F_x(s)$, see Eq. (4), a large number of influences can determine these random variables φ and s. A normal distribution is assumed for them in accordance with the central limit value theorem of the probability calculation, see Pfaffinger (1989). Therefore, mean value and standard deviation are sufficient for the determination of these distributions, which are included in the calculations from observations and empirical values. Thus, using a mean value of 0° and a standard deviation of $\sigma_{\varphi} = 10°$, it is taken into account that within a probability of 98% all damaged ships move in an angular fan of $\pm 20°$ towards the bridge component under consideration. The avoidance of a collision, probability of avoidance W2(s), depends substantially on the stopping ability of the ships, which can be determined due to permission tests in dependence of the drive power and the technical equipment. For this project a stopping distance between 300 m and 800 m with a mean value s = 500 m and a standard deviation $\sigma_s = 150$ m was taken into account.

The collision probabilities for the frontal and lateral collision of loaded ships with the piers were calculated. Due to the bundling of the seagoing vessels arriving from the north and the shielding effect of the bridge piers of the nearby Kattwyk Bridge, it is very unlikely that the Suederelbe Bridge crossing will be endangered from the north. For vessel traffic in the bridge passage, it is assumed that north and south passages travel centrally in the bridge passage, which is underlined by the bundling effect of the vessel passages through the Kattwyk Bridge. For the access of ships to the bridge from the north and from the south, a sailing path length of 1,000 m is taken into account in each case. The piers endangered by a possible ship collision are shown with the respective dimensions of about l x b = 25 m * 25 m.

The relevant collision rates are shown in Table 2 below (FF = frontal impact, FL = lateral impact, vN = from north, vS = from south). Thus, statistically, out of the reciprocal $(1/\Sigma\lambda)$ a ship collision every approx. 470 years on average is to be expected for pier east, while the value for pier west is approx. every 1,240 years on average.

The link between the impact load distribution function and the collision rate using the mathematical relationship

Table 2. Calculated collision rates [1/year]

Pier	$\lambda_{FF,vN}$	$\lambda_{FF,vS}$	λ_{FL}	$\Sigma\lambda$
Western pier	=0	=0	0.00081	0.00081
Eastern pier	=0	0.0016	0.00053	0.00213

$$\lambda * t_R = \frac{1}{1 - F_p(F)}, \tag{5}$$

Is based on the POISSON process, see Kunz (2011). It results in a dimensionless impact load distribution function by the product $\lambda * t_R$. from collision rate [1/year] and return period t_R [year] over the adjacent distribution function $F_p(F)$ of the dynamic impact load F_{dyn} in [MN], calculated for frontal impact, Fig. 7, and lateral impact. From this, a design impact load can be determined for a specified or economically optimised limiting risk, where t_R is the return period between the acceptable design impact events.

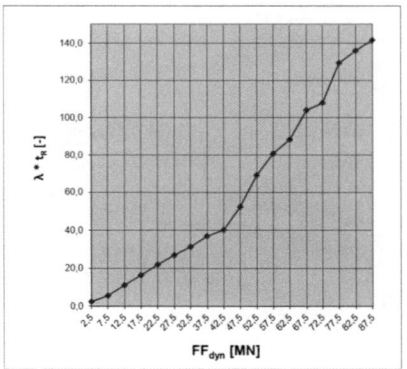

Fig. 7. Impact load distribution $\lambda * t_R. = f(F_P(F) = FF_{dyn})$ for frontal impact; Source: BAW (2016)

The representative value of this accidental impact is determined according to DIN EN 1991-1-7/NA (2010), with an exceedance probability of the impact energy or the impact load of $p = 10^{-4}$/year. This represents the above mentioned limiting risk. Out of this the return period t_R is determined as $1/(10^{-4}$ /year) = 10,000 years.

3.5 Impact Forces

With the return period of 10,000 years and the collision rates out of Table 2 and Fig. 7 the impact loads for the eastern pier is determined to $FF_{dyn} = 17.5$ MN for frontal impact, see Fig. 7, and $FL_{dyn} = 4.0$ MN for lateral impact. For the western pier there is only a lateral impact of $FL_{dyn} = 4.0$ MN. A dynamic calculation has been recommended. Guidance in this regard is given in EN 1991-1-7 (2006). The heights of the impact loads in relation to the water level and the impact area are to be applied according to EN. An impact load of 10% of the frontal impact load must also be applied to the superstructure in accordance with EN.

3.6   Implementation of Ship Impact

If necessary, the determined impact loads can be transferred from the bridge into the building ground. Bored piles with a diameter of 1.5 m are planned for the pile foundations, which will also have foot extensions. Additional safety in case of ship impact is provided by combined pipe sheet piles up to a height of +5.60 m + NN (sea level) in front of the pier foundations, which are also intended to counteract the sliding of ships. In order to dampen a ship's impact on the bank wall, the pipes are designed as composite piles and a solid concrete girder is provided as the upper termination.

Underwater concrete bases are required below the pile cap, which are also used to brace the walls against the soil behind the foundation. To counter water-side deformations at the wall head, 2 anchor rods to the pile cap are provided on each of the pipe piles. In the event of an impact, the anchors are supported in a sliding manner so that buckling can be ruled out.

4   Discussion and Outlook

According to the state of the art, which is currently represented by EN 1991-1-7 (2006) and DIN EN 1991-1-7/NA (2010), impact loads from ship collision were determined on a probabilistic basis for a verification of the piers of the planned Suederelbe Bridge crossing which piers will be located at the bank.

The determined loads are below the force values given in Table C.4 of EN 1991-1-7, even taking into account a harbour situation. On the one hand this is due to the generalised specifications in the code and, in this case, to the deterministically determined values for seagoing vessels in the EN. On the other hand, the site-specific individual case determination allows the special circumstances of the bridge pier location close behind the bank.

A plausibility check of the calculations was carried out und is suggested by Fig. 8, in which the distance of the east pier closer to the fairway axis was simulated almost to the "blockade" with the probabilistic model used. An increase of the model impact load

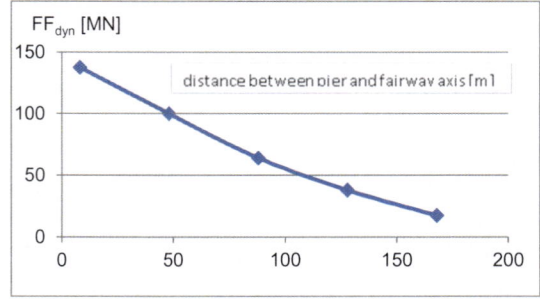

Fig. 8. Dynamic impact load dependent on the distance between eastern pier and fairway axis; Source: BAW (2016)

towards the deterministic value according to Eq. (1a/b) and also EN becomes clear, because the collision probability dwindles to zero. However, the deterministic values would not fully achieved, as ship displacements and speeds are still represented by variable distributions in the simulation. But, without these, one would come close to the deterministic value according to EN 1991-1-7.

References

BAW (2016) Ermittlung von Schiffsstoßlasten für die Süderelbequerung A26-Ost, Hamburg, Elbe-km 620. Expert opinion BAW-No. A395 100 10115, Karlsruhe (unpublished)

EN 1991-1-7 (2006) Eurocode 1: Basis of design and actions on structures – Part 1.7: Actions on structures – Accidental actions due to impact and explosions. CEN, Brussels

DEGES (2013) A26 Süderelbebrücke Moorburg – Dokumentation des Realisierungswettbewerbs. DEGES, Berlin, Juni 2013

DEGES (2016) Vorstellung des Wettbewerbsentwurfs für die Süderelbebrücke. Abgerufen unter http://www.deges.de/Startseite/A-26-Entwurf-fuer-neue-Suederelbebruecke-vorgestellt-E1215.htmam31.05.2016

DIN EN 1991-1-7/NA (2010) Eurocode 1: Einwirkungen auf Tragwerke - Teil 1–7: Allgemeine Einwirkungen, Außergewöhnliche Einwirkungen; Nationaler Anhang. Beuth-Verlag, Berlin. Meanwhile updated to version (2019)

Faltboot (2016) http://www.faltboot.org/wiki/index.php?title=S%C3%BCderelbe&printable=yes.Abgerufenam22.06.2016

Kunz C (2011) Probabilistic Modeling of Vehicle Impact as Accidental Action. In: International Conference on on Applications of Statistics and Probability in Civil Engineering 2011, Zuerich, Switzerland, 2011

Pedersen PT, Zhang S (1998) On Impact mechanics in ship collisions. In: Marine structures, vol 11, no 10

Pfaffinger D (1989) Tragwerksdynamik. Verlag Springer, Wien. https://doi.org/10.1007/978-3-7091-9026-5

Study on Design of Ship Lift Auxiliary Lock Water Filling/Emptying System and Operation Mode of Lock Gate

Xican Zhao[1], Chao Guo[2(✉)], Lei Wang[3], and Zhonghua Li[4]

[1] Yunnan Provincial Highway Engineering Supervision Consulting Co. Ltd, Kunming, China
1269349381@qq.com

[2] Nanjing Hydraulic Research Institute, Nanjing, China
cguo@nhri.cn

[3] Baise Hydro Project Navigation Investment Co. Ltd, Baise, China
30850955@qq.com

[4] Nanjing Hydraulic Research Institute, Nanjing, China
zhli@nhri.cn

Abstract. The docking between ship lift chamber and upstream/downstream is an important link in the whole process of ship lift operation. The wave and flow movement generated by the hub operating downstream can easily lead to the failure of ship lift docking or ship grounding. To deal with the problem of large amplitude and high frequency of water level fluctuations downstream of Baise ship lift, an auxiliary lock is set up at the lower lock head of the ship lift. The extreme lift height of Baise auxiliary lock is 5.64 m with a two-way hydraulic head, and the size of the chamber is large, therefore it is faced with the problems of water filling/emptying efficiency and ship navigation safety. In this paper, a kind of water F/E system combining water filling under the gate and grid energy dissipation type is designed, and a scale of 1:30 Baise auxiliary lock chamber physical model and three-dimensional mathematical model are established. The berthing conditions of ships in the auxiliary lock under different water level variations and lock gate operation modes are analyzed. Besides the hydraulic characteristics of the lock filling/emptying system are measured. The operation mode of lock gate under different water level variations is recommended.

Keywords: Auxiliary lock · Filling and emptying system · Berthing condition · Hydraulic characteristics · Operation mode of lock gate

1 Introduction

Ship lift and lock are two main types of navigation structures (Niu et al. 2007), among which ship lift has the advantages of small water consumption, fast operation speed and less technical restrictions on lifting height (Hu et al. 2016). It is especially suitable for navigation of high dams and has been widely used in navigation of mountainous rivers in central and western China in recent years.

Y. Li et al. (Eds.): PIANC 2022, LNCE 264, pp. 644–655, 2023.
https://doi.org/10.1007/978-981-19-6138-0_56

The docking between the ship chamber and the upper/lower approach is an important link in the whole process of ship lift operation (Hu et al. 2016). The hydro-junction operation such as flood discharge and powerplant generation can easily lead to large fluctuations of the water surface near the head and chamber of the ship lift. If the water level changes too fast during chamber docking, the chamber cannot be docked (Zhang Yong 2016). If the water level of the channel changes too much during chamber docking, the ship may hit bottom or the water level in the ship chamber exceeds the total height of the ship chamber structure and water overflow (Shang et al. 2020).

To deal with the problems of steep rise and fall of river water level in mountainous areas of central and western China and fluctuation of water level of hub operation, China invented a unique launching type vertical lifting ship lift with wire rope winch (Wang et al. 2013). The ship chamber of launching ship lift directly enters the water. If the ship chamber is large in scale, launching ship lift requires a large motor to enhance power (Li et al. 2016). At the same time, it will also face mechanical equipment processing manufacturing and layout problems.

China has conducted theoretical study and engineering practice on building up an auxiliary lock to cope with unsteady flow changes during ship lift docking, in addition to adopting the type of launching ship lift and warning measures of unsteady flow at the hydro-junction. The measures of setting auxiliary locks to block unsteady flow from downstream to ship chamber were proposed for the first time in the study on the influence of unsteady flow generated by the operation of TGP project on the docking operation of ship lift (Qi et al.2013), but they were not put into practical application. Also on the Yangtze River, the Xiangjiaba ship lift installs an auxiliary lock downstream of the ship lift to isolate the water level of the ship chamber from that of the downstream, avoiding the influence of water level changes, which is a first-time engineering approach for solving the problem (Hao et al. 2020; Mei et al. 2020).

In the navigation facilities to be built in Baise junction on the Youjiang River, the ship lift adopts the fully balanced steel wire rope winch lift. The Baise ship lift exit is 0.5 km below the Dongsun Power Station dam (Preliminary Design of Navigation Facilities Of Baise Water Conservancy Project in Guangxi (2021)). The Dongsun Power Station's discharge operation causes a huge variation and high frequency in the downstream water level of the Baise ship lift. An auxiliary lock is also set downstream of the ship lift to tackle the problem of water level variation downstream of the ship lift. The Baise ship lift auxiliary lock consists of a transition and berthing section that can accommodate two lines of vessels berthing inside the lock chamber, hence increasing the ship lift's traffic capacity., but the lock chamber area is increased to 2.39 times that of Xiangjiaba auxiliary lock chamber. Bearing the two-way hydraulic head, the extreme lift height of the Baise auxiliary lock is 5.64 m. The Xiangjiaba auxiliary lock's simple water filling/emptying (F/E) system under the gate cannot be employed. To increase water filling/emptying efficiency, minimize the transit time of ships through the ship lift as much as feasible, and maintain the safety of vessels berthing in the lock chamber, a reasonable water F/E system and gate operation mode must be determined.

2 Project Overview

The main buildings of navigation facilities of Baise Hydro Project Navigation include a saving-water lock, intermediate channel, navigable aqueduct, vertical ship lift, and so on. The design of the ship type is 2 × 500 ton fleet and 1000 ton single vessel.

The Baise ship lift auxiliary lock has a total length of 228 m, a progressive section length of 90 m, and an orifice width that gradually changes from 12.0 m to 34.0 m (Fig. 1).The straight part is 138 m long, and the lock's sill elevation is 109.7 m. The auxiliary lock's effective scale is 120 m × 34 m × 4.7 m (length, width, and sill depth). For ordinary operation, the lower head of the auxiliary lock uses a vertical lift gate for water retention.

This paper proposes a F/E system combining gate and grid energy dissipation type, and establishes a physical model of Baise auxiliary lock (scale 1:30) to evaluate the berthing condition of vessels in auxiliary lock and hydraulic characteristics of F/E system, then recommends the vertical gate operation mode of the auxiliary lock, which can provide a technical basis for design.

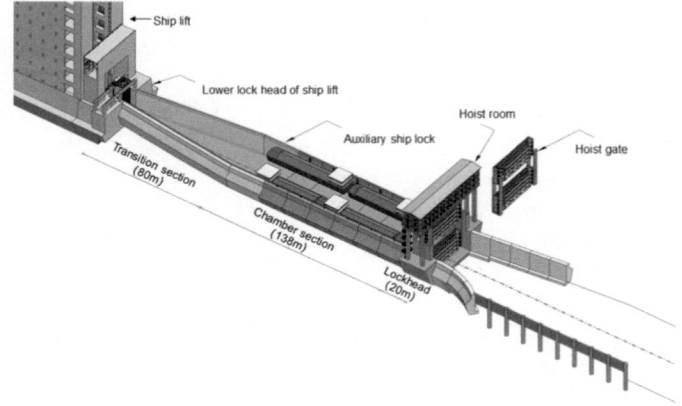

Fig. 1. Baise Ship lift and its auxiliary lock.

3 Water Filling/Emptying System Selection and Arrangment

3.1 Water Filling/Emptying System Selection

The filling/emptying systems can be divided into two main types. One is the filling and emptying"through the heads", and the other is the "through longitudinal culverts" system (PIANC report N 106 (2009)). According to Chinese Design Code for Filling and Emptying System of Shiplocks (JTJ 306-2001, China), the formula $m = T/\sqrt{H}$ can be used to select hydraulic system for inland navigation locks. In which, H(m) is the lift height of lock and T(min) the time to fill the chamber. According to the fluctuation frequency of downstream water level and traffic capacity of Baise ship lift, the design water filling/emptying time T is no more than 3.5 min when H = 1.5 m under normal condition, and T is no more than 5.0 min when H = 3.0 m under design condition for Baise ship lift auxiliary. Substitute the water filling time and lift height of Baise ship lift into the formula, $m = (3.5 \sim 5.0)/\sqrt{1.5 \sim 3.0} = 2.86 \sim 2.89$. According to the Chinese specification requirements,if $2.5 < m < 3.5$, the two types of water filling and emptying system can be selected.Considering the few days of extreme working condition of Baise ship lift every year, the normal lock lift is only 1.5 m, it is proposed to adopt "through the heads" system for downstream auxiliary lock.

3.2   Water Filling/Emptying System Layout

The Baise ship lift's auxiliary lock uses a through-the-head technology that combines a flat vertical gate with grid energy dissipation to ensure that the water energy is effectively dissipated and the water transit time is shortened once the ship enters the chamber. The grid energy dissipation chamber has a vertical gate at the bottom. Water flow energy dissipates as water through the grid energy dissipation when the vertical gate is lifted and the bottom of the vertical gate does not exceed the sill level, ensuring uniform flow and proper energy while entering the chamber. When the water level difference between upstream and downstream reaches a specific point, lift the vertical gate, and the water flow through the vertical gate and grid energy dissipation chamber significantly increases, reducing the time it takes for the water to fill and empty. This plan can not only adapt to the operation of a low lift height lock, but also the vertical gate of the lock is involved in water filling/emptying, eliminating the need for another culvert valve, allowing the project to save work while boosting the lock's water filling/emptying efficiency (Figs 2 and 3).

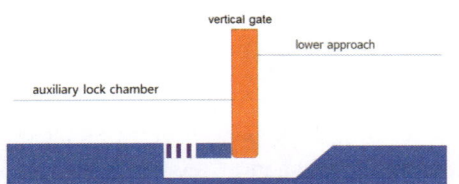
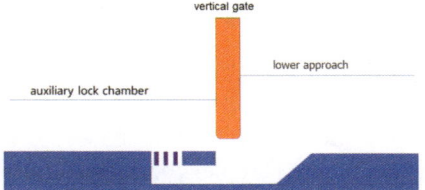

Fig. 2. The vertical gate is lifted to the sill and water go through the grid energy dissipation chamber.

Fig. 3. The vertical gate is lifted above the sill level and water go through the grid energy dissipation chamber and under the gate.

Two water inlets share one grid energy dissipation chamber, for a total of four grid energy dissipation chambers with a total volume of 196 m^3. The outlet grid is located at the top of the grid energy dissipation chamber; the length of the grid is 4.4 m and the width is 0.3 m; each grid energy dissipation chamber has 8 outlet grids, totaling 32 grids with a 42.24 m^2 area.

Two water inlets share one grid energy dissipation chamber, a total of four grid energy dissipation chambers, size of 7.0 m × 5.0 m × 1.4 m (length × width × height), total volume of 196 m^3. The top of the grid energy dissipation chamber is provided with the outlet grid, the length of the grid is 4.4 m, the width is 0.3 m, each grid energy dissipation chamber is provided with 8 outlet grids, a total of 32 grids and area of 42.24 m² (Fig. 4).

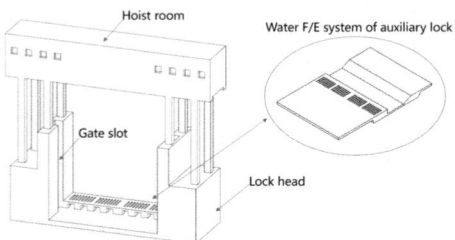

Fig. 4. Baise lock gate slot and its water F/E system.

The Baise ship lift auxiliary lock hydraulic physical model was created using the gravity similarity criterion and a length scale of L = 30. Auxiliary lock, water F/E system (containing vertical gate and grid energy dissipation chambers), and upper/lower approach are all included in the hydraulic model. The berthing conditions of the ship in the lock and the hydraulic characteristics of the water filling/emptying system are investigated using the physical model, the rationality of the water filling/emptying system design is confirmed, and the operation mode of the vertical gate of the auxiliary lock is recommended.

4   Berthing Conditions of Vessels in Lock Chamber

To evaluate the berthing circumstances of vessels in lock chamber, the Baise auxiliary lock chamber physical model and three-dimensional mathematical model are constructed. Because 2 × 500T fleets are uncommon in the Youjiang River, this paper focuses on the berthing conditions of 1000T vessels. The length, width, and full load draft of the 1000T vessels are 67.5 m × 10.8 m × 2.9 m, and its displacement is equivalent to the European CEMT-IV.

4.1   3D Mathematical Model Evaluation

The flow state when water fills the chamber and the mooring conditions of vessels in the chamber are computed first, using a three-dimensional mathematical model. The river's water level is 1.50 m higher than the auxiliary chamber (typical encountered), and the vertical gate bottom raises to sill elevation (109.7 m) at a speed of 2 m/min, according to the calculations.

Figure 5 depicts the typical flow pattern in the gate chamber while the auxiliary chamber is filled with water. The water flow in the chamber is rather stable during the auxiliary gate chamber filling process, as shown in the figure, and the water flow velocity in the inlet behind the gate exceeds 4m/s during the vertical gate raising procedure.

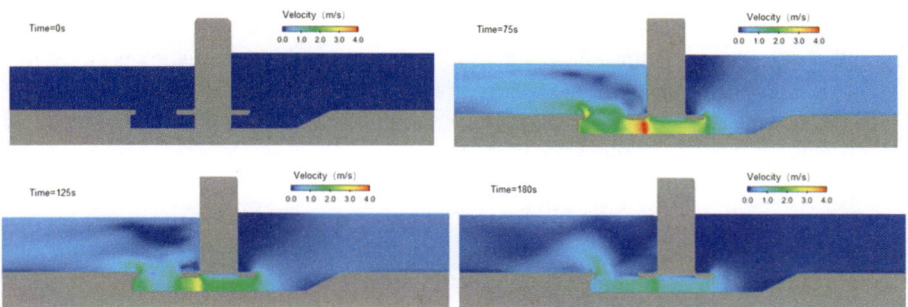

Fig. 5. Flow profile at the vertical gate in the chamber during water filling

The maximum longitudinal water surface slope and mooring force on the vessel when water filling the chamber calculated by the three-dimensional mathematical model are shown in Fig. 6. It can be seen from the figure that in normal conditions, when the vertical gate bottom is lifted to 109.7 m at a rate of 2 m/m, the maximum longitudinal water surface slope is 2.28‰. According to European Longitudinal force criteria for CEMT-IV class, the longitudinal water gradient of lock is required to be no more than 1.1‰ when locks filling through lock head (PIANC report N 106(2009)). So the lifting speed of the flat gate needs to be reduced to ensure the safety of ship berthing.

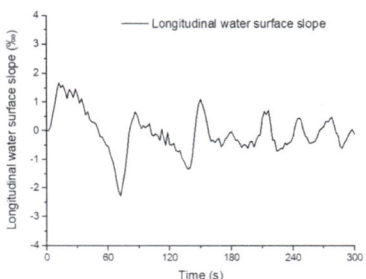

Fig. 6. Longitudinal water surface slope and mooring force on the vessel in the chamber.

4.2 Physical Model Evaluation

When the downstream river water level is higher than the lock chamber, the berthing circumstances of vessels in the lock chamber are termed extreme (H = 5.64 m), design (H = 3.00 m), and normal (H = 1.50 m) in physical model evaluation. The test vessels were of the 1000T class and carried a heavy load (67.5 m × 10.8 m × 2.9 m). For the test, the vessels were lined up in a single row in the calm part of the auxiliary lock. The vertical gate lifts at a rate of 2 m/min and 1 m/min, respectively. The lifting elevation of the vertical gate bottom in the physical model test at extreme and design conditions is the sill elevation (109.7 m) and 1m above the sill (110.7 m). The vertical gate is continuously elevating above the water in normal condition.

The maximum value of mooring force under the conditions of extreme head, design head and common head with different gate lifting speed and mode is shown in Table 1 and Table 2. Typical curves of mooring force of 1,000T vessel are shown in Fig. 7.

As shown in Table 1, the maximum longitudinal mooring force of a 1000T single vessel berthing in the upper lock is 70.93 kN, 47.14 kN, and 25.85 kN, respectively, when the vertical gate lifts to the sill elevation of 109.7 m at the speed of V = 2 m/min under three working conditions of extreme condition, design condition, and normal condition. Except for the normal condition, the mooring force values mentioned above are significantly higher than the standard requirements. The lifting speed of the vertical gate is too fast, given that the maximum longitudinal mooring force of the vessel in the chamber is mostly affected by the flow rate rise of $\Delta Q/\Delta T$ at the beginning stage of gate lifting (Huang et al. 2016), which leads to a large flow rate increase at the initial stage of water filling in the auxiliary chamber. Therefore, the lifting speed of the vertical gate shall be reduced.

The gate lifting speed is lowered to V = 1 m/min in extreme condition, and the increasing rate of beginning water flow is reduced when the vertical gate is raised to 109.7 m (i.e. sill elevation). A designed 1000T vessel's maximum longitudinal mooring forces are also reduced to 31.80 kN, which meets the China code's standards. If the vertical gate is lifted to 110.7 m (i.e., 1m higher than the sill elevation) at a speed of V = 1 m/min in order to improve the water traffic efficiency, not only does the energy dissipation grid chamber pass the flow, but additional flow flows beneath the vertical gate, increasing the maximum longitudinal mooring force of the specified 1000T single vessel moored in the chamber to 41.65 kN, which exceeds the code's standards. Therefore, it is recommended that the vertical gate be lifted to 109.7 m at a speed V = 1 m/min in extreme condition.

Under design condition, when the vertical gate lifts to 110.7 m at the speed V = 1 m/min, the maximum longitudinal and transverse mooring forces of the designed 1000T single vessel moored in the chamber are 24.54 kN and 8.73 kN respectively, which meet the requirements of the code. Therefore, in the design condition, it is recommended that the vertical gate be opened to 110.7 m at the speed V = 1m/min.

When the vertical gate lifts to 110.7 m at a speed V = 1 m/min under normal condition, the maximum longitudinal and transverse mooring forces of the designed 1000T single vessel mooring chamber are 14.39 kN and 3.57 kN, respectively. When the vertical gate is lifted above the water continuously, the maximum longitudinal and transverse mooring forces of the designed 1000T single vessel berthed in the chamber are 15.53 kN and 3.38 kN respectively. As the water level difference between upstream and downstream is small, the lifting speed of gate is slow. When the gate is lift to the sill evevation, the water level of upstream and downstream is basically flat, so the increase of ship mooring force when the gate is continuously lifted is small. The mooring forces under the two operation modes of the vertical gate both meet the requirements of the code. Considering the operation efficiency of the lock, it is recommended to lift the gate continuously above the water at a speed of V = 1 m/min in normal condition.

Table 1. Maximum mooring forces for 1000T vessels in lock chamber (energy dissipation way: Grid).

Lift Height H (m)	energy dissipation way	Elevation of gate stop position(m)	Longitudinal mooring force(kN)	Bow transverse mooring force (kN)	Stern transverse mooring force (kN)	Gate lifting speed (m/min)
5.64	Grid	109.7	70.93	18.71	10.49	2
3.00			47.14	11.32	3.49	
1.50			25.85	5.61	1.76	
5.64			31.80	8.33	8.59	1
3.00			21.71	6.21	4.08	

Note: The allowable value of longitudinal force for 1000T vessel is 32 kN, and that of transverse force is 16kN (Liu et al, 2014).

Table 2. Maximum mooring forces for 1000T vessels in lock chamber (energy dissipation way: Grid+gate).

Lift Height (H) (m)	energy dissipation way	Elevation of gate stop position (m)	Longitudinal mooring force (kN)	Bow transverse mooring force (kN)	Stern transverse mooring force (kN)	Gate lifting speed (m/min)
5.64	Grid + gate	110.7	41.65	11.31	6.71	1
3.00			24.54	8.73	3.50	
1.50			14.39	3.57	2.35	
1.50		Continuous	15.53	3.38	1.33	

Note: The allowable value of longitudinal force for 1000T vessel is 32 kN, and that of transverse force is 16 kN (Liu et al, 2014).

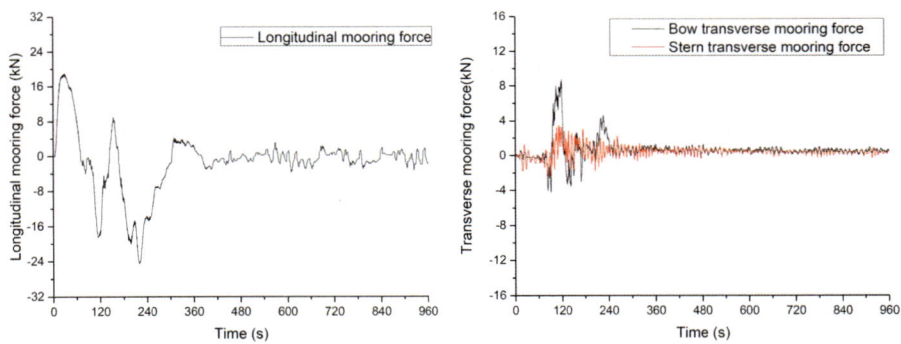

Fig. 7. Mooring force curves of 1000t vessel berthing in upper chamber. (H = 3.00 m, lift the vertical gate to 110.7 m at a speed V = 1 m/min)

5 Hydraulic Characteristics of Auxiliary Chamber

The physical model of the auxiliary lock of Baise ship lift adopts the recommended water filling and emptying system and gate operation mode. The measured hydraulic characteristic values of the water filling and emptying process are shown in Table 3, and the curves of the water filling flow process in typical working conditions are shown in Fig. 8 and Fig. 9.

As can be seen from Table 3, when the vertical gate is lifted at a speed of 1 m/min to 109.7 m at extreme condition, the water filling time of the chamber is 513 s, the maximum flow rate is 161.28 m^3/s, and the maximum water surface rising speed U_{max} of the chamber is 1.22 m/min.

When design condition, the vertical gate is lifted at the speed of 1 m/min to 110.7 m, the water filling time of the chamber is 239 s, which can meet the requirements of the design locking time T (not more than 5 min). The maximum flow rate is 162.41 m^3/s, and the maximum water surface rising speed U_{max} of the chamber is 1.20 m/min, which can meet the requirements of the specification.

By comparing the filling flow curves in the chamber under different conditions (Fig. 8 and Fig. 9), it can be seen that when the height of the gate stop elevation is 109.7 m, water only enters the lock through the grid energy dissipation chamber, and the flow curve shows a peak. When the height of the gate stop position is 110.7 m, the water flow first dissipates through the grid energy dissipation chamber, then the water flow enters the lock through the bottom of the gate and grid energy dissipation chamber as the height of the gate bottom exceeds 109.7 m, so there are two peaks in the curve of the flow rate.

When the vertical gate is continuously lifted at the speed of 1m/min at normal condition, the water filling time of chamber is 214 s, which basically meets the requirement that the design water filling/emptying time (no more than 3.5 min). The maximum water filling flow rate and the water surface rising speed of chamber are greatly dropped compared with the extreme and design conditions.

In extreme conditions, if the vertical gate is lifted at a speed of 1m/min and the gate stops at 109.7 m, the emptying time of the lock is 438 s, the maximum flow rate is 167.71 m^3/s, and the maximum water surface falling speed of chamber is 1.33 m/min.

When the design and normal conditions, the characteristic values of emptying time, maximum emptying flow rate and water surface falling speed of the chamber are greatly reduced compared with the extreme condition, which can meet the specification and design requirements.

Therefore, the recommended gate operation mode can meet the water filling/emptying time requirements of the lock, and the water filling and emptying system layout is reasonable.

Table 3. Hydraulic Characteristics of Filling and Emptying System in Chamber.

Conditions	H (m)	V (m/min)	Elevation of gate stop position (m)	energy dissipation way	T (s)	$Qmax$ (m³/s)	$Umax$ (m/min)
Filling	5.64	1.0	109.7	Grid	513	161.28	1.22
	3.00		110.7	Grid + gate	239	162.41	1.20
	1.50		Continuous	Grid + gate	214	85.57	0.63
Emptying	5.64	1.0	109.7	Grid	438	167.71	1.33
	3.00		110.7	Grid + gate	229	163.22	1.22
	1.50		Continuous	Grid + gate	194	86.70	0.64

Note: H is the lift height, V is the lifting speed of the gate, T is the water filling and emptying time of the chamber, Q_{max} is the maximum flow rate, U_{max} is the maximum chamber water surface rising (falling) speed.

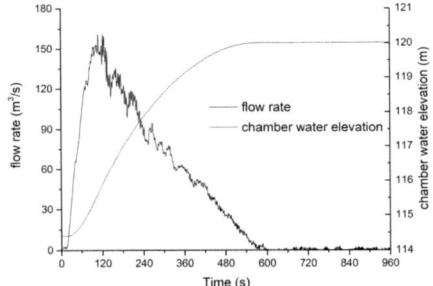

Fig. 8. Hydraulic characteristic curves of auxiliary chamber (H = 5.64 m, lift the gate to 109.7 m at a speed V = 1 m/min).

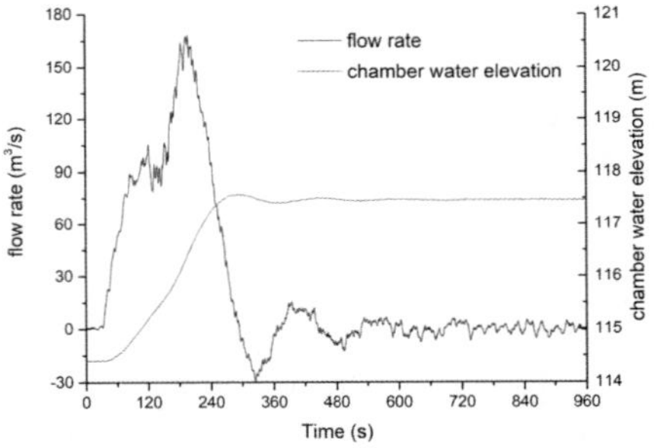

Fig. 9. Hydraulic characteristic curves of auxiliary chamber (H = 3.00 m, lift the gate to 110.7 m at a speed V = 1 m/min).

6   Conclusions

(1) The Baise ship lift's auxiliary lock uses an F/E system that combines gate + grid energy dissipation to satisfy the needs of rapid filling and emptying of the auxiliary chamber under various fluctuations in downstream water level. The hydraulic system is well-designed, and the use of an auxiliary lock to deal with changes in downstream water level is feasible.

(2) The recommended operation of the auxiliary lock gate under various lift heights is proposed based on model testing. The mooring force of the planned 1000T single vessel may meet the criteria of Chinese standard when the gate is operated as recommended.

(3) The physical model test measured the design condition (H = 3.00 m), the water filling time of the lock chamber is 239 s, and the maximum water filling flow is 162.41 m^3/s under the recommended lock gate operation mode. The water filling time of the lock chamber is 214 s at normal condition (H = 1.50 m), and the maximum water filling flow is 85.57 m^3/s. The water filling/emptying system structure is suitable, and the recommended gate operation mode can meet the lock's water filling/emptying time requirements.

Acknowledgements. The authors would like to thank Huang Yue, An Jiangfeng for their help and advise on building the physical model and data collection.

References

Hao Y, Huang R, Lu H (2020) Design introduction of Xiangjiaba Ship lift Auxiliary Chamber. China Water Transp 2020(9):73-75

Hu Ya 'an, Li Zhonghua, Li Y, Xuan G (2016) Research progress of large ship lift in China. Water Transp Eng 2016(12):10–19

Hu Y, Li Z, Lai D (2016) Influence of different tumble gate operation of ship lift chamber on vessels mooring force. Water Transp Eng 2016(12):148–152

Huang L, Tao G (2016) Channelization Engineering. People's Communications Press, Beijing

Li Q, Huang J, Wang M (2016) Design and application of large diameter drum in the wire rope-hoisting chamber-launching vertical ship lift. Mech Eng 2016(4):151–153

Liu Z, Wen X (2014) Hydraulic design manual part 2. China Water and Power Press, Beijing, p 2014

Mei X, Yang P (2020) Engineering measures and suggestions to realize the win-win situation of Xiangjiaba power station and shipping. China Water Transp (first half) 2020(9):88–89

Niu X, Song W (2007) Design of ship lock and ship lift. China Water and Power Press, Beijing

Nanjing Hydraulic Research Institute, Tianjin Water Transport Engineering Research Institute, 2001.Nanjing Hydraulic Research Institute, Tianjin Water Transport Engineering Research Institute (2001) Design Code for Filling and Emptying System of Shiplocks (JTJ306-2001), Beijing. People's Communications Press

Preliminary Design of Navigation Facilities of Baise Water Conservancy Project in Guangxi (2021) Zhongshui Pearl River Planning and Design Co., Ltd., Yangtze River Planning and Design Research Co., Ltd

Qi J, Zhang Y, Feng X, Zheng W (2013) Preliminary study on the influence of unsteady flow in lower approach channel on the operation of three gorges ship lift by setting auxiliary chamber. China Water Transp 2013(3):46–47

Shang., H, Li, R.: (2020) Influence of water level change on downstream docking operation of three gorges ship lift chamber. Water Transp Eng 2020(2):62–66

The World Association for Waterborne Transport Infrastructure (2009) Innovations in navigation lock design (Report n^0 106–2009)

Wang J, Hu Y, Li Z (2013) Review of hydraulic characteristics of the wirerope-hoisting chamber-launching vertical ship lift. Hydro-Sci Eng 2013(3):83–91

Zhang Y (2016) Preliminary study on improvement of overlimit surge at lower gate head of Three Gorges Ship lift. Waterw Port 2016(4):416–421

Study on the Operation Safety Evaluation System of Ship Lock Combined with Variation Coefficient Method and Matter-Element Extension Method

Junman Li[1,2(✉)], Yaan Hu[1(✉)], Xin Wang[1(✉)], Mingjun Diao[2(✉)], and Mingjun Diao[2(✉)]

[1] State Key Laboratory of Hydrology-Water Resources and Hydraulic Engineering, Nanjing Hydraulic Research Institute, 210029 Nanjing, China
411076959@qq.com
[2] State Key Laboratory of Hydraulics and Mountain River Engineering, Sichuan University, 610065 Chengdu, China

Abstract. Ship lock is the most widely used, the most promising and the most important type of navigation structure in the world at present. It is extremely necessary to evaluate the operation safety of ship lock in service, which has great social and economic benefits. The construction of safety evaluation system is the key step of safety evaluation of ship lock operation. Based on this, this paper systematically studies the evaluation system of ship lock operation safety, including indicator system, weighting method and evaluation model. The main work and conclusions are as follows: Firstly, a multi-indicator hierarchical indicator system including five first-class indicators and forty-seven second-class indicators for the safety evaluation of ship lock operation is established, and four safety evaluation grades of normal, deterioration, early warning and shutdown are divided. Secondly, the process and model of ship lock operation safety evaluation based on extension theory are put forward. Finally, the evaluation result shows that operation safety grade of ship lock belongs to the first grade – normal state, it is consistent with actual situation, indicating that the evaluation system is reliable. At the same time, the method can also be applied to other fields such as dam health evaluation, dam aging evaluation and rock mass quality evaluation, which provides a basis for safety evaluation.

Keywords: Ship lock · Operation safety · Indicator system · Extension evaluation model

1 The Necessity of Conducting Evaluation

Ship lock is a kind of navigable building that enables the ship to overcome the concentration drop of the water level of the channel, which is mainly composed of three basic parts such as pilot channel, head of gate and lock chamber and corresponding equipment. It integrates hydraulic structure, metal structure, hydraulic system, electrical system and ancillary facility, which is an open complex giant system (Yao 2003; Zhang 2001). As an important node project on the waterway - an integral part of the

© The Author(s) 2023
Y. Li et al. (Eds.): PIANC 2022, LNCE 264, pp. 656–667, 2023.
https://doi.org/10.1007/978-981-19-6138-0_57

canalization hub, its operation safety plays a decisive role in the safety of inland waterway shipping. Failure to function properly in the event of a malfunction will result in the obstruction of the entire route, or even the suspension of navigation (Changjiang Waterway Bureau 2004), and even catastrophic consequences.

2 The Current Status of Ship Lock Operation Safety Evaluation

In terms of industry standards, the Ministry of Transport issued "Technical Specifications for the Maintenance of Navigation Buildings" (JTS 320-2-2018, 2018) in 2018. The technical status grade standard of ship lock equipment and facility and the evaluation content of detection result of navigational water flow condition, hydraulic characteristic of water transmission system and gate and valve are proposed. This was followed in 2019 by "Technical Specification for Safety Testing and Evaluation of Shipping Hubs" (JTS 304-2-2019, 2019), which involves specific requirements related to ship lock safety assessment.

In terms of engineering applications, the main research on the safety evaluation of ship lock operation is: MA Kolosov (Kolosov 2002) summarized the number and causes of Russian ship lock accidents from 1985 to 1998, and introduced the typical types of lock room failures, and established an accident risk assessment model of gate and lock wall to analyze the safety of ship lock. Relying on the scientific research topic "Research on the Safety Evaluation System of Ship Lock Steel Structure" of Hohai University, and combining with the herringbone gate structure of the Siyang Ship Lock in northern Jiangsu province of the Beijing-Hangzhou Canal, Xu Haifeng (Xu 2007) carried out safety analysis of gate structure. YE Senitskiy and NY Kuzmin (Senitskiy and Kuzmin 2012) studied the dynamic characteristic of ship lock. The precise design relationship formula between the inherent vibration and forced vibration of the gate bottom is proposed, and the specific calculation example is given, and the accurate solution to the problem of self-vibration and forced vibration of the gate bottom is obtained. The frequency of the first five-order vibration is analyzed, and the significant effect of the liquid on the low-frequency part of the spectrum and the dynamic response of the gate bottom is determined. Zhang Yongen (Zhang 2013) took the miter gate at the lower lock head of the representative Gezhouba No.3 ship lock as the object to make qualitative and quantitative safety assessment with the safety assessment method based on the reliability theory.

3 Safety Evaluation System

3.1 Safety Grade

In this paper, referring to the division of safety status in pumping stations, sluices and other engineering fields, combined with the relevant regulations on the operation safety of ship lock in China, such as the "Technical Specifications for maintenance of navigable buildings", the safety of ship lock operation is divided into four grades:

"normal", "deterioration", "early warning" and "shutdown" (Lu 2019), the specific meaning of each safety grade is shown in Table 1.

Table 1. Ship lock operation safety grade and corresponding meaning.

Safety grade	Meaning
Grade 1 (Normal)	The actual state and function of ship lock meet the requirements of current relevant national regulation, norm and standard, the evaluation indicators are in a normal state, and the entire system can operate normally, the grade of safety is high
Grade 2 (Deterioration)	Some evaluation indicators show abnormal signs, reaching the deterioration threshold, the function and actual state of ship lock can't fully meet the requirements of the current national regulation, norm and standard, which may affect the normal use of ship lock project, and the failure is more frequent. The number of overhauls increases significantly and grade of operation safety is moderate
Grade 3 (Early warning)	Some evaluation indicators are in an abnormal state, reaching the early warning threshold, according to the current regulations, norms and standards, there are serious problems that endanger safety of ship lock, the number of major failures increases, and the grade of operation safety is low
Grade 4 (Shutdown)	Some evaluation indicators are in an abnormal state, reaching the shutdown threshold, the function and actual condition of ship lock can't meet the requirements of the current national regulations, norms and standards, and the project has serious safety problems and should be stopped immediately

3.2   Evaluation Indicator and Criteria

Ship lock is an open complex giant system, and its operation safety can be characterized by multiple subsystems and indicators (Wang and Lee 2001). In general, it can be divided into five subsystems: hydraulic structure, metal structure, hydraulic system, electrical system and hydraulic power, each subsystem is composed of multiple devices and facilities, and finally reflected by the corresponding indicators. The indicator system and evaluation criteria are shown in Table 2.

Table 2. Ship lock operation safety evaluation indicator system and criteria.

Target layer	Guideline layer	Indicator layer	Grade 1 (Normal)	Grade 2 (Deterioration)	Grade 3 (Early warning)	Grade 4 (Shutdown)
Operation safety of ship lock	Hydraulic structure	Ratio of damage degree to standard value (%)	[0, 33.33)	[33.33, 100)	[100, 140)	[140, + ∞)
		Deformation (mm)	[0, 1.5]	(1.5, 3]	(3, 6]	(6, + ∞)
		Ratio of crack width to standard value (%)	[0, 50)	[50, 100)	[100, 140)	[140, + ∞)
		Grinding depth (mm)	[0, 1)	[1, 2)	[2, 10)	[10, + ∞)
		Carbonization depth (mm)	[0, 1)	[1, 3)	[3, 6)	[6, + ∞)
		Ratio of stress to allowable value (%)	[0, 85]	(85, 100]	(100, 115)	(115, + ∞)
		Ratio of seepage flow to standard value (%)	[0, 41.67)	[41.67, 100)	[100, 140)	[140, + ∞)
		Ratio of strength to standard value (%)	(88.75, 100]	[70, 88.75]	[33.33, 70)	[0, 33.33)
		Cavitation depth (mm)	[0, 0.27)	[0.27, 2)	[2, 5)	[5, + ∞)
		Ratio of elastic modulus to standard value (%)	(90,100]	(75, 90]	(60, 75]	[0, 60]
	Metal structure	Ratio of static stress to allowable value (%)	[0, 75)	[75, 80)	[80, 90)	[90, + ∞)
		Fatigue	[0, 0.85)	[0.85, 1)	[1, 1.05)	[1.05, 2]
		Ratio of runout exceeding standard value (%)	[−100, 0]	(0, 100)	[100, 133)	[133, + ∞)
		Rust area ratio (%)	[0, 0.3)	[0.3, 10)	[10, 11)	[11, + ∞)
		Drift (mm)	[0, 3]	(3, 6)	[6, 9)	[9, 12]
		Deformation (mm)	[0, 1.5]	(1.5, 3]	(3, 6]	(6, + ∞)
		Amount of wear (mm)	[0, 2.5]	[2.5, 5)	[5, 7.5)	[7.5, 10]
		Lintel ventilation volume (m3/s)	(0.42, + ∞)	(0.37, 0. 42]	(0.33, 0. 37]	[0, 0.33]
		Ratio of pressure bar clearance exceeding standard value (%)	[−62.5, 0]	(0, 100)	[100, 133)	[133, + ∞)
		Average vibration displacement (mm)	[0, 0.0508)	[0.0508, 0.254)	[0.254, 0.508]	(0.508, 2]
		Crack area ratio (%)	[0, 0.15)	[0.15, 0.3]	(0.3, 1)	(1, + ∞)
		Friction ultrasound	No friction ultrasound	Mild friction ultrasound	Relatively severe friction ultrasound	Severe friction ultrasound
	Hydraulic system	System pressure (MPa)	[16, 20]	(14.1, 16)	(3, 14.1]	[0, 3]
		Piston rod deformation (mm)	[0, 1.5]	(1.5, 3]	(3, 6]	(6, + ∞)
		Running speed (m/min)	[0, 2)	[2, 4]	(4, 8)	[8, + ∞)
		Piston rod vibration acceleration extreme (g)	[0, 0.25]	[0.25, 0. 5]	(0.5, 1)	[1, + ∞)
		Ratio of opening and closing force to design value (%)	[0, 40)	[40, 70]	(70, 105)	[105, + ∞)
		Ratio of internal leakage amount to standard value (%)	[0, 40)	[40, 100]	(100, 140)	[140, + ∞)
		Aging of the pipeline	No aging	Slight aging	Noticeable aging	Severe aging
		Synchronization error (%)	[0, 5)	[5, 15]	(15, 20)	[20, + ∞)
	Electrical system	Power supply	Normal	Relatively normal	Relatively abnormal	Extremely abnormal
		Monitor latency (s)	[0, 2)	[2, 3)	[3, 6)	[6, + ∞)

(*continued*)

Table 2. (*continued*)

Target layer	Guideline layer	Indicator layer	Grade 1 (Normal)	Grade 2 (Deterioration)	Grade 3 (Early warning)	Grade 4 (Shutdown)
		Communication system stability	Good	Lower	Obviously lower	Significantly lower
		Electronic components failure rate (%)	[0, 5)	[5, 10)	[10, 30)	[30, 100]
		Sensor stability	Good	Lower	Poor	Extremely poor
		Navigation signal	Stable	Wane	Significantly weaken	Do not meet the requirements
		Aging of equipment and facility	Intact	Mild aging	Noticeable aging	Severe aging
		Insulation resistance (MΩ)	[5, + ∞)	[2, 5)	[0.5, 2)	[0, 0.5)
		Ground resistance (MΩ)	[0, 2)	[2, 4]	(4, 30)	[30, + ∞)
	Hydraulic power	Water transport characteristic	Good	Relatively good	Relatively poor	Extremely poor
		Cavitation noise of water flow (dB)	[0, 120)	[120, 140)	[140, 160)	[160, + ∞)
		Sonic vibration	Extremely weak	Weak	Strong	Extremely strong
		Siltation of the pilot channel	No siltation	Mild siltation	Significant siltation	Severe siltation
		Ratio of flow velocity in port area to standard value (%)	[0, 30)	[30, 100]	(100, 125)	[125, + ∞)
		Pilot channel water level fluctuation (m)	[0, 0.4]	(0.4, 0.45)	[0.45, 0. 5]	(0.5, + ∞)
		Amplitude of upstream and downstream water level pulsation (m)	[0, 0.1)	[0.1, 0.2)	[0.2, 0. 4]	(0.4, + ∞)
		Ratio of navigable water depth to standard value (%)	(150, + ∞)	[100, 150]	(47, 100)	[0, 47]

The grading criteria of quantitative indicator is determined by its own situation; qualitative indicator adopts a scoring system, and the corresponding score standards at all grades are as follows (Table 3):

Table 3. Corresponding score criteria at each grade of qualitative indicator.

Grade	Grade 1	Grade 2	Grade 3	Grade 4
Score	(90,100]	(75,90]	(60,75]	[0,60]

3.3   Extension Evaluation Method

The extension evaluation method is a method that takes the indicator and characteristic value as matter-element, and obtains the classic domain, the node domain and the correlation degree through the evaluation standard, so as to realize the qualitative and quantitative evaluation (Zeng 2014; Shen 2007; Sun et al. 2007). In this paper, according to the characteristic of ship lock and based on the system concept based on the overall situation, the extensibility theory (Zhang et al. 2013; Jia et al. 2003; Yang and Cai 2002; Hu 2001) is introduced to establish an extension evaluation model for the safety of ship lock operation.

3.3.1   Classic and Node Domain

$$
R = \begin{bmatrix} N_j & N_1 & N_2 & \cdots & N_m \\ C_i & V_{i1} & V_{i2} & \cdots & V_{im} \end{bmatrix}
$$

$$
= \begin{bmatrix} N_j & N_1 & N_2 & \cdots & N_m \\ C_1 & <a_{11},b_{11}> & <a_{11},b_{12}> & \cdots & <a_{1m},b_{1m}> \\ C_2 & <a_{21},b_{21}> & <a_{22},b_{22}> & \cdots & <a_{2m},b_{2m}> \\ \vdots & \vdots & \vdots & \vdots & \vdots \\ C_n & <a_{n1},b_{n1}> & <a_{n2},b_{n2}> & \cdots & <a_{nm},b_{nm}> \end{bmatrix} \quad (1)
$$

where, R is the classic domain matter-element of C_i about N_j; N_j is the jth safety grade divided (j = 1,2,3,4); m = 4 is the number of safety grade, that is, N = {N_1, N_2, N_3, N_4} = {normal, deterioration, early warning, shutdown}; C_i is the ith indicator (i = 1, 2, $\cdots$, n); n is the number of indicators; $V_{ij} = <a_{ij},b_{ij}>$ is the magnitude range of Ci on N_j, i.e. the classic domain

$$
R_N = [N \quad C_i \quad V_i] = \begin{bmatrix} N & C_1 & <a_1,b_1> \\ & C_2 & <a_2,b_2> \\ & \vdots & \vdots \\ & C_n & <a_n,b_n> \end{bmatrix} \quad (2)
$$

where, R_N is the node domain matter-element of C_i about N; N is all grades; $V_i = <a_i, b_i>$ is the magnitude range of C_i on N, that is, the node domain

3.3.2   Matter-Element to be Evaluated
The actual status of the evaluation indicator is expressed as

$$
R_P = [P \quad C_i \quad v_i] = \begin{bmatrix} P & C_1 & v_1 \\ & C_2 & v_2 \\ & \vdots & \vdots \\ & C_n & v_n \end{bmatrix} \quad (3)
$$

where, R_P is the matter-element to be evaluated of the evaluation object P; P is the evaluation object in criterion layer of ship lock operation safety evaluation indicator system; v_i is the value of the indicator C_i.

3.3.3   Single Indicator Correlation Degree

$$K_{ij} = \begin{cases} -\dfrac{\rho(v_i, V_{ij})}{|V_{ij}|} & v_i \in V_{ij} \\[2ex] \dfrac{\rho(v_i, V_{ij})}{\rho(v_i, V_i) - \rho(v_i, V_{ij})} & v_i \notin V_{ij} \end{cases} \tag{4}$$

$$\rho(v_i, V_i) = \rho(v_i, \langle a_i, b_i \rangle) = \left| v_i - \frac{a_i + b_i}{2} \right| - \frac{b_i - a_i}{2} \tag{5}$$

$$\begin{aligned} \rho(v_i, V_{ij}) &= \rho(v_i, \langle a_{ij}, b_{ij} \rangle) \\ &= \left| v_i - \frac{a_{ij} + b_{ij}}{2} \right| - \frac{b_{ij} - a_{ij}}{2} \end{aligned} \tag{6}$$

where, K_{ij} is the correlation degree of the ith evaluation indicator of the evaluation object P to the grade j; $\rho(v_i, V_i)$ is the distance between the point v_i and the interval V_i; $\rho(v_i, V_{ij})$ is the distance between the point v_i and the interval V_{ij} (Zhang et al. 2021; Li and Wang 2020).

3.3.4   Multi - indicator Comprehensive Correlation Degree
Combine the weight of the evaluation indicator with its correlation degree

$$K_j(P) = \sum_{i=1}^{n} W_i K_{ij} \tag{7}$$

where, $K_j(P)$ is the comprehensive correlation degree of the evaluation object P about the grade j; W_i is the weight of the ith indicator of the evaluation object, which satisfies $\sum_{i=1}^{n} W_i = 1$.

Then conduct a target layer evaluation

$$K_j = \sum_{i=1}^{n} W_i' K_j(P_i) \tag{8}$$

where, K_j is the comprehensive correlation degree of the evaluation target about the grade j; W_i' is the weight of the ith evaluation object, which satisfies $\sum_{i=1}^{n} W_i' = 1$; K_j (P_i) is the comprehensive correlation degree of the ith evaluation object P_i about the grade j.

3.3.5   Rating

$$K_{j'} = \max K_j \tag{9}$$

Then the evaluation target belongs to the grade j'.

3.4 Variation Coefficient Weighting Method

Among the objective weighting methods, the variation coefficient method avoids equal division of weight and makes result more reasonable. The steps are (Jiang 2011):

i. Calculate the variation coefficient of indicator

$$\delta_i = \frac{\sigma_i}{\overline{x}_i} \tag{10}$$

$$\sigma_i = \sqrt{\frac{\sum_{d=1}^{D}(x_{id} - \overline{x}_i)^2}{D}} \tag{11}$$

$$\overline{x}_i = \frac{\sum_{d=1}^{D} x_{id}}{D} \tag{12}$$

where, δ_i is the variation coefficient of the indicator C_i; σ_i is the mean variance of eigenvalue of C_i; $\overline{x}_i$ is the mean value of eigenvalue of C_i.

ii Calculate objective weight

$$W_i = \frac{\delta_i}{\sum_{i=1}^{n} \delta_i} \tag{13}$$

4 Instance Application

4.1 Calculate the Correlation Degree of a Single Indicator and Weight

The correlation degree of the single indicator and weight of second-class indicator are calculated according to formula (4) to (6) and (10) to (13). The final result is shown in Table 4.

Table 4. Weight of operation safety evaluation indicator based on the variation coefficient method.

Target layer	Guideline layer	Indicator layer						
		Indicator	N_1	N_2	N_3	N_4	Grade	Weight
Operation safety	Hydraulic structure	Ratio of damage degree to standard value	0.4236	−0.5764	−0.8588	−0.8991	1	0.0239
		deformation	−0.2414	0.4667	−0.2667	−0.6333	2	0.0253
		Ratio of crack width to standard value	0.4286	−0.5714	−0.7857	−0.8469	1	0.0297
		Grinding depth	0.3333	−0.6667	−0.8334	−0.9667	1	0.0102
		Carbonization depth	0.3333	−0.3333	−0.7778	−0.8889	1	0.0078
		Ratio of stress to allowable value	0	−1	−1	−1	1	0.0413
		Ratio of seepage flow to standard value	0.2001	−0.2001	−0.6667	−0.7619	1	0.0129
		Ratio of strength to standard value	−0.7183	−0.6429	−0.2499	0.2499	4	0.0104
		Cavitation depth	−0.5889	−0.5	−0.2	0.2	4	0.0202
		Ratio of elastic modulus to standard value	−0.9622	−0.9547	−0.9433	0.0567	4	0.0098
	Metal structure	Ratio of static stress to allowable value	−0.4	−0.25	0.5	−0.25	3	0.0046
		Fatigue	0.24	−0.76	−0.88	−0.9143	1	0.052
		Ratio of runout exceeding standard value	−0.1933	0.29	−0.3698	−0.4622	2	0.0132
		Rust area ratio	−0.4962	−0.3333	−0.3103	0.1011	4	0.0148
		Drift	−0.4118	−0.2857	0.3333	−0.1667	3	0.0153
		Deformation	0.1333	−0.1333	−0.5667	−0.7833	1	0.0245
		Amount of wear	0	−1	−1	−1	1	0.0293
		Lintel ventilation volume	−0.4048	−0.3243	−0.2424	0.2424	4	0.0175
		Ratio of pressure bar clearance exceeding standard value	−0.1212	0.1	−0.5538	−0.6292	2	0.0233
		Average vibration displacement	−0.4661	−0.2674	0.4252	−0.2126	3	0.0197
		Crack area ratio	−0.3684	−0.1429	0.0857	−0.64	3	0.0371
		Friction ultrasound	−0.25	0.3333	−0.4	−0.625	2	0.0154
	Hydraulic system	System pressure	−0.375	−0.2908	0.3694	−0.4118	3	0.0098
		Piston rod deformation	−0.2414	0.4667	−0.2667	−0.6333	2	0.0202
		Running speed	−0.5714	−0.5	−0.25	0.25	4	0.025
		Piston rod vibration acceleration extreme	−0.4269	−0.3288	0.04	−0.02	3	0.0151
		Ratio of opening and closing force to design value	0.125	−0.875	−0.9286	−0.9524	1	0.0448

(*continued*)

Table 4. (*continued*)

Target layer	Guideline layer	Indicator layer						
		Indicator	N_1	N_2	N_3	N_4	Grade	Weight
		Ratio of internal leakage amount to standard value	0.325	−0.675	−0.87		1	0.0237
		Aging of the pipeline	−0.3333	0.3333	−0.2	−0.5	2	0.0032
		Synchronization error	−0.3333	0.5	−0.3333	−0.5	2	0.0113
	Electrical system	Power supply	−0.7778	−0.7333	−0.6667	0.3333	4	0.0161
		Monitor latency	0.5	−0.5	−0.6667	−0.8333	1	0.0193
		Communication system stability	−0.4265	−0.2642	0.0667	−0.025	3	0.0069
		Electronic component failure rate	0	0	−0.5	−0.8333	2	0.0522
		Sensor stability	−0.4167	−0.2222	0.3333	−0.125	3	0.0135
		Navigation signal	−0.25	0.3333	−0.4	−0.625	2	0.0189
		Aging of equipment and facility	−0.3333	0.3333	−0.2	−0.5	2	0.0191
		Insulation resistance	−0.2	0.3333	−0.3333	−0.4667	2	0.0298
		Ground resistance	0	0	−0.5	−0.9333	2	0.0577
	Hydraulic power	Water transport characteristic	0.5	−0.5	−0.8	−0.875	1	0.0203
		Cavitation noise of water flow	0.4083	−0.4083	−0.4929	−0.5562	1	0.017
		Sonic vibration	−0.4556	−0.3467	−0.1833	0.1833	4	0.0342
		Siltation of the pilot channel	−0.25	0.3333	−0.4	−0.625	2	0.0028
		Ratio of flow velocity in port area to standard value	−0.4118	0	0	−0.2	3	0.0131
		Pilot channel water level fluctuation	0.25	−0.75	−0.7778	−0.8	1	0.0202
		Amplitude of upstream and downstream water level pulsation	−1	−1	−1	0	4	0.023
		Ratio of navigable water depth to standard value	0	−1	−1	−1	1	0.0246

4.2 Calculate the Comprehensive Correlation Degree of Multiple Indicators and Rating

The correlation degree of a single indicator and calculated weight in Table 4 are substituted into formula (7)and (8) to calculate the comprehensive correlation degree of multiple indicators. The final grade is evaluated, and the result is shown in Table 5.

Table 5. Extension evaluation result of the operation safety of a certain ship lock.

Item		N_1	N_2	N_3	N_4	Max	Grade
Guideline layer	Hydraulic structure	0.3122	−0.4526	−0.8115	−0.9722	0.3122	1
	Metal structure	−0.0661	−0.2884	−0.4873	−0.761	−0.0661	1
	Hydraulic system	−0.0252	−0.2	−0.5154	−0.7793	−0.0252	1
	Electrical system	0.0491	0.0808	−0.4188	−0.8098	0.0808	2
	Hydraulic power	0.1832	−0.5145	−0.7263	−0.8209	0.1932	1
Target layer	Operation safety	0.0971	−0.2872	−0.6005	−0.8349	0.0971	1

5   Conclusions

Operation safety evaluation of in-service ship lock is extremely essential and has significant social and economic benefits. In this paper, ship lock operation safety evaluation system, including the indicator system, evaluation method and weighting method, is systematically discussed. The main work done and the conclusions drawn are as follows:

i. The safety evaluation indicator system of ship lock is constructed, and the evaluation method of ship lock operation safety based on extension theory is proposed to provide a basis for safety evaluation;

ii. The evaluation result shows that operation safety grade of ship lock belongs to the first grade – normal state, and all the first-class indicators belong to the first grade except that electrical system belongs to the second grade – deterioration state. Among the second-class indicators, ratio of strength to standard value, cavitation depth and ratio of elastic modulus to standard value of hydraulic structure, rust area ratio and lintel ventilation volume of metal structure, running speed of hydraulic system, power supply of electrical system, sonic vibration and amplitude of upstream and downstream water level pulsation of hydraulic power belong to the fourth grade – shutdown state, which should be paid special attention to. The evaluation result is consistent with actual situation, indicating that the evaluation system is reliable.

Acknowledgements. This work was supported by National Key R&D Program of China (Grant No.2018YFB1600400).

References

Changjiang Waterway Bureau (2004) Waterway engineering manual (Fine). China Communications Publishing House

Hu BQ (2001) The interval extension assessment method of water environmental quality and its application. Eng Sci 3(6):53–56

Jia C, Xiao SF, Liu N (2003) Application of extenics theory to evaluation of tunnel rock quality. Chin J Rock Mech Eng 22(5):751–756

Jiang J (2011) Fuzzy comprehensive evaluation model based on entropy weight and variation coefficient combination weighting method. Capital Normal University

JTS 304-2-2019 (2019) Technical specification for safety detection and assessment of navigation Junction

JTS 320-2-2018 (2018) Technical code of maintenance for navigation structure

Kolosov MA (2002) Shipping lock (SL) safety

Li N, Wang H (2020) Analysis on the evaluation of sustainable utilization of water resources by extension evaluation method. Arab J Geosci 13(16):1–8. https://doi.org/10.1007/s12517-020-05845-2

Lu HY (2019) Research on safety evaluation method of urban rail transit operation. Manag Obs (16):88–92

Senitskiy YE, Kuzmin NY (2012) The oscillations of ship lock bottom. Mag Civ Eng 30(4):17–24

Shen B (2007) The application research of extension method on enterprise performance evaluation. Xi'an University of Architecture and Technology

Sun XL, Chu JD, Ma HQ, Cao SL (2007) Improvement and application of matter element extension evaluating method. J China Hydrol 27(1):4–7

Wang SJ, Lee CF (2001) Global quality of rock works for permanent shiplock of the Three Gorges Project on Yangtze River, China. Chin J Rock Mech Eng 20(5):589–596

Xu HF (2007) Analysis on space structure and research on safety evaluation of miter gate lock. Hohai University

Yang CY, Cai W (2002) Study on extension engineering. Eng Sci 2(12):90–96

Yao ZM (2003) Application of rock engineering system theory in the slope stability estimate. J Anhui Univ Sci Technol (Nat Sci) (4):23–27

Zeng JJ (2014) Diagnosis of vulnerability of urban water sources in Yunnan Plateau Basin. Yunnan Normal University

Zhang YE (2013) Safety test technology and evaluation method study on miter gates of ship lock. Zhejiang University

Zhang CH (2001) Numerical modelling of concrete dam-foundation-reservoir systems. Tsinghua University Publishing House

Zhang YD, Guo J, Dai XC, Zou B (2013) Operation safety risk evaluation of train control system based on multilevel extensible evaluation method. China Railw Sci 34(5):114–119

Zhang ZX, Wang WP (2021) Managing aquifer recharge with multi-source water to realize sustainable management of groundwater resources in Jinan, China. Environ Sci Pollut Res Int 28(9):10872–10888

Study on the Mechanism of Water Loss and Capsizing of Multi - point Suspension Ship Lift

Lin Chen[✉], Yaan Hu, Zhonghua Li, and Chao Guo

Key Laboratory of Navigation Structures, Nanjing Hydraulic Research Institute, Nanjing, China
xiaolinzi@tju.edu.cn

Abstract. The stability of ship lift chamber operation has always been the key factor restricting the development of ship lift, which has not been well solved for a long time. With the development of upsizing of ships, the super-huge fully balanced ship lift with hoisting steel wire rope over 10000–15000 t is the development trend in the future. The mechanism of water loss and capsizing of multi-point ship lift is the most key scientific problem, and the most core technical problem is the layout and operation control of ship lift system. A generalized physical model of ship lift with the model scale of 1:33 was established to study the influence of multiple factors on water loss stability of ship chamber, such as gravity counterweight/torque counterweight ratio, water leakage flow, the position of suspension points among the chamber, quantity of the suspension points, and so on. Secondly, the structural dynamics equations of hoisting system is deduced in detail, considering the water fluctuation in the process of water loss caused by longitudinal capsizing moment influence on balance system, and research on the transient dynamic characteristics of hoisting system of the filtration process, determine the ship chamber conditions of stability for the trim through the Lyapunov stability criterion. The optimization calculation model of ship lift mechanical suspension system layout under the action of multiple constraints was established, and the effective measures will be given to improve the pitch stability. The results can provide technical guidance for the upsizing of ship lifters.

Keywords: Wire rope hoist vertical ship lift · Chamber stability · Structural dynamic equation · Water loss · Layout optimization

1 Introduction

The wire rope hoist vertical ship lift is a navigable building that uses cables to raise the ship to overcome the water level difference between upstream and downstream (GB 51177 2016). During the suspension process, sloshing phenomenon in the ship chamber directly influences the ship lift operation safety. Especially under the pitching situation, the tremendous capsizing moments produced by sloshing may lead to catastrophic overturning accident. For the rectangular ship chamber with filling depth less than 0.05, it is essential to introduce a suitable engineering method to reasonably

© The Author(s) 2023
Y. Li et al. (Eds.): PIANC 2022, LNCE 264, pp. 668–679, 2023.
https://doi.org/10.1007/978-981-19-6138-0_58

describe the shallow water sloshing and analytically predict the capsizing moments. The classical Housner theory, originated from the physical intuition of fluid motion, is widely used in engineering by imagining a series of incompressible massless rigid membranes in fluid. (Ibrahim 2005; Dodge 2000; Faltinsen and Timokha 2010; Liao 2014; 1996; 1996) introduce the Housner theory (Housner 1957) in researching the capsizing moments of the ship lift firstly. For the ship chamber, the Housner theory is probably the most suitable engineering method to reveal the dynamic characteristics of the shallow water sloshing. (Zhang et al. 2019) introduce the linear modal theory to describe the linear shallow water sloshing in the ship lift chamber. The linear modal theory, as described by Faltinsen and Timokha (Faltinsen and Timokha 2009) has shown reasonable accuracy when applied to compute the surface waves and associated hydrodynamic loads with a few natural modes. It provides an effective analytical method to deal with linear sloshing problems. Furthermore, assuming a time-harmonic pitching excitation, a fluid dynamic model is presented to predict the capsizing moments based on the Housner theory (Zhang et al. 2019).

The issue of safety of full balanced hoist ship lift was originated by an event occurred in 1994. The ship chamber of the physical model of the ShuiKou ship lift, the first built full balanced hoist vertical ship lift in China (Wang and Hu 2008). Regarding the pitching stability of ship chamber, a discriminant of instability condition for ship chamber was derived based on the static equilibrium leading to the disappearing of tension of some hoist wire ropes, with respect to water leakage condition (Chen et al. 1996). The natural vibration characteristics and stability of full balanced hoist vertical ship lift, containing ship chamber, wire ropes, gravity counterweights, water and ship in ship chamber, was studied to develop the critical influential factors for natural vibration characteristics as well as the critical distance of suspension points maintaining stability (Chen and Ma 1996; Cheng et al. 2005; Ruan and Cheng 2003). However, the primary hoist subsystem is ignored, which plays an essential role in ensuring pitch stability and the influence of synchronous shaft in the main suspension system was neglected in all the above studies. By establishing a series of the multivariable nonlinear differential equations, and using numerical calculation based on the Runge–Kutta method, the stability of ship chamber subjected to the rigid vertical vibration was addressed (Shi et al. 2003; Pearlson 1985) A theoretical formulation was developed to analyze the critical distance of suspension points, by establishing a dynamical model without gravity counterweights (Nakayama and Washizu 1980). By building a coupling dynamics model of main hoist subsystem, ship lift chamber, and sloshing, but neglected the torque counterweight in the main hoist subsystem; the latter is an essential part in the fully balanced quality system. (Liao 2014; Cheng et al. 2018; Li et al. 2005). A semi-analytical scheme is proposed to analyze the pitching stability of the high-lift wire ropes vertical ship lift (Zhang et al. 2019). In this study, a new 9-DOF coupled dynamics model of the complete main hoist system, ship lift chamber motion and shallow water sloshing was established. Combining the Lyapunov motion stability theory, the critical distance of suspension points would be numerically calculated.

In a full balanced hoist vertical ship lift, with the closed rectangular synchronous shafts, long and narrow ship chamber, and the larger elasticity of wire ropes, it is difficult to accurately quantify the various influential factors for pitching stability and their influential degree. If the ship chamber capsizes due to instability, it will cause huge economic loss and intolerable social impact. Therefore, with the aim of distinguishing the parameters with great influence on pitching stability, it is very significant to study the accurate quantitative characterization for the influence of structural parameters on pitching stability.

In this paper, the influence of the distance between two suspension points, torque counterweight ratio, the number of suspension points et al will be researched in detail, and the effective measures will be given to improve the pitch stability.

2   Model

The stability of chamber is always a technical problem which is concerned about the safe and stable operation of ship lift. Through the physical model, the rationality of the hoist system and all kinds of torque weight ratio were proved.

2.1   Model Design

It is necessary to simplify the design of model with complex structure to study the characteristics of water sloshing in the chamber at the moment of pitch instability, that is to say, proposed a generalized physical model can achieve the research target, including Ignoring the fluctuation of water level in chamber, simplifying the main hoisting system greatly with the eight groups of torque counterweight to eight torque suspension point, arranging eight gravity counterweight suspension point on both ends of the ship chamber, crown blocks were configured to connect chamber and the balance weight. The size of chamber satisfied the similarity rule of geometric and gravity, rather than the stiffness similarity, and arranged multiple ear canals in the torque suspension point, so as to made it easy to adjust to the suspension point spacing. The generalized physical model is shown in Fig. 1.

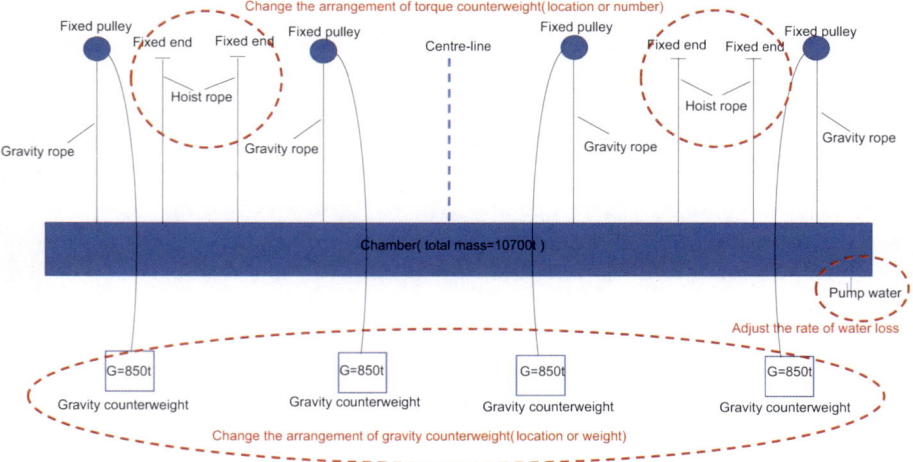

Fig. 1. Arrangement of the generalized physical model

As is shown in Fig. 1, the structure of full balanced hoist vertical ship lift is composed of ship chamber, hoisting system and tower column. Counterweights include gravity counterweights connected to ship chamber with wire ropes, and torque counterweights connected to winding drums with wire ropes.

The model scale is 1:33. in view of the relative small capsizing angel when the chamber lose its stability and in order to prevent the damage of model and avoid affecting the subsequent test, the chamber was placed about 50 cm above the ground and restricted the height of the gravity counterweight so as to limit the range of movement at the same time, The bottom of the gravity counterweight was about 40–50 cm above the ground. The chamber stopped when it fell to the ground. The size of the water area and the suspension point were refined and manufactured strictly according to the design. The tension sensors with 16 suspension points were used for weighing after the installation of the chamber, gravity counterweights were configured with electronic scales and the physical model after installation is shown in Fig. 2.

(a) chamber with water   (b) hoisting platform

Fig. 2. Photo of physical model

2.2   Layout of Measuring Points

Capsizing value of chamber: ultrasonic displacement meter was arranged on the deck of the four corners of the chamber to measure the elevation change. The initial level of the chamber was calibrated with a level instrument, and the levelness is adjusted by a suspension point rope receiver.

Tension of wire rope: Tension sensors were arranged at 16 simplified suspension points to measure the force during the capsizing process.

Depth and surface fluctuation: high precision wave probe was used to measure depth variation and surface fluctuation and it was arranged along one side to focus on the instability process and the changing characteristics of the surface in the chamber at the moment of instability.

The specific measuring points were arranged as shown in Fig. 3. There were 4 ultrasonic displacement meters (serial number: D1–D4) and 16 tension sensors (gravity counterweight suspension points: G1\G4\G5\G8\G9\G12\G13\G16; Torque counterweight suspension points: T2\T3\T6\T7\T10\T11\T14\T15 and 7 wave sensors (serial number: W1–W7).

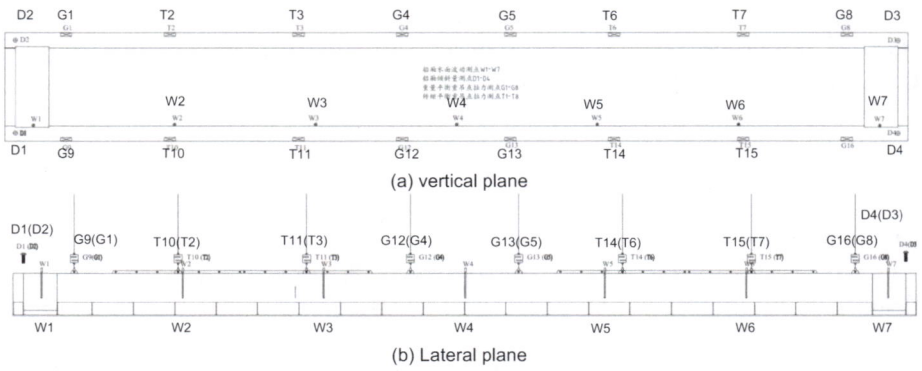

(a) vertical plane

(b) Lateral plane

Fig. 3. Layout of measuring points

3 Results

3.1 Water Loss Rate 9t/s

3.1.1 Whole Process of Overturning

The chamber was initially levelled in the horizontal state, and the water surface was stable. Submersible pumps (2 sets of different specifications and combination) were used at the end of the chamber to drain the water outward so as to simulate the process of the chamber water loss (leakage).

The chamber began to leak under the initial state of 3.9 m in depth, hoisting load of eight torque counterweights decreases continuously along with the reduction of chamber weight, until the force of torque counterweight suspension point was zero, at this time wire rope is flabby, chamber was pulled up at one end by the gravity counterweight, while the water flow to the opposite end. as a result, the load on this end increased and the chamber moved to the direction of ground, the entire process of capsizing is shown in Fig. 4.

(a) entire instability   (b) lose control

Fig. 4. Process of capsizing and destabilization of chamber

The critical instability of the non-leaking side started at the time about 402 s, however, the start-up of the leaking side lagged behind about 12–13 s, and the speed of capsizing changed from slow to sharp after the instability of the chamber, namely an accelerated capsizing process. When one end is pulled up about 15 m, it only took about 33 s, and the elevation of both sides also has certain deviation. The descending process of the other end was affected by the load generated by water flow, height, overflow and certain fluctuation, etc. The height difference between the two sides was relatively obvious.

Unstable motion process of chamber was analyzed in Fig. 5. It can be divided into two stages, In the first stage, with T7/T15 as the center of rotation, it rotates clockwise about 7.26°, and stops when the gravity counterweight in the non-leaking side reaches the preset boundary (G1/G4/G9/G12 falls to the ground, similar to braking). With water from the leaking side continues to leak outside, the leaking side is lifted clockwise with G1/G9 as the rotating center. After the completion of stage 2, the chamber is restored in the horizontal state. It is worth noting that, due to the preset boundary of the ground, if there is serious water loss of the chamber in the actual project, the chamber will be overturned.

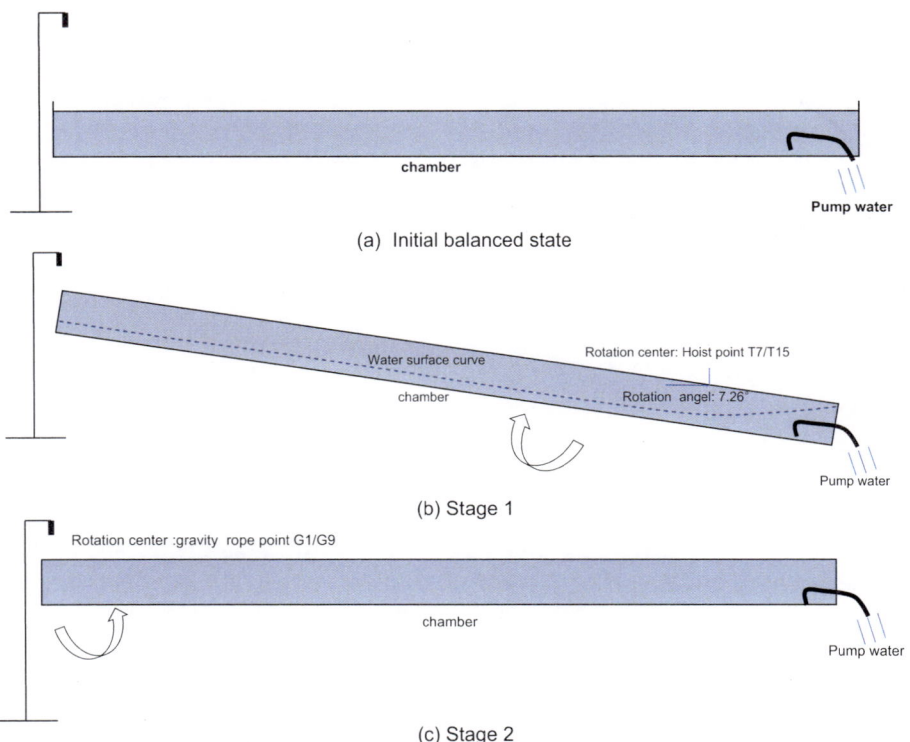

Fig. 5. Unstable motion process of chamber

3.1.2   Load Variation During Instability Process

As can be seen from Fig. 6, the weight of the chamber decreases uniformly in the process of leakage. When the total weight reaches about 7085t, that is to say, the loss weight is 3664t, the chamber begins to capsize and lose stability and a great impact load is generated on the side of the chamber.

Figure 7 showed the ship chamber capsized instability process within the complex movements of water, according to the seven wave height measuring point test data of the change of water depth available on one side during the capsized process. as shown in Fig. 8, in the process of water leakage, The water surface decreases evenly and maintain steady. with the decreasing of water, a small fluctuations occurs, finally the system reached the critical instability state, And then there's capsizing and instability. During the capsizing process, the water flows to one end, and the water flow also changes from the initial unsteady gradual to the sudden change flow, and forms the falling wave. The water surface decreases at the raised end, and the water surface is high at the lowered end, and the middle measuring point decreases rapidly after the short high.

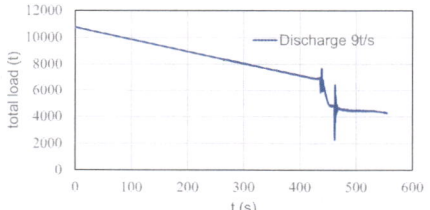

Fig. 6.   Total load variation

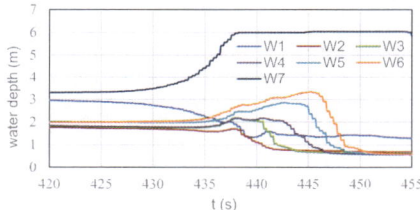

Fig. 7.   Water depth variation during instability process

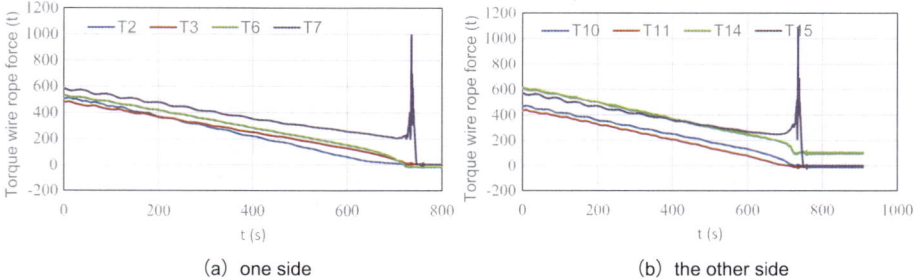

(a) one side   (b) the other side

Fig. 8.   Load variation line of torque counterweight load point

3.1.3   Determination of Instability Point

The critical instability process was so quickly, the determination of the critical moment is a huge problem, Take the derivative of the time line of 16 hoisting steel rope tension process, and get the time rate of change of load process line, Fig. 9 was gravity counterweight hoisting process of load change rate of time line, it can see that, before the critical instability point, load change rate was at 0 t/s, It showed that the suspension system always keeps the elastic vibration state in the process of water leakage, and the frequency of load variation is its own natural vibration frequency. Instability process can be seen that the theoretical analysis, it is because of the slack side of a lifting point stress reduce to zero, after ship reception chamber reaches instability state, analyze each side four torque counterweight process of the time rate of change of heavy load line, determine mutation point for instability point, as shown in Fig. 10, because of test error, the left and right side there is some deviation, but the overall rule is changeless, The instability process spreads gradually from non-water leakage measurement to water leakage measurement. The critical instability point is T = 402.12 s, and the instability occurs at T10 in the beginning.

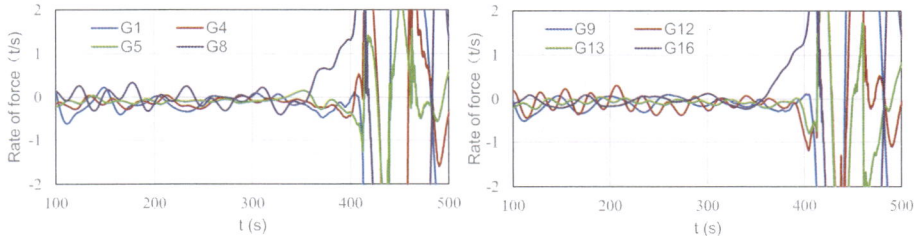

Fig. 9. Rate of load variability at gravity counterweight lifting point

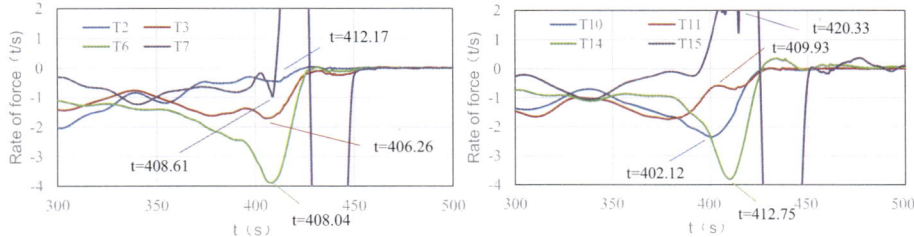

Fig. 10. Rate of load variability at torque counterweight point

3.2 Parallel Comparison of Working Conditions

From the theoretical analysis, it can be seen that the trim stability of the chamber is restricted by a variety of primary and secondary factors. In order to establish the quantitative relationship of the influence of multiple factors, multiple groups of parallel comparison tests of the leaking and capsizing are carried out. Considering the influence of gravity counterweight/torque counterweight ratio, water leakage flow, the layout of torque counterweight (number and position), and the layout of gravity counterweight on the critical instability, the statistics of test results are shown in Table 1.

Leakage flow has little effect on the critical instability, weight ratio, the layout of gravity counterweight is affected, the torque counterweight lifting points (quantity, location) was the main factors, compared to the number, spacing of hoisting position especially lifting point it is important to influence, while keeping the lateral lifting point, critical instability value of 0.945, For the layout of lifting points larger than 8, the relationship between the center distance of lifting points (double lifting points is the center of lifting points) and critical instability is established for convenient analysis, as shown in Fig. 11. The importance ranking is as follows: position of lifting points, number of lifting points, gravity counterweight arrangement, torque counterweight ratio, leakage flow rate. In order to achieve the maximum state, the lifting point and gravity counterweight should be closer to the end as far as possible. In order to delay the overturning time, torque counterweight ratio should be reduced as far as possible.

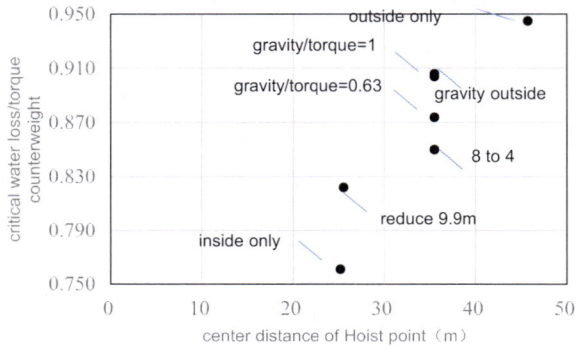

Fig. 11. Relationship between the distance between lifting points and critical instability

Table. 1. Test results under different working conditions

Factors	Set	Water loss flow	Gravity counterweight	Torque counterweight	Torque/ Gravity	Instability time (s)	Instability water loss/Torque	Aver– age
Gravity/torque	1	9.08	5350	5252	0.982	529.07	0.915	0.904
	2	5.02	5350	5363	1.002	953.54	0.893	
	3	4.00	5350	5320	0.994	1202.57	0.904	
	4	9.11	6600	4149	0.629	402.12	0.883	0.874
	5	5.23	6600	4176	0.633	692.85	0.867	
	6	4.10	6600	4221	0.639	897.88	0.872	
Remain (T2/T7/T10/T15)	7	8.64	5350	5236	0.979	574.40	0.948	0.945
	8	4.88	5350	5250	0.981	1014.83	0.944	
	9	3.85	5350	5291	0.989	1295.40	0.943	
Remain (T3/T6/T11/T14)	10	10.32	5350	5471	1.023	403.67	0.762	0.761
	11	5.51	5350	5507	1.029	756.16	0.757	
	12	4.46	5350	5484	1.025	942.91	0.766	
Distance between hoist decreases 9.9 m	13	8.89	6600	4150	0.629	372.76	0.798	0.822
	14	5.13	6600	4181	0.634	673.66	0.826	
	15	3.84	6600	4088	0.619	894.31	0.841	
Hoist points change from 8 to 4	16	8.46	6600	3921	0.594	392.81	0.848	0.850
	17	4.94	6600	4004	0.607	695.03	0.857	
	18	3.87	6600	4060	0.615	888.05	0.846	
RemainG1/G8/G9/G16, total gravity unchanged	19	9.16	5350	5337	0.998	517.01	0.887	0.906
	20	5.23	5350	5313	0.993	926.25	0.912	
	21	4.07	5350	5389	1.007	1215.78	0.918	

4 Discussion

(1) The influence of leakage flow: According to literature (Zhang et al. 2019), the overturning moment M under the pitch movement consists of four parts, the MIW of pulse pressure accounts for about 0.016%–0.056%, Pulse pressure action MIB accounts for about 0.447%–4.850%, both of which can be ignored basically. Due to the model of a smaller scale in three different leakage flow, the chamber almost no visible surface waves, under the condition of model test, wave high overturning moment caused by the difference is not large, the speed instability stage, the overturning moment performance for static and overturning moment, namely the water in front of the ship reception chamber instability a small fluctuation, instability capsized reception chamber in the process of flood wave is relatively small. At the moment of instability, the torque lifting point is zero, and the torque counterweight is in the relaxed state.

(2) Influence of lifting point spacing: The critical instability has a great relationship with the increase of the distance between lifting points. With the decrease of the distance between lifting points, the height of water surface swell increases, and the instability is easy to occur when the inclined emergency braking occurs.

(3) In fact, the process in manufacture and installment of the diameter of drum slot, the location of suspension point, chamber cannot guarantee absolute precision such as quality and size. a large number of ship lift construction and operation practice both at home and abroad shows that the change of the load hoisting height in the operation of the rise and fall, chamber will produce certain capsizing, due to the the free flow of water, thus causes the chamber offset center of gravity, Produce a capsizing moment; In addition, the water will be disturbed in the process of ship entering and leaving the chamber and the opening and closing of the flap gate, and the oscillating water wave will be formed, which will also produce an obvious overturning moment to the chamber. Compared with the static state, the force on the hoisting rope will change to some extent. The force on the hoisting rope will decrease in the process of falling, which is equivalent to reducing the control ability of the suspension system and reducing the anti-capsizing ability.

To ensure the stability and safety, it is suggested that before the balance drum safety brake is put into operation, the maximum water leakage must be less than 80% of the torque counterweight.

5 Conclusions

(1) The instability process of the chamber is a changing process from slow to rapid, which is the mutual promotion process of the intensified capsizing and the increase of eccentric load of the water. The initial stage of instability is the key stage of control, and the braking will produce a great impact load in the capsizing process.

(2) In the case that there is no obvious eccentric load, the more suspension points there are, the stronger the control ability of the chamber will be.

(3) In order to ensure the safety and reliability of the fully balanced ship lift, the weight of the torque should be increased as much as possible. In order to enhance the control ability of the main suspension mechanism to the suspension system, it is very important to increase the torque counterweight and correspondingly increase the braking capacity of the safety brake to improve the safety of the ship lift especially considering the leakage accident of the ship lift.

(4) The elastic rope and frictional resistance exist in the model test and the model scale effect, so a variety of factors have effects on critical state of instability, The model is almost rigid body, which cannot reflect the flexibility of the large scale carrier, and the initial eccentric load or disturbance factors such as the deformation of the carrier and the difference of water load are not obvious. through much of the contrast test in the subsequent and combining with the theoretical derivation and numerical simulation techniques was studied for the test results.

Acknowledgements. This research is supported by Fundamental Research Funds for The Central Public-Interest Scientific Research Institutes (Major Project, Grant No. Y121009; Youth Foundation, Grant No. Y122009), China Postdoctoral Science Foundation (General Programs, Grant No. 2021M701752).

References

Chen J, Ma G (1996) Chamber stability of hoisting fully balancing type vertical ship lift. Hydro-Sci Eng 4:301–308

Cheng G, Li H, Ruan S (2005) Free vibration characteristics and stability analysis of ship lift system. J Mech Strength 3:276–281

Cheng X, Shi D, Li H, Xia R, Zhang Y, Zhou J (2018) Stability and parameters influence study of fully balanced hoist vertical ship lift. Struct Eng Mech 66:583–594

Dodge FT (2000) The new dynamic behavior of liquids in moving containers. [Update of NASA SP-106]. Southwest Research Institute, San Antonio, Texas

Faltinsen OM, Timokha AN (2009) Sloshing. Cambridge University Press, Cambridge

Faltinsen OM, Timokha AN (2010) Book review. J Fluids Struct 26:1042–1043

GB 51177 (2016) Design Code for Shiplift. The Ministry of Water Resources of the People's Republic of China, Beijing, China (in Chinese)

Housner GW (1957) Dynamic pressures on accelerated fluid containers. Bull Seismo Soc Am 47:15–35

Lbrahim RA (2005) Liquid Sloshing Dynamics: Theory and Applications. Cambridge University Press, Cambridge

Liao LK (2014) Safety analysis and design of full balanced hoist vertical shiplifts. Struct Eng Mech 49:517–522

Liao LK, Shi DW (1996) Coupled vibration analysis of torsion of hoist and pitch of ship chamber of shiplift. Yangtze River 27:19–22

Liao LK, Shi DW (1996) Analysis of motion stability of shiplifts in suspension state. Water Conserv Electr Power Mach 6:9–12

Li HT, Cheng GD, Ruan SL (2005) Seismic response of ship lift system. J Dalian Univ Technol 473–479

Nakayama T, Washizu K (1980) Nonlinear analysis of liquid motion in a container subjected to forced pitching oscillation. Int J Numer Meth Eng 8:1207–1220

Pearlson DL (1985) Installation of syncrolift equipment in ports. In: Port engineering and operation, proceedings of the conference on British ports and their future, pp 177–189

Ruan S, Cheng G (2003) Calculation of ship-liquid-chamber coupled system in the ship lift with finite element method in time domain. China J Comput Mech 3:290–294

Shi D, Shong Z, Liao L (2003) Coupling analysis of ship-chamber of ship lift and parametric vibration of liquid. Eng J Wuhan Univ 1:77–80

Wang YX, Hu XW (2008) Overall design of 2× 500t vertical ship lift of ShuiKou hydraulic power station, Technology for Hydraulic Machinery Essay Collection in 2008

Zhang Y, Shi DW, Liao LK, Shi L, Cheng XH (2019) Pitch stability analysis of high-lift wire rope hoist vertical shiplift under shallow water sloshing–structure interaction. Proc Inst Mech Eng Part K: J Multi-body Dyn 223(4):942–955

Zhang Y, Shi DW, Shi L, Xia R, Cheng XH, Zhou J (2019) Analytical solution of capsizing moments in ship chamber under pitching excitation. Proc Inst Mech Eng Part C J Mech Eng Sci 233(15):5294–5301

Study on the Unsteady Flow of the Approach Channel's Entrance

Xiaodong Wang[1], Jinchao Xu[1,2(✉)], Long Zhu[1], Donghui Zhou[3], and Jun Zhao[1,2]

[1] Nanjing Hydraulic Research Institute, Nanjing, China
{xdwang,lzhu}@nhri.cn,
[2] Nanjing University of Information Science and Technology, Nanjing, China
jcxu@nuist.edu.cn
[3] Jiangsu Yangtze River Delta Smart Water Research Institute,
Nanjing University of Information Science and Technology, Nanjing, China

Abstract. The entrance of shiplock's approach channel always exist a mixing shear layer caused by the shear mixing layer, where is the junction of main river flow and the quiescent water of the approach channel. The flow structure of the turbulent mixing area presented as large-scale vortices frequently and periodically. And the fluctuations of the water surface and velocity induced by the separation of vortices may threat the navigation conditions, which should be considered during the engineering designing. The existing studies regard the mixing shear layer as steady flow and only take care of the average shear velocity, which may underestimate the harm of the shear flow. In this work, RNG k-ε model and LES model were adoped to study the hydraulic characteristics of the entrance. The vortex characteristics and the influence on the water level and velocity were analyzed. Results shown that LES model had better precision than RNG k-ε model in the unsteady characteristics. Then further studies about the recirculation flow were performed with LES model, The length and period of the recirculation flow was studied. It found that the vortex was generated at the upstream of shear zone, and then transferred with the recirculate flow until it was collapsed. All the above provide references for hydraulic characteristics of the entrance.

Keywords: Approach channel · Entrance · LES simulation · Hydraulic characteristics

1 Introduction

The entrance of port and lock approach is a junction of quiescent water and flow water, where is a turbulent mixing shear flow and circulation flow always exist (Kimura and Hosoda 1997). The flow structure of the entrance caused by turbulent mixing behave as large-scale vortexes generated frequently and periodically. The separation of large-scale vortexes causes the water surface and velocity fluctuation at the entrance and the approach channel, which affects the flow stability in the approach channel and threat to the safety of navigation (Sauida 2016). Many scholars have focus the flow characteristics. For example, Liu (1995) studied the mixing zone at the entrance of the dead

© The Author(s) 2023
Y. Li et al. (Eds.): PIANC 2022, LNCE 264, pp. 680–692, 2023.
https://doi.org/10.1007/978-981-19-6138-0_59

end and its recirculation characteristics with physical experiment, and explored the flow structure and mechanism of the entrance area. Li et al (2005) studied the flow conditions in the entrance area of the lower approach channel of the Gezhouba shiplock. The results shown that the flow velocity and direction of the recirculation flow at the entrance were extremely unstable, and the flow presented a periodic characteristic, which was similar to the oscillatory flow and had a significant influence on the navigation. Zhou et al. (2005) summarized the results of physical model test and prototype observation of some shiplocks in China, and analyzed the wave and discharge variation at the entrance caused by wind or the water releasing of nearby hydraulic engineering. The results shown that the fluctuation presented as short wave, and the period is less than 5 s with the amplitude could be reached more than 1 m. In addition, many scholars have used numerical models to study the hydraulic characteristics of the entrance area, such as transverse flow velocity, longitudinal flow velocity, recirculation velocity, and so on. However, existing studies mainly focusing on the flow structure and average flow filed of the entrance, in meanwhile ignored the unsteady characteristics of the vortex. Therefore, it is necessary to conduct further study on the vortex characteristics of the entrance, to explore the influence of unsteady characteristics on the approach channel.

The large eddy simulation (LES) and RNG k - ε turbulence model were used to study the flow characteristics in the entrance area. The flow field and periodic characteristics obtained by the two numerical simulation methods are compared and analyzed. In addition, the hydraulic characteristics of the recirculation zone in the entrance were discussed.

2   Simulation Model

RANS k - ε model and LES model were used to simulate the flow field of approach channel and its entrance respectively.

2.1   RANS k-ε Model

The control equations of RANS k - ε model can be expressed as (Girimaji 2006):

$$\frac{\partial k}{\partial t} + u_j \frac{\partial k}{\partial x_j} = \frac{\partial}{\partial x_j}\left[\left(\frac{v_t}{\sigma_k} + v\right)\frac{\partial k}{\partial x_j}\right] + G - \varepsilon \tag{1}$$

$$\frac{\partial \varepsilon}{\partial t} + u_j \frac{\partial \varepsilon}{\partial x_j} = \frac{\partial}{\partial x_j}\left[\left(\frac{v_t}{\sigma_\varepsilon} + v\right)\frac{\partial \varepsilon}{\partial x_j}\right] + C_{1\varepsilon}\frac{\varepsilon}{k}G - C_{2\varepsilon}\frac{\varepsilon^2}{k} \tag{2}$$

where, ρ is fluid density, t is time, u_j is the velocity vector in j direction, k is turbulent kinetic energy, ε is turbulent dissipation rate, σ_{ij} is the Kronecker function, the stress component is δ_{ij}, v_t is eddy viscosity, p is the pressure term. The other parameters are presented as:

$$\tau_{ij} = 2\left(v + \frac{k^2}{\varepsilon}\right)\sigma_{ij} - \frac{2}{3}k\delta_{ij}, \ \sigma_{ij} = \frac{1}{2}\left(\frac{\partial u_i}{\partial x_j} + \frac{\partial u_j}{\partial x_i}\right), \ v_t = C_d\frac{k^2}{\varepsilon}$$

$$G = 2v_t\sigma_{ij}\frac{\partial u_i}{\partial x_j}, \ C_d = 0.09, \ C_{1\varepsilon} = 1.44, \ C_{2\varepsilon} = 1.92, \ \sigma_k = 1.0, \ \sigma_\varepsilon = 1.3 \tag{3}$$

2.2 LES Model

In this work, Smagorinsky-Lilly dynamic model was used to solve the subgrid stress τ_{ij}, and the coefficients are obtained dynamically in the calculation process, rather than a pre-given value. Two filtering factors were introduced in the model, namely, grid filtering factor Δ and test filtering factor $\bar{\Delta}$. Generally, the value of test filtering factor is twice of the grid filtering factor. Then the subgrid stress was expressed with eddy viscosity model (Sagaut, P. 2006):

$$\tau_{ij} - \delta_{ij}\tau_{kk}/3 = -2C\bar{\Delta}^2\left|\bar{S}\right|\bar{S}_{ij} \tag{4}$$

in which, $\bar{S} = \sqrt{2\bar{S}_{ij}\bar{S}_{ij}}$, $\bar{S}_{ij} = \frac{1}{2}\left(\partial\bar{u}_i/\partial x_j + \partial\bar{u}_j/\partial x_i\right)$, $\bar{\Delta} = 2\Delta$, C is a coefficient and is considered to be constant between the two filter factors. The least square method was used to calculate the C value, as:

$$C = L_{ij}M_{ij}/\left(2M_{ij}M_{ij}\right) \tag{5}$$

where L_{ij} is calculated as:

$$L_{ij} = \overline{\bar{u}_i\bar{u}_j} - \overline{\bar{u}_i}\,\overline{\bar{u}_i}M_{ij}/\left(2M_{ij}M_{ij}\right) \tag{6}$$

$$M_{ij} = -2\left(\bar{\Delta}^2\left|\bar{\bar{S}}\right|\bar{\bar{S}}_{ij} - \bar{\Delta}^2\overline{\left|\bar{S}\right|\bar{S}_{ij}}\right) \tag{7}$$

2.3 Mesh and Boundary Conditions

The experiment of Kimura and Hosoda (1997) was used for model validation. The experiment was carried at a rectangular channel with a dead end of the length and width both 15 cm (B = L = 15 cm), the width of the rectangular channel is 10 cm, the bottom slope is 1: 500, and the water depth is 1 cm.

The structured mesh was used in numerical model. In order to capture the flow structure at the entrance, the refined mesh was adopted at the entrance. The minimum mesh size is 1.25 mm, and the number of mesh elements was 1138491. The velocity of inlet of main channel was 25.5 cm/s, and the outlet was set as pressure boundary. The free surface was treated by VOF method. The central difference scheme was used to solve the convection diffusion equations. The second-order implicit method was used to calculate the transient term, and the time step was set as 0.001 s. When the residual errors were less than 1×10^{-5}, the calculation was considered as stable.

Besides the condition of approach channel $L = 15$ cm, another 4 conditions had been presented to study the influence of length on the characteristics of recirculation flow, with $L = 30$ cm, 45 cm, 60 cm and 75 cm respectively, as shown in Table 1 (Fig. 2).

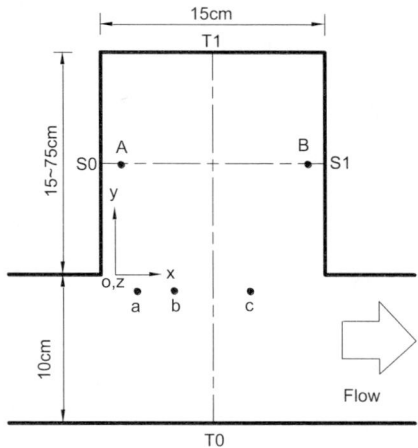

Fig. 1. Experiment model of in Kimura and Hosoda (1997)

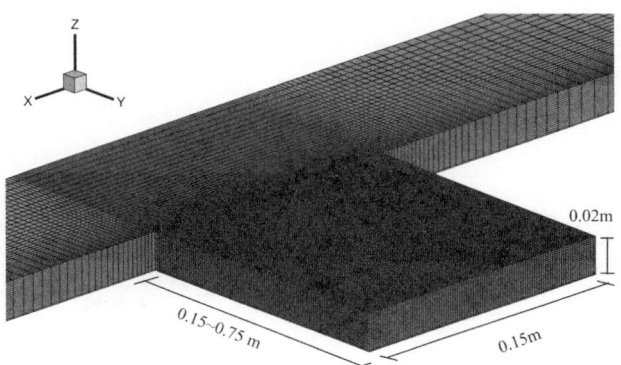

Fig. 2. Mesh of calculated zone

Table 1. Simulation conditions

Conditions	L (cm)
C-1	15
C-2	30
C-3	45
C-4	60
C-5	75

3   Results

3.1   Validation

The average flow velocities in the x direction of T0–T1 section (as shown in Fig. 1) calculated by RNG model and LES model are shown in Fig. 3. It can be seen that the time-average results of two models both close to the experiment results. For the instantaneous flow fields at three points a, b and c (the location as shown in Fig. 1), it can be seen that the instantaneous variation of flow velocity calculated by LES model also consistent with the experiment results. And the lowest order oscillation period of flow velocity is about 1 s. It was accorded with the calculation results of the wave period of approach channel $T = 2L/(n\sqrt{gh})$ (n = 1, 2, 3...). Meanwhile the results calculated by RNS model were only agree with the experiment results in amplitude, but failed with the period. The lowest order oscillation period of RNS model was about 2 s, which did not meet the variation of instantaneous flow field in the approach channel.

So with the RNG k-ε model, only the time-averaged characteristics of the flow field can be obtained. The results of LES model are agree with the experiment results which measured by Particle Image Velocimetry (PIV). With LES model, the spatial-temporal random distribution characteristics of the flow velocity can be captured, including the swing of the jet axis and the distribution of the vortex. The calculation results of LES model are more accurate than the RNG k-ε model in the recirculation zone of the entrance area (Fig. 4).

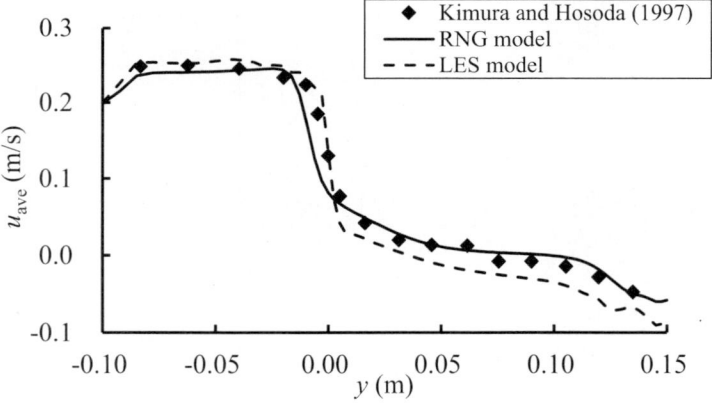

Fig. 3. Comparison of time-averaged velocity distributions along T0–T1

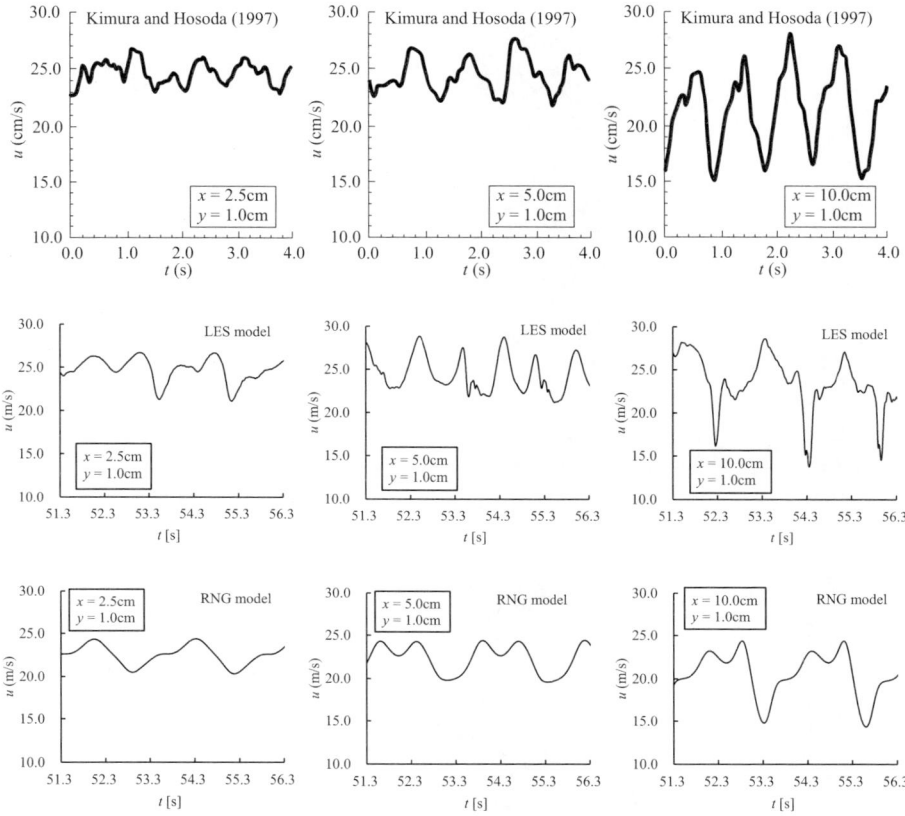

Fig. 4. Comparison of temporal variations in velocity in x-direction

3.2   Flow Characteristics

LES model was used to study the flow conditions of the approach channel under five conditions with different length of approach channels, as shown in Table 1.

3.2.1   Length of Recirculation Zone

In this work, the flow velocity greater than 1 cm/s was considered as effective zone when determined the recirculation zone. The length of recirculation zone was found equal the length of approach channel when the channel length smaller than 30 cm. if

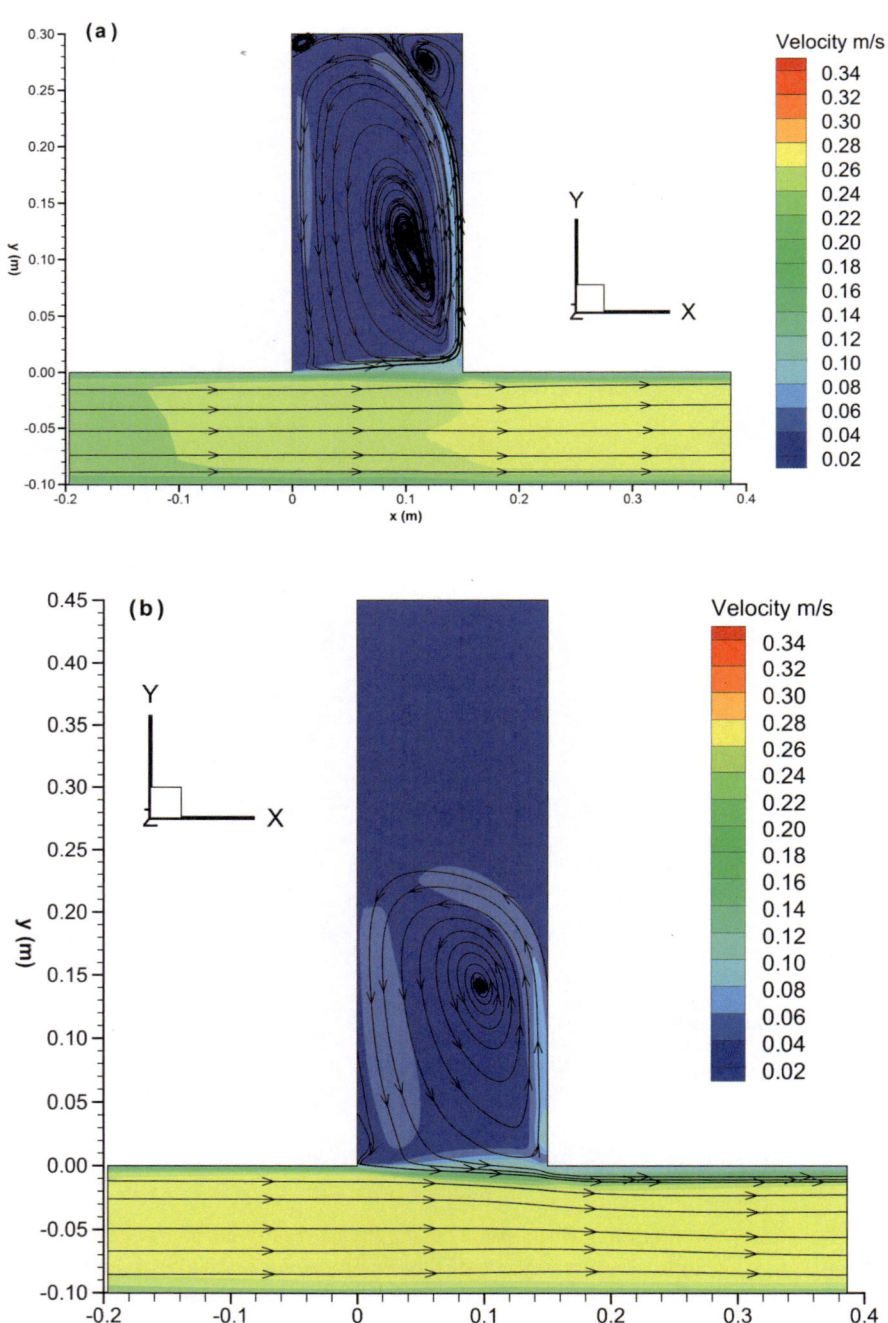

Fig. 5. The velocity field of recirculation zone in typical condition. (a) Condition C-2, (b) Condition C-3.

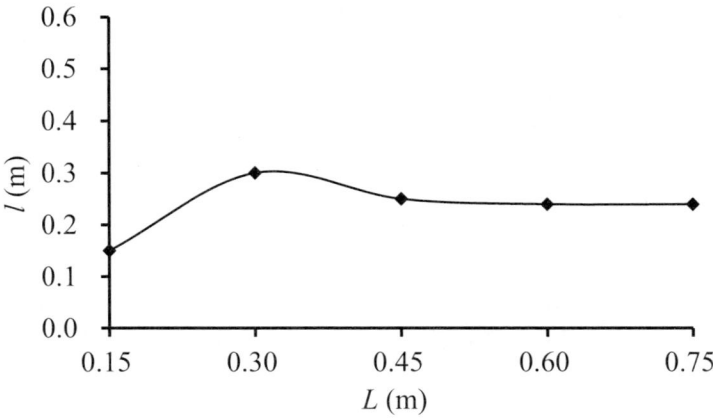

Fig. 6. Length of recirculation flow in different conditions

the length of approach channel larger than 45 cm, the length of recirculation was keep consistent with value of 25 cm, as shown in Fig. 5 and Fig. 6.

3.2.2 Period of Recirculation Zone

Results show that the water level on the T1-T0 axis in the recirculation zone maintained unchanged, and the water depth at the two sides of T1-T0 axis increases and decreases alternately. The period of recirculation zone of condition C-1 is 1.0 s, and 1.6 s in all

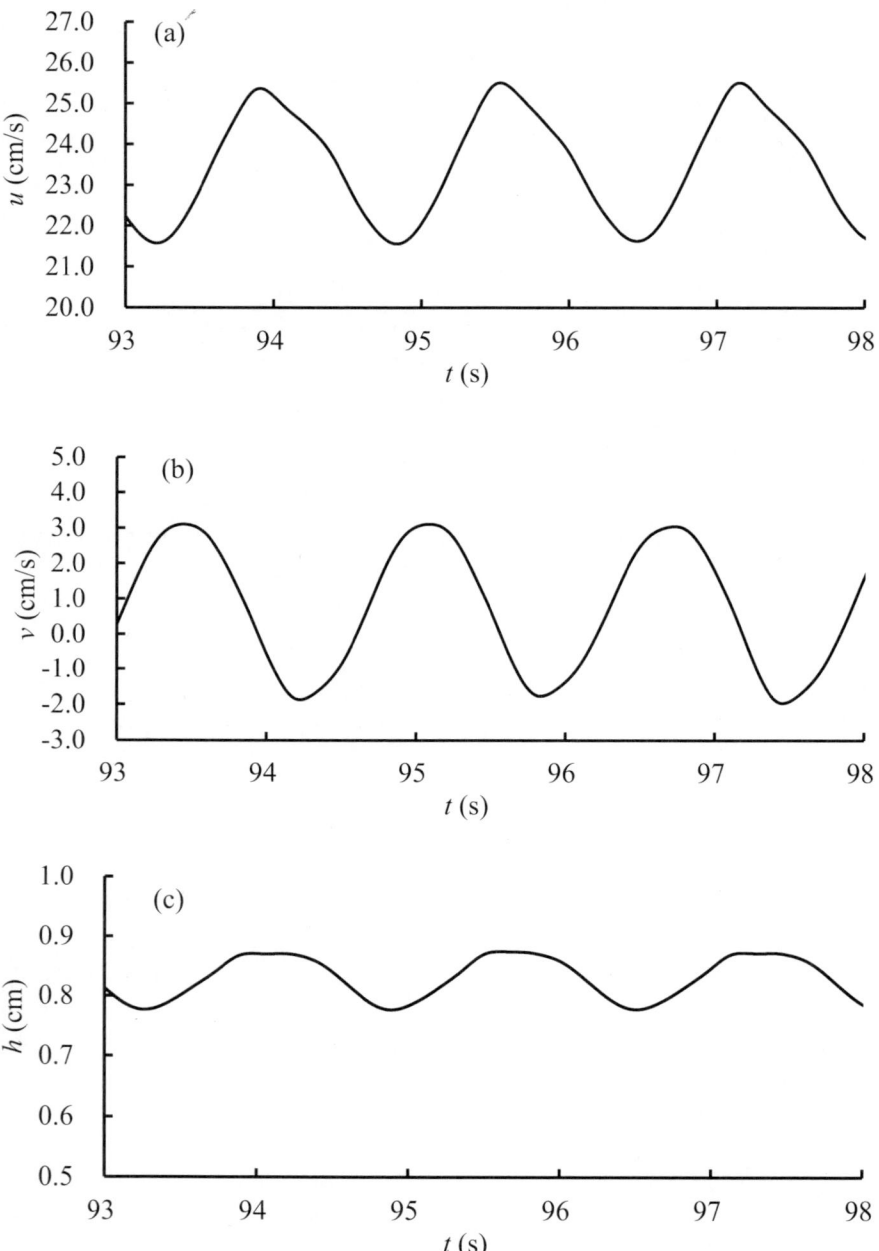

Fig. 7. Velocity variation at *c* point in typical conditions (condition of C-2) (a) velocity *u* at *x* axis, (a) velocity *v* at *y* axis (c) depth *h*

the other conditions, which in accordance with the results of the recirculation flow length, as shown in Fig. 7.

3.2.3   Vortex of the Entrance

The shear zone at the entrance of mainstream and approach channel is an area of vortex generation and collapse. The results shown that the vortex generated at the upstream corner, then transferred with the flow to the downstream and the approach channel, and

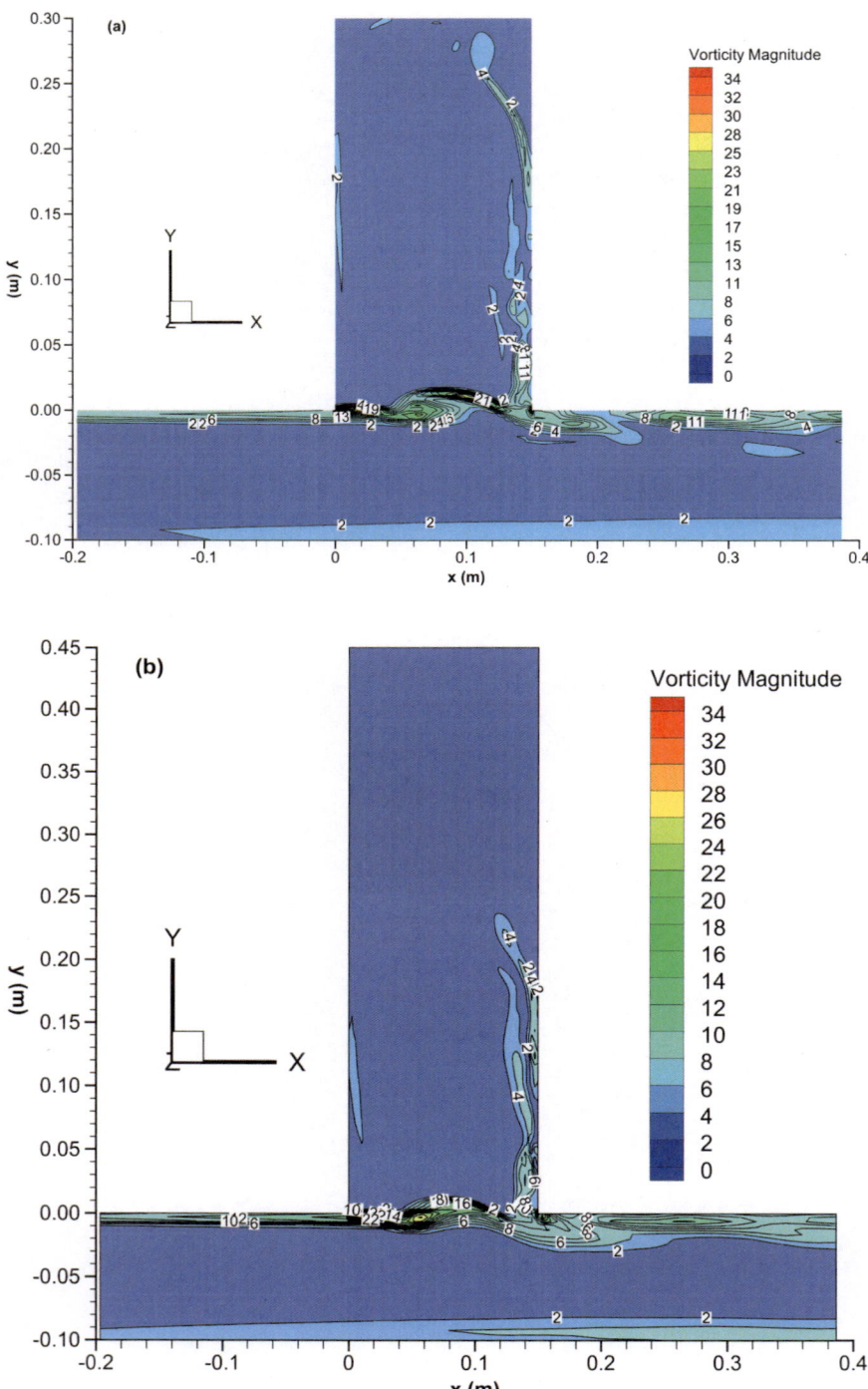

Fig. 8. Vorticity distribution at different time

collapsed at last. The velocity and water level fluctuation in the recirculation zone were shown in Fig. 8.

4 Conclusions

The flow characteristics of approach channel were studied with numerical model in this work, and the RNG k-ε and LES models were compared with Kimura and Hosoda (1997). The results showed that both of the two models could be used to study the time-average flow field at the recirculation zone of entrance. However, the simulation results of RNG k-ε model has ignored the eddy details, and cannot accurately presented the periodic characteristics of flow velocity and water level oscillation in the recirculation zone. Meanwhile, the results of LES model has a good precision for the simulation of velocity and water level fluctuation, as well as the period characteristics of recirculation zone. The LES model was adopted for the further studies of the effects of approach channel length on the recirculation zone. And the results shown that the vortex generates at the shear zone of entrance, and causes the period change in the oscillation of flow velocity and water surface, the length and period of the recirculation region is mainly decide by the dimensional parameters of approach channel and the flow conditions. All the above provide references for hydraulic characteristics of the entrance.

Acknowledgements. The authors appreciate the support of the National Key Research and Development Program of China (No. 2021YFC3201101), Project funded by China Postdoctoral Science Foundation (No. 2020T130309, 2019M651892), Jiangsu Water Resources Science and Technology Project (No. 2020022, 2021024), and Nanjing University of Information Science & Technology Research Foundation (No. 2017r097). The authors also want to thank the people for their helpful suggestions and corrections on the earlier draft of our study according to which we improved the content. There is no conflict of interest in our paper.

References

Girimaji SS (2006) Partially-averaged navier-stokes model for turbulence: a Reynolds-averaged Navier-stokes to direct numerical simulation bridging method. J Appl Mech 73(3):413–421

Kimura I, Hosoda T (1997) Fundamental properties of flows in open channels with dead zone. J Hydraul Eng 123(2):98–107

Li Y, Li F (2005) Discussion on improving measures for navigational conditions of entrance area of downstream approach channel in Sanjiang of Gezhouba. J Waterw Harbour 26(3):154–158

Liu Q (1995) The characteristics of water movement in cecum circulating flow. J Hydrodyn 10 (3):290–301

Sagaut P (2006) Large Eddy Simulation for Incompressible Flows: An Introduction. Springer Science & Business Media, Berlin, Heidelberg, pp 124. https://doi.org/10.1007/b137536

Sauida MF (2016) Dead zone area at the downstream flow of barrages. Ain Shams Eng J 7 (4):1053–1060

Zheng B, Zhou H, Meng X, Cao Y, Chen Z (2008) Discussion on flow structure of the entrance of lock and the range of junction block. J Waterw Harbour 29(3):199–204

Zhou H, Zheng B, Wang H (2005) Test on wave amplitude and slope during lock filling and emptying in approach channe. J Waterw Harbour 26(2):103–108

The Approach of Lock Hydraulic System Selection Based on Multiple Factors

Zhonghua Li[1]($\boxtimes$) and Guoxiang Xuan[2]

[1] Nanjing Hydraulic Research Institute, Nanjing, China
zhli@nhri.cn
[2] Key Laboratory of Navigation Structure Construction Technology,
Ministry of Transport, Nanjing, China

Abstract. There are many types of navigation lock filling and emptying systems, how to choose the appropriate hydraulic system is the key issues in the design of navigation lock. At present, there is no unified method and standard for the decision of the navigation lock filling and emptying system. In Chinese design codes, the traditional way to choose the lock hydraulic system types is based on the m coefficient which is related with the lock lift height (H) and leveling time(T). The influence of dimension of lock chamber is ignored in the method, including the length, width and initial water depth etc. which makes it difficult to choose an accurate and reasonable F/E system. In this paper, according to the energy dissipation mechanism of the filling and emptying system, combined with the influences of water head, filling and emptying time, chamber dimension and initial water depth, except the traditional discrimination index m, two dimensionless indexes m_L and m_C are introduced, which respectively represent the energy that needs to be eliminated in the horizontal, vertical and horizontal unit area of the F/E system. And the lock hydraulic system is classified into 7 categories based on the layout and energy dissipation characteristics of F/E system. The decision indexes of over 100 navigation locks with different characters are counted, then the relationship between the indexes and 7 lock hydraulic system categories is established using gray system theory. Finally, using the filling and emptying system selection method, the filling and emptying system of 13 typical ship locks are calculated and compared with the industry codes. The results show that this method quantifies the filling and emptying system selection process; the filling and emptying system selected by this method has high accuracy and good practical value.

Keywords: Navigation lock · Hydraulic system · Classification of F/E system · Decision indexes

1 Introduction

According to the layout of the water outlet of the lock chamber and the energy dissipation method, it is mainly divided into two categories: end filling and emptying system and longitudinal culvert filling and emptying system (NHRI 2005)˙ Both of them can be subdivided into many types depending on the range of application. Taking the longitudinal culvert filling and emptying system as a column, there are more than

Y. Li et al. (Eds.): PIANC 2022, LNCE 264, pp. 693–703, 2023.
https://doi.org/10.1007/978-981-19-6138-0_60

10 kinds of filling and emptying systems commonly used, such as the side orifice longitudinal culvert filling and emptying system, the bottom longitudinal culvert filling and emptying system with side ports and open ditches, and so on. At present, there is no unified method and standard for the decision of the navigation lock filling and emptying system at home and abroad. How to choose the appropriate filling and emptying system is the most concerned problem in the design of navigation lock.

In the Soviet Union, the key consideration in the design of lock is project investment. It is believed that when the water head is not large (the water head is below 15m), compared with the longitudinal culvert filling and emptying system, the engineering cost of the end filling and emptying system can be reduced by 10% to 60%. When $L \cdot H < 2000$ m^2 (L is the length of the lock chamber, H is the design head of the lock) and $H/h_k < 3$ (h_k is the water depth on the sill of the lock chamber), an end filling and emptying system can be used regardless of the demonstration. Only when the water head exceeds 18–20 m, consider using a longitudinal culvert filling and emptying system (В.д 1964). Choosing the water delivery system according to this standard often results in poor berthing conditions in the lock room, which is rarely used at present.

In the navigation lock design manual revised by the U.S. Army Corps of Engineers in 1985 (US Army Corps of Engineer 2006), the water head classification and the respective applicable filling and emptying systems are specified as follows, see Table 1. This method only considers the impact of the ship lock head and the decision of filling and emptying system is conservative. If the water head exceeds 12.2 m, it is recommended to use the inertial bottom longitudinal corridor filling and emptying system, which leads to large investment and poor economy in lock construction. The International Chamber of Shipping in 1986 (PIANC 1986) divided the lock head into three levels of less than 10.0 m, 10.0–15.0 m and greater than 15.0 m, and divided the filling and emptying system into three levels: simple, medium, and complex. Different water heads choose different filling and emptying system with different complexity.

Table 1. Water head classification and filling and emptying system of American ship locks

Water head classification	Applicable filling and emptying system
Extremely low water head: 0–3.5 m	End filling and emptying system
Low water head: > 3.5–12.2 m	Side orifice longitudinal culvert filling and emptying system; bottom horizontal corridor filling and emptying system
High water head: > 12.2–30.5 m	Inertial bottom longitudinal corridor filling and emptying system
Extremely high water head: > 30.5 m	Beyond the experience of the U.S. Army Corps of Engineers, high-head ship locks can be used in the preliminary design

In < Design Code for Filling and Emptying System of Ship locks > (JTJ306—2001) (referred to as "Code"), choose the form of filling and emptying system according to the discriminant coefficient m calculated by formula (1), when m > 3.5, use end filling and emptying system; when m > 2.4, the first type of longitudinal culvert filling and emptying system can be used; when 1.8 ≤ m ≤ 2.4, the second type of longitudinal culvert filling and emptying system can be used; when m < 1.8, the third type of longitudinal culvert filling and emptying system can be used; when 2.5 ≤ m ≤ 3.5 At the time, end filling and emptying system or a simple longitudinal culvert filling and emptying system can be used.

$$m = \frac{T}{\sqrt{H}} \tag{1}$$

In the formula: m is the discriminant coefficient; H is the design head (m); T is the design water delivery time (min).

Compared with the filling and emptying system decision for navigation lock abroad, in addition to the ship lock design head H, the Chinese code has increased the impact of the water delivery time T, and made a more detailed classification of the filling and emptying system. PIANC also recommended it in 2009 (PIANC 2009).This method also does not consider the influence of the ship lock's plane dimensions, the initial water depth of the lock chamber, and other factors, resulting in the filling and emptying system selected by the value of m that cannot meet the actual requirements, as shown in Table 2.It can be seen that the design head and water delivery time of Xijiang Changzhou 1# and 2# ship locks are exactly the same, m are both 2.54, and the effective plane dimensions of the ship lock are 200.0 m × 34.0 m and 190.0 m 23.0 m, respectively. The 1# ship lock selected the second type of longitudinal culvert filling and emptying system which is more complex and has better energy dissipation effect.

At the same time, the current standard m deviates from the actual selection of the filling and emptying system. For example, the head of the Qianjiang Datengxia ship lock reached 40.1 m, which is the single-stage ship lock with the largest water head in the world. The water delivery time is 15 min and the m value is only 2.37. According to the code, the second type of longitudinal culvert filling and emptying system can be selected, but the most complicated third type of longitudinal culvert filling and emptying system is actually selected. In addition, in the actual application process, there are many types of navigation lock filling and emptying systems. Only the second type of filling and emptying system in Chinese code has more than 10 forms. The code does not quantify the water delivery system and the discrimination coefficient. The above factors bring great difficulties to the selection of the ship lock water conveyance system.

Table 2. Some typical projects of selection deviation of filling and emptying system

Shiplock	Characteristic parameter					m	Code suggestion	Actual selected
	H/m	T/min	L/m	B/m	D/m			
Changzhou 1#	15.6	10.5	200.0	34.0	4.5	2.54	The first type/ the end	The second type
Changzhou 2#	15.6	10.5	190.0	23.0	4.0	2.54	The first type/ the end	The first type
Datengxia	40.1	15.0	280.0	34.0	5.8	2.37	The second type	The third type

2   Model of Filling and Emptying System Selection

2.1   Selection method

The selection of the filling and emptying system of the ship lock requires comprehensive evaluation of the influences of many factors such as the design head, the time of filling and emptying, the plane dimension and the initial water depth of the lock. At present, the comprehensive evaluation methods considering the influence of multiple factors mainly include analytic hierarchy process, fuzzy comprehensive evaluation, data envelopment analysis, artificial neural network evaluation and gray comprehensive evaluation (Du et al. 2008). Each of these evaluation methods has its own characteristics. Analytic hierarchy process (AHP) is mainly aimed at the decision problems of basic scheme determination and is generally only used for scheme optimization (Wang et al. 2020). The data envelopment analysis method is completely based on the objective information of the index data and eliminates the error caused by human factors. Artificial neural network (Ann) is a new evaluation method for complex systems. The grey comprehensive evaluation method quantitatively represents the degree of correlation between various factors, and is essentially a comparison of the characteristic parameters between the evaluation object and the reference object. The closer the characteristic parameters are, the closer the evaluation object is to the reference object (Gao 2012). The absolute correlation is a method to quantify the correlation between several factors in the grey comprehensive evaluation method, which does not require too much data. In this method, $H = \{X_i| i \in I = \{0,1,2, 3,\ldots,m\}\}$ is factor set (association space), $X_0 \in H$ is reference factor column, $X_0 = \{x_0(k) \mid k = 1,2,\ldots n\}$, $Y \in H$ is comparison factor column, $Y = \{y(k) \mid k = 1,2,\ldots n\}$, then the absolute correlation of comparison factor column Y to reference factor column X in H can be calculated as follows:

$$R(X_0, Y) = \frac{1}{n} \left(\sum_{k=2}^{n} \frac{|x_0(k) - x_0(k-1)| \wedge |y(k) - y(k-1)|}{|x_0(k) - x_0(k-1)| \vee |y(k) - y(k-1)|} + \frac{|x_0(n) - x_0(1)| \wedge |y(n) - y(1)|}{|x_0(n) - x_0(1)| \wedge |y(n) - y(1)|} \right)$$

(2)

In the formula: "∧"means take small and "∨" means take a big; $R\ (X_0,\ Y)$ means absolute correlation, the higher the value of $R\ (X_0,\ Y)$, the higher the correlation.

According to the above method, the lock filling and emptying system can be divided into i types according to certain characteristics. Through statistical analysis of the characteristic parameters of different types of filling and emptying system, establish the reference sequence $X_i = \{x_i(k) \mid k = 1,2,...n\}$ of influencing factors of class i, calculate the correlation $R(X_i, Y)$ of the ship lock between the comparison index sequence $Y = \{y(k) \mid k = 1,2,...n\}$ and the reference sequence. If the correlation between Y and X_i is the greatest, the i type of filling and emptying system is selected.

2.2 Classification of Filling and Emptying System

According to the layout and energy dissipation characteristics of different types of filling and emptying systems, based on the types in China codes, the commonly used forms of filling and emptying systems from simple to complex at present mainly include the following 7 categories:

1) End filling and emptying system. The water flow is fed into the chamber from the end of the chamber and flows from one head to another.
2) Partial longitudinal culvert filling and emptying system. The filling is "Through longitudinal culvert" system, the emptying is"through the heads" system.
3) The Type I filling and emptying system, such as the side orifice longitudinal culvert filling and emptying system.
4) The Type IIa filling and emptying system, such as the bottom longitudinal culvert filling and emptying system with side ports and open ditches, the bottom longitudinal culvert filling and emptying with side branch hole and so on, Its characteristic is that there is no branch corridor and no partition in the filling and emptying system.
5) The Type IIb filling and emptying system. It is characterized by transverse branch corridors arranged in the gate chamber and no longitudinal branch corridors.
6) The Type IIc filling and emptying system. It includes the filling and emptying system of the vertical and horizontal branches of the gate, and the filling and emptying system of the bottom branch of the horizontal diverter gate, etc. Its characteristics are that the filling and emptying system is divided in the gate chamber, the water flow enters through the middle corridor of the gate, and the longitudinal branch corridor is set up.
7) The Type III filling and emptying system is regulated by Chinese codes, such as Iso-inertial stereo distributary multistage decentralized filling and emptying system, characterized by the inertia arrangement of the system. The first shunt is three-dimensional.

2.3 Discriminant Index of Filling and Emptying System

From the perspective of the hydraulic filling and emptying system, the selection of filling and emptying system mainly affected by many factors like design head H, lock chamber length L and width W, the initial water depth S, lock filling time T. How to quantify the influence of various factors and establish the quantitative evaluation index of various types of filling and emptying system is the key to establish the selection model of filling and emptying system. The essence of the selection of the ship lock filling and emptying system is to select the appropriate form to eliminate the water flow energy entering the lock chamber. According to the equation of ship lock water conveyance energy, the maximum energy entering the lock chamber during the process of filling and emptying can be expressed as (Wang 1992):

$$E_{\max} = 0.95\rho g \frac{9.81 CH^2}{T\sqrt{K_V(2 - K_V)}} \qquad K_V \geq 0.25 \tag{3}$$

The water flow energy that needs to be eliminated per unit area of the lock water surface, longitudinal and transverse can be expressed as:

$$E_H = \frac{E_{\max}}{LB} = 0.95\rho g \frac{9.81 CH^2}{TLB\sqrt{K_V(2 - K_V)}} \propto \frac{CH^2}{TLB} \tag{4}$$

$$E_L = \frac{E_{\max}}{LD} = 0.95\,\rho g \frac{9.81 CH^2}{TLD\sqrt{K_V(2 - K_V)}} \propto \frac{CH^2}{TLD} \tag{5}$$

$$E_C = \frac{E_{\max}}{BD} = 0.95\,\rho g \frac{9.81 CH^2}{TBD\sqrt{K_V(2 - K_V)}} \propto \frac{CH^2}{TBD} \tag{6}$$

In the formula, $E_{\max}$ is maximum energy of water delivery; C is the calculation of water area of the gate chamber, the multi-stage lock takes half of the water area of the lock chamber(m^2); K_v is the ratio of valve opening time t_v to lock filling and emptying time T; E_H is the horizontal unit area energy of the lock chamber; E_L is the energy per unit longitudinal area of the lock chamber; E_C is the energy per unit area of the lock chamber.

Define three dimensionless variables $m_H = \frac{CH^2}{TLB}$, $m_L = \frac{CH^2}{TLD}$, $m_C = \frac{CH^2}{TBD}$ which include the influences of lock acting head H, lock filling and emptying time T, effective length L, effective width B, initial water depth D and water area C of lock chamber.

$$E_H \propto \frac{CH^2}{TLB} = m_H = \frac{1}{m^2} \tag{7}$$

$$E_L \propto \frac{CH^2}{TLD} = m_L \tag{8}$$

$$E_C \propto \frac{CH^2}{TBD} = m_C \tag{9}$$

It can be seen from the above that m_H, m_L, m_C, these three dimensionless variables, respectively represent the energy that needs to be eliminated in the horizontal, vertical and horizontal unit area of the filling and emptying system. In fact, the discriminant

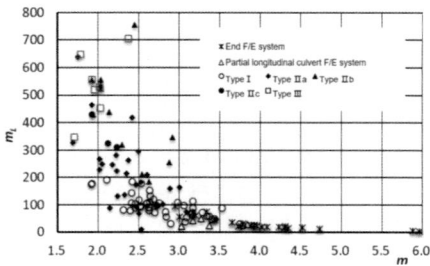

Fig. 1. Relationship between m and m_L for different filling and emptying systems

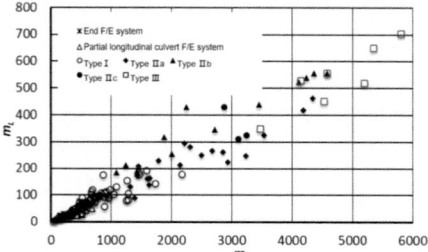

Fig. 2. Relationship between m_L and m_c for different filling and emptying systems

coefficient m of the ship lock filling and emptying system in China represents the energy of elimination per unit area of the water surface of the system. In order to be consistent with the current code, the value m is still used as the index of the energy per unit area of the water surface of the ship lock.

In order to analyze the actual elimination energy per unit area of different types of filling and emptying systems, the value of m, m_L, m_C, of 123 large ship locks (Li et al. 2020) at home and abroad are calculated and analyzed according to the above method, as shown in Fig. 1 and 2. Among them, there are 23 end filling and emptying systems with an average acting head of 6.7 m. There are 5 partial longitudinal culvert filling and emptying systems with an average acting head of 9.0 m. There are 45 TypeI filling and emptying system, with an average acting head of 11.5 m; 28 TypeIIa filling and emptying systems, with an average acting head of 18.6 m; 12 TypeIIb filling and emptying systems, with an average acting head of 25.7 m; 3 TypeIIb filling and emptying systems, with an average acting head of 28.4 m. The TypeIII filling and emptying system has 7 seats with an average acting water head of 37.0 m.

According to the classification standard of filling and emptying system, the mean values of m, m_L, m_C of 7 types of filling and emptying system are statistically analyzed, as shown in Table 3. According to the grey comprehensive evaluation method, the index reference sequence of influencing factors of filling and emptying system $X_i = \{m_i, m_{Li}, m_{Ci}\}, i = 1, ..., 7$, respectively represent the 7 types classified in Sect. 2.2.

Table 3. Index of Influencing factors of 7 types of filling and emptying system

Number	Type of filling and emptying system	m	m_L	m_C
1	End filling and emptying system	4.01	33	308
2	Partial longitudinal culvert filling and emptying system	3.33	58	504
3	Type I	2.80	100	870
4	Type II a	2.54	209	2093
5	Type II b	2.32	427	2816
6	Type II c	2.09	355	3075
7	Type III	1.97	537	4724

3 Applications

The parameters of 13 constructed or under construction ship locks with different water heads, plane dimension, initial water depth and lock filling and emptying time were selected for the decision of filling and emptying system (Li et al. 2020; Xuan et al. 2003; Li et al. 2005; Xuan et al. 2011; Li, et al., 2017).The acting head of the test lock is 7.0–45.2 m, the length of lock chamber varies from 120–300 m, the width of the lock chamber varies from 12 to 34 m, the initial water depth of the lock chamber is 3.0–10.0 m, and the lock filling and emptying time is 8.0–16.0 min. See Table 4 for the specific characteristic parameters of each test lock.

Table 5 shows the results of the calculated correlation between 13 ship locks and different types of filling and emptying system and is sorted according to the absolute correlation. The greater absolute correlation indicates that the lock is more suitable for the filling and emptying system. It can be seen from Table 6 that the 12 types of filling and emptying system selected by the method in this paper compared with the actual form used in the project are obviously superior to the current codes in terms of the type decision.

Table 4. Characteristic parameter of test lock and index of influencing factors

Number	Name	L/m	B/m	D/m	H/m	T/min	Discriminant coefficient		
							m	m_L	m_C
1	Anrenpu	230.0	23.0	4.0	7.0	10.0	3.78	31	310
2	Siyang3[#]	260.0	23.0	4.0	7.0	8.0	3.02	38	433
3	Xingan	180.0	23.0	3.5	10.1	8.0	2.52	94	732
4	Yuliang	190.0	12.0	3.5	12.5	10.0	2.83	57	895
5	Changzhou1[#]	200.0	34.0	4.5	15.6	10.0	2.54	214	1257
6	Changzhou2[#]	190.0	23.0	4.0	15.6	10.0	2.54	156	1288
7	Dongfengyan	200.0	34.0	10.0	16.0	10.0	2.50	102	599
8	Fuchunjiang	300.0	23.0	4.5	20.2	16.0	3.56	140	1832
9	Qiaogong	120.0	12.0	3.0	24.7	12.0	2.01	223	2228
10	Dahua	120.0	12.0	3.0	29.0	11.4	2.12	325	3246
11	Wanan2[#]	180.0	23.0	4.5	32.5	11.5	2.02	529	4143
12	Datengxia	280.0	34.0	5.8	40.1	15.0	2.37	705	5804
13	The Three Gorges	280.0	34.0	5.0	45.2	12.0	1.78	649	5346

Table 5. Correlation between the test lock and 7 types of filling and emptying system

Number	Name	Absolute correlation $R(X_i, Y)$						
		End	Partial longitudinal culvert	TypeI	TypeIIa	TypeIIb	TypeIIc	TypeIII
1	Anrenpu	**0.98**	0.58	0.33	0.14	0.10	0.09	0.06
2	Siyang3[#]	0.74	**0.80**	0.46	0.20	0.13	0.13	0.08
3	Xingan	0.39	0.66	**0.87**	0.38	0.25	0.24	0.16
4	Yuliang	0.40	0.69	**0.82**	0.38	0.26	0.25	0.16
5	Changzhou1[#]	0.21	0.36	0.63	**0.71**	0.46	0.46	0.30
6	Changzhou2[#]	0.22	0.38	**0.66**	0.65	0.43	0.42	0.28
7	Dongfengyan	0.45	0.76	**0.77**	0.34	0.22	0.22	0.14
8	Fuchunjiang	0.18	0.31	0.55	**0.81**	0.56	0.53	0.35
9	Qiaogong	0.13	0.23	0.40	**0.94**	0.72	0.70	0.45
10	Dahua	0.09	0.16	0.28	0.64	0.81	**0.93**	0.66
11	Wanan2[#]	0.07	0.12	0.20	0.47	0.71	0.72	**0.91**
12	Datengxia	0.05	0.08	0.15	0.34	0.52	0.52	**0.80**
13	The Three Gorges	0.05	0.09	0.16	0.37	0.56	0.57	**0.87**

Table 6. Selection of test ship lock filling and emptying system

Number	Name	Test results	Codes	Practical application
1	Anrenpu	End	End	End
2	Siyang3[#]	Partial longitudinal culvert	End or type I	Partial longitudinal culvert
3	Xingan	Type I	End or type I	Type I
4	Yuliang	Type I	End or type I	Type I
5	Changzhou1[#]	Type II a	End or type I	Type II c
6	Changzhou2[#]	Type I	End or type I	Type I
7	Dongfengyan	Type I	End or type I	Type I
8	Fuchunjiang	Type II a	Type II	Type II a
9	Qiaogong	Type II a	Type II	Type II a
10	Dahua	Type II c	Type II⁻	Type II c
11	Wanan2[#]	Type III	Type II	Type III
12	Datengxia	Type III	Type II	Type III
13	The three gorges	Type III	Type III	Type III

4 Conclusions

1) The non-dimensional discriminant coefficient of filling and emptying system m, $m_L = \frac{CH^2}{TLD}$, $m_C = \frac{CH^2}{TBD}$ can comprehensively reflect the influences of design head, plane scale of lock chamber, initial depth of water level and time of lock filling and emptying on the selection of filling and emptying system of ship lock.
2) The model is used to calculate the filling and emptying system of 13 typical ship locks, and the results show that the form of filling and emptying system calculated by the model is basically consistent with the form of filling and emptying system adopted in the engineering practice, which is not only the selection results are obviously better than the current specifications, but also the selection results can be quantified and good practicality.

Acknowledgements. This paper is supported by the National Key Research and Development Program of China (Grant No. 2016YFC0402001), the Science Foundation of NHRI (Grant No. Y120011).

References

Nanjing hydraulic research institute (2001) Code for design of ship lock filling and emptying system. JTJ306—2001

В.д. (1964) Hydraulics of ship lock

US Army Corps of Engineer (2006) Hydraulic design of navigation locks EM1110–2–1604. USACE

PIANC (1986) Final report of the international commission for the study of locks. PIANC

PIANC (2009) Innovations in navigation lock design NO. 106. PIANC

Huang Y, Cao FS (2019) Study on selection standard of ship lock filling and emptying system. China Water Transp. 19(05):114–116

Du D, Pang QH, Wu Y (2008) Modern comprehensive evaluation method and case selection

Wang DY, Cheng MY, Huang HJ (2020) Study on influencing factors of ship lock capacity based on analytic hierarchy process. Port Waterw Eng. 6:154–158

Gao L (2012) Research on application of multi-level grey evaluation method in ship recycling. Dalian Maritime University

Wang ZG (1992) The Design of Ship Lock. Water Conservancy and Electric Power Press

Li ZH, Xuan, GX (2020) Research report on selection of navigable buildings. NHRI

Li ZH, Xuan GX (2020) Report on key technology of filling and emptying system of 60m single stage giant lock. NHRI

Xuan GX, Li, ZH, Huang, Y. Study on model test of ship lock filling and emptying system of Guangxi Changzhou Hydro-junction. NHRI (2003)

Li ZH, Xuan GX, Huang Y: Hydraulic study on filling and emptying system of Qiaogong Lock, Hongshui River, Guangxi. NHRI (2005)

Xuan GX, Liu, BQ, Huang Y (2011) Study on the layout, hydraulic characteristics and model test of filling and emptying system of ship lock of Fuchun River reconstruction project. NHRI
Li J, Xuan GX, Huang Y (2017) Study on optimal layout and hydraulic physical model test of filling and emptying system of ship lock of Datengxia Hydro-junction. NHRI

The Finite Element Analysis of Dislocated Lock Heads in Double-Lane Ship Lock

Zhiguo Niu[1,2(✉)] and Yali Wang[3]

[1] State Key Laboratory of Hydrology-Water Resources and Hydraulic Engineering, Nanjing 210023, China
zgniu@nhri.cn
[2] Nanjing Hydraulic Research Institute, Nanjing 210023, China
[3] Teng Zhou Water Affairs Bureau, Tengzhou 277500, China

Abstract. In order to investigate the mechanical characteristics of dislocated lock heads, a three-dimensional finite element model of dislocated lock heads in double-lane ship lock was established, then the interaction between two middle piers was researched by the mixed finite element method, the influence of the new layout pattern on stress and deformation of lock heads was studied. The results showed this new pattern had little effect on structural stress, but it had obvious influence on subsidence and foundation reaction force of two middle piers. The anti-slide safety factor of transverse stability in design flood was smaller, this working condition was controllability in design.

Keywords: Lock head · Ship lock · The mixed finite element method · Stability safety factor · Subsidence

1 Introduction

The layout pattern of dislocated lock heads in double-lane ship lock is a new structural type, it has some advantages of less land occupation, environmental protection, low maintenance cost and reducing investment. This pattern has unique structural features, such as an expansion joint is built between two middle piers of lock heads, the axes of lock heads are not on same line, the earth pressure acting on side pier is dissymmetric.

Aiming at the support type of a double diaphragm wall, a finite element model was constructed to study the deformation and stresses of adjacent ship lock, the results showed larger wall spacing can cause greater reduction of the ship lock displacement (Carlos et al. 2020). Two layout schemes in accordance with Xi-Cheng canal ship lock project were proposed, the pattern of dislocated lock heads was considered to a more reasonable scheme through comparison and analysis (Chen et al. 2018). By the viscoelastic boundary and the dynamic contact model, the influence of contact nonlinearity on the structural stress and deformation under the action of earthquake was studied, the opening width of structural joints under the design earthquake and its influence on water seal were analyzed in Xin-Xia ship lock, the researches will improve the seismic design of dislocated lock heads (Ding et al. 2018).

Due to few practical engineering cases and above structural features, the stress and deformation behavior of dislocated lock heads in double-lane ship lock are unknown,

Y. Li et al. (Eds.): PIANC 2022, LNCE 264, pp. 704–711, 2023.
https://doi.org/10.1007/978-981-19-6138-0_61

the promotion and application of this structural pattern are limited. In order to investigate the mechanical characteristics of dislocated lock heads, a three-dimensional finite element model of dislocated lock heads in double-lane ship lock is established, then the interaction between two middle piers is researched by the mixed finite element method, the influence of this layout pattern on stability, structural stress , subsidence of lock heads is studied, these results in this paper will provide a theoretical basis for proving the design of the new structural pattern.

2   The Mixed Finite Element Method

Analysis on the mechanical mechanism of dislocated lock heads in double-lane ship lock is a key problem in the design, the mixed finite element method is used to solve contact problems between lock heads in order to research the interaction between two adjacent lock heads. Based on the characteristic of local nonlinearity, the forces acting on the contact bodies are divided into two parts: external forces and contact forces. The displacement of contact bodies is regarded as the basic variable, the nodal contact force in possible contact area is regarded as the iterative variable, so that the nonlinear iteration process was only limited in the possible contact surface. The complex contact nonlinearity is transformed into the variation of the contact forces, so the iterative procedure became easily to be carried out and highly efficient (Zhao et al. 2006).

The adjacent lock heads are denoted as Ω_1 and Ω_2 in Fig. 1, u and f are displacement and contact force of contact point pair on contact surface, respectively. ξ, η, ζ are the local coordinate system defined on the contact surface, respectively. The normal gap value between contact points can be expressed as:

$$w = (u - u)\vec{\xi} + w_0 \tag{1}$$

where w denotes gap value of contact points, w_0 denotes initial gap value of contact points.

When the contact process is regarded as a mechanical model, three contact conditions should be satisfied normal impenetrability condition, normal contact force is compressive stress condition or tangential friction condition. Three contact states can be summarized as follows:

$$\text{open state}: \quad f_1 = f_2 = 0, \quad w > 0 \tag{2}$$

$$\text{close state}: \quad w_0 = 0, \quad f_1 = -f_2, \quad \sqrt{f_\zeta^2 + f_\eta^2} < Ac - \mu f_\xi, \quad f_\xi < A\sigma_t \tag{3}$$

$$\text{slip state}: \quad w = 0, \quad \sqrt{f_\zeta^2 + f_\eta^2} = Ac - \mu f_\xi, \quad f_\xi < A\sigma_t \tag{4}$$

where f_ζ, f_η denote component of contact force along tangential directions, μ is friction coefficient, A is area controlled by contact points, c denotes cohesion, σ_t denotes tensile strength of contact surface, f_ξ denotes normal contact force component.

Under static conditions, the finite element equilibrium equation at the n + 1 load increment step can be expressed as follows:

$$K\Delta u_n = (F_{n+1} + f_n - \int B^T \sigma_n d\Omega) + \Delta f_n \tag{5}$$

where K is global stiffness matrix, Δu_n is displacement increment matrix, F_{n+1} is external load vector, f_n is the contact force vector at the previous one step, Δf_n is the contact force increment at the present step, B is strain matrix, σ_n is total stress increment step at the previous one step.

3   Engineering Example

A double-lane ship lock locates in Wuxi city, Jiangsu Province, and it is an integral part of the Xi-Cheng canal regulation project. The downstream head of this project is 1.5 km away from the main channel of the Yangtze River, and this ship lock project is a pivot engineering in Xi-Cheng canal.

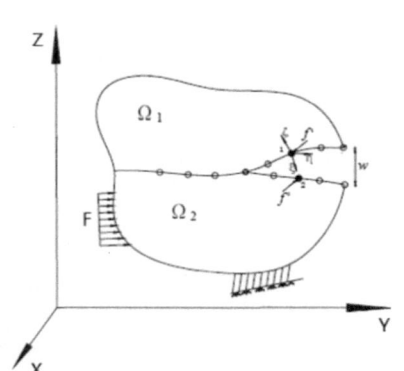

Fig. 1. Mechanical model for contact problem

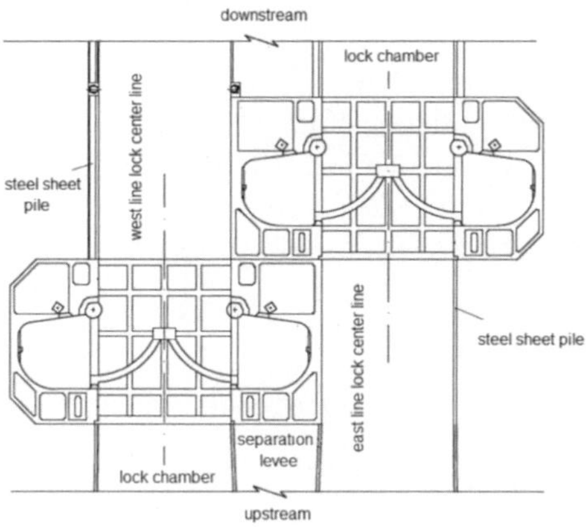

Fig. 2. The dislocation lock heads

The effective size of each single-lane ship lock is 180 m × 23 m × 4.0 m (length of chamber × width of chamber × minimum water depth), and the design water head is 3.47 m. the design navigability is 42 million tons per year. The chamber adopts a separate structural type, and the chamber walls on both side piers apply to single-anchored steel sheet pile. The shared middle chamber wall uses a counter-pulled sheet piles structural type.

The project adopts a new pattern of dislocated lock heads in double-lane ship lock, there is 28.8 m along the flow direction between two adjacent lock head axis, the arrangement is shown in Figure 2 and 3.

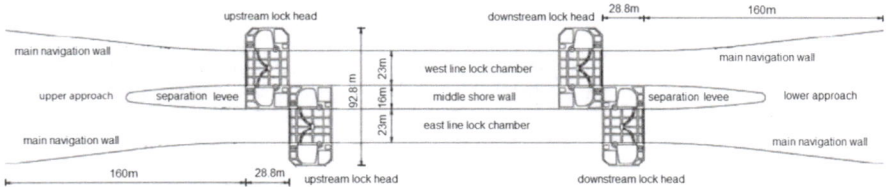

Fig. 3. The general layout of this double-lane ship locks

To account for deformation feature and mechanical mechanism of this new pattern, a three-dimensional finite element model is created. In the model, 8-node solid isoparametric and 6-node prism elements are adopted. the special contact elements are also used in the computational analysis. The whole FEM model consists of 417290 elements and 1154860 nodes, as seen in Figure 4. Dimensions of the simulations are 112.8 m (X axis) × 77.6 m (Y axis) × 30 m (Z axis), the transverse of this model is X axis and the along river direction is Y axis. In this study, we use GEIDP to calculate the deformations and stresses of this model, GEIDP is a finite element analysis software package developed by Li T.C. professor.

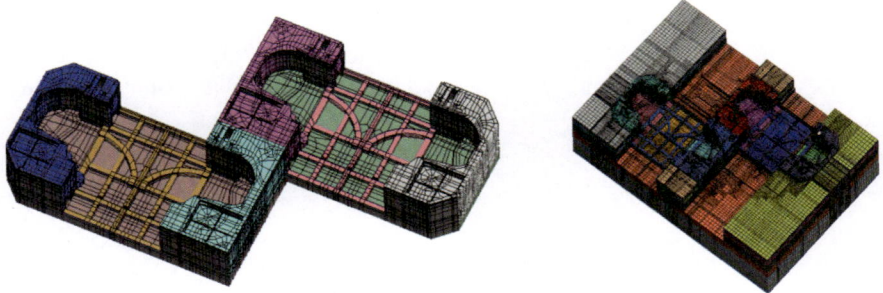

Fig. 4. Finite element model of dislocated lock heads

The main loads such as self-weight, water pressure, earth pressure, uplift pressure and so on are carried on the dislocated lock heads. The load combination is given in Table 1.

Table 1. Load combination

Working condition		Self-weight	Earth pressure	Uplift pressure	Water pressure	Ship load	Temperature
Normal operation	Positive direction	√	√	√	√	√	√
	Reverse	√	√	√	√	/	√
Construction completion		√	√	√		/	√
Maintenance condition		√	√	√		/	√
Design flood		√	√	√		/	√

4   Result Analysis

The maximum tensile stresses in the lock heads are shown in Table 2, the calculation proves the tensile stress is generally small in the dislocated lock heads, the maximum tensile stress is 1.69 MPa in the side pier, the maximum tensile stress is 1.73 MPa in the middle pier, the maximum tensile stress is 1.09 MPa in the lock floor, the maximum tensile stress does not exceed the concrete tensile strength for the whole structure.

The anti-slide safety factor of lock heads can reflect structural stability under various working conditions, the tangential force and normal force of each element on the sliding surface can be obtained according to the finite element results (Niu et al. 2009), the stability safety factor can be calculated by traditional formulas shown as following:

$$K_s = \frac{\sum \sigma_i f_i A_i + \sum c_i A_i}{\sum \tau_i A_i} \tag{6}$$

where σ_i is normal stress of element i, f_i is friction coefficient of element i, c_i is cohesion of element i, τ_i is tangential stress of element i, A_i is area of element i in the sliding direction. The stability safety factor is obtained in each working condition shown in Table 3.

Table 2. The maximum tensile stress unit: MPa

Working condition		Side pier	Middle pier	Floor
Construction completion		0.47	0.37	1.03
Normal operation	Positive direction	1.28	1.39	0.71
	Reverse	1.33	1.45	0.68
Maintenance condition		0.55	0.51	1.09
Design flood		1.69	1.73	0.62

From Table 3, it is obvious that each safety factor of shear fracture under four working conditions is more than 1.8 and the result satisfies the criterion, the structure can get enough safety reserve in overall stability. The anti-slide safety factor of transverse stability in design flood is smaller than other conditions, this working condition is controllability in design.

Table 3. Stability safety factor of gate head based on finite element method

Working condition		Transverse stability		Overall stability		
		Anti-slide safety factor	Anti-dip stability safety factor	Anti-slide safety factor	Anti-dip stability safety factor	Anti-floating stability safety factor
Normal operation	Positive direction	2.05	8.45	6.46	7.04	3.26
	Reverse	2.77	8.76	5.23	6.58	2.83
Construction completion		2.55	6.1	6.55	5.31	3.24
Maintenance condition		3.13	4.73	8.79	8.97	/
Design flood		1.82	9.27	7.54	7.49	2.06

Figure 5 shows the foundation reaction force distribution of lock heads under different working conditions. It is seen that the great foundation reaction force occurs in middle piers for each working condition. These forces are 74.45 kPa, 93.58 kPa, 98.04 kPa, respectively, under three working conditions. The foundation reaction force is largest under design flood condition, The foundation reaction forces of the middle-pier floors are greater due to the two middle piers interact with each other. When this ship lock is completed, the ratio of the maximum to the minimum of the foundation reaction force in the middle-pier floors is 4.13, which is slightly smaller than the allowable value 5.0 of the current ship lock design code. When the ship locks are operating, due to the action of water pressure in the lock heads, foundation reaction force can tend to be uniformly, the ratio of the maximum and minimum foundation reaction force is 2.40, which meets the requirements of the design code.

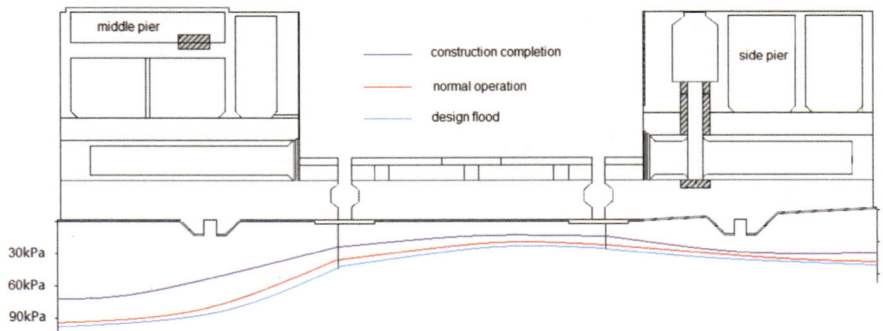

Fig. 5. The foundation reaction force distributions of lock heads

The subsidence of lock-head floor is shown in Fig. 6, we can see that there is a large subsidence on side floor, and the maximum value is 20.3 mm. Under different working conditions, the subsidence of lock-head floor is mainly rigid displacement, and this subsidence will not result in greater stress on the lock head structure.

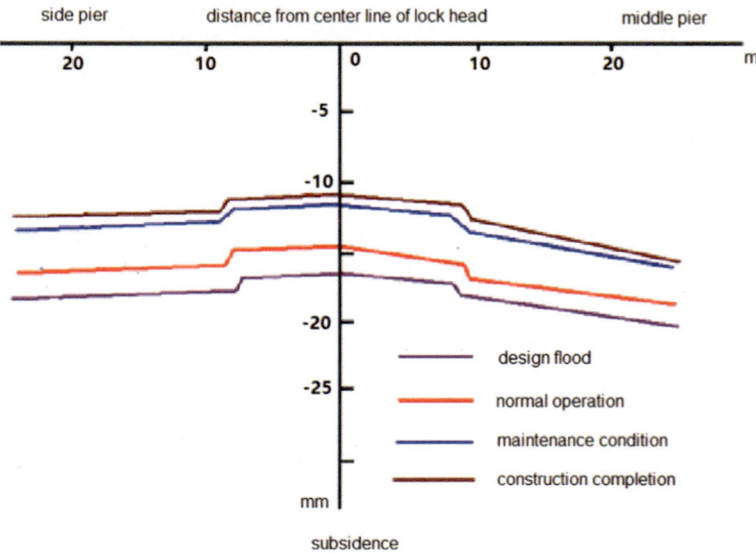

Fig. 6. Horizontal distribution of lock head floor subsidence

5   Conclusions

Based on the analysis above; the following conclusions could be reached:

(1) For the dislocated lock heads, the maximum tensile stress does not exceed the concrete tensile strength under different working conditions, this layout pattern has little effect on structural stress.

(2) This layout pattern has obvious influence on subsidence and foundation reaction force of middle piers. The subsidence of dislocated lock heads is mainly rigid displacement and large subsidence occurs in middle piers. The foundation reaction force of middle piers is quite large under design flood.

(3) For the dislocated lock heads, the anti-slide safety factor of transverse stability in design flood is smaller, this working condition is controllability in design.

Acknowledgements. The authors gratefully acknowledged the financial support from the National Natural Science Foundation of China (No. 51109142). This work in the article was supported by the National Key Research and Development Program of China (No. 2016YFC0401610). This research was supported by the Jiangsu provincial transportation science and technology project (No. 2015T47–1).

References

Carlos EJ (2020) Study on the Influence of wall spacing on double diaphragm wall supporting structure for ship lock expansion. In: 2020 3rd International Conference on Applied Mathematics, Modeling and Simulation

Chen EW, Wu LZ, Li SP (2018) Plane layout and structure optimization of Xinxia Port ship lock in Jiangyin. Port Waterw Eng 8(545):65–75

Ding TP, Niu ZG, Gao Y (2018) Dynamic response analysis of lock heads with dislocation layout in the double locks. Pearl River 39(12):129–134

Zhao LH, Li TC, Niu ZW (2006) Mixed finite element method for contact problems with friction and initial gaps. Chin J Geotech Eng 28(11):2015–2018

Niu ZG, Hu SW (2009) Optimization design of the powerhouse at the Dam Toe. In: Asia-pacific Power and Energy Engineering Conference. IEEE

The Type and Layout of the Lock Gate and the Scheduling for the Babao Lock on the Estuary with Strong Tidal Bore

Guoqiang Jin[1(✉)], Zhejiang Li[1], Yingbiao Shi[2], Jiming Zhai[1], and Runchen Ye[1]

[1] Zhejiang Institute of Communications Co. Ltd., Hangzhou, China
jingq@zjic.com
[2] Zhejiang Institute of Hydraulics and Estuary, Hangzhou, China
shiyb@zjwater.gov.cn

Abstract. The unique tidal bore of Qiantang River is called "the first tidal bore in the world". The outlet of Hangzhou Babao lock of Beijing-Hangzhou Canal is located in the strong tidal bore area of Qiantang River estuary. This paper focuses on the characteristics of Qiantang River tidal bore, the selection of lock gates under the action of strong tidal bore, and the navigation safety control mechanism of lock. The results show that during the spring tide period, the maximum height of the head of tidal bore at Babao lock of Qiantang River can reach 2.5 m, and the maximum tidal bore pressure on the front of the lock can reach 90–100 kPa. In order to ensure the safe operation of ship lock, the design put forward for the Qiantang river side lock head adopts the miter gates and tidal gate combination plan, namely to put a flat gate on the outside of miter gate, with gate mouth 23 m wide and height of 13 m, to withstand the enormous impact of tidal bore in a large tide by closing the tide gate before the arrival of the tide and opening the tide gate afterward, while only using the miter gate as the lock operation gate the rest of the time during which there are no tidal surges. According to the patterns of tidal bore propagation, the intelligent prediction and early warning mechanism of tidal bore, the operation rules, and the automatic control system of miter gate and tide gate are studied and established to ensure the safe and efficient operation of the lock.

Keywords: Babao lock · Tidal bore · Characteristics · Tide gate · Safe operation

1 Preface

With a total length of 1,794 km, the Beijing-Hangzhou Grand Canal is China's second "golden waterway", only after the Yangtze River. In order to address the conflict between the rapid growth of the Beijing-Hangzhou Canal transport volume in Zhejiang section and the incompatibility of channel capacity, a second channel connecting the Beijing-Hangzhou Canal with the Qiantang River was started in October 2017 and will be completed by the end of June 2022, resolving the grand canal navigation bottlenecks in consideration of the grand canal cultural heritage protection and utilization. Hangzhou Babao ship lock is a key project connecting the southern end of The Beijing-Hangzhou

© The Author(s) 2023
Y. Li et al. (Eds.): PIANC 2022, LNCE 264, pp. 712–723, 2023.
https://doi.org/10.1007/978-981-19-6138-0_62

Canal with the Qiantang River, which is capable of having 1000-ton ships navigating in the inland river. At the same time, a double-line lock is constructed, with the effective dimensions of the lock chamber being 300 m long, 23 m wide and 4.2 m threshold depth. The location of Babao Lock is shown in Fig. 1. The Qiantang River is famous for its unique tidal bore. The maximum height of tidal bore in tidal flood is 3 m, and the height of tidal bore on buildings can exceed 15 m. After tidal bore, the flow speed is generally 6–9 m/s, and the maximum is 10 m/s (Pan et al. 2008). The tidal bore at Sanbao lock is shown in Fig. 2. The outlet of Babao lock is located in the strong tide area at the mouth of Qiantang River. According to the research on the key technical issues of Babao lock gate and the in-depth research on the treatment scheme of Babao lock gate, the maximum tidal head height at the outlet of the lock can reach 2.5 m, and the maximum tidal bore pressure on the gate front can reach 90–100 kPa (Yang et al. 2016), which has a tremendous impact on buildings. Under such unusual circumstances, selection of lock gate is important as commonly used miter gate, triangle gate, and horizontal sliding gate cannot bear huge tidal impact. Therefore, in the project feasibility study and preliminary design stage, the key research is done on gate form and layout that can meet the safe operation conditions, and gate operation rules are formulated and intelligent safe operation control system of the lock is established according to the patterns of tidal propagation.

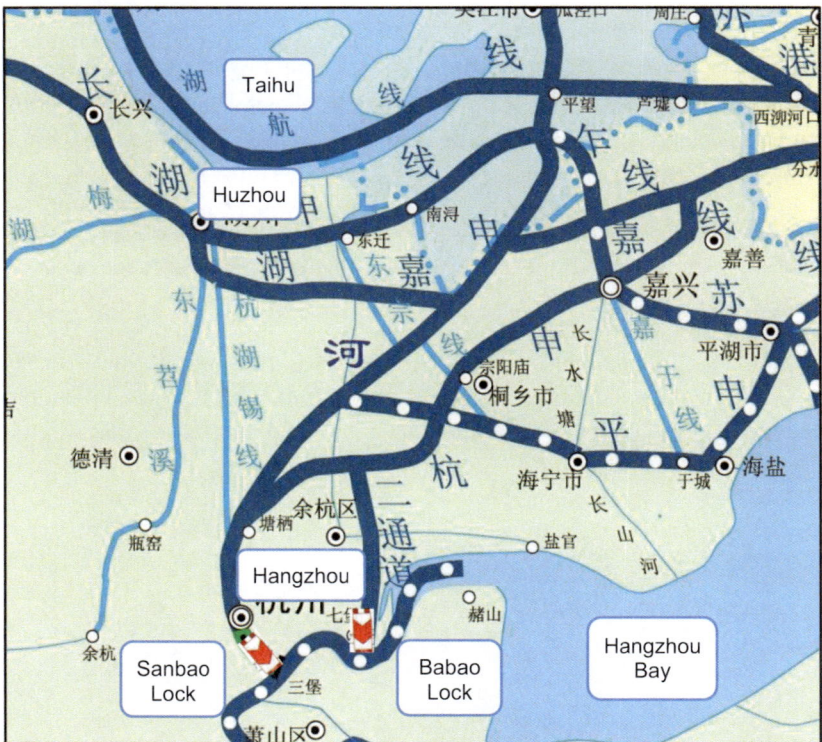

Fig. 1. Location of Babao lock of Beijing-Hangzhou Canal

Fig. 2. Tidal bore at Sanbao lock

2 Tidal Bore Characteristics

2.1 The Tides

The tides at the Qiantang River estuary are irregular semidiurnal tides. There are two tides every day, and there are two changes between spring tides and neap tides every month. The tides can reach the Fuchun River Hydropower Station about 120 km upstream of Babao. According to the long-term tidal level observation station near the gate of Babao lock, the measured perennial average high tide is 4.43 m (1985 national elevation benchmarks. Same below), the perennial average low tide is 3.65 m, the maximum tidal range is 4.28 m, and the perennial average tidal range is 0.8 m. The tidal current in the section near the entrance is a reciprocating current, lasting from 1 h 11 min to 4 h 10 min for a flood current and 7 h 42 min to 12 h 15 min for an ebb current. Except for the flood in Meiyu season, the flow velocity of floods is greater than that of ebbs, which is generally less than 1.5 m/s. When large tidal range occurs every five years, the average flow velocity of floods at perpendicular line can reach 3–4 m/s. In September 2021, the measured maximum surface flow velocity of floods was 2.71 m/s, and the maximum ebb tide flow velocity was 1.28 m/s. Figure 3 and Fig. 4 respectively show the validation diagram of secondary tide level measurement process and tidal current process near the lock entrance in September 2021.

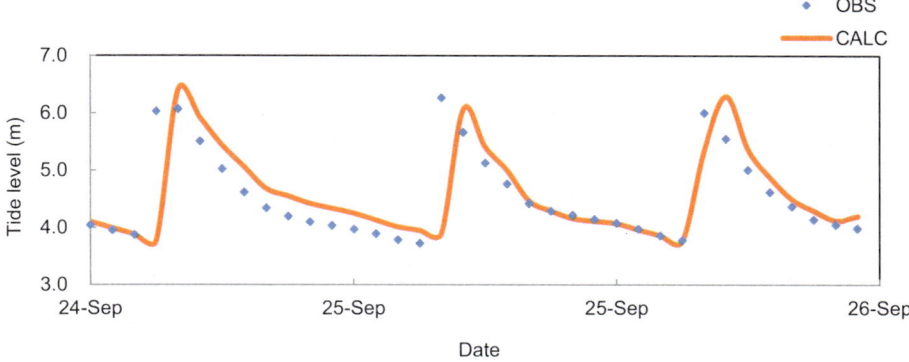

Fig. 3. Tide level process verification diagram

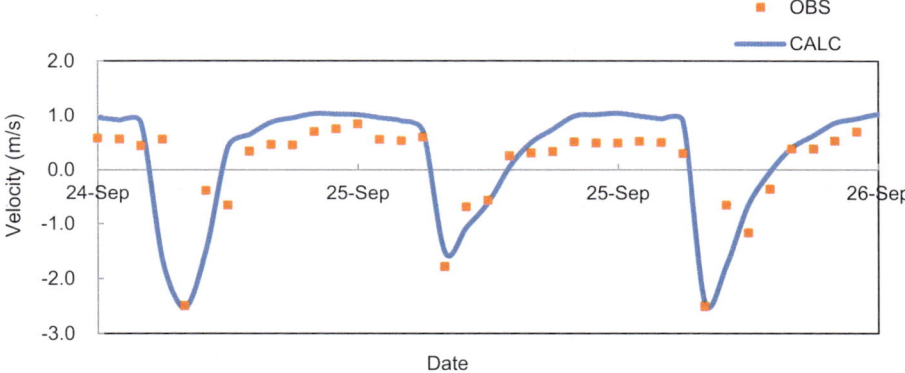

Fig. 4. Validation diagram of tidal flow proc

2.2 The Tidal Bore

The tidal bore of Qiantang River is formed in the area of Jianshan, and gradually increased in the upstream process, reaching the maximum in the area of Yanguan-Daquekou, and then gradually decreased in intensity. When the tide is strong, the head of the surge can reach Wenjiayan, and the whole journey is about 90 km. The strength of tidal bore is consistent with the size of tidal range: when the tidal range is large, tidal bore is strong; and when the tidal range is small, the tidal bore is weak. If the tidal range is less than 1.0 m, there will be almost no tidal bore.

2.2.1 The Height of the Tidal Bore
Traditionally, the height of the tidal head is used to represent the strength of the tidal bore. According to the observation data of tidal bore over the years and the relationship between tidal height and tidal range, the maximum height of tidal bore near Babao lock is about 2.5 m. According to the actual measurement, the velocity of the fast water measuring point following the head of the tidal bore is generally 6–9 m/s, with a

maximum of 10 m/s and a duration of about 15 min. According to the Babao data measured in September 2001, when the tidal range was 2.7 m and the height of the tidal bore was 1.8 m, the tidal bore velocity was about 7.34 m/s. In September 2021, when the measured tidal range was 2.5 m, the height of tidal bore was 1.3 m, and the propagation velocity of tidal bore was about 6.64 m/s.

2.2.2   The Pressure of Tidal Bore

The tidal bore propagates rapidly, and the impact force on the gate greatly exceeds the hydrostatic pressure. The pressure of tidal bore is closely related to tidal bore height. According to the physical model test of tidal bore pressure at Babao lock, the distribution of tidal bore pressure corresponding to different frequencies and tidal bore height is shown in Fig. 5.

It can be seen from Fig. 5 that the maximum pressure of tidal bore acting on the front of the gate is 92 kPa, and the action elevation is the low tidal level before the tidal bore, and the height decreases linearly upward and downward. The tidal bore pressure has also been observed in other projects on the Qiantang River and the results are similar.

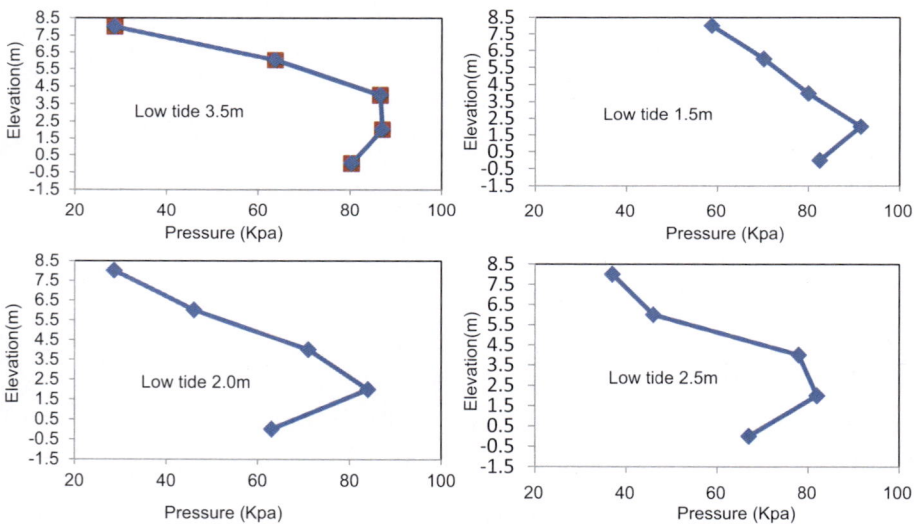

Fig. 5. Tidal action of gate at 2% tidal height

3   Lock General Layout and Gate Selection

In the stage of project feasibility study and preliminary design, detailed research and demonstration was carried out on the selection and layout of the gate of the Babao lock, as detailed below.

3.1   General Layout of Lock

The new double-line lock is arranged in parallel, the center distance of the lock chamber is 55 m, and the effective scale of the double-line lock chamber is the same at 300 m long, 23 m wide, and the threshold water depth is 4.2 m. The plane scale of the side gate head of Qiantang River is 43 m wide and 46.5 m long. The main ship lock is set up in the north of Xiasha Road, the main road of the city, and the boundary of the head is about 450 m away from the Qiantang River dike. The length of the straight section of the downstream approach channel is 438 m, among which the length of the navigation adjustment section is 138 m, the length of the ship section is 300 m, and the width is 110 m.The downstream approach channel entrance is set up with two guide dikes, with lengths of 200 m and 163.7 m respectively. The lock management area is set up on the west side of the lock, and auxiliary buildings such as the lock operation and management building, maritime management building and substation will be built in the lock area. The general layout of Babao Lock is shown in Fig. 6.

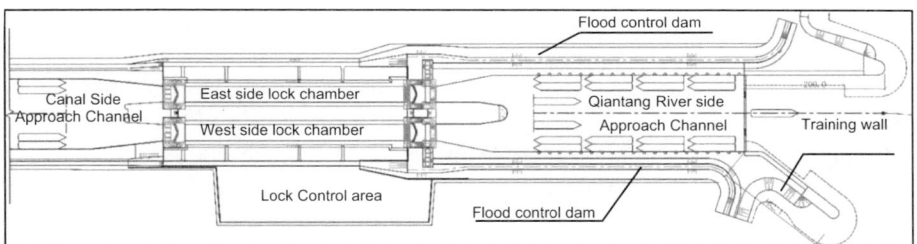

Fig. 6. General layout of the Babao lock

3.2   Lock Operating Conditions

1) Characteristic water level of the Babao Lock
    The designed maximum navigable water level of Qiantang River channel is 7.0 m, and the designed minimum navigable water level is 2.7 m. The water level is set at 9.3 m according to the high water level for a 300-year flood standard.
    The designed maximum navigable water level of the canal channel is 2.4 m, and the designed minimum navigable water level is 0.6 m. The water level is set at 3.5 m according to the high water level for a 100-year flood standard.
2) Lock working head
    The lock operates as a unidirectional head, and the maximum head is 6.4 m. At this time, the Qiantang River side is 7.7 m, and the canal side is 0.6 m. The maximum check head of the side gate of Qiantang River is 8.3 m, the side gate of Qiantang River is 9.3 m and the side gate of gate chamber is 1.0 m (the normal water level of the canal), so the flood control and tidal gate should be closed.

3) Annual navigating days

According to the meteorological, hydrological and sediment conditions, and taking the maintenance needs of the lock into consideration, the designed annual navigation time is set at 310 days.

4) Tidal bore effect

The cumulative rate of tidal range greater than 1.0 m was 40% on average for many years, with an average of 12 days of tidal bore per month and 2 semidiurnal tidal bores per day.

3.3   Selection and Layout of Lock Gate

3.3.1   Characteristics of Various Gate Types

Under the hydrologic conditions of the Babao lock, the main types of gate available are triangular gate, miter gate, flat lifting gate and sliding gate. Under normal circumstances, miter gate is most commonly used for unidirectional head lock, and triangular gate is most commonly used for bidirectional head lock. Both of them have the characteristics of economical and efficient operation, with a width of 12 m to 34 m in various specifications; Flat lifting gates and sliding gates can be adapted to bidirectional head and are used in some locks with widths ranging from 12 m to 23 m. Triangular gate, flat lifting gate and sliding gate can adapt to the action of bidirectional head and certain dynamic water load.

Facing the special hydrological and sediment conditions, the side gate of Qiantang River of the Babao lock is not ideal to be used as tide blocking gate because the miter gate and triangular gate can not bear the huge impact of tidal bore during the spring tide. For the sliding gate, when the gate is closed, it can bear a certain dynamic water effect. However, due to the high sediment content in Qiantang River during the spring tide or flood, the track groove is subject to silting, which needs to be cleaned regularly, and the underwater maintenance of the flat car at the bottom of the hoist is difficult, which affects the passing capacity of the lock, so the sliding gate scheme is not appropriate.

Flat lifting gate has the structural bearing capacity of tidal bore impact and can avoid the influence of gate groove silting, so it is a reasonable choice of tide blocking working gate. Located at about 7 km upstream of Qiantang River at the Babao lock, the Sanbao lock connects the Beijing-Hangzhou Canal to the Qiantang River with the width of the lock being 12 m, and the vertical lifting flat gate is adopted, which can withstand the impact of huge tidal bore. Thirty years of operation shows that this type of gate is safe. However, the net width of the gate of the Babao lock is 23 m, the height is nearly 13 m, the mass of the gate reaches 220,00 kg, and the lifting height is about 16 m. According to the designed capacity, the lifting times reaches 26 times a day, which is not ideal in terms of operational costs and work efficiency. Moreover, the raised gate is hoisted high in the air, resulting in poor visual and landscape effects.

3.3.2   Gate Selection and Layout

According to the characteristics of various feasible gate types and the use requirements of tide gate, the design adopts the combination scheme of "miter gate + tide gate". The gate head structure is lengthened, and a flat tide gate is set up on the outside of the miter gate, as shown in Fig. 7. Usually, when there is no tidal bore, only the miter gate is used. During the tidal bore, the tide gate is used. The tide gate is also used as a flood control and maintenance gate.

1) Miter gate

Gate head is 23 m wide, the miter gate scale is $13.58 \times 1.4 \times 9.0$ m (width × thickness × height), the gate top elevation is 7.0 m, the threshold elevation is −2.5 m, the designed working head is 6.4 m, bearing unidirectional head, and the panel is arranged towards the Qiantang River side. Miter gate open and close in still water, without considering the impact of tidal bore.

2) Tide gate

The width of the gate is 23 m, the height of the gate top is 11.5 m, and the height of the bottom sill is −1.5 m.The flat gate is 13 m high, 24.3 m wide and 2.81 m thick. When closed, both sides are stuck in the gate head slot. The structure of the tide gate meets the requirements of strength, stiffness and stability under the most adverse conditions, and the safety of the upper gate stuck in the gate slot is desirable. The tide gate is usually stored in the side gate storehouse of the head, and is hoisted into the working gate slot by the trolley hoist.

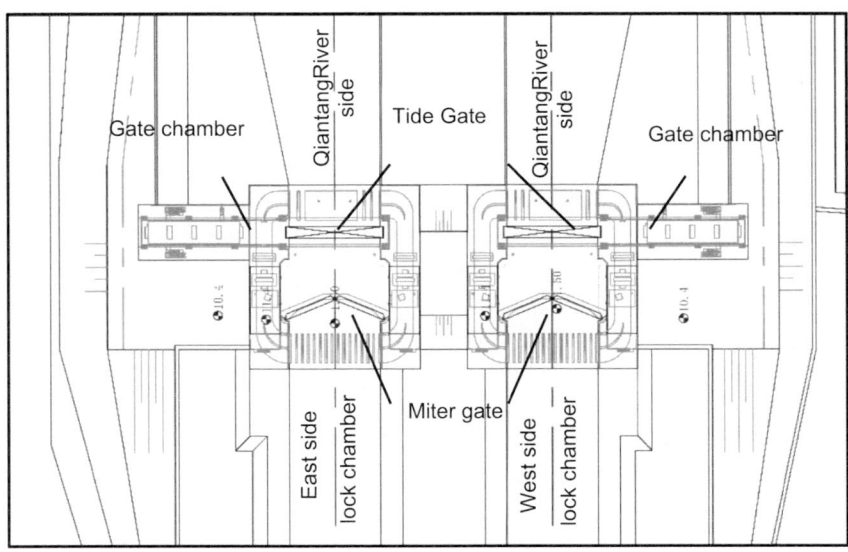

Fig. 7. Arrangement of gate along Qiantang River

3.3.3   Operation and use of Miter Gate and Tide Gate

During the normal operation of the lock without tidal bore, the miter gate is used as the working gate to block water. When the tidal range of Qiantang River is greater than 1 m and there is a tidal bore, the tide gate should be closed about 0.5 h before the arrival of the tidal bore, and the gate should be used to resist the impact of the tidal bore. The tide gate should be opened 1.5 h after the tide to make the Qiantang River navigable. Because the tide gate is only used in the middle and high tide, and is opened and closed only once per tide, and the use frequency is relatively not high, the height of the gate frame can be greatly reduced by lifting the gate to a certain height and moving horizontally to the gate chamber afterward. At the same time, the flat tide gate can also be used as the maintenance gate, and also as the accident gate when the miter gate is running, which further improves the safety of the project.

Considering the operation characteristics of the side gate of Qiantang River, that there are 12 days of tidal bore every month on average, and twice a day, the tide gate is used twice a day, and the gate is closed for 2 h each time. Only the miter gate is used during other times not affected by tidal bore. Under the circumstances that the lock is in operation 24 h a day, the number of ships passing locks is 26 times a day, and the miter gate is only required for the remaining 24 times except for the use of the tide gate twice. In addition, there is no tidal bore during 18 days of each month, so there is no need to use the tide gate.

4 Lock Safe Operation Control Measures

4.1 Navigation Control of Ship Entry and Exit

When the tidal range of Babao area is greater than 1.0 m, the ship entry and exit of the lock will be affected by tidal bore. According to the analysis of measured data, the accumulation rate of tidal range greater than 1.0 m is 39.7%. 280 tides out of 705 tides in a year have tidal bore phenomenon, during which ships need to take tide shelter measures for navigation. The tidal bore can travel 30 to 40 km upstream of Babao during strong tides, during which the ships passing the lock should shelter in the anchorage of the deep water area of the upstream, which is about 35 km from the Babao ship lock to the Zhijiang anchorage, and the tidal bore propagation time is about 70–80 min. In order to prevent ships from encountering tidal bore in the passage from Zhijiang to Babao lock, according to the operation experience of Sanbao lock in anti-tidal bore for more than 30 years or the operation rules of anti-tidal bore, ships entering and leaving the gate of Babao lock and navigation on Qiantang River are restricted for several hours before and after high tide. The navigation control time of ships passing the lock at Babao Lock is shown in Fig. 8.

1) Lock exit time: stop exit for a total of 5 h: 3.5 h before high tide (to ensure that the ship can safely reach the anchorage before the tide head given the single ship speed of 8–9 km/h) and within 1.5 h after high tide (flow rate > 2.0 m/s). Lock exit time within a tide (about 12.4 h) is during 1.5 h to 8.9 h after high tide, 7.4 h in total.

2) Lock entrance time: considering that entering the lock should be stopped 0.5 h before the tide, ships should be able to descend within 0.5–1 h after the tidal bore reaches the anchorage of Zhijiang (2–2.5 h after the tidal head of Babao ship lock). Based on the calculation of single ship speed of 10 km/h, the actual time for entering the lock of Babao ship is generally 5 h after the high tide, and the time for entering the lock within a tide is during 5 h to 11.9 h after the high tide, a total of 6.9 h.

According to the above control of gate entry and exit time, in theory, the ship lock can operate for about 20 h every day when there is tidal bore (tidal range > 1.0 m). During the small tide season when the tidal range is less than 1.0 m, as long as the water depth of the lock entrance and channel is allowed, ships entering and leaving the lock and sailing in the Qiantang River are not affected by the tidal bore, and they can theoretically pass for 24 h a day.

3) According to ship model sea trial and real ship sea trial, the safe navigable flow condition of ship lock entrance and exit area is recommended. The restricted flow rate of ship entrance and exit door is: single ship is no more than 2 m/s; The ebb tide velocity of the towing fleet shall not exceed 1.2 m/s, and the flood tide velocity shall not exceed 1.5 m/s.

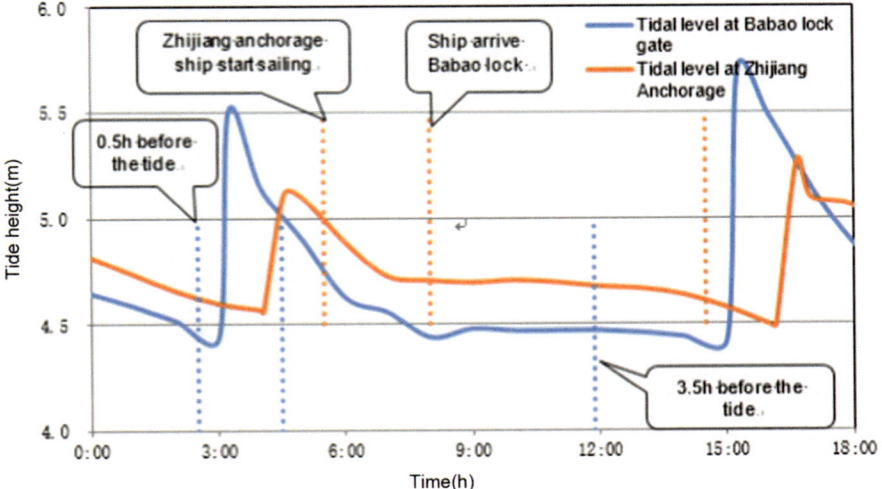

Fig. 8. Operation diagram of Babao lock

4.2 Intelligent Control of Lock Safe Operation

In order to ensure the operational safety of Babao lock and the navigation safety of ships, the project establishes a set of intelligent control system for the safe operation of the lock, which as the whole realizes the sharing of hydrological information in the upper and lower reaches of Qiantang River, accurate prediction of tidal bore arrival time, intelligent dispatching of ship crossing and safe control of gate operation.

1) Accurate prediction of tidal bore
   According to the historical hydrological data for many years from the anchorage of the Qiantang river estuary to upstream hydrological tide stations, compared with analysis of the measured data at Babao, correlation of the observed data at the Babao lock and. the hydrological stations of the Qiantang River has been established. Through short-term tidal prediction model algorithm, time and height of tides arrived at the Babao lock can be accurately predicted, and GIS (geographic

information system) technology is applied to dynamically display the real-time position, high tide and height of tidal bore in real time on the map.

2) Intelligent tidal bore warning

The system realizes the tidal bore warning function. When the tidal range is greater than 1 m and there is tidal bore, the system automatically sends out the tidal bore warning two hours and one hour before the tidal head reaches the gate of the Babao lock, and warns the lock management and control personnel to close the tide gate 0.5 h before. After the prediction and early warning system is associated with the lock control system, it can automatically control the closure of the tidal flat door.

3) Intelligent dispatching of ships passing the lock

Ships that plan to pass through the Babao lock shall apply in advance through the ship lock passing dispatching system platform of the channel network. AIS base stations are set up along the lock and channel, which can monitor and position each ship in real time and release tidal bore dynamic information for ships.The navigation dynamics of ships sailing on the upstream anchorage of Qiantang River and the time of their arrival at the Babao lock are monitored, and the risk of ships crossing is predicted by combining with the real-time dynamic information of tidal bore, so as to carry out intelligent dispatching for safe ship crossing.

4) Automatic control of gate operation

When the tidal range of Qiantang River is greater than 1 m and there is tidal bore, the tidal gate of the Babao lock should be closed 0.5 h in advance, and the opening and closing of the tidal gate should be intelligently linked with tidal bore prediction and warning. When the tidal range is less than 1 m and there is no tidal bore, only miter gate is used. The operation control model and control system of miter gate and tide gate are systematically studied to ensure the efficient and safe operation of lock.

5   Conclusion

1) The Babao lock exit is located in the area of strong tide of Qiantang river estuary area, and usually there is tidal bore in Qiantang river when tidal range is greater than 1 m. The average cumulative rate of total range greater than 1 m is 40% for many years, the largest tidal bore tide height can reach 2.5 m, and the maximum tidal bore pressure on the head on the building can reach 90–100 kPa, having a strong impact on the lock gate.

2) The head of the lock facing the Qiantang River adopts the combination scheme of "miter gate + tidal gate", that is, a flat tidal gate is set up on the outside of the miter gate to resist the huge impact of tidal surge. Usually when there is no bore, the tidal gate is placed in the head side door bank, and only the miter gate is used as the working gate. During the period of medium and large tides from June to November, there are 12 days affected by tidal bore on average every month, with 2 tidal bores every day. The tidal gate will be closed about 0.5 h before the arrival of tidal bores, and the tidal gate will be opened 1.5 h after the arrival of tidal bores, and the tidal gate will be used for navigation only twice a day.

3) In order to ensure the efficient and safe operation of the lock, accurate prediction of tidal bore arrival time, intelligent warning of tidal bore, and intelligent dispatching of ships, operation control system of double gates of miter gate and tide gate is studied and established, and integrated intelligent dispatching and safe operation control system of navigation lock in river section of strong tidal bore is constructed.

References

Jin G, Li Z, Zhai J (2017) Preliminary design report of babao lock in Hangzhou section of Beijing-Hangzhou Canal Zhejiang section waterway regulation engineering 73–83:154–172

Pan C, Lu H, Zeng J (2008) Characteristics and numerical simulation of tidal bore in Qiantang River. Hydro-Sci Eng No. 2, 1–9

Yang Y, Shi Y, Cao Y.(2016) Study on the regulation scheme of Babao gate in the third-class waterway regulation project of Zhejiang section of Beijing-Hangzhou canal. 12–20: 98–110

Transverse Mooring Forces Due to Asymmetrical Filling in a Lock with Longitudinal Culverts and Side Ports

C. Savary[1]($\boxtimes$), B. Bertin[2], M. Lenaerts[2], I. El Ouamari[2],
and D. Bousmar[1]

[1] Service Public de Wallonie, Hydraulic Research Laboratory, Châtelet, Belgium
{celine.savary, didier.bousmar}@spw.wallonie.be
[2] Université Catholique de Louvain, Louvain-la-Neuve, Belgium

Abstract. In lock design, several geometries are proposed for the levelling system with the aim of insuring a smooth and fast levelling operation. For high and medium lift lock, longitudinal culverts with side ports located in the lock walls are often chosen because this system distribute the flow along the entire length of the lock chamber. Nevertheless, when one of the culvert (valve) is out of order, the flow is asymmetrical what induced significant transverse mooring forces, especially during filling operations.

Field measurement on such a lock (225 m long, 25 m wide and 13.5 m lift) demonstrated transverse water slopes significantly larger than the admissible criteria during asymmetrical filling operation. Further investigations on a physical scale model highlighted the driving effect of the side port jets on the rolling flow. Detailed measurements were performed on the scale model covering: (1) water surface slope; (2) transverse velocity distribution; (3) transverse mooring forces and rolling angle of a vessel located in the lock chamber. It was concluded that the size of the outlets of the side ports has a major impact on the transverse forces. For smaller outlets at a given discharge, the flow velocity increases and more impulse is transferred to the rolling flow, resulting in larger forces on the vessel.

Keywords: Mooring forces · Lock · Physical model · Lock levelling

1 Introduction

The design of the levelling system of a navigation lock results from a compromise between three main constraints: (1) reducing the levelling duration; (2) minimizing the mooring forces acting on the vessel in the lock chamber; (3) keeping the lock culvert design simple and not too expensive.

Criteria for allowable mooring forces can be found in the literature and in some national regulations (PIANC WG 155 2015). These criteria cover in all cases the maximum longitudinal forces and sometimes also the transverse forces. Physical processes resulting to longitudinal forces are well documented. Notably, for through-the-head filling systems, different sources of force are identified: (1) translatory waves and longitudinal water slope; (2) impulse difference along the vessel length; (3) jet

Y. Li et al. (Eds.): PIANC 2022, LNCE 264, pp. 724–735, 2023.
https://doi.org/10.1007/978-981-19-6138-0_63

effect on the bow; and (4) friction on the vessel hull. On the other hand, only limited information on the physical processes involved in transverse forces can be found.

The present investigation focuses on locks filled through longitudinal culverts and side ports located in the lock walls. In normal operation, the filling discharge is distributed all along the lock chamber symmetrically between the two longitudinal culverts. The flow spreads from symmetrical side ports. Both jets face along the central axis of the lock and energy is efficiently dissipated (Fig. 1a). When one filling valve is out of order, e.g. due to maintenance operations, the lock is operated asymmetrically. The jets flowing from the ports on the operating side spread on the whole lock width. This generates a rolling flow in the chamber lock and significant transverse forces (Fig. 1b).

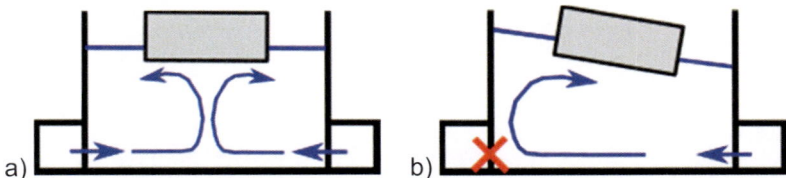

Fig. 1. Cross-section of the lock chamber - flow pattern for symmetrical (a) and asymmetrical (b) filling operation

To reduce transverse forces during asymmetrical filling operation, the valve opening time can be slowed down. However, it is usually not possible to find an opening schedule of the valve which allows decreasing the forces drastically with an acceptable duration for the leveling operation. Then, in addition, operational guidelines have to be applied: only one boat is allowed on the width of the lock chamber and it has to be moored on the side of the operational culvert (then the vessel is held against the wall due to transverse currents).

Figure 2 illustrates an incident that occurred at the lock of Lanaye in Belgium (225 m long, 25 m wide and 13.5 m lift), while these guidelines were note applied and the mooring conditions were not optimal. Ships A,D and G were moored on the side of the inoperative culvert and had to face significant difficulties. Site measurements revealed transverse slope of the free surface up to 3 to 5‰ depending on the valve opening schedule (Savary et al. 2019) and important gyratory movement in the lock chamber. The velocity profile measured at the middle of the lock chamber, centered on a port side is illustrated at Fig. 3, a transverse velocity higher than 2 m/s is observed near the free-surface.

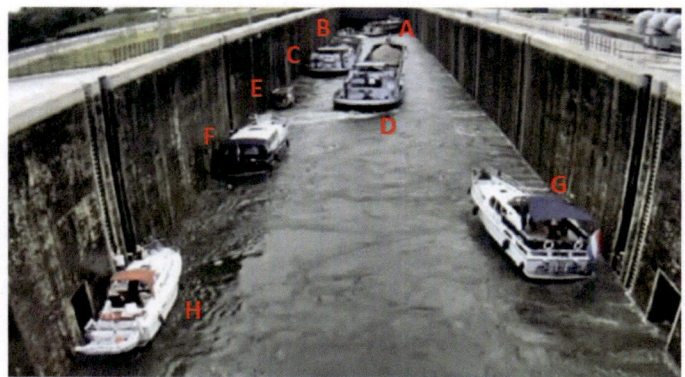

Fig. 2. Asymmetrical filling at the lock of Lanaye

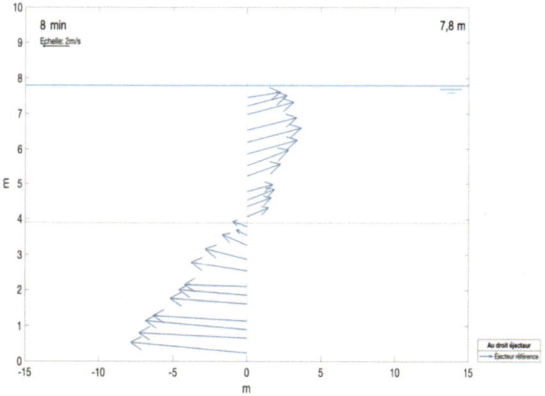

Fig. 3. Velocity profile at the lock of Lanaye

Additional investigations were realized on a physical model to measure the velocity field, the pressure around the boat, the transverse forces and the transverse slopes (water free-surface and ship) for different configurations. The measurements illustrate the influence of some parameters like the opening schedule of the valves, the position of the ship in the lock chamber and the size of the ports. The present paper focus on this experimental campaign and its results from the master thesis of El Ouamari I. and Lenaerts M. (2019) and Bertin B. (2021).

2   Physical Model and Experimental Set up

The physical model is located in the facilities of the Hydraulic Research Laboratory depending on Service public de Wallonie (Belgium). The model is a reproduction of a lock levelled through longitudinal culverts with side ports at a 1/25 scale (Fig. 4a). The lock chamber (9 m long, 0.72 m wide 0.58 m lift) is connected on each side to a longitudinal culvert (24 cm × 18 cm) through 20 ports (1 cm × 8 cm or 4 cm). The culverts are equipped with butterfly valves allowing to level the lock with different opening schedules.

The upstream and downstream reaches are equipped with large weirs to maintain a constant level during the lockage process.

Some experiments are realized with a vessel in the lock chamber. The boat dimensions are close to those of a Vb vessel at 1/25 scale (7.2 m long, 0.5 m wide, 0.3 m height), its total weight is 602.7 kg. The vessel is held in a fixed transverse position by means of two vertical bars crossing the structure (red arrows at Fig. 4a. The vessel is connected to the bars by means of a specific structure allowing rotation and vertical translation of the vessel (Fig. 4b). The mooring forces are deduced from the transverse and longitudinal forces applied by the vessel on the bars (deformation gauges). The system allows to consider elastic and unstretched mooring lines. In the present study, mooring lines are considered rigid and stretched.

a)   b)

Fig. 4. View from upstream of the physical model (a), view from above of the specific structure connecting the vessel to the vertical bars (b)

The gauges used for the measurements are:

- Ultrasonic water level gauges (Baumer Unam18U6903): measure the water level in the lock chamber and the reaches (US in Fig. 5). In the lock chamber, during lockage, they allow to calculate the evolution of the average water level and the discharge.

- Differential pressure gauges Yokogawa EJA110E): are used to estimate the free-surface slope in five cross-sections along the lock chamber (P_{dif} at Fig. 5). They are also used to estimate de hydrostatic pressure around the vessel. The range of the gauges is ± 50 mmH$_2$O with a 0.02 mmH$_2$O precision.
- Cable position sensors (ASM WS10SG): are placed at the bow and the stern to measure the trim and rolling angles of the vessel.
- Deformation gauges (TEDEA HUNTLEIGH): measurements from three deformation gauges (range 30 N, precision 0.006 N) are used to measure the forces on the vessel (F at Fig. 5).
- Electromagnetic velocity gauges (PEMS WL Delft Hydraulics): are used to measure the velocity field in cross-sections of the lock chamber. One probe (range ± 2.5 m/s, precision 0.01 m/s) is used, the same test is realized several times and the probe is displaced at each test to obtain the complete velocity field.

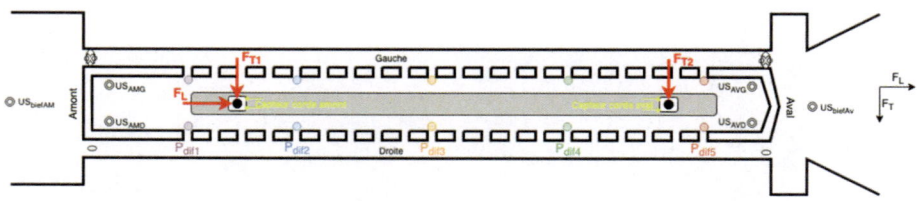

Fig. 5. Location of gauges

3   Forces, Slopes and Rolling Angle of the Vessel

Regarding the longitudinal forces during the levelling process, in many cases, the free-surface slope in the lock chamber is representative of the hawser forces because hydrostatic forces are the main components. However, the presence of a ship in the lock chamber during the levelling has an influence on the measured water slope. Longitudinal water slopes during levelling without a ship in the lock chamber will be lower than during levelling with a ship in the chamber (PIANC WG155 2015). As an order of magnitude, the presence of the ship increases the slope up to 30% for inland navigation locks depending on the ship and the size of the lock. Then the threshold value used as a reference to check that the levelling is safe will depend on what is measured or calculated (forces > free-surface slope with ship > free-surface slope without ship).

Regarding the transverse direction, in the specific case of asymmetrical filling, tests were realized on the scale model. Figure 6 illustrates the results and the difference between the measured free-surface slopes with and without ship in the lock chamber, force and rolling angle of the vessel. To be compared, all the values are expressed in ‰. As expected, the total transverse force is higher than the slope with a vessel. In this case, the transverse free-surface slope without ship is not representative of the transverse forces. The rolling angle is related to the transverse force, but of opposite sign as illustrated in Fig. 7.

The transverse free-surface slope presented in Fig. 6 is deduced from the difference in piezometric level measured against the lock walls. Additional tests were realized with a measurement made on both side against the vessel, the resulting slope is higher than the one plotted in Fig. 6 and closer from the measured forces.

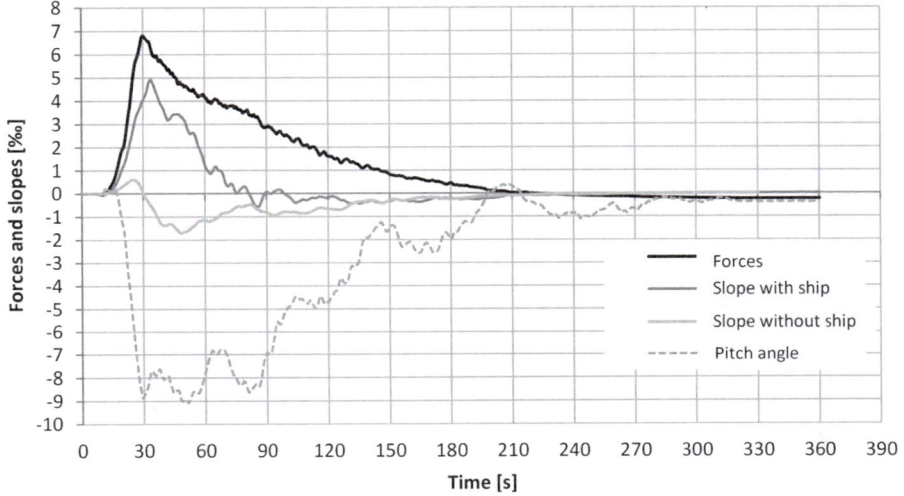

Fig. 6. Asymmetrical filling – comparison between measured transverse force, free-surface slope (with and without boat) and rolling angle (El Ouamari and Lenaerts 2019)

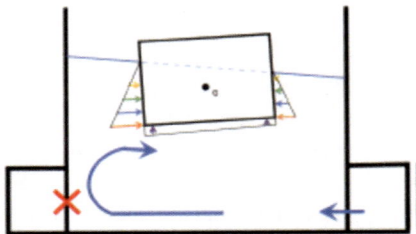

Fig. 7. Cross-section during asymmetrical filling (Bertin 2021)

4   Velocity Field

The velocity field was measured in two cross-sections, one centered on a side port and the other between two side-ports. Figure 8 illustrates the development of the gyratory movement during the asymmetrical filling process for the cross-section centered on a side port.

Tests were realized with and without a vessel in the lock chamber and for two ports size (1 cm × 8 cm and 1 cm × 4 cm). Figure 8b illustrate how the presence of the vessel influence the velocity field. Due to the dimension of the boat the depth of the gyratory movement is reduced.

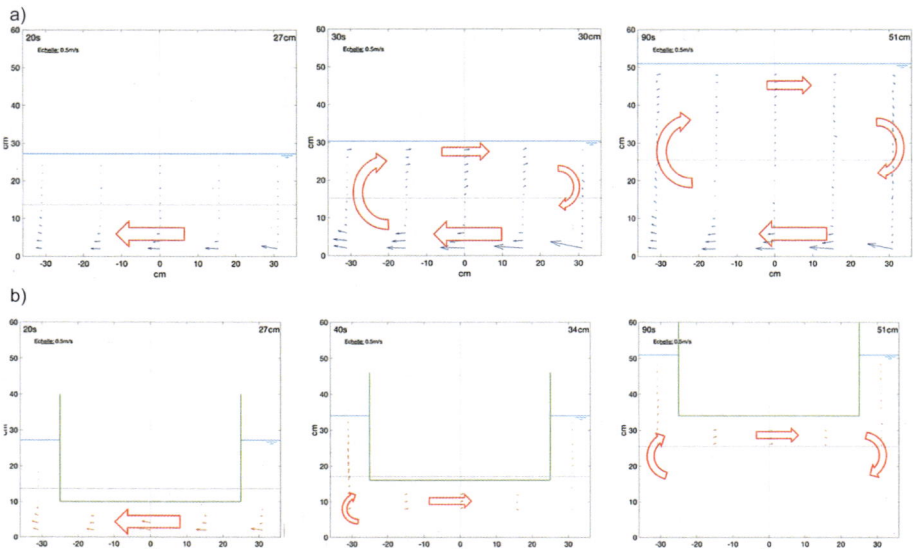

Fig. 8. Velocity field – cross section centered on a side port without (a) and with (b) a vessel (Bertin 2021)

The flow velocity coming from the side ports induces an impulse transfer to the static water of the lock chamber creating a rolling flow. To quantify this transfer, the discharge crossing different areas of the cross-section (Q1 to Q4 at Fig. 9) is roughly estimated from the integration of the velocity measurements and compared to the discharge coming from the ports Qp for the 2 geometries of the side ports. The results obtained for the maximum discharge are in Table 1. For a similar valve opening schedule, the discharge going through the ports at a given time is smaller for smaller outlets. To make possible the comparison between both geometries, the velocities measured for the smaller ports configuration was multiplied by a factor in order to have the same discharge at the ports. When the dimensions of the ports are smaller, the discharges are higher, showing that the impulse transfer is more important.

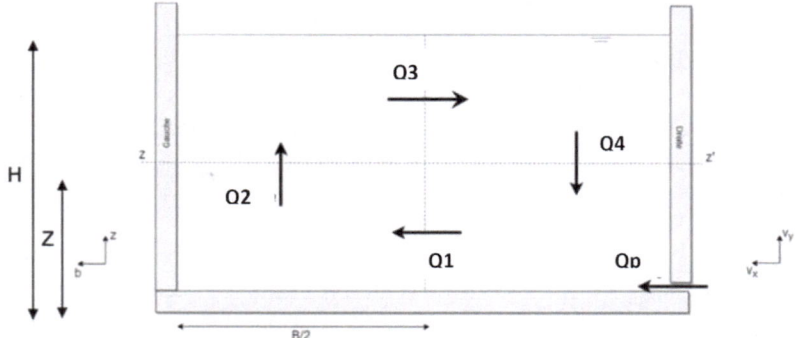

Fig. 9. Division of the cross-section (El Ouamari and Lenaerts 2019)

Table 1. Estimated discharge from integration of velocity measurements

Ports geometry	1 cm × 8 cm	1 cm × 4 cm
Qp (l/s)	1.2	1.2
Q1 (l/s)	8.3	11.5
Q2 (l/s)	8.5	11.4
Q4 (l/s)	4.5	7.8

5   Configurations Impacting the Transverse Forces

Among all the parameters that could have an impact on the transverse forces on the vessel, the influence of the valve opening schedule (3 schedules tested), the position of the vessel in the lock chamber (3 positions tested) and the dimensions of the side ports (2 width tested) was experimented.

5.1   Valve Opening Schedule

Figure 10 illustrates the measured hydrographs for 3 linear opening schedules of the valve (opening in 30 s, 60 s and 90 s).

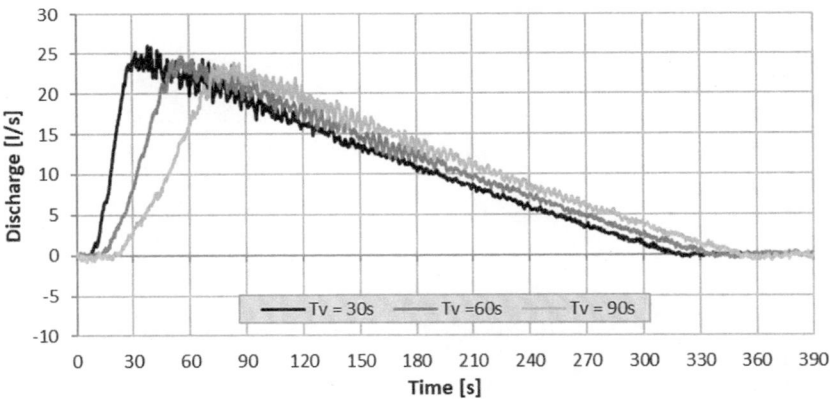

Fig. 10. Hydrographs for asymmetrical filling for 3 opening schedules of the valve (Tv) (El Ouamari and Lenaerts 2019)

Figure 11 illustrates the measured transverse force related to the different opening schedules. Opening the valve slower is quite efficient to reduce the transverse force with a limited impact, in this case, on the filling time. If the opening time of the valve is doubled (from 30 s to 60 s), the maximum transverse force decreases by 20% for a filling time 4% longer. If it is tripled (from 30 s to 90 s) the decrease in the transverse force is 30% and the increase of filling time is 11%. In this case, the efficiency of an increase of the valve opening time is related to the fact that initial value (30 s) is small regarding the filling time. If the valve opening time is further increased, the gain regarding the transverse force will become smaller and the cost on the filling time will become higher. An optimum has to be found.

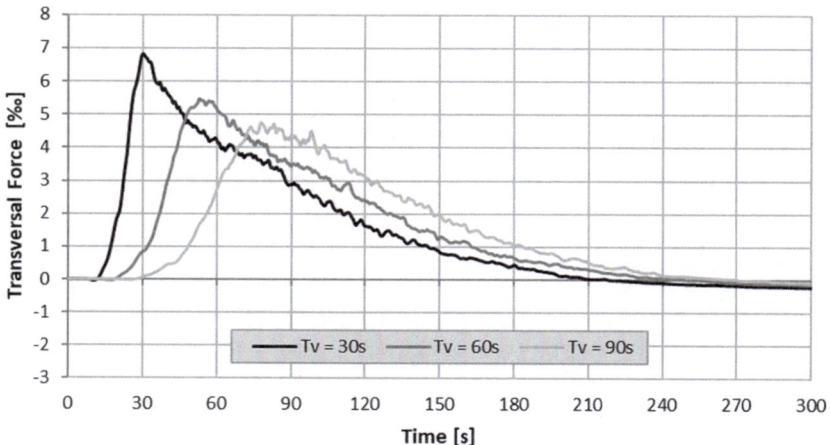

Fig. 11. Transverse force for 3 opening schedules of the valve (Tv) (El Ouamari and Lenaerts 2019)

5.2   Position of the Vessel

Figure 12 illustrates the measured transverse force for 3 positions of the vessel in the lock chamber. The results confirm the operational guidelines which recommend mooring the vessel on the wall located on the side of the functional culvert (valve). The transverse force is divided by 2 when the vessel is moored against the wall located on the side of the functional culvert in comparison with a mooring on the other wall.

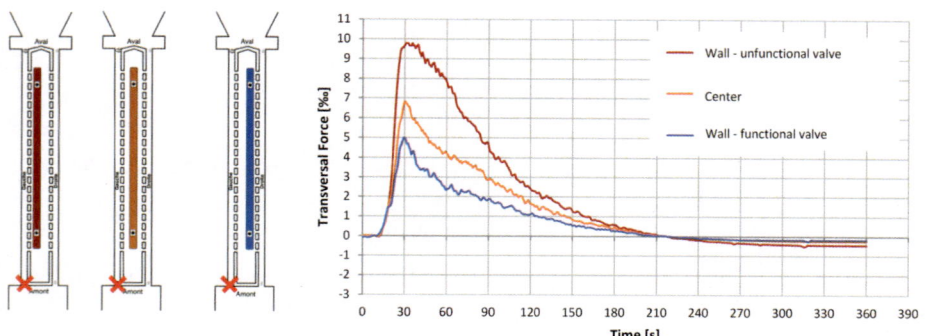

Fig. 12. Transverse force for 3 vessel positions in the lock chamber (El Ouamari and Lenaerts 2019)

5.3   Dimensions of the Side Ports

Figure 13 illustrates the hydrographs for an asymmetrical filling corresponding to 2 geometries for the outlet of the side ports (1 cm × 8 cm and 1 cm × 4 cm). As expected, for an identical opening schedule of the valve (linear opening in 30 s), due to an increased head loss, the filling operation is slower when the side ports are smaller.

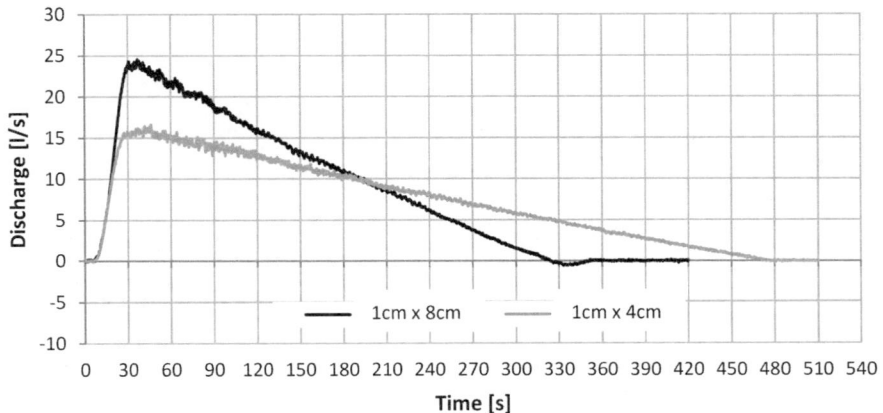

Fig. 13. Hydrographs for asymmetrical filling for 2 side ports geometries (El Ouamari and Lenaerts 2019)

As explained in Sect. 4, for smaller outlets at a given discharge, the flow velocity increases, and more impulse is transferred to the rolling flow, resulting in larger forces on the vessel. In the present case, when the width of the ports decreases, the discharge also decreases but it does not allow decreasing the transverse force significantly (Fig. 14) due to the increased impulse transfer.

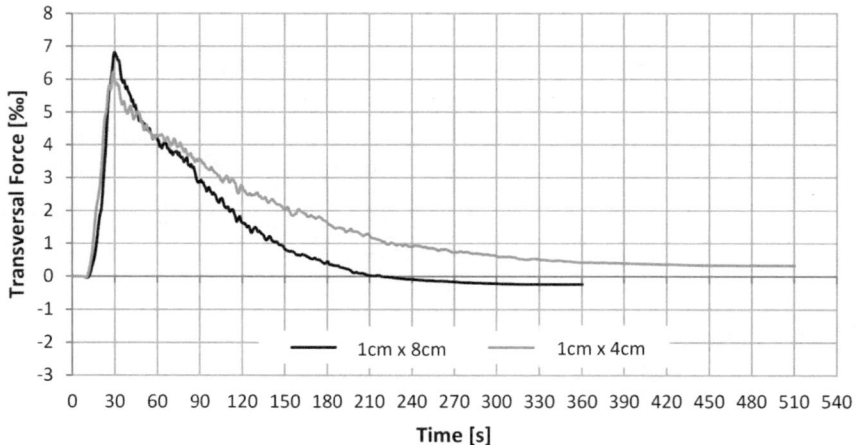

Fig. 14. Transverse force during asymmetrical filling for 2 side ports geometries (El Ouamari and Lenaerts 2019)

6 Conclusions

The experimental results highlight the behavior of the flow and the induced transverse forces applied on a vessel during the asymmetrical filling (one of the valve out of order) process of a lock with longitudinal culverts and side ports.

The results indicates that, especially in the transverse direction, one should be careful using the free-surface slope to estimate the force. Moreover, the free-surface slope is significantly impacted by the presence of a vessel in the lock chamber.

The transverse force can be decreased acting on the opening schedule of the valve, an optimization is necessary to avoid to long filling time. Usually it is not enough, it is important to apply the operational guidelines consisting in mooring the vessel at the wall located on the side of the operational culvert, it significantly limits the transverse forces. For safety reason, several boats must not be moored side by side. Reducing the area of the side ports is not efficient to reduce the transverse force.

Finally, as another mean to decrease transvers forces not treated in the present paper, some dissipation structures or baffles can be placed at the outlet of the ports to decrease the rolling effect of the flow.

Acknowledgements. The authors acknowledge the contribution of the technical staff of the hydraulic research laboratory for their participation in the preparation and realization of the experimental campaign. The authors also acknowledge the staff of the hydraulic research laboratory and the hydrological department for their participation in the realization of the site measurement campaign.

References

Bertin B. (2021) Efforts transversaux sur les convois lors du sassement en mode dégradé d'une écluse à aqueducs-larrons. Master thesis report. Ecole polytechnique de Louvain, Université catholique de Louvain. http://hdl.handle.net/2078.1/thesis:28310. (in French)

El Ouamari I, Lenaerts M (2019) Efforts d'amarrage latéraux lors du remplissage en mode dégradé d'une écluse à aqueducs larrons. Master thesis report. Ecole polytechnique de Louvain, Université catholique de Louvain, 2019. (in French)

PIANC (2015) Ship behavior in locks and lock approaches. PIANC Report WG155

Savary C, Bousmar D, Swartenbroekx C, Zorzan G (2019) Lanaye lock – perturbation in the lock chamber induced by asymmetrical filling. In: Smart Rivers 2019 Conference

Turbulent Flow Simulation of Bridge Piers and Navigation Safety of Ships in Curved River Sections with Variable Water Level

Zhirong Tan[1(✉)], Gang Xing[2], Xing Gao[1], and Xin Cui[1]

[1] School of Navigation, Wuhan University of Technology, Wuhan, China
tanzhi@whut.edu.cn
[2] Hubei Inland Shipping Technology Key Laboratory, Wuhan, China

Abstract. Some of the bridges across the Yangtze River in the Three Gorges Reservoir area are in the curved section, and the collision prevention of bridges is a hot issue in the current industry. The system of ships sailing across bridges is a complex system in the discipline of transportation engineering. It is affected by ship conditions, channel conditions, meteorology and hydrology, navigation management and human factors. In order to grasp the influence of turbulent flow near the piers of bridges in curved river sections and oblique flow during variable water level periods on ships crossing bridges, it is necessary to carry out numerical simulation analysis. Methods: By establishing a simple physical model of the water area of the bridge pier, the turbulent flow field of the bridge pier was simulated in two dimensions by Fluent software. Then the turbulent flow characteristics of the single pier and the tandem double pier were compared, and the flow velocity on the upstream side of the bridge pier was used as a parameter to carry out numerical simulation. Finally, the flow-induced drift of the ship is quantitatively calculated in combination with the flow velocity. Conclusion: The two-dimensional simulation confirms that the turbulent width of the bridge pier increases with the increase of the flow velocity. The turbulent width of the tandem double pier is larger than that of the single pier, but the vorticity extending downstream is smaller than that of the single pier. The deflection moment and flow-induced drift of the ship crossing the bridge confirm the risk of ship collision. It is necessary to add a multi-function navigation mark to collect water flow parameters in time, and introduce LED visual navigation in the bridge area to improve the reliability of ship navigation in the bridge area of the curved river section.

Keywords: Water level changes · Turbulent flow · Drift value · Multi-function navigation mark · LED

1 Introduction

In recent years, China's transportation industry has developed rapidly, and the demand for cross-river bridges has become more and more prominent. According to statistics, Chongqing Maritime Bureau has built (under) 56 bridges. Lots of wading bridges not only promote the economic development on both sides of the Yangtze River, but also change the original channel conditions, which has a certain impact on the safety of ships crossing the bridge. Therefore, it is necessary to study the ship crossing the bridge

Y. Li et al. (Eds.): PIANC 2022, LNCE 264, pp. 736–748, 2023.
https://doi.org/10.1007/978-981-19-6138-0_64

and the interaction between the ship and the bridge from the mechanism, so as to design the navigation safety guarantee facilities according to the analysis results (Zhang Dan 2012; Tan Zhirong 2011; Wu Changsheng 2020).

There is a certain range of turbulent flow in the flow field around the bridge pier. The complex flow state of turbulence leads to the change of the water current force when the ship sails on the water, which brings some safety hazards. In order to alleviate navigation risks, researchers mostly use physical model tests and numerical simulations to study the turbulent characteristics of water flow caused by water blocking by piers in the channel (Liu Zhigang 2020; Raudkivi 1983). Comparing the distribution of the flow field around different pier types, it is found that the average velocity, turbulence intensity and Reynolds shear stress around the composite pier are lower than those around the single cylindrical pier. This is due to the existence of pile caps that limit the formation of vortices around the piers (Johnson 2003; Cautam 2019; Das 2013; Han Min 2010).

The angle between the normal direction of the cross-river bridge in the curved river section and the water flow direction changes periodically with the change of the water level all the year round. As a result, when the water flow velocity is large, a large steering pressure will be generated when the ship sails across the bridge. Therefore, in addition to the analysis from the perspective of the flow field, the ship sailing across the bridge in the curved river section also needs to be studied from the direction of flow-induced drift.

In this paper, we select typical bridges in the curved river section of the Three Gorges Reservoir area, and the water flow parameters in the bridge area are also selected. Numerical simulation is carried out through Fluent to analyze the relationship between the turbulent flow range of the bridge pier and the flow velocity and other parameters, so as to explore the effect of the turbulent flow of the pier on the curved river section on the navigation safety of ships. Then, the influence of flow-induced drift of ships under the action of different water flow and ship speeds was analyzed, and the influence of flow-induced drift on the navigation safety of ships was explored. Finally, referring to the existing mature multi-functional navigation aids and LED technology (Allen 2009; Zhao 2011; Liu Dianhong 2015; Chen Chen 2014), the optimization suggestions for navigation aid induction in the bridge area are put forward.

2 Flow Field Simulation of Bridge Pier

2.1 Simple Physical Scene

Using the usual practice of planar physical scene and grid scale [1, 2], a physical relationship diagram of small-scale pier waters with flow from upstream left to right is constructed. The diameter of the cylinder is D, the distance between the centers of the double piers in series is L, L_I is the inflow scale, L_H is the vertical scale, and L_O is the outlet scale. The schematic diagram of the calculation grid of the single pier cylinder is shown as Fig. 1, and the calculation grid of the tandem double pier is shown as Fig. 2. Among them, the uniform free flow (its velocity is V) bypasses the cylinder from left to right, and the coordinates in the x-axis is along the flow direction, the y-axis is perpendicular to the incoming flow direction, and the coordinate origin is placed at the center of the cylinder.

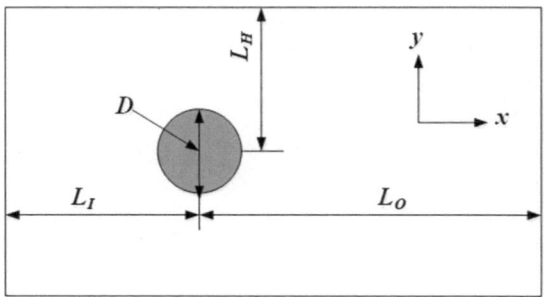

Fig. 1. Schematic diagram of flow around the single pier cylinder

2.2   Governing Equation

The fluid simulation uses the planar 2D flow conditions, the incompressible fluid continuity equation and the N-S equation:

Mass Conservation Equation:

$$\nabla \cdot \boldsymbol{u} = \frac{\partial u}{\partial x} + \frac{\partial v}{\partial y} = 0 \tag{1}$$

Momentum conservation equation:

$$\frac{\partial \boldsymbol{u}}{\partial t} + (\boldsymbol{u} \cdot \nabla)\boldsymbol{u} = -\frac{\nabla p}{\rho} + \frac{\mu}{\rho}\nabla^2 \boldsymbol{u} \tag{2}$$

In (2), ∇ and $\boldsymbol{u}$ are the gradient operator and the velocity vector. It includes two velocity components u and v, p is the pressure, and μ is the hydrodynamic viscosity coefficient.

2.3   Simulation Conditions

According to the data of the special report, the calculation conditions are taken from the observation data. The calculation conditions are simplified to:

Pier diameter: D = 10 m;
Bending radius: R = 800 m;
flow velocity: V = 1, 1.5, 2.0, 3.0 (m/s)

3   Comparison and Analysis of Numerical Simulation

3.1   Flow Field Analysis of Single Pier

Through the change of flow velocity parameters under the condition of variable water level, the flow velocity cloud map, pressure map, velocity streamline map and vorticity field around the bridge pier are compared between the single pier and the tandem double pier. We analyzed the relationship between the turbulent range of the pier and the flow velocity and other parameters, and explored the influence of the turbulence of the pier in the curved river section on the navigation safety of ships.

3.1.1   Flow Field Analysis

Velocity nephogram and vector analysis reflect the changing law of the turbulent range.

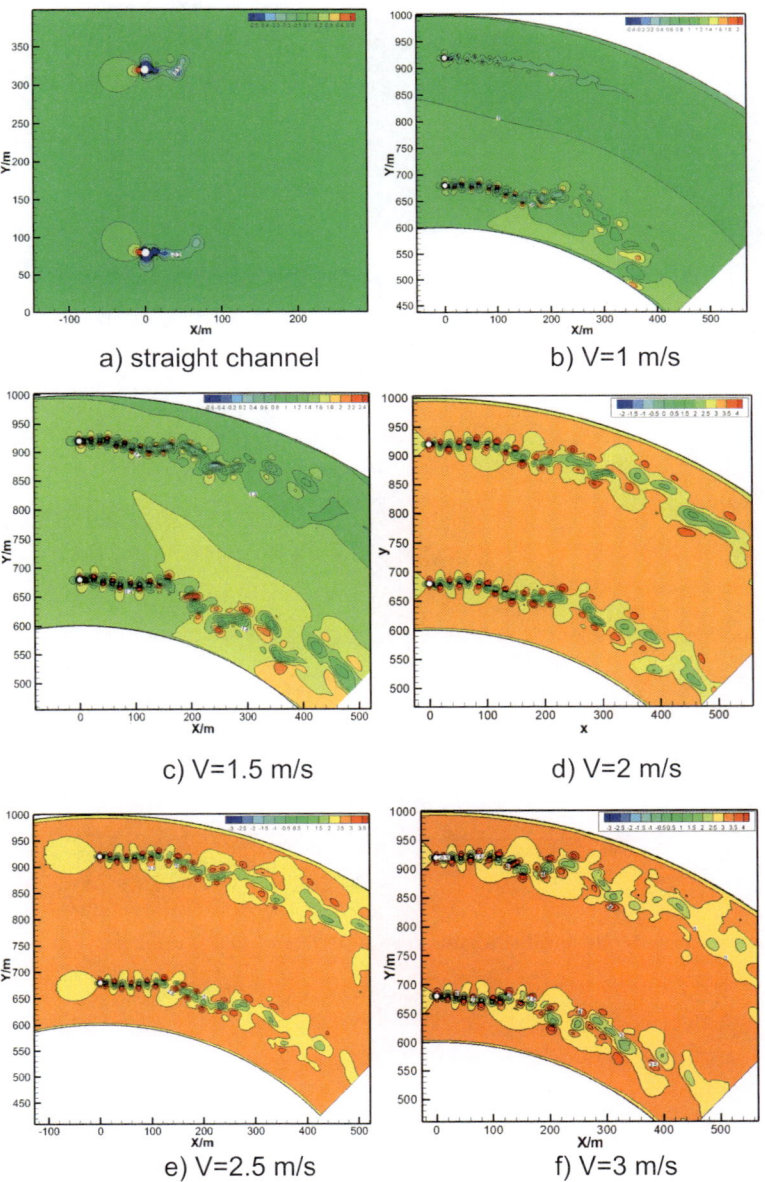

Fig. 2. Nephogram of different inflow velocities

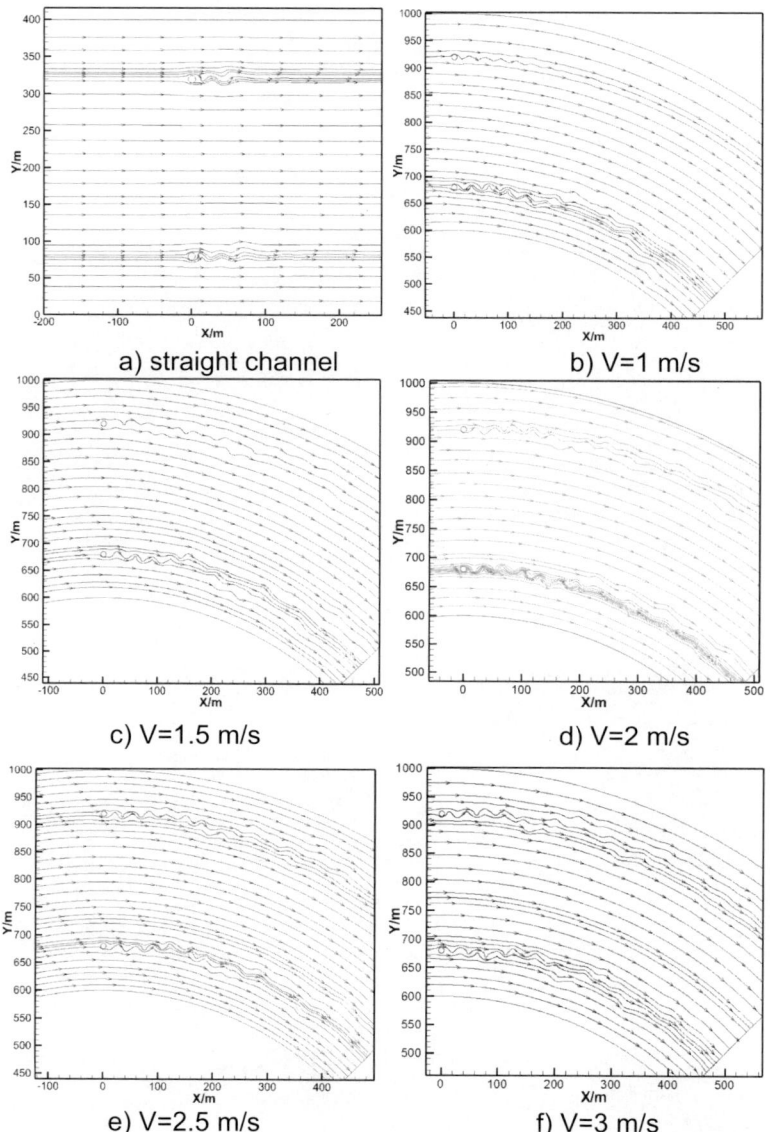

a) straight channel

b) V=1 m/s

c) V=1.5 m/s

d) V=2 m/s

e) V=2.5 m/s

f) V=3 m/s

Fig. 3. Vector distribution map of different inflow velocities

In Fig. 2 and Fig. 3, the overall trend of the influence of velocity on the flow field can be seen, and the characteristics of the velocity nephogram are as follows:

(1) With the increase of the incoming flow velocity, the phenomenon of flow around the pier becomes more obvious, the flow pattern in the downstream direction of the pier becomes more complex, and the turbulent flow width further increases. Under different working conditions, with the increase of the incoming flow velocity, the influence range of the two sides on the downstream also further increases.

(2) There are also differences in turbulent flow patterns on the concave and convex banks of the curved river reach. It can be analyzed by the quantification of the X/Y axis coordinates: When V = 1 m/s, the concave bank side piers have a significant impact on the flow regime in the downstream range of about 10D, and gradually weakened until disappearing at 35D. The pier on the side of the convex bank has a significant impact on the flow regime in the downstream 18D range, and the velocity isosurface is dense, indicating that in the river section from the bridge pier to the downstream 18D range, the flow regime is changeable and the ship's heading stability is poor. From the follow-up of the pier to 40D, the flow disorder gradually weakens to basically no effect.

(3) During the flood period, when the flow velocity increases, for example, under the conditions of V = 2 m/s and V = 3 m/s, the significant influence range of the concave bank side is 19D and 22D, and the significant influence range of the convex side is 23D and 36D, respectively. That is to say, with the increase of the incoming flow speed, the water flow field of the bridge pier changes more significantly, and the scope of influence on the ship's navigation also increases.

3.1.2   Pressure Field Analysis

The pressure field is a high-pressure area and a low-pressure area inside and outside the pier, which can easily lead to the deflection moment of the ship (Tan Zhirong 2022).

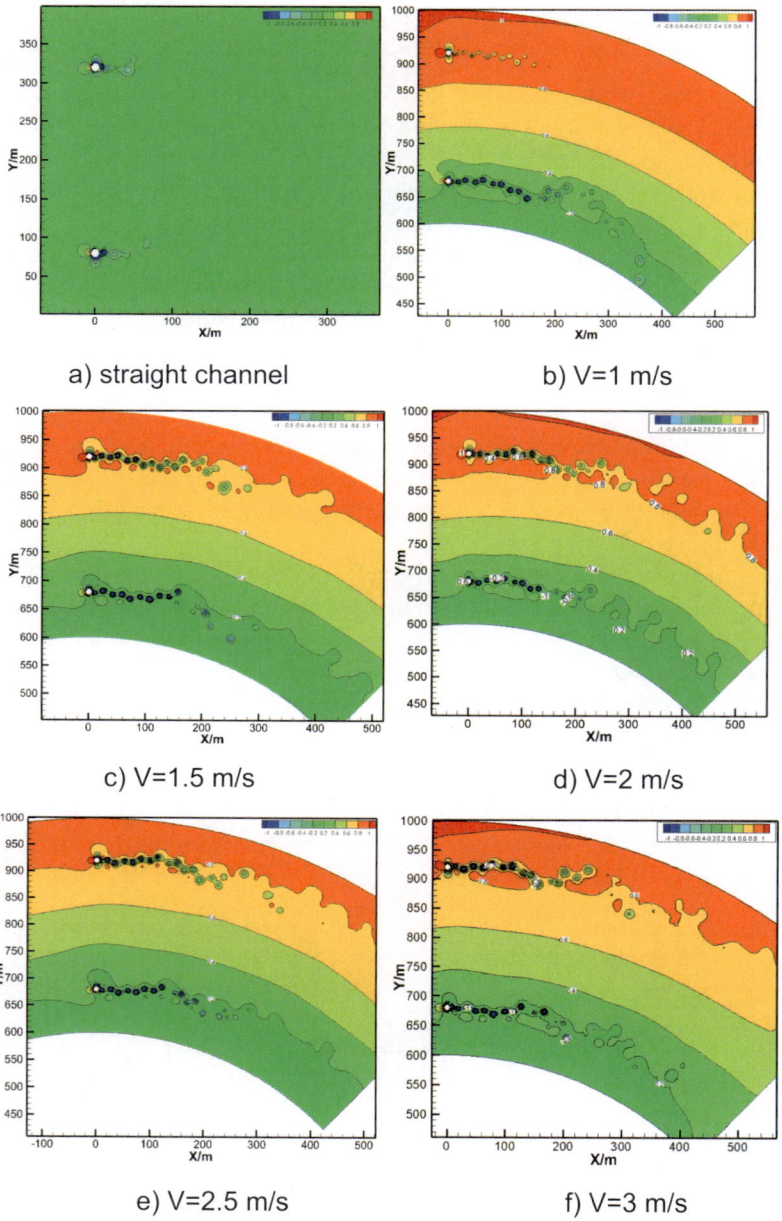

Fig. 4. Pressure distribution diagram of different inflow velocities

Figure 4 depicts the change in the distribution of the pressure field on the downstream side of the pier under different flow velocity conditions. With the increase of the flow velocity, the low pressure area under the bridge pier increases, and the spread is more obvious in the downstream direction. The pressure in the low-pressure area is relatively small, and when the ship travels in the continuous low-pressure area, it is easy to deflect to the side of the pier, resulting in the danger of touching the pier.

3.1.3   Vorticity Field Analysis

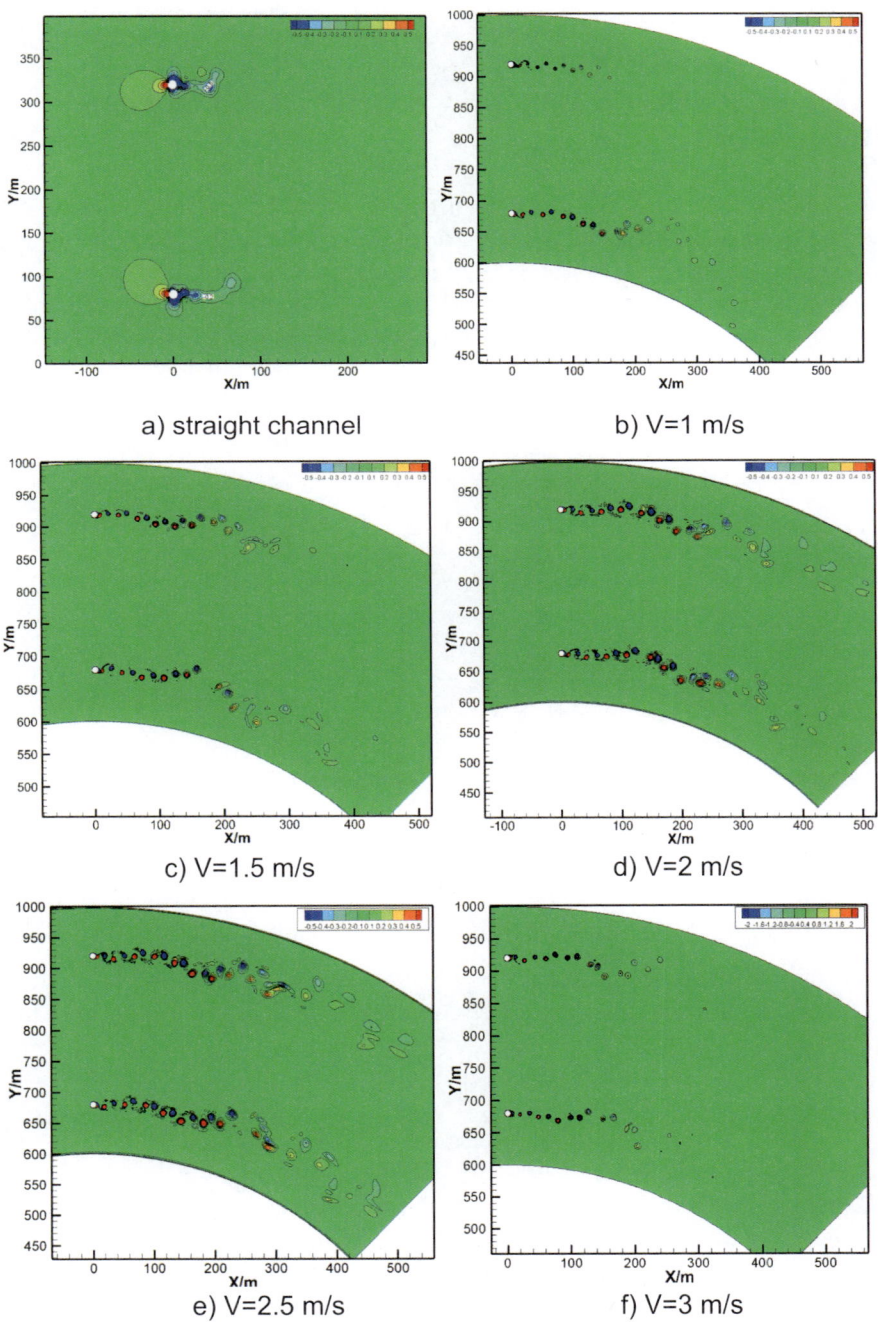

a) straight channel

b) V=1 m/s

c) V=1.5 m/s

d) V=2 m/s

e) V=2.5 m/s

f) V=3 m/s

Fig. 5. Vorticity distribution map of different inflow velocities

It can be seen from Fig. 5 that when D remains unchanged, with the increase of the flow velocity, the more vortices generated after the piers on the convex bank side and the concave bank side, the greater the vorticity, and the area extending to the downstream also further increases (vorticity legend in FIG. F increases to some extent).

4   Analysis of Ship Crossing Bridge Motion Based on Flow-Induced Drift

To study the flow-induced drift under the action of different water flows, a coordinate system needs to be established. Let the y-axis of the coordinates be parallel to the axis of the bridge, and the x-axis to be parallel to the normal of the bridge, as shown in Fig. 6 (a), (b).

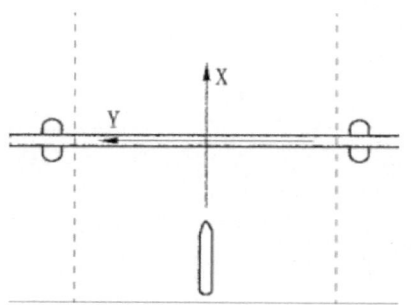
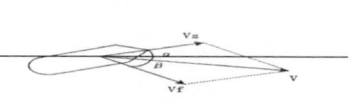

a) Flow-induced drift calculation axis schematic

b) Flow-induced drift modeling

Fig. 6. Schematic diagram of ships sailing across the bridge and drifting in the waters of the bridge area

The ship speed is V_s and the water flow velocity is V_f, according to the flow-induced drift formula in "The Navigation Standard of the Yangtze River Trunk Line 2020".

The time for the ship to cross the bridge is T:

$$T = S/(V_s \cos \alpha + V_f \cos \beta) \tag{3}$$

The flow-induced drift of the ship in the y-axis direction is B:

$$B = (V_s \sin \alpha + V_f \sin \beta) \times S/(V_s \cos \alpha + V_f \cos \beta) \tag{4}$$

It can be seen from Fig. 7 that when other conditions are constant, the greater the water velocity, the greater the flow-induced drift. At the same time, the greater the boat speed, the smaller the flow-induced drift. When the flow rate is high, the captain should keep sailing cautiously, increase the speed at an appropriate time, and adjust the sailing in real time, so as to reduce the amount of drift.

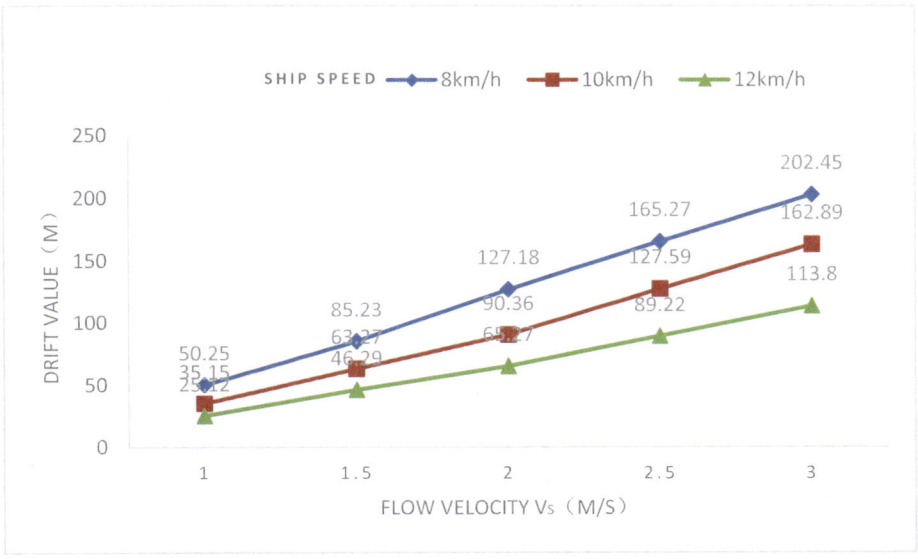

Fig. 7. Drift value at different speed and flow velocity (m)

The experimental conditions where the calculated river length is 1500 m, the water velocity is 3 m/s, and the ship speed is 14 km/h are selected to study the influence of different ship yaw angles on the drift. The results are shown in Fig. 8. It can be seen that when other conditions at a certain time, the larger the yaw angle of the ship is, the flow-induced drift will be reduced accordingly, and the larger the water-current included angle, the flow-induced drift will increase accordingly.

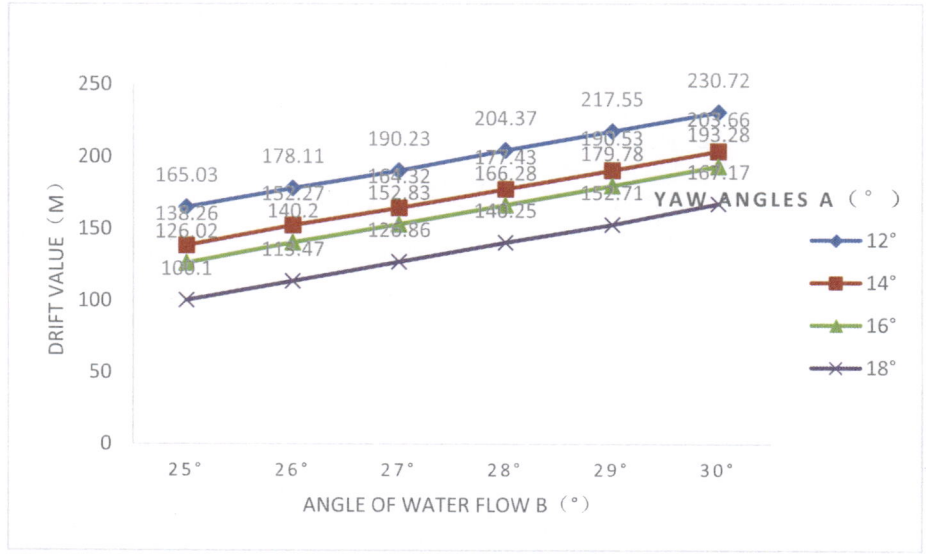

Fig. 8. Drift value due to different yaw angles and angles of water flow

5 Optimization Suggestions for Navigation Beacon Induction in Bridge Area

5.1 Multifunctional Beacon for Real-Time Flow Velocity Collection

In the waters of the bridge area, we can install some multi-function beacons. The real-time hydrological and meteorological data can be measured by various sensors on the multi-functional beacon. This data is then sent to passing ships, waterway authorities and maritime regulators by AIS equipment. After receiving the data sent by the multi-function beacon, the captain can judge the real-time status of the channel, provide information decision support for the ship's maneuvering, and ensure the safe navigation of the ship.

5.2 Recommended Route Guidance Based on LED Warning Signs

(1) Install 3 lamps of different colors above the bridge navigation holes, so as to illuminate a section of water before entering the water area of the bridge area. Ships traveling in different areas can observe the guiding lights of different colors, so as to judge the optimal route.

(2) ABCD is a parallelogram representing the water in the bridge area. The ship travels in the direction of Q, the crew on duty observes the waters ahead, and the visual navigation system is located at point P on the side of the bridge above the bridge navigation hole Q to illuminate the water surface. The illuminated area is EFGH and the light colors are red, yellow-green and blue. Among them, the most sensitive yellow-green light of the human eye at night represents the best driving

route. The red light is on the right side of the best route, the blue light is on the left. The red and blue lights flash quickly, and the yellow-green lights flash slowly.

(3) By observing the light color and flickering frequency, the crew can judge which area the ship is traveling in, and adjust the course in time so that they can travel to the correct route before entering the bridge area waters. So it can pass the bridge area waters safely according to the instructions of the bridge area beacon.

6   Conclusions

(1) From a microscopic perspective, this paper uses FLUENT to simulate the flow field of the bridge pier in two dimensions, and clarify the mechanism of the influence between the ship and the bridge when the ship crosses the bridge. Then, the relationship between the turbulent range and the flow velocity after the single pier and the tandem double pier under the condition of variable water level is obtained: with the increase of the incoming flow speed, the water flow field of the bridge pier changes more significantly, and the influence range of ship navigation gradually increases. Under the same flow velocity, the turbulent flow is more complex in the tandem double pier than the single pier, but the range is reduced.At the same flow rate, the turbulence of the tandem double pier is more complicated than that of the single pier, but the range is reduced.

(2) From a macro point of view, the flow-induced drift under different flow rates and different ship speeds is discussed. It is obtained that the greater the water flow speed, the greater the flow-induced drift, and the greater the ship speed, the smaller the flow-induced drift.

(3) From the perspective of navigation, it is proposed that the data such as the velocity of the bridge area during the high water level should be provided to the ship in time. Then, with the help of information technology, especially the popularized LED warning, which can be used as a directional light source to induce the recommended route.

Subsequent research can analyze the pier-water interaction under three-dimensional conditions.

References

Zhang D, Huang L, Chen L, Fan X (2012) Numerical analysis of the impact of bridge area waterway regulation project on large fleet bridge crossing. Water Transp Eng (03):15–118 +153

Tan Z (2011) Study on mechanism and risk assessment method of Yangtze River line (Doctoral Dissertation, Wuhan University of Technology)

Wu C, Gan L, Sun J, Fan J, Zhou C (2020) Data-driven ship path planning in lower bridge group waters. J Wuhan Univ Technol Transp Sci Eng Ed (06):1085–1090

Liu Z (2020) Numerical simulation of container ship movement characteristics in restricted water and safety navigation countermeasures (Master Dissertation, Zhejiang Ocean University)

Raudkivi AJ, Ettema R (1983) Clear water scour at cylindrical piers. J Hydraul Eng (03):338–350

Johnson KR, Ting FCK (2003) Measurements of water surface profile and velocity field at a circular pier. J Eng Mech (5):502–513

Cautam P, Eldho TI, Mazumder BS (2019) Experimental study of flow and turbulence characteristics around simple and complex piers using PIV. Exp Therm Fluid Sci (100):193–206

Das S, Das R, Mzumar A (2013) Comparison of charateristics of horseshoe vortex circular and square piers. Res J Appl Sci Eng Tech (17):4317–4387

Han M (2010) Study on turbulent characteristics of single cylindrical piers (Master Dissertation, Northwest A & F University)

Whang AJW, Chen YY, Teng YT (2009) Designing uniform illumination systems by surface-tailored lens and configurations of LED arrays. J Disp Technol (3):94–103

Zhao S, Wang K, Chen F, Wang K, Wu D (2011) Lens design of LED searchlight of high brightness and distant spot. JOSA A (5):815–820

Liu D, Zhang X, Chen C (2015) Free-form lens for rectangular illumination with the target plane rotating at a certain angle. JOSA A (11):1958–1963

Chen C, Zhang X (2014) Design of optical system for collimating the light of an LED uniformly. JOSA A 31(5):1118–1125

Tan Z, Wang Y, Wang H, Wang Z, Chen B (2022) Quantitative Analysis of the action of bridge pier turbulence on ship bow shaking moment. J Wuhan Univ Technol Transp Sci Eng Ed

Lecture Notes in Civil Engineering

Volume 264

Lecture Notes in Civil Engineering (LNCE) publishes the latest developments in Civil Engineering—quickly, informally and in top quality. Though original research reported in proceedings and post-proceedings represents the core of LNCE, edited volumes of exceptionally high quality and interest may also be considered for publication. Volumes published in LNCE embrace all aspects and subfields of, as well as new challenges in, Civil Engineering. Topics in the series include:

- Construction and Structural Mechanics
- Building Materials
- Concrete, Steel and Timber Structures
- Geotechnical Engineering
- Earthquake Engineering
- Coastal Engineering
- Ocean and Offshore Engineering; Ships and Floating Structures
- Hydraulics, Hydrology and Water Resources Engineering
- Environmental Engineering and Sustainability
- Structural Health and Monitoring
- Surveying and Geographical Information Systems
- Indoor Environments
- Transportation and Traffic
- Risk Analysis
- Safety and Security

To submit a proposal or request further information, please contact the appropriate Springer Editor:

- Pierpaolo Riva at pierpaolo.riva@springer.com (Europe and Americas);
- Swati Meherishi at swati.meherishi@springer.com (Asia—except China, Australia, and New Zealand);
- Wayne Hu at wayne.hu@springer.com (China).

All books in the series now indexed by Scopus and EI Compendex database!

More information about this series at http://www.springer.com/bookseries/15087

Yun Li · Yaan Hu · Philippe Rigo ·
Francisco Esteban Lefler ·
Gensheng Zhao
Editors

Proceedings of PIANC Smart Rivers 2022

Green Waterways and Sustainable Navigations

Set 2

 Springer

Editors
Yun Li
Nanjing Hydraulic Research Institute,
Ministry of Water Resources, Ministry of
Transport and National Administration of
Energy
Nanjing, China

Yaan Hu
Nanjing Hydraulic Research Institute,
Ministry of Water Resources, Ministry of
Transport, National Administration of
Energy
Nanjing, China

Philippe Rigo
Liege University
Liège, Belgium

Francisco Esteban Lefler
PIANC
Brussels, Belgium

Gensheng Zhao
Nanjing, China

Nanjing Hydraulic Research Institute This work was supported by Nanjing Hydraulic Research Institute

ISSN 2366-2557   ISSN 2366-2565   (electronic)
Lecture Notes in Civil Engineering
ISBN 978-981-19-6137-3   ISBN 978-981-19-6138-0   (eBook)
https://doi.org/10.1007/978-981-19-6138-0

This Springer imprint is published by the registered company Springer Nature Singapore Pte Ltd.
The registered company address is: 152 Beach Road, #21-01/04 Gateway East, Singapore 189721, Singapore

Contents

Contents

Inland Navigation Structure

Contents

River System Management

Logistics

Smart Shipping

A Study on Key Technologies for the Joint Scheduling and Control of Multi-level and Multi-line Ship Lock Groups

Wu Ning[1(✉)], Hantao Ye[1], Ning Lin[2], and Zhenyu Mo[3]

[1] Guangxi Beibu Guif International Port Group Co., Ltd., Nanning, China
Ningwu_xijiang@l63.com
[2] Guangxi Xijiang Development and Investment Group Co., Ltd.,
Wuzhou, China
[3] Fangcheng Port Group Co. Ltd., Fangcheng, China

Abstract. This paper aims at the scattered management system, low service level, multi-level and multi-line ship locks caused by the numerous ship locks in Guangxi and they belong to different owners. For outstanding problems such as lack of scheduling theory and methods, and weak security and intelligent management and control capabilities, the top-level design of the joint scheduling of multi-stage and multi-line ship lock groups in the basin is carried out, and the joint scheduling technology of parallel four-line ship lock groups with different scales and levels under complex navigation restrictions is carried out. Research and development of key technologies for intelligent scheduling platform for line ship lock groups, key technology research on management and control safety in multi-level and multi-line ship lock clusters, and research on joint dispatch model and dispatch theory and methods for busy multi-level ship lock groups, so as to realize joint dispatch and centralized management and control of watershed ship lock groups. The overall navigation efficiency of the basin will be greatly improved, and it will provide a reference for the innovation of the navigation management mode of the ship lock group in the basin.

Keywords: Joint scheduling · Centralized control

1 Background

Inland river shipping, with advantages of large traffic volume, small energy consumption, less land occupation and low cost, plays an important position in comprehensive transportation system in China. The Xijiang main line and its main tributaries are the important "one horizontal" in the "four vertical, four horizontal and two networks" of the national water transport main channel. The total navigable mileage of Xijiang main line reaches 5,873 km, which is an important part of China's modern comprehensive transportation system. Guangxi has developed hydrographic net and numerous locks, with four characteristics of "multi-line, multi-cascade, multiple elements and multiple owners". The "one trunk and three branches" shipping system includes 20 cascades and 25 locks, involving 13 enterprises and transportation, water conservancy, power generation and other administrative departments integrating navigation AIDS, water level, bridges, shoals, ships, docks, locks and other elements. The previous operation and

© The Author(s) 2023
Y. Li et al. (Eds.): PIANC 2022, LNCE 264, pp. 751–763, 2023.
https://doi.org/10.1007/978-981-19-6138-0_65

control mode of the ship lock cannot meet the needs of the rapid development of inland water transport in Guangxi. There are mainly prominent problems such as decentralized management system, low service level, insufficient research on multi-stage and multi-line lock scheduling technology, and relatively weak intelligent control ability of lock safety. The problems are: (1) the lack of effective management system between multiple owners, the prominent contradiction between navigation and electricity, and the great differences in the scheduling level of each lock, which affects the navigation efficiency; (2) The multi-stage and multi-line ship lock scheduling technology urgently needs a breakthrough. In particular, it is necessary to solve the scheduling problem of complex navigation restriction and super large lock group navigation restriction interwoven section after the expansion of Changzhou third and fourth line locks. Efficient navigation of river basin puts forward higher requirements for the joint scheduling of multi-stage and multi-line ship locks as well. (3) Repeated reports to the lock are needed when ships use the traditional gate mode to pass through the multi-cascade hub, which leads to low level of service. The rapid development of inland water transport in Guangxi urgently needs information technology. (4) It is urgent to improve the safety control ability of the lock after centralized scheduling and unattended lock. New technologies such as intelligent perception, remote monitoring and early warning need to be introduced to ensure the safe operation of the lock.

2   Research Content

Centering on the overall goal of Xijiang multi-cascade multi-line lock group scheduling integration, efficient navigation, intelligent control and intelligent service, We have carried out a lot of research and technology research and development from the aspects of multi-cascade and multi-line lock group management mode design, joint scheduling technology innovation, the establishment of lock intelligent scheduling platform, and the centralized control and safety of locks. We realize the "centralized control, unified dispatch, unified command, joint operation and remote monitoring" of the multi-cascade, multi-line and multi-owner ship lock of the Xijiang River, which improves the navigation efficiency and management level of the lock group, and ensures the safe operation of the lock group.

3   Concepts and Technical Structure

The project revolves around the efficient and safe navigation of multi-level, multi-line and multi-owner locks. Taking aspects of the top-level system design, multi-line lock scheduling technology, software and hardware research and development into consideration, we adopt the method of combining development and engineering application to construct the joint scheduling platform of multi-cascade and multi-line ship lock group, and innovated the lock group centralized control safety technology. We will learn from our experience and develop it independently. We follow the overall research principle of appropriate advanced and have overcome the key technical difficulties. The general idea and technical route of the research are shown in Fig. 1.

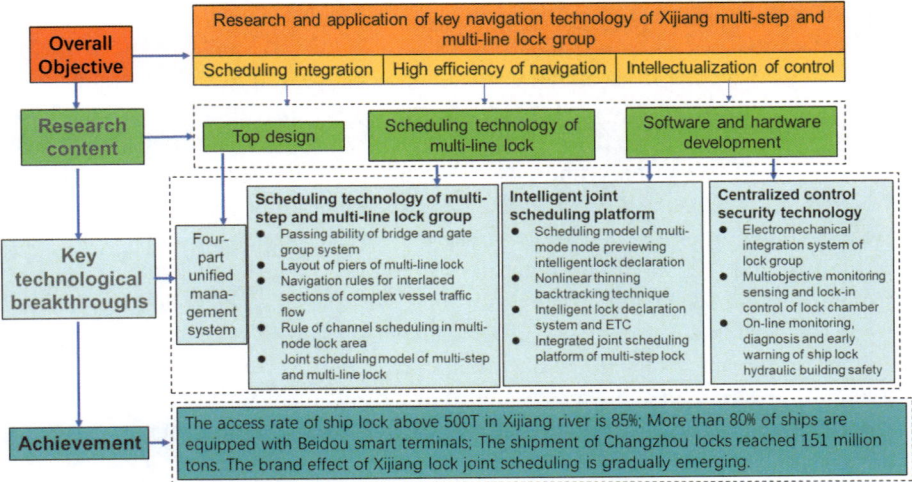

Fig. 1. General idea of project research

4 Research Findings

4.1 Innovation the Theory of Multi-cascade Multi-line Lock Group

(1) Based on operation rule of lock entrance area and traffic flow conflict theory, we propose a new method for the arrangement of complex multi-line lock (Fig. 2).

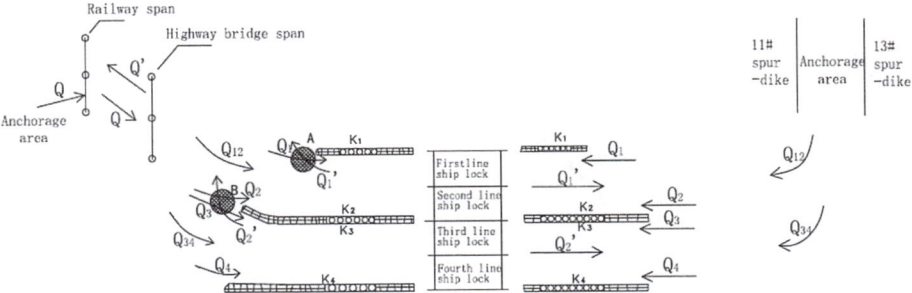

Fig. 2. Schematic diagram of ship traffic flow between upstream and downstream anchorage of Changzhou Junction

Changzhou four-lane ship lock project is taken as an example. In the scheme of double sides of piers, each side can dock ships for one lock operation time. By analyzing the traffic flow into and from the ship lock gate, we found that two-way docking piers were set on the inside of the first, second and third and fourth line sluice channels. During the operation of the lock, two conflict zones at the gate of the approach channel would be caused by the incoming and outgoing ship. That is, the zones A and B in Fig. 1. There are three possibilities: Q1 and Q1'; Q2 and Q2', and Q3 and Q2'.

If the single side pier is used, that is, the first, second and third and fourth line gate groups are only set on the right side pier, only one conflict zone at the gate of the channel would be caused, during the operation of the lock. That is, the area of area B in Fig. 1. There is only one possibility of crossover, namely, Q2 and Q2'. According to the analysis, the conflicts corresponding to the different landing pier setting methods are shown in Table 1. Therefore, it is safest to lay a single side pier in the first and second and third and fourth line lock diversion channels.

Combined with scheduling scheme research, We can draw the following conclusion. For double sides piers, the replenishment of ship to piers can still be performed during any bidirectional operation cycle, but the operating safety margin is relatively less. For the single side pier, the replenishment time is relatively more sufficient, and the safety and reliability are increased. The optimization scheme of piers layout of multi-line lock is put forward as follow: piers on the right side should be increased appropriately, while being retained on the left side, which is considered as a compromise scheme of above, so that more ships can park on the right side, or temporary docking. This arrangement makes it more like a single-sided scheme. At the same time, it is stipulated that the ship entering the approach channel have priority to dock on the right side of the pier. When the upper ship leaves the gate, the left ship on the ship pier can enter the gate first. At this point, the ship on the right side can move to the left side on the pier. In this way, it can operate according to the single-side dock pier scheme.

(2) The simulation and analysis technology of multi-line parallel lock joint scheduling was invented, and the traffic organization optimization method of multi-line lock group under limited and complex conditions was innovated.

Based on ship scheduling conditions and ship lock operation conditions, a mathematical model of ship lock operation state is established with ship lock operation conditions and ship scheduling conditions as constraints:

$$f(t_0 + t_1 + t_2) = \begin{cases} R(i) & \left(t_2 < t_{R(i)}\right) \\ R(i+1) & \left(t_2 > = t_{R(i)}, i+1 < = n\right) \\ R(0) & \left(t_2 > = t_{R(i)}, i = n\right) \end{cases} \tag{1}$$

where $f(t_0 + t_1 + t_2)$ is the state of lock at time $t_0 + t_1 + t_2$; t_0 is the time that the model has been running when the previous state ends; t_1 is the time to wait for the current state; t_2 is the running time of the current state; i Is the serial number of the state, n and is the maximum value of the serial number; $R(i)$ is the state of ship lock; $t_{R(i)}$ is the duration of the state; The ship lock operation process is divided into 10 states, the serial number of which is = 0, 1, 2, 3, 4, 5, 6, 7, 8, 9.

According to ship scheduling conditions and sailing routes, a mathematical model of fleet movement is established:

$$f(k) = \begin{cases} L(j) + k \times V(j) & (k < (L(j+1) - L(j))/V(j)) \\ L(j+1) & (k > = (L(j+1) - L(j))/V(j)) \end{cases} \tag{2}$$

where $f(k)$ is the position function of fleet movement at time k; k is the time when the fleet has moved; $L(j)$ is the distance from the starting position of the fleet to the lock; $L(j)$ is the distance between the target position of the fleet and the lock; $V(j)$ is the speed of the fleet in that segment; j is the sequence number of the target location.

According to the waiting time t_1 of the lock's current state in the simulation process, the time t_1 when the state is about to change is recorded as t_{1max}, which is the waiting time of each lock in each state. According to the waiting time t_{1max} of each lock in each state, the total waiting time of each lock in the simulated time is calculated as the following formula:

$$T_1 = \sum t_{1\,max} \tag{3}$$

where, T_1 is the sum of the waiting time of all locks in the simulated time. In the same simulation time, the larger the value is, the lower the operation efficiency of ship lock.

According to the position of the ships at each time and the position of the conflict zone, the traffic flow status of the two fleets that may have conflicts at each time can be judged as $p_1(t)$ and $p_2(t)$, respectively. When the fleet is located within the conflict zone at time t, it is 1; otherwise, it is 0.

Under the potential route crossing mode of each ship, whether there is a conflict in the conflict zone at time t can be determined by the following conditions:

$$P(t) = p_1(t) \times p_2(t) \tag{4}$$

(3) The existing scheduling model of lock chamber is improved by using the technology of scheduling point, and the utilization rate of the lock chamber is increased. According to the constraints of lock grade, main size, ship draft, ship tonnage and other factors, a complex multi-line joint scheduling model of lock is established (Fig. 3).

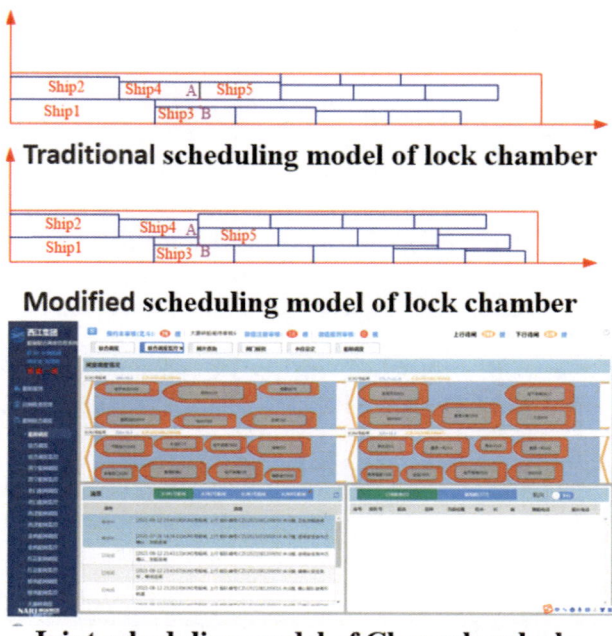

Traditional scheduling model of lock chamber

Modified scheduling model of lock chamber

Joint scheduling model of Changzhou locks

Fig. 3. Diagram of scheduling lock chamber model

In order to maximize the overall capacity and optimize the service level of the multi-step lock, an integrated scheduling model of the step lock was established. In view of the congestion of ships in the upstream and downstream of the hub under special conditions, a multi-step ship lock scheduling model of buffer sailing was developed to ease the traffic flow and avoid the large number of ships in a section (Fig. 4).

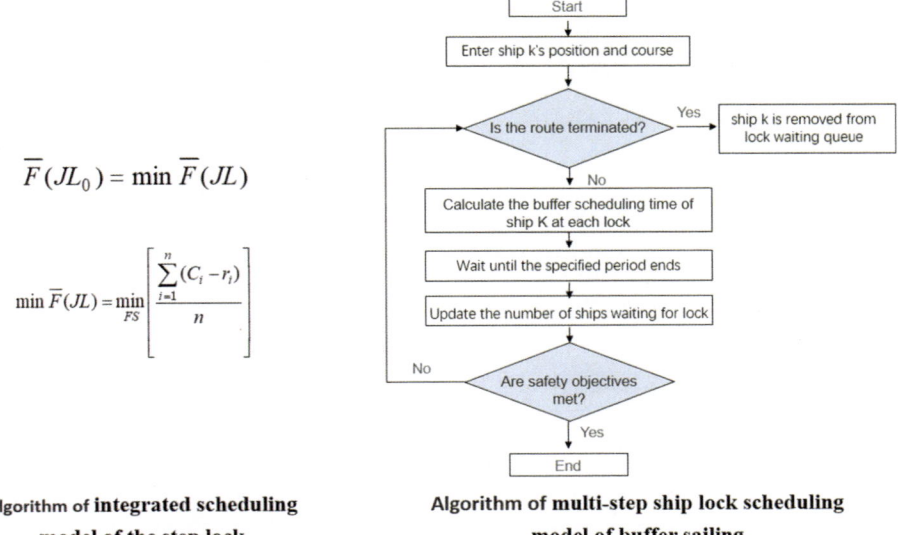

$$\overline{F}(JL_0) = \min \overline{F}(JL)$$

$$\min \overline{F}(JL) = \min_{FS} \left[\frac{\sum_{i=1}^{n}(C_i - r_i)}{n} \right]$$

Algorithm of **integrated scheduling**
model of the step lock

Algorithm of **multi-step ship lock scheduling**
model of buffer sailing

Fig. 4. Diagram of chain locks scheduling model

4.2 The Intelligent Joint Scheduling Technology of Multi-step and Multi-line Lock Group is Developed

Based on the developed inland river shipboard Beidou satellite navigation monitoring terminal and multi-mode node pre-viewing intelligent lock model, the shipboard intelligent lock declaration system was established, and the Xijiang multi-step and multi-line locks group intelligent joint scheduling platform was constructed, realizing multi-source information perception, interaction, processing, intelligent lock and centralized real-time scheduling.

(1) The high-precision Beidou positioning algorithm and black box safety warning model in complex environment were optimized, and the first Beidou satellite navigation monitoring terminal on inland river ships was developed, which solved the problem of multi-source information perception (Fig. 5).

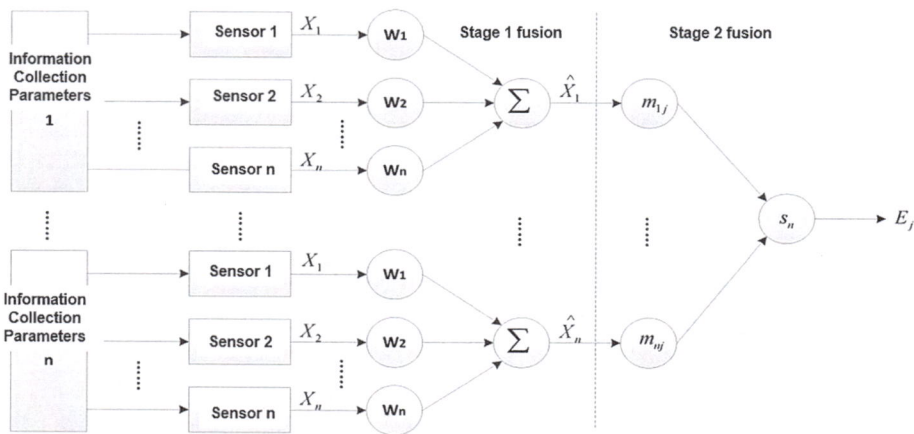

Fig. 5. Multi-information data fusion diagram of black box security warning for ship lock scheduling in complex environment

(2) Established multimode nodes in advance at intelligent brake model, innovation of nonlinear thin back mobile GIS technology, embedded mobile big data heterogeneous resource integration technology, developed ship intelligent brake system, to achieve "landed on ship registration, don't pay cost, a brake, broadly through", created the Xijiang river basin based on Beidou navigation "smart lock" application mode (Fig. 6).

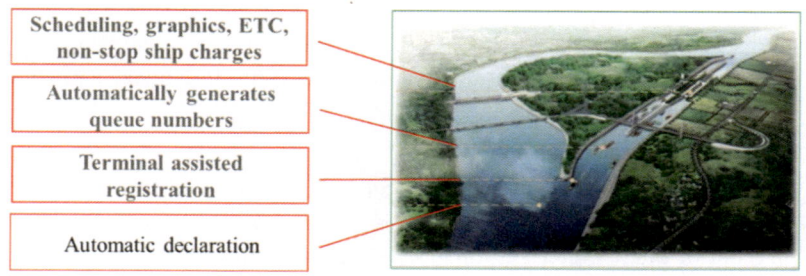

Fig. 6. Schematic diagram of ship automatic lock declaration

(3) By integrating multi-party communication command, cloud computing, Xijiang E-payment, Xijiang Tong APP, and video monitoring of the lock area, the integrated intelligent scheduling platform of multi-step and multi-line lock group has been built (Fig. 7).

Fig. 7. Xijiang lock joint scheduling platform

4.3 The Centralized Control and Safety Technology of Multi-step and Multi-line Lock Group is Innovated

The integrated mechanical and electrical monitoring and control of locks, multi-objective monitoring and sensing of lock rooms, online monitoring, diagnosis and early warning models for hydraulic structures of locks and other models have been established. The intelligent sensing and locking control technologies of anti-miter gate clamping ships and floating mooring bollards have been innovated. The centralized control system of multi-step and multi-line locks based on multi-source heterogeneous model has been developed. It has realized the safe, stable and efficient operation of Xijiang multi-step and multi-line lock group.

(1) Research and formulate data access specifications and data standards for multi-owner heterogeneous control system; The deep BP control model of mechatronics is established to realize the mechatronics control and automatic fault diagnosis of ship lock (Fig. 8).

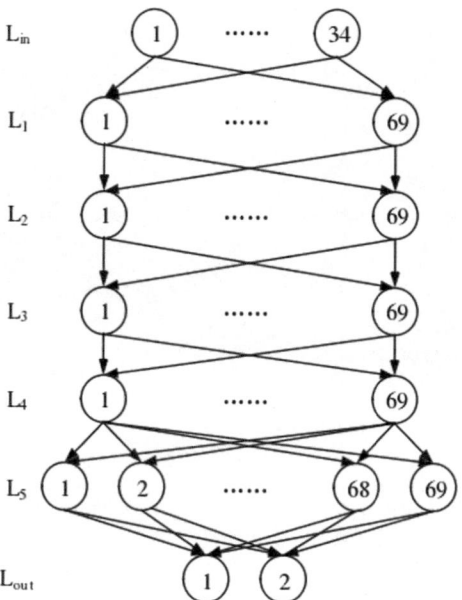

Fig. 8. Structure of deep BP control model

The training process of the complete model of deep BP neural network is as follows:

①The pre-trained weight matrix *[W]* and offset value vector *[B]* are used to initialize the corresponding parameters of each hidden layer.

②According to the current input sample, formula (5) is derived through BP algorithm to calculate the current output sample.

$$y_i = f\left(\sum_{j=0}^{n} W_{ij} x_j\right) \tag{5}$$

③The gradient term of neuron in output layer was calculated according to BP basic formula (6), and the error of output layer was calculated.

$$E = \frac{1}{2} \sum_l \left(t_l - f\left(\sum_j v_{ij} f\left(\sum_i w_{ij} x_i - \theta_j\right) - \theta_l\right)\right)^2 \tag{6}$$

④The error is propagated backward to the neuron of the upper hidden layer, and the gradient term of the neuron of the upper hidden layer is calculated.

⑤The error is transmitted to the next hidden layer, and the process of step (4) is repeated until all the errors of each hidden layer are calculated, and the ownership value matrix is adjusted and updated by formula (7) and (8).

$$v_{ij}(k+1) = v_{ij}(k) + \eta \delta_l y_j \tag{7}$$

$$W_{ji}(k+1) = W_{ji}(k) + \dot{\eta}\hat{\delta}_j x_i \tag{8}$$

Deep BP control model combined with specific scenes, after several rounds of personalized training, has the ability of automatic control for intelligent devices.

(2) A series of models of intelligent perception of ship line-crossing, intelligent perception of ship against miter door clamp, and intelligent perception of floating mooring bollards are established to innovate the intelligent perception of ship line-crossing and lock lockout control, and intelligent perception of floating mooring bollards and lock lockout control technologies (Fig. 9).

Fig. 9. Overline alarm system based on machine vision

(3) The on-line monitoring, diagnosing and early warning technology of lock hydraulic building safety were innovated. Piers and retaining wall structure monitoring, Beidou displacement monitoring of lock and other technologies integrated innovatively, which realize the online safety monitoring, diagnosing and automatic warning of ship lock hydraulic structures, and ensure the safe, stable and efficient operation of Xijiang multi-step and multi-line lock group.

In the sub-pixel displacement search algorithm for retaining wall structure position movement monitoring, its expression as follow:

$$w(\bar{s}) = \begin{cases} \frac{2}{3} - 4\bar{s}^2 + 4\bar{s}^3 & \bar{s} \le \frac{1}{2} \\ \frac{4}{3} - 4\bar{s} + 4\bar{s}^2 - \frac{4}{3}\bar{s}^3 & \frac{1}{2} < \bar{s} \le 1 \\ 0 & \bar{s} \le 1 \end{cases} \tag{9}$$

where $(\bar{s}) = \frac{s}{s_{max}}$ is the relative distance between a node in the supported domain and s_{max} is the supported radius. Then the residual sum function can be expressed as

$$J = \sum_{j=1}^{n} w(x - x_j) [f(x) - y_j]^2 = \sum_{j=1}^{n} w(x - x_j) [p^T(x)a(x) - y_j]^2 \quad (10)$$

where, y_j is the observed value. To the determining coefficient $a\,(x)$ and minimize the sum of residuals. The derivative of Eq. (11) with respect to a can be obtained:

$$\frac{\partial J}{\partial a} = a(x)A(x) - B(x)y = 0 \quad (11)$$

$$W(x) = diag(w_1(x), w_2(x), \ldots, w_n(x),) \quad (12)$$

It is the n-order diagonal matrix. Then $A(x)$ can be written in matrix form as $A(x) = p^T(x),\ W(x)p(x)$, which is an m-order symmetric matrix. $B(x)$ can be written in matrix form as $B(x) = p^T(x)W(x)$, which is an $m \times n$-order matrix. Finally, $a(x) = p^{-1}(x)B(x)y$ can be calculated by matrix multiplication (Fig. 10).

Fig. 10. Layout diagram of retaining wall structure displacement monitoring points

5   Conclusion

By researching on the key technologies of joint scheduling and control of multi-cascade and multi-line ship lock group, we have successfully realized the joint dispatch of multi-level, multi-line and multi-owner lock groups in the main line and branch lines of the Xijiang River Basin, as well as the centralized control and unmanned lock. We have created a new "four unified" management mode of Xijiang River navigation. Since the joint scheduling of cascade locks in Xijiang River Basin, it has played the advantages of unified coordination and unified management. This has greatly improved the navigation efficiency and service level of the Xijiang River Lock and provided a reference for the innovation of other navigable river management modes in China.

References

Zhang M, Feng XX (2015) Study on traffic organization rules of Changzhou four parallel ship locks. J Waterw Harb 36(4):329–333

Zhang M, Li M, Feng XX et al (2012) Causes and countermeasures of traffic jam in Changzhou double line ship lock. J Waterw Harb 33(2):56–60

Zhang M, Hao YY, Feng XX et al (2014) Application of ship maneuvering simulation technology for shipping route selection near reach of dam area. J Waterw Harb 35(2):175–179

Hao YY, Xue J, Zhang M (2013) Research on navigation encounter mode of double-line ship lock. J Waterw Harb 34(6):513–517

Jiang MF, Zhang M, Liao P, Feng XX et al (2013) Study on the recent lock capacity of the Changsha Hub. J Waterw Harb 4:359–363

Kong Z, Liao P, Yang CH et al (2017) Analysis of ship lock operation and throughput capacity based on traffic simulation model. Hydro-Sci Eng 1:73–79

Human impact on river channel changes downstream of Changzhou hydro-junction on the Xi River (Xijiang), China. 36th IAHR

Lin XZ (2016) Research on visual management of traffic logistics GIS road network. Popular Sci Technol 18(012):6–8

Lin XZ (2016) Design of Beidou satellite navigation technology transfer service platform. Sci-Tech Dev Enterp 2016(12):23–26

Lin XZ, Tang Y, Pan X (2014) Beidou CORS navigation accuracy continuous monitoring and evaluation model. Chin Satell Navig Positioning Assoc 2014:5

Lin XZ, Pan Y, Luo HP (2010) Construction of unified information service for information system under next Generation Network environment. J Guangxi Acad Sci 26(02):167–170

Shi HP (2019) Research on intelligent Ship Service System based on Beidou. Sci-Tech Dev Enterp 2019(03):54–56

Yang Q (2016) Transportation logistics Internet of Things Beidou terminal research and development. Sci-Tech Dev Enterp 2016(10):22–24

Pan X (2014) Optimization of transportation logistics early warning image transmission based on GIS. J Hechi Univ 34(02):65–70

Lin N (2019) Pan QB (2019) Study on the buffer sailing scheduling model of cascade ship lock in river Basin. W Chin Commun Sci Technol 11:168–171

Huang HZ (2016) Design and implementation of ship lock alarm management system based on Beidou Navigation. Sci-Tech Dev Enterp 2016(06):30–33

Ning W (2015) Research and development of Beidou Satellite navigation monitoring terminal for ship networking. Sci-Tech Dev Enterp 2015(24):20–22+25

A New Era for River Information Services (Wg125)

Piet Creemers[(✉)]

De Vlaamse Waterweg nv, Oostdijk 110, 2830 Willebroek, Belgium
piet.creemers@vlaamsewaterweg.be

Abstract. Inland Navigation and River Information Services (RIS) are facing in the coming decade many new challenges from various perspectives. All different transport modes are making the shift towards digitalization. Inland navigation is also moving towards a digital world under impulse of different worldwide initiatives. This means data and information services will become more and more crucial and critical for transport and traffic management in inland waterway transport (IWT).

Keywords: Digitalisation · Innovation · Multimodal · Synchromodal · Resilience

1 The Cycle of WG 125

The concept of River Information Services (RIS) was first initiated within the European Union in 1998. Since the first European RIS initiatives, this concept on information exchange to support traffic and transport management in inland navigation has found its way throughout the world. River Information Services are in an implementation stage in the different continents of the world, like America, Europe and Asia.

PIANC established a Working Group in 2002 that developed the Guidelines for River Information Services. These Guidelines are an important pillar in the implementation of River Information Services and have been updated in 2004, 2011 and 2019.

While the initial guidelines were initiated and inspired by academic research, the 2004 edition was more oriented on further developments of services and standards like Inland AIS (Automatic Identification System), Inland ECDIS (Electronic Chard & Display Information System), NtS (Notices to Skippers) and ERI (Electronic Reporting International). Initially these standards were addressed as RIS key technologies.

In 2009 PIANC established the *Permanent* Working Group 125 with the task to keep the Guidelines for River Information Services over the years to come up to date due their technical and innovative character.

Thanks to a number of small implementation projects by means of Proof-of-Concepts and Pilots, the guidelines in 2011 addressed the importance of having operational services within the domain of Traffic related Services. An academic approach is left behind and made room for an approach more based on experience and developments.

Y. Li et al. (Eds.): PIANC 2022, LNCE 264, pp. 764–768, 2023.
https://doi.org/10.1007/978-981-19-6138-0_66

In 2019 the guidelines aligned the initial RIS Key Technologies with the maritime e-Navigation concept, including the introduction of the e-Nav concept in which operational and technical services are defined. Developments in the area of corridor management were incorporated, primarily affecting the existing RIS Traffic Management Information service, but also highlights the needs to support Transport Logistics (Transport related Services). Finally, the 2019 guidelines were revised to move from a European-oriented perspective to guidelines optimized of use on a worldwide scale.

Finally the cycle of going from academic via Research to Developments is in 2022 with the concept of "synchromodal transport" back at the start of a new cycle, being an new academic era.

2 Synchromodal Transport

During the last years different emerging threats made a serious impression on mankind. Besides shocks of nature due to climate change or pandemics, also technological shocks due to automatization and digitalization are part of the world we live in today. Because of these shocks both public and private parties are aware they need to make their business processes not only more resilient but if possible also preventive.

A prominent statement towards climate change has been made by different governmental bodies by elaborating a green deal with actions defined in different domains like industry, energy, agriculture, ... and transport. To be able to meet the goals as defined in the policy documents (e.g. by the European Commission) mobility needs to be *Sustainable, Smart and Resilient* in which there is an important role to be fulfilled by Inland Water Transport (IWT) and River Information Services (RIS). To become:

- Sustainable, the share of transport by IWT needs to be increased;
- Smart, the full potential of data needs to be unleashed, freight transport needs to be paperless and automated mobility will be deployed on a large scale;
- Resilient, a fully multimodal transport network for sustainable and smart transport needs to be operational.

Recent years some new concepts were born in the field of transport and logistics, looking to optimize the whole transport chain and tackle the challenges as defined within the green deal. One of the remarkable ideas was the concept of the Physical Internet. If we are able to translate the principles of the Digital Internet, i.e. the transport of (data) containers aka packets over a network in an optimal and orchestrated way, to a Physical Internet (PI) we could have the same benefits in real life when transporting (goods) containers over a transport network of different transport modes. Since this concept requires a very high automated infrastructure together with an intelligent infostructure (aka digital infrastructure) to orchestrate the whole transport chain, a final implementation is foreseen by some organizations at around 2040. Although this horizon is quite far away, governmental and private bodies are already preparing themselves to make the transition as smooth as possible.

The principle behind the Physical Internet is based on the idea of synchromodal transport for which a description was elaborated by workgroup 125:

"The most efficient and appropriate transport solution in terms of sustainability, transport costs, duration and their reliability, in which the configuration of the transport chain is not static during transport, but is flexible, being able to adapt the mode of adequate transport according to the conditions in real time of infrastructure and capacity, through collaboration and the exchange of information in real time of all modes of transport, the terminal facilities and the actors involved in the transport logistics chain."

This description is fully compliant with the three objectives as defined within the green deal, i.e. looking towards a sustainable, smart and resilient transport of goods and people (aka mobility) and will demand a full cooperation between the different transport modes to realize these objectives.

While private companies are looking towards Industry 4.0, the governmental bodies are making investments in both digital and automatization initiatives like River Information Services (RIS), Intelligent Transport Systems (ITS), e-Navigation (e-Nav), ERTMS, RAILWAY 4.0, …

To make the inland waterways a valuable and trustful link within a synchromodal network different challenges need to be tackled:

- Geographical upscaling: River Information Services will need to interconnect different stakeholders of a waterway network on a continent by means of a (pan-) corridor management approach versus a national or regionally focussed approach like it is nowadays. This current narrow approach is hindering the growth of inland navigation.
- Synchromodal ready: Once these international waterway networks are able to offer operational and technical services, interconnection with other transport modes will be required. By exchange knowledge, experiences, ideas, technologies and lessons learned these different transport modes can converge to a similar approach and framework. This approach will finally result in a synchromodal network. The big challenge will be the coordination of such an approach covering and steering the different transport modes to act to fulfil a common goal and become synchromodal ready.
- Future Proof: new technologies will allow new applications in the area of infrastructure and vessels, including increasing requirements towards data and information service. At the same time, these new technologies will challenge the existing technical and operational services to evolve. These services must be designed to be scalable, trustful, of high quality and flexible enough to resist future demands. In the specific case of automated vessels, this need to be addressed within the concept and framework of Smart Shipping. Within this framework not only the vessel is taken into account, but also the interaction with the shore infrastructure (bridges, locks, traffic centres, …). Therefore RIS developments must be closely in line with the concept of Smart Shipping (WorkGroup 210).

3   The Approach of Pianc WG 125 (2019–2022)

How is the PIANC WG 125 going to deal with these complex challenges? The final goal is a valuable contribution of Inland Navigation to a well performing synchromodal transport network, therefore a good insight in other transport modes, together with a decent knowledge in different technology trends, is necessary.

Since River Information Services are covering a set of services on both traffic and transport management, RIS should be able to close the current gaps to reach the above mentioned goal. Related to other transport modes investigations will be necessary and done in:

- Intelligent Transport Systems (ITS) related to road traffic
- Telematics Applications within the EU rail system, mainly focused on Freight services (TAF)
- e-Navigation (e-Nav) related to Maritime Navigation

These investigations in intelligent transport systems of other transport modes will allow the PIANC WG 125 to look for similarities and connections with inland navigation and thus defining new or updating existing River Information Services (both operational and technical) to support practical use cases. E.g. once Smart Shipping is able to take place in a safe way with maximum respect of harmonization and standardisation, one of the next phases can be focused on efficiency within traffic management.

Finally these new or updated services need to be harmonized in a worldwide way, so inland waterway networks across continents can be interconnected via Maritime Services (i.e. concept of e-Navigation) allowing industry to benefit from it in a cost effective way stimulating the use and benefits of Inland Navigation.

The final report of PIANC WG 125 will elaborate this research and formulate guidelines useful for e.g. implementing countries, possible directives and regulations, Research & Development projects, implementation projects, etc.

4   Conclusions

One can conclude the role of RIS is still very important and focused on safety and efficiency, but faces the opportunities and challenges of new technologies and trends. RIS will be the facilitator for inland waterways to become a trustful link in a synchromodal transport network.

To be able to become such a trustful link, the first steps will be elaborated within the new PIANC RIS guidelines of 2022. It is important to know which (new) services for Inland waterways are needed to make inland navigation *synchromodal ready*.

Since the previous Smart Rivers in 2019, a lot of work has already been done. The progress of this work will be presented during the Smart Rivers Conference in 2022, providing the audience a good insight at the future guidelines.

A Web-Based Regional Economic Simulation Tool for U.S. Army Corps of Engineers' Civil Works Programs

Wen-Huei Chang[1](✉), Dena Abou-El-Seoud[2], and Kevin Knight[1]

[1] Institute for Water Resources, United States Army Corps of Engineers, Alexandria, USA
{Wen-Huei.Chang, kevin.p.knight}@usace.army.mil
[2] Great Lakes and Ohio River Division, United States Army Corps of Engineers, Chicago, USA
dena.abou-el-seoud@usace.army.mil

Abstract. The U.S. Army Corps of Engineers (USACE) is one of the world's largest public engineering, design, and construction management agencies. The major Civil Works (CW) mission areas include navigation, flood risk management, hydropower, ecosystem restoration, coastal storm damage reduction, hydropower, water supply and recreation. To capture the regional economic benefits of water infrastructure and programs, the USACE Institute for Water Resources (IWR) developed the Regional ECONomic System (RECONS). RECONS estimates short-term and long-term economic activity resulting from federal investments within CW mission areas. Estimates of economic activity are measured as industry output, employment, labor income and value added and are provided for three levels of geography: local, state, and national. This information supports federal investment decisions and stakeholder communication. USACE's navigation mission alone services 41 states. USACE operates and maintains 25,000 miles of waterways and 236 lock chambers for commerce. Short-term economic activity is attributed to federal spending on infrastructures and operations and maintenance (e.g., spending to replace a lock) while long-term economic activity is supported by infrastructure users (e.g., lock utilization by shippers). RECONS facilitates widespread application of input-output (I-O) analysis through a user-friendly, web-based simulation tool customized for CW mission areas. RECONS enables users to conduct valid and consistent economic impact analyses without the degree of knowledge required by comprehensive I-O models. As a result, over 200 USACE economists have used RECONS to conduct hundreds of analyses each year. This paper describes RECONS' capabilities and applications, and its implications for federal decision-making regarding investments in inland navigation infrastructure.

Keywords: Economic impact analysis · Web-based simulation tool · Inland navigation

Y. Li et al. (Eds.): PIANC 2022, LNCE 264, pp. 769–779, 2023.
https://doi.org/10.1007/978-981-19-6138-0_67

1 Usace Evaluation of Proposed Civil Works Projects

1.1 USACE Navigation Mission

The U.S. Army Corps of Engineers (USACE) is one of the world's largest public engineering, design, and construction management agencies. The major Civil Works (CW) mission areas include navigation, flood risk management, hydropower, ecosystem restoration, coastal storm damage reduction, hydropower, water supply and recreation (USACE 2000). Navigation is one of USACE's major missions. USACE maintains 12,000 miles of inland and intracoastal waterways (9- to 14-foot draft), 13,000 miles of channels of deep draft channels (greater than 14 feet) for commerce. In addition, the Corps also maintains 236 lock chambers at 191 sites on the coastal and inland navigation systems (USACE 2022). These navigation infrastructures affect all 50 states and many other nations in terms of their dependance on the economic activity supported by these navigation systems.

1.2 The Federal Objective

Improvements to the navigation system are funded by the federal government and implemented by USACE. The Economic and Environmental Principles and Guidelines for Water and Related Land Resources Implementation Studies (P&G) (U.S. Water Resources Council 1986) established the federal objective of water and related land resource planning: "to contribute to national economic development consistent with protecting the Nation's environment, pursuant to national environmental statutes, applicable executive orders, and other Federal planning requirements." Table 1 identifies the four accounts were identified to facilitate the evaluation and display of effects of alternative investment decisions (U.S. Water Resources Council 1986).

Table 1. Four accounts to facilitate the evaluation and display of effects of alternative investment decisions (U.S. Water Resources Council 1986).

Account	Acronym	Definition
National Economic Development	NED	Displays changes in the economic value of the national output of goods and services
Regional Economic Development	RED	Displays non monetary effects on significant natural and cultural resources
Environmental Quality	EQ	Registers changes in the distribution of regional economic activity that result from each alternative plan. Evaluations of regional effects are to be carried out using nationally consistent projections of income, employment, output, and population
Other Social Effects	OSE	Registers plan effects from perspectives that are relevant to the planning process, but are not reflected in the other three accounts

1.3   Contributions to National and Regional Economic Development

Improvements to the navigation system funded are by the federal government and implemented by USACE are subject to a detailed benefit-cost analysis which could include quantification of benefits and costs that accrue at both the national level. This evaluation facilitates an evaluation of national economic efficiencies (contributions to the NED account) to be realized by the navigation improvement. For navigation improvements, project benefits are generally defined as the reduction in transportation costs to shippers and other users of the waterways (USACE 2000).

In recent years, there has been a renewed emphasis on additional factors such as environmental justice, social equity, ecosystem services and local and regional economic effects of USACE projects. USACE must consider changes in the distribution of regional employment and regional income (RED account) is required as part of the formulation and evaluation of proposed CW projects (USACE 2021). Therefore, economic impact analysis is integral to USACE's evaluation of potential inland navigation improvements.

2   Regional Economic System (RECONS)

2.1   Purpose of RECONS

The USACE Institute for Water Resources (IWR) developed the Regional ECONomic System (RECONS) to provide accurate and defensible estimates of economic activity associated with USACE projects, programs, and infrastructure across all CW mission areas. Estimates of economic activity are measured as industry output, employment, labor income and value added and are provided simultaneously for three levels of geography: local, state, and national (IWR 2019). The RECONS modeling framework in conjunction with its accessible web-based interface enables USACE to readily leverage economic impact analyses to support decision making and stakeholder communication across all levels of USACE, whether it be to assist in the prioritization of which projects to fund through infrastructure bills or comparing alternative investment decisions for specific water resource problems.

RECONS automates calculations of the effect of Corps projects, programs, and infrastructure affect regional economies across the nation. This information helps demonstrate the importance of existing and potential Corps investments to decision-makers, project stakeholders and communities. RECONS was designed with four primary purposes:

1. Provide an efficient and consistent approach to conduct economic impact analyses;
2. Enable users to conduct valid and reliable economic impact analyses without having the degree of knowledge or experience of Input-Output analysis required by comprehensive I-O models;
3. Enable users to simultaneously estimate economic impacts at different geographic scopes (local, state, national) and to aggregate impacts across projects and regions; and

4. Increase rigor and consistency of economic impact analyses are conducted by USACE (IWR 2022).

RECONS is a certified model for use across USACE, a designation reflecting the model's compliance with the Agency's technical quality, system quality and usability requirements (USACE 2011).

2.2   Web-based Dynamic Simulation System

RECONS is a web-based dynamic simulation system and is developed with PHP/MySQL applications with the server hosted on Cloud Computing Services. RECONS utilizes a mapping function which allows users to portray the areas benefited by USACE programs. There are currently more than 1,500 built in regional I-O models that correspond to USACE's CW mission areas. Multipliers and other economic ratios and factors are from IMPLAN, US Bureau of Economic Analysis, and Bureau of Labor Statistics. It also allows frequent modifications/updates and new add-ons to be made through the server and instantly distributed.

The work USACE undertakes and funds through its CW budget is categorized into work activities. For each work activity, expenditures associated with that activity are mapped to IMPLAN sectors, and these allocations of expenditures are termed spending profiles. IMPLAN models were created for each geography, and multipliers, ratios, and regional purchase coefficients (RPCs) were extracted from the IMPLAN models and imported into RECONS. The extent to which an effect is captured within the impact area is represented by local purchase coefficients (LPCs), which are either based on the RPCs in the IMPLAN model or customized with information from USACE or industry experts (IWR 2019a).

2.3   Model Framework

RECONS is designed to provide decision makers a comprehensive understanding of how a region's economy is affected throughout the project life cycle. Short-term economic activity is attributed to direct federal spending on infrastructures and operations and maintenance (e.g., spending to replace an existing lock and dam on an inland waterway) while long-term economic activity results from the primary users of infrastructure constructed and maintained by USACE (e.g., long-term utilization of the lock by shippers to transport commodities). RECONS accounts for economic activity associated with directly affected industries (direct effects), backward linkages to the industries, businesses, and households supplying the goods and services used by the directly affected industries (indirect effects), as well as the household spending associated workers spending their income within the impact area (induced effects).

To facilitate the evaluation of short- and long-term economic activity supported by USACE CW projects, programs and infrastructure, RECONS comprises two sets of modules: Civil Works Spending Modules, and USACE Programs and Infrastructure Modules. The USACE CW spending models are used to estimate short-term regional economic impacts of project expenditures within the eight USACE CW business lines (e.g., construction or operation and maintenance expenditures). The USACE Programs and Infrastructure Modules are used to estimate the long-term regional economic impacts of activities induced by USACE programs and infrastructure (use of navigation infrastructure for shipping activities). These activities and expenditures support economic output, jobs, earnings, and value added. Results are shown for three levels of geography: local, state, and national impact areas (IWR 2022).

2.4   User Interface

RECONS is deployed across USACE offices for use by economists, planners, and others as an online tool accessible through the internet with a password. For each RECONS module, users are prompted through a series of screens to provide or verify the necessary inputs for the economic impact analysis. For example, to estimate the regional economic activity resulting from rehabilitation of a lock on an inland waterway, RECONS prompts the user through the following process (IWR 2019b):

Step 1: Select a Project
Step 2: Select a Business Line Work Activity and Year of Expenditure
Step 3: Review Economic Impact Area and Socioeconomic Information
Step 4: Confirm or Modify Project Expenditure
Step 5: Confirm or Modify Spending Profile
Step 6: Confirm or Modify LPCs
Step 7: Review Economic Impact Results.

Similarly, to estimate the regional economic activity resulting from long-term use of a lock by commercial shippers, RECONS prompts the user through the following process (IWR 2019b):

Step 1: Select an Inland Port or Harbor
Step 2: Select a Year of Expenditure
Step 3: Review Economic Impact Area and Socioeconomic Information
Step 4: Enter Shipping Volume per Commodity Type
Step 5: Confirm or Modify Spending Profile
Step 6: Confirm or Modify LPCs
Step 7: Review Economic Impact Results.

The results of each module are displayed in a similar manner, regardless of the module selected. Figure 1 provides a sample screenshot of RECONS mapping interface while Fig. 2 provides a screenshot of the RECONS 'Results' tab. The overall summary provides a snapshot of the local, state, and national economic impacts. Users can save their analysis with a title or download a report which provides documentation of all inputs and results from the evaluation.

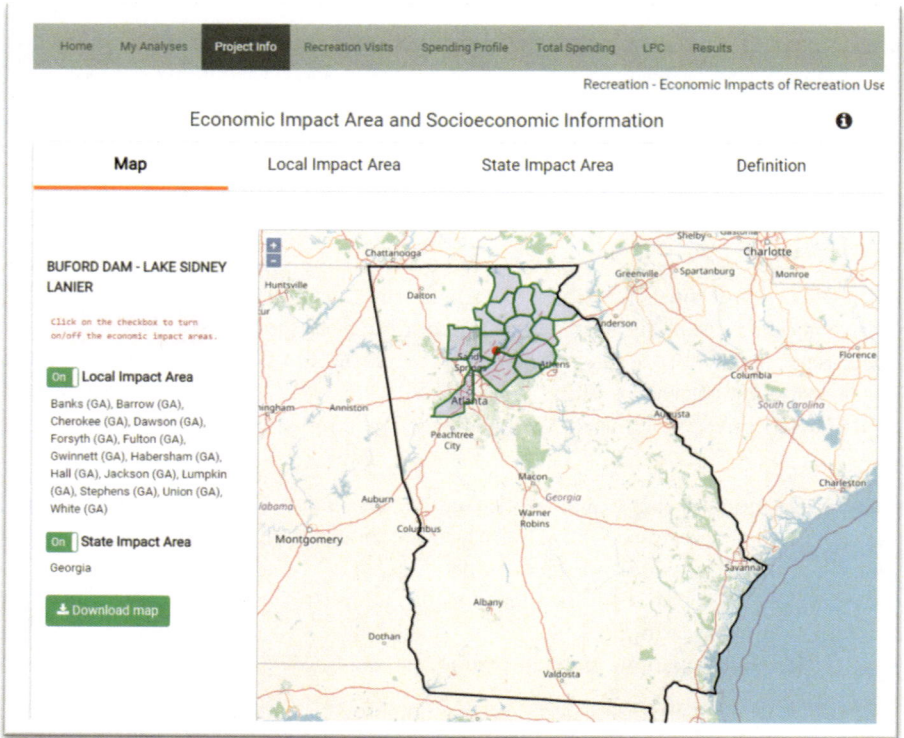

Fig. 1. RECONS Web Interface (IWR 2019b).

2.5   RECONS Applications for Enterprise-Level Reporting

RECONS is applied across all USACE CW mission areas on a regular basis. Over 200 USACE economists have used RECONS to conduct hundreds of analyses each year to support decision making and stakeholder communication across all levels of USACE. RECONS is utilized to communicate the value of USACE CW mission areas to nation. Although RECONS was initially developed to estimate the economic impacts of American Recovery and Reinvest Act in 2009 and was used as the agency tool to report how the government spending will support jobs and local economies, it has evolved into a more versatile and comprehensive regional economic impact analysis tool over these years. It has been utilized by the agency to estimate and report jobs supported by various federal government stimulus spending bills since its inception and is currently being used to estimate jobs supported by USACE's proposed construction projects for the 2021 US Infrastructure Investment and Jobs Act. Comprehensive regional economic impact assessments using RECONS assures USACE presents results consistently across business lines and geographic regions.

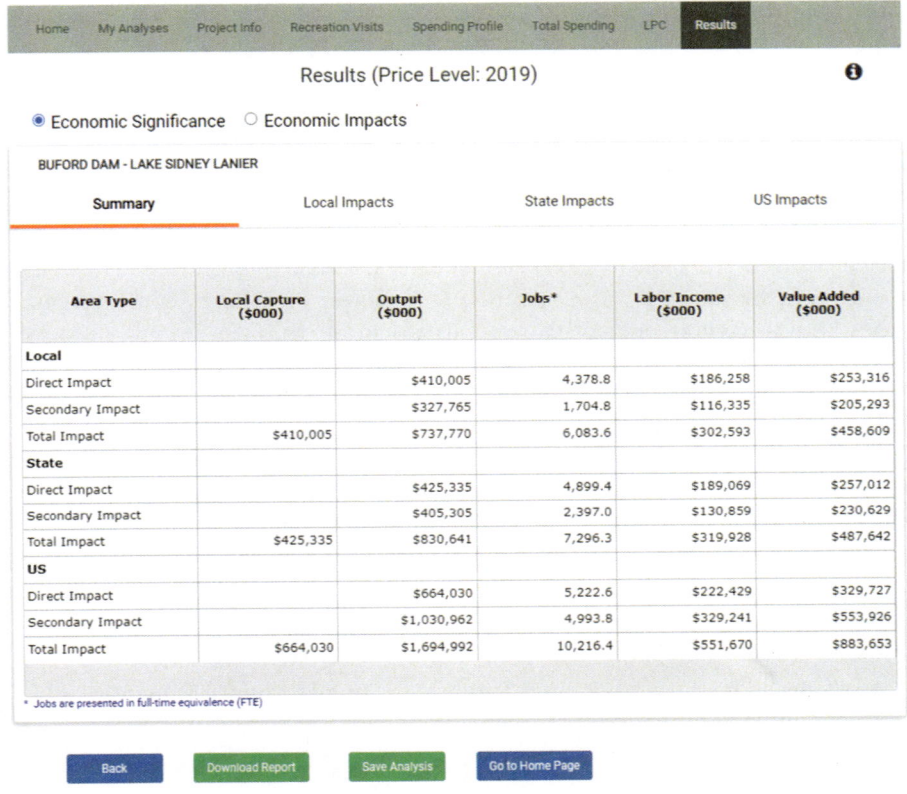

Fig. 2. RECONS 'Results' Screen (IWR 2019b).

2.6 RECONS Applications for Evaluating Civil Works Projects

RECONS is also used to address the RED account in the formulation and evaluation of proposed water resource projects, which supports the USACE decision-making process. The certification of RECONS for use in CW planning efforts in conjunction with the user-friendly interface facilitates use of the model at a low cost to projects. RECONS has been applied to facilitate navigation-related economic impact analyses across the nation.

RECONS is regularly used to support the evaluation of proposed navigation improvements. In 2017, RECONS was applied for an Economic Evaluation Study to look at the regional and national economic impact of short term and long term closure of the Soo Locks at Sault Ste Marie, Michigan, and the annual economic contribution of the Soo Locks to Great Lakes Basin industries that use the Soo Locks to ship commodities. RECONS has also been used to support the evaluation of economic and environmental tradeoffs. For example, RECONS was utilized in support of the Great Lakes and Mississippi River Interbasin Study – Brandon Road (USACE 2018), a study which sought to identify a solution to reduce the risk of aquatic invasive species transfer between the Great Lakes and Mississippi River Basins while minimizing impacts to waterway uses and users. RECONS was used estimate the long-term changes in regional economic activity arising from the potential implementation and subsequent operation and maintenance of aquatic nuisance species controls at Brandon Road Lock and Dam (BRLD) within the Chicago Area Waterway System (CAWS). With 11 million tons of cargo transit the lock on an average annual basis, it necessary to understand how industries the Chicago area would be impacted if modifications to BRLD hindered their access to markets. The evaluation considered local firms that ship goods through BRLD and those that receive shipments through the lock and dam. Expected changes to shipping costs were linked to changes in total sales. If shipping costs increased for industries using the BRLD to deliver or receive goods and commodities, then the selling price of those goods would be impacted. In turn, impacted firms would face a reduction in competitive advantage relative to other firms (USACE 2018). Estimated changes in employment and income in the Chicago area due to implementation of a GLMRIS-BR alternative plan was to help communicate the tradeoffs between maintaining and expanding uses of navigable waterways while reducing the risk of invasive aquatic species transfer. More recently, RECONS was applied to a study requested by International Joint Commission Great Lakes-St. Lawrence River Adaptive Management to estimate the economic impacts to commercial shipping industries of closure of St. Lawrence Seaway to Commercial Navigation (Chang et al. 2021).

2.7 Considerations for Developing Regional Economic Impact Modeling Tools for Federal Entities

The widespread and frequent application of RECONS in USACE suggests other Federal entities may benefit from developing customized regional economic impact modeling tools to better inform investment decisions and stakeholder communication. The successful development and implementation of RECONS provides insights as to what federal entities should consider prior to undertaking their own development of such a tool. Key considerations are listed below.

1. **Defensibility:** Federal entities should assure the modeling framework, data, methods, and results are defensible. To become eligible for USACE-wide application, RECONS underwent a rigorous review process to be a 'certified' model. The review assures a model's technical quality, system quality and usability requirements (USACE 2011). The resulting model certification ultimately serves as an agency-wide indicator that the model results can be readily utilized and reported with confidence. This increases the likelihood model will be used and helps assure consistent reporting across geographic regions and project types.
2. **Relevance:** Federal entities should assure a customized regional economic impact modeling tool is relevant, in that it accommodates a comprehensive array of applications. USACE implements a various types of Civil Works projects across numerous geographic regions. RECONS can be leveraged to estimate the economic

activity supported by federal spending across all Civil Works project types (inland navigation, deep-draft navigation, flood risk management, etc.). RECONS also provides capability to evaluate the long-term economic activity resulting from the primary users of infrastructure constructed and maintained by USACE. If needed, new work activities and impact areas may be requested. Offering diverse applications supports widespread application of the model and substantially reduces the need to obtain or develop additional regional economic impact modeling tools on an ad-hoc basis.

3. **Accessibility & Usability:** Federal entities should assure the regional economic impact modeling tool is accessible and user-friendly. RECONS is a web-based tool, making it readily available to any user with a username and password. This assures users are always accessing the current version of the model, and any technical improvements immediately benefit all users. The model framework is consistent across analysis types. Knowledge that users gain from a single RECONS application can be readily transferred to the other model applications. The RECONS user guide, methods manual and other support documents are available for download from within the RECONS model. This assures the support documents are readily accessible.

4. **Establishment of a Long-Term Model Support and Maintenance Plan:** Federal entities should assure a long-term model support structure and maintenance plan are identified and funded. The RECONS program is managed by the USACE Institute for Water Resources, which provides funding to a national technical support team on an annual basis. The support team develops and provides training, as are readily engaged by users across the enterprise for model support. A model development team is also funded on an annual basis, which provide continual support with model debugging and the periodic incorporation of new input data. Periodic virtual web-based training and in-person training workshops are scheduled, and ad-hoc trainings are provided upon request. These regular trainings also allow for users to provide feedback about potential model improvements and needs for additional model applications. Collectively, establishment of the model support structure and regular maintenance facilitates continued efficient and effective model application across the enterprise.

3 Conclusions

As one of the objectives of the USACE's Navigation mission is to provide safe, reliable, and efficient waterborne transportation systems, the USACE accomplishes this mission through a combination of capital improvements and the operation and maintenance of existing facilities and waterways through sound planning and best available data for decision making. This paper described the capabilities of the RECONS model, the assumptions and approach used for the model, and the use of the model to estimate existing and potential impacts on various harbors, ports and inland waterways across the USA. The results from RECONS can serve not only as a means to communicate with stakeholders of the importance and values associated with navigation infrastructure, also augment the traditional benefit cost analysis by providing more comprehensive benefits with regard to the return on investment for federal spending.

The development of RECONS enabled the defensible and consistent application of regional economic impact modeling across USACE. Estimates of jobs, revenues, income and value added help communicate the importance of USACE projects to decision makers, as well as non-federal project sponsors and stakeholders. RECONS is relied upon for an array of applications, whether it be to comprehensively evaluate of economic activity supported by USACE CW to support enterprise-level reporting or to serve as a readily accessible tool to characterize the regional economic effects of proposed navigation improvements. The widespread application of RECONS is encouraged by its web-based application and user-friendly interface. IWR continues to educate users on RECONS functionality and expand the model's capabilities. This continued investment in regional economic impact modeling is integral to USACE decision making and the effective communication of how its CW projects affect the geographic regions across the vast nation. In addition, RECONS also provides a best management practice and can be replicated by other public agencies to generate reliable and consistent statistics to assess policy decisions such as public investment across the nation. This consistent modeling approach across all government agencies is extremely important for federal budget planning and communication, especially for supplemental funding such as the 2021 US Infrastructure Investment and Jobs Act.

References

Chang WH, Stalikas S, Witherow L (2021) Economic impact assessment of halting commercial navigation on the montreal-to-lake ontario section of the St. Lawrence Seaway. Technical reports, in Press. Alexandria, VA: Institute for Water Resources, U.S. Army Corps of Engineers

United States Army Corps of Engineers (2018) The Great Lakes and Mississippi River Interbasin Study—Brandon Road Final Integrated Feasibility Study and Environmental Impact Statement— Will County, Illinois

United States Army Corps of Engineers (2022) Civil Works Navigation. Headquarters U.S. Army Corps of Engineers, U.S. Army Corps of Engineers. https://www.usace.army.mil/Missions/Civil-Works/Navigation/. Accessed 12 Apr 2022

United States Army Corps of Engineers (2000) Planning Guidance Notebook. US Army Corps of Engineers, Washington, DC

Water Resources Council (1983) Economic and Environmental Principles and Guidelines for Water and Related Land Resources Implementation Studies. Water Resources Council, Washington, DC

Institute for Water Resources (2019a) RECONS 2.0 Methods Manual. US Army Corps of Engineers, Washington DC

Institute for Water Resources (2019b) RECONS 2.0 User Guide. US Army Corps of Engineers, Washington DC

United States Army Corps of Engineers (2011)

US Army Corps of Engineers, 2011. Assuring the Quality of Planning Models., Washington, DC

Institute for Water Resources (2022) Regional Economic System (RECONS). US Army Corps of Engineers, Washington DC. https://www.iwr.usace.army.mil/missions/economics/regional-economic-system-recons/. Accessed 13 Apr 2022

United States Army Corps of Engineers (2021) Memorandum for Commanding General, U.S. Army Corps of Engineers, Policy Directive – Comprehensive Documentation of Benefits in Decision Document. US Army Corps of Engineers, Washington DC

Application and Prospect of Spatial Information Technology in Inland Waterway Resources Census

Pengpeng Jia[✉], Zhefei Jin, Honglin Feng, Qiong Yang,
and Wenwu Yang

Transport Planning and Research Institute Ministry of Transport, Beijing, China
{jiapp,jinzf,fenghl,yangqiong,yangww}@tpri.org.cn

Abstract. In recent years, the rapid development of GIS, remote sensing, big data and other spatial information technologies have provided a huge driving force for the information construction and development of many industries. As an important branch of transportation field, inland water transportation also has great development opportunities in information construction. Since the second National Inland Waterway Census, great changes have taken place in the current situation of China's inland waterway. It is urgent to carry out a new round of National Inland Waterway Census and comprehensively find out the current situation of China's inland waterway. This paper comprehensively discusses the current situation of the application of spatial information technology in the general census of resources in typical industries. This paper also summarizes the basic situation of Inland Waterway Census in China, and analyzes the key and difficult points of the application of spatial information technology in Waterway Census data acquisition, working base map compilation, application achievement display and so on. It initially puts forward the technical scheme of channel resources census based on information means and online system. Compared with the traditional offline census, the technical scheme proposed in this study has achieved a breakthrough in improving the efficiency of census and ensuring the quality of census results. The research can provide technical reserves and experience for the general survey of channel resources for the whole country and all the provinces.

Keywords: Inland water · Digitization · Database construction · GIS · Remote sensing

1 Application of Spatial Information Technology in Resource Census

1.1 Development Status of Spatial Information Technology

Spatial information technology is an emerging technology rising in the 1960s and developed rapidly in China around the 1970s. The core of spatial information technology includes "3S" technology, namely satellite positioning system, geographic information system and remote sensing technology (Liu et al. 2008). At present, spatial information technology not only includes the traditional "3S" technology, but also organically

© The Author(s) 2023
Y. Li et al. (Eds.): PIANC 2022, LNCE 264, pp. 780–788, 2023.
https://doi.org/10.1007/978-981-19-6138-0_68

combined with the Internet of things, mobile Internet, cloud computing, blockchain, BIM, big data, artificial intelligence and other new generation information technologies. It has been widely and deeply applied in the fields of transportation infrastructure life cycle management, infrastructure resource census, intelligent transportation platform construction, transportation planning and management and so on (Ming et al. 2017).

"3S" technology includes remote sensing (RS), global positioning system (GPS) and geographic information system (GIS). It is an important technology that runs through the whole process of spatial information data acquisition, analysis, management and update in the earth observation system. Among them, remote sensing is a means of observing the characteristics of long-range electromagnetic wave, which has the advantages and characteristics of remote detection, dynamic, high-resolution and multi-type sensors. The global positioning system includes GPS of the United States, GLONASS of Russia, Galileo of the European Union and Beidou satellite navigation system independently developed by China; At present, China's Beidou Positioning System which was developed independly has completed comprehensive networking and officially provided global services in 2020 (Wang et al. 2016). It plays an important role in the fields of land and resources, meteorological observation and prediction, earthquake prediction, transportation and so on. It is now called the world's four major satellite navigation systems together with the three major systems. Geographic information system is a computer software platform that specially manages and uses geographic information data. Its main feature is to classify, manage, combine, analyze and process geographic information data, and display data in two-dimensional and three-dimensional forms. Its advantage is that the obtained data can be mapped by computer. This technology has been widely used in various businesses in the transportation industry.

At present, the rapid development of emerging technologies represented by the Internet of things, mobile Internet, cloud computing, blockchain, BIM Technology, big data and artificial intelligence is triggering a new round of scientific and technological revolution all over the world (Yao et al. 2021; Yalcinkaya et al. 2015). The application of new generation information technology in the transportation industry will bring revolutionary influence and strongly promote the process of transportation modernization. The development of these emerging technologies has improved the ability of basic data collection, sorting, storage, analysis and sharing of traffic multi-source data and solving complex problems, and gave birth to new business forms for the development of the transportation industry (Hjelseth et al.; KIM et al. 2016).

1.2   Overview of Resource Census in Typical Industries

The first national water conservancy census conducted in 2010 is a full-scale census of basic water conservancy data of the water conservancy industry since the founding of the people's Republic of China. It is divided into three stages: inventory stage, general census stage, data collection and results display, development and utilization. It involves eight projects, including general census of basic conditions of rivers and lakes, general survey of basic conditions of water conservancy projects, survey of economic and social water use, general survey of development and protection of rivers and lakes, and general survey of water and soil conservation. Spatial information technology has been fully utilized in the three stages and eight projects. In particular, the general census of basic conditions of rivers and lakes and the general census of water and soil conservation mainly rely on spatial information technology and obtain relevant information by means of indoor acquisition and processing. In the aspect of achievement display, development and utilization, taking spatial information technology as the technical support, the "one map" of census results is realized, which intuitively shows the comprehensive and in-depth application of spatial information technology in water conservancy industry.

The national land census is a major census of national conditions and national strength for the purpose of verifying and clarifying land resources. It is the basis for comprehensively investigating the current situation of land resources utilization in China and accurately and comprehensively understanding the basic land data infor-mation. It is also the basis for building a modern and information society. It plays an important role in promoting the socialized service of land resources information. Compared with the second national land census, the third national land census puts forward more accurate and precise requirements for the basic land data. The remarkable feature of this task is the "refinement" and "informatization" of land resources, which requires that in the process of completing this work, we must make major technological breakthroughs, using the Internet, remote sensing technology, UAV measurement, global navigation Geographic information and other advanced modern information technology means to realize the accurate investigation of land resources.

2   General Census and Current Situation of Inland Waterway

2.1   Inland Waterway Census

In 2002, the Ministry of Communications issued the notice on carrying out the second National Inland Waterway Census, and carried out the census with the support and cooperation of the National Bureau of statistics (PARK et al. 2016). The general census mainly includes the overview of inland waterways, the current situation of waterways, hubs, river crossing structures, waterway management institutions and maintenance forces, the investigation of riverside facilities, the drawing of the current situation of waterways, etc. Based on the survey results, in 2007, the national development and Reform Commission and the Ministry of communications jointly issued the national inland waterway and port layout plan, which planned the layout of "two horizontal, one vertical, two networks and 18 lines" inland national high-grade waterway, with a total mileage of about 19000 km. Under the guidance of the plan, the state and local governments have successively carried out a series of waterway regulation projects such as the Yangtze River trunk line, the Xijiang River trunk line, the Beijing Hang-zhou canal, the Yangtze River Delta and the Pearl River Delta. By the end of 2020, the total mileage of high-grade inland waterways in China has reached 14500 km. Com-pared with 2002, the current situation of high-grade inland waterways in China has changed greatly. At the same time, the national inland waterway and port layout plan is currently being revised, and a new layout of "four horizontal, four vertical and two networks" inland high-grade waterway is planned. The total mileage of the waterway has reached 29000 km, which is nearly 10000 km compared with the previous version of the plan. In view of this, it is urgent for the competent department of transportation to carry out a new round of National Inland Waterway Census and comprehensively find out the current situation of China's inland waterway.

2.2   Current Situation of Waterway Census Research

In recent years, many provinces and cities across the country have successively carried out the general census of inland waterways. In addition to the traditional offline census, collection, sorting, review and summary, spatial information technologies such as Beidou, Internet, GIS and high-resolution remote sensing images are widely used, and information-based means are fully used to effectively improve the efficiency of the general survey and ensure the quality of the survey results.

In July 2021, the port and shipping census of Hubei Province was officially launched. The census covers all elements of water transportation, including waterway census, port census, ship census, general survey of navigation facilities (hubs, locks and ship lifts), general survey of river crossing channels (bridges and tunnels), general survey of Township ferries, general survey of waterway transportation enterprises, etc. At the same time, based on the census data, the provincial channel and port information system will be developed, mainly including the construction of basic database, platform function and mobile pad function.

In November 2020, the fourth general census of inland waterways in Jiangsu Province was officially launched, covering 3700 km of provincial trunk waterways, 6000 km of municipal trunk waterways and local general waterways. In view of the problems of insufficient coverage and weak applicability of the previous Waterway Census data, the fourth Inland Waterway Census in Jiangsu Province fully combines the construction of inland waterway electronic waterway map under way in Jiangsu Province. The census comprehensively covers the provincial and municipal planned trunk channels and local general channels to retain freight functions. On the basis of traditional attribute information such as channel, ship lock, navigation mark, water service area and bridge, the census objects added new unique attribute information such as tank washing station, gas filling station and ship traffic volume observation station, and established the data production standard of Jiangsu inland river electronic channel map. The census work makes full use of information-based means, and all channel census data collection adopts online system to realize the online input and review of census data, and further realize the query, statistical analysis and visualization of channel census results.

3   Application of Spatial Information Technology in Inland Waterway Census

3.1   Application of Spatial Information Technology in Census Technical Scheme

The general survey of inland waterways includes the general situation of waterways, navigation conditions of waterways, hubs, river crossing structures of navigable rivers, riverside structures of navigable rivers, river blocking structures of navigable rivers, channel management institutions and channel space vector data (Fig. 1 and Tables 1, 2 and 3).

Table 1. List of navigable waterway

No.	Name	Name of Start/End spot		Navigation mileage (KM)	Classification		Navigation structure		River-crossing structure	River side structure
		Start	End		Present	Plan	Lock	Lift	Bridge/Cable/Others	Port facilities
1	Han Jiang	Lujia miao	Zhang jiatao	57.6	Three level	second level	1	1	2	15

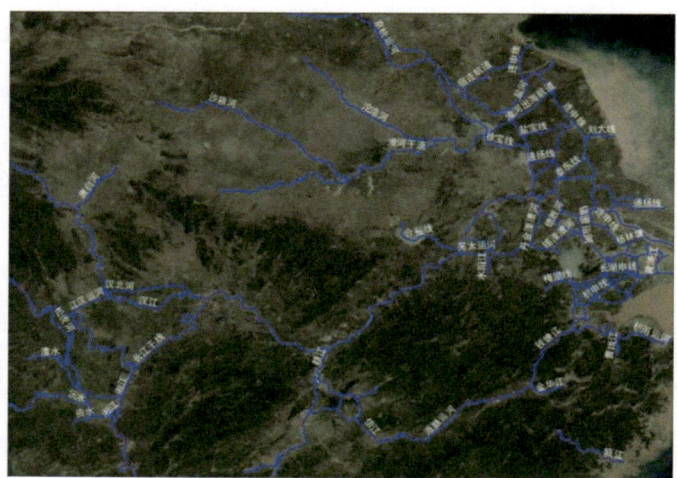

Fig. 1. Waterway space vector data

Table 2. List of navigation lock

No.	Water way name	Nav lock name	Mileage To Estuary (KM)	Completion time	Cap (10000 tons)	Effective dimension (M)	Navigable tonnage (10000 tons)	Chamber structure	Operation
1	Han Jiang	Gu Shan	797	under construction	404	120 × 23 × 3	500	Integral type	–

In the technical scheme of general census, the technical methods of each general census content adopt spatial information technology, especially the technical scheme of general census of spatial vector data of waterway, hub and navigation river flowing through river blocking structures, which has obvious characteristics of spatial information. Relying on spatial information and focusing on internal work, through high-resolution remote sensing image, digital elevation model (DEM), Digital Orthophoto Image (DOM) and digital line drawing (DLG), and using geographic information system software to directly obtain relevant census object information, which fully reflects the advantages of high-resolution remote sensing image in inland waterway resources census. Compared with traditional field survey methods, it can effectively reduce labor intensity and economic cost and improve work efficiency.

Table 3. List of river-crossing bridges

No.	Water way name	Bridge name	Completion time	navigation condition			Present	Plan	Anti-collision facilities
				Net height (M)	Net width (M)	Number of Nav Holes			
1	Yangtze River	BaDong Yangtze River Bridge	2004	30	22	2	One level	One level	Yes

Waterway overview and waterway space vector data adopt the methods of extracting data and field investigation and review. The basic characteristic parameters of inland waterway are analyzed and extracted by using 1:50000 DEM, DLG and DOM data and image data with resolution of 2.0 m and 0.5 m respectively. The main data sources include DEM, DLG and DOM in the 1:50000 national basic geographic information database, high-resolution satellite image data, 2.0 m high-resolution remote sensing image data covering the whole country and 0.5 m high-resolution remote sensing image data covering the national high-grade waterway.

Inland Waterway Survey is based on spatial information technology and organically integrates the relevant parts of RS, GPS and GIS technology to form a powerful technical system, which can realize the rapid, mobile, accurate and reliable collection, processing and updating of various spatial and environmental information.

3.2 Spatial Information Expression of Census Objects

The spatial data acquisition required for this waterway census includes: waterway space vector, hub space vector, navigation river crossing structure space vector, navigation river riverside structure space vector, navigation river blocking structure space vector, mainly collecting its spatial position and morphological characteristics. The spatial expressions of various census objects are shown in Table 4.

As shown in Table 4 above, all census objects are expressed in the form of spatial points, lines or surfaces. At the same time, under the condition of spatial expression of census objects, it is necessary to consider the spatial relationship between census objects and truly express the real relationship of census objects in the information system. The spatial relationship is as follows.

(1) Attachment Association: there are primary and secondary attachment relationships between different objects (the same life cycle). For example, the hub is attached to the river. When the river does not exist, the hub does not exist.

(2) Simple association: there is an association pointing relationship between different objects, such as the relationship between river water body and river sideline. The corresponding sideline can be associated and queried by river water body.

The location-based topological relationship is as follows.

Table 4. Spatial expression of census project

Project	Classification	Type
Navigable river	River system	Line (River Line) Face (River Face)
Waterway	Waterway	Line
Hinge	Navigation-power junction	Line
	Navigation lock	Line
	ship lift	Line
River-crossing structure	Bridge	Line
	Overhead pipeline	Line
	Underwater pipeline/Tunnel	Line
Riverside structure	Drain/Water intake	Point
	Culvert/Floodgate/Pumping Station	Point
River-blocking structure	rubber dam	Line
Hydrological station on navigable river	Hydrological station	Point

(1) Inclusion relationship: in space, one object belongs to a part of another object, such as between an avionics hub and a ship lock or ship lift.

(2) Crossing relationship: between water flows, an object crosses over or under another object to form an interchange. There is a crossing relationship between objects, such as between canals and rivers.

(3) Gland relationship: the boundary of the line and surface coincides on the point falling line and the bank line of the point falling surface. For example, the water intake and sewage outlet fall on the sideline of the river.

(4) Connection relationship: between objects on the same layer, in space, one end of one object is connected with one end of another object, and there is a connection relationship between objects, such as channel and upper and lower channels.

3.3 Application of Spatial Information Technology in the Base Map of General Census

The working base map of waterway census is the basic data guarantee for waterway census. The working base map is applied to the whole working process of waterway census. In the census stage, the role of the working base map is reflected in internal plotting and field census. In the data collection stage, the role of the working base map is reflected in the splicing of cross-border objects to realize the combination of spatial attributes and natural attributes. In the stage of achievement display, the role of working base map is more obvious, and the results of waterway census data will be fully and directly expressed in the form of spatial distribution.

The working base map is a digital composite product composed of national basic scale 1:50000 topographic data, remote sensing digital orthophoto mosaic image data with resolution no less than 2.0 m, administrative division ownership boundary, waterway survey object and other infrastructure.

3.4   Application of Spatial Information Technology in Achievement Display

The results of waterway census data are displayed in space in the form of "one map", that is, all kinds of objects in the results of waterway census are displayed in space, so as to realize the "classification, stratification and classification" display, comprehensive statistics and data management of waterway census data, realize the integrated management of waterway census spatial data and attribute data, and provide strong intelligent analysis and decision-making support for intuitively mastering the current situation of waterway. Using GIS, RS and virtual simulation technology, establish a "one map" management and service platform for waterway census results; On this basis, the functions of geographic information query, statistics and analysis are developed to realize the unified management, integration, exchange and sharing of spatial information of various waterway resources, and realize the "one-stop" Geographic Information Collaborative Service of waterway basic information. The display system realizes the effective integration and integration of channel basic data, application data and thematic data. The system not only includes national basic scale topographic map, administrative division ownership boundary and high-resolution remote sensing image, but also includes waterway overview, waterway navigation, hub, navigation river crossing buildings, navigation River adjacent buildings, navigation river blocking buildings, waterway management institutions and other census object achievement data, as well as their spatial and business relations. Attribute information is associated with spatial objects to realize the two-way query function of GIS and attribute data. The query results can be displayed in the form of hydrograph, histogram, pie chart and so on.

4   Summary and Prospect

In recent years, with the rapid development of economy and society, giving full play to the advantages of large inland shipping capacity, low energy consumption and low pollution, the construction and development of inland shipping in China has achieved remarkable results. In order to implement the requirements of the construction of a transportation power, carrying out a new round of Inland Waterway Census and comprehensively finding out the "family background" of China's Waterway resources has become a necessary basis for effectively promoting the high-quality development of China's inland waterway shipping.

Referring to the experience of resource census in various industries, spatial information technology can efficiently run through the whole workflow of inland waterway census, and will be widely used in the whole process cycle of inventory, census, data collection, achievement display, development and utilization. The application of spatial information technology has changed the traditional working mode of Waterway Census, and has the advantages of reducing labor intensity, accelerating work efficiency and saving economic cost.

At the same time, through the integration and sharing mechanism of waterway census and port census information resources, a standardized, unified and authoritative national water transport industry information platform will be formed, so as to provide reliable water transport industry information support and guarantee for the national economic and social development.

References

Liu Q, Yuan BS (2008) Techniques of GIS, GPS and RS for the development of intelligent transportation. Geomat Inf Sci Wuhan Univ 04:331–336

Ming FH, Li ZM (2017) Application and development of 3S technology in intelligent. J East China Jiaotong Univ 08:57–62

Wang XZ, Zhao LJ, Cao XF (2016) 3S, internet of things technology integration in the application of intelligent transportation. Geomat Spat Inf Technol 07:155–157

Yao HY, Xue TH, Qi Y (2021) Application of BIM technology in port planning under national spatial planning system. Port Waterw Eng 04:147–152

Yalcinkaya M, Singh V (2015) Patterns and trends in building information modeling (BIM) research: a latent semantic analysis. Autom Constr 59:68–80

Hjelseth E, Thiis TK: Use of BIM and GIS to enable climatic adaptations of buildings. ALAINZ. eWork and eBusiness in Architecture, Engineering and Construction

Kim HY (2016) Implementing a sustainable decision-making environment: cases for GIS BIM and big data utilization. J KIBIM 6(3):24–33

Park SH, Kim E (2016) Middleware for translating urban GIS information for building a design society via general BIM tools. J Asian Archit Build Eng 15(3):447–454

AIRIS-PS Project in the Port of Seville and in the Guadalquivir Waterway. Results of the Project Completed

Rodrigo García[1($\boxtimes$)], Angel Pulido[1], Isabel Navarro[2], Xavier Pascual[2], Cas Willems[3], and Alicia Yanes[4]

[1] Autoridad Portuaria de Sevilla, Sevilla, Spain
rgarcia@apsevilla.com
[2] SENER Ingeniería y Sistemas S.A., Barcelona, Spain
isabel.navarro@sener.es
[3] Smart Atlantis, Teteringen, The Netherlands
[4] Serviport Andalucía, Sevilla, Spain
ayanes@serviportandalucia.com

Abstract. The AIRIS-PS project has been a project financed by the European Commission within the funding instrument of the Connecting Europe Facility (CEF) for Transport under the priority of River Information Services. The objective of the priority has been to implement the River Information Services in the Port of Seville and in the Guadalquivir harmonized with the practices applied In the European Inland Waterway Network. River Information Services and Systems have been designed to provide support to improve traffic and transport management and enhance operational procedures. RIS contribute in an essential way to the improved competitiveness of waterborne transport in the supply chain. In this sense, the AIRIS-PS project is considered as a key project for the improvement of the accessibility and capacity of the river access to the Port of Seville. River Information Services lead to the increase of transport to the Port of Seville not only allowing the increase of number of ships calling for Seville but also bigger vessels access.

Keywords: RIS traffic & transport management · Fiware

1 River Information Services in the International Context

In 2005 the European Parliament and the European Council adopted Directive 2005/44/EC, dealing with harmonized River Information Services (RIS) on Inland Waterways of the Community. The definition of RIS, as stated in the RIS Directive (2005/44), is the following: "River information services means that the harmonized information services to support traffic and transport management in inland navigation, including, wherever technically feasible, interfaces with other transport modes".

A challenge in the port of Seville and the Guadalquivir is that RIS is implemented in an inland river with maritime traffic and transport. In this environment apart from the EU directive on River Information Services also maritime directives on monitoring and reporting of maritime vessels must be taken in account.

© The Author(s) 2023
Y. Li et al. (Eds.): PIANC 2022, LNCE 264, pp. 789–802, 2023.
https://doi.org/10.1007/978-981-19-6138-0_69

The RIS Guidelines are published by the European Commission as Commission Regulation n°. 2007/414/EC and contain guidelines on the development, implementation, and operation of River Information Services. The Guidelines and recommendations for river information services have been used in the AIRIS-PS project as guidance for the development of services and technologies for the Port of Seville and the Guadalquivir. The RIS guidelines define also the objectives of RIS and the information needs based on the objectives of the stakeholders. The RIS guidelines further provide an overview of the services as the essential core-principal of RIS and as given in the Fig. 1.

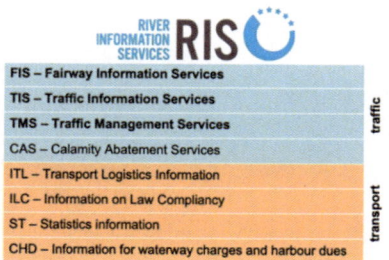

Fig. 1. River information services for transportation and traffic control

The RIS Guidelines provide a functional decomposition of RIS services into RIS information. RIS key technologies and RIS systems/applications are relevant for understanding the RIS concept and the implementation of the key technologies and systems and application in Seville and the Guadalquivir.

2   Implementation Strategy of RIS in the Seville

The study on applicability of RIS key technologies and RIS systems and applications in the port of Seville and the Guadalquivir was performed as a best practices paper-study. The structured approach of the AIRIS-PS implementation plan was based on the decomposition of the RIS services into RIS technologies and systems as specified in the RIS Guidelines.

The structured implementation of RIS contributes to the strategic target for safe, efficient in the Guadalquivir and the port and competitive transport through the Port of Seville by:

- Better understanding of the waterway: i.e., accurate bathymetry, accurate model of tidal, accurate measurement of the height of the water surface and waves, etc.,
- Better traffic information on the waterway and in the port: vessel characteristics - size and draft -, cargo characteristics, etc.,
- Better traffic management in the Guadalquivir Euroway: optimization of human (e.g., navigation pilots) and material resources, of anchorage areas, optimization of convoys entering and leaving the Euroway, etc.,

- Better management of emergencies,
- Better coordination of port operations, etc.
- Better transport information in a multimodal environment – maritime and inland waterway transport, rail and road transport.

The implementation plan takes in account the actual and predicted:

- Traffic intensity and complexity,
- The geographic infrastructural conditions and
- The transport performance and transport requirements.

3 River Information Services Implementation

The functional systems architecture as the basis for the AIRIS-PS implementation plan and implementation strategy was defined and three RIS pilots were defined, implemented and deployed in the Port of Seville and in the Guadalquivir Euroway, in order to demonstrate the feasibility of the solutions adopted in AIRIS-PS for RIS implementation:

- FIS - Fairway Information Services
- TIS – Traffic Information Services
- TMS – Traffic Management Services.

3.1   FIS-Fairway Information Services

3.1.1   Fiware IoT Platform

The following figure depicts the whole architecture of the AIRIS project where the different RIS services are interconnected and fully integrated through an advanced IoT Platform for service integration. This platform provides also access to IoT network deployed along the river and other external sources relevant to the domain (Fig. 2).

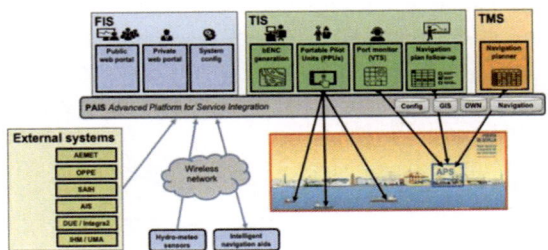

Fig. 2. Architecture of the AIRIS project

The Fiware IoT Platform constitutes the core of the project since it is the component that allows the integration of the different subsystems that are part of the functional architecture: sensor and actuator networks, the FIS web portal, TIS & TMS pilots. The platform provides a shared database for all RIS services that rely on it as a fundamental structural component. The platform has been developed with open-source technologies based on Fiware technical components The following figure depicts the components of the Fiware platform (Fig. 3):

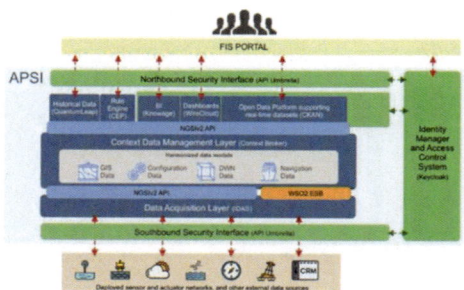

Fig. 3. Logical architecture of the FIS system solution

3.1.2 Meteorological and Hydrological Sensorization and Intelligent Maritime Signaling

The monitoring of the hydrological conditions of the channel allows obtaining information on water quality, turbidity, PH as well as detecting the presence of chemical and / or bacteriological discharges. In addition to water quality parameters, the hydrological monitoring of the channel also covers the measurement of quantities whose values are useful, such as real-time water level measurement, alteration of the sea surface or waves, or the speed and direction of flows of marine currents. On the other hand, access to real-time measurements or predictions of meteorological data allows to provide information on visibility, adverse weather conditions and other relevant data of interest in river navigation.

The signaling system allows creating warnings and alarms in cases of poor visibility on the waterway or dangerous magnitude and direction of the water current or flooding. The signaling luminous intensity does depend on the proximity of a ship and the luminosity existing in the environment. It also incorporates a siren that reinforces the functional operation by adding an alarm in the most critical case. This system was implemented in a zone where the change of currents between calm waters by the exit of the lock and the normal current of the river downstream, makes this environment a critical point in the navigation of the ships.

3.1.3 FIS Web Portal

The FIS Portal is a portal to provide in an integrated way the Fairway Information Services of the port of Seville and the Guadalquivir and the geography of the navigation area and their updates. The FIS Portal provides the stakeholders in a structured way information on:

- Static port and waterway information as with the main functions proposed in the RIS guidelines like:
  - Navigation aids and traffic signs

- • Status of the Guadalquivir, the lock in the port area
- • Physical limitations of the Guadalquivir and the lock
- • Lock operating times and opening hours of port facilities
- • Navigational rules and regulations
- (Semi) dynamic port and waterway information in accordance with the RIS guidelines like:
  - • Water depths in the Guadalquivir and the port
  - • Actual hydrological and meteorological information
  - • Water levels at gauges, present water level info for a certain stretch
  - • Actual and predicted restrictions caused by "flood"
- As Notices to Mariners the following information is provided:
  - • Actual and long-time term obstructions in the fairway
  - • Malfunctions of aids to navigation.

This Web Portal consists in an accessible webpage from the official website of the APS that allows the centralized publication of all the relevant FIS information of the channel so that all interested parties can easily consult public information. In addition, after authenticating and logging into the page, the user (portal administrator) can generate FIS information to publish it on the portal and access the user interface that interacts with the service platform in order to perform certain actions (Fig. 4).

Fig. 4. FIS web portal

3.1.4 Port Information Guide

The Port Information Guide is a detailed document available in the FIS Web Portal for all the users to provide them, in a harmonized way, reliable and accurate information about the Guadalquivir and the Port of Seville, ensuring that all information related to navigation, port and freight transport management is concentrated and off-line available.

3.2 TIS – Traffic Information Services

3.2.1 Vessel Traffic Monitoring

The Vessel Traffic Monitoring or 'Port Monitor', is a particularized VTS-system for the Port of Seville and the Guadalquivir Euroway. It provides the operator a complete overview on a real-time tactical traffic image of the operation area of the Seville Port

Authority on top of a Bathymetric electronic nautical chart (bENC) and allows displaying AIS information of the vessels inside the VTS area.

The Port Monitor has been configured to visualize the waterway as a whole and as different key sections. For AIRIS project, a specific screen layout has been chosen. In the upper part, the entire stretch of the Guadalquivir is displayed. In the lower part, two detail views are placed. The area displayed in these views can be switch to different preconfigured sections of the river by pressing hotkeys. The figure below shows an area near the estuary (left side) and the port of Seville (right side) (Fig. 5).

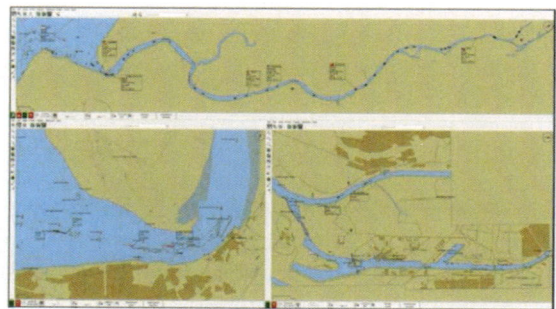

Fig. 5. Port Monitor's screen display with Guadalquivir overview map and detail views

The Port Monitor enables the operator to have special features for alarming and signaling of specific vessels within predefined rules. For example, in case of the local regulation states a maximum speed in an area, the Port Monitor can signal or even alarm at higher speed. Different types of alerting and signaling features are available, like speed and illegal anchoring alerts. The port monitor will also facilitate the operator with measuring tools like: CPA and TCPA (Time to Closest Point of Approach) between vessels and between vessel and shore, and SoG and CoG (speed and course over ground).

The Port monitor tool has been implemented in AIRIS project in two ways: as an internal application (inDTS), a fat client running on a dedicated workstation that will be used by the lock operators of the Port of Seville, and as a web application (inDTSweb) used for external/remote access to the traffic image. In addition, inDTSweb features two additional modules also developed in the AIRIS project and described below: Trip Planner and Trip Monitor.

3.2.2   Voyage Plan Tracking Application

The Voyage plan tracking application or 'Trip Monitor' shares the same graphical user interface as the 'Voyage Planner' (described below) but adds the AIS information in 2D representation to compare the planned voyage created with the 'Trip Planner', and the real one in order to see possible deviations or delays in the entry to the lock of Seville.

For each sailing plan the places or stretches with risk should be minimized to an acceptable level in order to get the agreed sailing plan. Based on an agreed sailing plan the pilot/vessel can execute the voyage. The agreed voyage plan is available for presentation on the pilot's PPU and the Port Monitor graphically in a time-way-diagram that represents the passage of the vessels through the Guadalquivir.

During the trip the AIS position of the vessels is represented over this time-way-diagram of the voyage plan in order to provide the pilot on the PPU-screen and the operator on the port Monitor Screen the facility to follow the progress and the deviations of the sailing plan.

The user can monitor that the course and duration of the actual itinerary carried out by the vessel correspond to the planned one. In this way, if needed measures can be taken on the basis of these deviation.

3.2.3   Bathymetric Chart Generation System

The Bathymetric chart generation system developed is based in a software application that enables the Port Authority to generate a bathymetric ENC, upload it to a cloud-based module to which different systems (PPUs and the VTS) can access to download the bENC.

The process of generating the bathymetric chart takes as input a standard nautical chart and the most recent bathymetry file generated by the port authority. The system is able to integrate the information content in a standardized nautical chart from these two data sources (Fig. 6).

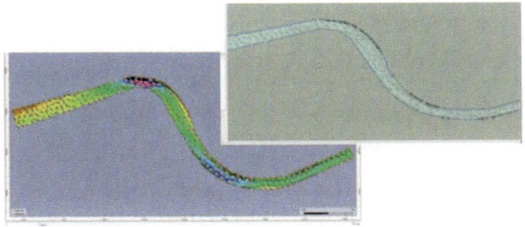

Fig. 6. Generation of bathymetric ENC's

The generated bathymetric chart complies with the international standards of nautical charts, so it is compatible with the navigation software installed in the navigation aid devices (Portable Pilot Units), that pilots use during river navigation and, as it was mentioned before, with the Port Monitor that shows the traffic image in real time of the Guadalquivir over the bathymetric nautical chart generated and available in the system.

3.2.4   Portable Pilot Units (PPUs)

AIRIS project has included the acquisition of navigation aid devices that pilots will use during river navigation. This Portable Pilot Units (PPUs), provided with the navigation software and the AIRIS tools developed, allow:

- To improve the safety and efficiency of waterborne traffic.

- To improve the fairway usability and accessibility for vessels with dimensions that are going to reach the minimum limits of the fairway by collecting, processing, and employing dynamic parameters of the vessel and the environment.
- Integration of the safety and efficiency related information exchange between the Port Monitor and the pilots.
- The preparation and exchange of the voyage plan with the ship master, the bridge team and port authorities, to reach a complete and transparent overview of all factors regarding the safety and efficiency of the intended passage.

The PPUs can be operated in:

- Information mode, where the PPU is using the pilot plug as information source for the navigation information. In this mode of use the traffic image presented on the PPU is for creating awareness of the traffic and navigation situation.
- Navigation Mode, where the PPU is connected to high precision positioning information and the pilot can use the PPU as a high qualified navigation and information system.

The PPU provides the pilot on board a navigation assistance tool during the work of piloting, docking and undocking the ship. During navigation tasks, the help consists of representing dynamic information of the navigable channel, of the vessel itself and of traffic around it through a traffic image overlaying the bENC. In the maneuvers of berthing and undocking of the ship, the PPU provides a medium-term view of the position of the vessel after starting the maneuver, allowing the pilot and the captain to take corrective measures in case of danger.

The PPU is also able to present the voyage planed for the individual ship allowing the pilot to compare the predict voyage with the actual one in order to allow the pilot to make decisions so that the expected plan is followed. The PPUs have installed a specific module used to provide the Trip Monitor to the pilots. During their voyage on the Guadalquivir, the Trip Monitor is disconnected from the VTS and receives AIS and GPS data from the PPU sensors, and allows to get monitor, measuring and alerting tools like already mentioned for the Port Monitor, as well as the logging of the complete traffic image and related information of a complete pilots-voyage (Fig. 7).

Fig. 7. Portable pilot units

3.3   TMS – Traffic Management Services Included

3.3.1   Voyage Planner with Tidal Window Prediction

In the Guadalquivir the period in which a vessel with a large draft can sail inward or outward bound is restricted by the tide. The tidal window for a vessel depends on the height of the tide and the ship's draft. Extending the tidal window or predicting in a more precise way increases the navigational safety of the Guadalquivir and the port of Seville more attractive for deep draft vessels or for vessels to carry more cargo. In addition, the avoidance of ship encounters or passings for inward bound and outward-bound vessels in critical areas of the river will improve the safety and efficiency of traffic in the Guadalquivir and thus the image of the port of Seville.

These two topics – the accurate prediction of a tidal window for vessels entering and leaving the port of Seville as well as avoiding encounters of vessels in the Guadalquivir – were the principal objective for the development and implementation of the voyage planner (or 'Trip Planner'). This software tool includes the tidal window planning for the development of sailing plans for the visiting vessels from sea to quay vice versa.

Principal stakeholders for this voyage planner are of course the vessels visiting the port of Seville, but the principal users of the voyage planner are going to be the pilots and the port authority APS. APS is the authority authorized to give permission to visiting vessels when they comply with the rules for entering or leaving the port.

The voyage planner takes in account:

- Fairway characteristics and information like:
  - Maximum speed allowed in sections of the fairway.
  - Sections where encounters of vessels are not allowed or not advisable.
  - Measured depths and thresholds of fairway.
  - Anchoring restrictions.
- Environmental conditions like:
  - Actual and predicted meteo information.
  - Actual and predicted tidal information.
- Ship characteristics like:
  - Ship details.
  - Place of destination/departure.
  - ETA for arriving. Vessels, ETD for departing vessels.
  - Actual draught.

With respect to the tidal information the University of Malaga has performed a study on the hydrographical model for the Guadalquivir. The results of this study have been used as input on the tidal prediction information on different positions in the Guadalquivir. The voyage planner calculates in an iterative process the sailing plan of each vessel that is heading for the port of Seville or is planned to departure from the port of Seville based on above mentioned information and characteristics. The resulting sailing plan of a vessel in addition complies with:

- Safety margin between 2 consecutive vessels.
- Required time of arrival at the lock for each vessel.
- Vessel passing and encounter restrictions.

As depicted below, the right part of the graphical user interface corresponds to the menu to configure the vessels available in a certain time whilst the left window provides the graphic image of the vessels entering/departing the port, the crossing areas and the tidal information represented in blue. The green vertical squares indicate areas where the crossing of vessels is safe (Fig. 8).

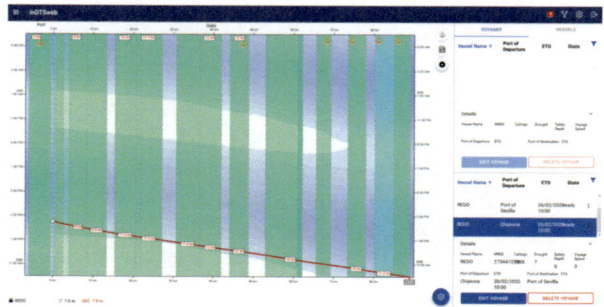

Fig. 8. Trip Planner graphical user interface

The Voyage Planner is a web-based application which enables the APS to enhance the planification of the vessels considering the best timeslot of the tidal conditions in the Guadalquivir estuary. The combination of the sailing plans of all vessels is presented in an overall voyage plan time-way-diagram that results in an integral and optimized voyage planification of all vessels that will sail through the Guadalquivir inward or outward bound of the Port area.

3.3.2   Digital Waterway Network

In special for the Voyage planner there is a need for a digitized waterway network. The Digital Waterway Network (DWN) is a set of logical objects that contain the topological, bathymetric, topographic, functional and operational information of the waterway (waterway). It consists of a logical abstraction of the waterway by fragmenting it into a set of finite segments joining certain reference nodes.

The segmentation of the river should be based on the characterizing conditions of a certain part of the river. So called nodes are connecting segments of the DWN, the nodes will be located in those locations in which there is a change in any of the features or attributes of the waterway. For the development of the DWN the RIS index with the ISRS code have been used as coding principle. By doing so the experience with the RIS index in the rest of Europe can be used and it results in a practical usable and consistent coding format.

For the voyage planner the segments characterize the topology of the waterway with respect to:

- Topography of the estuary.
- Areas with specific depth of the waterway.
- Areas with specific width of the waterway (e.g., area's where encounters are forbidden).

- Areas with speed restrictions.
- Location of specific infrastructure influencing navigation (Lock, Quays, and docks).
- Anchoring areas, etc.

4 Strategic View of the AIRIS Project's Results

The AIRIS Project is of high technological value and of a high impact on the port activities of the port of Seville and has allowed bringing the first steps to digitization of the Guadalquivir Euroway in order to monitor the entire estuary with sensors to obtain real time information about the state of the river, monitor and manage the vessel's traffic along the river. The substantial improvement of the services provided in the Guadalquivir Euroway through the RIS services implemented makes it possible to improve the traffic information and the management of the traffic of ships that use the Guadalquivir River as a waterway with origin or destination Seville. Applied technologies, such as that of the hydrological information service that can predict the height of the tide and the amount of water available throughout the route, jointly with traffic monitoring and travel's planification tools, allow the entry of ships with greater drafts and/or dimensions so that can perform their maneuvers safely. Also, another tool that helps to better understand the waterway is the one that integrates information on the bathymetries, giving exact results that allow a better performance in the dredging of the river or that can facilitate the ideal location to allow the passage of two or more ships simultaneously without putting the ship itself or the congestion that may arise in the Euroway itself at risk. All this become in a tool in decision-making and port management that favors the correct maneuverability of ships along the river and the prediction and correct management of hydrological and climatic conditions presented by the Euroway.

The implementation of this Project also supposes a reduction in the costs derived from the transport, since, by allowing the passage through the Euroway del Guadalquivir of ships of greater dimensions and drafts, it is allowing the passage to a greater capacity of movement of the load because said ships will have greater transport capacities, which means a reduction in the cost of transporting the merchandise.

The impact that this Project generates is high, as it allows improving coordination between different actors in the port community on the exchange of data and generating direct impact on the end customer of the port. Apart from these good results obtained, the AIRIS Project is an excellent example for Europe as the implementation of the intelligent RIS services and technologies will bring the port of Seville in the coming years to a higher performance level through:

- Optimize port resources (pilots, stevedores, moorings, tugboats).
- Improve under-keel-clearance predictions and real-time monitoring, the prediction of tides and windows, including some additional phenomena such as meteorological tides (storms and spring tides), river discharges, etc.
- Improve the identification and detection of the position of vessels in critical sections of the river.
- Improve ETA predictions by integrating with eNavigation developments.

- Improve cross-planning and real-time control of operations (Synchronization) between the water side (ships) and the land side (rail and road).
- Optimizing the speeds allowed in the waterway, as well as redefining the limitations for the crossing of vessels in the sections of the waterway and improve real-time control of joint operations (ship-train; ship-road).
- Improve document transfer by digitizing port processes. The FIS platform has a great capacity to interoperate both with systems - internal and external - as well as with physical devices.
- Optimize billing processes as the FIS platform allows the improvement of the quality of data mainly related to ship fees and, in the future, to the pricing of port services.

The strategic objective of the Port of Seville is to become the reference port in southern Spain for the management of intermodal traffic, mainly between the Canary Islands and the interior of the peninsula. This objective is part of a strategy aimed at guaranteeing sustainable growth to make the Port of Seville a key player in increasing economic activity and generating high-quality jobs in its immediate surroundings, especially in Seville and its area of influence. The application of RIS services allows to satisfy the strategic objectives that the port of Seville has proposed:

- Increase in domestic profits: this will be achieved through the AIRIS-PS project thanks to cost savings in transport, derived mainly from an improvement in relation to the ship (larger dimensions, load capacity), an improvement in relation to the state of the inland waterway (state of bathymetry, management of the passage of ships, greater control of incoming and outgoing ships), an improvement in the optimization of the navigation pilots (practical).
- Improvement in the management of emergencies (number of accidents, response time and resolution of the accident).
- Increase in social benefits, by reducing externalities on other types of transport and reinforcing its role as an accelerator of innovation in the region, being able to offer other value-added services such as: streamlining intermodality through the synchronization between modes of transport (reduction of waiting times for ships, merchandise on esplanades, etc.), and greater traceability of operations and services related to the ship and the merchandise (ETA/ATA, ETD/ATD, Improvement in the stay times of ships in the navigation channel and port, quality levels of basic port services, etc.).
- Increase in the capacity and competitiveness of the companies that operate in the port.

5   Conclusions

The substantial improvement of the services provided in the Eurovía del Guadalquivir with the RIS services implemented makes it possible to improve the management of the traffic of ships that use the Guadalquivir River as a waterway with origin or destination Seville. This means that certain applied technologies, such as that of the hydrological information service, can predict the height of the tide and the amount of water available throughout the route, allowing the entry of ships with greater drafts and / or dimensions so that can perform their maneuvers safely. Another tool that helps to better understand the waterway is the one that integrates information on the bathymetries, giving exact results that allow a better performance in the dredging of the Eurovía or that can facilitate the ideal location to allow the passage of two or more ships simultaneously without putting the ship itself or the congestion that may arise in the Eurovía itself at risk.

All this supposes a reduction in the costs derived from the transport, since, by allowing the passage through the Eurovía del Guadalquivir of ships of greater dimensions and drafts, it is allowing the passage to a greater capacity of movement of the load because said ships will have greater transport capacities, which means a reduction in the cost of transporting the merchandise.

In the future, services may be implemented that allow the ship operator to calculate optimal arrival times at the port of Seville, according to the conditions of the tide that will allow them to have a reduction of emissions in fuel consumption, or, from the point of view of accidents at sea, incidents can be detected and communicated more easily in order to respond as quickly as possible.

The implementation of the different RIS services generates and is expected to have a direct impact on the port management of the port of Seville. This impact is reflected at the European level, positioning Seville as a port that digitizes the entrance and exit channel of its port through sensors and technological tools that help to optimally manage the passage of ships from or to the sea, to the city of Seville, through its digital twin integrated into the FIS platform.

Based on the evaluation study it can be deducted that a great advance has been achieved in the digitization of the Guadalquivir Eurovía, from the physical point of view and applied in port management through the RIS services.

In the medium term, it is recommended to integrate information and services related to operations in the ship-port interface and towards intermodality, integrating data and processes related to other initiatives that the APS has currently underway or has planned in the near future.

Related with proposals for the evolution towards integrated management, the Port of Seville aims to evolve towards a benchmark multimodal node in the European transport network and become the engine of the main logistics and industrial cluster in the south of the Iberian Peninsula, as a valuable asset for Seville and a promoter of the Guadalquivir estuary.

To sum up, the RIS services generates and has a direct impact on the port management of the port of Seville. This impact is reflected at the European level, positioning Seville as a port that digitizes the entrance and exit channel of its port through technological tools that help to optimally manage the passage of ships from or to the sea, to the city of Seville. As the next step, after AIRIS I, APS is also involved in the AIRIS II Project, which addresses the implementation of RIS services associated with transport and intermodality management, through the development of real-time control and planning tools in order to improve the management of synchronization in the transport chain, considering the different modes of transport (rail, maritime and road trucks).

Acknowledgements. The AIRIS project has been co-financed with the Connecting Europe Facility (CEF) mechanism, which is a key EU funding instrument to support the development of high performing, sustainable and efficiently interconnected trans-European networks.

The authors acknowledge Portel Logistic Technologies S.A.U. and Electronic Engineering Department of the Seville University for their contribution and active participation in AIRIS.

References

Commission Regulation (EC) No 414/2007 (2007) Concerning the technical guidelines for the planning, implementation and operational use of river information services (RIS) referred to in Article 5 of Directive 2005/44/EC of the European Parliament and of the Council on harmonised river information services (RIS) on inland waterways in the Community. Official Journal of the European Union

Directive 2005/44/EC of the European parliament and of the council (2005) On harmonised river information services (RIS) on inland waterways in the Community. Official Journal of the European Union

Inland Transport Committee (2012) Guidelines and recommendations for river information services, revision 1. United Nations Economic Commission for Europe. https://unece.org/DAM/trans/doc/2012/sc3wp3/ECE-TRANS-SC3-165-Rev1e.pdf

Permanent Working Group 125 of the World Association for Waterborne Transport Infrastructure PIANC (2011) Guidelines and recommendations for river information services. Edition 3.0

Aids to Navigation Improvement to Optimize Ship Navigation

Raúl Redondo[✉], Raúl Atienza , Lourdes Pecharroman,
and Leandro Pires

Siport21, Madrid, Spain
{raul.redondo,raul.atienza,lourdes.pecharroman,
leandro.pires}@siport21.es

Abstract. Aids to navigation are established to assist vessels in the navigation with special importance in narrow channels, rivers, or areas with special dangers to navigation. Whether the design of the aids to navigation (AtoN) system is optimum depends on ship positioning, which is affected by several factors: waterway geometry, metocean conditions, design ship, pilotage, and the AtoN themselves, being a crucial part of the design of a waterway.

Optimizing the type of signals and the distance between them allows, not only for safer navigation, but to expand the limits of the waterway by assisting larger vessels to move safely along the waterway, as well as to minimize dredging costs and dredging area by optimizing simultaneously the design of both: the waterway and the Aids to Navigation. An equilibrium is required between the investment and maintenance costs, and therefore, this combined assessment and optimization allows to finally get both, technically and economically feasible projects without detriments on the safety of the waterway.

The use of fast time manoeuvring models and Real Time Simulations allows to check and verify the optimization and improvements in the waterway and in the aids to navigation system. Real Time Simulators and manoeuvring sessions are the perfect place for all stakeholders (Port Authorities, Pilots, Harbour Masters, waterway designers, shipowners, …) to interact, discuss, evaluate, and agree on the best option for the improvement of navigation conditions while preserving the safety of the operations, as it allows to verify the effectiveness of the improvements before implementing them in the reality.

Keywords: AtoN · Real time manoeuvre simulation · Waterway design · Dredging optimization · Navigation safety

1 Introduction

AtoN provide information on ships positioning along the waterway, increasing the safety of the navigation by showing the safe area for ships to navigate by using visual or instrumental references, either physical or virtual, PPUs… Nevertheless, the availability of AtoN implies a certain error positioning of the ship in a waterway, which varies (increases or decreases) depending on the type of aid, the local conditions (mainly related to visibility such as rain or fog), and the distance between consecutive signals.

Y. Li et al. (Eds.): PIANC 2022, LNCE 264, pp. 803–813, 2023.
https://doi.org/10.1007/978-981-19-6138-0_70

It is possible to improve the AtoN to optimize the navigation. This is done by minimizing the positioning error, allowing an increase in ship dimensions while keeping the fairway width constant, always maintaining the safety of the manoeuvres. Nevertheless, it is important to understand that there is a point in which by increasing the AtoN, the effective area of ships is not further reduced. An optimum design point is to be found, or at least, an adequate one.

Two examples of the optimization or improvement of the Aids to Navigation to allow the access of larger vessels or reduce dredging costs are presented. For the assessment of the optimization in navigation conditions by improving aids to navigation, both Fast Time manoeuvring models and Real Time Manoeuvring Simulators have been used, as those are the most adequate tools to analyse in detail manoeuvring conditions of ships for waterways design.

The importance of assessing together the waterway design with the Aids to Navigation, and their optimization to reduce dredging costs or increase ship sizes on already existing areas, without reducing the safety of the maritime operations, is presented. The use of fast time manoeuvring models and Real Time Simulations, to check and verify the optimization and improvements in the waterway and in the navigation conditions, is of essence in these assessments as it allows to verify the effectiveness of the modifications before implementing them in reality. Real Time Simulators and manoeuvring sessions are the perfect place for all stakeholders (Port Authorities, Pilots, Harbour Masters, waterway designers, shipowners, …) to interact, discuss, evaluate, and agree on the best option for the improvement of navigation conditions while preserving the safety of the operations.

2   Aton Positioning Errors

All available guidelines and recommendations for waterway design, national or international, consider that aids to navigation, which are placed for the safety of the manoeuvres by allowing ships to position themselves along waterways, produce a certain type of positioning error on the ships, and therefore this is to be considered for the waterway design.

AtoN might be fixed or floating. And this aspect is to be considered as well. Fixed systems have no error on their position, but floating buoys are subjected to external forces which swing the buoy at anchor, introducing a position error of the buoy itself.

For the purpose of this paper, the focus will be only on the positioning error that AtoN introduce on the ship navigation due to the type and number of the signals, but not on the intrinsic error of the buoys.

Positioning errors increase as the distance between buoys increases, and at reduced distances as reference, the next pair of buoys is to be considered, as some 200 m before reaching a pair of buoys, the Captain or Pilot is already considering the following pair as a reference. The following image shows how a pair of buoys concentrates the tracks of ships, reducing the required width to almost half of the width from previous buoy, which is only on one side of the track (Fig. 1).

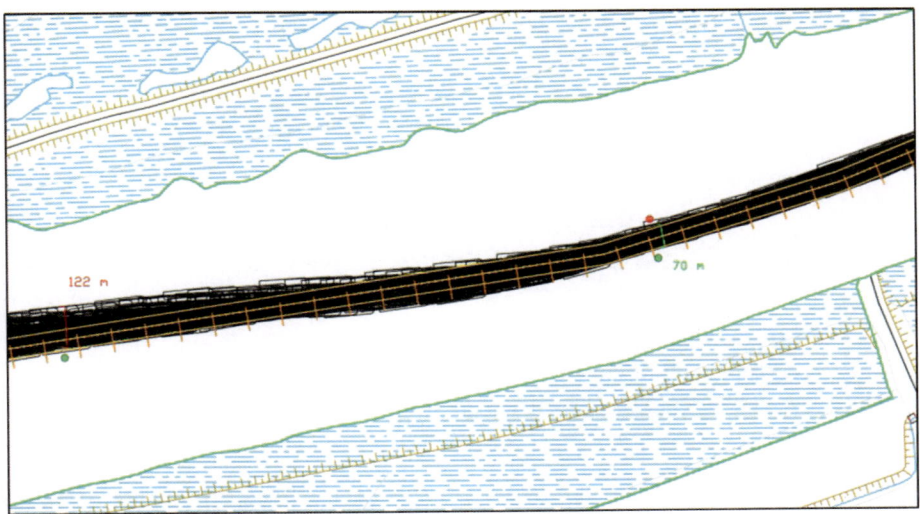

Fig. 1. Ships tracks along a waterway. Reduced dispersion at the pair of buoys

National and international guidelines and recommendations consider these positioning errors on the fairway design. The following tables and figures show different references that can be applied for waterway design, with distinct levels of detail (Tables 1, 2 and Fig. 2):

Table 1. Additional width due to positioning errors. PIANC 121 «Harbour Approach Channels Design»

Excellent	Good	Moderate
0.0 B of additional width	**0.2 B of additional width**	**0.4 B of additional width**
Channel: Paired lighted buoys with radar reflectors Lighted leading lines VTS, where applicable	**Channel:** Paired lighted buoys with radar reflectors Lighted leading lines	Anything less than mentioned on Excellent and Good
Availability: Pilots DGPS ECDIS	**Availability:** Pilots DGPS	

Table 2. Additional width due to positioning errors. ROM3.1-99 (Spanish national recommendation)

Type of Positioning	No experienced pilot/captain	Experienced pilot/captain
Visual Positioning		
Open estuaries, without navigation marking	100 m	50 m
Buoys or beacons in approach ways	50 m	25 m
Visual positioning between buoy or beacon alignments marking fairway limits	20 m	10 m
Leading lines	0.5°	0.5°
Radio electric systems (valid for locating on a nautical chart with no visual positioning)		
• Radio beacons	5.0°	5.0°
• Radar (aboard), S Band	1.5°	1.5°
• Radar (aboard), X Band	1.0°	1.0°
• RACON (distance/delay)	150 m/0.3°	150 m/0.3°
• TRANSIT Dual Frequency	25 m	25 m
GPS	100 m	100 m
DGPS	10 m	10 m

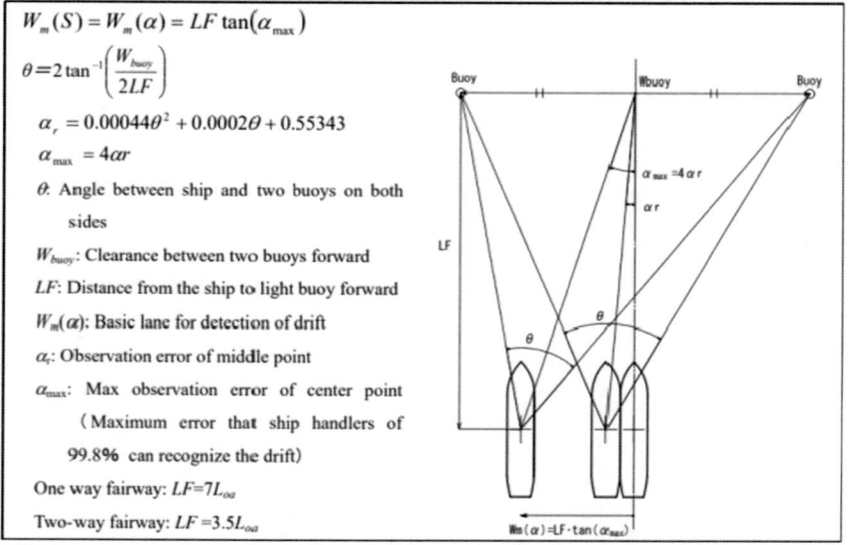

$$W_m(S) = W_m(\alpha) = LF \tan(\alpha_{max})$$

$$\theta = 2\tan^{-1}\left(\frac{W_{buoy}}{2LF}\right)$$

$$\alpha_r = 0.000044\theta^2 + 0.00002\theta + 0.55343$$

$$\alpha_{max} = 4\alpha r$$

θ: Angle between ship and two buoys on both sides

W_{buoy}: Clearance between two buoys forward

LF: Distance from the ship to light buoy forward

$W_m(\alpha)$: Basic lane for detection of drift

α_r: Observation error of middle point

α_{max}: Max observation error of center point (Maximum error that ship handlers of 99.8% can recognize the drift)

One way fairway: $LF = 7L_{oa}$

Two-way fairway: $LF = 3.5L_{oa}$

Fig. 2. Positioning errors derived from a pair of buoys. "Design standard for fairway in next generation"

When several positioning references are available in the same area, such a pair of buoys and leading lines, the one which provides the minimum error in each section is to be considered. Based on the recommendations on the calculation of positioning errors of the above references, the following errors are obtained for a waterway (Fig. 3):

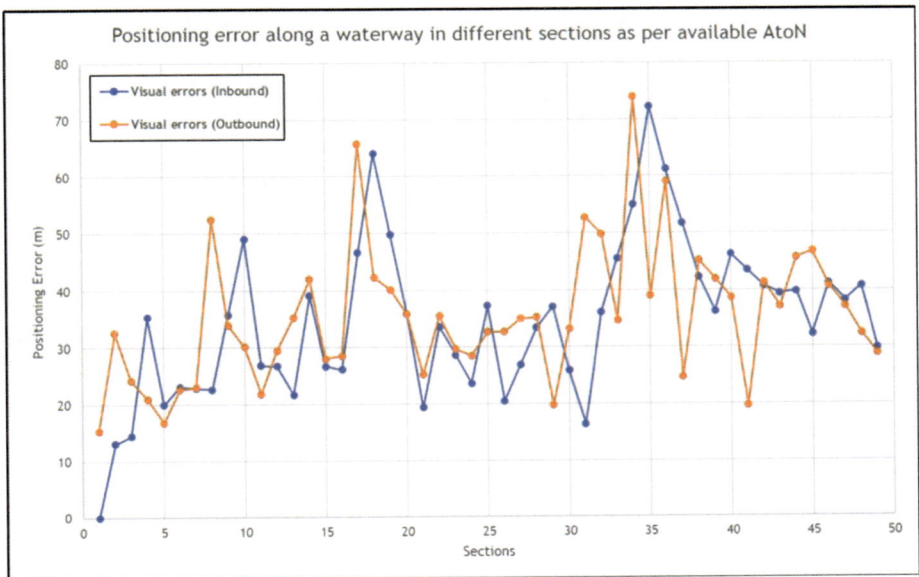

Fig. 3. Visual errors for different sections of a waterway

The area occupied by the ships in a waterway does not depend only on the AtoN, but other factors, such as wind, current, UKC, … also affect the required width. Conclusions are therefore derived based on all factors together. To evaluate the importance of AtoN visual error on the global width occupied by the ships the following image is presented, where this relation for both straight and curved section is shown (Fig. 4):

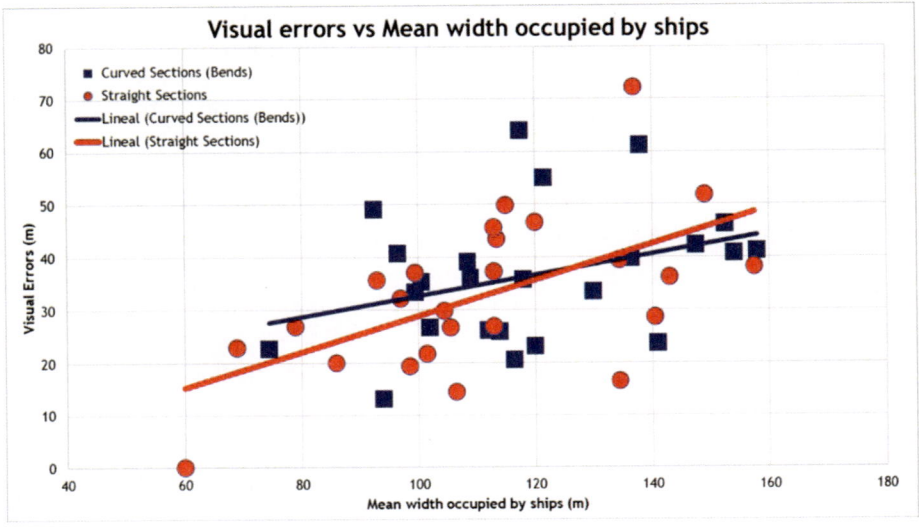

Fig. 4. Visual errors dispersion vs Mean occupied width

This graph shows a stronger relation for straight sections than for bends, which makes sense, as positioning the ship along a bend, and therefore the occupied area, is strongly linked to the response time and the human factor of the Pilot or Captain, and not the visual positioning when in a transitory phase.

3   Optimization by Fast Time Manoeuvre Simulations

Fast Time Manoeuvre Simulators allow to perform a large number of manoeuvres in a short time combining many different parameters. A methodology to evaluate AtoN improvements in large areas of waterways is established by means of performing randomized manoeuvre simulations with autopilot models combined with the statistical determination of the safe manoeuvring area as per PIANC 121.

Based on preliminary assessments, and as described in Sect. 2, the positioning error along a waterway due to the AtoN can be estimated.

With these positioning errors Fast Time Manoeuvre simulations are performed in a randomized way, in which the waypoints of the track-keeping autopilot are randomly varied, linked to a mean and standard deviation of the positioning errors in each stretch of the waterway.

Considering AtoN improvements, in the most critical areas identified, the positioning error can be reduced and therefore the randomized manoeuvres are repeated by reducing the dispersion in waypoints.

The comparison of the statistically determined safe navigable areas with and without AtoN improvements allows to preliminarily evaluate whether larger ships can be accommodated in already existing areas, or if an optimization of the required dredging areas is possible to allow safe navigation in narrow areas by increasing the positioning performance of the ships by improved AtoN.

Fast Time Manoeuvring Models allow to define the most critical areas for navigation, with larger potential for improvement in long waterways or rivers. These critical areas can be later evaluated in detail using Real Time Manoeuvre Simulators, where AtoN improvements are incorporated. Pilots and Captains performing the manoeuvres introduce the human factor, which is the missing factor when using Fast Time autopilot models.

4   Optimization by Using Real Time Manoeuvre Simulators

Real Time Manoeuvre Simulators are the best tool to assess the improvements of the AtoN. They allow to determine whether the modifications allow larger vessels to safely navigate through the fairway, as well as to verify the potential dredging optimizations due to an improvement of the AtoN.

Real Time Simulators involve the participation of expert Captains and Pilots which perform the manoeuvres in a virtual environment including all features and aids required for navigation, allowing therefore Captains and Pilots to evaluate AtoN improvement before implementing modifications. Most important, incorporating the human factor (perception and decision making) absent in the Fast Time Manoeuvring Models.

The following images show a narrow channel in the Real Time Simulator with the AtoN that allow large ships to be safely handled (Figs. 5, 6 and 7):

Fig. 5. Real Time Simulation view of an approach to a narrow channel, leading lights, and buoys

Fig. 6. Real Time Simulation visual image of a bulkcarrier approaching during night-time

Fig. 7. Real Time Simulation view of the departure of a bulkcarrier with the AtoN that mark channel limits

The next example shows a study conducted in phases to determine both the feasibility and the optimized dredging requirements to allow New-Panamax container vessels in an existing terminal, in a fairway initially designed for smaller ships. In this project a statistical assessment of manoeuvres based on PIANC WG121 report was performed to finally determine the feasibility of the manoeuvre strategy and the dredging requirements in combination with the AtoN design. Local conditions in the area are mild, which allows for a larger optimization of the waterway in combination with the AtoN. Initially the project consisted of the access of New-Panamax containerships of 366.0 m in length and 48.2 m in beam with restricted draught along a 1.5 nautical miles narrow channel leading to a swinging basin in front of the terminal, and with a predefined AtoN and compulsory use of PPU (Fig. 8).

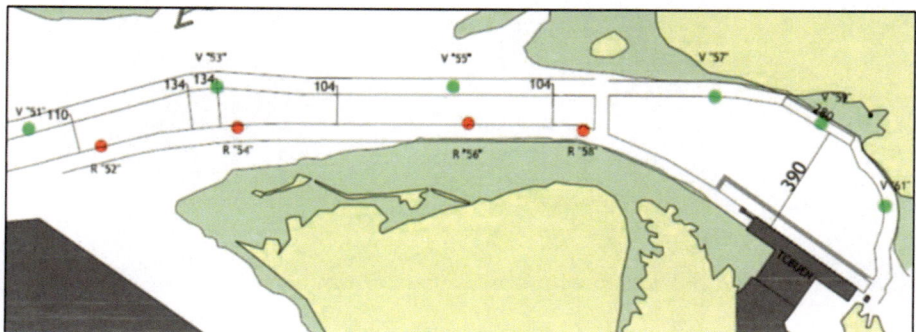

Fig. 8. Channel layout for the New Panamax containership assessment

Years later, the requirement to assess Maxi New Panamax containerships of 369.0 m in length and 52.1 m in beam with restricted draught was considered. During this assessment, the same methodology was applied, and it was found that, at the basin entrance, more space was required for the New Panamax container vessel compared to the Maxi New Panamax (with larger beam) for the same weather conditions and tug formation.

There was only one difference that justified the reduced area for the larger ships, and it consisted of the differences in the AtoN of the channel at the basin entrance, moreover when in this narrow channel buoys are the main visual reference to position the vessel during the transit. Considering the different results of navigable areas, it was concluded that buoys V-57 (in front of R-58) and V-59 are beneficial to reduce the required navigable areas by allowing the Pilot to position the vessel with a very low error (Fig. 9).

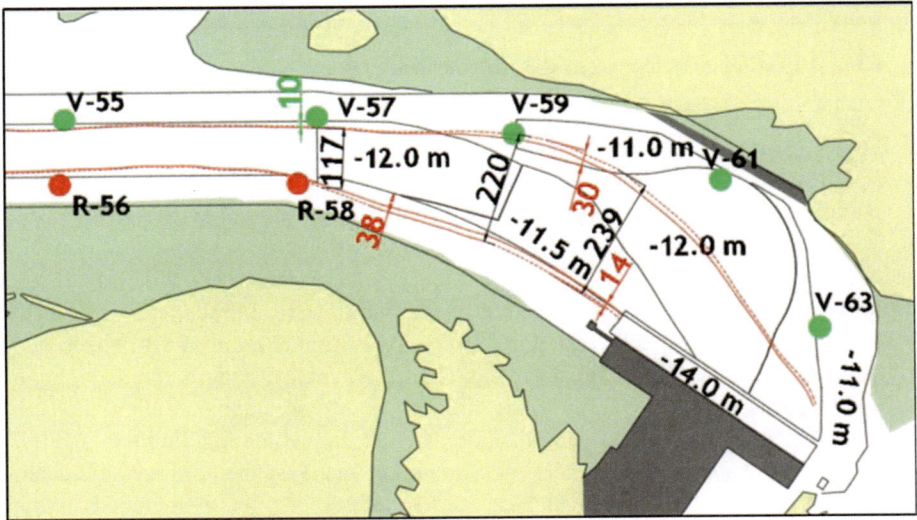

Fig. 9. Layout and required widths during the Maxi New Panamax containership assessment

Additionally, it was concluded that installing a new buoy in the south area of the basin entrance, paired with V-59, could reduce, or even avoid, the interference identified in this area for departures starboard side alongside (sailing astern) to help the vessel to position precisely and maintain a distance to this shallow area. This solution could avoid dredging this area to allow safe departures of this vessel under the strategy mentioned (Fig. 10).

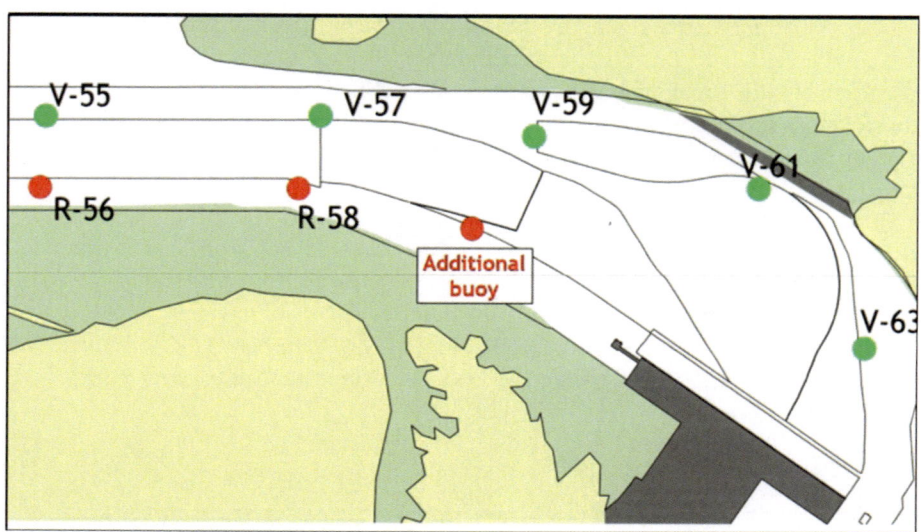

Fig. 10. Additional proposed AtoN to potentially reduce dredging requirements

5   Conclusions

This paper presents the importance of jointly assessing the waterway design with the Aids to Navigation system. This combined assessment allows to optimize navigable areas and therefore reduce dredging costs or increase ship sizes acceptable on already existing areas, without reducing the safety of the maritime operations.

The use of Fast Time and Real Time Manoeuvre Simulators allows to check and verify the optimization and improvements in the waterway and in the navigation conditions. These tools are of essence in these type of assessments as they allow to verify the effectiveness of the modifications before implementing them in reality. Unlike with Fast Time models, Real Time Simulations increase the accuracy and detail as the human factor (perception and decision making) is directly included in the assessment.

Real Time Simulators and manoeuvring sessions are the perfect place for all stakeholders (Port Authorities, Pilots, Harbour Masters, waterway designers, shipowners, …) to interact, discuss, evaluate, and agree on the best option for the improvement of navigation conditions while preserving the safety of the operations.

References

Ohtsu K, Yoshimura Y, Hirano M, Takahashi H, Tsugane M (2008) Design standard for fairway in next generation 20:6. https://doi.org/10.14856/kanrin.20.0_6

PIANC Report 30 (1997) Approach Channels. A Guide for Design

PIANC Report 121 (2014) Harbour Approach Channels Design Guidelines
Puertos del Estado, Spain (2000) ROM 3.1-99 Proyecto de la Configuración Marítima de los
    Puertos; Canales de Acceso y Áreas de Flotación

Container Barge Design to Optimize Hinterland Transport in Europe

Bianca Borca[1]([✉]), Lisa-Maria Putz[1], and Bernhard Bieringer[2]

[1] University of Applied Science Upper Austria, Steyr, Austria
{bianca.borca, lisa-maria.putz-egger}@fh-steyr.at
[2] Consulting Engineers for Naval Architecture, ZT Kanzlei Dipl.-Ing. Richard Anzböck, Vienna, Austria
bernhard.bieringer@anzboeck.com

Abstract. The aim of this paper is to develop and evaluate new design options for barges for the economically viable transport of 45' standard high-cube containers. Special attention was given to the potential performance of the new designs for navigation during low water periods on the Danube. To reach this aim, a four-step approach was conducted: first, a literature review was conducted, second, a scenario was selected leading to defined waterways and goods, third, the results were visualized, and fourth, the design options were evaluated. The research resulted in six design options for barges optimized for accommodating 45' standard high-cube containers. The barges should serve the Danube stretch between Enns/Austria and Giurgiu/Romania. A major conclusion is that which of the six barge designs fits best depends on the business case, since each of the options have several advantages and disadvantages which have to be investigated such as a different loading capacity. A barge being able to carry between 24 and 90' standard high cube containers.

Keywords: Inland waterway transport · Low water · Barge design · Container transport · Danube

1 Introduction

Inland waterway transport (IWT) is an environmentally friendly transport mode which is promoted to achieve the goal of the 2019 announced European Green Deal to reduce 90% of the transport related emissions until 2050 (European Commission 2019). IWT has, besides its eco-friendliness, other advantages such as minimal external costs and free capacities (Fastenbauer et al. 2019). To increase the share of inland navigation on the Danube and to promote a modal shift from road to IWT it is crucial to foster container transport. Until today, container transport on the Danube is limited and by far not as developed as on the Rhine. Containers have a lower relative density than bulk goods and can therefore navigate economically viable even during low water conditions. Moreover, the transport of containers allows new market opportunities for IWT on the Danube to meet the 50% increase goal of transport volume on IWT as announced in the Green Deal. (van Dorsser et al. 2020, CCNR 2020). In Europe 45 'standard high cube containers are highly relevant for the hinterland freight transport, in

Y. Li et al. (Eds.): PIANC 2022, LNCE 264, pp. 814–819, 2023.
https://doi.org/10.1007/978-981-19-6138-0_71

particular for multimodal transport chains, as they are broader than 40' standard containers. This width of 2.44 m allows a better utilization of the containers being loaded with standard euro pallets. Standard euro pallets are mostly used as a packing aid in European freight transport (Gronalt et al. 2010).

The aim of this paper is to develop and evaluate new design options for container barges, i.e. unmotorized vessels, on the Danube for an optimized hinterland freight transport in Europe. The following research questions (RQ) guides this paper: *RQ: Which barge designs are suitable for carrying 45' standard high cube containers on the Danube between Enns/Austria and Giurgiu/Romania?*

2 Methodology

For the development and evaluation of new design options for barges carrying 45' standard high cube containers we used a four-step approach. First, a data collection was conducted. We collected data about locks, bridges, and ports along the Rhine-Main-Danube-Corridor. This data was subsequently processed through a large-scale desktop and literature research. Second, we agreed on a scenario based on the business case of our partner Nothegger by defining the waterway and the type of goods the barge should serve. Third, we analyzed the existing barge options for container transport on the Danube and subsequently designed new barges, which are able to navigate safely and economically viable carrying 45' high-cube containers. All the six identified design options were finally visualized using the Naval Architecture software CAD. Fourth, we further analyzed and evaluated the identified design options for barges regarding their stability, required sight lines and possible construction materials as well as the nautical factors draught and air draught.

3 Findings and Discussion

3.1 Predefinitions

Based on the requirements of our business case for Nothegger, we agreed to design the barge options for the Danube River on the stretch between Enns in Austria and Giurgiu in Romania. Furthermore, the barges are designed to fit for carrying 45' standard high cube containers. 45' standard high cube containers are generally used in intermodal hinterland transport in Europe. The unique characteristic of these containers is that the size of the containers is adapted to the size of standard Euro pallets (dimensions: 120 cm × 80 cm). With the dimension of the container (13.56 m × 2.44 m × 2.7 m) euro pallets fit better in the container and therefore, increase their utilization. The reason for choosing container barges for intermodal hinterland transport is, that intermodal transport must be promoted, particularly on the Danube, since there hardly takes place container transport nowadays.

3.2 Designed Barge Options

To design new barge options, we started with existing barge design options for containers transport used on the Danube. Currently used barge designs on the Danube are the barges Europa 2b and Europa 3a, shown in Fig. 1. The new six barges were designed to allow a safe and economically viable transport of 45' high-cube containers on the Danube. Figure 1 presents the new design options for barges including the length, the breadth, the lightship displacement and the number of 45' standard high cube containers, which can be carried.

Each barge design can carry two or three layers of containers, which is the maximum on the Danube since there is insufficient bridge height on the Danube to carry more than three layers. A minimum of 24 45' high-cube containers (i.e. Barge Europa 2b in two layers) and a maximum of 90 45' high-cube containers (i.e. Barge iw-net – Container transverse v2 in three layers) can be carried using one barge. The barges differ in length and breadth, with each barge fitting into the maximum possible length and breadth of barges on the Danube according to the maximum fairway width and lock sizes. While the barge designs in general allow integration into usual pushed convoy formations with two barges side-by side, this is not the case for the broader IW-NET – Containers transverse barge type with a view to lock sizes available in particular on the upper Danube.

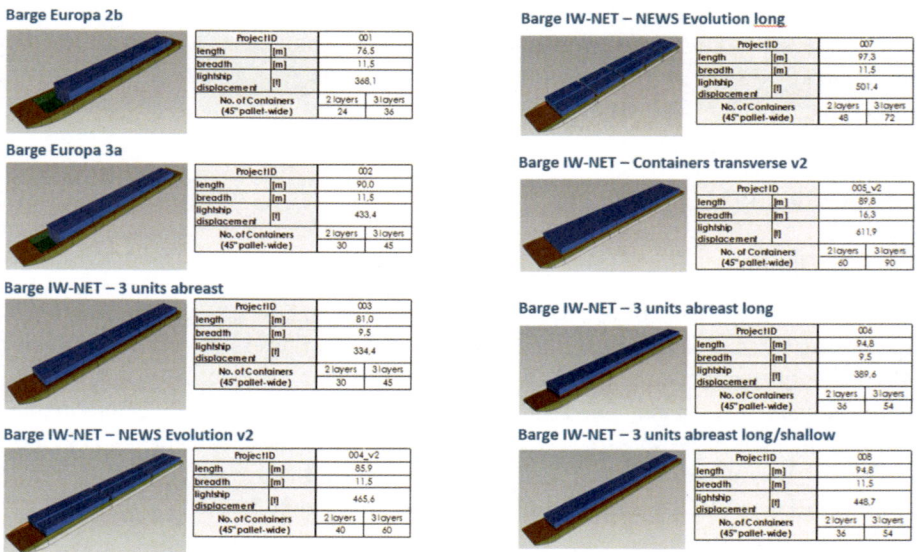

Fig. 1. Result: overview of the barge design options

3.3   Further Analysis of the Barge Design Options

The barge designs were further evaluated regarding their sight lines and stability resulting in several barge options being not feasible for transportation. Furthermore, different construction materials were compared and discussed.

The *stability* of barges with two layers of containers is basically given. Considering the barges with three container layers, the stability of three barge types is severely limited: for the "Barge iw-net - 3 units abreast" (Project ID 003), "Barge iw-net - 3 units abreast long" (Project ID 006) and "Barge iw-net - NEWS Evolution long" (Project ID 007) three layers of containers are limited to the transport of empty containers. Therefore, the barges with the Project ID 003, 006 and 007 cannot be used to transport loaded containers in three layers in most cases. However, the stability has only been assessed for pre-defined standard loading conditions and is always depending on the individual loading conditions of the barge. The stability of the barges was assessed against the requirements of Chapter 27 of ES-TRIN 2021 for non-secured containers. Figure 2 shows the design options and their stability in different loading conditions (CESNI 2021).

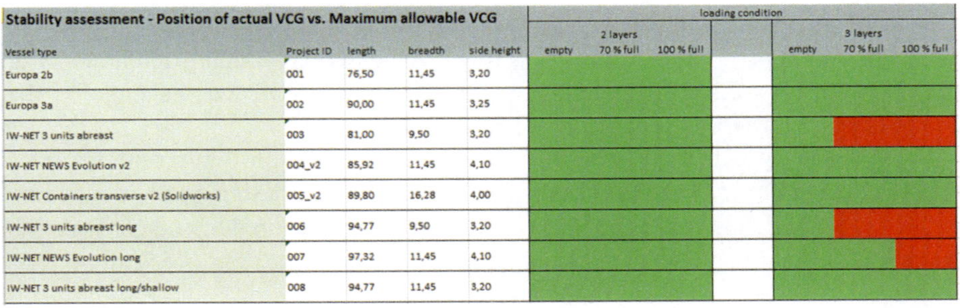

Stability assessment - Position of actual VCG vs. Maximum allowable VCG					loading condition						
						2 layers				3 layers	
Vessel type	Project ID	length	breadth	side height	empty	70 % full	100 % full		empty	70 % full	100 % full
Europa 2b	001	76,50	11,45	3,20							
Europa 3a	002	90,00	11,45	3,25							
IW-NET 3 units abreast	003	81,00	9,50	3,20							
IW-NET NEWS Evolution v2	004_v2	85,92	11,45	4,10							
IW-NET Containers transverse v2 (Solidworks)	005_v2	89,80	16,28	4,00							
IW-NET 3 units abreast long	006	94,77	9,50	3,20							
IW-NET NEWS Evolution long	007	97,32	11,45	4,10							
IW-NET 3 units abreast long/shallow	008	94,77	11,45	3,20							

Fig. 2. Result of the stability assessment

Regarding the *sight lines*, an elevating wheelhouse on the push-boat is necessary if more than one container layer is transported. In general, transportation with barges loaded with two container layers is feasible. Transportation with barges stacked with three container layers is restricted to loaded containers, as empty containers are too lightweight. Therefore, the barges do not immerse deep enough to allow an acceptable sight line. The assessment of sight lines was conducted against the requirements of Article 1.07 of CEVNI (United Nations Economic Commission for Europe 2021). Figure 3 shows the design options regarding the sight lines in different loading conditions.

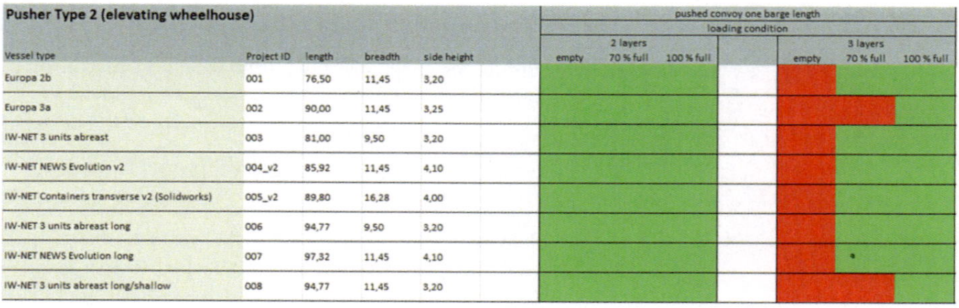

Pusher Type 2 (elevating wheelhouse)					pushed convoy one barge length						
					loading condition						
					2 layers				3 layers		
Vessel type	Project ID	length	breadth	side height	empty	70 % full	100 % full		empty	70 % full	100 % full
Europa 2b	001	76,50	11,45	3,20							
Europa 3a	002	90,00	11,45	3,25							
IW-NET 3 units abreast	003	81,00	9,50	3,20							
IW-NET NEWS Evolution v2	004_v2	85,92	11,45	4,10							
IW-NET Containers transverse v2 (Solidworks)	005_v2	89,80	16,28	4,00							
IW-NET 3 units abreast long	006	94,77	9,50	3,20							
IW-NET NEWS Evolution long	007	97,32	11,45	4,10						*	
IW-NET 3 units abreast long/shallow	008	94,77	11,45	3,20							

Fig. 3. Results of sight lines assessment

As *construction materials* for the barges we evaluated shipbuilding steel, aluminum and composites such as carbon. Composites are technically not feasible for the construction of cargo vessels in inland navigation, leading to an exclusion of composites for further evaluation. Shipbuilding steel succeeded as the most appropriate and affordable material. A major disadvantage of shipbuilding steel is its weight. In fact, shipbuilding steel is relatively heavy leading to unfavorable conditions for low water levels. Aluminum offers the advantage to be lightweight leading to a lower draught. Indeed, aluminum is less ductile as well as highly expensive in building and repair. The latest developments of the prices for steel and aluminum represent a challenge for the shipbuilding sector and the financing of future new barges.

A major finding is that all six design options are feasible for the transportation of containers during fluctuating water conditions. The six barge designs differ on the one hand in its size (i.e. in length or breadth) and on the other hand in the used construction material. It is hardly possible to determine the best suitable barge design for the transport of 45' standard high cube containers from Enns to Giurgiu, since this decision depends on the specific transport situation and requirements of the business case.

4   Conclusions

In our paper, we developed six barge designs which are suitable to carry 45' standard high cube containers on the Danube, specifically on the stretch between Enns/Austria and Giurgiu/Romania. The six barge designs differ in their size and their container carrying capacity. Each of the barges is able to carry between 24 and 90 45' standard high cube containers. Furthermore, we investigated each option regarding their stability and found, that three designs have major limitations if they are stacked in three layers. Considering the sightlines, it was found, that all barge options, loaded with more than one layer of containers need an elevating wheelhouse on the push-boat. For the construction materials of the barges, shipbuilding steel, aluminum and composites, such as carbon were investigated. The construction using composites is technically not feasible. Aluminum has the major advantage of being lightweight and therefore, highly suitable to navigate economically feasible on low water. Nevertheless, from today's perspective, steel has more advantages than aluminium in the construction of barges, such as the robustness of the material and the costs. One major output is, that it depends on the business case, which of the six designs fits best for the purpose of shippping 45' standard high cube containers between Enns and Giurgiu, as each of the designs have specific advantages and disadvantages. Therefore, the decision of which is the best suitable design should be made depending on the specific business case.

A limitation which leads to further research is that the barge designs are based on the infrastructural requirements of the Danube and cannot similarly be transferred to other waterways without certain adaptations. Furthermore, the new designed barges are specifically designed for container transport. Only a limited number of designs is also feasible for other goods, i.e. bulk goods. Several designs must be adapted for the transportation of other goods than containers. For further research, the research can be extended by optimizing the barges for the Danube stretch Enns-Giurgiu suitable for other goods, e.g., bulk goods or chemical products.

Acknowledgements. This research is part of the IW-NET project and has received funding from the European Union's Horizon 2020 research and innovation programme under grant agreement No. 861377.

References

CCNR (2020) "Act now!" on low water and effects on Rhine navigation. https://www.ccr-zkr. org/files/documents/workshops/wrshp261119/ien20_06en.pdf. Accessed 24 Nov 2020

CESNI (2021) European Standard laying down technical requirements for inland navigation vessels

European Commission (2019) The European Green Deal, Bruxelles

Fastenbauer M et al (2019) Manual on Danube Navigation, 4th edn. Viadonau, Vienna

Gronalt M, et al. (2010) Handbuch Intermodaler Verkehr. Kombinierter Verkehr: Straße - Schiene - Binnenwasserstraße. Bohmann

United Nations Economic Commission for Europe (2021) CEVNI European Code for Inland Waterways

van Dorsser C, Vinke F, van Hekkenberg R, Koningsveld M (2020) Effect of low water on loading capacity of inland ships. Eur J Transp Infrastruct Res 20(3):47–70. https://doi.org/10. 18757/ejtir.2020.20.3.3981

CEERIS (Central and Eastern European Reporting Information System) – SMART Electronic Reporting Platform for IWT

Introducing Reporting Only Once with Single Entering of Data in 8 Countries

Katrin Steindl-Haselbauer[✉], Mario Kaufmann,
and Thomas Zwicklhuber

via donau - Österreichische Wasserstrassen-Gesellschaft mbH,
1220 Wien, Austria
Katrin.Haselbauer@viadonau.org

Abstract. In order to increase the competitivity of Inland Waterway Transport (IWT) in Europe the concept of RIS enabled Corridor Management has been developed and was now implemented within the European flagship project RIS COMEX (https://www.riscomex.eu/.) in the 2 integrated platforms EuRIS (fairway information services, traffic information services, services for logistics) and CEERIS (retrieving of reporting requirements, electronic reporting services for skippers, receiving authority services). The novel centralized architecture enables CEERIS users of to fulfil all reporting obligations configured along their route with a central access point for one or more voyages by a single data entry, using his EuRIS data including access and reporting rights for claimed vessels. Thereby, the user does not have to take care about the complexity of reporting procedures and the further delivery of reports to the respective receiving authorities. The smart distribution of reports is managed by the system in the background proving all competent authorities with data based on the respective legal basis and configured means. Thus, reports can be received either through an authority dashboard at a central access point or email as XML (Extensible Markup Language.), PDF or as a graphical template. Data delivery and notifications using webhook is supported as well as the delivery of XML messages via API (Application Programming Interface.). The reporting platforms addresses authorities dealing with traffic management, customs, border police, statistics and ports in addition to the national RIS authorities. A smart user and identity management secures access authorization according to the consent agreement of reporting parties – consequently data can also be accessed and used by authorized third parties such as shippers, fleet managers and other logistics partners for the optimization of their services.

Keywords: River Information Services · Electronic Reporting International (ERI) · Data exchange · Corridor services

© The Author(s) 2023
Y. Li et al. (Eds.): PIANC 2022, LNCE 264, pp. 820–833, 2023.
https://doi.org/10.1007/978-981-19-6138-0_72

1   Introduction and Background

As described in Haselbauer et al. (2019) Inland Waterway Transports on the Danube waterway and its navigable tributaries were repeatedly facing major administrative barriers concerning the reporting of transported goods and passengers resulting in crucial obstacles for the efficiency and competitiveness of this transport mode. As a result, border crossing transports along the Danube have to transmit the required data to respective competent national authorities several times using different paper forms. These long-winded administrative processes are both time consuming and inefficient and lead to a prolongation of the total idle time, which is subsequently reflected in a prolongation of the total travel time of vessels. Since the costs for vessel operation make up a significant share of transport costs, transportation costs consequently increase as well, threatening the already low profit margins. River Information Services (RIS) can contribute to overcome outdated reporting processes by technologies such as Electronic Reporting (ERI), Vessel Tracking and Tracing (VTT). These River Information Services have been created to enhance Inland Waterway Transport (IWT) in terms of safety and efficiency of transport by means of telematics, with the overall goal to improve its modal split. However, the existing grown fragmentation of services was identified as setback hindering the development of new services towards synchro modality and competitivity of Inland Waterway Transport. Therefore, public authorities still cannot enjoy the full benefits of RIS for traffic management and the logistics users of RIS still have only a few benefits from RIS for transport management, most of them within the limits of national borders or of different quality in different countries. This conference contribution will introduce the implemented Electronic Reporting System CEERIS including a novel system architecture and will highlight benefits for different stakeholders together with further future development possibilities.

2   RIS COMEX

In order to tackle the existing fragmentation of services, the concept of RIS enabled Corridor Management was already established within the EU-project CoRISMa and has now been implemented in the EU-project RIS COMEX. The overarching goal of the CEF co-funded multi-beneficiary project RIS COMEX project was to provide harmonized services and functions through a centralized RIS platform aiming at improved traffic management by authorities and more efficient transport **management by the logistics sector and was based on a mutual agreement between waterway authorities along specific transport corridors.** RIS COMEX started in the course of 2016 and was finalized by June 2022. The project area covered altogether 13 different European countries having 14 partners joined their forces under the coordination of the Austrian Waterway Administration viadonau. As a result, a new common European RIS platform (EuRIS) was established, enabling the use of a variety of harmonized RIS services in all participating countries with a single access point. As further part of the implementation of new common RIS services, 8 countries with partially available national systems with a lacking interoperability have also developed the common Electronic Reporting platform CEERIS (Central and Eastern European

Reporting Information System), which is closely integrated with the services and security features offered in EuRIS.

3 COMEX Electronic Reporting System CEERIS

3.1 Scope

CEERIS comprises the services ILE.10 (provision of information for increased efficiency during vessel inspections), ILE.11a (reporting requirements) and ILE.11b (electronic report gateway service) as specified in RIS COMEX. The scope of the approach related to cross-border electronic reporting and exchange of information for logistics stakeholders was to create a single portal, supported by all competent RIS authorities and available for all types of transport such as dangerous goods transport, non-dangerous goods transport, container transport and passenger shipping. The entry of data is required only once, according to the reporting requirements along the respective transport route. In order to correspond to the prevalent principle of a single access platform and to further use synergies of already existing supporting services, a close integration with the EuRIS platform has been established. The platforms share their identity server for the user management. The reporting platform CEERIS covers a wide scope of receiving authorities, including authorities dealing with traffic management, customs, border police, statistics and ports in addition to the national RIS authorities. In order to be able to cover all responsibilities of these authorities, missing ERI XSDs such as INVRPT[1], WASDIS[2] and CUSCAR[3] were developed in the course of the implementation in order to be able to support reporting of ship stores, waste disposal and bill of lading using XML message transmission via webservice. For this purpose, an extended ERI data pool, the so called SoaD (sum of all data fields) including mapping to all ERI XSDs, was established. Furthermore, the smart user management secures access authorization according to the consent agreement of reporting parties – consequently data can also be used by authorised third parties such as shippers, fleet managers and other logistics partners, expanding the circle of beneficiaries.

3.2 CEERIS/EuRIS System Architecture and Integration

Figure 1 provides an overview of the high-level system architecture of the CEERIS platform, which comprises 3 system environments including development, testing and production and is hosted on a cloud platform in the EU. The production environment is based on redundant servers using load balancing. CEERIS uses the user account information stored within identity server provided by EuRIS for authentication and authorization purposes. Skippers and fleet operators can manage their vessel claims, privacy classes and reporting rights within the EuRIS platform. Furthermore, EuRIS

---

[1] Inventory Report Message.

[2] Waste Disposal Message.

[3] Customs Notification.

provides route calculations to CEERIS in order to derive route specific reporting requirements. The route information is derived from EuRIS as polyline through the route planner webservice based on basic transport data. Locations used for reporting including trigger locations for report delivery are synchronized via a RIS index API provided by EuRIS.

CEERIS transmits ERINOT and ERIVOY XML messages towards EuRIS to receive ETA updates and voyage information according to the movement of the vessel. CEERIS request ETAs from EuRIS along the remaining route. Recalculations are triggered when the vessel passes dedicated important objects. Active voyages in EuRIS and the related voyage and transport information can be used by authorised 3rd parties e.g. fleet operators, logistics operators or agents that have been granted access rights. Reporting parties with successful vessel claims can start the reporting for the planned voyage also directly from the EuRIS route planner using and redirection link to CEERIS, handing over route information and basic transport parameters.

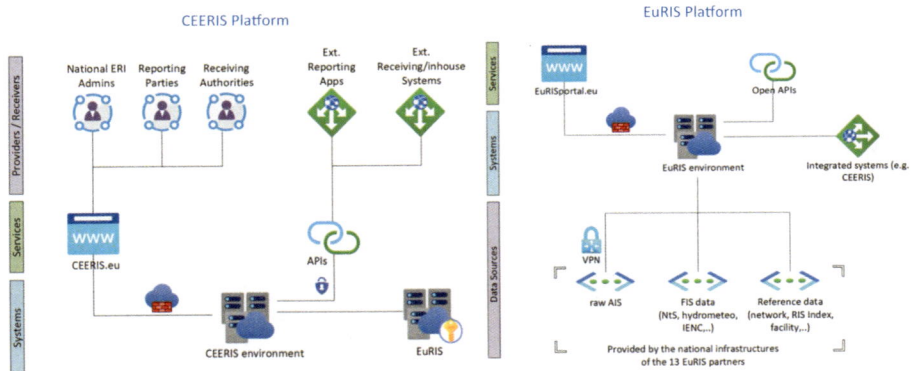

Fig. 1. High level system architecture of the CEERIS and EuRIS platforms

CEERIS address 3 user groups namely national ERI administrators (responsible for requirement configuration), reporting parties (skippers, fleet operators) and national receiving authorities (RIS, customs, statistics) and provides reporting services both through the user dashboards on the website and webservices. Users of external reporting applications as for example BICS can send ERINOT messages towards CEERIS through the interconnected Dutch message server Hermes. Skippers that are also registered in EuRIS and have approved vessel claims can access CEERIS to add missing data in case the reporting requirement along their specific rout cannot be fulfilled solely by the data covered by ERINOT.

Reporting party users can use API to publish voyages (new, update and cancel), retrieve information about his voyages, and retrieve reports generated for his voyages and responses for his reports following the workflow determined in Fig. 2. After publishing the voyage, reporting party can list all of his voyages for a specific vessel, filtered by minimum ETD. Information about the voyages contains common denominator which can be used to retrieve more information about a specific voyage. Using common denominator reporting party can retrieve list of reports generated for a specific voyage. Each report is identified by common denominator and reporting requirement ID. Using this information reporting party can retrieve details about a specific report which includes report

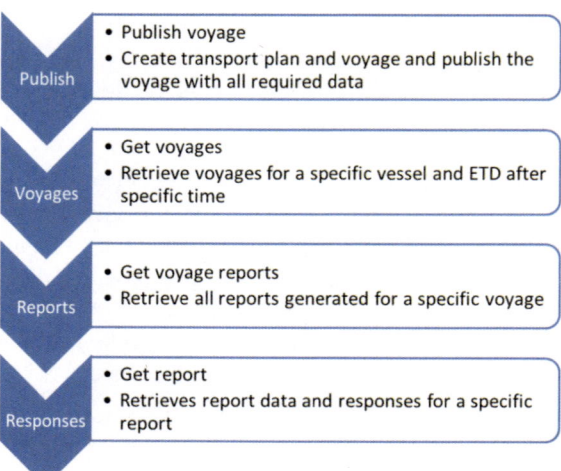

Fig. 2. Reporting party workflow via webservice

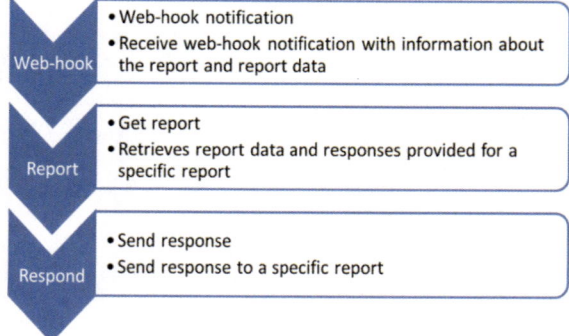

Fig. 3. Receiving authority workflow via webservice

data and all responses received from receiving authorities for that report in the form of ERIRSP.

National RIS systems and inhouse systems of national receiving authorities can interconnect to CEERIS though dedicated APIs following the workflow prescribe in Fig. 3. Web-hook notifications can be configured for authorities, which contain identification information of the report which can be used to retrieve details of the report. Web-hook notification also includes all report data in CEERIS specific format and if external receiving authority system uses that format get report step can be skipped. After getting report details, receiving authority can provide response in the form of ERIRSP if it has response permission in the reporting requirement for which report is provided. Receiving authorities can periodically check get reports API to retrieve a list of new reports, provided after a specific time. For each report, report identification is provided in the form of common denominator and reporting requirement id and can be used to retrieve derails of the report. After getting report details, receiving authority can provide response in the form of ERIRSP if it has response permission in the reporting requirement for which report is provided.

3.3   Access Control - The Principle of Privacy by Design

Administrators of organizations that have claimed vessel ownership determine by unique privacy classes the visibility of information for other users. Proprietors of claimed vessels can grant reporting rights for their vessel to other parties as charter companies, logistics operators or agents by setting a dedicated flag enabling these users to create transport plans for a vessel and submit data and electronic reports for this vessel. This feature support Economic Operators to provide cargo information for a transport. Vessel claims are examined and approved by national identity controllers. Vessel owners can grant access to their vessels with a number of insight privacy classes.

1. Position related:
- Class 1: fully anonymous
- Class 2: basic non-personal data (default)
- Class 3: standard AIS information (no SRM)
2. Voyage related:
- Class 4: voyage information (future). This is primarily the destination of the active voyage and the corresponding ETA.
- Class 5: detailed voyage information (future). All future waypoints and ETA is also available. The voyage can be plotted on a map.
- Class 6: detailed voyage information (past & future). All information about the current voyage with past (ATAs) & future (ETAs) times can be consulted.

4   CEERIS Implementation

The implementation of the system was finalized in December 2021 followed by a 6 months transition period to system operation, aiming at involving component national receiving authorities and interconnecting CEERIS to the final environment of the EuRIS platform. CEERIS comprises the corridor services ILE.10 (provision of information for increased efficiency during vessel inspections), ILE.11a (reporting requirements) and ILE.11b (electronic report gateway service) as specified in RIS COMEX.

4.1   Flexible Configuration of Reporting Requirements

CEERIS enables ERI national administrators to configure their national reporting requirements based on the requirements of the respective national and international legislation using an extended pool of data that is used in Electronic Reporting and mapped with all standardized or specified ERI messages.

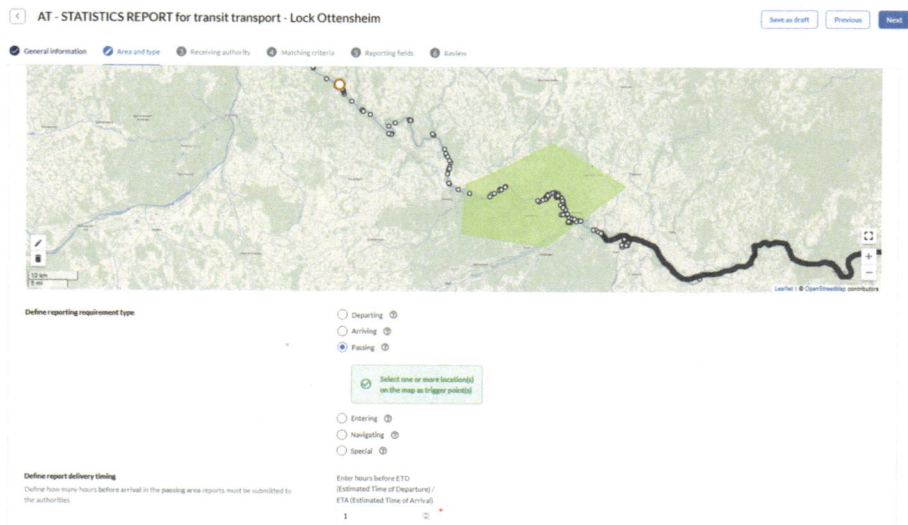

Fig. 4. Configuration dashboard enabling competent national administrators to configure national reporting obligations.

The configuration of reporting requirements includes a number of steps. In a first step the reporting rules and legal provision are defined. The geographical scope of a reporting requirement is determined by a configured polygon area (see Fig. 4). For reports that are triggered by entering a specific area ETA time triggers can be defined determining e.g. how many hours before entering/navigating in a specific a specified area reports must be submitted to the receiving authorities. Different types of reporting requirements reflect the nature of and processes behind a reporting requirement as arriving to a port or navigating in a specific area. The configuration of receiving authorities allows to set individual permissions for different types of receiving authorities (e.g. response permissions, update actual times permission, issue permit permission) and additional configure e-mail notifications with different reports formats attached (e.g. PDF, XML or graphical templates which appear identical to existing paper forms). In order to account for various different obligation types transport matching criteria are used to determine the transport type affected by an obligation (cargo, dangerous cargo, passenger transports), the travel type (domestic, international, both) and vessel specific criteria (e.g. LNG). In the main configuration steps the data fields required by an obligation are assigned. The reporting data pool is divided into logical groups aligned the ERI messages. Required fields are selected by a checkbox and can either be set to mandatory or optional. Further-more, more detailed information for a data item can be added as well as links to reference specifications.

As a result, the requirements finder service provides information about mandatory and voluntary reporting requirements, reporting require-ment type, receiv-ing authorities and data requirements of a reporting requirement along

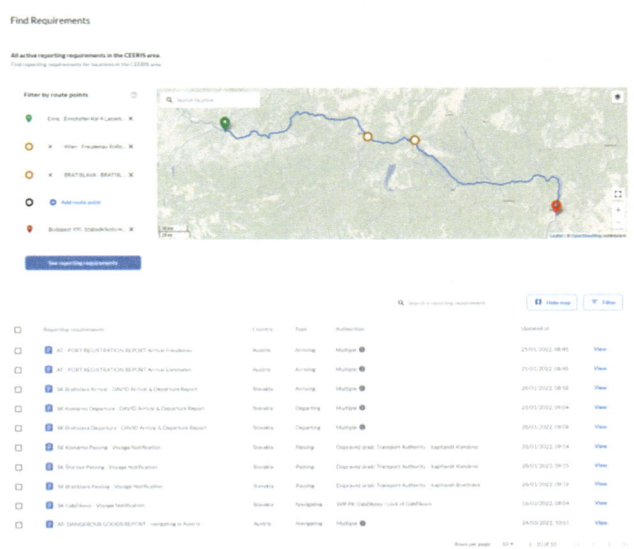

Fig. 5. Service providing an overview of reporting requirements along a specified route.

the route of a specific voyage to reporting parties such as skippers (see Fig. 5). Filters support the user in identifying requirements that are related to a specific authority, country, transport level, transport type or reporting requirement type. The user is enabled to retrace which authority receives which data based on which legal basis including rules and conditions for reporting. External systems can retrieve an overview of all configured reporting requirements using the web service infrastructure.

4.2 Electronic Reporting Services for Users

CEERIS enables users registered in EuRIS with approved ownership rights for a vessel or granted reporting rights to access the platform to fulfil reporting obligations for their transports and interact with the receiving authorities. The retrieval of applicable reporting obligations for a planned transport starts with the entry of the route (depar-ture, destination and route points) including estimated departure and arrival times (see Fig. 6). A transport plan can contain multiple voyages which are defined by events such as loading or unloading of goods or passenger movements and entail a change of data. Based on the provided voyage data route calculations are derived from the EuRIS platform using a dedicated webservice. By further defining the transport characteristics, the reporting obligation derived for a route are matched to a specific transport (e.g. transport of dangerous goods). Users receive a graphical (map with their voyages and requirements) and tabular overview of reporting obligations applicable for their voy-ages as shown in Fig. 8 and are enabled to obtain details on receiving authorities and the data submitted to these authorities. In order to guarantee data sovereignty users can optout individual requirements and fulfil obligations by conventional means. After confirmation of the reporting request, the reporting workflow is started. The overall design principle of "reporting only once with single entry of data" ensures that skippers or fleet operators have to fill-in all data only once, even if the related data field is required by different authorities in different countries and different reporting obligations. The system takes care of creating the individual reports and providing them to the related Receiving Authorities (reporting only once).

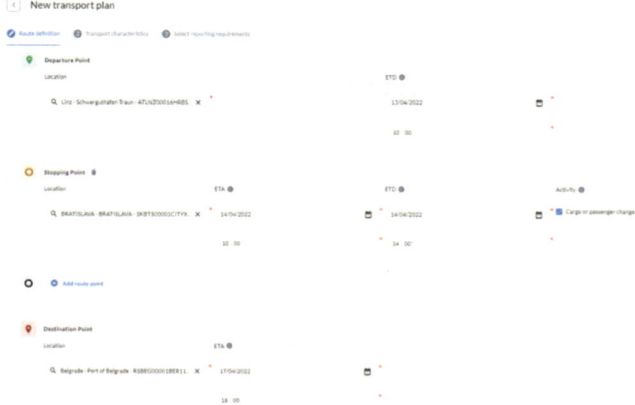

Fig. 6. Entry of departure, destination and route points including ETD and ETA for transport plans supporting multiple voyages with cargo discharge in between.

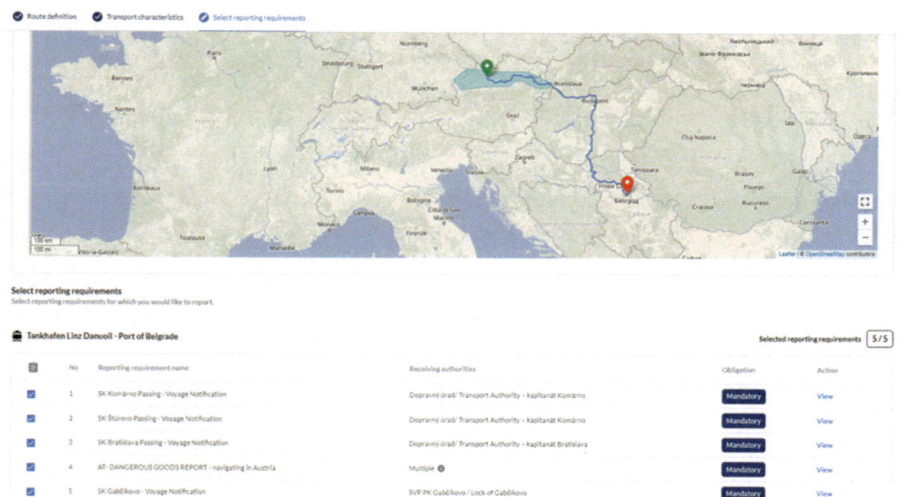

Fig. 7. Reporting GUI supporting users to fulfill all reporting requirement only once with single entry of data

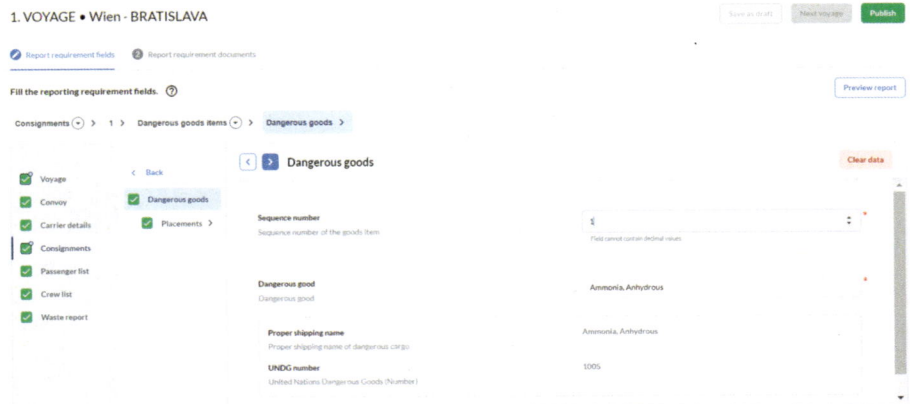

Fig. 8. Overview of reporting requirements and authorities for a specific route fulfilling the set transport matching criteria.

Based on smart data copy functions, the system allows users to complete or adapt only the part of the dataset that was changed by an intervening event. Data entry is supported by extended ERI reference data, which have been translated into all languages of the participating authorities. These reference data also include a mapping between different codes and classifications such as HS, NST and ADN. The entry starts a search query where all codes and names in the reference database are queried. Based on the selection of the user, all mapped data elements are assigned for a report reducing the number of reporting fields for reporting parties. This principle is also applied for locations where based on the assignment of an ISRS code further locations attributes such as hectometre, terminal name and location name are filled-in. By creating favourites for data sets on the main vessel, additional barges, parties such as the vessel operator, agents and invoicing parties, the reporting process can be conducted in increasingly shorter time spans. Additional documents such as certificates can be transmitted to authorities as attachments. During the fill-in workflow users are guided through the data groups via a navigation bar and continuously receive an overview of the overall reporting progress and status (see Fig. 7). Before reports are submitted, the user receives an overview of all created reports and data fields as well as the related receiving authorities to ensure full transparency of the process. After the handover of the dataset, the status per report (missing, prepared and sent) and authorities can be monitored under the respective voyage (see Fig. 9). This also includes the authority responses (pending, rejected or approved) as presented in Fig. 10.

Before the submission the status is indicated as draft and is changed to awaiting immediately after the submission. The data provided for a transport plan can be updated at any time by the vessel owner or 3rd party which has been granted reporting rights. Updates are highlighted both for the reporting party as well as for the receiving authority. Voyages can further be cancelled and discarded after cancellation. The reporting process for passenger vessels is facilitated by an excel import feature that support the import of long passenger list generated by inhouse solutions. Users as

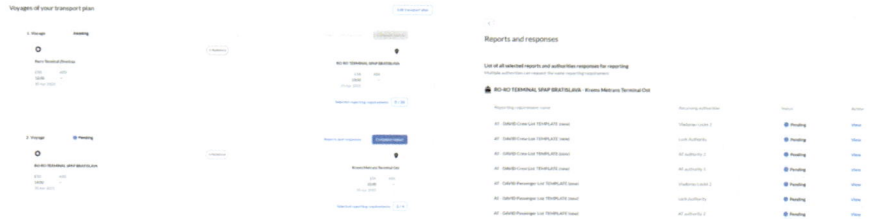

Fig. 9. Overview of all voyages assigned to a transport plan including response status, response details, ETA

Fig. 10. Overview and details of authority responses on submitted reports.

skippers or fleet operators can evaluate and export their data for the purpose of statistics and derive e.g. quarterly reports to Statistics authorities.

4.3   Services for 3rd Parties as Logistics

In addition to granting reporting rights for a vessel, vessel owners have the possibility to grant logistics users access to the voyage and transport information for their vessel in EuRIS either for an individual voyage or a certain time period or even unlimited. This allows logistic users to query and track the current position data of the vessel and transport information based on the ERI messages transmitted toward the EuRIS platform, depending on the granted privacy class. As presented in Fig. 11 the available information is grouped into general

Fig. 11. Voyage and transport information for 3rd parties as logistics operators based on access rights in the EuRIS platform.

information including loaded cargo, transport information as transport mode and transport dimensions, hull information (e.g. hull type) as well as travel plan and position information including ETA, ETD, ATA and ATD for the route. In the notification centre user can subscribe for specific events related to authorized vessels. These events comprise the receiving of ERI messages, ETA changes, stationary situations, the passage of bridges and locks as well as the arrival and departure at berths.

4.4 Receiving Authority Services

Implemented CEERIS receiving authority services aim at providing information to competent receiving authorities via a platform with a central access point enabling them to increase the efficiency of their task and thereby reducing the waiting times for users. The receiving authorities only have access to those data fields covered by their legal basis. The delivery and notification of the receipt of reports at the configured time are based on the respective authority needs and can be accessed either via the dashboard, email notification with report attachments as PDF or XML, webhook notifications or XML via API. In order to be able to support a broad variety of receiving authorities (traffic management, port authorities, statistics, customs and boarder police) also in making use of their inhouse systems and to be able to cover all data needs, the ERI data pool was extended. Furthermore, missing XSDs in the thematic area of customs (CUSCAR), waste (WASDIS) and ship stores (INVRPT) were implemented in addition to amendment of the existing specifications of standardized ERI messages as ERINOT for cargo and dangerous goods, ERIVOY for voyage information and PAXLST for passenger, crew and stowaway reporting. This will enable economic operators to cover the reporting of bill of lading, inland waterway BL and cargo manifest by electronic means in the future. The receiving authority dashboard (see Fig. 13) enables authority users to get an overview of all voyages and

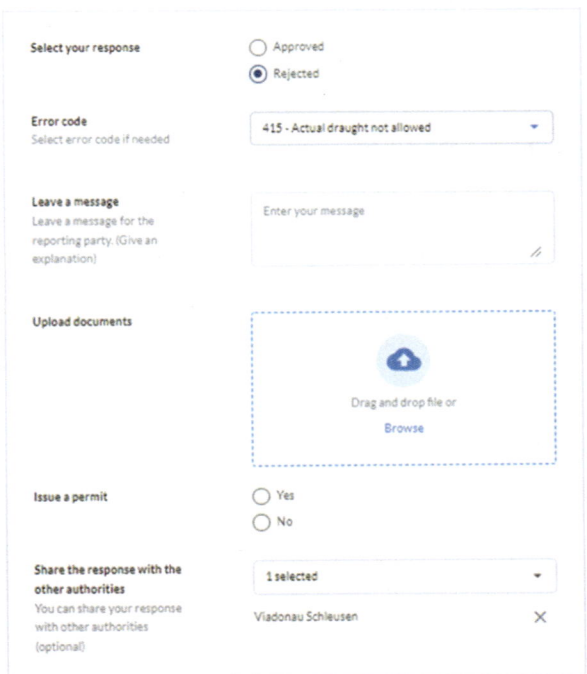

Fig. 12. Authority response service

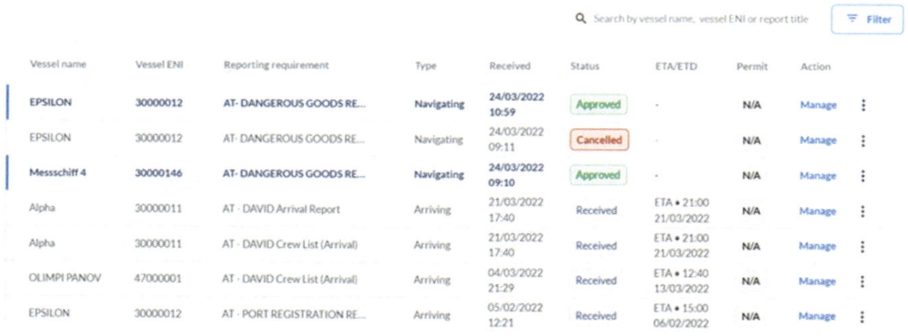

Vessel name	Vessel ENI	Reporting requirement	Type	Received	Status	ETA/ETD	Permit	Action
EPSILON	30000012	AT- DANGEROUS GOODS RE...	Navigating	24/03/2022 10:59	Approved	-	N/A	Manage ⋮
EPSILON	30000012	AT- DANGEROUS GOODS RE...	Navigating	24/03/2022 09:11	Cancelled	-	N/A	Manage ⋮
Messschiff 4	30000146	AT- DANGEROUS GOODS RE...	Navigating	24/03/2022 09:10	Approved	-	N/A	Manage ⋮
Alpha	30000011	AT - DAVID Arrival Report	Arriving	21/03/2022 17:40	Received	ETA • 21:00 21/03/2022	N/A	Manage ⋮
Alpha	30000011	AT - DAVID Crew List (Arrival)	Arriving	21/03/2022 17:40	Received	ETA • 21:00 21/03/2022	N/A	Manage ⋮
OLIMPI PANOV	47000001	AT - DAVID Crew List (Arrival)	Arriving	04/03/2022 21:29	Received	ETA • 12:40 13/03/2022	N/A	Manage ⋮
EPSILON	30000012	AT - PORT REGISTRATION RE...	Arriving	05/02/2022 12:21	Received	ETA • 15:00 06/02/2022	N/A	Manage ⋮

Fig. 13. Receiving authority dashboard providing an overview of reports submitted to a specific authority containing data based on the respective national legislation supporting the management of all received reports.

reports impacting the authority based on their area of competence and the applicable transport matching criteria. All details of a received report can be inspected and evaluated. Under the management features, user can react on the received report by either approving the report, requesting additional information or rejecting the report. Responses can further be shared together with additional notes with the other receiving authorities belonging to the same reporting requirement (see Fig. 12). In the response message ERIRSP error codes indicate different error types and additional explanations can be provided as free text. Based on national requirements authorities may receive a permission to update actual times and reported data. Access to position information can be configured for individual authorities within their area of responsibility and their legal mandate. Furthermore, reports can be exported in PDF, graphical templates, XML. All reports are anonymized after a maximum storage duration of 30 days and available for statistical evaluations.

5   Conclusions

The CEERIS system will be released in June 2022 and pursues the goal of gradually reducing existing reporting burdens with an increasing number of participating receiving authorities in the CEE region in the future. This conference contribution will introduce the implemented novel system architecture and highlight benefits for different stakeholders together with further future development possibilities.

Acknowledgements. The platforms EuRIS and CEERIS were developed and implemented within the CEF co-funded multi-Beneficiary project RIS COMEX aiming at the definition, specification, implementation and sustainable operation of Corridor RIS Services following the results of the CoRISMa study. The project area covered altogether 13 different European countries with 14 partners that joined their forces under the coordination of the Austrian Waterway Administration viadonau aiming at the common goal to realize Corridor RIS Services.

Reference

Haselbauer K (2019) Smart reporting - reducing reporting burdens along the Danube waterway. In: Smart Rivers Conference, Lyon

Development of S-401, Status of the New Standard for IECDIS

Gert Morlion[✉]

De Vlaamse Waterweg, Hasselt, Belgium
gert.morlion@vlaamsewaterweg.be

Abstract. Although ENCs (maritime) and IENCs (inland) are based on the same model, many differences have arisen over the years. An important issue with both chart systems is the lack of interoperability. In mixed zones as sea ports or estuary zones, the ECDIS viewer can't be used to visualize both chart types. This means that a vessel should be equipped with two different viewers.

IHO started with the development of a universal hydrographic framework that would be the base for all the 'S-products' which are in development. The framework makes it possible to interact between these different products in such a way that the visualisation of features and attributes is handled automatically. E. g. Notices to Marines (S-124) or Aids to Navigation Information (S-201) will be displayed correctly on the S-100 viewer). In other words, the S-100 framework will handle the interoperability between the different products without any interaction of the skipper.

The Inland ENC Harmonization Group (IEHG) decided to develop their own S-401 product. This new standard had to be in line with the maritime S-101 to make the use of inland and maritime charts possible on the same viewer onboard. The development of this new standard in replacement of the S-57, is mainly done within the European RIS COMEX project.

During this presentation, the status of all elements will be given together with the decisions the IEHG had to take.

Keywords: Navigational chart · IENCs · S-401 · S-100 · IHO

1 Introduction

The development of the S-100 framework is a major step in the collaboration of different worlds and products. Means for maritime and inland navigation will come closer to each other and information will be displayed in the same way in the wheelhouse. Notices to Mariners (S-124), AIS AtoNs (S-201), Water level information (S-104), Bathymetric information for Inland (S-402), … S-100 will handle them all in the same way and the interoperability will be set.

The development of the S-401 for inland electronic navigational charts is ongoing for several years and getting in a phase of testing. Although some elements have to be developed, the S-401WG is looking for manners to convert and test the product.

In the next presentation an overview will be given of the status of all elements within S-401.

© The Author(s) 2023
Y. Li et al. (Eds.): PIANC 2022, LNCE 264, pp. 834–837, 2023.
https://doi.org/10.1007/978-981-19-6138-0_73

2 Data Dictionary and Portrayal

2.1 Product Specification

One of the most important decisions the IEHG has taken concerned the compatibility between the maritime S-101 and the future inland S-401 standard. It had to possible to use both charts on the same S-100 viewer and therefore the S-401 has to be in line as much as possible with the maritime standard. This decision lead to the procedure that the development of S-401 would always run behind the development of S-101.

Once the first edition of the S-101 Product Specifications were published by IHO, the IEHG finalized their adapted S-401 Product Specification in December 2019. To make the publication possible, some issues had to be postponed and temporarily removed from the final document. The interoperability and validation were two of such issues which weren't solved yet.

Certainly the interoperability is a main issue for which a good solution is needed. The combination of S-products in the viewer must be handled in a correct way. Which feature must be visualized and which not. Which attribute is prior and must be displayed on top?

On the moment this paper is written, IHO is developing the S-98 standard which describes the interoperability for S-100. In the first draft, only S-101 has been taken into account as a base cell. This lead to many questions in the Inland ENC Harmonization Group:

- How will S-401 interact with the base cell of S-101?
- How will the S-100 viewer determine what to visualize?
- Can't the navigator chose between a maritime chart and an inland chart?
- ...

As a solution IHO has decided to amend S-401 as possible base cell as well. Hereby the skipper can chose between S-101 and S-401 as base layer on his screen.

2.2 Feature Catalogue

The Feature Catalogue is the overview of all features, attributes and enumerations that can be encoded into a chart. The existing Inland ENC feature catalogue is based on the S-57 format. Three type of can be distinguished:

- Specific Inland features (not used in maritime)
- Maritime features (also used in inland)
- Maritime features used in inland but with other attribute values

To get a complete list of the features, attributes, values and enumerations for the S-401, all maritime elements were copied into the new inland feature catalogue. The IHO Geo information Registry allows all domains (hydro, inland, ...) to register features and attributes into the registry database and avoids that different domains add or change features which are used in other domains. The Feature Catalogue Builder is connected to this database and allows the domain control body to add all available features (attributes) to a specific feature catalogue.

The whole process of amending, changing, updating the inland specific features, registering them in the IHO GI Registry and creating the feature catalogue has been finalized and lead to the S-401 FC edition 1.0.0.

2.3   Portrayal Catalogue

Once all features are described, they have to be visualized in the S-100 viewer. Lines, areas, points, colors and symbols need to be defined. The description of colors is done in different style sheets (day, dusk, night), while lines and area fills are composed in XML-files. On the other hand all symbols are in the SVG format. Some of the portrayal elements, were rather easy to create but others (e.g. symbols) needed some expertise.

For a lot of features the visualization is depending on the condition. Attribute values, combinations of features, lead to different visualization. In S-57, the conditions were described in the five look up tables. But S-401 isn't using ASCII look up tables anymore but separate LUA scripts.

Before the S-401 scripts could be developed, all the conditions needed to be determined. Therefore all the inland conditions were compared with the maritime conditions and as a result combined look up tables were created. The last step was to decide which S-101 LUA scripts could be used and which scripts needed to be developed from scratch.

At the moment this paper is written, the development of the S-401 LUA scripts is ongoing.

2.4   Data Classification and Encoding Guide

Although the most important elements for the S-401 are described in the previous overview, it will be very difficult to produce a chart without instructions and guidelines. For Inland ENCs the IEHG created the Encoding Guide which describes real world entities. For the manufacturer of charts, the instructions are a big help.

For S-100 products IHO has developed a Data Classification and Encoding Guide. The DCEG is a document with the same purpose as the Inland Encoding Guide but the big difference is that the DCEG is describing every feature separately instead of describing real world entities with the linked features.

The IEHG had to decide whether they would follow this approach or not. The existing Encoding Guide was a helpful document and without the possibility to amend these entities, the DCEG wouldn't be of any value. On the other hand the IEHG had decided to stay in line with S-101 as much as possible.

After the acknowledgment from IHO that it would be possible to add free text and pictures which wouldn't be lost by updating the DCEG, IEHG decided to use the DCEG in the future.

3   Next Steps/Issues

Before the S-401 can be implemented, a thorough testing has to be done. Testing in a viewer with the S-401 Feature Catalogue and Portrayal Catalogue and with S-401 charts. Converting existing IENCs will be made possible by the manufacturers of charting software and will be one of the next steps of the RIS COMEX team. But for the conversion of S-57 cells some rules and conversion guidelines have to be set. This will be done in the 'S-57 ENC to S-401 conversion guidance'. This document will describe the default values of attributes, how old values have to be translated into the new values and more.

At this moment the document is in preparation and hopefully finalized until the end of the year.

For the testing itself, the IEHG is relaying on the S1OOP (S-100 Open Online Platform) which is developed by KHOA. This online platform will enable the possibility to load different catalogues in the viewer and load charts of different products into the software. This will allow the project team to test not only the catalogues but also the charts.

One of the major issues the IEHG is facing in the development of S-401, is the interoperability. It must be possible to combine S-401 with other S-products in the S-100 viewer without losing important information. The displaying of features and attributes that come from other S-products may not interfere with the features and attributes of the S-401 base cell. This interoperability must be enforced and regulated. Even the interoperability with S-101 has to be described. In a first step IHO will describe this interoperability for S-101 in the product specification and the S-98. Once IHO has finalized the interoperability it's up to the IEHG to get in line with S-101.

4 Conclusions

The development of the S-401 for electronic inland navigational charts is a process that is ongoing and will take a couple of years before it can be implemented. Due to the compliance with S-101, the IEHG is relying on the development within IHO and therefore the IEHG is working together with the S-101 working groups. This will prevent any inconsistencies between the two worlds.

Exploration of Digitalization System and Technical Solutions for Inland Waterway

Jun Huang[✉], Haiyuan Yao, and Zhengyong Chen

Transport Planning and Research Institute, Ministry of Transport, Beijing, China
{huangjun,yaohy,chenzy}@tpri.org.cn

Abstract. The inland waterway is a crucial channel for external liaison in inland areas and is an essential carrier for inter-regional bulk commodities transportation. In general, the current mode of development and construction of inland waterway is still relatively traditional, and the level of waterway management cannot adapt to the requirements of information and digitization. To effectively improve the level and efficiency of inland waterway management, this paper is dedicated to using GIS, remote sensing, and other technologies to innovate the means of inland waterway management and explore the basic scheme suitable for the development of digital waterway in China. Firstly, it reviews the situation of digital waterway development and construction in China and abroad, and outlines the results achieved by existing research. Secondly, from the perspective of serving the information management of inland waterway, the basic structure of the digitalization system is proposed, including the object, scope, and specific contents of digitalization. At the same time, it outlines the main framework of the digital indicators for constructing a port big data center. Thirdly, it analyzes the development ideas to provide a basis for future planning, construction, and management of digital waterways. Finally, this paper puts forward relevant suggestions on data collection and standardization, technical solutions, and work implementation from the perspective of the entire life cycle development of inland waterway planning, construction, management, and maintenance. This paper can help the technical selection for the planning and construction of inland digital waterway and the development of port and shipping big data center.

Keywords: Inland · Digital waterway · Port · Standardization · Big data

1 Introduction

China is one of the most developed countries for inland waterway transportation. 2020, the country's inland waterways navigable mileage is 128 thousand kilometers, of which 67 thousand kilometers of graded waterways, accounting for 52.7% of the total mileage. The mileage of Class III and above waterways is 14 thousand kilometers, accounting for 11.3% of the total mileage. These waterways provide excellent support for the economic development of the areas through which they flow. They are essential to support China's inland areas' economic growth and significantly affect the gathering of various resource factors. For example, the GDP of the seven provinces and two cities along the Yangtze River waterway accounts for 40% of China's total GDP. Given the

Y. Li et al. (Eds.): PIANC 2022, LNCE 264, pp. 838–849, 2023.
https://doi.org/10.1007/978-981-19-6138-0_74

prominent role of inland waterways in developing inland areas, China has made the development of inland waterway shipping a strategic priority in building an integrated transport system. Technological innovation is the key to achieving the relevant policy goals. These policies are significant to exploit the comparative advantages of inland waterways, construct an integrated transport network, and build national strength in transportation.

As the scale of inland waterway development and modernization improves, the complexity of waterway construction, operation, and management increase continuously. Integrating business logic at all levels through digitalization means simplifying workflow and enhancing work efficiency are crucial to the current development of the inland waterway shipping industry. At the same time, with the deepening of national green and coordinated development concept, inland waterways are the critical link of transportation and the important content of territorial spatial planning. There is an increasing need for information exchange between parties in the inland navigation world, which puts forward an urgent demand for the development and construction of digital waterways. Only by continuously improving the level of digitalization and informatization of waterways can we adapt to the national big data strategy and the needs of regional collaborative development.

In recent years, new concepts such as smart waterway, digital waterway, and digitalization of waterway have emerged. This paper does not explore the conceptual differences of the above terms. From the perspective of engineering application, the digital waterway in this paper refers to the comprehensive management system of digitization, networking, and visualization of inland waterway facilities and related activities by comprehensively using various information technologies such as big data, the internet of things, mapping, GIS, etc. To this end, this paper will study the basic framework system of digital waterway construction and analyze the relevant technical paths.

2 Review of Digital Waterways Development

Digital waterway is of strategic importance to the development of inland waterways, and there are many practical engineering cases in the development and construction of digital waterways in China and abroad. On the whole, foreign digital waterway construction starts earlier and plays a vital role in inland waterway transportation.

In the late 1990s, several countries started to work on information systems for inland shipping, and the EU proposed River Information Services (RIS). RIS is the harmonized information service to support traffic and transport management in inland navigation. RIS has the goal of a safe and efficient transport process and thus contributes to intensive use of inland waterways, and is open for connections to the commercial sector. The functions of the RIS cover traffic information, waterway information, traffic supervision, emergency rescue, transportation and logistics information, law enforcement information, statistical information, fee collection and other service areas. RIS is a cross-regional, cross-department and cross-business system collaboration and resource integration for inland waterway shipping. It includes interfaces with other transport modes on sea, roads, and railways. It will collect, process, assess, and disseminate fairway, traffic and transport information.

European research projects paved the way towards full deployment of RIS. For example, the Central Commission for the Navigation of the Rhine (2002) adopted the guidelines and recommendations for river information services in 2002 and a subsequent revision in 2004. Then it was adopted by the Working Party on Inland Water Transport of Inland Transport Committee of United Nations Economic Commission for Europe (2005). After the publication of the Directive 2005/44/EC (European Parliament and The Council of EU 2005), comprehensive RIS deployment projects started in all European countries with connected inland waterways. These RIS Guidelines describe the principles and general requirements for planning, implementing and operational use of RIS and related systems, laying the foundation for the widespread use of the RIS in various countries.

Since the new century, the road of digital development of Chinese inland waterways has made significant progress. Especially the Yangtze River and Xijiang River as the representative of the two major trunk channels, have long attached great importance to the role of information technology to promote the cause of shipping and achieved remarkable results in the construction of digital waterways. The large-scale information work on the Yangtze River started in 2001. So far, several channel information projects have been carried out around the information network infrastructure, channel measurement equipment, electronic channel map, and business applications. With the implementation of the projects mentioned above, the Yangtze River waterway management department has realized the data collection and real-time monitoring of navigation aid, water level, and tugboat in the waterway, as well as the digitization of the relevant business management, which has comprehensively improved the waterway maintenance management and information public service capability. With the Yangtze River electronic navigation chart and related information service system and monitoring platform, 95% of the Yangtze River waterway mileage has been digitized for maintenance management. New technologies such as BIM, unmanned aerial vehicle (UAV), unmanned ship, and Beidou navigation have been widely used in engineering scheme design, topographic survey and monitoring management, etc. The information technology has been gradually expanded to the whole waterway from some business areas or local reach.

In the standardization of digital waterways, China has introduced several industry standards, covering the project construction of digital waterways (JTS/T185-2021), quality inspection (JTS/T267-2021), information exchange (JTS/T184-2021), electronic navigation chart (JTS 195-3-2019), remote monitoring and control for navigation aids (JT/T 788-2010) and other aspects. The introduction of these standards has standardized the basic methods of digital waterway construction from different perspectives, such as data collection, information exchange, platform construction, and network deployment.

3   The Basic Framework of Digitalization System for Inland Waterway

3.1   Review of Digital Waterway Research

With the expanding demand for information and digital applications in the development of inland waterways, research in digital waterways has made great progress in the past 20 years in China.

Li (2004) presented the theoretical system and the construction frame of digital navigation channel. Then the paper discussed the realization of digital navigation channel in Ganjiang River and developed the navigation channel management and maintenance model as well as the riverbed evolvement analytic model. It integrated the one and two plane dimension mathematic model of water flow and sediment with GIS, which can greatly help us to make a decision for the implementation of channel regulation engineering. Wan et al. (2004) analyzes the origin of the Yangtze River "digital waterway", as well as the meaning, aim and principle of building a digital waterway. The paper describes the overall frame, main contents and key technique of constructing the "digital waterway" in Yangtze River.

The above-mentioned studies put forward a preliminary concept of digital waterway construction, but they are constrained by the development stage, technical conditions, etc. In recent years, with the leap of modern science and technology, the pace of development of digitalization and informatization of inland waterway has been accelerated, and the research on digital waterway has been deepened and improved.

Li and Yan (2014) presented a new method based on the dynamic information from navigating ships to overcome the insufficient frequency based on the professional measurement in surveying and mapping, also the lack of space perception based on fixed point digital perception used Internet of things. The key technology was analyzed with multi-sensor integration, multi-source information quality evaluation, and multi-source data fusion. Then a comprehensive perception system with terminal perception, hydrographic survey, and navigating ships perception were reconstructed. This paper made an in-depth exploration of the data collection and processing involved in the construction of digital waterway. Lu (2016) proposed an overall framework of digital waterway for Yangtze River consisting of platform layer, data layer, application layer, service layer and support layer. It analyze the content of digital waterway construction plan, progress and achievements, and put forward the key requirements on the development of digital waterway for Yangtze River. Li (2017) summarized the technical research progress of inland digital waterway and briefly demonstrated the application of primary technological results in the planning, construction, maintenance, management and services of national high-class waterways, such as the Yangtze River, Xijiang River, Beijing-Hangzhou Grand Canal, etc. It pointed out the overall demands for the development of the technology from the prospective of orderly pushing forward the development of inland digital waterway. At the same, the paper discussed technical bottleneck in aspects of the perception, data, maintenance and management, public service of waterway information.

The established studies mainly combine with engineering cases related to the paper and study the ideas and ways of development and construction of digital waterways from many aspects such as main contents, technical conditions, etc. This paper attempts to build a framework for developing and constructing digital waterways in a universal sense based on the studies mentioned above, analyze its fundamental laws, and provide references for the long-term development of digital waterways in China's inland rivers.

3.2   Physical Objects of Digitization System

Digital waterway is the digital simulation and restoration of the natural world waterway development system. The construction of digital waterways is an inevitable requirement for the inland waterways themselves to improve their management level and efficiency and to realize the intelligent development of ports and shipping. The physical objects in the digital system of the waterway include not only the waterway itself but also the objects served by the waterway and the related activities, as well as the objects that have a limiting effect on water transportation activities.

Specifically, the physical objects of waterway digitization can be divided into 3 aspects, as shown in Fig. 1.

Fig. 1. Physical objects of the digital system.

First, the basic object, including the basic form of the river and the waterway infrastructure, is the primary carrier of the digital waterway. The basic form of the river is the spatial geographic information of the river, which involves critical indicators such as river basin, morphology, water depth, geology, water flow, and sediment movement. By carrying out the regular underwater topographic survey and studying the characteristics of water flow and sediment movement, we can provide the basic plan for building a digital waterway. The waterway infrastructure is mainly the position and maintenance scale of the channel and the navigation aids and revetment. The 2D digital channel models can be established by collecting and producing an electronic navigation chart. In turn, holographic navigation scene maps can be generated to expand the application area of the digital waterway combined with the elevation data.

Second, the service target, including ship, dock, anchorage, and water service facility. The ship data collection focuses on tracking the ship navigation trajectory and monitoring the loading and unloading activities at the quay berth. So it can realize the closed-loop tracking of the ship and cargo flow trajectory and create conditions for the logistics supply chain analysis based on port and shipping big data. The Port terminal, anchorage, and service area are the supporting facilities for ship arrival operation. Based on the electronic navigation chart, the information of wharf attributes is collected to enrich its content further, thus forming a big data resource pool for port and shipping.

Third, the conflict object, including ship lock, hydro-junction, bridge, and river-crossing pipeline, are external influencing factors for the development and construction of the waterway. Among them, bridges and pipelines have a specific constraint effect on the construction of waterway and ship navigation. Ship locks and hydro-junctions will significantly increase the complexity of ship navigation and extend the navigation time, but can improve the navigation conditions to a certain extent, thus creating the possibility for larger ships to navigate in the river. The construction of a digital waterway should comprehensively collect the information on the business attributes of the above hydraulic structures and integrate the development of the shipping industry with water conservancy and integrated transport to limit the negative impact of external factors and expand the positive effects.

In order to outline a complete digital portrait of physical object, this paper will describe each object in the following four dimensions: physical hardware, integrated management, production and operation, and time series, as shown in Table 1. These four indicators, which summarize the development and construction of various physical objects from different perspectives, can be used as the 1st level indicators of the data structure of physical objects. Combined with the characteristics of every entity, the corresponding 2nd level indicators can be further developed to form a complete indicator framework for a comprehensive description of physical objectives and as a basis for data collection and maintenance. It is important to note that each physical object contains only some of the selected metrics in Table 1, not all of them.

Table 1. Digitized content (classification of indicators)

	Dimension	Indicators
1	Physical	Location, space size, morphology, scale, physical conditions and laws
2	Management	Engineering solutions, technical grade, development positioning, main usage
3	Operation	Investment, consumption, product, emission
4	Time	History, current, future (near-term, medium-term, long-term)

3.3   Basic Framework

The construction of digital waterways should make full use of the existing advanced technology and adhere to the demand-oriented to meet the requirements of digital

China and intelligent transportation development. The construction process should focus on resource integration, take business applications as the entry point, and promote the integration of information technology and waterway construction, maintenance, and services. At the same time, the exchange and sharing of waterway data with port, integrated transport, water conservancy, and other related departments should be strengthened to achieve cross-river, cross-business, and cross-department work collaboration and give full play to the economic and social benefits of digitalization.

This paper divides the framework of digital waterway construction into six layers: hardware foundation layer, perception layer, communication layer, data layer, support layer, and business layer, as shown in Fig. 2. The leading roles of each layer are as follows.

The hardware base layer is the basis of the digital waterway, including three physical objects and information infrastructure. Among them, the latter includes the type and quantity of hardware facilities required to construct a digital waterway, which should be determined after a comprehensive assessment based on the specific content of digitization and the scale of data. At the same time, the equipment capacity should be left with the necessary surplus to ensure that the incremental demand for data and the possible functional expansion are met shortly.

The perception layer, communication layer, and data layer are mainly to build a big data center for port and shipping. Among them, the role of the perception layer is to sense and obtain various types of information by using appropriate data acquisition means, such as arranging sensors for the physical objects of the digital waterway. The communication layer is for data from different sources, using 5G, Internet, and other technical means to transmit it back to the big data center. At the same time, the relevant operating instructions of the system will be sent down to the entity objects to realize timely remote monitoring and control. The data layer will use appropriate storage and calculation methods, such as Hadoop, according to the structural characteristics of the data, to establish a big data center for port and shipping to realize efficient storage, calculation, and interaction of all collected data.

As a unified system platform, the support layer provides essential services for users, such as identity authentication, user management, application management, etc. At the same time, the support layer also provides a critical business support platform, including the GIS platform, electronic navigation chart, etc. Various geographic elements are packaged in map services within the GIS platform to achieve comparative analysis, unified management, and display.

The business layer includes various business application modules and the working interface for users to log in and access. According to the system's positioning, the responsibilities of the inland waterborne transport sector, and the availability of data, specific function modules can be developed and authorized for relevant users. It is worth noting that both the particular business workflow and the fusion analysis of multi-source heterogeneous data should simplify the operational logic and reduce complex human-computer interaction, thus increasing the interest and reliance of users in using the system.

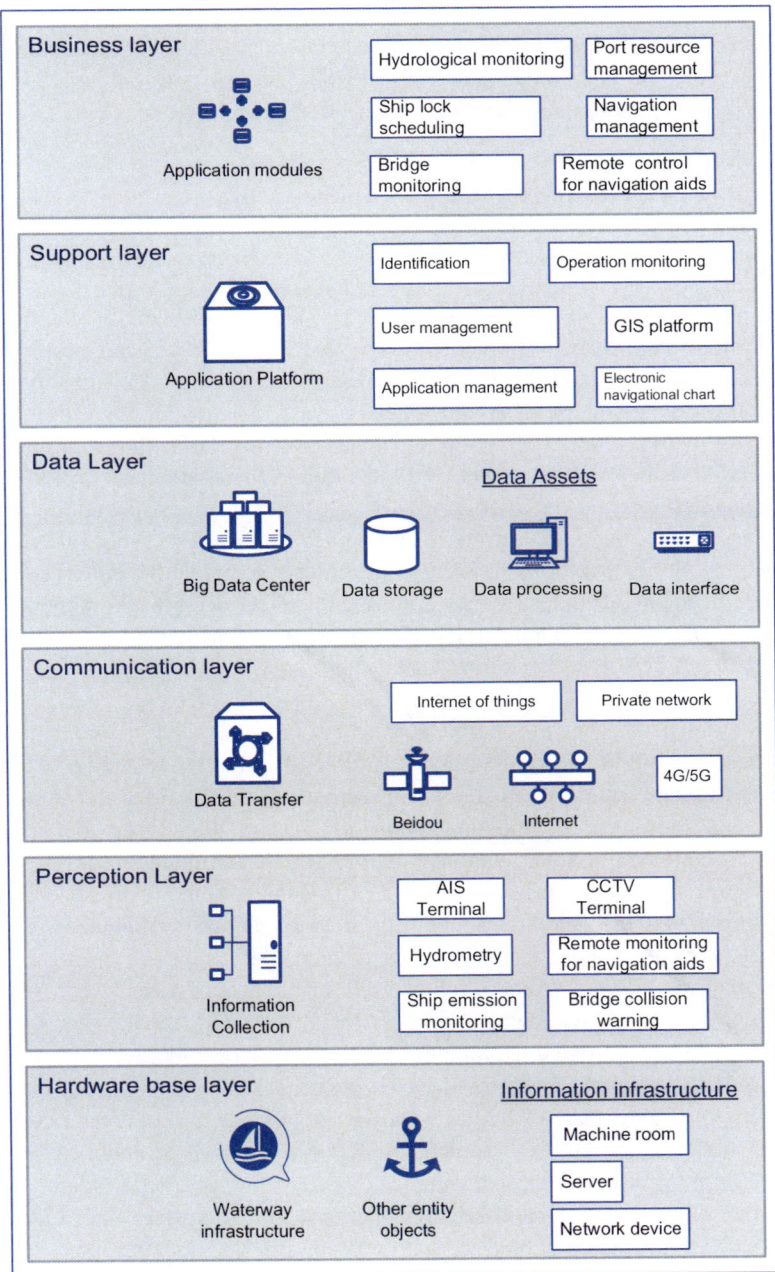

Fig. 2. Layered architecture of digital waterway.

The above layered architecture defines the basic hardware and software environment required for the construction of digital waterway. Usually, it should optimize the content of each layer by combining the specific needs and goals of data waterway construction. On this basis, relying on digital twins technology, the management of the waterway can realize interactive mapping from physical space to digital space, which provides an efficient and convenient information platform for the administration, decision-making, and simulation analysis of the waterway.

4 The Idea of Digital Waterway Construction

4.1 Implementation Path

The construction of a digital waterway is not just simple software development but complex system engineering. It should be oriented around goals and problems and choose the optimal path suitable for engineering characteristics. Usually, the implementation of a digital waterway should focus on the following four aspects.

Firstly, set the work target of the digital waterway, and clarify the main purpose of the information system and the object of digitization. Special attention should be paid to the target audience of the digital waterway. The specific application needs of enterprise users and government users of the digital waterway are very different. The former focuses on the operational aspects of the enterprise, while the latter is primarily concerned with social attributes such as investment and services. At the same time, the waterborne transport sector should fully sort out the business needs and focus on the problems that can be solved by information technology.

Next, develop the implementation plan of the digital waterway. The focus is on developing a detailed technical scheme, sorting out business logic, and designing the user interface. It should identify the possible problems in hardware and software, data collection, technical conditions, etc., and propose reasonable solutions. To facilitate the expansion and step-by-step implementation of system functions, the necessary redundancy of software and hardware capabilities should be retained.

Thirdly, the technical standards related to the construction of digital waterways should be formulated, the specific content of the data to be collected should be clarified, and a feasible collection scheme should be formulated. Data is the most critical content of an information system. Without enough rich, accurate, and timely data, the construction goal of a digital waterway is difficult to achieve. The above work should be carried out based on the existing standards and norms and combined with the characteristics of the project and application scenario to ensure that the relevant guidelines and programs meet the needs of digital waterway development and construction.

Fourth, after the system is completed, it should be well maintained and data updated. We should develop the maintenance, security emergency, and data updating mechanism to ensure the safe and reliable operation of the system. In addition, the waterborne transport sector should budget well to provide the necessary ongoing funding for system maintenance.

4.2 Implementation Suggestions

i. Attach great importance to data collection and continuous updating.

Data is the core asset of a digital waterway. Great importance should be attached to the collection and continuous acquisition of waterway, port, and related data. On the one hand, it needs to improve the hardware facilities of the waterway survey and sensing network, strengthen the dynamic monitoring of the operation of the critical navigable structure, and guarantee the stability of data sources. On the other hand, to realize the normalized and digitalized supervision of inland waterway, an effective mechanism for continuous updating of data should be established, and all stakeholders should be bound to share information promptly.

ii. Establish a unified digital waterway software system.

The system constitutes a complete business chain of data sensing, transfer, storage, analysis, display, and service, realizing the fusion processing of multi-source heterogeneous data and providing a one-stop service. The business modules will be developed around specific work requirements to serve waterway operation and management decisions and give full play to the value of data.

It is worth noting that, during the last two decades, a significant number of systems dealing with vessel traffic and transport management have been developed, and some are in operation. The inland waterborne transport sector is now faced with the challenge of integrating these building blocks into a common architecture.

iii. Increase the application of new technologies.

These new technologies are the basic guarantee for the construction of digital waterway, including but not limited to:

- the application of Beidou navigation system in the field of waterway survey, remote monitoring and control for navigation aids, ship supervision;
- the application of BIM and GIS in the design, construction and management of waterway engineering;
- the application of UAV and unmanned ship in the land and underwater topographic survey;
- the application of internet of things in the elements perception of waterway, and the big data technology in the fusion analysis of multi-source heterogeneous data, etc.

iv. Focus on information exchange and sharing.

The construction, management, and operation of the waterway are not the responsibility of a single department but involve the coordination of multiple departments. Therefore, the construction of the digital waterway, especially the management of data, should not be isolated. Under the premise of guaranteeing data security, the waterway data should be opened reasonably to achieve multi-sector and multi-level information exchange and sharing. Namely, it should provide diversified waterway information public services for ships, marine enterprises, management agencies, and the public.

5 Conclusions

The construction of digital waterways is to make full use of information technology to digitize all kinds of related physical objects such as waterways, wharves, and anchorage, and develop business modules to form a software system to realize dynamic simulation and effective control of the real world. It provides efficient, visualized, and accurate data analysis support and a business platform for a series of problems faced in the planning, construction, maintenance, and management of inland waterways. To effectively improve the level and efficiency of inland waterway management, this paper is dedicated to exploring the basic scheme of the digital waterway, which can help the technical selection for the planning and construction of inland digital waterway and the development of port and shipping big data center.

In general, information technology is becoming more mature and can effectively support the construction of digital waterways. Technology is no longer a constraint on the construction of the digital waterway. The key factors affecting the construction of a digital waterway are mainly the following: first, the accurate positioning of the use and target audience of the information system; second, how to effectively attract users to apply the digital waterway system to solve practical work problems. The former requires the inland waterborne transport sector to fully sort out the business needs and focus on the issues that can be solved by information technology. The latter requires that each business module design a relatively simple operation logic, reduce complex human-computer interaction, and avoid adding new burdens to users, thus increasing their interest and reliance on the use of the system.

References

Central Commission for the Navigation of the Rhine (2002) Guidelines and Recommendations for River Information Services. https://www.ccr-zkr.org/files/documents/ris/guidelines20_e. pdf

European Parliament and the Council of EU (2005) Directive 2005/44/EC of the European Parliament and of the Council on harmonized river information services (RIS) on inland waterways in the Community, Official Journal of the European Union, Strasbourg

Li TB (2004) Study on Construction of Digital Navigation Channel of Ganjiang River. Hohai University, Nanjing, pp 1–33

Li XX, Yan XP (2014) Overview of elements perception of Yangtze River waterway based on dynamic information from navigating ships. Port Waterway Eng. 2014(12):31–36

Li XX (2017) Summary and prospect of the technology of inland digital waterway. In: 2017 4th International Conference on Transportation Information and Safety (ICTIS), Banff, AB, Canada, pp. 163–171

Lu Y (2016) Progress and reflection of Changjiang digital waterway construction. Port Waterway Eng. 2016(1):12–14

United Nations Economic Commission for Europe (2005) Guidelines and recommendations for river information services. https://unece.org/fileadmin/DAM/trans/doc/finaldocs/sc3/TRANS-SC3-165e.pdf

Wan DB, Li GX, Yan CP (2004) A plan on constructing "digital waterway" in the Yangtze River. Port Waterway Eng. 2014(11):22–24

EURIS (European River Information Services System) – The Central European RIS Platform

Introducing a Joint RIS System Among 13 European Countries

Thomas Zwicklhuber[✉] and Mario Kaufmann

via donau – Österreichische Wasserstraßen-Gesellschaft mbH,
Donau-City-Straße 1, 1220 Vienna, Austria
{thomas.zwicklhuber,mario.kaufmann}@viadonau.org

Abstract. The development and implementation of the River Information Services (RIS) concept started in the end 1990s with various research projects followed by national or regional implementation projects in the first decade of this century. The resulting national RIS systems haven't been able to exploit the full potential of RIS when it comes to cross-border data exchange and interoperability. To overcome these gaps the concept of RIS Corridor Management was established aiming at linking the fragmented services together on a corridor to supply RIS along the complete route or network. The concept of RIS Corridor Management was taken up by the CEF (Connecting Europe Facility Programme) co-funded multi-beneficiary project RIS COMEX (www.riscomex.eu) with the goal to implement harmonized RIS services on European level. Within the RIS COMEX project the consortium of 13 countries realized a common and centralized single access point to Inland Waterway Information, the European River Information Services (EuRIS) System. EuRIS acts as European RIS platform fulfilling a great variety of information needs of inland waterway stakeholders like skippers, vessel and infrastructure operators, logistics and authorities. The system gathers relevant RIS information from the national systems in order to provide optimized fairway-, infrastructure- and traffic-related services in a single point of access for the users enabling reliable route- and voyage planning and sharing as well as traffic- and transport management on pan-European level. EuRIS provides access to its services via a user-friendly Graphical User Interface (GUI) or machine-readable Open Application Programming Interfaces (API).

In order to guarantee sustainable operation of EuRIS a legal, organizational and financial framework has been setup by the partners. The core aspects concern the joint governance of the system operation as well as the legal basis for RIS data exchange and usage. The full operation and further development of EuRIS is a major milestone in the sector enhancing attractiveness and competitiveness of Inland Waterway Transport in Europe and setting the basis for connectivity to other transport modes and synchro modal logistic operations.

Keywords: River Information Services (RIS) · RIS enabled corridor management · Single access point · Inland Waterway Transport (IWT) · European River Information Services System (EuRIS)

© The Author(s) 2023
Y. Li et al. (Eds.): PIANC 2022, LNCE 264, pp. 850–856, 2023.
https://doi.org/10.1007/978-981-19-6138-0_75

1   Introduction and Approach

Starting at the beginning of this century, the implementation of River Information Services (RIS) in Europe was focused mainly on national level resulting in a lack of interoperability between the national RIS systems realizing that the services to support transport management are limited and the potential benefits could not be utilized. Therefore, Corridor Management was considered as the next step in the development of RIS.

Within the EU-funded project CoRISMa (Corridor RIS Management) several EU member states defined the concept of "RIS enabled Corridor Management on Inland Waterways in Europe" under the coordination of Rijkswaterstaat (NL) in 2014–2015. RIS Corridor Management aims at linking the existing fragmented services together on a corridor in order to supply RIS not just locally but along the complete network.

In order to overcome the existing grown fragmentation of RIS as setback hindering the development of new services towards synchro modality and competitivity of Inland Waterway Transport, partners from 13 European countries (Austria, Belgium, Bulgaria, Croatia, Czech Republic, France, Germany, Hungary, Luxembourg, Netherlands, Romania, Serbia, Slovakia) joined their forces within the CEF co-funded multi-beneficiary project RIS COMEX starting in 2016 in order to realize the concept of RIS enabled Corridor Management along the European Inland Waterway Corridors.

In order to provide the specified RIS Corridor Services to the users, the RIS COMEX consortium agreed to realize a common and centralized RIS platform, the EuRIS system. EuRIS, an adapted clone of the existing Flemish VisuRIS system, was advanced to serve as European RIS platform fulfilling a great variety of information needs of inland waterway stakeholders.

2   System Architecture

The EuRIS platform consists of several core components as illustrated in Fig. 1. The central EuRIS environment depicts a virtualized processing hub gathering all required data from the national infrastructures of the 13 EuRIS partners.

The Data Sources provided by the national systems of the 13 EuRIS partners feed all required data into the central system. First, the reference data including the digital waterway network, the RIS Index objects or facility files of objects build the basic layer of the platform. Second, fairway information like Notices to Skippers (fairway-, traffic-, water-, weather-, ice related information), hydrometeo data (water levels, bridge clearance, depth information), Inland ENC or object status information is attached to gain actual data on the fairway and infrastructure. Third, raw AIS[1] data is provided via a secured VPN[2] connection to enable vessel tracking and sophisticated Estimated Time of Arrival (ETA) calculations.

---

[1] Automatic Identification System.

[2] Virtual Private Network.

Two main services are provided at the moment to the outside world. On the one hand the EuRIS web portal where all information can be retrieved via a GUI and on the other hand the open APIs where all data can be retrieved via machine readable interfaces.

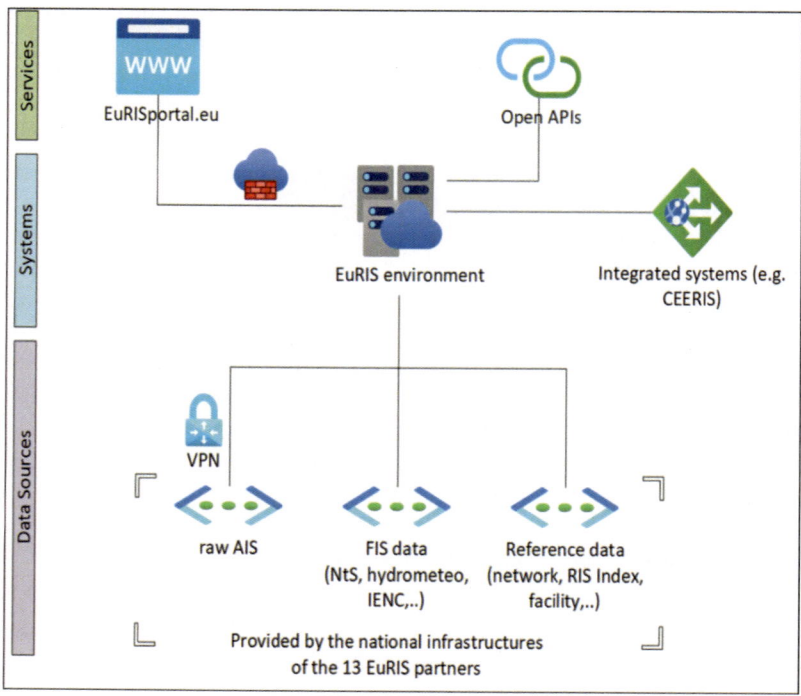

Fig. 1. System architecture of EuRIS

Data from the national infrastructure build the basis of the service provision in EuRIS. To cope with the various data categories to be implemented and interfaced, a variety of acceptable interfaces and data formats were defined on the central access point like the widely implemented Notices to Skippers Web Service but also de facto standards already known or tested.

3 Digital Waterway Network

The GIS based digital waterway network is the backbone of EuRIS representing the main interconnected European waterways. To establish such a digital network graph a Reference Network Model was newly specified defining all relevant parameters and data formats for the fairway network and objects. Together with the RIS Index data the information on the fairway objects was further enhanced.

The representation of the colored waterways based on the CEMT[3] class categorization in Fig. 2 highlights the exploit of the digital waterway network. All fairway objects like locks, bridges, harbors, terminal, berths or fairway hectometers are represented by a colored RIS Index dot depicted in the detailed picture representation.

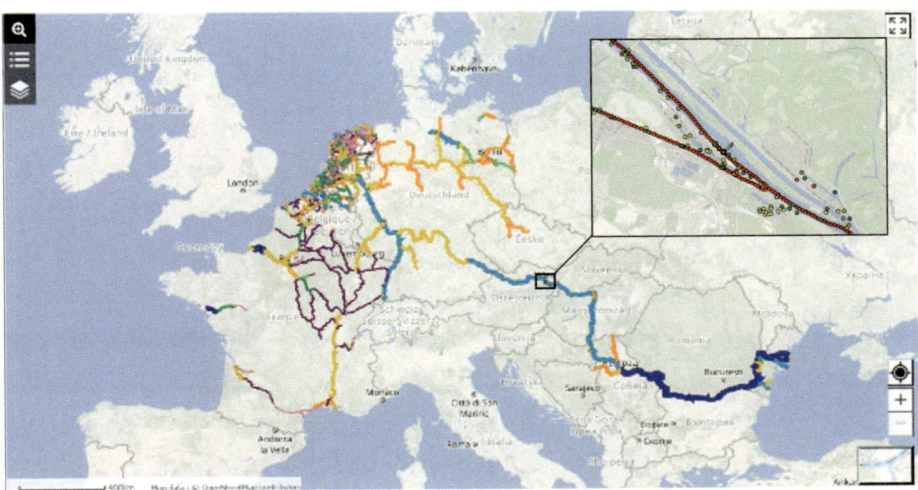

Fig. 2. Representation of the Digital Waterway Network and RIS Index data in EuRIS

4 Services Provided by Euris

By processing all input data, EuRIS provides a variety of information services to all kinds of users like skippers, vessel operators, logistics and authorities as single access point, preventing the users from having to gather all relevant information from many different websites and portals in a cumbersome approach.

The basic information layer comprises the static reference data as well as the dynamic fairway information providing most important information on limitations and blockages of the fairway or infrastructure. In addition, actual data on water levels, bridge clearance or water depth is crucial for any pre-trip planning.

Traffic related services are mainly derived from AIS position data enabling an anonymized traffic image of Europe as illustrated in Fig. 3 and access management by the data owner which is typically the vessel operator. Vessel- and voyage specific information are only available for authorized users, being authorities based on existing laws and regulations as well as skippers and vessel operators for their own vessels.

---

[3] Classification of European Inland Waterways.

Additional access permissions can be granted for third parties or logistics users by the vessel operator guaranteeing optimal transport management. A route and voyage planner combine all relevant information providing a voyage description and sophisticated arrival times throughout the European waterway network as a result.

Fig. 3. Visualisation of annonymised traffic image and berth occupation in EuRIS

5   Framework for Sustainable Operation

Cooperation between Member States is necessary to enable Corridor Management and the operation of EuRIS. In order to formalize and consolidate this cooperation in a sustainable way, a legal, organizational and financial framework has been established. Main goal of this framework was to guarantee on the one hand the legally-sound basis for the operation of the system and the related international exchange of RIS data. On the other hand, the partners had to agree on the organizational setup, the governance and financial aspects of the system operation and further developments. This solid and sustainable cooperation enables the transfer of EuRIS into sustainable operation.

The following frameworks were elaborated and signed:

- *European Corridor Management Agreement (EuRIS Cooperation Agreement):* Agreement among the EuRIS Parties towards the joint governance and operation of the EuRIS system and related Corridor Services
- *Core Arrangement 1:* Legal agreement on (GDPR compliant) data provision for Corridor RIS for vessel operators and logistics users concluded between national RIS Authorities/Providers
- *Core Arrangement 2:* Legal agreement on (GDPR compliant) data exchange between parties of the European Corridor Management Agreement in order to supply RIS enabled corridor management services and to share data with other national authorities/providers for the compliance with legal obligations

Accompanying the above-mentioned agreements and arrangements, specific data processing agreements with relevant contractors as well as data processing impact analysis were elaborated for the sake of data protection in terms of privacy related information.

6 Conclusions

With the joint introduction of the EuRIS system by 13 European countries a big step forward was achieved in providing harmonized River Information Services in Europe. Especially the formerly main drawback of locally provided services and information in a fragmented approach from user's point of view, is now solved as EuRIS serves as Single Access Point towards relevant fairway-, infrastructure-, traffic- and transport information.

By putting specific agreements and arrangements into force among the participating countries (EuRIS partners) a solid legal, organizational and financial framework was established for the sustainable joint operation and further development of the EuRIS system.

After the official launch of EuRIS by June 2022 the key focus is put on the integration of the users, interconnection of relevant systems and the further development of the provided services and functionality.

Acknowledgements. The EuRIS system was realized within the project RIS COMEX (2016–2022) which was co-funded by the European Commission in the frame of the CEF Programme (Connecting Europe Facilities). The project area covered altogether 13 European countries with 14 partners that joined their forces under the coordination of the Austrian Waterway Administration viadonau aiming at the common goal to realize Corridor RIS Services.

References

InCom PWG 125/I-2019: Guidelines and Recommendations for River Information Services
InCom PWG 125/II-2019: Technical report on the status of River Information Services
InCom PWG 125/III-2019: RIS Related Definitions
Reference is made to the Flemish VisuRIS system, operated by De Flaamse Waterweg nv, which was cloned and adapted, resulting in the EuRIS system. https://www.visuris.be/

Flow Analysis for Navigation Safety by Using iRIC Model Nays2DH

M. A. C. Niroshinie[1]($\boxtimes$), Nobuyuki Ono[1], Yasuyuki Shimizu[2], and Kazuya Egami[1]

[1] Ecoh Corporation, Tokyo, Japan
{c.niroshinie, n-ono, egami}@ecoh.co.jp
[2] Hokkaido University, Sapporo, Japan
yasu@eng.hokudai.ac.jp

Abstract. For safer navigation, it is important to reduce the shallow water effects and reduce the possibility of sedimentation. The shallow water effect is a vital factor in confluence areas where the depth of the navigation routes is subject to frequent changes during high flow and low flow times. This study summarizes the flow analysis during the process of improving the navigation channel around a confluence area by excavating shallow areas up to two possible bed elevation levels (3.3 OPm and 2.8 OPm). The analysis results, demonstrated that the depth has increased and the velocity has decreased in the excavated areas showing an improvement. Velocity has reduced from 1.1 m/s in the existing condition to 1.0 m/s for 3.3 OPm at excavation level and 0.8 m/s for 2.8 OPm at excavation level. Water depth is improved for 2.8 OPm at excavation level more than 3.3 OPm level. However, for both excavation levels water depth has been improved than the existing condition. Overall, the shallow water effect would be lessened after at least some of the proposed excavation plans.

Keywords: Flow analysis · Navigation safety · Nays2DH · Velocity field · Yodo river

1 Introduction

Inland navigation is a famous transport method since ancient times due to its cost-effectiveness and environmental friendliness. The main factors affecting navigation safety are navigable flow, water surface slope, water depth, and velocity (Zhang et al. 2019). For safer navigation in rivers, it is very important to reduce the shallow water effect which creates high velocities and turbulence under the hull increasing hull resistance. This increased resistance can cause instability of the vessels and more fuel consumption. Increased turbulence will result instability which might affect safety. Further, when the shallow areas are present, the reduced depth creates high velocities under the boat making low-pressure areas under the hull making boats squat. Therefore, the shallow water effects have to be reduced for safer navigation. This shallow water effect occurs mainly due to the presence of shallow areas on the navigation route. In inland navigation on rivers, these kinds of problems are very common especially

© The Author(s) 2023
Y. Li et al. (Eds.): PIANC 2022, LNCE 264, pp. 857–867, 2023.
https://doi.org/10.1007/978-981-19-6138-0_76

around the confluence areas as it is easier to form shallow areas due to the effect of flow separation zones at confluences. Therefore, it is vital to study possible dredging methods to improve the shallow points in navigation routes around confluences and their resulting flow conditions in the vicinity.

Yodo river in Osaka prefecture Japan, is a famous inland navigation area since ancient times for tourism and other purposes. Even though the most famous navigation area is the downstream areas of the river around Osaka bay, this study focuses on upstream areas of the river where the confluence of three rivers is present. Figure 1 shows the location of Yodo river, Yodo river basin, and navigation route in the area considered. This location is in a center of the tourism boat route from Hirakata to Yamasaki. On the navigation route, several shallow points exist near the confluence area and they have sometimes caused disturbances to the safety of the navigation in the area. Especially when two vessels cross their paths, the area may become critical. Figure 2(a) shows the bathymetry of the area and the navigation route with shallow points obtained from river bathymetric survey. Figure 2(b) shows the shallow points identified by interviewing the boat operators of the area (marked in purple and green marks according to the size) and bathymetric survey data. Figure 2(c) shows an aerial photograph taken during the green laser trial survey and the circled area shows high elevation rocky bottom. To improve navigation safety, several measures to improve the bed have been identified such as excavating the rocky bottom down to 3.3 OPm level (OP is with respect to a datum in Osaka called Osaka pale). However, when the rocky bottom is excavated to 3.3 OPm level, there are places that become deeper than the existing bed level. Therefore, the new flow field is required to analyze for velocity, depth and water surface elevations. This study summarizes the analysis of the existing condition and the conditions after the possible excavation options, for depth, velocity, and water surface elevations.

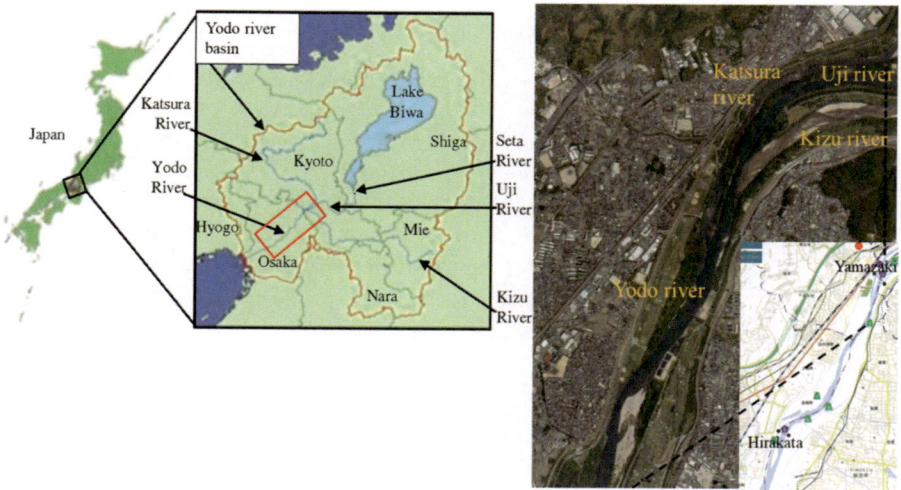

Fig. 1. The location of Yodo river, Yodo river basin and navigation route in the area considered

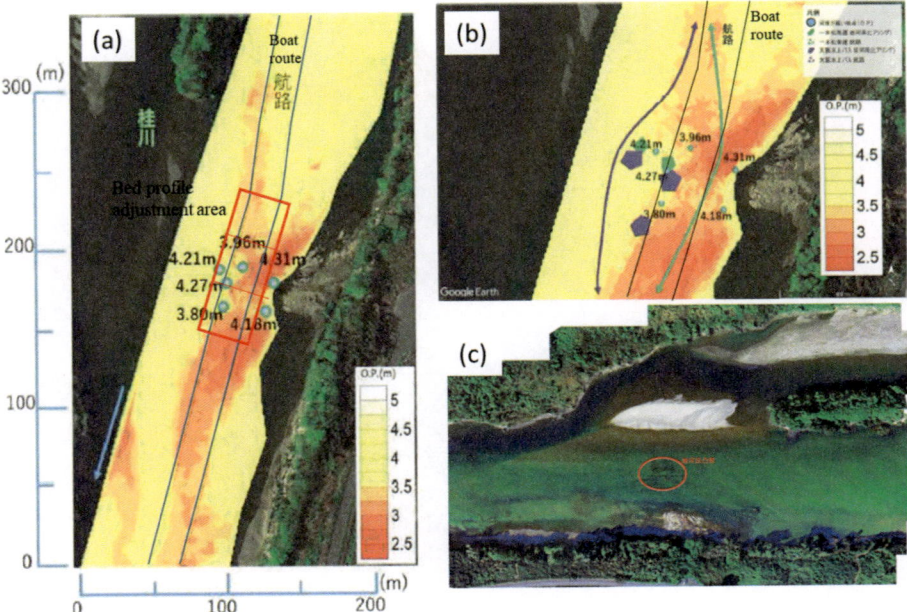

Fig. 2. (a) Elevation of the area and the boat route with shallow points (b) Shallow areas as per the interviews with boat operators (c) Arial photograph taken during green laser trial survey showing the rocky areas

2 Methodology

The basic test of improved navigation safety in rivers by excavation is to check the reduction of the shallow water effect. The reduction of the shallow water effect can be checked by the change in depth, velocity, and water surface elevation before and after the excavation along the channel in the vicinity of excavated area. Numerical simulations were carried out to identify the above effects for various possible excavation options proposed, considering the existing condition at the study location. For the numerical simulations, high-resolution bathymetry data were obtained with a high-density bathymetric survey using green laser about 1 m mesh size, and a calculation mesh of the area was then created with 5 m mesh in both x and y directions. Figure 3 shows the area modeled and river bed elevation plots of the area. Lines A, B and C are the lines along the navigation route. There are shallow points in the middle of the route as shown in the figure. The area inside the rectangle is the proposed excavation area which will be explained in Sect. 4.2.

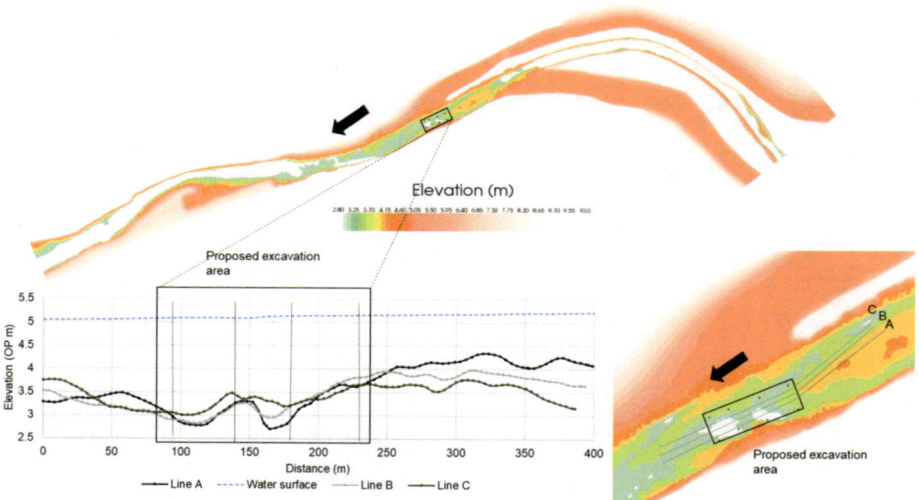

Fig. 3. Simulated area with proposed excavation area and the river bed elevation along the three lines and associated water surface elevation

3   Numerical Model

The flow analysis has been carried out with Nays2DH in iRIC GUI. Nays2DH, developed by Shimizu et al. 2014 is a computational model for simulating unsteady horizontal two-dimensional (2D) flow, sediment transport, and morphological changes of bed and banks in rivers using boundary-fitted coordinates within general curvilinear coordinates. This model has an established reputation for the calculation of unsteady flows accompanied by turbulence and laminar flow, and it is capable of dynamically showing the realistic motions of unsteady eddies.

3.1   Governing Equations

Equation of continuity and equations of motion of water are used as the governing equations. In order to consider the river shape accurately, a general coordinate system is used. The continuity and momentum equations in Cartesian coordinate system can be converted to the general coordinate system. The basic equations used in Cartesian coordinate system are shown in Eq. (1), (2) and (3).

$$\frac{\partial h}{\partial t} + \frac{\partial (hu)}{\partial x} + \frac{\partial (hv)}{\partial y} = 0 \tag{1}$$

$$\frac{\partial (uh)}{\partial t} + \frac{\partial (hu^2)}{\partial x} + \frac{\partial (huv)}{\partial y} = -hg\frac{\partial H}{\partial x} - \frac{\tau_x}{\rho} + D^x \tag{2}$$

$$\frac{\partial(uh)}{\partial t} + \frac{\partial(huv)}{\partial x} + \frac{\partial(hv^2)}{\partial y} = -hg\frac{\partial H}{\partial y} - \frac{\tau_y}{\rho} + D^y \tag{3}$$

Where, h is water depth, t is time, u is velocity in the x direction, v is velocity in the y direction, g is gravitational acceleration, and H is the total water depth. Here, the bed shear stresses in x and y directions (τ_x and τ_y) are expressed by using the friction coefficient of river bed C_f which can be given as in Eq. (4) and (5).

$$\frac{\tau_x}{\rho} = C_f u\sqrt{u^2 + v^2} \tag{4}$$

$$\frac{\tau_y}{\rho} = C_f v\sqrt{u^2 + v^2} \tag{5}$$

The friction coefficient of river bed C_f is estimated by Manning's roughness parameter n_m as in Eq. (6)

$$C_f = \frac{gn_m^2}{\sqrt[3]{h}} \tag{6}$$

The diffusion terms in x and y directions are expressed as in Eq. (7) and (8) respectively.

$$D^x = \frac{\partial}{\partial x}\left[v_t \frac{\partial(uh)}{\partial x}\right] + \frac{\partial}{\partial y}\left[v_t \frac{\partial(uh)}{\partial y}\right] \tag{7}$$

$$D^y = \frac{\partial}{\partial x}\left[v_t \frac{\partial(vh)}{\partial x}\right] + \frac{\partial}{\partial y}\left[v_t \frac{\partial(vh)}{\partial y}\right] \tag{8}$$

Here, v_t is the eddy viscosity coefficient. For the calculation of v_t zero equation model was used and v_t can be calculated by the Eq. (9).

$$v_t = \frac{\kappa}{6} A u_* h \tag{9}$$

Here κ iş von Karman constant (=0.4), u_* is the shear velocity.

4 Analysis of Flow Conditions Under Present and Proposed Conditions

4.1 Analysis of Existing Conditions

Navigable flow condition of 125 m³/s and low flow condition of 50 m³/s were considered for the initial analysis. Figure 4 shows the spatial variation of depth, velocity, and water surface elevation of the two cases. The area marked with a rectangle shows the excavation extent while the triangular marks show the shallow points. From Fig. 4, it is clearly visible that there are low depth areas along the navigation route near the excavation extent area and high velocities in the middle. Further, the velocity vectors show the concentrated flow occurring from the left bank side has increased the velocity on the navigation route. These low depth areas can cause the shallow water effect where high velocities and low pressure zones developed under the hull can cause the boats to squat depending on the speed of the boat and its specifications such as draft. Therefore, it is necessary to remove these kinds of shallow areas by excavation.

As shown on the right side of Fig. 4, the low flow condition has much shallower depths and high velocities in the middle. In this condition, it seems the depth is not sufficient for boat navigation in some locations.

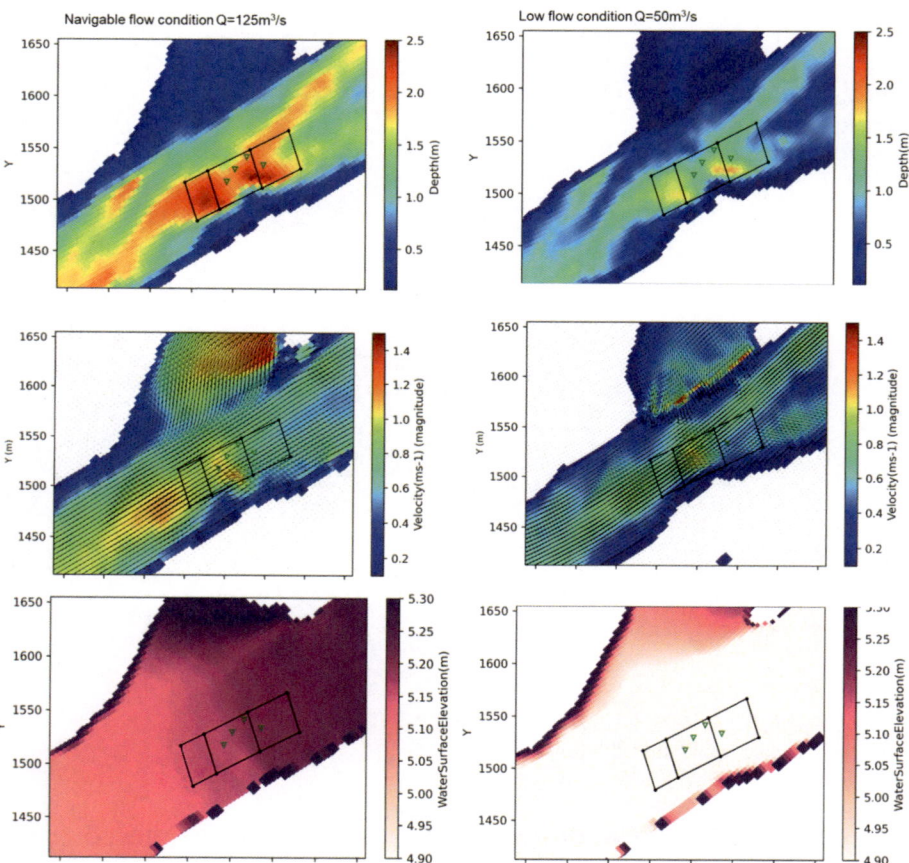

Fig. 4. Existing condition around the confluence area under navigable and low flow conditions (a) water depth, (b) Velocity and (c) Water surface elevation (Δ - shallow points)

4.2   Analysis of Proposed Excavation Options

Three possible excavation areas with two bottom levels were identified considering the navigable flow and the low flow situations. The excavation level is decided based on 200 day water level at Takahama which is 4.195 OPm. Assuming the water surface gradient between Takahama and the excavation area is the same as the average river bed gradient, the water surface elevation should be at around 4.6 OPm at the excavation area. Water depth at the location is targeted to be maintained as 0.8 m and a minimum safety margin of 0.5 m is allowed for the ship/boat passage. Therefore, the bed level should be, 3.3 OPm (=4.6 − 0.8 − 0.5 OPm). Therefore, in this study, one case is considered as 3.3 OPm bed elevation. Furthermore, according to harbor standards, a margin of another 0.5 m is required during low flow conditions. Accordingly, a second case with bed elevation of 2.8 OPm (=4.6 − 0.8 − 0.5 − 0.5 OPm) is also considered. Two main cases considering the two bed levels and extending excavation areas 6 possible scenarios are identified. Figure 5 shows the proposed excavation options and a cross-section which schematizes the excavation area in 2 different bed elevations (2.8 OPm and 3.3 OPm). Six different cases with proposed excavation options as shown in Fig. 5 were simulated.

Figure 6 shows the variation of flow velocity, the water surface, and bed elevation variation along the center line of the channel (line B in Fig. 3). From the 6 cases, it is visible that, flow velocity has reduced at the excavation area (Maximum velocity point) from 1.1 m/s to 1.0 m/s at the 3.3 OPm excavation level and 1.1 m/s to 0.8 m/s at the 2.8 OPm excavation level.

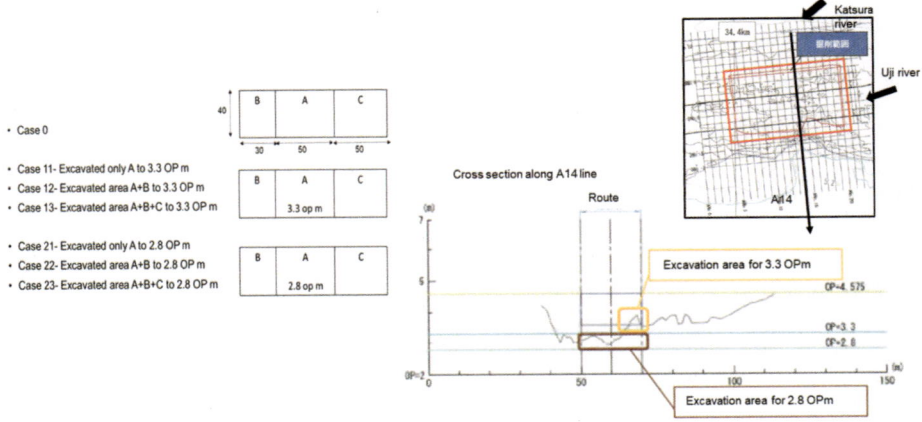

Fig. 5. Proposed excavation scenarios under navigable flow conditions and the areas to be excavated under each bed elevation along A14 line (This excavation area is different for each line)

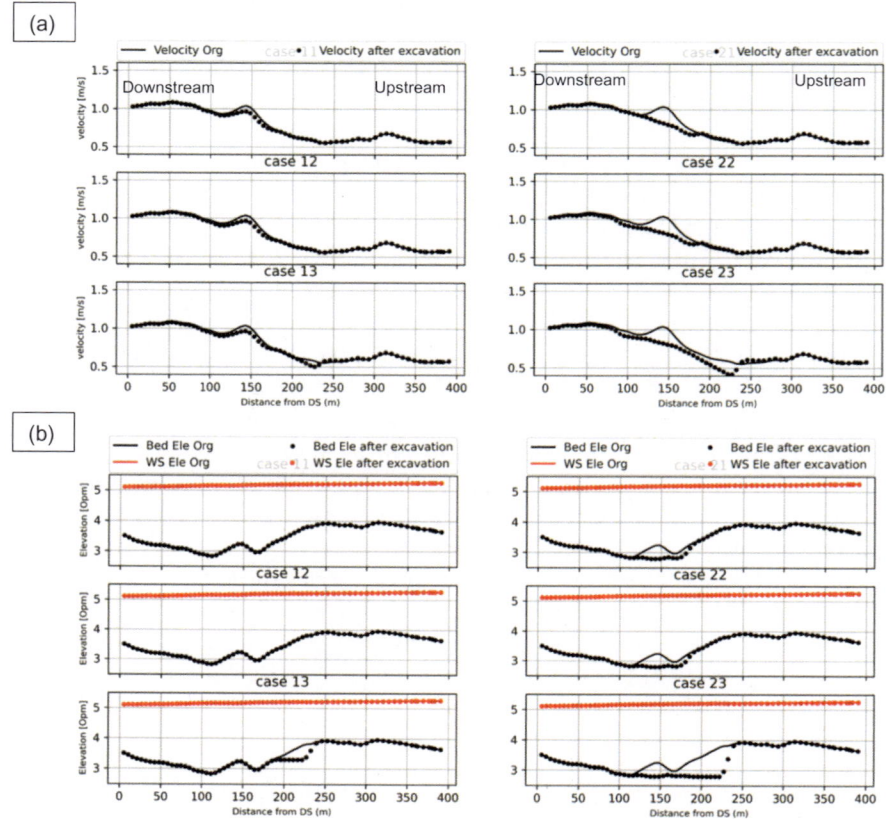

Fig. 6. Variations of (a) velocity and (b) water surface elevation for each excavation option along the center line of the channel

Comparing the elevation and velocity profiles along the navigation route for a total of seven cases of existing and with proposed excavation options, it is clearly visible that the water depth of the excavated area has increased and the velocity within the excavated area has decreased. During the excavation processes, it is important to confirm there is no significant change in water level and the velocities around the vicinity of excavated area. Otherwise, that may cause changes in sediment transport and the excavated areas may be easily filled up or some other areas might result in erosion. In this study, for all the excavation cases, no specific change in velocity expanded by the excavation effect in the downstream area was confirmed. Each case has variations in change in velocity and water depths. Out of all the cases, 2.8 OPm wider excavation extent (case 23) gives the best improved condition in depth and velocity.

Figure 7 is a spatial comparison between the case 0, case 13, and case 23 for depth, velocity, and water surface elevation. Case13 and 23 are the cases with largest excavation extent from 3 cases with 3.3 OPm level and 2.8 OPm level, respectively.

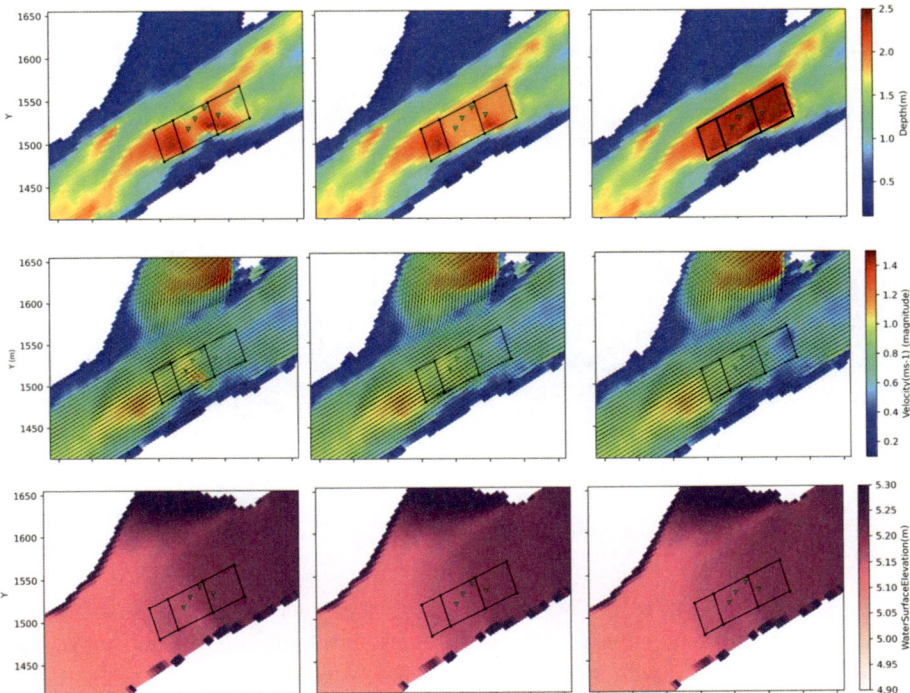

Fig. 7. Spatial variations of (a) depth, (b) velocity and (c) water surface elevation under existing condition (left), case 13 (middle) and case 23 (right)

According to the Fig. 7, it is clearly visible that the existing condition has been improved with the proposed excavation options. The turbulent area near the left bank has reduced in both cases (in the velocity plots; middle row in the Fig. 7). However, in case 23 where the excavation level is up to 2.8 OPm, the improvement is highly visible in both depth and velocity. The upstream areas and downstream areas away from the excavation extent area has no visible change. Water surface elevation shows no significant change with the existing conditions.

Figure 8 shows the variation of shear stress along the river center line which is related to sedimentation (and erosion) pattern. As seen from the figure, the shear stress has reduced along the excavated area and no specific change outside the excavated area. The shields parameter calculated assuming sandy grains with 0.55 mm–1 mm diameter shows that the sediment movement is bed load dominated. (The sand size was assumed, based on the survey of bed material in the Yodo river a few kilometers away by Azuma 2018). Therefore, the sediment entrainment would be even reduced than the present condition and thus this excavation would not increase the sediment movement which would have led to sedimentation within the study area.

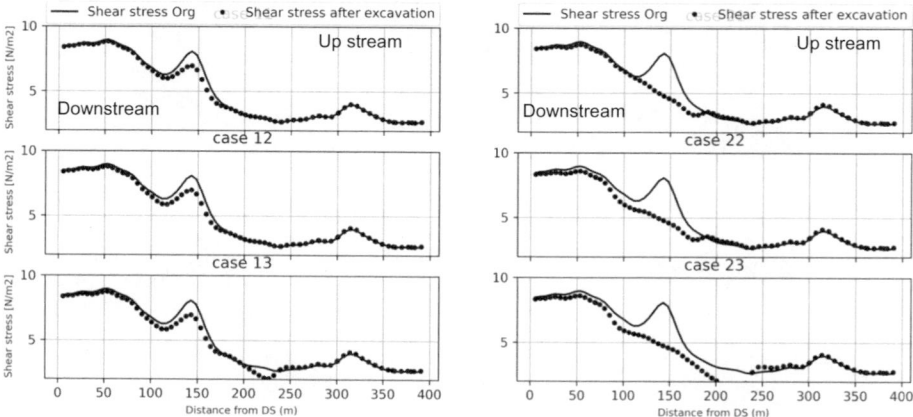

Fig. 8. Variation of shear stress for each excavation scenario along the center line of the channel

5   Conclusions

An area of 130 m × 40 m near a confluence of the Yodo river was analyzed for possible improvements to enhance navigation safety. From the bathymetric survey and the interviews conducted with local boat operators, existence of shallow areas and high velocity areas were identified. The flow analysis was carried out with 200 days of navigable flow condition of 125 m³/s compared with low flow condition of 50 m³/s to understand the existing condition. In the navigation route, high velocity areas, low depth areas and concentrated flow from the left bank side to the navigation route were identified.

As a measure to improve the safety of the navigation in the area, excavation of shallow rocky areas was identified. Flow analyses were carried out to understand the flow condition before and after the possible excavation options. Flow velocities, depths and water surface elevations were simulated with Nays2DH model (in iRIC GUI) for the 6 cases and compared with the existing condition.

The simulation results demonstrated that, the velocity reduction by increasing water depth and no significant change in water surface elevations were observed in the vicinity of the excavated areas. The current velocity is 1.1 m/s in the existing condition, and it decreases to 1.0 m/s at 3.3 OPm excavation level and 0.9 m/s at 2.8 OPm excavation level. In the case of the 2.8 OPm, the water depth is generally sufficient compared to the 3.3 OPm.

Though both excavation options were found to improve the shallow water condition, the bed level of 2.8 OPm shows more improvement. Since the shallow water effect would be reduced with the increased water depth, as per the analysis the adverse effect should be reduced with the excavation options.

To conclude, it has been demonstrated in this study that, Nays2DH 2D model in iRIC GUI can be used to analyze hydrodynamics of the river flows under various excavation options quite well in order to check the resulting improvements.

Acknowledgements. Authors would like to acknowledge the supports given from Yodo river office, Osaka and the members of the River and environmental department of ECOH corporation for their various supports. Dr Pailin Chatantavet is also acknowledged for her suggestions for the improvement of the paper.

References

Azuma T (2018) Distribution of anticancer drugs in river waters and sediments of the Yodo River basin, Japan. Appl. Sci. 8(11):2043

Shimizu Y, Takebayashi H, Inoue T, Hamaki M, Iwasaki T, Nabi M (2014) iRIC-Software. Nays2DH Solver Manual

Zhang S, Jing Z, Li W, Wang L, Liu D, Wang T (2019) Navigation risk assessment method based on flow conditions: a case study of the three river reach between Georges dam and the Gezhouba dam. Ocean Eng. 175:71–79

Marine Electrification is the Future: A Tugboat Case Study

Mark Hayton[✉]

United States Coast Guard, Buffalo, USA
mhayton9@gmail.com

Abstract. Increased emissions regulations, global volatility of petroleum supply chains, and a significant push to source energy from renewable and sustainable sources encourages companies and governments to move away from petroleum-based products. Research was conducted on the efficiencies and optimal operating parameters of internal combustion engines and electric motors, exposing situations where each would be best utilized given current energy infrastructure. To support the claim of partially electrified solutions for inland waterway vessels, an in-depth analysis was conducted for an inland waterway tugboat with a rated engine of 1800 kW. The unique operating parameters for tugboats make them prime candidates for plug-in-hybrid propulsion solutions. In this case, the 1800 kW rated tugboat operates at 360 kW or less 87% of the time. This means that most of the operating profile requires a very large engine to be running at low loads, wasting fuel. Proposing electric propulsion for operating modes that require 360 kW or less yields a 62% decrease in fuel consumption. Plug-in hybrid propulsion solutions allow for vessels to plug-in to charging stations after the completion of each voyage. Renewable sources like wind and solar, among others, directly feed the grid, permitting more flexibility in the move for sustainability. New developments in battery technology, require regulatory oversight to maintain safety compliance, specifically regarding the standardization of charging plugs and fire suppression systems for lithium-ion batteries. Implementing charging stations at frequented mooring locations will open the door for sustainable technology, like electrified propulsion solutions, to permeate the inland waterway infrastructure.

Keywords: Charging stations · Smart shipping · Efficiency · Safety compliance · Electric propulsion

1 Introduction

Energy diversification is closely tied to the independence of nations around the world. Countries that source energy from a multitude of renewable energy sources will prove resilient for generations. Global economic volatility shows the unpredictability of oil sourcing and distribution. During global crises, supply chains are largely disrupted, especially when sanctions are imposed on specific countries. To deal with economic turmoil and trade disruptions, energy diversification is the solution. Traditionally speaking, the marine industry is dominated by internal combustion engines and steam/gas turbine propulsion systems. Although slightly different in mechanical power

Y. Li et al. (Eds.): PIANC 2022, LNCE 264, pp. 868–879, 2023.
https://doi.org/10.1007/978-981-19-6138-0_77

transmission, both methods derive power from a single source: petroleum-based products. Combustion propulsion systems have been fine-tuned and optimized over the last century but are still limited by the theoretical efficiencies of the Otto, Rankine, and Carnot cycles. Over the last 40 years, new ship construction has been centered largely on internal combustion engines that have the ability to burn various petroleum-based products. Countries walk themselves into a geopolitical and economic trap when sourcing energy outside of their own territory.

In 2020, the International Maritime Organization (IMO) updated the Regulations for the Prevention of Air Pollution from Ships via MARPOL 73/78 Annex VI. IMO changed the allowable limits of sulfur content of fuel to not exceed 0.5% globally and 0.1% in emission control areas. As a result, vessels are forced to use more expensive, cleaner fuel or install costly exhaust gas scrubber systems. This updated regulation incentivizes companies to look for alternative means of energy. Hybrid or fully electric drives are the solution for the marine shipping industry. Although fully electric shipping for large-scale containerships and freighters is currently not feasible, electrification solutions for smaller operations are realistic and lucrative. Based on recent progress, it is evident that the maritime industry is moving towards a cleaner and more sustainable future.

The benefits of electric propulsion systems are numerous and have been implemented in some large-scale applications. For example, the *United States Naval Ship (USNS) T-AKE Class* ships are powered by a diesel electric propulsion system that boasts high maneuverability and quick start/stop capabilities, which greatly outperforms direct-link slow-speed propulsion engines. These engines need to be fully stopped, reversed, and started again to change the rotation of the propulsion shaft. Although the *USNS T-AKE* vessels use electric propulsion motors to actuate the shaft, the primary source of energy still resides with diesel fuel. Current technology and infrastructure deem fully electric large-scale vessels infeasible. Because of this, the commercial maritime industry is very reluctant to move away from fossil fuels. For example, it is very difficult to match the energy density of diesel fuel (42,800 J/g) when using batteries as the sole fuel source.

Throughout the world, governments will need to encourage electric propulsion in the maritime industry while still maintaining necessary standards for safety. Direct current (DC) charging stations at mooring locations give hybrid and electric commercial river tenders, recreational boaters, and transportation vessels the opportunity to charge after each voyage. Growing electrical applications across transportation and power generation industries have spurred an energy revolution backed by versatility and efficiency. However, large scale marine installation of electric or plug-in-hybrid technology remains to be implemented. Such applications are suited for vessels with shorter operating periods, an ideal situation for vessels operating on inland waters. Several small-scale applications of hybridized technology have resulted in reduced maintenance costs, increased maneuverability, and reduced consumption of petroleum-based fuel. Large-scale implementation of marine DC charging stations will yield a cleaner, lower maintenance, propulsion solution to steer the industry away from petroleum-based fuels. However, specific practices for the prevention and suppression of large-scale lithium-ion battery fires have not been published or standardized. It is essential that governments and regulatory agencies are at the forefront of battery developmental research to ensure that maritime safety remains a priority.

2 Propulsion Systems for Vessels

During the design process for new ship construction, engineers and naval architects conduct an in-depth analysis of the operating characteristics of newly constructed vessels. Sizing a propulsion system for a vessel largely relies on the needs of the customer, which is typically outlined in the customer's request for proposal (RFP) package. In the commercial maritime industry, sizing a propulsion plant for a standard vessel, like a containership or a bulk freighter, is relatively simple. The propulsion designer considers the expected tonnage and power requirements at maximum load parameters, while also meeting the speed requirement as specified by the RFP. In ideal situations, the vessel will be operating fully loaded with cargo for a majority of the time, which would maximize profits for the vessel owner. Internal combustion diesel engines are generally sized to operate at a high proportion of the maximum continuous rating (MCR), usually around 80%–90% MCR (Harrington 1992). That range of operation typically correlates to the best range for specific fuel consumption, meaning the engine operates most efficiently when fully loaded. Furthermore, smaller main engine load variations, while also maintaining the 80%–90% MCR, will lead to decreased fuel consumption. For operational characteristics of large vessels like containerships, diesel engines are functionally appropriate but retain high operational and maintenance costs.

3 A Closer Look, Inland Waterway Tugboat

3.1 Unique Operating Parameters of a Tugboat

Traditionally speaking, tugboat designers would size the rated power output of the engine according to the expected load at max towing capacity, which is usually around 80% of the maximum continuous rating (MCR). So far, this seems similar to the design profile of a containership, which is designed for optimal performance when operating at 80%–90% MCR.

To model the benefits of plug-in hybrid vessels, an in-depth review of an inland waterway tugboat will be examined. The vessel is a commercial tug with published operational data (Boyd and Macpherson 2014). See Table 1 and Table 2 below for the specific operating modes, conditions, and fuel specifications that will be referenced for the calculations and results.

Table 1. Engine and fuel data (Boyd and Macpherson 2014)

Engine data & fuel basis	
Rated RPM	1000
Rated power (kW)	1800
Density (kg/m^3)	840.00
Heating value (J/g)	42800

Table 2. Tugboat operating conditions and data (Boyd and Macpherson 2014)

Mode		Service	Speed (kt)	Time Percentage	Percent Engine Load	Engine Output (kW)	Mass flow rate of fuel (t/hr)
1	Standby	Idle	0	0.15	0.05	90	0.042
2	Transit low	Transit	6.6	0.3	0.024	43.2	0.019
3	Transit high	Transit	10	0.07	0.1	180	0.075
4	Assist	80% pull	1	0.01	0.8	1440	0.599
5	Assist	60% pull	1	0.01	0.6	1080	0.448
6	Assist	40% pull	1	0.09	0.4	720	0.305
7	Assist	20% pull	1	0.26	0.2	360	0.159
8	Barge Mv	60% pull	5	0.01	0.6	1080	0.448
9	Barge Mv	40% pull	5	0.01	0.4	720	0.305
10	Barge Mv	20% pull	5	0.09	0.2	360	0.159

Tugboats operate at 80% MCR only about 2% of the time underway. This is largely due to the small size of the vessel which requires high power output only when assisting larger vessels in maneuvering and mooring operations. This results in most tugboats operating the propulsion engines of the tug at an output much less than the MCR for 80%–90% of the time. Applying the above-mentioned procedure of sizing a diesel engine to a tugboat in order to maintain normal loading within 80%–90% of MCR creates a highly inefficient operating profile. The result is that most of the time the vessel is operating in a region of high inefficiency. For the purposes of this case study, a standard 8-h day was applied to the operating parameter percentages. Using the data from Table 2, the time bar chart was superimposed on the engine power output bar chart, as seen in Fig. 1.

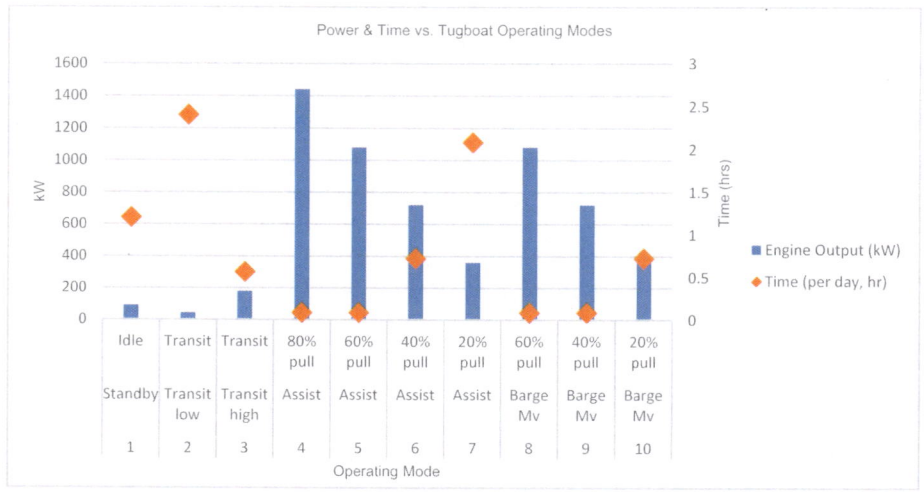

Fig. 1. Operating mode showing time allocations with coinciding power requirements

As seen above, for operational modes 4, 5, 8, and 9, a large amount of power is required for a short period of time. Conversely, for operational modes 1, 2, 3, 7, and 10, a much smaller amount of power is required for a significantly larger period of time. Considering both situations, a power threshold can be determined. The power threshold refers to the cut-off point where the vessel is operating at an abnormally low power output for a significant amount of time. The power threshold derived from this analysis is 360 kW. This means that modes that stay below 360 kW operate significantly outside of the range of peak efficiency for the rated 1800 kW engine, resulting in a high potential for fuel savings if alternative energy sources and power applications are used for these modes. Operating Mode 6 is an outlier because it requires moderate power for a moderate amount of time, approximately 518.4 kWh, with a peak power of 720 kW. In the specific case of this operating profile, the tug has a rated output power of 360 kW or less for 87% of the time underway.

3.2   Hybrid Propulsion

The sole focus of commercial enterprise in the maritime sector is centered on cost-saving measures with the goal of increasing profits. This includes maximizing fuel efficiency to decrease fuel costs, which account for a high percentage of the overall operating costs for commercial vessels. For vessels that have large fluctuations between high propulsion loads and low hotel loads, hybrid propulsion presents potential cost savings. Electric motors offer clear advantages in terms of versatility and maneuverability, namely, being able to operate at a larger range of loads while maintaining relatively high efficiency. While electric motors also tend to operate most efficiently at high loads, the efficiency taper at lower loads is much more gradual when compared to diesel engines (US Department of Energy 2014). This means the electric motors are able to retain a higher efficiency over the same operating conditions compared to diesel engines. Hybrid applications that boast large fuel savings are being implemented in various ways in industries that include ferry and tug vessels. One example of such benefits can be seen at Derecktor Shipyards of Mamaroneck, NY. Derecktor Shipyards has constructed a hybrid propulsion vessel with a 55% reduction in fuel usage by using two BAE Systems AC traction motors. Taking a closer look at Fig. 1 above reveals that electric propulsion from stored power in battery packs can be used to supplement the diesel engine for power conditions at 360 kW or less. Applying this analysis, Fig. 2 displays which operating modes should be powered by electricity.

Fig. 2. Operating modes differentiated by diesel and electric power in proposed hybrid propulsion system

The yellow-filled bars represent operating conditions to be powered by a fully electric system. The blue bars represent the operating conditions where the diesel engine will be turned on for high power output push-pull operations. Following this proposed power profile, the tugboat will use approximately 0.7 tons less fuel per 8-h working day when compared with a diesel-only vessel. Decreasing fuel usage by 0.7 tons per day equates to a 62% decrease in fuel consumption for this vessel. It should be noted that these calculations do not include the added weight of the electric propulsion system and storage batteries, which would require a capacity of just over 1320 kWh. Calculating new ship-specific power requirements and hull changes in order to implement hybrid technology are beyond the scope of this analysis. However, the estimated fuel savings shows the potential for hybridized electric propulsion in the tugboat industry and similar industries. Additionally, the 0.7 ton decrease in fuel consumption, the tugboat would only require 0.43 tons of fuel for an operational 8-h day, which would decrease the fuel storage tank capacity.

3.3   Battery Sizing

When designing a battery-electric or hybrid vessel, the primary limiting factor is the space required for battery storage. Even with modern advancements in battery technology, the volume required to store the needed watt-hours (W-HR) of energy needed for propulsion eventually limits the range and cargo capacity of the vessel. Balancing all three requires careful design of the battery storage areas.

Calculating the volume needed for battery storage first requires determining two design parameters: (1) the intended voltage (V) for the battery system, and (2) the total kilo-watt hours (kWh) needed for propulsion. With these two parameters determined, Eq. (1) can be used to calculate the Whr required for the system.

$$Watt - hours = Volt * Amp - hours = (kWh) * \left(\frac{1000\ W}{kW}\right) * \left(\frac{VA}{W}\right) \qquad (1)$$

With the required kWh for the system calculated, the volumetric energy density (kWh/L) can be used to determine the Liters of volumetric space needed for batteries. The gravimetric energy density, in kilowatt-hours per kilogram (kWh/kg) can also be used to determine the mass of batteries needed for the system.

A predictive model was constructed to approximate the weight of battery packs per kWh using the Tesla 5.3 kWh Battery Module as a standard due to the high energy density of the battery. Figure 3 below shows the trend of the weight of the battery as the kWh increases for larger applications, such as vessels. Included in Fig. 3 is the calculated kWh for the tugboat with corresponding weight of 6,216 kg.

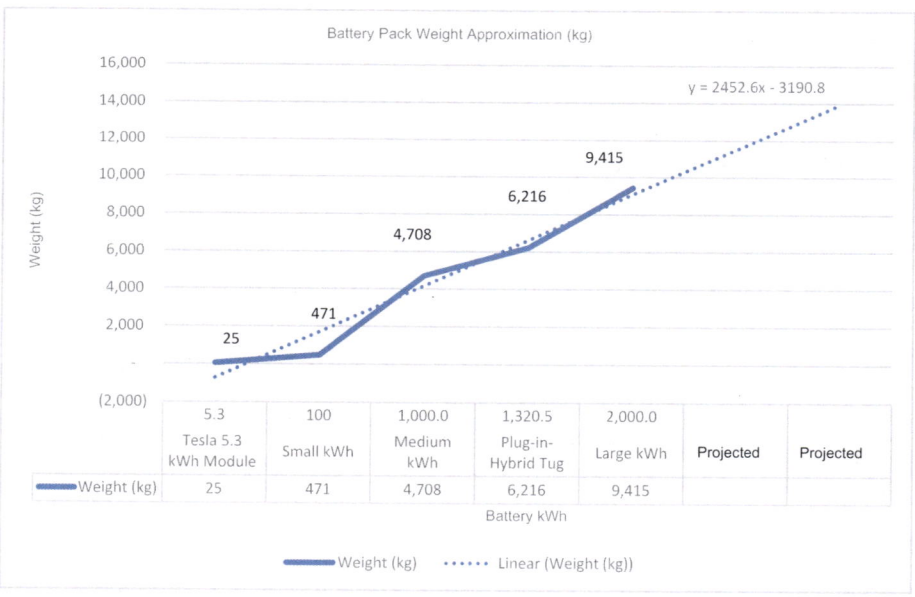

Fig. 3. Battery Pack Weight Approximation as kWh increases, based on Tesla 5.3 kWh Module

Having batteries on board a vessel that replace liquid fuel serves several advantages. As batteries are used for power, their mass effectively does not change, only about 0.5 mg of mass for a 1.7 MW battery system. When petroleum is used, the vessel becomes increasingly lighter, posing a ballast issue (Schurke 2021). As a result, batteries become solid ballast for the vessel, without the dangerous addition of the free surface effect from liquid ballast or fuel tanks.

Once the volumetric and mass requirements for the batteries are determined, several other factors must be accounted for. For example, batteries generate heat as they charge and release energy, so a cooling system must be designed to account for this heat load. Additionally, the heat load can require spacing batteries to allow space for the cooling system, which results in increased volume and mass requirements for the overall system. Fire suppression systems also account for additional space in the battery storage area.

4   Inland Waterway Infrastructure, Charging Stations

4.1   Applicability

The benefits of hybridized technology for an inland waterway tugboat operating profile are numerous, providing benefits to the vessel owner as well as the surrounding environment. These solutions are undercut by waterway electric grid infrastructure, which would take significant time and money to upgrade. Improving grid capacity for highly trafficked ports, to include recreational marinas, will give consumers the ability to explore plug-in hybrid or even fully electric propulsion options for vessels. These options are appealing to sailors and commercial operators who operate for limited hours per day, allowing for the time to recharge the battery pack(s) at night or during periods of non-operation. Current electric vehicle charging stations can be referenced when looking to install charging stations at marinas and commercial ports. Table 3 below depicts current electric vehicle charging stations with operating characteristics.

Table 3.   Typical battery charging stations (Palconit et al. 2018)

	Voltage (V)	Current (A)	Power (kW)	Type
Level 1	110	16	1.9	AC
Level 2	208/240	32	19	AC
Level 3	480	400	240	DC

For recreational sailors, installing lower-rated charging stations, like Level 1 or Level 2 chargers, would directly reduce the amount of fuel required at dockside refueling stations. Less frequent fueling operations directly correlate to a decrease in pollution incidents. For commercial operations that require significantly more power and larger capacity battery packs, like tugboats, Level 3 charging stations should be installed. If implemented, commercial enterprise and civilian sailors would have the opportunity to choose hybrid or fully electric boating options, further reducing the reliance on exclusively petroleum-based products.

4.2 Standardization

Bringing electric propulsion solutions to the maritime industry provides regulatory authorities and design engineers with an opportunity to learn from the previous implementation in the automotive transportation industry. Standardizing a grid approach is essential to streamline future upgrades to the system and accelerate widespread installation. Currently, there are three types of electric vehicle DC fast charging connections: CHAdeMO (developed by the Tokyo Electric Power Company for DC charging), SAE Combo (CCS), and Tesla Supercharger Plug. This is the result of regulatory authority lagging behind technological advancement. Simply put, private industry developed their own system and charger plugs, offering the public several options. Standardization was later implemented for Level 1 and Level 2 charging stations but remains to be updated as Level 3 charging stations become more abundant. The problem that this causes is inconvenience and inefficient charging station installment. This means, depending on the vehicle a customer purchases, only certain charging stations will be compatible. Learning from this, the marine industry has the opportunity to standardize two plugs for recreational boaters: regular charging and DC fast charging, resulting in base-line infrastructure widely available, regardless of manufacturer.

For commercial use, a higher rated standard should be developed based on the line voltage from the grid, to scale charging times for significantly larger battery packs. In the Port of Göteborg, medium voltage transformers provide 10 kV/6.6 kV 1250 kVA on the quay (Ericsson and Fazlagic 2008). Commercial ports have the available grid support to cater to larger battery packs used for propulsion, and a standard ship connection used for charging batteries, instead of supplying hotel loads, needs to be developed.

5 Solid-State Batteries

An in-depth discussion of the electrochemical properties of lithium-ion and solid-state batteries is beyond the scope of this paper. However, a brief overview and various benefits and drawbacks will be covered. Recent developments in solid-state lithium-ion battery production yield optimistic futures for recreational and commercial transportation industries. Limitations with liquid electrolytes in current lithium-ion batteries are extensive and have significant drawbacks in both performance and safety when compared to solid-state batteries. Therefore, further research in solid-state batteries may allow the maritime industry to decrease the added mass on a vessel when compared to traditional lithium-ion batteries.

The physical orientation of solid-state batteries allows for the cells to be arranged in series stacking and bipolar structures, thus allowing for less volume occupation on the vessel. As the total displaced volume of the battery decreases, ceteris paribus, the energy density increases. This is achieved by swapping the electrolyte material from an organic liquid to a solid electrolyte. This transition boasts higher electron transfer rates and further increases safety due to the inherently flammable nature of liquid organic electrolytes. Further, solid-state batteries are able to fully cycle thousands of times without losing capacity (Pistoia 2014).

Designing modular battery-pack systems in easily crane-accessible areas on the vessel would allow the battery storage pack to be upgraded over time, simply by swapping out the battery pack onto established battery mounts. Easily upgraded propulsion systems are the future of the marine industry.

6 Regulatory Oversight

6.1 Overview

Rapid advancements in technology require regulatory agencies to maintain safety compliance as a continuation of current standards, applying to shore-side infrastructure and waterborne transport. Electric propulsion systems and shore-side charging systems need to be designed and inspected to ensure tolerances for various load conditions that may arise from atypical operations. For example, systems need to be tested to withstand overloading conditions, under loading conditions, short circuit situations, and mechanical deformation of the battery packs, to name a few. The United States Coast Guard is working at the deck-plate level with manufacturers when constructing new systems, as well as reviewing plans for future builds to ensure safety compliance. Both levels of oversight are necessary to reduce the risk of failure at all system components.

6.2 Fire Prevention

Prevention of fire becomes a life or death scenario when applied to waterborne transport. There are several difficulties that present themselves when extinguishing a lithium-ion battery that will be covered in Sect. 6.3, but preventing fires in the first place is a priority. Lithium-ion batteries are at risk of thermal runaways. Thermal runaways occur when an exothermic reaction is caused by exposure to high temperatures (130 °C–150 °C) or when the battery is short-circuited. When this occurs, temperatures rise and gasses build up, which create an environment prone to fire or explosion. To prevent the occurrence of a fire or explosion, several safety measures must be installed and regulated. These measures include preventing thermal runaway using separators, preventing fire using flame retardant additives, and preventing the buildup of gas and pressure by cell venting.

Separators are components within the structure of the battery made of a semi-porous polymer in most cases. Placed in-between the positive and negative electrodes inside the battery, the separator prevents direct contact between the electrodes, which prevents an electrical short from occurring. Furthermore, as the temperature in the battery increases during a thermal runaway, the polymer material approaches its melting point, thus closing the pores and preventing the electrochemical reactions caused by the pathway between the electrodes. When this occurs, it is called a separator shutdown and is an inherently safe component of the battery structure.

Using flame retardants as an additive to the electrolyte or directly in the separator makes the batteries much safer. As the temperature in the battery rises during a thermal runaway, the electrolyte will not flash or combust when the flame retardant additive is used.

Safety vents are installed to release gasses from the battery in the event of a buildup. If the separator shutdown fails, thermal runaway will continue causing the internal venting mechanism to activate. This mechanism vents the battery abruptly, purging the gas buildup and drastically reducing the pressure to prevent a rupture or explosion. This will include toxic gas sensors and ventilation systems to remove toxic gasses created by batteries while charging/discharging. Items such as ductwork for ventilation and discharge piping and nozzles for the fire extinguishing system all must be accounted for when sizing a battery room for a vessel (Kong et al. 2018).

6.3   Fire Suppression

Subsequent systems for new technology must also be installed, and further research needs to be conducted to learn more about the thermal properties of lithium-ion batteries. For the safety of vessel operators, a fire detection and extinguishing system is necessary. When lithium batteries catch fire, they cause fires that range in classification. Additionally, thermal runaways make the extinguishing process slightly more complex. Instead of simply putting out the flame, extensive cooling is necessary to prevent a re-flash from occurring. Various agents can be used to extinguish lithium-ion batteries, including dry chemical, carbon dioxide, water, and halons to name a few. However, regulatory standards published by the Institute of Electrical and Electronics Engineers (IEEE) only focus on abuse and threshold testing of the batteries. Published data indicates that large-scale battery storage spaces require extended water hose stream application in order to fully extinguish the fire (Long and Misera 2019). There is a need to develop a system that adequately extinguishes lithium-ion batteries, cools the batteries sufficiently to prevent them from reigniting, and does not put the vessel in danger in terms of stability. There is no specific standard for extinguishing lithium-ion batteries, which leaves a major gap in the regulatory realm of safety compliance.

7   Conclusions

Electricity provides an opportunity to create a maritime industry more resilient to future energy changes. Rather than relying solely on petroleum-based energy, tapping into a diverse energy grid to source power from the growing number of renewable resources provides a flexible and secure maritime economy. There is an opportunity to start implementing electric or hybrid propulsion with smaller vessels that operate for relatively short periods of time while standardizing practices like charging station plugs. In addition to the energy diversification benefits, significant fuel savings are observed when compared to traditional internal combustion engines. It is common for vessels with large engines to operate under-loaded. These situations provide an opportunity to implement electric propulsion. A case study was conducted for an inland waterway tugboat with 10 distinct operating modes. The study yielded a 62% reduction in fuel consumption with hybridized technology. Applying this technology to vessels with similar operating modes has the potential to provide energy savings industry-wide. Progressing the grid in a more sustainable direction will take significant time and resources. As new battery technologies and applications continue to enter the maritime industry, governments need to be at the forefront of research and development in order to enforce new safety regulations that include fire prevention and suppression of lithium-ion battery fires. In order to get there, maritime authorities need to incentivize the installation of charging stations while maintaining regulatory oversight on emerging technologies to ensure safety compliance.

Acknowledgements. I would like to express my deepest appreciation to Lieutenant Commander Jon Benvenuto (USCG), PE, PMP, who offered steadfast mentorship and encouragement throughout the research process. I would also like to acknowledge the assistance of Ensign Kevin Reed (USN), Mercedes Walter, and John Hayton for providing thorough guidance and advice during the drafting of this paper.

References

Boyd E, Macpherson D (2014) Using detailed vessel operating data to identify energy-saving strategies. Hydrocompinc.com, UZMAR, Workboat and Tug Factory. https://hydrocompinc.com/wp-content/uploads/documents/Boyd%202014%20Using%20Detailed%20Vessel%20Operating%20Data%20to%20Identify%20Energy-Saving%20Strategies.pdf

Ericsson P, Fazlagic IF (2008) Shore-side power supply: a feasibility study and a technical solution for an on-shore electrical infrastructure to supply vessels with electric power while in port. Chalmers University of Technology, Department of Energy and Environment, Göteborg, Sweden, pp. 1–168

Harrington RL (1992) Marine Engineering. The Society of Naval Architects and Marine Engineers

Kong L, Li C, Jiang J, Pecht MG (2018) Li-Ion battery fire hazards and safety strategies. *Energies* 11(9):2191. https://doi.org/10.3390/en11092191

Long T, Misera AM (2019) Sprinkler protection guidance for lithium-ion based energy storage systems. Research Foundation, Research for the NFPA Mission, June 2019

Palconit EV, Abundo ML (2018) Electric ferry ecosystem for sustainable inter-island transport in the Philippines: a prospective simulation for Davao City – Samal Island Route. *Int J Sustain Energy* 38(4):368–381. https://doi.org/10.1080/14786451.2018.1512606

Pistoia G (2014) Lithium-Ion Batteries - Advances and Applications. Elsevier. https://app.knovel.com/hotlink/toc/id:kpLIBAA003/lithium-ion-batteries/lithium-ion-batteries

Schurke PW (2021) Battery-powered surf boats are not such a bad idea. Energy Fuel Matters 7–9

U.S. Department of Energy. Determining Electric Motor Load and Efficiency - Energy. https://www.energy.gov/sites/prod/files/2014/04/f15/10097517.pdf

Review of the Key Technology Research on Intelligent Locks

Maoming Xiang[1], Yaan Hu[1(✉)], and Xiaodong Wang[2]

[1] State Key Laboratory of Hydrology, Water Resources and Hydraulic Engineering, Nanjing Hydraulic Research Institute, Nanjing, China
xiangmm0209@163.com, yahu@nhri.cn
[2] Nanjing SURE TECH&DEV Co., Ltd., Nanjing, China
wxdgk@aliyun.com

Abstract. Owing to the high capacity, low costs, and environment-friendly characteristics, shipping has an important role in the sustainable development of the economy, society, and environment. However, with the development of the water transportation economy, the capacity of many navigation locks, such as the Three Gorges ship lock and the Beijing-Hangzhou Canal ship lock, has been hard to meet the increasing demand for passing through the locks. Thus, they have become the "bottleneck" limiting the navigation of the waterways. In the background of the vigorous development of intelligent transportation, the Intelligent Locks has been promoted through extensive research of the Intelligent Locks model, exclusive technology, and intelligent equipment, which is of great significance to improve the efficiency of vessels passing through the locks and enhance the function of waterways. This review starts from the current situation of the research on Intelligent Locks. We summarize the focus of the research on Intelligent Locks and the main tasks of Intelligent Locks. Moreover, the contents of Intelligent Locks are also introduced, such as the network communication platform, intelligent ship identification, positioning system, safety detection system, an intelligent guidance system for shipping, etc. Finally, we conclude that it is the informatization and intelligence that can improve the navigation ability and the management level of ship locks. At the same time, compared with the huge capital demand and the long-period construction of new ship locks, it will have obvious advantages of capital and time by attaching great importance to the automation, informatization, intelligent constructions, and applications of new technologies of existing ship locks.

Keywords: Intelligent locks · Safety and efficiency · Network communications

1 Introduction

According to the requirements of the "Outline of building China into a country with strong transportation network" issued by the Communist Party of China Central Committee and the State Council, China will be basically building the strength in transportation. As a water transportation hub, the inland navigation structure is also an important part of a country with a strong transportation network. By promoting the development of navigation structures, it can improve the efficiency of the locks and

© The Author(s) 2023
Y. Li et al. (Eds.): PIANC 2022, LNCE 264, pp. 880–893, 2023.
https://doi.org/10.1007/978-981-19-6138-0_78

enhance the navigability of waterways, which is of great significance to promote the development of the water transport economy and even the national economy. However, the existing large-scale navigation structures in China still have many prominent contradictions such as the mismatch between demand and supply, the mismatch between quantity, efficiency, and the lack of multi-node linkage. The intelligent operation management of navigation structures is still in its infancy, and the theoretical research has lagged behind the production practice. With the increasingly tight resource and environmental constraints faced by the construction of water transport infrastructure, as well as the Internet of Things and intelligentization of transportation infrastructure, it is urgent to re-engineer and optimize the process, and make use of emerging cross-border technologies such as 5G and artificial intelligence. The construction of Intelligent Locks is an important development trend and direction for navigation structures to improve the traffic supply capacity, safety guarantee capacity, and operation efficiency.

With the low social attention, the research on Intelligent Locks started relatively late. In the context of the vigorous construction of smart transportation, the concept of Intelligent Locks has gradually received attention in the process of theoretical research and practical development, attracting scholars to discuss it from different perspectives, but there is no unified standard definition for it yet. Chen (2015) demonstrated that the Intelligent Locks is a combination of sensor lock, digital lock, and palm lock. At the same time, it is also a scientific lock, a human lock, and an innovative lock. Intelligent Locks = interconnection of infrastructure and information + coordination of ship lock organization management and operation mechanism + smart thinking (Zhang and Qian 2018). Feng and Zou (2014) showed that through the means of the mobile terminal, video monitoring, programmable logic controller, and other means, the perception layer can be built to provide perception capabilities for Intelligent Locks. Wu (2011) believed that by improving the service level of supporting smart applications, more refined services can be provided for managers, crew members, and salesmen.

Under the requirements of safety, efficiency, and intelligence, the Intelligent Locks is based on lock informationization and intelligent transportation, relying on a more mature high-end technology system, a more coordinated system, and a more matching security system to achieve convenient and scientific management of ship locks. Ship lock informatization is the necessary premise and necessary stage of Intelligent Locks. And the Intelligent Locks is the advanced form and inevitable result of ship lock informatization.

2 Current Status and Focus of Intelligent Locks

At present, there are nearly 1,000 locks in China, and the proportion of medium and large locks is more than 30%. In terms of the lock control system, the Three Gorges Lock and the Gezhouba Lock have realized information-based operation and management, and are also the locks with the highest level of automation in China, realizing unmanned and centralized monitoring operation. In addition, the Three Gorges Lock control system is at the leading international level in terms of safety reliability. As the operation and management of the locks tend to be "unmanned or less manned", the computer monitoring system is more and more widely used in the daily operation of the locks to achieve monitoring and control of the locks, providing a strong guarantee for the safety and reliable operation of the locks. The computer monitoring system is more and more widely used in the daily operation of the locks to achieve monitoring and control of the locks, providing a strong guarantee for safe and reliable operation of the locks (Li and Kong 2013). It is worth noting that there are still a large number of ship locks in China with relatively primitive operation control methods and a low degree of informatization. They are still mainly based on manual experience management, and there is a huge gap between the intelligent operation and management of ship locks (Li 2001).

At the same time, many of the current intelligent concepts remain at the level of comprehensive information systems. They are only based on information systems, such as AI technology, big data technology, 5G, and other technologies. In addition, there is a lack of in-depth research on the concept and model of Intelligent Locks.

The study of Intelligent Locks must form a complete system from data collection to intelligent application, i.e., to establish an Intelligent Locks application model, using Intelligent Locks equipment, Intelligent Locks operation, and management technology to support the realization of the Intelligent Locks model. The criteria for judging the Intelligent Locks need to pay attention to the following four points:

(1) The efficiency of the locks;
(2) The level of security of the locks' operation;
(3) The level of maintenance, management, and joint scheduling of locks;
(4) The level of service to passing ships.

3   Research Objectives and Key Issues

The key to Intelligent Locks is to build an Intelligent Locks application model and innovative smart technology. Its research objects include the three main bodies of locks: "Lock", "Management" and "Ship". The "Lock" refers to the lock control system, monitoring system, sensing equipment, security inspection equipment, and fast crossing intelligent guidance equipment, including the lock entity. The "Management" refers to the comprehensive information system, the scheduling and charging system of the locks, etc., and the "Ship" refers to the ship's behavior of crossing the locks.

As shown in Fig. 1, the goal of Intelligent Locks is to build a trinity application model of "lock", "management" and "ship". The interconnection and data sharing among the three subjects can improve the navigation efficiency of locks, the security level of lock operation, the management level of locks, and the service level of passing ships.

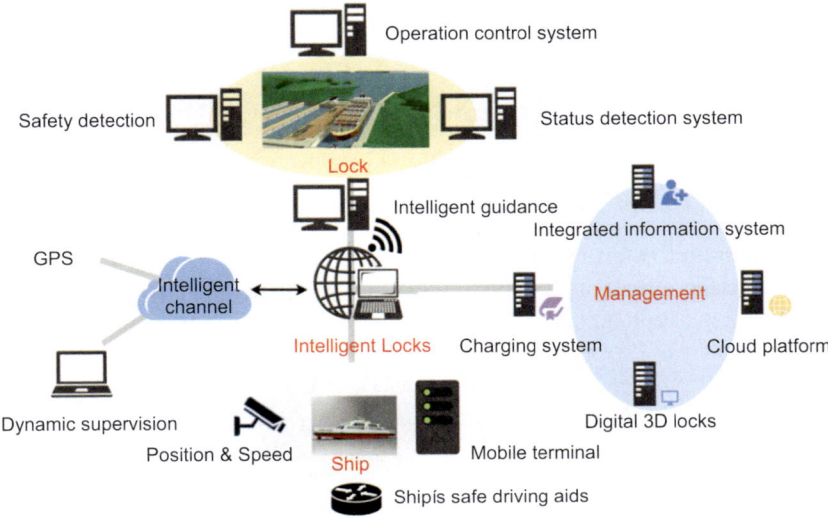

Fig. 1. Trinity application model of "lock", "management" and "ship"

The key issues to be realized in the Intelligent Locks are.

(1) Network communication platform: relying on the Internet, the trinity application model of "lock", "management" and "ship" is built to realize data interconnection and interoperability;

(2) Ship identification and positioning technology: to meet the functions of intelligent identification, precise positioning, and detecting the whole process of entering and leaving the lock;

(3) Ship safety detection technology: to complete the detection of dangerous goods, the load and draught of the ship, the superelevation of the ship, the prohibited parking area in the lock chamber, and the speed of the ship;

(4) Intelligent guidance equipment: to guide ships to quickly enter and leave the lock;

(5) Automatic ship berthing system equipment: to achieve automatic identification and mooring;

(6) Automatic monitoring system: to realize the automatic control of the gate;

(7) Intelligent dispatching system: to ensure the safe, fast, and orderly entry and exit of the lock.

4   Key Technology Research of Intelligent Locks

The research of Intelligent Locks can introduce cross-border technologies such as cloud services, big data analysis, AI (Artificial Intelligence), and Industry 4.0 (Chen 2018). The main research contents are as follows.

4.1 Network Communication Platform

The comprehensive information management system of the locks should be able to fully integrate the information of the systems such as scheduling plan, toll management, safety supervision, operation monitoring, status monitoring, and auxiliary navigation of the locks. To realize the integration of the whole information system of the lock, the platform shall meet the following requirements.

4.1.1 Information Interaction Between Lock and Ship

The ship's basic information, safety information (draft, pre-gear position, approach speed), berthing position and deviation, out-of-bounds status, and planned (remaining) lock-through time can be displayed at the lock control room end (see Fig. 2).

With the development of mobile business, to realize the intelligent guidance, the interactive system can show the information including approach position, navigation speed, actual speed, berthing position, planned crossing time, and the current status of the ship, to the ship through APP and cell phone client.

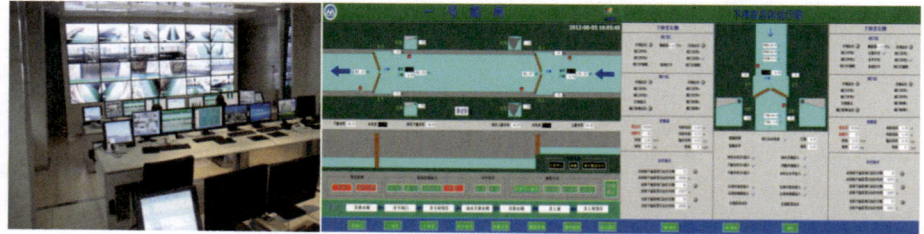

Fig. 2. The control room side of the lock

4.1.2 Active Optimization of Self-learning

The integrated information management system of locks can record and store the data of passing ships' size and gate times, perform unsupervised feature extraction on the pre-processed data, actively learn the optimal passing control parameters under various types of water level coordination relationships, establish a model under the real working parameters of locks and conduct training. The real-time data of the gate crossing are imported into the model and calculated to obtain the optimal fleet combination, as shown in Fig. 3.

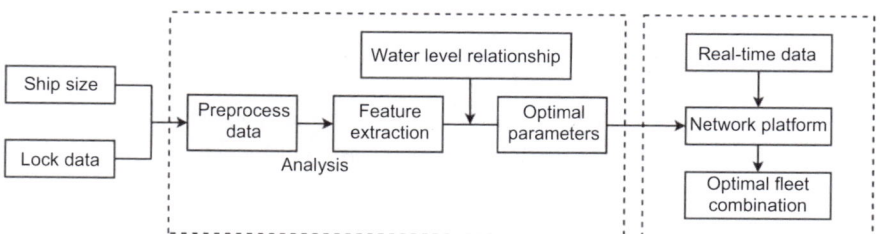

Fig. 3. Active optimization of self-learning model flow chart

4.2   Identification and Positioning System

As shown in Fig. 4, during the process of entering and leaving the lock, the identification and positioning system can obtain information about the ship and its precise position, which is essential to realize the intelligent operation of the lock.

- By locating and measuring the speed of ships, the efficiency and safety of passing the gates and moving berths can be improved.
- Through accurate localization of the ship's position in the lock chamber, it can be judged whether the ship's parking position meets the requirements and can give feedback on the real-time position information to the ships so that they can quickly complete the ship's parking and mooring in the lock chamber.
- Through the detection of the height limit and no-stopping area, it can ensure the safety of ships passing through the lock and the safety of the lock's hydraulic and metal structures (Lv 2020).
- By detecting the presence of ships or no ships in the lock, the automatic control system is realized in cooperation with the process of passing through the lock.

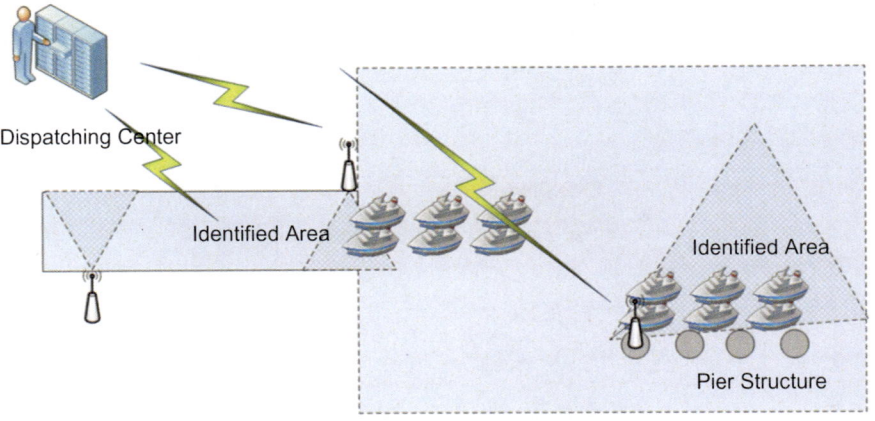

Fig. 4. Intelligent ship identification and positioning system

Ships can use GPS (Global Positioning System) for positioning and identification. However, when they enter the lock, it is impossible to use GPS for accurate positioning because of the water level drop, surrounding buildings blocking, and other factors. Combined with the current development of intelligent identification and positioning technology, the following technical research directions are mainly available:

- RFID (Radio Frequency Identification), as shown in Fig. 5, can be used for ship identification in the area of the pier structure, gate, and lock chamber. The error of identification can also be controlled within the acceptable limits according to the actual demand, and the accuracy of high-frequency recognition can reach more than 99.9% (Yang 2011).

- Within the detection distance of more than 500 m, millimeter-wave radar can reach the positioning accuracy of ±0.15 m and the speed measuring accuracy of ±0.15 m/s.
- With the detection of 2D and 3D, LiDAR (Light Detection and Ranging) has higher detection accuracy than millimeter-wave radar. However, the effective range of the LiDAR is currently between 50 and 100 m, which is shorter than millimeter-wave radar.
- The research of video recognition in ship locks started earlier than other technologies described above. Affected by light, environmental conditions, and other factors, the misjudgment rate of video recognition is also higher than other methods (Wang 2008).

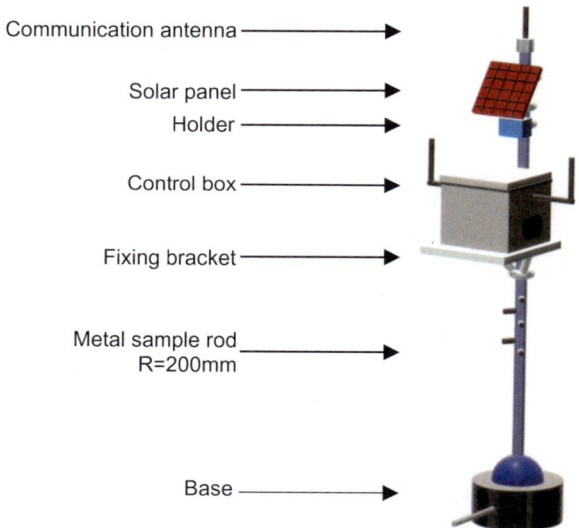

Communication antenna

Solar panel

Holder

Control box

Fixing bracket

Metal sample rod
R=200mm

Base

Fig. 5. The RFID equipment

Identification and positioning system is a comprehensive technology. In other words, the use of a single method often fails to meet the requirements of full regional coverage and detection accuracy. The methods, including image processing, pattern recognition, neuronal networks, and deep learning, need to be further explored. At the same time, the reliability of recognition still needs time to verify and continuously optimize.

4.3   Security Detection System

As an important part of intelligent services in the Intelligent Locks, security detection system, including the inspection of speed, dangerous goods, load, and draught of ships (see Figs. 6 and 7), can ensure the safety of ships when passing through the locks (Zhao 2020).

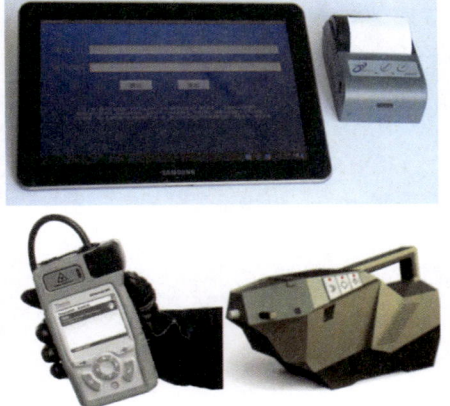

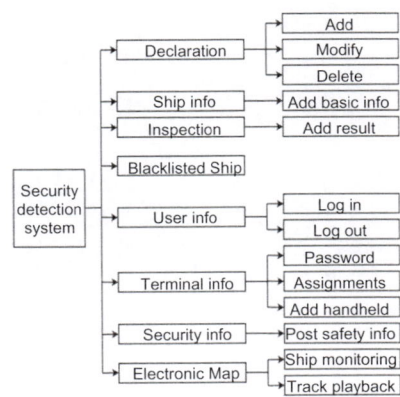

Fig. 6. Security check equipment for gates   **Fig. 7.** Gate-passing security check platform

Due to the large sizes of the ships and the variation in water level, the development and implementation of safety detection equipment have more technical and construction difficulties. For example, speed detection on the highway is a very simple technical application, but in the locks, the speed limit is only 0.8 m/s. Accompanied by a large variety of water levels, the different types of ships, and other factors, it requires a security detection system that has not only high accuracy but also intelligent identification capabilities.

4.4 Intelligent Guidance System

Based on ship identification and accurate positioning, an intelligent guidance system is to guide vessels to pass through the gate quickly using an intelligent ship navigation APP, fixed or mobile information board, lock chamber berth delineation guidance, and so on. There are the goals of an intelligent guidance system:

- Managing the order of ships entering the locks.
- Giving reasonable speed, and safe spacing to sailors.
- Notifying the berthing information for vessels and deviation alerts for parking.
- Offering feedback on the vessel's driving information to the guidance system.

As the bond between "ship", "gate" and "management", the intelligent guidance system reduces the manual work of ship dispatching and improves the efficiency and security of the management. Figure 8 shows that the 3D laser cloud point technology applied to car navigation will become the new direction of ship detection (Guo et al. 2020).

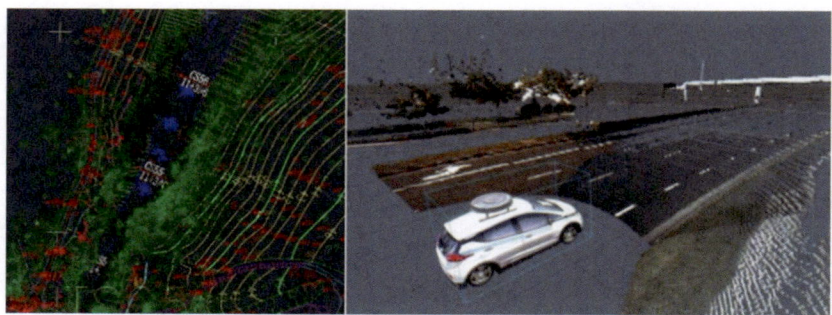

Fig. 8. 3D laser cloud point technology

4.5   Automatic Mooring and Parking System

In traditional locks, it takes a lot of time to enter, moor and exit the locks, which limits the operational efficiency of the locks. An automatic mooring and parking system will be a disruptive technology for the new locks.

4.5.1   Parking System

Automatic berthing includes two ways: self-propelled entry and exit of the lock and external force to pull in and out of the lock. To realize the automatic navigation of the ship, it is necessary to develop a self-propulsion system of the ship. However, due to the large differences between ships passing through the lock, the applicability is poor. Figure 9 shows an external force pulling in and out lock system suitable for large water heads, which can be implemented in the new construction or overall renovation of the ship lock. Alternatively, a six-degree-of-freedom intelligent robotic arm can be used, as shown in Fig. 10, to sample the target object with acquisition equipment such as a camera, and then combine pattern recognition, image processing, and other methods to analyze and process the acquired image data to obtain the spatial position of the ship. Finally, the obtained information is used to enable the robotic arm to automatically identify and grab the ship's bollard, and to quickly move the ship to the specified berthing position (Du et al. 2017).

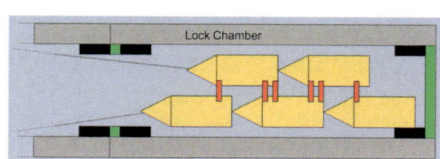

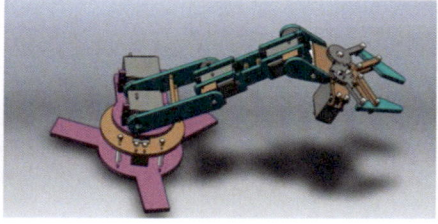

Fig. 9. The external traction   **Fig. 10.** The six-degree-of-freedom robotic arm

4.5.2   Mooring System

After the ship enters the lock, the ship needs to be moored reliably to resist the external impact caused by the water filling and draining. The vacuum-type automatic floating mooring system can also lift automatically with the change of water level in the lock chamber, as shown in Fig. 11. After the ship enters the lock, it should be able to identify with the berthing bollard system, and after the ship is accurately parked, the system should automatically connect to the bollard on the ship, and after the water filling and draining is completed, the bollard connection is automatically released.

Fig. 11. The vacuum-type automatic floating mooring system

4.6   Intelligent Monitoring System

The intelligent monitoring system can integrate the on-site sensor collection equipment, control equipment, centralized monitoring equipment, video monitoring equipment, broadcasting equipment, and navigation command guidance equipment of the locks into one whole (see Fig. 12). Compared with the common lock control system, it has the following features:

- Key sensors and components are redundantly configured and have self-diagnostic functions. On-site equipment should meet the standards for unattended operation, and be able to accurately locate, alarm, and automatically remove faulty equipment or emergency treatment when the system fails.
- The control system is interlinked with the video, broadcasting, and navigation command guidance equipment to achieve automatic monitoring.
- The system can fully guarantee the centralized monitoring operation's security, reliability, and stability.

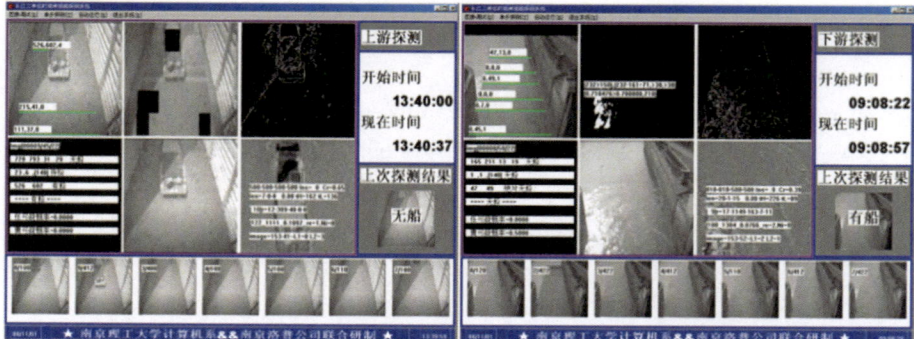

Fig. 12. The intelligent monitoring system

4.7   Intelligent Scheduling System

Scheduling is an important link to ensure the safe, fast, and orderly entry and exit of the ship lock. Through scientific and reasonable planning, the economic and social benefits of the ship lock can be fully utilized, and the transport conditions in a certain area can even be improved. In the existing ship lock scheduling methods, the shipping management department still mainly adopts the manual method, i.e., experience management and qualitative analysis (Wang 2021). Therefore, the efficiency of the ship lock will be limited to a certain extent by the technical level, work experience, working status, etc.

At present, an intelligent dispatching system can be developed, and advanced scheduling algorithms can be used to optimize the dispatching of ships passing through the locks and reduce the waiting time of the fleets passing through the locks. The development of shipping in foreign countries is more developed, and the standardization of ship locks and ships is higher than that in China (Shang et al. 2011). Bugarski et al. (2013) proposed a decision support system for ship lock management based on fuzzy theory, which can be used for decision support in the ship lock scheduling process and training of ship lock managers; Verstichel et al. (2014) believed that the ship lock scheduling problem can be divided into three interrelated sub-problems: ship layout, lock chamber allocation, and lock operation scheduling. Ji et al. (2017) established a multi-objective optimization model for the ship lock scheduling problem. Liu and Qi (2002) proposed a heuristic algorithm, which effectively solved the contradiction between the lock chamber area utilization and the priority of ships passing through the lock. Finally, the experimental data showed that the algorithm can improve the efficiency of the actual scheduling and scheduling of the Three Gorges Ship Lock.

The intelligent scheduling system should grasp the status of vessels passing through the gate in time, such as vessel tonnage, vessel size, personnel information, loading situation and arrival time, etc., and use advanced scheduling algorithms to automatically schedule the vessels to the gate according to the principle of maximizing the utilization rate of the gate space and minimizing the cost of time passing through the gate, with the following main functions and features:

- An intelligent scheduling model that meets the requirements of safety and efficiency.
- Supporting automatic visualization of scheduling information management.
- Supporting two-way information interaction management between ships and locks, such as APP to realize intelligent registration of ships waiting for locks and pushing of service information, etc., as shown in Fig. 13.
- Supporting the deepening application of AIS (Automatic Identification System).

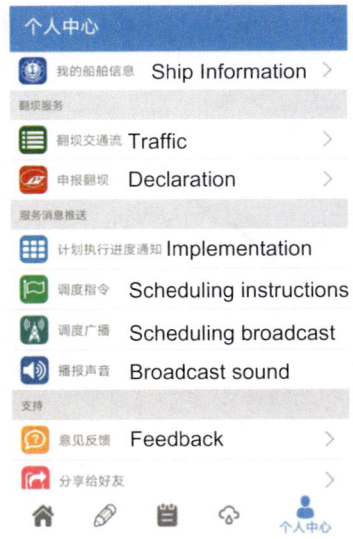

Fig. 13. "Intelligent dispatching" APP declaration interface (from Hunan General Aviation Dispatching APP)

5 Conclusions

Compared with the huge capital demand for new ship locks and their long construction period, the operation and management level of locks can be improved by fully tapping the existing lock resources, strengthening the team construction, improving the maintenance level of locks, attaching importance to the automation, informatization and intelligent construction of locks and the application of new technology, which has obvious financial and time efficiency advantages and should be the primary means to improve the navigation capacity of locks.

Under the guidance of informatization and intelligence, researching "Intelligent Locks" is an effective way to improve the navigation ability and the management level of ship lock operation. In the future, it is necessary to combine the basic data of the ship lock and the shipping information, use the big data platform to strengthen the top-level design and planning of "Intelligent Locks", promote the comprehensive integration of modern information technology with ship navigation management and services, and establish comprehensive management and maintenance.

Acknowledgements. The study is financially supported by the 333 High Level Talents Training Project of Jiangsu Province.

References

Bugarski V, Backalic T, Kuzmanov U (2013) Fuzzy decision support system for ship lock control. Expert Syst Appl 40(10):3953–3960

Ji B, Yuan X, Yuan Y (2017) Orthogonal design-based Nsga-Iii for the optimal lockage co-scheduling problem. IEEE Trans Intell Transp Syst 18(8):2085–2095

Verstichel J, De Causmaecker P, Spieksma F, Berghe GV (2014) The generalized lock scheduling problem: an exact approach. Transp Res Part E-Logist Transp Rev 65(SI):16–34

Chen C (2015) Intelligent transportation, 2nd edn. Tsinghua University Press. (in Chinese)

Chen C (2018) Liner operation optimization of the Yangtze River based on shipping big data. Wuhan University of Technology, p. 144. (in Chinese)

Du X, Cai Y, Lu T, Wang S, Yan Z (2017) A robotic grasping method based on deep learning. ROBOT 39(06):820–828 (in Chinese)

Feng F, Zou Y (2014) Study and implementation of IOT-based trinity ship supervision system for customs. Mod Electron Tech 37(06):83–87 (in Chinese)

Guo J, Liu C, Yu Z, Zheng M, He Z, Chu X (2020) Application and outlook of ship perception based on LiDAR. In: The 15th china intelligent transportation annual conference, Shenzhen. (in Chinese)

Li Z, Kong B (2013) Application of monitoring and control system for ship lock. Appl Traffic Eng Ind 10:33–36 (in Chinese)

Li Z (2001) A lock optimized scheduling and decision system based on the Internet. Nanjing Hydraulic Research Institute. (in Chinese)

Liu Y, Qi H (2002) The two-dimension optimization arranging heuristic algorithm and its application in the Yangtse Gorges Permanent Ship Lock decision system. Comput Modern. 01:1–3 (in Chinese)

Lv Y (2020) Research on intelligent integration of security inspection for ship passing through lock. Dalian Maritime University, p. 84. (in Chinese)

Shang J, Wu P, Tang Y (2011) On multiple-lane lock's joint scheduling plan based on computer simulation. Port Waterw Eng 09:199–204 (in Chinese)

Wang J (2008) The application of video tracking and identification technology in sea route video surveillance. Xiamen University, p. 58. (in Chinese)

Wang Y (2021) Multi-step ship locks chain scheduling of the Xijiang River shipping line based on dynamic binary tree. Port Waterw Eng 03:159–163 (in Chinese)

Wu X (2011) Research on solution for design and management to data layer of shipbuilding informatization service platform. Harbin Engineering University. (in Chinese)

Yang G (2011) Research on RFID localization algorithm. Henan University of Science and Technology, p. 75. (in Chinese)

Zhang J, Qian J (2018) Intelligent ship lock. Southeast University Press (in Chinese)
Zhao C (2020) Research on remote safety inspection system for navigable ships of Xiangjiaba ship lift. Dalian Maritime University, p 81. (in Chinese)

River Shoreline Project Management Based on BIM Technology: A Case Study of the Environmental Improvement Project of the Green Water Wetland in the Nanjing Reach of the Yangtze River

Tianzeng Huang[1], Haifeng Xu[1], Yanbo Wang[1], Huai Chen[2], Lei Zhang[3(✉)], and Hongxia Fan[2]

[1] Nanjing Water Planning and Designing, Institute Corporation Limited, Nanjing 210006, China
[2] State Key Laboratory of Hydrology-Water Resources and Hydraulic Engineering, Nanjing Hydraulic Research Institute, Nanjing 210029, China
chenhuai@nhri.cn, hxfan@hnri.cn
[3] Key Laboratory of Sediment Science and Northern River Training, The Ministry of Water Resources, China Institute of Water Resources and Hydropower Research, Beijing 100038, China
leizhang06@iwhr.com

Abstract. The Building information modeling (BIM) is one of the most promising developments in the architecture, engineering, and construction fields. It carries out the data management during the whole period from site analysis to later operation, and provides technical support and collaborative work platform for a built asset project. Based on the environmental improvement project of the Green-Water Wetland in the Nanjing reach of the Yangtze River, the BIM technology provides a fast and efficient communication platform for all partners involved in the construction period, and has been successfully and efficiently applied in the site design, model analysis, building design, and landscape design. Green-Water Wetland is located on the shoreline of the Nanjing reach of the Yangtze River. The main task of this project is to return the fishpond to the wetland, restore the forest, and improve the landscape of the whole wetland. The specific applications of BIM technology are as follows: (1) It provides a fast and efficient communication platform for all partners involved in the construction period, and couples with the application of GIS and other digital technologies; (2) The Revit and Civil3D software were carried out to realize 3D design of the real scene, and visually display the advantages and disadvantages of each scheme; (3) The preprocessing efficiency of data was greatly improved which lays the foundation for subsequent digital analog analysis; (4) The Mars software was used to render the design scheme in real time, intuitively express the design intention, and avoid repeated design.

Keywords: Shoreline management · Building information modeling (BIM) · Green-Water Wetland · Yangtze River

© The Author(s) 2023
Y. Li et al. (Eds.): PIANC 2022, LNCE 264, pp. 894–905, 2023.
https://doi.org/10.1007/978-981-19-6138-0_79

1   Introduction

BIM (building information modeling) technology was first proposed by Autodesk in 2002 and has been widely recognized by the industry all over the world. It can help realize the integration of building information, from the design, construction and operation of buildings to the end of the whole life cycle of buildings (Vysotskiy et al. 2015).

The core of BIM is to establish a virtual three-dimensional model of construction engineering and use digital technology to provide a complete and actual construction engineering information database for this model (Azhar 2011). The information base contains not only the geometric information, professional attributes and state information describing building components, but also the state information of non-component objects (such as space and motion behavior). Design teams, construction units, facility operation departments, and owners can work together based on BIM to effectively improve work efficiency, save resources, reduce costs, and achieve sustainable development.

BIM is widely used in industrial and civil construction. With the continuous advancement of the concept of smart city, the application of BIM in traditional fields, such as geotechnical and hydraulic engineering, is becoming more and more common, to cite a few (Ismail et al. 2017; Jin et al. 2015; Kalfa 2018; Kumar and Mukherjee 2009; Latiffi et al. 2013).

The model established by BIM has five characteristics: visualization, coordination, simulation, optimization and drawing. At present, the international mainstream BIM design software includes Autodesk Revit series (Demchak et al. 2009), Bentley Architecture series (Logothetis et al. 2015), Graphsoft's ArchiCAD series (Sulbaran et al. 2010), and Dassault's CATIA series (Yang et al. 2019). Taking the Autodesk Revit series as an example, the main softwares including in it are Revit, Civil 3D, Inventor, Naviswork, Infraworks and Lumion. In addition, the Autodesk 3D collaborative design platform Vault Professional provides a collaboration platform for designers. Collaborative design can be carried out on this platform to improve communication efficiency while avoiding collision of components and optimizing design results.

Relying on the Green Water Wetland Conservation and Environmental Improvement Project, this research provides a fast and efficient communication platform for all parties involved in the construction from multiple perspectives and levels. This study is a significant attempt to apply the BIM technology to the River shoreline project management, which has achieved social and economic benefits in the environmental improvement project of the Green-Water Wetland, and shows the prominent future of this technology.

2   Study Area and Methods

2.1   Study Area

Jiangbei New District is located in the north of the Yangtze River in Nanjing City. It is a national-level new district. In order to implement the protection of the Yangtze River,

Jiangbei New District has started the Yangtze River Wetland Protection and Environmental Improvement Project in Jiangbei New District. The Green Water Wetland has a total length of 23 km, and is included in the third-stage project. This project organically combines urban flood control, wetland protection and environmental improvement, and forms a hydrophilic landscape to serve the citizens. This project adopts a comprehensive upgrading method of "cleaning the shoreline, tidying up the river beach, and maintaining the wetlands" to carry out long-term protection of the Yangtze River shoreline in the Jiandbei New District (Tang et al. 2019).

The Green Water Wetland Park is located on the shoreline of the Yangtze River in Jiangbei New District, Nanjing, with a total length of about 12 km and an area of 15.82 km^2 (Fig. 1). It is an important ecological gateway of Jiangbei New District and an important interface in the spatial relationship among river, city and mountains. Under the overall goal of "the most beautiful shoreline and the nearest future", the protection of Green Water Wetland requires the regulation of water resources, water ecology, water environment, water safety, and water culture. And the goal of this project is to make Green Water Wetland Park the most representative National Park in China.

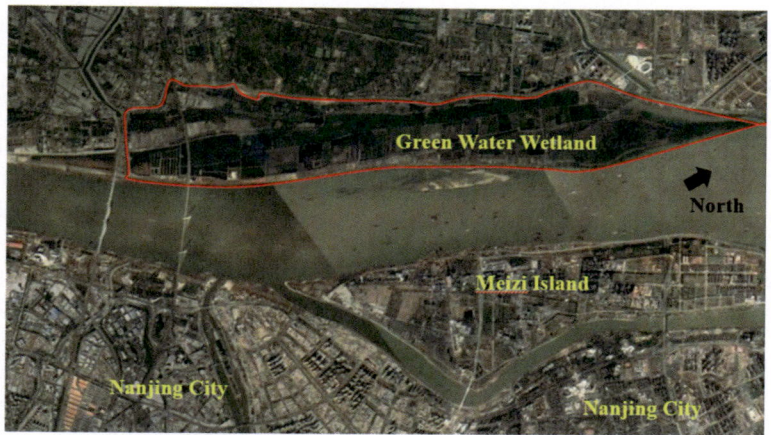

Fig. 1. The Green Water Wetland in Nanjing, China

2.2   Methods

The main design task of this project is to return fish ponds to wetland, and reorganize fish ponds, water pools, and beaches according to the requirements of habitat creation, water system connectivity and water-land transportation. To meet the requirements of biodiversity and environmental friendship, the entire wetland will be reconstructed and landscaped.

The Green Water Wetland Park is located in the Yangtze River shoreline. It covers a large area, which has complex and diverse regional planning, poor basic conditions, manifold survey contents, and complex water and underwater terrain, and brings great inconvenience to the designers. To decrease the environmental impact, the earthwork within the project scope needs to achieve the earthwork balance in sub-regions and the overall region. However, the layout of water network, water-land transportation and habitat creation are closely linked with each other, which brings great challenges to the calculation of earthwork balance and the analysis of hydrodynamic and water quality models. Many problems, such as, large design workload, low efficiency and poor verification, are difficult to solve in traditional design models. The landscape requirements of Green Water Wetland project are relatively high, the traditional design cannot verify the design effect in real time and cannot communicate effectively and timely with the construction party, which brings a huge modification workload to the design party.

Therefore, it is particularly necessary to actively explore the application of advanced technologies such as BIM and GIS in the project. By generating 3D digital models from DOM and DEM data, the on-site basic information can be provided intuitively and accurately; by opening up the data interface between 3D digital models and analysis models, the pre-processing workload of traditional analysis models can be greatly reduced, meanwhile the quality of basic data can be improved. Through the rendering and intuitive expression of the BIM model, the effect of landscape design in the project can be displayed and verified in real time.

3   Results

3.1   BIM Application in 3D Real Scene Model

The project covers a large area and is located along the shoreline. There are fish ponds all over the area. The basic conditions are poor and the terrain above and under the water is complex. In addition, the project requires a lot of investigation contents, which makes the site survey extremely difficult. It is impossible to accurately grasp and sort out the site situation only through photos and measurements. The aerial video shot by UAV only has on-site images without terrain data, so it is impossible to analyze the on-site situation thoroughly. Therefore, the study combines the UAV tilt photography and the unmanned ship underwater measurement technology to collect the on-site basic data, and build the three-dimensional real scene model, which lays a foundation for the implementation of subsequent projects and BIM application.

After completing the collection and processing of on-site Digital Orthophoto Map (DOM), Digital Elevation Model data (DEM) and underwater terrain data, the Autodesk RECAP and Context Capture software were used to fuse the data in the same coordinate system. Through the three-dimensional real scene model formed by data fusion (Fig. 2), the topographic and geomorphic patrol and roaming of large scenes and the measurement of elevation, distance and area can be realized. After the model is superimposed, the section cutting at any position is realized, and the section diagram is generated quickly.

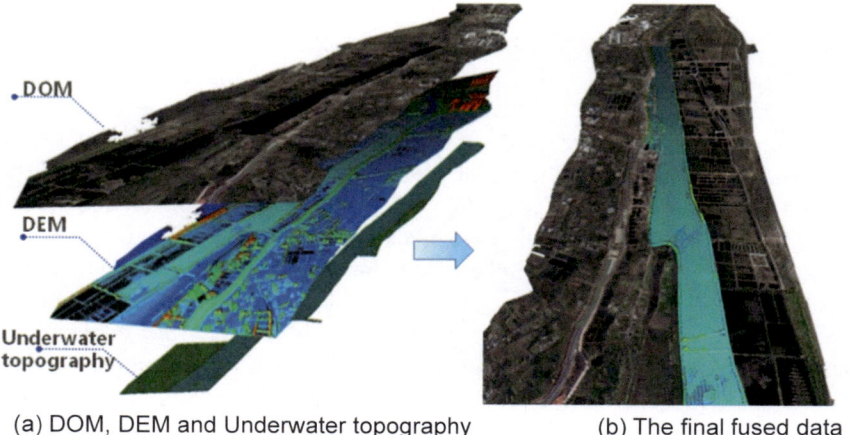

(a) DOM, DEM and Underwater topography   (b) The final fused data

Fig. 2. Data fusion

3.2   BIM Application in Site Design

There are many factors involved in the site layout in the design of this project, such as water network layout, water-land transportation, habitat creation, earthwork balance, etc. These factors all affect the layout of the site. Therefore, this design uses Autodesk Civil3D software for site layout design.

The DEM and underwater terrain data are directly imported into Civil3D to generate the original terrain surface (Fig. 3). Based on the original terrain, the layout of the water network and the design of water-land transportation were carried out. At the same time, the organic combination of habitat creation was taken into account, and the opening and closing changes of the water area were reasonably arranged, as well as the layout and shape of continents, bridges, streams, islands, and embankments.

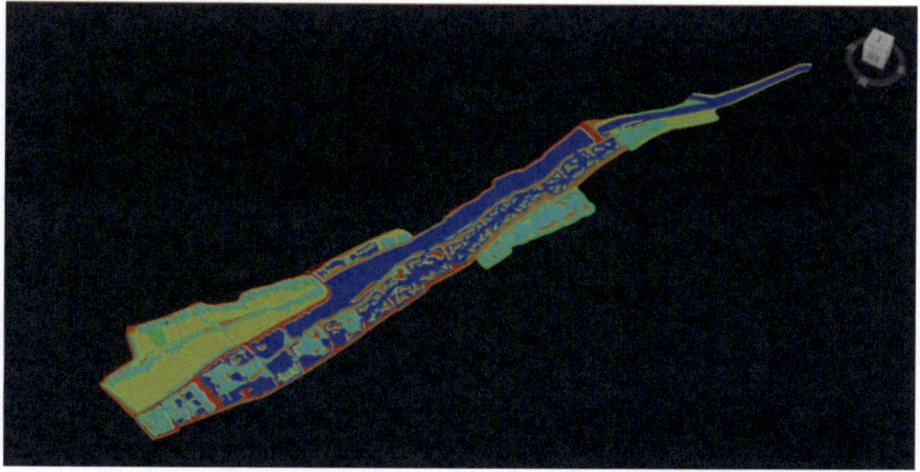

Fig. 3. The 3D terrain model of Green Water Wetland

After completing the design and creation of the entire site model, the elevation analysis function of the Civil3D software was used to analyze the submerged area and semi-submerged area and visually display the dynamic relationship between the submerged area and the non-submerged area, and to achieve water system connectivity to meet the needs of water transportation and vegetation (Fig. 4). At the same time, based on the requirements of the protection of the Yangtze River, the project area needs to achieve the basic balance of earthwork in each sub-area and the total balance of earthwork within the entire project scope. The volume panel function was used to quickly calculate the cut and fill volume of each sub area and total area, which not only meets the construction requirements, but also reduces the project investment.

Based on the site model, the current cross-section required for the design of water-land transportation can be quickly cut out. By utilizing the existing roads, the density and direction of the traffic road network can be reasonably set. The Subassembly Composer was used to parametrically design the logical relationship between the road section and the terrain surface, and then quickly create the road model (Fig. 5) according to the route designed in Civil3D, and quickly generate engineering quantities and 2D drawings.

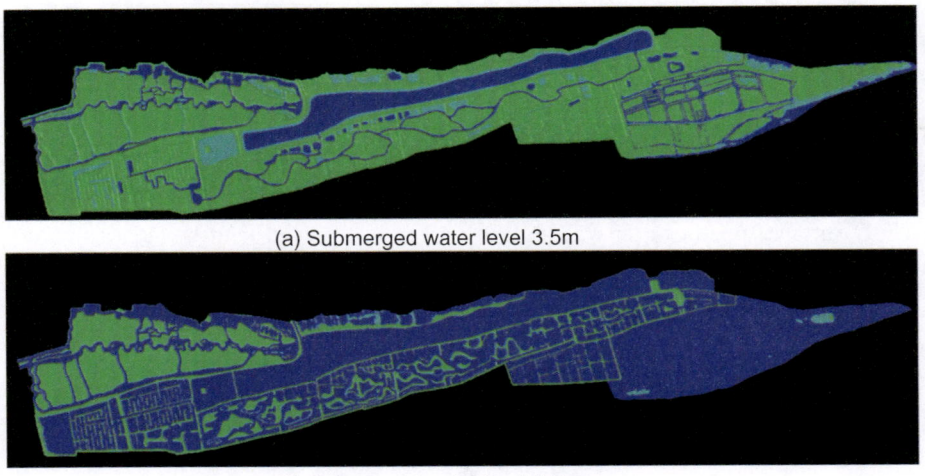

(a) Submerged water level 3.5m

(b) Submerged water level 5.5m

Fig. 4. Inundation analysis

Fig. 5. Model rendering

3.3 BIM Application in Model Analysis

This project has high requirements on hydrodynamic conditions and water quality conditions in the region, so verification and analysis of hydrodynamics, water quality and water depth are essential and especially important. There are many designed islands in the project area; the water network is crisscrossed; the deep grooves and shoals are densely distributed, which brings great difficulties to the modeling of analysis and verification. Different water levels will produce different land and water boundaries. The workload to find the boundary is rather cumbersome. The method of Civil3D surface superposition was used to find the water and land boundaries corresponding to different water level elevations, which significantly improves the efficiency (Fig. 6), and provided the basic support for model optimization analysis in the next step (Fig. 7).

3.4 BIM Application in Building Design

In order to solve the problem of water system connectivity, multiple bridges and culverts should be set up to improve the hydrodynamic and water quality conditions in the project area. Based on the Revit software platform, an application set of three-dimensional intelligent design system for culverts and gates was developed (Fig. 8). It can complete the structural layout and detailed structural design of small buildings. Combined with the standardized family library, it can automatically generate 3D models and output 2D structural drawings, which provides a fast, efficient and accurate design solution for small buildings (Fig. 9).

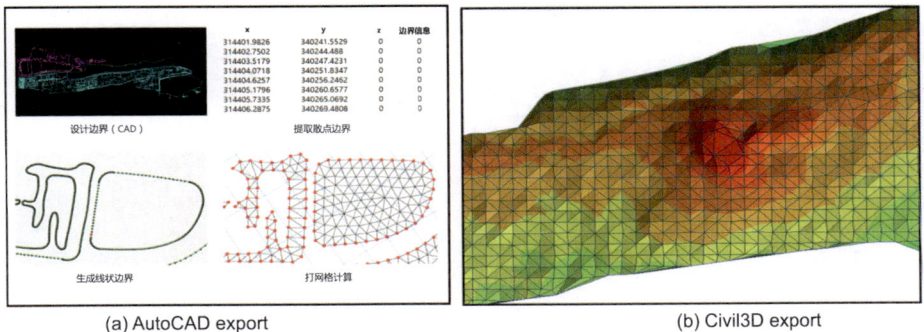

(a) AutoCAD export   (b) Civil3D export

Fig. 6. Export comparison between AutoCAD and Civil3D

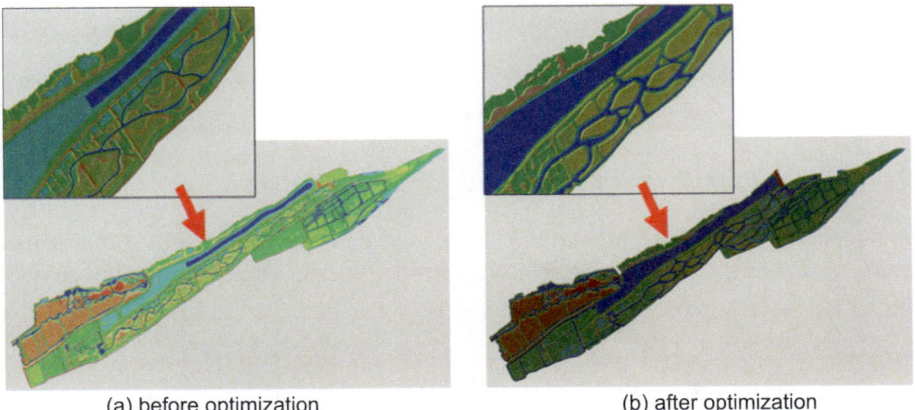

(a) before optimization   (b) after optimization

Fig. 7. Optimization of water system

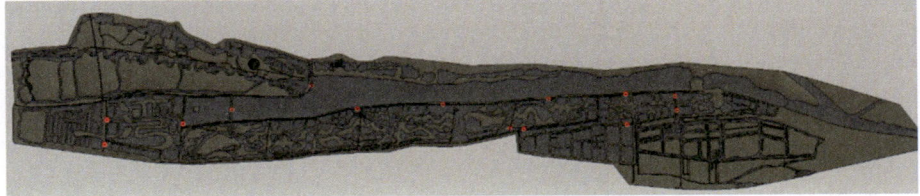

Fig. 8. Regional distribution of culverts and gates

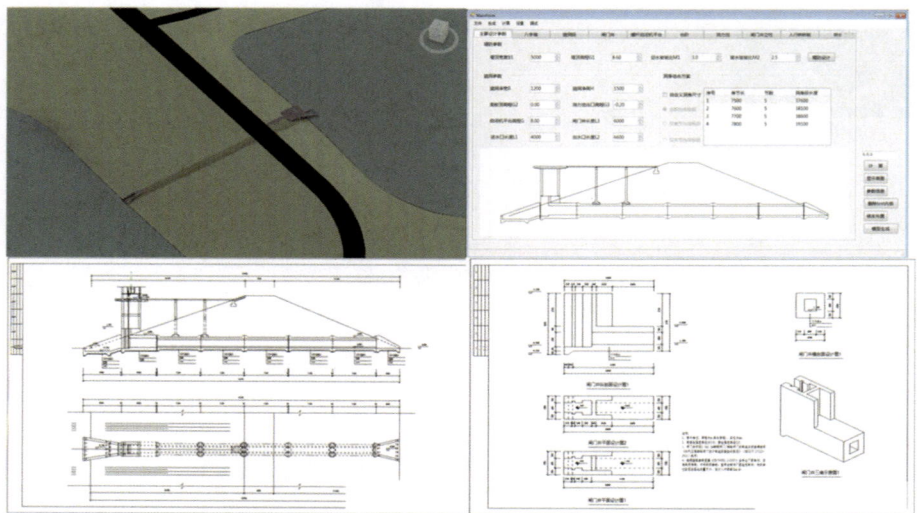

Fig. 9. 3D model of culvert and gate

3.5   BIM Application in Landscape Design

Combing BIM, GIS, and oblique photography technology, the designed landscape terrain and existing surrounding terrain were fused together. By loading multi-source heterogeneous data, such as, oblique photography, orthophoto, and BIM model in the Mars real-time rendering platform, the rapid restoration of real surrounding scenes was realized (Fig. 10).

The 3D visual planting in the software can intuitively adjust the landscape design scheme, which can realize what you see is what you get, simplify early design and reporting and reduce changes in the later stage of construction (Fig. 11).

Through the plant information model, the information about family, genus, ecological habits and other information of plants can be quickly checked, and tree species were reasonably selected according to the type of plant habitat to ensure the rationality of the design. Meanwhile, tree parameters such as plant height, crown width, and ground diameter at breast height can be adjusted conveniently. Landscape plants arranged in the scenario scene can be exported to EXCEL with a detailed resource list.

The scene can be analyzed from the perspective of human vision. Combined with the compass direction to simulate the real sunshine situation, the planting density and gradation can be adjusted in real time (Fig. 12).

Fig. 10. Real-time rendering of landscape design

Fig. 11. 3D visual planting with a vegetation resource list

Fig. 12. (a) Scene analysis from human perspective and (b) sunshine illumination simulation

4 Conclusions

Caused by complexity of the terrain of river shorelines and the diversity of regional planning, difficulties occur in on-site surveys and data collection, resulting in incomprehensible and inaccurate analysis, low design accuracy, and repeated plan revisions. Taking the Green Water Wetland project as an example, this study explores the application of digital technologies such as BIM and GIS. Revit and Civil3D software were used to carry out real 3D design, and intuitively display the advantages and disadvantages of each scheme. With the support of BIM technology, the pre-processing efficiency of data has been greatly improved, and unnecessary errors caused by manual data processing was also avoided, laying a foundation for subsequent digital-analog analysis. The Mars real-time rendering software was used to render the design plan in real time, express the design intent intuitively, and avoid design repetition.

Through the application of BIM technology in the Green Water Wetland project, the design quality can be effectively improved, and the design results can be displayed from multiple perspectives and dimensions, providing a fast and efficient communication platform for all parties involved in the construction. It is a useful attempt in the project, which has important guiding significance for the research and design in similar fields.

Acknowledgements. This work was funded by the National Natural Science Foundation of China (No. 52179072), the Innovative Team Project of Nanjing Hydraulic Research Institute of China (Grants Y220011), and the Open Research Fund of Key Laboratory of Sediment Science and Northern River Training, the Ministry of Water Resources, China Institute of Water Resources and Hydropower Research, Grant No. IWHR-SEDI-202107.

References

Azhar S (2011) Building information modeling (BIM): trends, benefits, risks, and challenges for the AEC industry. Leadersh Manag Eng 11(3):241–252

Demchak G, Dzambazova T, Krygiel E (2009) Introducing revit architecture 2009: BIM for beginners. Wiley, Indiana

Ismail NAA, Chiozzi M, Drogemuller R (2017) An overview of BIM uptake in Asian developing countries. In: Proceedings of the 3rd International Conference on Construction and Building Engineering (ICONBUILD), Palembang, Indonesia

Jin R, Tang L, Fang K (2015) Investigation into the current stage of BIM application in China's AEC industries. WIT Trans Built Environ 149:493–503

Kalfa SM (2018) Building information modeling (BIM) systems and their applications in Turkey. J Constr Eng Manage Innov 1(1):55–66

Kumar JV, Mukherjee M (2009) Scope of building information modeling (BIM) in India. J Eng Sci Technol Rev 2(1):165–169

Latiffi AA, Mohd S, Kasim N, Fathi MS (2013) Building information modeling (BIM) application in Malaysian construction industry. Int J Constr Eng Manage 2(4A):1–6

Logothetis S, Delinasiou A, Stylianidis E (2015) Building information modelling for cultural heritage: a review. In: 25th International CIPA Symposium, Taipei, Taiwan

Sulbaran T, Shiratuddin MF, Germany S (2010) Introduction to ArchiCAD: a BIM application. Delmar Cengage Learning, Kentucky

Tang S, Wang L, Wu X, Xu Z, Chen L (2019) Dynamic evaluation and spatial mapping of wetland ecosystem services value-a case study on Nanjing Jiangbei new area. Env Eng Manage J (EEMJ) 18(11):2519–2532

Vysotskiy A, Makarov S, Zolotova J, Tuchkevich E (2015) Features of BIM implementation using autodesk software. Procedia Eng 117:1143–1152

Yang T, Lu Y, Qin H (2019) CATIA-based BIM technology in highway tunnel design. Civil Eng J 3:469–482

River Project, An Innovative Way to Reduce Pollution on Riverboats

Abdel Aitouche[1(✉)], Raouf Mobasheri[1], Xiang Li[2], Jun Peng[3],
Chris Barnett[4], Uwe Bernheiden[5], Peter Dooley[6], Klaus Bieker[7],
Ahmed El Hajjaji[8], and Robin Pote[9]

[1] JUNIA, Lille, France
{abdel.aitouche, raouf.mobasheri}@junia.com
[2] University of Bedfordshire, Luton, UK
xiang.li@beds.ac.uk
[3] University of Lincoln, Lincoln, UK
jpeng@lincoln.ac.uk
[4] Canal River Trust, Crewe, UK
chris.barnett@canalrivertrust.org.uk
[5] Engine Control Electronics GmbH, Friedrichshafen, Germany
[6] CleanCarb, Luxemburg, Luxemburg
pdooley@pt.lu
[7] Development Centre for Ship Technology and Transport Systems, Duisburg,
Germany
bieker@dst-org.de
[8] University Picardie Jules Verne, Amiens, France
ahmed.hajjaji@univ-picardie.fr
[9] Counseling and Innovation in Logistic, Le Havre, France
ropte@circoe.fr

Abstract. Considering the EU environmental standards for non-road mobile machinery (NRMM), reducing pollutant emissions from inland waterway vessels is becoming increasingly important. The RIVER research project aims to find solutions to achieve nitrogen-free combustion in waterways transportation systems while also emitting zero CO_2 emission. RIVER addresses these issues using Carbon Capture and Storage (CCS) technology and Oxy-fuel combustion (OFC). The project is co-financed by the European Union, as part of the Interreg North-West Europe program. There are ten partners involved in this project (FR, UK, GE, NL, LU). In OFC technology, pure oxygen is used instead of air. Due to the absence of N_2 in the intake charge, NOx emissions will be eliminated. Consequently, the only products of combustion are CO_2 and water vapor. To have a stable combustion process and avoid overheating problems caused by using pure oxygen, some part of the exhaust CO_2 will be recirculated to the engine to create an oxygen-CO_2 mixture for being fed into the engine. A detailed CFD simulation carried out in this project has revealed that 21% oxygen and 79% carbon dioxide is the ideal mixture for the engine to run at maximum efficiency. The remaining CO_2 from the exhaust is collected. It is then condensed, compressed, and stored in a tank to be valorized later. It will be transformed into cosmetics, skincare products, and formic acid. These types of acids are used by the medical sector as an anti-rheumatic product. River's final demonstration will take place in Crewe, UK in July 2022.

Y. Li et al. (Eds.): PIANC 2022, LNCE 264, pp. 906–915, 2023.
https://doi.org/10.1007/978-981-19-6138-0_80

Keywords: Carbon Capture and Storage · Oxy-fuel combustion · Pollutant emissions · Riverboat · Diesel engine

1 Introduction

The global warming problem has been significantly exacerbated by greenhouse gas emissions (GHG) over the last few decades (Kanniche et al. 2010; Wang et al. 2015). The transportation sector consumes twenty percent of global fossil fuel production which puts it second in terms of emission of carbon dioxide (CO_2) and consequently, contributes to alarmingly increasing levels of atmospheric CO_2 and associated greenhouse gases. The European Union has set a target of less than 59 g/km of CO_2 tailpipe emissions in 2030, meaning that a 37.5% reduction is required through this decade. Discussions are underway to further reduce the target to align it with the European Green Deal, which is aiming for net-zero greenhouse gas (GHG) emissions by the year 2050. There have been several initiatives proposed to deal with the deteriorating climate crisis, including carbon neutrality (Figueroa et al. 2008; Li et al. 2021, 2022).

In 2017, the European Union adopted regulations requiring limits on carbon emissions and certification of internal combustion engines used in nonroad mobile machinery (Directive 97/68/EC). As a result, inland waterways (IW) vessels must now comply with more strict emission standards. In project RIVER, the goal is to find solutions to achieve nitrogen-free combustion in waterways transportation systems while also emitting zero carbon dioxide (Aitouche 2022).

It has been proposed that oxygen enriched combustion and oxy-fuel combustion (OFC) are an efficient way to increase engine efficiency and reduce pollutant emissions. OFC combustion uses pure oxygen for combustion instead of air. Due to the absence of N_2 in the intake charge, NOx emissions will be eliminated. As a result, carbon dioxide and water vapor are the only products of combustion. Studies to date have been mainly focused on applying oxy-fuel or oxygen enriched combustion technologies to gas turbines and coal-fired power plants. The utilization of OFC and CO_2 capture for IC engines has been gaining a lot of attention during the last few years. Research on oxygen-enriched combustion shows that a slight increase in oxygen reduces smoke emissions as well as the amount of CO and unburnt hydrocarbons but increases the amount of nitrogen oxides (NOx). Various technologies have been used to decrease NOx and particulates, such as exhaust gas recirculation (EGR) and optimum injection strategies. Research conducted recently has drawn attention to oxy-fuel and nitrogen-free combustion because of the benefits it brings to vehicles and is being used to make huge improvements to the efficiency of internal combustion engines and to achieved zero NOx emissions. As is known, over-high peak pressures and peak pressure increases can easily appear in engine cylinders in oxygen-enriched conditions. An increase in combustion flame temperature is expected if oxygen is used instead of air. To minimize overheating problems due to overheating, it is crucial that the fuel injection flow rate be accurately controlled in order to eliminate unexpected temperature rises in the premixed and diffusion combustion. Further, the diluent ratio and intake charge temperature have a meaningful impact on controlling in-cylinder temperature and combustion process (Giorgi et al. 2021).

Homogeneous Charge Compression Ignition (HCCI) is one of the low-temperature combustion regimes being researched for internal combustion engines. The HCCI concept has been widely studied as a promising concept since it produces emissions comparable to those of a SI engine while achieving thermal efficiency comparable to diesel engines with direct injection. As a result of the auto-ignition occurring nearly simultaneously over the entire combustion chamber, HCCI engines have limited power density. The rapid energy release leads to a large pressure rise in the combustion chamber under high load, causing pressure oscillations. Additionally, HCCI combustion often suffers from the lack of proper combustion phasing. HCCI combustion is initiated by chemical kinetics: the high propensity of diesel fuel to auto-ignite combined with the high compression of diesel engines results in combustion starting before top dead center, very rapid pressure raises rates, short combustion durations.

The current project, which is part of a European project called RIVER (funded by Interreg North-West Europe), aims to examine how different intake charge temperatures may affect oxy-fuel combustion in HCCI mode.

2 River Project Proposed Technology

Equation (1) and Eq. (2) present the chemical reaction process of conventional air combustion (CAC) and OFC, respectively. Unlike with the CAC, the main feature of OFC is oxygen replaces air to react with fuel directly, leading to the chemical products merely contain CO_2 and H_2O.

Moreover, compared to nitrogen of air, the main discrepancies in physicochemical properties for CO_2 can be found in Table 1, which would affect the combustion characteristics of OFC under some specific conditions (Wall et al. 2009; Chen et al. 2012). Regarding molecular weight, CO_2 is 57% higher than that of nitrogen. Hence, under the conditions of OFC, the combustion temperature will be adversely affected due to the higher heat capacity on mole basis of CO_2. In addition, under OFC, chemical reaction rates at early combustion stage would be potentially reduced owing to the low thermal diffusivity and oxygen diffusion of CO_2.

$$C_xH_yO_z + \left(x + \frac{y}{4} - \frac{z}{2}\right)(O_2 + 3.773N_2) \rightarrow xCO_2 + \frac{y}{2}H_2O + 3.773\left(x + \frac{y}{4} - \frac{z}{2}\right)N_2$$

$$(1)$$

$$C_xH_yO_z + \left(x + \frac{y}{4} - \frac{z}{2}\right)O_2 \rightarrow xCO_2 + \frac{y}{2}H_2O$$

$$(2)$$

Table 1. Physicochemical properties of CO2 and nitrogen at 1000 k and 0.1 MPa

Property	CO_2	Nitrogen
Molecular weight	44	28
Density (kg/m^3)	0.5362	0.3413
Kinematic viscosity (m^2/s)	7.69e−5	1.2e−4
Specific heat capacity (kJ/kg K)	1.2343	1.1674
Thermal conductivity (W/m K)	7.057e−2	6.599e−2
Thermal diffusivity (m^2/s)	1.1e−4	1.7e−4
Mass diffusivity of O2 (m^2/s)	9.8e−5	1.3e−4
Prandtl number	0.7455	0.7022
Emissivity and absorptivity	>0	∼0

The RIVER project's primary goal is to eliminate NOx emissions from inland boat engines as well as to capture and store carbon emissions from these engines. These issues are addressed through the use of Carbon Capture and Storage (CCS) technology and OFC in RIVER. A summary of the RIVER technology can be seen in Fig. 1.

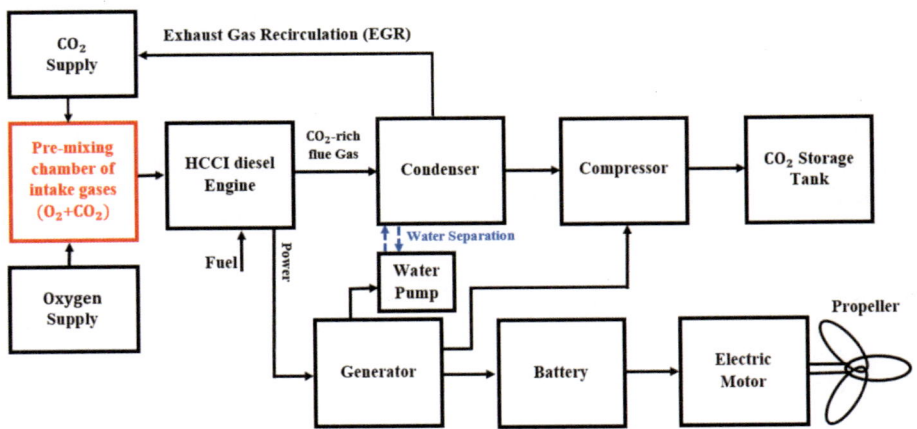

Fig. 1. An overview of RIVER technology

As shown in Fig. 1, power is supplied to the boat by a diesel generator. The engine operates in HCCI mode. There are three components involved in oxyfuel combustion: the oxygen supply system, the exhaust gas recirculation system, and the carbon capture system. Oxygen is provided by a commercial high-pressure oxygen cylinder. CO2-rich exhaust gases are condensed in the condenser, followed by the separation of water from it. Following recirculation, a portion of the remaining CO2 is recirculated back into the engine through EGR and mixed with oxygen prior to being used. Meanwhile, the rest of the CO2 is compressed and stored. It is predicted that this technology will eliminate NOx emissions while storing 100% of carbon dioxide. Optimizing oxy-fuel HCCI combustion requires using an appropriate oxygen concentration and EGR ratio. The authors examined how oxygen and carbon dioxide percentages affected combustion characteristics and engine operating conditions of OFC under HCCI mode. In the current study, the effect of intake charge temperature is discussed. This goal has been accomplished by studying the effects of five different intake charge temperatures on engine operating conditions for diluent strategies. The proposed idea in this study to apply oxy-fuel provides a significant advantage over previous studies in that the diluent strategy is integrated with HCCI combustion to control OFC process instead of water, which can eliminate problems associated with water injection such as lubrication and corrosion.

Because CO_2 is released at the outlet of the engine, no CO_2 is present during startup. It is being proposed to start the engine with an air mixture and wait a few cycles before switching to an oxygen/CO_2 mixture. Alternatively, install a tank in which CO_2 will be stored and solely used for starting. Both techniques have been tested, and both are effective. We will carry out tests under real conditions to see which works best. The ideal solution would be to use CO_2, as we don't take the risk of producing nitrogen oxide, which will mix with the CO_2.

Many boats in recent years have been equipped with exhaust gas recirculation (EGR) valves, making their installation of this technology much easier. However, this equipment must be added to older boats. All engines must be also equipped with oxygen supply valves. On smaller boats, oxygen can be provided by cylinders, but on larger boats, it should be produced on-site due to the large quantities needed. It must be noted the way to produce oxygen is much simpler than to produce hydrogen: Furthermore, oxygen does not have to be pure like in fuel cells, even 95% pure oxygen can be used.

The effect of different diluent strategies on in-cylinder cylinder pressure, and in-cylinder temperature are shown in Fig. 2 and Fig. 3, respectively, under a constant intake temperature and intake pressure.

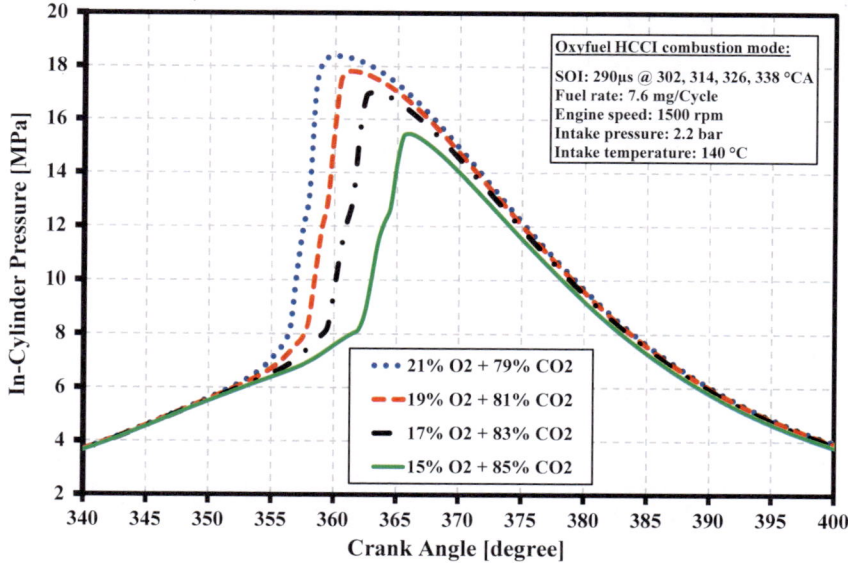

Fig. 2. Effect of different diluent strategies on in-cylinder pressure

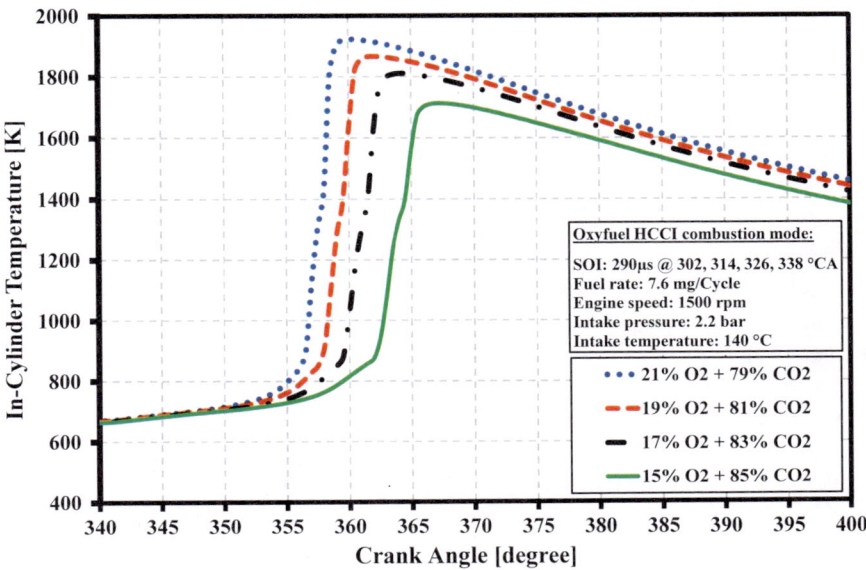

Fig. 3. Effect of different diluent strategies on in-cylinder temperature

As illustrated in Fig. 2 and Fig. 3, the increase of intake-air oxygen content from 15% to 21% results in a significant increase of peak in-cylinder pressure and peak in-cylinder temperature after TDC. In addition, it tends to advance peak pressure location and peak temperature location due to a significantly advanced main combustion process. Highest values of peak in-cylinder pressure and in-cylinder temperature are observed for the highest oxygen percentage (21%v/v). Applying oxy-fuel HCCI combustion instead of conventional HCCI combustion leads to acceleration of the combustion process which results in a shorter ignition delay period. In addition, premixed combustion is minimized while diffusion combustion is maximized. With the heat release rate dramatically increased, it takes a much shorter time to complete the entire heat release process. Subsequently, with such a high heat release rate, the in-cylinder temperature and in-cylinder pressure have increased.

In Table 2, it shows the amount of CO and PM emissions. The results indicate that the oxy-fuel HCCI combustion has brought the CO and PM emissions to a very ultra-low level while the NOx emission has been eliminated using the oxy-fuel combustion. In Table 1, Lambda$_{O2}$ is defined as follow:

$$\text{Lambda}_{O2} = \frac{\text{Actual}O_2 - \text{fuelratio}}{O_2 - \text{fuelratio for stoichiemetriccombustion}} \tag{3}$$

Table 2. Effects of different strategies on PM and CO emissions

Applied diluent strategy	Lambda$_{O_2}$ [-]	PM [gr/kg.fuel]	CO [gr/kg.fuel]
23% O_2 + 77%CO_2	2.38	1.92E−04	59.25
21% O_2 + 79%CO_2	2.18	1.17E−04	72.57
19% O_2 + 81%CO_2	1.97	3.66E−05	86.14
17% O_2 + 83%CO_2	1.77	3.88E−05	161.8

In addition, Fig. 4 and Fig. 5 shows the design and technical drawings of the whole after-treatment system in RIVER project. The system mainly involves a two-stage heat exchanger, water/gas separator, CO_2 compressors, CO_2 tank, required valves, controller, pipes, etc. The excess CO2 exhaust gas can be captured and stored in a storage tank to achieve zero carbon emissions. Furthermore, based on some initial simulation work in advance, the system would provide a good cooling capability, effectively reducing the exhaust temperature from around 800 K to 330 K before entering into the CO_2 tank.

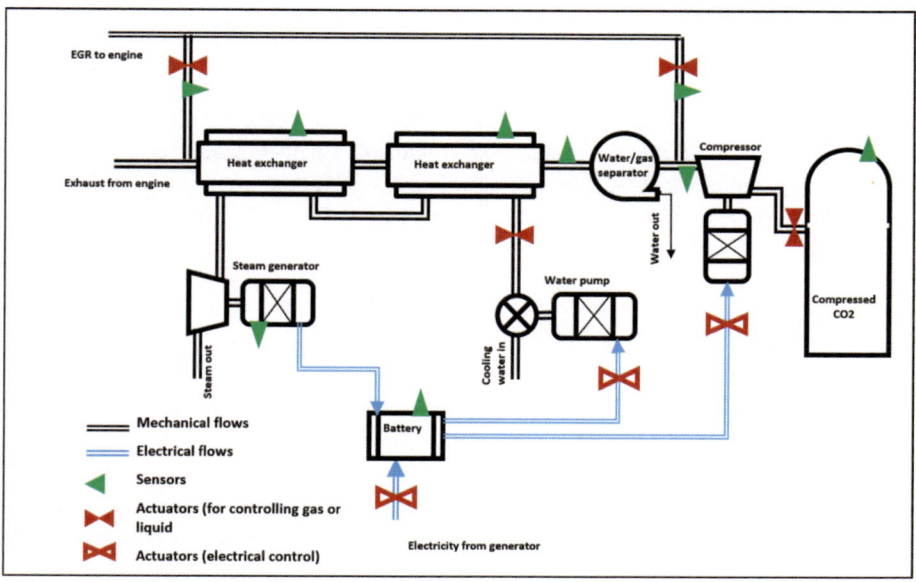

Fig. 4. Design of the after-treatment system in RIVER project

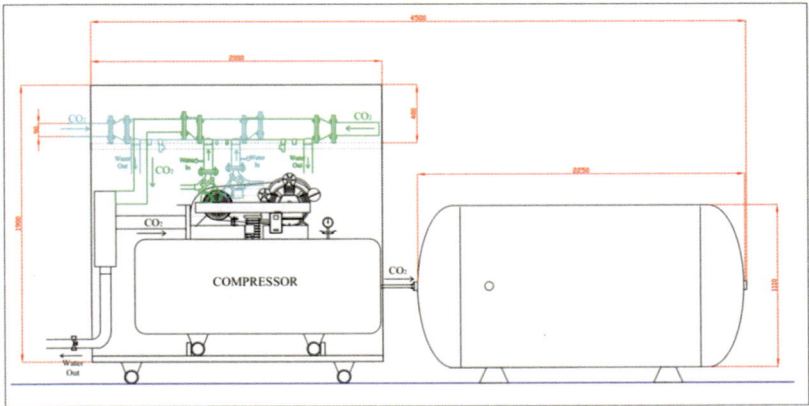

(a) Front view of the whole system

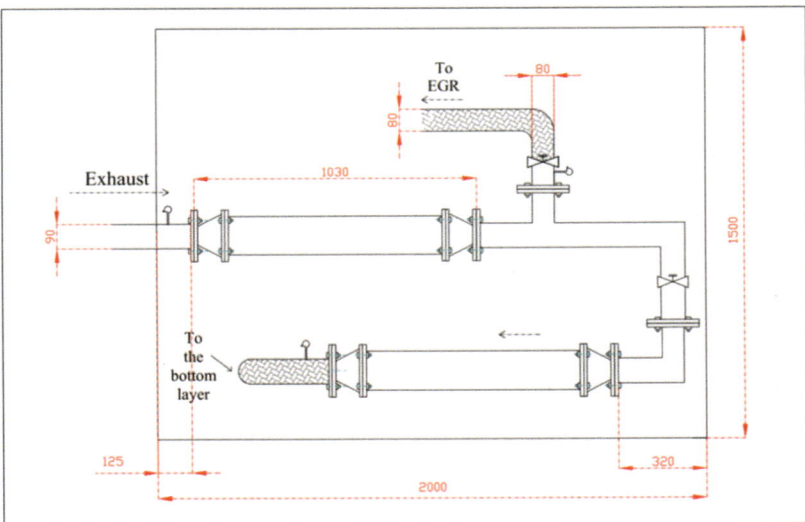

(b) Top view of the top layer

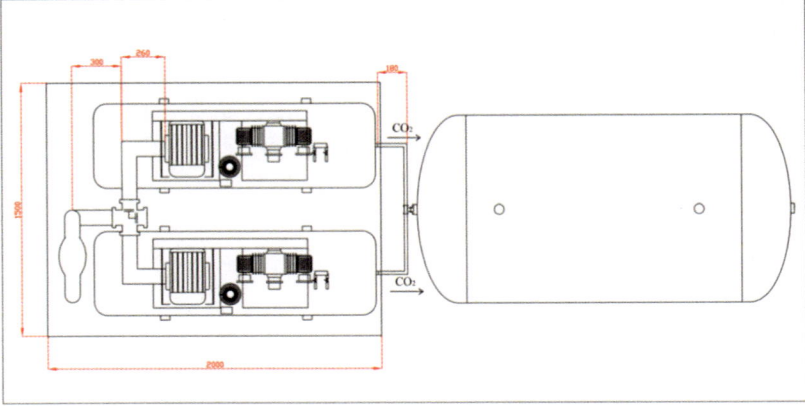

(c) Top view of the bottom layer and receiver

Fig. 5. Technical drawings of our after-treatment system

3 Conclusions

A technology has been proposed by the RIVER project to eliminate pollutant emission for the water transportation systems using CCS and OFC techniques. An advanced CFD simulation using detailed chemistry has been applied to explore the effect of different strategies on engine performance and the amount of pollutant emissions. Besides, a novel design and technical drawings of the after-treatment system of OFC has also been presented. The following conclusions may be drawn:

- Because no nitrogen is present in the intake charge of an oxy-fuel combustion, nitrogen oxides are eliminated. Consequently, combustion reactions produce CO_2-rich emissions. In all cases investigated, PM emissions were very low (<0.0004 gr/kg.fuel) and NOx emissions were eliminated when using OFC HCCI.
- The proposed application of oxy-fuel in this study offers a significant advantage over previous studies in that a diluent strategy is used to control oxy-fuel combustion instead of water, thus eliminating problems associated with water injection such as lubrication and corrosion.

Acknowledgements. The authors gratefully acknowledge the financial support of the Interreg North-West Europe (RIVER Project No. NWE553). Additionally, the authors would like to thank the AVL Company for their collaboration in this work to provide the AVL simulation software at JUNIA, France.

References

Aitouche A (2022) An overview of River Project. In: Workshop of prevention of pollution for inland waterway vessels and greening of the inland water transport. UECE, Division on sustainable transport, 16 February 2022, Genova, Switzerland

Chen L, Yong SZ, Ghoniem AF (2012) Oxy-fuel combustion of pulverized coal: characterization, fundamentals, stabilization and CFD modeling. Prog Energy Combust Sci 38(2):156–214

Figueroa JD, Fout T, Plasynski S, McIlvried H, Srivastava RD (2008) Advances in CO2 capture technology—the US Department of Energy's Carbon Sequestration Program. Int J Greenhouse Gas Control 2(1):9–20

Giorgi S, Vera-Quintana S, Mobasheri R, Aitouche A (2021) A CFD investigation into the effects of intake air oxygen enrichment on DI diesel engine combustion and emissions. In: 9th international conference on systems and control (ICSC), November, Caen, France, pp 24–26

Kanniche M, Gros-Bonnivard R, Jaud P, Valle-Marcos J, Amann JM, Bouallou C (2010) Precombustion, post-combustion and oxy-combustion in thermal power plant for CO2 capture. Appl Therm Eng 30(1):53–62

Li X et al (2022) Oxy-fuel combustion for carbon capture and storage in internal combustion engines–a review. Int J Energy Res 46(2):505–522

Li X, Pei Y, Ajmal T, Aitouche A, Mobasheri R, Peng Z (2021) Implementation of oxy-fuel combustion (OFC) technology in a gasoline direct injection (GDI) engine fuelled with gasoline–ethanol blends. ACS Omega ACS Publ 6(44):29394–29402

Wall T et al (2009) An overview on oxyfuel coal combustion—state of the art research and technology development. Chem Eng Res Des 87(8):1003–1016

Wang M, Joel AS, Ramshaw C, Eimer D, Musa NM (2015) Process intensification for post-combustion CO2 capture with chemical absorption: a critical review. Appl Energy 158:275–291

Status of Research and Application Cases in Intelligent Shipping

Jiayi Xu[1,2], Zixiang Li[1,2], Haifei Sha[1,2], and Shiqiang Wu[1,2(✉)]

[1] Nanjing Hydraulic Research Institute, Nanjing 210029, China
{jyx, sqwu}@nhri.cn
[2] State Key Laboratory of Hydrology Water Resources and Hydraulic Engineering, Nanjing 210098, China

Abstract. With the rapid development of information technology, the shipping industry is also experiencing an important period of change in the development of transition to intelligence and wisdom. Intelligent shipping is a service system that promotes inter-departmental cooperation, cross-regional information resource integration, transportation efficiency improvement and transportation cost reduction by using cloud computing, big data, Internet of Things, sensors and other technologies as a comprehensive application with the construction of big data center as a pioneer. Intelligent shipping is the innovative application of relevant advanced concepts and high technology in the shipping industry arising from the construction of intelligent society, however, the development of intelligent shipping is still in the primary stage, and relevant literature on the field of intelligent shipping is still lacking. In order to understand and develop intelligent shipping, this paper reviewed and summarized intelligent shipping based on the existing literature. This paper mainly introduced the background, connotation and development status of intelligent shipping at home and abroad, summarized the framework system of intelligent shipping from five aspects: intelligent ship, intelligent port, intelligent shipping insurance, intelligent supervision and intelligent service, summarized several key technologies in intelligent shipping, and gave examples of successful applications of intelligent shipping at home and abroad. Combined with the current research status of intelligent shipping, this paper put forward the shortcomings and challenges of intelligent shipping, and looks forward to the future development and research direction of intelligent shipping.

Keywords: Intelligent shipping · Intelligent ship · Intelligent port

1 Introduction

As an important part of the national economy and an indispensable part of production and life, transportation industry plays an important role. As an important cornerstone of the transportation industry, shipping is a transportation mode with large transportation volume, low energy consumption, low cost and environmental protection. It plays a very important role in the transportation of trade goods in China and the world. In recent years, with the rapid development of big data, artificial intelligence, machine learning and other technologies, new concepts such as smart city, smart water

Y. Li et al. (Eds.): PIANC 2022, LNCE 264, pp. 916–926, 2023.
https://doi.org/10.1007/978-981-19-6138-0_81

conservancy, smart maritime, smart robot and smart home have emerged one after another. The concept of "intelligent transportation" originated from land transportation and was later cited to water transportation, resulting in the concept of "intelligent shipping".

Intelligent shipping is a comprehensive application of cloud computing, big data, Internet of things, sensors and other technologies guided by the construction of big data center. It is a modern shipping service system that promotes cross departmental cooperation, cross regional information resource integration, improves transportation efficiency and reduces transportation costs. The basic components of intelligent shipping include five elements: intelligent ship, intelligent shipping insurance, intelligent port, intelligent supervision and intelligent service. The sustainable development of intelligent shipping requires the organic coordination and common development of all elements.

There are many key technologies used in intelligent shipping, such as cloud computing, big data, Internet of things, sensors and so on. Cloud computing is a new type of IT infrastructure with high performance, high speed and diversification. In the development of intelligent shipping, it will organically combine computing technology with traditional shipping, which is a collision combination of traditional technology and new technology in the development. "Cloud" refers to a virtual network. Cloud computing is a network that provides resources. Resources like cloud can be accepted and used in real time. In the early stage of the development of cloud computing, it is only simple distributed computing, and then the combination of computing results. Now it has developed to decompose the huge data in the cloud into infinite small tasks through computer processing, and then process and analyze these results through the server system and return them to users. Cloud computing technology is the core of making traditional shipping intelligent. Through multi data analysis, it can improve the synergy between ships and logistics and have efficient operation. With the "Internet plus" action and the national big data strategy advancing, data management, data management and data innovation will gradually influence the development of national governance.

In recent years, intelligent shipping has developed rapidly in China and around the world. In China, the proposal of the CPC Central Committee on formulating the 13th five year plan for national economic and social development issued in 2015 mentioned the need to speed up the improvement of water transport infrastructure network and the implementation of high-tech ship intelligent manufacturing; The guiding opinions on the development of intelligent shipping jointly issued by the Ministry of transport and other seven departments in 2019 put forward the strategic objectives and ten tasks for the development of intelligent shipping, carried out the top-level design of intelligent shipping development, and improved China's shipping competitiveness; In addition, the medium and long term development planning outline of scientific and technological innovation in the field of transportation (2021–2035) issued by the Ministry of transport in 2022 mentioned the need to strengthen the promotion of intelligent shipping technology innovation, so as to promote the application of global shipping service network based on blockchain. Internationally, the 98th meeting of the Maritime Safety Committee of the International Maritime Organization (IMO) (msc98) held in 2017 included "autonomous unmanned ships" as a new topic, and intelligent shipping has gradually become a research hotspot.

With the increase of the number of ships worldwide and the development of large ships, the shipping industry is facing many challenges, such as serious environmental pollution, increased labor cost, insufficient safety and so on. The development of intelligent shipping will effectively meet these challenges. Intelligent shipping can effectively improve shipping capacity, promote the development of green technology, reduce greenhouse gas and pollutant emissions, save energy and improve shipping safety. However, the development of intelligent shipping is still in its infancy, and the literature on the overall framework of intelligent shipping is relatively lacking. This paper mainly summarized the framework system of intelligent shipping from five aspects: intelligent ship, intelligent port, intelligent navigation insurance, intelligent supervision and intelligent service, summarized several key technologies of intelligent shipping, then gave examples of successful cases of intelligent shipping application in China and abroad, and finally put forward the current shortcomings and challenges of intelligent shipping. The future development and research direction of intelligent shipping will be prospected.

2   Intelligent Shipping System Framework

Intelligent shipping is mainly divided into five aspects: intelligent ship, intelligent port, intelligent aviation insurance, intelligent supervision and intelligent service. The overall framework of intelligent shipping is shown in Fig. 1. The following will mainly summarize the definitions and research status of these five aspects, and specifically introduce the two aspects of intelligent ship and intelligent port.

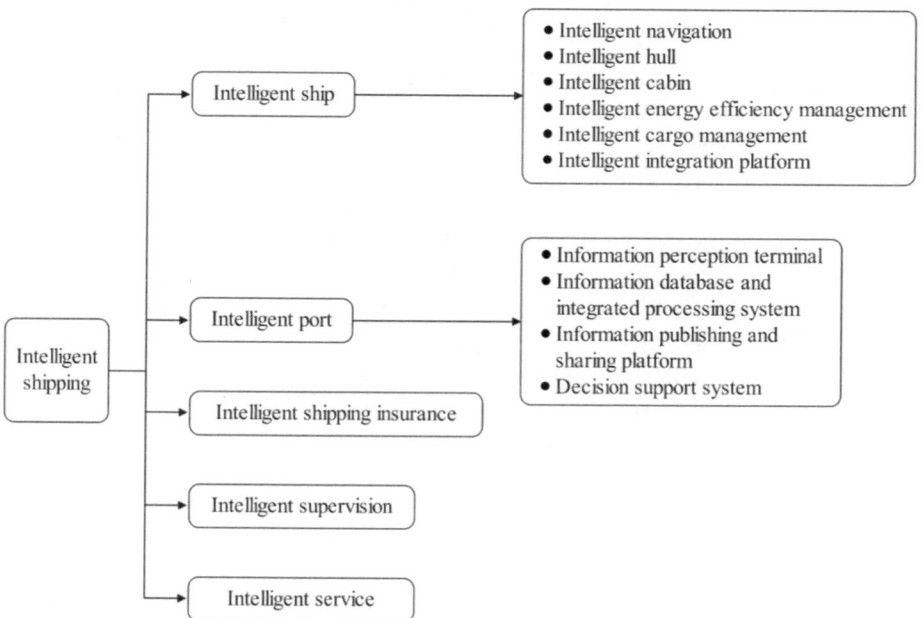

Fig. 1. Framework of intelligent shipping

2.1   Intelligent Ship

Intelligent ship refers to a ship that automatically senses and obtains the information and data of the ship itself, marine environment, logistics, port and other aspects by using technical means such as sensor, communication, Internet of things and Internet, and realizes intelligent operation in ship navigation, management, maintenance and cargo transportation based on computer technology, automatic control technology and big data processing and analysis technology, so as to make the ship safer, more environmentally friendly More economical and reliable (Li et al. 2021). Compared with traditional ships, intelligent ships have significant advantages in economy, safety, reliability, environmental protection and efficiency (Wang et al. 2021). At the same time, although ship intelligence reduces the impact of human factors on navigation safety, the research on its safety is still the focus of research in the shipbuilding industry (Zhang et al. 2021).

Internationally, Japan began to study AI ships with intelligent navigation function in the 1980s; South Korea launched the world's first intelligent ship in 2011. At present, its intelligent ship 2.0 plan is being promoted; Rolls Royce completed the world's first merchant ship remote operation in 2017, developed the world's first fully automatic ferry in 2018, and conducted automatic navigation test at the same time;, In 2019, Japan announced that it had completed the world's first sea test of intelligent ships and completed various test projects of the interim guide for autonomous ship test of the international maritime organization. At home, the action plan for the development of intelligent ships (2019–2021), prepared and issued by the Ministry of industry and information technology, the Ministry of transport and the Bureau of science, technology and industry for national defense in 2018, aims to promote the high-quality development of China's shipbuilding industry and keep the development of intelligent ships in China in step with the advanced level of the world. Some intelligent ships that have been developed are shown in Fig. 2.

Fig. 2. Examples of intelligent ship

China Classification Society (CCS) is an organization that provides the world's leading technical specifications and standards and classification inspection services for ships, marine facilities and related industrial products. At the same time, it also provides legal inspection, notarial inspection, certification and accreditation services in accordance with international conventions, rules and relevant regulations of authorized flag States or regions. Based on the scientific and technological research achievements of China Classification Society in recent years, and taking full account of the application experience of intelligent ships at home and abroad and the development direction of intelligent ships in the future, the intelligent ship system is composed of six functions: intelligent navigation, intelligent ship body, intelligent engine room, intelligent energy efficiency management, intelligent cargo management and intelligent integration platform. The operation diagram of intelligent ship is shown in Fig. 3.

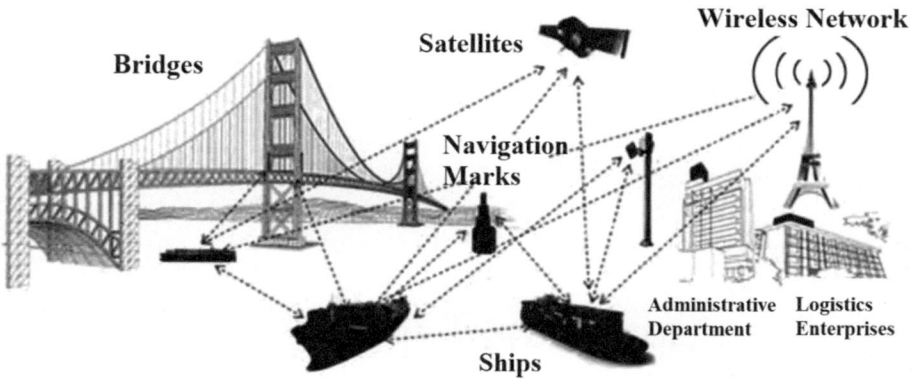

Fig. 3. Schematic diagram of intelligent ship operation (Yan and Liu 2016)

Intelligent navigation is to use technical means to analyze and process meteorological and logistics information, and design and optimize ship route and speed based on the analysis results. It can also have the function of autonomous navigation in open waters and advanced autonomous navigation ability of automatic berthing and disembarking from the wharf; Intelligent hull is to establish and maintain the hull database, and provide auxiliary decision-making for the safety and structural maintenance of the hull in the whole life cycle based on the database data; The intelligent engine room monitors the operation status of the main engine, auxiliary engine and shafting in the engine room, and analyzes and evaluates the operation status and health status of mechanical equipment according to the collected data; Intelligent energy efficiency is the on-line monitoring and automatic data collection of ship navigation status and energy consumption status. According to the collected data, the ship energy efficiency status, navigation and loading status are evaluated; Intelligent cargo management is an effective method to realize the monitoring, alarm and auxiliary decision-making of cargo hold and cargo, so as to optimize the stowage of cargo; The intelligent integration platform integrates the data of three systems: intelligent navigation, intelligent engine room and intelligent energy efficiency management. It can integrate the existing onboard information management system and subsequent new systems to realize the comprehensive monitoring and intelligent management of ships (Lei et al. 2018).

2.2   Intelligent Port

The development of the port is divided into four stages. The first generation is the "transportation center", which realizes the functions of port loading and unloading, transshipment, storage and goods receiving and dispatching; The second generation is "transportation center + Service Center", which provides some industrial and commercial value-added services based on the transportation center; The third generation is "international logistics center", which forms logistics services integrating goods, technology, information and capital; The fourth generation is the "supply chain center", which is an important link in the supply chain and has brand-new port characteristics.

As the latest fourth generation port form, intelligent port is a typical representative. Taking the information physical system as the framework, the intelligent port integrates the communication between the logistics supplier and the demander into the integrated system of collection, distribution and transportation through the innovative application of high and new technology; Greatly improve the comprehensive information processing capacity of the port and its related logistics parks and the optimal allocation capacity of related resources; Intelligent supervision, intelligent service and automatic loading and unloading have become its main forms, and can provide high safety, high efficiency and high-quality services for the modern logistics industry. The system structure diagram of intelligent port is shown in Fig. 4.

The world is setting off an upsurge in the development of intelligent ports. In China, the Ministry of transport issued the notice on carrying out intelligent port demonstration project in 2017, and a large number of Chinese scientific and technological innovation enterprises participated in it with artificial intelligence as the starting point. The schematic diagram or model diagram of some intelligent ports is shown in Fig. 5.

2.3   Intelligent Aviation Insurance

The core elements of intelligent navigation support system include shore based system, ship shore communication, navigation and positioning, information service and data standard. Ship shore communication is a channel that provides all information exchange and sharing. Intelligent ship navigation requires a large number of data exchange between ship and shore and between ships, which requires faster data communication bandwidth and more efficient and convenient data communication technology. Navigation and positioning provides basic location information, dynamic change information and unified reference time scale for all other technical elements and system cores. The data model includes data exchange standard protocol and data structure model. It integrates all marine related information, transmits according to the unified business model and data, and establishes a complete information service system to meet the needs of users (Gao and Wang 2019).

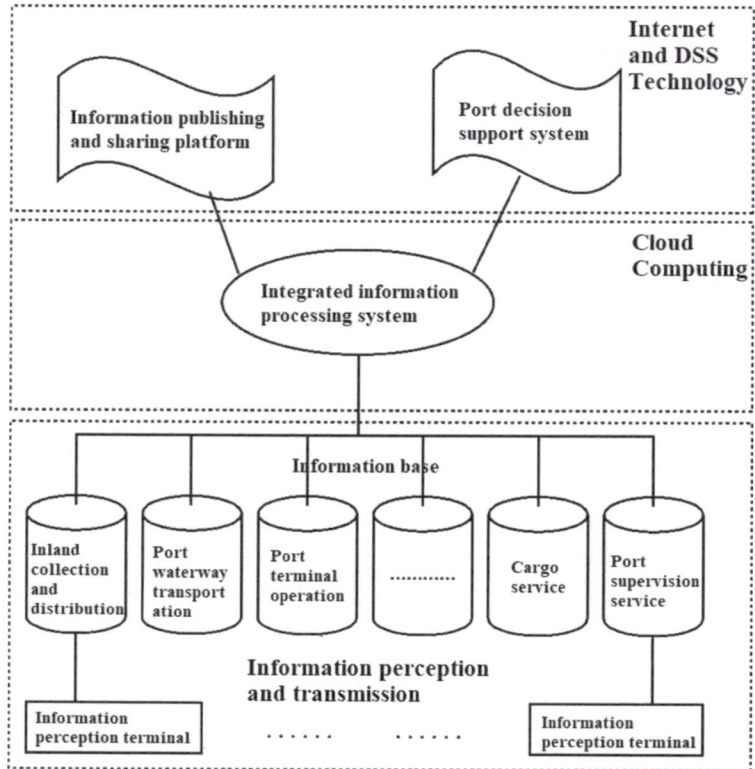

Fig. 4. System structure diagram of intelligent port (Bao 2013)

Fig. 5. Schematic diagram of intelligent port

2.4 Intelligent Supervision

With the intelligent development of ships and ports, shipping supervision must change and progress, realize intelligent supervision in ship inspection, safety supervision, health and epidemic prevention, border inspection and commodity inspection, promote the integration and sharing of information, realize a single window, comprehensively create a new pattern of shipping supervision with information and intelligent means, develop and utilize supervision resources and improve quality and efficiency.

2.5 Intelligent Services

Intelligent service is to realize intelligent shipping scheduling and shipping service through an open, standard and unified shipping comprehensive information service system based on multi-dimensional information resources. By exchanging and sharing data resources, various intelligent information services such as channels, ports, ships, logistics, pilotage, rescue, shipping management, shipping market and public services can be realized; Realize the release, exchange and sharing of information among shipping participants; Integrate various shipping business application systems, process and exchange relevant data and information, and provide government services, business services and public information services for all shipping participants.

3 Typical Cases of Intelligent Shipping

3.1 COSCO Intelligent Ship

Intelligent ship is an important part of intelligent shipping. The core of intelligent ship is ship energy efficiency system. Energy efficiency system is a cooperative system between ship and shore, and a control system for ship operation information and performance indicators. Ocean Shipping Group will cooperate with group A in the 20th century. According to the group's intelligent technology, the installation of its ships can monitor and detect the marine environment and ship performance. After mastering the data, it can monitor the operation of the ship and even the whole team. After half a year of trial operation, it will evaluate the economy before and after operation. The following table shows the economic improvement results between intelligent ships and traditional ships (Table 1).

Table 1. Benefit evaluation of single ship intelligent ship (Duan 2017)

Projects	Saving ratio	Amount (thousands of US dollars)
Route	1.5%	50
Speed	0.5%	150
Course	1.5%	50
Equipment	1.0%	75
Fuel	0.5%	150
Maintenance	1.0%	75
Testing	1.5%	50
Operation index	1.5%	50
Assets	1.0%	75
Other	2.5%	30
Total	12.5%	755

In terms of operation results, the installation of energy efficiency system can upgrade traditional ships to intelligent ships, which can save about US $7.855 million per ship every year, save human resources, save expenses and improve management level in daily operation. This result improves the feasibility of changing from single ship intelligence to fleet intelligence in the future, which not only saves expenses, but also saves resources, protects the environment and liberates the labor force.

3.2   Xiamen Ocean Automation Terminal

When artificial intelligence and 5g emerging technologies are organically combined with Xiamen ocean terminal, the automatic shipping terminal brings not only management transformation and efficiency improvement, but also a mileage of national marine terminal automation. In the upgrading process of traditional wharf, mechanical automation came first, and now it is a comprehensive intelligent simulation platform. Compared with traditional container terminals, fully automated container terminals can reduce operation and operation costs, save energy by more than 25% and reduce carbon emission by more than 16%.

Thanks to the help of intelligence, Xiamen ocean automation terminal saves more than 25% of energy, improves efficiency by 20% and reduces front-line operators by 70% compared with traditional terminals. The container throughput has increased from 300000 containers at the beginning of production in 2011 to more than 2 million containers in 2018. After seven years of production, the terminal throughput has increased nearly seven times. In 2019, Xiamen ocean automation terminal will continue to maintain double-digit growth. Since Xiamen ocean terminal fully automated wharf was put into commercial operation in March 2016, it has achieved "zero" safety accidents, and the container volume, work efficiency and economic benefits have been accelerated. The construction and completion of Xiamen ocean automation terminal is leading the new trend of terminal development in China in the future. Xiamen ocean automation terminal is shown in Fig. 6.

Fig. 6. Xiamen ocean automation terminal

4 Conclusion and Prospect

Starting from the concept and background of intelligent shipping and the current economic and social environment, this paper introduced the connotation of intelligent shipping and the research status in China and abroad, and gave examples of the excellent application of intelligent shipping. Intelligent shipping, as a new term with a short history and rapid development, has brought many conveniences and advantages. At the same time, there are still many problems in the development of intelligent shipping, such as the immature technology of intelligent system; relevant top-level design, policies and regulations are not perfect; network security risks and increased firewall security risks and so on. Compared with the problems, the greater impact must be positive. China's shipping industry has been at a low ebb since reaching its peak in the early 20th century, and its development speed is slow. The emergence of intelligent shipping is the industrial transformation of the shipping industry in the new era stage in combination with emerging technologies. It deepens the innovative ideas of intellectualization, modernization and efficiency into the shipping industry, promotes the transformation and upgrading of shipping, and intelligent shipping plays an important role in building a new shipping format It plays an important role in building a transportation power, realizing transportation modernization, improving service level and people's satisfaction. Promoting the upgrading of intelligent shipping in the direction of water conservancy is also one of the tasks in the new stage.

For the further development of intelligent shipping, some constructive suggestions are put forward:

(1) In terms of management methods, China should strengthen the construction of intelligent shipping laws and regulations, establish a set of basically perfect shipping process operation management system within the Ministry of transport, and realize shipping standardization and intelligence.
(2) In terms of technology, it is necessary to strengthen the research on intelligent shipping technology, actively learn from the advanced experience of foreign intelligent shipping, make it suitable for local conditions, and create a set of mature intelligent shipping system technology.
(3) In terms of talent training, we should strengthen the training of intelligent shipping related talents, so that young people can use more flexible ideas to contribute to the intelligent shipping system. We will vigorously carry out the integration and intersection of disciplines, so that different professional and technical personnel can develop intelligent shipping from different angles.
(4) In terms of top-level design, we should not copy the technologies of other countries, but firmly grasp the core technologies in our own hands that cannot be copied by other countries to realize the great rejuvenation of the Chinese nation.

Acknowledgements. This work is supported by the Natural Science Foundation of Jiangsu Province, China (Grant No. BK20200160).

References

Bao XG (2013) Concept of intelligent port and its systematics structure. Navig China 36 (02):120–123

Duan ZB (2017) Research on optimization of automatic container terminal operation based on big data analysis. JiMei University of Technology, Thesis of Master. (in Chinese)

Gao HZ, Wang YL (2019) Preliminary discussion on the construction of intelligent navigation guarantee system. China Maritime Safety 12:20–21 (in Chinese)

Lei JY, Chu XM, Jiang ZL et al (2018) Situation awareness system for vessel navigation based on visual analytics. Navig China 41(3):47–52 (in Chinese)

Li YJ, Zhang R, Wei MH et al (2021) State-of-the-art research and prospects of key technologies for ship autonomous navigation. Chin J Ship Res 16(1):32–44 (in Chinese)

Wang K, Hu WW, Huang LZ, et al (2021) Research progress and prospects of ship intelligent energy efficiency optimization key technologies. Chin J Ship Res 16(1):181–192, 199. (in Chinese)

Yan XP, Liu CG (2016) Review and prospect for intelligent waterway transportation system. CAAI Trans Intell Syst 11(6):807–817 (in Chinese)

Zhang D, Zhao YX, Cui YF, et al (2021) A visualization analysis and development trend of intelligent ship studies. J Transp Inf Saf 39:7–16 (in Chinese)

Study on Advanced Water Level Simulation Method for Inland Waterway Transport Based on the Extended Manning Formula

Junwei Zhou, Dianguang Ma$^{(\boxtimes)}$, Yu Duan, and Chao Ji

Tianjin Research Institute for Water Transport Engineering, No. 2618, Xinggang 2th Road, Binhai New Area, M.O.T, Tianjin, China
`zhoujw@tiwte.ac.cn`, `641683743@qq.com`

Abstract. The water level is a critical hydraulic parameter for inland ship safe navigation, as well as an important variable in inland waterway transport minoring and assistant systems. As a basic and traditional method, the one-dimensional (1D) hydrodynamic model is adapted to simulate river sections/waterway segments to obtain water levels numerically. However, the friction factor, i.e., Manning's coefficient n, is a sensitive parameter for the traditional 1D hydrodynamic model. Its calibration or identification is not only very time-consuming but also unpractical. Due to its sensitivity to the simulation results, usually, one identified parameter cannot be adopted into other flow scenarios. It has been concluded that the unfitness of the traditional empirical quasi-steady friction formulae leads to these consequences/phenomena. Besides finding advanced parameter calibration algorithms and updating friction parameters dynamically, employing a true unsteady friction formula to replace the quasi-steady friction formula is a thorough solution to the problem. In this study, we introduced a newly proposed 1D unsteady friction formula to the momentum equation of the Saint-Venant Equations, thus a modified 1D hydrodynamic model was developed. To validate its capability in simulating water levels, the modified model was adopted into the Xia-la-xian – La-he-lian section of Daying River; and compared with the traditional model with the Manning formula. Results showed that the modified hydrodynamic model performs better in both water level and cross-sectional average velocity simulation. The research results can be used to support the construction of intelligent water level warning systems, intelligent shipping, and digital waterway transportation platforms.

Keywords: Manning coefficient · Water level simulation · Smart shipping · Hydrodynamic model

1 Introduction

Building an accurate water level simulation model is a critical and fundamental technique for the development of digital waterways and smart shipping (Dafu et al. 2013; Zongjin 2019). Nevertheless, it is a difficult task for both hydraulic and hydrologic models. The hydrologic model usually focuses on the simulation of flow rates (Neal

© The Author(s) 2023
Y. Li et al. (Eds.): PIANC 2022, LNCE 264, pp. 927–937, 2023.
https://doi.org/10.1007/978-981-19-6138-0_82

et al. 2012). Due to the relationship between flow rate and the water level is complex and needs to overcome the difficulty in the simulation of anticlockwise looped shaped rating curve, the performance of the water level hydrologic model is not satisfactory (Bombar 2016). In terms of hydraulic simulation, the numerical simulation based on the two- or three-dimensional hydrodynamic models is time-consuming and needs excessive boundary and initial conditions. One dimensional (1D) hydrodynamic model is a light and optimal model for serving real-time water level simulation in a long river/inland waterway (Papanicolaou et al. 2004). The 1D hydrodynamic model is based on Saint-Venant Equations which are strictly deduced from the basis of assumptions and based on the laws of physics. However, the friction formula e.g. Chezy formula and Manning formula used in the model is proposed based on the combination of experimental data and experience in quasi-steady uniform flow circumstances (Chaudhry 1993). The real flow scenarios in the practical application are always unsteady flows. The traditional steady uniform friction formulas neglect the impact of flow unsteadiness and flow un-uniform on the friction resistance. They produce large errors in practical applications.

Recently, much research has been devoted to the development of more robust and appropriate friction-dependent models for improved estimations of unsteady friction-flow relationships in open channel/river systems. Some of these studies have focused on modifying the structure of the Manning formula to improve the prediction of unsteady open channel friction. For example, Tu and Graf (1993), Hsu et al. (2006), and Mrokowska et al. (2015) suggested that Manning's roughness coefficient fluctuates–throughout flood events, and Bellos et al. (2018) determined the roughness coefficient as a grey-box parameter and developed a new three-parameter friction model, which performs slightly better than the commonly used Manning equation. Bao et al. (2009) proposed a new formula to calculate the Manning roughness coefficient during flood events, which indicates that the unsteady open channel friction is closely linked to the historical river discharge or stage of its adjacent up-and down-channel sections. Other modifications have relied on adding more components to steady uni-form friction formulas to develop unsteady open channel friction models. For example, by adding the time derivative of flow rate to estimate the true unsteady friction slope, Ghimire and Deng (2011) proposed a hydrograph-based method to estimate the shear velocity during flood events. Additionally, after a careful correlation analysis based on a hydraulic experimental (flow-time) dataset, Bao et al. (2018) derived a linear unsteady friction model. These studies demonstrate the value of developing (incorporating) additional terms to account for the effects of unsteady flow in unsteady friction slope modeling. Zhou et al. (2022) developed a weighting friction model for unsteady open channel friction, which is also based on the structure of the Manning formula and so can be regarded as a modified or extended version of the Manning formula for unsteady open channel frictions. In contrast (or compared with) to previous parameter calibration algorithms and roughness coefficients updating methods such as (Bao et al. 2011; Hsu et al. 2006) and (Zeng and Huai 2009), these newly proposed unsteady frictional methods improve model performance without adjusting parameters dynamically. As a result, they do not suffer great deterioration in hydraulic forecasting, and can therefore help in providing a better understanding of the processes causing differences between steady (uniform) and unsteady friction slopes.

In this study, we introduced a newly proposed 1D unsteady friction formula, which the author's team proposed in (Zhou et al. 2022) namely the Extended Manning formula, to the momentum equation of the Saint-Venant Equations. Hence, a modified and more advanced water level simulation method for inland waterway transport was developed. To validate its capability in simulating water levels, the modified model was adopted into the Xia-la-xian – La-he-lian section of Daying River; and compared with the traditional model with the Manning formula. The advantages of the modified model are verified through the comparison study with the traditional 1D hydrodynamic model. And its potential support in further fields like the construction of intelligent water level warning systems, intelligent shipping, and digital waterway transportation platforms are discussed.

2 Methodology

2.1 Governing Equation

The proposed water level simulation method for inland waterway transport is based on the traditional 1D hydrodynamic model framework. Its governing equation is the Saint-Venant equation. The Saint-Venant equation consists of momentum and continuous equations, which can be expressed in terms of flow depth and cross-sectional averaged velocity using as follows:

$$\begin{cases} \frac{\partial D}{\partial s} + \frac{1}{g}\frac{\partial V}{\partial t} + \frac{V}{g}\frac{\partial V}{\partial s} + S_f - S_0 = 0 \\ \frac{\partial D}{\partial t} + D\frac{\partial V}{\partial s} + V\frac{\partial D}{\partial s} = 0 \end{cases} \tag{1}$$

where the upper equation is the momentum equation, and the lower equation is the continuous equation; D represents the water depth; V represents the cross-sectional averaged velocity; g is the acceleration of gravity; t is the time; s represents the longitudinal coordinates along the river; S_0 represents the bed slope; S_f is the frictional slope.

2.2 Preissmann Implicit Difference Method

To numerically solve the 1D hydrodynamic model, discretization algorithms are requested. In this paper, the Preissmann implicit difference method was used for discretization, in which the arithmetic average method and the weighted average method are applied in spatial partial derivative approximation and temporal partial derivative approximation, respectively.

$$F(s,t) = \frac{\theta}{2}\left(\Delta F_{j+1} + \Delta F_j\right) + \frac{1}{2}\left(F_{j+1}^k + F_j^k\right) \tag{2}$$

$$\frac{\partial}{\partial s}F = \theta\frac{\Delta F_{j+1} - \Delta F_j}{\Delta s} + \frac{F_{j+1}^k - F_j^k}{\Delta s} \tag{3}$$

$$\frac{\partial}{\partial t}F = \frac{\Delta F_{j+1} + \Delta F_j}{2\Delta t} \tag{4}$$

where F represents a certain variable such as flow depth and cross-sectional averaged velocity. $\Delta F_j = F_j^{k+1} - F_j^k$ and $\Delta F_{j+1} = F_{j+1}^{k+1} - F_{j+1}^k$. θ is the difference coefficient. Its ranges from 0.5 to 1. The superscripts k and $k+1$ represent time t. k and $k+1$ represent the current moment and the next moment, respectively. The subscripts j and $j+1$ represent the longitudinal coordinates along the river s. j and $j+1$ are the two adjoining cross sections, in which j^{th} cross section locates upstream of the $(j+1)^{\text{th}}$ cross section.

2.3   The Extended Manning Formula

In a classic 1D hydrodynamic model, the Manning formula is applied to calculate the friction slope/friction item in the Saint-Venant Equation. The Manning formula is an empirical friction equation derived and developed from the Chezy formula. Both of them are obtained and calibrated using some laboratory data of steady uniform flow. The Manning formula says:

$$S_f = n^2 \frac{V^2}{R^{4/3}} \tag{5}$$

where n is the Manning roughness coefficient; R is the hydraulic radius, in many engineering practices involving wide shallow rivers, $R \approx D$.

Since it has been concluded that the unfitness of the traditional empirical quasi-steady frictional formula like the Manning formula leads to significant simulation error in the 1D hydrodynamic model, the author's team proposed a modified version of the Manning formula in (Zhou et al. 2022). The most original article calls it a weighting function model for unsteady open channel friction, here we recognize its model structure as an extended version of the traditional Manning formula, and name it the Extended Manning formula. It goes:

$$S_f = n^2 \frac{V^2}{R^{4/3}} + \lambda^i \sum_{i=0}^N [w_{10}, w_{20}] \frac{V^2}{R^{4/3}} \left[\frac{R^{4/3}}{V^2} \frac{1}{V} \frac{\partial D}{\partial t}, \frac{R^{4/3}}{V^2} \frac{1}{g} \frac{\partial V}{\partial t} \right]_{t-i}^T \tag{6}$$

where n is the Manning roughness coefficient; λ is the weight decreasing coefficient; The weighting coefficient for the partial derivative of the velocity and the partial derivative of the water level are w_{10} and w_{20} respectively.

2.4   Advanced Water Level Simulation Method

By introducing Eq. (5) into the Saint-Venant Eq. (1), we can obtain the control equations of the traditional one-dimensional hydrodynamic model. Then, by discretizing it using the Preissmann implicit difference method, ones can obtain:

$$pa_{1j} \times \Delta V_j + pb_{1j} \times \Delta D_j + pc_{1j} \times \Delta V_{j+1} + pd_{1j} \times \Delta D_{j+1} = pe_{1j} \qquad (7)$$

$$pa_{2j} \times \Delta V_j + pb_{2j} \times \Delta D_j + pc_{2j} \times \Delta V_{j+1} + pd_{2j} \times \Delta D_{j+1} = pe_{2j} \qquad (8)$$

where Eqs. (7) and (8) are the two discretized equations derived from the momentum equation and the continuity equation between the jth and the $(j + 1)^{th}$ cross-sections, respectively. Discretization is a complex and basic mathematic tool for numerical simulation resolution. Its detailed process can refer to (Hsu et al. 2006). Subscript j and $j + 1$ represent the index of sections. ΔV_j, ΔD_j, ΔV_{j+1}, and ΔD_{j+1} are the increment of the next time relative to the current moment. pa_{1j}, pb_{1j}, pc_{1j}, pd_{1j}, pe_{1j}, pa_{2j}, pb_{2j}, pc_{2j}, pd_{2j}, and pe_{2j} are coefficients. They are determined by the current state of the hydraulic parameters and the Manning roughness coefficient, $(pa_{1j}... pe_{1j}, pa_{2j}... pe_{2j}) = f(\theta, V_j, D_j, V_{j+1}, D_{j+1}, n)$.

The core innovation of the proposed advanced water level simulation method is using a true unsteady friction formula, the extended Manning formula, to replace the quasi-steady friction formula, the traditional Manning formula. Hence, following the same logic, the numerical expression of the advanced water level simulation method can be obtained by following steps: First, by introducing Eq. (6) into the Saint-Venant Eq. (1), the control equations of the proposed method are obtained. Second, using the Preissmann implicit difference method, the discretized version of the proposed method is obtained:

$$ppa_{1j} \times \Delta V_j + ppb_{1j} \times \Delta D_j + ppc_{1j} \times \Delta V_{j+1} + ppd_{1j} \times \Delta D_{j+1} = ppe_{1j} \qquad (9)$$

$$ppa_{2j} \times \Delta V_j + ppb_{2j} \times \Delta D_j + ppc_{2j} \times \Delta V_{j+1} + ppd_{2j} \times \Delta D_{j+1} = ppe_{2j} \qquad (10)$$

where Eqs. (9) and (10) are the discretized equations for the advanced water level simulation method, in which Eq. (9) is the discretized momentum equation and Eq. (10) is the discretized continuity equation. Similar to the Eqs. (7) and (8), ppa_{1j}, ppb_{1j}, ppc_{1j}, ppd_{1j}, ppe_{1j}, ppa_{2j}, ppb_{2j}, ppc_{2j}, ppd_{2j}, and ppe_{2j} are coefficients. They are determined by the current state of the hydraulic parameters and the four parameters involved in the extended Manning formula, $(ppa_{1j}... ppe_{1j}, ppa_{2j}... ppe_{2j}) = f(\theta, V_j, D_j, V_{j+1}, D_{j+1}, n, \lambda, w_{10}, w_{20})$. Specifically, when $\lambda = 0$ or $w_{10} = w_{20} = 0$, the advanced water level simulation method degenerates into the traditional 1D hydrodynamic model; and $(ppa_{1j}... ppe_{1j}, ppa_{2j}... ppe_{2j}) = (pa_{1j}... pe_{1j}, pa_{2j}... pe_{2j})$.

3 Case Study

3.1 Study Area of Daying River

The proposed advanced water level simulation method is adopted in the Xia-la-xian – La-he-lian section of the Daying River. The Daying River is located in the southwest China of Yunnan province. The length of the Daying river is 204 km. It goes across Yingjiang County. Within Yingjiang County, the Daying River stretches 145.5 km. The Daying river basin has a mild climate, fertile land, rich specialties, and rich flora and fauna resources. In Yingjiang County, the riverbed is flat, with a maximum width of 1 km. The Xia-la-xian and La-he-lian are two national hydrometric stations. The Xia-la-xian – La-he-lian section of the Daying River stretches only 12.9 km, which is a perfect study case for 1D hydrodynamic models (Fig. 1).

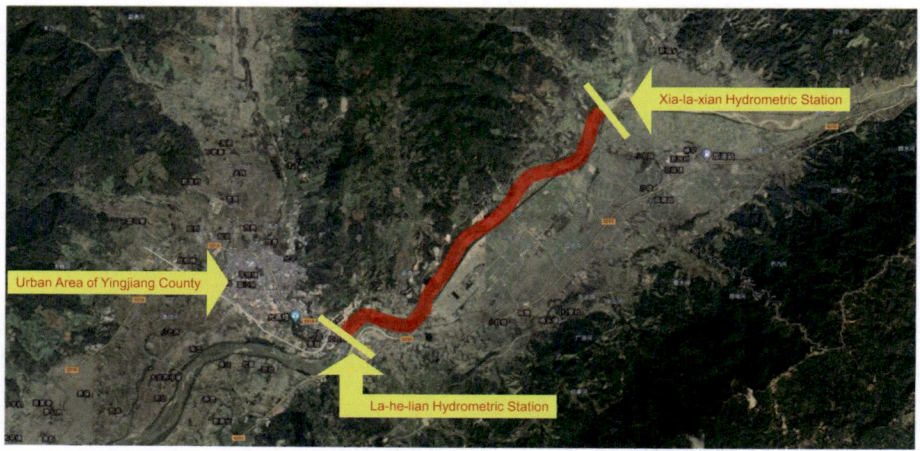

Fig. 1. Study Area (the Xia-la-xian – La-he-lian section) of Daying River and its relative location toward the urban area of Yingjiang county

3.2   Simulation Process

We collected the data of 9 Flood events in 1979 and 1980 at Xia-la-xian and La-he-lian two national hydrometric stations. Then, both the advanced water level simulation method and the traditional 1D hydrodynamic model are applied to simulate the water dynamics throughout the studied river section. Specifically, the discretized traditional 1D hydrodynamic model refers to Eqs. (7) and (8); The discretized advanced water level simulation method refers to Eqs. (9) and (10). In the study, the boundary conditions are designated using the cross-sectional averaged velocity at the upstream section (Xia-la-xian) and the water depth at the downstream section (La-he-lian). The initial condition is designated using the data when the flood event began. Both the two models are carefully calibrated using a Genetic Algorithm (GA) based parameter calibration algorithm. The details of this algorithm can be seen in a published paper by Zhou et al. (2018). The outputs of the two models are the same too, i.e., the water depth at the upstream section (Xia-la-xian) and the cross-sectional averaged velocity at the downstream section (La-he-lian). Finally, both the two models' results are compared to the observations to evaluate their performance.

3.3   Evaluation Criteria

For a better understanding of the performance of the advanced water level simulation method and the traditional 1D hydrodynamic model, the two models' simulation results are evaluated using the same 2 criteria, in which Mean Absolute Percentage Error (MAPE) is used to evaluate the model's overall performance.

$$\text{MAPE} = \frac{1}{N}\sum\nolimits_{i=1}^{N}\left(\frac{|cal_i - obs_i|}{obs_i} \times 100\%\right) \tag{11}$$

where *cal* represents model calculations; *obs* represents observations; *N* is the number of the outputs. It should be noted that both the outputs of water depth at the upstream section (Xia-la-xian) and cross-sectional averaged velocity at the downstream section (La-he-lian) need to be evaluated. Hence, $N = 2$ times the length of the flood time series.

Besides, Absolute Relative Error at Peak value (AREP) is adopted to evaluate the model performance in estimating flood peak values.

$$\text{REP\_D} = \frac{\max(cal\_D) - \max(obs\_D)}{\max(obs\_D)} \times 100\% \tag{12}$$

$$\text{REP\_V} = \frac{\max(cal\_V) - \max(obs\_V)}{\max(obs\_V)} \times 100\% \tag{13}$$

$$\text{AREP} = (|\text{REP\_D}| + |\text{REP\_V}|)/2 \tag{14}$$

where REP represents Relative Error at Peak value; REP_D and REP_V represent the calculated REPs for water depth and cross-sectional averaged velocity respectively.

4   Results and Discussion

4.1   Performance of the Traditional 1D Hydrodynamic Model

Table 1 summarizes the performance of the traditional 1D hydrodynamic model whose friction calculator is the Manning formula. It can be seen that MAPE of the traditional model ranges from 10% to 35% with a mean value of 23.4%, which indicates that the mean relative error is generally greater than 20%. The overall performance of the traditional 1D hydrodynamic model is not satisfactory. Besides, AREP measures the accuracy of peak value simulation. According to Table 1, the AREP of the traditional 1D hydrodynamic model ranges from 1% to 20% with a mean value of 6.5%, which is not good neither for a flood simulator.

Table 1. Summary of the performance of the traditional 1D hydrodynamic model using criteria MAPE and AREP

No.	MAPE (%)	AREP (%)
FloodEvent.01	12.7	3.6
FloodEvent.02	13.3	3.2
FloodEvent.03	17.7	5.4
FloodEvent.04	23.5	7.6
FloodEvent.05	30.9	17.9
FloodEvent.06	25.6	2.9
FloodEvent.07	32.2	7.2
FloodEvent.08	22.2	9.4
FloodEvent.09	33.2	1.3
In average	23.4	6.5

4.2   Performance of the Advanced Water Level Simulation Method

Table 2 summarizes the performance of our proposed advanced water level simulation method. The friction item in the proposed method is updated using the extended Manning formula to replace the traditional Manning formula. The results come out that MAPE of the modified version of 1D hydrodynamic model is much less than the traditional one. According to Table 2, the MAPE of the advanced water level simulation method ranges from 2% to 30% with a mean value of 9.4%. The overall relative error is reduced by $(23.4\% - 9.4\%)/23.4\% = 59.8\%$. Undoubtedly, it is a huge improvement in the lift of simulation performance. Besides, the AREP of the advanced water level simulation method ranges from 0% to 5% with a mean value of 1.8%. Compared to the performance of the traditional model, the modified model can estimate the peak values of flood events much better as well.

Table 2. Summary of the performance of the advanced water level simulation method using criteria MAPE and AREP

No.	MAPE (%)	AREP (%)
FloodEvent.01	4.3	1.4
FloodEvent.02	2.6	1.6
FloodEvent.03	3.9	2.1
FloodEvent.04	3.8	1.6
FloodEvent.05	5.8	1.4
FloodEvent.06	18.4	2.4
FloodEvent.07	8.6	1.1
FloodEvent.08	7.2	4.9
FloodEvent.09	30.5	0.1
In average	9.4	1.8

4.3 Result Comparison Between the Two Models

For better understand the advantages of the proposed advanced water level simulation method, we take two simulated flood event results as examples and draw the simulation results of the two models on a graph, sees Figs. 2 and 3.

Figure 2 illustrates a simulation example of single-peak flood hydrographs. It can be seen that compared to the traditional 1D hydrodynamic model (marked as SVN-TM on Fig. 2), the advanced water level simulation method (marked as SVN-EM on Fig. 2) performs better in both water level and cross-sectional average velocity simulation. Especially, in the peak of the flood event, the traditional model over-estimates the real peak, in contrast, the modified model can simulate the peak section really well.

Figure 3 illustrates a simulation example of multi-peaks flood hydrographs. Similar results can be seen in the Fig. 3 as that in Fig. 2. According to Fig. 3, the advanced water level simulation method (marked as SVN-EM on Fig. 3) performs better in both water level and cross-sectional average velocity simulation. In the peak of the flood event, the modified model can also better simulate the peak process compared to the traditional 1D hydrodynamic model (marked as SVN-TM on Fig. 3).

Therefore, the advantages of our proposed advanced water level simulation method are verified. The research results can be used to support the construction of intelligent water level warning systems, intelligent shipping, and digital waterway transportation platforms.

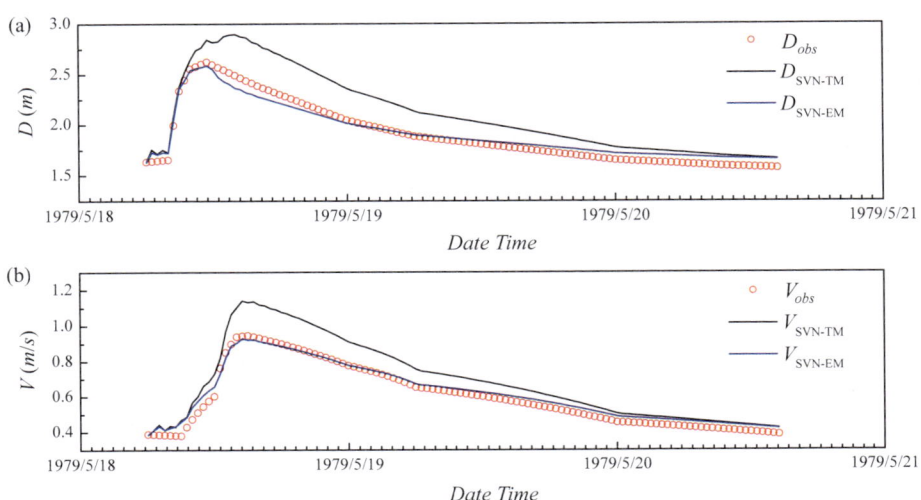

Fig. 2. Model performance comparisons between the proposed advanced water level simulation method (marked as SVN-EM) and the traditional 1D hydrodynamic model (marked as SVN-TM) in the studied segments (taking the 1st flood event as an example)

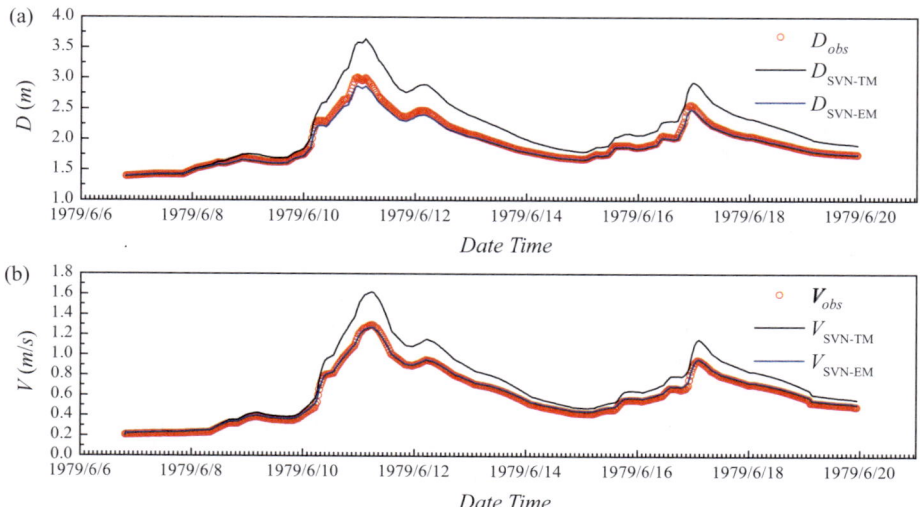

Fig. 3. Comparison of the simulation results between the proposed advanced water level simulation method (marked as SVN-EM) and the traditional 1D hydrodynamic model (marked as SVN-TM) in the studied segments (taking the 2nd flood event an example)

5 Conclusions

By introducing the extended Manning formula to the Saint-Venant equation to replace its traditional friction calculator (Manning formula), an advanced water level simulation method is proposed in this paper. The modified model is adopted to simulate 9 flood events in the Xia-la-xian – La-he-lian section of the Daying River to verify its advantages toward the traditional model.

The two models are used to simulate both single-peak flood hydrographs and multi-peaks flood hydrographs. For the traditional 1D hydrodynamic model, MAPE ranges from 10% to 35% with a mean value of 23.4%. AREP ranges from 1% to 20% with a mean value of 6.5%. For the advanced water level simulation method, MAPE ranges from 2% to 30% with a mean value of 9.4%. AREP ranges from 0% to 5% with a mean value of 1.8%.

In general, the modified hydrodynamic model performs better in both water level and cross-sectional average velocity simulation. The research find out can be used to support the construction of intelligent water level warning systems, intelligent shipping, and digital waterway transportation platforms.

Acknowledgements. This work was supported by the joint research on the ecological intelligent monitoring and impact assessment of inland waterway engineering (grant number: 2019YFE0121000); Research and application of key technologies for ecological inland waterway engineering and maintenance in the Xijiang River Basin (grant number: Guike AB22035084); Research and application of key technologies for the development of ultra-high head cascade water transport channels in Canyon rivers (grant number: [2018]3010).

References

Bao W, Zhou J, Xiang X, Jiang P, Bao M (2018) A hydraulic friction model for one-dimensional unsteady channel flows with experimental demonstration. Water 10(1):43

Bao WM, Zhang XQ, Qu SM (2009) Dynamic correction of roughness in the hydrodynamic model. J Hydrodyn Ser B 21(2):255–263

Bao WM, Zhang XQ, Yu ZB, Qu SM (2011) Real-time equivalent conversion correction on river stage forecasting with Manning's formula. J Hydrol Eng 16(1):1–9. https://doi.org/10.1061/(ASCE)HE.1943-5584.0000279

Bellos V, Nalbantis I, Tsakiris G (2018) Friction modeling of flood flow simulations. J Hydraul Eng 144(12):04018073

Bombar G (2016) The hysteresis and shear velocity in unsteady flows. J Appl Fluid Mech 9 (3):839–853

Chaudhry MH (1993) Open-channel flow 40(5):567–578

Dafu C et al (2013) Analysis on the development stages and characteristics of the Yangtze River channel. J Dalian Maritime Univ (Soc Sci Ed) 12(003):23–27

Ghimire B, Deng Z-Q (2011) Event flow hydrograph-based method for shear velocity estimation. J Hydraul Res 49(2):272–275

Hsu MH, Liu WC, Fu JC (2006) Dynamic routing model with real-time roughness updating for flood forecasting. J Hydraul Eng 132(6):605–619

Mrokowska MM, Rowiński PM, Kalinowska MB (2015) Evaluation of friction velocity in unsteady flow experiments. J Hydraul Res 53(5):659–669

Neal JC, Atkinson PM, Hutton CW (2012) Adaptive space–time sampling with wireless sensor nodes for flood forecasting. J Hydrol 414–415(2):136–147

Papanicolaou AN, Bdour A, Wicklein E (2004) One-dimensional hydrodynamic/sediment transport model applicable to steep mountain streams. J Hydraul Res 42(4):357–375

Tu H, Graf WH (1993) Friction in unsteady open-channel flow over gravel beds. J Hydraul Res 31(1):99–110

Zeng Y, Huai W (2009) Application of artificial neural network to predict the friction factor of open channel flow. Commun Nonlinear Sci Numer Simul 14(5):2373–2378

Zhou J, Bao W, Li Y, Cheng L, Bao M (2018) The modified one-dimensional hydrodynamic model based on the extended Chezy formula. Water 10:1743–1759

Zhou J et al (2022) A weighting function model for unsteady open channel friction. J Hydraul Res 60(3):460–475. https://doi.org/10.1080/00221686.2021.2004251

Zongjin J (2019) Build Jinma Cloud smart shipping logistics platform and create new opportunities for port and shipping enterprises to develop. China Water Transp (7):2

Study on the Economic Speed of the Grand Canal in North Jiangsu

W. Xie[1], S. D. Xu[1(✉)], N. N. Zhang[1], X. Yue[1], J. Liu[2], and S. H. Lu[1]

[1] Southeast University, Nanjing, China
sudongxu@seu.edu.cn
[2] North Jiangsu Grand Canal Administration of Navigational Affairs,
Nanjing, China

Abstract. The energy conservation and emission reduction have been the development trend of various industries. However, the carbon emissions of the shipping industry are increasing year by year. Optimizing ship energy consumption is an important task to develop environment-friendly shipping and reduce the operating cost of shipping companies. Speed is one of the most critical factors affecting ship energy consumption. Many studies confirm that it is an efficient way to save ship energy consumption by obtaining appropriate speed and reducing ship energy consumption. Compared with rivers and oceans, inland canals have the characteristics of shallow water depth, narrow width and low water velocity. This paper analyzed the low-energy operation scheme of canal ships with the research object of the canal ships in North Jiangsu. The main factors affecting the energy consumption and the ship resistance were analyzed. Accordingly, the mathematical model of ship energy consumption, ship speed and environmental factors was established. The optimized economic ship speed was then calculated by calling the optimization algorithm guided by the lowest ship energy consumption. It is verified by an example of ships in the North Jiangsu Grand Canal. Applying the optimized economic speed to actual ship navigation can effectively reduce the ship fuel consumption, which is of great significance to achieve the goal of energy conservation and emission reduction.

Keywords: Economic speed · Canal · Ship energy consumption · Genetic algorithm

1 Introduction

The International Maritime Organization (IMO) report pointed out that the carbon emission of the shipping industry in 2007 was nearly 1 billion tons. If the shipping industry does not control the carbon emission, the carbon emission of the shipping industry will double, accounting for 18% of the total global carbon emission in 2050. Due to the proposal of the "double carbon" goal, energy conservation and emission reduction are less and less valued by all walks of life, and the shipping industry is no exception. Jiangsu Province responded positively and continued to promote the adjustment of industrial and energy structure to realize the synergistic effect of pollution and carbon reduction. The realization of carbon emission reduction of shipping industry needs to start with reducing ship energy consumption.

Y. Li et al. (Eds.): PIANC 2022, LNCE 264, pp. 938–950, 2023.
https://doi.org/10.1007/978-981-19-6138-0_83

The researches on ship energy consumption optimization at home and abroad are mainly carried out in the following four directions: first, improve energy efficiency through the discovery and application of new energy; Second, the improvement of ship propulsion system to improve energy efficiency (Liang 2020; Yuan et al. 2017); Third, ship operation management and route scheduling to reduce energy consumption (Zhou 2020; Zhang 2020); Fourth, improve energy efficiency and reduce ship energy consumption through speed optimization (Li et al. 2020; Chang and Wang 2014; Kim et al. 2012). Ship speed optimization has the advantages of no need to modify the ship structure, convenient operation and less cost. There are many studies on ship speed optimization to reduce ship energy consumption. According to the measured data, Yuan et al. (2017) made a statistical analysis on the environmental factors of ship navigation. The data analysis shows that by optimizing the ship speed, the energy consumption efficiency of ships under different navigation conditions can be significantly improved. Sun (2019) studied the marine propulsion system and established the fuel consumption model of the marine diesel engine. The operation route of ship navigation is segmented, and the optimization of ship speed is completed by using genetic algorithm. However, there are few studies on the energy efficiency optimization of inland ships, and there is a relative lack of research on the energy consumption optimization of canal ships.

There are many factors affecting ship energy consumption, which are the main factors to be considered in ship energy consumption optimization. Through Spearman correlation analysis, Fan et al. (2017) proved that water flow and channel depth are the main factors affecting ship energy consumption. The energy efficiency of inland ships is significantly affected by the navigation environment including wind speed and direction, water depth and speed, as found in Yan et al. (2018). Therefore, this paper selects the channel water depth as the main influencing factor, establishes the mathematical model of ship fuel consumption, establishes the speed optimization algorithm, and calculates the economic speed corresponding to the water depth of the characteristic channel of the north Jiangsu Grand Canal under the guidance of the lowest ship energy consumption. The research results can be used to guide the navigation of ships in the northern Jiangsu canal section, and have important practical significance for reducing ship fuel consumption and realizing the "double carbon" standard.

2 Model Establishment

2.1 Ship Energy Consumption Model

2.1.1 Analysis of Ship Navigation Resistance

The basic resistance of the ship refers to the resistance of the water to the underwater part of the hull when the naked ship travels in calm water and deep water without accessories. The basic resistance of ship flow includes friction resistance and residual resistance, and the calculation formula is shown in formula (1).

$$R_0 = R_f + R_r \tag{1}$$

where, R_0 is the basic resistance of ship water flow (N); R_f is the friction resistance (N); R_r is differential pressure resistance, also known as residual resistance (N).

The basic flow resistance of inland ships can be calculated according to the formula of basic flow resistance of zvankov self-propelled ships, as found in Li (2002), as shown in formula (2).

$$R_0 = \left(0.17SV^{1.83} + C_r C_B A_M V^{17+4Fr}\right) \cdot g \tag{2}$$

where,

$$S = Ld\left[2 + 1.37(C_B - 0.274)\frac{B}{d}\right] \tag{3}$$

$$C_B = \frac{\Delta}{LBd} \tag{4}$$

$$A_M = C_M Bd \tag{5}$$

$$F_n = V/\sqrt{gL} \tag{6}$$

where, S is the wet surface area of the hull, which can be approximately estimated according to formula (3), (m^2); V is the speed of the ship to water (m/s); C_B is a square coefficient, which can be calculated according to formula (4); A_M is the water immersion area of the middle cross section, which can be calculated according to formula (5), (m^2); F_n is Froude number; C_r is the residual resistance coefficient; L is the waterline length of the ship (m); B is the width of the ship (m); d Is the draft of the ship (m); C_M is the mid cross section coefficient.

According to the analysis of natural environment data and field investigation, it can be seen that the water depth in most areas of the north Jiangsu section of the Beijing Hangzhou Grand Canal is relatively small, and the shallow water effect has a significant impact on the ship navigation resistance. The mathematical expression of ship navigation resistance considering the shallow water effect can be expressed by formula (7).

$$R_h = K_h R_0 \tag{7}$$

where,

$$K_h = 1 + \frac{0.0065V^2}{(h/d - 1)\sqrt{d}} \tag{8}$$

where, K_h is the conversion coefficient of shallow water navigation resistance, which can be calculated according to formula (8); R_h is the resistance of shallow water channel (N); h is the channel water depth (m); V is the ship speed (m/s).

2.1.2 Propulsion Characteristics of Diesel Engine

The working characteristics of marine diesel engine include speed characteristics, load characteristics and propulsion characteristics. Propulsion characteristics refer to the relationship between the performance parameters and the propeller speed and power when the diesel engine drives the propeller, as found in Zhang and Chen (2017).

According to the derivation of mathematical formula, the output power P_e of diesel engine is directly proportional to the third power of diesel engine speed n_e, as shown in formula (9). According to the multiple fitting results of measured data, as found in Wang (2014), it can be inferred that the diesel engine fuel consumption rate g_e and engine speed n_e can be fitted by quadratic polynomial, as shown in Eq. (10).

$$P_e = f_1\left(n_e^3\right) \tag{9}$$

$$g_e = f_2\left(n_e^2, n_e\right) \tag{10}$$

2.1.3   Open Water Characteristics of Propeller

The open water characteristics of propeller are mainly reflected by dimensionless coefficients such as thrust coefficient K_T, torque coefficient K_Q and propeller efficiency, and these coefficients are only related to advance coefficient J, the expression is shown in (11).

$$K_T = \frac{T}{\rho n_p^2 D^4} = f(J) \tag{11}$$

where,

$$J = \frac{h_p}{D} = \frac{V_p}{n_p D} \tag{12}$$

where, T is the ship thrust (N); Q is the propeller torque (N · m); ρ is the density of water (kg/m^3); n_p is the propeller speed (r/min); D is the propeller diameter (m); J is the advance coefficient, which is the ratio of process h_p to propeller diameter D; V_p is the propeller speed (m/s).

The functional relationship between thrust coefficient, torque coefficient, propeller efficiency and advance coefficient J can be drawn into the open water characteristic curve of propeller. The propeller open water characteristic curve can be fitted by inquiring the propeller open water characteristic map according to the propeller model.

In the actual navigation process, there are some errors, so it is necessary to correct the calculation of propeller open water characteristics.

Due to the viscous action of water itself, the ship will produce the wake effect of following the ship's hull during navigation, and the propeller advance speed V_p will be less than the ship's navigation speed V. Equation (13) is the relationship between them and the wake coefficient ω. The relationship between the corrected propeller speed V_p and the propeller speed coefficient J is shown in Eq. (14).

$$V_p = (1 - \omega) \cdot V \tag{13}$$

$$J = \frac{V_p}{n_p D} = \frac{(1 - \omega) \cdot V}{n_p D} \tag{14}$$

According to the relevant principles of hydrodynamics, the forward rotation of the propeller at the stern will reduce the stern pressure, resulting in the increase of the resistance of the ship during navigation. Therefore, the thrust of the propeller T should not only offset the navigation resistance of the ship R, but also offset the resistance ΔR increase caused by the rotation of the propeller. The drag increment ΔR is numerically equal to the thrust derating ΔT. Thrust derating fraction t is the ratio of thrust derating ΔT to ship thrust T, which can be used to express the relationship between ship thrust T and ship resistance R, as shown in Eq. (15).

$$t = \frac{T - R}{T} \tag{15}$$

2.1.4   Engine Speed and Ship Speed Relation

Equation (16) is the calculation of ship thrust, and Eq. (17) is the relationship between ship thrust and ship navigation resistance.

$$T = K_T \rho n_p^2 D^4 \tag{16}$$

$$T = R/(1 - t) \tag{17}$$

Substituting the ship navigation resistance $R = f(V)$ and Eq. (16) into Eq. (17), and substituting the known thrust derating fraction t, propeller thrust coefficient K_T, advance coefficient J and wake coefficient ω, Eq. (17) can be transformed into a univariate quadratic inequality of propeller speed n_p, and the expression of engine speed and speed can be obtained by using the root seeking formula.

2.1.5   Fuel Consumption Model of Diesel Engine

The fuel consumption of marine diesel engine in a voyage can be calculated according to Eq. (18).

$$W = g_e P_e t = f(V) \tag{18}$$

where, W is the fuel consumption of a voyage (g); g_e is the fuel consumption rate of diesel engine $(g/(kW \cdot h))$; P_e is the output power of diesel engine host (kW); t is the sailing time (h).

2.2   Speed Optimization Model

2.2.1   Genetic Algorithm

Genetic algorithm is an evolutionary algorithm based on the evolutionary law of "natural selection and survival of the fittest" in the process of biological evolution.

Because more friendly operation interface for non-professional personnel and graphical calculation results, this paper adopts the American Mathworks's GADST genetic algorithm and direct search toolbox brought by MATLAB. The main function of GADST genetic algorithm is *ga*. It is to find the optimal individual with the smallest fitness function. Functions can be called with the algorithm statements (19).

$$x = ga(@fitness, nvars, A, b, Aeq, beq, LB, UB, @nonlcon, options) \qquad (19)$$

$$[x, fval, exuitflag, output, population, scores] = ga(\cdots) \qquad (20)$$

where, the input parameters are *fitness, nvars, A, b, Aeq, beq, LB, UB, nonlcon, options*; the output parameter is *x, fval, exuitflag, output, population, scores*.

The main function *ga* of genetic algorithm can be expressed as Eq. (21). Equation (22) is constraint condition.

$$minf(x)(the\ number\ of\ x\ is\ nvars) \qquad (21)$$

$$\begin{cases} Ax \leq b \\ Aeqx = beq \\ LB \leq x \leq UB \\ nonlcon \end{cases} \qquad (22)$$

2.2.2   Algorithm Language for Speed Optimization Model

The expression of total fuel consumption of N segments is shown in Eq. (23). In genetic algorithm, Eq. (23) is the fitness function.

$$W = \sum_{i=1}^{n} W_i = \sum_{i=1}^{n} f(V_i)(i = 1, 2, \cdots n) \qquad (23)$$

where, W is the fuel consumption of the ship during the whole voyage (g); W_i is the fuel consumption of the ship in segment i; V_i is the speed of segment i; n is the number of segments of the whole voyage.

The upper and lower limit constraints of the speed optimization algorithm $[LB, UB]$ are the size range of the speed V, as shown in Eq. (24) and (25).

$$n_{min} \leq n_e \leq n_{max} \qquad (24)$$

$$V_{min} \leq V_i \leq V_{max} \qquad (25)$$

According to the above fitness value function expression and constraints, the genetic algorithm can be called with MATLAB statement. The algorithm is expressed as follows Eq. (26).

$$[x, fval] = ga(@fueloil\_total, n, [], [], [], [], LB, UB) \tag{26}$$

3 Ship Experimental Verification

3.1 Experimental Ship

In this study, a ship on the Beijing Hangzhou Grand Canal is selected as the research object, and its ship parameters are used to calculate the economic speed. The parameters of the ship are shown in Table 1.

Table 1. Ship related parameters.

Element	Parameter	Unit
Ship length	75.1	m
Waterline length	73.9	m
Ship width	14.8	m
Design draft	0.62	m
	3.50	m
Ship load	3041.5	t
Reduction ratio	4.33	/
Number of propeller blades	4	/
Disk ratio	0.48	/
Diameter	1.90	m
Pitch ratio	0.66	/

3.2 Fuel Consumption Model of Experimental Ship

The propulsion characteristic curve of the diesel engine is obtained from the operation manual of the Z6170ZLCZ diesel engine used by the research object ship. According to Eq. (9), the quadratic polynomial fitting of the diesel engine fuel consumption curve can obtain the mathematical expression of diesel engine fuel consumption and engine speed, as shown in Eq. (27). The curve is fitted with the third power function according to Eq. (10), and the mathematical expression of engine power is obtained, as shown in Eq. (28). The fitting curve is shown in Fig. 1.

$$g_e = 4.219 \times 10^{-4} \cdot n_{ei}^2 - 0.840 \cdot n_{ei} + 618.165 \tag{27}$$

$$P_e = 2.947 \times 10^{-7} \cdot n_{ei}^3 \tag{28}$$

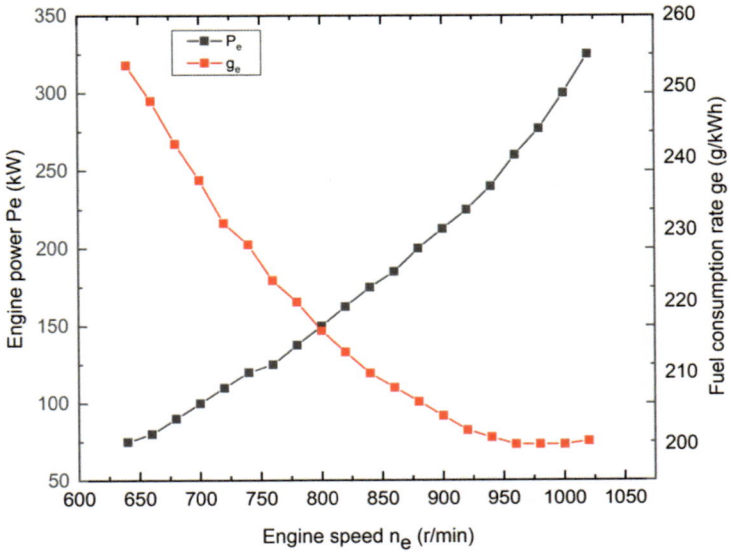

Fig. 1. Z6170ZLCZ Engine propulsion characteristic fitting curve.

According to the ship parameters and the previous calculation process, the mathematical model of fuel consumption of the research object ship on a certain leg can be obtained, as shown in Eq. (29).

$$W_i = \left(4.219 \times 10^{-4} \cdot n_{ei}^2 - 0.840 \cdot n_{ei} + 618.165\right)\left(2.947 \times 10^{-7} \cdot n_{ei}^3\right)\frac{L_i}{V_i} \quad (29)$$

where, n_{ei} is the engine speed of segment i; L_i is the mileage of segment i.

3.3   Experimental Verification

In order to verify the mathematical model of ship fuel consumption established in this paper, an experimental route with a total length of 47.9 km is designed. During the experiment, the working conditions of the ship following experiment were obtained by consulting the crew, recording the display data of relevant instruments and real-time monitoring. The specific contents are shown in Table 2.

Table 2. Experimental conditions.

Element	Parameter	Unit
Ship draft	1.0	m
Average ship speed	8.0	km/h
Water speed	2.0	km/h
Water depth	6.5	m
Ship status	No load	/
Ship load	1597	t
Sailing mileage	47.9	km
Measured fuel consumption	117.0	kg

The experimental conditions shown in Table 2 are substituted into the ship speed optimization model in this paper, and the optimal speed corresponding to the experimental conditions is calculated which is 6.96 km/h.

By substituting the empirical speed used in the experiment and the recommended speed calculated by the speed optimization model in this paper into the marine diesel engine fuel consumption model in this paper, the calculated fuel consumption consistent with the experimental conditions can be obtained. The calculated results and the real ship experimental results are listed in Table 3 for comparison and analysis. It can be seen that the error between the fuel consumption calculated by the fuel consumption model and the experimental fuel consumption is 1.26%, and the calculated fuel consumption corresponding to the recommended speed is 2.88% lower than the empirical speed.

Table 3. Model calculation and measured fuel consumption.

Situation	Ship speed (km/h)	Calculate fuel consumption (kg)	Measured fuel consumption (kg)
Experience	8.00	115.52	117.00
Economic	6.96	112.26	113.63

4 Economic Speed Calculation and Result Discussion

4.1 Calculation of Economic Speed

It can be seen from Sect. 2.1 that the main environmental impact factor considered in this paper is water depth when calculating ship navigation resistance. Due to the interception of multi-level ship locks in the northern Jiangsu canal, the water level before the two ship locks has little change. Therefore, according to the division principle of one navigation section between the two ship locks, the northern Jiangsu section of the Beijing Hangzhou canal is divided into 11 sections, namely $L_1, L_2, \cdots L_{11}$, and the mileage of each navigation section is shown in Table 4.

Taking the fuel consumption model of the whole voyage as the fitness function, the genetic algorithm is used to optimize the speed. According to the calculation flow in Sect. 2.4, the relationship between engine speed and speed of 11 segments is calculated according to different water depths of the segments, as shown in Eq. (28).

It is known that the engine speed of the research object ship during normal operation is controlled at 300–1200 (r/min). The upper and lower bound constraints of the speed optimization algorithm can be obtained through calculation, as shown in Eq. (29).

Table 4. Channel mileage and mean water depth of 11 segments of northern Jiangsu canal.

Segment	Mileage (km)	Water depth (m)
L_1	6.5	5.35
L_2	23.0	5.38
L_3	113.0	5.15
L_4	25.0	6.73
L_5	32.5	5.17
L_6	32.5	4.90
L_7	26.5	6.75
L_8	22.0	5.79
L_9	57.0	5.29
L_{10}	40.5	5.00
L_{11}	25.5	6.50

$$\begin{cases} n_{e1} = 141.21V_1 + 42.365 \\ n_{e2} = 140.65V_2 + 44.857 \\ n_{e3} = 146.23V_3 + 20.169 \\ n_{e4} = 122.03V_4 + 27.965 \\ n_{e5} = 145.82V_5 + 21.988 \\ n_{e6} = 154.18V_6 + 14.837 \\ n_{e7} = 121.82V_7 + 8.9210 \\ n_{e8} = 132.97V_8 + 78.922 \\ n_{e9} = 142.74V_9 + 35.611 \\ n_{e10} = 150.73V_{10} + 0.327 \\ n_{e11} = 124.09V_{11} + 38.694 \end{cases} \tag{30}$$

$$[LB, UB] = \begin{bmatrix} 1.825, 8.198 \\ 1.814, 8.213 \\ 1.914, 8.068 \\ 2.229, 9.605 \\ 1.901, 8.079 \\ 2.042, 7.879 \\ 2.389.9.777 \\ 1.662, 8.431 \\ 1.852, 8.157 \\ 1.988, 7.959 \\ 2.104, 9.359 \end{bmatrix} \tag{31}$$

According to the ship parameters and the above calculation process, the whole process fuel consumption mathematical model of the research object ship on the northern Jiangsu section of the Beijing Hangzhou Grand Canal can be obtained, as shown in Eq. (32). Taking the fuel consumption model of the whole voyage as the fitness function, the genetic algorithm is used to optimize the speed.

$$W = \sum_{i=1}^{11} W_i = \sum_{i=1}^{11} f(V_i)(i = 1, 2, \cdots 11) \tag{32}$$

Then, aiming at minimizing the total fuel consumption of the ship's voyage, the speed optimization algorithm is called to calculate the segmented economic speed.

4.2   Result and Discussion

According to the above calculating procedure, the optimized speed values based on the average water depth of each segment of the Beijing Hangzhou Grand Canal in northern Jiangsu are shown in Table 5.

Table 5. Segmented economic speed (unit: km/h).

Segment	L_1	L_2	L_3	L_4	L_5	L_6	L_7	L_8	L_9	L_{10}	L_{11}
Economic speed	6.713	6.724	6.616	7.871	6.623	6.475	8.023	6.889	6.682	6.534	7.665

It can be seen from Table 5 that the optimal solution of segmented speed in the north Jiangsu section of the Beijing Hangzhou canal is in the range of 6–8 (km/h). The deeper the water depth in the section, the smaller the impact of water depth on ship navigation. Under the same working condition, the corresponding reduction of ship fuel consumption and the greater the optimal solution of speed; On the contrary, the shallower the water depth of the section, the more obvious the influence of the shallow water effect on the ship's navigation resistance, and the larger the ship's fuel consumption under the same working conditions, so the smaller the optimal solution of the speed is obtained.

5   Conclusions

Through the analysis of the natural conditions of the canal in Northern Jiangsu, this study established a ship fuel consumption model and speed optimization model suitable for the canal environment. Taking a ship in the north Jiangsu canal as an example, the model established in this paper was experimentally verified. It can be seen that the economic speed calculated by the model can reduce the ship energy consumption by about 2.88% compared with the empirical speed. At the same time, the economic speed of the northern Jiangsu canal section was calculated, and the corresponding economic speeds of 11 sections are obtained. The following conclusions are drawn: the economic speed range of the north Jiangsu section of the Beijing Hangzhou canal is 6–8 km/h; Under the same working conditions, the deeper the water depth of the leg, the smaller the impact of the water depth on the ship's navigation, and the greater the calculated economy. These models and conclusions are helpful to the study of canal economic speed in the future and has reference significance for the canal to achieve low energy consumption and carbon emission reduction.

The current study only considered the influence of water depth and speed on ship energy consumption, ignoring the influence of canal width. At present, it mainly carried out theoretical research, so the applicability of the research has some limitations. More experiments will be carried out to verify the reliability of the ship model in the future. The model will also be used to calculate the ship energy consumption and economic navigation capital of the characteristic ship types of the northern Jiangsu canal. And the canal width will also be included in the study.

Acknowledgements. This study was sponsored by The Economic Speed Project of North Jiangsu Grand Canal Administration of Navigational Affairs, which we gratefully acknowledge.

References

Aris KM (2010) Particle Swarm Optimization for the optimal design of photovoltaic grid-connected systems. Solar Energy 84(12):2022–2033

Chang CC, Wang CM (2014) Evaluating the effects of speed reduce for shipping costs and CO_2 emission. Transp Res Part D 31:110–115

Fan AL, Yan XP, Yin QZ et al (2017) Clustering of the inland waterway navigational environment and its effects on ship energy consumption. Proc Inst Mech Eng Part M J Eng Maritime Env 231(1):57–69

Kim HJ, Chang YT, Kim KT et al (2012) An epsilon-optimal algorithm considering greenhouse gas emissions for the management of a ship's bunker fuel. Transp Res Part D 17(2):97–103

Li XH, Sun BZ, Guo CY et al (2020) Speed optimization of a container ship on a given route considering voluntary speed loss and emissions. Appl Ocean Res 94:1–10

Li YB (2002) Discussion on calculation method of navigation resistance of inland ships. Waterway Port 01:7–11

Liang ZX (2020) Research on ship energy efficiency optimization based on EEOI. Jimei University

Psaraftis HN, Kontovas CA (2014) Ship speed optimization: concepts, models and combined speed-routing scenarios. Transp Res Part C 44:52–69

Sun LK (2019) Research on ship energy consumption based on speed optimization. Harbin Engineering University

Wang ZZ (2014) Research on energy saving optimization of segmented speed of inland river ships. Dalian Maritime University

Yan XP, Wang K, Yuan YP et al (2018) Energy-efficient shipping: an application of big data analysis for optimizing engine, speed of inland ships considering multiple environmental factors. Ocean Eng 169:457–468

Yuan YP, Li ZX, Malekian R et al (2017) Analysis of the operational ship energy efficiency considering navigation environmental impacts. J Marine Eng Technol 16(3):52–69

Zhao CJ (2021) Collection and application of ship energy consumption data based on image recognition technology. Microprocess Microsyst 80:1–6

Zhao FY, Yang WM, Tan WW, et al (2016) Power management of vessel propulsion system for thrust efficiency and emissions mitigation. Appl Energy 161:124–132

Zhang TY, Chen JF (2017) Marine diesel engine. Harbin Institute of Technology Press, Harbin

Zhang Y (2020) Research on ship energy efficiency evaluation method improved by energy consumption model. Dalian Maritime University

Zhou WQ (2020) Research on ship track optimization based on energy consumption management. Zhejiang Ocean University

Smart Shipping on Inland Waterways

Ann-Sofie Pauwelyn[1]([⊠]) and Sim Turf[2]

[1] De Vlaamse Waterweg, Hasselt, Belgium
ann-sofie.pauwelyn@vlaamsewaterweg.be
[2] Flemish Department of Mobility and Public Works, Brussels, Belgium
sim.turf@mow.vlaanderen.be

Abstract. Inland shipping has been struggling with a shortage of skippers for several years. This means, among other things, that smaller vessels disappear and the smaller waterways are no longer used. In addition, it is also difficult for inland shipping to compete with road transport. In time, this will cause a reverse modal shift: cargo will be brought back from the waterway to the road. However, the road is already dealing with a lot of congestion while the potential of the waterway is being used less and less. This will lead to major mobility problems.

Over the years, an international consensus has grown that the automation of vessels can be a mean to solve a large part of the above problems and to revive transport via the waterways. In this way, the great pressure on our roads will also be reduced.

In order to gain a better insight in the possibilities of Smart Shipping, the PIANC WG 210 was established in 2019. The PIANC INCOM WG 210 Report on Smart Shipping on Inland Waterways has been published in March 2022. This report researches the impact of Smart Shipping developments on the physical and digital infrastructure and on traffic management, with focus on inland waterways. Smart shipping developments were viewed from the perspective of infrastructure providers and traffic managers of inland waterways to stimulate and maximize the deployment of Smart Shipping.

The report includes an analysis of the current (until 2019) Smart Shipping developments, what is currently lacking to stimulate Smart Shipping developments, as well as recommendations for the future that can be picked up in other PIANC working groups or research groups. This paper will highlight the findings of the WG and will zoom in on some more concrete examples of Smart Shipping in Belgium, where de Vlaamse Waterweg nv is monitoring a test area in which several 100s of test have taken place since 2019. Recent international legal initiatives will also be described.

Keywords: Smart Shipping · Automation · Automated inland vessels · Projects · Regulation

1 Introduction

Inland shipping has been struggling with a shortage of skippers for several years. This means, among other things, that smaller inland waterway vessels disappear and the smaller waterways are no longer used. In addition, it is also difficult for inland shipping

Y. Li et al. (Eds.): PIANC 2022, LNCE 264, pp. 951–958, 2023.
https://doi.org/10.1007/978-981-19-6138-0_84

to compete with road transport. In time, this will cause a reverse modal shift: cargo will be brought back from the waterway to the road. However, the road is already dealing with a lot of congestion while the potential of the waterway is being used less and less. This will lead to major mobility problems.

Over the years, an international consensus has grown that the automation of inland waterway vessels can be a mean to solve a large part of the above problems and to revive transport via the waterways. In this way, the great pressure on our roads will also be reduced.

In order to gain a better insight in the possibilities of Smart Shipping, this paper will go through the highlights of the PIANC INCOM WG 210 report on Smart Shipping on Inland Waterways. The PIANC WG 210 was established in 2019 and published its report in March 2022. In this paper an overview of the current regulations concerning automated shipping will be presented. Next to that some ongoing projects in Belgium will be discussed.

2   Pianc Findings on Smart Shipping

As digitalization broadens the possibilities for new business developments, Smart Shipping solutions are finding their way into the market, ranging from the development of inland waterway vessel trains, remote controlled ships to small(er) drone-like platforms for transportation of goods and people.

The new PIANC report focused on the interactions between autonomous vessels and the infrastructure, the role of the authorities and regulations with regard to Smart Shipping. Smart Shipping developments were viewed from the perspective of infrastructure providers and traffic managers of inland waterways to stimulate and maximize the deployment of Smart Shipping. An overview of recent Smart Shipping developments and use cases were analyzed in order to define the gaps that are prohibiting the further deployment of Smart Shipping developments. Possible solutions to cope with these gaps and recommendations for the future were described. These can be picked up and analyzed further in other PIANC working groups or research groups.

The biggest challenge as defined by PIANC WG 210 is the switch from human to machine, as many tasks, performed by humans on vessels as we know them today, might be executed by machines on automated vessels. Standards in communication today are human centered, but in the future they should also be machine-optimized, so that both human and machine can work with them. Therefore, collaboration with knowledge institutions and standardization organizations is necessary.

Next to that, more test areas should be created in order to create a safe space to test different technologies on their maturity. Collaboration between governments and private partners is necessary as the former can create the overall framework to make smart shipping possible but only the latter can develop specific technology and have in-depth expertise on specific topics. Collaboration is needed so that the overall framework answers the needs of the sector and the technology can be tested in a safe environment.

The inland waterway sector is not fully aware yet of all possibilities that Smart Shipping can offer. It is therefore important to raise awareness about the positive change Smart Shipping can bring and the problems it can solve. More awareness will increase the amount of R&D. Also learning from other sectors that work on automation is something that could be done more.

A last recommendation is to use more sailing simulators in order to train the crew for new ways of navigating an automated ship[1].

3   Regulation on Smart Shipping

The area of law surrounding Smart Shipping is both emerging and relatively untested. That is, the development of emerging Smart Shipping technologies is challenging current applications of legal regimes governing Smart Shipping operations. This, in turn, spurs significant debate in the domestic and international legal communities. Technology has outpaced the relevant regulations. Consequently, stakeholders and scholars continue to assess the use of Smart Shipping operations under the existing regulations, laws, treaties, and conventions and they have yet to reach universal consensus. Therefore, the PIANC report gives a general overview of the main international inland water transport legislation that could be relevant to Smart Shipping and that might require adaptation in the future, as the first step towards an international regulatory basis for the commercial use of automated inland navigation vessels (Smart Shipping).

The regulatory framework relevant to Smart Shipping for Europe, China and the U. S. has been reviewed based on the following nine policy areas: definitions; competences and crew qualifications; technical requirements for inland navigation vessels; presence of the boatmaster and crew members on board; responsibility and liability; communication between a vessel and a competent authority, and vessel to vessel communication; emergency situations; cybersecurity and inland waterway infrastructure. Those policy areas were chosen as they cover the main, most relevant aspects of Smart Shipping.

The analysis conducted in the report shows that there are important differences in the organisation of the regulatory and institutional framework between Europe, the U.S. and China. In Europe, there is a clear and strict separation between inland and maritime navigation regulations. Not only does Europe have a separate regulatory framework for inland and maritime navigation, there is also a strict separation between the competent institutions. Within Europe, several international institutions each have their own authority over inland navigation, such as the European Union, the United Nations Economic Commission for Europe, the Central Commission for the Navigation of the Rhine, the Danube Commission, the Moselle Commission, …At the national level, national/regional waterway authorities are responsible for managing inland waterways and drawing up inland navigation regulations. As a result, each member state has some kind of national inland navigation strategy. When looking at the way inland navigation is regulated in the U.S. and China, we find that there is no clear separation between inland and maritime navigation in terms of the competent institutions and the regulations themselves. Unlike in Europe, regulations are of a more hybrid nature, with

---

[1] PIANC INCOM Working Group 210, Report on Smart Shipping on Inland Waterways (2022), 41–48.

certain regulations applying to both inland and maritime navigation. The analyses also showed that the U.S. and China do not really have a clear structure of governmental institutions with exclusive responsibility for inland navigation. In the U.S. inland navigation is managed by a collection of federal, regional (states) and local agencies, each of which is partially responsible for inland navigation and partially for maritime navigation. As a result, the US doesn't really have a national inland navigation strategy. In China the roadmap is to gradually build up a guidance, standards, and rules in different levels for inland and maritime navigation particularly. In the meantime, the systems and platforms for testing and measurement of intelligent ships should be established with detailed steps and protocols. Without functional testing regulations, practical applications are hard to be issued by administration. All in all, China would like to establish the ability of testing in short term and a series of standards in the long term.

Owing to the difference in governance structure and the way inland navigation is regulated in Europe, China and the U.S he concept of automated navigation is also handled differently from a legal/regulatory point of view. When considering the situation in Europe, we find that it is first of all necessary to adapt existing inland navigation regulations and to subsequently consider the potential development of new regulations for automated inland navigation. In China, the China Classification Society (CCS) leads the regulations for ship building while the Maritime Safety Administration (MSA) leads the rules and regulations for crew-related training and management. Due to the differences between inland vessels and seagoing ships, some items of the existing rules and guidance may not be applicable for inland vessels.

In conclusion we can say that the analysis conducted in the report shows that there are important differences in the organisation of the regulatory and institutional framework between the three regions studied. This does not mean, however, that synergies regarding standardisation work should not be undertaken. For instance, in the field of River information services (RIS), similar technology blocks and associated standards (with sometimes regional adaptations) can be used. For Smart Shipping, it remains very relevant to continue the monitoring of the evolution of the regulatory frameworks in the three regions, because rules/standards developed in one of the regions could usefully inspire the others and then facilitate and speed up the development of Smart Shipping.[2]

4   Projects Findings on Smart Shipping

In Belgium several projects on Smart Shipping are already ongoing and some of them are finished. This Chapter gives an overview of the most important projects and their findings. Furthermore, attention is also drawn to an international project that will test in Belgian waters, the Horizon 2020 AUTOSHIP project.

---

[2] PIANC INCOM Working Group 210, Report on Smart Shipping on Inland Waterways (2022), 16–26.

Overall, no big incidents happened during testing. The companies also discuss needs for improvement regularly with the waterway authority. Thanks to the fact that communication is very transparent during testing, all tests have been successful so far.

4.1 Seafar

Seafar is a company that provides services to help operate unmanned and crew-reduced vessels. They support and control automated ships via their Control Centre in Antwerp. Seafar has already received several approvals for testing in the Flemish region of Belgium.

4.1.1 Testing Unmanned Ships

Since October 2019, tests have been carried out in the Westhoek region on the river Yzer and the Plassendale-Nieuwpoort canal with Watertruck X, a CEMT class II bulk carrier on behalf of company Decloedt. The project went through different stages of automation. The first few weeks were sailed with a full crew (phase 1). When the captain had sufficient contact with the ship, the captain was moved to the Remote Control Center (RCC) (phase 2). A second captain remained on board, along with the rest of the crew. The waterways were monitored from the RCC and more and more was controlled from the RCC itself. The captain on board had ultimate responsibility and intervened when necessary. In a third phase (phase 3a) the captain was removed from the ship and only a technical superintendent remained on the ship. This person could only intervene if the captain in the RCC gave him the command to intervene. The last phase (phase 3b) consists of complete control from the RCC with nobody on board of the ship anymore. The modalities for all phases were laid down in an experimental agreement in 2019. From 2019 until now (2021), 5 changes were made to this experimental agreement via addenda:

1. Since April 2020, additional permission has been given to deploy two additional Watertrucks on the same route: Watertruck VII and Watertruck VIII. In phase 2 there was a separate skipper on the SCC for each vessel.
2. Since July 2020, permission has been given to switch to the first part of phase 3, where there is no crew. In phase 3a, testing is carried out with the permanent presence of a technical superintendent on the ship. The responsibilities of the technical superintendent, as well as those of the skipper in the SCC were laid down in the addendum to the experimentation agreement. For example, the skipper always had ultimate responsibility and the technical superintendent was not allowed to sail longer sections and only bring the ship to safety on the instructions of the skipper.
3. After Seafar completed a full year of testing in October 2020, they were granted an extension to test for an additional year.
4. Since March 2021, Seafar has received additional permission to operate at night and move to phase 3b: testing without crew on board, but with full control from the SCC. Seafar has developed safety procedures for this and they have tested them during the phase with crew on board so that the ship can continue to sail safely once it sails unmanned.

5. On October 18, 2021, the test was extended for another year.

Seafar and the Vlaamse Waterweg nv are in regular contact and each addendum was prepared via a project change application, an updated risk analysis, gap analysis and ConOps and various evaluation meetings. Since April 2020, Seafar also received permission to sail with a (different) Watertruck on the Leuven-Dijle Canal (on behalf of Celis). The actual tests only started at the end of October/November 2021. The phases that will be followed here are the same as for the tests in the Westhoek. The Watertruck vessels are self-propelled barges, certified under Flemish regulations, in accordance with Article 24, second paragraph, of EU Directive 2016/1629 with regard to exemptions for vessels that travel limited routes of local importance or in port areas.

4.1.2   Testing with the Aim of Crew Reduction

Since June 2020, Seafar has also been testing the vessel Gamma, a CEMT class I bulk carrier, owned by Gitra BVBA, on the Bocholt-Herentals canal and the Brussels-Scheldt Sea Canal. This ship is manned: the captain is always on board. The ship therefore sails according to current laws and regulations. However, the ship is controlled from Seafar's RCC. The responsibility lies with the skipper on board. This project was completed at the end of 2021.

Since March 2021, the inland vessel Tercofin II (CEMT class Va - dry bulk/container) has been sailing between the Port of Antwerp and Liège, via the Albert Canal. There will always be crew on board, but control is with the RCC. The efficiency of the ship will be increased by supporting the ship's crew. This makes it possible to sail for longer with the same number of crew members on board, without exceeding the sailing and rest times. Work was also carried out in two phases. In phase 1, the crew on board will consist of 1 captain and 1 sailor instead of 2 captains and 2 sailors. The rest of the crew is located in the Seafar Remote Control Center. A team of 3 captains and 2 traffic controllers in the Remote Control Center will be in control of the ship. A Captain and Traffic Controller in the Remote Control Center work in 8 h shifts. The addendum to the experimental agreement allows the transition to phase 2 from March 2021. The ship is currently sailing with 1 mate and 2 sailors on board and 1 captain is present in the SCC, who is responsible for the ship. In March 2022 Seafar was granted an extension to test for an additional year.

Since February 2021, the container ship DESEO has been sailing between Zeebrugge and Antwerp. For this application, advice was given by the Vlaamse Waterweg nv, but the license itself was granted by the GNA, the authority for the Sea-Scheldt, which exists out of different governmental agencies from Belgium and the Netherlands. The ship is currently supported from the RCC in Antwerp with a full crew on board. Since August permission from GNA to sail the entire trajectory in control from the RCC, but the full crew still remains on board. This project is also working towards crew reduction.

4.2   DEME

DEME has been testing the autonomous vessel Marine Litter Hunter (MLH) at the Scheldt bridges Temse-Bornem from October 2020 until October 2021. In the first phase, testing was carried out with crew and since March 2021 a switch has been made to unmanned navigation. The MLH sailed autonomously and took certain mitigating actions itself in the event of problems. In case of unforeseen problems, a supervisor could help the MLH if necessary. Therefore, a responsible person always was designated as an "Autonomous Ship Supervisor", provided with a valid navigation license, to supervise the operations of the MLH from a distance.

The set-up consisted of a combination of a fixed installation that continuously removed "passive" floating waste from the water and a mobile system that "actively" collected larger floating debris, which can be harmful to shipping in the Scheldt. This mobile part was responsible for the part where shipping is allowed and focuses on larger floating debris (>200 mm) that can cause damage to shipping, such as ropes / mooring lines, fishing nets, wooden beams, pallets, and plastic items,…

The mobile system consisted of:

- a camera detection system by means of Artificial Intelligence (AI);
- the Marine Litter Hunter: The workboat that was fully electrically powered and equipped with an open push blade for actively catching floating debris and bringing it to the trap;
- a docking station, which was located at the Belgomine quay nearby the bridge.

Large floating waste and objects (such as tree trunks) were detected by smart cameras (AI) installed on the old Temse bridge at the height of the navigation channel. The waste was gathered in the collecting pontoon and was regularly transferred into a container by means of a crane equipped with a grab. The fixed crane was remotely controlled by an operator, using VR-3D vision technology. The container was placed on the workboat. When the container was full, the vessel autonomously took it to the docking station, where the container is unloaded by means of a transhipment crane on the Belgomine quay. The waste was transferred to a DVW waste container.

4.3   AUTOSHIP

AUTOSHIP is short for Autonomous Shipping Initiative for European waters. It is a project with several European partners and it is subsidized by the European Union under the Horizon 2020 program. The aim is to hold two demonstrations with ships equipped with Smart Shipping technology, with a focus on transport. One demonstration takes place in Norway and focuses on Short Sea shipping. Here there will be a limited crew on board of a fish feeder ship and the goal is crew reduction. The other demonstration takes place in Flanders and focuses on transport on inland waterways, with a ship from Zulu Associates. The route goes from the lock in Wintam to Willebroek and then back via the Rupel. During this demonstration there will be no crew on board and full control will therefore be at the shore control centre on shore.

The technology with which the boat will be converted is currently being developed by the Norwegian Partner Kongsberg. At the moment the shore control centre is being set up next to the lock in Wintam. Completion of the trajectory is planned for mid-2023 (this is still variable, depending on delays due to the corona crisis). The intention is to use the results from these 2 tests as optimally as possible. After all, they provide us with an enormous amount of information about legislation, security, socio-economic factors and cybersecurity and this will be analyzed. Furthermore, the aim is to develop a roadmap, standards and methods that can be used by future developers and thus further the commercialization of automated sailing.

5   Conclusions

This paper highlighted the findings of the WG and zoomed in on some more concrete **examples of Smart Shipping in Belgium**, where de Vlaamse Waterweg nv is monitoring a test area in which several 100s of test have taken place since 2019. Recent international legal initiatives have also be described.

Reference

(2022) PIANC INCOM working group 210, report on smart shipping on inland waterways. 16–26:41–48

SciPPPer: Automatic Lock-Passage for Inland Vessels – Practical Results Focusing on Control Performance

Alexander Lutz[✉] and Axel Lachmeyer

Argonics GmbH, Stollstr. 6, 70565 Stuttgart, Germany
{alexander.lutz, axel.lachmeyer}@argonics.de

Abstract. Navigating through locks is one of the most challenging tasks that skippers have to perform in inland navigation. Typical dimensions of a ship (width = 11.45 m) and a lock (width = 12 m) result in an error margin of less than 30 cm to the left and to the right of the ship when navigating within a lock chamber. Typical inland vessels on European waters have a length of 82 to 186 m. The wheel house on cargo vessels is located close to the stern of the vessel. This leads to low visibility of the bow in the lock chamber. In order to cope with this issue, a deck hand monitors the bow and announces distances to the skipper via radio. The quality of this information depends on the deck hand's ability to judge distances correctly and is prone to error. This highly demanding maneuver needs to be performed up to 15 times per day. Each lock passage can take up to 30 minutes. The research project SciPPPer aims at automating this complex navigational task.

The German acronym SciPPPer stands for Schleusenassistenzsystem basierend auf PPP und VDES für die Binnenschifffahrt – lock assistant system based on PPP and VDES for inland navigation. The idea is to fully automate the navigation into and out of a lock using high-precision GNSS (Global Navigation Satellite System) with PPP (precise point positioning) correction data which is transmitted from shore to ship using VDES (VHF Data Exchange System), an extension to AIS (Automatic Identification System). This absolute measurement data is complemented by relative measurement data using LiDAR and automotive RADAR and fused with inertial measurement data delivered by a mechanical gyro system. Apart from the challenge of precisely measuring the position and orientation of the vessel within the lock chamber, the control task poses an interesting problem as well. This contribution introduces both, the measuring and the control problem. However, the focus lies on the results of the control performance that was achieved on a full-bridge simulator as well as during real-world trials. A full-bridge simulator was used in order to test the control strategy and its algorithms safely. A number of different actuator configurations were investigated. Typical inland cargo vessels use one or two propellers with Kort nozzle and a twin rudder behind each propeller and a 360° turnable bow thruster. Typical inland passenger vessels use several (2–4) 360° turnable rudder propellers as main propulsion as well as a 360° turnable bow thruster or a classical tunnel thruster which can only apply forces to starboard or portside. These typical configurations were examined by simulation. The real-world trials were performed on a passenger vessel with three rudder propellers as main propulsion as well as a classical tunnel bow thruster acting left and right.

© The Author(s) 2023
Y. Li et al. (Eds.): PIANC 2022, LNCE 264, pp. 959–968, 2023.
https://doi.org/10.1007/978-981-19-6138-0_85

This contribution presents the results of the simulator study as well as the real-world trials in terms of control performance. It explains specific challenges due to the navigation within an extremely confined space. The contribution concludes with lessons learned as well as an outlook focusing on the potential of the introduction of such a system to the inland navigation market.

Keywords: Automation · Automatic lock-entering · Navigation · Control · Smart systems

1   Introduction

The project SciPPPer aims at automating the whole navigation task of passing a lock. It is comprised of several distinct challenges which are addressed within different work packages. The main aspects are the following:

1. Requirements and System Architecture
   The overall system is analyzed. Requirements for the individual components are determined. The architecture of the assistant system is developed in form of functional subsystems. The interfaces the subsystems are defined. In order to test the whole integrated system with all of its modules, a validation and demonstration concept is established.
2. Land-Based Technologies
   GNSS (Global Navigation Satellite System) is used whenever available during the lock passage. Due the extremely confined space, centimeter precision GNSS is necessary. Instead of using the well-established RTK (Real-time Kinematics) reference data, PPP (Precise Point Positioning) is investigated within this project. PPP data can be transmitted via broadcast with the same data for all receiving parties whereas RTK is calculated and transmitted to each vessel individually. The fact that broadcast is possible as well as the fact that the required data to be transmitted for PPP is a lot smaller than for RTK, this enables transmitting PPP correction data through VDES (VHF Data Exchange System), an addition to the well-known AIS (Automatic Identification System). PPP via VDES is established within SciPPPer for the first time as a means to enable centimeter precision together with integrity monitoring on several levels.
3. Communication
   In order to transmit the PPP data as well as data about the state of the lock (i.e. lock gate open or closed), VDES is employed. Prototypical transceivers are installed at the lock and on the ship to test the real-world behavior. Since the communication is carried out next to AIS antennas the collocation problem needs to be addressed. Otherwise, the normal AIS communication would be disturbed by the transmission of VDES data in close vicinity.
4. Onboard Technologies
   The PPP data being transmitted via VDES is used on board to generate high-precision centimeter-grade position accuracy. This GNSS data is fused with inertial measurement data in order to retain position and orientation information during GNSS outages. The well-known random walk when integrating accelerations and

rotation rates limits the use of inertial measurement data to short periods of time. In order to deal with this drawback, relative measurements between lock walls and the vessel are used. Three multi-beam lidar sensors, two at the bow and one at the stern, as well as three automotive radar sensors are employed for this task. The fusion of relative lidar and radar measurement together with absolute GNSS supported by inertial measurement data is part of this work package also.

5. Maneuvering Control and Simulation

This work package deals with the development, implementation and testing of a control strategy for the lock passage problem. This contribution focuses on this task and only touches on other aspects where necessary. In order to test in a risk-free environment, a full-bridge simulator is adapted such that it is able to mimic the real behavior of vessels during lock passage adequately.

6. System Integration and Demonstration

All the developed components are integrated into one fully-functional prototype. Some of the developed modules such as the control algorithms are tested on the full-bridge simulator. Trials on real vessels proof that the developed prototype works in practice. Its functionality is shown in a public demonstration.

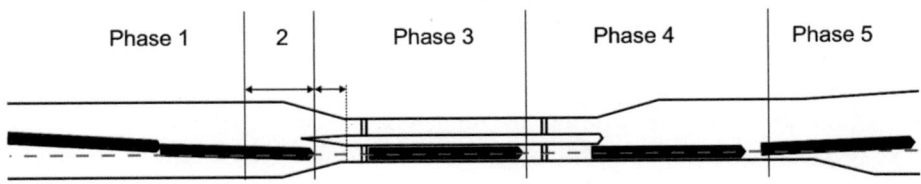

During the determination of requirements, the lock passage is divided into 5 different phases. Each phase has different requirements with respect to positioning accuracy and control performance. The 5 phases also differ with respect to sensor data availability. However, entering a lock at phase 2 is similar to leaving the lock at phase 4 whereas phase 1 is comparable to phase 5 with respect to positioning accuracies and sensor availability.

Phase 1: Approaching the lock. GNSS measurement available. Navigation radar still useful. Lidar, automotive radar not in range. Medium positioning requirements.

Phase 2: Lock chamber approach. GNSS measurement maybe available. Navigation radar not useful. Lidar, automotive radar starting to gather data. High positioning requirements for bow, medium requirements for stern.

Phase 3: Within the lock chamber. GNSS measurement not available. Navigation radar not useful. Lidar, automotive radar fully working. Highest positioning requirements.

Phase 4: Similar to phase 2. Lidar at bow may still work. High positioning requirements for stern, medium requirements for bow.

Phase 5: Similar to phase 1.

Section 1 gives a brief overview of the SciPPPer system and explains the requirements for the different sub-systems. Section 2 introduces the controller as well as the state estimator and all other functional modules that are part of the maneuvering control system. Section 3 shows the results of the full-scale trials.

2 System Overview

The SciPPPer projects comprises of land-based elements as well as onboard equipment. The communication infrastructure consists of antennas and transceivers for VDES both on shore and on the vessel. Servers on land receive PPP correction data, validate it and transmit it through the VDES communication channels. Lock status information is also processed and transmitted via VDES. The data stream is received onboard of the vessel, decoded and fed into GNSS receivers which are able to calculate centimeter-grade accuracies. The lock information is used to determine if the lock chamber is ready to be used.

Beside the GNSS compass there are 3 laser scanners (Lidar sensor) and 3 auto-motive radar systems to determine the relative position and orientation of the vessel with respect to the lock. One of the laser scanners is used to detect objects such as other vessels in front as well as to detect the lock gate when closed. The other two laser scanners are installed in order to find the walls of the lock chamber using a large horizontal field of view. The point clouds are processed such that the distance to the detected lock walls as well as the orientation between ship and lock can be computed.

All of this sensor information is shown to the skipper through dedicated displays. It is also fed into the control system which fuses the different sensor information into one state estimation using a mathematical ship model. The control system consists of a PLC-based processor with several interface modules which are able to read the sensor data as well as control the actuators by communicating with the propulsion control system. Several actuator configurations were evaluated. Figure 1 shows a simple overview of the SciPPPer system.

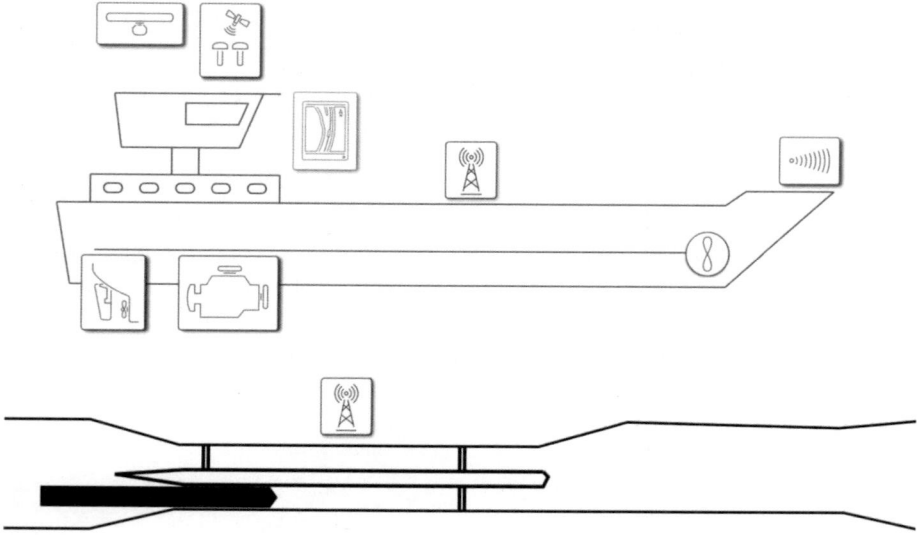

Fig. 1. Overview of SciPPPer system

2.1   Actuator Requirements

The employed control algorithm requires the independent setting of longitudinal and lateral forces as well as a moment about the vertical axis. This translates into a limitation of possible actuator configurations. Table 1 shows the configurations that fully satisfy the demand in green. Configurations that satisfy the constraint fully in theory but have limitations in the control performance due to other actuator constraints in yellow. As an example, vessels with one azimuth propeller in the front and a classical bow thruster fall into this category. This is because while this configuration may be able to apply forces in all directions, the rate of change of the forces is strictly limited by the turning speed of the azimuth thruster. Applying a forward force after applying a force backwards clearly illustrates this drawback. This limits the control performance dramatically and should be avoided. Configurations that are not taken into account are marked red. Apart from this, the continuous control of all actuators is preferred.

Table 1. Actuator configurations

Actuator Bow Actuator Stern	No Actuator at Bow	Classical Bow Thruster (90° oder - 90°)	360° Thruster	Bow Rudder
1x Propeller Fixed with Rudder				
2x Propellers Fixed with Rudder				
1x Propeller Fixed with Rudder and Flanking Rudder				
2x Propellers Fixed with Rudder and Flanking Rudder				
1x 360° Propeller (Azimuth)				
2x 360° Propellers (Azimuth)			Simulator Trials	
3x 360° Propellers (Azimuth)		Full-Scale Trials		
4x 360° Propellers (Azimuth)				

2.2   Position and Orientation Measurement Requirements

Typical locks on European waterways have a width of 12 m. The ship that are built to pass these locks are 11.45 m wide. This results in a lateral margin of less than 30 cm. The ships' lengths vary between 80 and 180 m. In order to be able to enter a lock automatically, the positioning accuracy as well as the control performance need to be well below 30 cm in lateral direction. Table 2 shows the requirements for the positioning and orientation accuracies for the different phases and different sensor systems. Note that phases 2, 3, and 4 are the most critical. GNSS may not be reliable during these phases due to low satellite visibility because of high lock walls. Accuracy is specified as 95% (2σ) of all measurement values being within the interval $\pm 2\sigma$ around the true value, σ being the standard deviation.

Table 2. Accuracy requirements for sensor systems

Phase	1	2	3	4	5
Positioning accuracy bow lateral GNSS [cm]	10	-	-	10	10
Positioning accuracy stern lateral GNSS [cm]	10	10	-	-	10
Positioning accuracy longitudinal GNSS [cm]	10	10	10	10	10
Resulting orientation accuracy [°]	0.11	0.11	-	-	0.11
(cm) for ship length L=100m	(20)	(20)			(20)
Longitudinal speed GNSS [cm/s]	10	10	-	-	10
Lateral speed GNSS [cm/s]	1	1	-	-	1
Time to alarm PNT unit [s]	10 (position)	10 (position)	10 (position)	10 (position)	10 (position)
	2 (heading)	2 (heading)	2 (heading)	2 (heading)	2 (heading)
Positioning accuracy bow lateral Lidar [cm]	-	1	1	1	-
Positioning accuracy bow longitudinal Lidar [cm]	-	-	10	-	-
Positioning accuracy stern lateral Lidar [cm]	-	-	1	1	-
Resulting orientation accuracy [°]	-	-	0,005	0,005	-
(cm) for ship length L=100m			(1)	(1)	
Longitudinal speed Lidar [cm/s]	-	-	1	-	-
Lateral speed Lidar [cm/s]	-	1	1	1	-
Time to alarm Lidar [s]	1s	1s	1s	1s	1s

2.3 Control Performance Requirements

The overall deviation consists of errors from position and orientation estimation as well as deviations from automatic control. Table 3 shows the accuracy requirements for the control performance for the lock passage phases 1–5. They are determined together with requirements for the pose determination task such that there is an extra margin for error of 10 cm for bow and stern for a ship of 100 m.

Table 3. Accuracy requirements for control systems

	Phase 1	Phase 2	Phase 3	Phase 4	Phase 5
Control deviation longitudinal position y [cm]	Not relevant	Not relevant	100	Not relevant	200
Control deviation lateral position x [cm]	50	10 (bow) 20 (stern)	10	20 (bow) 10 (stern)	50
Control deviation orientation [°] (cm for ship length L = 100)	0.23 (40)	0.11 (20)	0.057 (10)	0.11 (20)	0.23 (40)
Control deviation ROT [°/min]	2	2	2	2	2
Control deviation speed [cm/s]	10	10	10	10	10

3 Control System

The control system consists of several modules that are shown in Fig. 2. The sensor data is fed into a state estimator in the form of an extended Kalman filter. Once the final position within the lock has been chosen by the skipper the planning module plans a transition trajectory for the position and orientation as well as corresponding speeds from the initial state to the destination. This desired trajectory minus the current state from the Kalman filter forms the control error and is fed into a multi-input multi-output controller. The controller calculates forces in x and y as well as a moment about the vertical axis. This values are added to a values from a feedforward block which calculates nominal forces in x and y as well as a nominal moment using an inverted mathematical model of the vessel. A classical thrust allocation block computes RPMs and angles for the involved actuators. These signals are fed into the propulsion control system.

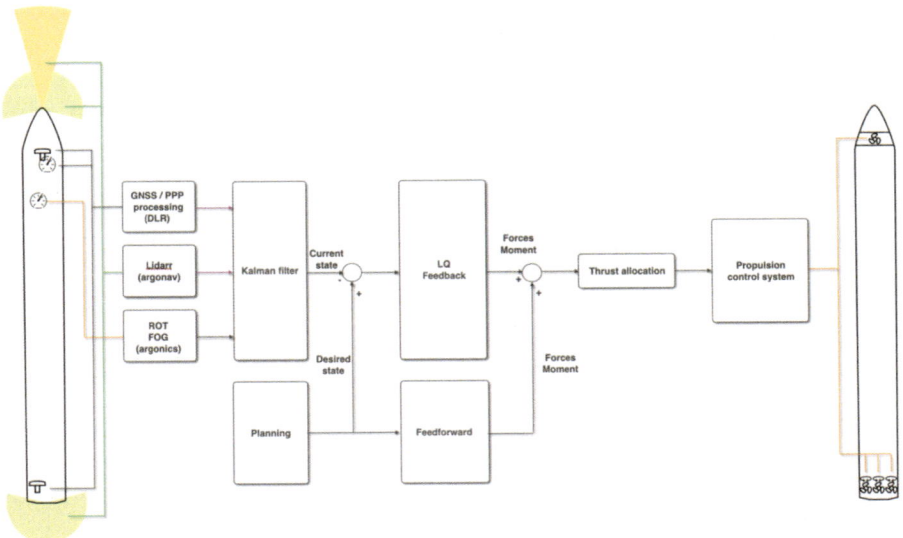

Fig. 2. Control system modules

3.1 Kalman Filter

The state estimator is implemented as an extended Kalman filter with the following dynamical states: N/S position, E/W position, N/S velocity, E/W velocity, with respect the center of gravity of the vessel as well as heading and rate of turn about the vertical axis. Thus, a classical 3DOF (degrees of freedom) model is used neglecting heave, pitch and roll. The inputs are three positive thruster RPMs as well as thruster angles in the case of 2 azimuth thrusters in the back and one in the front. In the case of a classical bow thruster, the third angle is fixed at $90°$ with the respective thruster RPM being positive and negative. Since the differential equations ask for forces and moments the actuator variables are transform using the inverse thrust allocation module.

3.2   Planning

The planning calculates desired values for all dynamical states. This is carried out in several phases: Phase 1 starts with the current state as initial value and plans a trajectory to a position in front of the lock with the middle of the ship being aligned with the middle of the lock chamber and the orientation being the orientation of the lock chamber. The respective speeds form at the desired position and orientation pose a design choice. The second phase starts with the last state of phase 1 as initial value and the desired position and orientation in the lock chamber as final state. Cubic splines in the velocity states ensure smooth transitions between the different poses.

3.3   Controller

The controller follows a two degree of freedom control strategy in the form of a feedforward as well as a feedback module. The feedforward module uses an inverted dynamical model which allows for the computation of nominal forces and moments from the desired trajectory. The employed mathematical model is not able to describe the dynamics perfectly, unmodeled disturbances also contribute to deviations when applying these nominal forces and moments. In order to compensate for these deviations, a feedback controller is implemented. This Riccati-type controller uses a static gain matrix and is therefore an infinite horizon linear quadratic optimal controller which is not taking constraints into account.

4   Results

This controller setup has been tested on the full-bridge simulator at Bundesanstalt für Wasserbau (BAW) in Karlsruhe on a cruise ship with 2 azimuth thrusters at the stern and one azimuth thruster at the bow during several simulation runs. The results of these simulator tests can be requested from the authors. Numerous simulator runs involving different types of actuator configurations have been performed at Argonics in Stuttgart at the in-house simulator station using a similar mathematical model as the one at BAW.

The first full-scale trials were carried out on the sister ship to the one in Karlsruhe at Hollands Diep. The final results were demonstrated again on a cruise ship which has three thrusters at the stern and a classical left/right bow thruster. This bow thruster was not continuously actuated but only offered four distinct RPM settings. This was compensated for by a special actuation strategy in order to not have to take the switching property into account explicitly. However, this limited actuation resulted in lower control performance.

Figures 3, 4, 5 show the desired values as well as the actual values for the lateral position, heading as well as longitudinal speed. At 12:30 h the lateral position jumps significantly. This is due to the stern Lidar sending faulty measurements. The state estimator weighs the different inputs but is unable to fully compensate for this error.

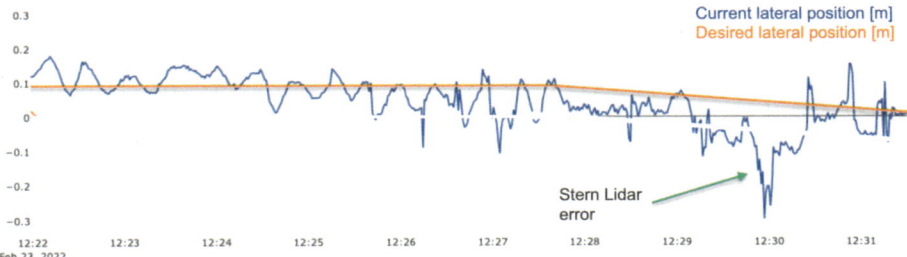

Fig. 3. Desired and estimated lateral positions

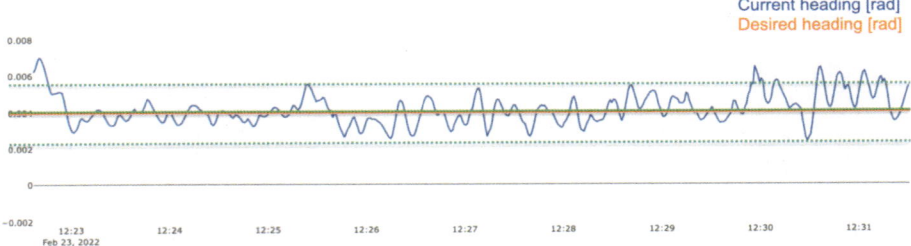

Fig. 4. Desired and estimated heading

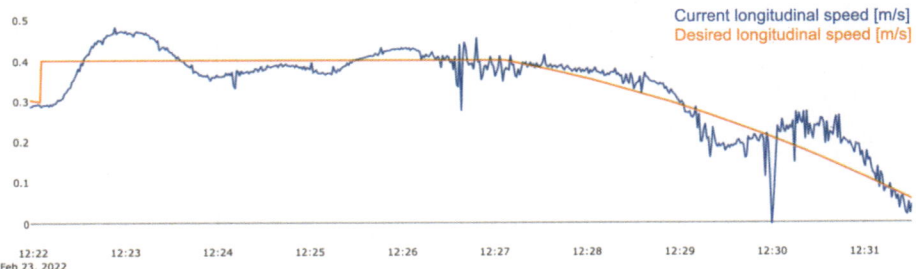

Fig. 5. Desired and estimated longitudinal speed

Table 4 shows the requirements next to the results of a typical trial run. The longitudinal position was not controlled very firmly due to the fact that the length of the lock basin was far greater than the ship length. Unfortunately, the tight requirements could not be met fully. This is mostly due to the fact that the bow thruster had the above mentioned switching property with only 4 distinct RPM settings as opposed to the required continuous operation. However, adding up the different errors laterally the total error adds up to 26 cm which is still below the overall requirement of a maximum of 30 cm.

Table 4. Requirements and achieved values

Control deviation	Requirements	Full-scale trial (Offset compensation)
Longitudinal position y [cm]	100	<250
Rate of turn [°/min]	2	2.23
Orientation [°]	5,73/L	
° at ship length L = 85m	0.06	0.11
cm at ship length L = 85m	10	16
Lateral position x [cm]	10	<10
Velocity [cm/s]	10	10

5   Conclusions

The research project SciPPPer was able to show that it is possible to automatically pass a lock with existing inland vessels using additional close proximity sensors such as Lidar as well as centimeter-grade GNSS receivers. PPP corrections distributed via VDES in broadcast mode showed to be a promising option for future high-precision GNSS services around existing infrastructure. The extremely tight constraints when passing a lock require serious calibration of all sensors. Small errors can already lead to a violation of these constraints. The automotive radar sensors were not precise enough to serve as a backup for the Lidar measurements. Error detection within the estimation algorithm still needs some further refinement. The control setup was able to achieve the requirements in the simulation runs even under undesirable environmental conditions. However, the desired requirements were not fully met during the full-scale trials due to the switching behavior of the bow thruster. However, several completely automatic runs were carried out successfully without intervention by the skipper. The control system stayed within the total bounds set by the difference of the lock basin width minus the width of a typical vessel.

Acknowledgements. This research has received funding from the Federal Ministry for Economic Affairs and Climate Action in Germany as part of the Maritime Research Program.

Ship Maneuvering Using a Ship Simulator in Search and Rescue Operation

Milan Kresojevic[1]([⊠]) and Vesna Ristic Vakanjac[2]

[1] University of Defense, Military Academy, Belgrade, Serbia
milan.kresojevic@va.mod.gov.rs
[2] Faculty of Mining and Geology, University of Belgrade, Belgrade, Serbia
vesna.ristic@rgf.bg.ac.rs

Abstract. In order to successfully steer a ship, a man, ship's watch officer, or commander of the ship must be physically ready and must know the maneuvering characteristics of the ship he steers, as well as the forces acting on him. For that, it is necessary for him to know the specifics of the ship because each ship has its own ``mood', its specific properties that depend on the type, and size of the ship. Therefore, maneuvering properties differ even in ships of the same type and therefore it is important to know the specifics of a particular ship. Also, the same ship will behave differently in different situations. When the ship finds itself in situations that require urgent reaction during the voyage, then the experience and skills of the ship's captain come to the fore. Then it is important that the captain, based on his experience, chooses the right maneuver, but also that he always has at least one reserve maneuver in his head. That decision in certain situations must be made in a very short period of time, and the future of the ship and human lives often depends on it, as well as the resources on board. Every skill is acquired through practice, and it is logical that the commander will react faster and more correctly in every new situation. Precisely because of this, the most intensive training is necessary for the formation of quality ship commanders. For that reason, the training of future ship commanders on ship simulators gives excellent results. On ship simulators, they can gain a lot of experience and go through countless scenarios. In this paper, the use-value of ship simulators from the aspect of the training was verified through the Search and Rescue (SAR) exercise realized on the ship simulator Wärtsilä Navigation Simulator NTPRO 5000. It has been shown that the simulator can successfully check and recognize the optimal SAR pattern, maneuvering characteristics of the ship, as well as practice, maneuvering the ship and resolving specific situations that ship commanders may encounter in real situations. Training on ship simulators cannot completely replace training in real situations, but it can be used to get acquainted with the maneuvering characteristics of the ship, train in working with navigation devices, and be a good starting point in preparing people who are trained to perform tasks in real situations. The use of hydrological data of relationships between different types of data using an autoregressive model (AR model) can contribute to the creation of more realistic scenarios on ship simulators. It is also possible to apply data in the modeling of the environment and connect them with the current hydrological situation of the waterway. In the exercise evaluation process, we can see if the ship's captain used the hydrological data in the right way.

© The Author(s) 2023
Y. Li et al. (Eds.): PIANC 2022, LNCE 264, pp. 969–977, 2023.
https://doi.org/10.1007/978-981-19-6138-0_86

Keywords: Ship maneuvering · Search and Rescue · Ship simulator ·
Hydrological data · Autoregressive model (AR model)

1 Introduction

The main objectives of this research are to define training scenarios on a ship simulator in a river environment using hydrometeorological data and the autoregressive model (AR model) of the river level to verify the performance of the SAR operation itself and to check the maneuvering characteristics of a given ship. If we look at the European Maritime Safety Agency (EMSA 2020) data on the total number of reported Marines victims and incidents during the period 2014–2020 we see that despite the new modern technology, a large number of accidents occur. The total number is 22532, and the annual average of the number of maritime casualties or incidents is 3218. In addition to the stated objectives, the training of the ship's captain (participant) in performing the SAR operation is checked, as well as his ability to use the available data.

The data at its disposal are defined in the task through the input parameters on the ship simulator (wind direction and speed, current direction and speed) while the water level is defined through the AR model of the river level for 6 (six) years. We have introduced this way of presenting the water level trend due to the possibility of reacting quickly to unforeseen situations and the ability of the participant to cope with the lack of current data on water levels. We also used a specific empirical formula to calculate wind pressure as a function of its speed.

Of course, all important parameters of wind direction and speed as well as the intensity and direction of the river current are entered into the simulator and assigned to each casualty (man in the water). From the moment of receiving the signal for the beginning of the exercise to the moment of the end of the exercise, the work of the participants is monitored and recorded if necessary for later analysis and making the most optimal decision possible.

In addition to monitoring the work of the participants in the SAR maneuver, the maneuvering characteristics of the ship are controlled. Specifically, we were based on monitoring the maneuvering characteristics of the ship's turning circle, given the fact that this maneuver is used to rescue a man in the water.

The maneuvering characteristics of the ship are shown on the basic ship information display, and we were using that defined data as controlled parameters. Basic data on the ship as well as data on the control maneuvering characteristics of the ship (turning circle) are shown in Fig. 1.

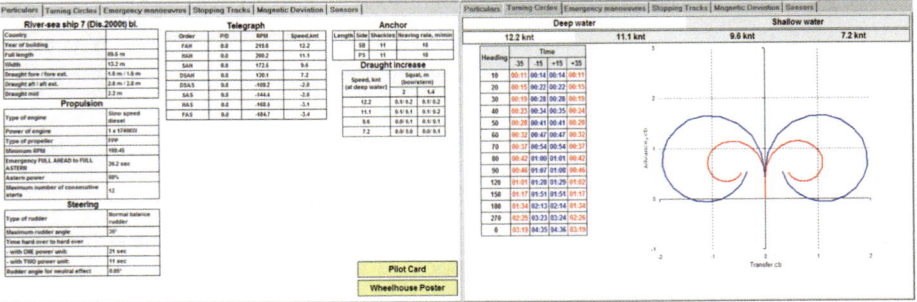

Fig. 1. Basic data of River-Sea ship, simulator Wärtsilä NT PRO 5000

Data on the ship's turning circle were checked during the SAR maneuver and the rescue operation in the water.

2 Simulated Scenario

Input environmental data for SAR operation were calculated and synchronized with real data based on SAR experience and a real incident in Budapest, Hungary, which occurred on 29th May 2019 (Némedi 2021).

We developed a scenario for SAR exercise and simulation environment (Table 1). The Hableany, a small sightseeing cruise collided from behind with a twenty-five-times bigger longship, MV Viking Sigyn capsized and sunk in seven seconds.

Table 1. SAR exercise simulation environment

Input parameters	Hableany disaster	Simulation parameters
Ship	Small sightseeing cruise	Small passenger cruise
Type of incident	Collision with a bigger ship	Collision with a sea-river ship
Wind	No data	N – 4,1 m/s
Temperature	Unusually cold	10 °C
Water temperature	No data	6,3 °C
Precipitation	Heavy rain	Heavy rain
Fog	No	No
Visibility	Low	Low (under 100 m)
Drift	Very strong	Very strong (
River current	Strong	Strong 110,5 °, 1,4 kn
Waves	Low	0,5–1 m
Flooding	Yes	Yes
Watercolor	No data	Moody
Number of persons onboard/rescued	37/7	30/7,41

Using this parameter the participant should predict the SAR area (Burciu, 2010). The initial position of the participant is close to the place of the accident, the distance is about 2 Nautical Mile (NM). We did not want the participant to waste a lot of time arriving at the scene of the accident.

The way these hydrometeorological parameters should be used by the Participants is by simply matching the wind pressure vector, which has its own direction and intensity, as well as the river current vector (Berawi, et al. 2019). The resultant of these two vectors in the interval of time should enable the participants to determine the optimized SAR region. The direction, and strength of the current we set in Table 1 Simulated parameters since we did not have accurate on-scene data values are assumed. What is important for the realization of the exercise is that the participant uses the data that he can see on the simulator or get in the task. It is very important that the participant knows how to determine the direction of the current, ie that he knows that the current flows in the direction defined in the table, unlike wind. It is assumed that the defined current represents the real value of the current, ie that for a defined time interval the starting point of the SAR region is moved by that value. In order to more accurately predict the impact of the wind itself, it is necessary to calculate its thrust using speed. The empirical formula by (Tesic 1971) we have used is:

$$P = 0.12 * V^2 \tag{1}$$

P – Wind pressure, v – the wind speed.

More accurate prediction of the search area is crucial for speed of response and saving lives (Xu et al 2011). Although we do not have current data on water level, participants need to take the assumed value on the basis of data obtained by the autoregressive model (AR model) done specifically for this exercise.

3   Search and Rescue Decision Making

Making a decision on the SAR operation itself is a complex activity based on the data, experience, and knowledge of the participants engaged in the operation (Baldauf et al 2015). An algorithm for making the most optimal decision in the operation is proposed in Fig. 2 (revised, Vidan et al 2010).

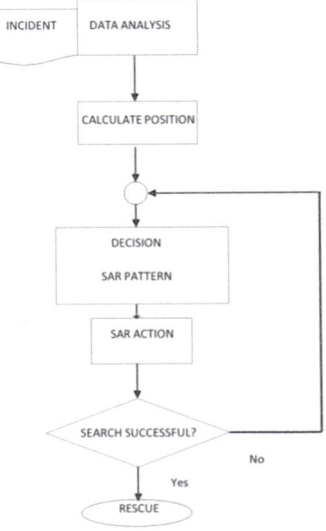

Fig. 2. SAR algorithm in the river environment

If the Search operation is successfully performed on the ship's simulator, participants start with maneuvering and rescuing people who are in the water.

Rescue should be conducted so that the ship reduces external hydrometeorological effects (IMO, 2019). The rescue act itself and the manner of its performance by the participant are monitored and recorded for later evaluation and conclusion on the exercise (Radojević and Kresojević, 2020).

At the end of this phase, the fall of a man into the water from the ship on the starboard side is simulated, the command bridge is alarmed and the work is monitored shown in Fig. 3.

Fig. 3. Saving man overboard, simulator Wärtsilä NT PRO 5000

When performing a full turn maneuver and rescuing a man from the water, the turkey movement of the ship is monitored in order to compare the turning circle with the defined parameters shown in Fig. 1.

4   The Autoregressive Model Applied to the Water Level

The autoregressive model (AR model) can be considered as one of the simplest regression models that can be used to simulate certain qualitative and quantitative parameters of surface flows, then karst spring water and springs and groundwater regime, where the dependent variable is Q_t - the predicted variable at a time t, and the independent variables Q_{t-1}, $Q_{t-2}, ... Q_{t-k}$ are values 1, 2,... k days before:

$$Q_t = a + b_1 \cdot Q_{t-1} + b_2 \cdot Q_{t-2} + ... + b_k \cdot Q_{t-k}$$

where a, b_1, b_2... b_k are model parameters.

In the first place, it is necessary to have a sufficiently long series of random variables (flow, water level, absolute elevations of the water mirror, nitrate concentrations, turbidity,...) to make the equation obtained as reliable as possible (Miladinović et al. 2015, Pešić et al. 2016, Ristić Vakanjac 2015, Ristić Vakanjac et al. 2018). For the purposes of this paper, the data on the absolute water levels of the Sava River observed in the gauging station (g.s.) Obrenovac was used. The average daily values of the water level of the Sava River observed in the Obrenovac profile were available to the authors for the period from the beginning of 1962 to the end of 2019. Due to the fact that there were frequent short-term interruptions in the data at this hydrological station, the longest continuous series was taken for the purpose of forming the simulation model (June 1st, 2014–December 31st, 2019). For different orders, the AR model coefficients were from 0.9917 (for order 1) to 0.9936 (for order 9) (Fig. 4).

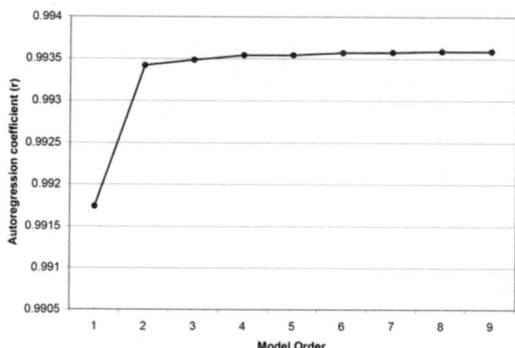

Fig. 4. The correlation coefficient for different orders regarding the AR model which was used for water level simulation

We can say that the correlation coefficients obtained for different orders do not differ significantly, so we will present only the results obtained for order 1. In this case, the equation obtained for order 1 is:

$$Z_{calculated,t} = 0.991862 \cdot Z_{measured,t-1} + 0.581253$$

The diagram of the dependence of the calculated and observed values of the water level of the Sava River in the Obrenovac profile is given in Fig. 5.

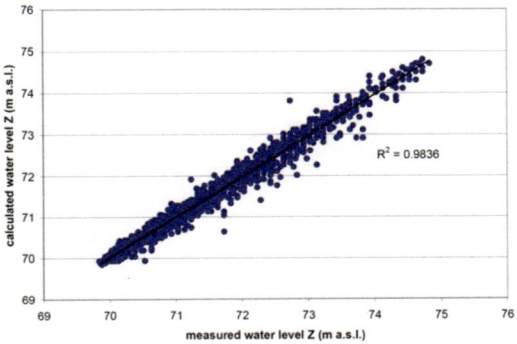

Fig. 5. Diagram of the dependence of the calculated values as a function of the measured values of the water level of the Sava River, g.s. Obrenovac

The comparative level diagram of the absolute levels is given in Fig. 6.

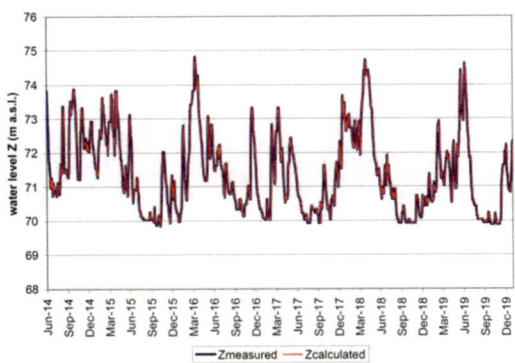

Fig. 6. Parallel water level diagram of observed and computed values, river Sava, g.s. Obrenovac

The AR model showed, as shown in Figs. 4, 5 and 6, that the assumed water level is almost identical to the measured levels. This allows us to assume with great accuracy the water level for the next day (which is most important to us from the point of view of navigation) if we do not have accurate data from the gauging station at the moment.

5 Conclusions

After the realization of the exercise, we noticed that the turning circle differs from the defined turning circle of the ship. However, taking into account the parameters of the environment, the deviation was expected, so we can conclude that the use of a ship simulator for maneuvering is useful in training shipmasters, i.e. the simulator faithfully shows the ship's behavior in defined conditions.

The simulator as a training tool cannot completely replace the sailing experience gained on a real ship, but it certainly represents a good basis and an excellent platform for training shipmasters. The two methods, conventional real ship training, and simulator training need to complement each other for the greatest training effects.

By using data on real accidents that happened on the waterway, their implementation through the simulator and later using a comparative method of exercises performed on the simulator with data on how the actual operation was performed, it is possible to obtain data that would be key to defining future training.

Using the autoregressive model (AR model), i.e. data on the ratio of the river level, for a specific river in which we navigate, can be of great importance to us, especially in situations where we do not have available data on current water levels. Although this is a little assumed scenario, monitoring the water level trend in this way can be useful when planning a voyage or performing specific tasks or specific maneuvers on the river.

In this specific exercise, we saw that the ship's captain successfully used data on the assumed water level, i.e. that he predicted the impact of water on his ship, which we confirmed by successfully positioning the ship in relation to the man in the water.

Acknowledgements. This work was supported by Project Operational and Functional Use of the Ship Simulator in Navigation University of Defence, Belgrade, Serbia.

References

Baldauf M, Schröder-Hinrichs J-U, Kataria A, Benedict K, Tuschling G (2015) Multi-dimensional simulation in team training for safety and security in maritime transportation. J Transp Saf Secur 8(3):197–213

Berawi MA et al (2019) Optimizing search and rescue personnel allocation in disaster emergency response using fuzzy logic. Int J Technol 10(7):1416–1426

Burciu Z (2010) Bayesian methods in the reliability of search and rescue action. Pol Marit Res 17:72–78. https://sciendo.com/article/10.2478/v10012-010-0039-7. Accessed 10 April 2022

European Maritime Safety Agency (2021) Annual Overview of Marine Casualties and Incidents 2020; European Maritime Safety Agency: Lisbon, Portugal, p 147. http://www.emsa.europa.eu/accident-investigation-publications/annualoverview.html. Accessed 14 April 2022

IMO. International Aeronautical and Maritime Search and Rescue Manual (2019). International Maritime Organization: London, UK, Vol 2

Miladinović B, Vakanjac VR, Bukumirović D, Dragišić V, Vakanjac B (2015) Simulation of mine water inflow: case study of the Štavalj coal mine (southwestern Serbia). Arch Min Sci 60(4):955–969. https://doi.org/10.1515/amsc-2015-0063

Némedi G, Petrétei D, Restás A (2021) The DVI hungary and its' first deployment, Védelem Tudomány – VI. évfolyam, 3. szám,. 7. Hó, pp 459–473

Pešić M, Ristić Vakanjac V, Vakanjac B, Kostadin J (2016) Turbidity simulation for short-term predictions: case study of the karst spring Surdup (Bor, Serbia), Comptes rendus de l'Académies bulgare des Sciences (ed. Todor Nikolov), Vol 69, pp 1183–1194

Radojević S, Kresojević M (2020) Saving migrants from the sea: improving training for search and rescue operations. TransNav: Int J Mar Navig Saf Sea Trans 14(1):129–133

Ristić Vakanjac V (2015) Forecasting long-term spring discharge. In: Monography: Karst Aquifers – Characterization and Engineering (Stevanović Z. ed), Series: Professional Practice in Earth Science, pp 435–454. ISBN 978–3–319–12849–8, https://doi.org/10.1007/978-3-319-12850-4, Springer International Publishing Switzerland

Ristić Vakanjac V, Čokorilo Ilić M, Papić P, Polomčić D, Golubović R (2018) AR, CR and ARCR modeling for simulations and analyses of karst groundwater quality parameters, Geološki anali Balkanskog poluostrva, Rudarsko geološki fakultet, pp 71–81, ISSN 0350–0608

Vidan P, Kasum J, Jolic N (2010) A proposal for the models and measures of search and rescue on inland waterways. Transport 25(2):178–185

Xu X, Turner CA, Santee WR (2011) Survival time prediction in marine environments. J Therm Biol 36(6):340–345

Tešić M (1971) Hidrometeorološki uslovi za spašavanje ljudstva u obalskim vodama Jugoslavije, Spasavanje ljudskih života na moru, Naučne rasprave, Pomorska biblioteka, sveska 23, Izdanje mornaričkog glasnika, Beograd

The Fairway to Corridor Management

Jan Gilissen$^{(\boxtimes)}$

De Vlaamse Waterweg nv, 3500 Hasselt, Belgium
jan.gilissen@vlaamsewaterweg.be

Abstract. This document is a description on the path we're working on in Flanders to implement corridor management. At this moment we're already providing services on European lever, but in other to be futureproof inland waterway transport has to take a next step to evolve to a reliable, modern, efficient and innovative transport mode. A brief description is given on the status today and the next phases in the program. Both the waterway authorities and the sector will have to do some efforts to realize the next phases. The document doesn't consists all the details because this is a work in process with all involved parties.

Keywords: Corridor management · ETA · RIS · Planning

1 Introduction

Since 2017 De Vlaamse Waterweg nv offers a certain amount of RIS services towards skippers and logistic parties by means of the VisuRIS portal. The provided services are meant to provide fairway information to the user or to share information with logistic partners on ETA's (estimated times of arrival) of certain vessels. The information is always a reflection on the status of the waterway network or the result of calculations made using collected voyage information and the status of the network, but it's always a snapshot of the situation at that moment.

In 2022 De Vlaamse Waterweg nv and 12 other European waterway authorities will launch these services on a larger scale towards the inland waterway users in Europe, as a result of the European RIS COMEX project. This will be done via the EuRIS platform where RIS services will be offered to facilitate corridor management. With this new portal users can already use the services on a European level and have one location for all information. Voyage information will also be available for cross border voyages.

This is a first, but indispensable, step towards green and efficient corridor management and planning in the future. But, to be able to facilitate a full-blown corridor planning some building blocks are missing in the current scope of EuRIS.

To get to the next level waterway authorities need to deliver an higher service level at locks and bridges. Therefore De Vlaamse Waterweg nv keeps building on the necessary features to facilitate full-blown corridor management.

Y. Li et al. (Eds.): PIANC 2022, LNCE 264, pp. 978–982, 2023.
https://doi.org/10.1007/978-981-19-6138-0_87

2   Traffic Management Today

Today inland vessel operators or captains have to estimate themselves at what time they have to leave their berth to be at the destination at the requested time. And for some vessels (for example container vessel going to a terminal in a seaport) it's not that easy because they only have a limited time window when they can more at this location.

The operator has to estimate, based on the information that's available on traffic density, possible blockages and operating hours at locks when they have to start their voyage.

During their voyage they have to take in account possible waiting times at locks. At the moment vessels are planned based on their arrival at the lock and first come first served. This means that they're not certain when they will be served at the moment they're making their voyage plan. This resolves in vessels racing each other in order the be first in line at the locks, waiting times at terminals because they're to early and last but not least in extra containers on a truck because the vessel had already left at the terminal to ensure his timeslot in the Harbour.

A first step to facilitate operational improvements in this use case DVW has already done some extra developments to support lock and bridge operators to plan traffic at their objects, but they can still only do this following the current, and outdated, rules.

3   Automated Voyage Plans and Traffic Image for Each Vessel

A first building block that's developed is the generation of the voyage plan. Combining information gathered through AIS and the voyage and cargo information of all vessels on our network, we're able to calculate detailed ETA's at all locks and bridges. These ETA's are constantly monitored and optimized, based on the AIS position, the voyage information and lock scheduling. This data is shared with the traffic planners on shore and the skippers.

Having the voyage plan on time is a major factor for success. To ensure this, skippers have to announce their voyage electronically. Based on this information, ETA's for locks and bridges are calculated automatically by the European routing software of EuRIS, taking in account the dimensions of the vessel, the network and the route points given by the skipper.

To avoid extra administrative burdens for the vessel operators we ask them to send us this info using reporting software, which can be integrated in the current software packages they currently use in the wheelhouse (iENC viewer, container planner, ...).

Also the process for collecting waterway charges is incorporated in this service by using the same information. Making use of the calculated voyage we will be able to automate and optimize the process both for waterway users and authorities.

When the voyage information is processed by VisuRIS it will constantly be monitored and recalculated when needed based on AIS and registrations in the lock planning software. But since planning is still only done within a certain perimeter from the lock the ETA at the end of the voyage is not certain.

At the moment the quality of the ETA-calculator is suboptimal due to different reasons:

- Missing information on interruptions of the voyage (staying overnight, delays at a berth...)
- ETD (Estimated Time of Departure) is not always correct or voyages are announced to late.
- Missing lock schedules at all locks
- Missing interaction with other transport modes (trains on movable bridges)
- Missing variables like, flood, weather conditions, traffic density, ...

Filling these gaps and improve data quality is one of the challenges in the future.

4 Lock and Bridge Management

Locks are, most of the time, a time-consuming factor during the voyage of a vessel, certainly in a country like Belgium where there are a lot of locks on the main waterways. Optimizing planning and considering RTA's (Requested Time of Arrival) of the skipper can have a huge impact on the reliability and carbon footprint of inland shipping.

Therefore we've provided our operators a lock planning software. Based on calculated ETA's, vessels are planned at the different locks and bridges. At this moment, each object (i.e. a lock or bridge) makes his own planning. This can be done based on an automated proposal that's made by the software or the operator can make it manually. When a vessel is planned at a certain lock the Estimated Time of Departure (ETD) will be added to the voyage plan.

Our lock operators have a good overview on the ETA of the vessels and plan traffic in both directions optimizing water usage. Today the planning is communicated through VHF but in the near future we'll also do this via an API (Application Programming Interface). Based on a validated schedule a digital message will be send to the each vessel. Through this message he can see where he must take position in the lock basin, the order that they have to follow to enter the lock and the dimension of the other vessel.

Thanks to the EuRIS system we're able to offer this information through a centralized European API towards the vessel operator, certainly in case of automated vessels this service will become indispensable.

To improve data quality and increase the usability, our scheduling software receives directly data from OPC and VHF systems. Some of these data like the position of a bridge, lock door, traffic light, ... will be offered via AIS and API's to the skippers on board or in the vessel control centers (in case of a remote operated vessel) to increase safety around the lock in all conditions.

Colleting all these data will enable us in the coming years to optimize our ETA calculation and predicted passage times in different situations (number of vessels in the lock, type of vessels, ...). This data is indispensable to reach automated corridor planning in the future.

5   Digital Handshake

To be able to move towards corridor planning, where we agree on passing times on all objects with the skipper not only legislation will have to change, but skippers will have to be convinced of this approach. We will have to convince them this will make inland shipping more efficient, reliable and green.

In the upcoming years we hope to evolve from a human traffic planner that operates one lock, to a traffic planner that agrees on planning with the users for the entire network.

This means that software on board of the vessel will have to communicate with lock planning software to agree on the time that the vessel has to be on each lock. This has to take in account the possible time windows at the lock, resting times of the vessel, speed of the vessel,....

A correct planning can only be made in collaboration between the vessel operator and the traffic planner. Based on a digital communication a timeslot has to be agreed upon for the different obstacles on their route. The vessel can optimize his time of departure and speed to be at the lock's at the foreseen time and the lock operator will be able to prepare his lock.

The most optimal scenario is that a all planning's are made before the voyage starts, but in a first phase it would be good to be able to offer these services on smaller corridors or within certain time windows and with locks that are managed by one traffic manager. But it's important to standardize communications and to foresee a single point of contact like is done in the EuRIS system. This will make it easier for suppliers of on board software to use these services.

6   Challenges

To provide a good corridor management service some challenges will have to be overtaken:

- Receiving voyage information of high quality & on time
- Reliable ETA calculation and lock passing times (cfr. supra)
- Convince waterway users of corridor planning and change legislation
- Connect software in the wheelhouse with lock planning software
- Modernize legislation focused on modern service levels

To achieve our goal both the waterway users and the authorities will have to some efforts. Therefore it's important to collect enough data to optimize the calculation and to do a GAP analysis to indicate and eliminate the blank spots.

Authorities also have to try to minimize the administrative burden for operators. This means that we first have to try the fill the gaps with automated processes and try not to ask more actions at the operator side.

But it will also be important to start with a realistic scope. Our dream is to offer a reliable planning and to strengthen the position of IWT. But it's not realistic to get there with a big bang. These services will have to grow starting with a smaller scope, for example a stretch with a manageable number of locks and within a certain time window, and evolving towards larger corridors and maybe in the future an entire voyage.

7   Opportunities

All services provided today are focusing on inland water transport but when corridor management is fully operational this can also bring opportunities for other transport modes. Sharing information when bridges will be open or closed with route planners will avoid certain traffic jams on certain roads. Communication with rescue services can be elaborated.

River System Management

Adaptive Regulation of Cascade Reservoirs System Under Non-stationary Runoff

Yu Zhang[1]([✉]), Xiaodong Wang[1], Zhixiang Min[2], Shiqiang Wu[1], Xiufeng Wu[1], Jiangyu Dai[1], Fangfang Wang[1], and Ang Gao[1]

[1] State Key Laboratory of Hydrology-Water Resources and Hydraulic Engineering, Nanjing Hydraulic Research Institute, Nanjing, China
yuzhang@nhri.cn
[2] College of Hydrology and Water Resources, Hohai University, Nanjing, China

Abstract. Under the influence of climate change and human activities, the spatial and temporal distribution of river runoff has changed. The statistical characteristics of runoff such as mean, variance and extreme values have changed significantly. Hydrological stationarity has been broken, deepening the uncertainty of water resources and their utilization. Hydrological stationarity is a fundamental assumption of traditional water resources planning and management. The occurrence of non-stationarity will undoubtedly have an impact on the operation and overall benefits of reservoirs, and may even threaten the safety of reservoirs and water resources. There is uncertainty as to whether reservoirs can operate safely and still achieve their design benefits under the new runoff conditions. Therefore, it is important to carry out adaptive regulation of reservoirs in response to non-stationary runoff. Based on the multi-objective theory of large system, a multi-objective joint scheduling model of the terrace reservoir group is constructed for adaptive regulation simulation. A set of combination schemes based on optimal scheduling, flood resource utilization, water saving is constructed. The adaptive regulation is validated using a real-world example of the Xiluodu cascade and Three Gorges cascade reservoirs system in Yangtze River, China. The adaptive regulation processes are analyzed by simulation and the adaptive regulation effects are evaluated. The results show that the non-stationary runoff in upper Yangtze River has had an impact on the comprehensive benefits of large hydropower projects. The use of non-engineering measures to improve flood resource utilization, adjust upstream water use behavior and optimize reservoir scheduling are effective means to reduce the negative impact of non-stationary runoff on cascade reservoirs system.

Keywords: Adaptive regulation · Cascade reservoirs system · Non-stationary runoff · Optimization and control · Hydropower system

1 Introduction

Climate change and human activities have led to significant changes in the global hydrological cycle, directly affecting precipitation, evapotranspiration, land use and land cover in watersheds, resulting in varying degrees of change in the spatial and temporal distribution of river runoff (Ye et al. 2020; Zhang et al. (2021a); Zhang et al.

© The Author(s) 2023
Y. Li et al. (Eds.): PIANC 2022, LNCE 264, pp. 985–1000, 2023.
https://doi.org/10.1007/978-981-19-6138-0_88

(2021b); Wang et al. 2021). The statistical characteristics of runoff such as mean, variance and extreme values have changed significantly. Hydrological stationarity, a fundamental assumption of traditional water resources planning and management, has been broken. This has increased the uncertainty in water resources development and use and has brought new challenges to water resources management (Dau et al. 2020).

Reservoirs usually have some storage capacity and play a key role in water resources management (Ahmad et al. 2014; Turner et al. 2020). The occurrence of non-stationarity will undoubtedly have an impact on reservoir operations and overall benefits, especially in large reservoir complexes (Qin et al. 2020). There is uncertainty as to whether the reservoir system can operate safely and achieve the design benefits under the new runoff conditions (Zhang et al. 2017; Chang et al. 2018). Therefore, how to mitigate the negative impacts of non-stationary runoff on reservoirs is a new challenge for water resources management under the current climate change and anthropogenic impacts (Alimohammadi et al. 2020).

There are two different ways to deal with this issue: one is to develop hydrological forecasting technologies that reduce the uncertainty of incoming water; the other is to develop theories and methods of adaptive operation under changing conditions (Liu et al. 2020). At present, under the first way, the effective period of the short-term hydrological forecast is extended, which improves the efficiency of reservoir operation (Stringer et al. 2020; Sun et al. 2020). The second idea is to adopt a dynamic learning method to propose control measures to guide the operation and management of the reservoir group based on the assessment of the impact of runoff change on the reservoir group, to overcome the design limitations under changing conditions.

In recent years, scholars have carried out adaptive research at the macro-level (framework and structural system) and micro-level (regulation strategy) in response to the changing environment. At the macro level, Edalat and Abdi (2018) explored a new framework of adaptive water management, described its concept, developed suitable decision support systems, and applied them to developing-country cities. At the micro-level, Brekke et al. (2009) analyzed the risks of climate change to reservoir operation and proposed strategies and methods for adaptive operation of reservoirs. Sowers et al. (2011) analyzed the adaptive strategies of water resources management in the Middle East and North Africa under climate change from the perspective of political, economic, and institutional decision-makers. Maran et al. (2014) focused on analyzing the impact of climate change on hydropower development and utilization in the Alpine Basin from the perspective of adaptive management strategies. Ahmadi (2015) formulated a dynamic optimal operation strategy of the reservoir for climate change scenarios, and the preliminary application in the Karoon-4 reservoir had shown that the adaptive dispatch strategy can effectively improve the reliability and reduce the vulnerability of hydropower generation; Alimohammadid et al. (2020) changed the operation policy of the Karaj hydropower dam reservoir (Iran) to mitigate the undesirable effects of climate change. The above-mentioned macro studies mostly focused on the description of control strategies and management frameworks; the micro studies mostly focused on the single reservoir and lacked research on the adaptive operation of cascade reservoirs.

In general, theoretical and methodological research on adaptive operation of reservoir complexes in changing environments is a hot topic in the field of water resources. However, there is a lack of general theories, models and methods, as well as studies that combine specific adaptive operation with quantitative control effect evaluation. Therefore, in this study, a multi-objective joint operation model of the cascade reservoir system for adaptive operation simulation is constructed, and an evolutionary algorithm is used for model solution. A set of combined schemes based on adaptive operation measures such as optimal adjustment of operation mode and utilization of flood resource is constructed. The adaptive operation effects of different schemes are evaluated by simulation. It is pointed out that non-engineering measures used to improve the utilization of flood resource and to optimize the operation mode of hydropower system are effective to reduce the impact of changed streamflow on the hydropower system.

2   Study Area

The Yangtze River is the river with the most abundant hydropower resources in the world, it also is the longest river in Asia, the third longest river in the world, which is about 6000 km in length. The Upper Yangtze River Basin (UYRB) has enriched nearly 90% of the hydropower resources in the entire basin, with a developable installed capacity of about 2 TW. In this study, we selected the giant cascade reservoirs in the UYRB as the research object, which is composed of Xiluodu, Xiangjiaba, Three Gorges, and Gezhouba. The spatial location of the cascade reservoirs is shown in Fig. 1, and the main control parameters of each reservoir are shown in Table 1.

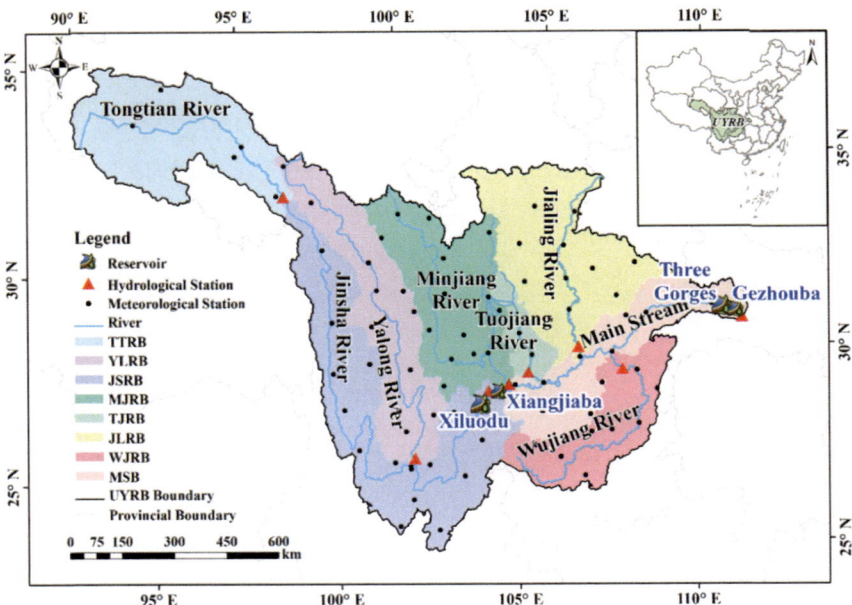

Fig. 1. Location of Xiluodu cascade and Three Gorges cascade reservoirs system.

Table 1. Main control parameters of the four reservoirs.

Reservoir	Normal water level (m)	Flood limited water level (m)	Dead water level (m)	Installed capacity (MW)
Xiluodu	600	560	540	12600
Xiangjiaba	380	370	370	6000
Three Gorges	175	146.5	–	22500
Gezhouba	66	–	63	2950

3 Methods

3.1 Multi-objective Joint Operation Model of Cascade Reservoirs

Currently, the way of single-purpose water resources development and utilization no longer exists, and it is replaced by the way of comprehensive development and utilization, which includes multiple goals such as flood control, water supply, power generation, and shipping. This paper took the restoration of power generation as the objective of adaptive regulation, and other objectives such as flood control, water supply, and shipping are transformed into constraints.

3.1.1 Objective Functions

$$\max E = \sum_{i=1}^{n} \sum_{t=1}^{T} N_{i,t} \cdot \Delta t \tag{1}$$

where, E is the maximum annual power generation of cascade hydropower plants; n is the number of hydropower plants; T is the numbers of time steps in the operation period; $N_{i,t}$ is the output of the ith reservoir at time t; Δt is the hours of the calculation period.

3.1.2 Constraint Conditions

(1) Water balance constraint

$$V_{i,t} = V_{i,t-1} + (Q_{i,t} - q_{i,t} - J_{i,t} - S_{i,t}) \cdot \Delta t \tag{2}$$

where $V_{i,t}$ and $V_{i,t-1}$ refer to the average storage of reservoir i in the tth and $(t-1)$th time steps, respectively; $Q_{i,t}$, $q_{i,t}$, $J_{i,t}$, and $S_{i,t}$ are the inflow, power generation flow, abandoned water flow, and loss flow of reservoir i at time t, respectively.

(2) Water level constraint

$$\underline{Z}_{i,t} \leq Z_{i,t} \leq \overline{Z}_{i,t} \tag{3}$$

where $Z_{i,t}$ is the average water level of the reservoir i at time t (m); $\underline{Z}_{i,t}$ and $\overline{Z}_{i,t}$ are the minimum and maximum water level limits of reservoir i at time t, respectively.

(3) Discharge flow constraint

$$q_{i,\min} \leq q_{i,t} \leq q_{i,\max} \tag{4}$$

where $q_{i,\min}$ and $q_{i,\max}$ are the minimum and maximum discharge from reservoir i at time t, respectively.

(4) Output constraints

$$\underline{N}_{i,t} \leq N_{i,t} \leq \min\left\{N_{i,t}^H, N_i^Y\right\} \tag{5}$$

where, $\underline{N}_{i,t}$, $N_{i,t}^H$, and N_i^Y are the minimum output, expected output, and installed capacity of reservoir i at time t, respectively.

3.1.3   Solution Method

The Particle Swarm Optimization algorithm (PSO) Zhang et al. (2021a) and Zhang et al. (2021b) was selected in this study to solve the above optimization model.

3.2   Adaptive Regulation

We constructed a series of scenarios of adaptive regulation which includes three control measures, i.e., optimization of the operation mode of the reservoir group, utilization of flood resource, and adjustment of water use behavior, to counteract the impacts of changes in river runoff on the reservoir group. The Control measures and scenarios are listed in Table 2.

Table 2. Control measures and scenarios of adaptive regulation

Scenario	Control measures	Description
A	Optimization of the operation mode of the reservoir group	The cascade four-reservoir joint operation
A	Utilization of flood resourc	The maximum control water level during the flood season is 146.5 m
B		150 m
C		153 m
A	Adjustment of water use behavior	Without considering water saving
D		10% of water-saving
E		20% of water-saving
F	Comprehensive control measures	The maximum control water level of the Three Gorges Reservoir is 150 m during flood season, and 10% water saving
G		150, 20%
H		153, 10%
I		153, 20%

3.2.1 Optimization of the Operation Mode of the Reservoir Group

The measures for the optimization of the operation mode of the reservoir group are through the joint operation of the four cascade reservoirs, changing the disorderly storage and release of the water to the coordinated way, to achieve the purpose of comprehensively controlling the water volume and head and restoring the power generation of the hydropower system. The scenario of the joint operation of cascade reservoirs (Scenario A) is simulated by setting the calculation conditions of the multi-objective joint operation model. At some time, the benefits of the adaptive control measure of joint optimal operation of reservoir group are analyzed by comparing the conventional operation.

3.2.2 Utilization of Flood Resource

The idea of regulating and controlling the utilization of flood resource is to dynamically control the flood limit water level of reservoirs without increasing system risks, improve the utilization of flood resource, and increase the effective runoff replenishment of the reservoir group, to control the water volume and restore the power generation of the hydropower system. The scenarios of flood resource utilization are simulated by setting the calculation conditions of the joint operation model. In the current calculation conditions, the flood limit water level of the Three Gorges reservoir is 146.5 m (Scenario A) during real-time operation. With that, two additional scenarios of flood resource utilization are set up by adjusting the maximum control water level of the Three Gorges reservoir to 150 m (Scenario B) and 153 m (Scenario C) during the flood season.

3.2.3 Adjustment of Water Use Behavior

The idea of adjustment of water use behavior is to reduce the water use outside rivers by adjusting the social and economic structure and promoting the construction of a water-saving society, thereby indirectly increasing the effective runoff supply of the water conservancy and hydropower system, to achieve the purpose of regulating the water volume of the reservoir group and restoring the power generation of the system. The scenarios of the water behavior adjustment are simulated by setting the calculation conditions of the joint operation model. Based on not considering the water-saving scenario (Scenario A), two additional scenarios of water behavior adjustment are set up by 10% (Scenario D) and 20% (Scenario E) of water-saving (accounting for the change in runoff).

3.2.4 Comprehensive Control Measures

The comprehensive control measures for the hydropower system to respond to changes in runoff are set up, which contains three aspects, i.e., optimization of the operation mode of the reservoir group, utilization of flood resource, and adjustment of water consumption behavior. The calculation conditions of the comprehensive control measures are listed in Table 2.

4 Results and Discussion

4.1 Runoff Change and Its Influence on Power Generation

The results show that the measured runoff at Yichang station, a control station on the upper Yangtze River, shows an obvious decreasing trend, with a speed of -0.35 km³/ year. The abrupt change of runoff in the upper Yangtze River occurred in 1993. Notably, the runoff data series used in the preliminary design of the Three Gorges Reservoir is as of 1990. The change will affect the operation mode and comprehensive benefits of the Three Gorges and other large hydropower projects, especially whether these hydropower projects can still play the design benefits under the new runoff conditions. Therefore, the study period is divided into the base period (1951–1993) and the change period (1994–2020). The amount of runoff change between the two periods is -17.2 km³, and the rate of change is -3.9%, as shown in Table 3.

Table 3 also shows the average power generation of the cascade hydropower plants in the base period and the change period under conventional operation. It can be seen that the power generation of the cascade hydropower plants decreases by 1.9 TWh (1.0%) compared with the base period during the change period due to the reduction in runoff. This means that the runoff change is not completely equivalent to the power generation change of hydropower plants. The power generation changes of each plant in the cascade hydropower plants are 0.8, 0.2, -2.8, and -0.1 TWh, respectively, with corresponding change rates of 1.4%, 0.6%, -3.0% and -0.5%. Among them, the power generation of Xiluodu and Xiangjiaba hydropower plants increase. Although it is verified that the measured runoff at Pingshan station has a weak decreasing trend during the study period (confidence is less than 80%), the average annual runoff in the changing period increases by 0.1 km³ compared to the base period, so the power generation of the two power stations has increased accordingly.

Table 3. Influence of runoff change on power generation of cascade reservoirs under conventional operation

Subperiod	Average annual power generation/TWh					Average annual runoff/km³
	Xiluodu	Xiangjiaba	Three Gorges	Gezhouba	Cascade hydropower plants	
Base period	56.57	30.56	91.74	16.49	195.37	438.2
Change period	57.35	30.74	88.98	16.41	193.49	420.9
Change value	0.78	0.18	−2.76	−0.08	−1.88	−17.2
Change rate	1.39%	0.59%	−3.00%	−0.47%	−0.96%	−3.9%

4.2 Processes and Effects of Adaptive Regulation Measures

4.2.1 Optimization of Reservoir Group Operation Mode

The year 2010 was selected as a typical year and focus on analyzing of the control mechanism of Xiluodu and Three Gorges reservoir, which have strong scheduling capacity, and draw the water level process line of joint operation and conventional operation of the reservoir group as shown in Fig. 3. It can be seen that:

(1) Falling period: The Xiluodu's centralized falling time is advanced in the joint operation compared with the conventional operation, and the water level process line is similar in the two operation modes; the Three Gorges's centralized falling time changes little in the joint operation compared with the conventional operation, but the water level process line of the Three Gorges Reservoir is slightly higher than that under the conventional dispatch due to the increase in discharge of Xiluodu during the joint operation.

(2) Flood period: To reduce the invalid abandoned water of the cascade hydropower plants, Xiluodu increases the outflow to maintain the water level at 540–560 m throughout the flood season while ensuring that it can be stored back to the flood limit water level (560 m) in the later stage of the flood season. The total power generation benefit of cascade hydropower plants increases due to the increase in the power generation flow of the Three Gorges. The water level of the Three Gorges maintains at the upper limit of the water level of 146.5 m during the flood season, and the water level process line has no significant difference under the two operation modes.

(3) Water storage period: The Xiluodu synchronously stores water with the Three Gorges under both joint and conventional operation, and the water level process lines almost overlap in the two operation modes. The Three Gorges mainly considers its power generation under the two different operation modes, so there is no obvious difference in the water storage process line.

Therefore, compared with the conventional operation, the joint operation is mainly embodied in reducing the reservoir water level in advance during the falling stage, increasing the discharge flow, thereby increasing the power generation of the downstream cascade power stations, and ultimately achieving the goal of maximum cascade power generation (Fig. 2).

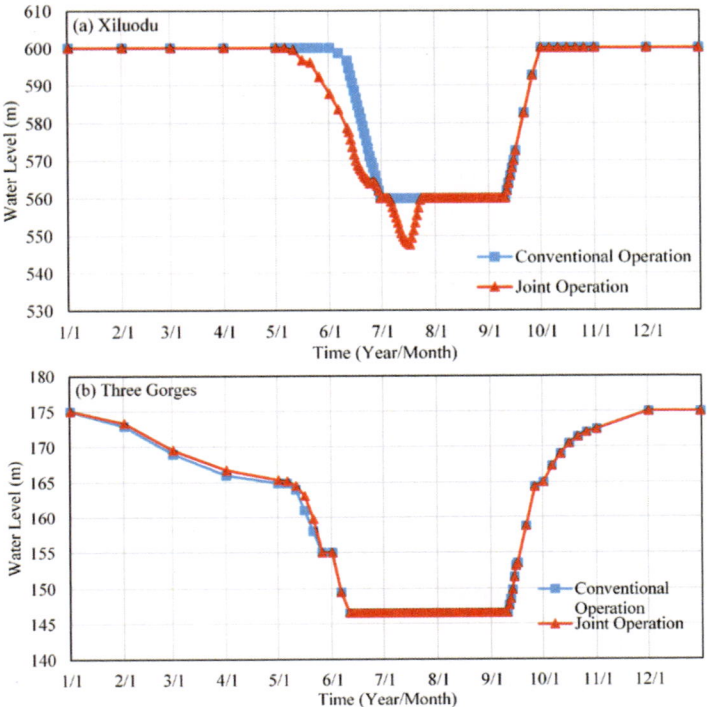

Fig. 2. Water level hydrograph of the Xiluodu reservoir and the three gorges reservoir under joint operation and conventional operation.

The adaptive operation effect is reflected in the recovery of power generation after the joint optimal operation of the reservoir group. Table 4 shows the average power generation in the change period under conventional operation and joint operation of the cascade hydropower plants. It can be seen that the power generation of the cascade hydropower plants increases by 0.59 TWh (0.31%) compared with the conventional operation for joint operation. Specifically, the power generation increase of each power station is -0.37, 0.47, 0.39, 0.10 TWh, respectively, with the increase rate of -0.65%, 1.52%, 0.44%, 0.62%. The power generation changes show that Xiluodu, as the first stage of the cascade hydropower plants, has sacrificed part of its power generation in exchange for increased power generation of other plants by falling in advance to achieve the overall power generation benefit of the cascade hydropower plants (Fig. 3a). The power generation is restored to a certain extent by adopting the mode of joint operation of cascade reservoirs, and the reduced power generation has been reduced from 1.88 TWh (0.96%) to 1.29 TWh (0.66%) due to the reduction in runoff. This means that the adaptive regulation of the reservoirs can alleviate the negative impacts of the reduction of river runoff on the power generation benefits of the cascade reservoirs.

Table 4. Power generation of cascade reservoirs under joint operation and conventional operation.

Scenario	Average annual power generation/TWh				
	Xiluodu	Xiangjiaba	Three Gorges	Gezhouba	Cascade hydropower plants
Conventional operation	57.35	30.74	88.98	16.41	193.49
Joint operation	56.98	31.20	89.38	16.52	194.08
Change value	−0.37	0.47	0.39	0.10	0.59
Change rate	−0.65%	1.52%	0.44%	0.62%	0.31%

4.2.2   Utilization of Flood Resource

Figure 3 shows the water level process line of the Three Gorges Reservoir under different flood limit water level scenarios (Scenario B-D). It can see that:

(1) Falling period: The Three Gorges reservoir water level process under scenarios C and D overlap with scenario B before June 1, and they fall to the flood limit water level from June 1 to On June 10.

(2) Flood period: The water level of the Three Gorges reservoir is maintained at the upper limit of the water level during the flood season under the three scenarios.

(3) Water storage period: The water level of the Three Gorges Reservoir under the three scenarios gradually returns from the flood limit water level to the control water level of 165 m at the end of September. Compared with Scenario A, the water level of the Three Gorges reaches 165 m earlier under scenarios B and C.

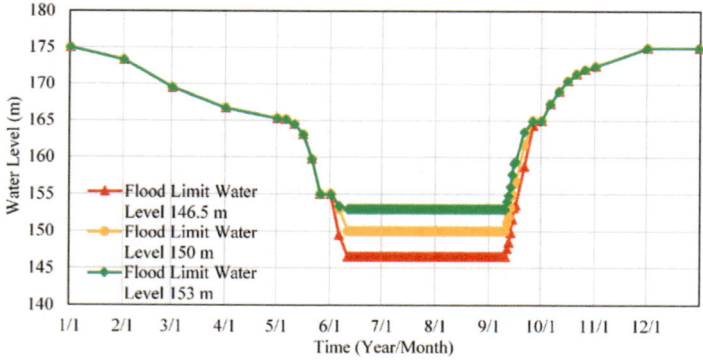

Fig. 3. Water level hydrograph of the Three Gorges Reservoir under scenario A, B, C.

Table 5 shows the calculation results of the average annual power generation of the cascade hydropower plants in the changing period under scenario A, B, C. As the maximum control water level of the Three Gorges Reservoir increases from 146.5 m (scenario A) to 150 (scenario B) m and 153 m (scenario C) during the flood season, the average annual power generation of the cascade hydropower plants during the change period increases from 194.08 to 195.26 and 196.71 TWh. Compared with scenario A, scenarios B and C have increased by 1.18 and 2.63 TWh respectively, and their growth rates are 0.61% and 1.34%. With the increase of the highest control water level in the flood season of the Three Gorges Reservoir, the power generation benefit of the cascade hydropower plants has increased significantly.

The regulation measures of raising the maximum control water level to 150 m (Scenario B) of the Three Gorges during the flood season can reduce the power generation reduction of the cascade hydropower plants due to the reduction of runoff from 1.88 TWh to 0.7 TWh. When the maximum control water level during the flood season of the Three Gorges is raised to 153 m (Scenario C), the power generation of the cascade hydropower plants has exceeded the 195.37 TWh of the conventional operation during the base period. This implies that the disadvantages of the reduction of runoff on the power generation of the cascade hydropower plants have been eliminated under scenario C. due to the increase of its power generation. For the power generation of the cascade hydropower plants, the increase in the maximum control water level of the Three Gorges reservoir during the flood season is mainly due to the increase of its power generation and has little impact on the Xiluodu, Xiangjiaba, and Gezhouba hydropower plants. From the results, it is clear that the utilization of flood resource can increase the effective runoff replenishment, regulate the water volume, and effectively increase the power generation of cascade hydropower plants (Fig. 4).

Table 5. Power generation under different flood control water level of the Three Gorges Reservoir.

Scenario	Average annual power generation/TWh					
	Xiluodu	Xiangjiaba	Three Gorges	Gezhouba	Cascade hydropower plants	Increase in cascade power generation
A	56.98	31.20	89.38	16.52	194.08	–
B	56.95	31.20	90.56	16.55	195.26	1.18 (0.61%)
C	56.96	31.21	91.27	16.58	196.71	2.63 (1.34%)

4.2.3   Adjustment of Water Use Behavior

In the long-term optimal operation, we focus on analyzing the regulation process of the Xiluodu and the Three Gorges Reservoir, which has strong regulation ability. Figure 5 shows the water level process line of the two reservoirs under scenario A, D, E. It can be seen that:

(1) Falling period: The water level process line of Xiluodu is slightly higher under the scenario E than scenario D and A due to the increase in incoming water, and it can be seen that the better water saving, the higher the water level process line. The

regulation process of the Three Gorges has the same characteristics, and its operating water levels of different scenarios overlap until the control water level of 155 m is reached on May 25.

(2) Flood period: The Xiluodu increases the outflow while ensuring that it can be stored to the flood limit water level of 560 m in the later stage of the flood season under scenario A, D, E. The water level is maintained within the range of 540–560 m during the entire flood period to reduce invalid abandoned water of other plants. When increasing the outflow of the Xiluodu, there is more water, and the drawdown depth is greater due to the control measures of water-saving. The water level of the Three Gorges reservoir is maintained at the upper limit of the water level during the flood season under scenario A, D, E.

(3) Water storage period: Due to the water-saving control measures, the upstream inflow is increased, and the speed of reservoir storage is faster. At the same time, the 20% of water saving scenario (scenario E) is faster than the 10% of water saving (scenario D) and no water saving scenario (scenario A). After Xiluodu reaches the normal water storage level on October 1, the water level process lines of different water-saving scenarios overlap. On the other hand, the water level of the Three Gorges reservoir reaches the control water level of 165 m before the end of September earlier and reaches the normal high water level on December 1 under water saving.

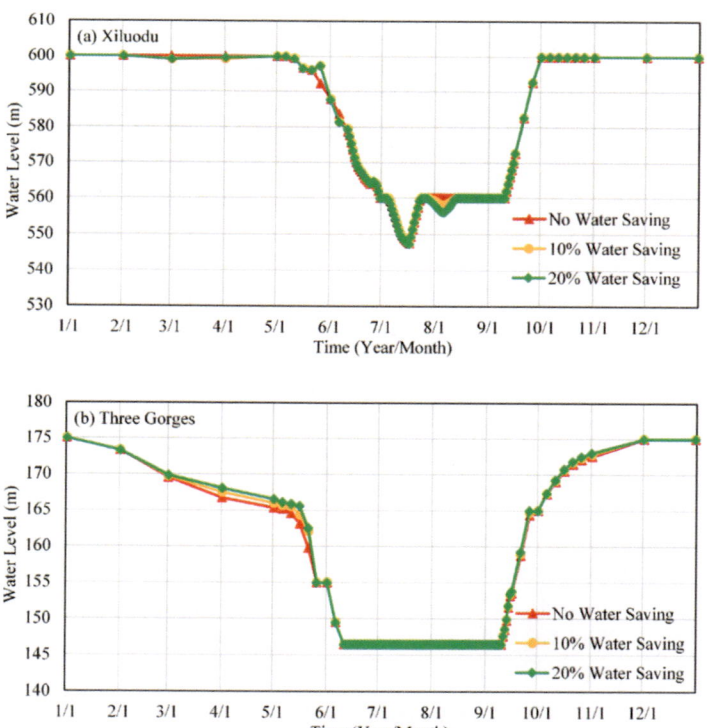

Fig. 4. Water level hydrograph of the Xiluodu and the Three Gorges Reservoir under scenario A, D, E.

Table 6 shows the average annual power generation of cascade hydropower plants during the changing period under scenario A, D, E. With the adjustment of water use behavior from no water saving (scenario A) to 10% water saving (scenario D) and 20% water saving (scenario E), the average annual power generation of cascade hydropower plants increases from 194.08 to 194.74 and 195.41 TWh during the change period. The increase in power generation is 0.66 and 1.32 TWh, respectively, and the corresponding increase rates are 0.34% and 0.68%. The regulation measures of 10% water saving (scenarios D) can reduce the power generation reduction of the cascade hydropower plants due to the reduction of runoff from 1.88 to 1.22 TWh. When we implement water saving 20% measures the power generation of the cascade hydropower plants has exceeded the 195.37 TWh of the conventional operation during the base period. This implies that the disadvantages of the reduction of runoff on the power generation of the cascade hydropower plants have been eliminated under scenario E.

In combination with Fig. 4, it is clear that the adjustment measures of water use behavior can increase the effective runoff replenishment and water volume of the hydropower system, thereby effectively increasing the power generation of the hydropower system. Together, reducing water outside the river (e.g., direct water consumption by humans) is one of the effective control measures to mitigate the impacts of upstream river runoff reduction on the hydropower system.

Table 6. Power generation of cascade reservoirs before and after the implementation of water-saving.

Scenario	Average annual power generation/TWh					
	Xiluodu	Xiangjiaba	Three Gorges	Gezhouba	Cascade hydropower plants	Increase in cascade power generation
A	56.98	31.20	89.38	16.52	194.08	–
D	57.16	31.28	89.77	16.53	194.74	0.66 (0.34%)
E	57.32	31.36	90.14	16.57	195.41	1.32 (0.68%)

4.2.4   Comprehensive Control Measures

Table 7 shows the average annual power generation of cascade hydropower plants during the changing period under scenarios F-I. The power generation of the cascade hydropower plants with the joint operation of scenarios F-I are 196.90, 197.40, 198.32, and 198.95 TWh, respectively. The power generation of the cascade hydropower plants under all scenarios can exceed the 195.37 TWh of the conventional operation during the base period. Combining with the results of utilization of rainwater and flood resources (scenario B and C) and implementation of water-saving (scenario D and E), it can be seen that the cascade power generation in Scenario B or C fails to recover to the 195.37 TWh of conventional operation during the base period. However, when combining these two measures (scenario F), the cascade power generation is 196.90 TWh (> 195.37 TWh), which can achieve the target of adaptive regulation. Therefore, the comprehensive control measures are effective control methods to counter the impacts of change in river runoff on the hydropower system.

Table 7. Power generation of cascade reservoirs with comprehensive control measures.

Scenario	Average annual power generation/TWh				
	Xiluodu	Xiangjiaba	Three Gorges	Gezhouba	Cascade hydropower plants
F	57.07	31.25	92.01	16.56	196.90
G	57.13	31.33	92.33	16.60	197.40
H	57.08	31.25	93.42	16.56	198.32
I	57.25	31.33	93.77	16.59	198.95

5 Conclusions

Due to the reduction of runoff, the annual power generation of the Xiluodu cascade and Three Gorges cascade reservoirs system reduces by 1.88 TWh during the change period using conventional operation. By adopting the joint operation of reservoir groups, Xiluodu sacrifices part of its own power generation in exchange for different degrees of increase in power generation from downstream power plants, enabling the reservoirs system to generate an additional 0.59 TWh, which restores the power generation of the system to a certain extent.

The utilization of rainwater and flood resources can increase the runoff recharge of the cascade reservoirs system to improve the system power generation. Raising the maximum control level of reservoirs during the flood season is beneficial for reservoirs to increase their own power generation and improve the overall power generation. When the maximum control water level during the flood season of the Three Gorges is raised to 153 m, the power generation of the cascade hydropower plants has exceeded the conventional operation during the base period, effectively eliminating the negative effect of reduced runoff on the power generation of the cascade reservoirs.

Developing water use control measures by adjusting the socio-economic structure, promoting water-saving can increase the runoff recharge of the cascade reservoirs system to improve the system power generation. When saving 20% of water use is implemented the negative effect of reduced runoff on the power generation is eliminated. Sometimes a single control measure cannot completely eliminate the negative effects of reduced runoff, but a combination of control measures is an effective way to hedge against the effects of runoff change on hydropower system.

Acknowledgements. This work is supported by the Natural Science Foundation of Jiangsu Province, China (Grant No. BK20200160).

References

Ahmad A, El-Shafie A, Razali SFM, Mohamad ZS (2014) Reservoir optimization in water resources: a review. Water Resour Manage 28(11):3391–3405

Ahmadi M, Bozorg-Haddad O, Loaiciga HA (2015) Adaptive reservoir operation rules under climatic change. Water Resour Manage 29(4):1247–1266

Alimohammadi H, Bavani ARM, Roozbahani A (2020) Mitigating the impacts of climate change on the performance of multi-purpose reservoirs by changing the operation policy from SOP to MLDR. Water Resour Manage 34(4):1495–1516

Brekke LD et al (2009) Assessing reservoir operations risk under climate change. Water Resour Res 45:W04411

Chang JX, Wang XY, Li YY, Wang YM, Zhang HX (2018) Hydropower plant operation rules optimization response to climate change. Energy 160:886–897

Dau QV, Kuntiyawichai K (2020) Identifying adaptive reservoir operation for future climate change scenarios: a case study in Central Vietnam. Water Resour 47(2):189–199

Edalat FD, Abdi MR (2018) Adaptive Water Management: concepts, principles and applications for sustainable development. Springer, Heidelberg

Liu HR, Sun YY, Yin XN, Zhao YW, Cai YP, Yang W (2020) A reservoir operation method that accounts for different inflow forecast uncertainties in different hydrological periods. J Clean Prod 256:120471

Qin PC et al (2020) Climate change impacts on three gorges reservoir impoundment and hydropower generation. J Hydrol 580:123922

Stringer N, Knight J, Thornton H (2020) Improving meteorological seasonal forecasts for hydrological modeling in European winter. J Appl Meteorol Climatol 59(2):317–332

Sowers J, Vengosh A, Weinthal E (2011) Climate change, water resources, and the politics of adaptation in the Middle East and North Africa. Clim Change 104(3–4):599–627

Sun Y, Bao W, Valk K, Brauer CC, Sumihar J, Weerts AH (2020) Improving forecast skill of lowland hydrological models using ensemble kalman filter and unscented kalman filter. Water Resour Res 56(8):e2020WR027468

Wang HN, Lv XZ, Zhang MY (2021) Sensitivity and attribution analysis based on the Budyko hypothesis for streamflow change in the Baiyangdian catchment, China. Ecol Ind 121:107221

Ye XC, Xu CY, Zhang ZX (2020) Comprehensive analysis on the evolution characteristics and causes of river runoff and sediment load in a mountainous basin of China's subtropical plateau. J Hydrol 591:125597

Zhang X, Liu P, Wang H, Lei X, Yin J (2017) Adaptive reservoir flood limited water level for a changing environment. Environ Earth Sci 76(21):1–14. https://doi.org/10.1007/s12665-017-7086-7

Zhang Y, Yu L, Wu SQ, Wu XF (2021a) A framework for adaptive control of multi-reservoir systems under changing environment. J Clean Prod 316

Zhang Y, Wang ML, Chen J, Zhong PA, Wu XF, Wu SQ (2021b) Multiscale attribution analysis for assessing effects of changing environment on runoff: case study of the Upstream Yangtze River in China. J Water Clim Change 12(2):627–646

Turner, S.W.D., Doering, K., Voisin, N.: Data-driven reservoir simulation in a large-scale hydrological and water resource model. Water. Resour. Res. **56**(10) (2020). e2020WR027902

Maran, S., Volonterio, M., Gaudard, L.: Climate change impacts on hydropower in an alpine catchment. Environ. Sci. Pol. **43**, 15–25 (2013). https://doi.org/10.1016/j.envsci.2013.12.001

Advances in Ecological and Environmental Effects of Mountain River Sediment

Longhu Yuan[1], Yongjun Lu[1,2(✉)], Jing Liu[1,3], Huaixiang Liu[1(✉)], Yan Lu[1], and Xiongdong Zhou[4]

[1] Nanjing Hydraulic Research Institute, Nanjing 210029, China
{yjlu,liuhx}@nhri.cn
[2] Yangtze Institute for Conservation and development, Nanjing 210029, China
[3] Department of College of Water Conservancy and Hydropower Engineering, Hohai University, Nanjing 210029, China
[4] State Key Laboratory of Hydroscience and Engineering-Tsinghua University, Beijing 10084, China

Abstract. Sediment is one of the main factors affecting the ecological environment of rivers, and its eco-environmental effect plays an important role in maintaining the balance of water environment and aquatic biodiversity. Sediment in mountain rivers has obvious characteristics such as wide gradation, which has unique impacts on the ecological environment. In addition, the increasingly intense human activities in mountain rivers, such as the construction and operation of large-scale cascade reservoirs, lead to further complicated changes in the ecological and environmental effects of sediment. In this paper, the environmental effects of mountain river sediment in adsorption, desorption and transport and the ecological effects on aquatic microorganisms, animals, plants and the entire food web were systematically reviewed. The problems existing in relevant researches were discussed, and the research prospects were presented, in order to provide guidance for the protection of mountain rivers.

Keywords: Mountain river · Cascade reservoir · Wide-graded sediment · Eco-environmental effect

1 Introduction

Mountain rivers are the pivotal corridors connecting mountains and plains, and are also the main source of supply, transport and storage of sediment. The unique characteristics of mountain river sediment are wide grain size distribution and large non-uniformity, that is, the wide-graded non-uniform sediment. The river bed composition is basically gravel, cobblesand or sand-gravel. In mountain rivers, the formation of wide-graded sediment depends on the thickness of bed cover and the non-uniformity of bed sediment. The non-uniform bed sediment is sorted and transported by the water flow, and the river bed is in an alternating process of bed armoring/stability. At the same time, fractural hillslope and weathered rock masses on both sides of the river provide a part of fine sediment, hence the mountain river sediment presents a unique wide-gradation feature (Lu and Zhang Hua-qing 1992).

© The Author(s) 2023
Y. Li et al. (Eds.): PIANC 2022, LNCE 264, pp. 1001–1016, 2023.
https://doi.org/10.1007/978-981-19-6138-0_89

The formation and movement of wide-graded sediment are diverse and intricate. In the process of interacting with water flow and riverbed, the impact of sediment on the water environment and aquatic biological community is complex (Lee and Ferguson 2002; Liu et al. 2014). For example, sediment in mountain rivers affects water environmental conditions through adsorbing and desorbing the nutrients and pollutants (Beauger et al. 2006). Wide-graded sediment plays a key role in controlling the density and diversity of benthic and fish populations by reshaping biological habitats through forming step-pool system, rib-like, cluster-like and other riverbed structures under the effect of water flow (Wang et al. 2009b). With the development and utilization of mountain rivers in China, the eco-environmental characteristics and influencing factors have become a new research hotspot. At present, studies on mountain rivers mainly focus on ecological flow estimation, ecological health assessment and ecological function restoration (Bockelmann et al. 2004; Meng et al. 2009), a systematic understanding of the ecological and environmental effects of sediment in mountain rivers are still required.

At present, mountain rivers are disturbed by more and more intense human activities, especially the cascade reservoir construction (Cheng et al. 2022; Yuan et al. 2021). Cascade reservoirs transform continuous, rapid-flowing natural river systems into intermittent, slow-flowing reservoir systems, resulting in dramatic changes in river landforms, flow patterns and biochemical cycles (BIRGITTA et al. 2010; Winemiller et al. 2016). The repeated cumulative effects of cascade reservoirs on river flow, sediment transport, and ecosystems significantly affect the overall river health status through processes such as dam blockage, intermittent discharge, and reservoir backwater (Yuan et al. 2021). Studies have shown that the cumulative impact of cascade reservoirs on the hydrodynamic process and flow regime of the lower reaches of the Jinsha River is significantly greater than that of a single reservoir. At present, studies on the impact of cascade on mountain rivers mostly focus on the destruction of river connectivity, the change of water and sediment transport process and the evolution of reservoir ecosystem (Huang et al. 2018; Yang et al. 2020), and lack of comprehensive analysis of the response process of water/sediment-environment-ecology of the mountain rivers under the influence of cascade reservoirs. This paper systematically reviewed and analyzed the ecological and environmental effects of sediment in mountain rivers, focusing on the impact caused by the cascade reservoirs, and put forward its research prospects, which can provide a guidance for the protection of mountain rivers.

2   Environmental Effects of Mountain River Sediment

With the increase in population and industrial development, water pollution has become a global problem (Yusuf et al. 2022). Mountain rivers are important reservoirs of water storages, and the enrichment and migration status of pollutants are related to water resources security. The transport of sediment is closely related to the migration of pollutants from source to convergence, such as adsorption, transport, deposition, and transformation of pollutants in water bodies, which play an important role in the distribution of pollutants in mountain rivers (Fig. 1). Studying on sediment environmental effects contributes to a scientific comprehending of the spatial and temporal pattern and variation of water pollution in mountain rivers (Sw et al. 2021; Zhao et al. 2021).

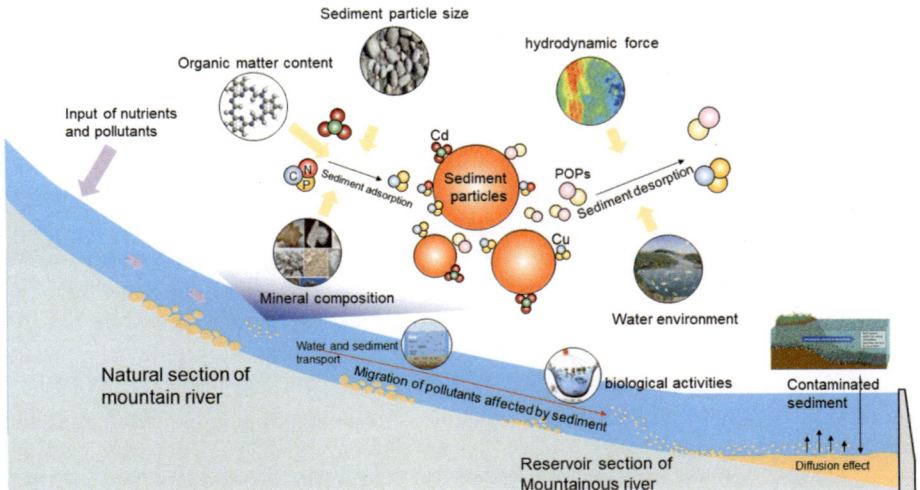

Fig. 1. Typical environmental effects related to sediment in mountain rivers

Under certain conditions, the transition of environmental substances from the dissolved state to adsorption state by sediment is called sediment adsorption. Sediment adsorption is a form of solid-liquid interface adsorption. The process of sediment adsorption is affected by its own characteristics and water environment. Factors affecting the adsorption of pollutants by sediment include sediment particle size, organic matter content, and mineral composition (Chen et al. 2021; Wang et al. 2009a). Finer particle size provides larger specific surface area and more adsorption sites, resulting in stronger adsorption capacity of sediment to heavy metals or organic pollutants. The coarse bedload in wide-graded mountain river sediment has limited adsorption capacity for pollutants. Therefore, when studying the sediment adsorption in mountain rivers, the impact of fine suspended load on pollutants should be considered.

Studies have shown that organic matter can promote the adsorption of pollutants by sediment (Xie et al. 2019). On the one hand, organic matter combines with minerals in the sediment to form complexes that absorb more pollutants through pore filling (Borggaard et al. 1990). On the other hand, organic matter promotes flocculation of fine particles and increases adsorption area (Gerbersdorf et al. 2008). For example, the adsorption capacity of fine-grained and high-organic-matter sediment in the urban section of the Yellow River is significantly higher than that of coarse-grained low-organic-matter sediment in the eroded loess section (Jiang et al. 2018). The surface of natural sediment in mountain rivers is wrapped by organic matter. Under the action of bridging, its specific surface area and pore size are significantly larger than those after artificial treatment, and the adsorption capacity is correspondingly enhanced (Wang et al. 2011). Moreover, the adsorption capacity of sediment is also related to the mineral composition of sediment. For example, the concentration of heavy metals such as Pb, Mn and Al in the deposited layer is positively correlated with the clay content, and the heavy metals exist in free form after separation from mineral oxides or as the surface coating of clay (Covelo et al. 2007; Miranda et al. 2022). The flow velocity of mountain rivers is fast, the clay content in the wide-graded sediment decreases, and the adsorption capacity for heavy metals decreases accordingly (Liao et al. 2017).

The kinetic model and isotherm adsorption model can quantitatively describe the process of sediment adsorption of pollutants, and are widely used in the studies of wide-graded sediment adsorption in mountain rivers (Liao et al. 2020). The kinetic model is used to describe the variation process of pollutants in the water-sediment phase with time, and the isothermal adsorption model is used to describe the distribution relationship of pollutants in the water-sediment phase under adsorption equilibrium state (Wu et al. 2021). The environment of mountain rivers is changeable, the temperature changes regularly with the altitude, the chemical conditions such as pH are affected by rainfall and human activities, and the properties of sediment are also distributed regionally due to different geological conditions, and the process of sediment adsorption is affected by the above factors. Combining the kinetic model with the isotherm adsorption model can explore the adsorption capacity of different particle sizes and different types of wide-graded sediments to various pollutants, and identify the key factors affecting the adsorption of wide-graded sediments.

The transition of pollutants from a sediment-based adsorption state to a dissolved state is called sediment desorption. The main influencing factors of sediment desorption include hydrodynamic force and water environment. In terms of hydrodynamic force, sediment desorption can be divided into two types according to the characteristics of water disturbance: 1) The water disturbance intensity is weak. The bed sediment does not move, and pollutants in the sediment gap migrate to the upper water through diffusion, which is affected by the pollutant diffusion flux, the pollutant content and the water depth; 2) When the water disturbance reaches a certain intensity, the bed sediment and suspended sediment exchange frequently, and the pollutants desorbed from them will affect the water quality. The seasonal distribution of water and sediment in mountain rivers is very uneven in the year. In dry seasons the general flow is relatively small and the bed surface is stable, so the desorption is mainly diffusion. While in flood seasons, the inflow and sediment increased sharply, the exchange between bed and suspended sediment strengthened, so the desorption was mainly caused by water disturbance. When discussing water environment, sediment desorption of pollutants is affected by pH, temperature and other factors. pH affects the desorption of sediment by changing the occurrence form of solutes. For example, strong alkali conditions promote phosphorus desorption, while strong acid conditions promote phosphorus adsorption. Under alkaline conditions, the phosphate ion in phosphate is replaced by OH^-, and desorption of phosphate by sediment is enhanced; under acidic conditions, phosphorus interacts with Fe and Al in sediment to generate insoluble phosphate; under neutral conditions, phosphorus mainly exists in the form of $H_2PO_4^-$ and $H_2PO_4^{2-}$ (Zhou et al. 2005). Temperature changes the intensity of sediment desorption of pollutants by affecting microbial and algal activity. For example, as the temperature increases, the respiration of bacteria and algae in the sediment is enhanced, and the consumption of dissolved oxygen increases to form a low-redox environment, which induces the reduction of ferric iron to ferrous iron, and the release of phosphorus in iron and aluminum phosphate (Xia et al. 2008). There are obvious temperature gradients in different reaches of mountain rivers due to height differences. Compared with high-altitude rivers, low-altitude rivers have higher temperatures, which intensifies the desorption of pollutants and nutrients by sediment and may lead to regional eutrophication happened.

The migration and transformation of pollutants in water depends on water flow and sediment movement, while in mountain rivers, the migration of pollutants is affected by water and sediment transport (such as sediment re-suspension) and biological activities. Studies have confirmed that more than 60% of sediment re-suspension is caused by wind and current (Ding et al. 2018). The flow velocity is positively correlated with the pollutant concentration in the overlying water, and the pollutant release in the sediment increases with the flow velocity and decreases with the water depth. The natural situation is that the mountain rivers have small water depth and fast flow speed, so the pollutant migration will be relatively frequent when there is more erosion and fine suspended matter in the flood season. The migration of pollutants in sediment is mainly affected by biological effects, which is mainly transmitted with biological enrichment. The pollutants in the sediment are absorbed by benthic organisms through ingestion, or desorption into the overlying water, absorbed by organisms through body surface contact or respiration, and then transferred along the food chain and enriched in higher order organisms. Benthic activity caused the local fine sediment re-suspension, but had limited effect on coarse sediment, and the effect of biological activity on sediment re-suspension decreased in mountain rivers dominated by wide sediment gradation (Nasermoaddeli et al. 2017).

3 Ecological Effects of Mountain River Sediment

Sediment is one of the important habitat conditions for aquatic organisms, and the ecological effect of sediment is closely related to the stability of the ecosystem. Mountain rivers have abundant hydraulic resources and complex natural environment, which breeds diverse biological habitats and rich aquatic organisms, and the interaction between sediment and aquatic organisms is frequent. The gradation of sediment in mountain rivers is wide, and the ecological effect of sediment varies with particle size. Coarse sediment provides habitat for organisms, and fine sediment absorbs nutrients and plays an important role in biological growth and reproduction. At present, the research scope of the mountain river sediment's ecological effects includes microorganisms, aquatic plants and aquatic animals (Bylak and Kukuła 2022; Mikuś et al. 2021).

Microbes is essential in the biochemical cycles of river ecosystems (Liu et al. 2018). As the material transformation medium and growth carrier of microorganisms, sediment affects the structure, function and diversity of microbial communities (Liu et al. 2017). River sediment provides a large number of attachment sites for microorganisms and provides energy sources for microorganisms by adsorbing nutrients. Studies have proven that the microbial community structure of mountain rivers is significantly affected by nitrogen and phosphorus content in deposited sediments. High nitrogen and phosphorus content promotes chemoheterotrophy and improves microbial community diversity (Wang et al. 2021). Meanwhile, the oxygen concentration in the hyporheic zone of the wide-graded bed of mountain rivers varies significantly. The fine sediment deposition zone has a larger specific surface zone, but the dissolved oxygen content is limited due to the low water flow rate. The coarse sediment deposition zone can provide more dissolved nutrients and oxygen to the microbial community due to the high flow rate caused by intergranular gaps. These factors promote the formation of diverse microbial communities (Fang et al. 2017).

Aquatic plants are the main producers of aquatic ecosystems, which are at the first trophic level and the beginning of the food chain. Water transparency, nutrient concentration, hydrodynamic conditions and other factors all affect aquatic plants. The effects of wide-graded sediment in mountain rivers on the growth of aquatic plants are multiform. Riverbed sediment can provide attachment points for attached algae, emergent plants, and submerged plants, which is conducive to plant reproduction. Suspended sediment is easy to deposit in and behind the vegetation area, and the nutrients attached by sediment can promote the growth of vegetation (Liu and Nepf 2016). However, excessive suspended sediment will block the light, reduce the transparency, inhibit the photosynthesis of phytoplankton, submerged plants and other categories, which is not conducive to the growth of vegetation. Sediment adsorption/desorption of nutrients has an impact on phytoplankton growth and root uptake of other plant species. Some phytoplankton, such as cyanobacteria, can resist extreme conditions such as low temperature, freezing, and weak light by dormancy in deposited sediments, resulting in their spatial distribution changing with the movement of sediments (Ouyang et al. 2021). Gravel beaches in mountain rivers can effectively intercept fine sediment and organic debris, provide a good substrate for the germination and growth of terrestrial and hygrophytic plants (Mikuś et al. 2013), which is conducive to increasing plant community diversity in mountainous rivers and coastal areas. Compared with the downstream plain rivers, mountain rivers generally have smaller water depth, lower sediment concentration, higher transparency level, and the bed rock is not easy to move, which is more beneficial to the reproduction of algae attached to the substrate (Zhao et al. 2020).

Fig. 2. Examples of typical benthic invertebrates with different particle sizes

Sediment in mountain rivers affects aquatic animals mainly by controlling the habitat environment. Studies have shown that there is a close relationship between the benthic community structure and the grain size in the riverbed. For example, the bedrock riverbed is suitable for aquatic insects with strong grasping force and flatworms (such as planarians) to survive; Ephemeroptera, Trichopterans and other aquatic insects are predominant in pebbles and gravel. There are many organisms such as Oligochaeta in the silt. In sandy riverbed, shrimp, bivalves and gastropods can inhabit (Fig. 2). Mountain rivers are diverse in sediments with wide gradation, which is suitable for a variety of organisms to inhabit. The coarse grain gap provides a large living space for benthic organisms and is not easy to be destroyed by erosion, so it has high biodiversity and biomass (Zhang 2009). The wide-graded sediment in mountain rivers can form a certain riverbed structure due to its high heterogeneity, providing differentiated habitat conditions and shelters for aquatic organisms. For example, mountain stream bed is mostly composed of alternately connected gentle slopes and stacked falls, which is called step-pool system (Changzhili 2004). At the beginning of this century, scholars began to explore the ecological effects of step-deep pool system. For example, Changzhili (2004) pointed out that the step-deep pool system improved the biodiversity and riverbed stability of mountain rivers. The composition of the system fully highlights the characteristics of wide-graded sediment distribution in mountain rivers, providing conditions for the formation of diverse aquatic habitats: cobbles accumulated as steps, fine sediment deposits formed pools between each step, the step and bed consists of different particle size of sediment, water flows above the cobble and through its pores, forming diverse habitat spaces. At the same time, the step-pool system forms water flow structures such as water falling and hydraulic jump with abundant dissolved oxygen, which is conducive to the survival of benthic animals, small fishes and small amphibians (Fig. 3).

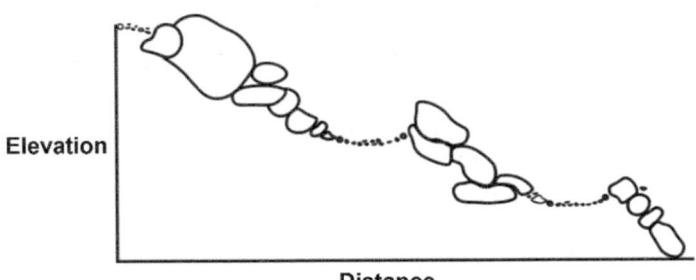

Fig. 3. Mountain River step-deep pool system typical longitudinal section

Overall, the wide-graded sediment in mountain rivers provides sufficient habitat conditions for microorganisms, aquatic animals and plants, and upstream water provides a continuous source of nutrients, thus forming a unique aquatic ecosystem (Zhang et al. 2021b). The structure and function of food web is one of the essential elements of ecosystem as a way of material and energy flux in between various organisms and habitat factors (Vesterinen et al. 2021). In mountain rivers, food web units mainly include various producers (phytoplankton, adherent algae, etc.) and consumers (benthic organisms, fish, etc.). The flux relationship of growth, death and predation of each unit is restricted by water flow, sediment, water quality and other indicators. As a nutrient carrier and a key habitat for organisms, the wide-graded sediment in mountain rivers undoubtedly has an important impact on the shaping of the structure and function of the food web. Understanding how the food web responds to the changes of water, sediment and water environment under the background of mountain rivers can provide new insights for the study of ecosystem stability.

4　Eco-Environmental Effects of Mountain River Sediment Under Influence of Reservoirs

The cascade development changes the continuous river system into a river-reservoir alternating discontinuous system, which changes the dynamic process of sediment in mountain rivers, and the sediment transformed from a single wide distribution to a spatial distribution of binary structure. The impact of cascade reservoirs on sediment is also different from the longitudinal changes of a single reservoir. The transport and siltation of sediment are affected by the overlapping effects of upstream dam discharge, downstream reservoir top support, and linkage regulation of upstream and downstream reservoirs, forming a step-by-step amplification cumulative effect (Yuan et al. 2021). The specific performance is that the sediment content of the water flow is greatly reduced, the particle size of the sand carried by the water flow is reduced, and the sediment transport capacity is decreased (Yan et al. 2021).

4.1　Changes in Environmental Effects of Sediment

The storage of cascade reservoirs results in the reduction of the total amount and grain size of sediment transported along mountain rivers. For example, after the construction of the cascade reservoirs along the lower reaches of the Jinsha River, the sediment load and sediment particle size discharged into the Three Gorges reach significantly decreased. the sediment amount from the Jinsha River into the Three Gorges Reach (in 2014) decreased by 147.6 million tons compared with that before (the average from 2003 to 2012). The proportion of sediment with particle size less than 0.062 mm (2014) increased by 19.6% compared with that before (1988–2012). Sediment retention by cascade reservoirs results in the retention of organic matter adsorbed by the sediment. From 1953 to 2016, the flux of particulate organic carbon transported from rivers to the ocean in China showed a downward trend, with an average annual decrease of 0.2 Tg C in total particulate organic carbon flux and the main reason for this phenomenon was sedimentation caused by reservoirs (Liu et al. 2020). The oxidative decomposition of particulate organic carbon in the silted sediment leads to the release of a large amount of carbon dioxide, which has an impact on the atmospheric environment. Due to the adsorption of sediment, the sedimentation of suspended sediment intercepted by the cascade reservoirs also leads to the retention of nutrients and reduces the supply of nutrients to the downstream. For example, due to the construction of the Three Gorges Reservoir, the transport of adsorbed phosphorus to the ocean in the upper reaches of the

Yangtze River decreased by about 22.5%. After a turbulent river is transformed into a reservoir, the environmental capacity is reduced, especially nutrients and pollutants are enriched in the reservoir area with sediment and gradually released to the water body, which will become a hidden danger to the river environmental safety in the long term. After the cascade development of the Lancang River Basin, the total phosphorus in each section of the main stream of the Lancang River generally exceeded the standard, and the water quality was Class V. Besides, the change of sediment particle size also affects the local nitrogen cycle in the reservoir area. Studies showed that the sediment with smaller particle size has larger specific surface area and higher organic carbon content, which leads to the growth of more denitrifying bacteria, resulting in a higher denitrification rate (Xia et al. 2017; Xia et al. 2018). Although a single reservoir also has the effect of decreasing particle size and transport capacity, the effect of cascade reservoir on sediment environment will be amplified step by step. For instance, cascade reservoir in Jinsha river reduced the concentration of suspended sediment in reservoir area and outlet, and significantly increased the concentration of particulate organic carbon in sediment. The concentration of particulate organic carbon in deposited sediment during the operation of cascade reservoir is 5 times that of single reservoir (Wu et al. 2020).

4.2   Changes in the Ecological Effect of Sediment

The effect of cascade reservoirs on sediment ecological effects is reflected in two aspects: regulating suspended sediment transport process and changing bed sediment conditions:

1. Control the suspended sediment transport process. In general, the sedimentation of suspended sediment in the reservoir area enhances the transparency of mountain rivers and, together with the reduced flow velocity, facilitates the growth of phytoplankton. However, the river habitat of cascade reservoirs in mountainous areas changes in a segmentalized manner due to multi-stage regulation. Taking the cascade reservoirs in the lower reaches of Jinsha River as an example, with the water storage of Xiluodu and Xiangjiaba reservoirs, the suspended sediment settled, and the phytoplankton and fish feeding on them increased. However, Wudongdong and Baihetan reservoirs still maintained normal flowing water habitat during the unfinished period, and the suspended sediment content was high, which was not conducive to the growth of phytoplankton, and the fish composition was dominated by the indigenous fluid-like water community. The temporal and spatial distribution of nutrients in mountain rivers changed with the change of suspended sediment transport process, which affected the community structure of algae and other aquatic plants. The sediment transported by rivers includes not only mineral particles, but also organic matter particles and bioclasts that can be eaten by organisms. The cascade reservoir, through the interception of multi-stage dams, flattens the downstream discharge, lowers the flow velocity of the reach, and weakens the sand-carrying capacity, which leads to the reduction of the nutrient and food contents carried by the water flow, thus significantly reducing the integrity of the river ecosystem(Duan et al. 2011). In addition, during the period of "Clear storage and muddy discharge" of some reservoirs, a large amount of sediment discharge also

has a negative impact on benthic organisms and fish, resulting in biomechanical damage, reduced food intake and individual death (Crosa et al. 2009).

2. Change the bed sediment characteristics. The decrease of the flow velocity in the reservoir area leads to the accumulation of sediment and nutrients in the riverbed and the acceleration of microbial growth and metabolism. Related metabolites (such as extracellular polymeric substances (EPS)) promote the formation of biofilm on the sediment surface, which affects the morphology, density and physical and chemical properties of sediment, leading to significant changes in sediment movement characteristics (Fang et al. 2017). The biofilm on the sediment surface can protect the sediment when it is eroded by water flow, and the network structure composed of EPS and other connected particle gaps enhances the anti-scour property of the sediment. Large cobbles in mountain river bed can provide shelter for benthic groups ranging from 2.5 mm to 10 mm, and small cobbles can also shelter scrapers, grazers and tearing of benthic animals, such as Rhithrogen, etc. With the sedimentation of reservoir sediment, cobble riverbed in reservoir area is covered by fine particles of sediment. The original benthic organisms lost a large number of shelters (Beauger et al. 2006), while the sediment was simplified and the riverbed structure in mountainous areas disappeared, resulting in a decrease in habitat conditions and the diversity of benthic organisms. For example, the species number and evenness index of benthic species in Xiluodu and Xiangjiaba reservoir areas of Jinsha River were lower than those in natural river channels, and the species with low oxygen tolerance and strong survival ability gradually became the dominant species in the reservoir areas. Therefore, some researchers believe that river cascade development will result in homogenized ecological environment and biome with obvious gradient (Petesse and Petrere 2012). In addition, fine-grained sediments in the mountainous river reservoir area will significantly reduce the porosity and dissolved oxygen content of the subsurface layer below the bed sediment, reduce the habitat depth of organisms, and threaten the subsurface microorganisms and benthic organisms. After scouring and armoring of the downstream riverbed of the reservoir, the gap fine sediment decreases, which may make the benthic species living in the original fine sediment, such as oligochaetes and chironomids, have no place to hide, and their biomass is greatly reduced. The process of riverbed evolution itself also has significant ecological effects, and the stronger riverbed evolution intensity will reduce the stability of the habitat and adversely affect the biological community. Erosion of river bed can wash out organic sediment and the benthic organisms in it, and rapid siltation can also reduce the quality of life of organisms, such as mayfly gills, affecting respiration and other functions (Duan et al. 2009). Furthermore, the destruction of benthic communities inevitably affects some mountain river fishes that feed on them. For example, fish that feed on benthic organisms, such as carp, are forced to consume food with low nutritional level due to the decrease of benthic organisms, leading to the decline of their nutritional level. The natural mountain rivers are in a relatively stable state of river evolution, but after the operation of the cascade reservoir, the changing backwater area, the perennial backwater area, and the downstream of the reservoir have the evolution process of alternating scouring and silting, continuous silting, continuous scouring, etc. Under these unstable states, the sediment activity will lead to the complex response of biological community.

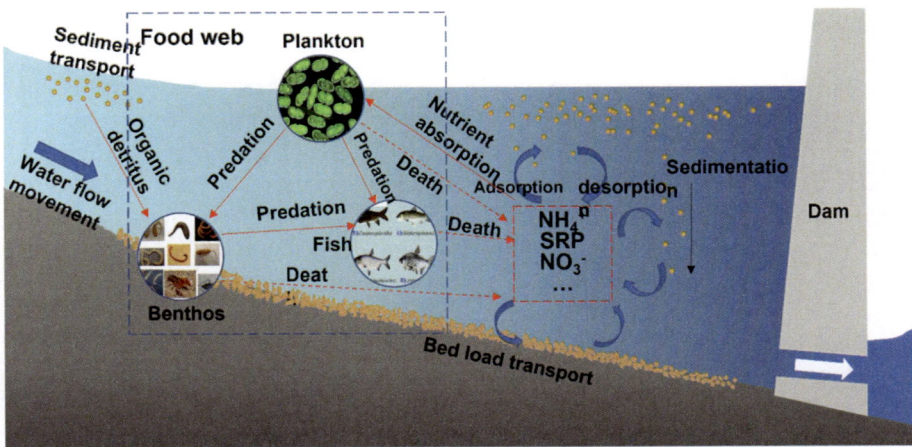

Fig. 4. Schematic diagram of the response mechanism of water/sediment-environment-ecology interaction (taking the reservoir area as an example)

4.3   Changes in Water/Sediment-Environment-Ecology Response Relationship

In summary, from the perspective of ecosystem as a whole, water and sediment regulation in cascade reservoirs of mountain rivers leads to the changes of various biological communities, and thus affects the interactions among species represented by food webs. The static water condition of the reservoir promotes the growth of algae, which leads to the transformation of the scavenged food web based on debris to the grazing food web based on algae (Mor et al. 2018). The operation of the reservoir results in the decrease of the habitats of original river organisms, which leads to the homogenization of biological communities in the reservoir and the decrease of food web diversity (Zhang et al. 2021a). At the same time, fine-grained sediment is deposited in the reservoir area, filling crevices of coarse-grained sediment in natural mountain rivers, resulting in the reduction of critical habitats and refuges, and the exposure of shallow benthic animals to water bodies, changing the predation relationship (Power et al. 2013). At present, studies on river food webs mainly focus on nutrient structure analysis and material/energy transfer process analysis [88, 89], while there are few studies on the relationship between food webs and hydrodynamic factors of water and sediment (Wootton et al. 1996). After the construction and operation of cascade reservoirs, a complete response process of water/sediment, water environment and water ecology (Fig. 4) are formed in the river. The material energy absorbed/desorbed by sediment can be absorbed by plankton, and then transmitted through the food web among benthic organisms, fish and other organisms. After the death of organisms, they are deposited and decomposed again, and become part of the sediment or migrate with water. Therefore, the changes of water flow and sediment transport brought by cascade reservoirs also transmit to the changes of biogenic substances, primary producers, consumers and other biological categories.

5 Conclusions and Future Perspectives

Mountain rivers play an important role in the conservation of water resources and biodiversity. With the implementation and promotion of sustainable development and ecological security strategy in China, ecological environmental effects of sediment in mountain rivers have gradually become an important research field. Sediment in mountain rivers significantly affects microbial, plankton, adherent algae and other species, especially the benthic community. The unique geomorphic characteristics and wide grain size distribution of mountain rivers contribute significantly to the formation of local microecosystem. The environmental effects of sediment adsorption and desorption are the key ways of pollutant migration and transformation. In recent decades, the construction and operation of cascade reservoirs in mountain rivers have caused great changes in the natural water and sediment processes, and the ecological and environmental effects of sediment have changed accordingly. On the one hand, the establishment of cascade reservoirs changes the characteristics and transport process of sediment, and greatly affects the pollutant migration and material circulation affected by sediment. On the other hand, the previous ecological process of mountain rivers is adjusted by the different response mechanisms of microorganisms, aquatic animals, plants and the whole food webs. At present, the research on the environmental effects of sediment adsorption and desorption in mountain rivers mainly focuses on laboratory simulation and mathematical models. Studies on sediment's ecological effects lack systematic analysis, which need massive data to form a complete knowledge framework. Recent studies can be further improved from the following aspects:

(1) The environmental characteristics of wide-graded sediment in mountain rivers. The sediment adsorption and desorption on pollutants are closely related to grain size and mineral composition of sediment. By improving the database of sediment environmental characteristics of major rivers and reservoirs, it can provide useful reference for understanding the law of pollutant migration and transformation, and provide corresponding basis for sediment pollution control in river basins.

(2) The combined approaches of ecology and environmental science with hydro-sediment dynamics to study the relationship between suspended/bed load transport rate and pollutant transport, especially the impact of wide-graded sediment on pollutants. Explore the whole series of pollutant migration and transformation processes including bedload transport and exchange with bed sediment, sediment adsorption/desorption, biological uptake, and enrichment.

(3) The response mechanism of water/sediment-environment-ecology interaction in cascade reservoirs of mountain rivers. The response process in river water/sediment, aquatic environment and aquatic ecology often involves the variation of multiple environmental factors and the succession of various biological communities. Especially under the impact of cascade reservoirs, the key environmental elements of the river and the structure and function of the food web may undergo major changes. It is necessary to conduct in-depth research on the complex comprehensive mechanism and carry out quantitative simulations.

Acknowledgements. The study is financially supported by the National Natural Science Foundation of China, the Yangtze River Water Research Joint Fund (U2040219).

References

Beauger A, Lair N, Reyes-Marchant P, Peiry JL (2006) The distribution of macroinvertebrate assemblages in a reach of the River Allier (France), in relation to riverbed characteristics. Hydrobiologia 571(1):63–76

RenÖFÄLt BM, Jansson R, Nilsson C (2010) Effects of hydropower generation and opportunities for environmental flow management in Swedish riverine ecosystems. Freshw Biol 55(1):49–67

Bockelmann BN, Fenrich EK, Lin B, Falconer RA (2004) Development of an ecohydraulics model for stream and river restoration. Ecol Eng 22(4–5):227–235

Borggaard OK, Jdrgensen SS, Moberg JP, Raben-Lange B (1990) Influence of organic matter on phosphate adsorption by aluminium and iron oxides in sandy soils. Eur J Soil Sci 41(3):443–449

Bylak A, Kukuła K (2022) Impact of fine-grained sediment on mountain stream macroinvertebrate communities: forestry activities and beaver-induced sediment management. Sci Total Environ 832:155079

Changzhili Z (2004) Development of step-pool sequence and its effects in resistance and stream bed stability. Int J Sedim Res 14(6):126–233

Chen Y, Huang L, Li X (2021) Quantifying the biofilm effects on phosphorus adsorption of sediment. J Soils Sediments 21(2):1302–1316. https://doi.org/10.1007/s11368-020-02851-5

Cheng Y, Zhao F, Wu J, Gao P, Wang Y, Wang J (2022) Migration characteristics of arsenic in sediments under the influence of cascade reservoirs in Lancang River basin. J Hydrol 606:127424

Covelo EF, Vega FA, Andrade ML (2007) Heavy metal sorption and desorption capacity of soils containing endogenous contaminants. J Hazard Mater 143(1):419–430

Crosa G, Castelli E, Gentili G, Espa P (2009) Effects of suspended sediments from reservoir flushing on fish and macroinvertebrates in an alpine stream. Aquat Sci 72(1):85

Ding W, Wu T, Qin B, Lin Y, Wang H (2018) Features and impacts of currents and waves on sediment resuspension in a large shallow lake in China. Environ Sci Pollut Res 25 (36):36341–36354. https://doi.org/10.1007/s11356-018-3471-3

Duan X-H, Wang Z-Y, Xu M-Z (2011) Effects of fluvial processes and human activities on stream macro-invertebrates. Int J Sedim Res 26(4):416–430

Duan X, Wang Z, Xu M, Zhang K (2009) Effect of streambed sediment on benthic ecology. Int J Sedim Res 24(3):325–338

Fang H, Chen Y, Huang L, He G (2017) Analysis of biofilm bacterial communities under different shear stresses using size-fractionated sediment. Sci Rep 7(1):1299

Gerbersdorf SU, Jancke T, Westrich BP, DM. (2008) Microbial stabilization of riverine sediments by extracellular polymeric substances. Geobiology 6(1):57–69

Huang X-R, Gao L-Y, Yang P-P, Xi Y-Y (2018) Cumulative impact of dam constructions on streamflow and sediment regime in lower reaches of the Jinsha river, China. J Mt Sci 15 (12):2752–2765. https://doi.org/10.1007/s11629-018-4924-3

Jiang Y, Yuan L, Liu L, Shi L, Guang A-L, Mu Z (2018) Bisphenol a in the yellow river: sorption characteristics and influential factors. J Hydrol 564:307–313 S0022169418305018-

Lee AJ, Ferguson RI (2002) Velocity and flow resistance in step-pool streams. Geomorphology 46(1–2):59–71

Liao J, Chen J, Ru X, Chen J, Wu H, Wei C (2017) Heavy metals in river surface sediments affected with multiple pollution sources, South China: distribution, enrichment and source apportionment. J Geochem Explor 176:9–19

Liao R, Hu J, Li Y, Li S (2020) Phosphorus transport in riverbed sediments and related adsorption and desorption characteristics in the Beiyun River. China. Environmental Pollution 266:115153

Liu B, Li Y, Zhang J, Zhou X, Wu C (2014) Abundance and diversity of ammonia-oxidizing microorganisms in the sediments of Jinshan Lake. Curr Microbiol 69(5):751–757. https://doi.org/10.1007/s00284-014-0646-0

Liu C, Nepf H (2016) Sediment deposition within and around a finite patch of model vegetation over a range of channel velocity. Water Resour Res 52(1):600–612

Liu D et al (2020) Changes in riverine organic carbon input to the ocean from mainland China over the past 60 years. Environ Int 134:105258

Liu R, Wu W, Zhou X, Yue Z, Zhao P (2017) Bacterioplankton community structure in Weihe river and its relationship with environmental factors. Huanjing Kexue Xuebao / Acta Scientiae Circumstantiae 37(3):934–944

Liu T et al (2018) Integrated biogeography of planktonic and sedimentary bacterial communities in the Yangtze river. Microbiome 6(1):16

Lu YJ, Hua-qing Z, T.R.I.o.W.T.E., Tianjin, P. R. China, (1992) A study on nonequilibrium transport of nonuniform bedload in steady flow. Adv Hydrodyn Res Engl Ed 2:8

Mikuś P et al (2021) Impact of the restoration of an incised mountain stream on habitats, aquatic fauna and ecological stream quality. Ecol Eng 170:106365

Mikuś P, Wyżga B, Kaczka RJ, Walusiak E, Zawiejska J (2013) Islands in a European mountain river: linkages with large wood deposition, flood flows and plant diversity. Geomorphology 202:115–127

Miranda LS, Ayoko GA, Egodawatta P, Goonetilleke A (2022) Adsorption-desorption behavior of heavy metals in aquatic environments: influence of sediment, water and metal ionic properties. J Hazard Mater 421:126743

Mor J-R, Ruhí A, Tornés E, Valcárcel H, Muñoz I, Sabater S (2018) Dam regulation and riverine food-web structure in a Mediterranean river. Sci Total Environ 625:301–310

Nasermoaddeli MH et al (2017) A model study on the large-scale effect of macrofauna on the suspended sediment concentration in a shallow shelf sea. Estuar Coast Shelf Sci 211(62–76):S0272771417300987

Ouyang W, Li Z, Yang J, Lu L, Guo J (2021) Spatio-temporal variations in phytoplankton communities in sediment and surface water as reservoir drawdown—a case study of Pengxi river in three gorges reservoir. China. Water 13(3):340

Petesse ML, Petrere M (2012) Tendency towards homogenization in fish assemblages in the cascade reservoir system of the Tietê river basin, Brazil. Ecol Eng 48:109–116

Power ME, Holomuzki JR, Lowe RL (2013) Food webs in Mediterranean rivers. Hydrobiologia 719(1):119–136. https://doi.org/10.1007/s10750-013-1510-0

Sw A, Rdv B, Jc C, Yan LA, Jf AX, A. (2021) Riverine flux of dissolved phosphorus to the coastal sea may be overestimated, especially in estuaries of gated rivers: Implications of phosphorus adsorption/desorption on suspended sediments - ScienceDirect. Chemosphere 287:132206

Vesterinen M, Perälä T, Kuparinen A (2021) The effect of fish life-history structures on the topologies of aquatic food webs. Food Webs 29:e00213

Wang CH, Gao SJ, Wang TX, Tian BH, Pei YS (2011) Effectiveness of sequential thermal and acid activation on phosphorus removal by ferric and alum water treatment residuals. Chem Eng J 172(2–3):885–891

Wang J et al (2021) Response of bacterial communities to variation in water quality and physicochemical conditions in a river-reservoir system. Glob Ecol Conserv 27:e01541

Wang Y, Shen Z, Niu J, Liu R (2009) Adsorption of phosphorus on sediments from the three-Gorges Reservoir (China) and the relation with sediment compositions. J Hazard Mater 162 (1):92–98

Wang ZY, Melching CS, Duan XH, Yu GA (2009) Ecological and hydraulic studies of step-pool systems. J Hydraul Eng 135(9):705–717

Winemiller KO, Mcintyre PB, Castello L, Fluet-Chouinard E, Saenz L (2016) Balancing hydropower and biodiversity in the Amazon, Congo, and Mekong. Science 351(6269):128–129

Wootton JT, Parker MS, Power ME (1996) Effects of disturbance on river food webs. Science 273(5281):1558–1561

Wu P, Wang N, Zhu L, Lu Y, Fan H, Lu Y (2021) Spatial-temporal distribution of sediment phosphorus with sediment transport in the three Gorges reservoir. Sci Total Environ 769:144986

Wu Y, Fang H, Huang L, Cui Z (2020) Particulate organic carbon dynamics with sediment transport in the upper Yangtze river. Water Res 184:116193

Xia J, Jin X, Yang Y, Li L, Wu F (2008) Effects of biological activity, light, temperature and oxygen on phosphorus release processes at the sediment and water interface of Taihu Lake. China. Water Res 42(8–9):2251–2259

Xia X, Jia Z, Liu T, Zhang S, Zhang L (2017) Coupled nitrification-denitrification caused by suspended sediment (SPS) in rivers: importance of SPS size and composition. Environ Sci Technol 51(1):212–221

Xia X et al (2018) The cycle of nitrogen in river systems: sources, transformation, and flux. Environ Sci Process Impacts 20(6):863–891

Xie F et al (2019) Adsorption of phosphate by sediments in a eutrophic lake: isotherms, kinetics, thermodynamics and the influence of dissolved organic matter. Colloids Surf, A 562:16–25

Yan H, Zhang X, Xu Q (2021) Variation of runoff and sediment inflows to the three Gorges reservoir: impact of upstream cascade reservoirs. J Hydrol 603:126875

Yang M et al (2020) Damming effects on river sulfur cycle in karst area: a case study of the Wujiang cascade reservoirs. Agr Ecosyst Environ 294:106857

Yuan Q-S, Wang P-F, Chen J, Wang C, Liu S, Wang X (2021) Influence of cascade reservoirs on spatiotemporal variations of hydrogeochemistry in Jinsha river. Water Sci Eng 14(2):97–108

Yusuf A et al (2022) Updated review on microplastics in water, their occurrence, detection, measurement, environmental pollution, and the need for regulatory standards. Environ Pollut 292:118421

Zhang H, Huo S, Cao X, Ma C, Zhang J, Wu F (2021) Homogenization of reservoir eukaryotic algal and cyanobacterial communities is accelerated by dam construction and eutrophication. J Hydrol 603:126842

Zhang M et al (2021) The aquatic benthic food webs: The determinants of periphyton biofilms in a diversion canal and its upstream reservoir. Ecol Eng 170:106363

Zhang X (2009) Effect of streambed sediment on benthic ecology. Int J Sedim Res 24(3):325–338

Zhao G et al (2020) Phytoplankton in the heavy sediment-laden Weihe River and its tributaries from the northern foot of the Qinling Mountains: community structure and environmental drivers. Environ Sci Pollut Res 27(8):8359–8370. https://doi.org/10.1007/s11356-019-07346-6

Zhao H et al (2021) Effect of extracellular polymeric substances on the phosphorus adsorption characteristics of sediment particles. Int J Sedim Res 36(5):628–636

Zhou A, Tang H, Wang D (2005) Phosphorus adsorption on natural sediments: modeling and effects of pH and sediment composition. Water Res 39(7):1245–1254

Analysis of Erosion and Deposition Characteristics in Hukou-Jiangyin Reach of the Lower Yangtze River After the Operation of the Three Gorges Project

Shuang Cao[1], Long Cheng[1(✉)], Nairu Wang[2], Qiang Li[1], and Hongyu Luo[1]

[1] Lower Changjiang River Bureau of Hydrology and Water Resources Survey, Changjiang Water Resources Commission, Nanjing, China
chenglong@whu.edu.cn
[2] Nanjing Hydraulic Research Institute, Nanjing, China

Abstract. With the operation of the Three Gorges Project (TGP) and the reservoirs in the upper reaches of the Yangtze River, the sediment transport reduced significantly in the lower reaches of the Yangtze River since 2003, leading to great changes in the relationship between flow and sediment. In order to study the variation characteristics of the river channel under new conditions, the variation characteristics of flow and sediment at the main hydrological stations, the amount and intensity of erosion/deposition in the Hukou-Jiangyin reach, and the distribution of erosion/deposition in the channel were analyzed. The results showed that the Hukou-Jiangyin reach experienced both erosion and deposition from 1998 to 2006, while unidirectional erosion in both benchland and channel after 2006. From 2006 to 2020, the average annual erosion intensities per kilometer below bankfull channel in Hukou-Datong reach and Datong-Jiangyin reach were 12.56×10^4 m³/km and 17.17×10^4 m³/km, increasing by 79.94% and 67.35%, respectively, compared with those during the period from 1998 to 2001, i.e., the times before the impoundment of TGP. The riverbed erosion intensity of Datong-Jiangyin reach was 37% higher than that of the upper Hukou-Datong reach. The erosion mainly occurred below bankfull channel, while the erosion between low water channel and bankfull channel only accounted for 10% – 20%. Specifically, the ratios of the erosion below low water channel and bankfull channel to those below flood channel accounted for 86.67% and 91.14% in Hukou-Datong reach, respectively. And those ratios in Datong-Jiangyin reach were 80.34% and 86.69%, respectively.

Keywords: Lower reaches of the Yangtze River · Variation of flow and sediment · The Three Gorges Project · Erosion and deposition

1 Introduction

Since 2003, with the completion and operation of the Three Gorges Project (TGP) and other eight reservoirs, including Liyuan, Ahai, Jin'anqiao, Longkaikou, Ludila, Guanyinyan, Xiangjiaba and Xiluodu in the upper reaches of the Yangtze River, the

Y. Li et al. (Eds.): PIANC 2022, LNCE 264, pp. 1017–1029, 2023.
https://doi.org/10.1007/978-981-19-6138-0_90

average runoff from Yichang to Datong of the lower reaches of the Yangtze River has slightly reduced, while the sediment transport has reduced sharply. The data of hydrological stations in the Yangtze river below Yichang showed that sediment transport decreased by 4/5 – 2/3 on average from upper to lower reaches. Besides, the sediment environment changed significantly, resulting in continuous long-distance erosion in the lower reaches of the Yangtze River (LRYR).

Many scholars have studied the erosion and deposition characteristics, as well as the evolution laws of the downstream river bed caused by clean water discharge after the completion of upstream reservoirs (Dai and Liu, 2013). The analysis of the erosion and deposition laws in the main stream from Yichang to Datong conducted by Xu et al. (2011) showed that the erosion and deposition characteristics of the middle and lower reaches of the Yangtze River had changed from "upper erosion and lower deposition" and "deposition in benchland and erosion in channel" to " erosion in both benchland and channel " after the application of TGP. And the erosion was almost along the entire lower reaches. Through the analysis of prototype observation data, Zhu et al. (2018) concluded that the downstream of TGP had entered a state of strong erosion. Li et al. (2018; 2015) analyzed the sedimentation characteristics of the Three Gorges Reservoir area by means of sediment transport rate method and section analysis method respectively. Shi et al. (2020) analyzed the variation of flow and sediment and the characteristics of erosion and deposition in the reach from TGP to Gezhouba Project, and found that the relationship between flow and sediment in the reach changed significantly. The main composition of sediment became coarser after the operation of the projects, and the activity in each sub-reach was spatially distinct.

At present, relevant studies are mainly focused on the Three Gorges Reservoir area and Yichang to Datong reach. There are few studies on the erosion and deposition characteristics of the entire Hukou-Datong -Jiangyin reach. Furthermore, the relevant studies have not gotten the data of the regional flood in 2020 involved. On the basis of flow and sediment characteristics variation data of the hydrological stations in LRYR since TGP construction, we used the past and the latest river terrain data of the Hukou to Jiangyin reach to analyze the erosion and deposition in LRYR. Riverbed erosion and deposition was calculated below 4 different water levels (i.e. low, medium, bankfull and flood water level). Temporal and spatial variations of channel erosion and deposition in LRYR since the completion of TGP were analyzed. It could provide the latest reference data for river regulation, waterway regulation, shoreline protection and exploitation in the new period.

2 Overview of the Reaches

The lower reaches of the Yangtze River, from Hukou to Jiangyin, flows through Hubei, Jiangxi, Anhui and Jiangsu provinces. The length of the main stream is about 639.2 km. The river is alternated in wide and narrow, with many sandbanks along the river, which are generally in the shape of lotus nodes. According to the nodes or narrowed sections or specific tributaries of the estuary, LRYR is divided into fifteen sub-reaches (Table 1).

3 Flow and Sediment Variation in LRYR Before and After the Impoundment of TGP

There are two perennial hydrological stations, i.e., Jiujiang hydrological station and Datong hydrological station, in LRYR. The spatial differences of flow and sediment characteristics were statistically analyzed before and after 2003, when TGP impounded.

After the impoundment of TGP, the average annual flow runoff of Jiujiang hydrological station slightly decreased by 4.6%, from 7.522×10^{12} m^3 to 7.177×10^{12} m^3, while the sediment runoff significantly decreased by 67.8%, from 2.774×10^8 t to 8.929×10^7 t (Fig. 1a). The average annual flow runoff of Datong hydrological station slightly decreased by 2.5%, from 9.046×10^{12} m^3 to 8.824×10^{12} m^3, while the sediment runoff significantly decreased by 69.5%, from 4.2686×10^8 t to 1.3225×10^8 t (Fig. 1b). Overall, the decrease of sediment runoff was significantly greater than that of flow runoff at both hydrological stations.

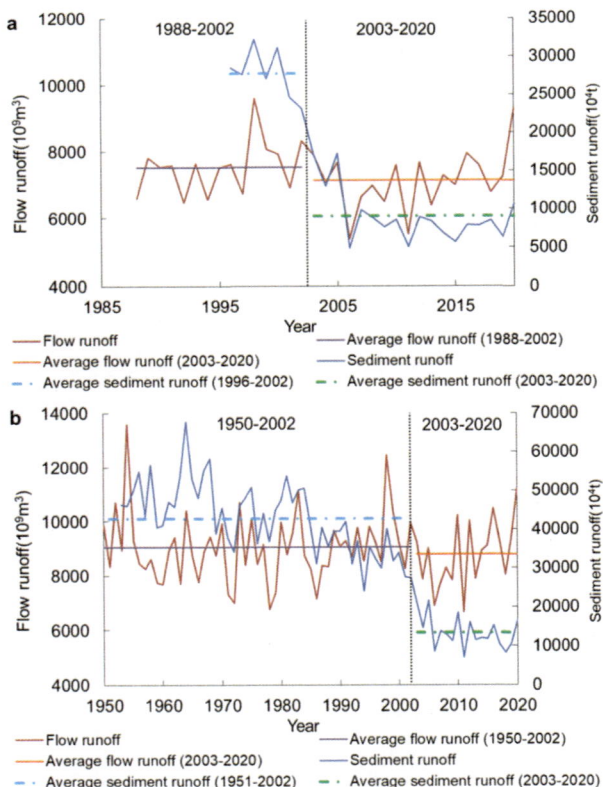

Fig. 1. Annual flow and sediment runoff of Jiujiang and Datong hydrological stations (a. Jiujiang hydrological station; b. Datong hydrological station)

4   Calculation and Analysis of Erosion and Deposition Changes in LRYR

4.1   Materials and Methods

Long range 1:10000 channel topographic maps of Hukou to Jiangyin in 1998 and every 5 years from 2001 to 2020 were collected. The the erosion and deposition of the river channel were calculated, using cross section method. The average interval of fixed cross sections was about 1.0 km, while that of curved reaches or reaches with large changes was reduced to about 500 m. Data extraction for cross sections was generally carried out according to 1:5000, i.e., the average horizontal point distance was 35 – 40 m. The horizontal point distance was encrypted to about 20m in steep slopes and areas with severe local terrain changes.

To understand the distribution of erosion and deposition in the river channel, water levels at the discharge of 10000, 30000, 45000 and 60000 m^3/s (i.e., low, medium, bankfull and flood·water level) were used to calculate the erosion and deposition. Since LRYR is as long as 639.2 km with a large drop of water head, the calculated water level of each sub-reach was determined by previous mathematical model results (Table 1).

4.2   Analysis of Calculation Results of Erosion and Deposition Characteristics in Hukou-Datong Reach

In 2003, TGP was completed and put into operation. From 2011 to 2016, 8 reservoirs in the upper reaches of the Yangtze River were successively completed and put into operation. The main stream reservoir group was basically completed and initial water storage has been achieved since 2016. According to the operation period of the reservoir group, the 23 years since 1998 were divided into 5 periods: i) before the operation of TGP (1998–2001); ii) shortly before and after operation of TGP (2001–2006); iii) initial period of TGP operation (2006–2011); iv) operation period of main stream reservoir group (2011–2016); v) after the operation of the main stream reservoir group (2016–2020). The variation characteristics of erosion and deposition in each reach were analyzed during each period.

Table 1. Water level values used for erosion and deposition in Hukou to Jiangyin reach

Sub-reaches	Length (km)	Water level (Initial cross section) (m)			
		Low level	Medium level	Bankfull level	Flood level
Shangxiasanhao	35.6	8.85	11.45	14.08	16.42
Madang	31.4	5.58	11.10	13.73	16.03
Dongliu	34.7	5.08	10.47	13.07	15.32
Guanzhou	29.6	4.53	9.75	12.34	14.52
Anqing	25.7	4.09	9.19	11.77	13.89
Taiziji	25.9	3.69	8.68	11.24	13.32
Guichi	23.3	3.32	8.20	10.75	12.77
Datong	21.8	2.92	7.69	10.23	12.20
Tongling	59.5	2.73	7.35	9.83	11.73
Heishazhou	33.8	2.41	6.46	8.68	10.32
Wuyu	49.8	2.23	5.94	8.02	9.51
Ma'anshan	30.6	1.95	5.18	7.01	8.29
Nanjing	95.5	1.80	4.70	6.34	7.46
Zhenyang	57.9	1.45	3.56	4.72	5.37
Yangzhong	84.1	1.29	2.94	3.84	4.26
Lower boundary	/	1.07	2.12	2.69	2.87

Table 2. Statistics of erosion/deposition amounts in Hukou-Datong reach

Period	Erosion/deposition amount ($10^4 m^3$) (deposition: +; erosion: −)			
	Low level	Medium level	Bankfull level	Flood level
1998–2001	−7012	859	4773	11509
2001–2006	2631	−4432	−7986	−15627
2006–2011	−5462	−6897	−7611	−12136
2011–2016	−21494	−20786	−21569	−20383
2016–2020	−11165	−11131	−10904	−11464
2006–2020	−38121	−38814	−40084	−43983
2001–2020	−35490	−43246	−48070	−59610

Table 3. Statistics of erosion/deposition intensities in Hukou-Datong reach

Period	Erosion/deposition intensity ($10^4 m^3$/km/a) (deposition: +; erosion: −)			
	Low level	Medium level	Bankfull level	Flood level
1998–2001	−10.25	1.26	6.98	16.83
2001–2006	2.31	−3.89	−7.01	−13.71
2006–2011	−4.79	−6.05	−6.68	−10.65
2011–2016	−18.85	−18.23	−18.92	−17.88
2016–2020	−12.24	−12.21	−11.96	−12.57
2006–2020	−11.94	−12.16	−12.56	−13.78

Table 4. Statistics of proportion of erosion/deposition in flood channel in Hukou-Datong reach

Period	Parameter	Low level	Medium level	Bankfull level	Flood level
2001–2020	Erosion/deposition amount ($10^4 m^3$)	−35490	−43246	−48070	−59610
	Proportion (%)	59.54	72.55	80.64	100.00
2006–2020	Erosion/deposition amount ($10^4 m^3$)	−38121	−38814	−40084	−43983
	Proportion (%)	86.67	88.25	91.14	100.00

As shown in Table 2, 3, 4 and 5 and Fig. 2, the accumulative performance of river bed in Hukou-Datong reach was erosion. However, before the impoundment of TGP, the river bed of Hukou-Datong reach was generally deposited except the low water level channel. After the impoundment of TGP in 2003, the river bed of the Hukou-Datong reach turned into erosion and then presented a unidirectional erosion trend. The specific characteristics were as followed.

i. Shortly before and after operation of TGP (2001–2006), the deposition mainly occurred in channel while the erosion in benchland. Both the amount and intensity of erosion increased with the elevation of water level, indicating that the erosion was mainly in the benchland during that period.

ii. During the initial period of TGP operation (2006–2011), the erosion, which occurred in both channel and benchland, was relatively slight.

iii. During 2011 to 2016, erosion gradually increased with the successively operation of the reservoirs upstream. The amounts and intensities of erosion were very close among 4 different water levels, which indicated that the erosion mainly occurred below the low water level.

iv. After the operation of the main stream reservoir group (2016–2020), erosion rate reduced, but the average erosion intensity was still higher than the pre-2011 level.

v. As shown in Table 4, erosion increased with the increase of calculated water level since 2006. The erosion amount below low, medium and bankfull water level channel accounted for 86.67%, 88.25% and 91.14% of flood water level channel, respectively. The data indicated that erosion occurred both in channel and benchland, and mainly below bankfull water level.

vi. In general, during the 14 years from 2006 to 2020, the average annual erosion intensity below bankfull water level increased by 79.94% compared with the period before the impoundment of TGP (1998–2001).

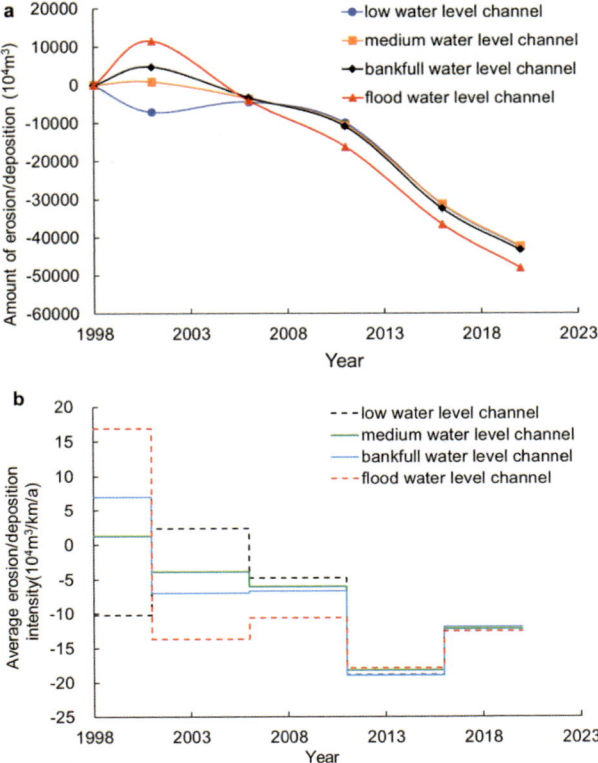

Fig. 2. Erosion/deposition amount and intensity in Hukou-Datong reach (deposition: +; erosion: -) (a. accumulative erosion/deposition amount; b. erosion/deposition intensity)

4.3   Analysis of Calculation Results of Erosion and Deposition Characteristics in Datong-Jiangyin Reach

As shown in Table 5 and 6 and Fig. 3, the performance of river bed in Datong-Jiangyin reach was unidirectional erosion. Before the impoundment of TGP, the river bed of Datong-Jiangyin reach experienced erosion in general. However, the erosion amount decreased with the elevation of water level, indicating that the benchland experienced slight deposition (Table 3, 1998–2001) (Table 6).

Table 5. Statistics of erosion/deposition amounts in Datong-Jiangyin reach

Period	Erosion/deposition amount ($10^4 m^3$) (deposition: +; erosion: −)			
	Low level	Medium level	Bankfull level	Flood level
1998–2001	−18739	−14743	−12653	−12113
2001–2006	−12485	−14357	−15087	−17540
2006–2011	−34029	−34885	−38150	−44446
2011–2016	−24819	−25580	−27109	−36372
2016–2020	−32770	−33926	−33601	−33215
2006–2020	−91618	−94390	−98860	−114033
2001–2020	−104103	−108747	−113947	−131573

Table 6. Statistics of erosion/deposition intensities in Datong-Jiangyin reach

Period	Erosion/deposition intensity ($10^4 m^3$/km/a) (deposition: +; erosion: −)			
	Low level	Medium level	Bankfull level	Flood level
1998–2001	−15.19	−11.95	−10.26	−9.82
2001–2006	−6.07	−6.98	−7.34	−8.53
2006–2011	−16.55	−16.97	−18.56	−21.62
2011–2016	−12.07	−12.44	−13.19	−17.69
2016–2020	−19.92	−20.63	−20.43	−20.19
2006–2020	−15.91	−16.40	−17.17	−19.81

Table 7. Statistics of proportion of erosion/deposition in flood channel in Datong-Jiangyin reach

Period	Parameter	Low level	Medium level	Bankfull level	Flood level
2001–2020	Erosion/deposition amount ($10^4 m^3$)	−104103	−108747	−113947	−131573
	Proportion (%)	79.12	82.65	86.60	100.00
2006–2020	Erosion/deposition amount ($10^4 m^3$)	−91618	−94390	−98860	−114033
	Proportion (%)	80.34	82.77	86.69	100.00

The river bed of the Datong-Jiangyin reach experienced unidirectional erosion, but the characteristics were different from the former period.

i. Shortly before and after operation of TGP (2001–2006), the amount and intensity of erosion increased with the elevation of water level. The erosion, which occurred in both channel and benchland, was relatively slight.

ii. During 2006 to 2011, erosion gradually increased to about 2.5 times of the former period.

iii. With the successively operation of the reservoirs upstream during 2011 to 2016, the erosion rate reduced, but still twice of that during 2001 to 2006.

iv. After the operation of the main stream reservoir group (2016–2020), the erosion sped up once again below all the 4 water levels.

v. As shown in Table 7, the erosion amount increased with the increase of calculated water level since 2006. The data indicated that erosion occurred both in channel and benchland, and mainly below bankfull water level.

vi. In general, during the 14 years from 2006 to 2020, the average annual erosion intensity below bankfull water level increased by 67.35%, compared with the period before TGP impoundment (1998–2001).

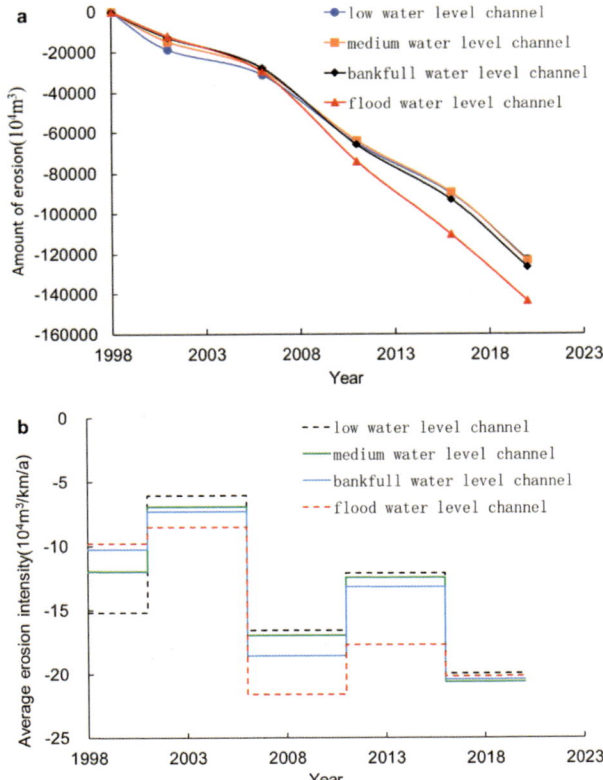

Fig. 3. Erosion/deposition amount and intensity in Datong-Jiangyin reach (deposition: +; erosion: -) (a. accumulative erosion/deposition amount; b. erosion/deposition intensity)

4.4   Temporal and Spatial Variations of Erosion/Deposition Intensity Along LRYR Below Bankfull Water Level

Based on the analysis in the previous section, erosion and deposition mainly occurred below bankfull water level. So, we set the calculation water level to bankfull level and analyzed the temporal and spatial variations of erosion/deposition intensity along LRYR.

The temporal characteristics of erosion/deposition were summarized as below. (Fig. 4).

  i. Before the operation of TGP, the riverbed of the Hukou-Datong reach experienced deposition, while that of Datong-Jiangyin reach experienced erosion. The extent of erosion/deposition was relatively small.

 ii. The riverbed in LRYR performed as slight erosion in general during 2001 to 2006.

iii. The erosion intensity gradually increased since 2006, and increased further with the successively operation of the reservoirs upstream since 2011.

 iv. During 2006 to 2020, the erosion in Hukou-Datong reach was about 1.37 times of that in Datong-Jiangyin reach.

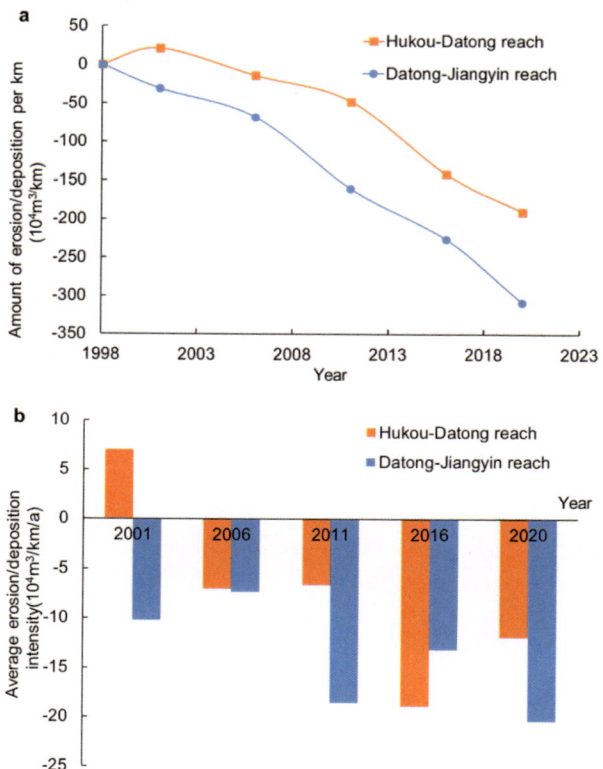

Fig. 4. Amount of erosion/deposition per km and average erosion/deposition intensity in LRYR (deposition: +; erosion: -) (a. amount of erosion/deposition per km; b. average erosion/deposition intensity).

Figure 5 and 6 showed the erosion and deposition of each sub-reach in LRYR below bankfull water level. The characteristics were summarized as blow (Fig. 6).

i. Upstream of Nanjing in LRYR, the annual average erosion/deposition intensity increased and decreased alternately. Accumulatively, the sub-reaches performed as erosion, with erosion intensity ranging from 3.74–26.8 × 10^4 m^3/km, except the Ma'anshan reach, Anqing reach and Taiziji reach.

ii. Three reaches, i.e., the Ma'anshan reach, Anqing reach and Taiziji reach, experienced deposition accumulatively, with relatively small deposition intensities of 8.40 × 10^4 m^3/km, 4.35 × 10^4 m^3/km and 7.39 × 10^4 m^3/km, respectively.

iii. The erosion intensities of Nanjing and its downstream reaches increased along the river, ranging from 7.81 × 10^4 m^3/km to 34.55 × 10^4 m^3/km.

iv. The erosion and deposition intensities of many reaches were close to that of the reaches about 100 km upstream, such as Zhenyang reach and Wuyu reach (126.1km upstream), Wuyu reach and Datong reach (93.3 km upstream), Nanjing reach and Heishazhou reach (80.4 km upstream), Guichi reach and Madang reach (115.9 km upstream), etc.

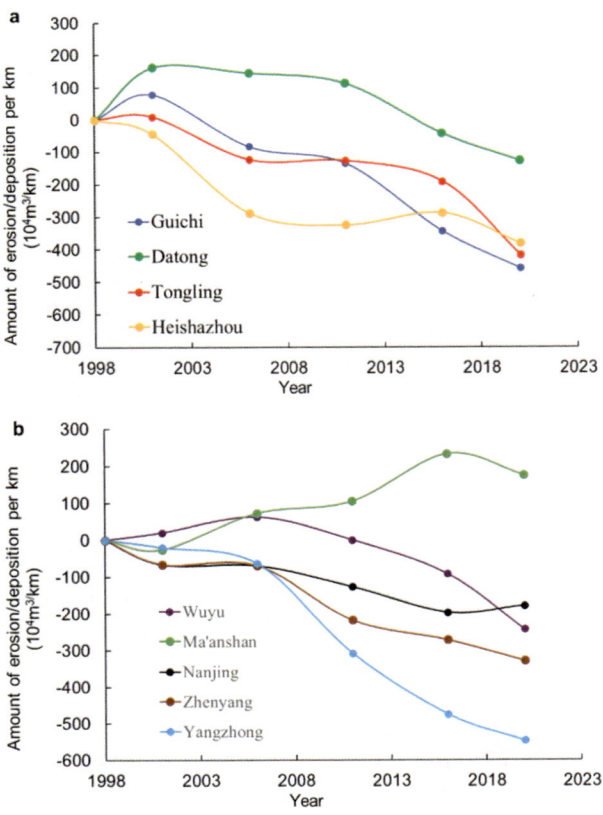

Fig. 5. Amount of erosion/deposition per km of each sub-reach in LRYR (bankfull channel; deposition: +; erosion: −)

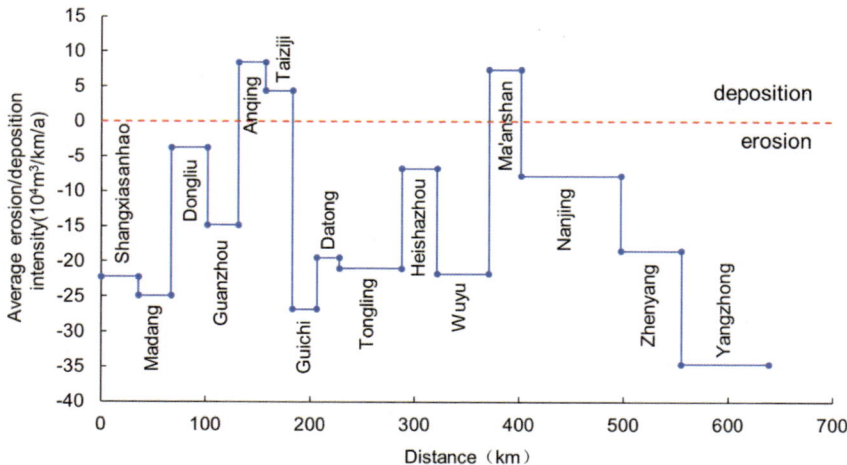

Fig. 6. Average annual erosion/deposition intensity along LRYR during 2006 to 2020 (bankfull channel; deposition: +; erosion: -)

5 Conclusions

i. The relationship between flow and sediment in LRYR has changed significantly since TGP and the reservoir group were put into operation. Annual flow runoff decreased slightly, and the process of runoff showed a reaction of flood peak cutting and valley filling. The sediment runoff decreased sharply, and was significantly planarized in flood season. The sediment runoff of Datong hydrological station has slightly increased during dry season, which was caused by discharge increase and erosion in the upstream reaches during dry season water. However, there was little difference in variation characteristics between Jiujiang hydrological station and Datong hydrological station. Overall, sediment variation far exceeded that of flow.

ii. Hukou-Jiangyin reach experienced both erosion and deposition the during 1998 to 2006, and unidirectional erosion in both benchland and channel after 2006. The average annual erosion intensity was significantly increased over time. The erosion in Hukou-Datong reach was about 1.37 times of that in Datong-Jiangyin reach. Erosion and deposition mainly occurred below bankfull channel, and the proportion of erosion between low water level channel and flood channel was only about 10–20%.

iii. During 2006 to 2020, the sub-reaches performed as erosion accumulatively except for the Ma'anshan reach, Anqing reach and Taiziji reach. Erosion mainly occurred below bankfull channel. The annual average erosion intensity fluctuated along the river upstream of Nanjing in LRYR, while that of Nanjing and its downstream reaches increased along the river. The erosion and deposition intensities of many reaches were close to that of the reaches about 100 km upstream.

Acknowledgements. This work was supported by Nanjing Jiangbei New District Public Engineering Construction Center (Hj221086).

References

Dai ZJ, Liu JT (2013) Impacts of large dams on downstream fluvial sedimentation: an example of the Tree Gorges Dam (TGD) on the Changjiang (Yangtze River). J Hydrol 480:10–18

Li WJ, Yang SF, Xiao Y, Fu XH, Hu J, Wang T (2018) Rate and distribution of sedimentation in the three gorges reservoir, upper Yangtze river. J Hydraul Eng 144(8):05018006

Li WJ, Yang SF, Fu XH, Xiao Y (2015) Sedimentation characteristics in the three Gorges reservoir during the initial operation stage. Adv Water Sci 26(5):676–685 (in Chinese with English abstract)

Shi CL, Niu LH, Zhao GL, Du LX (2020) Variation in water and sediment conditions and erosion and deposition characteristics in the reach between three Gorges dam and Gezhou dam. Adv Water Sci 31(6):875–884 (in Chinese with English abstract)

Xu QX, Yuan J, Wu WJ, Xiao Y (2011) Fluvial processes in middle Yangtze river after impoundment of three Gorges project. J Sedim Res 2:38–46 (in Chinese with English abstract)

Zhu LL, Xu QX, Chen ZH (2018) Extraordinary scour of Jingjiang reach downstream from Three Gorges project. J Basic Sci Eng 26(1):85–97 (in Chinese with English abstract)

Analysis of River Stability in the Middle Reaches of Huaihe River Based on Non-equilibrium Thermodynamicsins

Yu Duan[1,2](✉) and Guobin Xu[2]

[1] Tianjin Research Institute for Water Transport Engineering,
M.O.T, Tianjin, China
sqwu@nhri.cn
[2] State Key Laboratory of Hydraulic Engineering Simulation and Safety,
Tianjin University, Tianjin, China

Abstract. River stability is an important attribute of a river, which includes river pattern stability and river bed stability. The stability of the middle reaches of Huaihe River is one of the important problems concerned by the workers in Huaihe River regulation. The study of the stability of the middle reaches of Huaihe River is of great significance to the river regulation planning and flood prevention and control. To explore the stability of the middle reaches of Huaihe River, the research combined with hydrological data, trying to base on the theory of non-equilibrium thermodynamics system to determine the stability of river pattern, and using the unit stream power calculation formula to analyze the river stability. The research show that, the middle reaches of Huaihe River from Zhengyangguan to Fushan, the river pattern of each section is in a stable state, there is no possibility of conversion in the short time. The variation amplitude of unit stream power in each reach tends to decrease, the natural evolution of the riverbed is also in a stable state. Through the research, the applicability of the river stability analysis method based on the non-equilibrium thermodynamics theory in the Huaihe River is verified, and formed a set of analysis methods suitable for the stability judgment and development evolution trend of the Huaihe River. In this study, the stability of the middle reaches of the Huaihe River was judged theoretically, and the adjustment direction of the river and the evolution trend of the river bed were predicted.

Keywords: Middle reaches of Huaihe river · Excess entropy production · Unite stream power · River pattern · Stability judgment

1 Introduction

Huaihe River is located in the east of China, it is one of the seven great rivers of China, the total length is about 1000 km, with a total drop of 200 m. Historically, Huaihe River was a river flowing into the sea directly, later, the Yellow River encroached on the Huaihe River's waterway, diverting it into the Yangtze River, then the elevation of the ground formed Hongze Lake. Under the jacking of Hongze Lake, the water level of the middle Huaihe River rises, which seriously reduces the flood control and waterlogged elimination capacity. In addition, the middle Huaihe River stability is also a problem to

Y. Li et al. (Eds.): PIANC 2022, LNCE 264, pp. 1030–1040, 2023.
https://doi.org/10.1007/978-981-19-6138-0_91

be concerned about. River stability is an important property of a river, the study is of great significance to the regulation planning and flood disaster prevention of the middle Huaihe River. To explore the stability characteristics of the middle reaches of Huaihe River, system stability criterion of nonequilibrium thermodynamic theory is used to analyze river stability.

River stability includes river pattern stability and riverbed stability. River pattern stability means that under the influence of incoming water and sediment and riverbed boundary conditions, the river pattern will not change in the long run. And the riverbed stability mainly refers to the temporary, local, relative variation in the watershed development process, it is usually characterized by the riverbed stability coefficient.

In the existing river stability analysis research, they usually use the methods of measured data analysis and theoretical empirical formulas. For Huaihe River stability analysis, Yang et al. (2010) analyzed the evolution process of the Huaihe River from Bengbu to Fushan. Yu et al. (2011) taked Bengbu to Fushan reach as an example, analyzed the incoming water and sediment, bed-building discharge and riverbed stability coefficient. For other rivers, Tian et al. (2012) used the fractal theory to analysis the evolution of the lower Yellow River. Batalla et al. (2017) studied the morphology of the Ñuble River channel changing over the years based on aviation images, and quantitatively analyzed the channel morphology index. But these methods are all based on a large number of hydrological and topographic measured data, more detailed hydrological parameters need to be collected before the study. However, subject to technical conditions, some hydrological data are difficult to be collected. Thus, some scholars have applied the energy consumption rate extremum principle to river system based on basic hydrological data. Chang (1979) applied the energy consumption rate extremum to study the river pattern. Zhao and Xu (2015) based on the principle of super entropy generation, judged the stability of the river pattern in the lower Yellow River. Xu et al. (2016) applied minimum energy consumption rate principle, calculated the unit stream power of 3 river patterns in 6 reaches of the lower Yellow River. The existing studies have verified the feasibility of non-equilibrium thermodynamics in the river stability analysis.

Most of the studies on stability analysis of Huaihe River are based on historical measured data and based on traditional analysis methods. It is one of the problems worth studying to analyze river stability based on other theories and predict river development and evolution according to the analysis results. This paper is based on nonequilibrium thermodynamics theory, combined with hydrological data, using river pattern stability criterion and unit stream power calculation formula, analyzing the river pattern stability and riverbed stability of the middle reaches of Huaihe River.

2 Materials and Methods

2.1 Study Area and Data Description

The monthly flow time series data of Zhengyangguan, Bengbu and Fushan hydrological stations in the middle Huaihe River from 1985 to 2018 (408 months) are collected, the river reach between two adjacent hydrological stations is taken as the research object, each river reach is studied as a relatively independent system. Sine there are no major tributaries in the studied reaches, therefore, monthly series of river flow, water depth, river width, sediment concentration in the calculation are all used the average value of the inlet and outlet section data. The river surface slope taken as the ratio of water level difference between inlet and outlet section to river length. Figure 1 is the channel diagram of middle Huaihe River from Zhengyangguan to Fushan.

When analyze the river stability, using the monthly series data from 1985–2018 (408 months), quantitative judgment the river stability of Zhengyangguan-Bengbu (Zheng-Beng) reach and Bengbu-Fushan (Beng-Fu) reach respectively.

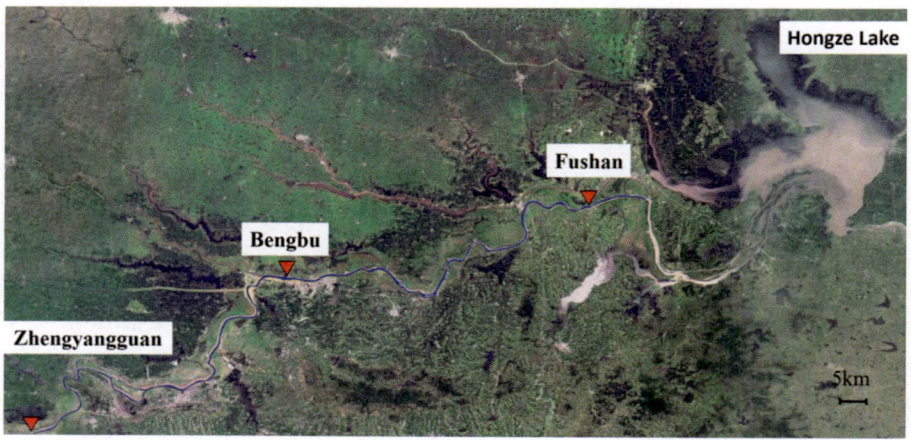

Fig. 1. Schematic diagram of the middle reaches of Huaihe River from Zhengyangguan to Fushan reach

2.2   River Stability Analysis

2.2.1   Channel Pattern Discrimination

Before the channel stability analysis, it is necessary to discrimination the channel pattern of the middle Huaihe River. As for the river type determination of plain rivers, scholars had carried out in-depth studies (Leopold,1957; Lane, 1957) and it showed that some discriminant methods are not applicable to the Huaihe River basin. In this paper, the channel pattern is discriminated by the method summarized by Leopold (1957). According to the study, the river development trend is determined by the river slope and river flow. Leopold counted nearly 50 rivers in the United States and India, and statistics the relationship between river slope and bankfull discharge, then a famous river pattern criterion for the relation between river slope and bankfull discharge is found, namely

$$\bar{J} = 0.0125 Q_d^{-0.44} \tag{1}$$

where, $\bar{J}$ is river slope, Q_d is bankfull discharge.

2.2.2   River Pattern Stability Analysis

The nonequilibrium thermodynamics principle is applied to open system, with the change of external constraints, the equilibrium state of an open system will deviate, while the lyapunov function can be used to judge the system stability. When the open system is in equilibrium, the lyapunov function of the system is entropy, this symbol can be used to determine the system stability. When the open system is in nonequilibrium, the lyapunov function of the system is entropy production, the results of entropy production compared with zero are used to judge the system stability(Nicllis and Prigogine, 1977). For an open system, the stability criteria can be applied, and the super entropy production symbol $\delta_X P$ in Eq. (2) is used to judge the system stability.

$$\begin{cases} \delta_X P > 0 & \text{System stability} \\ \delta_X P = 0 & \text{Critical stability} \\ \delta_X P < 0 & \text{System instability} \end{cases} \tag{2}$$

River act as an open system, also follow the open system stability criteria. Based on open system stability quantitative criterion Eq. (2), through theoretical derivation, obtained the channel pattern stability criterion Eq. (4) (Zhao and Xu, 2015).

$$\begin{cases} \frac{u}{g} \cdot \frac{du}{dl} < J & \text{River pattern stability} \\ \frac{u}{g} \cdot \frac{du}{dl} = J & \text{River critical stability} \\ \frac{u}{g} \cdot \frac{du}{dl} > J & \text{River pattern instability} \end{cases} \tag{3}$$

where, u is the average velocity, g is gravitational acceleration, $\frac{du}{dl}$ is the inlet and outlet flow rate of a reach varies with the river length, J is hydraulic gradient.

2.2.3   River Bed Stability Analysis

It can be known from the principle of minimum entropy production in nonequilibrium thermodynamics (Yang and Song 1979), the evolution of an open system is always in the direction of decreasing entropy production under the corresponding constraints, until the system reaches a nonequilibrium stationary state which is suitable for the constraint conditions, the entropy production of the system must be the minimum this moment. Based on the principle of minimum entropy production is equivalent to the principle of minimum energy dissipation rate, and combine the function expression of energy dissipation rate, deriving the mathematical expression of the principle of minimum energy dissipation rate (Xu and Lian, 2003a; Xu and Lian, 2003b). The mathematical expression of river energy dissipation rate per unit length can be expressed as:

$$\phi_l = \gamma Q J \tag{4}$$

where, Φ_l is energy dissipation rate per unit length, γ is water density, Q is river flow, J is hydraulic gradient.

According to Eq. (4), the expression of unit stream power can be deduced as(Yang, 1979):

$$\phi_N = uJ \tag{5}$$

where, Φ_N is unit stream power, u is average stream velocity, J is hydraulic gradient.

3   Application and Results

3.1   River Pattern

Drawing the coordinate points of bankfull discharge Q_d and rive slope $\overline{J}$ of Zheng-Beng and Beng-Fu in Fig. 2. It can be found that the coordinate points of the two river reaches are located in the meandering river reach. Therefore, the study river is determined as meandering river.

Through the research, the middle Huaihe River is mostly narrow and deep, and the water depth is increased by the jacking of Hongze lake. Under these conditions, the middle Huaihe River is seriously affected by the circumfluence, which is conducive to the river develop into the meandering river. However, restricted by the boundary conditions of the channel, the curved channel has not been fully developed.

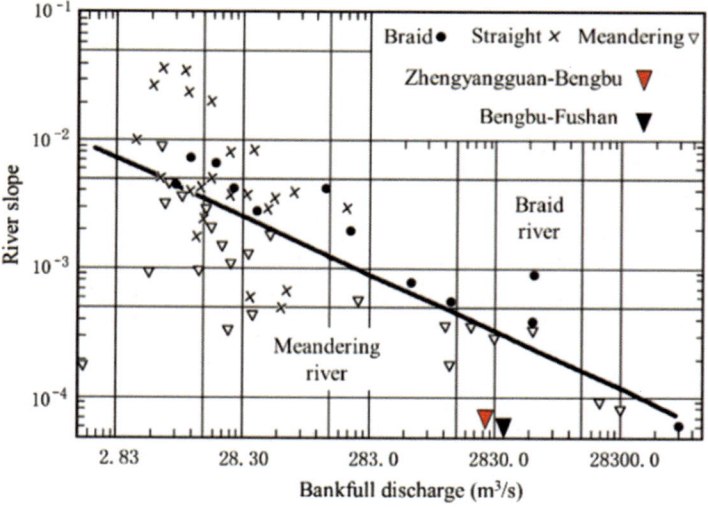

Fig. 2.   The relationship between Q_d and J in different river reaches

3.2   River Pattern Stability

According to the river stability criterion Eq. (3), the $\frac{u}{g} \cdot \frac{du}{dl}$ of Zheng-Beng reach and Beng-Fu reach from 1985 to 2018 are calculated respectively, plot the curve over time and add a linear trend line in Fig. 3. The calculated results show that the $\frac{u}{g} \cdot \frac{du}{dl}$ of the two river reaches are less than the specific drop J of the corresponding river reaches (The average ratio of zhengyangguan to Bengbu is 0.0364%, and that of Bengbu to Fushan is 0.0043%), which indicated that the river pattern of the two river reaches are in a stable state at present.

At the same time, the trend test of M-K nonparametric statistics for $\frac{u}{g} \cdot \frac{du}{dl}$ for two reaches was carried out, the statistical analysis results showed as Fig. 4. Among them, the trend statistic Z of Zheng-Beng reach is -4.18, which is less than 0 and exceeds the critical value -2.58. It indicates that the variation of $\frac{u}{g} \cdot \frac{du}{dl}$ for Zheng-Beng reach presents an abnormally significant decreasing trend. However, the trend statistic Z of the Beng-Fu reach is 3.53, which is greater than 0 and exceeds the critical value 2.58, indicating that the change of $\frac{u}{g} \cdot \frac{du}{dl}$ in the Beng-Fu reach shows an abnormal and significantly increasing trend.

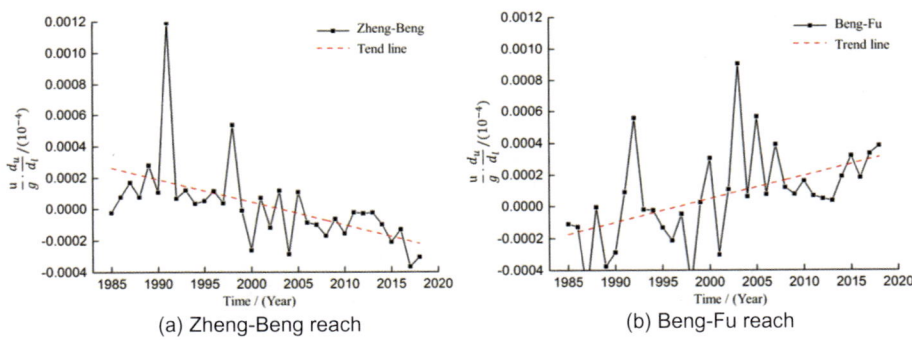

(a) Zheng-Beng reach   (b) Beng-Fu reach

Fig. 3. Variation trend of $\frac{u}{g} \cdot \frac{du}{dl}$ from Zhengyangguan to Fushan 1985–2018

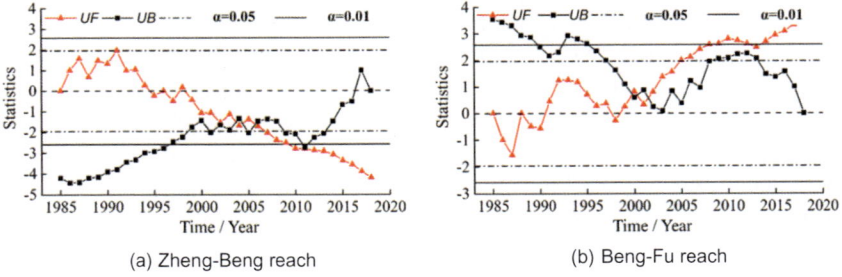

(a) Zheng-Beng reach   (b) Beng-Fu reach

Fig. 4. Variation of $\frac{u}{g} \cdot \frac{du}{dl}$ M-K statistics in different river reaches from 1985 to 2015

The factors affecting river pattern transformation mainly include natural factors and human factors, such as climate change, water and sediment inflow conditions, water conservancy projects construction and river control measures[Radoane, et al. 2013; Abate, et al. 2015; Morais, et al. 2016; Joshi and Jun, 2015]. According to the calculation results, the $\frac{u}{g} \cdot \frac{du}{dl}$ in Zheng-Beng reach shows a decreasing trend, which indicates that the river pattern of Zheng-Beng reach will not change in a short period of time. Although the $\frac{u}{g} \cdot \frac{du}{dl}$ in Beng-Fu reach shows a gradually increasing trend, but it is much smaller than the river slope, which indicates that the channel pattern is still in a stable state.

On the whole, the construction and control projects had a good effect on the river pattern control in the middle reaches of Huaihe River, but some of the river channels in the Bengbu to Fushan reach are affected by artificial sand mining, which will affect the river pattern stability to a certain extent. Therefore, the manual sand mining operation should be standardized and guided.

3.3   River Bed Stability

According to the Eq. (5), calculating the unit stream power of Zheng-Beng and Beng-Fu reaches from 1985 to 2018 respectively. Drawing the curve of uJ changing with time, and adding the linear trend line, as shown in Fig. 5.

The unit stream power of the two river reaches in the study shows serrated change with time due to the change of the inflow and sediment conditions and the river governance projects in different periods. This is because the stability of river bed is a dynamic erosion-deposition equilibrium, even when the unit stream power reaches the minimum value, it still fluctuates around the mean value of the minimum value. The linear trend lines of unit stream power show that the unit stream power of these two reaches decreases gradually with time, and the slope values of the trend lines are -0.00128 and -0.00063, respectively. The five-year mean variation line of unit stream power shows that the mean value of these two river reaches fluctuate with small increase or decrease every five years. Because the unit stream power will change with the change of external conditions, but from the point of long period of time. If the linear trend line per unit stream power is basically parallel to the horizontal axis, that means the riverbed was stable during the study period. While, the slope of the linear trend line of unit stream power of Zheng-Beng and Beng-Fu is very small. Therefore, it can be judged that the bed of the two reaches is in a stable state.

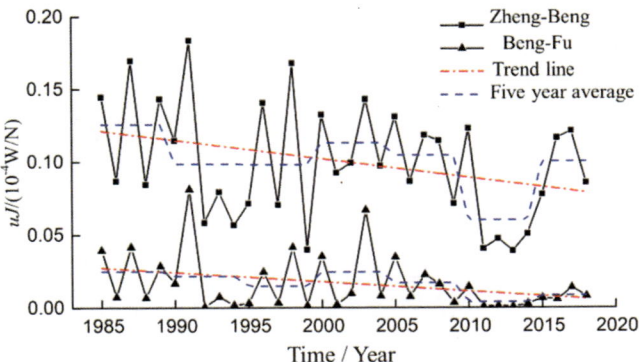

Fig. 5. Unite stream power of *uJ* from Zhengyangguan to Fushan 2005–2018

In order to explore whether there is a significant mutation point in the fluctuation of time series of unit stream power, the time series cumulative values of unit stream power time seires in different river sections are analyzed, and the change curve is obtained as shown in Fig. 6. The accumulative unit stream power values of Zheng-Beng and Beng-Fu are in good agreement with the trend line. It shows that the time series of unit stream power is in a relatively stable development and change, and there is no significant mutation point. Therefore, the fluctuations shown in the curves in Fig. 5 are all steady state fluctuations.

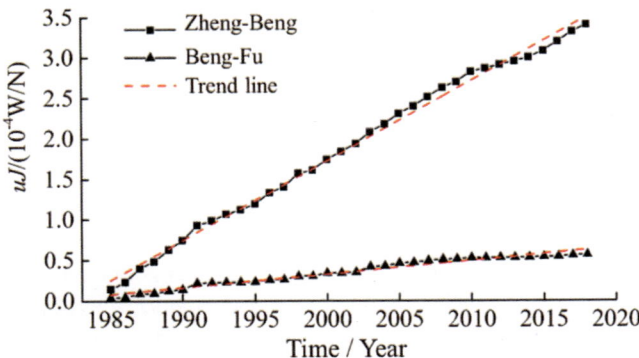

Fig. 6. Changing of time series accumulative value of unit stream power *uJ* in different reach

In order to explore the variation rule of unit stream power time series in these two reaches, M-K non-parametric statistical method was used for trend analysis and mutation point analysis, and the results of calculation and analysis were shown in Fig. 7. The trend statistics Z of unit stream power time series of Zheng-Beng and Beng-Fu reaches are −1.81 and −1.63, respectively, which are less than 0 but not exceeding the critical value −1.98. It is indicated that unit stream power does not decrease significantly with time. Meanwhile, no significant mutation point was found in the time series of unit stream power. The results of calculation and analysis are in agreement with the results of the unit stream power trend line and the time series cumulative curve.

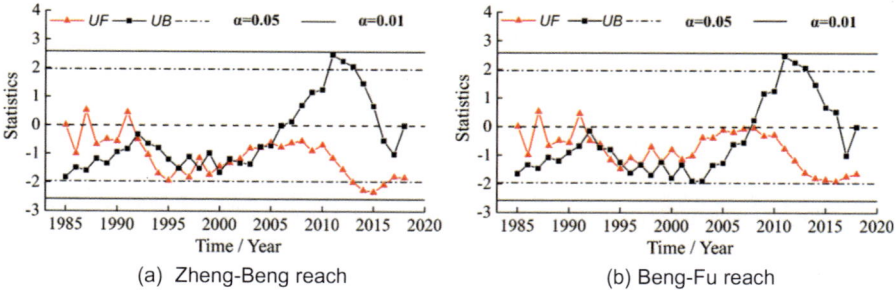

(a) Zheng-Beng reach   (b) Beng-Fu reach

Fig. 7. Variation of uJ M-K statistics of unit stream power in different reaches

Because the riverbed boundary of the middle reaches of the Huaihe River is mainly developed on the hard clay geological layer, and some reaches are also covered with soft sediments with a certain thickness, the riverbed erosion was restricted by the boundary conditions of the riverbed in the process of river evolution, so the riverbed of the middle reaches of the Huaihe River is maintained in a relatively stable state.

Based on the above analysis process and riverbed geology, it can be seen that the riverbed of Zheng-Beng reach is basically in a stable state and is slowly adjusting to a more stable direction. However, the riverbed of Beng-Fu reach has been in a stable state.

4 Conclusions

The river pattern stability is quantitatively determined by the super entropy generation discriminant, it is found that the river pattern in Zheng-Beng reach and Beng-Fu reach are both stable, and there is no river pattern transformation possibility in the near future. The river bed stability is quantitatively determined by using the unit stream power equation. The unit stream power in Zheng-Beng reach and Beng-Fu reach present a stable fluctuation state tending to decrease. It shows that although the incoming water and sediment conditions or riverbed geometry changing continuously during the studied years, but the riverbed evolution tends to be stable. By using the information entropy theory of river channel, it can be known that the influence of the boundary conditions on the riverbed evolution is greater than that of the incoming water and sediment. Therefore, strengthing the control of riverbed boundary conditions, especially the width depth ratio controlling, will more conducive to control the riverbed evolution.

The river pattern and riverbed evolution in the middle Huaihe River are in a stable state, and the results determined by the theories of nonequilibrium thermodynamics are consistent with the actual situation, which verifies the scientific of the method adopted, and proposed a new idea for exploring the basic characteristics of the river.

Acknowledgements. The study is financially supported by the National key R & D program of China (2017YFC0405602). The authors gratefully acknowledge the editors and the reviewers for their insightful and professional comments.

References

Abate M, Nyssen J, Steenhuis TS et al (2015) Morphological changes of Gumara river channel over 50 years, upper BlueNile basin, Ethiopia. J Hydrol 525:152–164

Batalla RJ, Iroumé A, Hernández M et al (2017) Recent geomorphological evolution of a natural river channel in a Mediterranean Chilean basin. Geomorphology 303:322–337

Chang HH (1979) Minimum stream power and river channel patterns. J Hydrol 41(3–4):303–327

Joshi S, Jun XY (2018) Recent changes in channel morphology of a highly engineered alluvial river-the lower Mississippi river. Phys Geogr 39(2):140–165

Lane EW (1957) A study of the shape of channels formed by natural streams flowing in erodible materia. U.S. Army Corps of Engineers, Missouri River Division, Sediment Series 9

Leopold LB, Wolmam MG (1957) River channel pattern: braided, meandering and straight. Palgrave Macmillan Publications, London

Morais ES, Rocha PC, Hooke J (2016) Spatiotemporal variations in channel changes caused by cumulative factors in a meandering river: the lower Peixe River, Bra-zil. Geomorphology 273:348–360

Nicolis G, Prigogine I (1977) Self-organization in non-equilibrium systems. Wiley Publications, New York, U.S.

Radoane M, Obreja F, Cristea I et al (2013) Changes in the channel-bed level of the eastern Carpathian rivers: climatic vs. human control over the last 50 years. Geomorphology 193:91–111

Tian SM, Su XH, Wang WH et al (2012) Application of fractal theory in the river regime in the lower yellow river. Appl Mech Mater 190–191:1238–1243

Xu GB, Zhao LN (2016) Yang CT (1957) Derivation and verification of minimum energy dissipation rate principle of fluid based on minimum entropy production rate principle. Int J Sediment Res 31(1):16–24

Xu GB, Lian JJ (2003) Theories of minimum rate of energy dissipation and the minimum entropy production of flow (I). J Hydraul Eng 5:35–40 (In Chinese)

Xu GB, Lian JJ (2003) Theories of minimum rate of energy dissipation and the minimum entropy production of flow(II). J Hydraul Eng 6:43–47 (In Chinese)

Yang CT (1976) Minimum unit stream power and fluvial hydraulics. J of Hydraul Div 102 (7):919–934

Yang CT, Song CCS (1979) Theory of minimum rate of energy dissipation. J Hydraul Div 105 (7):769–784

Yang XJ, Yu BY, Ni J (2010) Analysis on recent evolution of main stream of reach from Bengbu to Fushan of Huaihe river. Water Resour Hydropower Eng 41(10):70–72+86 (In Chinese)

Yu YQ, Zhang XJ, Zhang YS (2011) Analysis on flow change trend in Bengbu to Fushan section of mainstreams of the Huaihe river. Water Resour Plan Des 01:19–22 (In Chinese)

Zhao LN, Xu GB (2015) Discriminant of stability for channel pattern based on excess entropy production, J Hydraul Eng 46(10):1213–1221+1232. (In Chinese)

Application of Environmentally Active Concrete (EAC) for River Structure

Naozumi Yoshizuka[1]([✉]), Tomihiro Iiboshi[2], Hirokazu Nishimura[3], and Daisuke Kawashima[4]

[1] International Division, Nikken Kogaku Co., Ltd., Tokyo, Japan
yoshizuka@nikken-kogaku.co.jp
[2] Engineering Division, Nikken Kogaku Co., Ltd., Tokyo, Japan
iiboshi@nikken-kogaku.co.jp
[3] Marketing Division, Nikken Kogaku Co., Ltd., Tokyo, Japan
nishimura@nikken-kogaku.co.jp
[4] Environmental Symbiosis Research Division, Nikken Kogaku Co., Ltd.,
Tokyo, Japan
kawashima@nikken-kogaku.co.jp

Abstract. An increase in floods due to climate change has concern for causing enormous erosion damage. Concrete is demanded to play an even more important role in the development of river structures for disaster prevention and mitigation. In Japan, river structures using concrete have been actively developed to protect river banks and riverbeds from floods. Under these circumstances, in the River Act partial amendment in 1997, in addition to the previous concept of flood control and water utilization, the concept of the environment (improvement and conservation of river environment) was incorporated. Therefore, various efforts have been made, such as developing products and methods that utilize natural materials other than concrete and researching new environmentally friendly materials. The authors have developed "Environmentally Active Concrete (hereinafter referred to as EAC)" with environmental functions by mixing arginine, one of the amino acids, into the concrete and have put it into practical use. Demonstration experiments have confirmed the environmental performance of EAC in rivers. In river structures where EAC has been applied, the effects of promoting the growth of attached algae and habitat conservation for various organisms such as sweetfish, Japanese eel and Japanese giant salamander have also been confirmed. Utilizing EAC will make it possible to achieve river structures with both disaster prevention and environmental conservation functions.

Keywords: Environmentally Active Concrete · Amino acids · Working with nature · River and lake protection · Disaster prevention

Y. Li et al. (Eds.): PIANC 2022, LNCE 264, pp. 1041–1051, 2023.
https://doi.org/10.1007/978-981-19-6138-0_92

1   Introduction

The IPCC 6th Report (AR6) Working Group I (WG1) Report Summary for Policy-makers (SPM), released in August 2021, determined that it is "unequivocal" that human activities influence global warming. It also reported that global surface temperature will exceed 1.5 °C and 2.0 °C during the 21st century compared to pre-industrial levels unless deep reductions in CO_2 and other greenhouse gas emissions occur in the coming decades. There is a concern that the increases in annual mean precipitation, the rate of occurrence of extremely intense tropical cyclones, and peak wind speeds due to the rise in global surface temperature cause enormous erosion damage to riverbanks and lakeshores. In order to preserve the inland waterway and achieve sustainable operations of inland ports, countermeasure works to protect riverbanks and lakeshores are important.

In Japan, where heavy rains occur every year, especially during the rainy season and typhoon season, river structures such as revetment works, foot-protection works and riverbed protection works have been actively developed to protect riverbanks and riverbeds from floods. In the River Act partial amendment in 1997, in addition to the previous concept of flood control and water utilization, the concept of the environment (improvement and conservation of river environment) was incorporated. Under the circumstances, various efforts have been made, such as developing products and methods that utilize natural materials other than concrete and researching new environmentally friendly materials.

The authors have developed "Environmentally Active Concrete (hereinafter referred to as "EAC")" with environmental functions by mixing arginine, one of the amino acids, into the concrete and have put it into practical use. This paper introduces demonstration experiments on the environmental performance of EAC in rivers and specific application examples to river structures in Japan.

2   Environmentally Active Concrete (EAC)

EAC is a new concrete mixed with arginine, one of the amino acids, developed by Sato et al. (2011) and Nishimura et al. (2014). We have confirmed that arginine that slowly elutes from the concrete surface over a long period promotes the growth rate of microalgae and small-sized seaweeds attached to the concrete surface. EAC with attached microalgae is expected to contribute to the creation and improvement of habitat environment for various organisms and improvement of biodiversity, such as being used as habitats and feeding grounds for fish and benthos. In Japan, EAC has been applied to more than 500 projects. In 2018, the demonstration experiment at the Port of Wajima using EAC was certified by PIANC Working with Nature (Photo 1).

Photo 1. EAC Demonstration experiment in the construction of the 6th breakwater at the port of Wajima (PIANC Working with Nature Certificate of Recognition)

3 Demonstration Experiment of EAC in Rivers

3.1 Growth-Promoting of EAC in Rivers

In order to verify the growth-promoting effect of attached algae in rivers, a demonstration experiment was conducted in the middle basin of the Fushinogawa River, where sweetfish are naturally distributed (Photo 2). In the experiment, an EAC block mixed with 10% arginine by cement weight ratio (45 cm x 18 cm x 18 cm, hereinafter referred to as EAC), an ordinary concrete block of the same shape (hereinafter referred to as OC) and a natural stone collected from the river and stripped of attached algae on the surface (approx. 30 cm diameter) were used as test specimens for measurements. The amount of attached algae was evaluated by stripping off the adhering substances from within a 5 cm x 5 cm frame on the surface of each test specimen (2 locations/test specimen) and measuring the amount of chlorophyll-a.

Figure 1 shows the measurement results of the amount of chlorophyll-a in each test specimen. The amount of chlorophyll-a on the EAC surface is 1.7 to 2.2 times higher than that on the OC surface both 13 days and 41 days after installation. A comparison of EAC and natural stone shows similar values for 13 days and 41 days after installation. These results indicate that EAC can be expected to have the growth-promoting effect of attached algae equal to or higher than natural stone.

Fig. 1. Survey point of Fushinogawa river

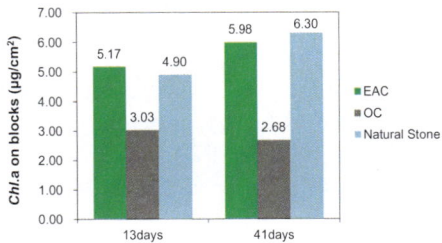

Photo 2. Comparison of chlorophyll-a on test piece and natural stone surfaces

3.2   Gathering Effect on Sweetfish

In order to verify the gathering effect of EAC on sweetfish, a demonstration experiment was conducted in an outdoor water tank (area of 60 m^2, water depth of 0.5 m) using water taken from the Fushinogawa River. Similar test specimens of EAC and OC as 3.1 were installed in the outdoor water tank. Two weeks later, 300 sweetfish were released, and the feeding times of sweetfish were observed with an underwater camera (Photo 3).

Figure 2 shows the measurement result of the feeding times of sweetfish for EAC and OC. The feeding times of sweetfish were remarkably higher for EAC in both the first and second times, with an average of 21.5 times/minute for EAC and 0.5 times/minute for OC. Since sweetfish have a taste response to arginine, we consider that many sweetfish gathered in EAC due to the taste and smell response to arginine eluted from concrete or arginine contained in algae.

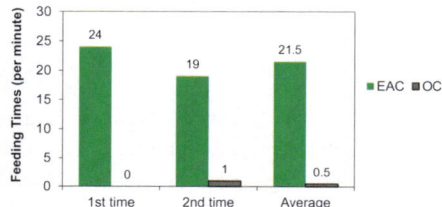

Photo 3. The moment when sweetfish feed on algae

Fig. 2. Comparison of feeding times by sweetfish

3.3   Gathering Effect on Japanese Eel

For verifying the gathering effect of EAC on Japanese eels, a demonstration experiment was conducted in the outdoor water tank used in 3.2. In the experiment, a total of four sets of five vinyl chloride pipes (7 cm diameter x 100 cm long) were prepared, of which two sets contained EAC (50 mm diameter x 50 mm long) and the remaining two sets contained OC (same shape as EAC) and installed in the outdoor water tank. Two weeks later, 10 Japanese eels were released into the outdoor water tank and the vinyl chloride pipes were collected at intervals of about two days to observe the gathering condition of the Japanese eels.

Photo 4 shows the gathering condition of the Japanese eels, and Fig. 3 shows the measurement results of the total number of Japanese eels in each pipe. As shown in Photo 4 and Fig. 3, Japanese eels gradually began to gather in the pipe where the EAC was installed as the day progressed. Similar results were also obtained by conducting experiments in which the installation positions of EAC and OC were replaced. As with the sweetfish mentioned above, we consider that Japanese eels also have a smell response to arginine eluted from EAC.

Photo 4. Gathering condition of Japanese Eels

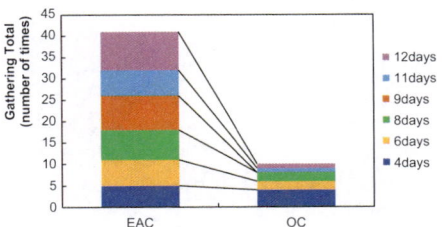

Fig. 3. Cumulative number of Japanese Eel gathering

4   Application Example of EAC for River Structures

4.1   Application Example for Riverbed Protection Work

At Katsuuri weir, which is an important intake facility for agricultural water in the Kinugawa River, Tochigi Prefecture, the riverbed protection work using concrete armor blocks (STONE-BLOCK 4-ton) are installed in the downstream area. At this site, four EAC panels (28 cm x 15 cm x 13 cm), which are expected to have an attractive effect on sweetfish, were attached to each of the 151 concrete armor blocks downstream area of the fishway facility in order to guide the sweetfish that run-up the Kinugawa River to the fishway attached to the riverbed protection work (Photo 5 and 6).

One month after the blocks were installed, many attached algae were observed growing on the EAC panels (Photo 7). As shown in Fig. 4, we also observed that the density of bite marks of sweetfish was up to 20 places/100 cm^2 downstream on the right bank side where the fishway is located. In addition, many sweetfish were found in the fishway (Photo 8).

We consider that sweetfish were guided to the fishway to run-up without straying into areas other than the fishway, based on the condition of the bite marks of sweetfish around the fishway. In addition to the effect of the EAC, the installation of a "Guide-flow waterway" that generates a strong current downstream entrance of the fishway is also thought to be a factor to improve the induction effect for sweetfish into the fishway (Fig. 4).

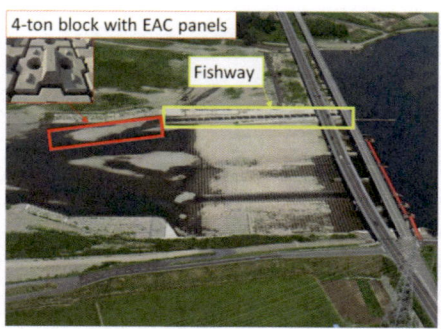

Photo 5. Aerial photograph of Katsuuri weir

Photo 6. Full view of the fishway (From downstream to upstream)

Photo 7. Attached algae on EAC panel

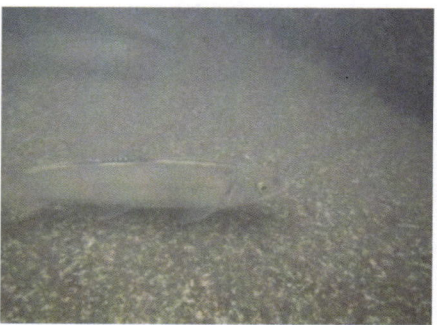

Photo 8. Sweetfish in the pool for fishway

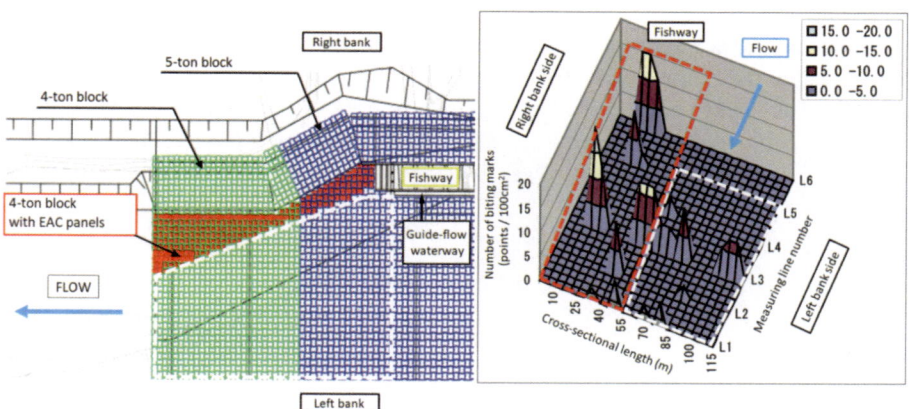

Fig. 4. Distribution chart of sweetfish bite marks

4.2   Application Example for Spur Dikes

In the Waragawa River, Gifu Prefecture, due to the restoration work of the revetment damaged by the heavy rain in July 2018, spur dikes with foot-protection blocks are being maintained. The Waragawa River is famous as a clear stream inhabited by many Japanese giant salamanders, which are designated as a special natural treasure. There is a locality that sweetfish grown in this river is also popular as a luxury food. Based on this background, 4-ton blocks (GASSHO-BLOCK) with high porosity and EAC panels (28 cm x 15 cm x 13 cm) were applied to the spur dikes aiming the habitat conservation for Japanese giant salamanders and sweetfish.

Six blocks with EAC panels were installed in each of 11 of the 19 total spur dikes (Fig. 5). The blocks with EAC panels were expected to attract fish by placing one upstream and one downstream of the center side of the stream at the first layer, where the running water always hits without stagnation (Fig. 6). In addition, for the second layer, the gathering effect was expected by using all four blocks with EAC panels.

Three months after the installation of the spur dikes, for each 4-ton block with EAC panels installed in spur dikes No.9, 11, and 15, the conditions of attached algae on the surface of the EAC panels (hereinafter referred to as EAC) and the surface of the ordinary concrete block (hereinafter referred to as OC) were compared. In each block, more attached algae were confirmed in EAC than in OC (Photo 9). Furthermore, the adhering substances in the 7 cm diameter frame on the EAC and OC surfaces were stripped off, and the amount of chlorophyll-a was measured and evaluated.

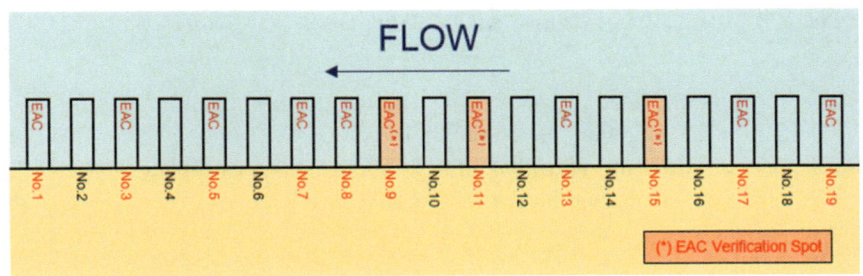

Fig. 5. Layout drawing of spur dikes (EAC applied to 11 out of 19 total spur dikes)

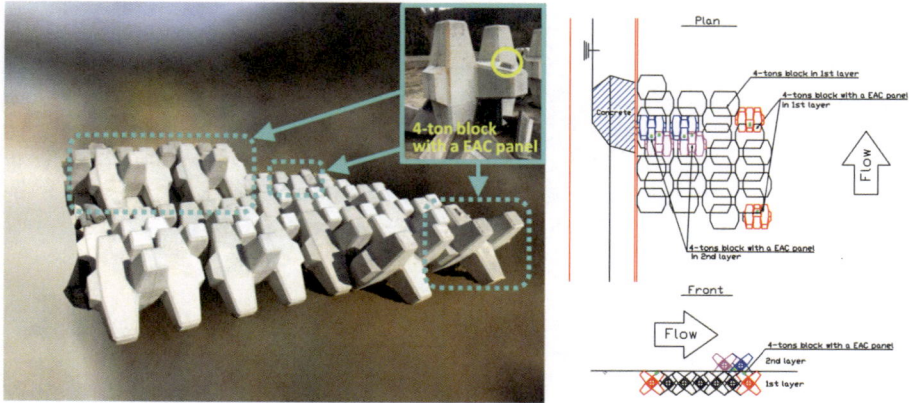

Fig. 6. Installation of 4-ton blocks with EAC panels

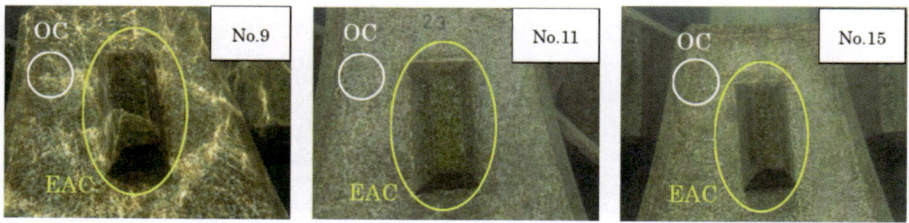

Photo 9. Attached algae on EAC and OC (Spur Dikes No.9, No.11, and No.15)

Figure 7 shows the measurement results of the amount of chlorophyll-a in the attached algae. The amount of chlorophyll-a was 0.12 μg/cm^2 for EAC and 0.035 μg/cm^2 for OC, confirming that EAC is about 3.4 times higher than OC. The number of appeared species and individuals of attached algae are shown in Fig. 8. The number of species was 30 in both EAC and OC. In regard to the composition of the species by class, the number of species of Bacillariophyceae was the most common in both EAC and OC, and the number of individuals was large in both Bacillariophyceae and Cyanophyceae. Among them, the Bacillariophyceae and Homoeothrix janthina of Cyanophyceae, the dominant species for both EAC and OC, are mainly used as food for sweetfish. The number of individuals was 147,863 for EAC and 49,984 for OC, resulting in a larger number of EAC. This result shows that EAC does not promote the growth of a specific species, but rather the growth of the entire algae species growing in the river.

In addition, seven months after the installation of the spur dikes, the growth of algae and the habitat conditions of sweetfish and Japanese giant salamander were investigated. The algae habitat conditions were good for the EAC, OC, and surrounding natural stones in each block of spur dike No.9, 11, and 15. Sweetfish were often found around the blocks with EAC panels (Photo 10). Bite marks of sweetfish were also confirmed everywhere on the surface of EAC, OC and the surrounding natural stones, indicating that blocks with EAC panels in the spur dikes contribute to creating the habitat for sweetfish.

A total of two individual adult Japanese giant salamanders were confirmed, one for each in the void parts of the 4-ton blocks of No. 9 and No. 15 (Photo 11). Both of them used the space created by 4-ton blocks as a habitat. Furthermore, many fishes such as sweetfish and Japanese dace lived in the void parts of the 4-ton block as well as Japanese giant salamanders. Since Japanese giant salamanders feed on fish, it is assumed that they prey on the fish around the block. In other words, benthos and sweetfish prey on the attached algae that grow on the surface of the EAC panels, which is thought to enrich the ecosystem base and the feeding environment of the Japanese giant salamanders, which are at the top of the ecological chain (Fig. 9).

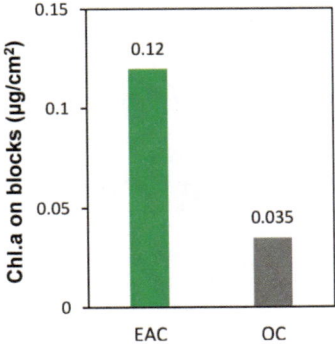

Fig. 7. Amount of chlorophyll-a in attached

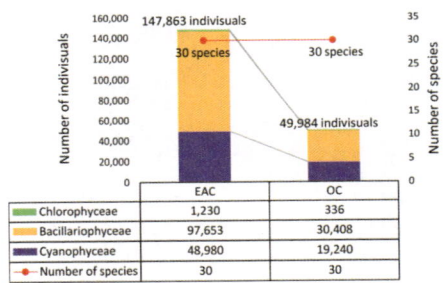

	EAC	OC
Chlorophyceae	1,230	336
Bacillariophyceae	97,653	30,408
Cyanophyceae	48,980	19,240
Number of species	30	30

Fig. 8. The number of species and individuals of attached algae

Photo 10. Sweetfish swimming around the 4-ton block and bite marks on the surface of EAC

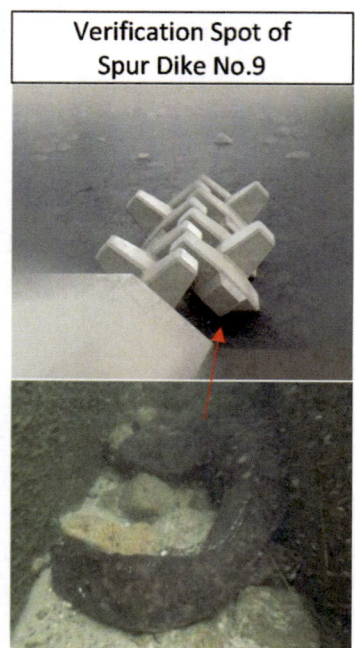

Photo 11. A Japanese giant salamander resting in the void part of 4-ton

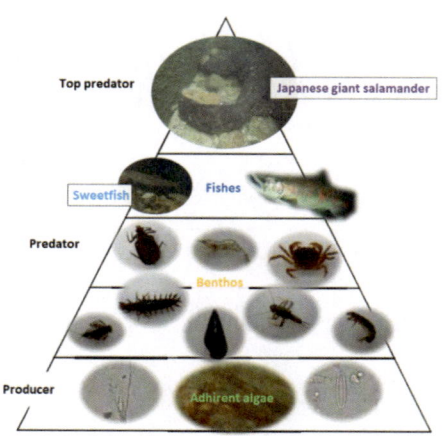

Fig. 9. Ecosystem pyramid in the Waragawa

5 Conclusions

This paper introduced the effectiveness of EAC and its application examples for river structures such as riverbed protection works and spur dikes. Among them, EAC contributed to promoting the growth of attached algae and was also confirmed to be effective in habitat conservation for various organisms such as sweetfish, Japanese eel and Japanese giant salamander.

Concrete is demanded to play an ever more important role in the development of river structures for disaster prevention and mitigation, as there is concern that the increase in floods due to climate change will cause extensive erosion in the future. Under these circumstances, adding environmental functions to concrete, such as EAC introduced in this paper, will make it possible to achieve river structures with both disaster prevention and environmental conservation functions. Hereafter, we hope that EAC will be widely used all over the world beyond Japan.

Acknowledgements. I would like to express my appreciation to Ms. Yuka Kizaki and Mr. Junya Inoue of Nikken Kogaku Co., Ltd. for their enormous cooperation in writing this paper.

References

Sato K, et al (2011) Characteristic of periphytic algae on concrete contained amino acid. In: 9th international conference on the environmental management of enclosed coastal seas

Nishimura H, Yamanaka R, Sato K, Tara C, Nakanishi T, Kozuki Y (2014) Material performance of "enviromentally active concrete" containing amino acids in water. In: PIANC world congress

An Integrated Approach to Define Estuarine System Resilience, Applied to the Upper Sea Scheldt, Flanders, Belgium

R. Adams[1](✉), G. van Holland[1], J. Vansteenkiste[1],
M. van Rompaey[1], M. de Beukelaer-Dossche[2], and S. Bosmans[2]

[1] International Marine and Dredging Consultants, Antwerp, Belgium
roeland.adams@imdc.be
[2] De Vlaamse Waterweg nv, Antwerp, Belgium
michael.debeukelaer-dossche@vlaamsewaterweg.be

Abstract. The Upper Sea Scheldt (Flanders, Belgium) is a part of the Scheldt estuary which extends from the North-Sea in the Netherlands to the shipping locks in Ghent with a total length of 160 km. Challenges on flood protection and nature development in this unique fresh water estuarine system are addressed in the Sigma-plan. Cumulative effects of i) the autonomous morphological development of the estuary, ii) the further evolution as a consequence of past realignments and dredging works, and iii) sea level rise, result in an increase of tidal dynamics and turbidity which affect both habitat and light climate, and finally disrupt the ecosystem functions.

In order to better understand the system functioning and to prepare for counteracting these undesired evolutions De Vlaamse Waterweg, the waterway manager, has launched a study programme to investigate solutions and to prepare a vision for the future management of the river. An extensive modelling instrument was developed coupling different state of the art modules into one model chain. The developed instrument proved to be highly effective to study (i) the interdependencies between the different river functions which allowed for an integrated analysis and evaluation of potential measures, and (ii) the robustness of the measures for climate change, and allowing the selection of a set of measures providing a desired level of system resilience. As such the results of the study form the backbone for the development of a future vision on estuary management, while the model instrument will continue to be used to study design alternatives and finetune measures for implementation.

Keywords: Adaptive river management · Climate resilience · Nature based solutions

1 Introduction

The Upper Sea Scheldt (Flanders, Belgium) is a part of the Scheldt estuary which extends from the North-Sea in the Netherlands to the shipping locks in Ghent, with a total length of 160 km. The upstream part presents a unique fresh water estuarine

© The Author(s) 2023
Y. Li et al. (Eds.): PIANC 2022, LNCE 264, pp. 1052–1068, 2023.
https://doi.org/10.1007/978-981-19-6138-0_93

system, facing challenges on flood protection and nature development, which are addressed in the Sigma-plan.

Cumulative effects of i) the autonomous morphological development of the estuary, ii) the further evolution as a consequence of past realignments and dredging works, and iii) sea level rise, result in an increase of tidal dynamics and turbidity which affect both habitat and light climate, and finally interfere with the ecosystem functions. Moreover these effects impacts the shipping function.

A study was commissioned by the Vlaamse Waterweg to better understand the functioning of the system, and to define a vision for the management of the river, improving its resilience and ensuring that it can still meet the requirements of the various river functions in the future.

To this end a state of the art modelling instrument was built to learn about the system response to changing boundary conditions and to measures. Measures have been combined into alternatives aiming at improving ecosystem functioning, creating habitat to provide better conditions for birds and fish, safeguarding (or even improving) the safety (against flooding) function, and creating better navigation conditions.

No specific ambition level was defined at the start of study (for example the creation of a specific habitat surface or rate of reduction of the tidal range), as the aim is to investigate what can be achieved when introducing measures, and what measures are required to achieve a certain goal. A reduction of the tidal dynamics, maintaining and improving boundary conditions for estuarine nature, protection against floods and improved navigation can be considered as general goals.

2   Integrated Study

In order to better understand the system functioning and to prepare for counteracting the undesired evolutions De Vlaamse Waterweg, the waterway manager, has launched a study programme to investigate solutions and to prepare a vision for the future management of the river.

An extensive modelling instrument was developed coupling different state of the art modules into one model chain (Fig. 1). This includes hydrodynamic, sediment transport modelling, water quality (ecosystem), habitat and higher trophic level models, and real time navigation simulations.

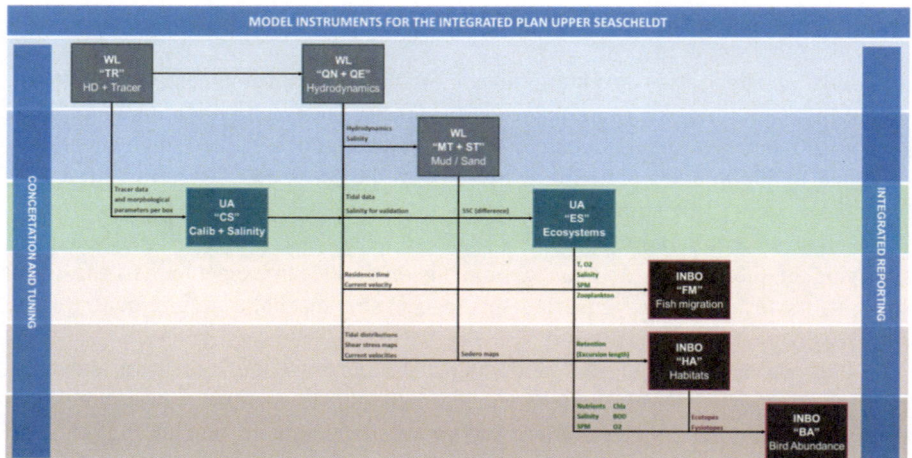

Fig. 1. Sequence and relations between the models of the model chain.

The models were used to study changing boundary conditions and modifications to the bathymetry as a part the measures. The changes in boundary conditions represent the results of climate change and cumulative effects including anthropogenic influences. For climate change, a time horizon of 2050 situation was considered. Besides a climate neutral scenario (CN), two climate change scenarios have been considered: a climate low (CL) and a climate high (CH) scenario, respectively assuming sea level rise of 15 and 40 cm with respect to 2013. Cumulative development concerns the unknown changes in hydrodynamic boundary conditions as a consequence from the further morphological development of past or yet undefined future dredging and sediment disposal activities in the Wester Scheldt and Lower Seascheldt, which may either lead to a reduction (Amin-) or an increase (Aplus) in tidal range. The reduction relates to a range of 500 cm in Schelle (corresponding to the 1960 situation). The increase implies a tidal range of 570 cm in Schelle, while in 2013 the tidal range is 550 cm. For changes in upstream discharge only one scenario was considered.

A detailed hydrodynamic model was constructed of the Scheldt estuary by Flanders Hydraulics Research (Smolders et al. 2016), from the North Sea to the upstream estuarine channels of the Scheldt and its tributaries using Telemac-3D. The model was calibrated for the 2013 situation using water levels from the existing tidal stations, velocity profiles across the river at several locations sampled during 13h measurement campaigns and a couple of permanent measurements at fixed locations. This was complimentend with salinity data over the entire length of the estuary. Measured wind data and measured upstream discharges were used as input.

The model is nested to the regional ZUNO model of the Southern North Sea (Maximova et al. 2016) from which downstream conditions can be taken for scenario analysis.

A mud transport model was constructed using SEDI3D (Smolders et al. 2020), the sediment transport module of Telemac-3D. The model was calibrated to available SSC measurements. For scenario analysis a delta-approach was used: the change of concentration obtained in the scenario models was transferred to the subsequent ecosystem and higher trophic level models rather than the absolute values and applied to the measurements to define the concentrations for the scenario models.

The existing quasi-2D Sigma model was modified to study the effect on safety against flooding (Coen et al. 2020).

The next model in the chain is the ecosystem model. It can be considered as a reduction of the original MOSES model (Soetaert et al. 1994), but extends further upstream and has a higher spatial resolution. It is developed by the Ecosystem Management Research Group of the University of Antwerp (Van Engeland et al. 2018). It is 1D tide-averaged model that resolves downstream volumetric transport of water and dissolved substances. The volumetric transport consists of a unidirectional advective component and a dispersive component. It simulates primary production, nutrient uptake by phytoplankton, phytoplankton loss, silica dissolution, zooplankton growth and loss, oxic remineralization, nitrification, denitrification.

For scenario analysis the temperature forcing is changed based upon literature study, while delta SSC and hydrodynamics are taken from the hydrodynamic model results. Hypsometric data are changed to study the measures. Dispersion is calibrated for these bathymetries using tracer tests results from the hydrodynamic models.

For the Habitat model (Van Braeckel et al. 2019) the MONEOS approach which is generally used in the estuary (Van Braeckel, 2013) was transformed to be able to work with model results and design bathymetry rather than observations. Boundary conditions are derived from the hydrodynamic model results, and geometric data from the design bathymetry. A major challenge proved to be the disequilibrium between the sea level rise and the bathymetry, as the alternative bathymetry was constructed from the existing bathymetry, and hence does not take into account the gradual adaptation of tidal marshes and mudflat geometry to the rising sea level. For this reason also a situation of the alternative bathymetry was simulated without sea level rise and without tidal amplification.

To study the higher trophic levels models were made to study habitat suitability for fish (using Twaite shad as research species[1]) (Vanoverbeke et al. 2019a) and bird abundance (referring to common teal) (Vanoverbeke et al. 2019b). The fish model uses temperature, salinity, oxygen, waterdepth and current velocity to determine a suitability index. The habitat suitability model uses the results of both the hydrodynamic and water quality models for the alternative bathymetries to study scenarios. As, therefore the model cannot be calibrated. The bird abundance model is a regression model comparing bird abundance to habitat: width, slope, dry up duration of mudflats. It uses the results of the hydrodynamic models for the different geometric alternatives to study scenarios.

The navigation simulations will not be discussed in this paper.

3 Measures Combined into Alternative Bathymetries

3.1 Alternative Development

Bathymetric alternatives have been composed after prior analysis of the hydrodynamic response of single measures, combinations of measures and previously studied alternatives aiming at improving the navigability of the river (IMDC, 2017). These

---

[1] Twait shad reappeared recently in the Scheldt as a result of improved water quality.

alternatives range from slightly widening the bends to creating a realigned waterway, however either respecting the current channel position or preserving the current channel as a side channel next to the new shipping channel. The latter alternatives have also been tested with the described model chain. The results of this analysis showed that although the interventions are beneficial to navigation, they do not create suitable environmental conditions when the valley is not included in the flow.

As it is not known how many measures are actually required to create a sufficiently resilient estuary a gradual approach in building up the alternatives was followed, in order to fully understand the extent to which measures of a certain scale respond. This allows to deduce from what kind of level of inserting singles measures on, one starts to have visible effects, and what is required additionally to achieve good conditions, whether the corresponding investment is reasonable, and what can be achieved with reasonable investments. From this level on one can start building up to arrive at what is either good or reasonable.

As such each of the investigated alternatives in this approach must considered as part of a further system exploration, and not as a final desired definition of the system.

3.2   Alternative Design Principles and Building Blocks

From the analysis of the measures and the so-called navigation alternatives, it was deduced that following design principles are considered favorable for the sustainable development of the estuary, they have been used as a guiding principle in designing the alternatives and identifying their building blocks:

- **Limiting flow velocity**: Limiting the flow velocity prevents (semi) planktonic organisms (e.g. the eggs / larvae of fish – in this case Twait shad) from being flushed out of the system. In addition, there is also a beneficial effect for the migration of adult fish and potentially for benthic density and feeding grounds for higher trophic levels. This can possibly be achieved by realizing additional room for the river through depoldering (pulling existing dikes land inwards or, creating additional depoldered areas by creating breaches in the existing Scheldt dikes, and/or connecting polders with the Scheldt through flood channels).
- **Limiting tidal dynamics**: By limiting tidal dynamics, beneficial effects are achieved for safety, navigability and sediment transport. This can be achieved both by (i) creating low-dynamic subtidal systems outside the fairway, (ii) depoldering and creating wide tidal flats with mild slopes, (iii) avoiding system shortening (cutting off bends).
- **Stimulating primary production**: Additional extra shallow zones (low-dynamic shallow water zones along the fairway, Flood Control Areas (FCA)/ with Controlled Reduced Tide function (CRT), depoldering) are considered to have a beneficial effect on primary production and the ecosystem.
- **Limiting turbidity**: By limiting turbidity, favorable effects are achieved for primary production and the development/viability of other organisms (e.g. Twait shad). Besides lowering the turbidity in the water by limiting dredging and dumping volumes in the estuary, extra depoldering is a measure that can limit turbidity by trapping sediment and lowering sediment transport.

- **Extra tidal flats, marshes and FCA/CRT**: An extension of tidal flat area is directly related to additional habitat for macrobenthos and resting and foraging avifauna (such as the common teal, research species for the study). Marshes and FCA/CRT are directly related to extra habitat for Natura 2000 habitat types habitat types (tidal willow shrubs and forest H91E0, and salt marsh meadows 1330) and reed beds with breeding avifauna (e.g. marsh harrier). This can be achieved by (i) depoldering - extra tidal flat area and (ii) creation of estuarine nature by FCA/CRT.

- **Improved water quality**: An additional connection with the valley and the creation of estuarine nature can result in improved water quality in the estuary. This can be achieved by creating connections to old meanders and other water bodies along the Scheldt. Note that even despite the decrease of water quality in these water bodies, this may have a positive effect on the water quality in the Scheldt.

- **Buffering** of peak flows. Due to climate change, more extreme events (longer dry periods in summer, more rainfall in winter) can be expected. The planned actions should contribute to the buffering of the expected peak discharges (additional FCA) and mitigate drought stress in the surrounding areas (FCA-CRT, depoldering).

3.3   Alternatives

Three different alternatives consisting of a combination of measures such as bend cut offs, depoldering with or without the introduction of flood or ebb channels, flood control areas with or without reduced tidal action, section widening and narrowing (by sediment disposal or groyne fields) have been composed and investigated (IMDC, 2021). The reference year 2050 was considered for a range of climate change and downstream tidal range scenarios in order to improve the understanding of the reaction of the estuarine system to such measures.

The investigated alternatives range from mild (only interventions in and along the navigation channel), over moderate (the latter complemented with a selection of measures in the valley) to extreme (the previous set with a maximum of measures in the valley).

The three alternatives have been developed with the following mindset:

- C1 alternative: Looking for opportunities in the river and redefining the Sigma-plan to improve habitat and to reduce the increase in tidal amplitude, and in the meantime to tackle the most prominent nautical bottlenecks;

- C2 alternative: Additional opportunities in the valley (depoldering, side channel) to improve habitat and to reduce increase in tidal amplitude are also included. Additionally also less prominent nautical bottlenecks are tackled, and additional measures for the most prominent ones have been defined. The additional nautical measures include a more extreme smoothening of the bends.

- C3-alternative: Additional measures (larger depolderings, additional depoldering at Weert, undeepening at Temse) aiming at providing added (climate) resilience while also improving habitat conditions, combined with further nautical measures (compared to C2) for a limited number of locations.

4   Effects

4.1   Tidal Range

The simulation results demonstrate that depoldering is a very effective measure to control the increase in tidal range (as a consequence of climate change and bend cut-offs). In addition, depoldering also has a very positive effect on the water quality and development of estuarine nature.

The effect of the alternatives on the hydrodynamics of the system increases from C1 to C3 (Bi at al. 2021a). The mean water level increases roughly from km 20–60 (from the upstream boundary (0 km) at Merelbeke, near the city of Ghent) in C2 and C3, and from km 40–60 in C1. Upstream of km 20, the mean water level decreases in C2 and C3. From km 0–40, the changes of mean water level in C1 is less than 2 cm, which is much smaller than the changes seen in C2 and C3. Further examination shows that the mean high water (HW) decreases and the mean low water (LW) increases in all the C-alternatives. As a result the tidal range decreases (Fig. 2), especially in the zones where the measures are implemented.

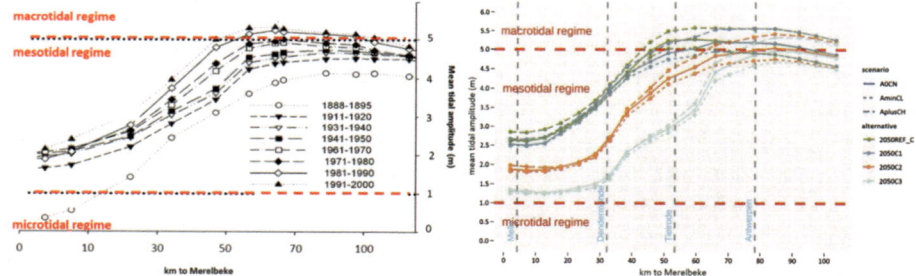

Fig. 2. Development of tidal range over the past century (left), resulting tidal ranges for the investigated alternatives (right). Source: Vanoverbeke et al. (2022).

The reduction is such that a historic mesotidal regime (based on the classification of Hayes, 1979) is restored over a large portion of the estuary. This is an interesting observation as tidal range increased significantly over the past century (Fig. 2). Given that an evolution of further increases in tidal amplitude is considered unfavorable, all three C-alternatives represent a favorable evolution of reduced tidal range. Given that reductions in tidal range also affect the availability and distribution of estuarine habitat (e.g. with a microtidal system the surface area of tidal mudflats and marshes along the river fairway can be strongly negatively affected) it needs to be considered what level of tidal amplitude is desirable to maintain a balance between safety, navigability, and the preservation of unique (freshwater) estuarine nature in the upper reaches of the Sea Scheldt. Overall, the evolution of the habitats in C1 is minor and mostly local (except for positive evolution of mudflat area in the upper reaches (< 20 km from Merelbeke)). In C2 and C3 subtidal area tends to undergo overall positive evolutions, due to the reduction in tidal amplitude and the conversion of mudflat area into subtidal area. However, mind that in the upstream reaches the regime tends to microtidal in the C3.

4.2   Habitat (Vanoverbeke et al. 2022)

Whereas in the downstream regions, depolderings largely create additional mudflat area (which partly drowns with sea level rise (but which in turn might be buffered by further autonomous evolution)), depolderings in the upstream area both create additional mudflat and marsh habitat (Fig. 3). As a result, in C2 and C3 the mudflats tend to undergo a positive evolution in surface area in almost the entire Upper Sea Scheldt, whereas, in contrast, the marshes only undergo a positive evolution in the upstream area, but undergo a negative evolution in the mid and downstream region (> 20 km from Merelbeke, km 0), with overall losses in habitat area. Even if changes in marsh area might be relatively small, strong changes in the location of the marshes may occur with a shift of marsh area from the originally existing marshes to newly created CRT areas. Indeed, due to the strong effects of measures (depolderings, CRT areas and channel adaptation) on the tidal range, evolutions in habitat area are often a combination of the creation of new habitat (depolderings and CRT areas) and the shift of one habitat type into another due to reduced tidal range. Due to these reductions in tidal amplitude, high mudflats will turn into marshes, low mudflats will turn into subtidal area, and higher marshes will turn into non-estuarine habitat. Furthermore, sea level rise will amplify the drowning of lower mudflats but will temper the desiccation of the higher marshes.

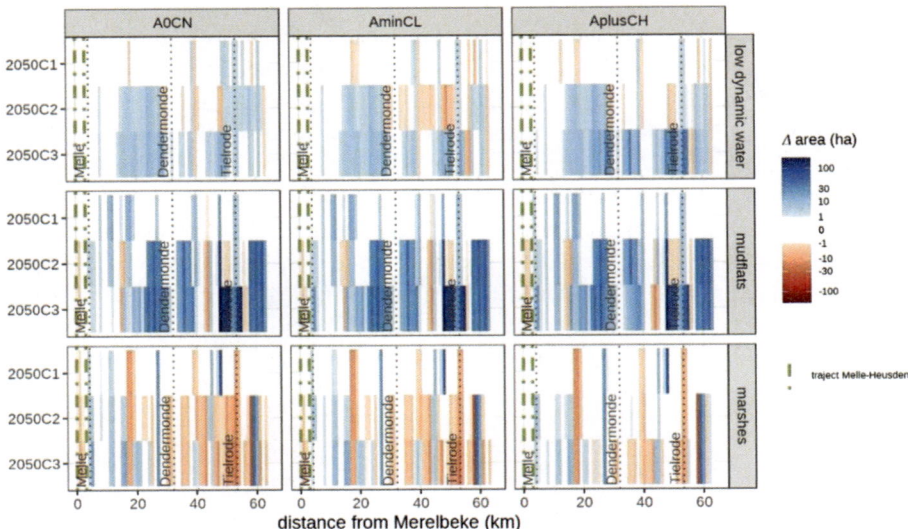

Fig. 3. Changes in surface area of ecologically important ecotopes. Blue indicates favorable evolution; red indicates unfavorable evolution. Source: Vanoverbeke et al. (2022).

The C-alternatives also affect the (mean cross-sectionally averaged) velocity. This also contributes to the habitat conditions. The effects are much stronger in C2 and C3 compared to those in C1. In all three C-alternatives, the effects on the ebb velocity are similar to the effects on the flood velocity. Based on the changes in the mean velocity, three zones can be distinguished in the Upper Sea-Scheldt: i) from 0 to about 28 km: the mean velocity in general becomes smaller, hence more favorable, as a consequence of depoldering between 24 and 28 km, the development of intertidal areas combined with alternative navigation channel development between 16 and 22 km and further smaller scale depoldering between km 5 and 12; ii) up to 55 km depolderings with side channels, and introducing CRT function in FCA areas, together with a major bend cut-off around 40 km result in variable velocity signals, either more favorable or unfavorable, but with sufficient lateral areas with low velocities; iii) the narrowing (and in C3 also undeepening) of the channel downstream of 55 km results in an (unfavorable) increase of the velocity in the main channel.

4.3   Safety Against Flooding

Another effect of the alternatives is the drop of waterlevels from C2 on, during flood events, particularly during storms. Despite the reduction of FCA at the cost of depoldering. This shows that depoldering can also have a significant contribution to safety against flooding. Differences may go up to 50 cm in some areas during storm events with return periods of 1000 years, hence even compensating the assumed level of sea level rise (Coen et al. 2021)).

4.4   Suspended Sediment

The suspended sediment concentration is barely affected in the C1 alternative as is shown by the results of the cohesive sediment transport model (Fig. 4). But is very prominent in the C2 and C3 alternatives (Bi et al. 2021b). The results show, as for most of the parameters, that the climate scenario only has a limited effect on the results compared to the effect of the alternatives (Fig. 4). Therefore, in general, conclusions will not be affected by the climate scenarios. In C2 and C3 the depoldered areas operate as sinks. Calculations based on observed settling velocities and suspended sediment concentrations indicate a long longevity (decades) of the measures, as the source of the mud is limited. The sand transport capacity aligns with the velocity hinting at the requirement of dredging in some areas upstream of 55 km to maintain the navigation channel, but definitely requiring protection to prevent erosion between 55 and 64 km where the channel is narrowed (and undeepened).

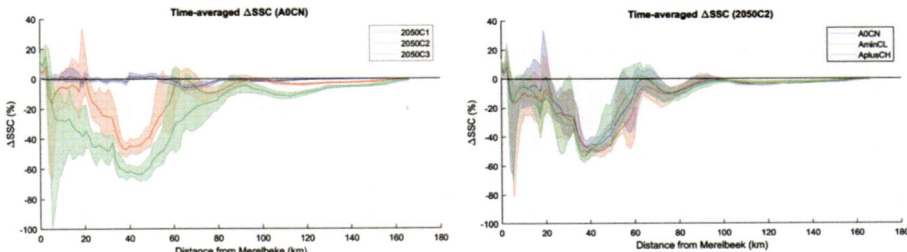

Fig. 4. Effect of alternative on SSC (left), effect of climate scenario on the result of one alternative (right, here showing C2). Source: Bi et al. 2021b.

4.5 Water Quality and Primary Production

The ecosystem model shows that the increase in shallow areas due to the addition of CRT areas in 2050 (Sigma-plan) positively influences primary production and oxygen concentration as a results of the generally lower water depths (Cox et al. 2021).

The measures taken in the C1, C2 and C3 alternatives and the reduction of suspended matter (better light penetration) lead to additional increase in primary production and oxygen concentrations (Cox et al., 2021). The increase in primary production and biomass are most clear in the C2 and C3 alternatives, as expected by the increasing scale and extent of the implemented measures, however with only relatively small changes between the C1 alternative and 2050 reference simulation (Fig. 5). The sharpest increases are detected in section 2 and section 3 (km 10 to 35 in Fig. 5). Primary production over Biomass (P:B) shows similar trends.The C1 alternatives only shows a clear difference with the reference in the depoldered areas in the upstream part (more primary production in C1).

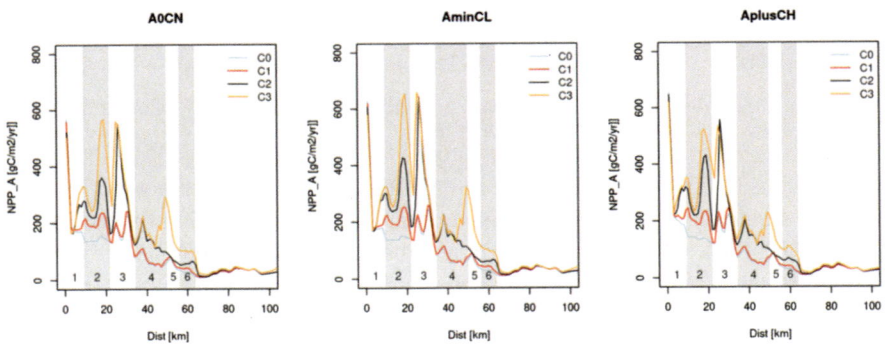

Fig. 5. Annual area-normalized net primary production (NPP) (gc/m2/yr). (Cox et al. 2021).

Patterns of increased primary production and oxygen upstream are followed by decay of organic material and lower oxygen concentrations downstream. This effect is strong downstream of the Rupel tributary, resulting in an "oxygen-dip" (which currently also occurs). Note that the impact of the current and future water quality of the Rupel tributary on the Scheldt is not modelled but may also be an important driver, as the discharge of the Rupel is relatively high compared to the one of the Scheldt. Generally oxygen concentrations show an increase in the direction of the 2050 reference situation (C0), C1, C2, C3 (Fig. 6). The C2 and C3 generally lead to improved oxygen conditions in the study area (zones 1–6 in Fig. 6), with the same patterns shown for the 3 climate scenarios. In the zone downstream from the Rupel the oxygen-dip effect for the C3 alternative is strong under the A0CN and AminCL scenario's, while the dip is smallest for the C2 alternative.

It is expected that improved water quality can further improve the situation. Also consider that large areas of mudflats and tidal marshes in the alternatives replace agricultural land. This probably will also lead to a reduction of nutrient input to the system, and will also influence the primary production. This effect is not modelled.

The impacts of measures installed for the different alternatives are generally robust under different climate scenarios, showing the same patterns among alternatives, with relatively small differences. However the AplusCH scenario (increase of tidal range and sea level rise of 40 cm) shows for some zones smaller Chlorofyll concentrations and primary production rates than the AminCL scenario.

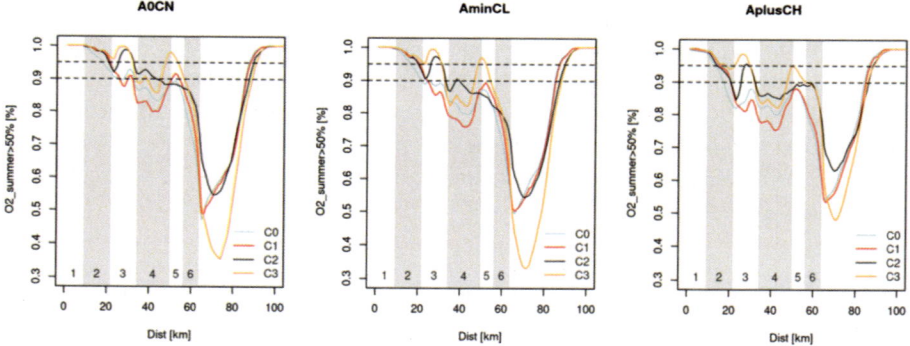

Fig. 6. % of time (Y-axis) summer oxygen concentrations are above 50% (Cox et al. 2021)

The predicted decrease in silica must be carefully assessed, suggesting an impact on diatoms, as the model is not designed for the simulation of population dynamics. The impact of silica release from implemented lateral areas in the different C alternatives is not modelled, and therefore the simulated silica depletion is probably overestimated.

4.6   Suitability for Twait Shad and Abundance of Common Teal

The reduced tidal influence in the upstream parts leads to reduced water dynamics and increased residence time. This not only results in higher primary production (as discussed before), but also leads to better (more sheltered) conditions for (juvenile) fish, that have no escape in the narrow, canalized riverbed.

Conditions are in general favorable for larval development upstream of Dendermonde (km 28) in the C-reference and alternative C1. In years with high suspended particle matter (SPM), however, suitability can strongly reduce in the region between Dendermonde and Tielrode (km 28 to km 50), which is the most important spawning area nowadays. In the C2 and C3 alternative, conditions between Dendermonde and Tielrode are clearly improved due to reduced levels of SPM.

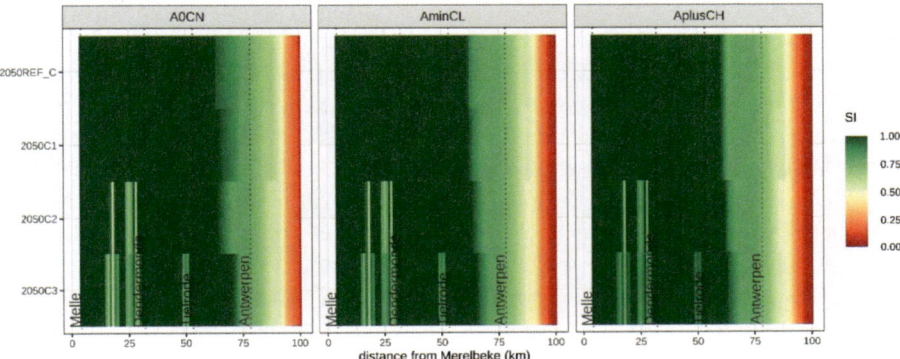

Fig. 7. Twait Shad suitability index for different alternatives (vertical), and different climate scenarios (horizontal). Suitability for spawning (mean value over modeled years), per kilometer from Merelbeke (0 km) to Antwerp (100 km). Source: Vanoverbeke et al. (2022).

Conditions for spawning are generally favorable in the Upper Sea Scheldt (Fig. 7), including the region between Dendermonde and Tielrode, where most of the spawning is observed today. Both for upstream migration (spawning) and downstream migration (developing juveniles) there are indications that between 50–90 km from Merelbeke in some years oxygen levels can be too low as a result of the earlier discussed decay of organic material, creating a barrier for migration. The C-alternatives do improve this, also today present, situation by narrowing the stretch where oxygen levels are unfavorable.There is a general positive evolution from C1 to C3 in the numbers of common teal when comparing the C-alternatives with the reference situation) due to the inclusion of large areas of newly created depolderings and CRTs, with positive evolutions on the area of mudflats.

4.7   Long Term Morphological Stability

The impact of the C-alternatives on the hydrodynamical and morphological stability has been assessed by means of an analysis of the relation between the wet cross-section, the tidal prism and the tidal range (IMDC, 2022). For this analysis it was assumed that the present state is in 'equilibrium'. This is debatable, as the river is largely impacted by human intervention, such as bank fixations, river straightening, (downstream) deepening, maintenance dredging, etc. The cumulative impact of the measures in the C-alternatives will have a strong global impact on in particular the tidal prism and the tidal range and a local impact on the wet cross-section. Especially the depolderings are dominant in the initial response, as they tend to dampen the tidal range and affect the tidal prism up- and downstream. Locally the wet cross-section will increase as a result of the added intertidal area. In the C-alternatives, straightening or removal of sharp bends in the river typically is combined with either depoldering, or the creation of intertidal area. While straightening would theoretically lead to a reduction of the convergence length and subsequent amplification, the creation of new intertidal area dominates the hydrodynamic response (damping).

Only as far as the increase in wet cross-section locally contributes to an increased flow capacity one can expect that (short-term) sedimentation may reduce the initial impact of the measures on the hydrodynamics. However, as most of the depolderings involve the lateral connection of large areas of polder land, local sedimentation will not lead to a diminution of the impact on the tidal prism. On the scale of in particular the C3 alternative, and in a lesser extent the C2-alternative, it is unlikely that sedimentation will be the driver in restoring the equilibrium on the longer term. Downstream erosion, as a result of increased tidal prism and reduced water levels, may result in an increase in tidal amplitude. But as a result of the logarithmic shape, it will require quite large modifications in wet cross-section in order to impact the tidal prism.

Another important aspect to consider is that different sediment types, i.c. sand and mud, will respond to the changes in different ways. Sand will respond quickly to the changed conditions, but may be limited to a local scale, because of limited availability. On the other hand, there is an abundance of mud in the system. But it requires large volumes of mud deposition before significant impact on the hydrodynamics will take place.

Deposition of mud is mainly expected in the low-dynamic areas, like the depoldered areas. These depoldered areas will therefore function as a sink, which allows mud transport against the net sediment transport direction that would be derived from the asymmetry of the tidal flow (eb-dominance). Given the large number of depolderings and the strong reduction in tidal dynamics, it may be expected that turbidity in the river will be reduced significantly. This in itself may pose a risk for drowning of the mudflats along the river, as less sediment may be available to adjust to changes in the average water level due to sea level rise. Further downstream, in the Lower Sea Scheldt, impact of the measures on the hydrodynamics are visible in the results. While it will take a long time to experience any impact on the sand transport, the impact on the mud concentrations may be noticeable. It may be expected that the proposed interventions (especially the C2 and C3 alternatives) may lead to a change in the dynamics in the estuarine turbidity maximum (ETM). On the one hand it is expected that the system will become more eb-dominant, but as a result of the tidal pumping a larger exchange of sediment to upstream may be expected

5   Discussion

Depolderings appear to be very efficient in reducing the tidal range, and are an important part of the measures. As we move further downstream, the measures must be larger to create the same hydrodynamic effect. On the other hand, large measures downstream have a major (and desired) impact on the tidal range upwards. A downward narrowing is therefore not considered necessary, and the undeepening of the access channel is perceived as negative. Not having to consider the latter has the advantage that expensive bank and soil protection costs can be avoided. Depoldering and CRT areas on the other hand, do require a redefinition of the functions of existing Sigma areas, or the development of additional areas, requiring the necessary compensations.

In addition to the system benefits caused by the reduction of the tidal range (better habitat conditions are created, and the reduction of suspended sediment, captured by the depoldered and CRT areas, leads to an improvement of the light climate, the Chlorophyll content and oxygen balance, and increased primary production), there are also significant indirect benefits from the avoided flood risk. In an evaluation of the safety level of the current Sigma-plan, and possible update taking into account the higher sea level rise associated with the current climate change scenarios, the investigated depolderings can occupy a prominent position.

The investigated depolderings compensate for the effects of bend straightening, which offers prospects for improving navigability. The nautical bottlenecks have been identified in the study, and although the measures examined for shipping may incur local but temporary impacts on habitat, but lead foremost to an improvement of navigability, without adverse system effects.

The measures seem to have a long life span. The depolderings capture sediment, but the sedimentation rate is slow, so that there is even a risk of drowning the system (turning mudflats into subtidal area), and the instantaneous loss of tidal marshes as a consequence of the drop of the high water level. A gradual construction (area by area) allows to catch sediment to gradually develop a balanced and diversified morphology, and to limit temporary damage by loss of habitat. The construction of the plan in a short period of time would also be a major challenge from a budgetary point of view, so the risk for such an unbalanced intervention is small.

In addition, it is far from certain what effects occur over the entire estuary. Redistribution of sand is expected soon after construction, however on the long term the morphological changes are difficult to predict, but it seems that the required adjustments of the downstream area are so large that it is unlikely to happen in the near future.

6   Conclusion: Towards a Vision and Bringing It into Practice

The wide range of simulated alternatives and achievable effects for different intervention levels, allow the selection of measures to obtain a desired level of system resilience. As such the results of the study form the backbone for the development of a future vision on estuary management. The vision aims at developing a sustainable and multifunctional estuarine system, compatible with the ambitions of both the waterway manager and its stakeholders.

The creation or restoration to a mesotidal regime should be aimed at. This implies the expansion of the estuarine domain by creating intertidal zones with controlled reduced tide function (CRT), and most notably additional depolderings.

Spreading the measures is considered advisable for fish (several well distributed low dynamic areas along the river axis), birds, but also for the ecosystem functioning (CRT ensures diatom development providing an important food source and building block of the food web). Regularly distributed measures can also be considered as stepping stones between larger habitat patches, each of which are necessary for a better functioning of the estuary.

The measures should be diversified according to location. The discharge dominated upstream section requires focus on the development of a fresh water tidal river landscape (with the preservation of wetlands). This means creating tidal flats to strengthen the pasture area next to the breeding area. This can be achieved by setting up depoldering, FCA (buffer effect), FCA/CRT (raising groundwater level) or side channels (impact on dynamics). Downstream, the estuarine nature can be improved (more qualitative area) by aiming at a well equilibrated tidal flat/marsh surface and more shallow subtidal area (flood channel). Therefore generally, we look for a different focus between the up and downstream section, upstream looking for opportunities to improve riverine and safety functions, and downstream to improve estuarine functions.

It is also considered important to keep the estuarine areas dynamic in order to increase the long term tidal mudflat function and limit sedimentation and vegetation succession by implementing large dike breaches when depoldering, direct connection to the main channel (dike withdrawal) or the introduction of creek dynamics (introducing side channels), in order to involve a larger part of the valley in the flow.

Already today, the gained insights from the modeling tool are put into practice by the installation and follow up of pilot projects, such as the depoldering in Wichelen, where a side channel has been created to improve the tidal dynamics and habitat. On the other hand the experience with such pilot projects will allow to gain insights to improve the design of new areas.

The developed model instrument proved to be highly effective in studying the interdependencies between the different river functions which allowed for an integrated analysis and evaluation of potential measures, and to test the robustness of the measures for climate change, and finally to allowing the selection of a set of measures providing a desired level of system resilience. The model instrument, and further improvements to it, will therefore continue to be used to study design alternatives and finetune measures for implementation.

Acknowledgements. We thank the partners of Flanders Hydraulics Research, the Research Institute for Nature and Forest, the Ecosystem Management research group of the University of Antwerp: Joris Vanlede, Qilong Bi, Leen Coen, Joost Vanoverbeke, Gunther Van Ryckegem, Alexander Van Braeckel, Tom Cox, Kristine De Schamphelaere for the scientific commitment, and the experts of the EGIPUS Expert Group Integrated Plan Upper Seascheldt: Han Winterwerp, Tom De Mulder, Peter Herman, Andreas Schöl, Mario Lepage, Patrick Meire, Erika Van den Bergh, for their critical comments on the used methods. Finally, and last but definitely not least, I also would like to thank Jürgen Roder for greatly improving the legibility of some figures, and being a most reliable colleague for over 25 years.

References

Bi Q, Vanlede J, Smolders S, Mostaert F (2021). Integraal plan Boven-Zeeschelde: sub report 16 – effect of the calternatives on the hydrodynamics. Version 1.0. FHR Reports, 13_131_16. Flanders Hydraulics Research: Antwerp

Bi Q, Vanlede J, Smolders S, Mostaert F. (2021). Integraal plan Bovenzeeschelde: sub report 18 – effect of the calternatives on mud transport. Version 3.0. FHR Reports, 13_131_18. Flanders Hydraulics Research: Antwerp

Coen L, Vanlede J, Mostaert F (2020). Integraal plan Boven-Zeeschelde - Veiligheidstoets B- en C-alternatieven: Deelrapport 3 - Veiligheidstoets C-alternatieven. Versie 0.1. Waterbouwkundig Laboratorium, Wl Rapporten, 14_176_3

Cox T,et al (2021). Evaluation of different management alternatives for the upper sea-scheldt. preliminary results – draft report research group ecosystem management

Hayes MO (1979) Barrier island morphology as a function of tidal and wave regime. In: Leatherman SP (ed) Barrier islands from the Gulf of Mexico to the Gulf of St. Lawrence. Academic Press, New York, pp 1–28

IMDC (2017) Integraal plan Boven Zeeschelde. Bouwstenenonderzoek. I/RA/11448/17.006/INE. De Vlaamse Waterweg

IMDC (2021) Integraal plan Boven Zeeschelde. definition of C alternatives. RA/11448/21.008/JVS. De Vlaamse Waterweg

IMDC (2022) Integraal plan Boven Zeeschelde. evaluation 1D equilibrium Sections for C-alternatives NO/11488/20223/BDH/JVS. De Vlaamse Waterweg

Maximova T, Vanlede J, Verwaest T, Mostaert F (2016). Vervolgonderzoek bevaarbaarheid Bovenzeeschelde: Subreport 4 – Modelling Train CSM – ZUNO: validation 2013. Version 3.0. WL Rapporten, 13_131. Flanders Hydraulics Research: Antwerp, Belgium

Smolders S, Maximova T, Vanlede J, Plancke Y, Verwaest T, Mostaert F (2016). Integraal plan Bovenzeeschelde: subreport 1 – SCALDIS: a 3D hydrodynamic model for the Scheldt estuary. Version 5.0. WL Rapporten, 13_131. Flanders Hydraulics Research: Antwerp, Belgium

Smolders S, Bi Q, Vanlede J, De Maerschalck B, Plancke Y, Mostaert F (2020). Integraal plan Boven-Zeeschelde: sub report 6 – scaldis mud: a mud transport model for the Scheldt Estuary. Version 4.0. FHR Reports, 13_131_6. Flanders Hydraulics Research

Soetaert K, Herman PMJ, Kromkamp J (1994) Living in the twilight: estimating net phytoplankton growth in the Westerschelde estuary (The Netherlands) by means of an ecosystem model (MOSES). J Plankton Res 16(10):1277–1301

Van Braeckel A, 2013. Geomorfologie – Fysiotopen - Ecotopen. p. 89–102. in Van Ryckegem, G. (red.). MONEOS – Geïntegreerd datarapport Toestand Zeeschelde INBO 2012. Monitoringsoverzicht en 1ste lijnsrapportage Geomorfologie, diversiteit Habitats en diversiteit Soorten. Rapport INBO.R.2013.26. Instituut voor Natuur-en Bosonderzoek, Brussel

Van Braeckel A, Vanoverbeke J, Elsen R, Van Ryckegem G (2019). Modelinstrumentarium voor het voorspellen van habitats in de Boven-Zeeschelde - Deelrapport voor het Integraal plan Boven-Zeeschelde. Rapporten van het Instituut voor Natuur- en Bosonderzoek 2019 (61)

Van Engeland T, Cox TJS, Buis K, Van Damme S, Meire P (2018). 1D ecosystem model of the Schelde estuary: model calibration and validation. Report research group Ecosystem Management ECOBE 018-R217

Vanoverbeke J, Van Ryckegem G, Van Braeckel A. Van den Bergh E (2019a). Modelinstrumentarium voor het voorspellen van habitatgeschiktheid van de Zeeschelde voor fint (Alosa fallax) - Deelrapport voor het Integraal plan Boven-Zeeschelde. Rapporten van het Instituut voor Natuur- en Bosonderzoek 2019 (18)

Vanoverbeke J, Van Reyckegem G, Van Braeckel A. Van den Bergh E (2019b). Modelinstrumentarium voor het voorspellen van overwinterende aantallen wintertaling (Anas crecca) in
  de Boven-Zeeschelde - Deelrapport voor het Integraal plan Boven-Zeeschelde. Rapporten van
  het Instituut voor Natuur- en Bosonderzoek 2019 (15)
Vanoverbeke J., Mertens A., Van Braeckel A. Van Ryckegem G (2022). Evaluation of the c-
  alternatives for habitats and higher trophic levels. INBO, preliminary report

Best Practice Approach for Layouting Technical-Biological Bank Protections for Inland Waterways – PIANC WG 128

Söhngen Benhard[1](✉), Dianguang Ma[2]
and Other Members of PIANC INCOM WG 128

[1] Federal Waterways Engineering and Research Centre, Karlsruhe, Germany
bernhard.soehngen@baw.de
[2] Tianjin Research Institute for Water Transport Engineering, M.O.T., Tianjin, People's Republic of China

Abstract. The worldwide increasing number of national guidelines and growing experience with realized green bank protections (constructions using insofar possible living or at least wooden construction material) in navigable waters, led to install a PIANC INCOM Working Group (WG) to collect and condense expert knowledge in this field of work and prepare it for practitioners for design purposes. The corresponding PIANC report, called "Technical-Biological Bank Protections for Inland Waterways", is foreseen to be released this year.

The report, whose structure, content, key findings and approach will be highlighted briefly in this contribution to the Smart Rivers Conference, tries to overcome the usual problems in design cases, which need knowledge and experience of civil engineers, eco-engineers and ecologists altogether and the way how the success of bank stabilization measures will be noticed and rated. The WG members had to notice that functionality assessment is not that simple, whereby partly huge differences between those who designed, realized and maintained measures and external parties as well as cultural differences occurred.

To overcome these problems and thus to objectivize the choice and layout of alternative solutions, which may help to convince people responsible for waterway development and maintenance to use green measures instead of traditional bank protections as riprap, a Best Practice Approach was developed, based on a catalogue of numerous realized measures, which are described e.g. in so-called Fact Files. The content of these descriptions, especially the local boundary conditions (BCs shortly in the following) and the balance between

Other members of PIANC INCOM WG 128—Katja Behrendt, BfG, Germany; Petra Fleischer, BAW, Germany; Kathrin Schmitt, BfG, Germany; Gonzalo Duro, Witteveen+Bos, Consulting, The Netherlands; Tetsunori Inoue, Port and Airport Research Institute, Japan; Kyle McKay, USACE, USA; Christophe Moiroud, CNR, France; Jeroen Verbelen, De Vlaamse Waterweg, Belgium; Jos Wieggers, Rijkswaterstaat, The Netherlands; Christian Wolter, IGB, Germany, Junwei Zhou, Tianjin Research Institute for Water Transport Engineering, China.

© The Author(s) 2023
Y. Li et al. (Eds.): PIANC 2022, LNCE 264, pp. 1069–1086, 2023.
https://doi.org/10.1007/978-981-19-6138-0_94

aims and achieved functionality issues, was used to assess the possible suitability of a chosen measure under generally different design conditions than those in the described realizations.

This was achieved inter alia by a scoring system, assessing differences between Design- (DC) and Analysis Cases (shortly ACs from the catalogue of measures), which is called Feasibility Check (it answers the question, whether experiences made with the AC-cases can be transferred to the DC) and differences between user-specified aims in the DC and expected performance issues from the ACs, called "Suitability Check" (answers the question, how far expected functionality issues may probably be achievable). This was done both for technical and ecological issues, whereby the scores were chosen and reviewed interdisciplinary and internationally to overcome the aforementioned assessment differences.

Keywords: Bank protection · Bioengineering · Multi-criteria analyses · Analytic hierarchy process

1   Introduction

The report of WG 128 offers a Best Practice Approach (BPA shortly in the following) to layout alternatives, so-called Technical-Biological Bank Protections (TBP shortly in the following). These measures use as far as possible vegetable construction elements as living plants or dead wood. Conventional construction components as sheet-piling or riprap (revetment made from heavy loose stones) shall be avoided as far as possible. This shall lead to bank protections, which are only as strong as really necessary and as weak or "green" as possible. Optimally, natural vegetation and succession shall take over the desired protection function short and long-term.

It is obvious that such "green" bank protections, whereby a collection is shown in Figs. 1, 3 and 5 cannot be designed in the same way as their technical counterparts. This is, as the latter need very much less information for design as TBPs. These are in principle the magnitude of ship-induced impacts and some properties of the bank to be protected as the slope inclination and of the friction angle only. These restricted design parameters allow also the standardisation of conventional bank protections to a large extent.

By contrast, the functionality of TBPs is depending generally on a huge number of influencing parameters.

Fig. 1. Visualisation of all measures described comprehensively in Part 2 of the WG 128 report: From DWA-M-519, over the UK Guide up to FFs for low water level changes (ΔW) and low ship-induced impact – the designations of the measures are abbreviated

These are e.g. the local BCs as the magnitude of water level changes, especially the duration of draught and flood periods, which influence the possible vegetation or the altitude difference between mean water (MW) and the highest shipping level (HSW), which determines the efficiency of wave-breaking pre-embankment constructions. Other important local boundary conditions determine the vitality of vegetation, e.g. the precipitation, the slope (inclination), the width of the vegetation zone or possible shading by buildings or large trees. Other impacts – besides those from navigation – such as frost heaving or ice drift are also relevant.

But also different planner's aims are design-relevant as the necessary stability, e.g. against the dominant navigation-related loads or the fulfilment of ecological demands as the enhancement of water-bound and terrestrial ecosystems, the creation of ecological stepping stones up to social or legal demands as the enabling of recreational activities.

This means, the design of TBPs is generally very much more complex compared to technical protections and led to the decision to develop a design approach, basing on experiences gained from numerous realized measures (collected in Part 2 of the report and visualized here in Figs. 1, 4 and 5). These experiences were analysed and extended by expert knowledge coming from the members of the PIANC WG, which are called "projections". They include the assessment of "worst case" BCs, where the measures may sill wok according to the expert's assessment and the higher functionality, if adaptions to the realized measures would be taken.

The challenge is, and this is the main purpose of the so-called Best Practice Design Approach, to "transfer" the experiences made e.g. concerning the achieved stability and sustainability, the potential ecological upgrade or other planner's aims at the site of realisation (called Analysis Case or AC shortly in the following), to the boundary conditions and planner's aims at the site, where a new measure is planned (called "Design Case or DC shortly), by tackling all relevant design criteria in a preferably quantitative way. This is a multi-criterion approach, which should be facilitated as far as possible for practitioners in a PIANC report.

2   Outline of the Best Practice Approach

This BPA bases on the following points, depending on the type of reader and applier (from a decision-maker in a waterway authority, which needs an overview of possible alternatives to riprap or sheet piling only, up to a planner in an engineering bureau) of the report and which is visualized in Fig. 2:

It is recommended to read Part 1 of the report first, maybe in different study depth as indicated in the coloured frames on the left on Fig. 2 according to the profession and interests of the reader. Also the application of the report may be more or less comprehensive for different appliers, as indicated by the dotted, coloured frames on the right hand side of Fig. 2.

It is recommended to read for this purpose at least

Chapter 2, reviewing relevant literature with focus on existing guidelines (Fig. 2), Chapter 3, offering short descriptions and common design features of TBPs and general design rules, Chapter 4, which discusses the content of Part 2 with descriptions of all the collected measures and explaining the "Screening Method" for pre-selection, which uses only a few characteristics of BCs and design features and which are collected in a big Overview Table.

Then it is recommended to have a first look into Part 2 next. This is very important, as the applier learns a lot about possible solutions with its various construction details and application ranges.

Then the applier is forced to go into details of Chapter 5, which presents the most important and very comprehensive pre-selection tools, developed from the WG and which are explained in the next paragraph of this paper in more detail. They tackle both technical and ecological issues as well as differences in BCs of the DC and the various AC(s), as well as possible discrepancies between demands and functionality issues of the measures considered. Social and legal aims are not tackled in the pre-selection tools as the focus is on answering the questions: What is feasible and what works! But these aims will be of course accounted for in the end while comparing selected variants by using the multi-criteria tool AHP (Analytic Hierarchy Process).

Then the applier is demanded to read Appendix B, which describes comprehensively an important part of a German Code of Practice, the DWA M-519 Bioengineering Approach, tackling a very restricted number of design criteria, that are all related to local boundary conditions and may be used as alternatives to other pre-selection tools and

Appendix C, which describes a reviewed and extended British Waterway Management Guide (1999), using a larger number of criteria than the DWA approach, but they are again mostly boundary-related. The UK Guide offers a very stringent, technological approach as the one of the DWA Guideline, whereby it is recommended to use it as a fourth alternative for pre-selection.

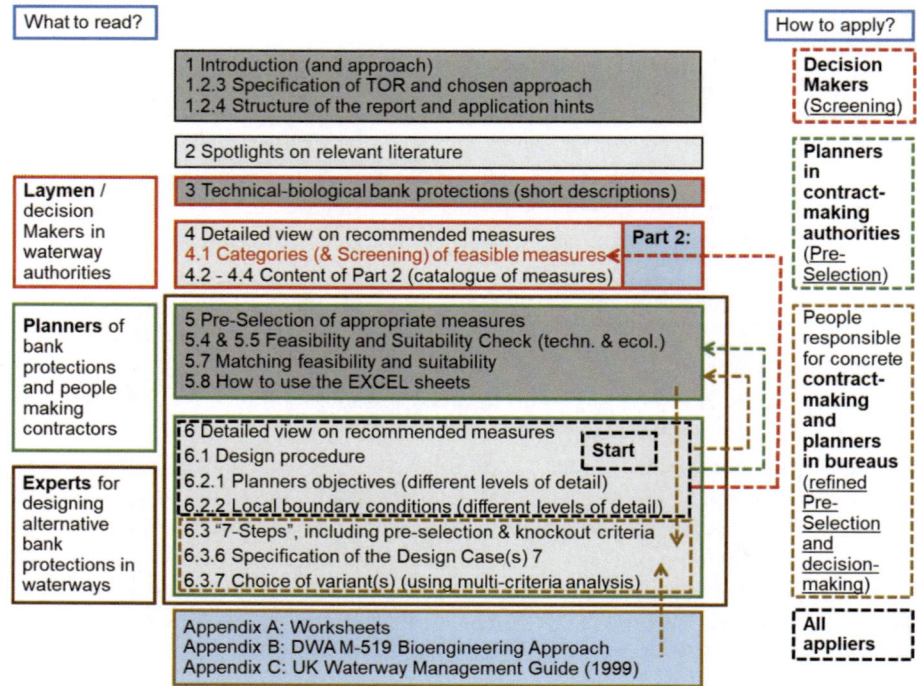

Fig. 2. How to read (hints on the left for laymen up to experts) and apply (from decision-makers up to planners on the right) the WG 128 report (Part 1) for different users – chapter names abbreviated

Then it is recommended to read and finally to apply the approach outlined in Chapter 6, which guides the user through the design process in the sense of process recommendations. Its kernel forms the procedure called "7 Steps" for design, which includes the aforementioned pre-selection tools.

The 7 Steps start with the discussion and specification of planner's aims (Step 1) and relevant local BCs (Step 2), whereby the applier is always forced to specify the boundary conditions and demands verbally in order to reduce the number of relevant design criteria as far as possible, whereby the report offers numerous tables with hints, on how the criteria will affect the design (also in Appendix A).

Then all the pre-selection tools (Screening, comprehensive Pre-Selection according to Chapter 5, DWA M-519 and UK Guide) shall be applied (Step 3), which end up e.g. in ranking lists of appropriate measures, which should be "matched", especially to account for both technical and ecological issues.

Then the applier should have a look, now more purposeful, into Part 2 (Step 4), e.g. concerning only the measures from the ranking lists, both from technical and ecological issues. A condensed overview on all the measures considered in the report provide, as mentioned earlier, Figs. 1, 4 and 5 in this paper. It is very important that the applier recognizes the various construction details, possible improvements to account for the various planner's aims and possible adaptations to relevant BCs.

Next, so-called "knockout criteria" as ice-effects in northern parts e.g. of Europe or the discussion of excess pore water pressure, which occurs if the drawdown speed of a passing vessels overtops the permeability and which demands for ballasting the bank slope e.g. with riprap or gabions, are discussed (Step 5). This important step may reduce the number of appropriate measures even more.

This leads to Step 6, where the applier is forced to specify "his" Design Case as far as possible, of course based on the collection of measures in Part 2. Again, the applier will be forced to specify the Design Case verbally, because it helps to concentrate on really important features.

The final design Step 7 is now to compare selected variants in a structured way. For this reason, the Analytic Hierarchy Process is recommended, see next paragraph for more details. The latter allows to match relevant technical design aspects as the congruence of BCs of DC and AC and the corresponding adaptability of the selected measures to different BCs, stability-related issues, whereby a first choice of the assessment may come from the pre-selection tools as well as those concerning the avoidance of efforts as to minimize the construction and maintenance costs, ecological aims as – again – the congruence of ecologically-relevant boundary conditions, the potential enhancement of selected taxa and aquatic as well as terrestrial habitats up to the consideration of social demands as those related to human activities, to landscape and preservation as well as legal demands, whereby all relevant design aspects will be supported by numerous tables, delivering comprehensive information, especially to knockout criteria (Appendix A).

It should be mentioned at this point that, designing of TBPs demands generally a "looped approach". This is because it is almost impossible to recognize all design aspects in the beginning. Therefore, after performing the pre-selection, including matching the technical and ecological ranking lists or at least after finally comparing variants, it is very probable that the chosen variants contradict the ideas from the beginning of the design. So, feedback on all planning results should be provided especially to decision makers in waterway authorities, if objectives sought and those which can realistically be achieved are not in agreement.

3 Selection Techniques and Decision Making

The AHP Multi-Criteria-Approach uses in principle a weighted average of several scores, which quantify the importance (by the weight) and the degree of fulfilment (by a score) of the different criteria, leading to a final score s_{AHP}, which is used to compare selected variants.

As its name implies, the AHP should base on hierarchically structured criteria. Here, three criteria groups were chosen, which are related to technical, social, and ecological issues. Each group contains three subgroups, which contain in principle a large number of single criteria that are tackled in the report by providing numerous scoring hints, but they will be tackled in applying the AHP as one sub-criterion only.

The subgroups of the technical and ecological issues describe in principle the accordance to boundary conditions and the assessed degree of fulfilment of functionality demands related to different aspects, e.g. to stability and economy in the technical group or to selected taxa and habitats in the ecological group (see Table 1), where the structure of the AHP and the scoring-system is shown.

The main purpose of the mathematical background of the AHP Approach is to objectify the weights. This is realized by pairwise comparisons of the importance of one criterion related to the other and supported by verbal specifications as "the first criterion is three times more important than the second one". This means mathematically that the weight of criterion 1 should be three times bigger than the weight of the second. Together with the reasonable rule that the sum of weights should be 1, this results generally in an overdetermined equation system for evaluating the weights, which is solved by the AHP algorithms. The advantage of this approach is that the most important criteria are carved out, so that the final weighted average result is not "diluted" as if the appliers assess the weights directly.

But even if the applier of the WG 128 approach gets assistance by choosing the weights, he is generally not an expert and needs at least assistance for choosing the scores. Besides several scoring hints in form of numerous tables, discussing all relevant criteria, the report offers the possibility to use results from the comprehensive pre-selection tools described in Chapter 5. This is, to get a first proposal (as the pre-selection uses of course a strongly reduced number of criteria only) for all the technical and ecological scores in Table 1, as they are directly related to those from the pre-selection – and the latter are the outcome of the evaluation of all the realized measures in Part 2 of the report – and this evaluation, including "projections", was performed from the experts of WG 128. So, the aim of the Best Practice Approach to "transfer" experience from realized measures to a planned measure will be supported in many ways in the report, inter alia by the comprehensive pre-selection schemes.

The applier of the BPA will also be supported by several EXCEL tools, tackling the Screening, the comprehensive Pre-Selection up to the application of AHP altogether, and delivers the aforementioned proposal for the AHP-scores by linking the results of pre-selection and AHP automatically.

Concerning the most important step for assessing the AHP-scores, the applying of the comprehensive pre-selection, he is forced to fill in four tables:

Table 1. Structure of the Analytic Hierarchy Process (AHP) for comparing TBP-variants with scoring and weighting hints. The weights of the criteria in each subgroup and those between groups are assessed by pairwise comparisons of the importance of the one criterion related to the other, using the AHP-algorithm.

Crite-rion group	Group weight	Single (subgroup) criterion	Weight related to (sub)group	Single (sub-group) score	Remarks concerning scoring	Weighted group score	Final score
1: Technical performance	$W1$ This weight quantifies the importance of criterion group 1 related to groups 2 and 3	Degree of fit to site and adap-tability	$W_{1.1}$ This weight quantifies the importance of criterion 1.1 related to 1.2 and 1.2	$S_{1.1}$	The score scales *differences* in BCs between the DC-site and those at the AC (where the example of the variant was already realized) - analogous to the scoring in Tables DFT and AFT, which should form the basis	Technical perfor-mance score: $S_1 = W_{1.1} \cdot S_{1.1} + W_{1.2} \cdot S_{1.2} + W_{1.3} \cdot S_{1.3}$	Overall perfor-mance score $S_{AHP} = W_1 \cdot S_1 + W_2 \cdot S_2 + W_3 \cdot S_3$
		Performance concerning stability and sustainability	$W_{1.2}$	$S_{1.2}$	The score assesses, up to which degree demands of the DC will be fulfilled by the variant - analogous to the scoring in Tables DST and AST, which should form the basis		
		Performance related to avoid technical efforts	$W_{1.3}$	$S_{1.2}$	The score assesses up to which degree demands of the DC will be fulfilled by the variant - analogous to the scoring in Tables DST and AST, which should form the basis		
2: Social and legal performance	W_2	Fulfilment of human demands	$W_{2.1}$	$S_{2.1}$	The score must be assessed from the applier directly, but in the sense of the degree of fulfilment of demands	Socical and legal performance score: $S_1 = W_{2.1} \cdot S_{2.1} + W_{2.2} \cdot S_{2.2} + W_{2.3} \cdot S_{2.3}$	
		Performance concerning landscape and preservation issues	$W_{2.2}$	$S_{2.2}$	The score must be assessed from the applier directly, but in the sense of the degree of fulfilment of demands		
		Fulfilment of legal demands and accep-tance	$W_{2.3}$	$S_{2.3}$	The score must be assessed from the applier directly, but in the sense of the degree of fulfilment of demands		
3: Ecological performance	$W_{1.1}$	Degree of fit to site conditions and adaptability	$W_{3.1}$	$S_{3.1}$	The score scales *differences* in BCs between the DC-site and those at the AC (where the example of the variant was already realized) - analogous to the scoring in Tables DFE and AFE, which should form the basis	Ecolo-gical perfor-mance score: $S_1 = W_{3.1} \cdot S_{3.1} + W_{3.2} \cdot S_{3.2} + W_{3.3} \cdot S_{3.3}$	
		Performance related to selected taxa	$W_{3.2}$	$S_{3.2}$	The score assesses, up to which potential degree demands of the DC will be fulfilled by the variant - analogous to the scoring in Tables DSE and ASE, which should form the basis		
		Performance related to selected habitats	$W_{3.3}$	$S_{3.3}$	The score assesses, up to which degree demands of the DC will be fulfilled by the variant - analogous to the scoring in Tables DSE and ASE, which should form the basis		

The first table, called DFT (Design Case, feasibility, technical issues), tackles the boundary conditions at the site of the planned measures with 8 selected single criteria: ship-induced impacts, slope inclination, erodibility, excess pore water pressure, hinterland space, water level changes (ΔW) between mean water MW (as vegetation with bank protection function can establish generally only above MW) and HSW (where navigational impact vanishes), vegetation growth conditions and ice-effects. The applier is demanded to assess the importance of each criterion, e.g. the ship-induced impacts in relation to the other criteria first. Then the scores have to be assessed between 0 (demanding for strong, more technical measures, e.g. very strong navigational impact) up to 1 (e.g. low ship-induced impact, speaking for "green" measures), whereby all scores will be explained by example of a selected Design Case, which is visualized in Fig. 5: A bank slope to be protected at the impounded Weser River. The scoring rules are explained comprehensively and supported by several tables, using easy-to-achieve data only that are e.g. the CEMT class or the fairway-bank distance to assess the ship-induced impact score. It is not necessary – even if of course helpful – to have e.g. wave data.

The scores in the DFT-table will be compared to analogous data in the AFT-Table, which contains the expert-rated scores (including projections) of all the realized measures in Part 2. The score of each criterion is calculated by using *differences* between the scores of DC and AC and leads to numbers between −1 (not feasible), 0 ("just O.K") and +1 (feasible). The final so-called Feasibility Score S_{FT}, which is a weighted average of all 8 differences, assesses therefore possible differences in BCs at DC and the various AC sites. If the differences are large (the final score is negative), even if the extended BCs from projections will be used, it is unlikely that the AC measure considered will "work" at DC site and the Feasibility Check fails.

The results of the Technical Feasibility Check will be offered in different ways by the EXCEL sheet, e.g. in form of ranking lists of the most appropriate measures, that are those with the highest feasibility scores.

The second table to be filled-in is called DST (DC, suitability, technical). The user is asked to specify his technical demands (by weights and scores) to the measure by 9 selected criteria. The latter concern the administrative support, access, erosion resistance, stability/durability, duration of initial phase, sustainability reflecting local conditions and materials, investment and maintenance expenses. The scores should be chosen again between 0 and 1, whereby a score of 0 means that e.g. the expenses may be very much higher than comparable technical solutions and 1 that the budget is low and thus low-cost solutions should be preferred.

The DST-scores will be compared with analogous scores from the various ACs (Table AST), where the experts in the WG assessed analogous functionality issues. If the demands are fulfilled entirely, the score is +1, if it totally fails it is −1, if the AC offers some positive properties, even if not demanded for, the score is 0.5 and so on. A weighted average of these comparison-scores is called "Suitability Score" and quantifies, up to which extent the diverse ACs will fulfil the demanded functionality issuers of the planned measure. Again, a "hit list" of the AC-measures with the best suitability scores will be provided from the EXCEL sheet.

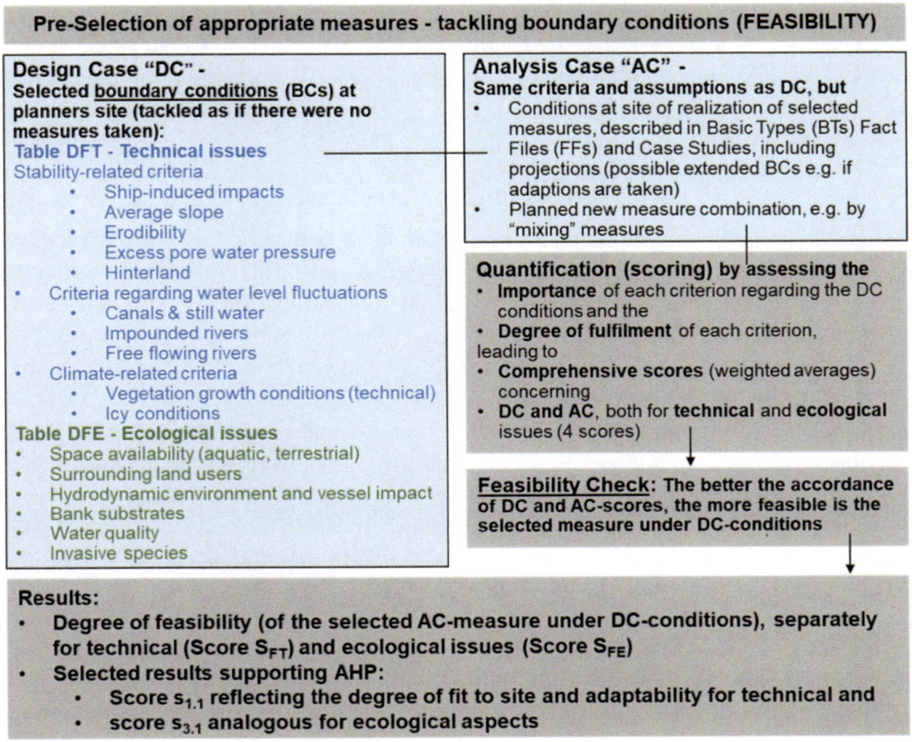

Pre-Selection of appropriate measures - tackling boundary conditions (FEASIBILITY)

Design Case "DC" -
Selected boundary conditions (BCs) at planners site (tackled as if there were no measures taken):
Table DFT - Technical issues
Stability-related criteria
- Ship-induced impacts
- Average slope
- Erodibility
- Excess pore water pressure
- Hinterland
- Criteria regarding water level fluctuations
  - Canals & still water
  - Impounded rivers
  - Free flowing rivers
- Climate-related criteria
  - Vegetation growth conditions (technical)
  - Icy conditions
Table DFE - Ecological issues
- Space availability (aquatic, terrestrial)
- Surrounding land users
- Hydrodynamic environment and vessel impact
- Bank substrates
- Water quality
- Invasive species

Analysis Case "AC" -
Same criteria and assumptions as DC, but
- Conditions at site of realization of selected measures, described in Basic Types (BTs) Fact Files (FFs) and Case Studies, including projections (possible extended BCs e.g. if adaptions are taken)
- Planned new measure combination, e.g. by "mixing" measures

Quantification (scoring) by assessing the
- **Importance** of each criterion regarding the DC conditions and the
- **Degree of fulfilment** of each criterion, leading to
- **Comprehensive scores** (weighted averages) concerning
- **DC and AC**, both for **technical** and **ecological** issues (4 scores)

Feasibility Check: The better the accordance of DC and AC-scores, the more feasible is the selected measure under DC-conditions

Results:
- **Degree of feasibility (of the selected AC-measure under DC-conditions), separately for technical (Score S_{FT}) and ecological issues (Score S_{FE})**
- **Selected results supporting AHP:**
  - **Score $s_{1,1}$ reflecting the degree of fit to site and adaptability for technical and**
  - **score $s_{3,1}$ analogous for ecological aspects**

Fig. 3. Definition sketch showing the main features of the Feasibility Check for tackling the importance of role of possibly different boundary conditions at planners site (Design Case DC) and at the site of realized measures (Analysis Case AC) for pre-selection.

In the same way as the technical aspects, also the ecological issues are tackled:

In Table DFE (DC, feasibility, ecological), the next table to be filled-in from the user, together with its AC counterpart AFE, the following boundary conditions were quantified and compared to each other to form the Ecological Feasibility Score S_{FE}: Space availability in the aquatic and terrestrial zone, land use behind the bank, hydraulic environment, vessel impacts (ecological view), bank substrate (e.g. according to vegetation growth conditions), water quality and the existence of invasive plant species.

The Ecological Suitability Check (Score S_{SE}) is based on the Tables DSE (DC, suitability, ecological), defining the *demands* to selected ecological issues at DC site (species: macrophytes, riparian vegetation, arthropods, benthic invertebrates, fish fauna, birds; selected habitats: bed and banks, riparian vegetation, connectivity), which are assessed by the approach itself (taking the maximum potentially achievable according to the boundary conditions), so that the user has to specify the *importance* of each criterion only and the Table ASE, containing all the expert-rated scores of the measures. Comparing and quantifying the potential degree of fulfilment of demands by the properties of the measures, leads to the Ecological Suitability Score.

The distinction between Feasibility and Suitability was methodologically neces-sary, even if both aspects are matched in numerous ways – as of course the func-tionality is linked to the local boundary conditions – because the number of cases to be distinguished in case of a combination of both criteria groups would be far too large. Therefore, the results of Feasibility and Suitability Checks have to be matched in some way, both for technical and ecological issues.

For this purpose, the EXCEL tools offer several matching techniques, e.g. the "Logical Matching", which offers the same result as an arithmetic "fifty-fifty-average" of both scores, but if only one of the both criteria is not fulfilled (the score is bad), meaning that either the boundary conditions are not comparable at all (despite of projections) or the AC doesn't fulfil the demands, the final score will be bad too.

From this matched results concerning feasibility and suitability, ranking lists can be constructed again, which can then be used, together with previous ranking lists, in Step 5 of the 7 Steps", to support a purposeful view into the collection of measures in Part 2 of the report. The aforementioned EXCEL tools allow also to match the technical and ecological results of pre-selection and construct according ranking lists. This is very useful, as technical and ecological aspects often contradict to each other, so that it can happen that there is no accordance at all in the different ranking lists.

Thus, before starting with the aforementioned Step 6, the concretion of the Design Cases (variants) to be compared by using AHP, the combined consideration of both technical and ecological issues, makes sense for these Design Cases, even if they are not tackled directly in the collection of measures, because the EXCEL sheets offer several "mixing tools" of variant properties. If e.g. ecological deficits of a preferred AC shall be upgraded, construction elements of another measure can be combined with the preferred solution. E.g. the example of willow brush mattresses protecting the bank slope above MW level (measure 4.3.3 in Fig. 5) could be upgraded by erecting a small dam on the riprap-covered bank in water depths below MW to create a shallow water zone as in the Rhone River close to Beauchastel (measure 3.1.1.2 in Fig. 1). This may be a good solution for the DC example at the impounded Weser River. These "mixing" algorithms, even if technically questionable, allow to check, which construction ele-ments may be appropriate to achieve the most positive effects and thus, to constrict the best solutions.

4   Detailed View on the Collection of Measures

Regardless of the different design approaches outlined above, the collection of mea-sures in Part 2 of the report, which are visualized here in Figs. 1, 4 and 5, has also an inestimable value in itself. Simply looking into Part 2 of the report and recognizing which solutions were used regarded the different conditions, helps to select and adapt appropriate measures for the DC considered.

Part 2 is divided into three categories of measures:

The first category is not related to a special measure with its unique environment. These measures are called "Basic Types" (BTs). There was no template prescribed to write the texts of the Basic Types as the sources of information varied widely. Thus, the authors were free to present the information in an appropriate way. Nevertheless, the scoring rules for filling-in the pre-selection tables were the same as for Fact Files and Case Studies with one exception: As the Basic Types don't reflect a special measure at a unique site as all the other measures described in FFs and CSs, the corresponding scores for boundary conditions and functionality issues must be assessed from the experts, that is, they could not extracted from the conditions at AC-site. The "scorer" considered therefore the measures as to be realized under "typical conditions", which were extracted from the sources of information. The same holds true for the "projections" concerning the applicability ranges of boundary conditions and functionality issues. Hence, the scores of the Basic Types are of course less accurate than those for FFs and CSs.

Nine BTs were extracted from the German DWA-M 519 code of practice on technical-biological bank protections for inland waterways. They are: Grass revetment, reed plantings, life willow, reed mat with vegetation roll as foot protection, pre-vegetated slope mat with grass, pre-vegetated slope mat with hardwood cuttings, willow weave, living fascine with hardwood cuttings, vegetated geotextile bodies with (hedge) brush (bush) layers, chamber revetment (vegetated stone mattresses, reed gabions), willow brush mattresses with heavy riprap as toe reinforcement and vegetated riprap, see Fig. 1.

Ten Basic Types were chosen from the numerous measures described in the UK Waterway Management Guide. As at least some ecological upgrade should be possible, "pure" technical measures were not selected. Note that all recommended measures are valid predominantly for UK or Middle Europe with a mild humid climate. They include: Grass revetment, reed planting, life willow on toe and bank, grass and geotextiles, timber and vegetation, rock and fibre rolls, toe-geotextile – fibre rolls, rock and fibre roll, vegetated open cell revetment, vegetated stone revetment and vegetated concrete units revetment, see again Fig. 1 for visualization.

The last group of Basic Types were extracted from the UK guideline "Green Approaches in River Engineering". Twelve case studies are presented that illustrate different types of measures and different types of rivers. The BTs (single measures) of the mostly described measure combinations were filtered out and used for the WG 128 report. Also measures that were implemented in tide-influenced river systems were not taken into account. The considered cases are: Willow spilling/bundles, turf/vegetated reinforced mattresses/seeded geotextile (e.g. seeded coir matting) with rock rolls as toe protection, life root wads and willow brush mattresses on coir matting with faggot bundles, rock rolls and logs as toe protection.

The second measure category contains so-called <u>Fact Files</u> (FFs).

In total, 34 Fact Files of realized measures from across Europe and China were collected and grouped according to the waterway type, where the measures were realized. This is, because the magnitude of water level fluctuations, which are associated with the waterway type (canals, impounded rivers, free flowing rivers), is generally the most important boundary condition to be considered in the design. A second categorization was chosen according to the strength of ship-induced impacts, which is the second most important criterion for design of TBPs according to the experiences of the WG 128 members. The lower the impact, the more "green" solutions are possible. Thus, the measures are categorized into those related to weak, average and strong impacts.

Each Fact File contains the designation and the affiliation to different measure types, as pre-embankment or direct measures or those using living and dead plants only (bioengineering measures) or bio-technical engineering measures and so on. Then, the boundary conditions and objectives of the measures are described. Next, construction details and the time course of realisation is reported, followed by the discussion of the achieved functionality in view of the boundary conditions and objectives. At last, the lessons learned, improvements, and advices for applications under different conditions are discussed.

The different measures can be seen again in the aforementioned Figures. It is obvious that most of the measures are realized in Channels with small, up to average water level changes. The problem with TBPs for free flowing rivers is, that long periods of draught or floods hinder the usage of plants as the only protection elements, which means that realized measures are often combined with technical components. One reason is that very high water level changes are generally not really natural, hence there is almost no experience available from real nature to be transferred to appropriate bank protections.

The third category of measures is described very comprehensively in form of Case Studies (CSs), containing the following points: Designation, objectives, brief description, realization in site, relevant local boundary conditions, evaluation, overall result and imprint. They are as the FFs categorized according to the magnitude of water level fluctuations.

For waterways with low water level fluctuations (canals, strongly canalized rivers and still water), two CSs are presented from the Brussels-Schelt canal: Shallow water zone behind a parallel dam and behind a sheet pile wall. The examples demonstrate impressively that the most important ecological deficits in canals arise from lacking shallow water zones. These are thus compensated by artificial ones, which are – as the impacts are large – sheltered by wave breakers. The challenge is to "balance" the necessary protection against ship impact with the need of water exchange between the channel and the shallows.

There is one example concerning moderate water level fluctuations between MW and the highest shipping level, as it is often the case in impounded rivers. It describes a *planned* measure, not a realized one, but it was derived on basis of a test area (measure 3.2.3.1 in Fig. 4) with similar boundary conditions and on numerous investigations. The measure creates a large shallow water zone behind a sheet pile wall. The most important design feature is that it offers adjustable top and under water openings ("fish windows"), because experiences showed that the vegetation in the artificially created shallows (planted and natural succession) has to be "controlled" in some way, whereby the openings have to be small in the initial phase and can be larger in the mature state.

Three Case Studies concern large water level fluctuations in free flowing rivers. They are situated in a German test section of TBPs on the Rhine River near Worms town and have been monitored up to now for more than 10 years. The first Case Study describes a vegetated riprap protection with a low-level dam ahead of the bank slope for improving fish ecology, the second measure protects the bank with vegetated gabions (reed gabions) and the third with living brush mattresses. The latter form the AC-example used everywhere in the report for explaining the various approaches.

Fig. 4. Visualisation of all measures described comprehensively in Part 2 of the WG 128 report: FFs for low water level changes (ΔW) and average ship-induced impact up to high ΔW and average ship-induced impacts.

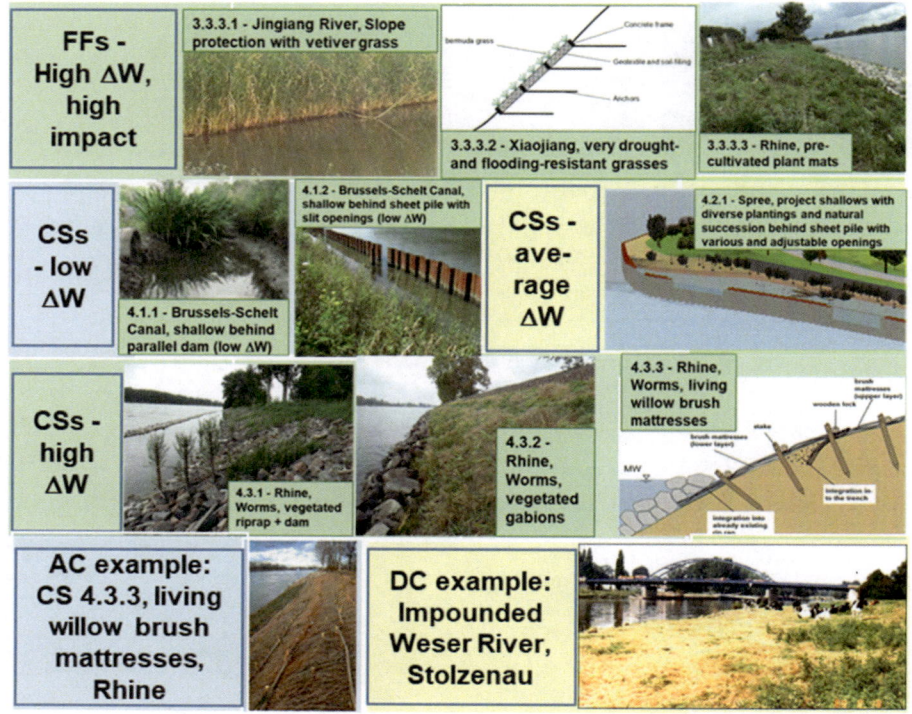

Fig. 5. Visualisation of all measures described comprehensively in the WG 128 report: Fact Files for high water level changes (ΔW) and high ship-induced impact, Case Studies and examples for applying the report.

5   Summary and Conclusions

The report of WG 128 is in its final stage at the time of writing this paper. It follows the reviewing phase, first by the members of the WG, then by INCOM. It is assumed that the report can be published by the end of 2022. After the publication, tutorials may start, which are very important, as the application of all the tools need some explanations and guidance by the authors.

The PIANC WG 128 has collected, comprehensively described and evaluated a lot of TBPs usable for different types of inland waterways with focus of those applicable under humid climate. To facilitate the selection of appropriate measures, the latter are grouped into different categories of water level changes and the magnitude of ship-induced impacts. These categorizations can be used in a first step to select generally applicable measures.

The report offers not only a collection of measures, but also several approaches to select appropriate measures on a rational, quantitative basis, based on a scoring system, whereby the scores of all the collected measures are provided from the experts in the WG. The selection tools use comparisons of boundary conditions of the Design Case and those of realized measures (ACs), called Feasibility Check as well as the balance of demands versus properties of the measures, which is called Suitability Check. These "Checks" are provided both for technical and ecological issues and concerning technical issues for the measures "as they are", but also "as they could be", e.g. up to which extent the measures may be appropriate even for worse boundary conditions. The ACs may be those from the collection, but also measure-combinations.

The pre-selection tools tackle selected technical and ecological criteria, but for the final selection all the relevant design criteria are considered, including social and legal design aspects, whereby these approaches are supported in numerous ways by EXCEL sheets. These concern various pre-selection tools up to the application of multi-criteria-analysis tools on the basis of a special Analytic Hierarchy Process.

Also two other, very stringent design tools, those derived and extended from the German Code of Practice DWA-M 519 and from the UK Waterway Management Guide (1999) were presented. They are recommended as two alternatives to the pre-selection tools.

The report offers finally process recommendations supporting the whole design process. This, especially for planners. But also decision-makers in waterway authorities or contract-makers, which need less comprehensive information and may be satisfied with the results from pre-selection. Therefore, the report allows a selective reading and application to satisfy all the usual users of the PIANC guidelines, which reach from technically interested laymen up to planners in engineering bureaus.

It is clear even to the authors, that the report will not solve all design problems and it may be necessary from case to case to perform additional studies. But it is obvious that the report provides a very useful and supportive tool for applying more nature-friendly bank protections by providing numerous design hints for solving the complex design problems associated with TBPs.

References

BAW, BfG (2019) Alternative technical-biological bank protection on Inland Waterways: Publications, Practical Information, Measures and Events. http://ufersicherung.baw.de/en

Ma D, Xing Y (2019) Theoretical research on the evaluation of the green ecological bank slope treatment for inland waterways. In: Smart rivers conference, Lyon

DWA (2016) Technisch-biologische Ufersicherungen an großen und schiffbaren Gewässern. DWA-M 519. Deutsche Vereinigung für Wasserwirtschaft, Abwasser und Abfall e. V. (DWA)

Gesing C, Söhngen B, Kauppert K (2016) Design of bank protection for inland waterways with GBBSoft+. ISRS 2016

Li J, Li Z (2013) Analysis on improving the economy of fabricated building by building industrialization technology progress. Housing Industry

Du L (2010) Study on compound revetment of restricted channel in inland river based on ecological engineering method. Southeast University

PIANC Report No. 99 (2008) Considerations to reduce environmental impact of vessels (results of PIANC INCOM WG 27)

PIANC (2019) Alternative technical-biological bank protections for inland waterways. In: Workshop in the framework of the smart rivers conference, Lyon

PIANC INCOM WG128 (2022, to appear) Technical-biological bank protections for inland waterways. INCOM WG Report n° 128 – 2021, 2021

Roca M, Escarameia, M, Gimeno O, et al (2017) Green approaches in river engineering - supporting implementation of green infrastructure. HR Wallingford, UK

Söhngen B, Fleischer P, Liebenstein H (2016) German Guidelines for designing alternative bank protection measures. ISRS 2016

UK (1999) Environment agency. Waterway bank protection: a guide to erosion assessment and management. R&D Publication, UK

UK (2019) Estuary edges: ecol. Design guidance. The River Restauration Centre, UK Environmental Agency

Verbelen J, Söhngen B (2019) Best practice approach of indirect technical-biological bank protection on waterways with strong ship-induced impacts. In: Smart rivers conference, Lyon

Wang Q (2010) Application of vetiver grass system in the treatment of medium and small rivers. Hunan Hydro & Power

Bao W (2018) Ecological effect of bermuda grass cynodon dactylon on the water-soil atmosphere system in Riparian zone of Pengxi River. Chongqin Jiaotong University, Chongqing

Xing Y, Ma D (2019) Design theories and technique of treatment of green ecological bank slopes for inland waterways. In: Smart rivers conference, Lyon

Changes of Channel Conditions of Zhangjiazhou Waterway for the 2020 Hydrological Year and Maintenance Countermeasures

Hui Xu[1(✉)], Qianqian Shang[1], and Xiangjun Xu[2]

[1] Key Laboratory of Port, Waterway and Sedimentation Engineering of the Ministry of Transport, Nanjing Hydraulic Research Institute, Nanjing, China
{huixu, qqshang}@nhri.cn
[2] Haian Water Conservancy Bureau, Nantong, China

Abstract. The 2020 hydrological year is featured with the high-water level and high flow period far exceeding the previous hydrological years. The sediment deposition in flood season shallowed and impeded the channel. Based on the hydrological analysis of 2020, the impact of the catastrophic flood year on the channel conditions in the upper shallow area of Nangang, and the reasons for the deterioration of the channel conditions in the shallow area are analyzed. Meanwhile, channel maintenance countermeasures, the necessity of emergency dredging, and the specific dredging scheme are proposed accordingly. The results show that: the upper shallow area of Nangang followed the law of silting in flood season and scouring in the dry season during a hydrological year. The channel conditions depend on the comparison between silting in the flood season and scouring after the flood season. In 2020, the flood lasted for a long time, so the water-falling period was relatively insufficient after flood season, and the upper shallow area was not sufficiently scoured, which resulted in navigation obstruction in the upper shallow area. Therefore, emergency dredging was necessary to be implemented on the channel, the dredging site was designed from #10 navigational mark to #9 navigational mark, the dredging bottom elevation was 4.2 m below the design water level, and the dredging width was 200 m. Considering the hysteresis of riverbed deformation, it is suggested to pay continuous attention to the water conditions and channel variation in the upper shallow area of Nangang in the future, and to formulate corresponding channel maintenance countermeasures.

Keywords: The 2020 hydrological year · Channel conditions · Countermeasures · Emergency dredging · Impact analysis

1 Introduction

In 2020, the Yangtze River Basin encountered continuous heavy rainfall, affected by which, this area experienced watershed flood which was second only to that in 1954 and 1998 since the founding of new China. The peak water level of Jianli to Datong

© The Author(s) 2023
Y. Li et al. (Eds.): PIANC 2022, LNCE 264, pp. 1087–1097, 2023.
https://doi.org/10.1007/978-981-19-6138-0_95

section in the middle and lower reaches ranked second to fifth in the measured records. Wherein, the peak water level of Jiujiang station and Hukou station ranked second in history, next to that in 1998. The tide level from Ma'anshan to the Zhenjiang section even exceeded the historical record (Feng 2020; Yao et al. 2020). The middle and lower reaches of the Yangtze River basin generally follow the law of "heavy water brought a heavy volume of sediment". A high volume of sediment carried by the flood in 2020 was deposited in the channel, resulting in shallow channel conditions after the flood season, which influenced the channel maintenance and dredging, and navigation in waterways with good conditions previously was also impeded by shallowing of channels. Taking Zhangjiazhou waterway as an example, this paper analyzed the impact of the flood in 2020 on the channel conditions of Zhangjiazhou waterway and put forward countermeasures on channel maintenance and dredging.

2   Research Region and Data Source

Located at the junction of Hubei, Jiangxi, and Anhui provinces, Zhangjiazhou water-way starts from Suojiang Building of Jiujiang City and ends at Bali River estuary, with a total length of about 30 km, belonging to a slightly curved double branch river. Zhangjiazhou divides the river into Nangang and Beigang, which converge at Bali at the end of the Zhangjiazhou, as shown in Fig. 1. Zhangjiazhou waterway is one of the key navigation obstruction shoal waterways in the lower reaches of the Yangtze River. At present, the channel maintenance scale is 5.0 m × 200 m × 1050 m (water depth × navigation width × bending radius).

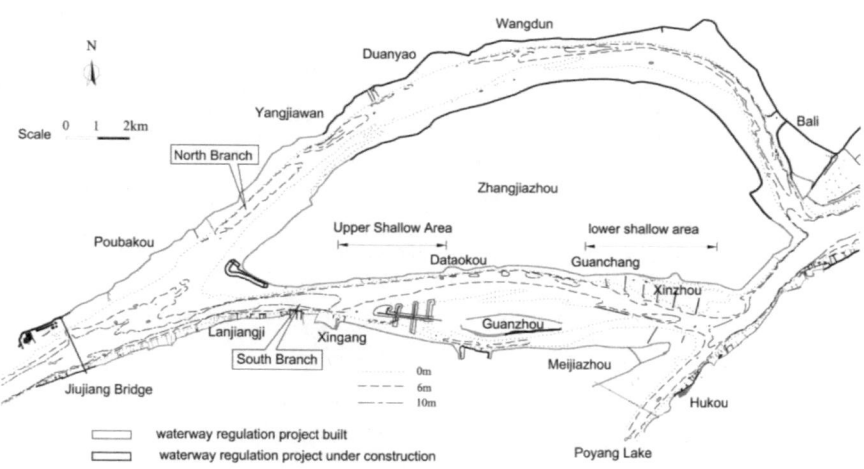

Fig. 1. River Regime of Zhangjiazhou waterway (March 2018)

The channel of Nangang (also named South Branch) is smooth and straight. As the main channel of the Yangtze River, Nangang Channel has silting bodies such as Guanzhou and Xinzhou, with Poyang Lake flowing in from the right bank in the lower section. There are two shallow areas in Nangang. Wherein, the upper shallow area is located in the widening section opposite to the new port. Due to the retreat of low shallow in front of Guanzhou, the thalweg swing increased, and the cross-section of the river in the shallow area was widened and shallowed (Xue et al. 2008), and the channel conditions deteriorated. In order to improve the channel conditions, Changjiang Waterway Bureau began to regulate the upper shallow area of Nangang of Zhangjiazhou in January 2009. The project passed the completion acceptance in February 2012. The regulation was successful, achieving the goal of 4.5 m × 200 m × 1050 m (Liu 2015; Li et al. 2009); The lower shallow area was in the transition section between Guanchang and Poyang Lake estuary. The widening river surface and the diffusing water flow in this area resulted in serious navigation obstruction. So, this area was the key to the smooth flow of the Zhangjiazhou waterway. Changjiang Waterway Bureau began to regulate the lower shallow area of Nangang in January 2001 (Huang 2002). In March 2006, the project passed the completion acceptance. After the regulation, the waterway conditions were significantly improved, and the channel condition was good (Li et al. 2010). The channel scale exceeded the goal of 4.5 m × 200 m × 1050 m.

The channel of Beigang (also named North Branch) channel is long, narrow, and bending. Before 1989, it was used as the main channel in dry seasons. However, there were many shallow areas, and revetment riprap rocks extending into the river along the banks of Chengjiaying to Bali River, so the disordered flow pattern, the bending flow as well as the low maintenance scale make it difficult to maintain the waterway. After 1989, with the increase of the diversion ratio of Nangang, the channel maintenance department mainly maintained Nangang and stopped using Beigang as the main channel of the Yangtze River (Wang 1989).

In order to open the 6-m-deep channel in Wuhan to Anqing section, the 6 m deep channel regulation project is being carried out for Zhangjiazhou waterway (Chen et al. 2019; Deng 2019), mainly to build a new bottom protection belt at the entrance of Guanzhoujia, a new beach protection belt with hook head in the upper section of the left side of Guanzhou; and repairing the #1 comb dam. See Fig. 1 and Table 1.

Table 1. Regulation works completed and under-construction in Zhangjiazhou waterway

	Project name	Project contents	Construction period	Waterway construction scale
1	Regulation works of the upper shallow channel of Nangang	Constructing a comb dam at the front of Guanzhou and a bottom protection belt at Guanzhoujia inlet	2009 to 2013	4.5 m × 200 m × 1050 m
2	Regulation works of the lower shallow channel of Nangang	Constructing 6 spur dikes on the left bank, and bank slope protection works between L3-L4# spur dikes; constructing 2 beach protection belts on Meijiazhou of the right bank; constructing revetment on the point bar of the end of Guanzhou	2002 to 2007	4.5 m × 200 m × 1050 m
3	6 m deep waterway regulation project in Wuhan to Anqing section	Constructing a bottom protection belt at Guanzhoujia inlet; a new beach protection belt with hook head in the upper section of the left side of Guanzhou Continent; and repairing #1 comb dam	2018 to now	6.0 m × 200 m × 1050 m

The data on which this paper is based mainly include three aspects, namely, river topography, local topography of channel, and flow and sediment data of hydrological stations. The topography data source from the survey data of the Changjiang Waterway Survey Center and Changjiang Jiujiang Channel Section; the flow and sediment data of hydrological stations are based on the flow and water level published by the Hydrographic Office of Changjiang Water Resources Commission.

3   Water Regime Analysis for 2020

There were nearly 20 large-scale heavy rainfall processes in southern China since the flood season of 2020. From June 1 to July 9, the cumulative precipitations of 85 stations in Hubei, Anhui, Jiangsu, Guizhou, Zhejiang, Chongqing, Hunan, Jiangxi, Shanghai, Guangxi, Sichuan, etc. have exceeded half of the average annual precipitations. The average precipitation in the Yangtze River Basin reached 369.9 mm, 54.8 mm more than the same period in 1998, and hit the highest in the same period since 1961. The rainstorm increased the flood control pressure in the Yangtze River Basin. The water level of Hankou station on the Yangtze River reached 28.77 m (frozen by Wu Song), ranking 4th since 1865, second only to 1954, 1998, and 1999.

From the flow process of Jiujiang Hydrometric Station in the upper reaches of the studied river section in 2020, the flow from January to April was basically stable in recent years and dropped sharply in late April; the flow was smaller than that in previous years from April to the end of May, and began to rise in early June, reaching the peak of 67,400 m³/s on July 9; the flow remained above 60,000 m³/s by the end of July; as of December 31, there were 103 days with a flow above 40,000 m³/s, 53 days above 50,000 m³/s days, and 22 days above 60,000 m³/s. The high-flow period was much longer than that in previous years. See Fig. 2.

Therefore, the water regime in 2020 was featured with a high-water level and high flow period far exceeding that in previous years.

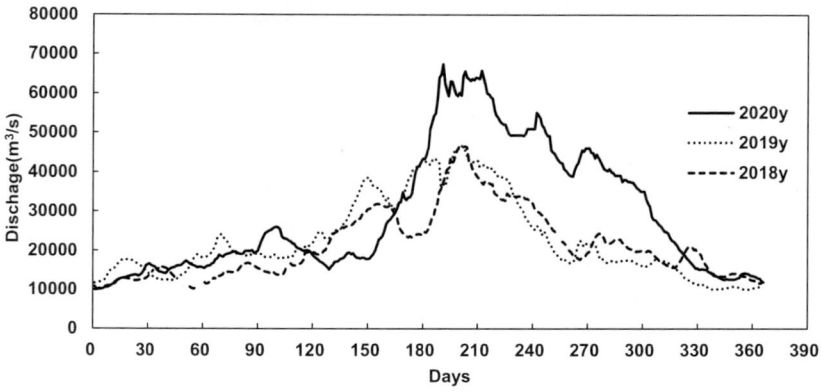

Fig. 2. Daily average flow process of Jiujiang hydrometric station in recent three years

4   Recent Riverbed Evolution

Zhangjiazhou waterway has been distributed into two branches for many years, and its river regime is relatively stable. Beigang is a slightly bending branch, which generally conforms to the general law of bend evolution, that is, silting on the convex bank and scouring on the concave bank. As bank protection works were imposed on the bank collapse section of the left bank of Beigang, the river coastline has been well controlled and was generally stable. The recent evolution is mainly manifested in the longitudinal scouring and silting deformation of the riverbed, the increasing and return of shoals due to siltation and scouring, and the reciprocating of the longitudinal change of thalweg. The plane shape of Nangang is relatively straight, but there is a secondary bifurcation in the middle and lower sections of the waterway, with Guangzhou in the upper section and Xinzhou in the lower section. Poyang Lake flows in from the right bank in the lower section. It generally conforms to the evolution law of a straight channel: the point bars on the sides of the river change alternately at the place where the waterway widens, and the mainstream swings indefinitely, which tends to form transition shoals.

After the 1990s, Nangang and the Beigang changed in an opposite way, with the diversion ratio of Nangang increasing gradually and that of Beigang reducing. With the increase of the diversion ratio of Nangang, the deep pool close to the right edge of Zhangjiazhou also began to develop scouring depth. At the same time, Guanzhou high beach gradually formed, and a relatively obvious branch was formed at Guangzhoujia. There is little change in the upper section of Nangang. In recent years, the thalweg in the upper shallow area has swung to the right, and the front of Guanzhou has retreated due to scouring; the bank lines of the lower section of Nangang is stable, and the evolution over the years is mainly manifested in the swing of thalweg and the growth and decline of upper- and lower-point bars.

From 2010 to 2016, the inlet channels of Nangang and Beigang of Zhangjiazhou were scoured and the low beach at the front of Zhangjiazhou was silted up. The river bed in Beigang was subject to both scouring and silting, mainly silting; Nangang showed both scouring and silting trends, with the low beach on the left edge and the tail of Guanzhou showing a scouring trend, and the inner part of Guanzhou showing slightly scouring and silting; From 2010 to 2016, back siltation with high volume was observed in the scouring area at the right edge of the front of Zhangjiazhou. The siltation area at the front of the regulation project of the lower shallow area moved downward. The low beach at the left edge of Guanzhou and the end showed a scouring trend. The deep pool gradually closed to the lower section of the left edge of Guanzhou, and there was no obvious scouring in Guanzhoujia.

From 2016 to 2019, the river channel generally showed a scouring trend. There was relatively obvious scouring in the inlet section of Beigang and the low beach area at the front of Zhangjiazhou, and the low beach at the front of Guanzhou and its left and right edges were dominated by scouring.

From July 2019 to August 2020, affected by the flood in 2020, the inlet section of Nangang was mainly formed by siltation, with a thickness of 1 to 3 m; the siltation thickness in the middle section was about 1 to 2 m; the region near L2 # and L3 # spur dikes in the lower section was subject to scouring of some extent, with an amplitude of about 2m. The scouring amplitude at the front of R1 # and R2 # beach protection belts was about 1m. The riverbed downstream of the confluence section of the lake mouth was generally silted up, with a maximum siltation thickness of about 5 m (Fig. 3).

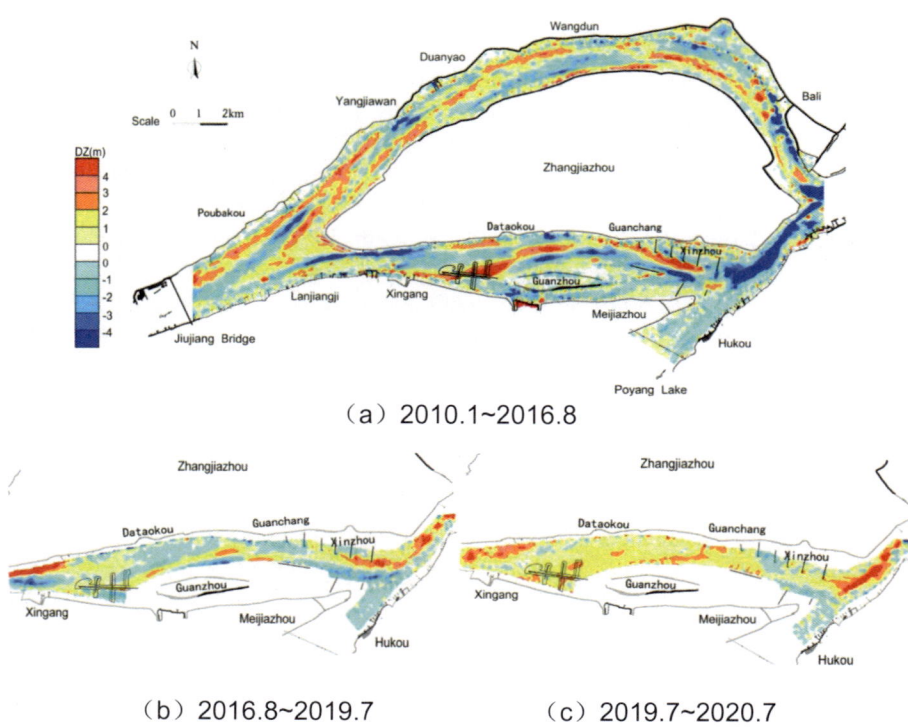

（a）2010.1~2016.8

（b）2016.8~2019.7   （c）2019.7~2020.7

Fig. 3. Recent channel scouring and silting of Zhangjiazhou waterway

5   Impact of Catastrophic Flood Year on Channel Conditions

5.1   Conditions of Shallow Channel

Comparing the channel conditions in the upper shallow area over the years, it was found that: in March 2013, the 5 m isobath at the narrowest part of the upper area of Zhangjiazhou South waterway was 310 m wide; in May 2018, the 5 m channel of Zhangjiazhou South waterway was penetrated with a width of more than 400 m; in April 2020, the 5 m isobath was penetrated with a minimum width of 400 m. By July 2020, due to a large volume of sediment deposition during the flood period, the 5 m isobath in the upper shallow area was disconnected, with a disconnection length of 370 m.

The large flow in the flood season in 2020 caused the flood plain of the main flow, the flow contraction capacity at the front of Guanzhou was weakened, the waterway in the upper shallow area of Zhangjiazhou was widened, and the sediment was deposited. In addition, the main flow transited from the right bank to the left bank in this section.

The flood in 2020 led to the staggering of the upper and lower deep grooves, and a shallow obstruction was formed in the middle.

In Comparison with the local mapping at the end of the water return period in 2019 and 2020 (Fig. 4), the 4 m isobath in November 2020 silted up about 330 m into the channel, compared with November 2019. The 4 m isobath in the channel was penetrated with a minimum width of only 25 m, while the 3 m isobath was penetrated with a minimum width of 215 m. The thalweg swung from the original north side to the middle position. The channel presented an "S" shape, which was difficult for the layout of navigation marks and navigation of vessels in the upper and lower area.

Since November 2020, according to the decline of water level and the change of scouring and silting of the channel, Jiujiang Channel Section has added conical side shore beacons for the upper shallow area of Nangang of Zhangjiazhou, Zhangnan 9-1 red float, and adjusted 25 navigation beacons. However, with the further decline of the water level, the channel cannot be effectively scoured, and dredging shall be implemented to meet the minimum maintenance scale of the channel.

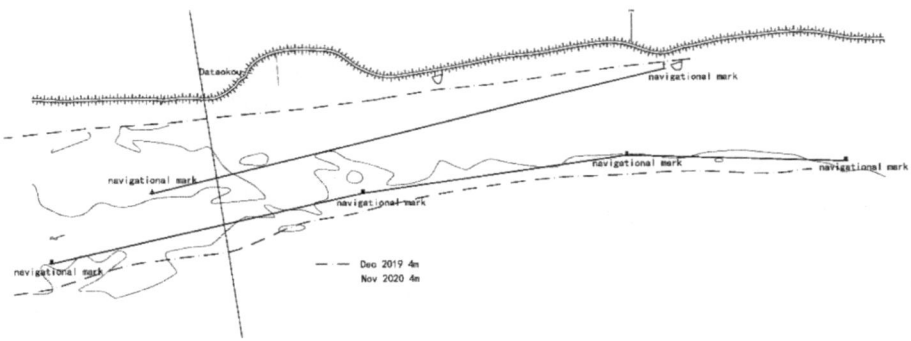

Fig. 4. Comparison of Isobaths between November 2019 and November 2020 in the upper shallow area of Nangang of Zhangjiazhou

5.2   Causes to Change of Channel in Shallow Area

Combined with the riverbed changes of Zhangjiazhou waterway over the years, the upper shallow area of Nangang followed the evolution law of "silting in flood season and scouring in the dry season" in the year. During the flood season, the point bar was silted up and the navigation channel was narrowed. With the recession of water after the flood season, the flow returned to the channel and scoured the channel, and the channel conditions were improved. Influenced by the flood in 2020, as in previous years, Nangang was still characterized by "silting in flood season and scouring in the dry season" (Fig. 5). In the dry season after the flood period in 2019 to April 2020, the inlet section and middle section of Nangang were mainly subject to scouring; in the flood season of 2020, due to the large flow and long flood period, the siltation range and amplitude in Nangang far exceeded those in previous years, wherein, the siltation amplitude in the inlet section exceeded 3 m locally.

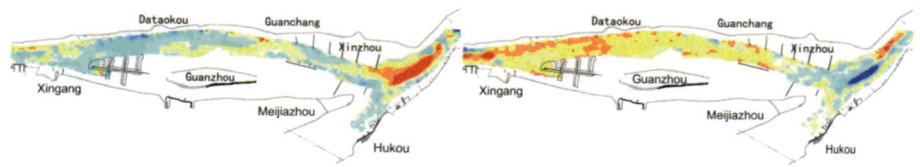

Fig. 5. Scouring and silting of Zhangjiazhou South Waterway in the year

According to the measured data from Jiujiang Channel Section, the water depth of shoal (below the navigational datum) in the upper shallow area of Zhangjiazhou Nangang changed with the inflow process of the year. When the water level was low, the shoal was scoured and the water depth increased, and when the water level was high, the shoal was silted up and the water depth decreased. Therefore, by comparing the days of high-water level period (above the flood land line water level) and days of backwater scouring in the years from 2017 to 2020 (Table 2), the relationship between channel conditions and inflow process in that year can be reflected. Research (Tang et al. 2012) showed that when the water level of Zhangjiazhou waterway retreated to about 6 m of the navigational datum, the shallow would be scoured. Therefore, the water level 6 m above the navigational datum was taken as the critical water level, the days that the shallow was lower than the critical water level shall be taken as the days of backwater scouring. When the ratio of the days of backwater scouring and the days of high water level was large, the scouring time would be relatively long, and the channel conditions were generally good in that year. For example, from 2017 to 2019, the ratio was between 0.39 and 0.87, so dredging maintenance was not required; when the ratio of the days of backwater scouring and the days of the high-water level was small, the scouring time will be relatively short. For example, in 2020, the days of backwater scouring after the flood season was only 0.26 of the days of high water level, the sediment deposited in the shoal couldn't be scoured sufficiently, resulting in insufficient channel width, and channel conditions shall be maintained through dredging (Fig. 6).

Table 2. Statistics of days of high-water level and days of backwater scouring from 2017 to 2020

Year	Days of high water level	Days of recession	Ratio	Channel conditions
2017	109	42	0.39	Good, no dredging
2018	84	42	0.50	Good, no dredging
2019	101	88	0.87	Good, no dredging
2020	150	39	0.26	Insufficient navigation width

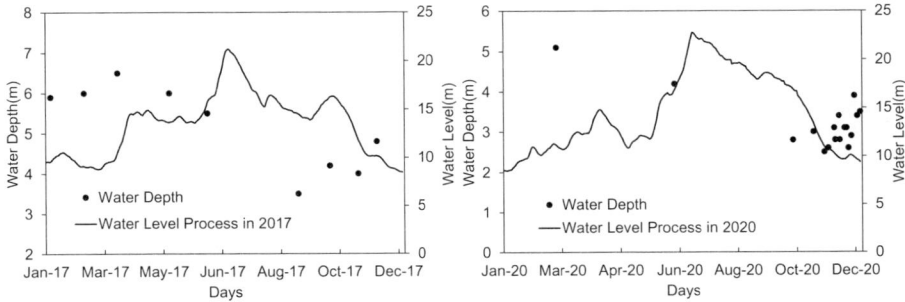

Fig. 6. Water depth and water level process of Shoal Ridge in the upper shallow area of Zhangjiazhou south waterway

6   Channel Maintenance Countermeasures

6.1   Maintenance Countermeasures for 2020

(1) Necessity of dredging

Influenced by the flood in 2020, the channel in the upper shallow area of Nangang was greatly silted up, with the shallowest point of only 2.7 m below the navigational datum. On December 2, the water level of Jiujiang station was only 3.37 m above the navigational datum, and the water depth in the channel was only 6.07 m. Considering that the rainfall would decrease significantly in December, it was very likely that the water level would fall further and rapidly. If the water level fell further, the water depth would not meet the channel maintenance scale (5 m 200 m, water depth × width) in the dry season. In addition, the shallow area of Zhangjiazhou was large, so maintenance volume and pressure were great. Therefore, according to the Measures for the Management of Maintenance and Dredging of the Yangtze River Trunk Line (Trial Implementation) and the Provisions on the Management of Channel Maintenance, when the channel scale was sharply reduced to near or even lower than the planned channel scale due to external conditions, emergency dredging and maintenance should be carried out immediately to ensure smooth navigation in the dry season.

(2) Emergency dredging scheme

Dredging depth: According to the statistics of water level in the dry season of Jiujiang Station in recent years, the lowest water level was 0.96 m above the navigational datum. To ensure the 5 m × 200 m maintenance scale of this waterway, the dredging bottom elevation was 4.24 m below the navigation foundation surface (considering the depth reserved for sedimentation of 0.2 m), and the design slope was 1:5.

Dredging width: According to the actual needs of channel maintenance, the dredging width shall not exceed 200 m.

Dredging site: The dredging site is located between Zhangnan #10 red float and Zhangnan #9 red float in the upper shallow area of Zhangjiazhou. The specific direction of the dredging site can be adjusted dynamically according to the dynamic changes of the shallow area during the dredging process, which shall be jointly determined by the maintenance unit, design unit, and the construction company according to the dynamic mapping during the construction.

Dredging volume: According to the mapping analysis on November 24, 2020, the dredging volume in the channel was expected to reach 290,000 m^3 in 2020.

Through the implementation of the above emergency dredging project, the minimum maintenance scale (5 m × 200 m, water depth × width) of Zhangjiazhou waterway in 2020 was guaranteed.

6.2   Future Concerns on Waterway Maintenance

Due to the impact of the catastrophic flood year in 2020, the siltation amplitude and range of the upper shallow area of the Nangang of Zhangjiazhou waterway were comparatively larger, which has an adverse impact on the channel pattern. At the end of the flood season, emergency dredging was carried out on some navigation channels to meet the minimum maintenance scale of the channel. Since the flow conditions change rapidly, a long time is required for the riverbed scouring and silting to adapt to the flow conditions, and the riverbed deformation has a certain lag in time, dredging may still be necessary to maintain channel conditions in the coming years. In dredging, attention should be paid to the action time of the flood in flood season and the effective scouring time in the recession period to reasonably determine the dredging scheme (intervention time, site, and dredging intensity) to meet the requirements of a smooth channel.

7   Conclusions

(1) The upper shallow area of Nangang of Zhangjiazhou follows the law of depositing in flood season and scouring in the dry season of the year. The channel conditions depend on the comparison between deposition in the flood season and backwater scouring after the flood season. Normally, the channels conditions were good if the backwater scouring period was long. In 2020, impacted by the special hydrologic year, the flood lasted for a long time, so the return time was relatively insufficient after flood season, and the upper shallow area was not sufficiently scoured, resulting in insufficient water depth and channel width in the shallow area, and hence larger navigation obstruction range.

(2) According to the channel mapping in November 2020 and the changing trend of the water level of Jiujiang station in the dry season, the necessity of emergency dredging was explained, and the dredging design parameters were put forward. Wherein, the dredging site was from Zhangnan #10 red float to Zhangnan #9 red float, the dredging elevation was 4.2 m below the navigational datum, and the dredging width was 200 m.

(3) In light of the lag in time of riverbed deformation, it is suggested to pay continuous attention to the channel changes in the upper shallow area of Nangang of Zhangjiazhou waterway and the water regime in the year to formulate corresponding channel maintenance countermeasures.

Acknowledgements. This research is supported by the Subsequent Work of the Three Gorges Project (No. SXHXGZ-2020-3) and Central Public-Interest Scientific Institution Basal Research Fund (No. Y222013).

References

Feng Y (2020) Analysis and reflection on characteristics of dikes dangers in middle and lower reaches of Yangtze River in 2020. Yangtze River 51(12):31–33

Yao SM, Lei WT, Qu G, Chai ZH, Luan HL (2020) Analysis on inundation condition in Poyang Lake during flood season of 2020 based on satellite image. Yangtze River 51(12):185–190

Xue XH, Huang ZB (2008) Analysis of riverbed evolution for Zhangjiazhou waterway downstream the Yangtze River. Port Waterw Eng 418:116–121

Liu T (2015) Waterway regulation works evaluation for Zhangjiazhou south branch up-shoal area of the Changjiang River. Hydro-Sci Eng 2:91–98

Li WQ, Deng XL, Lei JL, Zhang W (2009) Evolution analysis of the upper shallow area and study on waterway regulation measures of Zhangjiazhou Nangang downstream the Changjiang River. Hydro-Sci Eng 12:131–135

Huang ZB (2002) Research on the regulation scheme of Zhangjiazhou Nangang channel at downstream Yangtze River. Port Waterw Eng 3:39–45

Li WQ, Huang ZB, Deng XL et al (2010) Model experiment and effect analysis on waterway engineering of Zhangjiazhou Nangang downstream in the Changjiang River. Port Waterw Eng 3:97–102

Wang SQ (1989) Analysis of shallow area change and maintenance in dry season of Zhangjiazhou Beigang channel. Port Waterw Eng 2:12–17

Chen YJ, Jiang L (2019) Evaluation of waterway engineering construction and regulation effect in middle and lower reaches of the Yangtze River. Port Waterw Eng 1:6–11+34

Deng XL (2019) Regulation measures for 6 m-depth Zhangjiazhou waterway in lower reaches of the Yangtze River. Port Waterw Eng 7:155–161

Tang JW, Deng JY, You XY, et al (2012) Study of critical width of water retaining to channel in middle-reach channel of Yangtze River. Eng J Wuhan Univ 45(1):16–20+40

Do the Short-Term Water Diversion from Yangtze River Increase Phosphorus Bioavailability in the Water-Receiving Area?

Fuwei Tian[✉], Jiangyu Dai, Jiayi Xu, Xiufeng Wu, Shiqiang Wu,
Yu Zhang, Fangfang Wang, and Ang Gao

State Key Laboratory of Hydrology-Water Resources and Hydraulic
Engineering, Nanjing Hydraulic Research Institute, Nanjing 210029, China
2104803772@qq.com,
{jydai,xfwu,sqwu,yuzhang,ffwang,agao}@nhri.cn

Abstract. Water diversion projects have an important role in coping with water shortage and improving water quality, but they also have an impact on the ecological environment of lakes that cannot be ignored. As an important biogenic element for evaluating the primary productivity and eutrophication of lake water bodies, the influence of phosphorus by water diversion activities and its impact on the production and elimination of phytoplankton is lacking attention. In this study, we analyzed the phosphorus composition and bioavailability of the water channel and the Gonghu Bay of Lake Taihu under the influence of seasonal water diversion, revealed the spatio-temporal distribution patterns of phytoplankton communities in the water receiving river and lake, and analyzed the contribution of phosphorus to the variations of phytoplankton communities and their quantitative coupling relationships. The results showed that short-term water diversion in autumn and winter did not significantly increase the concentrations of particulate phosphorus in the receiving waters, but there was a risk of increasing the concentration of dissolved reactive phosphorus and dissolved organic phosphorus. The difference in total phosphorus concentrations between the diversion and non-diversion periods in Gonghu Bay was an important environmental factor influencing the phytoplankton community, and the bioavailable phosphorus could better fit the logarithm of algal cell density in all seasons, which was significantly and positively correlated with the phytoplankton cell densities. This study implies that the control of bioavailable phosphorus in the water channel can reduce the ecological risks of the water diversion project on the cyanobacterial blooms to some extent.

Keywords: Water diversion · Phosphorus bioavailability · Phytoplankton · Wangyu River · Lake Taihu

1 Introduction

Eutrophication of water bodies is one of the major water environment problems facing the world today. The eutrophication of rivers and lakes will become more and more severe due to factors such as changes in precipitation frequency and rapid human

© The Author(s) 2023
Y. Li et al. (Eds.): PIANC 2022, LNCE 264, pp. 1098–1112, 2023.
https://doi.org/10.1007/978-981-19-6138-0_96

urbanization in this century (Sinha et al. 2017), which will further aggravate the deterioration of water quality in rivers and lakes, the outbreak of water blooms and black odor of lake floods, and other disasters, and seriously threaten the regional economic development and ecological health of the basin (Qin et al. 2013; Cooke et al. 2016; Paerl et al. 2016; Monchamp et al. 2018). Therefore, how to quickly and effectively improve the water quality of eutrophic water bodies and mitigate cyanobacterial bloom disasters has become an extremely important research topic in the field of ecological restoration of eutrophic water bodies at home and abroad (Søndergaard et al. 2007; Vander Zanden et al. 2016; Zhang et al. 2016). The transfer and diversion of transit guest water to promote the flow of river and lake water bodies and strengthen the connectivity of river, river and lake systems within the watershed provides new ideas to mitigate cyanobacterial bloom disasters in eutrophic water bodies from an ecohydrological perspective.

While water diversion projects are effective in mitigating lake water hazard, the impact of exogenous phosphorus nutrient input from the diverted guest water on the water ecology of the receiving lake is also controversial (Wu et al. 2018; Yao et al. 2018; Qin et al. 2019; Dai et al. 2016; Yang et al. 2018). For example, the total phosphorus content of the Water Diversion from Yangtze River to Lake Taihu (WDYT) is higher than that of the Lake Taihu water body (Pan et al. 2015a), and the influx of phosphate and other pollutants along the diversion channel also makes the nutrient content under the Wangting Hdro-junction higher than that of the lake body (Dai et al. 2018; Dai et al. 2020). The migration and transformation of exogenous phosphorus nutrients from the passenger water in the receiving lake may have an impact on the phosphorus material cycling mechanism of the lake (Yao et al. 2018) as well as the growth and community structure of the primary producer planktonic algae (Amano et al. 2010; Huang et al. 2015), which in turn increases the aquatic ecological risk of the diversion (Zhai et al. 2010; Zhang et al. 2018; Pan et al. 2015b).

In fact, not all forms of phosphorus in nature contribute directly to lake primary productivity, depending on the biological effectiveness of phosphorus. Biologically active phosphorus (BAP) is the active phosphorus that can be used directly or indirectly by aquatic organisms (Wang et al. 2010). In general, BAP in lake waters is dominated by soluble reactive phosphorus (SRP) and enzymatically soluble phosphorus (EHP) (Gao et al. 2006; Chen et al. 2019), while BAP in sediments mainly includes DGT-P, water-soluble phosphorus (WSP), algal-available phosphorus (AAP), and $NaHCO_3$-extractable phosphorus (Olsen-P) types (Wang et al. 2016). Usually, phosphorus in river waters is predominantly in the particulate form due to the relatively high sediment content, and the potential biological effectiveness of particulate phosphorus is relatively high (Zhou et al. 2018). However, the impact of water diversion on the biological effectiveness of phosphorus in receiving rivers and lakes and its contribution to the production and elimination of planktonic algae in lakes are still not clarified, which is the primary issue limiting the scientific evaluation of the ecological and environmental impacts of rivers and lakes in water diversion projects such as WDYT.

Lake Taihu is one of the largest freshwater lakes in the middle and lower reaches of the Yangtze River, and the rapid economic development of the basin has led to serious pollution of the lake water bodies, with frequent occurrence of cyanobacterial blooms and "lake flooding" in the local lake area (Qin 2020). As an important emergency measure to alleviate the water bloom disaster in Lake Taihu and ensure water security in the basin, the river diversion project has been operating on a regular basis in recent years. The Taihu Basin is located in the subtropical monsoon region, and the ecological environment of rivers and lakes in the basin usually shows significant spatial and temporal variation (Wu et al. 2019). In this study, the spatial distribution and seasonal variation of phosphorus in the receiving areas of rivers and lakes during autumn and winter diversions were analyzed by means of field surveys, and the effects of autumn and winter diversions on the biological effectiveness of phosphorus and planktonic algae in the receiving areas of rivers and lakes were compared and studied to provide a basis for the prevention and control of exogenous phosphorus pollution in lakes by diversions of rivers and lakes. This study provides a basis for the diversion of water from the river to the lake.

2 Materials and Methods

2.1 Study Area

Since 2002, Taihu Basin has been implementing the "River Diversion Project" to transfer water from the Yangtze River, which has relatively good water quality, to Lake Taihu by using the Wangyu River, the backbone of the basin's water conservancy project, and to supply water from Lake Taihu to Shanghai and other downstream areas through the Taipu River Project. The water supply from Lake Taihu to Shanghai and other downstream areas through the Taipu River project, thus driving the optimal scheduling of many water conservancy projects in the basin and promoting the flow of water in Lake Taihu and the river network. The Wangyu River, which starts from the Yangtze River in the north and ends at Gonghu Bay in the south, has a total length of 60.8 km and a dense river network on both sides, and is one of the main diversion and drainage channels of the River Diversion and Tai project (Chu et al. 2014). Changshu water conservancy hub and Wangting water conservancy hub are the two control sections of the main stem section of the Wangyu River, of which Wangting water conservancy hub is the direct control gate for the diversion of water from the Wangyu River into the lake. The tributary gates on the east bank of the Wangyu River are basically controlled, while the tributary gates on the west bank are open for the drainage of flood water in the area, except for the section north of Fushan Pond near the Yangtze River in the north and some tributaries south of Jialing Dang in the south; in the process of water transfer, the diversion flow of the gates on the east bank of the Wangyu River is controlled by not exceeding 30% of the amount of river water transferred or the maximum diversion flow not exceeding 40 m³/s. The vast majority of the Yangtze River water diverted by the Changshu hub of the Wangyu River flows into Lake Taihu and the tributaries on the west bank (Ma et al. 2014).

Gonghu Bay is a large lake and bay type water in the northeast of Lake Taihu (30° 55′40″–31°32′58″ N, 119°52′32″–120°36′10″ E), with an area of about 150 km² and a year-round average water depth of 1.8 m in the bay (Zhong et al. 2012). The southwestern part of the bay is connected to Meiliang Bay and the center of Lake Taihu, and the northeastern corner is connected to the Wangyu River, which is the primary receiving lake for the diversion project of the Wangyu River. In recent years, with the continuous deterioration of the water environment in Meiliang Bay, Gonghu Bay has become the main water source of Wuxi City and one of the important water sources of Suzhou City. Gonghu Bay is a typical mixed grass-algae lake ecosystem, with both cyanobacterial bloom accumulation and a large amount of aquatic vegetation cover in the bay, but the aquatic vegetation is mostly concentrated in the east shore zone of Gonghu Bay (Zhao et al. 2015). The complexity of the ecological structure within Gonghu Bay makes its water environment and aquatic ecological elements show obvious spatial heterogeneity.

In this study, the river channel of Wangyu River into the lake is selected to study the characteristics of the water source of the diversion project into the lake, and the waters of Gonghu Bay are used as a sensitive receiving lake. The distance between the lake center of Lake Taihu and Gonghu Bay is far, and the open water is used as the control lake area of Gonghu Bay affected by the diversion. Nine monitoring points (Y, W-1–W-8, Fig. 1a) were placed along the entire route from the mouth of the Yangtze River to the Wangting Water Conservancy Hub in Lake Taihu to monitor the physicochemical parameters of the water bodies in the Wangyu River. The sampling sites in the lake area are as shown in Fig. 1b. Seven sampling sites are placed at equal distances along the axis of the bay of Gonghu Bay (G1–G7), and 3 monitoring points (C1–C3) are placed in the lake center area as reference points for the influence of water diversion in Gonghu Bay.

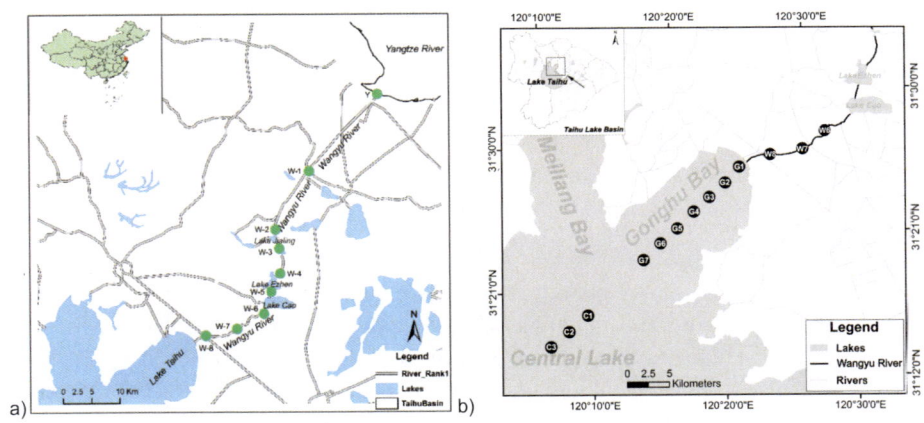

Fig. 1. Location of sampling sites in the Yangtze River, Wangyu River (a) and Lake Taihu (b).

2.2   Sampling and Physicochemical Parameters Measurement

According to the operation of the Wangyu River diversion project from 2014 to 2016, from November 2014 to January 2016, we selected the autumn and winter seasons, i.e. November and mid-January each year for field observation and sample collection, in which there were water diversions from the Wangyu River to the lake in November 2014 and mid-January 2015, and no water diversions in the rest of the monitoring time.

The spatial and temporal distribution patterns of phosphorus and phytoplankton community elements from the Wangyu River to the center of Lake Taihu were analyzed in different seasons, and the monitoring time started from 9:00 am and ended before 12:00 noon.

The 2014–2016 Wangyu River water inflow and Lake Taihu water level data were quoted from the Taihu Basin Authority Annual Report on River Diversion to Taihu (TBA 2014, 2015 & 2016). The wind direction at the sampling points was determined by a boat-mounted wind speed and direction meter, and weather and rain conditions were determined by the monitors on site. The water temperature and pH values at each sampling point were determined by HACH portable multi-parameter water quality meter HQ30d on site. Plexiglass column sampler was used to collect water samples from rivers and lakes 50 cm below the water surface, and one mixed sample was collected from each sampling point, totaling 1 L of water samples, collected in pre-washed transparent plastic bottles, placed in a holding tank with ice, transported to the laboratory, and total phosphorus (TP), total dissolved phosphorus (DTP) and soluble reactive phosphorus (SRP) and chlorophyll a (Chl-a) were measured within 24 h.

The planktonic algae samples were collected directly in the field by first collecting 1 L of mixed water samples (50 cm below the lake surface) from Lake Taihu, preserving them in washed plastic bottles, adding 15 ml of Lugol's reagent at 1% mass concentration to fix the algal cells, and transporting them to the laboratory for storage at room temperature and protected from light (Jin and Tu 1990).

TP and DTP were determined by referring to the relevant methods in the Specification for Lake Eutrophication Investigation (2nd edition), and DTP and SRP were determined by filtering the original water samples through 0.70 μm pore size GF/F glass fiber membrane beforehand and using the filtered water. The Chl-a content was determined by the hot ethanol method (Chen et al. 2003).

2.3   Identification of Phytoplankton Community

The 1 L of fixed sample was shaken and transferred to a partition funnel, and after 24 h of resting, 50 ml of the concentrated sample was separated from the bottom of the funnel and stored in an 80 ml glass vial at room temperature and protected from light for microscopic imaging and counting of planktonic algae (Jin and Tu 1990). The identification and enumeration of planktonic algal populations in water samples were carried out using automatic algal identification and enumeration software. 0.1 ml of the above static concentrated sample was taken in a 0.1 ml algal enumeration frame, imaged and enumerated under a $16 \times 40 \times$ biological microscope, and 100 available fields of view were read randomly and uniformly in the enumeration frame for each sample. The obtained field of view images were processed by the automatic algal identification and counting software for identification of planktonic algal populations and calculation of species numbers. The identification of planktonic algal species was carried out with reference to the literature (Hu and Wei 2006) method.

2.4   Statistical Analyses

Spatial differences in the monitoring indicators of the three waters of the Wangyu River, Gonghu Bay and the lake core area, as well as differences in the water environment parameters of Gonghu Bay during the diversion and non-diversion periods were compared using two-tailed t-tests in the numerical statistical software SPSS (IBM, Armonk, USA) 16.0. Data were mapped using Sigmaplot v14.0 scientific data mapping software (Systat Software Inc., London, UK).

The community composition of planktonic algae was based on the cell density of algae, and the percentage of the total cell density of each phylum or species of algae in the total cell density of a sample site was calculated. A ranking method was used to analyze the correlation between physicochemical parameters, phosphorus and planktonic algal community structure in the water column of Gonghu Bay during the diversion and non-diversion periods. The calculation procedure was performed by CANOCO 4.53 for Windows (Microcomputer Power) software (Ter Braak and Šmilauer 2002). Before analysis, the data of the planktonic algal community structure matrix based on algal density and the physicochemical parameters were transformed with open square roots to meet the requirements of statistical analysis. The data were selected by forward selection in the CANOCO software for those parameters that had a significant effect on the structure of the planktonic algal community. The significance of the calculated results ($p < 0.05$) was verified using 499 unrestricted Monte Carlo permutations.

3   Results and Discussion

3.1   Monitoring the Meteorological and Water Level Profile of the Lake

The wind direction, wind level, direction of lake flow and duration of water diversion in the waters of Gonghu Bay during the field survey are shown in Table 1. The wind direction in Gonghu Bay during the monitoring period in autumn and winter was mostly from the north, and the wind level was mainly 3, which had less influence on the lake flow. The duration of water diversions in both diversion periods up to the day of monitoring did not exceed one month.

Table 1. Wind speed, wind direction, water level and duration of water diversion in Gonghu Bay during the monitoring period.

Sampling days	Wind direction*	Wind speed	Lake water level (m)	Duration of water diversion (d)
2014-11-21	SW	2–3	3.2	29
2015-01-17	N	3–4	3.1	14
2015-11-27	N	0	3.5	3 d after diversion
2016-01-20	NE	3–4	3.4	/

* SW: South west, N: North, NE: North east; / represents the non-diversion periods.

3.2   Spatial Distribution of Phosphorus During Diversion and Non-diversion Periods

The spatial distribution of the values of the main physicochemical parameters of the water bodies in the Wangyu River, the bay center axis of Gonghu Bay and the lake center area in autumn and winter are shown in Fig. 2. The difference in water temperature between the diverted and non-diverted Wangyu River in autumn was significant (t-test, $p < 0.05$) and was significantly higher in the diverted period (November 2014) than in the non-diverted period (November 2015), and there was no significant difference in water temperature between the diverted and non-diverted Wangyu River in winter, but the difference was significant in Lake Taihu waters (t-test, $p < 0.05$). The pH of Lake Taihu was significantly lower during the diversion period than during the non-diversion period, influenced by the lower pH of Wangyu River waters, and the pH showed a significant increasing trend from Wangyu River to Lake Taihu. Similarly, Chl-a showed a significant increasing trend from the Wangyu River to Lake Taihu waters, but the values in the monitored waters during the diversion period were significantly lower than those in the non-diversion period (t-test, $p < 0.05$). The spatial variation patterns of the three phosphorus parameters were consistent, all showing a decreasing spatial trend from Wangyu River to Lake Taihu. The phosphorus concentrations in Gonghu Bay were significantly higher than those in the non-diversion period in winter (Fig. 2b, t-test, $p > 0.05$), while there was no significant difference between the DTP and SRP contents in the monitored waters in the diversion and non-diversion periods in autumn, except for TP (t-test, $p > 0.05$). The TP in Lake Taihu during the non-diversion period in autumn was significantly higher than that in the diversion period, which was related to the presence of cyanobacterial blooms in Lake Taihu during the non-diversion period, and this study also confirmed that the Chl-a content in Lake Taihu during the non-diversion period was significantly higher than that in the diversion period (Fig. 2a).

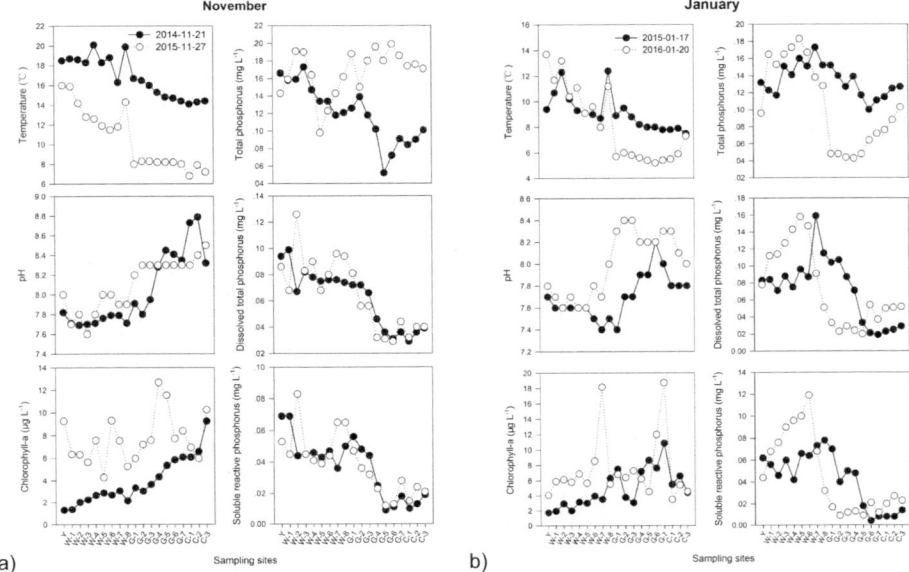

Fig. 2. Spatial distribution of physicochemical parameters of water bodies in the monitoring area in November (a) and January (b).

The proportion of particulate phosphorus in the water bodies of the Yangtze River and the Wangyu River was higher in the fall and winter diversion periods than in the non-diversion periods, but not higher than in the lake core area of Lake Taihu. Water diversions in autumn and winter still increase the biologically active phosphorus content in Gonghu Bay, but the impact area is limited to the inlet waters of the Wangyu River by the diversion flow and duration (Fig. 3). Studies have reported pollutants along the Wangyu River from tributaries into the Wangyu River, especially during non-diversion periods. In recent years, although some control measures have been implemented in the Wangyu River, further control of domestic and agricultural nonpoint pollution along the Wangyu River should be carried out in autumn and winter (Zhang et al. 2010). In autumn and winter, the water level of the Wangyu River during the diversion period is always higher than that of the non-diversion period due to low rainfall and the input of Yangtze River water. During the non-diversion period in winter, nutrients and organic pollutants from the tributaries can flow into the Wangyu River and increase the pollutant concentration in the main channel.

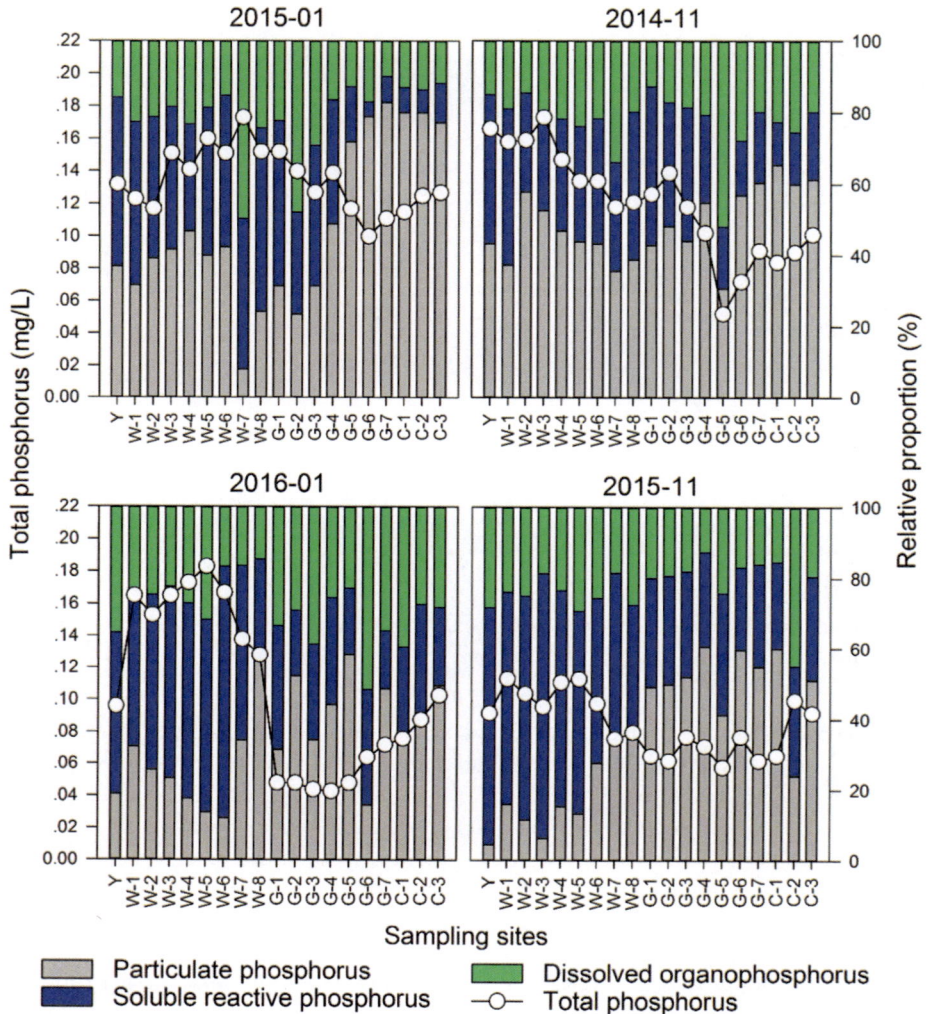

Fig. 3. Spatial distribution of phosphorus composition in the monitoring area during the diversion and non-diversion periods in autumn and winter.

3.3   Seasonal Variation of Phosphorus in Lake Taihu

The mean values of the three phosphorus concentrations in the monitored waters in different seasons of the diversion and non-diversion periods are shown in Fig. 4. The values of the three phosphorus concentrations in the winter diversion period were significantly higher than those in the non-diversion period in the same season (t-test, p < 0.05). On the contrary, the phosphorus concentration values in the autumn diversion period were significantly lower than those in the non-diversion period (t-test, p > 0.05), which was related to the appearance of cyanobacterial blooms in Lake Taihu in the autumn non-diversion period, which led to a significant increase in phosphorus content in the regional water column due to the accumulation of blooms. The values of all three phosphorus concentrations in the water coming from the Wangyu River in the fall and winter diversion periods were higher than those in the lake center area, and the values of each phosphorus in Gonghu Bay were between the Wangyu River and the lake center area, indicating that the diversion of water from the Wangyu River had the possibility of increasing the risk of phosphorus loading in Gonghu Bay.

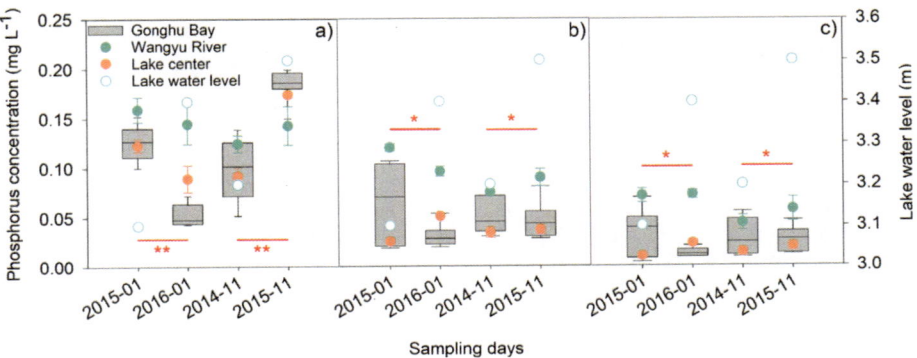

Fig. 4. Comparison of concentrations of total phosphorus (a), total dissolved phosphorus (b) and dissolved reactive phosphorus (c) in monitoring areas in autumn and winter.

3.4   Spatial and Temporal Distribution of Phytoplankton Community Composition

The composition of planktonic algal community in the monitoring area in autumn and winter is shown in Fig. 5. Most of the waters in the axis of the bay center of Gonghu Bay during the non-diversion period in winter are dominated by cyanobacteria, which is very similar to the composition of algal community in Wangyu River, and both the bay mouth and the lake center area are dominated by cyanobacteria. Diatoms dominated in the bay axis of Gonghu Bay during the winter diversion period, with relative proportions ranging from 70.6% to 94.3%, followed by green algae (relative proportions ranging from 2.2% to 13.2%), which is also very similar to the community composition of the incoming water from the Wangyu River, indicating the important influence of exogenous diatom input on the community composition of the receiving lake area of Lake Taihu. Cyanobacteria were still absolutely dominant in the bayou and core areas of Gonghu Bay.

The diatoms were dominant at the mouth of Wangyu River (G1–G5), while the cyanobacteria were dominant at the mouth of the bay and the center of the lake during the diversion period in autumn 2014, and the diatoms were dominant at the mouth of Wangyu River to G5 of Gonghu Bay during the non-diversion period in autumn 2015, while the cyanobacteria were dominant in all other waters (Fig. 5). The main reason for this is that the sampling date in November 2015 was on the third day after the end of the fall water diversion, and the effect of water diversion is continuous.

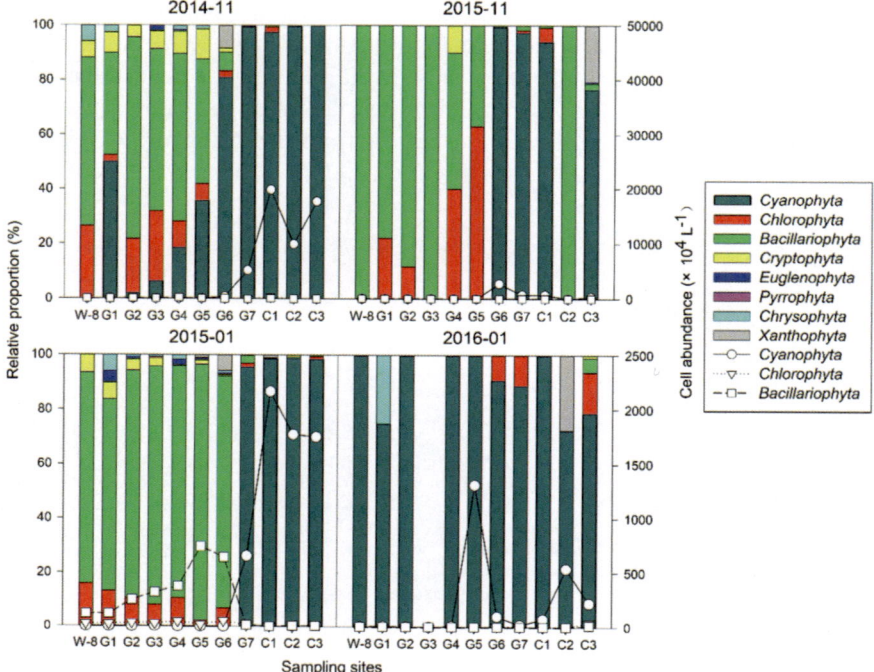

Fig. 5. Spatial distribution of phytoplankton community composition in the monitoring area in autumn and winter.

3.5   Contribution of Phosphorus Changes to Lacustrine Phytoplankton Communities

TP (p = 0.002) and DTP (p = 0.008) in autumn were significantly correlated with the difference in planktonic algal community structure between the diverted and non-diverted periods in Gonghu Bay in autumn, and the three dominant environmental factors explained a total of 20.7% of the spatial and temporal variation in algal community structure, with TP and DTP explaining 10.8% and 7.2% of the variation in algal community, respectively, and the highest values of TP and DTP were in the diverted period (Fig. 6a). The highest TP and DTP values belonged to the water diversion period (Fig. 6a). The results of CCA ranking of planktonic algal community structure and water column phosphorus during winter diversion and non-diversion periods in Gonghu Bay are shown in Fig. 6b, DTP (p = 0.002) was significantly correlated with the succession of algal community structure in Gonghu Bay. 11.8% of the differences in algal community structure.

The SRP content in the waters of Gonghu Bay during the non-diversion period in autumn was relatively higher than that in the diversion period, and the higher concentration of SRP in Gonghu Bay supplemented phosphorus nutrients for the proliferation and growth of planktonic algae in the non-diversion period. At the same time, the structure of algal community in Gonghu Bay during the diversion period was also importantly related to the higher DTP content in the incoming water of Wangyu River, and the relative proportion of diatoms and green algae in the bay during the autumn diversion period was higher than that in the non-diversion period. The study confirms that diatoms appear when the concentration of nitrate, phosphate and silicon oxide in the water column is high, and higher concentrations of phosphorus are instead more favorable for the growth of diatoms and other algae, which in turn replace the dominance of cyanobacteria.

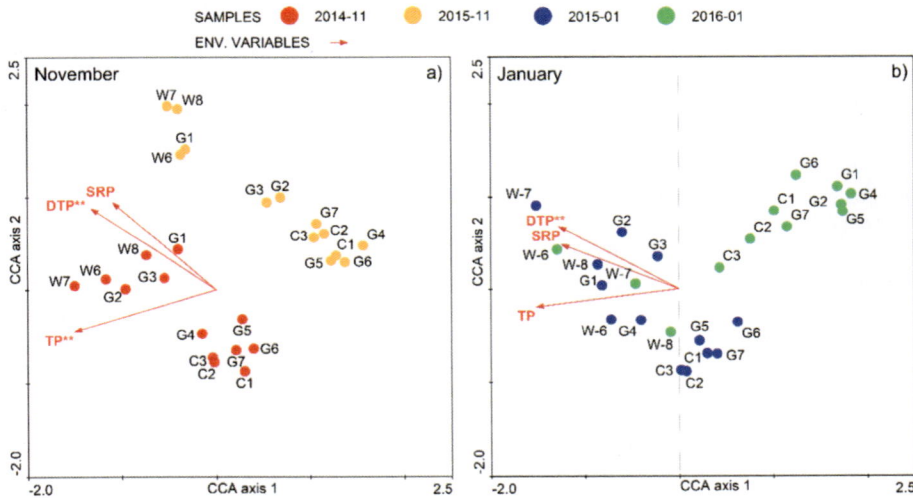

Fig. 6. Two-dimensional analysis of the paradigmatic correspondence between phosphorus and phytoplankton communities in autumn and winter.

4   Conclusions

The most significant effect of water diversion on phosphorus in Gonghu Bay in autumn and winter is at the mouth of the Wangyu River, and the phosphorus concentration decreases from the mouth of the Wangyu River to the mouth of Gonghu Bay during the water diversion period. In autumn and winter, water diversions still increased the biologically active phosphorus content in Gonghu Bay, but the impact area was limited to the mouth of the Wangyu River due to the flow and duration of diversions. The water diversion of Wangyu River significantly changed the community composition of planktonic algae in Gonghu Bay, and the water diversion in autumn and winter significantly promoted the proliferation of diatoms, making them the dominant species of planktonic algae. The difference in biologically effective phosphorus concentration between autumn and winter diversions and non-diversions in Gonghu Bay is an important environmental factor affecting the change of planktonic algae community, with the contribution of biologically effective phosphorus to the change of algal community ranging from 10.8 to 11.8%, and the contribution of phosphorus to the succession of planktonic algal community in Gonghu Bay in different seasons is also around 20%.

Acknowledgements. This work was jointly funded by the Project of the National Natural Science Foundation of China (52179073) and the Special Research Fund of the Nanjing Hydraulic Research Institute (Y120010) and the Natural Science Foundation of Jiangsu Province, China (Grant No. BK20200160).

We appreciate the cooperation and efforts of all authors in producing the Proceedings. This will be a great congress following the tradition of our sponsoring organizations.

References

Amano Y, Sakai Y, Sekiya T et al (2010) Effect of phosphorus fluctuation caused by river water dilution in eutrophic lake on competition between blue-green alga Microcystis aeruginosa and diatom Cyclotella sp. J Environ Sci 22(11):1666–1673

Chen J, Xu H, Zhan X et al (2019) Mechanisms and research methods of phosphorus migration and transformation across sediment-water interface. J Lake Sci 31(4):907–918

Chen Y, Fan C, Teubner K et al (2003) Changes of nutrients and phytoplankton chlorophyll-a in a large shallow lake, Taihu, China: an 8-year investigation. Hydrobiologia 506(1):273–279

Chu K, Kan L, Hua Z (2014) Construction and application of an indicator system for assessment of river ecosystem in plain tributary networks. J Hydroelectr Eng 33:138–144

Cooke GD, Welch EB, Peterson S et al (2016) Restoration and management of lakes and reservoirs. CRC Press, Boca Raton

Dai J, Wu S, Lv X et al (2016) Effect of water diversion on spatial-temporal dynamics of organic pollutants in Gonghu Bay, Lake Taihu. J Hydroecol 37(1):39–46

Dai J, Wu S, Wu X et al (2018) Effects of water diversion from Yangtze River to Lake Taihu on the phytoplankton habitat of the Wangyu River channel. Water 10(6):759

Dai J, Wu S, Wu X et al (2020) Impacts of a large river-to-lake water diversion project on lacustrine phytoplankton communities. J Hydrol 587:124938

Gao G, Zhu G, Qin B et al (2006) Alkaline phosphatase activity and the phosphorus mineralization rate of Lake Taihu. Sci China Ser D 49(1):176–185

Hu H, Wei Y (2006) Fresh algae in China - system, classification and ecology. China Science Press, Beijing

Huang J, Gao J, Zhang Y et al (2015) Modeling impacts of water transfers on alleviation of phytoplankton aggregation in Lake Taihu. J Hydroinform 17(1):149–162

Jin XC, Tu QY (1990) The standard methods for observation and analysis in lake eutrophication. Chinese Environmental Science Press, Beijing

Ma Q, Tian W, Wu C (2014) Total phosphorus and total nitrogen concentrations of the water diverted from Yangtze River to Lake Taihu through Wangyu River. J Lake Sci 26:207–212

Monchamp ME, Spaak P, Domaizon I et al (2018) Homogenization of lake cyanobacterial communities over a century of climate change and eutrophication. Nat Ecol Evol 2(2):317

Paerl HW, Gardner WS, Havens KE et al (2016) Mitigating cyanobacterial harmful algal blooms in aquatic ecosystems impacted by climate change and anthropogenic nutrients. Harmful Algae 54:213–222

Pan Y, Ma Y, Qin Y et al (2015a) Nutrients input characteristics of the Yangtze River and Wangyu River during the "water transfers on Lake Taihu from the Yangtze River." Environ Sci 36(8):2800–2808

Pan Y, Qin Y, Ma Y et al (2015b) Plasticizers input characteristics and environmental health risk assessment during the period of "water transfers on Lake Taihu from the Yangtze River." Acta Sci Circum 12:4128–4135

Qin BQ, Gao G, Zhu GW et al (2013) Lake eutrophication and its ecosystem response. Chin Sci Bull 58(9):961–970

Qin B, Paerl HW, Brookes JD et al (2019) Why Lake Taihu continues to be plagued with cyanobacterial blooms through 10 years (2007–2017) efforts. Sci Bull 64(6):354

Qin B, Zhou J, Elser JJ et al (2020) Water depth underpins the relative roles and fates of nitrogen and phosphorus in lakes. Environ Sci Technol 54(6):3191–3198

Sinha E, Michalak AM, Balaji V (2017) Eutrophication will increase during the 21st century as a result of precipitation changes. Science 357(6349):405–408

Søndergaard M, Jeppesen E, Lauridsen TL et al (2007) Lake restoration: successes, failures and long-term effects. J Appl Ecol 44(6):1095–1105

TBA (2014-2016) Annual report for water diversion from Yangtze River to Lake Taihu. Shanghai

Ter Braak CJF, Smilauer P (2002) CANOCO reference manual and canodraw for windows user's guide: software for canonical community ordination (version 4.5). Ithaca, New York

Vander Zanden MJ, Olden JD, Gratton C, et al (2016) Food web theory and ecological restoration. Foundations of restoration ecology. Island Press, Washington, DC

Wang J, Chen J, Yang H et al (2016) Bioavailable phosphorus in sediments from Lake Hongfeng, Southwestern China. Earth Environ 44(4):437–440

Wang S, Jin X, Bu Q et al (2010) Evaluation of phosphorus bioavailability in sediments of the shallow lakes in the middle and lower reaches of the Yangtze River region, China. Environ Earth Sci 60(7):1491–1498

Wu T, Qin B, Brookes JD et al (2019) Spatial distribution of sediment nitrogen and phosphorus in Lake Taihu from a hydrodynamics-induced transport perspective. Sci Total Environ 650:1554–1565

Wu Y, Dai R, Xu Y et al (2018) Statistical assessment of water quality issues in Hongze Lake, China, related to the operation of a water diversion project. Sustainability 10(6):1885

Yang Q, Wu S, Dai J et al (2018) Effects of short-term water diversion in summer on water quality and algae in Gonghu Bay, Lake Taihu. J Lake Sci 30(1):34–43

Yao X, Zhang L, Zhang Y et al (2018) Water diversion projects negatively impact lake metabolism: a case study in Lake Dazong, China. Sci Total Environ 613:1460–1468

Zhai S, Hu W, Zhu Z (2010) Ecological impacts of water transfers on Lake Taihu from the Yangtze River, China. Ecol Eng 36(4):406–420

Zhang L, Wang S, Han M et al (2010) Nitrogen and phosphorus pollution in the western district of Wangyu River and counter-measures, Taihu Basin. J Lake Sci 22(3):315–320

Zhang M, Wang S, Fu B et al (2018) Ecological effects and potential risks of the water diversion project in the Heihe River Basin. Sci Total Environ 619:794–803

Zhang Y, Liu X, Qin B et al (2016) Aquatic vegetation in response to increased eutrophication and degraded light climate in Eastern Lake Taihu: implications for lake ecological restoration. Sci Rep 6:23867

Zhao K, Li Z, Wei H et al (2015) The distribution of aquatic vegetation in Gonghu Bay, Lake Taihu, 2012. J Lake Sci 27(3):421–428

Zhong C, Yang G, Gao Y et al (2012) Seasonal variations of macrozooplankton community in Gonghu Bay of Lake Taihu. J Hydroecol 33:47–52

Zhou J, Zhang M, Li Z (2018) Dams altered Yangtze River phosphorus and restoration countermeasures. J Lake Sci 30(4):865–880

Ecological Evaluation of Waterways Based on Modified Neural Networks

Teng Wu, Jie Qin[✉], and Runzhuo Guo

College of Harbour, Coastal and Offshore Engineering, Hohai University,
Nanjing, China
{wuteng, jqin}@hhu.edu.cn

Abstract. The ecological condition of waterways has attracted increasing public attention in recent years, and the evaluation of the ecological status of waterways is of practical and scientific significance. To carry out an objective and credential evaluation of the ecological condition of waterways, this study compares the performance of two artificial neural networks (ANN) models, including the traditional Back Propagation (BP)-ANN model and the Particle Swarm Optimization (PSO)-BP-ANN model. The traditional BP-ANN model is characterized by a local search strategy and usually converges to local minima. This study combined the PSO algorithm with the BP-ANN model to improve the performance of the latter. An evaluation indicator system was established first based on the main features of waterway ecology, which includes 13 indicators in four aspects including the transportation function, the ecological function, the landscape function, and the economic function. A waterway ecology dataset was then established based on major waterways in China. The results show that the prediction accuracy of the BP-ANN is around 0.6 and unstable due to its local searching strategy. In contrast, the coefficient of determination of the PSO-BP-ANN model reaches 0.98, indicating the high prediction accuracy of the model. On this basis, this study further analyzed the relative importance of the four functions using the PSO-BP-ANN model. Results showed that the evaluation grades can be significantly improved by advancing the landscape function and the transportation function, and the improvement of these two scores can greatly reduce the probability of the lowest rating. The highest rating requires all four functions to be greater than certain threshold values.

Keywords: BP neural networks · PSO algarithm · Probability

1 Introduction

In China, the ecological states of waterways are receiving increasingly prevalent attention. Especially in recent years, emphasis has been placed on incorporating the concept of ecological development into waterway training projects in the Yangtze River and the Beijing-Hangzhou Canal. During the waterway regulation of the Jingjiang Reach in the middle Yangtze River, the concept of ecological and environmental development is a key component of the project. Waterway health changes in the Jingjiang Reach were assessed by Li et al. (2018) in the context of a waterway

© The Author(s) 2023
Y. Li et al. (Eds.): PIANC 2022, LNCE 264, pp. 1113–1120, 2023.
https://doi.org/10.1007/978-981-19-6138-0_97

regulation project. The results highlight potential mutual improvement between ecosystem health and navigation functions. In the Huai'an section of the Beijing-Hangzhou Canal in Jiangsu Province, the development of ecological waterways has been not just the protection of ecology, but also the integration of cultural heritage into the canal (Wang 2008). Based on the natural and socioeconomic attributes, Duan et al. (2021) constructed a theoretical framework and integrated index system to evaluate the ecological sensitivity, development potential, and current development patterns of riverfront resources. At present, a large number of projects focusing on the ecological development of waterways have been carried out, but much less research has been conducted related to the evaluation of the ecological situations of waterways.

Existing studies related to the ecological evaluation of waterways can be divided into two categories (Cai and Hu 2008): the first group focuses on specific ecological aspects and most of them relate to environmental factors such as pollutants, algae, invertebrates, and fishes. Methods in this group include the fish biotic integrity index proposed by Karr (1981) and the Riparian, Channel, and Environmental (RCE) Inventory developed by Prtersen (1992). The second group tries to carry out comprehensive evaluations. Li et al. (2018) took the Jingjiang section of the Yangtze River as an example, established an evaluation system, and used the hierarchical analysis method to evaluate the ecological situations of the Jingjiang reach in terms of the river navigation function, water quality, ecology, sand transport, and landscape recreation function.

In the last decade, neural network models have been widely used in various evaluation models (Chu et al. 2013; Hamida et al. 2020), but it has rarely been applied to evaluate ecological waterways. In addition, the quantitative analysis of the relative weights of each evaluation index is seldom reported. Therefore, the purpose of this study is to build ecological waterway evaluation indexes, evaluate ecological waterways by BP neural network model and PSO-BP neural network model, and then conduct a quantitative analysis of the relative weights of the indexes based on the ANN models.

2 Methodology

2.1 Data Collection

In China, eighty percent of waterways are distributed in five rivers or canals in terms of length of waterways: the Beijing-Hangzhou Canal, the Yangtze River, the Huai River, the Pearl River, and the Heilongjiang River. In addition, 97% of the throughput of China's waterways is contributed by the Yangtze River system and some southern waterways. Therefore, the dataset in this study was mainly collected from the four major waterway systems, namely, Yangtze River, Pearl River, Huai River, and Beijing-Hangzhou Grand Canal. Specifically, collected data includes the number of vessels, maximum vessel tonnage, water system connectivity, bank slope, the total number of vessel accidents, river width, coverage of ecological berms, bank vegetation coverage, bank plant richness, functional area water quality, riparian buffer zone width index, bank park area, and GDP per capita. Normalization is applied to all the collected data using the maximum-minimum method to convert all data to the range of (0, 1). Then the expert scoring method was used to determine the ecological rates of each waterway sample. A total of 30 experts provided their evaluation results and the final ecological rates are obtained by averaging these evaluation results. Finally, a dataset including 84 waterway samples was built for the following model establishment. This dataset can be found in Guo (2021) in which the measurement of each parameter was explained in detail.

2.2 Ecological Waterway Evaluation Model Based on Traditional BP Neural Network

The BP-ANN model for the evaluation of ecological waterways is established by MATLAB which contains a toolbox of neural networks. With the help of this toolbox, a BP-ANN model can be readily built by regulating relevant parameters according to specific requirements. Building the BP-ANN model includes the following steps: the construction of modeling structure, modeling training, and prediction. There are 13 input parameters and one output parameter for current BP-ANN model. The output parameter is the evaluation levels of waterways, including excellent, good, medium, poor, and bad, totally five grades. After testing different numbers of hidden units in the intermediate layer of the BP-ANN model, results show that the highest prediction accuracy obtained when the intermediate layer has 11 hidden units. To avoid overfitting, the dataset is divided into a training dataset (approximately 80% of the data) and a test dataset (approximately 20% of the data). The former is used to train the ANN model, and the latter is used to evaluate the predictive accuracy of the model.

2.3 Ecological Waterway Evaluation Model Based on PSO-BP Neural Network

The BP-ANN model randomly assigns connecting weights and thresholds during the model training process, which may fall into local optimal solutions and thus result in different outcomes every time. To address this issue, this study uses the PSO algorithm to improve the BP-ANN model which combines the particle swarm algorithm with the BP-ANN model. The specific steps of building PSO-BP-ANN model include: 1) use the particle swarm algorithm to obtain the position and velocity values of optimal particles; 2) assign the obtained values to the weights and thresholds in the BP-ANN model, respectively; 3) train the BP-ANN model based on the dataset and evaluate the neural network fitting effect by the decision coefficient R^2. This method avoids the random assignment of weights and thresholds in BP-ANN model, leading to a greater likelihood of obtaining a globally optimal solution.

3 Results

3.1 Performance of the BP-ANN Model and the PSO-BP-ANN Model

Three performance indexes of the BP-ANN model and the PSO-BP-ANN model are compared in Table 1. In this table, the R^2 of the traditional BP-ANN model is 0.62, while that of the PSO-BP-ANN model reaches 0.97, which indicates that the improved PSO-BP-ANN model has a higher prediction accuracy. However, the improved accuracy is accompanied by an increase in training time which reaches 25.29 s in comparison to the 1.3 s of BP-ANN model. This is attributed to the continuous iteration process in the PSO. This improvement in accuracy at the expense of time is a common feature of advanced algorithms. The number of iterations in the PSO-BP-ANN model is 8, while it is 15 in the conventional BP-ANN model indicating a significant reduction of the number of iterations of the former model. The PSO produces better estimates of weights and thresholds, which enables the neural network to reach the optimal solution quickly.

Table 1. The performance of the two ANN models

Model	R^2	Running time (s)	Iterations
Traditional BP-NNM	0.6209	1.3	15
PSO-BP-NNM	0.9895	26.59	5

3.2 Influence of the Variation of Parameters on Grading Levels

The relative importance of the evaluation parameters is carried out with the help of the PSO-BP-ANN model. In order to simplify the analysis, the evaluation indexes were categorized into four groups: transportation function (including the number of vessels, maximum vessel tonnage, water system connectivity, and the total number of vessel accidents), ecological function (including bank slope, river width, the coverage of ecological berms, bank vegetation coverage, and bank plant richness, and functional area water quality), landscape function (including riparian buffer zone width index and bank park area), and economic function (GDP per capita), and analyzes the influence of these four functions on the evaluation results accordingly. The scores of the functions are the average of the corresponding parameters. The specific steps include: 1) fix one function with a certain value, 2) randomly change the values of the other three functions, 3) calculate the probability of the five grades of the ecological evaluation, 4) change the value of the fixed function and repeat the previous steps to obtain the probability distribution of the five grades under different values of this function.

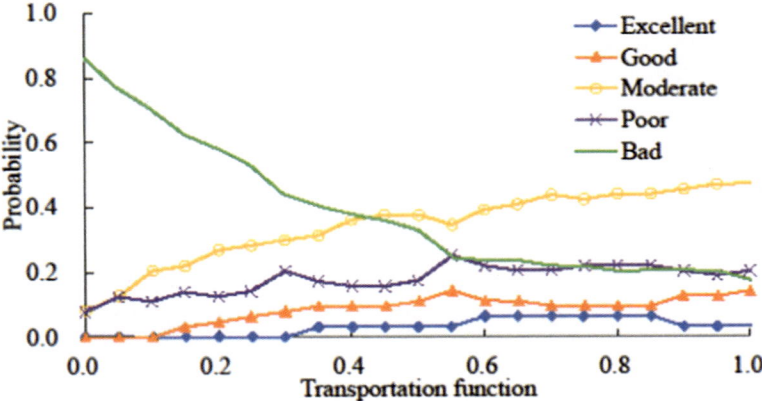

Fig. 1. The influence of transportation function on evaluating grades

As shown in Fig. 1, the abscissa indicates the score of the transportation function, and larger values indicate better conditions of navigation conditions. The vertical coordinate indicates the probability of the five grades. With the increase of the transportation function score (TFS), the probability of the excellent level increases first and then decreases when it is greater than 0.9. The overall probability of the excellent level is less than 0.1, which means the transportation function has limited influence on this grading level. When TFS is less than 0.3, the probability of the excellent level is 0. The probability of the good level varies in a limited range with the changing TFS, in an interval of 0 to 0.1. When the transportation function score is less than 0.1, the probability of the good level is 0. The probability of the poor level varies in a limited range as well, e.g., from 0.1 to 0.2. The possibility of the medium grade increases steadily from 0.1 to 0.5 with TFS increasing from 0 to 1. The bad grade changes most significantly compared with the other grades, and it decreases constantly for TFS increasing from 0 to 0.6 and remains stable thereafter. In summary, the enhancement of TFS can significantly reduce the probability of the bad level and increases the probability of the medium grade significantly. When the TFS score is higher than 0.6, the probability of all grading levels tends to be stable.

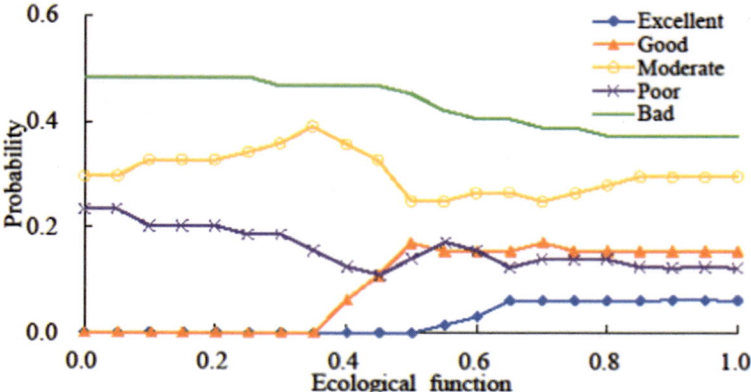

Fig. 2. The influence of the ecological function on the evaluating grades

As shown in Fig. 2, the abscissa indicates the ecological function score (EFS). The larger the EFS, the better the ecological conditions of waterways. With the increase of EFS, the probability of excellent and good grades shows an increasing trend. The probability of poor and bad grades shows a decreasing trend with increasing the EFS, while the probability of medium waterways shows a non-monotonical trend that it first increases and then decreases with the increasing EFS. The probability of a good grade abruptly increases from 0 to 0.5 when the EFS increases from 0.35 to 0.5. When the EFS exceeds 0.5, the probability of the excellent grade starts to increase from 0 and stabilized at 0.06 after the EFS is greater than 0.65. In summary, increasing the EFS can significantly increase the probability of the good grade given the EFS being greater than 0.35. Unlike the TFS, the EFS does not exhibit a monotonic effect on the middle, poor, and bad ratings, but they fluctuate with the EFS to some extent. After the EFS reaches 0.65, the probability of all grades tends to be stable.

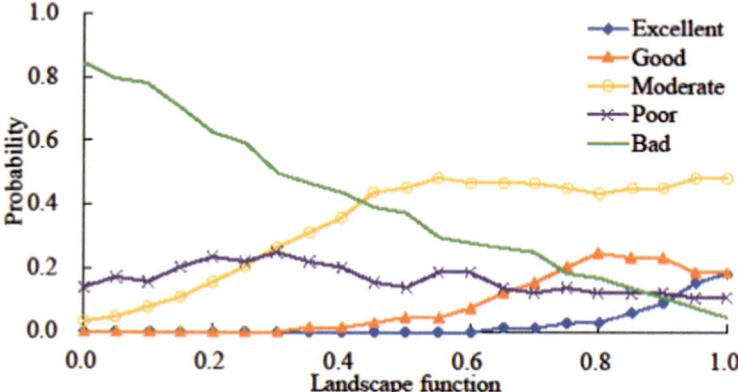

Fig. 3. The influence of landscape function on the evaluating grades

Figure 3 shows the different grading levels as a function of the landscape function score (LFS). In this figure, the probability of the excellent grade is zero until the LFS reaches 0.6 and then increases steadily to 0.2 with a further increase in the LFS. Similarly, the probability of the good grade is zero until the LFS reaches 0.4 and gradually increases with the LFS thereafter. The probability of the good grade stabilizes when the LFS reaches 0.8. The most significant change belongs to the probability of the bad grade that with the LFS increasing from 0 to 1, the poor grade constantly decreases from 0.9 to 0.1. In contrast, the probability of the poor level is almost invariant with the LFS. In summary, with the increase of the LFS, the probability of excellent, good, and moderate grades gradually increase, and the probability of the bad grade significantly decreases.

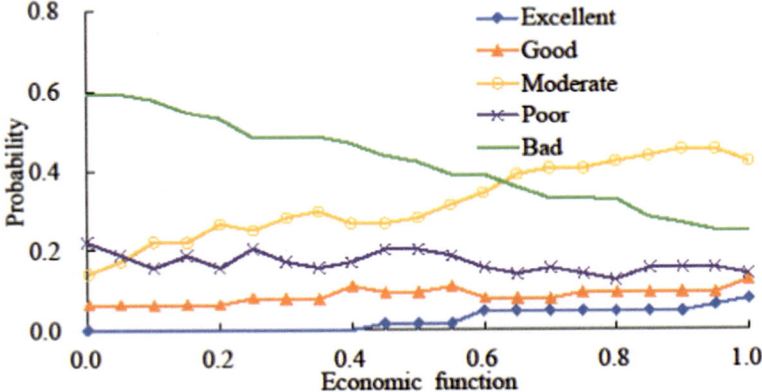

Fig. 4. The influence of the economic function on the evaluating grades

As shown in Fig. 4, the probability of the occurrence of the bad grade decreases constantly as the economic function score increases, i.e., it decreases from 0.6 to 0.25. The probability of the occurrence of the moderate grade increases substantially after the economic function score above 0.5, increasing from 0.28 to 0.48. The probability of occurrence of the excellent grade is 0 when the economic function score is below 0.4 and remains constant at 0.05 when the value lies between 0.6 and 0.9. It increases slightly after the score reaches 0.9. The probability of occurrence of poor and good levels remains relatively stable with the increase of the economic function score. In summary, the increase in the economic function score can greatly reduce the probability of the poor grade and increase the probability of the moderate grade, with no significant effect on the probability of remaining grades.

4 Conclusions

This study demonstrated the better performance of the PSO-BP-ANN model on the evaluation of the ecological waterway than the traditional BP-ANN model. Based on the former, the influence of four different function indicators including the transportation function, the ecological function, the landscape function, and the economic function on evaluating grades were analyzed. Among the four indicators, the landscape and transportation functions have the greatest influence on the probability of the bad level, and the increase of these two functions can significantly reduce the probability of this rating. The four indicators have an insignificant influence on the probability of the poor level. For the probability of the moderate level, the landscape function has the greatest impact. The ecological and landscape functions have a consequential influence on the probability of the good level. For the probability of the excellent level, all four indicators need to be greater than a certain value to reach this level.

Acknowledgements. This study was supported by the National Natural Science Foundation of China (52079043, 52179061).

References

Cai S, Hu X (2008) Concept of river health and index system for its evaluation. Adv Sci Technol Water Resour 28(1):23–27

Chu H, Lu W, Zhang L (2013) Application of artificial neural network in environmental water quality assessment. J Agric Sci Technol 15(2):343–356

Duan X, Zou H, Wang L, Chen W, Min M (2021) Assessing ecological sensitivity and economic potentials and regulation zoning of the riverfront development along the Yangtze River, China. J Clean Prod 291:125963

Guo R (2021) Ecological waterway evaluation based on PSO-BP neural network. Thesis, Hohai University, 86 p

Hamida A, Alsudairi A, Alshaibani K, Alshamrani O (2020) Environmental impacts cost assessment model of residential building using an artificial neural network. Eng Constr Archit Manag 28(10):3190–3215

Karr JR (1981) Assessment of biotic integrity using fish communities. Fisheries 6(6):21–27

Li T, Ding Y, Xia W (2018) An integrated method for waterway health assessment: a case in the Jingjiang reach of the Yangtze River. China. Phys Geogr 39(1):67–83

Prtersen RC (1992) The RCE: a riparian, channel, andenvironmrntal inventory for small streams in the agriculturel and scape. Freshw Biol 27(2):295–306

Wang J (2008) On delamination, protection and development of cultural heritage of the great canal. J Huaiyin Inst Technol 17(2):1–6

Environmental Impact Assessment of Watershed Plan Under the "Three Lines and One List" Environmental Governance

Runhe Cheng[✉] and Jinxiang Cheng

Institute of Environment Resources, Transport Planning and Research Institute
Ministry of Transport, Beijing 100028, China
754669651@qq.com

Abstract. Rapid and large-scale watershed development activities have imposed tremendous challenges to the sustainable development while driving economic prosperity in the areas along the watershed in China. Improving effectiveness of environmental impact assessment (EIA) of watershed planning has become a top priority for river ecological civilization. In this regard, the "three lines and one list" (TLOL) environmental governance, was proposed in the latest Chinese environmental management policy, consisting of an ecological conservation red line, an environmental quality bottom line and a resource utilization upper limit line and an environmental permit list are to be taken into account when assessing the potential effects of a watershed development plan. In this paper, an indicator system was established based on the TLOL requirements, and the rapid impact assessment matrix (RIAM) was adopt to asses watershed development alternatives. In an application of this methodology, the Jinjiang watershed development planning in Fujian province was taken as a case study to recommend an optimal alternative. Six alternatives were assessed by conducting a comprehensive comparison. The results showed that, the Alternative 2 is preferred because it has relative advantages in terms of allocating water resource in a reasonable way, safeguarding the ecological water use at downstream, controlling the small scaled hydropower generations. This research shows that the EIA of watershed planning on the basis of the TLOL governance policy is an effective way of integrating environmental management and river ecological civilization requirements into watershed development planning. It proposes not only a universal process for assessing watershed development alternatives, but also a feasible method of maximize the trade-off between water conservancy and hydropower, and other watershed development activities and river ecological protection.

Keywords: Three lines and one list · Watershed development plan · Environmental impact assessment · Rapid impact assessment matrix · Environmental governance

1 Introduction

China's watershed development has a long history, wide scope and high intensity, and the impact and consequences on the watershed ecosystem are extremely significant (Liu et al. 2021). Scientific planning of watershed development and strengthening

© The Author(s) 2023
Y. Li et al. (Eds.): PIANC 2022, LNCE 264, pp. 1121–1130, 2023.
https://doi.org/10.1007/978-981-19-6138-0_98

ecological protection has become an inevitable choice to deepen the construction of watershed ecological civilization (Huang et al. 2018). As a legal environmental management tool, environmental impact assessment (EIA) plays an important role in coordinating socio-economic development and ecological protection in China's watersheds. Domestic and international studies have shown that evaluating alternatives is a key step to improve the planning objectives and successfully implement the planning EIA (Huang et al. 2020). However, how to reasonably assess the ecological and environmental impacts of planning alternatives has been a prominent weakness in China's watershed planning EIA.

As China's latest environmental management policy of "three lines and one list" (TLOL) is a framework based on the threshold management of ecology, environment and resources (MEE 2016). It is composed of an ecological conservation red line, an environmental quality bottom line, a resource utilization upper limit, and an environmental permit list. With an aim of improving the quality of the environment, TLOL imposes explicit and rigid environmental constraints on economic development and resource utilization. Obviously, TLOL is in line with the watershed planning EIA of principle of weighing and maximizing the synergy between environmental governance requirements and development goals (Martínez-Bravo et al. 2019). Therefore, TLOL can provide a comprehensive framework for evaluating alternatives of watershed planning in EIA.

We also adopted Rapid impact assessment matrix (RIAM) to assess alternatives under the TLOL framework in this paper. The RIAM method is a semi-quantitative assessment method that combines standardized assessment scales with index calculation (Marcos et al. 2022). It has been regarded as an effective means to rapidly analyze the environmental impact of projects in EIA. In planning EIA studies, this method has been used for screening alternative of rehabilitation in small-scale water systems (Shakib-Manesh et al. 2014), evaluating urban industrial development planning (Phillips 2015), and selecting new site for municipal solid waste disposal (Seshagiri et al. 2016).

The remainder of this paper is organized as follows. Section 2 proposes an assessment indicator system amd RIAM method would be used to evaluate alternatives. Meanwhile, this section introduces an overview of development of Jinjiang watershed, which is used as our case study, and develops a set of alternatives for consideration. Section 3 presents results and discussion. Section 4 gives conclusion and implications.

2   Materials and Methods

2.1   Building up an Assessment Indicator System

The TLOL set ecological environment constraints on watershed development from three dimensions: ecological protection red line, environmental quality bottom line and resource utilization upper line. In this paper, the structural characteristics of the commonly used RIAM evaluation index system were considered, and four index sets were divided into ecological protection (EP), environmental improvement (EI), resource utilization (RU), and social economy (SE). Meanwhile, 34 specific indicators were selected as secondary indicators of this evaluation (Table 1).

Table 1. The indicator system for evaluating environmental impacts of alternatives of watershed development planning

Ecological Protection (EP)	Environmental Protection (EI)	Resource Utilization (RU)	Socio-Economic (SE)
EP1: Landscape fragmentation	EI1: Water quality standard attainment rate	RU1: Dry water level	SE1: Freshwater supply capacity
EP2: Landscape richness	EI2: Chemical oxygen demand	RU2: Annual average water level	SE2: Water energy supply capacity
EP3: River connectivity	EI3: Ammonia nitrogen	RU3: Flood water level	SE3: Supply capacity of aquatic products
EP4: River meandering	EI4: Total phosphorus concentration	RU4: Flood duration	SE4: Population health support capacity
EP5: Natural river scale	EI5: Total nitrogen concentration	RU5: Sediment content	SE5: Disaster regulation capability
EP6: Wetland retention rate	EI6: Dissolved oxygen	RU6: Water temperature	SE6: Leisure and recreation supply capacity
EP7: Riparian vegetation cover	EI7: Salinity	RU7: River dewatering section	SE7: Social Expenditure
EP8: Species diversity levels	EI8: Heavy metal content of substrate		
EP9: Habitat integrity of rare aquatic plants and animals			
EP10: Indigenous fish stocks			
EP11: Macroinvertebrate species, numbers and			

2.2 Assessing Alternatives by the RIAM Method

The assessment criteria that are usually employed in the RIAM approach are: importance of the condition (A1), magnitude of change/effect (A2), permanence (B1), reversibility (B2), and cumulative (B3). Against these criteria, both positive and negative impacts brought about by each alternative are assessed based on the environmental, ecological, societal, and economic indicators. Scores based on these assessments are calculated as follows:

$$(A1) \times (A2) = (AT) \tag{1}$$

$$(B1) + (B2) + (B3) = (BT) \tag{2}$$

$$(AT) \times (BT) = (BS) \tag{3}$$

The final result in this series of calculations, (ES), is the evaluated score for a given indicator. A1 and A2 are criteria that are important for the condition; for this reason a multiplication is required in Eq. (3). B1, B2, and B3 are criteria that are of value to the situation; to ensure that their collective importance is taken into account, addition is the appropriate operation in Eq. (4) (Pastakia and Jensen 1998).

To avoid placing undo significance on a specific number, a more efficacious system of assessment has been developed with the aid of range bands. Eleven range bands are usually employed: $-E$, $-D$, $-C$, $-B$, $-A$, N, $+A$, $+B$, $+C$, $+D$ and $+E$. An ES score of 0 implies that no impact or change occurs; in this case, the score falls in the range band designated as N. The range bands from $-A$ to $-E$ provide descriptive measures ranging from slightly negative to major negative impacts; those from $+A$ to $+E$ provide descriptive measures ranging from slightly positive impacts to major positive impacts (Table 2). Moving a step beyond the standard RIAM process, we have formulated an integrated environmental score (IES). Simply put, the IES for an alternative is summation of the ES values derived from the four indicator categories. It is calculated as follows:

$$IES = \sum_{i=1}^{4} (W_{c,i} \times ES_{c,i}) \tag{4}$$

where i represents an indicator category (PC, BE, SC, or EO) and $W_{c,i}$ and $ES_{c,i}$ are the assigned weight and the ES value, respectively, for category i. In this equation, $ES_{c,i}$ is derived from the formula:

$$ES_{c,i} = \sum_{j} (W_j ES_j) \tag{5}$$

where W_j and ES_j represent the weight and environmental score (ES), respectively, for indicator j in category i.

Table 2. The assessment criteria of RIAM

Criteria	Grade	Description	Criteria	Grade	Description
A1: importance of the condition	4	Important to national/international interests	B1: permanence	1	No change/not applicable
	3	Important to regional/national interests		2	Temporary
	2	Important to the local surrounding area		3	Permanent
	1	Only locally important	B2: reversibility	1	No change/not applicable
	0	Not important		2	Reversible
A2:magnitude of change/effect	+3	Strong positive impact		3	Irreversible
	+2	Significant improvement in the status quo	B3: cumulative	1	No change/not applicable
	+1	Improvement in the status quo			
	0	No change/status quo		2	Non-cumulative/independent action
	−1	Negative impact on the status quo			
	−2	Significant negative effect		3	Cumulative/synergistic effect
	−3	Strong negative impact			

By developing the impact level description system, the range of scores from −108 to 108 was divided into 11 environmental impact level segments, and the intervals where the scores of each index and the total scores of each alternative were located and the corresponding impact levels were found to obtain the ecological impact of the alternative on a certain index and on the whole watershed (Table 3).

2.3 Overview of the Jinjiang Watershed

Jinjiang watershed is located in the southeast coast of Fujian Province, bordering the Min River and Jiulong River. The administrative region includes the districts of Jinjiang, Nan'an, Anxi, Yongchun, Licheng, Fengze, and Luojiang. Jinjiang watershed covers an area of 5629 km², river length of 182 km, average river slope drop of 1.9%. Jinjiang watershed has experienced a rapid development of socio-economic, accounted for 1/4 of Fujian Province's GDP in 2016, and the area is known as the core of the Southern Fujian Golden Triangle (Yang et al. 2015).

The rapid development has led to ecological and environmental problems in the area of Jinjiang watershed. The utilization rate of water resources has reached 48.2%, which is higher than the average level in China. According to current trends, it is expected that the rate of hydro energy development in the basin can reach 60% by 2030 and cause a shortage of 1.33 billion m^3 of water supply (accounting for 37% of the total water demand) (Ma et al. 2015). Small hydropower is intensively developed in the Jinjiang watershed, especially the diversion type hydropower with less than 1000kw accounts for more than 85%, and most of them are not equipped with facilities for fish crossing and maintaining the ecological downstream flow. It has resulted in severe dewatering and a significant reduction in the pollution carrying capacity of the downstream reaches. It has resulted in severe dewatering and a significant reduction in the pollution carrying capacity of the downstream reaches. It also leads to the disappearance of aquatic organisms such as the original migratory fish in the basin. Optimizing the layout of hydropower development, improving the efficiency of water resources, controlling the total amount of water resources utilization, and strengthening ecological and environmental management are core issues of the sustainable development of the Jinjiang watershed.

Table 3. Conversions of the environmental scores to the range bands in RIAM

Score (ES)	Grade (indicated by letter)	Grade Description
108-72	E	Strong positive impact
71-36	D	Significant positive effect
35-19	C	Moderate positive impact
10-18	B	Positive Impact
1–9	A	Slight positive impact
0	N	No effect
−1–9	−A	Slight negative impact
−10−18	−B	Negative Impact
−19−35	−C	Moderate negative impact
−36−71	−D	Significant negative effect
−72−108	−E	Strong negative impact

2.4   Developing Alternatives of Jinjiang Watershed Planning

Due to the serious shortage of water resources and the over-exploitation of hydropower have seriously affected the ecological environment security of Jinjiang watershed and become the key restrictions of regional social and economic development. It is the top priorities to improve environmental sustainability of Jinjiang watershed development and promote trade-off synergy between watershed development activities and environmental protection.

To prevent the environmental problems mainly aggravated by irrational resource exploitation and watershed development activities, a new development plan of Jinjiang watershed was constructed.The planning and control objectives focus on solving the three most prominent problems of water resources utilization, hydropower development and ecological environmental protection in the Jinjiang watershed, and propose planning measures. Ecological environmental protection objectives are proposed in terms of water environment quality, ecological function improvement and ecological spatial integrity of the watershed; water resources development control objectives are clarified by controlling water resources development utilization rate and water saving efficiency; water energy development control objectives are proposed by water energy development access conditions and small hydropower cleanup. Based on the three control objectives, the Jinjiang watershed development plan proposes six alternatives (Table 4), combined with the requirements of the TLOL.

Table 4. Alternatives in the development planning of Jinjiang watershed

Alternatives	Water utilization		Hydropower development		Ecological protection
Alternative 1	Business as usual (BAU): no planning interventions; water utilization rate is expected to reach about 60% by 2030; hydropower development rate reaches 60%–70%				
Alternative 2	Water utilization rate is controlled within 40%, and the water intake of 10,000 Yuan GDP is reduced by 45%	Implement in-basin water diversion projects	① Delineate of prohibitions and restrictions construction areas fo hydropower development	Eliminate hydropower below 500kw, renovate hydropower below 1000kw, and expand 10% of total scale	① By 2030, the total amount of major pollutants entering rivers will be controlled within the carrying capacity, and the water quality of water function zones will meet 100% of the national standards ② Dam construction must ensure ecological and environmental water needs, over fish facilities or its alternative facilities must be built to ensure the stability of migratory fish populations
Alternative 3	Water utilization rate is controlled within 40%, and the water intake of 10,000 Yuan GDP is reduced by 48%	Expansion and renovation of existing reservoirs to take advantage of flooding	② Strictly control the addition of diversion-type small hydropower	Eliminate hydropower below 500kw, renovate hydropower below 1000kw, and expand 15% of total scale	
Alternative 4	Water utilization rate is controlled within 40%, and the water intake of 10,000 Yuan GDP is reduced by 50%	Implement inter-basin water transfer projects		Eliminate hydropower below 500kw, renovate hydropower below 1000 kw, and expand 20% of total scale	
Alternative 5	Water utilization rate is controlled within 40%, and the water intake of 10,000 Yuan GDP is reduced by 53%	Implementation of wastewater treatment reuse and desalination		Eliminate hydropower below 500 kw, renovate hydropower below 1000 kw, and expand 25% of total scale	
Alternative 6	Water utilization rate is controlled within 40%, and the water intake of 10,000 Yuan GDP is reduced by 55%	Strengthen water supply scheduling and reduce continuous downstream power generation		Eliminate hydropower below 500kw, renovate hydropower below 1000kw, and total scale is not expanded	

3 Results and Discussion

The assessment results of ES for alternatives are shown in Fig. 1. Results show that Alternative 1 would have negative impact on the four categories of "resource use (RU)", "environmental improvement (EI)", "ecological protection (EP)", and "social economy (SE)", which is significantly worse than the remaining alternatives. Alternative 2 would has a moderate positive impact on RU and EI, and a negative impact on EP. Alternative 3 would has no impact on RU, EI and EP, and a positive impact on SE. Alternative 4 would has a positive impact on all indicators of EI, and a positive impact on RU and SE, and a moderate negative impact on EP. Alternative 5 would has no impact on all four categories of indicators, with positive impact on EP and SE. Alternative 6 would has positive impacts on RU, EI, and EP, and no impact on SE.

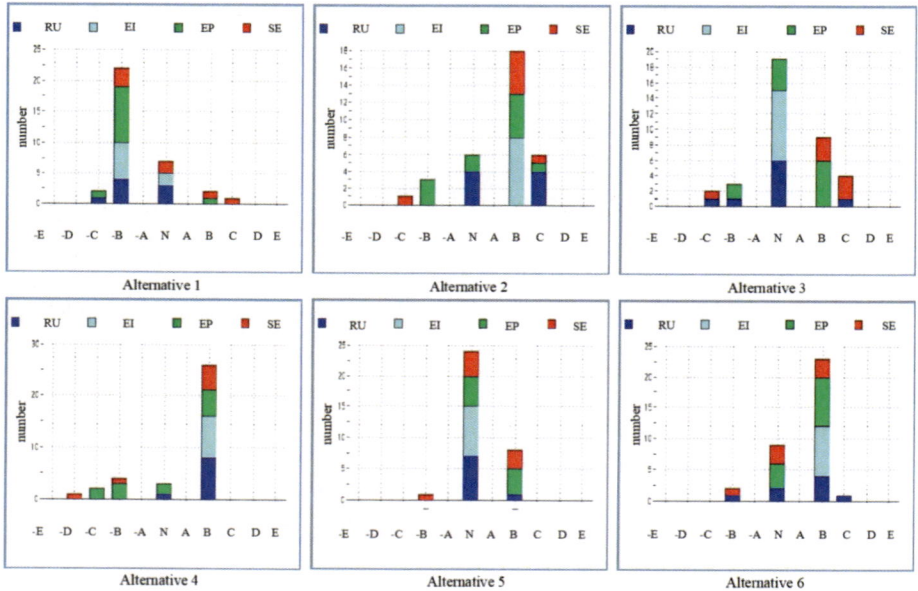

Fig. 1. The evaluating results of planning alternatives by RIAM.

Comprehensive assessment was conducted on the four index sets of RU, EI, EP and SE, and the results of IES are shown in Table 5. By comparing the results of the six alternatives, Alternative 2 has the best performance and Alternative 2 has the worst performance. .

Alternative 2 has a positive impact on most of the indicators, and the IES performance is significantly better than the remaining alternatives, so it can be judged that the implementation of Alternative 2 has a positive impact on EI, EP, RU and SE of the Jinjiang watershed. From the perspective of watershed planning EIA, the implementation of Alternative 2 is more advantageous in optimizing water resources allocation, ensuring ecological water use of downstream, and strictly controlling hydropower development. Therefore, Alternative 2 can be recommended for detailed assessment.

Table 5. IESs of the planning alternatives

Alternatives	RU	EI	EP	SE	IES
Alternative 1	−12	−10	−11	−2	−9
Alternative 2	16	14	6	11	11
Alternative 3	−2	0	5	16	4
Alternative 4	14	16	−2	2	7
Alternative 5	2	0	7	4	3
Alternative 6	9	14	9	4	9

4   Conclusion

This paper integrates the requirements of the "three lines and one list" environmental governance with the traditional RIAM method, and constructs the TLOL-RIAM method with an indicator system for watershed planning EIA including four dimensions of RU, EI, EP, and SE. Meanwhile, the traditional RIAM method is improved by introducing the Integrated Environmental Evaluation Score (IES). Applying the constructed method to a planning EIA of Jinjiang watershed in China as a case study. The results shows that, the IES performance of Alternative 2 is the best, and recommended it as the optimal one. Focusing on the measures of Alternative 2, the EIA proposed some suggestions to planning decision makers, including the implementation of water diversion projects in the watershed, strengthening the management of water resources dispatching, and controlling the scale of small hydropower.

The application case of the Jinjiang watershed shows that, the TLOL-RIAM method can help EIA to make a preliminary assessment of alternatives from a comprehensive perspective in the early stage, and propose timely optimization suggestions for planning. The finding in this study is expected to provide a useful reference for the study of watershed planning EIA in China.

References

Huang LH, Chen F, Jiang Y, Cheng HG, Yan DH (2018) Suggestions on solving the technical bottleneck of environmental impact assessment of comprehensive river basin planning. Environ Impact Assess 40(5):59–61

Liu H, Geng YZ, Dong Q (2021) Experiences of other countries in river basin development and protection and its enlightenment to China. China Water Resour 10:57–59+61

Ma L, Teng YG, Lin XY, Wang JS (2015) Spatial distribution of pollution load and critical source area identification in the Jinjiang River Basin. China Environ Sci 35(12):3679–3688

Martínez-Bravo M, Martínez-del-Río J, Antolín-López R (2019) Trade-offs among urban sustainability, pollution and livability in European cities. J Clean Prod 224:651–660

Ministry of Ecology and Environmental of the People's Republic of China (2016) The 13th Five-year environmental impact assessment Reform Implementation Scheme. http://www.mee.gov.cn/gkml/hbb/bwj/201607/t20160719_360949.htm

Pastakia CMR, Jensen A (1998) The rapid impact assessment matrix (RIAM) for EIA. Environ Impact Assess Rev 18(5):61–82

Phillips J (2015) A quantitative-based evaluation of the environmental impact and sustainability of a proposed onshore wind farm in the United Kingdom. Renew Sustain Energy Rev 49:1261–1270

Rodrigues MVC, Guimarães DV (2022) Urban watershed management prioritization using the rapid impact assessment matrix (RIAM-UWMAP), GIS and field survey. Environ Impact Assess Rev 94:106759

Shakib-Manesh TE, Hirvonenl KO (2014) Ranking of small scale proposals for water system repair using the rapid impact assessment matrix (RIAM). Environ Impact Assess Rev 49:49–56

Suthar S, Sajwan A (2014) Rapid impact assessment matrix (RIAM) analysis as decision tool to select new site for municipal solid waste disposal: a case study of Dehradun city, India. Sustain Cities Soc 13:12–19

Xiang J, Tang M (2021) Study of environment impact assessment for Beiliu River Basin comprehensive planning of Guangxi. Guangxi Water Resour Hydropower Eng 5:49–53

Yang LY, Li NB, Xu XY (2015) Study on water allocation and ecological compensation of water resources protection in Jinjiang River Basin. Yellow River 2:68–71

Knowledge About Sediment Transport Obtained Through Multiple Operations at Ports in Japan

Tomohiko Kachi[✉], Minoru Itui, Tatsunori Naruke, and Yukihiko Sugiura

IDEA Consultants, Inc, Yokohama, Japan
{kch20312, iminoru, ntatsuno, sugiura}@ideacon.co.jp

Abstract. At a port located at the estuary in Japan, 800,000 m^3/year of sediment is accumulated annually in navigation channel and basin. In order to consider effective countermeasures, it is necessary to elucidate sediment dynamics due to waves and tidal currents of the target port. Therefore, we are conducting a field observation of sediment transport and deposition. In addition, we have developed a numerical simulation model based on the results of the field survey for evaluated the countermeasure against the port sedimentation. We introduce the knowledge about sediment transport obtained through multiple operations.

Keywords: Port sedimentation · Field observation · Numerical simulation · Sediment disposal plan

1 Introduction

Some ports located at the estuary or inner bays in Japan suffer from river-derived sediment. For example, a large amount of sediment from rivers flows into estuary of Ariake Sea in Japan when the water rises due to seasonal rain or typhoon. Tideland at the mouth of the river stores sediment and contribute to the deposition of the port for a long term (refer to Fig. 1). To consider effective countermeasures, it is necessary to elucidate sediment dynamics due to waves and tidal currents of the target port. Therefore, field observation methods have been improved, and numerical simulation models for evaluating countermeasures have been developed for multiple ports in Japan. The first practical study of countermeasures against the port siltation in Japan was carried out in Kumamoto Port in Ariake sea. Tsuruya et al. (1990) developed the multi-layer numerical model that be able to predict the effect of submerged breakwater against the siltation. In the maintenance of navigation channels and basins, both to reduce dredging costs and to create disposal sites are issues. Furthermore, the procedure of environmental assessment for the construction of new disposal site takes long time. Therefore, it is necessary to predict medium- and long-term volume of dredging sediment, make a plan about the disposal of sediment and advance the preparation of construction of the disposal site.

Y. Li et al. (Eds.): PIANC 2022, LNCE 264, pp. 1131–1138, 2023.
https://doi.org/10.1007/978-981-19-6138-0_99

This paper introduces our knowledge about sediment transport obtained through multiple operations.

Fig. 1. Tideland at the mouths of Shirakawa River and Tsuboi River in Ariake Sea in Japan

2 Field Observation

2.1 Basic Surveys

It is classified the field survey as basic survey to grasp the present situation of sedimentation and application survey to specify the mechanism of sedimentation. For the basic survey, bottom sounding using a single-beam or multi-beam will be conducted to reveal the actual situation of navigation channels and basins. In recent years, the multi-beam sounding has been applying to the bathymetry survey for the progress of 3D visualization in the design and construction of facilities. Our company introduced Multibeam Echosounder (Sonic2024/Sonic2026), is able to carry out a wide range / high-resolution sounding depending on a purpose with an item (refer to Fig. 2).

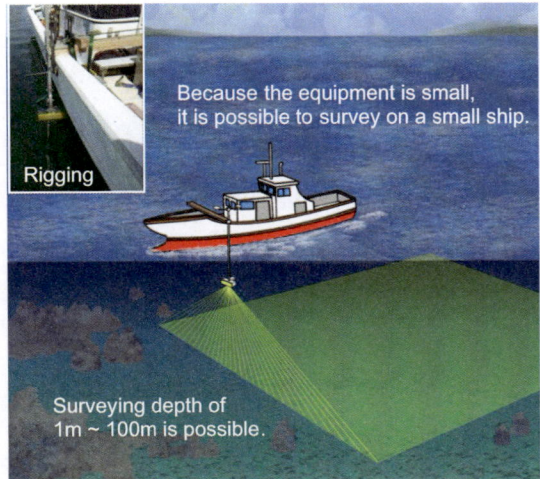

Features:

High-precision position measurement by GNSS post-analysis processing.

High-precision sway correction by motion sensor system.

A wide range can be measured in a short time.

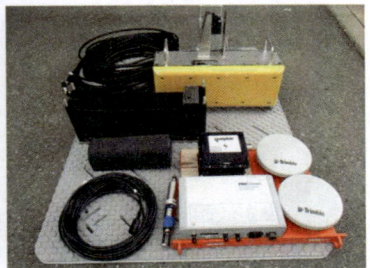

Fig. 2. Multibeam echosounder

2.2   Application Surveys

Application surveys are conducted in characteristic areas identified through the basic survey to elucidate suspended sediment transport and their deposition processes. Within a time limit and budget, it is important to make a plan of survey with building an appropriate hypothesis to get the high priority information.

Sediment transport is classified as suspended load and bed load. The suspended load travels a long distance and is often subject to sedimentation in navigation channels and basins. Current velocity/direction and SS concentration are investigated to know the suspended load transportation. Figure 3 shows survey method and Fig. 4 shows results of current velocity and SS concentration. It is recognized that the flood tide current includes a lot of sediment. Integrating temporally and spatially those, we can know sediment transport amount to inside the port.

Acoustic Doppler Current Profiler (ADCP) measures the current velocity and direction of multiple vertical or horizontal layers in the subsection. SS concentration is able to be calculated from echo intensity obtained by ADCP. Depending on a purpose and budget, single layer observation is conducted by using an electromagnetic velocity meter and a turbidity meter. Furthermore, there are some cases using sand level gauge to measure height of sedimentation.

There is a restriction in the selection of survey point because of precedence of using navigation channels and basins for usual activity in the port. If the survey point is not able to be located inside navigation channels and basins, it should be located on the course of sediment transport. Alternatively, it can be located in small pocket area which is dredged for the particular survey.

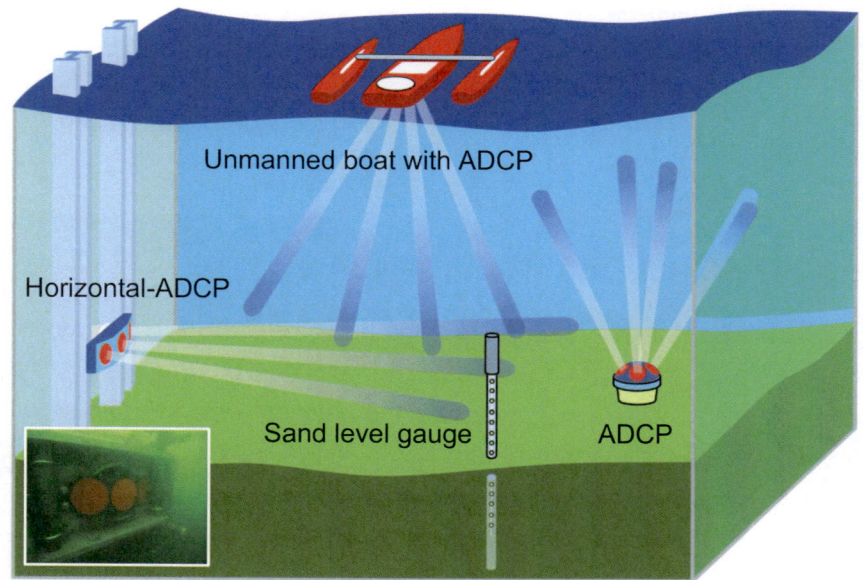

Fig. 3. Schematic diagram of current velocity survey method

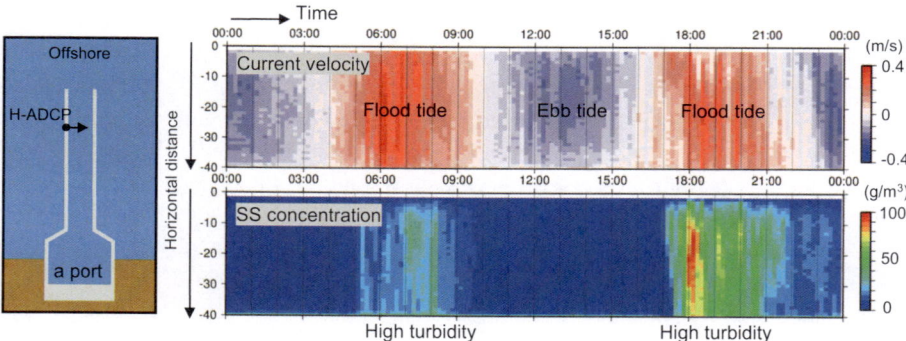

Fig. 4. An example of survey results

2.3   Mechanism of the Port Sedimentation

Sedimentation mechanism is considered by results of field survey and the hypothesis which has been built before is verified by observation data. At that time, the relationship between metrological phenomena and actual situation of sedimentation is important. In Japan, we can use some public information, NOWPHAS (wave data by MLIT (Ports and Harbors Bureau, Ministry of Land, Infrastructure, Transport, and Tourism)), AMEDAS (meteorological data by JMA (Japan Meteorological Agency)) and Water Information System (river data by MLIT).

3 Numerical Simulation

3.1 Building of the Practical Numerical Model

In practice, customers demand speed and comprehensibility. Therefore, we focus on simply expressing the phenomenon to get quick and easy-to-understand results.

Numerical simulation model of the port sedimentation has 3 main components, current, wave and topographical changes (refer to Fig. 5). This method is based on Tshuruya et al. (1990). When necessary, topographical changes component is taking account of fluid mud layer near the bottom. Modeling of the fluid mud based on Odd and Cooper (1989) has been improved by various field and laboratory survey.

Because the phenomenon of port sedimentation is very complex and to perfectly verify it by computing is very difficult, the calibration of model parameters is conducted focusing on the primary factor of sedimentation and the specified phenomenon.

We have been recently focusing on the pickup and diffusion of bottom sediment by ships on the waterway and improving the numerical model including those effects. Our numerical model has been already applied at several ports in Japan.

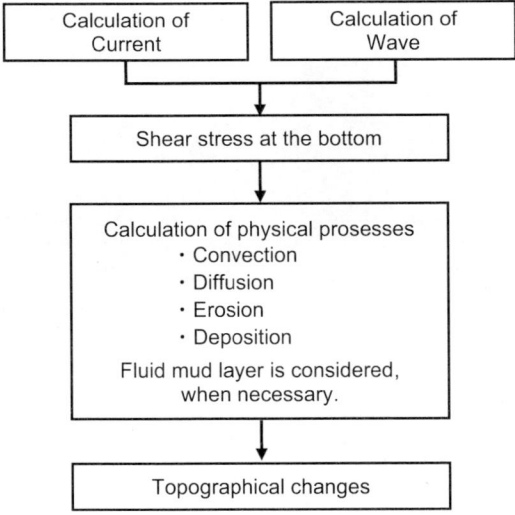

Fig. 5. Overview of numerical simulation model of the port sedimentation

3.2 Prediction and Estimation of the Effect of Countermeasure Against the Port Sedimentation

The effect of countermeasure against the port sedimentation is estimated by the difference between the case of existence of the installed countermeasure and the case of the uninstalled. The countermeasure is selected from the proposed in Fig. 6. In above-referenced Kumamoto port, the inverted-T submerged breakwater was applied as the countermeasure against the waterway sedimentation. The pocket dredging has been often applied in recent years because it doesn't give influence on the ship navigation. The pocket is constructed at the adjacent area of waterway or the upstream of the suspended load. It alternatively catches sediment and reduces sedimentation rate on the waterway.

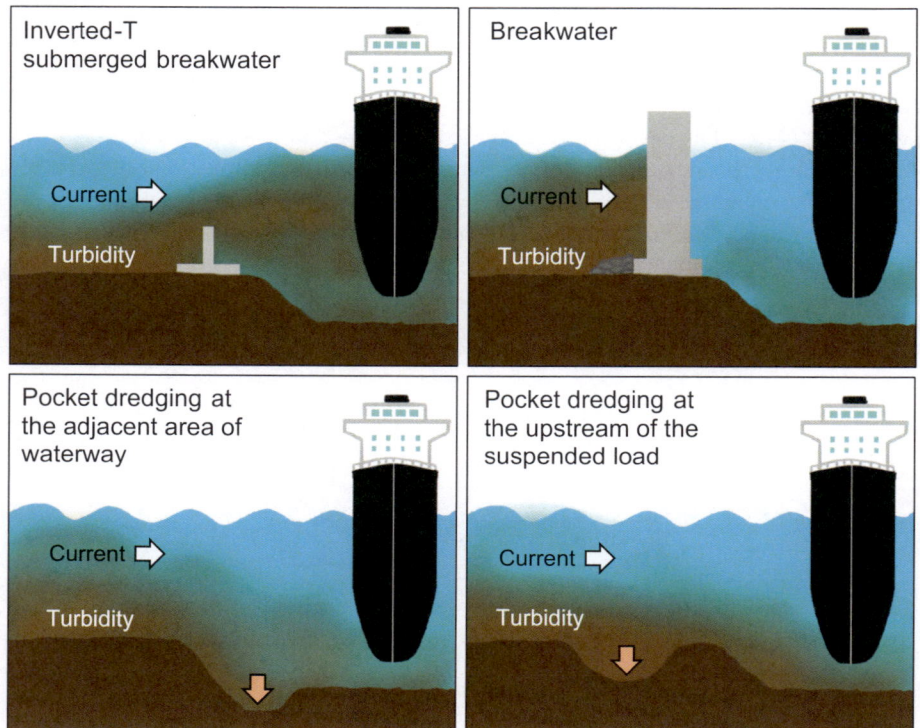

Fig. 6. Main Types of countermeasures

4   Dredging and Sediment Disposal Plan

4.1   Dredging Efficiency

To reduce the dredging cost, it is necessary not only to reduce the amounts also to improve the work efficiency. In case that the dredger ship can work on one spot, the dredging efficiency is good (Fig. 7). Therefore, the dredging at narrow pocket area is economically more reasonable than the slight dredging at wide area in the waterway or basin. However, it is necessary to estimate the effect with a long-term viewpoint due to the increase of sedimentation rate at pocket area which are deeper than surroundings. Therefore, we evaluate the effects of pocket dredging and countermeasure facilities using a numerical simulation model.

Fig. 7. Work status of dredging

4.2 Disposal Planning of Dredging Sediment

Dredging Sediments are deposited at the disposal site near the port and are appropriately treated. It is necessary to elaborate on a sediment disposal plan with estimation of sedimentation amount which will generate in future because the lack of disposal space affects activities in the port. But it is difficult to estimate long-term amount of dredging sediment because the sedimentation rate in the waterway or basin changes depending on typhoon coarse year by year. On that account, we have developed below methods.

   i. To estimate sedimentation amount from wave and bathymetry data (for approx. 30 years)
  ii. To estimate annual sedimentation rate by extreme value statistical analysis and draw out a probability density function
 iii. To calculate 10,000 cases of sedimentation rate by Monte Carlo method based on the probability density function
 iv. To estimate the expectation value of sedimentation rate as ensemble average of all cases.

Here, observation wave data by NOWPHAS can be used. When there is not it around the port, it is necessary to estimate wave height by using wind data.

4.3 Adaptive Management

In some cases, there is a gap between the sediment disposal plan and the actual results depending on the occurrences of meteorological event like a typhoon. Moreover, there is another case that the unexpected phenomenon which cannot be explained by assumed sedimentation mechanism occurs. Therefore, we recommend preparing "Clinical Chart to Consider Measures against Port Sedimentation" by each port. Historical records of field survey, dredging and countermeasure construction are organized in the chart. Moreover, the present situation about validation of the sedimentation mechanism and problem to be solved are also written in it. The aim of creation of the chart is to optimize maintenance of waterway and basin by means of handing over it from predecessor to successor for the maintenance.

5 Conclusions

For the consideration of measure against port sedimentation, to accurately grasp phenomenon which occurs on site, to estimate with numerical simulation based on the phenomenon and to plan an accommodative measure taking into account of uncertainty of the estimation are required. This paper introduces technical attempts from those 3 perspectives. A problem to be solved is to improve the measure against port sedimentation including domestic and overseas technical progress. We have been conducting overseas project about port sedimentation with technology which has been cultivated in Japan and continuing to contribute improvement of port accessibility in the world.

Acknowledgements. We give special thanks to the organizers which set the interesting theme, held the conference on the Coronavirus Crisis and accepted our paper.

References

Odd, NVM, Cooper, AJ (1989) Two-dimensional model of movement of fluid mud in a high energy turbid estuary. J Res Spec (5):185–193

Tshuruya H, Murakami K, Irie I (1990) Mathematical modeling mud transport in ports with a multi-layerd model. Rep Port Harbour Res Inst 29(1):3–51

Monitoring the Variation of Drought-Flood Abrupt Alternation and Its Response to Atmospheric Circulation at Multi-time Scales

Wuzhi Shi[1], Ke Zhang[1,2,3(✉)], Yuebo Xie[1], Lijun Chao[1], Tolossa Lemma Tola[1], and Xianwu Xue[4]

[1] State Key Laboratory of Hydrology-Water Resources and Hydraulic Engineering, and College of Hydrology and Water Resources, Hohai University, Nanjing 210024, Jiangsu, China
kzhang@hhu.edu.cn

[2] Yangtze Institute for Conservation and Development, Hohai University, Nanjing 210098, Jiangsu, China

[3] CMA-HHU Joint Laboratory for HydroMeteorological Studies, Hohai University, Nanjing 210098, Jiangsu, China

[4] Systems Research Group at Environmental Modeling Center, NOAA/NWS/NCEP, Washington, DC, USA

Abstract. As an emerging disaster, the drought-flood abrupt alternation (DFAA) may cause unprecedented socio-economic impacts under changing environment, which has attracted extensive attention in recent decades. DFAA involves drought to flood (DTF) and flood to drought (FTD). However, thus far, little effort has been made to identify DFAA with high spatial resolution. Moreover, few studies have fully revealed the driving mechanisms of DFAA by large-scale climate factors. Here, the Yellow River Basin (YRB) was selected as the research area, which is an important agricultural base in China. The spatiotemporal characteristics of DFAA at multiple time scales during flood season were analyzed using 0.25° grid precipitation from 1961 to 2020 in the YRB. Furthermore, the Pearson correlation method and cross wavelet method were used to investigate the relationship between circulation anomaly (such as Arctic oscillation (AO), Pacific decadal oscillation (PDO), El Niño Southern Oscillation (ENSO), and sunspot) and DFAA to explore the potential causes of DFAA in this region. The results demonstrated that: (1) FTD trend in the YRB is serious, and the short period of FTD trend is June-July > July-August > August-September; (2) spatially, the high-frequency long-period DFAA was distributed in the whole YRB, while the DFAA in June-July and July-August were concentrated in the center of the YRB; (3) AO and PDO are the key factors to induce DFAA in the YRB, especially the changes of AO and PDO phase. This study helps improve our understanding of the relationship between DFAA and large-scale climate factors and provides new insights for future disaster assessment.

Keywords: Drought-flood abrupt alternation · Spatiotemporal · Frequency · Driving mechanisms

© The Author(s) 2023
Y. Li et al. (Eds.): PIANC 2022, LNCE 264, pp. 1139–1151, 2023.
https://doi.org/10.1007/978-981-19-6138-0_100

1 Introduction

Increasingly severe climate change and human activities lead to more and more frequent global extreme climate disasters (Shi et al. 2020). The drought-flood abrupt alternation (DFAA) is caused by the complex relationship between drought and flood, which has developed into one of the most destructive disasters (Shi et al. 2021). It can cause enormous socioeconomic losses and seriously threaten agriculture, ecological environment, and human health. Therefore, understanding the spatiotemporal pattern of DFAA and its driving mechanism is of great significance for water resources management and early warning and mitigation of DFAA.

DFAA is quantified and monitored by the standardized precipitation index, long-/short-cycle drought-flood abrupt alternation index (LDFAI/SDFAI), and runoff abrupt change index. Among these indices, LDFAI and SDFAI were proposed in 2006 (Wu et al. 2006) and have the advantages of single input, namely only precipitation data is needed, and a simple calculation process, which become the most popular. The government and the public have paid great attention to when and where DFAA occurs. Numerous scholars have carried out multiple regional DFAA assessments. For instance, Li et al. (2017) defined the DFAA index using streamflow to identify DFAA of the Poyang Lake catchment in the past 50 years and found that drought to flood (DTF) and flood to drought (FTD) were concentrated in March to April and July to August, respectively. Spatiotemporal evolution characteristics of DFAA in Shanxi Province in summer were fully revealed using LDFAI (Liu et al. 2017).

The above studies mainly focused on the DFAA index with a single time scale to evaluate the changes of DFAA in a specific period. There was a lack of evaluation for DFAA in different periods. Therefore, it is necessary to strengthen the quantitative study of DFAA in different periods based on multi-time scale indicators. For example, Shan et al. (2015) comprehensively evaluated the spatiotemporal distribution of DFAA in the middle and lower reaches of the Yangtze River (MLYR) based on LDFAI and SDFAI and found that long-/short-cycle DFAA in this region was dominated by FTD. In addition, The change trends of LDFAI, SDFAI, and monthly precipitation in the growth period of tobacco were analyzed (Zhang et al. 2019).

DFAA is often affected by a variety of factors, such as precipitation, temperature, and evaporation, which directly affect DFAA. Moreover, many studies have shown that large-scale climate factors are closely related to DFAA. For instance, DFAA in the MLYR is related to phenomena of continuously low SST in the Nino 3.4 region, especially the La Nina phenomenon, which can provide an early warning for DFAA (Shan et al. 2018). Precipitation anomaly and its dynamic factors and water vapor sources during the DFAA period in the MLYR from May to June in 2011 were also comprehensively analyzed (Wang et al. 2014). However, these studies only focus on the influence of a single atmospheric index, or the causes of DFAA change in a particular year, while the relationship between different atmospheric indexes with DFAA and the driving mechanism of DFAA in historical/recent years has not been fully revealed. Therefore, the main objectives of this study are to: (1) analyze the spatiotemporal variation characteristics of long-/short-cycle DFAA; (2) comprehensively evaluate the potential driving mechanisms of various atmospheric factors on

DFAA. The results would help to understand the evolving pattern of DFAA under global warming and provide guidance for regional water resources management.

2   Study Area and Data

2.1   Study Area

The Yellow River basin (YRB), which is located between 96°– 119° E, 32° –42° N (Fig. 1), is surrounded by the Bohai Sea, Bayankara mountain, Qinling Mountains, and Yinshan Mountains in the east, west, south, and north, respectively (Ji et al. 2021). The YRB has a total drainage area of 795,000 km^2 including 42,100 km^2 of the inner basin, with a length of about 1900 km from east to west and a span of about 1100 km from north to south. The climatic characteristics of the YRB are complicated due to the monsoon and atmospheric circulation. The annual average precipitation and the annual average evaporation are concentrated in 200– 800 mm and 800 – 1800 mm, respectively, which is one of the typical climate-sensitive areas in China (Zhao et al. 2019). Under the influence of global warming and human activities, drought/flood disasters occur frequently in the YRB. Furthermore, the YRB is an important agricultural base in China, these frequent natural disasters seriously threaten the safety of agricultural products. According to the characteristics of the drainage system, the YRB is divided into the upper (UYRB) and middle and lower (MLYRB).

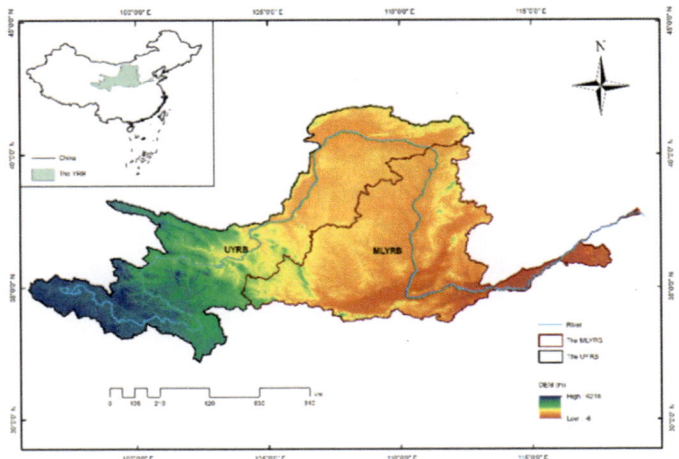

Fig. 1. The geographical position of the Yellow River basin.

2.2   Data Source

In this study, monthly precipitation data with the period for 1961–2020 was available from the Climate Research Unit (CRU) (https://www.cru.uea.ac. uk/data), which has a spatial grid of 0.5° × 0.5°. The nearest neighbor interpolation method was used to interpolate the precipitation data to a 0.25° grid. The monthly atmospheric circulation

factors over the same data, including Arctic Oscillation (AO), Pacific Decadal Oscillation (PDO), El Niño Southern Oscillation (ENSO), and sunspot, download from the
National Oceanic and Atmospheric Administration (NOAA) (https://psl.noaa.gov/
arctic/data/).

3 Methods

3.1 Long-Cycle Drought-Flood Abrupt Alternation Index

The precipitation in the YRB has obvious seasonal characteristics, and mainly concentrates from June to September (Xu and Zhang 2006). Therefore, the precipitation
during this period was selected as summer precipitation to quantitatively reveal the
evolution of the long-/short-cycle drought-flood abrupt alternation in this basin. The
time scale of the long-cycle drought-flood abrupt alternation index (LDFAI) is 2
months, namely, June-July is drought, August-September is flood, which is defined as
drought to flood (DTF) (Shi et al. 2021), and vice versa.

$$LDFAI = (R_{89} - R_{67}) \times (|R_{67}| + |R_{89}|) \times 1.8^{-|R_{67} + R_{89}|} \tag{1}$$

where, R_{67} and R_{89} are standardized precipitation from June-July and August-
September, respectively; $(R_{89} - R_{67})$ and $|R_{67}| + |R_{89}|$ represent intensity term of
DFAA and drought/flood, respectively; $1.8^{-|R_{67} + R_{89}|}$ is the weighting coefficient.
If LDFAI is greater than 1, it means DTF; if LDFAI is less than -1, it means flood to
drought (FTD), and the rest are normal (Yuan et al., 2021).

3.2 Short-Cycle Drought-Flood Abrupt Alternation Index

The short-cycle drought-flood abrupt alternation index (SDFAI), with a time scale of
1 month, is similar to the definition of LDFAI (Zhang et al. 2019). The specific formula
is as follows.

$$SDFAI = (R_j - R_i) \times (|R_i| + |R_j|) \times 3.2^{-|R_i + R_j|} \tag{2}$$

where, $j = i + 1$ $(i = 6, 7, 8)$; the evaluation criteria of SDFAI are consistent with
LDFAI.

4 Results and Discussion

4.1 Temporal Characteristics of Drought-Flood Abrupt Alternation

4.1.1 Temporal Pattern of Long-Cycle Drought-Flood Abrupt Alternation
LDFAI was calculated using the monthly precipitation of the YRB during 1961–2020.
Temporal variation of LDFAI is of great significance for understanding the long period of

DFAA events in the basin. Figure 2 presents the temporal variation of LDFAI in the UYRB and MLYRB, respectively. The red dotted line indicates the changing trend of LDFAI.

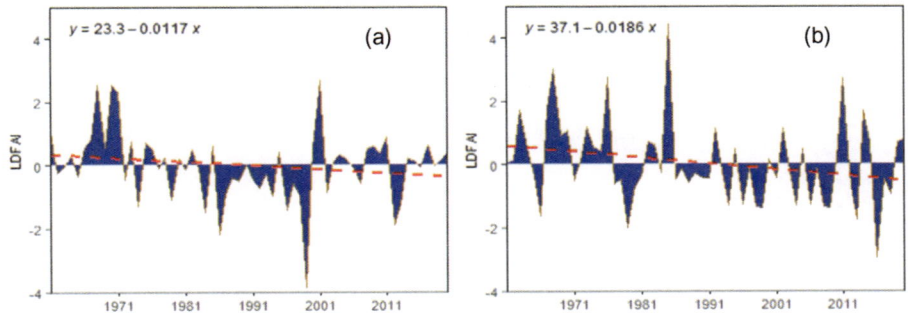

Fig. 2. Temporal variation of LDFAI in the UYRB (a) and MLYRB (b).

As can be seen from Fig. 2(a), the inter-annual variation of LDFAI was large, reaching a maximum value of 2.69 and a minimum value of −3.87 in 2001 and 1999, respectively. Overall, the slope of LDFAI is −0.117/10a in the UYRB, indicating that LDFAI presents a downward trend in the UYRB. This further illustrates that FTD tends to increase while DTF tends to decrease, which is consistent with the recent study by Shi et al. (2021). The temporal variation of LDFAI in the MLYRB is the same as that in the UYRB, with a maximum value of 4.44 in 1985 and a minimum value of −2.96 in 2016 (Fig. 2(b)). The LDFAI in the MLYRB decreased at a rate of 0.186/10a, indicating that the region was also prone to FTD. In addition, the decreasing rate of LDFAI in the MLYRB is significantly higher than that in the UYRB, which illustrates that FTD of the MLYRB is more frequent in the flood season. Liu et al. (2012) found that the increase and decrease of precipitation in the flood season of the YRB during 1961–2010 mainly occurred in June and July-September, respectively, and the increasing trend of precipitation in June was higher than the decreasing trend in July. This conclusion confirms that FTD is prone to occur in the YRB during flood season.

4.1.2   Temporal Pattern of Short-Cycle Drought-Flood Abrupt Alternation

This study quantified the change of SDFAI to further reveal the evolution characteristics of DFAA in the flood season of the YRB. Figures 3, 4 and 5 show the temporal variation of SDFAI in the YRB from June to July, July to August, and August to September, respectively. The SDFAI of the UYRB and MLYRB showed a slight downward trend from June to July, and the changing rates were 0.104/10a (Fig. 3(a)) and 0.0748/10a (Fig. 3(b)), respectively. The variation trend of SDFAI from July to August (Fig. 4) is the same as that from May to June, which indicates that FTD occurs frequently in the YRB from June to July and July to August. It is worth noting that the variation trend of SDFAI in the UYRB is greater than that in the MLYRB, indicating that FTD is more likely to occur in the UYRB during this period. Interestingly, the change rate of SDFAI in the UYRB from August to September was 0.107/10a (Fig. 5(a)), indicating that

SDFAI showed an upward trend, namely DTF occurred frequently in this region. However, SDFAI in the MLYRB decreased at a rate of 0.138/10a in this period, indicating that FTD in this region increased while DTF decreased. Meanwhile, DTF and FTD occurred before and after 1985 in the MLYRB, respectively. Xu and Zhang (2006) showed that precipitation in the lower of the YRB (LYRB) decreased significantly in September, which verified the reliability of this study. That is, the decline rate of SDFAI from August to September was significantly higher than that of other periods.

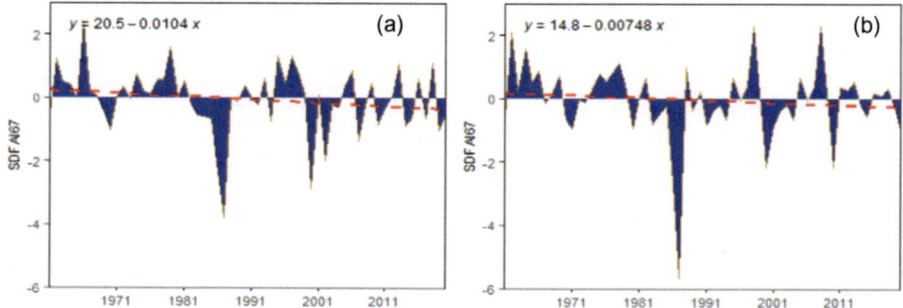

Fig. 3. Temporal variation of SDFAI67 in the UYRB (a) and MLYRB (b), SDFAI67 stands for SDFAI from June to July.

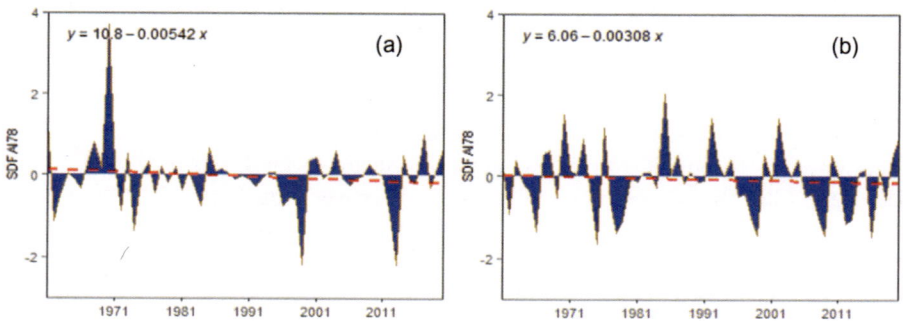

Fig. 4. Temporal variation of SDFAI78 in the UYRB (a) and MLYRB (b), SDFAI78 stands for SDFAI in July-August.

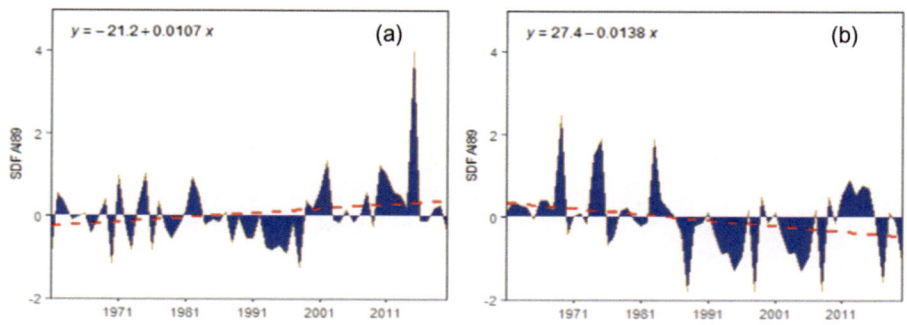

Fig. 5. Temporal variation of SDFAI89 in the UYRB (a) and MLYRB (b), SDFAI89 stands for SDFAI in August-September.

4.2 Spatial Characteristics of Drought-Flood Abrupt Alternation

4.2.1 Frequency of Drought-Flood Abrupt Alternation

The frequency of DTF and FTD in the YRB are shown in Fig. 6 and Fig. 7, respectively, and there are significant spatial differences in DFAA frequency in different periods. The spatial distribution of precipitation in the YRB is uneven and affected by topography and geomorphology (Li et al. 2016), which results in regional and complex distribution of DFAA. In general, the frequency of long-period DFAA in the YRB is more than that of short-period DFAA.

Figure 6(a) shows that DTF frequency is high in most areas of the MLYRB, especially the below Huayuankou, while it is slightly lower in the UYRB. For short DFAA, DTF is mainly concentrated in June to July (Fig. 6(b)), followed by August to September (Fig. 6(c)), and finally in July to August (Fig. 6(d)). The source region and the central region of the YRB were vulnerable to DTF impact from June to July, while sporadic DTF damage areas appeared on the southern edge of the MLYRB from July to August. Furthermore, from August to September, DTF with high frequency is mainly distributed in the inner flow region of the UYRB.

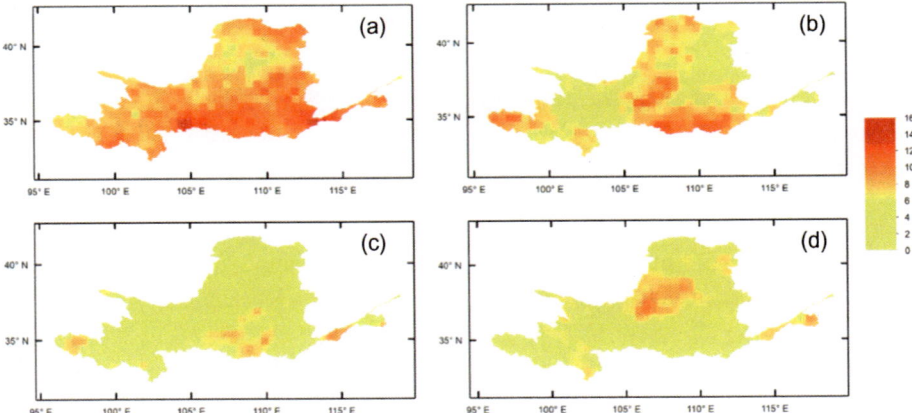

Fig. 6. Spatial distribution of DTF frequency in the YRB: (a) June-September, (b) June-July, (c) July-August, and (d) August-September.

For the long period (Fig. 7(a)), the spatial distribution pattern of FTD in the Yellow River is similar to that of DTF on the whole, but there was a slight difference locally. As can be seen from Fig. 7(b)-(d), the spatial distribution of FTD frequency is quite different in three periods of flood season. Among them, the high-frequency FTD in June-July and July-August was concentrated in the central region of the YRB, while the FTD in August-September was mainly distributed in the eastern edge of the MLYRB. Comparing Fig. 6(d) and Fig. 7(d), we can find DFAA types in the UYRB and MLYRB are opposite from August to September, namely DTF is dominant in the UYRB, while FTD is dominant in the MLYRB. Moreover, the frequency of FTD was highest from June to July, followed by July to August, and lowest from August to September. To sum up, short-cycle DFAA is mainly concentrated in June and July.

FTD is more frequent than DTF, which is consistent with the results in Sect. 4.1, namely LDFAI/SDFAI shows a declining trend overall.

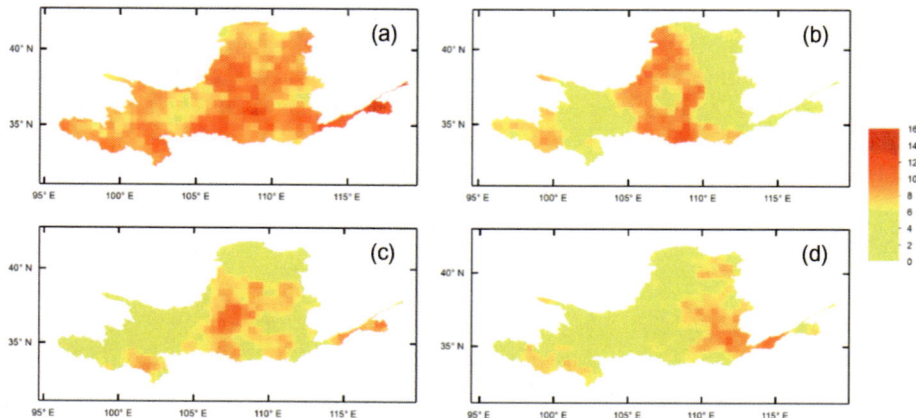

Fig. 7. Spatial distribution of FTD frequency in the YRB: (a) June-September, (b) June-July, (c) July-August, and (d) August-September.

4.2.2 The Trend of Drought-Flood Abrupt Alternation

The Mann-Kendall (MK) test was used to detect the existence of possible trends in DFAA. Figure 8 presented the spatial distribution of test statistic Z values at different time scales during the study period. $Z > 0$ indicates that LDFAI/SDFAI has an increasing trend, namely DTF is dominant in this region, on the contrary, FTD is dominant. The Z values of LDFAI (Fig. 8(a)) and SDFAI (Fig. 8(b)) from June to July are negative and distributed in a patchy pattern in the YRB, indicating that FTD occurred frequently in these periods. From July to August, negative Z was mainly found in the UYRB, showing speckled distribution, while a few positive Z appeared in the eastern part of the MLYRB (Fig. 8(c)). On the contrary, the positive and negative areas of SDFAI trend from August to September were concentrated in the UYRB and MLYRB, respectively (Fig. 8(d)), and there was a clear dividing line, indicating that the UYRB and MLYRB are DTF and FTD disturbance areas, respectively. This result is consistent with Sect. 4.2.1.

4.3 Impacts of Atmospheric Circulation on Drought-Flood Abrupt Alternation

After understanding the evolution of DFAA, it is necessary to reveal its possible driving mechanism. Regional precipitation anomalies are closely related to the large-scale circulation background (Liu et al. 2016). DFAA is no exception, mainly influenced by solar radiation and atmospheric circulation (Zhang et al. 2021). Solar radiation, as reflected by the sunspot index, is a key factor in climate formation (Hendon and Glick 1995). In addition, DFAA is not only affected by the atmospheric circulation of the same period but also affected by the earlier period, namely there is a lag correlation (Sun et al. 2006). Therefore, this study calculated the correlation coefficients between LDFAI/SDFAI and atmospheric circulation factors (such as AO, PDO, ENSO, and sunspot) based on the Pearson correlation method, considering simultaneity and lag correlation. The maximum correlation coefficients are shown in Table 1.

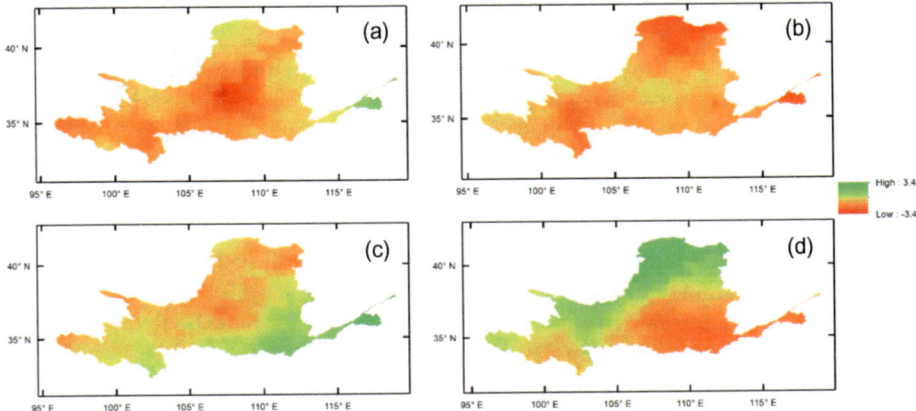

Fig. 8. Spatial distribution of DFAA trend in the YRB: (a) June-September, (b) June-July, (c) July-August, and (d) August-September.

Table 1. The correlation coefficient between LDFAI/SDFAI of YRB and atmospheric circulation factor.

	UYRB				MLYRB			
	AO	PDO	ENSO	Sunspot	AO	PDO	ENSO	Sunspot
LDFAI	−0.13	−0.18	−0.07	0.11	**0.38**	−0.19	−0.19	0.07
SDFAI67	−0.19	−0.20	−0.13	−0.07	0.13	**−0.29**	**−0.26**	−0.12
SDFAI78	0.23	0.21	0.13	0.14	**0.28**	0.22	0.15	0.19
SDFAI89	**0.31**	−0.16	**0.27**	−0.04	−0.22	**−0.26**	−0.16	0.24

Note: Bold indicates t significant at 95% confidence level

There was a significant positive correlation between LDFAI/SDFAI and AO in the YRB (Table 1). Previous study found that AO affects the middle and high latitudes of Eurasia mainly by influencing the Siberian high, which in turn changes the winter wind disturbance (Peng et al. 2021). In addition, a negative correlation was found for PDO between LDFAI/SDFAI. When PDO is in the cold phase, it tends to lead to more summer precipitation and shows a negative correlation (Setiawan et al. 2017). Compared with the UYRB, ENSO has a stronger correlation in the MLYRB. This is consistent with previous study (Dan et al. 2015), namely ENSO is weakly correlated with precipitation in the UYRB, while it is significantly negatively correlated with precipitation in the MLYRB, indicating that precipitation weakens when ENSO intensity increases. On the contrary, sunspot showed a weak negative correlation overall.

For long-cycle DFAA, LDFAI in the UYRB and MLYRB have the largest correlation coefficient with PDO and AO, respectively. For short-cycle DFAA, SDFAI67 and SDFAI78 of the YRB had the strongest correlation with PDO and AO, respectively. AO and PDO had the highest correlation with SDFAI89 in the UYRB and MLYRB, respectively. However, the interdecadal effect of atmospheric circulation on DFAA could not be derived using the correlation analysis, thus the cross-wavelet transform method was used to further investigate the driving process between them based on the maximum correlation coefficient.

According to Fig. 9(a), 1–3a antiphase resonance and 1-4a positive phase resonance existed in LDFAI and PDO during 1981–1986 and 1997–2005, respectively. Figure 9(b) shows a significant correlation between PDO and SDFAI from June to July, with periods of 14-18a and 0-2a during 1975–2004 and 2000–2006, respectively. AO had three significant resonance periods with SDFAI78, which were periodicities of 5a, 8-11a, and 1-4a during 1966–1975, 1997–1973, and 2012–2018, respectively (Fig. 9(c)). Moreover, there were three significant resonance cycles between AO and SDFAI from August to September, namely short-term oscillation of 1 and 1-5a (1968–1972 and 2012–2017) and long-term oscillation of 10-14a (2000–2012) (Fig. 9(d)). Previous research found that AO tended to increase and decrease during 1990–2000 and 2000–2014, respectively (Sun et al., 2018). Overall, AO showed a significant positive correlation with SDFAI78 and SDFAI89 after 2000, that is, AO can drive the decrease of SDFAI in the UYRB, indicating that AO is the key factor affecting the occurrence of FTD in this period.

In the MLYRB, LDFAI and AO showed significant oscillation cycles of about 8a, 5a, and 2-5a during 1969–1998, 1980–1986, and 2010–2017, respectively (Fig. 10 (a)). For short-scale DFAA, PDO had significant periods of 2-4a, 9a, and 3a during 1981–1991, 1990–2006, and 2008–2011, respectively, which passed the red noise test with a confidence level of 95% (Fig. 10(b)). There was a significant correlation between AO and SDFAI78 during 1984–1991 and 2008–2017, with a periodicity of about 17a and 4a, respectively (Fig. 10 (c)). Figure 10 (d) shows a significant positive correlation between PDO and SDFAI89 from 1980 to 1988, with a period of 3a. Moreover, a significant negative correlation with a periodicity of 12a was found during 1990–2014. Wang et al. (2016) found that PDO had a cold and warm phase exchange around 2000, which may be the main factor inducing short-cycle DFAA in the MLYRB. In addition, high index phase AO can weaken the summer monsoon and reduce the heat exchange between the north and the south, leading to regional precipitation anomaly and inducing DFAA of the YRB. Therefore, AO and PDO can be regarded as the dominant factor driving DFAA during the flood period in the YRB.

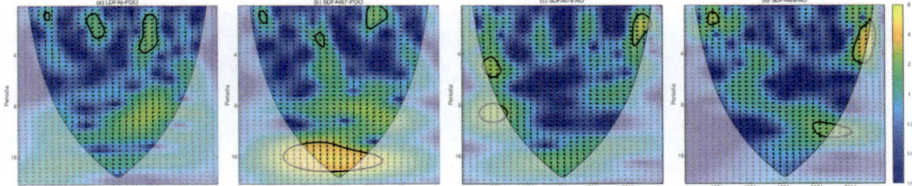

Fig. 9. Cross-wavelet transform between circulation factors with various indices in the UYRB, such as LDFAI (a), SDFAI67 (b), SDFAI78 (c), and SDFAI89 (d). The 95% confidence interval is represented by a thick black line; the relative phase relationship is represented by arrows (left anti-phase, right in-phase); the up and down arrows indicate drivers before or after DFAA, respectively.

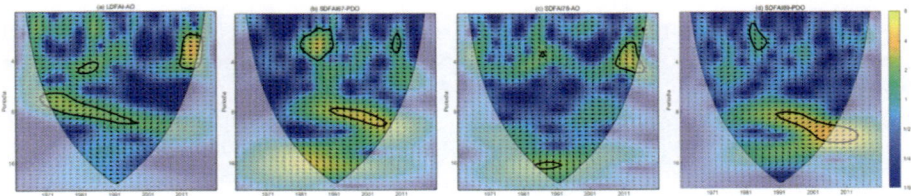

Fig. 10. Cross-wavelet transform between circulation factors with various indices in the MLYRB, such as LDFAI (a), SDFAI67 (b), SDFAI78 (c), and SDFAI89 (d).

5   Conclusions

This study evaluated the spatiotemporal pattern of DFAA at multiple time scales for flood season from 1961 to 2020 of the YRB. Furthermore, the relationship between large-scale climate drivers, including AO, PDO, ENSO, and sunspot, with DFAA was fully revealed using the Pearson correlation method and cross wavelet method. Results showed that the spatiotemporal distribution of DFAA was significantly different. First, LDFAI/SDFAI almost tended to decline, that is, FTD occurred frequently in the YRB from 1961 to 2020. The frequency of long-period DFAA is significantly higher than that of short-period DFAA, which is distributed in patches throughout the YRB. Second, based on the MK trend test, LDFAI and SDFAI67 tend to decrease in the YRB, indicating that FTD is prone to occur in this period. In addition, SDFAI78 tends to increase in the UYRB and decrease in the MLYRB, while SDFAI89 has the opposite trend. Third, results indicated a strong influence of AO and PDO in the occurrence of the DFAA in the YRB. Meanwhile, the cross-wavelet results showed that the occurrence of DFAA is closely related to the enhancement of AO and the phase change of PDO in the YRB since 2000.

As a complex emerging disaster, DFAA is induced by many factors and its actual physical driving mechanism is complex, such as topography, vegetation coverage, and human activities, which are not considered in this study. Therefore, the influence of these factors on DFAA should be analyzed from the perspective of the mechanism. Nonetheless, we provide new insights into how large-scale climate factors force DFAA and should help promote research in this area.

Acknowledgements. This study was supported by the National Natural Science Foundation of China (51879067) and Fundamental Research Funds for the Central Universities of China (B200204038, B220203051).

References

Dan Z, Bo Z, An M, Zhang Y, Jing L (2015) Responses of drought with different time scalese to the ENSO events in the Yellow River basin. J Desert Res 35(3):753–762

Hendon HH, Glick J (1995) Intraseasonal air-sea interaction in the tropical Indian and pacific oceans. J Clim 10(4):647–661

Ji G, Lai Z, Yan D, Wu L, Wang Z (2021) Spatiotemporal patterns of future meteorological drought in the Yellow River Basin based on SPEI under RCP scenarios. Int J Clim Chang Strateg Manag 14(1):39–53

Li Q, Yang M, Wan G, Wang X (2016) Spatial and temporal precipitation variability in the source region of the Yellow River. Environ Earth Sci 75(7):1–14. https://doi.org/10.1007/s12665-016-5583-8

Li X, Zhang Q, Zhang D, Ye X (2017) Investigation of the drought-flood abrupt alternation of streamflow in Poyang Lake catchment during the last 50 years. Hydrol Res 48(5):1402–1417

Liu Q, Yan C, Zhang Y, Yang J, Zheng S (2012) Variation of precipitation and temperature in yellow river basin during the last 50 years. Chin J Agrometeorol 33(4):475–480

Liu WB, Wang L, Chen DL, Tu K, Ruan C, Z. Y. (2016) Large-scale circulation classification and its links to observed precipitation in the eastern and central Tibetan Plateau. Clim Dyn 46 (11–12):3481–3497

Liu Y, Yuan Z, Guo L, Kong W, Zhang L, Wu L (2017) Characteristics of spatio-temporal variation of abrupt alternation of drought and flood in Shanxi province during summers in 1961–2013. J Ecol Rural Environ 33(4):332–340

Peng Y et al (2021) Spatial-temporal variations in drought conditions and their climatic oscillations in central Asia from 1990 to 2019. Chin J Eco-Agric 29(2):312–324

Setiawan AM, Lee W-S, Rhee J (2017) Spatio-temporal characteristics of Indonesian drought related to El Nino events and its predictability using the multi-model ensemble. Int J Climatol 37(13):4700–4719

Shan L, Zhang L, Chen X, Yang W (2015) Spatio-temporal evolution characteristics of drought-flood abrupt alternation in the middle and lower reaches of the Yangtze River basin. Resour Environ Yangtze Basin 24(12):2100–2107

Shan L, Zhang L, Zhang Y, She D, Xia J (2018) Characteristics of dry-wet abrupt alternation events in the middle and lower reaches of the Yangtze River Basin and their relationship with ENSO. Acta Geogr Sin 73(1):25–40

Shi W et al (2021) Drought-flood abrupt alternation dynamics and their potential driving forces in a changing environment. J Hydrol 597:126179. https://doi.org/10.1016/j.jhydrol.2021.126179

Shi W et al (2020) Dry and wet combination dynamics and their possible driving forces in a changing environment. 589:125211. https://doi.org/10.1016/j.jhydrol.2020.125211

Sun JH, He JH, Ren JZ, Zhong SS, Wang LJ (2006) Analysis of the relationship between the precipitation and the SST based on the TRMM data during the Asia monsoon season. In: conference on earth observing systems XI, San Diego, CA

Sun X et al (2018) A remarkable climate warming hiatus over Northeast China since 1998. Theoret Appl Climatol 133(1–2):579–594. https://doi.org/10.1007/s00704-017-2205-7

Wang C, Yang W, Zhou S, Hu Y (2014) Analysis on characteristics of atmospheric circulation and moisture around abrupt alternation of drought and flood in middle and lower reaches of the Yangtze River during May-June of 2011. Plateau Meteorol 33(1):210–220

Wang D, You Q, Jiang Z, Li Q (2016) Response of seasonal extreme temperatures in China to the global warming slow down. J Glaciol Geocryol 38(1):36–46

Wu Z, Li J, He J, Jiang Z (2006) Large-scale atmospheric singularities and summer long-cycle droughts-floods abrupt alternation in the middle and lower reaches of the Yangtze River. Sci Bull 51(16):2027–2034

Xu Z, Zhang N (2006) Long-term trend of precipitation in the Yellow River basin during the past 50 years. Geogr Res 25(1):27–34

Yuan Y, Gao H, Ding T (2021) Abrupt flood—drought alternation in southern China during summer 2019. J Meteorol Res 35(6):998–1011. https://doi.org/10.1007/s13351-021-1073-3

Zhang Y, Zhai L, Lin P, Cheng L, Wei X (2021) Variation characteristics and driving factors of drought and flood and their abrupt alternations in a typical basin in the middle reaches of Yangtze River. Eng J Wuhan Univ 54(10):887–897

Zhang ZZ, Yuan YJ, Shen DF, Fan H (2019) identification of drought-flood abrupt alternation in tobacco growth period in Xingren county under climate change in China. Appl Ecol Environ Res 17(5):12259–12269

Zhao Y, Wang M, Li J, Yang X, Zhang N, Chen H (2019) Diurnal variations in summer precipitation over the Yellow River basin. Adv Meteorol 2019:1–10

Morphological Evolution and Driving Factors of Tidal Flats in the Yangtze Estuary (China) During 1998–2019

Haifeng Cheng[1,2(✉)], Pei Xin[1], Jie Liu[2], Fengfeng Gu[2], Qi Shen[2], and Lu Han[2]

[1] State Key Laboratory of Hydrology-Water Resources and Hydraulic Engineering, Hohai University, Nanjing 210098, China
chenghaifeng_23@163.com
[2] Shanghai Estuarine and Coastal Science Research Center, Shanghai 201201, China

Abstract. This paper studies the morphological evolution and driving factors of the tidal flats in the Yangtze Estuary (YE), based on the bathymetric data over the last 20 years (1998–2019) and a three-dimensional numerical model (SWEM3D). The results show that: In the past two decades, the combined action of fluvial sediment decline and estuarine engineering has changed the morphological evolution trend of tidal flats in the YE. The fluvial sediment decline caused the decrease of suspended sediment concentration successively from the inner estuary to the mouth bar area (the outer estuary), which led to the erosion and steepening of the tidal flats in the YE, and the erosion of tidal flats in the inner estuary was earlier and more obvious than that in the mouth bar area. The estuarine engineering is the main controlling factor of the distribution and trend change of erosion-deposition in the adjacent tidal flat. The waterway regulation projects promoted the deposition of tidal flats within its sheltered area, while the reclamation and reservoir projects intensified the erosion of the lower tidal flats nearby. As for the remaining non-human-intervention tidal flats, those adjacent to the mainstream of ebb current in the inner estuary were significantly eroded, while those on the north side of the channel were slowly deposited due to the weaker hydrodynamics. In the future, the fluvial sediment supply may keep decreasing and maintain a lower level under the continued influence of anthropogenic activities in the Yangtze River basin, the unprotected tidal flats in the YE will face a risk of further erosion. It is necessary to take appropriate protection measures to improve the ecological service function of the tidal flats in the YE.

Keywords: Tidal flat · Morphological evolution · Fluvial sediment decline · Estuarine engineering · Yangtze Estuary

1 Introduction

Tidal flats in estuarine and coastal areas are the important natural resources and distributed with ecosystems such as salt marshes and mangroves, which have strong carbon sequestration capacity and play an important role in the global carbon cycle

Y. Li et al. (Eds.): PIANC 2022, LNCE 264, pp. 1152–1167, 2023.
https://doi.org/10.1007/978-981-19-6138-0_101

(Falkowski et al. 2000). However, most tidal flats in the world are facing serious challenges from both natural change and anthropogenic activities (Wei et al. 2015). Over the past century, there has been a net global reduction of sediment flux into coastal oceans resulting from human impacts (Syvitski et al. 2005). The fluvial sediment decline, which triggers the regression of the subaqueous delta (Syvitski et al. 2009; Yang et al. 2011), is primarily caused by sediment retention in inland reservoirs (Walling and Fang 2003). Moreover, estuarine engineering projects, including waterway regulation and reclamation, also play an important role in altering estuarine hydrodynamics and controlling estuarine tidal flat morphodynamics (Spearman et al. 1998; Antoine et al. 2009; Dai et al. 2013; Song and Wang 2014).

The Yangtze Estuary (YE), located in eastern China (Fig. 1), is one of the world's largest estuaries (Xie et al. 2009). It has been intensively influenced by human activities including altered river and sediment discharges in its catchment and local engineering projects in the estuary over the past half century (Luan et al. 2016). Due to the vulnerable natural conditions, intensive human activities and complex hydrodynamic and sediment systems, the tidal flats of the YE have attracted extensive attention as an ideal study area for studying on the temporal and spatial changes. Both sediment discharge and river flood events played important roles in the decadal morphological evolution of the YE (Luan et al. 2016), and the time-lag effect needs particular consideration in projecting future estuarine morphological changes under a low sediment supply regime and sea-level rise (Zhao et al. 2018). The Deep Waterway Project and reclamation projects were responsible for the polarization of seaward erosion and landward accretion, respectively (Wei et al. 2017). In addition, under favorable conditions (e.g., macrotidal range, strong tidal flow, flood dominance, sedimentary settling-lag/scour-lag effects, and increasing high-tide level), delta-front erosion of the YE can actually supply sediment to tidal flats, thereby maintaining the accretion rates in balance with relative sea-level rise (Yang et al. 2020). In general, previous studies mainly focused on the overall scouring and silting process of the YE riverbed and the evolution law and mechanism of some tidal flats, while the overall morphological evolution characteristics and driving factors of the YE tidal flats have not been systematically analyzed and discussed.

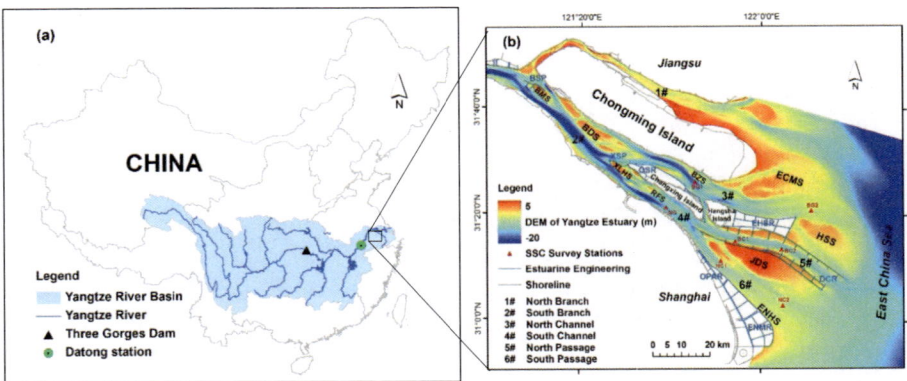

Fig. 1. Map of the Yangtze river basin (a) and Yangtze Estuary (b).

The physical conditions in terms of the prevailing hydrodynamic forcing, bed sediment composition and channel–shoal patterns in the YE varied from the inner estuary to the outer estuary (He et al. 2015). From this point of view, we divided the study area into two sections that was the inner estuary and the mouth bar area (the outer estuary) by taking Hengsha Island as the boundary. Tidal flats of the YE mainly include East Chongming Shoal (ECMS), Hengsha Shoal (HSS), Jiuduan Shoal (JDS, including Jiangya Shoal) and East Nanhui Shoal (ENHS) located in the mouth bar area, as well as Baimao Shoal (BMS), Biandan Shoal (BDS), Xinliuhe Shoal (XLHS), Ruifeng Shoal (RFS) and Baozhen Shoal (BZS) in the inner estuary (Fig. 1). Since 1998, many large-scale waterway regulation projects, hydraulic engineering and reclamation projects (Fig. 1) have been conducted in the YE, which have an increasingly significant influence on the riverbed evolution of the estuary, they mainly include the Baimao Shoal Protection Project (BSP, 2012–2013), XinLiuHe Shoal Protection Project (XSP, 2007–2009), Qingcao Shoal Reservoir Project (QSR, 2007–2009), Deepwater Channel Regulation Project (DCR, 1998–2010), and reclamation projects such as East Hengsha Shoal (EHSR, 2003–2019), Outside Pudong Airport (OPAR, 2008–2010) and East Nanhui Shoal (ENSR, 2013–2017).

Based on the long-term bathymetric data, this study systematically analyzes the morphological changes of tidal flats in the YE. Combined with the fluvial discharge and sediment flux as well as the construction of estuarine engineering projects, the driving factors of the morphological evolution of the tidal flats are discussed. The insights obtained from this study are helpful for the understanding and prediction of tidal flats evolution trend and the protection of wetlands in the YE, and can be useful reference to other estuaries and coastal areas.

2 Data and Methods

In order to analyze the topographic changes of tidal flats, we collected eight bathymetric maps (1998, 2002, 2010, 2013, 2016, 2017, 2018, 2019) of the study area (Fig. 1) from the Yangtze Estuary Waterway Administration Bureau, which were based on the chart datum. Using the GIS software package ArcGIS, all charts were digitized into Digital Elevation Model (DEM), and then spatial analysis and data statistics were performed. To assess the effects of river input, we collected annual runoff and suspended sediment discharge data at Datong station during 1950–2019 and the suspended sediment concentration (SSC) field survey data of fixed vertical line in the YE from 2002 to 2019. Meanwhile, the construction of estuarine engineering projects since 1998 had also been investigated.

Using a three-dimensional sigma coordinate coastal ocean model (SWEM3D), which was built based on unstructured C-grid and semi-implicit finite volume method by Shanghai Estuarine and Coastal Science Research Center, we simulated the changes of erosion and deposition trend of the tidal flats in the YE. By applying the semi-implicit method, SWEM3D can stay stable and efficient without the stability limitations of the surface gravity wave as well as the vertical diffusions. In addition, the finite volume method is used to calculate the surface gradient at each side which has no restriction on the grid orthogonality and makes the model more flexible for the complex coastal condition (Shen et al. 2014). The upstream boundary of the model is Datong Station of the Yangtze River, and the downstream boundary is −80 m deep area of the East China Sea. The total number of model triangular grids is 158,828, the minimum grid scale is about 30 m, the number of vertical layers is 10, and the time step length is 120 s. And the riverbed topography used in the model is a wide-range bathymetric map of the YE with a scale of 1:25000 in 1998. In order to discuss the influence of estuarine engineering projects and fluvial sediment decline on the erosion and deposition trend of the tidal flats, the following methods and indicators are used to calculate.

In this paper, the influence of the implementation of estuarine engineering projects on the morphological evolution of tidal flats is analyzed and calculated by using the indicators of erosion rate (f_e) and deposition rate (f_d), which are affected by hydrodynamic changes under fixed bed condition. f_e and f_d can indicate the probability of erosion and deposition of riverbed topography under certain hydrodynamic conditions, respectively (Cheng et al. 2020). The increase of f_e and decrease of f_d indicate that tidal flat tends to scour, and conversely, tidal flat tends to siltation. The formula of f_e and f_d are shown as:

$$f_e = \int_0^{T_1} m\left(\frac{\tau_b}{\tau_e} - 1\right) dt / 15 \qquad \tau_b > \tau_e \qquad (1)$$

$$f_d = \int_0^{T_2} \alpha\omega C_b\left(1 - \frac{\tau_b}{\tau_d}\right) dt / 15 \qquad \tau_b < \tau_d \qquad (2)$$

where, τ_e and τ_d are the critical deposition and starting stress at the bottom, which are 0.3 N/m^2 and 0.4 N/m^2 respectively; m is the erosion coefficient, set as 0.0001; α is the settling probability, which is 1; ω is the settling velocity of sediment at the bottom, take 0.2 mm/s; the above parameters are based on the field observation and laboratory test results of the YE (Qi et al. 2015). τ_b is actual bottom shear stress; C_b is the unit SSC, set at 1.0 kg/m^3, which means the influence of the changes of SSC is not taken into account; T_1 and T_2 are the duration of erosion and deposition, and the total calculation period is 15 d, which includes the full cycle process of spring-neap tide.

The upstream runoff process calculated by the model is the flow of Datong station from July 5 to 20, 2017, and the downstream boundary is the tide level of the offshore boundary in the corresponding period, which is calculated and given by the harmonic function.

The effect of fluvial sediment decline is calculated by using the indicator of deposition rate (f_d'), which is affected by SSC change under fixed bed condition, without considering the changes of riverbed topography and hydrodynamics. The decrease of f_d' indicates that tidal flat tends to scour, otherwise tends to siltation. f_d' is calculated as follows:

$$f_d' = \int_0^{T_2} \alpha\omega C_s\left(1 - \frac{\tau_b}{\tau_d}\right) dt / 365 \qquad \tau_b < \tau_d \qquad (3)$$

where, C_s is the actual SSC at the bottom calculated by the model, and the total calculation period is 365 d.

Fluvial sediment decline is represented by the following conditions: the SSC of Datong station decreased from 0.4 kg/m^3 in 1986–1997 to 0.2 kg/m^3 in 1998–2019. According to the calculation, the SSC decrease process in the YE caused by fluvial sediment decline tends to be stable in about one year under the condition of fixed bed. Therefore, the calculation results of the second year are selected to observe the changes of the erosion and deposition trend of tidal flats. The upstream runoff is the average discharge process of Datong station from 1998 to 2019, and the downstream tide is represented by the offshore boundary tide level calculated by harmonic function from 2018 to 2019.

3   Results

3.1   Area Changes of Tidal Flats

The calculation results of tidal flats area in the YE from 1998 to 2019 are shown in Fig. 2. During the last two decades, the tidal flats demonstrated different evolution processes. As mentioned above, the evolution characteristics of tidal flats were analyzed according to the upper and lower reaches of the YE, which were inner estuary and mouth bar area.

BMS, a typical central shoal (Fig. 1b), located in the main channel of the upper reaches of South Branch (SB), and the BSP was completed in 2013. Over the past 20 years, BMS shows a change of "scouring first, silting later, and then stabilizing" (Fig. 2a). The −5 m tidal flat of BMS was scoured from 1998 to 2013 with an area decrease rate of 0.7 km^2/yr, and turned into siltation from 2013 to 2016 with an area increase rate of 1.0 km^2/yr, which changed little from 2016 to 2019. The erosion of tidal flat above −2 m began in 2002, which was later than the lower tidal flat above −5 m. After that, the process of erosion and deposition was basically the same.

BDS is a relatively large tidal flat on the north side of the main channel of SB (Fig. 1b), which shows a general trend of siltation and expansion recently, especially the higher flat (Fig. 2b). During 1998–2019, the area of higher flat above 0 m and +2 m increased by 2 and 3 times respectively. However, the area of lower tidal flat above −5 m and −2 m only increased by 7% and 8%, and began to decrease in 2010 and 2016, respectively.

XLHS is a central shoal in the lower reaches of the SB (Fig. 1b), and the XSP was completed in 2009. Its +2 m higher flat developed and silted rapidly from 2010 to 2016, with an area growth rate of 0.5 km^2/yr, and tended to be stable from 2016 to 2019. The 0 m and −2 m tidal flats of XLHS were slowly silted from 1998 to 2010, and quickly silted from 2010 to 2016 with an area growth rate of 0.8 km^2/yr and 0.3 km^2/yr respectively, and remained stable from 2016 to 2019. However, the area of

−5 m lower tidal flat decreased at a rate of 0.2 km^2/yr from 1998 to 2010 and remained almost unchanged thereafter (Fig. 2c).

RFS and BZS are located in South Channel (SC) and North Channel (NC) of the inner estuary, respectively, and both of them are close to the thalweg of ebb channels (Fig. 1b). The area of the two tidal flats had declined significantly since 1998, and the lower tidal flats are the main ones (Fig. 2d and e). Take the −5 m isobath as an example, the area of RFS and BZS decreased at a rate of 0.9 km^2/yr and 0.4 km^2/yr, respectively.

It can be seen that each tidal flat in the inner estuary presents different erosion and accretion characteristics due to the different engineering layouts and local hydrodynamics and sediment conditions. Similarly, the variation of the tidal flats in the mouth bar area also shows the spatial difference.

ECMS is located on the north side of the mouth bar area, between the North Branch (NB) and North Channel (NC), and its upstream is covered by Chongming Island (Fig. 1b). Tidal flats above + 2 m and 0 m of ECMS expanded from 1998 to 2010, and then remained almost unchanged, while the tidal flats above −2 m and −5 m experienced a slowdown deposition process during 1998–2016, and slight erosion occurred during 2016–2019 (Fig. 2f).

HSS is located between NC and North Passage (NP), east of EHSR (Fig. 1b). The area of HSS at different elevations changed from expanding to shrinking, and the tidal flats above 0 m, −2 m and −5 m changed from deposition to erosion in 2016, 2016 and 2010, respectively (Fig. 2g). Among them, the area of –5 m lower tidal flat decreased by nearly 13% from 2010 to 2019.

JDS, the youngest tidal flat and the largest natural preservation zone in the YE (Fig. 1b), located between the NP and the South Passage (SP). The area of tidal flat above +2 m, 0 m and −2 m increased by 178%, 70% and 13% from 1998 to 2019, respectively. While the tidal flat above −5m changed from deposition to erosion in 2010, and its area decreased by about 4% from 2010 to 2019 (Fig. 2h).

ENHS is located on the northeast side of OPAR and ENSR, adjacent to the ebb channel of NP (Fig. 1b). It's obvious that the area of −5 m lower tidal flat continued to expand before 2010 and decreased significantly after 2010, when the reclamation project was constructed. Moreover, the erosion of −2 m lower tidal flats began in 2016 (Fig. 2i).

From the above, most tidal flats in the YE have presented an erosion trend successively in the past 20 years, which mainly occurred in lower tidal flats. Furthermore, the erosion of tidal flats in the inner estuary occurred earlier and more obvious than that in the mouth bar area.

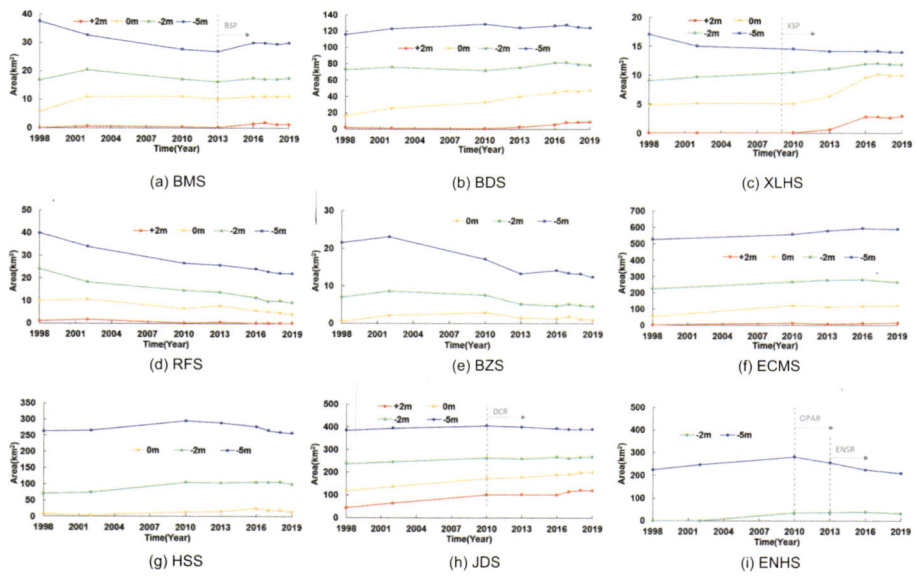

Fig. 2. Area changes of the tidal flats with respected to the benchmark water level.

3.2   Erosion and Deposition Distribution of Tidal Flats

The erosion and deposition distribution were obtained from the DEM by differencing the two years (Fig. 3). The results indicate that the YE experienced major changes over the 20 years from 1998 to 2019, and the spatial distribution of erosion and deposition varied considerably from the inner estuary to the mouth bar area. In general, the range of erosion and deposition in the inner estuary was significantly greater than that of the mouth bar area. From the perspective of spatial distribution, erosion mainly occurred in the thalweg of ebb channels and the areas of the bifurcation (the head of central shoal), while deposition mainly occurred in the areas of weaker hydrodynamics such as the sheltered waters of the estuarine engineering projects and the convex bank side of the channel.

It can be seen in Fig. 3 that BMS, XLHS and JDS, as typical central bar, had experienced different levels of erosion at the head of tidal flats and deposition at the higher mudflat in the past 20 years, what's more, the erosion of tidal flats in the inner estuary was more obvious. The southern edge of BDS, which is adjacent to the ebb channel of SB, showed severe erosion, and the scoured sediment transported downstream and deposited at the tail of BDS. Both RFS in the ebb channel of SC and BZS in the concave bank side of the NC were obviously eroded. ECMS on the north side of the mouth bar area had been mildly silted up during 1998–2019, with only the local area along the southern edge scoured slightly. The mid-lower part of HSS was slightly deposited, while the upper part and the north and south edges of the tidal flats were scoured. In addition to the head erosion mentioned above, JDS also showed obvious characteristics of northern edge deposition and southern edge erosion. ENHS located outside the OPAR and ENSR reclamation projects had been eroded slightly from 1998 to 2019.

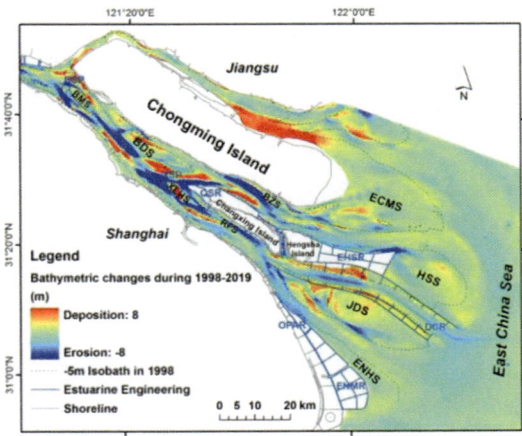

Fig. 3. Bathymetric changes of the Yangtze Estuary during 1998–2019 (positive values are for deposition and negative values are for erosion).

3.3   Shape Changes of Tidal Flats

The shape changes of tidal flat can not only directly reflect the vertical differences in the erosion and deposition process, but also indicate the characteristics of steepening or flattening. According to the shape changes of the tidal flats in the YE (Fig. 4), although the vertical erosion and deposition patterns of the tidal flats were obviously different in the past 20 years, they all showed a trend of gradual steepening.

BMS deposited in the higher mudflat and eroded in the lower mudflat during 1998–2002, then erosion and siltation occurred throughout the tidal flat, respectively, in 2002–2013 and 2013–2019 (Fig. 4a), the statistical results show that BMS steepened obviously before 2013, and leveled out slightly afterwards (Table 1). BDS has been steepening in the past 20 years, and the shape variation characteristics of BDS are as follows: the higher tidal flat above 0 m deposited continuously, the tidal flat between −1 m and −4 m changed from siltation to micro-erosion in 2016, and the −5 m lower tidal flat changed from deposition to erosion in 2010, which occurred earlier (Fig. 4b). XLHS steepened significantly from 1998 to 2019. Before 2010, the higher mudflat above 0 m of XLHS remained stable with rapid erosion of the lower mudflat below −4 m; after 2010, the higher mudflat above 0 m rapidly expanded with little change of the lower mudflat below −3 m (Fig. 4c). The significant erosion of RFS and BZS began in 1998 and 2002 respectively, among which the erosion of lower mudflat was more intense than that of the higher mudflat (Fig. 4d and e), therefore, the steepening characteristics of RFS and BZS were obvious.

Compared with the tidal flats in the inner estuary, the steepening trend of the tidal flats in the mouth bar area is relatively gentle. The entire tidal flat of ECMS continued to expand from 1998 to 2016, with slight erosion from 2016 to 2019 (Fig. 4f). Meanwhile, the steepness of ECMS remained stable (Table 1). HSS silted up and expanded before 2010, then the lower mudflat began to erode in 2010–2016, and the entire tidal flat scoured and shrank in 2016–2019 (Fig. 4g), with little change in the

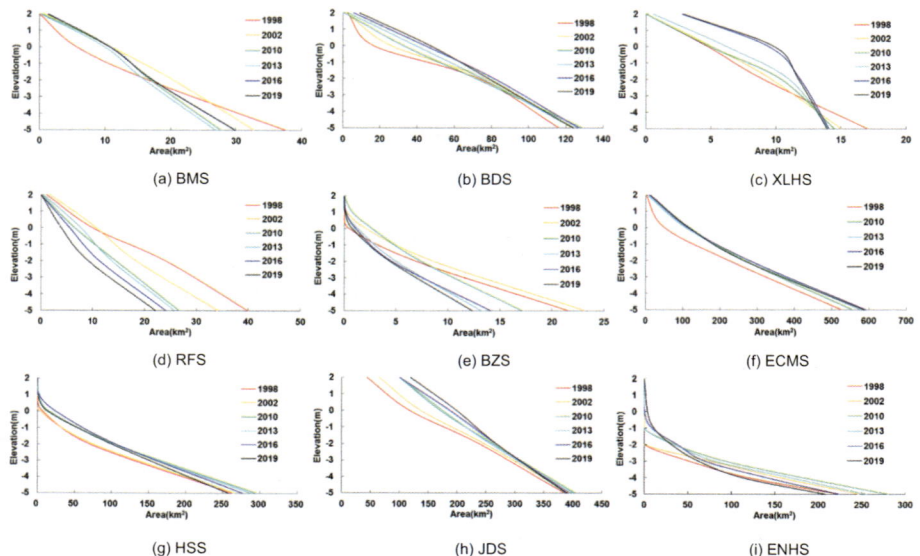

Fig. 4. Shape changes of the tidal flats.

Table 1. Steepness[a] of the tidal flats in the Yangtze Estuary.

Year	BMS	BDS	XLHS	RFS	BZS	ECMS	HSS	JDS	ENHS
1998	−0.178	−0.054	−0.412	−0.175	−0.285	−0.012	−0.023	−0.020	−0.023
2002	−0.219	−0.055	−0.464	−0.217	−0.276	–	−0.023	−0.021	−0.021
2010	−0.260	−0.056	−0.465	−0.261	−0.387	−0.013	−0.021	−0.023	−0.019
2013	−0.266	−0.058	−0.492	−0.278	−0.483	−0.012	−0.022	−0.023	−0.021
2016	−0.249	−0.058	−0.568	−0.287	−0.448	−0.012	−0.023	−0.024	−0.025
2019	−0.248	−.061	−0.573	−0.305	−0.511	−0.012	−0.024	−0.026	−0.027

[a] Steepness refers to the slope coefficient of the linear regression line of the "elevation-area relationship" in Fig. 4.

steepness. During the last 20 years, the higher mudflat of JDS was subjected to significant accretion, while erosion appeared in the lower mudflat after 2010, so the steepening trend of JDS was relatively obvious (Fig. 4h). ENHS silted up before 2010, and then the lower mudflat below −2 m began to erode (Fig. 4i) and the whole tidal flat steepened.

4   Discussions

4.1   Impact of Estuarine Engineering Projects

The large-scale estuarine engineering projects in the YE include waterway regulation projects such as BSP, XSP and DCR, reclamation projects such as EHSR, OPAR and

ENSR and hydraulic engineering such as QSR (Fig. 1b). The changes of erosion rate and deposition rate calculated by the numerical model (Fig. 5) show that the morphologic evolution of the tidal flats in the YE was significantly affected by the construction of large-scale wading projects.

Waterway regulation projects can not only control the river regime and form deep channel, but also play the role of protecting the shoal. BSP and XSP are the waterway regulation projects located at the SB-NB and SC-NC bifurcations respectively. Under the protection of BSP and XSP, the deposition rates of BMS and XLHS within the submerged breakwater of the projects increased and the erosion rates decreased, basically consistent with the actual erosion and deposition status of the two tidal flats (Fig. 3). DCR is located at the SP-NP bifurcation and its downstream. Due to the larger scale, DCR had obvious effects on HSS and JDS. Figure 5 shows that the deposition rates of HSS and JDS increased significantly, especially in the southern edge of HSS and the northern edge of JDS near DCR. Meanwhile, the erosion rate of the southern edge of HSS decreased slightly. However, the southern edge of the upper part of JDS tended to erosion under the influence of the increased diversion ratio of the SP after the implementation of DCR.

OPAR and ENSR narrow the width and enhance the hydrodynamic forces of SP (Cheng et al. 2020). Affected by the above changes, the erosion rate of ENHS outside the reclamation projects increased, while the erosion rate and the deposition rate of central southern edge of JDS increased and decreased, respectively. The above two areas showed an erosion trend, which is consistent with the actual erosion characteristics (Fig. 3). In addition, EHSR caused a slight increase in the erosion rate of the upper part of HSS near the dike. QSR is a large water source project located at the SC-NC bifurcation, which resulted in the increase of erosion rate on the north side (concave bank) of the main channel in the upper section of NC and intensified the erosion of BZS. The above changes show that the reclamation and reservoir projects occupy the higher tidal flat, and at the same time cause the erosion of the nearby lower tidal flat.

Due to the shielding effect of BSP extending downstream and the restriction of XSP and QSR on the flow from SB into NC, the deposition rate of BDS increased slightly and the erosion rate of the lower southern edge of BDS decreased (Fig. 5), which contributed to the stable deposition of BDS in the past 20 years (Fig. 3). It can also be seen from Fig. 5 that the deposition rate of ECMS generally increased, which should be related to the weakening effect of DCR and EHSR on the tidal dynamics in the northern waters of the projects. However, the impact of estuarine engineering projects on the erosion and deposition trend of RFS was relatively slight.

In summary, BSP and XSP were the key factors for BMS and XLHS to maintain the deposition of higher mudflats under the environment of significant erosion of tidal flats in the inner estuary, and DCR was the main cause of the siltation at the north side of JDS and southeast side of HSS. Meanwhile, the shielding effect of the estuarine engineering projects could promote the deposition of BDS and ECMS located on the north side of the estuary. However, QSR and EHSR, ENSR and other reclamation projects tended to cause erosion of adjacent lower tidal flats outside the projects.

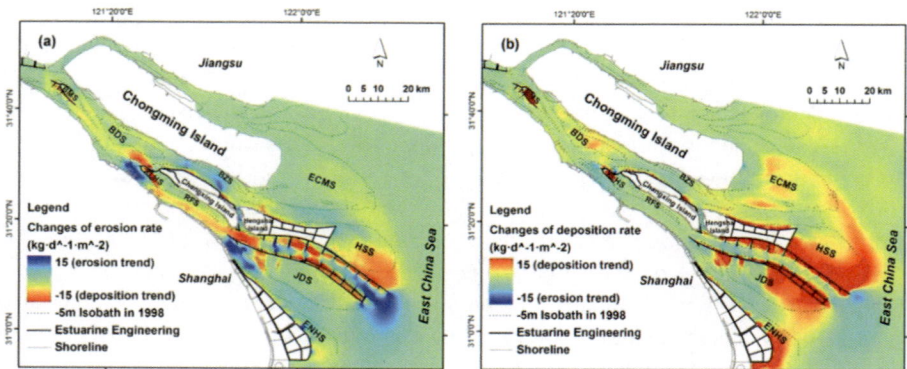

Fig. 5. Changes of erosion rate (a) and deposition rate (b) caused by estuarine engineering projects (results of numerical simulation, red represents the deposition trend and blue represents the erosion trend).

4.2   Impact of Fluvial Sediment Decline

The YE is rich in water and sediment. Over the years, the runoff of the Yangtze River basin has not changed much, but due to the impact of water and soil conservation and dam construction in the basin, the amount of sediment carried by runoff into the YE has decreased significantly (Liu et al. 2017). As shown in Fig. 6, the annual sediment discharge of Datong station began to decline after the mid-1980s, from an average of 471 million tons per year in 1955–1985 to 349 million tons per year in 1986–1997, with a reduction of 25.5%. Since 1998, the decline rate of sediment discharge has accelerated. The annual sediment discharge of Datong from 1998 to 2019 was 175 million tons per year on average, 48.6% less than that from 1986 to 1997. During this period, after the operation of the Three Gorges Project (2004–2019), the annual sediment discharge remained at a low value, with an average of only about 128 million tons per year.

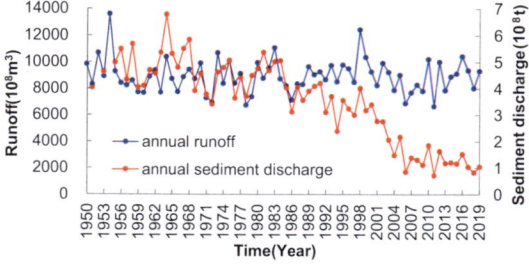

Fig. 6. Changes of the annual runoff and sediment discharge at the Datong station during 1950–2019.

Due to the fluvial sediment decline, the SSC in the lower reaches of the Yangtze River and the YE has been decreasing (Yang et al. 2013; Liu et al. 2017). The field data showed that the SSC of NG station in SC and BG1 station in NC decreased in fluctuation from 0.87 kg/m³ and 0.78 kg/m³ in August 2002 to 0.15 kg/m³ and 0.22 kg/m³ in August 2019, respectively (Fig. 7). From 2002 to 2009, 2010 to 2014, and 2015 to 2019, the average SSC of NC1 station in SP was 1.32 kg/m³, 1.06 kg/m³ and 0.30 kg/m³ respectively, while the average SSC of BC1 station in NP was 0.61 kg/m³, 0.48 kg/m³ and 0.26 kg/m³ respectively. The average SSC of NC2, BC2 and BG2 stations in the turbidity maximum zone decreased from 1.39 kg/m³, 1.44kg/m³ and 0.85 kg/m³ in 2012–2014 to 0.49kg/m³, 0.60kg/m³ and 0.39kg/m³ in 2016–2019, respectively. It can be seen that due to the time-lag effect (Zhao et al. 2018), the decrease of SSC in the mouth bar area is later than that in the inner estuary. In general, under the influence of the significant decrease of sediment discharge in the basin, the SSC in the YE had been declining successively from the inner estuary to the mouth bar area in the past 20 years.

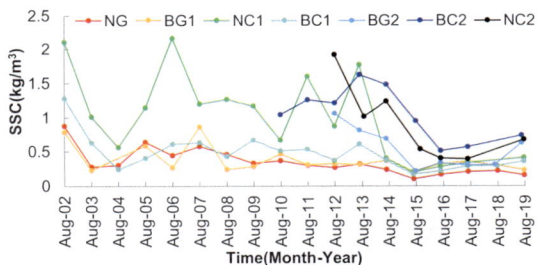

Fig. 7. Changes of SSC at typical survey stations in the Yangtze Estuary (Location is shown in Fig. 1)

For more than half a century, the tidal flats of the YE have been dominated by sedimentation. However, the decrease of SSC in the estuary would enhance the scouring capacity of the current, which would slow down the deposition rate of estuary and even make the transition to erosion (Liu et al. 2017). Studies have shown that when the annual sediment discharge of Datong is 260–280 million tons, the deposition and erosion of the YE subaqueous delta are in equilibrium (Yang et al. 2003; Yang et al. 2014). During the formation and development of the Yangtze River Delta, the average annual sediment discharge of Datong was 184–228 million tons (Li et al. 2004). The critical annual sediment discharge of Datong is about 254 million tons when the lower reaches of SB in the YE changes from silting to scouring (Cheng et al. 2020). From 1998 to 2019, the average annual sediment discharge of Datong is only 175 million tons, which is obviously lower than the critical annual sediment discharge proposed by previous scholars, and the estuary tends to be scoured on the whole (Liu et al. 2021).

According to the numerical model calculation results (Fig. 8), it can be seen that the decrease of SSC at Datong reduces the deposition rate of tidal flats in the YE, that is, the tidal flats tend to be scoured. Specifically, the erosion trend of the tidal flats in the inner estuary is more obvious than that of the mouth bar area. The erosion trend of BMS, BDS, XLHS and RFS in the inner estuary is generally more significant. At the same time, the erosion trend of the coastal side of the tidal flats in the mouth bar area is also obvious along the line of "ECMS-HSS-JDS". However, the erosion trend of tidal flats in the mouth bar area is relatively light, which is due to the time-lag effect of sediment accumulation in the turbidity maximum zone (Zhao et al. 2018). In addition, the erosion trend of lower tidal flats is generally more significant than that of higher tidal flats, especially that of tidal flats in the mouth bar area (Fig. 8), which accords with the law that the scouring of estuarine riverbed is first deep and then shallow under the condition of fluvial sediment decline (Liu et al. 2017; Cheng et al. 2020). It should be noted that the above characteristics of tidal flats erosion trend are only preliminary simulation results of the influence of fluvial sediment decline at Datong under fixed bed conditions, and the influence of complex coupling changes of "SSC, hydrodynamics and riverbed topography" on the tidal flat evolution trend needs to be further explored.

We speculate that, the erosion of tidal flats in the YE occurred from the inner estuary to the mouth bar area and from the lower mudflat to the higher mudflat successively, which was closely related to the successively decrease of SSC in the estuary caused by the fluvial sediment decline in the Yangtze River Basin.

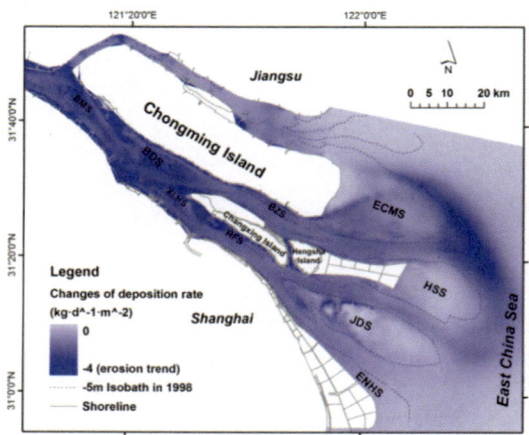

Fig. 8. Changes of deposition rate caused by fluvial sediment decline (results of numerical simulation, the darker the color and the more obvious the erosion)

4.3 Evolution Pattern of Tidal Flats

The above analysis results show that, under the fluvial sediment decline and estuarine engineering projects, the morphological evolution of tidal flats in the YE showed significant spatio-temporal differences in the past two decades, but it also followed a certain rule. The evolution pattern of tidal flats is summarized as follows.

In the first place, sediment decline in the basin had dominated the erosion of tidal flats in the YE from 1998 to 2019, the sequence of erosion was that the inner estuary preceded the mouth bar area, and the erosional location was mainly in the lower mudflat and gradually extended upward.

Secondly, different engineering boundaries and hydrodynamic conditions controlled the evolution trend of tidal flats and the distribution of erosion and deposition in the Yangtze Estuary under the macro environment of fluvial sediment decline. The tidal flats (BMS, XLHS) in the inner estuary protected by the projects were characterized by continuous erosion and obvious deposition, before and after the implementation of the projects, respectively; the tidal flats (RFS, BZS) without engineering protection and close to the mainstream of ebb current in the inner estuary were obviously eroded; the tidal flats (JDS, HSS) in the mouth bar area protected by the projects were characterized by first deposition and then erosion in the low mudflat and stable deposition in the high mudflat; the tidal flat (ENHS) in the mouth bar area without engineering protection silted up at first and then scoured as a whole; the tidal flats (BDS, ECMS) on the north side of the main channel changed from silting to scouring later and to a lesser degree.

5 Conclusions

The study reveals the morphological evolution characteristics of tidal flats in the YE under the influence of fluvial sediment continuous decline and the estuarine engineering projects in recent two decades.

Affected by the fluvial sediment decline, the SSC in the YE experienced a significant decline since 1998. However, due to the time-lag effect, which was caused by sediment accumulation in the turbidity maximum zone, the SSC in the mouth bar area decreased later than that in the upstream area. As a result, the tidal flats of the Yangtze Estuary presented a situation of erosion on the whole, Moreover, the tidal flats in the inner estuary were earlier and more obvious than those in the mouth bar area, and the tidal flats adjacent to the mainstream of ebb current were earlier and more obvious than those on the north side of the main channel (weaker hydrodynamics area). Meanwhile, the erosion place of tidal flats extended from the lower mudflat to the higher mudflat, resulting in the steepening of the tidal flats.

The estuary wading projects controlled the change of evolution trend and the distribution of scouring and silting position in the adjacent tidal flats. Among them, the estuarine waterway regulation projects were beneficial to the deposition of tidal flats while enhancing the erosion of deep channels. However, the reclamation and reservoir projects rented the higher tidal flats and caused the erosion of the lower tidal flats nearby.

We speculate that, under the influence of human activities such as dam construction and water-soil conservation, the sediment supply in the Yangtze River Basin may keep decreasing and maintain a lower level in the future, and without the implementation of targeted protection projects, the tidal flats in the YE will face a risk of further erosion. Therefore, we suggest that scientific and effective measures such as protection projects and ecological siltation-promotion should be taken as soon as possible to protect the tidal flats and improve the ecological service function of the wetlands in the Yangtze Estuary.

Acknowledgements. This study was supported by the National Natural Science Foundation of China (U2040204) and Shanghai Science and Technology Project (21DZ1201002). We thank the anonymous reviewers for their valuable comments and suggestions.

References

Antoine C, Julien D, Robert L, Christophe B (2009) Morphological responses of an estuarine intertidal mudflat to constructions since 1978 to 2005: the Seine estuary (France). Geomorphology 104:165–174

Cheng HF, Xin P, Liu J, Gu FF, Wang W, Han L (2020) Morphological evolution and dynamic mechanics of the Jiuduansha shoal (China) during 1959–2018. Adv Water Sci 31(4):19–29

Dai ZJ, Liu JT, Fu G, Xie HL (2013) A thirteen-year record of bathymetric changes in the north passage, Changjiang (Yangtze) Estuary. Geomorphology 187:101–107

Falkowski P et al (2000) The global carbon cycle: a test of our knowledge of earth as a system. Science 290:291–296

He Q, Guo L, Liu H, Wang Y (2015) Changjiang estuary sediment transport dynamics. In: Zhang J (ed) Ecological Continuum from the Changjiang (Yangtze River) Watersheds to the East China Sea Continental Margin. Springer International Publishing, Switzerland, pp 47–69

Li CX, Yang SY, Fan DD, Zhao J (2004) The change in Changjiang suspended load and its impact on the delta after completion of three-Gorges dam. Quat Sci 24(5):495–500

Liu J, Cheng HF, Han L, Wang ZZ (2017) Influence of fluvial sediment decline on the morphodynamics of the Yangtze estuary and adjacent seas. Adv Water Sci 28(2):249–256

Liu J, Cheng HF, Han L, Ye TT, Wang ZZ (2021) New trends of river channel evolution of the Yangtze river estuary under the influences of inflow and sediment variations and human activities. Hydro-Sci Eng 2:1–9

Luan HL, Ding PX, Wang ZB, Ge JZ, Yang SL (2016) Decadal morphological evolution of the Yangtze Estuary in response to river input changes and estuarine engineering projects (1958–2010). Geomorphology 265:12–23

Qi DM, Gu FF, Wang YY (2015) Waterway sedimentation mechanisms and near-bottom water and sediment monitoring techniques in the Yangtze Estuary. China Communications Press, Beijing, pp 122–132

Shen Q, Gu FF, Qi DM, Huang WR (2014) Numerical study of current and sediment variation affected by sea-level rise in the north passage of the Yangtze Estuary. J Coast Res 68:80–88

Song CC, Wang J (2014) Erosion-accretion changes and controlled factors of the submerged delta in the Yangtze Estuary in 1982–2010. Acta Oceanol Sin 69(11):1683–1696

Spearman JR, Dearnale MP, Dennis JM (1998) A simulation of estuary response to training wall construction using a regime approach. Coast Eng 33:71–89

Syvitski JPM et al (2009) Sinking deltas due to human activities. Nat Geosci 2:681–686

Syvitski JPM, Vörösmarty CJ, Kettner AJ, Green P (2005) Impact of humans on the flux of terrestrial sediment to the global coastal ocean. Science 308:376–380

Walling DE, Fang D (2003) Recent trends in the suspended sediment loads of the world's rivers. Glob. Planet. Chang. 39:111–125

Wei W, Dai ZJ, Mei XF, Liu JP, Gao S, Li SS (2017) Shoal morphodynamics of the Changjiang (Yangtze) estuary: influences from river damming, estuarine hydraulic engineering and reclamation projects. Mar Geol 386:32–43

Wei W, Tang ZH, Dai ZJ, Lin YF, Ge ZP, Gao JJ (2015) Variations in tidal flats of the Changjiang (Yangtze) estuary during 1950s–2010s: future crisis and policy implication. Ocean Coast Manag 108:89–96

Xie XP, Wang ZY, Charles SM (2009) Formation and evolution of the Jiuduansha Shoal over the past 50 years (1945–2001). J Hydraul Eng 135(9):741–754

Yang YP, Li YT, Fan YY (2014) Relationship between sediment elements of river basin and front sand islands evolution in Yangtze Estuary. Resour Environ Yangtze Basin 23(5):652–658

Yang Y, Li Y, Sun Z, Fan Y (2014) Suspended sediment load in the turbidity maximum zone at the Yangtze river Estuary: the trends and causes. J Geog Sci 24(1):129–142. https://doi.org/10.1007/s11442-014-1077-3

Yang SL, He SL, Xie WH (1998) The formation and evolution of the Jiuduansha tidal island as well as their relation to the development of the north and south passages in the Yangtze river estuary. Ocean Eng 16(4):55–65

Yang SL et al (2020) Role of delta-front erosion in sustaining salt marshes under sea-level rise and fluvial sediment decline. Limnol Oceanogr 9999:1–20

Yang SL, Milliman JD, Li P, Xu K (2011) 50,000 dams later: erosion of the Yangtze river and its delta. Glob Planet Chang 75:14–20

Yang SL, Du JL, Gao A, Li P, Li M, Zhao HY (2006) Evolution of Jiuduansha wetland in the Changjiang river estuary during the last 50 years. Sci Geogr Sin 26(3):335–339

Yang SL, Zhu J, Zhao QY (2003) A preliminary study on the influence of Changjiang river sediment supply on subaqueous delta. Acta Oceanol Sin 25(5):83–91

Zhao J, Guo LC, He Q, Wang ZB, van Maren DS, Wang XY (2018) An analysis on half century morphological changes in the Changjiang Estuary: spatial variability under natural processes and human intervention. J Mar Syst 181:25–36

Multi-purpose Management of the Walloon Waterways, from Local to Global Control of the Structures

Nathan Bertouille[1]([✉]), Gouverneur Ludovic[2], Franken Tim[3],
Dierickx Philippe[1], Savary Céline[1], and Meert Pieter[3]

[1] Service Public de Wallonie, Namur, Belgium
{nathan.bertouille, philippe.dierickx,
celine.savary}@spw.wallonie.be
[2] International Marine and Dredging Consultants, Antwerp, Belgium
ludovic.gouverneur@imdc.be
[3] Sumaqua, Leuven, Belgium
{tim.franken, pieter.meert}@sumaqua.be

Abstract. Given current issues such as climate change and traffic growth, Walloon Region, located at the heart of the European waterway network, decided to develop Orhyx, a tool that allows a harmonized, global and digital management of the waterways. The main goal of Orhyx is to optimize the management of an entire waterway network in order to enhance security and navigability, improve environmental safety and optimize energy production and consumption. This optimization tool is a modular web application built around three main modules being visualization and monitoring of the current state of the system, forecasting the future state of the system and optimization of the operational and real-time management of the system. A preliminary version was developed on a reduced area of the Walloon waterways network. Tests are planned on an operational point of view in order to upgrade Orhyx to cover the whole network. Extensive offline validation of the optimization and the web application shows encouraging results and outlines the benefits of using a global, centralized and optimized management as well as al the challenges that arise with such operational systems.

Keywords: Water Management · Optimization tool · Decision-support · Real-time · Digital river

1 Introduction

Located at the heart of the European waterway network, the Walloon Region from Belgium manages about 450 km navigable waters and more than 80 hydraulic structures including 6 weir reservoirs and has to deal with the increasing complexity of waterways and its current issues. Climate change increases the probability of both floods and extended dry spells, while the growing traffic adds additional stress to the system.

© The Author(s) 2023
Y. Li et al. (Eds.): PIANC 2022, LNCE 264, pp. 1168–1177, 2023.
https://doi.org/10.1007/978-981-19-6138-0_102

Last year in the Walloon Region, river transport of goods represented 7% of the total freight transport, after shipping by road (84%) and by train (9%). Considering that truck transportation is responsible for 99% of land shipping CO_2 emission, Wallonia aims to transfer good transports from trucks to ships[1]. For this purpose, the region wants to make waterways more attractive by modernizing the engineering structure of the waterways network and by improving its global management.

Alongside the current modernization and enlargement of the crossing structures, river dams allow to regulate water level for navigation. Nowadays, all those dams are controlled by local PLC - Programmable Logic Controller - based on a regulation loop. When the automatic mode is activated, those PLC can move the weir gates based on the upstream water level measurement and a specified water level setpoint. Given climate change and the complexity of the network, this local control does not prevent waves to propagate from reach to reach and is no longer considered the most optimal solution for a global management.

In the next few years, hydraulic structures of the Walloon Region will be remotely controlled from one and unique place, the recently inaugurated PEREX center in Namur. Structures will be connected to the center via optical fiber and a SCADA - Supervisory Control and Data Acquisition - system. The entire network could be visualized and controlled from PEREX. The project is currently underway. To have a global vision and an effective control of the network, Walloon Region wants to have a real-time decision support tool in order to optimize water management and to allow a harmonized, global and digital management of the waterways. The tool, working in close loop and supervised by an operator, will select the optimal action to be carried out on waterways hydraulic structures.

To develop this optimization tool - called Orhyx (Optimisation de la Régulation HYdraulique depuis pereX - Optimization of hydraulic control from perex) -, Walloon Region decided to get help from 2 Belgian compagnies with expertise in this field: International Marine and Dredging Consultants (IMDC) and Sumaqua. The different parties have therefore joined their efforts to create this tool which is currently under construction. The project started in 2019 and should be completed in 2028.

Orhyx is the start of a new intelligent and verry innovative system that optimizes the management of an entire waterway network in order to enhance security and navigability, improve environmental safety and optimize energy production and consumption.

A preliminary version was developed on two waterways of the Walloon network (Basse-Sambre and Canal Charleroi-Buxelles). Tests are planned on those two waterways on an operational point of view in order to test the system in real conditions. Next steps will consist in upgrading Orhyx to cover the whole network.

This paper aims to describe the tool and its characteristics as well as the optimization module behind it. The first available results will be presented as well as the improvement perspective for the rest of the project.

---

[1] Source: https://spw.wallonie.be/les-chiffres-du-transport-fluvial-et-de-l%E2%80%99intermodalit%C3%A9-en-wallonie

2 Description of the System

2.1 Information Flow

The optimization tool needs information as inputs to operate and produces results as outputs. Orhyx is therefore integrated in a complex flow of information. Figure 1 describes how the information loops through the global system and how the tool is integrated into the waterways manager physical and IT infrastructure.

Data coming from field measurement instruments (water levels, discharges, states of the equipments, regulation modes,) are transmitted in real-time to the back end of the tool where the conceptual model and the optimization algorithm are deployed. This calculation is triggered at regular intervals (30 min). The team of operators make sure the system is always up to date with latest known event occurring on the network (maintenance, heavy rainfall prediction) for the next 24 h. If approved by the operators, the new control strategy (setpoints for the equipments) is then sent to the network. Effects of these new states closes the information loop as they are recorded by the measurement instruments.

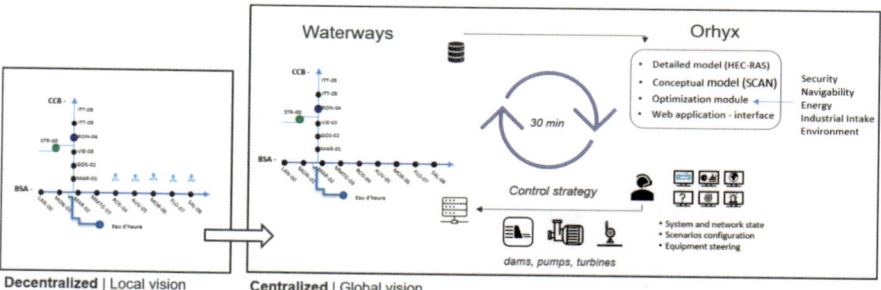

Fig. 1. Information flow throughout the physical and IT network.

2.2 Description of Orhyx

Orhyx is built as a modular web application (vue.js, django, influx and postgres databases) aimed to facilitate the management of such a complex system. The interface offers many functionalities:

Waterways Network Characteristics: all the relevant static information regarding infrastructures and equipment as well as waterway network (dams and reaches dimensions, pumps and turbines unitary flow,…) are grouped by operational site and river stretch. The operators can visualize them via a GIS interface or via a generic summary and export them in PDF.

Timeseries Exploration: for each river/channel of the network, the whole history (inputs and outputs of the scenarios) as well as the most recent data are displayed and compared on a flexible and customable dashboard. For instance, Fig. 2 shows water levels measurements for different sites on the graph above and regulation modes by site on the graph below (manual or automatic). Gates positions or water levels calculated as output by the tool can also be added on those graphs in order to visualize them.

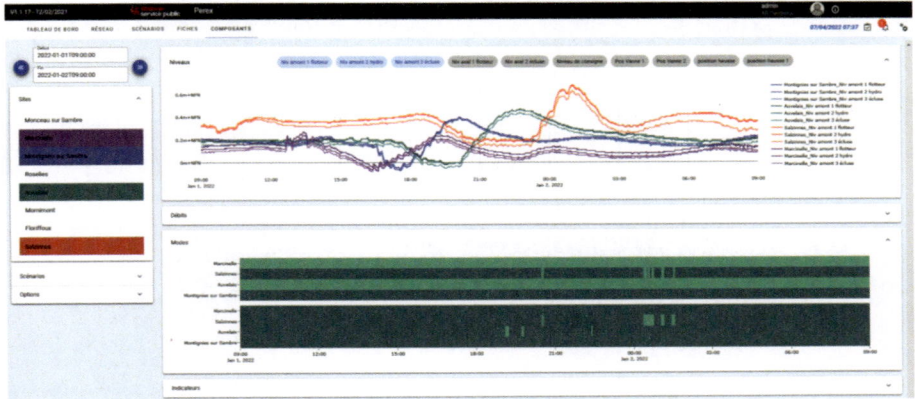

Fig. 2. Graphic dashboard in the Orhyx interface.

Scenarios - System Configuration: some events are not automatically caught by the system and must be manually implemented through the scenario editor. This component displays the input used by the optimization algorithm and are editable. It allows the operator to adapt the system according to reality on site (e.g. maintenance of gate) or to introduce a future event by modifying the relevant input and parameters. Once the configuration is adapted, all subsequent automatic simulations (running every 30 min) will take these changes into account.

Sensitivity Analysis and Remote Control: the scenario editor can also be used to study the influence parameters and run calculation beside the operational automatic simulations, or to simply steer the equipment remotely (change a gate level).

These components combined offer a wide and detailed view of the network and are meant to evolve in the coming years.

2.3   Optimization Module

The optimization module consists of a tailor-made reduced genetic algorithm that allows to optimize the future management of the system. The general structure of the optimization module is illustrated in Fig. 3.

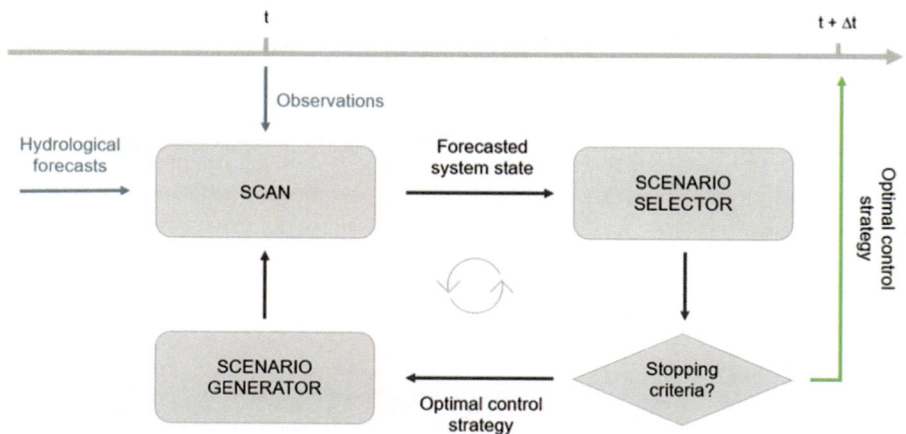

Fig. 3. Illustration the general structure of the optimization module.

The forecasting component combines the hydrological forecasts from the HYDROMAX model (Moens et al. 2018) with a surrogate hydraulic model named "SCAN" to forecast the future system states up to 24 h in advance (Wolfs et al. 2015). SCAN is a very fast simulating model due to the mix of data-driven and physical based modelling components. The model is calibrated against a detailed 1D hydrodynamic HEC-RAS model. It can simulate 24 h in 0.004 s for a waterway network of 88 km. The integrated data assimilation component ensures that the forecasts align with the latest observations and is an improved version of the assimilation procedure described in Vermuyten et al. (2018a).

The optimization module contains a scenario generator that generates a large set of potential control strategies for all controllable structures in the network within the physical realistic boundaries of the network and the structures. The scenario generator uses a tailor-made parallel reduced genetic algorithm that is built upon earlier work by Vermuyten et al. (2018b). New control strategies can be generated as the result of (a) mutation of an existing strategy, (b) a new randomized control strategy or (c) cross-over between parallel optimized control strategies. Mutations refine the control strategy and pushes the strategy towards the optimal solution. The randomized strategies and the cross-overs ensure that the scenario generator explores the entire solution space and does not get stuck in local optima.

The scenario selector evaluates the different control strategies and chooses the most optimal solution based on a transparent set of priorities and objective functions. Priorities group similar objective functions and allow to clearly prioritize certain objectives over others. Table 1 shows the different priorities that are implemented in Orhyx. The optimization works sequentially: it will first minimize the objective functions linked to the highest priority before moving towards lower priority objectives. Each priority contains a set of objective functions that mathematically describe the objectives of the priority. Soft constraints are used within the objective function to penalize the exceedance of certain thresholds. Figure 4 gives an example of the soft constraints linked to navigation and the resulting objective function. The soft constraints provide the upper and lower level in between water levels should remain to ensure navigation. The resulting objective function increases quadratically once the water level exceeds these boundaries, restraining the optimization to a solution that remains well within these thresholds. Next to the soft constraints, the objective function also contains weights that allow to flexibly shift the importance within a priority depending on the situation.

Table 1. List of priorities and objectives implemented in the optimization module.

Priority	Objective
1	Safety at the structures
2	General safety (floods)
3	Navigation
4	Discharge and water level variations, Ecological discharges, Economic costs and benefits

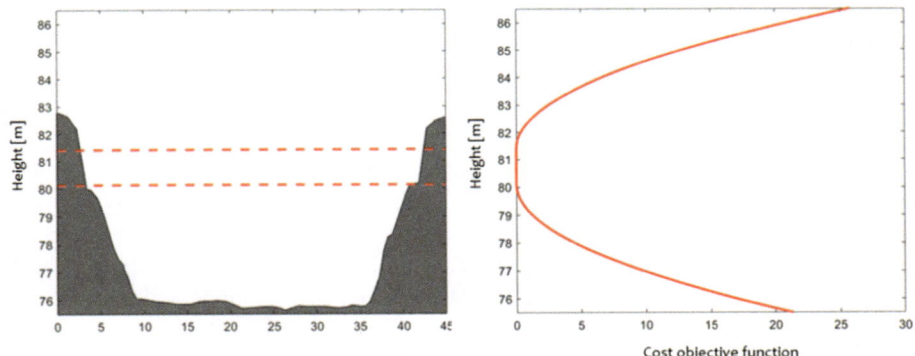

Fig. 4. Illustration of the soft constraints for navigation on a certain location (left) and the resulting objecitve function penalizing the exceedance of these thresholds (right).

Orhyx updates the optimal control strategy every 30 min based on the latest observations and forecasts. For each optimization 100.000 potential control strategies are simulated and evaluated. In the current setup 19 variables are simultaneously being controlled by the optimization module. The objective functions incorporate the model results of 87 variables along the study area.

3 Validation

The optimization module was evaluated on a broad range of historical events including extended dry spells in summer, high flow periods in winter, short high-intensity thunderstorms in summer and events with observed waves propagation through the system.

The optimization module produced a global control strategy for all structures for each of these events. These results were compared with a local control strategy where each of the structures respond individually to the disturbance. These local control strategies are used operationally and are based on years of experience with controlling the system manually. The results are evaluated for different indicators that summarize the results over the entire network. Figure 5 compares the navigability of the local current control strategy to the global control strategy (Orhyx tool). The navigability indicator gives the percentage of time where navigation on the channel can occur without problems. A value lower than 100% indicates a potential risk for navigation somewhere in the network for a certain period of time due to the undershoot of certain water level thresholds. Figure 5 shows that the global control strategy results in higher percentages of navigability for almost all events. Only for the first high flow event the percentage of navigability is lower in the global control strategy. This is linked to the fact that floods occur during this period and the global control strategy tries to avoid these floods at all costs given the general security has a higher priority than navigation (Table 1).

Figure 5 shows the comparison for the indicator linked to the variation of the water level. This indicator quantifies the short-term water level variations which could be linked to waves propagating through the network. The optimization module tries to minimize the water level fluctuations in the fourth priority. For most events the global control strategy results in lower variations of the water level. Only for the high flow events the global control strategy allows higher variations of the water levels to reduce flooding, which is a higher priority in the optimization.

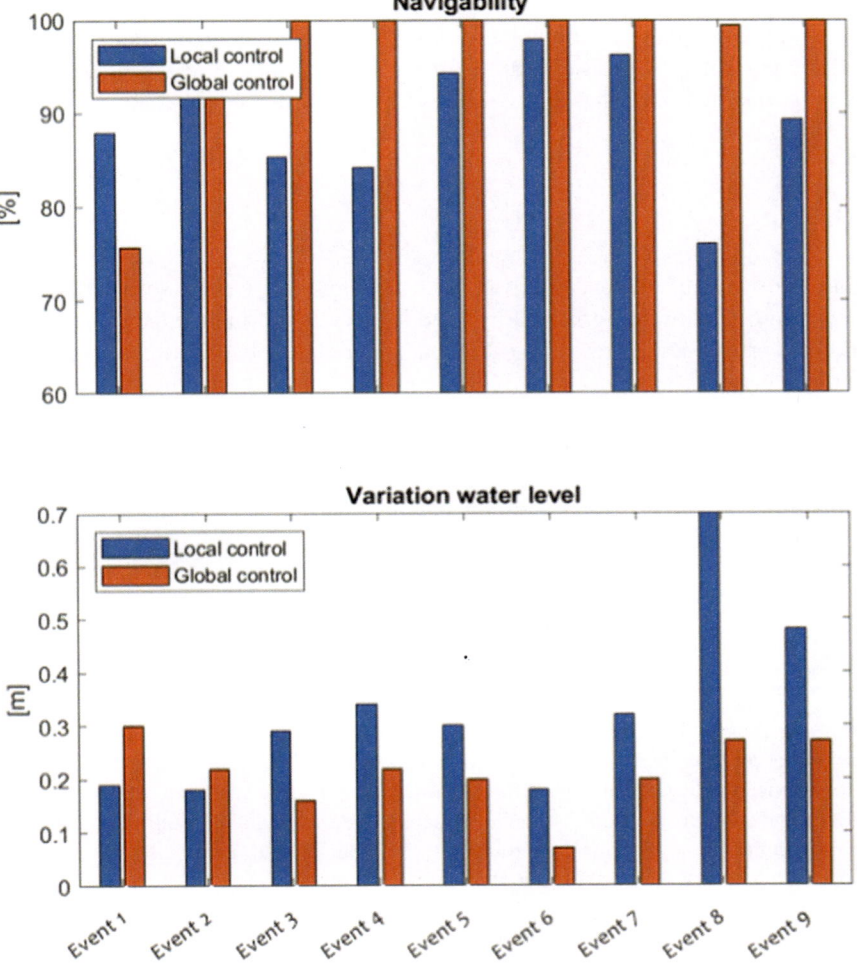

Fig. 5. Resulting time in which navigability is ensured (higher is better) and variations in water levels (lower is better) for different storms for the local and global control.

The results of the validation demonstrate the benefits of using a global control strategy compared to the existing local control strategy. This benefit is related to the increased temporal and spatial view of the global control strategy. The forecasting module allows the optimization to look into the future and anticipate on future disturbances with pro-active measures. Water levels in a reservoir (section of the river defined as a storage cell in the SCAN model) can e.g. be lowered to increase the available storage volume for future storm events thereby potentially reducing floods. The global control strategy also benefits from evaluating the entire system as a whole, allowing to take real time and reactive measures to e.g. reduce short term water level fluctuations.

4   Conclusions

Starting from scratch and developing an innovative tool like Orhyx is very challenging and rewarding. The framework of Orhyx and its operating principle have been studied and defined on a reduced area of the Walloon waterways network. Eighty-eight km of Walloon waterways have been therefore modeled and 14 hydraulic structures optimized with 48 optimization runs every day. The first off-line validation of Orhyx shows encouraging results and outlines the benefits of using a global, centralized and optimized management compared to the current local management. Next steps will consist in testing the outputs of the tool by implementing them in real conditions on site. For this purpose, water level setpoints and gates positions instructions will be sent to the sites. After this, Orhyx will be enlarge to the whole Walloon Region network. Challenges that arise in the next months will probably be the operational settlement and the interaction and communication with the hydraulic structure thanks to the SCADA system, also under construction.

Acknowledgements. Developing Orhyx asks a lot of multidisciplinary competences. A large number of people have been involved in this project and will continue to be whether they are from the Service Public de Wallonie or from IMDC and Sumaqua. Many hours have been spent in order to develop Orhyx, from hydraulic modeling, data assimilation, optimization development, operational set up or project coordination. We would like to thank all those people not mentioned in the article who have made the tool what it is today.

References

Moens L, Bastin G, Dierickx P, Thunus M (2018) PREVISIONS EN TEMPS REEL DES DEBITS AVEC HYDROMAX: 24 ANNEES D'EXPERIENCE.De la prévision des crues à la gestion de crise(SHF) (Avignon, du 14/11/2018 au 16/11/2018)
Vermuyten E, Meert P, Wolfs V, Willems P (2018) Model uncertainty reduction for real-time flood control by means of a flexible data assimilation approach and reduced conceptual models. J Hydrol 564:490–500. https://doi.org/10.1016/j.jhydrol.2018.07.033OpenAccess

Vermuyten E, Meert P, Wolfs V, Willems P (2018) Combining model predictive control with a reduced genetic algorithm for real-time flood control. J Water Resour Plan Manag 144(2):1–13. https://doi.org/10.1061/(ASCE)WR.1943-5452.0000859 Art.No. 04017083

Wolfs V, Meert P, Willems P (2015) Modular conceptual modelling approach and software for river hydraulic simulations. Environ Model Softw 71:60–77

Numerical Simulation of the Composite Bank Stability Process of the Songhua River

Jun Yang[1], Dongdong Jia[1,2(✉)], Lei Wu[1], Youzhi Hao[1],
and Zhuoying Cang[1]

[1] Waterway and Sedimentation Engineering of Ministry of Transport, Key
Laboratory of Port, Nanjing Hydraulic Research Institute, Nanjing, China
ddjia@nhri.cn
[2] Yangtze Institute for Conservation and Development, Nanjing, China

Abstract. The phenomenon of bank erosion and collapse is widely distributed in major rivers all over the world, and it is a kind of natural disaster with greater hazards. Bank erosion in seasonally frozen rivers (SFR) is subject to the coupling effects between hydrodynamic forces and freeze-thaw, and the mechanism is complex. Understanding of bank erosion mechanisms is of great significance for river bank protection and comprehensive river management. Taking the downstream near dam section of the Dadingzishan Navigation and Hydropower Project in the mainstream of the Songhua River as an example, the BSTEM model was used to analyze the degrees of riverbank stability in different periods considering the freeze-thaw effect. The results indicated that the degrees of riverbank stability are high during the periods of low water level and water-level rising before the flood season, and the slope toe scour is the main factor; and they are low during the periods of high water level and water-level recession with the continuous bank collapse occurring; The freezing and thawing effect has an important effect on the stability of river banks in the seasonal frozen region. so as to provide a certain reference for the study on the bank collapse of the river in seasonal freezing region and its simulation for the study of riverbank erosion in seasonally frozen rivers.

Keywords: Bank collapse · BSTEM · Seasonal frozen rivers · River bank stability

1 Introduction

From the perspective of fluvial evolution, bank collapse is the movement of a large number of sediment particles under the interaction between the river flow and the river bed boundary, which is caused by the unbalanced sediment transport (Yang et al. 2022). River bank collapse is widespread and is a natural disaster with great harm (Kimiaghalam et al. 2015) For the Heilongjiang River in Northeast, the main trunk reaches are important boundary rivers in our country, and the erosion of banks and beaches will directly cause the loss of farmland (Jia et al. 2021). There are significant differences in the process, and the characteristics of freezing and thawing are obvious.

© The Author(s) 2023
Y. Li et al. (Eds.): PIANC 2022, LNCE 264, pp. 1178–1185, 2023.
https://doi.org/10.1007/978-981-19-6138-0_103

Therefore, it is of great theoretical significance and practical guiding value to carry out research on the collapse of river banks and beaches in seasonally frozen regions.

For the problem of bank collapse, its formation mechanism and occurrence process are complex, and the existing research cannot fully reveal its evolution law. In response to such problems, many experts and scholars at home and abroad have carried out research on the mechanism of bank collapse, revealing different influencing factors of river bank collapse. At the same time, many scholars have constructed different collapse modes based on the theory of soil mechanics slope stability. Calculation model of river bank stability under. For example, in terms of revealing the influencing factors of river bank collapse analyzed the influence of curved circulation on bank erosion and its change process(Papanicolao et al. 2007). The interaction process of bank slope collapse and river bed erosion and deposition is also analyzed (Yu et al. 2016). The existing riverbank stability analysis models are usually based on the slope stability theory in soil mechanics. considered the influence of pore water and hydrostatic pressure, and made corresponding supplements and improvements to the previous viscous bank collapse model (Darby et al. 1996). However, such simulations do not take into account the effect of the freezing and thawing of seasonally frozen rivers, so further improvement is needed.

This paper takes the typical dual-structure bank of the Songhua River as the research object, considering the freezing and thawing of soil, lateral water pressure, and combining with the slope toe scour calculation module, the typical section near the dam in the downstream of the Dadingzishan Aviation and Power Project of the Songhua River is taken as the research object. Taking a river bank as an example, the BSTEM model was used to calculate the river bank stability in different periods of the entire hydrological year in 2009, and to analyze the change process of the corresponding safety factor F_s.

2   Materials and Methods

2.1   Study Area

The Songhua River is one of the seven major rivers in my country and the largest tributary on the right bank of the Heilongjiang River. The main stream of the Songhua River has a well-developed water system. There are many tributaries along the way. The main stream of the Songhua River is a typical plain alluvial river, with a total length of about 940 km and a catchment area of 186,400 km^2. Except for the hills and hills in the middle part, it is basically an alluvial plain. According to the terrain, topography and river nature, the main stream of the Songhua River can be divided into three sections: upper, middle and lower reaches. The vegetation in the basin is rich in humus. The surface layer on both sides of the river channel is black humus soil, 0.5–1.5 m below the surface layer is clay or sandy clay, the lower layer is silt or fine sand, and the river bed is composed of medium-fine sand and sandy loam. Composition, loose texture and poor impact resistance.

The phenomenon of river freezing is very common in the high latitudes of northern my country, among which river freezing can be divided into stable freezing and unstable freezing. Heilongjiang and its tributaries, Songhua River in the Northeast, due to the dominant low temperature in winter, although there is a large flow and flow rate in the river, it is frozen every year, which is a stable freezing, and the freezing period is generally 4 to 5 months. Therefore, such rivers are also called seasonally frozen rivers, or "seasonally frozen rivers" for short. For the non-freezing period (free flow period), the movement of water and sediment in the seasonally frozen area and its channel evolution are basically the same as those of conventional rivers. However, during the freezing and thawing period of the river, the physical and mechanical properties of the bank soil will change, and the river bank will be stable. Sex is also further affected.

The calculated section of the selected river section is located near the dam downstream of Dadingzishan Avionics Junction. The Dadingzishan Avionics Junction project started construction in 2004 and was put into use at the end of 2008. It is the eighth part of the overall plan for the cascade development of the Songhua River Channel One of the hubs. The operation of the junction and the reduction of upstream sediments have caused obvious erosion in the downstream section of the dam near the dam, deep groove swings, and beach avalanches occur from time to time.

2.2   Numerical Modeling

The BSTEM (bank stability and toe erosion model) model developed by the National Sediment Laboratory of the United States can simultaneously consider the effects of lateral water pressure, pore water pressure, soil matrix suction and different composi-tions of river bank soil layers. The scour and bank stability model simulates the process of bank collapse and is one of the most widely used models.

The model is mainly composed of the toe scour module (TEM) and the bank stability module (BSM). By inputting the typical river bank profile topography, channel water level, soil physical and mechanical parameters, etc., run the slope toe scour module and import it into the bank stability module to calculate the bank stability safety factor Fs according to the new terrain after scour.

2.3   TEM Modle

The lateral scour width acting on the bank soil is determined by the lateral scour rate and scour time of the river bank. Among them, the lateral scour rate of the river bank is mainly determined by the scour strength of the water flow and the scour resistance of the soil body. Only when the shear stress of the near-shore water flow applied to the river bank soil body is greater than the starting shear stress of the river bank soil body, the river bank soil body can start. The lateral scour width E (m) of the river bank is expressed as:

$$E = k \cdot \left(\tau_f - \tau_c \right) \cdot t \tag{1}$$

where E = erosion distance (m), k = erodibility coefficient (m^3/N s), Δt = time step (s), τ_f = average boundary shear stress (Pa), and τ_c = critical shear stress (Pa).

2.4   BSM Modle

The BSTEM model uses the limit equilibrium method to calculate the bank stability safety factor F_s, including the horizontal layer method, the vertical slice method and the cantilever shear collapse method. The main calculation formula of the stability safety factor is:

$$F_s = \sum_{i=1}^{I} \left(c_i' L_i + (\mu_a - \mu_w)_i L_i \tan \varphi_i^b + [W_i \cos \beta - \mu_{ai} L_i + P_i \cos(\alpha - \beta)] \tan \varphi_i' \right) / \sum_{i=1}^{I} (W_i \sin \beta - P_i \sin[\alpha - \beta])$$

(2)

where c_i' = effective cohesion of ith layer (kPa); L_i = length of the failure plane incorporated within the ith layer (m); W_i = weight of the ith layer (kN); P_i = hydrostatic-confining force due to external water level (kN/m) acting on the ith layer; b = failure-plane angle(degrees from horizontal); a = local bank angle (degrees from horizontal); and I = number of layers.

2.5   Field Data

(1) soil properties In order to obtain the calculation parameters of the soil composition and mechanical properties of the river bank required by the model, a field investigation was carried out on the typical section of the bank collapse, and layered sampling was carried out according to the composition, structure and properties of the soil at the section., according to the "Standards for Geotechnical Test Methods" (GB/T50123–2019), the shear test of the soil sample is carried out, and the shear strength of the corresponding soil body is obtained. The physical and mechanical properties of the riparian soil are shown in Table 1.

Table 1. Parameters for the bank soil properties at sections of DM1

Soil layer	Material	Layer thickness/m	Unit weight/ (kN/m³)	φ^b /°	Internal friction angle/(°)	Cohesion/ (kN/m²)
1	Clay	0.7	16.3	15	21.6	17.3
2	Clay	1.5	16.9	15	25.5	15.6
3	Silt	1.3	16.4	15	24.3	0.5
4	Silt	2	18.7	15	28.1	0.4
5	Silt	2	18.7	15	28.1	0.4

(2) Flow changes The annual average precipitation in the Songhua River Basin is generally around 500 mm. The precipitation from June to September in the flood season accounts for 60% to 80% of the whole year, and the precipitation from December to February in winter is only about 5% of the whole year. The model water level process during the simulation period is shown in Fig. 1.

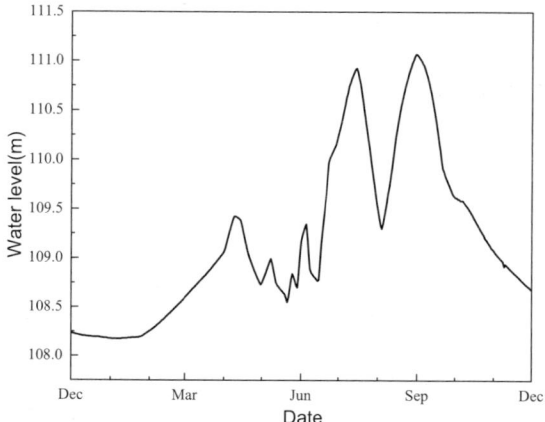

Fig. 1. Water level processes of model

3 Results and Discussion

In order to analyze the stability of the river bank in different periods, considering the effect of freezing and thawing, a typical section of DM1 at a distance of 3 km downstream of the Dadingzishan Aviation and Power Junction of the Songhua River was selected to calculate the river bank stability. According to the initial measured river bank shape, the hydrodynamic conditions at the cross section and the mechanical properties of the soil, the BSTEM model was used to calculate the stability of each period, and the change process of the corresponding safety factor Fs was obtained, as shown in Fig. 2, 3 and 4 The calculation results are as follows:

(1) During the dry season (12.15–3.15), the discharge flow of Dadingzishan Reservoir is relatively small and does not change much. The average water level through this section is less than 108.5 m, the water level in the channel is low, and the shear stress of near-shore water flow is relatively small. It is smaller, but still greater than the starting shear stress of the lower sandy soil layer. At this time, there is a certain scour in the lower sandy soil layer of the binary structure river bank. According to the calculation and statistics, the scouring amount of the slope foot in the dry season in 2009 was 13.87 $m^3 \cdot m^{-1}$; however, due to the gentle slope of the river bank and the small pore water pressure in this period, the river bank has a high degree of stability, and the stability safety factor Fs value is correspondingly large. The average F_s value is basically above 1.8, and the probability of bank collapse is very small.

(2) During the Level rising stage (3.16–5.31), the ice melts, the flow will increase accordingly, the water level in the river channel will rise, and the sandy soil layer on the lower part of the river bank will be further eroded. According to statistics, the scour amount of the sections in 2009 was 6.77 $m^3 \cdot m^{-1}$, and the value of the stability safety factor F_s was relatively small at this time. The existence of freeze-thaw effect affects the shear strength of soil, resulting in the reduction of river

bank stability. According to the calculation and statistics, the Fs value of the section at this time has dropped to about 1.32. In fact, during the freezing and thawing period around the beginning of April every year, a unique spring flood phenomenon will occur in a short period of time, and the resulting ice will also adversely affect the stability of the bank. In the situation, there is still the possibility of bank collapse during the rising water stage before the flood season. It belongs to the stage with strong shore collapse intensity.

(3) During the flood period (6.1–10.31), the flow in the river channel increases, the water level reaches the highest level in the whole year, the sandy soil layer in the lower part of the river bank is washed away, the bank slope becomes steep, and the water level in the river channel rises and falls during the flood peak period, and the diving level rises steadily, resulting in higher pore water pressure and lower overall bank stability. According to the calculation and statistics, the scour amount during the flood period in 2009 was 11.18 $m^3 \cdot m^{-1}$, and there was1 collapse, and the collapse amount was 26.06 $m^3 \cdot m^{-1}$. In general, during the flood period, the toe of the slope has the most severe erosion, the largest amount of collapse, and the poor stability of the river bank, which is the stage of frequent bank collapse.

(4) During the Recession stage (11.1–12.15), the water level in the river channel drops rapidly. As the water level decreases, the lateral water pressure acting on the river channel gradually decreases. As the angle increases, the safety factor decreases, increasing the possibility of river bank collapse. During the calculation process, the scour amount during the flood period in 2009 was 5.09 $m^3 \cdot m^{-1}$, and 1 collapses occurred, and the collapse amount was 16.75 $m^3 \cdot m^{-1}$; from this section, the scour amount during the 2009 flood period was less, but the collapse occurred large amount. It shows that through the cumulative effect of previous water current scouring and freeze-thaw effects, the possibility of bank collapse during the receding period is high, which belongs to the stage of frequent bank collapse.

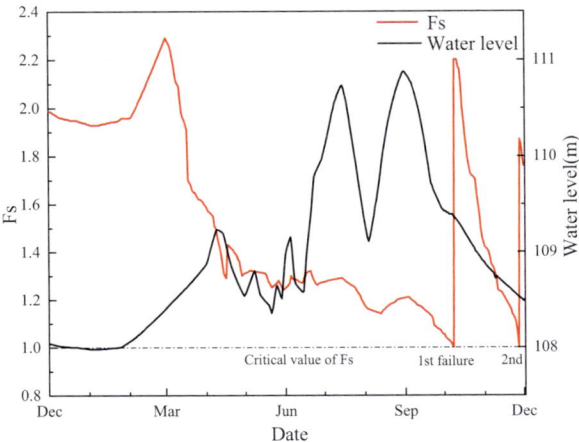

Fig. 2. Safety factors of bank satiability during different stages

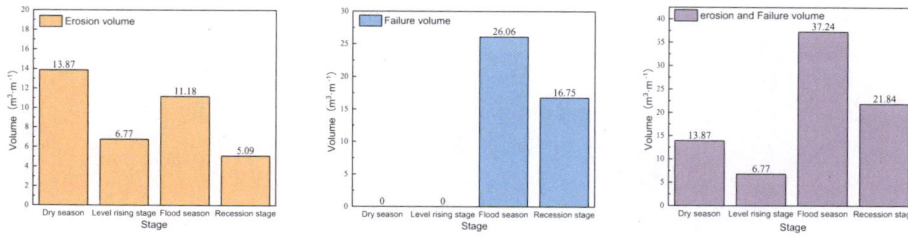

Fig. 3. Calculated bank erosion and failure volumes under different stages

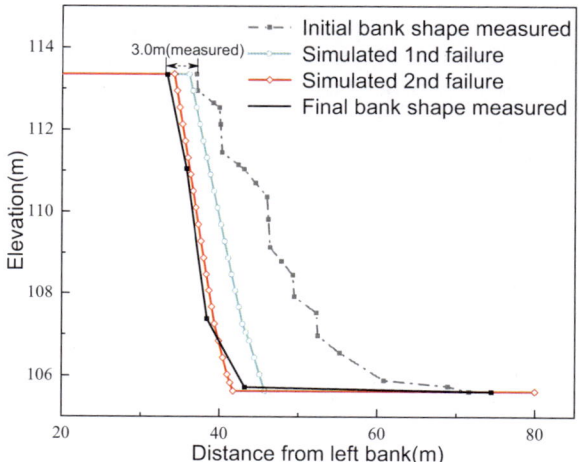

Fig. 4. Changes of bank profiles at different stages

4 Conclusions

Taking the typical section of the bank near the dam downstream of the Dadingzishan avionics project on the Songhua River as an example, considering the effect of soil freezing and thawing, combined with the scouring of the slope foot, the BSTEM model was used to calculate the river bank stability in the whole hydrological year of 2009, and quantitative analysis of the freezing and thawing effect was carried out. The effect of melting on river bank stability. The main research conclusions are:

(1) The stability of the bank in different periods of the typical dual structure river bank was calculated, and the results showed that the soil loss of the DM1 section in the dry period and the flood period was dominated by the erosion of the slope foot, and the collapse amount in the flood period and the receding period accounted for the majority. About 54% of the total scour and collapse volume is dominated by river bank collapse during this period.

(2) The river bank stability is relatively high in the dry and high water periods; the river bank stability is relatively low in the flood period and the receding water period, with 2 collapses in total, which is a period of frequent bank collapses.

(3) Taking into account the scouring of the slope toe, the effects of freezing and thawing, the stratification of soil mass and the accumulation of soil mass at the toe of the slope after the collapse, etc., the typical binary structure river bank was simulated, and the change of the bank safety factor F_s in the whole hydrological year was obtained. The simulation results of the river bank collapse law are in good agreement with the actual collapse results. The simulated method of river bank and beach collapse in seasonally frozen areas considering the effects of freezing and thawing provides an effective idea for in-depth study of river bank and beach collapse in seasonally frozen areas and its management of river banks.

Acknowledgements. The work presented in this paper is financially supported by the National Natural Science Foundation of China (Nos. U2040215 and 52079080).

References

Yang J, Jia DD, Gao J, Hao YZ, Wu L (2022) Study progress of mechanism and numerical simulation of bank collapse of river in seasonal freezing region. Water Resour Hydropower Eng 53(01):83–90

Kimiaghalam N, Goharrokhi M, Clark SP (2015) A comprehensive fluvial geomorphology study of riverbank erosion on the red river in Winnipeg, Manitoba. Can J Hydrol 529(3):1488–1498

Jia DD, Yang J, Cheng CY, Zhang XN, Ying Q (2021) Numerical simulation of bank erosion in a typical seasonally frozen river-case study of the Songhua river. China Adv Water Sci 32 (05):717–726

Papanicolaou AN, Elhakeem M, Hilldale R (2007) Secondary current effects on cohesive river bank erosion. Water Resour Res 43(431):497–507

Yu MH, Chen X, Wei HY, Hu CW, Wu SB (2016) Experimental of the influence of different near-bank riverbed compositions on bank failure. Adv Water Sci 27(02):176–185

Darby SE, Thorne CR (1996) Development and testing of riverbank-stability analysis. J Hydraul Eng ASCE 122(8):443–454

Research on Water Scour Conditions
of Wanjiazhai Reservoir, China

Kunhui Hong[2], Shouyuan Zhang[1], Wei Zhang[2($\boxtimes$)], and Teng Wu[2]

[1] State Key Laboratory of Hydrology-Water Resources and Hydraulic
Engineering, Ministry of Water Resources, Nanjing, China
oliviatungyur@outlook.com
[2] College of Harbour Coastal and Offshore Engineering, Hohai University,
Nanjing, China
Hongkunhui@yeah.net, wuteng@hhu.edu.cn

Abstract. Flood control, power generation and agricultural irrigation are the primary functions of reservoirs. The main reason for the serious deposition in the Yellow River reservoir is the high concentration of sediment in the water flow. Due to the establishment of the Yellow River cascade reservoir and the joint scheduling among reservoirs, the flow rate of the downstream reservoir decreases in flood season, which reduces the hydrodynamic flushing efficiency in the reservoir area. Wanjiazhai Reservoir is selected as the research sample in the present study, which is a first-level cascade reservoir in the connecting area of the upper and middle reaches of the Yellow River. In this study, a numerical model was built to analyze factors that influenced the operation and flushing of the upstream reservoir. And a formula was recommended to apply to the Wanjiazhai Reservoir on the basis of the sediment deposition pattern discrimination formula. Besides, various flushing methods and methods reducing the sediment deposition were carried out and analyzed. Results indicated that when the inflow rate increased, the maximum particle diameter of the incipient sediment was larger. When the inflow rate reached 1200 m^3/s, the diameter of the initial the incipient sediment varied from 0.015 mm to 0.094 mm. And the diameter range of the initial the incipient sediment was 0.014 mm–0.094 mm, when the inflow rate was 1800 m^3/s. When the inflow rate reached 2500 m^3/s, the diameter of the initial the incipient sediment varied from 0.010 mm to 0.094 mm. Notably, all sediment were discharged from the reservoir. Moreover, the sediment of the reservoir was flushed at a flow rate of 1200 m^3/s–1800 m^3/s under the premise of meeting the upstream water storage.

Keywords: Wanjiazhai Reservoir · Hydrodynamic simulation · Reservoir scour · Sediment starting size

1 Introduction

Wanjiazhai Reservoir is located in the canyon between the Tuoketuo section and the Longkou River section, which is lied in the upper section of the Yellow River main stream (Maria et al. 2015). The frequent torrential rain during summer and autumn, large flood peak flow, abundant sand source and high concentration of sediment in

© The Author(s) 2023
Y. Li et al. (Eds.): PIANC 2022, LNCE 264, pp. 1186–1194, 2023.
https://doi.org/10.1007/978-981-19-6138-0_104

water flow occurred in this area make it a typical sandy section (Medeiros et al. 2019). The Yellow River passes through the Inner Mongolia Plateau, where wind erosion and water erosion exist alternately and interact with each other (Lu and Siew 2006). With the improvement of the cascade reservoir system in the Yellow River mainstream, the annual average inflow rate of the Wanjiazhai Reservoir reduced significantly compared with how it was before implementing the joint regulation schedule of water and sediment. From 1998 to 2004, the annual average inflow rate was 1,278,000 m^3, which was far lower than the designed value. As a result, after the 13-years operation of the hydro-junction, the sediment deposition in reservoir area reached 4,278,000 m^3, which reaches the 47.7% of the total reservoir capacity. The sediment affects the flood control, irrigation and navigation seriously (Fu 2016).

Massive studies have investigated how to adjust the water and sediment outflow process, to reduce the sediment deposition, increase the efficiency of sediment flushing and the storage capacity. For example, Zhang and Zhang (1982) put forward methods of calculating the sediment deposition volume, location of sediment deposition and the variation of the reservoir storage capacity with time by analyzing multiple operations of reservoirs in flood season. Similarly, Zhang et al. (2018) compared more than 100 floods with high concentration of sediment in the Sanmenxia reservoir from 1961 to 2013, and analyzed their characteristics. Results revealed how floods with high concentration of sediment affected the reservoir using different sediment discharge methods. Besides, a calculation formula of sediment discharge rate in floods with high concentration of sediment were discharged using backwater. And another calculation formula of outflow sediment transport rate under open sediment discharge were adopted. Wei et al. (2005) analyzed all the inflow water and sediment data of XLD reservoir from 1999 to 2002, and calculated the quasi-two-dimensional sediment mathematical model of the Yellow River. It turned out that the model was in good agreement with the measured data in real reservoir area, including the scouring and deposition process of the riverbed, alterations in the river morphology, flow patterns in flood season and the change of water level. Tu (1980) concluded the discharge calculation formula of sediment under the discharge method of density current by investigating connections among the sediment discharge of large domestic reservoirs. Moreover, in Zhang's study (2004), fuzzy-neural network was used to calculate the scouring of sediment deposition in flood season. However, the research focusing on alterations in the along flow pattern under different water inflow volume and the particle diameter range of the incipient sediment under the influence of flow patterns in the reservoir area were still rare (Fig. 1).

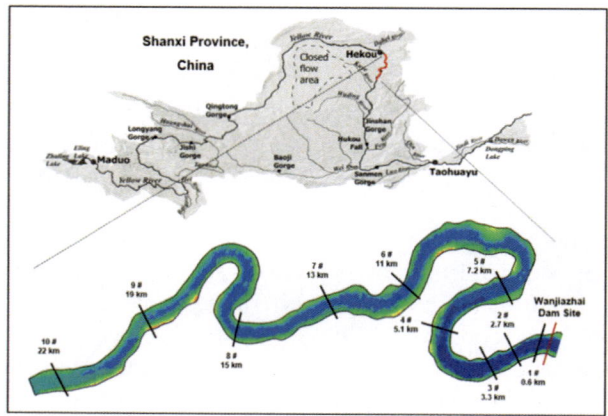

Fig. 1. Location of the Wanjiazhai Reservoir in China

2   Hydraulic Numerical Model

2.1   Numerical Model

Basing the incompressible Reynolds equation, the two-dimensional unsteady flow model which describes the unsteady flow of the open canal was built.. The water head is given by the following equations:

$$h = \eta + d \tag{1}$$

$$\frac{\partial h}{\partial t} + \frac{\partial (h\bar{u})}{\partial x} + \frac{\partial (h\bar{v})}{\partial y} = hS \tag{2}$$

$$\frac{\partial (h\bar{v})}{\partial t} + \frac{\partial (h\overline{uv})}{\partial x} + \frac{\partial (h\bar{v}^2)}{\partial y} = f\bar{u}h - gh\frac{\partial \eta}{\partial y} - \frac{h}{\rho_0}\frac{\partial p_a}{\partial y} - \frac{gh^2}{2\rho_0}\frac{\partial \rho}{\partial y} + \frac{\tau_{sy}}{\rho_0} - \frac{\tau_{by}}{\rho_0}$$
$$- \frac{1}{\rho_0}\left(\frac{\partial S_{yx}}{\partial x} + \frac{\partial S_{yy}}{\partial y}\right) + \frac{\partial}{\partial x}\left(hT_{xy}\right) + \frac{\partial}{\partial y}\left(hT_{yy}\right) + hv_sS \tag{3}$$

where $\bar{u}$ and $\bar{v}$ are the average flow velocity, m/s. t represents time, s. d represents the depth of still water, m. η represents the elevation of riverbed, m. h is the hydraulic head, m. u and v are components of velocity in x axis and y axis, m/s. g represents the gravitational acceleration, m/s^2. ρ is the density of water, kg/m^3. s_{xx}, s_{xy}, s_{yx} and s_{yy} are the components of radiation stresses. P_a is the atmospheric pressure, Pa. ρ_0 is the relative density of water, kg/m^3. u_s and v_s are the water flow velocity, m/s.

2.2   Model Establishment and Verification

The simulated area of the model is the front bend of the Yellow River midstream. The upstream boundary of the model is the first front bend, and the downstream boundary is based on the actual measured water level of Wanjiazhai reservoir. The roughness coefficient is 0.024. The model contains 9000 average-sized grids. The length of the biggest grid is less than 100 m, and its area is less than 2000 m².

The model was verified based on the actually measured hydrological data of the Hamaoer hydrometric station and the Wanmatou hydrometric station located in the flow entrance and the dam site from July 3 in 2011 to July 16 in 2022. The verification result is shown in Fig. 2. The calculated water level of the Wanjiazhai reservoir from the model is slightly different from the actually measured water level, and the curve trend of them are consistent. Among results, the maximum differential value between the actually measured water level in Hamaoer hydrometric station and the simulated water level is 0.10 m, with a maximum relative error rate of 1.23%. The maximum differential value in Wanmatou hydrometric is 0.04 m with a maximum relative error rate of 0.49%. Therefore, the topography and the roughness (0.025) established by the model are reasonable. It is feasible to use this model to simulate the evolution process of water and sediment from the front turning of the Yellow River midstream to the river section of the Wanjiazhai reservoir dam site.

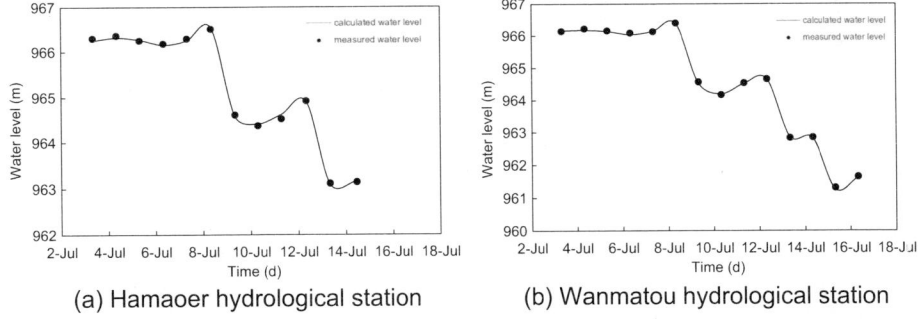

(a) Hamaoer hydrological station   (b) Wanmatou hydrological station

Fig. 2. Comparisons between calculated results and measured data

Considering factors such as the actual inflow rate, the annual precipitation, and the united dispatching, the outflow water level of the reservoir is set as 948 m, which can maintains the reservoir flushing. Owing to the united dispatching of the upstream reservoir, the inflow rate of the reservoir can be allocated by the upstream reservoir. The inflow rate of the upstream reservoir was set as five calculation groups, which were 500 m³/s, 800 m³/s, 1200 m³/s, 1800 m³/s and 2500 m³/s, respectively.

3   Variations of Hydraulic Parameters

3.1   Average Flow Velocity

The control variable method was applied to analyze the flow velocity under different inflow rates in this study. The average flow velocity and the water depth along the riverway in different flow rates are shown in Table 1. With the increasing inflow rate, the flow velocity in water area 7.2 km away from the dam axis is significantly affected by the inflow rate variations, and the flow velocity in water area 1 km away from the dam axis is slightly affected by the inflow rate variation. When the inflow rate increases from 1200 m^3/s to 1800 m^3/s, the increasing range of the average flow velocity is 0.17 m/s. When the inflow increases from 1800 m^3/s to 2500 m^3/s, the increasing range of the average flow velocity is 0.16 m/s. When the inflow increases from 800 m^3/s to 1200 m^3/s, the flow velocity variation is the most affected, and the increasing range of the average flow velocity is 0.21 m/s. Figure 3 demonstrates the alterative figure of the along flow velocity at different reservoir inflow rates. Under the same flow rate, the reservoir area 3.3 km away from the dam axis has the minimum flow velocity. And there is the maximum increasing range of the flow velocity within 4 km–7.2 km of the reservoir area. As the inflow rate goes up, the increasing range of the flow velocity gets larger. In water area 7.2 km away from the dam axis, the water flow velocity is significantly affected by the inflow rate, and its maximum increasing range of the average flow velocity reaches 211%. This hydrometrical section is located in the most front position of the delta deposition with a large channel gradient, whose average flow velocity is majorly affected by the inflow rate.

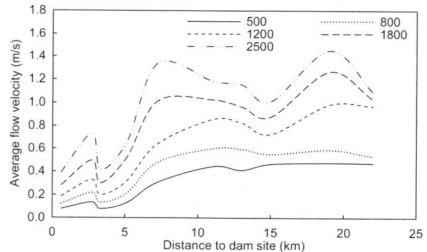

Fig. 3. Flow velocity in different inflow rates   **Fig. 4.** Water depth in different inflow rates

Table 1. Average flow velocity and water depth along the riverway in different inflow rates

Inflow rate (m^3/s)	500		800		1200		1800		2500	
Distance to dam site (km)	Average flow velocity (m/s)	Average water depth (m)	Average flow velocity (m/s)	Average water depth (m)	Average flow velocity (m/s)	Average water depth (m)	Average flow velocity (m/s)	Average water depth (m)	Average flow velocity (m/s)	Average water depth (m)
0.6	0.08	24.16	0.12	24.16	0.19	24.34	0.28	24.22	0.39	24.23
2.7	0.14	14.25	0.22	15.13	0.33	14.84	0.50	13.56	0.73	13.66
3.3	0.08	18.24	0.13	18.26	0.20	18.92	0.30	18.98	0.41	19.06
5.1	0.13	12.60	0.21	13.86	0.31	14.63	0.52	14.16	0.65	14.31
7.2	0.43	6.37	0.48	6.43	0.64	6.62	1.01	6.80	1.34	7.18
11	0.44	5.40	0.60	6.20	0.86	6.65	1.03	6.96	1.18	8.16
13	0.41	7.57	0.60	8.58	0.83	9.55	0.96	11.14	1.15	12.37
15	0.46	3.67	0.55	4.85	0.73	5.98	0.88	7.44	1.01	9.13
19	0.48	6.82	0.59	8.20	0.98	8.87	1.27	9.60	1.45	12.20
22	0.47	2.65	0.53	3.90	0.97	4.82	1.03	5.32	1.09	6.67

3.2 Average Water Depth

Based on the effect of different inflow rates to the water depth in reservoir area, it is found that when the reservoir inflow rate increases, the water depth in the reservoir area 19 km away from the dam axis is mostly influenced, and the water depth in the reservoir area within 1 km from the dam axis is minimally affected by the inflow rate variation. When the inflow rate increases from 1200 m³/s to 1800 m³/s, the increasing range of the average flow velocity in the reservoir area is the largest and reaches 0.90 m/s. When the inflow rate increases from 1800 m³/s to 2500 m³/s, the increasing range of the average flow velocity is 0.88 m/s. When the inflow rate increases from 500 m³/s to 800 m³/s, the increasing range of the average flow velocity is approximately 0.78 m. When the inflow rate increases from 800 m³/s to 1200 m³/s, the increasing range of the average water depth is 0.67 m.

Figure 4 shows alteration of the water depth at different inflow rates. In the same inflow rate, the variation trend of along water depth in the reservoir is basically consistent. And the water depth grows smaller with the increasing distance from the dam axis. Within 7.2 km of the dam axis, the water depth is basically not affected by the flow rate. However, in 7.2 km away from the dam axis, the influence of flow rate on the water depth is bigger, and the water gets deeper when the flow rate increases. Influenced by the topography, there are two sudden increases of the water depth in 13 km and 19 km away from the dam axis.

4 Diameter Analyses of the Reservoir Flushing Sediment

4.1 Calculation Formula of Incipient Sediment

It is considered that the reservoir sediment is mainly derived from the Yellow River midstream, and the sediment particle is smaller. Affected by the soil quality in the Yellow River midstream, the inflow sediment is finer and has certain agglutinating power among each other. Therefore, the Zhang's sediment incipient velocity formula is applied to calculate the diameter of the incipient sediment, which is set as criterion of the incipient sediment under certain flow rate. It is also used to initially estimate the diameter range of the incipient sediment in flood season. The sediment incipient velocity formula is displayed as follows:

$$U_c = \left(\frac{h}{d}\right)^{0.14}\left[17.6\frac{\rho_s - \rho}{\rho}d + 6.05 \times 10^7 \frac{10 + h}{d^{0.72}}\right]^{0.5} \tag{4}$$

where U_c is the velocity of the incipient sediment, m/s. h is the water depth, m. d is the particle diameter of sediment, m. ρ is the density of water, kg/m³. α is the effective volumetric weight coefficient.

4.2 Sediment Discharge Analysis

After studying the sediment deposition of the reservoir at different inflow rates, the connection between the diameter range of the incipient sediment and the flow velocity was clarified. According to the actually measured data, the diameter range of sediment in the reservoir is 0.007 mm–0.094 mm. The first 7.2 km of reservoir belongs to the front dam funnel area, where the hydrodynamic parameters are larger than other areas since the existence of the dam outfall and the sediment flushing outlet. Thus, sediment in this area can be discharged during the outflow process. The calculation result of the incipient sediment in the river section 7.2 km–22 km away from the dam axis is shown in Table 2.

Table 2. Minimum particle diameter of incipient sediment in different inflow rates (Unit: mm)

Distance to dam site (m)	Inflow rate (m³/s)				
	500	800	1200	1800	2500
7.2	No motion	0.100	0.046	0.017	0.010
11	No motion	0.051	0.022	0.017	0.014
13	No motion	0.067	0.034	0.028	0.021
15	0.070	0.053	0.032	0.025	0.022
19	0.108	0.065	0.019	0.015	0.013
22	0.029	0.050	0.015	0.014	0.015

When the inflow rate reaches 500 m³/s, the volume of the incipient sediment is little, and the flushing effect of water flow is limited. When the inflow rate reaches 800 m³/s, sediment with a diameter of 0.050 mm–0.094 mm is able to be started in the reservoir area. When the inflow rate is 1200 m³/s, sediment with a diameter of 0.015 mm–0.094 mm enables to start. When the inflow rate is 1800 m³/s, sediment with a diameter of 0.014 mm–0.094 mm enables to start. When the inflow rate is 2500 m³/s, sediment with a diameter of 0.010 mm–0.094 mm can be started, which can be all discharged from the reservoir.

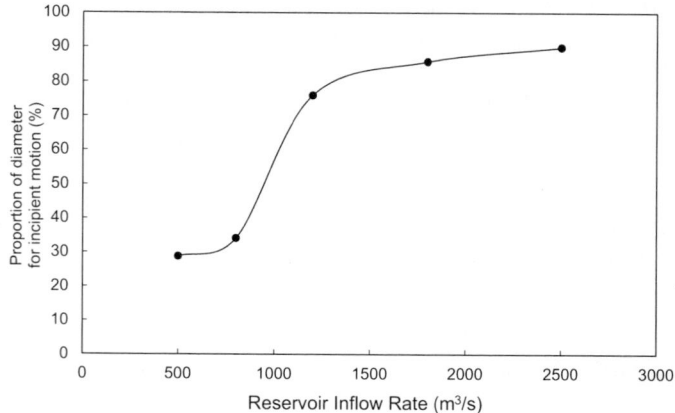

Fig. 5. Proportion of incipient sediment diameter

What's more, percentage of the incipient sediment diameter at different inflow rates is shown in Fig. 5. When the inflow rate increases from 500 m^3/s to 800 m^3/s, the percentage of the particle diameter range of the incipient sediment to the total particle diameter range of sediment increases from 28.73% to 34.02%. The increasing range is little, which is about 5.29%. When the inflow rate increases from 500 m^3/s to 800 m^3/s, the water flow is 0.5 time larger with an increasing range of 41.95%. The most effective sediment flushing effect appears under the largest increasing range of inflow rate. When the inflow rate increases from 1200 m^3/s to 1800 m^3/s, the increasing percentage is 9.77%. When the inflow rate increases from 1800 m^3/s to 2500 m^3/s, the increasing percentage is 4.13%, which is less. The diameter range of the flushable incipient sediment and the discharge flow rate both reaches the biggest when the flow rate is 2500 m^3/s. Compared with the flow rate of 1200 m^3/s, the discharge flow rate is 1.08 times larger. While the sediment flushing ability is limited with an increasing range of 18.30%. Besides, oversized outflow rate also affects the flushing of the downstream river course. In addition, the inflow rate of the reservoir has been few in recent years, and sediment flushing at a large flow rate has certain adverse impacts on the normal storage of the upstream reservoir. Therefore, it is inappropriate to choose 2500 m^3/s as the flow rate of sediment flushing in the reservoir. However, at a flow rate of 1200– 1800 m^3/s, the proportion of the incipient sediment particle is more than 75% and has a strong effect on sediment flushing. Besides, the discharge flow rate is moderate, which makes it reasonable for using this range as the recommended flow rate of flushing the sediment and deposition.

5   Conclusions

Based on the actually measured hydrology and topography data, a mathematical model was built in the present study to conclude characteristics of flow velocity and water depth along the Wanjiazhai Reservoir and clarify the particle diameter range of the incipient sediment by using the sediment starting velocity formula. Various flushing effects of sediment at different flow rates in flood season and characteristics of sediment particle diameter in the reservoir area were analyzed and studied. Results indicated that the riverbed gradient in the front position of the delta deposition was large, which is 7.2 km away from the dam axis. The flow velocity there was large and sensitive to the flow rate variation. However, the water depth within 7.2 km of the reservoir area was not sensitive to the flow rate, and the sensitivity within 7.2 km–22.0 km of the reservoir area increased with the flow rate. Additionally, when the flow rate increased, the maximum diameter of the initial sediment of movement was larger. And, when the inflow rate reached 1200 m^3/s, the sediment with a particle diameter of 0.015 mm– 0.094 mm can be started and discharged from the reservoir area. When the inflow rate reached 1800 m^3/s, the particle diameter range of the incipient sediment was 0.014 mm–0.094 mm. When the inflow rate reached 2500 m^3/s, the range was 0.010 mm–0.094 mm. When the inflow rate increased from 800 m^3/s to 1200 m^3/s, the proportion of the particle diameter of the incipient sediment increased the most, and the flushing effect was significantly enhanced. When the flow rate increased from

1800 m^3/s to 2500 m^3/s, the sediment flushing effect was limited. In summary, considering the flushing effect and downstream scouring, the appropriate flow rate is 1200 m^3/s–1800 m^3/s.

References

Fu WB (2016) The study on visualization of reservoir erosion and deposition of the three Gorges reservoir. D. Chongqing Jiaotong University, Chongqing

Lu XX, Siew RY (2006) Water discharge and sediment flux changes over the past decades in the Lower Mekong River: possible impacts of the Chinese dams. J Hydrol Earth Syst Sci 10 (2):181–195

Maria TC, Daniel M, Pablo P (2015) Potential accumulation of contaminated sediments in a reservoir of a high-Andean watershed: Morphodynamic connections with geochemical processes. J Water Resour Res 51(5):3181–3192

Medeiros IC, Costa SJ, Santos CAG (2019) Run off erosion modelling and water balance in the Epitácio Pessoa Dam river basin, Paraíba State in Brazil. J. Int J Environ Sci Technol 16 (7):3035–3048

Tu QH (1980) Calculation method of sediment erosion and sedimentation in large reservoirs. R. Yellow River Conservancy Commission of the Ministry of Water Resources, Zhengzhou

Wei ZL, Zhao LJ, Tan GM, Yu XM (2005) Scouring simulation of river bed downstream of the yellow river in the early operation period of Xiaolangdi reservoir. J Adv Eng Sci 42(2):10–15

Zhang QS, Zhang ZQ (1982) Calculation of reservoir scouring and sedimentation form and its process. J Sediment Res 7(1):1–13

Zhang JL, Liu YY, Lian JJ (2004) Calculation of sedimentation and scour of Sanmenxia reservoir in flood season by the fuzzy neural network. J Hydroelectr Eng 02:39–43

Zhang CP, Qu SJ, Shang HX, Chen Z (2018) Study on the characteristics of sediment release of hyper-concentrated flood in Sanmenxia reservoir. J North China Univ Water Resour Electr Power (Nat Sci Ed) 39(5):7–10

Research Progress on Fish Barrier Measures

Kaixiao Chen[1], Xiuyun Guo[2], Xiaogang Wang[1(✉)], Yun Li[1],
and Long Zhu[1]

[1] State Key Laboratory of Hydrology-Water Resources and Hydraulic
Engineering, Nanjing Hydraulic Research Institute, Nanjing, China
xgwang@nhri.cn
[2] Laluo Water Resources Hub and Irrigation District Administration of Tibet
Autonomous Region, Tibet, China

Abstract. Fishway is an artificial flume for fish migration through sluice gates or dams, which is widely used worldwide as an effective means to restore ecological connectivity of rivers. However, since the inlet of fishway is quite narrow compared to the width of the river, it is difficult for fish to find the inlet of fishway, and fish are easily attracted by the relatively high speed of water flow generated by turbines and cross the dam from turbines, resulting in a large number of fish casualties, so fish barrier measures play an important role in preventing fish from entering turbines, intakes and guiding fish into fishway inlets. This paper collects the main representative types of physical barrier and behavioral barrier, summarizes their advantages, disadvantages and application conditions, and analyzes their applications. It is found that most of the existing fish barrier measures are applicable to the downstream passage, while the upstream passage mainly attracts fish by setting up water auxiliary water systems at the inlet, and a few electric fence and acoustic system are also applied at present; in the near future, the hybrid fish barrier measures with physical barriers as the main body and behavioral barriers as the auxiliary will be the mainstream because of the more stable performance of physical barriers; behavioral barrier have the advantages of saving space, not affected by debris in water, and not causing head loss, but they are still in the experimental stage due to the unstable effect of fish interception, easy to be affected by water body or other external conditions, and high technical requirements. The efficient and stable behavioral barriers have certain research prospects.

Keywords: Fish barrier measures · Fishway · Physical barrier · Behavioral barrier

Y. Li et al. (Eds.): PIANC 2022, LNCE 264, pp. 1195–1208, 2023.
https://doi.org/10.1007/978-981-19-6138-0_105

1 Introduction

The construction of dams on rivers to make full use of water resources has brought great benefits, but also blocked the connectivity of rivers and led to the fragmentation of river water ecosystems. In order to compensate for the ecological impact on fish, the first fishway was built at the end of the 19th century to allow fish to cross the barrier and thrive, followed by various fishway arrangements such as Daniel fishway, vertical slot fishway and submerged orifice fishway (Katopodis and Williams 2012). According to the Hydraulic Design Manual (2nd Edition) (GIWP 2014), fish passage widths worldwide range from 1 m to 10 m, while river widths are often several hundred meters or even thousands of meters, for example, the narrowest point of the Yangtze River is 1100 m, so fishway projects are also "needle's eye". At the same time, considering that fishway is generally set next to the hydropower station, the fish are easily attracted by relatively high-speed water flow and cross the dam from turbines, causing fish death or injury (Silva et al. 2018); in some river areas, irrigation is the main water consumption project, and its water intake often causes a large number of fish entrainment and casualties.

Therefore, fish barrier measure has been more widely used as an effective artificial intervention, and according to the summary analysis of (Fisheries 2011), they can prevent fish from entering irrigation ditches, power plant tailwaters, streams with sudden changes in flow, turbine tailpipes, rivers with poor spawning gravels, poor water quality or insufficient water quantity, and guide them to fishway, bypass, etc. According to the characteristics of fish barrier measure, it can be divided into two categories: physical barrier and behavioral barrier, which are mainly represented by fish screen, porous dike, barrier net and guide wall. The interception of fish is mainly achieved through small apertures (spacing) or large heights; behavioral barriers mainly stimulate fish through flow, sound, light, electricity and biochemical means and cause avoidance effects (Noatch and Suski 2012). This paper will introduce the fish barrier measures from the aspects of the function principle, applicable conditions and application, and analyze the development trend of fish barrier measure.

2 Physical Barrier

2.1 Fish Screen

The role of the screen for fish mainly includes two parts: blocking and guiding. All fish screens usually have blocking function due to their relatively dense pores, while the guiding function of fish screen is mainly reflected in the fact that some of them can be arranged at a certain angle with the direction of water flow, which can guide fish into the bypass or other safe waters by generating sweeping flow.

2.1.1 Flat Plate Screen

A flat plate screen consists of a series of flat mesh grids that are fixed between support beams or rails and placed at an angle to the direction of the water flow. The water flow will pass through the grids, while fish and debris will be directed to the bypass. To minimize maintenance requirements and maintain efficient operation, any fish retention facility must include effective cleaning measures. For small grids and low debris loads, cleaning measures may require only manually operated rakes, brushes or scrapers. For larger systems, mechanically driven rakes, brushes, or scrapers may be required. Due to their excellent fish retention performance, low operating costs, reasonable cleaning measures and low water depth requirements, flat plate screens are now widely used in small and large irrigation diversions in Washington, Oregon and California, USA, where complete fish retention is required (USBR 2006). The main disadvantages are the accumulated cleaning and maintenance costs over time and the tendency for debris to enter the bypass along with fish and cause clogging.

2.1.2   Drum Screen

Drum screen consists of cylindrical frames covered with mesh (usually woven wire) that are placed at an angle to the direction of flow with the cylindrical axis placed horizontally. The installation can consist of individual screen at smaller diversion sites or, in wider river basins, a series of drum screens placed end-to-end.

The installed screen rotates slowly about its horizontal axis. As it rotates, the front surface of the drum rotates upward and out of water, while the rear surface rotates downward. The rotation picks up any debris on the drum and it is washed off at the back as the water passes through the screen. In order to provide sufficient screen area and optimize self-cleaning capability, drum screens must operate at 65% to 85% submersion and require high water level stability. Most commonly installed in rivers, drum screens are placed in the water at an angle along the axis to generate sweeping velocities to guide fish into the bypass, but are more costly than flat mesh screens due to their relatively complex construction.

2.1.3   Traveling Screen

Traveling screen is mainly composed of screen panels, ring chains, water injection systems and other structures, and are generally placed upstream of the intake to prevent debris and fish from entering the water. Traveling screen is usually placed parallel to the fluid or at a shallow angle, similar in principle to the two previous screens, providing guidance for fish by generating a good sweeping flow along the surface of the screen, thus reducing fish impingement. As the screen rotates out of the water, debris and fish impinging on the screen are removed by a high-pressure water spray generated by a water injection system to function as an autonomous cleaning screen.

(Black and Perry 2014) conducted more than 100 replicate tests on more than 13,000 fish at impact velocities of 0.3, 0.6 and 0.9 m/s and found that fish survival and interception rates exceeded 95%, demonstrating the efficiency and eco-friendliness of traveling screen. The effectiveness of traveling screen depends on many specific factors, such as fish size, flow rate, location and escape routes (Taft 2000). In addition, the overall cost is relatively high due to the complex structure and the high maintenance costs of the bearings, screen panels, conveyors and other structures for small and medium-sized waters, which require research trials to determine the solution in conjunction with specific detailed conditions when used.

2.1.4   Cylindrical Screen

Components of a cylindrical screen typically include a wire screen with a V-shaped or wedge-shaped cross-section having an internal baffle concept to produce a uniform velocity distribution, a water differential measurement system, and a cleaning system. Brushes on the outside or inside of the cylinder are used to clean debris from the screen surface. It is usually placed on a canal or riverbank track and acts primarily at the intake end of pumping or gravity diversion pipelines for irrigation, processing, cooling, and small hydroelectric supply. Current applications have shown that screens are biologically efficient in preventing entrainment and impingement of large fish and do not cause unusual maintenance problems (Veneziale 1992). However, as with any screening technology, the potential for clogging and biofouling is an issue that needs to be addressed in the design and operation of this technology (Smith and Ferguson 1979).

2.1.5   Inclined Screen

Inclined screen is a fence placed on the inverted slope of the channel, with the screen at an angle to the flow and submerged in the water. In addition to being arranged in the river, it can also be placed near the tailpipe to prevent fish from entering dangerous waters such as turbines. When the flow with fish and debris sweeps over the surface, the sweeping speed is generated along the direction of the surface due to the slope, while the water depth gradually decreases. The sweeping velocity will guide the fish and debris across the surface into the bypass, while the water flows down through the surface into the diversion pipe. The positive slope design tends to allow debris and branches to clog the screen, requiring regular cleaning; the negative slope design needs to ensure that the bypass has sufficient water depth for the target fish to migrate. Currently, some projects are installed with inclined screen in a removable support frame so that the downstream end of the frame can be raised and lowered to follow or adjust for changes in water surface elevation.

2.1.6   Eicher Screen

Eicher screen is a smooth-surfaced oval metal screen placed on a frame and angled toward the bypass, which can be cleaned by rotating and backflushing (EPRI 2013). The screen was designed in the late 1970s by biologist George Eicher to develop a better facility for fish to safely bypass turbines (Eicher 1982). The elliptical design of the fish screen allows it to be installed at an angle in a pressure pipe and to operate at flow rates of up to 2.4 m/s. The first Eicher screen was installed at the Sullivan Hydroelectric Plant in the United States in 1980, and in 1990, the American Electric Power Research Institute (Winchell 1992) conducted a two-year evaluation of the Eicher screen installed at the Elwha Hydroelectric Project in Washington State. The results showed that the survival rate of target fish, including juveniles, exceeded 98.7%.

2.1.7   Modular Inclined Screen

Modular inclined screen consists of an inlet with an bar rack, a stop log, an inclined screen at a small angle to the flow and bypass (EPRI 2005). Developed by EPRI in the 1990's, the fish screen was designed to accommodate any type of intake and was designed to operate at relatively high water velocities ranging from 0.6 m/s to 3 m/s. (EPRI 1994) conducted biological tests on the guidance efficiency of modular inclined screens at high water velocities range, and the results showed that the guiding efficiency was higher than 98% for most of the target fishes, and the salmon guiding efficiency was always 100%; based on the laboratory test results, (EPRI 1996) conducted a prototype test of this fish screen in the Green Island Hydroelectric Project to evaluate the effectiveness of fish guiding in the field, and the results showed similar to the laboratory results, with the guiding efficiency close to 100%.

2.1.8 Porous Dike

Porous dike is a short dike structure made of gravel with certain voids, which can prevent fish from entering power plant intakes and causing economic and ecological losses. (Anglin et al. 2012), the Wisconsin Electric Power Company built a porous dike around the cooling basin of the Port Washington Power Station to prevent the entry of fish, and the results showed that it performed beyond expectations, not only successfully limiting fish entrainment during intake withdrawal, but also reducing the downtime of the plant due to algae problems. Compared to fish screen, porous dike has the advantage of being structurally simple, stable and reliable, and the aperture size does not have to be smaller than the smallest local fish size, which are often afraid to pass through due to their dense and massive structural characteristics (Taft 2000).

However, the results of some experimental studies have shown that some fish can be trapped in the pore space or entrained in the pumping flow, algae and shellfish can clog the pore space and shellfish can grow in it, and when the water passes through the pore space, head loss occurs due to viscous dissipation, which is not conducive to the use of water energy. In view of these unfavorable aspects, its application is not widespread and needs to be verified in targeted tests before use.

2.1.9 Barrier Net

Barrier net is mainly composed of fixed anchors, net surface, support piles and floats. Data from the study results indicate that the fish interceptor nets were successful in reducing the number of fish entering the inlet (Michaud and Taft 2000; Reider et al. 1997). Although barrier net is more economical than other measures, they have problems such as the tendency of the mesh to be clogged and cause overall sinking, the difficulty of cleaning the net after algae accumulation, poor stability at high flow velocities, and ice clogging of the net in winter; therefore, the fish interception efficiency is higher when the flow velocity is low (less than 0.3 m/s), the temperature is suitable, and the debris load in the water is light (Taft 2000).

2.2 Guide Wall

Guide wall consists of a series of partial-depth panels or fish fences anchored to a river, reservoir or canal. These structures are designed for fish that are accustomed to surface migration in the water column (e.g., salmon). The concept of guide wall originated when dam operators observed fish aggregating along gravel fences, similar to the fences of guide walls. The angle between the guide wall and the incoming flow direction is generally less than 45° to ensure that the flow velocity along the guide wall direction, pointing to bypass, is greater than the vertical flow velocity, but because it is not a full-depth design, there is a large vertical flow velocity, which is prone to turbulence and affects the fish blocking effect. The influence can be eliminated by keeping the included Angle with incoming flow less than 15° and increasing the guide wall depth (Liedtke et al. 2009; Mulligan et al. 2018). The device is still in the experimental stage and is currently installed at Bonneville Dam, USA (Mulligan et al. 2017).

3   Behavioral Barrier

Behavioral barrier refer to fish barrier measures that produce stimuli to target fish through flow fields, sound, light, electric fields, bubble curtains, and biochemistry to provide a repellent or attraction effect. The use of behavioral barriers typically provides lower capital and operating costs, but because target fish are often diverse and lack adequate behavioral studies, most behavioral barriers are less effective than physical barriers to partially reduce fish entrainment.

3.1   Angled Bar Rack and Louver

Angled bar rack and louver consist of a series of vertical fence-like slats placed on a rack that span the channel in a diagonal pattern and end at the inlet of the fish bypass. The main difference between louver and angled bar rack is that the louver slats are oriented 90° to the flow while angled rack slats are angled 90° to the rack frame and their orientation to the flow will be dependent upon the angle of the entire rack structure (Shepherd et al. 2007). As fish approach the slats, turbulence will drive them to swim in the direction of the rack and thus into the bypass channel. Studies have shown that its efficiency in guiding fish reaches about 80–95%.

Since the angled bar rack and louver rely heavily on fish response to turbulent flow, its efficiency may be relatively low and unstable when river flow conditions change or when non-target fish are encountered, and it may not be effective in blocking juvenile fish when the slat spacing is too large, while too small spacing is likely to cause injury to weak fish that hit the fence.

3.2   Velocity Barrier

When the flow rate is within the range of fish's favorite flow rate, it can attract fish, and when the flow rate is too large and exceeds the limit of fish swimming speed, it can act as a velocity barrier (Noatch and Suski 2012). Currently, a velocity barrier is mainly used in the upstream fish passage to prevent fish from entering dangerous waters by mistake. Velocity barrier consists of a combination of concrete a weir and an apron, which prevent upstream fish from passing through by creating a smaller depth of water and a higher flow velocity on the apron, while fish cannot jump the height of the weir.

The advantage of the velocity barrier is that it does not have the problem of debris blockage like the others, and the debris and water can flow smoothly downstream through the weir. The disadvantage is that the presence of the weir creates a reservoir upstream, and the backwater effect may cause sediment accumulation upstream, resulting in loss of production and life, and a sufficient head difference is required to maintain good fish blockage. It is currently used at the Coleman National Fish Hatchery and the Vitale Tailrace in the United States.

3.3   Sound

Similar to positive and negative phototropism, fish also have positive and negative phonotropism. It is generally believed that fish are positively phonotropic to the swimming and feeding sounds of their baits and counterparts, and to the courtship sounds of the opposite sex, while they are negatively phonotropic to the negative and escape sounds of their counterparts, the feeding and swimming sounds of predators, and the abnormal sounds from fishing boats and fishing gear (Zhu 2007).

Sound decays slowly in the water, is highly directional, and is not affected by low light levels or water turbidity; light and electricity signals are less comprehensive than sound in these aspects, so sound fish guidance and attracting are more widely used and studied. (Murchy et al. 2017)'s experiments on the dispersal effect of boat engine sound on fish resulted in a combined deterrence efficiency of 90.5% for the target fish, indicating the feasibility of sound barriers.

Although sound has some advantages, it also has limitations and its effectiveness may be influenced by bottom morphology, hydrology and acoustic angle. (Maes et al. 2004) conducted field tests on the deterrence efficiency of infrasound fish barrier systems in power stations and found that their deterrence efficiency was only 60%, proving that they are not suitable to be used as a stand-alone fish barrier measure. Although infrasound has been an effective fish barrier measure in some studies, it does not propagate well in shallow water and hard substrates (Turnpenny and Nedwell 1994). Therefore, it should be used in combination with local specificities and multiple means.

3.4   Strobe Light

Fish are phototropic and produce directional movements after observing light, which can be classified as positive or negative phototropism depending on whether they are approaching or fleeing from the light source (Luo 1980). Using the negative phototropism of fish, they can be repelled to achieve the purpose of fish interception.

In the dark, strobe lights from power plants significantly alter light conditions in dangerous sites such as turbines, stimulating fish to take advantage of their negative phototropism and prompting them to seek bypass in the dark. Existing studies show that strobe lights are effective in stopping fish (Johnson et al. 2005; Noatch and Suski 2012), and fish do not become accustomed to continuous flash exposure, so strobes are effective in the long term, but studies to date have not demonstrated their full reliability as a single measure (Hamel et al. 2008).

In addition, taking advantage of the positive phototropism of fish to certain specific light, fish can be attracted to the inlet of the fish passage, thus increasing the efficiency of operation. Light attracting was first applied in fisheries, where early fishermen used vapor lamps to attract squid, and then gradually applied to fishway attracting systems.

For example, (Lin et al. 2019) evaluated the effect of different light colors on attracting fish into fish passage fish collection systems and found that warm white light had an important role in attracting fish. The study showed that since fish are selective to light (Zhang et al. 2019), a comprehensive and specific study of the target fish should be considered when making the design arrangement, otherwise the desired effect cannot be achieved, so a combination of measures should be used to improve the fish attraction effect.

3.5   Electric Screen

Electric screen is a suspended structure consisting of a main cable, sling, horizontal cable, electrodes and struts, and anchor piers that generate an electric field in the water (Wang et al. 2013).

Electric screen has been innovative in its delivery method, and the pulsed electric screen is about 1/40th of the power of the DC screen, while its health effects on fish are minimal and it has some prospects for application (Dong 2007). Pulsed electric screen has been used in China on a certain scale, mainly to prevent fish escape from reservoirs, and was tested with good results (Zhao et al. 2000).

Electric screen also has some limitations, for example, its cross-sectional flow velocity should be controlled below 0.7 m/s, if the flow velocity is too large, it is easy to increase the possibility of fish passage, and under bad weather conditions, fish riot may occur and "sudden change of fish escape", and there are limited action area, need for stable power supply, and the effect on small fish.

3.6   Bubble Curtain

Bubble curtain mainly produces bubbles by discharging compressed air in a large amount through holes in the tube. If the fish are far away from the bubble curtain, the fish tend to swim towards the bubble curtain for the purpose of luring (Qiao et al. 2011); however, when the fish are close to the bubble curtain, the "bubble curtain wall" has a visual effect on the fish, an auditory effect when the bubbles are strongly mixed with air and broken, and the bubbles cause a change in water pressure and mechanical pressure vibration (Mao 1985).

The efficiency of bubble curtain as a fish barrier measure is relatively low. Its practical application in power facilities was not effective, and since then, the bubble curtain in these power stations has been removed (Hocutt 1980); (Liu and He 1988) found through experiments that freshwater fish had obvious adaptation to the bubble curtain, and that the bubble curtain had a better deterrent effect in a short time and a poorer effect in a long time. Therefore, the bubble curtain can not fully achieve the effect of fish interception, and comprehensive measures are needed to improve efficiency.

3.7   Other Means

In addition to the above, water quality barriers, pheromones and magnetic fields are also viable experimental fish barrier measures.

Water quality barrier is a fish barrier formed by increasing the concentration of carbon dioxide in the ambient water, thereby preventing fish from breathing. (Theresa et al. 2008) suggest that the downstream migration of invasive fish can be stopped by discontinuing supplemental aeration in the river to create a hypoxic zone.

Pheromone is widely defined as a secreted chemical odor that elicits specific behavioral responses in fish (Sorensen and Stacey 2004). Pheromone plays an important role in reproduction and predator avoidance for fish. Attractive pheromone is usually secreted by sexually receptive individuals to attract the opposite sex. In contrast, alarm pheromone, is often released when the fish's skin is damaged and can trigger an escape response in conspecifics. Alarm pheromones can be used to exclude fish from specific locations, while attraction pheromones can be used to gather fish away from sources of danger. Study shows that chemicals released by decaying eels trigger avoidance behavior in their own kind (Wagner et al. 2011).

Some fishes are endowed with electro-sensory organs that are designed to provide fish with an advantage in locating prey. However, strong magnetic fields may over-stimulate these receptors, achieving deterrent or repulsive effects that may help guide the fish away from undesired locations. For example, a study by (O'Connell et al. 2010) showed that decoy areas containing magnets had a repulsive effect on fish.

All of the above measures are in the experimental stage, with limited application, and their efficiency and stability of fish interception need to be further studied.

4 Conclusion

Through the statistics in Table 1, it is easy to find that the physical barrier has the advantages of high efficiency and good stability, but also has the problems that it is mostly applicable to the waters above the dam, easy to encounter debris clogging and high maintenance cost; while the behavioral barrier has the advantages of space saving, no debris clogging, no head loss, etc., but the effect of fish interception is unstable, easily affected by the water body or other external conditions, and the research results are not yet mature.

Most of the fish barrier measures are currently applied in the upstream of the dam, but the downstream of the dam also has the possibility of fish entering the irrigation canal, tailwater canal, or non-upstream river channel and other dangerous waters, which need efficient fish barrier facilities to guide them into the upstream fish passage.

Although fish barrier measures are used in China, there are few data on their effectiveness and fish survival rate, and there is a lack of sufficient data for reference compared with other countries. In addition, there are few types of fish barrier measures applied in China, and there is a certain prospect of developing efficient fish barrier facilities that are specific to the actual situation in China.

Table 1. Overview of the main fish barrier measures

Name	Suitable situation	Range of flows	Advantages	Disadvantages	References
Flat plate screens	Allowing panels to be removed for cleaning	All	High optimum diversion efficiency, simple structure, low water depth requirements	High maintenance cost, debris blocking the bypass, large space occupation	(USBR 2006)
Drum screens	Stable elevations	0.1–30 m^3/s	High optimum diversion efficiency, good eco-friendliness and self-cleaning ability	Complex structure, high water depth requirements, needs continuous operation	(Neitzel et al. 1990)
Traveling screens	Stable flow conditions and low flow rates	0.1–2 m^3/s	Low workload, high optimum diversion efficiency, and self-cleaning ability	High acquisition costs, complex mechanics, high maintenance costs	(Black and Perry 2014; Taft 2000)
Cylindrical screens	Low velocity, pumped water diversion	0.1–5 m^3/s	No need for fish bypass, easy access for maintenance, low water depth requirements	No guidance function, clogging potential	(Smith and Ferguson 1979)
Inclined screens	Gravity flow diversion	0.1–20 m^3/s	High optimum diversion efficiency, simple structure, low costs	Clogging potential	(Harbicht et al. 2022)
Eicher screens	Hydro channels and penstocks	All	Space saving; low installation and running costs; fish friendly; unaffected by water level	Experimental phase	(EPRI 2013; Winchell 1992)
MIS	Hydro channels and penstocks	All	Space saving; low installation and running costs; fish friendly; unaffected by water level	Experimental phase	(EPRI 1994, 1996, 2005)
Porous dike	Stable flow conditions	All	Simple structure, stable and reliable	Easy to block and cause stagnant water	(Anglin et al. 2012)

(*continued*)

Table 1. (*continued*)

Name	Suitable situation	Range of flows	Advantages	Disadvantages	References
Barrier net	Suitable temperature and light debris load in water	<0.3 m/s	Simple and economical construction	Easy to clog, only suitable for slow flow and normal temperature environment	(Michaud and Taft 2000; Reider et al. 1997)
Guide wall	Suitable for surface fish	<1.2 m/s	Partial-depth, less influenced by water level	Juvenile can easily pass through the bottom and not achieve the effect of blocking fish	(Liedtke et al. 2009; Mulligan et al. 2017; Mulligan et al. 2018)
Angled bar rack and louver	Stable flow conditions	<0.9 m/s	Simple structure, low cost, and can pass through smaller debris	High bars spacing and maintenance requirement, relatively low efficiency	(Albayrak et al. 2018)
Velocity barrier	Stable flow conditions	All	Low costs, no risk of clogging and the ability to block fish that are weaker than target fish	Backwater effect, head difference requirement	(Noatch and Suski 2012)
Sound	Sufficient water depth, soft substrate	All	Slow decay in water; highly directional; unaffected by light levels or water turbidity	Not propagate well in shallow water and hard substrates, and should not be used alone	(Maes et al. 2004)
Strobe light	Dark environment, clear water	All	Long-term effectiveness; specific lights have a fish-attracting effect	General efficiency, different fish are selective to light types	(Noatch and Suski 2012)
Electric screen	Stable flow conditions and weather	<0.7 m/s	High fish blocking efficiency	Poor stability; requires stable power supply; not as effective on juveniles as adults	(Zhao et al. 2000)
Bubble curtain	Stable flow conditions	All	Better results in a short period of time	Not effective in blocking fish when used alone	(Hocutt 1980; Liu and He 1988)

(*continued*)

Table 1. (*continued*)

Name	Suitable situation	Range of flows	Advantages	Disadvantages	References
Other means	Stable flow conditions	All	Relatively new, with certain application prospects	Experimental phase	(O'Connell et al. 2010; Wagner et al. 2011)

Although some of the fish barrier measures can reach 100% efficiency for some specific fish in the test, considering the diversity of target fish and the variability of hydraulic conditions in the actual application, it is difficult to achieve efficient operation by using a certain fish barrier measure alone, therefore, a combination of hybrid fish barrier measures can be considered, for example, the fish trapping and replenishment spraying system combines the use of auxiliary water systems and acoustic fish trapping means, the bioacoustic fish screen uses a combination of strobe lights, sound barriers and bubble curtains.

Acknowledgements. This study is supported by Specific Funds for Basic Scientific Research of Central Public Scientific Institutes (Grant No. Y120009) and Monitoring and Evaluation of Fish Passage Operation Effectiveness of Tibetan Laluo Water Conservancy Hub and Supporting Irrigation District Project (XZLL-GC2021FW-002).

References

Albayrak I, Kriewitz CR, Hager WH, Boes RM (2018) An experimental investigation on louvres and angled bar racks. J Hydraul Res 56(1):59–75

Anglin D, MacIntosh K, Ryan L, Haubert P (2012) Modelling and design of a porous dike cooling water intake structure. In: Coastal structures 2011. World Scientific, pp 1210–1221

Black JL, Perry ES (2014) Laboratory evaluation of the survival of fish impinged on modified traveling water screens. North Am J Fish Manag 34(2):359–372

Dong Y (2007) Brief introduction to the application of pulse current in electric grid to block fish. Sci Technol China Rural Prosper (06):83

Eicher GJ (1982) A passive fish screen for hydroelectric turbines. Paper presented at the applying research to hydraulic practice

EPRI (1994) Biological evaluation of a modular inclined screen for diverting fish at water intakes. Palo Alto. https://www.epri.com/research/products/TR-104121

EPRI (1996) Evaluation of the Modular Inclined Screen (MIS) at the green island hydroelectric project. Palo Alto. https://www.epri.com/research/products/TR-106498

EPRI (2005) Chapter 11: Modular inclined screens, fish passage manual chapter updates. Palo Alto. https://www.epri.com/research/products/000000000001011428

EPRI (2013) Fish protection at cooling water intake structures: a technical reference manual–2012 update. Palo Alto. https://www.epri.com/research/products/000000003002000231

Fisheries N (2011) Anadromous Salmonid passage facility design. https://repository.library.noaa.gov/view/noaa/23894

GIWP (2014) Hydraulic design manual, 2nd edn. China WaterPower Press, Beijing

Hamel MJ, Brown ML, Chipps SR (2008) Behavioral responses of rainbow smelt to in situ strobe lights. North Am J Fish Manag 28(2):394–401

Harbicht AB et al (2022) Guiding migrating salmonid smolts: experimentally assessing the performance of angled and inclined screens with varying gap widths. Ecol Eng 174:106438

Hocutt CH (1980) Behavioral barriers and guidance systems. Power Plants 183–205

Johnson PN, Bouchard K, Goetz FA (2005) Effectiveness of strobe lights for reducing juvenile salmonid entrainment into a navigation lock. North Am J Fish Manag 25(2):491–501

Katopodis C, Williams JG (2012) The development of fish passage research in a historical context. Ecol Eng 48:8–18

Liedtke TL, Kock TJ, Ekstrom BK, Rondorf DW (2009) Behavior and passage of juvenile salmonids during the evaluation of a fish screen at Cowlitz Falls Dam, Washington, 2008. http://pubs.er.usgs.gov/publication/70182098

Lin C et al (2019) An experimental study on fish attraction using a fish barge model. Fish Res 210:181–188

Liu L, He D (1988) Response of five freshwater fishes to fixed bubble curtain. J Xiamen Univ (Nat Sci) (02):214–219

Luo H (1980) Phototactic behavior of fishes under light irradiation. Mar Fish (04):23–24+13

Maes J, Turnpenny AWH, Lambert DR, Nedwell JR, Parmentier A, Ollevier F (2004) Field evaluation of a sound system to reduce estuarine fish intake rates at a power plant cooling water inlet. J Fish Biol 64(4):938–946

Mao S (1985) Fish movement and fishing techniques. Ocean Press, Beijing

Michaud DT, Taft EP (2000) Recent evaluations of physical and behavioral barriers for reducing fish entrainment at hydroelectric plants in the upper Midwest. Environ Sci Policy 3:499–512

Mulligan KB, Towler B, Haro A, Ahlfeld D (2017) A computational fluid dynamics modeling study of guide walls for downstream fish passage. Ecol Eng 99:324–332

Mulligan KB, Towler B, Haro A, Ahlfeld DP (2018) Downstream fish passage guide walls: a hydraulic scale model analysis. Ecol Eng 115:122–138

Murchy KA et al (2017) Potential implications of acoustic stimuli as a non-physical barrier to silver carp and bighead carp. Fish Manag Ecol 24(3):208–216

Neitzel DA, Abernethy CS, Clune TJ (1990) Evaluation of rotary drum screens used to protect juvenile salmonids in the Yakima River Basin, Washington, USA. Pacific Northwest Lab., Richland, WA (USA)

Noatch MR, Suski CD (2012) Non-physical barriers to deter fish movements. Environ Rev 20 (1):71–82

O'Connell CP, Abel DC, Rice PH, Stroud EM, Simuro NC (2010) Responses of the southern stingray (Dasyatis Americana) and the nurse shark (Ginglymostoma cirratum) to permanent magnets. Mar Freshw Behav Physiol 43(1):63–73

Qiao Y, Huang H, Huang M, Chen S, Yin L (2011) Application of bubble curtains in fish behavior research. Fish Inf Strategy 26(12):29–32

Reider R, Johnson D, Latvaitis PB, Gulvas J, Guilfoos E (1997) Operation and maintenance of the Ludington Pumped Storage Project barrier net. Paper presented at the fish passage workshop, Milwaukee, Wisconsin

Shepherd D, Katopodis C, Rajaratnam N (2007) An experimental study of louvers for fish diversion. Can J Civ Eng 34(6):770–776

Silva AT et al (2018) The future of fish passage science, engineering, and practice. Fish Fish 19 (2):340–362

Smith L, Ferguson D (1979) Cleaning and clogging tests of passive intake screens in the Sacramento River, California. Paper presented at the proceedings of the passive intake screen workshop, Chicago

Sorensen PW, Stacey NE (2004) Brief review of fish pheromones and discussion of their possible uses in the control of non-indigenous teleost fishes. NZ J Mar Freshwat Res 38(3):399–417

Taft EP (2000) Fish protection technologies: a status report. Environ Sci Policy 3:349–359

Theresa MS, Verdel KD, Wendi L (2008) Effectiveness of piscicides for controlling round gobies (Neogobius melanostomus). J Great Lakes Res 34(2):253–264

Turnpenny AW, Nedwell JR (1994) The effects on marine fish, diving mammals and birds of underwater sound generated by seismic surveys: consultancy report

USBR (2006) Fish protection at water diversions–a guide for planning and designing fish exclusion facilities, Denver, Colorado

Veneziale EJ (1992) Fish protection with wedge wire screens at Eddystone Station, United States

Wagner CM, Stroud EM, Meckley TD (2011) A deathly odor suggests a new sustainable tool for controlling a costly invasive species. Can J Fish Aquat Sci 68(7):1157–1160

Wang Y, Chen X, Zhang T, Zhu S (2013) Study on layout scheme of fishway entrance. J Hydroecol 34(04):30–34

Winchell FC (1992). Evaluation of the Eicher Screen at Elwha Dam: 1990 and 1991 test results. Final report. United States. https://www.osti.gov/servlets/purl/10117448

Zhang N et al (2019) The effect of water flow on the phototaxis of juvenile grass carp. Acta Hydrobiol Sin 43(06):1253–1261

Zhao S, Nie R, Su Z (2000) The popularization and application of electric fish barrier project in reservoir. Curr Fish 25(4):12

Zhu C (2007) Advances in fish behavioral ecology. J Beijing Fish (01):20–24

River Regime Evolution of the South Channel in the Changjiang River Estuary, China, During Past 50 Years

Huiming Huang[1,2(✉)], Xiantao Huang[1,2], Yuliang Zhu[1,2], and Siqi Li[1,2]

[1] College of Harbour, Coastal and Offshore Engineering, Hohai University, Nanjing, China
hhm@hhu.edu.cn
[2] Key Laboratory of Coastal Disaster and Protection of Ministry of Education, Hohai University, Nanjing, China

Abstract. As the middle reach of the Changjiang River Estuary, China, the South Channel's evolution would directly affected upper and lower reaches. However in recent decades, fluvial water and sediment loads have appeared notable changes, consequently, the latest evolution of the South Channel would adjust either. Hence, topography data of the South Channel in different historical period during past 50 years were adopted, then geomorphology evolution were analyzed in detail. The result showed that river regime of the South Channel had undergone three stages. In the first stage among $1973 \sim 1981$, it deposited $1.0 \sim 9.7$ m in central shoal and eroded $-1.0 \sim -13.6$ m in two side deep channels. In the second stage during $1981 \sim 1994$, the erosion and sedimentation region turned around, wherein the central shoal eroded $-1.0 \sim -9.8$ m and two side channels deposited $1.0 \sim 10.6$ m. In the third stage during $1994 \sim 2016$, the central shoal partially deposited $1.0 \sim 8.1$ m and partially eroded $-1 \sim -6.1$ m, and both side channels eroded $-1 \sim -13.2$ m. Meanwhile, in the inlet cross section during $1973 \sim 2016$, location of main channel moved 275 m southward with average elevation changing -1.1 m and the central shoal eroded about -4.6 m. In the middle and outlet cross sections, their central shoal eroded $-1.3 \sim -5.8$ m and main channel respectively swung 450 m northward with elevation decreasing 0.4 m and 1775 m northward with elevation deepening 6.1 m. Moreover, average elevation of thalwegs changed from -14.0 m to -15.9 m during $1973 \sim 2016$. Furthermore, evolution characteristics of geomorphology and fluvial water and sediment load in past 50 years showed that there might be close relationship between them.

Keywords: The Changjiang Estuary · The South Channel · Topography evolution · Cross section · Thalweg

© The Author(s) 2023
Y. Li et al. (Eds.): PIANC 2022, LNCE 264, pp. 1209–1219, 2023.
https://doi.org/10.1007/978-981-19-6138-0_106

1   Introduction

The Changjiang River Estuary is the largest estuary in China which locates at north of Shanghai, its evolution has attracted much more attentions in past decades. Meanwhile, water load of the Changjiang River Basin into estuary is also the first one in China, and relevant sediment load is the second one that just only being less than Yellow River, in result, its water and sediment load always play an important role in evolution of the estuary. For many years, due to complexity and importance, evolution of the Changjiang Estuary is still the hot topic of researches till now.

In general, the Changjiang Estuary is divide into three branches. The first branch is the South Branch and North Branch, the second branch is the South Channel and North Channel, and the third branch is the South Passage and North Passage (Fig. 1). So we can see that the South Channel locates at the middle position of the estuary. It directly connects the upper South Branch and the lower South and North Passage. Hence, its evolution will directly impact the upper and lower reaches, especially the lower South and North Passage.

For many years, because of importance of the Changjiang Estuary, lots of researches have focused on its evolution characteristics on hydrodynamics (Wang et al. 2010; Shi and Lu 2011), suspended sediment transport (Shi et al. 1997), bed load movement (Wu et al. 2009), sand bodies variation (Yang et al. 2003), geomorphology (Zhao et al. 2018), engineering influence (Jiang et al. 2012; Chen et al. 2001) and so on (Qiu and Zhu 2013).

However, owing to afforestation after 1980s, the fluvial water and sediment load into the estuary had appeared variation after then. And till 2000s, owing to continuous construction of Three gorges Dam, South to North Water Diversion and many other conservancy projects in the Changjiang River Basin, the fluvial water and sediment load into the estuary have appeared more significant variation again in the latest 20 years. In result, the Changjiang Estuary also has inevitably emergent new evolution tendency (Chu et al. 2013), correspondingly, as an integral part of the estuary, the South Channel' geomorphology also appears new evolution characteristics.

In this background, based on historical underwater topography data in each 10 years during past half century, this paper tries to find out former decades and recent geomorphology evolution tendency and characteristics of the South Channel.

Fig. 1. Location of the South Channel in the Changjiang Estuary

2   Data and Method

In order to discover evolution characteristics of the South Channel in the Changjiang Estuary during the latest half century, underwater topography data in different period during the past 50 years were collected and edited. Their detail periods were respectively in 1973, 1981, 1994, 2007, 2016, and their datum were unified to the Theoretical Lowest Low Water Datum (Table 1).

Then, by the GIS software, their erosion and sedimentation between adjacent years were calculated through subtracting underwater topography of former year from that of later year.

Meanwhile, in order to further reveal regime evolution of the South Channel, thalwegs and cross sections in each period were also extracted and analyzed. Wherein, each thalweg represented the deepest location along the main channel in the South Channel during each period. And cross sections were respectively set at inlet, middle and outlet of the South Channel, and their locations were shown in Fig. 2a as pink line across the channel.

Table 1. Topography data information

Year	Map scale	Depth datum
1973	1:50000	Theoretical lowest low water datum
1981	1:50000	
1994	1:120000	
2007	1:50000	
2016	1:20000	

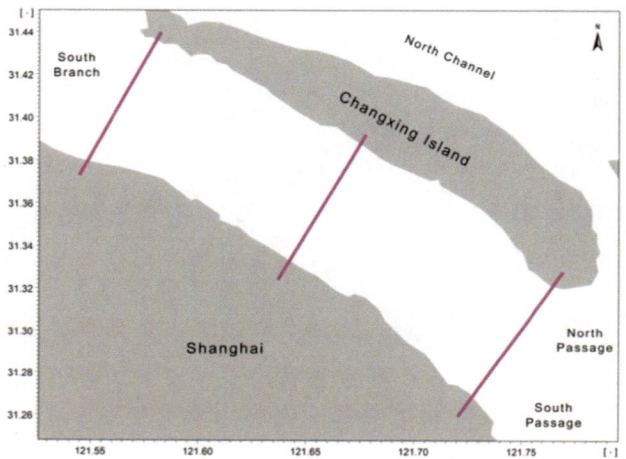

Fig. 2. Location of cross sections in the South Channel

3   Results and Discussion

3.1   Erosion and Sedimentation Characteristics

In order to reveal the evolution tendency of the South Channel, Fig. 3 gave out erosion and sedimentation characteristics of the South Channel during the past 50 years.

As the Fig. 3 shown, there were three stages during 1973~2016 for evolution of the South Channel. In the first stage among 1973~1981, the channel presented obvious sedimentation with value about 1.0~9.7 m, especially in the central region on longitudinal direction along the reach, but manifested erosion with value about −1.0~ −13.6 m along both side channels. Wherein, the erosion pattern ran through the whole main deep channel closed to south bank, but it was broken at inlet in the other channel closed to north bank.

In the second stage during 1981~1994, the erosion and sedimentation region turned around, wherein obvious erosion with value about −1.0~−9.8 m mainly appeared at the central region along the reach, and the maximum erosion mainly located at inlet and lower reach. Meanwhile, along both side channels, it presented notable sedimentation with value about 1.0~10.6 m, especially in the upper reach closed to south bank. But it should be concerned that in this stage the south and north bank all present local scouring

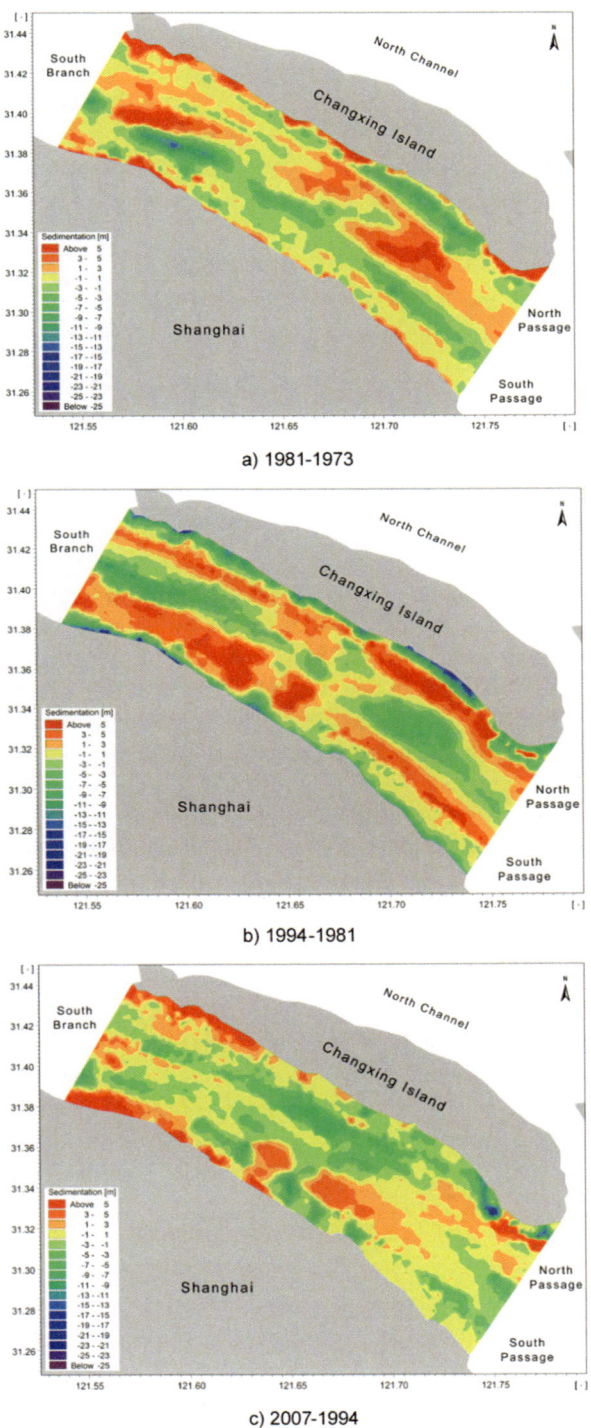

a) 1981-1973

b) 1994-1981

c) 2007-1994

Fig. 3. Erosion and sedimentation between adjacent years

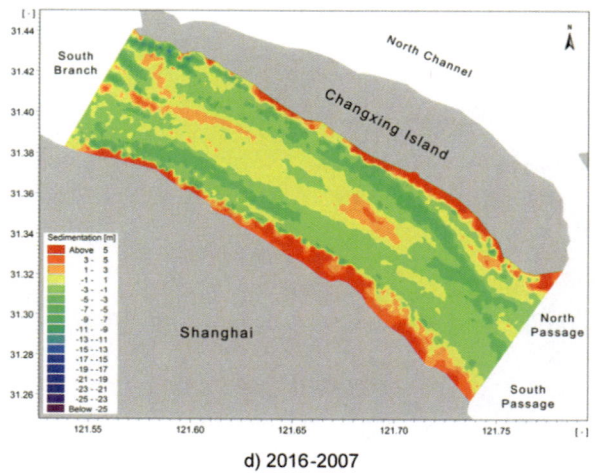

d) 2016-2007

Fig. 3. (*continued*)

along banks with maximum value nearly −15 m, which would threat the embankments on both banks. This phenomenon might be caused by building of ports along both banks, and decreasing of sediment load from upper basin due to afforestation in the Changjiang Basin after 1980s (Wang et al. 2008b).

In the third stage, till initial 2000s, the erosion and sedimentation regions varied again. The central region along the reach presented partial sedimentation with value about 1.0 ∼ 8.1 m, and both side channels presented partial erosion with value about −1 ∼ −11.5 m. In this stage, the sedimentation region was nearly equal to that of erosion, but in the upper and middle reach, the sedimentation played a dominant role. This may be caused by several big floods in 1990s, which transported large amount of sediment into lower reach and estuary and then deposited in the estuary gradually (Yu et al. 2009). After 2010s, following continuous impacts of constructions of the Three Gorges Dam and others water conservancy projects, fluvial sediment load into estuary continuously decreased. In result, the erosion and sedimentation pattern gradually varied in a certain extent but without qualitative change. The erosion in the central region along the reach increased to −1.0 ∼ −6.1 m, and the sedimentation decreased to 1.0 ∼ 4.1 m. Meanwhile, the erosion region along both side channels increased distinctly, wherein the erosion region ran through the whole reach with value about −1 ∼ −13.2 m, especially in the lower reach where the erosion took the dominant function.

3.2   Cross Sections Evolution

Evolution of cross sections in different period were shown in the Fig. 4.

As Fig. 4 shown, it could be found that every cross section had its own variation characteristics and they also undergone three stages for river regime. In the first stage during 1973 ~ 1981, the central shoal mainly presented sedimentation but erosion in both side channels. Then in the second stage during 1981 ~ 1994, the central shoal turned to distinct erosion especially in the upper reach but obvious sedimentation in both side channels. And in the third stage after 1990s, the central shoal still remained erosion trend but both side channels also changed into overall erosion.

In addition, it should be concerned that owing to implementation of afforestation and large scale conservancy projects, fluvial water and sediment load had appeared obvious variation during 1973 ~ 2016 especially the sediment load decreasing obviously after 1980s, hence the evolution of central shoal and both side channels appeared their respective characteristics in past 50 years.

For the cross section at inlet of the South Channel, the whole profile pattern was basically stable in past half century, but there were still many detailed changes. Firstly, in past 5 decades, its location of main deep channel moved about 275 m southward, but its elevation just only changed −1.1 m. Secondly, its central shoal had obvious erosion with average value about −4.6 m in past 50 years, which may be caused by continuous decreasing of sediment load from river basin due to afforestation and construction of large scale water conservancy projects.

And for the middle cross section, during 1973 ~ 2016, the initial pattern with obvious compound channel changed into inconspicuous compound channel pattern, where the central shoal became not very significant than that before. Meanwhile, the main southern channel moved about 450 m northward with elevation decreasing 0.4 m. Accordingly, the central shoal presented firstly sedimentation and then erosion tendency and lastly scoured about −1.3 ~ −5.7 m in the past decades.

Lastly, for the outlet cross section, being similar with middle cross section, its initial typical compound channel pattern in 1973 also changed into inconspicuous compound channel pattern in 2016. And the deepest location of the main southern channel moved about 1775 m northward with elevation deepening about 6.1 m. Meanwhile, the central shoal also presented firstly sedimentation and then erosion trend, in past 50 years, its elevation decreased about −5.8 m on average. Consequently, the central shoal also presented not significant at present.

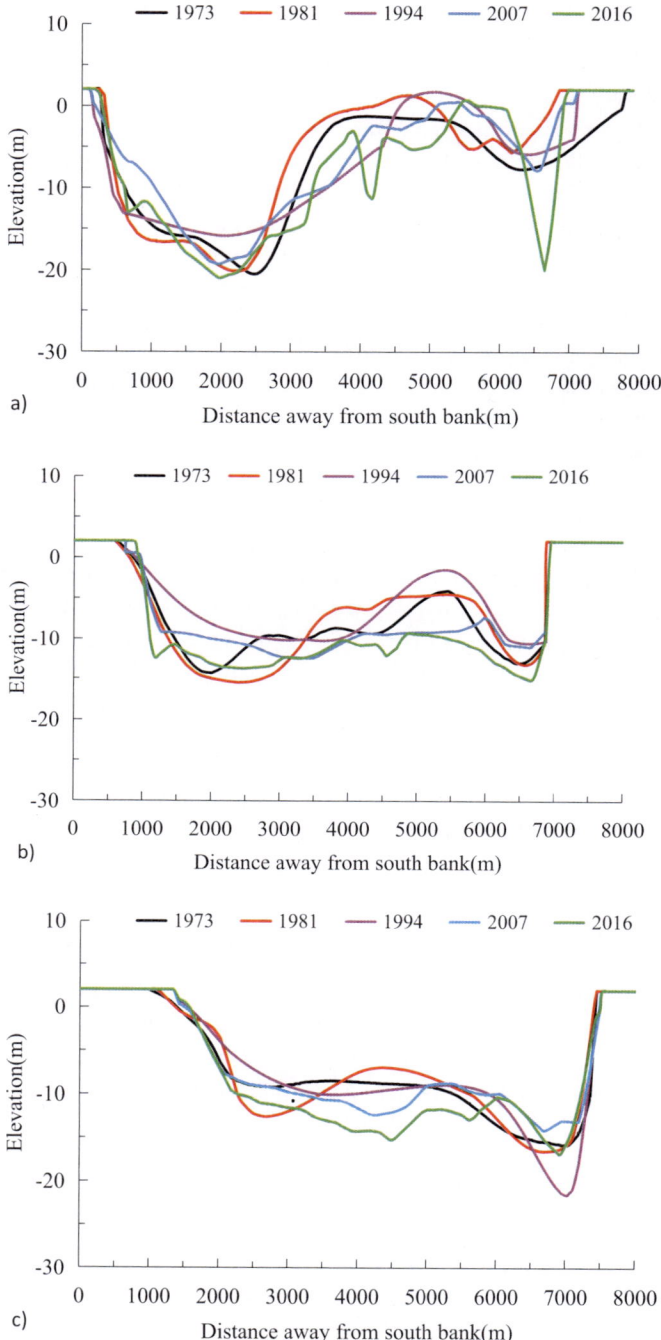

Fig. 4. Variation of cross sections in different years

3.3   Thalweg Evolution of the Main Channel

In order to reveal main southern channel evolution in the South Channel, Fig. 5 presented variation of thalwegs in different period.

As Fig. 5 shown, from 1973 to 2016, average elevation of thalwegs along the whole channel presented an overall fluctuating but slight decreasing trend. During 1970s to 1980s, average elevation of thalwegs decreased with value from −14.0 m to −16.6 m. Then till 1990s, owing to erosion of longitudinal shoals in the center region of the South Channel and returning deposition into channel, average elevation of thalgweg consequently increased to −13.4 m. And then following the construction of many large scale water conservancy projects such as the Three Gorges Dam, South to North Water Diversion and so on after 2000s, sediment load into estuary decreased obviously year by year but annual water load still remained stable. In result, average elevation of thalwegs started turning to decreasing trend in the whole till nowadays. Its value had changed to −15.9 m till 2016. But, integrating location of thalwegs in different period shown in Fig. 4, it could be found that elevation change of main channel in different period were always due to channel swing but not directly erosion.

In addition, comparing variation of central shoal and main channel, it could be found that variation of average elevation of thalwegs were all below 0.3 m/a during 1973 ∼ 2016, erosion and sedimentation range of main channel was distinctly smaller than that of central shoal.

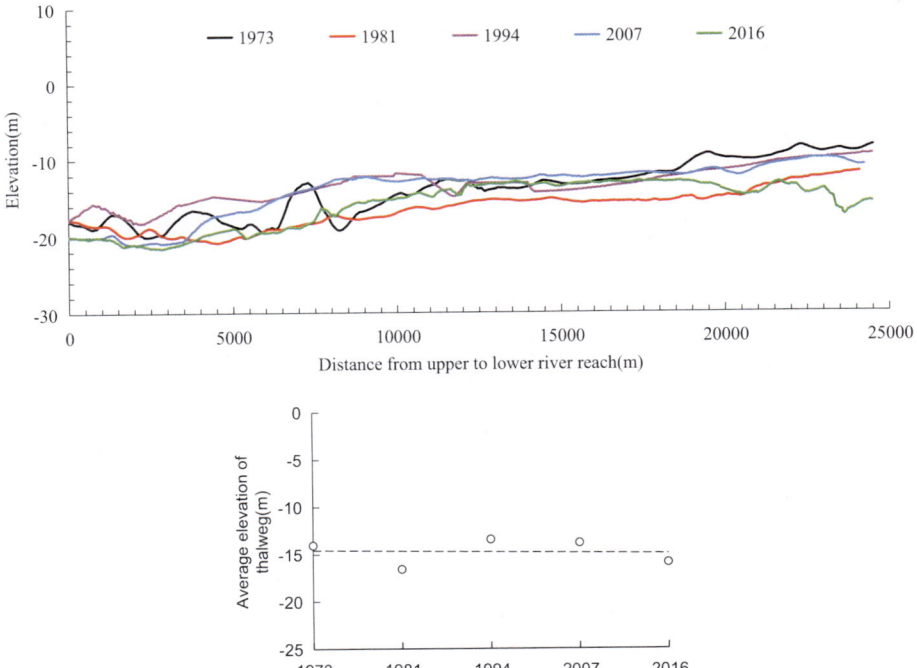

Fig. 5. Evolution of thalwegs in different period

4 Conclusion and Discussion

In sum, the evolution of the South Channel has experienced three stages in past 50 years, which mainly presented at exchange of erosion and sedimentation regions. Wherein, the variation of central shoal along the South Channel was the most distinct, and the main deep channel also presented variation in location but average elevation of its thalweg changed in smaller range than that of central shoal.

Meanwhile, according to fluvial water and sediment load variation characteristics in past decades, it could be preliminary found that there may be close relationship between the geomorphology evolution of the South Channel and the incoming water and sediment load from upper basin. From 1973 ~ 2016, the fluvial water load basically remained stable, but the fluvial sediment load presented obviously decreasing tendency after 1980s and then after 2000s its decreasing trend became more obvious. Correspondingly, the central shoal along the reach appeared erosion in general after 1980s, its elevation decreased obviously, especially in the middle and lower reach. Moreover, the main deep channel maintained a relatively stable state, though its location existed swing. Hence, to some extent, it could be preliminarily inferred that evolution of the central shoal along the South Channel might be mainly influenced by fluvial sediment load which had shown a decreasing tendency from 1980s till now, but evolution of the main deep channel might be mainly influenced by fluvial water load which always presented basically stable from 1970s till nowadays.

Acknowledgements. This study was supported by the National Natural Science Foundation of China (51979096), the National Natural Science Foundation of China (U2040203), the Key Laboratory of Coastal Disaster and Defence, Ministry of Education, Hohai University Program (The influence of the Deepwater Channel Regulation Project on saltwater intrusion in the Yangtze River Estuary), the Fundamental Research Funds for the Central Public Welfare Research Institutes, Nanjing Hydraulic Research Institute (YN912001).

References

Chen XQ, Zong YQ, Zhang EF et al (2001) Human impacts on the Changjiang Yangtze River basin, China

Chu ZX, Yang XH, Feng XL et al (2013) Temporal and spatial changes in coastline movement of the Yangtze delta during 1974–2010. J Asian Earth Sci 66:166–174

Jiang CJ, Li JF, Swart HED (2012) Effects of navigational works on morphological changes in the bar area of the Yangtze Estuary. Geomorphology 139–140:205–219

Qiu C, Zhu JR (2013) Influence of seasonal runoff regulation by the Three Gorges Reservoir on saltwater intrusion in the Changjiang River Estuary. Cont Shelf Res 71:16–26

Shi JZ, Lu LF (2011) A short note on the dispersion, mixing, stratification and circulation within the plume of the partially-mixed Changjiang River estuary, China. J Hydro-Environ Res 5:111–126

Shi Z, Ren LF, Zhang SY, Chen JY (1997) Acoustic imaging of cohesive sediment resuspension and re-entrainment in the Changjiang Estuary. Geo-Mar Lett East China Sea 17:162–168

Wang Y, Shen J, He Q (2010) A numerical model study of the transport timescale and change of estuarine circulation due to waterway constructions in the Changjiang Estuary. J Mar Syst China 82:154–170

Wang YH, Ridd PV, Wu HL, Wu JX, Shen HT (2008b) Long-term morphodynamic evolution and the equilibrium mechanism of a flood tide channel in the Yangtze Estuary (China). Geomorphology 99:130–138

(2001) With special reference to the impacts on the dry season water discharges into the sea. Geomorphology 41:111–123

Wu JX, Wang YH, Chen HQ (2009) Bedforms and bed material transport pathways in the Changjiang (Yangtze) Estuary. Geomorphology 104:175–184

Yang SL, Belkin IM, Belkina AI et al (2003) Delta response to decline in sediment supply from the Yangtze River: evidence of the recent four decades and expectations for the next half-century. Estuar Coast Shelf Sci 57:689–699

Yu FL, Chen ZY, Ren XY, Yang GF (2009) Analysis of historical floods on the Yangtze River, China: characteristics and explanations. Geomorphology 113:210–216

Zhao J, Guo LC, He Q et al (2018) An analysis on half century morphological changes in the Changjiang Estuary: spatial variability under natural processes and human intervention. J Mar Syst 181:25–36

Study on Wind Waves Similarity and Wind Waves Spectrum Characteristics in Limited Waters

Ang Gao[✉], Xiufeng Wu, Shiqiang Wu, Hongpeng Li, Jiangyu Dai, and Fangfang Wang

State Key Laboratory of Hydrology Water Resources and Hydraulic Engineering, Nanjing Hydraulic Research Institute, Nanjing 210029, People's Republic of China
{agao,xfwu,sqwu,jydai,ffwang}@nhri.cn

Abstract. Wind waves is an important factor affecting navigation safety and water environment in limited waters such as lakes and bays. Wind wave spectrum represents the frequency domain features of wind waves and has always been the focus of research. Based on the field observation and flume experimental method, the system analysis of similarity of two kinds of situations, discussed nonlinear response of the relationship of the spectral shape parameter of balance field α, β and wind waves basic frequency between factors like wind speed, wind blowing fetch and water depth. By means of wind tunnel flume and prototype observation data of nonlinear regression analysis, The relation formulas of wind wave frequency prediction considering the comprehensive influence of wind speed, wind blowing fetch and water depth is established. Relevant research is of great significance for revealing the evolution characteristics of wind waves in limited waters and guiding navigation safety and water environment management.

Keywords: Limited waters · Wind wave spectrum · Similarity · Wind-tunnel test · Prototype observation

1 Introduction

Wind waves are the product of the interaction between water and air and exist widely in natural waters. Because wind waves carry a certain amount of energy. Strong wind and wave may lead to shipwreck accident. Wave force is also one of the driving forces affecting the energy and material conversion of muddy-water interface. Therefore, wind waves research has been a hot spot in the field of water transportation and water environment. Among them, wind waves in lakes, harbors and other limited waters are not only restricted by wind speed, but also affected by wind speed, wind blowing fetch, and their morphological characteristics become more complex (Karimpour et al. 2017).

Field observation is the most direct method to study the characteristics of wind wave spectrum, but the field conditions are greatly affected by external factors, the observation is not controllable, and systematization of data is poor. The wind tunnel flume generalizes the field environment to a certain extent, and has the advantages of

© The Author(s) 2023
Y. Li et al. (Eds.): PIANC 2022, LNCE 264, pp. 1220–1235, 2023.
https://doi.org/10.1007/978-981-19-6138-0_107

being controllable and easy to measure, so that the evolution characteristics of wind wave spectrum can be systematically and carefully studied (Zakharov et al. 2015). Both methods play an important role in wind waves research, but there is no discussion on the similarity of wind waves under the two scenarios.

Because natural wind waves are affected by uneven and unstable wind field, the wind waves process at a certain point is the superposition of wind waves at different scales, and its total energy is the sum of the energy carried by wind waves at different scales. Wind wave spectrum is often obtained through energy spectrum analysis for further study of wind waves characteristics (Zakharov et al. 2015). The high frequency part of wind wave spectrum exists balance field. In this region, energy intake by wind and consumption by breakage of wind waves tends to balance (Grare et al. 2013), and the wind tunnel experimental research shows that balance and frequency domain energy relationship can use $S(\omega) = \alpha g^2 \omega^\beta$ to express ($\omega = 2\pi f$) (Hidy and Plate 1966). The α decreases with the increase of wind path and β characterization of balance between the different scale wave energy transfer efficiency (The smaller the absolute value, the faster the energy travels). For wind waves in shallow water, the effect of the bottom wall on wind and wave enhances its nonlinearity, inhibits the propagation of small-scale wind and wave to large-scale wind and wave, and results in significant changes in the spectral shape. For example, the spectrum width of wind waves increases, and the absolute value of β decreases (the energy transfer of wind and wave at different scales accelerates) (Young and Verhagan 1996; Nair and Kumar 2017; Fedele et al. 2019).

Wind waves dominant frequency (f_d) represents the size of the dominant wind wave scale. At present, the relationship between the dominant frequency and wind speed (U_{10}) and wind blowing fetch (F) is mostly established in the form of formula (1) (Hwang 2005). Statistical studies have found that when a is between $1.72 \sim 3.50$, and b is between $-0.24 \sim -0.33$ the two coefficients are highly discrete in different studies. Karimpour et al. (2017) found that for the wind waves in the fully developed shallow water area, the influence of relative wind blowing fetch F/d(D is water depth) on $f_d u_{10}/g$ and gF/u_{10}^2 should be considered, and the dominant frequency prediction formula (2) considering the influence of water depth is given. However, this formula is only applicable to fully developed wind waves, and it does not consider the wind tunnel experimental data.

$$f_d U_{10}/g = a\left(gF/U_{10}^2\right)^b \tag{1}$$

$$f_d U_{10}/g = 3.5\left(gF/U_{10}^2\right)^{-0.75(F/d)^{-0.1}} \tag{2}$$

In this study, the similarity of wind wave spectrum shape and spectrum parameters in the two scenarios was analyzed by combining field observation and wind tunnel flume experiment, and the response relationship between the spectrum parameters α and β in the equilibrium domain and wind speed, wind blowing fetch and water depth was further summarized. This paper analyzes the reasons for the large deviation of parameters a and b in the existing dominant frequency prediction relation, and puts forward a dominant frequency prediction method suitable for limited waters. Relevant research is of great significance for revealing the evolution characteristics of wind waves in limited waters and guiding navigation safety and water environment governance.

2   Research Methods

2.1   Prototype Observation

Lake Tai, a typical shallow water lake, is selected for prototype observation. The data of airflow, wind wave and water depth were collected from the monitoring platform at the baymouth of Zhushan Bay (Fig. 1). The wind speed and wind direction sensor was located 2.5 m above the water surface, and the water surface fluctuation data were measured by the pressure sensor located at the bottom of the lake (Gao et al. 2022).

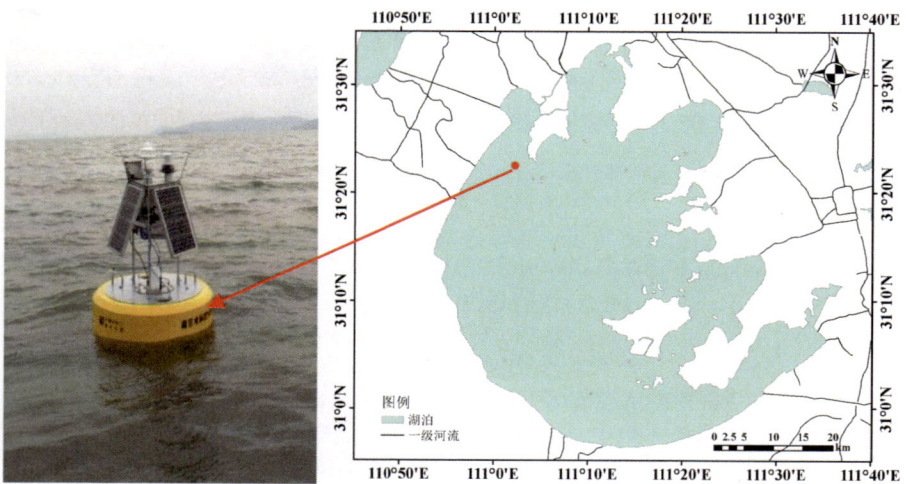

Fig. 1. Air flow and wind wave station arrangement

2.2   Flume Experiment

The experiment was carried out in the wind tunnel flume of Nanjing Hydraulic Research Institute (Fig. 2). Six measuring cross section (F4.5 ∼ F19.5) were arranged along the wind tunnel flume, and the cross section spacing was 3 m. A wind speed measuring line is arranged on the middle line of each measuring section, and 16 hot wire anemometers are arranged along each measuring line vertically. Accordingly, a pair of capacitive wave altimeters is arranged along the wind direction 10 cm from each measuring line, with a total of 12 (Gao et al. 2022).

Considering the influence of shallow lake water depth on wind waves, the experimental water depth d was 0.10 m, 0.15 m and 0.30 m respectively. A total of 7 levels of reference wind speed U_r were set (the average wind speed of wind tunnel cross section, which is carefully controlled by transducer), including the prevailing wind speed range of Lake Tai, as shown in Table 1. Two groups of tests were carried out for each wind speed, and the relative errors of the test results were all within 5%, and the average of the two results was the final result.

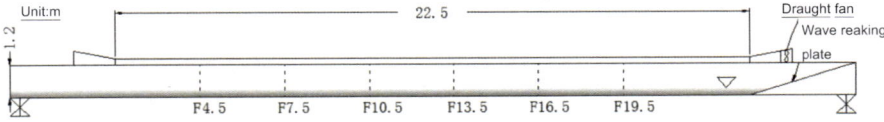

Fig. 2. Wind tunnel flume and instrument layout

Table 1. Test group design

Case number	Water depth d/m	Reference wind speed U_r/(m/s)						
Case1	0.10	3.2	4.7	6.2	7.7	9.3	10.8	12.3
Case2	0.15							
Case3	0.30							

3   Result Analysis

3.1   Characteristics of Wind Wave Spectrum

3.1.1   Prototype Observation

Figure 3 shows the relationship between wind wave spectrum and wind blowing fetch of Lake Tai at the same wind speed ($U_{2.5}$ = 2.9 m/s) (wind wave spectrum is obtained by fast Fourier transform method). The solid line in the figure is the smoothing fitted line of scattered data. It can be seen that with the increase of wind blowing fetch, the spectrum peak gradually rises and the dominant frequency shows a slight decreasing trend (the dominant frequency wind wave cycle increases). It reflects the characteristics of wind wave developing gradually with the increase of wind blowing fetch under the condition of limited wind blowing fetch.

Figure 4 shows the relationship between wind wave spectrum and wind speed in Lake Tai at the same wind blowing fetch (F = 33 km). It can be seen that with the increase of wind speed, the spectrum peak gradually rises and the dominant frequency gradually decreases, indicating that the wind wave scale of the dominant frequency increases. In addition, when the wind speed is high, the amplitude of the dominant frequency and spectrum peak energy decreases with the increase of wind speed, indicating that the wind wave tends to be saturated with the increase of wind speed. When the wind speed is 15.3 m/s, the dominant frequency of wind waves is about 0.31 Hz.

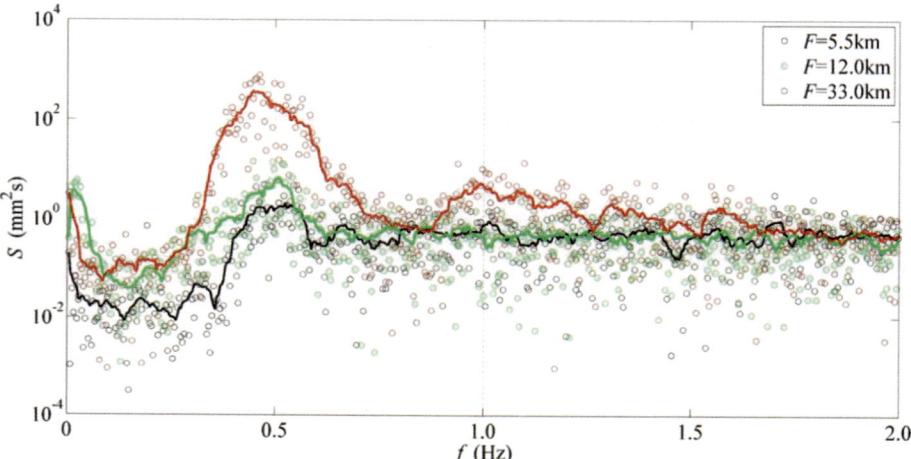

Fig. 3. Wind wave spectrum changes with wind blowing fetch in the field ($u_{2.5}$ = 2.9 m/s)

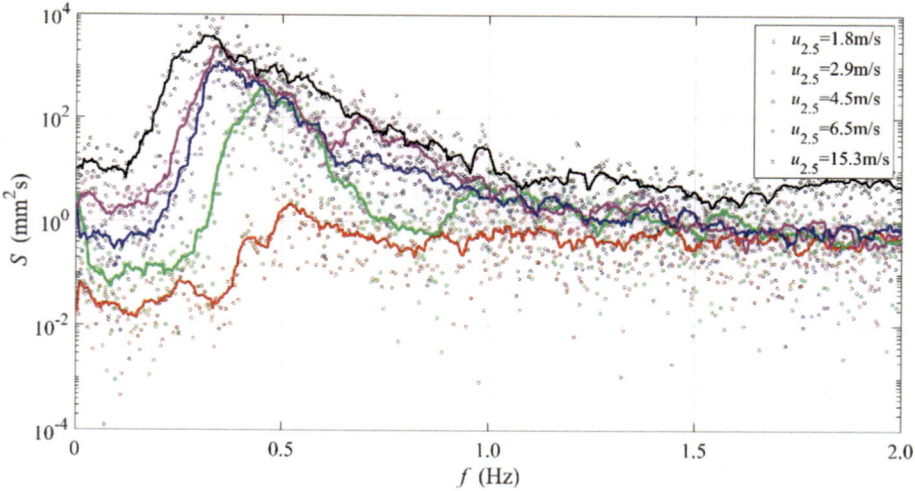

Fig. 4. Wind wave spectrum changes with wind speed in the field (F = 33 km)

3.1.2 Wind-Tunnel Test

Figure 5 shows the change of wind wave spectrum with wind blowing fetch when U_r = 7.7 m/s. The solid line in the figure is a smooth fitted line for scatter data. It can be seen that with the increase of wind blowing fetch, the dominant frequency gradually decreases and the peak value rises gradually. The spectral shape near the dominant frequency gradually changes from wide and low to sharp and narrow, indicating that with the increase of wind blowing fetch, there is a dominant wave scale which is relatively consistent with the dynamic and boundary conditions at this time, and the wave energy of other frequencies gradually shifts to the dominant frequency wave, and the spectral shape changes rapidly when the wind blowing fetch is small, but slowly when the wind blowing fetch is large, which is consistent with the law of measured wind wave spectrum.

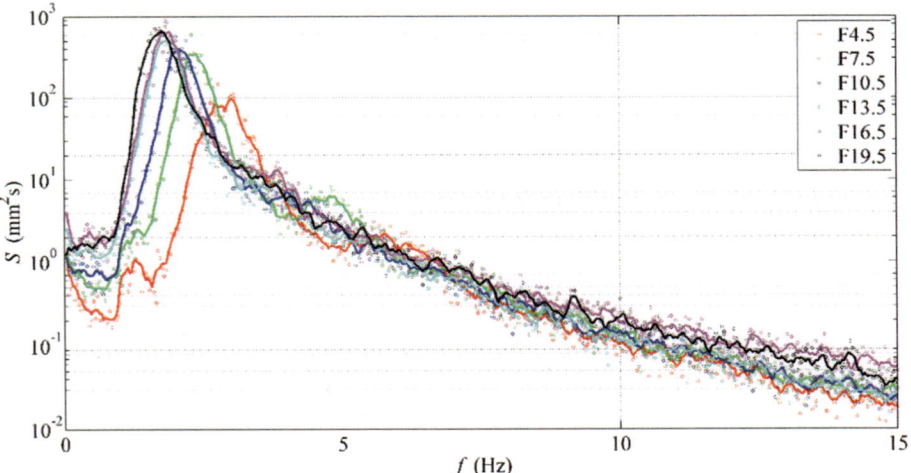

Fig. 5. Wind wave spectrum changes with wind blowing fetch in the flume

Figure 6 shows the change of spectral shape with the increase of wind speed when d = 0.15 and F = 13.5 m. It can be seen that the overall energy increases with the increase of wind speed, the spectral peak energy increases with the increase of wind speed, the spectral peak frequency decreases with the increase of wind speed, and the spectral shape near the main frequency gradually changes from wide and short to sharp and narrow with the increase of wind speed. It shows that the wind wave energy of other frequencies is gradually transferred to the dominant frequency wave, and the spectral shape changes rapidly with the increase of wind speed when the wind speed is small, while the spectral shape changes slowly when the wind speed is large, which is consistent with the response relationship between measured wind wave spectral shape and wind speed (Fig. 4).

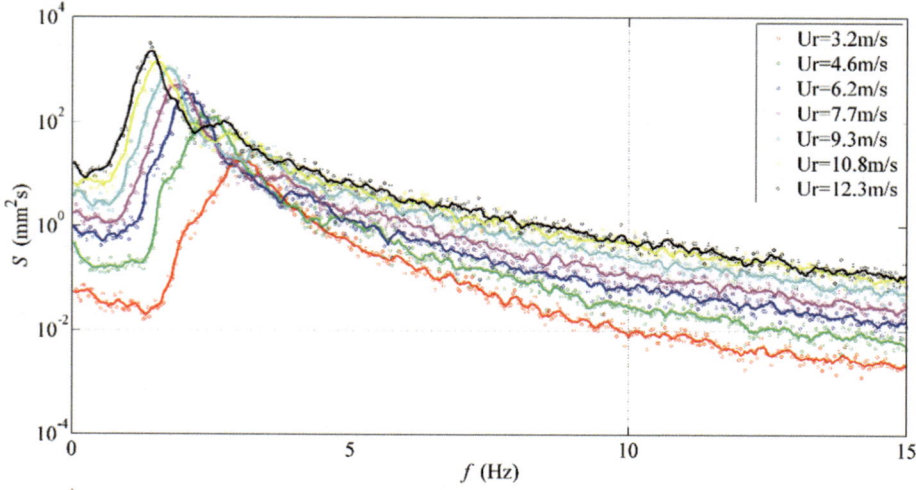

Fig. 6. Wind wave spectrum changes with wind speed in the flume

The influence of water depth on spectral shape is mainly reflected in the influence of bottom wall on wave shape and dynamic process when wave scale is large enough. Figure 7 shows the wind wave spectra at three water depths when F = 19.5 m U_r = 12.3 m/s. It can be seen that with the increase of water depth, the overall energy spectrum line increases, especially the increase of the main peak, and its spectral shape becomes more sharp and thin, indicating that under the same wind speed and wind blowing fetch, the greater the water depth, the more easily the energy is transferred to the main frequency wave, and the main frequency energy is larger.

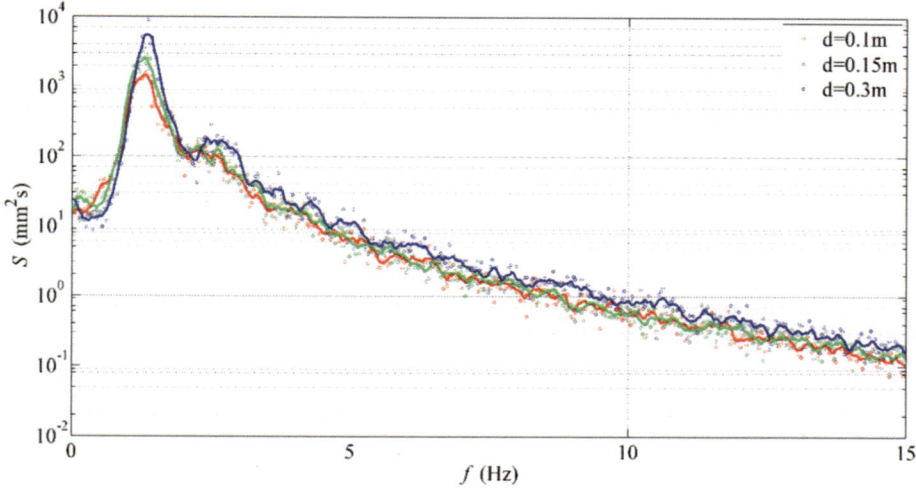

Fig. 7. Wind wave spectrum changes with water depth in the flume

3.1.3   Comparison Between Prototype Observation and Wind Tunnel Experiment

Figure 8 shows the comparison between the flume experiment and the measured wind wave spectrum in Lake Tai. For the convenience of comparison, the wind wave spectrum is normalized, in which frequency is normalized by dominant frequency (f_d) and energy is normalized by energy of dominant frequency (S_d). It can be seen that the spectral shapes obtained in field and flume are generally similar, but the dispersion degree of field measured data is significantly greater than that of flume data, which is believed to be caused by the variable external factors such as field wind speed and direction.

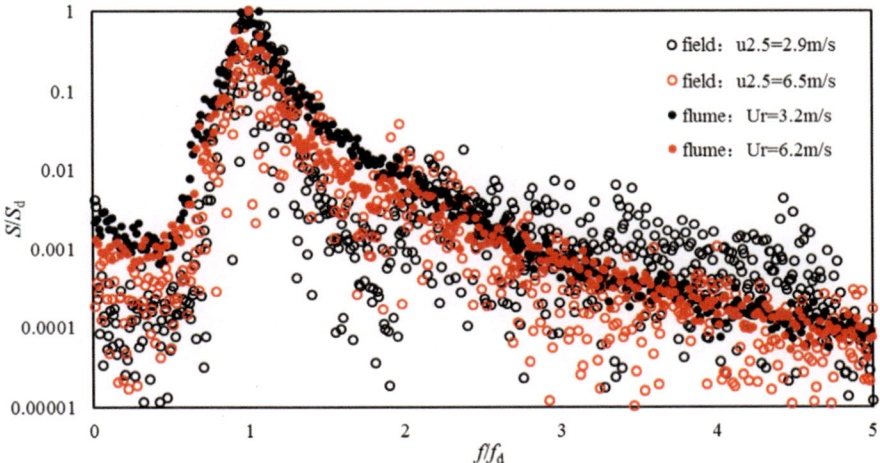

Fig. 8. Wind wave spectrum comparison between Lake Tai and flume experiment

3.2   Variation Rule of Parameters in Spectrum Equilibrium Domain

3.2.1   Prototype Observation

Figure 9 shows the relationship between the spectrum parameters α and β in the wind wave equilibrium domain observed at $U_{2.5} = 2.9$ m/s and the wind blowing fetch. It can be seen that α ranges from 1.6×10^{-3} to 7.3×10^{-3} and shows a decreasing power function trend with the wind blowing fetch.

Figure 10 shows the relationship between α and β with wind speed when F = 33 km. It can be seen that α is between 4.0×10^{-5} and 1.6×10^{-3}, and decreases with wind speed as a power function. β was between -8.76 and -3.20, and β showed a logarithmic increase trend with the increase of wind speed, indicating that the wave energy transfer efficiency in the equilibrium region tended to be saturated with the increase of wind speed, indicating that wind waves were broken under strong wind force and dissipated part of energy rapidly. In addition, the interaction between wind waves and bottom wall in limited water depth also dissipated part of energy.

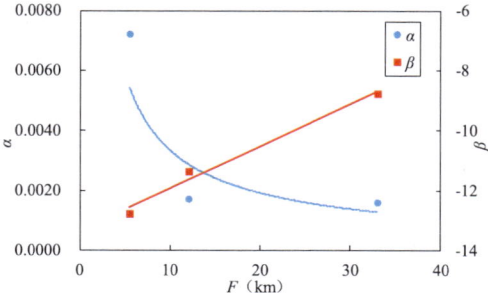

Fig. 9. Relationship between α and β with wind blowing fetch

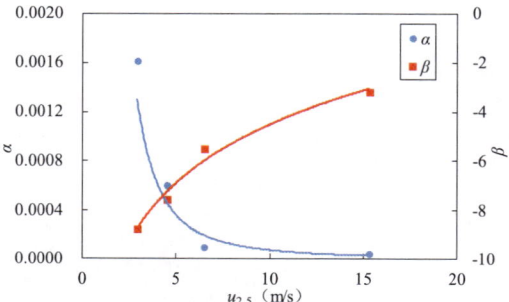

Fig. 10. Relationship between α and β with wind speed

3.2.2 Wind Tunnel Experiment

Figure 11 shows the relationship between the spectral parameters α and β and the wind blowing fetch when d = 0.15 m U_r = 7.7 m/s. It can be seen that α is between 2.5 10^{-4} and 2.1 10^{-3} and the wind blowing fetch shows a decreasing power function trend. β was in the range of $-4.53 \sim -3.89$, and β showed a linear increase trend with the increase of blowing distance, indicating that the wave energy transfer rate and blowing distance had a linear relationship with the measured wind wave law in Lake Tai.

Figure 12 shows the relationship between α and β and wind speed when d = 0.15 m   and   F = 13.5 m,   indicating   that   α   was   in   the   range   of $2.8 \times 10^{-4} \sim 1.0 \times 10^{-2}$, decreases as a power function with respect to wind speed; β was between -5.53 and -3.67, and β showed a logarithmic increase trend with the increase of wind speed, indicating that the wave energy transfer efficiency in the equilibrium region tended to saturation with the increase of wind speed, which was consistent with the measured wind wave law in Lake Tai.

Figure 13 (a) shows the relationship between α and β and water depth when F = 7.5 m and U_r = 4.6 m/s. It can be seen that α is positively correlated with water depth, while β is negatively correlated with water depth under test conditions. Figure 13 (b) shows the relationship between α and β when F = 19.5 m and Ur = 12.3 m/s. It can be seen that α is positively correlated with water depth under experimental conditions, while β is not significantly correlated with water depth. The relationship between β and water depth was different under the two test conditions. The analysis found that for the F = 7.5 m U_r = 4.6 m/s group, the corresponding F_c values (F_c = $g^{1.25}H^{0.5}T^{2.5}/d^{1.75}$) (Zhu 2015) were 9.8, 4.4 and 2.0 respectively under the three water depths (0.10 m, 0.15 m and 0.30 m), both less than 10, and wind wave nonlinear weak; The corresponding F_c values of F = 19.5 m U_r = 12.3 m/s at three water depths were 81.9, 45.4 and 27.4 respectively, both greater than 10, indicating weak wind wave nonlinearity. The comparison of the two cases shows that water depth only has an effect on the energy transfer efficiency in the balance domain when the nonlinear effect of wind and wave is weak, and when the nonlinear effect is strong, water depth has a weak effect on the energy transfer efficiency in the balance domain.

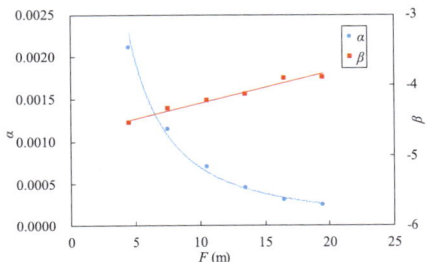

Fig. 11. Relationship between α and β with blowing distance

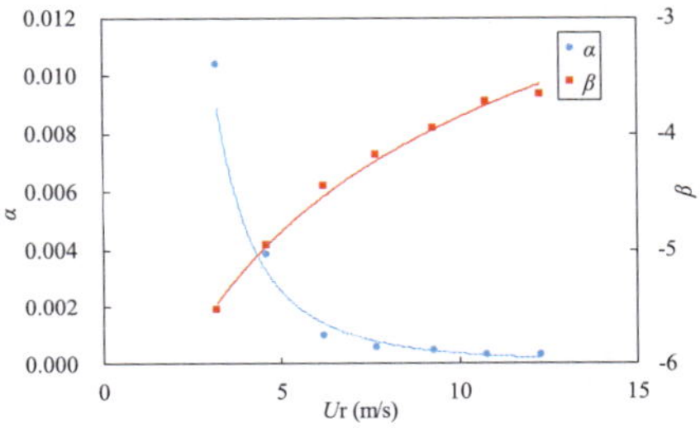

Fig. 12. Relationship between α and β with wind speed

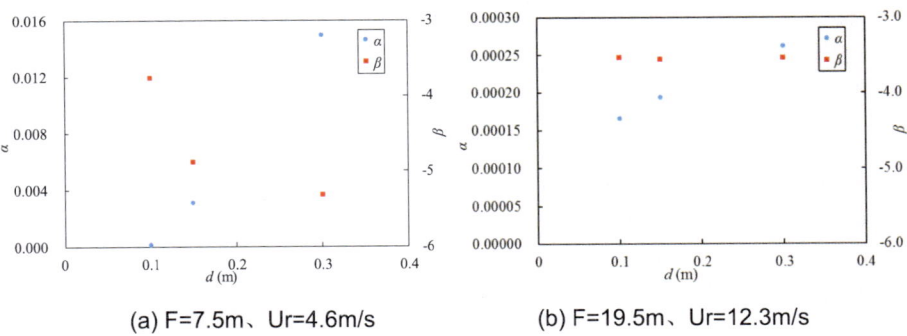

(a) F=7.5m、Ur=4.6m/s

(b) F=19.5m、Ur=12.3m/s

Fig. 13. Relationship between α and β with water depth

3.3 Variation Rule of Wind Wave Dominant Frequency

3.3.1 Prototype Observation

F_d reflects the scale of dominant wave groups in wave sequence and is an important parameter to quantify the shape of wind wave spectrum. The following formula is used to calculate f_d (Mo et al. 2018):

$$f_d = \frac{m_1}{m_0} \qquad (3)$$

where m_k (k = 0, 1) represents the k-order central moment of the spectrum.

Figure 14 shows the influence of measured wind speed (a) and wind blowing fetch (b) on the dominant frequency of wind waves in Lake Tai. It can be seen that with the increase of wind speed, the dominant frequency shows a non-linear decreasing trend, and the damping decreases with the increase of wind speed. The dominant frequency decreases linearly with the increase of wind blowing fetch.

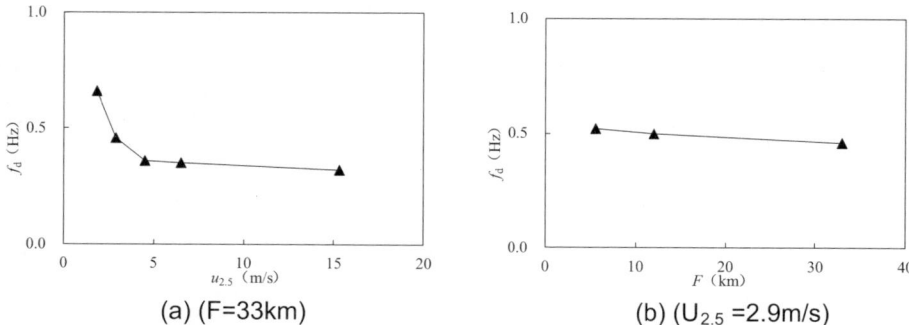

(a) (F=33km)   (b) (U$_{2.5}$ =2.9m/s)

Fig. 14. Influence of wind speed (a) and wind blowing fetch (b) on dominant frequency of wind waves in Lake Tai

3.3.2   Wind Tunnel Experiment

Figure 15 for three water depth under the experiment and the relationship between frequency and wind speed and fetch, obviously with the increase of wind speed and fetch the decrease trend of dominant frequency is nonlinear. When fetch and wind speed is small the damping of dominant frequency is larger, with the increase of fetch and wind speed, the damping gradually become smaller, trends related to the intensity of the nonlinear wave, and measured in Lake Tai is consistent. The change law of nonlinear frequency show that at the beginning of the interaction between water and gas, water surface changes in response to flow boundary and dynamic conditions and produce scale small waves and diversity is strong. With the increase of fetch both interaction, a concentrate towards the dominant wave after wave absorb wind energy scale increase gradually and the main frequency gradually decreases with the increase of wind blowing fetch. However, when large scale wind waves are generated under high wind speed and blowing range, the nonlinear effect caused by limited water depth inhibits the transfer of non-dominant frequency wave energy to dominant frequency wave to some extent, and the amplitude of dominant frequency decreases with the increase of blowing range and wind speed.

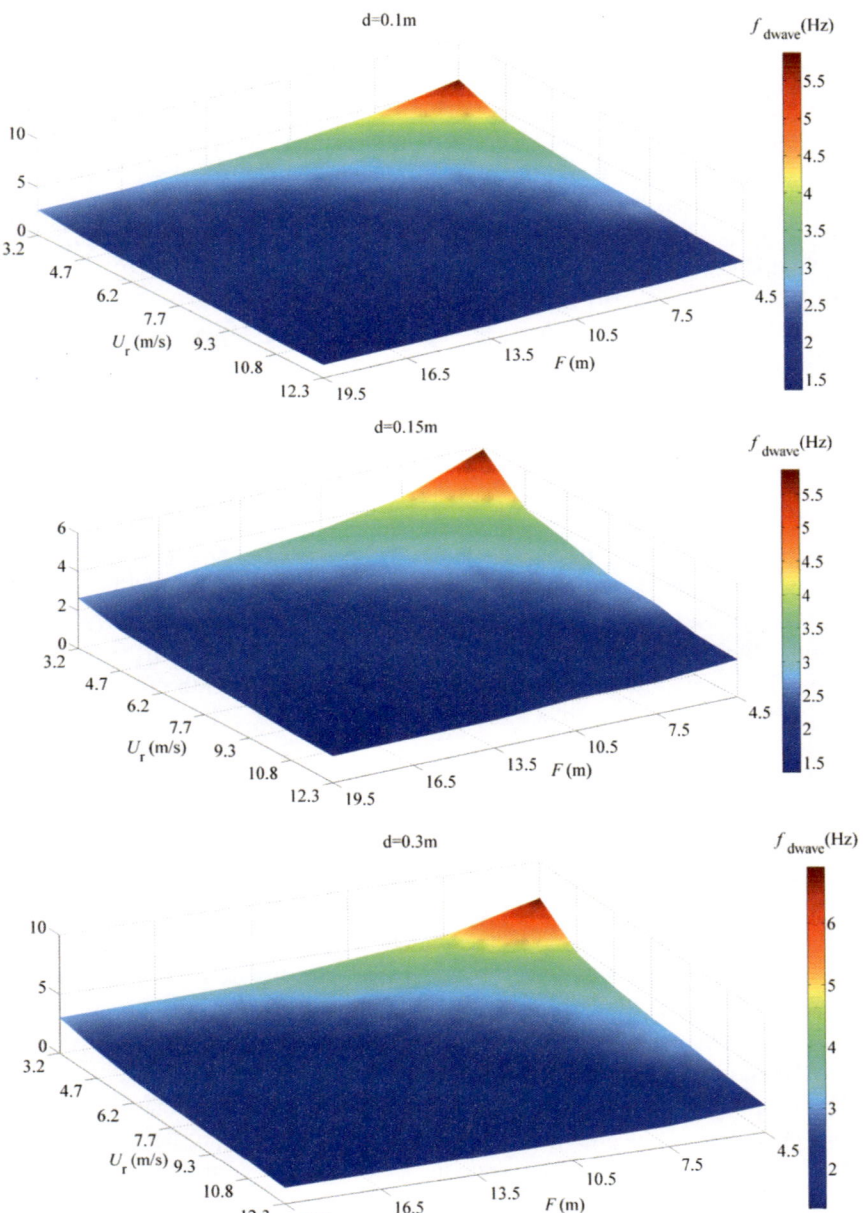

Fig. 15. Relationship between f_d with wind blowing fetch and wind speed under different water depths

3.4 Wind Wave Dominant Frequency Prediction

The relationship between experimental and prototype observation of wind wave dominant frequency and wind blowing fetch in Lake Tai was compared with previous studies (Formula (1)). The method of converting measured $U_{2.5}$ in Lake Tai to U_{10} was referred to (Gao et al. 2022), and the profile gradient method was used to extrapolate experimental and prototype observation for U_{10}. Figure 16 shows the relationship of $f_d U_{10}/g$ and gF/U_{10}^2 of the experimental and prototype observations and three fitting relationship lines of existing studies are also shown in the figure (CERC 1977; Donelan et al. 1985; Donelan et al. 1992). It can be seen that under experimental and prototype observation conditions, both of them are negatively correlated with power functions. Experimental and measured data of Lake Tai are basically located between the three fitting lines, indicating that the data are generally consistent with the trend of existing research. But there is a certain deviation between three fitting line. It is considered that one of the reasons for the difference is that the water depth of each research area is different and the water depth is not considered when establishing the dominant frequency relationship.

Based on the above results can be inferred formula (1) without considering the influence of the depth of the water, It has the problem of poor suitability in limited waters and we consider in establishing $f_d U_{10}/g$ and gF/U_{10}^2 depth was developed for the relation between the factors. The formula (4) is constructed by referring to the form of formula (2). When a and e are 0, the relation degenerates into the form of formula (1). According to the experimental and measured data of Lake Tai, relevant literature data (Ding 2005; Johnson et al. 1998; Donelan et al. 2006), 134 groups of wind tunnel test and 96 groups of field observation, the water depth (d) ranged from 0.1 to 4.0 m, the wind blowing fetch (F) ranged from 4.5 to 33000 m, the wind speed (U_{10}) ranged from 2.36 to 24.02 m/s, and the dominant frequency (f_P) ranged from 0.28 to 5.87 Hz. After nonlinear regression analysis of the relation, the relation Eq. (5) was obtained, with correlation coefficient 0.97 and determination coefficient 0.94. This equation shows that the dominant frequency is negatively correlated with the wind speed, wind blowing fetch and water depth, which is consistent with the previous results and the conclusion in reference (2017).

Figure 17 shows the comparison between the measured $f_d U_{10}/g$ and the calculated values by formula (5). It can be seen that the experimental and field observation data points are evenly and compact distributed on both sides of the 45° line. Figure 17(b) shows the measured $f_d U_{10}/g$ compared with the data calculated by formula (2). It can be seen that the prototype observation data points are evenly and compactly distributed on both sides of the 45° line, but the experimental data are at around $f_d U_{10}/g > 2$, data points and the 45° line deviation increases gradually, and show that type formula (5) does not apply to wind waves of the wind tunnel sink. However, both wind tunnel flume experiments and field observations, the interaction mechanism of water and wind is consistent with each other, thus suggests that there are some defects on the similarity between the wind tunnel experiment and field observation data in formula (2).

Formula (5) takes into account the wind wave similarity relationship between wind tunnel flume and shallow lake. By introducing water depth as a modulation factor, the prediction relationship of wind wave dominant frequency considering the comprehensive influence of wind speed and water depth is established, which makes up for the deficiency of Formula (2) without considering the wind tunnel experimental data.

$$f_{\rm d}U_{10}/{\rm g} = a + b\left({\rm g}F/U_{10}^2\right)^{c*(F/d)^e} \tag{4}$$

$$f_{\rm d}U_{10}/{\rm g} = -0.847 + 3.443\left({\rm g}F/U_{10}^2\right)^{-0.233*(F/d)^{-0.069}} \tag{5}$$

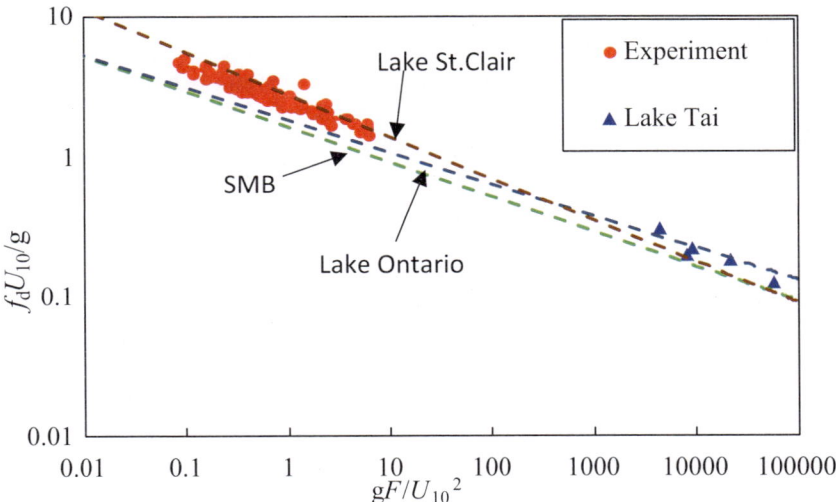

Fig. 16. Relationship between $f_{\rm d}U_{10}/{\rm g}$ and ${\rm g}F/U_{10}^2$

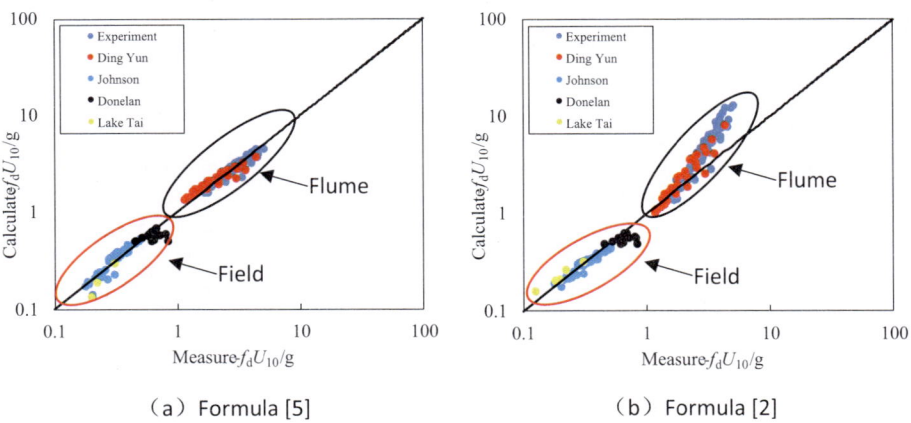

Fig. 17. Comparison of calculated and measured values

4 Conclusion

- The normalized spectrum of measured wind waves in Lake Tai is similar to that in flume experiment, but the dispersion degree of measured data in Lake Tai is greatly affected by unstable wind speed and direction. The measured parameters of wind wave spectrum in balance domain and the variation of dominant frequency, wind speed and wind blowing fetch in Lake Tai are consistent with the wind tunnel experiment.
- The spectrum parameter α decreases as a power function with the increase of wind speed and wind blowing fetch, and increases with the increase of water depth. The parameter β increases linearly with the increase of wind blowing fetch and logarithmically with the increase of wind speed. When the nonlinearity of wind wave is weak, the parameter β decreases with the increase of water depth. When the nonlinearity is strong, the parameter β does not change with the increase of water depth.
- The main frequency of wind wave is negatively correlated with wind speed and blowing range. The amplitude reduction is large at low wind speed and wind blowing fetch, and becomes smaller with the increase of wind speed and blowing range. With the increase of wind speed and wind blowing fetch, the dominant frequency energy increases first and then hold steady.
- By considering the similarity relationship between wind tunnel experiment and shallow lake, and by introducing water depth as the modulation factor, the dominant frequency prediction relationship of wind wave considering the comprehensive influence of wind speed, wind blowing fetch and water depth is established, which makes up for the deficiency that wind tunnel experiment data is not taken into account in existing studies.

References

Karimpour A, Chen Q, Twilley RR (2017) Wind wave behavior in fetch and depth limited estuaries. Sci Rep 7:40654

Zakharov VE, Badulin SI, Hwang PA et al (2015) Universality of sea wave growth and its physical roots. J Fluid Mech 780:503–535

Grare L, Peirson WL, Branger H et al (2013) Growth and dissipation of wind-forced, deep-water waves. J Fluid Mech 722:5–50

Hidy GM, Plate EJ (1966) Wind action on water standing in a laboratory channel. J Fluid Mech 26(4):651–687

Young IR, Verhagan L (1996) The growth of fetch limited waves in water of finite depth. Part 1: total energy and peak frequency. Coast Eng 27:47–78

Nair MA Kumar VS (2017) Wave spectral shapes in the coastal waters based on measured data off Karwar on the western coast of India. Ocean Sci Discuss 1–34

Fedele F, Herterich J, Tayfun A et al (2019) Large nearshore storm waves off the Irish coast. Sci Rep 9:15406

Hwang PA (2005) Duration- and fetch-limited growth functions of wind-generated waves parameterized with three different scaling wind velocities. J Geophys Res Atmos 111:C02005

Gao A, Wu XF, Wu SQ et al (2022) Experimental study on the evolution characteristics of near-surface airflow in limited waters. J Lake Sci 34(03):994–1005

Zhu XW (2015) Nonlinear characteristics of random waves in shallow water. Dalian University of Technology, Dalian

Mo D, Hou Y, Liu Y, Li J (2018) Study on the growth of wind wave frequency spectra generated by cold waves in the northern East China Sea. J Oceanol Limnol 36(5):1509–1526. https://doi.org/10.1007/s00343-018-7265-8

CERC (1977) Shore protection manual, 3 vols. U.S. Army Coastal Engineering Research Center, Washington, D.C.

Donelan MA, Hamilton J, Hui WH (1985) Directional spectra of wind-generated waves. Philos Trans R Soc Lond Ser A 315:509–562

Donelan MA, Skafel M, Graber H et al (1992) On the growth rate of wind-generated waves. Atmos Ocean 30:457–478

Ding Y (2005) Influence of wind wave state on wind stress at sea surface. Ocean University of China, Qingdao

Johnson HK, Højstrup J, Vested HJ et al (1998) On the dependence of sea surface roughness on wind waves. J Phys Oceanogr 28(9):1702–1716

Donelan MA, Babanin AV, Young IR et al (2006) Wave-follower field measurements of the wind-input spectral function. Part II: parameterization of the wind input. J Phys Oceanogr 36 (8):1672–1689

Shoreline Carrying Capacity Assessment Based on Satellite Remote Sensing Image: A Case Study of the Nanjing Reach of the Yangtze River

Huai Chen[1], Shan Wang[2], Suning Huang[3], Lei Zhang[4(✉)], Nairu Wang[1], and Lijun Zhu[1]

[1] State Key Laboratory of Hydrology-Water Resources and Hydraulic Engineering, Nanjing Hydraulic Research Institute, Nanjing 210029, China
{chenhuai, nrwang, ljzhu}@nhri.cn
[2] Water Service Management Station in Jiangning Street of Jiangning District, Nanjing 210000, China
[3] Qinhuai River Management Office in Nanjing City, Nanjing 210012, China
[4] Key Laboratory of Sediment Science and Northern River Training, The Ministry of Water Resources, China Institute of Water Resources and Hydropower Research, Beijing 100038, China
leizhang06@iwhr.com

Abstract. Research on shoreline carrying capacity is of great practical significance to promote the sustainable development of shoreline. Taking into account five aspects of shoreline health, resource supply, environmental pollution, ecological service and social service, the evaluation index system of shoreline carrying capacity was systematically established. Taking the Nanjing reach of the Yangtze River as a typical case, the variation trend of shoreline carrying capacity in recent 40 years was analyzed. The Landsat satellite remote sensing images from 1984 to 2020 were collected. The classification regression tree (CART) algorithm was used to classify the land use types in the remote sensing images, and then evaluation indexes of shoreline carrying capacity were calculated. The results show that the shoreline carrying capacity of the Nanjing reach was basically stable from 1984 to 2003. With the large-scale development and utilization of the shoreline since 2003, the carrying capacity of the Nanjing reach gradually decreased and approached the warning line. Due to the implementation of restrictive measures such as "action to clear the four chaos" and "the operation of responsibility system on river/lake leaders" by Chinese government after 2018, the carrying capacity of the Nanjing reach has rebounded rapidly. With the help of Mann-Kendall (MK) mutation analysis method, the mutation point of the time series of the shoreline carrying capacity of the Nanjing reach was found to occur in 1991, 2012 and 2018. The research results can help to discover unsustainably social and economic activities, put forward the productivity layout adjustment, and guide corresponding management measures in the reach.

Keywords: Nanjing reach of the Yangtze River · Landsat remote sensing image · Classification regression tree (CART) · Carrying capacity · Mann-Kendall (MK) mutation analysis

© The Author(s) 2023
Y. Li et al. (Eds.): PIANC 2022, LNCE 264, pp. 1236–1247, 2023.
https://doi.org/10.1007/978-981-19-6138-0_108

1   Introduction

Shoreline refers to the banded area where water and land intersect on both sides of the river or within a certain range around the lake. It is an important component of the natural ecological space of rivers and lakes. River shoreline not only has the natural attributes about flood discharge, regulating water flow and maintaining river ecological balance, but also has the resource attributes about development and utilization value and providing services for social and economic development.

Carrying capacity originally refers to the strength of the foundation and the load-bearing capacity of the building. Nowadays it has evolved into the most commonly used term to describe the limitation of development. At the end of the 19th century, the concept of carrying capacity was clearly put forward in the field of Applied Ecology, for example, the livestock carrying capacity management (Malthus 1970). The concept of carrying capacity has been widely used in the field of ecology since the 1950s (Park and Burgess 2019). After the 1960s, with the advancement of industrialization and urbanization, the contradiction between resource supply and human demand has become increasingly prominent. Many scholars have widely applied the carrying capacity to the research in the field of resources and environment, and successively put forward the concepts of resource and environment carrying capacity, such as land resource carrying capacity (Sun et al. 2020), water resource carrying capacity (Song et al. 2011), forest resources carrying capacity (Martire et al. 2015), mineral resources carrying capacity (Wang et al. 2016), regional environmental carrying capacity (Lane 2010), atmospheric environmental carrying capacity (Su and Yu 2020), urban environmental carrying capacity (Lu et al. 2017), coastal environmental carrying capacity (Wei et al. 2014), and etc. The internal essences of the concepts about different carrying capacities in all time stages are the same. It is the supporting capacity of resources and environment to population and economic development scale and the internal and external pressure on resources and environment. Carrying capacity can be described as the interactive coupling relationship between a material foundation and its receiving carrier, and its final manifestation is the quantitative characteristics of the receiving carrier that the material foundation can maintain (Price 1999).

Generally speaking, as long as the consumption speed of shoreline resources does not exceed its regeneration speed, the shoreline will always be able to recover itself and there will be no crisis. However, if the consumption speed of shoreline resources exceeds its regeneration speed, the regeneration basis of shoreline resources will be destroyed, making the recovery process of shoreline extremely slow and even impossible in serious cases. Therefore, although the shoreline is rich in resources, its carrying capacity for human economic activities is also limited. This degree is called the shoreline carrying capacity.

The carrying capacity of shoreline actually includes five aspects, namely, the stability of shoreline, resource supply capacity, environmental service capacity, ecological service capacity and social service capacity. Among them, the stability of the shoreline is the basic condition to ensure that the shoreline can exercise other capabilities. The resource supply capacity refers to the limit of human economic activities that shoreline resources can sustainably support. The environmental service capacity refers to the maximum pollution assimilative capacity of the shoreline. The ecological service capacity refers to the limit of ecological regulation, ecological support and ecological culture services that the shoreline system can provide. The social service capacity refers to the maximum degree of social services that shoreline can provide to human society.

In this study, the evaluation index system of shoreline carrying capacity was systematically established, and was use to evaluate the shoreline carrying capacity of the Nanjing reach of the Yangtze River. The research on shoreline carrying capacity is of great practical significance to promote the sustainable development of shoreline. As long as we get the carrying capacity threshold of a certain section of shoreline, we can correspondingly regulate human activities, so as not to cause damage to the shoreline ecology.

2   Study Area, Data and Methods

2.1   Study Area

Shoreline boundary line refers to the boundary line delimited along the riverbank or around the lake bank, which is divided into waterfront boundary line and outer boundary line. The waterfront boundary line is the boundary line in the waterfront zone defined along the flow direction on the waterfront side along the river or on the waterfront side around the lake (reservoir). The outer boundary line is the outer boundary line of the shoreline zone defined on the land side along the river or around the lake (reservoir) according to river and lake shoreline management, protection and maintenance requirement.

The Nanjing reach of the Yangtze River is a plain river, and its frontage boundary line is generally the intersection line between the water level corresponding to bed forming discharge or flat discharge with the land. In this study, the water land intersection line is used as the frontage boundary line. The outer boundary line can be the demarcated embankment line, as shown in Fig. 1.

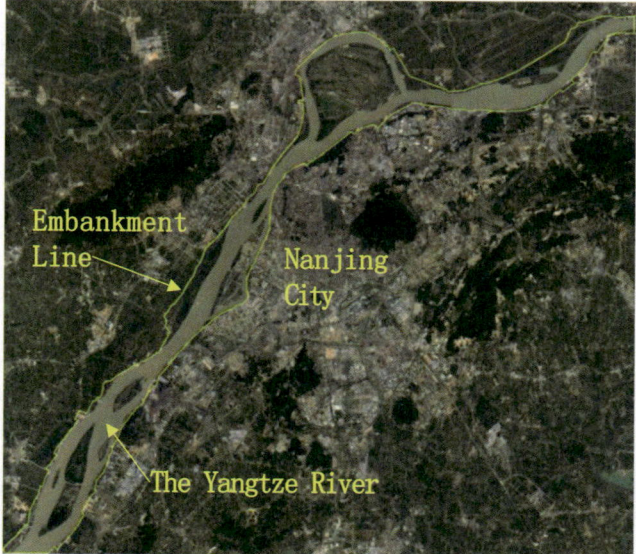

Fig. 1. Embankment in the Nanjing reach of the Yangtze River

2.2   Data

The remote sensing images of shoreline are mainly from landsat4, landsat5, Landsat7 and landsat8 satellites (Hansen and Loveland 2012). The relevant parameters are shown in the table below. About 400 remote sensing images of Nanjing reach from Landsat satellite from 1984 to 2021 were collected, and four bands (red, green, blue and NIR) with a resolution of 30 m were used (Table 1).

Table 1. Relevant parameters of satellites (Landsat 4, Landsat 5, Landsat 7 and Landsat 8)

Satellite	Sensor	Bands	wavelength (μm)	Resolution (m)	Period
Landsat4–5	TM	Red	0.45–0.52	30	Landsat4 1982–1993 Landsat5 1984–2012
		Green	0.52–0.60	30	
		Blue	0.63–0.69	30	
		NIR	0.76–0.90	30	
Landsat7	ETM+	Red	0.45–0.52	30	1999–2021
		Green	0.52–0.60	30	
		Blue	0.63–0.69	30	
		NIR	0.76–0.90	30	
Landsat8	OLI and TIRS	Red	0.43–0.45	30	2013–2021
		Green	0.53–0.59	30	
		Blue	0.64–0.67	30	
		NIR	0.85–0.88	30	

2.3   Methods

The evaluation indexes of shoreline carrying capacity of Nanjing reach of the Yangtze River were listed in Table 2. The evaluation indexes are river bank stability, river bank width, human disturbance index, shoreline utilization rate, pollution blocking function index, vegetation coverage rate, wetland retention rate and water area width. All of these eight indicators can be calculated directly from remote sensing images (Ministry of Water Resources of China 2020).

Table 2. Evaluation index system of shoreline carrying capacity based on remote sensing images

Target	Criterion	Weight	Sub-criterion	Index	Weight
Shoreline carrying capacity	Shoreline condition	0.20	Stability	River bank stability	0.10
			Spatial redundancy	River bank width	0.05
			Human disturbance	Human disturbance index	0.05
	Resource supply	0.25	Land resource	Shoreline utilization rate	0.25
	Pollution assimilation	0.25	Soil	Pollution blocking Function index	0.25
	Ecological service	0.16	Vegetation	Vegetation coverage rate	0.08
			Wetland	Wetland retention rate	0.08
	Social service	0.14	Traffic	Water area width	0.14

The overall carrying capacity of shoreline is calculated as follows:

$$C_{rl} = \sum_{n=1}^{8} w_n X_n, \tag{1}$$

where C_{rl} is the comprehensive index of shoreline carrying capacity; X_n is the evaluation index; w_n is the weight of corresponding evaluation index.

In this paper, three levels are adopted for the classification of carrying capacity, namely, underloaded, critical loaded, and overloaded. The specific classification standards are shown in the table below (Table 3).

Table 3. Levels of shoreline carrying capacity

Score	[75, 100]	[60, 75)	[0, 60)
Level	Underloaded	Critical loaded	Overloaded

The land use classification can be used to calculate the index of shoreline carrying capacity. The river bank stability, river bank width and water area width can be calculated from the water body information; the human disturbance index and shoreline utilization index can be calculated from the farmland and construction land information; the pollutant blocking function index can be calculated form the natural vegetation, wetland, farmland and construction land information; the vegetation coverage index can be calculated from vegetation information (grassland, farmland and forest land); the wetland retention index can be calculated from wetland information. It is concluded that the above eight indexes can be calculated from six types of land use (water body, forest land, construction land, grassland, farmland and wetland) information.

The classification regression tree (CART) algorithm is used to classify land use types of remote sensing images (Shao and Lunetta 2012). Figure 2(a) shows the remote sensing image of Landsat 8 satellite in the Nanjing reach of the Yangtze River on October 31, 2019 with corresponding CART classification training sample regions. Figure 2(b) shows the CART classification results. It is obvious that the classification of forest land, water body and construction land in the figure is accurate.

In order to further evaluate the accuracy of CART classification results, Fig. 3 shows the classification accuracy of remote sensing images about the Nanjing reach. It can be seen that the CART classification accuracy in recent 40 years is more than 90%, indicating that the classification results have high accuracy.

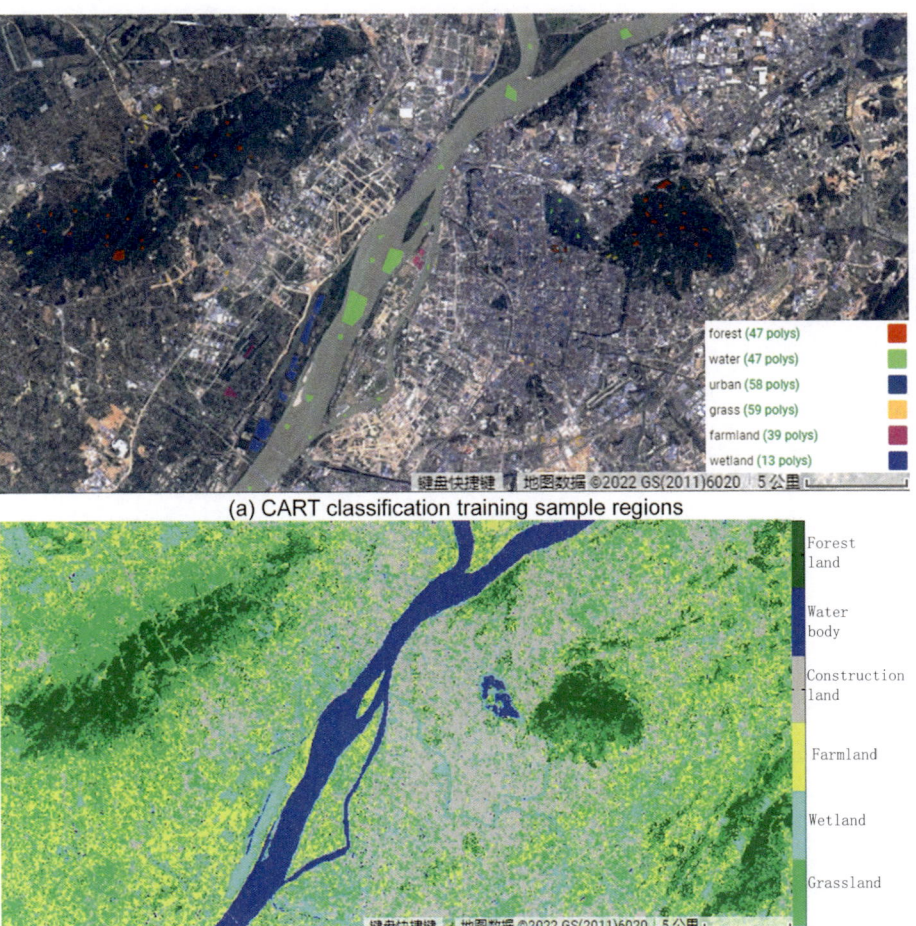

(a) CART classification training sample regions

(b) CART classification results of land use types

Fig. 2. CART classification of land use types of remote sensing images of the Nanjing reach (2019-10-31)

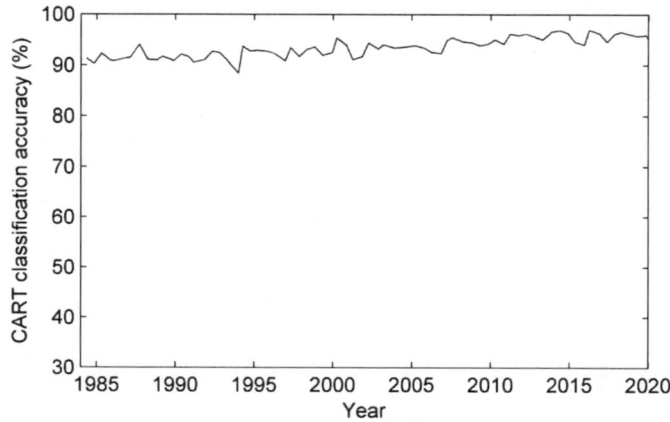

Fig. 3. CART classification accuracy of remote sensing images about the Nanjing reach

3   Results

3.1   Shoreline Morphology Evolution

The Nanjing reach of the Yangtze River is located in Jiangsu Province, starting from Maozi Hill to Sanjiangkou, with a total length of 92.3 km (Fig. 4). It is one of the 14 key reaches of the middle and lower reaches of the Yangtze River. The upper reaches of Nanjing reach are connected with Ma'anshan reach, and its lower reaches flow into Zhenyang reach. There are five major sandbars distributed in the Nanjing reach, namely, Bagua Sandbar, Meizi Sandbar, Xinqian Sandbar, Xinsheng Sandbar, and Xinji Sandbar.

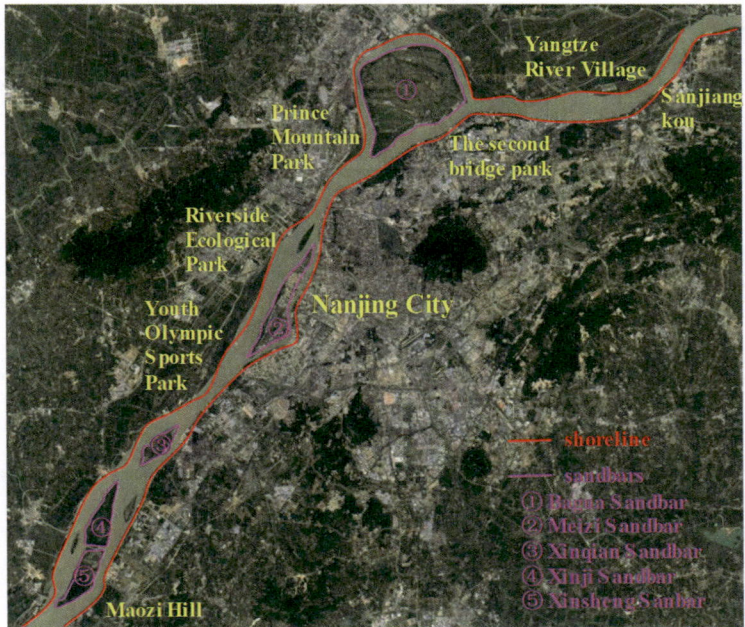

Fig. 4. Shoreline and sandbars in the Nanjing reach of the Yangtze River

Figure 5 shows the shoreline evolution in the Nanjing reach of the Yangtze River from October 30th, 1984 to November 18th, 2020. It can be seen that in the past 40 years, only local shorelines have change. The shoreline change near the Yangtze River Village is the fiercest, while the overall pattern of shoreline in the Nanjing reach has not changed greatly. The change degree of the shoreline in the right bank is more softer than that in the left bank, and its shoreline change is not violent.

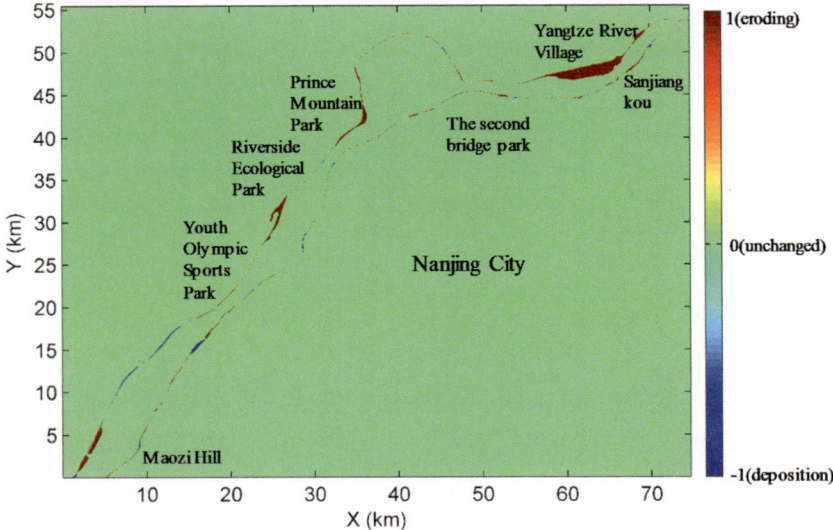

Fig. 5. Shoreline evolution in the Nanjing reach of the Yangtze River (1984–2020)

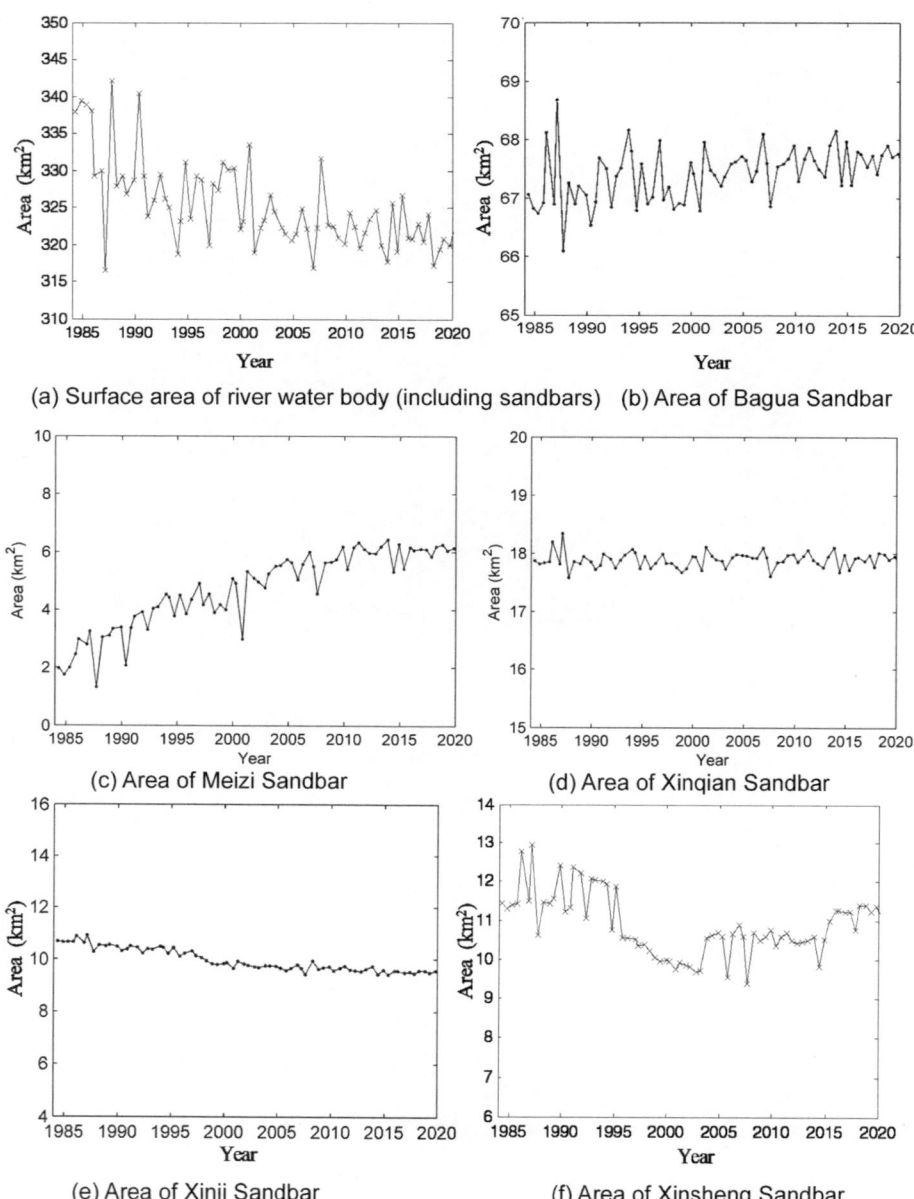

Fig. 6. Area variation of river water body and sandbars in the Nanjing reach in recent 40 years

Figure 6 shows the area variation of river water body and sandbars in the Nanjing reach in recent 40 years. It is worth noting that the water body area in Fig. 4(a) refers to the area surrounded by the shorelines in the Nanjing reach, and the sandbar area is also included. It can be seen that in the past 40 years, the water body area of the Nanjing reach has decreased slowly, from the largest 3.4 km^2 to 3.2 km^2 with a decrease of about 6%. This shows that human activities have been leading to the eroding of river shoreline in the past 40 years. Among the five large sandbars in the Nanjing reach, Bagua Sandbar, Meizi Sandbar and Xinji Sandbar are relatively stable and change little; The Xinsheng Sandbar changes greatly whose head degrades and tail grows, but the change degrees of the two evolutions are similar, resulting in a small change in its area. The deposition degree of the tail of the Xinqian Sandbar is serious, with an area increase of 200%. As there is almost no human activity there, the deposition of the tail of the Xinqian Sandbar is mainly caused by the change of water and sediment.

3.2   Shoreline Carrying Capacity

Figure 7 shows the comprehensive index of shoreline carrying capacity of the Nanjing reach of the Yangtze River in the last four decades. It can be seen that the shoreline carrying capacity in Nanjing reach has always been within the critical load zone, indicating that the development and utilization of the shoreline is still sustainable. The shoreline carrying capacity of the Nanjing reach was basically stable from 1984 to 2003. With the large-scale development and utilization of the shoreline since 2003, the carrying capacity of the Nanjing reach gradually decreased and approached the warning line. Due to the implementation of restrictive measures such as "action to clear the four chaos" and "the operation of responsibility system on river/lake leaders" by Chinese government after 2018, the carrying capacity of the Nanjing reach has rebounded rapidly.

Figure 8 shows the mutation analysis of the time series of the shoreline carrying capacity index detected by Mann-Kendall (MK) method. UF and UB are the statistical series calculated according to the order and reverse order of the time series respectively. If the value of UF or UB is greater than 0, it indicates that the series shows an increasing trend, and if it is less than 0, it indicates a decreasing trend. When they exceed the significance lines, it indicates that the increasing or decreasing trend is significant. If the UF and UB curves intersect and the intersection is between the significance lines, then the intersection is the mutation point.

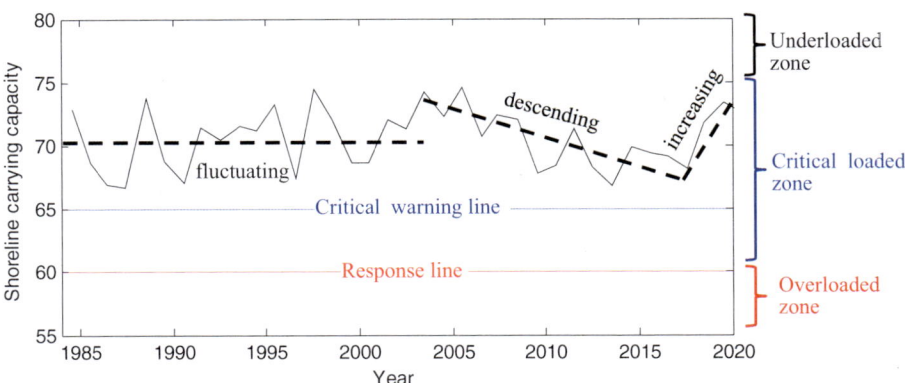

Fig. 7. Shoreline carrying capacity of the Nanjing reach of the Yangtze River in the last four decades

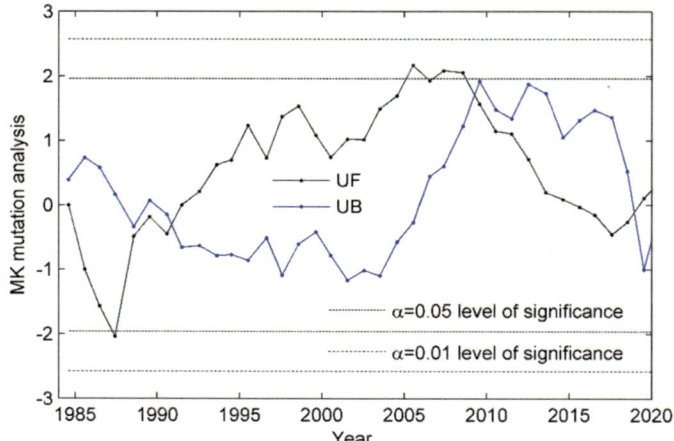

Fig. 8. MK mutation analysis of shoreline carrying capacity in the last four decades

In the past 40 years, the shoreline carrying capacity index of the Nanjing reach of the Yangtze River has shown an overall increasing trend with fluctuations, and the increasing trend exceeds the $\alpha = 0.05$ significance line, which means the mutation is significant. UF and UB curves intersected in 1991, 2012 and 2018, and the intersection point was between the $\alpha = \pm0.05$ significance line (±1.96), indicating that the shoreline carrying capacity of Nanjing reach mutated in 1991, 2012 and 2018.

4   Conclusions

Taking into account five aspects of shoreline health, resource supply, environmental pollution, ecological service and social service, the evaluation index system of shoreline carrying capacity was systematically established, and the comprehensive index of shoreline carrying capacity is classified into three levels: underloaded, critical loaded, and overloaded.

Based on the remote sensing images over the past 40 years, the variation of shoreline carrying capacity in Nanjing reach of the Yangtze River are quantitatively calculated. The results show that the comprehensive index of shoreline carrying capacity in Nanjing reach is within the critical loaded zone, indicating that the development and utilization of this reach is within the bearing range. Due to the implementation of restrictive measures such as "action to clear the four chaos" and "the operation of responsibility system on river/lake leaders" by Chinese government in recent years, the value of shoreline carrying capacity has increased.

The research results can help to discover unsustainably social and economic activities, put forward the productivity layout adjustment, and guide corresponding management measures in the reach.

Acknowledgements. This work was funded by the National Natural Science Foundation of China (No. 52179072) and supported by the Open Research Fund of Key Laboratory of Sediment Science and Northern River Training, the Ministry of Water Resources, China Institute of Water Resources and Hydropower Research, Grant NO. IWHR-SEDI-202107.

References

Hansen MC, Loveland TR (2012) A review of large area monitoring of land cover change using Landsat data. Remote Sens Environ 122:66–74

Lane M (2010) The carrying capacity imperative: assessing regional carrying capacity methodologies for sustainable land-use planning. Land Use Policy 27(4):1038–1045

Lu Y, Xu H, Wang Y, Yang Y (2017) Evaluation of water environmental carrying capacity of city in Huaihe River Basin based on the AHP method: a case in Huai'an City. Water Resour Ind 18:71–77

Malthus TR (1970) An essay on the principle of population. Penguin, New York

Martire S, Castellani V, Sala S (2015) Carrying capacity assessment of forest resources: enhancing environmental sustainability in energy production at local scale. Resour Conserv Recycl 94:11–20

Ministry of Water Resources of China (2020) Technical guidelines for river and lake health assessment (SL/T 793-2020). Printing Office of Ministry Of Water Resources, Beijing, China

Park RE, Burgess EW (2019) Introduction to the science of sociology. The University of Chicago Press, Chicago

Price D (1999) Carrying capacity reconsidered. Popul Environ 21(1):5–26

Shao Y, Lunetta RS (2012) Comparison of support vector machine, neural network, and CART algorithms for the land-cover classification using limited training data points. ISPRS J Photogramm Remote Sens 70:78–87

Song X, Kong F, Zhan C (2011) Assessment of water resources carrying capacity in Tianjin City of China. Water Resour Manag 25(3):857–873

Su Y, Yu Y (2020) Dynamic early warning of regional atmospheric environmental carrying capacity. Sci Total Environ 714:136684

Sun M, Wang J, He K (2020) Analysis on the urban land resources carrying capacity during urbanization–a case study of Chinese YRD. Appl Geogr 116:102170

Wang R, Cheng J, Zhu Y, Xiong W (2016) Research on diversity of mineral resources carrying capacity in Chinese mining cities. Resour Policy 47:108–114

Wei C, Guo Z, Wu J, Ye S (2014) Constructing an assessment indices system to analyze integrated regional carrying capacity in the coastal zones–A case in Nantong. Ocean Coast Manag 93:51–59

Simulation and Hazard Map of Flooding Caused by the Break of a Concrete Gravitational Dam

Qing Leng[1]([✉]), Ming Zhang[1], Gensheng Zhao[1], Senhao Mao[1], and Ang Jiang[2]

[1] State Key Laboratory of Hydrology, Water Resources and Hydraulic Engineering, Nanjing Institute of Hydraulic Sciences, Nanjing, China
{276998265,913833325}@qq.com, {zhangm,gszhao}@nhri.cn
[2] Hubei Province Hanjiang Yakou Shipping Hub Engineering Construction Headquarters, Xiangyang, China
1959261347@qq.com

Abstract. The simulation of Concrete Gravitational Dam burst floods is an important research content in the field of disaster prevention and mitigation in water conservancy projects. Due to the extreme rainfall, earthquake, structure failure and etc., the concrete gravitational dam usually breaks in a short time period. The dam break flood will give an extreme risk to the downstream communities. Taking the flood simulation of the Kaliwa Dam in the Philippines as an example, based on the downstream channel of the dam body and the measured terrain on both sides, a numerical simulation model of one-dimensional and two-dimensional coupled flood evolution is constructed, and the numerical simulation of the flood evolution process of the dam collapse is carried out, counts the inundation range, water depth, flow velocity, flood arrival time and other disaster causing factors in the downstream inundation area, and draws the flood hazard map of both banks downstream. The simulation results show that the KALIWA burst accident occurred, the total inundation area downstream is over 38 km^2 in the downstream of the Kaliwa Dam. The dam break flood peak takes 1.5 h to reach the downstream estuary, which is the shortest time. The General Nakar City and Infanta City are inundated completely with the depth of 1.0 m to 2.0 m. The terrain near the upper reaches of the lower estuary is open and flat, the downstream area will be affected seriously by the flood. It is proposed to build a flood warning system to give the people downstream of Kaliwa Dam. The results of the research will provide a scientific basis for dam-break flood risk analysis, disaster assessment.

Keywords: Dam-break flood · Concrete gravitational dam · Coupling model · Flood hazard map

1 Introduction

With the rapid economic development, all countries in the world have attached great importance to the safety of dams. While studying the possible dangers of dam-breaking floods to the downstream, they are also taking proactive preventive measures [1]. For example: the National Dam Safety Plan launched by the United States, the European

© The Author(s) 2023
Y. Li et al. (Eds.): PIANC 2022, LNCE 264, pp. 1248–1260, 2023.
https://doi.org/10.1007/978-981-19-6138-0_109

Community organized more than 10 member countries to carry out cooperative research on dam failure issues, and the International Committee on Dams integrated and coordinated on this basis, and released ICOLD No. 111 Breakthrough The dam flood analysis guidelines have a clear normative and guiding role for the analysis and calculation of dam break floods [2].

KALIWA Dam in the Philippines is located in Metro Manila, Philippines. From left to right, there are gravity dam, spillway dam with gate, free spillway dam and gravity dam. The length of the dam is 247.19 m. The height of the dam is 73.30 m. The elevation of the dam is 173.30 m and the dam foundation elevation is 100 m. The dam breach will happen due to the structural failure. The breach can be assumed into Instantaneous breach in second with a final breach opening. It is different from the breach process in the earthen dam with a breaching period. At the plane of EI 124.5 m, the upstream vertical dam body becomes a slope of 1:0.25. And there is gallery of 3 m × 3.5 m at EI 124.5. In the spillway dam, the elevation of the spillway is at around 124.5 m. The flow will give a lot of pressure on the spillway when the flow goes through it. So the force distributions are very complex at the plane of EI 124.5 m, and outside forces can destroy the plane and trigger a dam break here. The layout of the KALIWA dam is shown in Fig. 1.

Due to the occurrence of extreme events such as super-standard floods, earthquakes, and structural damage, concrete gravity dams have the risk of dam failure in a short period of time, and the dam-breaking floods will bring extreme disaster losses to the downstream. Therefore, it is necessary to carry out research work on the flood risk of dam break. According to the terrain characteristics of the KALIWA dam in the Philippines, this paper establishes a two-dimensional coupled hydrodynamic model, conducts numerical simulation for the proposed dam break condition, and counts the disaster-causing factors such as the submerged range, water depth, flow velocity, and flood arrival time in the downstream submerged area, draws a flood risk map on both sides of the downstream under normal water level dam failure, and provides technical support for the later concrete dam failure flood risk analysis and disaster loss assessment [3].

Fig. 1. Layout of Kaliwa dam.

2 Model Building

Dam-break flood propagation is an unsteady flow. According to ICOLD Bulletin 111, "Dam Break Flood Analysis," Mike FLOOD software is recommended for use in the Kaliwa Dam dam break flood simulation. Mike FLOOD dynamically links two separate software packages: Mike 11 (1D) and Mike 21 (2D) [4].

2.1 Model Range

In order to fully reflect the inundation caused by the collapse of the dam flood to the downstream area, the upper boundary of the simulation range is the dam site, and the lower boundary is the river estuary.

This time, two topographic maps were collected, namely the topographic map of the power station area and the topographic map of the Kaliwa Dam site. The topographic map of the power station area adopts the contour line format, and the elevation interval is 3.0 m. Its coverage area is large, and its lower boundary is about 39 km away from the dam site, which is determined as the dam failure simulation area. The topographic map of the dam site only covers the dam site area and has a small range. It is represented by a mixed format of contour lines and elevation points. The two topographic maps are corrected and merged and superimposed with the image map based on the GIS platform, as shown in Fig. 2.

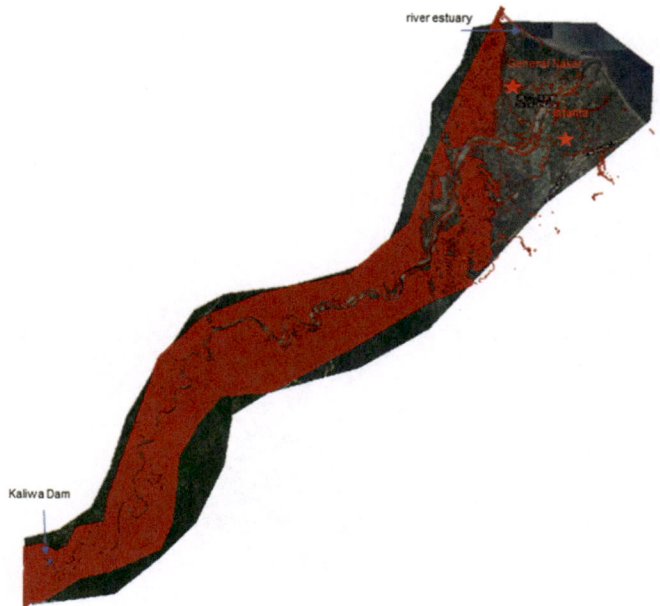

Fig. 2. Schematic diagram of the simulation range.

2.2   Model Establishment

Based on the measured cross-section and the topographic map of both sides of the strait, based on the MIKE FLOOD simulation platform, the one-dimensional and two-dimensional numerical simulation models of flood evolution downstream of the dam are constructed. Among them, the two banks of the river are steep mountains, using a one-dimensional hydrodynamic model, and the downstream terrain is relatively flat, which is a two-dimensional hydrodynamic model.

The establishment of the numerical simulation model of Kaliwa dam failure follows the principle from simple to complex, from point to line, and the model is formed by connecting the former set of specific single network objects.

(1)   Terrain model
Terrain data is the underlying data for the model. The creation and modification of cross-sections and submerged area meshes along the line are based on terrain data. Based on the GIS platform, a terrain model of the flood area on both sides of the river is generated. The 2D model simulates a range of about 492 km^2, with an average mesh size of 30 m, for a total of 615036 triangular grids.

(2)   Cross-section
Cross-sections are the most basic elements in model calculations. Therefore, the accuracy of the cross-sectional data directly affects the accuracy of the calculation. The last section is about 39 km from the dam, which is consistent with the topographic map. The one-dimensional model reflects the terrain features in the cross-section of the river channel, and the number of cross-sections entered in the model is 32 (Fig. 3).

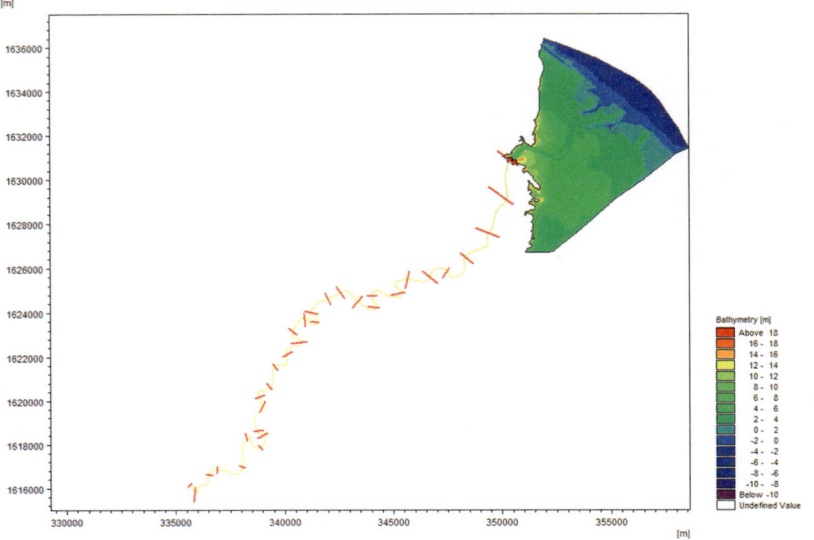

Fig. 3. Dam break flood model of Kaliwa dam.

2.3   Model Solving

The basic theoretical method for the numerical simulation of flood in complex basin is the hydrodynamic equation, namely the saint venant equations, describing the hydraulic characteristics of water bodies.

(1)   The basic principle of the one-dimensional flow mathematical model of the river reach [5]:

① Basic equation

Continuity equation:

$$B\frac{\partial Z}{\partial t} + \frac{\partial Q}{\partial x} = q \tag{1}$$

Momentum equation:

$$\frac{\partial Q}{\partial t} + \frac{\partial}{\partial x}\left(\alpha\frac{Q^2}{A}\right) + gA\frac{\partial Z}{\partial x} + gA\frac{Q|Q|}{K^2} - \left(V_x - \frac{Q}{A}\right)q = 0 \tag{2}$$

where $K = AC\sqrt{R}$ is modulus of discharge; $\alpha = \frac{\int_A v^2 dA}{\bar{v}^2 A}$ is momentum correction coefficients: where Q - is flow (m³/s); A - is area (m); V - is velocity (m/s); R - is hydraulic radius (m); C - is Chezy coefficient (m$^{1/2}$/s); B - is width of river channel (m); q - is lateral inflow (m²/s); Vx - is the velocity component of the lateral inflow in the direction of flow (m/s); $\bar{v}$ – is mean velocity of cross section (m/s).

② Definite condition

The control condition for the boundary of upstream and downstream are generally controlled in the forms of water level process, flow process and flow-water level relationship, which can be written in the following unified form:

$$\alpha(t)Z(x_b, t) + b(t)Q(x_b, t) + c(t) = 0 \tag{3}$$

When t = 0, the water level and flow of each cross section is known, that is,

$$\begin{aligned} Z(x,0) &= Z_0(x) \\ Q(x,0) &= Q_0(x) \end{aligned} \tag{4}$$

The definite solution of one-dimensional flood motion is composed by the basic Eq. (1), Eq. (2), boundary condition (3) and initial condition (4).

(2)   Two-dimensional numerical simulation model of protection zone along the channel.

① Basic equation

The conservation form of differential equation has the characteristics of maintaining the conservation nature of physical quantity, and has advantage of being easier in dealing with the nonlinear convection terms in the process of numerical discretization. Therefore, the conservation form of shallow water wave equation is adopted in this study as the control equation of two-dimensional flood motion.

$$\frac{\partial h}{\partial t} + \frac{\partial U}{\partial x} + \frac{\partial V}{\partial y} = s_0 \tag{5}$$

$$\frac{\partial U}{\partial t} + \frac{\partial}{\partial x}\left(\frac{U^2}{h}\right) + \frac{\partial}{\partial y}\left(\frac{UV}{h}\right) + gh\frac{\partial H}{\partial x} + gh(J_{fx} - J_{0x}) = 0 \tag{6}$$

$$\frac{\partial V}{\partial t} + \frac{\partial}{\partial x}\left(\frac{UV}{h}\right) + \frac{\partial}{\partial y}\left(\frac{V^2}{h}\right) + gh\frac{\partial H}{\partial y} + gh(J_{fy} - J_{0y}) = 0 \tag{7}$$

where h is the water depth; U and V are the single-width discharges in the $U = uh$ x and y *directions, respectively:* $V = vh$, u and the flow velocity in the v and direction, y respectively x; s_0 the source term; J_{ox} and J_{oy} the x bottom slope in the $J_{0x} = -\frac{\partial Z}{\partial x}$ and direction: y, $J_{0y} = -\frac{\partial Z}{\partial y}$, Z the river bottom elevation; J_{fx} and J_{fy} are respectively the x friction gradient in $J_{fx} = \frac{n^2 u\sqrt{u^2 + v^2}}{h^{4/3}}$ and direction: y, $J_{fy} = \frac{n^2 v\sqrt{u^2 + v^2}}{h^{4/3}}$, n is the Manning roughness coefficient.

② Definite condition
Inflow and outflow boundary conditions:

$$aU(x_b, y_b, t) + bV(x_b, y_b, t) + cH(x_b, y_b, t) + d = 0 \tag{8}$$

The solid boundary is the condition of slip boundary:

$$\vec{U}_n(x_b, y_b, t) = 0 \tag{9}$$

The initial conditions are:

$$\left.\begin{array}{l} U(x, y, 0) = U_0(x, y) \\ V(x, y, 0) = V_0(x, y) \\ h(x, y, 0) = h_0(x, y) \end{array}\right\} \tag{10}$$

3   Simulation Analysis of Dam Break Flood

3.1   Data Input

The study area in 2D model coupled 1D model is calculated from the upstream of Kaliwa Dam to the downstream with a length of 39 km. The computation mesh consists of the main river bed and river banks. The computation mesh in 2D model is set in an unstructured grid with the shape of triangles. Bathymetry of model domain in 2D model is collected from the measurement of the topographical map. The parameters of 2D model are set up as follows: Calculation time step is 5 s; Viscosity coefficient is 0.28 m²/s; Manning (M) coefficient arranges from 15–18 m^{1/3}/s. Because there are lack of flood data on the river channel and land, the Manning Coefficient in the land was used empirically. In the extreme flood simulations, the empirical Manning Coefficient performs very well. Testing the calculation results shows that Courant coefficient (Cr) is smaller than 1. It means that 2D model runs smoothly, and the detailed simulation the moving direction of flow.

Based on MIKE FLOOD, from upstream to downstream, lateral connection is adopted among adjacent sections of left and right bank and the coupling connection between the one-dimensional model of river channel and the two-dimensional model for both banks. Overflow weir is used to simulate the flood overflow on both banks. And the elevation of the dyke on left and right bank is reagred as the weir control elevation.

3.2   Dam Collapse Flood Simulation

According to ICOLD Bulletin 111 "Dam Break Flood Analysis", the breach of the Kaliwa dam can be assumed into full breach [6]. In the ICOLD bulletin, all dam-breaking flow formulas are intended to calculate the dam-breaking flow for earth dykes, not concrete gravity dam-breaking flows. Based on the rectangular gap weir formula, the maximum flow can be calculated using the weir formula [7].

$$Q_{max} = \frac{8}{27}\sqrt{g}\left(\frac{B}{b_m}\right)^{\frac{1}{4}}b_m H_0^{\frac{3}{2}} \tag{11}$$

which Q_{max} is the maximum dam break discharge, g is the gravitational acceleration, B is the length of the dam, b_m is the breach width, H_0 is the water depth in front of the dam.

According to formula (11), the maximum flow of Kaliwa dam failure under normal water level can be calculated to be 66953.09 m³/s. The parameters of the simulation scheme are set as shown in Table 1.

Table 1. Simulation scheme parameter settings.

Hydraulic boundary	Reservoir water level (m)	Maximum flow (m³/s)	Downstream water level (m)	Notch width (m)
Normal water level	168.50	66953.09	107.60	247.19

The capacity of Kaliwa reservoir needs to be designed to ensure that it will not reduce water supply of 600 MLD during a 1:10 year drought condition. The Kaliwa Dam reservoir volume – area – water level relationship are shown in Fig. 4.

The KALIWA Dam burst flooding process is shown in Fig. 5 below. This flow process serves as an upper boundary input condition calculated by the downstream flood evolution model. Lower boundary conditions take the sea level water level of the estuary, 0 m.

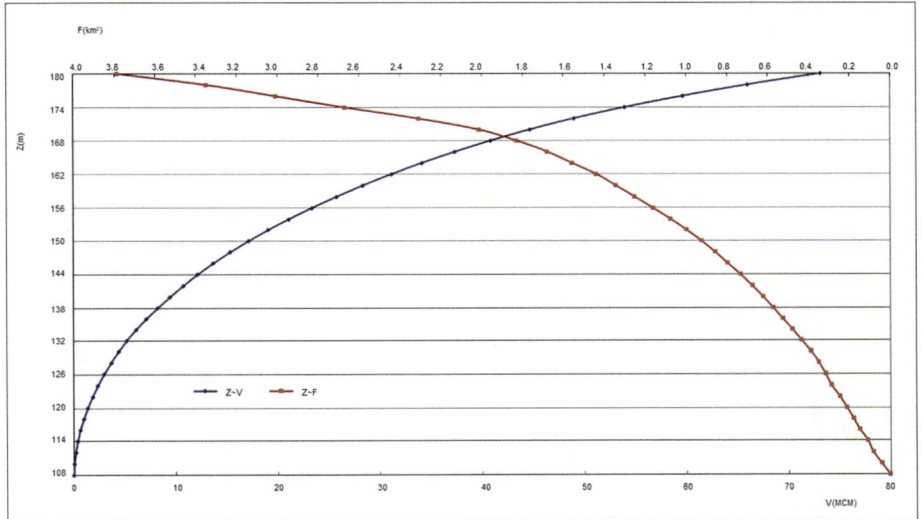

Fig. 4. Kaliwa dam reservoir volume – area – water level relationship.

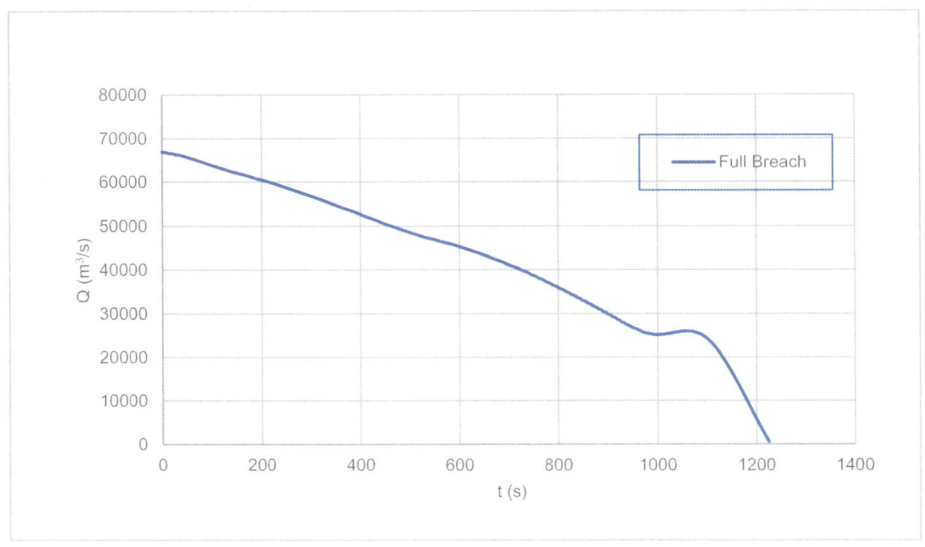

Fig. 5. Dam break discharge hydrograph in case of normal water level.

3.3   Simulation of Dam Collapse Flood Evolution

A numerical simulation model of one-dimensional and two-dimensional coupled flood evolution downstream of the dam body is used to simulate and calculate the above-mentioned dam collapse flood process.

3.3.1   Flood Inundation Results of the One - Dimensional Model

In the case with the full breach scenario in the condition of normal water level, the maximum dam break discharge is 66953 m^3/s. The cross-section downstream of the dam break reaches the highest water level 123 m after the dam break occurs 7 min. It takes 30 min for the flood to reach the last cross-section. The peak water level reduces to 20 m at the last cross-section when the dam break occurs 45 min, where the peak discharge is 23000 m^3/s. The flood last around 80 min. The water levels and discharges at different cross-sections are shown in Figs. 6 and Figs. 7.

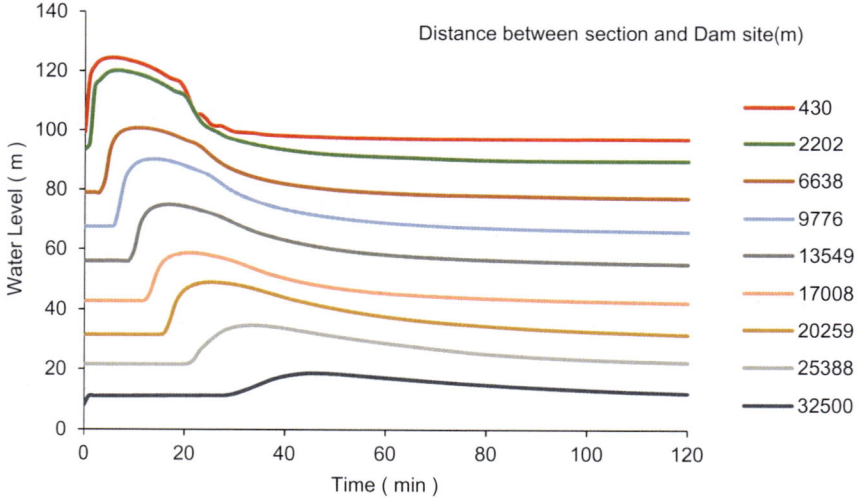

Fig. 6. Water level of different cross-sections.

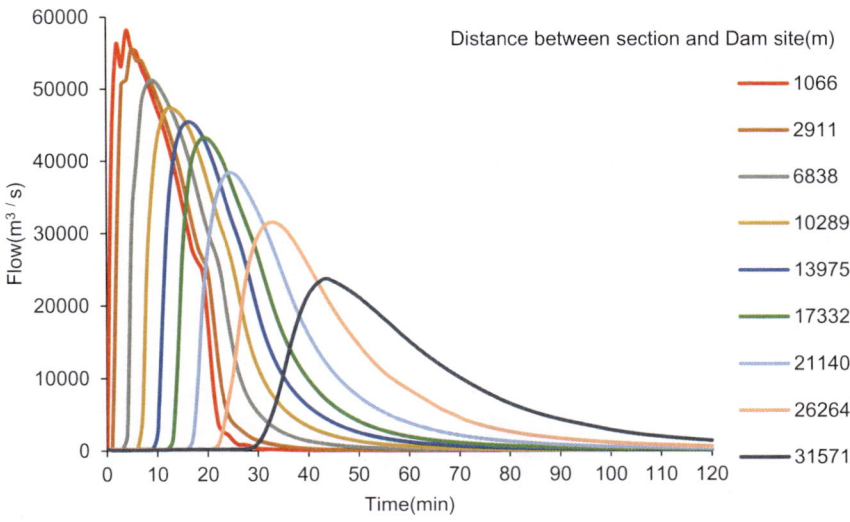

Fig. 7. Flood discharges in different cross-sections.

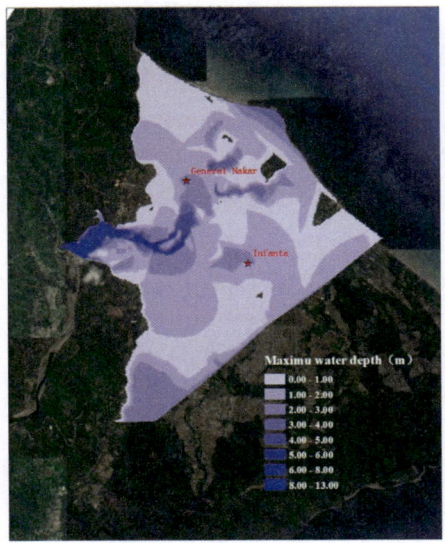

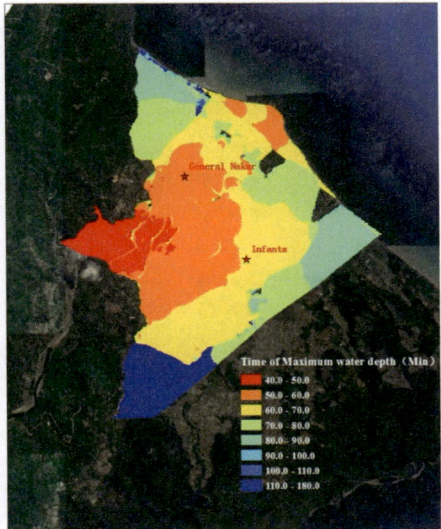

Fig. 8. Map of the maximum inundation depth of dam break flood at downstream.

Fig. 9. Map of time table for the maximum inundation depth of dam break flood at downstream.

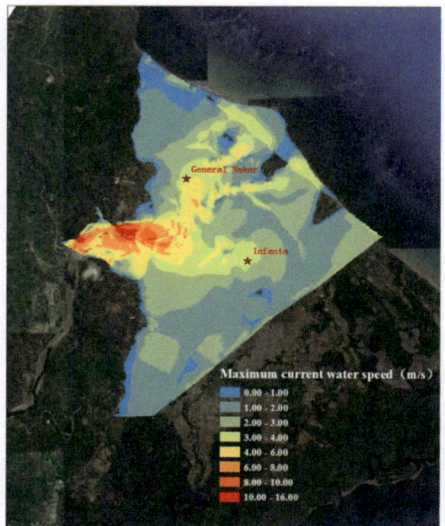

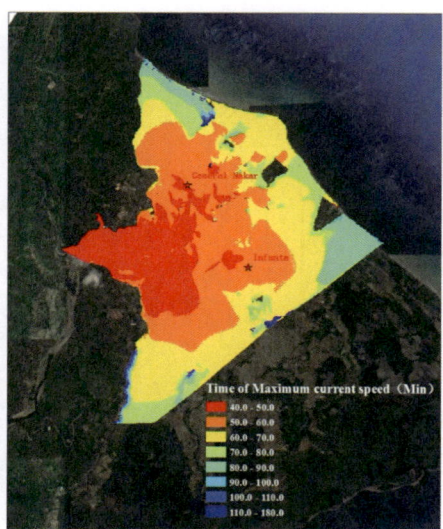

Fig. 10. Map of the maximum inundation velocity of dam break flood at downstream.

Fig. 11. Map of time table for the maximum inundation velocity of dam break flood at downstream.

3.3.2   2D Dam Break Flood Inundation Simulation

In the condition of the Normal Water Level, the maximum breach discharge is 66953 m^3/s when the full breach happens. The maximum inundation depth and the time to reach the maximum inundation depth, maximum inundation velocity and the time to reach the maximum velocity are shown in Figs. 8, Figs. 9, Figs. 10, Figs. 11, respectively.

The total inundation area downstream is about 38.77 km^2, with an average depth of 1.98 m and a maximum inundation depth of approximately 12.46 m in the downstream of the Kaliwa Dam. The dam break flood peak takes 1.5 h to reach the downstream estuary. In the upstream of the estuary, 7 km upstream of the estuary, the geometry becomes mild with low elevation. The dam break flood depth reduces to 1.0 m–5.0 m. The General Nakar City and Infanta City are inundated completely with the depth of 1.0 m to 2.0 m. The flood peak reaches General Nakar City and Infanta City around 50.0 min. The averaged velocities range from 1 m/s to 3 m/s in the flat estuary area.

4   Impact Analysis

The dam-break floods could affect the cities of Nakar, with a population of more than 30,000, and Infanta, with a population of more than 70,000. Both cities will be inundated by floodwaters with a depth of 1.0 m to 2.0 m. The estuary area will be flooded by dam-break floods. Farmland, roads, factories, houses will be destroyed by dam-breaking floods.

Since it takes around 50 min for the dam break flood peak to arrive at General Nakar City and Infanta City, it is proposed to build a flood warning system to give the people downstream of Kaliwa Dam. The waring time can be 80 min. People can be evacuated within 80 min if the Dam break happens.

Flood warning system is an important means to mitigate dam-break floods, and flood insurance may be another measure to mitigate dam-break floods. Flood insurance is recommended for people and property downstream of the Kaliwa Dam.

5   Conclusions

Kaliwa Dam is a concrete gravity dam, including gravity dam, spillway dam with gate, free spillway dam, and gravity dam. Concrete gravity dams usually burst within a short period of time due to extreme earthquakes, structural damage, etc. Dam-breaking floods will pose significant risks to downstream communities and infrastructure. In this paper, based on the one- dimensional and two-dimensional coupling technology, the numerical simulation analysis of the dam-break hydrological flow process and the flood inundation downstream of the dam-break is carried out. The main conclusions are as follows:

(1)   According to the ICOLD Bulletin 111 "DAM-BREAK FLOOD ANALYSIS", the breach of the Kaliwa dam is assumed into the full breach, the breach width is 247.19 m. The elevation of the breach is assumed as 124.5 m. The Mike flood model was used to simulate dam break flooding downstream of the Kaliwa Dam. The maximum flow can be calculated by the weir formula, and the hydrograph of the dam-breaking flood can be calculated by the water balance formula. The propagation of dam-break floods is simulated under normal water level hydraulic boundary conditions.

(2)   According to the simulation of dam-break flood cases, the total submerged area downstream of the Kaliwa Dam exceeds 38 km^2. It takes 1.5 h for the flood peak to reach the downstream estuary. Due to the limitations of the river bank geometry, floodwaters can pass through the channel with a width of 200 m-350 m. The main water depth in the channel can reach 12.46 m. In the upper reaches of the estuary, that is, 7 km upstream of the estuary, the terrain is flat and the elevation is low. The dam-break flood depth is reduced to 1.0 m–5.0 m. The cities of Nakar and Infanta were completely submerged at a depth of 1.0 m to 2.0 m. The flood peak reaches General Nakar City and Infanta City around 50 min. The average flow velocity in the flat estuary area drops to 1.0 m/s–3.0 m/s.

(3)   The downstream areas will be severely affected by flooding due to the failure of the Kaliwa Dam. The cities of Nakar and Infanta will be inundated with water depths exceeding 1.0 m. The estuary area will be inundated with a flood depth of more than 1.0 m, especially the area 7.0 km upstream of the estuary to the sea. It will take about 50 min for the dam break to reach the city of General Santos, and a flood warning system is recommended to provide information to people down-stream of the Kaliwa Dam. The minimum alert time is 80 min. If the dam breaks, people can evacuate within 80 min.

Acknowledgements. This study is financially supported by the National Key R&D Program of China (2021YFC3200302).

References

1. Zhan M, Guo Y, Yang Y (2022) Flood evolution simulation and impact analysis of face rockfill dam failure. Pearl River 43(1):11–18, 27
2. Cheng T, He W, Zhao H (2022) Research on dam-break flood risk analysis based on 1-D and 2-D hydraulic coupling calculation model. Water Power
3. Zhong QM, Wu WM, Chen SS (2016) Comparison of simplified physically based dam breach models. Nat Hazards 81(2):1385–1392
4. Lian Y (2016) Research on the compilation of flood risk map in Ankang city based on MIKE one-dimensional and two-dimensional coupling. Northwest A&F University
5. Song L (2012) Mathematical model and hydrodynamic characteristics of dam break flood. Huazhong University of Science and Technology
6. Wang L, Hu S (2007) A review of research on dam failure. Progr Water Resour Hydropower Sci Technol 27(1):80–85
7. Fread DL, DAMBRK (1984) The NWS dam break flood forecasting model. National Weather Service

Study on Air Bubble Plume in Open Channel with CFD-PBM Coupling Model

Jinchao Xu[1,2], Xiaodong Wang[2(✉)], Long Zhu[2], Donghui Zhou[3], and Jun Zhao[1,2]

[1] Nanjing University of Information Science and Technology, Nanjing, China
jcxu@nuist.edu.cn
[2] Nanjing Hydraulic Research Institute, Nanjing, China
{xdwang,lzhu}@nhri.cn
[3] Jiangsu Yangtze River Delta Smart Water Research Institute, Nanjing University of Information Science and Technology, Nanjing, China

Abstract. Air bubble plume flow has been applied widely in the dredging, ice breaking, and pollution control at navigation projects. But the interaction regimes among bubbles or between bubbles and water are not quite clear. Especially in open channels, the bubble plume flow are significantly affected by the separation phenomenon which is caused by the cross flow velocity. According to the existing research, the interaction force of gas-liquid and the distribution of bubble size are the key parameters to simulate the hydrodynamic characteristics of bubble plume flow. In order to explore the mechanism of air bubbles entrained plumes in open channels, an Eulerian-Eulerian approach for air-water flows numerical model was introduced, and the population balance model (PBM) was included to describe the distribution of bubble size. The cross flow velocity of open channels has been discussed in the proposed numerical model. It shows that the separation of bubble plume is strongly influenced by the cross flow velocity. The influence of these parameters on the movement characteristics of air bubbles is studied. The results indicate that the cross flow velocity has great impact on bubble plume as well as the lifting effectiveness of pneumatic sluicing. This research provides references for bubble plume in engineering applications.

Keywords: Bubbles plume · Cross flow · CFD-PBM · Breakup · Coalescence

1 Introduction

Pneumatic sluicing is an efficient dredging technology which can be used in approach channel, ports, or lake (Pan and Wu 2019). The air-injection of pneumatic sluicing can be employed to stir the bottom nutrients or sediments and take them to the downstream or other place (Ding et al. 2019). It is significant to study the rising process of bubble plume and air-water interaction in open channel in order to improve the transportation efficiency of sands or nutrients.

There are lots of scholars focusing on the movement of bubble plume in water, the gas-liquid characteristics such as gas holdup, bubble velocities, bubble size, and entrained plume flux (Qiang et al. 2018; Yao et al. 2019; Wan et al. 2017; Cheng et al.

© The Author(s) 2023
Y. Li et al. (Eds.): PIANC 2022, LNCE 264, pp. 1261–1270, 2023.
https://doi.org/10.1007/978-981-19-6138-0_110

2020). For example, Liang and Peng (2005) proposed a formula to calculate the velocity of rising bubble plume in quiescent water with gas discharge, the density of liquid-gas, and height. Song et al. (2011) studied the distribution characteristics of bubble void fraction by physical model experiment and image processing technology. The results show that when the aspect ratio of model's height and width is 1.0, the bubble plume structure is less affected by the pressure, and the plume structure is stable. With the increase of aspect ratio and pressure, the plume structure is unstable.

With the development of multi-phase numerical simulation, there are lots of researchers adopt air-water two phase flow to study the air plumes in water. And Euler-Euler model is used wildly in multiphase flow modeling with high void fraction (Duguay et al. 2021). For example, Liu and Li (2018) study the bubbly flows in water, and the sensitivity of different turbulence models and the scale-adaption of Euler-Euler model had been conducted. It found that the mesh size should be taken account in the model. Fleck and Rzehak (2019) studied the dynamic flow phenomena of bubble plume with Euler-Euler two-fluid model, and the periodically oscillating bubble plume was been simulated, which shown a good precision about the plume oscillation period. Godino et al. (2020) studied the air-water dispersed and segregated multiphase flows with experiments and Euler-Euler numerical model. Five cases with different flow regimes were simulated using the same set of interfacial force models, and the local rheology of the flow in different interfacial models was considered by a linear blending method. And so on.

Existing studies have gained many achievements in bubble plume and its application. Nevertheless, most of the studies only focus on the rising characteristics of bubble plume and its influence on surrounding waters in quiescent waters. And the study of bubble plume in open channels is insufficient, which the bubble plume is obviously affected by cross flow (Qiang et al. 2018). Based on numerical simulation, this paper uses Eulerian-Eulerian approach for air-water flows, and the population balance model (PBM) is applied to describe the distribution of bubbles, then the gas holdup, size distribution of bubbles, and the flow field of aerated area are studied.

2 Simulation Model

In CFD-PBM model, the turbulent dissipation rate, gas holdup and flow field are calculated by the two-fluid CFD model. And the results are used to calculate the bubble coalescence and breaking rate. Then the PBM equations are solved for the bubble size distribution, flow pattern, interphase force and the turbulence source term caused by bubble turbulence in the improved two-fluid model.

2.1 Euler-Euler Model

Euler-Euler model is used to simulate gas-liquid two phase flow. The model regards the bubble as a continuous phase and runs through water phase. It can simulates the distribution of gas holdup well, and greatly promotes the calculation speed. The continuity equations and momentum equations are represented as (Ranganathan and Sivaraman 2011):

$$\frac{\partial}{\partial t}(\alpha_i \rho_i) + \nabla \cdot (\alpha_i \rho_i \mathbf{u}_i) = 0, \; i = g, l \tag{1}$$

$$\frac{\partial}{\partial t}(\varepsilon_i \rho_i u_i) + \nabla \cdot (\alpha_i \rho_i \mathbf{u}_i \mathbf{u}_i) = -\nabla P' + \left(\alpha_i \mu_{eff}\left(\nabla \mathbf{u} + \nabla \mathbf{u}^T\right)\right) + F_{i,j} + \alpha_i \rho_i g \tag{2}$$

In these equations, ρ, α_i, $\mathbf{u}$, t stand for density, the ith group volume, fraction velocity vector, time, and the subscript i represents gas (g) or liquid (l). P', F, μ_{eff} and g, are modified pressure, interphase force, effective viscosity, and acceleration of gravity, respectively. P' can be calculated by Eq. (3):

$$P' = P + \frac{2}{3}\mu_{eff,l}\nabla \cdot \mathbf{u}_l + \frac{2}{3}\rho_l k_l \tag{3}$$

The effective viscosity of liquid phase $\mu_{eff,l}$ can be calculated as

$$\mu_{eff,l} = \mu_l + \mu_{t,l} + \mu_{b,l} \tag{4}$$

in which μ_l is the liquid viscosity, μ_{tl} and μ_{tg} represent the turbulence viscosity of liquid phase and the turbulence viscosity induced by gas phase, as given by Eq. (5) and (6):

$$\mu_{t,l} = \rho_l C_\mu \frac{k_l^2}{\varepsilon_l} \tag{5}$$

$$\mu_{b,l} = C_{\mu b}\rho_l \alpha_g d_b |u_g - u_l| \tag{6}$$

where k_l is turbulent kinetic energy, $C_{\mu b}$ is empirical parameters with a value of 0.6 (Ranganathan and Sivaraman 2011).

The k–ε model is used for turbulence model. Its advantage is that this model has the wide applicability and is verified useful in CFD-PBM model.

2.2   The Drag Force

During the pneumatic sluicing process, momentum exchange occurs at the interface between gas and liquid phases, and the interphase force plays an important role in the accuracy of simulation results. The gas-liquid interphase forces include the drag force, lift force, virtue mass force, turbulent dispersion force, and wall lubrication force. Compared the drag force, the other interphase forces can be neglected because of their less significance in the interaction. The drag force represents the interfacial momentum transfer caused by the gas-liquid phase velocity slip. It can be described by Schiller-Naumann formulas shown in Eq. (7) (Olmos et al. 2001):

$$F_D = \frac{3}{4}\frac{C_d}{d_b}\alpha_g \rho_l |u_g - u_l|(u_g - u_l)$$
$$C_D = \begin{cases} 24(1 + 0.15\text{Re}^{0.687})/\text{Re} & \text{Re} \le 1000 \\ 0.44 & \text{Re} > 1000 \end{cases} \tag{7}$$

where the C_D is the drag coefficient of a bubble size d_b.

2.3   PBM Model

PBM model is a general method to describe the distribution of dispersed phase size in multiphase flow system (Hulburt and Katz 1964). In gas-liquid multiphase flow system, PBM model can be used to consider the influence of bubble coalescence and breakup on the distribution of bubble size, so as to study the mechanism of two-phase interaction in gas-liquid system. For the gas-liquid system, the group equilibrium equation can be expressed as:

$$\frac{\partial n(v,t)}{\partial t} + \nabla \cdot [n(v,t)] = \underbrace{\frac{1}{2}\int_0^v a(v - v', v')n(v - v', t)n(v', t)dv'}_{\text{birth term due to coalescence}} -$$
$$\underbrace{\int_0^\infty a(v, v')n(v, t)n(v', t)dv'}_{\text{death term due to coalescence}} + \underbrace{\int_\Omega vb(v')\beta(v|v')n(v', t)dv' - b(v)n(v, t)}_{\text{birth term due to breakup}} - \underbrace{b(v)n(v, t)}_{\text{death term due to breakup}} \tag{8}$$

In which, v and v' is the bubble volume, $n(v, t)$ is the distribution function of bubble size, $a(v, v')$ is the bubble coalescence rate function, $b(v)$ is the bubble breaking rate function, $\beta(v|v')$ denotes the probability density function of bubble breaking up into daughter bubbles with the volume from v to v'. In this paper, the discrete method is employed, and transport equation of the bubbles in the kth group can be expressed by Eq. (9):

$$\frac{\partial}{\partial t}(\rho_g \alpha_k) + \nabla \cdot (\rho_g \alpha_k v_k) = \rho_g (B_{k,c} - D_{k,c} + B_{k,b} - D_{k,b}) \tag{9}$$

In which, ρ is the density of gas phase, α_k is the volume fraction of the bubbles in the kth group, as shown in Eq. (10). The source terms of bubble generation and extinction caused by coalescence and breakup are represented by Eq. (11)–(14).

$$\alpha_k = N_k V_k \qquad (k = 0, 1, ..., N - 1) \tag{10}$$

$$\overline{B}_{ag,k} = \frac{1}{2}\sum_{i=1}^N w_i \sum_{j=1}^N w_j (L_i^3 + L_j^3)^{k/3} a(L_i, L_j) \tag{11}$$

$$\overline{D}_{ag,k} = \sum_{i=1}^{N} L_i^k w_i \sum_{j=1}^{N} w_j a\left(L_i, L_j\right) \tag{12}$$

$$\overline{B}_{br,k} = \sum_{i=1}^{N} w_i \int_0^{\infty} L_k g(L_i) \beta(L/L_i) \tag{13}$$

$$\overline{D}_{br,k} = \sum_{i=1}^{N} w_i L_i^k g(L_i) \tag{14}$$

In this study, Luo's bubble breaking and coalescence model (Luo and Svendsen 1996) is used to investigate the distribution of bubble size in liquid.

The model mentioned above is employed in bubble columns and aeration tanks. Gas holdup, bubble breakup rate and the distribution of daughter bubble size are in good agreement with the tests data (Cheng et al. 2020).

2.4   Simulation Model

The sketch of simulation model is shown in Fig. 1. The model is 10 m long and 0.7 m high. The 0.1 m aeration zone is set in the middle bottom of the channel located at x is from 0.45 m to 0.55 m. The water depth is kept in 0.5 m. The left side of the model is velocity inlet and the right side is set as free outflow. The quadrilateral mesh is adopted with the size of 1 mm, and the total number is 70,000 (as shown in Fig. 1(a)). The gas enters from the inlet with the uniform velocity of 0.2 m/s and initial gas diameter of 0.03 mm, as shown in Fig. 1(b). The cross water flow velocity ranges from 0 to 1 m/s, as shown in Table 1. All the simulations are carried out on the ANSYS Fluent platform. SIMPLE algorithm is used for pressure-velocity equations. QUICK scheme is employed for momentum equations. The volume fraction equations are discretized by the first-order upwind format, and the relaxation factors are set with the default values. The time step in the calculation is 0.002s, and the process ends when air bubble plume is steady (Table 2).

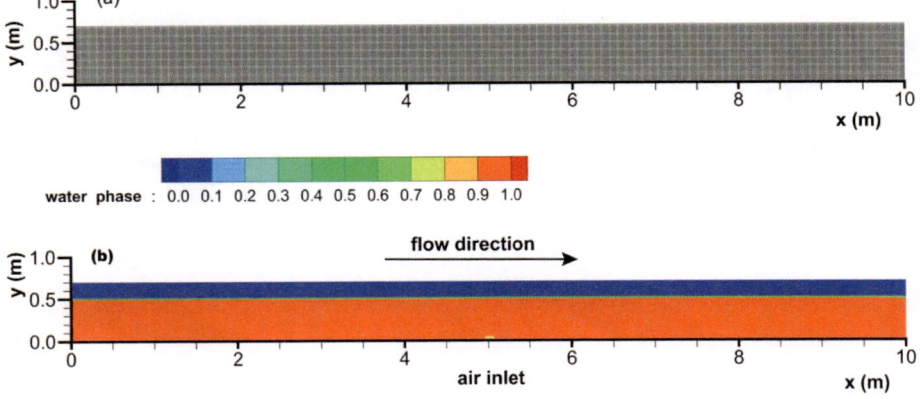

Fig. 1. Computational domain mesh and initial condition: (a) computational domain mesh, (b) initial condition

Table 1. Simulation conditions (unit: m/s)

Conditions	Velocity of water inlet	Velocity of air inlet
C-1	0.0	0.2
C-2	0.2	
C-3	0.4	
C-4	0.6	
C-5	0.8	
C-6	1.0	

Table 2. Size of discrete bubbles

Index	Size (mm)
1	1.19
2	1.89
3	3.00
4	4.76
5	7.56
6	12.00

3   Results

3.1   Gas Holdup

Figures 2 and 3 show the distribution contours of gas holdup at different flow velocities and gas holdup curves at different heights. It can be seen that gas holdup decreases gradually along the radial direction, and gas holdup in the middle of bubble plume is higher than that at the sides. With cross flow, the bubble plume is incline to the direction of velocity. And with the increase of cross velocity, the bubble plume is wider and gas hold up is lower at the same height than that with a small velocity.

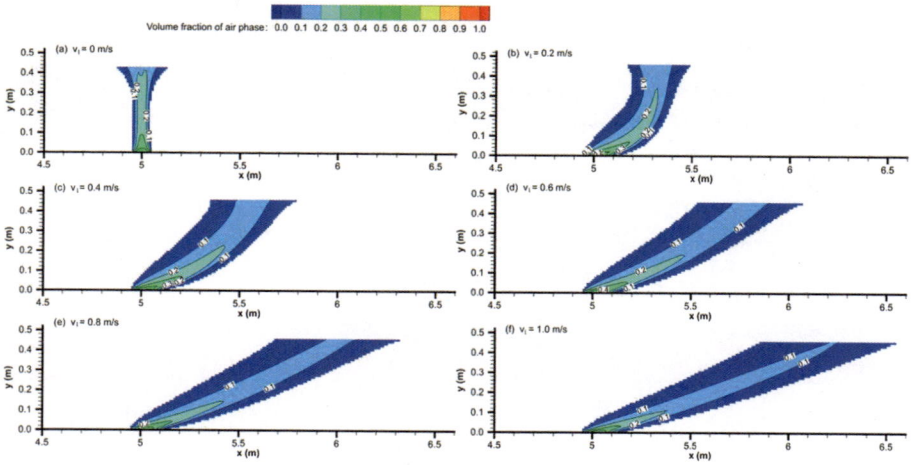

Fig. 2. Volume fraction of gas phase in different conditions

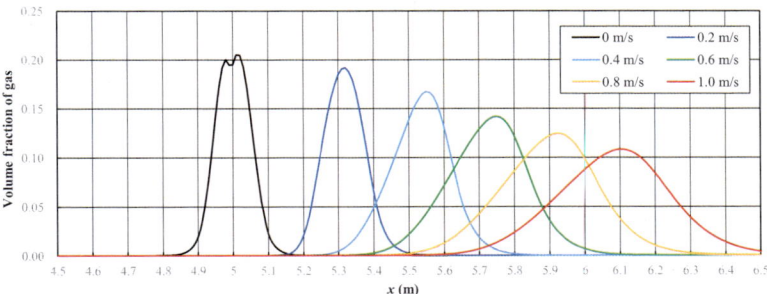

Fig. 3. Volume fraction of gas phase at $h = 0.4$ m ($v_l = 1.0$ m/s)

3.2   Bubble Size Distribution

Figure 4 shows the distribution of bubble size with different flow velocities. It can be seen that with the uplifting of the bubbles, the diameter of bubbles gradually increases, and the diameter reaches the maximum at the water surface. The coalescence of bubbles is effected by cross flow velocity, the smaller the flow velocity is, and the larger the diameter of bubble flow is.

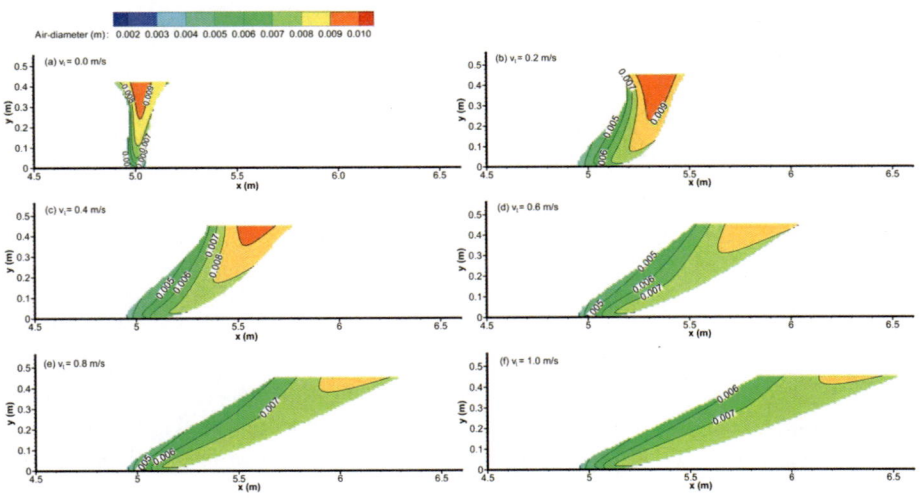

Fig. 4. Air diameters of different conditions

3.3   The Influence of Bubbles on Water Flow

The water velocity of the y direction in bubble plume is the main influencing factor for the pneumatic sluicing.

Figure 5 shows water vertical velocity of the y direction near bubbles at different heights with different flow velocities. Results show that in different conditions, the greater the cross flow velocity, the smaller the vertical velocity of water phase at the same height. Therefore, a smaller velocity of cross flow can bring the sediment or nutrients higher, but a larger flow velocity can take the masteries to the farther downstream.

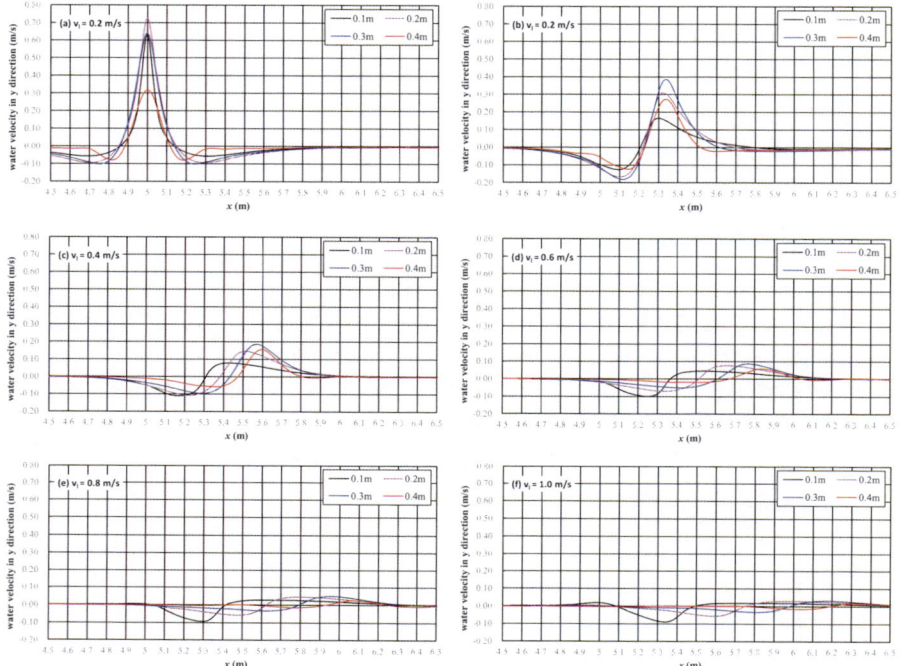

Fig. 5. Water velocity of the *y* direction in different conditions

4   Conclusions

The uplifting characteristics of bubble plume is a main influence factor of the pneumatic sluicing. There are little studies focus on the bubble plume with cross flow. Based on CFD-PBM model, this paper studied the bubble plume with different cross flow velocities, and the results show that:

(1) The width of bubble plume is increasing with the increase of cross velocity, and the gas hold up is reducing with the cross velocity.
(2) The bubble diameter is enlarged during the uplifting of bubbles and the bubble growth to maximum at the water surface. The larger the cross flow velocity, the smaller the bubble diameter at the same water level.
(3) The uplifting force is decreased with the cross flow velocity, and the greater the cross flow velocity, the smaller the vertical velocity of water phase.

The results indicate that the cross flow velocity has great influence on the bubble plume as well as the lifting effectiveness of pneumatic sluicing, which provides references for the further studies.

Acknowledgements. The authors appreciate the support of the National Key Research and Development Program of China (No. 2021YFC3201101), Project funded by China Postdoctoral Science Foundation (No. 2020T130309, 2019M651892), Jiangsu Water Resources Science and Technology Project (No. 2020022, 2021024), and Nanjing University of Information Science & Technology Research Foundation (No. 2017r097). The authors also want to thank the people for their helpful suggestions and corrections on the earlier draft of our study according to which we improved the content. There is no conflict of interest in our paper.

References

Cheng Y, Dong H, Yan R, Wei W (2020) Characteristics of bubble plume oscillation and the Coanda effect. J Chem Eng Chin Univ 34(04):904–911

Ding L, Luo Y, Dou X, Jiao J, Hu J (2019) A dynamic regulation and control technology of sediment based on hydraulic power and strong artificial measure–pneumatic sluicing. In: APAC 2019, Singapore

Duguay J, Lacey J, Masse A (2021) Evaluating the Euler-Euler approach for predicting a strongly 3D bubble-induced recirculatory flow with OpenFOAM. Chem Eng Sci 229:115982

Fleck S, Rzehak R (2019) Investigation of bubble plume oscillations by Euler-Euler simulation. Chem Eng Sci 207:853–861

Godino DM, Corzo SF, Ramajo DE (2020) Two-phase modeling of water-air flow of dispersed and segregated flows. Ann Nucl Energy 149:107766

Hulburt HM, Katz S (1964) Some problems in particle technology: a statistical mechanical formulation. Chem Eng Sci 19(8):555–574

Liang NK, Peng HK (2005) A study of air-lift artificial upwelling. Ocean Eng 32(5):731–745

Liu Z, Li B (2018) Scale-adaptive analysis of Euler-Euler large eddy simulation for laboratory scale dispersed bubbly flows. Chem Eng J 338:465–477

Luo H, Svendsen HF (1996) Theoretical model for drop and bubble breakup in turbulent dispersions. AIChE J 42(5):1225–1233

Olmos E, Gentric C, Vial C, Wild G, Midoux N (2001) Numerical simulation of multiphase flow in bubble column reactors. Influence of bubble coalescence and break-up. Chem Eng Sci 56 (21–22):6359–6365

Pan Q, Wu J (2019) A review on the research status of bubble plume and its applications. Mar Forecasts 36(02):97–104

Qiang Y et al (2018) Behaviors of bubble-entrained plumes in air-injection artificial upwelling. J Hydraul Eng 144(7):04018032

Ranganathan P, Sivaraman S (2011) Investigations on hydrodynamics and mass transfer in gas–liquid stirred reactor using computational fluid dynamics. Chem Eng Sci 66(14):3108–3124

Song C, Cheng W, Hu B, Cheng W (2011) Research on the calculation of void fraction of bubble plume and its instability pattern. J Hydraul Eng Shuili Xuebao 42(04):419–424

Wan H, Li R, Pu X, Zhang H, Feng J (2017) Numerical simulation for the air entrainment of aerated flow with an improved multiphase SPH model. Int J Comput Fluid Dyn 31(10):435–449

Yao Z et al (2019) Theoretical and experimental study on influence factors of bubble-entrained plume in air-injection artificial upwelling. Ocean Eng 192:106572

Study on Planning and Design of Ecological Pastoral Cultural Landscape Belt of Luliang River System in Yunnan Province, China

Hongzhuang Xu[1], Dean Wu[2(✉)], Shaofu Tang[1,3], Yuhong Huang[1], and Weiyi Qu[2]

[1] Yunnan Water Resources and Hydropower Engineering Co., Ltd., Kunming, China
[2] College of Harbor, Coastal and Offshore Engineering, Hohai University, Nanjing, China
wudeian@163.com
[3] China Water Resources Pearl River Planning, Surveying and Designing Co. Ltd., Guangzhou, China

Abstract. Luliang County of Yunnan Province has identified tourism as one of the four pillar industries, and established the new concept of large tourism, large resources, large market and large development, so as to make tourism a new growth point of Luliang County's national economy and the leader of the tertiary industry. Luliang will be integrated into a scenic spot with water as the core, integrating pastoral scenery with cultural landscape, combining modernity with tradition, beautiful and comfortable tourism environment, complete facilities and reasonable planning. Taking the opportunity of the national implementation of the river head system, Rural Revitalization and rural complex construction, taking the Xinpanjiang River, the Laopanjiang River and Yanfang River as the framework and aiming at "smooth river, clear water, green bank and beautiful scenery", the project fully excavates and makes use of Luliang's historical and cultural connotation and resources through flood control and drainage, sewage collection and treatment along the river, ecological green corridor, the waterfront landscape improvement and other measures shall be taken to comprehensively manage the three rivers, so as to create the waterfront landscape pattern of one heart and three belts of the wetland ecological tourism service core of the Xinpanjiang River and the Laopanjiang River Basin, the fast green tourism channel of the Xinpanjiang River, the ecological and cultural landscape belt of the Laopanjiang River and the ecological pastoral landscape belt of the Yanfang River, so as to improve the urban taste and the people's sense of obtaining a beautiful ecological environment. The project falls within the poverty-stricken area of fish. Rice and water township on the plateau, and its functional orientation is mainly ecological agricultural sightseeing, experience and poverty-stricken vacation. Therefore, the construction of waterfront landscape belt, park node construction and greening promotion along the Xinpanjiang River, the Laopanjiang River and the Yanfang River have beautified the environment of the dam area, created space for tourists and citizens to visit, visit and relax, and laid a solid foundation for the development of tourism in Luliang County.

© The Author(s) 2023
Y. Li et al. (Eds.): PIANC 2022, LNCE 264, pp. 1271–1284, 2023.
https://doi.org/10.1007/978-981-19-6138-0_111

Keywords: River channel · Ecology · Culture · Landscape belt · Planning and design

1   Introduction

Luliang County shown in Fig. 1 is situated on the east of Yunnan Province, known as the "Pearl of eastern Yunnan" (Wang et al. 2020), located in the upper reaches of the Nanpanjiang river and under the jurisdiction of Qujing City. Over the past few years, with the steady implementation of the Western national development strategy and the vital bridgehead protection strategy of opening up the southwest, Luliang County has made rapid economic and social development and urban construction. In the modern era, higher requirements are made for the advance of Luliang County. According to the "concept of green source construction of the Pearl River", Luliang water Township mainly uses rich water network and connects the five prominent natural resources of "mountain, water, forest, field and village" to form a new landmark of Qujing City. It will be the best mountain tourism, parent-child tourism, farm food, wedding photography and summer resort in Qujing City. According to the 13th five year plan for tourism in Luliang County, Qujing City, Luliang County defines tourism as one of the four pillar industries, establishes the new concept of large tourism, large resources, large market and large development, and makes tourism a new growth point of Luliang County's national economy and the dragon head of the tertiary industry. Luliang will be incorporated into a scenic spot with water as the core, rural scenery and cultural landscape integrated, modern and traditional, beautiful and comfortable tourism environment, complete facilities and reasonable planning (Yu 2018).

Depending on the area division in the plan, the construction project belongs to the poverty-stricken area of fish and rice water township on the plateau, and its functional orientation is mainly ecological agricultural sightseeing, experience and poverty-stricken Vacation (Zeng et al. 2013). Therefore, carry out waterfront landscape belt construction, park node construction, greening and upgrading along XinPan River, Laopan River and the Yanfang River, beautify the environment of the dam area, and create space for tourists and citizens to visit, visit and relax, so as to lay a solid foundation for the development of tourism in Luliang County.

Pastoral complex is the inevitable product of agrarian production and socio-economic development of a certain stage. It is the breakthrough of agrarian progress. It can effectively relieve the pressure of municipal functions, improve rural natural and living environment, and become the carrier of artistic creativity (Hu and Zheng 2018). The overall rural landscape planning is based on the investigation and evaluation of the rural landscape environment, takes the landscape science theory as the foundation, takes the landscape analysis as the core, takes the landscape planning and design technology system as the support, takes the rural human settlement environment construction as the center, and takes the rural sustainable development as the goal, carries out the landscape zoning and landscape planning of the rural landscape environment, and determines the overall characteristics, overall pattern and overall development direction of the rural landscape (Guo and Wang 2000).

Therefore, according to the action plan for the construction of beautiful and livable villages in Yunnan Province and the 13th five year plan for tourism in Luliang County, Qujing City formulated by Yunnan provincial government, the planning and design of ecological pastoral cultural landscape belt of main river system in Luliang, Yunnan Province is carried out (Tang 2020).

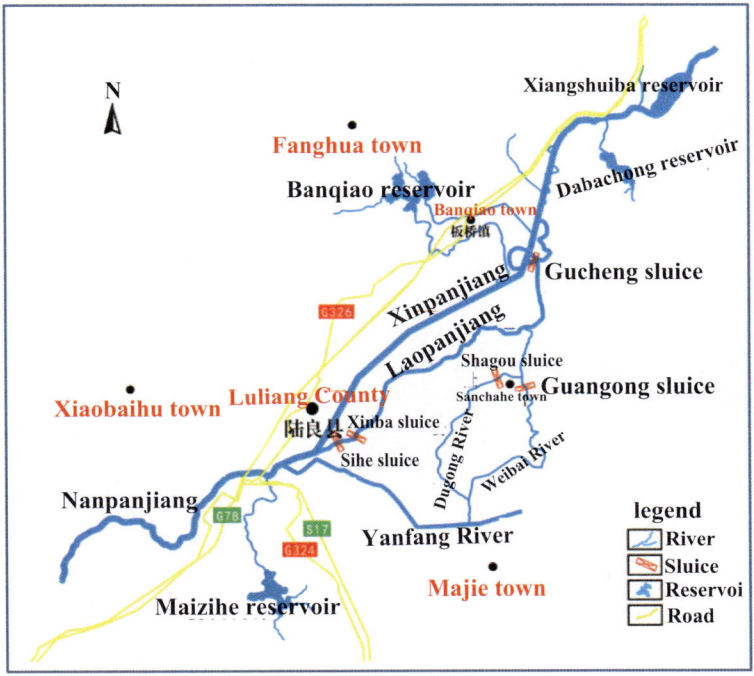

Fig. 1. The location and water system distribution of Luliang County

2 Slow Greenway and Facade Reconstruction of Buildings Along the River

2.1 Design of Slow Greenway

The dike tops road connecting lampooning along the whole line. On the one hand, a dike road integrating greenway function is added for the river section lacking dike road, so as to realize the greenway connection of lampooning; On the other hand, for the current 3–5 M wide rural road, the surrounding greening shall be improved to create a cool and comfortable greenway environment. At the same time, in order to better respond to the tourists on the greenway and residents on both sides of the Strait, a series of recreational viewing nodes are designed for the project, which is arranged at judicious intervals. Figure 2 is only an aerial view of the newly-built embankment section.

1) New embankment section

   For the construction of slow-moving Greenway in the green embankment section, a pleasant and comfortable greenway landscape is mainly created through the color asphalt pavement project, which is organically coupled with the existing rural roads to constitute a complete and unobstructed greenway tour system. On the basis of the allowable width of the bank slope, local secondary leisure walking paths and hydrophilic wood plank roads are added to create a diversified and multi-level slow-moving greenway system.

Fig. 2. Aerial view of new embankment

2) Current rural section

   The average width of the current pastoral road is 3–4 m, and most of the posterior slope is distributed with dense residential buildings, with large passenger flow. In view of the contemporary pastoral sections, we should concentrate on improving the greening on both banks. On the one hand, greening can be used to cover the residential buildings with larger volume and unsightly appearance. On the other hand, we can comprehensively improve the ecological environment of the green-way through rich greening landscape. At the same time, on the premise of sufficient bank slope width, secondary hydrophilic platforms are added locally to enrich the spatial hierarchy of greenways. Figure 3 shows in the aerial view of the new-built rustic road section.

Fig. 3. Aerial view of new-built rural road

The core service objects of the rest and viewing service shop are tourists and villagers on both sides of the Strait, which is designed to provide a convenient, comfortable and simple living waterfront place according to local conditions.

In view of the current situation, a large number of fragmented beaches and vegetable fields are distributed in the embankment defense line on both sides, and the landscape can be improved uniformly, which can be transformed into natural landscapes such as the lotus pond and wetland. Figure 4 is the new-built greenway rest and viewing service points.

Fig. 4. Greenway rest and viewing service point

A few greenways and pavilions, which can be combined with the cultural demand, can serve as a unified place for the reconstruction of the green road and the pavilion, and can be used as a place for the reconstruction of the green road and the abandoned Pavilion, which can greatly improve the cultural connotation of the building.

2.2   Facade Reconstructions of Buildings Along the River

There are a number of dense residential buildings on both sides of lampooning River, including many illegal residential buildings on the bank slope. These residential buildings lack overall planning and are mostly constructed by villagers spontaneously. Therefore, architectural features and styles on both sides of the Strait are clustered, and the overall building facade lacks aesthetic feeling. Therefore, through the unified building facade transformation, the project comprehensively improves the architectural style on both sides of the Laopanjiang River and is dovetailed with the overall style of the slow-moving Greenway of the Laopanjiang River. The architectural style of the facade of buildings along the river should follow the characteristics of local traditional villages and folk houses, partially retain the architectural characteristics of traditional tile roofed houses, extract the characteristic elements such as window flowers as unified symbols, start with Chinese style and use modern elements to create the facade image with traditional beauty and modern technology. Figure 5 shows the buildings along the river.

Fig. 5. Facade reconstruction of buildings along the river

3   Characteristic Zoning

In addition to the construction of slow-moving greenways along the whole line, the lampooning ecological and cultural belt also makes full use of the favorable landscape conditions such as idle ponds, wetlands and green spaces around the greenway, plans the overall resources, transforms and improves moderately, and turns the whole ecological and cultural landscape belt into a three-dimensional green corridor landscape belt with changeable space, different sizes and different moving scenery.

The ecological and cultural landscape belt is about 19 km long. According to the site characteristics and the distribution of surrounding cultural and natural resources, it can consist of three zones to form the landscape structure of "one river and three areas". Being the sequence of time in advance, the three areas from upstream to downstream are ancient village cultural sightseeing area, water village charm leisure area and water conservancy historical experience area. The upstream section highlights the origin of lampooning river and village, the midstream section highlights the change and integration of people, rivers and villages, and the downstream section highlights the harmony, prosperity and progress of people and water. Each of the three zones gets its characteristics. Tourists can experience the changing history of people and water from prehistoric times to contemporary times by visiting the slow greenway.

1) Ancient village cultural tourist area-Origin

Inspired by the local village history and local characteristic folk architecture (such as Guan Shenggong Temple), and guided by the allusions and deeds of ancestors and celebrities (such as the repeated achievements of Lu Liang Zhizhou and Du Zhen), the ancient village culture is highlighted through sightseeing and leisure design to explain the history of ancestors living in water and burying the seeds of civilization.

The ancient village is cultural tourism area mainly includes two landscape nodes: Dugong Kaiyuan and autumn scenery of the ancient village. The main construction contents are as follows: according to the current terrain, transform the current banded pond into five ecological reservoirs with the function of purifying water quality, form a continuous and changeable banded waterfront space combined with the facility layout of the entrance square, multiple hydrophilic platforms and wooden plank roads in the south, highlight the rich animal and plant community landscape in the wet land, and become the main hydrophilic leisure place for residents and tourists; Greening and upgrading the existing sparse forest and grassland to create a richer plant community, so as to have scenery in four seasons. At the same time, multiple rest and viewing platforms are arranged to facilitate residents to relax and visit outdoors; Reasonably arrange multiple theme sculptures, cultural landscape walls and other specific landscape signs to enrich the artistry and experience of the space and create a characteristic folk waterfront greenway landscape. Figure 6 shows in the aerial view of autumn nodes in the ancient village.

2) Shuixiang charm leisure area-change

Taking the literary history of recording the local natural mountains and rivers, characteristic features and cultural monuments as the clue, extract the relevant descriptions of the Laopanjiang River in literary biographies and oral legends, and combine the current environment of the Laopanjiang River, design a belt tour system that can be visited, played and appreciated, and use the ups and downs and rich and changeable three-dimensional landscape belt to metaphor the change history of the Laopanjiang river for many years, so as to create a water town charm leisure area with great regional characteristics. The layout of the leisure setting of the landscape belt is intimately combined with the surrounding cultural scenic spots. Through the efficient layout of the viewing platform, it can effectively connect with the surrounding cultural scenic spots, effectively collect and distribute the flow of people, and create a comfortable and convenient waterfront green belt.

Fig. 6. Aerial view of autumn nodes in ancient village

The water town charm leisure area mainly includes six nodes: Qingding Zixi, Delin Qingyin, tianjiangshenma, Feilai jade, Lingmeng flower fragrance and Biquan yingyue. The main construction contents are: transforming a few low-lying green spaces into the central landscape, arranging landscape corridors, wooden platforms and other rest facilities around the lake, focusing on optimizing the current greening environment, through rich plant planting design and semi private and semi open enclosed waterfront space design, Create a natural and comfortable gathering and leisure place connected with the greenway. Figure 7 shows in the aerial view of flower fragrance node named Lingmeng and Fig. 8 shows in the aerial view of spring reflecting the moon node.

Fig. 7. Aerial view of Lingmeng flower fragrance node

Fig. 8. Aerial view of Biquan yingyue node

3) Water conservancy development experience area – Development

Taking the history of frequent flood and drought disasters in the Nanpanjiang River and the climax of water conservancy construction after the founding of the people's Republic of China as the clue, make full use of the characteristic water conservancy buildings such as Xinba sluice retained in the current situation, highlight the theme of ecological water conservancy, and reflect the image of harmonious coexistence, prosperity and development between mankind and nature. The design transforms the existing fish pond into an ecological landscape lake with purification function, and combines a series of water education facilities to create a hydrophilic space integrating water conservancy schooling, popular science exhibition, sightseeing and rest.

The water conservancy development experience area mainly has four nodes: half river water color, legendary stage, colorful flower sea and water moistening all things. The main construction contents are: create an open and shared first impression through the design of open entrance square and entrance landscape wall logo, which can effectively attract popularity and create an open and comfortable green corridor entrance landscape; Transform the abandoned hydraulic structures into a Water Conservancy Museum, combined with the interactive installation of water art, to create an entrance space that can be learned, visited, played and rested; The landscape of the existing pond is improved through the layout of rainwater garden, hydrophilic steps, platforms and Wangjiang Pavilion. At the same time, the pond is turned into an ecological reservoir, which can effectively guide and treat the surface runoff of municipal roads, and then discharge it into the Laopanjiang River after purifying the water quality. At the same time, hydrophilic devices, leisure reclining chairs, hydrophilic square and other facilities with strong inter-action are arranged to create an education and leisure area for teaching and tourism; According to the needs of the surrounding intensive village and town activities, reasonably set up diversified children's facilities and create a dynamic place for villagers to personally leisure activities; Comprehensively optimize the greening environment of the site, design multiple color leaf plant forests, and create a natural ecological, ventilated and comfortable rest area. Figure 9 is very just an aerial view of the water moistening everything node.

Fig. 9. Aerial view of water moistening everything node

4   The Yanfang River-Ecological Pastoral Landscape Belt

4.1   Ecological Pastoral Greenway

The Yanfang River is located in the countryside of Luliang County. A large number of farmland are made available along both banks of the river, with a very broad vision of pastoral characteristics. TheYanfang River landscape belt takes full advantage of the pastoral characteristics of both banks and arranges ecological pastoral greenways along the river, so that tourists and villagers can enjoy the bucolic scenery of both sides while riding. The ecological pastoral greenway is mainly composed of four characteristic corridors, which uniformly upgrade the current and planned 3 M embankment Top Road into a 10 m long ecological greenway to create an ecological cycling greenway with diversion of people and vehicles, natural comfort and pastoral characteristics along the whole line of the Yanfang River.

Based on the basic construction of the unified corridor, the design content of each corridor will be added according to different themes. Viewing platform, Greenway post station and some outdoor fitness equipment will be in addition to the short corridor. Ginkgo veranda will be planted with Ginkgo bilobed on the green belt, and a water friendly wood platform will be in addition to make multiple best waterfront photography points. The fishing gallery is organized at the turning of the river, which is also the concentration of fish spawning. A number of hydrophilic fishing platforms and sunshade landscape kiosks will be centrally organised to facilitate the majority of fishing lovers. The forest language corridor will add ecological wooden stakes and other natural landscape structures that attract birds to maintain, and add landscape stone locally to make a leisure ecological corridor full of birds and flowers. Figure 10 displays the typical section of the Yanfang River corridor.

Fig. 10. Typical section diagram of the Yanfang River corridor

4.2   Design of Greenway Rests Service Point

Being reliant on the current characteristics of the site and the activity needs of future tourists. Three different types and scales of Greenway rest service points are designed: Scenic post station type, hydrophilic activity type and rest and leisure type. Figure 11 is the diagram of scenic post station service point.

1) Viewing post station type

    The scale of observation post service point is quite enormous, and the proportion of hard landscape is also bulky. It equips with landscape structures with sunshade function. This type of service point is mainly combined with greenway planning to build a leisure conversion point integrating bicycle post station and sightseeing and rest. Therefore, it can not respond the distribution needs of a large number of bicycle flow and people flow in a specified area.

Fig. 11. The scenic post station service point

2) Hydrophilic active type

The scale of hydrophilic active service points is relatively small. It mainly aims at the needs of some tourists for fishing, photography, hydrophilic and other activities, and creates a hydrophilic wood platform in combination with the existing embankment maintenance ladder. Figure 12 shows the hydrophilic active service points.

Fig. 12. The hydrophilic active service point

3) Leisure type

The scale of leisure service point is relatively unimportant, which owns the advantages of easy implementation and less land occupation. Construction content of the service point is mainly to appropriately add strip landscape stone or landscape wooden piles on the bank slope. Strip landscape stone or landscape wooden piles can not only provide tourists with a space to stay and rest, but also enrich the river landscape as stone scenes. Figure 14 shows the intention of leisure service points.

5 Conclusions

According to the action plan for the construction of beautiful and livable villages in Yunnan Province and the 13th five year plan for tourism in Luliang County, Qujing City formulated by Yunnan provincial government, this design carries out the planning and Design Research on the ecological pastoral cultural landscape belt of the main river system in Luliang, Yunnan Province. Main design results:

1) The dike top road connecting lampooning along the entire line. On the one hand, a dike road integrating greenway function is added for the river section lacking dike road, so as to realize the greenway connection of lampooning; On the other hand, for the current 3–5 M wide rural road, the surrounding greening shall be improved to create a cool and comfortable greenway environment.
2) Through the unified building facade transformation, the project comprehensively improves the architectural style on both sides of the Laopanjiang River and is dovetailed with the overall style of the Laopanjiang slow Greenway.
3) In addition to the construction of slow-moving greenways along the whole line, the lampooning ecological and cultural belt also makes full use of the favorable landscape conditions such as idle ponds, wetlands and green spaces around the greenway, plans the overall resources, transforms and improves moderately, and turns the whole ecological and cultural landscape belt into a three-dimensional green corridor landscape belt with changeable space, different sizes and different moving scenery.
4) The Yanfang River is located in the countryside of Luliang County. A large number of farmland are made available along both banks of the river, with a very broad vision of pastoral characteristics. The Yanfang River landscape belt makes extensive use of the pastoral characteristics of both banks and arranges ecological pastoral greenways along the river, so that tourists and villagers can enjoy the bucolic scenery of both sides while riding.

Acknowledgements. This study are supported by the scientific research projects entrusted by Yunnan water conservancy and Hydropower Engineering Co., Ltd. "Research on the regulation and harmless treatment technology of water hyacinth migration path in the comprehensive treatment of river channel in Luliang County, Qujing City, Yunnan Province" and "Research on the impact assessment and Countermeasures of river discharge into Luliang County, Qujing City, Yunnan Province on the regulation effect of river channel comprehensive treatment project"; This research is also supported by the National Natural Science Foundation of China (No. 41776024, No. 42176166).

References

Hu ZQ, Zheng Y (2018) Exploration on technical methods of landscape design of rural complex-taking the landscape design of rural complex in Surabaya, Shandong Province as an example. Special papers of the 21st academic annual meeting of China National Architecture Research Association. China National Architecture Research Association, Changsha, Hunan, pp 290–294. (in Chinese)

Guo HC, Wang YC (2000) Research on the development of sightseeing agriculture. Econ Geogr 2:119–124 (in Chinese)

Tang SF (2020) Feasibility study report on comprehensive regulation project of Luliang urban section of Nanpan River and the Yanfang River. China Water Resources Pearl River Planning, Surveying and Designing and Designing Co., Ltd. Guangzhou. (in Chinese)

Wang HY, Li QL, Wang T, Li YT, Lian C (2020) The cultural landscape characteristics and spatial distribution of geographical names in Luliang county based on GIS. J Guizhou Normal Univ (Nat Sci) 38(2):28–36 (in Chinese)

Yu W (2018) Trinity urban waterfront overall design-taking Hengli river ecological and cultural landscape project in Nanxiang Town, Shanghai as an example. Garden 2:66–69 (in Chinese)

Zeng Y, Zhu KX, Qi Jl, Li YQ, Pu YX, An TX (2013) Rural tourism situation and development strategies research of Luliang country. J Yunnan Agric Univ 7(S2):169–174. (in Chinese)

Study on the Design of Ecological Green Corridor Project for Comprehensive River Treatment in Luliang County, Yunnan Province

Shaolin Yi[1], Shaogan Sun[1], Guofang Huang[3], Dean Wu[2(✉)],
and Weiyi Qu[2]

[1] Yunnan Water Resources and Hydropower Engineering Co., Ltd.,
Kunming, China
[2] College of Harbor, Coastal and Offshore Engineering, Hohai University,
Nanjing, China
wudeian@163.com
[3] China Water Resources Pearl River Planning,
Surveying and Designing Co. Ltd., Guangzhou, China

Abstract. Luliang County is located in the east of Yunnan Province, knew as the "Pearl of eastern Yunnan". It is located in the upper reaches of the Nanpan River and is under the jurisdiction of Qujing City. In the modern era, higher requirements are made for the development of Luliang County. In order to improve the river flood control system, control the domestic pollution along the river, repair the damaged water ecosystem, create a waterfront landscape belt, and create basic conditions for the construction of "the most beautiful dam area", Luliang water authority has given priority to the comprehensive river regulation project of Luliang County, Qujing City, according to the principle of first trial and highlighting key points. According to the current conditions of the Laopanjiang River and the Yanfang River, based on the principles of safety, integrity, ecology, protection, context and hydrophilicity, taking the construction of "Luliang water town and the most beautiful dam area" as the overall concept, the rich water network is used to connect the five prominent natural resources of "mountain, water, forest, field and village", highlighting the local folk culture, water conservancy development history Cuan culture has three cultural characteristics. Coordinate the development of surrounding land, and activate the popularity of the site, promote local economic development, create a land of fish and rice with common prosperity of people and water, and reproduce the brilliance of "East Yunnan pearl, plateau granary" through reasonable functional layout and facility layout of river center, beach, Boulder, ecological wooden pile, aquatic plants and aquatic animals.

Keywords: Luliang County · River channel synthesis · Ecological green corridor · Engineering design

Y. Li et al. (Eds.): PIANC 2022, LNCE 264, pp. 1285–1301, 2023.
https://doi.org/10.1007/978-981-19-6138-0_112

1 Introduction

The dependence on rivers has changed from simple survival and economic utilization to diversified needs, paying attention to the return of river public ties, paying attention to the development of waterfront environment and the construction of corridor ecological greening, and improving the interactive relationship between people and rivers, which can not only reduce the misunderstanding and regret of development and construction, but also create a good carrier space for the city to shape rich urban culture, citizen spirit and Ecological Landscape (Zheng 2012). On May 4, 2018, the feasibility study report on the comprehensive river regulation project in Luliang County passed the review of the expert group (Tang 2018; Tang 2020), which is the first project in the overall concept of "Pearl River green source", and the river regulation in Luliang County has entered a new stage (Li et al. 2021). The task of river regulation is no longer part of the traditional dredging of river channels and reinforcement of river banks. "Smooth river, clear water, green back and beautiful scenery" have become the standard of comprehensive river regulation in the modern era. The provincial government also hopes to be able to revitalize the resources along and around the river through river regulation to drive the development of tourism and economy. Luliang water Township can make use of the rich water network to connect the five natural resources of "mountain, water, forest, field and village", take water as the core, integrate pastoral scenery with cultural landscape, and combine modern and traditional planning and governance concepts, improve the flood control capacity of river channels, control the domestic pollution along the river (Wang et al. 2020; Liu and Gu 2020), transform the surrounding environment, repair the damaged water ecosystem, ensure the ecological base flow, create a waterfront landscape belt and improve the river ecological landscape, Create a unique landscape of plateau water town and form a new landmark of Qujing City (Wu 2020). Create a river ecological landscape project integrating urban water environment landscape and ecological agriculture sightseeing, experience, leisure and vacation, fully tap and make use of Luliang's historical and cultural connotation and resources, and carry out environmental remediation of the moat and the Yanfang River through flood control and drainage, water diversion into the city, water ecological restoration and other measures, so as to create the Yanfang River ecological pastoral landscape and moat urban water landscape, and improve the urban quality from flood control, water environment, ecology Comprehensive treatment of landscape and other aspects.

The ecological green corridor uses the concept of landscape ecology to explore the application and practice of landscape ecology in the spatial planning and design of urban ecological green space from the perspective of landscape elements (Zhou 2021). This paper discusses how to make a localized and poetic urban ecological corridor landscape planning and design through the creation of scenic spot artistic conception, plant planting planning and landscape sketch design, integrating local historical context and regional characteristics in the process of urban ecological corridor landscape construction (Li 2015). Built on the analysis of ecological environmental base and ecological sensitivity, the landscape ecological pattern is the framework, and the distinctive landscape ecological planning is carried out from the perspective of landscape structure and function layout according to relevant laws, regulations and technical specifications. The ecological corridor is superimposed with the landscape corridor and recreation corridor to build a continuous landscape ecological pattern, and the multi-path corridor considering their own factors forms a more stable ecological network (Yan et al. 2016).

Therefore, according to the requirements of "minutes of Symposium on comprehensive river regulation project in Luliang County" and "design of feasibility study stage of comprehensive river regulation project in Luliang County", the design and research of Ecological Green Corridor Project for comprehensive river regulation in Luliang County is carried out. Build suitable habitat and environmental elements by artificial means, accelerate the restoration of ecosystem, and build a complete river ecological corridor, so as to provide assistance in regional ecological security of Luliang County.

2   Current Situation Analysis

2.1   The Waterfront Landscape Design is Single and Cannot Show the Regional Cultural Characteristics

As showed in Fig. 1, the current form of revetment is single, and the river with a length of more than ten kilometers basically adopts only one or two typical section forms. In the past, the traditional river planning scheme focused on the construction of water safety, with low requirements for water landscape and water culture, leading to the current situation of the remaining projects are mostly stereotyped and lack of beauty. Instancing River shown in Fig. 2 is close to the municipal road, and the section form is also a definite traditional hydraulic section, with a definite spatial level. Although the vegetation coverage on both banks is brought up, the plant planting is chaotic and lacks characteristics and reasonable plant color planning, resulting in a general overall landscape experience.

The residential buildings on both sides of the Laopanjiang River are densely distributed, and most of them hold the embankment, which belongs to illegal buildings. There are many folk cultural buildings on both sides, representing rich folk culture, such as Guansheng temple and Qinglong Temple (shown in Fig. 3), as well as many hydraulic structures with characteristics of the times, such as Xinba sluice (shown in Fig. 4).

As showed in Fig. 5, regardless of the fact that the cross-section form of the Yanfang River is single, large agricultural gardens and a small amount of wetlands are made available on both banks, with wide vision and good pastoral landscape resources.

These landscapes, which can fully reflect the history and regional characteristics of local water conservancy development and local characteristic folk culture, have not been fully respected, protected and utilized in previous planning projects, which is a major gap in the current comprehensive management of river courses.

Fig. 1. Current embankment

Fig. 2. Current embankment of the Laopanjiang River

Fig. 3. Guansheng temple and Qinglong temple on the Bank of the Laopanjiang River

Fig. 4. Xinba sluice on the Laopanjiang River

As showed in Fig. 5, despite the fact that the cross-section form of theYanfang River is single, large agricultural gardens and a small amount of wetlands are distributed on both banks, with wide vision and pleasant pastoral landscape resources.

These landscapes, which can fully reflect the history and regional characteristics of local water conservancy development and local characteristic folk culture, have not been fully respected, protected and utilized in previous planning projects, which is a major gap in the current comprehensive management of river courses.

Fig. 5. Rural scenery along the Yanfang River

2.2 Lack of Recreational Facilities and Green Leisure Places

At present, there are basically no landscape nodes arranged on both banks of the river for staying and shading, which is seriously missing in hydrophilicity. This not only blocks the relationship between man, nature and rivers, but also makes these three rivers lose their basic functions as important public recreational space in cities and towns, and can not provide a convenient green leisure place for the general public and villagers along the river.

2.3   The Watery Landscape System with Convenient Access and Organic Integration has not been Formed

The three rivers are the lack of connection, which can not create an organically integrated, accessible and convenient water landscape system for the city. At present, although there are intersections between instancing River and the Laopanjiang River, instancing River and cunningly River, the connection between rivers is weak and there is a lack of certain distribution and conversion space, resulting in the inability of these three rivers to effectively gather popularity and the dispersion of people and vehicles. At present, three relatively separated waterfront landscape space cannot form a whole, and the accessibility and convenience are poor. Figure 6 shows the intersection of the Xinpanjiang River and Yanfang River.

Fig. 6. Intersection of XinPan River and the Yanfang River

3   Design Principles and Concepts

3.1   Design Principles

1) Safety principle - flood control priority, on the premise of meeting the flood control function. 2) Overall principle - landscape design must fully consider the needs of urban planning, tourism planning and other relevant planning, and uniformly consider the buildings and land attributes around the land to reflect the image of the city. 3) Ecological principles - follow the principles and methods of "landscape ecology" to create rich and diverse river habitats. 4) Protection principle - protect the basic farmland within the design scope. On the basis of not requisitioning the basic farmland and not changing the land use nature of the basic farmland, highlight the farmland texture through reasonable landscape design methods, so as to achieve a win-win effect of reducing the cost of land requisition and showing the scenery of characteristic farmland. 5) Context Principle - respect and make rational use of regional cultural characteristics, pay attention to the communication and interaction between modern and traditional, and show the unique water culture of the river. 6) Hydrophilic principle - meet people's needs to be close to nature and water. 7) People oriented, pay attention to the construction of human feelings in urban and rural public space, and enhance the sense of belonging of local residents.

3.2 Design Concepts

In view of the three major problems existing in the current situation of the site, this landscape design will: 1) highly integrate natural and cultural resources and create a waterfront landscape space with regional cultural characteristics in combination with urban positioning. 2) Give full consideration to the leisure needs of local residents and tourists and create rich and diverse activity facilities. 3) Realize the effective connection of points, lines and surfaces, and create a waterfront landscape tourism system with clear theme, accessibility and convenience.

Taking the construction of "Luliang water town, the most beautiful dam area" as the overall concept, the rich water network is used to connect the five prominent natural resources of "mountain, water, forest, field and village", highlighting the three cultural characteristics of local folk culture, water conservancy development history and Cuan culture. Coordinate the development of surrounding land, activate the popularity of the site, promote local economic development, create a land of fish and rice with common prosperity of people and water, and reproduce the brilliance of "Pearl of East Yunnan and granary of Plateau" through reasonable functional layout and facility layout.

4 General Layout

4.1 Overall Spatial Layout

According to the different contemporary characteristics of the three rivers and the needs of urban development, the overall layout structure is "one heart and three belts", and seeks to create a waterfront landscape pattern suitable for living, tourism and industry. Of which:

1) One heart: the core of ecological service of Panjiang wetland

   The project proposes building Panjiang wetland at the core of the connection between the lower reaches of the Laopanjiang River and the central area. Panjiang wetland will make full use of its characteristic natural resources such as farmland, pond and wetland, unique Cuan cultural resources, convenient location and transportation, and coordinate the development needs of surrounding land and urban positioning on the basis of no expropriation of basic farmland, reasonable protection and utilization, so as to build a core of Panjiang wetland ecological tourism service integrating ecological tourism, leisure and entertainment and popular science education. Completion of Panjiang wetland can not only effectively improve the ecological and aesthetic characteristics of waterfront landscape, but also effectively stimulate the improvement of surrounding land value and the development of real estate industry, and stimulate provincial economic growth.

2) Three belts: three ecological landscape belts connecting the central area

   (1) The Xinpanjiang River- fast green sightseeing channel

       With the theme of "urban greenway, beautiful scenery", comprehensively optimize the greening on both sides of instancing River, select important landscape nodes and landscape belts for greening upgrading, and plant colored leaf plants to form a boulevard with rich corridor landscape color, so as to create a fast green sightseeing channel connecting Qujing urban area and Luliang center.

(2) The Laopanjiang River-ecological and cultural landscape belt

With the theme of "ancient water town, looking for mountains and asking for water", restore the crisscross water network of the Laopanjiang River, integrate the folk customs, legendary literature, water conservancy history and other historical resources of local villages and towns, take advantage of the undulating mountains in the distance, make full use of the current distribution of a large number of vegetable fields, beaches and local architectural layout, try to avoid demolition, integrate all favorable resources, and create water streets, water lanes Watercourses and other scenic spots full of the charm of the ancient capital.

The ecological and cultural landscape belt is about 19 km long. According to the site characteristics and the distribution of surrounding resources, it can consist of three zones to form the landscape structure of "one river and three areas". From the upstream to the downstream, the three areas are ancient village cultural tourism area, water village charm leisure area and water conservancy development experience area.

(3) The Yanfang River-ecological pastoral landscape belt

With the theme of "beautiful countryside and long green water flow", combined with the surrounding ecological agricultural scenic spots, create an ecological pastoral leisure atmosphere, integrate the slow greenway design, highlight the concept of pastoral and health, set up scenic spots such as characteristic fishing points, picking points and greenway post stations, and create an ecological pastoral landscape belt integrating light viewing, entertainment, ecological experience and leisure sports, Create a rich and diversified slow walking healthy greenway for local residents and tourists.

Being dependent on the current situation of "confusion" with a length of about 15 km, the landscape structure of "confusion" is proposed. "Three parts" refer to the folk culture park, cunningly ecological pastoral Park and cunningly Fengming Wetland Park. "Four corridors" to refer to cunningly ecological greenway, which is divided into four characteristic corridors, from downstream to upstream: sport corridor, ginkgo corridor, fishing corridor and Lin language corridor.

4.2   Layout of Ecological Restoration Work

The composition, structure and distribution pattern of natural riparian biological community are very different from those in dry land areas. It is a great transitional and interlaced zone for the exchange of material, energy and information between the river ecosystem and terrestrial ecosystem. Natural rivers have significant practical and potential values in controlling bank erosion, regulating microclimate, protecting river water quality, being home for aquatic and terrestrial animals and plants, maintaining river biodiversity and ecosystem integrity, improving river landscape quality and carrying out tourism activities. In fact, the water level, flow and velocity of the natural river channel that has not been greatly transformed by human beings change with the change of seasons, which fully reflect the operation law of nature. The rapids, slow flows, floodplains and deep pools generated by hydraulic and geological evolution in the natural river channel also provide rich and diverse habitats for aquatic organisms, and the local aquatic organisms also change their habitat and living rules, which is the result of the slow evolution of nature.

In the past, the main purpose of municipal and rural river regulation was to prevent water and soil loss and flood control. The vertical concrete and mortar masonry flood dike are the basic methods used for river revetment, resulting in the diversification of river habitat into simplification, the disappearance of some aquatic organisms with special ecological requirements and the reduction of species diversity. Therefore, reconstructing the diversity of river habitat is an urgent means in river regulation.

The ecological restoration task of this project is designed to transform the contemporary river based on its current status, provide more abundant habitats for aquatic organisms, and meet the habitat needs of river biological species as much as possible.

4.2.1 Zoning of Ecological Restoration Work

1) Upper reaches of the Laopanjiang River
   In the bucolic river section, there are basically no buildings on both sides, with little human interference, and there are many beaches on both sides of the river, which can make habitats with rich diversity. This section is used as an ecological restoration area, and the project includes the release of river course, beaches, boulders, aquatic plants and aquatic animals. Supplement emergent plants and repair the existing submerged plants.
2) Lower reaches of the Laopanjiang River
   In the village river section, there are dense buildings on both banks and large human interference, so small-area habitats can be built depending on regional conditions. This section serves as an ecological construction area, and the project content is to pile up boulders (a small amount), wooden piles, aquatic plants and aquatic animals.
3) Yanfang River-Zhongyuan Ze
   In the rustic river section, except for the most downstream river section, there are few buildings on both banks and little human interference. However, due to the recent regulation, it is not apt to undertake major actions in the river section to avoid disturbing the newly restored ecosystem. This section serves as an ecological maintenance area, and the project content is to pile boulders, wooden piles, emergent plants and aquatic animals.
   Natural rivers are composed of numerous habitat elements. In the comprehensive regulation of the Nanpan River, on the premise of not damaging the built river embankment project, through some relatively simple habitat construction, a variety of river morphological elements are brought together to make diversified habitats and promote the healthy development of the ecosystem. The river habitat construction elements mainly include the following types: River core, beach, Boulder, ecological stake, aquatic plant and aquatic animal food chain construction. In

addition, in view of the problem that the water and grass of the Laopanjiang River are too lush and hinder flood discharge, the water and grass trimming measures of the Laopanjiang River will be implemented.

4.2.2   Habitat Restoration Design

1) River shoal design

A pure river will form a river heartland owing to sediment deposition at the outlet of the torrent. Because it is very difficult for human beings to reach and less disturbed by human activities, the river heartland is a safe place for a variety of organisms to inhabit and survive.

The design of expansion is mainly used to restore the diverse forms of river channels, provide places apt for amphibians and birds that are difficult to be disturbed by human beings, change the regional flow pattern and form a richer flow pattern division. The design principles of expansion are as follows: relying on the micro- terrain at the bottom of the river, reduce the amount of earthwork. Combined with the existing siltation points to prevent hydraulic scouring and hollowing out.

In this habitat structure project, the river heartland is at the rate of the thick part of the country river section in the upper reaches of the Laopanjiang River, and Yanfang River does not set river heartland.

As showed in Fig. 7, No. 1 extension is in the chainage of the Laopanjiang River $0 + 350 \sim 0 + 600$, with an area of 4088 m^2. The river bottom elevation of this section is exceeded, so it has the inherent advantage of constructing river shoals.

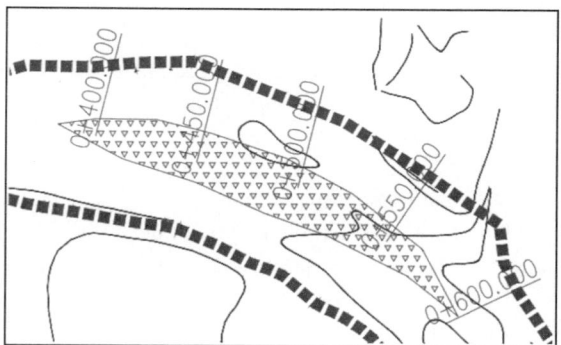

Fig. 7. Scope of No. 1 River shoal

As shown in Fig. 8, No. 2 River heartland is located between chainage $2 + 600$ and $2 + 800$, with an area of 1860 m^2. The original high convex land at the center of the river indicates that the hydrological situation here is conducive to sediment deposition and is suitable for the construction of River heartland.

Fig. 8. River shoals under construction

It is necessary to backfill the planting soil on the river shoal, with a backfill thickness of 0.3 m, and plant emergent plants, mainly local moist plants reed, Zizania latifolia, rush grass and water taro, so as to attract birds to stay and provide habitat for amphibians.

2) Beach land

Shoal is a commonplace river habitat structure in mountainous rivers, and it is the spawning ground of some fish and shrimp. The fine sand in the bank shoal is washed away by the water flow, and the gap between the gravels becomes larger, becoming the habitat of aquatic insects, attached algae and other organisms, which in turn attracts all kinds of fish for food, thus becoming the living space of all kinds of aquatic animals.

Depending on the field survey, there is a considerable area of beach habitat in the upper reaches of the Laopanjiang River. The project mainly relies on the existing beach in the river embankment to maintain the existing terrain as far as possible, protect and transform it. Engineering measures are mainly greening, and the quantities are integrated into the landscape greening project.

3) Boulder

As showed in Fig. 9, boulders can modify the direction of water flow, guide deposition, and from a variety of river environments such as river beach depression and still water area. Boulders in the river block the flow of water, change the flow pattern and oxygenate the river. The bottom of the boulder is another critical habitat for aquatic insects and molluscs. The gap between the boulders is a pretty good biological shelter, and the top of the boulder can also be utilized as a place for birds to stop.

In order to enrich the morphological diversity of the river channel and increase the habitat, boulders are given to the narrow upright, stable and steep river section according to the topographic characteristics of the river section. Among them, on the 4.5 km long river channel in the upper reaches of the Laopanjiang River, four ecological boulders with a volume of 0.2 m^3 are placed every 50 m, with a total of 72 m^3 boulders. On the 14.3 km long river channel in the lower reaches of the

Laopanjiang River, two ecological boulders with a volume of 0.2 m^3 are placed every 50 m, and a total of 114.4 m^3 boulders are placed. On the 15.2 km long river course of the Yanfang River, four ecological boulders with a volume of 0.2 m^3 are placed every 50 m, with a total of 243.2 m^3 boulders.

Fig. 9. Boulder and Wooden pile

4) Ecological stake

As showed in Fig. 10, the stump can also provide ecological water supply for vertebrates and birds, as showed in the singular ecological diagram. Since there are small houses on both sides of the upper reaches of the Laopanjiang River and there is relatively little human interference in the river, ecological wooden piles will not be added, but only in the lower reaches of the Laopanjiang River and Yanfang River.

Fig. 10. Wooden pile

In the 14.3 km long river channel in the lower reaches of the Laopanjiang River, a group of wooden piles are set every 100 m, each group is composed of 25 pine trees with a length of 4 m and a tail diameter of 0.12 M. In the Yanfang River, two groups of wooden piles are set every 100 m, each consisting of 25 pine trees with a length of 4 m and a tail diameter of 0.12 M. The wooden pile should be located 5–8 m offshore, and the length of the exposed part is about 0.5–1 m.

5) Aquatic plant

Aquatic plants are found and food for a variety of animals, and can also purify water quality and beautify the environment. Since there are many submerged plants in the regulated river section, the planting of submerged plants will not be aggregated, and only emergent plants will be placed. The planting range is the Laopanjiang River and Cunningly River, in which the planting length in the upper reaches of the Laopanjiang River accounts for 20%, the planting River in the lower reaches of the Laopanjiang River accounts for 40%, and the planting River in cunningly River accounts for 30%. Many kinds of plants are planted in landscape festivals around the point. The planting width is about 0.5 m, and each section can be adapted with singular theme colors. The color system is divided into yellow system (Acorus calamus and Canna flavor), red system (Canna safflower), pink system (calcium), white system (taro), variegated system (cattail and reed), etc. take a spot walk to create a rich plant landscape. Aquatic plants are given in Fig. 11.

Fig. 11. Aquatic plants

Depending on statistics, 900 m^2 emergent plants are positioned in the upper reaches of the Laopanjiang River, 5720 m^2 emergent plants are planted in the lower reaches of the Laopanjiang River, and 4560 m^2 emergent plants are planted in the Yanfang River. There are many submerged plants at the bottom of the Laopanjiang River, which affects flood discharge and is not hard to be covered with garbage. After the submerged plants die and decay, they will release nitrogen and phosphorus into the water body, causing seasonal deterioration of water quality. Therefore, it is necessary in order to locally harvest the submerged plants at the bottom of the river and clean up the garbage. The growth area of the ongoing submerged plants accounts for about 60% of the area at the bottom of the river. In order to properly preserve the current habitat, some submerged plants will be removed. The area ratio of 30% is always retained. The total cleaning scale is 169200 m^2.

6) Aquatic animals

Aquatic animals include zooplankton, benthos and fish. They feed on free bacteria, planktonic algae and organic debris in the water, which can effectively reduce child suspended solids in the water is enhanced the transparency of the water. Putting aquatic animals with an appropriate quantity and reasonable species ratio into the water body plays an obvious role in prolonging the food chain of the water environment ecosystem and improving the effect of biological purification.

The project plans to properly stocking zooplankton, benthos (mussels, snails, etc.) and fish (grass carp, carp) and other aquatic animals in the water body, extending the food chain, optimizing the food web structure and building a complete river ecosystem. Being dependent on statistics, 846 kg zooplankton, 5076 kg zoobenthos and 8460 kg fish were put into the Laopanjiang River. A total of 684 kg zooplankton, 4104 kg zoobenthos and 6840 kg fish were put into the Yanfang River.

4.3   Constructions of a Rapid Transit System

1) Analysis of current traffic system

According to the revision of Luliang County Urban Master Plan (2012–2030), except instancing, the red line of the project is not within the planning area of the urban master plan, and the overall traffic accessibility between the suburban area and the suburbs is poor.

The plan points out that in order to relieve the pressure of transit traffic on the central area of the city, it is planned to build the ring road and the ring transit line, and consider the ring east line on the east side of Nanpan River as the long term planning content. In the long term, Binjiang Avenue on the Jiangxi side of Nanpan River will be invoked as the ring of the east line.

Although the planned road can establish a basic road framework for the east of the urban area, it fails to meet the needs of the rapid development of the urban area, and fails to form an effective connection with the nodes such as characteristic towns, river landscape and ecological agriculture that have developed rapidly in the suburbs in recent years.

On the whole, there is a lack of effective momentary connection between suburban areas and principal urban areas, which limits the progress of urbanization to a certain extent.

2) Rapid transit system construction

According to the regional characteristics, urban development planning, current transportation system characteristics and future development direction of Luliang County, it is therefore proposed to add secondary roads, supplement rural road network, and set up multiple shared parking lots at the same time. The rapid transit system can effectively evacuate a large number of traffic and people from the inner urban area of Luliang County and the central urban area of Qujing, and quickly connect the inner urban area with the outlying area. The design of rapid transit system emphasizes "communication", which can provide tourists with fast sight-seeing channels. The slow traffic system of "one heart, three belts" emphasizes "reaching", which allows tourists to swim, slow down and stay in the slow traffic system with pleasant scale and scenery and multiple landscape nodes after arrival, so as to create a green ecological Waterfront Resort. The two will form an effective connection and form a characteristic sightseeing route for mutual benefit.

According to the revision of Luliang County Urban Master Plan (2012–2030), the planned construction of Ring Road and ring transit line is a long term plan, and the completion time has not intended to determine. Therefore, the newly added secondary roads and rural roads in the project are constructed according to the traffic road standard of 6 m wide agricultural logistics park, and the planned Ring Road along Laopanjiang River is also constructed according to the traffic road standard of 6 m wide agricultural logistics park, so as to ensure the smooth connection of the overall external traffic of the project. The aggregate length of the traffic load of the agricultural logistics park constructed by the project is about 41.6 km.

After the construction time of the planned construction of the ring road and the transit line around the city is determined, the pertinent departments can use the project to extend and expand the traffic load of the agricultural logistics park, so as to avoid repeating projects.

(1) New class II Highway

The newly added class II highway is mainly arranged along the Jiangxi Bank of the Laopanjiang River. The planned highway connecting the ring road and the transit line around the city can improve the main ring road network of the county, which owns the advantages of rapidity and convenience. Through the overall horizontal and vertical layout, the newly added class II highway can quickly guide and evaluate the traffic flow pressure from the principal urban area and fully activate the popularity of the site.

(2) Improve rural road network

On the basis of the current scattered rural road network, according to the river direction, important project nodes, surrounding important scenic spots and other resource conditions, the new rural road network connected with the secondary road network can effectively alleviate the traffic pressure of the secondary road network, improve the overall accessibility and hierarchy of the suburban area, and provide diversified choices for citizens and tourists.

(3) Set up green transfer service points

The green transfer service point is mainly arranged at the intersection of class II Highway and slow traffic system, and has the service functions of rapid transfer, parking, bicycle rental and so on. Tourists can quickly reach the service point through the ancillary highway, and quickly enter the lengthy travel system through the service functions such as parking and renting bicycles.

5 Discussion and Conclusions

The waterfront landscape design of the Xinpanjiang River, Laoanjiang River and Yanfang River are single, which can not show the regional cultural characteristics. The lack of recreational facilities makes it impossible to provide a green leisure place for citizens. In addition, the three rivers are lack of connection, which can not create an organically integrated, accessible and convenient water landscape system for the city.

Therefore, based on the current water system conditions of Luliang County, the design and research of the ecological green corridor project for comprehensive river treatment in Luliang County are carried out:

1) Based on the principles of safety, integrity, ecology, protection, context and hydrophilicity, build the overall concept of "Luliang water Township, the most beautiful dam area", use the rich water network to connect the five prominent natural resources of "mountain, water, forest, field and village", and highlight the three cultural characteristics of local folk culture, water conservancy development history and Cuan culture. According to the different current characteristics of the three rivers and the needs of urban development, the overall layout structure is "one heart and three belts", and strive to create a waterfront landscape pattern suitable for living, tourism and industry.

2) Based on the zoning of the ecological restoration project, the layout of the ecological restoration project is carried out, and then the habitat restoration design is carried out. On the basis of the current river channel, it is transformed, and the river habitat elements such as river core, beach, Boulder, ecological stake, aquatic plant and aquatic animal food chain are reasonably optimized and constructed, so as to provide a richer habitat for aquatic organisms and meet the habitat needs of river biological species as much as possible.

3) According to the regional characteristics, urban development planning, current transportation system characteristics and future development direction of Luliang County, it is suggested to add secondary roads, supplement rural road network, and set up multiple public parking lots at the same time. The rapid transit system can effectively evacuate a large number of traffic and people from the central urban area of Luliang County and the central urban area of Qujing, and quickly connect the central urban area with the suburban area. The design of rapid transit system emphasizes "communication", which can provide tourists with fast sightseeing channels. The slow traffic system of "one heart, three belts" emphasizes "reaching", which allows tourists to swim, slow down and stay in the slow traffic system with pleasant scale and scenery and multiple landscape nodes after arrival, so as to create a green ecological Waterfront Resort. The two will form an effective connection and form a characteristic sightseeing route for mutual benefit.

In a word, through the use of artificial means to build suitable habitats and environmental elements, accelerate the recovery of the ecosystem, and build a complete river ecological corridor, in order to contribute to regional ecological security of Luliang County.

Acknowledgements. This study are supported by the scientific research projects entrusted by Yunnan water conservancy and Hydropower Engineering Co., Ltd. "Research on the regulation and harmless treatment technology of water hyacinth migration path in the comprehensive treatment of river channel in Luliang County, Qujing City, Yunnan Province" and "Research on the impact assessment and Countermeasures of river discharge into Luliang County, Qujing City, Yunnan Province on the regulation effect of river channel comprehensive treatment project". This research is also supported by the National Natural Science Foundation of China (No. 41776024, No. 42176166).

References

Li H (2015) Planning and design of ecological corridor in Nanyang Xixia stork river road based on the regional characteristics. Mod Landsc Architect 12(6):462–468 (in Chinese)

Li LB, He XD, Yi SL (2021) Study on the present situation, problems and comprehensive treatment measures of the Laopanjiang river channel in Luliang County in Yunnan Province. Water Resour Dev Manag 9:19–24 (in Chinese)

Liu Y, Gu J (2020) Study on sewage treatment of old buildings along the river. Water Resour Dev Manag 6:39–42 (in Chinese)

Tang SF (2018) Feasibility study report on comprehensive river regulation project in Luliang County. China Water Resources Pearl River Planning, Surveying and Designing and Designing Co. Ltd., Guangzhou, China. (in Chinese)

Tang SF (2020) Preliminary design report of comprehensive river regulation project in Luliang County, Qujing City, Yunnan Province. China Water Resources Pearl River Planning, Surveying and Designing and Designing Co. Ltd., Guangzhou, China. (in Chinese)

Wang L, Yang MQ, Guo WZ (2020) Application of physical-ecological composite restoration technology in black-odor rivers regulation. Water Resour Dev Manag 7:1–7 (in Chinese)

Wu DP (2020) Analysis on comprehensive treatment of water environment in Luliang County, Yunnan Province. Flood Control Drought Relief China 30(4):43–47 (in Chinese)

Yan J, Xiang ZJ, Ling J (2016) Study on landscape ecological planning of Shijiu Lake and Ma'anshan Lake area. Resour Environ Yangtze River Basin 25(9):1375–1383 (in Chinese)

Zheng DF (2012) Exploration on city river's function returning from "edge line" to "public link"—A case study of ecological green corridor planning of Chuan Yang river. Shanghai Urban Plan 1:67–73 (in Chinese)

Zhou WW (2021) Research on space shaping method of urban ecological green corridor based on landscape ecology—Taking landscape planning and design of optical valley ecological corridor as an example. Urban Hous 28(12):96–98 (in Chinese)

Study on the Technical and Policy Pathway for Low-Carbon Development of the Water Transportation in Sichuan Province

Yonglin Zhang[1,2], Chaohui Zheng[1,2(✉)], Yue Li[1,2], Liguo Zhang[1,2], Jinxiang Cheng[1,2], and Mingjun Li[1,2]

[1] Transport Planning and Research Institute, Ministry of Transport, Beijing, People's Republic of China
zhengch@tpri.org.cn

[2] Laboratory of Transport Pollution Control and Monitoring Technology, Beijing, People's Republic of China

Abstract. Transport, one of the key sectors for the implementation of China's carbon peak and carbon neutrality goals, boasts a water transportation crucial to China's integrated transport infrastructure, which will face a rising demand with the continuous optimization of the transport system. Located upstream of the golden waterway of the Yangtze River, Sichuan province is one of the key regions for inland waterway transportation. In the context of the 'dual carbon' goals, the study proposes a conceptual framework for the low-carbon development pathway of water transport by analyzing the technical roadmap and policy recommendations for the Sichuan Province, which found that Sichuan can rely upon energy efficiency and energy substitution on the technical front, while upon the synergy of "government guidance + enterprise responsibility + market support + partnership". While focusing on slowing the growth of CO_2 emissions and improving energy efficiency with LNG as the main transition fuel in the near term, Sichuan should curb the total emissions and strategize around carbon-free solutions including electricity and hydrogen.

Keywords: Waterway transportation · Low-carbon development · Policy recommendation · Transition pathway

1 Introduction

With the national strategies of carbon peak and neutrality ambitions, challenging targets have been imposed on the transport sector, which is bound to transform towards a green and low-carbon trajectory. The waterway transportation, a pillar with emissions accouting for around 6.5% of the sectoral total (Li et al. 2021), will face an increasing demand with the continuous system optimization transferring bulk goods and mid-and-long-haul movements of goods 'from road to waterway' (Ministry of Transport of the People's Republic of China, 2021).

Waterway emissions mainly stem from vessels and ports. Chinese transport vessels, which are of large in volume, high in mobility and difficulty in management, mainly consume diesel and fuel oil. Despite generating less emissions than ships, ports must endeavor towards the same direction given its natural connections to the hinterland and maritime transport (Li 2021). Decarbonizing the waterway transport is a large undertaking with considerable implications and difficulties, thus requiring systematic and sophisticated research. While there have been studies analyzing the status quo of waterway emissions in China as well as the main challenges and solutions towards a less carbon-intensive waterway transportation (Hou 2017; Zheng et al. 2020; Ma et al. 2020), there is yet a comprehensive analysis on the technical and policy roadmap for its low-carbon development.

Taking Sichuan province as the subject, In the context of the 'dual carbon' goals, the study proposes a conceptual framework for the low-carbon development pathway of water transport by analyzing the technical roadmap and policy recommendations for the Sichuan province, and establishes a scientific baseline for the provincial measures towards carbon ambitions during the 14th Five Year Plan (2021–2025).

2   Research Scope

Located upstream of the golden waterway of the Yangtze River, Sichuan province is one of the key regions for inland waterway transportation, boasting strategic significance in the Yangtze River Economic Belt. The national strategy of "Yangtze Belt Powered by Golden Waterway" has brought unprecedented opportunities for the water transportation of Sichuan (Zhai 2014), a province with abundant resources including 4 high-grade waterways (Yangtze, Min, Jialing and Qu rivers) and numerous watercourses of other grades such as Tuo, Fu, Jinsha and Chishui, as well as exceptional shoreline resources and untapped potential (The People's Government of Sichuan Province 2012).

By 2020, Sichuan Province had 10,900 km of waterways, 4,718 transport vessels and 18 planned ports with 1,558 berths. In 2020, Sichuan's waterborne passenger traffic reached 9.54 million with a turnover of 104 million passenger kilometers, waterborne freight traffic 65.27 million tonnes with a turnover of 29.176 billion tonne-kilometers, port throughput 13.6 million tonnes and the annual container throughput 274,175 TEU (National Bureau of Statistics Zhai 2021). By 2020, the Sichuan Province had already embarked on its green journey including the introduction of 80 electric vessels for short-haul passenger transport, 15 sets of shore power facilities at 15 berths in Luzhou, Yibin and Nanchong ports, as well as over 80% of the province's official vessles using shore power at berth.

The 14th Five-Year Plan (2021–2025) for Comprehensive Transport Development in Sichuan Province stipulates that by 2025, the share of waterway transport in cargo turnover shall increase by 5 percentage points and CO_2 emissions per unit of transport turnover of operating vessels decrease by 5% (The People's Government of Sichuan Province 2021). Therefore, it is imperative for the provincial waterways to accelerate its green and low-carbon transformation.

Taking the ports and shipping of Sichuan as the subject, the study conducts an analysis of the technical roadmap and policy recommendations, addressing the direct CO_2 emissions from fossil fuel consumption by ships, cargo loading/unloading, and auxiliary operations at ports. Through an analysis of the key factors determining the low-carbon development of the waterway sector, the study reviews the technical roadmap from the perspectives of both energy efficiency and substitution, and provides policy recommendations on CO_2 emissions measurement, energy consumption monitoring, market mechanism, R&D (Research and Development) on low-carbon technologies, etc.

3   Key Factors of the Low-Carbon Water Transport Sector

The activity level of vessels and ports, mostly driven by regional demands, will remain an upward trend in large parts of China. For example, in 2021, China completed 15.55 billion tonnes of port cargo throughput with an increase of 6.8% year-on-year (Financial sector 2022), Sichuan province 20.44 million tonnes with a 50.29% bump (China Waterway Network 2022), Pearl River system 1.879 billion tonnes (China News Network 2022), and Ningbo-Zhoushan Port 1.224 billion tons, up 4.4% year-on-year, ranking the first in the world for the 13th consecutive year (Xin Min Evening News 2022).

The status quo analysis of the the water transport sector indicates that the low-carbon process may zero in on reducing emissions in the recent future. Table 1 shows the sectoral carbon emissions by transportation modes, and the formula variables already contain prospective approaches applicable to both ports and shipping including decreasing activity levels and unit energy consumption as well as using energy sources of lower emission factors.

At present, low-carbon shipping solutions concentrate on energy efficiency and cleaner fuels, which could also consider green marine engines and CCUS (Carbon capture, utilisation and storage) in the future. For example, improve energy efficiency in management and operations, adopt applications for less resistance and more propulsion, and prioritize cleaner ship fuels such as LNG, methanol, biodiesel, hydrogen and ammonia (China Classification Society 2021). The key to low-carbon port development involves production equipment, mobile machinery and other energy-consuming components, and mainly depends on energy substitution and efficiency improvement technologies, which are represented by process technologies and loading/unloading equipment powered by electricity or other clean energy, capacity matching as well as optimized energy efficiency and consumption during loading and unloading operations, facility upgrading by phasing out energy-and-emission-intensive and inefficient equipment (Hou 2017). Low-carbon development measures shall be tailored to the current and future specificalities of ports and shipping in the Sichuan province. The technology roadmap for this research is shown in Fig. 1.

Table 1. Calculation methods for carbon emissions by mode of waterborne transport

Mode of transportation	Method	Formula
Passenger	Turnover	CO_2 emissions $= \sum_i$(turnover × percentage of energy of type i× energy consumption per unit of passenger turnover for energy type i× CO_2 emission factor for the ith energy source)
Freight	Turnover	CO_2 emissions $= \sum_i$(turnover × percentage of energy of type i × energy consumption per unit of freight turnover for energy type i × CO_2 emission factor for the ith energy source)
Ports	Throughput	CO_2 emissions $= \sum_i$(throughput× energy consumption per unit of throughput ×percentage of energy of type ×CO_2 emission factor for the ith energy source)

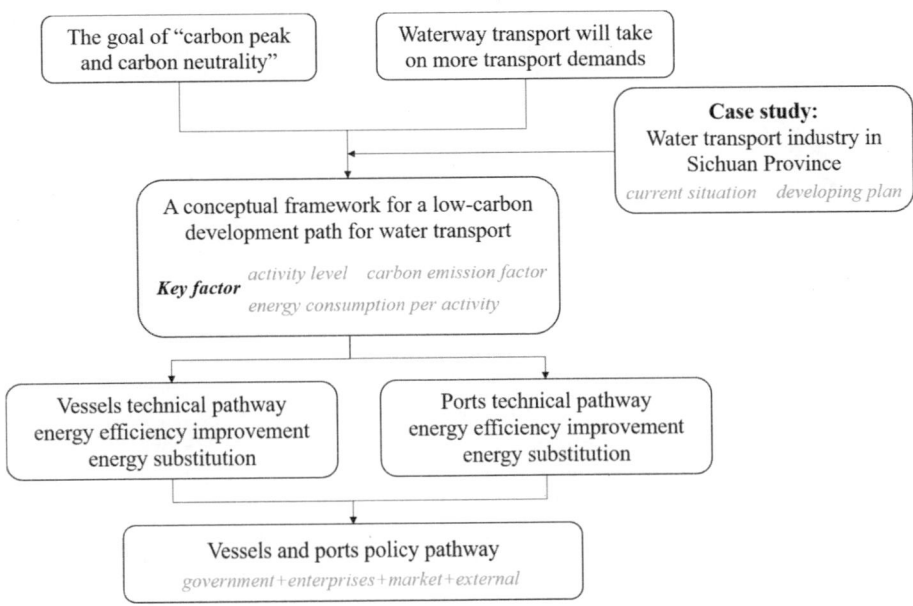

Fig. 1. Technology roadmap for this research.

4   Technical Options for Low Carbon Development of the Water Transport Industry in Sichuan Province

4.1   Energy Efficiency Improvement

Vessels

Vessels' energy efficiency could be improved at design stage and/or operation stage. Energy Efficiency Design Index (EEDI) of a vessel, which is defined by IMO's guidance, is an estimate of CO_2 emissions per freight volume in the stage of the vessel's specific design. Optimization of the propulsion system, hull form and additional drag reduction technologies are effective measures to improve EEDI. Energy Efficiency Operational Indicator (EEOI) is an indicator expressing the energy efficiency of the ship currently in service, which is significantly influenced by ship operations (including sailing speed, route and trimming). Lowering the speed is considered to be the most effective means to improve energy efficiency, reducing fuel consumption up to 60%, indicating a close linkage between vessel speed and fuel consumption. However, speed reduction is often subject to sailing conditions and the time sensitivity of shipping orders. Meanwhile, route optimization is a comprehensive solution that could decrease the total fuel consumption by planning optimal paths and taking into account wind currents (i.e. the shortest distance and voyage).

However, the obsolete vessels in the Chinese inland waterway transport fleet remain a major challenge. Given the fact that the Sichuan fleet is mostly dry bulk carriers and aging vessels for short-distance transshipment, and that the main navigable waterways in the province are of lower grades and varying levels, neither configuration modification nor operational optimization is optimal. Considering the present and future development of the Sichuan fleet, fleet renewal is generally recommended for better energy efficiency, replacing outdated vessels with new ones that are designed and built in accordance with the energy efficiency standards based on the latest technical specifications.

Ports

The energy efficiency of the port is reflected by energy consumption per unit of throughput. Generally speaking, substandard loading, unloading, storage and transfer of cargo may complicate or obstruct the operational processes, thus increasing the energy consumption of port activities. Therefore, terminal-specific optimization is particularly significant to improve the coordination, capacity-matching and efficiency of port operations.

Ports in Sichuan face challenges in terms of unbalanced geographical distribution and functional structure, inadequate infrastructure and supply capacity, and mismatched costs and benefits, painting a picture of 'small, weak and scattered'. Port modernization in Sichuan started fairly late with only 11% of modernization rate (Yu 2021) and 6 ports above designated scale, two of which (Luzhou Port and Yibin Port) are major ports along the Yangtze River. At present, besides the port groups of Luzhou-Yibin-Leshan in the southern economic zone, the northeastern Guangyuan-Nanchong-Guangan, and part of the Liangshan Port in the Panzhihua-Xichang region, other ports, mostly built upon natural slopes, still mainly transport sand and gravel without modern and scale-appropriate machinery.

Therefore, horizontal transportation for dry bulk cargo terminals can shift away from oil-powered to electricity-powered processes. General cargo terminals could apply suitable energy-efficient technologies to dry bulk cargo and breakbulk, and level up automation for the handling of containers or dry bulk in applicable terminals. From a local perspective, modernize the handling equipment and vehicles with higher energy efficiency. In the meantime, lower the equipment energy consumption through energy recovery of loading and unloading machinery and energy-saving renovation of power facilities.

The 5G smart port construction project in Zhicheng terminal of Yibin Port has been officially started. As the first 5G smart port project in Sichuan Province, it is an ideal pilot project for automatic and intelligent approach for low-carbon development.

In addition, the application of smart technologies is also recommended, which elevates operational accuracy and efficiency while reducing unit energy consumption. The Zhicheng 5G terminal in Yibin, as the first 5G smart port in Sichuan, serves as a blueprint for future demonstration projects.

4.2   Energy Substitution

Vessels

Research and practices worldwide have shown that there is yet a global consensus on the energy substitution in the realm of waterway transportation (Horvath et al. 2018; Brahim et al. 2019; The Oxford Institute for Energy Studies 2019; KPMG Germny 2021; Xu 2020; Huang 2021). The application of alternative fuels will likely be driven by regulations of international bodies, fuel prices, technological developments, availability of alternative fuels and development of infrastructure. Despite the consensus on the significance of electricity and hydrogen in achieving carbon neutrality, relevant technological means remain uncertain in terms of readiness, and never-ending debates still terrorize alternative fuels such as LNG, ammonia, methanol and biogas. The global LNG industry, despite a huge expansion of capacity with final investment decisions on projects potentially in excess of 100 bcm in 2019 and 2020, faces affordability and decarburization challenges for some reasons. Consider that ammonia is of high kindling point, low flame speed, low flammability limit, high heat of evaporation and produces nitrogen oxide emissions when used as fuel. Methanol, mainly extracted from coal and natural gas, is not a low carbon option according to the law of conservation of energy.

The Sichuan province, a key national base for clean energy and a national flagship on clean energy demonstration, has been committed to the implementation of Outline of Clean Energy Development Strategy. In 2020, the natural gas production in Sichuan Province reached 43.2 billion cubic meters, ranking first in the country, with an increase of 14% or 5.2 billion cubic meters year-on-year, accounting for 45% of the yearly incremental growth. Meanwhile, the abundance of hydropower resources implies the province's tremendous potential in "clean electricity". Recently, Sichuan and Chongqing are co-developing hydrogen economy, which has already produced the Chengdu-Chongqing hydrogen corridor for hydrogen vehicles, translating into favorable conditions for furthering the energy mix optimization of Sichuan's water transportation.

Considering the hydrogen endowment of Sichuan, LNG could be the transition fuel in the fleet renewal to achieve short-term carbon reduction. As electric and hydrogen power solutions mature up, new vessels can be fitted with cleaner engines, new container liners with chemical or hydrogen fuel cells, and local liner ships with hydrogen power system. However, Sichuan's lacking of LNG refuelling station is restricting the potential of LNG vessels. Therefore, it is urgent to redouble efforts in LNG terminals and LNG-powered vessels. Moreover, other measures such as green and environmentally-friendly technologies should also be included in waterways programs.

Ports

The energy consumption of port operations, relying on trucks, front cranes, single bucket loaders, excavators, tractors, etc., features mainly oil and electricity; Therefore, cleaning up the power of port machineries is instrumental to optimize fuel consumption at ports. At present, there are three container terminals (Luzhou, Yibin, and Nanchong) in the province that have achieved electrification. It is recommended to advance electrification of port equipment and LNG solutions for horizontal transport vehicles, which then gradually transition to battery and hydrogen power systems in the future. Self-sufficiency of wind and PV (Photovoltaic Power) is another stepping stone to achieve the two carbon goals. Advantaged ports can adopt an energy system of "distributed PV + energy storage + micro-grid", combined with smaller wind turbines and ground/air-sourced heat pumps to enable two-way power flow between the port and the grid.

Vessel-port coordination is important in the energy substitution process of the waterway transport sector, considering new vessel power such as shore power, LNG and hydrogen are based on the port. In accordance with relevant regulations and measures (The Standing Committee of the 13th National People's Congress 2020; Ministry of Transport of the People's Republic of China 2021), the renovation of receiving facilities of the shore power system shall be promoted with a focus on container ships, multi-purpose ships and dry bulk carriers of 600 GT and above. Meanwhile, increase the utilization rate of shore power facilities. In addition, build LNG terminals in the upstream of major waterways such as the Jialing River and Minjiang River, and provide LNG in the service areas of the waterways should be taken into consideration. Meanwhile, ports can also make preparations for vessel charging, battery swap and hydrogen refueling in advance.

5   Policy Recommendations

The central government has clearly pledged to enhance the policy environment for investment, green finance, fiscal and pricing framework, as well as the market mechanism. The Sichuan government can maximize the synergy of "government guidance + enterprise responsibility + market support + partnership" from the provincial baseline of the waterborne transportation. In the meantime, it should strengthen organizational leadership, coordination, accountability, supervision and review (The State Council the People's Republic of China 2021).

Firstly, tighten the grip on vessels and ports, reducing fuel consumption and emissions per unit to fulfill the latest standards. Conduct regular emission assessments to reflect the effectiveness of the decarbonizing efforts in order to make policy adjustments accordingly. It is recommendable to consider pilot projects on new energy vessels and zero carbon ports. *The Sichuan Province Five-Year Action Plan to Promote Green Water, Green Navigation and Green Development* strives to build two zero-carbon demonstration sites by 2025 where operations and facilities including HVAC will be powered by clean energy within the port area (Department of Transportation of Sichuan Province 2021).

Secondly, develop an energy efficiency improvement plan on the basis of a comprehensive survey on the fleet's energy efficiency, outlining the timetable for vessel renewal and introducing the energy efficiency standards. Establish a supervision system for ship pollution aligned with the vessel pollution regulation, the inland vessel energy consumption collection and reporting framework of the Yangtze River, as well as the intelligent supervision solution for sand and gravel vessels in Sichuan. Additionally, establish a life-cycle tracking and reporting system for port emissions based on energy usage records and statistics.

Thirdly, encourage the private sector to diversify funding sources, utilize the provincial emissions trading mechanism, and accelerate the establishment of the climate change investment and financing center. The Sichuan Joint Environmental Exchange is the first carbon emissions trading market outside the national pilot program in China. By the end of October 2020, the Sichuan Exchange had traded a total of 15,728,200 tonnes of voluntary greenhouse gas emission reductions (CCERs) with the unilateral transaction amount exceeding RMB 100 million. The government should provide informed guidance based on research on participating ports and vessels.

Fourthly, increase interconnection with the Chongqing Port while a shipping center in the Yangtze upper reach is being contemplated with aligned regulation, enforcement, execution, standards, as well as reward and penalty policies. Scale up low-carbon technologies in vessels by partnering up with ports in the middle and lower reaches of the Yangtze River. At the same time, competent authorities should provide platforms for scientific research and pilot projects for technology breakthroughs, which are often too complex and challenging for one actor to attain independently, so the government could weigh in as a coordinator and facilitator of cooperation. Sichuan Province boasts a solid foundation for R&D, and Chengdu has recently proposed to build a state-level key science and technology center. Such abundance of resources in science, technology and industrial application could further contribute to the low-carbon transformation of its water transport sector.

6 Conclusions and Prospect

In the context of the dual carbon goals, it is imperative for the water transport industry in Sichuan Province to accelerate its green and low-carbon transition. This study proposes a conceptual framework for a low-carbon development pathway for water transport with an analysis of technical roadmap and policy recommendations based on the current and future landscape of its ports and vessels. Sichuan can rely upon energy efficiency and energy substitution on the technical front, while upon the synergy of "government guidance + enterprise responsibility + market support + partnership". While focusing on slowing the growth of CO_2 emissions and energy efficiency with LNG as the main transition fuel in the near term, Sichuan should curb the total emissions through improved emissions monitoring and assessment, vessel fleet renewal, pilot projects on carbon-neutral ports while strategize around carbon-free solutions including electricity and hydrogen in the long term. Meanwhile, leverage the carbon trade and green finance in decarbonizing the sector.

For end-of-pipe treatment in waterway transportation, CCUS is not addressed in this research as the technology is far from application, which shall be considered based on a cost-benefit assessment of space and performance once it achieves maturity. Besides, it is recommended that new waterways, ports and terminals should maximize carbon sink in the landscaping phase, as well as ecological restoration and sponge capabilities.

Acknowledgements. This work was supported by Green Intelligent Inland Ship Innovation Programme & Prototype development of methane escape control device for Marine dual fuel/gas engines & Technology Project of Beijing Dongcheng District Bureau of Science, Technology and Information Technology (fund NO.2022-1-010).

References

Li XY et al (2021) Paths for carbon peak and carbon neutrality in transport sector in China. Strateg Study of CAE 23(6):7

Ministry of Transport of the People's Republic of China (2021) Notice on the Issuance of the "14th Five-Year Plan (2021–2025)" for the Development of Green Transport. https://xxgk.mot.gov.cn/2020/jigou/zhghs/202201/t20220121_3637584.html

Li QX (2021) Carbon emission status and carbon reduction path analysis of waterway transport in China. Transp Energy Conserv Environ Prot 17(2):5

Hou J (2017) Analysis of current situation and development trend of energy saving and emission reduction in waterway transport. Water Transp Eng 13(1):6

Zheng J, Liu CG, Lin ZQ (2020) Low-carbon development of green ships and related strategies. Strateg Study of CAE 22(6):9

Ma LY, Wu M (2020) Discussion on the current situation and application of energy saving and emission reduction technology in China's ports. Energy Conserv Environ Prot 10:2

Zhai XN (2014) Sichuan's golden waterway sails from here. China Marit 000(011):60–62

The People's Government of Sichuan Province (2012) Implementation Opinions on Accelerating the Development of Yangtze River and Other Inland Waterways Transport. https://www.sc.gov.cn/10462/10883/11066/2012/3/15/10203116.shtml

National Bureau of Statistics (2021) Chinese statistical yearbook 2020. *China Statistics Press*

The People's Government of Sichuan Province (2021) Notice on the Issuance of the "14th Five-Year Plan (2021–2025)" for Comprehensive Transport Development in Sichuan Province. https://www.sc.gov.cn/10462/zfwjts/2021/10/29/602da2a8aca9495fb114dc226a256ff7.shtml

Financial sector (2022) Ministry of Transport of the People's Republic of China: China to complete 15.55 billion tons of port cargo throughput in 2021. https://baijiahao.baidu.com/s?id=1726003942185388791&wfr=spider&for=pc

China Waterway Network (2022) Sichuan province aims to complete waterway transport investment of 5 billion yuan in 2022. https://new.qq.com/omn/20220224/20220224A0943Z00.html

China News Network (2022) Four provinces in the Pearl River system to reach 1.454 billion tonnes of waterway freight in 2021. https://baijiahao.baidu.com/s?id=1721114974762940299&wfr=spider&for=pc

Xin Min Evening News (2022) Ningbo Zhoushan Port to exceed 1.2 billion tonnes of cargo by 2021 for the first time. https://baijiahao.baidu.com/s?id=1721912929883870790w&wfr=spider&for=pc

China Classification Society (2021) Shipping Low Carbon Development Outlook 2021

Yu QB (2021) Problems and countermeasures of Sichuan inland waterway port development. China Ports 5:4

Horvath S, Fasihi M, Breyer C (2018) Techno-economic analysis of a decarbonized shipping sector: technology suggestions for a fleet in 2030 and 2040. Energ Convers Manage 164:230–241

Brahim TB, Wiese F, M, Münster. (2019) Pathways to climate-neutral shipping: A Danish case study. Energy 188:116009

The Oxford Institute for Energy Studies (2019) Challenges to the Future of LNG: decarbonisation, affordability and profitability

KPMG Germany (2021) The pathway to green shipping

Xu D (2020) Analysis of the advantages and shortcomings of hydrogen energy in port applications. China Water Transport 11:2

Huang B (2021) The impact of carbon neutrality on the port industry and the related development proposals and measures of Guangzhou Port. China Ports 12:5

The Standing Committee of the 13th National People's Congress (2020) Law of the People's Republic of China on the Protection of the Yangtze River. https://www.mee.gov.cn/ywgz/fgbz/fl/202012/t20201227_814985.shtml

Ministry of Transport of the People's Republic of China (2021) Decision on Amending the Measures for the Administration of Shore Power for Ports and Ships (Order of the Ministry of Transport of the People's Republic of China No. 31 of 2021). https://xxgk.mot.gov.cn/2020/jigou/fgs/202109/t20210922_3619619.html

The State Council the People's Republic of China (2021) Opinions of the Central Committee of the Communist Party of China (CPC) and the State Council on the complete and accurate implementation of the new development concept and the proper implementation of carbon peaking and carbon neutral work. http://www.gov.cn/zhengce/2021-10/24/content_5644613.htm

Department of Transportation of Sichuan Province (2021) Notice on the Issuance of the Five-Year Action Plan for the Promotion of Green Development of Green Water and Green Navigation in Sichuan Province

Sedimentary Process in Navigation Channel in an Estuarine Port, - A Case Study from the Port of Niigata, Japan

Yasuyuki Nakagawa[1]([✉]), Taichi Kosako[1], Hiroyuki Hayashi[2], and Tomohiro Watanabe[3]

[1] Port and Airport Research Institute, Yokosuka, Japan
{nakagawa, kosako}@p.mpat.go.jp
[2] Yokkaichi Port Authority, Yokkaichi, Japan
hayashi@yokkaichi-port.or.jp
[3] Hokuriku Regional Development Bureau, Ministry of Land, Infrastructure, Transport and Tourism, Niigata, Japan
watanabe-t84k2@mlit.go.jp

Abstract. The present study focused on the siltation process at the navigation channel in the Port of Niigata, Japan. A part of the port is located at the mouth of the Shinano River, which is the longest one in the country. Due to the discharged sediments through the river and their deposition in the port area, frequent dredging works are required for the safety navigations in the waterway and turning basins. The purpose of the present study is to get a better understanding of the specific features of the sedimentary process in the target area for the consideration of any appropriate countermeasures for the optimization of the dredging works. In the present study, we especially focus on the sedimentation processes around the dredged navigation channel, which need frequent dredging to keep the planned depth from −5.5 to −12 m, under the complicated estuarine hydraulic conditions and several field measurements have been carried out, including current measurements, bathymetric surveys by acoustic soundings with a narrow multibeam sonar system.

Keywords: Estuarine port · Sediment discharge · Siltation · Estuarine circulation · Fluid mud

1 Introduction

Many ports have been developed in estuarine areas of the rivers all over the world and played an important logistics function as a node between sea routes and inland transportation. While they have geographical advantages, such areas often experience some difficulties to keep the water depth required for ship navigation due to the accumulation of discharged sediments through rivers (e.g. PIANC 2008). Estuarine

---

H. Hayashi—Former Director of Niigata Port and Airport Office, Hokuriku Regional Development Bureau, MLIT).

Y. Li et al. (Eds.): PIANC 2022, LNCE 264, pp. 1312–1321, 2023.
https://doi.org/10.1007/978-981-19-6138-0_114

environments where freshwater and seawater encounter each other have several characteristics from the viewpoints of hydraulic and sedimentary processes. Current fields are controlled by the density distribution that could drive the estuarine circulation and sediment transport processes could become complicated due not only to physical but also to chemical processes, such as floc formation of suspended particles (e.g. Dyer 1997, Whitehouse et al. 2000).

Several ports and harbors have been also developed around the estuaries of Japanese major rivers and they support logistics transportation. However, in response to the increase in the depth of waterway for the growth of vessel size, the deposition of discharged sediments through rivers is an inevitable problem for the use and maintenance of them with securing appropriate disposal sites for dredged sediment. This paper shows some examples of field observations at the west part area of the port of Niigata, located at the mouth of the Shinano River, which have been carried out to elucidate the characteristics of sediment dynamics in the navigation channels.

2 Site Description

2.1 Location of the Study Site

The port of Niigata is one of the largest commercial ports on the west coast of the Japanese main island and the west part of the port has been developed around the mouth of the Shinano River, which is the longest one in the country, as shown in Figs. 1 and 2. An example of measured water depth along the channel is presented in Fig. 3 and the survey line is the dotted line in Fig. 5. As shown in the figure of the bottom topography, the channel is maintained with a depth from −5.5 to −7.5 m by using any type of dredger including grab type, pump type, and the trailing suction hopper dredger, Hakusan, which is a specially designed and operated for the port (e.g. Katoh 2018).

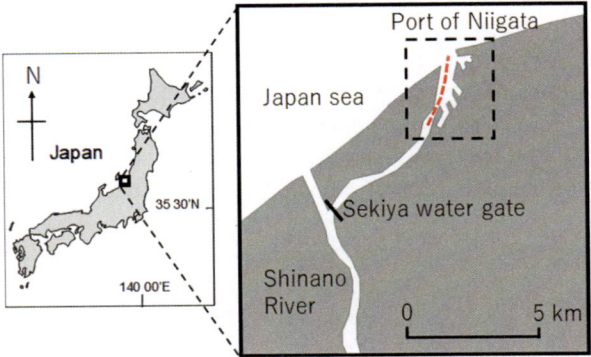

Fig. 1. Location of the study site.

Fig. 2. Aerial photo of the port area.

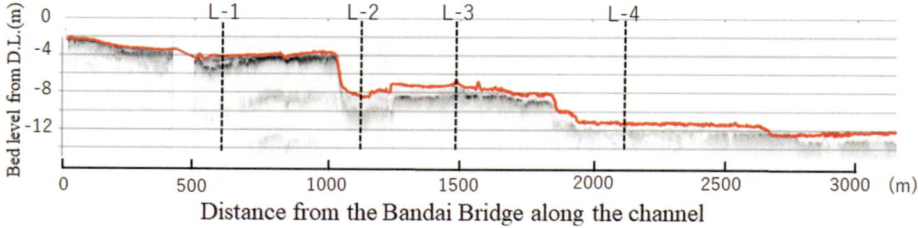

Fig. 3. Bottom topography along the channel measured in July 2013. (Locations of L-1 through L-4 are indicated in Fig. 6)

2.2   Condition of the River Discharge and Density Structure

As a condition of freshwater discharge, Fig. 4 shows temporal variations of the measured flow rate at the monitoring station located at the upper part of the Sekiya watergate in 2017 and 2020, which is operated by the water information system of the ministry of land, infrastructure, transport and tourism (MLIT). It is around 400 m³/s of the daily averaged rate during most of the period, but it could increase over 1,000 m³/s due to any high precipitation event in the upstream region. Since the freshwater inflow to the port area is controlled by the Sekiya water gate shown in Fig. 2, not all the freshwater flows into the port area. However, part of them flows through the gate with sediments and the annual maintenance dredging is required with an average amount of about 800,000 m³ per year.

According to the previous works (e,g, Nakagawa et al. 2017), we observed the stable pycnocline mainly due to the vertical salinity profiles formed by the near surface freshwater outflow and the deeper seawater in the dredged channel. The formation of the stable pycnocline is due to the micro-tidal condition with a tidal range of around 30 cm even during the spring tide period. The suspended sediment concentration, however, is highly variable depending on the upstream condition of the discharged flow. Since fluid muds are also detected in the upper area of the dredged channel, the sediment transport process in the target area is schematized in Fig. 5, which shows the

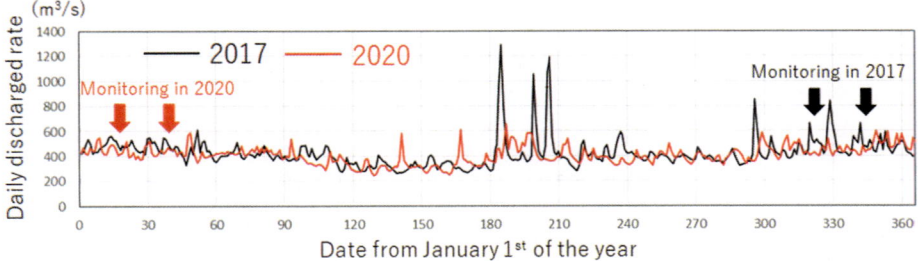

Fig. 4. Daily discharge measured at the upstream of the Sekiya water gate (Data source from the water information system of MLIT in 2017 and 2020).

suspended sediments in the near surface freshwater layer and fluid mud layer in the near bottom of the dredged area. In the present paper, additional field data are presented in terms of characteristics of the current field and bathymetry around the dredged channel, which have been obtained through monitoring surveys in 2017 and 2020.

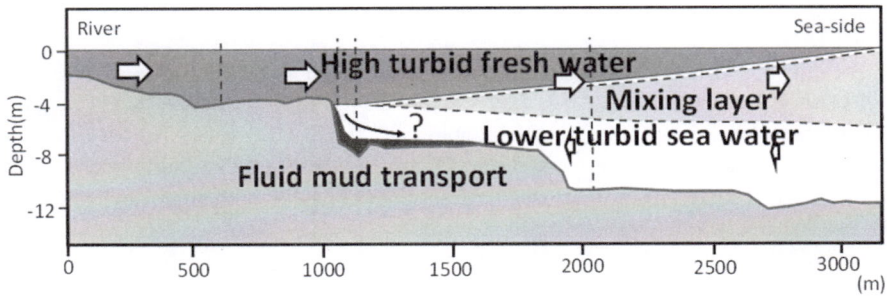

Fig. 5. Schematics of sediment transport process in the study site. (After from Nakagawa et al. 2017)

3   Field Measurements

Field measurements of the sediment fluxes at the boundary between a port area and the outer region are crucial to evaluate the siltation processes and their volume in the port area (e.g., Claeys et al. 2001). In the present study, an acoustic doppler current profiler or ADCP (Workhorse Sentinel, 1200 kHz, TRDI) was used to measure the vertical profiles of current velocities in the site in late November and December of 2017. The monitoring points are included near the upstream end of the dredging area as shown in Fig. 6. The ADCP was attached to the survey boat, which was anchored during the measurement period, and water quality parameters were also measured by casting a multi-sensor system (AAQ1183, JFE-Advantec Co.) for water temperature, salinity, and turbidity. The turbidity was measured by an optical backscatter sensor, and it was calibrated into the suspended sediment concentration (SSC) by using the relationship between the measured turbidity and analyzed SSC for the same samples. The turbidity

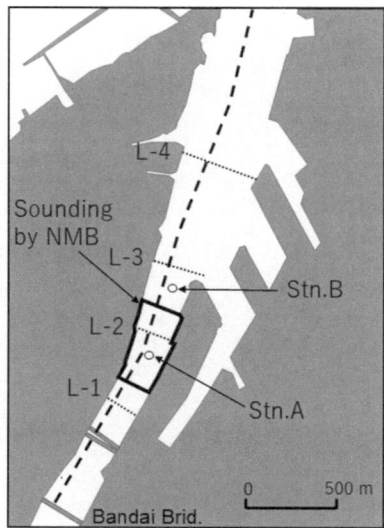

Fig. 6. Monitoring points of ADCP measurements, water qualities and sounding area by NMB system.

sensor can read up to around 2,500 mg/l before the saturation of the backscatter signal. A narrow multi-beam system or NMB (Sonic2022, Toyo Co.) was used for the measurement of detailed topography for the area shown in Fig. 6. The bathymetry observation was carried out twice at intervals of about two weeks in January and February 2020.

4   Results and Discussions

4.1   Measured Profiles and SSC Flux Analysis

Figures 7 and 8 show the observation results of the vertical profiles of measured velocity together with the water temperature, salinity, and calibrated suspended sediment concentration at the representative monitoring points of Stn. A and B in Fig. 6. The vertical resolutions of the plotted data are every 0.25 m for the velocity and 0.1 m for the temperature, salinity, and SSC. All the figures indicate the data measured on November 17 and December 11 in 2017. The current profiles are demonstrated as the current speed in the direction along the channel and the positive/negative value means the velocity in the downward/upward direction. As shown in Fig. 7, the current flow in the downward direction over the whole depth at Stn. A located in the upper river channel, where the freshwater is flowing with a salinity of almost zero. On the other hand, so-called gravitational estuarine circulations are observed at Stn. B as shown in Fig. 8(a) with slight compensatory flows to the upstream direction at the depth of around 4 m, which is just beneath the pycnocline. The circulation patterns are quite similar between the two measurements. The salinity profiles are also similar between the two data as shown in Figs. 7(b) and 8(b), although the water temperatures are slightly different. These density and current profiles are typical and can be categorized as a salt wedge estuary in the 2D regime (e.g., Dyer 1997). Secondary flow in the direction of the cross-section is generated when there is a curvature along the channel and the density structure also may modify the 3D flow field (e.g., Winterwerp et al. 2006). The channel in the present study between the sections L-1 and L-3 has a simple shape without a strong bend and the more 2D structure could be dominant here with a weak secondary current.

Although the observed current and density profiles are similar between the surveys, the SSC condition is quite different and turbid water with a higher SSC was discharged

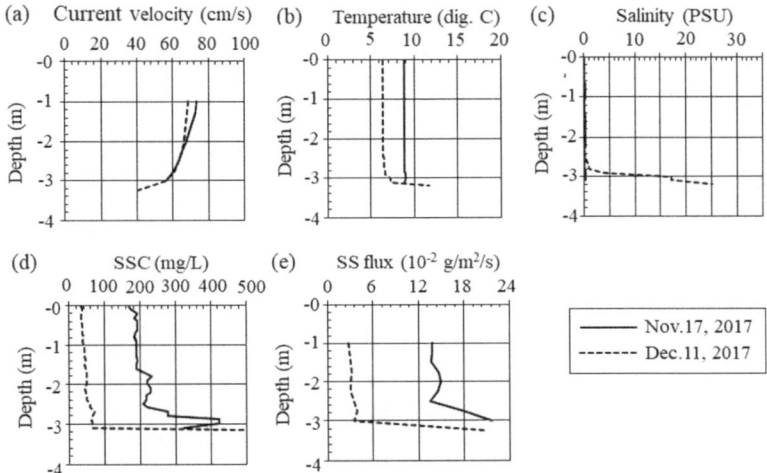

Fig. 7. Measured profiles at Stn. A: (a) current speed along the channel, (b) Water temperature, (c) salinity, (d) SSC calibrated from turbidity, and (e) calculated sediment flux.

on December 11 in 2017, showing the turbid fresh water flowing over the less turbid seawater as depicted in Fig. 5. Furthermore, SSC fluxes are also calculated with the measured current speed and calibrated SSC profiles (Figs. 7(e) and 8(e)).

4.2   Characteristics of Bathymetry Change

As mentioned above, the required water depth of each section of the navigation channel has been kept with frequent maintenance dredging works and monitored by soundings. In the present study, specific monitoring of the bathymetry by the NMB sonar system has been carried out at the uppermost domain of the dredged channel to elucidate the sedimentation process. The perspective view of the bottom topography measured on February 8, 2020, looking from the downward side is shown as an example in Fig. 9 (a), where the darker color means an area with a deeper water depth. Compared with the previous sounding survey on January 17, 2020, the bed level change between the surveys is demonstrated in Fig. 9(b). Blue colored substantial deeper region means

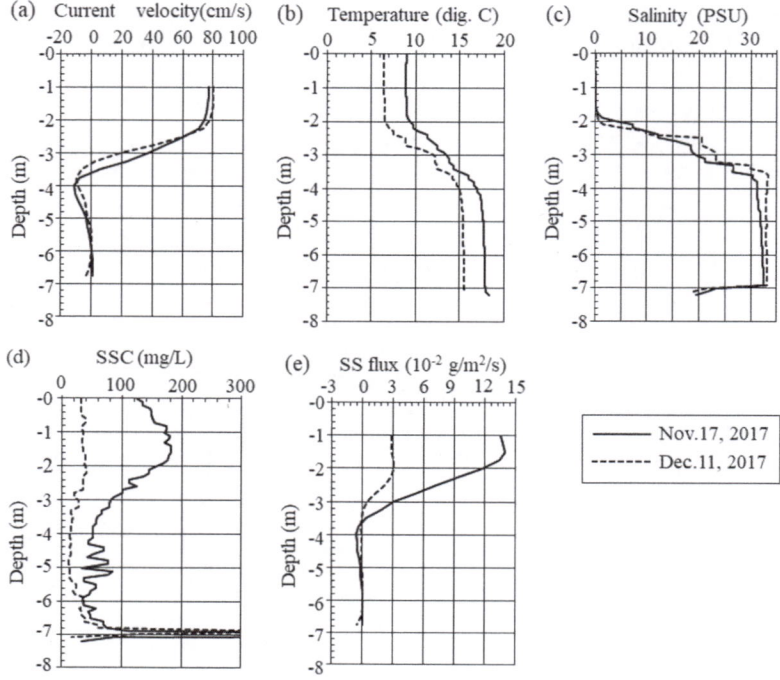

Fig. 8. Measured profiles at Stn. B: (a) current speed along the channel, (b) water temperature, (c) salinity, (d) SSC calibrated from turbidity, and (e) calculated sediment flux.

where dredging work was operated during the period and other warm-colored area means purely diminishing of water depth by the sedimentation in the period.

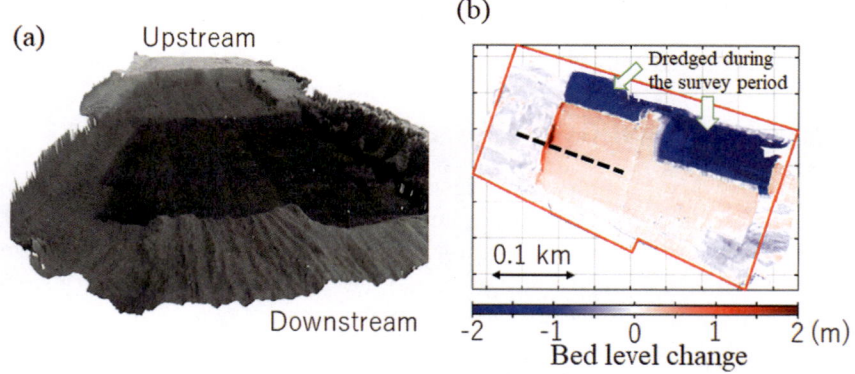

Fig. 9. Survey results of the multi narrow beam monitoring. (a) perspective view of the study area measured on Feb. 8 in 2020; (b) bathymetry change from the previous surveys on Jan. 17 in 2020. Nakagawa et al. 2021 (Modified from)

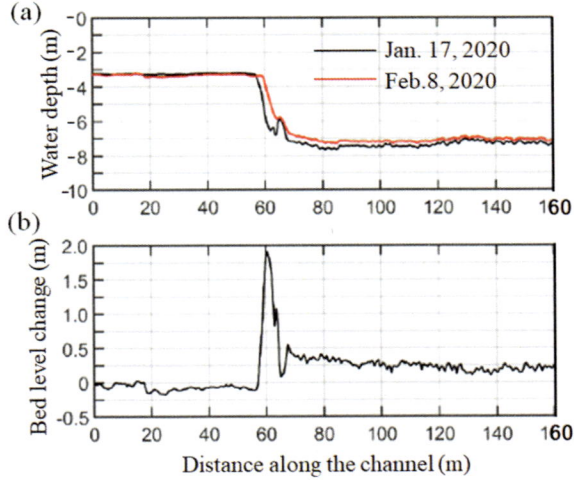

Fig. 10. Bathymetry change along the survey line indicated as dashed line in Fig. 9(b). (a) measured bathymetries along the line; (b) bed level change between the measurements.

The water depths measured along the dashed line in Fig. 9(b) are shown in Fig. 10 (a) and their difference is presented as the bed lave change in Fig. 10(b). Higher sedimentation is observed at the edge and slope region between the upstream shallow region and the downward dredged area with a maximum thickness of almost 2 m. One of the reasons for the higher sedimentation at the slope area could be the accumulation of the bedload transport from the upstream and it is clearly observed during the higher suspended sediment flux as shown in Fig. 7(e). Since the accumulation of so-called fluid mud with a higher water content of 300% has been observed around the bottom of the slope area in the previous monitoring works (Nakagawa et al. 2017), the downward fluid mudflow on the slope could be a key process of the sedimentation of the area. The return flows, furthermore, in the water column are clearly observed under the pycnocline in the deeper dredged area due to the gravitational estuarine circulation (Fig. 8(a)) and they could also accumulate suspended sediments around the upper edge of the dredged channel also from the downstream side as shown in Fig. 8(e). Since the flow pattern of the gravitational circulation just behind the gap of the water depth in an estuarine port is very sensitive to the topographic condition (e.g., Perillo et al. 2005), the deformation of the bottom topography by dredging should be also considered in the present site.

5   Conclusions

Siltation process in estuarine ports is often crucial to keep the safety of ship navigation and it consists of complicated transport dynamics due to the estuarine hydraulic and sedimentary dynamics. The present study demonstrates some field data sets of specific density and current structures around the navigation channel in an estuarine environment. The data was obtained through field surveys in the Port of Niigata located at the mouth of the Shinano River, including current field, bathymetry soundings, and in-situ bulk density for the fluid mud measurements.

In the study site, the density and current field structure can be categorized as a salt wedge estuary, where a stable pycnocline is formed by the freshwater outflow over the salty seawater in a dredged deeper channel. Return flows toward the upstream direction just beneath the pycnocline are observed and suspended sediments are also transported in the direction by the flow during a higher SSC period. Although the sediment transport rate by the return flow is smaller than the flux due to the near surface outflow, it could be important to estimate the accumulation or sedimentation rate at the upstream end of the dredged area with the rapid siltation observed by the sounding survey. The rapid siltation could be due to near-bed transport including fluid mud transport over the specific topography, in addition to the effect of return flow.

These findings are applied to the consideration of countermeasures to minimize the siltation as a future study. Furthermore, evaluation of the settling velocity of suspended particles is also crucial practically to estimate the overall volume to be dredged and the development of any reliable method for getting the parameters in situ is also an important challenge.

Acknowledgements. The authors appreciate ECOH corporation and MIKUNIYA corporation for their field measurement works in 2017 and 2020. The authors also appreciate valuable comments from the reviewers on the draft of the paper.

References

Claeys S, Dumon G, Lanckneus J, Trouw K (2001) Mobile turbidity measurement as a tool for determining future volumes of dredged material in access channels to estuarine ports. Terra et Aqua 84:8–16

Dyer KR (1997) Estuaries, A Physical Introduction. Wiley, Hoboken, p 195

Katoh H (2018) Introduction of dredging and oil recovery technology by drag suction dredging and oil recovery vessel. J JIME 53(6):95–101 (in Japanese)

Ministry of Land, Infrastructure, Transport and Tourism, Water Information System. http://www1.river.go.jp/

Nakagawa Y, Takashima N, Shinozawa T (2017) Field measurement and flume experiment studies on sedimentation in river mouth port, a case study in the Port of Niigata, Technical Note of Port and Airport Research Institute, no 1334, p 14

Nakagawa Y, Kosako T, Watanabe T (2021) Siltation processes of dredged navigation channel at estuarine port. In: Proceedings of INTERCOH2021, Delft, p 57

Perillo GME, Perez DE, Piccolo MC, Palma ED, Cuadrado DG (2005) Geomorphologic and physical characteristics of a human impacted estuary: Quequen Grande River Estuary. Argentina, Estuar Coast Shelf Sci 62:301–312

PIANC (2008) Minimizing harbour siltation, Working report, no 102, p 75

Whitehouse R, Soulsby W, Roberts W, Mitchener H (2000) Dynamics of Estuarine Muds. Thomas Telford Publishing, London, p 210

Winterwerp JC et al (2006) Flow velocity profiles in the lower Scheldt estuary. Ocean Dyn 56:284–294. https://doi.org/10.1007/s10236-006-0063-4

Logistics

Analysis of Knowledge Transfer of Inland Waterway Transport on the Danube Towards a Positive Attitude

Denise Beil[(⊠)] and Lisa-Maria Putz-Egger

Department of Logistics, University of Applied Sciences Upper Austria,
Steyr, Austria
denise.beil@fh-steyr.at

1 Introduction

Preventing serious climate change requires a reduction in global greenhouse gas emissions. Every year, about 50 billion tons of CO_2e emissions are generated worldwide, an increase of more than 40% compared to 1990. By comparing sectors in terms of emissions, it is evident that the transport sector is one of the largest polluters and contributes to approximately one quarter of all energy related greenhouse gas (GHG) emissions (Ritchie et al. 2020). Since road freight transport still has the highest share in the modal split and is significantly responsible for the high emission growth, a modal shift towards environmentally friendly transport modes such as inland navigation is necessary to reduce the environmental impact of the transport sector (European Commission 2019). Our present transportation system is a manifest of the decisions made by transportation professionals in our somewhat recent past. These decisions were influenced by the education that transport professionals received and by their experiences, set forth by a culture, that has shaped them throughout their professional lives. Being aware now of the negative environmental impacts our current transportation system has, it is increasingly important to explore new approaches to create awareness of sustainable transportation in education (Farris et al. 1969).

Inland navigation has by far the lowest GHG emissions, precisely because of its high mass efficiency, thus being one of the most eco-friendly modes of transport (Greene and Lewis 2019). Putz et al. (2018) demonstrated that knowledge about green freight transport is currently limited and that active learning can trigger knowledge retention. Therefore, there is a clear need to create a well-founded knowledge transfer and awareness-raising among young people about inland navigation. Since an effective positive anchoring of a topic takes place preferably in the early years of life, the target group of seven- to fourteen-year-old students was set in the present paper. (Dieleman and Huisingh 2006). Creating awareness at an early age leads to an initial understanding of the topic so that the waterway is anchored in the minds of young people as an environmentally friendly transport alternative.

Therefore, the goal of this paper is to analyze currently available teaching formats and didactics-methodical approaches guided by the following research question: *"Which formats and contents are suitable for imparting knowledge on the topic 'The*

Y. Li et al. (Eds.): PIANC 2022, LNCE 264, pp. 1325–1328, 2023.
https://doi.org/10.1007/978-981-19-6138-0_115

Danube waterway as an environmentally friendly mode of transport' for the target group of seven- to fourteen-year-old students?

2 Methodology

Teaching formats and didactics-methodical approaches for awareness raising of inland waterway transport were explored through a literature review. The literature review included research in the fields of sustainable inland navigation as well as pedagogical-didactic literature. The found formats were examined in terms of their degree of use and popularity. The output of the literature review built the basis for the execution of interviews with 20 people to obtain information regarding the integration of sustainable logistics and inland waterway transport into their work. The interview format was semi-structured and conducted with participants from various fields. In total 39 contacts requests were made: 22 contacts with teachers, schools and school administrations, 6 students, and 11 from the field of art and culture. Finally, 20 participants (n = 20) were successfully interviewed, which means that the self-imposed goal of obtaining information by means of interviews with different target groups was fulfilled beyond measure.

3 Result Analysis and Discussion

3.1 Literature Review on Existing Formats

The literature review was carried out on possible teaching formats and didactic-methodical, in particular also gamified, approaches for raising awareness focusing on the target group of seven- to fourteen-year-old students. The results of the literature research were divided into two sections: (1) Subject content regarding the Danube, inland navigation, climate and education, (2) Pedagogical-didactical literature (esp. with consideration of current findings from teaching/learning research and neuroscience). A strong focus was put on general possibilities for raising awareness methods. The review of the sources led to the first didactic preliminary considerations regarding the suitability of contents and methods and the design of formats for the young target group. Finally, an overview of common methodological approaches and didactic basic principles as well as contents, to be conveyed for awareness raising for sustainable freight transports on the Danube, was created.

3.2 Semi-structured Interviews

The aim of the interviews was to analyze the current status of teaching formats in the area of sustainable transportation. Furthermore, the methods and formats found within the literature review were evaluated together with the participants to identify suitable formats for the target group.

Regarding the current status of teaching formats and the role of climate protection in schools is described by both teachers and students as very large and extremely important. Likewise, the representatives of art and culture emphasize the absolute importance of this topic in the educational environment. However, the topic of sustainable freight transport is hardly dealt with at the schools of the respondents, although interest in the topic is expressed. About 70% of the teachers and 100% of the student's state that the topic was not dealt with, but interest is expressed by all. Consistently, all respondents stated that inland waterway transport as a sustainable way of transporting goods is not addressed in school, but interest in a future treatment is basically stated by all respondents in school. A large part of the survey also dealt with the evaluation of the identified teaching formats as well as further design of possible teaching formats and approaches. The results of the interviews combined with the output of the literature review built the basis for the creation of a spectrum matrix regarding awareness raising teaching formats in the field of sustainable inland navigation.

3.3   Matrix of Teaching Formats

The spectrum matrix represents an overview of the results and the findings from of the literature review and the 20 interviews conducted. Basically, the created offer matrix is a collection of results of possible teaching formats, materials, methods, and tools. It is designed as a table using MS Excel, i.e. with search and filter functions, and comprises 75 entries. As a result of the interviews conducted, it emerged that all of the methodological approaches and tools researched and discussed can be meaningfully assigned to one of the following main categories in terms of usability:

i. Offers for workshops (conducted by external persons at schools etc.),
ii. Teaching-learning materials for teachers (for their independent use),
iii. Online offers/-platforms
iv. Offers for events outside the school such as excursions and exhibitions
v. Offerings that primarily serve the purpose of dissemination/PR
vi. Offerings that appear to be suitable for all of the above areas, and
vii. Other (no meaningful assignment to one of the previously mentioned areas possible)

These seven categories are handled in the matrix with a filter function, and all entries are assigned to the corresponding category. Each entry contains information regarding usability (area, context), description (short explanation of the method), target group (estimated age range), added value (subjective-spontaneous expert indication) and information of the feasible applicability.

4   Conclusions

Through an extensive literature review and the execution of 20 interviews an overview of common methodological approaches, didactics basic principles and content for awareness raising for inland waterway transport on the Danube was found. As a result, a matrix of existing and desired materials was created including the formats and methods which were assessed and examined about its degree of use and popularity. Currently, a limited amount of up to date, diversified, playful materials and methods which is targeted for seven- to fourteen-year-old students about sustainable freight transport on the waterway are available. The matrix serves as a basis for further development of formats and materials suitable for the primary and secondary schoolers.

Due to a small amount of teaching materials further development of formats is necessary to enable a well-founded knowledge transfer. The research was focused on the Danube and thus, more research is needed to generalize the findings to other waterways. The findings underline, that there is an urge need to extend the existing materials and methods on inland waterway transport on the Danube for teachers and students.

Acknowledgement. This paper is part of the research project REWWay which was funded by viadonau and is art of the research field 'sustainable transport systems', which was funded by the State of Upper Austria as part of the research program 'FTI Struktur Land Oberösterreich'.

References

Dieleman H, Huisingh D (2006) Games by which to learn and teach about sustainable development. Exploring the relevance of games and experiential learning for sustainability. J Clean Prod 14(9–11):837–847. https://doi.org/10.1016/j.jclepro.2005.11.031

European Commission (2019) The European Green Deal. Bruxelles. https://ec.europa.eu/info/sites/info/files/european-green-deal-communication_en.pdf. Accessed 24 Nov 2020

Farris MT, Cochran DC, Davis GM, Gourley DR (1969) Transportation education-an interdisciplinary approach. Transp J 9(1):33–44. http://search.ebscohost.com/login.aspx?direct=true&db=buh&AN=7702732&site=ehost-live

Greene S, Lewis A (2019) Global Logistics Emissions Council Framework for Logistics Emissions Accounting and Reporting, version 2.0. Smart Freight Centre. Amsterdam. Accessed 10 July 2020

Putz L-M, Treiblmaier H, Pfoser S (2018) Field trips for sustainable transport education. Impact on knowledge, attitude and behavioral intention. Int J Logist Manag 9(3):235. https://doi.org/10.1108/IJLM-05-2017-0138

Ritchie H, Roser M, Rosado P (2020): CO$_2$ and Greenhouse Gas Emissions. OurWorldInData. org. https://ourworldindata.org/co2-and-other-greenhouse-gas-emissions. Accessed 1 Mar 2022

A Roadmap Towards Eliminating Greenhouse Gas Emissions and Air Pollutants of the Inland Navigation Sector by 2050 – How to Address the Related Economic, Financial, Technical and Regulatory Obstacles?

Raphael Wisselmann[✉], Laure Roux, and Benjamin Boyer

Central Commission for the Navigation of the Rhine, Strasbourg, France
{r.wisselmann, l.roux, b.boyer}@ccr-zkr.org

Abstract. The Ministers responsible for transport have tasked the Central Commission for the Navigation of the Rhine to develop a roadmap for reducing emissions in inland navigation. The ambition is to determine the transition pathways for developing the fleet and achieving "zero emission" inland navigation by 2050. After recalling the general context of climate change and its application to the inland navigation sector, the roadmap explains certain definitions and assumptions. It then identifies different technologies enabling alternatives to gasoil which could play a role in the energy transition of the inland navigation sector, in the form of two pathways for the fleet's development. Economic challenges are addressed, taking into account the huge financial gap to be bridged. It ends with a list of actions needed to achieve these objectives. Thus, this roadmap is intended as a public policy tool for European inland navigation.

Keywords: Energy transition · Climate change · Public policy

Preamble: The Central Commission for the Navigation of the Rhine (CCNR) is an intergovernmental organization that exercises an essential regulatory role in the navigation of the Rhine. It is active in the technical, legal, economic and environmental fields. In all its areas of action, its work is guided by the efficiency of transport on the Rhine, safety, social considerations, and respect for the environment. Many of the CCNR's activities now reach beyond the Rhine and are directly concerned with European navigable waterways more generally.

1 Initial Situation

Addressing the issue of climate change is a political priority both nationally and internationally. The Paris Agreement (2015), which aims to slow the pace of climate change (maximum 2 °C increase) by reducing greenhouse gases (GHG) emissions is one of its key components.

In the Declaration signed in Mannheim on 17 October 2018, the inland navigation Ministers of the Member States of the CCNR (Germany, Belgium, France,

© The Author(s) 2023
Y. Li et al. (Eds.): PIANC 2022, LNCE 264, pp. 1329–1337, 2023.
https://doi.org/10.1007/978-981-19-6138-0_116

Netherlands, Switzerland) reasserted the objective of largely eliminating GHG and other pollutants by 2050, a long-term vision which is also shared by the European Union (EU).

In addition, to further improve the environmental sustainability of navigation on the Rhine and inland waterways, the same Mannheim Declaration tasked the CCNR to develop a roadmap for:

- reducing GHG emissions by 35% compared with 2015 by 2035,
- reducing pollutant emissions by at least 35% compared with 2015 by 2035,
- largely eliminating GHG and other pollutants by 2050.

This energy transition must be seen as a crucial challenge for Rhine and European inland navigation. Based on today's knowledge, while innovations to reduce emissions from existing vessels and newbuilds have increased in recent years, they tend for the time being to be limited to pilot projects, which are however of utmost importance in gaining knowledge about new technologies, and addressing economic, financial, technical and regulatory obstacles to the deployment of relevant technologies.

Despite current uncertainties concerning especially the development, the cost, the level of maturity and the availability of the technologies contributing to the transition towards a zero-emission inland navigation sector, it is necessary to immediately start designing an approach towards this ambitious objective that can be sustained in the medium and long-term.

In this context, identifying and considering the measures enabling an accelerated transition towards zero-emissions (such as regulatory measures, emissions monitoring, financial support for the energy transition, …), together with the development of technology transition pathways for the fleet, are essential elements to be included when designing a realistic and sound roadmap.

In today's circumstances, air pollutants can be reduced to a large extent with internal combustion engines (ICE) equipped with modern aftertreatment, while the reduction of GHG emissions remains the most challenging part. Beyond the use of new energy carriers (like methanol or hydrogen) in combination with energy converters (like fuel cells or ICE), reduction of energy consumption by all possible means is an important lever to achieve the emission reduction objectives. This includes for example better use of vessels, increased efficiency by means of modern propulsion systems, the improvement of vessels' hydrodynamics, smart navigation with less waiting time at locks, and efficient integration of inland navigation into seaports logistics.

Wherever possible, careful attention should be paid to developments in other modes of transport, such as road, rail and short-sea shipping. Indeed, there is much to be learned from the experience gained by other modes regarding the energy transition.

Last but not least, the relatively small size of the European inland waterway vessel market implies that technological solutions designed specifically for the inland navigation sector alone might not be commercially viable. It is therefore unlikely that a technological solution will be developed for the inland waterway transport sector alone. From this perspective, synergies should be found with technologies developed for seagoing vessels and for non-marine applications whether in Europe or in other parts of the world.

In light of the above, largely eliminating both GHG and air pollutant emissions from inland navigation by 2050 is clearly no longer an option but a necessity if inland navigation wants to preserve and strengthen its position as a competitive, sustainable and environmentally friendly mode of transport.
In other words, the fleet modernisation and the energy transition are motivated by addressing climate change with reduction of GHG emissions, reducing health-related risks by improving air quality but also reducing the sector's operational costs (OPEX) by increasing efficiency of inland navigation.

2   Purpose of the Roadmap

The recently adopted roadmap aims primarily to deliver on the mandate conferred by the Mannheim Declaration in 2018 and to help address the crucial challenge of the energy transition for Rhine and European inland navigation.

Built on the CCNR study on the energy transition towards a zero-emissions inland navigation sector (2021)[1], the roadmap should be understood as the primary CCNR instrument for climate change mitigation and for giving effect to the energy transition. The objective is to reduce Rhine and inland navigation emissions by:

– setting transition pathways for the fleet (new and existing vessels),
– suggesting, planning, and implementing measures directly adopted or not by the CCNR,
– monitoring intermediate and final goals set by the Mannheim Declaration.

It goes without saying that many players will be involved in this energy transition, such as vessel owners, operators, shippers, and shipbuilders as well as representatives of the sector, classification societies, equipment manufacturers, infrastructure operators, service and energy providers, universities or research institutes, the scientific community with relevant organizations such as PIANC, international organisations including European Union institutions and river commissions[2], the CCNR, and national administrations of European States with inland waterways.

The CCNR hopes with its roadmap to contribute to the development of a shared vision of the energy transition and associated challenges within the European inland navigation sector, while also generating support and acceptance for related policy measures.

3   Scope and Assumptions of the Roadmap

To ensure a common understanding between all the actors involved in the energy transition of inland navigation, it was essential to agree on a scope for the roadmap and on key definitions. In particular, it was decided to:

– lay focus on inland navigation, meaning the transport of goods and the carriage of passengers by inland waterway vessels-recreational crafts, service vessels and floating equipment were not included at this stage-,
– define emissions as atmospheric pollutants and greenhouse gases (GHG) arising from the operation of an inland navigation vessel's propulsion and auxiliary systems,

---

[1] https://www.ccr-zkr.org/12080000-fr.html.
[2] Moselle Commission, Danube Commission, International Sava River Basin Commission.

– adopt a "tank-to-wake" approach, as an interim solution, until a "well-to-wake" approach is available for the relevant energy carriers. Application of this approach however implies making assumptions, notably concerning the upstream chains (emissions produced and fuel availability), which are idealised.

4 Two Transition Pathways for Inland Navigation Leading Up to 2050

In particular, the roadmap aims to outline, in addition to a business-as-usual scenario, two transition pathways for the fleet (new and existing vessels), meaning how the entire fleet will evolve by 2050.

A more conservative transition pathway, based on technologies that are already mature, is cost efficient in the short-term but with uncertainties on the availability on certain fuels, and a more innovative one, relying on technologies still in their infancy stage but providing more promising emission reduction potential in the long run.

The transition pathways also address the role which the different technological solutions will play in the energy transition challenge, assess their suitability according to the different fleet families in Europe and the sailing profiles of the vessels.

4.1 Technologies Considered in the Pathways

(See Table 1).

Table 1. Technologies considered in the pathways - energy carriers and converters (source: CCNR)

CCNR 2 or below, Diesel	Fossil diesel in an internal combustion engine which complies with the emission limits CCNR 2 or older engine
CCNR 2 + SCR, Diesel	Fossil diesel in an internal combustion engine which complies with the emission limits CCNR 2 and equipped with an additional Selective Catalytic Reduction system
Stage V, Diesel	Fossil diesel in an internal combustion engine which complies with the emission limits EU Stage V
LNG	Liquefied Natural Gas in an internal combustion engine which complies with the emission limits EU Stage V
Stage V, HVO	HVO in an internal combustion engine which compiles with the emission limits EU Stage V HVO stands for hydrotreated vegetable oil itself (without blending with fossil fuels) and all comparable drop-in biofuels (including e-fuels) as well as synthetic diesel made with captured CO_2 and sustainable electric power
LBM	Liquefied Bio Methane (or bio-LNG) in an internal combustion engine which compiles with the emission limits EU Stage V

(continued)

Table 1. (*continued*)

Battery	Battery electric propulsion systems, with fixed or exchangeable battery system
H_2, FC	Hydrogen stored in liquid or gaseous form and used in fuel cells
H_2, ICE	Hydrogen stored in liquid or gaseous form and used in internal combustion engines
MeOH, FC	Methanol used in fuel cells
MeOH, ICE	Methanol used in internal combustion engines

Remark 1: Regarding the energy converter, the mono-fuel engine is considered in the transition pathways for each fuel. In practice dual-fuel engines could also be applied, e.g. engines that run on LNG and gasoil but have significantly higher GHG emissions. This could also apply to MeOH and H_2 engines once these enter the market.

Remark 2: The stage "CCNR 2" refers to the emission limits adopted by the Resolution CCNR 2005-II-20[3]. The EU Stage V refers to emission limits adopted by Regulation (EU) 2016/1628 for non-road mobile machinery[4] (categories IWP, IWA, NRE or EURO VI marinised truck engines).

4.2 Transition Pathways Leading up to 2050

(See Fig. 1)

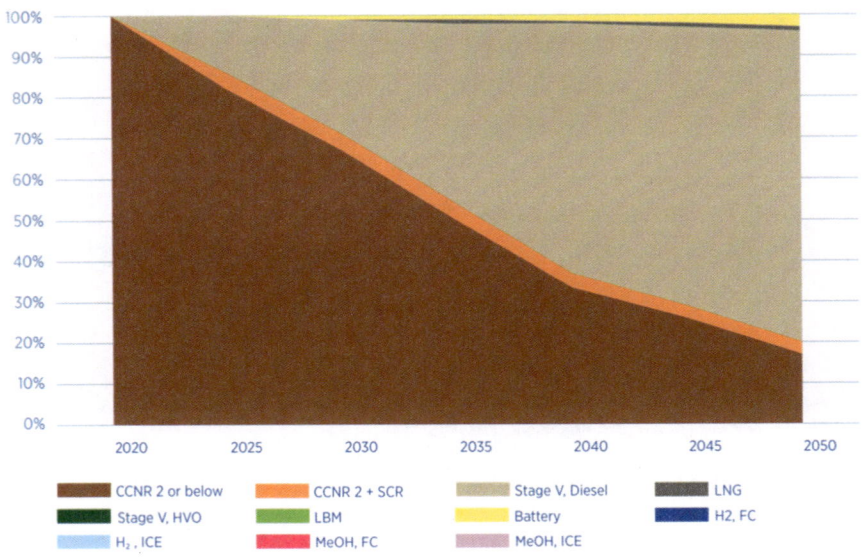

Fig. 1. Business-as-usual scenario: development of technologies by 2050 (source: CCNR)

---

[3] https://www.ccr-zkr.org/13020400-en.html.

[4] http://data.europa.eu/eli/reg/2016/1628/oj.

In summary, the outcome of this business-as-usual scenario is that the air pollutant and GHG emissions targets to be achieved in 2050 as provided for in the Mannheim Declaration cannot be achieved. Specific measures must be taken to achieve these objectives (Figs. 2 and 3).

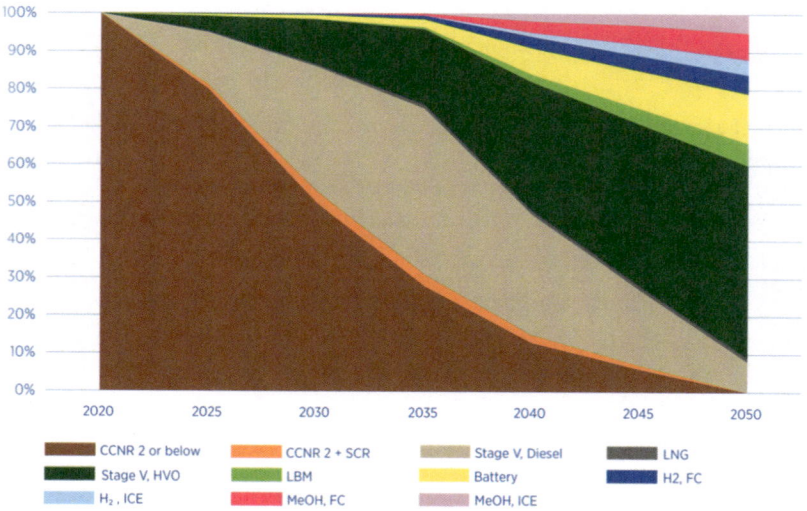

Fig. 2. Conservative transition pathway: development of technologies by 2050 (source: CCNR)

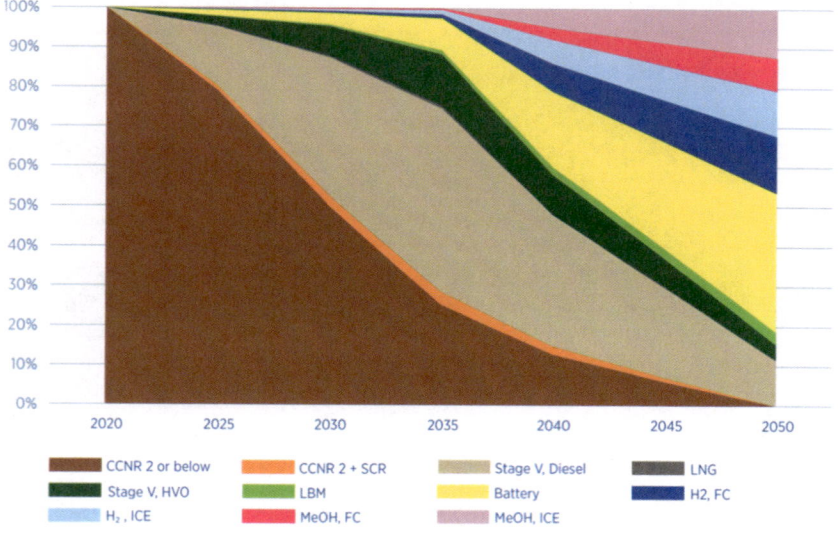

Fig. 3. Innovative transition pathway: development of technologies by 2050 (source: CCNR)

The two transition pathways are both sufficiently ambitious to achieve the emission reduction objectives of the Mannheim Declaration. A key conclusion points to the absence of a "one size fits all" technology solution adapted to all types of vessels and navigation profiles. A technologically neutral approach appears therefore relevant to achieve the energy transition.

5   The Energy Transition Financial Challenge

Initial estimates show that the financial challenge involved in achieving the aim of zero emissions by 2050 is considerable. Depending on the fleet transition pathway, the financial gap to be bridged to achieve the Mannheim Declaration emission reduction objectives varies significantly but amounts to several billions in any scenario. The energy transition-related costs will exceed the financial resources of the navigation profession, which will only be able to bear a part of the costs required to achieve this transition. Significant grants are needed to close this gap, and to make the transition pathways economically viable for the inland navigation industry, energy suppliers, and shore-side infrastructure operators. Strong public support is therefore necessary.

In order to support the energy transition of the inland waterway transport sector, the CCNR considers it opportune to pursue the idea of a European-wide financial support instrument for the energy transition of the inland waterway transport sector, based on mixed sources (public and private), including a sector contribution. To ensure a level playing field, such an instrument should be open to EU countries as well as Rhine and Danube riparian states which are not members of the EU (Switzerland, Serbia, Moldova and Ukraine, in particular). Easy access to such an instrument is paramount, as is administrative simplicity.

6   Implementation Plan and Next Steps

Economic, technical, social and regulatory aspects need to be considered to tackle the challenge of the energy transition towards zero emissions. How to address them through concrete policy measures was a guiding question when developing the implementation plan proposed in the roadmap, which aims at suggesting, planning and implementing measures to be adopted directly or not by the CCNR, as well as monitoring the intermediate and final objectives laid down by the Mannheim Declaration (Table 2).

Table 2. Implementation plan (source: CCNR)

Regulatory measures
Appropriate regulatory framework for the use of alternative fuels and batteries (vessel construction, crew, vessel operation, transport of dangerous goods, definition, fuel characteristics, blending and supply)
Scrutiny and where appropriate amendment of safety and statutory requirements for bunkering of alternative fuels in inland waterway transport
Possible phasing out of the most harmful technologies which appear inconsistent with the CCNR's and the EU's long-term emission reduction ambitions
Infrastructure requirements for alternative fuels and electricity for propulsion
Examination of the possibility of a sector contribution in the framework of a European funding and financing instrument
Voluntary measures
Label for environmental and climate protection
Carbon offsetting measures (carbon compensation)
Pilot vessel trials (all vessel types)
Innovative vessels (Database)
Innovation award
Situation reports: regularly analyse emissions reduction status and the effectiveness of measures
Financial measures
Examination of a European funding and financing instrument to support the inland navigation energy transition
EU Taxonomy – establishment of an EU classification system for sustainable activities
Stimulate research and innovation projects

The CCNR undertakes to

- report by 2025 on the progress in the implementation as well as the need to update the roadmap,
- at the latest in 2025 evaluate whether it is opportune to revise the CCNR's study, especially on the economic and technical evaluation of the technologies,
- review the "tank-to-wake" approach in a forthcoming revision of its roadmap,
- evaluate by 2025 whether it is opportune to extend the scope of the roadmap, for example to other greenhouse gases such as N_2O or to emissions associated with other aspects of the vessel's life-cycle, to the manufacturing and disposal of propulsion systems, to other types of vessels, or even to the technologies' safety,
- revise, if necessary, by 2030 the roadmap and the corresponding action plan.

Eventually, the CCNR aspires to this roadmap being of assistance in developing a shared vision of the energy transition and the concomitant challenges within the inland navigation sector. In particular, this publication in the framework of the Smart Rivers conference aims to broaden the scope of discussion at global level and allows possible exchange of views on the energy transition of the inland navigation sector in different parts of the world. It is desirable to deepen this cooperation to learn from each other and ensure that the policy measures are properly tailored to the opportunities and challenges of the inland navigation sector.

Reference

To learn more about the roadmap and the CCNR's contribution to emissions reduction:

- Dedicated webpage on reducing inland navigation emissions
- CCNR roadmap (English)
- Key points of the CCNR roadmap (English)
  CCNR study on energy transition towards a zero-emission inland navigation sector

Calculating Emissions Along Multimodal Transport Chains - Standards, Difficulties and Problems

Laura Hörandner[✉], Lisa-Maria Putz Egger, and Denise Beil

University of Applied Sciences Upper Austria,
Roseggerstraße 15, 4600 Wels, Austria
laura.hoerandner@fh-steyr.at

1 Introduction

The European Union has ambitious goals regarding decarbonization and the transport sector is of great importance in order to become climate neutral by 2050 (Haasza et al. 2018). To reach a higher level of transparency about the origin of greenhouse gas (GHG) emissions in a transport chain, carbon foot printing is an essential pre-requisite (Dobers et al. 2019). Users and providers of freight services need to be aware how much GHG they are emitting to be able to set and deliver carbon reduction targets (McKinnon 2021). This requires further insights into current emission levels and the development of an internationally accepted standard for the calculation of transport chain emissions to support comparisons, and with them the identification of potentials for reducing GHG emissions and best practice (Dobers et al. 2019).

During recent years substantial progress has been made in the measurement, calculation and reporting of transport chain emissions. The subject has generated a large literature of reports and journal papers, techniques for measuring emissions have been refined and there has been much greater harmonization of methodologies, particularly with the wide adoption of the EN 16258 standard and the Global Logistics Emissions Council (GLEC) framework (McKinnon 2021). Nonetheless, at present, a seemingly fragmented landscape of various programs with different methodological approaches exists. The quality of calculated emissions and emission intensities depends on availability, specification, quality and exchange of data. In many cases no complete real-world data are available for carbon footprint calculation. In such cases, it is possible to use aggregated average data (default factors) on emissions to substitute missing data (Dobers et al. 2019). These values, mostly given in gram per tkm vary greatly, especially for inland navigation. In fact, a comparison between available data of IWT, rail and road shows that the IWT sector is lacking behind (Schweighofer and Szalma 2014; Greene and Lewis 2019). For this paper, a literature review was conducted to focus on emission factors along multimodal transport chains and their effects on emission calculation results.

© The Author(s) 2023
Y. Li et al. (Eds.): PIANC 2022, LNCE 264, pp. 1338–1341, 2023.
https://doi.org/10.1007/978-981-19-6138-0_117

2   Result, Analysis and Discussion

Accounting for emissions from freight transportation requires an understanding of a diverse set of business models, modes of transport, regions and more. For the road sector there is a choice between many sources of reference data and default factors which can lead to comparability issues, whereas for rail and inland waterway transport there are limited data available, leading to very high-level assumptions being made, this time leading to uncertainty and potentially unrepresentative outputs being generated (Greene and Lewis 2019). Raising the GHG emission factors for inland waterway transport is quite extensive in comparison to other modes. The specific energy consumption of an inland vessel in relation to the weight of goods transported and the distance (grams of diesel per ton kilometer) can take on very different dimensions. The width and depth of the waterway including the distance between the side of the ship and the river bottom have a massive influence on the diesel consumption of a ship. Therefore, energy consumption data for individual waterways and the type of ship used are necessary (Bauer et al. 2011). If energy consumption data is not measured, but calculated using default values, then certain assumptions, e.g. about the load utilization of the vehicles, enter the calculation. These assumptions can have a considerable effect on the result but are not always correct (Schmied and Knörr 2012).

There exist large differences between fuel-based emission calculation and calculations using emission-intensity factors based on the volumes and kilometers transported. The use of intensity emission factors is often preferred due to unavailability of real fuel consumption values in companies. The use of these factors should be used with caution as for example in the calculation of a round trip, the distance without freight cannot be calculated. A reason for this is that the emission intensity factors already include a certain utilization and empty running factor. Thus, an upstream calculation with freight and a downstream calculation without freight with the emission intensity factors according to the GLEC Framework is not possible (Lomax 2022).

Another difficulty in calculation is that the vessel types in the GLEC Framework are not specified more precisely. As an example, it is not clear if the tanker vessel is a tank lighter or a motor tanker.

Carbon calculation tools such as EcoTransIT World or CarbonCare strive to simplify the process of calculating. EcoTransIT World is compliant with the EN 16258 and the GLEC framework, but a simple calculation example with a truck and an inland waterway vessel is emerging in different values compared with a calculation including the manual GLEC calculation. A major challenge are the different parameters for the emission calculation: The EcoTransIT World and the GLEC framework have varying vessel categories, differences in the calculation of the kilometers, the load factor, the empty trips and the emission standard of the vessel. To make the differences and obstacles regarding vessel types clearer, here is an example. To calculate the emissions of a tanker from Enns to Vienna with a cargo of 3,100 t, it is only possible to select a large inland cargo vessel Bulk V 1,500–3,000 t in the EcoTransIT vessel categories. However, with a cargo of 3,100 t, two ships would be needed, since no larger ship is available in EcoTransIT.

Simenc (2016) evaluated existing emission calculator that could be used for estimating emissions of inland waterway transport and concluded, that the range of available ready-to-use practical solutions is relatively narrow. There are few options available and even the estimation capabilities of existing ones could be thought of only as educated guesses, at best. They are only as good as the quality of emission factors and other parameters that are considered, over which the prospective users have no influence and are subject to uncertainties regarding the underlying calculation algorithms and ability to produce reliable results.

To further increase the accuracy of logistics emissions in multimodal transport chains, van Liere (2018) calculated the GHG emission factors for representative vessel classes in Europe on the basis of real-life data from barge operators. The data collected by barge owners/inland shipping lines has resulted in lower GHG emission factors in comparison to other recognized studies. As the dataset includes information on only approx. 1% of the vessels operating in Europe, they recommended to continue expanding the dataset with annual information on transport performance (distance covered, load factor, tones transported) and fuel consumption per representative vessel class.

Since 2018 the work is still in progress and some data collection programs are currently being undertaken to get a more accurate picture of the GHG emission per tkm from inland waterways. The work will be reflected in the update to the GLEC Framework by the end of the year 2022 and its resulting worldwide standard ISO 14083 (Quantification and reporting of greenhouse gas emissions arising from operations of transport chains) (Lomax 2022). The improved access to reliable data will help both business and governments make better decisions to collectively reach climate goals.

3 Conclusions

It is important, that the drive for consistency, transparency and comparability is strongly maintained in the future for the collection of freight emission data. This means that the future framework for calculating GHG emissions needs to be reliable, relevant, and accurate to enable adequate comparison of emissions from transport operations, thereby placing all modes and operations on equal footing. The GHG emission factors for inland waterway transport is quite extensive in comparison to other modes, so more effort is needed to ensure a broad database to stay competitive with other transport modes such as road and rail. The inland waterway sector is challenged to ensure a high level of accurate data collection among the GHG of vessels. Nevertheless, it should always be the goal to shift to primary data (primary fuel consumption data). This is the only way to get really reliable factors and to be able to track future progress. Future research should focus not only on the GHG emissions from the transportation, but also those emissions which are related to handling, where the data situation is currently inadequate.

Acknowledgement. This project received funding from the European Union's Horizon 2020 research and innovation program under grant agreement No 861377 for the IW-NET project as well as from the research cooperation REWWay which is funded by viadonau.

References

Bauer B, Kranke A, Rauser T (2011) CO2 Berechnung. Das Sonderheft zur Ermittlung von Treibhausgas-Emissionen in der Logistik. http://media1.verkehrsrundschau.de/fm/3576/VR-CO2_Spezial_2011_Juni.pdf?_gl=1*11qip1t*_ga*NTY1Mjg1OTkxLjE2NDkwMTU3NTU.*_ga_VZTE521QZ8*MTY0OTAxNTc1NS4xLjEuMTY0OTAxNTc3MS4w&_ga=2.77305362.649737176.1649015755-565285991.1649015755

Dobers K, Ehrler VC, Davydenko I, Rüdiger D, Clausen U (2019) Challenges to Standardizing Emissions Calculation of Logistics Hubs as Basis for Decarbonizing Transport Chains on a Global Scale. Transportation Research Record (Journal of the Transportation Research Board) 2673(9):502–513

Greene S, Lewis A (2019) Global Logistics Emissions Council Framework for Logistics Emissions Accounting and Reporting, version 2.0. Smart Freight Centre. Amsterdam

Haasza T et al (2018) Perspectives on decarbonizing the transport sector in the EU-28. Energ Strat Rev 20:124–132

Lomax M (2022) Emission factors inland waterway transport. E-mail message to Laura Hörandner, 2022

McKinnon A (2021) A Progress Report on the Measurement of European Freight Transport Emissions. Available online at https://fsr.eui.eu/a-progress-report-on-the-measurement-of-european-freight-transport-emissions/

Schmied M, Knörr W (2012) Calculating GHG emissions for freight forwarding and logistics services in accordance with EN 16258. Terms, Methods, Examples. Available online at https://www.clecat.org/media/CLECAT_Guide_on_Calculating_GHG_emissions_for_freight_forwarding_and_logistics_services.pdf

Schweighofer J, Szalma B (2014) Evaluation of a one-year operational profile of a danube vessel. Available online at https://www.researchgate.net/publication/266088659_EVALUATION_OF_A_ONE-YEAR_OPERATIONAL_PROFILE_OF_A_DANUBE_VESSEL

Simenc M (2016) Overview and comparative analysis of emission calculators for inland shipping. Int J Sustain Transp 10(7):627–637

van Liere R (2018) GHG emission factors for IWT. Final report. Available online at https://www.smartfreightcentre.org/pdf/GLEC-report-on-GHG-Emission-Factors-for-Inland-Waterways-Transport-SFC2018.pdf

Construction of Container Terminal in the Yangon River

Hiroki Kohno[1]([✉]), Masayuki Takahashi[2], Daisuke Niina[3],
and Satoshi Tokiwa[1]

[1] Civil Engineering Department, International Division,
Toyo Construction Co., Ltd., Tokyo, Japan
{kouno-hiroki, tokiwa-satoshi}@toyo-const.co.jp
[2] Mechanical Department, International Division,
Toyo Construction Co., Ltd., Tokyo, Japan
takahashi-masayuki@toyo-const.co.jp
[3] Civil Engineering Department, Toyo Construction Co., Ltd., Tokyo, Japan
niina-daisuke@toyo-const.co.jp

Abstract. The Port of Yangon is the largest port in Myanmar and a river port located approximately 40 km upstream from the Yangon estuary. The country's economic growth has led to the construction of a new container terminal in the Thilawa area, 16 km downstream from the Yangon port. Regardless of the difficult natural conditions in the area, project had to be completed in a short period at the request of the client. In order to shorten the construction period, construction of a jetty using the jacket method and soil improvement combined with the consolidation acceleration method were adopted. After repeated studies, the project was completed in a short period of 915 days. This paper describes the details of these studies.

Keywords: Jacket method · Pile driving · Transportation · Installation schedule · PVD

1 Introduction

Myanmar has experienced a rapid increase in logistics due to economic growth following the transfer of civil administration in March 2011. The existing Yangon port was less convenient due to its shallow water depth and the need to wait for high tide to enter the port. Therefore, a new port was to be built in Thilawa area, which is downstream from Yangon Port and has deeper water. Figure 1 shows the location of Yangon city and Thilawa area.

© The Author(s) 2023
Y. Li et al. (Eds.): PIANC 2022, LNCE 264, pp. 1342–1352, 2023.
https://doi.org/10.1007/978-981-19-6138-0_118

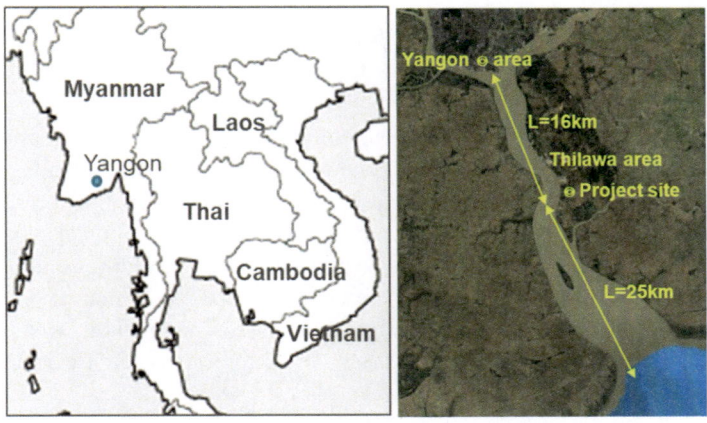

Fig. 1. Yangon City and Thilawa Area

The project contains the construction of a jetty with a length of 400 m and a planned water depth of C.D.L.-10 m. In the land area, 800 m long and 250 m wide (approximately 18 ha) of land was developed filling from the quay to the inland area as shown in Fig. 2. With the aim of constructing the terminal in a short time, the jacket method was adopted for the pier, and soil improvement was adopted in the terminal yard to prevent the ground settlement during terminal operation. Myanmar has a rainy season from June to October, during which a large amount of rain falls, approximately 3000 mm. In addition, the river has a maximum flow velocity of 6knots and tidal levels of H.W.L. +6.24 m and L.W.L. +0.33 m, with a large difference in tide. In order to implement the works under harsh natural conditions, various techniques were used and reported in this paper.

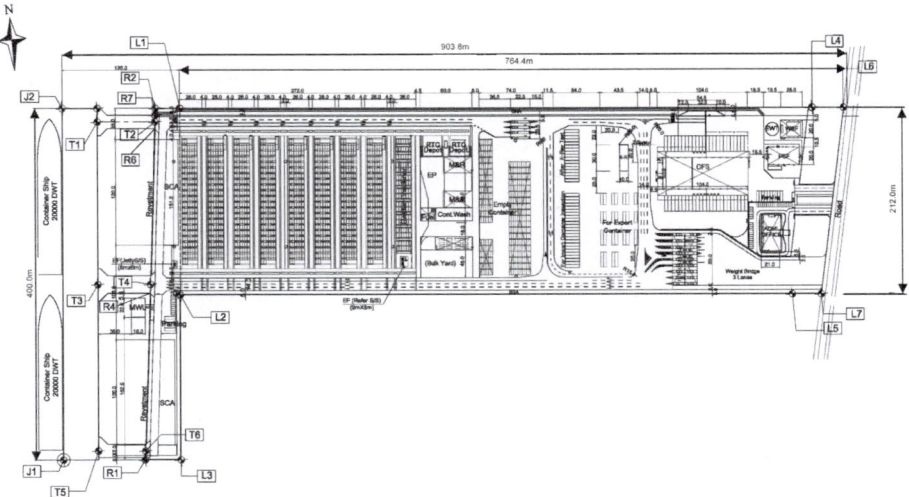

Fig. 2. General layout

2 Jetty Construction

2.1 Construction Outline

The jetty was constructed adopting jacket method, which is a three-dimensional truss structure called a jacket manufactured onshore and capped on steel pipe piles and then, superstructure was constructed using precast slabs. As the majority of the structure is fabricated on land, this method enables the fabrication of components of uniform quality without being affected by external factors, such as scaffoldings and climate. In addition, the construction period can be shortened because fabrication of precast slabs is carried out simultaneously with pile driving offshore. This method, which requires little underwater work, is useful at this site under conditions of high current velocity and very large tidal differences (Fig. 3).

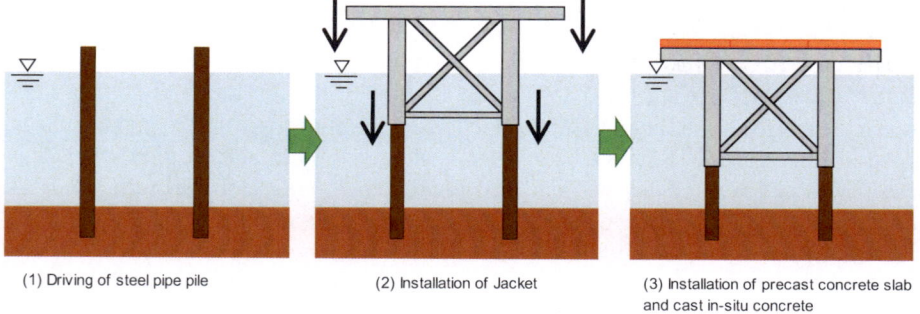

(1) Driving of steel pipe pile   (2) Installation of Jacket   (3) Installation of precast concrete slab
                                                                          and cast in-situ concrete

Fig. 3. Series of jacket method

2.2 Steel Pile Driving

The foundation of the jacket consists of 120 pieces of Ø1300 mm steel pipe piles. As the leg diameter of jacket is Ø1480 mm in relation to the diameter of the piles, the margins are 90 mm on both sides (Fig. 4). Considering the margins and the fabrication accuracy of the jackets, the steel pipe piles required high driving accuracy. Considering harsh conditions in the area, a 400-tonne lifting crane barge equipped with spuds for holding the vessel position and a pile keeper for adjustment of the steel pipe pile position were used and (Fig. 5).

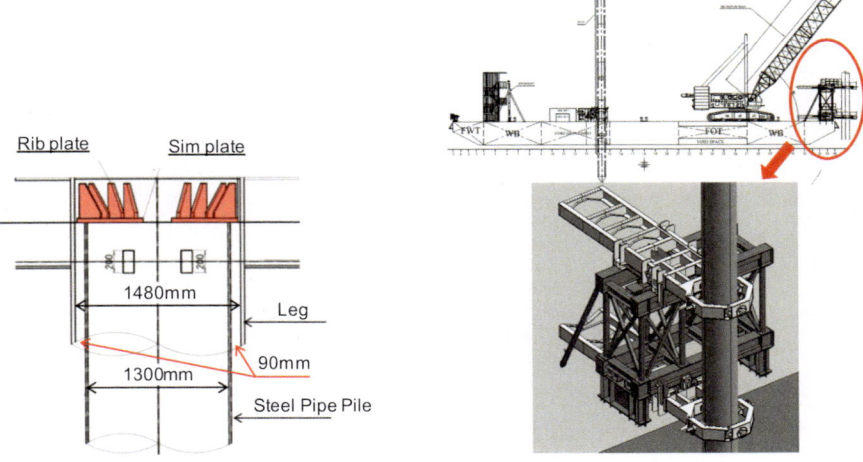

Fig. 4. Relation of Pile and Jacket Leg   **Fig. 5.** General View of Crane barge and pile keeper

The steel pipe piles were driven with a vibratory hammer for initial driving and a hydraulic hammer for final driving (Table 1). In order to improve the accuracy of driving, time was taken to ensure accuracy in the initial driving.

Table 1. Hammer specifications

Vibratory hammer (ICE 84C)			Hydrauric hammer (IHC S-200)		
Eccentric moment	kg-m	126	Blow energy	kN-m	200
Centrifugal force	kN	2,360	Blow rate	bl/min	45
Frequency	vpm	1,500	Ram mass	ton	13.6
Amplitude	mm	37	Total mass	ton	40.0
Total mass	ton	12.4			

The results of the pile driving are shown in Fig. 6. The average driving accuracy was 3.8 mm (standard deviation 18.1 mm) on the landward side and 0.04 mm (standard deviation 18.0 mm) downstream. More than 90% of the piles in both straight and perpendicular to the normal direction fell within ±30 mm of the design position, indicating high accuracy of driving.

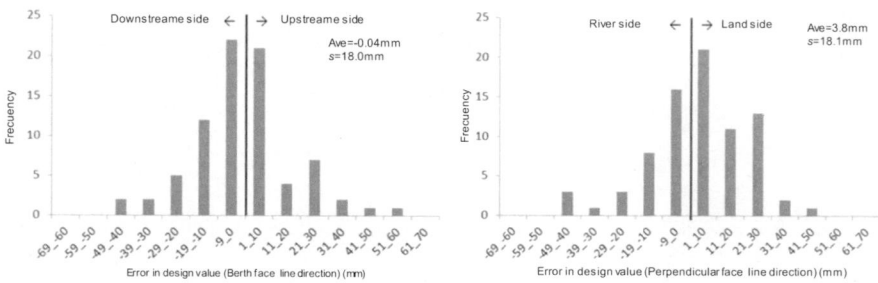

Fig. 6. Histogram of Piles Driving Results Analysis

2.3   Jacket Installation

The size of one jacket is 20 m in the direction normal to the pier and 40 m perpendicular to the pier, weighing 245 t/unit, for a total of 20 jackets (Fig. 7). The jackets were assembled in a temporary yard adjacent to the project site. The fabricated jackets were placed on rollers called tirtanks and transported by winch on a 600 m rail in the temporary yard to the temporary quay. The transported jackets were lifted by a 500t lifting crane barge. To hold the barge in the river, it was transferred by six anchors instead of boats (Fig. 8).

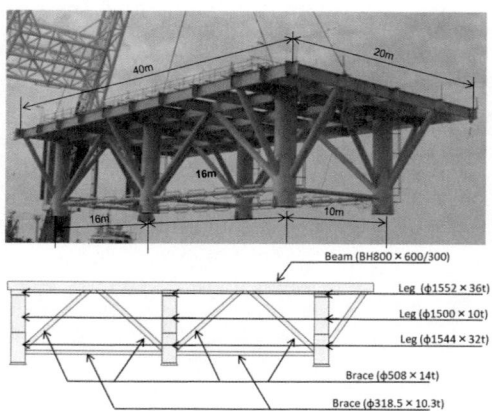

Fig. 7. Overview of Jacket

(a)Transportation on land

(b) Lifting by a crane barge

(c) Installation of Jacket

Fig. 8. Jacket construction

2.4 Installation Schedule

Jacket installation is affected by river conditions. As the draft of the barge is 2.5 m, the tide level must be at least C.D.L. +4.5 m at the time of jacket lifting. In addition, the top elevation of the steel pipe pile is C.D.L. +6.5 m and the pile head needs to be exposed from the water surface during jacket installation, so work must be carried out at a tide level of C.D.L. +5.5 m or less. Therefore, the jacket is lifted at high tide, and by the time the crane barge is moved to install the jacket, subsequently, the tide moves to low tide and the pile heads are protruding. Therefore, a construction plan was elaborated so that a series of operations could be completed in one day from the start of the jacket lifting, based on tide level observations. A cycle of the jacket transportation and installation is shown in Fig. 9. One jacket was successfully installed in a day following this cycle.

Fig. 9. Installation cycle of one Jacket

3 Soil Improvement

3.1 Overview

The plan view of reclamation and soil improvement works are shown in Fig. 10. The soft clay layer with a thickness of 20 m to 25 m is widely distributed on the site(see

Fig. 11), and the terminal will be constructed by reclaiming sand dredged from the Yangon River on top of this layer. Ground improvement was carried out to reduce the residual settlement of the terminal yard due to operational loads during the terminal operation. As it takes a long time for the underlying soil to achieve the required consolidation degree using loading fill only, the PVD (Prefabricated Vertical Drain) method was adopted as a consolidation accelerator.

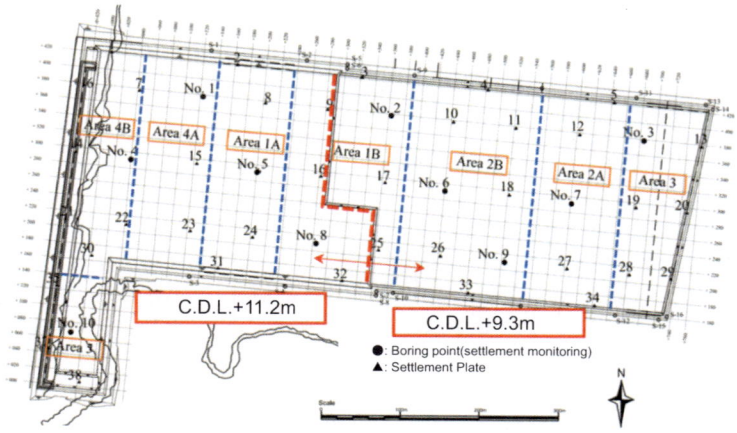

Fig. 10. Plan view of surcharge filling

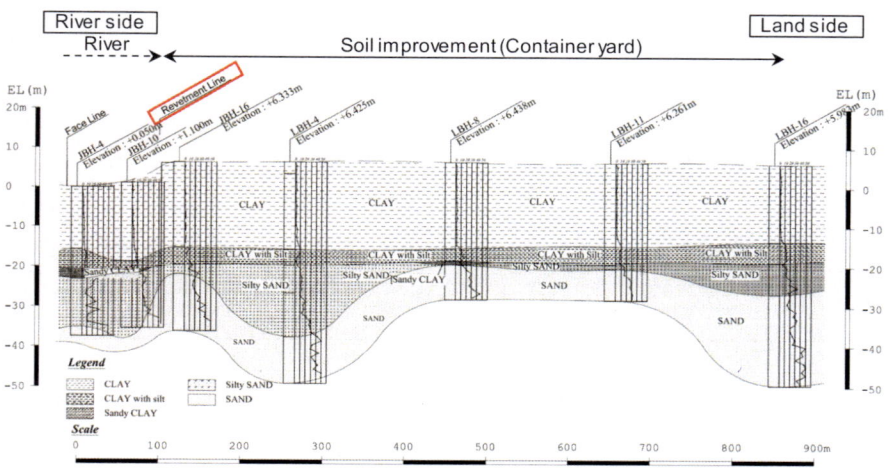

Fig. 11. Soil property along longitudinal direction

The series of construction procedures of soil improvement consisted of laying geotextiles on the local bedding, then reclaiming sand using the water spreading method to create a 1 m layer-thick sand mat, and driving PVD in a square arrangement of 1.1 m × 1.1 m. Subsequently, the container yard was filled to a height of C.D. L. +11.2 m and the building area to a height of C.D.L. +9.3 m (see Fig. 12). After leaving the soil to reach the required consolidation degree, the surcharge fill was removed and levelled at the finish grade.

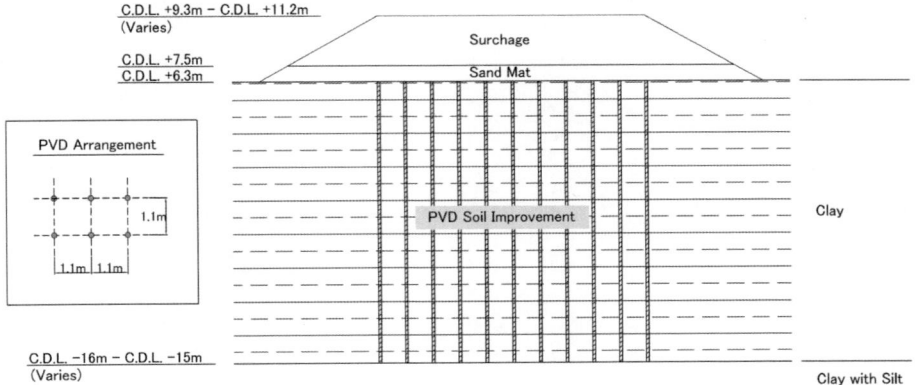

Fig. 12. Outline of soil improvement using PVD

3.2   Geotechnical Investigation and Construction Planning

Prior to construction, detailed soil investigations of the yard were carried out at ten locations to confirm soil properties and soil layer classification. Based on these data, the surcharge fill height and target leaving period were determined. The results of consolidation tests with specimens sampled at the borehole locations are shown in Table 2. The soil model set up based on this data is shown in Fig. 13. Settlement analysis was carried out using the finite element analysis program "DACSAR"[1] for the settlement observation points.

---

[1] Atushi IIZUKA and Hideki OHTA. (1987), "A determination procedure of input parameters in elasto-viscoplastic finite element analysis", Soils and Foundations, Vol.27, No. 3, pp. 71–87.

Table 2. Result of consolidation test and parameters set for design

Location	Soil Layer (Type)	Depth (m)	e_0	p_y (kN/m²)	Cc Value	Cc Set Value	Cv (cm²/day) Value	Cv (cm²/day) Set Value	p (kN/m²)	OCR Value	OCR Set Value
BH-01		2.9	1.360	98.1	0.494		26.2		28.85	3.400	
BH-03	Clay-II-1	3.4	1.910	78.5	0.926		34.4		32.50	2.415	3.0
BH-05	(CH)	3.4	1.230	93.6	0.406	0.456	31.4	30.0	32.50	2.880	(singular
BH-08		3.4	1.220	127.8	0.444		20.0		32.50	3.932	value
BH-09		3.4	1.420	230.5	0.478		24.0		32.50	7.092	deletion)
BH-04		5.9	1.400	61.0	0.508		63.1		50.75	1.202	
BH-02		6.4	1.700	64.2	0.628		40.2		54.40	1.180	
BH-06		6.4	1.510	74.9	0.548		43.6		54.40	1.377	
BH-07		6.4	1.610	88.3	0.679		35.8		54.40	1.623	
BH-10		6.4	1.640	78.5	0.617		36.5		54.40	1.443	
BH-01		8.9	1.680	78.5	0.669		55.0		72.65	1.081	
BH-03		9.4	2.000	93.9	0.850		70.3		76.30	1.231	
BH-05		9.4	2.000	112.0	0.772		70.6		76.30	1.468	1.4
BH-08		9.4	1.380	136.8	0.601		58.4		76.30	1.793	
BH-09		9.4	1.520	109.5	0.631		41.8		76.30	1.435	
BH-04		11.9	1.330	126.7	0.581		76.9		94.55	1.340	
BH-02	Clay-II-2	12.4	1.560	133.4	0.698	0.645	36.9	50.0	98.20	1.358	
BH-06	(CH)	12.4	1.440	126.7	0.628		44.3		98.20	1.290	
BH-07		12.4	1.710	142.7	0.750		47.3		98.20	1.453	
BH-10		12.4	1.450	142.8	0.610		46.5		98.20	1.454	
BH-01		14.9	1.420	155.5	0.624		50.8		116.45	1.335	
BH-03		15.4	1.520	138.6	0.745		45.2		120.10	1.154	
BH-05		15.4	1.310	138.0	0.517		56.0		120.10	1.149	
BH-08		15.4	1.280	170.9	0.522		49.1		120.10	1.423	
BH-09		15.4	1.630	142.8	0.688		40.1		120.10	1.189	1.2
BH-04		17.9	1.480	151.0	0.593		47.9		138.35	1.091	
BH-06		18.4	1.460	184.4	0.734		62.7		142.00	1.299	
BH-10		18.4	1.360	164.4	0.586		45.6		142.00	1.158	
BH-02		19.2	1.410	163.9	0.707		55.3		147.84	1.108	
Notation		e_0: initial void ratio, p_y: consolidation yield stress, Cc: compression index, Cv: coefficient of consolidation, p: overburden pressure, OCR: over consolidation ratio									

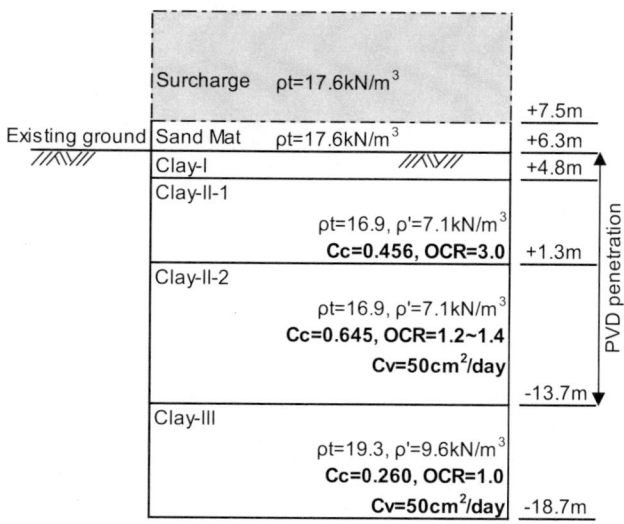

Fig. 13. Model of soil layer for design

3.3   Measurement Control

In deciding the removal of the surcharge fill after a period of leftover consolidation, final settlement predictions were made using the Asaoka method[2]. The final settlement prediction results for each observation point using the Asaoka method are shown in Fig. 14. Even though the loading conditions during construction were similar, the amount and trend of settlement differed at each observation point. The required consolidation ratio and expected residual settlement were obtained from the estimated final settlement. From this, it was confirmed that the removal conditions were met.

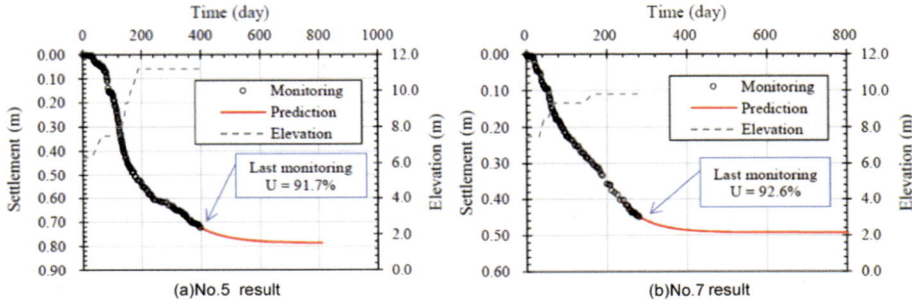

Fig. 14.   Final settlement prediction using the ASAOKA method

4   Conclusion

This project had to be completed in a short time under severe natural conditions. The jetty was built using a jacket method, which is a three-dimensional truss structure to be installed on six number of piles, requires high accuracy of piling. Considering harsh conditions in the area, a 400-tonne lifting crane barge equipped with spuds for holding the vessel position and a pile keeper for adjustment of the steel pipe pile position were used. Spending time to attain accuracy of pile positions during initial driving, more than 90% of the piles were placed within ± 30 mm of the design position, indicating high accuracy of driving.

The fabricated jackets were placed on rollers called tirtanks and transported by winch on a 600 m-rail in the temporary yard to the temporary quay. The transported jackets were lifted by a 500t lifting crane barge. To hold the barge in the river, it was transfered by six anchors instead of boats.

Jacket installation is affected by river conditions of the draft of the barge is 2.5 m. Therefore, the jacket is lifted at high tide, and by the time the crane barge is moved to install the jacket, subsequently, the tide moves to low tide and the pile heads are protruding. Therefore, a construction plan was elaborated so that a series of operations could be completed in one day.

2 Akira ASAOKA. (1978), "Observational procedure of settlement prediction", Soils and Foundations, Vol. 18, No. 4, pp. 87–101.

As the container terminal was to be built on soft ground, soil improvement using PVD was used to shorten the improvement period. Prior to construction, detailed soil investigations of the yard were carried out at ten locations to confirm soil properties and soil layer classification. Based on these data, the surcharge fill height and target leaving period were determined. Settlement analysis was carried out using the finite element analysis program "DACSAR" for the settlement observation points. The required consolidation ratio and expected residual settlement were obtained from the estimated final settlement using Asaoka method. From this, it was confirmed that the removal conditions were met.

Acknowledgements. We express our appreciation to the people concerned who gave us the opportunity to write this paper.

Reference

Murakami H, Tsuchida T, Yamada Y, Aoyama T (2015) Consideration of physical and mechanical characteristics of clayey soils in Myanmar. Japan Geotech Soc 10(1):163–172. (in Japanese)

Datasharing in Inland Navigation

Jef Bauwens[✉]

De Vlaamse Waterweg, Hasselt, Belgium
jef.bauwens@vlaamsewaterweg.be

1 Introduction

This paper describes the theme of sharing information with the focus on inland navigation. Looking at global actions needed to achieve the climate targets, inland shipping will play an important role in making transport more sustainable and to order to achieve a climate neutral logistic chain. De Vlaamse Waterweg (DVW) will play a pioneering role in this matter. Starting point is to consider inland shipping not only as climate neutral and resilient, but also to offer the necessary transparency to every stakeholder so that it can be used by e.g. shippers as a fully-fledged transport mode, alongside rail, road and air. The ultimate goal is to evolve from uni-modal thinking, where people make decisions based on habits, to a synchromodal (automated) mode by 2050. To achieve this, sharing qualitative data will be crucial. It is important to notice that DVW will set up a model to facilitate in sharing qualitative data and will only allow data to be used within legal constraints.

2 From Local Systems to Ecosystems

As of today, decisions to transport cargo on a particular mode of transport are often made a long time in advance. As a result, the most efficient mode of transport, let alone climate-neutral transport, is usually not used. There are many reasons: contractual obligations, limitations in the field of infrastructure, but often also ignorance. Biggest victim is the climate.

Taking into account that the transport sector is responsible for 21% of the total carbon emission as of 2021, a reduction of this seems to be able to be made quite easily by choosing the transport mode that has least impact on the climate at the right time at the right place. However, in practice, this is not as easy as written down. A lot of parameters can influence the decision to make the right choice for the right transport. This can result in a fast, polluting choice in some cases, but as we notice today, the choice is moistly made by the heart and not by the brain.

A lot of freight – most of them are parcels ordered by domestic users – in the logistics chain moves too fast, disrupting the regular "just in time" logistics chain. One of the reasons is that most home users receive their order the next day without even asking. As a result, only the fast modes (trucks, vans) make part of this chain. As long as this habit is not broken by ratio, an economical, more sustainable or slower mode such as inland shipping cannot stand a chance and thus cannot compete.

Y. Li et al. (Eds.): PIANC 2022, LNCE 264, pp. 1353–1356, 2023.
https://doi.org/10.1007/978-981-19-6138-0_119

Also to be taken into consideration is that a vast majority (some figures mention 70%) of the current fleet in inland navigation is not ready to meet the stringent requirements to meet the climate targets. This must be seen as an opportunity for investors as inland navigation can be an enabler towards a sustainable and climate neutral transport mode when new types of ships are build and provided. The challenge here is to provide accurate data to help investors make forward-looking decisions.

As looking to data in enabling stakeholders in the logistic chain in making decisions, we must also take into account the importance of the crossing point with other modes like smart hubs and data. The place to be build, size, etc. Can all be predicted by accurate data.

One of the objectives is to use the data we facilitate to offer ourselves and also third parties within an ecosystem the opportunity to develop innovative concepts, albeit always within a legal framework.

3   Step by Step Approach

In the future rational decisions will be made by supercomputers based on information from multiple sources. As a result we will notice that goods typically will no longer reach their destination via one mode of transport, but that the path of least resistance (the most climate-friendly in relation to the contract) will be chosen. In order to allow these decisions to be made in relation to the other transport modes (road, rail and air), it is therefore crucial that inland shipping can provide relevant transport and network data in a generic way. Inland navigation should act as one of these sources. We realize that it will take years to realize this goal. A step-by-step approach driven by use cases is proposed.

4   Use Case: Sharing Transport-Report and Location Information from Inland Vessels

As of 2022, all cargo movements in the Flanders region on inland shipping must be obligatory reported to the competent authority in a digital and standardized way. One stop shop principle is applied for the skipper. Today, this information is only used on an operational level by the competent authorities. As this information is also crucial as an enabler for a sustainable logistic chain in B2B -e.g. more efficient terminal planning leads to less congestion and less carbon pollution, a solution to share this data on a data sharing platform will be proposed. Conditions are a strict appointment system in relation with the legal requirements.

A project-based approach is proposed here. As a government organisation, we already have a solution of allowing data to flow through to all interested actors in the logistics chain. However, the authorization mechanism, albeit in accordance with the legal restrictions, is very strict and is experienced by all the actors involved as a barrier to integrating systems. A solution with an intermediate data platform seems appropriate. In this way the current actors, i.e. the shippers (and those who carry out the transports) can share data in a much simpler way with all other parties involved in the transport of the goods.

To achieve this goal, it is necessary to build up the certain levels of trust and keep also the technological solutions in line.

The first step is unlocking location information from inland vessels towards the endusers.

Important to keep in mind is that "De Vlaamse Waterweg" is only acting as an intermediate in this relation. "De Vlaamse Waterweg" has the allowance to capture location information from inland vessels. Under certain conditions, end users, other than the vessel owners themselves may make use of this information. In all cases, the vessel owner should know who is using its data and also for what purposes. Also, a mechanism to control should be provided.

Figure 1 shows the solution that will be set up: "De Vlaamse Waterweg" will provide data to a Data Sharing platform. Legal agreements and governance rules are agreed upon. Endusers will be able to capture the data only on the data sharing platform.

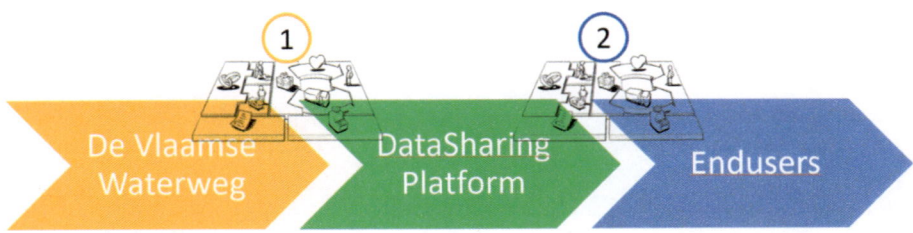

Fig. 1. A technological solution with an intermediate data platform

At the time the paper is submitted, a business case is in approval mode. Once approved, a project can be set up and planned.

5   Conclusion

Figure 2 the vision and ambition De Vlaamse Waterweg has on data sharing. As a waterway authority we are strong in building infrastructure to manage the actual vessels around the waterways. The next step is to build up a infostructure with all the relevant information we gather.

INFOSTRUCTURE

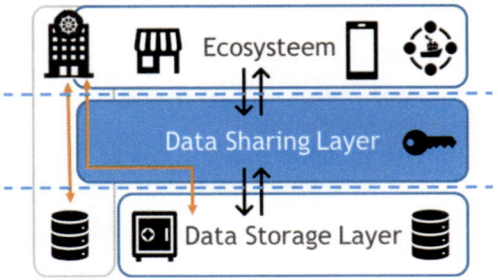

Fig. 2. A infostructure with all the relevant information

As De Vlaamse Waterweg is not the only dataprovider to a datasharing platform and as endusers will not be restricted to inland navigation, we are confident that this will be best practice to encourage creativity and disruptive innovation.

Examining the Logistics Performance of a Decentralized Waterborne Container Transportation Service in the West German Canal Network with the Help of Discrete-Event Simulation

Cyril Alias[1(✉)], Jonas Zum Felde[1], and Sven Severin[2]

[1] DST – Development Centre for Ship Technology and Transport Systems, Duisburg, Germany
{Alias,ZumFelde}@dst-org.de
[2] RIF Institut für Forschung und Transfer e. V., Dortmund, Germany
Sven.Severin@rt.rif-ev.de

Abstract. In order to alleviate road traffic congestion, a logistics concept based on small inland vessels in the West German canal network has been developed for the distribution of containers. The service concept rests on multiple transshipment points throughout the considered geographic area as potential origins and destinations of transport legs. After safeguarding the technical feasibility, the resulting service concept needs to be examined in order to assess its economic viability. Only with promising outcomes resulting from this assessment can potential operators be convinced to transfer the concept into reality, set up the decentralized waterborne container transportation service, and deploy actual inland vessels to be operated in the German waterways. Discrete-event simulation has established itself as an important analysis method in logistics and is suitable for the examination of logistics systems at an operational and tactical level. In the present setting, various scenarios have been selected as the most promising ones and simulated with appropriate models. Eventually, DES will help to determine the routes to be operated, the ports and transshipment points to be included in the respective routes, the vessels to be used, including their type and number, the manning regulation of the inland vessels to be selected, and the transshipment concept to be pursued. Ultimately, the logistics simulation reveals which scenarios turn out to be the most and least promising ones and allows overall statements on the expectable profitability of the service. Furthermore, it helps to identify utilization peaks of the examined inland vessels and transshipment points.

Keywords: Inland waterway transportation · Discrete-event simulation · Decentralized container transport · Transshipment · Logistics · Economic performance

Y. Li et al. (Eds.): PIANC 2022, LNCE 264, pp. 1357–1372, 2023.
https://doi.org/10.1007/978-981-19-6138-0_120

1 Introduction

Reliable, safe and environmentally compatible transport is an elementary prerequisite for a competitive and sustainable economy. To ensure that the transport sector does not become a bottleneck to economic development, the available resources must be used efficiently. However, the road and rail modes of transport are increasingly reaching their limits, and the heavy traffic load, especially on the roads, leads to high emissions of pollutants and greenhouse gases. At the same time, inland waterway transportation (IWT) is characterized by high reliability and energy efficiency and has considerable unused capacity reserves, especially in Northwestern and Central Europe[1]. It therefore lends itself to relieving the burden on road and rail.

Traditionally, IWT has focused on direct long-distance transports, such as bulk transports between two locations of up to several hundred kilometers. In view of the eroding volumes of bulk cargo in continental Europe, IWT is urgently required to open up new market segments and, in doing so, to exploit the opportunities offered by technological progress to the greatest possible extent for its benefit.

In order to utilize the unused transport capacities and develop new market segments for IWT, a decentralized waterborne container transportation service can be a viable solution. While the conception of such a service has already been described in previous publications, the configuration of the service in terms of the inland vessels to be used and the transshipment points to be served as well as an assessment of the expected economic key figures are still pending.

2 Transport Service Concept

The examined transport service concept encompasses a decentralized waterborne container transportation service operating in the hinterland of the European seaports in the range between Le Havre, France, and Hamburg, Germany. Specifically, it serves the hinterland areas of the ports in Hamburg and Bremerhaven (both Germany), Zeebrugge and Antwerp (both Belgium) as well as Rotterdam and Amsterdam (both the Netherlands).

The service foots on four sub-concepts, i.e., the logistics service design, the envisaged inland vessels, the designated transshipment operations, and – lastly – the integration of the envisioned service into the landscape of existing logistics services. Details of the four sub-concepts have been published earlier and are, thus, only briefly described in the following.

2.1 Vessel Concept

The inland vessels designed for the transport service have been developed specifically for the use in the West German canal network. As part of this, the decision in favor of

---

[1] An area between Eastern and Western Europe, generally accepted as comprising Austria, the Czech Republic, Germany, Hungary, Liechtenstein, Poland, Slovakia, Slovenia, and Switzerland (see Collins Dictionary).

small inland vessels was made deliberately to account for the different capacity demands expected for different stretches of the transport network. As such, the considered inland vessels can carry between eight and 30 TEU each. Due to structural restrictions in terms of bridge heights in the area of operation, the transport of only a single layer of containers can be considered realistically. Taking into account the remaining factors of the clearance profile, such as lock chamber dimensions and waterways drafts, the inland vessels measure between 50 and 95 m in length and 6.80 and 9.50 m in width and feature a draft of up to 1.12 m (Alias et al. 2021).

The previous examination failed to prove the economic viability of such a service due to the lack of freight volume to utilize the envisioned vehicle capacity fully. Inland barges of up to 96 TEU. In the RuhrCargo research project, a push train between the inland ports of Dortmund (at the end of the Dortmund-Ems Canal) and Duisburg (at the junction of the river Rhine and the Rhine-Herne Canal) was planned with a potential single-layer transport capacity of up to 96 TEU (DST 2013).

As the inland vessels are planned for operation in the canal network without any waterway currents to overcome, the power requirements for the vehicles can be covered by hybrid-electrical propulsion as the expected power consumption of the different inland vessels ranges between 43 and 196 kilowatt-hours, respectively, at a velocity of 12 km per hour, the maximum velocity permitted on the considered waterways. In addition, the inland vessels were hydrodynamically optimized with the help of CFD and model testing experiments (Ley et al. 2022). During the vessel design process, it was also considered to keep the construction as modular as possible and, thereby, allow for easily scalable and cost-effective production.

The inland vessels are also planned in a way that future modes of operation, such as one-person operation, remote control, and even fully automated operation, are enabled. By this, the potential for further cost reduction in the future is included. At the moment, such modes of operation are not recognized by the German waterway authority, while it is already permitted in the Netherlands.

As the inland vessels are designed for single-layer container transports only, they can be equipped with a mobile onboard crane to tie into the transshipment concept.

2.2   Transshipment Concept

The service area of the decentralized waterborne container transportation service extends over a geographic area between the river Rhine in the west and the river Weser in the northeast of the German state of North Rhine-Westphalia, the economically strongest and most densely populated region in the country. Apart from the existing container ports in the considered area, additional decentralized transshipment points have been taken into consideration. These additional transshipment points include active ports without any container handling business to date as well as abandoned transshipment points from the former era of coal, iron, and steel production in the considered geographic area (Alias et al. 2021).

Usually, either a container crane or a reach stacker is available at those transshipment points already engaged in the container business so that there is no further investment required to participate in the transport service. However, this applies to a small minority of the considered locations only. These remaining transshipment points will need to invest in container handling infrastructure, mainly reach stackers, once such an investment appears economically viable and rewarding. However, in order to develop a container business, even smaller investments may pose an obstacle for many locations, so a concept in which all required equipment will be provided by the inland vessel operator appears more promising. Hence, a mobile onboard crane can help include additional locations with initially low cargo volume but a sufficient growth potential and, thus, eliminate a potential entry barrier.

Designed as a mobile gantry crane running along the side passage of the inland vessel, the mobile onboard crane is operated by the inland vessel operator. In order to limit the usage of space, the control stand of the crane is co-located with the control station of the inland vessel. From its home position at the ship's bow, the crane is capable of traversing the inland vessel along its entire length. Additional details about the transshipment concept can be found in earlier publications.

2.3   Logistics Concept

In order to determine the logistics concept underlying the proposed transport service, both the demand and supply sides have been considered in the design process. The supply side is represented by the service area, which has been defined with the help of the waterways that are to be covered. The service area represents the aforementioned geographic area with five canals, i. e., the Datteln-Hamm Canal, the Dortmund-Ems Canal, the Mittelland Canal, the Rhine-Herne Canal, the Wesel-Datteln Canal, as well as the navigable part of the river Ruhr and a small portion of the river Weser (Alias et al. 2021).

The first level of transshipment point taken into consideration comprises those ports already actively participating in container transshipment. Only two ports, Minden and Dortmund, fit this criterion. As the next step, all active ports in the examined service area were added to the list of transshipment points – resulting in a total of 42 additional port locations. Lastly, yet another 61 formerly active ports and transshipment points have been identified as potential transshipment points on the third level. Many of these inactive and abandoned sites still fit the regulatory requirements for port operations. All three levels combined represent the complete set of 105 total destinations to be potentially used by the proposed service. To enable a comprehensive calculation of travel times, additional points of interest, such as locks, bridges, turning basins and waterway junctions, were included in the network (Alias et al. 2020).

The demand for such a service was derived in two ways. Firstly, the cargo already known to be transported with the proposed waterborne transport process was determined. In addition, a potential modal shift effect due to the introduction of the new service was quantified and added to the already determined volumes. The freight volumes were determined using publicly available information on container freight volumes and calculated at the municipal level before being allocated to ports (Alias et al. 2020). In addition, the determined volumes the values were validated on a sample basis by identifying real cases of potential users of the service.

Eventually, three service concepts were identified as possible solutions for the service: direct transport, scheduled liner service, and a hub-and-spoke system with a transshipment hub in the center of the service area.

2.4   Integration Concept

As has been mentioned before, the designated decentralized waterborne container transportation service operates in the hinterlands of the seaports in Belgium, Germany, and the Netherlands. More precisely, it is designed as serves as a feeder shuttle service between the hinterland hubs. The feeder shuttle bidirectionally serves the inland ports of Duisburg and Wesel on the Rhine and the port of Minden at the waterway junction of Mittelland Canal and the river Weser. In all three termini, the containers are transshipped onto the larger liner services to the seaports in Belgium and the Netherlands, and northern Germany, respectively.

Direct service to the seaports as a single inland vessel or as a push train had been considered earlier and is principally possible, although the reduced vehicle capacity compared to the existing liner services makes this option economically unattractive. Therefore, the options have been discarded during the examination and the service area of the envisioned transport service restricted to the above-mentioned geographic area.

In addition, the proposed decentralized waterborne container transport network must be designed with the international supply chains in mind in order to become an integral part of the intermodal transport chains. The need for this integration has been deliberately taken into account in the planning and booking processes for the proposed service, so that they are compatible with those used for the preceding and subsequent transport phases. This also applies to the control tools required to monitor and control the transport execution.

3   Research Methodology

For the detailed examination of the logistics performance of the decentralized waterborne container transportation service, discrete-event simulation (DES) has been chosen as the examination method of choice because it has established itself as a frequently adopted examination method in logistics when it comes to the multiple flows in complex technical systems. In a discrete-event simulation, the system under consideration is represented by events, activities and processes. As it uses random numbers, a sufficient number of replications is required to achieve statistical certainty of the results the DES study. DES has been selected to account for the complexity of the envisioned transport service operation as it has already been applied to a wide range of applications across multiple areas within the transportation and logistics domain (and beyond) (Kuhn et al. 2008, p. 73). Moreover, DES enables process transparency and, thereby, the comparative assessment of several decision alternatives (März et al. 2011, pp. 3–5).

In IWT, however, the application of DES is still a rarity and gains momentum only slowly. Therefore, a dedicated simulation tool and model had to be developed in order to examine and assess the predefined scenarios of the decentralized waterborne container transportation service in an effective manner.

3.1   Simulation Tool

A DES tool for the proposed waterborne transport service was created based on the VEROSIM toolbox to evaluate and compare different scenarios that could be envisioned as part of the decentralized waterborne container transportation service. VEROSIM is a software system originally designed for 3D simulations, virtual test environments, and simulation-based development in robotics and engineering. This foundation for the DES tool allowed for the abstraction of a two-dimensional model. Based on this, the necessary functionality for modeling an IWT simulation system had to be programmed to create a new tool tailored to the actual case (zum Felde et al., 2022). Although the development was strongly oriented towards the specifications of the particular case study, the flexibility of the toolbox allows for a complete representation of IWT in general, including further case studies.

3.2   Simulation Model

The discrete-event simulation model is based on the points of interest identified as relevant to the network. The points of interest, such as ports, locks and nodes, are located on a map covering the entire service area described above. Each point of interest is considered a node in the network connected by edges representing the waterway segments between two points of interest. Travel times are based on the distance of the waterways between each point of interest and the related inland vessel characteristics, such as maximum permissible speed and acceleration behavior. Each point is given the necessary attributes to represent its specific functionality. For instance, a lock node is provided with attributes that indicate the width and length of the chamber as well as the respective lock operation time, whereas a port node is defined by its number of available berths and handling facilities and the pertaining processing times.

In order to generate the container volumes to be processed in the model, the predetermined cargo volumes are fed into the simulation model based on the municipality of origin. The model determines the costs of road-based transportation on the first or last mile, i. e., between the respective municipality and the assigned canal port. In this study, only loaded containers were considered as the focus lay on load trips. In terms of inland vessel utilization, a significant increase can be expected once empty containers are included in the examination.

When creating a specific scenario, the DES model allows the creation of routes from the set of considered ports. The only restriction is that the route must serve a port with the ability to transfer containers onto a container liner service, i. e., Duisburg, Wesel, and Minden, as the transport service is designed as a feeder shuttle between the inland hubs. In addition to the freedom in route design, the DES model allows the specification of a minimum time between visits to improve the timing of the route. Further, it allows a combination of minimum utilization and maximum waiting time. While designing a route, the DES model automatically computes the length of the route and determines the smallest lock and lowest bridge that limit use by certain vessels. After a route is created, the required vehicles must be deployed on the route. Inland vessels can be added and configured according to available vessel types, crew specifications, and a home port for the vessel. The required vessel manning specifications are available according to both the Regulations for Rhine navigation personnel (RheinSchPersV) and German regulations for inland navigation personnel (BinSchPersV), which were both integrated in the German Inland Waterway Vessels Safety Ordinance (BinSchUO) formerly. Apart from lock size, waterway draft, and bridge height, there are no restrictive factors to the use of inland vessels so that any combination of type, crew and home port for each vessel used on a route is allowed.

In order to determine the resulting key performance indicators (KPI), additional data is required as input. This applies in particular to the calculation of the costs incurred by the envisaged transport system. Within the simulation model of the decentralized waterborne container transportation system, a large number of input parameters are variable. These input parameters include, but are not limited to, many of the costs associated with vessel operation, the capital investments of the ports, and the temporary storage of containers at the ports. The most detailed calculation is performed for vessel operations. Besides the wages for the crew required for the specific manning specification, further time-related costs, e. g., insurance and regular tasks of maintenance, repair, and overhaul (MRO) required to operate a vessel, and the distance-related cost of energy, which were determined by calculating accurate consumption based on vessel movements and collected from the experimental evaluation during the vessel design phase, are included.

To be precise, the energy costs caused by vessel movements are determined by calculating accurate values for the energy consumed. These values foot on previously collected data from the experimental results of the model tests of the designed inland vessels. The energy consumption determined for each barge is also used to calculate the CO_2 emissions caused by the service using a CO_2 conversion factor.

In addition, the costs of road-based pre-haul, i. e., from the consignor to the canal port, and the ones of the liner services between an inland hub and the seaport, e. g., between Duisburg and Rotterdam or between Minden and Hamburg, are included. By including these costs in the DES model, the costs incurred for the total transportation service between the consignor and the seaport can be determined. Thereby, it is possible to effectively compare the new decentralized waterborne container transportation service with its main competitor, i. e., the direct road transportation from the consignor to the seaport. In this way, both the costs per TEU kilometer of the new transport service on the West German canal network and the ones of the total transport leg can be determined and compared throughout the examined scenarios.

Furthermore, the DES model computes a port-related service level based on the container dwell time in a particular port. This interpretation of the service level differs from the classic definition of service level, which refers to shipments received on time and in full. Such an accurate calculation of the level of service cannot be determined because no order data is available to measure the punctuality factor and the damage freedom.

The DES model stores the events in an external database, which allows easy access to the simulation results for further analysis with external tools. For the most important KPI analyzed in each scenario, the DES model has a built-in option for evaluating and visualizing the KPI over time.

4 Simulation-Based Examination

4.1 Examined Scenarios

With the help of discrete-event simulation, the logistics performance of the envisioned decentralized waterborne container transportation service is examined. In total, the comparative examination comprises eight different scenarios, all of which are based on different configurations of the system. Figure 1 illustrates the different routes of the decentralized waterborne container transportation service as a network map, while Table 2 shows the number of routes served and inland vessels deployed for each of the ten considered scenarios.

Labelled as the baseline case, Scenario 1 represents the original configuration of the transport service, i. e., a feeder service of the hinterland hubs Duisburg and Wesel serving the seven biggest ports in the canal network with liner services operating on seven different routes with 28 inland vessels of different types and capacities. The assignment of the inland vessels to the routes is determined by the volume of cargo to be hauled and handled. The inland vessels are crewed following the BinSchPersV regulations in such a way that 24-h operation is allowed by regular shift changes of the crew.

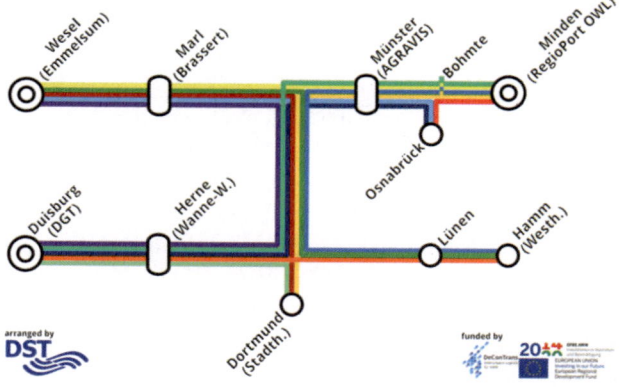

Fig. 1. Network map of the decentralized waterborne container transportation service with termini in Duisburg (DUI), Wesel (WES), Dortmund (DTM), Hamm (HMM), and Minden (MID).

Scenario 2 is a slight modification of the first one, with one stop at the Mittelland Canal being replaced by another one. Precisely, the inland port of Osnabrück has replaced the container terminal in Bohmte and led to the inclusion of additional operational challenges as the inland port of Osnabrück is situated at the end of a branch canal, which again is limited by two small locks restricting the vessel size to a maximum length of 82 m. This fact results in two additional routes being serviced by two inland vessels each, elevating the total number of routes to nine, and the number of ships to 32. Based on the baseline scenario, the two subsequent scenarios observe the impact of an increase in transshipment points to be served. Scenarios 3 and 4 increase the number of ports to 44 and 105, respectively. In both scenarios, the seven routes are served by a total of 39 inland vessels in order to compensate for the higher number of stops. Scenario 5 examines the effect of the baseline scenario operating in shuttle mode and results in a configuration of 16 routes with 37 inland vessels. An examination of more ports served by a shuttle service is refrained from because the number of required inland vessels and routes would exceed any realistic expectations. Table 1 shows an exemplary overview of the configuration results for the Scenarios 1 to 4.

Table 1. Overview of served routes, deployed vessels and allocated capacities for the scenarios 1 to 4. (colors and termini correspond to the lines in the network map in Fig. 1).

		Route 1	Route 2	Route 3	Route 4	Route 5	Route 6	Route 7	Route 8	Route 9
Scenario 1	transport relation [start - end]	DUI - WES	DUI - DTM	DUI - HMM	DUI - MID	WES - DTM	WES - HMM	WES - MID		
	no. of vessels	7	3	3	5	3	2	5		
	vessel type (TEU)	VI (30)	VI (30)	IV (24)	I (8)	III (18)	III (18)	I (8)		
Scenario 2	transport relation [start - end]	DUI - WES	DUI - DTM	DUI - HMM	DUI - MID	DUI - OSN	WES - DTM	WES - HMM	WES - MID	WES - OSN
	no. of vessels	7	3	3	4	3	3	2	4	3
	vessel type (TEU)	VI (30)	VI (30)	IV (24)	I (8)	I (8)	III (18)	III (18)	I (8)	I (8)
Scenario 3	transport relation [start - end]	DUI - WES	DUI - DTM	DUI - HMM	DUI - MID	WES - DTM	WES - HMM	WES - MID		
	no. of vessels	9	3	5	8	3	3	8		
	vessel type (TEU)	VI (30)	VI (30)	III (18)	I (8)	II (12)	III (18)	I (8)		
Scenario 4	transport relation [start - end]	DUI - WES	DUI - DTM	DUI - HMM	DUI - MID	WES - DTM	WES - HMM	WES - MID		
	no. of vessels	9	3	5	8	3	3	8		
	vessel type (TEU)	VI (30)	VI (30)	III (18)	I (8)	II (12)	III (18)	I (8)		

Apart from liner services and direct haulage, a third traffic concept, i. e., a hub-and-spoke system, is examined in Scenarios 6 and 7. Both scenarios assume a fictional hub in the center of the West German canal network, precisely at the transshipment in Datteln, and differ in the inclusion of Bohmte and Osnabrück, respectively. While six routes are required for the earlier case, seven are required for the latter.

Another variation of the scenarios is the extension of the service area to the hinterland of North German seaports and, thus, the consideration of northbound cargo volumes towards the ports of Hamburg and Bremerhaven. This variation considers Minden as an additional inland hub because it is already served by regular liner services to the seaports. This variation is examined with three scenarios: Scenario 8 is based on the baseline scenario, while Scenario 9 foots on Scenario 2. Scenario 8 results in nine routes and 30 inland vessels, while Scenario 9 requires twelve routes and 35 inland vessels. Scenario 10, designed analogously to Scenario 5, requires an additional seven routes, leading to 23 routes with 46 inland vessels in total. All increases in routes and inland vessels are the necessary result of serving a third destination port and connecting the complete network.

4.2   Simulation Study

The results of the aforementioned scenario configurations are presented as average values of the results collected from over a hundred simulation runs of each scenario. This is done to account for the statistical variation induced by the model. Four key performance indicators will be observed to compare the scenarios and determine the quality of each of them: the costs per TEU-kilometer, the service level, the capacity utilization of the inland vessels, and the productive time shares of each inland vessel.

The most relevant performance indicator is the costs per TEU-kilometer as it is needed to determine the economic viability of the service and to compare it with existing transport alternatives. The costs are analyzed for both the IWT leg with the outlined service as well as the complete transport leg between seaport and consignor (and vice versa).

Additionally, a service level is calculated for each port. Subsequently, both the minimum service level at each location and an aggregated average service level are determined. The minimum service level is set up to act as a threshold that all ports need to accomplish in order to ensure that containers are not being stored in ports for excessive amounts of time. Else, this would lead consignors to refrain from using such a service. To ensure all ports have a decent level of connectivity within the new transport service, a minimum service level of 75% has been established. This translates into 50% of the containers being picked up within the first 24 h after their arrival at a port.

From the operator's perspective, the major performance indicator is the capacity utilization of the inland vessels used. With respect to the DES model, this metric guarantees that the port-related service level is not increased by simply running a multitude of underutilized inland vessels and, thereby, minimizing the container dwell times in the canal ports. It is to be noted again that vessel utilization refers to load runs of containers only. The empty container cycle has not been simulated and needs to be considered as extra volumes on top of the given capacity utilization.

Eventually, the productive time shares of each vessel include the time of transport and locking operations as well as loading and unloading processes. They are used to compare traffic flow in each scenario.

4.3   Simulation Results

After running a hundred simulation runs of each of the ten scenarios, the results have been collected and compiled. Table 2 shows an overview of the results of all scenarios considered in this article.

The baseline scenario, Scenario 1, reaches a service level of 90.78% while just about achieving the required minimal service level of merely 75.3%. Moreover, the baseline scenario yields an average vessel utilization of 47.69% and a productive time share of 66.39%. The costs for the new service amount to 1.07 EUR per TEU-kilometer for the canal leg, and 0.68 EUR for the total leg, respectively.

The other scenarios, including westbound liner service concepts, i. e., Scenarios 2 to 4, feature similar results in terms of port-related service level and vessel utilization albeit costs per TEU-km do increase quite significantly with an increasing number of transshipment points to be served. The service levels vary between 87.03% and 91.61% and vessel utilization ranges between 43.69% and 48.19% whereas costs per TEU-kilometer for the canal leg grow from 1.09 EUR (Scenario 2) to 1.31 EUR (Scenario 4). All three scenarios miss the minimum service level threshold. While the results of Scenario 2 are close to the one of the baseline scenario, the differences become more pronounced as the number of transshipment increases. This is mostly the result of the increased numbers of stops resulting in longer roundtrip times for routes while boosting the productive time shares.

The first scenario with a direct traffic concept, Scenario 5, almost reaches the service level of the baseline scenario with 90.45%. However, the minimum service level, on the other hand, only reaches 71.77% and, thus, misses the threshold of 75%. The productive time share of the vessels is on the same level as the baseline scenario as well, averaging at 66.7%. Vessel utilization comes in a little short with only 44.94%. In essence, the transport costs are quite comparable with 1.09 EUR per TEU-kilometer for the canal leg and 0.72 EUR for the total leg, respectively.

Table 2. Overview of logistics performance results of the scenarios 1 to 10.

	Scen 1	Scen 2	Scen 3	Scen 4	Scen 5	Scen 6	Scen 7	Scen 8	Scen 9	Scen 10
no. of routes served	7	9	7	7	16	6	7	9	12	23
no. of inland vessels deployed	28	32	39	39	37	13	14	30	35	46
service level (in percent)	90.78	91.07	91.61	87.03	90.45	25.19	16.28	90.95	91.27	93.57
minimum service level (in percent)	75.30	74.63	66.42	62.42	71.77	0.24	0.17	82.25	82.58	78.18
productive time share (in percent)	66.39	66.42	69.14	74.33	66.70	51.53	61.44	71.54	71.59	72.58
vessel utilization (in percent)	47.69	46.85	43.69	48.19	44.94	82.98	80.38	48.41	46.00	39.62
annual emissions per vessel (in tons CO_2)	168.13	150.81	133.55	119.69	126.23	161,70	164,07	165.05	148,36	121.88
costs per TEU-km (canal only) (in EUR)	1.07	1.09	1.29	1.31	1.09	1.03	0.93	1.09	1.11	1.14
costs per TEU-km (total leg) (in EUR)	0.68	0.70	0.74	0.75	0.72	2.00	2.02	0.68	0.71	0.77

The scenarios with the hub-and-spoke concept, Scenarios 6 and 7, return remarkable results of capacity utilization with 82.98% and 80.38%, respectively. This result is only possible at the expense of the service level with only 25.19% and 16.28%, respectively, as well marginal results for the minimum service level with less than 0.5% for either scenario. Likewise, the productive time shares dropped in either scenario with 51.53% and 61.44%, respectively. The main reason behind these results is the limited transshipment capacity at the hub port (Datteln). This again leads to disadvantageous costs per TEU-kilometer because the penalties for long storage times portside drive these upward.

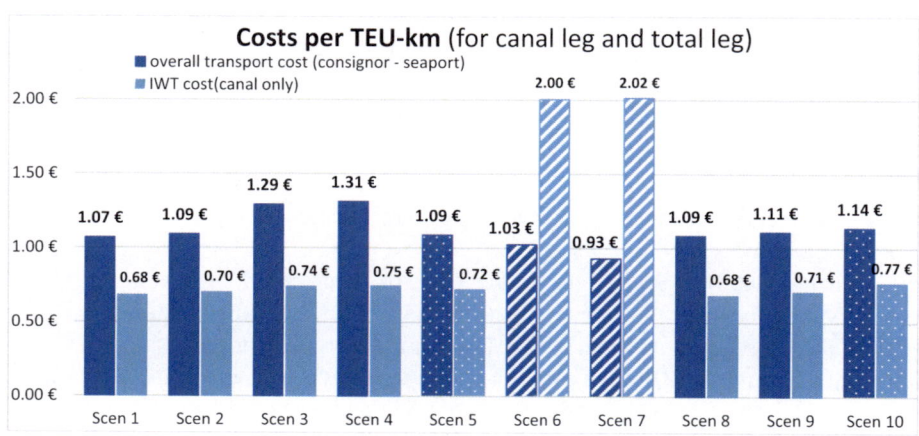

Fig. 2. Comparison of costs per TEU-kilometer for the canal leg and the total leg in the scenarios 1 to 10

When extending the westbound service area by a northbound orientation, as represented in Scenarios 8 to 10, the additional container volumes lead to a stabilization of the service level and particularly the minimum service level. On comparing Scenario 8 with the baseline scenario, the delta between average and minimum service level has shrunk to 8.7% (as opposed to 15.49% in the baseline scenario). Similarly, the vessel utilization and the productive times shares profited from the extension of the service area. The costs per TEU-kilometer remained stable with a slight increase by 0.02 EUR to 1.09 EUR for the canal leg. For the total leg, the costs per TEU-kilometer amount to 0.68 EUR – exactly the same as in the baseline scenario. Similar to the comparison of Scenario 2 with the baseline scenario, Scenarios 8 and 9 feature similar results with 0.71 EUR per TEU-kilometer. This leads to the assessment that the selection of Bohmte or Osnabrück has only a minor impact on the logistics performance of the transport service. The effect of the northbound extension of the service concept is reflected in the results of Scenario 10 as well. However, the scenario with the most routes served and inland vessels used among all observed scenarios shows that the costs per TEU-kilometer are 0.77 EUR for the overall transport and 1.14 EUR for the envisaged IWT service only. In a direct comparison with Scenario 5, the port-related results improve due to the additional freight volume, whereas the vessel-oriented indicators see a downturn with reduced utilization rates (39.62%) and lower productive time shares.

Figure 2 provides an overview of the costs per TEU-kilometer computed for each scenario, while Fig. 3 exhibits the economic KPI of the examined ten scenarios.

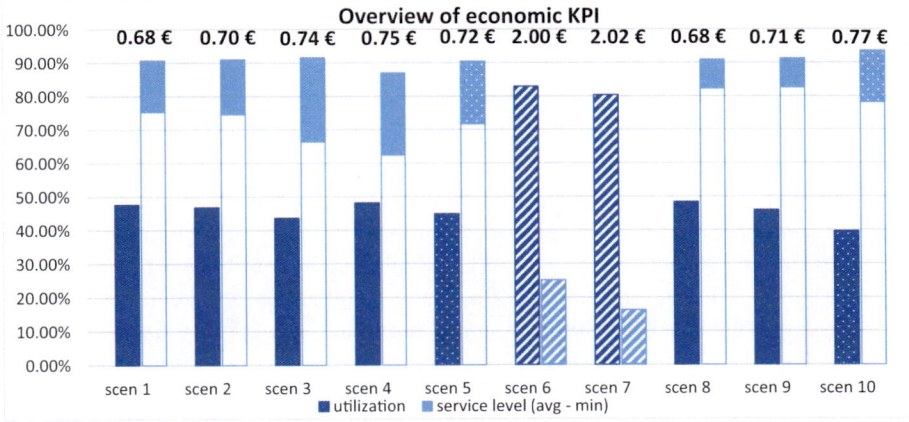

Fig. 3. Comparison of vessel utilization and port service levels in the scenarios 1 to 10.

4.4 Discussion

First, the costs determined in the considered scenarios need to be put into perspective. On the one hand, the designated transportation service still improves on the costs per TEU-kilometer of 2.30 EUR for road transportation, the main competitor for the considered transport leg. The only major deviation from these observations results from the hub-and-spoke scenarios. On the other hand, the costs for operating such a service remain significantly higher than the liner services operated on the river Rhine which can achieve costs as low as 0.18 EUR per TEU-kilometer through its significantly larger economies of scale.

Considering the different service types, the liner services prove to be the best performing and most cost-efficient service type (see Fig. 2 and Fig. 3). Even throughout the synopsis of all considered scenarios, the liner serves yield the best results in comparison with the residual service types.

However, the direct traffic scenarios prove to be competitive despite their relatively high number of routes and the massive increase in the number of inland vessels for the northbound extension of the service area. It is capable of offering high average service levels while remaining economically competitive enough with a noticeable advantage over road transportation (see Fig. 2 and Fig. 3).

The hub-and-spoke system, on the contrary, is severely limited in its operation. A realistic potential hub in the center of the canal network has only limited availability of berths as the majority of transshipment points in the central region have not been designed for the number of simultaneous transshipment operations required of the hub. The identified hub location is realistically capable of handling merely three vessels at a time and, thus, cannot provide the needed transshipment capacity. Additionally, the routes between the hub and the destination ports at the Rhine need to handle almost the entire transport volume as they are limited by the vehicle capacity and the high number of inland vessels. This again puts additional strain on the central hub. All these reasons lead to a reversed structure of the key performance indicators for the hub-focused scenarios, as can be seen in Fig. 3. This is reflected in the disadvantageous cost structure. Consequently, the overall costs per TEU-kilometer rise significantly as penalty charges are applied for long storage times in ports.

Another aspect worth discussing is the effect of accommodating both westbound and northbound containers (instead of westbound containers only). While the service with westbound container flows already shows good results, the service improves even further with the expansion. Thus, the introduction of additional liner services toward the inland hub of Minden results in a performance stabilization or even improvement, e. g., regarding vehicle capacity utilization and minimum service level. In Scenarios 8 and 9, this improvement is achieved through the introduction of additional routes and vessels as the additional container volume is sufficient to offset the additional costs and eventually result in a small increase in the costs of the canal leg and no effect on the costs of the total leg.

5 Conclusion

The article presents the case of decentralized waterborne container transportation service in the West German canal network designed as a feeder shuttle between the inland hubs in the hinterlands of the seaports of Belgium, the Netherlands, and Germany. The service concept needs to be examined in order to determine the best-performing scenario among multiple conceivable scenarios and to assess its respective economic viability. With the help of discrete-event simulation, various scenarios have been selected as the most promising ones and examined with an appropriate DES model of the envisioned transport service in order to determine operational parameters, such as routes and ports to be served, vessel types and related capacities to be used, manning regimes to be applied, and transshipment concepts to be pursued.

Various alternative configurations have been examined and compared with each other so that the effects of the variations of service type, i. e., liner service, direct traffic, and hub-and-spoke system, service area, i. e., only for westbound containers or also for northbound ones, operation modes, or manning regimes on the vessel can be derived from the multitude of simulation runs.

The results reveal that a bidirectional liner service between the destination ports at the Rhine and the Weser serving, ideally for both west- and northbound containers, with a 24-h operation appears to be the best scenario from an economic perspective.

6 Outlook

Based on the results, the scenarios need further variation in order to move away from the exclusive consideration of a particular service type on all routes and destinations and to replace the approach with a blend of a different, relation-specific configuration of the transport routes.

Furthermore, the effect of future operation modes, such as one-person operation, remote control, and fully automated operation, needs to be examined as they appear as promising next evolutionary steps of the concept.

Likewise, the integration of waterborne container transportation with more sustainable and emission-reduced modes of transport on the first or last leg in order to achieve a better ecological footprint deserves further attention and examination. For instance, the use of heavy-duty electric vehicles could be an option.

Acknowledgments. The authors express their gratitude towards Mr. Frank Eduardo Alarcón Olalla, Mr. Anatolij Buchhammer, Mr. Guido Giesen, Mr. Dieter Gründer, Mr. Hergen Hanke, Mr. Henk Heuvelman, Mr. Sebastian Jezek, and Mr. Andreas Stolte as well as those colleagues at the respective organizations that have supported this work actively by means of inspiring discussions, fruitful collaboration, and careful proof-reading of the manuscript.

The research leading to these results has received funding from the EFRE.NRW (2014–2020) Joint Research Funding Programme of the European Union (EFRE) and the Ministry of Economy, Innovation, Digitalization, and Energy of the German Federal State of North Rhine-Westphalia (NRW) under grant agreement EFRE-0801222 (ML-2-1-010A, DeConTrans).

References

Alias C, Broß H, zum Felde J, Gründer D (2021) Enabling decentralized transshipment in waterborne container transportation. In: Jahn C Kersten W Ringle CM (eds) proceedings of the hamburg international conference of logistics: vol. 32, adapting to the future: maritime and city logistics in the context of digitalization and sustainability, 1st edn., pp. 137–166. epubli GmbH. https://doi.org/10.15480/882.3997

Alias C et al (2020) Identifying suitable transshipment points for a decentralized waterborne container transportation network. In: 2020 IEEE international conference on industrial engineering and engineering management (IEEM): proceedings, pp. 799–806. IEEE. https://doi.org/10.1109/IEEM45057.2020.9309903

Alias C, Gründer D, Dahlke L, zum Felde J, Pusch L (2020) Determining the freight volumes for a decentralized waterborne container transportation service. In: 2020 IEEE international conference on industrial engineering and engineering management (IEEM): proceedings, pp. 786–793. IEEE. https://doi.org/10.1109/IEEM45057.2020.9309828

Alias C et al (2021) Designing a decentralized waterborne container transportation service using small inland vessels. In: Proff H (ed) Making connected mobility work: technische und betriebswirtschaftliche aspekte, 1st edn., pp. 573–601. Springer Gabler. https://doi.org/10.1007/978-3-658-32266-3_36

DST (2013) RUHRCARGO: Entwicklung eines Systems zum Transport von Containern (DST-Report No. 2044). Duisburg, Germany. DST - Entwicklungszentrum für Schiffstechnik und Transportsysteme   e.V.   http://www.dst-org.de/wp-content/uploads/2016/01/DST-Bericht-2044.pdf

Kuhn A, Wenzel S (2008) Simulation logistischer systeme. In: Arnold D, Kuhn A, Furmans K, Isermann H, Tempelmeier H (eds) (2008). VDI. Handbuch Logistik, 3rd edn., pp 73–94. Springer, Berlin Heidelberghttps://doi.org/10.1007/978-3-540-72929-7

Ley J, Broß H, Kämmerling E (to appear) Experimental investigation of passing ship effects on moored ships in a canal port. In MASHCON 2022. Symposium conducted at the meeting of University of Strathclyde, Glasgow, United Kingdom

März L, Krug W, Rose O, Weigert G (2011) Simulation und optimierung in produktion und logistik: praxisorientierter leitfaden mit fallbeispielen. VDI-Buch. Springer Berlin Heidelberg. https://doi.org/10.1007/978-3-642-14536-0

zum Felde J, Alias C, Goudz A (to appear) Comparing generic and dedicated tools of discrete event simulation for examining inland waterway transportation services. In: 2022 IEEE international conference on logistics operations management (GOL): proceedings. IEEE

Free Zones as Booster of Growth of Ports

Rodrigo Nicolás Benítez Leto[1(✉)] and Savarese Ariel Jose[2]

[1] Administración Portuaria Puerto Barranqueras, Barranqueras, Argentina
rodrigo.benitez.leto@gmail.com
[2] Empyria STM/Administración Portuaria Puerto Barranqueras,
Buenos Aires, Argentina
ariel.savarese@empyriastm.com

Abstract. Free zones are defined as: "a field where the goods are not subject to the usual control from the customs service and its introduction and extraction are just only taxed with service tax which can be created and there is no economic prohibition on it".

In Argentina, Law N° 24.331 says that the free zones can be of storage, commercial, services or industrial. The tax benefits that this legal regime provides are that the goods in this area are not subject to export or import taxes, created or to be created. Despite the fact that the free zones are a key tool in developing economically deprived areas in a country or a region, those can not do it by themselves. Free zones must be integrated to the logistic and industrial system of the region, making, in this way, the necessary synergy in order to succeed.

Port of Barranqueras which is located on the right bank of the Paraguay - Paraná Waterway in the province of Chaco, Argentina, at around 1,000 km North of Buenos Aires city. Historically, this is a marginal area of Argentina in terms of investment and development, with large infrastructure deficiencies. However, the Port of Barranqueras has unique advantages like its location on the geographic center of MERCOSUR with good road (RN11 & RN14) and railway (Belgrano Cargas) connections; unique possibilities for sustainable growth as it is not constrained by large urban areas and for being a multimodal port with facilities for container and general cargo handling as well as grain storage. Those characteristics make it ideal to set up a free zone.

The objective of this paper is to analyze how a free zone integrated to the Port of Barranqueras will enhance it until converting it into an industrial and logistic center, making more competitive all the export industries in the zone. This analysis includes a map that will help others in similar conditions.

Keywords: Paraguay · Parana · Waterway · Sustainable growth · Inland terminals · Free zone

1 Argentinian Legal Regime

Argentinian free zones are defined in customs code, which article 590 says: "Free zone is an ambit where goods are not bringed under by the usual custom service and the introduction and extraction are not taxed, unless service fee, and those goods would not be reached by economic prohibitions.

Y. Li et al. (Eds.): PIANC 2022, LNCE 264, pp. 1373–1378, 2023.
https://doi.org/10.1007/978-981-19-6138-0_121

On the other hand, Law N° 24.331 (Free Zones Law) establish that free zones be of storage, commercial, services or industrial, this last one with the unique objective of exports goods to thirds countries and they can produce capital goods wich have not been produce in argentinian territory, in order to import it from the free zones.

This custom regime that is proposed by Free Zones Law, establishes among other things, tax exemptions on goods imported to the free zone that are actually or going to be created. Also, tax exemptions are on escentials services in the free zone.

Another interesting incentive that this Law establishes is the suspension of the destination of export/import, that means that you must nationalize the goods when they are going to get out from the free zone. This situation applies to all the formalities that must be done in order to export a good. Lastly, you can introduce all classes of goods or services which are not included in imports list, created or to create, unless guns, ammunition and other things that are against moral, health, animal and vegetal health, security and environment preservation.

The Decision N° 33/15 from the Common Market Council of the Common Market of the Southern Cone (MERCOSUR), establishes a positive landmark for the free zones, because it allows that the goods which are originative from a Member State (from MERCOSUR), or a third party state that have the same originative rules than Member State, under commercial agreements subscribed by MERCOSUR, not going to lose their characteristic of originative, when, in the lapse of they transport or storage, use a special custom zone, an export prosecution zone or a free zone, every time these areas are under custom control of any Member State and it only can be subject of operations aimed to secure they commercialization, conservation, division or another operations, as long as it tariff item would not be change nor it originative character.

In those terms, the Argentinian State made the Resolution N° 669/21, which regulates the Certificado de Origen Derivado, which certifies that goods which transited or had been stored in argentinians free zones, accomplishes all requirements in Decision N° 33/15 from the Common Market Council of the Common Market of the Southern Cone (MERCOSUR), and enables that a specialized organism (Cámara Argentina de Concesionarios de Zonas Francas) issue those certificates.

2   Successful Experiences

Examples that are going to be analyzed were selected because their creation answers the objectives of this paper, that justify a free zone in Barranqueras.

2.1   Free Zone Manaos

Manaos is a city located in the heart of Brazilian west amazonia. Founded in 1669 by Portuguese colonizers. It has a golden age called "caucho feber". This golden age lasted from 1890 to 1915, when asian caucho and synthetic caucho appeared in the market. Unless the caucho industry did not disappear, it had a short upturn in the Second World War, it was clear that to develop that asylee zone, that is far away from the main economics center, was necessary a state impulse. This impulse was made in 1957, with the creation of a free port that, ten years later, had been turned into a free zone by means of Law 288.

The Manaos free zone creation has these pillars: reduccion or exemption of federals, states and municipal taxes, symbolic prices on land acquisition inside industrial park and the construction of a wide range of infrastructure in health, energy and communication services. This last point was made with the availability of ships, which frequency and storage in Amazona waterway, was a pillar of its development. The free zone has three sectors, commercial, industrial and agricultural. If an enterprise wants to settle in the free zone, it must present a project and it must be approved by the SUFRAMA, an collegiate organism, which has managed the free zone since its creation.

Fifty years after its creation, it can be told that this experience was a success. Since then, Manaos population growth from 175 thousand inhabitants to more than 2.5 millions, being the largest city of Amazonas and the seventh most inhabited city of Brazil. Just in the last decade, the annual average growth of Manaos PBI was 9%, more or less three times the growth of Brazil itself. The industrial park has more than 500 enterprises settled, most of then multinationals, which employs 120 thousand of people directly and about 400 thousand indirectly. It is remarkable that the aim of the free zone has variegated since its creation. At the beginning, it did the process of import substitution for the internal market but it bringed serious problems to the balance of payments. It was for that, in the year 2007, incentives were created in order to settle export enterprises.

2.2   La Plata's Free Zone for General Purpose

Created in 1997, located in bonaerense city of Ensenada, in a strategic location just 60 km long from Buenos Aires City, neuralgic center of consumption in Argentina. It has a surface of 700000 Km^2, with fluvial and road access and a service infrastructure like an industrial park, but with the typical tax advantage of a free zone. Nowadays, it has about 100 direct users and more than 3400 indirect users.

It is important to highlight that in practice this free zone works only as a logistic support area for the nearby ports of Tec-Plata (La Plata) and Exolgan (Dock Sud-Avellaneda). This function allows it to get about 90% of free zone commerce in Argentina. This is an example of success, of how to develop a free zone supporting a port community which has already been established in the area.

3   Barranqueras Free Zone

With all the information that we have senn, we can note some similarities with Manaos and La Plata free zones potentialities. As we have been told in the introduction, Port of Barranqueras is located in a strategic point in Argentina, in the heart of Paraná-Paraguay waterway and the Bi-Oecanic corridor, near medium center of consumption in argentine territory and a medium distance from big center of consumption in Brazil, Paraguay, Bolivia and Chile. At the same time, it has excellents lines of communication with those centers of consumption and production, how to be Paraná-Paraguay waterway, nationals routes which connects north (Asunción-Paraguay), south (Rosario-Buenos Aires) and east (Brasil) and west (NOA-Chile) and railway connection with ROSAFE, Buenos Aires, NOA region and Chile.

This brings the possibility that a free zone can be a truly logistic center for the export of goods because of its tax benefits, NOA region goods can be direct exported to consumptions center in Paraguay and Brazil via, in principle, by one of the mos economic and ecologic ways of transport, river transport, by the waterway. Inside these goods, we can find different types of minerals, use them as fertilizer and give lower cost to production from Center and NEA region to export to Brazil Southern Region and Paraguay, and to achieve to asians markets via chilean's pacific ports via railways and road connections.

Setting up a free zone near the Port of Barranqueras, would create a virtuous circle, generating quality employment and improve all service provided in both cities, Barranqueras and Resistencia, because of the growth of commercial, export and logistics activities. Transforming the economic matrix of this region, from a mostly primary and residually industrial and service, to a logistic and export powerhouse, just as we saw in Manaos free zone success example.

It is important to highlight that the land proposed as study hypothesis, has excellent road conexion through bypass avenues which directly connects with national routes. At the same time, we can find to a few meters, C3 railway which is operated by Belgrano Cargas and from the Port of Barranqueras, which has the necessary equipment to attend the river freight (Figs. 1, 2 and 3).

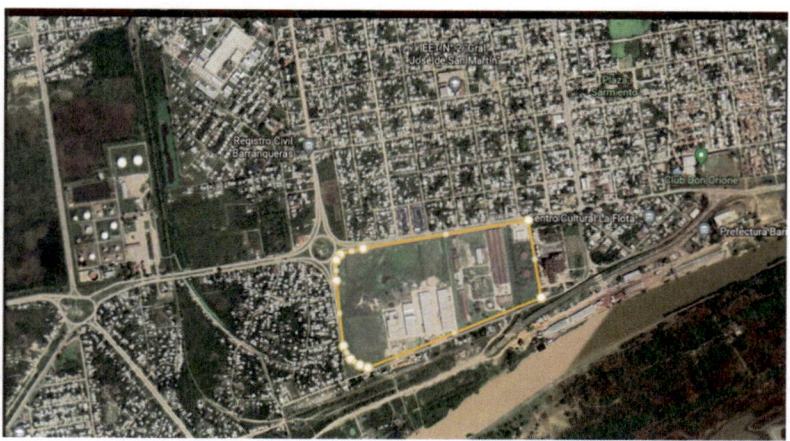

Fig. 1. The green land proposed for the free zone with a 0.29 Km² land near Port of Barranqueras.

We also know that there are a lot of difficulties in order to settle a free zone in the proximity of the Port of Barranqueras.

Firstly, the bureaucracy in order to enable a free zone in Argentina, could generate delayings in the completion of the project. On the other hand, the lack of culture in

Fig. 2. The railway connection and its proximity with the land proposed.

Fig. 3. Road access through its connection with national routes.

Argentina about free zones and its benefits, could extend the deadlines of the project because of the time it would require to convince the authorities and the society in order to achieve the political approval in order to get the lands to settle a free zone.

Secondly, there is a possibility that the purpose of the free zone will be modify, limit or remove most of the comparative benefits that it had, because of the change of the rules, affecting in this way, the predictability and the financial economic equation of the project, jeopardize the probabilities of the project.

Lastly, the economic context of Argentina will always add a risk factor to every project or investment that someone wants to make, because of the macroeconomics instability.

Although the existence of problems like other projects, those are not insurmountable. Through a round of negotiation between authorities and the investor, those problems could be solutioned, transforming in this way, in a great opportunity to generate an invaluable improvement for the region. There is a background of success, in similar conditions to this hypothesis, the infrastructure exists, the location and the legal changes makes this a unique opportunity.

References

https://normas.mercosur.int/simfiles/normativas/
https://proexpansion.com/en/articles/29
https://www.bazflp.com/
http://www.cotia.com.ar/
https://www.cronista.com/. (Transport & cargo supplement)
https://www.gov.br/suframa/pt-br
https://www.tecplata.com/
Zonas francas en Brazil: análisis de la legislación vigente Martin Lautaro Vega Correa

Grain Flow Through the Northern Arch of Brazil

Ana Paula Harumi Higa[1,2](✉), José Gonçalves Moreira Neto[3],
Rodrigo Guimarães Trajano[3], and Herbert Koehne De Castro[3]

[1] Brazilian National Waterway Transportation Agency, São Paulo, Brazil
ana.higa@antaq.gov.br
[2] Graduate Program in International Relations San Tiago Dantas (UNESP,
UNICAMP, PUC-SP), São Paulo, Brazil
[3] Brazilian National Waterway Transportation Agency, Brasília, Brazil
{jose.moreira, rodrigo.trajano,
herbert.castro}@antaq.gov.br

Abstract. Brazil is the world's largest producer of soybean followed by the United States. These two countries accounting for roughly 68.7% of world production. The largest soybean producing region in Brazil is the Midwest, followed by the southern region. Between 2010 and 2020, the exported volume of soybean rose from 29.1 million tons to 83 million tons, respectively, representing 185% rise. The soybean exports are drained by nine logistic corridors and recently it calls attention the expansion of the logistics corridors of the Northern Arch of Brazil. It is expected that Ferrogrão, a railway project connecting the 933 km from Sinop (Mato Grosso state) to Itaituba (Pará state), will consolidate the Tapajós River Corridor as one of the most important corridors to the grain flows and it will help to reduce freight prices. The objective of this paper is to present the recent change in the grain exports flows through the Northern Arch of Brazil and discuss the potential impact of Ferrogrão rail in the Brazilian logistics of grain exports. For it, at the first item will be presented the importance of Brazilian soybean production in the world market and the main producing region in Brazil; in the second item will be presented the main logistic corridors for grain exports; in the third item will be presented the Northern Arch logistic corridors; in the fourth item will be presented the Ferrogrão railway project and in the fifth item will be presented the final considerations.

Keywords: Logistic corridors · Grain flows · Tapajós river · Madeira river · Northern Arch logistic corridors

1 Brazil Soybean Production

Brazil is the world's largest producer of soybean followed by the United States. These two countries accounting for roughly 68.7% of world production. The United States Department of Agriculture estimates a world production of 363.3 million tons of soybean in the 2020/2021 harvest, which represents an increase of 6.9% against the previous period. According to the US Department of Agriculture, Brazil is expected to produce 137 million tons of soybean in the 2020/2021 harvest, representing a share of 37.7% of the world production (Table 1).

© The Author(s) 2023
Y. Li et al. (Eds.): PIANC 2022, LNCE 264, pp. 1379–1387, 2023.
https://doi.org/10.1007/978-981-19-6138-0_122

Table 1. World soybean production – select countries – million metric tons

Countries	2019/20	2020/21*	Change	Participation
Brazil	128.5	137.0	6.6%	37.7%
United States	96.7	112.6	16.4%	31.0%
Argentina	48.8	46.0	−5.7%	12.7%
China	18.1	19.6	8.3%	5.4%
India	9.3	10.5	12.4%	2.9%
Paraguay	10.1	9.9	−2.0%	2.7%
Canada	6.2	6.4	3.4%	1.8%
Russia	4.4	4.3	−1.1%	1.2%
Ukraine	4.5	3.0	−33.3%	0.8%
Others	13.2	14.1	6.4%	3.9%
World Production	**339.7**	**363.3**	**6.9%**	**100.0%**

* Preliminary estimate – September 2021
Source: United States Department of Agriculture (Sept. 2021)

The largest soybean producing region in Brazil is the Midwest, followed by the Southern region. Between 2010 and 2020, the exported volume of soybean rose from 29.1 million tons to 83 million tons, respectively, representing a raise of 185% (Fig. 1).

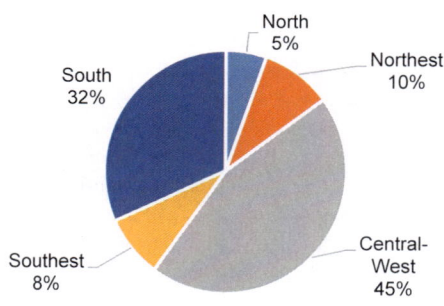

Source: CONAB (2021)

Fig. 1. Participation in the soybean production by country region

2 Main Logistic Corridors for Grain Exports

The expansion of grain production and exports was also accompanied by a significant increase in Brazilian ports operation. In 2010, 42.5 million tons of grain were handled in Brazilian ports. In 2015, the volume handled almost doubled to 84 million tons and in 2020, the volume handled jumped to 113.5 million tons, representing an increase of 35% in relation to the movement of 2015 and of 167.1% in relation to the 2010 movement.

This growth of grain movement in Brazilian ports occurred along with the raise in grain movement in ports located in the North Region, called Northern Arch and composed by ports in the states of Rondônia, Amazonas, Pará, Amapá and Maranhão (Fig. 2).

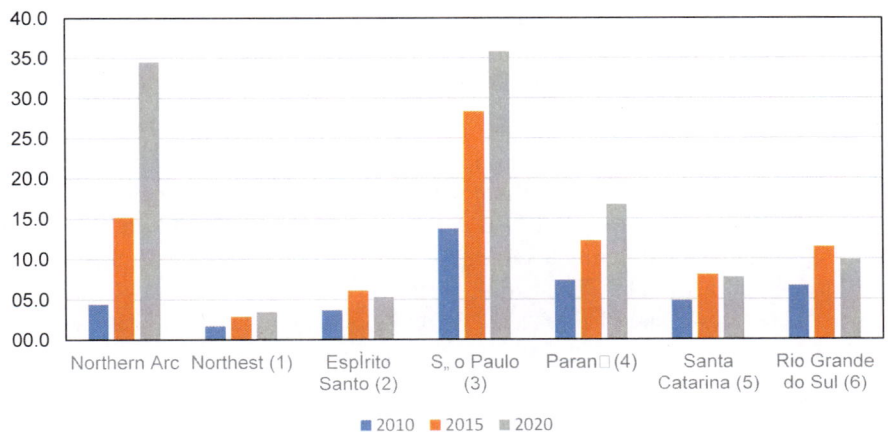

Source: Brazilian National Waterway Transportation Agency

Fig. 2. Grain exports movement in brazilian ports – million metric tons

Between 2017 and 2018, the Brazilian Ministry of Transports, Ports and Civil Aviation, released a series of reports on the project "Strategic Logistics Corridors", which aimed to present a diagnosis and a panoramic view of the transport infrastructure focused on the flow of goods and services of the country's main cargoes, namely: soy and corn; iron ore; auto-vehicles; sugar and ethanol and fuels.

Regarding grain flow, the Brazilian Ministry of Transports considered "Strategic Logistics Corridors" as being the complex system of modal and intermodal routes through which cargo from soybean and corn complex converge (MINISTÉRIO DOS TRANSPORTES, PORTOS E AVIAÇÃO CIVIL, 2017).

The mentioned report identified and detailed the main routes for the flow of grain production and analyzed the options for routes that connect the production hub (origin) to the export hubs (destiny). According to the methodology adopted in the report, the point of origin of each route was defined as being a city close to a federal highway, chosen to represent the producing region close to this city. As for the destination, in the case of export, the city where the destination port complex is located was considered.

As a resulted it was identified nine logistic corridors used for the transport of soybeans and corn correspond to approximately 37 thousand km of transport routes, divided between the modes: road (23.6 thousand km), rail (9.2 thousand km) and waterway (4 thousand km), with a predominance of highways.

Considering export flows, nine logistical corridors were defined, namely:

- Northern Logistics Corridor – Madeira Axis;
- Northern Logistics Corridor – Tapajós Axis;

- Northern Logistics Corridor – Tocantins Axis;
- Northeast Logistics Corridor – São Luís Axis;
- Northeast Logistics Corridor – Salvador Axis;
- Southeast Logistics Corridor – Vitória Axis;
- Southeast Logistics Corridor – Santos Axis;
- Southern Logistics Corridor – Paranaguá Axis;
- Southern Logistics Corridor – Rio Grande Axis.

The North Corridor includes the Madeira Axis, the Tapajós Axis and the Tocantins Axis. Besides the highways, this corridor encompasses the following waterways: Madeira river, Amazonas river and Tapajós river; the following port complexes: Itacoatiara (AM), Santarém (PA), Belém/Barcarena (PA), Santana (PA) and São Luis (MA); and transshipment terminals in the city of Porto Velho and the city of Itaituba/district of Miritituba. The North Corridor also counts with two railways: "Ferrovia Norte-Sul – Tramo Norte" and "Ferrovia Estrada de Ferro Carajás". A more detailed description of this corridor will be done in the next section.

The Northeast Corridor includes de São Luis Axis and Salvador Axis. It has only highways and contains the port complexes of São Luis and Salvador. It is important to note that the São Luis port complex is part of both the North and Northeast corridors, as it receives cargo from routes that come from both the Tocantins Axis and the São Luis Axis (Fig. 3).

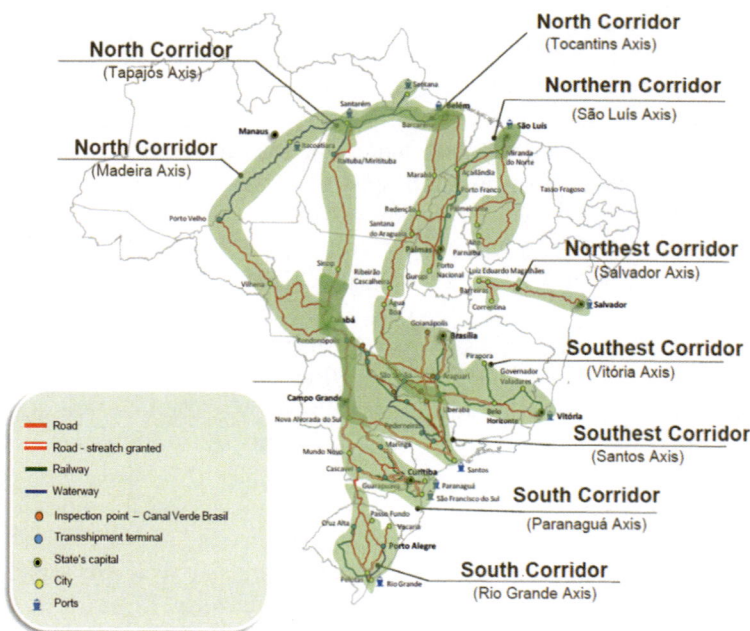

Source: Ministério dos Transportes, Portos e Aviação Civil (2017)

Fig. 3. Logistics corridors for grain flow

The three modes of transports (road, railway and waterway) are used in the Southeast Corridor which includes Vitoria Axis and Santos Axis. Besides de highways, the Southeast Corridor encompass the following railways: "Ferrovia América Latina Malha Norte (ALLMN)", "Ferrovia América Latina Paulista (ALLMP)"; "MRS Logística", "Ferrovia Centro Atlântica (FCA)" e "Ferrovia Estrada de Ferro Vitória Minas (EFVM)". Further, this corridor includes the main Brazilian waterway, Tietê-Paraná, and the largest Brazilian port, the Santos port complex. The Vitória's port complex is also part of the Southeast Corridor. There are also 9 transshipments terminals.

The South Corridor includes de Paranaguá Axis and the Rio Grande Axis. This corridor has three modes of transport, with two sections of railways – "Ferrovia America Ltina Malha Sul (ALLMS)" and "Estrada de Ferro Paraná-Oeste (EFPO)" - and one waterway – "Hidrovia do Sul". Furthermore, three ports complexes are part of this corridor: "Paranaguá", "São Francisco" and "Rio Grande". There are also 5 transshipments terminals (Table 2).

Table 2. Number of routes by modes of transports at the exports corridors

	Road	Rail	Road-rail	Road-water	Road-rail-water	Total
Corridor North	5	–	3	7	–	15
Corridor Northeast	4	–	–	–	–	4
Corridor Southeast	3	4	4	–	2	13
Corridor South	5	1	2	1	–	9

Source: Ministério dos Transportes, Portos e Aviação Civil (2017)

The aforementioned report identified 41 different routes. All corridors have roads as mode of transports and the Northeast Corridor is the only one that has only roads. Although the North and South Corridors have the three modes of transport (road, rail and water), only the Southeast Corridor has integration between all of them for the transport of grain.

3 Northern Arch

We consider ports of Northern Arch those located in the state of Rondônia, Amazonas, Pará, Amapá and Maranhão. In 2010, 4.4 million tons were exported through the ports of the Northern Arch. In 2015, this volume increased by 242% and reached 15.4 million tons of grains. In 2020, 34.4 million tons of grains were handled in that corridor to be sent abroad.

Considering the figures above, it calls attention the development occurred in Tapajós River. Moreover, the Northern Arch encompasses the North Logistics Corridor and part of the Northeast Logistics Corridor, that is, the Madeira Axis, Tapajós Axis, Tocantins Axis and São Luis Axis.

More specifically on the cargo route through the Tapajós Logistics Corridor, the grain harvest is transported from the north region of the state of Mato Grosso by trucks along the BR-163 highway and unloaded at transshipment terminals located in the city of Itaituba. Afterwards, the cargo is transported by barges across Rio Tapajós with destination to Santarém or to Barcarena/Vila do Conde (Fig. 4).

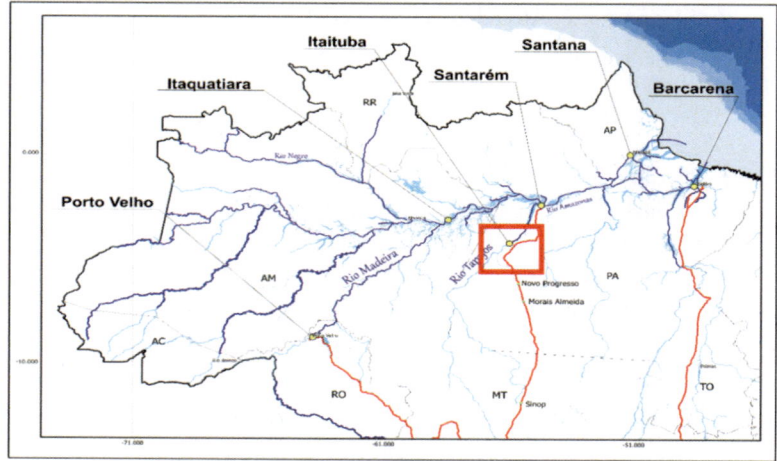

Source: Brazilian National Waterway Transportation Agency

Fig. 4. Main waterways in Amazon region

In 2021, with the aim of capturing the perception of large agribusiness companies that operate mainly in the export of grains, about multimodal transport operations, the Brazilian National Waterway Transportation Agency held meetings with some of the largest companies in this sector, namely: Archer Daniels Midland Company (ADM), Amaggi Group and Louis Dreyfus Company (LDC). In addition to these companies, a meeting was also held with Hidrovias do Brasil, in order to learn more about the operationalization of the flow of the grain harvest through Northern Arch.

The grain harvest arrives at Brazilian ports through different logistic corridors, depending on the region of production. Apart from the location where the soybean or corn is produced, the domestic freight cost is also important for the choice of a port, in addition to the shorter distance. In this sense, Brazil's competitiveness in the international grain market depends directly on the costs related to the production and to the transportation.

The agribusiness companies confirmed that, regarding the logistics corridor and the port used for grain exports, in general, the choice depends on the freight market conditions. The companies also highlighted that they consider as being very important to have different modal options and logistic corridors available for the flow of grains to the ports, as it increases competition between modes and also among logistics corridors, contributing to the freight costs reduction.

As large agribusiness companies have a verticalized market structure, it means that they are holders of assets along the logistics chain. They were the largest investors in construction of port terminals in the city of Itaituba and Miritituba district.

Since 2014, the National Waterway Transportation Agency (ANTAQ) authorized 6 grain terminals to operate in the city of Itaituba (Fig. 5).

Source: Movimento Pró-Logística

Fig. 5. Transshipment cargo station – Itaituba/Miritituba (Pará)

The construction of Ferrogão railway has the potential to consolidate the Northern Arc as the main logistic alternative to the grain flows from the Midwest of Brazil.

4   Ferrogrão Railway Project

The Ferrogrão railway is a project that intends to connect Sinop in Mato Grosso up to Itaituba in Pará. It has an extension of 933 km and will be constructed in parallel to BR-163 road. It will allow an integration between rail and water and it is expected to consolidate the Northern Arch as the main logistic corridor for grain exports.

The Ferrogrão would allow the flow of grain production from the Midwest region of Brazil (municipality of Sinop) to the city of Itaituba, Miritituba district, where the cargo would be loaded on barges that would go on Tapajós river to the Santarém port or Barcarena port where cargo would be transshipped to long-haul ships.

The paving of the road BR-163, which connects the city of Sinop in the state of Mato Grosso to Miritituba in the state of Pará, gives a good idea of the potential that a logistic connection can have on the freight costs. The freight costs between Sinop and Miritituba fell by 26% after the conclusion of the 51 km paving that remained until Miritituba, in 2019. It is expected that with Ferrogrão railway, the freight costs fell between 30% to 40% (Fig. 6).

According to the Brazilian Nacional Land Transport Agency (ANTT) the Ferrogrão railway should take 10 years to operate and it will be granted for the private sector for 69 years. The Brazilian National Agriculture Confederation estimates that 20 million tons of grains could be transported by Ferrogrão railway, changing the logistic axis from the Southeast to the North region of Brazil and in the end period of the concession the transported volume could be around 48 million tons.

Nevertheless, there are some important steps to follow before Ferrogrão railway could be conceded the private sector and it is still under analysis of the Brazilian Federal Court of Accounts (TCU). Among the main obstacles, the Ferrogrão railway crosses environmental protection area in the Amazon region and also indigenous land.

Source: Brazilian Nacional Land Transport Agency (ANTT)

Fig. 6. Transshipment cargo station – Itaituba/Miritituba (Pará)

5   Conclusions

Brazil as the largest world producer of soybeans counts with important and diversified logistic corridors to bring the grain production from the Midwest of Brazil to the coast, to be exported. An important factor of competitiveness in the international markets of grains is the freight costs, which demands a low internal freight cost. As a result an efficient logistic transportation which combines different modes of transports is essential to make Brazilian grain production competitive in the international markets.

Recently, the logistics corridors of the Northern Arch gained importance, specially the Tapajós Axis, with the expansion of port terminals in the region of the Itaituba/Miritituba. Now, the ports of the Northern Arch disputes importance with port of Santos related to the grain exports. The Ferrogrão railway is expected to consolidate the importance of the ports of the Northern Arch, but it could take some time to be concluded and prior to it, this railway has to deal with important questions related to environmental protection area and indigenous land.

Acknowledgements. We would like to express our special thanks of gratitude to the Board of Directors of the Brazilian National Waterway Transportation Agency, to our colleagues of the Superintendency of Performance, Development and Sustainability and to our colleagues of the Development and Studies Management.

We also would like to express our gratitude to the representatives of Archer Daniels Midland Company (ADM), Amaggi Group and Louis Dreyfus Company (LDC).

References

ANTT – Agência Nacional de Transportes Terrestres. Plano de Outorga para Concessão da EF-170 Ferrogrão: Trecho ferroviário compreendido entre os municípios de Sinop/MT e Itaituba/PA. Brasília: ANTT, maio de (2019)

CONAB – Companhia Nacional de Abastecimento. Acompanhamento da safra brasileira de grãos, Brasília, DF, vol. 8, safra 21 2020, no. 12 décimo segundo levantamento, setembro. (2021) https://www.conab.gov.br/info-agro/safras/graos/boletim-da-safra-de-graos. Accessed 14 April 2022

Filassi M, Oliveira ALR, Makiya IK (2017) Logística de exportação da soja brasileira: uma avaliação do corridor intermodal Centro-Norte. Revista Espacios, Venezuela, **38**(7).https://www.revistaespacios.com/a17v38n07/17380721.html. Accessed 14 April 2022

Subtil M (2021) Ferrogrão: entenda sobre o projeto de ferrovia que promete impulsionar o escoamento de grãos pelo Norte, mas enfrenta impasse legal. G1, Amazônia, 11 July 2021. https://g1.globo.com/natureza/amazonia/noticia/2021/07/11/ferrograo-entenda-sobre-o-projeto-de-ferrovia-que-promete-impulsionar-o-escoamento-de-graos-pelo-norte-mas-enfrenta-impasse-legal.ghtml. Accessed 14 April 2022

Ministério dos Transportes, Portos e Aviação Civil. Corredores logísticos estratégicos. vol. I – Complexo de soja e milho. Brasília: MTPA (2017)

Rodrigues J (2018) O Arco Norte e as Políticas Públicas Portuárias para o Oeste do Estado do Pará (Itaituba e Rurópolis): Apresentação, Debate e Articulações. Revista Nera, Presidente Prudente, ano 21(42), 202–228. https://doi.org/10.47946/rnera.v0i42.5693

United States Department of Agriculture. World Agriculture Production. [Washington]: USDA (2021). https://downloads.usda.library.cornell.edu/usda-esmis/files/5q47rn72z/0z709w03d/bk129951s/production.pdf. Accessed 14 April 2022

Handling of Inland Vessels in Seaports – Necessary Actions and Additional Options to Support Container Transport on Inland Waterways

Laure Roux[1](✉), Iven Krämer[2], and Rob Konings[3]

[1] Central Commission for the Navigation of the Rhine, Strasbourg, France
L.Roux@ccr-zkr.org
[2] Ministry of Science and Ports, Free Hanseatic City of Bremen, Bremen, Germany
iven.kraemer@swh.bremen.de
[3] Rijkswaterstaat, Den Bosch, The Netherlands
rob.konings@rws.nl

Abstract. Deep sea container transport is despite the existence of various crisis situations a worldwide steady growing business and as a result seaports need to handle ever growing container volumes. Transporting these containers with as little as possible emissions, efficiently and in a reliable manner from the seaport into the hinterland has strategic importance for the attractiveness and hence the competitiveness and growth potential of a container seaport. For ports having access to the hinterland by inland waterways, inland shipping is a mode to contribute to hinterland accessibility and to provide a more environmental-friendly and safer solutions for hinterland transport compared to truck transport, which is still very dominant.

The physical as well as the informational interfaces between the seaport and the hinterland modalities (truck, train and inland vessel) appear to have significant importance for the overall cost and quality performance of the hinterland transport chain, in particular on relatively short transport distance. Several container seaports world-wide have a well-developed hinterland transport system for container barges, but are facing inefficient handling processes of inland vessels in the seaport which cause tempered growth of container inland shipping.

This paper discusses different causes for inefficient handling of container inland vessels in seaports and evaluates initiatives, proposals and necessary actions for improvement. These matters are discussed in the context of the major Northwest-European container seaports. Solutions can be found both in organisational, information-based and technical improvements. These improvements are jointly needed to realise the growth ambitions for container inland shipping.

Keywords: Container transport · Seaports · Inland shipping

© The Author(s) 2023
Y. Li et al. (Eds.): PIANC 2022, LNCE 264, pp. 1388–1397, 2023.
https://doi.org/10.1007/978-981-19-6138-0_123

1   Introduction

The quality of hinterland transport has become increasingly important for the competitiveness of a seaport. Shippers and carriers value the attractiveness of a port not only on the performance of the seaport, but also on its hinterland accessibility and thus consider a ports competitive position not any more as a single important argument but include the overall port activities into a combined supply chain performance approach. In fact, large seaports are important transshipment places, not only for maritime vessels, but also for inland vessels, as they represent the interface between maritime trade and hinterland transport. This hold for the container transport market in particular. Containerisation has changed liner shipping spectacularly and affected seaports and their hinterland transport systems. Ports are much more in competition to serve the same hinterland areas. Especially in Northwest-Europe, where the distance of container ports to major hinterland areas is not very different. This has made hinterland accessibility a strategic matter for the competitiveness of seaports.

In the largest container seaports in Europe, inland container shipping has developed as a major mode for hinterland transport. In Rotterdam 3 million TEU were transported by barge in 2020 with a market share of 32% in hinterland transport. Antwerp recorded 2,5 million TEU with a market share of 35%. The German ports of Hamburg and Bremerhaven recorded significantly lower shares between 2% and 3%, as their focus is due to their geographical position and traditional development put on rail transport. In view of increasing global container traffic, accessibility and achieving their sustainability goals, most seaports are targeting larger volumes to be handled by barge and aim at increasing its modal share (e.g. Rotterdam targets a market share of 45% at Maasvlakte area in 2033). This is also driven by the ambitious national and European modal shift to inland waterways targets with dedicated organisations and specific strategic targets in almost all European countries. Over the last 15 years, however, while the modal split share of barge in the main seaports of Rotterdam, Antwerp and Hamburg has increased slightly (Table 1), efforts are still needed to achieve their modal split targets.

Table 1. Development of modal split in container hinterland traffic in port of Rotterdam, Antwerp and Hamburg, 2005–2020 (x 1.000 TEU and %)

	Rotterdam		Antwerp		Hamburg	
	2005	2020	2005	2020	2005	2020
Truck	4.056	5.296	3.897	4.214	2.991	2.490
Barge	2.056	3.026	2.312	2.491	102	128
Rail	634	978	540	484	1.420	2.808
Total	6.746	9.300	6.749	7.189	4.513	5.426
Truck	60%	57%	58%	58%	66%	46%
Barge	31%	33%	34%	35%	2%	2%
Rail	9%	10%	8%	7%	32%	52%
Total	100%	100%	100%	100%	100%	100%

Source: port authorities.

Studies (e.g. Shobayo and Van Hassel, 2019) show that the handling of barges in the seaport has a great impact on the performance and reliability of barge services to the hinterland (dependent on transport distance easily 30% of the total hinterland chain costs) and that the current handling processes in the seaport suffer from different problems. Improvement of the handling operations in the seaports is therefore an important avenue to increase competitiveness and support growth of container inland shipping. In fact, in order to exploit the potential of container transport even further, a constant improvement of inland navigation's integration into logistics chains is a cornerstone, in particular to adapt to changes affecting global trade flows. In addition, recent events such as the Covid-19 crisis, the Suez Canal incident (ship blockage) as well as the Brexit, led to increased congestion issues at the level of seaports, highlighting further their vulnerability and the need to find solutions to address this challenge. In the next section the details of the inefficient handling process are unravelled. Next different possible solutions to improve the process are discussed. The paper ends with conclusions and future perspectives.

2   Current Processes and Problems with Handling Container Barges in Seaports

Inefficient handling of container barges in seaports is not a new problem and seem to remain a persistent problem due to multiple mutually related factors:

– Barges very often have to call many (easily 6 to 9) deep-sea terminals in the seaport: this is time consuming and leads to many small call sizes that also lowers the overall terminal productivity.
– Many barges call at the same terminal: it causes congestion and waiting times.
– Sea vessels in general have priority over barges since there is no contractual relation between terminal operators and barge operators: this causes additional congestion and waiting times, especially when the handling equipment for deep-sea vessels is used to handle barges.
– The technology used for the handling of deep-sea vessels becomes in combination with the continuous rise in ship sizes less competitive if this technology is also used for barge handling.
– A delay at one terminal can easily lead to missing agreed time windows for handling at the following terminals, i.e. causing a domino-effect. For instance, in Rotterdam in 2019 and in Antwerp in 2020, delays amounted up to 20–30 h on average, but peaks of 60 h were sometimes registered.
– Lack of coordination in the planning of handling barges since the planning takes place bilaterally between the terminal operator and barge operator: both actors may be responsible for disturbing the planning.
    The implication of these process characteristics is that barge services become more expensive (because the turnaround time of barges in the seaport is too long, that is to say making the barges less productive), less reliable (because of the unpredictable delays) and may increase the transit time of individual containers. Ultimately, this can lead to loss of competitiveness for barge compared to other modes.
    The negative impacts on the barge service performance are further enhanced by the following circumstances and developments (CCNR, 2022):

- Growth in container traffic in general causing more pressure on the handling capacity of seaports, although ports may have expansion plans.
- Increasing scale of deep-sea vessel and call size that create greater peaks in demand for handling capacity, and consequently may lead to longer waiting times for barges. The capacity of the largest deep-sea vessels calling at the north western European Ports exceeds 24.000 TEU. The average call size on such vessel i.e. has continuously increased to 8.300 TEU in Rotterdam.
- Lack of dedicated handling capacity for barges at the individual seaport terminals. Moreover, using expensive cranes to handle deep-sea vessels also to handle barges makes the handling of barges not only unnecessary more expensive, but also slower.
- Unreliable deep-sea container shipping schedules that make the planning of handling barges less reliable.
- The strong commercial position of maritime actors compared to inland operators, strengthened further by a wave of market consolidation also took place in the last decades in the global container shipping industry, creating further imbalances in the global container trade.
- Increasing tightening of demurrage and detention conditions by shipping companies which puts more pressure on the timeframe in which containers can be supplied or disposed of free of charge around a deep-sea call and further increases the peak load.

To conclude, to overcome the inefficiencies of the current handling process of barges in the seaport solutions should contribute to a more reliable and faster process.

3   Measures and Best Practices to Improve the Handling of Container Barges in Seaports

As shown, the inefficiencies in barge handling have different dimensions: organisational, technological and information-based. This observation is directional for promising solutions. However, politics and regulation can also play a role.

3.1   Organisational Developments

The most common pattern of container barge services consists of one or a very few terminal visits in the hinterland and a rather large number of terminal visits in the seaport. The need to visit many seaport terminals arises from the fact that a shipment of containers of one inland terminal consists of shipments of several deep-sea lines calling at different seaport terminals. Together these shipments can fill a barge. A different way of bundling container flows can lead to a reduction in the number of terminal visits in seaports and an increase in call size. Both results are beneficial for the barge operator and deep-sea terminal operator. Alternative bundling of flows can take place in the seaport and in the hinterland.

Hub Bundling in the Seaport
Hub bundling is a very effective approach to realise benefits from network operations (economies of scale, density and scope). The decoupling of barge services on the seaport side and hinterland side enables several advantages:

Performance improvement of barge hinterland operations:
- Improvement of the hinterland vessel turnaround time (higher productivity);
- Improvement of the costs and reliability performance of barge hinterland services.

More efficient performance of deep-sea terminals:
- A higher crane/quay productivity, because the average productivity of handling containers in large call sizes is higher;
- A better utilisation of space at deep-sea terminals as the dwell time of containers at deep-sea terminals can be reduced if the hub terminal can also facilitate a storage function for (empty) containers.

Moreover, port accessibility can be improved if the hub terminal can act as an 'extended gate' for container trucks that drop and pick up their containers at the hub instead visiting the deep-sea terminals. In this way the congestion on roads in the port region will be reduced.

The economic advantages of hub bundling, however, should be large enough to compensate a main disadvantage, i.e. the costs of the extra transhipment operation in the hub.

In the port of Antwerp, the concept of bundling was first introduced through the implementation of a five-year project in 2018 that was submitted to the European Commission. The successful results of this test led to the full application of the bundling concept in the port and since the beginning of January 2021, barge operators WeBarge and Contargo Transbox have been operating container shuttle operations together in the port of Antwerp. They are now bundling their volumes at consolidation hubs which organise the transport to and from the maritime terminals at a fixed transport rate per hub.

Since 2018, the Port of Rotterdam also implemented this concept of container bundling. In the port, containers are bundled at Maasvlakte, Waal-Eemhaven and Alblasserdam and transported directly by inland vessels to and from the deep-sea terminals according to a fixed schedule.

These initiatives in both ports have resulted notably in call sizes that are two to four times bigger than previously and in shorter port calls (approximately - 40%).

Hub bundling is so far mainly noticeable in these seaport regions, but could also be applied at greater distance from the seaport. The inland port of Duisburg (at 250 km distance from Rotterdam) can also perform the hub function very well (Fig. 1).

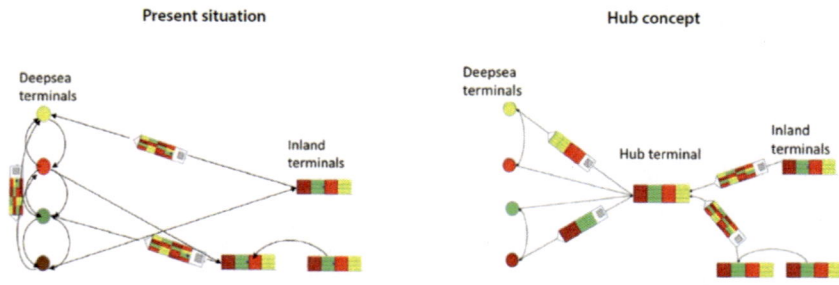

Fig. 1. The hub bundling concept (source: CE Delft, 2020)

Corridor Bundling in the Hinterland

Bundling taking place in the hinterland with services to the port of Rotterdam are organized in the form of corridor partnerships, that is to say, it is based according the bundling principle of line bundling along a corridor.

Various barge operators and inland terminals located along a hinterland route have decided to work together to bundle the flow of containers destined for specific deep-sea terminals. As a result, large volumes of containers can be moved between the different deep-sea and inland terminals using inland vessels that sail according to a fixed schedule. In addition, the barge operators and the deep-sea terminals have agreed to load and unload containers during specific time slots, and to put through a minimum number of containers (moves) per inland vessel. These agreements have made the handling of inland container shipping flows in the port of Rotterdam both more reliable and more efficient.

The first corridor partnerships set up in 2018 were the West Brabant Corridor (WBC) and the Ruhr Express and they have been successful. Some results from the WBC: the call size tripled, 30% shorter duration times of vessels in the seaport, 30% reduction in number of barge visits at deep-sea terminals, 75% fewer deviations in visit appointments and a modal shift from truck to barge of 20%. Soon the North West central corridor and the Limburg Express were also established.

In Belgium, the government is going to launch a subsidy program to support the development of corridor bundling (Logistiek Magazine, 2022) (Fig. 2).

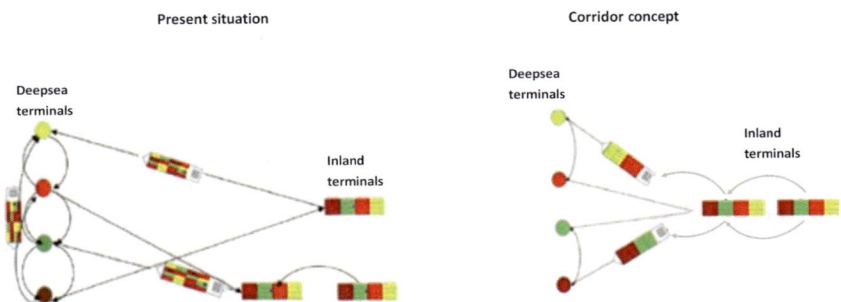

Fig. 2. The corridor bundling concept (source: adapted from CE Delft, 2020)

Push Barge 'LEgo' System

This proposed concept, having similarities with the hub bundling concept but without applications so far, is aimed at optimisation of the utilisation rate of the deep-sea terminal quays, by decoupling the loading and unloading process at the terminal. The system is based on the use of unmanned push barges that are buffered in the port area and called for (un-)loading whenever capacity is available (Fig. 3).

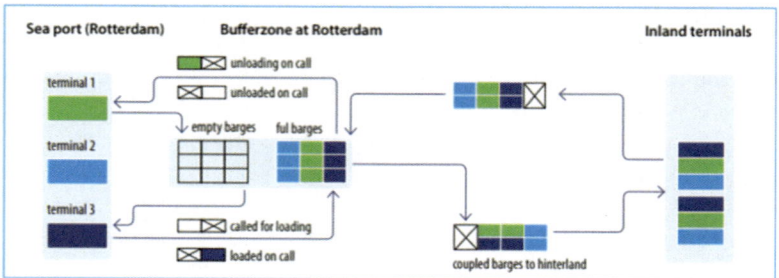

Fig. 3. The push barge 'Lego' system (source: CE Delft, 2020)

3.2   Cooperation and Exchange of Information

In the current planning process of handling barges, planners make a plan for their own vessels and quays and hence only know their own plan and available capacity. Therefore, it is often impossible to achieve a match between demand and supply. However, for the short-term, it is clear that cooperation and exchange of information between the different actors in the transport chains and port operations can generate "quick-wins". Many examples can be found in main seaports regarding how to improve information sharing. However, a key challenge is to make information systems compatible with each other so that they can be used by all those involved in the process. A prerequisite for this is also the willingness of all process participants to share their information with each other.

At the port of Rotterdam, Nextlogic is a tool and information platform that offers an integral, port-wide and neutral planning of the handling of container barges in the port of Rotterdam. The planning is feasible and appropriate for all parties based on the delivered information, pre-defined KPIs and calculation rules. The system continuously optimises taking into account real-time changes. Sharing correct and complete information in time by all relevant actors is a key requirement for the well-functioning of the planning system.

Following a pilot five barge operators and three terminals were using Nextlogic at the end of 2021. More participants are envisaged and also required to make the system functioning truly well (Fig. 4).

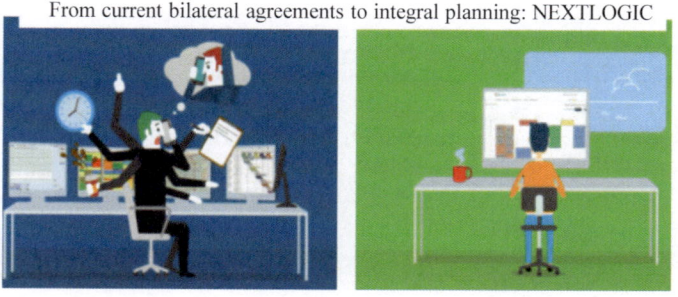

Fig. 4. The NEXTLOGIC planning system (source: Nextlogic)

A similar system also exists in the Port of Antwerp, where the platforms "C-POINT" and NextPort offer a complete package of applications to promote digital communication between all actors present in and around the port.

3.3   Technology: Dedicated Infrastructure and Innovative Transport and Transhipment Technologies

Investments in dedicated barge handling infrastructure in the seaports is without doubt an effective solution. Small cranes could be used to save handling costs of the barges and dedicated handling space for inland barges can reduce the waiting time of the barges. Nevertheless, outstanding issues remain, notably as to which actor would bear the responsibility of these investments and what the cost-benefit analysis of these investments would be.

Except for just creating more handling capacity it remains interesting to consider new vessel, terminal and transhipment technologies that enable a cost-efficient and fast (un-)loading process of vessels, benefitting from the possibilities of further increasing the scale of vessels and automation and robotisation.

Over the last three decades, and mainly in the nineties of last century, a lot of concepts with a technological component were launched and developed (see Binsbergen et al, 2009). Although studies showed their technical feasibility it appeared difficult to bring these concepts to market realisation.

Inspired by the terminal automation of ECT in Rotterdam the Barge Express-concept (1996) contained deployment of very large push-barge units (624 and 2x280 TEU capacity) to reduce sailing costs and automated loading/unloading to reduce handling costs. Quay transport could be performed by the also well-known AGVs. Except for the automated crane the concept was based on proven technology and hence the investment costs and risks were rather limited. Although technically feasible it was economically not, because the transport volumes were insufficiently large at that time. It would be worthwhile to reconsider this and other technologically advanced concepts again in the current spirit of the age (Fig. 5).

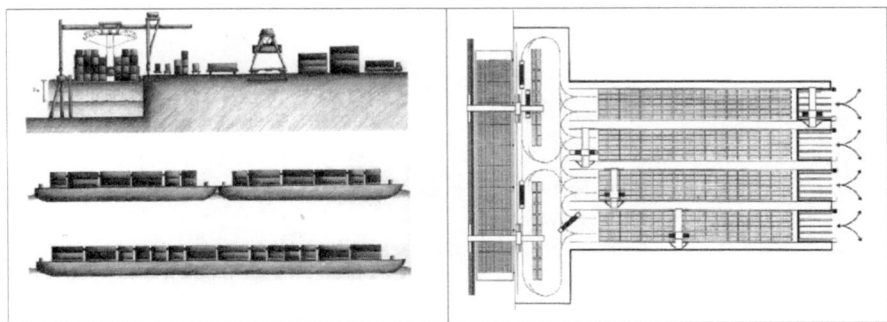

Fig. 5. Barge Express: perspective for large-scale container barging (source: TRAIL Onderzoekschool, 1996)

4   Conclusions and Perspectives

In view of increasing global container traffic, port-hinterland accessibility and sustainability goals main container seaports in Western Europe are targeting larger volumes to be handled by inland shipping. Although the volumes transported by barge have generally significantly increased during the last two decades it has remained a big challenge to increase the market share of barge transport in hinterland container traffic. Hinterland container barge transport is suffering from inefficiencies in the handling of inland vessels in the seaport. Considering the great impact of this handling process in the overall performance of barge services to the hinterland solving the inefficiencies in container barge handling in the seaport is an important avenue to improve the attractiveness and create growth potential for barge transport in hinterland container traffic.

After a long struggle in solving this problem, some concrete promising initiatives have now been implemented or are envisaged. Important actions that have been undertaken to improve the handling of barges are the reorganisation of barge hinterland services and the improvement of the information process regarding the handling of barges in the seaport. The first experiences seem positive. Although these actions tackle important dimensions of the inefficiency problem more actions are needed. Historically, barge has never been a standalone transport mode as it has always had to use the infrastructure of deep-sea vessels. In view of the huge ambitions in terms of the modal share of barge transport the role of the inland shipping mode at deep-sea terminals should be revalued. This revalue requires more than the current organisational and information-based improvements: for instance (more) own facilities (quays and cranes), but also use of innovative technologies (automation and robotisation), that are already being rolled out to handle deep-sea vessels, need also become available to handle barges.

Overall, inland container handling inefficiencies in seaports are considered to be strongly linked to commercial issues which might not necessarily be solved by additional regulatory measures. Nevertheless, national public authorities can have a non-negligible capacity to influence this issue. Indeed, the fact that the problem of handling inefficiencies in seaports is a long lasting issue; seem to demonstrate that such inefficiencies cannot be solved by the market alone. This would ultimately lead to difficulties in achieving the ambitious European and national modal shift to inland waterways targets. This justifies possible interventions on the side of national public authorities.

References

Binsbergen AJ van, MGM van der Horst, A Veenstra and R Konings (2009) containerlogistiek in de binnenvaart, delft: trail onderzoeksschool

CE Delft (2020) outlook hinterland and continental freight 2020, delft

CCNR (2022) assurer un chargement/déchargement plus rapide et efficace des conteneurs en navigation intérieure dans les ports maritimes - analyse du Secrétariat, Strasbourg (forthcoming)

Logistiek Magazine (2022) Vlaamse regering: 'modal shift naar binnenvaart stimuleren' [Flemish government: encouraging modal shift to inland shipping] https://www.logistiek.nl/182630/vlaamse-regering-modal-shift-naar-binnenvaart-stimuleren

Shobayo P, E van Hassel (2019). container barge congestion and handling in large seaports: a theoretical agent-based modelling approach. J Shipping Trade 4 26

TRAIL onderzoekschool, business development international, Arthur D. little international, vuyk engineering rotterdam (1996) barge express: perspectief voor grootschalige binnenvaart, Delft

Regulation for the Decarbonisation of IWT in Europe

Gernot Pauli[1(✉)] and Benjamin Boyer[2]

[1] Berlin, Germany
gernot.pauli@outlook.de
[2] Central Commission for the Navigation of the Rhine, Strasbourg, France
b.boyer@ccr-zkr.org

Abstract. To ensure its future, the response of IWT to climate change must be twofold: it must adapt to the changing climate and mitigate its carbon footprint. The former will make IWT resilient against adverse effects of climate change, the latter will bring IWT in line with the Paris Agreement and decarbonise IWT. PIANC, in its 2019 Declaration on Climate Change, stated that "… PIANC and its members will strive to develop approaches to decarbonise the operation of port and navigation infrastructure (i.e. move to net zero emissions), whilst at the same time enabling the reduction of greenhouse gas (GHG) emissions from vessels by providing the necessary facilities, infrastructure and, where appropriate, incentives…" The European Union (EU) and the Central Commission for the Navigation of the Rhine (CCNR) more explicitly aim at zero-emission vessels and eliminating GHG emissions from inland navigation vessels by 2050. For doing so, EU and CCNR will go beyond the measures foreseen in PIANC's declaration and will employ regulations and standards as well, since regulations and standards are important policy instruments to facilitate the transition towards carbon neutral IWT. This paper will present respective regulations and standards which are already in place, currently under development or whose development is foreseen to effectively support the transition towards a zero-emission IWT fleet in Europe. Vessel technical requirements are at the core of the contribution, but requirements concerning vessel operation and crew training are also considered. The paper analysis the basic content of the regulations and standards and provides general recommendations for the way forward.

Keywords: Decarbonisation · IWT · Regulation · CCNR · CESNI

1 Policy Context

The European Commission's Green deal for Europe (European Commission, 2019) and Smart and Sustainable Mobility Strategy (European Commission, 2020a) lay out priority actions to be realised for achieving a GHG emission reduction target of roughly 55% by 2030 compared with 1990 (for all sectors), and a GHG reduction target of 90% by 2050 (for the transport sector). In line with the above, the European Commission's NAIADES III Action plan (European Commission, 2021) includes the core objective

---

G. Pauli—Self-Employed.

© The Author(s) 2023
Y. Li et al. (Eds.): PIANC 2022, LNCE 264, pp. 1398–1408, 2023.
https://doi.org/10.1007/978-981-19-6138-0_124

of facilitating the transition to zero-emission vessels by 2050. In the Declaration signed in Mannheim in 2018 (Central Commission for the Navigation of the Rhine, 2018), the inland navigation ministers of the Member States of the CCNR, namely Belgium, Germany, France, The Netherlands and Switzerland, defined similar targets of largely eliminating GHG emissions by 2050.

Regulations and standards are important policy instruments to enable and stimulate the transition towards carbon neutral IWT. Regulations and standards provide for legal certainty, which in turn facilitates investments in new technologies. Indeed, legal uncertainties and long administrative procedures could be a bigger obstacle to decarbonising the fleet than strictly technological issues. Furthermore, regulations and standards ensure safe deployment as well as public support and confidence in the new technologies and energy carriers, which are all needed to overcome the many challenges arising with the decarbonisation of IWT.

Presently, IWT in Europe views battery electric propulsion as well as fuel cells using hydrogen or methanol as the most promising solutions for decarbonisation of the fleet. They are therefore at the centre of the current work on new regulations and standards.

2   Scope

This paper considers regulations and standards adopted by the EU and the CCNR, which can be seen as having established a shared governance for IWT in Europe. The paper takes into account this shared governance in order to provide analysis and recommendations aiming at a smooth and coordinated transition of regulatory frameworks towards decarbonized IWT in Europe.

The scope of this paper is limited to fleet related regulations: vessel construction and equipment, vessel operations including bunkering as well as crew qualification and manning. These regulations have significant effects on the total cost of ownership and the legal certainty.

There are numerous technologies which can be applied in possible pathways for the decarbonisation of inland navigation vessels. However, when developing regulations

Table 1. TRLs defined within the Horizon 2020 Programme (European Commission, 2020b)

	DEFINITION
TRL 1	basic principles observed
TRL 2	technology concept formulated
TRL 3	experimental proof of concept
TRL 4	technology validated in lab
TRL 5	technology validated in relevant environment (industrially relevant environment in the case of key enabling technologies)
TRL 6	technology demonstrated in relevant environment (industrially relevant environment in the case of key enabling technologies)
TRL 7	system prototype demonstration in operational environment
TRL 8	system complete and qualified
TRL 9	actual system proven in operational environment (competitive manufacturing in the case of key enabling technologies; or in space)

and standards for new technologies, only those technologies can be considered, which are already sufficiently developed and understood. An accepted method for expressing their development stage is the technology readiness level (TRL). Table 1 depicts the TRL as defined within the Horizon 2020 Programme.

The CCNR commissioned an assessment of technologies in view of zero-emission IWT, using these TRL definitions. Figure 1 summarizes suitable technologies and the respective TRLs.

The technologies listed in Fig. 1 reflect the current state of knowledge. The CCNR decided to focus on a set of technologies with a TRL of 5 and above. Other technologies like lithium-air batteries or LOHC (Liquid Organic Hydrogen Carrier) could be studied at later stage. Even if ammonia seems to be a serious candidate for seagoing vessels, it presents major safety issues, especially while bunkering (Heitink et al. 2021), and is therefore excluded from this analysis.

Technologies considered in the pathways	Description	TRL (1-9) vessel application	TRL (1-9) fuel / energy production and supply
Stage V, Diesel	Fossil diesel in an internal combustion engine which complies with the emission limits EU Stage V.	9	9
LNG	Liquefied Natural Gas in an internal combustion engine which complies with the emission limits EU Stage V.	9	9
Stage V, HVO	HVO in an internal combustion engine which complies with the emission limits EU Stage V. HVO stands for hydrotreated vegetable oil itself (without blending with fossil fuels) and all comparable drop-in biofuels (including e-fuels) as well as synthetic diesel made with captured CO_2 and sustainable electric power.	9	9
LBM	Liquefied Bio Methane (or bio-LNG) in an internal combustion engine which complies with the emission limits EU Stage V.	9	8
Battery	Battery electric propulsion systems, with fixed or exchangeable battery systems.	8	7
H_2, FC	Hydrogen stored in liquid or gaseous form and used in fuel cells.	7	7
H_2, ICE	Hydrogen stored in liquid or gaseous form and used in internal combustion engines.	5	7
MeOH, FC	Methanol used in fuel cells.	7	6
MeOH, ICE	Methanol used in internal combustion engines.	5	6

Fig. 1. Technologies on possible pathways for the decarbonisation of inland navigation vessels

with TRLs for application on the vessels as well as TRLs for the fuel / energy production and supply.
(Dahlke-Wallat et al. 2021).

3 Vessel Design and Propulsion System

In the legal context of the EU and the CCNR the vessel technical requirements and regulations applicable to engines or energy converters determine the legal feasibility of the use of alternative energies for the propulsion of IWT vessels.

3.1 Vessel Technical Requirements

A vessel operating on EU waterways or the Rhine must carry either a Union inland navigation certificate or a Rhine vessel inspection certificate. Both certificates are issued by the competent national authorities (inspection bodies) and confirm full compliance of the vessel with the European Standard laying down Technical Requirements for Inland Navigation vessels (ES-TRIN) (European Committee for drawing up Standards in the field of Inland Navigation, 2021a). This standard contains provisions on inland navigation vessel construction and equipment as well as special provisions for certain categories of vessels such as passenger or container vessels. The objective of these technical requirements is to guarantee a high level of safety in inland navigation, thereby also protecting the environment and the people on board. ES-TRIN is updated every two years by the European Committee for drawing up standards in the field of inland navigation (CESNI). References to ES-TRIN are included in the legal frameworks of the EU and the CCNR, Directive (EU) 2016/1629 (EU, 2016a) and Rhine Vessel Inspection Regulations (RVIR) (Central Commission for the Navigation of the Rhine, 2020) respectively.

ES-TRIN foresees that vessels use conventional diesel as fuel. Its Article 8.01(3) states that "Only internal-combustion engines burning fuels having a flashpoint of more than 55 °C may be installed." Article 8.05 includes safety requirements for diesel tanks and piping. However, in 2017 the first step was taken to recognize alternative fuels in ES-TRIN, by including general provisions for low flash point fuels (Chapter 30) and an annex dedicated to liquefied natural gas (LNG) (Annex 8). But at present, ES-TRIN does not permit the use of other low flash point fuels (such as hydrogen). Edition 2021 of ES-TRIN also regulates lithium-ion batteries (especially the design of rooms where such batteries are stored) (Article 10.11).

Gaining knowledge and experience from pilot projects facilitates the development of regulations and standards for innovative technologies. This is one reason, why the regulatory frameworks of the EU and the CCNR foresee derogations from ES-TRIN. For example, on 17 June 2021, the CCNR granted a derogation for the motor vessel MAAS which will use hydrogen as fuel and a fuel cell as energy converter (Central Commission for the Navigation of the Rhine, 2021). In 2019 CESNI published a guidance (European Committee for drawing up Standards in the field of Inland Navigation, 2019a) describing the procedure for derogations with the objective to facilitate administrative procedures and actively support the greening of the fleet.

In 2020, CESNI established a temporary working group to prepare amendments to ES-TRIN allowing the use of fuel cells but also the storage of methanol and hydrogen on board of inland navigation vessels. The composition of the group reflects the CCNR's experience that intensive stakeholder involvement is an important success factor for regulatory and standardisation work. Members of the group are therefore not only drawn from the member states' administrations, but also from relevant industries (European Committee for drawing up Standards in the field of Inland Navigation, 2021b). The group has chosen an approach based on a combination of prescriptive rules and risk analysis (like the LNG rules currently in ES-TRIN). The amendments prepared by the group foresee a revision of Chapter 30 and reorganisation of Annex 8 of ES-TRIN to distinguish between different energy converters (engine or fuel cell), but also between energy converters and fuel storage. This allows for a stepwise integration of new fuels and energy converters in the future. But most importantly, the amendments contain technical requirements for fuel cells installed in inland navigations vessels. It is expected that the amendments will be included in ES-TRIN 2023 which could enter into force in January 2024 (European Committee for drawing up Standards in the field of Inland Navigation, 2022). CESNI expects to finalise the standardisation work on methanol storage and use in combustion engines by end of 2022. The temporary working group is now deliberating requirements for the storage of hydrogen, which introduces complex technical and regulatory issues and requires therefore more time. Because almost all hydrogen pilot projects rely on swappable fuel tanks, the group will consider such arrangements in future amendments of ES-TRIN.

3.2   Regulations Applicable to Engines or Energy Converters

Since 2003, the engines of inland navigation vessels are subjected to specific requirements regarding emissions of air pollutants. Indeed, first limits for air pollutants were introduced in the RVIR in 2003 (so called CCNR stage I). These limits only applied to newly installed engines onboard inland navigation vessels. Just four years later, the CCNR and the EU introduced more stringent emission limits in the RVIR and the Directive 2004/26/EC (so called CCNR II or EU IIIa). Taking effect as of 1 January 2019, new emission limits (so called EU Stage V limits) were introduced by the "NRMM Regulation" (EU) 2016/1628 (EU, 2016b). Seeing no further need for parallel emission requirements within its own regulatory framework, the CCNR decided to align it with that of the EU and to refer to this regulation. A summary of the successive limits of air pollutants emissions is given in Fig. 2.

As pointed out in the EU funded project Prominent (Stichting Projecten Binnenvaart, 2016), most vessels of the European inland navigation fleet are equipped with engines installed before 2003. These engines are not subject to any emission limits because the emission requirements apply only to newly installed engines. The reason for the slow renewal of engines used for inland navigation is their long technical lifetime.

To reflect LNG becoming a fuel for inland navigation vessels, the NRMM Regulation includes specific provisions on total hydrocarbon (HC) limits for fully and partially gaseous-fuelled engines. The objective was notably to limit the emission of unburned methane (which is a potent GHG) from the combustion engine, the so-called methane slip. Indeed, on-board measurements suggest that marine engines using LNG show a significant methane slip, particularly at low loads (Ushakov et al. 2019), which are typical for the operation of inland navigation vessels. Thus, the reduction of pollutant emissions by using LNG may lead to higher GHG emissions in comparison with conventional diesel engines. At least, the reduction of GHG emission associated in theory with the use of LNG does not translate in IWT practice.

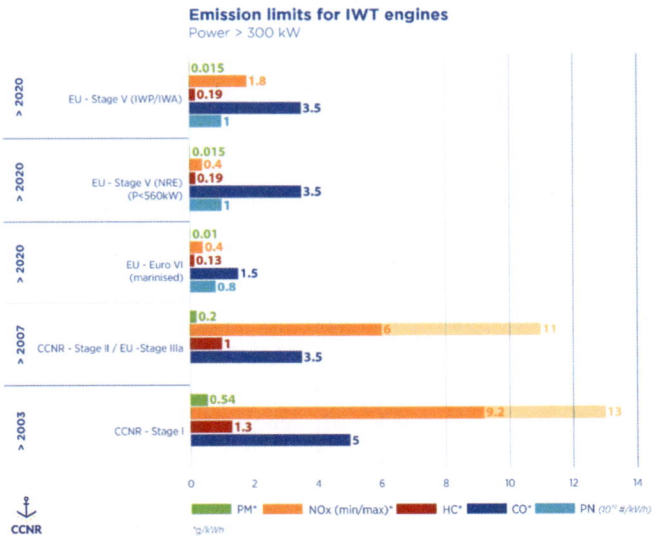

Fig. 2. Summary of mandatory limits of air pollutants emissions (source: CCNR)

4   Vessel Operation

In Europe, vessel operations are generally covered by national and international police regulations. In the following, the police regulations applicable for the Rhine and equivalent recommendations adopted by the United Nations Economic Commission for Europe (UNECE) are considered. There is no EU legal instrument in the field of police requirements.

The Police Regulations for the Navigation of the Rhine (RPR) (Central Commission for the Navigation of the Rhine, 1995) contain the core rules applicable to the traffic on the Rhine such as visual and sound signals, radiotelephony, waterway signs and markings, as well as navigational, crossing and berthing rules. It also regulates for example the electronic reporting or mandatory use of the Automatic Identification System (AIS) equipment and provides a list of paper or electronic documents required on board.

The RPR constitute the legal basis for the international navigation of the Rhine and inspire significantly national regulations of the CCNR member states.

After having temporarily authorised 15 vessels using LNG as fuel for trial purposes, the CCNR updated in 2015 its core regulations, among them the RPR, to allow the use of LNG as regular fuel for vessels sailing on the Rhine (Central Commission for the Navigation of the Rhine, 2015). In close cooperation with the inland navigation profession and technical experts who already had experience in the use of LNG, the CCNR incorporated in the RPR requirements concerning markings, passing through locks, electronic reporting, safekeeping and surveillance as well as bunkering of LNG. For the latter, a dedicated checklist was drawn up based on a list developed for maritime ports. More generally, the LNG requirements contained in the RPR were an important source of inspirations for national or international instruments.

To implement the objectives of the Mannheim Declaration, the CCNR included in its work programme 2022–2023 updating the RPR to facilitate the use of alternative fuels other than LNG, in particular hydrogen and methanol (Central Commission for the Navigation of the Rhine, 2021). Here again, the experience gained with pilot vessels will be of major importance. The main challenge will be to determine, whether additional rules are necessary at all to ensure an equivalent of safety when using these new fuels. The design of possibly necessary rules could be a lesser challenge, as the already existing rules for vessels using LNG as a fuel provide a suitable template. The best outcome for the navigation industry and also for accelerating the decarbonisation of IWT would probably be a scientifically sound reasoning, that no additional rules are needed.

The European Code for Inland Waterways (CEVNI) contains the core rules applicable to the traffic on inland waterways in the UNECE region such as marks and draught scales on vessels, visual and sound signals on vessels, waterway signs and markings, as well as navigational, crossing and berthing rules.

CEVNI is not legally binding but constitutes a technical basis to facilitate the harmonisation of the police requirements for inland navigation in Europe. Indeed, CEVNI is regularly updated to follow the evolution of the navigation regulations of River Commissions or national rules.

The sixth revised edition of CEVNI (United Nations Economic Commission for Europe, 2021) takes into account best practices from the existing up-to-date traffic regulations. In particular, this edition contains the provisions for the safety of vessels using LNG.

5   Crew Qualification and Manning

The Regulations on Navigation Personnel on the Rhine (RPN) (Central Commission for the Navigation of the Rhine, 2022) lay down the rules for personnel on board vessels sailing on the Rhine. The regulations' primary objective is safety of navigation. They cover the qualifications of personnel, particularly the skills required of boat-masters and the manning requirements of crews (i.e. number and qualification of the crew for specific vessels types). In 2015, the CCNR added provisions concerning the expertise of crew members of inland navigation vessels fuelled by LNG (Chapter 4a) (Central Commission for the Navigation of the Rhine, 2015). It stipulates that all crew members involved in the bunkering procedure shall have sufficient expertise. It lays down also the content of training courses and examinations.

The respective regulations of the EU and the CCNR are currently reviewed with the aim of referring in the future to CESNI's standards related to professional qualifications in inland navigation (ES-QIN) (European Committee for drawing up Standards in the field of Inland Navigation, 2019a). The latter lays down details of a new competence-based approach for deck crew members to improve navigational safety with competent, well-trained personnel, creating interesting career opportunities and easing job mobility within Europe. The CESNI work programme for 2022–2024 (European Committee for drawing up Standards in the field of Inland Navigation, 2021b) also includes the development of "competence standards for new and innovative technologies including the use of relevant alternative fuels, batteries and electric propulsion systems".

Moreover, the EU and the CCNR strive for a sweeping modernisation of the manning regulations. Taking into account the organisations' policy objectives (Central Commission for the Navigation of the Rhine, 2018; European Commission, 2021) and specifically commissioned research work (Henn et al. 2019), the CESNI drafted a roadmap with the aim to investigate all important subjects for developing standards for manning regulations, including innovation and technology changes. A sector (stake-holder) consultation event dedicated to the roadmap for European manning regulations was organised in December 2021 (European Committee for drawing up Standards in the field of Inland Navigation, 2021c). The sector' comments on the roadmap will feed into the future work of CESNI under its new work programme 2022–2024.

6   Conclusions

Table 2 indicates the progress of the regulatory work for the decarbonisation of IWT in Europe. For vessel technical requirements, necessary standards for the deployment of LNG, hydrogen and methanol as well as electric batteries are already drafted or at least under development. For police and crew related requirements, necessary work is foreseen in the respective work programmes, but still in the phase of needs evaluation.

Table 2. Summary of the status of regulations and standards for the use of new technologies for the propulsion of IWT vessels in Europe

Energy carrier for IWT propulsion	Vessel technical requirements	Police requirements	Qualification and manning requirements
Diesel (or Bio-Diesel)	Ready	Ready	Ready
LNG (or Bio-LNG	Ready	Ready	Ready
Battery	Ready (for Lithium-ion batteries but ongoing work for swappable battery containers)	Evaluation of the need of safety requirements	Evaluation of the need of safety requirements / Reseach work
Hydrogen (combustion engines or fuel cells)	Ready for fuel cells. On-going for storage and use in combustion engine	Evaluation of the need of safety requirements	Evaluation of the need of safety requirements / Research work
Methanol (combustion engines or fuel cells)	Ready for fuel cells. On-going for storage and use in combustion engine	Evaluation of the need of safety requirements	Evaluation of the need of safety requirements / Research work

Nevertheless, the authors, being involved in this work already for many years chiefly regarding vessel technical requirements, are convinced, that the remaining work will also achieve the objectives stipulated in the respective work programmes. Important research projects, often funded by the EU, and numerous pilot vessels, initiated by forward-looking shipowners, are creating knowledge and experience, from which the CCNR and the EU can draw. Possibly the most important success factor for this work is the direct involvement of experts from equipment manufacturers, shipyards and shipping companies as well as classification societies in the development of necessary standards and regulations, not only consulting them on draft documents, but rather involving them already early on in the drafting process.

IWT needs to decarbonise not only in Europe, but everywhere. The authors hope that sharing the results of the ongoing regulatory and standardisation work as well as the lessons learned in Europe will be useful for experts involved in the development of standards and regulations elsewhere in the world and will eventually kickstart a worldwide exchange on this existential challenge for IWT.

References

Central Commission for the Navigation of the Rhine (1995). Règlement de police pour la navigation du Rhin (RVBR). Strasbourg

Central Commission for the Navigation of the Rhine (2015). CCNR Plenary session – Spring 2015. Press release. CC/CP (15)02, Strasbourg

Central Commission for the Navigation of the Rhine (2018). Mannheim Declaration: "150 years of the Mannheim Act – the driving force behind dynamic Rhine and inland navigation". Strasbourg

Central Commission for the Navigation of the Rhine (2020). Règlement de visite des bateaux du Rhin (RVBR). Strasbourg

Central Commission for the Navigation of the Rhine (2021). Session d'automne 2021 – Résolutions adoptées. CC/R (21) 2 final, Strasbourg

Central Commission for the Navigation of the Rhine (2022). Règlement relatif au personnel de la navigation sur le Rhin (RPN). Strasbourg

Dahlke-Wallat F, Friedhoff B, Karaarslan S, Martens S, Quispel M (2021). CCNR Study Research question C: Assessment of technologies in view of zero-emission IWT (Edition 2). Central Commission for the Navigation of the Rhine, Strasbourg

European Committee for drawing up Standards in the field of Inland Navigation (2019a). Leaflet on deliberation on derogations and equivalences of technical requirements of the ES-TRIN for specific craft, Strasbourg

European Committee for drawing up Standards in the field of Inland Navigation (2019b). European Standard for Qualifications in Inland Navigation (ES-QIN). Edition 2019, Strasbourg

European Committee for drawing up Standards in the field of Inland Navigation (2021a). European Standard laying down Technical Requirements for Inland Navigation vessels (ES-TRIN). Edition 2021/1, Strasbourg

European Committee for drawing up Standards in the field of Inland Navigation (2021b). Collection of CESNI resolutions and decisions - Meeting of 28 October 2021. CESNI (21) 37 final, Strasbourg

European Committee for drawing up Standards in the field of Inland Navigation (2021c). Sector consultation event: one step further towards harmonised European manning regulations - 14 December 2021. Strasbourg

Committee E, for drawing up Standards in the field of Inland Navigation, (2022) CESNI Meeting on 12 April 2022. Press release, Strasbourg

European Commission (2019). Communication from the Commission to the European Parliament, the Council, the European Economic and Social Committee and the Committee of the Regions. The European Green Deal. COM(2019) 640 final, Brussels

European Commission (2020). Communication from the Commission to the European Parliament, the Council, the European Economic and Social Committee and the Committee of the Regions. Sustainable and Smart Mobility Strategy – putting European transport on track for the future. COM(2020) 789 final, Brussels

European Commission (2020b). H2020, TRL levels. Retrieved 15 April 2022, https://ec.europa. eu/research/participants/data/ref/h2020/wp/2014_2015/annexes/h2020-wp1415-annex-g-trl_ en.pdf

European Commission (2021). Communication from the Commission to the European Parliament, the Council, the European Economic and Social Committee and the Committee of the Regions. NAIADES III: Boosting future-proof European inland waterway transport. COM(2021) 324 final, Brussels

EU (2016a). Regulation (EU) 2016/1629 of the European Parliament and of the Council of 14 September 2016 laying down technical requirements for inland waterway vessels, amending Directive 2009/100/EC and repealing Directive 2006/87/EC. Official Journal of the European Union, L 252, Volume 59. Brussels. 118-

EU (2016b). Regulation (EU) 2016/1628 of the European Parliament and of the Council of 14 September 2016 on requirements relating to gaseous and particulate pollutant emission limits and type-approval for internal combustion engines for non-road mobile machinery, amending Regulations (EU) No 1024/2012 and (EU) No 167/2013, and amending and repealing Directive 97/68/EC. Official Journal of the European Union, L 252, Volume 59, Brussels, 53–117

Henn R, Holtmann B, Schreibers K, Van der Weide R, Turnbull P (2019). TASCS study (Towards A Sustainable Crewing System). European Social Partners Organisations EBU, ESO, ETF, Brussels

Heitink J, Mentink L.M.A. (2021). Safety aspects of new energy sources inland navigation – Report for the Ministry of Infrastructure and Water Management. Adviesgroep AVIV BV, Enschede

Stichting Projecten Binnenvaart (2016). Prominent, D1.1 List of operational profiles and fleet families Identification of the fleet, typical fleet families & operational profiles on European inland waterways

Ushakov S, Stenersen D, Einang PM (2019) Methane slip from gas fuelled ships: A comprehensive summary based on measurement data. J Mar Sci Technol 24:1308–1325

United Nations Economic Commission for Europe (2021). European Code for Inland Waterways CEVNI (Sixth revised edition). ECE/TRANS/SC.3/115/Rev.6, New York

Strategies for High Quality Development of Smart Inland Shipping in Zhejiang Province Based on "Four-Port Linkage"

Jianan Zhou[✉], Wanfeng Liu, and Jianwei Wu

Zhejiang Institute of Communications Co. Ltd., Hangzhou, China
{zhoujn, liuwf, wujw}@zjic.com

Abstract. Currently Zhejiang ports have gradually changed from large-scale expansion to high-quality upgrading development, and gradually pay attention to the integrated development with railway, aviation, highway and other transportation modes. In 2019, Zhejiang Province proposed a new development pattern of "Four-port Linkage" with seaports (including inland ports) as the leader, dry ports as the basis, airports as the characteristics and infoport as the link, in which improving the basic service system of sea-river intermodal transport and optimizing the inland waterway network for sea-river intermodal transport are important contents, and cooperate with the requirements of the national outline for the development of inland shipping. With the navigation of a series of high-grade channels such as Hangpingshen Canal, Qiantang River, Hangzhou-Ningbo Canal and Oujiang River, the goal of inland shipping revival of "Upgrading north, unblocking south, opening east and revitalizing west" in Zhejiang Province has been basically realized, and all 11 cities in Zhejiang province are connected to the river and the sea. With the emergence of modern information technologies such as big data, IoT and 5G and the rapid development of multimodal transport, the development of smart inland shipping in Zhejiang Province greets new opportunities. This paper discusses the strategy of "Four-port Linkage" to help the high-quality development of Zhejiang smart inland shipping, from the perspectives of developing a comprehensive big data platform for inland shipping, developing a sea-river intermodal transport service information system, promoting the linkage of inland shipping information systems across provinces and cities, and suggesting smart inland shipping typical projects.

Keywords: Four-port linkage · Smart inland shipping · Sea-river intermodal transport · Transportation information systems

1 Introduction

In the past ten years, Zhejiang ports have experienced rapid development. In 2011, the total cargo throughput of ports in Zhejiang Province was 1.22 billion tons, of which the inland ports cargo throughput was 350 million tons. By 2021, the cargo throughput of ports in the whole province had reached 1.93 billion tons, of which the inland ports cargo throughput was 440 million tons and ranked the fourth in China (China Ports &

© The Author(s) 2023
Y. Li et al. (Eds.): PIANC 2022, LNCE 264, pp. 1409–1418, 2023.
https://doi.org/10.1007/978-981-19-6138-0_125

Harbors Association 2021). After nearly ten years of large-scale expansion, with the continuous improvement of the requirements of internal and external freight transport quality, the port gradually expanded from scale expansion to transformation and upgrading, and entered a new stage of integration with various modes of transportation, such as railways, aviation, and highways (Liu 2014; Li 2021; Liu 2021). Under this background, the Zhejiang Province has issued the Plan for Accelerating the Development and Construction of the "Four-port Linkage" of the Seaports, Dry ports, Airports and Infoport, and proposed the mechanism of the "Four-port Linkage" with the seaports (including inland ports) as the leader, dry ports as the basis, airports as the characteristics and infoport as the link. Important contents include improving the basic service system of sea-river intermodal transport and optimizing the inland waterway network for sea-river intermodal transport. "Four-port Linkage" puts forward higher level requirements for Zhejiang inland shipping development in the fields of smart port, smart channel, smart logistics, smart service, and smart supervision (Mao 2019; Li 2021).

In May 2020, the Ministry of Transport issued the Outline for Inland River Shipping Development, pointing out that it is necessary to improve the service level of sea-river intermodal transport, strengthen sharing resources and information, and vigorously develop multimodal transport centering on ports and one-stop logistics. Also, it requires the use of big data, block chain and other information technology to realize efficient and convenient services of intelligent and comprehensive information, and establish a port economic ecosystem with multiple resources. With the navigation of a series of high-grade channels such as Hangpingshen Canal, Qiantang River, Hangzhou-Ningbo Canal and Oujiang River, the goal of inland shipping revival of "Upgrading north, unblocking south, opening east and revitalizing west" in Zhejiang Province has been basically realized, and all 11 cities in Zhejiang province are connected to the river and the sea, providing powerful facilities guarantee for the development of smart inland shipping in Zhejiang Province. From the view of industry development, with the rapid development of new technologies such as 5G and IoT, transportation industry will be deeply integrated with modern information technology. New changes will be presented in the big data platform for inland shipping business, the information level of inland port terminals and shipping infrastructure, and the innovation of intelligent shipping technology.

2   Integrating "Four-Port Linkage" Information to Develop a Comprehensive Big Data Platform for Zhejiang Inland Shipping

2.1   Develop Comprehensive Big Data Platform for Inland Shipping Across the Province

Actively plan the construction of "four-port linkage" comprehensive transportation big data basic exchange and service platform, develop comprehensive big data platform for inland shipping across the province, and serve as an important sub platform for port and shipping information management system in the province. This platform 1) covers the port and waterway infrastructure, monitoring and control, production operation and logistics services of the inland river ports in the whole province, 2) promotes the digital management of planning, investment, construction, maintenance and approval of inland river shipping facilities in the province, 3) improves the informatization monitoring and control level of inland shipping enterprises operation quality and safety risk across the province, 4) promotes the resource coordination and dispatching cooperation, business linkage and information sharing, 5) strengthens the connectivity of public logistics information of inland river ports, trade, logistics enterprises, highway-railway-waterway intermodal transportation networks in the province and 6) and forms a comprehensive inland river shipping data warehouse across the province.

2.2 Develop a Dynamic Information Collection Mechanism Shared by the Government and Enterprises

The platform is used to 1) collect the dynamic port-related logistics data, 2) develop a long-term and stable data collection cooperation relationship between port related enterprises, 3) cooperate and guide the government to formulate a data sharing mechanism, 4) collect the information of port-related logistics resource and 5) obtain the information of terminal operation, yard status, ships, sources of goods, customs clearance, market dynamics, industry news and ship lock status etc. provided by the port, shipping company, cargo owner, government, ship lock and other subjects, assisting cargo owner, freight forwarders and carriers to achieve demand supply matching. Form a unified, stable and structured inland shipping data management scheme, so that the historical data in multiple fields can be queried and tracked.

2.3 Develop a Dynamic Application Mechanism for Inland Shipping Data

According to the needs of management and monitoring, the platform is used for 1) directional analysis and research of inland shipping big data and 2) applying artificial intelligence, cloud computing, data mining, data visualization and other technologies to carry out massive data modeling and calculation, so as to provide data and information support for production planning, industry supervision, decision analysis, etc., and carry out data innovation services for port planning, path optimization, flow direction analysis, capacity analysis, safety early warning, precision marketing, credit investigation services, etc. (Fig. 1).

3 Developing a Multimodal Transport Information System for Inland Ports Focusing on Sea-River Intermodal Transport

3.1 Develop Sea-River Intermodal Transport Information System

Participate in the construction of "four-port linkage" bulk cargo and container intermodal transport information exchange hub with seaports (including inland ports) as the core, highlight container and bulk cargo logistics, and develop sea-river intermodal

Fig. 1. The structure of comprehensive big data platform for Zhejiang inland shipping based on integrating "four-port linkage" information.

transport service information system. Based on the information interconnection standards such as interconnection document and data query standards formed by seaports, inland ports and freight forwarders, the system 1) integrates document flow, cargo logistics and information flow, 2) carries out sea-river intermodal transport business handling services, and 3) promotes the online handling of inland river transit shipping agency, container and bulk cargo transit agency, inland river cargo agency, barge handling, pre-allocation manifest, customs declaration and clearance, so as to promote the sharing of business information and data between seaports and inland ports, and provide one-stop inquiry service of whole process logistics information.

3.2 Promote Comprehensive Connectivity of Inland Ports with Seaports, Dry Ports, and Airports

Carry out multimodal transport services in inland ports, and provide users with "door-to-door" whole process logistics and transportation services such as railway-watery, highway-waterway, sea-river and air-waterway intermodal transport, and inland river transit. Based on the comprehensive big data platform of inland shipping across the province and the Intelligent Cloud Platform of Zhejiang Comprehensive Transportation, the data of inland ports, seaports, dry ports and airports are fully connected, combining the transportation demand and supply and logistics data with cloud computing, machine learning, data mining and other technologies to provide real-time whole process multimodal transport logistics solutions for shippers, freight forwarders and carriers, so as to fully expand users' right to know and choose when purchasing multimodal transport services. It promotes the sharing of logistics entrustment orders, lading bills and other business documents by all parties in the process of multimodal transport, and forms a "one-stop" lading bill for the whole process logistics of multimodal transport with high-level real rights. It promotes the visual and dynamic tracking of the whole process logistics, provides users with one-stop information query of the whole process logistics with multiple information sources and multiple entrances, and improves the convenience of users for information query of the whole process logistics.

3.3   Provide Convenient Commercial and Financial Information Services

Promote the application of smart trading e-commerce for bulk commodity transactions such as metal ore, coal, oil, grain and chemicals. Through cooperation with financial and insurance institutions, develop information platforms such as logistics finance and pledge supervision, trading platforms such as shipping insurance and shipping space, as well as shipping trust and ship financing, so as to realize the integration of cargo import and export trade, cross-border e-commerce and commercial logistics finance. Guided by market demand and user experience, build an online inland port logistics mall and related mobile applications to improve the convenience of users' multimodal transport business handling, ordering, payment and settlement (Fig. 2).

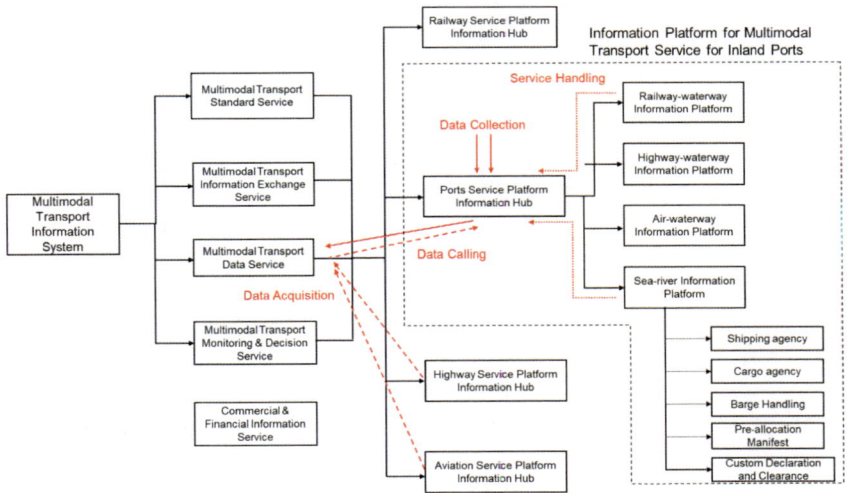

Fig. 2. The structure of multimodal transport information system focusing on inland ports based on integrating "four-port linkage" information and sea-river intermodal transport.

4   Promoting the Connection with Inland Shipping Information Systems in Other Provinces and Cities in the Yangtze River Delta

4.1   Advocate the Integrated Development of Inland Shipping Information System in the Yangtze River Delta

Under the background of the integrated development of Yangtze River Delta, actively promote the development of docking standards of inland river shipping information system among Zhejiang Province, Shanghai City, Jiangsu Province and Anhui

Province in terms of operation management, logistics transaction, document mutual recognition and port management, and advocate the synchronous development of the integration and standardization of the inland river shipping information systems of all provinces and cities in the Yangtze River Delta. Actively promote the electronization of main business documents of inland ports, strengthen the digital circulation and information sharing of inland port operation documents, and promote the sharing of business information among inland shipping enterprises in the Yangtze River Delta, so as to make important support for promoting the integration of inland shipping resources, promoting the integrated development of ports and shipping in the Yangtze River Delta, and forming a world-class inland port cluster with reasonable labor division and mutual cooperation.

4.2   Plan to Share Inland Shipping Information and Develop Information Platform Across Administrative Regions

It may 1) connect the Zhejiang inland river shipping information system with Shanghai, sharing sea-river intermodal transport information and cooperating between inland ports (*e.g.* Hangzhou, Jiaxing and Huzhou Port) and coastal ports (*e.g.* Jiaxing Zhapu Port, Shanghai Waigaoqiao Port and Yangshan Port), 2) connect the Zhejiang inland river shipping information system with Jiangsu and Anhui, sharing shipping information and cooperating between Zhejiang inland ports and the ports along the Yangtze River (*e.g.* Wuhu, Nanjing, Nantong and Taicang Port), especially sharing sea-river intermodal transport information between the inland river ports in northern and western Zhejiang and the ports in southern Jiangsu, and 3) promote sharing sea-river intermodal transport information between Ningbo Zhoushan Port and the ports along the Yangtze River, and finally assisting expanding the service scope of Zhoushan River-sea Intermodal Service Center. Plan to develop public port and shipping logistics information platform across administrative regions, promote the exchange and sharing of multi-modal transport information across administrative regions with reference to LOGINK China and the shipping electronic data interchange (EDI) center, and actively develop public information platforms for railway-highway-waterway intermodal transport of bulk cargo and sea-river intermodal transport of containers, explore the "four-port linkage" mechanism across administrative regions under the background of integrated development of Yangtze River Delta from the perspective of inland shipping (Fig. 3).

5   Promoting Typical Projects of Smart Inland River Shipping

5.1   Promote Typical Applications of "Four-Port Linkage" Information Interconnection Covering Inland Shipping

Promote developing reservation and pick-up system for the container yard of inland ports, promote interconnection and data sharing between fourth party logistics platforms and the infoports of "four-port linkage", open up the information transmission channels of terminals, freight forwarders, storage yards, fleets and truck drivers, and

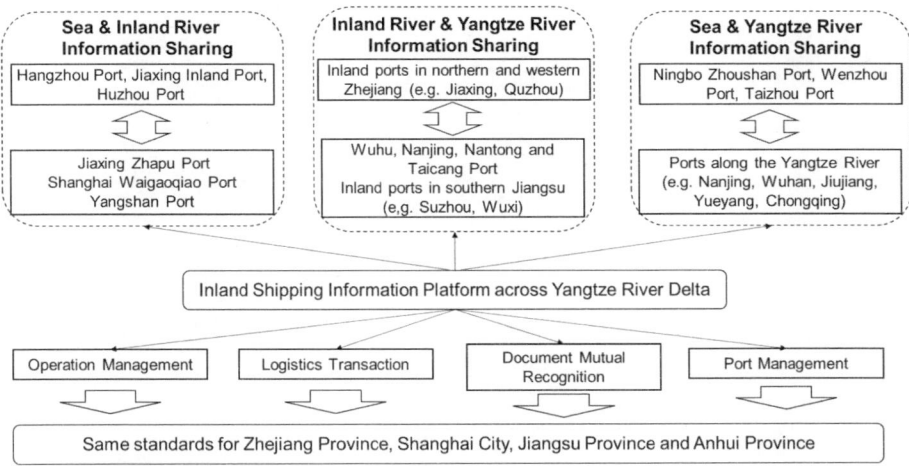

Fig. 3. Suggested framework for the development of inland shipping information platform across Yangtze River Delta.

form an efficient collaborative system for container and bulk cargo highway-waterway intermodal transport of inland ports. Promote the typical application of inland ports railway-waterway intermodal transport information interconnection, promote the unification of freight tickets, waybills and other standard messages between important inland ports (*e.g.* Hangzhou, Jiaxing and Huzhou Port) and Shanghai Railway Bureau, overcome the information barriers between inland ports and China Railway e-commerce system and dry ports, and realize the information interconnection of container status in the whole process of railway-waterway intermodal transport. Promote the typical application of information interconnection between Zhoushan River-sea Intermodal Service Center and Yangtze River port, and expand the service scope of Zhoushan river-sea intermodal transport public information platform. Focusing on bulk cargo, promote the information interconnection between Ningbo Zhoushan port and Yangtze River port and Commodity Futures Exchange. Focusing on one-stop logistics, develop intelligent service platform for container sea-river intermodal transport in Northern Zhejiang, connect with the information systems of offshore routes, barge branch lines, sea-railway intermodal transportation and highway transportation of Ningbo Zhoushan Port and Jiaxing Port, realize the information interconnection of container multimodal transport, and provide one-stop, and door-to-door services (Fig. 4).

5.2   Promote the Construction of Automatic Terminals in Inland Ports

Taking Linping, Xiasha container operating areas in Hangzhou Port and Xucun container operating areas in Jiaxing Inland Port as pilots, promote the construction of automated container terminals in inland ports, utilize digital management and automated operating equipment, improve the intelligence level of operation and comprehensive management of inland ports by using digital management and automatic

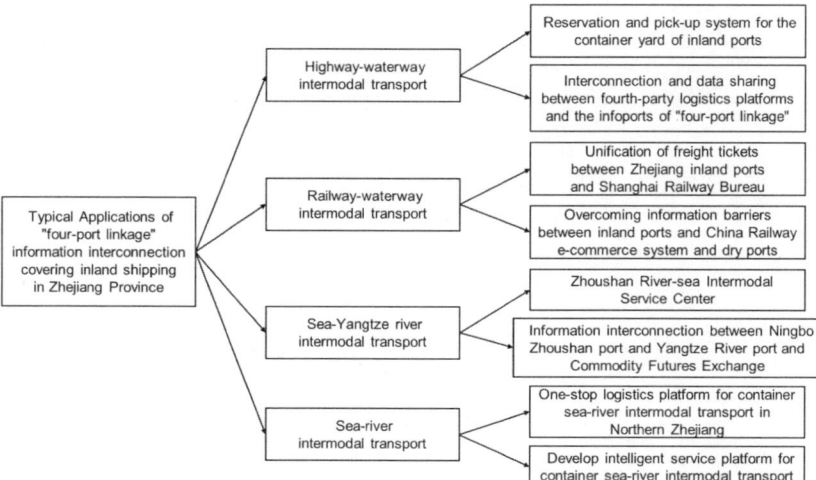

Fig. 4. Typical applications of "four-port linkage" information interconnection covering inland shipping.

operating equipment, and realize the improvement of port capacity and operational efficiency. The key construction includes intelligent remote control of inland port equipment, intelligent management system of production process, intelligent tallying system, and 5G technology applied in inland port. Taking Linpu operating zone in Hangzhou Port and Luoyang operating zone in Jinhua Port as pilots, promote preliminary exploration and project implementation of automated bulk terminals in inland port, and explore the application of automatic stacker-reclaimer, digital yard, automatic loading and unloading ship machine, intelligent monitoring and production information management system.

5.3   Promote Smart Channel Construction Projects

Strengthen the construction of intelligent channels, and accelerate the digitization of infrastructure and the development of and operation management system for higher-grade channel network in northern Zhejiang, Qiantang River, Hangzhou-Ningbo Canal and Oujiang River. Carry out construction and maintenance combined with intelligent control in Xinba second line ship lock, and explore driverless shipping in Heshangtang channel. Use information technology to comprehensively enhance the perception, pre control and adaptability of port and shipping operation, and realize the dynamic real-time monitoring of water transportation. Develop the provincial construction, operation and maintenance and network security standard system for port and shipping informatization. Take Zhezhatong (Zhejiang Lock Management System) as the platform to develop a unified intelligent dispatching platform for ship locks and inland river shipping in the whole province, all ship locks in the whole province will be incorporated into the system, and further improve the functions including electronic

Fig. 5. Layout of smart channels for the Container Transportation Channel of High-grade Channel Network in North Zhejiang.

channel map, ship trajectory and dynamic supervision, positioning perception etc. (Fig. 5).

6   Concluding Remarks

This paper discusses the strategy of "Four-port Linkage" to help the high-quality development of Zhejiang smart inland shipping, from the perspectives of developing a comprehensive big data platform for inland shipping, developing a sea-river intermodal transport service information system, promoting the linkage of inland shipping information systems across provinces and cities, and suggesting smart inland shipping typical projects. Key measures include developing a dynamic information collection and monitoring mechanism for inland shipping shared by governments and enterprises, promoting the comprehensive connection of multimodal transport information between inland ports and sea, land and air ports, and planning the development of inland shipping information platform across administrative regions. This paper provides reference for the development of smart inland shipping in Zhejiang Province.

Acknowledgements. This paper is supported by *Zhejiang Provincial Seaport investment & operation Group Co., Ltd.* Through the research project Research on Promoting the Development of "Four-port Linkage" by Zhejiang Provincial Seaport Investment & Operation Group Co., Ltd., and this support is gratefully acknowledged.

References

China Ports & Harbors Association (2021) China Ports Year Book 2021

Li H, Wang D (2020) Liu T (2020) Problems and suggestions for the construction of smart ports in China. Ship Manage 1:23–25

Li Y, Wang F, Lian J, Wang J (2021) Suggestions on developing Jinhua multimodal transport hub. Ship Manage 2021(12):13–14, 19

Liu C, Li X, Ji D (2014) Research on the development model of sea-river intermodal transport in Zhejiang Province. China Ports 2014(1):49–51, 54

Liu W (2021) Wang J (2021) "Golden decade" for the development of inland water transportation in Qiantang River. China Water Transport 10:13–15

Mao J (2019) Accelerate the construction of first-class ports in the world and world-class port clusters. China Ports 2019(12):1–4

The Ministry of Transport of the People's Republic of China (2020) The outline for inland river shipping development

Zhejiang Provincial Leading Group for Comprehensive Transportation Reform and Development (2019) The plan for accelerating the development and construction of the "four-port linkage" of the seaports, dry ports, airports and infoport

The Linear Regression Model for Estimate the Price in Crossing Transport Services: Methodology and Application

Isaac Monteiro do Nascimento[✉] and Eduardo Pessoa de Queiroz

National Agency for Waterway Transportation, Brasília, Brazil
{isaac.nascimento,eduardo.queiroz}@antaq.gov.br

Abstract. The Brazilian's urbanization process, allied to the expansion of the economic frontier, which have occurred concomitantly in the last five decades, are responsible for the increase in the number urban settlements and for the increase in the country's arable area. However, in many cases, transport infrastructure did not keep pace with the process of territorial occupation. In this sense, many highways were provided without the proper instruments for transposing watercourses and many crossing shipping companies emerged. Nowadays, there are ninety federal crossing lines in operation. One of the main issues inherent to these services is the prices charged and the respective affordability. Considering this scenario, the present work selected crossing sample to estimate prices using the Linear Regression Model. The selected model was applied to forecast crossing price in 8 (eight) crossing lines. As result, the price charged in all of them were out of the prediction interval model, indicating that the Regulatory Agency should investigate the reason why they are dissonant to market practices.

Keywords: Forecast · Crossing · Price

1 Introduction

The crossing waterway transport is carried out in all Brazilian macro-regions. The main characteristic of this transport is the interconnection of road stretches, usually in a short journey, integrating road systems or also as an element of urban/metropolitan transport. Nevertheless, this type of transport, for thousands of people, is the main or the only option for crossing a body of water. This essentiality reflects the amount paid for the service. Often, it is not modest, due to the lack of a logistical alternative. In this sense, the present work seeks to present a model for forecasting the price of the crossing transport service, especially in the crossing of vehicles. However, before entering the proposed model, it is necessary to understand, in general, the territorial process that explains the number of crossings in Brazil, a movement intensified from the second half of the 20th century (Fig. 1).

Y. Li et al. (Eds.): PIANC 2022, LNCE 264, pp. 1419–1431, 2023.
https://doi.org/10.1007/978-981-19-6138-0_126

Fig. 1. Crossing navigation: Manaus - Careiro route in Amazon river

2 Historical Context

Brazil has undergone major social, economic and territorial transformations in the last decades of 20th century. Can be listed the construction of a new capital; the agricultural and urban frontier expansion; increased of the industrial sector in the Gross Domestic Product (GDP) and; the intensification of the urbanization process.

Regarding the territorial occupation process, I highlight that in the period between 1950 and 2000, the Brazilian population jumped from 51 to 169 million inhabitants; now 73% of the population is urban, compared to 31% in the middle of the last century (Table 1).

Table 1. Brazil - number of people living in urban and rural areas (1950–2021[1]).

Year	1950	1960	1970	1980	1991	2000	2010	2021
Total	51.784.325	70.608.046	94.508.642	121.150.573	146.917.459	169.590.693	190.755.799	211.846.681
Urban	16.278.109	28.540.752	47.540.991	72.066.333	98.518.438	123.460.941	143.792.645	
Rural	35.506.216	42.067.294	46.967.651	49.084.240	48.399.021	46.129.752	46.963.154	
Urban (%)	31	40	50	59	67	73	75	
Rural (%)	69	60	50	41	33	27	25	

Source: IBGE – demographic survey

In the same decades, there was an increase in municipal units throughout the national territory and, consequently, an increase in the number of cities. In 1950 there were 1,889 municipal units. Currently, the country has 5,570 municipalities (Table 2).

---

[1] Population Estimation available: https://www.ibge.gov.br/estatisticas/sociais/populacao/9103-estimativas-de-populacao.html?=&t=resultados.

[2] Numbers from the Population Estimation Research, available from: https://www.ibge.gov.br/estatisticas/sociais/populacao/9103-estimativas-de-populacao.html?=&t=resultados.

Table 2. Number of municipalities (1950–2021[2]).

Year	1950	1960	1970	1980	1991	2000	2010	2021
Municipalities	1.889	2.766	3.952	3.991	4.491	5.507	5.565	5.570

Source: IBGE – demographic survey

These numbers emphasize the territorial transformation that took place in the last century, with the outlining of new development vectors. As already pointed out by several authors, including Steinberger (1999), Becker (2005) and Queiroz (2007), it is a process of internalization of productive activities, especially agriculture, which generated poles of population attraction for the regions previously considered as "demographic voids", mainly in the North and Midwest of Brazil.

This process was called "March to the West", initiated in the Getúlio Vargas government (1930–1945). The creation of the new capital in the Central Plateau of the country, called Brasília, intensified this process. In parallel, with the change in the political axis, there was the construction of several road sections connecting the capital to the other regions of the country, a fact that accelerated the process of expansion of economic and urban borders, demonstrated in a certain way in the data above and in the maps below.

This movement was not always consistent with the occupation of land for agricultural exploitation. The interconnection of the new territorial divisions in the different regions of the country allowed for a greater participation of these in global markets, even with evident inefficiencies and gaps in infrastructure. The crossing lines, responsible for interconnecting road sections without bridges, are partially explained by this context (Figs. 2 and 3).

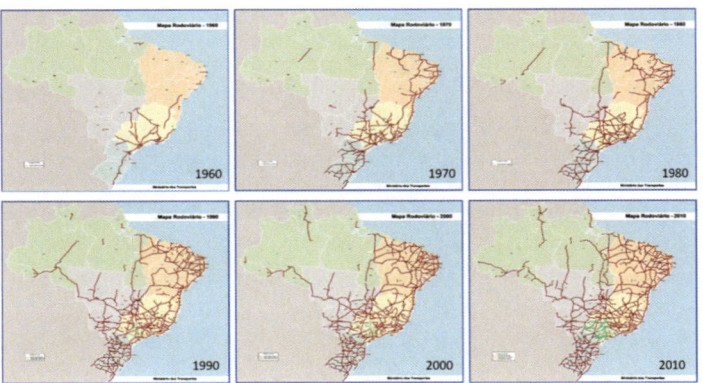

Fig. 2. Brazilian roadways development: 1960–2010

Currently, Brazil has 90 crossing transport lines under the regulatory authority of the National Waterway Transport Agency – ANTAQ (Fig. 4). Approximately 300 shipping companies authorized by ANTAQ operates the federal lines. Almost 50% of them have a tax framework for individual microentrepreneurs (MEI). The MEI are exempted from submitting bookkeeping, annual survey of balance sheet and economic result. They are also not obliged to issue an invoice when providing the service to final consumer. Such tax benefits have resulted in great informality in the sector.

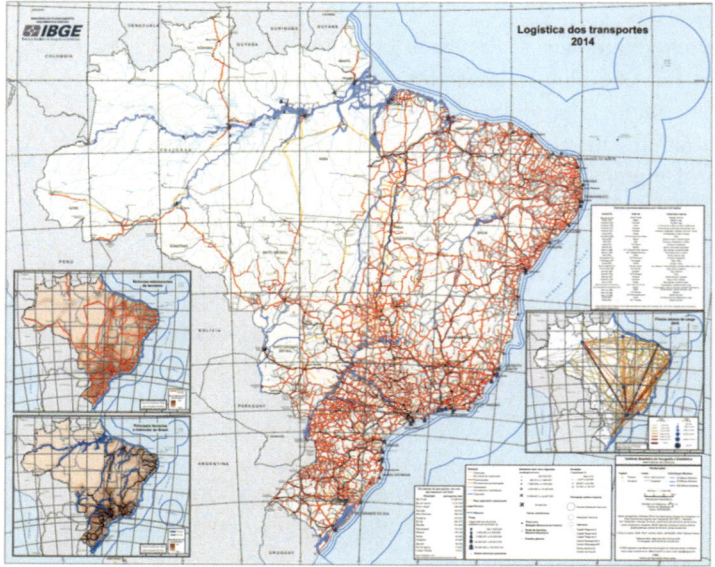

Fig. 3. Brazilian multimodal network (IBGE)

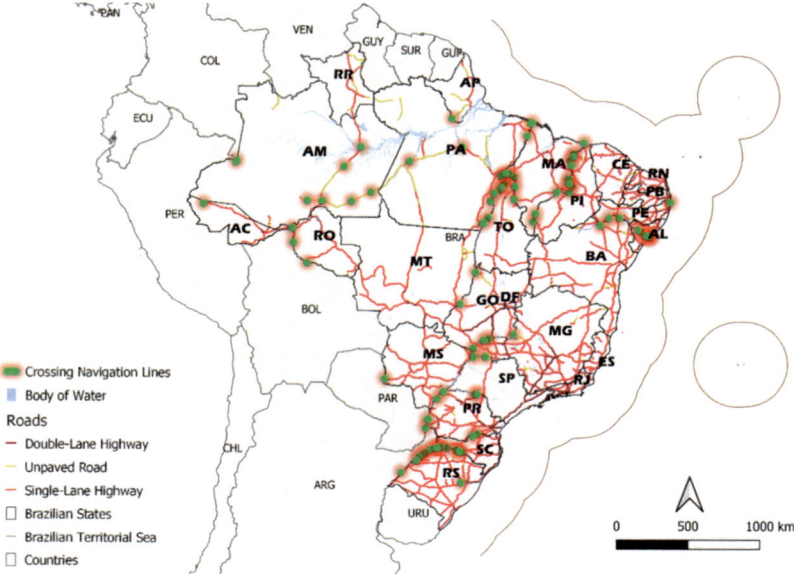

Fig. 4. Brazilian crossing navigation lines.

3   Service Affordability Price

After understand de crossing historical context, it should be highlight that only in 2015 the transport was sign up as a fundamental right in the Brazilian Federal Constitution (CF). The text compiled from art. 6 of the CF says: "These are social rights: education, health, food, work, housing, transport, leisure, security, social security, maternity and childhood protection, assistance to the destitute, in the form of this Constitution".

The Brazilian Waterway Transport Regulatory Framework[3] establishes that the State can crossing transport service under the authorization regime. Subsequently, the Law determines freedom prices in free and competitive markets.

The Law[4] also establishes that the provision of transport services must be carried out in an appropriate manner, satisfying the conditions of regularity, efficiency, safety, topicality, generality, courtesy in the provision of the service, and affordability prices.

Public passenger transport has functional characteristics that differentiate it from the private services regime. Public services are directly linked to the concept of essentiality in order to meet the appropriate service attribute that satisfies the price affordability requirement[5]. In practice, most crossing lines fit the public service characteristics.

In this context, the challenge is to establish practical parameters for the attribute of reasonable prices, considering the essentiality of the service to the community and the guarantee of freedom of prices and tariff services provided for by Law.

4   Monitoring Regulated Prices

In most of the crossing transport markets in inland navigation can be find a few or only one company operating on the lines authorized by ANTAQ. In 2020, more than 65% of the lines in operation had only one authorized shipping company. In some cases, technical and operational line's characteristics do not allow a significant number of operators. In other cases, there are no companies interested in competing for service's provision.

The ANTAQ does not yet have a methodology for measuring the affordability price. The topic will be developed in the 2022–2024 Regulatory Agenda. Until the study is completed, this work proposes a linear regression model to predict prices. The model can be useful both in the investigation of current authorizations and for new crossing line authorizations.

In short, it is proposed to compare the intended price with that practiced in the transport market, based only on the length of the crossing line. If the target price is within the model's prediction range, then the hypothesis that it is statistically similar to

---

[3] Law n° 10,233, of June 5, 2001.

[4] Art. 28, item I, of Law 10.233/2001.

[5] Art. 175, sole paragraph, item IV, of CF/1988; arts. 6 and 7 of the Concessions Law (Law n°. 8,987/1995); art. 6 of the Consumer Defense Code, Law n°. 8,078, of September 11, 1990; and art. 4 of the Public Services Law, Law n°. 13,460, of June 26, 2017.

the market price is accepted. If it is below or above, the regulatory agency may request a detailed explanation of the proposed pricing.

5 Methodology

A random sample of 42 lines of vehicle crossings was selected from the federal crossing line's database (Table 3). The table of prices charged in December 2021 was checked for each line. Considering the variety of vehicles listed in the operators' price list, the passenger car was selected as standard for the modelling.

At least one ANTAQ authorized shipping company was operating the crossing lines selected. The typical sample vessel has a steel hull, 12 m of beam, 50 m length and 505 deadweight tons, in average. The type of engine was not considered. The operational schedule programmed foresaw at least five daily trips. Finely, Google Earth was used to estimate the average length (in kilometers) of each crossing line.

In a preliminary approach, the scatter plot of price per kilometer by line extension indicated a logarithmic pattern. Under the assumption of association between variables, the Linear Regression method was adopted to assess the effect of an explanatory variable (extension) on the response variable (price per kilometer). The method also allows forecasting values for the response variable.

The linear regression model expressed by Johnson (2007, p. 362) was represented by Eq. (1):

$$Y = Z * B + \varepsilon, \qquad E(\varepsilon) = 0 \ and \ Cov(\varepsilon) = \sigma^2 I \tag{1}$$

where: Y is dependent variable (or response variable); Z is the matrix of the independent variable (or predictor variable); B is the matrix of unknown parameters to be estimated by the n independent observations; ε is the random residual that assumes zero mean, constant variance and zero covariance (uncorrelated).

When the residuals have a normal distribution, the prediction interval (IP) for a future observation can be written as:

$$IP(Y_f; \gamma) = \widehat{y}_f \pm t_\gamma S_e \sqrt{1 + \frac{1}{n} + \frac{(x_f - \bar{x})^2}{\sum (x_i - \bar{x})^2}} \tag{2}$$

Considering the complexity of the calculations involved, the entire process of model selection, statistics calculation and model checking was supported by the Statistica software.

Table 3. Sample selected of Brazilian vehicle crossing lines.

N°	Crossing line	Line extend	Price charged for car
1	Alto Parnaíba (MA)/Santa Filomena (PI)	100 m	R$ 16,99
2	Amarante (PI)/São Francisco do Maranhão (MA)	250 m	R$ 19,41
3	Apuí - Lado direito do Rio Aripuanã (AM)/Diretriz de Rodovia - 230	320 m	R$ 25,00
4	Apuí - Lado Direito do Rio Sucunduri (AM)/Diretriz de Rodovia - 230	120 m	R$ 25,00
5	Araguacema (TO)/Santa Maria das Barreiras (PA)	850 m	R$ 33,35
6	Belém de São Francisco - Porto da Barra (PE)/Chorrochó - Barra do Tarrachil (BA)	1 km	R$ 30,00
7	Belo Monte - Barra do Ipanema (AL)/Porto da Folha - Ilha do Ouro (SE)	1,5 km	R$ 25,00
8	Borba (AM)/Diretriz de Rodovia – 319	200 m	R$ 30,00
9	Brejo (MA)/Matias Olímpio (PI)	450 m	R$ 19,41
10	Canutama - Diretriz da BR 230 - KM 752,8 a 753,0 (AM)/ Lábrea	100 m	R$ 30,00
11	Carolina/MA – Filadélfia/TO	900 m	R$ 18,36
12	Concórdia (SC)/Mariano Moro (RS)	750 m	R$ 25,00
13	Duque Bacelar (MA)/Miguel Alves (PI)	220 m	R$ 19,40
14	Floresta do Araguaia (PA)/Pau D Arco (TO)	2,9 km	R$ 29,13
15	Imperatriz (MA)/São Miguel do Tocantins (TO)	700 m	R$ 24,25
16	Itaituba - Diretriz BR 230 (PA)/Itaituba - Miritituba (PA)	2,8 km	R$ 25,00
17	Itapiranga (SC)/Pinheirinho do Vale (RS)/Barra do Guarita (RS)	820 m	R$ 28,12
18	Machadinho (RS)/Capinzal (SC)	260 m	R$ 24,00
19	Manaus - Terminal CEASA (AM)/Careiro - Terminal Careiro (AM)	12 km	R$ 30,00
20	Marcelino Ramos (RS)/Alto Bela Vista (SC)	350 m	R$ 21,50
21	Mondaí (SC)/Vicente Dutra (RS)	500 m	R$ 25,00
22	Nova Iorque (MA)/Porto Alegre do Piauí (PI)	880 m	R$ 29,13
23	Palestina do Pará (PA)/Ananás (TO)	1 km	R$ 24,25
24	Pão de Açúcar (AL)/Porto da Folha - Niterói (SE)	2 km	R$ 30,00
25	Parnarama (MA)/Palmeirais - Porto Mangueira (PI)	450 m	R$ 16,87
26	Penedo (AL)/Neópolis - Passagem (SE)	2,6 km	R$ 32,00
27	Piçarra (PA)/Araguanã (TO)	2,1 km	R$ 29,13
28	Porto (PI)/Buriti - Buriti de Inácia Vaz (MA)	600 m	R$ 15,00
29	Porto Franco (MA)/Tocantinópolis (TO)	800 m	R$ 24,25
30	Porto Velho - Abunã - Diretriz da Rodovia Federal Br - 364 (RO)	1,5 km	R$ 20,00
31	Porto Vera Cruz (RS) /Porto Panambi- Missiones (ARGENTINA)	1,11 km	R$ 40,00
32	Sampaio (TO)/Cidelândia (MA)	630 m	R$ 33,35

(continued)

Table 3. (*continued*)

N°	Crossing line	Line extend	Price charged for car
33	Santa Albertina (SP)/Carneirinho - Estrela da Barra (MG)	2,7 km	R$ 40,00
34	Santa Helena (PR)/Puerto Índio - Sanga Funda (PARAGUAI)	8,44 km	R$ 45,00
35	Santa Mariana (PR)/Florínia (SP)	500 m	R$ 30,00
36	São Francisco de Sales (MG)/Riolândia (SP)	2,3 km	R$ 25,00
37	São Geraldo do Araguaia (PA)/Xambioá (TO)	1,6 km	R$ 19,40
38	São João do Araguaia (PA)/Esperantina (TO)	860 m	R$ 24,25
39	São Pedro do Paraná - Porto São José (PR)/Bataiporã - Porto São João (MS)	1,25 km	R$ 40,00
40	São Sebastião do Tocantins (TO)/Vila Nova dos Martírios (MA)	800 m	R$ 24,25
41	Tasso Fragoso (MA)/Santa Filomena (PI)	100 m	R$ 14,77
42	Xinguara (PA)/Santa Fé do Araguaia (TO)	2,8 km	R$ 29,13

Source: ANTAQ database

6   The Linear Regression Model

It was selected the model that met all the assumptions of linear regression at a significance level of 5%. The equation of the adjusted regression model for crossing a car was y = 3.2672 − 0.8827 * x, where, y = Ln (unit price of car in R$/km) and x = Ln (line length in km). The R^2 statistic, that measures the variability explained by the model, presented a value equivalent to 94.96% (Table 4). Both regression coefficient, the intercept and the explanatory variable, had a significance level lower than 0.05 (Table 5). The following tables presents the summary of the analysis of variance.

Table 4. Summary ANOVA statistics

Statistic	Value
N	42
R	0,97448523
R^2	0,94962146
Ajusted R^2	0,948362
F (1,40)	753,99
p	<0,0000
Std. error of estimate	0,22892

Table 5. Regression summary statistics

Coefficient	b*	Std. err. of b*	b	Std. err. of b	t(40)	p-value
Intercept	–	–	3,267182	0,036034	90,6701	0,00001
Ln (crossing line extend)	−0,974485	0,035489	−0,882721	0,032147	−27,4589	0,00001

Considering the selected sample, it was concluded that the length of the line influence significantly the price charged for crossing a car, explaining about 95% of its variability. The model makes it possible to make predictions about the market crossing price. The Table 6 presents as example the forecast interval (with 95% of confidence) generated by the model for crossing lines with 100 m, 500 m, 1 km and 1.5 km in length.

Table 6. Forecast interval for some crossing lines extension.

Crossing line extension (meters)	Forecast range for price (reais)	
	IP (−95%)	IP (+95%)
100	R$ 12,30	R$ 32,60
500	R$ 15,13	R$ 38,67
1,000	R$ 16,43	R$ 41,91
1,500	R$ 17,20	R$ 44,02

In a crossing line of 500 m in length, it is expected that the cost of crossing a car varies from R$ 15.13 to R$ 38.67. If the practiced (or intended) price is outside this range, then the Agency may analyze the case seeking for additional information to identify dissonant market practices.

6.1   Model Checking

If the model is valid, each residual is an estimate of the error, which is assumed to be normal random variable with mean zero and variance S^2 (Johnso 2007, p. 381).

The normality of residuals examined graphically showed residuals that tend to align with the quantiles of a theoretical normal distribution (Fig. 5). Although some asymmetry is observed, the hypothesis of normality was accepted.

Still in relation to the analysis of the residuals of the regression model, their independence was tested. The Durbin-Watson statistic (d) that measures the correlation between each residue and the residue corresponding to the immediately previous observation indicated the value of 1.975, confirming the hypothesis of independent residues (Table 7).

Table 7. Durbin-Watson test independent residuals

Autocorrelation test	Statistics
Durbin-Watson (d)	1,975
p-value	0,0059
N	42
d (minimum)	1,246
d (maximum)	1,344
Result	Accept independent residuals hypothesis

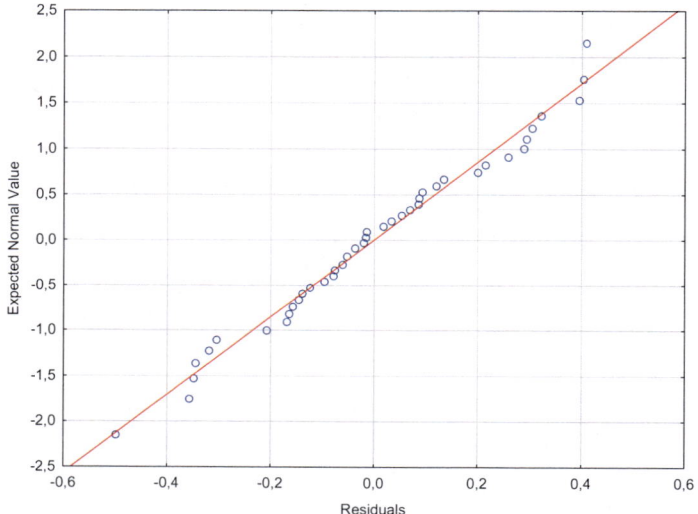

Fig. 5. Normal probability plot of residuals from Eq. (1).

The homoscedasticity (or constant variance) of the residuals examined by graphic visual analysis (Bussa 2003) indicated residuals scattered randomly around zero. In this case, the hypothesis of constant variance was accepted (Fig.6).

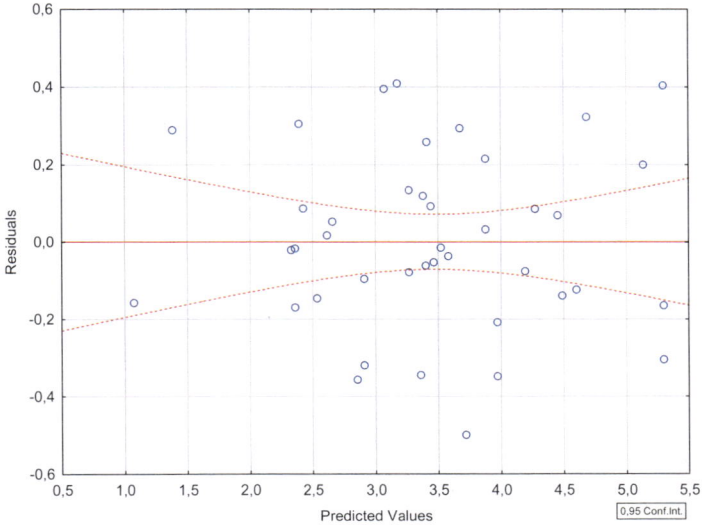

Fig. 6. Predicted vs residual scores.

7   Application

The crossing price forecast model was applied to another 8 (eight) crossing lines regulated by ANTAQ. They prices charged were outside the model's forecast range. Three of them were below the lower limit and the last five were above the upper limit of the forecast range. In other words, these line are statistically significant different from the model market (Table 8).

This result indicates that the regulatory agency should get additional information to identify why the price in these lines are asymmetry to the modeled market. The following table presents the crossing lines taken from the model.

The posterior analysis identified that in the case of the crossing line Anapu (PA)/ Vitória do Xingu (PA) - Directive BR-230 (Belo Monte) the operator practices a discount policy for cars to the detriment of large vehicles. For the other crossings, the ongoing analysis requested operational and economic-financial information from the companies.

Table 8. Forecasting statistics for eight crossing lines applied.

Crossing line	Line extension (meters)	Car price charged (reais)	Predicted value (reais)	Foracast range for price (reais)	
				IP (−95%)	IP (+95%)
Anapu (PA)/Vitória do Xingu (PA) - Diretriz BR-230 (Belo Monte)	450	R$ 11,00	R$ 23,89	R$ 14,94	R$ 38,21
Cachoeira do Piriá (PA)/Centro Novo do Maranhão (MA)	150	R$ 50,00	R$ 21,00	R$ 12,99	R$ 33,97
Canoinhas - Distrito de Paulo Pereira (SC)/São Mateus do Sul (PR)	220	R$ 10,00	R$ 21,72	R$ 13,49	R$ 34,99
Inaciolândia (GO)/Ipiaçu - Porto Gouveinha (MG)	1,350	R$ 65,00	R$ 27,18	R$ 17,00	R$ 43,46
Paula Freitas (PR)/Irineópolis (SC)	200	R$ 6,00	R$ 21,72	R$ 13,49	R$ 34,99
Porto Mauá (RS)/Alba Posse - Província de Missiones (ARGENTINA)	550	R$ 48,00	R$ 24,46	R$ 15,31	R$ 39,09
Porto Xavier (RS)/San Javier - Mnes (ARGENTINA)	550	R$ 50,00	R$ 24,46	R$ 15,31	R$ 39,09
Porto Murtinho (MS)/Colônia Carmelo Peralta (PARAGUAI)	350	R$ 170,00	R$ 34,47	R$ 21,51	R$ 55,21

8   Conclusions

The linear regression model selected for forecasting the transport market price is statistically valid. For the crossing lines surveyed, there is a strong correlation between the price of a car crossing and the length of the line, so this information was enough to generate price market estimates.

The application of the regression model in other lines in operation identified eight cases in which the practiced price is outside the forecast range. In one of these cases, it was identified a practice of discounts for automobiles. For the other cases, the ongoing analysis requested additional information from the authorized companies.

The statistical method applied in this work does not replace the study of price formation for regulatory purposes. In order to curb practices harmful to competition and the abuse of economic power, the regulatory Agency may use methods of price determination such as price-cap, yardstick competition, revenue-cap, or also use the legal framework of competition law.

References

Becker BK (2005) Geopolítica da Amazônia. Dossiê Amazônia Brasileira I. Estud av 19(53). https://doi.org/10.1590/S0103-40142005000100005. Accessed 07 Mar 2022

Bussab WO, Morettin PA (2003) Estatística Básica. 5 edn, Saraiva, São Paulo, pp 436–470

Grotti DAM (2019) Apontamentos sobre os serviços públicos e serviços privados de transporte. https://revistas.pucsp.br/red/issue/download/1594/6. Accessed 04 Jan 2019

Instituto brasileiro de geografia e estatística (2022) Número de municípios e População nos Censos Demográficos por tamanho da população. Sistema IBGE de Recuperação Automática – SIDRA. https://sidra.ibge.gov.br/Tabela/1290#resultado. Accessed 07 Mar 2022

Instituto brasileiro de geografia e estatística (2022) Número de cidades e População nas cidades nos Censos Demográficos por tamanho da população. Sistema IBGE de Recuperação Automática – SIDRA. https://sidra.ibge.gov.br/Tabela/1294#resultado. Accessed 07 Mar 2022

Johnson RA, Wichern DW (2007) Applied Multivariate Statistical Analysis. 6 edn, New Jersey: Pearson Prentice Hall, pp 360–417

Queiroz EP (2007) A formação histórica da região do Distrito Federal e entorno: dos municípios-gênese à presente configuração territorial. https://repositorio.unb.br/handle/10482/2354. Accessed 08 Mar 2022

Steinberger M (1999) Formação do Aglomerado Urbano de Brasília no contexto nacional e regional. In: Paviani A (org) Brasília, Gestão urbana: conflitos e cidada

A Study on the Han-Hai Fleet Slot Mutual Chartering Model of Han-Shen Line Based on Linear Alliance

Yi Zhang, Cheng Peng, Yike Li, Jinling Li, and Jinshan Dai[✉]

Wuhan University of Technology, Wuhan, China
zhangyi_logistic@whut.edu.cn

Abstract. To raise the income of the container liner company and improve the shipping space utilization rate, an integer programming model is constructed in view of the seaworthiness characteristics of each ship route along Hanshen line of the Yangtze river, OD container transport demand, and the container rental business under the alliance. Aiming to maximize the gross profit of the Hanhai liner shipping fleet series, the container was optimized route network along the Yangtze river waterway and determined how to choose the port call for each route and the types of vessels. While optimizing, the model is established according to the characteristics of liner multi-port attachment and cargo nonstop transportation routes, comprehensively considering various factors including freight demand, maximum cargo capacity of a single ship, minimum voyage frequency, and freight rate level. The numerical analysis of the Hanhai fleet turns out that the model can perfectly simulate the situation in reality and optimize the existing routes and boost the profits of liner transportation companies.

Keywords: Liner alliance · Multi-port call · Slot allocation · Slot leasing · Optimization model

1 Introduction

The liner company needs to meet the most favorable basic conditions both technically and economically. Since the 1990s, scholars at home and abroad have been exploring route allocation and fleet planning for large liner fleets, seeking to establish a set of scientific mathematical methods and techniques to solve this problem. Perakis and Jaramillo were the first to define the route allocation problem, and they proposed a preliminary model and available solutions, which to some extent provided the basis for subsequent research has laid the foundation (Jaramillo 1991). The works of Wu Changzhong and Yang Qiuping are the earlier works on the route allocation problem in China, which established a linear model with operating cost and opportunity cost as the objective function and capacity and volume limitation as the constraints from the perspective of shipping companies, which laid a good foundation for the subsequent studies (Wu 1992; Yang 2011). XIE X and others proposed a nonlinear model for fleet planning (Xinlian 1993). Zhao Gang explored a more realistic nonlinear impact of route allocation (Zhao 1997). The studies of An Fen, Huang Yong have greater significance for the research of this paper, and these are more targeted studies for the route optimization of inland river container liner system in China (An 2014; Huang2007). Huang Yong initially studied the optimization of inland river container routes in Yangtze River, and achieved the lowest total cost of the whole system through route search and model solution (Huang 2007). An Fen analyzed the container capacity of single vessel and the container loading capacity of each port of call on the basis of Huang Yong's studies (An 2014). The empty container transfer problem caused by the imbalance of two-way transportation was also considered. By the 21st century, the study by Maxim A. Dulebenets extended the cooperation agreement between liner carriers and sea container terminal operators by considering, for example, the negotiation of vessel time windows (TW) and loading and unloading rates between liner carriers and sea container terminal operators (Dulebenets 2018; Dulebenets 2019). New forms of cooperative agreements will improve liner shipping and ocean container terminal operations. Chen et al. provide a systematic overview of research on liner shipping alliance management, including forming alliances, selecting partners, and designing cooperative mechanisms (Chen et al. 2022). This has similarities with Dulebenets' research on alliances.

Mamedio noted that alliances can facilitate the effective integration of information and resources among members, and are a channel for sharing technology and other capabilities, so they can effectively integrate resources and capabilities among members (Mamedio 2019). Cariou and Cuillotreau's study argues that alliances can be effective in addressing the problem of overcapacity, noting that alliances can play a greater role in reducing overcapacity when the number of competitors is limited. They point out that when the number of competitors is limited, alliances can play a greater role in reducing overcapacity (Cariou 2021).

Based on the above studies, there is an end to the study of alliances between inland shipping and fleet companies for liner shipping issues, from the initial maritime to inland waterways, and from the initial individual interests to the current alliance interests. The study in this paper is inspired by the above-mentioned studies, but focuses more on the impact of company alliances on benefits for particular vessel types, examining the overall fleet benefits under alliances and non-alliances. It also conducts a numerical analysis on the Han-Hai series of the 1,000-case fleet of the Han-Shen Line.

This study focuses on the ship scheduling of the Wuhan-Shanghai Hanshin line, and focuses on the 1,000 container fleet "Hanhai Series". The five thousand container ships belong to three companies, among which Hanhai 2 and 3 are operated by Wuhan COSCO Shipping Container Transport Company Limited, Hanhai 7 is operated by Wuhan Changwei International Shipping Industry Co. The five vessels operate the direct river-sea service from Yangluo Port to Yangshan Port. The ports (in the order of visiting ports) through which the service passes include Yangluo Port, Maanshan Port, Jiujiang Port, Taicang Port, Yangshan Port and other series of ports. The containers transported have a corresponding starting port (Original Port) and destination port (Destination Port). Between each two ports can form a port pair (O,D). Linear arrangement of any two ports on the inland waterway trunk line may have a demand for container traffic between them, called OD traffic (An 2014). Han Shen Line container liner route network consists of several liner routes. Considering the alliance situation, each route may be operated by different carriers. However, its initial starting port is Yangluo port and destination port is Shanghai port, as shown in Fig. 1. The ports of call for each ship on a route are not necessarily the same for the outbound and return trips, as shown in Fig. 2.

2   Inland Container Liner Problem Description

This section describes in detail the liner shipping routes to be analyzed and presents the main modeling assumptions.

2.1   Network Planning Problem Description

This study focuses on the ship scheduling of the Wuhan-Shanghai Hanshin line, and focuses on the 1,000 container fleet "Hanhai Series". The five thousand container ships belong to three companies, among which Hanhai 2 and 3 are operated by Wuhan COSCO Shipping Container Transport Company Limited, Hanhai 7 is operated by Wuhan Changwei International Shipping Industry Co. The five vessels operate the direct river-sea service from Yangluo Port to Yangshan Port. The ports (in the order of visiting ports) through which the service passes include Yangluo Port, Maanshan Port, Jiujiang Port, Taicang Port, Yangshan Port and other series of ports. The containers transported have a corresponding starting port (Original Port) and destination port (Destination Port). Between each two ports can form a port pair (O,D). Linear arrangement of any two ports on the inland waterway trunk line may have a demand for container traffic between them, called OD traffic. Han Shen Line container liner route network consists of several liner routes. Considering the alliance situation, each route may be operated by different carriers. However, its initial starting port is Yangluo port and destination port is Shanghai port, as shown in Fig. 1. The ports of call for each ship on a route are not necessarily the same for the outbound and return trips, as shown in Fig. 2.

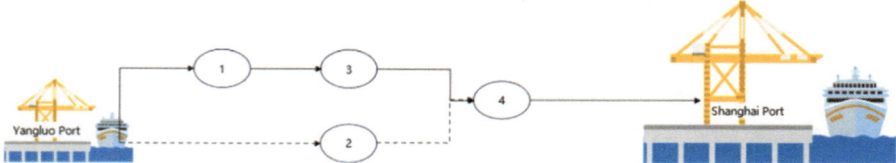

Fig. 1. Liner alliance network

Under liner alliances, there is a need to consider the issue of slot chartering. Which refers to the act of liner companies chartering slots from cargo ship operators on one or more routes so as to operate the relevant routes without increasing capacity input, and paying slot chartering fees regularly. The space lessor can earn rental income and effectively solve the problem of surplus capacity due to declining demand.

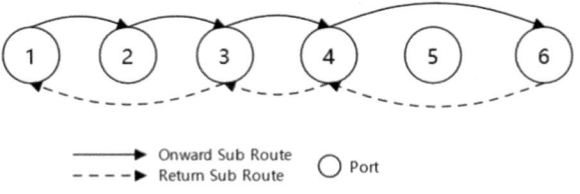

Fig. 2. Route structure

Therefore, assuming that the Hanseatic series of operating companies reach an alliance, different companies can trade their own ships as resources by signing a charter agreement stipulating the basic amount of chartered space. At this point, the problem changes to the Hanseatic liner fleet with a certain scale, through the cargo source survey to grasp the amount of freight demand and the level of freight rates between any ports on its operating routes during the study period, and design the optimal alliance fleet plan to achieve the maximum overall return of the alliance. Each carrier can also maximize its own interests when operating according to the scheme of the maximum overall return of the alliance, so as to maintain the willingness of the alliance cooperation.

2.2 Network Planning Problem Assumptions

Since the route network planning of inland container liner transportation system involves numerous factors and details, if all the details are considered, the model is complicated and difficult to study. For the convenience of the study, the following hypotheses are proposed.

(1) The study period is one year, one quarter, one month or other unit of time. The line is operating normally and there is no port leakage problem.

(2) The volume of traffic between any two port pairs is deterministic and can be predicted from prior data, and the volume of traffic is essentially constant over a period of time; the level of tariffs has been determined, and the number of chartered slots and chartered tariffs in the liner alliance are also determined.

(3) The container type is unique, and only TEU containers, i.e. 20-foot containers and only full container shipments are considered in this study, without considering the case of LCL shipments.

(4) The optimal speed of each type of vessel on each route during the study period is predetermined; the time when the vessel enters and leaves the port is fixed and does not vary according to the type of vessel, and no congestion and delay problems occur.

(5) The fixed costs of containers during the study period are not considered separately, i.e. the depreciation, repair, storage, insurance, chartering and other costs of containers are incorporated into the ship's costs; at the same time, it is assumed that the fixed costs of ships are fixed and can be determined in advance according to the actual operation.

3   Inland Container Liner Mathematical Problem

3.1   Illustration of Symbols and Parameters

See Table 1.

Table 1. Nomenclature

Nomenclature	
OD	Set of ports to be visited
Decision variables	
q_{od}^{rm}	transport volume of M ship on THE OD port pair of R line during the study period
β^r	= 1 if Liner companies lease space to alliance partners at route r
f^r	Number of cabins leased by alliance partners
Auxiliary variables	
$p_{o,d}^r$	Unit freight rate between OD port pairs on the r route
$q_{o,d}^r$	The volume between OD port pairs at route r
e^r	The charter price of an alliance partner at route r
$Fuel_r$	Fuel consumption at voyage
$Port_r$	Berth cost at port p route r
$Cont_r$	Handing cost at port p route r
α_k	Fuel consumption per unit time of m type ship on r route
T_k^r	The total time of a single voyage for m vessel on r route
C_p	Unit handling cost at port p route r
TN_k^r	Sailing time of single voyage of type M vessel at route r
TEO_k^r	Total inbound and outbound time for a single voyage of m vessel on r route
d_r	Distance per voyage of m type ship at route r
S_{rk}	Speed per voyage of m type ship at route r
PT	The total time for a ship to enter or leave the port once within the planning period
Parameters	
U^r	M vessel maximum cargo capacity per vessel
z^r	Maximum number of shipping space leased by liner company
b^r	Maximum number of shipping spaces leased by alliance company
N	The boat number of m type at route r
cl	The average cost of loading and unloading a container
Q_k	Other expenses in the fixed cost of daily operation at route r
H_k	Fixed costs of daily operations at route r, among other expenses
P_r	revenue of ships at route r during planning period
C_r	fixed cost of ships at route r during planning period
S_r	operation cost of ships at route r during planning period

3.2 Modeling

3.2.1 Model Objective Function

To solve the above problems, a mathematical model of shipping space leasing and allocation is established to maximize the total fleet profit for Liner Companies during the study period, as in Eq. (1) (total revenue minus total operating cost)]

$$Max\, Z_r = P_r - S_r - C_r \tag{1}$$

(1) Revenue of Ships at route r during planning period refers to the total profit of shipping containers transported by liner companies during the actual transportation of goods on THE OD port pair at route r which is affected by freight rate and volume, number of cabins rented and unit price of each cabin. Provided that it is assumed that freight rate and volume are known in advance, so the profit is expressed as in Eq. (2):

$$P_r = \sum_{o,d \in N} p_{o,d}^r q_{o,d}^r + e^r f^r \tag{2}$$

(2) Operation cost of Ships at route r during planning period refers to the total cost of shipping containers transported by liner companies during the actual transportation of goods which is affected by sailing distance, speed, time, number of attached ports, Berthing time in port fuel consumption and container loading influence. The operating cost of ships at route r in the planning period mainly contains of the fuel cost, including fuel, moisture and materials; port charges, including port charges for sluice and dam turning, which are not considered in this paper; as well as container handling fee. It is expressed as in Eq. (3)

$$S_r = Fuel_r + Port_r + Cont_r \tag{3}$$

Fuel consumption could be calculated as in Eq. (4):

$$Fuel_r = \sum_k (\alpha_k \cdot T_k^r \cdot N) \tag{4}$$

Berth cost at port p route r could be calculated as in Eq. (5):

$$Port_r = \sum_k C_p \tag{5}$$

Handling cost at port p route r could be calculated as in Eq. (6):

$$Load_r = cl \sum_{o \in N} \sum_{d \in N} q_{od}^r \tag{6}$$

It's assumed that there is no LCL transport situation, so in this system all THE OD port pair for container loading and unloading expenses are only related to the total number of containers, namely, the number of loading and unloading is constant, so we can equal the container handling fee and the product of unit price and the amount of container. The container handling fee doesn't change by optimization scheme. Thus, it could be treated as a constant value.

(3) Fixed cost of ships at route r during planning period

Fixed cost of ships at route r during planning period refers to ship once the leaves, is not affected by the container transport costs, including the crew wages, employee welfare, staff insurance premium, labor protection, ship maintenance fees, transportation and regulation fees, ship repair and other fees. For the simple model, we simply divided the fixed cost into two parts. It could be calculated as in Eq. (7):

$$C_r = \sum_k \frac{T_k^r \cdot N}{24}(H_k + Q_k) \tag{7}$$

At present, there is one variable to be processed. It contains round-trip sailing time and ship in-and-out time, which is calculated as in Eq. (8):

$$T_k^r = TN_k^r + TEO_k^r \tag{8}$$

As for the sailing time, it could be calculated as in Eq. (9):

$$TN_k^r = \frac{d_r}{S_{rk}} \tag{9}$$

As for the ship in-and-out time, it could be calculated as in Eq. (10):

$$TEO_k^r = \sum_{o \in N}\sum_{d \in N}(x_{od}^r \cdot PT) \tag{10}$$

Combined with what has been stated above, the objective function can be integrated into

$$Z_r = \sum_{o,d \in N^*} p_{o,d}^r q_{o,d}^r + e^r f^r - \sum_k \left(S_k^r \cdot \alpha_k \cdot T_k^r \cdot N \right)$$
$$- \sum_k C_p \cdot T_k^r - cl\sum_{o \in N}\sum_{d \in N}q_{od}^r - \sum_k \frac{T_k^r \cdot N}{24}(H_k + Q_k)) \tag{11}$$

3.2.2   Model Constraints

(1) Ship cargo carrying capacity constraint, the sum of all shipping Spaces starting from the same port O and those leased to alliance partners does not exceed the maximum carrying capacity of the ship:

$$\sum_d^D q_{od}^{rm} + f^r \leq U^r \forall o \in N \tag{12}$$

(2) Ship cargo carrying capacity constraint, the sum of all shipping Spaces terminating at the same D port does not exceed the sum of shipping Spaces leased from alliance partners and the maximum carrying capacity of ships:

$$\sum_{o}^{O} q_{od}^{rm} \leq U^r + f^r \forall d \in N \tag{13}$$

(3) Lease constraints (the number of slots rented to alliance partners on self-operated routes does not exceed the maximum shortage of alliance partners):

$$f \leq \beta^r z^r \tag{14}$$

(4) Lease constraints (the number of slots rented from the alliance partner on other routes does not exceed the number of slots available for leasing by the Alliance partner)

$$f^r \leq (1 - \beta^r) b^r \tag{15}$$

(5) Variable constraints (variables are non-negative and integers)

$$f, f^r, q_{od}^r \in N \tag{16}$$

4 Inland Container Liner Numerical Experiments

On condition that Wuhan Changhai international shipping agency co., LTD. (hereinafter referred to as "company A"), Wuhan Changwei international shipping industrial co., LTD. (hereinafter referred to as company B), Wuhan Chinese transportation maritime container transport co., LTD. (hereinafter referred to as "company C) The three companies have opened a total of 3 multi-port docking circular routes on both ends of Yangluo port, Shanghai port, a total of six port area of liner alliance. Company A operates R2, Company B R3, and Company C R4, while R1 is the direct route from Yangluo to Yangshan of Han-Shen Line. The route structure of the four routes involved is shown in the graph (Figs. 3, 4, 5 and 6).

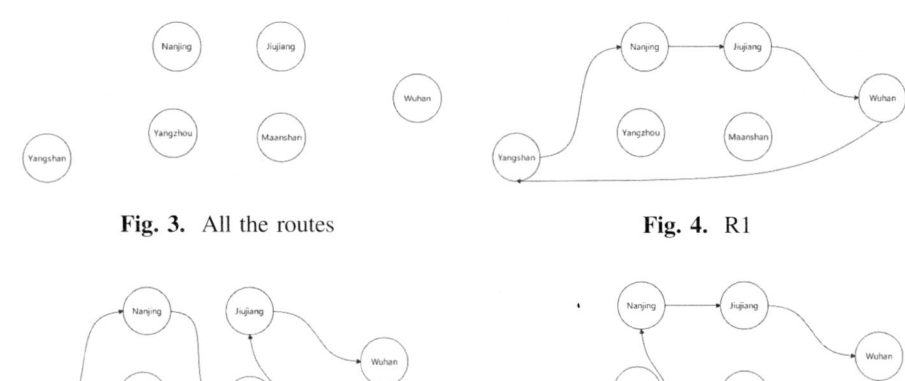

Fig. 3. All the routes

Fig. 4. R1

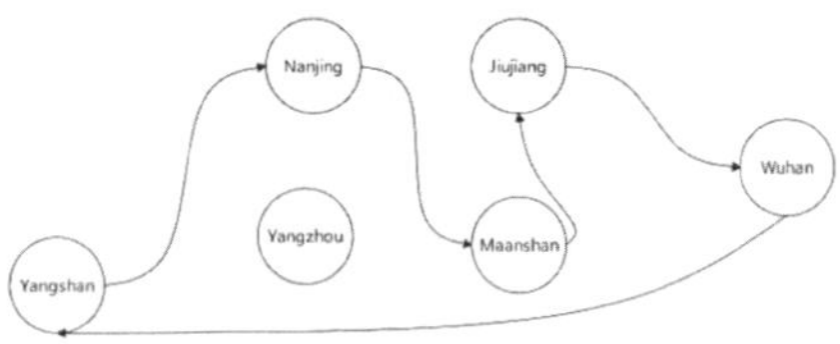

Fig. 5. R2

Fig. 6. R3

In addition, the transport capacity of company A on the R1 route is 2280TEU. According to the initial alliance agreement, the maximum number of cabins that company A can rent in R2 is 500TEU, and the maximum number of cabins that company B and Company C can rent in R3 and R4 are both 500TEU. For the convenience of listing, the port names are replaced as shown in the Table 2:

Table 2. Port code description

Port	Symbol
Yangshan port	A1
Yangzhou port	B1
Nanjing port	B2
Maanshan port	B3
Jiujiang port	B4
Wuhan port	B5

Due to the particularity of this route, the time from Yangluo port to Yangshan port is the optimal target, and port affiliation is almost not involved. Therefore, this study focuses on the return route affiliation. The research focus is on the shipping space leasing of inland river container liner. Provided that the known transport demand between ports, the research assumes that the transport volume between any two ports is determined, which is not the research content of this paper. Based on the assumption of the growth trend of container transport volume between ports returning from Han-Shen Line, the transport demand between ports operated by each company during the planning period (one week) is shown in the Table 3.

Table 3. Route related data.

R1	A1-B2	A1-B4	A1-B5	B2-B4	B2-B5	B4-B5	B5-A1	B1-A2	B2-A1	B2-A2	B3-A1	B3-A2
Unit Price	164	241	185	200	201	167	182	0	0	0	0	0
$\bar{D}$/TEU	210	430	93	382	407	183	221	0	0	0	0	0
δ/TEU	10	15	15	12	17	14	10	0	0	0	0	0
Volume limit	35	25	23	31	20	25	18	0	0	0	0	0
R2	A1-B2	A1-B3	A1-B4	A1-B5	B2-B3	B2-B4	B2-B5	B3-B4	B3-B5	B4-B5	B5-A1	B3-A2
Unit Price	164	179	241	185	218	200	201	189	211	167	182	0
$\bar{D}$/TEU	121	404	272	118	538	210	475	550	269	171	412	0
δ/TEU	18	15	11	11	15	15	19	14	14	18	17	0
Volume limit	21	20	28	20	22	25	29	25	15	24	16	0
R3	A1-B1	A1-B2	A1-B4	A1-B5	B1-B2	B1-B4	B1-B5	B2-B4	B2-B5	B4-B5	B5-A1	B3-A2
Unit Price	195	164	241	185	166	216	160	200	201	167	182	0
$\bar{D}$/TEU	207	376	282	203	463	426	512	434	463	331	493	0
δ/TEU	18	18	11	14	15	13	17	19	14	15	15	0
Volume limit	31	24	32	33	24	27	24	19	19	16	20	0

MATLAB was used to calculate and optimize respectively the profit of company A, B and C, while the other two companies were taken as constraints, and the mutual leasing results were obtained as shown in the Tables 4 and 5.

Table 4. Result of space exchange between routes

Route	R1	R2	R3
A	500	0	0
B	0	500	0
C	0	0	500

Table 5. Results of seat allocation for each route of Company A

R1	A1-B2	A1-B4	A1-B5	B2-B4	B2-B5	B4-B5	B5-A1	B1-A2	B2-A1	B2-A2	B3-A1	B3-A2
F20	183	258	93	382	364	183	221	0	0	0	0	0
R2	A1-B2	A1-B3	A1-B4	A1-B5	B2-B3	B2-B4	B2-B5	B3-B4	B3-B5	B4-B5	B5-A1	B3-A2
F20	500	0	0	0	500	0	0	500	0	500	500	0
R3	A1-B1	A1-B2	A1-B4	A1-B5	B1-B2	B1-B4	B1-B5	B2-B4	B2-B5	B4-B5	B5-A1	B3-A2
F20	500	191	0	0	309	0	0	500	0	500	500	0

According to the calculation, company A, B and C can obtain additional potential customers on other lines and increase their income after implementing the shipping space leasing of liner lines. On the basis of the model, variables and constraints related to mutual leasing are removed and the same method is used to calculate. Taking Company A as an example, its total revenue is reduced by 980,000 yuan. In order to verify the robustness of the model, on the basis of the above calculation example, two groups of different demand values are taken respectively, and other factors and data are controlled unchanged. The benefits are calculated and compared with the non-reciprocal leasing strategy, as shown in the Table 6:

Table 6. Comparison of benefits under different circumstances

Requirement taking value	Cabin swap	No cabin swap	Difference in value
D_{ij}	2564367	1577096	987271
$D_{ij} + \delta_{ij}$	2579910	1623442	956468
$D_{ij} - \delta_{ij}$	2550140	1522807	1027333

It can be seen from the table that adopting the mutual leasing strategy can bring more benefits to the shipping company through multiple groups of different demand values, which fully proves the robustness of the algorithm.

It can be seen from the above calculation results that the optimal model of space allocation for Wuhan thousand-container fleet of Han-Shen line based on alliance mutual leasing strategy constructed in this paper can not only expand the route coverage of the liner company, but also improve the space utilization rate of the liner company and bring additional operating benefits to the liner company. Therefore, the inter-leasing strategy of inland line liner alliance is a long-term win-win strategy of benefit sharing and risk sharing, and its optimal decision of inter-leasing and allocation is of vital importance to improve the market share and competitiveness of liner companies.

5 Conclusions

In the light of the characteristics of inland container liner shipping alliance space rent, each strategy is studied under the spandex optimization allocation problem of shipping space, the integer programming model is built to maximize the profit of the liner companies and example of Hanhai series is used to verify. Results of shipping space and rent and space distribution optimization decisions are got. It has certain reference value and guiding significance for liner alliance cooperation.

This study considers the space rent and space distribution of the decision, but only single optimal routes of a company operating decisions as the goal, while at this time of the other company are seen as a whole, without considering the influence of the individual. In the future study, we can build the model which considers the interests of the global optimization model from the perspective of the liner alliance as a whole in order to better realize the goal of win-win cooperation of liner alliance. In the meantime, this study ignores the difference of the two companies' operating cycles and has not added the constraint of shipping cycle. Time cycle can be added in the future to get closer to the actual operation situation.

References

An F (2014) Network planning and design for inland container liner shipping routes. Shanghai Jiao Tong University, Shanghai

Cariou P, Guillotreau P (2021) Capacity management by global shipping alliances: findings from a game experiment. Marit Econ Logist 24(1):41–66. https://doi.org/10.1057/s41278-021-00184-9

Chen J et al (2022) Liner shipping alliance management: overview and future research directions. Ocean Coast Manage 219:106039

Dulebenets MA (2018) A comprehensive multi-objective optimization model for the vessel scheduling problem in liner shipping. Int J Prod Econ 196:293–318

Huang Y (2007) The optimization of Yangtzi freshwater liner shipping routes. Shanghai Jiao Tong University, Shanghai

Jaramillo DI, Perakis AN (1991) Fleet deployment optimization for liner shipping Part 2. Implementation and results. Marit Policy Manage 18(4):235–262

Dulebenets MA, Pasha J, Abioye OF, Kavoosi M (2019) Vessel scheduling in liner shipping: a critical literature review and future research needs. Flex Serv Manuf J 33(1):43–106. https://doi.org/10.1007/s10696-019-09367-2

Mamédio D, Rocha C, Szczepanik D, Kato H, (2019) Strategic alliances and dynamic capabilities: a systematic review. J Strategy Manage 12(1):83–102

Xinlian X, Zhuoshang J, Yu Y (1993) Nonlinear programming for fleet planning. Int Shipbuild Prog 40(421):93–103

Yang Q, Xinlian X, Guangshi P (2011) Modeling and simulation of fleet planning for liner shipping. J Southwest Jiaotong Univ 46(6):1046–1054

Wu Z (1992) Shipping management. Dalian Mariner Transport College, Dalian

Zhao G (1997) Analysis and improvement on the vessel allocation model of liner service. J Syst Eng 12(1):80–86

Special Session

Analysis of Water Level Fluctuations in Bifurcating Approach Channels Under the Flow Regulation of Reservoirs

Zhiyong Wan[1,2(✉)], Yun Li[1(✉)], Jianfeng An[1], Xiaogang Wang[1], and Xiujun Yan[1]

[1] State Key Laboratory of Hydrology-Water Resources and Hydraulic Engineering, Nanjing Hydraulic Research Institute, Nanjing, China
zhiyongwan@whu.edu.cn, yli@nhri.cn
[2] School of Water Resources and Hydropower Engineering, Wuhan University, Wuhan, China

Abstract. An approach channel is a restricted channel connecting the lock head with the upstream and downstream navigation waterway, characterized by a narrow navigation channel with a closed-end, small coefficient of cross-section, and low speed for vessel navigating. Due to the uncertain impact of the waves induced by the flow regulation of reservoirs on water-level fluctuations in bifurcating approach channels, the river reach between the Three Gorges Dam (TGD) and the Gezhouba Dam (GZD) is selected as the study area and the variations of water-level fluctuating amplitude in the approach channel located on the left bank downstream of the TGD are revealed under the joint regulation of the Three Gorges and Gezhouba reservoirs based on numerical simulations. Furthermore, the relationship between water level variation at the lower lock head of the ship lift and the flow regulation of reservoirs is investigated in detail. Results indicated that water level variation of the ship lift approach channel exhibits an amplifying effect and the lower lock head of the ship lift is a key concern for navigable flow conditions. In addition, the flow variation rate should be controlled within the range of 2,000 m³/15 min in the context of joint regulation of two reservoirs, which enables the safe docking of vessels at the lower lock head of the ship lift. The findings in this study may contribute to the designing and planning of the newly-planned Three Gorges ship lock approach.

Keywords: Approach channel · Flow regulation · Reservoirs · Ship lift · Water-level fluctuations

1 Introduction

Reservoirs have been deployed as long-term investments to ensure water and energy security. However, the high transient flow triggered by the discharge regulation of reservoirs alters the hydrodynamic conditions of the natural river, which also exerts an impact on the navigation safety of vessels in the waterway to some extent (Liu et al. 2012). To satisfy the requirements of water supply, power generation, shipping and

© The Author(s) 2023
Y. Li et al. (Eds.): PIANC 2022, LNCE 264, pp. 1447–1459, 2023.
https://doi.org/10.1007/978-981-19-6138-0_128

flood control, the discharge regulation of reservoirs generally involves a combination of regulation schemes (Tefs et al. 2021); the differences in hydraulic parameters such as river flow variation rate and flow amplitude during the regulation of reservoirs throughout the year result in periodic rising and falling water along the river.

In recent years, most of the studies on wave propagation triggered by the flow regulation of reservoirs have been studied for hydraulic elements such as wave period and hydraulic gradient, with prototype observations, physical model tests, and numerical simulation techniques being the general technical tools employed to investigate the hydrodynamic phenomenon. Although prototype observation techniques intuitively reflect the real-world hydraulic variation pattern of the river, the hydrodynamic field coupled with external complicated environmental factors (e.g., ship-generated waves) in the large scale plane makes it costly and difficult to obtain accurate data; on the other hand, although the physical model addresses the issue of a larger planar scale, the scale effect between the physical model and prototype as well as the accuracy control of experimental apparatus greatly restrict the improvement of the agreement between the model test and prototype measuring results. In light of this, to overcome the unfavourable factors such as long observation periods and high costs, numerical modelling techniques have become an accepted technical aid for the shipping community to investigate navigable flow conditions. Nowadays, the famous shallow water equations have been widely used in the field of river hydrodynamics, and many methodologies upon the solutions of shallow water equations have been investigated (Guinot 2003; Yoon and Kang 2004). Simultaneously, the hydrodynamic processes have been reported in engineering applications using commercial software (Nguyen and Zheng 2012; Muñoz et al. 2021). However, for a multi-branched channel, in recent years, several literature has reported the relationship between sediment distribution and flow rate (Das et al. 2022; Du et al. 2016) rather than the navigational flow conditions in the approach channel.

Previous research shows that gravity wave flows driven by reservoir scheduling have a greater impact on safe navigation in the river reaches (Bravo and Jain 1991; Shang et al. 2017). Due to the limitations of the sampling frequency of the prototype data, there are many research results involving steady flow, while the unsteady flow is mainly focused on the analysis of the flow field and wave characteristics (Wan et al. 2020). When the wave generated by the joint operation of the hub evolves into a lock approach connecting the lock head with the upstream and downstream navigation channels, due to the small cross-section coefficient of the channel and the significant shallow water effect (PIANC 1997; Vantorre 2003; Kazerooni and Seif 2013), gravity long-wave and ship hull coupling may affect the comfort of vessels and additional safety risks for navigating vessels. More importantly, considering the operation of the ship lock and ship lift, the safe docking of vessels at the lower lock head of the ship lift and the constraint of the reverse hydraulic head of the mitre gate further hinder the enhancement of the navigation guarantee rate (Xu et al. 2020). Therefore, in the context of the development of inland navigation, previous studies on the navigable flow conditions of the approach channel have been conducted (Zhao et al. 2012; Ioan et al.

2016; Xie et al. 2016). However, these studies have mainly focused on single or double approach channels under single-stage reservoir operations, and the driving factor of water-level fluctuations has not been highlighted, nor have the results of the available analyses been in-depth. With the reconstruction or expansion of inland navigation structures (Duviella et al. 2018; Li et al. 2021; Yan et al. 2019), the arrangement of multi-lane navigation structures may emerge with multiple channels coupled, and their hydrodynamic processes threaten the safety of ship navigation.

With the increasing demand for hydropower industry and inland waterway transport freight, water-level fluctuations in multiple approach channels under the joint operation of the two reservoirs have become a hot topic, and its specificity is manifested in two aspects: (i) the scale, number of branches and layout type of approach channels; and (ii) the flow regulation mode under the joint operation of the two reservoirs. In light of this, to investigate the evolution of the waves in bifurcating approach channels driven by the joint regulation of the two reservoirs, and considering the flow regulation of the dual reservoirs and the typical layout of the bifurcating approach channels, this study attempts to explore the influence of the discharge regulation mode of the two reservoirs on water-level fluctuations in bifurcating approach channels based on a two-dimensional (2-D) hydrodynamic model and to reveal the intrinsic link between the hourly variation of water level in the approach channel and flow regulation mode of the two reservoirs.

2   General Description of the Study Area

The TGD and the GZD located along the Yangtze River are nearly 38.0 km apart, and the Three Gorges and the Gezhouba reservoirs constitute a typical dual-reservoir. The river section is mostly presented as U- or V-shaped in cross-section. To be specific, the river reach from the TGD site to Letianxi is a wide valley section around 9.6 km in length, which the river reach from Letianxi to Nanjinguan Station is 26.49 km with several bends such as Letianxi, Liantuo, Shipai, and Nanjinguan. For reservoir functioning, the Gezhouba reservoir serves as a counter-regulation reservoir for the TGD with a limited regulating capacity of merely 0.86 billion m^3, and the water level in front of the GZD varies within the range of 63.0–66.5 m.

In the waterway, the typical bifurcating approach channels consist of a ship lift approach, an existing lock approach, and a newly-planned lock approach. The ship lift approach channel merges into an existing lock approach 1,100 m downstream of the lock head and enters the main channel of the Yangtze River. The bottom width of the lock approach is around 180–200 m, and the elevation of the riverbed is 56.5 m. Several sampling sites in bifurcating approach channels are shown in Fig. 1.

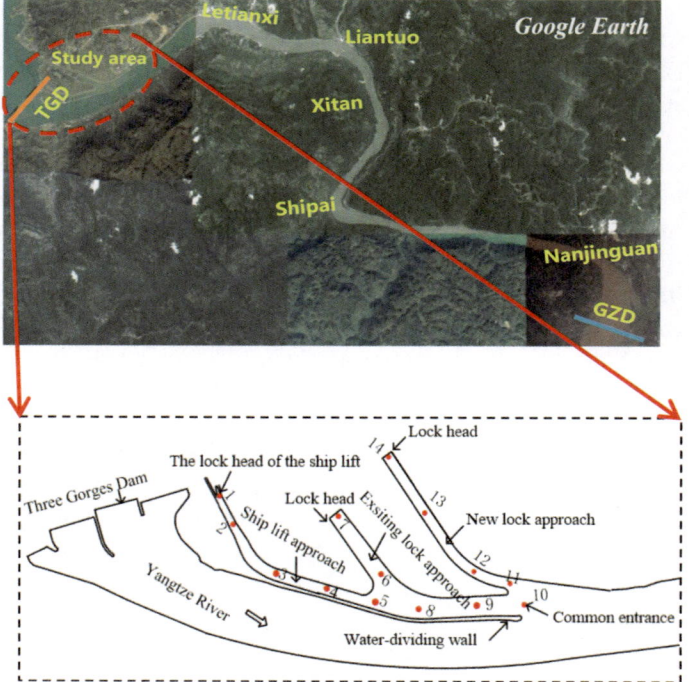

Fig. 1. Sampling locations of the study area.

3 Methodology

3.1 Hydrodynamic Model

In modelling the waterway flow, due to the horizontal scale being much larger than the vertical scale, the variation in water depth and velocity along the vertical direction is much smaller than that along the horizontal direction. Therefore, the three-dimensional Navier-Stokes equations can be averaged vertically to derive a set of depth-averaged two-dimensional equations, resulting in the following well-known two-dimensional St. Venant equations, including the continuity equation and momentum equation (without considering the effect of the Coriolis force and wind stress):

$$\frac{\partial h}{\partial t} + \frac{\partial hu}{\partial x} + \frac{\partial hv}{\partial y} = 0 \tag{1}$$

$$\frac{\partial uh}{\partial t} + \frac{\partial uuh}{\partial x} + \frac{\partial uvh}{\partial y} = \frac{\partial}{\partial x}\left(\gamma_t h \frac{\partial u}{\partial x}\right) + \frac{\partial}{\partial y}\left(\gamma_t h \frac{\partial u}{\partial x}\right) - gh\frac{\partial z}{\partial x} - g\sqrt{u^2 + v^2}/C^2 \tag{2}$$

$$\frac{\partial vh}{\partial t} + \frac{\partial uvh}{\partial x} + \frac{\partial vvh}{\partial y} = \frac{\partial}{\partial x}\left(\gamma_t h \frac{\partial v}{\partial x}\right) + \frac{\partial}{\partial y}\left(\gamma_t h \frac{\partial v}{\partial y}\right) - gh\frac{\partial z}{\partial y} - g\sqrt{u^2 + v^2}/C^2 \tag{3}$$

Where t is time; x and y are horizontal Cartesian coordinates; h is water depth; u and v are depth-averaged velocity components in x and y directions, respectively; g is gravitational acceleration; z is bed elevation; C is the Chezy coefficient; γ_t is the horizontal eddy viscosity coefficient.

The spatial discretization of the equations is performed using a cell-centred finite volume method, and the second-order spatial accuracy is achieved via employing a linear gradient reconstruction technique. The time integration of the governing equations is performed using the second-order Runge-Kutta method. This study focuses merely on water levels without considering turbulence effects and the code used is programmed based on the finite volume method. In addition, the integrated hydraulic roughness of the model is determined by the water level of the river at steady-state conditions.

3.2   Mesh Generation and Boundary Conditions

The computation domain of the river reach between the TGD and the GZD is gridded using triangular mesh, with an encrypted grid number of 51,717 in the approach channel and a grid scale of 5–10 m; and a grid scale of 20–50 m in the river channel. The number of nodes and cells is 90,643 and 174,970 in the model, respectively. The layout of the grid within the approach channel is shown in Fig. 2.

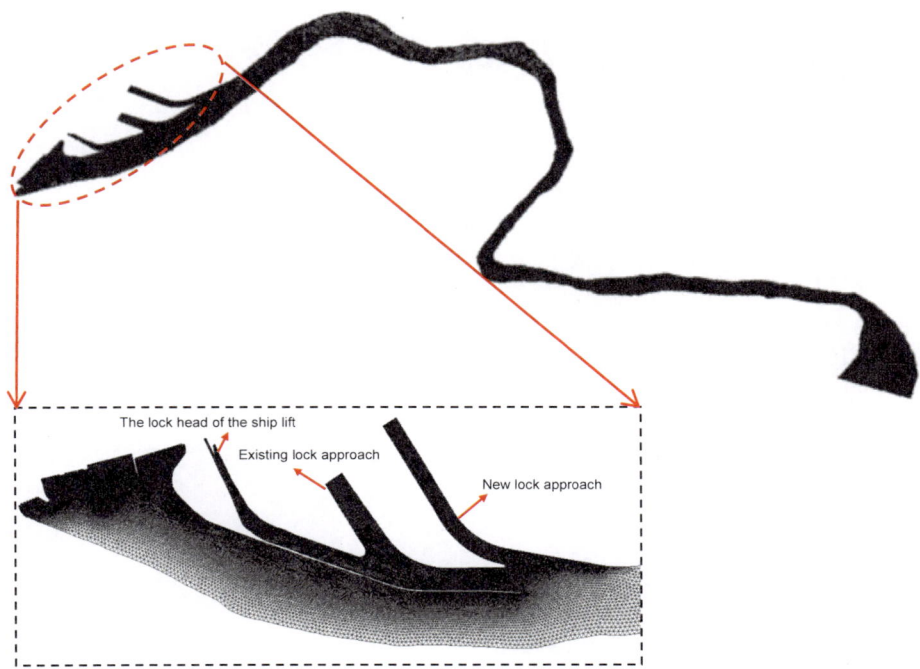

Fig. 2. Mesh generation in bifurcating approach channels.

Boundary conditions adopted for the model are set as follows: (i) the outflow boundary of the Three Gorges reservoir is set as the flow boundary; and (ii) the outflow boundary of the Gezhouba reservoir can be set as the flow boundary and water level boundary according to the specific scenarios investigated in this case study. The remaining boundaries are set as the land boundaries; normal fluxes were forced to zero for all variables along the closed boundaries.

3.3   Performance Evaluation Indices

To check and compare the performance of modelling approaches, two performance evaluation indices, including root mean square error (*RMSE*) and determination coefficient (R^2), are selected to assess the adaptability and feasibility of the simulation results. The expressions of these performance evaluation indices are listed as follows:

$$RMSE = \sqrt{\frac{1}{N} \sum_{i=1}^{N} (M_i - S_i)^2} \tag{4}$$

$$R^2 = \left[\sum_{i=1}^{N} (M_i - \overline{M})(S_i - \overline{S}) \right]^2 / \sum_{i=1}^{N} (M_i - M)^2 \sum_{i=1}^{N} (S_i - \overline{S})^2 \tag{5}$$

Where M_i and S_i are the measured values and calculated ones, respectively; while $\overline{M}$ and $\overline{S}$ are the mean of the measured and simulated values, respectively; N represents the number of observations.

4   Results and Discussion

4.1   Model Validation

To assess the reliability and stability of the established hydrodynamic model, an operational scenario is assumed as follows: the discharge flow of the upstream reservoir is set to 6,000 m³/s and the outflow variation rate of the downstream reservoir is set to 2,000 m³/s/15 min. As shown in Fig. 3, two indices, R^2 and *RMSE*, are utilized to evaluate the simulated and measured relative water depths corresponding to the initial water depths at the lower lock head of the ship lift and the entrance of the common approach channel.

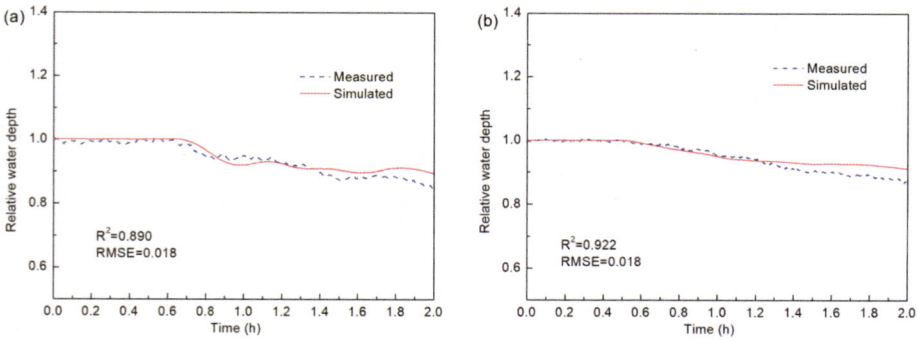

Fig. 3. Numerical simulations versus physical model tests. (a) The lower lock head of the ship lift; and (b) The entrance of the common approach channel.

A good correlation between the measured values and calculated ones can be found in Fig. 3. The simulated relative depth at the lower lock head of the ship lift is in accordance with that of the measured results from the physical model test, and the RMSE can reach 0.018. Overall, the established numerical model in this study can simulate the wave propagation in the approach channel.

4.2   Uncertainty Analysis of the Outflow Boundary of Reservoirs

Generally, the upstream reservoir is set as the flow boundary to satisfy the flow regulation of the dual reservoirs, whereas the boundary conditions of the downstream reservoir subjected to counter-regulation functioning can be divided into three types: constant water level, constant flow, and the outflow variation of the downstream reservoir is in accordance with that of the upstream reservoir. Therefore, assuming the following scenarios: the discharge flow of the Three Gorges reservoir is set to 6,000 m^3/s and the variation rate is set to 2,000 m^3/s/15 min; the outflow boundary at the GZD is set at the water surface elevation of 63.0 m, a constant flow of 6,000 m^3/s, and the same flow variation rate as that of the Three Gorges reservoir. Figure 4 shows the amplitude of water-level fluctuations in bifurcating approach channels.

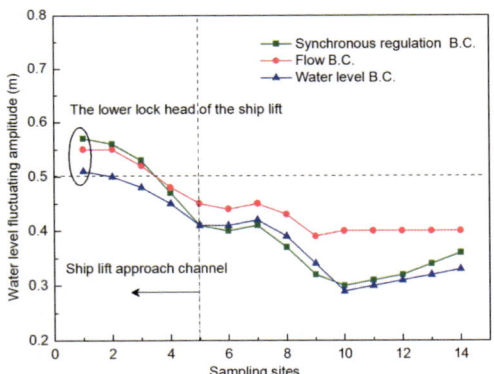

Fig. 4. The amplitude of water-level fluctuations under different boundary conditions

The retrograde wave generated by the constant discharge flow of the downstream reservoir enables the water level fluctuating amplitude at the lower lock head of the ship lift to rise continuously, nevertheless, the amplitude at the lower lock head of the ship lift under the constant water level boundary oscillates periodically. Furthermore, the wave evolves from the entrance to the channel, with a small variation of water level at the entrance and a greater fluctuating amplitude at the lock head. In addition, the water level variation at the lower lock head of the ship lift under the constant water level and flow boundary increases by around 25% from the entrance of the ship lift approach channel to the lower lock head, whereas the consistency of the flow variation process of the two reservoirs increases water level fluctuating amplitude at the lower lock head of the ship lift by 42.5%, which highlights the amplification effect of the water level variation at the lower lock head of the ship lift. Overall, under the flow variation rate of 2,000 m³/s/15 min, the maximum hourly variation of the water level at the lower lock head of the ship lift under three types of boundary conditions exceeds 0.5 m (MOT 2018), therefore, the safe docking of the vessels at the lower lock head of the ship lift fails to meet the engineering application requirements.

4.3   Synchronous Variation of Flow Processes at the Two Reservoirs

The flow conditions in the river reach between the TGD and the GZD should meet the requirements for safe navigation of vessels as well as the maximum hourly variation of the water level at the vicinity of the lower lock head of the ship lift. Considering the influence of flow variation and variation rate on the amplitude of water-level fluctuations in bifurcating approach channels, under the scenario that the outflow process of the dual reservoirs is consistent, the initial outflows from the upstream and downstream reservoirs are set to 6,000 m³/s, and the various rates of the flow increment are set to 2,000 m³/s/5 min, 2,000 m³/s/15 min, and 2,000 m³/s/30 min, respectively. The water level fluctuating amplitudes are shown in Fig. 5.

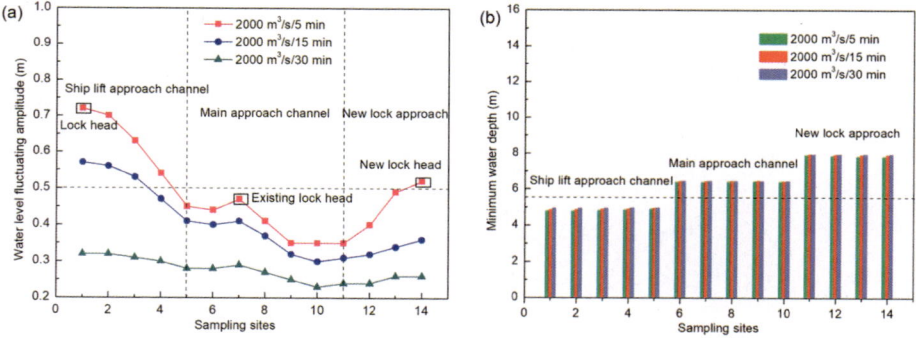

Fig. 5. The maximum hourly amplitude of water-level fluctuations at various flow variation rates and the minimum water depth in bifurcating approach channels. (a) The maximum hourly amplitude of water-level fluctuations, and (b) Minimum water depth in the approach channel.

As shown in Fig. 5, the water level variation in the approach channel increases with the growth of the flow variation rate. The water level variation at the lower lock head of the ship lift under the flow variation rate of 2,000 m³/s/5 min is 2.25 times that under the flow variation rate of 2,000 m³/s/30 min, and thus flow variability has a significant effect on the water-level fluctuation at the lower lock head of the ship lift. Notably, the amplification effect of wave amplitude is more obvious from the entrance of the ship lift approach channel to the lower lock head, especially when the flow variation rate is 2,000 m³/s/5 min, and the water level variation at the lower lock head of the ship lift under this scenario increases by 1.06 times. Furthermore, the minimum water depth corresponding to the trough in the approach channel meets the requirement of ship draft. In general, in the case of the flow variation rate of 2,000 m³/s/30 min, the amplitude of water level variation at the lower lock head of the ship lift is within the threshold value.

4.4   Time-Delay Effect of Flow Regulation

The flow regulation process of the two reservoirs has both synchronous and lagging characteristics, and in exceptional regulation situations there may be a lag in the flow regulation time of the downstream reservoirs affecting the water-level fluctuations in bifurcating approach channels. To clarify this issue, this subsection takes the river reach between the TGD and the GZD as an example and assumes that there is a hysteresis in regulation time for the two reservoirs. The base flow from the upstream reservoir is set to 6,000 m³/s, the initial water level of the downstream reservoir is set to 63.0 m, and the outflow variation rates of the reservoirs vary by 2,000 m³/s within 15 min. The propulsive wave reaches the GZD site for around 31 min. For this reason, 0, 31, 40, and 45 min following the occurrence of a flow pulse from the upstream reservoir are selected as the lag periods for the analysis, the amplitudes of water-level fluctuations at the entrance of the approach channel and the lower lock head of the ship lift at the flow variation rate of 2,000 m³/s/15 min are shown in Fig. 6.

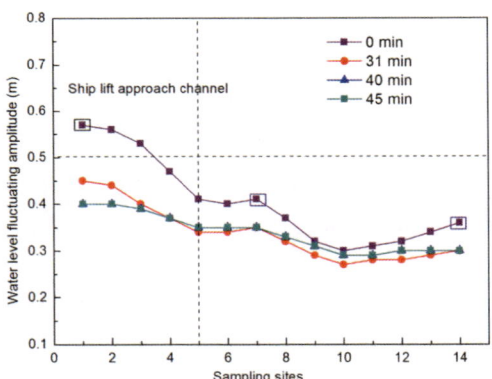

Fig. 6. Water level fluctuating amplitude considering flow regulation lag periods.

The linearized increment of pulse flow triggers the nonlinear process of water level variation in the river channel and approach channel. When the flow variation rates of the reservoirs increase by 2,000 m³/s/15 min synchronously, as presented in Fig. 6, the positive wave triggered in the upstream reservoir interacts with the negative falling wave triggered in the downstream reservoir, which leads to an increase in wave amplitude in the approach channel. As the flow regulation time of the downstream reservoir is extended, the initial base flow from the upstream reservoir increases from 6,000 to 8,000 m³/s, and the water level fluctuating amplitude decreases due to the increment of base flow. Hence, the lag of flow regulation could turn the scenario of the dual reservoirs regulation into a single-step regulation mode, and the water level fluctuating amplitude in the approach channel can be reduced to some extent.

4.5   The Constant Flow Operation Mode of Reservoirs

4.5.1   Three Gorges Reservoir with Constant Outflow

To explore the effect of flow amplitude on the water-level fluctuations in the approach channel, let the base flow from the Three Gorges reservoir be 6,000 m^3/s and kept constant, the initial water level of the waterway is set to 64.0 m, and the outflow of Gezhouba reservoir is increased by 1,000 m^3/s, 2,000 m^3/s, and 3,000 m^3/s respectively within 15 min, then the maximum hourly variation of water level in the approach channel under different flow amplitudes is shown in Fig. 7(a). Additionally, considering the effect of the retrograde wave triggered by the flow variation rate of the downstream reservoir outflow on the water-level fluctuations in the approach channel, the variation rates of the flow regulation of Gezhouba reservoir are set to 1,000 m^3/s/5 min, 1,000 m^3/s/15 min, and 1,000 m^3/s/25 min, respectively, then the maximum hourly variation of water level at typical monitoring points in the approach channel under various flow variation rates are shown in Fig. 7(b).

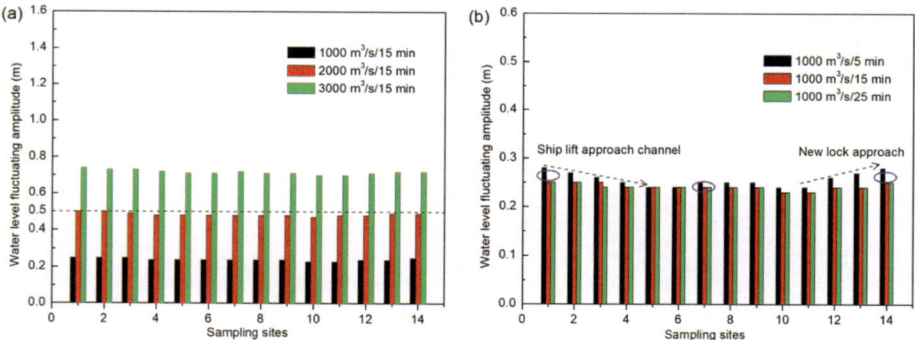

Fig. 7. Maximum hourly variation of water level at various flow amplitude and variation rates. (a) Flow amplitude, and (b) Flow variation rate.

The differences in flow amplitude and flow variation rate during the regulation of the downstream reservoir lead to the retrograde falling waves propagating upstream and evolving into the bifurcated approach channel, and the flow pulse magnitude and variation rate directly affect the amplitudes of water-level fluctuations in the approach channel. As shown in Fig. 7(a), the flow amplitude and maximum hourly water level fluctuating amplitude show a linear correlation, in particular, for every additional 1,000 m^3/s, the water level variation at the lower lock head of the ship lift increases by 0.23–0.25 m; however, the water level fluctuating amplitude at various flow variation rates under the flow amplitude of 1,000 m^3/s satisfies the requirements for safe docking at the lower lock head of the ship lift (Fig. 7(b)).

4.5.2   Gezhouba Reservoir with Constant Outflow

Under the scenario of a constant discharge flow of 6,000 m³/s from the Gezhouba reservoir, the following scenarios are assumed to analyze the amplitude of water-level fluctuations in a multi-bifurcated approach channel: (i) the discharge flow from the Three Gorges reservoir increases by 1,000 m³/s, 2,000 m³/s, and 3,000 m³/s within 15 min, respectively; and (ii) the flow variation rates are 2000 m³/s/5 min, 2,000 m³/s/15 min, and 2,000 m³/s/25 min, respectively. The maximum hourly variation of the water level at typical sampling sites in the approach channel under these scenarios is shown in Fig. 8.

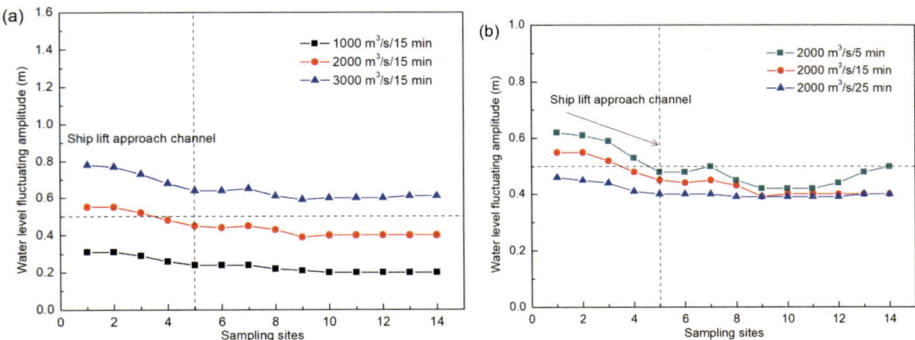

Fig. 8. Maximum hourly variation of water level at various flow amplitudes and variation rates. (a) Flow amplitude, and (b) Flow variation rate.

When the discharge flow of the GZD remains constant, the operation of the Three Gorges reservoir results in a prograde rising wave in the river and approach channel. Compared with the water level variations in other stations in the approach channel, it is found that flow transient regulation leads to a significant amplification effect of the water level fluctuating amplitude at the lower lock head of the ship lift. In particular, when the flow variation rate exceeds 2,000 m³/s/15 min, the lower lock head of the ship lift fails to meet the safety docking requirements of the ship-bearing compartment. Additionally, in comparison with the constant flow conditions of the TGD, the retrograde wave generated by the flow regulation of the Gezhouba reservoir has a larger wave energy loss due to the boundaries reflection by the bends such as Nanjinguan, Shipai, Liantuo, and Letianxi. With the same flow variation rate, the rising wave amplitude at the lower lock head of the ship lift triggered by the outflow of the upstream reservoir is larger than the falling wave amplitude caused by the operation of the downstream reservoir.

5   Conclusions

In this study, a hydrodynamic model is applied to investigate the influence of unsteady flow induced by the flow regulation of the two reservoirs on the water-level fluctuations in bifurcating approach channels, and combined with the typical operating conditions in the course of the regulation of reservoirs, flow amplitude and variation rate suitable for the safe docking of vessels at the lower lock head of the ship lift is explored, which may provide quantitative control for shipping safety. The following conclusions can be drawn from this study.

(1) In the navigational scheduling system including the ship lift approach and the multi-lane lock approach, wave energy transport in bifurcating approach channels is driven by various boundary conditions, which leads to the most significant wave amplitude at the lower lock head of the ship lift. In particular, the special narrow hydraulic section of the ship lift approach channel causes the wave height to be amplified at the lock head during the evolution of gravity waves.

(2) The difference in the scheduling mode of the dual reservoirs triggers high transient flow, which breaks the volumes of water balance between the two dams, and the flow amplitude and variation rate have a great impact on the water-level fluctuations in bifurcating approach channels.

(3) When the variation rate of the reservoir outflow is 2,000 m^3/s/15 min, the maximum hourly variation of the water level at the lower lock head of the ship lift exceeds the critical threshold value (0.5 m). The actual operation of the project should be avoided. However, the maximum hourly variation of water level at the lower lock head of the ship lift can be reduced provided that the simultaneous regulation of the dual-stage reservoir is generalized to interval scheduling.

Finally, it should be noted that the effect of flow variation rate and the flow amplitude on the water-level fluctuations in the approach channel is merely discussed from the perspective of the typical flow regulation modes, whereas the propagation processes of the gravity wave in the approach channel under more complex conditions are further enhanced in the subsequent study.

Acknowledgements. This research was jointly funded by the National Key Research and Development Program of China (2016YFC0401906) and the Three Gorges Follow-up Work Research Project (SXXTD-2018-8).

References

Bravo HR, Jain SC (1991) Flow fields in lower lock approaches induced by hydro-plant releases. J Waterw Port Coast Ocean Eng 117:369–389

Das A, Barman BC, Nandi N (2022) On some aspects of flow characteristics of the bifurcated channel–an experimental approach. ISH J Hydraul Eng 11:1–7

Du Q, Tang HW, Yuan SY, Xiao Y (2016) Predicting flow rate and sediment in bifurcated river branches. In: Proceedings of the Institution of Civil Engineers-Water Management, vol 169, no 4. Thomas Telford Ltd, pp 156–167

Duviella E, Doniec A, Nouasse H (2018) Adaptive water-resource allocation planning of inland waterways in the context of global change. J Water Resour Plan Manag 144(9):04018059

Guinot V (2003) Riemann solvers and boundary conditions for two-dimensional shallow water simulations. Int J Numer Meth Fluids 41(11):1191–1219

Ioan V, Beilicci R, Beilicci E, Visescu M (2016) approach channel modelling with advanced hydroinformatic tool, case study: small hydro power plant Huta Certeze, Romania. Procedia Eng 161:898–903

Kazerooni MF, Seif MS (2014) Experimental evaluation of ship squat in shallow waters. J Teknol (Sci Eng) 66(2):15–20

Li WJ, Wang DW, Yang SF, Yang W (2021) Three Gorges project: benefits and challenges for shipping development in the upper Yangtze river. Int J Water Resour Dev 37(5):758–771

Liu ZH, Ma AX, Cao MX (2012) Shuifu-Yibin channel regulation affected by unsteady flow released from Xiangjiba hydropower station. Procedia Eng 28:18–26

MOT (Ministry of Transport). Changjiang River Administration of Navigational Affairs (2018) Three Gorges-Gezhouba Water Conservancy Hub Navigation Scheduling Regulations, Wuhan, China. (In Chinese)

Muñoz DF, Yin D, Bakhtyar R, Moftakhari H, Xue Z, Mandli K, Ferreira C (2021) Inter-model comparison of Delft3D-FM and 2D HEC-RAS for total water level prediction in coastal to inland transition zones. JAWRA J Am Water Resour Assoc 1–16

Nguyen VT, Zheng JH (2012) Preliminary study of regulation schemes for navigation channel in Dinh An Estuary, Vietnam. Appl Mech Mater 212–213:117–122

Pianc P (1997) Approach channels—a guide for design. Report of working group II-30, 95, 14–28

Shang YZ, Li XF, Gao XR, Guo YX, Ye YT, Shang L (2017) Influence of daily regulation of a reservoir on downstream navigation. J Hydrol Eng 22(8):05017010

Tefs AAG et al (2021) Simulating river regulation and reservoir performance in a continental-scale hydrologic model. Environ Model Softw 141(2021):105025

Vantorre M (2003) Review of practical methods for assessing shallow and restricted water effects, In: International Conference on Marine Simulation and Ship Maneuverability (MARSIM 2003), Kanazawa, Japan, pp WS-4-1/11

Wan ZY, Li Y, Wang XG, An JF, Dong B, Liao YP (2020) Influence of unsteady flow induced by a large-scale hydropower station on the water level fluctuation of multi-approach channels: a case study of the Three Gorges Project, China. Water 12(10):2922

Xie MF, Zhou JZ, Li CL, Lu P (2016) Daily generation scheduling of cascade hydro plants considering peak shaving constraints. J Water Resour Plan Manag 142(4):04015072

Xu JC, Li Y, Xuan GX, Melville BW, Macky GH (2020) Numerical simulation of turbidity current in approach channels with a closed end. J Waterw Port Coast Ocean Eng 146(5):04020036

Yan T, Yang YP, Li YB, Chai YF, Cheng XB (2019) Possibilities and challenges of expanding dimensions of waterway downstream of Three Gorges dam. Water Sci Eng 12(2):136–144

Yoon TH, Kang S-K (2004) Finite volume model for two-dimensional shallow water flows on unstructured grids. J Hydraul Eng 130(7):678–688

Zhao J, Zhao SQ, Cheng XZ (2012) Impact upon navigation conditions of river reach between the two dams by peak shaving at Three Gorges hydropower station. Procedia Eng 28:152–160

Analysis on the Characteristics of Channel Scour and Deposition in the Nanjing Reach of the Yangtze River After Impoundment of the Three Gorges Reservoir

Nairu Wang[1(✉)], Suning Huang[2], Shuang Cao[3], Hongyu Zhang[4], and Taotao Zhang[5]

[1] Nanjing Hydraulic Research Institute, Nanjing, China
nrwang@nhri.cn
[2] Nanjing Qinhuai River Administration Office, Nanjing, China
[3] Survey Bureau of Hydrology and Water Resources of Lower Reaches of Changjiang River, Nanjing, China
[4] Yangzhou Water Conservancy Bureau, Yangzhou, China
[5] Hohai University, Nanjing, China

Abstract. The construction and operation of the Three Gorges reservoir (TGR) alters the water and sediment conditions in the downstream channels, rendering the sediment transport capacity of water flow in these downstream channels at an unsaturated state, subsequently leading to scour and affecting flood prevention. Therefore, this study statistically analyzed the variations of channel scour and deposition in the Nanjing reach since the operation of the TGR, using the measured hydrologic and channel topographic data. The results showed that, the scour in the Nanjing reach was weak during 2001–2006, but prominently intensified during 2006–2020, with the multi-year average bankfull channel scour amount being 745.5×10^4 m³/a for the entire Nanjing reach. For the period of 2020.11–2021.03 after the basin-scale big flood in 2020, the scour was further intensified, with the total bankfull channel scour amount reaching 2677×10^4 m³. Since 2006, the Nanjing reach demonstrated an overall trend of "scour in both main channel and floodplain", and bankfull channel scour was dominant, with the scour amount being 85% of that of the flood channel scour amount. Furthermore, scour was more intense on the left bank than on the right bank. The left branch of Xinjizhou reach and its downstream reaches were all erosional, demonstrating a pattern of scour amount being larger in the upper reach and smaller in the lower reach. Also, the multi-year average scour amount was the greatest at the Longtan and Yizheng reach, while the multi-year average scour intensity was the strongest at the main reach upstream the Meizizhou bifurcated reach.

Keywords: The lower reaches of the Yangtze River · The Nanjing reach · Channel scour-deposition · The Three Gorges reservoir (TGR) · Measurement data

© The Author(s) 2023
Y. Li et al. (Eds.): PIANC 2022, LNCE 264, pp. 1460–1470, 2023.
https://doi.org/10.1007/978-981-19-6138-0_129

1 Introduction

Since the operation of 8 reservoirs (including Three Gorges reservoir, Liyuan reservoir, Ahai reservoir, Jinanqiao reservoir, Longkaikou reservoir, Ludila reservoir, Guanyinyan reservoir, Xiangjiaba reservoir, and Xiluodu reservoir) in 2003, the sediment runoff in the middle and lower reaches of Yangtze River has considerably declined. As a consequence, riverbed scour and deposition of the reaches has long been in the process of adaptive adjustment (Xu et al. 2019). Due to that scour and deposition of river channels downstream the TGR directly affects flood prevention, shipping and socio-economic development on both riverbanks (Yuan et al. 2014), researchers have paid great attention to investigating the characteristics of riverbed scour and deposition in the lower reach of the TGR has broken the original relative scour-deposition balance of river channels in the middle and lower reaches of Yangtze River, shifting the status from pre-impoundment "scour upstream, deposition downstream" and "scour in main channel, deposition in floodplain" to post-impoundment "scour in both main channel and floodplain", and such all-the-way scour has developed to the Datong reach. The research of Hu et al. (2017) showed that, after the operation of the TGR, river channel scour kept developing downstream and parts of the reach swung frequently, leading to more complex evolution of scour and deposition. The Nanjing reach is in the lower reach of the Yangtze River, within which the bottomlands are quite well-developed and the evolutions of river channel scour and deposition are complex (Ling et al. 2021; Zang et al. 2021). Around 1985, the Nanjing reach shift from depositional to erosional. Then, after channel regulation for many years and under the effects of natural nodes and bank-protection projects, the overall river regime of the Nanjing reach was generally stable. However, parts of the channel still experienced intense scour and deposition variations (Qu et al. 2008; Wang et al. 2007). Associated with the construction and operation of the TGR and other reservoirs along the main stream, the adjustive characteristics of scour and deposition of the Nanjing reach would experience certain new changes, due to the impact of altered sediment content in upstream water. Therefore, there is an urgent need to re-evaluate the recent scour-deposition evolution pattern of Nanjing reach.

At present, most relevant studies mainly focus on the overall scour-deposition characteristics of Yichang-Datong reaches in the middle and lower reaches of the Yangtze River, while little attention has been paid to the spatiotemporal scour-deposition variation characteristics of the Nanjing reach across the 20 years. In the meantime, relative research has seldom included the data after the basin-scale big flood of the Yangtze River in 2020. This study will, based on analysis of the water and sediment characteristics of the Nanjing reach, use successive and the newest channel topographic data to calculate scour and deposition via four water levels (flooding, bankfull, medium, low), thereby analyzing the spatiotemporal scour-deposition variation characteristics of the Nanjing reach after the operation of the TGR. Research findings from this study would provide the latest scientific evidence for the protection and exploitation of the Yangtze River bank line in the new era.

2 Overview of the Nanjing Reach of Yangtze River

The Nanjing reach of the Yangtze River extends from the upstream Cihu stream outlet to the downstream Dadao stream outlet (Fig. 1), with a total length of 95.5 km. Within the Nanjing reach, the bottomlands are quite well developed, with the plane form being alternatively wide and narrow, lotus rhizome knot shaped, and bifurcated. From upstream to downstream, there are Xinjizhou branching channel, Meizizhou branching channel, Baguazhou branching channel, and Longtan bend channel segment (Wang et al. 2020). The mainstream direction exhibits a continuous wave shape, flowing through Xinshengzhou and Xinjizhou right branches, reaching the Qiba node on the left bank via Jiqian watercourse, turning right towards the head of Meizizhou at Dashengguan, entering Xiaguan and Pukou narrow sections via Meizizhou and Qianzhou left branch, turning to Yanziji on the right bank via the right border of the head of Baguazhou, transiting to Tianhekou on the left bank, bending towards the Xinshengyu port area on the right bank, turning to the Xiabatou node at the confluence segment of Baguazhou, finally turning to the Longtan bend on the right bank and entering the Yizheng watercourse via Sanjiangkou node.

3 Water and Sediment Characteristic Variations of the Nanjing Reach

Variations in the conditions of incoming water and sediment are directly linked to the adjustment of scour and deposition. The Nanjing reach is a tidal reach, and the multi-year measured data of Nanjing tidal level station show that the runoff effect of the upper reach of Yangtze River is the main factor affecting the riverbed scour-deposition evolution of Nanjing reach. The Datong hydrometric station is the control station of Yangtze River runoff. According to statistics, the amount of water inflow into the Yangtze River from main stream downstream the Datong hydrologic station accounts for approximately 3% of the total water discharge of the station. Therefore, the amount of incoming water and sediment in the lower reach of Yangtze River is determined using the data of Datong hydrologic station (Table 1).

Before the impoundment of the TGR, the multi-year average runoff of Datong station was $9051 \times 10^8 \text{ m}^3$, and the sediment runoff was 4.27×10^8 t. After impoundment of the TGR, the incoming water and sediment from upper reach of Yangtze River declined in certain years. Compared to the pre-impoundment average values, the multi-year average average runoff between 2003–2020 for the Datong station varied slightly, being approximately 3.1% lower. However, since the reduced incoming sediment from upstream of the reservoir and sediment detention by the TGR, the sediment load of downstream of the reservoir declined greatly. Between 2003–2020, the multi-year average sediment load of Datong station was 1.34×10^8 t, falling by 68.6% in comparison to the pre-impoundment value. Meanwhile, the multi-year average sediment concentration decreased to 0.151 kg/m^3.

In 2020, another basin-scale big flood took place in the Yangtze River, following previous two in 1954 and 1998. The big flood in 2020 led to historical high water levels in the Nanjing reach of the Yangtze River. In 2020, the maximum water discharge of Datong station was 83800 m^3/s, and the multi-year average runoff was $11180 \times 10^8 m^3$, increasing by 23.5% and 27.5% compared to the pre-impoundment and post-impoundment multi-year average runoffs, respectively. Of the of entire year 2020, there were 48 days when the water level of Nanjing Xiaguan station exceeded the warning water level. While of the 48 days, there were 6 days when the water level was higher than the highest water level ever recorded, reaching a maxima of 10.39 m (on 21 July), which was 0.17 m higher than the previous historic high. The multi-year average sediment load of Datong station in 2020 was 1.64×10^8 t, remarkably decreasing by 61.6% compared to the pre-impoundment multi-year average value, but increasing by 22.4% in comparison to the post-impoundment multi-year average sediment load (Ling et al. 2021).

Table 1. Statistical table of runoff and sediment characteristics of Datong hydrologic station.

Item	Period of statistics	Characteristic value
Runoff/$10^8 m^3$	Pre-impoundment multi-year average (1950–2002)	9051
	Post-impoundment multi-year average (2003–2020)	8767
	Rate of change	−3.1%
Sediment load/10^4t	Pre-impoundment multi-year average (1950–2002)	42700
	Post-impoundment multi-year average (2003–2020)	13400
	Rate of change	−68.6%
Sediment concentration/kg·m^{-3}	Pre-impoundment multi-year average (1950–2002)	0.473
	Post-impoundment multi-year average (2003–2020)	0.151

4 Computational Analysis of Scour-Deposition Variations of the Nanjing Reach

4.1 Study Area, Data Source and Computing Method

This paper chose the section of the Nanjing reach as the study area, being the segment from Xinjizhou (Xinshengzhou) branch to Longtan-Yizheng watercourse (Fig. 1), with a total length of 95.5 km. To facilitate computational analysis, this study divided the Nanjing reach into 9 sub-segments, namely left branch of the Xinjizhou (Xinshengzhou) reach (L1), right branch of the Xinjizhou (Xinshengzhou) reach (R1), main reach upstream the Meizizhou bifurcated reach (M1), left branch of the Meizizhou reach (L2), right branch of the Meizizhou reach (R2), main reach upstream the Baguazhou bifurcated reach (M2), left branch of the Baguazhou reach (L3), right branch of the Baguazhou reach (L3), and Longtan and Yizheng reach (M3). Based on the segmentation, this study analyzed the scour-deposition conditions of the entire Nanjing reach and the 9 sub-segments.

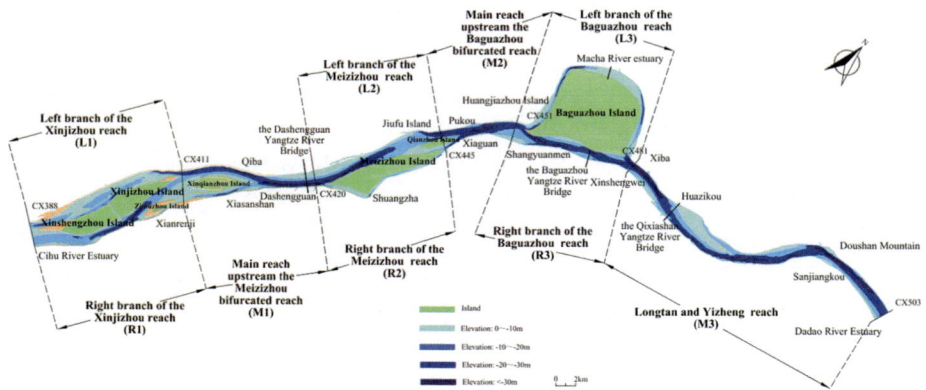

Fig. 1. Schematic of the study area.

The data used include 1:10000 fluvial geomorphology and observational data of fixed sections of the Nanjing reach in 1998, 2001−2020 at a 5-years interval, and 2021. This study adopted the commonly used cross-section method to calculate the scour and deposition of the Nanjing reach. The interval between two fixed sections is about 1.0 km, and reduced to 500 m for bending sections and sections with great changes. In total, there are 128 sections. Generally, the river cross-sections are extracted using the ratio of 1:5000, namely that the mean transverse point-to-point distance is 30−40 m, and reduced to 20 m for scarp and local areas with dramatic topographical changes.

To understand the scour and deposition conditions of different parts of the river channel, the current study uses water levels corresponding to water discharge of 10000 m³/s, 30000 m³/s, 45000 m³/s and 60000 m³/s to calculate the sediment scour and deposition amount in low-water channel, medium-water channel, bankfull water channel and flood water channel, respectively. Since the analyzed reach is quite long and the upstream-downstream water level gradient is large, this study refers to previous numerical simulation results to determine the calculated water levels. Table 2 lists the calculated water levels for different segments.

Table 2. Calculated water levels of control sections for computing scour and deposition of Nanjing reach.

Channel segment numbers	Channel segments	Channel length /km	Calculated water levels (m)							
			Discharge of Datong/m³·s⁻¹ (10000)		Discharge of Datong/m³·s−1 (30000)		Discharge of Datong/m³·s⁻¹ (45000)		Discharge of Datong/m³·s⁻¹ (60000)	
			Initial section	End section	Initial section	End section	Initial section	End section	Initial section	End section
L1	Left branch of the Xinjizhou reach	14.8	1.8	1.72	4.68	4.48	6.32	6.03	7.43	7.08
R1	Right branch of the Xinjizhou reach	14.7	1.8	1.72	4.68	4.48	6.32	6.03	7.43	7.08
M1	Main reach upstream the Meizizhou bifurcated reach	13.2	1.72	1.65	4.48	4.29	6.03	5.77	7.08	6.76
L2	Left branch of the Meizizhou reach	13.4	1.65	1.58	4.29	4.11	5.77	5.53	6.76	6.46
R2	Right branch of the Meizizhou reach	14.8	1.65	1.58	4.29	4.11	5.77	5.53	6.76	6.46
M2	Main reach upstream the Baguazhou bifurcated reach	7.3	1.58	1.56	4.11	4.03	5.53	5.41	6.46	6.3
L3	Left branch of the Baguazhou reach	23.3	1.56	1.53	4.03	3.9	5.41	5.21	6.3	6.03
R3	Right branch of the Baguazhou reach	12.4	1.56	1.53	4.03	3.9	5.41	5.21	6.3	6.03
M3	Longtan and Yizheng reach	31.2	1.53	1.45	3.9	3.56	5.21	4.72	6.03	5.37
N1	the entire Nanjing reach	95.5	1.8	1.45	4.68	3.56	6.32	4.72	7.43	5.37

4.2 Variations of Channel Scour and Deposition

To facilitate description, this study, according to the operations of the TGR and the reservoir group at main stream of upstream Yangtze River, divides the period into 6 stages: pre-impoundment (1998.10–2001.10), impoundment of TGR (2001.10–2006.10), early operation of TGR (2006.10–2011.10), successive operation of main stream reservoirs (2011.10–2016.10), utilization of the main stream reservoir group (2016.10–2020.11), and the period after the basin-scale big flood (2020.11–2021.03). The spatiotemporal variation characteristics of channel scour and deposition for each stage are described below.

After impoundment, the TGR had intercepted over 70% of the inbound sediment, thereby greatly reducing the amount of outbound sediment and leading to prominent scour-deposition adjustment of main stream in middle and lower reaches of Yangtze

River. In particular, since 2006, the Nanjing reach has experienced rather intense scour. During 2006–2021, the multi-year average scour amount of bankfull channel for the entire reach reached 874.3 × 104 m³/a, being approximately 16 times of the multi-year average value for 2001–2006. Scour has become dominant for the whole Nanjing reach, especially for the Longtan and Yizheng reach. Figure 2 and Table 3 compare the amount of scour or deposition of Nanjing reach in different periods.

Prior to the impoundment of the TGR (1998.10–2001.10), owing to the channel-forming effect of the Yangtze River flood in 1998, the Nanjing reach experienced intense scour. For the whole reach, the total sediment scour amount corresponding to low-water channel, bankfull channel and flood channel was 6405 × 104 m³, 6346 × 104 m³, and 7384 × 104 m³, with the multi-year average scour intensity being 2.36 × 104 m³/(km·a), 122.15 × 104 m³/(km·a) and 25.77 × 104 m³/(km·a), respectively. The overall scour-deposition status was "scour in both main channel and floodplain" and dominated by low-flow channel scour.

Since impoundment of the TGR (2001.10–2021.03), the total bankfull channel scour of the entire Nanjing reach was 13368 × 104 m³, corresponding to a multi-year average scour amount of 668.4 × 104 m³/a and a multi-year average scour intensity of 7.00 × 104 m³/(km·a). Along the Nanjing reach, scour amount was higher at the main reach upstream the Meizizhou bifurcated reach, the left branch of the Meizizhou reach, and Longtan-Yizheng reach, with the total scour amount of the three reaches accounting for 80.3% of the total scour amount of the whole Nanjing reach. Of this period:

During the cofferdam impoundment period (2001.10–2006.10), the scour of Nanjing reach slowed down, with the total bankfull channel scour amount being merely 254 ×104 m³, corresponding to the scour intensity of 0.53 × 104 m³/(km·a). According to the distribution along the Nanjing reach, the scour is the most intense at the right branch of the Xinjizhou reach.

Since 2006, the riverbed scour of Nanjing reach has prominently increased (Figs. 2c–2f, Table 3). During 2006–2021, the total bankfull channel scour amount of Nanjing reach reached 13114 × 104 m³, and the multi- year average scour intensity was 9.15 × 104 m³/(km·a), being approximately 16 times of the multi-year average scour intensity between 2001–2006. From the early operation of the TGR to the successive operation of reservoirs at the main stream of Yangtze River (2006.10–2016.10), the total bankfull channel scour amount was 12150 × 104 m³, while the multi-year average scour intensity was 12.72 × 104 m³/(km·a), being much greater than that of the impoundment period. Meanwhile, the scour amount was the highest at Longtan and Yizheng reach, while the scour intensity was the strongest at the Main reach upstream the Meizizhou bifurcated reach. After the utilization of the main stream reservoir group (2016.10–2020.11), almost the riverbed along the entire Nanjing reach has experienced deposition, with the total bankfull channel deposition about being 1713 × 104 m³ and the multi-year average deposition intensity being 4.48 × 104 m³/(km·a). Also, the deposition amount was the highest at the left branch of the Xinjizhou reach, accounting for 57.9% of the deposition of the whole Nanjing reach. After the basin-scale big flood in 2020 (2020.11–2021.03), the Nanjing channel section again experienced intense scour, the total bankfull channel scour amount reached 2677 × 104 m³, the multi-year average scour intensity was 28.03 × 104 m³/(km·a),

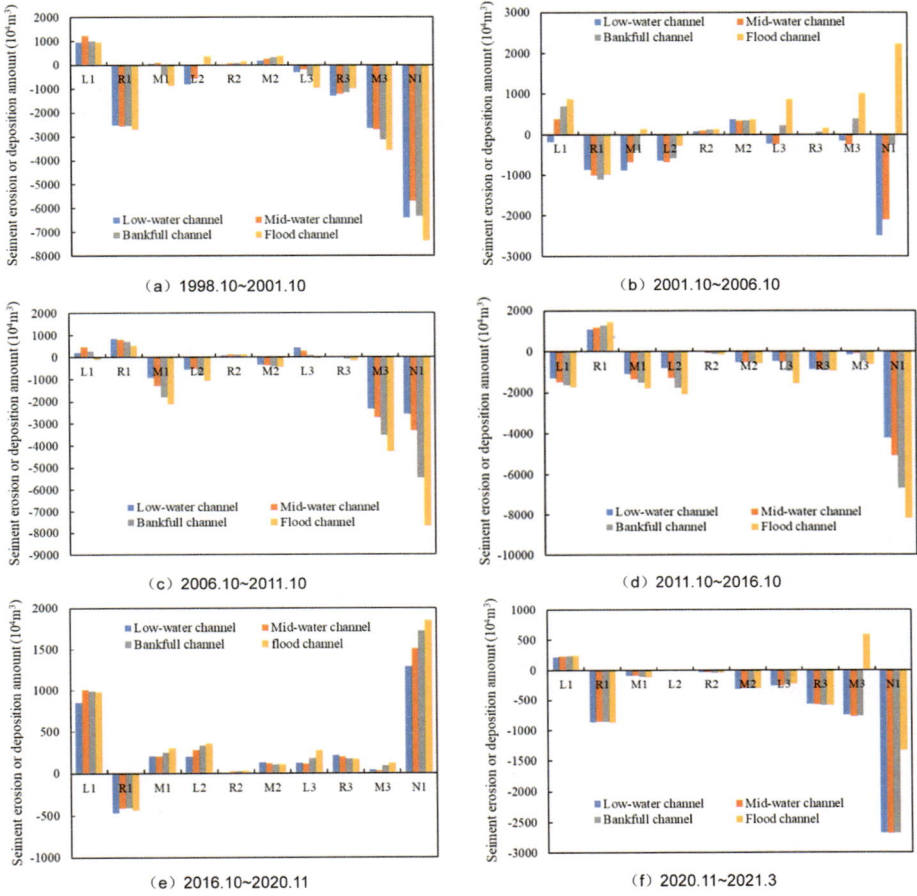

Fig. 2. Scour and deposition conditions of Nanjing reach during 1998–2021.

and the scour was the strongest at the right branch of the Xinjizhou reach. It is clear from these figures that the operation of the TGR has gradually intensified the riverbed scour of the Nanjing reach. Such continuous scour would threaten the stability of wading projects on both sides of the riverbank and the security of flood prevention. Therefore, it is necessary to enhance the monitoring.

According to the spatial distribution of scour of Nanjing reach (Fig. 3, Table 3), it is clear that during 2006–2021, when the riverbed scour was quite prominent, the river cross-sections demonstrated an overall trend of "scour in both main channel and floodplain", with bankfull channel scour being dominant, accounting for 85% of flood channel scour amount. Meanwhile, scour was more intense on the left river bank, and the corresponding mean value of the multi-year average scour intensity of each river reach was 8×10^4 m^3/(km·a), being about 1.14 times of that of the right river bank. Along the Nanjing reach, under the effect of perennial deposition of upstream Maan-shan reach, the right branch (main branch) of Xinjizhou reach showed an overall trend of deposition. However, the left branch of Xinjizhou reach and the downstream reaches demonstrated an scour pattern, with multi-year average scour being larger in the upper reach and smaller in the lower reach. Also, the multi-year average scour was the greatest at the Longtan-Yizheng reach (313×10^4 m^3/a), and the multi-year average scour intensity was the strongest at Main reach upstream the Meizizhou bifurcated reach (16.13×10^4 m^3/(km·a)).

Table 3. Comparison of bankfull channel scour amount of different parts of the Nanjing reach during 1998–2021.

Item	Period	Reach L1	Reach R1	Reach M1	Reach L2	Reach R2	Reach M2	Reach L3	Reach R3	Reach M3	Reach N1
Amount of scour or deposition/ 10^4 m^3	1998.10–2001.10	1014	−2547	−420	−38	92	319	−475	−1152	−3139	−6346
	2001.10–2006.10	701	−1111	−396	−585	117	342	227	60	391	−254
	2006.10–2011.10	275	707	−1803	−797	115	−421	99	−93	−3545	−5462
	2011.10–2016.10	−1632	1271	−1536	−1790	−127	−532	−956	−918	−466	−6688
	2016.10–2020.11	992	−408	248	329	23	98	176	169	86	1713
	2020.11–2021.03	233	−860	−110	5	−41	−307	−240	−587	−770	−2677
	2006.10–2021.03	−132	710	−3201	−2253	−30	−1162	−921	−1429	−4695	−13114
	2001.10–2021.03	569	−401	−3597	−2838	87	−820	−694	−1369	−4304	−13368
Multi-year average intensity of scour or deposition/ 10^4 m^3.km^{-1}. a^{-1}	1998.10–2001.10	22.89	−57.85	−10.58	−0.95	2.08	14.62	−6.81	−31.07	−33.58	−22.15
	2001.10–2006.10	9.49	−15.14	−5.98	−8.75	1.59	9.40	1.95	0.97	2.51	−0.53
	2006.10–2011.10	3.73	9.64	−27.25	−11.92	1.56	−11.58	0.85	−1.50	−22.75	−11.44
	2011.10–2016.10	−22.11	17.32	−23.22	−26.77	−1.72	−14.64	−8.22	−14.86	−2.99	−14.01
	2016.10–2020.11	16.80	−6.95	4.69	6.15	0.40	3.36	1.89	3.41	0.69	4.48
	2020.11–2021.03	15.78	−58.60	−8.31	0.37	−2.78	−42.21	−10.32	−47.49	−24.71	−28.03
	2006.10–2021.03	−0.60	3.22	−16.13	−11.23	−0.14	−10.66	−2.64	−7.71	−10.04	−9.15
	2001.10–2021.03	1.93	−1.37	−13.59	−10.61	0.29	−5.64	−1.49	−5.54	−6.91	−7.00

Note: positive values represent deposition, and negative values represent scour.

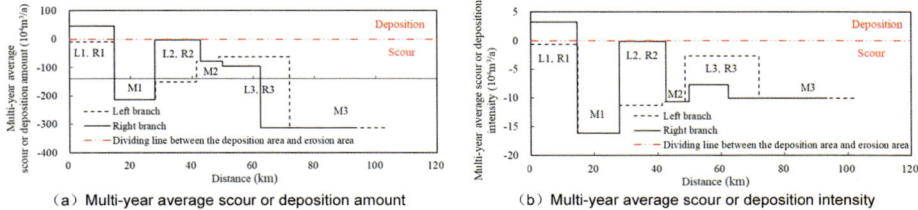

(a) Multi-year average scour or deposition amount   (b) Multi-year average scour or deposition intensity

Fig. 3. Comparison of bankfull channel scour along the Nanjing reach after the operation of the TGR (2006.10–2021.03).

5   Conclusions

(1) Since the impoundment of the TGR, the discharge of clean water has gradually intensified the scour of the Nanjing reach. During the cofferdam impoundment period of the TGR (2001–2006), the scour was relatively weak, with the total bankfull channel scour amount of the entire Nanjing reach being 254×10^4 m^3, and the multi-year average scour intensity being 50.8×10^4 m^3/a. Since 2006, the scour apparently became more intense. In particular, between 2006.10–2020.11, the total bankfull channel scour amount reached 10437×10^4 m^3, and the multi-year average scour amount was 745.5×10^4 m^3/a, which was approximately 14 times of the multi-year average scour amount for 2001–2006. After the basin-scale flood in 2020, during the period of 2020.11–2021.03, the scour amount of the Nanjing reach further increased to 2677×10^4 m^3. The continuous scour may potentially threat the stability of wading projects on both river banks of Nanjing reach and the security of flood prevention, so monitoring should be strengthened in the future.

(2) During 2006–2021, the Nanjing reach demonstrated an overall trend of "scour in both main channel and floodplain", and bankfull channel scour was dominant, with the scour amount being 85% of that of the flood channel scour amount. Along the Nanjing reach, the scour on the left bank was relatively intense, with the mean value of the multi-year average scour intensity of each river reach being about 1.14 times of that on the right bank. Moreover, the left branch of Xinjizhou (Xinshengzhou) reach and its downstream reaches demonstrated an scour pattern, with multi-year average scour being larger in the upper reach and smaller in the lower reach. Also, the multi-year average scour was the greatest at the Longtan and Yizheng reach, while the multi-year average erosion intensity was the strongest at the main reach upstream the Meizizhou bifurcated reach.

Acknowledgements. This work was supported by the National Natural Science Foundation of China (52179072; 52079080) and Nanjing Jiangbei New District Public Engineering Construction Center (Hj221086).

References

Hu CH, Fang CM (2017) Research on the solution to the sediment problem and operation effect of the three gorges project. Sci China 47(08):832–844

Ling Z, Zhang P, Lyu XY, Luo LH, Yuan WX (2021) Overview on the monitoring and analysis of the key banks of the Jiangsu Yangtze river in 2020. Jiangsu Water Resour (S02):13–15, 33

Qu GX, Wang J, Gao ZR, Bai SB, Cao GJ (2008) Effect of shoal head revetment on fluvial process of the Meizizhou reach of the Yangtze river based on GIS. Resour Environ Yangtze Basin 17(6):927–931

Wang J, Liu P, Gao ZR, Bai SB, Cao GJ, Qv GX (2007) Temporal- spatial variation of the channel in Jiangsu reach of the Yangtze river during the last 44 years. Acta Geogr Sin 62 (11):1185–1193

Wang YK, Wang XJ, Zhu CG (2020) Practices and thoughts on the river regulation project of Xinji continent and the costal flood control safety along the Yangtze river in Nanjing. Jiangsu Water Resour (1):69–72

Xu V, Cheng HQ, Zheng SW, Wang SP, Chen G, Yuan XT (2019) Evolution of Nanjing channel in the Yangtze river and its response to human activities during the last 20 years. Sci Geogr Sinica 39(4):663–670

Yuan J, Hu GY, Hu LL (2014) The research of flow and sediment transportation disciplinarian in middle and downstream of Changjiang river in past 50 years. In: 9th National Symposium on Fundamental Theory of Sediment Research, Hangzhou, China

Zang YP, Li TZ, Zhu CG, Chen L, Sun XZ (2021) Analysis on river regime change in Nanjing reach of the Yangtze river. Jiangsu Water Resour (S02):86–88

Experimental Study on Navigation Flow Condition of Downstream Approach Channel of Navigation Facilities of Baise Water Conservancy Project

Kaiwen Yu[1(✉)], Changhai Han[1], Kang Han[1], Jianjun Zhao[1], and Zhiguang Yu[2]

[1] State Key Laboratory of Hydrology Water Resources and Hydraulic Engineering, Nanjing Hydraulic Research Institute, Nanjing, China
kaiwyu@163.com
[2] Hubei Institute of Water Resources Survey and Design, Wuhan, China

Abstract. Baise Water Conservancy Project is the second cascade in the Yujiang River planning, and its navigation facilities are one of the key projects to get through Yunnan and Guizhou. The exit of the downstream approach channel of the navigation facilities of the Baise Water Conservancy Project is located about 700 m downstream of the Dongsun Hydropower Station, so the operation of Dongsun Hydropower Station has a direct impact on the navigation flow condition of the approach channel. In addition, the topography of the river also has an obvious influence on the navigation conditions at the entrance area of the approach channel. So, based on the overall hydraulic model with a scale of 1:80 and the self-navigating ship simulation test, the navigation flow conditions at the entrance area of the approach channel and the characteristics of the ship entering and exiting the ship lock were studied. The test results show that: (1) Under the condition that the discharge of Dongsun Hydropower Station is less than1500 m^3/s at full power, the flow pattern in the downstream channel of the hydropower station and the entrance area of the approach channel is relatively smooth. The flow pattern, velocity and wave height can meet the specification requirements. The maximum rudder angle and drift angle at the entrance area of the approach channel do not exceed the requirements, and the ship can enter and exit the downstream approach channel smoothly; (2) The maximum navigation discharge of 1700 m^3/s can be achieved by adjusting the route of the connecting section of the downstream approach channel to the left bank and dredging the two convex points on the right bank of the entrance area of the downstream approach channel.

Keywords: Baise Water Conservancy Project · Downstream approach channel · Navigation flow condition · Self-navigating ship test · Optimal operation

1 Introduction

The entrance area of the approach channel is usually located in the boundary sudden change, mainstream diffusion, dynamic and static water junction area, and it has been thought of as the throat of ships entering and exiting the approach channel (Li et al.

© The Author(s) 2023
Y. Li et al. (Eds.): PIANC 2022, LNCE 264, pp. 1471–1480, 2023.
https://doi.org/10.1007/978-981-19-6138-0_130

2016).In the entrance area, there exist oblique flow, backflow, bubble vortex, high-intensity eddy current and other complex flow patterns (Yang et al. 2016; Huang et al. 2017;Liu et al. 2021), which directly affect the navigation safety of ships. From the review of the current literature, it can be seen that the flow conditions at the entrance area of the approach channel are jointly restricted by factors such as flow rate, river trend, terrain, scheduling modes and anti-regulation (Han et al. 2014; Yu et al. 2014; Chen et al. 2008; Wu et al. 2016). There are important achievements on how to improve the complex navigation flow conditions and ensure the safety of ship navigation in past studies (Li et al. 2016; Yang et al. 2016; Wang et al. 2019). While the actual situation of each project varies greatly, and the problems of the navigation flow are different, it is necessary to improve and optimize the navigation flow conditions according to the actual situation of projects. The exit of the downstream approach channel of the navigation facilities of the Baise Water Conservancy Project is located about 700 m downstream of the Dongsun Hydropower Station, so the operation of Dongsun Hydropower Station has a direct impact on the navigation flow conditions of the approach channel. And because the entrance area of the approach channel is located in a narrow river area, the navigation conditions are also affected by the river terrain. According to *Code for Master Design of Shiplocks*, for IV ship lock, in the entrance area, the longitudinal flow velocity, transverse velocity and back flow velocity should be less than 2.0 m/s, 0.3 m/s, and 0.4 m/s, respectively. Therefore, based on the overall hydraulic model test of the large-scale and the simulation test of the self-propelled ship, the flow conditions at the entrance area of downstream approach channel are studied to propose the reasonable layout schemes of downstream approach channel from the perspective of hydraulics.

2   Methodology

2.1   Overall Physical Model Test

Navigation facilities of Baise Water Conservancy Project adopt the combination scheme of ship lock and ship lift, which consist of ship lock, intermediate channel, vertical ship lift and downstream approach channel (Fig. 1). The navigation route is arranged on the left bank of Baise Water Conservancy Project with total length of 4245 m. The designed navigation scale of the project is 2×500 t class and reconciling the 1000 t class single ship. The Dongsun Hydropower Station is located 6.5 km downstream of the Baise Water Conservancy Project, and the exit of the downstream approach channel of the navigation facilities of the Baise Water Conservancy Project is located about 700 m downstream of the Dongsun Hydropower Station, and it has an intersection angle of 20.203° with the downstream river. The downstream approach channel is arranged at the downstream of the auxiliary lock. The size of downstream approach channel in the preliminary plan is about 188 m long in a straight section, 34 m–60 m wide at the bottom, 110.0 m in the bottom elevation, 555 m in the centerline turning radius, 20.203° in the center angle, and 60 m wide at the entrance area. The right side of the downstream approach channel is provided with a permeable navigation embankment, the top elevation, length and insert plate height of which are121.04 m, 117 m and 6.94 m, respectively.

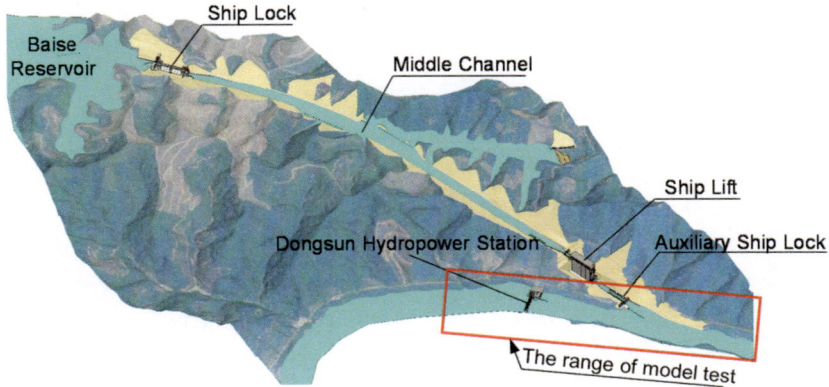

Fig. 1. The layout of Baise Water Conservancy Project

The geometric scale of the model is 1:80. The upstream boundary of the model is taken to the dam site of Dongsun Hydropower Station, the downstream boundary is taken to 2000 m downstream of Dongsun Hydropower Station, and the upstream boundary of the approach channel is taken to the lower lock head of auxiliary lock. In this paper, an experimental study on the navigation flow conditions of the downstream approach channel was carried out according to the operation and scheduling modes of Dongsun Hydropower Station and the designed maximum navigation discharge (1500 m³/s).

The bottom flow velocities were measured by a DPJ propeller current meter with an incipient velocity of 1 cm/s and an accuracy of 0.01 cm/s.

2.2   The Simulation Test of the Self-propelled Ship

The ship simulation meets the requirements of gravity similarity and the scale is the same as that of the overall model to ensure the ship draft and the coordination of speed. The trajectories of the ship model are measured in real time by the multilateral wireless positioning method. Four wireless base stations are arranged on the shore, and one wireless tag module is respectively arranged at the bow and the stern of the ship model. The distance between the bow, stern and each base station is collected in real time and calculated by the computer to obtain the navigation trajectories and drift angles. The roll and trim of the ship model are transmitted to the computer by the wireless transmitter module. As shown in Fig. 2.

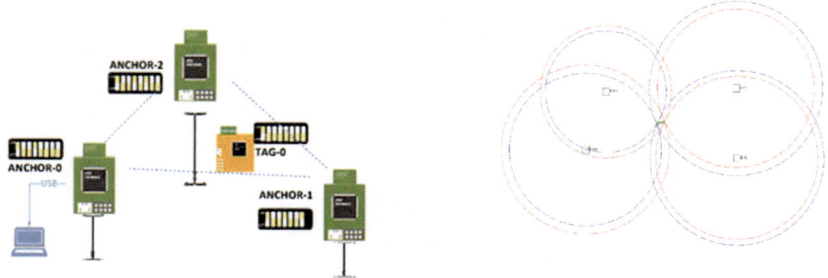

Fig. 2. Principle of multilateral measurement and positioning method

A single ship is simulated in self-propelled ship test, and representative size of the 1000 t class single ship design is 67.5 m × 10.8 m × 2.6 m (length × width × full load draft), corresponding model size is 84.4 cm × 13.5 cm × 3.3 cm.

3   Results and Analysis

3.1   Velocity Distribution at the Approach Channel Entrance Area and Connecting Section

Velocity distribution in the approach channel entrance area and connecting section can be seen from Table 1 and Fig. 3. For condition 1–3, the flow velocities at the entrance area of downstream approach channel meet the specification requirements. For condition 4, the longitudinal and transverse flow velocity of individual measuring points on the right edge of the curved section of the downstream approach channel entrance area slightly exceeds the standard, but the flow velocities of the channel centerline and other areas can meet the specification requirements. For condition 5, the over-standard range of the longitudinal and transverse flow velocity in the local area on the right side of the curved section of the downstream approach channel entrance area is larger than that in condition 4, the backflow range on the left side of the entrance area also increased significantly from about 150 m in condition 4 to 350 m, and the flow velocity at the entrance area of downstream approach channel exceeded the standard obviously.

Therefore, under the condition that the total discharge of Dongsun Hydropower Station not exceed 1500 m³/s and the power station is fully powered, the flow conditions at the entrance area of the downstream approach channel can meet navigation requirements.

Table 1. The velocity at the entrance area of the downstream approach channel in the preliminary scheme

Operating condition	Discharge (m³/s)	Overcurrent flow of power station (m³/s)	Sluice gate scheduling method	Maximum measured flow velocity (m/s)		
				Longitudinal	Transverse	Backflow
1	600	Full/600	Closed	0.99	0.16	0.10
2	1000	Full//600	Evenly partial open	1.50	0.24	0.04
3	1500	Full//600	Evenly partial open	1.95	0.17	0.21
4	1500	Closed/0	Full open	2.08	0.37	0.12
5	2000	Closed /0	Full open	2.20	0.38	0.25
Control index				≤2.0	≤0.30	≤0.40

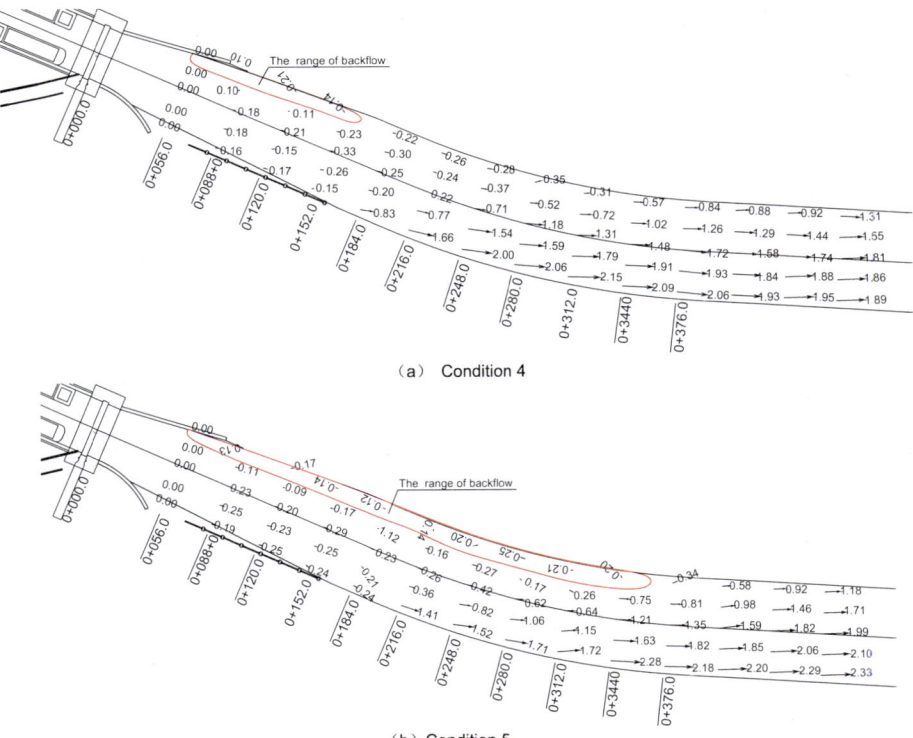

(a) Condition 4

(b) Condition 5

Fig. 3. Velocity distribution at the entrance area of approach channel

3.2   Wave Characteristics of River Channel and the Entrance Area

In the model test, the fluctuation of the water surface of downstream was measured. A total of 6 measurement points were arranged, including 4 points in the main channel which are located in sequence: 0 + 64 m, 0 + 384 m, 0 + 640 m, 0 + 960 m; and 2 in the entrance area of approach channel which are 120 m and 250 m away from the auxiliary ship lock respectively (Fig. 4). The measurement results are shown in Table 2.

It can be seen that the wave height in the entrance area has exceeded the requirement under the conditions of 4 and 5.

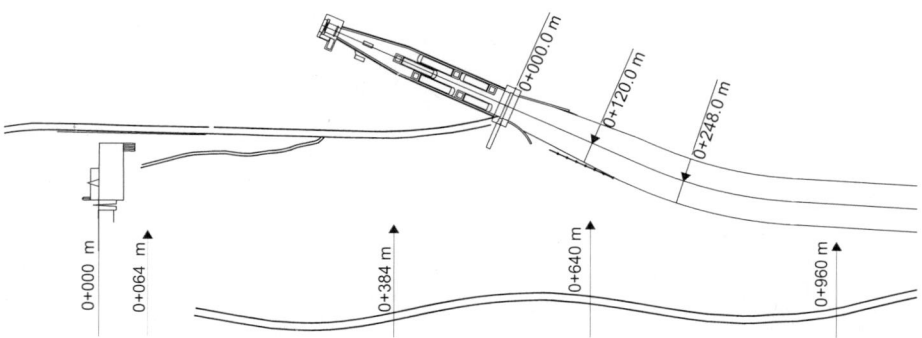

Fig. 4. Arrangement of wave height measuring points

Table 2. The wave height of the downstream channel and the entrance area in the preliminary scheme

Operating condition	0 + 64 m		0 + 384 m		0 + 640 m		0 + 960 m		0 + 120 m		0 + 248 m	
	RMS	Max (m)	RMS	Max (m)	RMS	Max (m)	RMS	Max (m)	RMS	Max (m)	RMS	Max (m)
2	0.1096	0.38	0.0334	0.12	0.0084	0.03	0.0055	0.02	0.0253	0.00	0.0144	0.06
3	0.1409	0.40	0.1041	0.30	0.0302	0.12	0.0221	0.04	0.0193	0.04	0.0174	0.05
4	0.1428	0.45	0.1026	0.30	0.0589	0.18	0.0248	0.04	0.0243	0.04	0.0317	0.12
5	0.2407	0.80	0.2556	0.80	0.1406	0.45	0.0315	0.12	0.0296	0.11	0.0491	0.20
Control index	/	/	/	/	/	/	/	/	/	0.10	/	0.10

(RMS represents root mean square value of wave height, Max represents the maximum wave height).

3.3   The Navigation Parameters of the Ship Model

A simulation test of a self-propelled ship was carried out, and navigation parameters can be seen from Table 3. Under all test conditions, the ship model can smoothly pass through the entrance area to enter and exit the approach channel. While in condition 5, the maximum drift angle of the whole journey has exceeded the requirements, which occurs about 360 m downstream of the gate. Besides, when the ship passes through the river channel shaped "S", the track of the ship is affected by the contracted flow, so it is necessary for steer in time to overcome the transverse flow and straighten the course.

Table 3. Navigation parameters of the ship model

Operating condition	Calm water speed (m/s)		Ship speed against bank (m/s)		Steering angle of the entrance area (°)		The maximum steering angle of the whole journey (°)		Drift angle of the entrance area (°)		The maximum drift angle of the whole journey (°)	
	Entering	Exiting	Entering	Exiting	Entering	Exiting	Entering	Exiting	Entering	Exiting	Entering	Exiting
2	2.8–3.2	2.0–3.0	1.8–2.3	3.0–4.4	−4–12	−10–−6	−10–12	−10–10	−6–2	−1–4	−6–8	−8–5
3	3.2–3.6	3.0	1.8–2.7	2.0–5.3	−4–13	−10–4	−10–13	−10–6	0–4	0–6	−0–7	−10–6
4	2.5–4.0	2.0–2.5	2.0	2.0–4.0	−8–10	−10–5	−10–10	−10–12	0–4	4–7	−10–10	−10–8
5	3.6–4.2	2.0–2.7	1.9–2.5	2.0–4.2	0–12	−8–2	−8–12	−8–10	1–6	4–8	−9–10	−14–8
Control index	/		/		−20–20				−10–10	/		

What can be learned from navigation flow condition test and self-propelled ship model test is that the preliminary layout scheme of approach channel fully meets the navigation requirements under the designed maximum navigation discharge of 1500 m^3/s, and the only shortcoming is that it is necessary for steer in time to overcome the transverse flow due to the influence of the protruding rock point on the right bank downstream of the entrance area.

4   Optimized Layout of Approach Channel and Its Navigation Conditions

In order to solve the above problems in the preliminary scheme and improve the navigation capacity of the downstream approach channel, the optimized layout schemes were proposed by adjusting the course and dredging. The flow velocity of each optimized scheme is shown in Table 4.

In Optimized Scheme 1: Adjust the downstream route under the condition that the axis of the ship lift and the auxiliary ship lock remain unchanged. Compared with the original scheme, the straight section of the approach channel is offset to the left bank by 3.7° starting from 470 m from the lower gate of the auxiliary ship lock, and the route of the optimized scheme is laterally shifted to the left by 23 m at about 780 m away from the lower gate, shown in Fig. 5. After the adjustment of the route of downstream approach channel, the characteristic flow velocity of the entrance area of downstream approach channel in condition 3 meets the requirements of the navigation specification, but the local characteristic flow velocity on the right side in conditions 5 and 6 exceeds specification requirements.

In Optimized Scheme 2: On the basis of optimized scheme 1, excavation and dredging were carried out at the rocky point on the right bank about 500 m downstream of the entrance area, as shown in Fig. 6. After the dredging, for condition 4, the longitudinal and transverse velocity of individual measuring points on the right edge of the bend section of the downstream entrance area slightly exceeds the standard, but the characteristic velocity of the channel centerline and other areas can meet the specification requirements, so it is considered that the navigation conditions are met according to the use experience of the built navigation facilities. For conditions 5 and 6, the flow velocity in a large range on the right side of the center line exceeds the specification requirements.

In Optimized Scheme 3: In order to explore the possibility of increasing the maximum navigable discharge to 1700 m³/s which is once in 5 years, on the basis of optimized scheme 2, excavation of the protruding rocky point in upstream of the right bank was carried out, as shown in Fig. 7. After the excavation, only the velocity index of local area of the outer boundary of the bend section of the downstream entrance area exceeds the standard, and the area is small. In addition, the maximum wave height in the entrance area is not greater than 0.1 m. Although the maximum drift angle of the whole journey slightly exceeds the requirements (Table 5), the ship can enter and exit the entrance area smoothly, and the number of steering times per minute is less than 2 times. Therefore, it is reasonable to take 1700 m³/s as maximum navigable discharge.

Table 4. The velocity at the entrance area of the downstream approach channel in the optimized scheme

Optimized scheme	Operating condition	Discharge (m³/s)	Overcurrent flow of power station (m³/s)	Sluice gate scheduling method	Maximum measured flow velocity (m/s)		
					Longitudinal	Transverse	Backflow
1	3	1500	Full/600	Evenly partial open	2.02	0.18	0.13
	5	2000	Closed/0	Full open	2.39	0.42	0.14
	6	2500	Closed /0	Full open	2.61	0.44	0.19
2	4	1500	Closed /0	Full open	2.20	0.39	0.20
	5	2000	Closed/0	Full open	2.27	0.40	0.27
	6	2500	Closed /0	Full open	2.52	0.42	0.22
3	7	1700	Closed /0	Full open	2.28	0.40	0.23
Control index					≤2.0	≤0.30	≤0.40

Table 5. Navigation parameters of the ship model with the discharge of 1700 m³/s

Operating condition	Direction	Ship speed against bank (m/s)	Steering angle of the entrance area (°)	The maximum steering angle of the whole journey (°)	Drift angle of the entrance area (°)	The maximum drift angle of the whole journey (°)	Transverse swing angle
7	Entering	1.0–2.2	10.5	−8.0–12.0	4.5	−7.5–11.4	0.6–0.7
	Exiting	1.8–4.8	−15–3	−18–10.1	−10.0–3.4	−10.0–8.0	<2.0
Control index		/	−20–20		−10–10		/

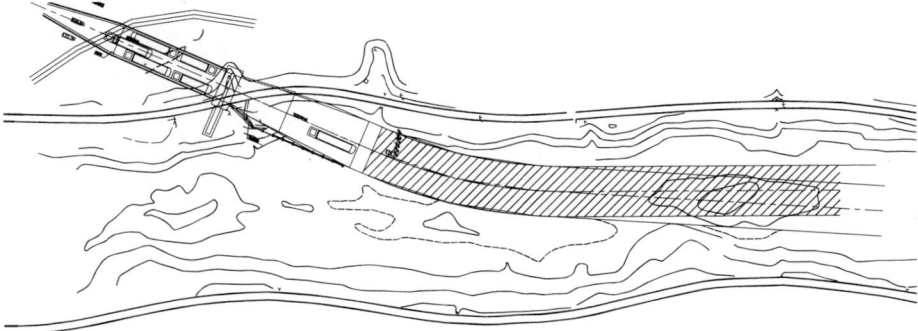

Fig. 5. Comparison of the route of preliminary plan and optimized scheme 1 (shaded)

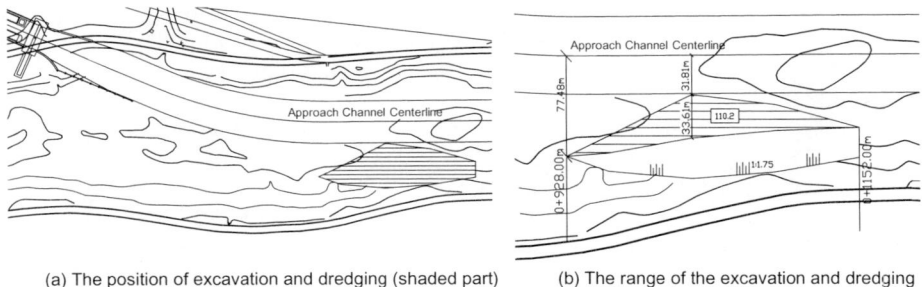

(a) The position of excavation and dredging (shaded part)   (b) The range of the excavation and dredging

Fig. 6. Excavation and dredging at the rocky point on downstream of the right bank

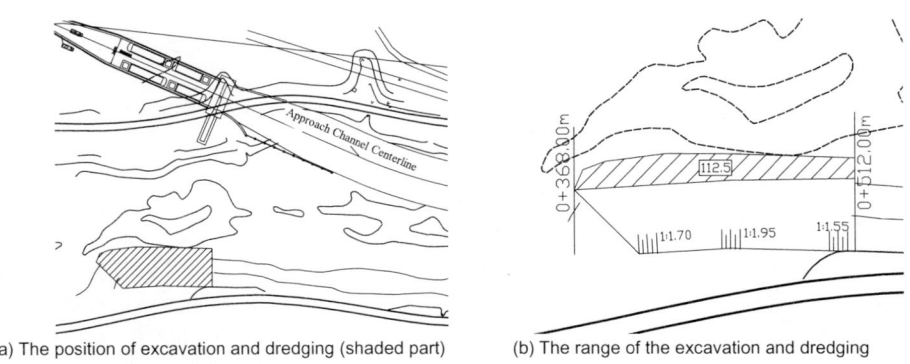

(a) The position of excavation and dredging (shaded part)   (b) The range of the excavation and dredging

Fig. 7. Excavation and dredging at the rocky point on upstream

5 Conclusions

From the review of the test result, the following can be concluded:

(1) In the preliminary scheme, under the condition that the discharge of the power station is 1500 m³/s at full power, the flow pattern, velocity, wave height, and navigation parameters all can meet the requirements of the specification, and the

ship can smoothly enter and exit the downstream approach channel. The only shortcoming is that it is necessary for steer in time to overcome the transverse flow due to the influence of the protruding rock point on the right bank downstream of the entrance area.

(2) After excavation and dredging of two rock points on the right bank, the maximum navigable discharge can be increased from 1500 m³/s to 1700 m³/s, so as to meet the requirements of navigable flow once in five years.

References

Li J, Zhao, Ji J, Hong J (2016) Flow condition optimization for upstream and downstream approach channels when ship lock lies on concave bank. Port WaterwEng 12:101–105

Yang Y, Li YF, Han CH (2016) Experimental study on navigation flow condition of Chengjing hydro-power station in Guizhou Qingshui river. Port Waterw Eng 12:126–131

Huang J, Zhao JJ, Gu JD (2017) Navigation flow condition at great angle bend. Port Waterw Eng 4:146–150

Liu JM, Wang X, Huang Q (2021) Study on navigation flow condition of downstream approach channel of Jinghong ship lift. Port Waterw Eng 1:6–11

Han CH, Yang Y, Li YF (2014) Layout of navigable channel of navigation-power the convex bank with slightly bended river junction on section. Port Waterw Eng 10:121–125

Yu ZG, Han CH, Peng CB (2014) Overview of bent waterway navigation hub layout research. Pearl River 35(06):102–105

Chen YN, Hu XY, Li B (2008) Influence of re-regulation hydro-project on navigable flow condition. Hydro-Sci Eng 4:66–70

Wu ZY, Jiang CB, Chen J (2016) Influence of sluice gate opening mode on navigation flow condition. Adv Sci Technol Water Resour 36(3):73–77, 82

Wang JP, Xing FL, Chen YF (2019) Optimization of navigable flow condition of ship lock entrance area in curved river. Port Waterw Eng 2019(11):86–91

Flow and Sedimentation Characteristics of Tidal Waterways – with the Kouanzhi Waterway in the Lower Yangtze River as an Example

Jie Qin[1(✉)], Ye Jing[1], Xueting Lei[2], Teng Wu[1], and Elikplim Agbemafle[1]

[1] College of Harbour, Coastal and Offshore Engineering, Hohai University, Nanjing, China
{jqin,wuteng}@hhu.edu.cn
[2] Changjiang Waterway Planning, Design and Research Institute, Wuhan, China

Abstract. River flow in the Lower Yangtze River (LYR) is influenced by the combined effect of runoff and tides, and the complex flow conditions tend to cause sediment deposition in waterways, which in turn affects navigation conditions. In order to improve the understanding of the river mechanism of tidal reaches, this study selected a typical reach in the LYR – the Kouanzhi Waterway (KW) – as an example to investigate the flow characteristics that affect sediment transport processes. Sedimentation annually occurs at the entrance of the KW and a large sidebar constantly increases in size, causing the reduction of navigation depth and width, but the causes of the sedimentation are still unclear. In this study, a three-dimensional (3D) model of the KW was established based on the Delft3D, and the k-epsilon model was chosen to simulate the turbulent flow. The model simulates the flow processes during flood tides in flood and dry seasons. The results of the numerical simulation show that a significant difference exists in the large-scale flow structures between the flood and the dry seasons. In the flood season, flow at the entrance of the KW is extremely turbulent and a large-scale vortex shedding phenomenon is formed downstream of the channel entrance, which causes a great sediment transport rate at the entrance. In addition, because the river width widens in the middle part of the KW, a large-scale circulation flow structure is developed near the Sanyiqiao sidebar. The circulation is supposed to trap and deposit sediment on the sidebar. In the dry season, the intensity of flow turbulence is greatly reduced, which causes the sediment transport rate at the entrance decreases accordingly. In addition, the circulation flow structure at the Sanyiqiao disappeared, and thus the sidebar is supposed to be under erosion during this period. The results of this study provide a vital reference for the engineering works of waterway regulations in tidal river reaches.

Keywords: Sidebar · Sedimentation · Turbulence · Circulation

Y. Li et al. (Eds.): PIANC 2022, LNCE 264, pp. 1481–1491, 2023.
https://doi.org/10.1007/978-981-19-6138-0_131

1 Introduction

Most of the existing studies on numerical simulation of tidal rivers are carried out on a large spatial scale, and thus most of these models are 2-dimensional (2D) which can be used to investigate depth-averaged velocity distribution or the topographic evolution of rivers (Xie et al. 2018). These 2D models can estimate large-scale sediment erosion and deposition, but can hardly reproduce complex local flow structures. During the routine maintenance of waterways, local scouring and deposition are of concern to waterway management departments. In order to study the local sediment transport and topographic changes, detailed 3D numerical simulations are needed (Bever and MacWilliams 2016).

Hu et al. (2009) carried out a numerical simulation near the estuary of the Yangtse river using a coarse grid with the minimum grid spacing inside the estuary being about 300 m, and the average grid spacing reaching 2,000 m. A later study carried out by Ding (2011) used a much finer grid in comparison to the former with a grid spacing around 30 m in the tidal reach of the Yangtse river. Because this study used a mixing length model to calculate the eddy viscosity in the Reynolds equations, the applicability of this simple model based on the mixing length concept is limited by the need to prescribe mixing-length variations for different flow conditions (Pourahmadi and Humphrey 1983). In addition, in spite of a finer grid being used in Ding et al. (2011), the grid is still too coarse to reveal 3D flow structures for a river reach characterized by complicated topography, e.g., bifurcation sections in a tidal river.

The KW is a typical bifurcation section in the downstream Yangtze River. Sedimentation frequently occurs at the entrance of the KW, which reduced the navigation depth of the waterway. Previous studies have found that the flow velocity tends to decrease in the center of the channel and thus causes sediment deposition (Cao et al. 2011, Chen et al. 2012, Wang 2016). However, the mechanism that causes the decreased flow velocity is unclear. As a consequence, this study aims at studying the detailed flow characteristics in the KW using a fine-grid 3D turbulence model to simulate flow processes and explains the observed sedimentation issue in the waterway.

2 Numerical Model

The 3D model in this study is built based on the Delft3D by employing multiple vertical layers. In the Delft3D-FLOW module, four turbulent stress closure models can be used, and the widely used k - ε model is selected. Previous studies have demonstrated that the k - ε model can well simulate the flow characteristics in most cases in river simulations (Fischer-Antze et al. 2008). The upstream boundary of the model was located far from the entrance of the KW (10 km upstream) in order to accurately simulate the inlet flow conditions considering the complex topography upstream of the KW. A curved orthogonal grid was generated with the number of cells being around 800,000. The grid size of the planar grid ranges from 10 m to 50 m. By comparing the simulation results of models with 5, 10, 15, and 20 vertical layers, the secondary flow structures can be well reproduced when the number of layers reaches 10. Therefore, the 10-layer model was chosen for the subsequent simulation.

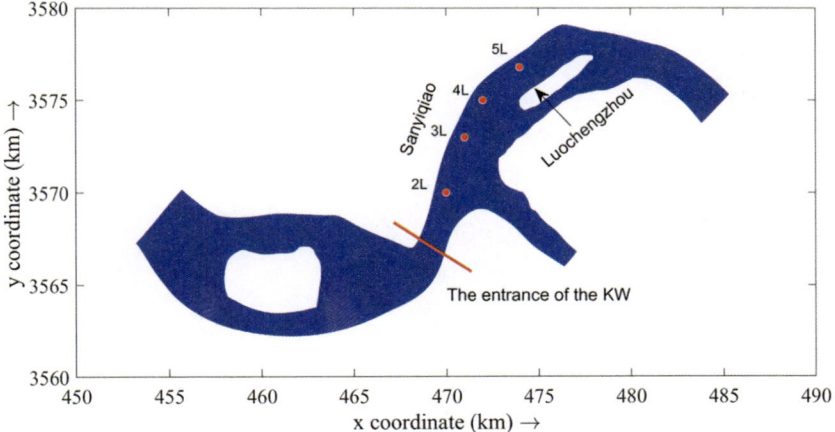

Fig. 1. The computational grid of the KW

The boundary conditions of the 3D model are provided by another large 2D model. The upstream and downstream boundaries of the 2D model are located at the Datong and the Xuliujing, respectively, and these two stations have long-term measured flow and water level data. The 2D model is calibrated and validated, and then provides the boundary conditions for the current 3D model, i.e., the flow discharge and water level temporal processes at the upstream and the downstream, respectively. To compare the flow characteristics under different hydrological conditions, the flow processes during flood tides in flood and dry seasons are simulated.

The initial value of the model roughness (the Manning coefficient) was 0.025, which was adjusted during the calibration process according to the measured flow processes and the calibrated roughness coefficient ranges from 0.02 to 0.03. The turbulent eddy viscosity is calculated based on the zero-equation turbulent flow model, $U_t = \alpha u_* h$, where u_* is the frictional flow velocity and α is a constant value of 1. The calculation time step is taken as $\Delta t = 6$ s.

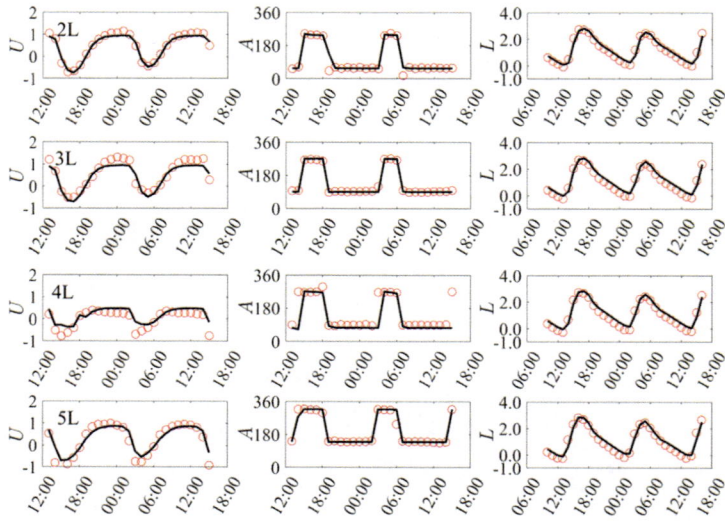

Fig. 2. Validation results of the 3D model

The 3D model was validated using the data measured during flood tides from February 20 to February 21, 2019, and data of ebb tides from February 15 to February 16, 2019. Due to a large number of measurement points, several locations shown in Fig. 1 are selected for validation. The validation results are shown in Fig. 2, in which only the magnitude and orientation of the surface velocities were displayed. This figure shows the simulated velocity magnitude, flow direction, and water level during the flood tide compared with the measured data. The model well reproduced the water level process at the validation positions and the tidal level error is less than 0.10 m. The flow velocity and direction are in good agreement with the measured results as well.

3   Results

The planar large-scale flow characteristics in the KW are studied based on the depth-averaged velocity field and vortex distribution. The secondary flow structures are analyzed by extracting typical cross-sections from the 3D velocity field, and four cross-sections are extracted for each experimental run, including CX1, CX2, CX3, and CX4 as shown in Fig. 3. The vertical coordinates of all cross-sections were stretched (magnified by 30 times) for a clearer view of the cross-sectional flow structures.

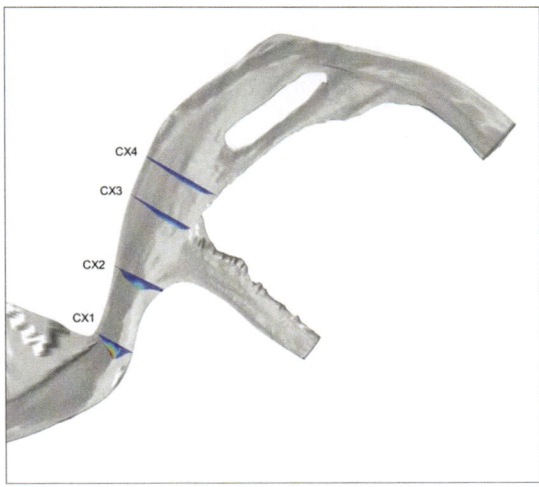

Fig. 3. The locations of the four typical cross-sections in the KW

3.1   Flow Characteristics During the Flood Season

Figure 4(a) shows the distribution of the depth-averaged flow field, from which significant large-scale flow structure can be readily observed. Firstly, at the entrance of the KW, the strong recirculation structures on both sides of the channel occur due to the upstream narrowing topography and sequential secondary circulations develop. In addition to the circulations at the entrance of the KW. Because of the flow separation of the mainstream near the entrance, circulation structures of different sizes are distributed on the left side of the river channel from the entrance to the vicinity of the Sanyiqiao sidebar, specifically, the biggest circulation being developed covering the sidebar. The large-scale structures indicate that this area is prone to slow flow velocity where sediment is easily deposited during the flood season, which explains the development of the sidebar near the Sanyiqiao.

Figure 4(b) shows the vorticity distribution at the KW, from which the development and shedding of large-scale vortices near the entrance of the KW can be observed. The chaotic flow pattern at the entrance contrasts with the flow pattern downstream, and the turbulent flow explains the origin of the high shear stress and high turbulent kinetic energy at the entrance. The formation of these large and small vortices is influenced by two factors, one is the strong turbulence flow caused by the upstream topography, and the strong contraction at the entrance of the KW. A strong wake area developed downstream the entrance of the KW where vortices are continuously generated, grew, and shed. The dislodged vortex can be transported to the middle area of the Sanyiqiao sidebar.

Figure 5(a) and (b) show the velocity distribution, streamline, and turbulent kinetic energy (TKE) distribution at the CX1 cross-section. The most evident feature of this section is a strong and narrow vortex on the right side of the river. To the left of this vortex is the core of the main flow characterized by high velocities, in contrast to the low velocities on the right side of the vortex. Such a flow distribution is due to the fact that this region is located downstream of a topographic constricted section, and a planar circulation flow is formed behind this contraction. Thus, a strong secondary flow structure is formed at the intersection of this circulation and the main flow, similar to the cross-sectional flow near a groin field.

On the left side of this vortex, the flow consists of two large vortices. The right vertex is a typical secondary flow structure observed in bend flow. However, this vortex is limited to the center of the river compressed by vortices on two sides. The maximum flow velocity of the CX1 occurs on the left side of the river instead of the deepest portion, which is different from normal bend flow. This is attributed to the presence of a strong lateral vortex on the right side, which squeezes the mainstream and makes the core of the main flow to the left.

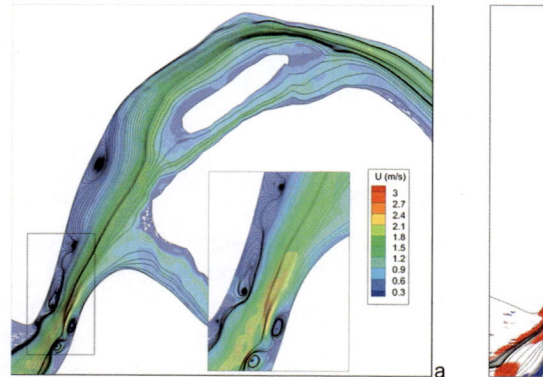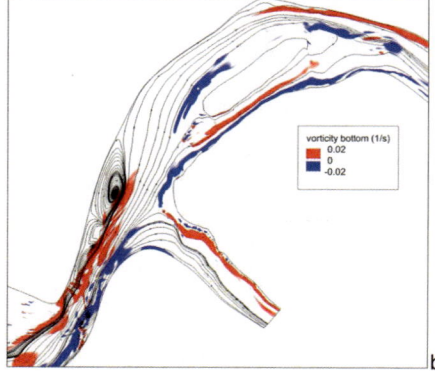

Fig. 4. Depth-averaged flow velocity and vorticity distribution in the KW during the flood season

As the maximum flow velocity in CX1 shifts to the left, the maximum gradient of flow velocity moves to the left accordingly, which leads to the maximum TKE near the bed being located on the left side of CX1. Such a distribution has a substantial impact on downstream sediment transport. The bedload transportation is often consistent with the bed shear stress distribution, and the TKE is usually positively correlated to bed shear stress. Thus, the largest sediment transport zone in this section should be located on the left side of the river. As a consequence, sediment advances along the left bank and deposits when flow velocity reduces, which represents a major sediment source for the sedimentation on the downstream sidebar.

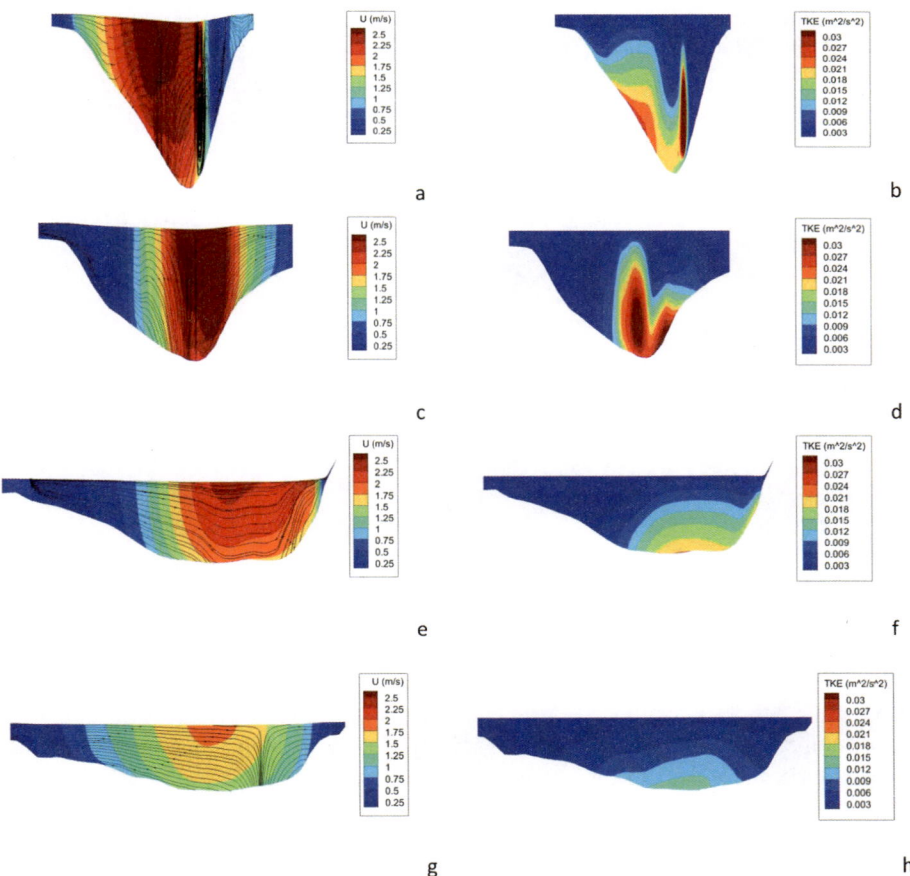

Fig. 5. The velocity distribution (left column) and the TKE (right column) of the four cross-sections during the flood season

On the right side of the channel, due to the presence of a strong lateral vortex, a strong turbulence region is formed near the vortex. The intensity of the TKE in this region exceeds that of the bed. This intense turbulence contributes to the suspension of the sediment and the transported sediment may deposit downstream where flow velocities decrease.

3.2   Flow Characteristics During the Dry Season

Figure 6(a) shows the depth-averaged flow velocity and vorticity distribution in the KW during the dry season. The main difference from the flood season is the disappearance of the large-scale circulation structures near the Sanyiqiao. In Fig. 6, streamlines pass smoothly through the Sanyiqiao sidebar. This phenomenon means that the Sanyiqiao sidebar is prone to sedimentation during the flood season, whereas scouring occurs in the dry season.

In addition, the flow near the entrance of the KW changed significantly. Due to the reduced flow discharge, the separation of streamlines is not as evident as in the flood season, and the resulting horizontal circulations on both sides shrink accordingly. There is a single circulation structure on each side, in contrast to a series of secondary circulation structures in the flood season.

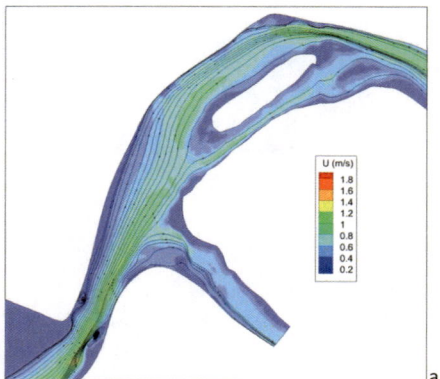 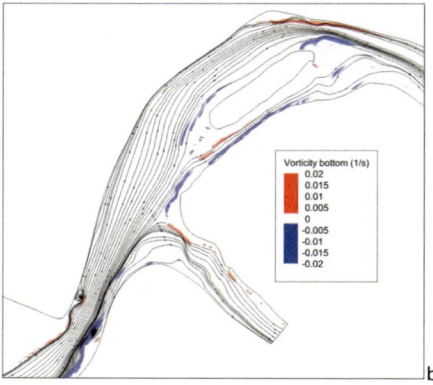

Fig. 6. Depth-averaged flow velocity and vorticity distribution in the KW during the dry season

Figure 6(b) shows the distribution of vorticity during the flood tide of the dry season, from which it can be seen that the development of vortices is significantly reduced compared to that during the flood season. Especially at the entrance of the KW, the magnitude and coverage of vortices are significantly reduced compared with that during the flood season, and no large-scale vortices mixing phenomenon occurs. However, the topographic contraction on the right side of the channel still causes small-scale vortices, and the shedding vortices move downstream along the right bank.

Figure 7 shows the flow velocity, streamlines, and the TKE distribution of the cross-sections. In the CX1 cross-section, the secondary flow structure caused by the topographic contraction at the right bank is obvious, thus forming a low flow velocity area on the right side. Due to the existence of the secondary flow structure which squeezes the high-velocity flow to the left side of the river. The flow velocity distribution is thus similar to that during the flood season, with the large velocity gradient being located on the left bank. As a result, the greatest TKE occurs on the left bank of the riverbed. The high TKE at the junction between the slow flow on the right and the main flow can be observed, but it is much weaker compared to the TKE on the left side of the riverbed. The high TKE indicates that intense sediment transport is located on the left bank, rather than the deepest area.

The flow velocity in the CX2 cross-section is much lower compared to that during the flood season. The flow separation is greatly reduced, as shown by the shrinking low flow velocity area on both sides compared with that during the flood season. The circulation is significantly weakened and the maximum velocity occurs at the deepest area, and the maximum velocity area shifts to the left to a certain extent in comparison to the flood season. Similar changes occur in the cross-sectional variation of the TKE. In addition, due to the lack of strong mixing during the flood, the TKE gradually decreases with increasing distance from the bed. Intense turbulence still exists in the center of the riverbed.

The flow at the cross-section CX3 can be divided into three secondary flow structures, whereas there no such structures can be observed during the flood season. On the left side, due to the disappearance of large-scale circulation on the Sanyiqiao sidebar, no slow flow area is found on the left side. The distribution of the TKE is consistent with that of the flow velocity gradient, which is similar to the distribution during the flood season but with reduced intensity.

Due to the enlarged area at the CX4 cross-section and the vanish of the large circulation at the Sanyiqiao, the lateral flow structure at this cross-section was well adapted to the local topographic features, showing two large secondary flow structures on the left and right sides corresponding to the inlet of the downstream left and right branches. The maximum TKE of the cross-section is located at the deepest area, which means that the intensive sediment transport zone is located near this area and readily enters the downstream left branch.

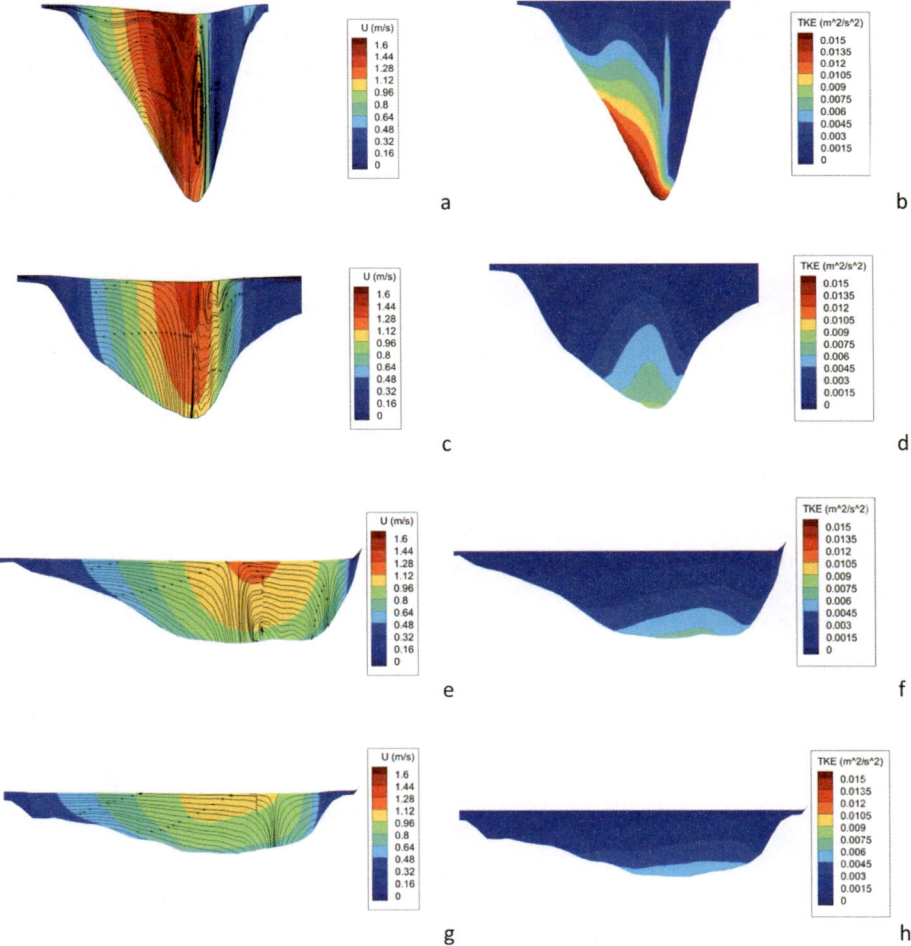

Fig. 7. The velocity distribution (left column) and the TKE (right column) of the four cross-sections during the dry season

4   Conclusions

In this study, flow characteristics of the tidal reach, the KW which is a typical bifurcation reach in the lower Yangtze River, were studied based on a 3D numerical model. With the help of the simulation results, this study demonstrated the existence of large-scale flow structures in both planar and vertical directions and these structures vary from the flood season to the dry season. In the flood season, the flow at the entrance of the KW is characterized by intense turbulence due to the upstream complex topography and the topographic constraint at the entrance. Due to the suddenly increased width of flow downstream the constraint, large-scale circulation structures are developed on the Sanyiqiao sidebar. In addition, the intense turbulent flow at the inlet increases the sediment transport rate, which makes this circulation area highly susceptible to sedimentation. Therefore, in the flood season, sediment accumulation is likely to occur at the Sanyiqiao sidebar, which causes the decreased depth of the waterway and often channel dredging is usually required. In the dry season, the above two factors are greatly weakened, and the circulation structure at the Sanyiqiao sidebar is disappeared. The combination of the influence of the above factors causes sediment erosion during the dry season, and the navigation conditions of the channel improve accordingly.

Acknowledgements. This study was supported by the National Natural Science Foundation of China (52079043, 52179061).

References

Pourahmadi F, Humphrey JA (1983) Prediction of curved channel flow with an extended k-epsilon model of turbulence. AIAA J 21(10):1365–1373

Ding Y, Jia Y, Wang SS (2011) Three-dimensional numerical simulation of tidal flows in the Yangtze River Estuary. In: World environmental and water resources congress 2011: bearing knowledge for sustainability, pp 2135–2144

Hu K, Ding PX, Wang ZB, Yang SL (2009) A 2D/3D hydrodynamic and sediment transport model for the Yangtze Estuary, China. J Mar Syst 77:114–136

Xie Q, Yang J, Lundström S, Dai W (2018) Understanding morphodynamic changes of a tidal river confluence through field measurements and numerical modeling. Water 10(10):1424

Bever AJ, MacWilliams ML (2016) Factors influencing the calculation of periodic secondary circulation in a tidal river: numerical modelling of the lower Sacramento River, USA. Hydrol Process 30(7):995–1016

Chen Z, Zhang X, Zhang S (2012) Study of regulation schemes of Sanyiqiao shallow in Lower Yangtze River. J Sediment Res 5:65–69

Wang J (2016) Regulation ideas and project layout in Kou'anzhi waterway. Port Waterw Eng. 2:1–9

Cao M, Li Q, Cai G, Yuan D Wang X (2011) On local scour deformation flume generalization experiment of Manyusha shoal head protection engineering in straight reach of Yangtze River estuary: I. Design of flume generalized and protection engineering building simulation. Port Waterw Eng 7:106–112

Fischer-Antze T, Olsen NRB, Gutknecht D (2008) Three-dimensional CFD modeling of morphological bed changes in the Danube River. Water Resour Res 44(9):W09422

Hydraulic Research on Filling and Emptying System of Water-Saving Ship Lock for Navigation-Power Junction in Mountainous River

Benqin Liu[1(✉)], Jin Yang[2], Yue Huang[1], and Lei Wang[2]

[1] Key Laboratory of Navigation Structure Construction Technology, Ministry of Transport, Nanjing Hydraulic Research Institute, Nanjing, China
`bqliu@nhri.cn`
[2] Baise Hydro Project Navigation Investment Co. Ltd, Baise, China

Abstract. Usually the ship lock has the characteristics of high water head and large variation of navigable water level in mountainous area. It is suitable to construct the water-saving ship lock, which can not only save the water consumption but also reduce the working head of each stage. Therefore, it is also beneficial to solve the hydraulic problems of high head ship lock. Taking Baise ship lock as an example of ship locks for navigation-power junctions in mountainous river, we studied the water saving layout scheme and suggested that a high water saving pool and a low one should be set on both sides of the lock chamber. And each pool adopts a new type of trapezoidal transverse section in order to make full use of the topographic conditions of the project and reduce the excavation volume. The water level classification of water-saving ship lock is calculated and analyzed. The elevations of the water saving pools are determined. The layout of the filling and emptying system of the water-saving ship lock is put forward. The hydraulic characteristic indexes, the pressures of the culverts near the valves and the ship berthing conditions in lock chamber under different operating conditions of the water-saving ship lock are obtained through physical model research. Furthermore, the opening and closing modes of the valves are recommended. The results show that the water saving scheme of Baise ship lock is reasonable and feasible in hydraulics.

Keywords: Navigation-power junction in mountainous river · Water-saving ship lock · Water saving pool with trapezoidal transverse section · Filling and emptying system · Operation mode

1 Introduction

The water level difference is large between the upstream and downstream of navigation-power junction in the mountainous river. And the terrain on both banks is also complex. Usually the ship lock has the characteristics of high water head and large variation of navigable water level in mountainous area. The water-saving ship lock can not only save the water consumption of the ship lock, but also reduce the working head of each stage, which is conducive to solving the hydraulic problems of the high head

Y. Li et al. (Eds.): PIANC 2022, LNCE 264, pp. 1492–1501, 2023.
https://doi.org/10.1007/978-981-19-6138-0_132

ship lock. Therefore, it has been applied to several ship locks in China (Liu and Xuan 2019; Zhu and Xuan 2019). Baise ship lock in Guangxi is a typical case that adopts this water-saving scheme.

Baise water control project in Guangxi is located in the upper reaches of Youjiang River. It is a junction that focuses on flood control and has comprehensive utilization functions such as power generation, irrigation, navigation and water supply. The navigation buildings are arranged separately from the main buildings of the dam. They are located in Nalu ditch on the left bank of the main dam of the hydro project, including ship lock, intermediate channel and ship lift from upstream to downstream. The total length of the navigation route is about 4384 m. The total maximum navigation head of Baise hydro project is 113.6 m and the maximum design head of the ship lock is 25.0 m.

The effective dimension of Baise ship lock is 130 m × 12 m × 4.7 m (length × width × threshold depth). The designed ship types are 2 × 500t fleet and 1000t single ship. The upstream and downstream characteristic water levels of the lock during water saving operation are as follows: (1) The water level of upstream reservoir is 228.0 m to 214.0 m with the variation of 14.0 m. (2) The water level of downstream intermediate channel is 203.2 m to 203.0 m with the variation of 0.2 m.

2 Layout of the Water Saving Pools and Water Level Calculation

2.1 Construction Scheme of the Water Saving Pools

In combination with the project layout conditions, ship lock scale, operating head and navigation capacity requirements, it is determined that Baise ship lock is equipped with two-stage water saving pools through calculations and analyses. The topographic features of Baise ship lock site are low in the middle and high on both banks. Therefore, the water saving pools adopt decentralized layout scheme. And the high and low pools are respectively arranged on the left and right sides of the lock chamber. In order to make full use of the topographic conditions on both banks and reduce the amount of excavation and backfilling, the section shape of both pools is designed to be the new trapezoidal section form that is shown in Fig. 1.

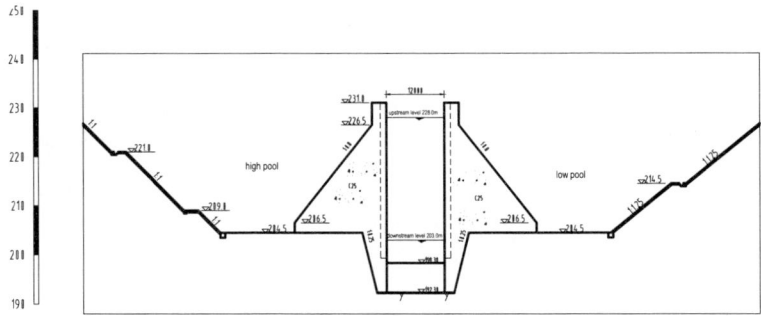

Fig. 1. Typical cross section of the water saving pools and lock chamber

2.2   Characteristic Water Level of the Pools

The classification of water level is very important for the operation of water saving ship lock. And many studies have been carried out by relevant scholars (Yang et al. 2021; Li and Xu 2020; He et al. 2020).

In this paper, the dynamic water valve closing mode is adopted in order to save water filling and emptying time. And there is a small amount of residual head between the pool and the chamber when the valves have been closed completely. Under the design condition with 0.15 m residual head, the characteristic water levels during water filling and emptying processes are calculated and analyzed for the trapezoidal section water saving pool scheme. The results are shown in Table 1 and Table 2. And Fig. 2 shows the initial and the end water levels of the water saving pools under three typical water level combinations. We can see that the characteristic water levels in the pools are obviously different because of the large variation range of the water level in upstream.

Table 1. Characteristic water levels during water filling from pools to chamber (unit: m)

H_u	H_d	From low pool to chamber					From hogh pool to chamber				
		H_{pb}	H_{pe}	H_{cb}	H_{ce}	h_s	H_{pb}	H_{pe}	H_{cb}	H_{ce}	h_s
228.00	203.00	213.09	210.14	203.00	209.99	0.15	220.08	217.91	209.99	217.76	0.15
226.00	203.00	212.36	209.53	203.00	209.38	0.15	218.74	216.64	209.38	216.49	0.15
224.00	203.00	211.63	208.92	203.00	208.77	0.15	217.40	215.37	208.77	215.22	0.15
222.00	203.00	210.88	208.32	203.00	208.17	0.15	216.06	214.12	208.17	213.97	0.15
220.00	203.00	210.13	207.73	203.00	207.58	0.15	214.71	212.87	207.58	212.72	0.15
218.00	203.00	209.37	207.14	203.00	206.99	0.15	213.36	211.63	206.99	211.48	0.15
216.00	203.00	208.59	206.56	203.00	206.41	0.15	212.00	210.41	206.41	210.26	0.15
214.00	203.00	207.81	205.99	203.00	205.84	0.15	210.64	209.19	205.84	209.04	0.15

Notes: H_u, H_d, H_{pb}, H_{pe}, H_{cb}, H_{ce} and h_s respectively represent the upstream and downstream water level, the initial and end water level of the pool, the initial and end water level of the chamber and the residual head between the pool and the chamber after the valves are closed. And the same below

Table 2. Characteristic water levels during water emptying from chamber to pools (unit: m)

H_u	H_d	From chamber to high pool					From chamber to low pool				
		H_{pb}	H_{pe}	H_{cb}	H_{ce}	h	H_{pb}	H_{pe}	H_{cb}	H_{ce}	h
228.00	203.00	217.91	220.08	228.00	220.23	0.15	210.14	213.09	220.23	213.24	0.15
226.00	203.00	216.64	218.74	226.00	218.89	0.15	209.53	212.36	218.89	212.51	0.15
224.00	203.00	215.37	217.40	224.00	217.55	0.15	208.92	211.63	217.55	211.78	0.15
222.00	203.00	214.12	216.06	222.00	216.21	0.15	208.32	210.88	216.21	211.03	0.15
220.00	203.00	212.87	214.71	220.00	214.86	0.15	207.73	210.13	214.86	210.28	0.15
218.00	203.00	211.63	213.36	218.00	213.51	0.15	207.14	209.37	213.51	209.52	0.15
216.00	203.00	210.41	212.00	216.00	212.15	0.15	206.56	208.59	212.15	208.74	0.15
214.00	203.00	209.19	210.64	214.00	210.79	0.15	205.99	207.81	210.79	207.96	0.15

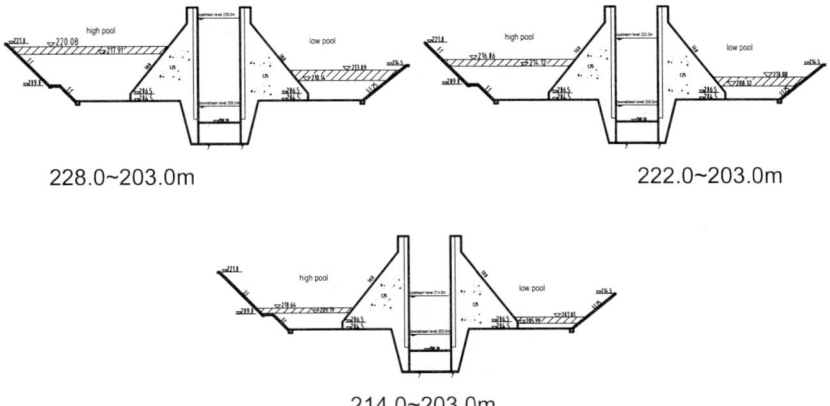

228.0~203.0m   222.0~203.0m

214.0~203.0m

Fig. 2. The initial and end water levels of the poos under three typical water level combinations

3   Layout of Filling and Emptying System

3.1   Type of the Filling and Emptying System

Under the design maximum head of 25.0 m, when the ship lock adopts the water saving operation mode, the initial water head between the lock chamber and the pool is 10.09 m, and the initial water head between the lock chamber and the upstream or downstream approach channel is 10.24 m. The designed total water filling or emptying time should be no more than 16 min. Through the analysis of data indicators, it is found that the hydraulic index of the ship lock is high.

According to "Design Code for Filling and Emptying System of Shiplocks" (JTJ306-2001), it is calculated that the type discrimination coefficient of filling and emptying system is 1.87. So the distributed type should be selected. Furthermore, combined with the structural form of the chamber, lock bottom long-culvert filling and emptying system is adopted after comprehensive analyses.

3.2   Layout of the Filling and Emptying System

According to the research and calculation results, the section size of the culvert at the valves is 2.0 m wide and 3.0 m high. It can save water filling & emptying time and facilitate the operation management and maintenance of the ship lock that the culverts of the lock heads and the pools have the same control section in size.

The section size of the lock bottom culvert is 5.2 m wide and 3.0 m high. And 22 outlet branch holes are arranged on each side of the bottom culvert. The orifice size is 0.4 m wide and 0.9 m high. And the length of all the side branch holes is 1.2 m. In order to ensure the smooth flow and reduce the shape resistance of the branch holes, both ends of the inlet and outlet of the holes are rounded on three sides with a radius of 0.25 m. The center distance of adjacent holes is 4.0 m. In this way, the total length of chamber outlet section is 88.0 m, accounting for 67.7% of the effective length of lock chamber.

Two energy dissipation open ditches are arranged along the length direction of the lock chamber outside the outlet holes. The width of the open ditch is 2.2 m and the height of the retaining sill is 3.0 m. In order to guide the water flow to the middle of the chamber, the upper part of the open ditch is inclined to the center of the chamber according to the slope of 1:1.

Three inlet holes are arranged in each side culvert of the upper lock head. Each inlet hole is 3.0 m wide and 3.0 m high. The submerged water depth of the inlet holes is from 26.2 m to 12.2 m.

Considering that the downstream of the ship lock is connected with the intermediate channel and the ship lift, in order to meet the safety requirements of the ship lift, the water level fluctuation of the intermediate channel is limited. Therefore, for weakening the fluctuation in downstream approach channel and intermediate channel caused by the unsteady flow during lock emptying, an innovative layout of water outlet is proposed. Specifically, the outlet of the left culvert is arranged in the lower lock head and the right one is arranged in the downstream approach channel, with a spacing of 60.0 m. The outlet type is top branch holes of the energy dissipation chamber. Two stilling sills are set in the energy dissipation chamber to adjust the flow distribution and flow pattern in the holes and downstream approach channel. Moreover, the problem of uniform outflow will be also solved while emptying through single side culvert.

The outlet section at the top of the culverts of the pools is 4.0 m long and 2.0 m wide. And it is expanded around with a radius of 1.0 m to form a horn shaped outlet, which is connected with the bottom elevation of the water saving pool.

Table 3 shows the characteristic sectional areas of the filling and emptying system. The overall layout is shown in Fig. 3.

Table 3. The characteristic sectional areas of the filling and emptying system

Section	Upstream inlet	Filling valve	Lock bottom culvert	Branch holes in chamber	Pool culvert	Pool outlet	Emptying valve	Downstream outlet
Area/m^2	54.0	12.0	15.6	15.84	12.0	48.0	12.0	69.0
Ratio to valve	4.5	1.0	1.3	1.32	1.0	4.0	1.0	5.75

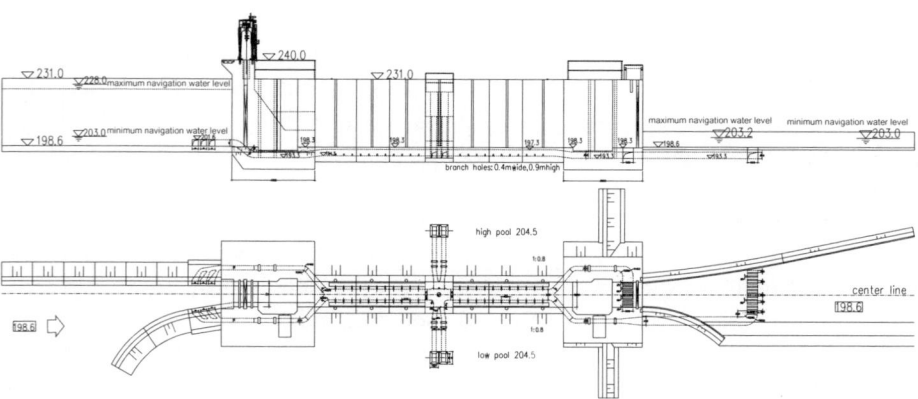

Fig. 3. Overall layout of the filling and emptying system

4 Physical Model Test

4.1 Physical Model

According to the gravity similarity criterion, the overall physical model of the filling and emptying system of the ship lock with trapezoidal section pools is designed and established. The geometric scale between the model and the prototype is 1:20. The scope of the model includes the upstream and downstream reservoirs, lock chamber, the upper and lower lock heads, a full set of filling and emptying system, two water saving pools and some upstream and downstream approach channels, which is shown in Fig. 4.

Fig. 4. The overall physical model of filling and emptying system of ship lock with trapezoidal section pools

4.2 Hydraulic Characteristics

Through the physical model test, the hydraulic characteristic values of the ship lock under different operating conditions are obtained in Table 4. All hydraulic indexes meet the code requirements and the average filling and emptying time meets the design requirement. The discharge coefficient is 0.772 while lock filling from upstream, 0.701 emptying to downstream, 0.801 filling from pools and 0.722 emptying to pools. The results show that the hydraulic performance is superior. The theoretical water saving rate of the ship lock is 54.9% to 59.0% when the upstream water level changes between 214.0 m to 228.0 m.

The minimum pressure is 4.34 m water column during water filling process, which appears behind the water filling valve of the low pool. And it is 1.95 m during water

Fig. 5. Flow pattern of downstream approach channel

emptying process, which appears behind the water emptying valve beside the lower lock head. So the culvert pressure conditions can meet the safety requirements of lock operation.

It is observed in the test that the innovative layout of the downstream outlet plays a certain role in weakening the water level fluctuation caused by watering emptying. The fluctuation is small in downstream approach channel and the outlet flow is evenly distributed along the transverse direction. The flow pattern of downstream approach channel is shown in Fig. 5.

Table 4. Hydraulic characteristic values under different operating conditions

Operating conditions	$t_{vl}/t_{vh}/t_{vc}$ (min)	T (min)	Q_{maxl} (m³/s)	Q_{maxh} (m³/s)	Q_{maxc} (m³/s)	U_{max} (m/min)	v_{maxl} (m/s)	v_{maxh} (m/s)	v_{maxc} (m/s)
228.0 m–203.0 m Filling	1.0/1.0/2.0	12.77	121	121	105	4.03/4.03/3.50	10.08	10.08	8.75
	1.5/1.5/2.0	13.43	111	112	105	3.70/3.73/3.50	9.25	9.33	8.75
	2.0/2.0/2.0	14.05	102	103	105	3.40/3.43/3.50	8.50	8.58	8.75
228.0 m–203.0 m Emptying	1.0/1.0/3.0*	17.85	105	105	43	3.50/3.50/1.43	8.75	8.75	3.58
	1.5/1.5/3.0*	18.68	100	100	43	3.33/3.33/1.43	8.33	8.33	3.58
	2.0/2.0/3.0*	19.55	91	92	43	3.03/3.07/1.43	7.58	7.67	3.58
222.0 m–203.0 m Filling	1.0/1.0/2.0	11.24	102	102	92	3.40/3.40/3.07	8.50	8.50	7.67
	1.5/1.5/2.0	12.05	93	95	91	3.10/3.17/3.03	7.75	7.92	7.58
	2.0/2.0/2.0	12.59	86	86	91	2.87/2.87/3.03	7.17	7.17	7.58
222.0 m–203.0 m Emptying	1.0/1.0/3.0*	16.51	91	93	39	3.03/3.10/1.30	7.58	7.75	3.25
	1.5/1.5/3.0*	17.12	84	85	38	2.80/2.83/1.27	7.00	7.08	3.17
	2.0/2.0/3.0*	17.72	75	77	38	2.50/2.57/1.27	6.25	6.42	3.17
214.0 m–203.0 m Filling	1.0/1.0/2.0	9.37	73	73	64	2.43/2.43/2.13	6.08	6.08	5.33
	1.5/1.5/2.0	9.95	63	64	64	2.10/2.10/2.13	5.25	5.33	5.33
	2.0/2.0/2.0	10.48	54	56	64	1.80/1.87/2.13	4.50	4.67	5.33
214.0 m–203.0 m Emptying	1.0/1.0/3.0*	14.10	66	67	30	2.20/2.23/1.00	5.50	5.58	2.50
	1.5/1.5/3.0*	14.82	56	59	30	1.87/1.97/1.00	4.67	4.92	2.50
	2.0/2.0/3.0*	15.16	51	53	30	1.70/1.77/1.00	4.25	4.42	2.50

Notes: t_{vl}, t_{vh} and t_{vc} respectively refer to the valve opening time of the lower pool, the higher pool and the lock head; T refers to the water filling or emptying time; Q_{maxl}, Q_{maxh} and Q_{maxc} respectively refer to the maximum flow rate between the lock chamber and the lower pool, the higher pool and the upstream or downstream; U_{max} refers to the maximum rising and falling speed of the water surface in lock chamber; v_{maxl}, v_{maxh} and v_{maxc} respectively refer to the maximum section flow velocity of the culvert of the lower pool, the higher pool and the lock head. 3.0* represents that the opening time of the valve is 3 min intermittently.

4.3   Berthing Conditions in Lock Chamber

The mooring force characteristics of 2 × 500t fleet and 1000t single ship in lock chamber are measured while water filling. Among them, the 1000t single ship berths at three typical positions: the upper, the middle and the lower part of the lock chamber. It can fully reflect the berthing conditions in this way.

Figure 6 shows the change process of mooring force of fleet and single ship in lock chamber under typical operating conditions. For 228 m–203 m water level combination condition, the maximum longitudinal mooring force of $2 \times 500t$ fleet is 14.33 kN to 16.50 kN and the maximum transverse mooring force is 4.98 kN to 7.45 kN when the opening time of the pool valves is 1 min to 2 min. The maximum longitudinal and transverse mooring forces of the 1000t ship with a draft of 2.9 m are 11.19 kN to 21.23 kN and 4.56 kN to 10.60 kN respectively. The maximum mooring force of the ship is less than the allowable value of the code, and there is a certain surplus. The results indicate that the flow energy dissipation in lock chamber is good and the flow distribution is relatively uniform.

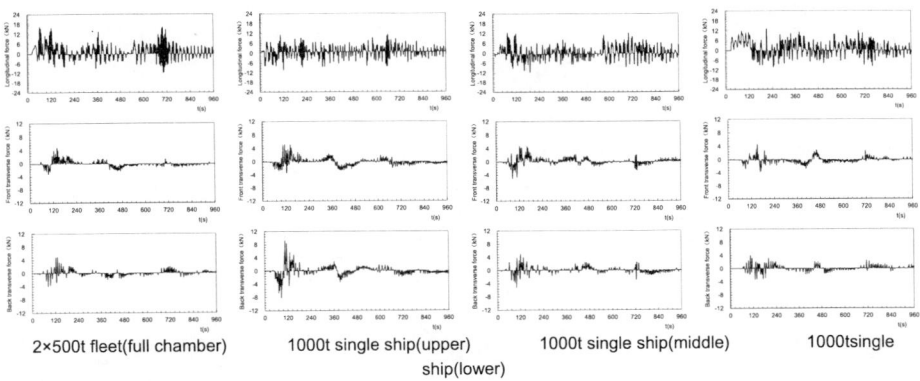

| 2×500t fleet(full chamber) | 1000t single ship(upper) | 1000t single ship(middle) | 1000tsingle |

ship(lower)

Fig. 6. The change process of mooring force of fleet and single ship in lock chamber

4.4 Valve Operating Mode

According to the hydraulic characteristics, the culvert pressure conditions, the ship berthing conditions in lock chamber and the flow conditions in downstream approach channel, based on comprehensive analyses, the valve operating modes are recommended as follows: (1) The opening time of valves for water saving pools is 2.0 min, the remaining head is 1.5 m when they begin to close, and the closing time is 1.0 min; (2) The opening time of valves at the upper lock head is 2 min, and it is 3 min intermittently at the lower lock head. See Fig. 7 for details.

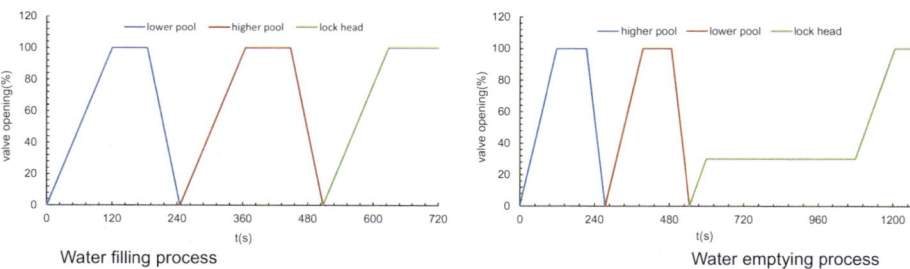

Fig. 7. Recommended valve operating modes

5   Conclusions

In view of the characteristics of large water level drop, complex topography on both banks, high lock head and large variation of navigable water level of navigation-power junction in mountainous rivers, a two-stage new trapezoidal section water saving pool scheme of ship lock is studied and proposed. It can make full use of mountainous terrain conditions and reduce excavation and backfilling. On this basis, the water level classification of the ship lock is calculated, and the type and specific layout of filling and emptying system are studied and determined. The innovative scheme of decentralized layout of downstream outlets can meet the requirements of downstream intermediate channel flow conditions.

Through the physical model test, the hydraulic characteristics, culvert pressure characteristics, ship berthing conditions and key hydraulic indexes of the water saving ship lock are studied, and the operation modes of the valves are recommended.

The research shows that the hydraulic performance of the filling and emptying system is superior, and the hydraulic indexes and flow conditions under the recommended valve opening mode meet the requirements of the safe operation of the ship lock. The scheme of trapezoidal section water saving pools proposed in the study is reasonable and feasible in hydraulics. Within the range of design water level variation, the theoretical water saving rate of the ship lock is 54.9%–59.0%.

References

Liu B, Xuan G (2019) Hydraulic model test study on filling and emptying system of Bajiangkou ship lock reconstruction and expansion project in Guilin. Research report, NHRI

Zhu L, Xuan G (2019) Experimental study on key technologies of Baishi ship lock in Qingshui River. Research report, NHRI

Yang J, Liu B, Wang L (2021) Study on water level classification of water saving ship lock under large water level variation. Port Waterw Eng (11):63–69

Li Z, Xu D (2020) Water level calculation and influencing factors of single-step lock with water saving basins. Port Waterw Eng (11):7–11

He S, Yu B, Ge G (2020) Study on several problems of integrated water saving locks. China Harbour Eng (4):6–10

Impact of Three Gorges Project (TGP) on Riverbed Revolution and Navigation Conditions in the Taipingkou Waterway

Qilin Yang[1(✉)], Qianqian Shang[1], Hui Xu[1], and Min Xu[2]

[1] Nanjing Hydraulic Research Institute, Nanjing, China
1246062029@qq.com, {qqshang,huixu}@nhri.cn
[2] Yangtze River Channel Planning and Design Institute, Wuhan, China

Abstract. Taipingkou channel is the first sandy shoal channel in the lower reaches of the Three Gorges Project (TGP). The change of river boundary is basically stable under human control, but the change of beach trough in the channel is complex, which has adverse effects on navigation. In particular, the shrinkage of Sanbatan is harmful to the design of navigation bridge hole of Jingzhou Yangtze River Bridge in dry season. By analyzing the hydrological and sediment characteristics of Taipingkou waterway before and after the impoundment of the Three Gorges Project, and studying the changes of river regime, erosion and deposition adjustment, beach evolution and diversion ratio, the influence of the impoundment of the Three Gorges Project on the riverbed evolution of Taipingkou waterway is discussed. The results show that the Three Gorges Project has changed the inflow and sediment process of Taipingkou Waterway, resulting in the frankness of annual runoff process and the reduction of sediment concentration. Under the condition of new water and sediment, the tidal flats of Taipingkou waterway show a general erosion trend. The branch with large drawdown and good inflow conditions is easier to develop, and the peak regulation greatly limits the back siltation of the convex bank side beach in the recession period. It is believed that the root cause of the current navigation obstruction of Taipingkou waterway is that the navigation bridge hole setting of Jingzhou Yangtze River Bridge cannot adapt to the trend of river evolution, and dredging is an effective measure to ensure the smooth flow of the waterway.

Keywords: Three Gorges Project (TGP) · The Taipingkou waterway · Riverbed evolution · Flow-sediment regime · Waterway

1 Introduction

The Three Gorges Project (TGP) has brought comprehensive benefits since its impoundment in 2003. Nevertheless, the water impoundment also changed the flow-sediment regime of downstream rivers, leading to general scour, and variations in transects, longitudinal slope, riverbed composition as well as river regime.

The Taipingkou waterway, located at the Jingjiang reaches of the Yangtze River, is the first sandy-shoal waterway in the downstream of TGP project. Variations in erosion and deposition there are uncertain, and thus, the navigation conditions are changing all

© The Author(s) 2023
Y. Li et al. (Eds.): PIANC 2022, LNCE 264, pp. 1502–1511, 2023.
https://doi.org/10.1007/978-981-19-6138-0_133

the time, involving a variety of research. Liu Ka et al. (2018) established a two-dimensional simulation, and illustrated that northern distributaries of Yanglinji would be silted and thus, the navigation condition would become severe; Jia Dongdong et al. (2017) built a three-dimensional dynamic model to verify that slide collapse of river-banks could lead to scouring; Cao Xiaoqin et al. (2017) proposed that TGP was difficult to maintain because the bridge anti-collision setting cannot meet the year-round navigation requirements; Kong Xianwei et al. (2020) analyzed the risks of the Jingzhou Yangtze River Bridge based on actual situation, and created possible design proposals for new bridges.

2   Overview of the Taipingkou Waterway

The Taipingkou waterway is located in the Jingjiang reach, starting from Chenjiawan and down to Yuheping, about 475 km to 492 km from Hankou waterway. The north bank of the Taipingkou waterway lies in Jingzhou city, which belongs to straight, meandering, and anabranching channels controlled by ecological revetment. The Taipingkou waterway consists of two bifurcation reaches, with Yanglinji point bar as the boundary. Its upstream part is divided into north and south channels by the Taipingkou mid-channel bar, and the downstream part is divided into north and south branches by the Sanba mid-channel bar, where the Jingzhou Yangtze River Bridge is located.

Both north branch and south branch of the downstream part are used as navigable channel, the north one being the main branch. In recent years, however, the main flow during low water shifted from the north branch to the south branch. The south branch has developed into the main channel, which moved left to the middle and formed a new navigable channel. The original channel gradually lost their functions due to siltation. The dimension of the Taipingkou channnel is 3.5 m × 150 m × 1000 m (Figs. 1 and 2).

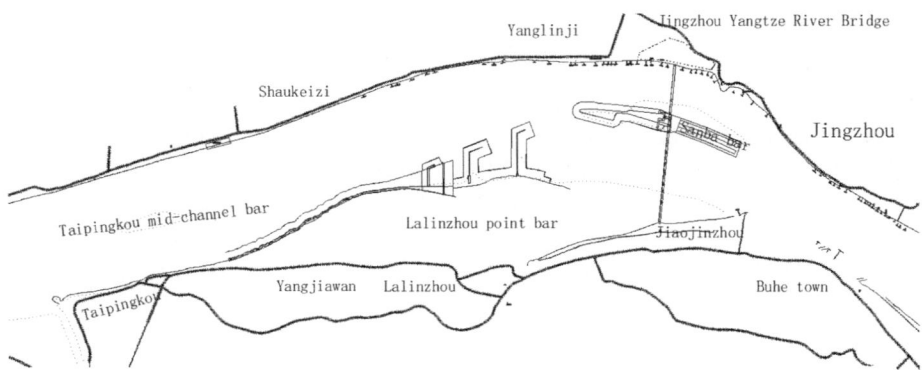

Fig. 1. River regime of the Taipingkou waterway.

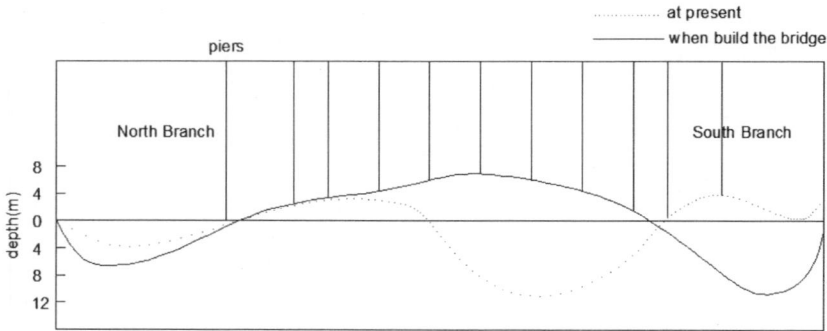

Fig. 2. Schematic diagram of pier setting of the Jingzhou Yangtze River Bridge.

After the impoundment of the TGP project, the Sanba mid-channel bar shrinks rapidly, facing the threat of disintegration again. In this regard, several river training projects were undertaken to ensure the good navigation, including the urgent protection measures of the Sanba mid-channel bar and the reinforcement of the urgent protection projects (or the first-stage training project of the Shashi waterway), and the protection project for the middle-north and upper-north parts of the Lalingzhou point bar.

3 Varation Flow-Sediment Regime

After TGP impoundment, the flow-sediment regime downstream has changed significantly, and the Taipingkou waterway has been greatly affected. In this section, we took Yichang station and Shashi station as study cases to illustrate the changes of flow-sediment regime in the Taipingkou waterway.

3.1 Flow Variations

The interannual flow is changeable due to the randomness of the hydrological process, but there is no change in trend, and the commissioning of the upstream terrace reservoir does not have a continuous effect on the downstream annual discharge. As to yearly distribution, as the TGP project advances, the discharge decreases gradually, especially during dry period (i.e., January to April), with discharge increases slowly, and recession period (i.e., September and October), with discharge decreases significantly (Fig. 3).

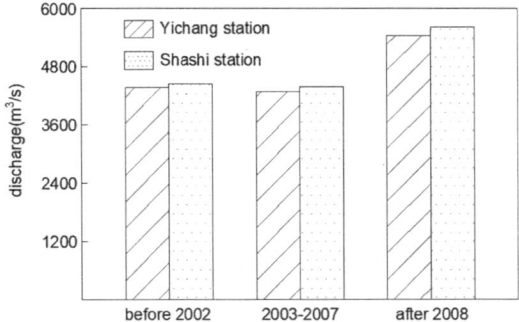

Fig. 3. Annual average discharge changes in major hydrological stations.

3.2 Variations in Sediment Load

Under TGP project, sediment load has been reduced significantly, and it reduced more after the impoundment. Influenced by upstream cascade reservoirs, sediment load from downstream further decreases. Even though annual discharge is high, sediment load from the downstream remains unchanged. (Fig. 4).

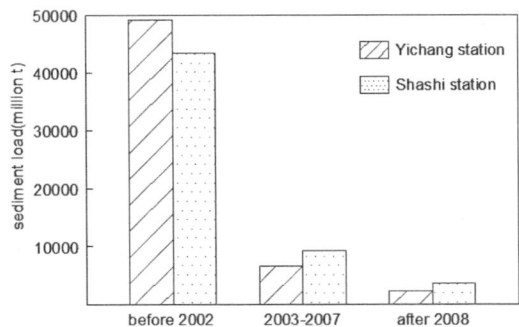

Fig. 4. Variations in sediment load of major hydrological stations.

4 Riverbed Evolution and Analysis

4.1 Before TGP Impoundment

Before the 1980s, the Taipingkou waterway was close to the left bank due to the large areas of Lalinzhou point bar and Sanba mid-channel bar. Its upstream had only one channel, while the downstream was divided into two branches. However, after the 1980s, the operation of the Gezhouba Dam changed the form of the Taipingkou waterway. Taipingkou mid-channel bar gradually silted up, while Lalinzhou point bar and Sanba mid-channel bar gradually shrank, with two branches in both its upstream and downstream. The downstream was often in a scattered multi-channel form due to 1998 China floods.

4.2   After TGP Impoundment

The Taipingkou waterway shows a planar form of compound channels with artificial nodes. At present, the waterway is suffering cumulative, intense scouring, but the river regime is stable.

After TGP impoundment, the north channel in the upper part rarely changes, while the south changed deeply, with increasing divided water ratio, and thus, has become the main channel. For the upper part, changes are obvious and intense: the scour and deposition ranges in the south and north branches are relatively large, but the plane position of the north branch is relatively stable. In recent years, the south branch has moved northward and widened year by year, becoming the main branch. In general, the Taipingkou main channel maintains'a south-south (upstream-downstream) direction. The variations in divided water ratio of the north-south branches as well as the north-south channels of the Taipingkou waterway are shown in Fig. 5.

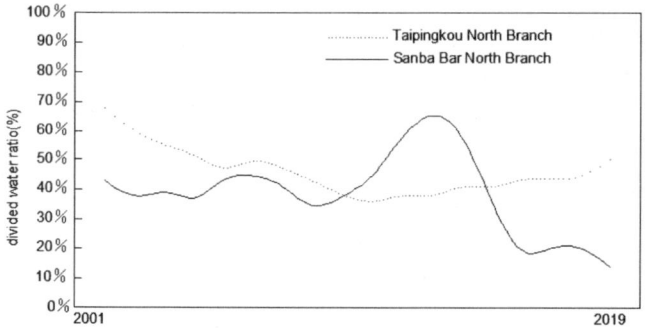

Fig. 5. Variations in divided water ratio.

4.3   Analysis of the Adjustment

The reduction of sediment load intensifies the scouring of the convex bank during flood seasons. After the impoundment of TGP project, the mid-year discharge has long retention and short recession period, which greatly limits the siltation of the convex bank. The Yanglinji point bar does not match the sediment transport capacity of the north channel. Thus, the hydrodynamic conditions for generating the Yanglinji point bar still exist in the inlet of the north branch.

Under the action of unsaturated water flow, the development of the branching channel in straight channels depends on the good headwater conditions, and the branches with good headwater conditions tends to develop. The development of the two branches is repeated. Under the action of unsaturated water flow, when slope of the south branch is greater, the south branch would develop with strong discharge, which can be transformed into hydrodynamic conditions.

5 Navigation Obstruction and Dredging Projects

5.1 Characteristics of Navigation-Obstruction

The current characteristics of Navigation-Obstruction in the Taipingkou waterway are closely related to the construction of the Jingzhou Yangtze River Bridge. The 1998 China Floods and the operation of the TGP project has greatly influenced Sanba channel bar. It was rapidly washed away and shrank, making it difficult to restore. Accordingly, the main flow in the north branch is very important for good navigational conditions because bridge piers are very closely distributed in the south branch, thus limiting navigation. However, after 2012, the main flow during low water shifted from the north branch to the south branch, bringing great challenges to navigation. It is obvious that the north branch cannot be used as main waterway anymore.

The cause of the current obstruction in the Taipingkou waterway is that the bridge abutment setting of Jingzhou Yangtze River Bridge cannot adapt to the natural evolution of rivers. The designed navigational channel in the south branch have been severely deposited, while the non-designed navigation bridge channel limit the utilization of better natural water depth conditions.

5.2 Recent Maintenance

During the initial period of the TGP project (2003–2006), the Taipingkou waterway was frequently changing, with unstable channels and adverse navigation conditions. Dredging and maintenance work were urgently needed during the dry season every year.

Since 2016, the dredging and maintenance of north branch of downstream has been continued. The dredging area is mainly concentrated in the transition area from the branch section of Taipingkou mid-channel bar to the branch section of Sanbatan mid-channel bar, and there are two transitional forms: Taipingkou South Branch-Sanbatan North Branch (2# channel) and Taipingkou North Branch-Sanbatan North Branch (1# channel), which can be divided into three stages: (1) North-South transition, centralized maintenance of 2# channel; (2) 1# channel and 2# channel are constructed alternately, navigable alternately, and the two channels are equally important; (3) North-north transition, drainage and stabilization of channels, and efforts to dredge 1# channels.

5.3 Variations in Navigation Conditions

In 2018, due to the more unfavorable water conditions, the Yanglinji point bar shifted down significantly and the Sanba bar mid-channel bar were connected into one, making the navigational conditions in the north branch very bad after the floods, and channel 2# was fully blocked. In 2018–2019, the maintenance method of alternate construction and alternate navigation of channel 1# and channel 2# was adopted to achieve year-round ship passage in the northern branch, and the survey map in March 2019 showed that the 4 m line of channel 1# was close to penetration and the width of the upper section 3 m line reached more than 180 m; the 5 m line of channel 2# was penetrated and the narrowest width was about 180 m. In July 2019, the 3.5 m line under the most depleted water level of channel 1# was penetrated, and the width could reach about 200 m. Channel 2# was slowly silted up after dredging, and the width of the 3.5 m line under the most depleted water level in the shallow area was less than 150 m. Before the flood in 2019, the channel maintenance department will change the main branch to channel 1#, mainly to maintain channel 1#, channel 2# gradually silt waste. By March 2020, after many years of maintenance and dredging, the diversion ratio of channel 1# increased to more than 20%, and the wandering beach body - Yanglinji point bar was gradually moved out. Even if the impact of the unfavorable water period in 2020 is encountered, the impact on the channel conditions of the Taipingkou waterway is not significant due to the low upstream sand source, and the amount of dredging and maintenance in 2020 is comparable to that of the previous year.

In July 2021, the width of the 3.5 m line under the driest water level of the north branch reaches more than 200 m, but from the comparison of August 12–28, 2020 and September 16–17, 2020 flushing and siltation, channel 1# to the north branch is dominated by siltation, with about 1.08 million square meters of siltation and about 300,000 square meters of flushing, with a net siltation of 780,000 square meters. Considering that the flooding period from July to September is the period of the largest siltation in the northern branch of Taipingkou, according to the large scale measurement map of the shallow area in late August 2021, the shallow area of the northern branch has increased significantly compared with July, and the change trend of the development of the southern branch and the shrinking of the northern branch has not been reversed, and the northern branch is still in the process of slow siltation.

5.4   Analysis of the Effect of Dredging Project

i. In the context of relatively limited engineering management measures, maintenance dredging is an effective measure to ensure the smooth flow of waterways. The external conditions of the Taipingkou waterway are complex, with various constraints as well as difficult implementation. In this context, dredging has become a necessary tool to ensure the smooth flow of waterways.

ii. Adjusting the river flushing and siltation area by "digging and throwing" is one of the effective ways to cope with the adverse effects of TGP storage on the middle and lower reaches of the waterway. By dredging the navigable channel and throwing the dredged soil to the beach body, non-navigable branches and other areas that need to be controlled, it enhances the undercutting of the target channel on the one hand, and helps to alleviate the unfavorable phenomena such as beach scouring and branch development on the other hand, which is one of the effective ways to cope with the adverse effects of TGP clear water discharge on the navigation channel.

iii. The trenching design idea of the diversion plastic tank has a certain contribution to the formation of a stable navigation path. From the last five years, under the initial control of the remediation project, after continuous dredging, the northern branch of the Taipingkou waterway diversion overall showed a small increase in the trend, the channel conditions have all shown some improvement, dredging maintenance volume also generally decreased slightly, the difficulty of channel maintenance has also been reduced.

6 Forecast of the Evolutionary Trend

Under the combined effect of the guarding project, shore protection project and dredging project, the river potential of the Taipingkou waterway is basically stable, but with the change of water conditions, bars and branches are still subject to change. Combined with the historical flow-sediment regime and the evolution of the riverbed, the future evolution trend of the Taipingkou waterway is predicted as follows.

i. The Taipingkou mid-channel bar: Since 2018, with the increase in the fight against illegal sand mining, the current river form has the trend of repairing the impact of sand mining, mainly in the form of local sand mining deep pits gradually silt flat; in the past two years only the remaining beach body still maintain an obvious scouring trend.

ii. Transition section and Sanba mid-channel bar: In recent years, the evolution of the transition section and Sanba mid-channel bar has actually been strongly influenced by dredging activities. The dredging project for the north branch is essentially a process of gradually removing the Yanglinji point bar, which not only stabilizes the diversion of the northern branch to a certain extent, but also retards the further development of the south branch. Combined with the dredging in recent years, continuous dredging can avoid the complete decline of the northern branch; considering the weakening of the upstream sand recharge conditions, the possibility of generating large-scale Yanglinji point bar is small, and the north branch has the conditions for maintenance.

iii. From the changes in the river during the maintenance and dredging in the past two years, the maintenance and dredging has ensured the overflow capacity of the north branch and maintained the current channel scale, but the trend of siltation in the north branch and development in the southern branch is still difficult to reverse, and the dredging only eases the degree of siltation in the north branch.

iv. From the variations in the yearly divided water ratio of the Taipingkou waterway, it can be seen that before 2014, the ratio of the north branch was mostly above 30%, and the channel conditions were better; since 2014, the ratio of the north branch decreased sharply and has been maintained at less than 30%. Therefore, it can be inferred that when the divided water ratio of the north branch is greater than 30%, the channel conditions will improve significantly.

7 Conclusions

This paper presents a comprehensive analysis of the Taipingkou waterway based on the new water and sand conditions after impoundment of TGP, comparison of riverbed evolution before and after impoundment, and changes in channel conditions, with the following main conclusions.

i. After TGP is impounded, the average annual flow of the Taipingkou waterway remains basically unchanged, but the distribution of flow changes within the year, limiting the re-siltation of the beach during the receding period. At the same time, the sand content of the Taipingkou waterway has been significantly reduced, resulting in the general scouring of the channel, which currently maintains the mainstream orientation of the Taipingkou south branch-Sanba bar south branch.

ii. The obstruction factor of the Taipingkou waterway is that the navigable bridge piers of Jingzhou Yangtze River Bridge cannot adapt to the evolution trend of rivers. Since the impoundment of the TGP project, a series of dredging projects have been carried out and the channel conditions have stabilized. In the current situation, dredging is still a feasible measure to keep the waterway open. Although the north branch, where the main navigable bridge piers is located, continues to shrink, it still has the conditions for maintenance.

iii. The river can basically remain stable in the future period. Under the strong influence of the dredging project, the divided water ratio of north branch was stabilized and the channel conditions were improved, but the trend of siltation in the north branch remained irreversible. In addition, maintaining the divided water ratio of the north branch at 30% or more is conducive to the improvement of waterway conditions.

Acknowledgements. This research is supported by the Subsequent Work of the Three Gorges Project (No. SXHXGZ-2020-3).

References

Cao XQ, Zhan JY, Xu YF (2017) Analysis on the evolution trend of Taipingkou waterway after the impoundment of 175m in the Three Gorges and analysis of waterway maintenance and safeguard measures. China Water Transp. Channel Technol (02)

Jia DD, Xia HF, Chen CY, Zhang XN (2017) Three-dimensional numerical simulation of the influence of shore erosion on channel conditions—taking the Taipingkou waterway in the middle reaches of the Yangtze River as an example. Adv Water Sci 28(02)

Kong XW, Zhang QH et al (2020) Research on the influence of river bed evolution on water navigation safety in Xiaqiao district. Chin J Saf Sci 31(03)

Li RB (2014) Comprehensive management of Taipingkou waterway in the middle reaches of the Yangtze River. China Water Transp (09)

Liu K, Chen L, Gao CL, Guo Q (2018) The evolution trend of the Taipingkou waterway in the middle reaches of the Yangtze River and its governance countermeasures. Water Transp Eng (10)

Xia JQ, Zhou MR et al (2020) The characteristics of riverbed adjustment and bank collapse in the middle reaches of the Yangtze River after the Three Gorges Project was used. People's Yangtze River 51(01)

Yao SM, Hu CW, Qu G (2021) Research on the evolution and ecological governance of the lower reaches of the Three Gorges Project. Proc Yangtze River Acad Sci 38(10)

On Characterizing Flow Resistance in a Tidal Reach

Ye Jing[1]([✉]), Xueting Lei[2], Jie Qin[1], Teng Wu[1],
and Elikplim Agbemafle[1]

[1] College of Harbour, Coastal and Offshore Engineering, Hohai University,
Nanjing, China
liaozihu@qq.com, {jqin,wuteng}@hhu.edu.cn
[2] Changjiang Waterway Planning, Design and Research Institute, Wuhan, China

Abstract. The tidal reaches are characterized by unsteady and non-uniform flow (UNF), which is significantly different from the commonly assumed steady and uniform flow (SUF) in hydraulics. The SUF shows invariant temporal and spatial flow characteristics, and thus flow acceleration is absent in a prismatic channel. However, for the UNF, the variation of flow velocity and depth in both temporal and spatial scales causes the loss of flow energy, and thus increases the flow resistance. In order to clarify the variation of flow resistance and its influencing factors in tidal reaches, this study investigates the flow resistance characteristics under UNF conditions. In this study, a typical tidal section of the Lower Yangtze River (LYR) – Kouanzhi Waterway (KW) – was selected as the study area, where the temporal variation of water surface along the river course at different tide levels, the bathymetry of multiple cross-sections, the distribution of cross-sectional flow velocity and its temporal variation were measured in detail. Based on these field measurement data, the contribution terms to the energy slope were calculated and evaluated, by decomposing the momentum equation. The calculated contributing terms include water surface gradient, local acceleration, and convective acceleration. The results showed that the local acceleration and convective acceleration have a substantial impact on the energy slope during specific time periods, which was found to be more significant than the findings in previous studies. The results show that the local acceleration term is more significant than the convective acceleration term except when the water surface slope is close to zero, and its contribution is significant throughout the flood tide and the initial ebb tide periods. The above research results are of great significance for the investigation of flow resistance mechanisms and numerical simulations in tidal rivers.

Keywords: Momentum equation · Flow resistance · Energy slope · Tidal river

1 Introduction

Resistance coefficients of rivers vary with flow conditions and physical boundaries, such as bed roughness, cross-sectional geometry, river planar forms, the spatial distribution of roughness, and other factors that cause resistance changes (Yen 2002); Pagliara and Chiavaccini (2006) studied the hydraulics of rough channels with or

© The Author(s) 2023
Y. Li et al. (Eds.): PIANC 2022, LNCE 264, pp. 1512–1521, 2023.
https://doi.org/10.1007/978-981-19-6138-0_134

without the insertion of protruding boulders and influencing parameters such as slope, and relative submergence, and blocks disposition were systematically investigated. Rubol et al. (2018) explored the effect of canopy morphology on vegetated channel flow structure and resistance by treating the canopy as a porous medium characterized by an effective permeability. In addition to the roughness influence, Hohermuth and Weitbrecht (2018) found that bed-load transport significantly decreased flow resistance and increased near-bed velocity for the conditions investigated.

2 Literature

Most existing studies related to flow resistance assumed steady and uniform flow conditions, without considering the unsteadiness of flow. For steady flow in a prismatic channel, there is no temporal and spatial variation of flow velocity and flow depth, indicating no temporal and spatial gradient for velocity and depth. In contrast, the variation in the velocity and depth introduces extra loss of flow energy in comparison to steady and uniform flow, e.g., the flow processes in tidal rivers (Brown and Richard 1980). Specifically, there is a lack of knowledge and data to reliably estimate the flow resistance in tidal reaches which is a vital component in calibrating numerical models. Mrokowska et al. (2015) analyzed friction slope, friction velocity, and Manning coefficient in unsteady flow and they found that the former two parameters are more sensitive to applying simplified formulas than the Manning coefficient. Note that their measurements were obtained from artificial dam-break flood waves in a small lowland watercourse. Bao et al. (2018) investigated a modified formula aimed at improving the prediction of unsteady flow resistance. They proposed an equation including ten terms relating to the first- and second-order temporal and spatial partial derivatives of hydraulic parameters. Their results showed that the optimal number of additional terms is three, and extra terms can barely improve the unsteady friction estimation. These three items correspond to the three first-order derivatives in previous research (Rowiński et al. 2000; Mrokowska et al. 2015).

This study aims at analyzing the effect of unsteady flow characteristics on the energy slope in a large tidal river based on measured data and then compares the results with previous findings obtained from small channels. The method and data acquisition are given in Sect. 2, the results of the analysis are provided in Sect. 3, and this study is concluded in Sect. 5.

3 Methodology

For a one-dimensional channel, the momentum equation for unsteady flow is (Knight 1981):

$$-S_f = \frac{\partial \eta}{\partial x} + \frac{U}{g}\frac{\partial U}{\partial x} + \frac{1}{g}\frac{\partial U}{\partial t} + \frac{h + \eta}{2\rho}\frac{\partial \rho}{\partial x} \tag{1}$$

where η is the water level; U is the average cross-sectional flow velocity; x is the distance; t is the time; h is the depth; ρ is the fluid density, and g is the acceleration of gravity. With the help of the Sf, resistance coefficients can be calculated by resistance formulas, such as the Manning equation. The four terms on the right-hand side in Eq. (1) represent the slope of water surfaces, the convective acceleration, the local acceleration, and the density gradient, respectively. The last term is not considered in this study given the invariant flow density. For unsteady flow, the first three terms on the right-hand side determine the magnitude of the energy slope. Therefore, the main task of this study is to study the contribution of the three terms to the energy slope.

The KW is located in the LYR (latitude and longitude: 119°41′36.6″, 32°15′6.1″). The Datong hydrological station, about 350 km upstream of the KW, is the last runoff observation station of the Yangtze River. This station has an annual average flow discharge of 28,700 m^3/s and an annual average sediment concentration of 0.461 kg/m^3. The KW is a tidal reach and its flow process is influenced by both runoff and tides. Normally, there is no reversing flow in the KW during the moderate flood and flood periods and occurs only during the spring tide in dry seasons. Therefore, the main driving force in the river section is the upstream runoff and is affected by the downstream tidal processes (Cao et al. 2011).

To study the resistance caused by the temporal and spatial variation of flow processes, the following data were collected: the temporal variation of water surface along the river course, the cross-sectional profiles along the river course, and the temporal cross-sectional distribution of flow velocity. These data were collected during spring tides in February 2019 (dry season) and May-June 2014 (flood season). The streamwise variation of the water surface was measured by temporary tidal stations, indicated by the solid red points in Fig. 1. The cross-section profiles and flow velocity were obtained by ADCP measurements indicated by the black solid lines. Due to the long distance among measured cross-sections, this study spatially interpolated the upstream and downstream cross-sections (red solid lines) near the measured cross-sections, and the convective acceleration terms are calculated based on these interpolated cross-sections and the accordant measured cross-sections. In addition, the interval of measured water level was 1 h, which was interpolated into an interval of 15 min. The water level was recorded relative to the 1956 Yellow Sea Height Datum of China. By spatially and temporally interpolating the measured data, the time series of water level along the course and the spatial distribution of cross-sectional flow velocity were obtained. Based on these data, the water level and flow velocity were spatially differenced to obtain the terms in Eq. (1).

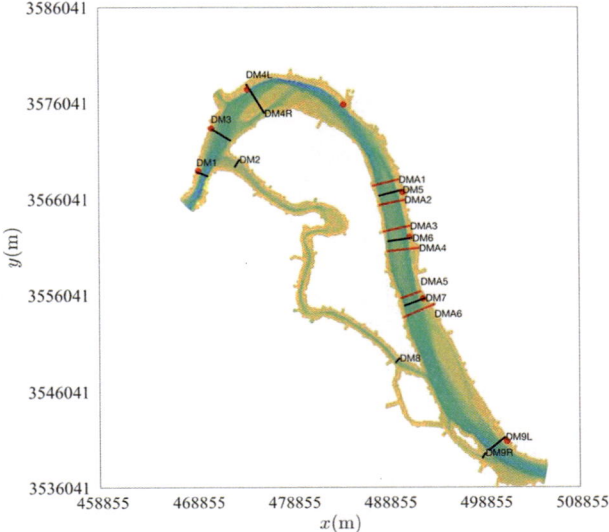

Fig. 1. Locations of water surface and ADCP measurements in the KW. The black and red cross-sections indicate the measured and interpolated locations, respectively.

4   Results

4.1   Variation of the Terms of Energy Slope in the Flood Season

Figure 2 shows the calculated results of the terms of the energy slope for the cross-sections DMA1 and DMA2 during a complete tidal cycle. The results of the other cross-sections are similar to those of the two sections. In each plot in Fig. 2, the upper subplot shows the temporal process of the contribution of each term in Eq. (1), and the lower subplot shows the corresponding temporal tide level (Lw) and depth-averaged velocity (U) process. The temporal variation of the components is similar in magnitude, in spite of the fact that more fluctuations occur in DMA2 because of the more evident temporal variation of the water surface gradient in this cross-section. The Sf continuously changes with the variation of tide level. It gradually varies during the long ebb tide., however, during the flood tide and the initial stage of the ebb tide, Sf varies more evidently in magnitude. In general, the main contribution of Sf comes from the slope of the water surface, $\partial h/\partial x$. The local acceleration term and the convective acceleration terms contribute much less in comparison to the gradient of the water surface for most of the time. Only during flood tide and the initial period of the ebb tide, do the acceleration terms increase significantly. The magnitude of the convective acceleration term is small compared to the gradient of water surface and the local acceleration term, with a small contribution only near the flood tide period.

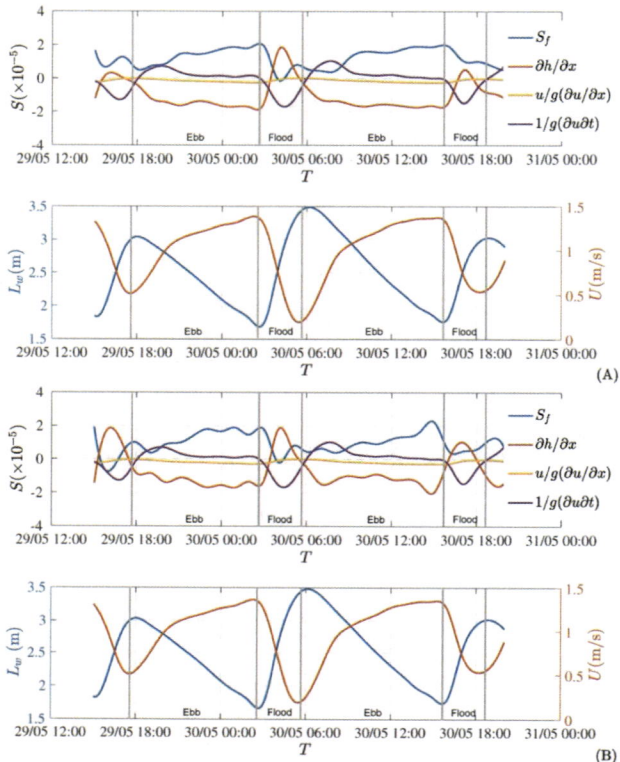

Fig. 2. Variation of different components of the energy slope during a complete tide cycle in the spring tide of a flood season. Plot (A) and (B) correspond to the DMA1 and DMA2 cross sections, respectively.

4.2   Variation of the Terms of Energy Slope in the Dry Season

Figure 3 shows the distribution of the magnitude of each term of the energy slope in a complete tidal cycle during the dry season. The gradient of water surface in DMA2 shows a more significant contribution to the total energy slope in comparison to the DMA1 which in combination with the weaker local acceleration terms in DMA2 makes the energy slope in DMA2 show a more gentle variation with time in comparison to that in DMA1. Compared to the flood season, the tidal magnitudes become smaller, decreasing from a tidal difference of around 2 m in the flood season to around 1 m in the dry season. In addition, the depth-averaged velocity also decreases from a range of 0.1 to 1.4 m/s during the flood season to 0.1 to 0.8 m/s during the dry season. The above variation decreases the magnitude of the items during the dry season. The trends shown in Fig. 3 are similar to the results in Fig. 2 to some extent, whereas differences can also be readily observed. The energy slope varies with the water level, and the main contribution comes from the water surface slope which is the same as the findings in the flood season. However, the relative contribution of local acceleration increases in comparison to the values in the flood season. During flood tide and initial ebb tide, the magnitude of the local acceleration is comparable to that of the water surface slope, which causes a significant change in Sf. The decrease in the magnitude of each item makes the temporal change in Sf more significant and fluctuations more evident. Due to the vital influence of the water surface slope on the energy slope, it is supposed that the relation between the water level and the resistance coefficient will be improved in comparison to the widely used relation between resistance coefficient and flow depth.

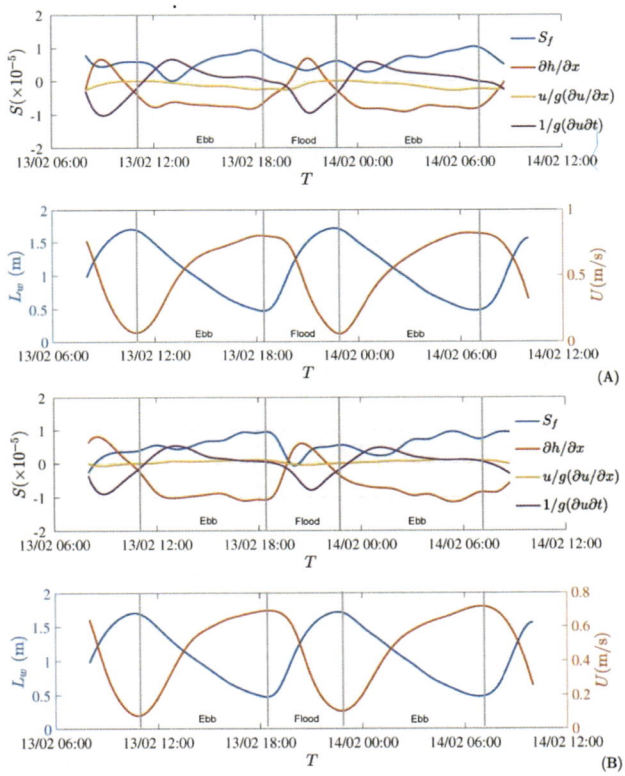

Fig. 3. Variation of different components of the energy slope during a complete tidal cycle in the spring tide of the dry season. Plot (A) and (B) correspond to the DMA1 and DMA2 cross-sections, respectively.

5 Discussion

In Knight (1981), local and convective accelerations were significant only when the water surface slope is around zero, and most of the time the energy slope consistently varies with the water surface slope. This differs to some extent from the results of this study. The contribution of local acceleration and convective acceleration terms to the energy slope compared to the water surface slope is much more evident than the results found by Knight (1981). Although the results in this study show that the local and convective accelerations are certainly more significant when the water surface slope is zero, which is consistent with Knight's findings, their contributions remain significant throughout the flood tide and the initial ebb tide periods. This difference is mainly due to the fact that the magnitude of water surface variation in the straight section of the KW is smaller compared to the reach studied by Knight (1981); therefore, the change in water surface slope is less significant than that observed in Knight (1981). The KW is farther away from the Yangtse estuary (240 km) and thus has a smaller change in water surface gradient, compared to the Tal-y-Cafn reach studied by Knight (1981), which is only 10 km from the estuary. In addition, the Tal-y-Cafn section has a straight planar shape with little variation in cross-section along the course, while the cross-section morphology of the KW reach varies more evidently compared to the Tal-y-Cafn river, resulting in increased local and convective acceleration. These twofold impacts make the energy slope variation of the KW more complex than that observed by Knight (1981).

The effect of water depth on the resistance coefficient has been the major concern in river flow dynamics irrespective of tidal or non-tidal rivers. Many empirical formulas have been proposed to establish relations between water depth and flow coefficient (Yen 2002). A general form of such a relation can be obtained by combining the logarithmic equation of friction factor with the Manning equation. The logarithmic flow resistance for rough flow conditions is:

$$\frac{U}{u_*} = 6.25 + 5.75 \log_{10}\left(\frac{R}{k_s}\right) \tag{2}$$

where U and u_* are depth-averaged flow velocity and shear velocity, respectively. R is the hydraulic radius, and k_s is the roughness length scale. The Manning equation can be written as:

$$\frac{U}{u_*} = \frac{R^{1/6}}{n\sqrt{g}} \tag{3}$$

where n is the Manning coefficient and g is the gravity acceleration. Substitution of Eq. (2) into Eq. (3), the influence flow depth or hydraulic radius on the Manning coefficient can be obtained given a known k_s value:

$$n = \frac{\kappa R^{1/6}}{\sqrt{g}[\ln(R/k_s) - 1]} \tag{4}$$

where κ is the von Karman constant. This equation reflects the variation of flow resistance with hydraulic radius. Because the Manning coefficient varies with flow depth, mathematical models prefer to use Darcy-Weisbach factor to account for the flow resistance. However, it is difficult to accurately estimate ks, because it changes with the variation of sediment transport conditions.

Figures 4 and 5 show the relations between Manning coefficient, n, with the water level, Lw, and hydraulic radius, R, respectively. The relation between n and Lw in Fig. 4 is much better than that between n and R. The most evident difference between them is that the n is uniformly distributed with the variation of the water level, i.e., the variance of the scattered points rarely changes with the varying water level. In contrast, the relation between n and the hydraulic radius shows an irregular distribution of the scattering points, that the variance of the points may increase or decrease with the varying hydraulic radius.

Equation (3) based on the logarithmic law was applied in Fig. 5 to compare with the measured data. Since the size of sand waves on the river bed is unknown, different ks values from 0.05 to 0.2 m were applied to the equation. As shown in Fig. 5, the formula generally agrees with the measured data, but there is a significant deviation when the hydraulic radius is greater than 18 m. In addition, all the relations in Fig. 5 appear to be parallel straight lines, indicating that the formula predicts almost invariant Manning coefficients. However, as shown in Fig. 4, a trend of decreasing Manning coefficient as increasing water level can be readily observed. This is attributed to the fact that the shape of sand waves on the river bed changes with varying water levels, which further affects the resistance coefficient. Therefore, the relation between water level and resistance coefficient obtained based on the measured data is more suitable for the estimation of resistance coefficient in tidal rives in comparison to the relation based on the hydraulic radius. More importantly, there is no effective method to determine the ks parameter, and thus the application of Eq. (3) is of great limitation. An empirical function was fitted between the Manning coefficient with the water level, and the correlation coefficient, R2, reaches 0.72, which can be used to estimate the resistance coefficient in the KW.

$$n = L_w^{-0.0047} \tag{5}$$

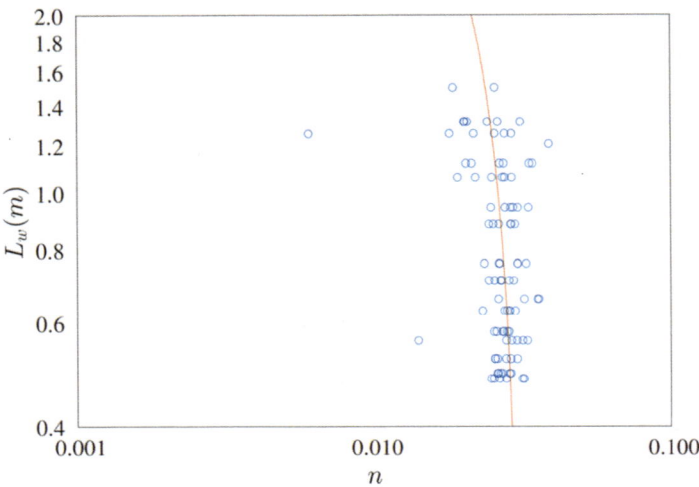

Fig. 4. Relation between water levels and the Manning coefficient

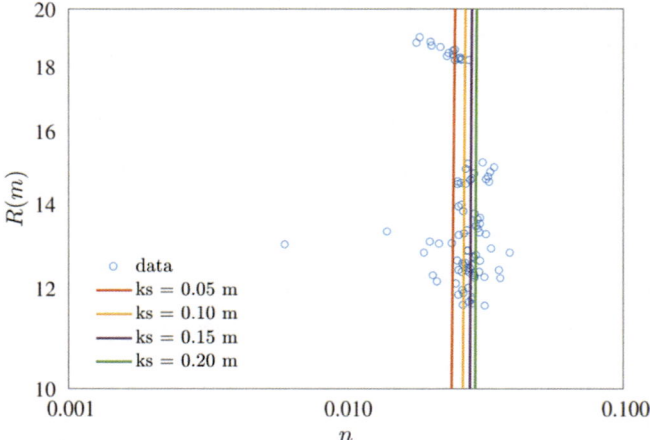

Fig. 5. Relation between the hydraulic radius and the Manning coefficient. The lines correspond to Eq. (4) with different k_s values.

6 Conclusions

By calculating the energy slope components of the KW during the flood and dry seasons, the contributions of different terms to the energy slope are compared during a complete tidal cycle. The following conclusions can be drawn from the results:

1. the water surface gradient is the main contribution to the energy slope irrespective of the flood or dry seasons.
2. the local acceleration and convective acceleration have a significant contribution to the total energy slope during the flood tide and initial stage of the ebb tide.
3. the relation between the resistance coefficient and the hydraulic radius during a complete tidal cycle is no better than the relation between the resistance coefficient and water levels. Therefore, in the lower reaches of the Yangtze River, the latter is preferred for estimating the resistance coefficient.
4. the coefficient of resistance varies in a limited range during tidal cycles, and thus applying a fixed Manning coefficient is an acceptable treatment if no prediction equation is available.

These results provide a reference for the resistance research in other tidal river reaches, whereas due to the empirical nature of the formula in this study, it is necessary to obtain a similar formula based on measured data.

Acknowledgements. This study was supported by the National Natural Science Foundation of China (52079043, 52179061).

References

Bao W, Zhou J, Xiang X, Jiang P, Bao M (2018) A hydraulic friction model for one-dimensional unsteady channel flows with experimental demonstration. Water 10(1):43

Brown WS, Richard PT (1980) A study of tidal energy dissipation and bottom stress in an estuary. J Phys Oceanogr 10(11):1742–1754

Cao M, Li Q, Cai G, Yuan D, Wang X (2011) On local scour deformation flume generalization experiment of Manyusha shoal head protection engineering in straight reach of Yangtze River estuary: I. Design of flume generalized and protection engineering building simulation. Port WaterwEng 7:106–112

Hohermuth B, Weitbrecht V (2018) Influence of bed-load transport on flow resistance of step-pool channels. Water Resour Res 54(8):5567–5583

Knight WD (1981) Some field measurements concerned with the behaviour of resistance coefficients in a tidal channel. Estuar Coast Shelf Sci 12(3):303–322

Mrokowska MM, Rowiński PM, Kalinowska MB (2015) A methodological approach of estimating resistance to flow under unsteady flow conditions. Hydrol Earth Syst Sci 19:4041–4053

Rowiński PM, Czernuszenko W, Marc J (2000) Time-dependent shear velocities in channel routing. Hydrol Sci J 45(6):881–895

Rubol S, Ling B, Battiato I (2018) Universal scaling-law for flow resistance over canopies with complex morphology. Sci Rep 8(1):1–15

Pagliara S, Chiavaccini P (2006) Flow resistance of rock chutes with protruding boulders. J Hydraul Eng 132(6):545–552

Yen BC (2002) Open channel flow resistance. J Hydraul Eng 128(1):20–39

Research on the Non-constant Navigable Water Flow Conditions of the River Channel Downstream of the Navigable Facilities of Baise Water Control Project

Mingfang Guan[1], Ming Zhang[2(✉)], Changhai Han[2], Leng Qing[2], and Yu Hang[1]

[1] Zhejiang Institute of Communications Co. LTD., Hangzhou, China
[2] Key Laboratory of Navigation Structures, Nanjing Hydraulic Engineering Institute, Nanjing, China
zhangm@nhri.cn

Abstract. The exit of Baise ship lift on the Right River is only about 700 m away from Dongsun Hydropower Station. Its dispatching operation, especially the unsteady flow, has a direct impact on the navigable water flow conditions of the approach channel gate area and river channel downstream of the Baise ship lift. According to some characteristics of Baise Junction and Dongsun Junction and the measured topography of Baoai River, a plane two-dimensional numerical model of the research river section is constructed, and the water flow changes in the mouth area under different working conditions are simulated and calculated. The research results show that the safety of navigation in the entrance area can be guaranteed from two aspects: optimizing the operation mode of the Baise and Dongsun cascade hydropower stations, and rationally formulating the downstream ship navigation management mode. Under the operating conditions of the four units of Baise Power Station, it is recommended that the ship wait for 0.5 h after the two units in Dongsun are started, so as to avoid the maximum water level increase period in the first half hour.

Keywords: Baise ship lift · Peak shaving power station · Unsteady flow · Navigation safety

1 Introduction

The exit of Baise ship lift on Youjiang River is only about 700 m away from Dongsun hydropower station. As the reverse regulation reservoir of Baise hydropower station, Dongsun hydropower station only has the function of daily regulation. The dispatching operation of Dongsun hydropower station, especially the unsteady flow, has a significant and direct impact on the navigable water flow conditions at the entrance of the approach channel and the river channel downstream of Baise ship lift.

According to the partial characteristics of Baise hub and Dongsun hub projects and the measured topography of TiAI river channel, the plane two-dimensional numerical model of the research river section is constructed. Combined with the operation

Y. Li et al. (Eds.): PIANC 2022, LNCE 264, pp. 1522–1534, 2023.
https://doi.org/10.1007/978-981-19-6138-0_135

conditions of Baise power station and Dongsun power station, different simulation conditions are designed, and the water rheology in the entrance area under different working conditions is simulated and calculated. The calculation results of the scheme show that the unsteady discharge process of peak shaving in Baise power station has a significant impact on the navigable flow conditions in the gate area of the downstream Baise ship lift. Due to the weak regulation capacity of Dongsun power station, when four units of Baise power station are full, the maximum water level variation in the entrance area is too large, which affects the navigation safety of ships in the entrance area.

2 Project Overview

The Baise Water Control Project is the second cascade in the Yujiang River planning. It is located in the upper reaches of the Right River, the main stream of the Yujiang River, 22 km away from Baise City. It is a key project of the Right Channel on the southern route, one of the three main channels leading to Yungui in the Right Shipping Plan.

The general layout plan of the navigable buildings of Baise Water Conservancy Project adopts the combination plan of ship lock and ship lift, and the navigation route is arranged in Nalugou on the left bank of Baise Water Conservancy Project. The ship-passing facilities are composed of ship locks, intermediate channels, retaining earth dams, navigable aqueducts, flood-retaining and maintenance locks, and vertical ship lifts. The total length of the line is about 4245 m. The navigable buildings are divided into three relatively independent areas, namely the approach channel upstream of the ship lock, the intermediate channel, and the approach channel downstream of the ship lift. The overall layout of the ship-passing facility project is shown in the figure (Fig. 1).

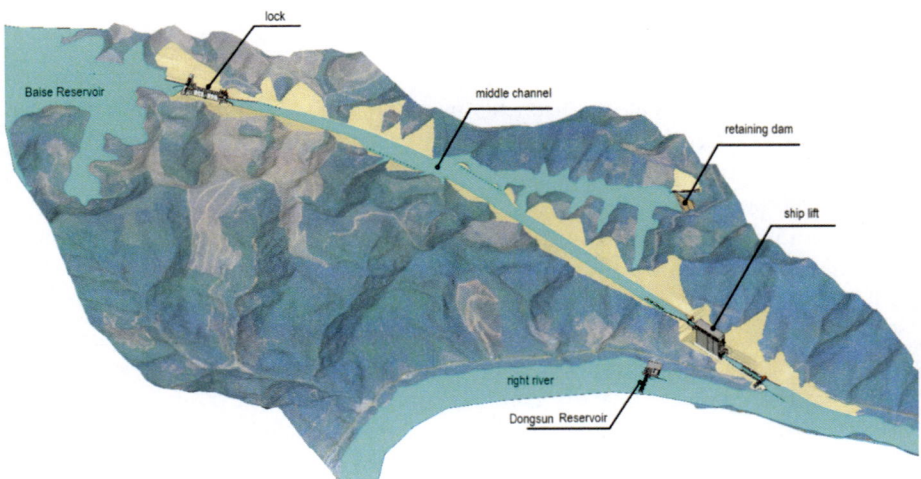

Fig. 1. The general layout of the ship-passing facilities of Baise water control project.

The ship-passing facility project of Guangxi Baise Water Conservancy Project involves three cascade hubs including Baise Water Conservancy Project, Dongsun Hydropower Station and Naji Shipping Hub.

(1) Baise Water Conservancy Project

The main buildings of Baise Water Control Project include dam, power plant, sluice gate and navigation buildings, etc. The normal water level is 228 m, the dead water level is 203 m, and the regulating storage capacity is 2.62 billion m^3. It is an incomplete multi-year regulating reservoir; the installed capacity of the power station is 540 MW, The maximum output is 580 MW; the flood control storage capacity is 1.64 billion m^3, the limited water level during the flood season is 214 m; the daily average minimum discharge is 100 m^3/s.

The Baise Water Control Project is affected by the anti-regulation effect of Dongsun Hydropower Station and Naji Shipping Junction.

(2) Dongsun Hydropower Station

Dongsun Hydropower Station is located 6.5 km downstream of Baise Hydropower Station. It is a water conservancy project mainly based on Baise Reservoir's anti-regulation reservoir for peak regulation, and also has comprehensive utilization benefits such as power generation and water supply. The normal water level is 122.5 m, and the corresponding storage capacity is 6.34 million m^3; the check flood level is 132.68 m, and the corresponding storage capacity is 18.15 million m^3. The available flow is 600 m^3/s, and the installed capacity of the power station is 2 × 12 MW.

The specific layout from left to right along the axis of the dam is: powerhouse on the right bank, regulating gate, 9 m open overflow dam (practical weir), 5-hole 14m 6.5 m overflow gate dam (overflow weir surface is wide top weir), 9 m open overflow dam (utility weir), thorn wall on the left bank. The net width of the overflow front of the gate dam is 70 m, the elevation of the weir crest is 116.5 m, the net width of the overflow front of the practical weir is 18 m, and the elevation of the weir crest is 122.5 m.

The operating principle of Dongsun Hydropower Station: when the incoming water flow Q ≤ 300 m^3 /s, the daily adjustment operation of the reservoir can be carried out according to whether the upstream Baise Reservoir is peak-regulated; when 300 < Q ≤ 600 m^3/s, the comprehensive utilization of the reservoir needs, the water level of the reservoir is maintained at the normal storage level of 122.5 m; when 600 < Q ≤ 1500 m^3/s, in addition to the turbine overflow of 600 m^3/s, the excess water is discharged from the overflow gate and dam, and the water level on the dam slightly It is 122.5 m higher than the normal water storage level; when Q > 1500 m^3/s, the unit stops generating electricity, and the water level of the reservoir basically returns to the state of natural river course.

Dongsun Hydropower Station is about 700 m away from the exit of Baise ship lift, so Dongsun Hydropower Station has a direct impact on the navigable water flow conditions in the approach channel gate area and connecting section downstream of Baise ship lift.

(3) Naji Shipping Hub

The Naji Shipping Hub is located in the downstream of Dongsun Hydropower Station, and its backwater range reaches the exit of Baise Ship Lift. It is a large-scale shipping hub project focusing on shipping, with comprehensive utilization benefits such as power generation, irrigation, flood control, and water supply. The reservoir The normal water level is 115 m, the corresponding storage capacity is 103 million m³, and the dead water level is 114.4 m. It has daily regulation performance. The Naji Shipping Hub is the anti-regulation reservoir of Baise Reservoir. Its main task is to regulate the peak regulation process of Baise Power Station and ensure the downstream navigation. The flow rate is not less than 140 m³/s, and at the same time, the channel of the Naji Reservoir area can be channeled to improve the navigation standard of the Right River.

As mentioned above, the exit of Baise ship lift is only about 700 m away from Dongsun Hydropower Station. Dongsun Hydropower Station, as the counter-regulating reservoir of Baise Power Station, only has daily adjustment function. It has a direct impact on the navigable flow conditions of the area and the river. According to the actual situation of the daily variation of the water level downstream of the Dongsun Hydropower Station, the daily variation of the hourly water level is often greater than 1.0 m, and in extreme cases it is greater than 5.0 m, which causes great harm to the safety of downstream navigation. Therefore, it is very necessary to carry out research on the influence of unsteady flow in the downstream channel of the navigation facilities of Baise Hub, and by optimizing the operation and scheduling mode of Dongsun Hydropower Station, so that the characteristics of the unsteady flow in the downstream channel can meet the navigation safety requirements, and it is necessary to minimize the influence of unsteady flow.

Using the two-dimensional mathematical model simulation method, the influence of the unsteady flow in the downstream channel of the navigation facilities of the Baise Hub and the improvement measures are studied. Navigation safety requirements.

3 Numerical Model of Two-Dimensional Flow in the Downstream Channel of the Hub

3.1 Model Construction

According to the characteristics of Baise Junction and Dongsun Junction and the measured topography of the Baoai River, a two-dimensional numerical model of the river reaches is constructed.

(1) Model range
According to the research object, from Baise Junction along the Poai River to the downstream Baise (III) Hydrological Station of Dongsun Hydropower Station. Among them, the section from Baise Junction to Dongsun Power Station is about 6.2 km long, and the length from Dongsun Junction to Baise (III) Hydrological Station is about 17.6 km, see Fig. 2 below.

(2) Model building

This time, taking Dongsun Power Station as the boundary, two-dimensional plane hydraulic models were established for calculation respectively, as shown in Figs. 3 and 4 below.

Fig. 2. Model calculation area.

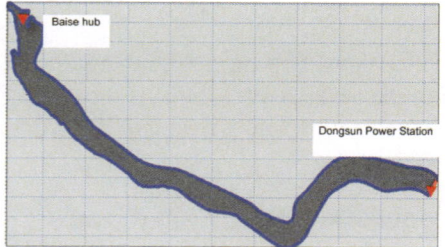

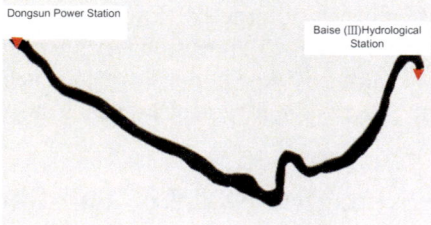

Fig. 3. The 2D model of the dam site of Baise Project-Dongsun project.

Fig. 4. The 2D model of Dongsun project Dam Site-Baise (III) Hydrological station

(3) Boundary conditions

Upper boundary condition: The model calculates the upper boundary using the corresponding flow process.

Lower boundary conditions: the upstream model, the corresponding design of the power generation flow process of Dongsun Power Station; the downstream model, the water level-flow relationship of the Baise (III) hydrological station (Fig. 5).

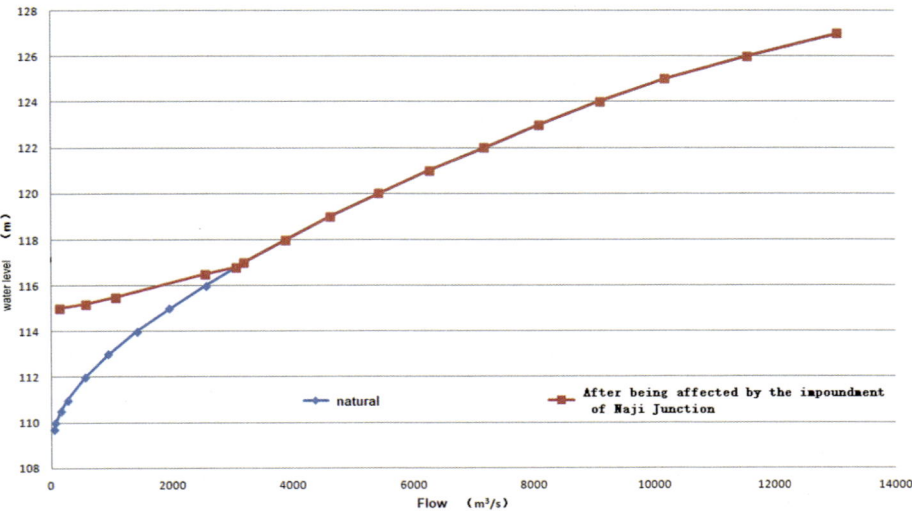

Fig. 5. The relationship between water level and flow at Baise Hydrology (III) Station.

(4) Roughness rate determination

Due to the lack of on-site measured hydrological data, this time, the relationship between the design water level and flow in the mouth area was used to calibrate the parameters of the model. The results show that the average roughness can be in the range of 0.016 to 0.014, and the roughness decreases slightly when the flow rate is large.

3.2   Simulation Conditions

The navigable water flow conditions of Baise ship lift gate area and section channel are affected by the dispatching operation of Baise, Dongsun and Naji hubs. Based on the constructed two-dimensional hydrodynamic simulation model of the study area, the operating characteristics of the three cascade hubs are comprehensively considered, and different simulation conditions are designed. The characteristics of the water level change process in the Entrance area under flow conditions are statistically and analyzed to provide data support for the navigation management of ships downstream of Dongsun Hydropower Station under unsteady flow conditions (Table 1).

Table 1. Simulation conditions.

Working condition	Baise power station			Dongsun power station				Naji junction
	Boot time (min)	Power generation flow (m³/s)	Boot Number of units	Boot time (min)	Power generation flow (m³/s)	Before the dam initial water level (m)	Boot Number of units	Before the dam initial water level (m)
1	5	348	2	5	300	122.5	1	114.4
2	5	348	2	5	300	122.5	1	115.0
3	5	348	2	30	300	122.5	1	114.4
4	5	348	2	30	300	122.5	1	115.0
5	10	696	4	10	600	122.5	2	115.0
6	10	696	4	10	600	121.5	2	115.0
7	10	696	4	10	600	122.5	2	114.4
8	10	696	4	10	600	121.5	2	114.4
9	5-60-5	348-348	4	30-60-30	300-300	121.5	2	114.4
10	5-60-5	348-348	4	30-60-30	300-300	121.5	2	115.0
11	5	174	1	5	174	122.5	1	114.4

The operation modes of Baise and Dongsun power stations under various working conditions are described as follows:

(1) Simulation scheme 1. Baise Power Station: Start 2 units within 5 min, reach the rated power generation flow of 348 m³/s, the water level in front of the dam of Dongsun Power Station at the lower boundary is 122.5 m, and the maximum water level in front of the dam is controlled at 122.8 m. Dongsun Power Station: Simultaneously with the start-up of Baise Power Station, Dongsun Power Station starts one unit within 5 min, reaching a rated power generation flow of 300 m³/s, and the initial water level downstream of the dam site is 114.4 m.

(2) Simulation scheme 2. Baise Power Station: Start 2 units within 5 min, reach the rated power generation flow of 348 m³/s, the water level in front of the dam of Dongsun Power Station at the lower boundary is 122.5 m, and the maximum water level in front of the dam is controlled at 122.8 m. Dongsun Power Station: Simultaneously with the start-up of Baise Power Station, Dongsun Power Station starts one unit within 5 min, reaching a rated power generation flow of 300m³/s, and the initial water level downstream of the dam site is 115.0 m.

(3) Simulation schem 3. Baise Power Station: Start 2 units within 5 min, reach the rated power generation flow of 348m³/s, the water level in front of the dam of Dongsun Power Station at the lower boundary is 122.5 m, and the maximum water level in front of the dam is controlled at 122.8 m. Dongsun Power Station: Simultaneously with the start-up of Baise Power Station, Dongsun Power Station starts one unit within 30 min, reaching a rated power generation flow of 300m³/s, and the initial water level downstream of the dam site is 114.4 m.

(4) Simulation scheme 4. Baise Power Station: Start 2 units within 5 min, reach the rated power generation flow of 348m^3/s, the water level in front of the dam of Dongsun Power Station at the lower boundary is 122.5 m, and the maximum water level in front of the dam is controlled at 122.8 m. Dongsun Power Station: In synchronization with the start-up of Baise Power Station, Dongsun Power Station will start one unit within 30 min, with a rated power generation flow of 300 m^3/s and an initial water level of 115.0 m at the lower boundary.

(5) Simulation scheme 5. Baise Power Station: 4 units are started within 10 min, reaching a rated power generation flow of 696 m^3/s, the water level in front of the dam of Dongsun Power Station at the lower boundary is 122.5 m, and the maximum water level in front of the dam is 122.8 m. Dongsun Power Station: Simultaneously with the start-up of Baise Power Station, Dongsun Power Station starts 2 units within 10 min, reaching a rated power generation flow of 600 m^3/s, and the initial water level downstream of the dam site is 115.0 m.

(6) Simulation scheme 6. Baise Power Station: 4 units are started within 10 min, reaching a rated power generation flow of 696 m^3/s, the water level in front of the dam of Dongsun Power Station at the lower boundary is 121.5 m, and the maximum water level in front of the dam is 122.8 m. Dongsun Power Station: Simultaneously with the start-up of Baise Power Station, Dongsun Power Station starts 2 units within 10 min, reaching a rated power generation flow of 600 m^3/s, and the initial water level downstream of the dam site is 115.0 m.

(7) Simulation scheme 7. Baise Power Station: 4 units are started within 10 min, reaching a rated power generation flow of 696m^3/s, the water level in front of the dam of Dongsun Power Station at the lower boundary is 122.5 m, and the maximum water level in front of the dam is 122.8 m. Dongsun Power Station: Simultaneously with the start-up of Baise Power Station, Dongsun Power Station starts 2 units within 10 min, reaching a rated power generation flow of 600 m^3/s, and the initial water level downstream of the dam site is 114.4 m.

(8) Simulation scheme 8. Baise Power Station: 4 units are started within 10 min, reaching a rated power generation flow of 696 m^3/s, the water level in front of the dam of Dongsun Power Station at the lower boundary is 121.5 m, and the maximum water level in front of the dam is 122.8 m. Dongsun Power Station: Simultaneously with the start-up of Baise Power Station, Dongsun Power Station starts 2 units within 10 min, reaching a rated power generation flow of 600 m^3/s, and the initial water level downstream of the dam site is 114.4 m.

(9) Simulation scheme 9. Baise Power Station: start 2 units within 5 min, reaching the rated power generation flow of 348 m^3/s; after an interval of 1.0 h, start two units within 5 min, totaling 696 m^3/s; the water level in front of the dam of Dongsun Power Station at the lower boundary is 121.5 m, the maximum water level in front of the dam is 122.8 m. Dongsun Power Station: In synchronization with the start-up of Baise Power Station, Dongsun Power Station starts one unit within 30 min, reaching the rated power generation flow of 300 m^3/s; after an interval of 1.0 h, starts another unit within 30 min, with a total of 600 m^3/s; The initial water level downstream of the dam site is 114.4 m.

(10) Simulation scheme 10. Baise Power Station: Start 2 units within 5 min, reaching the rated power generation flow of 348 m³/s; after an interval of 1.0 h, start two units within 5 min, totaling 696 m³/s; the water level in front of the dam of Dongsun Power Station at the lower boundary is 121.5 m, the maximum water level in front of the dam is 122.8 m. Dongsun Power Station: In synchronization with the start-up of Baise Power Station, Dongsun Power Station starts one unit within 30 min, reaching a rated power generation flow of 300 m³/s; after an interval of 1.0 h, starts another unit within 30 min, reaching a total of 600 m³/s; dam The initial water level downstream of the site is 115.0 m.

(11) Simulation scheme 11. Baise Power Station: start one unit within 5 min, the rated power generation flow is 174 m³/s, the water level in front of the dam of Dongsun Power Station in the lower boundary is 122.5 m, and the maximum water level in front of the dam is 122.8 m. Dongsun Power Station: Simultaneously with the start-up of Baise Power Station, Dongsun Power Station starts one unit within 5 min, the power generation flow is 174 m³/s, and the initial water level downstream of the dam site is 114.4 m.

In working conditions 9 and 10, Baise Power Station started 2 units within 5 min, reaching the rated power generation flow of 348 m³/s; after an interval of 1.0 h, started two units within 5 min, totaling 696 m³/s. Simultaneously with the start-up of Baise Power Station, Dongsun Power Station starts one unit within 30 min, reaching a rated power generation flow of 300 m³/s; after an interval of 1.0 h, starts another unit within 30 min, reaching a total of 600 m³/s.

4   Statistics and Analysis of Water Flow Conditions in 3entrance Area

4.1   Analysis of Water Flow Conditions in the Mouth Area

Table 2 for the statistics of water flow simulation results in the gate area and anchorage under the above 11 working conditions.

Table 2. Each simulated working condition and corresponding simulation result table.

Working condition	Simulation conditions						Simulation results		
	Baise power station		Dongsun power station			Naji hub	Water level in the entrance area		Anchorage water level
	Boot time (min)	Power generation flow (m³ /s)	Boot time (min)	Power generation flow (m³ /s)	Before the dam initial water level (m)	Before the dam initial water level (m)	Maximum time-varying amplitude (m /h)	Maximum downhill (‰)	Maximum time-varying amplitude (m/h)
1	5	348	5	300	122.5	114.4	1.12	1.12	0.73
2	5	348	5	300	122.5	115.0	0.88	1.10	0.59
3	5 .	348	30	300	122.5	114.4	1.08	0.90	0.73
4	5	348	30	300	122.5	115.0	0.83	0.88	0.56
5	10	696	10	600	122.5	115.0	1.70	1.98	1.17
6	10	696	10	600	121.5	115.0	1.70	1.93	1.17
7	10	696	10	600	122.5	114.4	2.07	2.06	1.41
8	10	696	10	600	121.5	114.4	2.07	2.01	1.41
9	5-60-5	348-348	30-60-30	300-300	121.5	114.4	1.08	0.97	0.72
10	5-60-5	348-348	30-60-30	300-300	121.5	115.0	0.83	0.93	0.56
11	5	174	5	174	122.5	114.4	0.67	0.91	0.56

The analysis of water flow conditions in the entrance area is as follows:

(1) The increase of the water level in the downstream port area of the unsteady flow of Dongsun Hydropower Station is affected by the initial downstream water level and the start-up time of the unit. Hourly maximum increase decreases.

(2) If two units of Baise Power Station are fully powered, and one unit of Dongsun Power Station is fully powered, the initial downstream water level is 114.4 m, the dead water level of Naji Power Station. The maximum hourly increase in the Entrance area is 1.12 m; if the initial downstream water level is 115.0 m, the normal water level of the Naji Junction, the maximum hourly increase in the Entrance area is 0.88 m. Under the operating conditions of the cascade power station, it can basically meet the requirements of the navigable water flow conditions in the Entrance area.

When the two units of Baise Power Station are fully powered, the maximum hourly increase in the downstream anchor area is about 0.73 m, and the water level after stabilization is about 115.8 m–115.9 m.

(3) If the four units of Baise Power Station generate electricity normally, limited by the weak storage capacity of Dongsun Reservoir, the maximum hourly increase of water level in the downstream Entrance area will exceed 2.0 m.

The analysis of the results also shows that the water level has the largest increase rate in the first half hour, and within one hour after half an hour, that is, in the period of 0.5 h–1.5 h, even under the most unfavorable working conditions (Condition 7, Dongsun Hydropower Station is not considered. Adjustment and storage, the downstream water level is 114.4 m), the water level increases from 115.92 m to 116.71 m, and the maximum hourly increase is about 0.79 m.

When the four units of Baise Power Station are fully powered, the maximum hourly increase in the downstream anchor area is about 1.41 m (the water level in front of Naji Dam is 114.4 m) or 1.17 m (the water level in front of Naji Dam is 115.0 m), and the water level after stabilization is about 116.7 m–117.1 m.

(4) By properly adjusting the operation settings of the cascade power station, the flow field characteristics in the entrance area can be improved. The adjustment plan is: Baise Power Station starts 2 units within 5 min, and starts 2 units within 5 min after an interval of 60 min. Dongsun Power Station starts one unit within 30 min simultaneously, and starts another unit within 30 min after an interval of 60 min. Under this operation plan, even if the initial water level downstream of Dongsun Hydropower Station is unfavorable at 114.4 m, the maximum hourly increase in the Entrance area is about 1.08 m, which can basically meet the navigation requirements.

Under this plan, the maximum hourly increase of the water level in the anchorage area is about 0.72 m (the water level in front of Naji Dam is 114.4 m) or 0.56 m (the water level in front of Naji Dam is 115.0 m), and the water level after stabilization is about 116.7 m–117.1 m.

In the case of Baise Power Station with only one unit generating power (174 m^3/s), even if the initial downstream water level is 114.4 m, the maximum hourly increase of the water level in the Entrance area is 0.67 m.

When the four units of Baise Hydropower Station are in full power normally, the maximum water surface slope within 600 m of the downstream mouth of Dongsun Hydropower Station is about 2‰. By implementing the operation plan adjustment of Baise and Dongsun power stations (conditions 9 and 10), the maximum water surface slope within 600 m of the Entrance area is reduced to less than 1‰.

The above water level analysis and statistical results can provide strategic support for the safety of ship navigation in the mouth area.

4.2 Ship Operation Plan in the Port Area and Waterway

The unsteady discharge process of Baise Power Station's peak regulation has a significant impact on the navigable water flow conditions in the gate area of the downstream Baise ship lift. Due to the weak regulation capacity of Dongsun Power Station, when the four units of Baise Power Station are fully powered, the maximum water level variation in the Entrance area exceeds 2.0 m/s, which seriously affects the navigation safety of ships in the Entrance area. It is necessary to ensure the safety of navigation in the entrance area from two aspects: optimizing the operation mode of Baise and Dongsun cascade hydropower stations, and the management mode of downstream ship navigation.

For the operation of ships in the gate area and waterway downstream of Dongsun Power Station, it is necessary to make corresponding navigation operation plans based on the actual operating conditions of Baise Power Station and the specific conditions of the downstream initial water level:

(1) If the upstream Baise Power Station generates electricity for one unit, the operation of the ship will not be controlled;

(2) The upstream Baise Power Station generates power for 2 units, and the initial water level downstream of Dongsun Power Station is not lower than 115.0m, then one unit of Dongsun Power Station starts to operate normally, and the navigation of ships in the downstream port area will not be affected.

(3) The upstream Baise Power Station generates power for 2 units, and the initial water level downstream of Dongsun Power Station is 114.4 m, so one unit of Dongsun Power Station is recommended to start operation within 30 min, and the flow field in the downstream port area can meet the navigation requirements.

(4) Baise Power Station is powered by 4 units. It is recommended that the ship sail for 0.5 h after the two units in Dongsun are started, so as to avoid the maximum water level increase period in the first half hour.

(5) If the ship suspension (0.5 h) management plan is not adopted, the operation plan of Baise and Dongsun cascade power stations can be adjusted to meet the navigation requirements of the Entrance area.

5 Conclusions

In this paper, the unsteady navigable water flow conditions of the river channel downstream of the navigable facilities of the Baise Water Control Project are studied, combined with the operating characteristic parameters of the Baise Power Station and Dongsun Power Station, the corresponding simulation conditions are set, and the numerical model is used to simulate and analyze various unsteady flow schemes. The research results show that it is necessary to optimize the operation mode of the Baise and Dongsun cascade hydropower stations, and to rationally formulate the downstream ship navigation management mode to ensure the safety of navigation in the Entrance area: 1) By properly adjusting the operation settings of the cascade hydropower stations, the flow field in the Entrance area can be improved. characteristic. Baise Power Station started 2 units within 5 min, and after 1 h interval, started 2 units within 5 min. Dongsun Power Station starts one unit within 30 min synchronously, and then starts another unit within 30 min after 1 h interval. Under the operation mode of this cascade power station, the non-constant water flow conditions in the downstream gate area and the connecting section can meet the navigation requirements. Require. 2) From the formulation of the ship's operation management mode, under the operating conditions of the four units of Baise Power Station, it is recommended that the ship wait for 0.5h after the start of the two units in Dongsun to sail, so as to avoid the maximum water level increase period of the first half hour.

Acknowledgements. This study is financially supported by the National Key R&D Program of China (2021YFC3200302).

References

Huang GD (2017) Initial operation and regulation optimization of Baise multi-purpose dam project. Guangxi Water Resour Hydropower Eng 6:94–97

Sun XP (2014) Water resource bearing capacity study for Xijiang river economic belt in Guangxi. Guangxi Water Resour Hydropower Eng 5:30–33

Zhang X (2003) Feasibility analysis of Dongsun as reverse regulation reservoir of Baise water control project. Pearl River 6:7–9

Huang K, Pan ZH, Chen SL (2022) Analysis of water imbalance of hydropower stations on Yujiang river. China Rural Water Hydropower 4:30–36

Standard for inland navigation (GB50139-2014)

Hydrological Specification for Inland Navigation Engineering (JTS145-1-2011)

Technical specification for waterway improvement engineering (JTJ312-2003)

Overall design code for river port engineering (JTJ212-2006)

Report on the Model Test Results of the Downstream Approach Channel in the Feasibility Study Stage of the Ship Passing Facility Project of the Baise Water Control Project in Guangxi (2021) Nanjing Institute of Water Resources

Spatiotemporal Evolution Characteristics and Influencing Factors of Incoming Water and Sediment in Three Gorges Reservoir

Peng Chen[1(✉)], Jinyou Lu[1], Zhongwu Jin[1], Yinjun Zhou[1],
Rouxin Tang[2], Zhaoxi Liu[1], and Qiuba Han[3]

[1] Key Laboratory of River Regulation and Flood Control of Ministry of Water Resources, Changjiang River Scientific Research Institute, Wuhan, China
ChenPeng5211@outlook.com

[2] State Key Laboratory of Water Resources and Hydropower Engineering Science, Wuhan University, Wuhan, China

[3] Chongqing Water Conservancy and Electric Power Construction Survey, Design and Research Institute Co., Ltd., Chongqing, China

Abstract. Based on the hydrological and sediment observation data of the main stream and main stream of the upper Yangtze River from 1956 to 2018, using M-K test, Wavelet analysis, Approximate entropy and Lyapunov exponent, this paper analyzes the spatiotemporal evolution characteristics from the perspectives of catastrophe, periodicity, complexity and chaos of incoming water and sediment in The TGR (TGR), defines the connotation of "new water and sediment conditions" in the TGR, and discusses the main influencing factors of spatiotemporal variation of incoming water and sediment in the TGR. The results show that: (1) Except Pingshan station of Jinsha River, Fushun station of Tuojiang River and Wulong station of Wujiang River, the runoff of other main and tributaries and TGR in the upper Yangtze River has no significant change, but the sediment discharge has obvious mutation, and the runoff and sediment have periodic changes in different degrees; (2) The complexity of annual runoff and sediment transport in the TGR increases gradually, and the complexity of sediment transport is greater than that of runoff. Both annual runoff and sediment transport are chaotic, and the complexity of runoff and sediment transport sequence increases gradually from the upper reaches to the lower reaches; (3) The connotation of "new water and sediment condition" shows that there is no obvious mutation in runoff and sediment transport in time scale, the main cycle of water and sediment change is basically the same, and the complexity and chaos of sediment transport in space scale are significantly greater than runoff, and the change degree of sediment transport is more significant than runoff; (4) From 1991 to 2002, climate change (rainfall) was an important factor affecting the sediment discharge in the upper Yangtze River. From 2003 to 2018, the impact of climate change was relatively insignificant, and human activities became the most important factor.

Keywords: The TGR · Water and sediment characteristics · Spatiotemporal evolution · Climate change · Human activities

© The Author(s) 2023
Y. Li et al. (Eds.): PIANC 2022, LNCE 264, pp. 1535–1552, 2023.
https://doi.org/10.1007/978-981-19-6138-0_136

1 Introduction

The Yangtze River is the third largest river in the world and the largest river in China. From the source of the Yangtze River to Yichang, Hubei Province, the Yangtze River is about 4500 km long, with a total catchment area of about 1 million km^2 (Wang et al. 2020; Liu et al. 2019; Tan et al. 2019). The runoff in the upper Yangtze River mainly comes from Jinsha River, Minjiang River, Tuojiang River, Jialing River and Wujiang River, and the sediment mainly comes from Jinsha River and Jialing River. The change of water and sediment conditions in the upper Yangtze River has a great impact on the sediment deposition and regulation in The TGR area, as well as the flood control and ecological environment in the middle and lower Yangtze River. Since the 1990s, with the construction of a large number of reservoirs, the largest reservoir group has been built in the upper Yangtze River in the world. The operation and joint operation of the reservoir group in the upper Yangtze River with The TGR as the core has changed the temporal and spatial distribution of sediment macroscopically (Liu et al. 2020; Tan et al. 2019).

Since 1950s, many experts and scholars have studied the characteristics of runoff and sediment in the upper reaches of the Yangtze River. Based on the analysis of the changes of runoff and sediment transport at Yichang station, Pingshan station and Beibei station of Jialing River in the upper Yangtze River since the 1950s, it is considered that the obvious differences in the way and degree of human activities, vegetation destruction and restoration, soil and water loss control, sediment retention by water conservancy projects and sediment increase by engineering construction are the main reasons for the different changes of runoff and sediment in Jialing River and Jinsha River since the 1980s reason (Zhang 1999; Zhang and Wen 2002). Based on the measured data from 1954 to 1992, Spearman rank correlation test, Kendell rank correlation test and linear regression test are used to analyze the discharge and sediment discharge of Pingshan station on Jinshajiang River, and the variation trend of water and sediment is analyzed and tested. It is concluded that there is no obvious trend of water and sediment variation (Huang et al. 2002). Based on the measured data of runoff and sediment in the Yangtze River Basin in recent 50 years, the variation trend of runoff and sediment in the main stations of the main stream and tributaries is analyzed macroscopically. It is found that the average annual sediment discharge of Hankou station, Datong station and Yichang station of the main stream has a significant decreasing trend (Renshou et al. 2003). The variation of runoff and sediment in the upper reaches of the Yangtze River in recent decades is analyzed by using the statistical analysis method of double cumulative curve of measured annual runoff and annual sediment discharge and Spearman rank correlation test (Dai et al. 2007). Based on the analysis of recent runoff and sediment conditions and their changing trend in the upper reaches of the Yangtze River, it is found that there is no obvious change in the composition of runoff and sediment in the upper reaches of the Yangtze River, but the composition of sediment has changed greatly (Quanxi et al. 2004; Xu et al. 2005, Xu et al. 2009). Based on the measured runoff and sediment data of Pingshan, Cuntan, Yichang and other control stations in the main stream of the Yangtze River from 1950 to 2005, the temporal, spatial and correlation variation characteristics and trends of runoff and sediment in the main stream of the Yangtze River are studied by using the methods of time interval analysis, regression analysis, correlation analysis, moving average, frequency analysis, modulus coefficient comparison and cumulative curve (Dong et al. 2008). By means of wavelet analysis, Mann Kendall test, runoff concentration degree and concentration period, the interannual and interannual variations of runoff in Yichang station in recent decades are studied. The results show that the operation of The TGR aggravates the trend of runoff decrease in flood season and increase in dry season in the upper reaches of the Yangtze River, and reduces the difference of annual distribution (Zhao et al. 2012). Based on the 60 year series of runoff and sediment data from the main stream of the upper reaches of the Yangtze River and its main tributaries, Minjiang River and Jialing River Control hydrological stations, and based on the M-K test method and R/S analysis method, the variation characteristics and development trend of runoff sediment transport time series are preliminarily analyzed by calculating the characteristic curve (Xiao et al. 2017). In addition, Zhu (2000) discussed the changes of sediment transport and concentration in the upper Yangtze River and the ways of sediment reduction, and Xiang (1993) studied the source, transport and sedimentary characteristics of sediment in the Yangtze River, Li (2001) studied the temporal and spatial distribution characteristics of water and sediment transport in the Yangtze River, Chen (2001) and Chen (2001) studied the variation characteristics of water and sediment transport in the Yangtze River, and analyzed the variation of water flow into the sea in the dry season, Lu (2003) also analyzed the variation characteristics of sediment transport in the upper Yangtze River, Su (2005) preliminarily predicted the variation trend of water and sediment in the Yangtze River Basin, and Yang (1993) respectively predicted the Yangtze River movement under the influence of human activities In addition, Hu (1993) also made a preliminary study on the water and sediment variation of Jialing River.

At present, in view of the research on the variation characteristics of water and sediment in the upper Yangtze River, the relevant scholars only recognize the "new water and sediment conditions" from the time level (time point, mutation, periodicity, etc.), and ignore the analysis and discussion from the spatial level to a certain extent. However, the movement of water and sediment not only changes with the passage of time, but also changes in different spatial categories. As an objective natural phenomenon, the movement of water and sediment can be explored more comprehensively only by fully covering its spatial and temporal characteristics. Therefore, based on the hydrological and sediment observation data of the upper Yangtze River and the main stream from 1956 to 2018, this paper analyzes the spatiotemporal evolution characteristics of water and sediment in Sanxin reservoir from four aspects of mutation, periodicity, complexity and chaos, and defines the "new water and sediment conditions" of the TGR. The main factors affecting the temporal and spatial variation of water and sediment in the TGR are discussed.

2   Data and Methods

2.1   Data

Taking Zhutuo and Cuntan in the main stream of upper Yangtze River, Beibei in the Jialing River and Wulong in the Wujiang River as representative stations, the data of Cuntan + Wulong (1956–2002) and Zhutuo + Beibei + Wulong (2003–2018) are used for The TGR. The runoff and sediment data of the upper Yangtze River from 1956 to 2018 provided by the Bureau of hydrology of the Yangtze River Water Resources Commission and the Yangtze River Sediment bulletin are used as the basic data for the study. The river system boundary and hydrological stations in the upper Yangtze River are shown in Fig. 1.

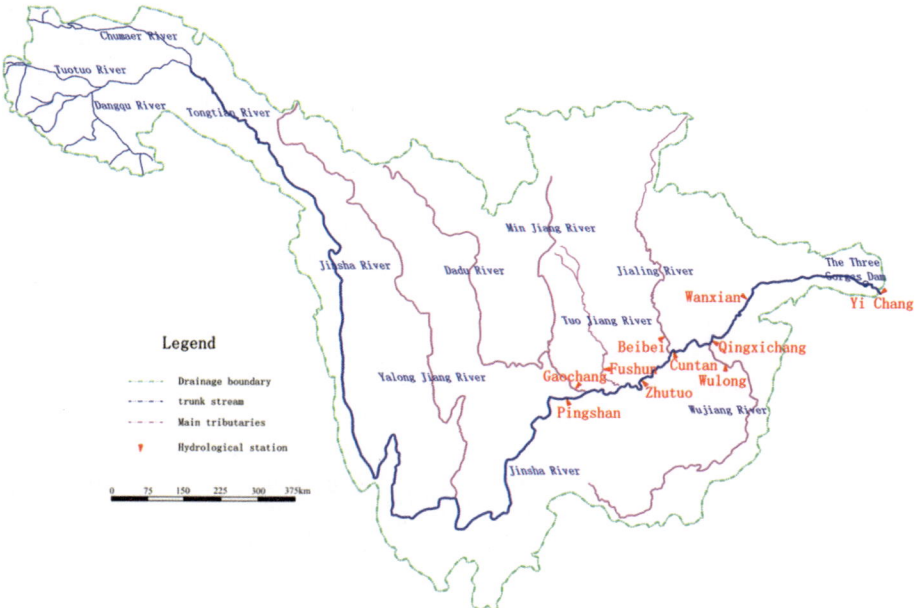

Fig. 1. Drainage boundary and hydrological control stations in the upper Yangtze River

2.2   Methods

(1) M-K test method: It is very effective for the change test of change elements from a relatively stable state to another state. M-K test method is widely recognized by hydrologists because it does not require the original data to obey a specific probability distribution and only satisfies the random independence of time series (Chai et al. 2017).

(2) Wavelet analysis method: The basic idea is to use a cluster of wavelet function system to represent or approximate a signal or function (Wu et al. 2016). Water and sediment time series are non-stationary signals and have multi time scale characteristics. Wavelet analysis has multi-resolution function, which is suitable for analyzing water and sediment time series, realizing the localization of time-frequency domain and revealing the variation characteristics of water and sediment time series.

(3) Approximate entropy (ApEn) method: ApEn is a method to quantify the complexity of time series based on edge probability distribution statistics. The calculation process is simple, and the requirements for the length of time series and other conditions are relatively low, so it has high practical value (Chen et al. 2019; Yang et al. 1997). It is a new way to understand the dynamic characteristics of runoff and sediment transport from the perspective of approximate entropy.

(4) Lyapunov exponent method: Lyapunov exponent can quantitatively describe the separation or contraction direction of trajectories between two adjacent points at exponential rate. In chaos identification, only the maximum Lyapunov exponent λ_{max} is usually calculated. As long as $\lambda_{max} > 0$, chaos exists. The larger the value is, the stronger the chaotic characteristics of the system are; the smaller the value is, the weaker the chaotic characteristics of the system are (Xu et al. 2017).

3   Results

3.1   Mutation

From 1956 to 2018, the results of M-K test of runoff and sediment discharge in the upper Yangtze River and the TGR (Fig. 2) show that: (1) There is no significant change in the annual runoff of the TGR control station, but the annual runoff of the TGR mutated in 2003. (2) The annual sediment discharge of the upper Yangtze River and the TGR has obvious abrupt change. The annual sediment discharge of Zhutuo station changed abruptly in 2008; Cuntan station changed abruptly in 2001; Beibei station changed abruptly in 1989; Wulong station changed abruptly in 2002; The TGR changed abruptly in 2003. (3) The M-K test of abrupt change significance of runoff and sediment discharge in the upper Yangtze River and the TGR has been passed (Table 1).

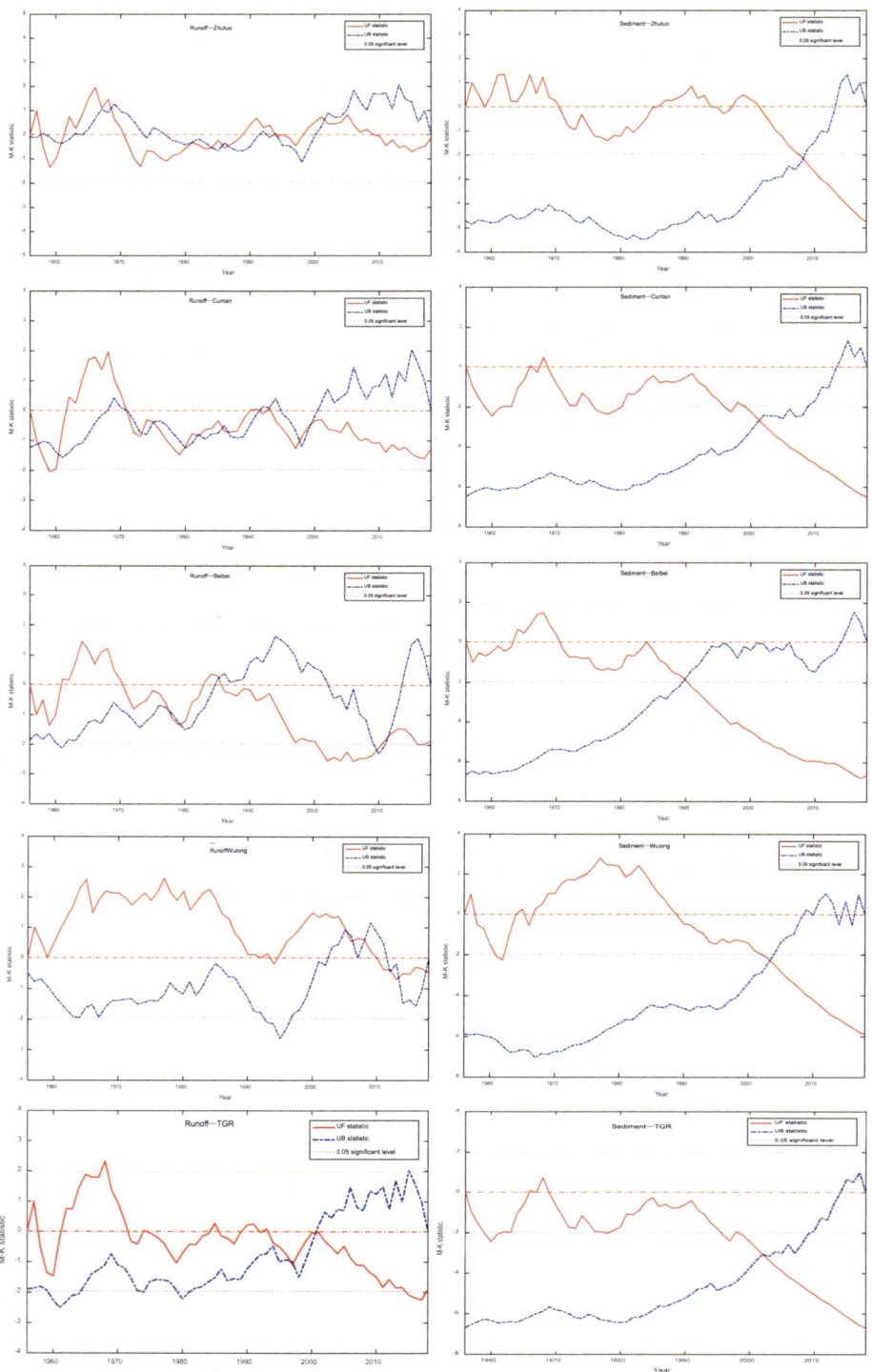

Fig. 2.   M-K test results of annual runoff and sediment series in the upper Yangtze River and the TGR

Table 1. Significance test of M-K mutation of water and sediment in Upper Yangtze River and TGR in recent 60 years

	Zhutuo		Beibei		Cuntan		Wulong		TGR	
	Z	P	Z	P	Z	P	Z	P	Z	P
Runoff	−0.459	>0.05	−0.857	>0.05	−0.227	>0.05	−1.315	>0.05	−2.132	<0.05
Sediment	−3.511	<0.05	−4.581	<0.05	−3.327	<0.05	−5.686	<0.05	−5.947	<0.05

3.2 Periodicity

Morlet wavelet is used as the base wavelet to transform the annual runoff and sediment of Zhutuo, Beibei, Wulong and The TGR respectively, and the wavelet transform coefficients Wf(a,b) are obtained. Then, the time-frequency distribution of Wf(a,b) wavelet coefficients is drawn with b as abscissa and a as ordinate. Finally, the wavelet variance is obtained from the wavelet transform coefficients, and the wavelet variance diagram is drawn according to the variation of wavelet variance with a. According to the time-frequency distribution of wavelet coefficients, the characteristics of hydro-logical time series can be analyzed. The wavelet coefficients under different time scales can reflect the variation characteristics of water and sediment in corresponding time scales, that is, if the wavelet coefficient is positive, it indicates that the period is wet (or sandy), if the wavelet coefficient is not negative, it indicates that the period is dry (or sandy), if the wavelet coefficient is zero, it corresponds to the mutation point. The larger the absolute value of the wavelet coefficient is, the more significant the time scale change is. According to the wavelet variance diagram, the main period of each sequence can be found out. The larger the wavelet variance is, the stronger the time scale oscillation is and the more significant the period is. Figure 3 and Fig. 4 show the time-frequency distribution of wavelet coefficients of annual runoff and annual sediment of each station respectively. The solid line indicates that the wavelet coefficients are positive, and the dotted line indicates that the wavelet coefficients are negative.

It can be seen that the structure of runoff variation at Zhutuo station is different at different time scales, and more complex small-scale oscillations are nested in large-scale oscillations. In the larger scale of 8–9 years, the periodic oscillation of annual runoff is very obvious, there are six alternations of wet and dry. Before 1960, 1967–1976, 1987–1995 and 2003–2011, the wavelet coefficients were positive, indicating the wet season, while the other years were the dry season. In the scale of 4–6 years, there are 12 alternations of wet and dry in the past 60 years, and the location and distribution of mutation points are relatively clear. There are many abrupt changes in the 2–3 year scale(Fig. 3). There are four peaks in wavelet variance, which indicates that the first, second, third and fourth main periods of annual runoff change at Zhutuo station are 16 years, 12 years, 9 years and 4 years respectively. The wavelet characteristics of annual sediment discharge at Zhutuo station are similar to and different from those of annual runoff. On the scale of 6–8 years, the sediment discharge presents a cycle of "more-less", with abrupt changes in 1965, 1976, 1982, 1990 and 2004. The smaller the scale, the more complex the phase change and the more mutation points. The main periods of annual sediment discharge at Zhutuo station are 12 years and 4 years, which corresponding to the result of real frequency distribution (Fig. 4).

It can be seen that the scales of annual runoff change at Beibei station are mainly 10–12 years, 5–6 years and 2–3 years. On the scale of 10–12 years, the mutation point is clear, and there are six alternations of wet and dry. The mutation points of wet and dry are 1976, 1985, 1996 and 2006. On the scale of 5–6 years, there are 11 alternations of wet and dry in the past 60 years. On the scale of 2–3 years, there are many abrupt changes and frequent phase transformation (Fig. 3). The main periods of annual runoff change of Beibei station are 10 years, 8 years and 2 years. On the scale of 9–11 years, the annual sediment discharge of Beibei station also presents a cycle of "more-less", with abrupt changes in 1976, 1990 and 2006. With the decrease of time scale, the mutation points increase. The main periods of annual sediment discharge at Beibei station are 14 years, 8 years and 4 years, which corresponding to the result of real frequency distribution (Fig. 4).

It can be seen that the scales of annual runoff change at Wulong station are mainly 8–10 years and 3–4 years. On the scale of 8–10 years, the mutation points were clear, and there were 7 alternations of wet and dry seasons. The mutation points of wet and dry seasons were 1965, 1973, 1982, 1990, 1998 and 2006. On the scale of 3–4 years, there are many abrupt changes and frequent phase transformation (Fig. 3). The main periods of annual runoff change at Wulong station are 18 years, 10 years and 4 years. It can be seen that the annual sediment discharge of Wulong station also presents a cycle of "more-less" on the scale of 7–9 years, with abrupt changes in 1978, 1984 and 2002. With the decrease of time scale, the mutation points increase. The main periods of annual sediment discharge at Wulong station are 7 years and 4 years, which corresponding to the result of real frequency distribution (Fig. 4).

It can be seen that the scale of annual runoff variation of the TGR is mainly 15–17 years, 8–9 years and 4–5 years. On the scale of 15–17 years, the mutation points were clear, and there were 8 alternations of wet and dry seasons. The mutation points were 1961, 1970, 1979, 1988, 1998, 2003 and 2010. On the scale of 4–5 years, there are many abrupt changes and frequent phase transformation(Fig. 3). The main periods of annual runoff variation of the TGR are 17 years, 9 years and 3 years. On the scale of 7–9 years, the annual sediment discharge of the TGR also presents a cycle of "more-less", with abrupt changes in 1964, 1971, 1979, 1987, 1995, 2003 and 2008. With the decrease of time scale, the mutation points increase. The main periods of annual sediment discharge are 17 years, 9 years and 2 years, which corresponding to the result of real frequency distribution(Fig. 4).

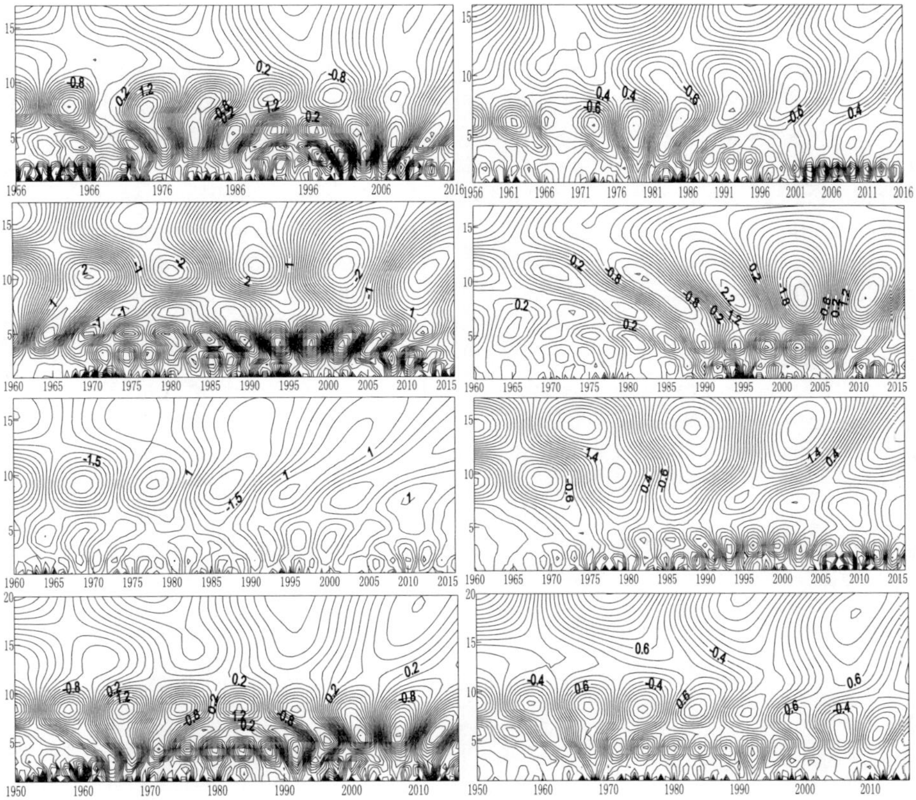

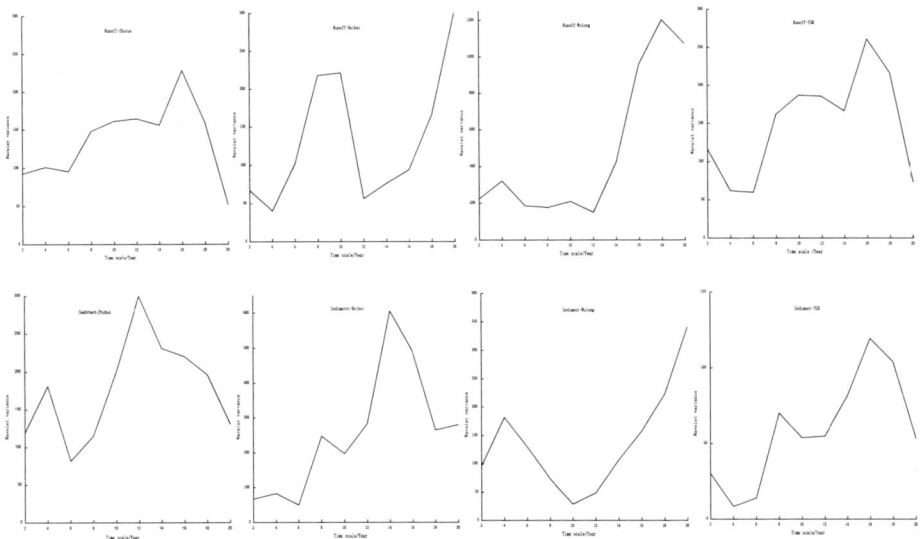

Fig. 3. Real frequency distribution of annual runoff and sediment of Zhutuo, Beibei, Wulong and TGR

Fig. 4. Wavelet variogram of annual runoff and sediment of Zhutuo, Beibei, Wulong and TGR

3.3   Complexity

The complexity of runoff and sediment transport time series in the upper Yangtze River and the TGR is studied based on ApEn. Firstly, the approximate entropy values of the original runoff and sediment transport time series of each hydrological station are calculated; secondly, the approximate entropy values of the runoff and sediment transport time series of each hydrological station are compared, and the temporal and spatial variation laws of the annual runoff and sediment transport complexity of each hydrological station are analyzed qualitatively and quantitatively. Based on the runoff and sediment transport data of Zhutuo, Beibei, Wulong and the TGR from 1956 to 2018, this paper uses ApEn method to calculate the approximate entropy of each station, and uses the approximate entropy to express the complexity of annual runoff and sediment transport. It can be seen that both runoff and sediment transport in the upper Yangtze River and the TGR are complicated. The complexity of water and sediment transport in Zhutuo station of the main stream is greater than that in Beibei station and Wulong station, and the complexity of sediment transport in each station is greater than that of runoff, which indicates that the variation of sediment transport in the upper Yangtze River and the TGR is more obvious (Table 2).

Table 2. ApEn values of runoff and sediment in the upper Yangtze River and the TGR

	Zhutuo	Beibei	Wulong	TGR
Runoff	0.437	0.268	0.206	0.535
Sediment	0.522	0.344	0.240	0.865

3.4   Chaos

Based on the chaos theory, the chaotic characteristics of runoff and sediment transport time series in the upper reaches of the Yangtze River and the TGR are quantitatively analyzed by calculating the saturation correlation dimension, and the chaotic characteristics are further verified by using the maximum Lyapunov exponent method.

3.4.1   Delay Time

The variation of autocorrelation function with delay time of annual average runoff time series of TGR is shown in Fig. 5. According to the numerical test results, the time corresponding to the first zero crossing of the correlation function curve is taken as the best delay time. The best delay time of reconstructing the phase space of the TGR annual average runoff time series is $T = 3$.

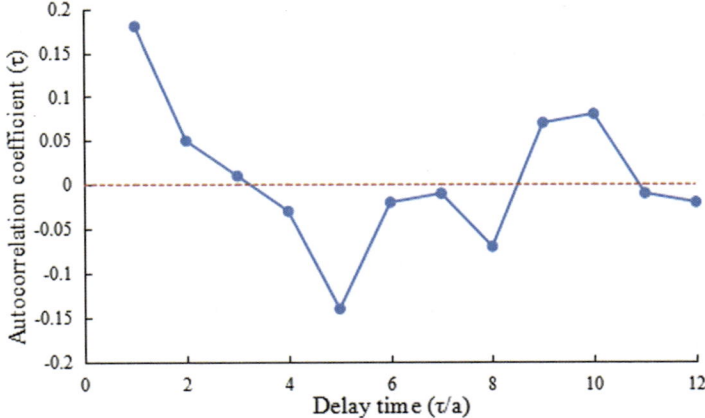

Fig. 5. Autocorrelation function of annual runoff in the TGR

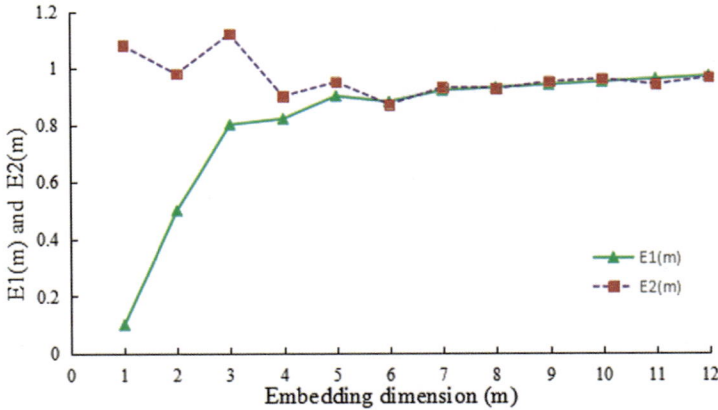

Fig. 6. Embedding dimension of annual runoff in the TGR

3.4.2   Embedding Dimension

According to the delay time τ of the annual average runoff time series of The TGR, the Cao method is used to calculate the embedding dimension m, and the relationship curves between E1(m) and E2(m) and the embedding dimension m are obtained (Fig. 6). When m = 7, E1(m) changes little. At this time, M = 8 is the embedding dimension of the phase space reconstruction of The TGR annual average runoff time series. At the same time, the simulation results show that E2(m) fluctuates up and down at 1, so it can be determined that the average annual runoff time series of The TGR is chaotic time series.

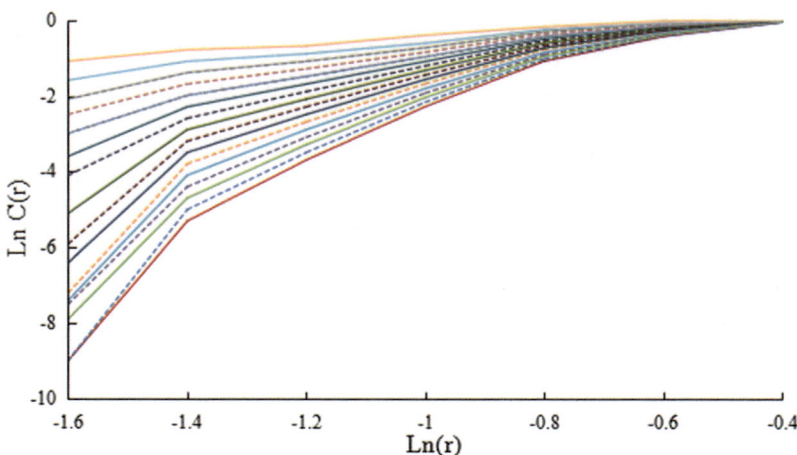

Fig. 7. LnC(r)-Lnr relationship of annual runoff in the TGR

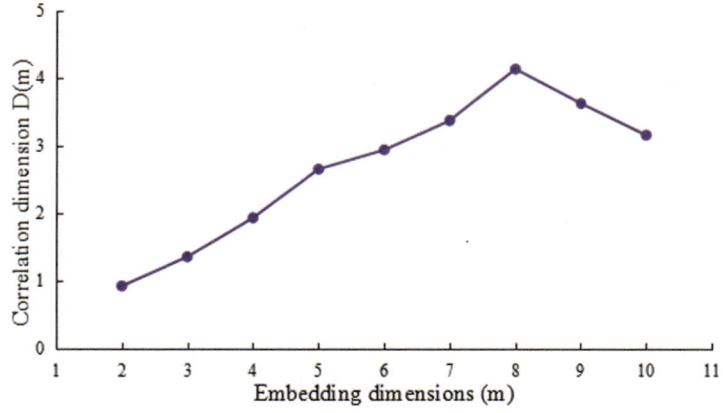

Fig. 8. D(m)-m relationship of annual runoff in the TGR

3.4.3   Saturation Correlation Dimension

According to the delay time of the TGR annual average runoff time series T = 3, the G-P algorithm is used to obtain the lnC(r)-lnr curves under different embedding dimensions m (Fig. 7) and the relationship between correlation dimension D and different embedding dimensions m (Fig. 8).

Under different embedding dimensions m, each curve in lnC(r)-lnr graph has a straight line segment, and the slope of this segment is the correlation dimension d corresponding to each embedding dimension M (Fig. 7). This process is illustrated in Fig. 10 to obtain the relationship between the embedding dimension m and the correlation dimension D. It can be seen that the correlation dimension reaches a saturation value of D2 = 4.135 when the embedding dimension M = 8. This means that when the embedded phase space of the annual mean runoff time series reaches dimension 8, the system will have a stable attractor dimension. Therefore, the annual average runoff series of The TGR has chaotic characteristics.

3.4.4   Largest Lyapunov Exponents

The existence of saturation correlation dimension indicates that the system has a sin-

Table 3. Comparison of embedding dimension and maximum Lyapunov exponent of annual runoff in the TGR

Embedding dimension	3	4	5	6	7	8	9	10
Maximum Lyapunov exponent	0.3621	0.3035	0.2638	0.2234	0.1946	0.1747	0.1738	0.1732

gular attractor, that is, the average annual runoff series of the TGR has chaotic characteristics. In order to further verify its chaotic characteristics, the maximum Lyapunov exponent method is used to identify the chaotic characteristics of the TGR annual average runoff series when the delay time is determined to be 3. The maximum Lyapunov exponent under different embedding dimensions is obtained as shown in the table below (Table 3).

It can be seen that when the embedding dimension increases to 8, the maximum Lyapunov exponent no longer changes greatly with the increase of m value, with L_E = 0.1747. The maximum predictable scale is 5.72 according to $t0 \approx \frac{1}{L_E}$. The actual

Table 4. Chaos analysis results of annual runoff and sediment in the TGR

Station		Delay time	Embedding dimension	Maximum Lyapunov exponent	Saturation correlation dimension
Zhutuo	Runoff	3	6	0.1096	3.2643
	Sediment	3	7	0.1467	3.4679
Beibei	Runoff	3	7	0.1248	3.5784
	Sediment	3	8	0.1659	3.6897
Wulong	Runoff	2	7	0.1316	3.8649
	Sediment	3	8	0.1832	4.1026
TGR	Runoff	3	8	0.1747	4.1348
	Sediment	4	9	0.2018	4.3869

physical meaning is that when using the actual data of the average annual runoff of the TGR to predict, the maximum prediction time is at most six years under the condition of less serious loss of accuracy. In addition, $L_E > 0$ indicates that the annual mean runoff of Yichang station is chaotic. Combined with the results of Cao method and saturation correlation dimension, it can be confirmed that the average annual runoff of the TGR has chaotic property.

It can be seen that both the annual average runoff and sediment of the TGR have saturation correlation dimension, and the maximum Lyapunov exponent is greater than 0. Therefore, it can be confirmed that the annual average runoff and sediment of the TGR have chaotic characteristics. In addition, the state space dimension of runoff and sediment of representative stations in the upper Yangtze River gradually increases from upstream to downstream, and the corresponding saturation correlation dimension also increases from upstream to downstream (Table 4). This shows that the characteristics of runoff and sediment in the upper Yangtze River are more and more complex, and the complexity of runoff and sediment sequence increases from upstream to downstream, which is consistent with the results of the previous complexity analysis.

3.5   Connotation of "New Water and Sediment Conditions"

This paper analyzes the temporal and spatial evolution characteristics of water and sediment in the TGR from four aspects of mutation, periodicity, complexity and chaos. The connotation of "new water and sediment conditions" is as follows: After the impoundment of the TGR, the runoff does not change much and the sediment discharge decreases sharply; from the perspective of time scale, there is no obvious mutation in runoff and sediment discharge, and the main cycle of water and sediment change is basically the same; from the perspective of space scale The results show that the complexity and chaos of sediment transport are significantly greater than that of runoff, and the change degree of sediment transport is more significant than that of runoff.

4   Influencing Factors of "New Water and Sediment Conditions"

The main factors affecting the change of water and sediment in the TGR can be divided into two aspects: natural and man-made. Natural factors mainly refer to climate (rainfall), soil and vegetation conditions, geology and geomorphology, while human factors mainly include soil and water conservation, water conservancy project sand blocking, river sand mining, etc. (Ding et al. 2008). The results show that: (1) Climate change is mainly composed of precipitation, regional distribution of rainfall and rainfall intensity, which has a great impact on sediment yield and transportation in the basin, and its randomness is strong, and generally has a certain periodicity; (2) The sediment reduction caused by the reservoir intercepting sediment also has a certain long-term effect. The initial sediment retention benefit is very significant, but when the reservoir sedimentation is balanced, it will no longer have the sediment retention benefit, and the upstream sediment inflow will return to the natural state. The main reason is that when there is no obvious change in the amount of erosion in the basin, the sediment transport ratio will be greatly reduced in a certain period of time due to the blocking effect of the reservoir, but when the upstream reservoir is full, the sediment transport ratio will return to the previous situation or greater, and the sediment entering the TGR will increase; (3) The sediment reduction caused by soil erosion control has a certain long-term nature, but it will reach the limit value of relative equilibrium in a certain period of time; (4) The impacts of other human activities, such as increasing or reducing sediment in engineering construction, sand mining in river channel, water use in industry and agriculture, will change with the development of social economy, and also affect the change of water and sediment in river basin.

The relationship between precipitation and sediment discharge and the relationship between runoff and sediment discharge are power functions: $W_s = a \times P^b$ and $W_s = a \times R^b$. Where, W_s is annual sediment discharge, P is annual rainfall and R is annual runoff, a and b are fitting coefficients and indexes respectively. Generally, the relationship can be used to roughly estimate the effect of precipitation/runoff on sediment transport. The relationship between runoff and sediment discharge in the upper Yangtze River is complex, and the correlation varies greatly in different periods. Based on the results of M-K mutation test, the contribution of climate change and human activities to sediment transport in the upper Yangtze River was further separated and quantified.

According to the preliminary estimation of the relationship between runoff and sediment discharge, the actual sediment reduction in the upper Yangtze River from 1991 to 2002 is 0.108 billion t/a, and the runoff is 7 billion m³/a less than that before 1990. Among them, the sediment reduction caused by the change of precipitation/runoff (under the underlying surface conditions before 1990, the same below) is 0.067 billion t/a, and the sediment change caused by rainfall accounts for 79.8%. From 2003 to 2018, the actual sediment reduction was 0.206 billion t/a and the runoff decreased by 19.5 billion m³/a compared with that before 1990. Among them, the sediment reduction caused by precipitation/runoff (under the underlying surface conditions before 1990, the same below) was 0.043 billion t/a, and the sediment change caused by rainfall accounted for 20.9% (Fig. 9).

Table 5. Statistics of rainfall (mm), runoff (billion m3) and sediment (billion t) in the upper Yangtze river

Time interval	Rainfall (mm)	Runoff (billion m³)	Sediment (billion t)
① 1956–1990	867	3957	4.831
② 1991–2002	852	3869	3.537
③ 2003–2018	833	3645	1.540
②–①	−1.7%	−2.2%	−26.7%
③–②	−2.2%	−5.8%	−56.5%

To sum up, climate change (rainfall) was an important factor affecting the sediment discharge in the upper Yangtze River from 1991 to 2002. From 2003 to 2018, the impact of climate change on the sediment discharge in the upper Yangtze River was relatively insignificant, and human activities became the most important factor (Table 5).

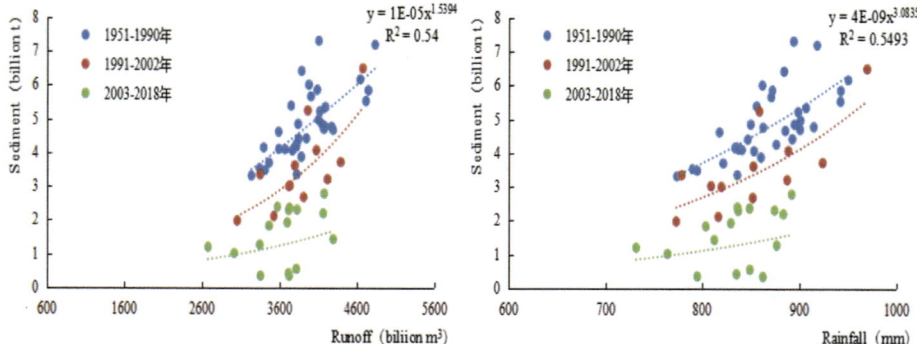

Fig. 9. Annual runoff-sediment and rainfall- sediment at different periods in the upper Yangtze River

5   Conclusion

Based on the hydrological and sediment observation data of the upper Yangtze River and the main stream from 1956 to 2018, this paper analyzes the spatiotemporal evolution characteristics of the incoming water and sediment of the TGR from four aspects of mutation, periodicity, complexity and chaos based on the M-K test method, wavelet analysis method, ApEn method and Lyapunov exponent method, and defines the connotation of "new water and sediment conditions" of the TGR, The main factors affecting the temporal and spatial variation of water and sediment in theTGR are discussed. The main conclusions are as follows:

(1) Except Pingshan station of Jinsha River, Fushun station of Tuojiang River and Wulong station of Wujiang River, the runoff of other main and tributaries and The TGR in the upper Yangtze River has no significant change, but the sediment discharge has obvious mutation, and the runoff and sediment have periodic changes in different degrees;

(2) The complexity of annual runoff and sediment transport in the upper Yangtze River increases gradually, and the complexity of sediment transport is greater than that of runoff. Both annual runoff and sediment transport are chaotic, and the complexity of runoff and sediment transport sequence increases gradually from the upper reaches to the lower reaches;

(3) The connotation of "new water and sediment condition" shows that there is no obvious mutation in runoff and sediment transport in time scale, the main cycle of water and sediment change is basically the same, and the complexity and chaos of sediment transport in space scale are significantly greater than runoff, and the change degree of sediment transport is more significant than runoff;

(4) From 1991 to 2002, climate change (rainfall) was an important factor affecting the sediment discharge in the upper Yangtze River. From 2003 to 2018, the impact of climate change was relatively insignificant, and human activities became the most important factor.

Acknowledgements. We are grateful to the Changjiang Water Resources Commission (CWRC) for the access to the valuable data sets. This study was supported by the Open Research Fund of State Key Laboratory of Simulation and Regulation of Water Cycle in River Basin (China Institute of Water Resources and Hydropower Research) (Grant No. IWHR-SKL-202111), the Fundamental Research Funds of the Central Government (Grant No. CKSF20211743/HL), the Hubei Provincial Natural Science Foundation Program Youth Project (Grant No. 2020CFB322). We gratefully acknowledge the anonymous reviewers and editors for their reviews and suggestions.

References

Chen XQ, Zong YQ, Zhang EF, Xu JG, Li SJ (2001) Human impacts on the Changjiang (Yangtze) river basin, China, with special reference to the impacts on the dry season water discharge into the sea. Geomorphology 41:111–123

Chen ZY, Li JF, Shen HT, Wang ZH (2001) Yangtze river of China: historical analysis of discharge variability and sediment flux. Geomorphology 41:77–91

Chen P, Tan G, Deng J, Xu Q, Tang R (2019) Mutation analysis of annual sediment discharge at Wu Long station in Wu Jiang river basin from 1960 to 2016. PLoS ONE 14(12):e0225935

Tan G, Chen P, Deng J, Xu Q, Tang R, Feng Z, Yi R (2019) Review and improvement of conventional models for reservoir sediment trapping efficiency. Heliyon 5(9):e02458

Xu G, Zhao L (2017) Study on identifying chaotic characteristics of river based on multiple time series. J Sediment Res 42(03):7–13

Huang C, Lou X, Liu Y (2002) Analysis of the sediment development and trend in Jinshajiang river drainage area. J Chongqing Univ (Nat Sci Ed) 25:21–23

Dai H, Wang L, Jiang D (2007) Near term water flow and silt concentration variation trend of Yangtze river before and after impounding of Three Gorges reservoir. Shuili Xuebao 10 (S1):226–231

Zhao J, Li J, Dai Z et al (2012) Analysis the runoff variotion of Yangtze river in Yichang. Resour Sci 32(12):2306–2315

Zhu J (2000) Variation of sediment transportation in the Yangtze river and the way for its reduction. J Hydroelectr Eng 3:38–48

Liu J, Yang S, Shen Y (2019) Impact of runoff and sediment from the Upper Yangtze river on deposition in the Three Gorges reservoir. J Sediment Res 44(06):33–39

Jie L, Ying S, Shuqing Y (2020) Influence of non-balanced water and sediment transport in the Upper Yangtze river on sediment siltation in the TGR area. Resour Environ Yangtze Basin 29 (06):1333–1342

Lu XX, Ashmore P, Wang J (2003) Sediment load mapping in a large river basin: the Upper Yangtze, China. Environ Model Softw 18:339–353

Xu Q, Shi G, Chen Z (2004) Analysis of recent changing characteristics and tendency runoff and sediment transport in the upper reach of Yangtze river. Adv Water Sci 15(4):420–426

Xu Q, Chen S, Xiong M (2005) Analysis on runoff and sediment characteristics and affecting factors of the Jialing river basin. J Sediment Res

Xu Q, Zhang X, Yuan J (2009) Resources and environment in the Yangtze basin. Study of frequency jump phenomenon of sediment transport time series in the Upper Changjiang river

Fu R, Yu Z, Liu J, Fang H (2003) Variation trend of runoff and sediment load in Yangtze river. Shuili Xuebao (11):21–30

Su BD, Xiao B, Zhu DM, Jiang T (2005) Trends in frequency of precipitation extremes in the Yangtze river basin, China: 1960–2003. Hydrol Sci J 50(3):479–492

Hu S, Wang Z, Wang G, Liu X (2004) Effects of watershed management on the reduction of sediment and runoff in the Jialing river, China. J Int Sediment Res 02:63–69

Tan G, Chen P, Deng J et al (2019) Estimations and changeṣ of the dominant discharge in Three Gorges Reservoir channel. Arab J Geosci 12:82

Wang Z, Mao H, Shen J, Tang X, Chen X (2020) Analysis of water and sediment characteristics in the main tributaries of the Yangtze river and their associated influence factors. J East China Norm Univ (Nat Sci) (01):126–138

Yang W, Li Z, Li X (1997) A new method of analysing climate jump and its application. Q J Appl Meteorol 81(1):119–123

Ding W, Zhang P, Ren H (2008) Quantitative analysis on evolution characteristics and driving factors of annual runoff and sediment transportation changes for Jialing river. J Yangtze River Sci Res Inst 25(30):23–27

Wu X, Na L, Wang L (2016) Characteristics of runoff and sediment discharge in Yangtze river in recent 60 years. J Sedim Res (05):40–46

Zhang X (1999) Status and causes of sediment change in the upper Yangtze river and sediment reduction measures—comparison of Jialing river with Jinsha river. Soil Water Conserv China 2:22–24

Zhang X, Wen A (2002) Variations of sediment in upper stream of Yangtze river and its tributary. Shuili Xuebao (4):56–59

Li X, Yang J, Chen Z (2001) Characteristics of discharge and sediment transportation in Yangtze river. J East China Norm Univ (Nat Sci) (4):88–95

Dong Y, Hui X, Lin Q (2008) Preliminary analysis on characteristics and changing tendency of annual runoff and sediment load of Changjiang river main channels. J Yangtze River Sci Res Inst 25(02):16–20

Xiao Y, Yang S, Shao X (2017) Trends of temporal variation of the flow-sediment discharges into the TGR. J Sediment Res 42(06):22–27

Yang SL, Zhao QY, Belkin IM (2002) Temporal variation in the sediment load of the Yangtze river and the influences of human activities. J Hydrol 263:56–71

Yang SL, Belkin IM, Belkina AI, Zhao QY, Zhu J, Ding PX (2003) Delta response to decline in sediment supply from the Yangtze river: evidence of the recent four decades and expectations for the next half-century. Estuar Coast Shelf Sci 57:689–699

Chai Y, Li Y, Li S, Zhu B, Wang J (2017) Analysis of recent variation trend and cause of runoff and sediment load variations in the Yangtze river basin. J Irrig Drain 36(03):94–101

Xiang Z, Zhou G (1993) Characteristics of sediment transport in the Yangtze river. J China Hydrol (6):8–13

Study on Classification Arrangement and Hydraulic Characteristics of Water-Saving Ship Lock Under Ultra-high Head

Long Zhu$^{(\boxtimes)}$, Xiaodong Wang, Yue Huang, Benqin Liu, and Zhonghua Li

Nanjing Hydraulic Research Institute, Nanjing, China
zhulong@nhri.cn

Abstract. The ship lock with water saving basins can store part of water during the operation, which has the outstanding advantage of saving water resources. At the same time, the water-saving ship lock can divide the total water head into stages, significantly reduce the operating head of each stage, which can provide technical conditions for simplifying the ship lock's filling & emptying system and improving the operating conditions of lock's valves. For ship lock with high head and large water level variation, the layout of the saving basins and the division of water level are very important. They determine the overall layout, operation efficiency and safety performance of the ship lock. With the increase of water head and water level variation, the classification arrangement of water-saving ship lock will become very complex. Therefore, based on the Baishi water-saving ship lock project (60 m-class, ultra-high head), the analysis and calculation research are carried out in this paper. First of all, the influence laws of key factors such as water-saving rate, number of water saving basins, area of water saving basins and operating head on water level division are obtained. On this foundation, a reasonable water level classification scheme for 60m-class ship lock is proposed, and the hydraulic characteristics of ship lock under different operating conditions are analyzed, and the feasibility of ship lock layout scheme is demonstrated. The research results can provide direct technical support for Baishi ship lock project, and provide reference for the design and construction of similar water-saving ship lock projects.

Keywords: Water-saving ship lock · Ultra-high head · Classification arrangement · Hydraulic characteristics

1 Introduction

The water-saving ship lock is a special type of ship lock construction. Usually, water saving basins are arranged on one or both sides of the ship lock. Through the continuous water exchange between the lock chamber and the water saving basins, some water bodies are reused to achieve the effect of saving water resources. In addition, another important function of the water-saving ship lock is to divide the total head into stages, so as to significantly reduce the operating head of each stage, provide technical conditions for simplifying the filling & emptying system and improving the operating

© The Author(s) 2023
Y. Li et al. (Eds.): PIANC 2022, LNCE 264, pp. 1553–1564, 2023.
https://doi.org/10.1007/978-981-19-6138-0_137

conditions of lock's valves, and have the advantages of improving the flow conditions of the upstream and downstream approach channels of the ship lock. It provides an idea for solving the problem of high head ship lock construction in alpine and gorge areas.

2 Layout and Engineering Background of Water-Saving Ship Lock

2.1 Construction, Application and Layout Type of Water-Saving Ship Lock

As a special type of ship lock construction, the water-saving ship lock is widely used in foreign canals, such as the three-stage water-saving ship lock of the Panama Canal, the series ship locks on the Rhine-Main-Danube in Germany, the cascade ship locks on the Seine Nordic canal (under planning and construction), etc. According to statistics, Germany is the country with the largest number of water-saving ship locks in the world, and has rich experience in design and construction. 《Innovations in navigation lock design》 published by the PIANC summarizes the characteristics and key technical problems of water-saving ship locks (PIANC 2009).

There are relatively few examples of water-saving ship locks in China. So far, almost no water-saving ship lock has been built and put into operation. However, in recent years, with the planning and construction of high head navigation buildings in southwest mountainous areas and the demonstration and implementation of major cross river canal projects, many water-saving ship lock design schemes have been put forward and experimental studies have been carried out (Zhu and Xuan 2019; Wu and Cao 2013; Wang and Liu 2013), such as the Yinpan ship lock on Wujiang River, the Baishi ship lock on Qingshui River, the Jinjiayan ship lock on Xiaoqing River, the Baima ship lock on Xiujiang River, the Bajiangkou ship lock on Guijiang River, etc.

According to the layout type of the water saving basins of the ship lock, the water-saving ship lock can be divided into two typical schemes (Wu and Xuan 2009): closed integral type and open decentralized type. Figure 1 shows the schematic diagram of two schemes. In the closed integrated scheme, the water saving basins are overlapped on one or both sides of the ship lock, which is relatively concentrated and covers a small area, but its structure is complex and the water level division of the water saving basins is difficult. In the open decentralized type, The water saving basins are arranged on one or both sides of the lock, which is scattered and covers a large area. However, this layout mode has simple structure and relatively flexible water level division.

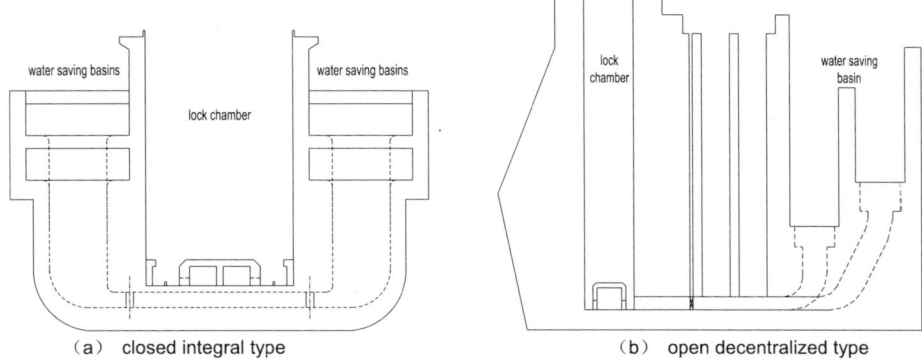

(a) closed integral type   (b) open decentralized type

Fig. 1. Typical layout of water saving ship lock

2.2   The Engineering Background

The Baishi hydropower station on Qingshui River is the last cascade of Qingshui River in Guizhou Province. It is a project focusing on power generation and comprehensive utilization benefits such as flood control, shipping and aquaculture. From left to right, the existing buildings in the riverbed of the dam site are: connecting dam section, vertical ship lift, overflow dam, power plant and connecting dam section. The existing navigation structures do not meet the national standards for 500t ships. Therefore, it is proposed to build new navigation structures on the right bank of the hub and adopt the water-saving ship lock scheme. The effective dimension of the new ship lock is 140 m × 12 m × 4.0 m (long × wide × depth), and the maximum design water head is 59.6 m. The water saving basins are arranged on the right side of the ship lock, and the total water filling/emptying time is about 25 min.

The highest navigable water level in the upper reaches of Baishi ship lock is 300 m and the lowest navigable water level is 294 m; The highest navigable water level in the downstream is 251.44 m and the lowest navigable water level is 240.4 m. It is the single-stage water-saving ship lock with the highest water head in China (60 m class, which is a veritable ultra-high water head ship lock), and the water levels in the upstream and downstream vary greatly. It is very necessary to study its classification arrangement and relevant hydraulic characteristics.

3   Analysis on Key Influencing Factors of Classification Arrangement of Water-Saving Ship Lock

There are many factors affecting the classification arrangement of water-saving ship lock (Dong et al. 2020; Li et al. 2020; Liu et al. 2016). The following factors need to be paid attention to when arranging the water saving basins and dividing the water level of the ship lock: ① operating head; ② water saving rate; ③ number of stages; ④ area of water saving basins. The above influencing factors are not single variables, but multi-objective variables coupled with each other. In order to analyze each factor in detail, the schematic diagram of the filling and emptying process with three-stages water saving basins is drawn, as shown in Fig. 2.

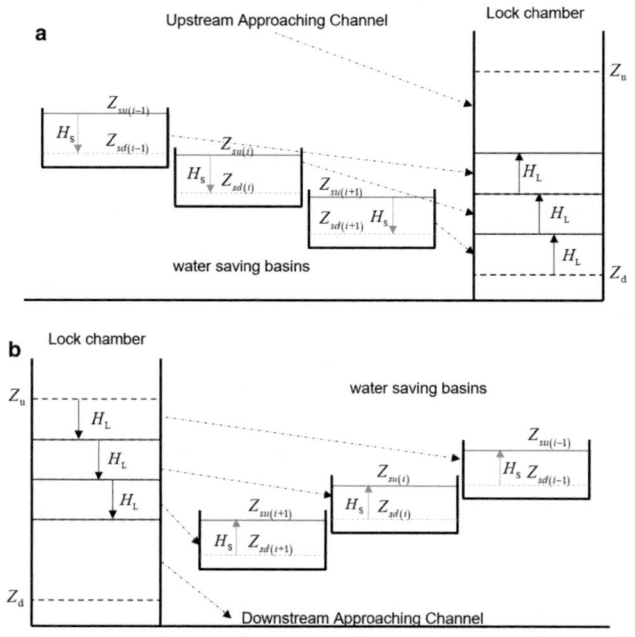

Fig. 2. The schematic diagram of the filling and emptying process

According to the Fig. 2, the following definitions are given in this paper. Z_u is the navigable water level upstream of the ship lock (m); Z_d is the navigable water level downstream of the ship lock (m); $Z_{u\,max}$ is the highest navigable water level upstream (m); $Z_{u\,min}$ is the lowest navigable water level upstream (m); $Z_{d\,max}$ is the highest navigable water level downstream (m); $Z_{d\,min}$ is the lowest navigable water level downstream (m); Z_{sui} is the high water level of 'stage i' water saving basin (m); Z_{sdi} is the low water level of 'stage i' water saving basin (m); ΔH_u is the variation of upstream water level(m); ΔH_d is the variation of downstream water level(m); H_L is the change of water level in the lock chamber during each stage of water-saving operation (m); H_S refers to the change of water level in the water saving basin during each stage of water-saving operation (m); H_n refers to the water head during each stage of water-saving operation (m); A_L is the water area of lock chamber (m^2); A_S is the water area of the water saving basin (m^2), we assume that all water saving basins have the same area; m is the water area ratio between the water saving basins and the lock chamber; E_S is the water saving rate of ship lock; n is the number of stages of the water saving basins.

According to the operation principle and mass conservation equation of the ship lock, the parameter variables in the process of water saving of the ship lock meet the following basic relations:

$$H_S \times A_S = H_L \times A_L \tag{1}$$

$$H_L = H_S \times m \tag{2}$$

3.1 Relationship Between Water Head H_n, Area Ratio m and Stage Number n

For a specific upstream and downstream water level combination, the operating head of each stage of the ship lock can be deduced, as shown in Eq. (3). When the ship lock is under the maximum design head condition, the operating head of each stage of water-saving operation is also the maximum value. The analysis shows that the water head (H_n) during the water-saving operation is related to the total head of the ship lock (($Z_u - Z_d$)),the stage number n and the area ratio m of the water saving basin to the lock chamber. In order to master the response law of water head (H_n) and the key characteristic parameters, the dimensionless number $H_n/(Z_u - Z_d)$ is introduced, which means the ratio percentage of actual water head during water-saving operation to the total design water head of the ship lock. The relationship between $H_n/(Z_u - Z_d)$, m and n is shown in Fig. 3. It can be seen from the formula and figure that with the increase of the water-saving stage number and the area ratio of water saving basin to lock chamber, the ratio percentage of actual water head to the total head shows an obvious downward trend, but the downward range gradually slows down.

$$H_n = H_L + H_S = \frac{(m+1)(Z_u - Z_d)}{m(n+1)+1} \tag{3}$$

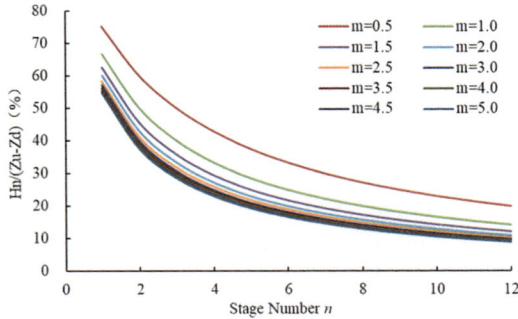

Fig. 3. The relationship between $H_n/(Z_u - Z_d)$, m and n

Further statistics on the relationship between the decline range (The decline range refers to the ratio of the change value of the dependent variable to the original value as the value of the independent variable increases. For example, if m = 1. When n is 1, the value of $H_n/(Z_u - Z_d)$ is 66.67%. When n changes from 1 to 2 and the value of $H_n/(Z_u - Z_d)$ changes to 50%, which means that the decline range is about 25%) of $H_n/(Z_u - Z_d)$, m and n are shown in Fig. 4. The research shows that: ① When the area ratio m between the water saving basin and the ship lock is fixed; If the number of water-saving stages is greater than or equal to 4 (n $\geq$ 4), all the decline range may be less than 20%; If n > 8, the decline range begins to be less than 10%. ② When the stage number n of the water saving basins is determined, with the change of the area ratio m between water saving basin and the ship lock, the decline range does not change significantly; Especially when m $\geq$ 2.0, it has little effect on the change of water head.

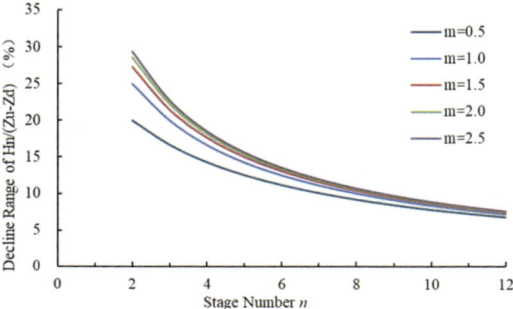

Fig. 4. Correlation between $H_n/(Z_u - Z_d)$, m and n

3.2 Relationship Between Water Saving Rate E_S, Area Ratio m and Stage Number n

The water saving rate of the ship lock measures the ability of the ship lock to save water in each operation, that is, the ratio between the water saved by the ship lock and the water body required for a single operation of the ship lock, which is recorded as E_S. According to the operation process of the ship lock, it can be deduced:

$$E_S = \frac{n \times V_S}{V_L} = \frac{mn}{mn + m + 1} \tag{4}$$

Based on this, the relationship between water saving rate E_S and area ratio m and stage number n is drawn as shown in Fig. 5. The research shows that the water saving rate of ship lock is closely related to the number of water saving stages and the area ratio of water saving basin to lock chamber. In order to quantitatively understand the influence degree of each parameters on the water saving efficiency, the variation law of the growth rate(Similar to the 'decline range', the growth rate of E_S refers to the ratio of the change value to the original value. For example, if m = 1. When n is 2, the value of E_S is 50%. When n changes from 2 to 3 and the value of E_S changes to 60%, which means that the growth rate is about 20%) with m and n is shown in Fig. 6. The comprehensive analysis shows that: ① On the premise that the area of the water saving basin remains unchanged, with the increase of the number of water saving stages, the water saving efficiency gradually increases, but the upward trend gradually slows down. When the number of water saving stages is greater than 4, the growth percentage of the water saving efficiency gradually decreases to less than 10%; ② On the premise that the number of water saving stages remains unchanged, with the increase of the area of water saving basin, which means that m increases, the water saving efficiency of the ship lock increases. Similarly, the upward trend gradually slows down with the increase of area. When m is greater than 2.0, the growth of water saving efficiency is very slow.

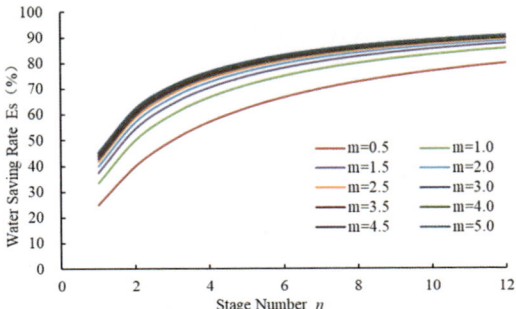

Fig. 5. Correlation between E_S, m and n

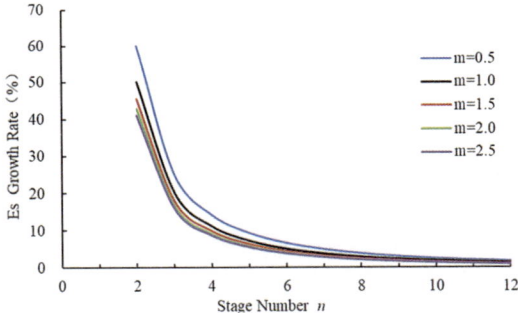

Fig. 6. Correlation between E_S growth rate, m and n

3.3 Relationship Between Water Level $Z_{su(i)}$, $Z_{sd(i)}$ and Area Ratio m, Stage Number n

It is defined that the uppermost water saving basin is the first stage, and the lowest water saving basin is the n-stage. According to the formula, the high water level of the i-stage water saving basin can be deduced as follows:

$$Z_{su(i)} = Z_d + [m(n - i + 1) + 1] \times \frac{Z_u - Z_d}{m(n + 1) + 1} \tag{5}$$

The low water level of the i-stage water saving basin is:

$$Z_{sd(i)} = Z_d + m(n - i + 1) \times \frac{Z_u - Z_d}{m(n + 1) + 1} \tag{6}$$

According to the derivation of the above formula, after determining the water saving stage and the area of the water saving basins, the water level of the water saving basins at all stages can be analyzed and calculated. The results show that the water level of each water saving basin is closely related to the variation of upstream and downstream water level.

Generally, in practical engineering, we can approximate that when the water levels in the upstream and downstream are high water levels, the water levels in the water saving basins are also high water levels; When the water levels in the upstream and downstream are low water levels, the lowest water levels is in water saving basins.

4   Classification Arrangement of Water-Saving Ship Lock Under Ultra-high Head and Large Water Level Variation

In the design scheme, the maximum head of Baishi ship lock is 59.6 m (60 m-class). Table 1 shows the typical cases of high head large ship lock at home and abroad. The comparison shows that the current maximum head of the ship lock is only about 40 m (40 m-class). For the Baishi ship lock with the maximum water head of 60 m-class, if the conventional single-stage ship lock is directly adopted, the hydraulic index is expected to be very high, and there are still many technical problems to be solved. If the layout of water-saving ship lock is adopted, the working head can be significantly reduced and its working index can be within the application range of existing mature technology.

Table 1. Statistics of some high head and large-scale ship locks in the world (m)

Number	Name	Country	Dimensions (length × width)	Maximum head	Type
1	Walter Bouldin*	U.S.A	192.0 × 25.6	39.6	Single-step
2	Zaporojie	Russia	290.0 × 18.0	39.2	Single-stage
3	Lajeado	Brazil	210.0 × 25.0	37.3	Single-stage
4	Tucurui (upstream)	Brazil	210.0 × 33.0	36.5	Single-stage
5	Tucurui (downstream)	Brazil	210.0 × 33.0	35.0	Single-stage
6	John Day	U.S.A	205.7 × 26.2	34.5	Single-stage
7	Pak Beng	Laos	120.0 × 12.0	32.38	Single-stage
8	Lower Granite	U.S.A	205.7 × 26.2	32.0	Single-stage
9	Little Goose	U.S.A	205.7 × 26.2	30.8	Single-stage
10	Three Gorges	China	280.0 × 34.0	113.0(45.2)	Dual-way five-stage
11	Datengxia	China	340.0 × 34.0	40.25	Single-stage
12	Wan'an (second line)	China	180.0 × 23.0	32.5	Single-stage
13	Letan	China	120.0 × 12.0	29.1	Single-stage
14	Gezhouba	China	280.0 × 34.0	27.0	Single-stage
15	Baishi*	China	140.0 × 12.0	59.6	Water saving basin

Note:*For the research scheme.

According to the analysis results of H_n, m and n, the relationship between the operating head of Baishi ship lock, stage number and area ratio can be drawn in Fig. 7. It can be seen that the effect of increasing the area ratio m on reducing the operating head is relatively not obvious. Under the condition of the same stage number n, if you want to reduce the operating head by 5 m, the area ratio m needs to be 2.5 or more; Increasing the stage number n has an obvious effect on reducing the operating head of ship lock, but when the stage number is greater than 4, the effect slows down.

Combined with the specific conditions of Baishi ship lock, the relatively suitable regional scope is shown in the figure. This refers to reducing the operating head of each stage to 20 m–25 m. The technology of ship lock construction is mature at this level, and the hydraulic problems of the filling & emptying system are easy to solve. Therefore, it can be preliminarily determined that it is reasonable to adopt the layout scheme of three-stage water saving basins for Baishi ship lock.

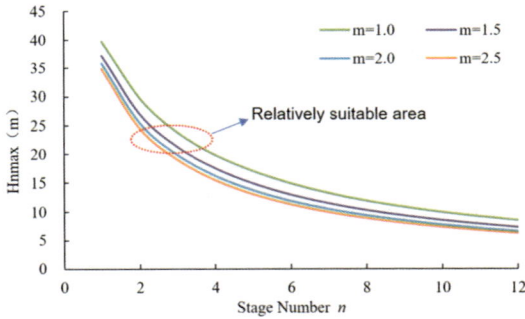

Fig. 7. Relationship between maximum operating head of Baishi ship lock, m and n

The green transportation development strategy is being implemented in China. The water-saving ship lock has achieved relatively good water-saving effect while significantly reducing the actual operating head of the ship lock. According to Eq. (4), the change trend of water saving rate of Baishi ship lock is shown in Fig. 8 under the arrangement condition of three-stage water saving basins. It can be seen from the figure that when the area ratio m changes from 0.5 to 1.0, the theoretical water saving rate E_S increases from 50.0% to 60.0%, and the effect is obvious; When the area ratio m changes from 1.0 to 1.5, the theoretical water saving rate increases from 60.0% to 64.3%; When it changes from 1.5 to 2.0, the theoretical water saving rate increases from 64.3% to 66.7%, and the growth rate is only 3.7%.

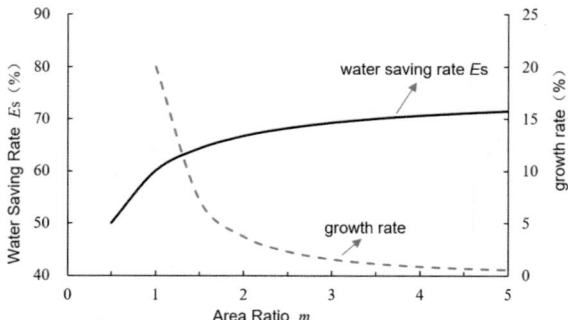

Fig. 8. Relationship between the change trend of water saving rate and m of Baishi ship lock

Based on the comprehensive consideration of spatial layout conditions, investment economy and other aspects, in Baishi ship lock project, it is preliminarily determined to adopt the open decentralized type of three-stage water saving basins, which are arranged on one side of the ship lock, and the area ratio of water saving basin to lock chamber is 1.0:1.0. Based on this, the water level is divided according to Eq. (5) and Eq. (6), and the results are shown in Table 2. The layout type is shown in Fig. 9.

Table 2. Water level classification scheme of Baishi ship lock

	300 ~ 251.44 (m)		300 ~ 240.4 (m)		294 ~ 240.4 (m)	
	High water level	Low water level	High water level	Low water level	High water level	Low water level
The first stage	290.29	280.58	288.08	276.16	283.28	272.56
The second stage	280.58	270.86	276.16	264.24	272.56	261.84
The third stage	270.86	261.15	264.24	252.32	261.84	251.12

Note: $n = 3$; $m = 1$; the theoretical water saving rate $E_S = 60.0\%$

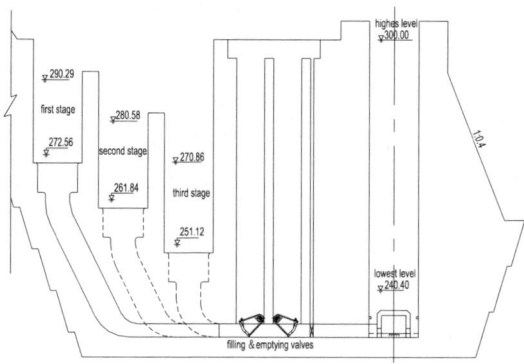

Fig. 9. The three-stage water saving scheme and water level classification of Baishi ship lock

5 Hydraulic Characteristics of Ship Lock Based on Water Saving Layout

According to the three-stage water saving scheme of Baishi ship lock determined in this paper, when all filling & emptying valves of the ship lock adopt the operation mode of opening time $t_{v1} = 1.5$ min and closing time $t_{v2} = 1.0$ min, the hydraulic characteristics of the whole process are estimated, and the typical hydraulic indexes such as filling & emptying time, discharge process and water flow velocity are obtained, as shown in Table 3. It can be seen that under the condition of this classification arrangement, the filling & emptying time at the maximum head of the ship lock is about 24 min, which can meet the design requirements; The maximum flow rate from the water saving basins of the ship lock is about 106.01 m³/s–112.33 m³/s, and the maximum flow rate from the lock head is about 216.82 m³/s–244.70 m³/s; Under this condition, the maximum velocity of the culvert section at each valve of the ship lock is about 10.59 m/s–11.95 m/s; The maximum lifting speed of water surface in the lock chamber is about 6.30 m/min–7.12 m/min.

The above indexes can meet the relevant requirements of the design code for filling and emptying system of ship locks. At the same time, compared with the ship lock projects at home and abroad, the hydraulic indexes are relatively good, which shows that the layout scheme of ship lock proposed in this study is reasonable and achieves the expected effect.

Table 3. Summary of hydraulic characteristic indexes of Baishi ship lock

Filling & emptying	Water level combination (m)	Filling & emptying time (min)	Maximum flow rate from WSBs (m³/s)	Maximum flow rate from lock head (m³/s)	Maximum lifting speed of water surface (m/min)	Maximum velocity of the culvert section at each valve (m/s)
Filling	$300.00 \sim 251.44$	21.48	106.01	216.82	6.30	10.59
	$300.00 \sim 240.40$	23.63	119.36	244.27	7.12	11.93
	$294.00 \sim 240.40$	22.52	112.26	229.65	6.68	11.21
Emptying	$300.00 \sim 251.44$	21.47	106.04	217.09	6.30	10.60
	$300.00 \sim 240.40$	23.60	119.39	244.70	7.12	11.95
	$294.00 \sim 240.40$	22.43	112.33	230.14	6.69	11.24

6 Conclusion and Prospect

(1) This paper makes a theoretical calculation and analysis of the key influencing factors, establishes the calculation models of water saving operating head, water saving rate and the water level of water saving basin. The mutual influence and variation laws of various parameters are put forward. According to the results, it is preferred to adopt: the number of stages of the water saving basins $n \leq 4$; the area ratio between the water saving basin and the lock chamber $m \leq 2.0$.

(2) Aiming at the 60 m class ultra-high head ship lock (Baishi ship lock), the overall layout of the water saving basins and a reasonable water level classification are carried out. The hydraulic characteristics of the filling & emptying process are analyzed and calculated. The research results can provide technical reference for the determination of the final scheme of the project.

(3) By setting multi-stage water saving basins, the utilization efficiency of water resources has been significantly improved. At the same time, the layout type of water-saving ship lock divides the filling & emptying process into stages, so as to reduce the actual operating head of the ship lock, which provides a solution to the hydraulic problems of filling & emptying system of ship lock under ultra-high head. Therefore, it has good applicability and application prospects in the navigation field of high dams.

References

PIANC (2009) Innovations in Navigation Lock Design NO 106, Brussels

Zhu L, Xuan G (2019) Study on key technology of Baishi ship lock on Qingshui river. Nanjing Hydraulic Research Institute, Nanjing (in Chinese)

Wu P, Cao F (2013) Construction technology and development trends of water-saving lock. In: Proceedings of the 167th China engineering science and technology forum, pp 188–193 (in Chinese)

Wang X, Liu C (2013) Development and study status of water-saving ship lock. Chongqing Archit. 10:52–54 (in Chinese)

Wu P, Xuan G (2009) New advances in navigation lock design. Hydro-Sci Eng 4:122–127 (in Chinese)

Dong S, Wang Q, Zhang N et al (2020) Calculation method of water saving rate of multistage ship lock. China Water Transp 12:87–89 (in Chinese)

Li Z, Xu D, An J (2020) Water level calculation and influencing factors of single-step lock with water saving basins. Port Waterw Eng 11:7–11 (in Chinese)

Liu B, Li Y, Hu Y et al (2016) Water-saving layout and hydraulic simulation of high head and large scale ship lock. Port Waterw Eng 12:42–46 (in Chinese)

Technical Status Evaluation of River Training Works Based on the Improved DS Evidence Theory

Zhonglian Jiang[1]([✉]), Xiao Chu[1], Zhen Yu[2], Jianqun Guo[1], and Xiumin Chu[1]

[1] National Engineering Research Center for Water Transport Safety, Wuhan University of Technology, Wuhan, China
{z.jiang, cx330}@whut.edu.cn
[2] Changjiang Waterway Institute of Planning and Design, Wuhan, China
chingyue@yeah.net

Abstract. A large number of river training works have been built in the inland waterway regulation projects to improve ship navigation conditions. However, water damages to river training works happen frequently in practice. Technical status evaluation of river training works is regarded as a fundamental content of inland waterway maintenance. Due to the various influencing factors and complex mechanisms, the content involved in the standard files is recognized as qualitative and no quantitative evaluation method is recommended so far. The technical status of river training works is currently evaluated through on-site investigation which is time-consuming and individual-dependent. By means of multi-source sensors, massive status data of river training works could be obtained instantaneously. Study on the technical evaluation model based on the multi-source information fusion theory attracts more attention in recent years. The classical DS evidence theory could fail as evidence conflict occurs. Thus, the Pearson's correlation coefficient is calculated and utilized to update the probability distribution in the present study. A novel technical status evaluation model based on the improved DS evidence theory is established. The model is further verified through three case studies of traditional river training works (spur dike and flexible mattress belt) in the Yangtze River, China. The model outputs are consistent with the technical survey reports as well as the published research article. Quantitative and accurate evaluation of river training works could be accomplished by applying the proposed evaluation model. Moreover, the model could be embedded in the Inland Electronic Chart Display and Information System. The present study would provide theoretical basis for inland waterway maintenance and infrastructure monitoring in the future.

Keywords: River training works · DS evidence theory · Technical status evaluation · Pearson correlation coefficient · Inland waterway

Y. Li et al. (Eds.): PIANC 2022, LNCE 264, pp. 1565–1576, 2023.
https://doi.org/10.1007/978-981-19-6138-0_138

1 Introduction

River training works have been widely applied in the inland waterway regulation projects to provide better navigation conditions. Common river training works including dams (Macfarlane et al. 2017), spur dikes (Kiani et al. 2017), dredging works (Mendes et al. 2016), etc. would facilitate navigation conditions improvement and transport safety. In the upper reach of Yangtze river, the stochastic characteristics of turbulent flows and gravel bedload transport are highlighted (Cui et al. 2021). Severe river bank erosion results in the riverbed declination as well as the riverbank retreat. Meanwhile, the flood discharge capacity would decrease as siltation occurs. Except for some water conservancy projects to promote social economy and human activities (Ren et al. 2020), majority of river training works are performed for the purpose of maintaining the navigable conditions and the riverbank stability. Since the 21st century, a large number of river training projects have been conducted in China, e.g. the Yangtze estuary deep water waterway regulation project and Jingjiang waterway regulation project, among which spur dikes and flexible mattress belts have been widely applied. Due to the nonlinear and complex flow conditions, the failure mechanism differs for river training works (Crotti and Cigada 2019). Failures of river training works would lead to riverbank retreat and deteriorate navigation conditions. In view of aforementioned situation, reasonable maintenance strategies and solutions are required. Following the annual survey report of Wuhan Waterway Bureau (*Ministry of Communication*, China) of Yangtze River, 5 sites of river training works are seriously damaged and 17 sites are damaged to a certain degree. Intuitive evaluation of training works would be achieved through in situ investigation. However, it is known as subjective and time-consuming. The common qualitative evaluation methods include analytic hierarchy process and expert judgment method (Fan et al. 2016), which are usually straightforward but arbitrary in some cases. Quantitative evaluation methods include fuzzy comprehensive evaluation method and principal component analysis method (Gu et al. 2020; Wu et al. 2019), which require numerous fundamental information and calculations. The DS evidence theory-based model combines characteristics of both qualitative and quantitative methods, and provides a comprehensive evaluation of river training works through multi-source information fusion.

Since 1967, DS evidence theory (Dempster 1967; Shafer 1967) has been widely used in different fields, such as information fusion, fault diagnosis, expert identification, and target identification (Gao et al. 2021; Khan and Anwar 2019; Wickramarathne et al. 2013). In the absence of prior information, the DS evidence method could complete multi-sensor data fusion and judgment. Since information conflicts occur between different sensors, the information fusion results would be greatly affected. Therefore, DS evidence conflicts need to be properly resolved in the case of technical status evaluation. Interval assignment method is introduced to evaluate impacts of different factors. The improved DS evidence theory is therefore applied and validated in the technical status evaluation of river training works.

2 Technical Evaluation Model

2.1 Framework of Technical Evaluation Model

The technical evaluation framework of river training works presented by Li et al. (2020) are adopted and reproduced in Fig. 1. A total number of 14 indexes are classified as functional and structural indicators.

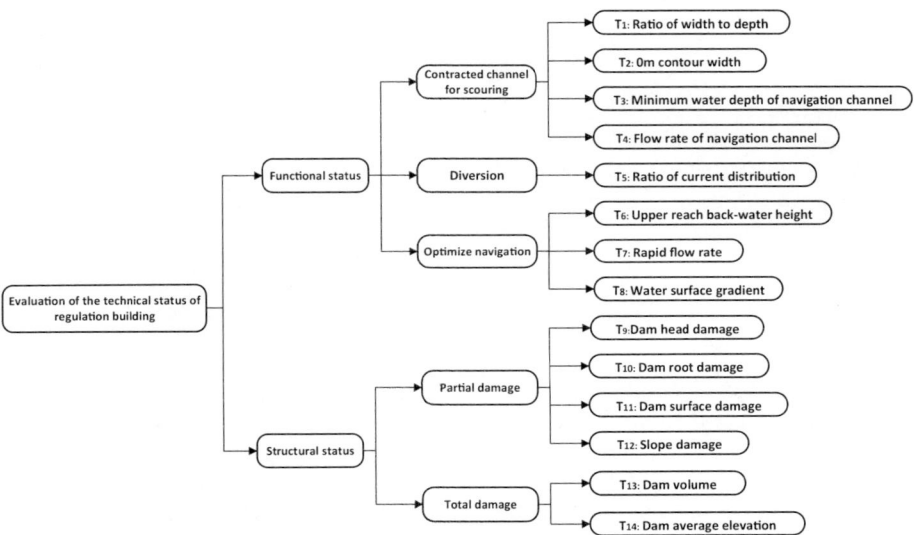

Fig. 1. Index evaluation system for technical status of inland river training works

In the middle reaches of the Yangtze River, the river training works include flexible mattress belts, riverbank revetments, spur dikes, etc. Enrockment method has been used to protect river training works from severe erosion. Following the Technical Code of Waterway Maintenance (JTS/T 320–2021, China), the technical status of river training works is categorized as five levels and the corresponding maintenance recommendations are tabulated (Table 1).

Table 1. Evaluation level definition for river training works

Evaluation level	Score	Maintenance recommendations
Level 1	100	No countermeasure is required
Level 2	85	Follow-up monitoring and analysis is necessary
Level 3	70	Timely maintenance is required
Level 4	55	Countermeasures are required urgently
Level 5	40	No countermeasure is needed

2.2   Improved DS Evidence Theory

DS evidence theory is actually an evidence combination method. The basic probability distribution directly affects the credibility of the evidence theory function (mass function). The identification framework expressed as: $\Theta = \{\theta_1, \theta_2, \ldots, \theta_3\}$ is noted as the foundation of model establishment. $\theta_i (i = 1, 2, \ldots, n)$ represents a single element in the frame, and n denotes the number of elements. For $\forall A \subseteq \Theta$, the Dempster composition rule of the two basic probability distribution functions m_1, m_2 on Θ is written as:

$$m_1 \oplus m_2(A) = \frac{1}{K} \sum_{B \cap C = A} m_1(B) \cdot m_2(C) \tag{1}$$

For A, the composite basic probability distribution function is the sum of the products of all two hypotheses that intersect with A. As the two basic probability distribution functions are calculated, they are divided by the normalization coefficient K. The K algorithm is defined as follows:

$$K = \sum_{B \cap C \neq \emptyset} m_1(B) \cdot m_2(C) = 1 - \sum_{B \cap C = \emptyset} m_1(B) \cdot m_2(C) \tag{2}$$

Although DS evidence theory has been widely applied in the field of information fusion, evidence conflict would occur when two evidences are interrelated. There are four common evidence conflict types named as complete conflict, 0 confidence conflict, 1 confidence conflict, and high conflict. The composition rule is highly sensitive to the hypothetical probability (i.e. the evidence is 0 and 1) which often causes conflicts of evidence and inaccurate evaluation results.

The evidence conflict in DS evidence theory is caused when two evidence units are not independent with each other. The main reason of evidence conflicts is that the certain hypothetical probability with greater difference occupies a dominant position when the evidence is fused. Taking 0 confidence conflict as an example, the result of the synthesis with a probability of 0 is only due to the 1 probability factor of 0 in the hypothesis, which accounts for 100% of the weight ratio of the fusion result of the evidence unit. The Pearson correlation coefficient is therefore used to calculate the weight of each evidence body in the overall weight. The correlation degree between two evidence bodies is clarified and the BPA value is updated through modifying the weight ratio. An improved DS evidence theory model based on Pearson's correlation coefficient is proposed, which solves the problem of evidence conflict in the classic DS synthesis rules. The model can be further utilized in a multi-index fusion evaluation system for the purpose of technical status evaluation of river training works in the Yangtze river. The flowchart of the proposed improved DS evidence theory-based model is shown in Fig. 2.

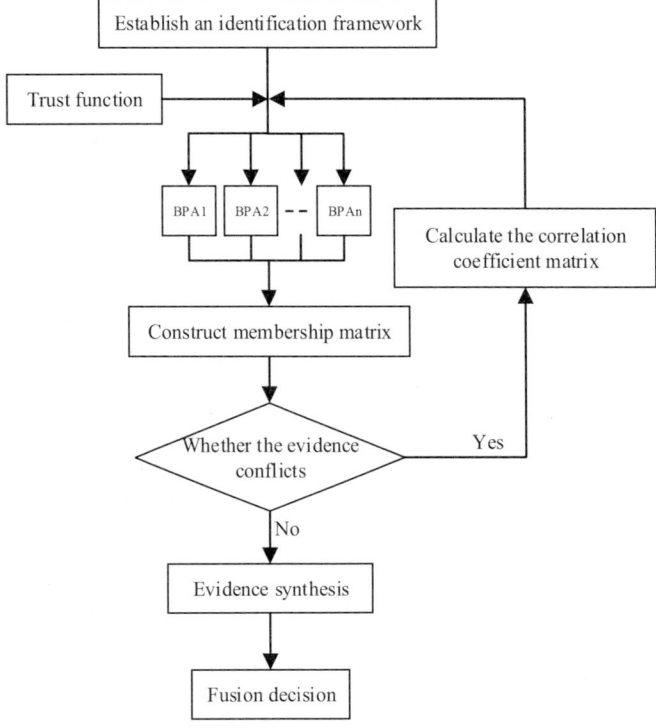

Fig. 2. Flowchart of improved DS evidence theory-based model.

The membership function is the basis of fuzzy statistics, which calculates the probability of each indicator under different evidence units and denotes the membership degree. The correlation coefficient matrix is constructed to modify the probability of the membership matrix of each indicator, combining with the zero-factor correction. A threshold value of 0.001 is introduced to replace the zero-factor in the matrix, and later subtracted from the evidence body with the highest probability.

For improved DS evidence theory based model, the correlation coefficient between the two evidence units is calculated by Eq. (3).

$$s_{ij} = \frac{\mathrm{cov}(m_i, m_j)}{\sigma_{m_i}\sigma_{m_j}} = \frac{E((m_i - \mu_{m_i})(m_j - \mu_{m_j}))}{\sigma_{m_i}\sigma_{m_j}} \tag{3}$$

E is the mathematical expectation; μ and σ denote the mean and variance respectively. Their relationship with the mathematical expectation E is demonstrated as follows:

$$\mu_{m_i} = E(m_i) \tag{4}$$

$$\sigma_{m_i} = \sqrt{E(m_i^2) - E^2(m_i)} \tag{5}$$

By calculating the correlation coefficient of evidence units m, the evidence correlation coefficient matrix S_{ij} could be constructed as follows:

$$S_{ij} = \begin{bmatrix} s_{11} & s_{12} & \cdots & s_{1n} \\ s_{21} & s_{22} & \cdots & s_{2n} \\ \vdots & \vdots & & \vdots \\ s_{n1} & s_{n2} & \cdots & s_{nn} \end{bmatrix} \tag{6}$$

To avoid zero confidence factor in the DS synthesis rules, a zero factor correction is adopted. Then the initial credibility $cred$ of the evidence body m_i can be obtained by Eq. (7).

$$cred(m_i) = \frac{\sum\limits_{j=1}^{n} s_{ij}(m_i, m_j)}{\sum\limits_{i=1}^{n} \sum\limits_{j=1, i\neq,}^{n} s_{ij}(m_i, m_j)} \tag{7}$$

Considering the influence of the decimal point of the credibility value in the basic probability of the evidence unit, $m_i^*(X)$ is derived through Eq. (8):

$$m_i^*(X) = m_i(X) * cred(m_i) \tag{8}$$

The reasonableness of each evidence $D[m_i(A)]$ is calculated after updating the evidence value.

$$D[m_i(A)] = \frac{2m_i(A)\overline{m}(A)[1 - dm_i(A)]}{m_i^2(A) + \overline{m}^2(A)} \tag{9}$$

$d[m_i(A)]$ represents the BPA distance between each evidence and the average value as defined in Eq. (10):

$$d[m_i(A)] = |m_i(A) - \overline{m}(A)| \tag{10}$$

$\overline{m}(A)$ is the average evidence obtained by multiplying the credibility and the original evidence value, as shown in Eq. (11):

$$\overline{m}(A) = \sum\limits_{i=1}^{n} [m_i * cred(m_i)] \tag{11}$$

The credibility value of the fusion evidence is finally obtained:

$$F_{cred}[m_i(A)] = \frac{D[m_i(A)]}{\sum\limits_{q=1}^{N} D[m_i(A_q)] + D[m_j(A_q)]} \tag{12}$$

3   Case Study

Case study has been performed for technical status evaluation of river training works in the Yangtze River, China. Two types of river training works (spur dike and flexible mattress belt, as shown in Fig. 3) are considered in the present study. Spur dikes are frequently applied to protect river banks while the flexible mattress belts are used to combat beach erosion. Following the aforementioned evaluation index system and index scoring rules, the evaluation index scores of three cases are calculated and presented in Table 2.

Fig. 3. Classic river training works (left: spur dike; right: flexible mattress belt).

Table 2. Evaluation index calculations for case studies in the Yangtze River, China.

Evaluation index	Dongxikou spur dike	Taipingkou flexible mattress belt	Majiazui flexible mattress belt
Ratio of width to depth (T_1)	100	100	50
0 m contour width (T_2)	100	100	100
Minimum water depth of navigation channel (T_3)	100	100	100
Flow rate of navigation channel (T_4)	80	80	80
Ratio of current distribution (T_5)	100	50	60
Upper reach back-water height (T_6)	100	80	80
Rapid flow rate (T_7)	60	60	60
Water surface gradient (T_8)	100	100	100
Dam head damage (T_9)	80	100	40
Dam root damage (T_{10})	60	100	80
Dam surface damage (T_{11})	60	100	50
Slope damage (T_{12})	100	100	60
Dam volume (T_{13})	80	100	80
Dam average elevation (T_{14})	100	100	80

4 Results

4.1 Dongxikou Spur Dike

The evaluation results of Dongxikou spur dike are shown in Table 3. The results showed that the technical status of spur dike is generally well, which is consistent with conclusions drawn by Li et al. (2020). The authors presented a coupled model of fuzzy mathematical theory and Bayesian network for river training works in the upper reaches of Yangtze River. The uncertain nature of evidence is characterized by probability, which is generally similar to the improved DS evidence theory in this study. However, the probability calculation between different levels is not remarkable (approximately 0.06) in the literature (Li et al. 2020). Therefore, misinterpretation might occur in the technical status evaluations. The present study demonstrated that quantitative evaluation of river training works could be accomplished by both improved DS evidence theory-based model and Bayesian evaluation model. Considering the probability calculations, the improved DS evidence theory-based model is more recommended in the practices.

Table 3. Probability calculations for Dongxikou spur dike

Evaluation level	Probability value
Level 1	0.64
Level 2	2.79×10^{-5}
Level 3	0.36
Level 4	2.79×10^{-5}
Level 5	3.36×10^{-11}

4.2 Taipingkou Flexible Mattress Belt

The Taipingkou channel located at the middle reaches of the Yangtze River, China. The north riverbank is straight and protected by the artificial structures. The Lalinzhou shoal locates at the south riverbank. Three flexible mattress belts have been deployed to protect Lalinzhou shoal from erosion. The length of flexble mattress belt is approximately 360 m. Following the technical survey report after the flood season of year 2019, the sediment has been silted up in the lower beaches of the Lalinzhou shoal. No significant variations have been observed in the downstream areas of the revetment works. The flexible mattress belt is recognized as functional well and structural complete. Therefore, the technical status is evaluated as level 1 in the field survey of year 2020. The BPA values are calculated and presented in Table 4. It is noted that the probability values of each evidence are not independent and evidence conflict could occur in the application of classic DS theory. The calculated reliability value is shown in Table 5. Based on the evaluation model established in the Sect. 2.2, the Pearson correlation coefficient matrix is calculated and the basic probability distribution updated. Finally, a diagonal matrix of correlation coefficients is obtained. The revised BPA matrix is presented in the Table 6.

Table 4. BPA calculations of Taipingkou flexible mattress belt

Evaluation index	Level 1	Level 2	Level 3	Level 4	Level 5
T_1	1	0	0	0	0
T_2	1	0	0	0	0
T_3	1	0	0	0	0
T_4	0	0.667	0.333	0	0
T_5	0	0	0	0.667	0.333
T_6	0	0.667	0.333	0	0
T_7	0	0	0.333	0.667	0
T_8	1	0	0	0	0
T_9	1	0	0	0	0
T_{10}	1	0	0	0	0
T_{11}	1	0	0	0	0
T_{12}	1	0	0	0	0
T_{13}	1	0	0	0	0
T_{14}	1	0	0	0	0

The technical status evaluation by the improved DS evidence theory is shown in Table 7. The technical status of the Taipingkou flexible mattress belt is evaluated as Level 1, which is consistent with the in-site survey report (Changjiang Waterway Institute of Planning and Design 2020).

Table 5. Calculation results of index reliability value

Evaluation index	*Fcred*
T_1	0.046
T_2	0.046
T_3	0.046
T_4	0.155
T_5	0.115
T_6	0.155
T_7	0.115
T_8	0.046
T_9	0.046
T_{10}	0.046
T_{11}	0.046
T_{12}	0.046
T_{13}	0.046
T_{14}	0.046

Table 6. The revised BPA value of Taipingkou flexible mattress belt

Evaluation index	Level 1	Level 2	Level 3	Level 4	Level 5
T_1	0.001	0.001	0.001	0.001	0.046
T_2	0.001	0.001	0.001	0.001	0.046
T_3	0.001	0.001	0.001	0.001	0.046
T_4	0.001	0.001	0.052	0.104	0.001
T_5	0.038	0.077	0.001	0.001	0.001
T_6	0.001	0.001	0.052	0.104	0.001
T_7	0.001	0.077	0.038	0.001	0.001
T_8	0.001	0.001	0.001	0.001	0.046
T_9	0.001	0.001	0.001	0.001	0.046
T_{10}	0.001	0.001	0.001	0.001	0.046
T_{11}	0.001	0.001	0.001	0.001	0.046
T_{12}	0.001	0.001	0.001	0.001	0.046
T_{13}	0.001	0.001	0.001	0.001	0.046
T_{14}	0.001	0.001	0.001	0.001	0.046

Table 7. Evaluation results of Taipingkou flexible mattress belt

Evaluation level	Probability value
Level 1	≈ 1
Level 2	2.77×10^{-15}
Level 3	2.55×10^{-15}
Level 4	2.65×10^{-15}
Level 5	2.26×10^{-15}

4.3   Majiazui Flexible Mattress Belt

According to the technical report of river training works status evaluation, obvious deformation has been observed at the slope foot of the Majiazui flexible mattress belt. The riverbank slope has become even steep. The overall trend is recognized as scouring and undercutting, which seriously affects the stability of the shoal revetment. The experimental data obtained by the field survey indicate that the technical status of Majiazui falls in the category of level 3. The detailed scoring results are shown in the Table 8.

Table 8. Evaluation results of the Majizui flexible mattress belt

Evaluation level	Probability value
Level 1	5.99×10^{-13}
Level 2	0.001
Level 3	0.998
Level 4	4.61×10^{-4}
Level 5	4.31×10^{-11}

It can be seen from the table that the technical status of the Majiazui flexible mattress belt are classified as level 3, which are consistent with the in situ technical survey report. As a result, timely maintenance is required to prevent structural damages of Majiazui flexible mattress belt.

5   Conclusions

The technical status evaluation of inland river training works is necessary for infrastructure maintenance. Owing to the complicated hydrodynamic conditions, their evolutionary characteristics have not yet been revealed. The wide applications of IoT sensors produce massive data of river training works which provide an effective way of big data analysis as well as multi-information fusion.

In view of the evidence conflict of applying classic DS evidence theory, the Pearson's correlation coefficient has been introduced to describe the relationship between two evidence units. The basic probability distribution is thus updated and BPA matrix is revised for multi-information fusion. The improved DS evidence theory-based model is further validated through case studies in the Yangtze River, China. Two types of river training works (spur dike and flexible mattress belt) are discussed in the present study. The results demonstrate that quantitative evaluations of river training works could be accomplished by the improved DS evidence model. Both Dongxikou spur dike and Taipingkou flexible mattress belt fall in the category of level 1, which are consistent with published research article (Li et al. 2020) and in situ survey reports. Due to the complex flow conditions, the technical status of Majiazui flexible mattress belt is evaluated as level 3. Timely maintenance is required to protect the flexible mattress belt from erosion damages. Moreover, the improved DS evidence theory could also be embedded with the Inland Electronic Chart Display and Information System which is the fundamental service platform of water transport. In the future, further studies on the quantification of indicators should be carried out to achieve more reliable evaluations of river training works.

Acknowledgements. The present study is supported by the National Natural Science Foundation of China (52071250, 51709220), the Fundamental Research Funds for the Central Universities (2018IVB078).

References

Changjiang Waterway Institute of Planning and Design (2020) Technical report of river training works status evaluation (Year 2020), Wuhan Waterway Bureau, 1–239

Crotti G, Cigada A (2019) Scour at river bridge piers: real-time vulnerability assessment through the continuous monitoring of a bridge over the river Po, Italy. J Civ Struct Heal Monit 9 (4):513–528. https://doi.org/10.1007/s13349-019-00348-5

Cui G, Su XS, Liu Y, Zheng SD (2021) Effect of riverbed sediment flushing and clogging on river-water infiltration rate: a case study in the Second Songhua River Northeast China. Hydrogeol J 29(2):551–565

Dempster AP (1967) Upper and lower probabilities induced by a multiple valued mapping. Ann Math Stat 38(2):325–339

Fan GC, Zhong DH, Yan FG, Yue P (2016) A hybrid fuzzy evaluation method for curtain grouting efficiency assessment based on an AHP method extended by D numbers. Expert Syst Appl 44:289–303

Gao XE, Jiang PL, Xie WX, Chen YF, Zhou SB, Chen B (2021) Decision fusion method for fault diagnosis based on closeness and Dempster-Shafer theory. J Intell Fuzzy Syst 40(6):12185–12194

Gu H, Fu X, Zhu YT, Chen YJ, Huang LX (2020) Analysis of social and environmental impact of earth-rock dam breaks based on a fuzzy comprehensive evaluation method. Sustainability 12:6239

Khan N, Anwar S (2019) Paradox elimination in Dempster-Shafer combination rule with novel entropy function: application in decision-level multi-sensor fusion. Sensors 19(21):4810

Kiani A, Masjedi A, Pourmohammadi MH, Heidarnejad M, Bordbar A (2017) Experiment of local scour around a series of spur dikes in river bend. Fresenius Environ Bull 26(8):5331–5339

Li WJ, Zhang HY, Zhang W, Yang SF, Hu J (2020) Technical status evaluation of regulation buildings in upper reaches of Yangtze River channel based on Fuzzy Bayesian Network. J Chongqing Jiaotong Univ (Nat Sci) 39(9):112–118

Macfarlane WW et al (2017) Modeling the capacity of riverscapes to support beaver dams. Geomorphology 277:72–79

Mendes DS, Fortunato AB, Pires-Silva AA (2016) Assessment of three dredging plans for a wave-dominated inlet. Proc Inst Civil Eng-Maritime Eng 169(2):64–75

Ren JR, Mao GX, Zhang F, Wei YH (2020) Research on investment decision-making in waterway engineering based on the hub economic index. Sustainability 12(4):1–19

Shafer G (1967) A mathematical theory of evidence. Princeton University Press, Princeton

Wickramarathne TL, Premaratne K, Murthi MN (2013) Toward efficient computation of the Dempster-Shafer belief theoretic conditionals. IEEE Trans Cybern 43(2):712–724

Wu Q, Wang L, Jin N, Li N, Hu X (2019) Social stability risk evaluation of major water conservancy projects in fragile eco-environment regions. Appl Ecol Environ Res 17(4):9097–9111

The Review and Prospect of the Development of Guangdong-Hong Kong-Macao Greater Bay Area Port Cluster and Its River-Sea Intermodal Transport System

Wentao Ding[✉], Zhengyong Chen, Rui Wang, Tianhan Xue, and Haiyuan Yao

Transport Planning and Research Institute, Ministry of Transport, Beijing, China
dwt0404@126.com

Abstract. Guangdong-Hong Kong-Macao Greater Bay Area is one of the most developed regions in China and a world-class manufacturing base. China has built a world-class port cluster with Hong Kong Port, Shenzhen Port and Guangzhou Port as the core members and a river-sea intermodal transport system based on the Pearl River Delta high-grade waterway network which features delivery of containers, coal, and grain. Promoting the high-quality development of the Guangdong-Hong Kong-Macao Greater Bay Area, achieving carbon peaking by 2030, and achieving carbon neutrality by 2060 are major strategies for China's future development. To realize these goals, China will further promote the construction and development of the Guangdong-Hong Kong-Macao Greater Bay Area port cluster, and will also pay more attention to environmental advantages such as high inland water transport capacity, low energy consumption and low pollution. Meanwhile, the government will make further reforms in planning, construction and institutional mechanism to accelerate the improvement of the river-sea intermodal transport system around the core ports in the Bay Area. This paper systematically reviews the layout, construction, and development of ports in the Guangdong-Hong Kong-Macao Greater Bay Area, the Pearl River Delta high-grade waterway network, and the river-sea intermodal transport system around seaports since the 1980s. Based on the "Outline of the Development Plan of Guangdong-Hong Kong-Macao Greater Bay Area", this paper foresees the future development trend of the port cluster and its river-sea intermodal transport system from the perspectives of inter-port relationship, high-grade waterway network and river-sea intermodal transport hub.

Keywords: Guangdong-Hong Kong-Macao Greater Bay Area port cluster · Pearl River water system · River-sea intermodal transport system · Review and prospect

© The Author(s) 2023
Y. Li et al. (Eds.): PIANC 2022, LNCE 264, pp. 1577–1588, 2023.
https://doi.org/10.1007/978-981-19-6138-0_139

1   Introduction

The Guangdong-Hong Kong-Macao Greater Bay Area is located in the southern coastal region of China, including nine cities in Guangzhou, Shenzhen, Zhuhai, Foshan, Huizhou, Dongguan, Zhongshan, Jiangmen and Zhaoqing (the "PRD Region") and two special administrative regions in Hong Kong and Macau, with a total area of 56,000 square kilometers (Fig. 1). It plays an important role in the construction of the "Maritime Silk Road" (Wang et al. 2022). In 2021, the total population will reach 86.56 million; GPD will be 12.6 trillion yuan; foreign trade import and export volume will be 16.6 trillion yuan, accounting for 6.1%, 10.8% and 34.7% of the country respectively.

The region is connected by rivers and seas, with important ports connecting internationally and high-grade waterways radiating inland areas, and has been a more developed area for water transportation in China in recent years. In 2021, the ports in the Guangdong-Hong Kong-Macao Greater Bay Area completed throughput of 1.61 billion tons and 78.55 million TEU of container throughput, with three ports ranking among the top ten in the world in terms of container throughput, forming a river-sea intermodal transport system featuring by containers and coal, which is an important port cluster region in the world. The ports play an important role in linking international markets, transferring energy and raw materials, and driving economic and trade development in the Guangdong-Hong Kong-Macao Bay Area and the Xijiang River Basin hinterland.

For the development of Bay Area-type ports, China's experience is: making the plan first and playing the role of government macro-control, opening up the market and relying on the market to allocate factor resources, keeping pace with the times and highlighting different development priorities at different times, and promoting together by the central government and local authorities, and utilizing the enthusiasm of both the central and local.

Fig. 1. Distribution of ports and inland waterways in the Guangdong-Hong Kong-Macao Greater Bay Area.

2 Evolution and Development of Ports in Guangdong, Hong Kong and Macao Bay Area

2.1 Evolutionary History

2.1.1 1978–1990

In 1978, China implemented reform and opening up, and established four special economic zones, including Shenzhen, Zhuhai and Shantou in 1980, and 14 coastal open cities, including Guangzhou and Zhanjiang, in 1984. During this period, Hong Kong's advantages in capital, technology, management and shipping were combined with Guangdong's advantages in policy, geography, land and labor, forming the economic cooperation model of "front store and back factory" between Guangdong and Hong Kong. Hong Kong's processing manufacturing industry began to transfer to the Pearl River Delta region, especially in Shenzhen and Dongguan on the east coast of the Pearl River Estuary, where the "three to come and one to make up" industry (processing with materials, processing with samples, assembly with parts and compensation trade) took the lead in development. From 1978 to 1990, the average annual speed of economic development in Guangdong Province was as high as 12.7%, and the GDP and foreign trade import and export volume accounted for 8.3% and 36.3% of the country respectively, ranking first in the country.

Guangzhou port is the oldest port in South China. In order to meet the energy demand brought by economic development, coal terminals such as Guangzhou West Base and Dongguan Shajiao Power Plant were built during this period, and bulk cargo represented by coal and crude oil was the main cargo category of the PRD port. Along with the transfer of manufacturing industry, the container production place was transferred from Hong Kong to the PRD region, and the containers produced in the mainland were mainly transported to Hong Kong by road.

However, compared with waterway transport, the transportation by road takes longer time and longer distance. The goods by waterway can be transported to Hong Kong in only more than ten hours, thus forming the front of the construction of the terminal, the rear of the construction of the "processing zone" mode and can enjoy the preferential policies for Hong Kong by using the length of 50 m or less "Hong Kong and Macao line" container ship. Pearl River Delta ports almost do not have routes for container foreign trade, and more than 90% of the containers transit through Hong Kong. Hong Kong is the only container trunk port for Pearl River Delta region to connect with the world. In 1970, Hong Kong's container throughput was still outside the world's 30th place, but with the abundant cargo supply in the Pearl River Delta region, it jumped to the world's 4th place in 1975. Hong Kong then overtook Kobe and New York, and in 1987 overtook Rotterdam as the world's No. 1 in terms of container throughput, rapidly developing into an international shipping center (Wang 2007; Wang 1997). See Fig. 2 for details.

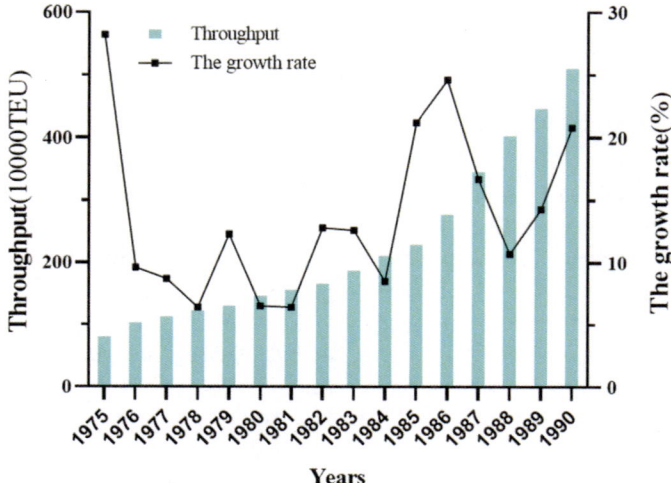

Fig. 2. Hong Kong container throughput 1975–1990.

2.1.2   1991–2000

After Deng Xiaoping's Southern Tour speech in 1992, the Pearl River Delta region further accelerated the transfer of industries, and the economy and foreign trade showed a new round of rapid development. The lack of energy transportation and container transportation capacity was the prominent problem at that time. Therefore, the demand for water transportation grew faster, and port construction was accelerated continuously. In order to rationalize the use of port shoreline resources and improve investment efficiency, the central government and Guangdong Province jointly formulated the "Guangdong Coastal and Pearl River Delta Port Layout Plan", which was the first regional port layout plan in China and determined the basic shape of port development in the Pearl River Delta region.

In terms of container transportation, the plan selects Shenzhen Port as the future container trunk port in the Pearl River Delta region, Guangzhou Port, Zhuhai Port as feeder ports, and other small and medium-sized ports as feeding ports. In terms of coal transportation, Guangzhou Port and Zhuhai Port are selected as transition ports, and it is confirmed that the power plants in the Pearl River Estuary are transported by bulk carriers with a 35,000-ton class, and coastal power plants are transported by 50,000-ton or l00,000-ton bulk carriers. During this period, the planning of a high-grade waterway network in the Pearl River Delta was taken seriously and the government began to formulate the "General Layout Plan for Inland Waterways in Guangdong Province", which proposed a planning pattern of "three horizontal and three vertical" for the region's inland waterway network.

Severely constrained by land, roads, environment, costs, etc., Hong Kong's major terminal operators have begun to turn to the mainland for new development. With its close proximity to Hong Kong and its natural harbor advantages, Shenzhen has become the preferred investment destination for terminal operators such as Hutchison Whampoa. In 1994, Shenzhen Yantian International Container Terminal (Phase I) was put into operation, and the container transport of Shenzhen port began to develop rapidly. In 2000, Shenzhen port completed container throughput of 3.99 million TEU, becoming the second largest container port in mainland China. The pattern that foreign trade containers in the Pearl River Delta region could only go to Hong Kong for transit began to change, but Hong Kong is still the most important container trunk port in the region.

During this period, Guangzhou port was still the distribution center for energy, raw materials and other materials in the region, and began to expand from Huangpu to Xinsha, with the construction of Xinsha coal receiving and unloading terminal and inland river transshipment terminal in 1995. Other cities in the PRD became more aware of the importance of water transport and set up factories along the water, and built a large number of inland container terminals feeding Hong Kong, and river-sea intermodal transport developed rapidly. The government attaches great importance to the construction of the high-grade waterway network in the PRD and has improved the waterway from Zhaoqing to Humen and Hengmen in the lower reaches of Xijiang River according to the standard of navigable 3000t class sea vessels, and the inland waterway in the PRD region has been developed rapidly.

2.1.3   Since 2000

Entering the new century, the Guangdong-Hong Kong-Macao Greater Bay Area has developed into a world-class city cluster and manufacturing base. The 2006 National Coastal Port Layout Plan and the 2007 National Inland Waterway and Port Layout Plan further refined the layout of ports and high-grade waterway networks in the PRD region.

The general plan of Shenzhen Port, Guangzhou Port, Zhuhai Port and Dongguan Port refines the distribution of labor and resource utilization for each port. A number of large deep-water berths such as Yantian (Phase III Project) and Da Chan Bay in Shenzhen, Nansha in Guangzhou, Gaolan in Zhuhai and Tsuen Wan in Huizhou were put into operation one after another, and the 3,000-tonne inland waterway extended to Zhaoqing. During this period, the ports in the PRD began to develop the container river and sea intermodal transport mode, and the scope of the port hinterland was further subdivided, gradually narrowing the gap between the ports of Hong Kong, Shenzhen and Guangzhou in the container transport field (Wang 2013; Chen and Wang 2015; Wu et al. 2013).

Selecting the ports of Hong Kong, Shenzhen and Guangzhou as the core team, and the ports of Zhuhai, Dongguan, Huizhou, Foshan and Zhongshan as the supportive members, a world-class port cluster in the Guangdong-Hong Kong-Macao Greater Bay Area has begun to take shape. Additionally, the "three longitudinal and three horizontal" high-grade waterway network has been completed, strongly supporting the development of the Greater Bay Area.

Hong Kong Port: In 2013, the container throughput of Hong Kong Port was surpassed by Shenzhen Port, but with the advantage of the convenient "free port" customs clearance system, well-developed shipping services, financial services and trade functions, Hong Kong Port is still the most preferred port for high value containers and plays an irreplaceable role.

Shenzhen port: Shenzhen port continues to promote the construction of container terminals, and the integration of the western container terminals. It has formulated the joint development pattern of two wings composed of the eastern Yantian, the western Nanshan and Dachan Bay; The eastern Yantian port area is mainly dominated by North American and European ocean routes; the western port area is dominated by near-ocean routes, which has the natural advantage of developing river-sea intermodal transport. After more than 20 years of high-speed expansion, Shenzhen port encounters the same problems with Hong Kong which are the land, resources, population and environmental constraints. After the global financial crisis in, 2008, it began to focus on container transport featuring coal, grain and other bulk cargo to the surrounding ports out.

Guangzhou Port: Guangzhou Port continues to play an important role in the intermodal transportation of coal and grain by river and sea, and started to plan to build the Nansha Port Area. 2004 saw the completion of the first phase of the Nansha Port Area, and through joint ventures with multinational companies such as COSCO, as well as local enterprises such as Foshan and Zhongshan, the first to fourth phases of the Nansha Container Terminal and the supporting inland river vessel transfer area were built one after another, while the water depth of the sea channel was continuously dredged to 17 m. The construction of Nansha port area has changed the situation of lacking large terminals in Guangzhou port and it also shortens the distance to the international routes. Starting from developing domestic trade containers and foreign trade routes to Southeast Asia and Africa, Guangzhou port gradually has become a new container trunk port in the region. See Fig. 3 for details.

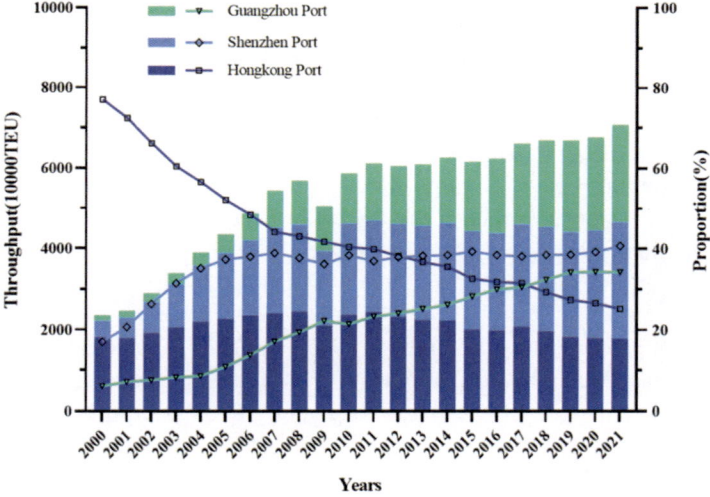

Fig. 3. Changes in container throughput of the ports of Hong Kong, Shenzhen and Guangzhou 2000–2021.

Zhuhai Port: Zhuhai Port focused on developing Gaolan port area and constructing harbor-dredging railroad. Therefore, it has become a bulk cargo transshipment hub for coal, iron ore, oil and gas on the west bank of the Pearl River Estuary.

Dongguan Port: Starting with Shajiao Power Plant coal terminal and as a feeder line to Hong Kong, Dongguan Port has gradually taken over the increased bulk cargo in the region and the bulk cargo transferred from Shenzhen Port. It plays an important role in coal, grain and domestic trade container transportation.

Huizhou Port: Relying on the 300,000-ton crude oil terminal at Mabianzhou, Huizhou Port has driven the development of Daya Bay world-class refining and chemical base.

The process of port development in the Guangdong-Hong Kong-Macao Greater Bay Area is shown in Fig. 4.

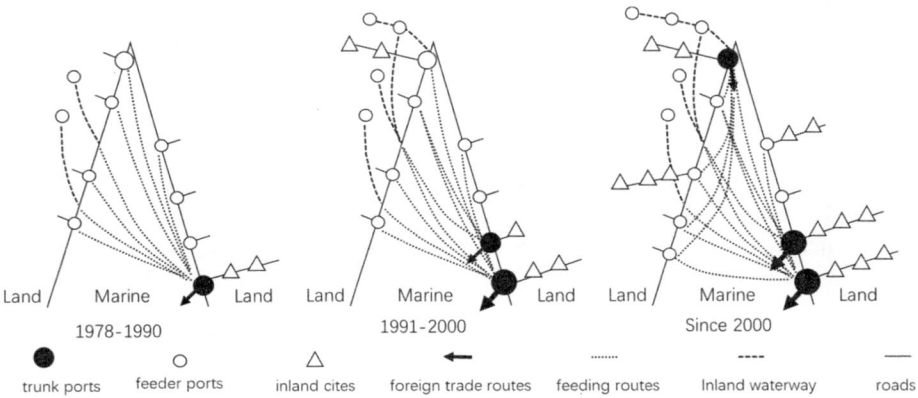

Fig. 4. Schematic diagram of the evolution of the port pattern in the Guangdong-Hong Kong-Macao Greater Bay Area.

2.2   Current Situation

2.2.1   Guangdong, Hong Kong, Macao and the Greater Bay Area is One of the Busiest Regions in the World in Terms of Container Transportation

In 2021, the ports in Guangdong, Hong Kong and Macao Greater Bay Area finished a total of 78.55 million TEU of container throughput, accounting for 26.1% of China's container throughput; more than 400 foreign trade container routes were opened, ranking 1st in China in terms of route density; the international container transshipment volume exceeded 10 million TEU, ranking first in the world's four major bay areas; the container throughput of Shenzhen Port, Guangzhou Port and Hong Kong Port ranked 4th, 5th and 10th respectively in the global container port ranking respectively.

2.2.2   A World-Class Port Cluster with Hong Kong Port, Shenzhen Port and Guangzhou Port as Hubs Has Been Formed

Hong Kong Port: It maintained the scale of container transportation at 20 million TEU/year, and 17.8 million TEU in 2021. It not only focused on import routes and the development of international trade, finance, shipping services and other service industries, but also took into account the transportation of some high-value export goods.

Shenzhen Port: Focusing on container transportation, about 60% of foreign trade containers in the Pearl River Delta region are transported through Shenzhen Port. In addition, an LNG energy receiving and unloading base has been built in Dapeng Bay.

Guangzhou Port: It has various transportation functions such as containers, coal, grain, refined oil products and commercial vehicles. The container ocean shipping line has been initially developed; the scale of domestic trade container and near-ocean shipping line has been expanded; and the river and sea intermodal transport network has been improved.

Zhuhai Port and Dongguan Port: It plays an important role in container feeder transportation and domestic trade transportation.

Huizhou Port: Relying on Daya Bay Refining and Chemical Base, it has developed oil transportation and port-side industries.

Foshan Port, Zhongshan Port, Jiangmen Port, Zhaoqing Port and other ports: It mainly serves processing and manufacturing enterprises in the cities where they are located, and is an important feeding port for container trunk ports.

2.2.3 A Convenient and Efficient River-Sea Intermodal Transport System Has Been Formed

For containers. Guangzhou Nansha, Shenzhen Nanshan and Hong Kong Kai Hong Kong complete the loading and unloading of river vessels at the Tuen Mun River Trade Terminal. Enterprises such as China Merchants Group, Hutchison Whampoa, Guangzhou Port Group and Guangdong Shipping Group attach much importance to container river-sea intermodal transport business, and invest in river terminals in the Pearl River Delta region through acquisition, equity participation and new construction in various ways, cultivating the formation of a comprehensive container river-sea intermodal transport network.

As for coal, Guangzhou Nansha and Xinsha, Dongguan MaChong and Zhuhai Gaolan are the hubs in the Guangdong-Hong Kong-Macao Greater Bay Area. Coal from northern Chinese ports and Indonesia and Australia is picked up and unloaded in these port areas and transported to the Pearl River Delta and cities along the West River via inland vessels.

3 Experience Learned

3.1 Establish a Sound Water Transport Planning System and Guide the Development of the Industry Through Planning

The government attaches importance to the macro-control role of public policy and has established a perfect planning system for ports and waterways. Through the national layout plan and provincial layout plan, the development direction of each port and the navigable grade of each waterway can be determined. Through the port master plan and the five-year development plan, the terminal layout and the development priorities within a certain period of time can be determined. See Fig. 5 for details.

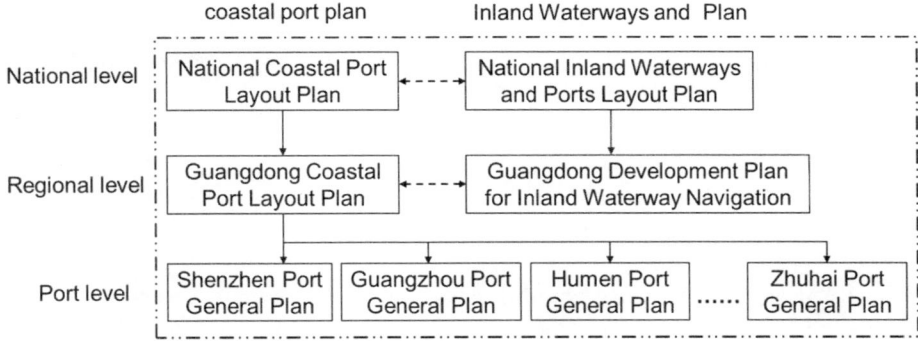

Fig. 5. Diagram of the port and waterway planning system.

3.2 Opening Up the Port Construction and Operation and Forming a Market Pattern of Diversified Equity

After Hutchison Whampoa invested in Yantian Container Terminal in Shenzhen, multinational terminal operators and shipping companies such as Modern Terminals and COSCO Shipping have invested in container terminals in Shenzhen and Guangzhou. Joint ventures and cooperation have also been carried out among mainland enterprises, forming a diversified pattern of equity in the port market of Guangdong, Hong Kong and Macao Greater Bay Area. Under the market regulation, a benign competition and cooperation relationship has been established. See Table 1, 2 for details.

Table 1. Shenzhen port container terminal shareholding structure. (Unit: %)

Company name	Yantian			Shekou & Chiwan	Mawan	DaChan Bay
	Phase I & II	Phase III	Phase III extension			
Hutchison Whampoa Limited	56.4	51.6	51.6	-	-	-
China COSCO SHIPPING	14.6	13.7	-	-	-	-
China Merchants Port Holdings Company	-	-	-	Holdings	67	-
Modern Terminals Limited	-	-	-	Participation	-	65
Yantian Port Group	29	35	48.4	-	-	35
Sinotrans Limited	-	-	-	-	33	-

Table 2. Guangzhou port Nansha container terminal shareholding structure. (Unit: %)

Company name	Phase I	Phase II	Phase III	Phase IV
China COSCO SHIPPING	40	59	-	-
Guangzhou Port Group	51	41	100	65
Nansha Infrastructure Investment Co	9	-	-	-
Foshan Public Utilities Holdings	-	-	-	19
Zhongshan City Construction Group Company	-	-	-	16

3.3   Adjusting the Focus of Port and Waterway Development According to the Economic and Trade Development Needs of the Hinterland at Different Times

Since the development stage of the hinterland determines the scale and structure of cargo transported by the port, China has therefore continuously adjusted the focus of port and waterway development in the PRD region. In the 80's, due to power and energy constraint, the main development of coal terminals was container transportation depended on Hong Kong; in the 90's, taking advantage of the transfer of Hong Kong's manufacturing and shipping industries to the mainland, the development of container terminals was accelerated, and the river-sea intermodal transport system for bulk cargoes such as coal began to be improved; since 2000, joint ventures and cooperation between enterprises have become increasingly close, and the government has increased its efforts to improve inland waterways, so the river-sea intermodal transport system for containers and coal has become more complete. The intermodal transport system has become more perfect (Guangdong Provincial Department of Transport 2022).

3.4   The Central Government and Local Governments Have Formed a Joint Effort to Promote the Construction and Development of Port Clusters in the Greater Bay Area

The central and local governments jointly formulate regional port layout plans, waterway development plans, and master plans for major ports, and jointly invest in the construction of waterways, breakwaters and other port infrastructure. The cumulative investment since 2000 is about 20 billion yuan (RMB) for inland waterway construction.The central government has given localities full autonomy in development, and localities attach importance to water transportation and develop port industries.

4   Future Outlook

In 2019, China announced the Outline of the Development Plan of Guangdong, Hong Kong and Macao Greater Bay Area to support the integration of Hong Kong and Macao into the overall development of the country. The outline of the 14th Five-Year Plan for National Economic and Social Development proposes to build a world-class port cluster in the Guangdong-Hong Kong-Macao Greater Bay Area. These policies will have a significant impact on the development of ports and the river-sea intermodal transport system in the Guangdong-Hong Kong-Macao Greater Bay Area.

4.1 Port Cluster

At present and for a period of time in the future, the ports in the Guangdong-Hong Kong-Macao Greater Bay Area will cooperate more closely and their functions will be further subdivided. The port group with Hong Kong as the international shipping center, Shenzhen Port and Guangzhou Port as the container trunk ports, and Dongguan Port, Zhuhai Port, Foshan Port and Zhongshan Port as the feeder ports will be continuously improved. The level of sharing public resources such as the Pearl River Estuary Sea channel and anchorage is expected to be further enhanced.

4.2 High-Grade Waterway

China attaches more and more importance to stimulate the domestic market demand, and will gradually form a "dual cycle" economic pattern, the prospects of the inland areas of the Pearl River basin will be very positive. The government plan to further enhance the navigational level of the main channel of Xijiang River, promote the construction of tributary channels such as Dongjiang River, Beijiang River, Zuojiang River and Duliu River, improve the connectivity between the main channel and tributary channels, and realize the extension of high-grade channels to the north and east of Guangdong Province, Guangxi, Yunnan, Guizhou and other regions.

4.3 River-Sea Intermodal Hub

The coupling between transportation, space and industrial layout will be further enhanced. Important hubs with river and sea intermodal functions, such as Shenzhen Shekou Port Area, Guangzhou Nansha Port Area, Zhuhai Gaolan Port Area, Dongguan Machong Port Area, etc., will guide inland waterways, railroads, highways, and other transportation resources to converge into a network at the port to form a comprehensive transportation system, which will provide important physical support for strengthening cooperation in the Guangdong-Hong Kong-Macao Greater Bay Area.

4.4 Potentiality

China has set the goal of achieving carbon peaking by 2030 and carbon neutrality by 2060. In the context of sustainable development, increasing the proportion of water transport carriers and giving full play to its advantages of large capacity, small land area, low energy consumption and light pollution is an effective way to relieve the pressure of land constraints and carbon emissions in the Guangdong-Hong Kong-Macao Greater Bay Area.

5 Conclusion

In the 40 years of development of Guangdong-Hong Kong-Macao Greater Bay Area since the reform and opening up, port transportation has played an important role in economic development and urban construction. Through a series of development plans such as the "Guangdong Coastal and Pearl River Delta Port Layout Plan" and "General Layout Plan for Inland Waterways in Guangdong Province" and other development plans set by the government, the ports in Guangdong-Hong Kong-Macao Greater Bay

Area have achieved rapid development. A world-class port group with Hong Kong Port, Shenzhen Port and Guangzhou Port as hubs has been formed, and a convenient and efficient river-sea intermodal transport system has been established. It has now become one of the busiest areas for container transportation in the world. Through sorting out and summarizing, establishing a sound water transport planning system, opening up port construction and operation, timely adjusting the development focus of port and waterway, and forming a joint force between the central and local governments are important experiences for the rapid development and construction of Guangdong-Hong Kong-Macao Greater Bay Area. Looking back at the development of Guangdong-Hong Kong-Macao Greater Bay Area in the past 40 years, port transport has played an important role. Currently, the development of ports in the Guangdong-Hong Kong-Macao Greater Bay Area is facing historic opportunities as well as challenges at the same time. The formation of an efficient and collaborative port cluster, the development of river-sea intermodal transport, and the realization of low-carbon, sustainable and intelligent development are the important paths as well as key solutions for the Guangdong-Hong Kong-Macao Greater Bay Area in the future.

References

Cheng J, Wang C (2015) Evolution and dynamic mechanism of container port system in the Pearl River Delta. Acta Geogr Sin 70(8):1260–1264

Guangdong Provincial Department of Transport (2022) History of Water Transport in Guangdong Province, 440–500

Wang J (2013) Research on evolution mechanism and cooperation development of the Pearl River Delta port system. South China University of Technology, Guangzhou

Wang J (2007) Hong Kong: from modern shipping center towards postmodern logistics management center in Asia. Maritime China (7):40

Wang J (1997) Hong Kong container port: the south China load center under treat. J Eastern Asia Soc Transp Stud 1(2):101–114

Wang Q, Chen W, Wei C (2022) Participation of the global shipping network in the Guangdong Hong Kong-Macao Greater Bay area. Trop Geogr 42(02):236–246

Wu Q, Zhang H, Ye Y et al (2013) Port system evolution model in Pearl River Delta. Trop Geogr 33(2):171–177

Three-Dimensional Hydrodynamic Analysis and Early Warning of Ω-Collapse in the Lower Reaches of the Yangtze River Based on Experimental Study on Generalized Model

Menghao Jia[1,2], Fanyi Zhang[1], Xinyi Lyu[3], Yuncheng Wen[1(✉)], and Hua Xu[1]

[1] Nanjing Hydraulic Research Institute, Nanjing, China
ycwen@nhri.cn
[2] Hohai University, Nanjing, China
[3] Jiangsu Water Conservancy Project Planning Office, Nanjing, China

Abstract. Under the background of the construction of cascade reservoir group in the main stream of the Yangtze River, the lower reaches of the Yangtze River are faced with new water and sediment situation, which leads to the increased risk of bank collapse in the lower reaches of the Yangtze River. Therefore, the purpose of this study is to have a certain supporting significance for the prevention and control of riverbank collapse disasters. The study takes the Jiangsu section of the lower reaches of the Yangtze River as the research object. Firstly, according to the measured data over the years, the macroscopic characteristics of the collapsed bank in the Jiangsu section are analyzed, and the temporal and spatial distribution characteristics are mainly analyzed. The research results show that the collapse of the Jiangsu section of the Yangtze River is dominated by Ω-collapse, which mostly occurs in the flood season and the post-flood receding water period. The frequency of bank collapse in the Yangzhong Reach is relatively the highest among all river segments, and there are more bank collapses on the north bank than the south bank. According to statistics, the average collapse width of the collapse in the Jiangsu section can reach 130 m, the depth of the collapse can reach 60 m, and the ratio of the average bank collapse to the depth can reach 2.15. Then, aiming at the characteristics of the main bank collapse type in the Jiangsu section is the Ω-collapse, the experimental investigation and numerical calculation are used to conduct in-depth research, combined with the measured data, probability and statistical analysis and theoretical analysis. Three-dimensional hydrodynamic analysis is carried out on the mainstream area near the collapse area and outside the collapse area, focusing on the analysis of its nearshore velocity and shear stress and other factors. Combined with the water tank test, the water flow in the inner surface, middle and bottom layers of the Ω-type nest was studied under different flow levels, and it was concluded that the water flow in the nest had a counterclockwise backflow, and the backflow intensity gradually weakened from the side wall to the center. And the phenomenon that the surface layer and the bottom water flow are separated, and this phenomenon becomes more prominent with the increase of the flow rate.

Keywords: Riverbank stability · Cavity reflow · The lower reaches of the Yangtze River · 3D Hydrodynamics · Physical model

© The Author(s) 2023
Y. Li et al. (Eds.): PIANC 2022, LNCE 264, pp. 1589–1603, 2023.
https://doi.org/10.1007/978-981-19-6138-0_140

1 Introduction

The bank collapse patterns in the lower reaches of the Yangtze River can be divided into nest collapse, strip collapse and wash collapse. Nest collapse and strip collapse are the most common forms of bank collapse in the lower reaches of the Yangtze River, with a greater degree of harm. Washing collapse is generally caused by waves, ship waves, etc. with less intensity and less harm, and generally occurs in the estuary area. Collapses mostly occur on the banks of the binary structure. When the height difference of the shoal and groove is large, the nearshore is washed away, the bank slope becomes steep and unstable, and large-scale collapses may occur. Washing avalanche refers to the sliding or slumping of the surface layer of the bank slope or local small-scale soil scoured by water flow, so it is called "washing avalanche". Strip collapse refers to the collapse of the beach soil with a length of tens of meters, hundreds of meters or even a few kilometers along the direction of the river flow. Strip, so called "strip collapse". Nest collapse refers to the collapse or collapse of a large area of soil on the river bank or floodplain, and the phenomenon of soil collapse develops into the bank slope. The collapsed soil volume can reach hundreds of thousands of cubic meters, or even millions of cubic meters, and its appearance forms an inwardly concave nest, so it is called "nest collapse".

The collapsed banks of the Jiangsu section of the Yangtze River are mainly nest collapses. According to incomplete statistics (since 1949, there have been 315 bank collapses of different scales in the Jiangsu section of the Yangtze River. Among them, 191 times of nest collapse occurred, accounting for 60.63% of the total number of collapses, 64 times of strip collapse, accounting for 20.32% of the total number of collapses, including 60 times of washing collapse, accounting for 19.05% of the total number of collapses. It is mainly guarded with the bank slopes of the Jiangsu section of the Yangtze River, and most of its continental beaches are also guarded. The occurrence of strip collapse and wash collapse is relatively small. Due to the large height difference between the riverbed and the beach, the scale of the generated nest collapse is generally large. Therefore, this study conducted an in-depth analysis of the Ω-type nest collapse (Fig. 1).

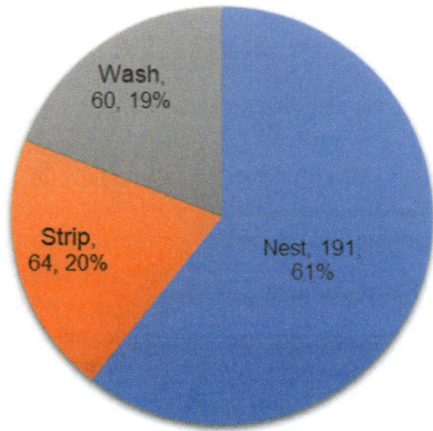

Fig. 1. Pie chart of landslide type distribution.

2   Research Progress

River bank collapse is an important subject in the field of river development and governance. From the 1950s to the 1980s, river bank slope erosion was defined as the process of separating, entraining, and transporting river bank soil in the form of particles or aggregates due to water flow and geological action. River bank collapse is defined as partial or complete collapse of the river bank due to soil instability. River bank collapse is the product of the comprehensive action of many factors, including water erosion, seepage erosion, mechanical properties of river bank soil, weathering, etc. Among them, water erosion and seepage erosion will directly lead to the occurrence of river bank collapse. In the mid-1970s, S.frydmom et al. conducted a centrifugal simulation test of river bank instability, and analyzed the soil slip surface and the displacement and strain of the bank slope soil. In the early 1980s, the U.S. Army Corps of Engineers Waterway Test Station began the corresponding experimental and theoretical research work in the lower Mississippi River in the United States. Jingjiang Riverbed Experiment Station, Hankou Hydrological Station and Nanjing Riverbed Experiment Station analyzed the bank collapse of the Jingjiang River, Chenglingji to Jiujiang and Nanjing rivers according to the measured data of the Yangtze River. The main influencing factors are the water flow intensity, the shape of the river bend and the soil conditions of the river bank. Based on the measured data, Yu and Zeng (1986) analyzed the process of forming a "pocket-type" collapse, and believed that for a bend strongly affected by the top-scouring action of the water flow, the near-shore water flow encounters a bank with weaker shock resistance, and it will rush into the bank to form a rotational property. With the strong backflow, with the expansion of the size of the collapsed nest, the river bank eventually forms a "pocket-type" planar shape that is compatible with the backflow intensity.

After the 1980s, Osman and Thorne (1988), Thorne and Osman (1988) and Darby and Thorne (1996), Darby et al. (2015) proposed relatively complete physical models of various bank slope collapses. Osman and Thorne (1988) analyzed the river bed scouring and river bank erosion, and believed that the most common cause of bank collapse was that the lateral erosion process of the river bank widened the river channel and made the bank slope steeper. The ratio reflects whether the river bank is safe or not. On this basis, Darby and Thorne (1996). Added the relationship between the resistance and the dynamic force of the soil, and further analyzed the slope stability. In the 1990s, a number of Dutch scholars also conducted special research on the stability of river bank slopes. Li (1992) analyzed and proposed that the dynamic factor of the collapse of the Nanjing reach of the Yangtze River is the large-scale longitudinal axis (water flow direction) spiral flow. At the same time, the collapse occurs on the river bank with a binary structure where the upper layer of the river bank is sub-clay or silty clay and the lower layer is a silty-fine sand layer, the time is generally high in the flood season. The continuous deepening of the deep groove and the approach to the shore can be used as a condition and signal for the occurrence of the collapse.

At the beginning of this century, Dapporto et al. (2001, 2003) conducted long-term observations on the pore water pressure on both sides of the Sieve River. The river bank of the river is a typical binary structure bank. The observation results found that the variation of matrix suction inside the bank slope is affected by a combination of factors such as seasonal changes, precipitation, flow, evaporation, and changes in the water surface. Huang et al. (2002) pointed out that the main reason for the instability of the beach is the change of the physical properties, state indicators and strength indicators of the soil itself, and introduced the slope stability analysis and seepage calculation methods, and carried out a sensitivity analysis of the main influencing factors. Leng (1993) believes that groundwater movement only inhibits or promotes the collapse of the bank, and the collapse of the lower reaches of the Yangtze River mostly occurs in the post-flood or dry season. Zhang et al. (2006a, b) and others have classified the collapse of the Yangtze River in detail according to the manifestations, failure modes and formation mechanism of the collapse of the Yangtze River. According to the manifestations of the bank collapse of the Yangtze River, the river bank collapse of the Yangtze River is classified into washing collapse, strip collapse and nest collapse, and the characteristics and common river sections of wash collapse, strip collapse and nest collapse are analyzed in detail. Zhang et al. (2006a, b) believes that the process of river bank collapse includes key physical processes such as water flow erosion and seepage erosion. Research by Chen (2000) and others pointed out that the reasons for the collapse of the Yangtze River can be attributed to the infiltration force, the water current scouring the slope foot, the infiltration of the rainstorm into the bank slope soil and the wave dynamic water pressure.

In recent years, with the improvement of computer calculation level, a method of simulating the bank collapse process based on mathematical model calculation has appeared. Xia et al. (2020) combined the plane two-dimensional water and sediment model with the viscous bank collapse mechanics model to simulate the river bank collapse process of the generalized straight reach under different erosion and deposition conditions. Based on the mechanical mechanism of river bank collapse, Jia (2010) deduces and establishes a viscous bank and a dual-structure river bank collapse mechanics model considering the influence of adjacent soil, and establishes a three-dimensional flow and sediment mathematical model to simulate the lateral swing process of the river channel caused by the heterogeneous river bank collapse.

To sum up, there are certain differences in the existing analysis of the causes of river bank collapse in the middle and lower reaches of the Yangtze River, and there is no convincing unified theory for the key influencing factors of river bank collapse, especially the mechanism of collapse. The complexity is directly still not fully understood. Therefore, in this study, the Ω-type nest collapse was studied in depth, and the three-dimensional hydrodynamics and turbulent structure of the collapsed nest were analyzed.

3   Research Method

3.1   Analysis of Bank Collapse Data

The size of the nest collapse is related to the geological conditions of the bank slope, the height difference of the beach and groove, etc. The width of the collapse can reach tens to hundreds of meters, and the depth can also reach tens to hundreds of meters. The average width of the collapse scale in the Jiangsu section of the Yangtze River is about 150 m, and the depth is about 80 m. The typical bank collapse scale of the Jiangsu section of the Yangtze River is shown in Fig. 2.

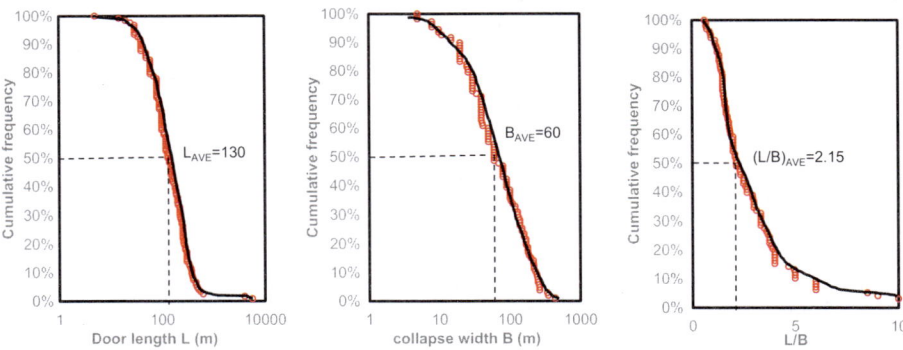

Fig. 2. The width of the collapsed door, the depth of the collapse and the ratio distribution.

This study counted the frequency distribution of the width of the mouth of the nest, the depth of the collapse and the corresponding ratios when the collapse occurred since 1949. The results show that the width of the entrance door of the collapsed nest is generally between 50 m–600 m, and the depth of collapse is generally between 10 m–350 m. The average gate width is about 130 m, the average collapse depth is about 60 m, and the median ratio is 2.15. In this physical model experiment, the length of the semi-circular door is 160 m, the width of the collapsed nest is 80 m, and the collapsed nest with a ratio of 2 is used to simulate the collapsed nest, which is in line with the typicality of nest collapse. The selected Ω-type collapsing nests are mainly generalized with reference to typical collapsing nests.

3.2   Analysis of Bank Collapse Data

3.2.1   Model Design

The length of the established test water tank is 40 m, the height is 0.5 m, the width of the water tank is 1.9 m except for the collapsed nest, the radius of the Ω-type collapsed nest is 47 cm, the length of the door is 66.93 cm, and the opening angle is 90°. The axial surface of the collapsed nest is 12.8 m away from the entrance. The upstream of the entire tank is controlled by the water weir, the downstream is controlled by the tailgate, and the slope of the bottom of the tank is 1:1000. A total of 38 measuring points are arranged around the collapsed nest and outside the collapsed nest. The measurement contents of the model test include: water level measurement along the way, three-dimensional flow velocity measurement in the collapse nest and main flow area (Fig. 3 and Fig. 4).

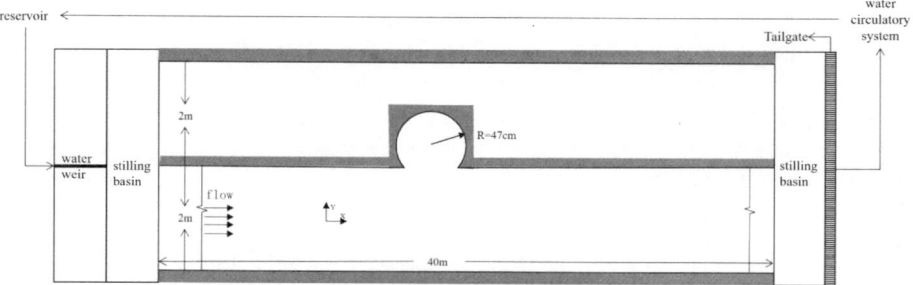

Fig. 3. Schematic diagram of sink layout.

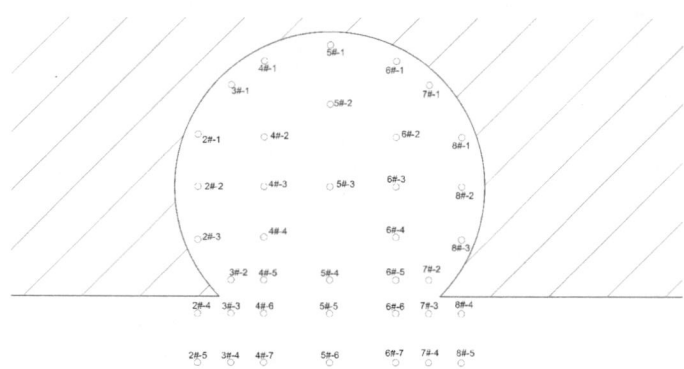

Fig. 4. Local map of measuring points in the collapsed nest area.

The three-dimensional flow velocity of the measuring point is measured by the Vectrino Xiaoweilong Acoustic Doppler Point Velocity Meter (ADV) developed by Nortek in Norway. The probe of ADV consists of one transmitting transducer and four receiving transducers. The working principle of ADV is based on the Doppler effect. The transmitting transducer sends out a pulse signal of a certain frequency, and the flow velocity information is obtained by measuring the phase difference of the pulse signal by the receiving transducer. The flow velocity measuring point is 5 cm away from the transmitting transducer, the sampling frequency is 25 Hz, and the accuracy is ± 1 mm. The ADV probe has two forms, one is a downward-looking probe, and the other is a side-looking probe. This test uses a downward-looking probe for measurement.

3.2.2   Model Parameters

Referring to the nearshore flow velocity and water depth conditions of the natural bank collapse risk section of the Jiangsu section of the Yangtze River, considering that the natural water depth is generally 20–40 m, and the flow rate is 1.5–3 m/s, the combination of water depth and flow rate is selected as the selection range of the fixed bed test. The model similarity conditions are as follows:

It is necessary to meet the gravity similarity criterion and the resistance similarity requirement at the same time.

Flow rate scale: $\lambda_u = \lambda_h^{1/2}$

Flow scale: $\lambda_Q = \lambda_h^{5/2}$

Time scale: $\lambda_t = \lambda_h^{1/2}$

Roughness scale: $\lambda_n = \lambda_h^{1/6}$

The turbulence limiting conditions and the surface tension limiting conditions must be met.

$Re_m > 1000$ and $h_m > 1.5$ cm

The scale chooses $\lambda_h = \lambda_l = 100$, then $\lambda_u = 10, \lambda_n = 2.1544$.

The test groups are as follows (Table 1):

Table 1. Test groups.

Groups	H	V	Q
1	0.40 m	0.30 m/s	0.228 m³/s
2	0.30 m	0.22 m/s	0.1254 m³/s
3	0.20 m	0.15 m/s	0.057 m³/s

4   3D Hydrodynamic Model Study

4.1   The Distribution Characteristics of Plane Velocity and Flow Field in the Collapse Nest Area

Based on the results of the water tank model test, the plane flow velocity and flow field distribution in the Ω-shaped collapse nest were analyzed, and the water surface ratio was 0.96‰ under heavy water conditions. Under various water flow conditions, the water surface line is slightly high at the entrance of the collapsed nest, which is because the backflow in the collapsed nest area at the entrance merges into the main flow and causes local stagnant water (Fig. 5).

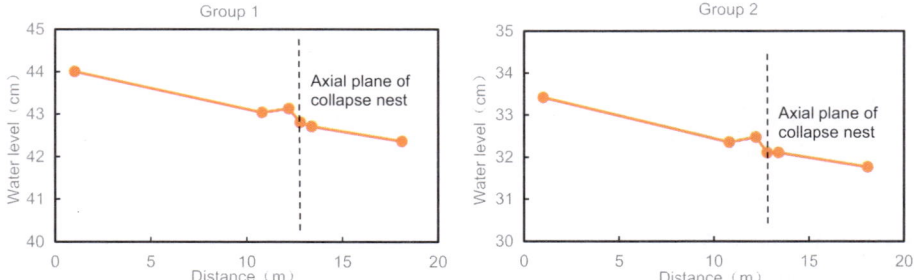

Fig. 5. Variation diagram of water surface line along the route.

Figure 6 is the test site map of the collapsed nest area under three water flow conditions. The surface velocity distribution is observed by using paper flowers. The surface velocity distribution under the three water flow conditions is basically similar, and there is a counterclockwise movement along the side wall near the collapsed nest. In the center of the collapsed fossa, there is a slow flow area, and there is a local vortex in a long transition area near the mouth.

Fig. 6. Test site map of the collapsed nest area.

Figure 7 and Fig. 8 is the flow velocity distribution diagram of the water tank test. Under the condition of high water, the flow velocity of the upper and lower sections of the collapsed nest and the outer area of the collapsed nest is about 0.26–0.33 m/s. The internal velocity of the collapsed nest increases with the distance from the center of the collapsed nest. The flow velocity at the side wall and the mouth of the collapsed nest is about 0.1–0.13 m/s, which is about 40% of the velocity of the upper and lower sections of the collapsed nest and the outer area of the collapsed nest. The flow velocity in the external area is about 0.21–0.26 m/s, and the inside of the collapsed nest is still the largest at the sidewall and the mouth of the collapsed nest, between 0.09–0.11 m/s, which is about the upper and lower sections of the collapsed nest and the outside of the collapsed nest. About 43% of the regional flow velocity, the flow velocity of the upper and lower sections of the collapsed nest and the outer area of the collapsed nest is about 0.15–0.19 m/s under small water conditions, and the flow velocity at the side wall and the mouth of the collapsed nest is between 0.06–0.07 m/s. It is about 39% of the flow velocity of the upper and lower sections of the collapsed nest and the outer area of the collapsed nest. It can be seen that the flow velocity at the inner wall of the collapse nest is generally about 40% of the flow velocity in the mainstream area. Figure 9 shows the relationship between the flow velocity in the mainstream area and the flow velocity near the wall of the collapsed nest under the condition of three water flows. Combined with the results of Yu and Su (2007) measured in the collapsed nest area at Yangzhou Port, the correlation between the flow velocity in the mainstream area and the maximum flow rate in the collapsed nest is calculated. The fitting of the relationship shows that the linear relationship is better, and the regression coefficient can reach 0.95. For the axial plane of the collapsed nest, the size of the external and anterior and posterior sections of the collapsed nest is basically similar, but slightly different. The flow velocity into the collapsed nest decreased sharply, and the flow velocity approached to zero at the center of the collapsed nest. Although the flow direction was opposite to the side wall, the flow velocity gradually increased. The flow velocity inside the collapsed nest basically presents a "V"-shaped flow velocity distribution, and the flow velocity at the top of the bend is basically the same as the flow velocity at the mouth, and the flow direction is opposite.

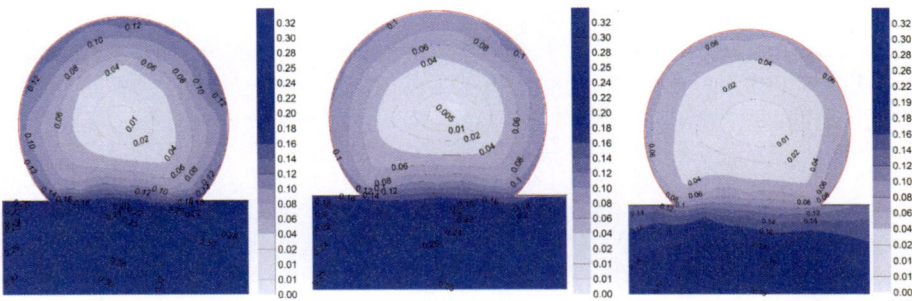

Fig. 7. Velocity distribution in the collapsed nest area.

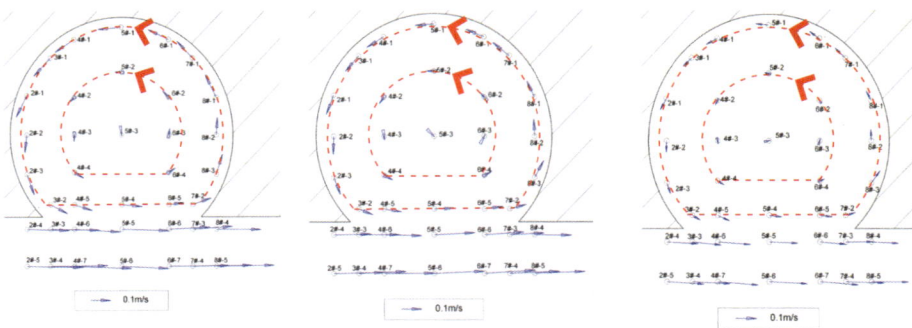

Fig. 8. Distribution of flow field in the collapsed nest area.

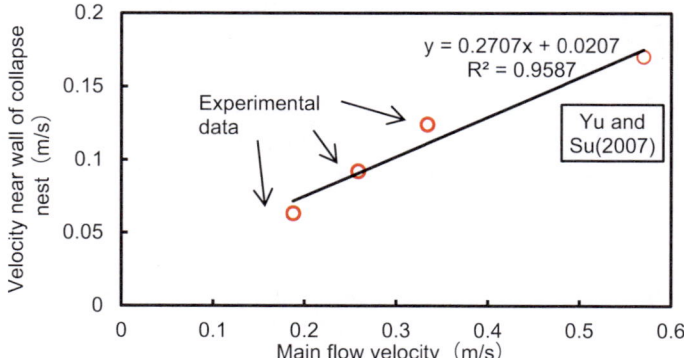

Fig. 9. The relationship between the flow velocity in the main flow area of the Ω-type nest collapse and the maximum flow velocity near the wall in the collapse nest area.

4.2   Distribution Characteristics of Vertical Flow Velocity in the Collapse Nest

Figures 10 are the flow field diagrams of the vertical velocity distribution at the surface, middle and bottom layers at the side wall of the collapse nest under three water flow conditions. Under high water conditions and under moderate water conditions, the difference in the flow velocity of the middle and bottom layers near the wall of the collapse nest is basically very small, and the flow direction near the wall is shown in Fig. 10. The flow rate of the middle bottom layer basically follows the boundary of the collapse nest, and the surface flow velocity appears obvious in some areas. The vertical flow velocity of the stratification is significantly larger than the flow velocity of the middle and bottom layers, and its direction is always biased towards the side wall direction, and has a normal sub-velocity relative to the side wall. Through analysis, the phenomenon of separation of flow velocity in the middle and bottom of the table is basically collapsed. On the chord of the nest (ie from 45° to 225° relative to the center of the collapse), the dispersion of the vertical flow velocity will cause brushing at the side walls of the collapse. This is because the stronger the separation characteristics of the surface and the bottom layer of the water flow, the more conducive to the diffusion and energy consumption, causing scouring, and there is a secondary backflow.

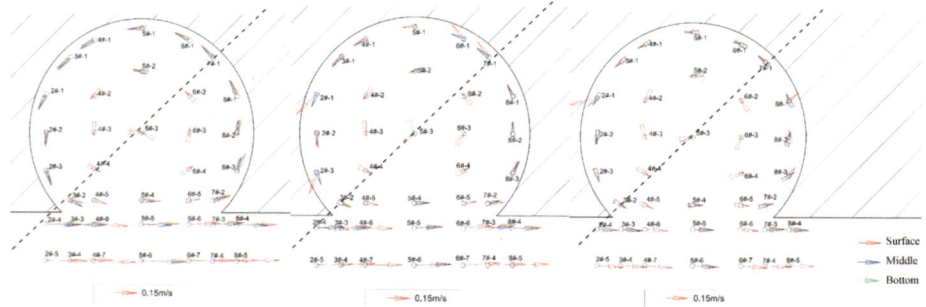

Fig. 10. Distribution of vertical stratified flow velocity in the collapsed nest area.

Figure 11 establishes the relationship between the relative deflection angle and the flow velocity of the bottom layer of the collapsed nest under the condition of reclaimed water. Combining with Fig. 10, it can be seen that when the water flow enters the collapsed nest (that is, after passing through 7#-2 and 8#-3), the flow rate begins to decrease. But the flow velocity is still at a relatively large level in the collapse nest, and the surface and bottom declination angles of about 8° and 5° are generated. It enters the collapse nest and continues to the vicinity of 8#-1, and the flow velocity continues to decrease. There is no obvious deviation from the bottom layer of the surface, and the deflection angle is only about 1°. The water flow velocity fluctuates along the process from 7#-1 to 2#-3. The flow velocity is relatively large at 0.1–0.13 m/s, and the declination angle of the surface and bottom water flow is relatively large and basically between 15–45°. The above shows that the water flow inside the collapse nest exists with the increase of the flow velocity during the swirling motion. The factor of water flow diffusion and energy consumption, under the action of water flow diffusion, the risk of digging and brushing of the side wall of the collapsed nest increases.

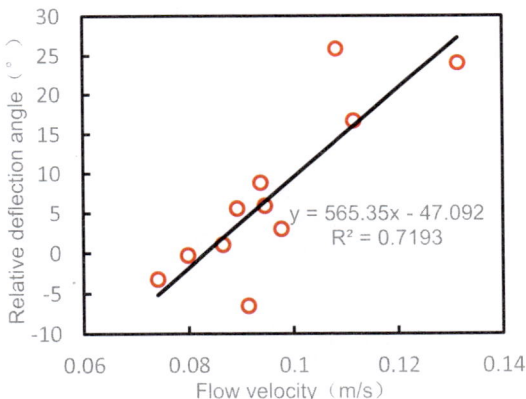

Fig. 11. The relationship between the relative deflection angle of the surface and bottom flow velocity of the collapse nest and the flow velocity.

4.3   The Distribution Characteristics of Turbulent Kinetic Energy and Shear Stress in the Collapsed Nest Area

Turbulence kinetic energy is the kinetic energy possessed by fluid turbulence, and is an important indicator to measure the degree of water turbulence. Except for a small part of the turbulent energy that is directly dissipated as heat energy due to viscous deformation in the mainstream area, most of the turbulent energy is transferred to the near-wall area through shear force, and converted into turbulent kinetic energy there. A vortex with a certain size and rotation speed is generated and rises into the mainstream area, and the vortex decomposes and transfers energy step by step until the vortex is so small that the viscous effect dominates, and finally it is dissipated as heat through the viscous deformation of the water flow. It is this part of the energy that makes the turbulent flow diffuse and forms a special shear field to suspend the sediment. The calculation formula of turbulent kinetic energy k_T is as follows:

$$k_T = \frac{1}{2}\left(\overline{u'^2} + \overline{v'^2} + \overline{w'^2}\right) \tag{1}$$

where u', v', w' are the velocity fluctuations in the x, y, and z directions after the flow velocity undergoes Reynolds decomposition.

Three different kinds of water flow conditions under the disorder kinetic energy as shown in Fig. 12 (unit cm^2/s^2), the turbulent kinetic energy of the mainstream area under large water conditions is generally about 2, the near wall of the collapse nest is at entrance and exit and in the lower part of the mouth. The turbulence kinetic energy is significantly higher than that of other near the wall, generally above 10 cm^2/s^2, the reclaimed water condition is similar to the water, the disorder kinetic energy of the mainstream area is generally about 2 cm^2/s^2, and the near wall of the collapse nest is at the entrance and exit and in the lower part of the mouth. Regional disorder kinetic energy is significantly higher than other near-wall, generally above 4 cm^2/s^2, under small water conditions and the former is slightly different, the turbulent kinetic energy of the flow zone is generally about 2 cm^2/s^2, but at $-30°$ the disorder kinetic energy is large, can reach more than 10 cm^2/s^2.

Fig. 12. Turbulent kinetic energy distribution under different flow conditions.

Shear stress is the most intuitive expression reflecting the action of water flow. In this study, the shear stress was solved by the turbulent kinetic energy method. The three-dimensional turbulent kinetic energy method, also known as the turbulent energy method, uses the fluctuating flow velocity in three dimensions to calculate the bed shear stress. Its formula is as follows:

$$\tau_b = C_1\rho 0.5\left(\overline{u'^2} + \overline{v'^2} + \overline{w'^2}\right) \tag{2}$$

where τ_b is the turbulent shear stress, C_1 is an empirical constant, generally taken as 0.19, McLelland and Nicholas (2000) uses a new method to calculate the total measurement error for the inherent error of ADV measurement, which mainly includes sampling error, ADV noise, and single-point ADV caused by shear flow velocity. The error between the relevant sampling quantities is measured, and the results show that the ADV can characterize the turbulent velocity at the frequency of the maximum sampling rate with very small error. Therefore, it can be considered that the position close to the bed surface is measured by ADV, and the shear stress calculated from the obtained data is the bed surface shear stress. The distribution of small turbulent shear stress for three different water flow conditions is similar to the distribution of turbulent kinetic energy.

5   Conclusions

In this study, the Ω-type nest collapse was studied by using technical methods such as field investigation, theoretical analysis, probability and statistical analysis, and generalized water tank test. The main conclusions are as follows:

1. There is a circulation that moves counterclockwise along the side wall in the collapsed nest area, while in the center of the collapsed nest is a slow flow area, and there is a local vortex in a long transition area near the mouth. For the axial plane of the collapsed nest, the flow velocity into the collapsed nest decreases sharply, and the flow velocity at the center of the collapsed nest has approached zero, and the flow direction near the side wall is reversed, but the flow velocity gradually increases. The flow velocity inside the collapse nest basically presents a "V"-shaped flow velocity distribution, and the flow velocity at the top of the bend is basically the same as the flow velocity at the mouth, and the flow direction is opposite. The backflow in the collapsed nest area at the entrance merges into the main stream, causing local backwater and raising the water surface line. The flow velocity at the inner wall of the collapse nest is generally about 40% of the flow velocity in the mainstream area.
2. The vertical flow velocity inside the collapse nest will be inconsistent in direction. The phenomenon of separation of flow velocity at the bottom of the table is basically the upper string of the collapse nest (that is, the area from 45° to 225° relative to the center point of the collapse nest). The dispersion of the vertical flow velocity will cause brushing at the side walls of the collapsed nest. When the water flow inside the collapsing nest is doing gyratory motion, there is a factor of water flow diffusion and energy consumption with the increase of the flow velocity.

3. Under the conditions of high water and medium water, the turbulent kinetic energy of the near wall of the collapse nest at the inlet and outlet and the attachments at the lower part of the mouth door is significantly higher than that of other near walls. The shear stress distribution has a good positive correlation with the turbulent kinetic energy distribution.

Acknowledgement. This research was funded by the Jiangsu Water Conservancy Science and Technology Project (2020010, 2020002).

References

Thorne CR, Osman AM(1988) Riverbank stability analysis. II: Applications. J Hydraul Eng 114 (2):151–172

Osman AM, Thorne CR (1988) Riverbank stability analysis. I: Theory. J Hydraul Eng 114 (2):134–150

Darby SE, Gessler D, Thorne CR (2015) Computer program for stability analysis of steep, cohesive riverbanks. Earth Surf Proc Land 25(2):175–190

Darby SE, Thorne CR (1996) Development and testing of riverbank-stability analysis. J Hydraul Eng 122(8):443–454

Dapporto S, Rinaldi M, Casagli N et al (2003) Mechanisms of riverbank failure along the Arno River, central Italy. Earth Surf Proc Land 28(12):1303–1323

Dapporto S, Rinaldi M, Casagli N (2001) Failure mechanisms and pore water pressure conditions: analysis of a riverbank along the Arno River (Central Italy). Eng Geol 61(4):221–242

Huang BS, Bai YC, Wan YC (2002) Model for dilapidation mechanism of riverbank. J Hydraul Eng 09:49–54+60

Zhang XN, Li CH et al (2006a) Generalized model test of bank collapse and its mechanism research.Nanjing Hydraulic Research Institute, Nanjing, China

Zhang XN, Li CH, et al (2006b) Analysis and research on the formation mechanism and law of bank collapse. Nanjing Hydraulic Research Institute, Nanjing, China

Xia JQ, Zhou MR, Xu QX et al (2020) Bank collapse and river bed adjustment in middle Yangtze River after operation of Three Gorges Project. Yangtze River 51(01):16–27

McLelland SJ, Nicholas AP (2000) A new method for evaluating errors in high-frequency ADV measurements. Hydrol Process 14:351–366

Yu WC, Su CC (2007) The formation process and water flow structure of the "pocket type" collapse nest in the middle and lower reaches of the Yangtze River. Yangtze River 08:156–159

Yu WC, Zeng JX(1986) Research on local scour and protection of spur dams on the Yangtze River Revetment. The Second China-Japan River Engineering and Dam Engineering Conference, Beijing

Li BZ (1992) Talking about the causes and protection of the collapse of the Nanjing Reach. Yangtze River 23(11):26–28

Leng K (1993) Preliminary study on the formation conditions and protective measures of cave collapse in the lower reaches of the Yangtze River. Adv Water Sci 4(4):281–287

Chen ZY, Sun YS(2000) Discussion on the collapse mechanism and engineering measures of the Yangtze River embankment. China Water Resour 15(2):28–29,4

Jia DD (2010) 3D numerical simulation of heterogeneous riparian channel oscillation. Tsinghua University, Beijing

Author Index

© The Editor(s) (if applicable) and The Author(s) 2023
Y. Li et al. (Eds.): PIANC 2022, LNCE 264, pp. 1605–1609, 2023.
https://doi.org/10.1007/978-981-19-6138-0